THE HUMAN BODY IN HEALTH & DISEASE

7th Edition

Kevin T. Patton, PhD
Founding Professor of Life Science, Emeritus Faculty
St. Charles Community College
Cottleville, Missouri
Professor of Human Anatomy and Physiology Instruction
New York Chiropractic College
Seneca Falls, New York

Gary A. Thibodeau, PhD
Chancellor Emeritus and Professor Emeritus of Biology
University of Wisconsin—River Falls
River Falls, Wisconsin

ELSEVIER

ELSEVIER

3251 Riverport Lane
St. Louis, Missouri 63043

THE HUMAN BODY IN HEALTH & DISEASE,
SEVENTH EDITION

ISBN: 978-0-323-40211-8 (Softcover)
ISBN: 978-0-323-40210-1 (Hardcover)

Notices

Knowledge and best practice in this field are constantly changing. As new research and experience broaden our understanding, changes in research methods, professional practices, or medical treatment may become necessary.

Practitioners and researchers must always rely on their own experience and knowledge in evaluating and using any information, methods, compounds, or experiments described herein. In using such information or methods they should be mindful of their own safety and the safety of others, including parties for whom they have a professional responsibility.

With respect to any drug or pharmaceutical products identified, readers are advised to check the most current information provided (i) on procedures featured or (ii) by the manufacturer of each product to be administered, to verify the recommended dose or formula, the method and duration of administration, and contraindications. It is the responsibility of practitioners, relying on their own experience and knowledge of their patients, to make diagnoses, to determine dosages and the best treatment for each individual patient, and to take all appropriate safety precautions.

To the fullest extent of the law, neither the Publisher nor the authors, contributors, or editors, assume any liability for any injury and/or damage to persons or property as a matter of products liability, negligence or otherwise, or from any use or operation of any methods, products, instructions, or ideas contained in the material herein.

Previous editions copyrighted 2014, 2010, 2005, 2002, 1997, and 1992.

Library of Congress Cataloging-in-Publication Data

Names: Patton, Kevin T., author. | Thibodeau, Gary A., 1938- author.
Title: The human body in health & disease / Kevin T. Patton, Gary A. Thibodeau.
Other titles: Human body in health and disease
Description: 7th edition. | St. Louis, Missouri : Elsevier, [2018] | Includes bibliographical references and index.
Identifiers: LCCN 2016050640| ISBN 9780323402118 (pbk. : alk. paper) | ISBN 9780323402101 (hardcover : alk. paper)
Subjects: | MESH: Physiological Phenomena | Anatomy | Pathologic Processes
Classification: LCC QP34.5 | NLM QT 104 | DDC 612--dc23 LC record available at https://lccn.loc.gov/2016050640

Executive Content Strategist: Kellie White
Senior Content Development Manager: Laurie Gower
Senior Content Development Specialist: Karen C. Turner
Publishing Services Manager: Jeffrey Patterson
Book Production Specialist: Carol O'Connell
Design Direction: Ashley Miner

Printed in China
Last digit is the print number: 9 8 7 6 5

ABOUT THE AUTHORS

Kevin Patton has taught anatomy and physiology to high school, community college, university, and graduate students from various backgrounds for more than three decades. He has earned several citations for teaching anatomy and physiology (A&P), including the Missouri Governor's Award for Excellence in Teaching. His teaching experience has helped him produce a text that will be easier to understand for all students. "One thing I've learned," says Kevin, "is that most of us learn scientific concepts more easily when we can *see* what's going on." His talent for using imagery to teach is evident throughout this edition, with its improved illustration program.

Kevin found that the work that led him to a PhD in vertebrate anatomy and physiology instilled in him an appreciation for the "big picture" of human structure and function. He also has a keen interest in the science of learning, which is reflected in the enhanced pedagogical design of this edition.

Kevin's interest in promoting excellence in teaching anatomy and physiology has led him to take an active role in the Human Anatomy and Physiology Society (HAPS). He serves as HAPS President Emeritus and was the founding Director of HAPS Institute (HAPS-I), a professional continuing education program for anatomy and physiology teachers. As a founding faculty member of a Master of Science in Anatomy & Physiology Instruction, he currently mentors those who are preparing to teach A&P or improve their skills. Kevin also produces several online resources for A&P students and teachers, including *theAPstudent.org* and *theAPprofessor.org*.

Kevin is also a leader among textbook authors, serving many roles in the Textbook & Academic Authors Association (TAA) and mentoring other textbook authors in a variety of disciplines. In 2016, TAA recognized Kevin's service to the profession with the Norma Hood Award.

To my family and friends, who never let me forget the joys of discovery, adventure, and good humor.

To the many teachers who taught me more by who they were than by what they said.

To my students, who help me keep the joy of learning fresh and exciting.

Kevin T. Patton

Gary Thibodeau has been teaching anatomy and physiology for more than three decades. This new edition of *The Human Body in Health & Disease* is a logical extension of his interest and commitment to education. Gary's teaching style encourages active interaction with students using a variety of teaching methodologies. He is considered a pioneer in the introduction of collaborative learning strategies to the teaching of anatomy and physiology.

Gary has been active in numerous professional organizations, including the Human Anatomy and Physiology Society (HAPS), the American Association of Anatomists, the American Association of Clinical Anatomists, the American Pharmaceutical Association, the American Society for Reproductive Medicine (ASRM), and the American Association for the Advancement of Science (AAAS). His biography is included in numerous publications, including *Who's Who in America, Who's Who in American Education, Outstanding Educators in America, American Men and Women of Science,* and *Who's Who in Medicine and Healthcare.*

While earning master's degrees in both zoology and pharmacology, and a PhD in physiology, Gary says that he became "fascinated by the connectedness of the life sciences." That fascination has led to this edition's unifying themes, which focus on how each concept fits into the "big picture" of the human body.

To my parents, M.A. Thibodeau and Florence Thibodeau, who had a deep respect for education at all levels and who truly believed that you never give up being a student.

To my wife, Emogene, an ever-generous and uncommonly discerning critic, for her love, support, and encouragement over the years.

To my children, Douglas and Beth, for making it all worthwhile.

To my grandchildren, Allan Gary Foster and Johanna Lorraine Foster, for proving to me that you really can learn something new every day.

Gary A. Thibodeau

CONTRIBUTOR PANEL

Lead Contributors

Rhonda J. Gamble, PhD
Professor of Physiology and Life Sciences
Mineral Area College
Park Hills, Missouri

Linda Swisher, RN, EdD
Suncoast Technical College
Chair of Health Sciences Division (retired)
Advanced Trauma Life Support
Instructor
Course Educator for the American College of Surgeons
Sarasota, Florida

Contributors

Ed Calcaterra, BS, MEd
Instructor
DeSmet Jesuit High School
Creve Coeur, Missouri

Jeff Kingsbury, MD
Professor, Life Sciences
Mohave Community College
Associate Professor, Department of Biological Sciences
Northern Arizona University
Flagstaff, Arizona

Daniel J. Matusiak, Ed D
Life Science Instructor
St. Dominic High School
O'Fallon, Missouri
Adjunct Professor
St. Charles Community College
Cottleville, Missouri

Amy L. Way, PhD
Professor of Health Science
Lock Haven University of Pennsylvania
Clearfield, Pennsylvania

SCIENTIFIC REVIEW PANEL

Reviewers of Current Edition

Frank Bell, DC, MSHAPI
SUNY Adirondack
Queensbury, New York

Angela Erickson, MSN, RN
Mineral Area College
Park Hills, Missouri

Elizabeth G. F. Granier, PhD
St. Louis Community College
St. Louis, Missouri

Heiko Heisermann, PhD
Winona State University
Winona, Minnesota

Ann L. Henninger, PhD
Wartburg College
Waverly, Iowa

Virginia Johnson, MSHAPI, RN, LMT
Stark State College
North Canton, Ohio

Fiona A. Murray, PhD, BSc (hons)
Swedish Institute
New York, New York

Tina K. Putman, CST, CRST
Lord Fairfax Community College
Middletown, Virginia

Paula D. Silver, BS, PharmD
ECPI University
Newport News, Virginia

Jennifer Swann, PhD
Lehigh University
Bethlehem, Pennsylvania

Barbara L. Westrick, AAS, CPC, CMA (AAMA)
Ross Medical Education Center
Brighton, Michigan

Peggie Williamson, MS, MSHAPI
Central Texas College
Killeen, Texas

Dawn Zuidgeest-Craft, RN
Grand Rapids Community College
Grand Rapids, Michigan

Reviewers of Previous Editions

Bert Atsma
Union County College
Cranford, New Jersey

Janis A. Baker
School of Vocational Nursing
Harlingen, Texas

Rachel Venn Beecham
Mississippi Valley State University
Itta Bena, Mississippi

Christi A. Blair
Holmes Community College
Goodman, Mississippi

Andrew Case
Southeast Community College
Lincoln, Nebraska

Deborah Cipale
Des Moines Area Community College
Ankeny, Iowa

Erin Clason
Spokane Community College
Spokane, Washington

Virginia Clevenger
Mercer County Vocational School
Trenton, New Jersey

Mentor David
Barton County Community College
Great Bend, Kansas

Leslie Day
Northeastern University
Boston, Massachusetts

Judith Diehl
Reid State Technical Campus
Atmore, Alabama

Paul Ellis
St. Louis College of Health Careers
Saint Louis, Missouri

Judy Fair
Sandusky School of Practical Nursing
Sandusky, Ohio

John Finnegan
Cortiva Institute
Somerset, New Jersey

Beth A. Forshee
Freeman Health System
Joplin, Missouri

Linda Fulton
North Hills School of Health Occupations
Pittsburgh, Pennsylvania

Christy Gee
South College
Asheville, North Carolina

Natalie Greene
Macoupin County Health Department
Carlinville, Illinois

Sharon Harris-Pelliccia, BS, RPA
Mildred Elley College
Albany, New York

Beulah Hoffman
Indiana Vocational Technical College
Terre Haute, Indiana

Rita Hoots
Yuba College
Woodland, California

Marilyn Hunter
Daytona Beach Community College
Daytona Beach, Florida

Jon-Phillippe Hyatt
Georgetown University
Washington, DC

Pablo Irusta
Georgetown University
Washington, DC

Tanys Gene James
North Central Texas College
Gainesville, Texas

Michelle Kennedy
Morgan County High School
Madison, Georgia

Brian H. Kipp
Grand Valley State University
Allendale, Michigan

Kathy Korona
Community College of Allegheny County
West Mifflin, Pennsylvania

Anne Lilly
Santa Rosa Junior College
Santa Rosa, California

Caleb Makukutu
Kingwood Junior College
Kingwood, Texas

Susan Caley Opsal
Illinois Valley Community College
Oglesby, Illinois

Darrell Pietarila
Flint Hills Technical School
Emporia, Kansas

Henry M. Seidel
The Johns Hopkins University School of Medicine
Baltimore, Maryland

Donna Silsbee
SUNY Institute of Technology
Utica, New York

Gerry Silverstein
University of Vermont
Burlington, Vermont

Greg K. Sitorius
Minden High School
Minden, Nebraska

Sharon Spalding
Mary Baldwin College
Staunton, Virginia

William Sproat
Walters State Community College
Morristown, Tennessee

Deborah Sulkowski
Pittsburg Technical Institute
Oakdale, Pennsylvania

Karen Tvedten
School of Radiologic Technology
Madison, Wisconsin

Patricia A. West
Independent A&P and Massage Therapy Consultant
Waterville, Ohio

Rebecca S. Wiggins
West Florida High School of Advanced Technology
Pensacola, Florida

Shirley Yeargin
Rend Lake College
Ina, Illinois

Nina Zanetti
Siena College
Loudonville, New York

PREFACE

This book about the human body represents the latest and best information available. *The Human Body in Health & Disease* is a guide for future health professionals who are just beginning their exploration of the complex human organism. It not only presents introductory material on the elegance and efficiency of the healthy human body but also shows what happens when things go wrong. To truly understand the human body, one must appreciate both normal and abnormal structure and function.

As we prepared this newest edition, each decision regarding how concepts were to be presented in our book was evaluated by teachers actually working in the field—teachers currently helping students learn about human structure, function, and disease for the first time. We also consulted closely with working health professionals and medical writers to ensure that our references to disease processes and related topics are current, accurate, and clearly summarized. We also paid careful attention to what research tells us about how the brain reads, processes, and learns new information, enabling us to present information in a way the promotes student success. The result is a text that students will read with enthusiasm—one designed to help the teacher teach and the student learn.

This book is particularly suited to introductory courses about the human body in relation to various health professions. *The Human Body in Health & Disease* emphasizes concepts that are required knowledge for entry into more advanced courses, completion of professional licensing examinations, and success in a practical, work-related environment.

Instructional Design

Unifying Themes

Anatomy, physiology, and introductory pathology encompass a body of knowledge that, because of its sheer magnitude, can easily discourage and overwhelm the new student. There is no question, however, that competency in these fields is essential for student success in almost every health-related or science curriculum. If a textbook is to be successful as a teaching tool in such a complex and important learning environment, it must help unify information, stimulate critical thinking, and motivate students to master a new vocabulary as they learn about the beauty and "connectedness" of human structure and function and the "disjointedness" of human disease.

The Human Body in Health & Disease is dominated by two unifying conceptual themes: the *complementarity of normal structure and function* and *homeostasis.* In every chapter of the book the student is shown how organized anatomical structures of a particular size, shape, form, or placement serve specific functions. Emphasis of this principle encourages students to integrate otherwise isolated factual information into a cohesive and understandable whole. The breakdown of normal integration of form and function is identified as the basis for many disease processes.

The integrating principle of homeostasis is used to show how normal structure and function are maintained by dynamic counterbalancing of forces within the body. Failures of homeostasis are shown as basic mechanisms of disease—a concept that reinforces understanding of the regulatory systems of the human body.

The Human Body in Health & Disease is dominated by two unifying pedagogical themes: the *language of science and medicine* and a *multisensory approach* to learning.

To successfully learn and apply the concepts of human science, students must first master the complex terminology and usage of scientific language. This edition again features expanded word lists that start at the beginning of each chapter to assist both native speakers and English language learners. Embedded hints encourage students to read and say new terms out loud before encountering them in the context of learning concepts—a proven reading strategy based on how we naturally learn new language and process reading in the brain. Included pronunciation guides help students "get it right" without guessing and included translations of word parts help students see how scientific language is constructed. Online audio chapter summaries and an audio glossary add an auditory dimension to the proper pronunciation and usage of essential scientific language, while the suggested "saying terms out loud" provides a helpful kinesthetic/motor experience.

Our multisensory approach is further enhanced by the many intriguing visual elements such as detailed drawings, photographs, medical imaging, and flow charts that help students "picture" the concepts described in the text. Unique "anatomical compass" rosettes help students develop a sense of anatomical direction. Expanded legends and embedded "steps" provide a "walkthrough" of key diagrams that reinforce essential concepts of each chapter. The digital coloring activities on the Evolve website reinforce visual learning with kinesthetic experiences.

Organization and Content

The 25 chapters of *The Human Body in Health & Disease* present the core material of anatomy, physiology, and pathology *most important* for introductory students.

The sequence of chapters in the book follows that most commonly used in courses taught at the undergraduate level. Basic concepts of human biology—anatomy, physiology, biochemistry, cytology, histology, and pathology—are presented in Chapters 1 through 6. Chapters 7 through 25 present material on more specialized topics, such as individual organs or systems, the senses (Chapter 11), immunity (Chapter 16), and genetics and genetic diseases (Chapter 25). Because each chapter is self-contained, instructors are given the flexibility to alter the sequence of material to fit personal teaching preferences or the special content or time constraints of their courses.

In this edition, we continue developing our popular "conversational" style of narrative with additional "chunking" of content for better comprehension. We broke apart longer sentences and paragraphs and labeled subtopics with additional descriptive headings.

An equally important goal for us in designing this text is to present information using a conceptual framework on which the student can build an understanding of the human body. Rather than simply listing a set of facts, each chapter outlines the broad concepts that allow students to relate the facts to one another in a meaningful way. For example, rather than presenting diseases in a disjointed scattering of definitions or descriptions typically seen in other texts, we explain disease conditions within a framework of patterns that facilitates a more complete understanding of the *process* of disease and allows the student to compare and contrast related disorders easily.

Instructors who teach courses with less emphasis on concepts of pathology may wish to examine an alternate text with a similar instructional design: *Structure & Function of the Body,* also available from Elsevier.

Illustrations and Page Design

A major strength of *The Human Body in Health & Disease* is the exceptional quality, accuracy, and beauty of the illustration program and page design. Many illustrations have been revised or updated and several new illustrations have been added. We have continued to use a consistent color scheme and illustration style to enhance student understanding.

Our popular directional rosettes appear inconspicuously in all anatomical illustrations. These rosettes, like the compass rosettes found on all modern maps, orient the user, pointing which way is left and which way is right—directions that in anatomy may appear "backward" to the beginning student. Many illustrations feature expanded legends and embedded "steps" that provide a helpful "walk through" describing the concepts represented and pointing out important elements of the illustration.

Advancing an approach to page design first attempted in the previous two editions, we have made additional improvements in our intuitive internal design that integrates the illustrations and other learning tools more closely with the text narrative. Summary tables are used extensively to visually organize concepts in a way that help students compare and contrast them to better understand the relationships among structures and functions of the body. This has produced a more usable and more attractive platform for student learning.

Clear View of the Human Body

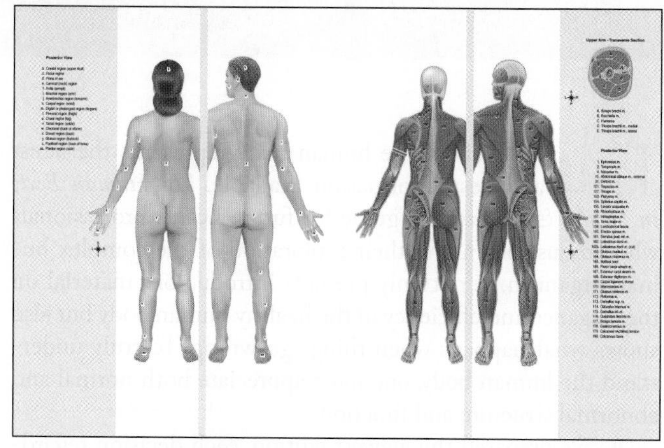

A full-color, semitransparent model of the body called the *Clear View of the Human Body* is located between Chapters 3 and 4 in the textbook. This revised feature provides a handy "always on" virtual dissection of male and female human bodies along several different planes. A student favorite, this tool helps learners use both visual and kinesthetic experiences to assimilate their knowledge of the complex, three-dimensional nature of the human body. It also helps students visualize human anatomy in the manner of today's clinical and athletic body-imaging technology.

Embedded Learning Tools

The Human Body in Health & Disease is a student-oriented text. Written in a friendly style accessible to both expert and challenged students, it has many learning aids within the text that maintain interest and motivation. Every chapter contains the following elements, each of which facilitates teaching and learning.

Hints: Hints embedded within learning aids—marked with a bold icon—highlight strategies in using the provided learning tools effectively. These hints are based on strategies recommended by learning and reading experts.

Hint

Chapter Outline: An overview outline introduces each chapter and enables the student to preview the content and flow of the chapter at the major concept level before embarking on the detailed reading.

Chapter Objectives: Each chapter opening page contains several measurable learning objectives. Each clearly identifies for the student, before he or she reads the chapter, what the key goals should be and what concepts should be mastered.

Language of Science and Medicine: Key terms (boldface in the text) are listed starting at the beginning of each chapter so that students can say them out loud using the provided pronunciation guides—a strategy based on how the brain processes reading and recommended by reading experts. Word parts that form each term are identified and translated to help students slowly build a foundation in understanding the structure of scientific language and medical terminology.

Quick Check questions: Brief sets of questions appearing at intervals throughout each chapter encourage students to pause and reflect on what they just read. This strategy improves reading comprehension and retention by practicing the retrieval of recently learned information and ideas—also preparing them for an eventual end-of-chapter retrieval practice. Answers to all Quick Check questions are on Evolve.

Active Concept Maps: New to this edition are animated, narrated flow charts of selected concepts that many students find difficult to understand are now available at the Evolve website. Called out at appropriate locations within the chapters, they use sight and sound to virtually walk students through conceptual relationships.

Boxed sidebars: Brief boxed sidebars appear in every chapter. These boxes include information ranging from clinical applications of the information to highlights of recent research or related topics to relevant discussions of exercise and fitness. Pathological conditions are sometimes explained in essay format to help students better understand the relationship between normal structure and function. All sidebars are highlighted with an easily recognized symbol so that students can see at a glance whether the box contains wellness, clinical, research, or science application information. In this edition, the featured boxes cover four categories:

- The *Health & Well-Being* boxes contain information about wellness, fitness and exercise, athletics, public health, and related issues and problems.

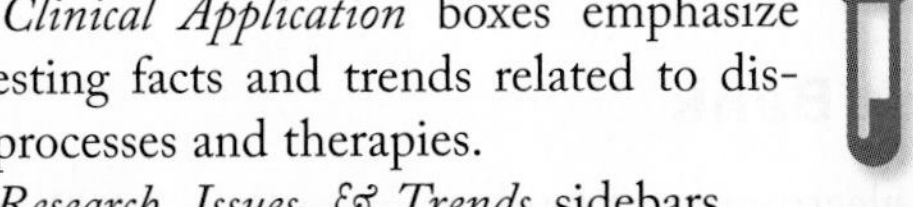

- The *Clinical Application* boxes emphasize interesting facts and trends related to disease processes and therapies.

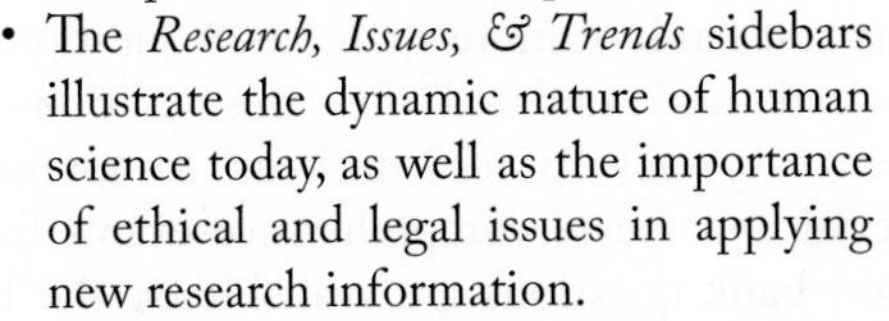

- The *Research, Issues, & Trends* sidebars illustrate the dynamic nature of human science today, as well as the importance of ethical and legal issues in applying new research information.

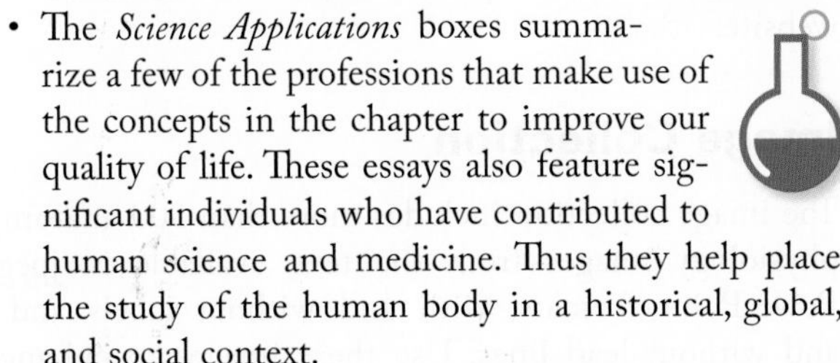

- The *Science Applications* boxes summarize a few of the professions that make use of the concepts in the chapter to improve our quality of life. These essays also feature significant individuals who have contributed to human science and medicine. Thus they help place the study of the human body in a historical, global, and social context.

Connect It! online articles: New to this edition are a collection of brief articles that illustrate, clarify, and apply concepts encountered in the text. Embedded within the text narrative, small boxes connect students with specific illustrated articles available online at *evolve.elsevier.com.* These articles stimulate thinking, satisfy the natural curiosity of students, and help integrate concepts, so that each student better understands the "big picture" of human structure, function, and disease.

AnimationDirect: Each chapter has small boxes that point the reader to animations of important principles. These are available in AnimationDirect, which is included on Evolve. The brief animated sequences are designed to demonstrate concepts that are not easily illustrated in static diagrams. In effect they help put a student's understanding in motion and thus help solidify learning using a multisensory approach.

Outline Summaries: Many students are "low structure builders," meaning they have trouble building a complex conceptual framework on their own. Extensive and detailed end-of-chapter summaries in outline format provide excellent guides for students as they build the conceptual structures needed to understand the content of each chapter. They also help review the text materials when preparing for examinations. Many students also find these detailed guides to be useful as a chapter preview—the conceptual blueprint—in conjunction with the chapter outline.

Audio Chapter Summary boxes: Found at the Evolve website and called out with an icon at the start of each chapter outline summary, brief audio (mp3) summaries can be played on a variety of digital devices. These summaries are useful for both previews and reviews of chapter content, enabling students to use multiple sensory modalities in their learning.

Active Learning tools: A complete set of end-of-chapter challenges allow students to test their own learning to find weak spots that require additional study and provide opportunities to try their hand at analyzing and evaluating questions and cases to apply the concepts they have learned.

- *Study Tips.* A list of specific active study strategies to most effectively study the concepts presented in the chapter. By participating in suggested study activities, students not only master the concepts of a specific chapter but also build their overall competence as self-directed learners.
- *Review Questions.* Subjective review questions at the end of each chapter allow students to use a narrative format to discuss concepts and also serve to synthesize important chapter information that can then be reviewed to assess comprehension of the material.
- *Critical Thinking Questions.* Review questions that encourage students to use critical thinking skills are highlighted at the end of the Review Questions section.
- *Chapter Tests.* Objective-type Chapter Test questions included at the end of each chapter serve as self-tests for the recall and mastery of important subject matter. They also provide practice needed to increase the re-

tention of information and build confidence before an in-class quiz or test.
- *Case Studies.* Each chapter ends with a few case studies that ask students to practice their skills of analysis, evaluation, and application by solving specific, practical problems.
- Answers for all Active Learning sections are on the Evolve website.

Glossary: An extensive listing of key terms, pronunciations, and definitions serves as a handy reference for students as they progress through the course.

Index: A comprehensive index aids in locating information anywhere in the book quickly and easily.

Digital Learning Tools

A wide variety of multisensory learning tools are available at *evolve.elsevier.com/PattonThibodeau/humanbody:*

- **Audio chapter summaries** for each chapter in the book are a student favorite. These concise, narrated overviews are called out at the start of each chapter Outline Summary and can be accessed on the Evolve website. Some students find that these audio reviews improve their retention of chapter concepts when used immediately after reading the chapter.
- Throughout the text, small boxes callout the **Active Concept Map** and **AnimationDirect** features that are found on Evolve. These provide audio and visual coverage of a wide range of topics and body systems.
- **Connect It!** articles, which are called out throughout the text, expand on topics in the text to stimulate thinking beyond the coverage of the textbook.
- The **Body Spectrum Electronic Anatomy Coloring Book** is one of our most popular interactive features on Evolve. Using a visual-kinesthetic approach, this tool simplifies the way students learn anatomy and medical terminology by offering more than 70 detailed anatomy illustrations that can be colored online or printed out to color and study offline.
- Not sure how a term is pronounced? Listen to the proper pronunciation with the **Audio Glossary**!
- **FAQs**—Frequently Asked Questions—by students, along with answers from the authors.
- The **Self-Test** and **Matching Exercises** activities on Evolve allow for interactive practice with immediate feedback, providing an excellent tool for gauging comprehension.
- **Appendixes** that were formerly in the print book have been moved to the Evolve site for easier reference online. These provide detailed information on selected pathological conditions, medical terminology, and clinical and laboratory values and conversion factors.

Online Resources for Instructors

TEACH Instructor Resource Manual

TEACH has been updated and revised for this edition. The TEACH lesson plans help instructors prepare for class and make full use of the rich array of ancillaries and resources that come with the textbook. The content covered in each textbook chapter is divided across one or more lesson plans, each designed to occupy 50 minutes of class time. Lesson plans are organized into easily understandable sections that are each tied to the chapter learning objectives:

- **Instructor Preparation:** This section provides a checklist of all the things you need to do to prepare for class, including a list of all the items that you need to bring to class to perform any activity or demonstration included in the lesson plan, and all pertinent key terms covered in that lesson.
- **Student Preparation:** Textbook readings, study guide exercises, online activities, and other applicable homework assignments for each lesson are provided here along with an overall estimated completion time.
- **The 50-Minute Lesson Plan:** A lecture outline that reflects the chapter lecture slides that come as part of TEACH is included, as well as classroom activities and online activities, one or more critical thinking questions, and time estimates for the classroom lecture and activities.
- **Assessment Plan:** To ensure that your students have mastered all the objectives, the new TEACH includes a separate "Assessment Plan" section. An easy-to-use table maps each assessment tool to the lesson plans and chapter objectives so you can see all your assessment options—by chapter, by lesson, and by objective—and choose accordingly.

Test Bank

An electronic test bank of more than 3,600 questions with answers—revised and updated for this edition—gives instructors an easy way to test students' comprehension of text material and create comprehensive exams for students. The test bank questions are available on the Evolve website.

Image Collection

The image collection includes more than 500 anatomy and physiology images from the text, available in jpeg and PowerPoint formats, with and without labels and with and without lead lines. Use these images to enhance the visual elements of your lectures, discussions, case studies, quizzes, tests, handouts, and more. The image collection is available on the Evolve website.

Update Guide

Compiled by the lead author during the revision process of this textbook, the *Changes to the 7th Edition* document is a detailed list including updates in the textbook since the last edition. These include global, chapter-by-chapter, and section-by-section changes. This document will help in planning as you upgrade from the 6th edition to the new 7th edition of *The Human Body in Health & Disease*. Updates and teaching tips are also posted at *PattonHD.org*.

Supplements

The supplements package has been carefully planned and developed to assist instructors and to enhance their use of the text. Each supplement has been thoroughly reviewed by many of the same instructors who reviewed the text.

- *Study Guide.* Written by Linda Swisher, it provides students with additional self-study aids, including chapter overviews, topic reviews, and application and labeling exercises (such as matching, crossword puzzles, fill in the blank, and multiple choice), as well as answers in the back of the guide.
- *Anatomy and Physiology Online for The Human Body in Health & Disease.* This optional tool is a 24-module online course that brings A&P to life and helps you understand the most important concepts presented in *The Human Body in Health & Disease*. It includes over 125 animations, 300 interactive exercises, and quizzes and exams to assess student comprehension. Available at the Evolve website.
- *Elsevier Adaptive Learning (EAL).* This personalized and interactive tool enables students to learn faster and remember longer. It's fun; it's engaging; and it's constantly tracking student performance and adapting to deliver content precisely when it's needed to ensure information is transformed into lasting knowledge.
- *Survival Guide for Anatomy & Physiology.* An entertaining and easy-to-read set of tips, shortcuts, and advice, this "survival kit" helps students achieve success in anatomy and physiology.

Summary of Changes to the Seventh Edition

A comprehensive detailed list of changes appears in the *Changes to 7th Edition* guide. Look for it in the *Instructor Resources* on Evolve. A few selected revisions for this edition are listed here.

- Chapter and section headings have been simplified for clarity
- New *Connect It!* boxes embedded in the text point students to illustration articles that help them integrate and apply concepts.
- Additional *chunking* and reorganization of sections, paragraphs, and sentences to improve reading comprehension.
- New animated, narrated *Active Concept Maps* have been added to the Evolve website resources and called out where appropriate in each chapter
- *Language of Science* and *Language of Medicine* word lists now start on each chapter-opener page, then continue in the end-of-chapter resources. Each boldface term from the text narrative is listed with a pronunciation guide and word parts (with literal translations).
  - Nearly 350 words have been added to the word lists. Mostly words that already appear in the book, the additions to the word lists ensure that students have immediate access to pronunciation guides and translation of word parts.
  - Pronunciation guides are more complete, providing help with standard vocabulary that may be challenging to English language learners and underprepared students. Pronunciations for unusual plurals forms have also been added.
- Cross references to the *Clear View of the Human Body* and to the appendixes on Evolve have been embedded in the text to assist student learning
- Specific learning *Hints* throughout each chapter tell students how best to use the included learning tools have been clarified for ease of use.
- The career choices mentioned in the *Science Applications* boxes are now emphasized in boldface or italic and have been added to the word lists, along with pronunciations and word part translations; photographs of health professionals in action also have been added to some boxes.
- Numerous new illustrations were added or replaced previous illustrations
  - A number of illustrations were redrawn or revised to include explanatory "step boxes," often with matching numerals embedded within an illustration's detail, help walk students through complex concepts.
  - Many illustrations were revised or redrawn for clarity.
- The *Clear View of the Human Body* has additional new layers. Some existing layers have additional structures labeled.
- The human *microbiome*—an emerging concept of rapidly growing importance in health care—is now introduced in Chapter 1 and integrated into many of the remaining chapters.
- The terms "lower arm" and "upper arm" replaced with the more precise terms "forearm" and "arm."
- Section on passive transport mechanisms (diffusion and osmosis) has been reorganized and osmosis definition clarified to distinguish it from diffusion.

- Additional skull and other bone illustrations help students better understand the skeleton.
- New body movement photographs show how muscles move joints.
- Content on hearing and equilibrium reorganized and revised.
- Discussion of blood clotting revised for clarity.
- Discussion and illustrations of the heart's conduction system revised for clarity.
- Clarified that the primary principle that drives the flow of any fluid is a pressure gradient.
- Introduced the term hemodynamics as a heading to better frame the discussion of the role of blood pressure in blood circulation.
- Added a boxed discussion of arterial blood gases (ABGs) to illustrate how acid/base balance can be assessed.
- New content covers how the body regulates food intake.

A Word of Thanks

Many people have contributed to the development and success of *The Human Body in Health & Disease.* We extend our thanks and deep appreciation to the various students and classroom instructors who have provided us with helpful suggestions following their use of the earlier editions of this text.

For this edition, we would like to thank the following experts for their contributions: Rhonda Gamble, for contributing many of the *Connect It!* articles and editing all of them. Linda Swisher, who helped us improve the learning opportunities in every chapter of the book—as well as producing a very useful *Study Guide.* Janie Corbitt, Virginia Johnson, and Angela Palmier for their excellent work reviewing and updating learning tools on the Evolve website; Jennifer Bertucci for the *TEACH Instructor Resource Manual;* Dan Matusiak for reviewing and revising the online course; and Peggie Williamson for updating the test bank on Evolve. We are grateful for the prior contributions of Ed Calcaterra, Jeff Kingsbury, and Amy Way.

At Elsevier, thanks are due to all on the talented and creative team that produced this 7th edition. We wish especially to acknowledge the support, effort, and occasional crisis management of Kellie White, executive content strategist; Karen Turner and Heather Bays, content development specialists; Jeffrey Patterson, publishing services manager; Carol O'Connell, book production specialist; Doris Cadd, copy editor; and Ashley Miner, book designer, all of whom were instrumental in bringing this edition to successful completion. We also wish to thank all the many others of the Elsevier family who support our ongoing efforts, from animation creation, to online resource creation and support, to the hard-working marketing and sales professionals.

Kevin T. Patton
Gary A. Thibodeau

HOW TO USE THIS BOOK

You might think that it's obvious how to use a textbook—you just open it up and read it! But that won't get you very far in your learning. You need a *textbook strategy* to make the best use of this tool. Here, we outline a three-step plan that you can adapt to your own learning goals and study preferences.

1. Get Your Head Into It

- **Regularly think about how you learn best**—and notice what is working for you and what is not. Learning coaches call this *metacognition,* a term that means "thinking about your thinking." Surprisingly, this alone will improve your success! The many *Hints* embedded in each chapter will help you focus on how you are thinking about your reading.
- **Avoid distractions while your use your book.** This means a quiet environment and no interruptions. This includes *turning off your music*. Even if you think it's helping you, research shows that listening to music nearly always reduces your ability to focus fully on what you are reading.
- **Realize that deep learning takes time and effort.** This recommended strategy will not work if you don't put significant work into it. So make sure you've scheduled a lot of short study sessions (20-45 minutes) to get it all done. Space out your study sessions, interspersing other activities between them, or your study time will be wasted.

2. Gain Familiarity

- **Familiarity is recognition of the terminology and basic ideas in each chapter.** This is only the first step to learning, but it's the necessary foundation for later *mastery* of the content.
- **First, take a look at the Chapter Objectives.** This tells you what you need to master by the time you've finished learning the chapter. That will help you focus on the "big ideas" as you read.
- **Then slowly read through the *Chapter Outline* at the beginning of each chapter** to get your brain familiar with how the "story" of the chapter is laid out. This helps you "sort out" the ideas in a chapter as you read and build a mental framework of new concepts.
- **Next, read each term in the *Language of Science* and *Language of Medicine* out loud.** This book has as many new words as a foreign-language textbook, so recognizing that *you must learn a new language* is an important step to success. As you say each term aloud, use the provided *pronunciation guide*—so you won't trip over complex terms when you read the chapter. Take a moment to glance at the *word parts* that make up each term. This helps you start recognizing how scientific terminology works—therefore, it helps you learn additional new terms faster and more accurately. More than one pass at this step works even better.
- **Read the chapter.** Read each section separately, skimming the subheadings before you begin your reading—to help you build your mental framework of concepts. By reading in sections, it won't matter that you can't read the whole chapter in one sitting.
- **Answer the embedded *Quick Check* questions** before you move ahead in your reading. This helps you make sure you have paused to think about what you've read. And it helps you see if you have really understood the main ideas. If you can't correctly answer the *Quick Check,* consider reading that part of the chapter again, so you "get it" before moving to the next section.
- **Use the embedded resources to help you understand the text narrative.** Take the time to study the illustrations called out in the reading. If there are *Animation-Direct* videos or *Active Concept Maps,* watch and listen to them. Read all the related *Connect It!* articles and boxed sidebars.

3. Seek Mastery

- **Move beyond familiarity to mastery.** Many students don't realize that *familiarity*—which is incomplete and easily forgotten—is not the same as *mastery.* Mastery is a more complete understanding of the concepts, how they relate to one another, and how they can be applied to "real world" situations. Mastery is something that stays with you for years—when you really need it—rather than being forgotten in a few days or weeks.
- **Review the Outline Summary.** Start thinking about how it all fits together and consider how familiar you are with each major idea. You may need to go back and review a section that you have forgotten.
- **Work through the Active Learning process.** Work through the *Review Questions,* then *Critical Thinking* items, the *Chapter Test,* and *Case Studies* at the end of the chapter. Write out your answers—don't just answer them "in your head." Check your answers and correct any weaknesses before moving on to the next

set of items. Then do this again in a few days. You'll have forgotten some things, but that's expected—and that's why continuing, spaced repetition of knowledge-retrieval practice is so important!

- **Use your Evolve Resources for additional retrieval practice.** Your Evolve resources include practice questions in the "Prepare for Exams" section.
- **Combine this textbook strategy with the other course components.** Your instructor has carefully planned a whole course around the textbook—so this strategy is just a part of what you must be doing to learn and master all the essential concepts. Make sure you *participate fully* in your course and use effective study strategies to maximize your learning.

CONTENTS

6 Mechanisms of Disease, 112

7 Skin and Membranes, 144

8 Skeletal System, 174

9 Muscular System, 218

13 Blood, 348

14 Heart, 378

15 Circulation of Blood, 402

16 Lymphatic System and Immunity, 428

17 Respiratory System, 458

18 Digestive System, 492

19 Nutrition and Metabolism, 532

20 Urinary System, 544

21 Fluid and Electrolyte Balance, 584

22 Acid-Base Balance, 600

23 Reproductive Systems, 616

THE HUMAN BODY IN HEALTH & DISEASE

Introduction to the Body

OUTLINE

Scan this outline before you begin to read the chapter, as a preview of how the concepts are organized.

OBJECTIVES

Before reading the chapter, review these goals for your learning.

After you have completed this chapter, you should be able to:

1. Define the terms *anatomy, physiology,* and *pathology.*
2. Describe the process used to form scientific theories.
3. List and discuss in order of increasing complexity the levels of organization of the body.
4. Define the terms *anatomical position, supine,* and *prone.*
5. List and define the principal directional terms and sections (planes) used in describing the body and the relationship of body parts to one another.
6. Do the following related to body cavities and body regions:
   - List and discuss the major cavities of the body and the subdivisions of each.
   - List the nine abdominopelvic regions and the abdominopelvic quadrants.
   - Discuss and contrast the axial and the appendicular subdivisions of the body. Identify a number of specific anatomical regions in each area.
7. Do the following related to the balance of body functions:
   - Define and discuss *homeostasis.*
   - Compare and contrast negative and positive feedback loops and give an example of each in the body.

CHAPTER 1

There are many wonders in our world, but none is more wondrous than the human body. This is a textbook about that incomparable structure. It deals with two very distinct and yet interrelated sciences: **anatomy** and **physiology.**

As a science, anatomy is often defined as the study of the structure of an organism and the relationships of its parts. The word *anatomy* is derived from two word parts that mean "cutting apart." Anatomists learn about the structure of the human body by cutting it apart. This process, called **dissection,** is still the principal technique used to isolate and study the structural components or parts of the human body.

Physiology, on the other hand, is the study of the functions of living organisms and their parts. Physiologists use scientific experimentation to tease out how each activity of the body works, how it is regulated, and how it fits into the complex, coordinated operation of the whole human organism.

In the chapters that follow, you will see again and again that anatomical parts have structures exactly suited to perform specific functions. Each has a particular size, shape, form, or position in the body related directly to its ability to perform a unique and specialized activity. This principle—that *structure fits function*—is the key to understanding all of human biology.

Although an understanding of the normal structure and function of the body is important, it is also important to know the mechanisms of **disease.** Disease conditions result from abnormalities of body structure or function that prevent the body from maintaining the internal stability that keeps us alive and healthy. **Pathology,** the scientific study of disease, uses principles of anatomy and physiology to determine the nature of particular diseases. The term *pathology* comes from *pathos,* the Greek word for "disease." Chapter 6 provides an overview of the basic mechanisms of disease, such as infection and cancer.

Throughout the rest of this textbook, explanations of normal structure and function are supplemented by discussions of related disease processes. By knowing the structure and function of the healthy body, you will be better prepared to understand what can go wrong to cause disease. At the same time, having knowledge of disease states will enhance your understanding of normal structure and function.

LANGUAGE OF SCIENCE

Hint Before reading the chapter, say each of these terms out loud. This will help you to avoid stumbling over them as you read.

abdominal
(ab-DOM-ih-nal)
[*abdomin-* **belly,** *-al* **relating to**]

abdominal cavity
(ab-DOM-ih-nal KAV-ih-tee)
[*abdomin-* **belly,** *-al* **relating to,** *cav-* **hollow,** *-ity* **state**]

abdominopelvic cavity
(ab-DOM-ih-noh-PEL-vik KAV-ih-tee)
[*abdomin-* **belly,** *-pelv-* **basin,** *cav-* **hollow,** *-ity* **state**]

abdominopelvic quadrant
(ab-DOM-ih-noh-PEL-vik KWOD-rant)
[*abdomin-* **belly,** *-pelv-* **basin,** *quadran-* **fourth part**]

abdominopelvic region
(ab-DOM-ih-noh-PEL-vik REE-jun)
[*abdomin-* **belly,** *-pelv-* **basin,** *-ic* **relating to**]

anatomical position
(an-ah-TOM-ih-kal poh-ZISH-un)
[*ana-* **apart,** *-tom-* **cut,** *-ical-* **relating to,** *posit-* **place,** *-tion* **state**]

anatomist
(ah-NAT-oh-mist)
[*ana-* **apart,** *-tom-* **cut,** *-ist* **agent**]

anatomy
(ah-NAT-oh-mee)
[*ana-* **apart,** *-tom-* **cut,** *-y* **action**]

antebrachial
(an-tee-BRAY-kee-al)
[*ante-* **front** *-brachi-* **arm,** *-al* **relating to**]

anterior
(an-TEER-ee-or)
[*ante-* **front,** *-er-* **more,** *-or* **quality**]

Continued on p. 17

1

Scientific Method

What we often call the **scientific method** is merely a systematic approach to discovery. Although there is no single method for scientific discovery, some scientists follow the steps outlined in **Figure 1-1** to discover the concepts of human biology discussed in this textbook.

First, one makes a tentative explanation, called a **hypothesis.** A hypothesis is a reasonable guess based on previous informal observations or on previously tested explanations.

After a hypothesis has been proposed, it must be tested—a process called **experimentation.** Scientific experiments are designed to be as simple as possible to avoid the possibility of errors. Often, **experimental controls** are used to ensure that the test situation itself is not affecting the results.

For example, if a new cancer drug is being tested, half the test subjects will get the drug and half the subjects will be given a harmless substitute. The group getting the drug is called the *test group,* and the group getting the substitute is called the *control group.* If both groups improve, or if only the control group improves, the drug's effectiveness has not been demonstrated. If the test group improves, but the control group does not, the hypothesis that the drug works is tentatively accepted as true. Experimentation requires accurate measurement and recording of data.

If the results of experimentation support the original hypothesis, it is tentatively accepted as true, and the researcher moves on to the next step. If the data do not support the hypothesis, the researcher tentatively rejects the hypothesis.

RESEARCH, ISSUES, AND TRENDS

METRIC SYSTEM

Scientists, many government agencies, and increasing numbers of American industries are using or moving toward the conversion of our system of English measurements to the metric system. The metric system is a decimal system in which measurement of length is based on the meter (39.37 inches) and weight or mass is based on the gram (about 454 grams equal a pound).

A micrometer is one millionth of a meter. (Micron is another name for micrometer.) In the metric systems, the units of length are as follows:

1 kilometer = 1000 meters
1 meter (m) = 39.37 inches
1 centimeter (cm) = 1/100 m
1 millimeter (mm) = 1/1000 m
1 micrometer (μm) or micron (μ) = 1/1,000,000 m
1 nanometer (nm) = 1/1,000,000,000 m
1 Angstrom (Å) = 1/10,000,000,000 m

Approximately equal to 1 inch:

- 2.5 cm
- 25 mm
- 25,000 μm
- 25,000,000 nm

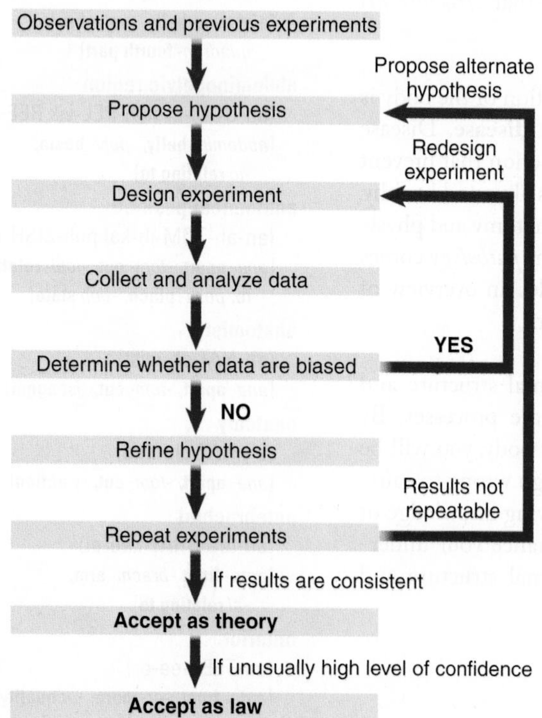

FIGURE 1-1 Scientific method. In this classic example, initial observations or results from other experiments may lead to formation of a new hypothesis. As more testing is done to ensure that outside influences and biases are eliminated, results become more consistent and scientists begin to have more confidence in the tested principle, which can then be called a theory or law.

Knowing which hypotheses are incorrect is as valuable as knowing which hypotheses are valid. Scientists can thus focus on the ideas shown to have merit and avoid wasting time with disproven hypotheses.

Initial experimental results are published in scientific journals so that other researchers can benefit from them and verify them. If experimental results cannot be reproduced by other scientists, then the hypothesis is not widely accepted. If a hypothesis withstands this rigorous retesting, the level of confidence in the hypothesis increases. A hypothesis that has gained a high level of confidence is called a **theory** or **law.**

Why is it important to know the steps of experimentation and developing theories if your main interest is a career in science applications—such as a health career? If you do not understand how concepts are discovered and how they can change after additional experimentation, it is hard to fully grasp them.

The facts presented in this textbook are among the latest theories of how the body is built and how it functions. As methods of imaging the body and measuring functional processes improve, we find new data that cause us to replace old theories with newer ones.

Levels of Organization

Before you begin the study of the structure and function of the human body and its many parts, it is important to think about how those parts are organized and how they might logically fit together into a functioning whole.

Examine **Figure 1-2**. It illustrates the differing **levels of organization** that influence body structure and function. Note that the levels of organization progress from the least complex (chemical level) to the most complex (organism level).

Because you already know that "structure fits function," it should not surprise you that the highly complex and coordinated functions of the whole body can be understood by discovering the many basic processes that occur in the smaller parts, such as organs, tissues, and cells.

Organization is one of the most important characteristics of body structure. Even the word *organism*, used to denote a living thing, implies organization.

Although the body itself is considered a single structure, it is made up of trillions of smaller structures. Atoms and molecules are often referred to as the **chemical level** of organization. The existence of life depends on the proper levels and proportions of many chemical substances in the cells of the body.

Many of the physical and chemical phenomena that play important roles in the life process are reviewed in Chapter 2.

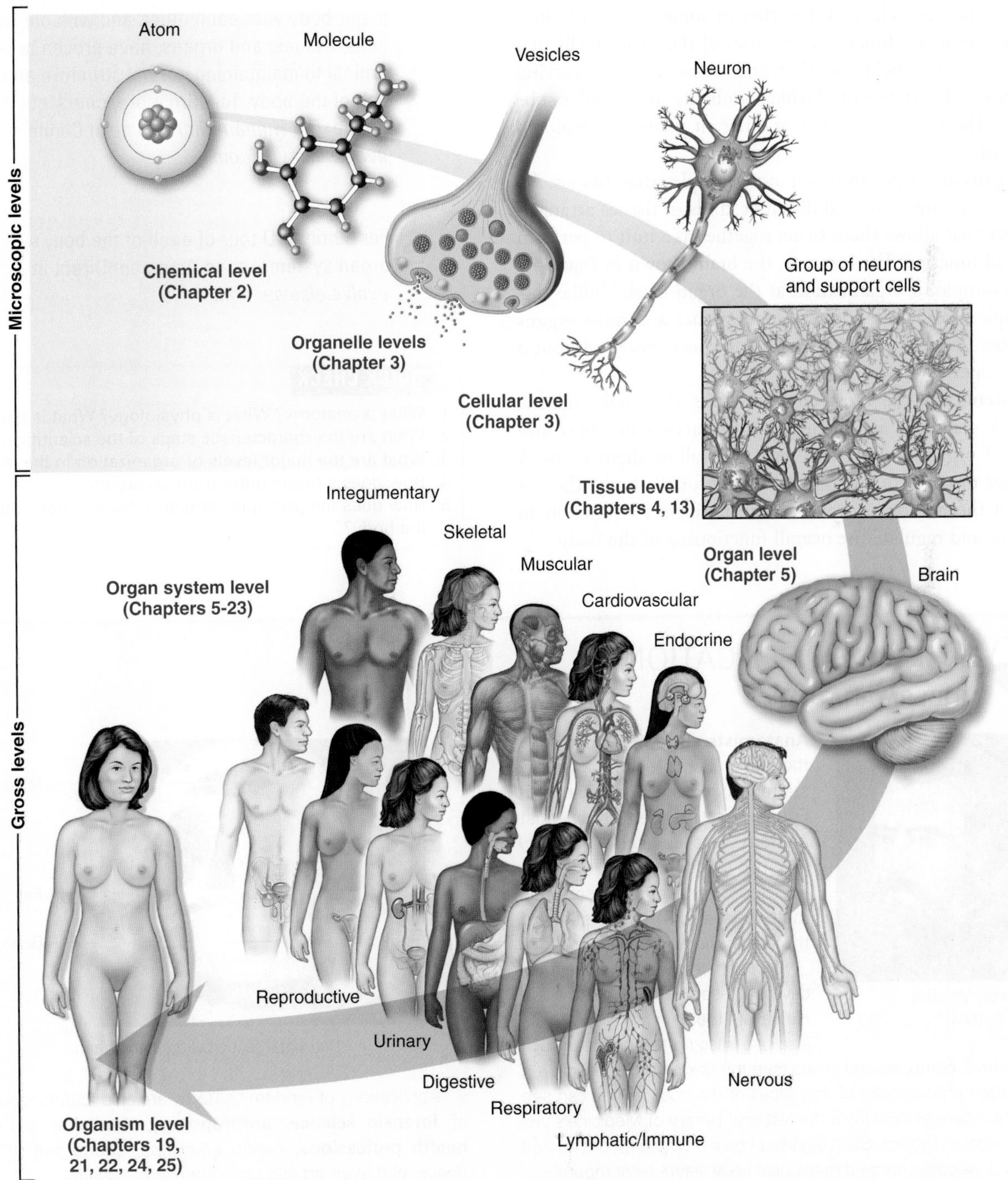

FIGURE 1-2 Levels of organization in the body. Atoms, molecules, and cells ordinarily can be seen only with a microscope, but the gross (large) structures of tissues, organs, systems, and the whole organism can be seen easily with the unaided eye.

Such information provides an understanding of the physical basis for life and for the study of the remaining levels of organization that are so important in the study of anatomy and physiology—cells, tissues, organs, and systems.

Cells are considered to be the smallest "living" units of structure and function in our bodies. Although long recognized as the simplest units of living matter, cells are far from simple. They are extremely complex, a fact you will discover in Chapter 3.

Tissues are somewhat more complex than cells. By definition a tissue is an organization of many cells that act together to perform a common function. The cells of a tissue may be of several types, but all work together in some way to produce the structural and functional qualities of the tissue. Cells of a tissue are often held together and surrounded by varying amounts and varieties of gluelike, nonliving intercellular substances. The varied properties of different tissues are explored in Chapter 4.

Organs are larger and even more complex than tissues. An organ is a group of several different kinds of tissues arranged in a way that allows them to act together as a unit to perform a special function. For instance, the brain shown in **Figure 1-2** is an example of organization at the organ level. Unlike microscopic molecules and cells, some tissues and most organs are gross (large) structures that can be seen easily without a microscope.

Systems are the most complex units that make up the body. A system is an organization of varying numbers and kinds of organs arranged in ways that allow them to work together to perform complex functions for the body. The organs of the nervous system shown in **Figure 1-2** function to monitor and regulate the overall functioning of the body.

The *body as a whole*—the human **organism**—is all the atoms, molecules, cells, tissues, organs, and systems that you will study in subsequent chapters of this text. Although capable of being dissected or broken down into many parts, the body is a unified and complex assembly of structurally and functionally interactive components, each working together to ensure healthy survival.

The many important roles of the *microbial systems* of the body, or human **microbiome,** have come to the forefront of human biology. The complex interactions of microorganisms (such as bacteria) in our body with each other, and with our own cells, tissues, and organs, have proven to be critical to maintaining normal structure and function of the body. To learn more, check out the article *The Human Microbiome* at Connect It! at *evolve.elsevier.com.*

For a brief 3-D tour of each of the body's organ systems, go to AnimationDirect at *evolve.elsevier.com.*

QUICK CHECK

1. What is anatomy? What is physiology? What is pathology?
2. What are the characteristic steps of the scientific method?
3. What are the major levels of organization in the body?
4. How does a tissue differ from an organ?
5. How does the principle "structure fits function" relate to the body?

SCIENCE APPLICATIONS

MODERN ANATOMY

Andreas Vesalius (1514–1564)

Anatomists study the structure of the human body. Modern anatomy started during the Renaissance in Europe with the Flemish scientist Andreas Vesalius *(shown at left)* and his contemporaries. Vesalius was the first to apply a scientific method (see p. 4) to the study of the human body.

Most anatomists still dissect cadavers (preserved human remains). However, today many anatomists also use imaging technologies such as x-rays, computerized scans (see the box on p. 132, and even digitized photographs of thin slices of the body as you can see in the *figure at right* from the National Library of Medicine's Visible Human Project. Such digitized images can be reconstructed into dissectible, three-dimensional body views by computers.

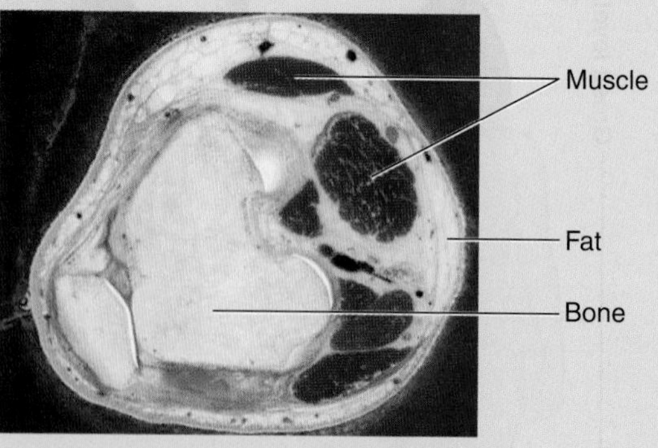

Horizontal section of the human arm

Applications of modern anatomy are also found in the fields of **forensic science, anthropology, medicine** and **allied health professions,** *health education, sports* and *athletics, dance,* and even *art* and *computerized animation.*

Anatomical Position

Discussions about the body, the way it moves, its posture, or the relationship of one area to another assume that the body as a whole is in a specific position called the **anatomical position.** In this reference position (Figure 1-3), the body is in an erect, or standing, posture with the arms at the sides and palms turned forward. The head also points forward, as do the feet, which are aligned at the toe and set slightly apart.

The broken line along the middle, or *median,* of the body demonstrates that the body has external **bilateral symmetry**—that is, the left and right sides of the body roughly mirror each other.

The anatomical position is a reference position that gives meaning to the directional terms used to describe the body parts and regions. In other words, you need to know the anatomical position so that you know how to apply **directional terms** correctly regardless of the particular position of the body being described.

Supine and **prone** are terms used to describe the position of the body when it is not in the anatomical position. In the supine position the body is lying face upward, and in the prone position the body is lying face downward.

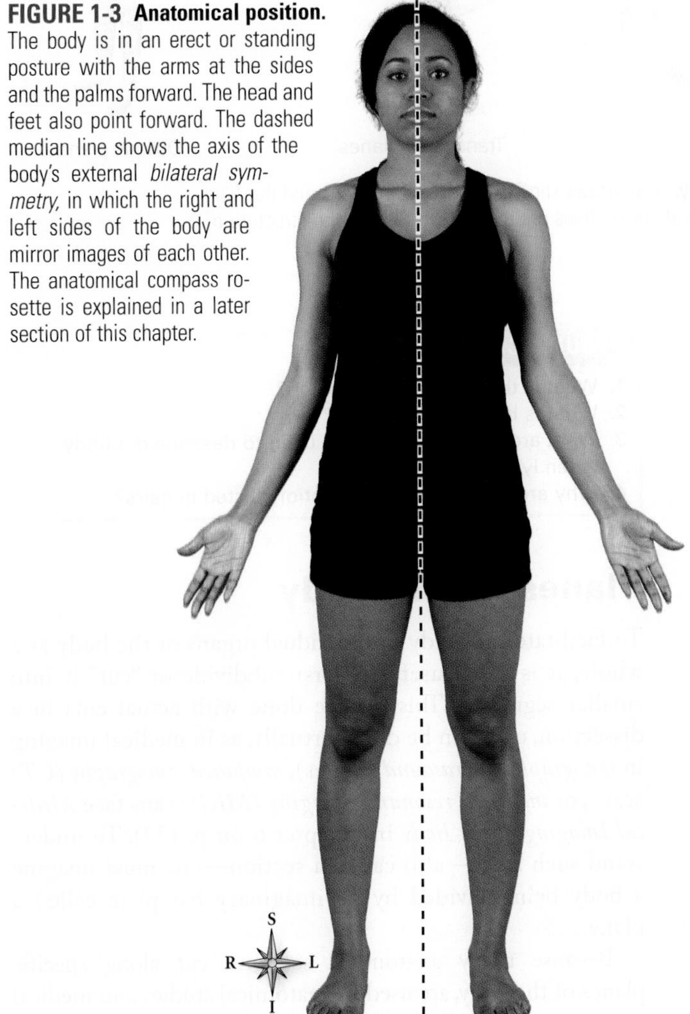

FIGURE 1-3 Anatomical position. The body is in an erect or standing posture with the arms at the sides and the palms forward. The head and feet also point forward. The dashed median line shows the axis of the body's external *bilateral symmetry,* in which the right and left sides of the body are mirror images of each other. The anatomical compass rosette is explained in a later section of this chapter.

Anatomical Directions

Directional Terms

When studying the body, it is often helpful to know where an organ is in relation to other structures. The following directional terms are used in describing relative positions of body parts. To help you understand them better, they are listed here in sets of opposite pairs:

1. **Superior** and **inferior** (Figure 1-4). *Superior* means "toward the head," and *inferior* means "toward the feet." *Superior* also means "upper" or "above," and *inferior* means "lower" or "below." For example, the lungs are located superior to the diaphragm, whereas the stomach is located inferior to the diaphragm (refer to Figure 1-8 if you are not sure where these organs are located). The simple terms *upper* and *lower* are sometimes used in professional language as well. For example, the term "upper respiratory tract" and "lower gastrointestinal tract" are used commonly by anatomists and health professionals.
2. **Anterior** and **posterior** (see Figure 1-4). *Anterior* means "front" or "in front of." *Posterior* means "back" or "in back of." In humans, who walk in an upright position, *ventral* (toward the belly) can be used in place of anterior, and *dorsal* (toward the back) can be used for posterior. For example, the nose is on the anterior surface of the body, and the shoulder blades are on its posterior surface.
3. **Medial** and **lateral** (see Figure 1-4). *Medial* means "toward the midline of the body." *Lateral* means "toward the side of the body or away from its midline." For example, the great toe is at the medial side of the foot, and the little toe is at its lateral side. The heart lies medial to the lungs, and the lungs lie lateral to the heart.
4. **Proximal** and **distal** (see Figure 1-4). *Proximal* means "toward or nearest the trunk of the body, or nearest the point of origin of one of its parts." *Distal* means "away from or farthest from the trunk or the point of origin of a body part." For example, the elbow lies at the proximal end of the forearm, whereas the hand lies at its distal end. Likewise, the distal portion of a kidney tubule is more distant from the tubule origin than is the proximal part of the kidney tubule.
5. **Superficial** and **deep.** *Superficial* means nearer the surface. *Deep* means farther away from the body surface. For example, the skin of the arm is superficial to the muscles below it, and the bone of the arm is deep to the muscles that surround and cover it.

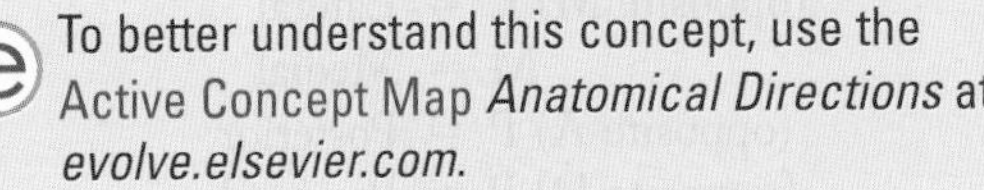
To better understand this concept, use the Active Concept Map *Anatomical Directions* at *evolve.elsevier.com.*

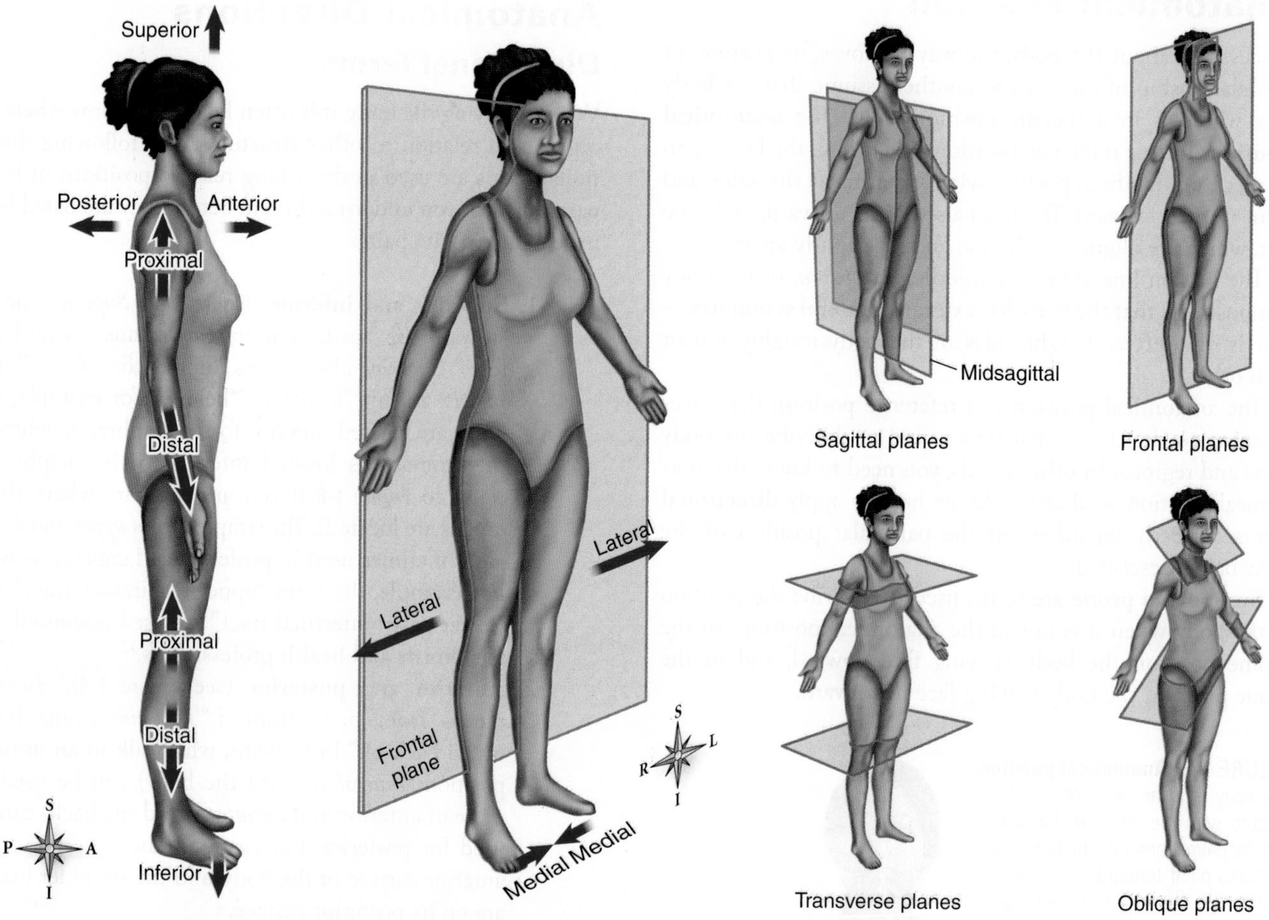

FIGURE 1-4 Directions and planes of the body. The *arrows* show anatomical directions and the *blue plates* show examples of body planes along which cuts or sections are made in visualizing the structure of the body.

Anatomical Compass Rosette

To make the reading of anatomical figures a little easier for you, we have used an anatomical compass rosette throughout this book. On many figures, you will see a small compass rosette like you might see on a geographical map. Instead of being labeled **N, S, E,** or **W,** the anatomical compass rosette is labeled with abbreviated anatomical directions.

For example, in **Figure 1-3** (p. 7), the rosette is labeled S (for *superior*) on top and I (for *inferior*) on the bottom. Notice that in **Figure 1-3** the rosette shows **R** *(right)* on the subject's right—not your right. Now look at the rosettes in **Figure 1-4** and compare them to the body positions shown.

Here are the directional abbreviations used with the rosettes in this book:

A = Anterior
D = Distal
I = Inferior
(opposite R) L = Left
(opposite M) L = Lateral
M = Medial
(opposite A) P = Posterior
(opposite D) P = Proximal
R = Right
S = Superior

QUICK CHECK

1. What is the anatomical position?
2. What is bilateral symmetry?
3. What are two terms that are used to describe the body when lying down?
4. Why are the anatomical directions listed in pairs?

Planes of the Body

To facilitate the study of individual organs or the body as a whole, it is often useful to first subdivide or "cut" it into smaller segments. This can be done with actual cuts in a dissection, or it can be done virtually, as in medical imaging in *sonography* (*ultrasound* images), *computed tomography (CT)* scans, or *magnetic resonance imaging (MRI)* scans (see *Medical Imaging of the Body* in Chapter 6 on p. 132). To understand such a cut—also called a section—one must imagine a body being divided by an imaginary flat plate called a plane.

Because many anatomical sections, cut along specific planes of the body, are used in anatomical studies and medical imaging, we describe them here. As you read the following descriptions, identify each type of plane in **Figure 1-4**.

This chapter continues on p. 10, following the Clear View insert.

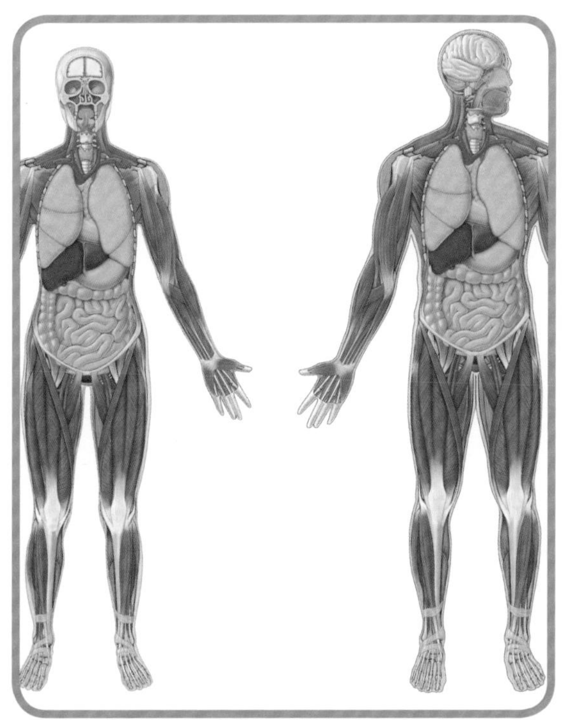

Clear View of the Human Body

Developed by
KEVIN PATTON and PAUL KRIEGER
Illustrated by
Dragonfly Media Group

Introduction

A complete understanding of human anatomy and physiology requires an appreciation for how structures within the body relate to one another. Such appreciation for anatomical structure has become especially important in the twenty-first century with the explosion in the use of diverse methods of medical imaging that rely on the ability to interpret sectional views of the human body.

The best way to develop your understanding of overall anatomical structure is to carefully dissect a large number of male and female human cadavers—then have those dissected specimens handy while reading and learning about each system of the body. Obviously, such multiple dissections and constant access to specimens are impractical for nearly everyone. However, the experience of a simple dissection can be approximated by layering several partially transparent, two-dimensional anatomical diagrams in a way that allows a student to "virtually" dissect the human body simply by paging through the layers.

This **Clear View of the Human Body** provides a handy tool for dissecting simulated male and female bodies. It also provides views of several different parts of the human body in a variety of cross sections. The many different anterior and posterior views also give you a perspective on body structure that is not available with ordinary anatomical diagrams. This Clear View is an always-available tool to help you learn the three-dimensional structure of the body in a way that allows you to see how they relate to each other in a complete body. It will always be right here in your textbook, so place a bookmark here and refer to the Clear View frequently as you study each of the systems of the human body.

Hints for Using the Clear View of the Body

1. Starting at the first page of the Clear View, slowly lift the page as you look at the anterior view of the male and female bodies. You will see deeper structures appear, as if you had dissected the body. As you lift each successive layer of images, you will be looking at deeper and deeper body structures. A key to the labels is found in the gray sidebar.
2. Starting with the second section of the Clear View, notice that you are looking at the posterior aspect of the male and female body. Lift each layer from the edge to reveal body structures in successive layers from the back to the front. This very unique view will help you understand structural relationships even better.
3. On each page of the Clear View, look at the transverse section represented in the sidebar. The section you are looking at on any one page is from the location shown in the larger diagram as a red line. In other words, if you cut the body at the red line and tilted the upper part of the body toward you, you would see what is shown in the section diagram. Notice that each section has its own labeling system that is separate from the labels used in the larger images.

KEY

1. Epicranius m.
2. Temporalis m.
3. Orbicularis oculi m.
4. Masseter m.
5. Orbicularis oris m.
6. Pectoralis major m.
7. Serratus anterior m.
8. Basilic vein
9. Brachial fascia
10. Cephalic vein
11. Rectus sheath
12. Linea alba
13. Rectus abdominis m.
14. Umbilicus
15. Abdominal oblique m., external
16. Abdominal oblique m., internal
17. Transverse abdominis m.
18. Inguinal ring, external
19. Fossa ovalis
20. Fascia of the thigh
21. Great saphenous vein
22. Parietal bone
23. Frontal bone
24. Temporal bone
25. Zygomatic bone
26. Maxilla
27. Mandible
28. Sternocleidomastoid m.
29. Sternohyoid muscle
30. Omohyoid muscle
31. Deltoid m.
32. Pectoralis minor m.
33. Sternum
34. Rib (costal) cartilage
35. Rib
36. Greater omentum
37. Frontal lobe
38. Parietal lobe
39. Temporal lobe
40. Cerebellum
41. Nasal septum
42. Brachiocephalic vein
43. Superior vena cava
44. Thymus gland
45. Right lung
46. Left lung
47. Pericardium
48. Liver
49. Gallbladder
50. Stomach
51. Transverse colon
52. Small intestines
53. Biceps brachii m.
54. Brachioradialis m.
55. Adductor longus m.
56. Sartorius m.
57. Quadriceps femoris m.
58. Patellar ligament
59. Tibialis anterior m.
60. Sup. extensor retinaculum
61. Inf. extensor retinaculum
62. Cerebrum of brain
63. Cerebellum
64. Brainstem
65. Maxillary sinus
66. Nasal cavity
67. Tongue
68. Thyroid gland
69. Heart
70. Hepatic veins
71. Esophagus
72. Spleen
73. Celiac artery
74. Portal vein
75. Duodenum
76. Pancreas
77. Mesenteric artery
78. Ascending colon
79. Transverse colon
80. Descending colon
81. Sigmoid colon
82. Mesentery
83. Appendix
84. Inguinal ligament
85. Pubic symphysis
86. Extensor carpi radialis m.
87. Pronator teres m.
88. Flexor carpi radialis m.
89. Flexor digitorum profundus m.
90. Quadriceps femoris m.
91. Extensor digitorum longus m.
92. Thyroid cartilage
93. Trachea
94. Aortic arch
95. Right lung
96. Left lung
97. Pulmonary artery
98. Right atrium
99. Right ventricle
100. Left atrium
101. Left ventricle
102. Coracobrachialis m.
103. Inferior vena cava
104. Descending aorta
105. Right kidney
106. Left kidney
107. Right ureter
108. Rectum
109. Urinary bladder
110. Prostate gland
111. Iliac artery and vein
112. Uterus
113. Parietal bone
114. Frontal sinus
115. Sphenoidal sinus
116. Occipital bone
117. Palatine process
118. Cervical vertebrae
119. Corpus callosum
120. Thalamus
121. Trapezius m.
122. Acromion process
123. Coracoid process
124. Humerus
125. Subscapularis m.
126. Deltoid m. (cut)
127. Triceps m.
128. Brachialis m.
129. Brachioradialis m.
130. Radius
131. Ulna
132. Diaphragm
133. Thoracic duct
134. Quadratus lumborum m.
135. Psoas m.
136. Lumbar vertebrae
137. Iliacus m.
138. Gluteus medius m.
139. Iliofemoral ligament
140. Sacral nerves
141. Sacrum
142. Coccyx
143. Femur
144. Vastus lateralis m.
145. Femoral artery and vein
146. Adductor magnus m.
147. Patella
148. Fibula
149. Tibia
150. Fibularis longus m.
151. Spinal cord
152. Nerve root
153. Platysma m.
154. Splenius capitis m.
155. Levator scapulae m.
156. Rhomboideus m.
157. Infraspinatus m.
158. Teres major m.
159. Lumbodorsal fascia
160. Erector spinae m.
161. Serratus post. inf. m.
162. Latissimus dorsi m.
163. Gluteus medius m.
164. Gluteus maximus m.
165. Iliotibial tract
166. Flexor carpi ulnaris m.
167. Extensor carpi ulnaris m.
168. Extensor digitorum m.
169. Carpal ligament, dorsal
170. Interosseous m.
171. Gluteus minimus m.
172. Piriformis m.
173. Gemellus sup. m.
174. Obturator internus m.
175. Gemellus inf. m.
176. Quadratus femoris m.
177. Biceps femoris m.
178. Gastrocnemius m.
179. Calcaneal (Achilles) tendon
180. Calcaneus bone
181. Subcutaneous fat
182. Corpus spongiosum
183. Corpora cavernosa
184. Umbilical ligaments
185. Epigastric artery and vein
186. Right testis
187. Transverse thoracic m.
188. Parietal pleura
189. Common bile duct
190. Lesser omentum
191. Flexor digitorum profundus
192. Epiglottis

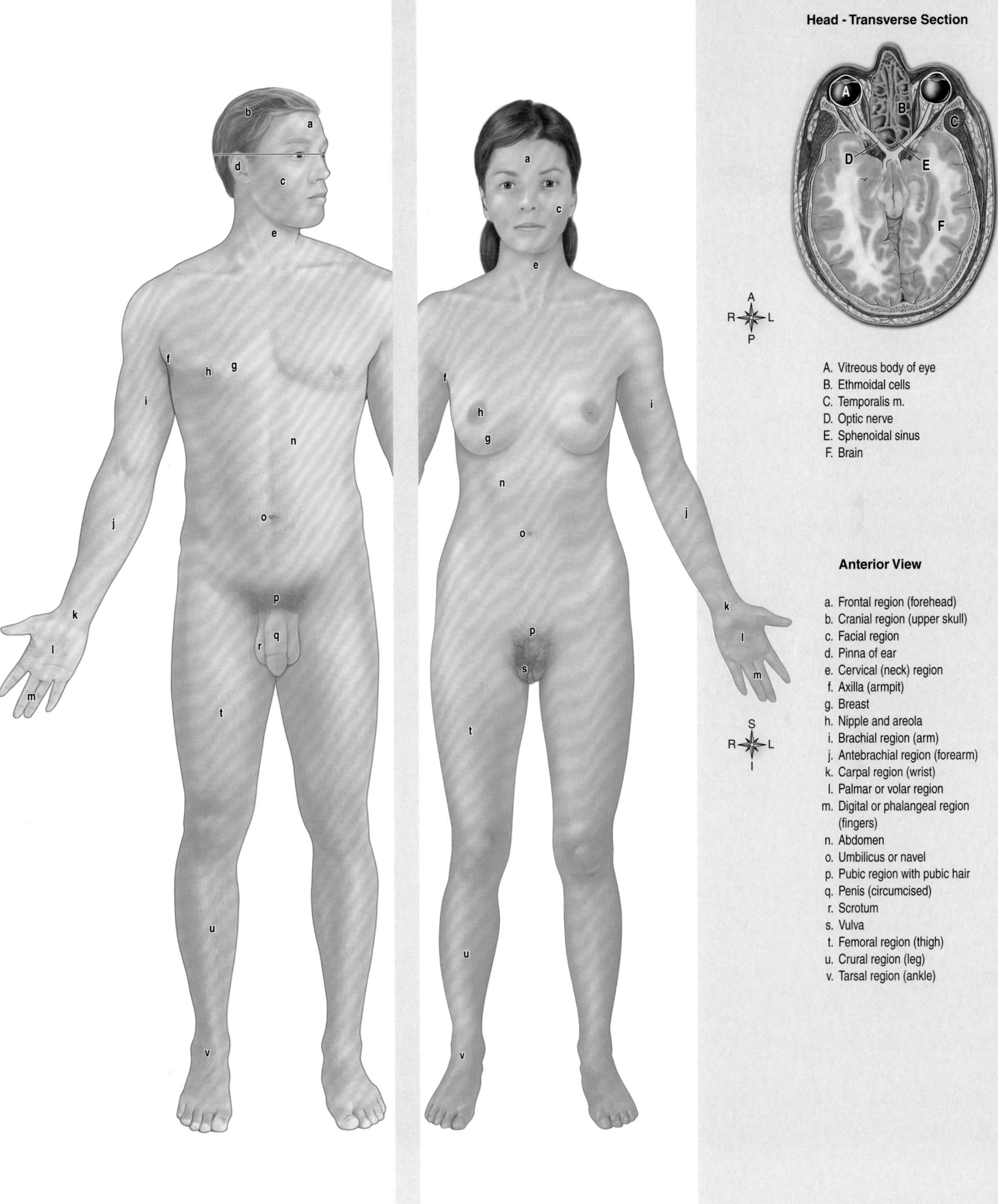
Head - Transverse Section
A. Vitreous body of eye
B. Ethmoidal cells
C. Temporalis m.
D. Optic nerve
E. Sphenoidal sinus
F. Brain
Anterior View
a. Frontal region (forehead)
b. Cranial region (upper skull)
c. Facial region
d. Pinna of ear
e. Cervical (neck) region
f. Axilla (armpit)
g. Breast
h. Nipple and areola
i. Brachial region (arm)
j. Antebrachial region (forearm)
k. Carpal region (wrist)
l. Palmar or volar region
m. Digital or phalangeal region (fingers)
n. Abdomen
o. Umbilicus or navel
p. Pubic region with pubic hair
q. Penis (circumcised)
r. Scrotum
s. Vulva
t. Femoral region (thigh)
u. Crural region (leg)
v. Tarsal region (ankle)

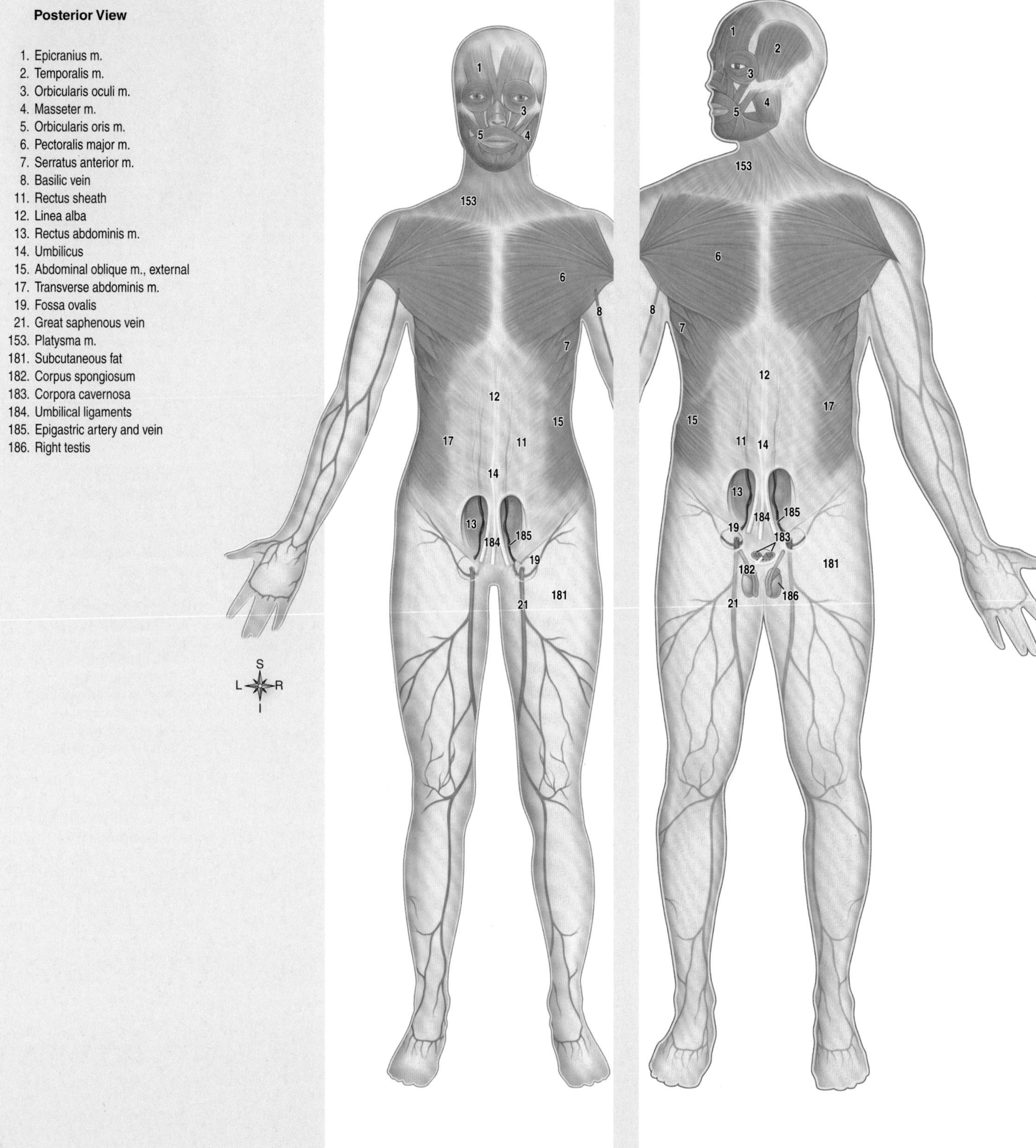
Posterior View
1. Epicranius m.
2. Temporalis m.
3. Orbicularis oculi m.
4. Masseter m.
5. Orbicularis oris m.
6. Pectoralis major m.
7. Serratus anterior m.
8. Basilic vein
11. Rectus sheath
12. Linea alba
13. Rectus abdominis m.
14. Umbilicus
15. Abdominal oblique m., external
17. Transverse abdominis m.
19. Fossa ovalis
21. Great saphenous vein
153. Platysma m.
181. Subcutaneous fat
182. Corpus spongiosum
183. Corpora cavernosa
184. Umbilical ligaments
185. Epigastric artery and vein
186. Right testis
S
L
R
I
1
3
5
4
153
6
8
7
12
15
17
11
14
13
184
185
19
181
21
1
2
3
4
5
153
6
8
7
12
17
15
11
14
13
184
185
19
183
182
181
186
21

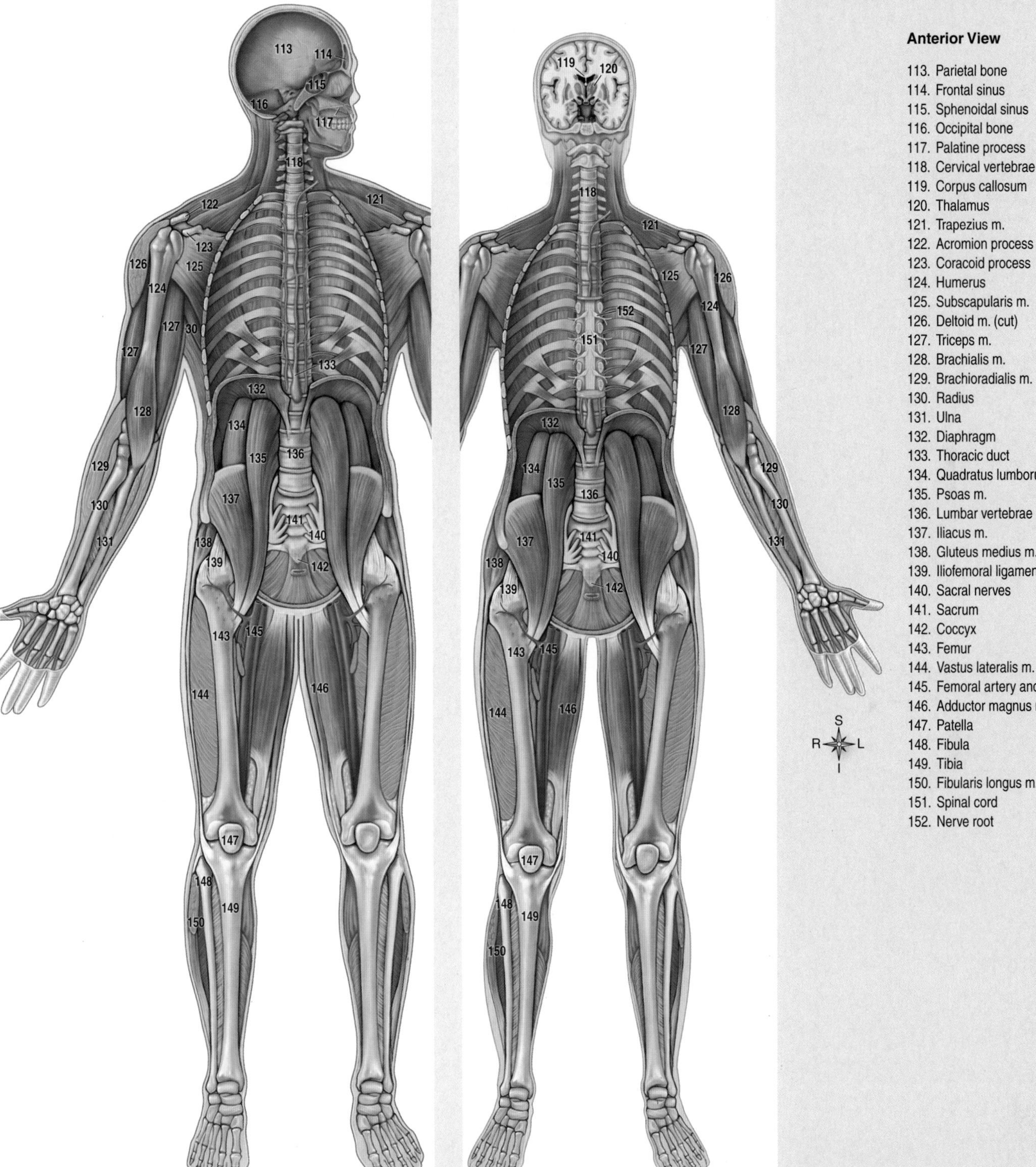

Anterior View

113. Parietal bone
114. Frontal sinus
115. Sphenoidal sinus
116. Occipital bone
117. Palatine process
118. Cervical vertebrae
119. Corpus callosum
120. Thalamus
121. Trapezius m.
122. Acromion process
123. Coracoid process
124. Humerus
125. Subscapularis m.
126. Deltoid m. (cut)
127. Triceps m.
128. Brachialis m.
129. Brachioradialis m.
130. Radius
131. Ulna
132. Diaphragm
133. Thoracic duct
134. Quadratus lumborum m.
135. Psoas m.
136. Lumbar vertebrae
137. Iliacus m.
138. Gluteus medius m.
139. Iliofemoral ligament
140. Sacral nerves
141. Sacrum
142. Coccyx
143. Femur
144. Vastus lateralis m.
145. Femoral artery and vein
146. Adductor magnus m.
147. Patella
148. Fibula
149. Tibia
150. Fibularis longus m.
151. Spinal cord
152. Nerve root

Posterior View

b. Cranial region (upper skull)
c. Facial region
d. Pinna of ear
e. Cervical (neck) region
f. Axilla (armpit)
i. Brachial region (arm)
j. Antebrachial region (forearm)
k. Carpal region (wrist)
m. Digital or phalangeal region (fingers)
t. Femoral region (thigh)
u. Crural region (leg)
v. Tarsal region (ankle)
w. Olecranal (back of elbow)
x. Dorsal region (back)
y. Gluteal region (buttock)
z. Popliteal region (back of knee)
aa. Plantar region (sole)

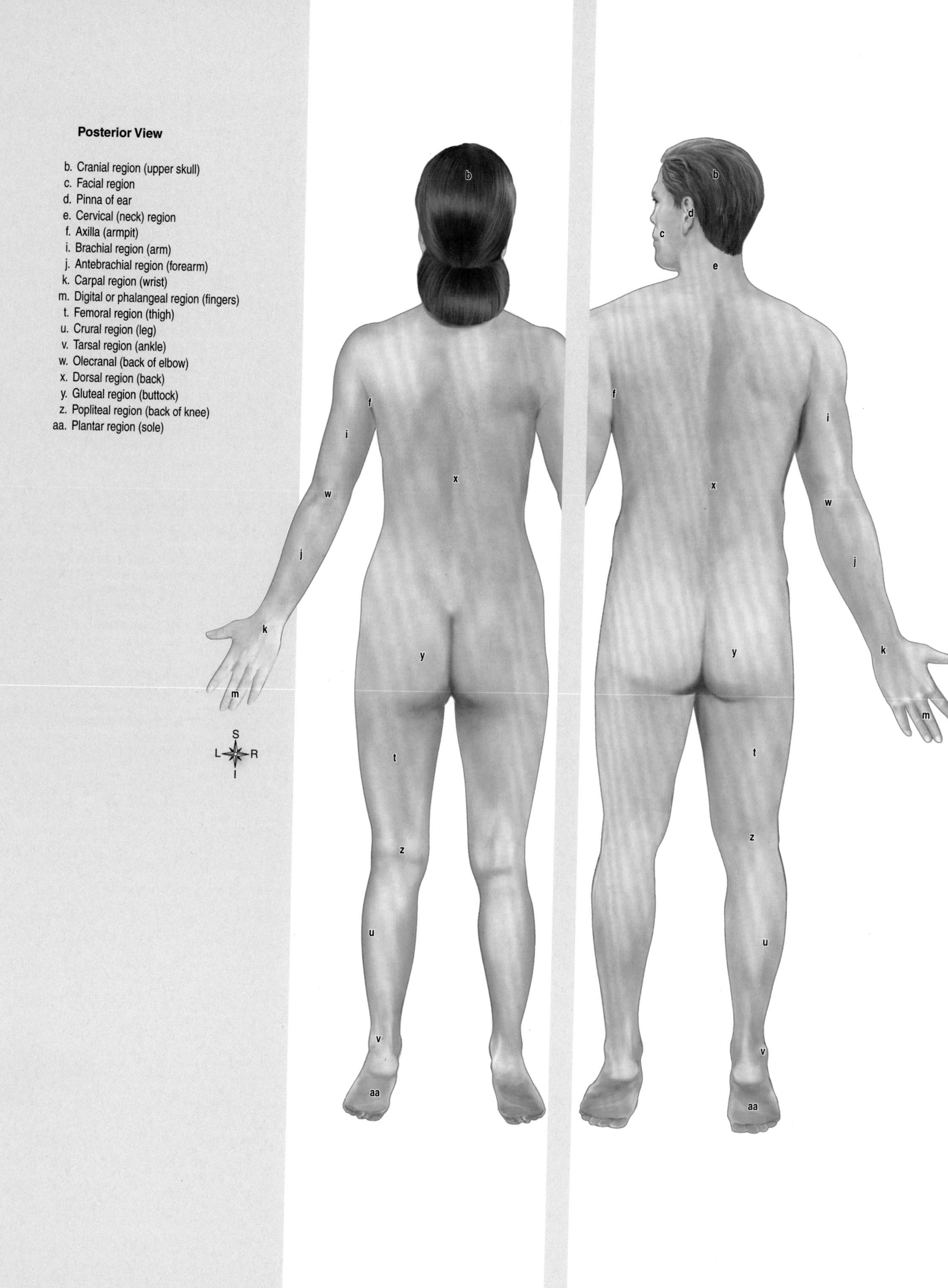

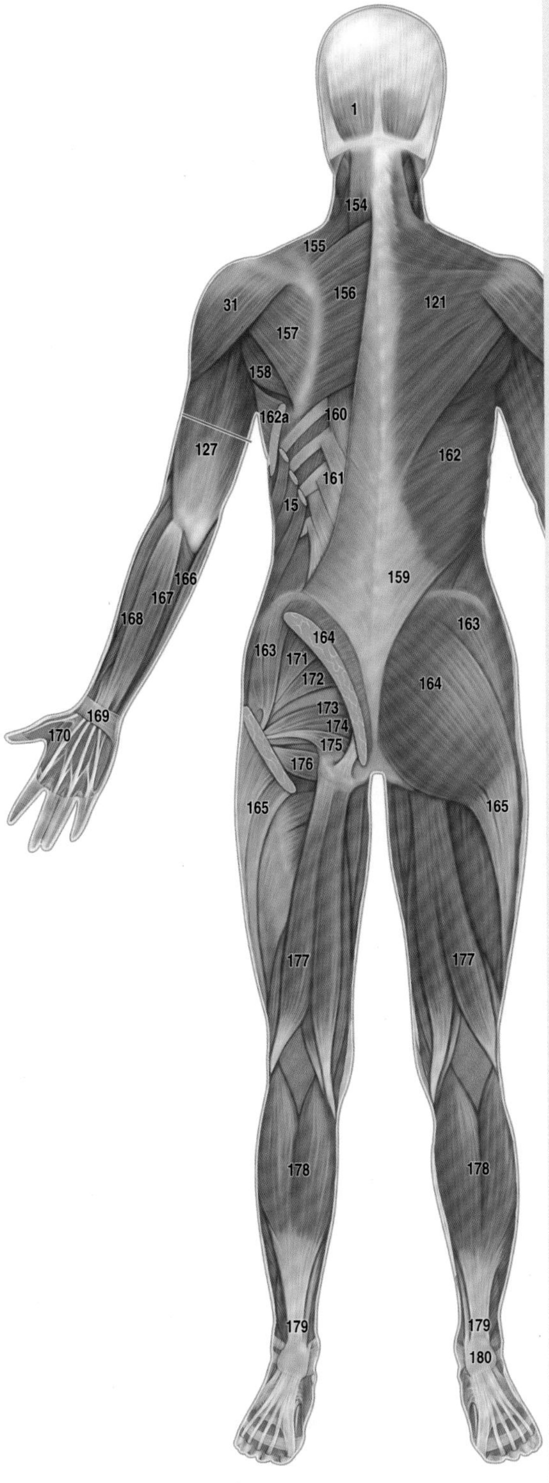

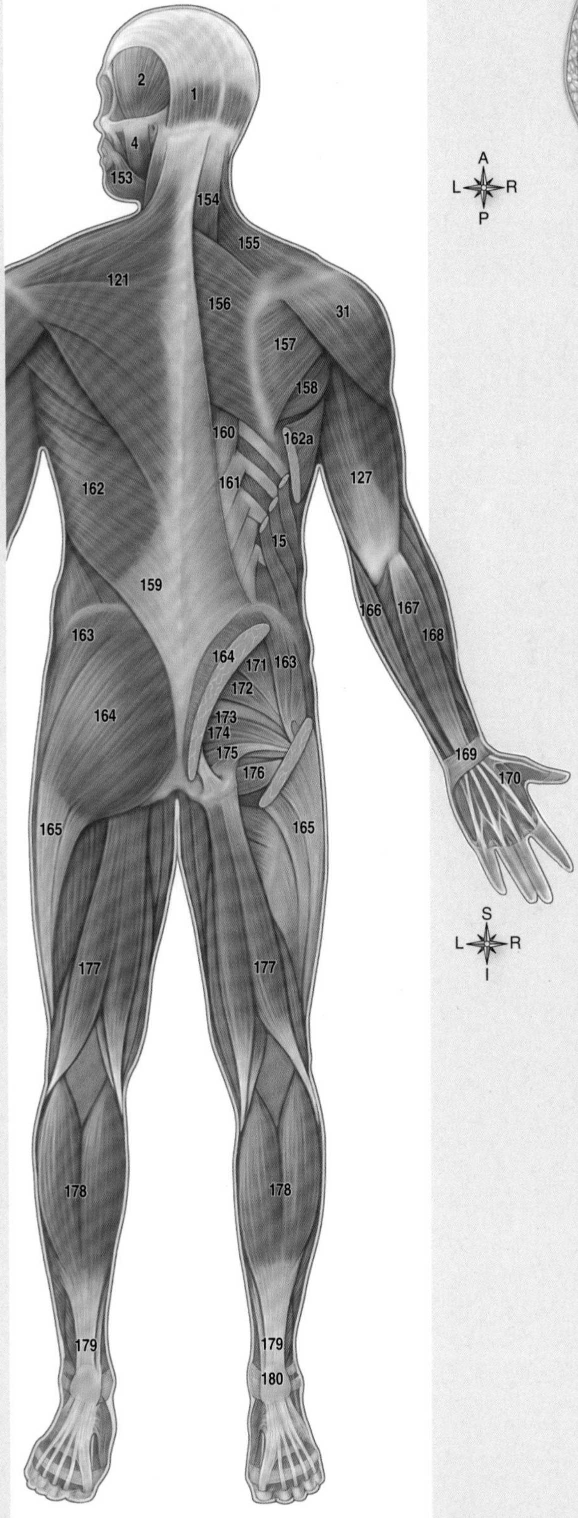

Upper Arm - Transverse Section

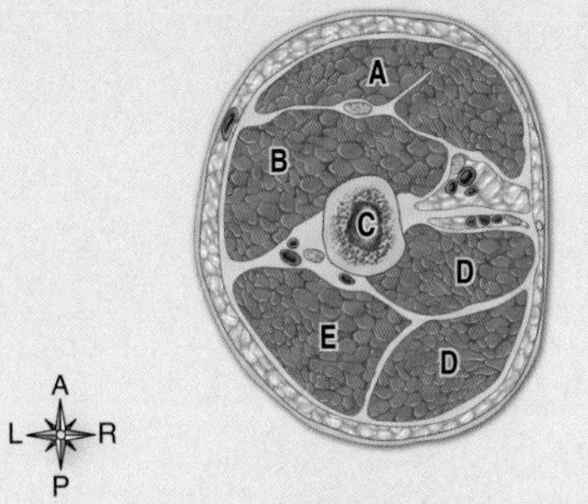

A. Biceps brachii m.
B. Brachialis m.
C. Humerus
D. Triceps brachii m., medial
E. Triceps brachii m., lateral

Posterior View

1. Epicranius m.
2. Temporalis m.
4. Masseter m.
15. Abdominal oblique m., external
31. Deltoid m.
121. Trapezius m.
127. Triceps m.
153. Platysma m.
154. Splenius capitis m.
155. Levator scapulae m.
156. Rhomboideus m.
157. Infraspinatus m.
158. Teres major m.
159. Lumbodorsal fascia
160. Erector spinae m.
161. Serratus post. inf. m.
162. Latissimus dorsi m.
162a. Latissimus dorsi m. (cut)
163. Gluteus medius m.
164. Gluteus maximus m.
165. Iliotibial tract
166. Flexor carpi ulnaris m.
167. Extensor carpi ulnaris m.
168. Extensor digitorum m.
169. Carpal ligament, dorsal
170. Interosseous m.
171. Gluteus minimus m.
172. Piriformis m.
173. Gemellus sup. m.
174. Obturator internus m.
175. Gemellus inf. m.
176. Quadratus femoris m.
177. Biceps femoris m.
178. Gastrocnemius m.
179. Calcaneal (Achilles) tendon
180. Calcaneus bone

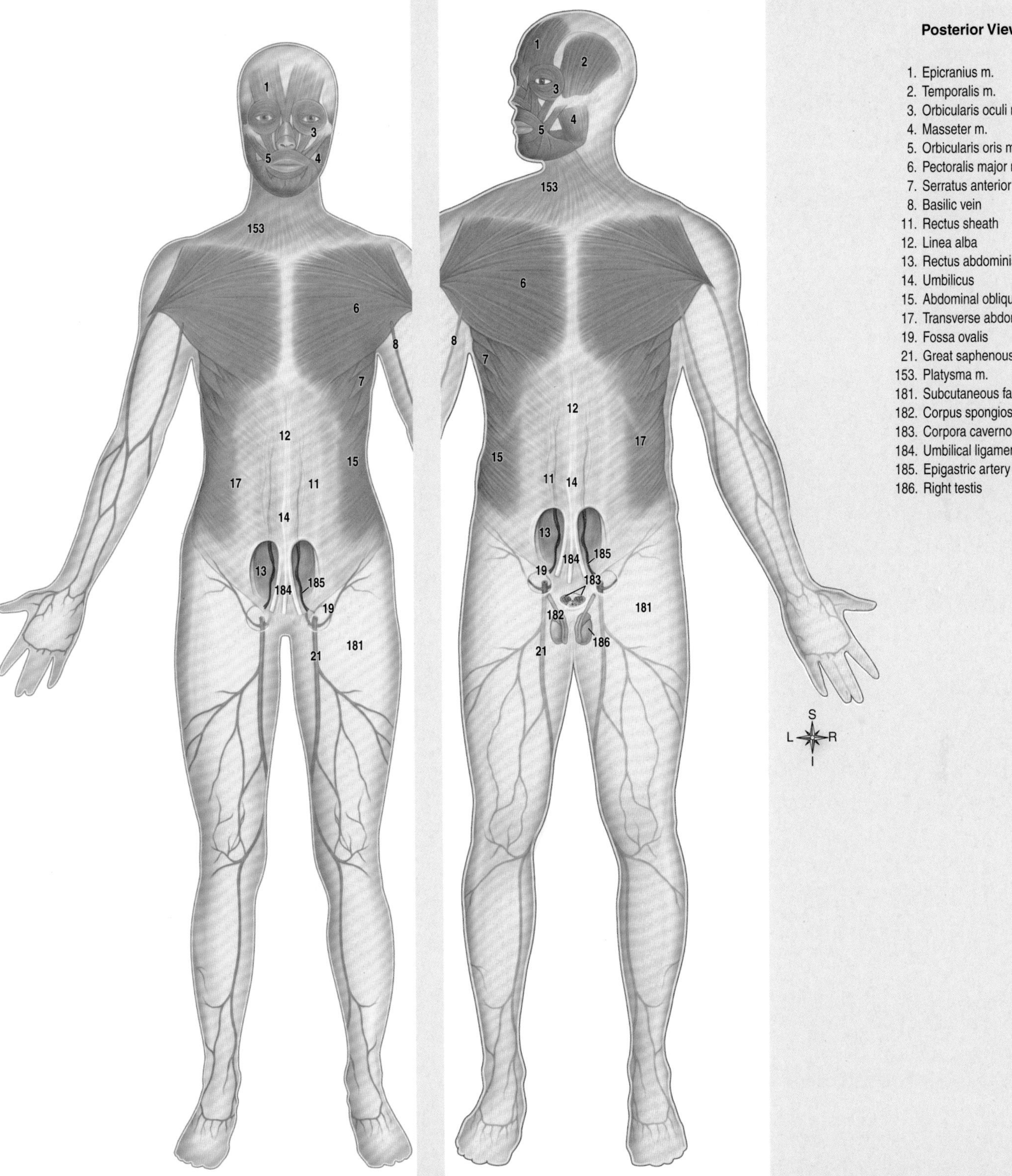

Posterior View

1. Epicranius m.
2. Temporalis m.
3. Orbicularis oculi m.
4. Masseter m.
5. Orbicularis oris m.
6. Pectoralis major m.
7. Serratus anterior m.
8. Basilic vein
11. Rectus sheath
12. Linea alba
13. Rectus abdominis m.
14. Umbilicus
15. Abdominal oblique m., external
17. Transverse abdominis m.
19. Fossa ovalis
21. Great saphenous vein
153. Platysma m.
181. Subcutaneous fat
182. Corpus spongiosum
183. Corpora cavernosa
184. Umbilical ligaments
185. Epigastric artery and vein
186. Right testis

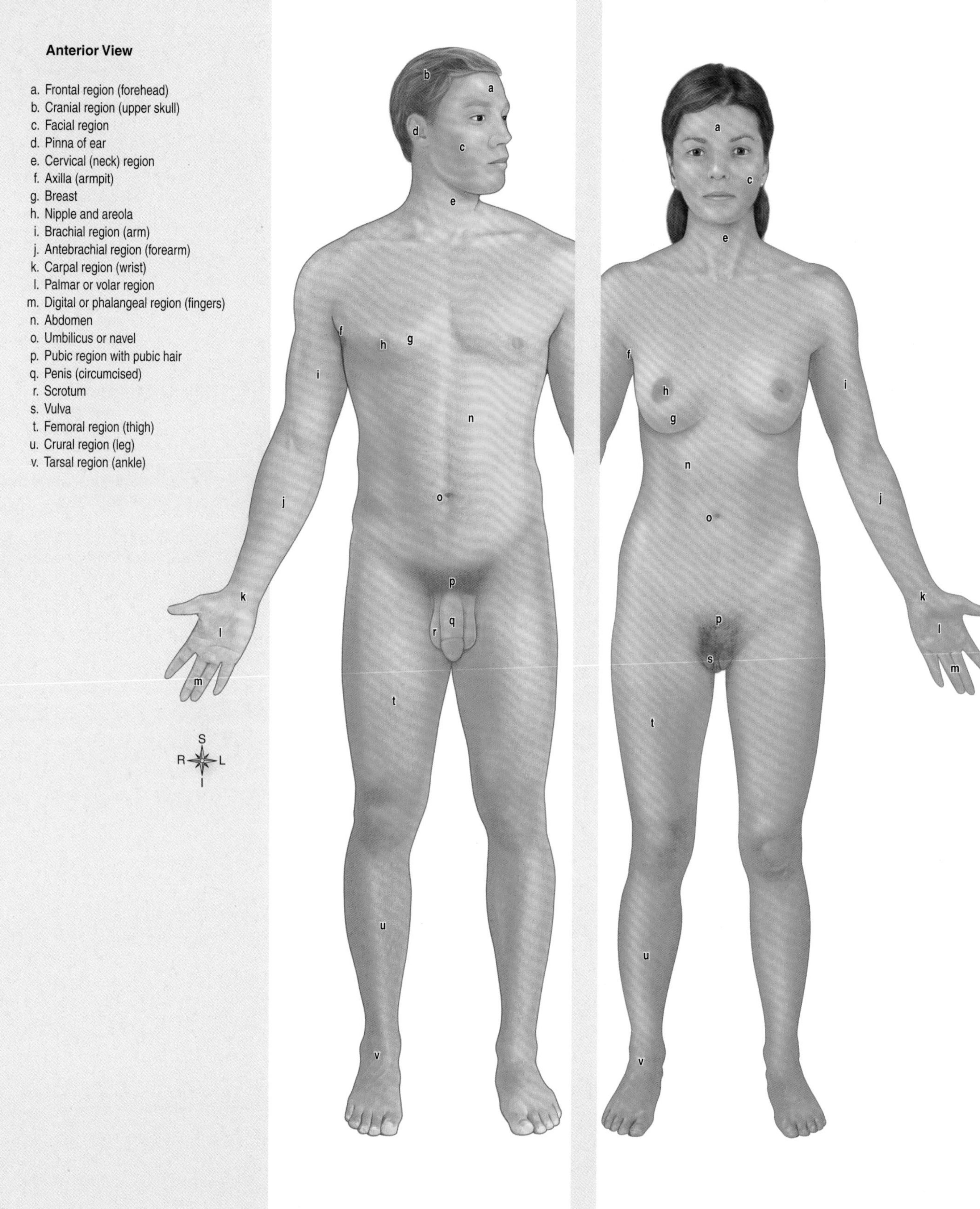
Anterior View
a. Frontal region (forehead)
b. Cranial region (upper skull)
c. Facial region
d. Pinna of ear
e. Cervical (neck) region
f. Axilla (armpit)
g. Breast
h. Nipple and areola
i. Brachial region (arm)
j. Antebrachial region (forearm)
k. Carpal region (wrist)
l. Palmar or volar region
m. Digital or phalangeal region (fingers)
n. Abdomen
o. Umbilicus or navel
p. Pubic region with pubic hair
q. Penis (circumcised)
r. Scrotum
s. Vulva
t. Femoral region (thigh)
u. Crural region (leg)
v. Tarsal region (ankle)
S
R
L
I

1. **Sagittal plane**—a sagittal cut or section that runs along a lengthwise plane running from anterior to posterior. It divides the body or any of its parts into right and left sides. The **midsagittal plane** shown in **Figure 1-4** is a unique type of sagittal plane that divides the body into two *equal halves.*
2. **Frontal plane**—a frontal plane *(coronal plane)* is a lengthwise plane running from side to side. As you can see in **Figure 1-4**, a frontal plane divides the body or any of its parts into anterior and posterior (front and back) portions.
3. **Transverse plane**—a transverse plane is a crosswise or *horizontal plane.* Such a plane (see **Figure 1-4**) divides the body or any of its parts into superior and inferior portions.

Understanding planes of the body is essential to being able to interpret medical images.

Sometimes it is helpful to make a cut along a plane that is not parallel to the planes we have already mentioned. Such diagonal cuts are made along **oblique planes,** which you can see illustrated in **Figure 1-4**.

Besides using planes to cut the body into various sections, we sometimes use planes to describe movement. For example, one rotates the head in a transverse plane, and one can move a finger along both a sagittal plane and along a frontal plane.

Explore the *Clear View of the Human Body* insert located just prior to this page. Note that the larger transparency images show the body and its organs sectioned along frontal planes. However, the smaller images in the margins show transverse sections at specific locations in the body.

To explore some of the major types of medical images that use sectional views of the body, see the article *Medical Imaging of the Body* at Connect It! at *evolve.elsevier.com*.

Body Cavities

Contrary to its external appearance, the body is not a solid structure. It is made up of open spaces or cavities that in turn contain compact, well-ordered arrangements of internal organs. The major body cavities are categorized as the **dorsal body cavities** and **ventral body cavities.** The location and outlines of the major body cavities are illustrated in **Figure 1-5**.

Dorsal Cavities

The dorsal cavities shown in **Figure 1-5** include the space inside the skull that contains the brain. It is called the **cranial cavity.** The space inside the spinal column is called the **spinal cavity.** It contains the spinal cord. The cranial and spinal cavities are *dorsal cavities because they are located in a dorsal position in the body.*

Ventral Cavities

The *ventral cavities* are located in a ventral position in the body.

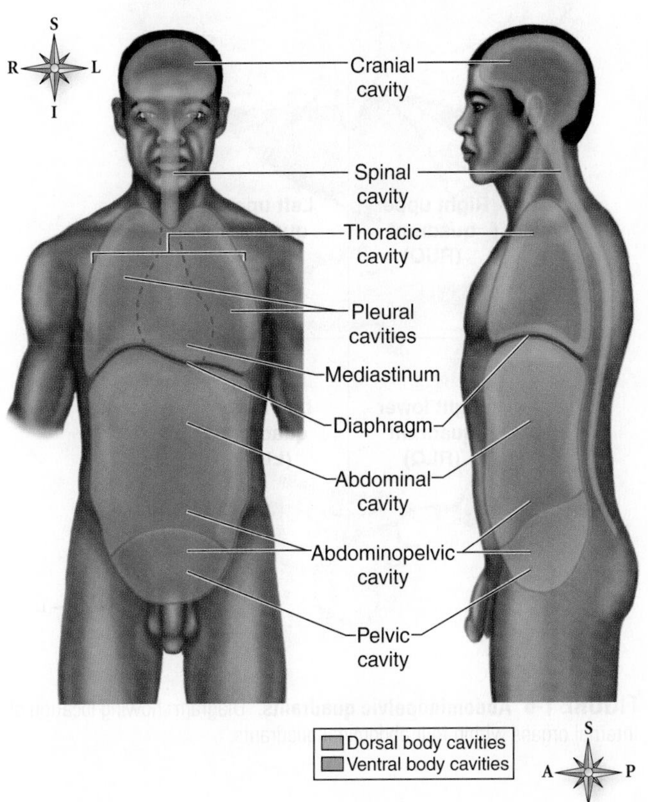

FIGURE 1-5 Body cavities. Location and subdivisions of the dorsal and ventral body cavities as viewed from the front (anterior) and from the side (lateral).

Thoracic and Abdominopelvic Cavities

The upper ventral cavities include the **thoracic cavity,** a space that you may think of as your chest cavity. Its midportion is a subdivision of the thoracic cavity, called the **mediastinum.** The lateral subdivisions of the thoracic cavity are called the right and left **pleural cavities.**

The lower ventral cavities in **Figure 1-5** include an **abdominal cavity** and a **pelvic cavity.** Actually, they form only one cavity, the **abdominopelvic cavity,** because no physical partition separates them. In **Figure 1-5** a faint line shows the approximate point of separation between the abdominal and pelvic subdivisions. Notice, however, that an actual physical partition separates the thoracic cavity above from the abdominopelvic cavity below. This muscular sheet is the **diaphragm.** It is dome-shaped and is the most important muscle for breathing.

Abdominopelvic Quadrants and Regions

Abdominopelvic Quadrants

To make it easier to locate organs in the large abdominopelvic cavity, anatomists have divided the abdominopelvic cavity into four **abdominopelvic quadrants:**

1. *Right upper quadrant* or *RUQ* (right superior quadrant)
2. *Right lower quadrant* or *RLQ* (right inferior quadrant)
3. *Left upper quadrant* or *LUQ* (left superior quadrant)
4. *Left lower quadrant* or *LLQ* (left inferior quadrant)

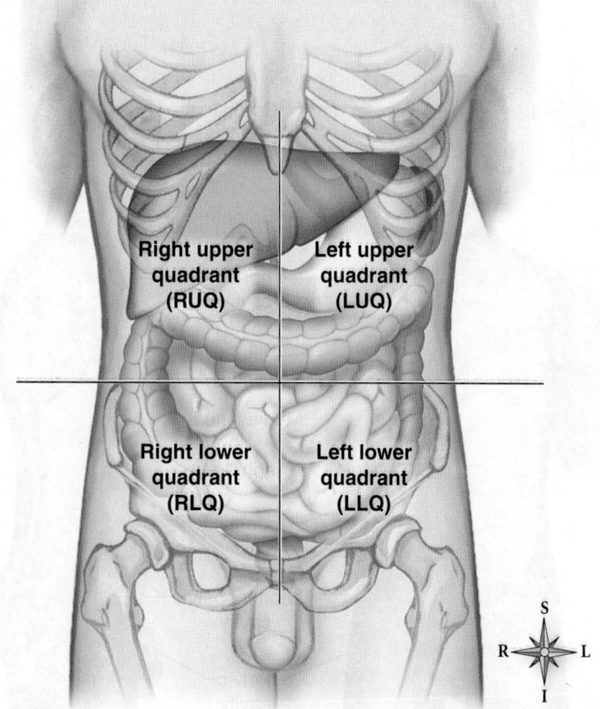

FIGURE 1-6 Abdominopelvic quadrants. Diagram showing location of internal organs within four abdominal quadrants.

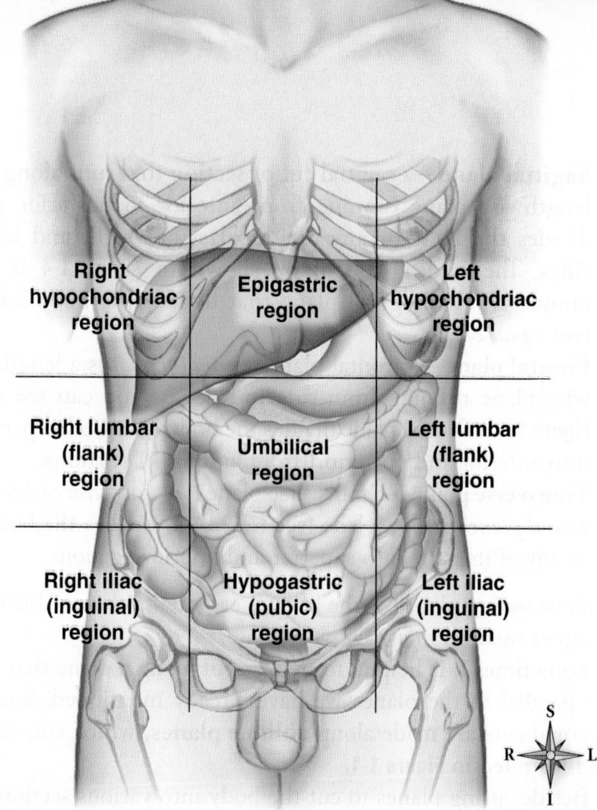

FIGURE 1-7 Abdominopelvic regions. The most superficial organs are shown. Look at **Figure 1-8** (p. ***)—can you identify the deeper structures in each region?

As you can see in **Figure 1-6**, the midsagittal and transverse planes, which were described in the previous section, pass through the navel (umbilicus) and divide the abdominopelvic region into the four quadrants. This method of subdividing the abdominopelvic cavity is frequently used by health professionals and is useful for locating the origin of pain or describing the location of a tumor or other abnormality.

You may notice that terms like *upper* and *lower* are often used to name quadrants, which may seem overly informal compared with the more technical terms *superior* and *inferior*. However, this practice reflects the usage found in many clinical environments, where one commonly encounters a mix of informal and formal terminology.

Abdominopelvic Regions

Another and perhaps more precise way to divide the abdominopelvic cavity is shown in **Figure 1-7**. Here, the abdominopelvic cavity is subdivided into nine **abdominopelvic regions** defined as follows:

1. *Upper abdominopelvic regions*—the **right hypochondriac region, left hypochondriac region,** and the **epigastric region** lie above an imaginary line across the abdomen at the level of the ninth rib cartilages.
2. *Middle abdominopelvic regions*—the **right lumbar region, left lumbar region,** and the **umbilical region** lie below an imaginary line across the abdomen at the level of the ninth rib cartilages and above an imaginary line across the abdomen at the top of the hip bones.
3. *Lower abdominopelvic regions*—the **right iliac region, left iliac region** (also called **inguinal regions**), and the **hypogastric region** lie below an imaginary line across the abdomen at the level of the top of the hip bones.

Some of the organs in the largest body cavities are visible in **Figure 1-8** and are listed in **Table 1-1**. Find each body cavity in a model of the human body if you have access to one. Try

TABLE 1-1 Body Cavities

BODY CAVITIES	ORGAN(S)
Dorsal Body Cavities	
Cranial cavity	Brain
Spinal cavity	Spinal cord
Ventral Body Cavities	
Thoracic Cavity	
Mediastinum	Heart, trachea, esophagus, thymus, blood vessels
Pleural cavities	Lungs
Abdominopelvic Cavity	
Abdominal cavity	Liver, gallbladder, stomach, spleen, pancreas, small intestine, parts of large intestine
Pelvic cavity	Lower (sigmoid) colon, rectum, urinary bladder, reproductive organs

to identify the organs in each cavity, and try to visualize their locations in your own body. Study **Figure 1-5** and **Figure 1-8** and explore the layers of the *Clear View of the Human Body* insert located in this book after p. 8.

QUICK CHECK

1. What is meant by a *section* of the body?
2. What are the two major sets of cavities of the body?
3. What is the difference between the *abdominal cavity* and the *abdominopelvic cavity*?
4. What is the difference between *right upper quadrant* and *right superior quadrant*?

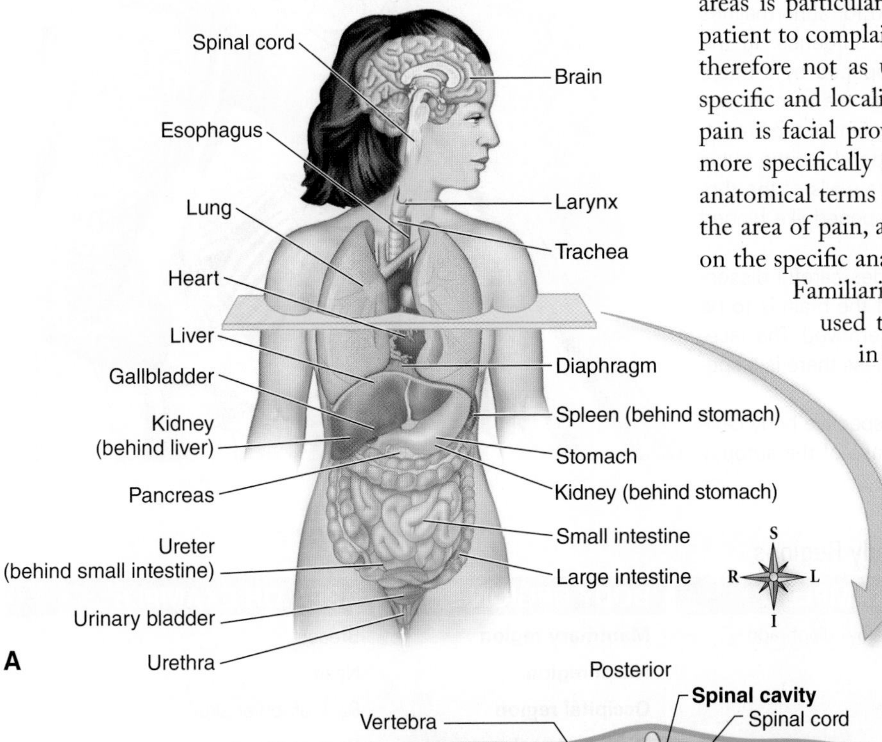

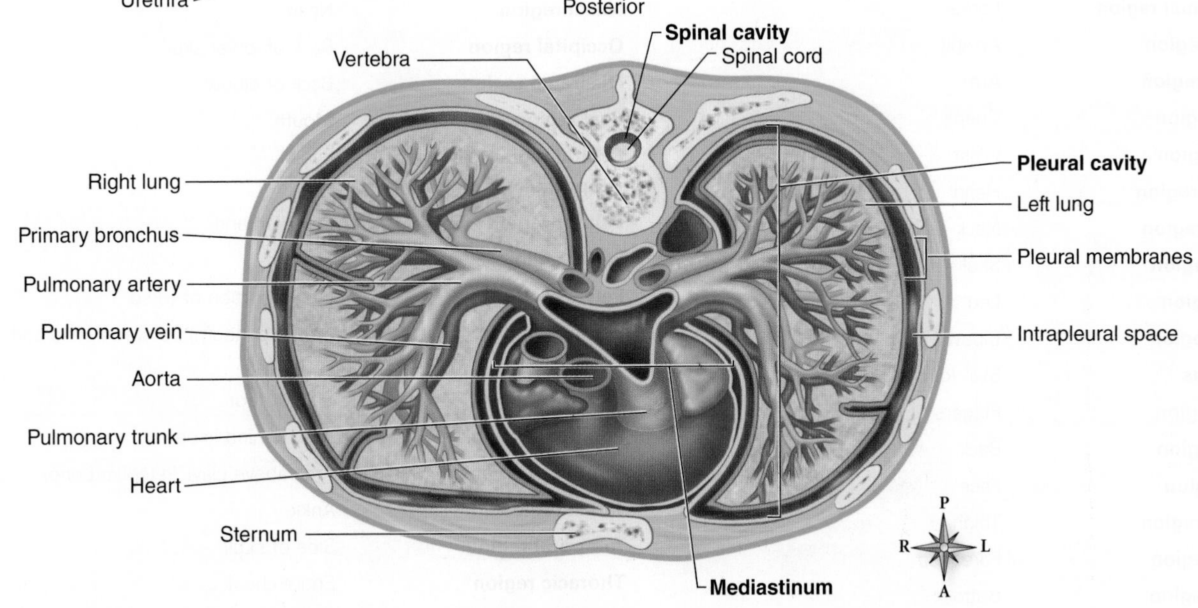

FIGURE 1-8 Organs of the major body cavities. A, A view from the front. **B,** Transverse section viewed from above.

Body Regions

To recognize an object, you usually first notice its overall structure and form. For example, a car is recognized as a car before the specific details of its tires, grill, or wheel covers are noted. Recognition of the human form also occurs as you first identify overall shape and basic outline. However, for more specific identification to occur, details of size, shape, and appearance of individual body areas must be described. Individuals differ in overall appearance because specific body areas such as the face or torso have unique identifying characteristics. Detailed descriptions of the human form require that specific regions be identified and appropriate terms be used to describe them.

The ability to identify and correctly describe specific body areas is particularly important in the health sciences. For a patient to complain of pain in the head is not as specific, and therefore not as useful to a health professional, as a more specific and localized description would be. Saying that the pain is facial provides additional information and helps to more specifically identify the area of pain. By using correct anatomical terms such as forehead, cheek, or chin to describe the area of pain, attention can be focused even more quickly on the specific anatomical area that may need attention.

Familiarize yourself with the more common terms used to describe specific body regions identified in **Figure 1-9** and listed in **Table 1-2**. Explore the *Clear View of the Human Body* insert located in this book after p. 8 to find the major body regions.

CLINICAL APPLICATION

AUTOPSY

Knowledge of human anatomy is important in conducting an **autopsy** or postmortem examination. The term *autopsy* is built from the words *auto* (meaning "self") and *opsis* (meaning "view"). Autopsies are procedures in which a human body is examined after death to accurately determine the cause of death, to confirm the accuracy of diagnostic tests, to discover previously undetected problems, and to assess the effectiveness of surgeries or other treatments. Medical and allied health students often attend autopsies to improve their knowledge of human anatomy and pathology.

Autopsies are usually performed in three stages. In the first stage, the exterior of the body is examined for abnormalities such as wounds or scars from injuries or surgeries. In the second stage, the ventral body cavity is opened by a deep, Y-shaped incision. The arms of the Y start at the anterior surface of the shoulders and join at the inferior point of the breastbone (sternum) to form a single cut that extends to the pubic area. After the rib cage is sawn through, the walls of the thoracic and abdominopelvic cavities can be opened like hinged doors to expose the internal organs.

The second stage of the autopsy includes careful dissection of many or all of the internal organs. If the brain is to be examined, a portion of the skull must be removed. The face, arms, and legs are usually not dissected unless there is a specific reason for doing so.

After the organs are returned to their respective body cavities, and the body is sewn up, the third phase of the autopsy begins. It is a microscopic examination of tissues collected during the first two phases. Tests to analyze the chemical content of body fluids or to determine the presence of infectious organisms also may be performed.

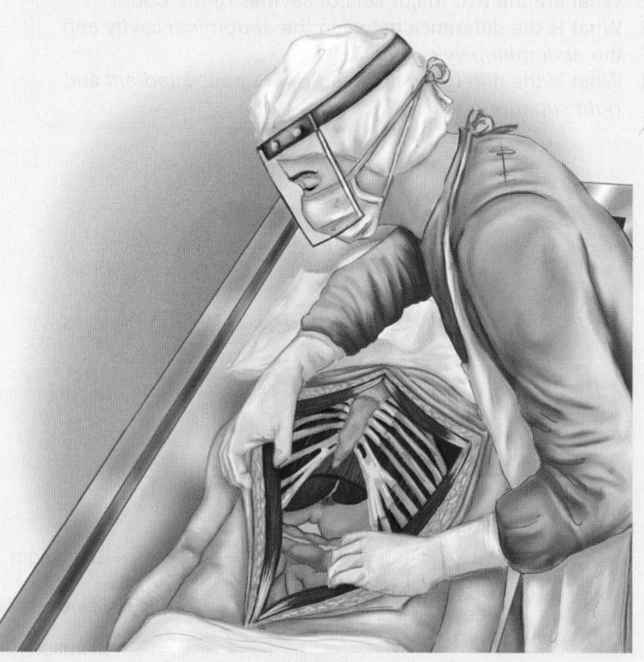

TABLE 1-2 Descriptive Terms for Body Regions

BODY REGION	AREA OR EXAMPLE	BODY REGION	AREA OR EXAMPLE
Abdominal region	Anterior torso below diaphragm	**Mammary region**	Breast
Antebrachial region	Forearm	**Nasal region**	Nose
Axillary region	Armpit	**Occipital region**	Back of lower skull
Brachial region	Arm	**Olecranal region**	Back of elbow
Buccal region	Cheek	**Oral region**	Mouth
Carpal region	Wrist	**Orbital region** *or* **ophthalmic region**	Eyes
Cephalic region	Head		
Cervical region	Neck	**Palmar region**	Palm of hand
Cranial region	Skull	**Pedal region**	Foot
Crural region	Leg	**Pelvic region**	Lower portion of torso
Cubital region*	Elbow	**Perineal region**	Area (perineum) between anus and genitals
Cutaneous	Skin (or body surface)		
Digital region	Fingers or toes	**Plantar region**	Sole of foot
Dorsal region	Back	**Popliteal region**	Area behind knee
Facial region	Face	**Supraclavicular region**	Area above clavicle (collar bone)
Femoral region	Thigh	**Tarsal region**	Ankle
Frontal region	Forehead	**Temporal region**	Side of skull
Gluteal region	Buttock	**Thoracic region**	Entire chest
Inguinal region	Groin	**Umbilical region**	Area around navel or umbilicus
Lumbar region	Lower back between ribs and pelvis	**Volar region**	Palm or sole
		Zygomatic region	Upper cheek

*The term *cubital* may also refer to the forearm.

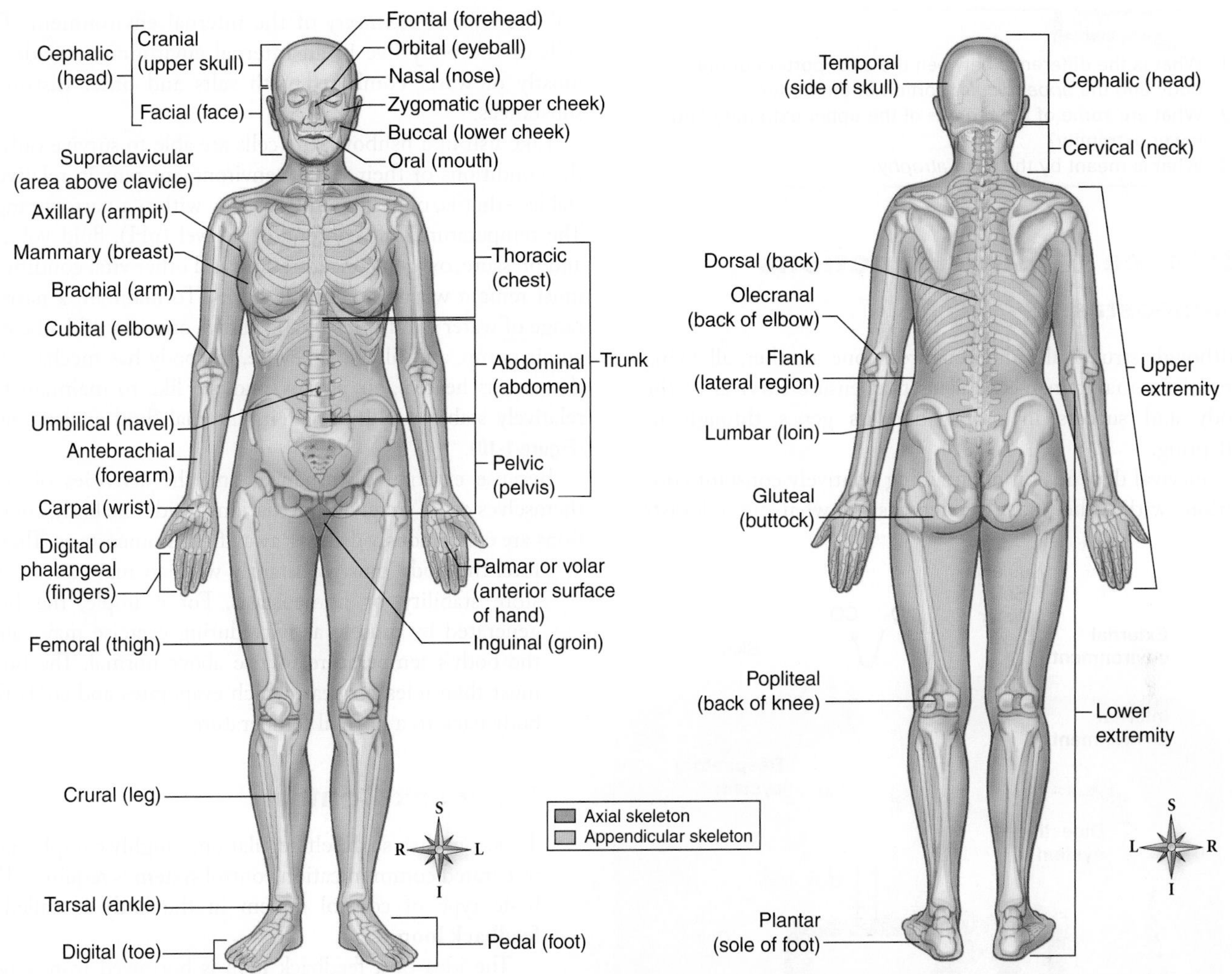

FIGURE 1-9 Axial and appendicular divisions of the body. Specific body regions are labeled (examples in parentheses). For example, the cephalic region includes the head. Notice how the axial and appendicular regions of the body frame are distinguished by contrasting colors.

The body as a whole can be subdivided into two major portions or components: **axial** and **appendicular.** The axial portion of the body consists of the head, neck, and torso or trunk. The appendicular portion consists of the upper and lower extremities (or limbs). Each major axial and appendicular area is subdivided as shown in **Figure 1-9**. Note, for example, that the torso is composed of thoracic, abdominal, and pelvic areas, and the upper extremity is divided into arm, forearm, wrist, and hand components.

Although most terms used to describe gross body regions are well understood, misuse is common. The word *leg* is a good example: it refers to the area of the lower extremity between the knee and ankle and not to the entire lower extremity.

The structure of each person's body is unique. Even identical twins have some variations in the size, shape, and texture of various tissues and organs.

The structure of the body also changes in many ways and at varying rates during a lifetime. Before young adulthood, the body develops and grows. After young adulthood, the body gradually undergoes changes related to aging. For example, with the reduced activity of the body as one advances through older adulthood, body organs and tissues decrease in size and therefore change in their functions. A degenerative process that results from disuse is called **atrophy.** In many cases, atrophy can be reversed with therapy. Some tissues simply lose their elasticity or ability to regenerate as we get older. Nearly every chapter of this book refers to a few of the changes that occur through the life cycle.

Before moving ahead, we pause to consider what seems like an overwhelming number of scientific terms introduced in the preceding sections. It is important to know that such terminology is a "new language" that you must learn as you continue your studies. Now is a good time to review the introduction to this new language in *Appendix B* at *evolve.elsevier.com.* Then, in upcoming chapters, make it a habit to read through the new terms in the chapter word lists—pausing to pronounce each term out loud and glance at its word parts—before starting your reading. Such a strategy will help you slowly and comfortably build a mastery of scientific language.

1

QUICK CHECK

1. What is the difference between the *axial* portion of the body and the *appendicular* portion of the body?
2. What are some of the regions of the upper extremity and lower extremity?
3. What is meant by the term *atrophy*?

Balance of Body Functions

Homeostasis

Although structurally different from one another, all living organisms maintain mechanisms that ensure survival of the body and success in propagating its genes through its offspring.

Survival depends on maintaining relatively constant conditions within the body. **Homeostasis** is what physiologists call the *relative constancy* of the internal environment. The cells of the body live in an internal environment made up mostly of water combined with salts and other dissolved substances.

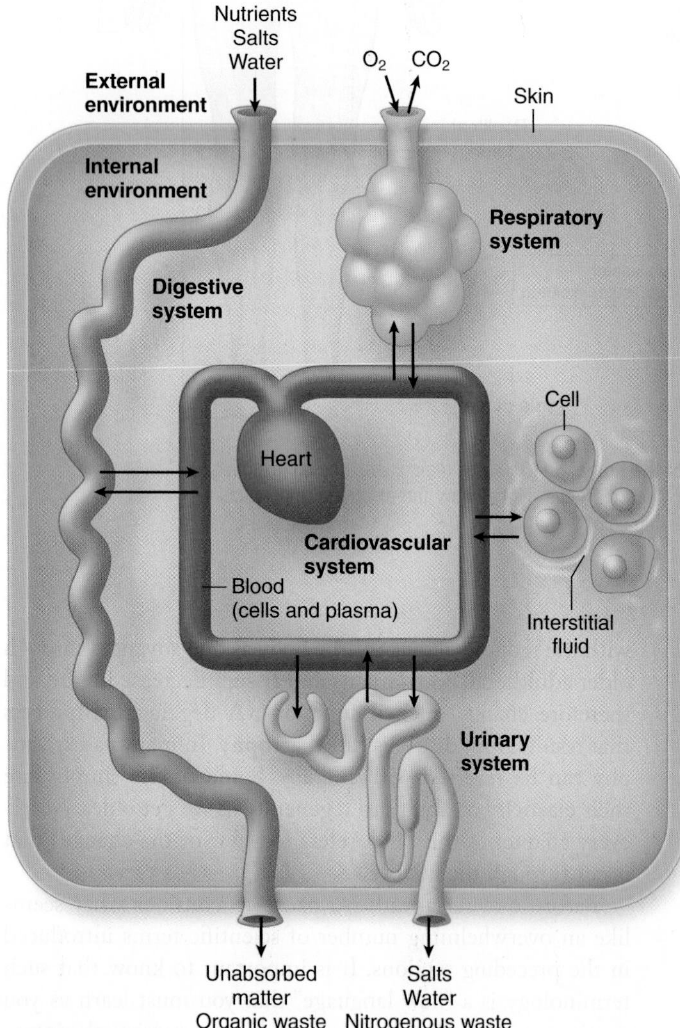

FIGURE 1-10 Diagram of the body's internal environment. The human body is like a bag of fluid separated from the external environment. Tubes, such as the digestive tract and respiratory tract, bring the external environment to deeper parts of the bag where substances can be absorbed into the internal fluid environment or excreted into the external environment. All the organs and systems somehow help maintain a constant environment inside the bag that allows survival of the cells that live there.

Like fish in a fishbowl, the cells are able to survive only if the conditions of their watery environment remain relatively stable—that is, only if conditions stay within a narrow range. The temperature, salt content, acid level (pH), fluid volume and pressure, oxygen concentration, and other vital conditions must remain within acceptable limits. To maintain a narrow range of water conditions in a fishbowl, one may add a heater, an air pump, and filters. Likewise, the body has mechanisms that act as heaters, air pumps, and the like to maintain the relatively stable conditions of its internal fluid environment (**Figure 1-10**).

Because external disturbances and the activities of cells themselves cause frequent fluctuations inside the body, conditions are continuously drifting away from homeostasis. Therefore, the body must constantly work to maintain or restore stability, or homeostasis. For example, the heat generated by muscle activity during exercise may cause the body's temperature to rise above normal. The body must then release sweat, which evaporates and cools the body back to a normal temperature.

Feedback Control

To accomplish such self-regulation, a highly complex and integrated communication control system is required. The basic type of control system in the body is called a **feedback loop.**

The idea of a feedback loop is borrowed from engineering. **Figure 1-11**, *A,* shows how an engineer would describe the feedback loop that maintains stability of temperature in a building. Cold winds outside a building may cause the building temperature to drop below normal. A **sensor,** in this case a thermometer, detects the change in temperature. Information from the sensor *feeds back* to a **control center**—a thermostat in this example—that compares the actual temperature to the normal temperature and responds by activating the building's furnace. The furnace is called an **effector** because it has an effect on the controlled condition (temperature). Because the sensor continually feeds information back to the control center, the furnace will be automatically shut off when the temperature has returned to normal.

As you can see in **Figure 1-11**, *B,* the body uses a similar feedback loop to restore body temperature when we become chilled. Nerve endings that act as temperature sensors feed information to a control center in the brain that compares actual body temperature to normal body temperature. In response to a chill, the brain sends nerve signals to muscles that cause rapidly repeated contractions. This shivering produces heat that increases our body temperature. We stop shivering when feedback information tells the brain that body temperature has increased to normal.

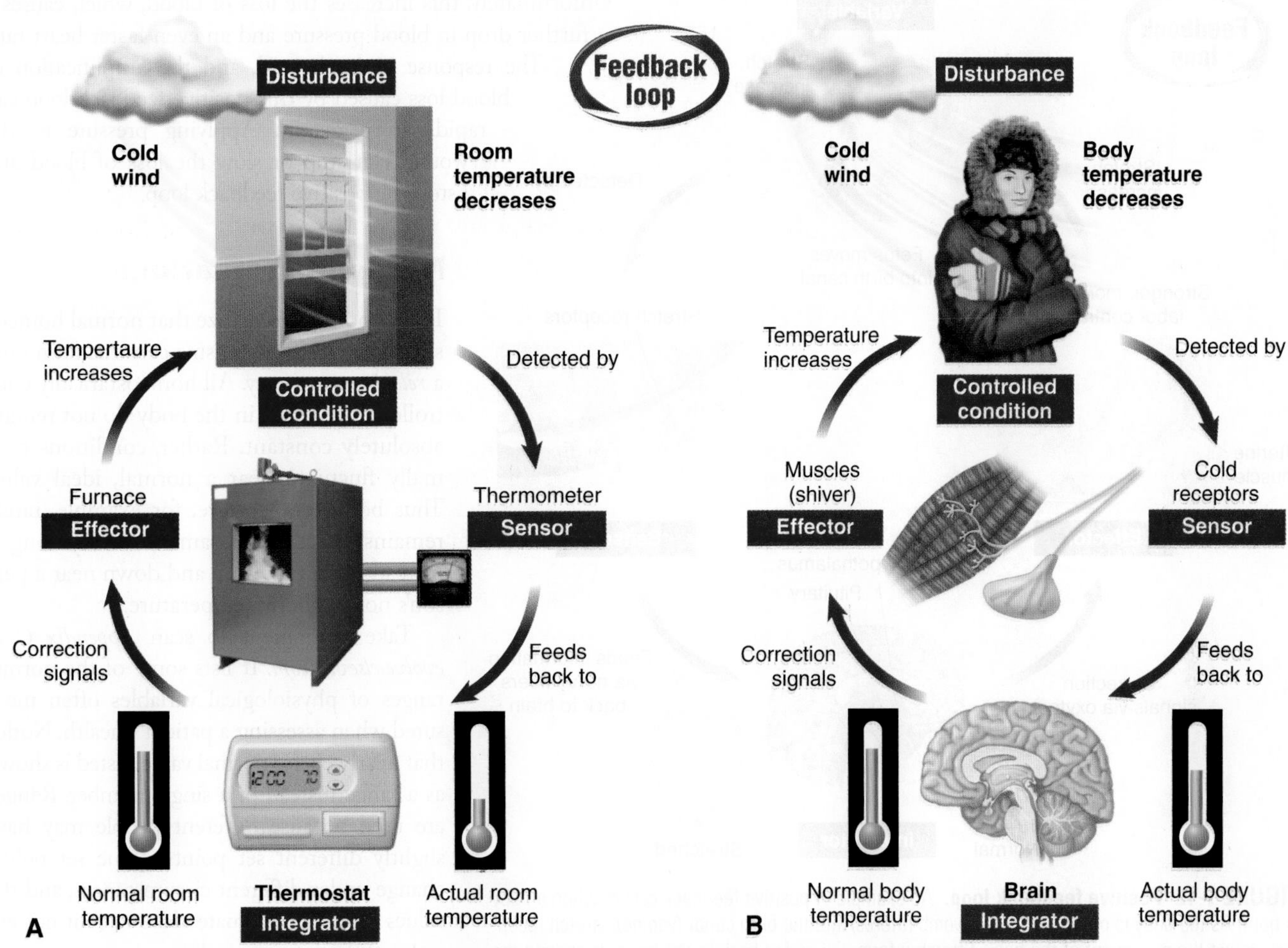

FIGURE 1-11 Negative feedback loops. A, An engineer's diagram showing how relatively constant room temperature (controlled condition) can be maintained. A thermostat (control center) receives feedback information from a thermometer (sensor) and responds by counteracting change from normal by activating a furnace (effector). **B,** A physiologist's diagram showing how a relatively constant body temperature (controlled condition) can be maintained. The brain (control center) receives feedback information from nerve endings called *cold receptors* (sensors) and responds by counteracting a change from normal by activating shivering by muscles (effectors).

Negative Feedback

Feedback loops such as those shown in Figure 1-11 are called **negative feedback** loops because they oppose, or negate, a change in a controlled condition. Most homeostatic control loops in the body involve negative feedback because reversing changes back toward a normal value tends to stabilize conditions—exactly what homeostasis is all about.

Think about the opposite circumstance of that shown in Figure 1-11, as when we become overheated during hot weather. Temperature receptors detect a body temperature higher than normal, and the brain sends signals to the sweat glands to cool us down through evaporation. Thus the conditions are reversed and balance is restored.

Another example of a negative feedback loop occurs during exercise. As muscles contract, they produce additional CO_2 that is transported by blood. This increase in blood CO_2 levels is detected by sensory receptors that transmit the information to respiratory control centers in the brain. This triggers an increase in breathing rate that brings the blood CO_2 level back down toward normal.

An additional example is the excretion of larger than usual volumes of urine when the volume of fluid in the body is greater than the normal, ideal amount.

Positive Feedback

Although not common, **positive feedback** loops do exist in the body and are sometimes also involved in normal function. Positive feedback control loops are stimulatory. Instead of opposing a change in the internal environment and causing a "return to normal," positive feedback loops temporarily amplify or reinforce the change that is occurring. This type of feedback loop causes an ever-increasing rate of events to occur until something stops the process. An example of a positive feedback loop includes the events that cause rapid increases in uterine contractions before the birth of a baby (Figure 1-12).

1

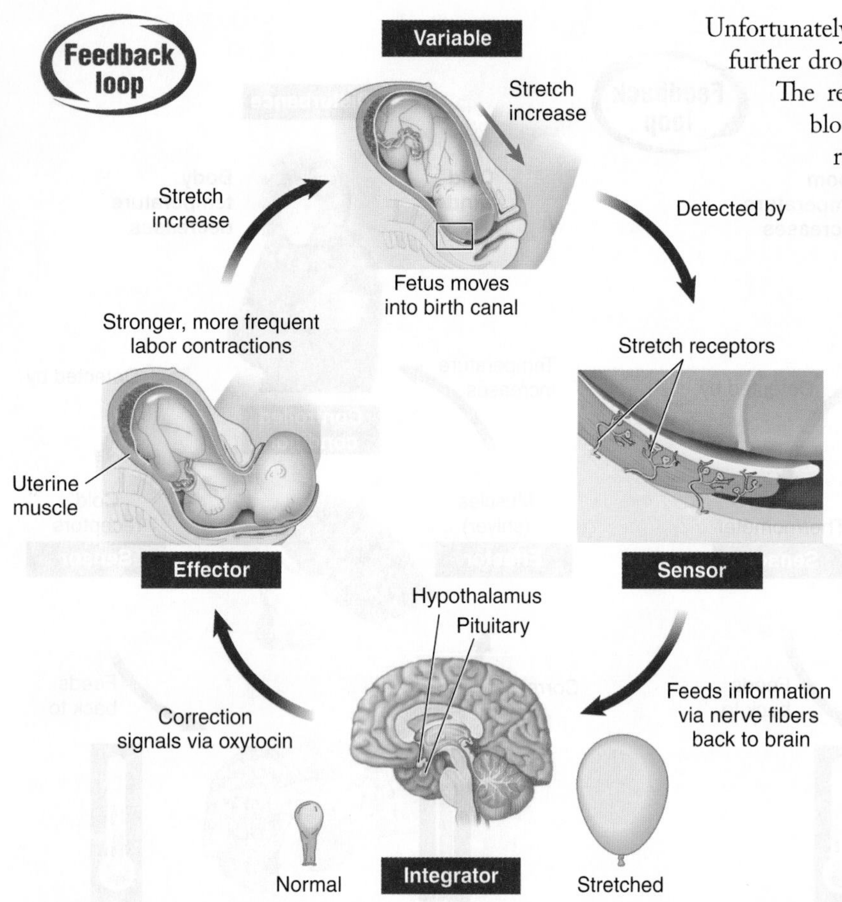

FIGURE 1-12 Positive feedback loop. An example of positive feedback occurs when a baby is born. As the baby is pushed from the womb *(uterus)* into the birth canal *(vagina)*, stretch receptors detect the movement of the baby. Stretch information is fed back to the brain, triggering the pituitary gland to secrete a hormone called *oxytocin* (OT). OT travels through the bloodstream to the uterus, where it stimulates stronger contractions. Stronger contractions push the baby farther along the birth canal, thereby increasing stretch and stimulating the release of more OT. Uterine contractions quickly get stronger and stronger until the baby is pushed out of the body, and the positive feedback loop is broken. OT also can be injected therapeutically by a physician to stimulate labor contractions.

Another example of normal positive feedback regulation in the body is the activity of blood cells called *platelets,* which become increasingly sticky in response to damage to a blood vessel. Circulating platelets rapidly cling to the damaged area and release chemicals that attract additional platelets that accumulate at the site of damage to form a blood clot. The blood clot forms to control bleeding.

In each of these cases, the process rapidly increases until the positive feedback loop is stopped suddenly by the birth of a baby or the formation of a clot. In the long run, such normal positive feedback events also help maintain constancy of the internal environment.

However, negative feedback can abnormally turn into positive feedback, possibly causing a deadly shift in body function.

For example, consider the role of blood pressure and the effect that severe bleeding may have on blood pressure. A normal blood pressure is necessary to ensure that blood flows through blood vessels at an appropriate rate. When blood is lost, as occurs with severe bleeding, blood pressure drops. To compensate, the heart beats faster to try to restore normal pressure. Unfortunately, this increases the loss of blood, which causes a further drop in blood pressure and an even faster heart rate. The response is accelerated, and the amplification of blood loss caused by this positive feedback loop can rapidly turn deadly. Applying pressure to the wound can stop or slow the loss of blood and stop the positive feedback loop.

Normal Fluctuations

It is important to realize that normal homeostatic control mechanisms can maintain only a *relative* constancy. All homeostatically controlled conditions in the body do not remain absolutely constant. Rather, conditions normally fluctuate near a normal, ideal value. Thus body temperature, for example, rarely remains exactly the same for very long—instead it fluctuates up and down near a person's normal body temperature.

Take a moment to scan *Appendix C* at *evolve.elsevier.com.* It lists some of the normal ranges of physiological variables often measured when assessing a patient's health. Notice that nearly every "normal value" listed is shown as a range instead of a single number. Ranges are used because different people may have slightly different set points, some set points change under different circumstances, and the values normally fluctuate close to (but not exactly at) the set point value.

Because all organs function to help maintain homeostatic balance, we discuss negative and positive feedback mechanisms often throughout the remaining chapters of this book.

Before leaving this brief introduction to physiology, we must pause to state an important principle: the ability to maintain the balance of body functions is related to age. During childhood, homeostatic functions gradually become more and more efficient and effective. They operate with maximum efficiency and effectiveness during young adulthood. During late adulthood and old age, they gradually become less and less efficient and effective.

Changes and functions occurring during the early years are called *developmental processes.* Changes occurring after young adulthood are called *aging processes.* In general, developmental processes improve efficiency of functions. Aging processes, on the other hand, often diminish efficiency of body functions.

QUICK CHECK

1. Why is *homeostasis* also called "balance" of body function?
2. What is a *feedback loop* and how does it work?
3. How does *negative feedback* differ from *positive feedback*?
4. How can negative feedback abnormally turn into positive feedback?

LANGUAGE OF **SCIENCE** *(continued from p. 3)*

anthropology
(an-thro-POL-oh-jee)
[*anthropo-* **human,** *-log-* **words (study of),** *-y* **activity**]

appendicular
(ah-pen-DIK-yoo-lar)
[*append-* **hang upon,** *-ic-* **relating to,** *-ul-* **little,** *-ar* **relating to**]

axial
(AK-see-al)
[*axi-* **axis,** *-al* **relating to**]

axillary
(AK-sil-layr-ee)
[*axilla-* **wing,** *-ary* **relating to**]

bilateral symmetry
(bye-LAT-er-al SIM-eh-tree)
[*bi-* **two,** *-later-* **side,** *-al* **relating to,** *sym-* **together,** *-metr-* **measure,** *-ry* **condition of**]

brachial
(BRAY-kee-al)
[*brachi-* **arm,** *-al* **relating to**]

buccal
(BUK-al)
[*bucca-* **cheek,** *-al* **relating to**]

carpal
(KAR-pul)
[*carp-* **wrist,** *-al* **relating to**]

cavity
(KAV-ih-tee)
[*cav-* **hollow,** *-ity* **state**]

cell
(sel)
[*cell* **storeroom**]

cephalic
(seh-FAL-ik)
[*cephal-* **head,** *-ic* **relating to**]

cervical
(SER-vih-kal)
[*cervic-* **neck,** *-al* **relating to**]

chemical level
(KEM-ih-kal LEV-el)
[*chem-* **alchemy,** *-ical* **relating to**]

control center
(kon-TROHL SEN-ter)

cranial
(KRAY-nee-al)
[*crani-* **skull,** *-al* **relating to**]

cranial cavity
(KRAY-nee-al KAV-ih-tee)
[*crani-* **skull,** *-al* **relating to,** *cav-* **hollow,** *-ity* **state**]

crural
(KROOR-al)
[*crur-* **leg,** *-al* **relating to**]

cubital
(KYOO-bih-tal)
[*cubit-* **elbow,** *-al* **relating to**]

cutaneous
(kyoo-TAYN-ee-us)
[*cut-* **skin,** *-aneous* **relating to**]

deep

diaphragm
(DYE-ah-fram)
[*dia-* **across,** *-phrag-* **enclose**]

digital
(DIJ-ih-tal)
[*digit-* **finger or toe,** *-al* **relating to**]

directional term
(dih-REK-shun-al term)

dissection
(dis-SEK-shun)
[*dis-* **apart,** *-sect-* **cut,** *-tion* **process**]

distal
(DIS-tal)
[*dist-* **distance,** *-al* **relating to**]

dorsal
(DOR-sal)
[*dors-* **back,** *-al* **relating to**]

dorsal cavity
(DOR-sal KAV-ih-tee)
[*dors-* **back,** *-al* **relating to,** *cav-* **hollow,** *-ity* **state**]

effector
(ef-FEK-tor)
[*effect-* **accomplish,** *-or* **agent**]

epigastric region
(ep-ih-GAS-trik REE-jun)
[*epi-* **upon,** *gastr-* **stomach,** *-ic* **relating to**]

experimental control
(eks-payr-ih-MENT-el KON-trohl)
[*ex-* **out of,** *-peri-* **tested,** *-ment-* **thing,** *-al* **relating to**]

experimentation
(eks-payr-ih-men-TAY-shun)
[*ex-* **out of,** *-peri-* **tested,** *-ment-* **thing,** *-tion* **process**]

facial
(FAY-shal)
[*faci-* **face,** *-al* **relating to**]

feedback loop
(FEED-bak loop)

femoral
(FEM-or-al)
[*femor-* **thigh,** *-al* **relating to**]

frontal
(FRUN-tal)
[*front-* **forehead,** *-al* **relating to**]

frontal plane
(FRUN-tal playn)
[*front-* **forehead,** *-al* **relating to,** *plan-* **flat surface**]

gluteal
(GLOO-tee-al)
[*glut-* **buttocks,** *-al* **relating to**]

homeostasis
(hoh-mee-oh-STAY-sis)
[*homeo-* **same or equal,** *-stasis* **standing still**]

hypochondriac region
(hye-poh-KON-dree-ak REE-jun)
[*hypo-* **under or below,** *-chondr-* **cartilage,** *-ac* **relating to**]

hypogastric region
(hye-poh-GAST-rik REE-jun)
[*hypo-* **under or below,** *gastr-* **stomach,** *-ic* **relating to**]

hypothesis
(hye-POTH-eh-sis)
pl., hypotheses
(hye-POTH-eh-seez)
[*hypo-* **under or below,** *-thesis* **placing or proposition**]

iliac region
(IL-ee-ak REE-jun)
[*ilia-* **loin or gut (ileum),** *-ac* **relating to**]

inferior
(in-FEER-ee-or)
[*infer-* **lower,** *-or* **quality**]

inguinal
(ING-gwih-nal)
[*inguin-* **groin,** *-al* **relating to**]

lateral
(LAT-er-al)
[*later-* **side,** *-al* **relating to**]

law

levels of organization
(LEV-elz ov or-gan-ih-ZAY-shun)

lumbar
(LUM-bar)
[*lumb-* **loin,** *-ar* **relating to**]

lumbar region
(LUM-bar, REE-jun)
[*lumb-* **loin,** *-ar* **relating to**]

mammary
(MAM-mah-ree)
[*mamma-* **breast,** *-ry* **relating to**]

medial
(MEE-dee-al)
[*media-* **middle,** *-al* **relating to**]

mediastinum
(MEE-dee-as-TYE-num)
[*mediastin-* **midway,** *-um* **thing**]

Continued on p. 18

1

LANGUAGE OF **SCIENCE** *(continued from p. 17)*

microbiome
(my-kroh-BYE-ohm)
[*micro-* **small,** *-bio-* **life,** *-ome* **entire collection**]

midsagittal plane
(mid-SAJ-ih-tal playn)
[*mid-* **middle,** *-sagitta-* **arrow,** *-al* **relating to**]

nasal
(NAY-zal)
[*nas-* **nose,** *-al* **relating to**]

negative feedback
(NEG-ah-tiv FEED-bak)
[*nega-* **deny,** *-tive* **relating to**]

oblique plane
(oh-BLEEK playn)
[*obliq-* **slanted,** *plan-* **flat surface**]

occipital
(awk-SIP-it-al)
[*occipit-* **back of head,** *-al* **relating to**]

olecranal
(oh-LEK-rah-nal)
[*olecran-* **elbow,** *-al* **relating to**]

ophthalmic
(of-THAL-mik)
[*oph-* **eye or vision,** *-thalm-* **inner chamber,** *-ic* **relating to**]

oral
(OR-al)
[*or-* **mouth,** *-al* **relating to**]

orbital
(OR-bih-tal)
[*orbi-* **circle,** *-al* **relating to**]

organ
(OR-gan)
[*organ* **tool or instrument**]

organism
(OR-gah-niz-im)
[*organ-* **instrument,** *-ism* **condition**]

palmar
(PAHL-mar)
[*palm-* **palm of hand,** *-ar* **relating to**]

pedal
(PEED-al)
[*ped-* **foot,** *-al* **relating to**]

pelvic
(PEL-vik)
[*pelvi-* **basin,** *-ic* **relating to**]

pelvic cavity
(PEL-vik KAV-ih-tee)
[*pelvi-* **basin,** *-ic* **relating to,** *cav-* **hollow,** *-ity* **state**]

perineal
(payr-ih-NEE-al)
[*peri-* **around,** *-ine-* **excrete (perineum),** *-al* **relating to**]

physiology
(fiz-ee-OL-oh-jee)
[*physio-* **nature (function),** *-o-* **combining vowel,** *-log-* **words (study of),** *-y* **activity**]

plane
(playn)
[*plan-* **flat surface**]

plantar
(PLAN-tar)
[*planta-* **sole of foot,** *-ar* **relating to**]

pleural
(PLOOR-al)
[*pleura-* **rib,** *-al* **relating to**]

pleural cavity
(PLOOR-al KAV-ih-tee)
[*pleura-* **rib,** *-al* **relating to,** *cav-* **hollow,** *-ity* **state**]

popliteal
(pop-lih-TEE-al)
[*poplit-* **back of knee,** *-al* **relating to**]

positive feedback
(POZ-it-iv FEED-bak)
[*posit-* **to place or amplify,** *-tive* **relating to**]

posterior
(pohs-TEER-ee-or)
[*poster-* **behind,** *-or* **quality**]

prone
[*prone* **lying face down**]

proximal
(PROK-sih-mal)
[*proxima-* **near,** *-al* **relating to**]

sagittal plane
(SAJ-ih-tal playn)
[*sagitta-* **arrow,** *-al* **relating to,** *plan-* **flat surface**]

scientific method
(sye-en-TIF-ik METH-od)

section
(SEK-shun)
[*sect-* **cut,** *-ion* **process or state**]

sensor
(SEN-sor)
[*sens-* **feel,** *-or* **relating to**]

spinal cavity
(SPY-nal KAV-ih-tee)
[*spin-* **backbone,** *-al* **relating to,** *cav-* **hollow,** *-ity* **state**]

superficial
(soo-per-FISH-al)
[*super-* **over or above,** *-fici-* face, *-al* **relating to**]

superior
(soo-PEER-ee-or)
[*super-* **over or above,** *-or* **quality**]

supine
(SOO-pyne)
[*supin-* **lying on the back**]

supraclavicular
(soo-prah-klah-VIK-yoo-lar)
[*supra-* **above or over,** *-clavi-* **key,** *-ul-* **little,** *-ar* **relating to**]

system
(SIS-tem)
[*sy(n)-* **together,** *-stem* **standing**]

tarsal
(TAR-sal)
[*tars-* **ankle,** *-al* **relating to**]

temporal
(TEM-poh-ral)
[*tempora-* **temple (of head),** *-al* **relating to**]

theory
(THEE-ah-ree)
[*theor-* **look at,** *-y* **act of**]

thoracic
(thoh-RAS-ik)
[*thorac-* **chest (thorax),** *-ic* **relating to**]

thoracic cavity
(thoh-RAS-ik KAV-ih-tee)
[*thorac-* **chest (thorax),** *-ic* **relating to,** *cav-* **hollow,** *-ity* **state**]

tissue
(TISH-yoo)
[*tissu-* **fabric**]

transverse plane
(TRANS-vers playn)
[*trans-* **across or through,** *-vers* **turn,** *plan-* **flat surface**]

umbilical
(um-BIL-ih-kul)
[*umbilic-* **navel,** *-al* **relating to**]

ventral
(VEN-tral)
[*ventr-* **belly,** *-al* **relating to**]

ventral cavity
(VEN-tral KAV-ih-tee)
[*ventr-* **belly,** *-al* **relating to,** *cav-* **hollow,** *-ity* **state**]

volar
(VOH-lar)
[*vola-* **hollow of hand,** *-ar* **relating to**]

zygomatic
(zye-goh-MAT-ik)
[*zygo-* **union or yoke,** *-ic* **relating to**]

LANGUAGE OF **MEDICINE**

allied health professions
(AL-ayed helth pro-FESH-unz)

atrophy
(AT-roh-fee)
[*a-* **without,** *-troph-* **nourishment,** *-y* **state**]

autopsy
(AW-top-see)
[*auto-* **self,** *-ops-* **view,** *-y* **procedure**]

disease
(dih-ZEEZ)
[*dis-* **without,** *-ease* **comfort**]

forensic science
(foh-REN-zik SYE-ens)
[*forens-* **public forum,** *-ic* **relating to,** *scienc-* **knowledge**]

medicine
(MED-ih-sin)
[*med-* **heal,** *-ic-* **relating to,** *-ine* **of or like**]

pathology
(pah-THOL-oh-jee)
[*patho-* **disease,** *-o-* **combining vowel,** *-log-* **words (study of),** *-y* **activity**]

☐ **OUTLINE** SUMMARY

To download a digital version of the chapter summary for use with your device, access the ***Audio Chapter Summaries*** *online at evolve.elsevier.com.*

Scan this summary after reading the chapter to help you reinforce the key concepts. Later, use the summary as a quick review before your class or before a test.

Scientific Method

A. Science involves logical inquiry based on experimentation and can use a variety of methods (**Figure 1-1**)
   1. Hypothesis—idea or principle to be tested in experiments
   2. Experiment—series of tests of a hypothesis; a controlled experiment eliminates biases or outside influences
   3. Theory or law—a hypothesis that has been supported by experiments and thus shown to have a high degree of confidence
B. The process of science is active and changing as new experiments add new knowledge

Levels of Organization

A. Organization is the most important characteristic of body structure
B. The body as a whole (organism) is a unit constructed of the following smaller units (**Figure 1-2**):
   1. Atoms and molecules—chemical level
   2. Cells—the smallest structural units; organizations of various chemicals
   3. Tissues—organizations of similar cells
   4. Organs—organizations of different kinds of tissues
   5. Systems—organizations of many different kinds of organs
   6. Organism—organization of all systems together, forming a whole body
C. Microbiome—set of interacting communities of bacteria and other microorganisms that inhabit the human body; influences normal body function

Anatomical Position

A. Reference position in which the body is standing erect with the feet slightly apart and arms at the sides with palms turned forward (**Figure 1-3**)
B. Anatomical position gives meaning to directional terms
C. Terms that describe the body not in anatomical position
   1. Supine—lying face upward
   2. Prone—lying face downward

Anatomical Directions

A. Commonly used directional terms
   1. Superior—toward the head, upper, above
   2. Inferior—toward the feet, lower, below
   3. Anterior—front, in front of (same as ventral in humans)
   4. Posterior—back, in back of (same as dorsal in humans)
   5. Medial—toward the midline of a structure
   6. Lateral—away from the midline or toward the side of a structure
   7. Proximal—toward or nearest the trunk, or nearest the point of origin of a structure
   8. Distal—away from or farthest from the trunk, or farthest from a structure's point of origin
   9. Superficial—nearer the body surface
   10. Deep—farther away from the body surface
B. Anatomical compass rosette—indicator of anatomical directions in an illustration that uses abbreviated directional terms

1

Planes of the Body (Figure 1-4)

A. Sagittal plane—lengthwise plane that divides a structure into right and left sections
B. Midsagittal—sagittal plane that divides the body into two equal halves
C. Frontal (coronal) plane—lengthwise plane that divides a structure into anterior and posterior sections
D. Transverse plane—horizontal plane that divides a structure into upper and lower sections
E. Oblique plane—any plane that is not parallel to any of the planes listed above, thus producing a slanted section

Body Cavities (Figure 1-5, Table 1-1)

A. Dorsal cavities
   1. Cranial cavity contains brain
   2. Spinal cavity contains spinal cord
B. Ventral cavities
   1. Thoracic cavity
      a. Mediastinum—midportion of thoracic cavity; heart and trachea located in mediastinum
      b. Pleural cavities—right lung located in right pleural cavity, left lung in left pleural cavity
   2. Abdominopelvic cavity
      a. Abdominal cavity contains stomach, intestines, liver, gallbladder, pancreas, and spleen
      b. Pelvic cavity contains reproductive organs, urinary bladder, and lowest part of intestine
      c. Abdominopelvic subdivisions
         (1) Four abdominopelvic quadrants (**Figure 1-6**)
         (2) Nine abdominopelvic regions (**Figure 1-7**)
C. Organs of the major body cavities can be seen in **Figure 1-8** and are also referenced in Table 1.1

Body Regions (Figure 1-9, Table 1-2)

A. Axial region—head, neck, and torso or trunk
B. Appendicular region—upper and lower extremities (limbs)
C. Body structure and function vary among individuals and also throughout an individual's lifespan; atrophy (decrease in size) occurs when an organ is not used

Balance of Body Functions

A. Survival of the individual and of the genes that make up the body is of the utmost importance
B. Survival depends on the maintenance or restoration of homeostasis (relative constancy of the internal environment)
   1. The internal environment is a fluid that must be kept stable by the operation of various organ systems (**Figure 1-10**)
   2. The body uses stabilizing negative feedback loops (**Figure 1-11**) and, less often, amplifying positive feedback loops (**Figure 1-12**) to maintain or restore homeostasis
   3. Feedback loops involve a sensor, a control center, and an effector
   4. Negative feedback loops can turn into positive feedback loops during injury or disease, possibly causing a deadly shift in body function
C. All organs function to maintain homeostasis
D. Ability to maintain balance of body functions is related to age; peak efficiency occurs during young adulthood, diminishing efficiency occurs after young adulthood

❑ ACTIVE LEARNING

STUDY TIPS

Hint *Consider using these tips to achieve success in meeting your learning goals.*

1. A number of topics are introduced in this chapter that will be important throughout the rest of the course.
2. One of your first steps should be mastering the new terminology of each chapter. Read the new terms listed at the beginning of each chapter out loud before attempting to read or learn each new topic. Use the pronunciation guides provided, saying each term several times to "get it into" your working memory. Pay attention to word parts, too—they'll help you master the terminology of science and medicine more quickly. (For more terminology tips, see *my-ap.us/fsboS2*.)
3. The most important concept is probably homeostasis. The word itself tells you what it means: *homeo* means "the same," *stasis* means "staying." Homeostasis is the balance the body tries to maintain by making sure its internal environment "stays the same." Make sure you understand this concept. (For more tips on homeostasis, see *my-ap.us/rs3KqV*.)
4. Another important topic introduced in this chapter is the structural levels of organization. The lower levels are the building blocks on which the upper levels depend. As various disease processes are explained in later chapters, notice how many of these processes cause failure at the chemical or cellular level and how this failure affects organs, systems, and even the body as a whole.
5. Become familiar with the directional terms—you will see them in almost every diagram in the text. The terms also are used in naming several body structures (for example, superior vena cava, distal convoluted tubule). The terms are fairly easy to learn because they are presented in opposite pairs, so if you learn one term, you almost always automatically know its opposite. Flash cards will help you learn them. (For more on using flash cards effectively, see *my-ap.us/LzuowE*. See *my-ap.us/K9GtVc* for more tips on learning directions.)
6. **Table 1-2** and Appendix B (at *evolve.elsevier.com*) are helpful resources to keep in mind when you see an unfamiliar term.
7. In your study group, try to come up with examples of negative feedback loops that help maintain a balance. Be creative—and try to use something other than the furnace example. Go over your directional-term flash cards or photocopy **Figure 1-4** and then blacken out the terms so you and your fellow students can use the illustration to quiz each other. Go over the questions at the end of the chapter and discuss possible test questions.

Review Questions

Write out the answers to these questions after reading the chapter and reviewing the Chapter Summary. If you simply think through the answer without writing it down, you will not retain much of your new learning.

1. Define anatomy, physiology, and pathology.
2. Disease results from what general conditions in the body?
3. Describe the process used to form scientific theories.
4. List and explain the levels of organization in the human body.
5. Describe the anatomical position.
6. Name and describe the three planes or sections of the body.
7. List two organs of the mediastinum, two organs of the abdominal cavity, and two organs of the pelvic cavity.
8. List the nine regions of the abdominopelvic cavity, beginning at the upper left region and ending at the lower right region.
9. Name the main areas of the axial skeleton.
10. Name the two subdivisions of the dorsal cavity. What structure does each contain?
11. Explain the difference between the terms lower extremity, thigh, and leg.
12. List four conditions in the cell that must be kept in homeostatic balance.
13. List the three parts of a negative feedback loop and give the function of each.

Critical Thinking

After finishing the Review Questions, write out the answers to these more in-depth questions to help you apply your new knowledge. Go back to sections of the chapter that relate to concepts that you find difficult.

14. Identify a structure that is inferior to the heart, superior to the heart, anterior to the heart, posterior to the heart, and lateral to the heart.
15. The maintenance of body temperature and the birth of a baby are two body functions that are regulated by feedback loops. Explain the different feedback loops that regulate each process.
16. If a person complained of pain in the epigastric region, what organs could be involved?
17. Give an example of a negative feedback loop that occurs during exercise. Explain the physiology involved during the process.

Chapter Test

Hint

After studying the chapter, test your mastery by responding to these items. Try to answer them without looking up the answers.

1. _______ is a term derived from two Greek words meaning "cutting apart."
2. _______ means the study of the function of living organisms and their parts.
3. _______ is the scientific study of disease.
4. A hypothesis that has been rigorously tested and has gained a high level of confidence is called a _______ or _______.
5. _______, _______, _______, _______, and _______ are the five levels of organization in a living thing.
6. _______ and _______ are terms used to describe the body position when it is not in anatomical position.
7. A _______ section cuts the body or any of its parts into upper and lower portions.
8. A _______ section cuts the body or any of its parts into front and back portions.
9. A _______ section cuts the body or any of its parts into left and right portions.
10. If the body is cut into equal right and left sides, the cut is called a _______ section or plane.
11. The body portion that consists of the head, neck, and torso is called the _______ portion.
12. The body portion that consists of the upper and lower extremities is called the _______ portion.
13. _______ is the term used to describe the fact that the left and right sides of the body appear alike or mirror each other.
14. An _______ plane is an imaginary flat plane that runs diagonally to an axis of the body or one of its parts, producing a slanted section or cut.
15. The two major cavities of the body are the:
    a. thoracic and abdominal
    b. abdominal and pelvic
    c. dorsal and ventral
    d. anterior and posterior
16. The structure that divides the thoracic cavity from the abdominal cavity is the:
    a. mediastinum
    b. diaphragm
    c. lungs
    d. stomach
17. The epigastric region of the abdominopelvic cavity is:
    a. inferior to the umbilical region
    b. lateral to the umbilical region
    c. medial to the umbilical region
    d. none of the above
18. The hypogastric region of the abdominopelvic cavity is:
    a. inferior to the umbilical region
    b. lateral to the left iliac region
    c. medial to the right iliac region
    d. both a and c
19. Which of the following is an example of a positive feedback loop?
    a. Maintaining a constant body temperature
    b. Contractions of the uterus during childbirth
    c. Maintaining a constant volume of water in the body
    d. Both a and c
20. Which of the following is an example of a negative feedback loop?
    a. Maintaining a constant body temperature
    b. Contractions of the uterus during childbirth
    c. Maintaining a constant volume of water in the body
    d. Both a and c

Match each directional term in column B with its opposite term in column A.

Column A	Column B
21. _______ superior	a. posterior
22. _______ distal	b. superficial
23. _______ anterior	c. medial
24. _______ lateral	d. proximal
25. _______ deep	e. inferior

Case Studies

To solve a case study, you may have to refer to the glossary or index, other chapters in this textbook, and other resources.

1. Mrs. Miller was referred to the clinic by her regular physician to have a mole on her skin examined. The referral form states that the mole is on the left upper quadrant of her abdomen. Give a more detailed description of its location. In preparing Mrs. Miller for the examination, how would you position her? Should you ask her to assume a supine or prone position? During the examination, the physician notices that the referral form states that Mrs. Miller has a similar mole in the occipital region. What position should she assume so that the physician can examine that mole?
2. Mr. Sanchez has just severed the distal tip of the fourth digit on his upper extremity. Describe in layman's terms which body part he has injured. As blood poured out of the injured tissue, his blood pressure dropped. His heart then pumped faster to restore normal pressure. What effect would this response have on Mr. Sanchez's homeostasis? Would such a response be an example of negative or positive feedback?
3. Mrs. Tipps is a high school algebra teacher who thinks that she might have high blood pressure (hypertension). She also thinks that it might be stress related. She experiences symptoms such as a flushing of her face and headaches at certain times during the day. She notices them particularly during her fifth period class, which has been particularly challenging. She also seems to be symptom free on the weekends. She has a regular appointment already scheduled with her doctor in 3 weeks. What data could she take with her that might help her doctor with a diagnosis and possibly affirm her thinking?

Answers to Active Learning Questions can be found online at evolve.elsevier.com.

Chemistry of Life

OUTLINE

Hint *Scan this outline before you begin to read the chapter, as a preview of how the concepts are organized.*

OBJECTIVES

Before reading the chapter, review these goals for your learning.

After you have completed this chapter, you should be able to:

1. Describe the structure of an *atom.*
2. Define and discuss the terms *element, molecule,* and *compound.*
3. Compare and contrast the major types of chemical bonding.
4. Do the following related to inorganic chemistry:
   - Distinguish between organic and inorganic chemical compounds.
   - Discuss the chemical characteristics of water.
   - Discuss acids, bases, and salts.
   - Explain the concept of pH.
5. Discuss the structure and function of the following types of organic molecules: carbohydrate, lipid, protein, and nucleic acid.

CHAPTER 2

Life is full of chemistry and the more we learn about chemicals and their structures, the better we can understand chemical processes in the human body. The digestion of food, the formation of bone tissue, and the contraction of a muscle are all chemical processes. Thus the basic principles of anatomy and physiology are ultimately based on principles of chemistry. A whole field of science, **biochemistry,** is devoted to studying the chemical aspects of life. To truly understand the human body, it is important to understand a few basic facts about biochemistry, the chemistry of life. The best place to begin is with the building blocks of matter.

Levels of Chemical Organization

Matter is anything that occupies space and has mass. Biochemists classify matter into several levels of organization for easier study. The smallest unit of matter is the **atom.** Atoms are used to build more complicated substances in the body. In the body, most chemicals are in the form of molecules. **Molecules** are particles of matter that are composed of one or more atoms. Atoms are considered to be the basic units of matter. So a good place to start is with the atom.

Atoms

Atoms are so small they can be observed only with very sophisticated equipment. For example, *tunneling microscopes* and *atomic force microscopes (AFM)* can produce pictures of individual atoms (**Figure 2-1**). Atoms are composed of several kinds of *subatomic particles:* **protons, electrons,** and **neutrons.**

At the core of each atom is a **nucleus** composed of positively charged protons and uncharged neutrons. The number of protons in the nucleus is an atom's **atomic number.** The number of protons and neutrons combined is the atom's **atomic mass.**

Negatively charged electrons surround the nucleus at a distance. If an atom is neutral (carries no electrical charge), there is one electron for every proton. Electrons do not stay still. Instead, electrons keep darting about within certain limits called **orbitals.** Each orbital can hold two electrons. Even though atomic models and the name orbital imply that electrons move around in confined elliptical orbits, electrons actually move about in chaotic, unpredictable paths.

Orbitals are arranged into **energy levels** (shells), depending on their distance from the nucleus. The farther an orbital extends from the nucleus, the higher its energy level. The energy level

LANGUAGE OF **SCIENCE**

Hint Before reading the chapter, say each of these terms out loud. This will help you to avoid stumbling over them as you read.

acid
(AS-id)

adenine
(AD-eh-neen)
[*aden-* **gland**]

adenosine diphosphate (ADP)
(ah-DEN-oh-seen dye-FAHS-fayt [ay dee pee])
[*adenos-* shortened from *adenine-ribose, -ine* **chemical,** *di-* two, *-phosph-* **phosphorus,** *-ate* **oxygen**]

adenosine triphosphate (ATP)
(ah-DEN-oh-seen try-FOS-fayt [ay tee pee])
[*adenos-* shortened from *adenine-ribose, -ine* **chemical,** *tri-* **three,** *-phosph-* phosphorus, *-ate* **oxygen**]

alkaline
(AL-kah-lin)
[*alkal-* **ashes,** *-ine* **relating to**]

amino acid
(ah-MEE-no AS-id)
[*amino* NH_2, *acid* **sour**]

aqueous solution
(AY-kwee-us suh-LOO-shun)
[*aqu-* **water,** *-ous* **relating to,** *solut-* **dissolved,** *-ion* **process**]

atom
(AT-om)
[*atom* **indivisible unit**]

atomic mass
(ah-TOM-ik mas)
[*atom-* **indivisible unit,** *-ic* **relating to**]

atomic number
(ah-TOM-ik NUM-ber)
[*atom-* **indivisible unit,** *-ic* **relating to**]

base
(bays)
[*base* **foundation**]

Continued on p. 36

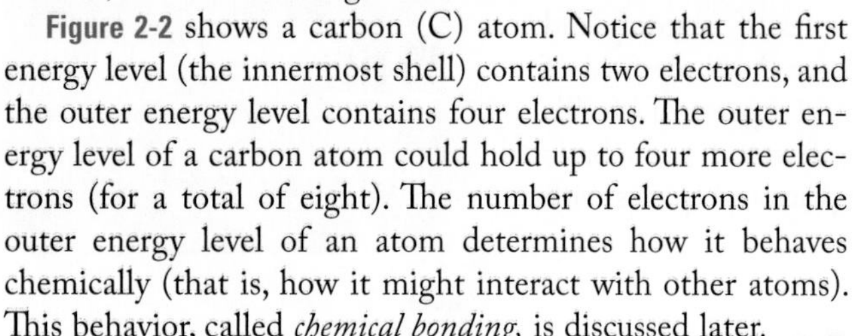

FIGURE 2-1 Atoms. A group of cloudlike atoms in a crystal as pictured by atomic force microscopy (AFM). Added colors highlight different kinds of atoms.

closest to the nucleus has one orbital, so it can hold two electrons. The next energy level has up to four orbitals, so it can hold eight electrons.

Figure 2-2 shows a carbon (C) atom. Notice that the first energy level (the innermost shell) contains two electrons, and the outer energy level contains four electrons. The outer energy level of a carbon atom could hold up to four more electrons (for a total of eight). The number of electrons in the outer energy level of an atom determines how it behaves chemically (that is, how it might interact with other atoms). This behavior, called *chemical bonding,* is discussed later.

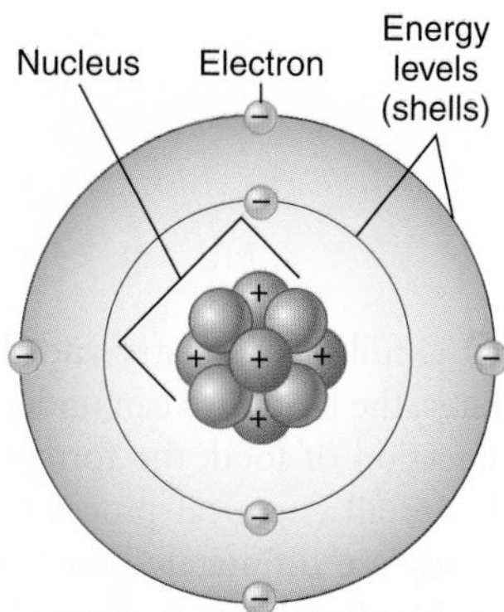

FIGURE 2-2 A model of the atom. The nucleus—protons (+) and neutrons—is at the core. Electrons inhabit outer regions called energy levels. This is a carbon atom, a fact that is determined by the number of its protons. All carbon atoms (and only carbon atoms) have six protons. Because there are only four electrons in the outer energy level, which can hold up to eight electrons, this carbon atom will share electrons with other atoms so that its outer energy level becomes full. (One proton and two neutrons in the nucleus are not visible in this illustration.)

Elements, Molecules, and Compounds

Substances can be classified as **elements** or **compounds.** Elements are pure substances, composed of only one of more than a hundred types of atoms that exist in nature. Only four kinds of atoms (**oxygen, carbon, hydrogen,** and **nitrogen**) make up about 96% of the human body. There are traces of about 20 other elements in the body. **Table 2-1** lists some of the elements in the body and also gives their universal chemical *symbols*—the abbreviations used by chemists worldwide.

Atoms usually unite with each other to form larger chemical units called *molecules.* Some molecules are made of several atoms of the same element. **Compounds** are substances whose molecules have more than one element in them.

CLINICAL APPLICATION
RADIOACTIVE ISOTOPES

Each element is unique because of the number of protons it has. In short, each element has its own *atomic number.* However, atoms of the same element can have different numbers of neutrons. Two atoms that have the same atomic number but different atomic masses are **isotopes** of the same element.

An example is hydrogen. Hydrogen has three isotopes: ^{1}H (the most common isotope), ^{2}H, and ^{3}H. The accompanying figure shows that each different isotope has only one proton but different numbers of neutrons.

Some isotopes have unstable nuclei that radiate (give off) particles. Radiation particles include protons, neutrons, electrons, and altered versions of these normal subatomic particles. An isotope that emits radiation is called a **radioactive isotope.**

Radioactive isotopes of common elements are sometimes used to evaluate the function of body parts. Radioactive iodine (^{125}I) that is put into the body and then taken up by the thyroid gland gives off radiation that can be easily measured. Thus the rate of thyroid activity can be determined using this method.

Images of internal organs can be created by radiation scanners that plot out the location of injected or ingested radioactive isotopes. For example, radioactive technetium (^{99}Tc) is commonly used to image the liver and spleen. The radioactive isotopes ^{13}N, ^{15}O, and ^{11}C are often used to study the brain by way of a technique called the *PET (positron emission tomography) scan.*

Radiation can damage cells. Exposure to high levels of radiation may cause cells to develop into cancer cells. Higher levels of radiation completely destroy tissues, causing **radiation sickness.** Low doses of radioactive substances are sometimes given to cancer patients to destroy cancer cells. The side effects of these treatments result from the unavoidable destruction of normal cells along with the cancer cells.

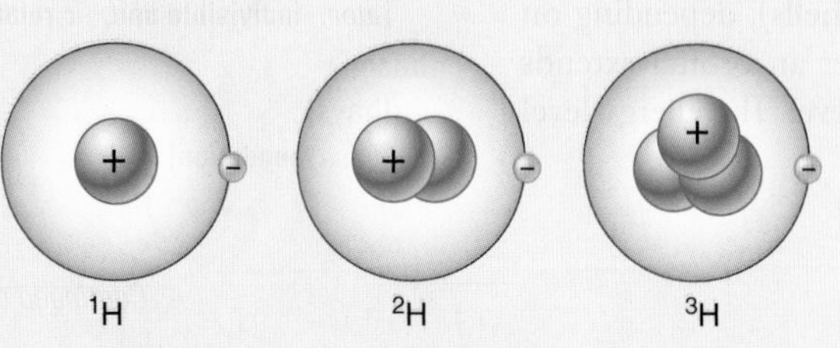

To learn more about the health applications of nuclear radiation and to see medical images made with radiation techniques check out the articles *Radiation, Medical Imaging of the Body,* and *Bone Scans* at Connect It! at *evolve.elsevier.com.*

TABLE 2-1 Important Elements in the Human Body

ELEMENT	SYMBOL	NUMBER OF PROTONS IN NUCLEUS	NUMBER OF ELECTRONS IN OUTER SHELL*
Major Elements (Greater Than 96% of Body Weight)			
Oxygen	O	8	6
Carbon	C	6	4
Hydrogen	H	1	1
Nitrogen	N	7	5
Trace Elements (Examples of More Than 20 Trace Elements Found in the Body)			
Calcium	Ca	20	2
Phosphorus	P	15	5
Sodium (Latin *natrium*)	Na	11	1
Potassium (Latin *kalium*)	K	19	1
Chlorine	Cl	17	7
Iodine	I	53	7

*Maximum is eight, except for hydrogen. The maximum for that element is two.

In order to describe which atoms are present in a compound, a chemical **formula** is used. The formula for a compound contains symbols that represent each element in the molecule. The number of atoms of each element in the molecule is expressed as a subscript after the elemental symbol. For example, each molecule of the compound **carbon dioxide** has one carbon (C) atom and two oxygen (O) atoms—thus its molecular formula is CO_2.

To learn more about molecule formation, go to AnimationDirect at *evolve.elsevier.com*.

QUICK CHECK

1. What are the three main subatomic particles of an atom?
2. What is *matter*?
3. Describe an energy level.
4. What is a *compound*? An *element*? A *molecule*?

Chemical Bonding

Chemical bonds form to make atoms more stable. An atom is said to be chemically stable when its outer energy level is "full" (that is, when its energy shells have the maximum number of electrons they can hold). All but a handful of atoms have room for more electrons in their outermost energy level. A basic chemical principle states that atoms react with one another in ways that make their outermost energy level full. To create this full energy level, atoms can share, donate, or borrow electrons.

For example, a hydrogen atom has one electron and one proton. Its single energy shell has one electron but *can* hold two—so it is not full. If two hydrogen atoms "share" their single electrons with each other, then both will have full energy shells, making them more stable *as a molecule* than either would be as an atom. This is an example of how atoms **bond** to form molecules. Other atoms may instead donate or borrow electrons until the outermost energy level is full and then form crystals.

Ionic bonds

One common way in which atoms make their outermost energy level full is to form **ionic bonds** with other atoms. Such a bond forms between an atom that has only one or two electrons in the outermost level (which would normally hold eight) and an atom that needs only one or two electrons to fill its outer level. The atom with one or two electrons simply "donates" its outer shell electrons to the one that needs one or two.

2

For example, as you can see in **Table 2-1**, the sodium (Na) atom has one electron in its outer level and the chlorine (Cl) atom has seven. Both need to have eight electrons to fill their outer shell. **Figure 2-3** shows how sodium and chlorine form an ionic bond when sodium "donates" the electron in its outer shell to chlorine. Now both atoms have full outer shells (although sodium's outer shell is now one energy level lower).

Because the sodium atom lost an electron, it now has one more proton than it has electrons. This makes it a positive **ion,** an electrically charged atom. Chlorine has "borrowed" an electron to become a negative ion called the *chloride ion.* Because oppositely charged particles attract one another, the sodium and chloride ions are drawn together to form a sodium chloride (NaCl) crystal—common table salt (**Figure 2-3**, *B*). The crystal is held together by *ionic bonds.*

Ionic compounds usually dissolve easily in water because water molecules are attracted to ions and wedge between the ions—thus forcing them apart. When this happens, we say the compounds **dissociate** to form free ions. Compounds that form ions when dissolved in water are called **electrolytes.**

The formula of an ion always shows its charge by a "+" or "−" superscript after the chemical symbol. Thus the sodium ion is Na^+, and the chloride ion is Cl^-. Calcium (Ca) atoms lose two electrons when they form ions, so the calcium ion is written as Ca^{++}.

Because the body's internal environment is mostly water, we find many dissolved ions in the body. Specific ions have important roles to play in muscle contraction, nerve signaling, and other vital functions. **Table 2-2** lists some of the more important ions present in body fluids. Many of these ions are discussed in later chapters. Chapter 21 describes mechanisms that maintain the homeostasis of the electrolytes throughout the body.

Covalent Bonds

Atoms also may fill their energy levels by sharing electrons rather than donating or receiving them. When atoms share electrons, a **covalent bond** forms. **Figure 2-4** shows how two

TABLE 2-2 Important Ions in Body Fluids

NAME	SYMBOL
Sodium (Latin *natrium*)	Na^+
Chloride	Cl^-
Potassium (Latin *kalium*)	K^+
Calcium	Ca^{++}
Hydrogen	H^+
Magnesium	Mg^{++}
Hydroxide	OH^-
Phosphate	$PO_4^{\equiv}$
Bicarbonate	HCO_3^-

hydrogen atoms may move together closely so that their energy levels overlap. Each energy level contributes its one electron to the sharing relationship. That way, both outer levels have access to both electrons.

Because atoms involved in a covalent bond must stay close to each other, it is not surprising that covalent bonds are not easily broken. Covalent bonds normally do not break apart in water.

Carbon, nitrogen, oxygen, and hydrogen almost always share electrons to form covalent bonds, making this type of bonding important in the human body. Covalent bonding is used to form all of the major organic compounds found in the body.

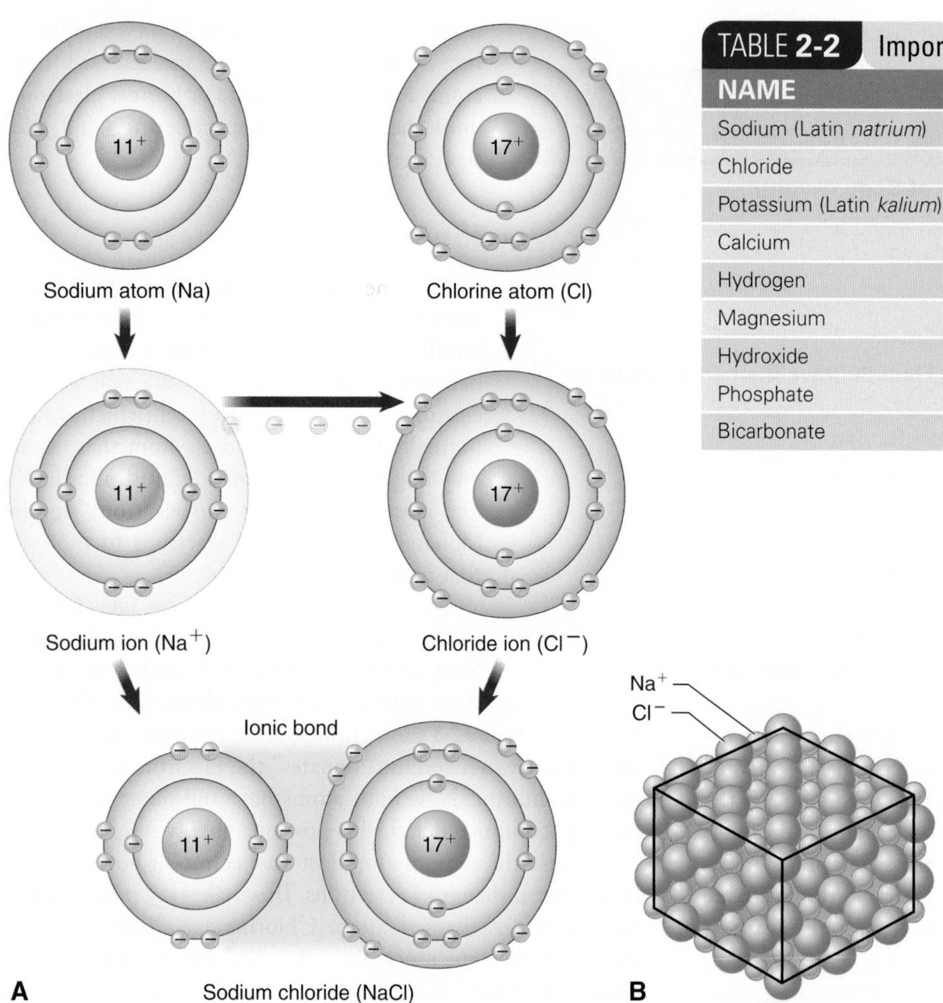

FIGURE 2-3 Ionic bonding. A, The sodium atom donates the single electron in its outer energy level to a chlorine atom having seven electrons in its outer level. Then both have eight electrons in their outer shells. Because the electron/proton ratio changes, the sodium atom becomes a positive sodium ion. The chlorine atom becomes a negative chloride ion. The positive-negative attraction between these oppositely charged ions is called an *ionic bond.* **B,** A cube-shaped crystal of sodium chloride (table salt).

Hydrogen Bonds

A kind of weak attraction that helps hold your body's substance together is the **hydrogen bond.** Slight electrical charges may develop in different regions of a molecule when tiny hydrogen atoms are not able to equally share their electrons in a covalent bond. Oppositely charged ends of various molecules then electrically attract one another (**Figure 2-5**).

Hydrogen bonds do not form new molecules, but instead provide subtle forces that help a large molecule stay in a particular shape. They also may help hold together neighboring molecules. For example, hydrogen bonds help maintain the complex folded shapes of proteins (see **Figure 2-12** on p. 34). Hydrogen bonds also keep water molecules loosely joined together—giving water a weak gluelike quality that helps hold your body together (see **Figure 2-5**).

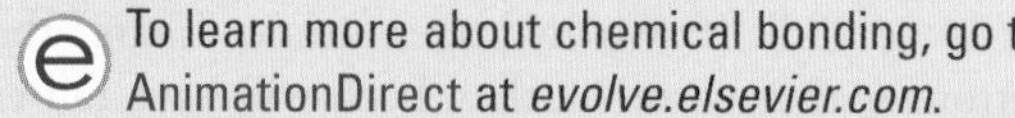

To learn more about chemical bonding, go to AnimationDirect at *evolve.elsevier.com.*

FIGURE 2-4 Covalent bonding. Two hydrogen atoms move together, overlapping their energy levels. Although neither gains nor loses an electron, the atoms share the electrons, thereby forming a covalent bond.

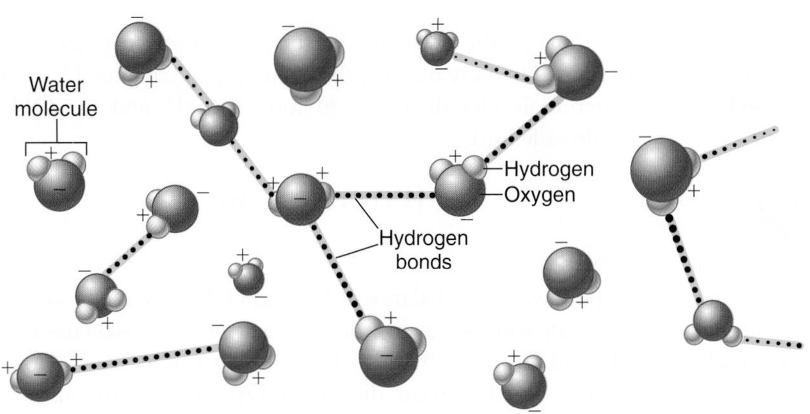

FIGURE 2-5 Hydrogen bonds. Because the tiny hydrogen atoms in water cannot share their electrons equally with a large oxygen atom, the water molecule develops slightly different charges at each end. Like weak magnets, the water molecules form temporary attachments (hydrogen bonds) that give liquid water its slightly gluelike properties.

QUICK CHECK

1. What is an *ionic* bond? What is a *covalent* bond?
2. What is meant by an electrolyte *dissociating* in water?
3. How does hydrogen bonding differ from ionic and covalent bonding?
4. Why is the symbol for calcium expressed C^{++}?

Inorganic Chemistry

All compounds in living organisms can be classified as either **organic** or **inorganic.** Organic compounds are composed of molecules that contain carbon-carbon (C—C) covalent bonds or carbon-hydrogen (C—H) covalent bonds—or both kinds of bonds. Few inorganic compounds have carbon atoms in them and none have C—C or C—H bonds. Organic molecules are generally larger and more complex than inorganic molecules. The human body has both kinds of compounds because both are equally important to the chemistry of life. We will discuss the chemistry of inorganic compounds first, and then move on to some of the important types of organic compounds.

Water

Though **water** is an inorganic compound, it is essential to life. Found in and around each cell, water is the most abundant compound in the body. Its slightly gluelike properties help hold the tissues of the body together.

Solutions

Water is the **solvent** in which most other compounds or **solutes** are dissolved. When water is the solvent for a *mixture* (a blend of two or more kinds of molecules), the mixture is called an **aqueous solution.**

An aqueous solution containing common salt (NaCl) and other molecules forms the "internal sea" of the body. Water molecules not only compose the basic internal environment of the body but also participate in many important *chemical reactions.* Chemical reactions are interactions among molecules in which atoms regroup into new combinations.

Water Chemistry

Dehydration synthesis is a common type of chemical reaction in the body. In any kind of synthesis reaction, the **reactants** combine to form a larger **product.** In dehydration synthesis, reactants combine only after hydrogen (H) and oxygen (O) atoms are removed. These removed H and O atoms combine to form H_2O, or water. As **Figure 2-6** shows, the result of a dehydration synthesis reaction is both the large product molecule and a water molecule.

Just as dehydration of a cell is a loss of water from the cell and dehydration of the body is loss of fluid from the entire internal environment, dehydration synthesis is a reaction in which water is lost from the reactants.

Hydrolysis is another common reaction in the body that involves water. In this reaction, water *(hydro)* disrupts the bonds in large molecules, breaking them down into smaller molecules *(lysis).* Hydrolysis is virtually the reverse of dehydration synthesis, as **Figure 2-6** shows.

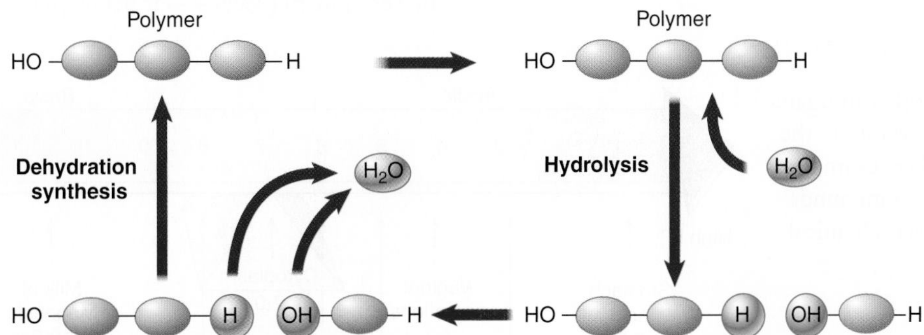

FIGURE 2-6 Water-based chemistry. Dehydration synthesis *(on the left)* is a reaction in which small molecules are assembled into large molecules by removing water (H and O atoms). Hydrolysis *(on the right)* operates in the reverse direction—H and O from water are added as large molecules are broken down into small molecules.

Not only is water the medium in which all major types of organic compounds are formed and broken down; it is also a product (dehydration synthesis) or reactant (hydrolysis) in these types of reactions. Clearly, water is an important substance in the body!

Chemical reactions always involve energy transfers. Energy is required to build the molecules. Some of that energy is stored as potential energy in the chemical bonds. The stored energy can then be released when the chemical bonds in the molecule are later broken apart. For example, a molecule called *adenosine triphosphate (ATP)* breaks apart in the muscle cells to yield the energy needed for muscle contraction (see Figure 2-15 on p. 35).

Chemists often use a *chemical equation* to represent a chemical reaction. In a chemical equation, the reactants are separated from the products by an arrow (→) showing the "direction" of the reaction. Reactants are separated from each other, and products are separated from each other by addition, or plus, signs (+). Thus the reaction *potassium and chloride combined to form potassium chloride* can be expressed as the following equation:

$$K^+ + Cl^- \rightarrow KCl$$

The single arrow → is used for equations that occur in only one direction. For example, when hydrochloric acid (HCl) is dissolved in water, all of it dissociates to form H^+ and Cl^-.

$$HCl \rightarrow H^+ + Cl^-$$

The double arrow ↔ is used for reactions that happen in "both directions" at the same time. When carbonic acid (H_2CO_3) dissolves in water, some of it dissociates into H^+ (hydrogen ion) and HCO_3^- (bicarbonate), but not all of it. As additional ions dissociate, previously dissociated ions bond together again, forming H_2CO_3.

$$H_2CO_3 \leftrightarrow H^+ + HCO_3^-$$

In short, the double arrow indicates that at any instant in time both reactants and products are present in the solution at the same time.

Acids, Bases, and Salts

Besides water, many other inorganic compounds are important in the chemistry of life. For example, acids and bases are compounds that profoundly affect chemical reactions in the body, and as such are closely regulated. As explained in more detail at the beginning of Chapter 22, a few water molecules dissociate to form the H^+ and the OH^- (hydroxide ion):

$$H_2O \leftrightarrow H^+ + OH^-$$

Acids

In pure water, the balance of H^+ and OH^- is equal. However, when an acid such as hydrochloric acid (HCl) dissociates into H^+ and Cl^-, it shifts this balance in favor of excess H^+ ions. In the blood, carbon dioxide (CO_2) forms carbonic acid (H_2CO_3) when it dissolves in water. Some of the carbonic acid then dissociates to form H^+ ions and HCO_3^- (bicarbonate) ions, producing an excess of H^+ ions in the blood. Thus high CO_2 levels in the blood make the blood more acidic.

Bases

Bases, or **alkaline** compounds, on the other hand, shift the balance in the opposite direction. For example, sodium hydroxide (NaOH) is a base that forms OH^- but not H^+.

Looking at it simply, **acids** are compounds that produce an excess of H^+ ions, and bases are compounds that produce an excess of OH^-. Since OH^- can bind to H^+, bases actually decrease H^+ concentration of a solution.

pH

The relative H^+ concentration is a measure of how acidic or basic a solution is. The H^+ concentration is usually expressed in units of **pH.** The formula used to calculate pH units assigns a value of 7 to pure water. A higher pH value indicates a low relative concentration of H^+—a base. A lower pH value indicates a higher H^+ concentration—an acid.

Figure 2-7 shows a scale of pH from 0 to 14. Notice that when the pH of a solution is less than 7, the scale "tips" toward the side marked "high H^+." When the pH is more than 7, the scale "tips" toward the side marked "low H^+." pH units increase or decrease by factors of 10. Thus a pH 5 solution has 10 times the H^+ concentration of a pH 6 solution. A pH 4 solution has 100 times the H^+ concentration of a pH 6 solution.

A *strong acid* is an acid that completely, or almost completely, dissociates to form H^+ ions. Strong acids are indicated by very low pH values—far below pH 7. A *weak acid,* on the

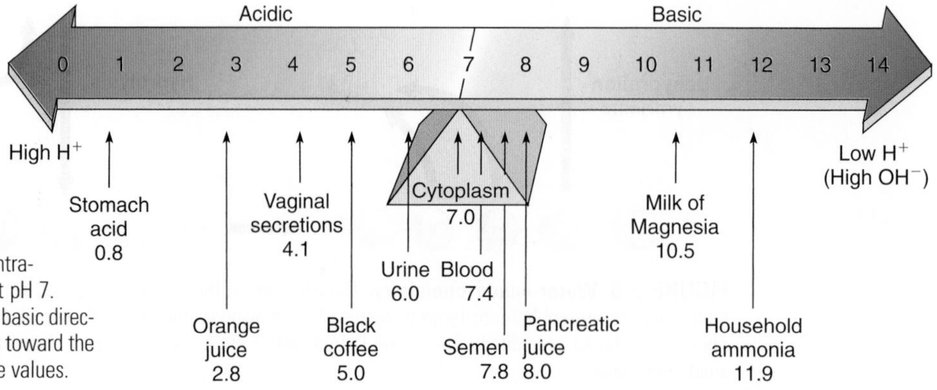

FIGURE 2-7 The pH scale. The H^+ concentration is balanced with the OH^- concentration at pH 7. At values above 7 (low H^+), the scale tips in the basic direction. At values below 7 (high H^+), the scale tips toward the acidic side. Examples given are normal, average values.

other hand, dissociates very little and therefore produces few excess H^+ ions in solution. Weak acids have a pH value just below 7.

Likewise, strong bases produce a very low relative H^+ concentration and have a very high pH value—far above 7. Weak bases produce a H^+ concentration a bit lower than pure water and thus have a pH value just a bit higher than 7.

To better understand this concept, use the Active Concept Map *Concept of pH* at *evolve.elsevier.com.*

Salts

When a strong acid and a strong base mix, excess H^+ may combine with the excess OH^- to form water. That is, they may *neutralize* each other. The remaining ions usually form neutral ionic compounds called **salts.** For example:

$$\underset{\text{acid}}{HCl} + \underset{\text{base}}{NaOH} \rightarrow H^+ + Cl^- + Na^+ + OH^- \rightarrow \underset{\text{water}}{H_2O} + \underset{\text{salt}}{NaCl}$$

Homeostasis of pH

The pH of body fluids affects body chemistry so greatly that normal body function can be maintained only within a narrow range of pH of about 7.35 to 7.45. **Acidosis** (low blood pH) and **alkalosis** (high blood pH) are equally dangerous and thankfully rarely occur because of the homeostatic mechanisms of the body.

The body can remove excess H^+ ions by excreting them in the urine (see Chapter 22). Another way to remove acid is by increasing the loss of CO_2 (an acid) by way of the respiratory system (see Chapter 17).

A third way to adjust the body's pH is by using **buffers**—chemicals in the blood that maintain pH. Buffers maintain pH balance by preventing sudden changes in the H^+ ion concentration. Buffers do this by forming a chemical system that neutralizes acids and bases as they are added to a solution.

The mechanisms by which the body maintains pH homeostasis, or acid-base balance, are discussed further in Chapter 22.

QUICK CHECK

1. Define an organic compound.
2. What is the difference between dehydration synthesis and hydrolysis?
3. What determines whether a solution is an acid, a base, or neutral?
4. How do buffers adjust the body's pH?
5. What are the chemical characteristics of water?

Organic Chemistry

Organic compounds are much more complex than inorganic compounds. In this section, we describe the basic structure and function of each major type of organic compound found in the body: *carbohydrates, lipids* (fats), *proteins,* and *nucleic acids.*

Table 2-3 summarizes the structure and the function of each major type of organic compound in the body. Refer to this table as you read through the descriptions that follow.

All four of these organic compounds are formed by dehydration synthesis reactions. Conversely, their bonds can be broken by hydrolysis.

Carbohydrates

The name **carbohydrate** literally means "carbon (C) and water (H_2O)," signifying the types of atoms that form carbohydrate molecules.

The basic unit of many carbohydrate molecules is called a **monosaccharide** (Figure 2-8). Glucose (dextrose) is an important monosaccharide in the body—cells use it as their primary source of energy (see Chapter 19).

A molecule made of two saccharide units is a double sugar, or **disaccharide.** The disaccharides sucrose (table sugar) and lactose (milk sugar) are important dietary carbohydrates. After they are eaten, the body breaks them down, or digests them, to form monosaccharides that can be used as cellular fuel.

Many saccharide units joined together form **polysaccharides.** Examples of polysaccharides are **glycogen** and *starch*. Glycogen is the polysaccharide of glucose that the human body stores. Plants store glucose as starch. Each glycogen molecule is a chain of glucose molecules joined together. When there is excess glucose in the blood, liver cells and muscle cells pull glucose out of the blood and store it as glycogen for later use. When we eat plants, we can break apart their starch molecules to get glucose.

Carbohydrates have potential energy stored in their bonds. When the bonds are broken in cells, they release energy that can then be used to do work. Chapter 19 explains more about the process by which the body extracts energy from carbohydrates and other nutrient molecules.

To better understand this concept, use the Active Concept Map *Metabolism of Glucose to Generate ATP* at *evolve.elsevier.com.*

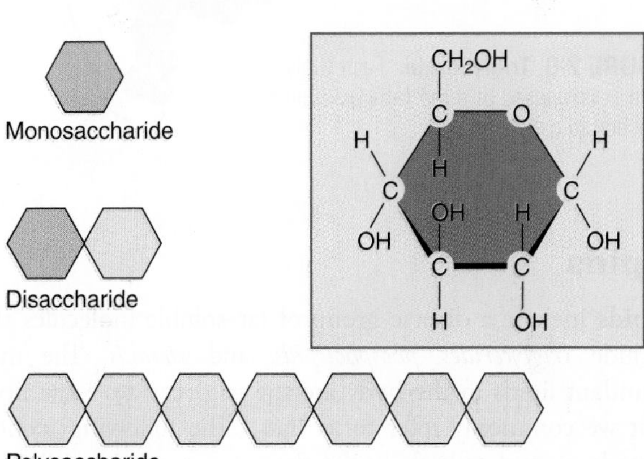

FIGURE 2-8 Carbohydrates. Monosaccharides are single carbohydrate units joined by dehydration synthesis to form disaccharides and polysaccharides. The detailed chemical structure of the monosaccharide *glucose* is shown in the inset.

2

TABLE 2-3 Major Types of Organic Compounds

EXAMPLE	COMPONENTS	FUNCTIONS
Carbohydrate		
Monosaccharide (glucose, galactose, fructose)	Single monosaccharide unit	Used as source of energy Unit used to build other carbohydrates
Disaccharide (sucrose, lactose, maltose)	Two monosaccharide units	Can be broken into monosaccharides
Polysaccharide (glycogen, starch)	Many monosaccharide units	Used to store monosaccharides (thus to store energy)
Lipid		
Triglyceride	One glycerol head, three fatty acid tails	Stores energy Provides protective or structural padding
Phospholipid	One glycerol/phosphate head, two fatty acid tails	Forms cell membranes
Steroid	Four carbon rings at core	Stabilizes cell membranes Category of hormones
Protein		
Structural proteins (fibers)	Amino acids	Form structures of the body
Functional proteins (enzymes, hormones)	Amino acids	Facilitate chemical reactions Carry signals Regulate functions
Nucleic Acid		
Deoxyribonucleic acid (DNA)	Nucleotides (contain deoxyribose)	Contains information (genetic code) for making proteins
Ribonucleic acid (RNA)	Nucleotides (contain ribose)	Serves as a copy of a portion of the genetic code during protein synthesis
Adenosine triphosphate (ATP)	Modified nucleotide (ribose, adenine, and three phosphates)	Transfers energy from nutrient molecules to power work in the cell

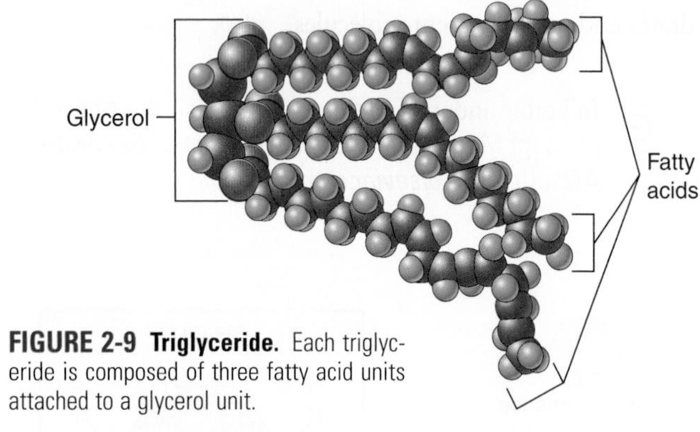

FIGURE 2-9 Triglyceride. Each triglyceride is composed of three fatty acid units attached to a glycerol unit.

Lipids

Lipids include a diverse group of fat-soluble molecules that include *triglycerides, phospholipids,* and *steroids.* The most abundant lipids in the body are the triglycerides—the lipids that we commonly refer to as "fats." The following sections provide more details about the three major groups of lipids.

Triglycerides

Triglycerides, or fats, are lipid molecules formed by a **glycerol** unit or "head" joined to three **fatty acid** "tails" (**Figure 2-9**). The fatty acid components can be classified as *saturated* or *unsaturated.* Saturated fatty acids tend to be solid at room temperature and are found in butter and lard. Unsaturated fatty acids tend to be liquids and are found in oils like corn oil and olive oil.

Like carbohydrates, the bonds in triglycerides can be broken to yield energy (see Chapter 19). Thus triglycerides are useful in storing energy in cells for later use.

Triglycerides stored in fat tissue also provide helpful "padding" around organs and under the skin to stabilize and protect body structures.

Phospholipids

Phospholipids are similar in structure to triglycerides but instead of having three fatty acid chains attached to a glycerol, they have two fatty acid chains and a phosphorus-containing unit called a *phosphate* attached to a glycerol. The phosphate at the base of the glycerol "head" attracts water. The two fatty acid "tails" repel water.

Figure 2-10, *A* shows the head and tail of the phospholipid molecule. This structure allows them to form a stable double layer—a **bilayer**—in water. A phospholipid bilayer forms the foundation for the cell membrane. In **Figure 2-10,** *B* the water-attracting heads face the water and the water-repelling tails face away from the water (and toward each other).

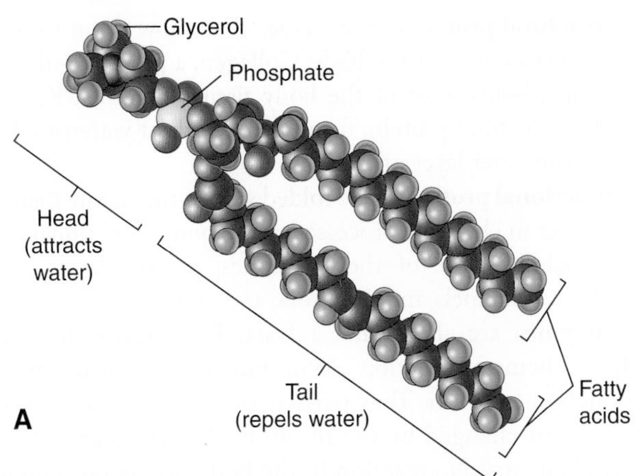

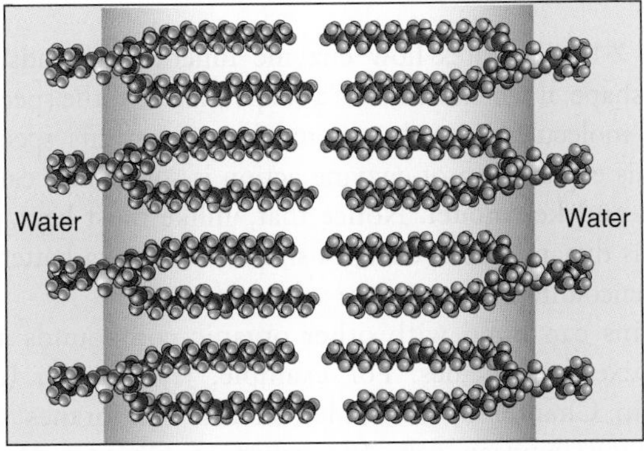

B

FIGURE 2-10 Phospholipids. A, Each phospholipid molecule has a phosphorus-containing "head" that attracts water and a lipid "tail" that repels water. **B,** Because the tails repel water, phospholipid molecules often arrange themselves so that their tails face away from water. The stable structure that results is a bilayer sheet forming a small bubble. The membrane around each cell of the body is formed of such a structure.

Steroids

Steroid molecules have multiple-ring structures, as shown in Figure 2-11.

Cholesterol is an important steroid lipid that performs several critical functions in the body. For example, it is embedded within the cells to help stabilize its bilayer structure. As Chapter 12 explains, the body also uses cholesterol as a starting point in making *steroid hormones* such as estrogen, testosterone, and cortisone.

Cholesterol

Cortisol
(a steroid hormone)

FIGURE 2-11 Steroids. Cholesterol *(left)* has a steroid structure, represented here as four colored rings. Changes to the side groups can convert cholesterol to cortisol (shown) or other steroid hormones.

CLINICAL APPLICATION

BLOOD LIPOPROTEINS

A lipid such as cholesterol can travel in the blood only after it attaches to triglycerides and is wrapped with a layer of phospholipids studded with protein molecules (see figure). This large, complex **lipoprotein** capsule thus becomes a "transport vehicle" for cholesterol in the blood.

Some of these lipoprotein structures are called *high-density lipoproteins (HDLs)* because they have a high density of protein (more protein than lipid). Another type of molecule contains less protein (and more lipid), so it is called *low-density lipoprotein (LDL)*.

LDL carries cholesterol *to cells*, including the cells that line blood vessels. If a large amount of cholesterol builds up in artery walls, a condition called **atherosclerosis** may develop. Atherosclerosis may progress to life-threatening arterial blockages, especially when they occur in the heart or brain (see Figure 15-5 on p. 407).

HDL, on the other hand, carries cholesterol *away from cells* and toward the liver for elimination from the body. A high proportion of HDL in the blood is associated with a low risk of developing atherosclerosis—and may even reduce existing buildup of cholesterol in arterial walls.

Factors such as cigarette smoking decrease HDL levels and thus contribute to the risk of atherosclerosis. Factors such as exercise increase HDL levels and thus decrease the risk of atherosclerosis.

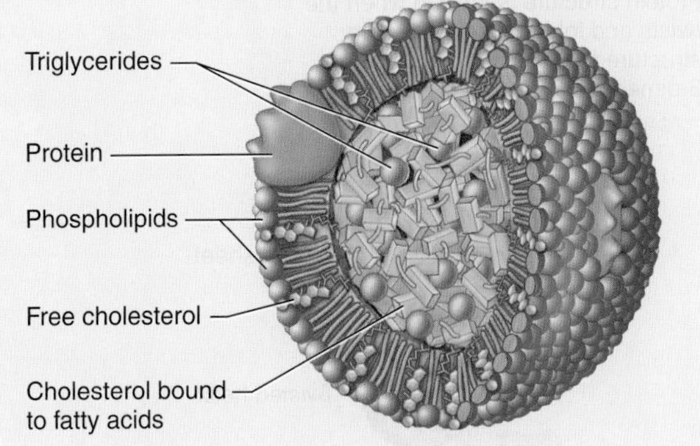

Proteins

Proteins are very large molecules composed of basic units called **amino acids.** In addition to carbon, hydrogen, and oxygen, all amino acids contain nitrogen (N). Many different amino acids are linked together in a particular sequence to form all of the proteins in cells. The process that joins amino acids by **peptide bonds** is fully discussed in Chapter 3.

Attractions between positively charged and negatively charged regions along the long amino acid chain cause it to fold over on itself and maintain its unique shape. The complex, three-dimensional molecule that results is a protein molecule (Figure 2-12). The folded shape of a protein molecule determines its role in body chemistry.

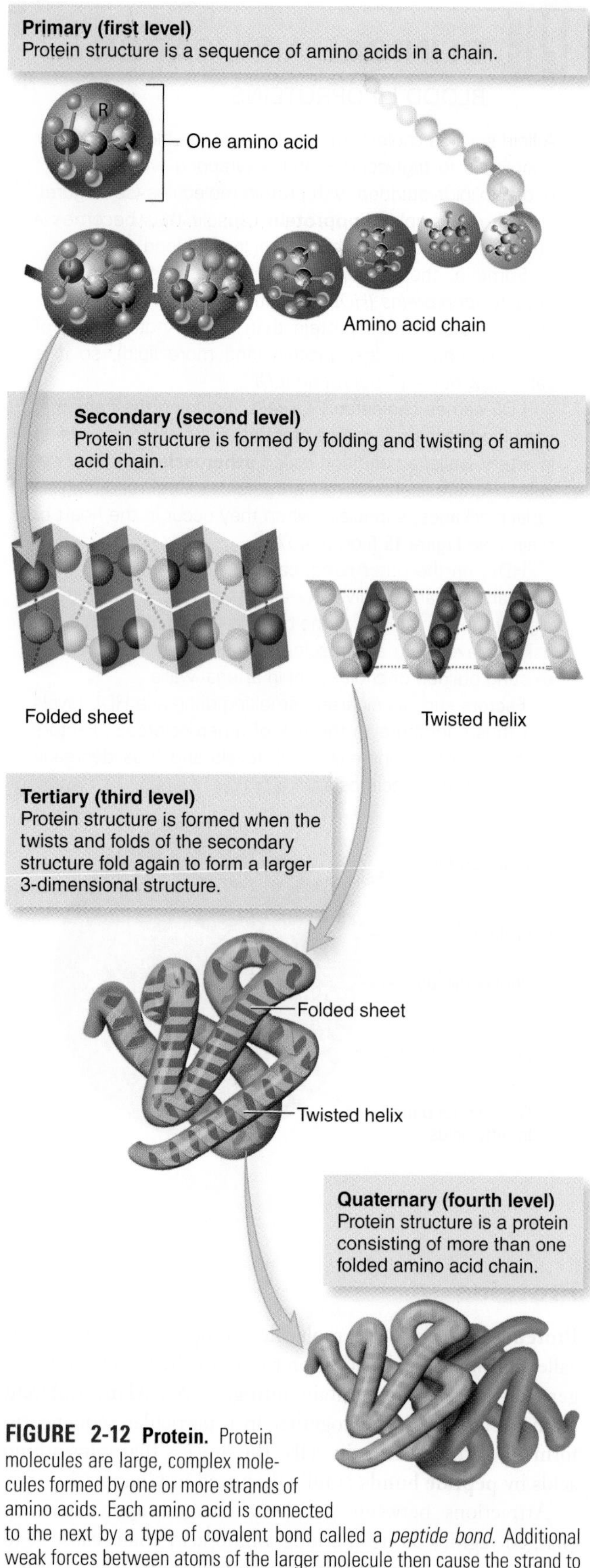

FIGURE 2-12 Protein. Protein molecules are large, complex molecules formed by one or more strands of amino acids. Each amino acid is connected to the next by a type of covalent bond called a *peptide bond.* Additional weak forces between atoms of the larger molecule then cause the strand to twist or fold into a *secondary* (second-level) protein structure. New relationships among the atoms then cause the molecule to fold again on itself to form a three-dimensional *tertiary* (third-level) protein structure. Several tertiary proteins may join to form a *quaternary* (fourth-level) protein structure.

Structural proteins have shapes that allow them to form essential structures of the body. Collagen, a protein with a fiber shape, holds most of the body tissues together. Keratin, another structural protein, forms a network of waterproof fibers in the outer layer of the skin.

Functional proteins have folded shapes that allow them to participate in chemical processes of the body. Functional proteins include some of the hormones, growth factors, cell membrane channels and receptors, and enzymes.

Enzymes are chemical catalysts. This means that they help a chemical reaction occur but are not reactants or products themselves. They participate in chemical reactions but are not changed by the reactions. Enzymes are vital to body chemistry. No reaction in the body occurs fast enough unless the specific enzymes needed for that reaction are present.

Figure 2-13 illustrates how enzyme function depends on enzyme shape. Each enzyme has a shape that "fits" the specific *substrate* molecules it works on, much as a key fits specific locks. This explanation of enzyme action is sometimes called the **lock-and-key model.** Notice that, unlike most keys, the enzyme is dynamic, so it changes shape to ensure a better fit when it encounters one or more substrates.

Proteins can bond with other organic compounds and form "mixed" molecules. For example, *glycoproteins* (described in Chapter 3) embedded in cell membranes and *proteoglycans* between cells (described in Chapter 4) are proteins with sugars attached. *Lipoproteins* are lipid-protein combinations, as described in the Clinical Application box on p. 33.

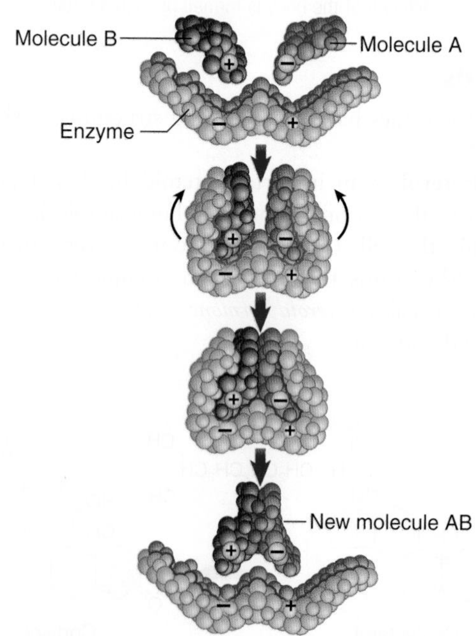

FIGURE 2-13 Enzyme action. Enzymes are functional proteins whose molecular shape and ability to alter shape allow them to catalyze chemical reactions. Substrate molecules *A* and *B* are brought together by the enzyme to form a larger molecule, *AB.*

TABLE 2-4 Components of Nucleotides

NUCLEOTIDE	DNA	RNA
Sugar	Deoxyribose	Ribose
Phosphate	Phosphate	Phosphate
Nitrogen base	Cytosine	Cytosine
	Guanine	Guanine
	Adenine	Adenine
	Thymine	Uracil

Nucleic Acids

The two forms of **nucleic acid** are **deoxyribonucleic acid (DNA)** and **ribonucleic acid (RNA).** As outlined in Chapter 3, the basic building blocks of nucleic acids are called **nucleotides.**

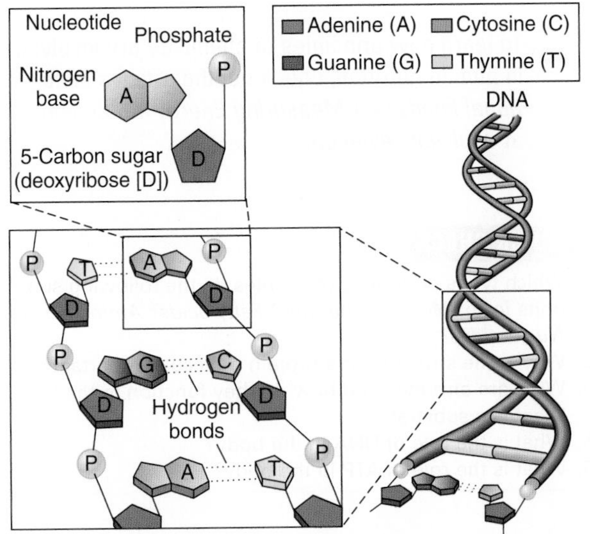

FIGURE 2-14 DNA. Deoxyribonucleic acid (DNA), like all nucleic acids, is composed of units called *nucleotides.* Each nucleotide has a phosphate, a sugar, and a nitrogen base. In DNA, the nucleotides are arranged in a double helix formation.

Each nucleotide consists of a *phosphate unit,* a sugar (*ribose* or *deoxyribose*), and a *nitrogen base.* DNA nucleotide bases include **adenine, thymine, guanine,** and **cytosine.** RNA uses the same set of bases, except for the substitution of **uracil** for thymine (Table 2-4).

Nucleotides bind to one another to form strands or other structures. In the DNA molecule, nucleotides are arranged in a twisted, double strand called a **double helix** (Figure 2-14).

The sequence of different nucleotides along the DNA double helix is the "master code" for assembling proteins and other nucleic acids. *Messenger RNA (mRNA)* molecules have a sequence that forms a temporary "working copy" of a portion of the DNA code called a gene. The genetic code in nucleic acids ultimately directs the entire "symphony" of living chemistry.

A modified nucleotide called **adenosine triphosphate (ATP)** plays an important energy-transfer role in the body. As Figure 2-15 shows, adenosine (a base and a sugar) has not just one phosphate, as in a standard nucleotide, but instead has three phosphates. The "extra" phosphates are attached to the molecule with unstable "high-energy" bonds that require a great amount of energy (from nutrients) to form. Therefore, they release a large amount of energy when broken. When a phosphate breaks off of ATP—forming **adenosine diphosphate (ADP)**—the energy released is used to do work in cells.

ATP thus acts as a sort of energy-transfer "battery" that picks up energy from nutrients and then quickly makes the energy available to cellular processes. Chapter 19 outlines details of how ATP works in the cells.

To learn more about how energy in the body is measured, including examples of the energy content of food and the energy cost of common activities, check out the article *Measuring Energy* at Connect It! at *evolve.elsevier.com.*

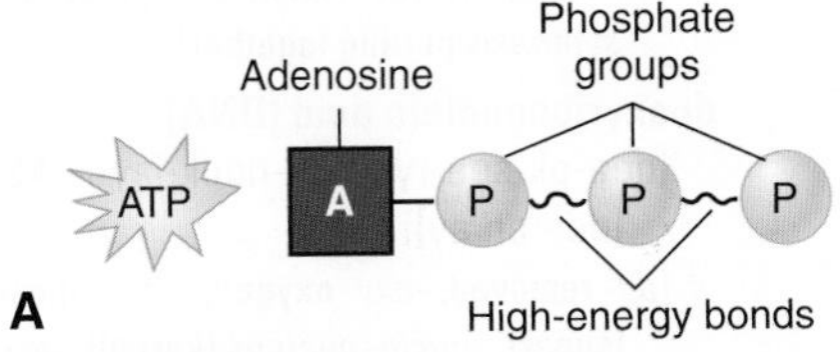

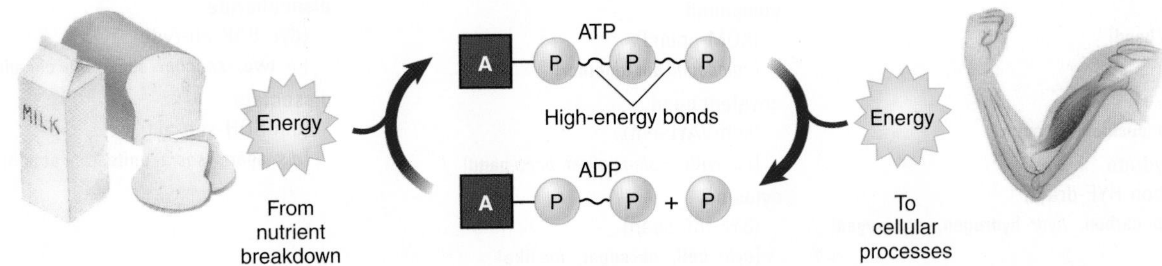

FIGURE 2-15 ATP. A, Structure of adenosine triphosphate (ATP). Because the adenosine group is made up of a sugar (ribose) and a base (adenine), ATP is really a nucleotide with added phosphates. **B,** The role of ATP in transferring energy from nutrient molecules to cellular processes. ADP, Adenosine diphosphate.

SCIENCE APPLICATIONS

BIOCHEMISTRY

Rosalind Franklin (1920–1958)

British scientist Rosalind Franklin was one of the leading **biochemists** of the modern age. Franklin used x-rays to cast shadows through DNA to analyze its structure. When she was only 32 years old, she discovered the unusual helical (spiral) structure of the DNA molecule and how the sugars and phosphates form an outer backbone for the molecule (**Figure 2-14**).

Her breakthrough helped James Watson, Francis Crick, and Maurice Wilkins to finally work out the structure and function of DNA in 1953 and thus crack the "code of life." The three men received a Nobel Prize for their achievement in 1962, but Franklin's early death from cancer in 1958 prevented her from sharing in the credit for one of the greatest discoveries of all time.

Biochemists continue to make important discoveries that increase our understanding of human structure and function. Aided by **laboratory technicians** and *lab assistants*, biochemists also find ways to help other professionals apply biochemistry to solve everyday problems. For example, **clinical laboratory technicians** analyze samples from the bodies of patients for signs of health or disease. Others who use biochemistry as a basis for their work include **nuclear medicine technologists, pharmacists** and **pharmacy technicians, dietitians,** *forensic investigators,* **genetic counselors,** and even *science journalists*.

Clinical Applications of Chemistry

In addition to the clinical applications already mentioned in this chapter—such as pH imbalances—there are many yet to come as we move through each remaining chapter. Take a moment to scan through Appendix C (at *evolve.elsevier.com*), which lists normal concentrations of various chemicals found in the blood, urine, or other body fluids of a healthy person. You can see that our bodies are, in a way, chemical systems where each chemical must be maintained in a healthy balance.

Continue to watch for the important roles played by water, oxygen, carbon dioxide, ions, pH, carbohydrates, lipids, proteins, and nucleic acids as you progress through your course. This will help you understand the "big picture" of human structure and function. Do not hesitate to return to this chapter later on in your studies when you need a quick "refresher" on one of these chemistry topics.

To learn how principles of chemistry are involved in human nutrition, check out the articles *Functional Foods* and *Measuring Energy* at Connect It! at *evolve.elsevier.com*.

QUICK CHECK

1. Which types of organic molecules do the following subunits form: *Monosaccharides? Fatty acids? Amino acids? Nucleotides?*
2. Why is the structure of a protein molecule important?
3. What are enzymes and how do they function in the body?
4. What is a substrate?
5. What is the role of DNA in the body?
6. What is the role of ATP in the body?

LANGUAGE OF **SCIENCE** *(continued from p. 25)*

bilayer
(BYE-lay-er)
[*bi-* **two, layer**]

biochemist
(bye-oh-KEM-ist)
[*bio-* **life,** *-chem-* **alchemy,** *-ist* **agent**]

biochemistry
(bye-oh-KEM-is-tree)
[*bio-* **life,** *-chem-* **alchemy,** *-ist* **agent,** *-ry* **practice of**]

bond
[*bond* **band**]

buffer
(BUF-er)
[*buffe-* **cushion,** *-er* **actor**]

carbohydrate
(kar-boh-HYE-drayt)
[*carbo-* **carbon,** *-hydr-* **hydrogen,** *-ate* **oxygen**]

carbon
(KAR-bun)
[*carbon* **coal**]

carbon dioxide (CO_2)
(KAR-bun dye-AHK-syde [see oh too])
[*carbon* **coal,** *di-* **two,** *-ox-* **sharp (oxygen),** *-ide* **chemical**]

cholesterol
(koh-LES-ter-ol)
[*chole-* **bile,** *-stero-* **solid,** *-ol* **oil**]

compound
(KOM-pound)
[*compound* **put together**]

covalent bond
(koh-VAYL-ent)
[*co-* **with,** *-valen* **power,** *bond* **band**]

cytosine
(SYE-toh-seen)
[*cyto-* **cell,** *-os-* **sugar,** *-ine* **like**]

dehydration synthesis
(dee-hye-DRAY-shun SIN-the-sis)
[*de-* **from,** *-hydrat-* **water,** *-tion* **process,** *synthesis* **putting together**]

deoxyribonucleic acid (DNA)
(dee-ok-see-rye-boh-nook-lay-ik AS-id [dee en ay])
[*de-* **removed,** *-oxy-* **oxygen,** *-ribo-* **ribose (sugar),** *-nucle-* **nucleus (kernel),** *-ic* **relating to,** *acid* **sour**]

disaccharide
(dye-SAK-ah-ryde)
[*di-* **two,** *-sacchar-* **sugar,** *-ide* **chemical**]

dissociate
(dih-SOH-see-ayt)
[*dis-* **apart,** *-socia-* **unite,** *-ate* action]

LANGUAGE OF SCIENCE (continued from p. 36)

double helix
(DUB-el HEE-lix)
pl., helices
(HEE-lis-eez)
[*helix* **spiral**]

electrolyte
(eh-LEK-troh-lyte)
[*electro-* **electricity,** *-lyt* **loosening**]

electron
(eh-LEK-tron)
[*electr-* **electric,** *-on* **unit**]

element
(EL-eh-ment)
[*element* first principle]

energy level
(EN-er-gee LEV-el)
[*en-* **in,** *-erg-* **work,** *-y* **state**]

enzyme
(EN-zyme)
[*en-* **in,** *-zyme* **ferment**]

fatty acid
(FAT-tee AS-id)
[*fat-* **fat,** *-ty* **state,** *acid* **sour**]

formula
(FOR-myoo-lah)
[*form-* **form,** *-ula* **little**]

functional protein
(FUNK-shen-al PROH-teen)
[*function-* **perform,** *-al* **relating to,** *prote-* **primary,** *-in* **substance**]

glycerol
(GLIS-er-ol)
[*glyce-* **sweet,** *-ol* **alcohol**]

glycogen
(GLYE-koh-jen)
[*glyco-* **sweet,** *-gen* **produce**]

guanine
(GWAH-neen)
[*guan-* **guano,** *-ine* **like**]

hydrogen
(HYE-droh-jen)
[*hydro-* **water,** *-gen* **produce**]

hydrogen bond
(HYE-droh-jen bond)
[*hydro-* **water,** *-gen* **produce,** *bond* **band**]

hydrolysis
(hye-DROHL-ih-sis)
[*hydro-* **water,** *-lysis* **loosening**]

inorganic compound
(in-or-GAN-ik KOM-pownd)
[*in-* **not,** *-organic* **natural,** *compound* **to assemble**]

ion
(AYE-on)
[*ion* **to go**]

ionic bond
(aye-ON-ik bond)
[*ion* **to go,** *-ic-* **relating to,** *bond* **band**]

isotope
(AYE-soh-tohp)
[*iso-* **equal,** *-tope* **place**]

lipid
(LIP-id)
[*lip-* **fat,** *-id* **form**]

lipoprotein
(lip-oh-PROH-teen)
[*lipo-* **fat,** *prote-* **primary,** *-in* **substance**]

lock-and-key model
(lok and kee MAHD-el)

matter
(MAT-er)
[*matter* **something from which something is made**]

molecule
(MOL-eh-kyool)
[*mole-* **mass,** *-cul* **small**]

monosaccharide
(mon-oh-SAK-ah-ryde)
[*mono-* **one,** *-sacchar-* **sugar,** *-ide* **chemical**]

neutron
(NOO-tron)
[*neuter-* **neither,** *-on* **unit**]

nitrogen
(NYE-troh-jen)
[*nitro-* **soda,** *-gen* **produce**]

nucleic acid
(noo-KLAY-ik AS-id)
[*nucle-* **kernel,** *-ic* **relating to,** *acid* **sour**]

nucleotide
(NOO-klee-oh-tyde)
[*nucleo-* **nut or kernel,** *-ide* **chemical**]

nucleus
(NOO-klee-us)
pl., nuclei
(NOO-klee-aye)
[*nucleus* **kernel**]

orbital
(OR-bih-tal)
[*orb-* **circle,** *disk* **ring,** *-al* **relating to**]

organic compound
(or-GAN-ik KOM-pownd)
[*organ-* **tool or instrument,** *-ic* **relating to,** *compound* **to assemble**]

oxygen (O_2)
(AHK-sih-jen [oh too])
[*oxy-* **sharp,** *-gen* **produce**]

peptide bond
(PEP-tyde)
[*pept-* **digest,** *-ide* **chemical,** *bond* **band**]

pH
(pee aych)
[abbreviation for *potenz* **power,** *hydrogen* **hydrogen**]

phospholipid
(foys-foh-LIP-id)
[*phospho-* **phosphorus,** *-lip-* **fat,** *-id* **form**]

polysaccharide
(pahl-ee-SAK-ah-ryde)
[*poly-* **many,** *-sacchar-* **sugar,** *-ide* **chemical**]

product
(PROD-ukt)

protein
(PROH-teen)
[*prote-* **primary,** *-in* **substance**]

proton
(PROH-ton)
[*proto-* **first,** *-on* **unit**]

radioactive isotope
(ray-dee-oh-AK-tiv AYE-soh-tohp)
[*radio-* **send out rays,** *iso-* **equal,** *-tope* **place**]

reactant
(ree-AK-tant)
[*re-* **again,** *-act-* **act,** *-ant* **agent**]

ribonucleic acid (RNA)
(rye-boh-noo-KLAY-ik AS-id [ar en ay])
[*ribo-* **ribose (sugar),** *-nucle-* **nucleus,** *-ic* **relating to,** *acid* **sour**]

salt
(sawlt)

solute
(SOL-yoot)
[*solut* **dissolved**]

solvent
(SOL-vent)
[*solv-* **dissolve,** *-ent* **agent**]

steroid
(STAYR-oid)
[*ster-* **sterol,** *-oid* **like**]

structural protein
(STRUK-shur-al PROH-teen)
[*structura-* **arrangement,** *-al* **relating to,** *prote-* **primary,** *-in* **substance**]

thymine
(THYE-meen)
[*thym-* **thymus,** *-ine* **like**]

triglyceride
(try-GLIH-ser-ayed)
[*tri-* **three,** *-glycer-* **sweet,** *-ide* **chemical**]

uracil
(YOOR-ah-sil)
[*ura-* **urea,** *-il* **chemical**]

water
(WAyT-er)

2

2

LANGUAGE OF **MEDICINE**

acidosis
(as-ih-DOH-sis)
[*acid-* **sour,** *-osis* **condition**]

alkalosis
(al-kah-LOH-sis)
[*alkal-* **ashes,** *-osis* **condition**]

atherosclerosis
(ath-er-oh-skleh-ROH-sis)
[*ather-* **porridge,** *-sclero-* **harden,** *-osis* **condition**]

clinical laboratory technician
(KLIN-ih-kal LAB-rah-tor-ee tek-NISH-en)
[*clin-* **sickbed,** *-ic* **relating to,** *-al* **relating to,** *labor-* **work,** *-tory* **place of activity,** *techn-* **art or skill,** *-ic* **relating to,** *-ian* **practitioner**]

dietitian
(dye-eh-TISH-en)
[*diet-* **way of living,** *-itian* **practitioner**]

genetic counselor
(jeh-NET-ik KOWN-se-ler)
[*gene-* **produce,** *-ic* **relating to,** *counsel-* **advise or plan,** *-or* **agent**]

laboratory technician
(LAB-rah-tor-ee tek-NISH-en)
[*labor-* **work,** *-tory* **place of activity,** *techn-* **art or skill,** *-ic* **relating to,** *-ian* **practitioner**]

nuclear medicine technologist
(NOO-klee-ar MED-ih-sin tek-NOL-oh-jist)
[*nucle-* **nut or kernel,** *-ar* **relating to,** *techn-* **art or skill,** *-log-* **words (study of),** *-ist* **agent**]

pharmacist
(FAR-mah-sist)
[*pharmac-* **drug,** *-ist* **agent**]

pharmacy technician
(FAR-mah-see tek-NISH-en)
[*pharmac-* **drug,** *-y* **location of activity,** *techn-* **art or skill,** *-ic* **relating to,** *-ian* **practitioner**]

radiation sickness
(ray-dee-AY-shun SIK-nes)
[*radiat-* **send out rays,** *-ion* **process**]

❑ OUTLINE SUMMARY

To download a digital version of the chapter summary for use with your device, access the ***Audio Chapter Summaries*** *online at evolve.elsevier.com.*

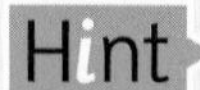

Scan this summary after reading the chapter to help you reinforce the key concepts. Later, use the summary as a quick review before your class or before a test.

Levels of Chemical Organization

A. Atoms (**Figures 2-1** and **2-2**)
   1. Nucleus—central core of atom
      a. Proton—positively charged particle in nucleus
      b. Neutron—uncharged particle in nucleus
      c. Atomic number—number of protons in nucleus
      d. Atomic mass—number of protons and neutrons combined
   2. Energy levels—orbital regions surrounding atomic nucleus that contain electrons
      a. Electron—negatively charged particle
      b. May contain up to eight electrons in each level
      c. Energy level increases the farther it is from the nucleus
B. Elements, molecules, and compounds
   1. Element—a pure substance; made up of only one kind of atom
   2. Molecule—a group of atoms bound together in a group
   3. Compound—substances whose molecules have more than one kind of atom

Chemical Bonding

A. Chemical bonds form to make atoms more stable
   1. Atoms react with one another in ways that make their outermost energy level full
   2. Atoms may share electrons or donate or borrow them to become stable
B. Ionic bonds (**Figure 2-3**)
   1. Ions form when an atom gains or loses electrons in its outer energy level to become stable
      a. Positive ion—has lost electrons; indicated by superscript positive sign(s), as in Na^+ or Ca^{++}
      b. Negative ion—has gained electrons; indicated by superscript negative sign(s), as in Cl^-
   2. Ionic bonds form when positive and negative (oppositely charged) ions attract each other
   3. Electrolyte—compound that dissociates (breaks apart) in water to form individual ions; an ionic compound
C. Covalent bonds (**Figure 2-4**)
   1. Covalent bonds form when atoms try to complete their outer energy levels by sharing electrons
   2. Covalent bonds do not easily dissociate in water
   3. Covalent bonding is used to form all of the major organic compounds found in the body
D. Hydrogen bonds (**Figure 2-5**)
   1. Hydrogen bonds are relatively weak bonds that do not create new molecules
   2. Hydrogen bonds form when partially charged regions of neighboring molecules attract one another
   3. Hydrogen bonds are present in water, DNA, and proteins

Inorganic Chemistry

A. *Organic* molecules contain carbon-carbon covalent bonds and/or carbon-hydrogen covalent bonds; *inorganic* molecules do not
B. Organic molecules are generally larger and more complex than inorganic molecules
C. Water
 1. Water is essential to life
 2. Water is a solvent (liquid in which solutes are dissolved), forming aqueous solutions in the body
 3. Water is involved in chemical reactions (**Figure 2-6**)
  a. Dehydration synthesis—chemical reaction in which water is removed from small molecules so they can be strung together to form a larger molecule
  b. Hydrolysis—chemical reaction in which water is added to a large molecule to break it into smaller molecules
  c. All the major organic molecules are formed through dehydration synthesis and broken apart by hydrolysis
  d. Chemical reactions always involve energy transfers, as when energy is used to build ATP molecules
  e. Chemical equations show how reactants interact to form products; arrows separate the reactants from the products
D. Acids, bases, and salts
 1. Water molecules dissociate to form equal amounts of H^+ (hydrogen ion) and OH^- (hydroxide ion)
 2. Acid—substance that shifts the H^+/OH^- balance in favor of H^+; opposite of base
 3. Base—substance that shifts the H^+/OH^- balance against H^+; also known as an *alkaline;* opposite of acid
 4. pH—mathematical expression of relative H^+ concentration in an aqueous solution (**Figure 2-7**)
  a. A pH value of 7 is neutral (neither acid nor base)
  b. pH values above 7 are basic; pH values below 7 are acidic
 5. Neutralization occurs when acids and bases mix and form salts
 6. pH imbalance occurs when blood pH is too high (alkalosis) or too low (acidosis); homeostasis restores and maintains pH balance in the body
 7. Buffers form chemical systems that neutralize excess acids or bases and thus maintain a relatively stable pH

Organic Chemistry

A. Carbohydrates—sugars and complex carbohydrates (**Figure 2-8**)
 1. Contain carbon (C), hydrogen (H), oxygen (O)
 2. Monosaccharide—basic unit of carbohydrate molecules (e.g., glucose)
 3. Disaccharide—double sugar made up of two monosaccharide units (e.g., sucrose, lactose)
 4. Polysaccharide—complex carbohydrate made up of many monosaccharide units (e.g., glycogen; stored by the body)
 5. Function of carbohydrates is to store energy for later use
B. Lipids—diverse group of fat-soluble molecules
 1. Triglycerides (**Figure 2-9**)
  a. Formed by a glycerol unit joined to three fatty acids
   (1) Saturated fatty acids are usually solid at room temperature
   (2) Unsaturated fatty acids are usually liquid at room temperature
  b. Store energy for later use
  c. Provide padding around organs and under the skin
 2. Phospholipids
  a. Similar to triglyceride structure, but have phosphorus-containing units—each with a head and two tails (**Figure 2-10**)
  b. The head attracts water and the double tail does not, thus forming stable double layers (bilayers) in water
  c. Form membranes of cells
 3. Steroids
  a. Molecules have a structure made up of multiple rings (**Figure 2-11**)
  b. Cholesterol is an important steroid
  c. Cholesterol stabilizes the phospholipid tails in cellular membranes and is also converted into steroid hormones by the body
C. Proteins
 1. Very large molecules made up of amino acids held together in long, folded chains by peptide bonds (**Figure 2-12**)
 2. Structural proteins
  a. Form various structures of the body
   (1) Collagen is a fibrous protein that holds many tissues together
   (2) Keratin forms tough, waterproof fibers in the outer layer of the skin
 3. Functional proteins
  a. Participate in chemical processes of the body
  b. Examples include hormones, cell membrane channels and receptors, and enzymes
 4. Enzymes—chemical catalysts
  a. Help chemical reactions occur
  b. Enzyme action sometimes called lock-and-key model because enzymes fit their substrates much like a key fits into a lock (**Figure 2-13**)
 5. Proteins can combine with other organic molecules to form "mixed" molecules such as glycoproteins or lipoproteins
D. Nucleic acids
 1. Made up of nucleotides that include:
  a. A phosphate unit
  b. A sugar (ribose or deoxyribose)
  c. A nitrogen base (adenine [A], thymine [T] or uracil [U], guanine [G], cytosine [C])
 2. DNA (deoxyribonucleic acid)
  a. Used as the cell's "master code" for assembling proteins
  b. Uses deoxyribose as the sugar and A, T (not U), C, and G as bases
  c. Forms a double helix shape (**Figure 2-14**)

3. RNA (ribonucleic acid)
   a. Used as a temporary "working copy" of a gene (portion of the DNA code)
   b. Uses ribose as the sugar and A, U (not T), C, and G as bases
4. By directing the formation of structural and functional proteins, nucleic acids ultimately direct overall body structure and function
5. ATP (adenosine triphosphate) is a modified nucleotide used to transfer energy from nutrients to cellular processes, thus acting as an energy-transfer "battery" (**Figure 2-15**)

Clinical Applications of Chemistry

A. Appendix C (at *evolve.elsevier.com*) illustrates the normal values of various chemicals in the body
B. The human body is a chemical system in which all chemicals must remain in a healthy balance

2

❑ ACTIVE LEARNING

STUDY TIPS

Consider using these tips to achieve success in meeting your learning goals.

This chapter introduces you to some basic chemical concepts that are used later in other chapters to describe structures and functions of the body. First of all, it is important that you can read and understand a handful of important chemical symbols and equations.

1. Practice by putting the chemical symbols found in **Tables 2-1** and **2-2** on flash cards, and then pair up with a classmate and quiz each other on what the symbols stand for. Also learn to identify whether each one *is* or *is not* an ion.
2. If your instructor requires you to know the parts of the atom, make your own labeled diagram of an atom or make a three-dimensional model out of household items such as marshmallows, toothpicks, and string. There are also many online resources that illustrate the parts of the atom. Using multiple senses will help you learn and remember the information.
3. It is important that you learn the concept of pH value, which will be an integral part of later discussions. Develop a *T-chart* that lists the pH values (1 to 14) and give examples (besides those listed in your text) of substances and their appropriate pH value. (To learn about T-charts, go to *my-ap.us/Lzxuko*).
4. **Table 2-3** summarizes some important concepts of the structure and function of the major organic compounds that you will be using later in the course. Make your own version of the table on a poster-sized piece of paper and add simple pictures of the different molecules. Then make flash cards or use online flash cards and practice identifying which category different molecules belong to: protein, carbohydrate, lipid, or nucleic acid. Practice identifying which function each compound performs.

Review Questions

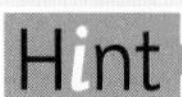

Write out the answers to these questions after reading the chapter and reviewing the Chapter Summary. If you simply think through the answer without writing it down, you will not retain much of your new learning.

1. Define the terms *element, compound, atom,* and *molecule.*
2. Name and define three kinds of particles within an atom.
3. What is an energy level?
4. What is a chemical bond?
5. What are the major types of chemical bonds?
6. What is an electrolyte? An ion?
7. Define the terms *organic compound* and *inorganic compound.*
8. What is a solvent? A solute?
9. Explain the concept of pH.
10. What is an acid? A base?
11. Briefly describe the structure of each of the following: protein, lipid, carbohydrate, nucleic acid.
12. Briefly state the principal functions of each of the following: carbohydrate, protein, lipid, nucleic acid.
13. What are the three main parts of nucleotides?
14. What is alkalosis?
15. What organic compound is associated with atherosclerosis?
16. Describe atherosclerosis and give an example of a habit that may increase and one that can decrease your risk of developing atherosclerosis.
17. What is an aqueous solution?

Critical Thinking

After finishing the Review Questions, write out the answers to these more in-depth questions to help you apply your new knowledge. Go back to sections of the chapter that relate to concepts that you find difficult.

18. Compare and contrast how ionic bonds and covalent bonds solve the problem of achieving stability in atoms.
19. A particular protein molecule is hydrolyzed by an enzyme. How would you explain that process to someone unfamiliar with chemical terminology?
20. Your blood normally has a pH of around 7.4. Is your blood alkaline, acid, or neutral?
21. If a newly discovered protein was found to regulate how hormones influence the functions of cells in the body, would the protein be a structural protein or a functional protein?
22. Describe how DNA regulates all of the body's structures and functions?
23. How would you explain the difference between ^1H, ^2H, and ^3H?

Chapter Test

Hint *After studying the chapter, test your mastery by responding to these items. Try to answer them without looking up the answers.*

1. ________ is anything that occupies space and has mass.
2. Molecules are made up of particles called ________.
3. Positively charged particles within the nucleus of an atom are called ________.
4. Electrons inhabit regions of the atoms called ________ levels.
5. Substances with molecules having more than one kind of atom are called ________.
6. A(n) ________ chemical bond occurs when atoms share electrons.
7. The symbol K^+ represents the potassium ________.
8. A compound that dissociates in water to form ions is called a(n) ________.
9. Molecules that have a carbon-carbon bond in them are classified as ________ compounds.
10. In salt water, salt is the solute and water is the ________.
11. When water is used to build up small molecules into larger molecules, the process is called ________.
12. ________ are solutions that have an excess of hydrogen ions.
13. The blood contains chemicals called ________ that maintain a stable pH.

Match each term in column B with its related term in column A.

Column A	Column B
14. ________ glycogen	a. salt
15. ________ collagen	b. acid
16. ________ RNA	c. base
17. ________ cholesterol	d. carbohydrate
18. ________ NaCl	e. lipid
19. ________ NaOH	f. protein
20. ________ HCl	g. nucleic acid

21. An ion is formed when:
    a. electrons are shared
    b. electrons remain in place
    c. electrons are gained or lost
    d. neutrons are added to the nucleus
22. In the equation $H_2O + CO_2 \rightarrow H^+ + HCO_3^-$, which of the compounds is a reactant?
    a. CO_2
    b. HCO_3^-
    c. O_2
    d. $\rightarrow$
23. Which of these chemical subunits is found in DNA?
    a. Uracil
    b. Ribose
    c. Amino acid
    d. Deoxyribose
24. Which of these values represents an acid?
    a. pH 7.5
    b. pH 6.1
    c. pH 9.0
    d. pH 7.0
25. Steroid hormones are:
    a. carbohydrates
    b. proteins
    c. lipids
    d. nucleic acids

Case Studies

To solve a case study, you may have to refer to the glossary or index, other chapters in this textbook, and other resources.

1. Grania knows that the pH of blood is normally 7.35 to 7.45. She sees that her blood test results show 7.57 as her blood plasma pH. Is Grania's blood too acid or too alkaline—or is her blood pH within normal range?
2. Baraka has adopted a "high carb" dieting strategy to help him prepare for an upcoming athletic event. What category of organic compound will Baraka be eating in higher proportions than usual? What are some examples of this type of compound that might be found in Baraka's food? What role does this type of organic compound play in Baraka's body? Why might this be an advantage in an athletic event?
3. Sinead's husband, Shane O'Shaunessey, just received the results from his annual physical examination. Shane sheepishly reported to Sinead that his HDL cholesterol levels have increased significantly. Sinead smiled and told Shane not to worry. Why would Sinead not be troubled by this increase in Shane's HDL level?

Answers to Active Learning Questions can be found online at evolve.elsevier.com.

Cells

OUTLINE

Scan this outline before you begin to read the chapter, as a preview of how the concepts are organized.

OBJECTIVES

Before reading the chapter, review these goals for your learning.

After you have completed this chapter, you should be able to:

1. Identify three major components of a cell and discuss the basic structure and function of the plasma membrane and cytoplasm.
2. List and briefly discuss the functions of the primary cellular organelles.
3. Discuss the basic structure and function of the nucleus.
4. Compare the major passive and active transport processes that act to move substances through cell membranes, as well as discuss the relationship of cell transport to diseases.
5. Compare and discuss DNA and RNA and their functions in protein synthesis.
6. Discuss the stages of mitotic cell division and explain the importance of normal cellular reproduction.
7. Explain how changes in cell growth and reproduction allow the cell to adapt to its ever-changing environment.

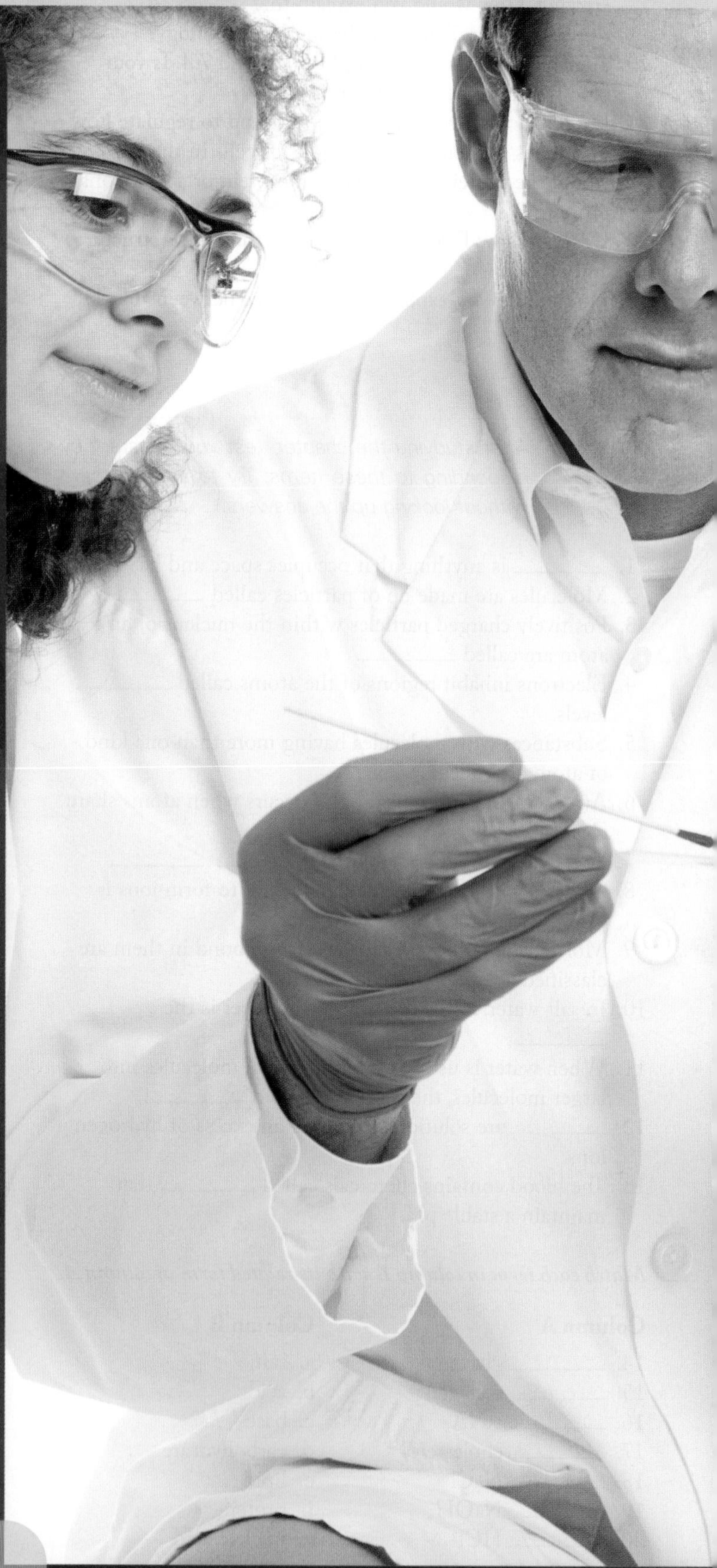

CHAPTER 3

About 350 years ago, Robert Hooke looked through his microscope—one of the very early, somewhat primitive ones—at some plant material. What he saw must have surprised him. Instead of a single magnified piece of plant material, he saw many small pieces. Because they reminded him of miniature monastery cells, that is what he called them—cells. Since Hooke's time, thousands of individuals have examined thousands of plant and animal specimens and found them all, without exception, to be composed of cells.

This fact, that cells are the smallest structural units of living things, has become the foundation of modern biology. Many living things are so simple that they consist of just one cell. The human body, however, is so complex that it consists not of a few thousand or millions or even billions of cells but of many trillions of them. This chapter explores the structure and function of cells.

Overview of Cells

Size and Shape

Human cells are microscopic in size—that is, they can be seen only when magnified by a microscope. However, the different types of human cells vary considerably in

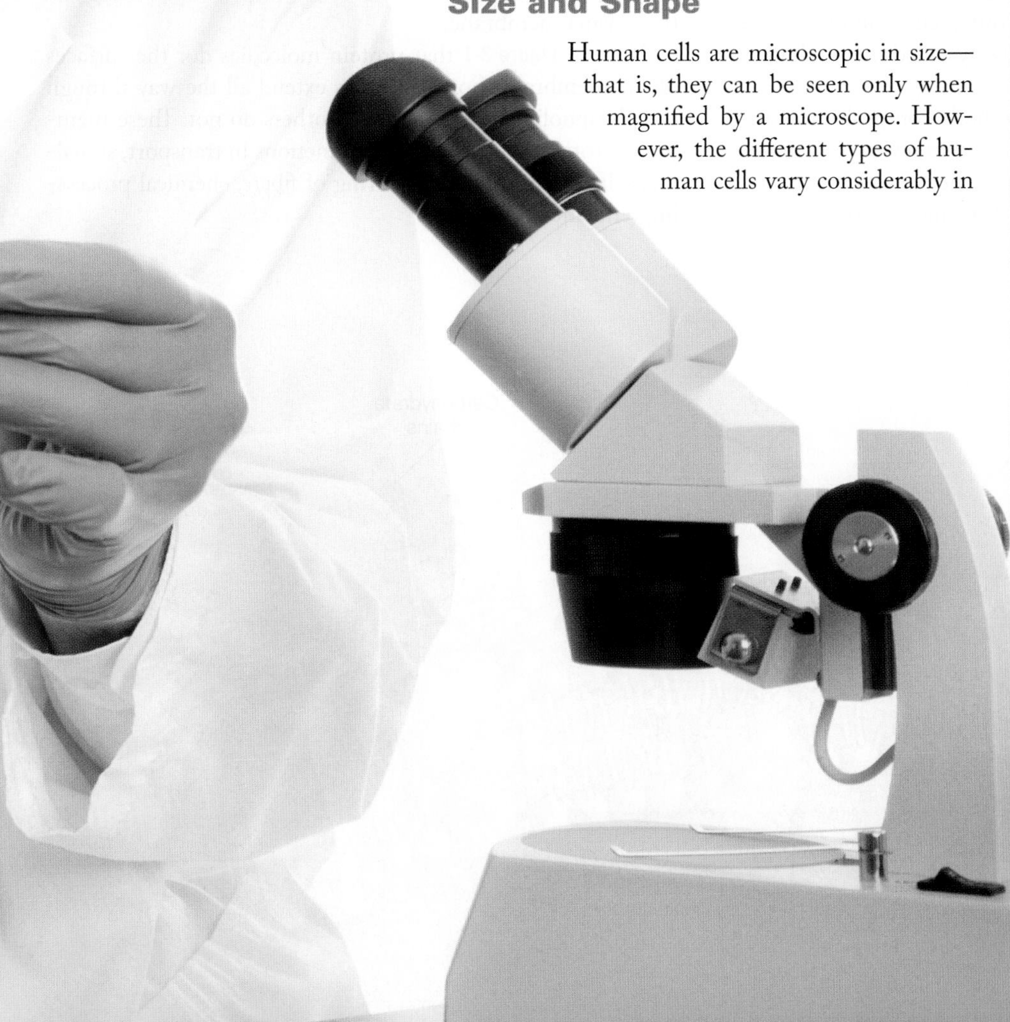

LANGUAGE OF **SCIENCE**

Hint Before reading the chapter, say each of these terms out loud. This will help you to avoid stumbling over them as you read.

active transport
(AK-tiv TRANZ-port)
[*act-* **move,** *-ive* **relating to,** *trans-* **across,** *-port* **carry**]

adenosine triphosphate (ATP)
(ah-DEN-oh-seen try-FOS-fayt [ay tee pee])
[*adenos-* shortened from *adenine-ribose, -ine* **chemical,** *tri-* **three,** *-phosph-* **phosphorus,** *-ate* **oxygen**]

anaphase
(AN-ah-fayz)
[*ana-* **apart,** *-phase* **stage**]

apoptosis
(ap-oh-TOH-sis *or* ap-op-TOH-sis)
[*apo-* **away,** *-ptosis* **falling**]

centriole
(SEN-tree-ohl)
[*centr-* **center,** *-ole* **small**]

centromere
(SEN-troh-meer)
[*centr-* **center,** *-mere* **part**]

centrosome
(SEN-troh-sohm)
[*centr-* **center,** *-som* **body**]

chromatid
(KROH-mah-tid)
[*chrom-* **color,** *-id* **structure or body**]

chromatin granule
(KROH-mah-tin GRAN-yool)
[*chrom-* **color,** *-in* **substance,** *gran-* **grain,** *-ule* **little**]

chromosome
(KROH-meh-sohm)
[*chrom-* **color,** *-som-* **body**]

cilium
(SIL-ee-um)
pl., cilia
(SIL-ee-ah)
[*cili-* **eyelid,** *-um* **thing**]

Continued on p. 63

size. An ovum (female sex cell), for example, has a diameter of about 150 micrometers, but red blood cells have a diameter of only 7.5 micrometers.

Cells differ even more notably in shape than in size. Some are flat, some are brick shaped, some are threadlike, and some have irregular shapes.

Advanced microscopes can picture human cells and their internal structures with detail never before possible. Check out these methods and the dramatic images they produce in the article *Tools of Microscopic Anatomy* at Connect It! at *evolve.elsevier.com.*

Composition

Cells contain **cytoplasm**—the living substance that exists only in cells. The term *cyto-* is a word part that means "cell." Each cell in the body is surrounded by a thin membrane, the **plasma membrane.** This membrane separates the cell contents from the salty solution called **interstitial fluid (IF),** or simply tissue fluid, that bathes every cell in the body. Numerous specialized structures called **organelles,** which are described in subsequent sections, are contained within the cytoplasm of each cell. A small, circular body called the **nucleus** is also inside the cell.

Important information related to body composition is included in Chapter 2. You are encouraged to review this material, which includes a discussion of the chemical elements and compounds important to body structure and function.

Parts of the Cell

The three main parts of a cell are

1. Plasma membrane
2. Cytoplasm
3. Nucleus

The plasma membrane surrounds the entire cell, forming its outer boundary. The cytoplasm is all the living material inside the cell (except the nucleus). The nucleus is a large, membrane-bound structure in most cells that contains the genetic code.

Plasma Membrane

As the name suggests, the **plasma membrane** is the membrane that encloses the cytoplasm and forms the outer boundary of the cell. It is an incredibly delicate structure—only about 7 nm (nanometers) or 3/10,000,000 of an inch thick! Yet it has a precise, orderly structure (**Figure 3-1**).

Two layers of phosphate-containing lipid molecules called **phospholipids** form a fluid framework for the plasma membrane. Another kind of lipid molecule called *cholesterol* is also a component of the plasma membrane. Cholesterol helps stabilize the phospholipid molecules to prevent breakage of the plasma membrane.

Note in **Figure 3-1** that protein molecules dot the surfaces of the membrane. Some proteins extend all the way through the phospholipid framework and others do not. These membrane proteins have a variety of functions in transport, signaling, self-identification, anchoring of fibers, chemical processing (enzymes), and more.

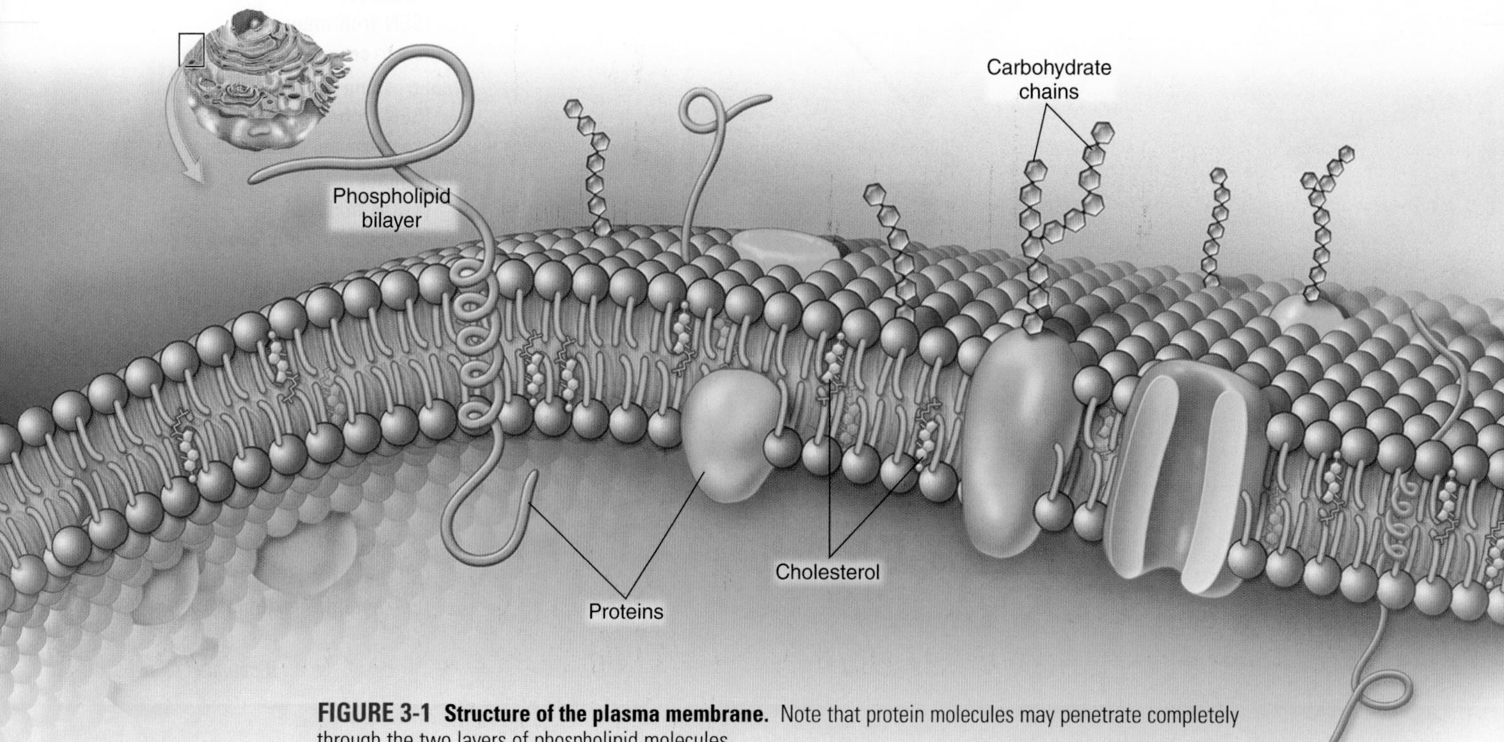

FIGURE 3-1 Structure of the plasma membrane. Note that protein molecules may penetrate completely through the two layers of phospholipid molecules.

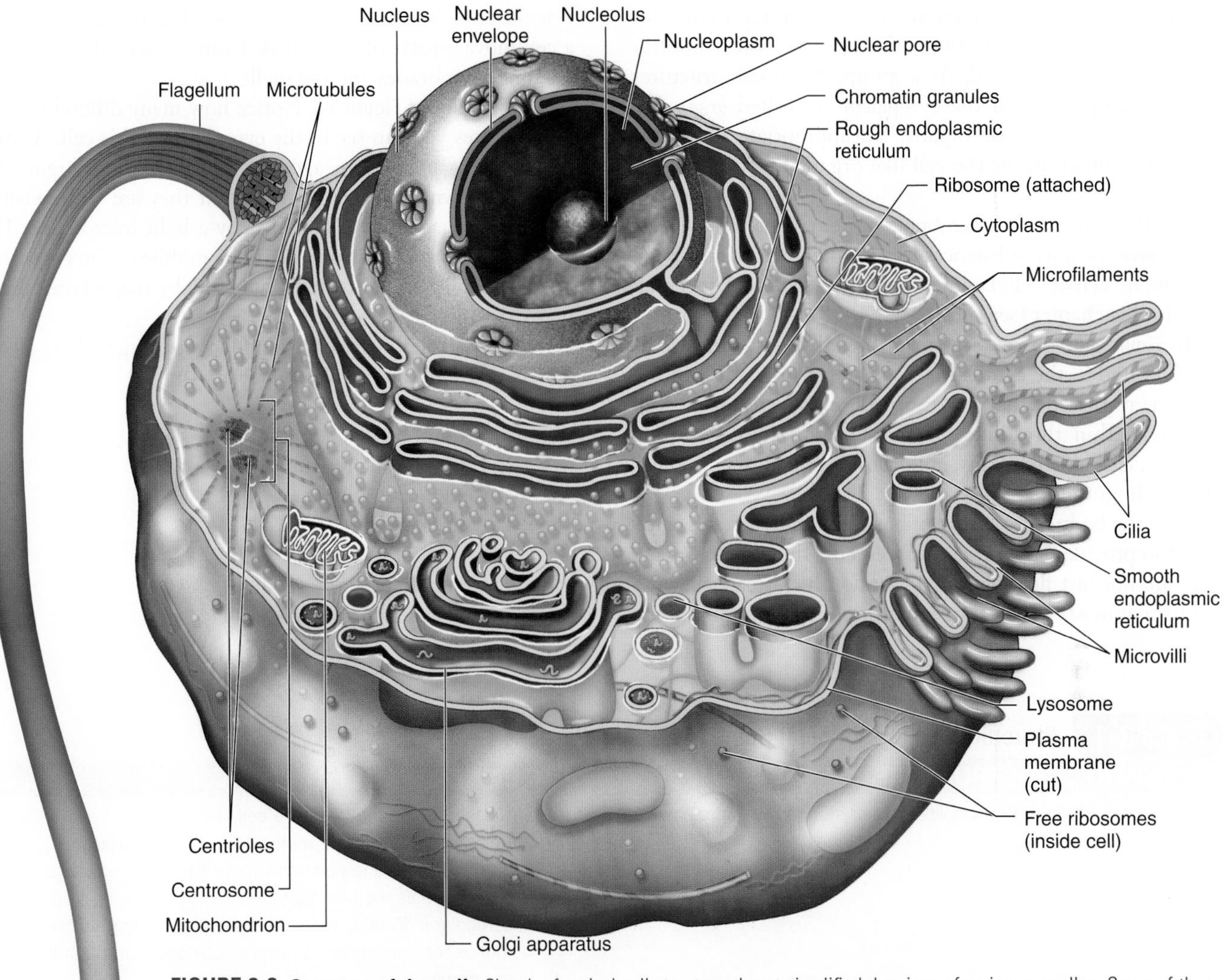

FIGURE 3-2 Structure of the cell. Sketch of typical cell structure shows simplified drawings of major organelles. Some of these structures, such as a flagellum or groups of cilia, are present only in certain types of cells.

3

Despite its seeming fragility, the plasma membrane is strong enough to keep the cell whole and intact and also performs other life-preserving functions for the cell. It serves as a well-guarded gateway between the fluid inside the cell and the fluid around it. Certain substances can move through the membrane by way of transporter channels and carriers, but other substances are barred from entry.

The plasma membrane even functions as a communication mechanism. In what way, you may wonder? Some of the proteins on the membrane's outer surface serve as receptors for certain other molecules when these other molecules contact the proteins. In other words, certain molecules bind to certain receptor proteins. For example, some hormones (chemicals secreted into blood from ductless glands) bind to membrane receptors, and a change in cell functions follows. We might therefore think of such hormones as carriers of chemical messages that are communicated to cells by way of binding to their receptors in the membrane.

The plasma membrane also identifies a cell as being part of one particular individual. Some of the surface proteins serve as positive identification tags because they occur only in the cells of that individual. Carbohydrate chains and hybrid molecules attached to the surface of cells also may play a role in the identification of cell types. A practical application of this fact is made in *tissue typing*, a procedure performed before an organ from one individual is transplanted into another.

Cytoplasm

Cytoplasm is the internal living material of cells. It fills the space between the plasma membrane and the nucleus, which can be seen in **Figure 3-2** as a round or spherical structure in the center of the cell. Numerous small structures are part of the cytoplasm, along with the fluid that serves as the interior environment of each cell. As a group, the small structures that make up much of the cytoplasm are called *organelles.* This name means "little organs," an appropriate name because they function for the cell like organs function for the body.

Another function of membrane proteins is as transporters that move various substances across the membrane. Such movement across cellular membranes is discussed in detail later in this chapter (see p. 50).

In **Figure 3-2** you can see small threadlike structures scattered around in the cytoplasm. You can see only a few of the very many threads that make up the **cytoskeleton** or "cell skeleton." Thin threadlike filaments in this framework are called microfilaments. Tiny, hollow tubes called microtubules also are important.

Like the body's framework of bones and muscles, the cytoskeleton provides support and movement. The various organelles are not just floating around randomly. Instead, they are held (or moved) by the fibers and molecular motors of the cytoskeleton. When a cell moves—or when organelles within a cell move—parts of the cytoskeleton are actually pulling or pushing membranes and organelles.

Look again at **Figure 3-2**. Notice how many different kinds of structures you can see in the cytoplasm of this cell. A little more than a generation ago, almost all of these organelles were unknown. They are so small that they are still invisible even when magnified 1000 times by a light microscope. The advent of electron microscopes in the middle of the twentieth century finally brought them into view by magnifying them many thousands of times.

Next we briefly discuss the following organelles, all of which are found in cytoplasm (**Table 3-1**):

1. Ribosomes
2. Endoplasmic reticulum
3. Golgi apparatus
4. Mitochondria
5. Lysosomes
6. Centrosome
7. Cell extensions

3

TABLE 3-1 Structures and Function of Some Major Cell Parts

CELL PART	STRUCTURE	FUNCTION(S)
Plasma membrane	Phospholipid bilayer studded with proteins	Serves as the boundary of the cell Protein and carbohydrate molecules on outer surface of plasma membrane perform various functions—for example, they serve as markers that identify cells as being from a particular individual, receptor molecules for certain hormones, or transporters to move substances through the membrane
Ribosomes	Tiny particles, each made up of rRNA subunits	Synthesize proteins—a cell's "protein factories"
Endoplasmic reticulum (ER)	Membranous network of interconnected canals and sacs, some with ribosomes attached (rough ER) and some without attachments (smooth ER)	Rough ER receives and transports synthesized proteins (from ribosomes) Smooth ER synthesizes lipids and certain carbohydrates
Golgi apparatus	Stack of flattened, membranous sacs	Chemically processes, then packages substances from the ER
Mitochondria	Membranous capsule containing a large, folded internal membrane embedded with enzymes; contains its own DNA molecule	Adenosine triphosphate (ATP) synthesis—a cell's "power plant" or "battery charger"
Lysosome	"Bubble" of hydrolysis enzymes encased by membrane	A cell's "digestive bag," it breaks apart large molecules
Centrioles	Pair of hollow cylinders at right angles to each other, each made up of tiny tubules within the centrosome	Help organize and move chromosomes during cell reproduction
Cilia	Hairlike cell surface extensions supported by an internal cylinder made of microtubules (longer than microvilli)	Sensory "antennae" to detect conditions outside the cell; some cilia also move substances over surface of the cell
Flagella	Long whiplike projection on the sperm; similar to a cilium but much longer	The only example in humans is the "tail" of a sperm cell, propelling the sperm through fluids
Nucleus	Double-membraned, spherical envelope containing DNA strands	Contains DNA, which dictates protein synthesis, thereby playing an essential role in other cell activities such as transport, metabolism, growth, and heredity
Nucleolus	Dense region of the nucleus	Makes subunits that form ribosomes

rDNA, Ribosomal RNA.

Ribosomes

Organelles called **ribosomes,** shown as dots in **Figure 3-2**, are very tiny particles found throughout the cell. They are each made up of two tiny subunits constructed mostly of a special kind of RNA called *ribosomal RNA (rRNA).*

Some ribosomes are found temporarily attached to a network of membranous canals called *endoplasmic reticulum (ER).* Ribosomes also may be free-floating in the cytoplasm. Ribosomes perform a very complex function—they make enzymes and other protein compounds. Thus they are aptly nicknamed "protein factories."

Endoplasmic Reticulum

An **endoplasmic reticulum (ER)** is a system of membranes forming a network of connecting sacs and canals that wind back and forth through a cell's cytoplasm. The ER extends from the nucleus almost to the plasma membrane. The tubular passageways or canals in the ER carry proteins and other substances through the cytoplasm of the cell from one area to another.

There are two types of ER: *rough* and *smooth.*

Rough ER gets its name from the many ribosomes that are attached to its outer surface, which gives it a rough texture similar to sandpaper. As they make their proteins, some ribosomes attach to the rough ER and insert the protein into the interior of the ER.

As the proteins begin folding insides, the ER transports them to areas where chemical processing takes place. These areas of the ER are so full of enzymes and other molecules that ribosomes have no room into which they can pass their proteins, and so they do not attach. The absence of attached ribosomes gives this type of ER a smooth texture. Fats, carbohydrates, and proteins that make up cellular membrane material are manufactured in smooth ER. Thus the smooth ER makes new membrane for the cell.

To sum up: rough ER receives, folds, and transports newly made proteins, and smooth ER manufactures new membrane.

Golgi Apparatus

The **Golgi apparatus** consists of tiny, flattened sacs stacked on one another near the nucleus. Little bubbles, or sacs, pinch off the ER and carry new proteins and other compounds to the sacs of the Golgi apparatus (**Figure 3-3**). These little sacs, also called **vesicles,** fuse with the Golgi sacs and allow the contents of both to mingle.

The Golgi apparatus chemically processes the molecules from the ER by further folding, sorting, and modifying proteins and combining them with other molecules to form quaternary proteins (see **Figure 2-12** on p. 34) or combinations such as glycoproteins (carbohydrate/protein molecules).

The Golgi apparatus then packages the processed molecules into new little vesicles that pinch off and pull away from the Golgi apparatus, moving slowly outward to the plasma membrane. Each vesicle then fuses with the plasma membrane, opens to the outside of the cell, and releases its contents. The wall of each vesicle then becomes incorporated into the plasma membrane—a mechanism for adding new membrane components.

An example of a Golgi apparatus product is the slippery substance called *mucus.* If we wanted to nickname the Golgi apparatus, we might call it the cell's "chemical processing and packaging center."

Mitochondria

The **mitochondrion** is another kind of organelle found in all cells except red blood cells. Mitochondria are so tiny that a lineup of 15,000 or more of them would fill a space only about 2.5 cm (1 inch) long.

Two membranous sacs, one folded tightly inside the other, compose a single mitochondrion. The folds of the inner membrane look like incomplete partitions. Within a mitochondrion's fragile membranes, complex, energy-releasing chemical reactions occur continuously. Because these reactions supply most of the power for cellular work, mitochondria have been nicknamed the cell's "power plants."

Enzymes (molecules that promote specific chemical reactions), found in mitochondrial membranes and the mitochondrial fluids, break down products of glucose and other nutrients to release energy. The mitochondrion uses this released energy to "recharge" **ATP (adenosine triphosphate)** molecules—the "batteries" required for cellular work. This process, which requires oxygen and releases carbon dioxide, is called *cellular respiration.*

Each mitochondrion has its own tiny, ring-shaped DNA molecule, sometimes called a *mitochondrial chromosome,* that contains information for building and running the mitochondrion.

e To better understand this concept, use the Active Concept Map *Metabolism of Glucose to Generate ATP* at *evolve.elsevier.com.*

e To learn more about how energy in the body is measured, including examples of the energy content of food and the energy cost of common activities, review the article *Measuring Energy* at Connect It! at *evolve.elsevier.com.*

Lysosomes

The **lysosomes** are membrane-walled organelles that in their active stage look like small sacs, often with tiny particles in them (see **Figure 3-2**). Because lysosomes contain enzymes that promote hydrolysis, they can break apart (digest) large nutrient molecules. Therefore, they have the nickname "digestive bags."

Lysosomal enzymes can also digest substances other than nutrients. For example, they can digest and thereby destroy microbes that invade the cell. Thus lysosomes can protect cells against destruction by microbes.

Formerly, scientists thought lysosomes were involved in programmed cell death. Now, however, we know a different set of mechanisms is responsible for "cell suicide," or **apoptosis,** which makes space for newer cells. When apoptosis does not occur normally, the cell may remain and cause overgrowth of the tissue—possibly producing a tumor.

Why learn about organelles and their functions? In addition to helping us learn about normal body structure and function, these concepts help us understand disease mechanisms that involve organelles. To learn about some of these diseases, check out the article *Organelle Diseases* at Connect It! at *evolve.elsevier.com.*

Centrosome

The **centrosome** is a region of cytoplasm near the nucleus of each cell. It serves as the *microtubule-organizing center* of the cell, thus playing an important role in organizing and moving the structures within the cell.

Centrioles are paired organelles found within the centrosome. Two of these rod-shaped structures exist in every cell. They are arranged so that they lie at right angles to each other (see Figure 3-2). Each centriole is composed of microtubules that form a tapered framework or "spindle" that moves chromosomes during cell division, as we shall see later in this chapter.

The centrosome also plays a role in forming and organizing the cell's cytoskeleton, including some of a cell's outward extensions.

Cell Extensions

Most cells have various indentations and extensions that serve many different functions. Here we describe three of the major types of cell extensions (Figure 3-4).

Microvilli

Microvilli are small, fingerlike projections of the plasma membrane of some cells (Figure 3-4, *A*). These projections

FIGURE 3-3 The cell's protein export system. The Golgi apparatus processes and packages protein molecules delivered from the endoplasmic reticulum (ER) by small vesicles. Some vesicles migrate to the plasma membrane to secrete the final products, and other vesicles remain inside the cell for a time and serve as storage vessels for the substance to be secreted.

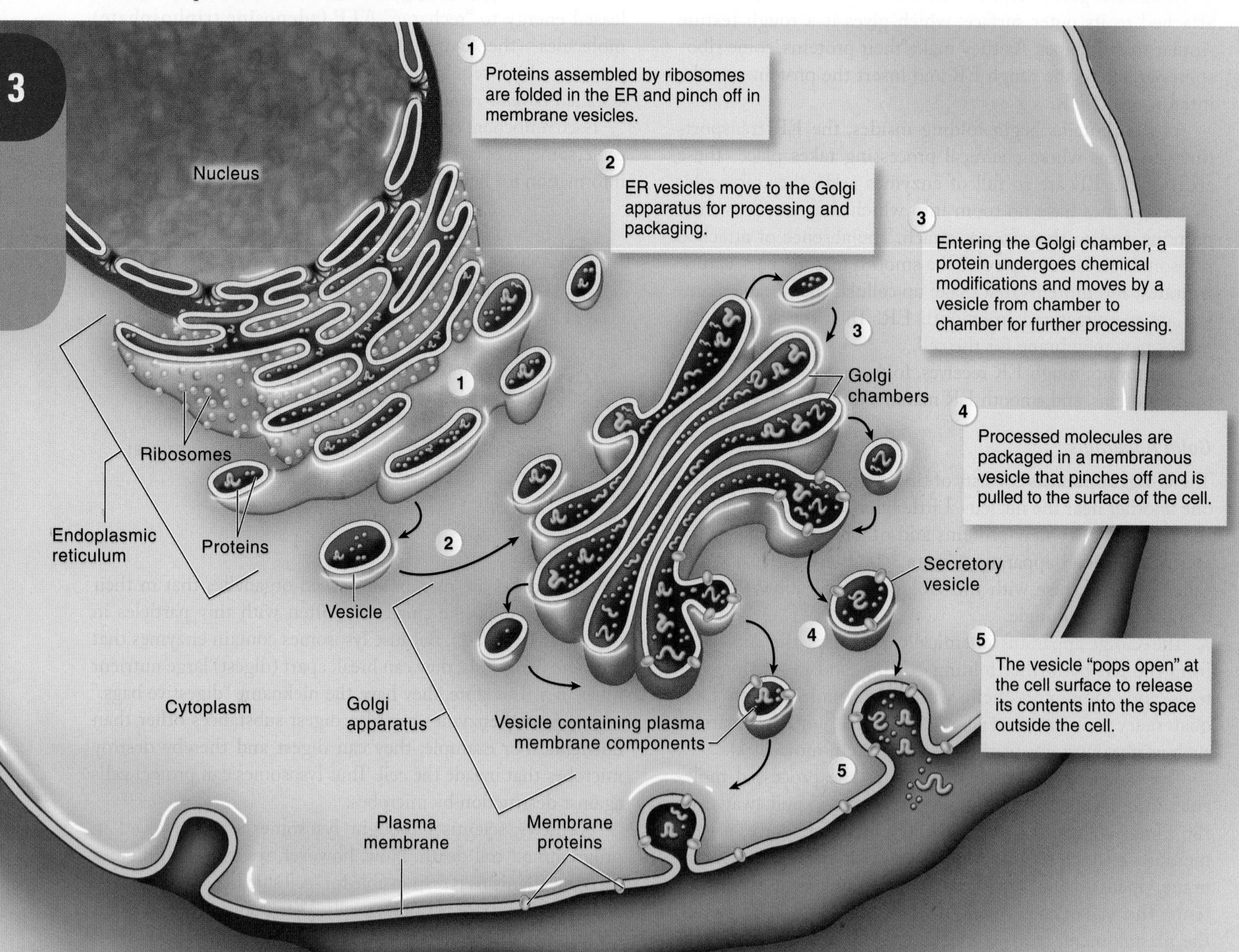

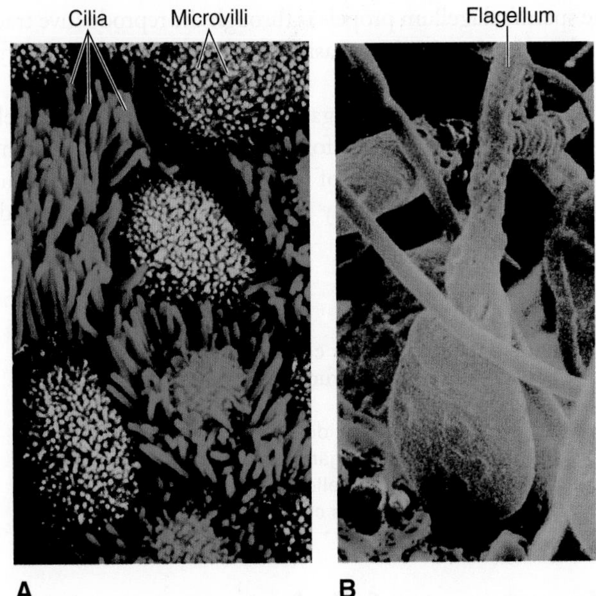

FIGURE 3-4 Cell extensions. A, Microvilli *(light blue)* are small, fingerlike extensions of the plasma membrane that increase the surface area for absorption. Cilia *(darker blue)* are longer than microvilli and move back and forth, pushing fluids along the surface. **B,** The tail-like flagellum that propels each sperm cell is so long that it does not fit into the photograph at this magnification.

increase the surface area of the cell and thus increase its ability to absorb substances. For example, cells that line the small intestine are covered with microvilli that increase the absorption rate of nutrients into the blood. Microvilli have microfilaments inside them that produce wobbly movement and thus make absorption more efficient.

Cilia

Cilia are extremely fine, hairlike extensions on the exposed or free surfaces of cells (see **Figure 3-4**, *A*). Cilia are larger than microvilli and possess inner microtubules that support and enable them to move. Every cell has at least one cilium.

All cilia act like an insect's antennae, allowing the cell to sense its surroundings. For example, the hairlike cilia in the taste buds of the mouth can detect different chemicals dissolved in saliva.

Some cells have hundreds of cilia capable of moving together in a wavelike fashion over the surface of a cell (**Figure 3-5**). By moving as a group in one direction, they propel mucus over the cells that line the respiratory or reproductive tubes.

> Cilia make one of the most important mechanisms helping protect the delicate tissues of the body's airways. To preview these strategies, check out the article *Protective Strategies of the Respiratory Tract* at Connect It! at *evolve.elsevier.com.*

Flagella

A **flagellum** is a single projection extending from the cell surface. Flagella are structurally similar to cilia but much longer. Like cilia, flagella can move. The cylinder of microtubules inside the flagellum moves in a way that whips the flagellum around allowing it to act like a propeller that pushes the cell forward (see **Figure 3-5**).

In the human, the only example of a flagellum is the "tail" of the male sperm cell (see **Figure 3-4**, *B*). Wiggling movements of the flagellum make it possible for sperm to "swim" or move toward the ovum after they are deposited in the female reproductive tract.

Nucleus

Central Structure of a Cell

Viewed under a light microscope, the nucleus of a cell looks like a very simple structure—just a small sphere usually near the center of the cell. In certain specialized cells, the nucleus may be pushed to one side and perhaps slightly compressed into a more flattened shape.

However, its simple appearance belies the complex and critical role the nucleus plays in cell function. The nucleus contains most of the cell's genetic information, which ultimately controls every organelle in the cytoplasm. It also controls the complex process of cell reproduction. In other words,

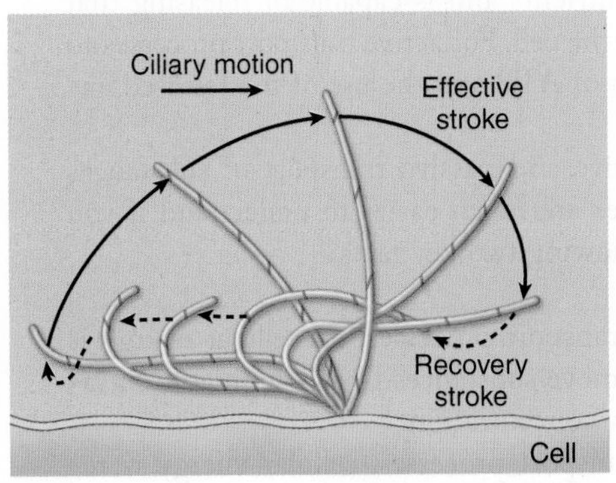

Cilium

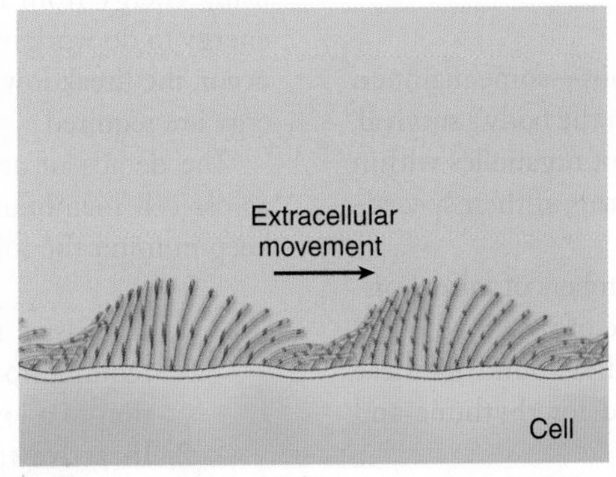

Cilia

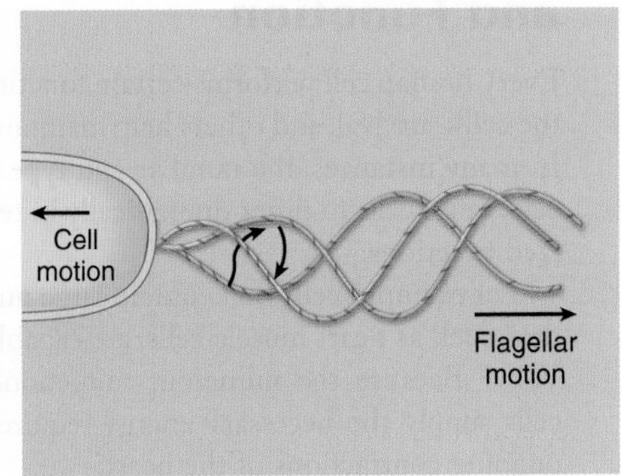

Flagellum

FIGURE 3-5 Movement patterns. In humans, cilia *(left* and *middle)* found in groups on stationary cells beat in a coordinated oarlike pattern to push fluid and particles in the extracellular fluid along the outer cell surface. A flagellum *(right)* produces wavelike movements, which propels a sperm cell forward—like the tail of an eel.

the nucleus must function properly for a cell to accomplish its normal activities and be able to duplicate itself.

Note that the cell nucleus in **Figure 3-2** is surrounded by a **nuclear envelope,** a structure made up of two separate membranes. The nuclear envelope has many tiny openings called *nuclear pores* that permit large molecules to move into and out of the nucleus. The nuclear envelope encloses a special type of cell material within the nucleus called **nucleoplasm.**

Nucleoplasm contains a number of structures. Two of the most important structures are the *nucleolus* and the *chromatin granules*—both pictured in **Figure 3-2.**

Nucleolus

The **nucleolus** is a dense region of the nuclear material that is critical in protein formation because it is where the cell makes the subunits that form ribosomes. The ribosome subunits then migrate through the pores of the nuclear envelope into the cytoplasm of the cell where they assemble into ribosomes, the protein-making machinery of the cell.

Chromatin and Chromosomes

3

Chromatin granules in the nucleus are made of proteins around which are wound segments of the long, threadlike molecules called **DNA,** or **deoxyribonucleic acid.** DNA is the genetic material often described as the chemical "cookbook" of the body. Because it contains the code for building both structural proteins and functional proteins, DNA determines everything from gender and metabolic rate to body build and hair color in every human being.

During cell division, DNA molecules become tightly coiled. They then look like short, rodlike structures and are called **chromosomes.**

Each cell of the body contains a total of 46 different DNA molecules in its nucleus and one copy of a 47th DNA molecule in each of its mitochondria. The importance and function of DNA are explained in greater detail in the section on cell reproduction later in this chapter.

Relationship of Cell Structure and Function

Every human cell performs certain functions—some maintain the cell's survival, and others help maintain the body's survival. In many instances, the number and type of organelles within cells cause cells to differ dramatically in terms of their specialized functions.

For example, cells that contain large numbers of mitochondria, such as heart muscle cells, are capable of sustained work. Why? Because the numerous mitochondria found in these cells supply the necessary energy required for rhythmic and ongoing contractions of the heart.

Movement of the flagellum of a sperm cell is another example of how each type of organelle has a particular function. The sperm's flagellum propels it through the reproductive tract of the female, thus increasing the chances of successful fertilization.

This is how and why organizational structure at the cellular level is so important for function in living organisms. Examples in every chapter of the text illustrate how structure and function are intimately related at every level of body organization.

QUICK CHECK

1. What are the three major components of a cell?
2. What is the molecular structure of the cell *plasma membrane*?
3. What is *cytoplasm*? How does it serve the body?
4. What are the primary organelles of the cell? What are the functions of these organelles?
5. Which two cell structures contain DNA?

Movement of Substances Through Cell Membranes

Types of Membrane Transport

The plasma membrane in every healthy cell separates the contents of the cell from the tissue fluid that surrounds it. At the same time, the membrane must permit certain substances to enter the cell and allow others to leave. Heavy traffic moves continuously in both directions through cell membranes. Molecules of water, nutrients, gases, wastes, and many other substances stream in and out of all cells in endless procession.

A number of different **transport processes** allow this mass movement of substances into and out of cells. These transport processes are classified under two general headings:

1. Passive transport processes
2. Active transport processes

As implied by the name, active transport processes require the expenditure of energy by the cell, and passive transport processes do not. The energy required for active transport processes is obtained from **ATP.** ATP is produced by the cell using energy from nutrients and is capable of releasing that energy to do work in the cell. For active transport processes to occur, the breakdown of ATP and the use of the released energy are required.

The details of active and passive transport of substances across cell membranes are much easier to understand if you keep in mind the following two key facts:

1. In passive transport processes, no cellular energy is required to move substances from a high concentration to a low concentration.
2. In active transport processes, cellular energy is required to move substances from a low concentration to a high concentration.

Passive Transport Processes

The primary **passive transport** processes that move substances through the cell membranes include the following:

1. Diffusion
2. Osmosis
3. Dialysis
4. Filtration

Scientists describe the movement of substances in passive systems as going "down a concentration gradient." This means that substances in passive systems move from a region of high concentration to a region of low concentration until they reach equal proportions on both sides of the membrane. As you read the next few paragraphs, refer to **Table 3-2**, which summarizes important information about passive transport processes.

FIGURE 3-6 Diffusion. The molecules of a lump of sugar are very densely packed when they enter the water. As sugar molecules collide frequently in the area of high concentration, they gradually spread away from each other—toward the area of lower concentration. Eventually, the sugar molecules become evenly distributed.

Diffusion

Diffusion—a good example of a passive transport process—is the process by which substances scatter themselves evenly throughout an available space. The system does not require additional energy for this movement. Diffusion can thus be described as a trend of movement of particles down a concentration gradient—that is, net movement from an area of high concentration toward an area of lower concentration.

To demonstrate diffusion of particles throughout a fluid, perform this simple experiment the next time you pour yourself a cup of coffee or tea (**Figure 3-6**). Place some sugar on a teaspoon and lower it gently to the bottom of the cup. Let it stand for 2 or 3 minutes, and then, holding the cup steady, take a sip off the top. It will taste sweet. Why? Because some of the sugar molecules will have diffused from the area of high concentration near the mound of sugar at the bottom of the

TABLE 3-2 Passive Transport

PROCESSES	DESCRIPTION	EXAMPLES
Diffusion	Movement of particles through a membrane from an area of high concentration to an area of low concentration—that is, down the concentration gradient	Movement of carbon dioxide out of all cells; movement of sodium ions into nerve cells as they conduct an impulse
Osmosis	Passive movement of water through a selectively permeable membrane in the presence of at least one nonpenetrating solute	Movement of water into and out of cells to correct imbalances in water concentration
Filtration	Movement of water and small solute particles, but not larger particles, through a filtration membrane; movement occurs from area of high pressure to area of low pressure High pressure Low pressure	In the kidney, water and small solutes move from blood vessels but blood proteins and blood cells do not, thus beginning the formation of urine

cup to the area of low concentration at the top of the cup—thus sweetening the entire solution.

Assume that the tea is brewed using a tea bag made of shredded tea leaves inside a pouch of porous filter paper. One can easily watch the diffusion of dark pigment particles from a concentrated area inside the tea bag to the less concentrated area in the water outside the tea bag. Thus, the pigment particles moved through a membrane (the paper) by diffusion—the tendency to spread out and create a uniform concentration or **equilibrium.**

The key to diffusion across a membrane is the presence of pores big enough for the particles to pass through. In cell membranes, most molecules cannot pass through the membrane unless there are gateways that permit it. Various protein channels act as gated doorways that permit certain molecules to diffuse through them. Other protein structures act as carriers that bind to the particles and carry them through to the other side of the membrane. Without these transporters, most **solutes** (substances dissolved in the water) could not diffuse through cell membranes.

The process of diffusion is shown in **Figure 3-7**. Note that both substances diffuse rapidly through the porous membrane in both directions. However, as indicated by the purple arrows, more of the solute (dissolved substance) moves out of the 20% solution, where the concentration is higher, into the 10% solution, where the concentration is lower, than in the opposite direction. This is an example of movement down a concentration gradient.

The result? *Equilibration* or balancing of the concentrations of the two solutions after an interval of time. After this equilibrium is reached, equal amounts of solute will diffuse in both directions.

The plasma membrane of a cell is said to be *selectively permeable* because it permits the passage of certain substances but not others. Put another way, the membrane has specific *channels* and *carriers* to allow diffusion of specific kinds of molecules. This necessary property permits some substances, such as nutrients, to gain entrance to the cell while excluding others.

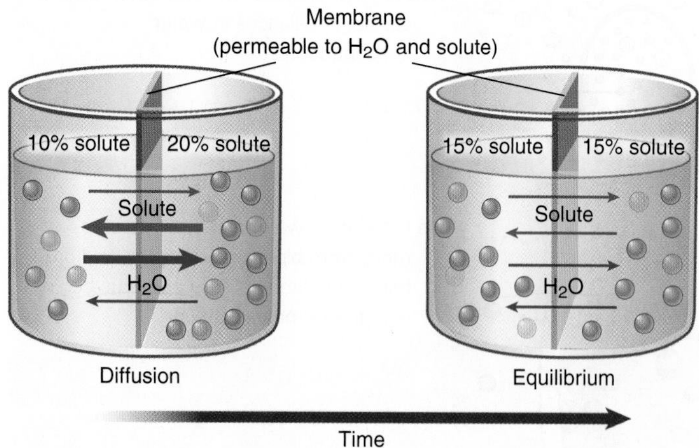

FIGURE 3-7 Diffusion through a membrane. Note that the membrane is permeable to solute and water and that it separates a 10% solution of solute particles from a 20% solution. The container on the left shows the two solutions separated by the membrane at the start of diffusion. The container on the right shows the result of diffusion after some time has passed.

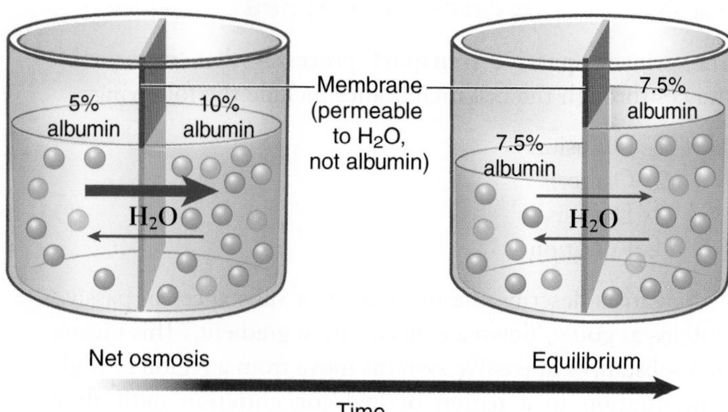

FIGURE 3-8 Osmosis. The solute albumin cannot cross the semipermeable membrane, but water can. The resulting movement of water (only) produces equilibration of the solutions, as water moves away from the side where it is most abundant and toward the solution with more solute particles. Osmosis also causes a shift in fluid volume and pressure (osmotic pressure).

Osmosis

Osmosis is a special case of passive transport. It is in many ways similar to diffusion, but is thought to involve unique mechanisms at the pores of cell membranes.

Osmosis is the passive movement of *water molecules* through water channels in a selectively permeable membrane when some of the *solute* cannot cross the membrane (because there are no open channels or carriers for that solute). **Figure 3-8** shows that osmosis moves water in a direction that results in dilution of solution to a type of equilibrium called *osmotic balance.*

In osmosis, because water moves into a space but there is no exchange of solutes, a change in fluid pressure may result. Such fluid pressure is called *osmotic pressure.*

Dialysis

In a process called **dialysis,** some solutes move across a selectively permeable membrane by diffusion and other solutes do not (**Figure 3-9**). Thus, dialysis results in an uneven distribution of various solutes.

Dialysis is often used as a medical procedure in which blood is pumped through membranous tubing bathed in a solution that mimics normal body fluids. Because the small waste molecules normally removed by the kidney diffuse into the bath solution, but the larger proteins in the blood cannot diffuse, such dialysis can safely "clean" the blood of waste.

Another strategy is to instead pump the bath solution into the fluid space of the abdominopelvic cavity to accept the blood's wastes by dialysis. After some time, the "dirty" solution is then pumped back out of the body.

These dialysis procedures can be used when the kidney is not functioning efficiently.

CLINICAL APPLICATION

OSMOTIC BALANCE

The internal fluid environment of the body is mostly a weak solution of salts such as NaCl and other solutes—as is the solution inside each cell of the body. Osmotic balance is maintained through homeostasis. However, disruptions of homeostasis can cause potentially dangerous movement of water and resulting shifts in pressure. Here, we explore examples of what can go wrong.

A NaCl solution is said to be **isotonic** (*iso* = equal) if it contains the same concentration of salt normally found in a living red blood cell, which measures 0.9% NaCl. Salt particles (Na^+ and Cl^- ions) do not cross the plasma membrane easily, so salt solutions that differ in concentration from the cell's fluid will promote the osmosis of water one way or the other. A solution that contains a higher level of salt than the cell (above 0.9%) is said to be **hypertonic** (*hyper* = above) to the cell and one containing less (below 0.9%) is **hypotonic** (*hypo* = below) to the cell.

With what you now know about filtration, diffusion, and osmosis, can you predict what would occur if red blood cells were placed in isotonic, hypotonic, and hypertonic solutions?

Examine the figures. Note that red blood cells placed in isotonic solution remain unchanged because there is no effective difference in salt or water concentrations. The movement of water into and out of the cells is about equal. This is not the case when red cells are placed in hypertonic salt solution. In this case, the cells immediately lose water from their cytoplasm into the surrounding salty solution—and the cells shrink. This process is called **crenation** because under a microscope, these shriveled cells appear to have a crenated (scalloped) border.

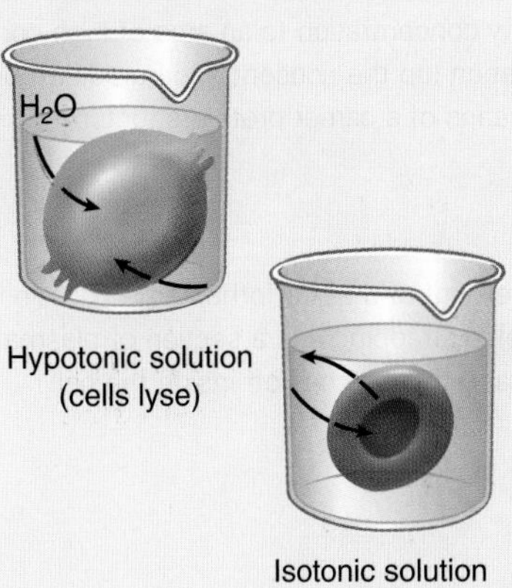

The opposite occurs if red cells are placed in a hypotonic solution—the cells swell as water enters from the surrounding dilute solution. Eventually the cells break or **lyse,** and the hemoglobin they contain is released into the surrounding solution.

Filtration

Filtration is the movement of water and solutes through a membrane as a result of a pushing force that is greater on one side of the membrane than on the other side. The force is called *hydrostatic pressure,* which is simply the force or weight of a fluid pushing against some surface (an example is blood pressure, in which blood pushes against vessel walls).

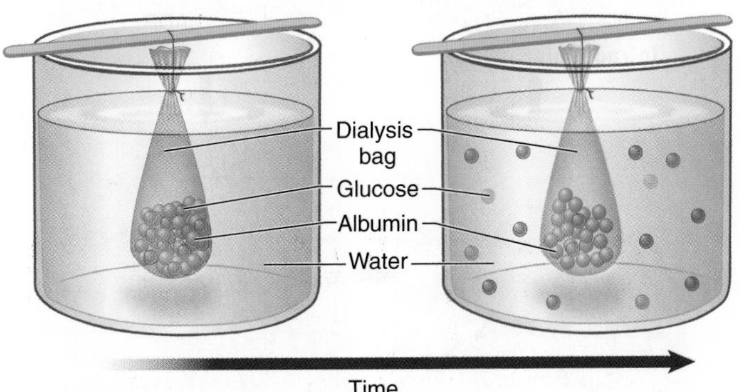

FIGURE 3-9 Dialysis. A membrane bag containing glucose, water, and albumin (protein) molecules is suspended in pure water. Over time, the smaller solute molecules (glucose) diffuse out of the bag. The larger solute molecules (albumin) remain trapped in the bag because the bag is impermeable to them. Thus dialysis results in separation of small and large solute particles.

A principle concerning filtration that is of great physiological importance is that it always occurs *down* a hydrostatic pressure gradient. This means that when two fluids have unequal hydrostatic pressures and are separated by a membrane, water and diffusible solutes or particles (those to which the membrane is permeable) will filter out of the solution that has the higher hydrostatic pressure into the solution that has the lower hydrostatic pressure.

Filtration is partly responsible for moving water and small solutes from blood into the fluid spaces of the body's tissues. Filtration is one of the processes responsible for urine formation in the kidney. Dissolved waste particles are filtered out of the blood into the kidney tubules because of a difference in hydrostatic pressure.

Active Transport Processes

Active transport is the uphill movement of a substance through a living cell membrane. *Uphill* means "up a concentration gradient" (that is, from a lower to a higher concentration). The energy required for this movement is obtained from ATP. Because the formation and breakdown of ATP require cell activity, active transport mechanisms occur only through living membranes. **Table 3-3** summarizes active transport processes.

TABLE 3-3 Active Transport Processes

PROCESS	DESCRIPTION	EXAMPLES
Ion pump	Movement of solute particles from an area of low concentration to an area of high concentration (up the concentration gradient) by means of a carrier protein structure	In muscle cells, pumping of nearly all calcium ions to special compartments—or out of the cell
Phagocytosis	Movement of a cell or other large particle into a cell by trapping it in a section of plasma membrane that pinches off inside the cell	Trapping of bacterial cells by phagocytic white blood cells
Pinocytosis	Movement of fluid and dissolved molecules into a cell by trapping them in a section of plasma membrane that pinches off inside the cell	Trapping of large protein molecules by some body cells

3

Ion Pumps

A complex membrane component called the *ion pump* makes possible a number of active transport mechanisms. An ion pump is a protein structure in the cell membrane called a *carrier.* The ion pump uses energy from ATP to actively move ions across cell membranes *against* their concentration gradients. "Pump" is an appropriate term because it suggests that active transport moves a substance in an uphill direction just as a water pump does, that is, moves water uphill.

An ion pump is specific to one particular ion. Therefore, different ion pumps are required to move different types of ions. For example, sodium pumps move sodium ions only. Likewise, calcium pumps move calcium ions and potassium pumps move potassium ions.

Some ion pumps are "coupled" to one another so that two or more different substances may be moved through the cell membrane at one time. For example, the **sodium-potassium pump** shown in **Figure 3-10** pumps sodium ions out of a cell while it pumps potassium ions into the cell. Because both ions are moved against their concentration gradients, this pump creates a high sodium concentration outside the cell and a high potassium concentration inside the cell. Such a pump is required to remove sodium from the inside of a nerve cell after it has rushed in as a result of the passage of a nerve impulse.

Some ion pumps are coupled with other specific carriers that transport glucose, amino acids, and other substances. However, there are no transporter pumps for moving water—it can move only passively by osmosis.

Phagocytosis

Phagocytosis is another example of how a cell can actively move an object or substance through the plasma membrane and into the cytoplasm. The term *phagocytosis* comes from a Greek word meaning "to eat." The word is appropriate because this process permits a cell to engulf and literally "eat" relatively large particles.

Certain white blood cells can use phagocytosis to destroy invading bacteria and chunks of debris from **tissue** damage. During this process the cytoskeleton extends the cell's plasma membrane to form a pocket around the particles to be moved

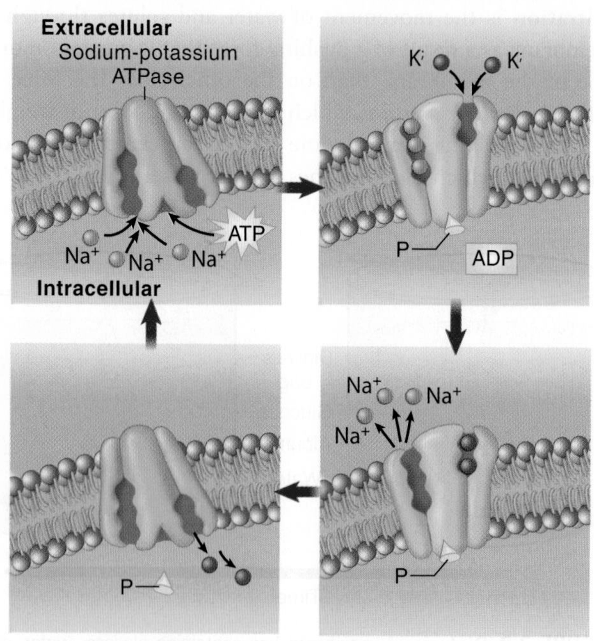

FIGURE 3-10 Sodium-potassium pump. Three sodium ions (Na^+) are pumped out of the cell and two potassium ions (K^+) are pumped into the cell during one pumping cycle of this carrier molecule. Adenosine triphosphate (ATP) is broken down in the process so that the energy freed from ATP can be used to pump the ions. ADP, Adenosine diphosphate.

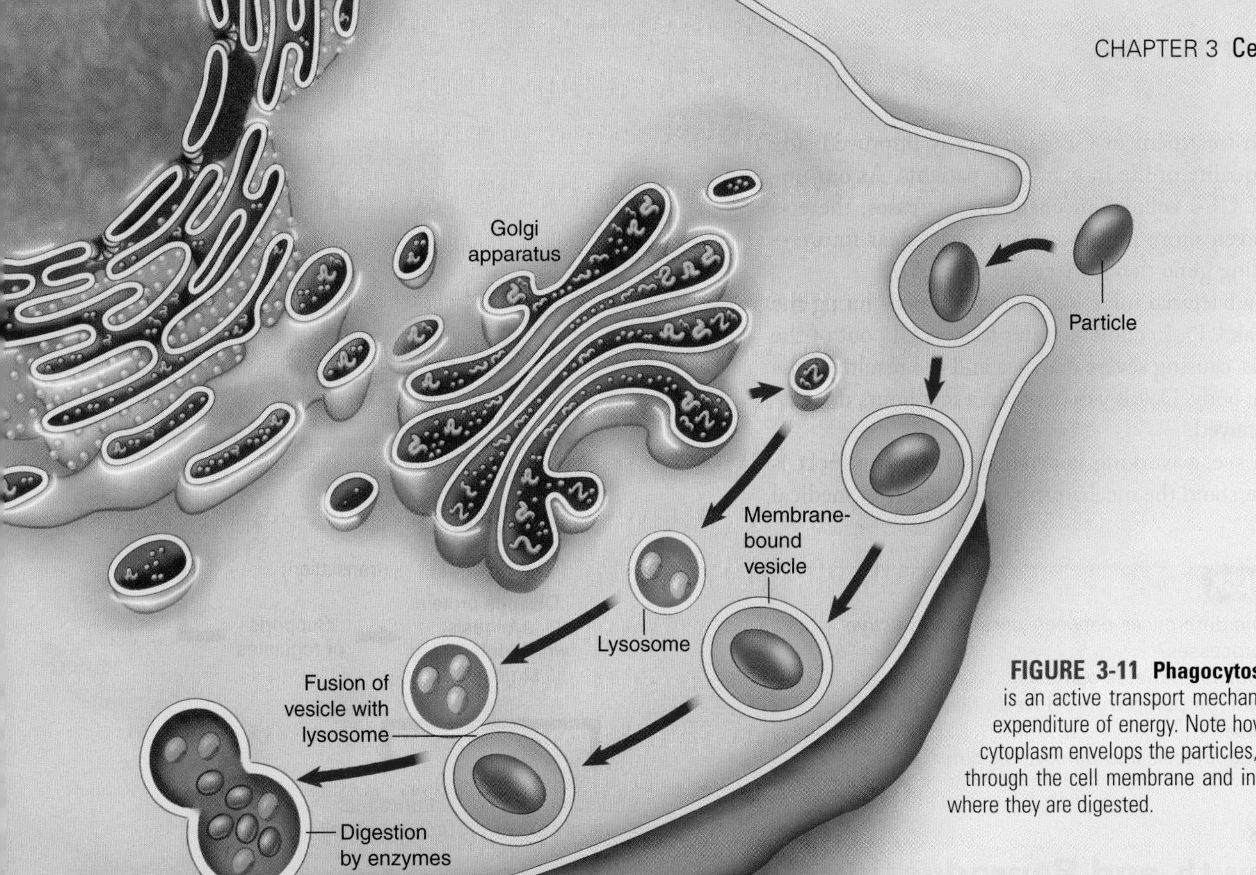

FIGURE 3-11 Phagocytosis. Phagocytosis is an active transport mechanism that requires expenditure of energy. Note how an extension of cytoplasm envelops the particles, which are drawn through the cell membrane and into the cytoplasm, where they are digested.

into the cell and thus encloses the material in a vesicle. Movements of the cytoskeleton pull the vesicle deeper into the cell.

Once inside the cytoplasm, the phagocytic vesicle fuses with a lysosome containing digestive enzymes and the particles are broken apart (**Figure 3-11**).

Pinocytosis

Pinocytosis is an active transport mechanism used to incorporate fluids or dissolved substances into cells by trapping them in a pocket of plasma membrane that pinches off inside the cell. Again, the term is appropriate because the word part *pino* comes from the Greek word meaning "drink."

Because the cytoskeleton uses energy from ATP to produce the movements of both pinocytosis and phagocytosis, these processes are active transport mechanisms.

Cell Transport and Disease

Considering the importance of active and passive transport processes to cell survival, you can imagine the problems that arise when one of these processes fails. Several very severe diseases result from damage to cell transport processes.

Cystic fibrosis (CF), for example, is an inherited condition in which chloride ion (Cl^-) pumps in the plasma membrane are misfolded and not functioning properly. Movement of negative Cl^- ions attracts positive sodium ions (Na^+), which in turn osmotically attract water molecules. Therefore, transport of Cl^- out of cells results in the release of water as well. In CF, the mucus and other watery secretions of cells get very thick because they contain very little water. In the lungs, this thick mucus impairs normal breathing and frequently leads to recurring lung infections.

Figure 3-12 shows a newborn with severe CF. Because of the difficulty with breathing and digestion and other problems caused by the disease, the affected child has not developed normally and has a bloated abdomen. Digestion is compromised by thick pancreatic secretions that may plug the duct leading from the pancreas and thereby prevent important digestive juices from flowing into the intestines. Thickened mucus can also cause intestinal blockage and disrupt normal absorption of nutrients.

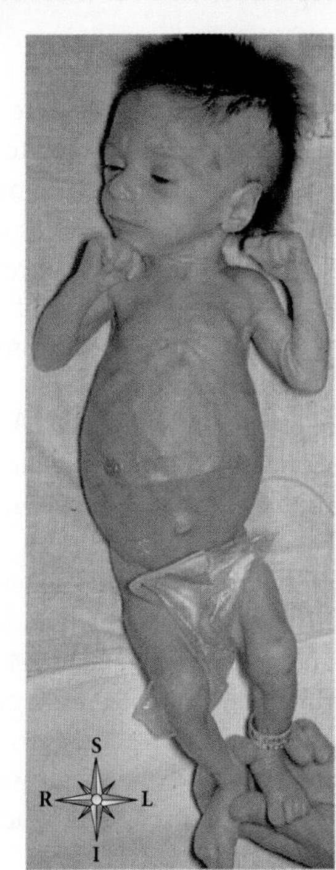

FIGURE 3-12 Cystic fibrosis. In cystic fibrosis (CF), the absence of chloride ion pumps causes thickening of watery secretions in the body. Because thickened secretions block airways, intestines, and digestive ducts, children born with this disease often become weakened and bloated—and, without treatment, may die before adulthood. Recent availability of advanced CF therapies has increased the quality and length of life in CF patients.

Advances in treatment of CF have greatly improved survivability and quality of life in many CF patients. As our understanding of CF's cellular mechanisms increases, there is real hope for even more improvements in the near future—perhaps including gene therapy (see Chapter 25).

Cholera is a bacterial infection that causes cells lining the intestines to leak Cl^-. In cholera, water follows Cl^- out of the cells by osmosis, causing severe diarrhea and the resulting loss of water by the body. Death can occur in a few hours if treatment is not received.

As you can see, a working knowledge of cell transport is needed to understand the mechanisms of a number of medical conditions.

QUICK CHECK

1. What are the differences between *passive* and *active* transport processes?
2. What is *diffusion*? What is *osmosis*?
3. How does an *ion pump* work? How do faulty ion pumps cause disease?
4. How do *phagocytosis* and *pinocytosis* differ?

3

Cell Growth and Reproduction

Cell Growth

For normal growth and maintenance, the cell must continually produce the many diverse structural and functional proteins needed for human life. The functional proteins then synthesize carbohydrates and lipids and help regulate all cell functions.

The two *nucleic acids* **deoxyribonucleic acid (DNA)** and **ribonucleic acid (RNA)** play crucial roles in directing protein synthesis in each cell. We start our story of cell growth and reproduction with these amazing molecules.

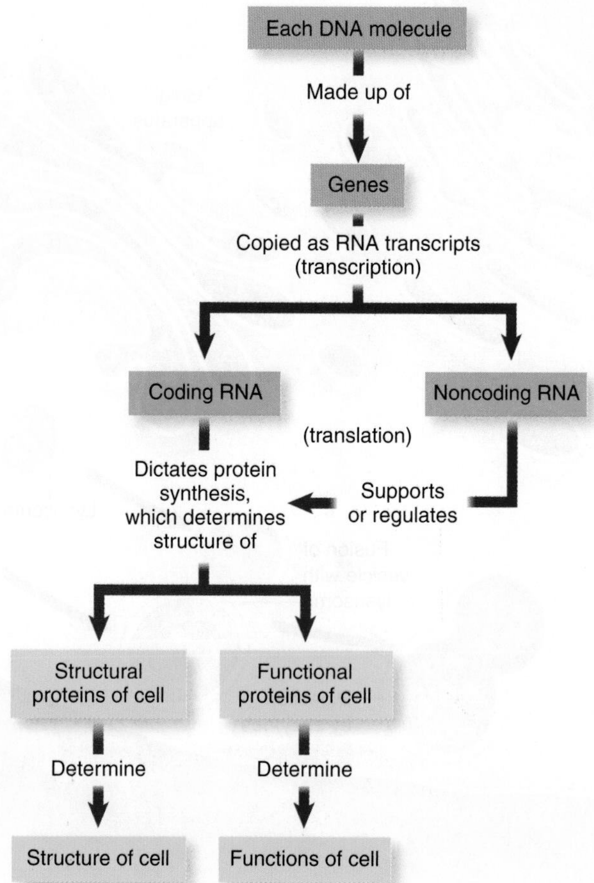

FIGURE 3-13 Function of genes. Genes copied from deoxyribonucleic acid (DNA) are copied to ribonucleic acid (RNA) in a process called *transcription.* The RNA transcripts are then used in a process called *translation,* in which a code that determines the sequence of amino acids is translated to form a protein. The structure of the resulting protein determines the role of the protein in body structure and function—and ultimately, the structure and function of the body.

DNA

Chromosomes, which are composed largely of DNA, contain the information needed to make all the proteins of the cells—the information that allows a cell to live and function normally. The *genetic code* contained in segments of the DNA molecules that are called **genes** ultimately determines the structure and function of all cells (**Figure 3-13**). This coded information can be transmitted to generations of cells and eventually to offspring.

Structurally, the DNA molecule resembles a long, narrow ladder made of a pliable material. It is twisted round and round its axis, taking on the shape of a double helix (see **Figure 2-14**, p. 35).

Each DNA molecule is made of many smaller units called *nucleotides*. Each nucleotide is made up of a sugar, a phosphate, and a base (see **Table 2-4** on p. 35). The bases are adenine, thymine, guanine, and cytosine. These nitrogen-containing chemicals are called *bases* because by themselves they have a high pH and chemicals with a high pH are called "bases" (see p. 30 for a discussion of acids and bases).

As you can see in **Figure 2-14** (p. 35), each step in the DNA ladder consists of a pair of bases. Only two combinations of bases occur, and the same two bases *always* pair off with each other in a DNA molecule. Adenine always binds to thymine, and cytosine always binds to guanine. This characteristic of DNA structure is called **complementary base pairing.**

A *gene* is a specific segment of base pairs in a chromosome. Although the types of base pairs in all chromosomes are the same, the order or *sequence* of base pairs is not the same. This fact has tremendous functional importance because it is the sequence of base pairs in each gene of each chromosome that determines the genetic code.

Most genes direct the synthesis of at least one kind of protein molecule. Each protein may function, for example, as an enzyme, a structural component of a cell, or a specific hormone. Or it may combine with other protein molecules—or even with carbohydrates or lipids—to form any number of large, complex molecules such as quaternary proteins, glycoproteins, proteoglycans, or lipoproteins.

The enzymes and other functional molecules produced by protein synthesis facilitate and regulate cellular chemical reactions that drive all the functions of cells—and thereby all the functions of the body.

In humans having 46 nuclear chromosomes and one kind of mitochondrial chromosome in each body cell, DNA has a content of genetic information totaling about 3 *billion* base pairs in perhaps 19,000 or so protein-coding genes. Sections of DNA that do not code for protein structure have other functions, which include regulation of turning genes on and off and regulating protein synthesis. This means that over a billion bits of information are inherited from each of our two biological parents. Is it any wonder, then, with all of this genetic information packed into each of our cells, that we are such complex organisms?

RNA

The genetic information contained in protein-coding genes is capable of "directing" the synthesis of a specific protein. Some genes instead contain information needed to build regulatory types of RNA molecules.

Regulatory RNA molecules act as functional molecules that affect some of the chemical processes in a cell. For example, *ribosomal RNA (rRNA)* molecules form most of the ribosome's protein-synthesizing structure and other RNA molecules that serve as temporary working copies of genetic code.

Most of the DNA, with its genetic code that dictates directions for protein synthesis, is contained in the nucleus of the cell. The actual process of protein synthesis, however, occurs at ribosomes in the cytoplasm and on ER. Another nucleic acid, RNA, copies this genetic information from the nucleus and carries it to the cytoplasm. RNA also may be an end product formed in the nucleus using the DNA code and transported out to the cytoplasm, where it regulates various functions of the cell.

Both RNA and DNA are composed of nucleotide subunits made up of a sugar, a phosphate, and one of four bases. RNA subunits, however, contain a different sugar and base component. In RNA nucleotide subunits, the base uracil substitutes for the base thymine. The types of RNA discussed here are all single-stranded molecules—not double-stranded like DNA. However, short double-stranded RNA molecules also exist in nature.

Table 3-4 lists the major types of RNA involved in protein synthesis.

TABLE 3-4 Types of RNA*

ACRONYM	NAME	DESCRIPTION	ROLE IN CELL FUNCTION
mRNA	Messenger RNA	Single, unfolded strand of nucleotides	Serves as working copy of one protein-coding gene
rRNA	Ribosomal RNA	Single, folded strand of nucleotides	Component of the ribosome (along with proteins); attaches to mRNA and participates in translation
tRNA	Transfer RNA	Single, folded strand of nucleotides; has an anticodon at one end and an amino acid–binding site at the other end	Carries a specific amino acid to a specific codon of mRNA at the ribosome during translation

*Cells contain other types of RNA that perform complex functions beyond the scope of this book.

Protein Synthesis

The process of transferring genetic information from the nucleus into the cytoplasm, where proteins are actually produced, requires completion of two steps called **transcription** and **translation.**

Transcription

During transcription the double-stranded DNA molecule separates or unwinds, and a special type of RNA called **messenger RNA (mRNA)** is formed (**Figure 3-14**, *Step 1*). Each strand of mRNA is a complementary copy of a particular gene sequence along one of the newly separated DNA spirals. The messenger RNA is said to have been "transcribed" or copied from its DNA mold or template. The mRNA then functions as a temporary "working copy" of the genetic information in a gene from DNA.

The mRNA transcripts pass from the nucleus to the cytoplasm to direct protein synthesis in the ribosomes (**Figure 3-14**, *Step 2*).

Translation

Translation is the process of "translating" the genetic code in the mRNA transcript to synthesize a protein. Translation occurs within ribosomes, which attach around the mRNA strands in the cytoplasm. The ribosomes move along the mRNA transcript and "read" the information encoded there to direct the choice and sequencing of the appropriate chemical building blocks called *amino acids.*

First, the two subunits of a ribosome attach at the beginning of the mRNA molecule (**Figure 3-14**, *Step 3*). Recall that ribosomes are themselves made mostly of RNA—ribosomal RNA (rRNA). The ribosome then moves down the mRNA strand as amino acids are assembled into their proper sequence (**Figure 3-14**, *Step 4*).

Transfer RNA (tRNA) molecules assist the process by bringing specific amino acids in to "dock" at each **codon** along the mRNA strand. A codon is a series of three nucleotide bases—a "triplet"—that acts as a code representing a specific amino acid. Each gene encoded in the mRNA is made up of a series of codons that tell the cell the sequence of amino acids to string together to form a protein strand. Each tRNA includes an anticodon segment at one end, which is a complementary sequence of three bases that allows the tRNA to recognize the

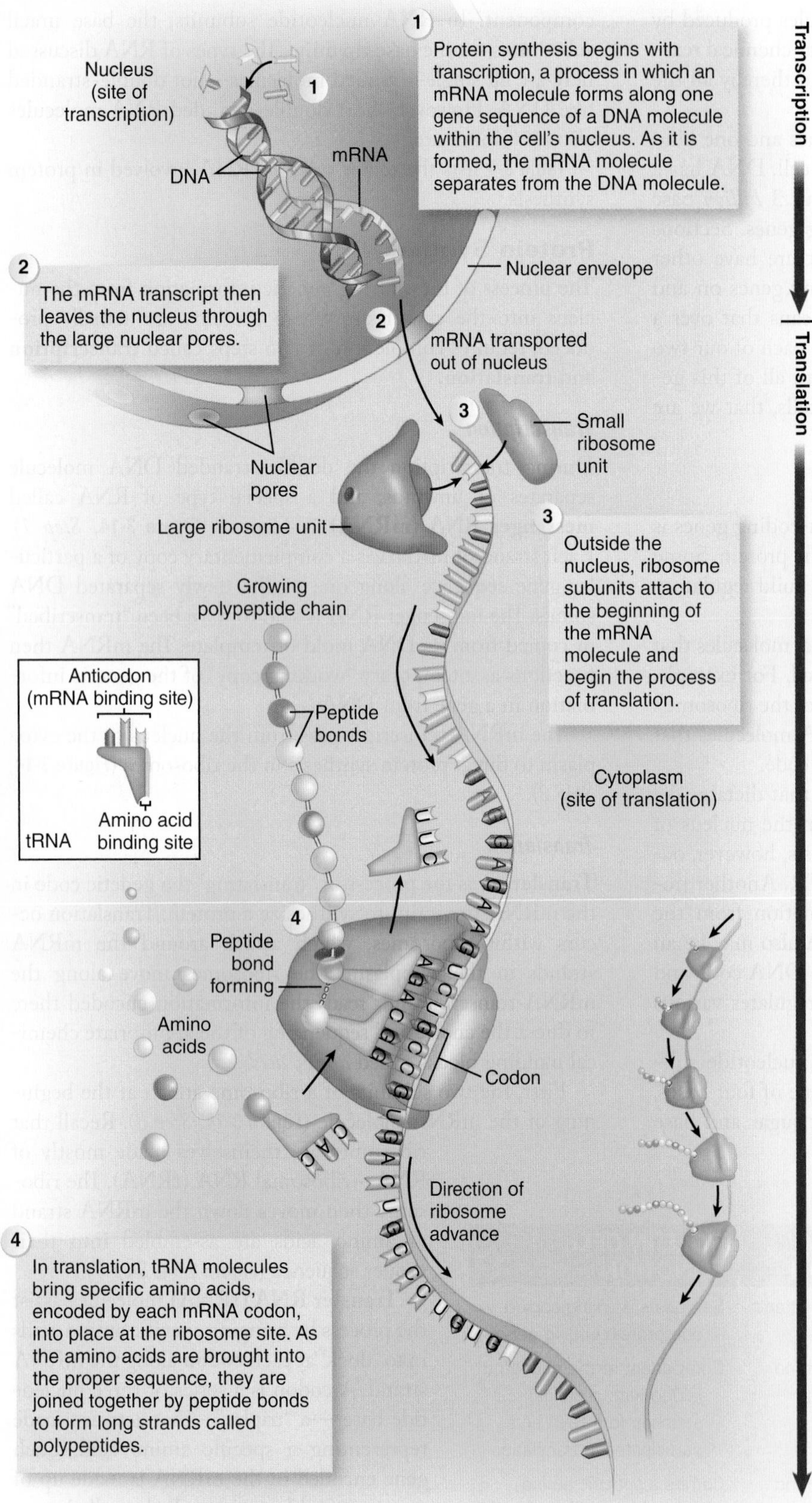

FIGURE 3-14 Protein synthesis. Steps show transcription of the DNA code to mRNA and subsequent translation of the mRNA at the ribosome to assemble a polypeptide. Several polypeptide chains may be needed to make a complete protein molecule. DNA, Deoxyribonucleic acid; mRNA, messenger RNA (ribonucleic acid); tRNA, transfer RNA.

particular codon for the type of amino acid carried by that tRNA molecule (see **Figure 3-14**, *inset*).

The strand of amino acids formed during translation then folds on itself and perhaps even combines with another strand to form a complete protein molecule (see **Figure 2-12**, p. 34). The specific, complex shape of each type of protein molecule allows the molecule to perform specific functions in the cell. It is clear that because DNA directs the shape of each protein, DNA also directs the function of each protein in a cell (see **Figure 3-13**).

Protein Synthesis and Disease

Many diseases have a cellular basis. That is, they are basically cell problems even though they may affect the entire body. Because individual cells are members of an interacting "community" of cells, it is no wonder that a problem in just a few cells can have a "ripple effect" that influences the entire body. Most of these cell problems can be traced to abnormalities in the DNA itself or in the process by which DNA information is transcribed and translated into proteins.

In individuals with inherited diseases, abnormal DNA from one or both parents may cause production of dysfunctional proteins in certain cells or prevent a vital protein from being synthesized. For example, DNA may contain a mistake in its genetic code that prevents production of normal blood-clotting proteins. Deficiency of these essential proteins results in excessive, uncontrollable bleeding—a condition called *hemophilia* (see Chapters 13 and 25).

Chemical or mechanical irritants, radiation, bacteria, viruses, and other factors can directly damage DNA molecules and thus disrupt a cell's normal function. For example, the *human immunodeficiency virus (HIV)* eventually inserts its own genetic codes into the DNA of certain cells. The viral codes trigger synthesis of viral molecules, detouring raw materials intended for use in building normal human products. This does two things: it prevents human white blood cells from performing their normal functions, and it provides a mechanism by which the virus can

RESEARCH, ISSUES, AND TRENDS

HUMAN GENOME

The sum total of all of the DNA in each cell of the body is called the **genome.** Intense, coordinated efforts by scientists recently mapped all of the gene locations in the human genome. Efforts at reading the different genetic codes possible at each location are still underway.

Much of the work of mapping the human genome was done as part of the Human Genome Project (HGP), which was started in 1990. Besides producing a human genetic map and developing tools of genetic mapping, a field called *genomics*, the HGP also addresses the ethical, legal, and social issues that may arise—a notable first for such a massive scientific research effort.

With the human genome already mapped, many scientists are working now to fill in the details concerning the many genes and gene variants found in the human genome. An offshoot of the HGP is *ENCODE, The Encyclopedia of DNA Elements.* ENCODE scientists have mapped large regions of DNA between the genes that contain a rich and complex assortment of "switches" that regulate gene activity.

Many scientists are also working in the emerging field of *proteomics*—the study of all the proteins encoded by each of the genes of the human genome.

reproduce itself and spread to other cells. When enough cells of the human immune system are affected, they can no longer protect us from infections and cancer—a condition that may eventually lead to death.

The genetic basis for disease discussed briefly in Chapter 6 is more fully explained in Chapter 25.

Cell Reproduction

Cell Life Cycle

The process of cell reproduction is one part of the cell's life cycle. It involves the division of the cell into two genetically identical daughter cells. Cell reproduction thus requires division of the nucleus—a process called **mitosis**—and division of the cytoplasm.

As you can see in **Figure 3-15**, when a cell is not dividing, but instead going about its usual functions, it is in a period of its life cycle called **interphase.**

Interphase includes the initial growing stages of a newly formed cell, in which a cell is busy with protein synthesis and other growth and maintenance functions. This initial growth period of interphase is followed by a period during which the cell prepares for possible cell division.

During interphase, the cell is said to be "resting." However, it is resting only from the standpoint of active cell division. In all other aspects it is exceedingly active. During interphase and just before mitosis begins, the DNA of each chromosome makes an identical copy of itself. The cell then enters another growth period of interphase before it begins to actively divide.

DNA Replication

DNA molecules are somewhat unusual in that, unlike most molecules in nature, they can make identical copies of themselves—a process called **DNA replication.** Before a cell divides to form two new cells, each DNA molecule in its nucleus forms another DNA molecule just like itself.

When a DNA molecule is not replicating, it has the shape of a tightly coiled double helix. As it begins replication, short segments of the DNA molecule uncoil and the two strands of the molecule pull apart between their base pairs. The separated strands therefore contain unpaired bases.

Each unpaired base in each of the two separated strands attracts its complementary base (in the nucleoplasm) and binds to it. Specifically, each adenine attracts and binds to a thymine, and each cytosine attracts and binds to a guanine. These steps are repeated over and over throughout the length of the DNA molecule. Thus each half of a DNA molecule becomes a whole DNA molecule identical to the original DNA molecule.

After DNA replication is complete, the cell continues to grow until it is ready for the first phase of mitosis.

Mitosis

Mitosis is the process of dividing the replicated genetic material—the DNA—of the nucleus in an orderly way so that each resulting daughter cell has a complete identical set.

Prophase

Look at **Figure 3-15** and note the changes that identify the first stage of mitosis, **prophase.** The chromatin becomes "organized." Chromosomes in the nucleus have formed two strands called **chromatids.** Note that the two chromatids are held together by a beadlike structure called the **centromere.** In the cytoplasm the centrioles are moving away from each other as a network of tubules called **spindle fibers** forms between them. These spindle fibers serve as "guidewires" and assist the chromosomes to move toward opposite ends of the cell later in mitosis.

Metaphase

By the time **metaphase** begins, the nuclear envelope and nucleolus have disappeared. Note in **Figure 3-15** that the chromosomes have aligned themselves across the center of the cell.

3

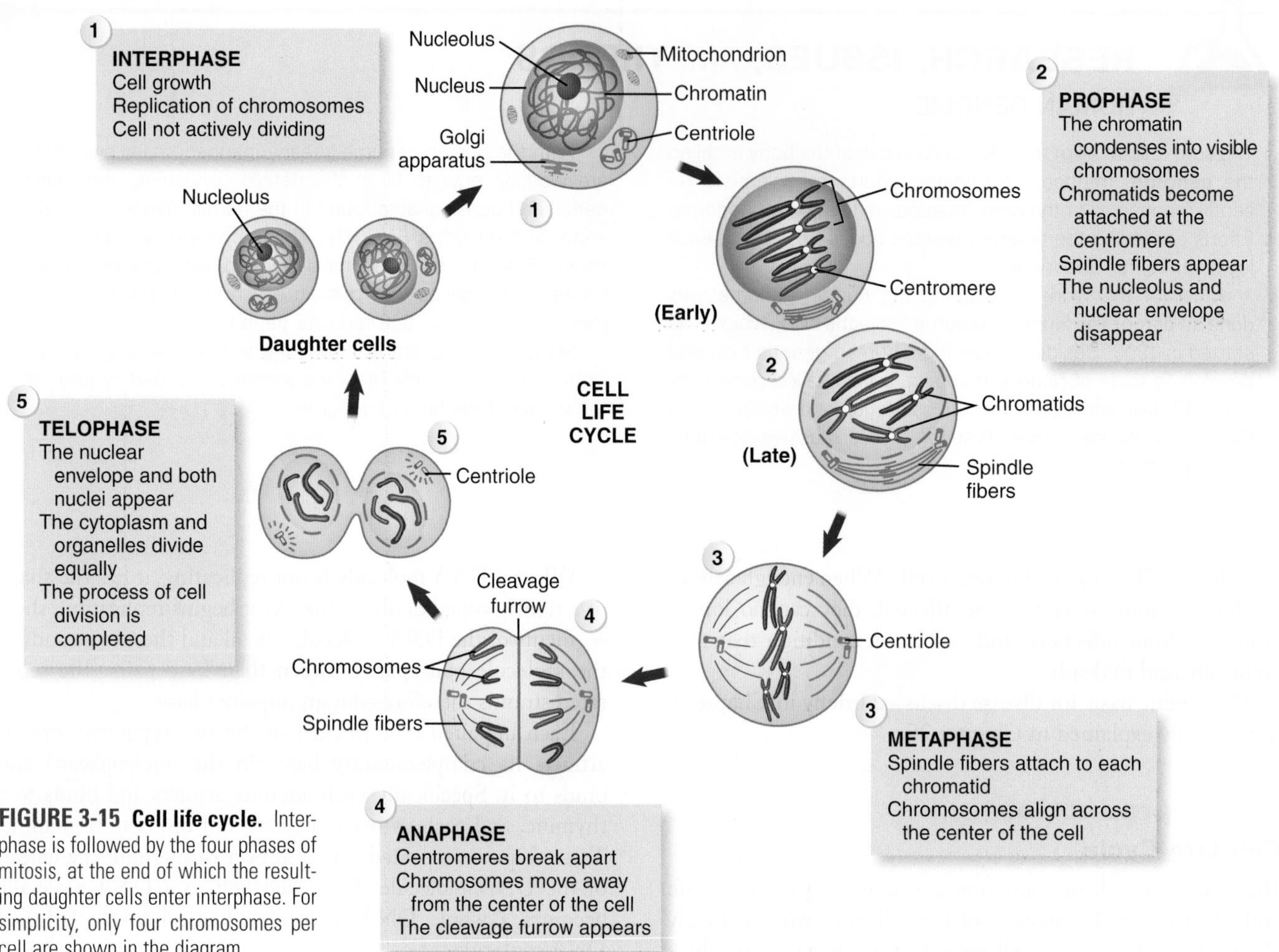

FIGURE 3-15 Cell life cycle. Interphase is followed by the four phases of mitosis, at the end of which the resulting daughter cells enter interphase. For simplicity, only four chromosomes per cell are shown in the diagram.

Also, the centrioles have migrated to opposite ends of the cell, and spindle fibers are attached to each chromatid.

Anaphase

As **anaphase** begins, the beadlike centromeres, which were holding the paired chromatids together, break apart. As a result, the individual chromatids, identified once again as chromosomes, move away from the center of the cell. Movement of chromosomes occurs along spindle fibers toward the centrioles. Note in **Figure 3-15** that chromosomes are being pulled to opposite ends of the cell.

A **cleavage furrow** that begins to divide the cell into two daughter cells can be seen for the first time at the end of anaphase.

Telophase

During **telophase,** cell division is completed. Two nuclei appear, and chromosomes become less distinct and appear to break up. As the nuclear envelope forms around the chromatin, the cleavage furrow completely divides the cell into two parts. The division of the plasma membrane and cytoplasm surrounding the nucleus is called **cytokinesis.**

Before division is complete, each nucleus is surrounded by cytoplasm in which organelles have been equally distributed. By the end of telophase, two separate daughter cells, each having identical genetic characteristics, are formed. Each daughter cell is now in interphase, is fully functional, and will perhaps itself undergo mitotic cell division (cell reproduction) in the future.

Now is a good time to review again the stages of mitosis summarized in **Figure 3-15**.

Results of Cell Division

Mitotic cell division results in the production of identical new cells. During developmental years, the addition of cells helps tissues and organs grow in size. During such periods of body growth, mitosis also allows groups of similar cells to **differentiate** or develop into different *tissues.* The next chapter explores the major types of tissues in the human body that result from differentiation.

In the adult, mitosis replaces cells that have become lost or less functional with age, as well as cells damaged or destroyed by illness or injury.

Changes in Cell Growth and Reproduction

Cells have the ability to adapt to changing conditions. Cells may alter their size, reproductive rate, or other characteristics to adapt to changes in the internal environment.

CLINICAL APPLICATION

STEM CELLS

Scientists all over the world are currently engaged in intensive research efforts to unravel the biological secrets of a special kind of undifferentiated cell called a **stem cell.** As the illustration shows, stem cells produce daughter cells with specific characteristics during the process of **differentiation.** Stem cells may also produce additional daughter stem cells, which may then later produce differentiated daughter cells.

Embryonic stem cells, which are obtained from a developing embryo, can be isolated and cultured in the laboratory. Using complex research methods, these primitive cells can then be stimulated to produce additional stem cells or be "directed" to produce many different kinds of differentiated daughter cell types including nerve, blood, muscle, and various types of glandular tissue.

Adult stem cells are undifferentiated cells found scattered within mature tissues throughout the body. Current research suggests that all adult tissues have some of these undifferentiated cells that are capable of producing any of the specialized cell types within its particular tissue.

Injecting adult bone marrow stem cells is a therapy now being used to treat patients with leukemia or bone marrow damaged by toxins or high-dose x-ray. Current research suggests that some adult stem cells, like embryonic stem cells, can be coaxed into producing a variety of different types of cells.

We will revisit the role of stem cells in the body—and in therapy—in later chapters.

Such adaptations usually allow cells to work more efficiently. However, sometimes cells alter their characteristics abnormally—decreasing their efficiency and threatening the health of the body.

Common types of changes in cell growth and reproduction are summarized in Figure 3-16, and in Table 3-5.

Cells may respond to changes in function, hormone signals, or availability of nutrients by increasing or decreasing in size. The term **hypertrophy** refers to an increase in cell size, and the term **atrophy** refers to a decrease in cell size.

Either hypertrophy or atrophy can occur easily in muscle tissue. When a person continually uses muscle cells to pull against heavy resistance, as in weight training, the cells respond by increasing in size. Bodybuilders thus increase the size of their muscles by hypertrophy—increasing the size of muscle cells.

Atrophy often occurs in underused muscle cells. For example, when a broken arm is immobilized in a cast for a long period, muscles that move the arm often atrophy. Because the muscles are temporarily out of use, muscle cells decrease in size. Atrophy also may occur in tissues whose nutrient or oxygen supply is diminished.

Sometimes cells respond to changes in the internal environment by increasing their rate of reproduction—a process called **hyperplasia.** The word part *-plasia* comes from a Greek word that means "formation"—referring to formation of new cells. Because *hyper* means "excessive," *hyperplasia* means excessive cell reproduction.

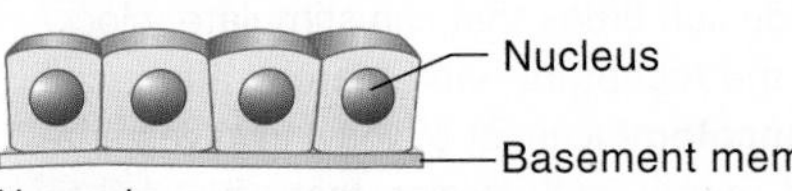

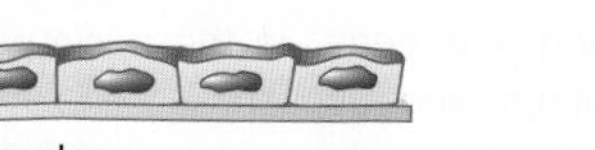

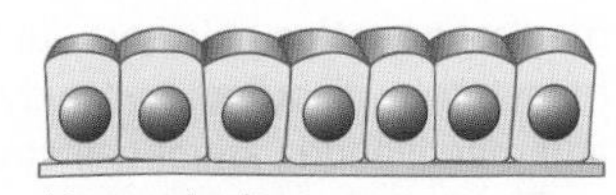

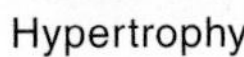

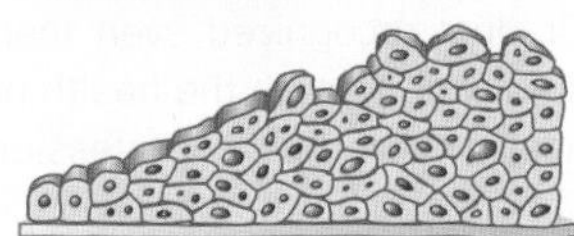

FIGURE 3-16 Alterations in cell growth and reproduction.

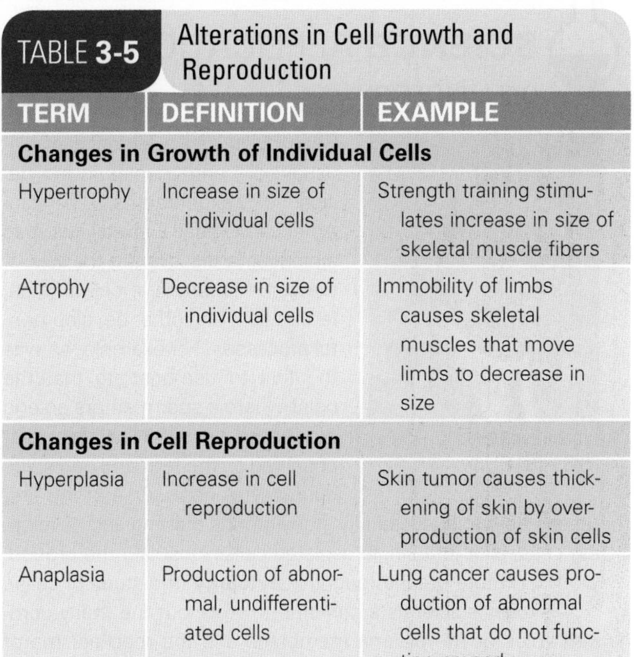

TABLE 3-5 Alterations in Cell Growth and Reproduction

TERM	DEFINITION	EXAMPLE
Changes in Growth of Individual Cells		
Hypertrophy	Increase in size of individual cells	Strength training stimulates increase in size of skeletal muscle fibers
Atrophy	Decrease in size of individual cells	Immobility of limbs causes skeletal muscles that move limbs to decrease in size
Changes in Cell Reproduction		
Hyperplasia	Increase in cell reproduction	Skin tumor causes thickening of skin by overproduction of skin cells
Anaplasia	Production of abnormal, undifferentiated cells	Lung cancer causes production of abnormal cells that do not function properly

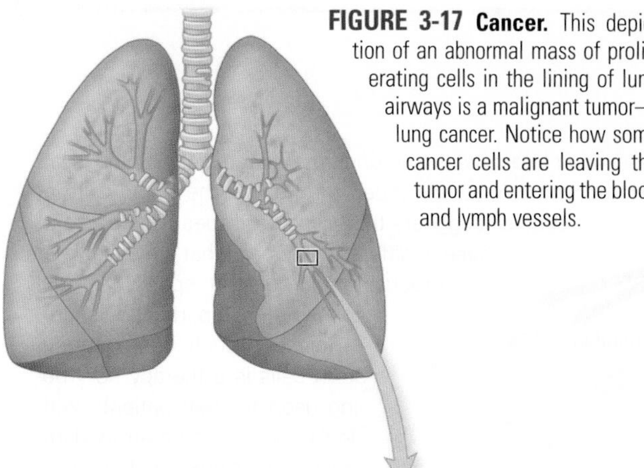

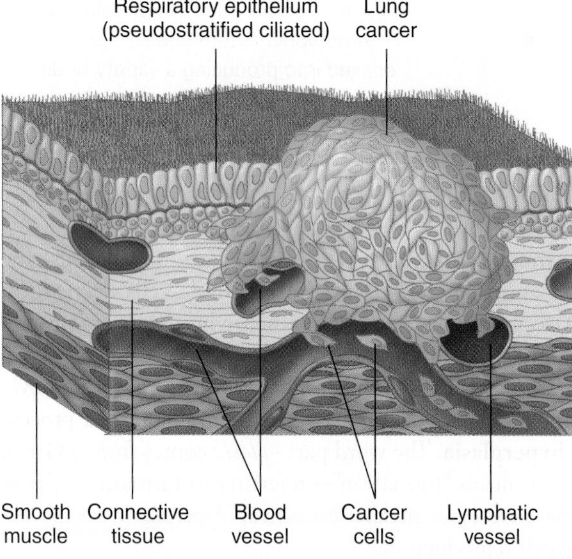

FIGURE 3-17 Cancer. This depiction of an abnormal mass of proliferating cells in the lining of lung airways is a malignant tumor—lung cancer. Notice how some cancer cells are leaving the tumor and entering the blood and lymph vessels.

Like hypertrophy, hyperplasia causes an increase in the size of a tissue or organ. However, hyperplasia is an increase in the *number of cells* rather than an increase in the size of each cell. A common example of hyperplasia occurs in the milk-producing glands of the female breast during pregnancy. In response to hormone signals, the glandular cells reproduce rapidly, preparing the breast for nursing.

If the body loses its ability to control the cell life cycle—cell growth, reproduction, differentiation, and death—abnormal hyperplasia may occur. The new mass of cells thus formed is a tumor or **neoplasm.**

Many neoplasms also exhibit a characteristic called **anaplasia.** Anaplasia is a condition in which cells change in orientation to each other and fail to mature normally—that is, they fail to differentiate into a specialized cell type and appear disorganized.

Neoplasms may be relatively harmless growths called **benign tumors.** If tumor cells can break away and travel through the blood or lymphatic vessels to other parts of the body (**Figure 3-17**), the neoplasm is a **malignant tumor** or **cancer.** Neoplasms are discussed further in Chapter 6.

QUICK CHECK

1. How do *genes* determine the structure and function of the body?
2. What are the main steps in making proteins in the cell?
3. What are the four phases of mitotic cell division?
4. Describe *hypertrophy, hyperplasia,* and *atrophy.*

SCIENCE APPLICATIONS

CELL BIOLOGY

Ernest Everett Just (1883-1941)

Recognized as "a biologist of unusual skill" and "genius in the design of experiments," African-American Ernest Everett Just was a pioneer in discovering the role of the cell membrane in cell division, fertilization, and other developmental processes. For example, he was the first to demonstrate that the point where a sperm enters an egg cell becomes the cleavage point observed as the fertilized egg splits during cytokinesis. One of the first researchers to successfully study groups of living and developing cells—not just single, isolated cells—Just laid the groundwork for future breakthroughs in **cytology** (the study of cells).

As today's scientists continue to work out the many complex roles of the plasma membrane and the mechanisms of cell division, they provide clues to the prevention and treatment of disease. For example, **pharmacologists** study the role of membrane receptors that regulate cell functions and use that information to design drugs that can stimulate, block, or otherwise influence the receptors—and thereby affect cell function. The study of **oncology** (cancer biology) involves analyzing the processes of mitotic cell division that often fail to function properly in cancer. Professionals who specialize in medical **genetics** help sort out the mechanisms of cancer and many other disorders by studying how genes in the nucleus and mitochondria affect cell structure and function.

The holistic view of biology championed a century ago by Dr. Just recognized even then that nearly every function we study and treat in the health professions is *cell* function. Thus, nearly every health profession directly or indirectly relies on understanding the basic principles of cell biology.

LANGUAGE OF SCIENCE *(continued from p. 43)*

cleavage furrow
(KLEE-vij FUR-oh)
[*cleav-* **split,** *-age* **state,** *furrow* **trench**]

codon
(KOH-don)
[*cod-* **book,** *-on* **thing**]

complementary base pairing
(kom-pleh-MEN-tah-ree bays PAYR-ing)
[*comple-* **complete,** *-ment-* **process,** *-ary* **relating to**]

cytokinesis
(sye-toh-kin-EE-sis)
[*cyto-* **cell,** *-kinesis* **movement**]

cytology
(sye-TOL-oh-jee)
[*cyto-* **cell,** *-log-* **words (study of),** *-y* **activity**]

cytoplasm
(SYE-toh-plaz-em)
[*cyto-* **cell,** *-plasm* **to mold**]

cytoskeleton
(sye-toh-SKEL-eh-ton)
[*cyto-* **cell,** *-skeleto-* **dried body**]

deoxyribonucleic acid (DNA)
(dee-ok-see-rye-boh-nook-lay-ik AS-id [dee en ay])
[*de-* **removed,** *-oxy-* **oxygen,** *-nucle-* **nucleus (kernel),** *-ic* **relating to,** *acid* **sour**]

dialysis
(dye-AL-ih-sis)
[*dia-* **apart,** *-lysis* **loosening**]

differentiate
(dif-er-EN-shee-ayt)
[*different-* **difference,** *-ate* **action**]

differentiation
(dif-er-EN-shee-AY-shun)
[*different-* **difference,** *-ation* **process**]

diffusion
(dih-FYOO-zhun)
[*dis-* **apart,** *-fus-* **flow,** *-tion* **process**]

DNA replication
(D N A rep-lih-KAY-shun)
[*re-* **again,** *-plic-* **fold,** *-ation* **process**]

endoplasmic reticulum (ER)
(en-doh-PLAZ-mik reh-TIK-yoo-lum [ee ar])
[*endo-* **inward or within,** *-plasm-* **to mold,** *-ic* **relating to,** *ret-* **net,** *-ic-* **relating to,** *-ul-* **little,** *-um* **thing**]

equilibrium
(ee-kwih-LIB-ree-um)
[*equi-* **equal,** *-libr-* **balance,** *-um,* **thing**]

filtration
(fil-TRAY-shun)
[*filtr-* **strain,** *-ation* **process**]

flagellum
(flah-JEL-um)
pl., flagella
(flah-JEL-ah)
[flagellum **whip**]

gene
(jeen)
[*gen-* **produce or generate**]

genetics
(jeh-NET-iks)
[*gene-* **produce,** *-ic* **relating to**]

genome
(JEE-nohm)
[*gen-* **produce (gene),** *-ome* **entire collection**]

Golgi apparatus
(GOL-jee ap-ah-RA-tus)
[*Camillo Golgi* Italian histologist]

hypertonic
(hye-per-TON-ik)
[*hyper-* **excessive,** *-ton-* **tension,** *-ic* **relating to**]

hypotonic
(hye-poh-TON-ik)
[*hypo-* **under or below,** *-ton-* **tension,** *-ic* **relating to**]

interphase
(IN-ter-fayz)
[*inter-* **between,** *-phase* **stage**]

interstitial fluid
(in-ter-STISH-al FLOO-id)
[*inter-* **between,** *-stit-* **stand,** *-al* **relating to**]

isotonic
(aye-soh-TON-ik)
[*iso-* **equal,** *-ton-* **tension,** *-ic* **relating to**]

lyse
(lyze)
[*lysis* **loosening**]

lysosome
(LYE-soh-sohm)
[*lyso-* **dissolution,** *-som* **body**]

messenger RNA (mRNA)
(MES-en-jer ar en ay [em ar en ay])
[*RNA-* **ribonucleic acid**]

metaphase
(MET-ah-fayz)
[*meta-* **change (place),** *-phase* **stage**]

microvillus
(my-kroh-VIL-us)
pl., microvilli
(my-kroh-VIL-aye or my-kroh-VIL-ee)
[*micro-* **small,** *-villus* **shaggy hair**]

mitochondrion
(my-toh-KON-dree-on)
pl., mitochondria
(my-toh-KON-dree-ah)
[*mito-* **thread,** *-chondrion* **granule**]

mitosis
(my-TOH-sis)
[*mitos-* **thread,** *-osis* **process**]

nuclear envelope
(NOO-klee-ar)
[*nucle-* **nucleus (kernel),** *-ar* **relating to**]

nucleolus
(noo-KLEE-oh-lus)
[*nucleo-* **nucleus (kernel),** *-olus* **little**]

nucleoplasm
(NOO-klee-oh-plaz-em)
[*nucleo-* **nucleus (kernel),** *-plasm* **substance**]

nucleus
(NOO-klee-us)
[*nucleus* **kernel**]

organelle
(or-gah-NELL)
[*organ-* **tool or instrument,** *-elle* **small**]

osmosis
(os-MOH-sis)
[*osmos-* **push,** *-osis* **process**]

passive transport
(PAS-iv TRANZ-port)
[*pass-* **submit,** *-ive* **relating to,** *trans-* **across,** *-port* **carry**]

phagocytosis
(fag-oh-sye-TOH-sis)
[*phago-* **eat,** *-cyte-* **cell,** *-osis* **process**]

phospholipid
(fos-foh-LIP-id)
[*phospho-* **phosphorus,** *-lip-* **fat,** *-id* **form**]

pinocytosis
(pin-oh-sye-TOH-sis)
[*pino-* **drink,** *-cyto-* **cell,** *-osis* **process**]

plasma membrane
(PLAZ-mah MEM-brayn)
[*plasma* **substance,** *membrane* **thin skin**]

prophase
(PROH-fayz)
[*pro-* **first,** *-phase* **stage**]

ribonucleic acid (RNA)
(rye-boh-noo-KLAY-ik AS-id [ar en ay])
[*ribo-* **ribose (sugar),** *nucle-* **nucleus,** *-ic* **relating to,** *acid* **sour**]

ribosome
(RYE-boh-sohm)
[*ribo-* **ribose or RNA,** *-som-* **body**]

sodium-potassium pump
(SOH-dee-um poh-TAS-ee-um)
[*sod-* **soda,** *-um* **thing or substance,** *potas-* **potash,** *-um* **thing or substance**]

Continued on p. 64

3

LANGUAGE OF **SCIENCE** *(continued from p. 63)*

solute
(SOL-yoot)
[*solut* **dissolved**]

spindle fiber
(SPIN-dul FYE-ber)

stem cell
(stem sel)
[*stem* **tree trunk,** *cell* **storeroom**]

telophase
(TEL-oh-fayz or TEE-loh-fayz)
[*telo-* **end,** *-phase* **stage**]

tissue
(TISH-yoo)
[*tissu* **fabric**]

transcription
(trans-KRIP-shun)
[*trans-* **across,** *-script-* **write,** *-tion* **process**]

transfer RNA (tRNA)
(TRANS-fer ar en ay [tee ar en ay])
[*RNA-* **ribonucleic acid**]

translation
(trans-LAY-shun)
[*translat-* **bring over,** *-tion* **process**]

transport process
(TRANS-port PROH-ses)
[*trans-* **across,** *-port* **carry**]

vesicle
(VES-ih-kul)
[*vesic-* **blister,** *-cle* **little**]

LANGUAGE OF **MEDICINE**

3

anaplasia
(an-ah-PLAY-zhah)
[*ana-* **without,** *-plasia* **shape**]

atrophy
(AT-roh-fee)
[*a-* **without,** *-troph-* **nourishment,** *-y* **state**]

benign tumor
(bee-NYNE TOO-mer)
[*benign* **kind**]

cancer
(KAN-ser)
[*cancer* **crab or malignant tumor**]

cholera
(KAHL-er-ah)
[*chole-* **bile,** *-a* **state**]

crenation
(kreh-NAY-shun)
[*crenat-* **scalloped or notched,** *-ation* **process**]

hyperplasia
(hye-per-PLAY-zhah)
[*hyper-* **excessive,** *-plasia* **shape**]

malignant tumor
(mah-LIG-nant TOO-mer)
[*malign* **bad,** *-ant* **state;** *tumor* **swelling**]

neoplasm
(NEE-oh-plaz-em)
[*neo-* **new,** *-plasm* **formation**]

oncology
(ong-KAHL-oh-jee)
[*onco-* **tumor,** *-log-* **words (study of),** *-y* **activity**]

pharmacologist
(far-mah-KAHL-oh-jist)
[*pharmaco-* **drug,** *-log-* **words (study of),** *-ist* **agent**]

❑ OUTLINE SUMMARY

To download a digital version of the chapter summary for use with your device, access the ***Audio Chapter Summaries*** *online at evolve.elsevier.com.*

Scan this summary after reading the chapter to help you reinforce the key concepts. Later, use the summary as a quick review before your class or before a test.

Cells

A. Size and shape
   1. Human cells vary considerably in size
   2. All are microscopic
   3. Cells differ notably in shape

B. Composition
   1. Cells contain cytoplasm—substance found only in cells
   2. Organelles are specialized structures within the cytoplasm
   3. Cell interior is surrounded by a plasma membrane

C. Parts of the cell
   1. Plasma membrane (**Figure 3-1**)
      a. Forms outer boundary of cell
      b. Composed of a thin, two-layered membrane of phospholipids containing proteins
      c. Proteins and other molecules embedded in membrane can function in transport, signaling, self-identification, anchoring of fibers, chemical processing (enzymes), and more

2. Cytoplasm (**Figure 3-2**)—internal cell fluid and numerous organelles
   a. Ribosomes
      (1) Made of two tiny subunits of mostly ribosomal RNA
      (2) May attach to rough ER or lie free in cytoplasm
      (3) Manufacture enzymes and other protein compounds
      (4) Often called *protein factories*
   b. Endoplasmic reticulum (ER)
      (1) Network of connecting sacs and canals
      (2) Carry substances through fluid cytoplasm
      (3) Two types—rough and smooth
      (4) Rough ER collects, folds, and transports proteins made by ribosomes
      (5) Smooth ER synthesizes chemicals; makes new membrane
   c. Golgi apparatus (**Figure 3-3**)
      (1) Group of flattened sacs near nucleus
      (2) Collect chemicals into vesicles that move from the smooth ER outward to plasma membrane
      (3) Called the *chemical processing and packaging center*
   d. Mitochondria
      (1) Composed of inner and outer membranous sacs
      (2) Involved with energy-releasing chemical reactions
      (3) Often called *power plants* of the cell
      (4) Contain one DNA molecule
   e. Lysosomes
      (1) Membranous-walled organelles
      (2) Contain digestive enzymes
      (3) Have protective function (engulf and destroy microbes)
   f. Centrioles
      (1) Paired organelles that lie at right angles to each other near the nucleus
      (2) Function in cell reproduction
   g. Microvilli (**Figure 3-4**)
      (1) Small, fingerlike extensions of the plasma membrane
      (2) Increase absorptive surface area of the cell
   h. Cilia (**Figure 3-4**)
      (1) Fine, hairlike extensions found on free or exposed surfaces of some cells
      (2) Some are found in groups and capable of moving in unison in a wavelike fashion (**Figure 3-5**)
      (3) Single, nonmoving cilia in some cells serve sensory functions
   i. Flagella (**Figure 3-4**)
      (1) Single projections extending from cell surface; much larger than cilia
      (2) Propel a cell through its fluid environment (**Figure 3-5**)
      (3) "Tails" of sperm cells only example of flagella in humans
3. Nucleus
   a. Controls cell because it contains DNA, the genetic code—instructions for making proteins, which in turn determine cell structure and function
   b. Component structures include nuclear envelope, nucleoplasm, nucleolus, and chromatin granules
   c. DNA molecules become tightly coiled chromosomes during cell division
   d. Each cell has 46 chromosomes in the nucleus

D. Relationship of cell structure and function
1. Every human cell has a designated function—some help maintain the cell; others regulate life processes of the body itself
2. Specialized functions of a cell differ depending on number and type of organelles

Movements of Substances Through Cell Membranes

A. Types of membrane transport
1. Transport processes move substances into and out of cells
2. Types of transport
   a. Passive transport—does not require the cell to expend energy
   b. Active transport—requires the cell to expend energy (from ATP)

B. Passive transport processes
1. Passive transport processes do not require added energy and result in movement "down a concentration gradient"
2. Diffusion
   a. Substances scatter themselves evenly throughout an available space, the particles moving from high to low concentration (**Figure 3-6**)
   b. Solute particles may thus move through channels or carriers in a membrane to reach an equilibrium (equality of concentration) of solution on both sides of the membrane (**Figure 3-7**)
   c. Passive process—it is unnecessary to add energy to the system
3. Osmosis (**Figure 3-8**)
   a. Passive movement of water molecules when some solutes cannot cross the membrane
   b. Similar to diffusion, water moves in a direction that produces an equilibrium
   c. Because water moves, but not all the solutes, osmotic pressure may change across the membrane
4. Dialysis—some solutes move across a selectively permeable membrane by diffusion and other solutes do not, thus resulting in uneven distribution of solute types (**Figure 3-9**)
5. Filtration—movement of water and solutes caused by hydrostatic pressure on one side of membrane

C. Active transport processes
   1. Active transport processes occur only in living cells; movement of substances is "up the concentration gradient"; requires energy from ATP
   2. Ion pumps (**Figure 3-10**)
      a. An ion pump is protein complex in cell membrane
      b. Ion pumps use energy from ATP to move substances across cell membranes against their concentration gradients
      c. Examples: sodium-potassium pump; calcium pump
      d. Some ion pumps work with other carriers so that glucose or amino acids are transported along with ions
   3. Phagocytosis and pinocytosis
      a. Both are active transport mechanisms because they require cell energy
      b. Phagocytosis is a protective mechanism often used to destroy bacteria (**Figure 3-11**)
      c. Pinocytosis is used to incorporate fluids or dissolved substances into cells

D. Cell transport and disease
   1. Cystic fibrosis, characterized by abnormally thick secretions in the airways and digestive ducts, results from failed Cl^- transport (**Figure 3-12**)
   2. Cholera is a bacterial infection that causes Cl^- and water to leak from cells lining the intestines, resulting in severe diarrhea and water loss

3

Cell Growth and Reproduction

A. Cell growth
   1. Proteins determine the structure and function of cells
   2. Protein synthesis is directed by two nucleic acids: deoxyribonucleic acid (DNA) and ribonucleic acid (RNA)
   3. DNA
      a. Make up 46 chromosomes contained in cell nucleus
      b. Large molecule shaped like a spiral staircase; sugar (deoxyribose) and phosphate units compose sides of the molecule; base pairs (adenine-thymine or guanine-cytosine) compose "steps" (see **Figure 2-14**, p. 35)
      c. Base pairings always the same (complementary base pairing), but the sequence of base pairs differs in different DNA molecule
      d. A gene is a specific sequence of base pairs within a DNA molecule
      e. Genes dictate formation of enzymes and other proteins by ribosomes, thereby indirectly determining a cell's structure and functions (**Figure 3-13**)
   4. RNA (**Table 3-4**)
      a. RNA molecules are made from genes that do not code directly for proteins
      b. RNA molecules regulate cell processes, such as protein synthesis
      c. RNA subunits are made up of nucleotides, but have ribose as their sugar and have the base uracil instead of thymine
         (1) mRNA—messenger RNA; transcribed working copy of one gene
         (2) rRNA—ribosomal RNA; component of ribosome
         (3) tRNA—transfer RNA; carries specific amino acid to its location on a ribosome during translation
   5. Protein synthesis—occurs in cytoplasm; thus genetic information must pass from the nucleus to the cytoplasm (**Figure 3-14**)
      a. Transcription
         (1) Double-stranded DNA separates, and one strand copied to form messenger RNA (mRNA)
         (2) Each strand of mRNA is a copy (transcript) of a particular gene (base-pair sequence) from a segment of DNA
      b. mRNA molecules pass from the nucleus to the cytoplasm where they direct protein synthesis in ribosomes and ER
      c. Translation
         (1) Translation of code in mRNA transcript allows ribosomes to synthesize proteins
         (2) Codon—a series of three nucleotide bases in mRNA that acts as a code for a specific amino acid
         (3) tRNA—carries a specific amino acid and has an anticodon, which is a three-base sequence that complements the mRNA codon that signifies that amino acid
         (4) tRNA brings amino acids into place along the mRNA strand where it is held by a ribosome, thus forming a strand of amino acids

B. Protein synthesis and disease
   1. Abnormal DNA that is inherited or that results from damage is often the basis of disease
   2. Factors that cause damage to DNA molecules include chemical or mechanical irritants, radiation, bacteria, and viruses

C. Cell reproduction
   1. Cell life cycle—includes reproduction of cell involving division of the nucleus (mitosis) and the cytoplasm (**Figure 3-15**)
      a. Two daughter cells result from the division
      b. Interphase—period of life cycle when the cell is not actively dividing
   2. DNA replication—process by which each half of a DNA molecule becomes a whole molecule identical to the original DNA molecule; precedes mitosis

3. Mitosis—process in cell division that distributes identical nuclear chromosomes (DNA molecules) to each new cell formed when the original cell divides (**Figure 3-15**)
   a. Prophase—first stage
      (1) Chromatin granules become organized
      (2) Chromosomes (pairs of linked chromatids) appear
      (3) Centrioles move away from nucleus
      (4) Nuclear envelope disappears, freeing genetic material
      (5) Spindle fibers appear
   b. Metaphase—second stage
      (1) Chromosomes align across center of cell
      (2) Spindle fibers attach themselves to each chromatid
   c. Anaphase—third stage
      (1) Centromeres break apart
      (2) Separated chromatids now called *chromosomes*
      (3) Chromosomes are pulled to opposite ends of cell
      (4) Cleavage furrow develops at end of anaphase
   d. Telophase—fourth stage
      (1) Cell division is completed
      (2) Nuclei appear in daughter cells
      (3) Nuclear envelope and nucleoli appear
      (4) Cytoplasm is divided (cytokinesis)
   e. Mitosis ends as daughter cells become fully functional and enter interphase
4. Results of cell division
   a. Two identical cells result from cell division, growing tissues or replacing old or damaged cells
   b. Differentiation—process by which daughter cells can specialize and form different kinds of tissue

D. Changes in cell growth and reproduction (**Figure 3-16** and **Table 3-5**)
   1. Hypertrophy—increase in size of individual cells; increasing size of tissue
   2. Atrophy—decrease in size of individual cells; decreasing size of tissue
   3. Hyperplasia—increase in cell reproduction, increasing size of tissue
   4. Anaplasia—production of abnormal, undifferentiated cells
   5. Uncontrolled cell reproduction results in formation of a benign or malignant neoplasm (tumor) (**Figure 3-17**)

3

❑ ACTIVE LEARNING

STUDY TIPS

Hint *Consider using these tips to achieve success in meeting your learning goals.*

Chapter 3 should be a review of your general biology course—most of what is in this chapter should be familiar.

1. The section on cell structure begins with the plasma membrane. It is made up mostly of phospholipids, but the most important part of the membrane structure is the proteins embedded in the phospholipids. They play important roles in a number of systems in the body such as the nervous and endocrine systems.
2. The organelles may seem to have strange-sounding names. Use the vocabulary list with the word origins in this chapter to help you determine the meaning of each organelle name. Flash cards would be helpful in learning these new terms.
3. The transport processes of osmosis and dialysis are special cases of diffusion—osmosis with water and dialysis with solutes. Filtration uses a pressure rather than a concentration difference to move substances.
4. Phagocytosis and pinocytosis are descriptions of what the cell is doing. *Phago* means "to eat," *pino* means "to drink," *cyto* means "cell," and *-sis* means "condition."
5. When studying protein synthesis, keep the goal of the process in mind. The cell wants a protein made, the DNA has the plans, but the ribosome is the factory. The DNA needs to tell the ribosome what to build (transcription), and the factory needs to put the protein together in the correct order (translation).
6. Use flash cards to study the phases of mitosis—remember that the phases are based on what is happening to the chromosomes. There are many online resources that illustrate the phases of mitosis. These resources include animations that can help you better understand what occurs in each phase of mitosis.
7. Make and use flash cards to learn the terms used to describe changes in cell growth and reproduction.
8. Link the diseases or conditions described in the chapter by constructing a *T-chart* with the cell structure or function that is abnormal.
9. In your study group, review the flash cards for the organelles, mitosis, and changes in cell growth and reproduction. Be sure to discuss steps in protein synthesis and the cell transport processes. Go over the questions at the end of the chapter and discuss possible test questions.
10. Check *my-ap.us/JEdgof* for the latest tips on studying cells.

Review Questions

Hint *Write out the answers to these questions after reading the chapter and reviewing the Chapter Summary. If you simply think through the answer without writing it down, you will not retain much of your new learning.*

1. Describe the structure of the plasma membrane, cytoplasm, and nucleus.
2. List the functions of the plasma membrane, cytoplasm, and nucleus.
3. List the functions of each of the following organelles: ribosome, endoplasmic reticulum, Golgi apparatus, mitochondria, lysosome, centrosome, centrioles, microvilli, cilia, and flagella.
4. Explain the function of the nucleus and nucleolus.
5. Explain the difference between chromatin and chromosomes.
6. Describe the processes of diffusion and filtration.
7. Describe the functioning of the ion pump.
8. Explain the process of phagocytosis.
9. Describe the process of transcription.
10. Describe the process of translation.
11. List the four stages in active cell division (mitosis) and briefly describe what occurs in each stage.
12. What important event in mitosis occurs during interphase?

Critical Thinking

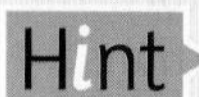 *After finishing the Review Questions, write out the answers to these more in-depth questions to help you apply your new knowledge. Go back to sections of the chapter that relate to concepts that you find difficult.*

13. Explain what would happen if a cell containing 97% water were placed in a 10% salt solution.
14. If one side of a DNA molecule had the following base sequence: adenine-adenine-guanine cytosine-thymine-cytosine-thymine, what would the sequence of bases on the opposite side of the molecule be?
15. If a molecule of mRNA was made from the DNA base sequence in question 14, what would the sequence of bases be in the RNA?
16. If an intravenous solution contains 1.1% salt, would it be hypertonic, hypotonic, or isotonic to cells?

Chapter Test

Hint *After studying the chapter, test your mastery by responding to these items. Try to answer them without looking up the answers.*

1. ________ and ________ are two lipid-based molecules that make up part of the structure of the plasma membrane.
2. ________ is a term that refers to small structures inside the cell—it means "little organs."
3. ________ is the movement of substances across a cell membrane using cell energy, whereas ________ is the movement of substances across a cell membrane without using cell energy.
4. ________ refers to the movement of fluids or dissolved molecules into the cell by trapping them in the plasma membrane.
5. ________ is a disease caused by the inability of cells to transport Cl^- ions.
6. ________ occurs when enzymes in the mitochondria use oxygen to break down glucose and other nutrients to release energy needed for cellular work.
7. ________ is the process in protein synthesis that forms the mRNA molecule.
8. A ________ is a segment of base pairs in a chromosome.
9. ________ is the total genetic information packaged in a cell.
10. Translation occurs within the ________.
11. Another name for "cell suicide" is ________.
12. The disease caused by an inherited mistake in the genetic code that prevents production of normal blood clotting proteins is:
    a. cystic fibrosis
    b. hemophilia
    c. Duchenne muscular dystrophy
    d. AIDS
13. The synthesis of a protein by ribosomes begins with:
    a. translation
    b. transcription
    c. transfer RNA
    d. complementary base pairing
14. During what stage of mitosis do the chromosomes move away from the center of the cell?
    a. Interphase
    b. Metaphase
    c. Anaphase
    d. Telophase
15. Atrophy refers to a(n):
    a. increase in cell size
    b. decrease in cell size
    c. increase in use
    d. condition that results from overextending a muscle

16. During what stage of mitosis do the chromosomes align in the center of the cell?
    a. Interphase
    b. Metaphase
    c. Prophase
    d. Telophase
17. During what stage of mitosis does the chromatin condense into chromosomes?
    a. Interphase
    b. Metaphase
    c. Prophase
    d. Telophase
18. During what stage of mitosis do the nuclear envelope and nuclei reappear?
    a. Interphase
    b. Prophase
    c. Anaphase
    d. Telophase
19. Which of the following terms refers to an increase in cell size?
    a. Hyperplasia
    b. Hypertrophy
    c. Anaplasia
    d. Atrophy
20. Which of the following terms does *not* belong?
    a. Sugar
    b. Phosphate
    c. Nitrogen base
    d. Lipid

Match the cell structure in column A with its corresponding description in column B.

Column A	Column B
21. ________ ribosome	a. long cell projections used to propel sperm cells
22. ________ endoplasmic reticulum	b. bags of digestive enzymes in the cell
23. ________ Golgi apparatus	c. tubelike passages that carry substances throughout the cell
24. ________ mitochondria	d. short hairlike structures on the free surfaces of some cells
25. ________ lysosomes	e. chemically processes and packages substances from the endoplasmic reticulum
26. ________ flagella	f. directs protein synthesis, contains DNA—the "brain" of the cell
27. ________ cilia	g. "protein factories" in the cell, made of RNA
28. ________ nucleus	h. small structure in the nucleus that helps in the formation of ribosomes
29. ________ nucleolus	i. "powerhouse" of the cell—where most of the cell's ATP is formed

Case Studies

To solve a case study, you may have to refer to the glossary or index, other chapters in this textbook, and other resources.

1. One form of the inherited condition *glycogen storage disease*—a form called *Pompe disease*—results in accumulation of excessive glycogen in cells of the heart, liver, and other organs. The accumulation of glycogen can disrupt cell function, causing heart and other problems that can progress to death. Glycogen is a large carbohydrate molecule formed by linking numerous glucose molecules into a branched chain (see Chapter 2). Glycogen formation is normal in the affected cells. The accumulation of excessive glycogen results from the failure of enzymes that are supposed to break apart the glycogen so that the cell can use the glucose subunits. In what organelle would you expect to find these glycogen-digesting enzymes? Explain how the presence of nonfunctional enzymes could have been inherited.
2. Malignant tumors are sometimes treated with drugs that halt mitosis, and thus stop the production of new cancer cells. Two such drugs, vincristine sulfate and vinblastine sulfate, interfere with the formation of spindle fibers. How could this action halt mitosis? Antibiotics such as mitomycin C and inorganic compounds such as *cis*-platinum also can be used to stop the growth of tumors. These drugs interfere with DNA synthesis in treated cells. How could this action halt mitosis?
3. Lauren is a 2-year-old girl with cystic fibrosis (CF). Because Lauren has this condition, her mother frequently turns her over, cups her hand, and quickly but firmly pats her sharply on the back between the shoulder blades. How could this help Lauren's condition?
4. Elizabeth is a senior in high school and has selected the field of proteomics to study when she enters college. She has been accepted at a prestigious institution and has even received scholarship money. You are excited for her and are planning to contact her and congratulate her on her choice. When discussing Elizabeth's selection and good fortune with your parents, they ask you questions such as: What is the field of proteomics? What are ENCODE scientists? How do you answer them?

Answers to Active Learning Questions can be found online at evolve.elsevier.com.

Tissues

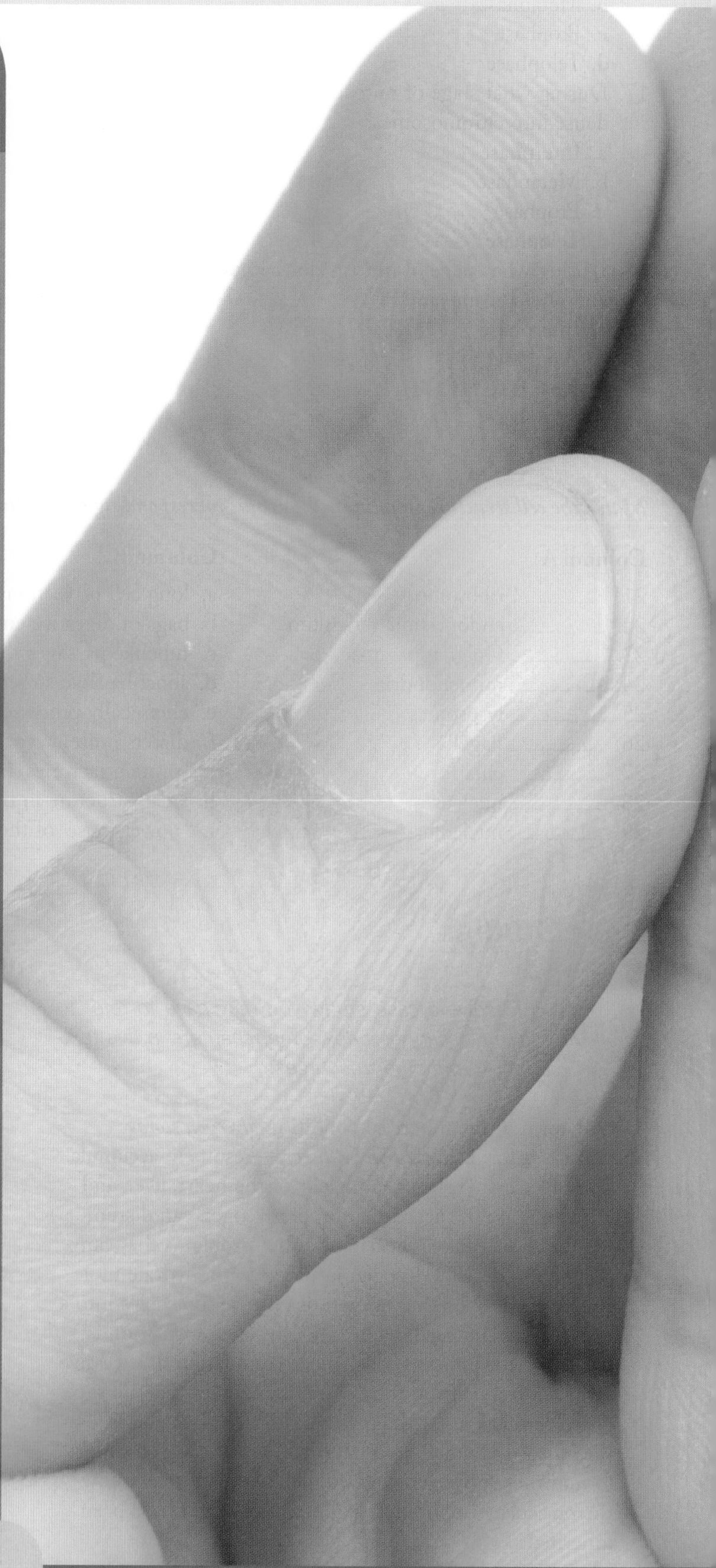

OUTLINE

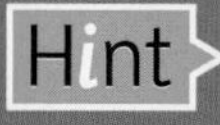

Hint ▶ *Scan this outline before you begin to read the chapter, as a preview of how the concepts are organized.*

OBJECTIVES

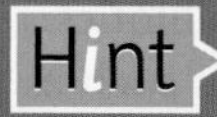

Hint ▶ *Before reading the chapter, review these goals for your learning.*

After you have completed this chapter, you should be able to:

1. Explain how epithelial tissue is grouped according to shape and arrangement of cells.
2. List and briefly discuss the major types of connective and muscle tissue.
3. Discuss nervous tissue and list the three structural components of a neuron.
4. Describe how injured tissues regenerate in each of the four major types of tissue.

CHAPTER 4

We explored cells in the previous chapter and now we turn our attention to the various *groups* of cells called **tissues.** The arrangement of cells in one tissue may form a thin sheet only one cell deep, whereas the cells of another tissue may form huge masses containing millions of cells. Tissues are the "fabric" of the body, and like the various fabrics that make up a garment, each tissue of an organ specializes in performing unique functions that help the organ do its job. This collaborative functioning of tissues within our body's organs maintains homeostatic balance and thus is vital to our survival.

In this chapter, we briefly survey the major kinds of tissues that form the organs of the body. As we progress through later chapters, we will revisit each of these tissue types and explore more detail about their locations, structures, and functions.

Introduction to Tissues

Tissue Types

Tissues differ from each other in the size and shape of their cells, in the amount and kind of material between the cells, and in the special functions they perform to help maintain the body's survival. In **Tables 4-1** through **Table 4-3**, you will find a listing of the four major tissues and the various subtypes of each. The tables also include the structure of each subtype along with examples of the location of the tissues and a primary function of each tissue type.

The four main kinds of tissues that compose the body's many organs follow:

1. Epithelial tissue—forms sheets that cover or line the body
2. Connective tissue—provides structural and functional support
3. Muscle tissue—contracts to produce movement
4. Nervous tissue—senses, conducts, and processes information

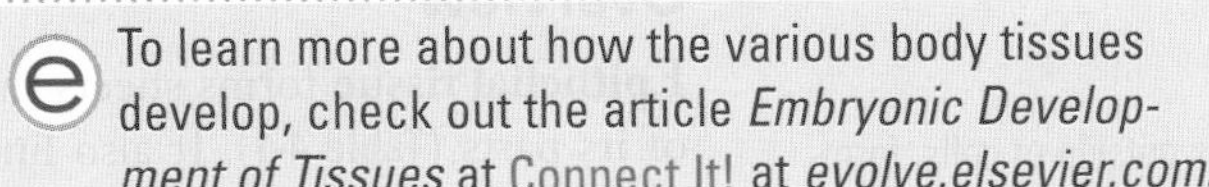

To learn more about how the various body tissues develop, check out the article *Embryonic Development of Tissues* at Connect It! at *evolve.elsevier.com.*

LANGUAGE OF SCIENCE

Hint Before reading the chapter, say each of these terms out loud. This will help you to avoid stumbling over them as you read.

adipose
(AD-ih-pohs)
[*adipo-* **fat,** *-ose* **full of**]

antigen
(AN-tih-jen)
[*anti-* **against,** *-gen* **produce**]

areolar tissue
(ah-REE-oh-lar TISH-yoo)
[*are-* **open space,** *-ola-* **little,** *-ar* **relating to,** *tissue* **fabric**]

axon
(AK-son)
[*axon* **axle**]

basement membrane
[*base-* **base,** *-ment* **thing,** *membrane* **thin skin**]

blood
(blud)

bone
(bohn)

cardiac muscle tissue
(KAR-dee-ak MUS-el TISH-yoo)
[*cardi-* **heart,** *-ac* **relating to,** *mus-* **mouse,** *-cle* **small,** *tissue* **fabric**]

cell body
(sel BOD-ee)
[*cell* **storeroom,** *body* **body**]

chondrocyte
(KON-droh-syte)
[*chondro-* **cartilage,** *-cyte* **cell**]

collagen
(KAHL-ah-jen)
[*colla-* **glue,** *-gen* **produce**]

columnar
(koh-LUM-nar)
[*column-* **column,** *-ar* **relating to**]

Continued on p. 85

TABLE 4-1 Epithelial Tissues

TISSUE	STRUCTURE	LOCATION(S)	FUNCTION(S)
Simple squamous	Single layer of flattened cells	Alveoli of lungs Lining of blood and lymphatic vessels	Diffusion of respiratory gases between alveolar air and blood Diffusion, filtration, and osmosis
Stratified squamous	Many layers; outermost layer(s) is flattened cells	Surface of lining of mouth and esophagus	Protection
Simple cuboidal	Single layer of cells that are as tall as they are wide	Glands; kidney tubules Surface of skin (epidermis)	Secretion, absorption Protection
Simple columnar	Single layer of tall, narrow cells	Surface layer of lining of stomach, intestines, parts of respiratory tract; lining of uterine (fallopian) tubes	Protection; secretion; transport (absorption)
Pseudostratified	Single layer of tall cells that wedge together to appear as if there are two or more layers	Lining of portions of the respiratory tract; portion of the male reproductive tract	Protection
Stratified transitional	Many layers of varying shapes, capable of stretching	Urinary bladder	Protection

Matrix

Recall from Chapter 1 that a central principle of human physiology is homeostasis—the relative constancy of the internal fluid environment. This fluid environment fills the spaces between the cells of the body. Tissues differ in the amount and kind of fluid material between the cells—the **matrix.** It is also called the *extracellular matrix (ECM)* to emphasize its location between cells.

The matrix varies in amount and composition among the various tissues—which reflects the variety of functions among tissue types. Epithelial tissues have very little matrix because the cells are so closely connected to each other. Connective tissues, on the other hand, are mostly matrix—with the cells few and far between.

Matrix is like jelly, made up of mostly water with various interlocking fibers that thicken it (**Figure 4-1**). The kinds and amounts of fibers can produce a variety of matrix types—all with different functions.

The thin watery matrix of blood—plasma—has no fibers at all (except when forming a blood clot), which allows it to remain free-flowing. The tissue of tendons and ligaments is dense with strong, twisted fibers that give the matrix a thick, ropelike quality. Bone's matrix fibers are encrusted with mineral crystals to give it the characteristics of reinforced concrete.

Collagen is a protein that forms microscopic twisted ropes within the matrix of many tissues. Collagen gives a tissue flexible strength. **Elastin** is present in some tissues, and its rubbery quality gives tissues the ability to stretch and rebound easily.

The matrix contains various *polysaccharides* and **proteoglycans** that provide tissues with specialized properties. For example, these molecules link cells, absorb shock, regulate function, and offer lubrication.

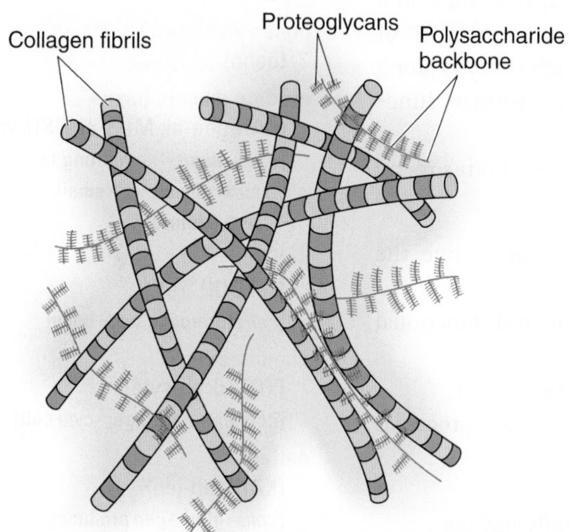

FIGURE 4-1 Extracellular matrix. The matrix is outside of cells, forming a connecting gel that contributes to the overall function of the tissue. This example illustrates thick, ropelike collagen fibrils interspersed with proteoglycan (protein-carbohydrate) structures attached to polysaccharide backbones—all surrounded by water. The collagen gives tissue strength, and the polysaccharide-proteoglycan structures absorb shocks. (Collagen is naturally white, but is often stained pink to make it more visible in microscopic studies.)

4

Epithelial Tissue

Introduction to Epithelial Tissues

Overview

Epithelial tissue forms sheets that cover the body and many of its parts (**Table 4-1**). It also lines various parts of the body and forms ducts or tubes.

Protein connectors in the plasma membranes of adjacent cells firmly hold epithelial cells together. Because epithelial cells are packed close together with little or no **interstitial** fluid or other matrix between them, they form continuous sheets that contain no blood vessels.

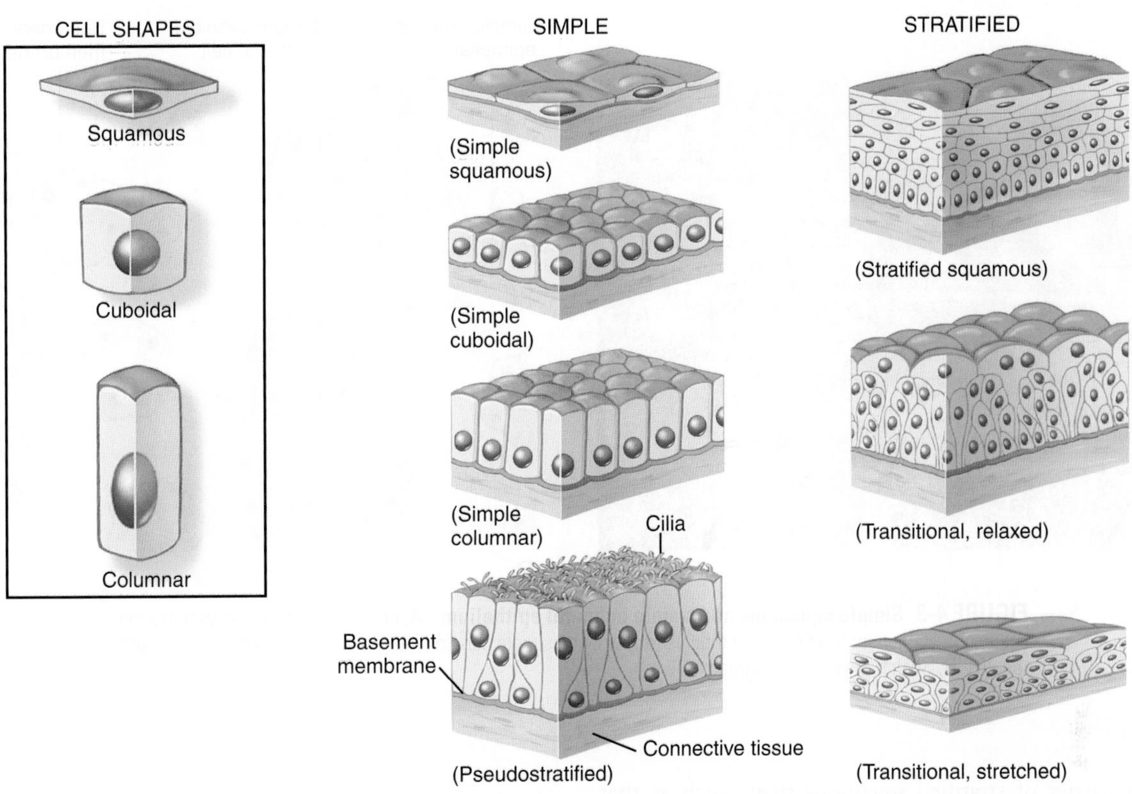

FIGURE 4-2 Classification of epithelial tissues. The tissues are classified according to the shape and arrangement of cells.

Because epithelium lacks blood vessels, epithelial cells must get their oxygen and nutrients from nearby blood vessels in the connective tissue that always underlies sheets of epithelial cells. A gluelike layer called the **basement membrane** connects epithelial tissue to its connective tissue foundation.

Examine Figure 4-2. It illustrates how this large group of tissues can be subdivided according to the *shape* and *arrangement* of the cells found in each type.

Shape of Cells

If classified according to shape, epithelial cells are identified as:

1. **Squamous**—flat and scalelike
2. **Cuboidal**—cube-shaped
3. **Columnar**—taller than they are wide
4. **Transitional**—varying shapes that can stretch

Arrangement of Cells

If classified according to *arrangement* of cells, epithelial tissue can be labeled as one of the following:

1. **Simple**—a single layer of cells of the same shape
2. **Stratified**—many layers of cells, named for the shape of cells in the outer layer

Several types of epithelium are described in the paragraphs that follow and are illustrated in Figures 4-2 to 4-8.

The translucent, microscopic structures that distinguish different types of epithelial tissue are visible in light microscopes only when stained. Many of the pink and purple colors in many of the micrographs in this chapter reflect common stains used to visualize tissues. For more on microscopic tools, review the article *Tools of Microscopic Anatomy* at Connect It! at *evolve.elsevier.com*.

Squamous Epithelium

Simple Squamous Epithelium

Simple squamous epithelium consists of a single layer of very thin and irregularly shaped cells. Because of its structure, substances can readily pass through simple squamous epithelial tissue, making transport its special function. Absorption of oxygen into the blood, for example, takes place through the simple squamous epithelium that forms the tiny air sacs in the lungs (Figure 4-3).

Stratified Squamous Epithelium

Stratified squamous epithelium (Figure 4-4) consists of several layers of closely packed cells, an arrangement that makes this tissue especially adept at protection. For instance, stratified squamous epithelial tissue protects the body against invasion by microorganisms. Most microbes cannot work their way

A

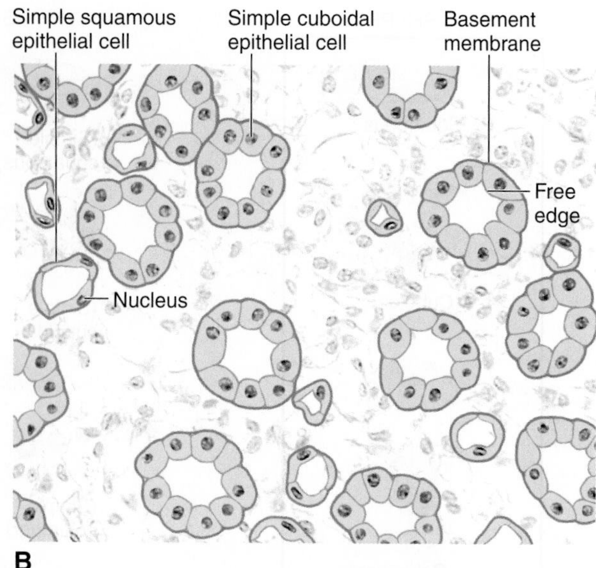

B

FIGURE 4-3 Simple squamous and simple cuboidal epithelium. A, Photomicrograph shows thin simple squamous epithelium forming some tubules *(arrows)* and simple cuboidal epithelium forming the walls of other tubules. **B,** Sketch of photomicrograph.

through a barrier of stratified squamous tissue such as that composing the surface of skin and of mucous membranes.

One way of preventing infections, therefore, is to take good care of your skin. Prevent it from becoming cracked from chapping, and guard against cuts and scratches.

Knowledge of the tissues that make up the skin is critical to understanding the different types of skin cancer. Check out the article *Skin Cancer* at Connect It! at *evolve.elsevier.com.*

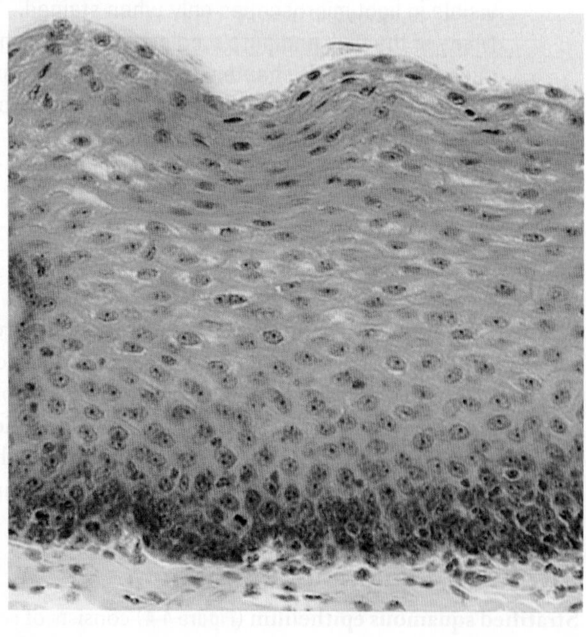

A

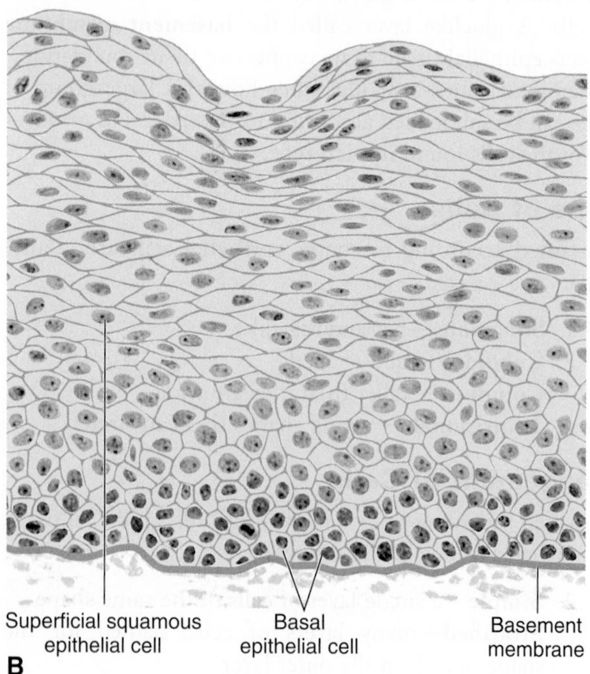

B

FIGURE 4-4 Stratified squamous epithelium. A, Photomicrograph. **B,** Sketch of the photomicrograph. Note the many layers of epithelial cells and the flattened (squamous) cells in the outer layers.

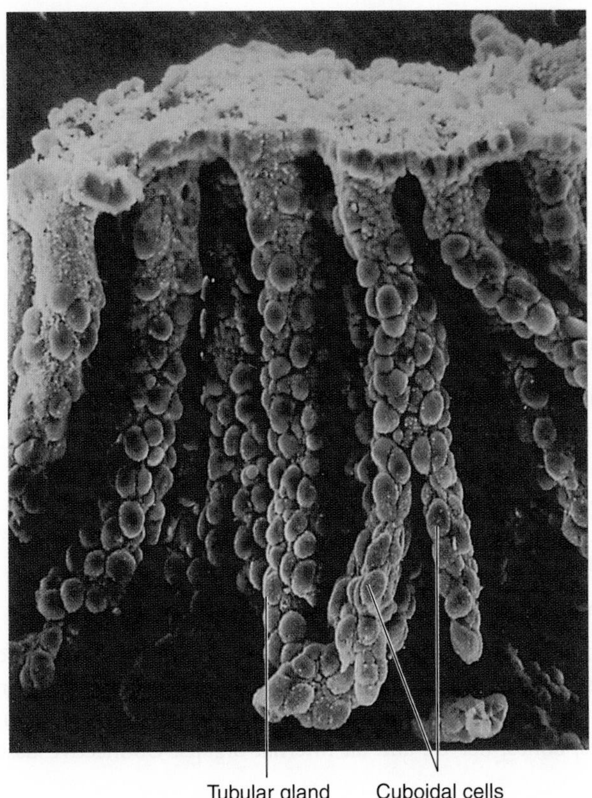

FIGURE 4-5 Simple cuboidal epithelium. This scanning electron micrograph shows how a single layer of cuboidal cells can form glands. The secreting cells arrange themselves into single or branched tubules that open onto a surface—the lining of the stomach in this case.

Cuboidal Epithelium

Simple cuboidal epithelium is a single layer of cells that are typically about as tall as they are wide—thus resembling a cube shape. This tissue does not form protective coverings. Instead, it forms tubules or other groupings adapted for secretory activity (**Figure 4-5**), which is why they appear to form ringlike arrangements in cross section (see **Figure 4-3**). These secretory cuboidal cells usually function in tubes or clusters of secretory cells commonly called **glands.**

Glands of the body may be classified as **exocrine** if they release their secretion through a duct or as **endocrine** if they release their secretion directly by diffusion into the bloodstream. Examples of glandular secretions include saliva, digestive juices, sweat, and hormones such as those secreted by the pituitary or thyroid glands. Simple cuboidal epithelium also forms the tubules that form urine in the kidneys.

In some glands, cuboidal epithelium occurs in more than one layer. Such *stratified cuboidal epithelium* may be found in the sweat gland ducts.

Simple Columnar Epithelium

Simple columnar epithelium can be found lining the inner surface of the stomach, intestines, and some areas of the respiratory and reproductive tracts. In **Figure 4-6** the simple columnar cells are arranged in a single layer lining the inner surface of the colon or large intestine. These epithelial cells are taller than they are wide, and the nuclei are located toward the bottom of each cell. The "open spaces" seen among the cells are specialized **goblet cells** that produce mucus. The regular columnar-shaped cells specialize in absorption.

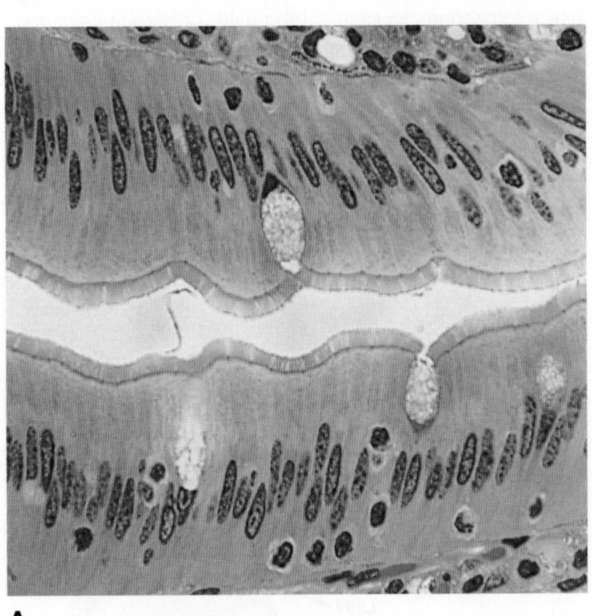

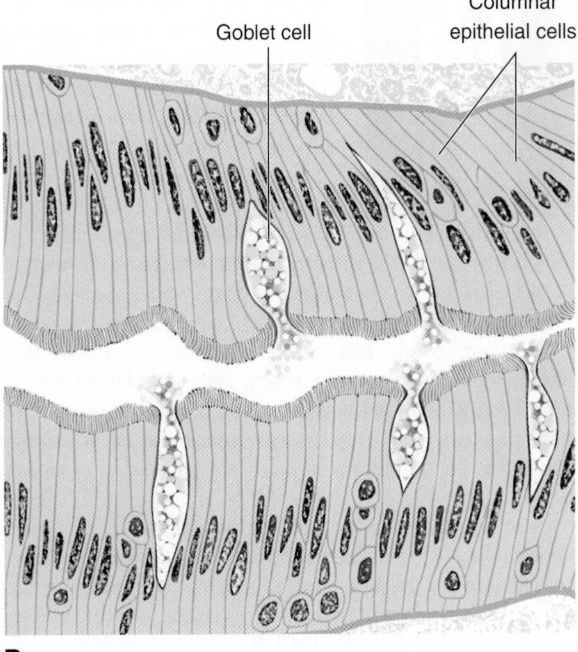

FIGURE 4-6 Simple columnar epithelium. A, Photomicrograph. **B,** Sketch of the photomicrograph. Note the oblong nuclei in all the cells and the goblet or mucus-producing cells that are present.

Pseudostratified Epithelium

Pseudostratified epithelium, illustrated in **Figures 4-2** and **4-7**, is typical of that which lines the trachea or windpipe. Look carefully at the illustrations. Note that each cell actually touches the gluelike basement membrane that lies under all epithelial tissues. Although the epithelium in **Figure 4-7** appears to be several cell layers thick, it is not. This is the reason it is called *pseudo* (or false) stratified epithelium.

The cilia that extend from the cells are capable of moving in unison (see **Figure 3-5**, p. 49). As they do so, they move mucus along the lining of the trachea, thus protecting the lungs against entry of dust or other foreign particles.

Transitional Epithelium

Stratified transitional epithelium is typically found in body areas subjected to stress and must be able to stretch. An example would be the lining of the urinary bladder. In many instances, up to 10 layers of differently shaped cells of varying sizes are present when the tissue is not stretched. When stretching occurs, the epithelial sheet expands, the number of cell layers decreases, and cell shape changes from roughly cuboidal to nearly squamous (flat) in appearance.

In the wall of the urinary bladder, the ability of transitional epithelium to stretch easily without damage keeps the bladder wall from tearing as urine fills the space inside. Stratified transitional epithelium is shown in **Figures 4-2** and **4-8**.

QUICK CHECK

1. List the four main types of tissues found in the body.
2. What is matrix?
3. Name the types of epithelial tissue according to their shape.
4. What is the difference between simple epithelium and stratified epithelium?
5. Which type of epithelial tissue is found in the wall of the urinary bladder?

Connective Tissue

Introduction to Connective Tissue

Overview

Connective tissue is the most abundant and widely distributed tissue in the body. It also exists in more varied forms than any of the other tissue types. It is found in skin, membranes, muscles, bones, nerves, and all internal organs. Connective tissue

SCIENCE APPLICATIONS

MICROSCOPY

Antonie van Leeuwenhoek (1632-1723)

Until the very hour of his death in 1723, the Dutch drapery merchant Antonie van Leeuwenhoek *(left)* spent most of his 91 years pursuing adventures with the hundreds of microscopes he had built or collected. Using what were, even then, very simple lenses or combinations of lenses, van Leeuwenhoek discovered a whole world of tiny structures he called "animalcules" in body fluids. Although scientists a century later would declare that all living organisms are made up of cells, van Leeuwenhoek was the first to see and describe human blood cells (see **Figure 4-15** on p. 81), human sperm cells (see **Figure 3-4**, *B*), and many other cells and tissues of the body. He was also the first to observe many microscopic organisms that live on or in the human body—many of which are capable of producing disease.

Scientists today use light microscopes that are much more advanced than those of van Leeuwenhoek's time. Some of the most modern microscopes, called *electron microscopes*, use electron beams instead of light to produce images of very high magnification (see **Figure 3-4**). Both cell biologists and **histologists** (tissue biologists) use microscopes to research the fine structure and function of the human body. A wide variety of professions have found practical applications for microscopy. Most health professionals use microscopes, or the images produced with microscopes, to perform routine duties. For example, *clinical laboratory technicians* and **pathologists** use microscopes to assess the health of human cells and tissues. Outside of the health sciences, professionals such as law enforcement *investigators, archaeologists, anthropologists,* and **paleontologists** often use microscopes to further their study of human and animal tissues.

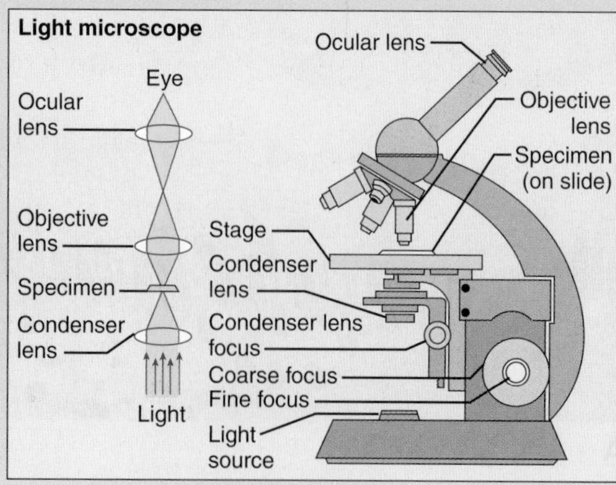

Modern compound light microscope.

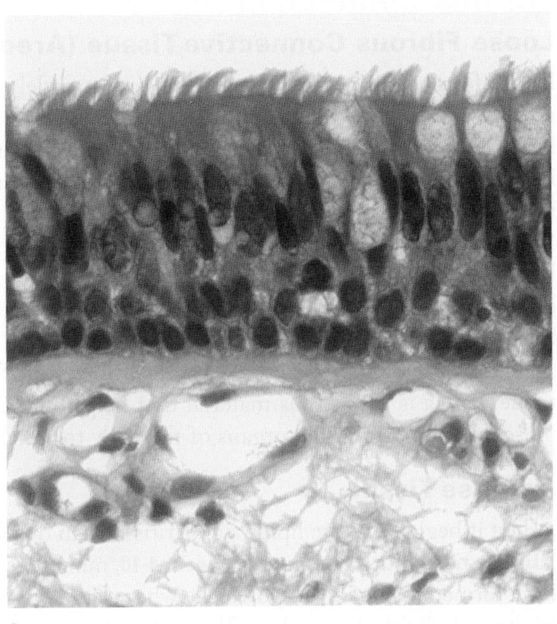
A

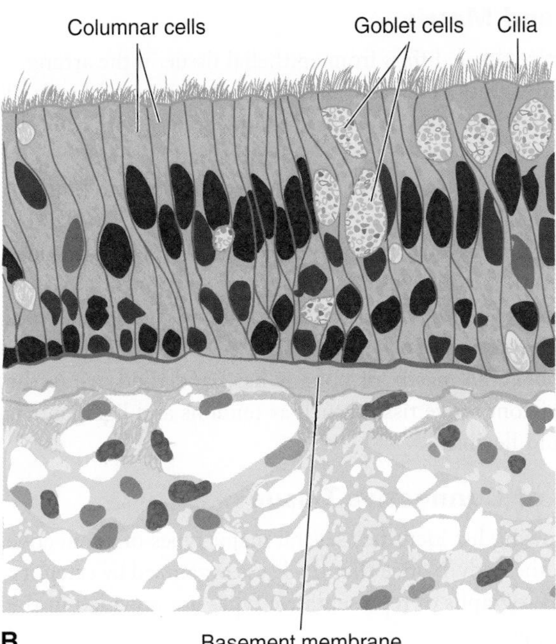

B

FIGURE 4-7 Pseudostratified columnar epithelium. A, Photomicrograph. The arrangement of nuclei makes this specimen seem stratified, but it is not because each cell reaches the basement membrane, thus forming just one layer. **B,** Sketch of the micrograph. Note the presence of goblet cells and cilia.

exists as delicate, paper-thin webs that hold internal organs together and give them shape. It also exists as strong and tough cords, rigid bones, and even in the form of a fluid—blood.

The functions of connective tissue are as varied as its structure and appearance. It connects tissues to each other and forms a supporting framework for the body as a whole and for its individual organs. As blood, it transports substances throughout the body. Several other kinds of connective tissue function to defend us against microbes and other invaders.

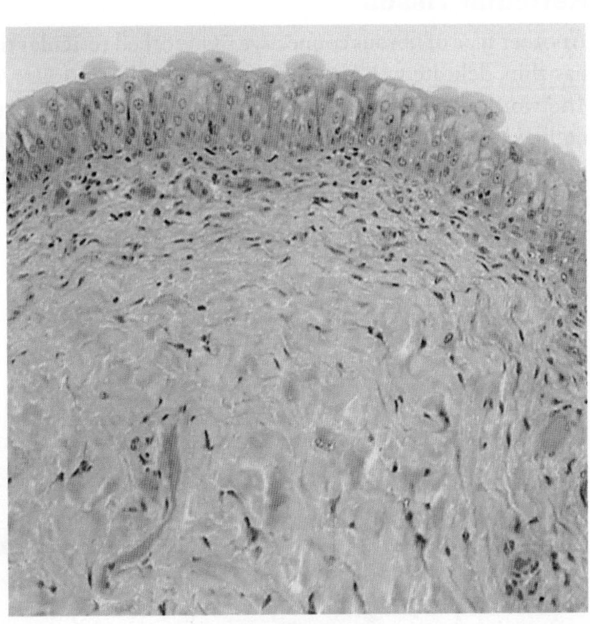
A

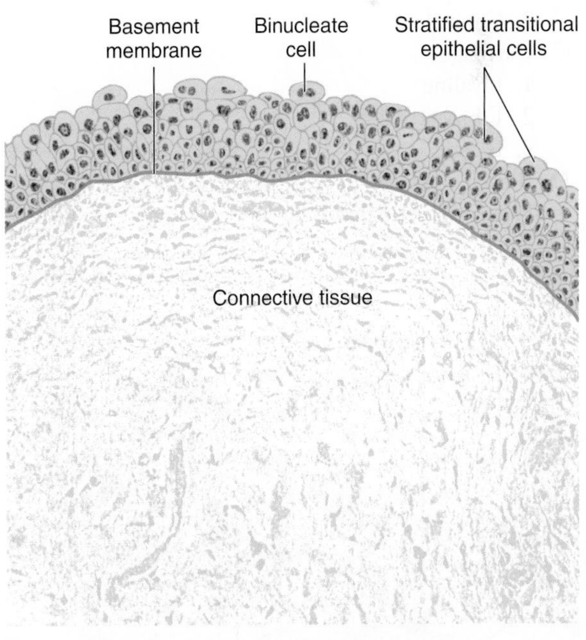

B

FIGURE 4-8 Stratified transitional epithelium. A, Photomicrograph of tissue lining the urinary bladder wall. **B,** Sketch of the photomicrograph. Note the many layers of epithelial cells of various shapes in this relaxed (unstretched) specimen.

Cells and Matrix

Connective tissue differs from epithelial tissue in the arrangement and variety of its cells and in the amount and kinds of intercellular material, or *matrix,* found between its cells. In addition to the relatively few cells embedded in the matrix of most types of connective tissue, varying numbers and kinds of fibers are also present.

The structural quality and appearance of the matrix and fibers determine the qualities of each type of connective tissue. The matrix of blood, for example, is a liquid, but other types of connective tissue, such as cartilage, have the consistency of firm rubber. The matrix of bone is hard and rigid, while the matrix of connective tissues such as tendons and ligaments is strong and flexible.

Types of Connective Tissue

The following list identifies several major types of connective tissue in the body. Notice that the list is organized by category. Photomicrographs of representative types are provided in the following pages.

A. Fibrous (connective tissue proper)
  1. Loose fibrous (areolar)
  2. Adipose (fat)
    a. White
    b. Brown
  3. Reticular
  4. Dense fibrous
    a. Regular
    b. Irregular
B. Bone
  1. Compact
  2. Cancellous
C. Cartilage
  1. Hyaline
  2. Fibrocartilage
  3. Elastic
D. Blood
E. Hematopoietic tissue

Fibrous Connective Tissue

Loose Fibrous Connective Tissue (Areolar)

Loose fibrous connective tissue is the most widely distributed of all connective tissue types. It is the "glue" that helps keep the organs of the body together. Also called **areolar tissue,** it consists of webs of fibers and of a variety of cells embedded in a loose matrix of soft, sticky gel (**Figure 4-9**).

Some of the fibers are made of **collagen,** a strong but flexible fibrous protein. Some are stretchy fibers made of rubbery **elastin** proteins. These **elastic fibers** help tissues return to their original length after having been stretched.

The **fascia** of the body is primarily composed of areolar tissue. Fascia is the fibrous material that helps bind the skin, muscles, bones, and other organs of the body together.

Adipose Tissue

When it begins to store lipids, areolar tissue can develop into **adipose tissue,** or **fat tissue.** In **Figure 4-10**, numerous vesicles have formed inside the adipose cells where large quantities of triglyceride lipids accumulate. These clear lipid-storage vesicles scatter light like so many snowflakes, giving ordinary adipose tissue a whitish appearance. This is why it is sometimes called *white fat.* Excess triglycerides move into storage after a meal and out of storage as energy-producing nutrients are needed by other tissues.

A special kind of adipose tissue called *brown fat* actually burns its fuel when the body is cold to produce heat. This heat, along with shivering by muscles, helps restore homeostasis of body temperature (see **Figure 1-11**, p. 15).

All types of adipose tissue also secrete hormones that help regulate metabolism and fuel storage in the body.

Reticular Tissue

Another type of fibrous connective tissue called **reticular tissue** has thin, delicate webs of collagen fibers called *reticular fibers.* The word *reticular* means "netlike," and it aptly describes the netlike structure of this tissue, as you can see in **Figure 4-11**.

Reticular tissue is found in bone marrow, for example, where it helps support cells of the blood-forming *hematopoietic tissue.*

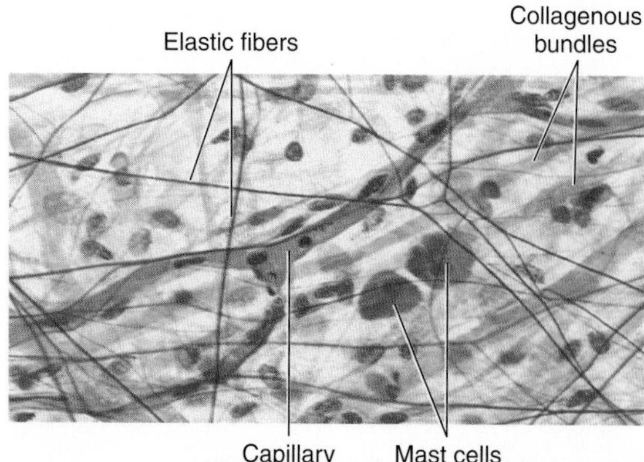

FIGURE 4-9 Loose fibrous (areolar) connective tissue. Notice how the staining used renders the bundles of collagen *pink* and the elastin fibers *dark purple.* Compare the loose arrangement of fibers here with those in **Figure 4-12**.

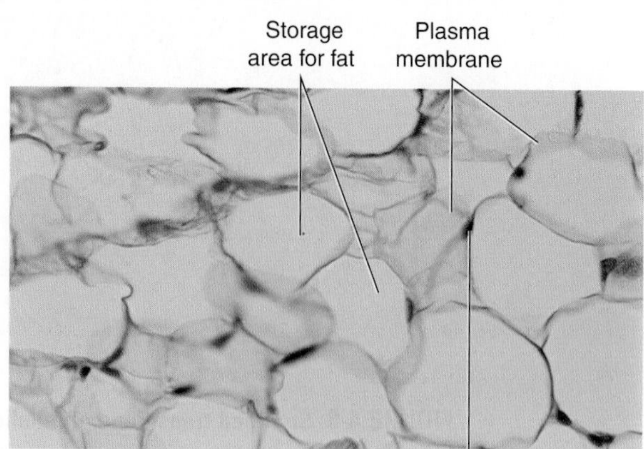

FIGURE 4-10 Adipose tissue. Photomicrograph showing the large lipid storage spaces inside the adipose cells of white fat.

TABLE 4-2 Connective Tissues

TISSUE	STRUCTURE	LOCATION(S)	FUNCTION(S)
Loose fibrous (areolar)	Loose arrangement of collagen fibers, elastic fibers, and cells	Area between other tissues and organs (fascia)	Connection
Adipose (white and brown fat)	Cells contain triglyceride vesicles	**White Fat**	
		Area under skin; padding at various points	Protection, insulation, support, nutrient reserve, regulation
		Brown Fat	
		Pockets within white fat of neck and torso	Heat production, regulation
Reticular	Network of fine collagen (reticular) fibers	Bone marrow, spleen, lymph nodes, cancellous bone cavities	Supports blood-producing cells and immune cells
Dense fibrous (regular and irregular)	Dense arrangement of collagen fiber bundles forming straps or sheets	Tendons, ligaments, skin (deep layer), fascia, scar tissue	Flexible but strong connection
Bone (compact and cancellous)	Hard, calcified matrix arranged in osteons (compact) or network of beams (cancellous)	Skeleton	Support, protection
Cartilage (hyaline, fibrocartilage, and elastic)	Hard but somewhat flexible gel matrix with embedded chondrocytes	**Hyaline**	
		Part of nasal septum, area covering surfaces of bones at joints, larynx wall, rings in trachea, and bronchi	Firm but flexible support
		Fibrocartilage	
		Disks between vertebrae and in knee joint	Withstands pressure
		Elastic	
		External ear	Flexible support
Blood	Liquid matrix with flowing red and white cells	Blood vessels	Transportation
Hematopoietic	Liquid matrix with dense arrangement of blood cell–producing cells	Red bone marrow	Blood cell formation

It is also found in the spleen and lymph nodes, where it supports developing cells of the immune system.

Dense Fibrous Connective Tissue

Dense fibrous connective tissue consists mainly of thick bundles of strong, white **collagen** fibers that are packed closely together. A few fiber-producing cells are scattered among the bundles.

Regular dense fibrous connective tissue has its collagen fiber bundles arranged in roughly parallel rows (Figure 4-12). This type of connective tissue makes up tendons—the strong straps that connect muscle to bone. It provides great strength and

FIGURE 4-11 Reticular connective tissue. The supportive framework of reticular fibers is stained *black* in this section of a lymph node. Note also the faint blood cells and blood-forming (hematopoietic) cells within the network of fibers.

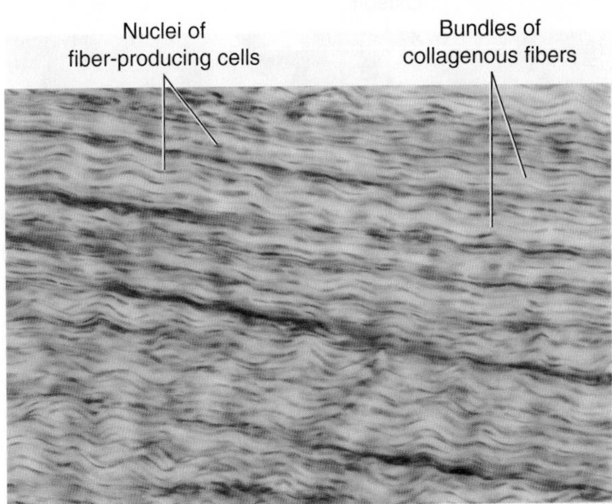

FIGURE 4-12 Dense fibrous connective tissue. Bundles of wavy collagen fibers are roughly parallel to one another in dense regular tissue. Dark nuclei of fiber-producing cells are also visible in this sample from a tendon.

flexibility, but it cannot stretch. Such characteristics are ideal for these structures that anchor our muscles to our bones.

Irregular dense fibrous connective tissue has its collagen arranged in a chaotic swirl of tangled bundles. This type of tissue forms the tough sheets in the deepest layer of the skin. It forms a tough, flexible support to the epithelial superficial layer of the skin. Although the swirled pattern of fiber bundles allows the skin to stretch a little, overstretching the skin often causes tears in the irregular fibrous tissue called *stretch marks*.

Bone

The matrix of **bone** is hard because it has a dense packing of collagen bundles encrusted with mineral crystals containing calcium. Bones are a storage area for calcium and provide support and protection for the body.

Compact bone is the solid form of bone that makes up the outer walls of bones in the skeleton. Compact bone is made up of numerous structural building blocks called **osteons** or **haversian systems.** When compact bone is viewed under a microscope, we can see these circular arrangements of calcified matrix and cells that give bone its characteristic appearance (**Figure 4-13**).

Inside each bone is a type of tissue called *cancellous bone* or *spongy bone.* The term *cancellous* refers to something that is made up of a lattice. The term applies to this bone type because it is a chaotic lattice of branching beams. Like a bath sponge, the lattice forms many, interconnected hollow spaces—giving this bone type the name *spongy.* These beams are nearly as hard as compact bone, but spongy bone cannot be compressed like a wet bath sponge. In fact, the crisscrossing pattern of the bony lattice *adds* rigidity—just like the crossed beams that often support roofs of buildings.

The spaces within cancellous bone are filled with blood-forming hematopoietic tissue or adipose tissue.

4

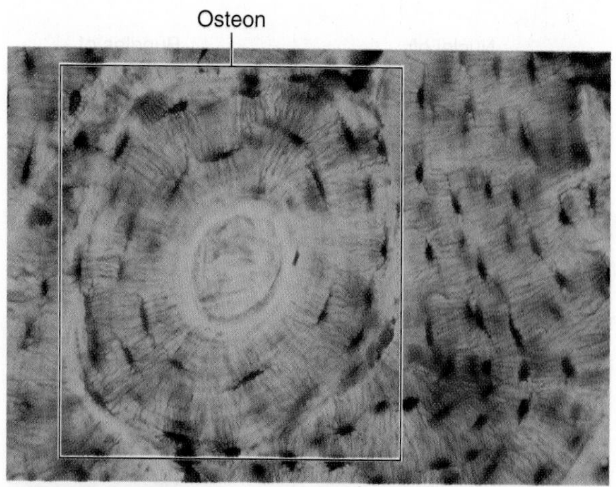

FIGURE 4-13 Bone tissue. Photomicrograph of a chip of compact bone. A cylindrical structural unit of bone, known as an *osteon* (haversian system), is seen in this cross section.

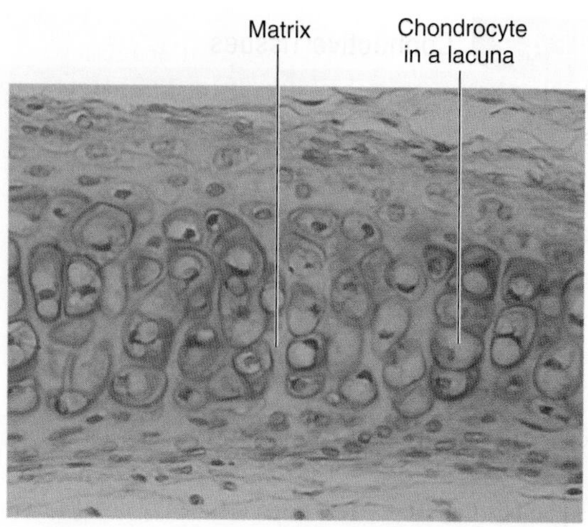

FIGURE 4-14 Cartilage. Photomicrograph showing the chondrocytes distributed throughout the gel-like matrix of hyaline cartilage.

Cartilage

Overview

Unlike bone, the collagen bundles of the cartilage matrix are not encrusted with hard minerals. Instead, cartilage matrix has the consistency of a firm plastic or a gristlelike gel. Cartilage cells, which are called **chondrocytes,** are located in many tiny spaces distributed throughout the matrix—giving this tissue the appearance of Swiss cheese (**Figure 4-14**). There are three major types of cartilage.

Hyaline Cartilage

Hyaline cartilage has only a moderate amount of collagen in its gel matrix, giving it a translucent, glasslike appearance. The name *hyaline* means "glassy." This is the most common type of cartilage in the body. It is found in the support rings of the respiratory tubes and covering the ends of bones that form movable joints.

Fibrocartilage

Fibrocartilage is the strongest and most durable type of cartilage. The matrix is rigid and filled with a dense packing of strong collagen fibers. Fibrocartilage disks serve as shock absorbers between adjacent vertebrae and in the knee joint.

Elastic Cartilage

Elastic cartilage contains few collagen fibers but large numbers of very fine elastic fibers that give the matrix material a high degree of flexibility. This type of cartilage is found in the external ear and in one of the components of the voice box, or larynx.

Blood Tissue

Blood is perhaps the most unusual form of connective tissue because its matrix—blood *plasma*—is liquid. It has transportation and protective functions in the body. Blood contains *red blood cells, white blood cells,* and *platelets* (**Figure 4-15**).

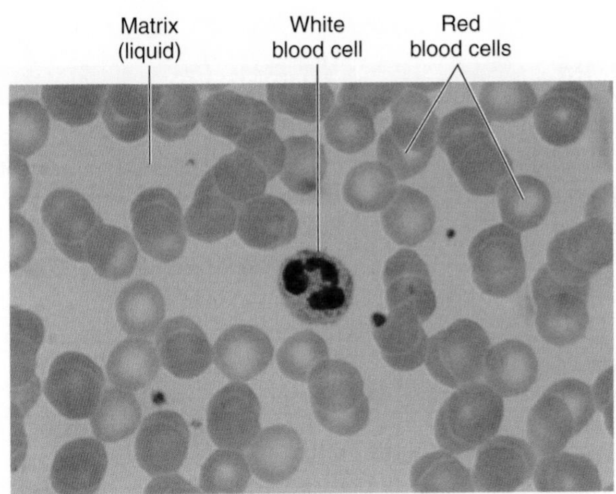

FIGURE 4-15 Blood. Photomicrograph of a human blood smear. This smear shows a white blood cell surrounded by a number of smaller red blood cells and tiny platelets. The liquid matrix of this tissue is also called *plasma*.

Hematopoietic Tissue

Hematopoietic tissue is the bloodlike connective tissue found in the red marrow cavities of bones and in organs such as the spleen, tonsils, and lymph nodes (see **Figure 4-11**). This type of tissue is responsible for the formation of blood cells and lymphatic system cells important in our defense against disease (**Table 4-2**).

QUICK CHECK

1. What are the functions of connective tissue?
2. List the five major classifications of connective tissue.
3. Which type of connective tissue composes tendons?
4. How does cartilage differ from bone?

Muscle Tissue

Introduction to Muscle Tissue

Muscle cells are the movement specialists of the body. They have a higher degree of contractility (ability to generate force or contract) than any other tissue cells. Besides producing movement, muscle tissue can also provide stability—and even produce body heat. Unfortunately, injured muscle cells are sometimes slow to heal and often replaced by fibrous scar tissue if injured.

There are three kinds of muscle tissue: skeletal muscle tissue, cardiac muscle tissue, and smooth muscle tissue (**Table 4-3**).

Skeletal Muscle Tissue

Skeletal muscle or *striated muscle* is called voluntary because willed or *voluntary* control of skeletal muscle contractions is possible. Note in **Figure 4-16** that, when viewed under a microscope, skeletal muscle is characterized by many cross striations and many nuclei per cell. Individual cells are long and threadlike and are often called *fibers*.

CLINICAL APPLICATION

SCREENING DONATED ORGANS AND TISSUES

Tissue typing is a screening process in which cell markers in a donated organ or tissue are identified so that they can be matched to recipients with similar cell markers. Cell markers are specific protein molecules (called **antigens** [see Chapter 16]) on the surface of plasma membranes. If the cell markers in donated tissue are different from those in the recipient's normal tissue, the recipient's immune system will recognize the tissue as foreign. When the immune system mounts a significant attack against the donated tissue, a **rejection reaction** occurs. The inflammation and tissue destruction that occur in a rejection reaction not only destroy or "reject" the donated tissue but also may threaten the life of the recipient. Although drugs such as cyclosporine can be used to inhibit the immune system's attack against donated tissue, crossmatching of tissues by their cell markers is the primary method of preventing rejection reactions.

If you know your blood type, you already know some of your tissue markers. In the ABO system (see Chapter 13), type A blood has the A marker, type B blood has the B marker, type AB blood has both A and B markers, and type O blood has neither A nor B markers. It is important to type and crossmatch blood before a blood transfusion takes place to prevent a rejection reaction that could kill the recipient. The American Red Cross and other agencies that coordinate procurement of organs and tissues for transplantation are developing computer networks to monitor availability of organs with specific cell markers. Such high-speed computer networks allow physicians to immediately locate organs or tissues for emergency transplants or transfusions.

Another procedure used to screen potential donor organs and tissues involves checking for the presence of infectious agents, especially viruses. Because viruses are difficult to find, most screening tests screen for the presence of specific antibodies. Antibodies are protein molecules produced by some white blood cells upon exposure to a virus or other infectious agent. Each type of virus triggers production of a specific kind of antibody, so the presence of a specific antibody type means that the tissue may have the corresponding virus. For example, a test called *ELISA (enzyme-linked immunosorbent assay)* is used to test for the presence of antibodies produced in response to HIV (human immunodeficiency virus). HIV may progress to *acquired immunodeficiency syndrome (AIDS)*, a fatal disease that can be transmitted through HIV-contaminated tissues or body fluids. Routine screening for viral hepatitis antigens and other viral antibodies is also done by most tissue and blood banks. Because there is some lag time between infection by a virus and the resulting production of antibodies, such screening tests may fail to identify a virus-contaminated tissue from a recently infected donor.

4

TABLE 4-3 Muscle and Nervous Tissue

TISSUE	STRUCTURE	LOCATION(S)	FUNCTION(S)
MUSCLE			
Skeletal (striated voluntary)	Long, threadlike cells with multiple nuclei and striations	Muscles that attach to bones Eyeball muscles Upper third of esophagus	Maintenance of posture; movement of bones; produces body heat Eye movements Involved in first part of swallowing
Cardiac (striated involuntary)	Branching, interconnected cylinders with faint striations	Wall of heart	Contraction of heart
Smooth (nonstriated involuntary or visceral)	Threadlike cells with single nuclei and no striations	Walls of tubular viscera of digestive, respiratory, and genitourinary tracts Walls of blood vessels and large lymphatic vessels Ducts of glands Intrinsic eye muscles (iris and ciliary body) Arrector muscles of hairs	Movement of substances along respective tracts Changing of diameter of vessels Movement of substances along ducts Changing of diameter of pupils and shape of lens Erection of hairs (goose pimples)
NERVOUS			
	Nerve cells with large cell bodies and thin fiberlike extensions; supportive glial cells also present	Brain; spinal cord; nerves	Irritability; conduction

Skeletal muscles are commonly attached to bones and, when contracted, can produce voluntary and controlled body movements.

Cardiac Muscle Tissue

Cardiac muscle forms the walls of the heart, and the regular but involuntary contractions of cardiac muscle produce the heartbeat. Under the light microscope (**Figure 4-17**), cardiac muscle fibers have faint cross striations (like skeletal muscle) and thicker dark bands called *intercalated disks.*

4

Cardiac muscle fibers branch and connect to various other cardiac fiber branches to produce a three-dimensional, interlocking mass of contractile tissue. Tubelike membrane proteins link the muscle fibers end-to-end and thus allow them to function as if they are one large unit. This arrangement allows the impulse that triggers contraction to move along all the fibers quickly—producing a nearly simultaneous contraction in the wall of heart chambers.

Smooth Muscle Tissue

Smooth muscle *(visceral muscle)* is said to be involuntary because it is not under conscious or willful control. Under a microscope (**Figure 4-18**), smooth muscle cells are seen as long, narrow fibers but not nearly as long as skeletal fibers.

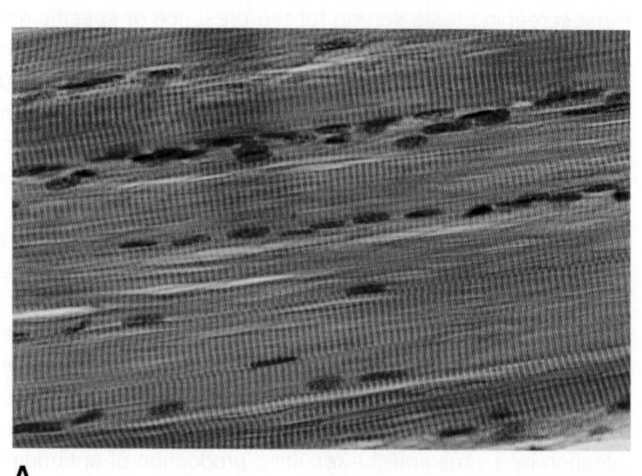

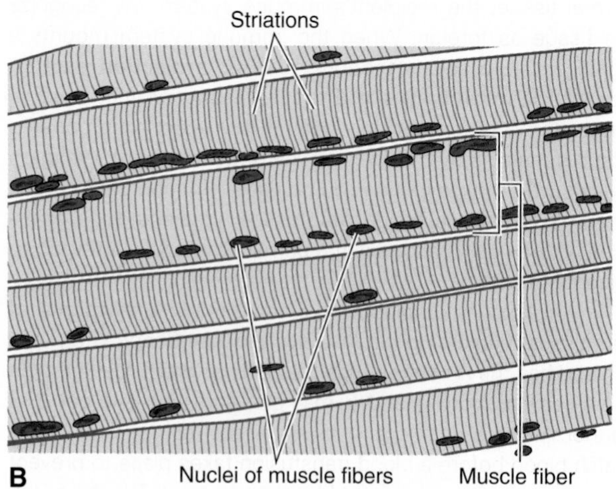

FIGURE 4-16 Skeletal muscle. A, Photomicrograph. **B,** Sketch of the photomicrograph. Note the striations of the muscle cell fibers seen in this longitudinal section.

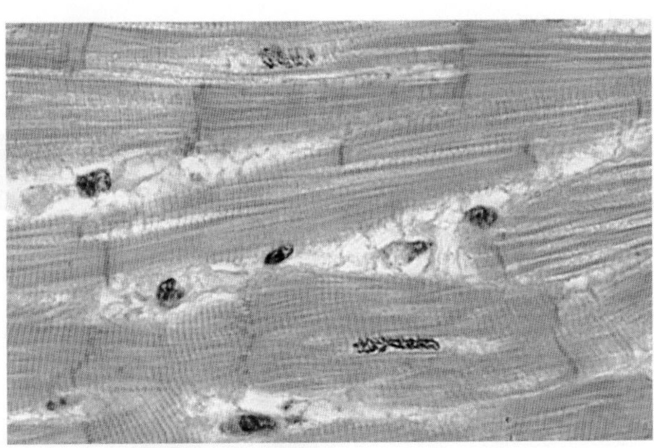

A

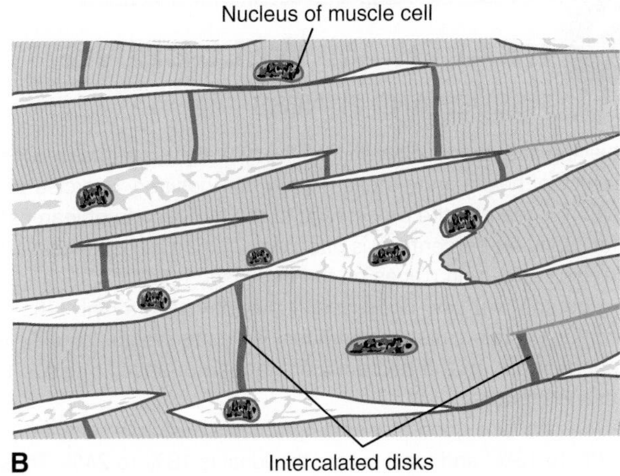

B

FIGURE 4-17 Cardiac muscle. A, Photomicrograph. **B,** Sketch of the photomicrograph. Note the branched, lightly striated fibers. The darker bands, called *intercalated disks,* which are characteristic of cardiac muscle, are easily identified in this tissue section.

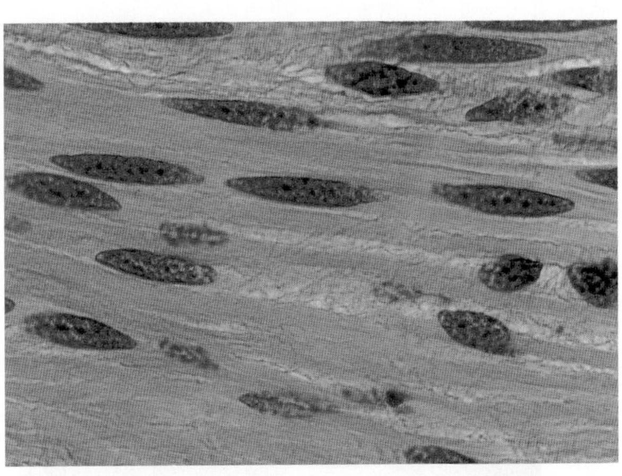

A

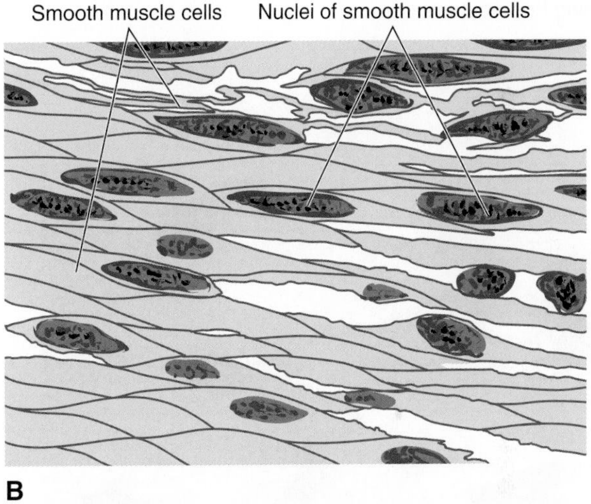

B

FIGURE 4-18 Smooth muscle. A, Photomicrograph. **B,** Sketch of the photomicrograph. Note the central placement of nuclei in the spindle-shaped smooth muscle fibers in this longitudinal section.

Individual smooth muscle cells appear smooth (that is, without cross striations) and have only one nucleus per fiber.

Smooth muscle helps form the walls of blood vessels and hollow organs such as the intestines and other tube-shaped structures in the body. Contractions of smooth (visceral) muscle propel material through the digestive tract and help regulate the diameter of blood vessels. Contraction of smooth muscle in the tubes of the respiratory system, such as the bronchioles in the lungs, can impair breathing and result in asthma attacks and labored respiration.

QUICK CHECK

1. Name the three major classifications of muscle tissue.
2. Which types of muscle tissue are under involuntary control?
3. Where would you find smooth muscle tissue in the body?

Nervous Tissue

Nervous tissue controls body functions and coordinates rapid communication between body structures (see **Table 4-3**). Nervous tissue consists of two kinds of cells: **neurons** are the impulse conducting units of the system, while **glia** or **neuroglia** are special connecting and supporting cells.

All neurons are characterized by a **cell body** and two types of processes: (1) one **axon,** which transmits a nerve impulse away from the cell body, and (2) one or more **dendrites,** which carry impulses toward the cell body. The large neurons in **Figure 4-19** have many dendrites extending from the cell body.

Tissue Repair

When damaged by mechanical or other injuries, tissues have a varying capacity to repair themselves. Damaged tissue will either regenerate or be replaced by tissue known as *scars.*

HEALTH AND WELL-BEING

TISSUES AND FITNESS

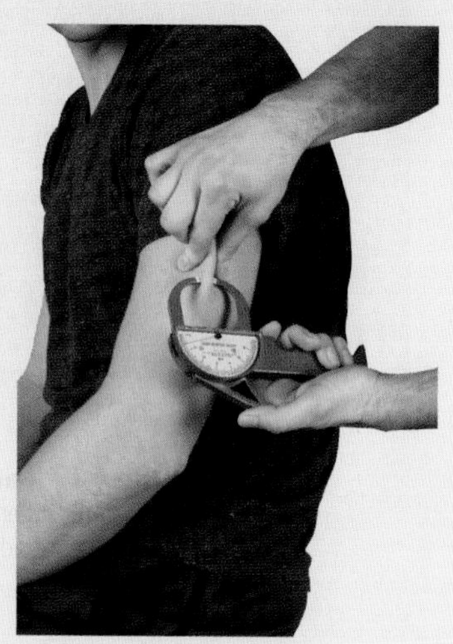

Achieving and maintaining an ideal body weight is a health-conscious goal. However, a better indicator of health and fitness is *body composition.* Exercise physiologists assess body composition to identify the percentage of the body made of lean tissue and the percentage made of fat. Body-fat percentage is often determined by using calipers to measure the thickness of skin folds at certain places on the body (see figure).

A person with low body weight may still have a high ratio of fat to muscle, an unhealthy condition. In this case the individual is "underweight" but "overfat." In other words, fitness depends more on the percentage and ratio of specific tissue types than the overall amount of tissue present.

One goal of a good fitness program is a desirable body-fat percentage. For men, the ideal is 12% to 18%, and for women, the ideal is 18% to 24%. These ranges can vary as a result of various other health factors and should be discussed with a physician.

Because fat contains stored energy (measured in calories), a low fat percentage means a low energy reserve. High body-fat percentages are associated with several life-threatening conditions, including diabetes and cardiovascular disease. A balanced diet and an exercise program can ensure that the ratio of fat to muscle tissue stays at a level appropriate for maintaining homeostasis.

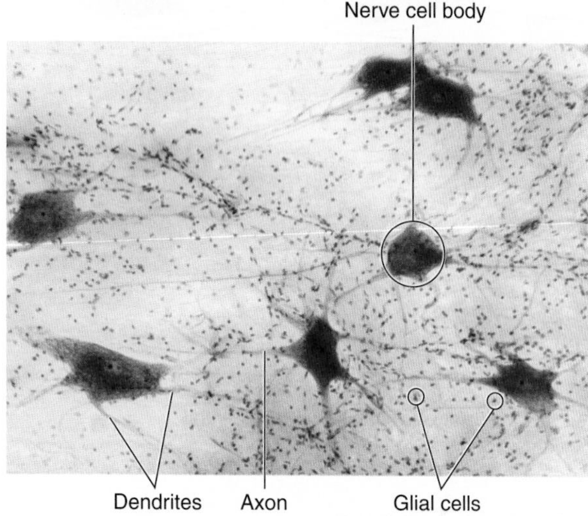

FIGURE 4-19 Nervous tissue. Photomicrograph of neurons and glia in a smear of the spinal cord. The neurons in this slide display characteristic cell bodies and multiple cell processes. Nuclei of glia are visible as *dark dots* surrounding the neuron.

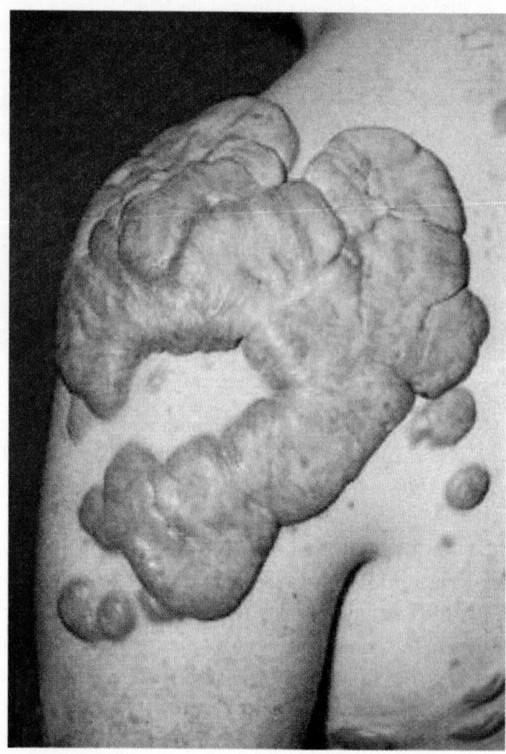

FIGURE 4-20 Keloid. Keloids are thick scars that form in the lower layer of the skin in predisposed individuals.

During repair, phagocytic cells remove dead or injured cells and allow new tissue cells to fill in the gaps that are left. This growth of new tissue is called **regeneration.**

Epithelial and connective tissues have the greatest capacity to regenerate. When a break in an epithelial membrane occurs, as in a cut, cells quickly divide to form daughter cells that fill the wound.

In connective tissues, cells that form collagen fibers become active after an injury and fill in a gap with an unusually dense mass of fibrous connective tissue. If this dense mass of fibrous tissue is small, it may be replaced by normal tissue later. If the mass is deep or large, or if cell damage was extensive, it may remain a dense fibrous mass called a **scar.** An unusually thick scar that develops in the lower layer of the skin, such as that shown in Figure 4-20, is called a **keloid.**

Skeletal muscle tissue often regenerates itself when injured. Cardiac and smooth muscle seem to have less ability to regenerate—especially when the damage is severe.

> Crushing injuries of skeletal muscle can release massive amounts of intracellular substances into the bloodstream that can have life-threatening consequences. Learn more in the article *Rhabdomyolysis* at Connect It! at *evolve.elsevier.com*.

Nerve tissue had been viewed as having a limited capacity to regenerate, but accumulating evidence shows that these limitations are not as great as once thought.

Neurons outside the brain and spinal cord can sometimes regenerate on their own, but the process is very slow and only occurs if certain neuroglia are present to "pave the way." In the normal adult brain and spinal cord, neurons may not always grow back when injured. Thus brain and spinal cord injuries often result in permanent damage. Fortunately, the discovery of *nerve growth factors* produced by neuroglia offers the promise of treating brain damage by stimulating release of these factors.

QUICK CHECK

1. What are the two main cell types found in nervous tissue?
2. What is meant by the term *regeneration* of tissue?

LANGUAGE OF SCIENCE *(continued from p. 71)*

connective tissue
(koh-NEK-tiv TISH-yoo)
[*con-* **together,** *-nect-* **bind,** *-ive* **relating to,** *tissue* **fabric**]

cuboidal
(KYOO-boyd-al)
[*cub-* **cube,** *-oid* **like,** *-al* **relating to**]

dendrite
(DEN-dryte)
[*dendr-* **tree,** *-ite* **part (branch) of**]

dense fibrous connective tissue
(dens FYE-brus koh-NEK-tiv TISH-yoo)
[*dense* **thick,** *fibro-* **fiber,** *-ous* **relating to,** *con-* **together,** *-nect-* **bind,** *-ive* **relating to,** *tissue* **fabric**]

elastic cartilage
(eh-LAS-tik KAR-tih-lij)
[*elast-* **drive or propel,** *-ic* **relating to,** *cartilag-* **cartilage**]

elastin
(e-LAS-tin)
[*elast-* **strike or beat out,** *-in* **substance**]

endocrine
(EN-doh-krin)
[*endo-* **within,** *-crin* **secrete**]

epithelial tissue
(ep-ih-THEE-lee-al TISH-yoo)
[*epi-* **on or upon,** *-theli-* **nipple,** *-al* **relating to,** *tissue* **fabric**]

exocrine
(EK-soh-krin)
[*exo-* **outside or outward,** *-crin* **secrete**]

fascia
(FASH-ee-ah)
[*fascia* **band or bundle**]

fat tissue
(fat TISH-yoo)
[*tissue* **fabric**]

fibrocartilage
(fye-broh-KAR-tih-lij)
[*fibr-* **thread or fiber,** *-cartilag-* **cartilage**]

gland
(gland)
[*gland* **acorn**]

glia
(GLEE-ah)
[*glia* **glue**]

goblet cell
(GOB-let sel)
[*goblet* **small bowl,** *cell* **storeroom**]

haversian system
(hah-VER-zhun or HAV-er-zhun SIS-tem)
[*Clopton Havers* **English physician,** *-ian* **relating to**]

hematopoietic tissue
(hee-mah-toh-poy-ET-ik TISH yoo)
[*hema-* **blood,** *-poie-* **to make,** *-ic* **relating to,** *tissue* **fabric**]

hyaline cartilage
(HYE-ah-lin KAR-tih-lij)
[*hyal-* **glass,** *-ine* **of or like,** *cartilag-* **cartilage**]

interstitial
(in-ter-STISH-al)
[*inter-* **between,** *-stit-* **stand,** *-al* **relating to**]

loose fibrous connective tissue
(LOOS FYE-brus kon-NEK-tiv TISH-yoo)
[*fibr-* **thread or fiber,** *-ous* **relating to,** *con-* **together,** *-nect-* **bind,** *-ive* **relating to,** *tissu-* **fabric**]

matrix
(MAY-triks)
[*matrix* **womb**]

nervous tissue
(NER-vus TISH-yoo)
[*nerv-* **nerves,** *-ous* **relating to,** *tissue* **fabric**]

neuroglia
(noo-ROG-lee-ah or noo-roh-GLEE-ah)
sing., neuroglial cell
(noo-ROG-lee-al or noo-roh-GLEE-al sel)
[*neuro-* **nerve,** *-glia* **glue**]

neuron
(NOO-ron)
[*neur-* **string or nerve,** *-on* **unit**]

osteon
(AHS-tee-on)
[*osteo-* **bone,** *-on* **unit**]

proteoglycan
(PRO-tee-oh-GLYE-kan)
[*proteo-* **protein,** *-glycan* **polysaccharide** (from *-glyc-* **sweet**)]

pseudostratified epithelium
(SOOD-oh-STRAT-ih-fyed ep-ih-THEE-lee-um)
[*pseudo-* **false,** *-strati-* **layer,** *-fied* **made,** *epi-* **on or upon,** *theli-* **nipple,** *-um* **thing**]

reticular tissue
(reh-TIK-yoo-lar TISH-yoo)
[*ret-* **net,** *-ic-* **relating to,** *-ul-* **little,** *-ar* **characterized by,** *tissue* **fabric**]

simple
(SIM-pel)
[*simple* **not mixed**]

simple columnar epithelium
(SIM-pel koh-LUM-nar ep-ih-THEE-lee-um)
[*simple* **not mixed,** *column-* **column,** *-ar* **relating to,** *epi-* **on,** *-theli-* **nipple,** *-um* **thing**]

simple cuboidal epithelium
(SIM-pel KYOO-boyd-al ep-ih-THEE-lee-um)
[*simple* **not mixed,** *cub-* **cube,** *-oid* **like,** *-al* **relating to,** *epi-* **on,** *-theli-* **nipple,** *-um* **thing**]

simple squamous epithelium
(SIM-pel SKWAY-mus ep-ih-THEE-lee-um)
[*simple* **not mixed,** *squam-* **scale,** *-ous* **characterized by,** *epi-* **on,** *-theli-* **nipple,** *-um* **thing**]

skeletal muscle tissue
(SKEL-et-tal MUS-el TISH-yoo)
[*skelet-* **dried body,** *-al* **relating to,** *mus-* **mouse,** *-cle* **small,** *tissue* **fabric**]

Continued on p. 86

4

LANGUAGE OF SCIENCE (continued from p. 85)

smooth muscle tissue
(smoothe MUS-el TISH-yoo)
[*smooth* **smooth,** *mus-* **mouse,** *-cle* **small,** *tissue* **fabric**]

squamous
(SKWAY-mus)
[*squam-* **scale,** *-ous* **characterized by**]

stratified
(STRAT-ih-fyde)
[*strati-* **layer,** *-fied* **made**]

stratified squamous epithelium
(STRAT-ih-fyde SKWAY-mus ep-ih-THEE-lee-um)
[*strati-* **layer,** *-fied* **made,** *squam-* **scale,** *-ous* **characterized by,** *epi-* **on,** *-theli-* **nipple,** *-um* **thing**]

stratified transitional epithelium
(STRAT-ih-fyde tran-ZISH-en-al ep-ih-THEE-lee-um)
[*strati-* **layer,** *-fied* **made,** *trans-* **across,** *-tion-* **process,** *-al* **relating to,** *epi-* **on,** *-theli-* **nipple,** *-um* **thing**]

tissue
(TISH-yoo)
[*tissue* **fabric**]

transitional
(tranz-IH-shen-al)
[*trans-* **across,** *-tion-* **process,** *-al* **relating to**]

LANGUAGE OF MEDICINE

histologist
(hih-STOL-oh-jist)
[*histo-* **tissue,** *-log-* **words (study of),** *-ist* **agent**]

keloid
(KEE-loyd)
[*kel-* **claw,** *-oid* **like**]

paleontologist
(pay-lee-un-TOL-oh-jist)
[*paleo-* **ancient,** *-onto-* **being,** *-log-* **words (study of),** *-ist* **agent**]

pathologist
(pah-THOL-oh-jist)
[*patho-* **disease,** *-log-* **words (study of),** *-ist* **agent**]

regeneration
(ree-jen-er-AY-shun)
[*re-* **again,** *generat-* **produce,** *-tion* **process**]

rejection reaction
(reh-JEK-shun ree-AK-shun)
[*re-* **again,** *-ject* **to throw,** *-tion* **process of,** *re-* **again,** *-act-* **to do,** *-tion* **process of**]

rhabdomyolysis
(RAB-doh-mye-OL-ih-sis)
[*rabdo-* **rod,** *-myo-* **muscle,** *-lysis* **loosening**]

scar
(skahr)
[*scar* **scab**]

tissue typing
(TISH-yoo TYE-ping)
[*tissue* **fabric,** *type-* **kind**]

4

❑ OUTLINE SUMMARY

To download a digital version of the chapter summary for use with your device, access the ***Audio Chapter Summaries*** *online at evolve.elsevier.com.*

Scan this summary after reading the chapter to help you reinforce the key concepts. Later, use the summary as a quick review before your class or before a test.

Introduction to Tissues

A. Four main tissue types
  1. Epithelial tissue—forms sheets that cover or line the body
  2. Connective tissue—provides structural and functional support
  3. Muscle tissue—contracts to produce movement
  4. Nervous tissue—senses, conducts, and processes information

B. Matrix—also called extracellular matrix (ECM)
  1. Internal fluid environment of the body, surrounding cells of each tissue
  2. Mostly water, but also often contains fibers and other substances that give it thick, jellylike consistency (**Figure 4-1**)
     a. Collagen—protein that forms twisted ropelike fibers that provide flexible strength to tissue
     b. Elastin—rubbery protein that provides elastic stretch and recoil in tissues
     c. Polysaccharides and proteoglycans help link cells, absorb shock, regulate function, and lubricate

Epithelial Tissue

A. Introduction to epithelial tissue (**Table 4-1**)
   1. Covers body and lines body cavities; forms tubes and ducts
   2. Strongly connected cells are packed closely together with little matrix; contains no blood vessels
   3. Gluelike basement membrane holds epithelium to underlying connective tissue
   4. Classified by shape of cells (**Figure 4-2**)
      a. Squamous—flat and scalelike
      b. Cuboidal—cube-shaped
      c. Columnar—taller than they are wide
      d. Transitional—varying shapes that can stretch
   5. Also classified by arrangement of cells into one or more layers: simple or stratified
B. Squamous epithelium
   1. Simple squamous epithelium—single layer of scalelike cells adapted for transport (e.g., absorption) (**Figure 4-3**)
   2. Stratified squamous epithelium—several layers of closely packed cells specializing in protection (**Figure 4-4**)
C. Cuboidal epithelium
   1. Simple cuboidal epithelium—single layer of cubelike cells often specialized for secretory activity; may secrete into ducts, directly into blood, and on body surface (**Figure 4-3** and **Figure 4-5**)
   2. Stratified cuboidal epithelium—two or more layers of cubelike cells, sometimes found in sweat glands and other locations
D. Columnar epithelium
   1. Simple columnar epithelium—tall, columnlike cells arranged in a single layer; contain mucus-producing goblet cells; specialized for absorption (**Figure 4-6**)
   2. Pseudostratified epithelium—single layer of distorted columnar cells; each cell touches basement membrane (**Figure 4-7**)
E. Transitional epithelium
   1. Stratified transitional epithelium—up to 10 layers of roughly cuboidal cells that distort to squamous shape when stretched (**Figure 4-8**)
   2. Found in body areas that stretch, such as urinary bladder

Connective Tissue

A. Introduction to connective tissue (**Table 4-2**)
   1. Most abundant and widely distributed tissue in body, with many different types, appearances, and functions
   2. Relatively few cells in extracellular matrix between tissue cells
   3. Types
      a. Fibrous—loose fibrous (areolar), adipose (fat), reticular, dense fibrous
      b. Bone—compact and cancellous (spongy)
      c. Cartilage—hyaline, fibrocartilage, elastic
      d. Blood
      e. Hematopoietic tissue
B. Fibrous connective tissue
   1. Loose fibrous connective tissue (areolar)—fibrous glue (fascia) that holds organs together; collagenous and elastic fibers, plus a variety of cell types (**Figure 4-9**)
   2. Adipose (fat) tissue—white fat stores lipids (triglycerides); brown fat produces heat; both types regulate metabolism (**Figure 4-10**)
   3. Reticular tissue—delicate net of collagen fibers, as in bone marrow (**Figure 4-11**)
   4. Dense fibrous tissue—bundles of strong collagen fibers, densely packed
      a. Regular—parallel collagen bundles; example is tendon (**Figure 4-12**)
      b. Irregular—chaotic, swirling collagen bundles; example is deep layer of skin
C. Bone tissue—matrix is collagen bundles encrusted with calcium mineral crystals
   1. Compact bone—made up of cylindrical osteons (haversian systems); forms outer walls of bones
   2. Cancellous bone—made up of thin, crisscrossing beams of bone; found inside bones; also called *spongy bone*
   3. Bone functions in support and protection (**Figure 4-13**)
D. Cartilage tissue—matrix has consistency of gristlelike gel; chondrocyte is cell type (**Figure 4-14**)
   1. Hyaline cartilage—moderate amount of collagen in matrix; forms a flexible gel
   2. Fibrocartilage—matrix is very dense with collagen; forms very tough, hard gel
   3. Elastic cartilage—matrix has some collagen with elastin; forms a soft, elastic gel
E. Blood tissue—matrix is fluid plasma; contains red blood cells, white blood cells, and platelets; functions are transportation and protection (**Figure 4-15**)
F. Hematopoietic tissue—blood-forming tissue with a liquid matrix

Muscle Tissue

A. Muscle tissue contracts to provide movement or stability; produces body heat (**Table 4-3**)
B. Skeletal muscle tissue—attaches to bones; also called *striated* or *voluntary;* control is voluntary; striations apparent when viewed under a microscope (**Figure 4-16**)
C. Cardiac muscle tissue—also called *striated involuntary;* composes heart wall; ordinarily cannot control contractions (**Figure 4-17**)
D. Smooth muscle tissue—also called *nonstriated (visceral)* or *involuntary;* no cross striations; found in blood vessels and other tube-shaped organs (**Figure 4-18**)

4

Nervous Tissue

A. Function—rapid communication between body structures and control of body functions (**Table 4-3**)
B. Neurons (**Figure 4-19**)
  1. Conduction cells
  2. All neurons have cell body and two types of processes: axon and dendrite
    a. Axon (one) carries nerve impulse away from cell body
    b. Dendrites (one or more) carry nerve impulse toward the cell body
C. Glia (neuroglia)—supportive and connecting cells

Tissue Repair

A. Usually accomplished by means of regeneration of tissue
B. Epithelial and connective tissues have greatest capacity for self-repair
  1. Scar—dense fibrous mass sometimes forms if damage was extensive
  2. Keloid scars are unusually thick (**Figure 4-20**)
C. Muscle and nervous tissue can regenerate under favorable conditions

❑ ACTIVE LEARNING

STUDY TIPS

Consider using these tips to achieve success in meeting your learning goals.

1. Tissue identification may seem a bit overwhelming at first glance. But if you look for "key characteristics" as you study each one, it becomes easier. See *my-ap.us/learntissues* for advice.
2. Tissue types are additional topics that could be learned using flash cards. It may be helpful to remember that epithelial tissues are covering and protective tissues, and the important thing about connective tissue is the matrix surrounding the cells.
3. To understand the shape of epithelial cells, you can use a soda can analogy. Imagine a soda can that has been completely smashed. This would represent a squamous-shaped cell. A soda can that has only been smashed half way would represent a cuboidal-shaped cell. A soda can that has not been smashed would represent a columnar-shaped cell. Finally, soda cans arranged in all of these shapes would represent stratified transitional epithelium.
4. Because membranous epithelium covers the body or lines a cavity, there is always an exposed space. After you have identified the exposed space, classify the shape (squamous, cuboidal, or columnar) of the cells. Then determine the number of layers (simple or stratified). Develop a concept map that depicts the different epithelial tissues. Use **Table 4-1** and include the locations of these tissues in the body.
5. When classifying connective tissues, pay close attention to the matrix. Identify whether the matrix is fibrous protein, protein that is ground substance, or fluid. Use available resources (textbook, lab manual, atlas, or Internet sources) to familiarize yourself with the difference among these matrices. Develop a *concept map* that depicts the different connective tissues. Use **Table 4-2** and include the locations of these tissues in the body. (To learn about concept mapping, go to *my-ap.us/MExHCf.*)
6. Familiarize yourself with the unique characteristics that define each type of muscle tissue. Refer to **Table 4-3**. Construct a *T-chart* that lists the different muscle tissues and their locations.
7. The use of flash cards or review cards is an excellent strategy to learn the various types of tissues. There are many online sources that have tissue images. Obtain photos or illustrations of the different types of tissues. Place the photo or illustration on one side of an index card. On the opposite side of the card, put the name of the tissue. You can also add additional information such as unique characteristics or location in the body.
8. Check out the following websites for interactive strategies that will enhance your understanding of the different tissues: *my-ap.us/JuM3p3* and *my-ap.us/KR4tvs.*

Review Questions

Write out the answers to these questions after reading the chapter and reviewing the Chapter Summary. If you simply think through the answer without writing it down, you will not retain much of your new learning.

1. Define the term *tissue* and identify the four principal tissue types.
2. Name and describe three epithelial tissues.
3. Classify epithelium according to the layers of cells present.
4. Where can stratified squamous cells be found?
5. What is the special function of simple columnar epithelium?
6. How does pseudostratified epithelium differ from stratified epithelium?
7. What are some examples of substances secreted through glands or tubules made up of simple cuboidal epithelial cells?
8. Name and describe three connective tissues.
9. Where is connective tissue found?
10. What type of connective tissue is the most widely distributed throughout the body?
11. What type of connective tissue provides great strength and flexibility, but no stretch?
12. What is the function of hematopoietic tissue?
13. Name and describe two muscle tissues.
14. Describe the structure and distinctive traits of skeletal muscle.
15. Give some examples of smooth muscle tissue.
16. Name the two types of nervous tissue. Which is functional nerve tissue and which is support tissue?
17. Give a general description of a neuron.
18. What is the function of an axon? How do dendrites serve the nervous system?
19. What tissues have the greatest capacity to regenerate?
20. Name the tissues that do not regenerate.
21. What is the fluid material between cells?
22. What is *brown* fat?
23. What is cancellous bone?
24. Name three types of cartilage.

Critical Thinking

After finishing the Review Questions, write out the answers to these more in-depth questions to help you apply your new knowledge. Go back to sections of the chapter that relate to concepts that you find difficult.

25. Explain what is meant by tissue typing. Why has this become so important in recent years?
26. You are working in a pathology lab and have been given an epithelial tissue sample to identify. What steps would you take to determine what type of epithelial tissue you are examining?
27. If a small, but deep cut involving skin and muscle occurs, predict which tissue will probably heal first and which will heal more completely. Explain your answer.
28. Compare and contrast tissue repair in epithelial, connective, muscle, and nervous tissue.

Chapter Test

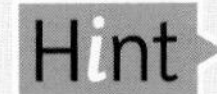

After studying the chapter, test your mastery by responding to these items. Try to answer them without looking up the answers.

1. ________, ________, ________, and ________ are the four main tissues in the body.
2. ________ tissue covers the body and many of its parts.
3. Epithelial cells that vary in shape that can stretch are classified as ________ epithelium.
4. The "open spaces" seen among epithelial cells are specialized ________ cells that produce mucus.
5. The most abundant and widely distributed tissue in the body is ________ tissue.
6. Connective tissue differs from epithelial tissue in the arrangement and variety of its cells and in the amount and kinds of intercellular material, called ________.
7. Cardiac muscle fibers have faint cross striations and thicker dark bands called ________.
8. All neurons are characterized by a ________ ________ and two processes: one ________ and one or more ________.
9. The growth of new tissue is called ________.
10. An unusually thick scar that develops in the lower layer of the skin is called a ________.

4

11. Epithelial tissue that contains cells that are flat and scalelike are classified as:
    a. stratified transitional
    b. squamous
    c. cuboidal
    d. columnar
12. The type of epithelial tissue that protects the body from invasion by microorganisms is:
    a. simple squamous
    b. stratified squamous
    c. pseudostratified
    d. simple columnar
    e. stratified transitional
13. Which epithelial tissue forms tubules or other groupings for secretory activity?
    a. Simple squamous
    b. Pseudostratified
    c. Stratified transitional
    d. Cuboidal
14. What are the functions of connective tissue?
    a. Connects tissue to each other
    b. Forms a supporting framework for the body
    c. Transports substances throughout the body
    d. All of these choices are correct
15. What type of tissue is the most widely distributed of all connective tissue?
    a. Areolar
    b. Adipose
    c. Dense fibrous
    d. Reticular
16. Tendons are examples of which connective tissue type?
    a. Dense fibrous
    b. Areolar
    c. Adipose
    d. Reticular
17. Which of the muscle tissue types are involuntary?
    a. Skeletal
    b. Smooth
    c. Cardiac
    d. Both b and c
18. This type of muscle tissue can be found in the walls of blood vessels and intestines.
    a. Skeletal
    b. Smooth
    c. Cardiac
    d. Both a and b
19. A neuron process that carries the impulse away from the cell body is called a(n):
    a. axon
    b. dendrite
    c. glia
    d. neuroglia
20. Which tissue is least likely to regenerate itself?
    a. Simple squamous epithelium
    b. Dense fibrous connective tissue
    c. Smooth muscle tissue
    d. Stratified squamous epithelium

Match the tissue type in column A with its corresponding description in column B.

Column A	**Column B**
21. ______ Simple squamous epithelium	a. Found in the walls of the intestines
22. ______ Pseudostratified epithelium	b. Glia cells
23. ______ Dense fibrous connective tissue	c. Absorption of oxygen into the blood
24. ______ Cartilage	d. Intercalated disks
25. ______ Adipose	e. Lines the trachea
26. ______ Reticular	f. Composes tendons
27. ______ Skeletal muscle tissue	g. Fat tissue
28. ______ Smooth muscle tissue	h. Delicate webs of collagen fibers
29. ______ Cardiac muscle tissue	i. Chondrocytes
30. ______ Nervous tissue	j. Striated, voluntary

4

Case Studies

To solve a case study, you may have to refer to the glossary or index, other chapters in this textbook, and other resources.

1. Faye was brought to the health clinic by her granddaughter. Faye sustained a burn on her right foot from burning trash. The area is blistered, swollen, hot, and tender to the touch. The area surrounding the burn has an increased redness. Based on the structure and function of types of tissue, what type of injury do you suspect Faye has sustained? Because the basement membrane was not completely destroyed in Faye's injury, what type of tissue repair would you expect to occur?
2. Lauren is a bodybuilder who is obsessed with her physique. She exercises daily and eats a very low-fat diet. Her sister Ellen is a personal fitness trainer and she assessed Lauren's body fat at 12%. Determine whether Lauren is too lean or too fat. Explain the relationship between her body-fat percentage and lifestyle.
3. Dennis is a sedentary, middle-aged smoker who is complaining of chest pain. Ultimately, he is diagnosed with lung cancer. What type of tissues would be involved?

Answers to Active Learning Questions can be found online at evolve.elsevier.com.

4

Organ Systems

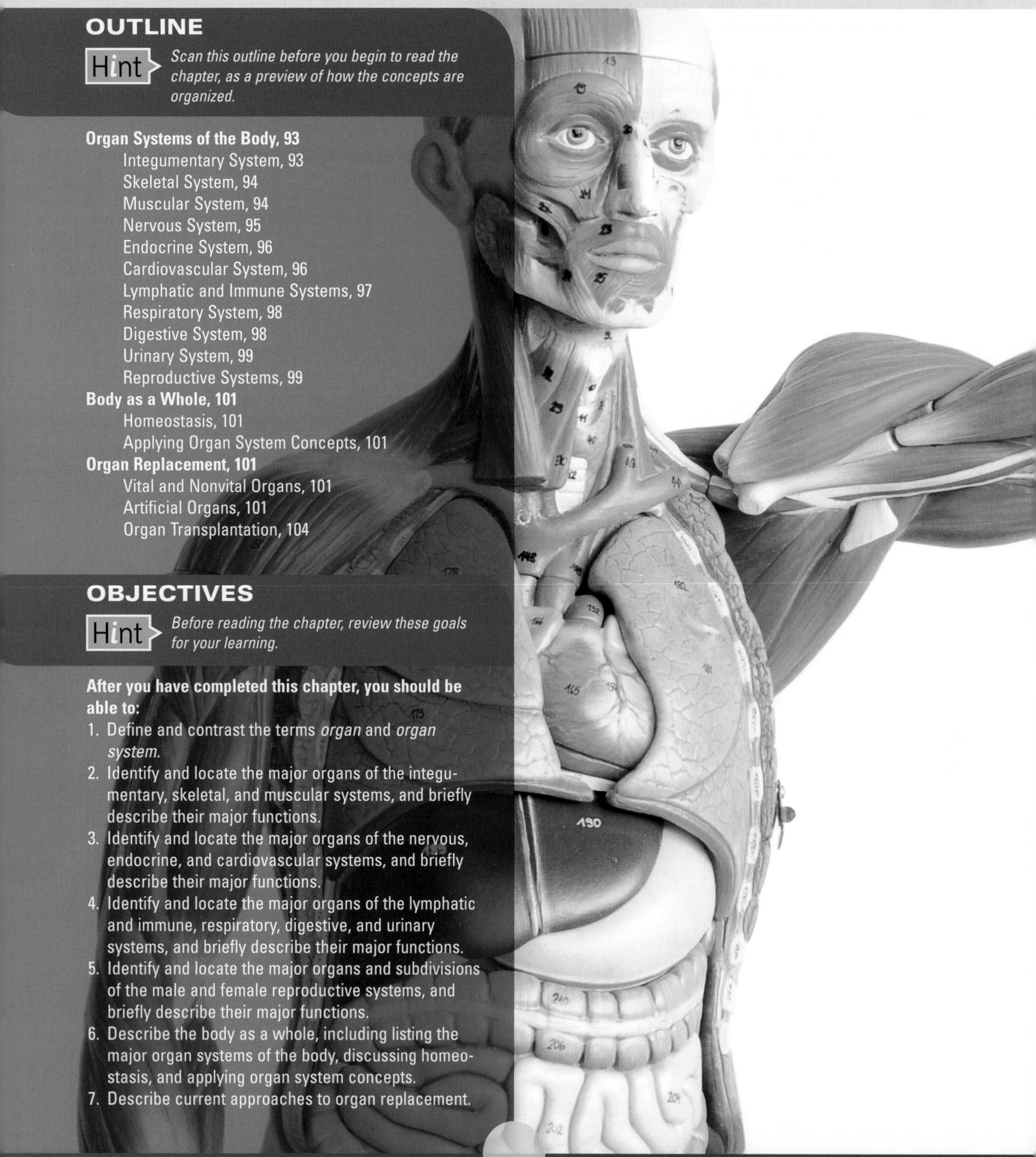

OUTLINE

Scan this outline before you begin to read the chapter, as a preview of how the concepts are organized.

OBJECTIVES

Before reading the chapter, review these goals for your learning.

After you have completed this chapter, you should be able to:

1. Define and contrast the terms *organ* and *organ system.*
2. Identify and locate the major organs of the integumentary, skeletal, and muscular systems, and briefly describe their major functions.
3. Identify and locate the major organs of the nervous, endocrine, and cardiovascular systems, and briefly describe their major functions.
4. Identify and locate the major organs of the lymphatic and immune, respiratory, digestive, and urinary systems, and briefly describe their major functions.
5. Identify and locate the major organs and subdivisions of the male and female reproductive systems, and briefly describe their major functions.
6. Describe the body as a whole, including listing the major organ systems of the body, discussing homeostasis, and applying organ system concepts.
7. Describe current approaches to organ replacement.

CHAPTER 5

After exploring cells and tissues in the previous chapters, we are ready to look at the organs and systems of the body. An **organ** is a structure made up of two or more kinds of **tissue** and is organized to perform a more complex function compared to a single tissue. A **system** is a group of organs that together perform a more complex function than does one organ. This chapter gives a brief overview of the major organ systems of the body.

An overview of body systems provides the foundation needed to see the "big picture" of human structure and function as we later reveal the details of each system. As you progress through your detailed study of the major organ systems in the chapters that follow, it will be possible to view the body not just as an assembly of individual parts but as an integrated and functioning whole.

Organ Systems of the Body

Integumentary System

The **integumentary system** includes only one organ: the skin (**Figure 5-1**). In most adults, the skin alone weighs 20 pounds or more—accounting for about 16% of total body weight and making it the body's heaviest organ.

Although the integumentary system has only one organ, that one organ, the skin or **integument,** has many millions of *appendages* (structures attached to a main part) and glands. These skin structures include the hair, nails, and sweat- and oil-producing glands. The skin includes many microscopic sense receptors, making it the largest sensory organ of the body. Skin sense receptors permit the body to respond to pain, pressure, touch, texture, vibration, and changes in temperature.

LANGUAGE OF **SCIENCE**

Hint Before reading the chapter, say each of these terms out loud. This will help you to avoid stumbling over them as you read.

adrenal gland
(ah-DREE-nal gland)
[*ad-* **toward,** *-ren-* **kidney,** *-al* **relating to,** *gland* **acorn**]

alimentary canal
(al-eh-MEN-tar-ee kah-NAL)
[*aliment-* **nourishment,** *-ary* **relating to,** *canal* **channel**]

alveolus
(al-VEE-oh-lus)
pl., alveoli
(al-VEE-oh-lye)
[*alve-* **hollow,** *-olus* **little**]

antibody
(AN-tih-bod-ee)
[*anti-* **against,** *-body* **body**]

artery
(AR-ter-ee)
[*arteri* **vessel**]

bronchus
(BRONG-kus)
pl., bronchi
(BRONG-kye)
[*bronchus* **windpipe**]

capillary
(KAP-ih-layr-ee)
[*capill-* **hair,** *-ary* **relating to**]

cardiac muscle
(KAR-dee-ak MUS-el)
[*cardi-* **heart,** *-ic* **relating to,** *mus-* **mouse,** *-cle* **little**]

cardiovascular system
(kar-dee-oh-VAS-kyoo-lar SIS-tem)
[*cardio-* **heart,** *-vas-* **vessel,** *-ular* **relating to,** *system* **organized whole**]

Continued on p. 106

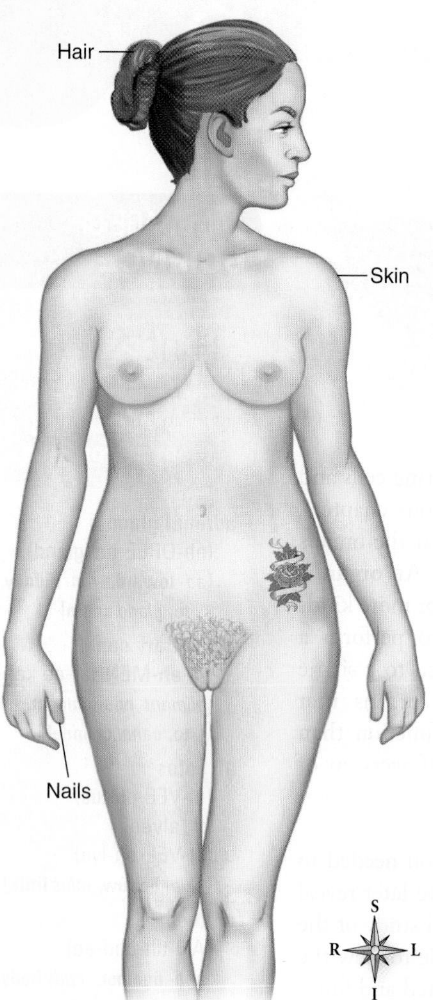

FIGURE 5-1 Integumentary system.

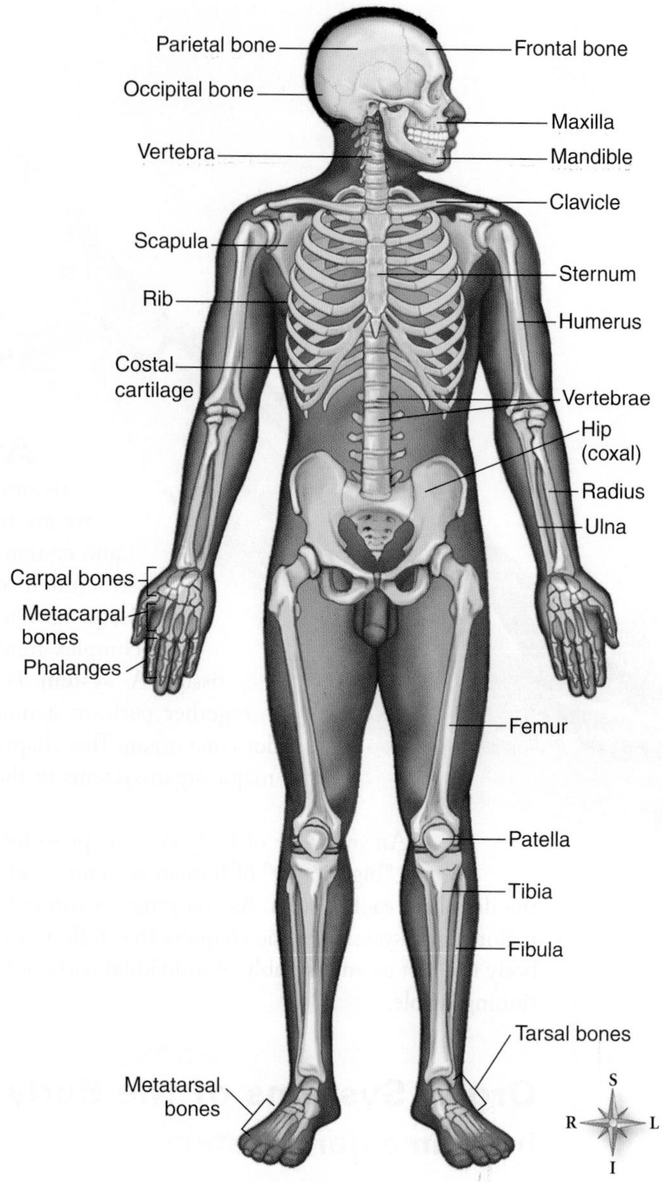

FIGURE 5-2 Skeletal system.

The integumentary system is crucial to survival. Its primary function is *protection.* The skin protects underlying tissue against invasion by harmful bacteria, bars entry of most chemicals, and minimizes the chances of mechanical injury to underlying structures. In addition, the skin regulates body temperature by sweating and by controlling blood flow and therefore heat loss at the body surface. The skin also synthesizes important chemicals, such as vitamin D, and functions as a sophisticated sense organ for temperature, touch, pressure, pain, vibration, and more.

5

Skeletal System

Bones are the primary organs of the skeletal system. **Figure 5-2** shows examples of the 206 individually named bones found in the **skeletal system.** Each individual also has some variable bones that differ from person to person and do not have specific names.

The skeletal system includes not only bone but also related tissues such as cartilage. Cartilage can cushion bones that are linked together and can act as the connection between one bone and another. Look at the large cartilage bands (costal cartilage) that connect the ribs to the sternum in **Figure 5-2**.

Ligaments are bands of fibrous connective tissue that help hold bones together. Connections between two or more bones are called **joints.** The moveable joints between bones make various movements of individual body parts possible. Without movable joints, our bodies would be rigid, immobile hulks.

The skeleton provides protection and a supporting framework for the brain and other internal organs. Bones also serve as storage areas for important minerals such as calcium and phosphorus.

The formation of blood cells in the red marrow of certain bones is another crucial function of the skeletal system.

Muscular System

Skeletal Muscles

Individual skeletal muscles are the organs of the **muscular system.** Muscles are made up of mostly **skeletal muscle** tissue. Also called **voluntary muscle,** this tissue has the ability to contract when stimulated by conscious nerve regulation. Although movement of the body is the primary function of the muscular system, it also maintains stability of our posture (body position) and provides heat to maintain our body temperature.

A **tendon** is a dense strap or sheet of regular dense fibrous connective tissue. A tendon is part of a muscle organ that attaches the muscle to a bone (or to another muscle). The anterior tibialis tendon of the leg labeled in Figure 5-3 shows how tendons attach muscles to bones.

When stimulated by a nervous impulse, skeletal muscle tissue shortens or contracts. Voluntary movement occurs when skeletal muscles shorten—a function of the way muscles are attached to bones and the way bones articulate (join) with one another in joints. Sometimes it is useful to think of this cooperative functioning of the bones and muscles as the *skeletomuscular system* or *musculoskeletal system.*

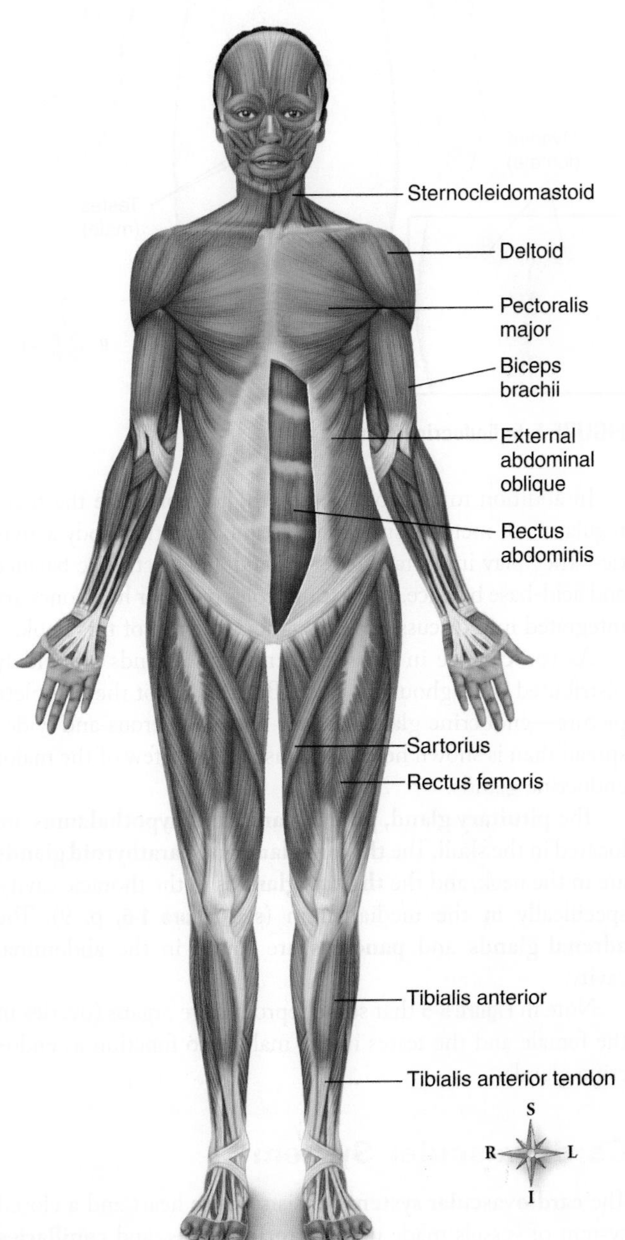

FIGURE 5-3 Muscular system.

Muscles of Other Systems

In addition to organs of the muscular system, the body contains other types of muscle tissue that form parts of organs in other body systems.

For example, **smooth muscle** tissue is found in the walls of hollow organs such as the stomach and small intestine. Smooth muscles help move fluids through organs and often form valves that regulate when fluids may move from one section of a hollow organ to another.

A third type of muscle tissue is the **cardiac muscle** in the wall of the heart. By contracting, it pumps blood through the circulatory system. Some cardiac muscle cells in the heart generate the rhythm of the heartbeat.

Smooth and cardiac muscle tissues are **involuntary** because they are regulated by subconscious mechanisms.

Nervous System

The brain, spinal cord, and nerves are the organs of the **nervous system** (Figure 5-4). The brain and spinal cord make up the **central nervous system (CNS).** These two organs provide the central control of the whole nervous system.

The *cranial nerves* extend from the brain and the *spinal nerves* extend from the spinal cord. The cranial and spinal nerves, and all their branches, make up the **peripheral nervous system (PNS).** The word *peripheral* means "around the boundary," an apt term for the nerve branches that extend all the way to the farthest boundaries of the body.

The extensive networking of the components of the nervous system makes it possible for this complex system to perform its primary functions. These include the following:

1. Communication between body organs
2. Integration of body functions
3. Control of body functions
4. Recognition of sensory stimuli

These functions are performed by signals called **nerve impulses.** In general, the functions of the nervous system result in rapid activity that lasts usually for a short duration. For example, we can chew our food normally, walk, and perform coordinated muscular movements only if our nervous system functions properly. The nerve impulses permit the rapid and precise control of diverse body functions. Other types of nerve impulses cause glands to secrete hormones or other fluids.

In addition, elements of the peripheral nervous system can recognize certain **stimuli,** such as heat, light, sound, pressure, or temperature, that affect the body. When stimulated, these **sense organs** (discussed in Chapter 11) generate nerve impulses that travel to the brain or spinal cord where analysis or relay occurs and, if needed, appropriate action is initiated.

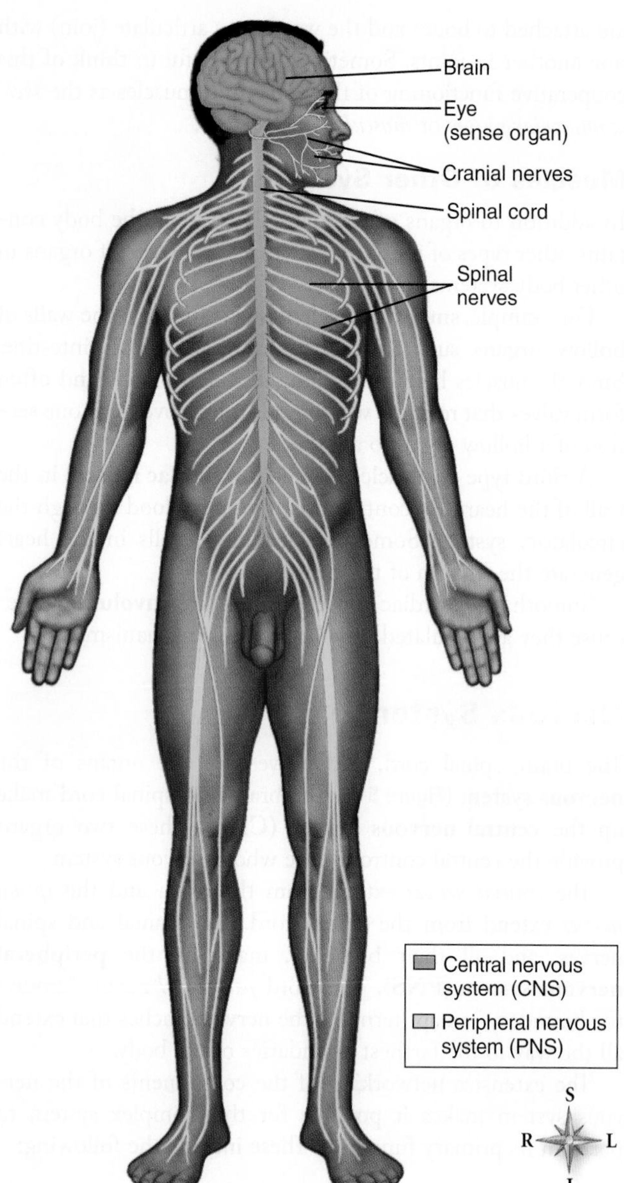

FIGURE 5-4 Nervous system.

Endocrine System

The **endocrine system** is composed of glands that secrete chemicals known as **hormones** directly into the blood. Sometimes called *ductless glands,* the organs of the endocrine system perform the same general functions as the nervous system: communication, integration, and control. The nervous system provides rapid, brief control by fast-traveling nerve impulses. The endocrine system provides slower but longer-lasting control by hormone secretion. For example, secretion of growth hormone controls the rate of development over long periods of gradual growth.

Because of their influence over activities throughout the body, it is no wonder that the nervous and endocrine systems are sometimes thought of as one large regulatory system—the *neuroendocrine system.*

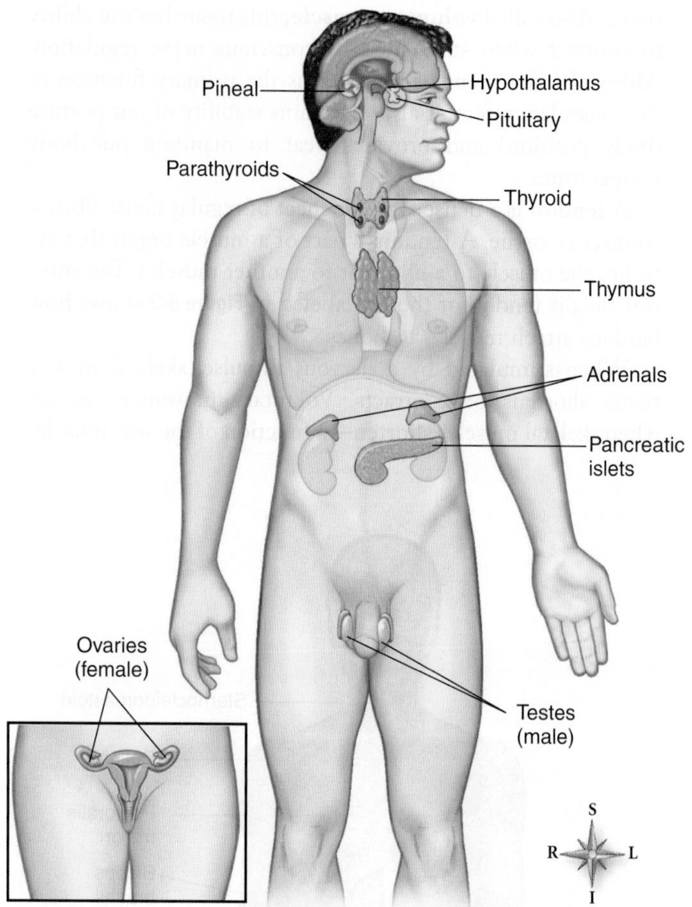

FIGURE 5-5 Endocrine system.

In addition to controlling growth, hormones are the main regulators of metabolism, reproduction, and other body activities. They play important roles in fluid and electrolyte balance and acid-base balance. The various roles of major hormones are integrated into discussions throughout the rest of this book.

As you can see in **Figure 5-5**, endocrine glands are widely distributed throughout the body. But this is not the complete picture—endocrine glands are far more numerous and widespread than is shown here. We consider just a few of the major endocrine glands.

The **pituitary gland, pineal gland,** and **hypothalamus** are located in the skull. The **thyroid gland** and **parathyroid glands** are in the neck, and the **thymus gland** is in the thoracic cavity, specifically in the mediastinum (see **Figure 1-5**, p. 9). The **adrenal glands** and **pancreas** are found in the abdominal cavity.

Note in **Figure 5-5** that some reproductive organs (ovaries in the female and the testes in the male) also function as endocrine glands.

Cardiovascular System

The **cardiovascular system** consists of the heart and a closed system of vessels made up of **arteries, veins,** and **capillaries** (**Figure 5-6**). As the name implies, blood contained in this

5

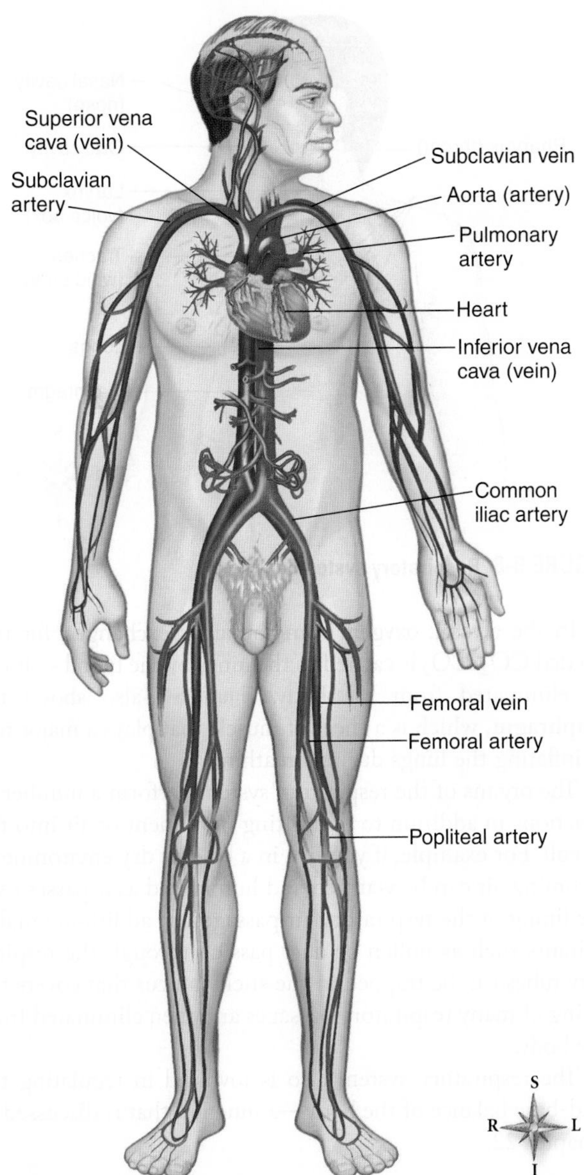

FIGURE 5-6 Cardiovascular system.

system is pumped by the heart around a closed circle, or circuit, of vessels as it passes through the body. The cardiovascular system is sometimes called the **circulatory system.**

The primary function of the cardiovascular or circulatory system is *transportation.* The need for an efficient transportation system in the body is critical. Transportation needs include continuous movement of oxygen (O_2) and carbon dioxide (CO_2), nutrients, hormones, and other important substances. Wastes produced by the cells are released into the bloodstream on an ongoing basis and are transported by the blood to the excretory organs.

The cardiovascular system also helps regulate body temperature by distributing heat throughout the body and by assisting in retaining or releasing heat from the body by regulating blood flow near the body surface. Some cells of the cardiovascular system also function in defense by way of immunity.

QUICK CHECK

1. What is the *integument,* and what are its functions?
2. Give examples of organs of the skeletal system.
3. What are the major functions of the nervous system?
4. What organs make up the cardiovascular system?

Lymphatic and Immune Systems

The two systems we describe next work as partners to provide defense of the body's internal environment against harmful agents such as pathogens and cancer.

Lymphatic System

The **lymphatic system** is composed of **lymphatic vessels** together with other lymphatic organs made up of masses of defensive cells often called *lymphoid tissue.* These lymphoid organs include the **lymph nodes, tonsils, thymus gland,** and **spleen** (Figure 5-7). Note that the thymus functions as an endocrine gland and as a lymphatic organ. Although it is part of the skeletal system, *red bone marrow* is also often considered to be a lymphoid structure.

Instead of containing blood, the lymphatic vessels are filled with **lymph,** a watery fluid that contains lymphocytes, proteins, and some fatty molecules, but no red blood cells. The lymph is formed from the fluid around the body cells that diffuses into the lymph vessels.

Unlike blood, lymph does not circulate repeatedly through a closed circuit, or loop, of vessels. Instead, lymph flowing through lymphatic vessels eventually enters the cardiovascular, or circulatory, system by passing through large ducts, such as the **thoracic duct,** which in turn connect with veins in the upper thoracic cavity. Many biologists consider the lymphatic system to be part of the cardiovascular system.

The functions of the lymphatic system include moving interstitial fluids and small particles back into blood vessels and transporting fats absorbed from the digestive tract to the blood.

Lymph nodes and other lymphoid structures act as small filters that trap and destroy bacterial cells, cancerous cells, and other debris that are carried by the lymph fluid as it flows through the tissues. As such, the organs of the lymphatic system play a role in immunity. Because of this overlap of functions, the lymphatic system and immune system are often discussed together.

Figure 5-7 shows groupings of lymph nodes in the axillary (armpit) and in the inguinal (groin) areas of the body.

Immune System

All of the body's defense systems together make up the **immune system.** It protects us from disease-causing microorganisms, harmful toxins, transplanted tissue cells, and any of our own cells that have turned malignant or cancerous.

The immune system is composed of protective cells (such as phagocytes) and various types of defensive protein molecules (produced by secretory immune cells). Some immune system cells have the ability to attack, engulf, and destroy

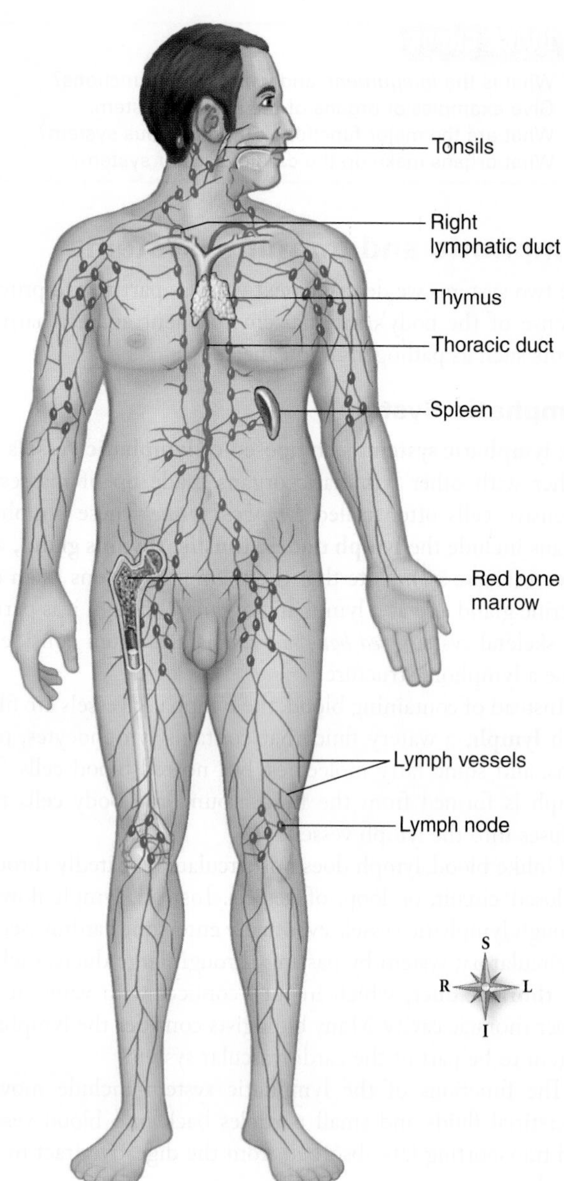

FIGURE 5-7 Lymphatic and immune systems.

harmful bacteria directly by phagocytosis. Other more numerous immune system cells secrete protein compounds called **antibodies** and **complements.** These substances produce chemical reactions that help protect the body from many harmful agents.

5

The lymphatic and immune systems, which are linked to each other and to the cardiovascular system, are discussed in Chapter 16.

Respiratory System

The major organs of the **respiratory system** include the **nose, pharynx** (throat), **larynx** (voice box), **trachea** (windpipe), **bronchi,** and **lungs** (Figure 5-8). Together these organs facilitate the movement of air into the tiny, thin-walled sacs of the lungs called **alveoli.**

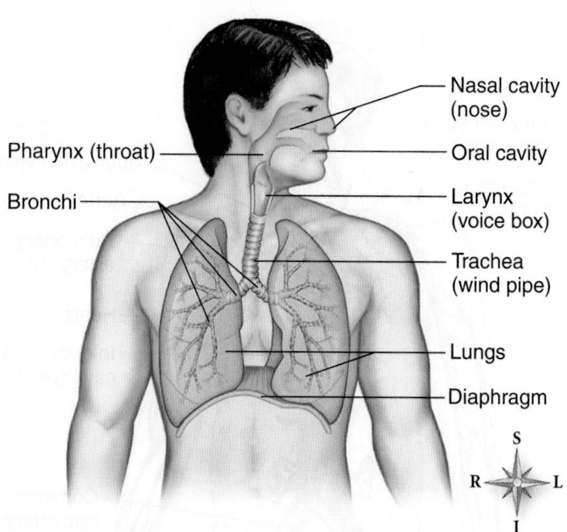

FIGURE 5-8 Respiratory system.

In the alveoli, oxygen from the air is exchanged for unneeded CO_2. CO_2 is carried to the lungs by the blood so it can be eliminated from the body. Figure 5-8 also shows the **diaphragm,** which is a sheet of muscle that plays a major role in inflating the lungs during breathing.

The organs of the respiratory system perform a number of functions in addition to permitting movement of air into the alveoli. For example, if you live in a cold or dry environment, incoming air can be warmed and humidified as it passes over the lining of the respiratory air passages. In addition, inhaled irritants such as pollen or dust passing through the respiratory tubes can be trapped in the sticky mucus that covers the lining of many respiratory passages and then eliminated from the body.

The respiratory system also is involved in regulating the acid-base balance of the body—a function that is discussed in Chapter 22.

To protect the delicate, vital alveoli deep inside the lungs, the respiratory tract has many complex mechanisms that guard against injury and disease. Check out the article *Protective Strategies of the Respiratory Tract* at Connect It! at *evolve.elsevier.com.*

Digestive System

The organs of the **digestive system** (Figure 5-9) are often separated into two groups: the *primary organs* and the *secondary* or *accessory organs*. They work together to ensure proper digestion and absorption of nutrients—and elimination of waste.

The primary organs of digestion form the digestive tract. They include the mouth, pharynx, esophagus, stomach, small intestine, large intestine, rectum, and anal canal. The accessory organs of digestion may attach to the digestive tract (or be inside it). Accessory digestive organs include the teeth, salivary glands, tongue, liver, gallbladder, pancreas, and appendix.

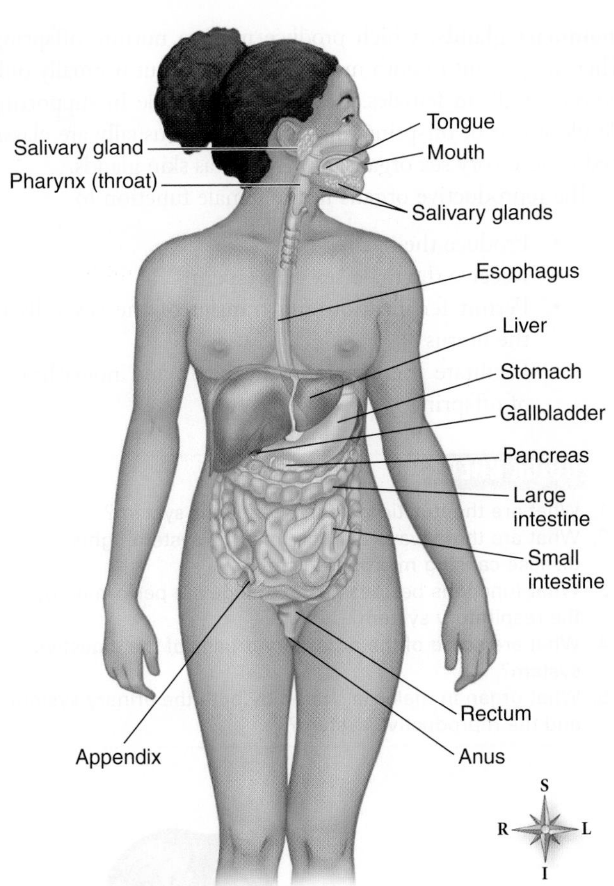

FIGURE 5-9 Digestive system.

The digestive tract is a tube, open at both ends. It is also called the **alimentary canal,** a major part of which is the **gastrointestinal (GI) tract.** Food that enters the alimentary canal is digested, its nutrients are absorbed, and the undigested residue is eliminated from the body as waste material called **feces.**

Urinary System

The organs of the **urinary system** include the **kidneys, ureters, urinary bladder,** and **urethra.**

The kidneys (Figure 5-10) filter out, or "clear," the blood of the waste products continually produced by the metabolism of nutrients in the body cells. The kidneys also play an important role in maintaining the electrolyte, water, and acid-base balances in the body.

The waste product produced by the kidneys is called **urine.** After it is produced by the kidneys, urine flows out of the kidneys, through the ureters, and into the urinary bladder where it is temporarily stored. Urine passes from the bladder to the outside of the body through the urethra.

In the male the urethra passes through the penis and has a double function—it transports both urine and semen (seminal fluid). Therefore it has urinary and reproductive purposes. In the female the urinary and reproductive passages are completely separate, so the urethra performs only a urinary function.

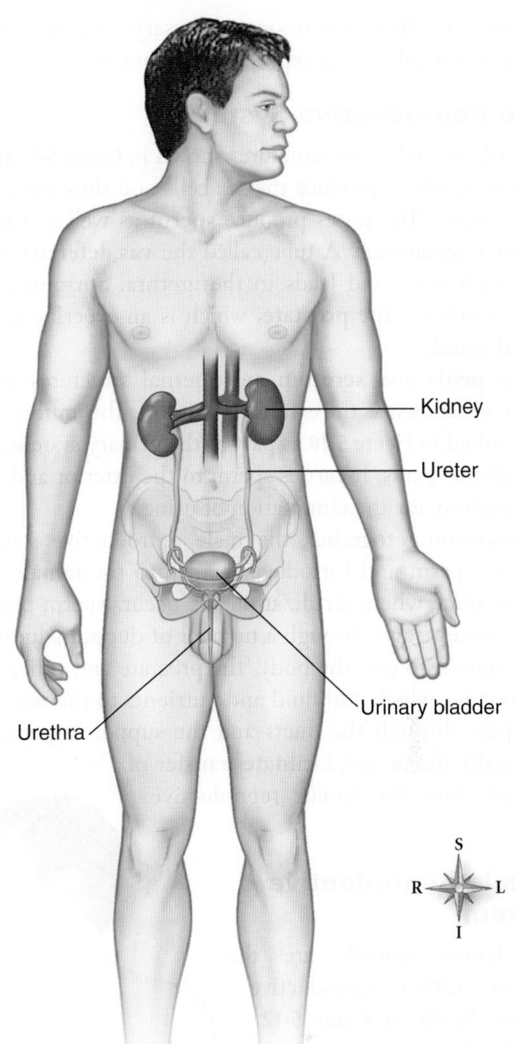

FIGURE 5-10 Urinary system.

In addition to the organs of the urinary system, other organs also help eliminate body wastes. Undigested food residues and metabolic wastes are eliminated from the intestinal tract as feces, and the lungs rid the body of carbon dioxide. The skin also serves an excretory function by eliminating water and some salts in sweat.

Reproductive Systems

Human Reproduction

The normal function of the **reproductive system** is different from the normal function of other organ systems of the body. The proper functioning of the reproductive systems ensures survival, not of the individual but of the genes. In addition, production of the hormones that permit the development of sexual characteristics also affects other structures and functions of the body.

Humans reproduce sexually (two-parent reproduction) so the male and the female reproductive systems must work together to produce offspring. Both systems have **gonads** that

5

produce sex cells and hormones necessary for producing offspring and regulation of reproductive functions.

Male Reproductive System

The male reproductive structures shown in **Figure 5-11** include the **testes,** which produce the sex cells and thus serve as the male gonads. The testes produce **sperm** as well as the male hormone *testosterone.* A tube called the **vas deferens** extends from each testis and leads to the urethra. Surrounding the upper urethra is the **prostate,** which is an exocrine gland or ducted gland.

The **penis** and **scrotum** are external structures and together are known as the external **genitalia.** The urethra, which is identified in **Figure 5-10** as part of the urinary system, passes through the penis. It carries sperm to the exterior and acts as a passageway for the elimination of urine.

Functioning together, the male reproductive structures produce sperm and introduce them into the female reproductive tract, where fertilization can occur. Sperm produced by the testes travel through a number of ducts, including the vas deferens, to exit the body. The prostate and other accessory organs, which add fluid and nutrients to the sex cells as they pass through the ducts and the supporting structures (especially the penis), facilitate transfer of sex cells into the female reproductive tract.

Female Reproductive System

The female gonads are the **ovaries.** Other reproductive organs shown in **Figure 5-12** include the **uterus, uterine tubes** or **fallopian tubes,** and the **vagina.** In the female the term **vulva** is used to describe the external genitalia.

Eggs, or **ova,** are sex cells produced by the ovaries. Ova travel through the uterine tubes, where they may be fertilized by sperm. As the offspring formed by the union of sperm and ovum matures, it moves down the uterine tube to the uterus, where it implants and forms a connection with the mother's blood vessels. After about 9 months, the offspring is delivered through the *cervix* (neck) of the uterus and through the vagina.

The breasts are fatty extensions of the skin that house the **mammary glands,** which produce milk to nurture offspring. They are present in both males and females, but normally only produce milk in females. Because of their role in supporting development of offspring, mammary glands usually are classified as accessory sex organs, rather than as skin glands.

The reproductive organs in the female function to

- Produce the ova (eggs)
- Receive the male sex cells (sperm)
- Permit fertilization and transfer of the sex cells to the uterus
- Facilitate the development, birth, and nourishment of offspring

QUICK CHECK

1. What are the functions of the lymphatic system?
2. What are three ways that the immune system fights disease-causing microorganisms?
3. What functions besides gas exchange are performed by the respiratory system?
4. What are some of the accessory organs of the digestive system?
5. What organ in males is shared by both the urinary system and the reproductive system?

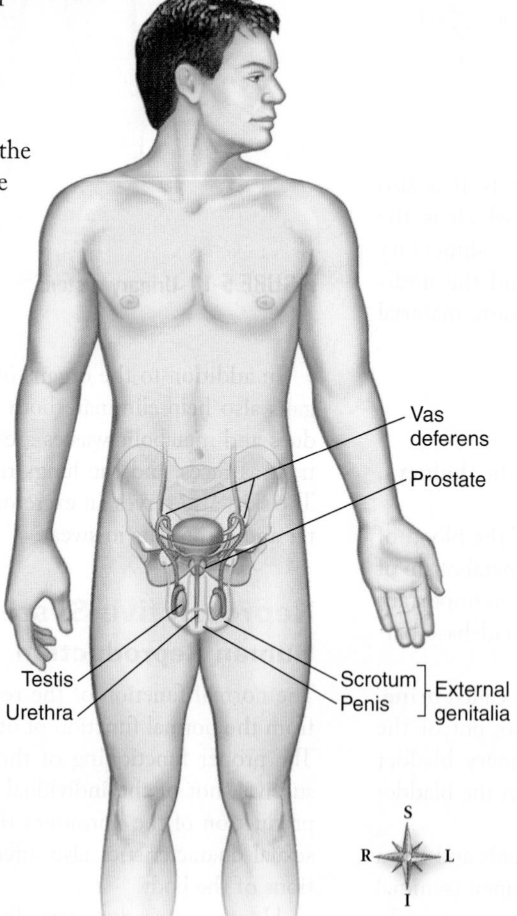

FIGURE 5-11 Male reproductive system.

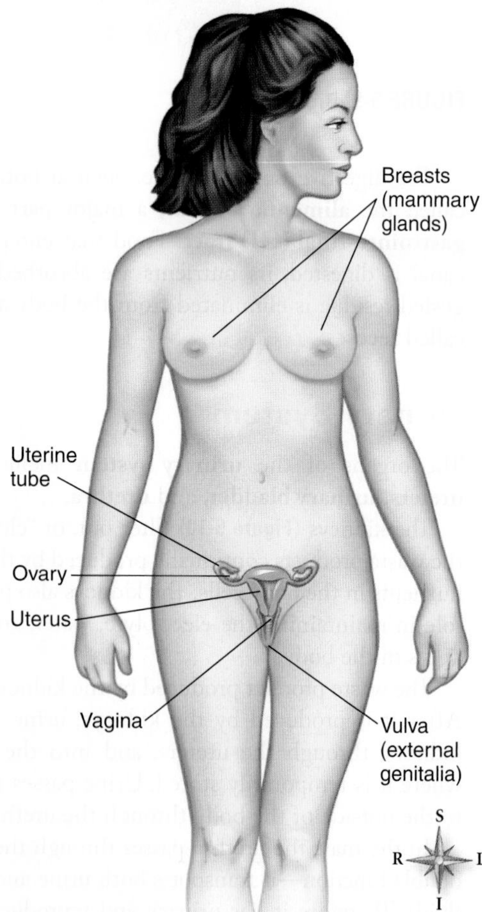

FIGURE 5-12 Female reproductive system.

5

HEALTH AND WELL-BEING

CANCER SCREENING

Knowledge of the structure and function of the organ systems is a critically important "first step" in understanding and using information that empowers us to become more sophisticated guardians of our own health and well-being.

For example, a better understanding of the reproductive system helps individuals participate in a more direct and personal way in cancer screening. Breast and testicular self-examinations to detect cancer are two important ways that women and men can participate directly in protecting their own health.

Likewise, awareness of normal nervous system functions can alert a person of a possible brain tumor when experiencing unexplained headaches or changes in vision or other senses. Unusual immune system functions, such as night sweats, also can alert physicians to the possibility of cancer.

Information on cancer screening is available from the American Cancer Society and from most hospitals, clinics, and health care providers.

Body as a Whole

As you study the structure and function of the various organ systems in the chapters that follow, it is important that you focus on how each system and its component organs relate to other systems and to the body as a whole.

Homeostasis

Table 5-1 lists major organs of each system and identifies the function of each system in the context of *homeostasis*. The concept of homeostasis, introduced in Chapter 1, explains how the body maintains or is able to restore relative constancy to its internal environment even when faced with changing external surroundings or internal needs.

For example, muscle contraction can produce a specific body movement only if it is attached appropriately to a bone in the skeletal system. In order for contraction to begin, muscles must first be stimulated by nerve impulses generated in the nervous system. Then, in order to continue contracting, muscles must receive both oxygen from the respiratory system and nutrients absorbed from the digestive system. Numerous wastes produced by contracting muscles must be eliminated by the urinary and respiratory systems. The cardiovascular system provides transportation for the respiratory gases, nutrients, and waste products of metabolism.

No one body system functions entirely independently of other systems. Instead, you will find that they are structurally and functionally interrelated and interdependent. Homeostasis can be maintained only by the coordinated and carefully regulated functioning of all body organ systems.

Applying Organ System Concepts

Notice that **Table 5-1** groups the major organ systems into larger systems or divides some of them into smaller systems. We group or split the major organ systems when it makes a particular situation easier to understand—or in certain clinical applications where it helps clarify professional communication.

For example, physical therapists sometimes find it most useful to use the concept of a *skeletomuscular system* rather than thinking of the skeletal and muscular systems separately. The nervous system is so complex that often it is easier to understand if it is split into *central* and *peripheral* nervous systems.

Notice also in the table that some organs belong to more than one system. For example, the hypothalamus is part of the brain and therefore is in the nervous system, but it also secretes hormones, so it is in the endocrine system as well.

Organ Replacement

Vital and Nonvital Organs

As we all know, disease and injury are sometimes unavoidable. Therefore it is common to suffer damage that renders an organ incapable of proper function.

By definition, a nonvital organ is not required for life to continue—a vital organ is. If a *nonvital organ* is damaged, a person's health may be in some peril, but even permanent loss of that organ will not result in death. For example, we can survive easily without the use of our spleen, appendix, and tonsils. We can also survive, although less easily, without the use of our eyes, arms, and legs. However, if the functions of a *vital organ* are lost, we are in immediate danger of dying. For example, when the heart or brain ceases to function, death will result.

Over the past few decades, health science professionals have made great advances in the ability to replace or repair lost or damaged organs. In the case of nonvital organs, these techniques have improved the quality of life for many patients. In the case of vital organs, these techniques have saved and extended life.

Artificial Organs

Nonvital Organ Replacement

A nonvital organ is often successfully replaced or enhanced by an artificial organ or **prosthesis.** The term *prosthesis,* in the broadest sense, refers to any device used to replace a body part or improve its function. **Figure 5-13** shows examples of many types of prostheses now in use.

TABLE 5-1 Organ Systems of the Body

GROUPED SYSTEM*	MAJOR BODY SYSTEMS (by chapter)†	GENERAL FUNCTIONS	SPLIT SYSTEM*	PRINCIPAL ORGANS	
Skeletomuscular (musculo-skeletal)	Integumentary (7)	Separates internal environment from external environment		Skin (includes hair, nails, glands)	
	Skeletal (8)	Supports, protects, and moves body Stores minerals	Axial	Bones of skull, spine, thorax (many)	Ligaments and joints of skull, spine, thorax (many)
			Appendicular	Bones of extremities (many)	Ligaments and joints of extremities (many)
	Muscular (9)	Powers and directs skeletal movement Stabilizes the skeleton to maintain posture Generates heat		Muscles (many)	
Neuroendocrine	Nervous (10, 11)	Major regulatory system of the internal environment Senses changes, integrates information, and sends signals to effectors (muscular organs, glands)	Central	Brain	Spinal cord
			Peripheral‡	Cranial nerves (and branches) Peripheral nerves (and branches)	Sense organs (many)
	Endocrine (12)	Regulates internal environment by secreting hormones that travel through bloodstream to target areas		Pituitary gland Pineal gland Hypothalamus Thyroid gland Adrenal glands	Pancreatic islets Ovaries Testes Other glands
Circulatory	Cardiovascular (13-15)	Transports nutrients, water, oxygen, hormones, wastes, and other materials within the internal environment		Heart Arteries (many)	Veins (many) Capillaries (many)
	Lymphatic/immune (16)	Drains excess fluid from tissues, cleans it, and returns it to the blood Defends internal environment from injury by abnormal cells, foreign particles, and other irritants	Lymphatic§	Lymph nodes Lymph vessels Thymus	Spleen Tonsils
			Immune	Lymph nodes	All other lymphoid organs
	Respiratory (17)	Exchanges O_2 and CO_2 between the internal and external environment		Nose Pharynx Larynx	Trachea Bronchi Lungs
	Digestive (18, 19)	Breaks apart nutrients from the external environment and absorbs them into the internal environment		PRIMARY Mouth Pharynx Esophagus Stomach Small intestine Large intestine Rectum Anal canal	ACCESSORY Teeth Salivary glands Tongue Liver Gallbladder Pancreas Appendix

*Some systems are grouped or split into other systems when needed; a few examples are given here.
†Numerals in parentheses after each system name refer to the chapter numbers where that system is discussed.
‡The nervous system often is split in other ways, such as sensory/motor or somatic/autonomic.
§The lymphatic system includes both lymphoid organs and an extensive network of lymph vessels, whereas the immune system includes only lymphoid organs with defensive functions.

TABLE 5-1 Organ Systems of the Body—cont'd

GROUPED SYSTEM*	MAJOR BODY SYSTEMS (by chapter)†	GENERAL FUNCTIONS	SPLIT SYSTEM*	PRINCIPAL ORGANS	
Urogenital	Urinary (20-22)	Adjusts internal environment by excreting excess water, salt, wastes, acids, and other substances		Kidneys Ureters	Urinary bladder Urethra
	Reproductive (23-25)	Produces sex cells that form offspring, ensuring survival of genes Female system is also site of fertilization and early offspring development	Male	Testes (gonads) Vas deferens Urethra	Prostate Penis Scrotum
			Female	Ovaries (gonads) Uterus Uterine (fallopian) tubes	Vagina Vulva Mammary glands (breasts)

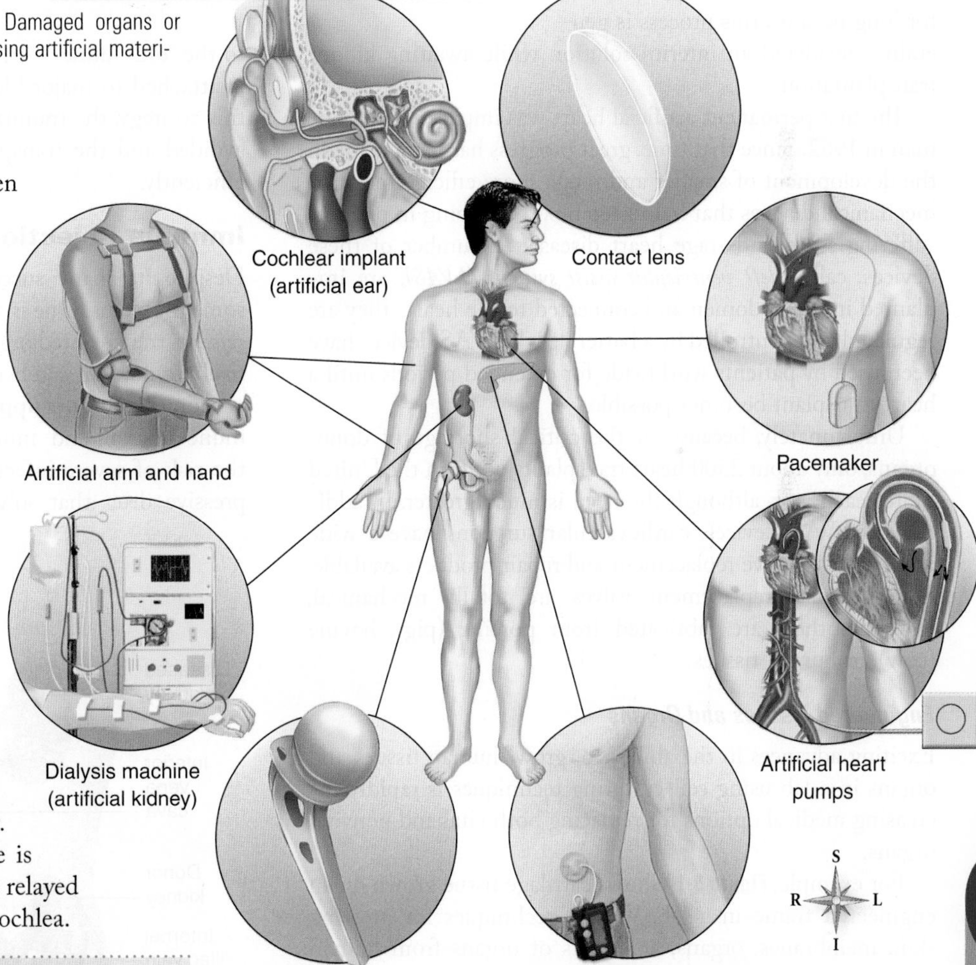

FIGURE 5-13 Examples of prostheses. Damaged organs or tissues often can be replaced or repaired by using artificial materials or devices.

Crude artificial limbs have been used for centuries, but the availability of new materials and advanced engineering have made more efficient types possible. For example, new computer-assisted arm and hand replacements can manipulate small objects with amazing dexterity. Artificial sense organs have even been able to restore sight to the blind and hearing to the deaf.

For example, many people suffering from deafness have had their hearing partially restored by "artificial ears" called **cochlear implants.** In cochlear implants, a miniature microphone is surgically implanted under the skin near the outer ear and wired to an electrode in the inner ear, or *cochlea.* Sound picked up by the microphone is converted to electrical signals that are relayed directly to the auditory nerve in the cochlea.

If you would like to see diagrams that show how cochlear implants work, check out the article *Cochlear Implants* at Connect It! at *evolve.elsevier.com.*

Vital Organ Replacement

The use of artificial materials—transplanted animal or human tissues or mechanical devices—to partially or completely replace vital organs or to augment their functions is occurring more and more frequently in modern medical practice.

Medical Machines

One of the earliest devices to augment vital functions was the "artificial kidney," or *dialysis machine* (see Chapter 20).

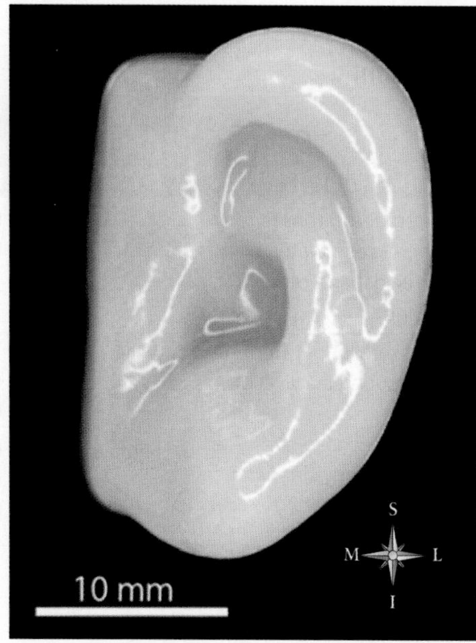

FIGURE 5-14 Tissue-engineered human ear cartilage. Photo showing cartilage tissue grown from human cells on an engineered frame in a lab.

Kidney dialysis machines pump blood through permeable tubes in an external apparatus, allowing waste products to diffuse out of the blood and into a salt-water type of electrolyte fluid that surrounds the semipermeable dialysis tubes. Patients with kidney failure must be "hooked up" to the machine, generally for 2- to 4-hour periods 2 or 3 times each week. Although ongoing machine-based dialysis treatments can extend the lives of kidney failure patients for long periods, this process is generally considered an interim solution while awaiting kidney transplantation.

The first permanent artificial heart was implanted in a human in 1982. Since that time, great progress has been made in the development of smaller and much more efficient electromechanical devices that help keep blood pumping in patients suffering from end-stage heart disease. A number of these devices, called *left ventricular assist systems (LVAS),* are implanted in the abdomen and connected to the heart. They are regulated and controlled by a battery pack. LVAS devices have been used by patients worldwide for extended periods until a heart transplant becomes possible.

Unfortunately, because of the critical shortage of donor organs, only about 2500 heart transplants occur in the United States each year, although the need is much greater. In addition to LVAS devices, cardiovascular surgeons have a wide array of heart valve replacement and repair products available. Some cardiac replacement valves are totally mechanical, whereas others are fabricated from porcine (pig), bovine (cow), or human tissues.

Engineered Tissues and Organs

Exciting advances in the ability to "grow" human tissues and organs in a lab using cell-culturing techniques is rapidly increasing medical options for replacing both vital and nonvital organs.

For example, **Figure 5-14** shows cartilage tissue grown on an engineered frame in a lab. Various techniques for growing skin, membranes, organs, and parts of organs from human cells have been developed. When grown from a patient's own cells, these engineered tissues and organs have a better chance of success than structures received from donors.

Although new materials and bioengineering advances are encouraging, many challenges remain for successful, long-term transplantation of many organs. Total artificial replacement for *vital organs,* if possible at all, is generally employed only to ensure survival until a more permanent solution (most often organ transplantation) can occur.

Organ Transplantation

Surgical Transplants

One approach that offers the hope of a permanent solution to loss of vital organ function is *organ transplantation.* In this technique, a normal living organ from a donor is surgically transplanted into the recipient. Kidney, liver, pancreas, lung, small intestine, and heart transplants are now done at many hospitals throughout the world.

When a new organ is transplanted into the body, the old organ may or may not be removed. For example, failed kidneys are often left in place at the posterior of the ventral body cavity. As **Figure 5-15** shows, the "new" kidney is nestled inferior to the "old" kidney—in the curve of the pelvic bone, where it is attached to major blood vessels and to the bladder. Using this strategy, the trauma of removing the damaged kidneys is avoided and the transplanted kidney can still process blood efficiently.

Immune Rejection of Transplants

Despite its many successes, organ transplantation still has some problems. One is that a recipient's immune system often *rejects* a transplanted organ. The immune response may be relatively minor, but can become life-threatening in some cases.

Some **immunosuppressive drugs** that suppress the immune system and inhibit rejection reactions also increase the risk of severe infection. Cyclosporine is an immunosuppressive drug that solves this problem to some degree by

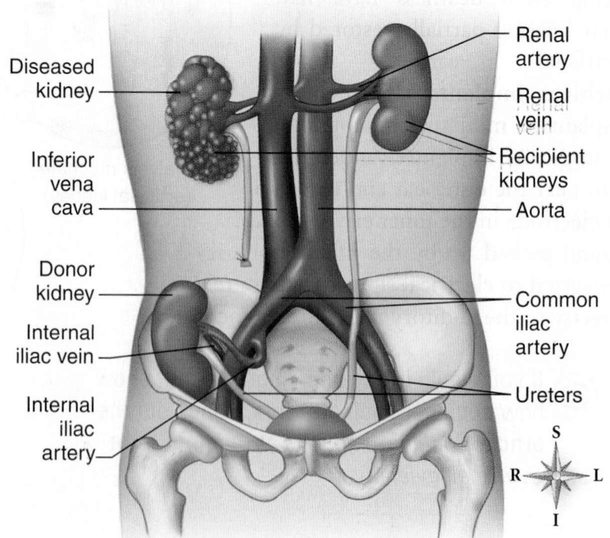

FIGURE 5-15 Kidney transplantation. In kidney transplantations, the diseased organs are left in place, and the donated organ is nestled in another part of the body.

5

suppressing rejection reactions without severely inhibiting infection control.

In addition, better tissue-typing procedures, continued development of new and more effective antirejection drugs, and the possibility of using a patient's own cells to build an engineered organ, also offer hope of reducing organ-rejection problems.

Another way to solve the rejection problem is to build "new" organs from a patient's own tissues. For example, in a method called *free-flap surgery,* pieces of tissue from one part of the body are surgically remodeled and then grafted to a new part of the body.

After cancerous breasts are removed, "new" breasts can be formed from skin and muscle tissue taken from the thighs, buttocks, or abdomen. Parts of the intestine can be used to repair the urinary bladder. Toes can even be transplanted to the hand to replace missing fingers. The advantage of using a patient's own tissues is that the possibility of rejection is greatly reduced.

Another major problem with organ transplants is the limited availability of donor organs. One solution is to eliminate or reduce the need for human donors. Researchers are now working on a variety of methods by which new organs or tissues can be "grown" in a tissue culture or in a patient's body. For example, it is hoped that one day, normal liver cells can be safely removed from a healthy donor and implanted in a plastic sponge that will be placed in the recipient's body. The transplanted cells may then reproduce and form a mass capable of performing some liver function. Researchers are also culturing colonies of healthy nervous tissue in laboratory dishes in the hope that it can someday be used to repair damaged sections of the brain or spinal cord.

QUICK CHECK

1. How is a nonvital organ often replaced?
2. What are two examples of medical machines that are used when organ transplantation is not available or an option?
3. What is the greatest concern regarding organ transplantation? What are some possible options to avoid this problem?
4. What is free-flap surgery?

SCIENCE APPLICATIONS

RADIOGRAPHY

Wilhelm Röntgen (1845–1923)

In 1895, the German physicist Wilhelm Röntgen (RENT-gun) made one of the most important medical discoveries of the modern age: radiographic imaging of the body.

Radiography, or x-ray photography, earned Röntgen a Nobel Prize and is the oldest and most widely used method of noninvasive imaging of internal body. While studying the effects of electricity passing through gas under low pressures, Röntgen accidentally discovered x-rays when they caused a plate coated with special chemicals to glow. Not long after that, he showed that they could produce shadows of internal organs such as bones on photographic film. His first, and most famous, radiograph was of his wife Bertha's hand. Although a little fuzzy, it clearly showed Bertha's finger bones and the outline of her ring. When this radiograph was published by a Vienna newspaper, the entire world became instantly aware of his breakthrough discovery.

The figure at the right shows how radiography works. A source of waves in the x band of the radiation spectrum beams the x-rays through a body and to a piece of photographic film or phosphorescent screen. The resulting image shows the outlines of bones and other dense structures that absorb the x-rays. As the figure shows, one way to make soft, hollow structures such as digestive organs more visible is to use radiopaque contrast material. For example, barium sulfate (which absorbs x-rays) can be introduced into the colon to make it more visible in a radiograph.

Today, many variations of Röntgen's invention are used to study internal organs without having to cut into the body. For example, computed tomography (CT) scanning is a modern, computerized type of x-ray photography. **Radiological technologists** are health professionals whose chief responsibility is to make radiographs, and **radiologists** are responsible for interpreting these images. Many *medical, veterinary*, and *dental professionals* rely on these images and interpretations in their diagnosis, assessment, and treatment of patients. In addition, radiography is used in many *industrial* and *investigative* settings—even by *archaeologists* studying mummies.

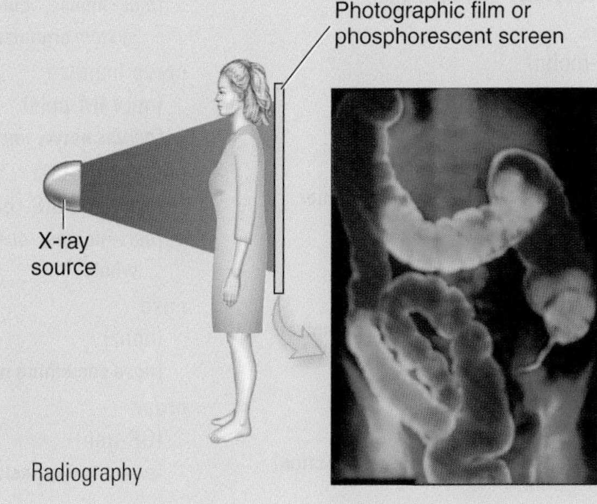

Radiography

5

LANGUAGE OF SCIENCE (continued from p. 93)

central nervous system (CNS)
(SEN-tral NER-vus SIS-tem [see en es])
[*centr-* **center,** *-al* **relating to,** *nerv-* **nerves,** *-ous* **relating to,** *system* **organized whole**]

circulatory system
(SER-kyoo-lah-tor-ee SIS-tem)
[*circulat-* **go around,** *-ory* **relating to,** *system* **organized whole**]

complement
(KOM-pleh-ment)
[*comple-* **complete,** *-ment* **result of action**]

diaphragm
(DYE-ah-fram)
[*dia-* **across,** *-phrag-* **enclose,** *-(u)m* **thing**]

digestive system
(dye-JES-tiv SIS-tem)
[*digest-* **break apart,** *-tive* **relating to**]

endocrine system
(EN-doh-krin SIS-tem)
[*endo-* **inward or within,** *-crin* **secrete,** *system* **organized whole**]

fallopian tube
(fal-LOH-pee-an toob)
[*Gabriele Fallopio* **Italian anatomist**]

feces
(FEE-seez)
[*feces* **waste**]

gastrointestinal tract (GI tract)
(gas-troh-in-TES-tih-nal trakt [jee aye trakt])
[*gastr-* **stomach,** *-intestin-* **intestine,** *-al* **relating to,** *tract* **trail**]

genitalia
(jen-ih-TAYL-yah)
sing., genital
(JEN-ih-tul)
[*gen-* **produce,** *-al* **relating to**]

gonad
(GOH-nad)
[*gon-* **offspring,** *-ad* **relating to**]

hormone
(HOR-mohn)
[*hormon* **excite**]

hypothalamus
(hye-poh-THAL-ah-mus)
[*hypo-* **under or below,** *-thalamus* **inner chamber**]

immune system
(ih-MYOON SIS-tem)
[*immun* **free (immunity),** *system* **organized whole**]

integument
(in-TEG-yoo-ment)
[*in-* **on,** *-teg-* **cover,** *-ment* **result of action**]

integumentary system
(in-teg-yoo-MEN-tar-ee SIS-tem)
[*in-* **on,** *-teg-* **cover,** *-ment-* **result of action,** *-ary* **relating to,** *system* **organized whole**]

involuntary (smooth) muscle
(in-VOL-un-tayr-ee MUS-el)
[*in-* **not,** *volunt-* **will,** *mus-* **mouse,** *-cle* **little**]

joint
(joynt)

kidney
(KID-nee)
[*kidney* **womb (shape)**]

larynx
(LAYR-inks)
[*larynx* **voice box**]

ligament
(LIG-ah-ment)
[*liga-* **bind,** *-ment* **result of action**]

lung
(lung)

lymph
(limf)
[*lymph* **water**]

lymph node
(limf nohd)
[*lymph* **water,** *nod* **knot**]

lymphatic system
(lim-FAT-ik SIS-tem)
[*lymph-* **water,** *-atic* **relating to,** *system* **organized whole**]

lymphatic vessel
(lim-FAT-ik VES-el)
[*lymph-* **water,** *-atic* **relating to,** *vessel* **container**]

mammary gland
(MAM-mah-ree gland)
[*mamma-* **breast,** *-ry* **relating to,** *gland* **acorn**]

muscular system
(MUS-kyoo-lar SIS-tem)
[*mus-* **mouse,** *-cul-* **little,** *-ar* **relating to,** *system* **organized whole**]

nerve impulse
(nerv IM-puls)
[*nervus* **nerve,** *impulse* **to drive**]

nervous system
(NER-vus SIS-tem)
[*nerv-* **nerves,** *-ous* **relating to,** *system* **organized whole**]

nose
(nohz)
[*nose* **something obvious**]

organ
(OR-gan)
[*organ* **tool or instrument**]

ovary
(OH-var-ee)
[*ov-* **egg,** *-ar-* **relating to,** *-y* **location of process**]

ovum
(OH-vum)
pl., ova
(OH-vah)
[*ovum* **egg**]

pancreas
(PAN-kree-as)
[*pan-* **all,** *-creas* **flesh**]

parathyroid gland
(payr-ah-THYE-royd gland)
[*para-* **beside,** *-thyr-* **shield,** *-oid* **like,** *gland* **acorn**]

penis
(PEE-nis)
[*penis* **male sex organ**]

peripheral nervous system (PNS)
(peh-RIF-er-al NER-vus SIS-tem [pee en es])
[*peri-* **around,** *-phera-* **boundary,** *-al* **relating to,** *nerv-* **nerves,** *-ous* **relating to,** *system* **organized whole**]

pharynx
(FAYR-inks)
[*pharynx* **throat**]

pineal gland
(PIN-ee-al gland)
[*pine-* **pine,** *-al* **relating to,** *gland* **acorn**]

pituitary gland
(pih-TOO-ih-tayr-ee gland)
[*pituit-* **phlegm,** *-ary* **relating to,** *gland* **acorn**]

prostate (gland)
(PROS-tayt gland)
[*pro-* **before,** *-stat-* **set or place,** *gland* **acorn**]

reproductive system
(ree-proh-DUK-tiv SIS-tem)
[*re-* **again,** *-produc-* **bring forth,** *-tive* **relating to,** *system* **organized whole**]

respiratory system
(RES-pih-rah-tor-ee SIS-tem)
[*re-* **again,** *-spir-* **breathe,** *-tory* **relating to,** *system* **organized whole**]

scrotum
(SKROH-tum)
[scrotum **bag**]

sense organ
(sens OR-gan)
[*organ* **tool or instrument**]

skeletal system
(SKEL-eh-tal SIS-tem)
[*skeleto-* **dried body,** *-al* **relating to,** *system* **organized whole**]

5

LANGUAGE OF **SCIENCE** *(continued from p. 106)*

smooth muscle
(smoothe MUS-el)
[*smooth* **smooth,** *mus-* **mouse,** *-cle* **small]**

sperm
(spurm)
[*sperm* **seed]**

spleen
(spleen)

stem cell
(stem sel)
[*stem* **tree trunk,** *cell* **storeroom]**

stimulus
(STIM-yoo-lus)
pl., stimuli
(STIM-yoo-lye)
[*stimulus* **incitement]**

system
(SIS-tem)
[*system* **organized whole]**

tendon
(TEN-don)
[*tend-* **pulled tight,** *-on* **unit]**

testis
(TES-tis)
pl., testes
(TES-teez)
[*testis* **witness (male gonad)]**

thoracic duct
(thoh-RAS-ik dukt)
[*thorac-* **chest (thorax),** *-ic* **relating to,** *duct* **to lead]**

thymus gland
(THY-mus gland)
[*thymus* **thyme flower,** *gland* **acorn]**

thyroid gland
(THY-royd gland)
[*thyro-* **shield,** *-oid* **like,** *gland* **acorn]**

tissue
(TISH-yoo)
[*tissue* **fabric]**

tonsil
(TAHN-sil)
[*tons-* **goiter,** *-il* **little]**

trachea
(TRAY-kee-ah)
[*trachea* **rough duct]**

ureter
(YOOR-eh-ter)
[*ure-* **urine,** *-ter* **agent or channel]**

urethra
(yoo-REE-thrah)
[*ure-* **urine,** *-thr-* **agent or channel]**

urinary bladder
(YOOR-ih-nayr-ee BLAD-er)
[*urin-* **urine,** *-ary* **relating to,** *bladder* **blister, pimple]**

urinary system
(YOOR-ih-nayr-ee SIS-tem)
[*urin-* **urine,** *-ary* **relating to,** *system* **organized whole]**

urine
(YOOR-in)
[*ur-* **urine,** *-ine* **chemical]**

uterine tube
(YOO-ter-in toob)
[*uter-* **womb,** *-ine* **relating to,** *tube* **pipe]**

uterus
(YOO-ter-us)
[*uterus* **womb]**

vagina
(vah-JYE-nah)
[*vagina* **sheath]**

vas deferens
(vas DEF-er-enz)
pl., vasa deferentia
(VAS-ah def-er-EN-chah)
[*vas* **duct or vessel,** *deferens* **carrying away]**

vein
(vayn)
[*vena* **blood vessel]**

voluntary (skeletal) muscle
(VOL-un-tayr-ee MUS-el)
[*volunt-* **will,** *mus-* **mouse,** *-cle* **little]**

vulva
(VUL-vah)
[*vulva* **wrapper]**

LANGUAGE OF **MEDICINE**

cochlear implant
(KOHK-lee-ar IM-plant)
[*cochlea-* **snail shell,** *-ar* **relating to,** *im-* **into,** *-plant* **plant]**

immunosuppressive drug
(ih-myoo-noh-soo-PRES-iv drug)
[*immuno-* **free (immunity),** *-suppress-* **press down,** *-ive* **relating to]**

kidney dialysis
(KID-nee dye-AL-ih-sis)
[*dia-* **apart,** *-lysis* **loosening]**

radiography
(ray-dee-OG-rah-fee)
[*radio-* **ray, frequency,** *-graphy* **drawing]**

radiological technologist
(ray-dee-oh-LOJ-ih-kul tek-NOL-oh-jist)
[*radi(at)-* **emit rays,** *-log-* **words (study of),** *-ic-* **relating to,** *-al* **relating to,** *techn-* **art or skill,** *-log-* **words (study of),** *-ist* **agent]**

radiologist
(ray-dee-AHL-oh-jist)
[*radi(at)-* **emit rays,** *-log-* **words (study of),** *-ist* **agent]**

5

❑ OUTLINE SUMMARY

*To download a digital version of the chapter summary for use with your device, access the **Audio Chapter Summaries** online at evolve.elsevier.com.*

Scan this summary after reading the chapter to help you reinforce the key concepts. Later, use the summary as a quick review before your class or before a test.

Definitions and Concepts

A. Organ—a structure made up of two or more kinds of tissues that can together perform a more complex function than a single tissue

B. Organ system—a group of organs that perform a more complex function than can any organ alone

C. Knowledge of individual organs and how they are organized into groups improves the understanding of how a particular organ system functions as a whole

Organ Systems of the Body

A. Integumentary system (**Figure 5-1**)
   1. Structure
      a. Only one organ, the skin, but has many appendages (attached structures)
      b. Skin appendages
         (1) Hair
         (2) Nails
         (3) Microscopic sense receptors
         (4) Sweat glands
         (5) Oil glands
   2. Functions
      a. Protection—primary function
      b. Regulation of body temperature
      c. Synthesis of chemicals
      d. Sense organ

B. Skeletal system (**Figure 5-2**)
   1. Structure
      a. Bones—organs of the skeletal system
         (1) 206 named bones in the skeleton
         (2) Additional variable bones occur in each individual
      b. Cartilage connects and cushions joined bones
      c. Ligaments—bands of fibrous tissue that hold bones together
      d. Joints—connections between bones that make movement possible
   2. Functions
      a. Supporting framework for entire body
      b. Protection of brain and internal organs
      c. Movement (with joints and muscles)
      d. Storage of minerals
      e. Formation of blood cells

C. Muscular system (**Figure 5-3**)
   1. Structure
      a. Muscles are the primary organs
         (1) Voluntary or striated skeletal muscle
         (2) Involuntary or smooth muscle tissue in walls of some organs
         (3) Cardiac muscle in wall of the heart
   2. Functions
      a. Movement
      b. Maintenance of body posture
      c. Production of heat
   3. Skeletomuscular system—combination of the skeletal and muscular systems; also called *musculoskeletal system*

D. Nervous system (**Figure 5-4**)
   1. Structure
      a. Central nervous system (CNS)
         (1) Brain
         (2) Spinal cord
      b. Peripheral nervous system (PNS)
         (1) Cranial nerves and their branches
         (2) Spinal nerves and their branches
         (3) Sense organs
   2. Functions
      a. Communication between body organs
      b. Integration of body functions
      c. Control of body functions
      d. Recognition of sensory stimuli

E. Endocrine system (**Figure 5-5**)
   1. Structure—ductless glands that secrete signaling hormones directly into the blood
   2. Functions
      a. Same as nervous system—communication, integration, control
      b. Control is slow and of long duration
      c. Neuroendocrine system—combination of nervous and endocrine systems
      d. Examples of functions regulated by hormones
         (1) Growth
         (2) Metabolism
         (3) Reproduction
         (4) Fluid and electrolyte balance

F. Cardiovascular system (also called *circulatory system*) (**Figure 5-6**)
   1. Structure
      a. Heart
      b. Blood vessels
   2. Functions
      a. Transportation of substances throughout the body
      b. Regulation of body temperature
      c. Immunity (body defense)

5

G. Lymphatic and immune systems (**Figure 5-7**)
  1. Lymphatic system
    a. Structure
      (1) Lymphatic vessels
      (2) Lymph nodes and tonsils
      (3) Thymus
      (4) Spleen
    b. Functions
      (1) Transportation of lymph
      (2) Immunity
  2. Immune system
    a. Structure
      (1) Unique cells
        (a) Phagocytes
        (b) Secretory cells
      (2) Defensive protein compounds
        (a) Antibodies
        (b) Complements
    b. Functions
      (1) Phagocytosis of bacteria
      (2) Chemical reactions that provide protection from harmful agents

H. Respiratory system (**Figure 5-8**)
  1. Structure
    a. Nose
    b. Pharynx
    c. Larynx
    d. Trachea
    e. Bronchi
    f. Lungs
  2. Functions
    a. Exchange of waste gas (carbon dioxide or CO_2) for oxygen (O_2) in the alveoli of the lungs
    b. Filtration of irritants from inspired air
    c. Regulation of acid-base balance

I. Digestive system (**Figure 5-9**)
  1. Structure
    a. Primary organs
      (1) Form a tube called the alimentary canal, or GI tract
      (2) Include mouth, pharynx, esophagus, stomach, small intestine, large intestine, rectum, anal canal
    b. Accessory organs
      (1) Assist the digestive process
      (2) Include teeth, salivary glands, tongue, liver, gallbladder, pancreas, appendix
  2. Functions
    a. Mechanical and chemical breakdown (digestion) of nutrients
    b. Absorption of nutrients
    c. Elimination of undigested waste product—referred to as *feces*

J. Urinary system (**Figure 5-10**)
  1. Structure
    a. Kidneys
    b. Ureters
    c. Urinary bladder
    d. Urethra (part of both urinary and reproductive systems in males)
  2. Functions
    a. "Clearing," or cleaning, blood of waste products—excreted from the body as *urine*
    b. Electrolyte balance
    c. Water balance
    d. Acid-base balance

K. Reproductive systems
  1. Structure
    a. Male (**Figure 5-11**)
      (1) Gonads—testes
      (2) Other structures—vas deferens, urethra, prostate, external genitalia (penis and scrotum)
    b. Female (**Figure 5-12**)
      (1) Gonads—ovaries
      (2) Other structures—uterus, uterine (fallopian) tubes, vagina, external genitalia (vulva), mammary glands (breasts)
  2. Functions
    a. Survival of genes
    b. Production of sex cells (male: sperm; female: ova)
    c. Transfer and fertilization of sex cells
    d. Development and birth of offspring
    e. Nourishment of offspring
    f. Production of sex hormones

Body as a Whole

A. Homeostasis
  1. No one body system functions entirely independently of other systems (**Table 5-1**)
  2. All body systems are structurally and functionally interrelated and interdependent

B. Applying organ system concepts
  1. Systems can be grouped or split for better understanding (**Table 5-1**)
    a. Skeletomuscular system made up of both skeletal and muscular systems
    b. Nervous system can be split into central and peripheral nervous systems
  2. Some organs, such as the hypothalamus, can be in more than one system

Organ Replacement

A. Vital and nonvital organs
  1. Loss of function in nonvital organs is not immediately life-threatening
  2. Loss of function in vital organs *is* immediately life-threatening
  3. Loss of function in organs can be treated by organ replacement

B. Artificial organs (**Figure 5-13**)
1. Nonvital organ replacement
a. Prosthesis—any device that replaces a body part or improves its function
b. Example—cochlear implant to improve hearing
2. Vital organ replacement
a. Medical machines—for example, kidney dialysis (artificial kidney) machine, artificial heart pumps
b. Engineered tissues and organs—for example, trachea (windpipe) grown from cultured human cells (**Figure 5-14**)

C. Organ transplantation
1. Surgical transplants—organs from donors are surgically transferred to a recipient; sometimes the original is left in place (**Figure 5-15**)
2. Immune rejection of transplants is sometimes treated with immunosuppressive drugs; new organs engineered with compatible cells may avoid rejection

❑ ACTIVE LEARNING

STUDY TIPS

Consider using these tips to achieve success in meeting your learning goals.

Chapter 5 is the "big picture" chapter. It is a preview of the systems discussed in the remaining chapters.

1. Put the names of the systems on one side of a flash card and the functions of the systems and their organs on the other side. Notice how each organ contributes to the functioning of the system. Use **Table 5-1** as a resource.
2. Review the various types of artificial organs, transplants, and some of the problems in transplantation.
3. Before you begin the chapter dealing with a particular system, it would be helpful to get an overview of that system by reviewing the synopsis of that system in this chapter. That will give you a quick look at its major functions and the organs in that system.
4. In your study group, review the body system flash cards you have made. Discuss how several systems need to be involved in accomplishing one function in the body, such as getting nutrients or oxygen to the cells. Go over the questions at the back of the chapter and discuss possible test questions.
5. Consider starting some *running concept lists* for each of the systems and organs that you will encounter in this course. Then, each time you learn something new about each one, you can add your new knowledge to the appropriate concept list. See *my-ap.us/JlLFb6* to learn more about how to use *running concept lists.*
6. Make use of the many online resources that provide an overview of the body's systems. Examples include: *my-ap.us/JmMkpi*, *my-ap.us/K9GtVc*, and *my-ap.us/Lzv45j*.

Review Questions

Write out the answers to these questions after reading the chapter and reviewing the Chapter Summary. If you simply think through the answer without writing it down, you will not retain much of your new learning.

1. Define *organ* and *organ system*.
2. What is the function of skin sense receptors?
3. How is the skin able to assist in the body's ability to regulate temperature?
4. What is the costal cartilage?
5. What is a tendon and describe what function it serves.
6. What structure is part of the skeletal system, but is often considered to also be a lymphoid structure?
7. Name the organs that help rid the body of waste. What type of waste does each organ remove?
8. With the exception of bone, what other types of tissue are included in the skeletal system, and what function do they serve for the body?
9. List the eleven organ systems discussed in this chapter.
10. Most of the organ systems have more than one function. List two functions for the following systems: integumentary, skeletal, muscular, lymphatic and immune, respiratory, and urinary.
11. What is unique about the reproductive system?
12. Name three artificial organs or prostheses. What organs do they replace or assist?
13. What is the role of drugs such as cyclosporine in organ transplantation?

Critical Thinking

After finishing the Review Questions, write out the answers to these more in-depth questions to help you apply your new knowledge. Go back to sections of the chapter that relate to concepts that you find difficult.

14. Explain the difference between the nervous and endocrine systems. Include what types of functions are regulated and the "message carriers" for each system.
15. The term *balance* is used throughout this chapter. This is another term for *homeostasis*. Review the functions of the body systems and list the homeostatic functions of each.
16. What explanation would you give to a friend who has recently had a kidney transplantation but can't believe it when a nurse told him that his old kidneys were left in place?
17. Explain how the use of stem cells will have a profound impact on human health.

Chapter Test

After studying the chapter, test your mastery by responding to these items. Try to answer them without looking up the answers.

1. The primary organs of the digestive system make up a long tube called the ________.
2. ________ is another name for voluntary muscle.
3. ________ is another name for involuntary muscle.
4. The nervous system can generate special electrochemical signals called ________.
5. The ________, ________, and ________ are called accessory structures of the skin.
6. The ________ gland is part of both the lymphatic and endocrine systems.
7. The ________ is part of both the male reproductive system and urinary system.
8. The gonads for the male reproductive system are the ________; the gonads for the female reproductive system are the ________.
9. The skeletal system is composed of bone and what two related tissues: ________ and ________.
10. A ________ is an "artificial ear" used to improve hearing.
11. ________ are undifferentiated cells taken from embryonic tissue or cord blood and can be used in "organ engineering."

Match each system in column A with its corresponding function in column B.

Column A	Column B
12. ________ integumentary	a. Provides movement, body posture, and heat
13. ________ skeletal	b. Uses hormones to regulate body function
14. ________ muscular	c. Transports fatty nutrients from the digestive system into the blood
15. ________ nervous	d. Causes physical and chemical changes in nutrients so they can be absorbed into the blood
16. ________ endocrine	e. Cleans the blood of metabolic waste and regulates water and electrolyte balance
17. ________ cardiovascular	f. Protects underlying structures, has sensory receptors, and regulates body temperature
18. ________ lymphatic	g. Responsible for the transport of substances from one part of the body to another
19. ________ respiratory	h. Ensures the survival of the species rather than the individual
20. ________ digestive	i. Uses electrochemical signals to integrate and control body functions
21. ________ urinary	j. Exchanges oxygen and carbon dioxide and helps regulate acid-base balance
22. ________ reproductive	k. Provides a rigid framework for the body and stores minerals

Case Studies

To solve a case study, you may have to refer to the glossary or index, other chapters in this textbook, and other resources.

1. Tommy has been diagnosed as having irreversible kidney failure. Which system of the body is involved in this condition? What functions has Tommy lost? What options do his physicians have in treating Tommy's condition?
2. Mr. Davidson was referred to a urologist for diagnosis and treatment of an obstruction in his urethra. What bodily functions may be affected by Mr. Davidson's conditions?
3. Tom complained of pain in his abdomen one night and decided to go to the emergency department. His blood work was negative and a flat screen x-ray of his abdomen revealed nothing abnormal. The doctor, however, still suggested that Tom be admitted to the hospital. He advised Tom that he would be ordering some additional tests and x-rays. What are some x-ray options that the doctor might order, and what are the individual advantages of these x-rays?

Answers to Active Learning Questions can be found online at evolve.elsevier.com.

Mechanisms of Disease

OUTLINE

Scan this outline before you begin to read the chapter, as a preview of how the concepts are organized.

OBJECTIVES

Before reading the chapter, review these goals for your learning.

After you have completed this chapter, you should be able to:

1. Explain the study of disease, including disease terminology and patterns of disease.
2. List and describe the basic mechanisms of disease and risk factors associated with disease.
3. List and describe six categories of pathogenic organisms and explain how they cause disease.
4. List and describe the ways pathogens can be spread, as well as prevention and control measures.
5. Do the following related to tumors and cancer:
   - Distinguish between the terms *benign* and *malignant.*
   - Describe the pathogenesis of cancer.
6. Outline the events of the inflammatory response and explain its role in disease.

CHAPTER 6

The title of this book uses the words *health* and *disease.* We use these words all the time, but what do they really mean? In scientific study, health is defined as physical, mental, and social well-being—not merely the absence of disease. Disease can be described literally as "lack of ease," or a physiological disturbance that threatens well-being. A named disease is a specific set of structural or functional abnormalities, as defined by characteristic signs and symptoms. In this chapter we explore these basic ideas about disease and how disease disrupts normal function.

Studying Disease

Disease Terminology

Signs, Symptoms, and Disease

Pathology is the study of disease. Researchers want to know the scientific basis of abnormal conditions. Health practitioners want to know how to prevent and treat a wide variety of diseases. When we suffer from the inevitable "cold" or something more serious, we all want to know what is going on and how best to deal with it.

Pathology has its own terminology, as in any specialized field. Most of these terms are derived from Latin and Greek word parts.

LANGUAGE OF **SCIENCE**

Hint Before reading the chapter, say each of these terms out loud. This will help you to avoid stumbling over them as you read.

age
(ayj)

ameba
(ah-MEE-bah)
pl., amebas or amebae
(ah-MEE-bahz or ah-MEE-bee)
[*amoeba* **change**]

archaea
(ARK-ee-ah)
sing., archaeon
(ARK-ee-ahn)
[*archae* **ancient**]

arthropod
(AR-throh-pod)
[*arthro-* **jointed,** *-pod* **foot**]

autoimmunity
(aw-toh-ih-MYOO-nih-tee)
[*auto-* **self,** *-immun-* **free,** *-ity* **state**]

bacillus
(bah-SIL-us)
pl., bacilli
(bah-SIL-aye)
[*bac-* **staff,** *-ill-* **small,** *-us* **thing**]

bacterium
(bak-TEER-ee-um)
pl., bak-TEER-ee-ah
[*bacter-* **rod,** *-ium* **thing**]

chemotaxis
(kee-moh-TAK-sis)
[*chemo-* **chemical,** *-taxis* **movement or reaction**]

ciliate
(SIL-ee-at)
[*cili-* **eyelid,** *-ate* **of or like**]

coccus
(KOK-kus)
pl., cocci
(KOK-sye or KOK-see)
[*coccus* **grain, berry**]

Continued on p. 136

For example, *patho-* comes from the Greek word for "disease" *(pathos)* and is used to form many terms, including *pathology* itself. If you are unfamiliar with word parts commonly used in medical science, refer to Appendix B at *evolve.elsevier.com*.

Many disease-causing organisms are known by their scientific names, which are Latin names that are often italicized to show that they are non-English terms.

Disease conditions are usually *diagnosed* or identified by signs and symptoms. **Signs** are objective abnormalities that can be seen or measured by someone other than the patient, whereas **symptoms** are the subjective abnormalities felt only by the patient.

Although *sign* and *symptom* are distinct terms, we often use them interchangeably. A **syndrome** is a collection of different signs and symptoms, usually with a common cause that presents a distinct picture of a pathological condition. The condition or syndrome, as defined by a characteristic set of signs and symptoms, is what we commonly refer to as a **disease.**

Disease Progression

When signs and symptoms appear suddenly, persist for a short time, then disappear, we say that the disease is **acute.** On the other hand, diseases that develop slowly and last for a long time (perhaps for life) are labeled **chronic** diseases. The term *subacute* refers to diseases with characteristics somewhere between acute and chronic.

The study of all factors involved in causing a disease is referred to as **etiology.** The etiology (causes or origin) of a skin infection often involves a cut or abrasion and subsequent invasion and growth of a bacterial population. Diseases with undetermined causes are said to be **idiopathic. Communicable,** or **infectious,** diseases can be transmitted from one individual to another.

The term *etiology* refers to the theory of a disease's cause, but the actual mechanism of a disease's development is called its **pathogenesis.** The common cold, for example, begins with a *latent,* or "hidden," stage during which the cold virus establishes itself in the patient. No signs of the cold are yet evident at this stage. In infectious diseases, the latent stage is also called **incubation.** After incubating, the cold may then manifest itself as a mild nasal drip, triggering a few sneezes. It then progresses to its full fury and continues for a few days. After the cold has run its course, *convalescence,* or recovery, occurs. During this stage, body functions return to normal.

Those who develop a chronic disease such as cancer may exhibit a temporary reversal of signs and symptoms that seems to be a recovery. Such reversal of a chronic disease is called a **remission.** If a remission is complete and permanent, we say that the person is "cured."

Patterns of Disease

Introduction to Epidemiology

Epidemiology is the study of the occurrence, distribution, and transmission of diseases in humans. Epidemiologists are physicians or medical scientists who study patterns of disease occurrence in specific groups of people. For example, a hospital may employ a staff epidemiologist who is responsible for infection-control programs within the hospital. Many governments and other agencies employ epidemiologists who track the spread of disease through a local community or even the world at large.

A disease that is native to a local region is called an **endemic** disease. If the disease spreads to many individuals at the same time within a defined geographic region, the situation is called an **epidemic.**

Pandemics are epidemics that spread throughout the world. HIV (human immunodeficiency virus) is now considered a pandemic because it is found worldwide. Because of the speed and availability of modern air travel, pandemics are more common than they once were. Almost every flu season, we see a new strain of influenza virus quickly spreading from continent to continent.

RESEARCH, ISSUES, AND TRENDS

CENTERS FOR DISEASE CONTROL AND PREVENTION

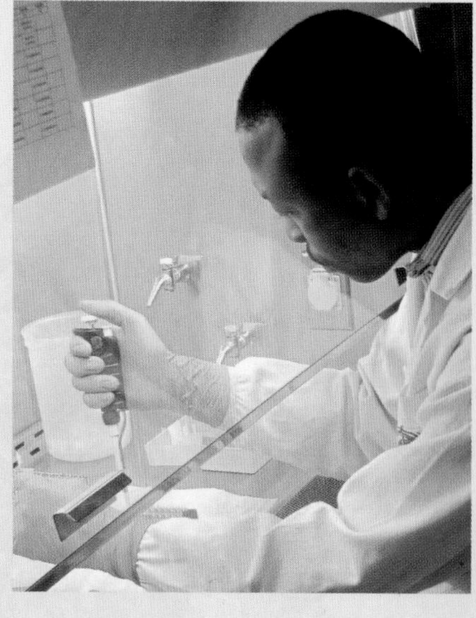

Epidemiology is a major concern of the *Centers for Disease Control and Prevention (CDC)*. Scientists and health professionals at CDC headquarters in Atlanta, Georgia, and around the world continually track the incidence and spread of disease in this country and worldwide.

Much of the CDC's tracking information is published in the *Morbidity and Mortality Weekly Report (MMWR)*. Available to physicians and other health professionals, this report provides recent information on disease rates in specific populations (**morbidity**) and the numbers of deaths caused by specific diseases (**mortality**).

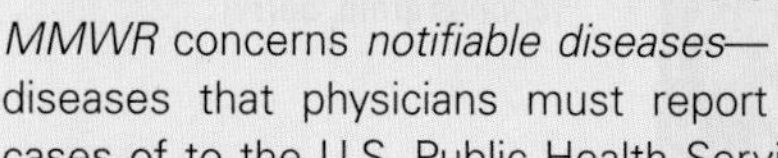

Much of the information in the *MMWR* concerns *notifiable diseases*—diseases that physicians must report cases of to the U.S. Public Health Service. Gonorrhea, measles, HIV, Zika virus, Lyme disease, anthrax, tuberculosis, and tetanus are examples of notifiable diseases.

Tracking Disease

Tracking the cause of a disease and its pattern of spread through a population can be very difficult. One reason is that there are so many different factors involved in the spread of disease. Nutrition, age, gender, sanitation practices, and socioeconomic conditions may play a role in the spread of disease.

Infectious agents, for example, can spread quickly and easily through an unsanitary water supply. Likewise, accumulation of untreated sewage or garbage can harbor disease-causing organisms or chemicals. Infectious agents or other contaminants in food also may spread disease to a large number of people. Crowded conditions may often play a role in spreading disease because more people come in close contact with one another. In crowded regions with poor sanitation and food-handling practices, disease may spread quickly.

The pattern of a disease's spread may be difficult to explain because of the different kinds of agents that can cause disease. For example, imagine that the majority of students in your class became ill with headaches and nausea (upset stomach) at about the same time. One would have to investigate many possible causes and modes of transmission before an explanation could be offered. Is it food poisoning? Is it an outbreak of the "flu" or another virus? Is the water supply for the drinking fountain contaminated? Is there a leak of toxic fumes in the building? Is there radioactive material nearby? Because any of these can cause the situation that is described, a thorough investigation is needed to distinguish the *causal* relationships from the *coincidental* relationships.

Causal relationships establish the cause of a disease outbreak (any of the possibilities listed in the previous paragraph are potentially causal). Coincidental relationships are events that coincide by chance. Using the example above, the professor may have worn a particularly unattractive sweater on the day the students became ill. However, it is much more likely that is a coincidental relationship, than a causal one. Only when all possible causal factors have been investigated can a reasonable answer be proposed that would explain the etiology of the disease outbreak.

Stopping the Spread of Disease

Epidemiologists study the spread of disease so that ways of stopping it can be found. The two most obvious strategies for combating disease are *prevention* and *therapy*.

Therapy or treatment of diseases was perhaps the first strategy used by humans to fight disease. The continued search for therapeutic treatments for almost all known diseases is evidence that we still value this strategy.

However, we have always known that an even more effective disease-fighting strategy is prevention. Only recently have we understood many diseases well enough to know how to prevent them.

Although the war on human disease will probably never end, we have had some dramatic successes. The often fatal viral infection, *smallpox*, once caused catastrophic epidemics, but natural outbreaks have been eliminated at this point in history because of successful prevention strategies such as worldwide *vaccination* and education (**Figure 6-1**).

FIGURE 6-1 The last smallpox patient. Ali Maow Maalin of Somalia contracted the last known naturally occurring case of smallpox in 1977. Successful disease prevention techniques completely eradicated natural outbreaks of this disease that once killed millions worldwide and dramatically affected human history. The World Health Organization (WHO) considers naturally occurring cases to be eradicated—thus the vaccine for smallpox is no longer required in the United States. Unfortunately, the potential of smallpox being used as a biological weapon remains a threat.

Unfortunately, smallpox and other pathogens that rarely if ever now produce natural outbreaks of disease may nonetheless become available for use as weapons. Such biological weapons could produce epidemics in local regions and would thus not only generate alarm but also would severely burden public health resources. See the Research, Issues, and Trends box on the facing page.

QUICK CHECK

1. What is the difference between a *sign* and a *symptom*?
2. How does an acute disease differ from a chronic disease?
3. Define *pathogenesis*. What are the stages of a cold virus?
4. What is the difference between an *epidemic* and a *pandemic*?
5. How do morbidity and mortality differ?
6. What are the two primary strategies for combating disease?

Pathophysiology

Mechanisms of Disease

Disturbances of Homeostasis

Pathophysiology is the study of the underlying physiological processes associated with disease. Pathophysiology is a branch of pathology, the general study of disease. Pathophysiologists attempt to understand the mechanisms of a disease and its pathogenesis. Although pathophysiologists uncover information that leads to the discovery of strategies of prevention and treatment, developing and applying these strategies is left to other professionals.

Many diseases are best understood as disturbances of homeostasis, the relative constancy of the body's internal environment. Under normal physiological conditions, if homeostasis is disturbed, a variety of feedback mechanisms returns the body to normal. Negative and positive feedback, or feedback loops, were introduced in Chapter 1. When a disturbance of homeostasis goes beyond normal fluctuations, a disease condition exists.

In acute conditions, the body recovers its homeostatic balance quickly. In chronic diseases, a normal state of balance may never be restored. If the disturbance keeps the body's

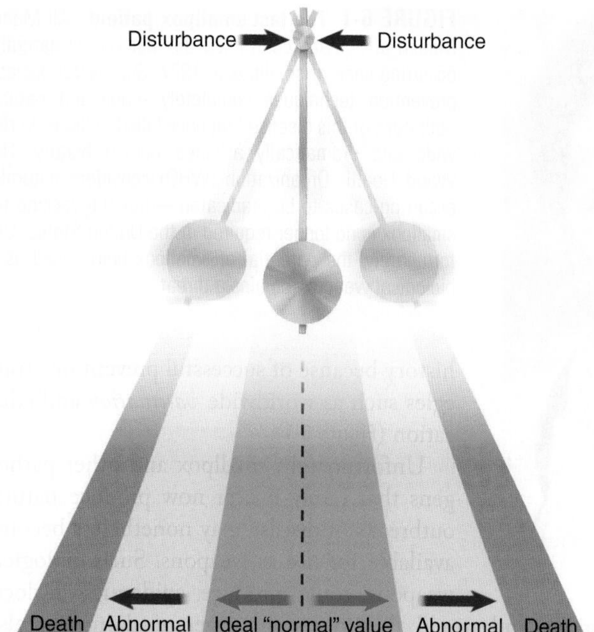

FIGURE 6-2 Model of homeostatic balance. Movement of the parameter in question, away from the ideal normal value, is depicted as normal fluctuations. Sometimes a physiological disturbance pushes the body beyond its capacity to maintain homeostasis and into the abnormal range for a given physiological parameter—a disordered condition. Disturbances in the extreme range may result in death.

internal environment too far from normal for too long, death may result (**Figure 6-2**).

Disturbance of homeostasis and the body's responses to that disturbance are the basic mechanisms of disease. Because of the variety of disease mechanisms, they are easier to study if categorized as in the following subsections.

Types of Disease Mechanisms

Genetic Mechanisms

A **genetic** mechanism occurs when altered or *mutated* genes cause production of abnormal proteins. These abnormal proteins often simply do not perform their originally intended function, resulting in the absence of an essential function. On the other hand, such proteins may perform an abnormal, disruptive function instead. Either case may be a threat to the constancy of the body's internal environment.

The basis for genetic diseases is discussed in Chapter 25, and important genetic conditions are summarized in Appendix A at *evolve.elsevier.com*.

Infectious Mechanisms

An **infectious** mechanism occurs when *pathogenic* (disease-causing) organisms or particles damage the body in some way.

An organism that lives in or on another organism to obtain its nutrients is called a **parasite.** The presence of microscopic-size or larger parasites may interfere with normal body functions of the *host* and thereby cause disease. Organisms other than parasites can poison or otherwise damage the human body to cause disease.

Some of the major pathogenic organisms are listed later in this chapter and in Appendix A at *evolve.elsevier.com*.

Neoplastic Mechanisms

A **neoplastic** mechanism occurs when abnormal tissue growths or **neoplasms** develop. Neoplasms such as benign tumors or cancers (malignant) can cause a variety of physiological disturbances. Many such mechanisms are described later in this chapter.

Traumatic Mechanisms

A **traumatic** mechanism involves injury by physical or chemical agents such as toxic or destructive chemicals, extreme heat or cold, mechanical injury (trauma), or radiation that can affect the normal homeostasis of the body.

Examples of pathological conditions caused by physical agents are summarized in Appendix A at *evolve.elsevier.com*. These conditions include injuries such as fractures and lacerations caused by physical trauma or poisoning caused by chemical agents.

Trauma to skeletal muscles—especially crushing injuries—can have catastrophic bodywide effects. Review the article *Rhabdomyolysis* at Connect It! at *evolve.elsevier.com*.

Metabolic Mechanisms

Metabolic mechanisms include malnutrition or endocrine imbalances that cause insufficient or imbalanced intake of nutrients.

A variety of diseases caused by metabolic mechanisms are outlined in Chapters 12, 19, 22, and other chapters. Some are also described in Appendix A at *evolve.elsevier.com*.

Inflammatory Mechanisms

Inflammatory mechanisms are commonly occurring disease mechanisms that can be involved along with other mechanisms to produce disease—or it may be the primary mechanism of a disease. **Inflammation** refers to the set of reactions of the immune system that often includes pain, redness, swelling, and warmth. It is a common response of the body to disturbances.

The *inflammatory response* is a normal mechanism that usually speeds recovery from an infection or injury. However, when the inflammatory response occurs at inappropriate times or is abnormally prolonged or severe, normal tissues may be damaged. Thus some disease symptoms are *caused by* the inflammatory response.

Autoimmunity is a faulty response or overreaction of the immune system that causes it to attack the body. Autoimmunity, literally "self-immunity," is discussed in Chapter 16 along with other immune system disturbances. Examples of autoimmune conditions are listed in Appendix A at *evolve.elsevier.com*.

Degeneration

Breaking apart, or **degeneration,** of tissues occurs by means of many still unknown processes. Although an expected consequence of aging, degeneration of one or more tissues resulting from disease can occur at any time. The degeneration of tissues associated with aging is discussed in Chapter 24.

Risk Factors

Other than direct causes or disease mechanisms, certain *predisposing conditions* may make the development of a disease more likely to occur. Usually called **risk factors,** they often do not actually cause a disease but may put one "at risk" for developing it.

Types of Risk Factors

There are many different types of risk factors. We describe some of the major categories of risk factors in the following sections.

Genetic Factors

There are several types of **genetic** factors that increase risk of certain diseases. In such a case, an inherited trait puts a person at a greater-than-normal risk for developing a specific disease.

For example, light-skinned people are more at risk for developing certain forms of skin cancer than are dark-skinned people. This occurs because light-skinned people have less pigment in their skin to protect them from cancer-causing ultraviolet radiation (see Chapter 7).

Membership in a certain ethnic group or *gene pool* involves the "risk" of inheriting a disease-causing gene that is common in that gene pool. For example, certain Africans and their descendants are at a greater-than-average risk of inheriting *sickle cell anemia*—a deadly blood disorder (see Chapter 13).

Age

A person's **age** can be a risk factor for certain diseases. Biological and behavioral variations inherent during different phases of the human life cycle put us at greater risk for developing certain diseases at certain times in life.

For example, middle ear infections are more common in infants than in adults because of the difference in ear structure at different ages. A person is at a higher risk for certain types of arthritis and bone fractures during the later adult years.

Lifestyle

The way we live and work—our **lifestyle**—can put us at risk for some diseases. For example, people whose work or personal activity puts them in direct sunlight for long periods have a greater chance of developing skin cancer because they experience more exposure to ultraviolet radiation from the sun.

Research has shown that the high-fat, low-fiber diet common among people in developed nations increases their risk of developing certain cancers such as colon cancer. Obesity is another risk factor for disease and has been shown to lead to increased incidence in type 2 diabetes and high blood pressure. Using tobacco is another important risk factor for disease.

Stress

Physical, psychological, or emotional **stress** can put one at risk of developing problems such as headaches, chronic high blood pressure (hypertension), depression, heart disease, and cancer. Conditions caused by psychological factors are sometimes called **psychogenic** (mind-caused) disorders.

Environmental Factors

Although **environmental factors** such as climate and pollution can cause injury or disease, some environmental situations simply put us at greater risk for getting certain diseases. For example, because some parasites survive only in tropical environments, we are at risk for diseases caused by those particular organisms only if we live in or travel to that climate.

Preexisting Conditions

A **preexisting condition,** such as an infection, can adversely affect our capacity to defend ourselves against further attack. Thus a *primary* (preexisting) condition can put a person at risk of developing a *secondary* condition. For example, blisters from a preexisting burn may break open and thus increase the risk of a bacterial infection of the skin.

Combined Risk Factors

Combined risk factors can increase a person's chances of developing a specific disease even more. For example, a light-skinned person can add to the genetic risk of developing skin cancer by spending a long time in the sun without skin protection—a lifestyle risk added to a genetic risk.

As you may have guessed, many of these categories of risk factors overlap. For example, stress can be a component of lifestyle, or it could be considered a preexisting condition. Sometimes a high-risk group is identified by epidemiologists, but the exact mechanism that puts them at high risk may be uncertain. For example, a high incidence of heart disease in a small ethnic group may point to a genetic risk factor but also could result from some aspect of a shared lifestyle.

Avoiding Risk of Disease

Risk factors for many deadly diseases can be reduced or avoided. For example, risk of heart disease, diabetes, cancer, infections, and other types of disease can be decreased by making informed choices about lifestyle and personal health management. Such choices may influence diet and exercise, stress management, the environment, and treatment of preexisting conditions.

QUICK CHECK

1. Define pathophysiology.
2. List seven general mechanisms that may cause disease.
3. List six risk factors that may cause disease.
4. How does a *primary* condition differ from a *secondary* condition?

Pathogenic Organisms and Particles

Many kinds of organisms and particles can cause disease in humans. Even humans can cause human disease through accidental or intentional injury to themselves or others. In pathophysiology, the pathogenic organisms most often studied are microscopic or just barely visible to the unaided eye. Microscopic organisms, also called **microbes,** include *bacteria, fungi,* and *protozoa.* Larger organisms, the pathogenic *animals,* are also medically important.

The smallest of all pathogens, microscopic nonliving particles called *viruses* and *prions,* lead our list of important disease-causing agents.

Viruses

Introduction to Viruses

Viruses are intracellular parasites that consist of a nucleic acid (DNA or RNA) core surrounded by a protein coat and sometimes a lipoprotein envelope.

Biologists hold that viruses are not technically living organisms because they are not made up of cells. However, because they infect living cells and contain their own unique genetic code, they remain the subject of biological study and are classified into groups as if they are organisms.

Virus particles can multiply—but only by using the mechanisms of their host cell. They invade cells and insert their own genetic code into the host cell's genetic processes, causing the host cell to produce viral DNA or RNA and protein coats. They thus pirate the host cell's nutrients and organelles to produce more virus particles. These newly formed viruses may leave the cell to infect other cells by way of vesicles or by bursting the cell membrane (**Figure 6-3**).

FIGURE 6-3 HIV. The human immunodeficiency virus, or HIV (*blue* in this electron micrograph), is released from infected white blood cells and soon spreads over neighboring cells, infecting them in turn. The individual viruses are very small—more than 200 million would fit on the period at the end of this sentence.

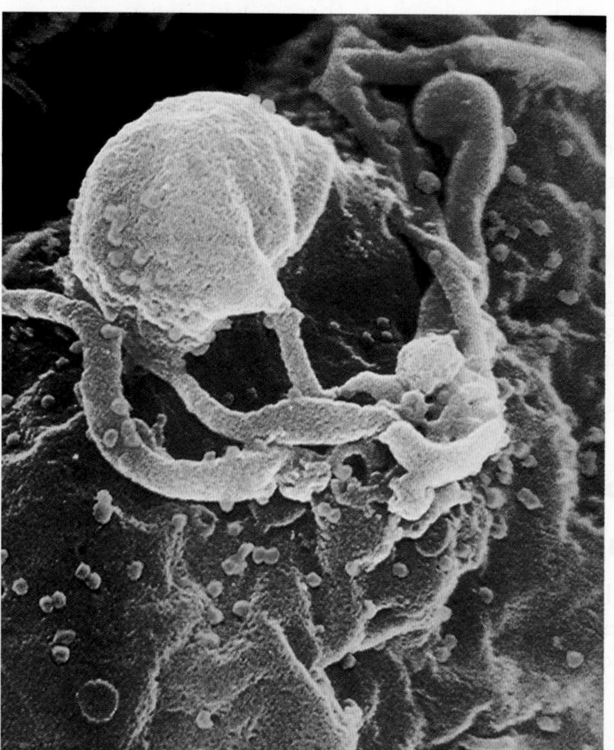

The symptoms of viral infections may not appear right away. The viral genetic code may not become active for some time, or viral multiplication may not immediately cause significant cellular damage. In any case, the effects of the intracellular viral parasite may eventually take their toll and thus produce symptoms of disease.

Viruses are a very diverse group, as illustrated in **Figure 6-4**. They are usually classified according to their shape, DNA or RNA content, and their method of multiplying. Some examples of medically important viruses are listed in **Table 6-1**. Many of these and some other viral diseases are discussed in detail in later chapters.

Examples of Viruses

There are many types of viruses that infect humans—and more are being discovered or are newly appearing in the human population all the time. Here, we discuss just a few of the many interesting examples of human viruses.

Human Immunodeficiency Virus

The most discussed virus in recent history is **human immunodeficiency virus (HIV).** HIV is an RNA-containing **retrovirus.** A retrovirus uses its RNA to "transcribe backward"

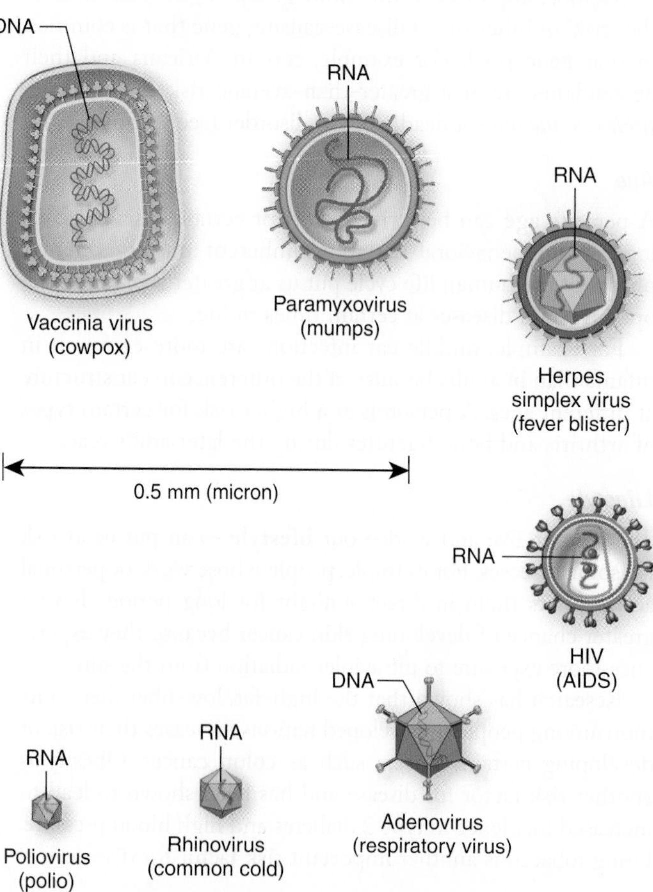

FIGURE 6-4 Diversity of pathogenic viruses. Some viruses are relatively large; others are extremely tiny. A human hair would be 8 meters (over 26 feet) thick if drawn at the same scale as the particles depicted here.

TABLE 6-1 Examples of Pathogenic Viruses

VIRAL TYPE	VIRUS	DISEASES CAUSED
DNA virus	Human papillomavirus (HPV)	Warts
	Hepatitis B	Hepatitis B (viral liver infection)
	Herpes simplex 1 and 2	Fever blisters and genital herpes
	Epstein-Barr virus (EBV)	Mononucleosis
RNA virus	Influenza A, B, and C	Various influenza infections
	Human immunodeficiency virus (HIV)	Acquired immunodeficiency syndrome (AIDS)
	Flavivirus	West Nile virus (WNV), yellow fever, dengue, Zika, St. Louis encephalitis
	Paramyxovirus	Measles, mumps, and parainfluenza
	Rhinovirus	Common cold and upper respiratory infections
	Coronavirus (CoV)	Respiratory infections, including common cold, severe acute respiratory syndrome (SARS), and Middle East respiratory syndrome (MERS)
	Hepatitis A and C *Ebolavirus*	Hepatitis A and C (viral liver infections) Ebola virus disease (EVD or Ebola hemorrhagic fever)

See Table 2 in Appendix A at *evolve.elsevier.com* for a list of viral diseases and their descriptions.

to produce the virus's primary genetic code and insert it into the host's DNA genome.

HIV attacks the immune system, thus rendering the host organism susceptible to a variety of infections. The immune system gives our bodies the ability to fight infections. HIV finds and destroys a type of white blood cell (a variety of T cells called CD4 cells) that the immune system must have to fight disease (see **Figure 6-3**). If untreated, an HIV infection may progress to stage 3, more commonly known as **acquired immunodeficiency syndrome (AIDS).**

HIV infection was first identified in the United States in 1981 after a number of homosexual men started getting sick with a rare type of cancer. It took several years for scientists to develop a test for the virus, to understand how HIV was transmitted between humans, and to determine what people could do to protect themselves against being infected. During the early 1980s, as many as 150,000 people became infected with HIV each year in the United States. By the early 1990s, this rate had dropped to about 40,000 to 50,000 each year, where it remains today.

The spread of HIV infection in regions of the world that are economically disadvantaged remains a major global health concern. This persists as a consequence of lack of education about disease prevention and lack of resources needed to treat the infected individuals with antiviral therapy.

It is also important to note that antiviral therapy prevents the immune system collapse characteristic of AIDS, but does not eliminate HIV from the infected individual, who thus remains contagious. HIV is a fragile virus, and cannot live for very long outside the body. The virus is not transmitted through day-to-day activities such as shaking hands, hugging, or a casual kiss. HIV is not spread from a toilet seat, drinking fountain, doorknob, dishes, drinking glasses, food, or pets. Unlike other infections discussed later, HIV infection is not an arthropod-borne disease and cannot be spread by mosquitoes or other biting arthropods.

HIV is primarily found in the blood, semen, or vaginal fluid of an infected person. HIV is transmitted in three main ways:

- Sexual contact (anal, vaginal, or oral) with someone infected with HIV
- Sharing needles and syringes with someone infected with HIV
- Being exposed (fetus or infant) to HIV before or during birth or through breastfeeding

Coronaviruses

Coronaviruses are RNA viruses characterized by a "crown" of surface projections when viewed under an electron microscope. The word part *corona-* means "crown." They use their RNA in host cells to produce their own viral enzymes and structural proteins needed to replicate.

Coronaviruses are found nearly everywhere in our environment and are the second leading cause of the common cold—after *rhinoviruses* (see **Table 6-1**). Coronavirus infections are spread when virus particles are shed by an infected body by way of respiratory fluids or other body fluids, and these particles then come in contact with another person's body fluids. A person is most likely to pick up shed viruses in the moist mucous membranes of the mouth, nose, eyes, or genitals.

Some coronavirus infections can be very serious. An example is infection by the **SARS-associated coronavirus (SARSCoV),** the cause of **severe acute respiratory syndrome (SARS).**

Flaviviruses

Flaviviruses are RNA viruses that are transmitted less directly than coronaviruses. Flaviviruses move from an infected bird or other animal to a mosquito or other biting insect and then finally to the human host. Such viruses cannot move directly from an infected bird to a human—they require the insect to carry the virus to humans. This role of animals, including biting insects, in disease transmission is discussed later in this chapter.

Various types of flaviviruses (literally "yellow" viruses) cause **yellow fever, dengue, West Nile virus (WNV)** infection, **Zika virus** disease, and other potentially serious infections.

There are many types of viruses that affect humans. Review **Table 6-1** for more examples.

To learn more about virus replication, go to AnimationDirect online at *evolve.elsevier.com.*

Prions

The word **prion** is a shortened form of the phrase "PROteinaceous INfectious particle." Prions are pathogenic protein molecules that can cause misfolding of other proteins in the infected cell. Review normal protein folding in **Figure 2-12** on p. 34.

The misfolding induced by prions converts normal proteins of the body into abnormal proteins, causing abnormalities of function (**Figure 6-5,** *A*). The abnormal form of the protein also may be inherited by offspring of an affected person.

We do know that prions can affect proteins in the nervous system and cause diseases such as **bovine spongiform encephalopathy** (BSE or "mad cow disease") and **variant Creutzfeldt-Jakob disease (vCJD)** (**Figure 6-5,** *B*). Both of these diseases are very rare, fatal conditions characterized by degeneration of brain tissue and progressive loss of nervous system function.

Many scientists believe that prions from infected cattle were consumed as beef by humans with these diseases, but there are many unanswered questions about the exact mechanisms of transmission of prion diseases.

Bacteria

A **bacterium** (*pl.,* bacteria) is a tiny, primitive cell without a nucleus. Bacteria produce disease in a variety of ways. They can secrete toxic substances that damage human tissues, they may become parasites inside human cells, or they may form populations in the host body that disrupt normal human function.

Like viruses, bacteria also are a diverse group of *pathogens* ("disease-producers"). There are several ways to classify bacteria:

1. Growth requirements—Bacteria can be categorized according to whether they need oxygen to grow. For example, they can be categorized as *aerobic* (requiring oxygen for their metabolism) or *anaerobic* (requiring an absence of oxygen).
2. Staining properties—Bacteria stain differently, depending on the compounds in their walls. For example, *gram-positive* bacteria are stained purple by the **Gram staining technique,** whereas

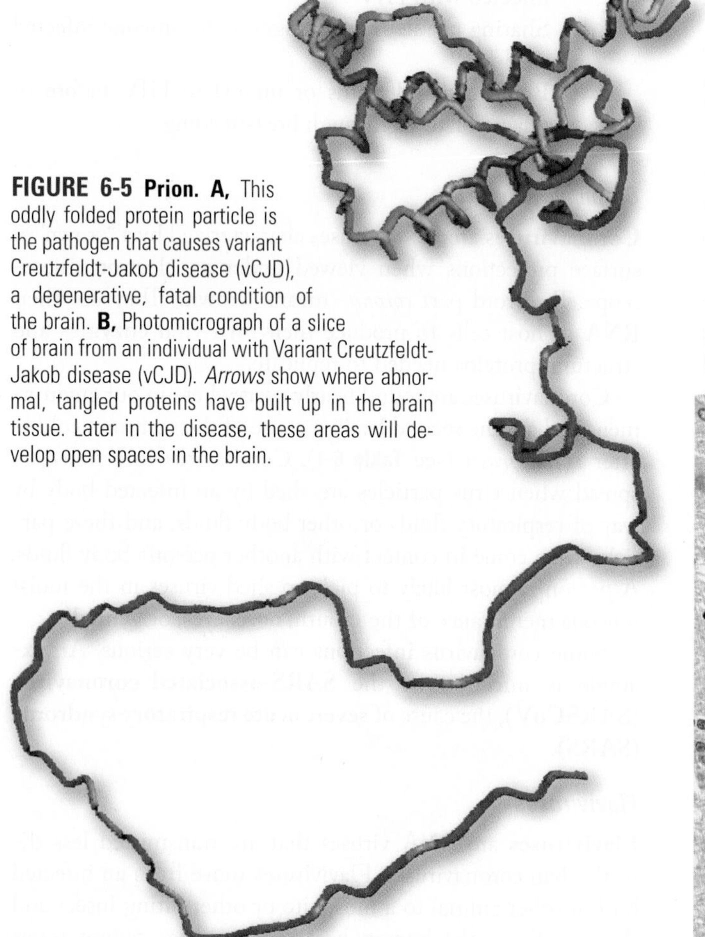

A

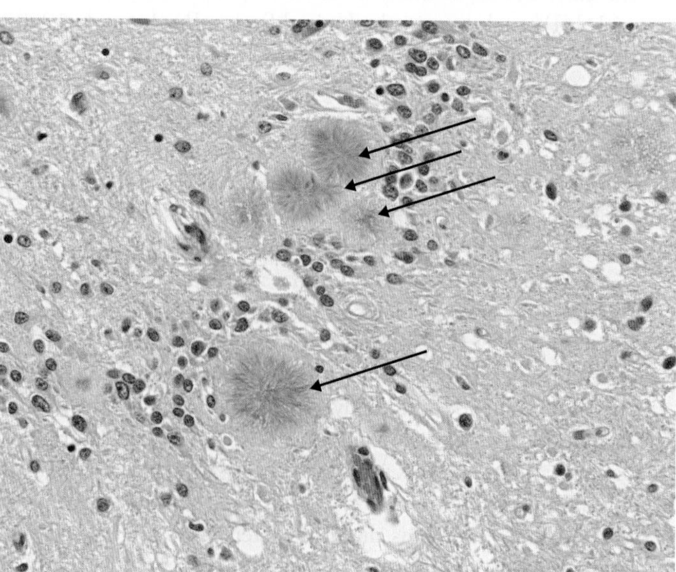

B

FIGURE 6-5 Prion. A, This oddly folded protein particle is the pathogen that causes variant Creutzfeldt-Jakob disease (vCJD), a degenerative, fatal condition of the brain. **B,** Photomicrograph of a slice of brain from an individual with Variant Creutzfeldt-Jakob disease (vCJD). *Arrows* show where abnormal, tangled proteins have built up in the brain tissue. Later in the disease, these areas will develop open spaces in the brain.

6

TABLE **6-2** Examples of Pathogenic Bacteria

STRUCTURAL CLASSIFICATION	GRAM STAIN CLASSIFICATION	BACTERIUM	DISEASES CAUSED
Bacillus (rod)	Gram-positive	*Bacillus* organisms	Anthrax and gastroenteritis
	Gram-positive	*Clostridium* organisms	Botulism, tetanus, and soft tissue infections
	Gram-negative	*Enterobacteriaceae* organisms	*Salmonella* diseases and gastroenteritis
	Gram-negative	*Pseudomonas* organisms	External otitis (swimmer's ear), endocarditis, and pulmonary infections
Coccus (sphere)	Gram-positive	*Staphylococcus* organisms	Staphylococci infections, food poisoning, urinary tract infections, and toxic shock syndrome
	Gram-positive	*Streptococcus* organisms	Throat infections, pneumonia, sinusitis, otitis media, rheumatic fever, and dental caries
	Gram-negative	*Neisseria* organisms	Meningitis, gonorrhea, and pelvic inflammatory disease
Curved or spiral rod	Gram-negative	*Vibrio* organisms	Cholera, gastroenteritis, and wound infections
	Gram-negative	*Campylobacter* organisms	Diarrhea
	Gram-negative	Spirochetes	Syphilis and Lyme disease
Small bacterium	Gram-negative	*Rickettsia* organisms	Rocky Mountain spotted fever and Q fever
	Gram-negative	*Chlamydia* organisms	Genital infections, lymphogranuloma venereum, pelvic inflammatory disease, conjunctivitis, and parrot fever

gram-negative bacteria are not (see Clinical Application box on p. 122).

3. Shape and size—Bacteria are most commonly classified by their varied shapes (**Table 6-2**). Medically significant bacteria range in size from less than 0.5 μm to more than 5 μm, making size a useful characteristic for classification. The abbreviation μm represents *micrometers* or *microns,* one millionth of a meter. Some major groupings based on shape and size follow:
   a. **Bacilli**—large, rod-shaped cells found singly or in groups.
   b. **Cocci**—large, round bacteria found singly, in pairs *(diplococci),* in strings *(streptococci)* as shown in **Figure 6-6**, or in clusters *(staphylococci).*
   c. Curved or spiral rods—curved rods arranged singly or in strands, or large curved or spiral cells arranged singly or in cell colonies.
   d. Small bacteria—round or oval bacteria that are so small that some of them were once thought to be viruses. They can reproduce only inside other living cells, so they are sometimes called *obligate intracellular parasites.* ***Rickettsia*** and ***Chlamydia*** are two types of small bacteria.

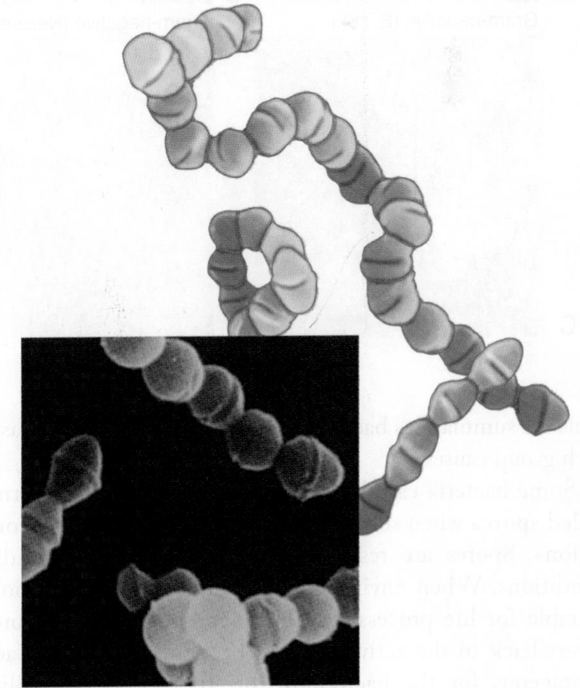

FIGURE 6-6 Bacteria. As the scanning electron micrograph and drawing show, individual spherical bacteria (cocci) may adhere to each other to form chains.

6

CLINICAL APPLICATION

LABORATORY IDENTIFICATION OF PATHOGENS

Often the evident signs or symptoms of a disease caused by bacteria or other pathogens provide enough information for a health professional to make a diagnosis. To be sure that the correct course of treatment is given, laboratory tests are often required to positively identify a pathogen.

Sometimes pathogens can be observed in specimens of blood, feces (stool), cerebrospinal fluid (CSF), mucus, urine, or other substances from the body. To view a microscopic pathogen, a portion of the collected specimen is smeared on a microscope slide and then stained to enhance visibility.

Certain stains color only certain types of cells. For example, only gram-positive bacteria retain the violet stain used in the Gram staining technique (Figures **A** and **B**). Gram-negative bacteria do not retain the violet stain; they retain only a red counterstain (Figures **C** and **D**). Thus gram-positive (violet) bacteria can be distinguished from gram-negative (red) bacteria using the Gram method. The staining properties, shape, and size of a pathogen are a few of the characteristics sometimes used to identify pathogens in specimen samples.

Pathogens are sometimes identified by **cultures** (propagation of microorganisms in special media conducive to their growth) that originate from specimens taken from a patient. These populations of bacteria can be grown only on certain *media* (liquid or *agar* gel–containing nutrients). Thus pathogenic bacteria are often identified by the type of medium in which they grow best.

For example, mucus swabbed from a sore throat and placed in a medium that contains blood may produce pinpoint-sized colonies of pathogenic streptococci bacteria. The streptococci bacteria that cause "strep throat" typically have a distinct, transparent ring around each colony. The rings result from hemolysis—bursting of red blood cells in the surrounding medium. A few viruses also can be cultivated but only within living cells.

Gram-positive (Listeria)

A

Gram-positive (Streptoccus)

B

Gram-negative (E. coli)

C

Gram-negative (Neisseria)

D

Some infections can be diagnosed on the basis of immunological tests that check for antibodies against a particular pathogen. If antibodies are found, it is assumed that the patient has been exposed to a pathogen; a large number of antibodies usually indicates an active infection. An example is the test for anti-HIV antibodies used to identify HIV infections. Recall from Chapter 5 that such tests are often used to screen donated tissues and organs for pathogenic organisms.

Even newer tests use rapid biological sensors that borrow the recognition mechanisms from immune cells sensitive to particular pathogens and link them to special proteins from jellyfish that cause the sensor to glow when the pathogen is present. A wide variety of different immunological tests are now available for bacterial and viral infections and other new technologies are on the horizon.

The recent explosion of knowledge in genetics and genomics (see Chapter 25) has led to many newer methods of rapidly and accurately identifying a variety of pathogens. For example, patterns of DNA or RNA code that are unique to a particular virus, bacterium, or other pathogen can be detected by special sensors or laboratory tests.

Table 6-2 summarizes bacterial types and some of the diseases each group causes.

Some bacteria can develop into resistant dormant forms called **spores** when subjected to adverse environmental conditions. Spores are resistant to chemicals, heat, and dry conditions. When environmental conditions become more suitable for life processes such as reproduction, the spores revert back to the active form of bacterium. Although advantageous for the bacterium, this transformation ability often makes it difficult for humans to destroy pathogenic bacteria.

Microbes of another type that are similar to bacteria are the **archaea.** They differ from bacteria in their chemical makeup and metabolism. Also, unlike bacteria, many archaea thrive in extremely harsh environments that are very hot, very acid, or very salty. Although archaea are found as normal residents in the human body, in the mouth for example, none have yet been proven to cause disease. However, they may play an important role in the human *microbiome*.

The presence of bacteria and other microorganisms in and on the body is normal and necessary for normal function. To learn more, review the article *The Human Microbiome* at Connect It! at *evolve.elsevier.com*.

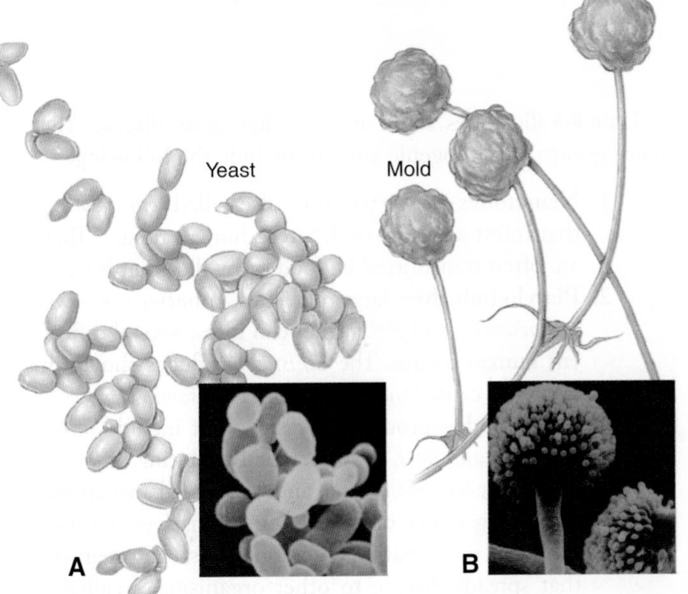

FIGURE 6-7 Examples of pathogenic fungi. Electron micrographs and drawings. **A,** Scanning electron micrograph of yeast cells. Yeasts commonly infect the urinary and reproductive tracts. **B,** This electron micrograph shows *Aspergillus* organisms, a mold that can infect different parts of the body where it forms characteristic "fungus balls."

Fungi

Fungi (*sing.*, fungus) are simple organisms that are similar to plants but without chlorophyll (green pigment). Without chlorophyll, pathogenic fungi cannot produce their own food, so they must consume or parasitize other organisms.

Most pathogenic fungi parasitize tissue on or near the skin or mucous membranes, as in athlete's foot and vaginal yeast infections. A few systemic (body-wide) fungal infections, such as San Joaquin fever, can disrupt the entire body.

Figure 6-7 shows that *yeasts* are small, single-celled fungi and *molds* are large, multicellular fungi. Fungal infections, or **mycotic infections,** often resist treatment, so they can become a quite serious health problem. **Table 6-3** lists some of the important pathogenic fungi and the diseases that they cause.

TABLE 6-3 Examples of Pathogenic Fungi

FUNGUS	DISEASES CAUSED
Candida organisms	Thrush and mucous membrane infections (including vaginal yeast infections)
Epidermophyton and *Microsporum* organisms	Tinea infections: ringworm, jock itch, and athlete's foot
Histoplasma organisms	Histoplasmosis
Aspergillus organisms	Aspergillosis and pneumonia
Coccidioides organisms	Coccidioidomycosis (San Joaquin fever)

See Table 4 in Appendix A at *evolve.elsevier.com* for a list of mycotic diseases and their descriptions.

Protozoa

Protozoa are *protists,* one-celled organisms that are larger than bacteria and whose DNA is organized in a nucleus. **Table 6-4** illustrates some of the pathogenic protozoa. Protozoa can infect human fluids and cause disease by parasitizing cells or directly destroying them (**Figure 6-8**).

TABLE 6-4 Examples of Pathogenic Protozoa

CLASSIFICATION	PROTOZOAN	DISEASES CAUSED
Ameba	*Entamoeba* organisms	Diarrhea, amebic dysentery, and liver and lung infections
	Naegleria fowleri	Brain infections
Flagellate	*Giardia* organisms	Giardiasis, diarrhea, and malabsorption syndrome
	Trichomonas organisms	Trichomoniasis, vaginitis, and urinary tract infections
Ciliate	*Balantidium* organisms	Gastrointestinal disturbances, including pain, nausea, and anorexia
Sporozoan (coccidium)	*Isospora* organisms	Isosporiasis infection of gastrointestinal tract, diarrhea, and malabsorption syndrome
	Plasmodium organisms	Malaria
	Toxoplasma organisms	Toxoplasmosis and congenital damage to fetus

See Table 5 in Appendix A at *evolve.elsevier.com* for a list of diseases caused by protozoa.

Some major groups of pathogenic protozoa include the following:

1. **Amebas**—large cells of changing shape; amebas extend their membranes to form *pseudopodia* ("false feet") that pull themselves along.
2. **Flagellates**—protozoa that are similar to amebas but move by wiggling long, whiplike extensions called **flagella.**
3. **Ciliates**—protozoa that move by means of many short, hairlike projections called cilia.
4. **Sporozoa**—protozoa with unusual organelles at their tips that allow them to enter host cells; also called

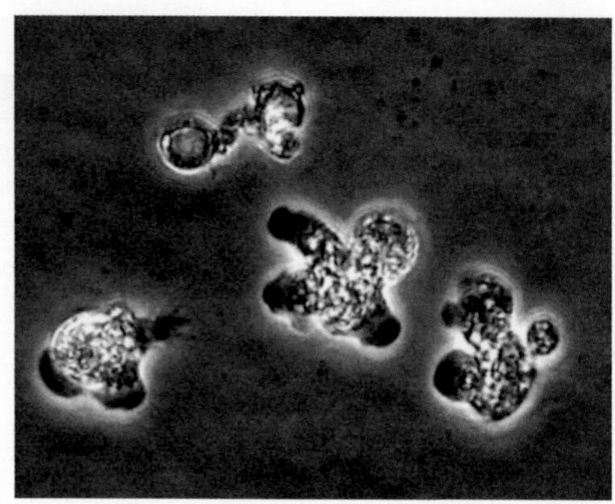

FIGURE 6-8 Pathogenic amebas. The *Naegleria fowleri* organism seen in this light micrograph is an emerging pathogen in the southern United States. It is an ameba found in warm, fresh-water ponds, lakes, streams, and warm springs. The organism enters the body by swimming up the nose and invading the brain through the thin ethmoid bone. It can cause a fatal central nervous system infection.

coccidia. They often oscillate between two different hosts, having two different stages in their life cycle. The sporozoa that cause malaria exhibit this pattern.

Pathogenic Animals

Pathogenic animals sometimes called **metazoa** are large, multicellular organisms. Animals can cause disease by parasitizing humans or causing injury in some other way.

Table 6-5 illustrates some animals that cause disease. The major groups of pathogenic animals include the following:

1. **Nematodes**—large parasites, also called *roundworms,* that infest a variety of different human tissues. They are often transmitted by food or by flies that bite.
2. **Platyhelminths**—large parasites, otherwise known as *flatworms* and *flukes,* that can infest several different human organs. The *Schistosoma* flukes shown in Figure 6-9 cause "snail fever," or *schistosomiasis.*
3. **Arthropods**—group of parasites that include *mites, ticks, lice,* and *fleas.* Also included are biting or stinging *wasps, bees, mosquitoes,* and *spiders.* All are capable of causing injury or infestation themselves but also can carry other pathogenic organisms. An organism that spreads disease to other organisms is called a **vector** of the disease.

Table 6-5 summarizes some of the major health problems associated with selected pathogenic animals.

QUICK CHECK

1. Name a disease caused by a *virus* and a disease caused by a *prion.*
2. What is a retrovirus? Name a disease that is caused by a retrovirus.
3. How do bacteria cause disease?
4. Name a common fungal infection.
5. What are protozoa, and how do they cause disease?
6. List the major groups of pathogenic animals.

TABLE 6-5 Examples of Pathogenic Animals

CLASSIFICATION	ANIMAL	DISEASES CAUSED
Nematode	*Ascaris* organisms	Intestinal roundworm infestation, gastrointestinal obstruction, and bronchial damage
	Enterobius organisms	Pinworm infestation of the lower gastrointestinal tract, itching around the anus, and insomnia
	Trichinella organisms	Trichinosis, fever, and muscle pain
Platyhelminth	*Schistosoma* organisms	Schistosomiasis (snail fever)
	Fasciola organisms	Liver fluke infestation
	Taenia organisms	Pork and beef tapeworm infestation
Arthropod	*Arachnida* organisms	Infestation by mites and ticks; toxic bites by spiders, scorpions; and transmission of other pathogens
	Insecta	Infestation by fleas and lice; toxic bites by wasps, mosquitoes, and bees; and transmission of other pathogens; ticks (Lyme disease)

See Table 6 in Appendix A at *evolve.elsevier.com* for a list of diseases caused by pathogenic animals.

6

FIGURE 6-9 Pathogenic animals. This light micrograph shows both male and female *Schistosoma* flukes mating in the human bloodstream (the male is the larger of the two).

Prevention and Control

The key to preventing many diseases caused by pathogenic organisms is stopping them from entering the human body. This sounds simple enough but is often very difficult to accomplish.

Mechanisms of Transmission

A key to stopping pathogens is to understand the mechanisms by which they spread—mechanisms that potentially can be disrupted. The following sections can be used as a partial list of the ways in which pathogens can spread.

Person-to-Person Contact

Small pathogens often can be carried in the air from one person to another. Also, direct contact with an infected person or with contaminated materials handled by the infected person is a common mode of transmission. The *rhinoviruses* and *coronaviruses* that cause the common cold are often transmitted in these ways.

Some viruses, such as those that cause *hepatitis B*, *hepatitis C*, and *AIDS*, are instead transmitted when infected blood, semen, or another body fluid enters a person's bloodstream.

Preventing the spread of these diseases often involves educating people about avoiding certain types of contact with individuals known or suspected of carrying the disease. Another strategy, called **aseptic technique,** involves killing or disabling pathogens on surfaces before they can spread to other people. **Table 6-6** summarizes the major approaches taken when using aseptic technique.

Environmental Contact

Many pathogens are found throughout the local environment—in food, water, soil, and on assorted surfaces. Under normal conditions, these pathogens infect only individuals who happen to come across them or who are already weakened by some other condition. If improper sanitation practices create an environment that promotes increased growth and spread of pathogens, an epidemic could result.

Disease caused by environmental pathogens can often be prevented by avoiding contact with certain materials and by maintaining safe sanitation practices.

Opportunistic Invasion

Some potentially pathogenic organisms are found on the skin and mucous membranes of nearly everyone. However, they do not cause disease until they have the opportunity. That is, they do not create a problem until and unless conditions change or they enter the body's internal environment.

For example, the fungi that cause athlete's foot are often present on the skin of people who do not have symptoms of this infection. Only when the skin is kept warm and moist for prolonged periods can the fungus reproduce and create an infection.

Preventing opportunistic infection involves avoiding conditions that could promote infections. Changes in the pH (acidity), moisture, temperature, or other characteristics of skin and mucous membranes often promote opportunistic infections. Cleansing and aseptic treatment of accidental or surgical wounds also can prevent these infections.

TABLE 6-6 Common Aseptic Methods That Prevent the Spread of Pathogens*

METHOD	ACTION	EXAMPLES
Sterilization	Destruction of all living organisms; does not usually affect prions	Pressurized steam bath, extreme temperature, gas (ethylene oxide), or radiation used to sterilize surgical instruments and garments or other surfaces
Disinfection	Destruction of most or all pathogens on inanimate objects but not necessarily all harmless microbes	Chemicals such as iodine, chlorine, alcohol, phenol, and soaps
Antisepsis	Inhibition or inactivation of pathogens	Chemicals such as alcohol, iodine, quaternary ammonium compounds (quats), and dyes
Isolation	Separation of potentially infectious people or materials from noninfected people	Quarantine of affected patients; protective apparel worn while giving treatments; and sanitary transport, storage, and disposal of body fluids, tissues, and other materials

*Spores (special bacterial forms) may resist methods that would ordinarily kill active bacterial cells.

SCIENCE APPLICATIONS

PUBLIC HEALTH

Robert Koch (1843–1910)

Robert Koch astounded his parents when, at the age of 5 (in 1848), he showed his parents that he had taught himself to read. His determination and his methodical use of newspapers in his home not only helped young Robert teach himself to read, it also foreshadowed a brilliant career as an investigative scientist.

Koch became a physician, and while still a young man, he proved that the anthrax bacillus (bacterium) causes the anthrax infection (see *Disease as a Weapon* box, p. 128). Thus he was the first to prove that specific bacteria cause specific diseases. He later went on to do similar ground-breaking work with wound infections, tuberculosis, cholera, and many other infections. Perhaps more importantly, he laid the groundwork for the laboratory study of bacteria and the control of individual infections as well as epidemics. In so doing, Robert Koch laid the groundwork for modern **public health,** the field that strives to prevent and control disease and promote good health in the human population.

Public health is a field that includes many different endeavors, all aimed at promoting the health and wellness of us all. For example, medical and allied health professionals treat disease and work to prevent and control epidemics. *Pathologists* and *laboratory technicians* help us better understand diseases that affect the human population, and *researchers* help develop effective prevention and treatment.

Many *public health advisors,* **environmental health scientists,** and *public health activists* work to help us understand and resolve issues related to exposure to pollutants, the effects of our lifestyle, technological advances, and social choices that affect our health. *Public health administrators* and staff, including volunteers, help organize and support the worldwide effort to promote public health. The photo shows a nurse in the commissioned corps of the *United States Public Health Service (USPHS)* assessing a patient.

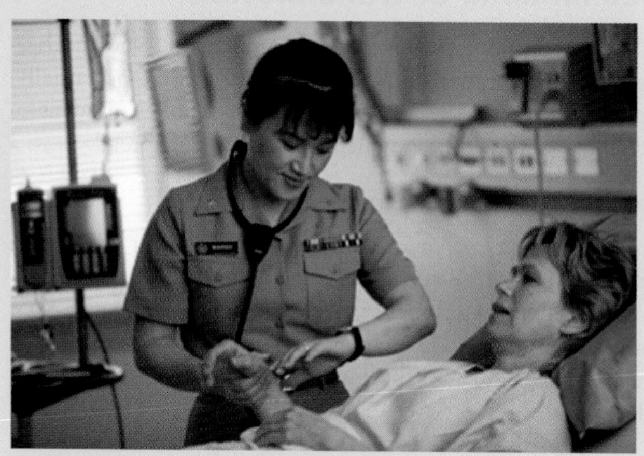

Transmission by a Vector

As stated previously, a vector such as an arthropod acts as a carrier of a pathogenic organism. For example, the spirochete bacterium that causes *Lyme disease* is not usually transmitted directly from human to human. Instead, a vector such as the deer tick carries it from one person to another or between animals and humans. **Table 6-7** gives examples of several tick-borne illnesses.

The most effective way to stop vector-borne diseases from spreading is a combination of reducing the population of vectors and reducing the number of contacts with vectors. *Malaria,* still a major killer in some parts of the world, was virtually eliminated from North America in this way. Many mosquitoes that transmit the malaria organism were destroyed with pesticides, and at the same time, people were educated about ways to prevent mosquito bites. Consistent use of both strategies resulted in the collapse of the pathogen population in the vector and host.

The fact that sporadic cases of malaria still occur in North America—and the emergence of other vector-borne diseases—demonstrates the need for ongoing monitoring of vector populations and the incidence of each disease.

Prevention and Treatment Strategies

Vaccination

A prevention strategy that has worked with some bacterial and viral pathogens has been the **vaccine.** A vaccine is a killed or attenuated (weakened) pathogen or part of a pathogen that is given to a person to stimulate immunity. *Vaccination* is a preventive method that stimulates a person's own immune system in a way that promotes development of resistance to a particular pathogen.

Although vaccines are very safe, they can cause mild side effects such as temporary flu-like symptoms, mild pain, and fainting; they rarely cause more severe side effects such as allergic reactions or febrile seizures. Thorough research has disproven a widely held belief that childhood vaccines (including preservatives) causes *autism spectrum disorder (ASD),* a difference in brain function. This dangerous myth still persists, however, reducing vaccination rates and increasing the incidence of dangerous childhood diseases such as *measles, mumps, rubella, whooping cough, and polio.*

More discussion of vaccination and other immune system strategies of disease prevention is found in Chapter 16.

To learn more about vaccination, go to AnimationDirect online at *evolve.elsevier.com*.

Drug Therapy

After an infection has begun, there are several ways to treat the patient and attempt to gain control of the disease. One common approach is the use of drug therapy to destroy pathogens or inhibit their growth.

Antibiotic Drugs

Antibiotics are compounds produced by certain living organisms or in a laboratory that kill or inhibit bacterial pathogens. *Penicillin*—produced by a fungus—and *streptomycin*—produced by a bacterium—are well-known antibiotics. A few synthetic chemicals are now also used to treat bacterial infections.

Antibiotic resistance is an important consideration in the treatment of infectious diseases today. As we continue to use antibiotics to treat bacterial disease, new generations of bacteria eventually develop resistance to antibiotic drugs. Some of this occurs as a natural consequence of bacterial adaptation. However, we accelerate this process by prescribing antibiotics for diseases that are not of bacterial etiology or by using the medications inappropriately after they are prescribed.

New antibiotics are needed when bacteria develop resistance to the drug or when the drug treatment cycle is difficult to complete. A bacterial disease called tuberculosis (TB) gives us an example of both of these issues. Current TB drugs are taken over a 6- to 9-month period, and there are strains of the TB bacteria that have become resistant to those drugs.

Several new drug combinations to treat TB are currently in clinical trials. Adding one of the newer anti-TB drugs to the standard therapy may reduce the usual treatment time required by current TB antibiotics and may be able to kill drug-resistant strains of TB.

The World Health Organization (WHO) estimates that around one fifth of the world's population is infected with TB—and that 1.5 million people died of the disease in 2014. The good news is that this is an 18% reduction in TB mortality rate since the start of this century.

TABLE 6-7 Examples of Tick-Borne Diseases

BACTERIAL PATHOGEN (DISEASE)	VECTOR	SIGNS AND SYMPTOMS*
Ehrlichia ewingii (Ehrlichiosis or human granulocytic ehrlichiosis)	*Amblyomma americanum* (Lone Star tick)	Fever, headache, muscle pain, nausea, and vomiting Rash is rare Low white blood cell count, low platelet count, anemia, elevated liver enzymes, kidney failure, and respiratory insufficiency
Borrelia burgdorferi (Lyme disease) Bull's eye rash.	*Ixodes scapularis* (deer tick or blacklegged tick)	Fever, headache, muscle pain, joint pain (arthritis), and swollen lymph nodes Red, expanding rash called erythema migrans (EM) or "bull's eye rash" in about 70% of cases Arthritis (pain and swelling) in the large joints (such as knees)
Rickettsia rickettsii (Rocky Mountain spotted fever)	*Dermacentor variabilis* (American dog tick)	Fever, headache, muscle pain, nausea, vomiting, and loss of appetite Rash after fever in 50% of adults and 90% of children Rash may involve palms and soles

*Symptoms vary among individual patients.

6

Antibiotic resistance occurs when some pathogenic bacteria in our microbiome survive antibiotic therapy and then reproduce. Over time, this shifts the ecological balance in favor of antibiotic-resistant strains. In a person with a well-functioning microbiome, antibiotic use can also cause short-term imbalances of pathogenic and nonpathogenic organisms that disrupt our microbial ecosystem and cause some of the "side effects" of antibiotics, including indigestion, opportunistic infection, and inflammation. Please review the article *The Human Microbiome* at Connect It! at *evolve.elsevier.com.*

Antiviral Drugs

Antiviral drugs, especially when used in carefully formulated combinations (often called "drug cocktails"), do not stop viral infections entirely. Instead, they inhibit viral reproduction and thus slow down the progression of viral infections. This strategy may reduce acute episodes of some virus infections or prevent the development of serious, perhaps life-threatening, viral disease and other complications.

Among the growing list of synthetic antiviral agents are *oseltamivir (Tamiflu)* for treating influenza A and B, *acyclovir (ACV)* for treating herpes infections, and *efavirenz* for treating HIV infections.

Antiviral agents used against HIV—especially when used in carefully formulated combinations called "drug cocktails"—may inhibit viral activity enough to slow or prevent HIV infection from progressing to stage 3 (AIDS).

QUICK CHECK

1. What are four ways that a disease can be transmitted?
2. What is *aseptic technique?*
3. How is an *antibiotic* drug different from an *antiviral* drug?
4. Why are new antibiotics often needed in the treatment of bacterial disease?

Tumors and Cancer

Neoplasms

Benign and Malignant Tumors

The term **neoplasm** literally means "new matter" and refers to an abnormal growth of cells. Neoplasms, also called **tumors,** can take the form of distinct lumps of abnormal cells or, in blood tissue, can be diffuse.

Neoplasms are often classified as **benign** or **malignant** (**Table 6-8**). Benign tumors remain localized within the tissue from which they arose. Malignant tumors tend to spread to

RESEARCH, ISSUES, AND TRENDS

DISEASE AS A WEAPON

World events have shown us that the intentional transmission of disease can be used as a weapon of terror. **Anthrax,** a bacterial infection caused by *Bacillus anthracis*, is an example of a pathogen that has been intentionally distributed to otherwise healthy people in acts of **bioterrorism.**

The anthrax bacterium ordinarily affects mainly plant-eating animals such as sheep, cattle, and goats, often resulting in their death.

The anthrax bacterium can assume the form of a **spore** that is resistant to heat, drying, and chemicals and then later becomes active to cause infection. Rarely, humans inhale some anthrax spores or get the spores in an open cut when handling infected animals or their hides. The inhaled form may be fatal if not treated quickly with antibiotics. The cutaneous (skin) form is less serious, characterized by a reddish brown patch on the skin that ulcerates and then forms a dark, nearly black scab (see figure), followed by muscle pain, internal hemorrhage (bleeding), headache, fever, nausea, and vomiting.

Anthrax causes disease by releasing a toxin that latches onto receptors on the cells of the host, then punching a hole in the cell's membrane, and inserting a portion of the toxin called "lethal factor" that destroys proteins in the cell and kills it.

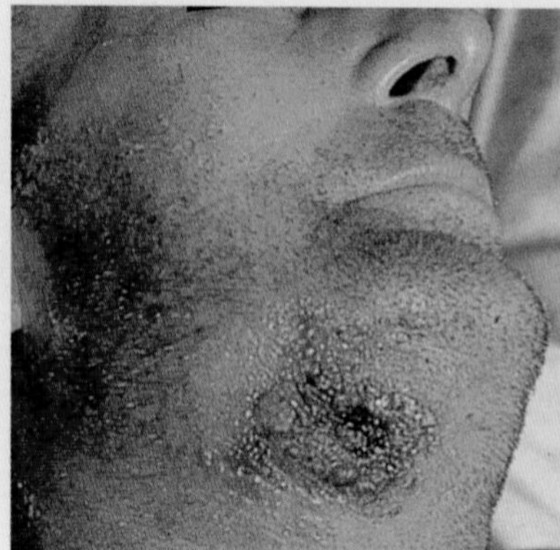

Cutaneous anthrax

If the infection is discovered before the anthrax bacteria have time to make large amounts of toxin, antibiotics such as *doxycycline* and *ciprofloxacin* can cure anthrax. Scientists also are working to perfect drugs that imitate the cell's receptors and would thus "gum up" the toxin on fake receptors before it can attack cells. Vaccines are available, but these must be given long before possible exposure to the spores.

Anthrax spores have been refined for military purposes, even though this is outlawed by various global treaties, and have been used by terrorists to attempt to intimidate or disrupt civilian populations. Such a situation occurred in the United States when anthrax-contaminated packages were sent through the mail in the fall of 2001.

Other bacteria such as *Yersinia pestis* (plague), viruses such as smallpox, and a variety of genetically engineered forms of known pathogens continue to be added to the potential arsenal of terrorists.

6

other regions of the body. **Cancer** is another term for a malignant tumor.

Benign tumors are called such because they do not spread to other tissues and they usually grow very slowly. Their cells are often well differentiated, unlike the undifferentiated cells typical of malignant tumors. Cells in a benign tumor tend to stay together, and they are often surrounded by a capsule of dense tissue. Benign tumors are usually not life threatening but can be if they disrupt the normal function of a vital organ (**Figure 6-10**).

Malignant tumors, on the other hand, are not encapsulated and do not stay in one place. Their cells tend to fall away from the original neoplasm and may start new tumors in other parts of the body. For example, cells from malignant breast tumors usually form new (secondary) tumors in bone, brain, and lung tissues. The cells migrate by way of lymphatic or blood vessels. This manner of spreading is called **metastasis.** Cells that do not metastasize still can spread, but in another way: they grow rapidly and extend the tumor into nearby tissues. Malignant tumors may replace part of a vital organ with abnormal, undifferentiated tissue—a life-threatening situation (**Figures 6-10** and **6-11**).

TABLE 6-8 Comparison of Benign and Malignant Tumors

CHARACTERISTIC	BENIGN TUMOR	MALIGNANT TUMOR
Rate of growth	Slow	Rapid
Structure	Encapsulated	Nonencapsulated (infiltrates surrounding tissue)
Pattern of growth	Expanding but not spreading to other tissues	Metastasizing (spreading) to other tissues
Cell type	Well differentiated (similar to normal tissue cells)	Undifferentiated (abnormal in structure and function)
Mortality rate	Low	High if condition remains untreated

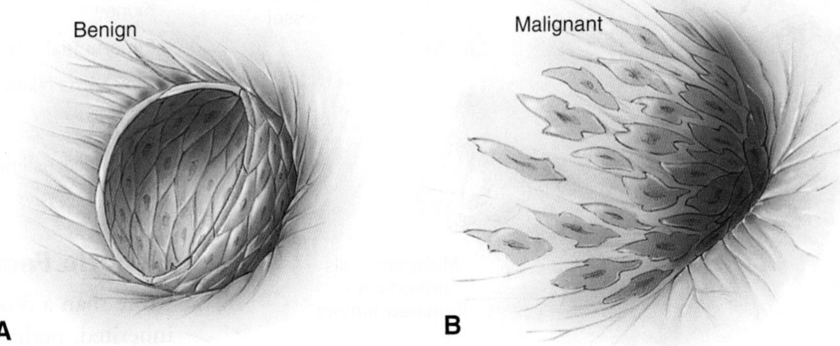

FIGURE 6-10 Types of neoplasms. A, Benign neoplasms (tumors) are usually encapsulated and grow slowly. **B,** Malignant neoplasms or cancers are not encapsulated. They grow rapidly, extending into surrounding tissues. Some cells metastasize, that is, they fall away from the original tumor and form tumors in other parts of the body.

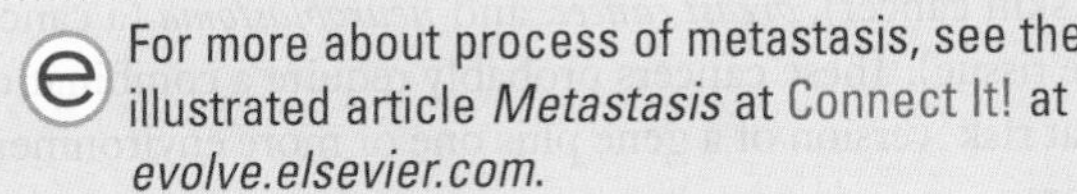

For more about process of metastasis, see the illustrated article *Metastasis* at Connect It! at *evolve.elsevier.com.*

Types of Tumors

Benign and malignant neoplasms are classified into subgroups, depending on appearance and the location where they originate. Benign and malignant tumors can be divided into three types—epithelial tissue, connective tissue, and miscellaneous tumors. Some examples of each follow:

1. Benign tumors that arise from epithelial tissues
   a. **Papilloma**—a type of tumor that forms a finger-like projection, as in a wart
   b. **Adenoma**—a general term for benign tumors of glandular epithelium
   c. **Nevus**—a variety of small, pigmented tumors of the skin, such as moles
2. Benign tumors that arise from connective tissues
   a. **Lipoma**—a tumor arising from adipose (fat) tissue
   b. **Osteoma**—a tumor that involves bone tissues
   c. **Chondroma**—a tumor of cartilage tissue
3. Malignant tumors that arise from epithelial tissues, generally called **carcinomas**
   a. **Melanoma**—a type of cancer that involves melanocytes, the pigment-producing cells of the skin
   b. **Adenocarcinoma**—the general term for malignant tumor of glandular epithelium
4. Malignant tumors that arise from connective tissues, generally called **sarcomas**
   a. **Lymphoma**—a term used to describe a cancer of lymphatic tissue
   b. **Osteosarcoma**—a term that refers to a malignant tumor of bone tissue
   c. **Myeloma**—a type of malignant bone marrow tumor
   d. **Fibrosarcoma**—a general term used to describe cancers involving fibrous connective tissues.

Miscellaneous tumors do not fit any of the other categories. For example, an **adenofibroma** is a benign neoplasm formed by epithelial and connective tissues. Another example is **neuroblastoma,** a malignant tumor that arises from nerve tissue.

Cancers can be further classified by their location. For example, malignant tumors may be labeled *skin cancer, stomach cancer,* or *lung cancer* according to the location of the affected

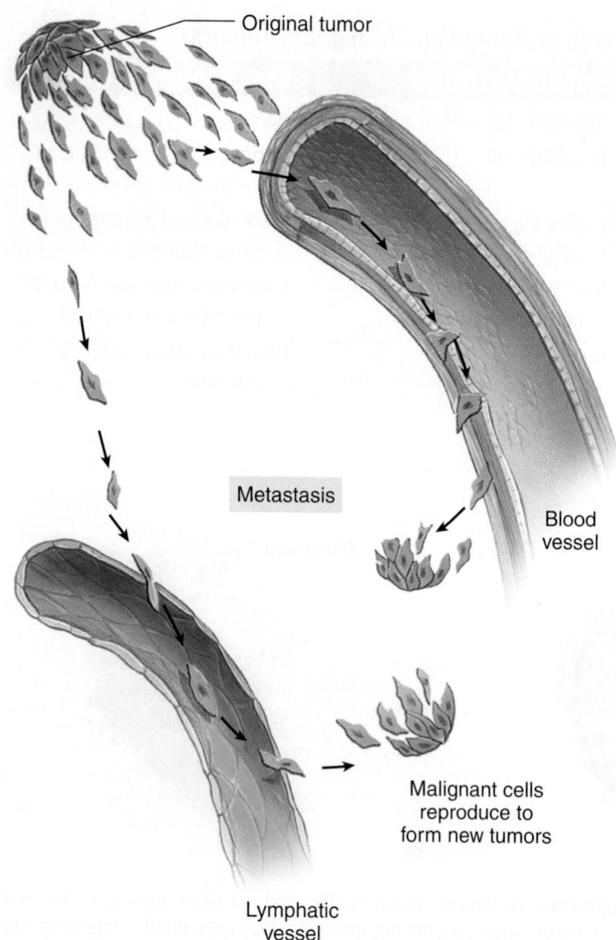

FIGURE 6-11 Metastasis. Abnormal cells from malignant tumors fall away from the original neoplasm and travel along lymphatic vessels, through which they can enter and exit easily. Malignant cells also can travel through the bloodstream and burrow through a blood vessel wall to invade other tissues.

tissues. The more common cancer types (by location) in the United States are listed in **Table 6-9** and are described in later chapters.

Causes of Cancer

Abnormal Cell Division

The etiologies of various forms of cancer puzzle researchers no less today than they did 100 years ago. The more we know about how cancer develops, the more questions we have. Currently, the best answer to the question "What causes cancer?" is "Many different things."

We know that cancer is a type of neoplasm, which means that it involves uncontrolled cell division. A process called **hyperplasia** produces too many cells. Also, abnormal, undifferentiated tumor cells are often produced by a process called **anaplasia.** Thus the mechanism of all cancers is a mistake or problem in cell division. However, science remains uncertain of all the possible triggers of the abnormal cell division.

Currently, the factors listed in the following sections are known to play a role.

TABLE 6-9 Common Forms of Cancer*

TYPE (by location)	NEW CASES (per year)	DEATHS (per year)
Lung	224,390	158,080
Colorectal	134,490	49,190
Pancreatic	53,070	41,780
Breast	249,260	40,890
Female	246,660	40,450
Male	2,600	440
Prostate	180,890	26,120
Leukemia (blood cancer)	60,140	24,400
Non-Hodgkin lymphoma (lymphoid tissue cancer)	72,580	20,150
Bladder	76,960	16,390
Kidney	62,700	14,240
Endometrial (uterus)	60,050	10,470
Melanoma (skin)	76,380	10,130
Thyroid	64,300	1,980

*2016 annual estimates in the United States, listed in order of death rate.

Genetic Factors

More than a dozen forms of cancer are known to be directly inherited, perhaps involving abnormal "cancer genes" called **oncogenes.** The way in which every known oncogene works is not yet clearly understood and is likely to involve a number of different mechanisms.

Other cancers may develop primarily in those people with genetic predispositions to specific forms of cancer. Cancers with known genetic risk factors include *basal cell carcinoma* (a type of skin cancer), *breast cancer,* and *neuroblastoma* (a cancer of nerve tissue). These cancers probably require a combination of the "at risk" version of a gene plus one or more environmental factors.

Carcinogens

Carcinogens ("cancer makers") are chemicals that affect genetic activity in some way, causing abnormal cell reproduction. Some carcinogens are **mutagens** ("mutation makers"). Mutagens cause changes in a cell's DNA structure. Although many industrial products such as benzene are known to be carcinogens, a wide variety of natural vegetable and animal materials are also carcinogenic. Tobacco, for example, contains carcinogens.

Age

Certain cancers are found primarily in young people (for example, leukemia) and others primarily in older adults (for example, colon cancer). The age factor may result from changes in the genetic activity of cells over time or from accumulated effects of cell damage.

Environment

Exposure to damaging types of radiation or chronic mechanical injury can cause cancer. For example, sunlight can cause

TABLE 6-10 The Warning Signs of Cancer*

Sores that do not heal
Unusual bleeding
A change in a wart or mole
A lump or thickening in any tissue
Persistent hoarseness or cough
Chronic indigestion or difficulty swallowing
A change in bowel or bladder function
Bone pain that wakes one at night and is located on only one side

*Any chronic change in body structure or function could be a sign of cancer and should be investigated by a physician.

skin cancer, and breathing asbestos fibers can cause lung cancer. Also, exposure to high concentrations of certain metals such as nickel or chromium can cause tumors to develop.

Viruses

Several cancers have now been identified as having a viral origin. This makes sense because we know that viruses often change the genetic machinery of infected cells. For example, *human papillomaviruses (HPVs)* have been found to have a causal relationship in some cases of cervical cancer in women and penile cancer in men. Thus, HPV vaccines can prevent both the viral infections and the resulting cancers.

Pathogenesis of Cancer

Detecting Cancer

Signs of cancer include those a person would expect of a malignant neoplasm—the appearance of abnormal, rapidly growing tissue. Cancer specialists, or *oncologists,* have summarized some major signs of early stages of cancer. These signs are listed in **Table 6-10**.

Early detection of cancer is important because in the early stages of development of primary tumors, before metastasis and the development of secondary tumors has begun, cancer is most treatable. Some methods currently used to detect the presence of cancer include those described in the following sections.

Self-Examination

Examining one's self for the early signs of cancer is a surprisingly effective method of detection. For example, women are encouraged to perform a monthly breast self-examination. Likewise, men are encouraged to perform a monthly testicular self-examination. Self-examination of the skin and other accessible organs or tissues is also recommended by cancer specialists.

If an abnormality is found, it can be further investigated with one of the methods described later.

Diagnostic Imaging

A variety of methods are available for forming images of internal body organs to detect tumors without exploratory surgery.

Radiography is the oldest and still the most widely used method of noninvasive imaging of internal body structures. Radiography is the use of *x-rays* to form a still or moving picture of some of the internal tissues of the body. A *mammogram,* for example, is an x-ray photograph of a breast. Potentially cancerous lumps show up as small, white areas on the mammogram (**Figure 6-12**, *A*).

Computed tomography (CT) scanning is a type of radiography in which x-rays produce a cross-sectional image of body regions (**Figure 6-12**, *B*).

Magnetic resonance imaging (MRI) is a type of scanning that uses a magnetic field to induce tissues to emit radio waves. Different tissues can be distinguished because each emits different signals. With MRI, tumors can then be visualized on a computer screen in cross sections similar to those produced in CT scanning (**Figure 6-12**, *C*). MRI is also sometimes called *nuclear magnetic resonance (NMR)* imaging.

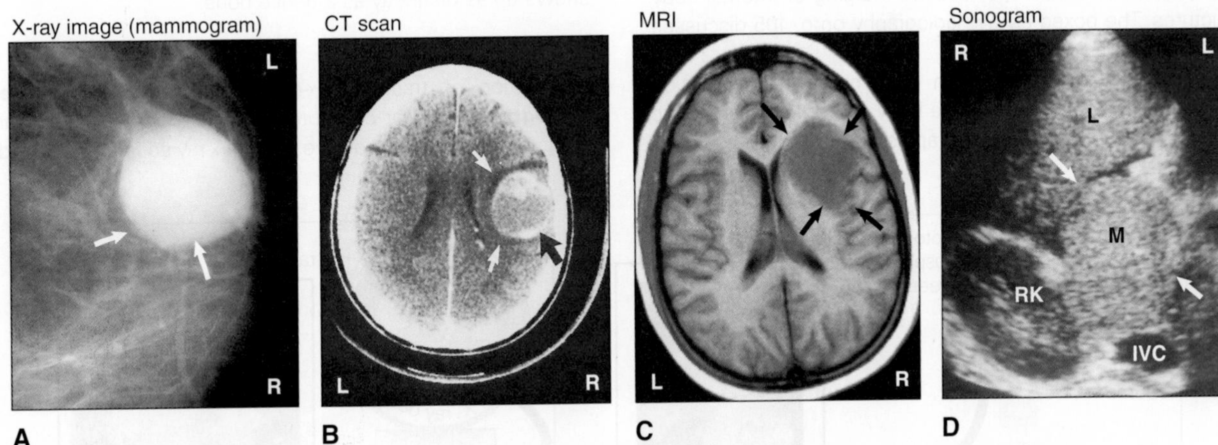

FIGURE 6-12 Medical images of tumors. A, A mammogram (x-ray image) showing carcinoma of a breast duct. **B,** CT scan of the brain showing a tumor in the left hemisphere. **C,** MR image of the brain showing a tumor in the left hemisphere. **D,** Sonogram showing a transverse view of an abdominal tumor. L, Left; R, right; IVC, inferior vena cava; L, liver; M, mass; RK, right kidney.

6

In **ultrasonography,** high-frequency sound waves can be reflected off internal tissues to produce images, or *sonograms,* of tumors (**Figure 6-12,** *D*). For more detailed information regarding medical imaging, refer to the Clinical Application box below.

Review additional information and examples of medical images in the article *Medical Imaging of the Body* at Connect It! at *evolve.elsevier.com.*

Biopsy

After a neoplasm has been identified with one of the previously mentioned techniques, a **biopsy** of the tumor may be done. Biopsy of an accessible tumor may precede or even eliminate the need for extensive medical imaging.

A biopsy is the removal and examination of living tissue. Microscopic examination of tumor tissue removed surgically or through a needle sometimes reveals whether the tissue is malignant or benign.

A very simple, noninvasive type of biopsy used to detect some types of cancer involves simply scraping or brushing cells from an exposed surface and smearing them on a glass microscope slide. For example, the **Papanicolaou test** or *"Pap smear"* is a common screening procedure in which cells from the neck of the uterus (cervix) are examined (see Chapter 23).

To see an example of a needle biopsy, see the article *Kidney Biopsy* at Connect It! at *evolve.elsevier.com.*

Blood Test

Changes in the concentration of normal blood components, such as ions or enzymes, sometimes can indicate cancer.

CLINICAL APPLICATION

MEDICAL IMAGING OF THE BODY

Cadavers (preserved human bodies used for scientific study) can be cut into sagittal, frontal, or transverse sections for easy viewing of internal structures, but living bodies, of course, cannot. This limitation has hampered medical professionals who strive to determine whether internal organs are injured or diseased. In some cases, the only sure way to detect a lesion or variation from normal is by performing extensive *exploratory surgery.* Fortunately, advances in medical imaging allow physicians to visualize internal structures of the body without risking the trauma or other complications associated with extensive surgery. Some of the more widely used techniques are briefly described here.

Radiography

Radiography, or x-ray photography, is the oldest and still most widely used method of noninvasive imaging of internal body structures. The boxed essay Radiography on p. 105 discusses the origins of radiography.

With this method, energy in the x band of the radiation spectrum is beamed through the body and onto photographic film (Figure **A**). The x-ray photograph shows the outlines of the bones and other dense structures that partially absorb the x-rays.

In *fluoroscopy,* a phosphorescent screen sensitive to x-rays is used instead of photographic film. A visible image is formed on the screen as x-rays passing through the subject cause the screen to glow. Fluoroscopy allows a medical professional to view the internal structures of the subject's body in real time as it moves.

Without enhancement aids, radiography works best as a tool to view solid objects such as bones. One way to make soft, hollow structures such as blood vessels or digestive organs more visible is to use *radiopaque* contrast media. Substances such as barium sulfate that absorb x-rays are injected into or swallowed by the patient to fill the hollow organ of interest. As the screen in Figure **A** shows, the hollow organ then shows up as distinctly as a dense bone.

Computed Tomography

A variation of traditional x-ray photography is **computed tomography (CT)** or *computed axial tomography (CAT)* scanning. In this method, a device with an x-ray source on one side

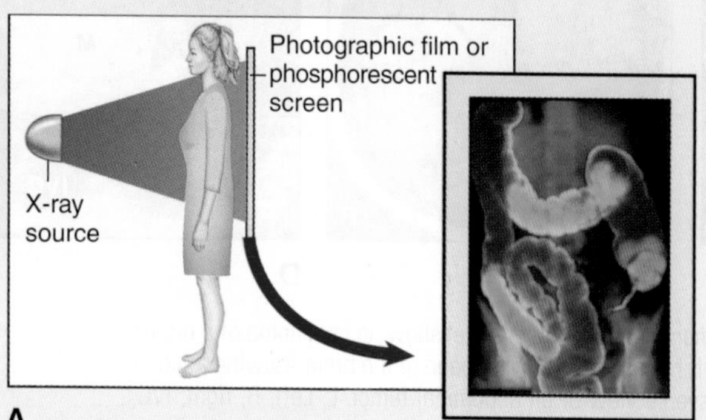

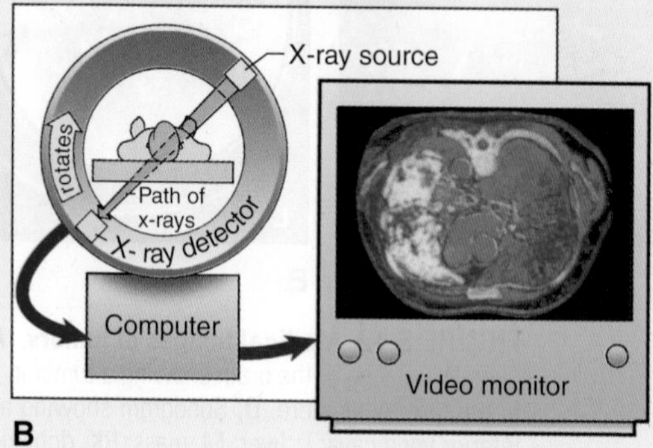

Cancer cells also may produce or trigger production of abnormal substances—substances often referred to as *tumor markers*. For example, bone cancer and some other malignancies can elevate the blood concentration of calcium ions (Ca^{++}) above normal levels. Blood tests to help detect tumor markers of prostate and other cancers are being developed and introduced.

Stages and Grades of Cancer

The information gained from these and other techniques can be used to *stage* and *grade* malignant tumors. *Staging* involves classifying a tumor based on its size and the extent of its spread. *Grading* is an assessment of what the tumor is likely to do, based on the degree of cell abnormality. Grading is a useful basis for making a **prognosis,** or statement of the probable outcome of the disease.

Without treatment, cancer may result in death. The progress of a particular type of cancer depends on the type of cancer and its location. Many cancer patients suffer from **cachexia,** a syndrome involving loss of appetite, severe weight loss, and general weakness. The cause of cachexia in cancer patients is uncertain. A variety of anatomical or functional abnormalities may arise as a result of damage to particular organs. The ultimate causes of death in cancer patients include secondary infection by pathogenic microbes, organ failure, hemorrhage (blood loss), and in some cases, undetermined factors.

Cancer Treatment

Of course, after cancer has been identified, staged, and graded, every effort is made to treat it and thus prevent or delay its development.

Surgical removal of cancerous tumors is preferable; although for anatomic reasons, that is not always possible. Even with surgical removal, the possibility that malignant cells have been left behind must be addressed.

Chemotherapy, or "chemical therapy," using *cytotoxic* ("cell-killing") compounds or *antineoplastic* drugs can be used

of the body and an x-ray detector on the other side is rotated around a central axis of the subject's body (Figure **B**). Information from the x-ray detectors is interpreted by a computer, which generates a video image of the body as if it were cut into anatomical sections.

The term *computed tomography* literally means "picturing a cut using a computer." Because CT scanning and other recent advances in diagnostic imaging produce images of the body as if it were actually cut into sections, it has become especially important for students of the health sciences to become familiar with *sectional anatomy,* which is the study of the structural relationships visible in anatomical sections. Find examples of sectional anatomy in margins of the *Clear View of the Human Body* (follows p. 8).

Magnetic Resonance Imaging

Magnetic resonance imaging (MRI) is a type of scanning that uses a magnetic field to induce tissues to emit radiofrequency (RF) waves. An RF detector coil senses the waves and sends the information to a computer that constructs sectional images similar to those produced in CT scanning (Figure **C**). Different tissues can be distinguished from each other because each emits different radio signals. MRI, also called *nuclear magnetic resonance (NMR)* imaging, avoids the use of potentially harmful x radiation and often produces sharper images of soft tissues than other imaging methods.

Ultrasonography

During **ultrasonography,** high-frequency (ultrasonic) waves are reflected off internal tissues to produce an image called a *sonogram* (Figure **D**).

Because it does not involve x radiation, and because it is relatively inexpensive and easy to use, ultrasonography has been used extensively—especially in studying maternal or fetal structures in pregnant women. However, the image produced is not as clear or sharp as those produced by MRI, CT scanning, or traditional radiography.

Variations of these and other technological advances that have improved the ability to study the structure and functions of the human body are discussed more in later chapters.

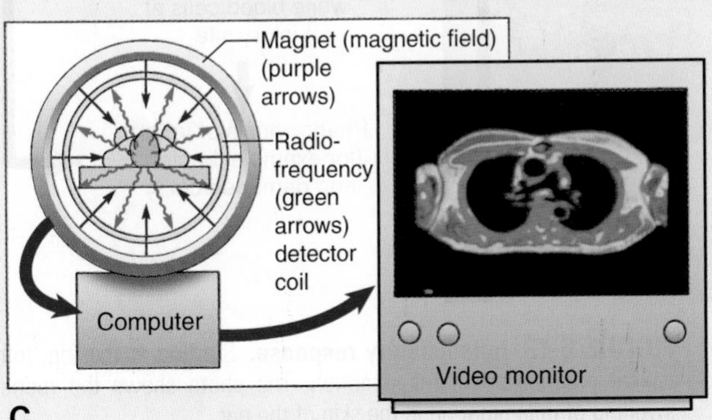

C

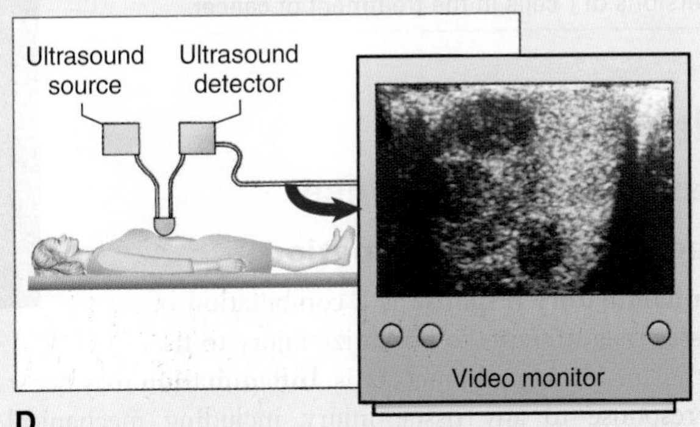

D

after surgery to destroy any remaining malignant cells. A more recent approach is the use of *rational drugs* in chemotherapy. Rational drugs are those that target only specific molecules, enzymes, or receptors unique to cancer cells or tumor growth, thereby affecting only the cancer and sparing the normal cells. Rational drugs thus increase the efficiency of chemotherapy and reduce its side effects.

Radiation therapy, also called *radiotherapy,* using destructive x-ray or gamma radiation may be used alone or with chemotherapy to destroy remaining cancer cells. Chemotherapy and radiation therapy may have severe side effects because normal cells are often killed along with the cancer cells.

Laser therapy, in which an intense beam of light destroys a tumor, is also sometimes performed in addition to chemotherapy or radiation therapy.

Immunotherapy, a newer type of cancer treatment, bolsters the body's own defenses against cancer cells. Because viruses cause some types of cancer, oncologists hope that vaccines against certain forms of cancer will be developed.

There has been some success with the treatment of people with cancers such as advanced *chronic lymphocytic leukemia (CLL)* by using genetically engineered versions of their own T cells. The newly programmed T cells have been shown to remove specially targeted cancerous cells throughout the patients' bodies for several months. Genetic engineering of the host immune system is hoped to provide ongoing protection against recurrence of several types of cancers.

Although new and different approaches to cancer treatment are being investigated, many researchers are concentrating on improving existing methods and promoting cancer prevention. Despite progress in reducing cancers in developed countries, the steep rise in smoking in developing regions threatens to make cancer the major killer worldwide.

QUICK CHECK

1. What is metastasis?
2. Give examples of benign and malignant tumors that arise from epithelial tissue and benign and malignant tumors that arise from connective tissue.
3. Name five general causes of cancer.
4. What are four methods that are used to detect cancer?
5. How is cancer treated?
6. Describe the use of genetically engineered versions of T cells in the treatment of cancer.

Inflammation

Inflammatory Response

Introduction to Inflammation

The **inflammatory response** is a combination of processes that attempts to minimize injury to tissues, thus maintaining homeostasis. **Inflammation** may occur as a response to any tissue injury, including mechanical injuries such as cuts and burns or damage caused by many other irritants such as chemicals, radiation, or toxins released by bacteria. The processes of inflammation eventually eliminate the irritant, after which tissue repair can begin.

As you learned in Chapter 4, tissue repair is the replacement of dead cells with living cells. In a type of tissue repair called *regeneration,* the new cells are similar to those that they replace. Another type of tissue repair is *replacement.* In replacement, the new cells are different from those that they replace, resulting in a scar. Often, *fibrous* tissue replaces the old tissue, a condition called **fibrosis.** Most tissue repairs are a combination of regeneration and replacement.

Inflammation also may accompany specific immune system reactions (which are discussed in Chapter 16). First described by a Roman physician almost 2000 years ago, the inflammatory response has four primary signs—redness, heat, swelling, and pain. These signs are indicators of a complex process that is summarized in the following paragraphs and in Figures 6-13 and 6-14.

Mechanisms of Inflammation

As tissue cells are damaged, they release **inflammation mediators** such as **histamine, prostaglandins (PGs),** and compounds called **kinins.**

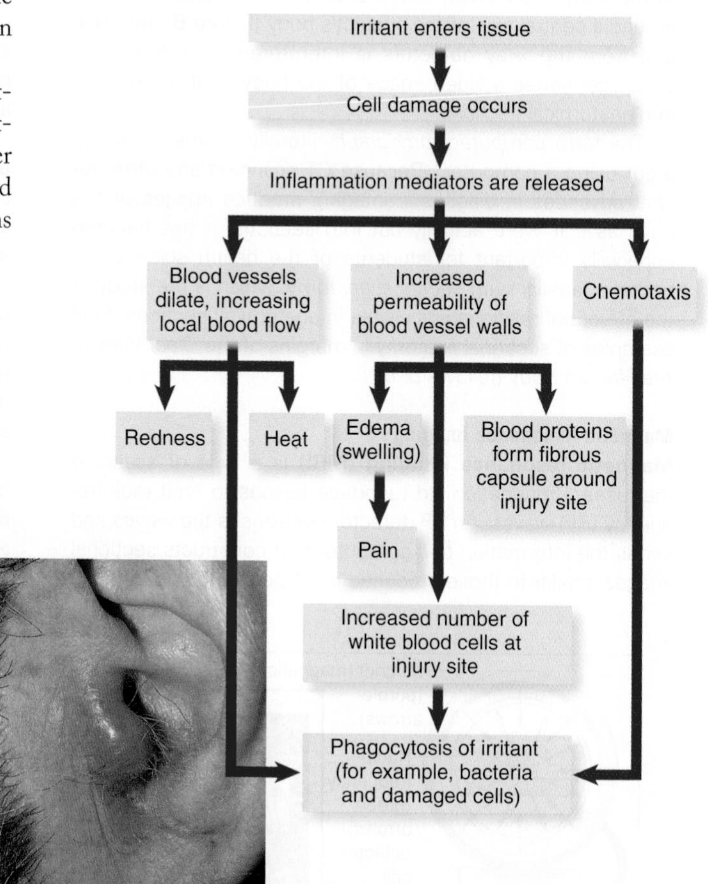

FIGURE 6-13 Inflammatory response. Starting at the top, follow the typical progression of inflammation. The photo shows the redness and swelling of inflammation in the skin of the ear.

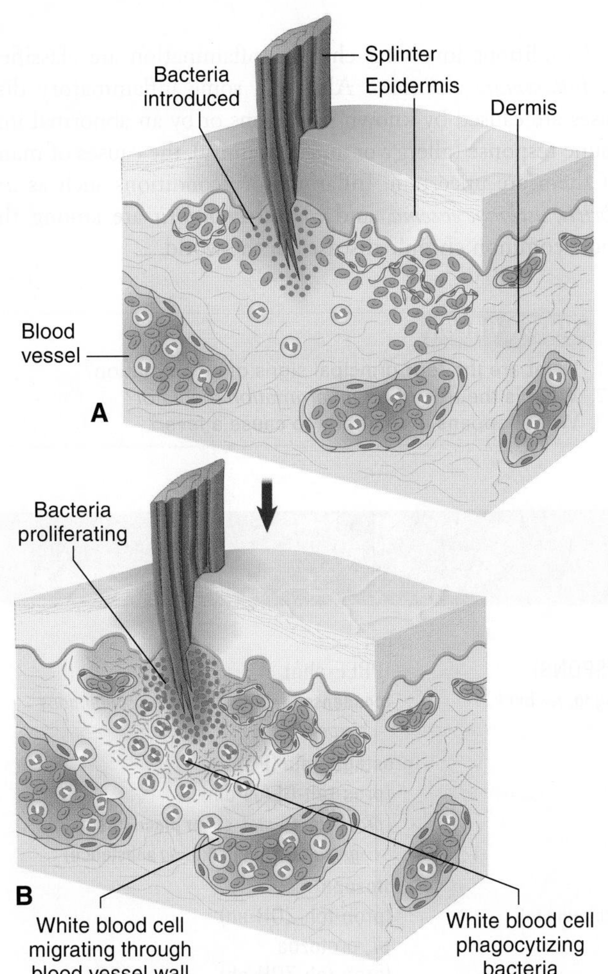

FIGURE 6-14 Typical inflammatory response to a mechanical injury. A, A splinter damages tissue and carries bacteria into the body. Blood vessels dilate and begin leaking fluids, causing swelling and redness. **B,** White blood cells are attracted to the injury site and begin to consume bacteria and damaged tissue cells. A fibrous capsule separates the injury site from surrounding tissue.

Some inflammation mediators cause blood vessels to dilate (widen), increasing blood volume in the tissue. Increased blood volume produces the redness and heat of inflammation. This response is important because it allows immune system cells (white blood cells) in the blood to travel quickly and easily to the site of injury.

Some inflammation mediators increase the permeability of blood vessel walls. This allows immune system cells and other blood components to move out of the blood vessels easily where they can deal directly with injured tissue. As water leaks out of the vessel, tissue swelling, or **edema,** results. The pressure caused by edema triggers pain receptors, consciously alerting an individual of the damage. The excess fluid often has the beneficial effect of diluting the irritant.

The fluid that accumulates in inflamed tissue is called **inflammatory exudate.** Blood proteins that leak into tissue spaces begin to clot within a few minutes. The clot forms a fibrous capsule around the injury site, preventing the irritant from spreading to nearby tissues.

Inflammatory exudate is slowly removed by lymphatic vessels and is carried to lymph nodes, which act as filters. Bacteria and damaged cells trapped in the lymph nodes are acted on by white blood cells in each lymph node. In some cases, lymph nodes enlarge when they are processing a large amount of infectious material.

Inflammation mediators also can act as signals that attract white blood cells to the injury site. The movement of white blood cells in response to chemical attractants is called **chemotaxis.**

Once in the affected tissue, white blood cells often consume damaged cells and pathogenic bacteria by means of *phagocytosis.* When the inflammatory exudate becomes thick with the accumulation of white blood cells, dead tissue and bacterial cells, and other debris, **pus** is formed.

Occasionally, the inflammatory response is more intense or prolonged than desirable. In such a case, inflammation can be suppressed by drugs such as antihistamines or aspirin. **Antihistamines** block the action of histamine, as their name implies. *Aspirin* disrupts the body's synthesis of PGs, a group of inflammation mediators.

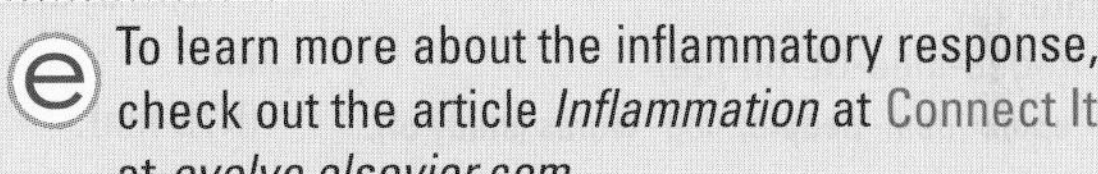
To learn more about the inflammatory response, check out the article *Inflammation* at Connect It! at *evolve.elsevier.com.*

To learn more about acute inflammatory response, go to AnimationDirect online at *evolve.elsevier.com.*

Inflammatory Disease

Local and Systemic Inflammation

Although many inflammation events are *local,* some affect the entire body, producing *systemic* inflammation. Local inflammation occurs when damage caused by an irritant remains isolated in a limited area, as in a small cut that becomes infected. Systemic inflammation occurs when the irritant spreads widely throughout the body or when inflammation mediators cause changes throughout the body.

Fever

One example of a systemic (body-wide) manifestation of the inflammatory response is a **fever.** The irritant or inflammation mediators can cause the "thermostat" of the brain to reset at a higher temperature. Instead of the normal body temperature, the body achieves and maintains a new, higher temperature.

Increased temperature often kills or inhibits pathogenic microbes. Some pathophysiologists also believe that the higher temperature enhances the activity of the immune system. Fevers usually subside or "break" after the irritant has been eliminated. Fevers also can be reduced by drugs that block the fever-producing agents.

The fever response in children and in the elderly often differs from that in the normal adult. Young children often develop very high temperatures in response to mild infections as

6

compared with adults, sometimes causing **febrile seizures**—abnormal brain activity caused by fever.

Elderly people often have reduced or absent fever responses during infections, which may reduce their ability to resist the infectious agent.

Acute and Chronic Inflammation

Acute inflammation is an immediate, protective response that promotes elimination of an irritant and subsequent tissue repair. See the photo inset in **Figure 6-13**.

Occasionally, chronic inflammatory conditions occur. Chronic inflammation, whether local or systemic, is always damaging to affected tissues.

Conditions involving chronic inflammation are classified as *inflammatory diseases.* Although some inflammatory diseases are caused by known pathogens or by an abnormal immune response (allergy or autoimmunity), the causes of many of them are uncertain. Inflammatory conditions such as *arthritis, asthma, eczema,* and *chronic bronchitis* are among the most common chronic diseases in the world.

QUICK CHECK

1. What are the four principal signs of inflammation?
2. What is the role of an *inflammation mediator*?
3. What happens in the body to cause a *fever*?

LANGUAGE OF SCIENCE (continued from p. 113)

coronavirus
(koh-ROHN-ah-vye-rus)
[*corona-* **crown,** *-virus* **poison]**

epidemiology
(EP-ih-dee-mee-OL-oh-jee)
[*epi-* **upon,** *-dem-* **people,** *-log-* **words (study of),** *-y* **activity]**

flagellate
(FLAJ-eh-lat)
[*flagell-* **whip,** *-ate* **of or like]**

flagellum
(flah-JEL-um)
pl., flagella
(flah-JEL-ah)
[*flagellum* **whip]**

flavivirus
(FLAY-vih-vye-rus)
[*flav-* **yellow,** *-virus* **poison]**

fungus
(FUNG-gus)
pl., fungi or funguses
(FUN-jye or FUN-gye or FUN-gus-ez)
[*fungus* **mushroom]**

Gram staining technique
(gram STAYN-ing tek-NEEK)
[Hans Christian Joachim Gram **Danish pathologist]**

histamine
(HIS-tah-meen)
[*hist-* **tissue,** *-amine* **ammonia compound]**

inflammation
(in-flah-MAY-shun)
[*inflam-* **set afire,** *-tion* **process]**

inflammation mediator
(in-flah-MAY-shun MEE-dee-ay-tor)
[*inflam-* **set afire,** *-tion* **process,** *mediat-* **intervene,** *-or* **condition]**

inflammatory exudate
(in-FLAM-ah-toh-ree EK-soo-dayt)
[*inflam-* **set afire,** *-ory* **relating to,** *exud-* **sweat out,** *-ate* **thing]**

inflammatory response
(in-FLAM-ah-toh-ree ree-SPONS)
[*inflam-* **set afire,** *-ory* **relating to,** *re-* **back,** *-sponse* **answer]**

kinin
(KYE-nin)
[*kin-* **move,** *-in* **substance]**

metabolic
(met-ah-BOL-ik)
[*meta-* **over,** *-bol-* **throw,** *-ic* **relating to]**

metazoan
(met-uh-ZOH-an)
pl., metazoa
(met-uh -ZOH-ah)
[*meta-* **over,** *-zoan* **animal]**

microbe
(MY-krobe)
[*micro-* **small,** *-b(io)* **life]**

nematode
(NEM-ah-tohd)
[*nema-* **thread,** *-ode* **like]**

oncogene
(ON-koh-jeen)
[*onco-* **swelling or mass (cancer),** *-gen-* **produce or generate]**

parasite
(PAYR-ah-syte)
[*para-* **beside,** *-site* **food]**

pathogenesis
(path-oh-JEN-e-sis)
[*patho-* **disease,** *-genesis* **origin]**

pathophysiology
(path-oh-fiz-ee-OL-oh-jee)
[*patho-* **disease,** *-physio-* **nature (function),** *-log-* **words (study of),** *-y* **activity]**

platyhelminth
(plat-ih-HEL-minth)
[*platy-* **broad, flat,** *-helminth* **worm]**

prion
(PREE-ahn)
[condensed from *proteinaceous infectious particle***]**

prostaglandin (PG)
(pros-tah-GLAN-din)
[*pro-* **before,** *-stat-* **set or place (prostate),** *-gland-* **acorn (gland),** *-in* **substance]**

protozoan
(proh-toh-ZOH-an)
pl., protozoa
(proh-toh-ZOH-ah)
[*proto-* **first,** *-zoan* **animal]**

psychogenic
(sye-koh-JEN-ik)
[*psycho-* **the mind,** *-gen-* **production,** *-ic* **relating to]**

pus
(pus)
[*pus* **rotten]**

retrovirus
(ret-roh-VYE-rus)
[*retro-* **backward,** *-virus* **poison]**

risk factor
(risk FAK-tor)
[*risk* **run into danger,** *factor* **agent]**

spore
(spor)
[*spore* **seed]**

sporozoan
(spor-oh-ZOH-an)
pl., sporozoa
(spor-oh-ZOH-ah)
[*spor-* **seed,** *-zoan* **animal]**

vector
(VEK-tor)
[*vector* **carrier]**

virus
(VYE-rus)
[*virus* **poison]**

LANGUAGE OF **MEDICINE**

acquired immune deficiency syndrome (AIDS)
(ah-KWY-erd ih-MYOON deh-FISH-en-see SIN-drohm [aydz])
[*immun-* **free**, *-de-* **down**, *-fic-* **perform**, *-ency* **state**, *syn-* **together**, *-drome* **running or (race) course**]

acute
(ah-KYOOT)
[*acut-* **sharp**]

adenocarcinoma
(ad-eh-noh-kar-sih-NOH-mah)
[*adeno-* **gland**, *-carcin-* **cancer**, *-oma* **tumor**]

adenofibroma
(ad-eh-noh-fye-BROH-mah)
[*adeno-* **gland**, *-fibr-* **fiber**, *-oma* **tumor**]

adenoma
(ad-eh-NOH-mah)
[*adeno-* **gland**, *-oma* **tumor**]

anaplasia
(an-ah-PLAY-zhah or an-ah-PLAY-zee-ah)
[*ana-* **without**, *-plasia* **shape**]

anthrax
(AN-thraks)
[*anthrax* **boil**]

antibiotic
(an-tih-by-OT-ik)
[*anti-* **against**, *-bio* **life**, *-ic* **relating to**]

antihistamine
(an-tih-HIS-tah-meen)
[*anti-* **against**, *-histo-* **tissue**, *-amine* **ammonia compound**]

antiviral drug
(an-tee-VYE-ral [or an-tih-VYE-ral] drug)
[*anti-* **against**, *-vir-* **poison**, *-al* **relating to**]

aseptic technique
(ay-SEP-tik tek-NEEK)
[*a-* **without**, *-septi-* **putrid**, *-ic* **relating to**, *techn-* **method**]

benign
(bee-NYNE)
[*benign* **kind**]

biopsy
(BYE-op-see)
[*bio-* **life**, *-ops-* **view**, *-y* **act of**]

bioterrorism
(bye-oh-TAYR-or-iz-em)
[*bio-* **life**, *-terror-* **fear**, *-ism* **condition**]

bovine spongiform encephalopathy (BSE)
(BOH-vyne SPUN-jeh-form en-sef-uh-LOP-uh-thee [bee es ee])
[*bovi-* **ox or cow**, *-ine* **of or like**, *spongi-* **sponge**, *-form* **shape**, *-en-* **within**, *-cephalo-* **head**, *-path-* **disease**, *-y* **state**]

cachexia
(kah-KEKS-ee-ah)
[*cache-* **bad**, *-(h)exia* **state**]

cancer
(KAN-ser)
[*cancer* **crab or malignant tumor**]

carcinogen
(kar-SIN-oh-jen)
[*carcino-* **cancer**, *-gen* **produce**]

carcinoma
(kar-sih-NOH-mah)
[*carcin-* **cancer**, *-oma* **tumor**]

chemotherapy
(kee-moh-THAYR-ah-pee)
[*chemo-* **chemical**, *-therapy* **treatment**]

Chlamydia
(klah-MID-ee-ah)
[*chlamyd-* **short mantle**, *-ia* **condition**]

chondroma
(kon-DROH-mah)
[*chondr-* **cartilage**, *-oma* **tumor**]

chronic
(KRON-ik)
[*chron-* **time**, *-ic* **relating to**]

communicable
(kom-MYOO-nih-kah-bil)
[*communic-* **share**, *-able* **capacity**]

computed tomography (CT)
(kom-PYOO-ted toh-MOG-rah-fee [see tee])
[*com-* **together**, *-pute-* **think**, *tomo-* **cut**, *-graph-* **draw**, *-y* **process**]

culture
(KULT-chur)
[*cultur-* **till land**]

degeneration
(dih-jen-uh-RAY-shun)
[*de-* **down**, *-generat-* **produce**, *-tion* **condition**]

dengue
(DENG-gay or DENG-gee)
[*dengue* **seizure or cramp**]

disease
(dih-ZEEZ)
[*dis-* **opposite of**, *-ease* **comfort**]

edema
(eh-DEE-mah)
[*edema* **swelling**]

endemic
(en-DEM-ik)
[*en-* **in**, *-dem-* **people**, *-ic* **relating to**]

environmental factor
(en-VYE-ron-ment-al FAK-tor)
[*environ-* **surround**, *-ment-* **condition**, *-al* **relating to**, *factor* **agent**]

environmental health
(en-VYE-ron-ment-al helth)
[*environ-* **surround**, *-ment-* **condition**, *-al* **relating to**]

epidemic
(ep-ih-DEM-ik)
[*epi-* **upon**, *-dem-* **people**, *-ic* **relating to**]

etiology
(ee-tee-AHL-oh-jee)
[*eti-* **cause**, *-o-* **combining vowel**, *-log-* **words (study of)**, *-y* **activity**]

febrile seizure
(FEB-ril SEE-zhur)
[*febri-* **fever**, *-ile* **characterized by**, *seiz-* **grasp suddenly**, *-ure* **action**]

fever
(FEE-ver)
[*fev-* **heat**]

fibrosarcoma
(fye-broh-sar-KOH-mah)
[*fibro-* **fiber**, *-sarco-* **flesh**, *-oma* **tumor**]

fibrosis
(fye-BROH-sis)
[*fibr-* **fiber**, *-osis* **condition**]

genetic
(jeh-NET-iks)
[*gene-* **produce**, *-ic* **relating to**]

human immunodeficiency virus (HIV)
(HYOO-mon ih-myoo-noh-deh-FISH-en-see VYE-rus [aych aye vee])
[*immuno-* **free (immunity)**, *-de-* **down**, *-fic-* **perform**, *-ency* **state**, *virus* **poison**]

hyperplasia
(hye-per-PLAY-zhah or hye-per-PLAY-zee-ah)
[*hyper-* **excessive**, *-plasia* **shape**]

idiopathic
(id-ee-oh-PATH-ik)
[*idio-* **peculiar**, *-path-* **disease**, *-ic* **relating to**]

immunotherapy
(im-yoo-noh-THAYR-ah-pee)
[*immuno-* **free**, *-therapy* **treatment**]

incubation
(in-kyoo-BAY-shun)
[*in-* **in or on**, *-cubat-* **lie**, *-tion* **process**]

infectious
(in-FEK-shus)
[*infect-* **stain**, *-ous* **relating to**]

inflammatory
(in-FLAM-ah-toh-ree)
[*inflam-* **set afire**, *-ory* **relating to**]

Continued on p. 138

LANGUAGE OF **MEDICINE** *(continued from p. 137)*

laser therapy
(LAY-zer THAYR-ah-pee)
[*laser* **acronym for light amplification by stimulated emission of radiation**]

lifestyle
(LYFE-style)

lipoma
(lih-POH-mah)
[*lip-* **fat,** *-oma* **tumor**]

lymphoma
(lim-FOH-mah)
[*lymph-* **water,** *-oma* **tumor**]

magnetic resonance imaging (MRI)
(mag-NET-ik REZ-ah-nens IM-ah-jing [em ar aye])
[*magnet-* **lodestone,** *-ic* **relating to,** *re-* **again,** *-sona-* **sound,** *-ance* **state**]

malignant
(mah-LIG-nant)
[*malign-* **bad** *-ant* **state**]

melanoma
(mel-ah-NOH-mah)
[*melan-* **black,** *-oma* **tumor**]

metastasis
(meh-TAS-tah-sis)
[*meta-* **change,** *-stasis* **standing**]

morbidity
(mor-BID-ih-tee)
[*morbid-* **sick,** *-ity* **condition**]

mortality
(mor-TAL-ih-tee)
[*mortal-* **subject to death,** *-ity* **condition**]

mutagen
(MYOO-tah-jen)
[*mutat-* **change,** *-gen* **producer**]

mycotic infection
(my-KOT-ik in-FEK-shun)
[*myco-* **fungus,** *-ic* **relating to,** *infect-* **stain,** *-tion* **process**]

myeloma
(my-eh-LOH-mah)
[*myel-* **marrow,** *-oma* **tumor**]

neoplasm
(NEE-oh-plaz-em)
[*neo-* **new,** *-plasm* **formation**]

neoplastic
(nee-oh-PLAS-tik)
[*neo-* **new,** *-plas(m)-* **formation,** *-ic* **relating to**]

neuroblastoma
(noo-roh-blas-TOH-mah)
[*neuro-* **nerve,** *-blast-* **germ,** *-oma* **tumor**]

nevus
(NEE-vus)
[*nevus* **birthmark**]

osteoma
(os-tee-OH-mah)
[*oste-* **bone,** *-oma* **tumor**]

osteosarcoma
(os-tee-oh-sar-KOH-mah)
[*osteo-* **bone,** *-sarc-* **flesh,** *-oma* **tumor**]

pandemic
(pan-DEM-ik)
[*pan-* **all,** *-dem-* **people,** *-ic* **relating to**]

Papanicolaou test
(pah-peh-nik-oh-LAH-oo)
[*George N. Papanicolaou* **Greek physician**]

papilloma
(pap-ih-LOH-mah)
[*papill-* **nipple,** *-oma* **tumor**]

preexisting condition
(pree-ig-ZIST-ing kon-DISH-un)
[*pre-* **before,** *-existing* **being**]

prognosis
(prog-NOH-sis)
[*pro-* **before,** *-gnosis* **knowledge**]

pathology
(pah-THOL-oh-jee)
[*patho-* **disease,** *-o-* **combining vowel,** *-log-* **words (study of),** *-y* **activity**]

psychogenic
(sye-koh-JEN-ik)
[*psycho-* **the mind,** *-gen-* **production,** *-ic* **relating to**]

public health
(PUB-lik helth)

radiation therapy
(ray-dee-AY-shun THAYR-ah-pee)
[*radiat-* **send out rays,** *-ion* **process,** *therapy* **treatment**]

radiography
(ray-dee-OG-rah-fee)
[*radi-* **rays, frequency,** *-graphy* **drawing**]

remission
(ree-MISH-un)
[*re-* **back or again,** *-miss-* **send,** *-sion* **condition of**]

Rickettsia
(rih-KET-see-ah)
[*H.T. Ricketts* **U.S. pathologist,** *-ia* **condition**]

sarcoma
(SAR-koh-mah)
[*sarco-* **flesh,** *-oma* **tumor**]

SARS-associated coronavirus (SARSCoV)
(SARZ-as-ohs-ee-AYT-ed koh-ROHN-ah-vye-us [sarz koh vee])
[*SARS-* **severe acute respiratory syndrome,** *associa-* **unite,** *-ate-* **process,** *corona-* **crown,** *-virus* **poison**]

severe acute respiratory syndrome (SARS)
(seh-VEER ah-KYOOT res-pir-ah-TOR-ee SIN-drohm [sarz])
[*acut-* **sharp,** *re-* **again,** *-spir-* **breathe,** *-tory* **relating to,** *syn-* **together,** *-drome* **running or (race) course**]

sign
(syne)

stress
(stres)
[*stress* **tighten**]

symptom
(SIMP-tum)
[*sym-* **together,** *-tom* **fall**]

syndrome
(SIN-drohm)
[*syn-* **together,** *-drome* **running or (race) course**]

traumatic
(truh-MAT-ik)
[*trauma-* **wound,** *-atic* **relating to**]

tumor
(TOO-mer)
[*tumor* **swelling**]

ultrasonography
(ul-trah-son-OG-rah-fee)
[*ultra-* **beyond,** *-sono-* **sound,** *-graph-* **draw,** *-y* **process**]

vaccine
(VAK-seen)
[*vaccin-* **cow (cowpox),** *-ine* **relating to**]

variant Creutzfeldt-Jakob disease (vCJD)
(VAYR-ee-ant KROYTS-felt YAH-kohb dih-ZEEZ)
[*Hans G. Creutzfeldt* **German neurologist,** *Alfons M. Jakob* **German neurologist,** *dis-* **opposite of,** *-ease* **comfort**]

West Nile virus (WNV)
(west nyle VY-rus [DUB-el-yoo en vee])
[*West Nile* **region of Africa,** *virus* **poison**]

yellow fever
(YEL-oh FEE-ver)
[*fev-* **heat**]

Zika virus
(ZEE-kah VYE-rus)
[*Zika* **forest in Uganda,** *virus* **poison**]

❑ OUTLINE SUMMARY

To download a digital version of the chapter summary for use with your device, access the ***Audio Chapter Summaries*** *online at evolve.elsevier.com.*

Scan this summary after reading the chapter to help you reinforce the key concepts. Later, use the summary as a quick review before your class or before a test.

Studying Disease

A. Disease terminology
 1. Health—physical, mental, and social well-being—not merely the absence of disease
 2. Disease—an abnormality in body function that threatens health
 3. Etiology—the study of the factors that cause a disease
 4. Idiopathic—refers to a disease with an unknown cause
 5. Signs and symptoms—the objective and subjective abnormalities associated with a disease
 6. Pathogenesis—the pattern of a disease's development
B. Patterns of disease
 1. Epidemiology is the study of occurrence, distribution, and transmission of diseases in human populations
 2. Endemic diseases are native to a local region
 3. Epidemics occur when a disease affects many people at the same time
 4. Pandemics are widespread, perhaps global, epidemics
 5. Discovering the cause of a disease is difficult because many factors affect disease transmission
 6. Disease can be fought through prevention and therapy (treatment) (**Figure 6-1**)

Pathophysiology

A. Mechanisms of disease
 1. Pathophysiology—the study of underlying physiological aspects of disease
 2. Many diseases are best understood as disturbances of homeostasis (**Figure 6-2**)
 3. Genetic mechanism
 4. Infectious mechanism (pathogenic organisms and particles)
 5. Neoplastic mechanism (tumors and cancer)
 6. Traumatic mechanism (physical and chemical agents)
 7. Metabolic mechanism (endocrine imbalances or malnutrition)
 8. Inflammatory mechanism
  a. Inflammation
  b. Autoimmunity
 9. Degeneration
B. Risk factors (predisposing conditions)
 1. Types
  a. Genetic factors
  b. Age
  c. Lifestyle
  d. Stress
  e. Environmental factors
  f. Preexisting conditions
 2. Some risk factors can combine or overlap
 3. Risk can be managed in some cases

Pathogenic Organisms and Particles

A. Viruses (**Table 6-1** and **Figure 6-4**)
 1. Introduction
  a. Not "alive" because they are not made of cells, but are still studied by biologists because they infect cells and have a genetic code; viruses are classified into groups like living organisms
  b. Virus particles are microscopic, intracellular parasites that consist of a nucleic acid core with a protein coat
  c. Invade host cells and pirate organelles and raw materials
  d. May be transmitted directly from human to human, or may be transmitted indirectly through a biting insect
  e. Classified by shape, nucleic acid type, and method of reproduction
 2. Examples of viruses
  a. Human immunodeficiency virus (HIV) (**Figure 6-3**)
   (1) Retrovirus that can transcribe its RNA "backwards" to produce DNA that becomes part of the host cell's genome
   (2) If untreated, can progress to acquired immune deficiency syndrome (AIDS)
  b. Coronaviruses
   (1) RNA viruses that have a "crown" of surface projections and make their own proteins inside the host cell
   (2) Found everywhere; second-leading (after rhinoviruses) cause of common cold
   (3) Some can cause serious infections such as severe acute respiratory syndrome (SARS)
  c. Flaviviruses
   (1) RNA viruses transmitted by mosquitoes
   (2) Cause yellow fever, dengue, West Nile virus (WNV) infection, Zika virus disease, and other potentially serious infections
B. Prions (**Figure 6-5**)
 1. Pathogenic protein molecules
 2. Convert normal proteins to abnormal proteins by inducing misfolding, causing abnormal functions that produce disease; may be passed on to offspring
 3. Cause rare, degenerative disorders of the nervous system such as BSE (bovine spongiform encephalopathy) and vCJD (variant Creutzfeldt-Jakob disease)

C. Bacteria (**Table 6-2** and **Figure 6-6**)
   1. Tiny cells without nuclei
   2. Secrete toxins, parasitize host cells, or form colonies
   3. Classification
      a. By growth requirements
         (1) Aerobic—require oxygen
         (2) Anaerobic—require no oxygen
      b. By staining properties (depend on composition of cell wall)
         (1) Gram-positive
         (2) Gram-negative
      c. By shape and size
         (1) Bacilli—rod-shaped cells
         (2) Cocci—round cells
         (3) Curved or spiral rods
         (4) Small bacteria—obligate parasites
      d. Spores—nonreproducing forms of bacteria that resist unfavorable environmental conditions
      e. Archaea are similar to bacteria but have a different chemical makeup and different metabolism (allowing them to survive harsh conditions); none found that infect humans, but are nonpathogenic residents found in the human body

D. Fungi (**Table 6-3** and **Figure 6-7**)
   1. Simple organisms similar to plants but lacking chlorophyll
   2. Yeasts—small, single-celled fungi
   3. Molds—large, multicellular fungi
   4. Mycotic infections—often resist treatment

E. Protozoa (**Table 6-4** and **Figure 6-8**)
   1. Large, one-celled organisms having organized nuclei
   2. May infest human fluids and parasitize or destroy cells
   3. Major groups
      a. Amebas—possess pseudopodia
      b. Flagellates—possess flagella
      c. Ciliates—possess cilia
      d. Sporozoa (coccidia)—enter cells during one phase of a two-part life cycle; borne by vectors (transmitters) during the other phase

F. Pathogenic animals (**Table 6-5**)
   1. Large, complex multicellular organisms
   2. Parasitize or otherwise damage human tissues or organs
   3. Major groups
      a. Nematodes—roundworms
      b. Platyhelminths—flatworms and flukes (**Figure 6-9**)
      c. Arthropods
         (1) Parasitic mites, ticks, lice, fleas
         (2) Biting or stinging wasps, bees, mosquitoes, spiders
         (3) Are often vectors of disease (**Table 6-7**)

Prevention and Control

A. Mechanisms of transmission
   1. Person-to-person contact
      a. Can be prevented by education
      b. Can be prevented by using aseptic technique (**Table 6-6**)
   2. Environmental contact
      a. Can be prevented by avoiding contact
      b. Can be prevented by safe sanitation practices
   3. Opportunistic invasion
      a. Can be prevented by avoiding changes in skin and mucous membranes
      b. Can be prevented by cleansing of wounds
   4. Transmission by a vector
      a. Can be prevented by reducing the population of vectors and reducing contact with vectors

B. Prevention and treatment strategies
   1. Vaccination—stimulates immunity; may produce mild side effects, but does not cause autism (ASD)
   2. Drug therapy—destroy or inhibit pathogens
      a. Antibiotics—natural compounds derived from living organisms; synthetic antibacterial drugs are also available
      b. Antiviral drugs—synthetic compounds such as Tamiflu, ACV, and efavirenz; slow progression of viral infections, but do not cure them; may be given in combination (drug cocktails)

Tumors and Cancer

A. Neoplasms (tumors)—abnormal growths of cells (**Table 6-8** and **Figure 6-10**)
   1. Benign tumors remain localized
   2. Malignant tumors spread, forming secondary tumors
   3. Metastasis—cells leave a primary tumor and start a secondary tumor at a new location (**Figure 6-11**)
   4. Classification of tumors
      a. Benign, epithelial tumors
         (1) Papilloma—fingerlike projection
         (2) Adenoma—glandular tumor
         (3) Nevus—small, pigmented tumor
      b. Benign, connective tissue tumors
         (1) Lipoma—adipose (fat) tumor
         (2) Osteoma—bone tumor
         (3) Chondroma—cartilage tumor
      c. Carcinomas (malignant epithelial tumors)
         (1) Melanoma—involves melanocytes
         (2) Adenocarcinoma—glandular cancer
      d. Sarcomas (connective tissue cancers)
         (1) Lymphoma—lymphatic cancer
         (2) Osteosarcoma—bone cancer
         (3) Myeloma—bone marrow tumor
         (4) Fibrosarcoma—cancer of fibrous tissue
      e. Can also be classified by location in the body (**Table 6-9**)

B. Causes of cancer—varied and still not clearly understood
   1. Cancer involves hyperplasia (growth of too many cells) and anaplasia (development of undifferentiated cells)
   2. Factors known to play a role in causing cancer
      a. Genetic factors (for example, oncogenes—cancer genes)
      b. Carcinogens—chemicals that alter genetic activity
      c. Age—changes in cell activity over time or accumulated effects of cell damage

d. Environment—chronic exposure to damaging substances
e. Viruses—cause change in genetic "machinery"

C. Pathogenesis of cancer
1. Signs of cancer (**Table 6-10**)
2. Methods of detecting cancers (**Figure 6-12**)
a. Self-examination
b. Diagnostic imaging—radiography (for example, mammogram and CT scan), magnetic resonance imaging (MRI), ultrasonography
c. Biopsy (for example, Pap smear)
d. Blood tests
3. Staging—classifying tumors by size and extent of spread
4. Grading—assessing the likely pattern of a tumor's development
5. Cachexia—syndrome including appetite loss, weight loss, and general weakness
6. Causes of death by cancer—secondary infections, organ failure, hemorrhage, and undetermined factors
7. Treatments
a. Surgery
b. Chemotherapy (chemical therapy)
c. Radiation therapy (radiotherapy)
d. Laser therapy
e. Immunotherapy
f. New strategies (for example, rational drugs that target specific molecules, enzymes, or receptors)

Inflammation

A. Inflammatory response—reduces injury to tissues, thus maintaining homeostasis (**Figures 6-13** and **6-14**)
1. Signs—redness, heat, swelling, and pain
2. Inflammation mediators (histamine, prostaglandins, and kinins)
a. Some cause blood vessels to dilate, increasing blood volume (redness and heat)—white blood cells travel quickly to injury site
b. Some increase blood vessel permeability (causing swelling, or edema, and pain)—white blood cells move easily out of vessels, irritant is diluted, and exudate accumulates
c. Some attract white blood cells to injury site (chemotaxis)

B. Inflammatory diseases
1. Inflammation can be local or systemic (body-wide)
2. Fever—high body temperature caused by a resetting of the body's "thermostat"—destroys pathogens and enhances immunity
3. Acute inflammation is an immediate response that often rids the body of an irritant
4. Chronic inflammation can constitute a disease itself because it causes damage to tissues

❑ ACTIVE LEARNING

STUDY TIPS

Consider using these tips to achieve success in meeting your learning goals.

This is a very challenging chapter. It presents a great deal of information, much of which may be new to you.

1. Divide the chapter into parts: disease terminology, mechanisms and risk factors, pathogenic organisms, tumors and cancer, and inflammation. In each of these sections, go over the terms in bold print. You may be surprised at how many you already know. Put the ones you do not know on flash cards.
2. Use flash cards to learn the mechanisms of disease; most of them are self-explanatory. Divide the pathogenic organisms into viruses, bacteria, fungi, protozoa, and pathogenic animals. Use flash cards for each group. Write a brief description of each type of organism.
3. Methods of disease spread are fairly self-explanatory. Make sure you know the distinction between prevention (e.g., vaccination) and treatment (e.g., antibiotic). In the cancer section, be sure you understand the difference between a carcinoma and a sarcoma.
4. Check out the *Centers for Disease Control and Prevention* website: *cdc.gov*. This site contains information on a variety of diseases.
5. The causes of cancer are also self-explanatory, as are the methods of detection and types of treatment. When studying the types of treatment, do not forget about surgery; it is not in bold type and therefore could be missed.
6. As you study inflammation, make flash cards for the four primary signs and their causes. Be sure you understand what chemotaxis is. Learn the positive effects of fever and the different effects fever has on the young and the elderly.
7. Meet with your study group early and often. This is not material you can master in one night. You may want to go over only one or two parts of the chapter per session. Review definitions, flash cards, major concepts, online resources, questions at the end of the chapter, and possible test questions. Keep your study material for this chapter handy; you may wish to refer to it as you study future chapters.

6

Review Questions

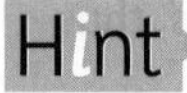

Write out the answers to these questions after reading the chapter and reviewing the Chapter Summary. If you simply think through the answer without writing it down, you will not retain much of your new learning.

1. Define or explain the following terms: etiology, idiopathic, communicable, and latent or incubation period.
2. What is the difference between an *epidemic* and a *pandemic*? What factor makes pandemics increasingly common in modern times?
3. List four factors involved in the spread of disease.
4. List seven mechanisms of disease.
5. What is a risk factor?
6. List the six risk factors discussed in the chapter.
7. Describe a virus. How does a virus damage a cell?
8. Define a flavivirus. List four examples of flaviviruses.
9. Briefly describe a bacterium. List the ways in which bacteria produce disease.
10. List three ways to classify bacteria.
11. Distinguish between anaerobic and aerobic bacteria.
12. Name the shapes and sizes used to classify bacteria. Which of these include the obligate parasites?
13. Describe a spore.
14. Describe fungi. Distinguish between yeasts and molds.
15. Describe protozoa. List the four major groups of protozoa.
16. Name and give an example of each of the pathogenic animals. Which of the arthropods are parasitic? What is a vector?
17. List the four ways disease can be spread.
18. Distinguish between *malignant* and *benign* tumors.
19. List the three benign tumors that arise from epithelial tissue.
20. List the three benign tumors that arise from connective tissue.
21. What are sarcomas? List the four sarcomas discussed in the chapter.
22. List the five factors that are known to play a role in the development of cancer. What is a mutagen?
23. List the four methods used to detect the presence of cancer.
24. List six methods of cancer treatment.
25. What are the four primary signs of inflammation? What causes each of them?
26. What is chemotaxis?
27. What are two positive effects of fever?

Critical Thinking

After finishing the Review Questions, write out the answers to these more in-depth questions to help you apply your new knowledge. Go back to sections of the chapter that relate to concepts that you find difficult.

28. The doctor noticed a rash on a boy's arm. The boy complained that the rash itched. Which of these was a sign? Which was a symptom? What is the difference between the two?
29. Of the risk factors listed in the text, which can you change, and which can't you change?
30. Why do bacteria that form spores present a greater health risk than those that do not form spores?
31. You have been given an antibiotic as a treatment for a disease. Why were you not given a vaccine instead of the antibiotic?

Chapter Test

After studying the chapter, test your mastery by responding to these items. Try to answer them without looking up the answers.

1. ________ are objective abnormalities that can be seen or measured.
2. ________ are subjective abnormalities felt only by the patient.
3. A disease with an undetermined cause is said to be ________.
4. A ________ affects a larger geographical region than does an epidemic.
5. A ________ is an attenuated pathogen given to a person to stimulate immunity.
6. A ________ tumor tends to spread to other regions of the body.
7. ________ is a process by which cancer cells are spread by lymphatic or blood vessels.
8. ________ are malignant tumors that arise from connective tissue.
9. ________ are malignant tumors that arise from epithelial tissue.
10. A ________ is a cause of cancer that damages or changes DNA structure.
11. The four primary signs of inflammation are ________, ________, ________, and ________.
12. ________ are not technically "living organisms" because they are not made up of cells. They do, however, infect living cells.
13. Stage 3 HIV infection is also known as ________.
14. The Zika virus is in a category of viruses known as ________ that move from an infected bird or other animal to a mosquito or biting insect and then to a human.

15. Which of the following is not a risk factor for disease?
    a. Stress
    b. Genetic factors
    c. Age
    d. Autoimmunity
16. Which of the following is not a means by which pathogens can spread?
    a. Environmental contact
    b. Vectors
    c. Person-to-person contact
    d. All of the above can spread pathogens
17. Treatment for cancer includes everything except:
    a. biopsy
    b. surgery
    c. chemotherapy
    d. genetic engineering
18. Which of the following is not an inflammation mediator?
    a. Prostaglandins
    b. Edema
    c. Histamine
    d. Kinins

Match the descriptions in Column A with the corresponding pathogenic organism in Column B. Some items in Column B will be used more than once.

Column A

19. ________ intracellular parasites made up of DNA or RNA and surrounded by a protein coat
20. ________ causes mycotic infections
21. ________ roundworms
22. ________ can be gram-positive or gram-negative
23. ________ causes malaria
24. ________ vector for Lyme disease
25. ________ insert their genetic code into the host's genetic code
26. ________ one-celled organism with a nucleus; can be a ciliate
27. ________ similar to plants but with no chlorophyll
28. ________ mites, lice, and fleas
29. ________ tiny primitive cells without nuclei; can be rod shaped
30. ________ flatworms and flukes
31. ________ can be bacilli or cocci shaped

Column B

a. virus
b. bacteria
c. fungus
d. protozoa
e. nematodes
f. platyhelminths
g. arthropods

Case Studies

To solve a case study, you may have to refer to the glossary or index, other chapters in this textbook, and other resources.

1. Without warning, Mr. Lee begins to feel sick. His most obvious symptom is a high fever. Within 24 hours, everyone in the Lee household also feels sick and has a high temperature. Before long, nearby households have the same experience—many people in the community are now sick. The local health department would probably call on what type of health professional to investigate this situation? Would the health professional label this situation an epidemic or a pandemic? If the symptoms are caused by a bacterial infection, list some ways the pathogen could have been transmitted to so many people within a short span of time.
2. Sandy is a nurse at the local university hospital. One of her patients has a severe staphylococcal infection. What would the pathogen responsible for this infection look like under a microscope? Sandy's patient is taking a newly developed antibiotic in the hope that it will cure the infection. Do you think that this drug is natural or synthetic?
3. Fred is a first-year medical student. He received a minor scrape during a basketball game on the parking lot outside his dorm. He has cleansed the wound and applied an antibiotic as a preventive measure. The affected area is red, swollen, and mildly painful. How do you explain these symptoms? Fred's roommate suggested applying an anti-inflammatory drug to the wound, but Fred refuses. What advantage might Fred see in avoiding such treatment?
4. Lola came home from preschool and was not feeling well. She had not been feeling well for the past few days. Her mother noticed that she had not been sleeping well and she complained about itching in the region of her anus. She was not eating well and could not even be tempted with her favorite foods. Her mother took her to the pediatrician who told Mrs. Byers that it was possible that Lola had a round worm. With the information that you have from your text, could you be more specific with that possible diagnosis?

Answers to Active Learning Questions can be found online at evolve.elsevier.com.

Skin and Membranes

OUTLINE

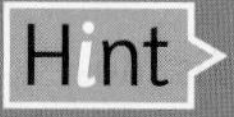

Hint *Scan this outline before you begin to read the chapter, as a preview of how the concepts are organized.*

OBJECTIVES

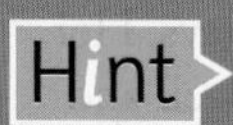

Hint *Before reading the chapter, review these goals for your learning.*

After you have completed this chapter, you should be able to:

1. Classify, compare the structure of, and give examples of each type of body membrane.
2. Describe the structure and function of the epidermis and dermis.
3. List and briefly describe each accessory organ of the skin.
4. List and discuss the five primary functions of the skin.
5. Do the following related to disorders of the skin:
   - List and describe major categories of skin disorders, and give examples of each.
   - Describe the classification of burns and how to estimate the extent of a burn injury.
   - List and briefly describe the three most common types of skin cancer.

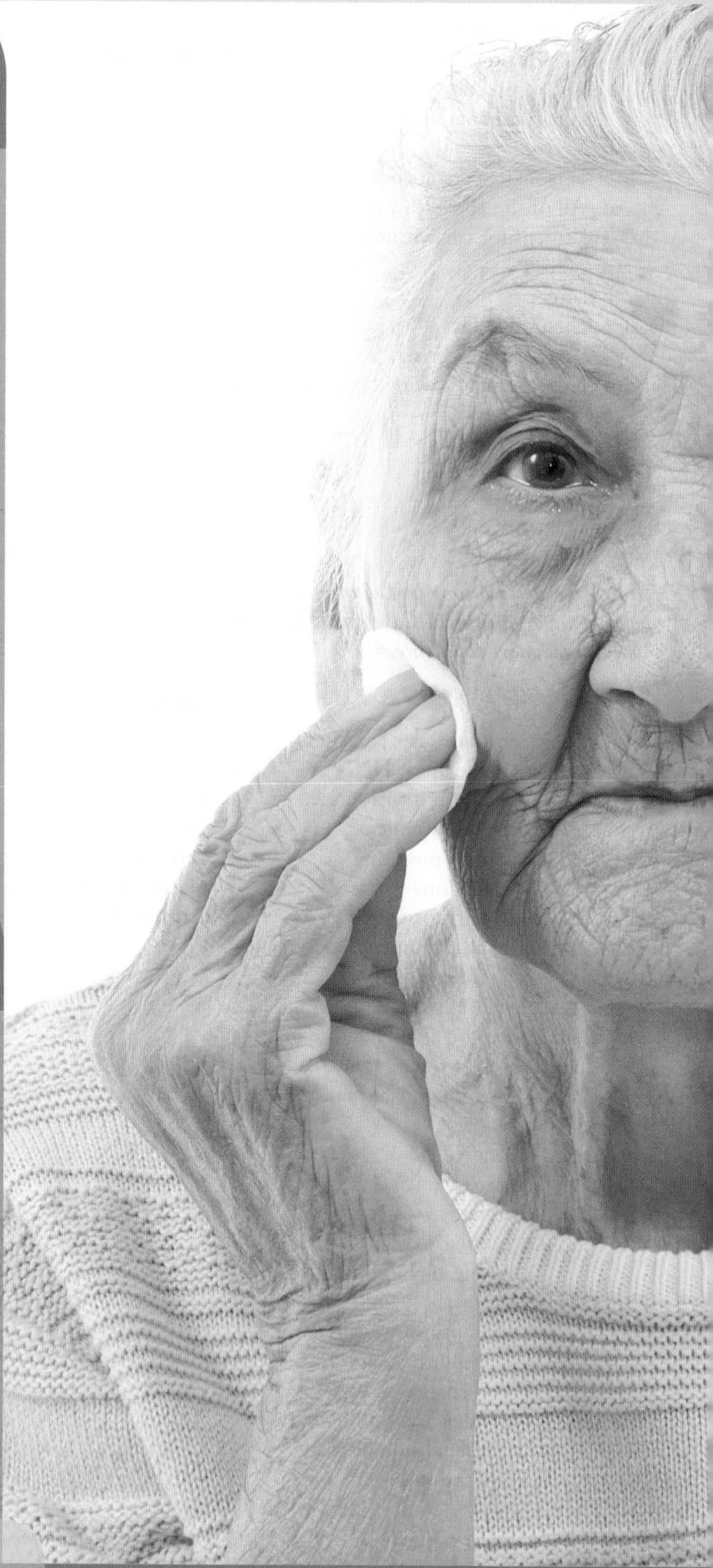

CHAPTER 7

In Chapter 1 the concept of progressive organization of body structures from simple to complex was established. Complexity in body structure and function progresses from cells to tissues and then to organs and organ systems. This chapter discusses the skin and its **appendages**—the hair, the nails, and the skin glands—as an organ system. This system is called the **integumentary system. Integument** is another name for the skin, and the skin itself is the principal organ of the integumentary system. The skin is one of a group of anatomically simple but functionally important sheetlike structures called **membranes.** This chapter begins with classification and discussion of the important body membranes. Study of the integument follows—our first exploration of how the structure and function of a body system are interrelated.

To get an overview of the integumentary system, go to AnimationDirect online at *evolve.elsevier.com.*

Body Membranes

Classification of Membranes

The term *membrane* refers to a thin, sheetlike structure that may have many important functions in the body. Membranes cover and protect the body surface, line body cavities, and cover the inner surfaces of the hollow organs such as the digestive, reproductive, and respiratory passageways.

Some membranes anchor organs to each other or to bones, and others cover the internal organs. In certain areas of the body, membranes secrete lubricating fluids that reduce friction during organ movements such as the beating of the heart or lung expansion and contraction. Membrane lubricants also decrease friction between bones in joints.

LANGUAGE OF **SCIENCE**

Hint Before reading the chapter, say each of these terms out loud. This will help you to avoid stumbling over them as you read.

apocrine
(AP-oh-krin)
[*apo-* **from,** *crin-* **secrete**]

apocrine sweat gland
(AP-oh-krin swet gland)
[*apo-* **from,** *-crin-* **secrete,** *gland* **acorn**]

appendage
(ah-PEN-dij)
[*append-* **hang upon,** *-age* **related to**]

areola
(ah-REE-oh-lah)
pl., areolae
(ah-REE-oh-lee)
[*are-* **area or space,** *-ola* **little**]

arrector pili
(ah-REK-tor PYE-lye)
[*arrector* **raiser,** *pili* **of hair**]

basement membrane
(BAYS-ment MEM-brayn)
[*base-* **base,** *-ment* **thing,** *membrane* **thin skin**]

bursa
(BER-sah)
pl., bursae
(BER-see or BER-say)
[*bursa* **purse**]

connective tissue membrane
(kon-NEK-tiv TISH-yoo MEM-brayn)
[*con-* **together,** *-nect-* **bind,** *-ive* **relating to,** *tissu-* **fabric,** *membran-* **thin skin**]

cutaneous membrane
(kyoo-TAYN-ee-us MEM-brayn)
[*cut-* **skin,** *-aneous* **relating to,** *membrane* **thin skin**]

cuticle
(KYOO-tih-kul)
[*cut-* **skin,** *-icle* **little**]

Continued on p. 165

There are two major categories or types of body membranes:

1. **Epithelial membranes** are composed of epithelial tissue and an underlying layer of fibrous connective tissue
2. **Connective tissue membranes** are composed exclusively of various types of connective tissue; no epithelial cells are present in this type of membrane

Epithelial Membranes

Types of Epithelial Membrane

There are three types of epithelial tissue membranes in the body:

1. Cutaneous membrane
2. Serous membranes
3. Mucous membranes

Cutaneous Membrane

The **cutaneous membrane,** or **skin,** is the primary organ of the integumentary system. It is one of the most important and one of the largest and most visible organs of the body. In most individuals the skin composes some 16% of the body weight.

The skin fulfills the requirements necessary for an epithelial tissue membrane in that it has a superficial layer of epithelial cells and an underlying layer of supportive connective tissue. Its structure is uniquely suited to its many functions. The skin is discussed in depth later in the chapter.

Serous Membranes

Serous membranes are found only on surfaces within closed cavities. Like all epithelial membranes, a serous membrane is composed of two distinct layers of tissue. The epithelial sheet is a thin layer of simple squamous epithelium. The connective tissue layer forms a very thin, gluelike **basement membrane** that holds and supports the epithelial cells.

The serous membrane that lines body cavities and covers the surfaces of organs in those cavities is in reality a single, continuous sheet of tissue covering two different surfaces. This arrangement results in two distinct layers of serous membranes. One serous membrane layer lines body cavities and the other layer covers the organs within those cavities.

The serous membrane layer that lines the walls of a body cavity, much like wallpaper covers the walls of a room, is called the **parietal** layer. The other serous membrane layer instead folds inward to cover the surface of organs found within a body cavity and is called the **visceral** layer.

Two serous membranes of the thoracic and abdominal cavities are identified in Figure 7-1. In the thoracic cavity the serous membranes are called **pleura,** and in the abdominal cavity, they are called **peritoneum.**

Look again at Figure 7-1 to note the placement of the **parietal pleura** and **visceral pleura** and the **parietal peritoneum** and **visceral peritoneum.** In both cases the parietal layer forms the lining of the body cavity, and the visceral layer covers the organs found in that cavity.

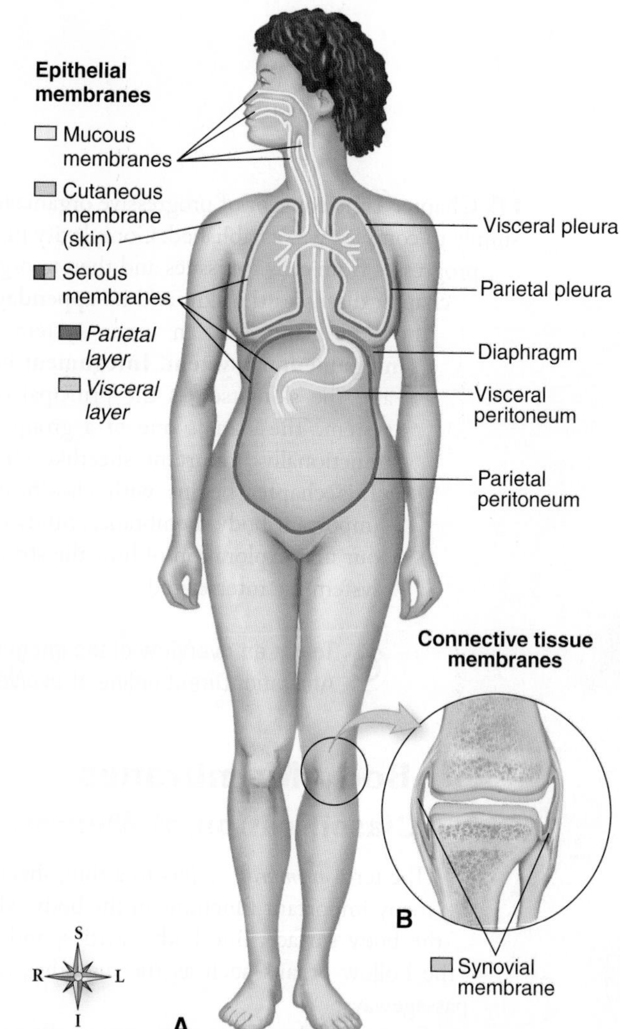

FIGURE 7-1 Types of body membranes. A, Epithelial membranes, including cutaneous membrane (skin), serous membranes (parietal and visceral pleura and peritoneum), and mucous membranes. **B,** Connective tissue membranes, including synovial membranes.

Serous membranes secrete a thin, watery fluid that helps reduce friction and serves as a lubricant when organs rub against one another and against the walls of the cavities that contain them.

The heart is surrounded by a fibrous sac lined with a thin, slippery membrane that doubles back on itself to form a lubricating, fluid-filled pocket around the heart. Figure 7-2 shows how the serous membrane around the heart—the **pericardium**—resembles a water-filled balloon with a fist thrust into it.

Pleurisy or *pleuritis* is a very painful pathological condition characterized by inflammation of a serous membrane (pleura) that lines one side of the chest cavity and covers a lung. Pain is caused by irritation from friction as the visceral pleura on the lungs rub against the parietal pleura lining the walls of the chest cavity. The parietal pleura is much more sensitive to pain than the visceral pleura. In severe cases the inflamed surfaces of the pleura fuse, and permanent damage may develop.

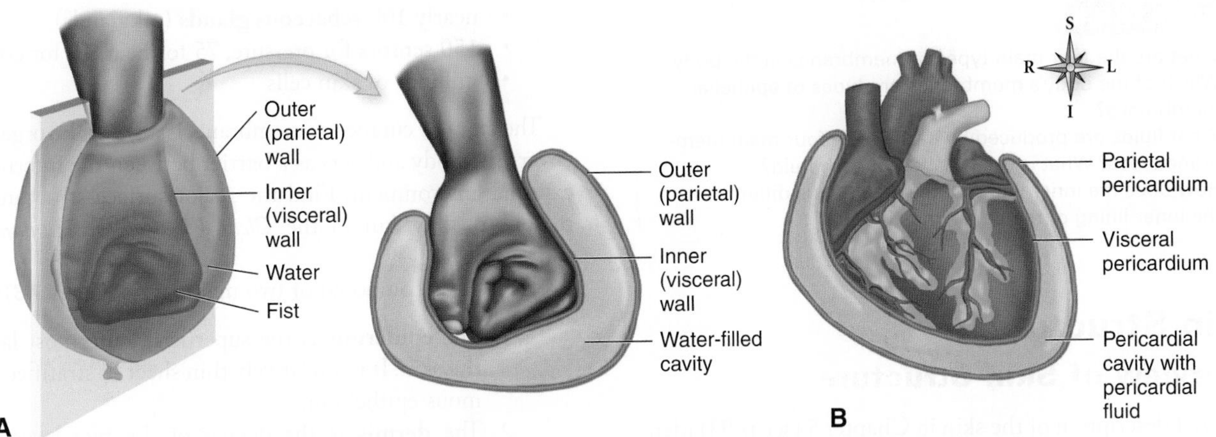

FIGURE 7-2 Serous membranes. A, The analogy of a fist thrust into a water-filled balloon demonstrates how a serous membrane forms a double-walled structure containing a thin pocket of fluid. **B,** The heart is surrounded by the serous pericardium, which forms a parietal and visceral layer filled with lubricating serous fluid called *pericardial fluid.*

The term **peritonitis** is used to describe inflammation of the serous membranes in the abdominal cavity. Peritonitis is sometimes a serious complication of an infected appendix.

To learn more about serous membranes, go to AnimationDirect online at *evolve.elsevier.com.*

Mucous Membranes

Mucous membranes are epithelial membranes that contain both an epithelial layer and a fibrous or connective tissue layer. These membranes line body surfaces opening directly to the exterior of the body.

Examples of mucous membranes include those lining the respiratory, digestive, urinary, and reproductive tracts. The epithelial component of a mucous membrane varies, depending on its location and function. In most cases the cell composition is either stratified squamous, simple columnar, or pseudostratified epithelia.

In the esophagus, for example, a tough, abrasion-resistant stratified squamous epithelium is found. This is a good example of the "structure fits function" principle. Without the protection of a tough epithelial lining, ingested food that is coarse, like popcorn, might cause injury to the esophageal wall when swallowed, resulting in irritation or even infection and hemorrhage.

A thin layer of simple columnar epithelium lines the walls of the lower segments of the digestive tract. In the stomach and small intestine, ingested food undergoes digestion and is changed into a smooth, liquefied material that is no longer abrasive. The single layer of lining epithelial cells in these segments of the intestinal tract is well suited to a primary function: nutrient and water absorption.

The epithelial cells of most mucous membranes secrete a thick, slimy material called **mucus** that keeps the membranes moist and soft. The fibrous connective tissue underlying the epithelium in mucous membranes is called the **lamina propria.**

Note that the term *mucous* identifies the type of membrane while *mucus* refers to the secretion produced by that membrane.

The term **mucocutaneous junction** is used to describe the transitional area that serves as a point of "fusion" where skin and mucous membranes meet. Such junctions lack accessory organs such as hair or sweat glands that characterize the skin. These transitional areas are generally moistened by mucous glands within the body orifices, or openings, where these junctions are located. The eyelids, lips, nasal openings, vulva, and anus have mucocutaneous junctions that may become sites of infection or irritation.

To learn more about mucous membranes, go to AnimationDirect online at *evolve.elsevier.com.*

Connective Tissue Membranes

Unlike cutaneous, serous, and mucous membranes, connective tissue membranes do not contain epithelial components. The **synovial membranes** lining the joint capsules that surround and attach the ends of articulating bones in movable joints are classified as connective tissue membranes (see **Figure 7-1**, *B*, and **Figure 8-28** on p. 198).

These membranes are smooth and slick and secrete a thick and colorless lubricating fluid called **synovial fluid.** The membrane itself, with its fluid that resembles egg white, helps reduce friction between the opposing surfaces of bones in movable joints. Synovial membranes also line the small, cushionlike sacs called **bursae** found between moving body parts.

To learn more about connective tissue and synovial membrane, go to AnimationDirect online at *evolve.elsevier.com.*

7

QUICK CHECK

1. What are the four main types of membranes in the body?
2. Which of the body's membranes are types of epithelial membranes?
3. What fluids are produced by each of the four main membrane types? What is the function of each fluid?
4. How does the inner lining of the esophagus differ from the inner lining of the small intestine?

Skin Structure

Overview of Skin Structure

The brief description of the skin in Chapter 5 (see p. 93) identified it as not only the primary organ of the integumentary system but also as the largest and one of the most important organs of the body. Architecturally, the skin is a marvel. Consider the incredible number of structures fitting into 1 square inch (6.5 cm^2) of skin:

- 500 sweat glands
- more than 1000 nerve endings
- yards of tiny blood vessels
- nearly 100 **sebaceous glands** (oil glands)
- 150 sensors for pressure, 75 for heat, 10 for cold
- millions of skin cells

The skin, or cutaneous membrane, is a sheetlike organ that covers the body and acts as a barrier between the internal and external environment. Find the major surface structures and regions of the skin in the *Clear View of the Human Body* (follows p. 8).

The skin is composed of two main layers (**Figure 7-3**):

1. The **epidermis** is the superficial, outermost layer of the skin. It is a relatively thin sheet of stratified squamous epithelium.
2. The **dermis** is the deeper of the two layers. It is thicker than the epidermis and is made up mostly of connective tissue.

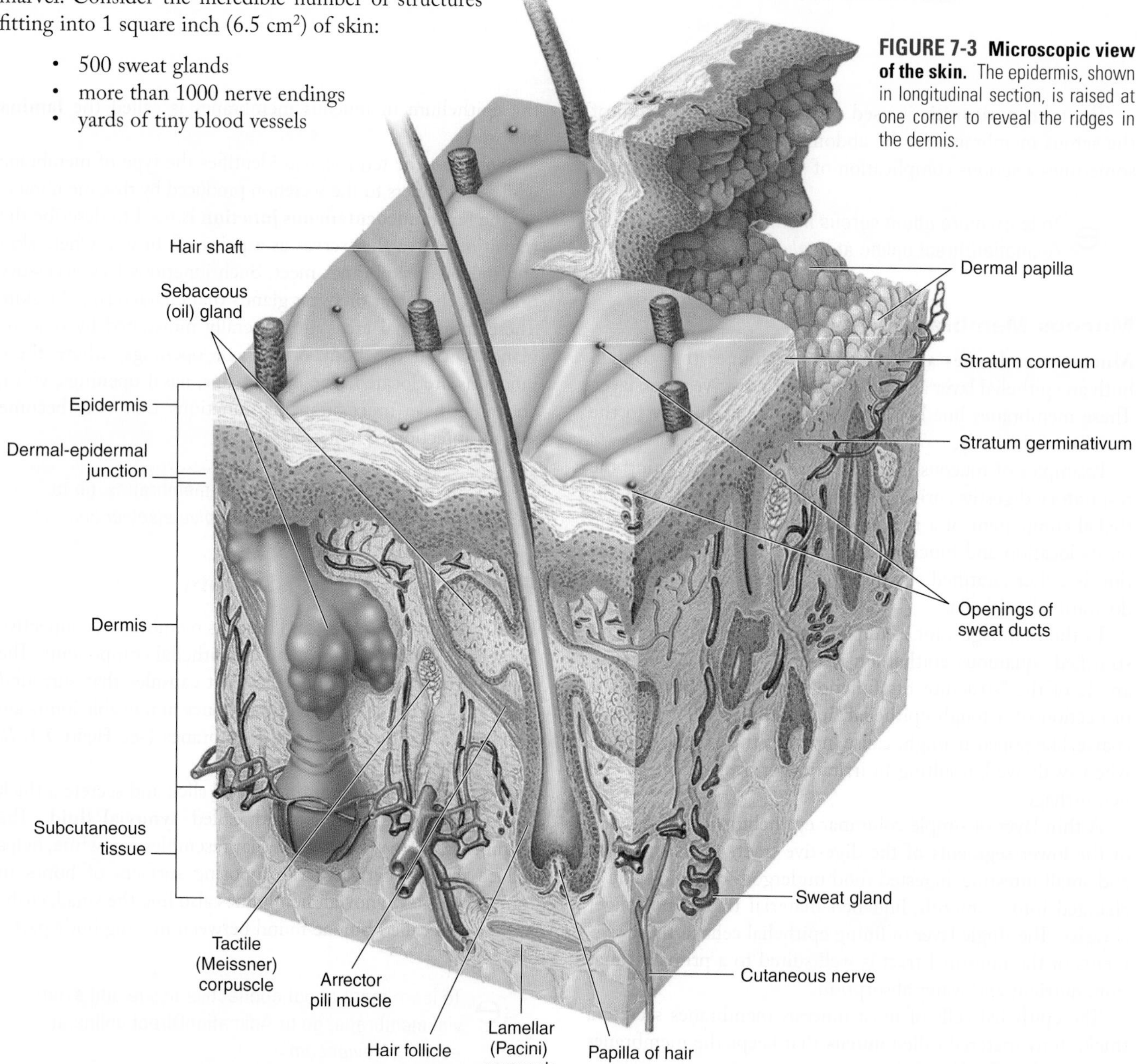

FIGURE 7-3 Microscopic view of the skin. The epidermis, shown in longitudinal section, is raised at one corner to reveal the ridges in the dermis.

As you can see in Figure 7-3, the layers of the skin are supported by a thick layer of loose connective tissue and adipose tissue called **subcutaneous tissue,** or the **hypodermis.**

Fat in the adipose tissue of the subcutaneous layer insulates the body from extremes of heat and cold. It also serves as a stored source of energy for the body and can be used as a nutrient source if required. In addition, the subcutaneous tissue acts as a shock-absorbing pad and helps protect underlying tissues from injury caused by bumps and blows to the body surface.

Epidermis

Epidermal Structure

The tightly packed epithelial cells of the epidermis are arranged in up to five distinct layers.

The basal cells of the innermost layer, called the **stratum germinativum,** undergo mitosis and reproduce themselves (see Figure 7-3). As new cells are produced in the deep layer of the epidermis, they are pushed upward through additional layers, or "strata," of cells.

As they approach the surface, the epidermal cells die and their cytoplasm is replaced by one of nature's most unique proteins, a substance called **keratin.** Keratin is a tough, waterproof material that provides cells in the outer layer of the skin with a horny, abrasion-resistant, and protective quality. The tough, keratinized outer layer of the epidermis is called the **stratum corneum.**

In the photomicrograph of the skin shown in Figure 7-4, many of the surface cells of the stratum corneum have been dislodged. These dry, dead cells filled with keratin "flake off" by the thousands onto our clothes and bedding, into our bathwater, and onto things we handle. Millions of epithelial cells reproduce daily to replace the millions shed—just one example of the work our bodies do without our knowledge, even when we seem to be resting.

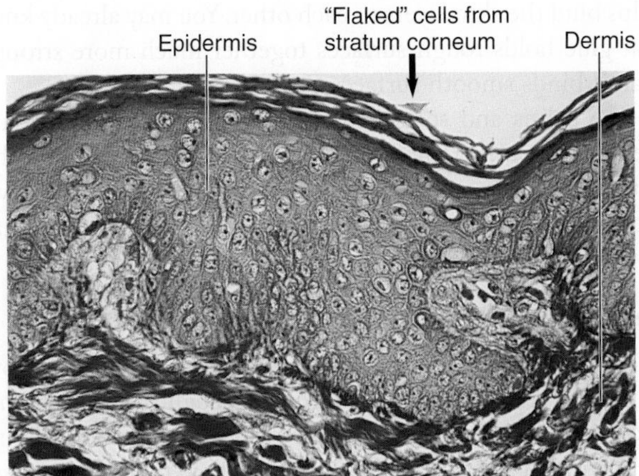

FIGURE 7-4 Photomicrograph of the skin. Many dead cells of the stratum corneum have flaked off from the surface of the epidermis. Note that the epidermis is very cellular. The dermis has fewer cells and more connective tissue.

Skin Pigment

Melanin

The deepest cell layer of the stratum germinativum identified in Figure 7-3 is responsible for the production of a **pigment** that gives color to the skin. The term *pigment* comes from a Latin word meaning "paint."

The brown pigment **melanin** is produced by cells in the basal layer called **melanocytes.** Melanocytes package the melanin in vesicles and distribute it to the surrounding epithelial cells, making them a darker color. The higher the concentration of melanin distributed in the layers of epithelial cells, the deeper is the color of skin. The primary function of melanin is to absorb harmful ultraviolet (UV) radiation from sunlight before it reaches tissues below the outer layers of the epidermis.

Skin Color Changes

The amount and type of melanin in your skin depends first on the skin color genes you have inherited. That is, heredity determines how dark or light your basic skin color is. However, other factors such as sunlight exposure can modify this hereditary effect. Prolonged exposure to sunlight in light-skinned people darkens the exposed area because it leads to increased melanin deposits in the epidermis—a protective mechanism that keeps deeper tissues safe from UV radiation.

If the skin contains little melanin, as under the nails where there is no melanin at all, a change in color can occur if the volume of blood in the skin changes significantly or if the amount of oxygen in the blood is increased or decreased. In these individuals, increased blood flow to the skin or increased blood oxygen levels can cause a pink flush to appear. However, if blood oxygen levels decrease or if actual blood flow is reduced dramatically, the skin turns a blue-gray color—a condition called **cyanosis.**

In general, the less abundant the melanin deposits in the skin are, the more visible will be the changes in color caused by the change in skin blood volume or oxygen level. Conversely, the richer the skin's pigmentation is, the less noticeable such changes will be.

The term **vitiligo** is used to describe a condition characterized by patchy looking areas of light skin resulting from the acquired loss of epidermal melanocytes. The term *vitiligo* is derived from the Greek word for calf. Early physicians compared the white spots caused by the loss of pigment to the light patches often seen on calves.

Although not as apparent in light-skinned individuals, vitiligo may be very obvious in those with darker skin. About 50% of cases begin before age 20 and progress slowly over a period of years. The hands, face, genitalia, and body folds, including the axillae are often involved (Figure 7-5).

Most cases of vitiligo are apparently genetic in origin and occur in individuals who have no other associated findings. Occasionally, the condition is related to autoimmune- or endocrine-related diseases, especially thyroid disorders. Some success has been achieved in darkening depigmented skin areas by using drugs and steroid hormones and by transplantation of skin epidermis containing melanocytes.

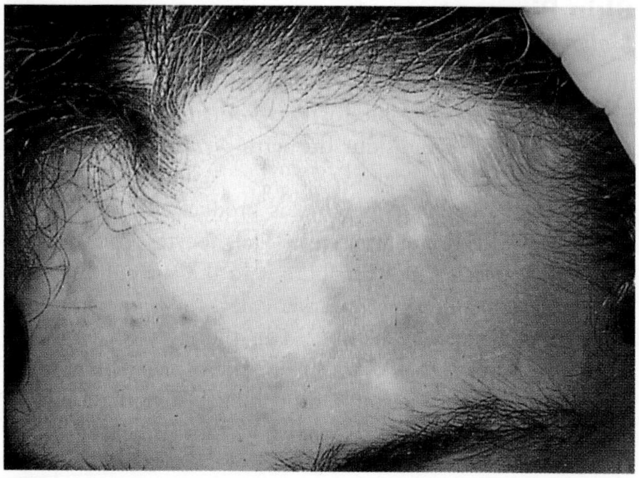

7

FIGURE 7-5 Vitiligo. Note the patchy loss of pigment on the forehead.

A hereditary condition called **albinism** is characterized by a partial or total lack of melanin pigment in the skin and eyes (see Chapter 25, pp. 682–683). Affected individuals are subject to eye damage and sunburn if exposed to direct sunlight.

A normal increase in skin pigmentation caused by hormonal changes is almost universal in pregnant women. It is most common in the genital area, nipples, and the **areola** surrounding each nipple. In addition, about 70% to 75% of pregnant women develop blotchy areas of brown pigmentation over the forehead, cheeks, nose, upper lip, and chin. It is sometimes called the "mask of pregnancy." The pigmented areas gradually fade after delivery.

One common variant of normal skin pigmentation is the small light brown or red **freckle** (look ahead to **Table 7-1** on p. 158). Freckles are small flat **macules** that most often occur as a genetic trait in light-skinned individuals and are usually confined to the face, upper extremities, and back.

In chronically sun-exposed areas of the skin, especially in older adults, brown-colored *age spots* are common. Incorrectly called "liver spots," these flat, pigmented lesions become more numerous with advancing age. They may develop into malignant lesions and should be monitored carefully for changes in size and appearance.

Dermal-Epidermal Junction

The junction that exists between the thin epidermal layer of skin above and the dermal layer below forms a type of basement membrane called the **dermal-epidermal junction.**

The deeper cells of the epidermis are packed tightly together. They are held firmly to one another by cellular junctions between the membranes of adjacent cells, sometimes described as "spot welds." They are also held firmly to the dermis below by a unique type of gel that serves to "glue" the two layers of the skin together. The thick dermis is thus able to provide support for the thin epidermis attached to its upper surface.

Small nipplelike bumps that project upward from the dermis into the epidermis, called **dermal papillae**—which are discussed in the next section—also play an important role in stabilizing the dermal-epidermal junction (see **Figure 7-3**).

If the dermal-epidermal junction is weakened or destroyed, the skin falls apart. When this occurs over a limited area because of burns, friction injuries, or exposure to irritants, **blisters** may result. Any widespread detachment of a large area of epidermis from the dermis is an extremely serious condition that may result in overwhelming infection and death.

QUICK CHECK

1. What are the two major layers of the skin?
2. What is keratin and where is it located?
3. How is the condition known as *vitiligo* related to melanin?
4. Give two examples each of elevated, flat, and depressed skin lesions.
5. What is the dermal-epidermal junction?
6. What is the primary function of melanin?

Dermis

Overview of Dermis

The dermis is the deeper of the two primary skin layers and is much thicker than the epidermis.

The mechanical strength of the skin is in the dermis. It is composed largely of connective tissue. Instead of cells being crowded close together like the epithelial cells of the epidermis, they are scattered far apart, with many fibers in between. Some of the fibers are tough and strong (collagen or white fibers), and others are stretchable and elastic (elastic or yellow fibers).

Papillary Layer

Dermal Papillae

The upper region, or *papillary layer,* of the dermis is characterized by parallel rows of peglike projections called *dermal papillae,* which are visible in **Figure 7-3**. The papillary layer takes its name from the papillae on its surface.

The papillary layer and its papillae are composed essentially of loose connective tissue elements and a fine network of thin collagenous and elastic fibers. The dermal papillae increase the surface area of the gluelike dermal-epidermal junction that helps bind the skin layers to each other. You may already know that glue holds rough surfaces together much more strongly than it binds smooth surfaces.

The palms and soles (and palmar surfaces of fingers and toes) possess *thick skin,* which is a special category of skin that is thick, hairless, and deeply ridged (**Figure 7-6**, *A*). However, most of the skin is *thin skin,* which has hair and irregular, shallow grooves (**Figure 7-6**, *B*).

Dermal Ridges

The thick skin on the palms and soles have large, distinct rows of dermal papillae that help form the roughly parallel *friction ridges* seen in **Figure 7-6**, *A.* Friction ridges help us to walk upright without slipping and to make and hold tools. These ridges also help us sense textures on surfaces in our environment.

You can observe these ridges on the tips of the fingers and on the skin covering the palms of your hands. Observe in **Figure 7-3** how the epidermis follows the contours of the dermal papillae. These ridges develop sometime before birth. Not

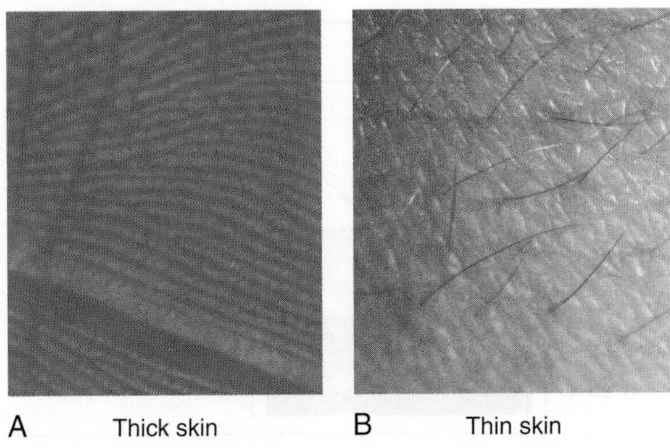

FIGURE 7-6 Thick and thin skin. A, Thick skin is hairless and has roughly parallel friction ridges. **B,** Thin skin has hairs and shallow, irregular grooves.

only is the pattern unique in each individual but also it never changes except to grow larger—two facts that explain why our fingerprints or footprints positively identify us. Many hospitals identify newborn babies by footprinting them soon after birth.

Reticular Layer

Fibrous Network

The deeper area, or *reticular layer,* of the dermis is filled with a dense network of interlacing fibers. Most of the fibers in this area are collagen that gives toughness to the skin. However, elastic fibers are also present. These make the skin stretchable and elastic (able to rebound).

During pregnancy, the skin over a woman's abdomen may stretch beyond the ability of the elastic and connective tissue elements in the dermis to rebound. The result is creation of "stretch marks" called **striae.** Although they fade after delivery, they never completely disappear (see **Table 7-1** on p. 158). As we age, the number of elastic fibers in the dermis decreases, and the amount of fat stored in the subcutaneous tissue is reduced. Wrinkles develop as the skin loses elasticity, sags, and becomes less soft and pliant.

In addition to connective tissue elements, the dermis contains an extensive network of nerves and nerve endings to detect sensory information such as pain, pressure, touch, and temperature. At various levels of the dermis, there are muscle fibers, hair follicles, sweat and oil glands, and many blood vessels.

Birthmarks

Developmental malformation of dermal blood vessels can result in pigmented *birthmarks* in significant numbers of newborns. One of the most common (**Figure 7-7**) is a collection of dilated vessels that may initially appear as a bruise at birth and then grow rapidly during the first year into a bright red nodule called a **strawberry hemangioma.**

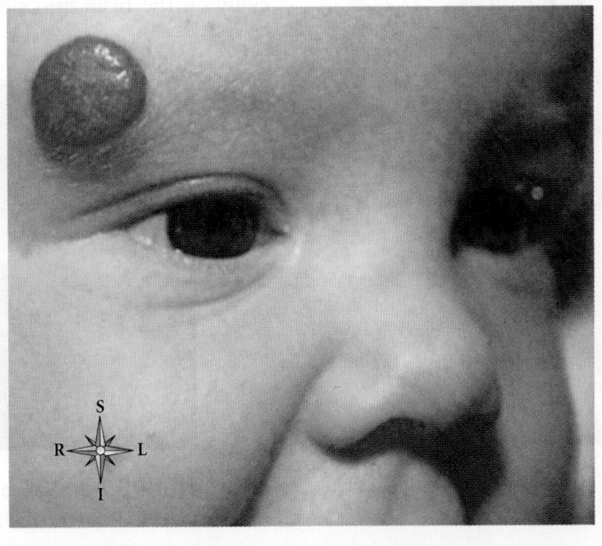

FIGURE 7-7 Strawberry hemangioma. This birthmark resembles a strawberry because of a mass of dilated dermal blood vessels.

A majority of these birthmarks shrink, fade, and disappear without treatment of any kind by age 7. Unfortunately, some pigmented vascular birthmarks, such as the **port-wine stain,** are permanent and do not fade with age. In these cases several types of laser-based therapy or use of opaque makeup can often provide effective cosmetic treatment.

Dilation of dermal capillaries at the nape of the neck in a baby that occur during development results in a birthmark called a **"stork bite"** in 50% to 70% of newborns. Although they often persist for life, these birthmarks generally are covered by hair and are inconspicuous.

Subcutaneous Tissue

The subcutaneous tissue (hypodermis) is often called the **superficial fascia** by anatomists. It is not a layer of the skin. Instead, it lies deep to the dermis and forms a connection between the skin and the underlying structures of the body such as muscle and bone.

Loose fibrous and adipose tissues are prominent in subcutaneous tissue, and in obese individuals, fat content in this layer may exceed 10 cm or more in thickness (see box on p. 84). The loose, spongy nature of subcutaneous tissue allows sliding movement of the skin over the muscles and bones as our body parts move. If not for this sliding, our skin would tear when we move our arms or legs.

Liquid medicines such as insulin are often administered by *subcutaneous injection* with a hypodermic needle into this area (see the Clinical Application box).

Hair, Nails, and Skin Receptors

Hair

Location of Hair

The human body is covered with millions of hairs. Indeed, at the time of birth most of the pocketlike **follicles** that are required for hair growth are already present. They develop early in fetal life and by birth are present in most parts of the skin.

The hair of a newborn infant is extremely fine and soft; it is called **lanugo** from the Latin word meaning "down." In premature infants, lanugo may be noticeable over most of the body, but soon after birth the lanugo is lost and replaced by new hair that is stronger and more pigmented.

Although only a few areas of the skin are hairless—notably the lips, the palms of the hands, and the soles of the feet—most body hair remains almost invisible. Hair is most dense—and

CLINICAL APPLICATION

SUBCUTANEOUS INJECTION

Although the subcutaneous layer is not part of the skin, it carries the major blood vessels and nerves that supply the skin above it. The rich blood supply and loose, spongy texture of the subcutaneous layer make it an ideal site for the rapid and relatively pain-free absorption of injected material. Liquid medicines such as insulin and pelleted implant materials such as synthetic hormones are often administered by **subcutaneous injection** into this spongy and porous layer beneath the skin. Needles used to inject materials into the hypodermis are called *hypodermic* needles.

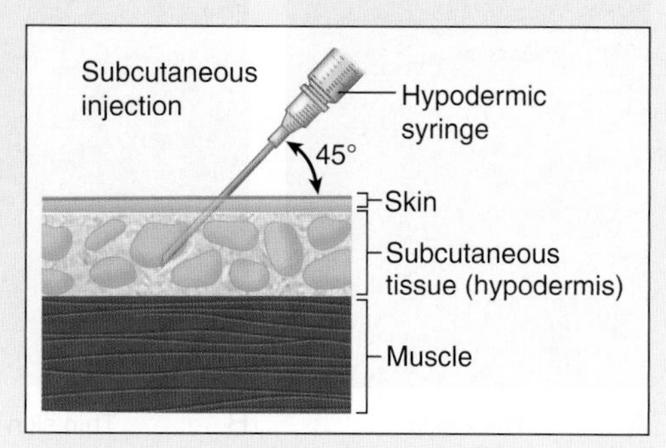

therefore, most visible—on the scalp, eyelids, and eyebrows. The coarse hair that first appears in the pubic and axillary regions at the time of puberty develops in response to the secretion of hormones.

Hair Growth

Hair growth begins when cells of the epidermal layer of the skin grow down into the dermis, forming a small tube called the **hair follicle.** The relationship of a hair follicle and its related structures to the epidermal and dermal layers of the skin is shown in **Figure 7-8**. Note in **Figure 7-8** that part of the hair, namely the *hair root,* lies hidden in the follicle. The visible part of a hair is called the *shaft.*

Hair growth begins from a small bump called the **hair papilla,** which is located at the base of the follicle. The papilla is nourished by dermal blood vessels and covered with a form of *stratum germinativum*—the epidermal growth layer. As in other areas of the skin, when new cells are formed, the older cells are pushed outward and become filled with keratin—producing a strong, keratinized cylinder of hair. The type of keratin in hair is a bit more rigid than the softer, more flexible keratin of stratum corneum.

As long as stem cells in the papilla of the hair follicle remain alive, new hair will replace any that is cut or plucked. Contrary to popular belief, frequent cutting or shaving does not make hair grow faster or become coarser. Why? It is because neither process affects the epithelial growth cells that form the hairs.

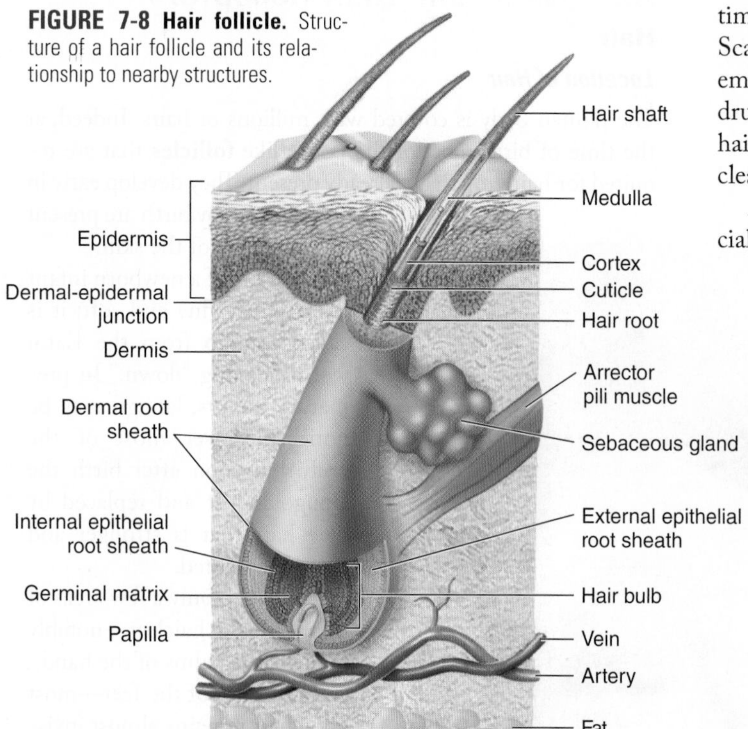

FIGURE 7-8 Hair follicle. Structure of a hair follicle and its relationship to nearby structures.

Hair Loss

Hair loss of any kind is called **alopecia.** Some forms of alopecia, such as *male pattern baldness,* are not diseases but are simply inherited traits. Alopecia also may be a normal consequence of aging. Sudden loss of hair in round or oval "exclamation point" patches on the scalp, such as that seen in **Figure 7-9**, is called *alopecia areata (AA). Areata* means "bald spot."

Alopecia can occur without a known cause but is sometimes associated with certain metabolic or endocrine diseases. Scalp infections, chemotherapy, radiation treatment, severe emotional or physical stress, and reactions to various types of drugs also can cause rapid hair loss. In most cases, regrowth of hair begins in 1 to 3 months, and the condition generally clears completely in 1 year without treatment.

A significant number of women experience hair loss, especially on the front and sides of the scalp, 1 to 4 months after

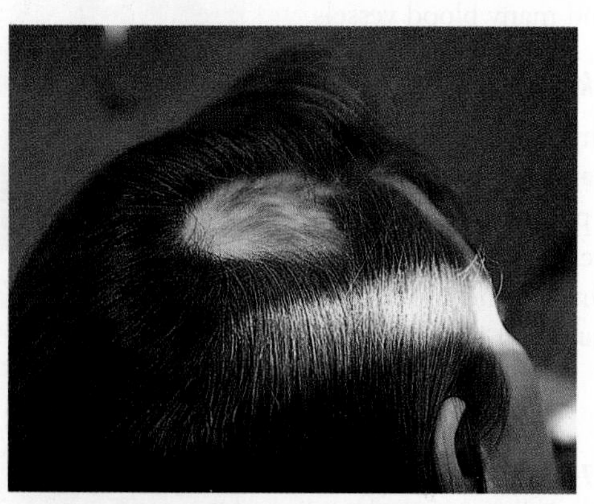

FIGURE 7-9 Alopecia areata (AA). A sudden, abnormal loss of a patch of hair, often of unknown cause.

childbirth. The condition is called *postpartum alopecia* (*post-*, after; *-partum*, birth). As in alopecia areata, full regrowth of hair generally occurs in less than a year.

Occasionally, and most often in young people, total loss of scalp hair occurs without apparent cause and in the absence of other findings. The condition, called *alopecia totalis* may be accompanied by cycles of partial hair regrowth and loss, but the chances for significant long-term regrowth are poor.

Arrector Pili Muscle

A tiny, smooth (involuntary) muscle can be seen in **Figure 7-8**. It is called an **arrector pili** muscle. It is attached to the base of a dermal papilla above and to the side of a hair follicle below. Generally, these muscles contract only when we are frightened or cold.

When contraction occurs, each muscle simultaneously pulls on its two points of attachment (that is, up on a hair follicle but down on a part of the skin). This produces little raised places, called *goose pimples,* between the depressed points of the skin and at the same time pulls the hairs up until they are more or less straight. The name *arrector pili* describes the function of these muscles. It is Latin for "erectors of the hair."

We subconsciously recognize these facts in expressions such as "I was so frightened my hair stood on end."

QUICK CHECK

1. What are dermal papillae? Why are they important?
2. What are birthmarks? List three types.
3. Identify the prominent tissues that compose subcutaneous tissue (superficial fascia).
4. How is hair formed?
5. What is alopecia?
6. What causes *goose pimples?*

Nails

Nail Growth and Structure

Nails are classified as accessory organs of the skin and are produced by cells in the epidermis. They develop when epidermal cells over the terminal ends of the fingers and toes fill with keratin and become hard and platelike.

The components of a typical fingernail and its associated structures are shown in **Figure 7-10**. In this illustration the fingernail of the index finger is shown in a posterior view and in a sagittal section. (Recall that a sagittal section divides a body part into right and left portions.)

Look first at the posterior view of the nail in **Figure 7-10**, *A*. The visible part of the nail is called the **nail body.** The rest of the nail, namely, the **root,** lies in a groove and is hidden by a fold of skin called the **cuticle.** In the sagittal section you can see the nail root from the side and note its relationship to the cuticle, which is folded back over its upper surface.

The nail body nearest the root has a crescent-shaped white area known as the **lunula,** or "little moon." You should be able to identify this area easily on your own nails; it is most noticeable on the thumbnail.

Now look at the sagittal section of the nail in **Figure 7-10**, *B*. Under the nail lies a layer of epithelium called the *nail bed.* Because it contains abundant blood vessels, it appears pink through the translucent nail bodies. If blood oxygen levels drop and cyanosis develops, the nail bed will turn blue.

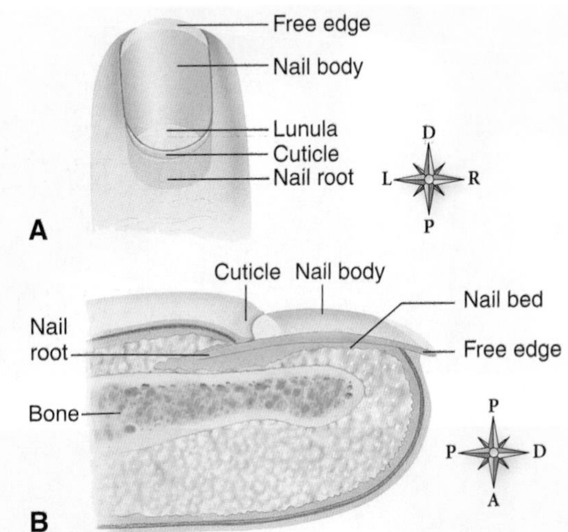

FIGURE 7-10 Structure of nails. A, Posterior view of fingernail. **B,** Sagittal section of fingernail and associated structures.

Variations in Nail Structure

Age and ethnicity can influence the normal shape and appearance of the nails. For example, longitudinal ridges are common in those with light skin of all ages and are especially prominent in the elderly, whereas pigmented bands are a normal finding in dark-skinned individuals (**Figure 7-11**).

Pathologic changes in the appearance of the nails often occur as a result of certain diseases and because of trauma. For

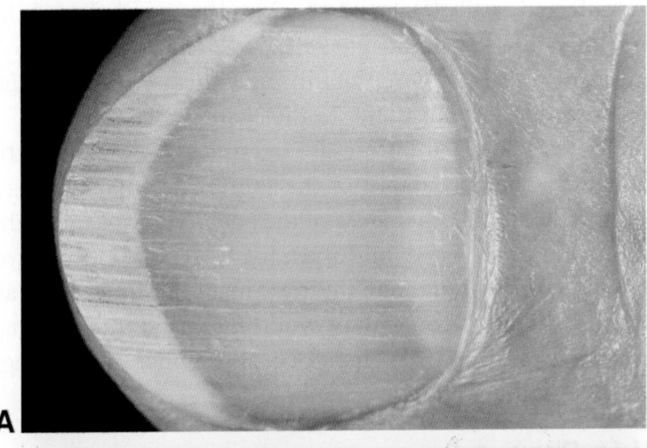

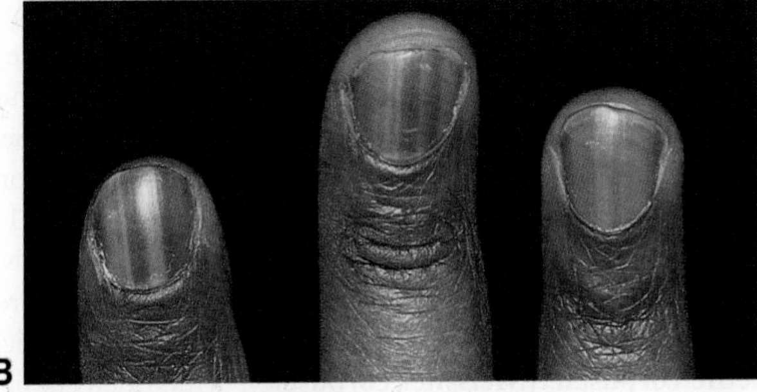

FIGURE 7-11 Normal variations in nail structure. A, Longitudinal ridges in light-skinned people are common. **B,** Pigmented bands are a normal finding in dark-skinned individuals.

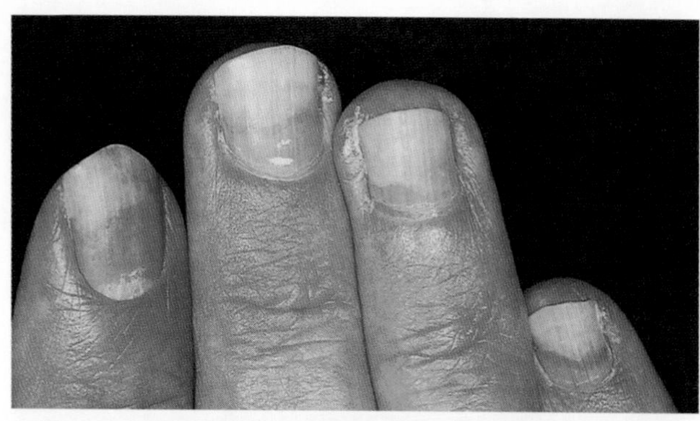

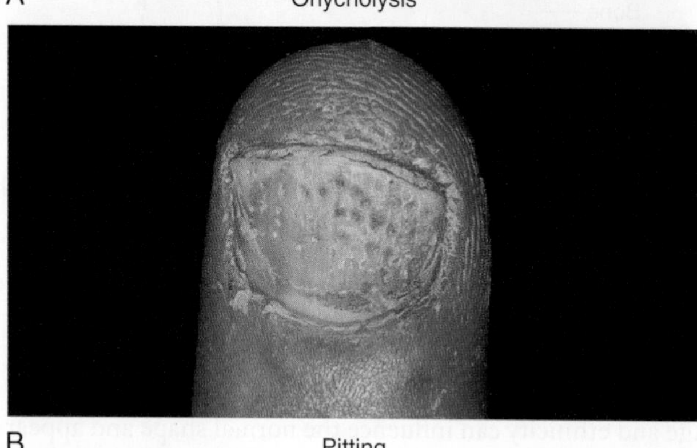

FIGURE 7-12 Abnormal nail structure. A, Onycholysis. Separation of nail from the nail bed begins at the free edge. **B,** Nail pitting. A common finding in persons with psoriasis.

example, even minor trauma to long fingernails can sometimes result in a loosening of the nail from the nail bed with a resulting separation that starts at the distal or free edge of the affected nail (Figure 7-12, A). The condition, called **onycholysis,** is common. Figure 7-12, *B,* shows pitting of the nail. Nail pitting often occurs in individuals with *psoriasis,* a skin disorder described later in the chapter (see p. 162).

Cyanosis and nail pitting are examples of how distinctive changes in the appearance of the skin or its appendages can point to disease in other areas of the body. Many of the pathologic conditions listed in Appendix A at *evolve.elsevier.com* and described throughout the text, including infectious and internal diseases, congenital syndromes, and tumors, have symptoms that appear as changes in appearance of the integumentary system and body membranes.

Skin Receptors

Receptors in the skin make it possible for the body surface to act as a sense organ, relaying messages to the brain concerning sensations such as touch, pain, temperature, and pressure. Sensory receptors, which differ in structure from the highly complex to the very simple, are discussed in detail in Chapter 11.

Two skin receptors are visible in Figure 7-3. One is a **lamellar corpuscle (Pacini corpuscle),** which detects pressure deep in the dermis. The other is the more superficial **tactile corpuscle (Meissner corpuscle),** which detects light touch. Other receptors mediate sensations such as crude touch, vibration, temperature, and pain.

Burn injuries, which are discussed later in the chapter, destroy skin receptors. By doing so, they also may destroy the ability of the burned skin to function as a sense organ.

Skin Glands

The skin glands include the two varieties of **sweat glands** and the tiny **sebaceous glands.**

Sweat Glands

Sweat glands, also called **sudoriferous glands,** are the most numerous of the skin glands (see Figure 7-3). They can be classified into two groups—**eccrine** and **apocrine**—based on type of secretion and location.

Eccrine Glands

Eccrine sweat glands are by far the more numerous, important, and widespread sweat glands in the body. They are quite small and, with few exceptions, are distributed over the total body surface. Throughout life they produce a transparent, watery liquid called **perspiration,** or **sweat.**

Sweat assists in the elimination of waste products such as ammonia and uric acid. In addition to elimination of waste, sweat plays a critical role in helping the body maintain a constant temperature.

Anatomists estimate that a single square inch of skin on the palms of the hands contains about 3000 eccrine sweat glands. With a magnifying glass you can locate the pinpoint-size openings on the skin that you probably call **pores.** The pores are outlets of small ducts from the eccrine sweat glands.

Apocrine Glands

Apocrine sweat glands are found primarily in the skin of the axilla (armpit) and in the pigmented skin areas around the genitals. They are larger than the eccrine glands, and instead of watery sweat, they secrete a thicker, milky secretion.

The odor associated with apocrine gland secretion is not caused by the secretion itself. Instead, it is caused by the contamination and decomposition of the secretion by skin bacteria. Apocrine glands enlarge and begin to function at puberty.

Sebaceous Glands

Sebum

Sebaceous glands secrete oil for the hair and skin. Oil glands, or sebaceous glands, are found where hairs grow. Their tiny ducts open into hair follicles (see Figure 7-3) so that their secretion, called **sebum,** lubricates the hair and skin. Someone aptly described sebum as "nature's skin cream" because it prevents drying and cracking of the skin.

Sebum secretion increases during adolescence, stimulated by the increased blood levels of the sex hormones. Frequently sebum accumulates in and enlarges some of the ducts of the sebaceous glands, forming white pimples. This sebum often darkens, forming a **blackhead** or **comedo.** Sebum secretion decreases in late adulthood, contributing to increased wrinkling and cracking of the skin.

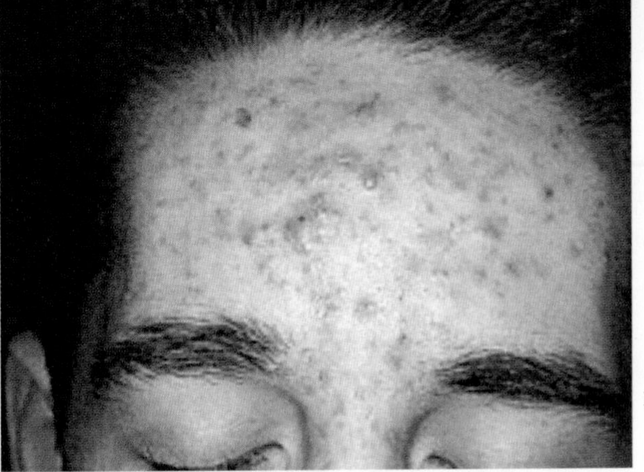

FIGURE 7-13 Acne. Acne vulgaris results from blocked sebaceous glands that become inflamed or infected.

Acne

The most common kind of acne, **acne vulgaris** (Figure 7-13), occurs most frequently during adolescence. This condition results from the more than fivefold increase in sebum secretion between the ages of 10 and 19.

The oversecretion of sebum results in blockage of the sebaceous gland ducts with sebum, skin cells, and bacteria. The inflamed lesions that result are called **papules.** Pus-filled pimples called **pustules** often develop and then rupture, resulting in secondary infections in the surrounding skin.

Formation of acne lesions can be minimized by careful cleansing of the skin to remove sebaceous plugs and to inhibit anaerobic skin bacteria.

Combinations of topical (external) medications are now used to effectively treat many types of acne. Topical use of vitamin A acid (tretinoin) speeds up mitosis in the hair follicle, thus preventing sebum from building up as the hair moves quickly out of the follicle. It is often combined with drying and peeling agents such as benzoyl peroxide and externally applied antibiotics. In addition, physicians or other trained health professionals may use a special surgical instrument called an *extractor* to remove compacted sebum ("blackheads") and the contents of acne pustules ("whiteheads") to hasten healing.

More severe cases of acne may require additional treatment with oral antibiotics, which reduce infection and inflammation.

Functions of the Skin

The skin, or cutaneous membrane, serves many important functions that contribute to survival. The most important functions are as follows:

1. Protection
2. Temperature regulation
3. Sense organ activity
4. Excretion
5. Synthesis of vitamin D

Protection

The skin as a whole is often described as our "first line of defense" against a multitude of hazards. It protects us against the daily invasion of deadly microbes. The tough, keratin-filled cells of the stratum corneum also resist the

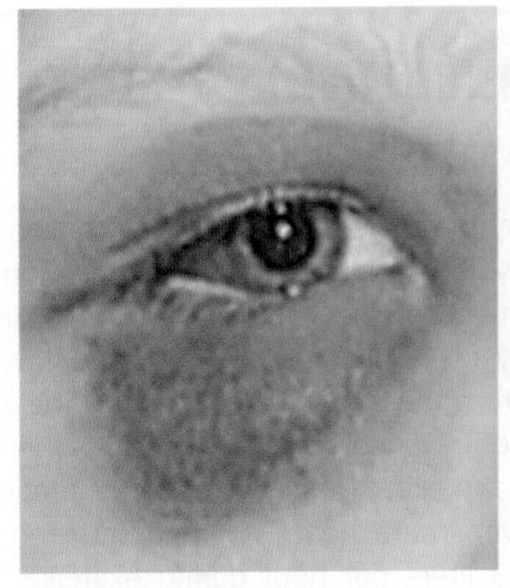
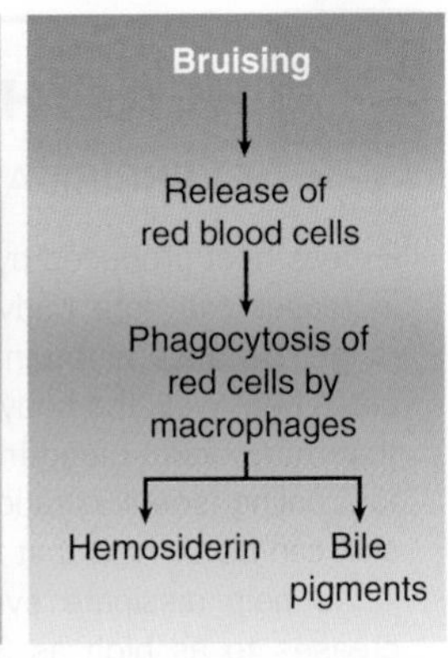

FIGURE 7-14 Bruising. Color changes caused by the deoxygenation, clotting, and breakdown of blood are easily seen in light-skinned individuals.

entry of harmful chemicals and protect against physical tears and cuts. Because it is waterproof, keratin also protects the body from excessive fluid loss. Melanin in the pigment layer of the skin prevents the sun's harmful UV rays from penetrating the interior of the body.

Physical damage can cause bruising of the skin when blood vessels in the skin break open (Figure 7-14). Release of red blood cells initially produces a reddish color in the skin. It then begins to darken and produce bluish colors when hemoglobin loses oxygen. As the blood clots, it may begin to appear darker blue or even black. Cells break down the hemoglobin into iron-containing *hemosiderin* (a yellow-brownish pigment) and several iron-free *bile pigments* that are greenish and yellowish.

Skin grafts may be required to provide some degree of protection to areas of the body that are no longer covered by skin because of burns or to replace skin destroyed by disease or trauma. Figure 7-15 shows a skin graft covering a burned hand.

Temperature Regulation

The skin plays a key role in regulating the body's temperature. Incredible as it seems, on a hot and humid day the skin can

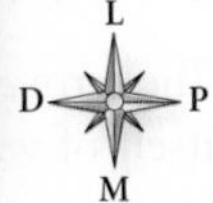

FIGURE 7-15 Skin graft. Photograph shows a skin graft covering a severe burn to the hand. Multiple slits allow the grafted piece of skin to stretch over a larger area than would otherwise be possible.

7

HEALTH AND WELL-BEING

EXERCISE AND THE SKIN

Excess heat produced by the skeletal muscles during exercise increases the core body temperature far beyond the normal range. Because blood in vessels near the skin's surface dissipates heat well, the body's control centers adjust blood flow so that more warm blood from the body's core is sent to the skin for cooling (see illustration). During exercise, blood flow in the skin can be so high that the skin takes on a redder coloration.

To help dissipate even more heat, sweat production increases to as high as 3 liters per hour during exercise. Although each sweat gland produces very little of this total, more than 3 million individual sweat glands are found throughout the skin.

Sweat evaporation is essential to keeping body temperature in balance, but excessive sweating can lead to a dangerous loss of fluid. Because normal amounts of drinking may not replace the water lost through sweating, it is important to increase fluid consumption during and after any type of exercise to avoid **dehydration**—excessive water loss that disrupts homeostasis.

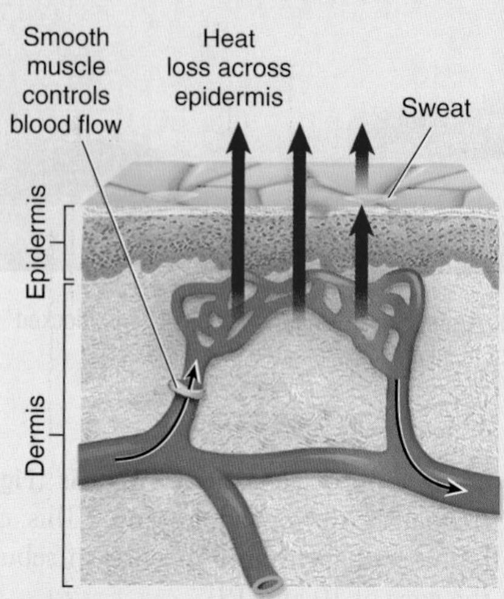

Heat loss during exercise. Excess heat produced by working muscles can be lost from blood through the skin to the air. Sweat on the skin can also absorb some of the heat and evaporate—cooling the body further. These mechanisms help maintain homeostasis of body temperature.

serve as a means for releasing almost 3000 calories of body heat—enough heat energy to boil more than 20 liters of water! It accomplishes this feat by regulating sweat secretion and by regulating the flow of blood close to the body surface. When sweat evaporates from the body surface, heat is also lost. The principle of heat loss through evaporation is basic to many cooling systems.

When increased quantities of blood are allowed to fill the vessels close to the skin, heat is also lost by radiation. Blood supply to the skin far exceeds the amount needed by the skin. The overabundant blood supply primarily enables the regulation of body temperature.

Sensation

The skin functions as an enormous sense organ. Its millions of nerve endings serve as antennas or receivers for the body, keeping it informed of changes in its environment. The sensory receptor types shown in **Figure 7-3** make it possible for the body to detect sensations of light touch (tactile corpuscles) and pressures (lamellar corpuscles). Other receptors make it possible for us to respond to the sensations of pain, heat, and cold.

Excretion

The term *excretion* refers to any process in which the body rids itself of waste or surplus substances. Excretion of substances in sweat can influence the amounts of certain ions (such as sodium) and waste products (such as uric acid, ammonia, and urea) that are present in the blood. Excess vitamins, drugs, and even hormones in the blood can also be excreted onto the skin by sweat.

Synthesis of Vitamin D

Synthesis of vitamin D is another important function of the skin. It occurs when the skin is exposed to *UV light*—usually from the sun. When UV light penetrates the skin, a precursor substance in skin cells forms, then is transported to the liver and kidneys where it is converted into an active form of vitamin D. Research shows that vitamin D affects many different functions in the body, thus emphasizing the importance of this skin function.

QUICK CHECK

1. Identify the structural components of a nail.
2. How do eccrine and apocrine sweat glands differ?
3. How is sebaceous gland function related to acne?
4. List the major functions of the skin.

Disorders of the Skin

Any disorder of the skin can be called a **dermatosis,** which simply means "skin condition." Many dermatoses involve inflammation of the skin, or **dermatitis.** Only a few of the many disorders of the skin are discussed here.

Review concepts of the inflammatory response in the article *Inflammation* at Connect It! at *evolve.elsevier.com.*

Skin Lesions

A **lesion** is any measurable variation from the normal structure of a tissue. Lesions are not necessarily signs of disease—they instead may be benign, ordinary variations. For example, freckles are considered lesions but are not signs of disease.

Almost all diseases affecting the skin are discovered and diagnosed after observing the nature of the lesions present. Lighting the skin from the side with a penlight is a method used to determine the category of a lesion: elevated, flat, or depressed. Elevated lesions cast shadows outside their edges; flat lesions do not cast shadows, and depressed lesions cast shadows inside their edges.

Important examples of each type of lesion are summarized in **Table 7-1**.

Lesions are often distinguished by abnormal density of tissue or abnormal coloration. Overgrowth or deficient growth of skin cells, calcification, and edema can cause changes in skin density. Discoloration can result from overproduction or underproduction of skin pigments such as the increase in melanin seen in a mole. A decrease in blood flow or oxygen content can give the skin a bluish cast (cyanosis), whereas an increased blood flow or oxygen content can give a red or darker hue to the skin. Discoloration of the affected area is associated with most skin lesions.

Some of the most common lesions result from scrapes and cuts that our skin often endures in its role of protection.

Figure 7-16 shows the way in which such injuries typically repair themselves. First, clotting of blood stops blood loss. Then cells of the stratum germinativum produce more epithelial cells to rebuild the epidermis as the clot dissolves. At the same time, fiber-producing cells of the dermis replace torn collagen fibers.

Often, the fibrous tissue replaced during skin repair is denser than the original tissue—providing extra strength in the case of further injury but also sometimes producing a scar.

Burns

Burns constitute one of the most serious and frequent problems that affect the skin. Typically, we think of a burn as an injury caused by fire or by contact of the skin with a hot surface. However, overexposure to UV light (sunburn) or contact of the skin with an electric current or a harmful chemical such as an acid also can cause burns.

Severity of Burns

The seriousness of a burn injury, as well as appropriate treatment and the possibility for recovery, are determined by three major factors:

1. Depth and number of tissue layers involved
2. Total body surface area affected
3. Type of homeostatic mechanisms that are damaged or destroyed, such as respiratory and blood pressure control or fluid and electrolyte balance

The age and general state of health of the individual at the time of injury also are important. A "moderately severe" burn in an otherwise healthy young adult may become a life-threatening "major" burn in an infant or an elderly individual with preexisting respiratory problems or heart disease.

Depth Classification of Burns

Burns can be classified in a variety of ways, including how deeply the tissues are damaged (**Figure 7-17**).

First-degree Burns

A **first-degree burn** (for example, a typical sunburn) causes minor discomfort and some reddening of the skin. Although the surface layers of the epidermis may peel in 1 to 3 days, no blistering occurs, and actual tissue destruction is minimal.

Second-degree Burns

A **second-degree burn** (**Figure 7-18,** *A*) involves the deep epidermal layers and always causes injury to the upper layers of the dermis. Although deep second-degree burns damage sweat glands, hair follicles, and sebaceous glands, complete

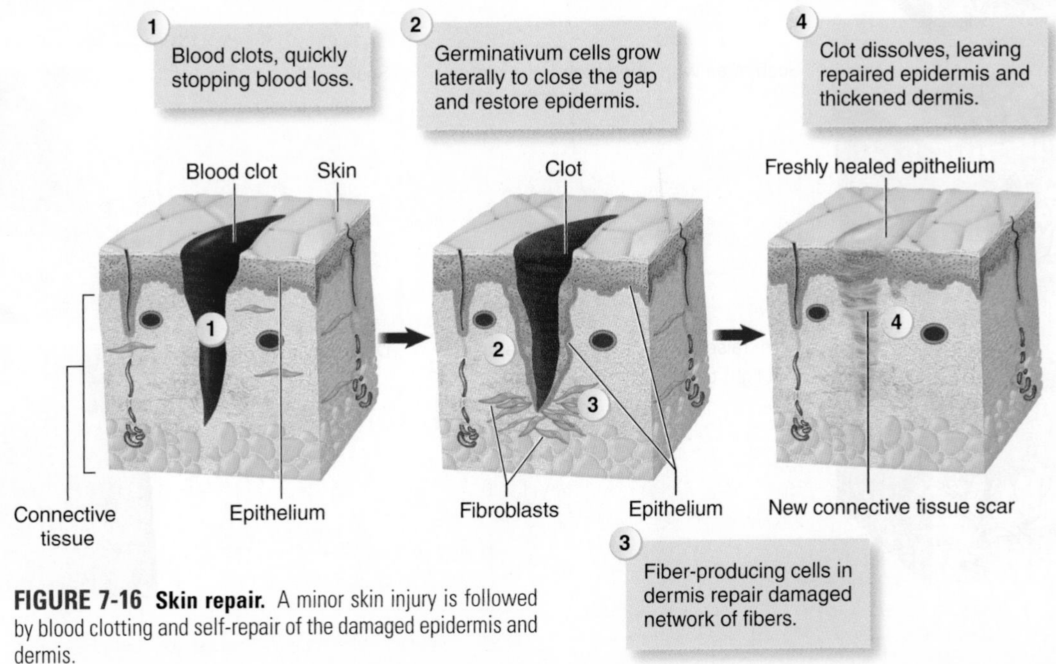

FIGURE 7-16 Skin repair. A minor skin injury is followed by blood clotting and self-repair of the damaged epidermis and dermis.

7

TABLE 7-1 Common Skin Lesions

LESION	DESCRIPTION	EXAMPLE
Elevated		
Papule	Firm, raised lesion (less than 1 cm in diameter)	Warts
Plaque	Large, raised lesion (greater than 1 cm in diameter)	Plaque caused by friction
Vesicle	Thin-walled blister filled with fluid that is smaller than 1 cm (a vesicle larger than 1 cm is a bulla)	Non-genital herpes vesicles
Pustule	Elevated lesion filled with pus	Acne
Crust	Scab; area with dried blood or exudate	Scab
Wheal (hive)	Firm, raised area of irregular shape with a light center	Drug-sensitivity hives

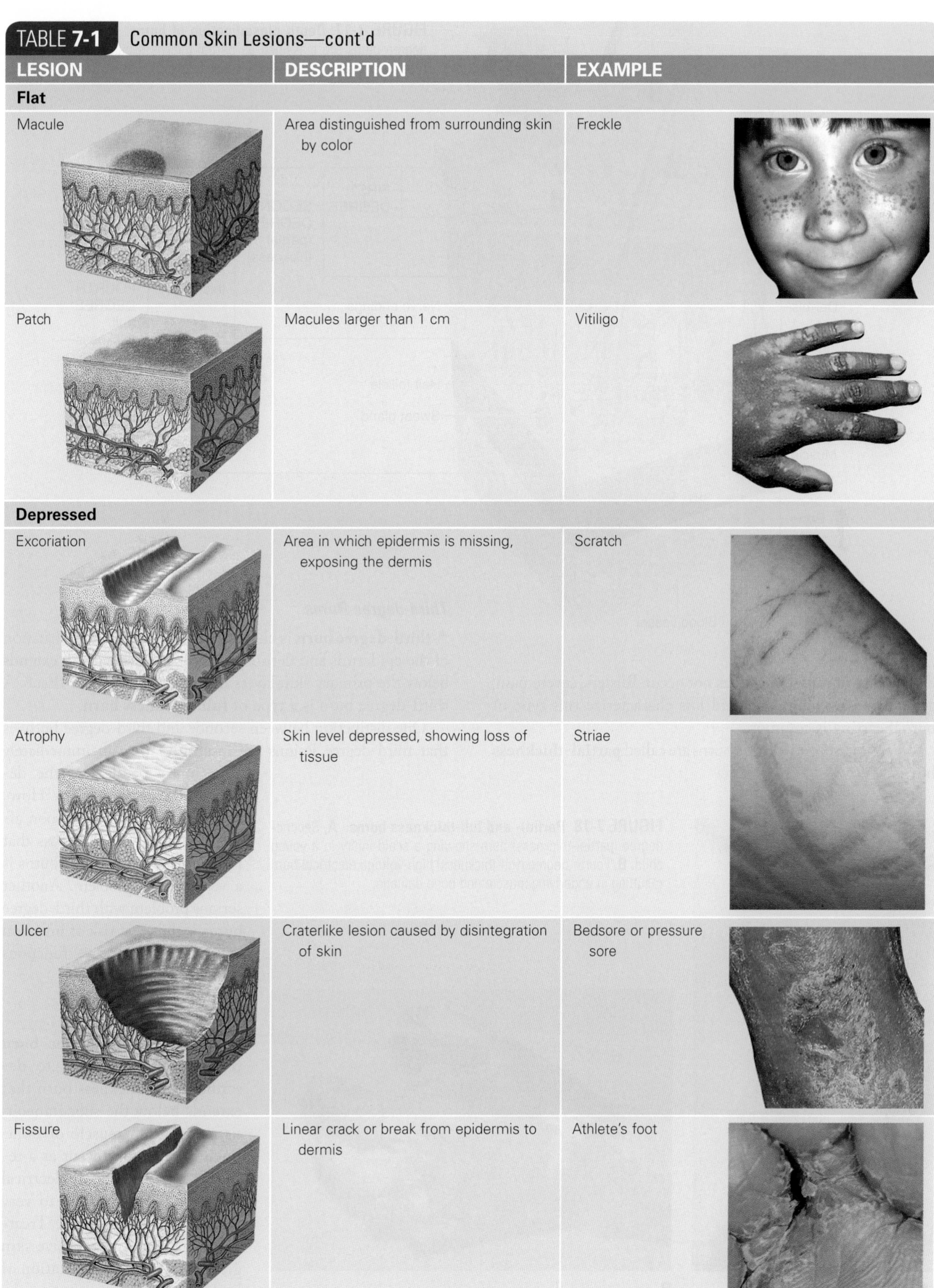

TABLE 7-1 Common Skin Lesions—cont'd

LESION	DESCRIPTION	EXAMPLE
Flat		
Macule	Area distinguished from surrounding skin by color	Freckle
Patch	Macules larger than 1 cm	Vitiligo
Depressed		
Excoriation	Area in which epidermis is missing, exposing the dermis	Scratch
Atrophy	Skin level depressed, showing loss of tissue	Striae
Ulcer	Craterlike lesion caused by disintegration of skin	Bedsore or pressure sore
Fissure	Linear crack or break from epidermis to dermis	Athlete's foot

FIGURE 7-17 Depth classification of burns. First- and second-degree burns are classified as partial-thickness burns and third- and fourth-degree burns as full-thickness burns.

destruction of the dermis does not occur. Blisters, severe pain, generalized swelling, and fluid loss characterize this type of burn. Scarring is common.

First- and second-degree burns are called **partial-thickness burns.**

Third-degree Burns

A **third-degree burn** is characterized by complete destruction of the epidermis and dermis. In addition, tissue death extends below the primary skin layers into the subcutaneous tissue. A third-degree burn is a type of **full-thickness burn.**

One distinction between second- and third-degree burns is that third-degree lesions are insensitive to pain immediately after injury because of the destruction of nerve endings. However, intense pain occurs soon after the injury. The fluid loss that results from third-degree burns is a very serious problem. Another serious problem with third-degree burns is the great risk of infection because the protective functions of the skin are lost.

Fourth-Degree Burns

The term **fourth-degree burn** (**Figure 7-18**, *B*) is used to describe a full-thickness burn that extends below the subcutaneous tissue to reach muscle or bone. Such injuries may occur as a result of high-voltage electrical burns or from exposure to very intense heat over time. Treatment may require extensive skin grafting and even amputation of limbs.

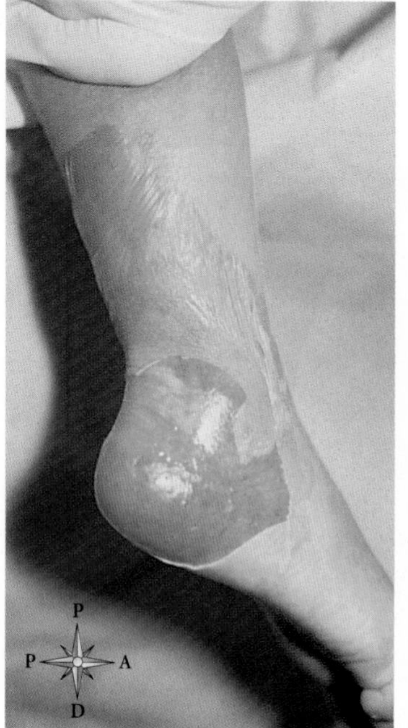

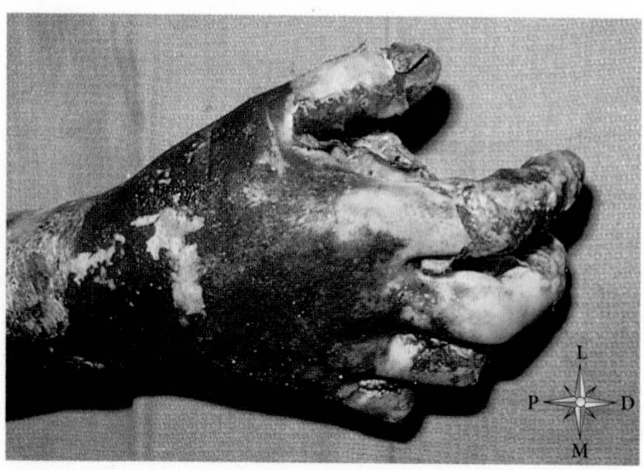

FIGURE 7-18 Partial- and full-thickness burns. A, Second-degree (partial-thickness) burn showing a scald injury in a young child. **B,** Fourth-degree (full-thickness) high-voltage electrical burn resulting in underlying muscle and bone damage.

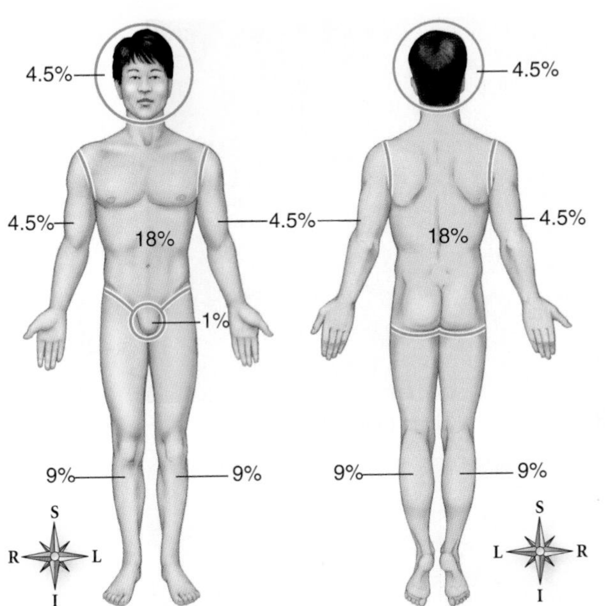

FIGURE 7-19 Rule of nines. Dividing the adult body into 11 areas of 9% each helps in estimating the amount of skin surface burned in an adult.

Estimating Body Surface Area

When burns involve large areas of the skin, treatment and the possibility for recovery depend in large part on the *total area involved* and the *severity of the burn.* The severity of a burn is determined by the depth of the injury, as well as by the amount of body surface area affected.

The **"rule of nines"** is one of the most frequently used methods of determining the extent of a burn injury in adults. With this technique (Figure 7-19), the body is divided into 11 areas of 9% each, with the area around the genitals representing the additional 1% of body surface area.

As you can see in Figure 7-19, in the adult 9% of the skin covers the head and each upper extremity, including anterior and posterior surfaces. Twice as much, or 18%, of the total skin area covers the anterior and posterior of the trunk and each lower extremity, including all surfaces.

To learn more about burns, go to AnimationDirect online at *evolve.elsevier.com.*

Skin Infections

The skin is the first line of defense against microbes that might otherwise invade the body's internal environment. So the skin is a common site of infection. Viruses, bacteria, fungi, or larger parasites cause skin conditions such as those listed here. Refer to Appendix A at *evolve.elsevier.com* for more information on these and other skin infections.

Impetigo

Impetigo is a highly contagious condition that results from staphylococcal or streptococcal infection and occurs most often in young children. It starts as a reddish discoloration, or **erythema,** but soon develops into vesicles and yellowish **crusts** (Figure 7-20, *A*). Occasionally, impetigo becomes systemic (body-wide) and thus life threatening.

Tinea

Tinea is the general name for many different *mycoses* (fungal infections) of the skin. Ringworm, jock itch, and athlete's foot are classified as tinea.

Signs of tinea include erythema, scaling, and crusting. Occasionally, **fissures,** or cracks, develop at creases in the epidermis. Figure 7-20, *B*, shows a case of ringworm, a tinea infection that typically forms a round rash that heals in the center to form a ring.

Antifungal agents usually stop the acute infection. Recurrence can be avoided by keeping the skin dry because fungi require a moist environment to grow.

Warts

Caused by a papillomavirus, **warts** are a type of benign neoplasm of the skin. However, some warts do transform and become malignant.

The nipplelike projections characteristic of this contagious condition are shown in Table 7-1, p. 158. Transmission of warts generally occurs through direct contact with lesions on the skin of an infected person. Warts can be removed by freezing, drying, laser therapy, or application of chemicals.

Boils

Also called **furuncles,** boils are most often local staphylococcal infections of hair follicles and are characterized by large, inflamed pustules (Figure 7-20, *C*). A group of untreated boils may fuse into even larger pus-filled lesions called **carbuncles.**

Scabies

Scabies is a contagious skin condition caused by the itch mite *(Sarcoptes scabiei).*

Transmitted by skin-to-skin contact, as in sexual activity, the female mite digs under the hard stratum corneum and forms a short, winding burrow where she deposits her eggs (Figure 7-20, *D*). Young mites called *larvae* hatch, forming tiny, red papules. After a month or so, a hypersensitivity reaction (see Chapter 16) may cause a rash characterized by erythema and numerous papules.

As the name of the culprit indicates, infestation of the skin by itch mites causes intense itching. **Excoriation** that results from scratching the itchy infested areas may lead to secondary bacterial infections.

Vascular and Inflammatory Skin Disorders

Just a few examples of the many vascular and inflammatory skin disorders are described in the following sections.

Decubitus Ulcers

Every caregiver should be aware of the causes and nature of pressure sores or **decubitus ulcers** (Figure 7-21, *A*). *Decubitus* means "lying down," a name that hints at a common cause of

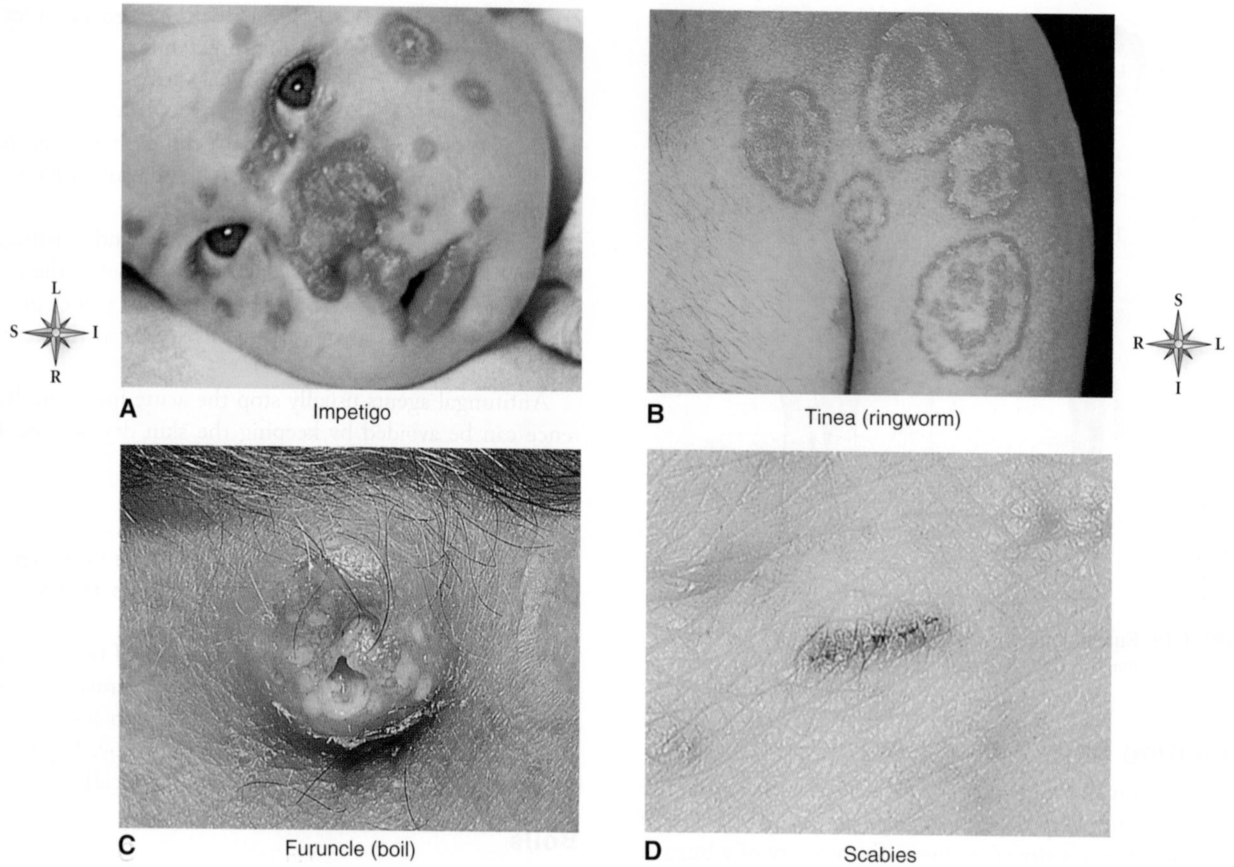

FIGURE 7-20 Skin infections.

pressure sores: lying in one position for long periods. Also called *bedsores,* these lesions appear after blood flow to a local area of skin slows or is obstructed because of pressure on skin covering bony prominences such as the heel. Ulcers form and infections develop because lack of blood flow causes tissue damage or death.

Frequent changes in body position and soft support cushions help prevent decubitus ulcers.

Hives

Hives, or **urticaria,** is a common condition characterized by raised red lesions called **wheals** (Figure 7-21, *B*). Urticaria is often associated with severe itching. Hives are generally short lived, lasting from a few hours to a few weeks.

The lesions are caused by leakage of fluid from the skin's blood vessels. They change in size and shape over time; new lesions erupt as old ones disappear when fluid in the raised wheals is reabsorbed by the body. Hypersensitivity or allergic reactions to drugs or food, physical irritants, and systemic diseases are common causes of urticaria.

Scleroderma

Scleroderma is an autoimmune disease that affects the blood vessels and connective tissues of the skin.

The name *scleroderma* comes from the word parts *sclera,* which means "hard," and *derma,* which means "skin." Hard skin is a good description of the lesions characteristic of scleroderma. Scleroderma begins as an area of mild inflammation that later develops into a patch of yellowish, hardened skin.

Scleroderma most commonly remains a mild, localized condition. Very rarely, localized scleroderma progresses to a systemic form, affecting large areas of the skin and other organs. Persons with advanced systemic scleroderma seem to be wearing a mask because skin hardening prevents them from moving their mouths freely. Both forms of scleroderma occur more commonly in women than in men.

Psoriasis

Psoriasis is a common, chronic, and often lifelong skin disease that affects 1% to 3% of the population. It is characterized by silvery white, scalelike **plaques** that may remain fixed on the skin for months (Figure 7-21, *C*).

Psoriasis is thought to have a genetic basis and tends to affect skin on the elbows, knees, and scalp most often. Individuals with psoriasis often show pitting of the nails (see Figure 7-12, *B*). The scales or plaques associated with the disease develop from an excessive rate of epithelial cell growth.

Eczema

Eczema is the most common inflammatory disorder of the skin. This condition is characterized by inflammation that is often accompanied by papules, vesicles, and crusts.

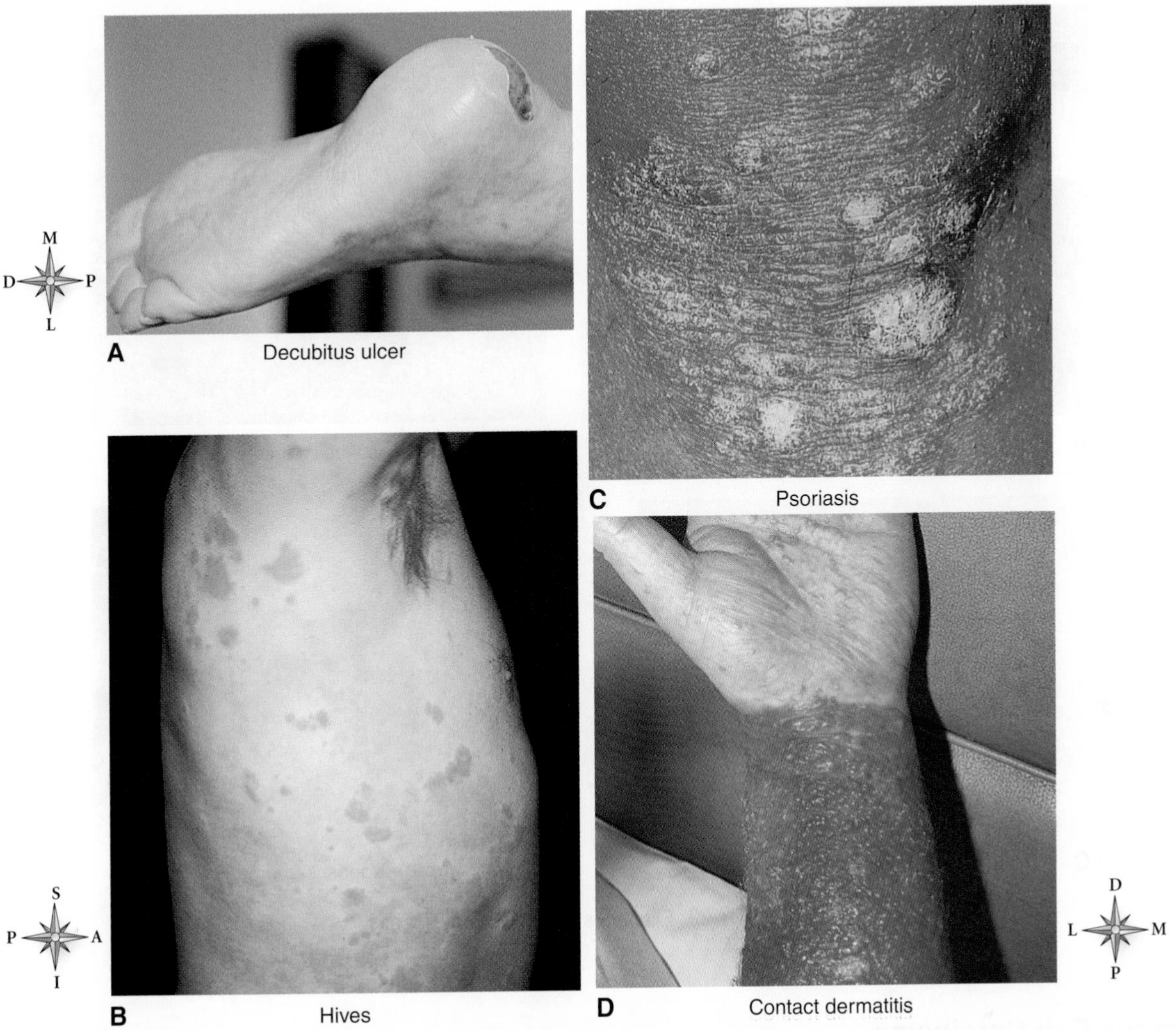

A Decubitus ulcer

B Hives

C Psoriasis

D Contact dermatitis

FIGURE 7-21 Vascular and inflammatory skin disorders.

Eczema is not a distinct disease but rather a sign or symptom of an underlying condition. For example, an allergic reaction called *contact dermatitis* can progress to become eczematous. The blisters and marked redness on the arm shown in Figure 7-21, *D* are the result of contact dermatitis caused by soap used in laundering a long-sleeved shirt.

Skin Cancer

Role of Ultraviolet Radiation

Of the many types of skin cancer, the most common are **squamous cell carcinoma, basal cell carcinoma,** and malignant **melanoma.**

Although genetic predisposition also plays a role, many pathophysiologists believe that exposure to the sun's UV radiation is the most important causal factor in the three most common skin cancers. UV radiation damages the DNA in skin cells, causing the mistakes in mitosis that produce cancer.

Skin cells have a natural ability to repair UV damage to the DNA, but in some people, this inherent mechanism may not be able to deal with a massive amount of damage. People with the rare, inherited condition **xeroderma pigmentosum** cannot repair UV damage at all and almost always develop skin cancer.

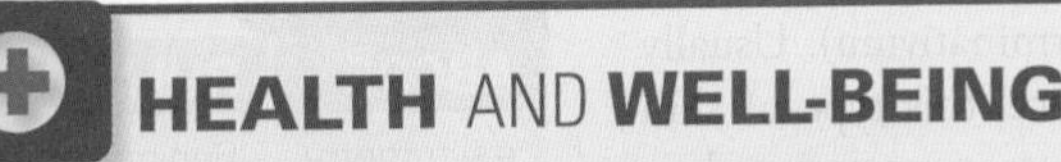

HEALTH AND WELL-BEING

SUNBURN AND SKIN CANCER

Burns caused by exposure to harmful ultraviolet (UV) radiation in sunlight are commonly called *sunburns.* As with any burn, serious sunburns can cause tissue damage and lead to secondary infections and fluid loss. Cancer researchers have recently theorized that blistering (second-degree) sunburns during childhood may trigger the development of malignant melanoma later in life. Epidemiologic studies now show that adults who had more than two blistering sunburns before the age of 20 have a much greater risk of developing melanoma than someone who experienced no such burns. This theory helps explain the dramatic increase in skin cancer rates in the United States observed in recent years. Those who grew up sunbathing and experienced sunburns in their youth are now, as older adults, exhibiting melanoma at a much higher rate than those in previous generations.

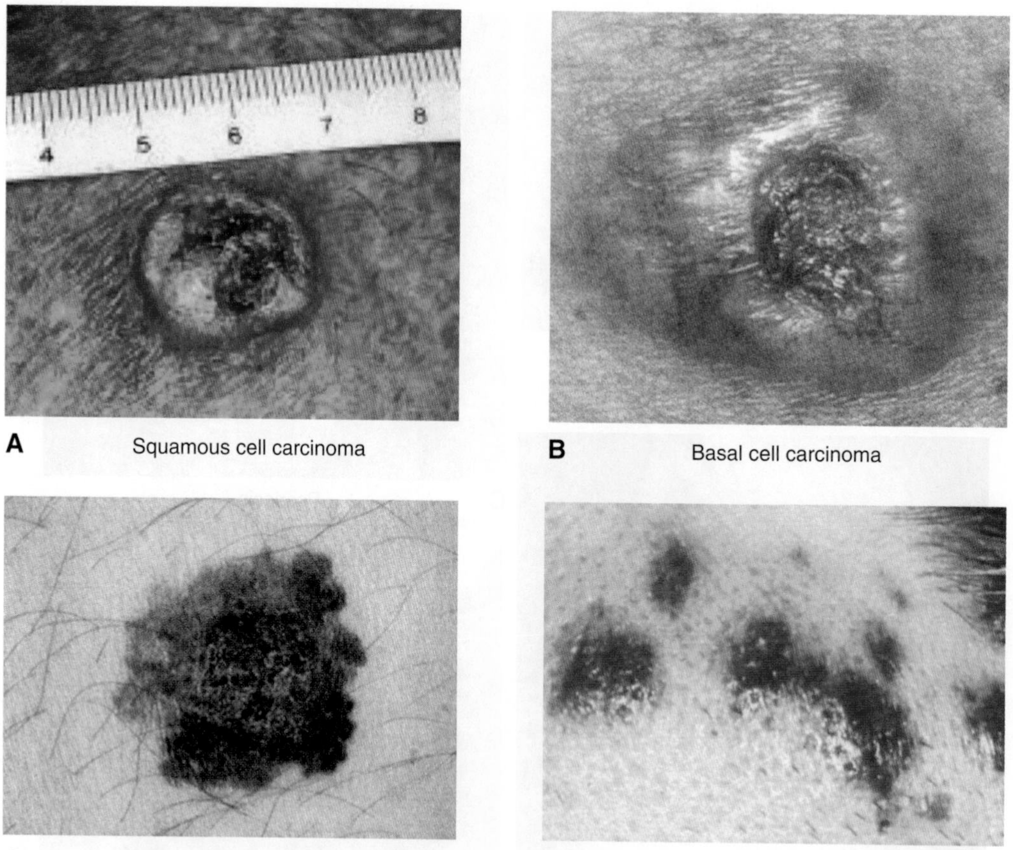

A Squamous cell carcinoma

B Basal cell carcinoma

C Malignant melanoma

D Kaposi sarcoma

FIGURE 7-22 Examples of skin cancer lesions.

Squamous Cell Carcinoma

Squamous cell carcinoma is a slow-growing malignant tumor of the epidermis. Lesions typical of this form of skin cancer are hard, raised nodules that are usually painless (**Figure 7-22**, A). If not treated, squamous cell carcinoma may metastasize, invading other organs.

Basal Cell Carcinoma

As its name implies, **basal cell carcinoma** begins in cells at the base of the epidermis (the basal layer of stratum germinativum). Usually occurring on the upper face, this type of skin cancer is much less likely to metastasize than other types. It is the most common type of skin cancer. Basal cell carcinoma lesions typically begin as papules that erode in the center to form a bleeding, crusted crater (**Figure 7-22**, *B*).

Melanoma

Malignant **melanoma** is the most serious form of skin cancer. Unfortunately, the incidence of melanoma in the U.S. population is increasing. In the absence of early treatment, it causes death in about one in every four cases.

This type of cancer sometimes develops from a pigmented **nevus (mole)** and transforms into a dark, spreading lesion (**Figure 7-22**, *C*). Benign moles should be checked regularly for warning signs of melanoma because early detection and removal are essential in treating this rapidly spreading cancer. The "ABCDE" rule of self-examination of moles is summarized in **Table 7-2**.

TABLE 7-2 Warning Signs of Malignant Melanoma

ABCDE	RULE
Asymmetry	Benign moles are usually symmetrical; their halves are mirror images of each other. Melanoma lesions are asymmetrical or lopsided.
Border	Benign moles are outlined by a distinct border, but malignant melanoma lesions are often irregular or indistinct in shape.
Color	Benign moles may be any shade of brown but are relatively evenly colored; melanoma lesions tend to be unevenly colored, exhibiting a mixture of shades or colors.
Diameter	By the time a melanoma lesion exhibits characteristics A, B, and C, it also is probably larger than 6 mm (¼ inch)
Evolving	Moles that continue to evolve, or change over time, may be cancerous. Besides the changes noted above, melanoma lesions may begin to itch, form an ulcer, or bleed.

Kaposi Sarcoma

Kaposi sarcoma (KS) is caused by *Kaposi sarcoma–associated herpes virus (KSHV)*, also known as *human herpes virus 8 (HHV8)*. Once associated mainly with certain ethnic groups, a form of this cancer now also appears in many cases of AIDS and other immune deficiencies.

Kaposi sarcoma, first appearing as purple papules (**Figure 7-22**, *D*), quickly spreads to the lymph nodes and internal organs.

For more information, including additional photographs, review the article *Skin Cancer* at Connect It! at *evolve.elsevier.com*.

QUICK CHECK

1. How are burns classified?
2. How can the skin surface area damaged by a burn be estimated?
3. Identify five skin infections and four vascular and inflammatory skin disorders.
4. List the three major types of skin cancer.
5. What are the warning signs of malignant melanoma?

SCIENCE APPLICATIONS

SECRETS OF THE SKIN

Dr. Joseph E. Murray (1919-2012)

The skin is our most visible organ, so it is no wonder that observing the structure and function of skin has generated sparks that have lit the fires of scientific discovery through the ages. The ancient Romans outlined the process of inflammation in detail after observing it first in the skin. In the twentieth century, Joseph Murray (see figure) noticed that skin he grafted onto burned soldiers he treated during World War II would eventually be rejected by the body (see **Figure 7-15** on p. 155).

After the war, Murray tried to understand the body's immune reactions to transplanted tissues and his work led to the first successful kidney transplants. His breakthroughs in transplanting kidneys not only earned him a Nobel Prize in 1990, it also paved the way for all the different types of tissue and organ transplantation that we see today.

Many scientists continue to study the secrets of the skin, and many physicians and other health-care professionals are also pioneers in developing new methods of skin care and treatment in the fields of **dermatology, allergy, immunology,** *burn medicine,* **reconstructive surgery,** and **cosmetic surgery.**

Additional practical applications of some of this skin science are practiced by people working with cosmetics and other skin, nail, and hair treatments. For example, *industrial researchers, product developers,* **cosmeticians,** *spa specialists,* and *hair stylists* all require some knowledge of current skin science to do their jobs effectively.

The photo shows a physician and medical assistant using a medical laser to remove a tattoo from the skin of a patient.

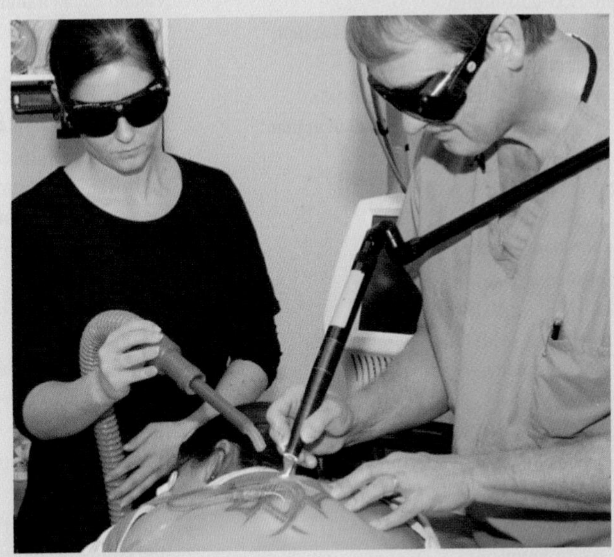

LANGUAGE OF SCIENCE *(continued from p. 145)*

dermal papilla
(DER-mal pah-PIL-ah)
pl., papillae
(pah-PIL-ee)
[*derma-* **skin,** *-al* **relating to,** *papilla* **nipple**]

dermal-epidermal junction
(DER-mal EP-ih-der-mal JUNK-shun)
[*derma-* **skin,** *-al* **relating to,** *epi-* **on or upon,** *-derma-* **skin,** *-al* **relating to,** *junc-* **join,** *-tion* **condition**]

dermis
(DER-mis)
[*dermis* **skin**]

eccrine
(EK-rin)
[*ec-* **out,** *-crin-* **secrete**]

eccrine sweat gland
(EK-rin swet gland)
[*ec-* **out,** *-crin-* **secrete,** *gland* **acorn**]

Continued on p. 166

7

LANGUAGE OF **SCIENCE** *(continued from p. 165)*

epidermis
(ep-ih-DER-mis)
[*epi-* **on or upon,** *-dermis* **skin**]

epithelial membrane
(ep-ih-THEE-lee-al MEM-brayn)
[*epi-* **on or upon,** *-theli-* **nipple,** *-al* **relating to,** *membran-* **thin skin**]

follicle
(FOL-lih-kul)
[*foll-* **bag,** *-icle* **little**]

freckle
(FREK-uhl)

hair follicle
(hayr FOL-lih-kul)
[*foll-* **bag,** *-icle* **little**]

hair papilla
(hayr pah-PIL-ah)
[*papilla* **nipple**]

hypodermis
(hye-poh-DER-mis)
[*hypo-* **under or below,** *-dermis* **skin**]

integument
(in-TEG-yoo-ment)
[*in-* **on,** *-teg-* **cover,** *-ment* **result of action**]

integumentary system
(in-teg-yoo-MEN-tar-ee SIS-tem)
[*in-* **on,** *-teg-* **cover,** *-ment-* **result of action,** *-ary* **relating to**]

keratin
(KER-ah-tin)
[*kera-* **horn,** *-in* **substance**]

lamellar corpuscle
(lah-MEL-ar KOR-pus-ul)
[*lam-* **plate,** *-ella-* **little,** *-ar* **relating to,** *corpus-* **body,** *-cle* **little**]

lamina propria
(LAM-in-ah PROH-pree-ah)
[*lamina* **thin plate,** *propria* **proper**]

lanugo
(lah-NOO-go)
[*lanugo* **down**]

lunula
(LOO-nyoo-lah)
[*luna-* **moon,** *-ula* **small**]

Meissner corpuscle
(MYZ-ner KOR-pus-ul)
[*George Meissner* **German physiologist,** *corpus-* **body,** *-cle* **little**]

melanin
(MEL-ah-nin)
[*melan-* **black,** *-in* **substance**]

melanocyte
(MEL-ah-noh-syte)
[*melan-* **black,** *-cyte* **cell**]

membrane
(MEM-brayn)
[*membran-* **thin skin**]

mucocutaneous junction
(myoo-koh-kyoo-TAY-nee-us JUNK-shun)
[*muco-* **slime or mucus,** *-cut-* **skin,** *-aneous* **relating to,** *junc-* **join,** *-tion* **condition**]

mucous membrane
(MYOO-kus MEM-brayn)
[*muc-* **slime,** *-ous* **characterized by,** *membran-* **thin skin**]

mucus
(MYOO-kus)
[*mucus* **slime**]

nail body
(nayl BOD-ee)

Pacini corpuscle
(pah-CHEE-nee KOR-pus-ul)
[*Filippo Pacini* **Italian anatomist,** *corpus-* **body,** *-cle* **little**]

parietal
(pah-RYE-ih-tal)
[*pariet-* **wall,** *-al* **relating to**]

parietal peritoneum
(pah-RYE-ih-tal payr-ih-TOH-nee-um)
[*parie-* **wall,** *-al* **relating to,** *peri-* **around,** *-tone-* **stretched,** *-um* **thing**]

pericardium
(pair-ih-KAR-dee-um)
pl., pericardia
(pair-ih-KAR-dee-ah)
[*peri-* **around,** *-cardi-* **heart,** *-um* **thing**]

peritoneum
(payr-ih-toh-NEE-um)
[*peri-* **around,** *-tone-* **stretched,** *-um* **thing**]

perspiration (sweat)
(per-spih-RAY-shun [swet])
[*per-* **through,** *-spire* **breathe,** *-ation* **process**]

pigment
(PIG-ment)
[*pigment* **paint**]

pleura
(PLOO-rah)
pl., pleurae
(PLOO-ree)
[*pleura* **side of body (rib)**]

pore
(por)

root

sebaceous gland
(seh-BAY-shus gland)
[*seb-* **tallow (hard animal fat),** *-ous* **relating to,** *gland* **acorn**]

sebum
(SEE-bum)
[*sebum* **grease**]

serous membrane
(SEE-rus MEM-brayn)
[*sero-* **watery body fluid,** *-ous* **characterized by,** *membran-* **thin skin**]

stratum corneum
(STRAH-tum KOR-nee-um)
[*stratum* **layer,** *corneum* **horn**]

stratum germinativum
(STRAH-tum jer-min-ah-TIV-um)
[*stratum* **layer,** *germinativum* **something that sprouts**]

subcutaneous tissue
(sub-kyoo-TAY-nee-us TISH-yoo)
[*sub-* **beneath,** *-cut-* **skin,** *-ous* **relating to,** *tissu-* **fabric**]

sudoriferous gland
(soo-doh-RIF-er-us gland)
[*sudo-* **sweat,** *-fer-* **bear or carry,** *-ous* **relating to,** *gland* **acorn**]

superficial fascia
(soo-per-FISH-al FAH-shah)
[*super-* **over or above,** *-fici-* **face,** *-al* **relating to,** *fascia* **band**]

sweat gland
(swet gland)
[*gland* **acorn**]

synovial fluid
(sih-NOH-vee-al FLOO-id)
[*syn-* **together,** *-ovi-* **egg (white),** *-al* **relating to**]

synovial membrane
(sih-NOH-vee-al MEM-brayn)
[*syn-* **together,** *-ovi-* **egg (white),** *-al* **relating to,** *membran-* **thin skin**]

tactile corpuscle
(TAK-tyle KOR-pus-ul)
[*tact-* **touch,** *-ile* **relating to,** *corpus-* **body,** *-cle* **little**]

visceral
(VIS-er-al)
[*viscer-* **internal organ,** *-al* **relating to**]

visceral peritoneum
(VIS-er-al payr-ih-TOHN-ee um)
[*viscer-* **internal organ,** *-al* **relating to,** *peri-* **around,** *-tone-* **stretched,** *-um* **thing**]

LANGUAGE OF **MEDICINE**

acne vulgaris
(AK-nee vul-GAR-is)
[*acne* **point,** *vulgaris* **common**]

albinism
(AL-bih-niz-em)
[*alb-* **white,** *-in-* **characterized by,** *-ism* **state**]

allergy
(AL-er-jee)
[*all-* **other,** *-erg-* **work,** *-y* **state**]

alopecia
(al-oh-PEE-sha)
[*alopec-* **fox,** *-ia* **condition**]

basal cell carcinoma
(BAY-sal sel kar-sih-NOH-mah)
[*bas-* **base,** *-al* **relating to,** *cell* **storeroom,** *carcin-* **cancer,** *-oma* **tumor**]

blackhead
(BLAK-hed)

blister
(BLIS-ter)

burn
(bern)

carbuncle
(KAR-bung-kul)
[*carbun-* **coal,** *-cle* **little**]

comedo
(KOM-ee-doh)
[*comedo* **glutton (secretions resemble body-devouring worms)**]

cosmetic surgery
(koz-MET-ik SUR-jeh-ree)
[*cosmet-* **adorned,** *-ic* **relating to,** *surger-* **hand,** *-y* **activity**]

cosmetician
(koz-meh-TISH-un)
[*cosmet-* **adorned,** *-ic* **relating to**]

crust
(krust)

cyanosis
(sye-ah-NOH-sis)
[*cyan-* **blue,** *-osis* **condition**]

decubitus ulcer
(deh-KYOO-bih-tus UL-ser)
[*decubitus* **lying-down position,** *ulcer* **sore**]

dehydration
(dee-hye-DRAY-shun)
[*de-* **remove,** *-hydro* **water,** *-ation* **process**]

dermatitis
(der-mah-TYE-tis)
[*derma-* **skin,** *-itis* **inflammation**]

dermatology
(der-mah-TOL-uh-jee)
[*derma-* **skin,** *-log-* **words (study of),** *-y* **activity**]

dermatosis
(der-mah-TOH-sis)
[*derma-* **skin,** *-osis* **condition**]

eczema
(EK-zeh-mah)
[*eczema* **to boil over**]

erythema
(er-ih-THEE-mah)
[*erythem-* **become red,** *-a* **condition**]

excoriation
(eks-koh-ree-AY-shun)
[*ex-* **off,** *-cori-* **skin,** *-ation* **process**]

first-degree burn
(furst dih-GREE bern)

fissure
(FISH-ur)
[*fiss-* **split,** *-ure* **thing**]

fourth-degree burn
(fohrth dih-GREE bern)

full-thickness burn
(ful THIK-nis bern)

furuncle (boil)
(FUR-un-kul [boyl])
[*fur-* **thief,** *-uncle* **little**]

immunology
(im-yoo-NOL-uh-jee)
[*immuno-* **free (immunity),** *-logy* **study of**]

impetigo
(im-peh-TYE-go)
[*impet-* **attack**]

kaposi sarcoma (KS)
(KAH-poh-see sar-KOH-mah [kay es])
[*Moritz K. Kaposi* **Hungarian dermatologist,** *sarco-* **flesh,** *-oma* **tumor**]

lesion
(LEE-zhun)
[*les-* **hurt,** *-ion* **condition**]

macule
(MAK-yool)
[*macul(a)-* **spot**]

melanoma
(mel-ah-NOH-mah)
[*melan-* **black,** *-oma* **tumor**]

nevus
(NEE-vus)
pl., nevi
(NEE-vye)
[*nevus* **birthmark**]

onycholysis
(ahn-ik-oh-LYE-sis)
[*onycho-* **nail,** *-lysis* **loosen**]

papule
(PAP-yool)
[*papul(a)* **pimple**]

partial-thickness burn
(PAR-shal THIK-nis bern)

peritonitis
(payr-ih-toh-NYE-tis)
[*peri-* **around,** *-ton-* **stretch (peritoneum),** *-itis* **inflammation**]

plaque
(plak)
[*plaque* **patch**]

pleurisy
(PLOOR-ih-see)
[*pleur-* **side of body (rib),** *-isy* **condition**]

port-wine stain
(port wyne stayn)
[*port wine* **type of dark red wine**]

psoriasis
(soh-RYE-ah-sis)
[*psor-* **itching,** *-iasis* **condition**]

pustule
(PUS-tyool)
[*pus-* **pus (rotten),** *-ule* **little**]

reconstructive surgery
(ree-kon-STRUK-tiv SUR-jeh-ree)
[*re-* **again,** *-con-* **with,** *-struct-* **build,** *-ive* **relating to,** *surger-* **hand,** *-y* **activity**]

rule of nines
(rool ov nahynz)

scabies
(SKAY-beez)
[*scabies* **mange or itch**]

scleroderma
(skleer-oh-DER-mah)
[*sclero-* **hard,** *-derma* **skin**]

second-degree burn
(SEK-und dih-GREE bern)

skin graft
(skin grahft)
[*graft* **shoot inserted into another plant**]

squamous cell carcinoma
(SKWAY-mus sel kar-sih-NOH-mah)
[*squam-* **scale,** *-ous* **characterized by,** *cell* **storeroom,** *carcin-* **cancer,** *-oma* **tumor**]

stork bite
(stork byte)
[*stork* **large bird related to birth mythology**]

strawberry hemangioma
(STRAW-ber-ee hem-an-jee-OH-mah)
[*hem-* **blood,** *-angi-* **vessel,** *-oma* **tumor**]

striae
(STRYE-ee)
sing., stria
(STRYE-ah)
[*stria-* **furrow or flute of a column**]

Continued on p. 168

LANGUAGE OF **MEDICINE** *(continued from p. 167)*

subcutaneous injection
(sub-kyoo-TAY-nee-us in-JEK-shun)
[*sub-* **under,** *cut-* **skin,** *-aneous* **relating to,** *in-* **in,** *-ject-* **throw,** *-tion* **process**]

third-degree burn
(third dih-GREE bern)

tinea
(TIN-ee-ah)
[*tinea* **gnawing worm**]

urticaria (hives)
(er-tih-KAYR-ee-ah [hyvez])
[*urtica-* **nettle,** *-ia* **condition**]

vitiligo
(vit-ih-LYE-go)
[*vitiligo* **blemish**]

wart
(wort)
[*wart* **swelling**]

wheal
(weel)
[*wheal* **welt or whip mark**]

xeroderma pigmentosum
(zeer-oh-DER-mah pig-men-TOH-sum)
[*xero-* **dry,** *-derma* **skin,** *pigment-* **paint,** *-osum* **characterized by**]

7

OUTLINE SUMMARY

*To download a digital version of the chapter summary for use with your device, access the **Audio Chapter Summaries** online at evolve.elsevier.com.*

Scan this summary after reading the chapter to help you reinforce the key concepts. Later, use the summary as a quick review before your class or before a test.

Body Membranes

A. Classification of body membranes (**Figure 7-1**)
  1. Epithelial membranes—composed of epithelial tissue and an underlying layer of connective tissue
  2. Connective tissue membranes—composed exclusively of various types of connective tissue

B. Epithelial membranes
  1. Cutaneous membrane—the skin
  2. Serous membranes—simple squamous epithelium on a connective tissue basement membrane
    a. Layers (**Figure 7-2**)
      (1) Parietal—line walls of body cavities
      (2) Visceral—cover organs found in body cavities
    b. Examples
      (1) Pleura—parietal and visceral layers line walls of thoracic cavity and cover the lungs
      (2) Peritoneum—parietal and visceral layers line walls of abdominal cavity and cover the organs in that cavity
      (3) Pericardium—parietal and visceral layers line a fibrous sac around the heart and a visceral layer covers the heart wall
    c. Diseases
      (1) Pleurisy—inflammation of the serous membranes that line the chest cavity and cover the lungs
      (2) Peritonitis—inflammation of the serous membranes in the abdominal cavity that line the walls and cover the abdominal organs
  3. Mucous membranes
    a. Line body surfaces that open directly to the exterior
    b. Produce mucus, a thick secretion that keeps the membranes soft and moist

C. Connective tissue membranes
  1. Do not contain epithelial components
  2. Produce a lubricant called *synovial fluid*
  3. Examples are the synovial membranes in the spaces between joints and in the lining of the bursal sacs

Skin Structure

A. Overview (**Figure 7-3**)—two primary layers
  1. Epidermis—superficial layer
  2. Dermis—deep layer

B. Epidermis
  1. Epidermal structure
    a. Outermost and thinnest primary layer of skin
    b. Composed of several layers of stratified squamous epithelium
    c. Stratum germinativum—innermost (deepest) layers of cells; basal layer continually reproduces, pushing older cells toward the surface
    d. As cells approach the surface, they become filled with a tough, waterproof protein called *keratin* and eventually flake off (**Figure 7-4**)
    e. Stratum corneum—outermost layer of keratin-filled cells

2. Skin pigment
   a. Melanin—brown skin pigment
      (1) Basal layer of stratum germinativum has pigment-producing melanocyte cells
      (2) The brown pigment melanin produced by melanocytes is distributed to other epithelial cells, giving skin a darker color
      (3) Amount and type of melanin, determined by genes, helps determine basic skin color
   b. Skin color changes
      (1) Sunlight promotes additional pigmentation
      (2) Pink flush indicates increased blood volume or increased blood oxygen
      (3) Cyanosis—bluish color of skin indicates decreased blood oxygen level
      (4) Vitiligo—patchy light skin areas resulting from acquired loss of epidermal melanocytes (**Figure 7-5**)
      (5) Increased skin pigmentation caused by hormonal changes in pregnant women
      (6) Freckles—small, flat macules; common normal skin pigment variation

C. Dermal-epidermal junction
1. Gluelike layer between the dermis and epidermis
2. Small bumps called *dermal papillae* help stabilize the junction
3. Blisters—caused by breakdown of union between cells or primary layers of skin

D. Dermis
1. Deeper and thicker of the two primary skin layers; composed largely of connective tissue
2. Papillary layer
   a. Dermal papillae—parallel rows of tiny bumps that characterize the upper area of dermis
      (1) Thick skin has parallel friction ridges in dermis and no hairs (**Figure 7-6**)
      (2) Thin skin has irregular, shallow grooves and hair
   b. Dermal ridges
      (1) Ridges and grooves in dermis form pattern unique to each individual
      (2) Improves grip for tool use and walking; helps in sensing textures on surfaces
      (3) Basis of fingerprint identification
3. Reticular layer—deeper area of dermis is filled with network of tough collagenous and stretchable elastic fibers
   a. Number of elastic fibers decreases with age and contributes to wrinkle formation
   b. Striae—"stretch marks"; elongated marks caused by overstretching of skin
   c. Dermis also contains nerve endings, muscle fibers, hair follicles, sweat and sebaceous glands, and many blood vessels
   d. Birthmarks—malformation of dermal blood vessels
      (1) Strawberry hemangioma (**Figure 7-7**)
      (2) Port-wine stain
      (3) Stork bite

E. Subcutaneous tissue
1. Also called *hypodermis* or *superficial fascia*
2. Mostly loose fibrous connective tissue and adipose tissue
3. Connecting layer
   a. Deep to dermis, between the skin and the underlying structures such as bone and muscle
   b. Not part of the skin
   c. Allows sliding movement of skin as body parts move

F. Hair, nails, and skin receptors
1. Hair (**Figure 7-8**)
   a. Location of hair
      (1) Hair grows from pocketlike hair follicles in the epidermis
      (2) Soft hair of fetus and newborn called *lanugo*
      (3) Hair is distributed widely (except lips, palms, soles) but varies in density and coarseness
   b. Hair growth
      (1) Hair growth occurs within the tube of the hair follicle
      (2) Hair-forming cells reproduce on bump at the bottom of each follicle called the *hair papilla*
      (3) Hair cells become keratinized like the epidermis, but in a cylinder form and having a tougher type of keratin
      (4) Hair root lies hidden in follicle; visible part of hair called *shaft*
   c. Hair loss
      (1) Alopecia (**Figure 7-9**)—any type of hair loss
      (2) Can result from normal aging, various disorders or treatments, pregnancy, or have unknown causes
   d. Arrector pili—smooth muscle of the skin that produces "goose pimples" and causes hair to stand up straight
2. Nails (**Figure 7-10**)
   a. Produced by epidermal cells over terminal ends of fingers and toes
   b. Visible part called *nail body*
   c. Root lies in a groove and is hidden by cuticle
   d. Crescent-shaped area nearest root called *lunula*
   e. Nail bed may change color with change in blood flow
   f. Normal variations in nail structure (**Figure 7-11**)
      (1) Longitudinal ridges in light-skinned individuals
      (2) Pigmented bands in dark-skinned individuals
   g. Abnormal variations in nail structure (**Figure 7-12**)
      (1) Onycholysis—separation of nail from nail bed
      (2) Pitting—common in psoriasis

3. Receptors (**Figure 7-3**)
   a. Sensory nerve endings—make it possible for skin to act as a sense organ
   b. Tactile (Meissner) corpuscle—capable of detecting light touch
   c. Lamellar (Pacini) corpuscle—capable of detecting pressure

G. Skin glands—two main types: sweat and sebaceous glands
1. Sweat glands; also called *sudoriferous glands*
   a. Eccrine sweat glands
      (1) Most numerous, important, and wide-spread of the sweat glands
      (2) Produce perspiration or sweat, which flows out through pores on skin surface
      (3) Function throughout life and assist in body heat regulation
   b. Apocrine sweat glands
      (1) Found primarily in axilla and around genitalia
      (2) Secrete a thicker, milky secretion quite different from eccrine perspiration
      (3) Breakdown of secretion by skin bacteria produces odor
2. Sebaceous glands
   a. Sebum
      (1) Oily secretion for hair and skin, lubricates and prevents drying, cracking
      (2) Level of secretion increases during adolescence
      (3) Amount of secretion regulated by sex hormones
      (4) Sebum in sebaceous gland ducts may darken to form a blackhead (comedo)
      (5) Acne vulgaris (**Figure 7-13**)—inflammation of sebaceous gland ducts

Functions of the Skin

A. Protection—first line of defense
1. Against infection by microbes
2. Against ultraviolet (UV) rays from sun
3. Against harmful chemicals
4. Against cuts and tears
5. Bruising can cause discoloration of skin as blood release from damaged vessels breaks down (**Figure 7-14**)
6. Skin grafts may be needed to replace skin destroyed by disease or trauma (**Figure 7-15**)

B. Temperature regulation
1. Skin can release almost 3000 calories of body heat per day
2. Mechanisms of temperature regulation
   a. Regulation of sweat secretion
   b. Regulation of flow of blood close to the body surface

C. Sensation
1. Receptors serve as receivers for the body, keeping it informed of changes in its environment
   a. Skin can detect sensations of light touch, pressure, pain, heat, and color

D. Excretion—sweat excretes waste products such as uric acid, ammonia, urea

E. Synthesis of vitamin D, which requires UV light available in the skin

Disorders of the Skin

A. Skin lesions (**Table 7-1**)—any measurable variation from the normal structure
1. Elevated lesions—cast a shadow outside their edges
   a. Papule—small, firm raised lesion
   b. Plaque—large raised lesion
   c. Vesicle—blister
   d. Pustule—pus-filled lesion
   e. Crust—scab
   f. Wheal (hive)—raised, firm lesion with a light center
2. Flat lesions—do not cast a shadow
   a. Macule—flat, discolored region
3. Depressed lesions cast a shadow within their edges
   a. Excoriation—missing epidermis, as in a scratch wound
   b. Ulcer—craterlike lesion
   c. Fissure—deep crack or break
4. Some lesions are produced by scrapes and cuts—the skin can repair itself (**Figure 7-16**)

B. Burns
1. Treatment and recovery or survival depends on total area involved and severity or depth of the burn
2. Depth classification of burns (**Figure 7-17**)
   a. First-degree (partial-thickness) burns—only surface layers of epidermis involved
   b. Second-degree (partial-thickness) burns—involve the deep epidermal layers and always cause injury to the upper layers of the dermis
   c. Third-degree (full-thickness) burns (**Figure 7-18**)—characterized by complete destruction of the epidermis and dermis
      (1) Lesion is insensitive to pain because of destruction of nerve endings immediately after injury—intense pain is experienced after the initial injury
      (2) Risk of infection is increased
   d. Fourth-degree burns—full-thickness burns that extend to muscle or bone
3. Estimating body surface area using the "rule of nines" (**Figure 7-19**) in adults
   a. Body divided into 11 areas of 9% each
   b. Additional 1% of body surface area around genitals

C. Skin infections (**Figure 7-20**)
1. Impetigo—highly contagious staphylococcal or streptococcal infection
2. Tinea—fungal infection (mycosis) of the skin; several forms occur
3. Warts—benign neoplasm caused by papillomavirus
4. Boils—furuncles; staphylococcal infection in hair follicles
5. Scabies—parasitic infection

D. Vascular and inflammatory skin disorders (**Figure 7-21**)
   1. Decubitus ulcers (bedsores) develop when pressure slows down blood flow to local areas of the skin
   2. Urticaria or hives—red lesions caused by fluid loss from blood vessels
   3. Scleroderma—disorder of vessels and connective tissue characterized by hardening of the skin; two types: localized and systemic
   4. Psoriasis—chronic inflammatory condition accompanied by scaly plaques
   5. Eczema—common inflammatory condition characterized by papules, vesicles, and crusts; not a disease itself but a symptom of an underlying condition

E. Skin cancer (**Figure 7-22, Table 7-2**)
   1. The most important causative factor in common skin cancers is exposure to sunlight
   2. Three common types
      a. Squamous cell carcinoma—characterized by hard, raised tumors
      b. Basal cell carcinoma—characterized by papules with a central crater; rarely spreads; the most common type
      c. Melanoma—malignancy in a nevus (mole); the most serious type of skin cancer
   3. Kaposi sarcoma—caused by a virus and characterized by purple lesions, is associated with certain ethnic groups, as well as AIDS and other immune deficiencies

❑ ACTIVE LEARNING

STUDY TIPS

Hint *Consider using these tips to achieve success in meeting your learning goals.*

1. The body membranes are either epithelial or connective. The epithelial membranes cover or protect. The difference between mucous and serous membranes is their location in the body. If the membrane is exposed to the environment, it is a mucous membrane. Connective tissue membranes cover joints.
2. The skin is divided into two parts: epidermis and dermis. *Epi-* means "on," so the epidermis is *on* the dermis. The job of the epidermis is protection. The dermis contains most of the skin appendages: nails, sense receptors, hair, and glands.
3. The functions of the skin are related to its location: protection, sensation, heat regulation, excretion, and synthesis of vitamin. Develop a concept map that details the specific functions of the skin.
4. Burns are classified by how much damage has been done and how deep the damage goes.
5. Take a photocopy or use your mobile phone to take a photo of the illustrations of the membranes, the microscopic view of the skin, the hair, and the nails. Blacken out the labels on the photocopy and quiz each other in your study group on the location and function of various structures.
6. Make a T-chart of the different types of skin disorders. Your chart will be most helpful if you organize it according to *mechanisms*. Group the diseases by pathogenic organisms and by internal or external conditions.
7. Go over the questions at the end of the chapter and discuss possible test questions with your study group. Use online resources that provide tutorials and diagrams. One example is *studyblue.com*. This is a free online site that allows you to create flash cards, and download apps for all academic disciplines. Other flash card sites and tips are found at *my-ap.us/LzuowE*
8. Review the Language of Science and Language of Medicine terms and their word origins to help you better understand the meaning of the terms in this chapter.
9. Review the outline at the end of this chapter. This outline provides an overview of the material and would help you understand the general concepts of the chapter.

Review Questions

Write out the answers to these questions after reading the chapter and reviewing the Chapter Summary. If you simply think through the answer without writing it down, you will not retain much of your new learning.

1. Define *membrane*.
2. List the two major categories of body membranes.
3. Explain the structure of a serous membrane, including the difference between the visceral and parietal membranes.
4. What is responsible for the pigment of the skin?
5. Explain the structure of a mucous membrane, including an explanation of the mucocutaneous junction.
6. Explain the structure of a synovial membrane. What is the function of synovial fluid?
7. Name and briefly describe the layers of the epidermis.
8. Explain the structure of the dermis.
9. What is the structural advantage of the loose, spongy nature of subcutaneous tissue?
10. Differentiate between the hair papilla, the hair root, and the hair shaft.
11. Explain what happens when the arrector pili contracts.

12. Name two receptors of the skin. To what stimuli does each respond?
13. Give the location of eccrine glands, their function, and what type of fluid they produce.
14. Give the location of apocrine glands, their function, and what type of fluid they produce.
15. Give the location of sebaceous glands, their function, and what type of fluid they produce.
16. What is sebum? Why is sebum sometimes described as "nature's skin cream"?
17. Explain the difference between second- and third-degree burns. Which is considered a full-thickness burn?
18. List the three most common forms of skin cancer and explain the factors involved in their development.
19. List the five types of skin infections and list the cause as viral, bacterial, fungal, or arthropod.
20. What is the cause of a decubitus ulcer, and what are methods of prevention?

Critical Thinking

Hint *After finishing the Review Questions, write out the answers to these more in-depth questions to help you apply your new knowledge. Go back to sections of the chapter that relate to concepts that you find difficult.*

21. An individual running a marathon expends a great deal of energy. Much of this energy generates heat. What is the role of the skin in balancing body temperature during this strenuous exercise?
22. Concern about skin cancer is reducing the amount of time people spend in the sun. If this caution is carried to the extreme, how would you explain the impact on skin function?
23. If a person burned all of his back, the posterior of his right arm, and the posterior of his right thigh, approximately what percent of his body surface area would be involved? How did you determine this?
24. A coroner was discussing a recent autopsy and "cause of death" to a group of students. He related to the group that the organ of trauma was lined with tough, abrasive-resistant stratified squamous epithelium that had been destroyed in areas throughout the lining of the organ. What most likely was the organ? Can you suggest a possible explanation for the areas of destruction?
25. Why are subcutaneous injections given by a hypodermic needle?

Chapter Test

Hint *After studying the chapter, test your mastery by responding to these items. Try to answer them without looking up the answers.*

1. ________, ________, and ________ are the three types of epithelial membranes.
2. Epithelial membranes are usually composed of two distinct layers: the epithelial layer and a supportive layer called the ________.
3. The membrane lining the interior of the chest wall is called the ________.
4. The membrane covering the organs of the abdomen is called the ________.
5. The connective tissue membrane that lines the space between bone and joint capsule is called ________.
6. The two main layers of the epidermis of the skin are the ________ and the ________.
7. As new skin cells approach the surface of the skin, their cytoplasm is replaced by a unique waterproof protein called ________.
8. The upper region of the dermis forms projections called ________ that form unique fingerprints.
9. The ________ are the sweat glands found in armpits; they produce a thicker secretion.
10. The ________ are the sweat glands found all over the body; they produce a transparent, watery liquid.
11. Sebaceous glands secrete an oil called ________.
12. ________, ________, ________, ________, and ________ are the five primary functions of the skin.
13. The "rule of nines" is used to estimate body surface involved in ________.
14. ________ are pressure sores caused by reduced blood flow to local areas of the skin.
15. The most common type of skin cancer is ________ carcinoma.
16. ________ often results from a fivefold increase in sebum secretions that occur between the ages of 10 and 19.

17. The receptors in the skin that detect pressure deep in the dermis are the:
    a. Meissner corpuscles
    b. lamellar corpuscles
    c. free nerve endings
    d. Krause end bulbs
18. The receptors in the skin that respond to light touch are:
    a. Meissner corpuscles
    b. lamellar corpuscles
    c. free nerve endings
    d. Krause end bulbs
19. The fine, soft hair of a newborn is referred to as:
    a. alopecia
    b. lanugo
    c. vitiligo
    d. striae
20. Which of the following is *not* a birthmark?
    a. Strawberry hemangioma
    b. Port wine stain
    c. Tinea
    d. Stork bite

Match each structure in column A with its description of the part of the hair in column B.

Column A	Column B
21. ________ hair follicle	a. the part of the hair hidden in the follicle
22. ________ hair papilla	b. the growth of the epidermal cells into the dermis forming a small tube
23. ________ hair root	c. the part of the hair that is visible
24. ________ hair shaft	d. a bump at the base of the follicle where hair growth begins

Match each skin condition in column A with its description in column B.

Column A	Column B
25. ________ furuncle	a. an autoimmune skin condition
26. ________ urticaria	b. skin cancer that can develop from a mole; the most serious form of skin cancer
27. ________ excoriation	c. another name for hives
28. ________ melanoma	d. skin lesion caused by a shallow scratch
29. ________ scleroderma	e. another name for a skin boil
30. ________ Kaposi sarcoma	f. a virus-caused skin cancer that sometimes develops in immune-deficient individuals

Case Studies

To solve a case study, you may have to refer to the glossary or index, other chapters in this textbook, and other resources.

1. Dana is an intern who has just been assigned to the burn unit at Mercy Hospital. One patient in the unit has burns covering the lower half of each arm (front and back). What is Dana's estimate of the total percent of skin surface area affected by the burn?
2. Uncle Ed, a light-skinned older man, has just learned from his physician that the spot on his forehead is skin cancer. Of course, dark-skinned Aunt Gina is very upset. Before Uncle Ed explains to the family what type of skin cancer he has, you examine the lesion and notice that it is a papule with an ulcer in the center. What type of skin cancer do you think Uncle Ed has? What do you know about this type of cancer that may help comfort Aunt Gina?
3. During your shift at the clinic, a young man arrives with a red, scaly rash formed into rings. What is this patient's diagnosis likely to be? What causes this condition? How can he avoid this rash in the future?
4. Christy is at high risk for melanoma. She wants to be proactive. What measures would you suggest that she include in her daily routine?

Answers to Active Learning Questions can be found online at evolve.elsevier.com.

Skeletal System

OUTLINE

Scan this outline before you begin to read the chapter, as a preview of how the concepts are organized.

OBJECTIVES

Before reading the chapter, review these goals for your learning.

After you have completed this chapter, you should be able to:

1. List and discuss the generalized functions of the skeletal system.
2. Identify the types of bones, the major anatomical structures found in a typical long bone, and the structure of flat bones.
3. Discuss the microscopic structure of bone and cartilage, including the identification of specific cell types and structural features.
4. Explain how bones are formed, how they grow, and how they are remodeled.
5. Identify the two major subdivisions of the skeleton and list the bones found in each area.
6. Do the following related to skeletal variations:
   - Compare and contrast the differences between a man's and a woman's skeleton.
   - Discuss age differences and environmental factors.
7. List and compare the major types of joints in the body and give an example of each.
8. Name and describe major disorders of bones and joints.

CHAPTER 8

The primary organs of the skeletal system, that is, bones, lie buried beneath the muscles and other soft tissues, providing a rigid framework and support structure for the whole body. In this respect the skeletal system functions like steel girders in a building; however, unlike steel girders, bones can be moved. Bones are also living organs. They can change and help the body respond to a changing environment. This ability of bones to change allows our bodies to grow and change.

Our study of the skeletal system begins with an overview of its function. The bones are classified by their structure and described by identifying characteristics of a typical bone. After discussing the microscopic structure of skeletal tissues, we briefly outline bone growth and formation. Having this information makes the study of specific bones and the way they are assembled in the skeleton more meaningful. The chapter ends with a discussion of skeletal variations and disorders and an overview of joints between bones.

An understanding of how bones articulate with one another in joints and how they relate to other body structures provides a basis for understanding the functions of many other organ systems. Coordinated movement, for example, is possible only because of the way bones are joined to one another and because of the way muscles are attached to those bones. In addition, knowing where specific bones are in the body will assist you in locating other body structures that are discussed in later chapters.

Functions of the Skeletal System

Support

The skeleton provides the internal framework of the body much like tent poles help maintain the structure of a tent. Skeletal muscles are attached to the bones, and internal organs are found in the cavities surrounded by the bones and skeletal muscles. The skeletal system can provide this support only when the composition of the bone is strong enough to hold the body weight and yet flexible enough to withstand twisting forces.

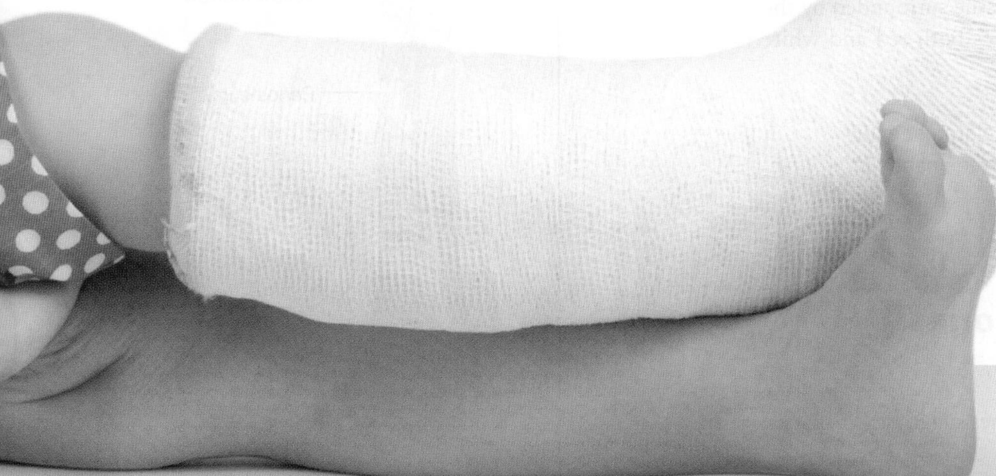

LANGUAGE OF **SCIENCE**

Hint Before reading the chapter, say each of these terms out loud. This will help you to avoid stumbling over them as you read.

abduct
(ab-DUKT)
[*ab-* **away,** *-duct* **lead**]

abduction
(ab-DUK-shun)
[*ab-* **away,** *-duct-* **lead,** *-tion* **process**]

acetabulum
(as-eh-TAB-yoo-lum)
[*acetabulum* **vinegar cup**]

adduct
(ad-DUKT)
[*ad-* **toward,** *-duct* **lead**]

adduction
(ad-DUK-shun)
[*ad-* **toward,** *-duct-* **lead,** *-tion* **process**]

amphiarthrosis
(am-fee-ar-THROH-sis)
pl., amphiarthroses
(am-fee-ar-THROH-seez)
[*amphi-* **both sides,** *-arthr-* **joint,** *-osis* **condition**]

appendicular skeleton
(ah-pen-DIK-yoo-lar SKEL-eh-ton)
[*append-* **hang upon,** *-ic-* **relating to,** *-ul-* **little,** *-ar* **relating to,** *skeleton* **dried body**]

arch

articular cartilage
(ar-TIK-yoo-lar KAR-tih-lij)
[*artic-* **joint,** *-ul-* **little,** *-ar* **relating to,** *cartilage* **gristle**]

articulation
(ar-tik-yoo-LAY-shun)
[*artic-* **joint,** *-ul-* **little,** *-ation* **state**]

athletic trainer
(ath-LET-ik TRAY-ner)
[*athlet-* **prize contender,** *-ic* **relating to**]

Continued on p. 209

8

Protection

The skeletal system protects the soft tissues that are located inside of bony cavities. The skull protects the brain, and the ribs and breastbone protect vital organs in the chest (heart and lungs). Bone also protects the vital blood cell–forming tissue inside the bones themselves. The bones of the extremities can also be moved and used in the defense of the body from outside forces.

Movement

The firm attachments between bones and muscles make body movement possible. As muscles contract and shorten, they pull on bones and thereby move them.

Storage

Bones play an important part in maintaining homeostasis of blood calcium, a vital substance required for normal nerve and muscle function. They serve as a safety-deposit box for calcium. When the amount of calcium in blood increases above normal, calcium moves out of the blood and into the bones for storage. Conversely, when blood calcium decreases below normal, calcium moves in the opposite direction. It comes out of storage in bones and enters the blood.

The balance of calcium deposits and withdrawals to and from the skeleton is regulated by a balance of hormones. For example, **calcitonin (CT)** from the thyroid gland increases mineralization of bone and thus reduces blood calcium. **Parathyroid hormone (PTH)** from the *parathyroid glands* counterbalances the effects of calcitonin by decreasing calcium in the bone and thus increasing blood calcium.

The medullary cavities inside of long bones also store fat.

Hematopoiesis

The term **hematopoiesis** is used to describe the process of blood cell formation. It is a combination of two word parts: *hemato* meaning "blood" and *poiesis* meaning "making." Blood cell formation is a vital process carried on in red bone marrow. **Red bone marrow** is soft connective tissue surrounded by the hard walls of some bones that produces both red and white blood cells.

To see exactly where in the skeleton hematopoiesis takes place, check out the images in *Sites of Hematopoiesis* at Connect It! at *evolve.elsevier.com.*

Gross Structure of Bones

Types of Bones

There are four types of bones. Their names suggest their shapes:

- *Long bones*—for example, humerus or arm bone
- *Short bones*—for example, carpals or wrist bones
- *Flat bones*—for example, frontal or skull bone
- *Irregular bones*—for example, vertebrae or spinal bones

Some scientists recognize an additional category called *sesamoid* ("like a sesame seed"), or *round,* bones. These small bones may develop within a tendon. An example of a large sesamoid bone is the kneecap (patella), which develops within the patellar tendon.

Many important bones in the skeleton are classified as long bones, and all have several common characteristics. By studying a typical long bone, you can become familiar with the structural features of the entire group of bones.

Structure of Long Bones

Figure 8-1 will help you learn the names of the main parts of a long bone. Identify each of the following:

1. **Diaphysis** or *shaft*—hollow tube made of hard compact bone, hence a rigid and strong structure light enough in weight to permit easy movement

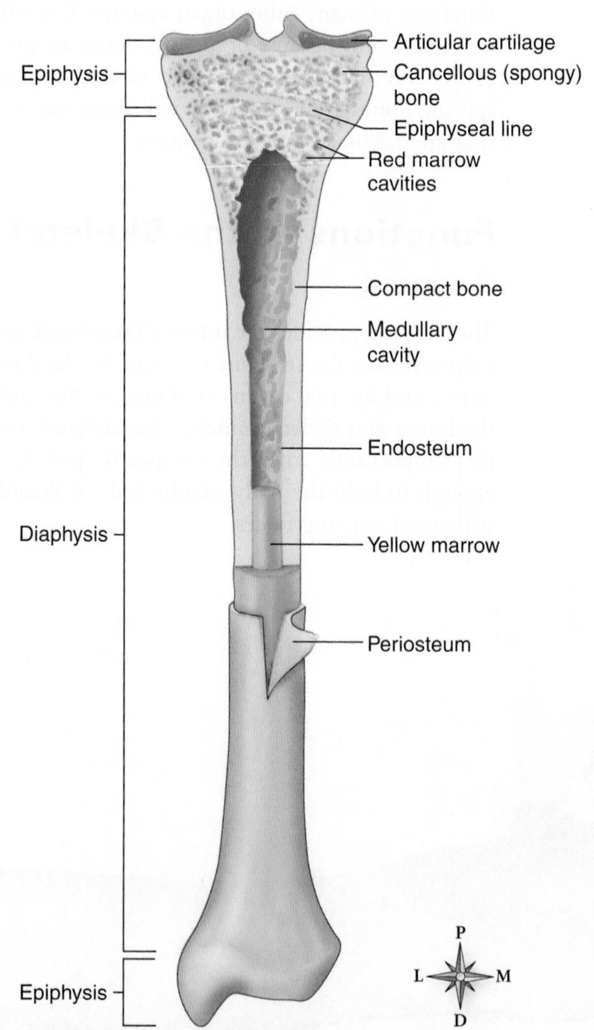

FIGURE 8-1 Long bone. Frontal section (partial) of a tibia.

2. **Medullary cavity**—hollow area inside the diaphysis of a bone; contains soft, **yellow bone marrow,** an inactive, fatty form of marrow found in the adult skeleton
3. **Epiphyses**—the ends of a long bone; red bone marrow fills in small spaces in the spongy bone inside the epiphyses; some yellow marrow may appear as a person ages
4. **Articular cartilage**—thin layer of hyaline cartilage covering each epiphysis; functions like a thin, smooth rubber cushion would if it were placed over the ends of bones where they form a joint
5. **Periosteum**—strong membrane of dense fibrous tissue covering a long bone everywhere except at joint surfaces, where it is covered by articular cartilage
6. **Endosteum**—thin membrane that lines the medullary cavity

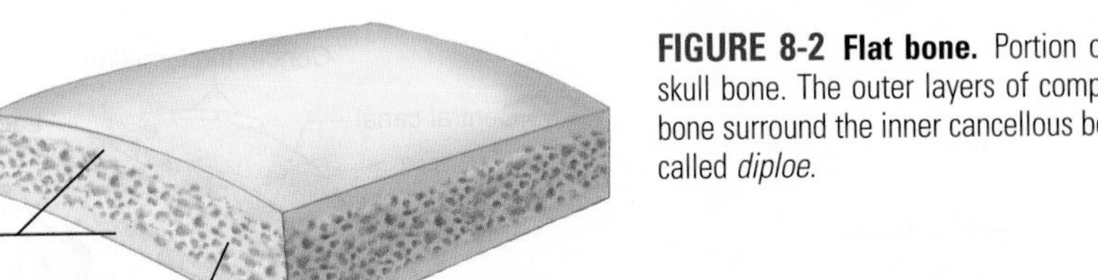

FIGURE 8-2 Flat bone. Portion of a skull bone. The outer layers of compact bone surround the inner cancellous bone called *diploe.*

QUICK CHECK

1. Name some of the organs of the skeletal system.
2. What are the five major functions of the skeletal system?
3. What are the four categories of bones in the skeleton?
4. Describe the main features of a long bone.

Structure of Flat Bones

Flat bones, such as the sternum (breastbone), the ribs, and many of the skull bones have a simpler structure than most long bones. As Figure 8-2 shows, flat bones have a layer of cancellous bone between outer layers of compact bone. The cancellous bone layer is called the **diploe.**

Microscopic Structure of Bones

The bones of the skeletal system contain two major types of connective tissue: *bone* and *cartilage.*

Bone Tissue Structure

Bone tissue has different microscopic structures, depending on its location and function. In Figure 8-3, *A,* the outer layer of bone is hard and dense. Bone of this type is called **compact bone.** Compact bone appears solid to the naked eye. The porous bone tissue on the inside of individual bones is called **cancellous bone** or **spongy bone.**

Cancellous Bone (Spongy Bone)

As the name implies, spongy bone contains many spaces—like a bath sponge. The cavities are filled with red or yellow marrow. The beams that form the lattice of spongy bone are called **trabeculae.** Figure 8-3, *B,* shows the microscopic appearance of cancellous bone.

Compact Bone

As you can see in Figure 8-3 and Figure 8-4, compact bone does not contain a network of open spaces. Instead, the extracellular matrix is organized into numerous structural units called **osteons** or **haversian systems.** Each circular and tubelike osteon is composed of calcified matrix arranged in multiple layers that resemble the rings of an onion. Each ring is called a **concentric lamella.**

The circular lamellae surround the **central canal,** or *haversian canal,* which contains blood vessels and nerves. The central canals are connected to each other by **transverse canals,** sometimes called **Volkmann canals.**

Bones are not lifeless structures. Within their hard, seemingly lifeless matrix are many living bone cells called **osteocytes.** Osteocytes are mature bone cells that were formerly active bone-making osteoblast cells, but which have now become dormant. These osteocytes lie between the hard layers of the lamellae in little spaces called **lacunae.**

In Figure 8-3, *B,* and Figure 8-4, note that tiny passageways, or canals, called **canaliculi** connect the lacunae with one another and with the central canal in each osteon. Nutrients pass along cell extensions of the osteocytes from the blood vessel in the central canal through the canaliculi and are distributed to all osteocytes of the osteon.

Note also in Figure 8-3, *B,* that numerous blood vessels from the outer **periosteum** enter the bone and eventually pass through transverse canals—and eventually to central canals.

Cartilage Tissue Structure

Cartilage both resembles and differs from bone. As with bone, it consists more of intercellular substance than of cells. Innumerable collagenous fibers reinforce the matrix of both tissues. However, in cartilage the fibers are embedded in a firm gel instead of in a calcified cement substance like they are in bone. As a result, cartilage has the flexibility of a firm plastic rather than the rigidity of bone.

Cartilage cells, called **chondrocytes,** as with the osteocytes of bone, are located in lacunae (Figure 8-5). In cartilage, lacunae are suspended in the cartilage matrix much like air bubbles in a block of Swiss cheese. Because there are no blood vessels in cartilage, nutrients must diffuse through the matrix to reach the cells. Because of this lack of blood vessels, cartilage rebuilds itself very slowly after an injury.

Bone Development

Making and Remodeling Bone

When the skeleton begins to form in a baby before its birth, it consists not of bones but of cartilage and fibrous structures shaped like bones. Gradually these cartilage "models" become

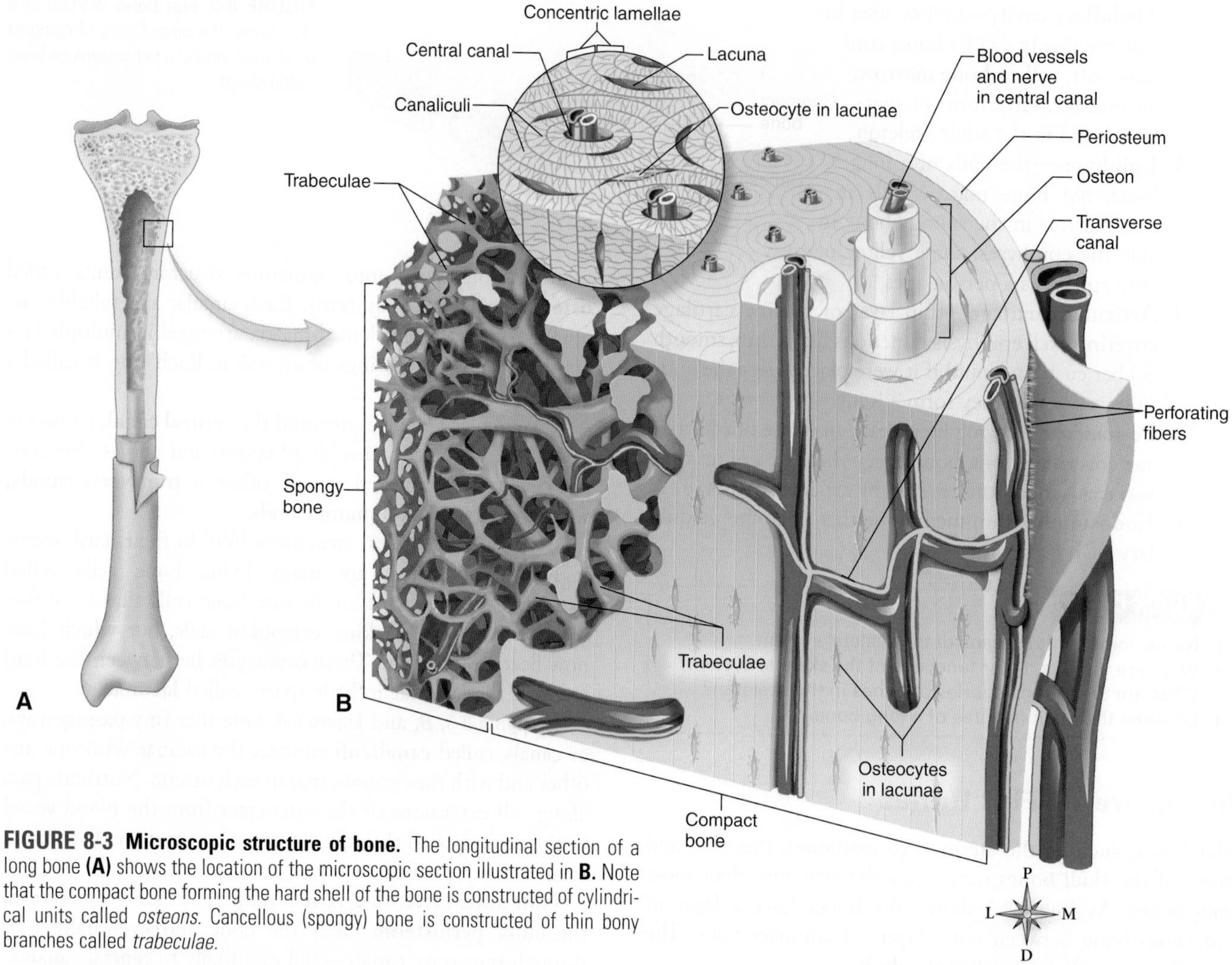

Concentric lamellae
Central canal
Lacuna
Blood vessels and nerve in central canal
Canaliculi
Osteocyte in lacunae
Periosteum
Osteon
Trabeculae
Transverse canal
Perforating fibers
Spongy bone
Trabeculae
Osteocytes in lacunae
Compact bone
A
B
P
L
M
D

FIGURE 8-3 Microscopic structure of bone. The longitudinal section of a long bone **(A)** shows the location of the microscopic section illustrated in **B.** Note that the compact bone forming the hard shell of the bone is constructed of cylindrical units called *osteons.* Cancellous (spongy) bone is constructed of thin bony branches called *trabeculae.*

Osteon (haversian system)
Canaliculi
Central (haversian) canal
Lacunae (containing osteocytes)

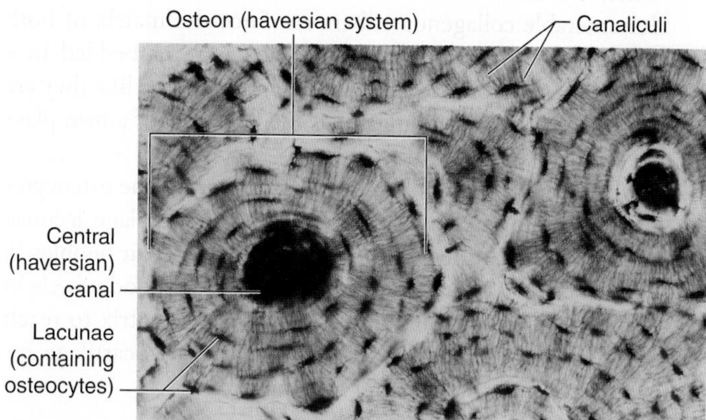

FIGURE 8-4 Compact bone. Photomicrograph shows circular cross section of a cylindrical osteon.

Chondrocytes (in lacunae)
Matrix

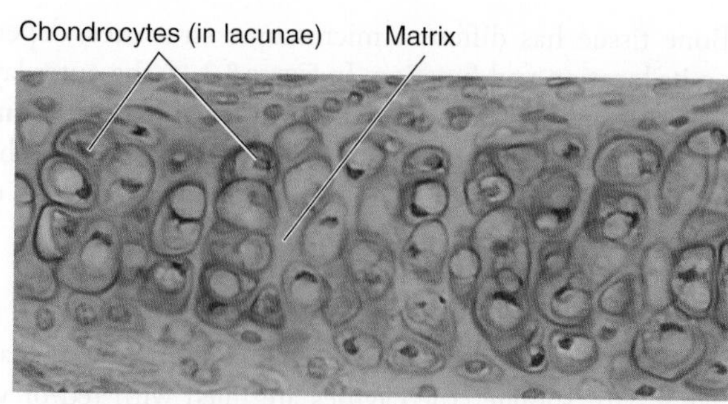

FIGURE 8-5 Cartilage tissue. Photomicrograph shows chondrocytes scattered around the tissue matrix in spaces called *lacunae.*

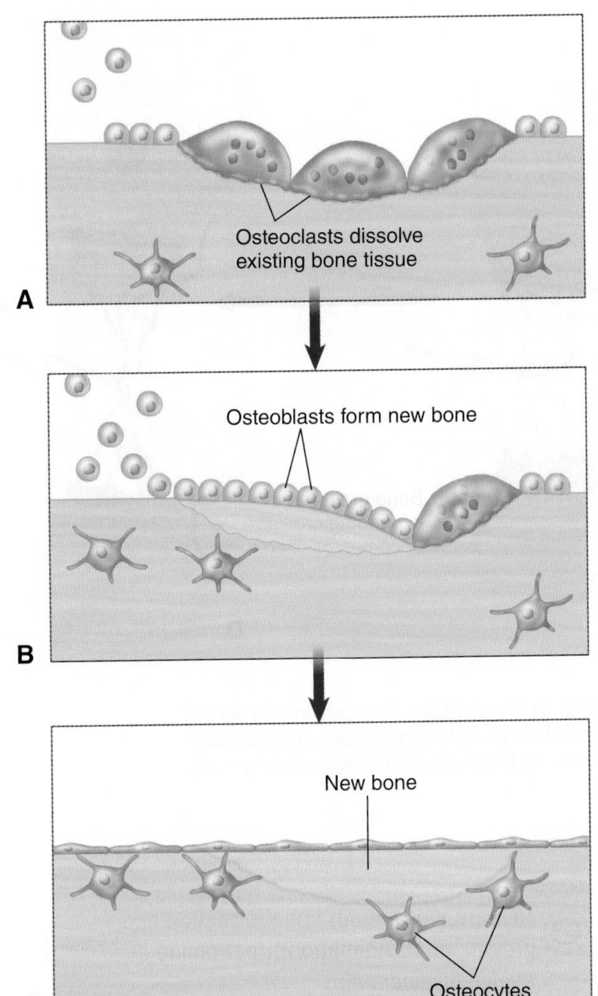

FIGURE 8-6 Bone remodeling. During remodeling of bone, bone-dissolving osteoclasts remove the hard calcium salts in bone matrix (**A**). Osteoblasts then form new bone matrix in the area (**B**) until they eventually become surrounded and "trapped" by hard bone and are then called *osteocytes* (**C**).

transformed into real bones when the cartilage is replaced with calcified bone matrix. This process of constantly "remodeling" a growing bone as it changes from a small cartilage model to the characteristic shape and proportion of the adult bone requires continuous activity by bone-forming cells called **osteoblasts** and bone-dissolving cells called **osteoclasts,** both seen in Figure 8-6.

The laying down of bone matrix is an ongoing process. Osteoblasts first lay down organic collagen fibers if needed. They also release a solution of inorganic calcium salts that crystallize on the fibers. The fibers reinforce the matrix to withstand twisting forces, and the mineral crystals calcify the bone to make it as "hard as bone."

When osteoblasts eventually become "trapped" between lamellae of hard bone matrix, they stop forming bone and are called *osteocytes*. Osteocytes resume their bone-making activity when osteoclasts (or an injury) remove the surrounding bone.

Osteoclasts release acids that dissolve the calcium crystals. This has two effects: the hard bone matrix is removed, and the calcium ions are released from bone tissue to diffuse into the bloodstream.

The combined breaking-building actions of the osteoblasts and osteoclasts remodel bones into their adult shapes (Figure 8-7). The process of "sculpting" by the bone-forming and bone-reabsorbing cells allows bones to respond to stress or injury by changing size, shape, and density.

When a bone is mechanically stressed from the pull of a muscle, the osteoblasts are stimulated to strengthen the bone at that location to resist the stress of pulling muscle. For this reason, athletes or dancers may have denser, stronger bones than less active people.

To learn more about bone remodeling, go to AnimationDirect online at *evolve.elsevier.com.*

Endochondral Ossification

Many bones of the body are formed from cartilage models, as illustrated in Figure 8-7 and Figure 8-8. This process is called **endochondral ossification,** meaning "formed in cartilage."

As you can see in Figure 8-7, a long bone grows and ultimately becomes "ossified" from small centers within a developing bone. These centers of ossification are located in the *epiphyses* at the ends of a long bone and from a larger center located in the *diaphysis* **(shaft)** of the bone.

An area of cartilage called an **epiphyseal plate** or "growth plate" remains between the epiphyses and the diaphysis as long as growth continues. Growth ceases when all epiphyseal cartilage is transformed into bone. All that remains is a faint **epiphyseal line** that marks the location where the two centers of ossification have fused together.

Physicians sometimes use concepts of bone development to determine whether a child is going to grow any more. They have an x-ray study performed on the child's wrist. If it shows a layer of epiphyseal cartilage, they know that additional growth will occur. However, if it shows no epiphyseal cartilage, they know that growth has stopped and that the individual has attained adult height.

Intramembranous Ossification

Some bones, such as the skull bones illustrated in Figure 8-8, are formed by calcification of fibrous membranes in a process called **intramembranous ossification.**

The soft spots, or **fontanels,** on a newborn baby's skull are areas of fibrous membrane that have not yet fully ossified (see Figure 8-8). As intramembranous ossification progresses, a hard bone plate forms a complete flat bone.

QUICK CHECK

1. What is the basic structural unit of compact bone tissue called?
2. What are osteocytes? Where would you find them in bone tissue?
3. How does cartilage differ from bone?
4. What is ossification? What is the role of the osteoblast?

8

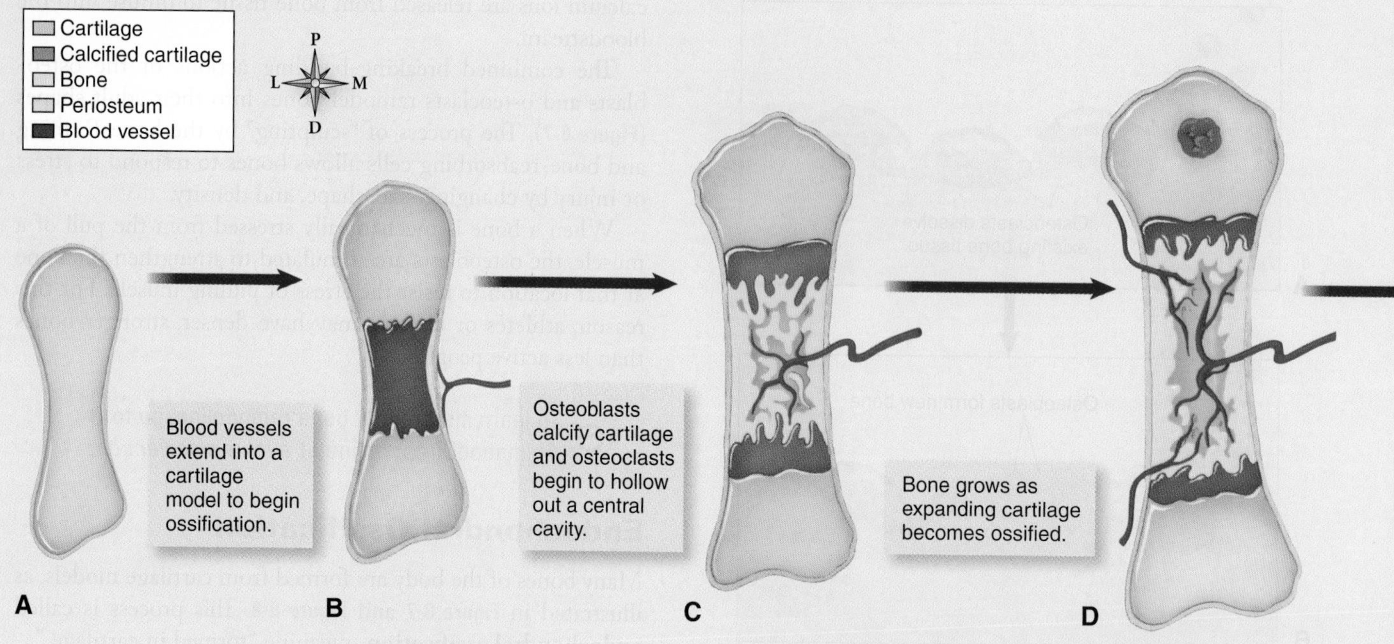

FIGURE 8-7 Endochondral ossification. As the cartilage of an immature long bone expands by normal growth, it is "invaded" by blood vessels carrying bone cells. Osteoblasts calcify cartilage as it becomes available. As the bone grows, osteoclasts hollow out the medullary cavity. Eventually, ossification overtakes cartilage expansion and further growth is not possible.

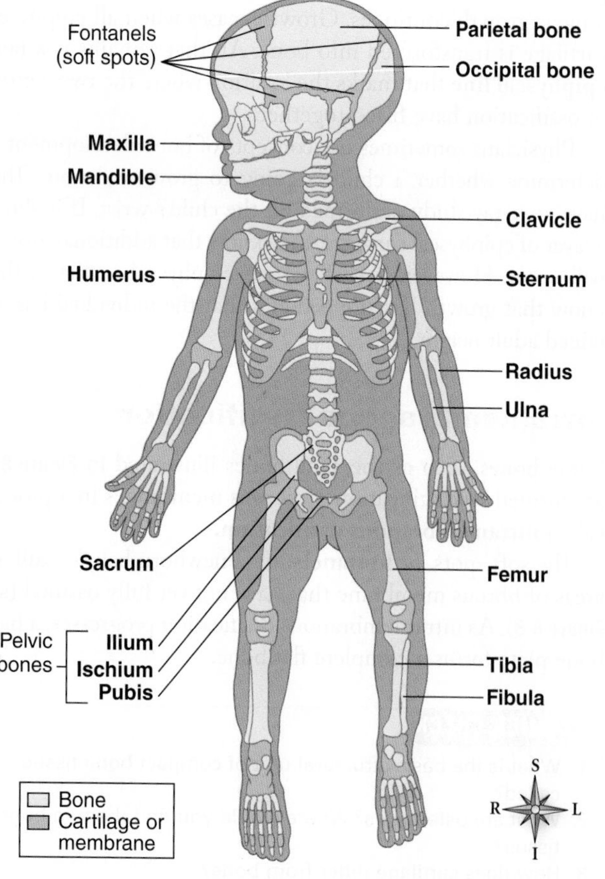

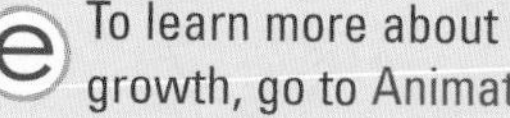

To learn more about bone formation and growth, go to AnimationDirect online at *evolve.elsevier.com*.

Axial Skeleton

The human skeleton has two divisions: the **axial skeleton** and the **appendicular skeleton.** Go back to **Figure 1-9** on p. 13 to review the axial-appendicular division of body regions.

Bones of the center, or axis, of the body make up the axial skeleton. The bones of the skull, spine, and chest and the hyoid bone in the neck are all in the axial skeleton. The bones of the upper and lower extremities or appendages make up the appendicular skeleton. The appendicular skeleton consists of the bones of the upper extremities (shoulder or pectoral girdle, arms, forearms, wrists, and hands) and the lower extremities (hip or pelvic girdle, thighs, legs, ankles, and feet) (**Table 8-1**).

Locate the various parts of the axial skeleton and the appendicular skeleton in **Figure 8-9**.

To better understand this concept, use the Active Concept Map *Organization of the Skeleton* at *evolve.elsevier.com*.

FIGURE 8-8 Bone development in a newborn. An infant's skeleton has many bones that are not yet completely ossified.

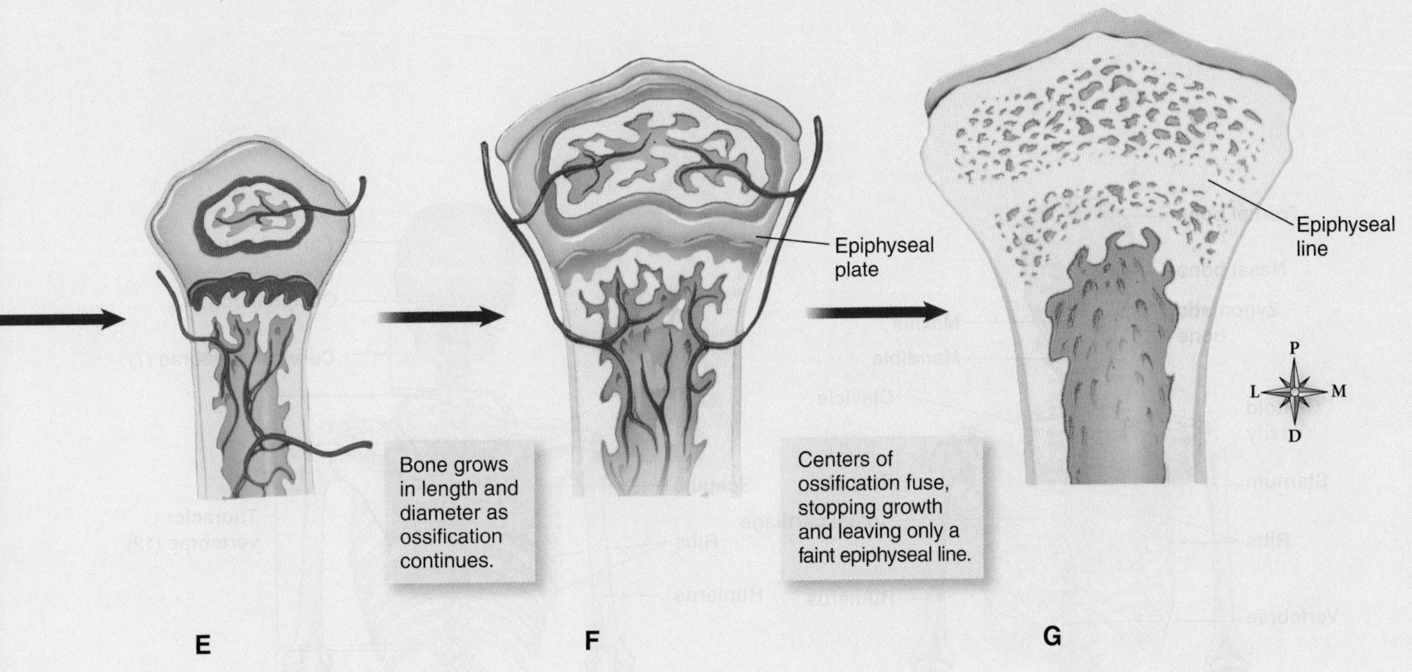

TABLE 8-1 Main Parts of the Skeleton (206 bones)

AXIAL SKELETON (80 bones)	APPENDICULAR SKELETON (126 bones)
Skull	Upper extremities
Cranial bones	Pectoral girdle
Ear bones	Arm and forearm bones
Face bones	Wrist bones
Spine	Hand bones
Vertebrae	Lower extremities
Thorax	Pelvic girdle
Ribs	Thigh and leg bones
Sternum	Ankle bones
Hyoid bone	Foot bones

Skull

Regions of the Skull

The **skull** consists of 8 bones that form the **cranium,** 14 bones that form the **face,** and 6 tiny bones in the **middle ear.** You can learn the names and locations of these bones by studying **Table 8-2**. Find as many of them as you can on **Figure 8-10**. Feel their outlines in your own body where possible. Examine them on a medical skeleton or skull model if you have access to one.

Sinuses

Sinuses are spaces or cavities inside some of the cranial bones. Four pairs of them (those in the **frontal, maxilla, sphenoid,** and **ethmoid bones**) have openings into the nose and thus are referred to as **paranasal sinuses** (**Figure 8-11**).

"My sinuses give me so much trouble." Have you ever heard this complaint or perhaps uttered it yourself? Sinuses cause trouble when the mucous membrane that lines them becomes inflamed, swollen, and painful. For example, inflammation in the frontal sinus *(frontal sinusitis)* often begins as a result of a nasal infection. The word part *-itis* added to a word means "inflammation of."

Mastoiditis, inflammation of the air spaces within the mastoid portion of the **temporal bone,** can produce very serious medical problems if not treated promptly (**Figure 8-12**). Locate the *mastoid process* in **Figure 8-10**, *A*.

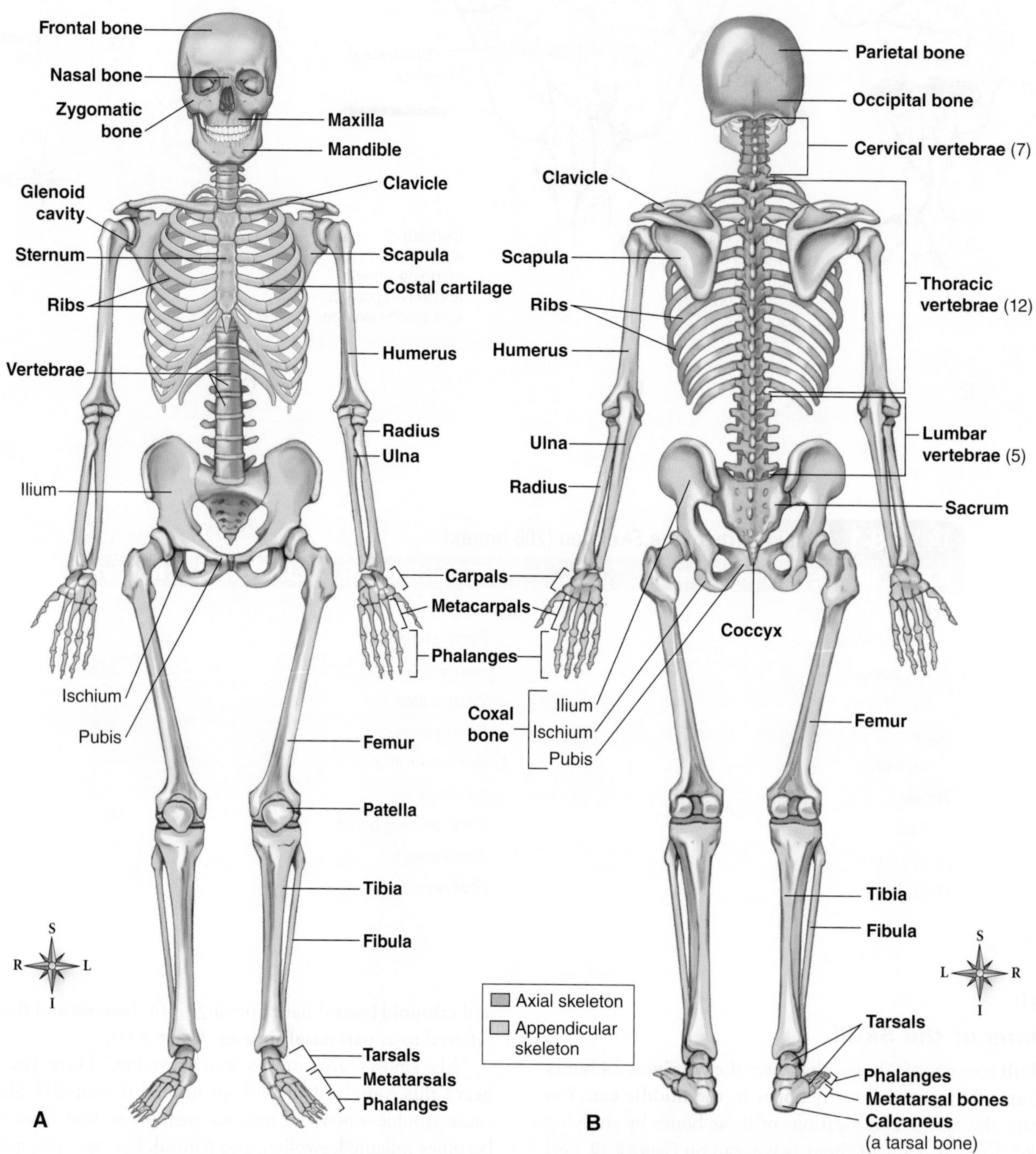

FIGURE 8-9 Human skeleton. The axial skeleton is distinguished by a blue tint. **A,** Anterior view. **B,** Posterior view.

TABLE 8-2 Bones of the Skull

NAME	NUMBER	DESCRIPTION
Cranial Bones		
Frontal	1	Forehead bone; also forms front part of floor of cranium and most of upper part of eye sockets; cavity inside bone above upper margins of eye sockets (orbits) called *frontal sinus;* lined with mucous membrane
Parietal	2	Form bulging topsides of cranium
Temporal	2	Form lower sides of cranium; contain *middle and inner ear structures; mastoid sinuses* are mucosa-lined spaces in *mastoid process*, the protuberance behind ear; *external auditory canal* is a tube leading into temporal bone; muscles attach to *styloid process*
Occipital	1	Forms posterior of skull; spinal cord enters cranium through large hole *(foramen magnum)* in occipital bone
Sphenoid	1	Forms central part of floor of cranium; pituitary gland located in small depression in sphenoid called *sella turcica (Turkish saddle);* muscles attach to pterygoid process
Ethmoid	1	Complicated bone that helps form floor of cranium, side walls and roof of nose and part of its middle partition (*nasal septum*—made up of the ethmoid's *perpendicular* plate and the *vomer* bone), and part of orbit; contains honeycomblike spaces, the *ethmoid sinuses; superior* and *middle conchae* are projections of ethmoid bone; form "ledges" alongside wall of each nasal cavity
Face Bones		
Nasal	2	Small bones that form upper part of bridge of nose
Maxilla	2	Upper jawbones; also help form roof of mouth, floor, and side walls of *nasal cavity* and floor of orbit; large cavity in *maxillary* bone is *maxillary sinus*
Zygomatic	2	Cheek bones; also help form orbit
Mandible	1	Lower jawbone articulates with temporal bone at *condyloid process*; small anterior hole for passage of nerves and vessels is the *mental foramen*
Lacrimal	2	Small bones; help form medial wall of eye socket and side wall of nasal cavity
Palatine	2	Form posterior part of roof of mouth and floor and side walls of nose and part of floor of orbit
Inferior nasal concha	2	Form curved "ledge" along inside of side wall of nose, below middle concha
Vomer	1	Forms lower posterior part of nasal septum
Ear Bones		
Malleus	2	Malleus, incus, and stapes are tiny bones in middle ear cavity in temporal bone; *malleus* means "hammer"—shape of bone
Incus	2	*Incus* means "anvil"—shape of bone
Stapes	2	*Stapes* means "stirrup"—shape of bone
Hyoid Bone		
Hyoid bone	1	U-shaped bone in neck; not joined to any other bone (not part of skull); between mandible and upper edge of larynx

8

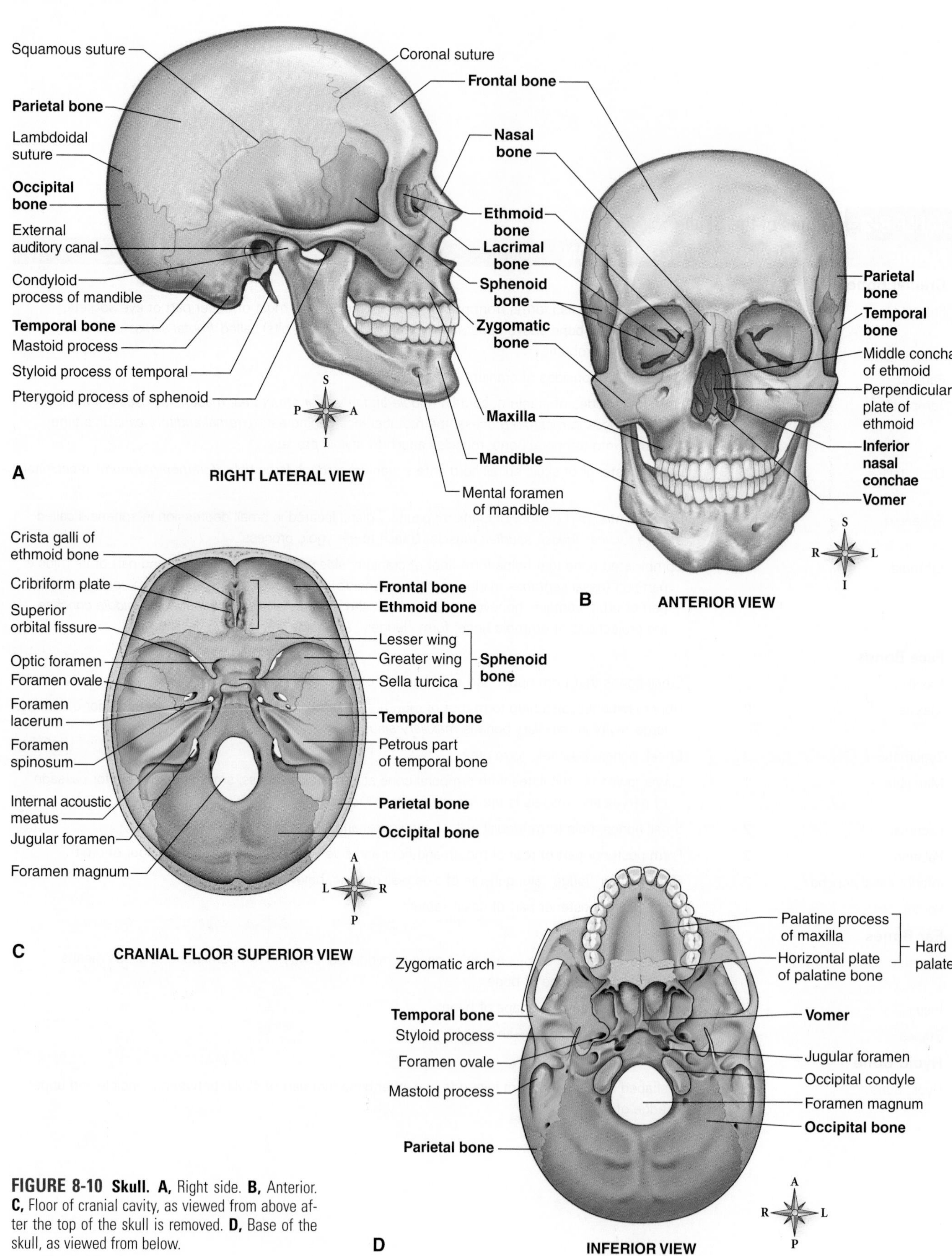

FIGURE 8-10 Skull. **A,** Right side. **B,** Anterior. **C,** Floor of cranial cavity, as viewed from above after the top of the skull is removed. **D,** Base of the skull, as viewed from below.

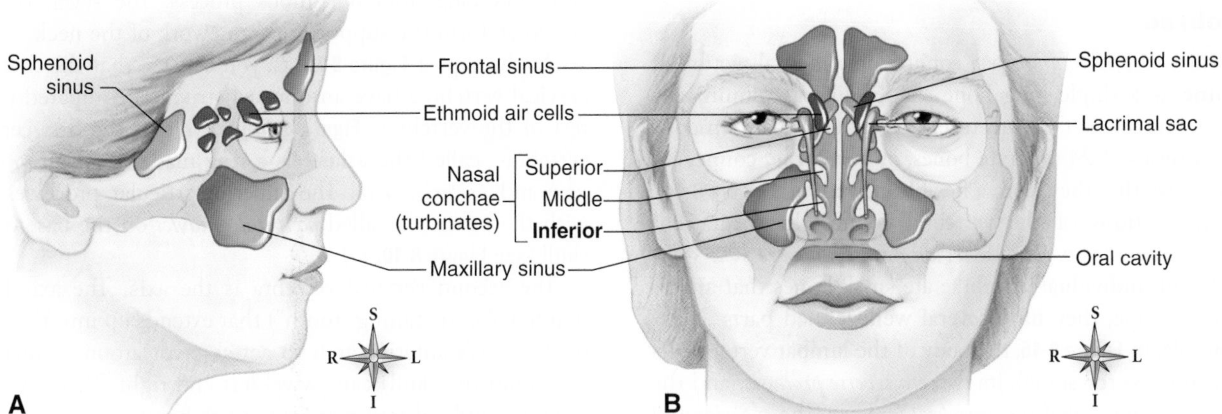

FIGURE 8-11 Paranasal sinuses. A, Lateral view of the face shows the position of the paranasal sinuses. **B,** Anterior view shows the relationship of the sinuses to each other and the nasal cavity.

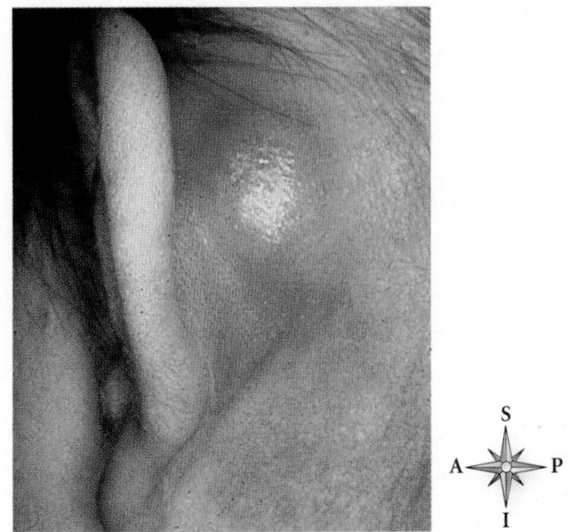

FIGURE 8-12 Mastoiditis. Note the redness and swelling over the mastoid process of the temporal bone behind the left ear.

Infectious material from middle ear infections sometimes finds its way into the mastoid air cells. These air cells do not drain into the nose like the paranasal sinuses do. Thus infectious material that accumulates may damage the thin, bony partition that separates the air cells from the brain. If this occurs, the infection may spread to the brain or the membranes covering the brain, a life-threatening situation.

If antibiotic therapy fails, chronic mastoiditis may be treated by surgically removing the affected tissue, including internal parts of the ear—rendering the individual deaf in the affected ear.

Sutures and Fontanels

Note in Figure 8-10 that two **parietal bones,** which give shape to the bulging topside of the skull, form immovable joints called **sutures** with several bones.

- *Lambdoidal suture*—joins posterior margins of parietal bones to the **occipital bone**
- *Squamous sutures*—joins lateral margin of each parietal bone with the superior margin of the temporal bone and to the lateral part of the sphenoid bone
- *Coronal suture*—joins the anterior margins of parietal bones with the posterior margin of the **frontal bone**
- *Sagittal suture*—joins the medial margins of the parietal margins to each other (not visible in Figure 8-10)

You may be familiar with the "soft spots" on a baby's skull. The soft areas are six **fontanels,** or areas where intramembranous ossification remains incomplete at birth. You can see them depicted in Figure 8-8. Fontanels allow some compression of the skull during birth without much risk of breaking the skull bones. They also may be identified by a clinician as an important diagnostic indication of the position of the baby's head before delivery.

The fontanels fuse to form sutures before a baby is 2 years old.

8

Hyoid Bone

The odd little **hyoid bone** resembles the Greek letter upsilon (γ or ν). Unlike other bones, it does not form a joint with any other bone of the skeleton. As you can see in Figure 8-13, the hyoid bone is located in the neck, where it serves as an anchor for tongue muscles and helps support the larynx (voice box).

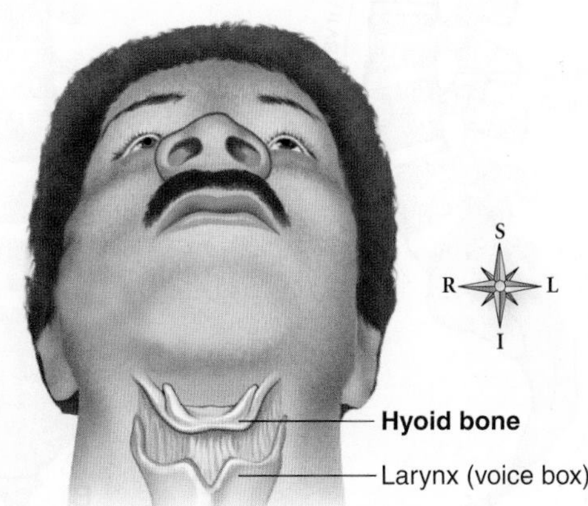

FIGURE 8-13 Hyoid bone. The U-shaped hyoid bone is unique because it does not attach to any other bone of the skeleton.

Vertebral Column (Spine)

Vertebrae

The term *vertebral column* may conjure up a mental picture of the **spine** as a single long bone shaped like a column in a building, but this is far from true. The vertebral column consists of a series of 24 separate bones, or **vertebrae,** connected in such a way that they form a flexible curved rod (**Figure 8-14**). Different sections of the vertebral column have different names: *cervical region, thoracic region,* and *lumbar region.*

Although individual vertebrae are small bones that are irregular in shape, they have several well-defined parts. Note, for example, in **Figure 8-15**, the body of the lumbar vertebra, its spinous process (or spine), its two *transverse processes,* and the hole in its center, called the *vertebral foramen.* The *superior* and *inferior articular processes* permit limited and controlled movement between adjacent vertebrae.

To feel the tip of the spinous process of one of your vertebrae, simply bend your head forward and run your fingers down the back of your neck until you feel a projection of bone at shoulder level. This is the tip of the seventh cervical vertebra's long, forked spinous process. The seven cervical vertebrae form the supporting framework of the neck.

At the top of **Figure 8-14**, *C*, you can see that the first two cervical vertebrae have an unusual structure compared to the rest of the vertebrae. **Figure 8-16** shows that the first cervical vertebra—called the **atlas**—is a ring made up of an *anterior arch* and *posterior arch.* The superior articular processes join with the processes called *occipital condyles* on the base of the skull (see **Figure 8-10**, *D*).

The second cervical vertebra is the **axis.** The axis has a pointed *dens* (meaning "tooth") that extends up into the curve of the atlas's anterior arch to act as pivot around which the atlas (and the skull) can swivel left and right. This is yet another example of *structure fits function* because rotation of the neck would be very limited without this unique structure.

Sacrum and Coccyx

The **sacrum** and **coccyx** are two additional bones of the vertebral column located just below the 24 vertebrae. In infants, the sacrum exists as five separate vertebrae that start to fuse

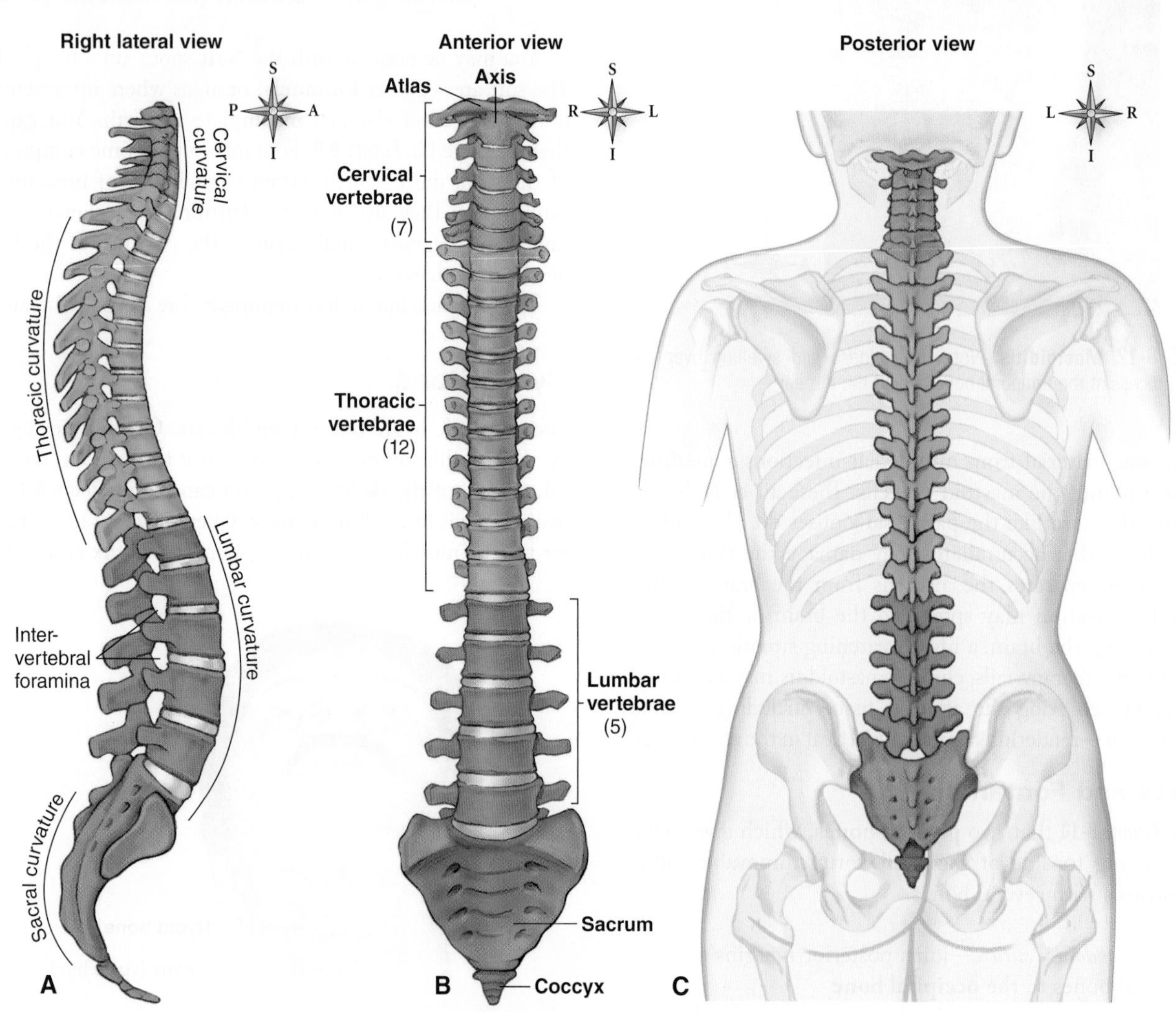

FIGURE 8-14 The vertebral column.

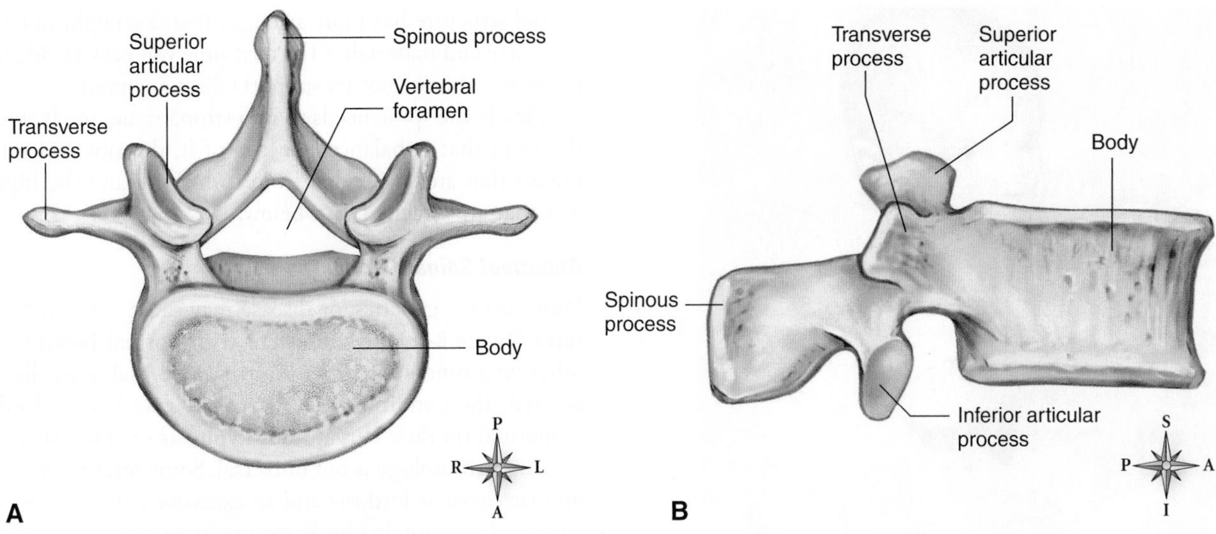

FIGURE 8-15 Typical vertebra. Third lumbar vertebra. **A,** From above (superior view). **B,** From the side (lateral view).

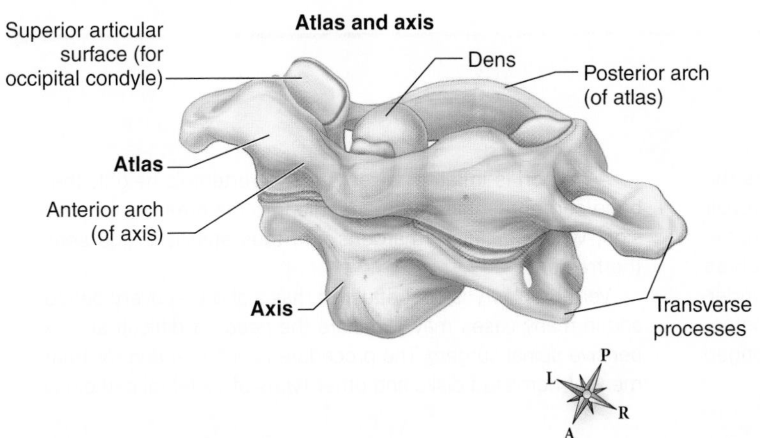

FIGURE 8-16 Atlas and axis. The toothlike dens of the axis (second cervical vertebra) extends along the inside of the anterior arch of the atlas (first cervical vertebra) to act as a pivot. Because the atlas supports the entire skull, this arrangement allows the head to rotate.

TABLE 8-3 Bones of the Vertebral Column

NAME	NUMBER	DESCRIPTION
Cervical	7	Upper 7 vertebrae, in neck region; first cervical vertebra called *atlas;* second, *axis*
Thoracic vertebrae	12	Next 12 vertebrae; ribs attach to these
Lumbar vertebrae	5	Next 5 vertebrae; in small of back
Sacrum	1	In child, 5 separate vertebrae; in adult, fused into one
Coccyx	1	In child, 3 to 5 separate vertebrae; in adult, fused into one

together about age 18 and are completely fused by 25 to 33 years of age. Likewise, three to five tiny "tail" vertebrae fuse to form a single coccyx by early adulthood.

All 26 bones of the vertebral column are illustrated in **Figure 8-14** and described in **Table 8-3**.

Spinal Curvatures

Have you ever noticed the four curves in your spine? Your neck and the small of your back curve slightly inward or forward, whereas the chest region of the spine and the lowermost portion curve in the opposite direction (see **Figure 8-14**).

Convex and Concave Curvatures

When you look at the spine from the rear, you will see the thoracic and sacral curves, called **convex curvatures** because they round outward. The cervical and lumbar curves of the spine are called **concave curvatures** because they curve inward.

This is not true, however, of a newborn baby's spine. It forms a continuous convex curve—called the *primary curvature*—from top to bottom (**Figure 8-17**). Gradually, as the baby learns to hold up his or her head, a reverse or concave curve develops in the neck (cervical region). Later, as the baby learns to stand, the lumbar region of his or her spine also becomes concave. The concave cervical and lumber curvatures are sometimes called *secondary curvatures* because they appear later in development than the primary (convex) curvatures.

The normal curves of the spine have important functions. They give it enough strength to support the weight of the rest of the body. These curves also make it possible to balance the weight of the body, which is necessary for us to stand and walk on two feet instead of having to crawl on all fours. A

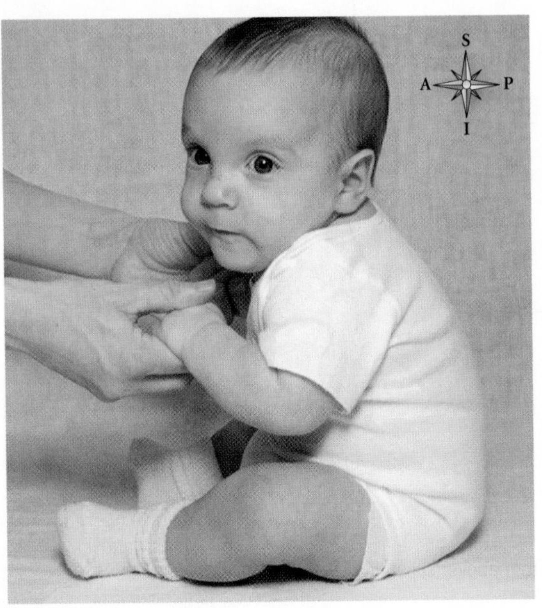

FIGURE 8-17 Spinal curvature of an infant. The spine of the newborn baby forms a continuous convex curve.

curved structure has more strength than a straight one of the same size and materials. (The next time you pass a bridge, look to see whether or not its supports form a curve.)

Clearly the spine needs to be a strong structure. It supports the head that is balanced on top of it, the ribs and internal organs that are suspended from it in front, and the hips and legs that are attached to it below.

Abnormal Spinal Curvatures

Poor posture, genetics, or disease may cause abnormal curvatures (**Figure 8-18**) that interfere with normal breathing and other vital functions. If the lumbar curve is abnormally exaggerated, the condition may be called *swayback* or **lordosis.** Abnormal thoracic curvature is **kyphosis** or "hunchback."

This terminology is not universal. Some refer to the normal lumbar curve as lordosis and to excessive curvature as *hyperlordosis.* Likewise, kyphosis may refer to the normal thoracic curve and *hyperkyphosis* to excessive curvature.

Abnormal side-to-side curvature is **scoliosis.** Scoliosis is a relatively common condition that appears before adolescence, usually of unknown cause.

8

CLINICAL APPLICATION

VERTEBROPLASTY

Vertebroplasty is an orthopedic procedure that involves the injection of a "super glue" type of bone cement to repair fractured and compressed (collapsed) vertebrae (see illustration). In these patients the body of one or more vertebrae (generally lower thoracic and/or lumbar segments) has undergone a compression fracture due to osteoporosis, trauma, tumors, or prolonged use of steroid drugs.

A balloon is inflated to restore the vertebra's height; then bone cement is injected by needle into the area of compression, where it quickly hardens and thus stabilizes and seals the fracture.

Vertebroplasty is cost effective, has a short recovery period, and in many cases may eliminate the need for difficult and expensive spinal surgery. The procedure is not intended for treatment of herniated disks and other types of vertebral pathology.

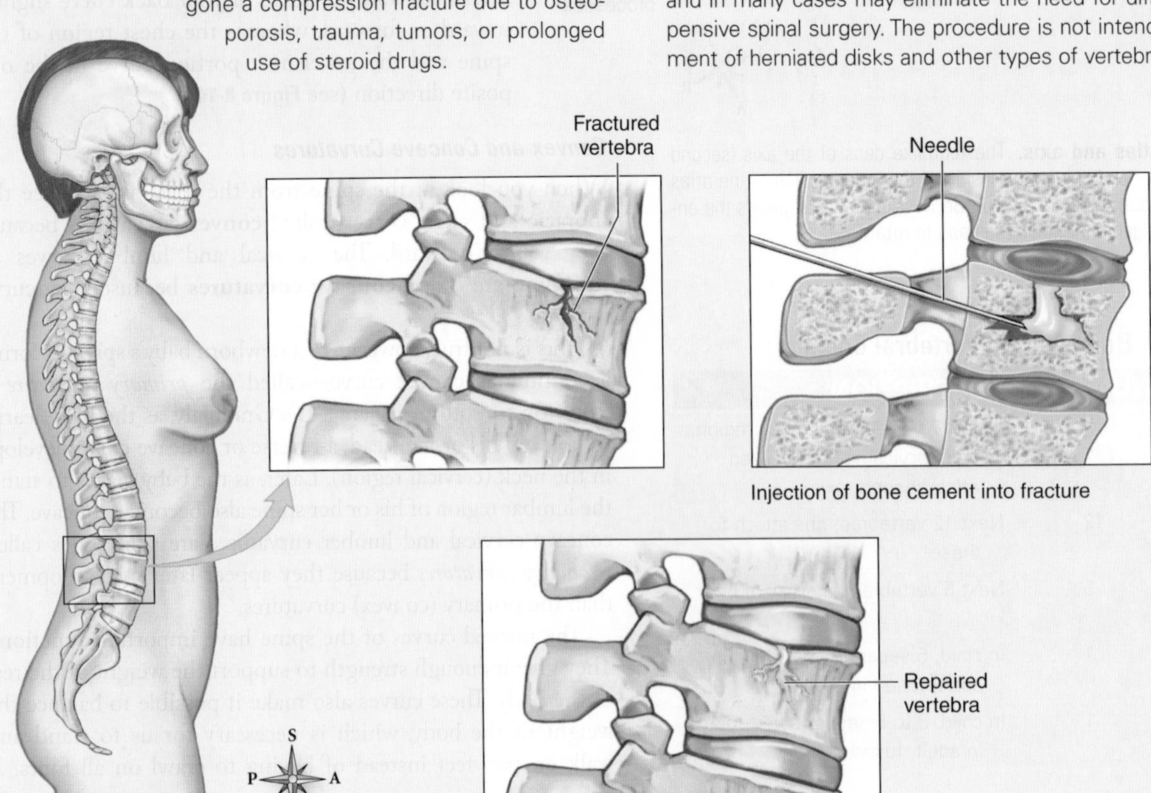

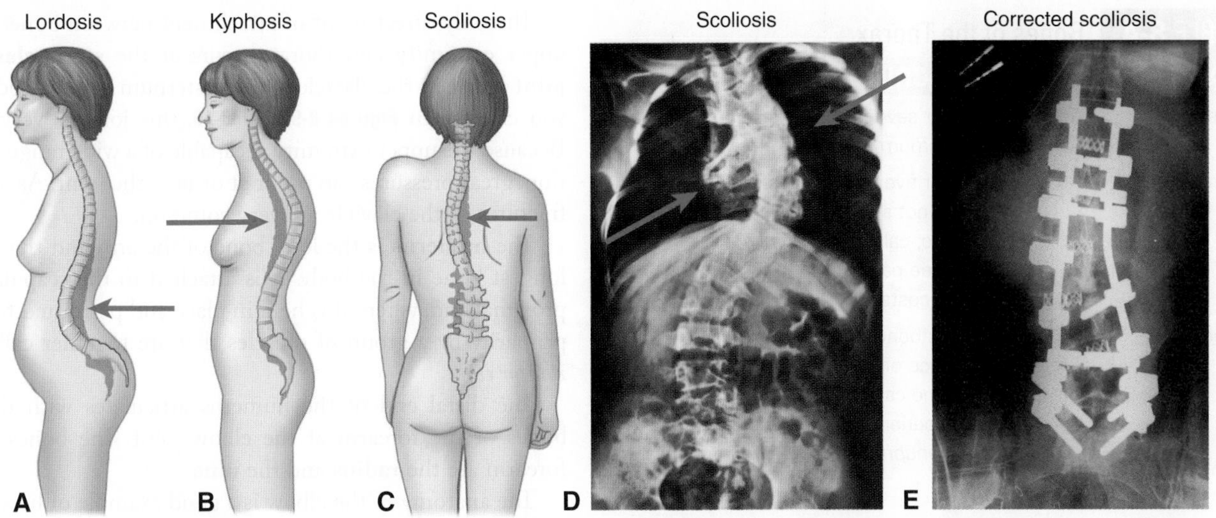

FIGURE 8-18 Abnormal spinal curvatures. A, Lordosis (hyperlordosis). **B,** Kyphosis (hyperkyphosis). **C,** Scoliosis. **D,** An x-ray showing pronounced scoliosis and **E,** an x-ray of an 11-year-old girl after corrective surgery for scoliosis.

Scoliosis treatments vary depending on the degree of lateral curvature and resulting deformity of individual vertebrae. Traditional treatments for scoliosis include long-term use of supportive braces, transcutaneous ("through-the-skin") muscle stimulation, and surgery. Electrical stimulation of muscles on one side of the spine over time helps pull the vertebrae into a more normal position. Surgical procedures to straighten the spine may involve bone grafts and the insertion of internal metal rods.

Thorax

Twelve pairs of **ribs,** the **sternum** (breastbone), and the thoracic vertebrae form the bony cage known as the **thorax** or **chest.**

Each of the 12 pairs of ribs is attached posteriorly to a vertebra. Also, all the ribs except the lower two pairs are attached to the sternum and so have anterior and posterior anchors.

Look closely at **Figure 8-19** and you can see that the first seven pairs of ribs (sometimes referred to as the *true ribs*) are attached to the sternum by *costal cartilage.* The remaining ribs are called *false ribs* because they do not attach directly to the sternum.

False rib pairs 8, 9, and 10 attach to the cartilages of rib pair 7. The last two pairs of false ribs, in contrast, are not attached to any costal cartilage but seem to float free in front, hence their descriptive name, *floating ribs* (**Table 8-4**).

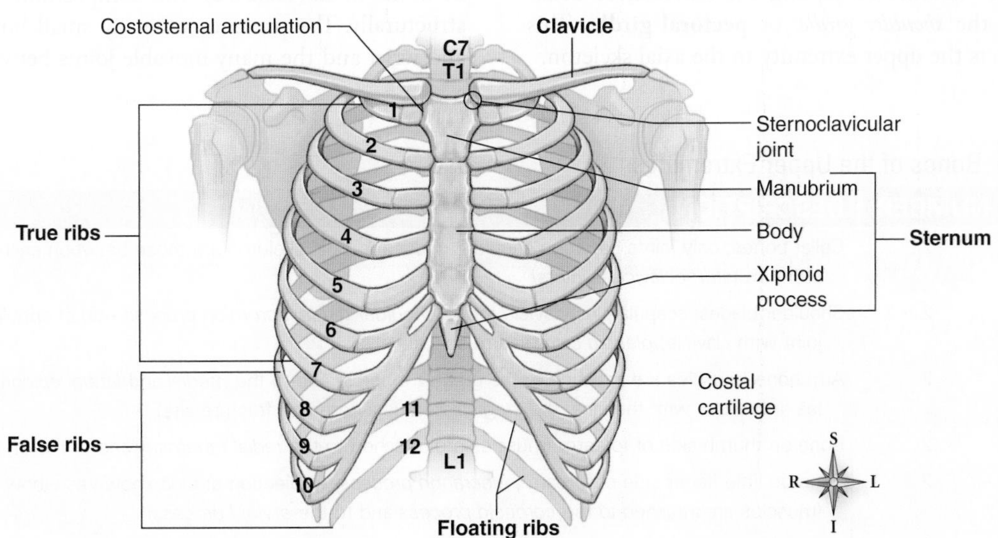

FIGURE 8-19 Thorax. Rib pairs 1 through 7, the *true ribs,* are attached by cartilage to the sternum. Rib pairs 8 through 10, the *false ribs,* are attached to the cartilage of the seventh pair. Rib pairs 11 and 12 are called *floating ribs* because they have no anterior cartilage attachments.

TABLE 8-4 Bones of the Thorax

NAME	NUMBER	DESCRIPTION
True ribs	14	Upper seven pairs; attached to sternum by *costal cartilages*
False ribs	10	Lower five pairs; lowest two pairs do not attach to sternum; therefore, called *floating ribs;* next three pairs attached to sternum by costal cartilage of seventh ribs
Sternum	1	Breastbone; shaped like a dagger; piece of cartilage at lower end of bone called *xiphoid process;* superior portion called the *manubrium*

Appendicular Skeleton

Of the 206 bones that form the skeleton as a whole, 126 are contained in the appendicular subdivision (see **Figure 8-9**).

Note that the bones in the shoulder, or *pectoral girdle,* connect the bones of the arm, forearm, wrist, and hands to the axial skeleton of the thorax, and the hip, or *pelvic girdle,* connects the bones of the thigh, leg, ankle, and foot to the axial skeleton of the pelvis.

8

QUICK CHECK

1. What is the difference between the axial skeleton and the appendicular skeleton?
2. What is a suture? A fontanel? A sinus?
3. What is the *hyoid bone* and where is it located?
4. Name the sections of the vertebrae? How many bones are in each section?
5. How does a *false rib* differ from a *true rib*?

Upper Extremity

The **scapula,** or shoulder blade, and the **clavicle,** or collar bone, compose the *shoulder girdle,* or **pectoral girdle.** This structure connects the upper extremity to the axial skeleton.

The only direct point of attachment between bones of the upper extremity and thorax occurs at the **sternoclavicular joint** between the clavicle and the sternum or breastbone. As you can see in **Figures 8-9** and **8-19**, this joint is very small. Because the upper extremity is capable of a wide range of motion, great pressures can occur at or near the joint. As a result, fractures of the clavicle are very common.

The **humerus** is the long bone of the arm and the second longest bone in the body. It is attached to the scapula at its proximal end, where it is held in place and permitted to move primarily by a group of muscles that are together called the *rotator cuff.*

The distal end of the humerus articulates with the two bones of the forearm at the elbow joint. The bones of the forearm are the **radius** and the **ulna.**

The anatomy of the elbow is a good example of how structure is related to function. Note in **Figure 8-20** that the rounded *trochlea* of the humerus fits into the *trochlear notch* of the ulna to form a hingelike structure that allows the elbow to bend or *flex.*

Notice also that the large bony process of the ulna, called the **olecranon,** fits nicely into a large depression on the posterior surface of the humerus, called the **olecranon fossa.** This arrangement prevents the "hinge" of the elbow from extending beyond a straight-arm position—a stability needed to hold objects efficiently.

The radius and the ulna of the forearm articulate with each other and with the distal end of the humerus at the elbow joint. In addition, they also touch each another distally where they articulate with the bones of the wrist. In the anatomical position, with the arm at the side and the palm facing forward, the radius runs along the lateral side of the forearm, and the ulna is located along the medial side of the forearm.

The wrist and the hand have more bones in them for their size than any other part of the body—8 **carpal** or wrist bones, 5 **metacarpal** bones that form the support structure for the palm of the hand, and 14 **phalanges,** or finger, bones—27 bones in all (**Table 8-5**). This composition is very important structurally. The presence of many small bones in the hand and wrist and the many movable joints between them makes

TABLE 8-5 Bones of the Upper Extremities

NAME	NUMBER	DESCRIPTION
Clavicle	2	Collar bones; only joints between shoulder girdle and axial skeleton are those between each clavicle and sternum *(sternoclavicular joints)*
Scapula	2	Shoulder blades; scapula plus clavicle forms *shoulder girdle; acromion process*—tip of shoulder that forms joint with clavicle; *glenoid cavity*—arm socket
Humerus	2	Arm bone (muscles are attached to the greater tubercle and to the *medial* and *lateral epicondyles;* the trochlea articulates with the ulna; the *surgical neck* is a common fracture site)
Radius	2	Bone on thumb side of forearm (muscles are attached to the *radial tuberosity* and to the *styloid process*)
Ulna	2	Bone on little finger side of forearm; *olecranon process*—projection of ulna known as elbow or "funny bone" (muscles are attached to the *coronoid process* and to the *styloid process*)
Carpal bones	16	Irregular bones at upper end of hand; anatomical wrist
Metacarpals	10	Form framework of palm of hand
Phalanges	28	Finger bones; three in each finger, two in each thumb

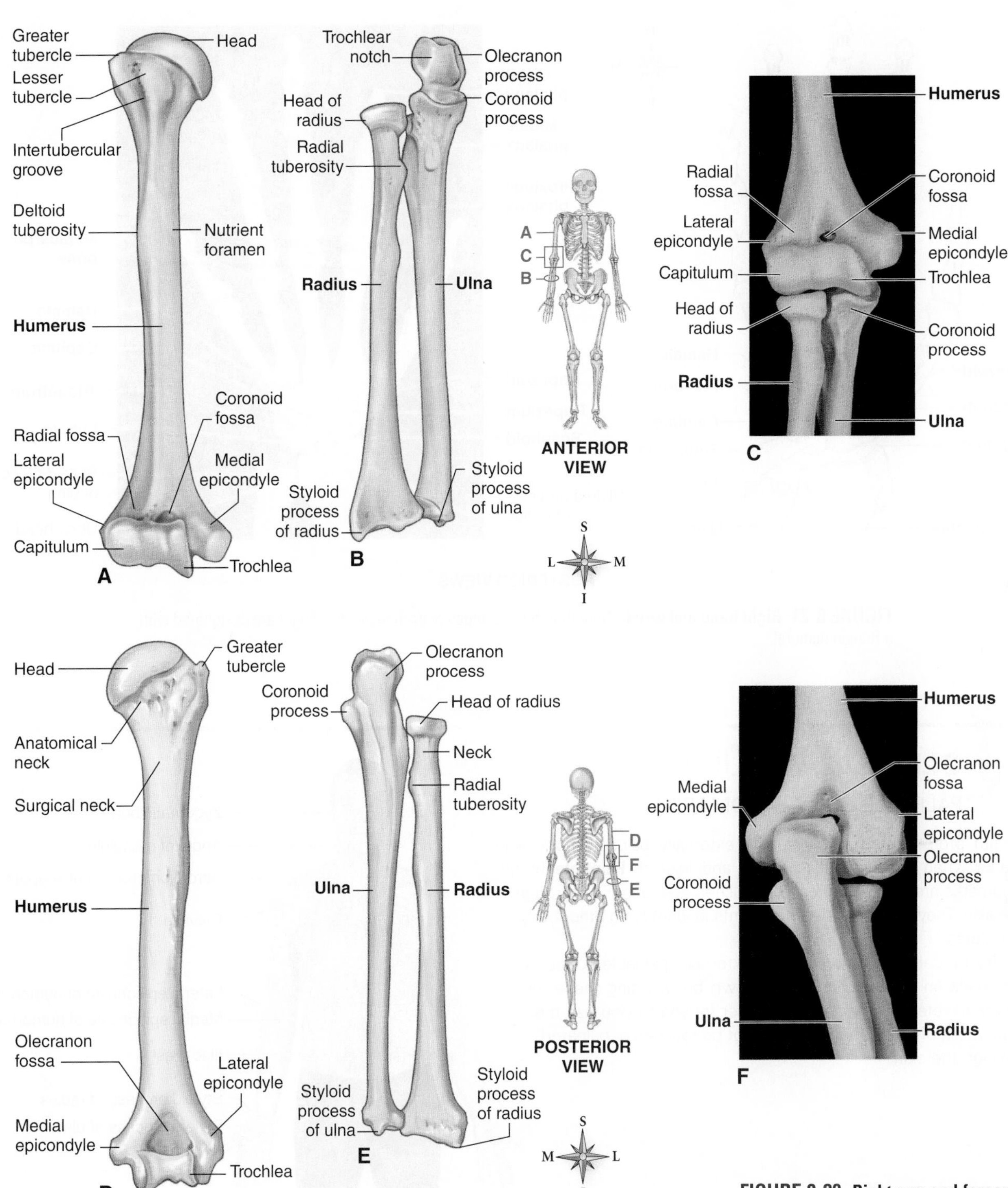

FIGURE 8-20 Right arm and forearm.

the human hand highly maneuverable—allowing us to easily make and use tools.

Figure 8-21 shows the relationships between the bones of the wrist and hand.

Lower Extremity

The *hip girdle,* or **pelvic girdle,** is the part of the lower extremity that connects to the trunk. The pelvic girdle as a whole consists of two large **coxal bones,** one located on each side of the pelvis, attached inferiorly to the *sacrum* of the vertebral column. This ringlike arrangement of bones provides a strong base of support for the torso and connects the lower extremities to the axial skeleton.

In an infant's body, each coxal bone consists of three separate bones—the **ilium,** the **ischium,** and the **pubis** (see **Figure 8-8**). These bones grow together to become one bone in an adult (see **Figures 8-9** and **8-25** on pp. 182 and 195.

Just as the humerus is the only bone in the arm, the **femur** is the only bone in the thigh (**Figure 8-22**). It is the longest bone

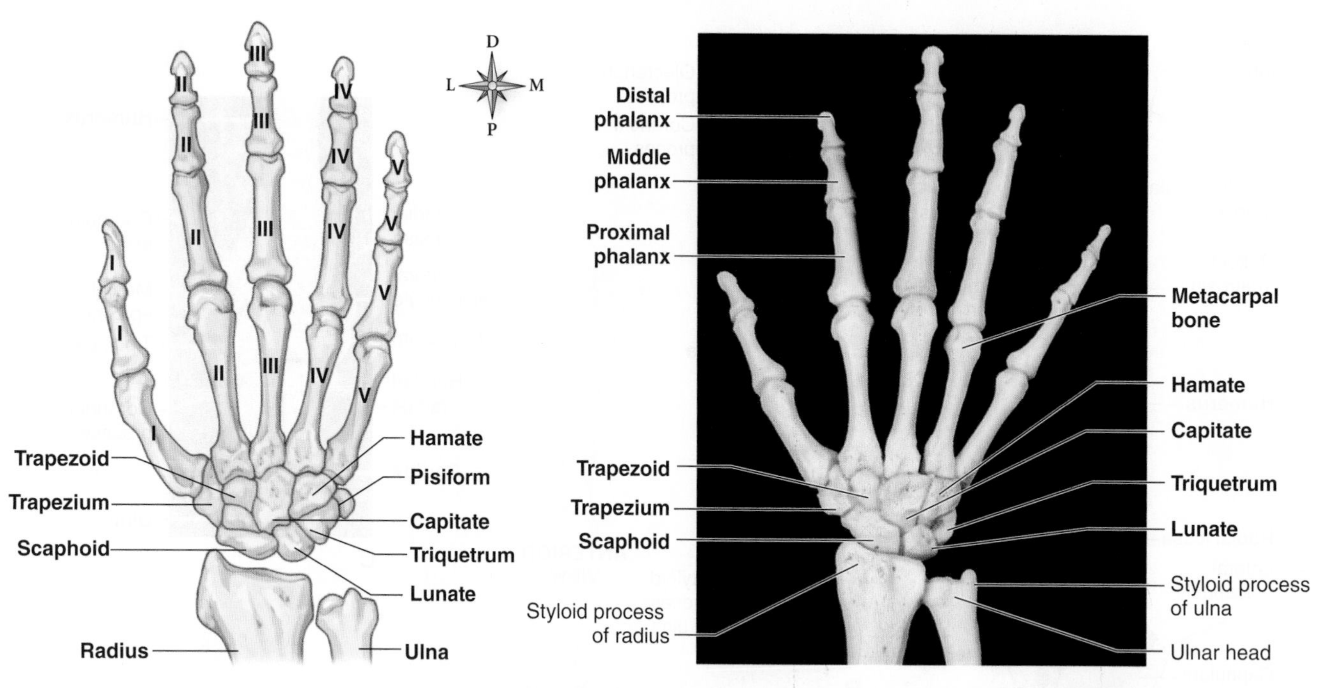

D
L
M
P
I
II
III
IV
V
Trapezoid
Trapezium
Scaphoid
Radius
Hamate
Pisiform
Capitate
Triquetrum
Lunate
Ulna
Distal phalanx
Middle phalanx
Proximal phalanx
Trapezoid
Trapezium
Scaphoid
Styloid process of radius
Metacarpal bone
Hamate
Capitate
Triquetrum
Lunate
Styloid process of ulna
Ulnar head
POSTERIOR VIEWS

FIGURE 8-21 Right hand and wrist. Note that the phalanges of each respective finger are designated with a Roman numeral.

CLINICAL APPLICATION

PALPABLE BONY LANDMARKS

Health professionals often identify externally palpable bony landmarks when dealing with the sick and injured. **Palpable** bony landmarks are bones that can be touched and identified through the skin. They serve as reference points in identifying other body structures.

Try to identify as many of the externally palpable bones of the skeleton as possible on your own body. Using these as points of reference will make it easier for you to visualize the placement of other bones that cannot be touched or palpated through the skin.

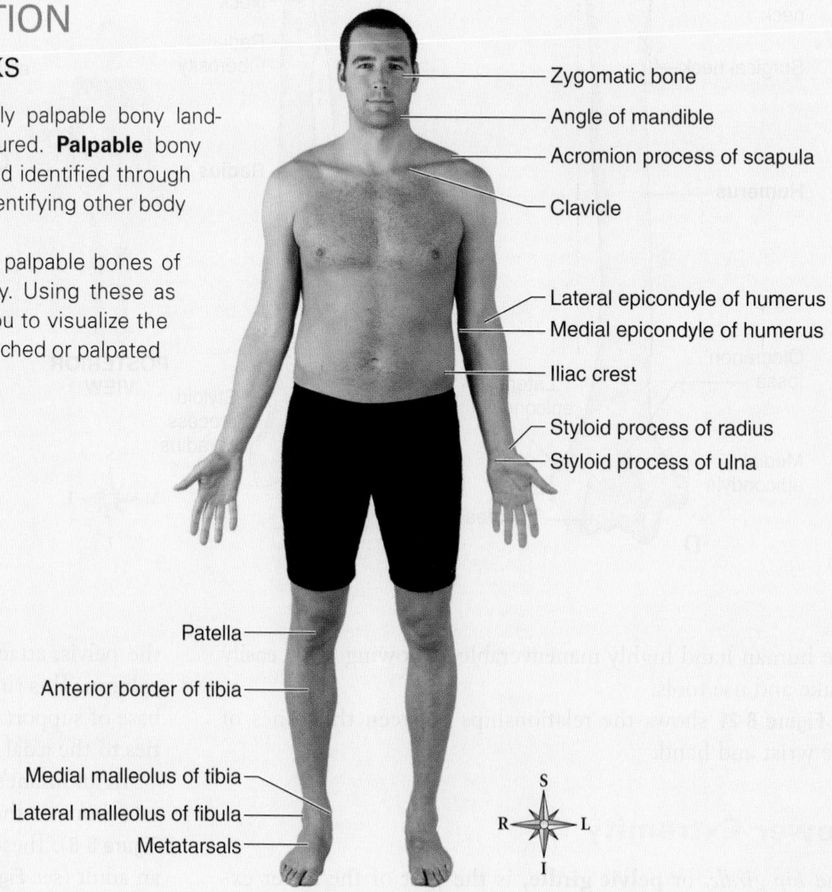

Zygomatic bone
Angle of mandible
Acromion process of scapula
Clavicle
Lateral epicondyle of humerus
Medial epicondyle of humerus
Iliac crest
Styloid process of radius
Styloid process of ulna
Patella
Anterior border of tibia
Medial malleolus of tibia
Lateral malleolus of fibula
Metatarsals
S
R
L
I

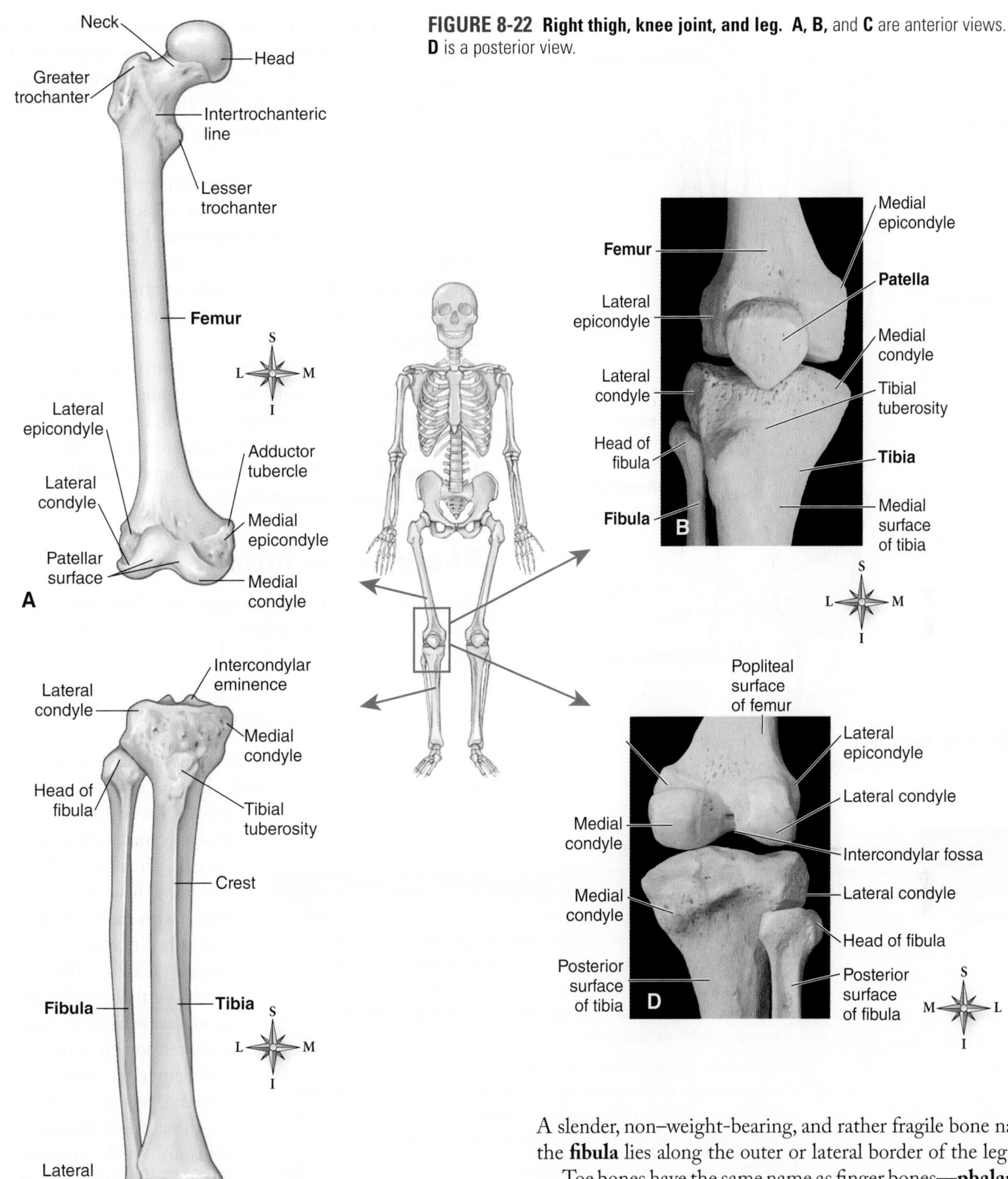

FIGURE 8-22 Right thigh, knee joint, and leg. A, B, and **C** are anterior views. **D** is a posterior view.

in the body and articulates proximally (toward the hip) with the coxal bone in a deep, cup-shaped socket called the **acetabulum.** The articulation of the head of the femur in the acetabulum is more stable than the articulation of the head of the humerus with the scapula in the upper extremity. As a result, dislocation of the hip occurs less often than does disarticulation of the shoulder. Distally, the femur articulates with the kneecap, or **patella,** and the **tibia,** or "shinbone." The tibia forms a rather sharp edge or crest along the anterior of the leg. A slender, non–weight-bearing, and rather fragile bone named the **fibula** lies along the outer or lateral border of the leg.

Toe bones have the same name as finger bones—**phalanges.** There is the same number of toe bones as finger bones, a fact that might surprise you because toes are shorter than fingers.

Foot bones comparable to the metacarpals and carpals of the hand have slightly different names. They are called **metatarsals** and **tarsals** in the foot (**Figure 8-23**). Just as each hand contains five metacarpal bones, each foot contains five metatarsal bones. However, the foot has only seven tarsal bones, in contrast to the hand's eight carpals. The largest tarsal bone is the **calcaneus,** or heel bone. The **talus** is the second largest tarsal bone and articulates with the tibia at the ankle joint. The bones of the lower extremities are summarized in **Table 8-6**.

8

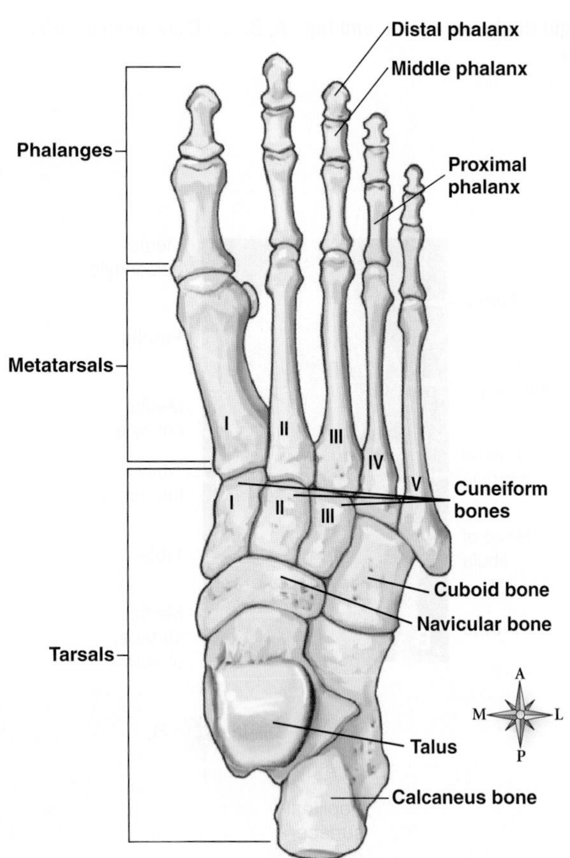

FIGURE 8-23 Right foot. Compare the names and numbers of foot bones (viewed here from above) with those of the hand bones shown in **Figure 8-21**.

8

NAME	NUMBER	DESCRIPTION
Coxal bone	2	Hipbones; *ilium*—upper flaring part of pelvic bone; *ischium*—lower posterior part; *pubic bone*—lower front part; *acetabulum*—hip socket; *symphysis pubis*—joint in midline between two pubic bones; *pelvic inlet*—opening into *true pelvis* or pelvic cavity; if pelvic inlet is misshapen or too small, infant skull cannot enter true pelvis for natural birth
Femur	2	Thigh bones; *head of femur*—ball-shaped upper end of bone; fits into acetabulum (muscles are attached to the greater and *lesser trochanters* and to the *lateral* and *medial epicondyles;* the *lateral* and *medial condyles* form articulations at the knee)
Patella	2	Kneecap
Tibia	2	Shinbone; *medial malleolus*—rounded projection at lower end of tibia commonly called *inner anklebone;* muscles are attached to the *tibial tuberosity*
Fibula	2	Long slender bone of lateral side of leg; *lateral malleolus*—rounded projection at lower end of fibula commonly called *outer anklebone*
Tarsal bones	14	Form heel and posterior part of foot; anatomical ankle; largest is the *calcaneus*
Metatarsals	10	Form part of foot to which toes are attached; tarsal and metatarsal bones arranged so that they form three arches in foot; medial *(inner) longitudinal arch* and *lateral (outer) longitudinal arch,* which extend from front to back of foot, and transverse or *metatarsal arch,* which extends across foot
Phalanges	28	Toe bones; three in each toe, two in each great toe

TABLE **8-6** Bones of the Lower Extremities

You stand on your feet, so it is important that certain features of their structure make them able to support the body's weight. The great toe, for example, is considerably more solid and less mobile than the thumb. The foot bones are held together in such a way as to form springy lengthwise and crosswise **arches.**

Two arches extend in a lengthwise direction in the foot (**Figure 8-24**, *A*). One lies on the inside part of the foot and is called the **medial longitudinal arch.** The other lies along the outer edge of the foot and is named the **lateral longitudinal arch.** Another arch extends across the ball of the foot; this arch is called the **transverse arch,** or **metatarsal arch** (**Figure 8-24**, *C*).

These arches provide great supporting strength and a highly stable base. Strong ligaments and leg muscle tendons normally hold the foot bones firmly in their arched positions. Frequently, however, the foot ligaments and tendons weaken. The arches then flatten, a condition appropriately called *fallen arches* or **flatfeet** (**Figure 8-24**, *B*).

Skeletal Variations

Many different factors cause each individual's skeleton to vary from all other human skeletons. In this section, we explore a few of those factors.

Male-Female Skeletal Differences

A man's skeleton and a woman's skeleton differ in several ways. If you were to examine a male skeleton and a female skeleton placed side by side, you would probably first notice the difference in their sizes. Most male skeletons have bones that are larger, with more distinct bumps and other markings, than most female skeletons. This difference results partly from the difference in muscle tension on bones—the more tension on bone, the bigger and denser the bone gets at the points of muscle attachment.

These male-female distinctions are visible in nearly every bone of the body, so it is no wonder that forensic scientists can often accurately determine the sex of human remains using just a few bones. Sex differences are also important in sports physiology and medicine, where it is useful in improving athletic performance in some sports and in avoiding certain injuries.

Perhaps the most obvious of the many structural differences between the male and female skeletons are in the *pelvic girdle* or **pelvis**—the ring formed by the two coxal bones and sacrum. The word *pelvis* means "basin." The wide structure of the female pelvis allows the body of a fetus to be cradled in it before birth, and its wide opening allows the baby to pass through it during birth.

Although the individual male coxal bones are generally larger than the individual female coxal bones, together the male coxal bones form a narrower structure than do the female coxal bones. A man's pelvis is shaped more like a funnel than the broad, shallow basin of the female pelvis (**Figure 8-25**).

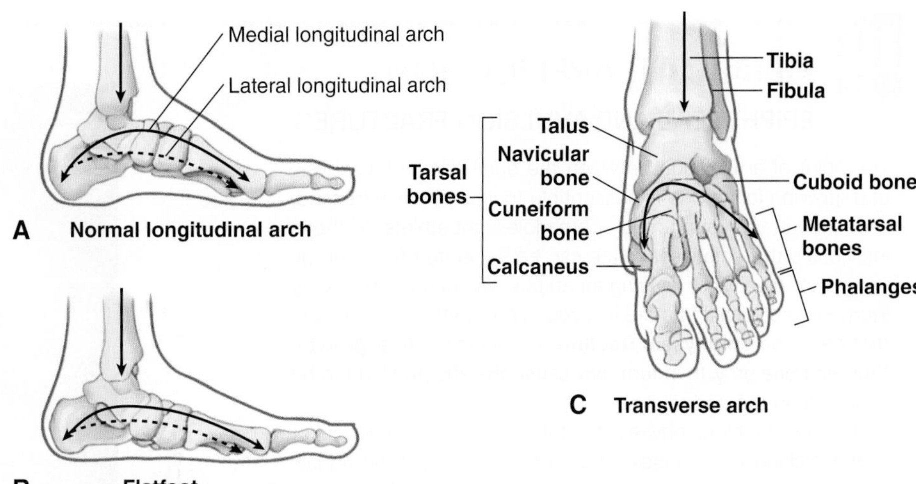

FIGURE 8-24 Arches of the foot. A, Medial and lateral longitudinal arches. (*Arrows* show direction of force.) **B,** "Flatfoot" occurs when tendons and ligaments weaken and the arches fall. **C,** Transverse arch.

You can also see in **Figure 8-25** that the openings from the abdomen into and through the pelvis—the pelvic inlet and pelvic outlet—are both normally much wider in the female than in the male. This effect is partly because the angle at the front of the female pelvis where the two pubic bones join is wider than in the male. Such an arrangement allows more room for a fetus's head to move through during childbirth.

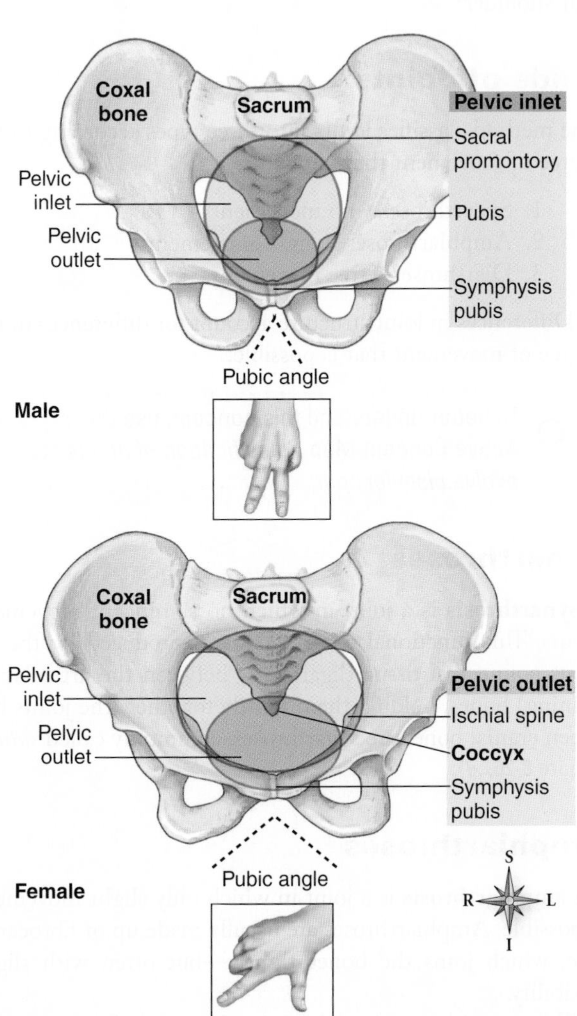

FIGURE 8-25 Comparison of the male and female pelvis. Notice the narrower width of the male pelvis, giving it a more funnel-like shape than the female pelvis. The pubic angle is narrower in males than females.

Find the major structures of the skeletal system in the *Clear View of the Human Body* (follows p. 8) and compare the male and female structures, which are illustrated side-by-side.

Age Differences

As we learned earlier in the chapter, during childhood and adolescence the bones of the skeleton enlarge and become more ossified. The human skeleton is considered to reach its mature state around age 25. From then until about age 50 or so, the skeleton is in a state of active maintenance, continually remodeling—dissolving and rebuilding—bone tissue.

After age 50, the density of bone often decreases slowly because of a shift in the remodeling activity. An elderly person's skeleton often weighs much less than it did when they were in their 30s.

Environmental Factors

Among the many factors that can cause variations in one's skeleton is nutrition. Without enough calcium and vitamin D, especially during the developmental years, the skeleton may not reach its full potential of growth or it may show signs of early degeneration.

Load-bearing, or mechanical stress, of using the skeleton affects how bone tissue is remodeled. Exercise has a profound effect on the skeleton. An active older person can reverse much or all of the bone loss associated with aging. Scientists can sometimes tell a person's occupation by which bones—or which parts of bones—are more developed. For example, a person who works with heavy loads on their right arm every day will have denser bones in the right arm and shoulder than in the left arm.

Breaks and repairs similarly cause individual variations in the skeleton.

CLINICAL APPLICATION

EPIPHYSEAL AND AVULSION FRACTURES

The point of articulation between the epiphysis and diaphysis of a growing long bone is susceptible to injury if overstressed, especially in the young child or preadolescent athlete. In these individuals the epiphyseal plate can be separated from the diaphysis or epiphysis, causing an epiphyseal fracture. This x-ray study shows such a fracture in a young boy. Without successful treatment, an **epiphyseal fracture** may inhibit normal growth. Stunted bone growth in turn may cause the affected limb to be shorter than the normal limb.

In addition to epiphyseal fractures, violent contraction or overstretching of a muscle in skeletally immature individuals also can cause a fragment of bone under the point of attachment to break away from the bone as a whole. The result is called an **avulsion fracture** (see p. 205).

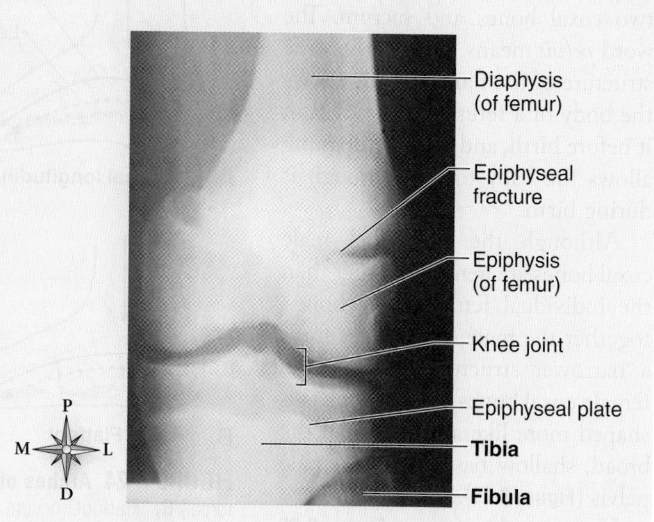

To see images of skeletal variations, check out the article *Skeletal Variations* at Connect It! at *evolve.elsevier.com.*

8

QUICK CHECK

1. Name some of the bones of the upper and lower extremities.
2. Identify the three bones that fuse together to form the coxal bone of an adult.
3. Name the arches that form the foot and explain their significance.
4. How does the female pelvis differ from the male pelvis?

Joints

Articulation of Bones

The term *joint* is borrowed from carpentry, where it refers to the structure formed when pieces of wood are joined together. In anatomy, a joint is the structure formed when bones join together. Joints are also called **articulations**—a term based on the word part *arthro,* which means "joint."

Every bone in the body, except one, connects to at least one other bone. In other words, every bone but one forms a joint with some other bone. The exception is the hyoid bone in the neck, to which the tongue anchors. Most of us probably never think much about our joints unless something goes wrong with them, and they do not function properly. Then their tremendous importance becomes painfully clear.

Joints hold our bones together securely and at the same time make it possible for movement to occur between the bones—between most of them, that is. Without joints we could not move our arms, legs, or any other of our body parts. Our bodies would, in short, be rigid, immobile hulks.

Try, for example, to move your arm at your shoulder joint in as many directions as you can. Try to do the same thing at your elbow joint. Now examine the shape of the bones at each of these joints on a skeleton or in **Figures 8-9** and **8-20**. Looking at the anatomy, do you understand why you cannot move your arm at your elbow in nearly as many directions as you can at your shoulder?

Kinds of Joints

One method classifies joints into three types according to the degree of movement they allow:

1. Synarthroses—no movement
2. Amphiarthroses—slight movement
3. Diarthroses—free movement

Differences in joint structure account for differences in the degree of movement that is possible.

To better understand this concept, use the Active Concept Map *Classification of Joints* at *evolve.elsevier.com.*

Synarthroses

A **synarthrosis** is a joint in which no significant movement occurs. This functional characteristic is produced by the fibrous connective tissue (ligaments) between the articulating (joining) bones, holding them tightly together. The joints between cranial bones are synarthroses, commonly called *sutures* (**Figure 8-26,** *A*).

Amphiarthroses

An **amphiarthrosis** is a joint in which only slight movement is possible. Amphiarthroses are usually made up of fibrocartilage, which joins the bones tightly—but often with slight flexibility.

The symphysis pubis, the joint between the two pubic bones, is an amphiarthrosis (**Figure 8-26,** *B*). It normally only flexes late in pregnancy when movement of the pelvic girdle is helpful during delivery of an infant.

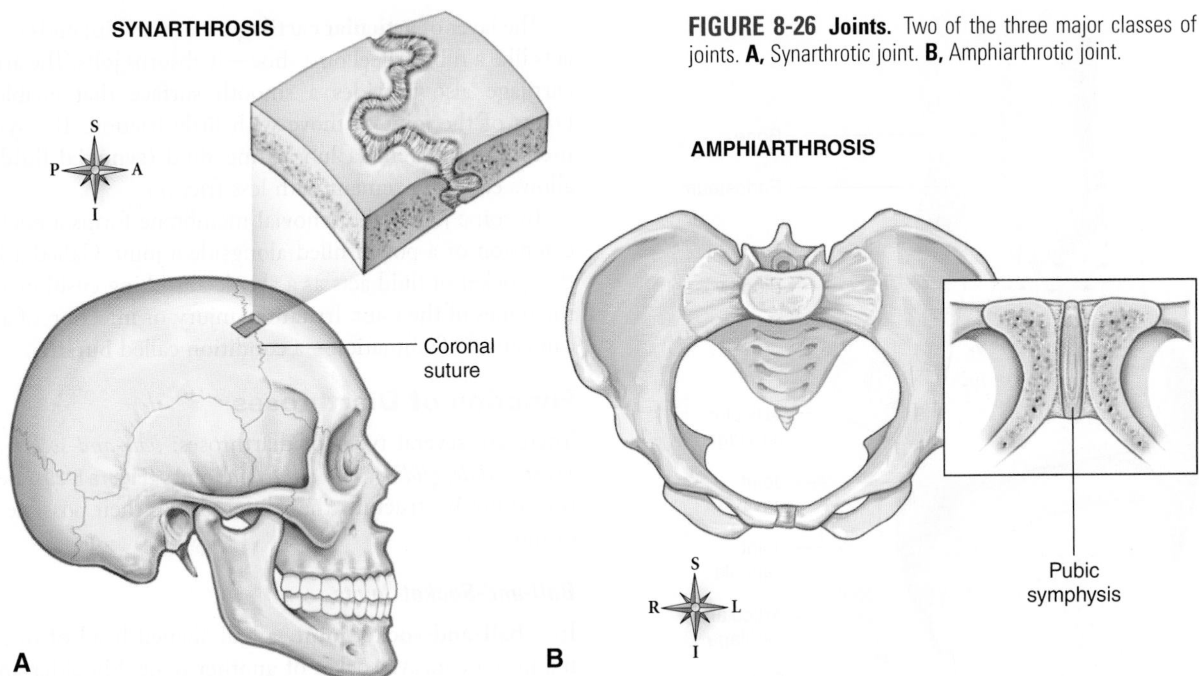

FIGURE 8-26 Joints. Two of the three major classes of joints. **A,** Synarthrotic joint. **B,** Amphiarthrotic joint.

Another example is the *vertebral disk* that connects each vertebral body to the next in the spinal column. Damage to a disk caused by the pressure of sudden exertion or injury may push its wall into the spinal canal (Figure 8-27). Severe pain may result if the disk presses on the spinal cord or spinal nerve. Popularly known as a *slipped disk,* this condition is known to health professionals as a **herniated disk.**

Diarthroses

The vast majority of our joints are diarthroses. Such joints allow considerable movement, sometimes in many directions and sometimes in only one or two directions.

Structure of Diarthroses

Diarthroses (freely movable joints) are made alike in certain ways. All have a joint capsule, a joint cavity, and a layer of hyaline cartilage (articular cartilage) over the ends of two joining bones (Figure 8-28).

The **joint capsule** is made of the body's strongest and toughest material, fibrous connective tissue, and is lined with a smooth, slippery synovial membrane. The capsule fits over the ends of the two bones somewhat like a sleeve. Because it attaches firmly to the shaft of each bone to form its covering (*periosteum*), the joint capsule holds the bones securely together but at the same time permits movement at the joint.

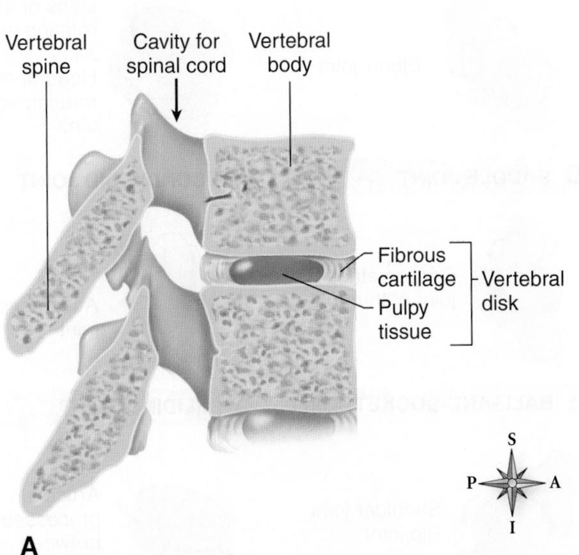

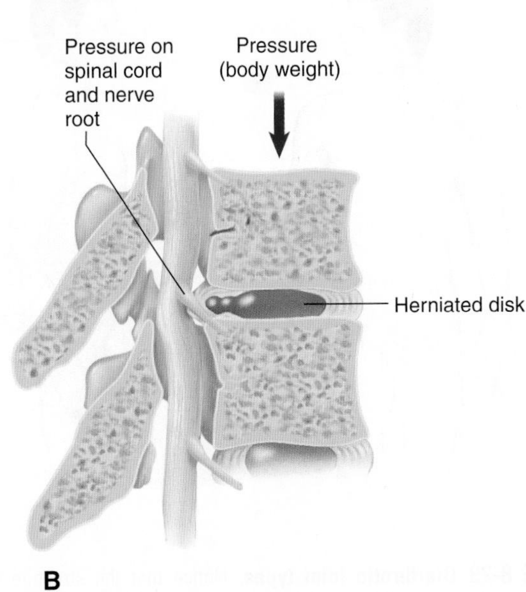

FIGURE 8-27 Herniated disk. Sagittal section of vertebrae showing **(A)** normal and **(B)** herniated disks.

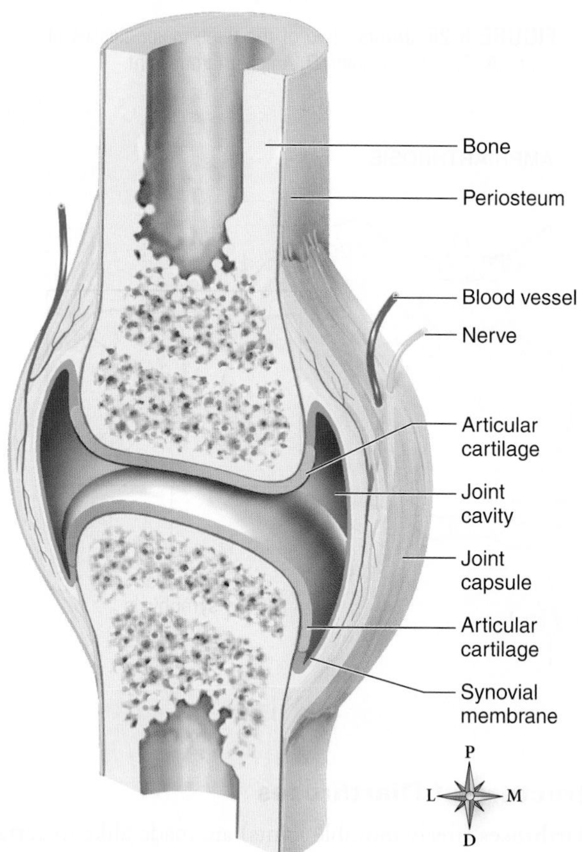

FIGURE 8-28 Diarthrotic joint structure. Each diarthrosis has a joint capsule, a joint cavity, and a layer of cartilage over the ends of the joined bones.

The structure of the joint capsule, in other words, helps make possible the joint's function.

Ligaments (cords or bands made of the same strong fibrous connective tissue as the joint capsule) also grow out of the periosteum and lash the two bones together even more firmly.

The layer of **articular cartilage** over the joint ends of bones acts like a rubber heel on a shoe—it absorbs jolts. The articular cartilage also provides a smooth surface that enables the bones of the joint to move with little friction. The **synovial membrane** secretes a lubricating fluid **(synovial fluid)** that allows easier movement with less friction.

In some joints, the synovial membrane forms a pocketlike extension or a pouch filled alongside a joint. Called a **bursa,** this pocket of fluid acts as a shock-absorbing cushion around the bones of the joint. Irritation, injury, or infection of a bursa can cause inflammation—a condition called **bursitis.**

Function of Diarthroses

There are several types of diarthroses: *ball-and-socket, hinge, pivot, saddle, gliding, and condyloid* joints (**Figure 8-29**). Because they differ in structure, they differ also in their possible range of movement.

Ball-and-Socket Joints

In a **ball-and-socket joint,** a ball-shaped head of one bone fits into a concave socket of another bone. Shoulder and hip joints, for example, are ball-and-socket joints. Of all the joints in our bodies, these permit the widest range of movements. Think for a moment about how many ways you can move your arms. You can move them forward, backward, away from the sides of your body, and back down to your sides. You can also move them around so as to make a circle in the air with your hands.

Hinge Joints

Hinge joints, like the hinges on a door, allow movements in only two directions, namely, flexion and extension. **Flexion** is bending a joint; **extension** is straightening it out (**Table 8-7**). Elbow and knee joints and the joints in the fingers are hinge joints.

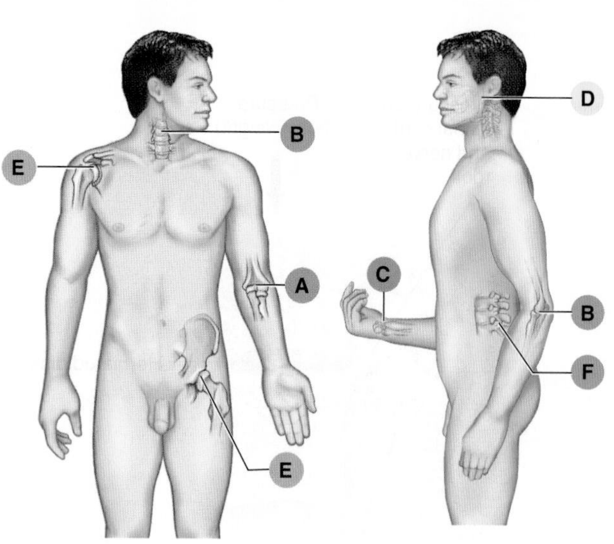

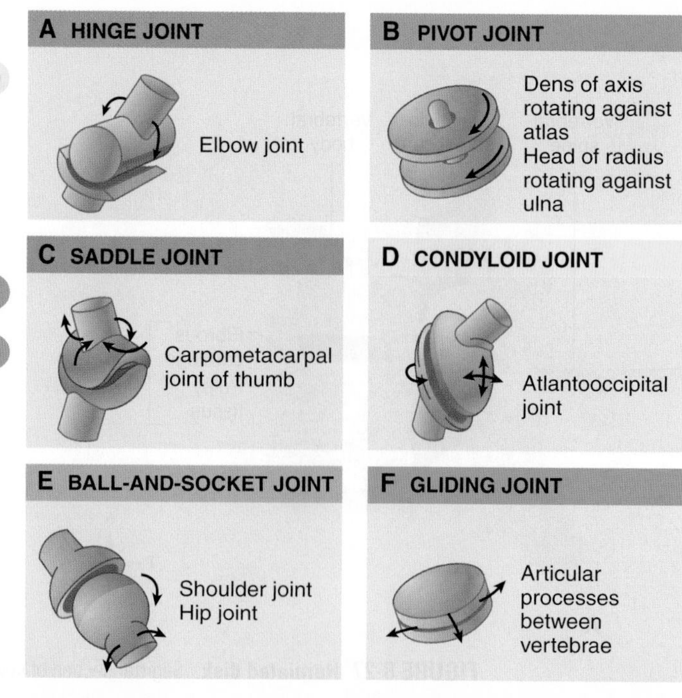

FIGURE 8-29 Diarthrotic joint types. Notice that the structure of each type indicates its function (movement). The mechanical diagrams represent the type of action at the highlighted anatomical joints.

TABLE 8-7 Types of Joint Movements

MOVEMENT	EXAMPLE	MOVEMENT	EXAMPLE
Flexion (to *flex* a joint) Reduces the angle of the joint, as in bending the elbow	Flexion	**Extension** (to *extend* a joint) Increases the angle of a joint, as in straightening a bent elbow	Extension
Abduction (to *abduct* a joint) Increases the angle of a joint to move a part away from the midline, as in moving the arm to the side and away from the body	Abduction	**Adduction** (to *adduct* a joint) Decreases the angle of a joint to move a part toward the midline, as in moving the arm in and down from the side	Adduction
Rotation (to *rotate* a joint) Spins one bone relative to another, as in rotating the head at the neck joint	Rotation	**Circumduction** (to *circumduct* a joint) Moves the distal end of a bone in a circle, while circumducting a joint, keeping the proximal end relatively stable, as in moving the arm in a circle and thus circumducting the shoulder joint	Circumduction

8

Pivot Joints

Pivot joints are those in which a small projection of one bone pivots in an arch of another bone. For example, recall from Figure 8-16 (on p. 187) that a projection of the *axis* (second cervical vertebra) is a point around which an arch of the *atlas* (first cervical vertebra) can pivot. This pivoting motion is referred to as **rotation**. Because the skull rests on the atlas, this action *rotates* the head.

Saddle Joints

Only one pair of **saddle joints** exists in the body—between the metacarpal bone of each thumb and a carpal bone of the wrist (the name of this carpal bone is the *trapezium*). Because the articulating surfaces of these bones are saddle-shaped, they make possible the human thumb's great mobility, a mobility no animal's thumb possesses. We can **flex, extend, abduct, adduct,** and **circumduct** our thumbs, and most important of all, we can move our thumbs to touch the tip of any one of our fingers. (This movement is called *opposing the thumb to the fingers.*)

Without the saddle joints, we could not do simple acts such as picking up a pin or grasping a pencil between thumb and forefinger.

Gliding Joints

Gliding joints are the least movable diarthrotic joints. Their flat articulating surfaces allow limited gliding movements, such as that at the superior and inferior articulating processes between successive vertebrae.

Condyloid Joints

Condyloid joints are those in which a condyle (an oval projection) fits into an elliptical socket. An example is the fit of the distal end of the radius into depressions in the carpal bones.

To learn more about types of joint movement, go to AnimationDirect online at *evolve.elsevier.com.*

CLINICAL APPLICATION

TOTAL HIP REPLACEMENT

Because total hip replacement (THR) is one of the more common orthopedic operations performed on older persons (more than 300,000 procedures per year in the United States), home health care professionals often are engaged to work with patients recovering from THR surgery. It is a type of **arthroplasty**—that is, a procedure that partially or totally replaces a diseased joint with an artificial device *(prosthesis)*.

The THR procedure involves replacement of the femoral head by a metal prosthesis and the acetabular socket by a polyethylene cup. The prosthesis is usually coated with a porous material that allows natural growth of bone to mesh with the artificial material. Such meshing of tissue and prosthesis ensures stability of the parts without the loosening that the use of glues in the past often allowed.

First introduced in 1953, THR technique has advanced to the state that it is now one of the most successful surgical procedures in adults.

As patients recover at home after THR surgery, they can expect to progress through normal surgical healing and recovery, which includes stabilization of the prosthesis as new tissue grows into the porous coating that was applied to the prosthesis. Typically, THR patients also can expect almost immediate pain relief and to regain some previously lost function in the affected hip—including improved weight-bearing and walking movements.

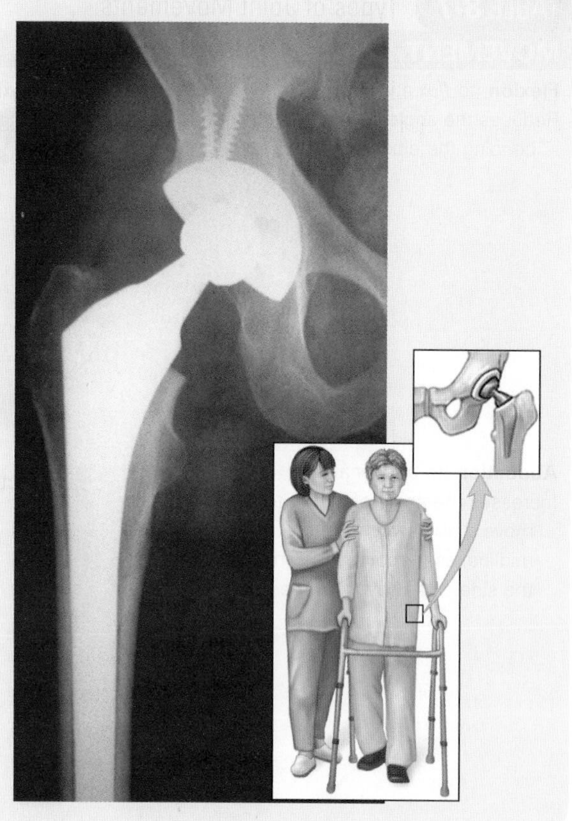

QUICK CHECK

1. Identify the three major types of joints in the skeleton. Name an example of each.
2. What membrane in a diarthrotic joint provides lubrication for movement?
3. What is a ligament?
4. Name three examples of a hinge joint and describe the movement a hinge joint allows.

Skeletal Disorders

Tumors

Two of the most serious skeletal disorders involve malignant tumors of bone tissue and cartilage.

Bone Tumors

The most common and devastating malignant neoplasm of bone is called **osteosarcoma.** Twenty-five years ago nearly all patients diagnosed with this disease died within 3 years. Although still considered a very aggressive and destructive type of cancer, earlier diagnosis and newer treatment options are increasing survival rates and decreasing the need for immediate and complete amputation of affected limbs.

Osteosarcomas occur most often in the distal femur (**Figure 8-30**, *A*) and proximal areas of the tibia and humerus. Nearly twice as many males are affected as females and most cases occur between 20 and 40 years of age.

These tumors are characterized by severe, unrelenting pain. Treatment involves surgical removal of the tumor and both presurgical and postsurgical chemotherapy.

Cartilage Tumors

Chondrosarcoma is cancer of skeletal hyaline cartilage tissue and is the second most common type of cancer affecting bones.

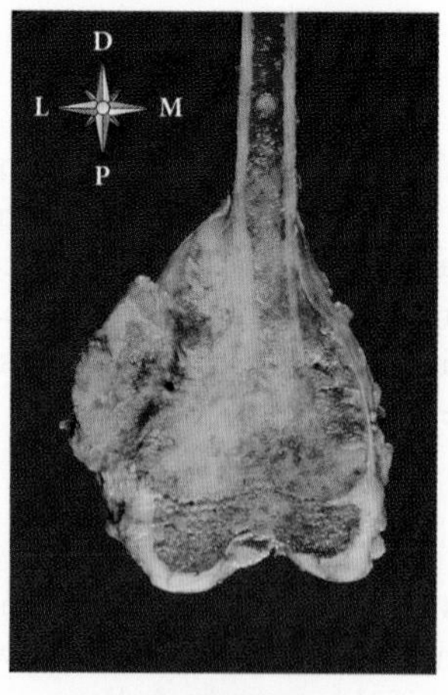

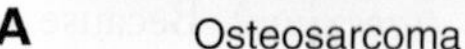

A Osteosarcoma

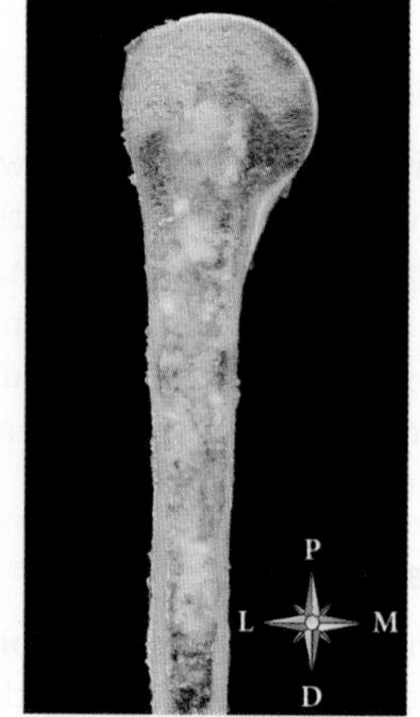

B Chondrosarcoma

FIGURE 8-30 Tumors. Surgical specimens sectioned longitudinally. **A,** Osteosarcoma of distal femur. The tumor has broken out of the medullary cavity and is growing on the surface of the bone. **B,** Chondrosarcoma of proximal humerus. Note the glistening appearance of the hyaline cartilage tumor in the medullary cavity.

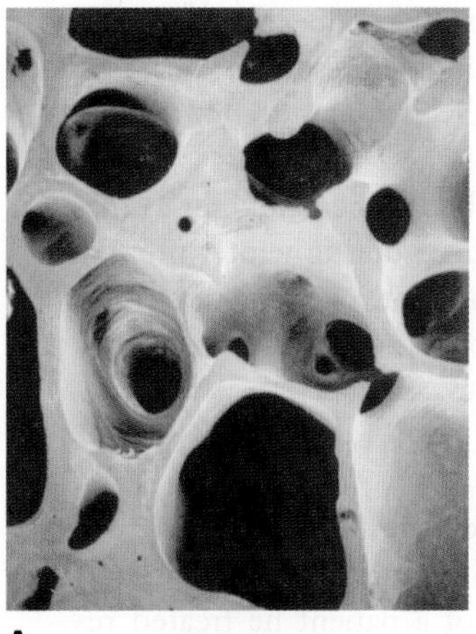

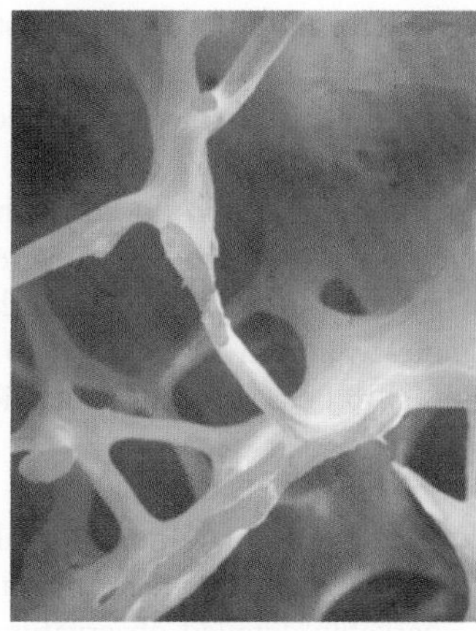

FIGURE 8-31 Osteoporosis. Scanning electron micrograph (SEM) of **(A)** normal bone and **(B)** bone with osteoporosis. Note the loss of trabeculae and appearance of enlarged spaces in the osteoporotic bone.

The most common tumor sites involve the medullary cavity of the humerus, femur, ribs, and pelvic bones (**Figure 8-30**, *B*). Chondrosarcomas occur most often in adults between 40 and 70 years of age. Incidence is slightly higher in males.

Pain is a common but not a universal symptom and when present is generally less severe than in osteosarcoma. Treatment is surgical removal of the lesion. Chemotherapy is not effective in treating chondrosarcoma.

Tumors and cancers of bone and cartilage are often difficult to detect early. Medical imaging can help—check out the article *Bone Scans* at Connect It! at *evolve.elsevier.com.*

Metabolic Bone Diseases

Osteoporosis

Osteoporosis is the name of the disorder in which bones lose minerals and become less dense, as evidenced on scanning electron micrograph studies (**Figure 8-31**). It is one of the most common and serious bone diseases.

Although the cause remains unknown, genetics play a part in the etiology, as do low estrogen levels and postmenopausal status in women. Certain drugs, a diet low in calcium-containing foods, lack of weight-bearing exercise, and smoking are also risk factors. Osteoporosis occurs most often in elderly white women.

The disease is characterized by excessive loss of calcified bone matrix. A reduction in the number of branching trabeculae in spongy bone is particularly noticeable. The name *osteoporosis* means "condition of bone pores" referring to the holes or pores formed as bone tissue is lost. Compare the appearance of normal and osteoporotic bone specimens in **Figure 8-31**.

A progressive increase in bone porosity causes the bones in people with osteoporosis to become brittle and easily broken. Fractures may be completely "spontaneous" or occur with even minor trauma or during routine activities. The most common fracture sites are the wrists, hips, and vertebrae. Compression fractures of the vertebrae result in a shortened stature and the classic kyphosis of the thoracic spine called **"dowager's hump"** in elderly women suffering from the disease.

Treatment or preventive measures may include drug therapy, weight-bearing exercise, and dietary supplements of calcium and vitamin D to replace deficiencies or to offset intestinal malabsorption.

Rickets and Osteomalacia

Rickets in young children and **osteomalacia** in adults are metabolic skeletal diseases that affect significant numbers of individuals worldwide. Both diseases are characterized by demineralization, or loss of minerals, from bone related to vitamin D deficiency. Vitamin D helps the intestines absorb calcium from the diet. The loss of minerals is coupled to an increased production of unmineralized matrix.

Rickets involves demineralization of developing bones in infants and young children before skeletal maturity. In osteomalacia, mineral content is lost from bones that have already matured.

In rickets, the lack of rigidity caused by demineralization of developing bones results in gross skeletal changes including a classic "bowing of the legs" symptom (**Figure 8-32**). The

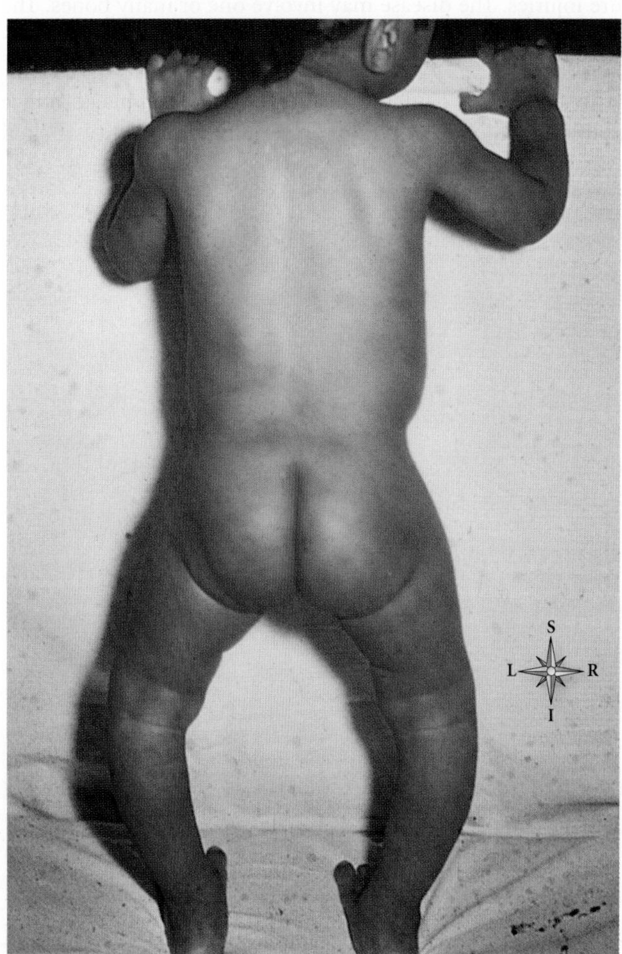

FIGURE 8-32 Rickets. Bowing of legs in this toddler is due to poorly mineralized bones

demineralization of bones in osteomalacia does not generally affect overall skeletal contours but does result in increased susceptibility to fractures, especially in the vertebral bodies and femoral necks.

Vitamin D is produced by the body when sunlight strikes the skin and its production is reduced during winter months where people wear warmer clothing that covers more of their bodies. Vitamin D is added to some foods (milk and some juices) to help reduce this deficiency.

Paget Disease

Paget disease of bone, also called **osteitis deformans,** was first described by British surgeon Sir James Paget in 1882. His remarkably detailed observations of a patient he treated repeatedly over a 20-year period for a deforming bone disease are considered classic in the annals of surgery.

The disease Paget described is characterized by localized, intermittent, and uncontrolled episodes of almost "frenzied" osteoclastic (bone resorbing) and osteoblastic (bone forming) activity. The faulty remodeling process results in bones that are deformed, unstable, and easily fractured (**Figure 8-33**).

Paget disease is often asymptomatic early in its course but becomes painful as the weak but often thickened areas of defective bone cause deformity, arthritic symptoms, and fracture injuries. The disease may involve one or many bones. The spine, skull, pelvis, and long bones of the extremities are common sites.

In addition to pain, the location of the lesion may produce additional unique symptoms. For example, deformity of skull

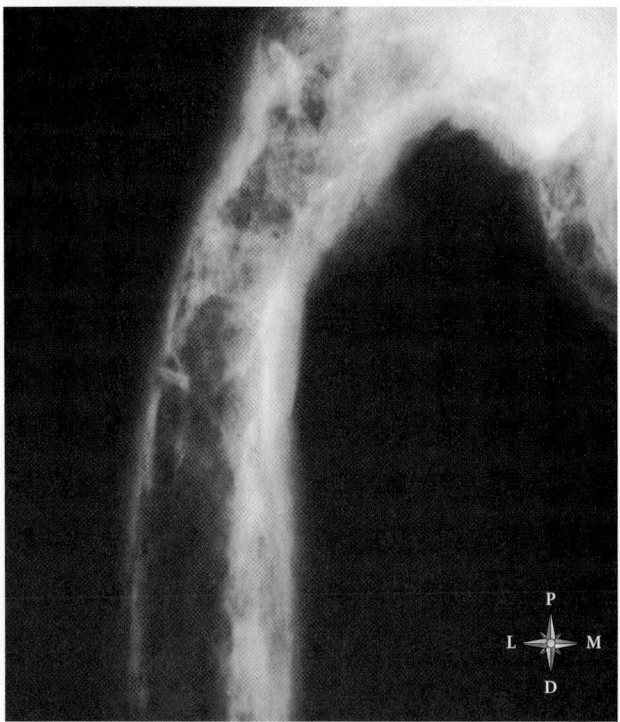

FIGURE 8-33 Paget disease (osteitis deformans). The fuzzy, disorganized appearance of this femur results from weakened and irregular cancellous bone overgrowth.

bones may compress cranial nerves causing deafness, blindness, headaches, and facial paralysis. Unfortunately, areas of diseased bone develop into osteosarcomas in about 1% of affected individuals.

Paget disease affects about 3% of people over 50 years of age. The disease has a genetic tendency and may be triggered by viral infections. Disease treatment includes mostly pain control and drugs that improve the strength of the bone.

Osteogenesis Imperfecta

Osteogenesis imperfecta is a genetic disease that can affect 1 in 30,000 births and is also called *brittle-bone disease.* The bones are brittle as the result of a lack of production of the fibrous matrix of bone.

The fibrous material (mostly collagen) in bone gives it the ability to withstand twisting and compression forces without breaking. The same concept applies to the practice of placing metal rods inside of concrete in bridge or road construction. A bridge without metal rods would not be able to withstand the weight and vibration of traffic. Bones without organic fibrous material are thus very fragile and easily fractured.

The diagnosis of osteogenesis imperfecta usually follows increased fracture rates and a blood test for the enzyme alkaline phosphatase. Treatment may include splinting of the bone to reduce fracture during growth and treatment with drugs that decrease the activity of cells that break down bone (**Figure 8-34**).

Bone Infection

Osteomyelitis is the general name for bacterial infections of bone and marrow tissue. *Staphylococcus* bacteria are the most common pathogens found in this condition. They may reach bone via the bloodstream or from an adjacent soft tissue infection such as an abscess, or be introduced directly into the bone as a result of open fractures, penetrating wounds, or by flawed surgical aseptic techniques.

On rare occasions, infections also may occur after insertion of infected donor tissues into bones or joints or by contaminated joint prostheses. Besides bacterial infections, bone tissue is also susceptible to damage by viruses, fungi, and other pathogens.

Any bone infection is difficult to treat because of the density of the bone and the slowness of the healing process compared with that of other tissues. Osteomyelitis produces persistent and severe pain, muscle spasm, swelling, and fever. Pus collecting in the confined space of the bone increases pressure, decreases blood flow, and in time will cause necrosis or death of bone tissue. **Figure 8-35** shows a segment of infected bone in a severe case of chronic osteomyelitis.

Severe cases of osteomyelitis are often treated with intensive and repeated drug therapy including **parenteral**—that is, **intravenous (IV)**—administration of antibiotics. Difficult cases of osteomyelitis may become chronic and reappear months or years after an assumed cure.

QUICK CHECK

1. Name two types of malignant tumors that affect bones.
2. What metabolic bone disease is characterized by kyphosis of the thoracic spine called "dowager's hump"?
3. Identify the metabolic bone disease characterized by loss of bone minerals and vitamin D deficiency in children. What is the disease called in adults?
4. What is the general name for bacterial infections of bone and marrow tissue?

Bone Fractures

Excessive mechanical stress on bones can result in breaks or fractures. Sometimes bone cancer or metabolic bone disorders weaken a bone to the point that it fractures with very little stress. **Figure 8-36** shows some of the major types of fractures summarized in the following paragraphs.

Open fractures, or *compound fractures,* in which bone pierces the skin, invite the possibility of infection or osteomyelitis. **Closed fractures,** also known as *simple fractures,* do not pierce the skin and so do not pose an immediate danger of bone infection.

In **complete fractures** the bone fragments separate completely, whereas in incomplete fractures the bone fragments are still partially joined. **Incomplete fractures** in which a bone is bent but broken only on the outer curve of the bend are often called *greenstick fractures.* Greenstick fractures, common in children, usually heal rapidly.

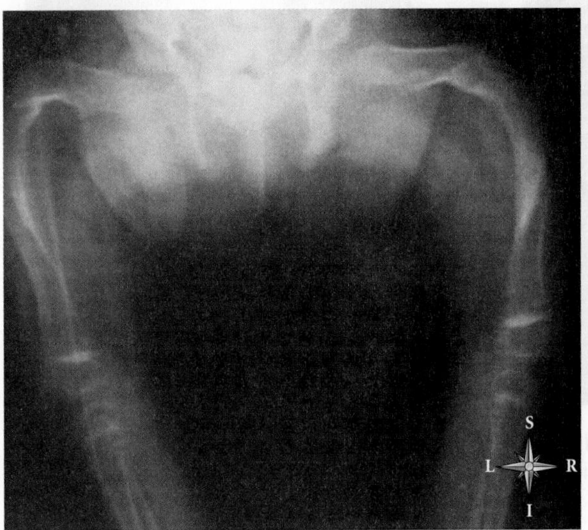

A

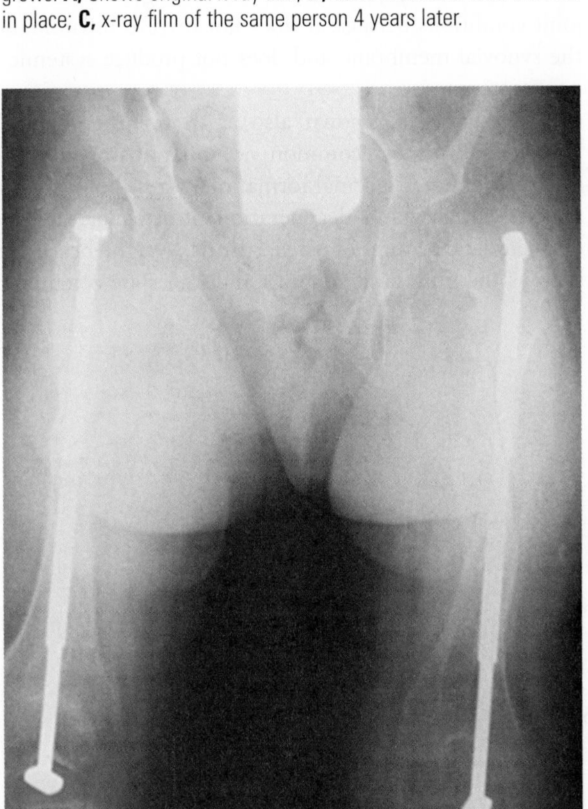
B

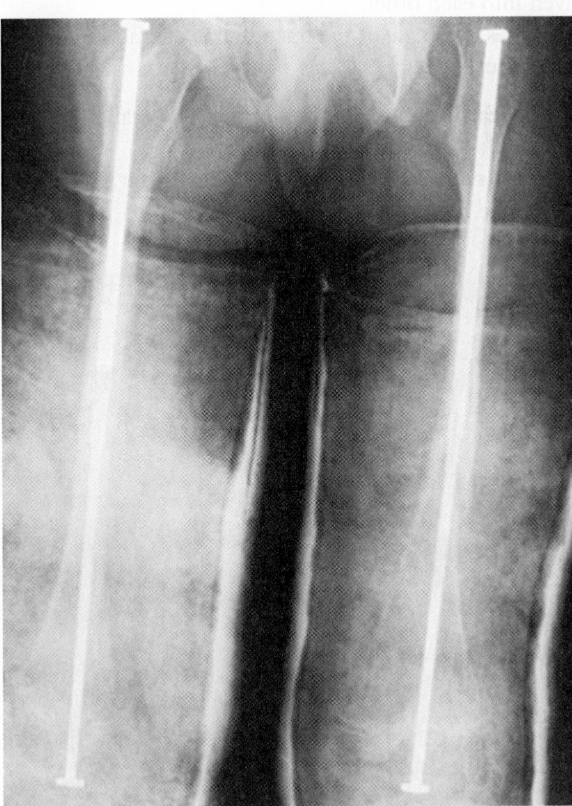
C

FIGURE 8-34 Osteogenesis imperfecta. These x-ray studies show the progression of treatment of femurs in the same individual with *brittle-bone disease.* A telescoping rod is inserted into each medullary cavity to provide rigidity, lengthening as the individual grows. **A,** Shows original x-ray film; **B,** same individual with the rods in place; **C,** x-ray film of the same person 4 years later.

8

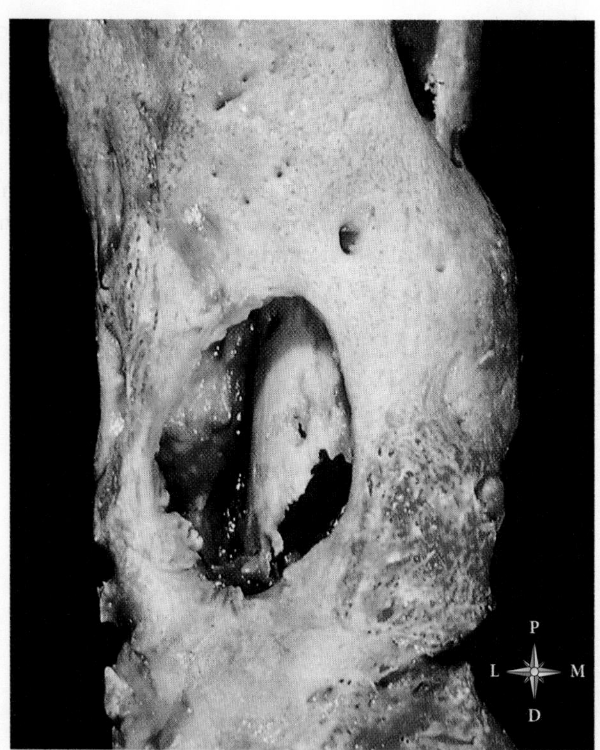

FIGURE 8-35 Osteomyelitis. Segment of femur from a patient with chronic osteomyelitis showing extensive damage from infection.

Comminuted fractures are breaks that produce many fragments. **Impacted fractures** occur when bone fragments are driven into each other.

Sometimes the angle of the fracture line or crack is used in labeling fracture types:

- **Linear fracture**—fracture line is parallel to the bone's long axis.
- **Transverse fracture**—fracture line is at a right angle to the bone's long axis.
- **Oblique fracture**—fracture line is diagonal to the bone's long axis. If the oblique fracture line seems to spiral around a bone like the stripe on a candy cane, the fracture may be called a *spiral fracture.*

After a fracture occurs, a bone usually bleeds, becomes inflamed, and then forms a bony framework called a **callus** around the injury (Figure 8-37). The callus tissue stabilizes the bone fragments and thus aids in the long healing and remodeling process.

X-rays are commonly used in diagnosing bone fractures. To see examples and get a new visual perspective on skeletal anatomy, check out the article *Skeletal Radiography* at Connect It! at *evolve.elsevier.com.*

To learn more about bone fracture and repair, go to AnimationDirect online at *evolve.elsevier.com.*

Joint Disorders

Joint disorders can be classified as noninflammatory joint disease or inflammatory joint disease.

Noninflammatory Joint Disease

Noninflammatory joint disease is distinguished from other joint conditions because it does not involve inflammation of the synovial membrane and does not produce systemic signs or symptoms.

Osteoarthritis, known also as *degenerative joint disease (DJD),* is the most common noninflammatory disorder of movable joints. Abnormal formation of new bone *(bone spurs)* at joint surfaces and degeneration of articular cartilage are characteristic features of osteoarthritis. Weight-bearing joints, such as the hips, lumbar spine, and knees are often involved.

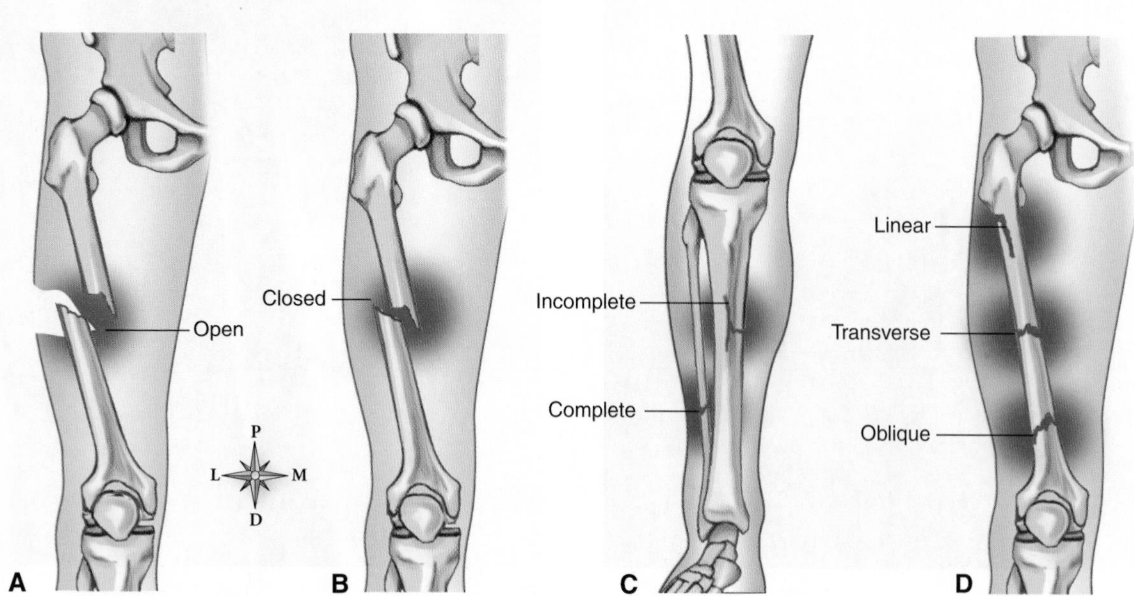

FIGURE 8-36 Bone fractures. A, Open. **B,** Closed. **C,** Incomplete and complete. **D,** Linear, transverse, and oblique.

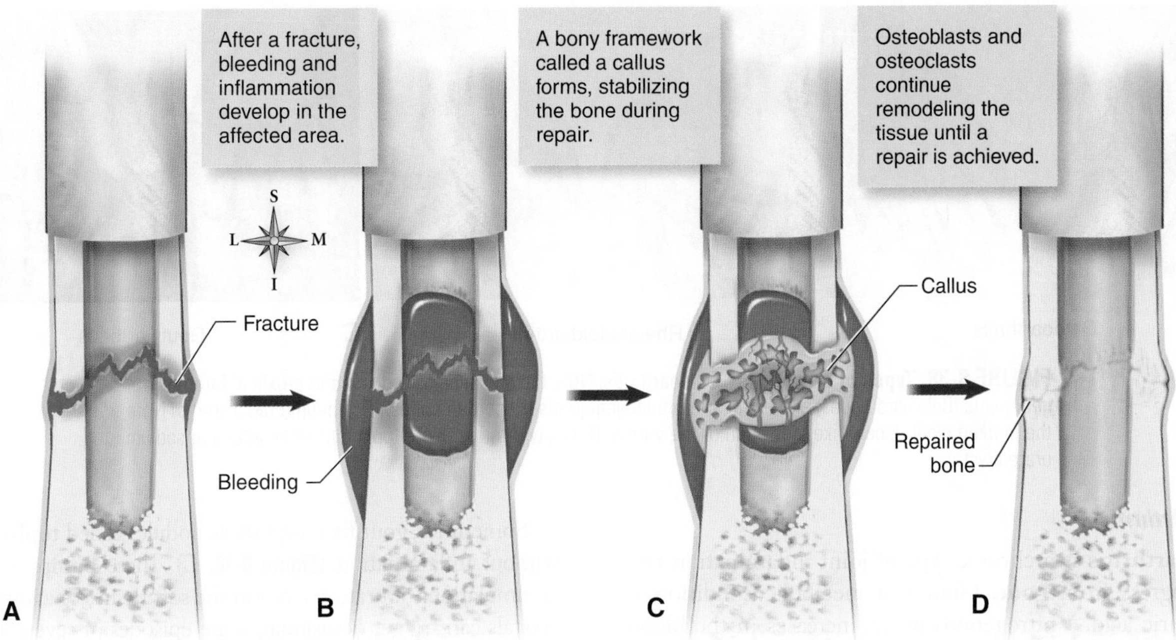

FIGURE 8-37 Bone repair. After a fracture **(A)**, there is bleeding and inflammation around the affected area **(B)**. Special tissue forms a bony framework called a *callus* **(C)** that stabilizes the bone until the repair is complete **(D)**.

Localized tenderness over affected joints, morning stiffness, and pain on movement are frequent symptoms.

Figure 8-38, *A*, shows another common sign of osteoarthritis at the **interphalangeal joints** of the fingers—the joints between phalanges. The fingers of individuals with this form of DJD often show bony bumps (nodes) at both the proximal interphalangeal joints (**Bouchard nodes**) and the distal interphalangeal joints (**Heberden nodes**).

The etiology of most cases of DJD is unknown, but advanced age, joint damage caused by "wear and tear," and obesity are known risk factors. Advanced cases of osteoarthritis are the most common cause for partial or total hip and knee replacements.

Traumatic injury is often the cause of noninflammatory joint problems. **Dislocation** occurs when the articular surfaces of bones forming the joint are no longer in proper contact with each other. A **subluxation** is a partial or minor dislocation in which the bones are slightly misaligned. A **sprain** is an acute injury to the ligaments around a joint. A common cause of sprains is a twisting or wrenching movement often associated with "whiplash"-type injuries.

The term **strain** is used to describe an injury involving the **musculotendinous unit** and may involve the muscle, the tendon, and the junction between the two, as well as their attachments to bone. Most strains occur in muscle tissue, often because of overstretching, or violent types of muscle contraction (see Chapter 9, p. 235).

However, the portion of the musculotendinous unit injured depends on which component is weakest. In preadolescent athletes with incompletely ossified bones, the muscle component of the unit or the point of attachment of the muscle to the bone may be stronger than the developing bone itself or the union between its epiphysis and diaphysis. In these cases, violent muscle contractions can cause an **avulsion fracture** (often near a joint), in which a piece of bone is pulled free, or an **epiphyseal fracture** between the epiphysis and diaphysis of the involved bone (see box on p. 196).

Inflammatory Joint Disease (Arthritis)

Arthritis is a general name for many different inflammatory joint diseases. Arthritis can be caused by a variety of factors, including infection, injury, genetic factors, and autoimmunity. We now explore a brief list of major types of arthritis.

Rheumatoid arthritis

Believed to be a type of autoimmune disease, **rheumatoid arthritis (RA),** involves chronic inflammation of connective tissues. It begins in the synovial membrane and spreads to cartilage and other tissues, often causing severe crippling.

A characteristic deformity of the hands in rheumatoid arthritis is **ulnar deviation** of the fingers. As you can see in Figure 8-38, *B* ulnar deviation involves an "elbow-like" bending of the finger joints.

The autoimmune nature of the disease may cause damage to many body organs, such as the blood vessels, eyes, lungs, and heart. Because it is a systemic disease, fever, anemia, weight loss, and profound fatigue are common.

Juvenile rheumatoid arthritis (JRA) is usually more severe than the adult form but involves similar deterioration and deformity of joints. The joint inflammatory process often destroys growth of cartilage, and growth of long bones is arrested. This form begins during childhood and is more common in girls.

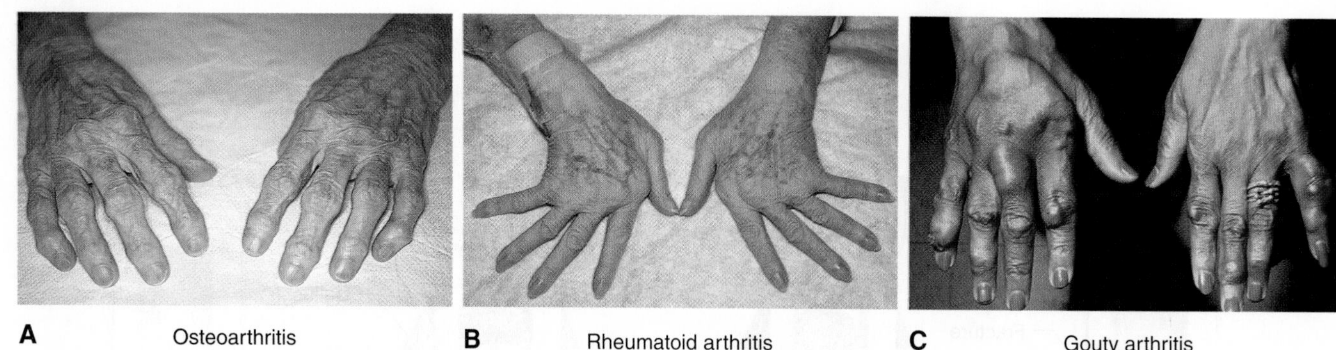

FIGURE 8-38 Types of arthritis. A, Osteoarthritis. Note the presence of nodes in the proximal interphalangeal joints (Bouchard nodes) and distal interphalangeal joints (Heberden nodes). **B,** Rheumatoid arthritis. Note the marked ulnar (elbow-like) deviation of the wrists. **C,** Gouty arthritis. Note tophi (stones) containing sodium urate crystals.

Gouty Arthritis

Gouty arthritis is a chronic type of joint inflammation that can progress from **gout.** Gout is a metabolic condition in which uric acid, a nitrogenous waste, increases in the blood. Excess uric acid is deposited as sodium urate crystals in distal joints and other tissues, causing inflammation.

Sometimes, gout first appears as nodules called **tophi** with or without inflammation (Figure 8-38, *C*). These bumps are gritty accumulations of urate crystals in the soft tissues of a joint. These crystals can trigger occasional, acute episodes of severe arthritis. However, without treatment it may progress to the chronic inflammation and tissue damage characteristic of gouty arthritis.

HEALTH AND WELL-BEING

KNEE JOINT INJURY

The knee is the largest movable joint in the body—but also one of the most vulnerable to injury. Because the knee is very moblie, but also often subjected to sudden, strong forces during athletic activity, knee injuries are among the most common type of athletic injury.

Sometimes, the concave disks of articular cartilage called **menisci** on the tibia tear when the knee twists while bearing weight. The ligaments holding the tibia and femur together can also be injured in this way.

Figure A shows tears in the lateral and medial ligaments outside the joint cavity, as well as tears in the crossed **cruciate ligaments** inside the joint. Knee injuries may also occur when a weight-bearing knee is hit by another person or a moving object.

Women have a higher risk of knee injuries than men. There have been several proposed causes, including the following:

- The wider pelvis of a woman results in the femur angling toward the knee at a greater angle, called the *Q angle* (Figure B). This results in the pull on the patella (kneecap) by anterior lateral muscles of the thigh, which pulls the patella out of alignment and may weaken the knee.
- The *intercondylar notch* of the femur is narrow in females compared to the males. This narrow notch may rub and weaken the *anterior cruciate ligament (ACL)* that stabilizes the knee.
- An increase in the female hormone estrogen may cause the ligaments to be more flexible and more prone to overstretching and rupturing. This may increase flexibility of ligaments when estrogen levels are higher, as occurs during ovulation or pregnancy.

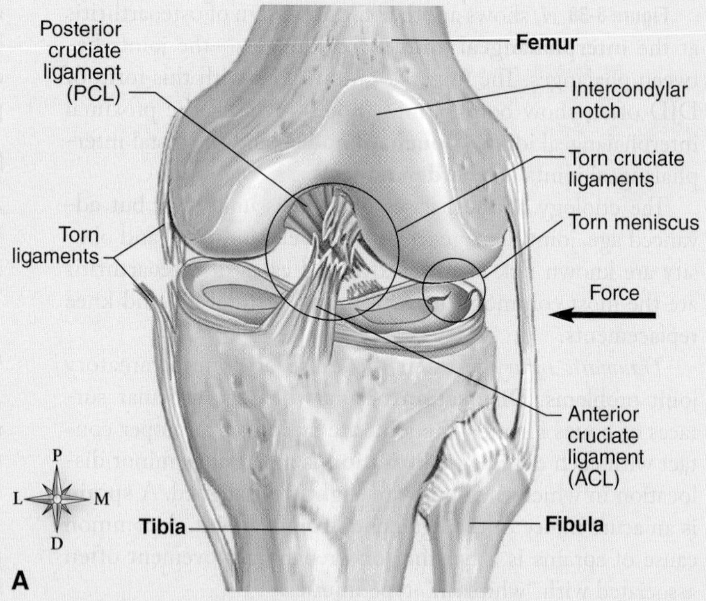

- Females tend to have stronger *quadriceps* (anterior thigh muscles) than hamstrings (posterior thigh muscles). The hamstring muscles keep the tibia in place in the knee when jumping or making sudden stops.

In order to avoid injury to the knee, it is recommended that women

- Wear shoes that prevent the twisting of the foot and knee

Acute episodes of gout are often successfully treated with *nonsteroidal anti-inflammatory drugs (NSAIDs)* such as *ibuprofen* or *naproxen*. However, *aspirin* can make the gout worse by changing uric acid levels in the blood. If the episodes become frequent, gout may be treated with *allopurinol,* which reduces the production of uric acid in the body—thus preventing or lessening acute episodes.

Review the article *Inflammation* at Connect It! at *evolve.elsevier.com.*

Infectious Arthritis

A variety of pathogens can infect synovial membrane and other joint tissues and thereby cause **infectious arthritis.**

One form of infectious arthritis, **Lyme arthritis,** or **Lyme disease,** has become a problem throughout most of the United States in only the last few decades. Lyme disease was identified in Old Lyme, Connecticut, in 1975 and is caused by a spirochete bacterium carried by deer ticks (**Figure 8-39**). This condition is characterized by inflammation in the knees or other joints accompanied by a variety of systemic signs and symptoms.

Another group of bacteria called *Ehrlichia* is also carried by ticks and includes the agents that cause the various forms of **ehrlichiosis.** This bacterial infection has some of the same symptoms as Lyme disease, but is more prevalent than Lyme arthritis in some parts of the United States. One form of ehrlichiosis is more prevalent in the midwestern United States and another form is found mostly in the southern United States.

Both Lyme arthritis and ehrlichiosis are treated by the use of antibiotics. See **Table 6-7**, p. 127, for more on tick-borne illness.

QUICK CHECK

1. What type of bone fracture is most likely to result in infection or osteomyelitis?
2. What is the most common noninflammatory disorder of movable joints?
3. "Ulnar deviation" of the fingers is a common sign in what type of inflammatory joint disease?
4. What form of arthritis is a metabolic condition characterized by an increase in uric acid in the blood?

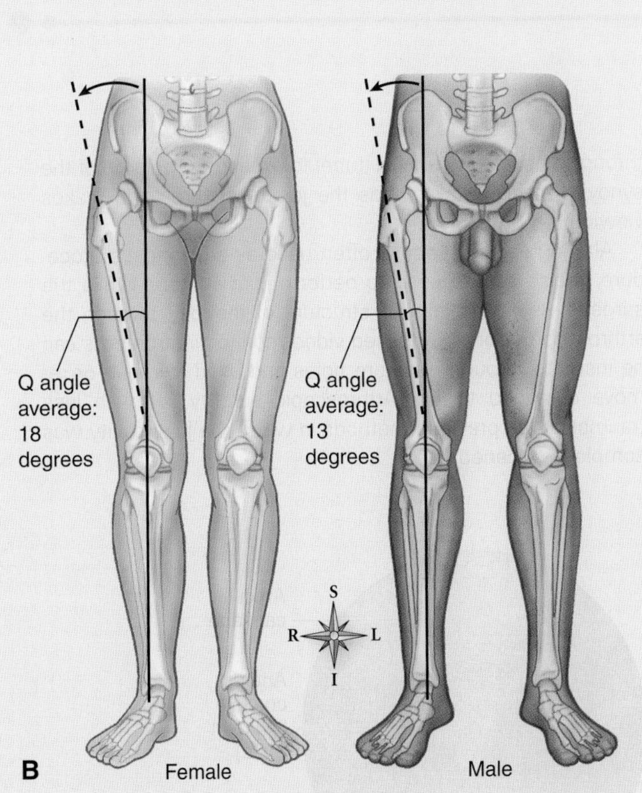

- Exercise both the hamstrings and the quadriceps muscles
- Have a knowledgeable, certified athletic trainer teach them how to land after jumping and how to make quick turns without twisting the knee.

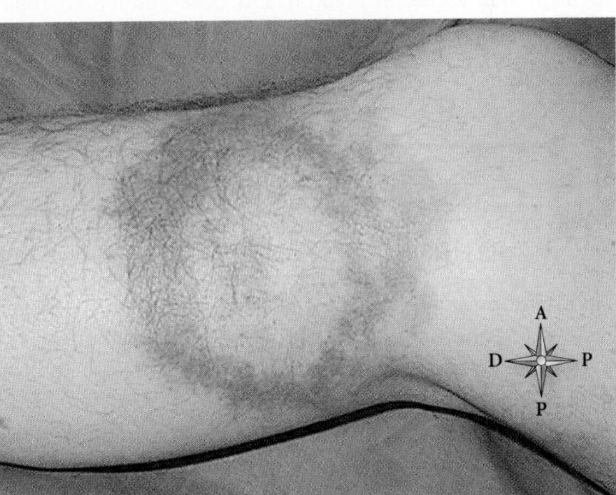

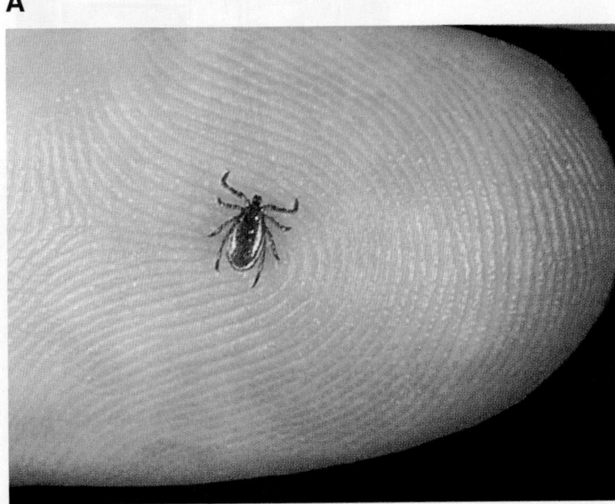

FIGURE 8-39 Lyme disease. A, Circular, expanding rash resembling a bull's-eye target caused by the spirochete bacteria *Borrelia burgdorferi.* **B,** Deer tick, vector for transmission of Lyme disease.

SCIENCE APPLICATIONS

BONES AND JOINTS

Hippocrates (460-377 BC)

Ever since 400 B.C.E., when Hippocrates (the Greek physician often regarded as a founder of the medical profession) first described treatments of human bone and joint disorders and injuries, many approaches to treating the human skeleton have been taken. **Physical therapists** and **occupational therapists** help patients regain movement in joints through physical exercises and **orthopedic surgeons** help their patients by means of surgical operations.

Because the skeleton, with its many bones and joints, is the framework of the entire body, it is not surprising to learn that many different health professionals work directly with the skeleton. **Podiatrists** work with the bones and joints of the foot and ankle, **athletic trainers** and **sports physicians** work with many parts of the skeleton, and **chiropractors** often work to align the vertebral column and other bones to maintain health. The photo shows a sports physician and athletic trainer assessing an injured athlete on the field.

Radiographic technologists and *radiologists* are often called upon to make medical images of the bones and joints and interpret the meaning of these images.

8

CLINICAL APPLICATION

ARTHROSCOPY

Arthroscopy is an imaging technique that allows a physician to examine the internal structure of a joint without the use of extensive surgery. The photos show arthroscopy of the left knee. As Figure **A** shows, a narrow tube with lenses and a fiber-optic light source is inserted into the joint space through a small puncture. Isotonic saline (salt) solution is injected through a needle or cannula (tube) to expand the volume of the synovial space. This spreads the joint structures and makes viewing easier (**B**).

Although arthroscopy is often used as a diagnostic procedure, it also can be used to perform joint surgery. While the surgeon views the internal structure of the joint through the arthroscope or on an attached video monitor, instruments can be inserted through puncture holes and used to repair or remove damaged tissue. Arthroscopic surgery is much less traumatic than previous methods in which the joint cavity was completely opened.

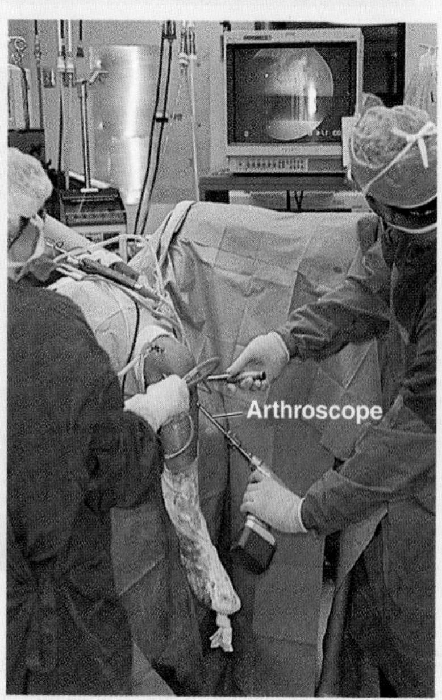

A

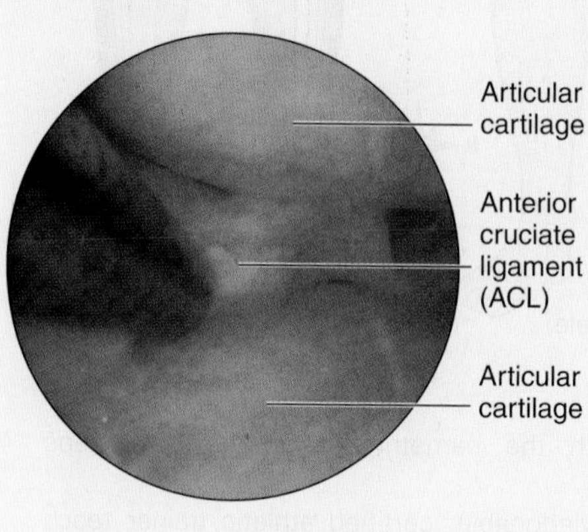

B

LANGUAGE OF SCIENCE (continued from p. 175)

atlas
(AT-lis)
[*Atlas,* **Greek mythical figure who supports the world**]

axial skeleton
(AK-see-al SKEL-eh-ton)
[*axi-* **axis,** *-al* **relating to,** *skeleton* **dried body**]

axis
(AK-sis)
pl., axes
(AK-seez)

ball-and-socket joint

bursa
(BER-sah)
pl., bursae
(BER-see or BER-say)
[*bursa* **purse**]

calcaneus
(kal-KAY-nee-us)
[*calcane-* **heel,** *-ous* **having to do with**]

calcitonin (CT)
(kal-sih-TOH-nin)
[*calci-* **lime (calcium),** *-ton-* **tone,** *-in* **substance**]

canaliculus
(kan-ah-LIK-yoo-lus)
pl., canaliculi
(kan-ah-LIK-yoo-lye)
[*canal-* **channel,** *-uculus* **little**]

cancellous bone
(KAN-seh-lus bohn)
[*cancel-* **lattice,** *-ous* **characterized by**]

carpal
(KAR-pul)
[*carp-* **wrist,** *-al* **relating to**]

cartilage
(KAR-tih-lij)
[*cartilage* **gristle**]

central canal
(SEN-tral kah-NAL)
[*centr-* **center,** *-al* **relating to,** *canal* **channel**]

chest

chiropractor
(KYE-roh-prak-ter)
[*chiro-* **hand,** *-pract-* **practical,** *-or* **agent**]

chondrocyte
(KON-droh-syte)
[*chondro-* **cartilage,** *-cyte* **cell**]

circumduct
(ser-kum-DUKT)
[*circum-* **around,** *-duct* **lead**]

circumduction
(ser-kum-DUK-shun)
[*circum-* **around,** *-duct-* **lead,** *-tion* **process**]

clavicle
(KLAV-ih-kul)
[*clavi-* **key,** *-cle* **little**]

coccyx
(KOK-sis)
[*coccyx* **cuckoo (beak)**]

compact bone
(kom-PAKT bohn)

concave curvature
(kon-KAYV KUR-vah-chur)
[*con-* **together,** *-cave* **hollow,** *curvat-* **bend,** *-ure* **state**]

concentric lamella
(kon-SEN-trik lah-MEL-ah)
pl., lamellae
(lah-MEL-ee)
[*con-* **together,** *-centr-* **center,** *-ic* **relating to,** *lam-* **plate,** *-ella* **little**]

condyloid joint
(KON-dih-loyd joynt)
[*condylo-* **knuckle,** *-oid* **like**]

convex curvature
(kon-VEKS KUR-vah-chur)
[*con-* **together,** *-vex* **convey,** *curvat-* **bend,** *-ure* **state**]

coxal bone
(kok-SAL bohn)
[*coxa-* **hip,** *-al* **relating to**]

cranium
(KRAY-nee-um)
[*cranium* **skull**]

cruciate ligament
(KRU-shee-ayt)
[*cruci-* **cross,** *-ate* **of or like**]

diaphysis
(dye-AF-ih-sis)
pl., diaphyses
(dye-AF-ih-seez)
[*dia-* **through or apart,** *-physis* **growth**]

diarthrosis
(dye-ar-THROH-sis)
pl., diarthroses
(dye-ar-THROH-seez)
[*dia-* **between,** *-arthr-* **joint,** *-osis* **condition**]

diploe
(DIP-loh-EE)
[*diploe* **folded over (doubled)**]

endochondral ossification
(en-doh-KON-dral os-ih-fih-KAY-shun)
[*endo-* **inward or within,** *-chondr-* **cartilage,** *-al* **relating to,** *oss-* **bone,** *-fication* **to make**]

endosteum
(en-DOS-tee-um)
[*endo-* **within,** *-osteum* **bone**]

epiphyseal line
(ep-ih-FEEZ-ee-al lyne)
[*epi-* **upon,** *-phys-* **growth,** *-al* **relating to**]

epiphyseal plate
(ep-ih-FEEZ-ee-al playt)
[*epi-* **upon,** *-phys-* **growth,** *-al* **relating to,** *plate* **flat**]

epiphysis
(eh-PIF-ih-sis)
pl., epiphyses
(eh-PIF-ih-seez)
[*epi-* **upon,** *-physis* **growth**]

ethmoid bone
(ETH-moyd bohn)
[*ethmo-* **sieve,** *-oid* **like**]

extend
(ek-STEND)
[*ex-* **outward,** *-tend-* **stretch**]

extension
(ek-STEN-shun)
[*extend* **to stretch out,** *-ion* **process**]

face
(fays)

femur
(FEE-mur)
[*femur* **thigh**]

fibula
(FIB-yoo-lah)
[*fibula* **clasp**]

flex
(FLEKS)
[*flex* **bend**]

flexion
(FLEK-shun)
[*flex-* **bend,** *-ion* **process**]

fontanel
(FON-tah-nel)
[*fontan-* **fountain or source,** *-el* **little**]

frontal bone
(FRUNT-al bohn)
[*front-* **forehead,** *-al* **relating to**]

gliding joint
(GLY-ding joynt)

hematopoiesis
(hee-mah-toh-poy-EE-sis)
[*hemo-* **blood,** *-poiesis* **making**]

hinge joint
(hinj joynt)

humerus
(HYOO-mer-us)
[*humerus* **arm**]

hyoid bone
(HYE-oyd bohn)
[*hy-* **Greek letter upsilon (γ or ν),** *-oid* **like**]

Continued on p. 210

8

LANGUAGE OF SCIENCE (continued from p. 209)

ilium
(IL-ee-um)
[*ilium* **flank**]

interphalangeal joint
(in-ter-fah-LAN-jee-al joynt)
[*inter-* **between,** *-phalang-* **rows of soldiers (finger bones),** *-al* **relating to**]

intramembranous ossification
(in-trah-MEM-brah-nus os-ih-fih-KAY-shun)
[*intra-* **within,** *-membran-* **thin skin,** *-ous* **characterized by,** *os-* **bone,** *-fic-* **make,** *-ation* **process**]

ischium
(IS-kee-um)
[*ischium* **hip joint**]

joint capsule
(joynt CAP-sool)

lacuna
(lah-KOO-nah)
pl., lacunae
(lah-KOO-nee)
[*lacuna* **pit**]

lateral longitudinal arch
(LAT-er-al lawnj-ih-TOOD-in-al)
[*later-* **side,** *-al* **relating to,** *longitud-* **length,** *-al* **relating to**]

ligament
(LIG-ah-ment)
[*liga-* **bind,** *-ment* **result of action**]

maxilla
(mak-SIH-lah)
pl., maxillae
(mak-SIH-lee)
[*maxilla* **upper jaw**]

medial longitudinal arch
(MEE-dee-al lon-jih-TOO-dih-nal)
[*medi-* **middle,** *-al* **relating to,** *longitud-* **length,** *-al* **relating to**]

medullary cavity
(MED-oo-lar-ee KAV-ih-tee)
[*medulla-* **marrow,** *-y* **related to,** *cav-* **hollow,** *-ity* **state**]

meniscus
(meh-NIS-kus)
pl., menisci
(meh-NIS-aye or meh-NIS-kye)
[*meniscus* **crescent**]

metacarpal
(met-ah-KAR-pal)
[*meta-* **beyond,** *-carp-* **wrist,** *-al* **relating to**]

metatarsal
(met-ah-TAR-sal)
[*meta-* **beyond,** *-tars-* **ankle,** *-al* **relating to**]

middle ear
(MID-ul eer)

musculotendinous unit
(mus-kyoo-loh-TEN-din-us YOO-nit)
[*mus-* **mouse,** *-cul-* **little,** *-tend-* **pulled tight,** *-in-* **unit,** *-ous* **relating to**]

occipital bone
(ok-SIP-it-al bohn)
[*occipit-* **back of head,** *-al* **relating to**]

occupational therapist
(ak-yoo-PAY-shun-al THAYR-ah-pist)
[*occup-* **occupy,** *-tion-* **process,** *-al* **relating to,** *therap-* **treatment,** *-ist* **agent**]

olecranon
(oh-LEK-rah-nohn)
[*ole-* **elbow,** *-cran-* **head,** *-on* **unit**]

olecranon fossa
(oh-LEK-rah-non FOS-ah)
[*ole-* **elbow,** *-cran-* **head,** *-on* **unit,** *fossa* **ditch**]

orthopedic surgeon
(or-thoh-PEE-dik SUR-jen)
[*ortho-* **straight or normal,** *-ped-* **feet,** *-ic* **relating to,** *surg-* **hand,** *-eon* **practitioner**]

osteoblast
(OS-tee-oh-blast)
[*osteo-* **bone,** *-blast* **bud or sprout**]

osteoclast
(OS-tee-oh-klast)
[*osteo-* **bone,** *-clast* **break**]

osteocyte
(OS-tee-oh-syte)
[*osteo-* **bone,** *-cyte* **cell**]

osteon (haversian system)
(AHS-tee-on [hah-VER-zhun or HAV-er-zhun SIS-tem])
[*osteo-* **bone,** *-on* **unit,** *Clopton Havers,* **English physician**]

palpable
(PAL-pah-bul)
[*palp-* **touch gently,** *-able* **capable**]

paranasal sinus
(payr-ah-NAY-zal SYE-nus)
[*para-* **beside,** *-nas-* **nose,** *-al* **relating to,** *sinus* **hollow**]

parathyroid hormone (PTH)
(pair-ah-THYE-royd HOR-mohn)
[*para-* **besides,** *-thyr-* **shield,** *-oid* **like,** *hormon-* **excite**]

parietal bone
(pah-RYE-ih-tal bohn)
[*parie-* **wall,** *-al* **relating to**]

patella
(pah-TEL-ah)
[*pat-* **dish,** *-ella* **small**]

pectoral girdle
(PEK-toh-ral GIRD-el)
[*pector-* **breast,** *-al* **relating to,** *girdle* **belt**]

pelvic girdle
(PEL-vic GIRD-el)
[*pelvi-* **basin,** *-ic* **relating to,** *girdle* **belt**]

pelvis
(PEL-vis)
pl., pelves or pelvises
(PEL-veez or PEL-vis-ez)
[*pelvis* **basin**]

periosteum
(payr-ee-OS-tee-um)
[*peri-* **around,** *-osteum* **bone**]

phalanges
(fah-LAN-jeez)
sing., phalanx
(fah-LANKS)
[*phalanx* **formation of soldiers in rows**]

physical therapist
(FIS-ik-al THAYR-ah-pist)
[*physic-* **medicine,** *-al* **relating to,** *therap-* **treatment,** *-ist* **agent**]

pivot joint
(PIV-it joynt)

podiatrist
(poh-DYE-ah-trist)
[*pod-* **foot,** *-iatr-* **treatment,** *-ist* **agent**]

pubis
(PYOO-bis)
[*pubis* **groin**]

radius
(RAY-dee-us)
[*radius* **ray**]

red bone marrow
(red bohn MAR-oh)

rib

rotation
(roh-TAY-shun)
[*rot-* **turn,** *-ation* **process**]

sacrum
(SAY-krum)
[*sacr-* **holy,** *-um* **thing**]

saddle joint
(SAD-el joynt)

scapula
(SKAP-yoo-lah)
[*scapula* **shoulder blade**]

shaft

sinus
(SYE-nus)
[*sinus* **hollow**]

skull
(skuhl)
[*skull* **head**]

8

LANGUAGE OF SCIENCE *(continued from p. 210)*

sphenoid bone
(SFEE-noyd bohn)
[*spheno-* **wedge,** *-oid* **like**]

spine
(spyne)
[*spine* **thorn**]

spongy bone
(SPUN-jee bohn)
[*spong-* **sponge,** *-y* **characterized by**]

sports physician
(sports fih-ZISH-un)
[*physic-* **medicine,** *-ian* **practitioner**]

sternoclavicular joint
(ster-noh-klah-VIK-yoo-lar joynt)
[*sterno-* **breastbone (sternum),** *-clavi-* **key,** *-ular* **relating to**]

sternum
(STER-num)
pl., sterna or sternums
(STER-nah or STER-numz)
[*sternum* **breastbone**]

suture
(SOO-chur)
[*suture* **seam**]

synarthrosis
(sin-ar-THROH-sis)
pl., synarthroses
(sin-ar-THROH-seez)
[*syn-* **together,** *-arthr-* **joint,** *-osis* **condition**]

synovial fluid
(sih-NOH-vee-al FLOO-id)
[*syn-* **together,** *-ovi-* **egg (white),** *-al* **relating to**]

synovial membrane
(sih-NOH-vee-al MEM-brayn)
[*syn-* **together,** *-ovi-* **egg (white),** *-al* **relating to,** *membran-* **thin skin**]

talus
(TAY-lus)
pl., tali
(TAY-lye)
[*talus* **ankle**]

tarsal
(TAR-sal)
[*tars-* **ankle,** *-al* **relating to**]

temporal bone
(TEM-poh-ral bohn)
[*tempora-* **temple (of head),** *-al* **relating to**]

thorax
(THOR-aks)
[*thorax* **chest**]

tibia
(TIB-ee-ah)
[*tibia* **shin bone**]

trabecula
(trah-BEK-yoo-lah)
pl., trabeculae
(trah-BEK-yoo-lee)
[*trab-* **beam,** *-ula* **little**]

transverse arch (metatarsal arch)
(TRANS-vers arch [met-ah-TAR-sal])
[*trans-* **across or through,** *-vers-* **turn** (*meta-* **beyond,** *-tars-* **ankle,** *-al* **relating to**)]

transverse canal (Volkmann canal)
(tranz-VERS kah-NAL [VOLK-man])
[*trans-* **across,** *-vers-* **turn,** *Richard von Volkmann,* **German surgeon**]

ulna
(UHL-nah)
[*ulna* **elbow**]

vertebra
(VER-teh-bra)
pl., vertebrae
(VER-teh-bree or VER-teh-bray)
[*vertebra* **joint or turning part**]

yellow bone marrow
(YEL-oh bohn MAR-oh)

8

LANGUAGE OF MEDICINE

arthritis
(ar-THRY-tis)
[*arthr-* **joint,** *-itis* **inflammation**]

arthroplasty
(AR-throh-plas-tee)
[*arthr-* **joint,** *-plasty* **surgical repair**]

arthroscopy
(ar-THROS-skah-pee)
[*arthr-* **joint,** *-scop-* **see,** *-y* **activity**]

avulsion fracture
(ah-VUL-shun FRAK-chur)
[*avuls-* **pull away,** *-ion* **process,** *fracture* **a breaking**]

Bouchard node
(boo-SHAR nohd)
[*Charles J. Bouchard* **French physician,** *nod-* **knot**]

bursitis
(ber-SYE-tis)
[*burs-* **purse,** *-itis* **inflammation**]

callus
(KAL-us)
[*callus* **hard skin**]

chondrosarcoma
(KON-droh-sar-KOH-mah)
[*chondro-* **cartilage,** *-sarco-* **flesh,** *-oma* **tumor**]

closed fracture
(klohzd FRAK-chur)
[*fracture* **a breaking**]

comminuted fracture
(kom-ih-NOO-ted FRAK-chur)
[*commin-* **break into pieces,** *-ute* **perform action,** *fracture* **a breaking**]

complete fracture
(kom-PLEET FRAK-chur)
[*fracture* **a breaking**]

dislocation
(dis-loh-KAY-shun)
[*dis-* **apart,** *-locat-* **locate,** *-tion* **condition**]

dowager's hump
(DOW-ah-jerz hump)
[*dowager* **widow with a dower (gift or wealth)**]

ehrlichiosis
(ur-lik-ee-OH-sis)
[*Paul Ehrlich* **German microbiologist**]

epiphyseal fracture
(ep-ih-FEEZ-ee-al FRAK-chur)
[*epi-* **on,** *-phys-* **growth,** *-al* **relating to,** *fracture* **to break**]

flatfeet
(flat FEET)

gout
(gowt)
[*gout* **drop**]

gouty arthritis
(gow-TEE ar-THRY-tis)
[*gout-* **drop,** *-y* **of or like,** *arthr-* **joint,** *-itis* **inflammation**]

Heberden node
(HEB-er-den nohd)
[*William Heberden* **English physician,** *nod-* **knot**]

herniated disk
(HER-nee-ayt-ed disk)
[*hernia-* **rupture,** *-ate* **act of**]

impacted fracture
(im-PAK-ted FRAK-chur)
[*impact-* **push into,** *fracture* **a breaking**]

Continued on p. 212

LANGUAGE OF **MEDICINE** *(continued from p. 211)*

incomplete fracture
(in-kom-PLEET FRAK-chur)
[*fracture* **a breaking**]

infectious arthritis
(in-FEK-shus ar-THRY-tis)
[*infect-* **stain,** *-ous* **relating to,** *arthr-* **joint,** *-itis* **inflammation**]

intravenous
(in-trah-VEE-nus)
[*intra-* **within,** *-ven-* **vein,** *-ous* **relating to**]

juvenile rheumatoid arthritis (JRA)
(JOO-veh-ney-el ROO-mah-toyd ar-THRY-tis)
[*juven-* **youth,** *-ile* **of or like,** *rheuma-* **flow,** *-oid* **like,** *arthr-* **joint,** *-itis* **inflammation**]

kyphosis
(kye-FOH-sis)
[*kypho-* **hump,** *-osis* **condition**]

linear fracture
(LIN-ee-ar FRAK-chur)
[*linea-* **line,** *-ar* **relating to,** *fracture* **a breaking**]

8

lordosis
(lor-DOH-sis)
[*lordos-* **bent backward,** *-osis* **condition**]

Lyme arthritis
(lyme ar-THRY-tis)
[*Lyme* **city in Connecticut,** *arthr-* **joint,** *-itis* **inflammation**]

mastoiditis
(mas-toyd-AYE-tis)
[*mast-* **breast,** *-oid-* **like,** *-itis* **inflammation**]

oblique fracture
(oh-BLEEK FRAK-chur)
[*obliq-* **slanted,** *fracture* **a breaking**]

open fracture
(OH-pen FRAK-chur)
[*fracture* **a breaking**]

osteoarthritis
(os-tee-oh-ar-THRY-tis)
[*osteo-* **bone,** *-arthr-* **joint,** *-itis* **inflammation**]

osteogenesis imperfecta
(os-tee-oh-JEN-eh-sis im-per-FEK-tah)
[*osteo-* **bone,** *-gen-* **produce,** *-esis* **process,** *im-* **not,** *-perfecta* **perfect**]

osteitis deformans
(os-tee-AYE-tis deh-FOR-manz)
[*oste-* **bone,** *-itis* **inflammation,** *deformans* **deforming**]

osteomalacia
(os-tee-oh-mah-LAY-shah)
[*osteo-* **bone,** *-malacia* **softening**]

osteomyelitis
(os-tee-oh-my-eh-LYE-tis)
[*osteo-* **bone,** *myel-* **marrow,** *-itis* **inflammation**]

osteoporosis
(os-tee-oh-poh-ROH-sis)
[*osteo-* **bone,** *-poro-* **pore,** *-osis* **condition**]

osteosarcoma
(os-tee-oh-sar-KOH-mah)
[*osteo-* **bone,** *-sarc-* **flesh,** *-oma* **tumor**]

Paget disease
(PAJ-et dih-ZEEZ)
[*Sir James Paget* **British surgeon and pathologist,** *dis-* **without,** *-ease* **comfort**]

parenteral
(pah-REN-ter-al)
[*par-* **beside,** *-enter-* **intestine,** *-al* **relating to**]

rheumatoid arthritis (RA)
(ROO-mah-toyd ar-THRY-tis [ar ay])
[*rheuma-* **flow,** *-oid* **like,** *arthr-* **joint,** *-itis* **inflammation**]

rickets
(RIK-ets)
[unknown origin]

scoliosis
(skoh-lee-OH-sis)
[*scolio-* **twisted or crooked,** *-osis* **condition**]

sprain
(sprayn)
[unknown origin]

strain
(strayn)
[*strain* **stretch**]

subluxation
(sub-luks-AY-shun)
[*sub-* **below or near,** *-luxat-* **locate,** *-tion* **condition**]

tophus
(TOH-fus)
pl., tophi
(TOH-fye)
[*tophus* **porous rock**]

transverse fracture
(TRANS-vers FRAK-chur)
[*trans-* **across,** *-verse* **turn,** *fracture* **a breaking**]

ulnar deviation
(UL-nur dee-vee-AY-shun)
[*ulna-* **elbow,** *-ar* **relating to,** *de-* **out of,** *-via-* **road or path,** *-ation* **process**]

vertebroplasty
(ver-tee-broh-PLAS-tee)
[*vertebr-* **joint or backbone,** *-plasty* **surgical repair**]

❑ OUTLINE SUMMARY

To download a digital version of the chapter summary for use with your device, access the ***Audio Chapter Summaries*** *online at evolve.elsevier.com.*

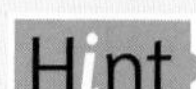

Scan this summary after reading the chapter to help you reinforce the key concepts. Later, use the summary as a quick review before your class or before a test.

Functions of the Skeletal System

A. Provides internal framework that supports and gives shape to the body
B. Protects internal organs and tissues
C. Makes movement possible when bones at movable joints are pulled by muscles
D. Storage of vital substances
   1. Calcium—hormones regulate calcium storage: calcitonin (CT) increases storage and parathyroid hormone (PTH) reduces stores of calcium
   2. Fat—stored in cavities of some bones

E. Hematopoiesis—blood cell formation in the red bone marrow

Gross Structure of Bones

A. Four major types, according to overall shape of the bone
   1. Long—example: humerus (arm)
   2. Short—example: carpals (wrist)
   3. Flat—example: frontal (skull)
   4. Irregular—example: vertebrae (spinal bones)
   5. Some also recognize a sesamoid (round) bone category—example: patella (kneecap)

B. Structure of long bones (**Figure 8-1**)
   1. Diaphysis, or shaft—hollow tube of hard compact bone
   2. Medullary cavity—hollow space inside the diaphysis that contains yellow marrow
   3. Epiphyses—ends of the bones made of spongy bone that contains red bone marrow
   4. Articular cartilage—thin layer of hyaline cartilage that covers each epiphysis; provides a smooth cushion
   5. Periosteum—strong, fibrous membrane covering bone everywhere except at joint surfaces
   6. Endosteum—thin membrane that lines medullary cavity

C. Structure of flat bones (**Figure 8-2**)
   1. Spongy bone layer sandwiched between two compact bone layers
   2. Diploe—spongy bone layer of a flat bone

Microscopic Structure of Bones

A. Bone tissue structure (**Figure 8-3**)
   1. Cancellous (spongy) bone
      a. Texture results from needlelike threads of bone called *trabeculae* surrounded by a network of open spaces
      b. Found in epiphyses of bones
      c. Spaces contain red bone marrow
   2. Compact bone
      a. Structural unit is an osteon—calcified matrix arranged in multiple layers or rings called concentric lamella (**Figure 8-4**)
      b. Bone cells are called *osteocytes* and are found inside spaces called *lacunae,* which are connected by tiny tubes called *canaliculi*

B. Cartilage (**Figure 8-5**)
   1. Cell type called *chondrocyte*
   2. Matrix is gel-like and lacks blood vessels

Bone Development

A. Making and remodeling bone tissue (**Figure 8-6**)
   1. Early bone development (before birth) consists of cartilage and fibrous structures
   2. Osteoblasts
      a. Form new bone matrix by encrusting collagen fibers with calcium crystals
      b. Osteocytes are inactive osteoblasts
   3. Osteoclasts dissolve bone, releasing calcium ions for reabsorption into the bloodstream
   4. Remodeling is a combined action of making and dissolving bone matrix that eventually sculpts bone into the adult shape

B. Endochondral ossification—cartilage models gradually replaced by calcified bone (**Figure 8-7** and **Figure 8-8**)

C. Intramembranous ossification—fibrous membranes are ossified into hard bone plates; fontanels are soft, not-yet-ossified regions

Axial Skeleton

A. Skeleton can be divided into central *axial* and peripheral *appendicular* regions (**Figure 8-9** and **Table 8-1**)

B. Axial skeleton includes 80 bones:
   1. Skull (**Figure 8-10** and **Table 8-2**)
      a. Bones of the cranium (8), face (14), and middle ear (6)
      b. Includes spaces called *paranasal sinuses* (**Figure 8-11**)
      c. Mastoiditis is inflammation of mastoid process of temporal bone (**Figure 8-12**)
   2. Hyoid bone (**Figure 8-13**)
   3. Vertebral column (spine) (**Figure 8-14** and **Table 8-3**)
      a. 24 vertebrae: cervical (7), thoracic (12), lumbar (5), sacrum, coccyx
      b. Vertebrae are irregular bones with well-defined parts, such as body, spine, transverse process, vertebral foramen, and articular processes (**Figure 8-15**)
      c. Atlas and axis—first two cervical vertebrae form a unique pivoting structure (**Figure 8-16**)
      d. Spinal curvatures support the body, but can become abnormally exaggerated (**Figure 8-17** and **Figure 8-18**)
   4. Thorax—ribs (24), sternum (**Figure 8-19** and **Table 8-4**)

Appendicular Skeleton

A. Bones of the upper and lower extremities (126)

B. Upper extremity (64) (**Table 8-5**)
   1. Pectoral (shoulder) girdle—scapula (2), clavicle (2)
   2. Arm and forearm—humerus (2), radius (2), ulna (2) (**Figure 8-20**)
   3. Wrist and hand—carpal bones (16), metacarpal bones (10), phalanges (28) (**Figure 8-21**)

C. Lower extremity (62) (**Table 8-6**)
   1. Pelvic (hip) girdle—coxal bone (2) (**Figure 8-25**)
   2. Thigh and leg—femur (2), patella (2), tibia (2), fibula (2) (**Figure 8-22**)
   3. Ankle and foot
      a. Tarsal bones (14), metatarsal bones (10), phalanges (28) (**Figure 8-23**)
      b. Arched structure of foot provides dynamic support for entire skeleton (**Figure 8-24**)

Skeletal Variations

A. Male and female skeletal differences
   1. Size—male skeleton generally larger
   2. Shape of pelvis—male pelvis deep and narrow, female pelvis shallow and broad
   3. Size of pelvic inlet—female pelvic inlet generally wider, normally large enough for baby's head to pass through it (**Figure 8-25**)
   4. Pubic angle—angle between pubic bones of female generally wider
B. Age differences
   1. Bones enlarge and become more ossified until maturity at age 25
   2. Bones actively remodel (dissolve and rebuild) in middle adulthood
   3. Bones become less dense during elderly years
C. Environmental factors
   1. Nutrition affects growth and maintenance of bone tissue
   2. Mechanical stress, including exercise, affects bone remodeling

8

Joints

A. Articulation—a joint between two or more bones
B. Every bone except the hyoid (which anchors the tongue) connects to at least one other bone.
C. Kinds of joints
   1. Synarthroses (no movement)—fibrous connective tissue (ligaments) grows between articulating bones; for example, sutures of skull (**Figure 8-26**)
   2. Amphiarthroses (slight movement)
      a. Fibrocartilage connects articulating bones; for example, symphysis pubis (**Figure 8-26**)
      b. Another example is the vertebral disk forming each joint between vertebral bodies, which can become herniated (**Figure 8-27**)
   3. Diarthroses (free movement)—most joints belong to this class
      a. Structure (**Figure 8-28**)
         (1) Structures of freely movable joints—joint capsule and ligaments hold adjoining bones together but permit movement at joint
         (2) Articular cartilage—hyaline cartilage covers joint ends of bones where they form joints with other bones
         (3) Synovial membrane—lines joint capsule and secretes lubricating synovial fluid
         (4) Joint cavity—space between joint ends of bones
         (5) Bursa—fluid-filled pouch that absorbs shock; inflammation of bursa is called *bursitis*
      b. Functions of freely movable joints—ball-and-socket, hinge, pivot, saddle, gliding, and condyloid—allow different kinds of movements determined by the structure of each joint (**Figure 8-29** and **Table 8-7**)

Skeletal Disorders

A. Tumors
   1. Bone tumors—osteosarcoma (**Figure 8-30,** *A*)
      a. Most common and serious type of malignant bone neoplasm
      b. Frequent sites include distal femur and proximal tibia and humerus
   2. Cartilage tumors—chondrosarcoma (**Figure 8-30,** *B*)
      a. Cancer of skeletal hyaline cartilage
      b. Second most common cancer of skeletal tissues
      c. Frequent sites include medullary cavity of humerus, femur, ribs, and pelvic bones
B. Metabolic bone diseases
   1. Osteoporosis (**Figure 8-31**)
      a. Characterized by loss of calcified bone matrix and reduction in number of trabeculae in spongy bone
      b. Bones fracture easily—especially in wrists, hips, and vertebrae
      c. Treatment includes drug therapy, exercise, and dietary supplements of calcium and vitamin D
   2. Rickets and osteomalacia—both diseases characterized by loss of bone minerals related to vitamin D deficiency
      a. Rickets (**Figure 8-32**)
         (1) Loss of bone minerals occurs in infants and young children before skeletal maturity
         (2) Lack of bone rigidity causes gross skeletal changes (bowing of legs)
         (3) Treated with vitamin D
      b. Osteomalacia
         (1) Mineral content is lost from bones that have already matured
         (2) Increases susceptibility to fractures
         (3) Treated with vitamin D
   3. Paget disease (osteitis deformans) (**Figure 8-33**)
      a. Faulty remodeling results in deformed bones that fracture easily
      b. Cause may be genetic or triggered by viral infections
   4. Osteogenesis imperfecta (also called *brittle-bone disease*) (**Figure 8-34**)
      a. Bones are brittle because of a lack of organic matrix
      b. Treatment may include splinting to reduce fracture and drugs that decrease cell activity
C. Bone infection
   1. Osteomyelitis (**Figure 8-35**)
      a. General term for bacterial (usually staphylococcal) infection of bone
      b. Treatment may involve surgery, drainage of pus, and parenteral (IV) antibiotic treatment—often over prolonged periods
D. Bone fractures (**Figure 8-36**)
   1. Open (compound) fractures pierce the skin and closed (simple) fractures do not

2. Fracture types include complete and incomplete, linear, transverse, and oblique
3. Bone repair—bleeding and inflammation, followed by formation of a callus (supportive framework), and finally remodeling of bone (**Figure 8-37**)

E. Joint disorders
1. Noninflammatory joint disorders—do not usually involve inflammation of the synovial membrane; symptoms tend to be local and not systemic
   a. Osteoarthritis, or *degenerative joint disease* (DJD) (**Figure 8-38**, *A*)
      (1) Most common noninflammatory disorder of movable joints—often called "wear and tear" arthritis
      (2) Symptoms include joint pain, morning stiffness, and appearance of Bouchard nodes (at proximal interphalangeal joints) and Heberden nodes (at distal interphalangeal joints) of the fingers
      (3) Most common cause for partial and total hip and knee replacements
   b. Traumatic injury
      (1) Dislocation—articular surfaces of bones in joint are no longer in proper contact; *subluxation* is a minor dislocation or misalignment
      (2) Sprain—acute injury to *ligaments* around joints (example: whiplash-type injuries)
      (3) Strain—acute injury to any part of the *musculotendinous unit* (muscle, tendon, junction between the two, and attachments to bone)
2. Inflammatory joint disorders—the term *arthritis* is a general name for several types of inflammatory joint diseases that may be caused by infection, injury, genetic factors, and autoimmunity. Inflammation of the synovial membrane occurs, often with systemic signs and symptoms.
   a. Rheumatoid arthritis (**Figure 8-38**, *B*)—systemic autoimmune disease—chronic inflammation of synovial membrane with involvement of other tissues such as blood vessels, eyes, heart, and lungs
   b. Gouty arthritis (**Figure 8-38**, *C*)—synovial inflammation caused by chronic gout, a condition in which sodium urate crystals form in joints and other tissues, sometimes forming accumulations called *tophi*
   c. Infectious arthritis (**Figure 8-39**)—arthritis resulting from infection by a pathogen, as in Lyme arthritis and ehrlichiosis, caused by two different types of bacteria that are transmitted to humans by tick bites.

8

❑ ACTIVE LEARNING

STUDY TIPS

Hint *Consider using these tips to achieve success in meeting your learning goals.*

Before starting your study of Chapter 8, go back to Chapter 5 and review the synopsis of the skeletal system.

1. There are several terms in this chapter that use prefixes or suffixes that help explain their meaning. The prefixes *epi-* (upon) and *endo-* (within) were discussed earlier. *Peri-* means "around," *osteo-* or *os-* refers to bone, *chondro-* refers to cartilage, *-cyte* means "cell," *-blast* means "young cell" or "building cell," and *-clast* means "to destroy." Knowing the meaning of these prefixes or suffixes makes most of the terms self-explanatory.
2. When studying the microscopic structure of the skeletal system, remember that bone tissue heals fairly well, whereas cartilage does not. This is because there are many blood vessels throughout the bone; this is not so in cartilage. The cells must have a way of receiving nutrients and oxygen and a way to get rid of waste products. The structure of the osteon allows this to occur.
3. Reviewing the figures of the full skeleton and the skull may be the best way to learn the bones of the skeleton. Use flash cards or online resources to supplement the text material. One such online resource is: *getbodysmart.com*. This site has excellent illustrations, tutorials, and quizzes. Additional online tips are found at *my-ap.us/JJEEMF*.
4. The joints are named based on the amount of movement they allow (*arthro* means "joint"). The joint capsule is an example of a synovial membrane discussed in Chapter 7.
5. In your study group you can use flash cards to learn the terms in the bone structure and joints. Discuss the formation and structure of the osteon. A photocopy or a cellphone picture of the skeleton figures with the labels blackened out will help you learn the names and characteristics of the bones. There is no real shortcut to learning the names and locations of the bones, it is simply a memorization task, but quizzing each other will help you learn them faster.
6. Construct a table to help yourself learn the bone and joint disorders. Again, as in previous chapters, it would be helpful to organize them by mechanism or cause. The bone cancers should be easy to remember because they use the prefixes; *osteo-* for bone and *chondro-* for cartilage.
7. Review the Language of Science and Language of Medicine terms and their word origins to help you better understand the meaning of the names of the bones.
8. Review the outline at the end of this chapter. This outline provides an overview of the material and would help you understand the general concepts to the chapter.

Review Questions

Write out the answers to these questions after reading the chapter and reviewing the Chapter Summary. If you simply think through the answer without writing it down, you will not retain much of your new learning.

1. List and briefly describe the five functions of the skeletal system.
2. Describe the structure of the osteon.
3. Describe the structure of cartilage.
4. Explain briefly the process of endochondral ossification, including the function of the osteoblasts and osteoclasts.
5. Explain the importance of the epiphyseal plate.
6. In general, what bones are included in the axial skeleton? The appendicular skeleton?
7. The vertebral column is divided into five sections based on location. Name the sections and give the number of vertebrae in each section.
8. Distinguish between true, false, and floating ribs. How many of each is in the human body?
9. Describe and give an example of a synarthrotic joint.
10. Describe and give an example of an amphiarthrotic joint.
11. Describe and give two examples of a diarthrotic joint.
12. Briefly describe a joint capsule.
13. Describe open, closed, and comminuted fractures.
14. Describe the three types of arthritis.
15. Describe *bursa* and define *bursitis*.
16. Describe an avulsion fracture.

Critical Thinking

After finishing the Review Questions, write out the answers to these more in-depth questions to help you apply your new knowledge. Go back to sections of the chapter that relate to concepts that you find difficult.

17. When a patient receives a bone marrow transplant, what vital process is being restored?
18. Explain how the canaliculi allow bone to heal more efficiently than cartilage.
19. Based upon what you know, what would happen to bone tissue if one of the three types of bone cells were missing, but the other types of bone cells were present in the tissue?
20. What effect does the task of childbearing have on the difference between the male and female skeleton?
21. Explain why a bone fracture along the epiphyseal plate may have serious implications in children and young adults.
22. Compare and contrast the causes and changes associated with osteoporosis, osteomalacia, and Paget disease.
23. Why is mastoiditis potentially more dangerous than a paranasal sinus infection?
24. Explain how the anatomy of the elbow is a good example of how "structure fits function."
25. Explain how "palpable bony landmarks" are used in the medical professions.

Chapter Test

After studying the chapter, test your mastery by responding to these items. Try to answer them without looking up the answers.

1. The thin layer of cartilage on the ends of bones where they form joints is called the ________.
2. The hollow area in the shaft of a long bone where marrow is stored is called the ________.
3. The needlelike threads of spongy bone are called ________.
4. The structural units of compact bone are called either osteons or ________.
5. Osteocytes and chondrocytes live in small spaces in the matrix called ________.
6. Bone-resorbing cells are called ________.
7. Bone-forming cells are called ________.
8. The process of forming bone from cartilage is called ________.
9. If an ________ remains between the epiphysis and diaphysis, bone growth can continue.
10. The two major divisions of the human skeleton are the ________ skeleton and the ________ skeleton.
11. The three types of joints, named for the amount of movement they allow are ________, ________, and ________.
12. The ________ are cords or bands made of strong connective tissue that hold bones together.
13. Abnormal side-to-side curvature of the vertebral column is called ________.
14. The skeletal disorder, common in elderly white women and characterized by excessive loss of calcified matrix and collagen fibers is called ________.
15. Microbial infection of the bone is called ________.
16. A ________ fracture invites the possibility of infection because the skin is pierced.
17. Degenerative joint disease, or ________, involves wearing away of articular cartilage.
18. The ________ bone serves as an anchor for tongue muscles and helps support the larynx.
19. The cervical and lumbar curves of the spine are called ________ curvatures.
20. A ________ is an orthopedic procedure that involves the injection of a bone cement to repair fractured and compressed vertebrae for those who have experienced a compression fracture due to osteoporosis, trauma, tumors, or prolonged use of steroid drugs.
21. Tophi are often the first indication of ________ arthritis.
22. Which of the following is not a function of the skeleton?
    a. Mineral storage
    b. Blood formation
    c. Heat production
    d. Protection

23. The strong fibrous membrane covering all of a long bone except the joint is called the:
    a. endosteum
    b. periosteum
    c. ligament
    d. tendon
24. The fibrous lining of the hollow cavity in the long bone is called the:
    a. endosteum
    b. periosteum
    c. medullary cavity
    d. ligament
25. The end of a long bone is called the:
    a. endosteum
    b. periosteum
    c. diaphysis
    d. epiphysis
26. The shaft of a long bone is called the:
    a. endosteum
    b. periosteum
    c. diaphysis
    d. epiphysis
27. Cancer of the cartilage is called:
    a. osteosarcoma
    b. chondrosarcoma
    c. fibrosarcoma
    d. osteomalacia
28. The inflammatory joint disease that is caused by an increase of uric acid is:
    a. osteoarthritis
    b. rheumatoid arthritis
    c. gouty arthritis
    d. infectious arthritis

Match the bones in Column A with their locations in Column B.

Column A	Column B
29. ________ ulna	a. skull
30. ________ mandible	b. upper extremity (arm, forearm, wrist, and hand)
31. ________ humerus	c. trunk
32. ________ metatarsals	d. lower extremity (thigh, leg, ankle, and foot)
33. ________ tibia	
34. ________ rib	
35. ________ fibula	
36. ________ sternum	
37. ________ scapula	
38. ________ femur	
39. ________ metacarpals	
40. ________ frontal bone	
41. ________ patella	
42. ________ zygomatic bone	
43. ________ clavicle	
44. ________ occipital bone	
45. ________ carpals	
46. ________ maxilla	

Case Studies

To solve a case study, you may have to refer to the glossary or index, other chapters in this textbook, and other resources.

1. Andrew is a young boy who loves to climb trees. While attempting to climb his favorite oak tree, Andrew fell and fractured his humerus. The radiologist described Andrew's injury as a "greenstick fracture." What does this label tell you about the appearance of the fracture? If given proper medical care, is the injury likely to heal rapidly or slowly?
2. Christine is a young music student at the local college. One of her professors suggested that she analyze her conducting technique by videotaping herself as she conducts the choir. As she replays the tape, Christine notices that her hips and shoulders seem awkwardly bent—even when she is in a formal standing position. What condition might cause this abnormal curve of the trunk? What are some ways of treating this condition?
3. Agnes is an elderly woman with osteoporosis. She recently sustained a severe bone fracture for no apparent reason (she did not fall or otherwise injure herself). The fracture was not treated for some time, and as a result, Agnes developed osteomyelitis. Based on what you know of osteoporosis, how do you explain her mysterious fracture? How can a fracture progress to osteomyelitis?
4. Alice is 10 years old and over the past few years has had several fractures even though she is not an active child. A recent blood test indicated that her alkaline phosphatase level was highly elevated. What diagnosis is possible with her history and lab values? What are some treatment options if your diagnosis is correct?

Answers to Active Learning Questions can be found online at evolve.elsevier.com.

Muscular System

OUTLINE

Scan this outline before you begin to read the chapter, as a preview of how the concepts are organized.

OBJECTIVES

Before reading the chapter, review these goals for your learning.

After you have completed this chapter, you should be able to:

1. List, locate in the body, and compare the structure and function of the three major types of muscle tissue.
2. Discuss the structure and function of skeletal muscle.
3. Describe the role of other body systems in movement.
4. Discuss the role of the motor unit in muscle stimulation and how a muscle fiber contracts.
5. Compare the major types of skeletal muscle contractions.
6. Describe the primary effects of exercise on skeletal muscle.
7. List and explain the most common types of movement produced by skeletal muscles.
8. Name, identify on a model or diagram, and give the function of the major muscles of the body.
9. Name and describe the major disorders of skeletal muscles.

CHAPTER 9

Although we initially review three types of muscle tissue introduced earlier (see Chapter 4), the plan for this chapter is to focus on skeletal, or voluntary, muscle—those muscle masses that attach to bones and actually move them about when contraction or shortening of muscle cells, or muscle fibers, occurs.

If you weigh 120 pounds, about 50 pounds of your weight comes from your skeletal muscles, the "red meat" of the body that is attached to your bones. The soft muscular tissue that lies between your skin and your bones is often called "flesh"—the site of injury in a "flesh wound."

Muscular movement occurs when chemical energy from nutrient molecules is transferred to protein filaments in each muscle fiber and converted to mechanical energy that shortens (contracts) the muscle. As the muscle fibers in a muscle contract, they pull on the bones to which they are attached and thus produce movement of the body.

Movements caused by skeletal muscle contraction vary in complexity from blinking an eye to the coordinated, powerful, and fluid movements of a gifted athlete. Not many of our body structures can claim as great an importance for happy, active living as can our voluntary muscles, and only a few can boast of greater importance for life itself. Our ability to survive often depends on our ability to adjust to the changing conditions of our environment to maintain homeostasis. Movements frequently constitute a major part of this homeostatic adjustment.

LANGUAGE OF SCIENCE

Hint Before reading the chapter, say each of these terms out loud. This will help you to avoid stumbling over them as you read.

abduct
(ab-DUKT)
[*ab-* **away,** *-duct* **lead**]

abduction
(ab-DUK-shun)
[*ab-* **away,** *-duct-* **lead,** *-tion* **process**]

acetylcholine (ACh)
(as-ee-til-KOH-leen [ay see aych])
[*acetyl-* **vinegar,** *-chole-* **bile,** *-ine* **made of**]

actin
(AK-tin)
[*act-* **act or do,** *-in* **substance**]

adduct
(ad-DUKT)
[*ad-* **toward,** *-duct* **lead**]

adduction
(ad-DUK-shun)
[*ad-* **toward,** *-duct-* **lead,** *-tion* **process**]

adductor muscle
(ad-DUK-tor MUS-el)
[*ad-* **toward,** *-duct-* **lead,** *-or* **condition,** *mus-* **mouse,** *-cle* **small**]

antagonist
(an-TAG-oh-nist)
[*ant-* **against,** *-agon-* **struggle,** *-ist* **agent**]

biceps brachii
(BYE-seps BRAY-kee-aye)
[*bi-* **two,** *-cep* **head,** *brachii* **related to the arm**]

brachialis
(bray-kee-AL-is)
[*brachi* **arm,** *-al-* **relating to,** *-is* **thing**]

Continued on p. 238

Muscle Tissue

Skeletal Muscle

Under the microscope, threadlike and cylindrical **skeletal muscle** cells appear in bundles. They are characterized by many crosswise stripes and multiple nuclei (Figure 9-1, *A*). Each fine thread is a muscle cell—usually called a **muscle fiber.**

This type of muscle tissue has three names: *skeletal muscle,* because it attaches to bone; *striated muscle,* because of its cross stripes or striations; and *voluntary muscle,* because its contractions can be controlled voluntarily.

Cardiac Muscle

In addition to skeletal muscle, the body also contains two other types of muscle tissue: *cardiac muscle* and *smooth muscle.*

Cardiac muscle composes the bulk of the heart. Cells in this type of muscle tissue are also cylindrical, branch frequently (Figure 9-1, *B*), and then recombine into a continuous mass of interconnected tissue. As with skeletal muscle cells, they have cross striations. They also have unique dark bands called **intercalated disks** where the plasma membranes of adjacent cardiac fibers come in contact with each other.

Cardiac muscle tissue demonstrates the principle that "structure fits function." The interconnected nature of cardiac muscle fibers helps the tissue contract as a unit and increases the efficiency of the heart muscle in pumping blood.

Smooth Muscle

Smooth muscle fibers are tapered at each end and have a single nucleus (Figure 9-1, *C*). Because they lack cross stripes, or striations, they are sometimes called *nonstriated* muscle cells. They have a smooth, even appearance when viewed through a microscope. They are called *involuntary* because we normally do not have control over their contractions.

Smooth, or involuntary, muscle forms an important part of blood vessel walls and of many hollow internal organs (viscera) such as the gut, urethra, and ureters. Because of its location in many visceral structures, it is sometimes also called *visceral muscle.*

Although we cannot willfully control the action of smooth muscle, its contractions are highly regulated, which facilitates, for example, the passage of nutrients through the digestive tract or urine through the ureters into the bladder.

All three muscle cell types—skeletal, cardiac, and smooth—specialize in contraction or shortening. Every movement we make is produced by contractions of skeletal muscle cells. Contractions of cardiac muscle cells pump blood through the heart, and smooth muscle contractions help pump blood and other substances through our other hollow organs.

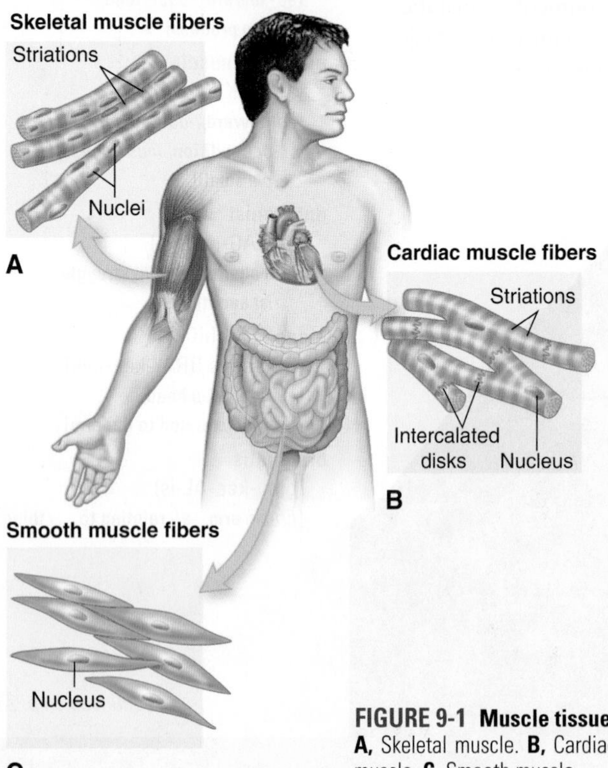

FIGURE 9-1 Muscle tissue. **A,** Skeletal muscle. **B,** Cardiac muscle. **C,** Smooth muscle.

Structure of Skeletal Muscle

Muscle Organs

A skeletal muscle is an organ composed mainly of skeletal muscle fibers and connective tissue.

Fibrous connective tissue wraps around each individual muscle fiber. The tissue continues as it wraps around groups of muscle fibers called *fascicles* and finally forms a "wrapper" around the entire muscle organ.

Fascia is the loose connective tissue outside the muscle organs that forms a flexible, sticky "packing material" between muscles, bones, and the skin.

Most skeletal muscles attach to two bones that have a movable joint between them. In other words, most muscles extend from one bone across one or more joints to another bone. Also, one of the two bones is usually more stationary during a given movement than the other. The muscle's attachment to this more stationary bone is called its **origin.** Its attachment to the more movable bone is called the muscle's **insertion.**

The rest of the muscle (all except its two ends) is called the *body* of the muscle (Figure 9-2).

Tendons anchor muscles firmly to bones, being made of dense, fibrous connective tissue that extends from the muscle "wrappers" described earlier. In the shape of heavy cords or broad sheets, tendons have great strength. They do not tear or pull away from bone easily. Yet any emergency room nurse or physician sees many tendon injuries—severed tendons and tendons torn loose from bones.

Small fluid-filled sacs called **bursae** lie between some tendons and the bones beneath them. These small sacs are made of connective tissue and are lined with **synovial membrane.** The synovial membrane secretes a slippery lubricating fluid—**synovial fluid**—that fills the bursa. Like a small, flexible cushion, a bursa makes it easier for a tendon to slide over a bone when the tendon's muscle shortens.

Tendon sheaths enclose some tendons. Because these tube-shaped structures are also lined with synovial membrane

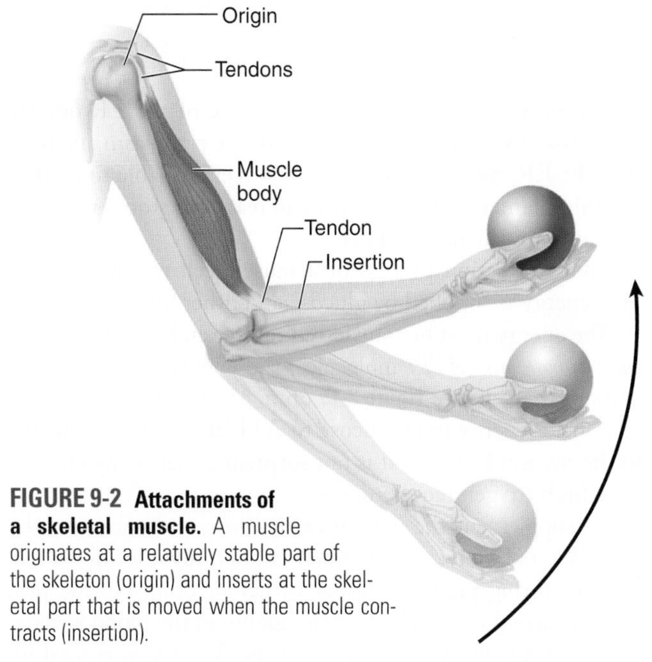

FIGURE 9-2 Attachments of a skeletal muscle. A muscle originates at a relatively stable part of the skeleton (origin) and inserts at the skeletal part that is moved when the muscle contracts (insertion).

and are moistened with synovial fluid, they, like the bursae, facilitate body movement.

Muscle Fibers

Structure of Muscle Fibers

Skeletal muscle tissue consists of elongated contractile cells, or muscle fibers, that look like long, tapered cylinders. Their flexible connective tissue wrappings hold them together in parallel groups, allowing the muscle fibers to all pull together in the same direction—as a team.

Each skeletal muscle fiber has a very unique cytoskeleton structure. The fiber's internal framework is organized into many long cylinders, each made up of two kinds of threadlike microfilaments called **thick** and **thin myofilaments.** The thick myofilaments are formed from a protein called **myosin,** and the thin myofilaments are composed mainly of the protein **actin.**

Each shaftlike myosin molecule has a "head" that sticks out toward the actin molecules. At rest, the actin is blocked from connecting with the myosin heads by small proteins attached to the actin. During contraction, however, the blocking proteins release actin and the myosin heads connect to form *crossbridges* between the thick and thin filaments.

Find the label **sarcomere** in Figure 9-3. Think of the sarcomere as the basic functional, or *contractile,* unit of skeletal muscle. The submicroscopic structure of a sarcomere consists of numerous thick and thin myofilaments arranged so that when viewed under a microscope, dark and light stripes or

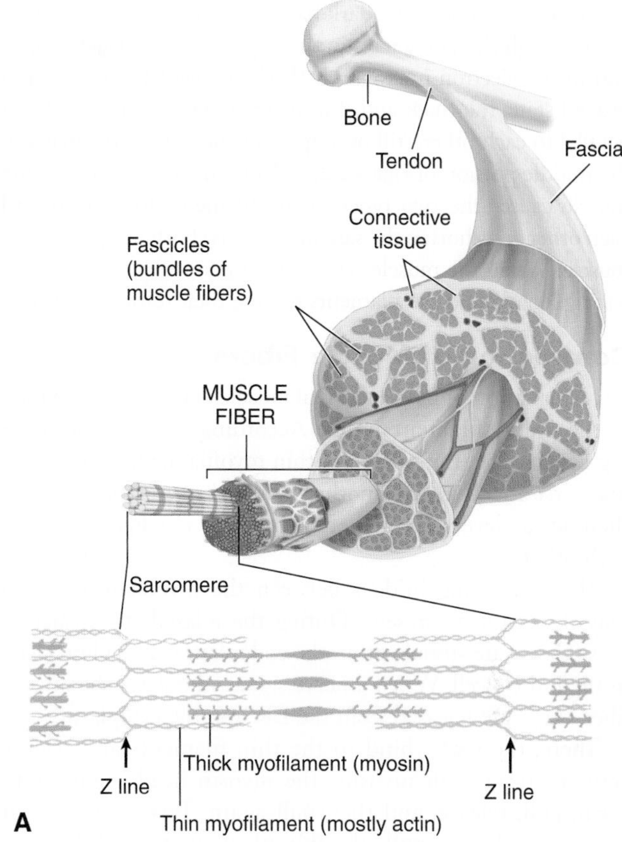

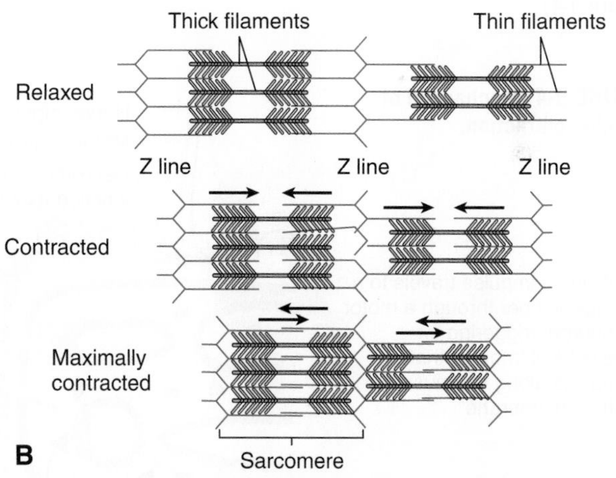

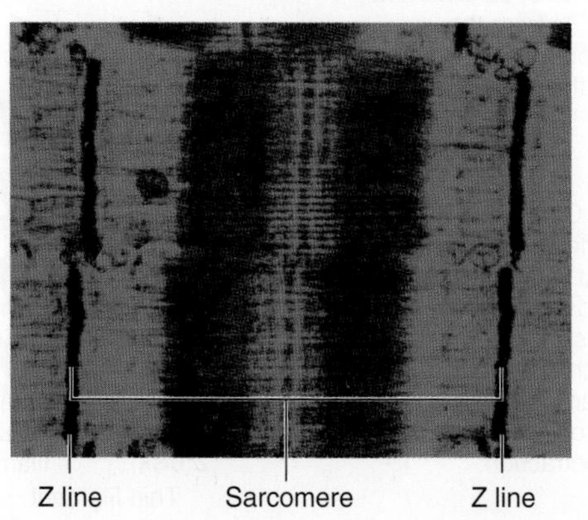

FIGURE 9-3 Structure of skeletal muscle. A, Each muscle organ has many muscle fibers, each containing many bundles of thick and thin myofilaments. The diagrams show the overlapping thick and thin filaments arranged to form adjacent segments called *sarcomeres.* **B,** During contraction, the thin filaments are pulled toward the center of each sarcomere, thereby shortening the whole muscle. **C,** This electron micrograph shows that the overlapping thick and thin filaments within each sarcomere create a pattern of dark striations in the muscle. The extreme magnification allowed by electron microscopy has revolutionized our concept of the structure and function of skeletal muscle and other tissues.

9

striations are seen. The repeating units, or sarcomeres, are separated from each other by dark bands called Z *lines* or Z *disks.*

Although the sarcomeres in the upper portions (**Figure 9-3**, *B*) and in the electron micrograph (EM) of **Figure 9-3**, *C*, are in a relaxed state, the thick and thin myofilaments, which are lying parallel to each other, still overlap. Now look at the diagrams in the middle portion of **Figure 9-3**, *A*. Note that contraction of the muscle causes the two types of myofilaments to slide toward each other and shorten the sarcomere, thus shortening the entire muscle. When the muscle relaxes, the sarcomeres can return to resting length, and the filaments resume their resting positions.

Contraction of Muscle Fibers

An explanation of how a skeletal muscle contracts is provided by the **sliding filament model.** According to this model, during contraction, the thick and thin myofilaments in a muscle fiber first attach to one another by forming crossbridges that then act as levers to ratchet or pull the myofilaments past each other.

The connecting bridges between the myofilaments form only if calcium is present. During the relaxed state, calcium ions (Ca^{++}) are stored within the endoplasmic reticulum (ER) in the muscle cell. When a nerve signal stimulates the muscle fiber, Ca^{++} are released from the ER into the cytoplasm.

There, the Ca^{++} bind to the thin filaments and release actin to react with myosin. The myosin heads connect to actin, pull, release, and then pull again. This ratcheting of myosin heads thus pulls the thin filaments toward the center of the sarcomere—producing the muscle contraction (**Figure 9-4**).

Calcium ion pumps in the membrane of the ER quickly pull most of the free Ca^{++} back out of the cytoplasm and back into the ER. Because Ca^{++} are no longer available to bind to the thin filaments, the actin-myosin reactions stop—restoring relaxation in the muscle fiber.

The contraction process of a muscle cell requires energy. This energy is supplied by glucose and other nutrients.

The energy must be transferred to myosin heads by *adenosine triphosphate (ATP)* molecules, the energy-transfer molecules of the cell (see **Figure 2-15** on p. 35). Oxygen is required to transfer energy from glucose to ATP and make it available to the myosin heads, so it is not surprising that many muscles have high oxygen requirements.

To supplement the oxygen carried to muscle fibers by the *hemoglobin* of blood, muscle fibers contain **myoglobin**—a red, oxygen-storing pigment similar to hemoglobin. During rest, oxygen carried to muscles by hemoglobin in the blood is taken up by myoglobin within muscle fibers. As oxygen is used up quickly during muscle contractions, oxygen from myoglobin adds to the oxygen from hemoglobin—thereby allowing maximum "recharging" of energy-containing ATP molecules.

We discuss the processes of transferring energy to ATP to cellular processes in Chapter 19.

> To learn more about how energy in the body is measured, including examples the energy cost of common muscular activities, check out the article *Measuring Energy* at Connect It! at *evolve.elsevier.com.*

FIGURE 9-4 Mechanism of muscle contraction.

1. A nerve impulse travels to a muscle fiber through a motor neuron, triggering an electrical impulse that travels along the muscle fiber membrane.
2. The impulse triggers the release of calcium ions (Ca^{++}) from the endoplasmic reticulum and into the cytoplasm.
3. The Ca^{++} ions bind to thin filaments and permit actin to react with myosin. Myosin heads form ratcheting crossbridges with actin, which pull the thin filaments toward the middle of the sarcomere—thus producing a contraction.

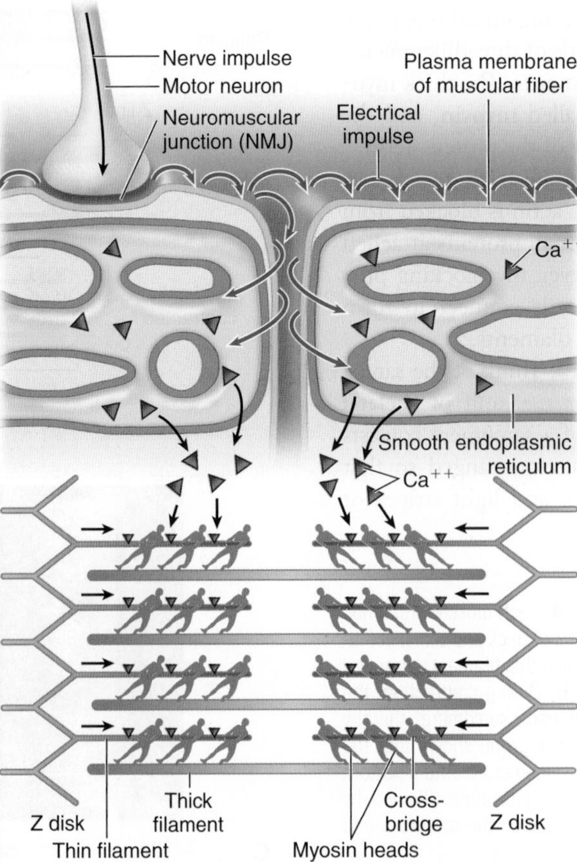

QUICK CHECK

1. What are the three main types of muscle tissue? How do they differ?
2. Distinguish between *fascicles* and *fascia*.
3. What is a muscle's *origin*? Its *insertion*?
4. How do a muscle's myofilaments provide the mechanism for movement?

> To better understand this concept, use the Active Concept Map *How a Skeletal Muscle Fiber Contracts* at *evolve.elsevier.com.*

Function of Skeletal Muscle

The functions of the muscular system are many. Most obviously, this system *produces movement* of the skeleton and permits us to move our whole body, as well as our

CLINICAL APPLICATION

RIGOR MORTIS

The term **rigor mortis** is a Latin phrase that means "stiffness of death." In a medical context, the term *rigor mortis* refers to the stiffness of skeletal muscles sometimes observed shortly after death.

What causes rigor mortis? At the time of death, stimulation of muscle cells ceases. However, some muscle fibers may have been in mid-contraction at the time of death—when the myosin-actin crossbridges are still intact. ATP is required to release the crossbridges and "energize" them for their next attachment. Because the last of a cell's ATP supply is used up at the time it dies, many crossbridges may be left "stuck" in the contracted position. Thus muscles in a dead body may be stiff because individual muscle fibers ran out of the ATP required to "turn off" a muscle contraction.

individual limbs. By producing continuous tension on the skeleton—**muscle tone**—this system also helps maintain a stable body position, or *posture*. As we learned in Chapter 1, skeletal muscles also *produce heat* and thus help us maintain homeostatic balance of body temperature.

Movement

Muscles move bones by pulling on them. Because the length of a skeletal muscle becomes shorter as its fibers contract, the bones to which the muscle attaches move closer together. As a rule, only the insertion bone moves. Look again at **Figure 9-2**. As the ball is lifted, the shortening of the muscle body pulls the insertion bone toward the origin bone. The origin bone stays put, holding firm, while the insertion bone moves toward it. Remember this simple rule: a muscle's insertion bone moves toward its origin bone. It can help you understand muscle actions.

Shortening of a muscle is the primary example of muscle action in this chapter, but it is important to remember that muscles can also produce tension as they extend. This occurs when muscles lengthen under tension, as the muscle insertion is pulled by a load away from the origin. For example, when you lower a heavy bowling ball from your shoulder the muscles in your arm produce tension as they lengthen and allow you to gently lower it—otherwise the ball would suddenly fall and possibly cause injury. Tension during muscle lengthening is often called *eccentric contraction*.

Voluntary muscular movement is normally smooth and free of jerks and tremors because skeletal muscles generally work in coordinated teams, not individually. Several muscles contract while others relax to produce almost any movement that you can imagine.

Of all the muscles contracting simultaneously, the one that is mainly responsible for producing a particular movement is called the **prime mover** for that movement. The other muscles that help in producing the movement are called **synergists.** As prime movers and synergist muscles at a joint contract, other muscles, called **antagonists,** relax. When antagonist muscles contract, they produce a movement opposite to that of the prime movers and their synergist muscles.

Locate the biceps brachii, brachialis, and triceps brachii muscles in **Figure 9-10** on p. 231. All of these muscles are involved in bending and straightening the forearm at the elbow joint. Each may play a different role, depending on the starting positions of the bones and from which direction the load is resisting the movement.

Imagine the elbow is slightly flexed and forearm is slightly twisted so that the hand is "thumb up," as in holding a large bottle of water upright. The biceps brachii may act as the prime mover as you flex the elbow to lift the bottle to drink. In this case, the brachialis may act as its helper or synergist muscle.

If you instead lifted a heavy object from a palm-up position, the brachialis and biceps brachii muscles would likely reverse roles, with the brachialis becoming the prime mover. This happens because the direction of the load has changed—so muscle control will adapt accordingly.

SCIENCE APPLICATIONS

MUSCLE FUNCTION

Andrew Huxley (1917–2012)

The British physiologist Andrew F. Huxley is largely responsible for explaining how muscle fibers contract. After making pioneering discoveries in how nerves conduct impulses, a feat for which he shared the 1963 Nobel Prize in Medicine or Physiology, Huxley turned his attention to muscle fibers. It was he who, in the 1950s, proposed the sliding filament model along with its mechanical explanation of muscle contraction.

Today, **exercise physiologists** and other researchers continue to find out more about how muscle fibers and muscle organs work. These discoveries are being applied in many different professions. For example, **nutritionists** use this information in advising athletes and others concerning what and when to eat to maximize muscular strength and endurance. *Athletes* themselves, along with their coaches and athletic trainers, use current concepts of muscle science in helping them improve their performance.

Health professionals such as **physicians, nurses,** *physical therapists,* and *occupational therapists* use information about muscular problems such as myasthenia gravis and muscular dystrophy to help clients improve their mobility and quality of life. Many other professions such as **chiropractic, massage therapy, ergonomics, physical education,** *fitness, dance, art,* and **biomechanical engineering** also rely on up-to-date information on muscle structure and function for optimal performance.

In either case, when the biceps brachii and brachialis muscles lift the forearm, the triceps brachii relaxes to allow the movement—thus acting as the antagonistic muscle.

When the forearm straightens, these three muscles continue to work as a team. However, during straightening, the triceps brachii becomes the prime mover and the biceps brachii and brachialis become the antagonistic muscles. This combined and coordinated teamlike activity is what makes our muscular movements smooth and graceful.

To learn more about movement of the muscles, go to AnimationDirect online at *evolve.elsevier.com*.

Posture

We are able to maintain our body position because of a specific type of skeletal muscle contraction called *muscle tone* or **tonic contraction.** Because relatively few of a muscle's fibers shorten at one time in a tonic contraction, the muscle as a whole does not shorten, and no movement occurs. Consequently, tonic contractions do not move any body parts. They do hold muscles in position, however. In other words, muscle tone maintains **posture.**

Good posture is the definition of body positioning that favors best functioning of all body parts. Such positioning balances the distribution of weight and therefore puts the least strain on muscles, tendons, ligaments, and bones.

Skeletal muscle tone maintains posture by counteracting the pull of gravity. Gravity tends to pull the head and trunk downward and forward, but the tone in certain back and neck muscles pulls just hard enough in the opposite direction to overcome the force of gravity and hold the head and trunk erect.

Heat Production

Healthy survival depends on our ability to maintain a constant body temperature. A fever, or elevation in body temperature, of only a degree or two above 37° C (98.6° F) is almost always a sign of illness. Just as serious is a fall in body temperature. Any decrease below normal, a condition called **hypothermia,** drastically affects cellular activity and normal body function. The contraction of muscle fibers produces most of the heat required to maintain body temperature.

Energy required to produce a muscle contraction is obtained from ATP. Some of the energy transferred to ATP and released during a muscular contraction is used to shorten the muscle fibers; however, much of the energy is lost as heat during its transfer to ATP. This heat helps us to maintain our body temperature at a constant level.

Sometimes the heat from generating ATP during heavy muscle use can produce too much heat, and we have to sweat or shed layers of clothing to cool back down to our setpoint temperature.

Fatigue

If muscle fibers are stimulated repeatedly without adequate periods of rest, the strength of the muscle contraction decreases, resulting in **fatigue.** If repeated stimulation occurs, the strength of the contraction continues to decrease, and eventually the muscle loses its ability to contract.

9

HEALTH AND WELL-BEING

SLOW AND FAST MUSCLE FIBERS

Sports physiologists know there are three basic skeletal muscle fiber types in the body: slow, fast, and intermediate fibers. Each type is best suited to a particular style of muscular contraction—a fact that is useful when considering how different muscles are used in various athletic activities.

Slow fibers are also called "red fibers" because they have a high content of oxygen-storing myoglobin (a red pigment similar to hemoglobin). Slow fibers are best suited to endurance activities such as long-distance running (pictured) because they do not fatigue easily. Muscles that maintain body position—posture—have a high proportion of slow fibers.

Fast fibers are also called "white fibers" because they have a low red myoglobin content. Fast fibers are best suited for quick, powerful contractions because even though they fatigue quickly they can produce a great amount of ATP very quickly. Fast fibers are well suited to sprinting and weight-lifting events. Muscles that move the fingers have a high proportion of fast fibers—a big help when playing computer games or musical instruments.

Intermediate fibers have characteristics between the extremes of slow and fast fibers. This muscle type is found in muscles such as the calf muscle (gastrocnemius) that is used both for posture and occasional brief, powerful contractions such as jumping.

Each muscle of the body is a mixture of varying proportions of slow, fast, and intermediate fibers.

During exercise, the stored ATP required for muscle contraction becomes depleted. Formation of more ATP results in rapid consumption of oxygen and nutrients, often outstripping the ability of the muscle's blood supply to replenish them. When oxygen supplies run low, the muscle fibers switch to a type of energy conversion that does not require oxygen. This process produces lactic acid that may contribute to a burning sensation in muscle during exercise.

The simple term **oxygen debt** describes the continued increased metabolism that must occur in a cell to remove excess lactic acid that accumulates during prolonged exercise. Thus the depleted energy reserves are replaced. Labored breathing after the cessation of exercise is required to "pay the debt" of oxygen required for the metabolic effort.

The technical name for oxygen debt used by exercise physiologists is *excess post-exercise oxygen consumption (EPOC),* a term that more directly describes what happens after exercise.

The oxygen debt mechanism is a good example of homeostasis at work. The body returns the cells' energy and oxygen reserves to normal, resting levels.

Interested in learning more about EPOC? Check out the illustrated article *Oxygen Debt* at Connect It! at *evolve.elsevier.com.*

Integration with Other Body Systems

Remember that muscles do not function alone. Other structures such as bones and joints must function along with them. Most skeletal muscles cause movements by pulling on bones across movable joints.

However, the respiratory, cardiovascular, nervous, muscular, and skeletal systems all play essential roles in producing normal movements. This fact has great practical importance. For example, a person might have perfectly normal muscles and still not be able to move normally. He or she might have a nervous system disorder that shuts off impulses to certain skeletal muscles, which results in **paralysis.** *Multiple sclerosis (MS)* creates paralysis in this way, but so do some other conditions such as a brain hemorrhage, a brain tumor, or a spinal cord injury.

Skeletal system disorders, especially arthritis, have disabling effects on body movement.

Muscle functioning, then, depends on the functioning of many other parts of the body. This fact illustrates a principle that is repeated often in this book. It can be simply stated: Each part of the body is one of many components in a large, interactive system that maintains homeostasis. The normal function of one part depends on the normal function of the other parts.

QUICK CHECK

1. What are the three primary functions of the muscular system?
2. When a *prime mover* muscle contracts, what does its *antagonist* do?
3. How would you define the term *posture?*
4. How does muscle function affect body temperature?
5. What is *oxygen debt?*

Motor Unit

Before a skeletal muscle can contract and pull on a bone to move it, the muscle must first be stimulated by nerve impulses. Muscle cells are stimulated by a nerve fiber called a **motor neuron** (see Figures 9-4 and 9-5). The junction between the nerve ending and the muscle fiber is called a **neuromuscular junction (NMJ).**

Signal chemicals called *neurotransmitters* are released by the motor neuron in response to a nervous impulse. The type of neurotransmitter operating in each NMJ is called **acetylcholine (ACh).** The released ACh diffuses across a tiny gap at the NMJ and triggers events within the muscle cell that stimulate the contraction process illustrated in Figure 9-4. A single motor neuron, with the muscle cells it innervates, is called a **motor unit** (see Figure 9-5).

Muscle Stimulus

In a laboratory setting, a single muscle fiber can be isolated and subjected to stimuli of varying intensities so that it can be studied. Such experiments show that a muscle fiber does not contract until an applied stimulus reaches a certain level of intensity. The minimal level of stimulation required to cause a fiber to contract is called the **threshold stimulus.**

A muscle is composed of many muscle cells that are controlled by different motor units and that have different threshold-stimulus levels. Consider also that different numbers of motor units can be activated simultaneously, thus affecting the overall strength of contraction. This fact has tremendous importance in everyday life. It allows you to pick up either a small bottle of water or a two-gallon jug because different numbers of motor units can be activated for different loads.

9

Types of Muscle Contraction

In addition to the tonic contraction of muscle that maintains muscle tone and posture, other types of contraction occur as well. These additional types of muscle contraction include the following:

1. Twitch contraction
2. Tetanic contraction
3. Isotonic contraction
4. Isometric contraction

Twitch and Tetanic Contractions

A **twitch** is a quick, jerky response to a stimulus. Twitch contractions can be seen in isolated muscles during research, but they play a minimal role in normal muscle activity. To accomplish the coordinated and fluid muscular movements needed for most daily tasks, muscles must contract in a smooth and sustained way—not in the jerky manner of a twitch.

A **tetanic contraction** is a more sustained and steady response than a twitch. It is produced by a series of stimuli bombarding the muscle in rapid succession. Contractions

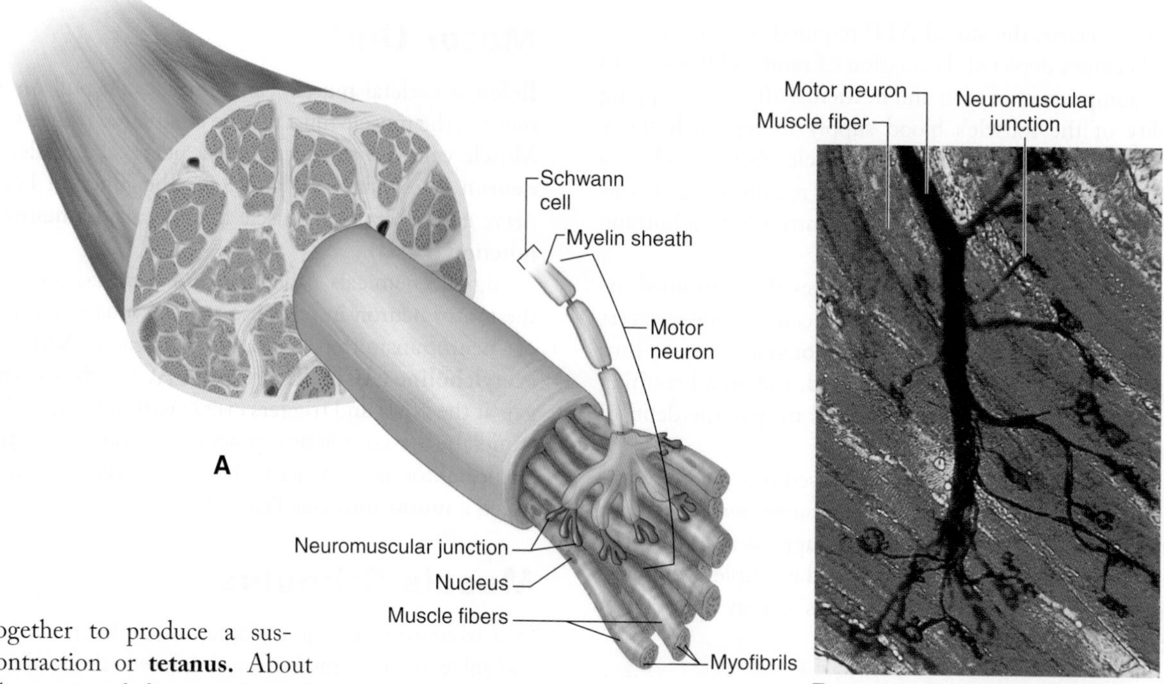

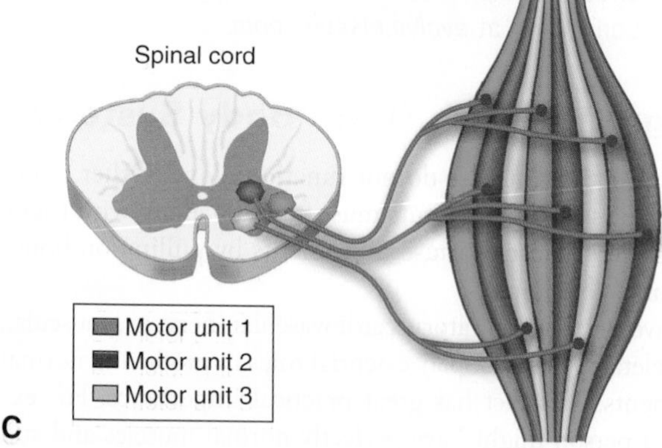

FIGURE 9-5 Motor unit. A, A motor unit consists of one motor neuron and the muscle fibers supplied by its branches. **B,** Micrograph of a motor unit. **C,** Diagram of several motor units, each controlled by its own motor neuron.

"melt" together to produce a sustained contraction or **tetanus.** About 30 stimuli per second, for example, evoke a tetanic contraction in certain types of skeletal muscle. Tetanic contraction is not necessarily a maximal contraction in which each muscle fiber responds at the same time. In most cases, only a few motor units undergo contractions at any one time.

Isotonic Contraction

In most cases, **isotonic contraction** of muscle produces movement at a joint. With this type of contraction, the muscle changes length, and the insertion end moves relative to the point of origin (**Figure 9-6,** *A*).

There are two types of isotonic contraction. One is **concentric contraction,** in which the muscle shortens. The other is **eccentric contraction,** in which the muscle lengthens but still provides work.

For example, lifting this book requires concentric contraction of the biceps muscle that flexes your elbow. Lowering the book slowly and safely requires eccentric contraction of the biceps muscle. Thus, what we call muscle "contraction" really means any pulling of the muscle whether it shortens or not.

Walking, running, breathing, lifting, twisting, and most body movements are examples of isotonic contraction.

Isometric Contraction

Contraction of a skeletal muscle does not always produce movement. Sometimes, it increases the tension within a muscle but does not change the length of the muscle. When the muscle contracts and no movement results, it is called an **isometric contraction.** The word *isometric* comes from Greek words that mean "equal measure." In other words, a muscle's length during an isometric contraction and during relaxation is about equal.

Although muscles do not shorten (and thus produce no movement) during isometric contractions, tension within them increases (**Figure 9-6,** *B*). Because of this, repeated isometric contractions make muscles grow larger and stronger. Pushing against a wall or other immovable object is a good example of isometric exercise. Although no movement occurs and the muscle does not shorten, its internal tension increases dramatically.

Effects of Exercise

We know that exercise is good for us. Some of the benefits of regular, properly practiced exercise are greatly improved muscle tone, better posture, more efficient heart and lung function, less fatigue, and looking and feeling better.

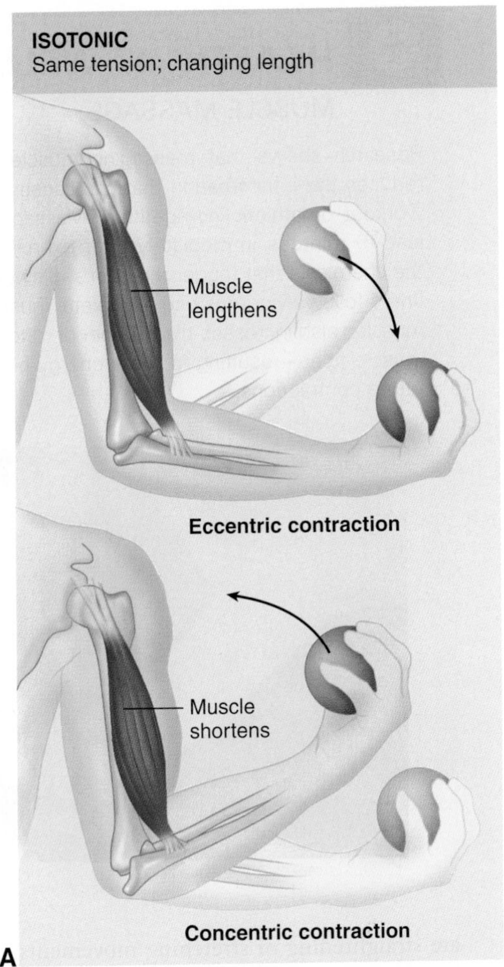

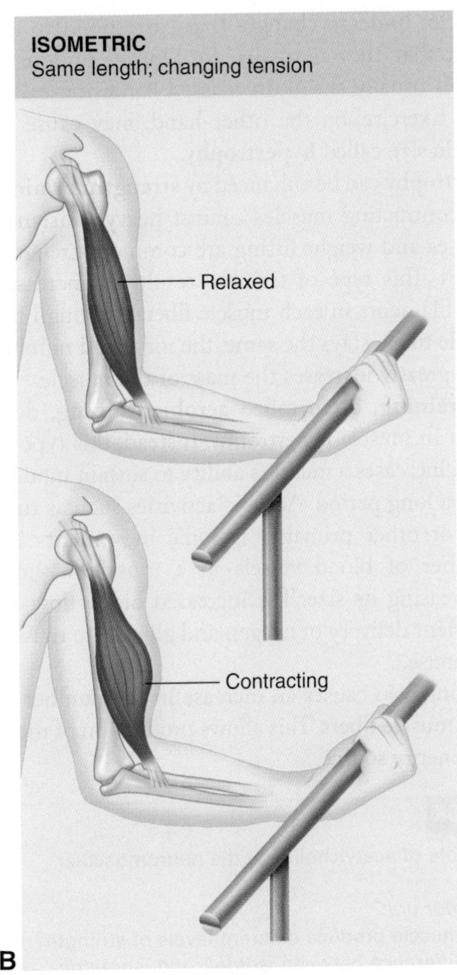

FIGURE 9-6 Types of muscle contraction. A, In isotonic contraction the muscle changes length, producing movement either by eccentric contraction (muscle lengthens) or concentric contraction (muscle shortens). **B,** In isometric contraction the muscle pulls forcefully against a load but does not shorten.

RESEARCH, ISSUES, AND TRENDS

ENHANCING MUSCLE STRENGTH

The most obvious and effective way of increasing skeletal muscle strength is by strength training, that is, regularly pulling against heavy resistance. The maximal amount of muscular strength one can achieve is determined mainly by genetics. However, there are a number of chemical enhancements athletes have tried over the centuries to improve strength. An early fad among athletes in the twentieth century was the overuse of vitamin supplements. Although moderate vitamin supplementation will ensure adequate intake of vitamins necessary for good muscle function, overuse may lead to *hypervitaminosis* and possibly serious consequences.

Another type of chemical often abused by athletes is *anabolic steroids.* Anabolic steroids are usually synthetic derivatives of the male hormone *testosterone.* As with testosterone, they do in fact stimulate an increase in muscle size and strength, making them dangerously attractive to coaches and athletes wanting to win their events. However, use of these hormones can cause serious, even life-threatening, hormonal imbalances. For this reason, anabolic steroids are strictly banned from organized sports.

Sports physiologists are now investigating a whole variety of chemicals such as *creatine phosphate* and various coenzymes (enzyme helpers) that are reported to enhance strength or endurance. Always *carefully* review the latest research findings on such substances with the help of your reference librarian and discuss them with your physician before using them yourself, or you may suffer serious health consequences.

Skeletal muscles undergo changes that correspond to the amount of work that they normally do. During prolonged inactivity, muscles usually shrink in mass, a condition called **disuse atrophy.** Exercise, on the other hand, may cause an increase in muscle size called **hypertrophy.**

Muscle hypertrophy can be enhanced by **strength training,** which involves contracting muscles against heavy resistance. Isometric exercises and weight lifting are common strength-training activities. This type of training results in increased numbers of myofilaments in each muscle fiber. Although the number of muscle fibers stays the same, the increased number of myofilaments greatly increases the mass of the muscle.

Endurance training, often called **aerobic training,** does not usually result in muscle hypertrophy. Instead, this type of exercise program increases a muscle's ability to sustain moderate exercise over a long period. Aerobic activities such as running, bicycling, or other primarily isotonic movements increase the number of blood vessels in a muscle without significantly increasing its size. The increased blood flow allows a more efficient delivery of oxygen and glucose to muscle fibers during exercise.

Aerobic training also causes an increase in the number of mitochondria in muscle fibers. This allows production of more ATP as a rapid energy source.

QUICK CHECK

1. What is the role of acetylcholine at the neuromuscular junction?
2. What is a *motor unit*?
3. How does a muscle produce different levels of strength?
4. What is the difference between *isotonic* and *isometric* muscle contraction?
5. How does strength training differ from *endurance training*?

HEALTH AND WELL-BEING

MUSCLE MASSAGE

Research shows that massaging muscles after exercise reduces the inflammation that can cause soreness after a workout or athletic competition. Massaging muscles for just 10 minutes immediately after exercise can decrease the chemicals that trigger the pain and swelling that sometimes follows vigorous use of skeletal muscles. Massaging muscles also increases the number of mitochondria in each muscle fiber—resulting in more energy being available for future contractions.

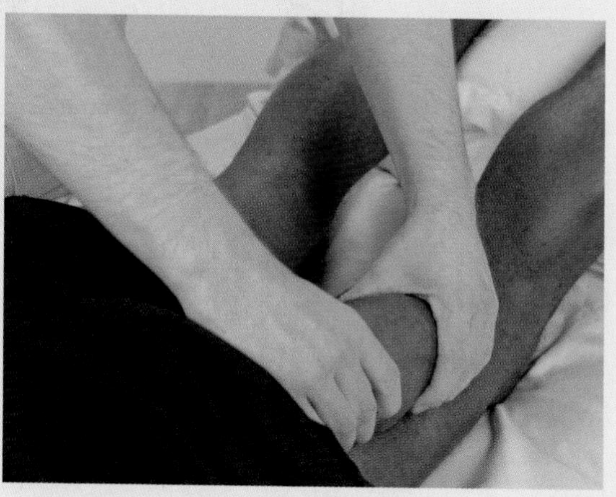

9

Movements Produced by Muscles

The particular type of movement that may occur at any joint depends on the muscles acting at that joint, on their origin and insertion points, on the shapes of the bones involved, and the joint type (see Chapter 8). Muscles acting on some joints produce movement in several directions, whereas only limited movement is possible at other joints. The terms most often used to describe body movements are described in the following sections.

Angular Movements

Flexion is a movement that makes the angle between two bones at their joint smaller than it was at the beginning of the movement. Most flexions are movements commonly described as bending. If you bend your elbow or your knee, you **flex** it.

Extension movements are the opposite of flexions. They make the angle between two bones at their joint larger than it was at the beginning of the movement. Therefore, extensions are straightening or stretching movements rather than bending movements. When you straighten your elbow or knee, you **extend** it.

Figure 9-7 and **Figure 9-9**, *A*, show flexion and extension of the elbow. **Figure 9-8** illustrates flexion and extension of the knee.

Abduction means moving a part away from the midline of the body, such as moving your arm out to the side.

Adduction means moving a part toward the midline, such as bringing your arms down to your sides from an elevated position. When you move your arm to the side to wave, you **abduct** it. When you move your arm back toward your body, you **adduct** it. **Figure 9-9**, *B*, shows abduction and adduction.

Circular Movements

Rotation is movement around a longitudinal axis. You **rotate** your head and neck by moving your skull from side to side as in shaking your head "no" (**Figure 9-9**, *C*).

Circumduction moves a part so that its distal end moves in a circle (**Figure 9-9**, *D*). When a pitcher winds up to throw a ball, she **circumducts** her arm.

Supination and **pronation** refer to hand positions that result from rotation of the forearm. (The term *prone* refers to the body as a whole lying face down. *Supine* means lying face up.) Supination results in a hand position with the palm turned to the anterior position (as in the anatomical position) and pronation occurs when you turn the palm of your hand so that it faces posteriorly (**Figure 9-9**, *E*).

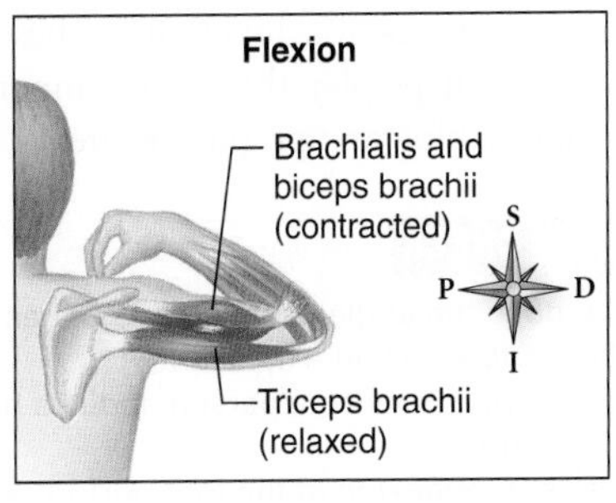

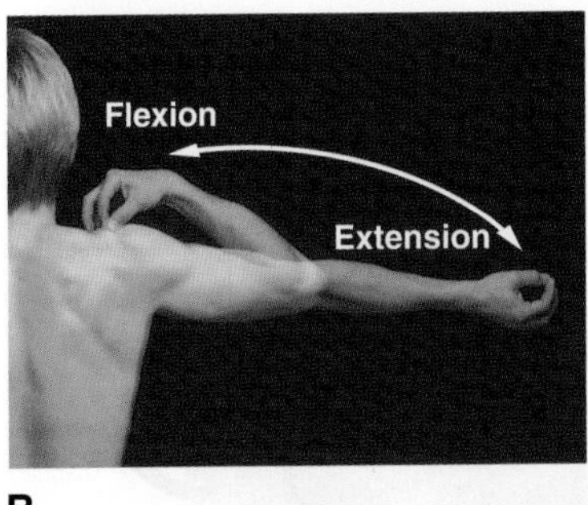

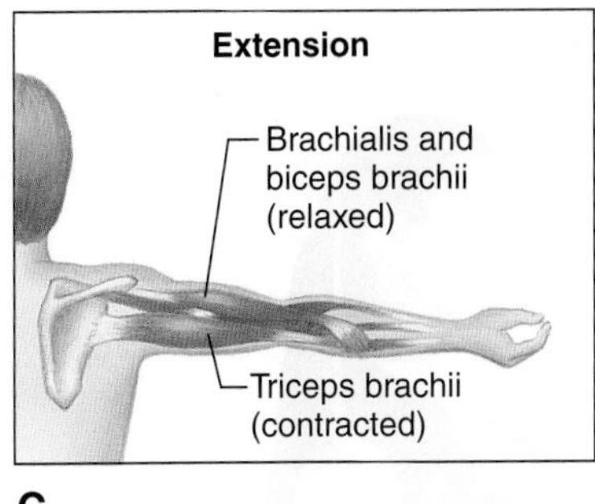

FIGURE 9-7 Flexion and extension of the forearm. **A** and **B,** When the forearm is flexed at the elbow, the brachialis and biceps brachii contract while an antagonist, the triceps brachii, relaxes. **B** and **C,** When the forearm is extended, the brachialis and biceps brachii relax while the triceps brachii contracts.

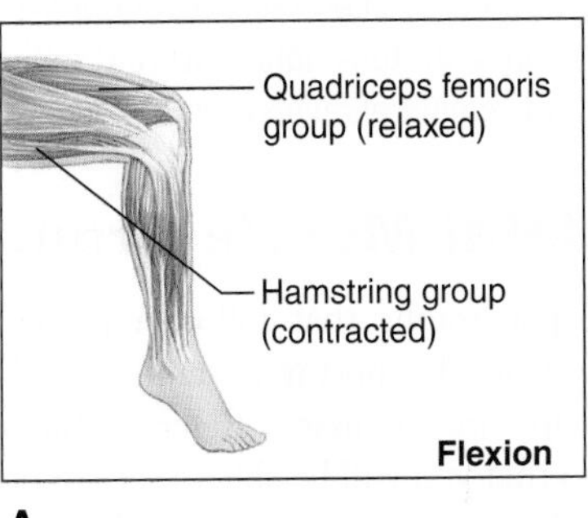

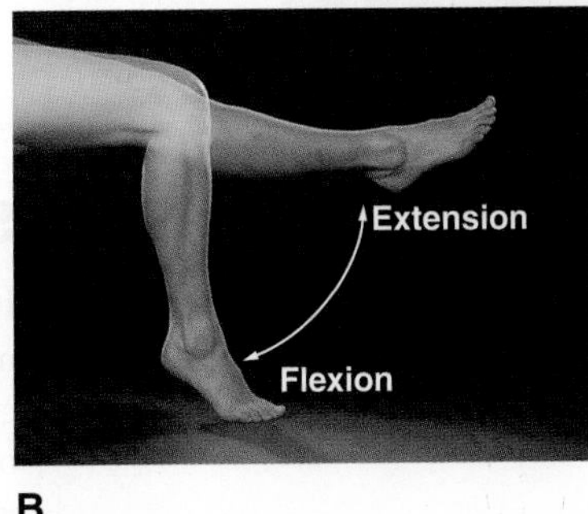

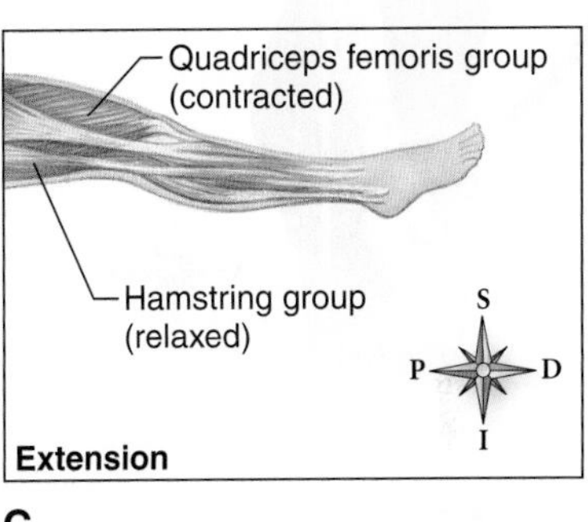

FIGURE 9-8 Flexion and extension of the leg. **A** and **B,** When the leg flexes at the knee, muscles of the hamstring group contract while their antagonists in the quadriceps femoris group relax. **B** and **C,** When the leg extends, the hamstring muscles relax while the quadriceps femoris muscle contracts.

When you ask for change, you **supinate** as you twist the forearm outward and you **pronate** as you twist the forearm inward to put the change in your pocket.

Special Movements

Some body parts, such as the foot, are difficult to describe with ordinary terms, so special terms are often used to describe their unique movements.

Dorsiflexion and **plantar flexion** refer to ankle movements. To **dorsiflex** the ankle, the dorsum or top of the foot is elevated with the toes pointing upward—as when standing on your heel. To **plantar flex** the ankle, the bottom of the foot is directed downward so that you are in effect standing on your toes (**Figure 9-9**, *F*).

Inversion and **eversion** are also ankle movements. Inversion moves turn the ankle so that the bottom of the foot faces toward the midline of the body (**Figure 9-9**, *G*). You **invert** your foot when the sole moves inward. Eversion turns the ankle in the opposite direction, so that the bottom of the foot faces toward the side of the body (**Figure 9-9**, *H*). You **evert** your foot when your sole moves outward.

As you study the illustrations and learn to recognize the muscles discussed in this chapter, you should attempt to group them according to function, as in **Table 9-1**. Some muscle names include the type of movement the muscle produces, such as the muscle named *adductor longus.* You will note, for example, that flexors produce many of the movements used for walking, sitting, swimming, typing, and many other

TABLE 9-1 Muscles Grouped According to Function

PART MOVED	FLEXORS	EXTENSORS	ABDUCTORS	ADDUCTORS
Arm	Pectoralis major	Latissimus dorsi	Deltoid	Pectoralis major and latissimus dorsi contracting together
Forearm	Biceps brachii	Triceps brachii	None	None
Thigh	Iliopsoas Sartorius Rectus femoris	Gluteus maximus Hamstrings	Gluteus medius	Adductor group
Leg	Hamstrings	Quadriceps group	None	None
Foot	Tibialis anterior	Gastrocnemius Soleus	Fibularis longus Fibularis brevis	Tibialis anterior Fibularis tertius

9

ANGULAR

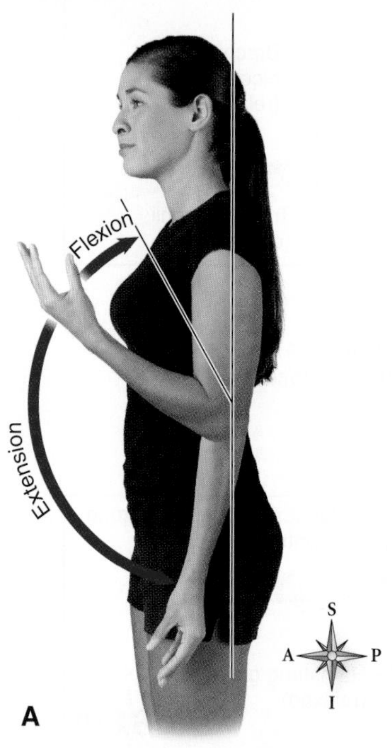

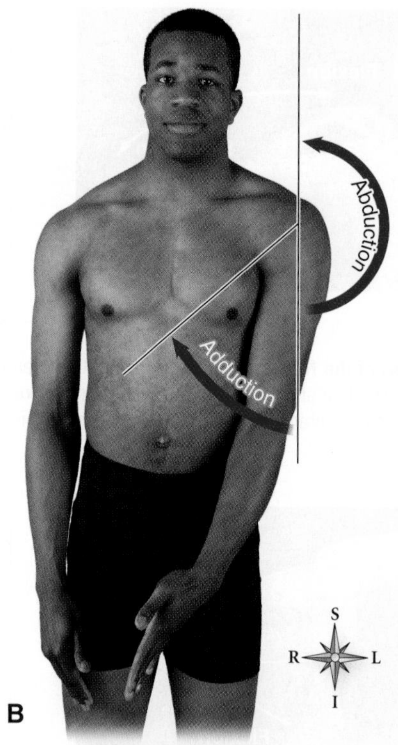

activities. Extensors also function in these activities but perhaps play their most important role in maintaining an upright posture.

QUICK CHECK

1. How does the angle between two bones differ in *flexion* and *extension*?
2. What happens when a person abducts his or her arm?
3. How is *dorsiflexion* of the foot performed?
4. Flexors and extensors function in many of the same activities; however, the extensors play what important role?

To learn more about movement produced by skeletal muscle contractions, go to AnimationDirect online at *evolve.elsevier.com.*

Skeletal Muscle Groups

In the paragraphs that follow, representative muscles from the most important skeletal muscle groups are discussed. Refer to **Figure 9-10** often so that you will be able to see a muscle as you read about its placement on the body and its function.

CIRCULAR

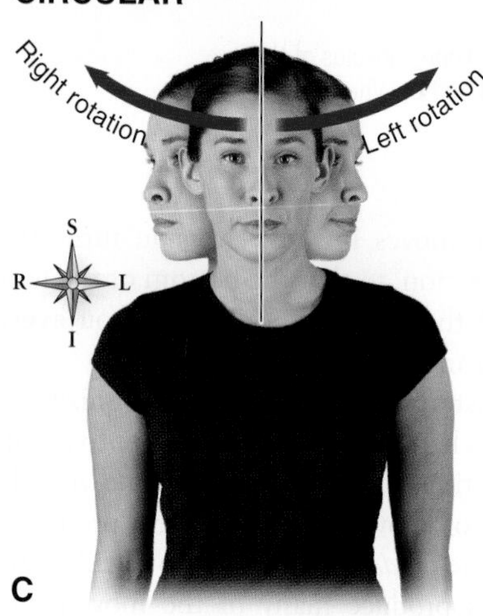

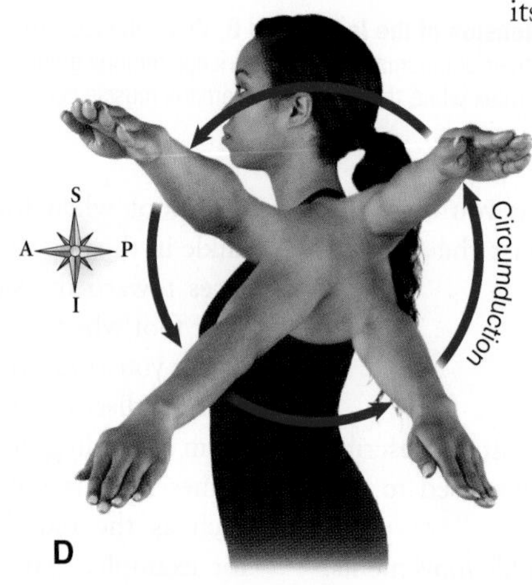

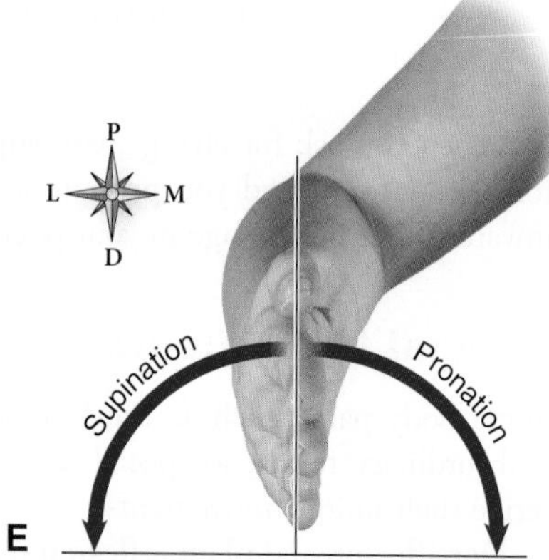

9

SPECIAL

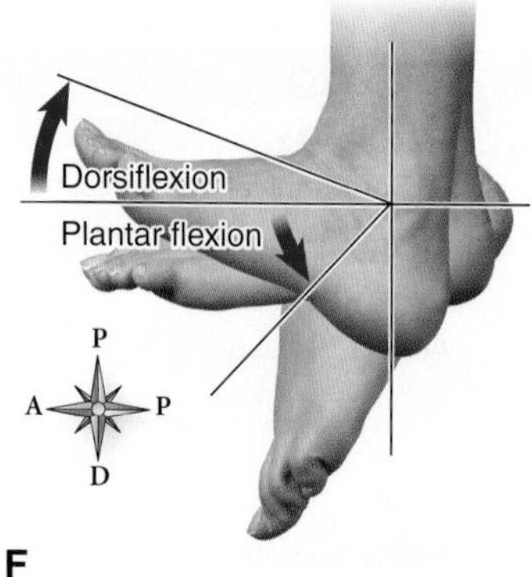

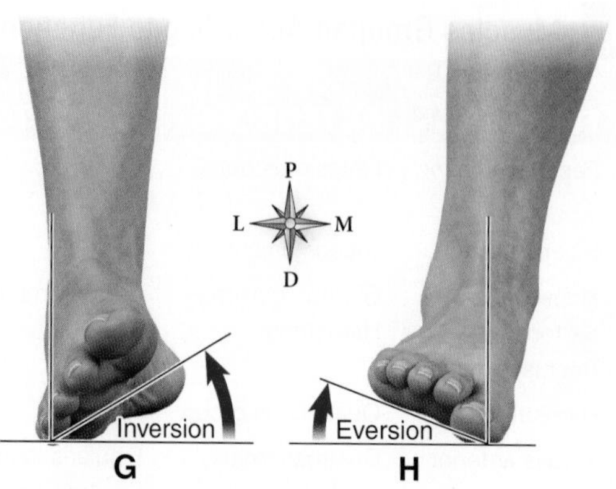

FIGURE 9-9 Examples of body movements. **A,** Flexion and extension. **B,** Adduction and abduction. **C,** Rotation. **D,** Circumduction. **E,** Pronation and supination. **F,** Dorsiflexion and plantar flexion. **G,** Inversion. **H,** Eversion.

Table 9-2 through **Table 9-5** identify and group muscles according to function and provide information about muscle action and points of origin and insertion. Keep in mind that muscles move bones, and the bones that they move are their insertion bones.

As you study the major muscles of the body, try to find them in the *Clear View of the Human Body* (follows p. 8).

Muscles of the body do not work as isolated engines of movement, but instead act in functional teams to produce effective movement. Check out the illustrated article *Whole Body Muscle Mechanics* at Connect It! at *evolve.elsevier.com*.

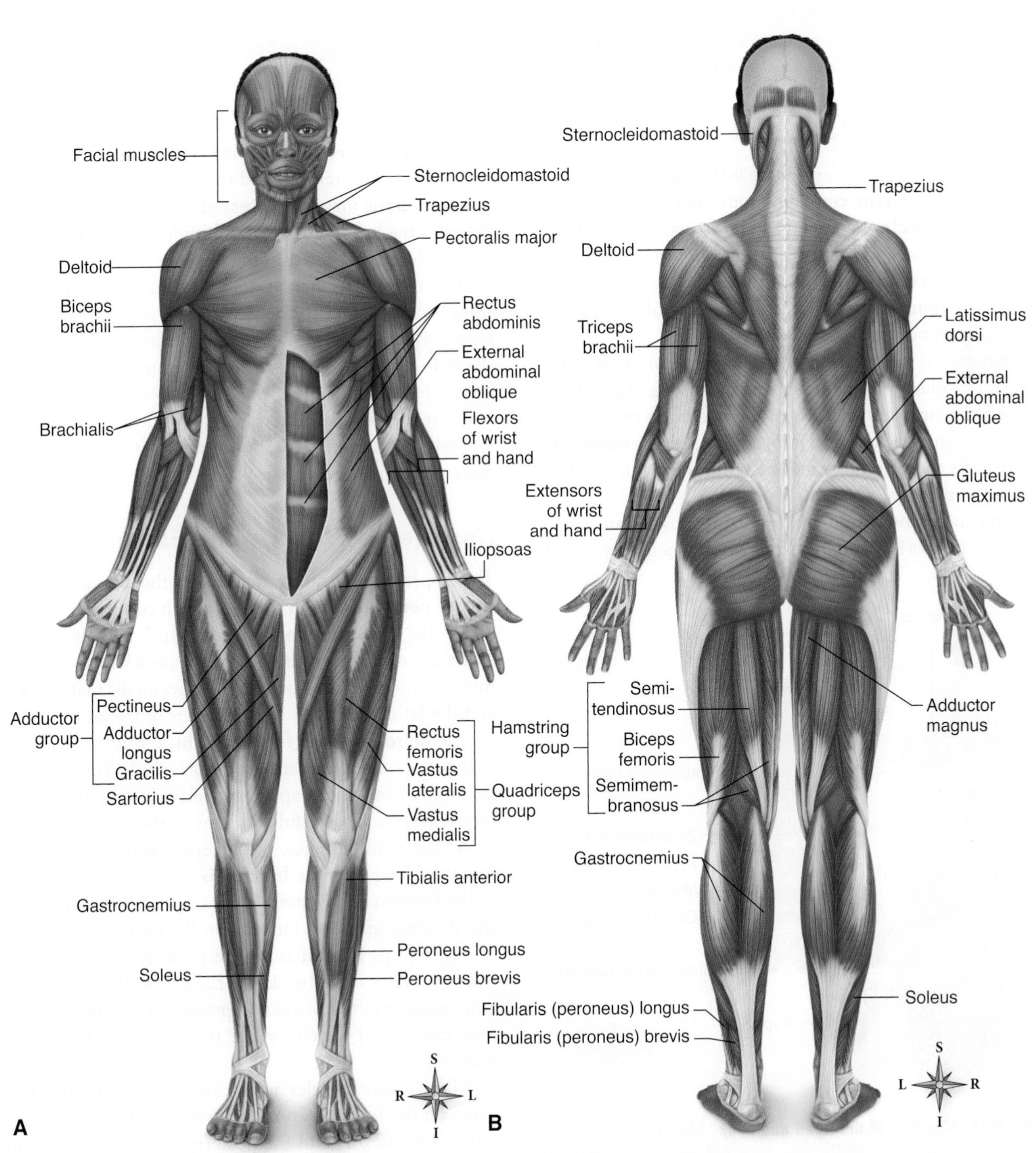

FIGURE 9-10 Overview of muscles of the body. A, Anterior view. **B,** Posterior view. Both views show an adult female.

TABLE 9-2 Muscles of the Head and Neck

MUSCLE	FUNCTION	INSERTION	ORIGIN
Frontal	Raises eyebrow	Skin of eyebrow	Occipital bone
Orbicularis oculi	Closes eye	Maxilla and frontal bone	Maxilla and frontal bone (encircles eye)
Orbicularis oris	Draws lips together	Encircles lips	Encircles lips
Zygomaticus	Elevates corners of mouth and lips	Angle of mouth and upper lip	Zygomatic
Masseter	Closes jaws	Mandible	Zygomatic arch
Temporal	Closes jaws	Mandible	Temporal region of the skull
Sternocleidomastoid	Rotates and flexes head and neck	Mastoid process	Sternum and clavicle
Trapezius	Extends head and neck Moves or stabilizes scapula	Scapula	Skull and upper vertebrae

Muscles of the Head and Neck

The *muscles of facial expression* (Figure 9-11) allow us to communicate many different emotions nonverbally. Contraction of the **frontal muscle,** for example, allows you to raise your eyebrows in surprise and furrow the skin of your forehead into a frown. The **orbicularis oris,** called the *kissing muscle,* puckers the lips. The **zygomaticus** elevates the corners of the mouth and lips and has been called the *smiling muscle.*

The muscles of **mastication** are responsible for closing the mouth and producing chewing movements. As a group, they are among the strongest muscles in the body. The two largest muscles of the group, identified in Figure 9-11, are the **masseter,** which elevates the mandible, and the **temporal,** which assists the masseter in closing the jaw.

The **sternocleidomastoid** and **trapezius** muscles are easily identified in Figures 9-10 and 9-11. The two sternocleidomastoid muscles are located on the anterior surface of the neck. They originate on the sternum and then pass up and cross the neck to insert on the mastoid process of the skull. Working together, they flex the head on the chest. If only one contracts, the head is both flexed and tilted to the opposite side.

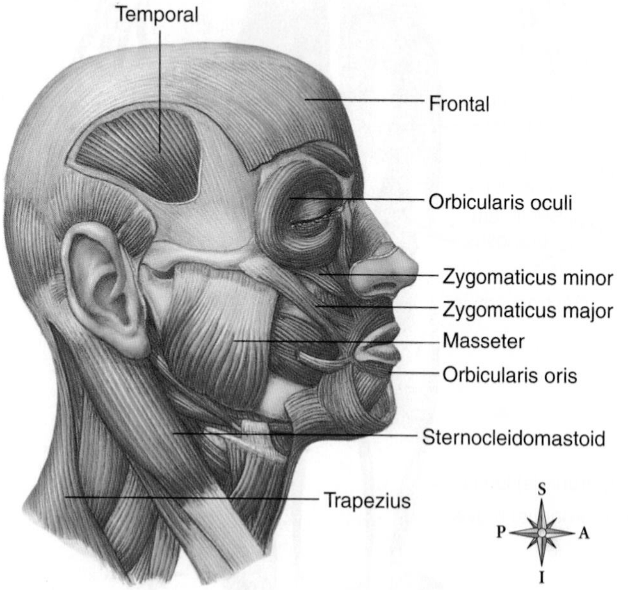

The triangular-shaped trapezius muscles form the line from each shoulder to the neck on its posterior surface. They have a wide line of origin extending from the base of the skull down the spinal column to the last thoracic vertebra. When contracted, the trapezius muscles help elevate the shoulders and extend the head backward.

Table 9-2 summarizes important facts about the major muscles of the head and neck.

Temporal
Frontal
Orbicularis oculi
Zygomaticus minor
Zygomaticus major
Masseter
Orbicularis oris
Sternocleidomastoid
Trapezius
S
P
A
I

FIGURE 9-11 Muscles of the head and neck. Muscles that produce most facial expressions surround the eyes, nose, and mouth. Large muscles of mastication stretch from the upper skull to the lower jaw. These powerful muscles produce chewing movements. The neck muscles connect the skull to the trunk of the body, rotating the head or bending the neck.

Muscles of the Upper Extremities

The upper extremity is attached to the thorax by the fan-shaped **pectoralis major** muscle, which covers the upper chest, and by the **latissimus dorsi** muscle, which takes its origin from structures over the lower back (see Figures 9-10 and 9-12). Both muscles insert on the humerus. The pectoralis major is a flexor, and the latissimus dorsi is an extensor of the arm.

The **deltoid** muscle forms the thick, rounded prominence over the shoulder and arm (see Figure 9-10). The muscle takes its origin from the scapula and clavicle and inserts on the humerus. It is a powerful abductor of the arm.

As the name implies, the **biceps brachii** is a two-headed muscle. Along with the **brachialis** muscle, it serves as a flexor of the forearm (see Figure 9-10). It originates from the bones of the shoulder girdle and inserts on the radius in the forearm.

The **triceps brachii** is on the posterior or back surface of the arm. It has three heads of origin from the shoulder girdle and inserts into the olecranon process of the ulna. The triceps is an extensor of the elbow and thus performs a straightening function.

Many of the muscles acting on the wrist and hand are located on the forearm. In most instances, the anterior forearm muscles are flexors and the posterior forearm muscles are extensors (see Figure 9-10).

Table 9-3 summarizes important facts about muscles of the upper extremities.

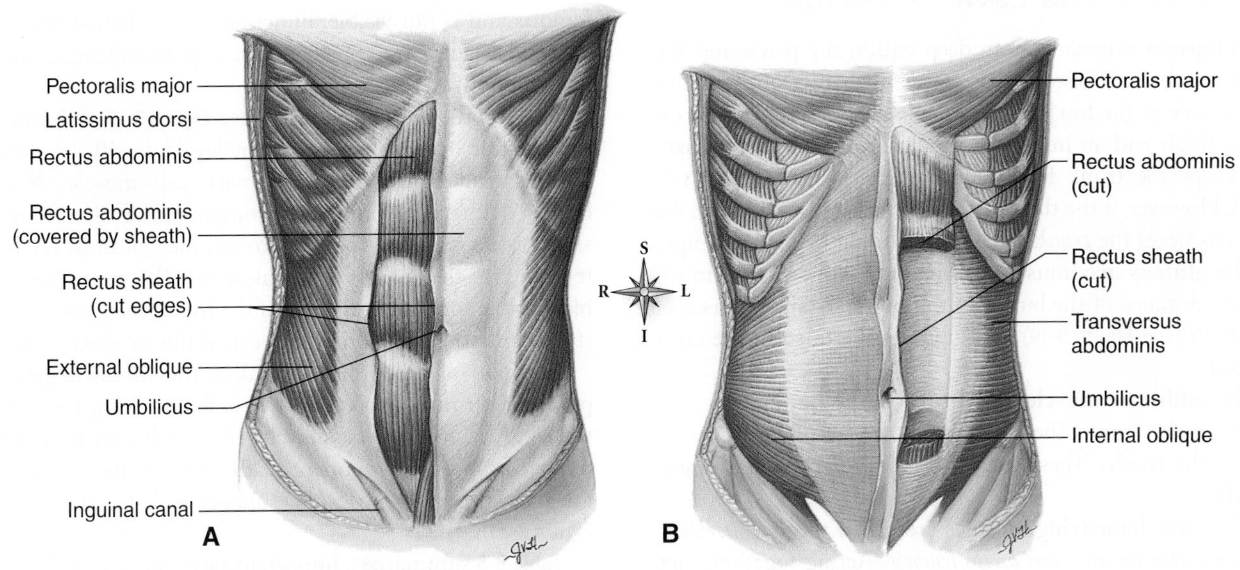

FIGURE 9-12 Muscles of the trunk. A, Anterior view showing superficial muscles. **B,** Anterior view showing deeper muscles.

TABLE 9-3 Muscles of the Upper Extremities

MUSCLE	FUNCTION	INSERTION	ORIGIN
Pectoralis major	Flexes and helps adduct arm	Humerus	Sternum, clavicle, and upper rib cartilages
Latissimus dorsi	Extends and helps adduct arm	Humerus	Vertebrae and ilium
Deltoid	Abducts arm	Humerus	Clavicle and scapula
Brachialis	Flexes elbow	Humerus	Ulna
Biceps brachii	Flexes elbow; pronates/supinates wrist	Radius	Scapula
Triceps brachii	Extends elbow	Ulna	Scapula and humerus

Muscles of the Trunk

The muscles of the anterior, or front, side of the abdomen are arranged in three layers, with the fibers in each layer running in different directions much like the layers of wood in a sheet of plywood (**Figure 9-12**). The result is a very strong "girdle" of muscle that covers and supports the abdominal cavity and its internal organs.

The three layers of muscle in the anterolateral (side) abdominal walls are arranged as follows: the outermost layer or **external oblique;** a middle layer or **internal oblique;** and the innermost layer or **transversus abdominis.** In addition to these sheetlike muscles, the band- or strap-shaped **rectus abdominis** muscle runs down the midline of the abdomen from the thorax to the pubis. In addition to protecting the abdominal viscera, the rectus abdominis flexes the spinal column. The muscles of the trunk can be seen in **Figure 9-12**.

The *respiratory muscles* are discussed in Chapter 17. **Intercostal muscles,** located between the ribs, and the sheetlike **diaphragm** separating the thoracic and abdominal cavities change the size and shape of the chest during breathing. As a result, air is moved into or out of the lungs.

Table 9-4 summarizes important facts about muscles of the trunk.

9

TABLE 9-4 Muscles of the Trunk

MUSCLE	FUNCTION	INSERTION	ORIGIN
External oblique	Compresses abdomen	Midline of abdomen	Lower thoracic cage
Internal oblique	Compresses abdomen	Midline of abdomen	Pelvis
Transversus abdominis	Compresses abdomen	Midline of abdomen	Ribs, vertebrae, and pelvis
Rectus abdominis	Flexes trunk	Lower rib cage	Pubis
Diaphragm	Expands thoracic cavity during inspiration	Circumference of lower rib cage	Fibrous tissue (central tendon) at center of diaphragm

Muscles of the Lower Extremities

The **iliopsoas** originates from deep within the pelvis and the lower vertebrae to insert on the lesser trochanter of the femur and capsule of the hip joint. It is generally classified as a flexor of the thigh and an important postural muscle that stabilizes and keeps the trunk from falling over backward when you stand. However, if the thigh is fixed so that it cannot move, the iliopsoas flexes the *trunk*. An example would be doing sit-ups.

The **gluteus maximus** forms the outer contour and much of the substance of the buttock. It is an important extensor of the thigh (see Figure 9-10) and supports the torso in the erect position.

The **adductor muscles** originate on the bony pelvis and insert on the femur. They are located on the inner or medial side of the thighs. These muscles adduct or press the thighs together.

The three **hamstring muscles** are called the *semimembranosus, semitendinosus,* and *biceps femoris.* Acting together, they serve as powerful flexors of the leg and extensors of the thigh (see Figure 9-10). They originate on the ischium and insert on the tibia or fibula.

The **quadriceps femoris** muscle group covers the upper thigh. The four thigh muscles—the *rectus femoris* and three *vastus* muscles—extend the leg (see Figure 9-10 and Table 9-2). One component of the quadriceps group has its origin on the pelvis, and the remaining three originate on the femur; all four insert on the tibia. Only two of the vastus muscles are visible in Figure 9-10. The vastus intermedius is covered by the rectus femoris and is not visible. Functionally, the hamstrings (flexors) and quadriceps (extensors) act as powerful antagonists in movement of the leg.

The **tibialis anterior** muscle (see Figure 9-10) is located on the anterior, or front, surface of the leg. It dorsiflexes the foot. The **gastrocnemius** is the primary calf muscle. Note in Figure 9-10 that it has two fleshy components arising from both sides of the femur. It inserts through the calcaneal (Achilles) tendon into the heel bone or calcaneus. The gastrocnemius is responsible for plantar flexion of the foot; because it is used to stand on tiptoe, it is sometimes called the *toe dancer's muscle.*

A group of three muscles called the **fibularis group** or **peroneus group** (see Figure 9-10) is found along the sides of the leg. As a group, these muscles plantar flex the foot. A long tendon from one component of the group—the *fibularis longus* muscle tendon—forms a support arch for the foot (see Figure 8-24).

Table 9-5 summarizes important facts about muscles of the lower extremities.

QUICK CHECK

1. What function do the *muscles of mastication* make possible?
2. What muscles are responsible for changing the size and shape of the chest during breathing?
3. What three muscles compose the anterolateral (side) abdominal walls?
4. What action do the *hamstring muscles* perform?

TABLE 9-5 Muscles of the Lower Extremities

MUSCLE	FUNCTION	INSERTION	ORIGIN
Iliopsoas	Flexes thigh or trunk	Femur	Ilium and vertebrae
Sartorius	Flexes thigh and rotates leg	Tibia	Ilium
Gluteus maximus	Extends thigh	Femur	Ilium, sacrum, and coccyx
Adductor Group			
Adductor longus	Adducts thigh	Femur	Pubis
Gracilis	Adducts thigh	Tibia	Pubis
Pectineus	Adducts thigh	Femur	Pubis
Hamstring Group			
Semimembranosus	Flexes knee; extends thigh	Tibia	Ischium
Semitendinosus	Flexes knee; extends thigh	Tibia	Ischium
Biceps femoris	Flexes knee; extends thigh	Fibula	Ischium and femur
Quadriceps Group			
Rectus femoris	Extends knee	Tibia	Ilium
Vastus lateralis Vastus intermedius Vastus medialis	Extend knee	Tibia	Femur
Tibialis anterior	Dorsiflexes ankle	Metatarsals (foot)	Tibia
Gastrocnemius	Plantar flexes ankle	Calcaneus (heel)	Femur
Soleus	Plantar flexes ankle	Calcaneus (heel)	Tibia and fibula
Fibularis Group			
Fibularis longus Fibularis brevis Fibularis tertius	Everts and plantar flexes ankle (tertius dorsiflexes ankle)	Tarsal and metatarsals (ankle and foot)	Tibia and fibula

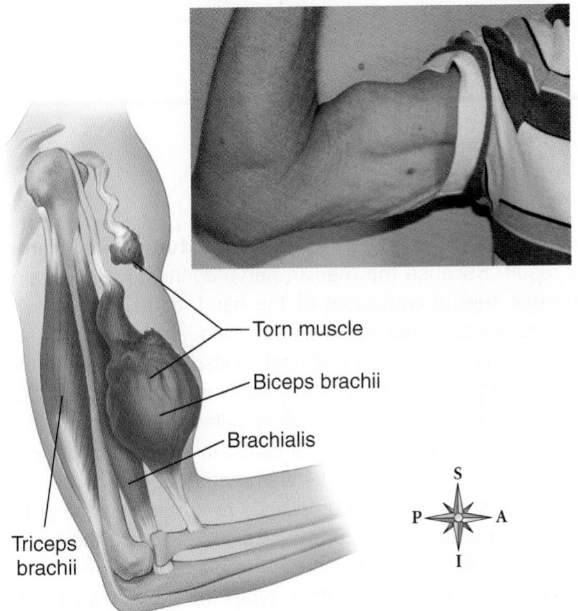

FIGURE 9-13 Muscle strain. Severe strain of the biceps brachii muscle. When a muscle is severely strained, it may break in two pieces, causing a visible gap in muscle tissue under the skin. Notice how the broken ends of the muscle reflexively contract (spasm) to form a knot of tissue.

Muscular Disorders

As you might expect, muscle disorders, or **myopathies,** generally disrupt the normal movement of the body. In mild cases, these disorders vary in degree of discomfort from merely inconvenient to slightly troublesome. Severe muscle disorders, however, can impair the muscles used in breathing—a life-threatening situation.

Muscle Injury

Injuries to skeletal muscles resulting from overexertion or trauma usually result in a **muscle strain** (Figure 9-13). Muscle strains are characterized by muscle pain, or **myalgia,** and involve overstretching or tearing of muscle fibers. If an injury occurs in the area of a joint and a ligament is damaged, the injury may be called a **sprain.**

Any muscle inflammation, including that caused by a muscle strain, is termed **myositis.** If tendon inflammation occurs with myositis, as when one experiences a *charley horse,* the condition is termed **fibromyositis.** Although inflammation may subside in a few hours or days, it usually takes several weeks for damaged muscle fibers to repair themselves. Some damaged muscle cells may be replaced by fibrous tissue, forming scars. Occasionally, hard calcium is deposited in the scar tissue.

Cramps are painful muscle spasms (involuntary twitches). Cramps often result from mild myositis or fibromyositis, but can be a symptom of any irritation or of an ion and water imbalance.

Minor trauma to the body, especially a limb, may cause a muscle *bruise* or **contusion.** Muscle contusions involve local internal bleeding and inflammation. Severe trauma to a skeletal muscle may cause a *crush injury.* Crush injuries not only greatly damage the affected muscle tissue but also cause the release of muscle fiber contents into the bloodstream, which can be life threatening. For example, the reddish muscle pigment *myoglobin* can accumulate in the blood and cause kidney failure.

Review the article *Rhabdomyolysis* at Connect It! at *evolve.elsevier.com.*

Stress-induced muscle tension can result in myalgia and stiffness in the neck and back and is thought to be one cause of "stress headaches." Headache and back-pain clinics use a variety of strategies to treat stress-induced muscle tension. These treatments include massage, biofeedback, and relaxation training.

Muscle Infections

Several bacteria, viruses, and parasites are known to infect muscle tissue—often producing local or widespread myositis. For example, in *trichinosis* (see Appendix A at *evolve.elsevier.com*), widespread myositis is common. The muscle pain and stiffness that sometimes accompany influenza is another example.

Once a tragically common disease, **poliomyelitis** is a viral infection of the nerves that control skeletal muscle movement. Although the disease can be asymptomatic, it often causes paralysis that may progress to death. Eliminated in the United States as a result of a comprehensive vaccination program, it still affects individuals in other parts of the world.

Another infection that affects muscle and is effectively prevented by vaccination is infection by the *Clostridium tetani* bacterium, which is found nearly everywhere in our environment. This condition is often called *tetanus* because the toxin released into the body by *C. tetani* bacteria can produce involuntary, sustained (tetanic) contractions throughout the body (Figure 9-14). If not successfully treated, tetanus infections can be fatal. Tetanus vaccinations, usually combined with other vaccines, are recommended throughout one's lifetime to remain protected.

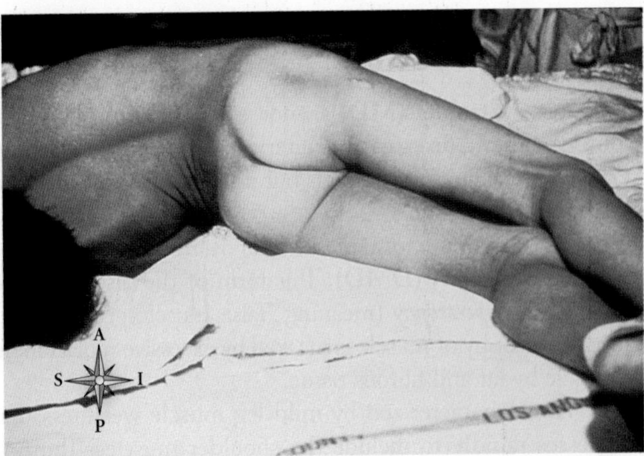

FIGURE 9-14 Tetanus infection. Severe muscle cramping caused by the involuntary, sustained (tetanic) contractions caused by toxins released into the body from *C. tetani* bacteria can produce this tense, twisted body posture.

CLINICAL APPLICATION

OCCUPATIONAL HEALTH PROBLEMS

Some epidemiologists specialize in the field of *occupational health,* the study of health matters related to work or the workplace. Many problems seen by occupational health experts are caused by repetitive motions of the wrists or other joints. Word processors (typists) and meat cutters, for example, are at risk of developing conditions caused by repetitive motion injuries.

One common problem often caused by such repetitive motion is **tenosynovitis**—inflammation of a tendon sheath. Tenosynovitis can be painful, and the swelling characteristic of this condition can limit movement in affected parts of the body. For example, swelling of the tendon sheath around tendons in an area of the wrist known as the *carpal tunnel* can limit movement of the wrist, hand, and fingers.

The figure shows the relative positions of the tendon sheath and median nerve within the carpal tunnel. If this swelling, or any other lesion in the carpal tunnel, presses on the median nerve, a condition called **carpal tunnel syndrome** may result. Because the median nerve connects to the palm and radial side (thumb side) of the hand, carpal tunnel syndrome is characterized by weakness, pain, and tingling in that part of the hand. The pain and tingling also may radiate to the forearm and shoulder. The median nerve and muscles that flex the fingers pass through a concavity called the *carpal tunnel.*

Prolonged or severe cases of carpal tunnel syndrome may be relieved by injection of anti-inflammatory agents. A permanent cure is sometimes accomplished by surgically cutting the fibrous band called the *flexor retinaculum* enclosing the carpal tunnel—thus relieving pressure on the median nerve.

Repetitive motion and other types of trauma also may cause inflammation of a bursa, known as **bursitis.** For example, carpet layers, roofers, and others who work on their knees are prone to bursitis involving the knee joints. Bursitis is most often treated with anti-inflammatory agents.

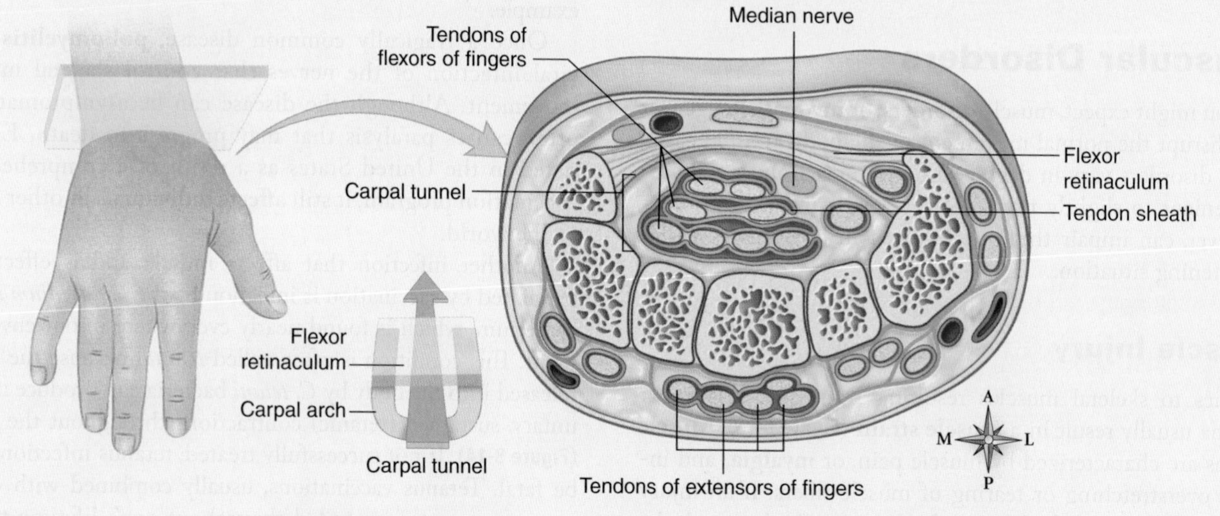

The median nerve and tendons of flexor muscles pass through a concavity called the carpal tunnel.

Muscular Dystrophy

Muscular dystrophy (MD) is not a single disorder but a group of genetic diseases characterized by atrophy (wasting) of skeletal muscle tissues. Some, but not all, forms of muscular dystrophy can be fatal.

The most common form of muscular dystrophy is **Duchenne muscular dystrophy (DMD).** This form of the disease is also called *pseudohypertrophy* (meaning "false muscle growth") because the atrophy of muscle is masked by excessive replacement of muscle by fat and fibrous tissue.

DMD is characterized by mild leg muscle weakness that progresses rapidly to include the shoulder muscles. The first signs of DMD are apparent at about 3 years of age, and the stricken child is usually severely affected within 10 to 12 years. Severe respiratory or cardiac muscle weakness often occurs by the time the patient is in their 20s.

DMD is caused by a missing gene in the X chromosome. DMD occurs primarily in boys. Because girls have two X chromosomes and boys only one, genetic diseases involving X chromosome abnormalities are more likely to occur in boys than in girls.

This is true because girls with one damaged X chromosome may not exhibit an "X-linked" disease if their other X chromosome is normal (see Chapter 25). The missing gene codes for a cell protein called *dystrophin.* In DMD, dystrophin is missing from muscle cells, which then become too weak to hold together during muscle contraction. Immune system responses may add to the damage. Gene therapy (see Chapter 25) is now being tested for use in DMD patients, with the hope

CLINICAL APPLICATION

INTRAMUSCULAR INJECTIONS

Many drugs are administered by **intramuscular injection (IM)**—that is, into the skeletal muscle tissue.

If the amount to be injected is 2 mL or less, the deltoid muscle is often selected as the site of injection. Note in Figure **A** that the needle is inserted into the muscle about two fingers' breadth below the acromion process of the scapula and lateral to the tip of the acromion.

If the amount of medication to be injected is 2 to 3 mL, the gluteal area shown in Figure **B** is often used. Injections are made into the gluteus medius muscle near the center of the upper outer quadrant—sometimes called the *ventrogluteal (VG)* site—as shown in the illustration.

Another technique of locating the VG site is to place the base of the palm over the patient's greater trochanter (femur) using your left palm on the patient's right thigh (and vice versa). Then feel for the anterior superior iliac spine at the end of the iliac crest with the tip of your index finger. Separate your next finger to form a V or "peace sign" and inject into the area in the middle of the V (see Figure B, photo inset). It is important that the sciatic nerve and the superior gluteal blood vessels be avoided during the injection. Proper technique requires knowledge of the underlying anatomy.

The vastus lateralis muscle of the thigh is another common site for IM injections.

In addition to intramuscular injections, which are generally administered by a health care provider in an institutional setting, many individuals must self-administer injections of needed medications on a regular basis in their homes. Educating these patients or their caregivers on how to correctly administer medication by injection is an important issue in the delivery of home health care services. Topics that must be covered include instruction on proper injection techniques, selection of needle length and gauge, identification of important anatomical landmarks when making injection site selections, and the preparation and rotation of selected injection sites.

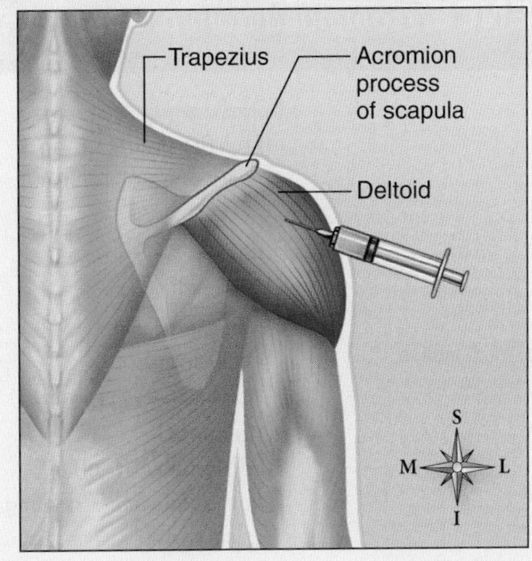

A

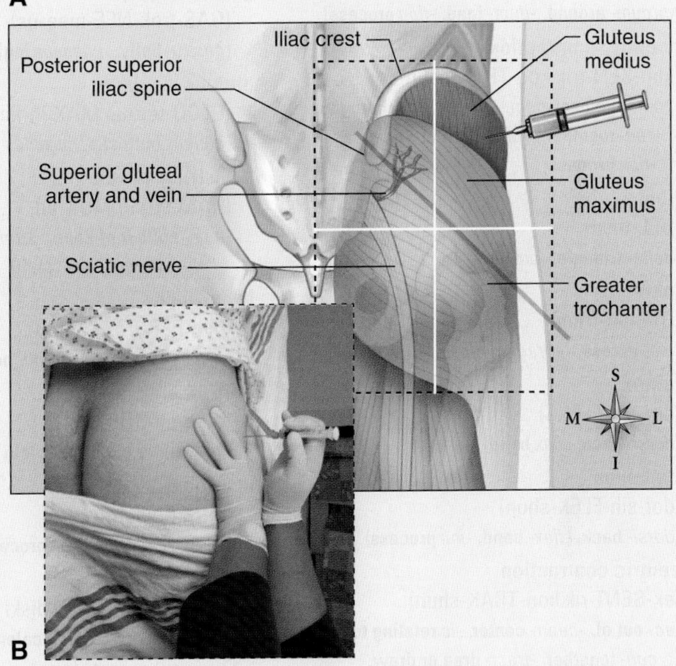

B

that the gene for dystrophin can be replaced by cells from a healthy donor.

Myasthenia Gravis

Myasthenia gravis (MG) is a chronic disease characterized by muscle weakness, especially in the face and throat. Most forms of this disease begin with mild weakness and chronic muscle fatigue in the face, then progress to wider muscle involvement.

When severe muscle weakness causes immobility in all four limbs, a myasthenic crisis is said to have occurred. A person in *myasthenic crisis* is in danger of dying from respiratory failure because of weakness in the respiratory muscles.

Myasthenia gravis is an autoimmune disease in which the immune system attacks ACh receptors on muscle cells at the NMJ (see **Figure 9-4**). Nerve impulses from motor neurons are then unable to fully stimulate the affected muscle. Treatment is individualized for each patient, but may include drugs that block the breakdown of ACh (so more is available at the NMJ) and immunosuppressant drugs that reduce the autoimmune damage to the NMJ.

QUICK CHECK

1. How does myositis differ from fibromyositis?
2. What is poliomyelitis and what effect does it have on skeletal muscle?
3. What is the cause of *Duchenne muscular dystrophy (DMD)*?
4. Describe the characteristics of myasthenia gravis.

9

LANGUAGE OF **SCIENCE** *(continued from p. 219)*

bursa
(BER-sah)
pl., bursae
(BER-see or BER-say)
[*bursa* **purse**]

cardiac muscle
(KAR-dee-ak MUS-el)
[*cardi-* **heart,** *-ac* **relating to,** *mus-* **mouse,** *-cle* **small**]

circumduct
(SER-kum-DUKT)
[*circum-* **around,** *-duct* **lead**]

circumduction
(ser-kum-DUK-shun)
[*circum-* **around,** *-duct-* **lead,** *-tion* **process**]

concentric contraction
(kon-SEN-trik kon-TRAK-shun)
[*con-* **together,** *-centr-* **center,** *-ic* **relating to,** *con-* **together,** *-tract-* **drag or draw,** *-tion* **process**]

deltoid
(DEL-toyd)
[*delta-* **triangle,** *-oid* **like**]

diaphragm
(DYE-ah-fram)
[*dia-* **across,** *-phrag-* **enclose,** *-(u)m* **thing**]

dorsiflex
(dor-sih-FLEKS)
[*dorsi-* **back,** *-flex* **bend**]

dorsiflexion
(dor-sih-FLEK-shun)
[*dorsi-* **back,** *-flex-* **bend,** *-ion* **process**]

eccentric contraction
(ek-SENT-rik kon-TRAK-shun)
[*ec-* **out of,** *-centr-* **center,** *-ic* **relating to,** *con-* **together,** *-tract-* **drag or draw,** *-tion* **process**]

eversion
(ee-VER-shun)
[*e(x)-* **outward,** *-ver-* **turn,** *-sion* **process**]

evert
(ee-VERT)
[*e(x)-* **outward,** *-ver-* **turn**]

extend
(ek-STEND)
[*ex-* **outward,** *-ten* **stretch**]

extension
(ek-STEN-shun)
[*ex-* **outward,** *-tens-* **stretch,** *-sion* **process**]

external oblique
(eks-TER-nal oh-BLEEK)
[*extern-* **outside,** *-al* **relating to,** *obliq-* **slanted**]

fatigue
(fah-TEEG)

fibularis group
(fib-YOO-lay-ris groop)
[*fibula-* **clasp** (fibula bone), *-aris* **relating to**]

flex
(fleks)
[*flex* **bend**]

flexion
(FLEK-shun)
[*flex-* **bend,** *-ion* **process**]

frontal muscle
(FRUN-tal MUS-el)
[*front-* **forehead,** *-al* **relating to,** *mus-* **mouse,** *-cle* **small**]

gastrocnemius
(GAS-trok-NEE-mee-us)
[*gastro-* **belly,** *-cnemius* **leg**]

gluteus maximus
(GLOO-tee-us MAX-ih-mus)
[*gluteus* **buttocks,** *maximus* **greatest**]

hamstring muscle
(HAM-string MUS-el)
[*ham-* **hollow of knee,** *-string* **(tendon),** *mus-* **mouse,** *-cle* **small**]

hypertrophy
(hye-PER-troh-fee)
[*hyper-* **excessive,** *-troph-* **nourishment,** *-y* **state**]

iliopsoas
(il-ee-oh-SOH-as)
[*ilio-* **loin or gut,** *-psoas* **loin muscle**]

insertion
(in-SER-shun)
[*in-* **in,** *-ser-* **join,** *-tion* **process**]

intercalated disk
(in-TER-kah-lay-ted disk)
[*inter-* **between,** *-cala-* **calendar,** *-ate* **act of**]

intercostal muscle
(in-ter-KOS-tal MUS-el)
[*inter-* **between,** *-costa-* **rib,** *-al* **relating to,** *mus-* **mouse,** *-cle* **small**]

internal oblique muscle
(in-TER-nal oh-BLEEK MUS-el)
[*intern-* **inside,** *-al* **relating to,** *obliq-* **slanted,** *mus-* **mouse,** *-cle* **little**]

inversion
(in-VER-shun)
[*in-* **in,** *-ver-* **turn,** *-sion* **process**]

invert
(in-VERT)
[*in-* **in,** *-ver-* **turn**]

isometric contraction
(aye-soh-MET-rik kon-TRAK-shun)
[*iso-* **equal,** *metric* **relating to measure,** *con-* **together,** *-tract-* **drag or draw,** *-tion* **process**]

isotonic contraction
(aye-soh-TON-ik kon-TRAK-shun)
[*iso-* **equal,** *-ton-* **tension,** *-ic* **relating to,** *con-* **together,** *-tract-* **drag or draw,** *-tion* **process**]

latissimus dorsi
(lah-TIS-ih-mus DOR-sye)
[*latissimus* **broadest,** *dorsi* **relating to back**]

masseter
(mah-SEE-ter)
[*masseter* **chewer**]

mastication
(mas-tih-KAY-shun)
[*mastica-* **chew,** *-ation* **process**]

motor neuron
(MOH-ter NOO-ron)
[*mot-* **movement,** *-or* **agent,** *neuron* **nerve**]

motor unit
(MOH-ter YOO-nit)
[*mot-* **movement,** *-or* **agent,** *unit* **single**]

muscle fiber
(MUS-el FYE-ber)
[*mus-* **mouse,** *-cle* **little,** *fibr-* **thread**]

muscle tone
(MUS-el tohn)
[*mus-* **mouse,** *-cle* **little,** *tone* **stretching**]

myofilament
(my-oh-FIL-ah-ment)
[*myo-* **muscle,** *-fila-* **thread,** *-ment* **thing**]

myoglobin
(my-oh-GLOH-bin)
[*myo-* **muscle,** *-glob-* **ball,** *-in* **substance**]

myosin
(MY-oh-sin)
[*myos-* **muscle,** *-in* **substance**]

neuromuscular junction (NMJ)
(noo-roh-MUS-kyoo-lar JUNK-shun [en em jay])
[*neuro-* **nerve,** *-mus-* **mouse,** *-cul-* **little,** *-ar* **relating to,** *-junc-* **join,** *-tion* **condition**]

orbicularis oris
(or-bik-yoo-LAYR-is OR-is)
[*orbi-* **circle,** *-cul-* **little,** *-aris* **relating to,** *oris* **relating to mouth**]

origin
(OR-ih-jin)
[*origin* **source**]

oxygen debt
(AHK-sih-jen det)
[*oxy-* **sharp,** *-gen* **produce,** *debt* **thing owed**]

pectoralis major
(pek-teh-RAH-liss MAY-jor)
[*pector-* **breast,** *-alis* **relating to,** *major* **greater**]

LANGUAGE OF SCIENCE *(continued from p. 238)*

peroneus group
(per-on-EE-uss groop)
[*pero-* **boot,** *-us* **thing**]

plantar flex
(PLAN-tar fleks)
[*planta-* **sole,** *-ar* **relating to,** *flex* **bend**]

plantar flexion
(PLAN-tar FLEK-shun)
[*planta-* **sole,** *-ar* **relating to,** *flex-* **bend,** *-ion* **process**]

posture
(POS-chur)
[*postur-* **position**]

prime mover
(pryme MOO-ver)
[*prime* **first order**]

pronate
(PROH-nayt)
[*prona-* **bend forward,** *-ate* **process**]

pronation
(PROH-nay-shun)
[*prona-* **bend forward,** *-ation* **process**]

quadriceps femoris
(KWOD-reh-seps feh-MOR-is)
[*quadri-* **four,** *-ceps* **head,** *femoris* **related to the thigh (femur)**]

rectus abdominis
(REK-tus ab-DOM-ih-nus)
[*rectus* **straight,** *abdominis* **related to the abdomen**]

rigor mortis
(RIG-or MOR-tis)
[*rigor* **stiffness,** *mortis* **of death**]

rotate
(roh-TAYT)
[*rot-* **turn,** *-ate* **process**]

rotation
(roh-TAY-shun)
[*rot-* **turn,** *-ation* **process**]

sarcomere
(SAR-koh-meer)
[*sarco-* **flesh,** *-mere* **part**]

skeletal muscle
(SKEL-et-al MUS-el)
[*skelet-* **dried body,** *-al* **relating to,** *mus-* **mouse,** *-cle* **small**]

sliding filament model
(SLY-ding FIL-ah-ment MAH-del)
[*slide-* **glide,** *-ing* **action,** *fila-* **thread,** *-ment* **thing,** *model* **standard**]

smooth muscle
(smoothe MUS-el)
[*smooth* **smooth,** *mus-* **mouse,** *-cle* **small**]

sternocleidomastoid
(STERN-oh-KLYE-doh-MAS-toyd)
[*sterno-* **breastbone (sternum),** *-cleid-* **key (clavicle),** *-masto-* **breast (mastoid process),** *-oid* **like**]

supinate
(soo-pih-NAYT)
[*supin-* **turned backward (belly up),** *-ate* **process**]

supination
(soo-pih-NAY-shun)
[*supin-* **turned backward (belly up),** *-ation* **process**]

synergist
(SIN-er-jist)
[*syn-* **together,** *-erg-* **to work,** *-ist* **agent**]

synovial fluid
(sih-NOH-vee-al FLOO-id)
[*syn-* **together,** *-ovi-* **egg (white),** *-al* **relating to**]

synovial membrane
(sih-NOH-vee-al MEM-brayn)
[*syn-* **together,** *-ovi-* **egg (white),** *-al* **relating to,** *membran-* **thin skin**]

temporal
(TEM-poh-ral)
[*tempora-* **temple (of head),** *-al* **relating to**]

tendon
(TEN-don)
[*tend-* **pulled tight,** *-on* **unit**]

tendon sheath
(TEN-don sheeth)
[*tend-* **pulled tight,** *-on* **unit**]

tetanic contraction
(teh-TAN-ik kon-TRAK-shun)
[*tetanus* **tension,** *-ic* **relating to,** *con-* **together,** *-tract-* **drag or draw,** *-tion* **process**]

tetanus
(TET-ah-nus)
[*tetanus* **tension**]

threshold stimulus
(THRESH-hold STIM-yoo-lus)
[*stimul-* **to excite,** *-us* **thing**]

tibialis anterior
(tib-ee-AL-is an-TEER-ee-or)
[*tibia-* **shinbone,** *-alis* **relating to,** *ante-* **front,** *-er-* **more,** *-or* **quality**]

tonic contraction
(TAHN-ik kon-TRAK-shun)
[*ton-* **to stretch,** *-ic* **relating to,** *con-* **together,** *-tract-* **drag or draw,** *-tion* **process**]

transversus abdominis
(tranz-VERS-us ab-DAH-min-us)
[*trans-* **across,** *-vers-* **turn,** *abdomin-* **belly**]

trapezius
(trah-PEE-zee-us)
[*trapezius* **small table (irregular 4-sided shape)**]

triceps brachii
(TRY-seps BRAY-kee-aye)
[*tri-* **three,** *-ceps* **head,** *brachii* **related to the arm**]

twitch
(twich)
[*twitch* **quick jerk**]

zygomaticus
(zye-goh-MAT-ik-us)
[*zygo-* **union or yoke,** *-ic-* **relating to,** *-us* **thing**]

9

LANGUAGE OF MEDICINE

aerobic training
(ayr-OH-bik TRAYN-ing)
[*aer-* **air,** *-bi-* **life,** *-ic* **relating to**]

biomechanical engineering
(bye-oh-meh-KAN-ik-al en-juh-NEER-ing)
[*bio-* **life,** *-mechan-* **machine,** *-ic-* **relating to,** *-al* **relating to,** *engin-* **devise or design,** *-eer* **practitioner**]

bursitis
(ber-SYE-tis)
[*burs-* **purse,** *-itis* **inflammation**]

carpal tunnel syndrome
(KAR-pul TUN-el SIN-drohm)
[*carp-* **wrist,** *-al* **relating to,** *syn-* **together,** *-drome* **running or (race) course**]

chiropractic
(kye-roh-PRAK-tik)
[*chiro-* **hand,** *-practic* **practical**]

contusion
(kon-TOO-zhun)
[*contus-* **bruise,** *-sion* **result**]

cramps
(kramps)

Continued on p. 240

LANGUAGE OF **MEDICINE** *(continued from p. 239)*

disuse atrophy
(DIS-yoos AT-roh-fee)
[*dis-* **absence of,** *a-* **without,** *-troph-* **nourishment]**

Duchenne muscular dystrophy (DMD)
(doo-SHEN MUS-kyoo-lar DIS-troh-fee)
[*Guillaume B.A. Duchenne de Boulogne* **French neurologist,** *muscul-* **little mouse (muscle),** *-ar* **relating to,** *dys-* **bad,** *-troph-* **nourishment,** *-y* **state]**

endurance training
(en-DUR-ance TRAYN-ing)
[*en-* **in,** *-dur-* **harden,** *-ance* **state]**

ergonomics
(er-go-NOM-iks)
[*ergo-* **work,** *-nom-* **arrangement,** *-ic* **relating to]**

exercise physiologist
(EK-ser-syze fiz-ee-OL-oh-jist)

fibromyositis
(fye-broh-my-oh-SYE-tis)
[*fibro-* **fiber,** *-myos-* **muscle,** *-itis* **inflammation]**

intramuscular injection (IM)
(in-trah-MUS-kyoo-lar in-JEK-shun)
[*intra-* **within,** *mus-* **mouse,** *-cle* **little,** *-ar* **relating to,** *in-* **in,** *-ject-* **throw,** *-tion* **process]**

hypothermia
(hye-poh-THER-mee-ah)
[*hypo-* **under or below,** *-therm-* **heat,** *-ia* **abnormal condition]**

massage therapy
(mah-SAHJ THAYR-ah-pee)
[*mass-* **handle,** *-age* **process,** *therap-* **treatment,** *-y* **activity]**

muscle strain
(MUS-el strayn)
[*mus-* **mouse,** *-cle* **small,** *strain* **stretch]**

muscular dystrophy (MD)
(MUS-kyoo-lar DIS-troh-fee [em dee])
[*muscul-* **little mouse (muscle),** *-ar* **relating to,** *dys-* **bad,** *-troph-* **nourishment,** *-y* **state]**

myalgia
(my-AL-jee-ah)
[*my-* **muscle,** *-algia* **pain]**

myasthenia gravis
(my-es-THEE-nee-ah GRAH-vis)
[*my-* **muscle,** *-asthenia* **weakness,** *gravis* **severe]**

myopathy
(my-OP-ah-thee)
[*myo-* **muscle,** *-path-* **disease,** *-y* **state]**

myositis
(my-oh-SYE-tis)
[*myos-* **muscle,** *-itis* **inflammation]**

nurse
(nurs)
[*nurs-* **nourish or nurture]**

nutritionist
(noo-TRISH-en-ist)
[*nutri-* **nourish,** *-tion-* **process,** *-ist* **agent]**

paralysis
(pah-RAL-ih-sis)
[*para-* **beside,** *-lysis* **loosening]**

physical education
(FIS-ik-al ed-yoo-KAY-shun)
[*physic-* **medicine,** *-al* **relating to]**

physician
(fih-ZISH-en)
[*physic-* **medicine,** *-ian* **practitioner]**

poliomyelitis
(pol-ee-oh-my-eh-LYE-tis)
[*polio-* **gray,** *-mye-* **marrow,** *-itis* **inflammation]**

sprain
(sprayn)

strength training
(strengkth TRAYN-ing)
[*strength* **power,** *train-* **instruct,** *-ing* **action]**

tenosynovitis
(ten-oh-sin-oh-VYE-tis)
[*teno-* **pulled tight (tendon),** *-syn-* **together,** *-ovi-* **egg white (joint fluid),** *-itis* **inflammation]**

9

❑ OUTLINE SUMMARY

To download a digital version of the chapter summary for use with your device, access the ***Audio Chapter Summaries*** *online at evolve.elsevier.com.*

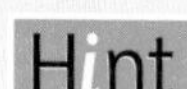

Scan this summary after reading the chapter to help you reinforce the key concepts. Later, use the summary as a quick review before your class or before a test.

Introduction

A. Muscular tissue enables the body and its parts to move
   1. Movement caused by ability of muscle cells (called *fibers*) to shorten or contract
   2. Muscle cells shorten by converting chemical energy (obtained from nutrients) into mechanical energy, which causes movement
   3. Three types of muscle tissue exist in the body (see Chapter 4)

Muscle Tissue

A. Skeletal muscle—also called *striated* or *voluntary* muscle (**Figure 9-1**, *A*)
   1. Is 40% to 50% of body weight ("red meat" attached to bones)
   2. Microscope reveals crosswise stripes, or striations
   3. Contractions can be voluntarily controlled
B. Cardiac muscle—composes bulk of heart (**Figure 9-1**, *B*)
   1. Cardiac muscle cells branch frequently
   2. Characterized by unique dark bands called *intercalated disks*
   3. Interconnected nature of cardiac muscle cells allows heart to contract efficiently as a unit
C. Smooth muscle—also called *nonstriated, involuntary,* or *visceral* muscle (**Figure 9-1**, *C*)
   1. Lacks cross stripes, or striations, when seen under a microscope; appears smooth
   2. Found in walls of hollow visceral structures such as digestive tract, blood vessels, and ureters
   3. Contractions not under voluntary control; movement caused by contraction is involuntary
D. Function—all muscle cells specialize in contraction (shortening)

Structure of Skeletal Muscle

A. Muscle organs—mainly skeletal muscle cells and connective tissue
   1. Connective tissue forms "wrappers" around each muscle fiber, around fascicles (groups) of muscle fibers, and around the entire muscle; fascia surrounds muscle organs and nearby structures
   2. Most skeletal muscles extend from one bone across a joint to another bone
   3. Regions of a skeletal muscle (**Figure 9-2**)
      a. Origin—attachment to the bone that remains relatively stationary or fixed when movement at the joint occurs
      b. Insertion—point of attachment to the bone that moves when a muscle contracts
      c. Body—main part of the muscle
   4. Muscles attach to bone by tendons—strong cords of fibrous connective tissue; some tendons are enclosed in synovial-lined tubes and are lubricated by synovial fluid; tubes are called *tendon sheaths*
   5. Bursae—small synovial-lined sacs containing a small amount of synovial fluid; located between some tendons and underlying bones
B. Muscle fibers (**Figure 9-3**)
   1. Contractile cells, or muscle fibers—grouped into bundles and intricately arranged
   2. Fibers of the cytoskeleton form cylinders made up of myofilaments
      a. Thick myofilaments contain myosin
      b. Thin myofilaments contain mainly actin
      c. Basic functional (contractile) units called *sarcomeres,* separated from each other by dark bands called *Z lines*
   3. Sliding filament model explains mechanism of muscle fiber contraction
      a. Thick and thin myofilaments slide past each other as a muscle contracts
      b. Contraction requires calcium and energy-rich ATP molecules (**Figure 9-4**)

Function of Skeletal Muscle

A. Movement
   1. Muscles produce movement by pulling on bones as a muscle contracts
      a. The insertion bone is pulled closer to the origin bone
      b. Movement occurs at the joint between the origin and the insertion
   2. Groups of muscles usually contract to produce a single movement
      a. Prime mover—muscle whose contraction is mainly responsible for producing a given movement
      b. Synergist—muscle whose contraction helps the prime mover produce a given movement
      c. Antagonist—muscle whose action opposes the action of a prime mover in any given movement
B. Posture
   1. A type of muscle contraction, called *tonic contraction,* enables us to maintain body position
      a. In tonic contraction, only a few of a muscle's fibers shorten at one time
      b. Tonic contractions produce no movement of body parts
      c. Tonic contractions maintain muscle tone called *posture*
   2. Good posture (optimum body positioning) favors best body functioning
   3. Skeletal muscle tone maintains posture by counteracting the pull of gravity

9

C. Heat production
  1. Survival depends on the body's ability to maintain a constant body temperature
    a. Fever—an elevated body temperature—often a sign of illness
    b. Hypothermia—body temperature below normal
  2. Contraction of muscle fibers produces most of the heat required to maintain normal body temperature

D. Fatigue
  1. Reduced strength of muscle contraction
  2. Caused by repeated muscle stimulation without adequate periods of rest
  3. Repeated muscular contraction depletes cellular ATP stores and outstrips the ability of the blood supply to replenish oxygen and nutrients
  4. Contraction in the absence of adequate oxygen produces lactic acid, which contributes to muscle burning
  5. *Oxygen debt*—term used to describe the metabolic effort required to burn excess lactic acid that may accumulate during prolonged periods of exercise
    a. Labored breathing after strenuous exercise is required to "pay the debt"
    b. This increased metabolism helps restore energy and oxygen reserves to pre-exercise levels

E. Integration with other body systems
  1. Muscle functioning depends on the functioning of many other parts of the body
    a. Most muscles cause movements by pulling on bones across movable joints
    b. Respiratory, cardiovascular, nervous, muscular, and skeletal systems play essential roles in producing normal movements
  2. Multiple sclerosis, brain hemorrhage, and spinal cord injury are examples of how pathological conditions in other body organ systems can dramatically affect movement

9

Motor Unit

A. Stimulation of a muscle by a nerve impulse is required before a muscle can shorten and produce movement

B. A motor neuron is the nerve cell that transmits an impulse to a muscle, causing contraction

C. A neuromuscular junction (NMJ)
  1. Junction between a nerve ending and the muscle fiber it innervates
  2. Chemicals called neurotransmitters cross a small gap at the NMJ to trigger contraction in the muscle
  3. Acetylcholine (ACh) is the neurotransmitter operating at each NMJ

D. Motor unit—combination of a motor neuron and the muscle cell or cells it innervates (**Figure 9-5**)

Muscle Stimulus

A. A muscle will contract only if an applied stimulus reaches a certain minimal level of intensity—called a *threshold stimulus*

B. Different muscle fibers in a muscle are controlled by different motor units having different threshold-stimulus levels

C. Differing numbers of motor units can be activated simultaneously to execute contractions of graded force

Types of Muscle Contraction

A. Twitch and tetanic contractions
  1. Twitch contractions—quick, jerky responses to a stimulus—are laboratory phenomena and do not play a significant role in normal muscular activity
  2. Tetanic contractions are sustained and steady muscular contractions caused by a series of stimuli bombarding a muscle in rapid succession

B. Isotonic contractions (**Figure 9-6**)
  1. Contraction of a muscle that produces movement at a joint
  2. During isotonic contractions, the muscle changes length, causing the insertion end of the muscle to move relative to the point of origin
  3. Concentric contractions shorten muscles
  4. Eccentric contractions allow muscles to increase in length
  5. Most types of body movements such as walking and running are caused by isotonic contractions

C. Isometric contractions (**Figure 9-6**)
  1. Isometric contractions are muscle contractions that do not produce movement; the muscle as a whole does not shorten
  2. Although no movement occurs during isometric contractions, tension within the muscle increases

Effects of Exercise

A. Exercise, if regular and properly practiced, improves muscle tone and posture, results in more efficient heart and lung functioning, and reduces fatigue

B. Specific effects of exercise on skeletal muscles
  1. Muscles undergo changes related to the amount of work they normally do
    a. Prolonged inactivity causes *disuse atrophy*
    b. Regular exercise increases muscle size, called *hypertrophy*

2. Strength training involves contraction of muscles against heavy resistance
   a. Strength training increases the numbers of myofilaments in each muscle fiber, and as a result, the total mass of the muscle increases
   b. Strength training does not increase the number of muscle fibers
3. Endurance training increases a muscle's ability to sustain moderate exercise over a long period; it is sometimes called *aerobic training*
   a. Endurance training allows more efficient delivery of oxygen and nutrients to a muscle via increased blood flow
   b. Endurance training does not usually result in muscle hypertrophy

Movements Produced by Muscles
(Figures 9-7 through **9-9)**

A. Angular movements
   1. Flexion—decreases an angle
   2. Extension—increases an angle
   3. Abduction—away from the midline
   4. Adduction—toward the midline
B. Circular movements
   1. Rotation—around an axis
   2. Circumduction—move distal end of a part in a circle
   3. Supination and pronation—hand positions that result from twisting of the forearm
C. Special movements—those not easily described with general terms
   1. Dorsiflexion and plantar flexion—foot movements (upward and downward ankle movement)
   2. Inversion and eversion—foot movements (sideways)
D. Muscles can be named or grouped according to function (movement) (**Table 9-1**)

Skeletal Muscle Groups (Table 9-2)

A. Muscles of the head and neck (**Figures 9-10** and **9-11; Table 9-2**)
   1. Facial muscles
      a. Orbicularis oculi
      b. Orbicularis oris
      c. Zygomaticus
   2. Muscles of mastication
      a. Masseter
      b. Temporal
   3. Sternocleidomastoid—flexes head
   4. Trapezius—elevates shoulders and extends head
B. Muscles of the upper extremities (**Table 9-3**)
   1. Pectoralis major—flexes arm
   2. Latissimus dorsi—extends arm
   3. Deltoid—abducts arm
   4. Biceps brachii—flexes forearm at elbow
   5. Brachialis—flexes forearm at elbow
   6. Triceps brachii—extends forearm
   7. Flexor muscles in forearm—flex wrist and hand
   8. Extensor muscles in forearm—extend wrist and hand
C. Muscles of the trunk (**Figure 9-12** and **Table 9-4**)
   1. Abdominal muscles
      a. Rectus abdominis
      b. External oblique
      c. Internal oblique
      d. Transversus abdominis
   2. Respiratory muscles
      a. Intercostal muscles
      b. Diaphragm
D. Muscles of the lower extremities (**Table 9-5**)
   1. Iliopsoas—flexes thigh
   2. Gluteus maximus—extends thigh
   3. Adductor muscles—adduct thighs
   4. Hamstring muscles—flex leg and extend thigh
      a. Semimembranosus
      b. Semitendinosus
      c. Biceps femoris
   5. Quadriceps femoris group—extend leg
      a. Rectus femoris
      b. Vastus muscles
   6. Tibialis anterior—dorsiflexes foot
   7. Gastrocnemius—plantar flexes foot
   8. Fibularis (peroneus) group—flexes foot

Muscular Disorders

A. Myopathies—muscle disorders; can range from mild to life threatening
B. Muscle injury
   1. Strain—injury from overexertion or trauma; involves stretching or tearing of muscle fibers (**Figure 9-13**)
      a. Often accompanied by myalgia (muscle pain)
      b. May result in inflammation of muscle (myositis) or of muscle and tendon (fibromyositis)
      c. If injury is near a joint and involves ligament damage, it may be called a *sprain*
   2. Cramps are painful muscle spasms (involuntary twitches)
   3. Crush injuries result from severe muscle trauma and may release cell contents that ultimately cause kidney failure
   4. Stress-induced muscle tension can cause headaches and back pain

C. Muscle infections
   1. Several bacteria, viruses, and parasites can infect muscles
   2. Poliomyelitis is a viral infection of motor nerves that ranges from mild to life threatening
   3. Tetanus infections are caused by *C. tetani* bacteria, which are found everywhere in our environment and release into the body a toxin that causes involuntary, sustained (tetanic) muscle contractions; can be fatal (**Figure 9-14**)

D. Muscular dystrophy (MD)
   1. A group of genetic disorders characterized by muscle atrophy
   2. Duchenne muscular dystrophy (DMD) is the most common type; also called *pseudohypertrophic*
      a. Characterized by rapid progression of weakness and atrophy
      b. X-linked inherited disease, affecting mostly boys

E. Myasthenia gravis—autoimmune muscle disease characterized by weakness and chronic fatigue

❑ ACTIVE LEARNING

STUDY TIPS

Consider using these tips to achieve success in meeting your learning goals.

Go back to Chapter 5 and review the synopsis of the muscular system. The three types of muscle tissue were covered in Chapter 4.

1. There are two prefixes that refer to muscle. One is *myo-* which means "muscle" and the other is *sarco-*, which means "flesh" (the soft muscular tissue between skin and bone). Several terms in this chapter have one or the other of these prefixes.
2. Movement is one of the functions of the muscular system. In order to create movement, muscle cells must get shorter. The sarcomere is the structure in the muscle that actually shortens. The sliding filament model explains how this shortening occurs. The shortening of the sarcomere requires energy. ATP supplies this energy. Check online resources with animations of muscle contraction.
3. The names of the muscles are probably less familiar to you than the names of the bones. But muscle names can give you information about the muscle. Muscles are named for their shape: *deltoid, trapezius.* They are named for the number of origins they have: *triceps* brachii, their points of attachment: *sternocleidomastoid,* their size: gluteus *maximus,* and the direction of the muscle fibers: *rectus* abdominis (*rectus* means the muscle has fibers running parallel to the midline of the body). When you are learning the muscles, try to look for meaning in the muscle names. Review the Language of Science and Language of Medicine terms and their word origins to help you better understand the meaning of the muscle names. Check out tips online at *my-ap.us/LnDZ2U*
4. Most of the terms for muscle movement are fairly straightforward. One way to remember the difference between *supination* and *pronation* is to picture your hand holding a bowl of soup—that's supination (silly, but an effective memory aid).
5. Draw a chart showing the mechanisms of muscular disorders by type: injury, infection, dystrophy, and myasthenia gravis.
6. Prepare flash cards and refer to online resources to help you learn the terms in this chapter. Review them in your study group. Also discuss the process of contraction and fatigue, and be sure you understand the movement terms. If you are asked to learn the names and locations of the muscles, a photocopy of the muscle figures with the labels blackened out can be used to quiz each other. There are many online labeling exercises *(getbodysmart.com)* that you can use as tutorials. If you are asked to learn the function, origin, and insertion of the muscles, prepare and use flash cards along with the figures.
7. Go over the questions at the end of the chapter and discuss possible test questions in your study group. Review the outline at the end of this chapter. This outline provides an overview of the material and would help you understand the general concepts to the chapter.

Review Questions

Write out the answers to these questions after reading the chapter and reviewing the Chapter Summary. If you simply think through the answer without writing it down, you will not retain much of your new learning.

1. Briefly describe the structure of cardiac muscle.
2. Briefly describe the structure of smooth muscle.
3. Describe the function of tendons, bursae, and synovial membranes.
4. Explain the sliding filament model of muscle contraction.
5. Explain why it is necessary to maintain good posture.
6. Suggest an example of how two body systems other than the muscular system contribute to the movement of the body.
7. Explain *twitch* and *tetanic* contractions.
8. Explain *isotonic* contractions.
9. Explain *isometric* contractions.
10. Describe the following movements: flexion, extension, abduction, adduction, and rotation.
11. Name two muscles in the head or neck and give the origin, insertion, and function of each.
12. Name two muscles that move the upper extremity and give the origin, insertion, and function of each.
13. Name two muscles of the trunk and give the origin, insertion, and function of each.
14. Name three muscles of the lower extremity and give the origin, insertion, and function of each.
15. What signs and symptoms are likely to accompany a moderate muscle strain?
16. What causes the signs and symptoms of myasthenia gravis?

Critical Thinking

After finishing the Review Questions, write out the answers to these more in-depth questions to help you apply your new knowledge. Go back to sections of the chapter that relate to concepts that you find difficult.

17. Explain the role of the biceps brachii and triceps brachii in terms of prime mover and antagonist in flexion and extension.
18. Explain the interaction of the prime mover, the synergist, and the antagonist in efficient movement.
19. Describe the condition that causes a muscle to develop an "oxygen debt." How is this debt paid off?
20. Why can a spinal cord injury be followed by muscle paralysis?
21. Can a muscle contract very long if its blood supply is shut off? Give a reason for your answer.
22. Briefly explain changes that gradually take place in bones, joints, and muscles in a person who habitually gets too little exercise.
23. Using fiber types, describe a muscle best suited for a marathon runner and different muscle for a 100-meter-dash sprinter. Explain your choices.
24. Briefly explain the progression of Duchenne muscular dystrophy.
25. Explain the need for myoglobin, in addition to hemoglobin, when oxygenating muscle fibers.
26. Explain how cardiac tissue demonstrates the principle that "structure fits function."

9

Chapter Test

After studying the chapter, test your mastery by responding to these items. Try to answer them without looking up the answers.

1. ________ is another name for muscle cell.
2. Cardiac muscle makes up the bulk of the tissue of the ________.
3. The muscle attachment to the more movable bone is called the ________.
4. The muscle attachment to the more stationary bone is called the ________.
5. ________ is the protein that makes up the thin myofilament.
6. ________ is the protein that makes up the thick myofilament.
7. The ________ is the basic functional unit of contraction in a skeletal muscle.
8. The three functions of the muscle system are ________, ________, and ________.
9. The molecule ________ supplies energy for muscle contraction.
10. ________ is the waste product produced when the muscle must switch to an energy-supplying process that does not require oxygen.
11. A single motor neuron with all the muscle cells it innervates is called a ________.
12. ________ is the minimal level of stimulation required to cause a muscle to contract.
13. ________ is a type of muscle contraction that produces movement in a joint and allows the muscle to shorten.
14. ________ is a type of muscle contraction that does not produce movement and does not allow the muscle to shorten but increases muscle tension.
15. ________ is a term that describes movement of a body part away from the midline of the body.
16. ________ is a term used to describe the movement that is opposite inversion.
17. ________ describes the hand position when the body is in anatomical position.
18. Excessive stretching or tearing of muscle fibers is called ________.
19. Inflammation of muscle and tendon is termed ________.
20. ________ is a viral infection of motor nerves that may progress to life-threatening paralysis of the respiratory muscles.
21. ________ is a group of muscle disorders characterized by muscle atrophy and that often progresses to death before age 21.
22. ________ is an autoimmune muscle disease characterized by weakness and chronic fatigue.
23. Skeletal muscles can also be called:
    a. visceral muscle
    b. voluntary muscle
    c. cardiac muscle
    d. all of the above
24. Smooth muscle can also be called:
    a. visceral muscle
    b. voluntary muscle
    c. cardiac muscle
    d. all of the above
25. What action does the hamstring group provide?
    a. Flexion
    b. Adduction
    c. Abduction
    d. Eversion
26. Which of the following is *not* a type of muscle fiber?
    a. Red fibers
    b. Fast fibers
    c. White fibers
    d. Gray fibers
27. Acetylcholine:
    a. stimulates muscle cell contraction
    b. allows muscle cells to relax
    c. is the fluid within bursae
    d. is an oxygen-storing pigment similar to hemoglobin
28. Walking, running, and breathing are examples of:
    a. isometric contraction
    b. isotonic contraction
    c. tetanic contraction
    d. twitch contraction

Match each muscle in Column A with its corresponding location or function in Column B.

Column A		Column B
29. ________	temporal muscle	a. muscles of the head or neck
30. ________	biceps brachii	b. muscles that move the upper extremity
31. ________	sartorius	c. muscles of the trunk
32. ________	masseter	d. muscles that move the lower extremity
33. ________	gastrocnemius	
34. ________	pectoralis major	
35. ________	external oblique	
36. ________	gluteus maximus	
37. ________	sternocleidomastoid	
38. ________	rectus abdominis	
39. ________	rectus femoris	
40. ________	triceps brachii	

Case Studies

To solve a case study, you may have to refer to the glossary or index, other chapters in this textbook, and other resources.

1. Your 3-year-old nephew, Tom, has just been diagnosed with *pseudohypertrophic muscular dystrophy.* How did he get this disease? Is his twin sister, Geri, likely to develop the same condition? Are you likely to get this disease? (HINT: See Chapter 25.) You know that muscular dystrophy typically causes atrophy or wasting of muscle tissue, yet Tom's leg muscles seem particularly well developed. Tom's physician said that the appearance of Tom's legs is typical for this form of muscular dystrophy. Can you explain this apparent contradiction?
2. Your friend Elena is suffering from a strain of her gastrocnemius muscle. What type of injury is this, and where in Elena's body is it located? What symptoms are likely to accompany Elena's injury? What movements should Elena avoid to prevent further injury to the gastrocnemius muscle?
3. Robert has decided to improve his appearance by exercising. He would like to build up his chest and shoulder muscles so that he looks better in the T-shirts he is so fond of wearing. He has decided to play racquetball every day as his primary training program because he knows that he uses his upper body muscles in this sport. After his first game of racquetball, you ask him how he likes his new sport and he can hardly answer you—he seems out of breath. Is Robert's plan likely to help him meet his goal? How do you explain his breathing difficulties?
4. Jessica was in the hospital for an extended period with a severe respiratory infection. During that period she received multiple IM injections and was discharged with orders to continue the injections daily until the doctor advised her to discontinue them. The nursing staff had been injecting the medication into the gluteal area, but this area was becoming sore from all of the prior injections. What options might she use for the future to avoid pain in the area of the injections?

Answers to Active Learning Questions can be found online at evolve.elsevier.com.

9

Nervous System

OUTLINE

Scan this outline before you begin to read the chapter, as a preview of how the concepts are organized.

OBJECTIVES

Before reading the chapter, review these goals for your learning.

After you have completed this chapter, you should be able to:

1. List the organs and divisions of the nervous system and describe the generalized functions of the nervous system as a whole.
2. Do the following related to the cells of the nervous system:
   - Identify the major types of cells in the nervous system and discuss the function of each.
   - Describe the function, size, and shape of glia.
   - Discuss various disorders of nervous tissue.
3. Identify the structure of a nerve.
4. Identify the anatomical components of a reflex arc and explains its function.
5. Explain the mechanisms of transmission of a nerve impulse along a nerve fiber and across a synapse.
6. Identify and discuss the major anatomical components of the brain and spinal cord, their functions, and brain disorders.
7. Identify and discuss the coverings and fluid spaces of the brain and spinal cord.
8. Compare and contrast cranial nerves and spinal nerves, and identify peripheral nerve disorders.
9. Discuss the structure and function of the two divisions of the autonomic nervous system, and identify disorders of these two divisions.

CHAPTER 10

The human body must accomplish a gigantic and enormously complex job—keeping itself alive and healthy. Each one of its trillions of cells performs some activity that is a part of this function. Control of the body's trillions of cells is accomplished mainly by two communication systems: the nervous system and the endocrine system. Both systems transmit information from one part of the body to another, but they do it in different ways. The nervous system transmits information very rapidly by nerve impulses conducted from one body area to another. The endocrine system transmits information more slowly by chemicals secreted by ductless glands into the bloodstream and then circulated to other parts of the body.

Nerve impulses and hormones communicate information to body structures, increasing or decreasing their activities as needed for healthy survival. In other words, the communication systems of the body are also its control and integrating systems. They combine the body's hundreds of functions into its one overall function of keeping itself alive and healthy.

Recall that homeostasis is the balanced and controlled internal environment of the body that is basic to life itself. Homeostasis is possible only if our physiological control and integration systems function properly. Our plan for this chapter is to name the cells, organs, and divisions of the nervous system and then explain how these impulses move between one area of the body and another. We not only discuss the major structures of the nervous system, such as the brain, spinal cord, and nerves, but also learn how they function to maintain and regulate homeostasis. In Chapter 11, we consider the senses.

Organs and Divisions of the Nervous System

The organs of the nervous system as a whole include the brain and spinal cord, the numerous nerves of the body, the special sense organs such as the eyes and ears, and the microscopic sense organs such as those found in the skin. The system as a whole consists of two principal divisions: the central nervous system and the peripheral nervous system (**Figure 10-1**). Because the brain and spinal cord occupy a midline or central location in the body, together they are called the **central nervous system,** or **CNS.** Similarly, the usual designation for the nerves of the body is the **peripheral nervous system,** or **PNS.** The term *peripheral* is appropriate because nerves extend to outlying or peripheral parts of the body. A subdivision of the peripheral nervous system, called the **autonomic nervous system,** or **ANS,** consists of structures that regulate the body's automatic or involuntary functions (for example, the heart rate, the

LANGUAGE OF SCIENCE

Hint Before reading the chapter, say each of these terms out loud. This will help you to avoid stumbling over them as you read.

acetylcholine (ACh)
(as-ee-til-KOH-leen [ay see aych])
[*acetyl-* **vinegar,** *-chole-* **bile,** *-ine* **made of**]

action potential
(AK-shun poh-TEN-shal)
[*act-* **moving,** *-ion* **condition,** *poten-* **power,** *-ial* **relating to**]

adrenergic fiber
(ad-ren-ER-jik FYE-ber)
[*ad-* **toward,** *-ren-* **kidney,** *-erg-* **work,** *-ic* **relating to,** *fibr-* **thread**]

afferent neuron
(AF-fer-ent NOO-ron)
[*a[d]-* **toward,** *-fer-* **carry,** *-ent* **relating to,** *neur-* **string or nerve,** *-on* **unit**]

antidiuretic hormone (ADH)
(an-tee-dye-yoo-RET-ik HOR-mohn [ay dee aych])
[*anti-* **against,** *-dia-* **through,** *-uret-* **urination,** *-ic* **relating to,** *hormon-* **excite**]

arachnoid mater
(ah-RAK-noyd MAH-ter)
[*arachn-* **spider(web),** *-oid* **like,** *mater* **mother**]

astrocyte
(AS-troh-syte)
[*astro-* **star shaped,** *-cyte* **cell**]

autonomic effector
(aw-toh-NOM-ik ef-FEK-tor)
[*auto-* **self,** *-nom-* **rule,** *-ic* **relating to,** *effect-* **accomplish,** *-or* **agent**]

autonomic nervous system (ANS)
(aw-toh-NAHM-ik NER-vus SIS-tem [ay en es])
[*auto-* **self,** *-nom-* **rule,** *-ic* **relating to,** *nerv-* **nerve,** *-ous,* **relating to**]

Continued on p. 280

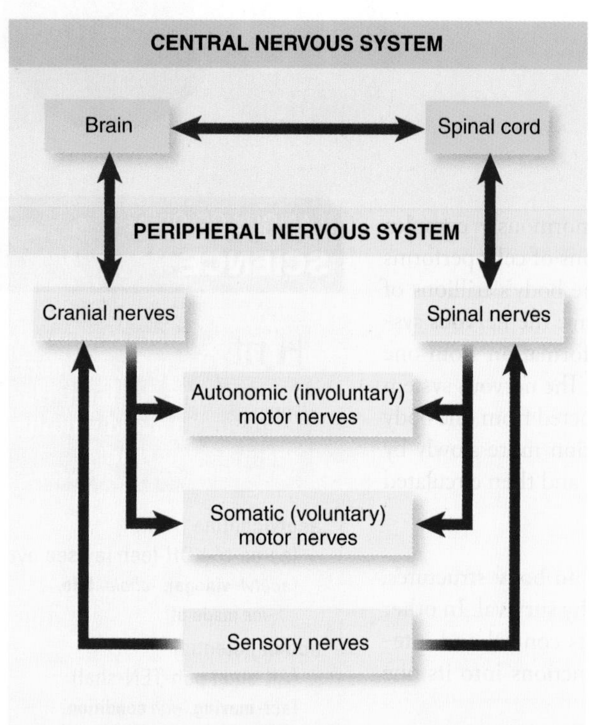

FIGURE 10-1 Divisions of the nervous system.

contractions of the stomach and intestines, and the secretion of chemical compounds by glands).

As you study the brain spinal cord, and nerves, try to find their location in the *Clear View of the Human Body* (follows p. 8) and note their relationships to surrounding structures.

To better understand this concept, use the Active Concept Map *Organization of the Nervous System* at *evolve.elsevier.com.*

To get an overview of the nervous system, go to AnimationDirect online at *evolve.elsevier.com.*

Cells of the Nervous System

In Chapter 4, we learned that nervous tissue is the major component of the nervous system. And we learned two major types of cells are found in nervous tissue: **neurons** or nerve cells and **glia,** which are support cells (see Figure 4-19 on p. 84). Neurons conduct impulses, whereas glia support neurons.

10

Neurons

Neuron Structure

Each neuron consists of three parts: a main part called the neuron **cell body,** one or more branching projections called **dendrites,** and one elongated projection known as an **axon.**

The axon shown in Figure 10-2, *B*, is surrounded by a segmented wrapping of a material called **myelin.** Myelin is a white, fatty substance formed by *Schwann cells* that wrap around some axons outside the central nervous system. Such fibers are called **myelinated fibers.** In Figure 10-2, *B*, one such axon has been enlarged to show additional detail. **Nodes of Ranvier** are gaps between adjacent Schwann cells.

The outer wrapped layer of a Schwann cell is called the **neurilemma.** It is clinically significant that axons in the brain and spinal cord have no neurilemma, because neurilemma plays an essential part in the regeneration of cut and injured axons. Therefore, the potential for regeneration in the brain and spinal cord is far less than it is in the peripheral nervous system.

Identify each part on the neuron shown in Figure 10-2. Dendrites are the processes or projections that carry impulses to the neuron cell bodies, and axons are the processes that carry impulses away from the neuron cell bodies.

Types of Neurons

There are three major types of neurons classified according to the direction in which they carry impulses.

Sensory Neurons

Sensory neurons carry impulses to the spinal cord and brain from all parts of the body. Sensory neurons are also called **afferent neurons.**

Motor Neurons

Motor neurons carry impulses in the opposite direction—away from the brain and spinal cord. They do not conduct impulses to all parts of the body—only to two kinds of tissue—muscle and glandular epithelial tissue. Motor neurons are also called **efferent neurons.**

Interneurons

Interneurons conduct impulses from sensory neurons to motor neurons. They also often connect with each other to form complex, central networks of nerve fibers. Interneurons are sometimes called *central* or *connecting* neurons.

Glia

Function of Glia

Glia—or **neuroglia**—do not specialize in transmitting impulses. Instead, they are special types of supporting cells. Their name is appropriate because it is derived from the Greek word *glia* meaning "glue." One function of glial cells is to hold the functioning neurons together and protect them.

We now know that glia perform many different functions, including the regulation of neuron function. Therefore, they not only act as "glue" in the physical sense but also help bring the various functions of nervous tissue together into a coordinated whole.

An important reason for discussing glia is that one of the most common types of brain tumor—called **glioma**—develops from them.

Central Glia

Glia vary in size and shape (Figure 10-3). Some are relatively large cells that look somewhat like stars because of the threadlike extensions that jut out from their surfaces. These glial cells are called **astrocytes,** a word that means "star cells" (see **Figure 10-3**, *A*). Their threadlike branches attach to neurons and to small blood vessels, holding these structures close to each other.

Along with the walls of the blood vessels, astrocyte branches form a two-layer structure called the **blood-brain barrier (BBB).** As its name implies, the BBB separates the blood tissue and nervous tissue to protect vital brain tissue from harmful chemicals that might be in the blood.

Microglia are smaller than astrocytes (see **Figure 10-3**, *B*). They usually remain stationary, but in inflamed or degenerating brain tissue, they enlarge, move about, and act as microbe-eating scavengers. They surround the microbes, draw them into their cytoplasm, and digest them. They likewise help

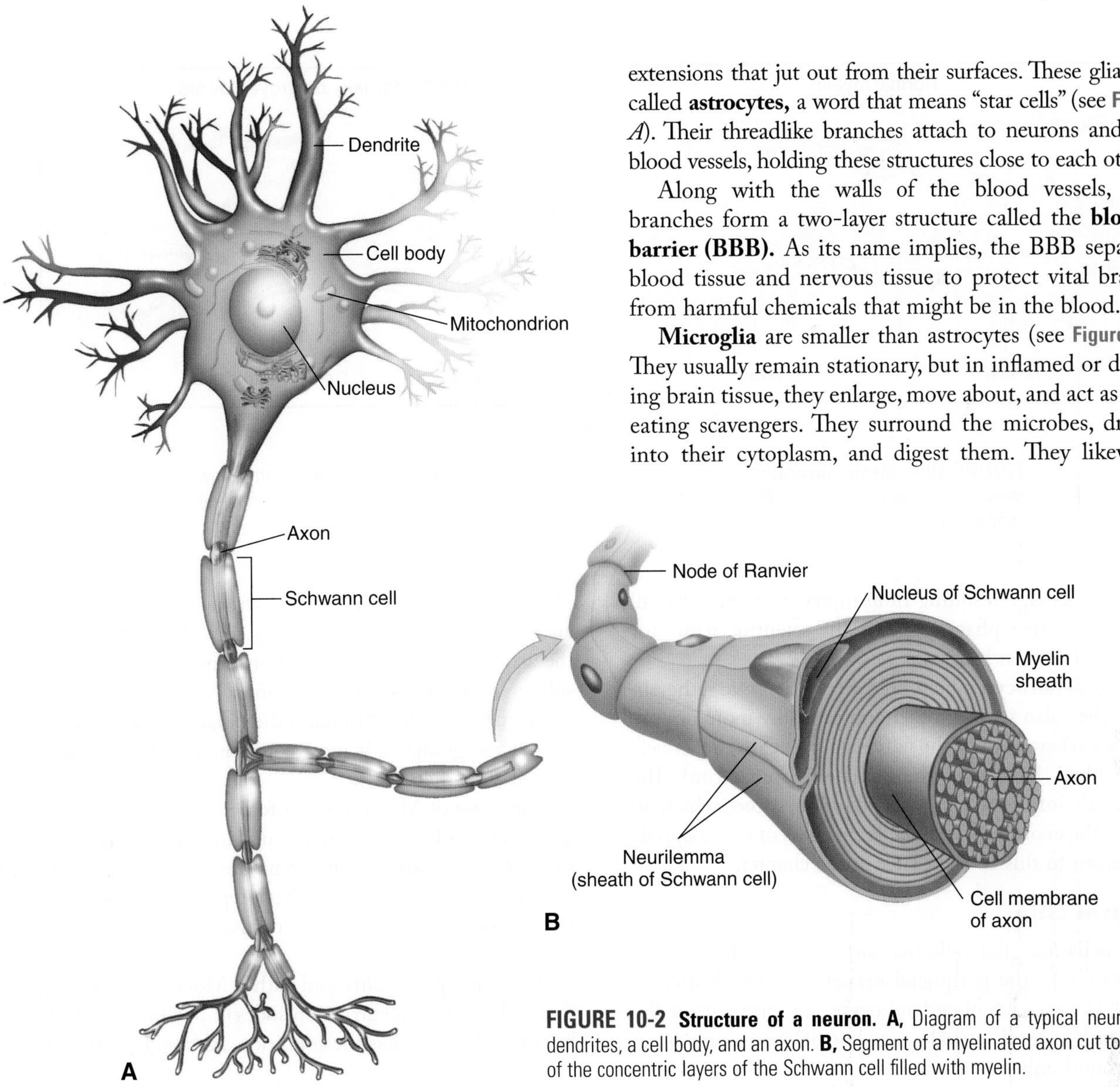

FIGURE 10-2 Structure of a neuron. A, Diagram of a typical neuron showing dendrites, a cell body, and an axon. **B,** Segment of a myelinated axon cut to show detail of the concentric layers of the Schwann cell filled with myelin.

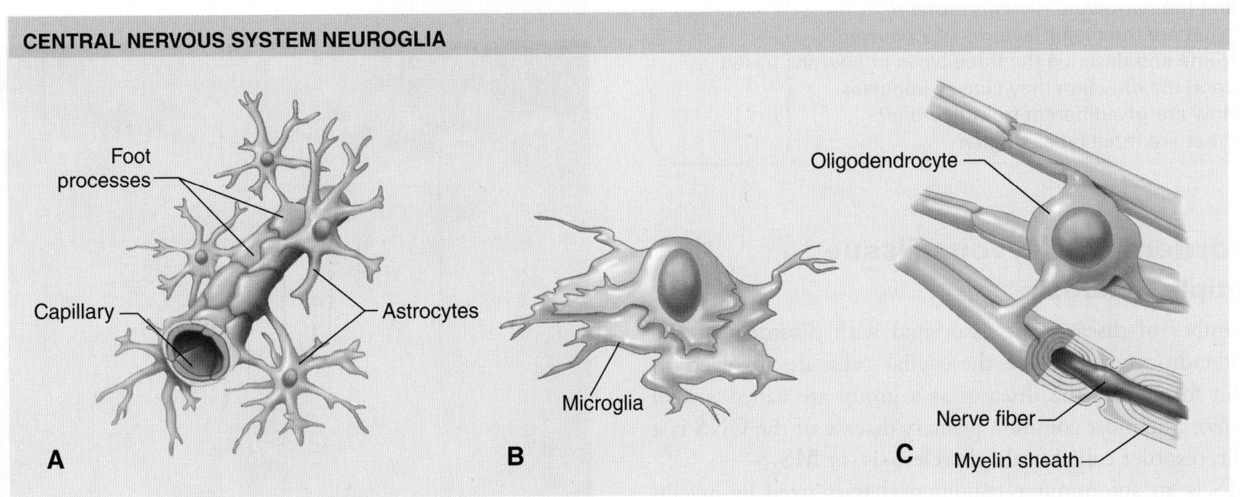

FIGURE 10-3 Central glia. A, Astrocytes have extensions attached to blood vessels in the brain. **B,** Microglia within the central nervous system can enlarge and consume microbes by phagocytosis. **C,** Oligodendrocytes have extensions that form myelin sheaths around axons in the central nervous system. Glia are not shown to scale.

10

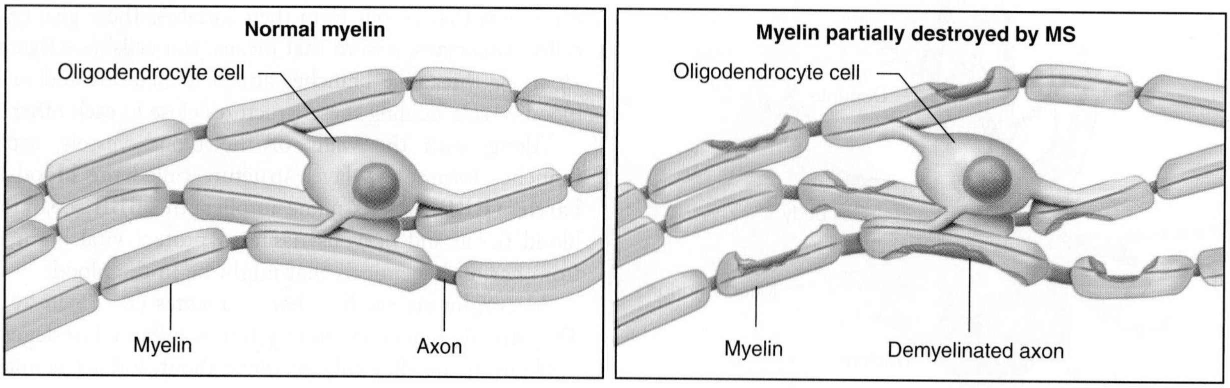

FIGURE 10-4 Myelin disorders. Diagram showing effects of multiple sclerosis (MS). **A,** A normal myelin sheath allows rapid conduction. **B,** In those with MS, the myelin sheath is damaged, disrupting normal nerve conduction.

clean up cell damage resulting from injury or disease. Recall from Chapter 3 that phagocytosis is the scientific name for this important cellular process.

The **oligodendrocytes** help hold nerve fibers together in the CNS. They also serve another and probably more important function: they produce the fatty myelin sheath that envelops nerve fibers located in the brain and spinal cord. The myelin sheath influences nerve conduction speed, which in turn affects the coordination and effectiveness of nerve signaling. We return to this concept later in the chapter.

Peripheral Glia

Schwann cells are glial cells that also form myelin sheaths but do so only in the peripheral nervous system. Notice in Figure 10-3, *C,* that each oligodendrocyte can form part of the myelin sheath around several axons but Schwann cells wrap entirely around only one axon.

QUICK CHECK

1. What is the difference between the *central nervous system* and the *peripheral nervous system*?
2. What are the major features of a *neuron*?
3. Name and describe the three types of neurons based upon the direction they convey impulses.
4. How are *glia* different from *neurons?*
5. What are three types of *glia*?

Disorders of Nervous Tissue

Multiple Sclerosis

A number of diseases are associated with disorders of the oligodendrocytes. Because these glial cells are involved in myelin formation, the diseases as a group are called *myelin disorders.* The most common primary disease of the CNS is a myelin disorder called **multiple sclerosis,** or **MS.**

MS is an autoimmune condition characterized by myelin loss and destruction accompanied by varying degrees of oligodendrocyte injury and death (Figure 10-4). The result is areas of demyelination throughout the white matter of the CNS. Hard, plaquelike lesions replace the destroyed myelin, and affected areas are invaded by inflammatory cells. As the myelin around the axons is lost, nerve conduction is impaired, potentially resulting in weakness, incoordination, visual impairment, and speech disturbances. Although the disease occurs in both sexes and all age-groups, it is most common in women between the ages of 20 and 40 years.

The cause of MS is thought to be related to autoimmunity and to viral infections in some individuals. MS is characteristically relapsing and chronic in nature, but some cases of acute and unremitting disease have been reported. In most instances the disease is prolonged, with remissions and relapses occurring over many years.

Television personality and author Montel Williams reports that he lived with recurring episodes of MS for 20 years before he realized that he has the condition. Although there is not yet

FIGURE 10-5 Multiple neurofibromatosis. This photo shows multiple tumors of Schwann cells in the nerves of the skin that are characteristic of this inherited condition.

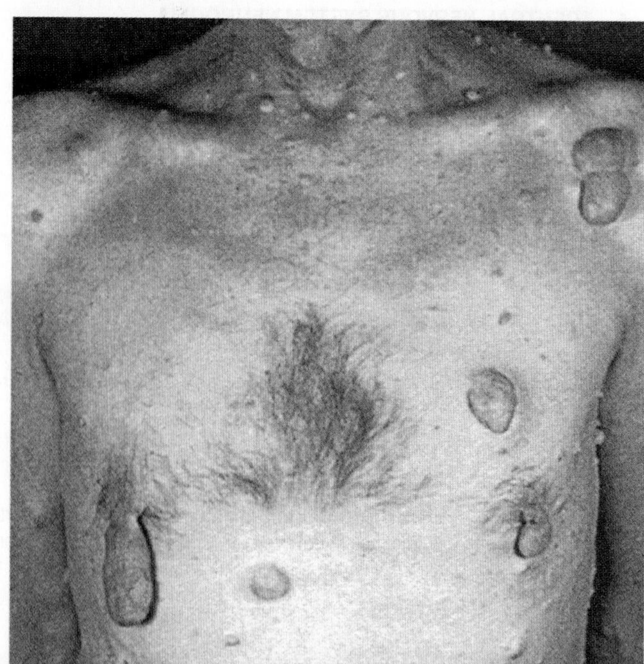

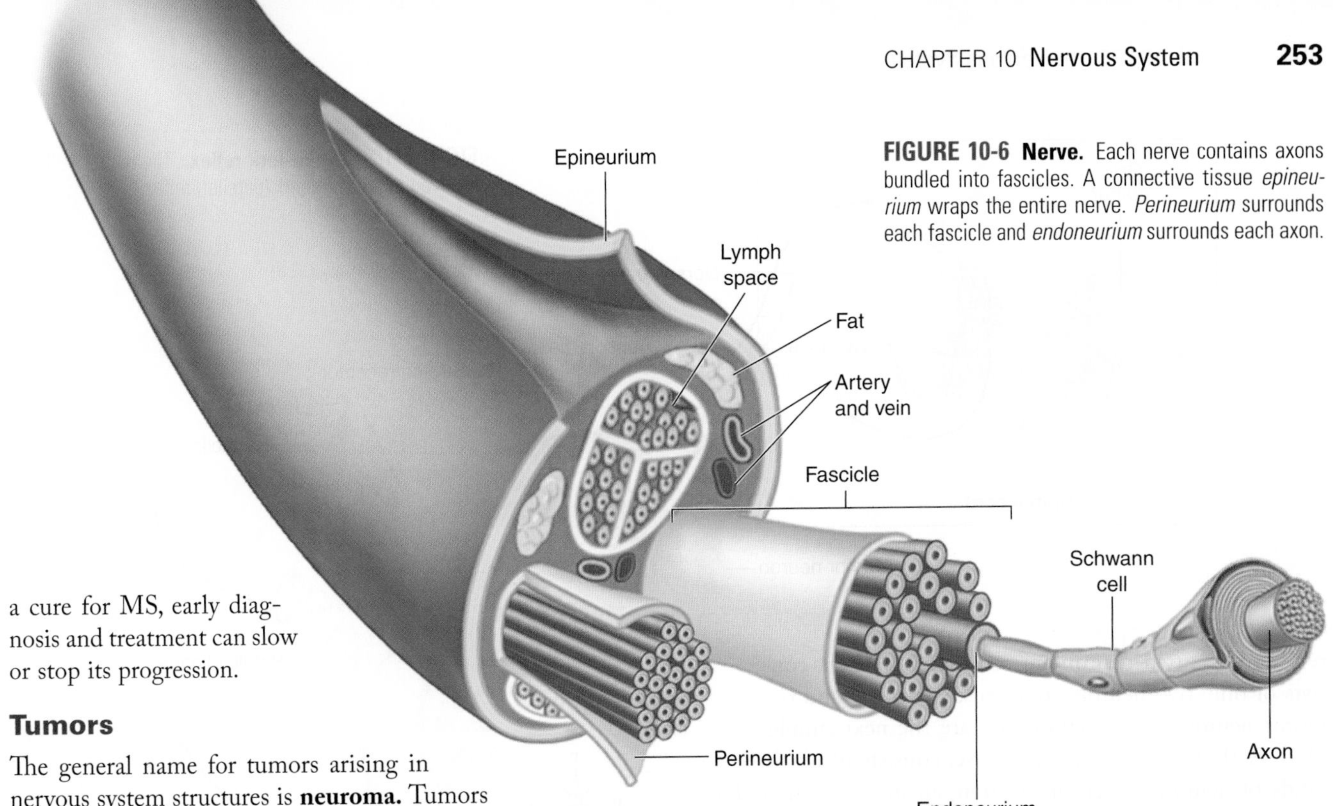

FIGURE 10-6 Nerve. Each nerve contains axons bundled into fascicles. A connective tissue *epineurium* wraps the entire nerve. *Perineurium* surrounds each fascicle and *endoneurium* surrounds each axon.

a cure for MS, early diagnosis and treatment can slow or stop its progression.

Tumors

The general name for tumors arising in nervous system structures is **neuroma.** Tumors do not usually develop directly from neurons but instead from glia, membrane tissues, and blood vessels.

As stated earlier, a common type of brain tumor—*glioma*—occurs in glia. Gliomas are usually benign but may still be life threatening. Patients usually show deficits reflecting damaged function of the area in which the tumor is located (Figure 10-4, *B*). Because these tumors often develop in deep areas of the brain, they are difficult to treat. Untreated gliomas may grow to a size that disrupts normal brain function, perhaps leading to death.

Multiple neurofibromatosis is an inherited disease characterized by numerous fibrous neuromas throughout the body (Figure 10-5). The tumors are benign, appearing first as small nodules in the Schwann cells of cutaneous nerves. In some cases, involvement spreads as large, disfiguring fibrous tumors appear in many areas of the body, including muscles, bones, and internal organs.

Most malignant tumors of glia and other nervous tissues do not originate there but instead are secondary tumors resulting from metastasis of cancer cells from the breast, lung, or other organs.

QUICK CHECK

1. What is a *myelin disorder*? How does a myelin disorder disrupt nervous system function?
2. What is a *neuroma*?
3. What is a common disorder that is characterized by myelin loss and destruction of varying degrees of the oligodendrocytes?

Nerves and Tracts

A **nerve** is a group of peripheral nerve fibers (axons) bundled together like the strands of a cable. Peripheral nerve fibers usually have a myelin sheath and because myelin is white, peripheral nerves often look white.

Figure 10-6 shows that each axon in a nerve is surrounded by a thin wrapping of fibrous connective tissue called the **endoneurium.** Groups of these wrapped axons are called **fascicles.** Each fascicle is surrounded by a thin, fibrous **perineurium.** A tough, fibrous sheath called the **epineurium** covers the whole nerve.

Bundles of axons in the CNS, called **tracts,** also may be myelinated and thus form the **white matter** of the brain and cord. Brain and cord tissue composed of cell bodies and unmyelinated axons and dendrites is called **gray matter** because of its characteristic gray appearance.

Nerve Signals

Reflex Arcs

Neuron Pathways

During every moment of our lives, nerve impulses speed over neurons to and from our spinal cords and brains. If all impulse conduction ceases, life itself ceases.

Only neurons can provide the rapid communication between cells that is necessary for maintaining life. Hormones are the only other kind of signal the body can send, and they travel much more slowly than nerve signals. Hormones can move from one part of the body to another only via circulating blood. Compared with nerve impulse conduction, hormone circulation is a very slow process.

Nerve impulses, often called *action potentials,* can travel over trillions of routes—routes made up of neurons because they are the cells that conduct impulses. Hence the routes traveled by nerve impulses are sometimes spoken of as *neuron pathways.*

A basic type of neuron pathway, called a **reflex arc,** is important to nervous system functioning. The simplest kind of

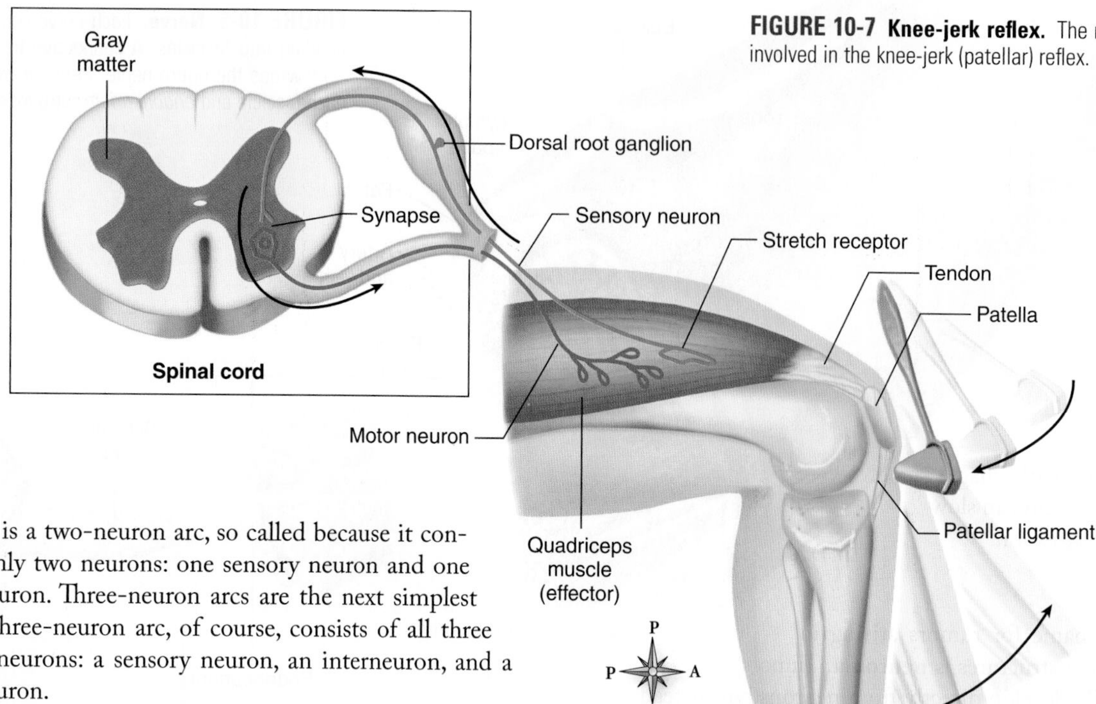

FIGURE 10-7 Knee-jerk reflex. The neural pathway involved in the knee-jerk (patellar) reflex.

reflex arc is a two-neuron arc, so called because it consists of only two neurons: one sensory neuron and one motor neuron. Three-neuron arcs are the next simplest kind. A three-neuron arc, of course, consists of all three kinds of neurons: a sensory neuron, an interneuron, and a motor neuron.

Structure of Reflex Arcs

Reflex arcs are like one-way streets; they allow impulse conduction in only one direction. The next paragraph describes this direction in detail. Look frequently at Figure 10-7 as you read it.

Impulse conduction normally starts in receptors. **Receptors** are the beginnings of dendrites of sensory neurons. They are often located far from the spinal cord (in tendons, skin, or mucous membranes, for example).

In Figure 10-7 the sensory receptors are located in the quadriceps muscle group—which acts to extend the leg. In the reflex that is illustrated there, stretch receptors are stimulated when muscles are stretched as a result of a tap on the patellar ligament from a rubber hammer used by a physician to elicit a reflex during a physical examination. The nerve impulse that is generated, its neurological pathway, and its ultimate "knee-jerk" effect provide an example of the simplest form of a two-neuron reflex arc.

In the knee-jerk reflex, only sensory and motor neurons are involved. The nerve impulse that is generated by stimulation of the stretch receptors travels along the length of the sensory neuron's dendrite to its cell body located in the *dorsal root ganglion*. A **ganglion** is a group of nerve-cell bodies located in the PNS. This ganglion is located near the spinal cord. Each dorsal root ganglion contains not one sensory neuron cell body as shown in Figure 10-7, but hundreds of them.

The axon of the sensory neuron travels from the cell body in the dorsal root ganglion and ends near the dendrites of another neuron located in the gray matter of the spinal cord. A microscopic space separates the axon ending of one neuron from the dendrites of another neuron. This gap serves as a junction between nerve cells called a **synapse.** The nerve impulse stops at the synapse, chemical signals are sent across the gap, and then the new impulse continues along the dendrites, cell body, and axon of the motor neuron.

The motor neuron axon forms a synapse with a structure called an **effector,** an organ that puts nerve signals "into effect." Effectors are usually muscles or glands, and muscle contractions and gland secretion are the kinds of reflexes operated by these effectors.

Reflex Responses

An involuntary response to impulse conduction over a reflex arc is called a **reflex.** In short, impulse conduction by a reflex arc causes a reflex to occur. In our example reflex, the nerve impulses that reach the quadriceps muscle (the effector) result in the "knee-jerk" response.

Some reflexes involve three rather than two neurons. In these more complex types of responses, an interneuron, in addition to the sensory and motor neurons, is involved. In three-neuron reflexes, the end of the sensory neuron's axon synapses first with an interneuron before chemical signals are sent across a second synapse, resulting in conduction through the motor neuron.

For example, application of an irritating stimulus to the skin of the thigh initiates a three-neuron reflex response that causes contraction of flexor muscles to pull the leg away from the irritant—a three-neuron arc reaction called the **withdrawal reflex.**

All interneurons lie entirely within the gray matter of the brain or spinal cord. Gray matter forms the H-shaped inner core of the spinal cord. Because of the presence of an interneuron, three-neuron reflex arcs have two synapses. A two-neuron reflex arc, however, has only a sensory neuron and a motor neuron with one synapse between them.

Identify the motor neuron in Figure 10-7. Observe that its dendrites and cell body, like those of an interneuron, are located in the spinal cord's gray matter. The axon of this motor neuron, however, runs through the ventral root of the spinal nerve and terminates in a muscle.

To explore applications of neural pathway concepts, check out the article *Pain Control Areas* at Connect It! at *evolve.elsevier.com*.

QUICK CHECK

1. How is white matter different from *gray matter*?
2. Can you explain the function of a *reflex arc*?
3. What is a *sensory receptor*? How does it relate to the reflex arc?
4. What is an *effector*? How does it relate to the reflex arc?
5. What is a *ganglion*?

Nerve Impulses

Definition of a Nerve Impulse

What are nerve impulses? Here is one widely accepted definition: a nerve impulse is a self-propagating wave of electrical disturbance that travels along the surface of a neuron's plasma membrane. You might visualize this as a tiny spark sizzling its way along a fuse.

Nerve impulses do not continually race along every nerve cell's surface. First they have to be initiated by a stimulus, a change in the neuron's environment. Pressure, temperature, and chemical changes are the usual stimuli.

Mechanism of a Nerve Impulse

Figure 10-8 depicts a simplified summary of the mechanism of a nerve impulse.

The membrane of each resting neuron has a slight positive charge on the outside and a negative charge on the inside, a state

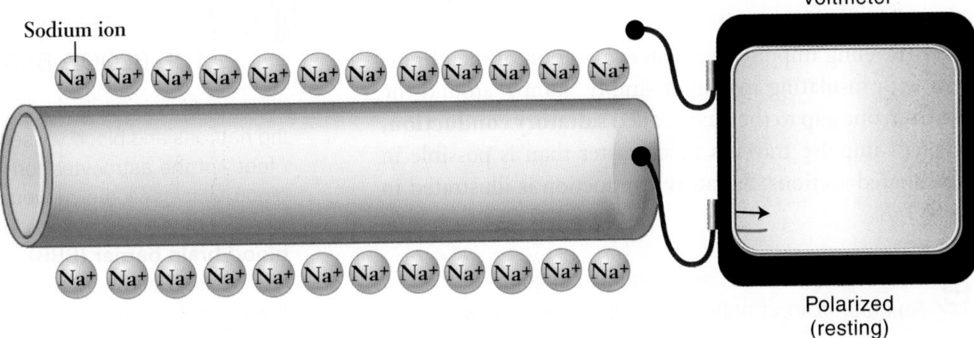

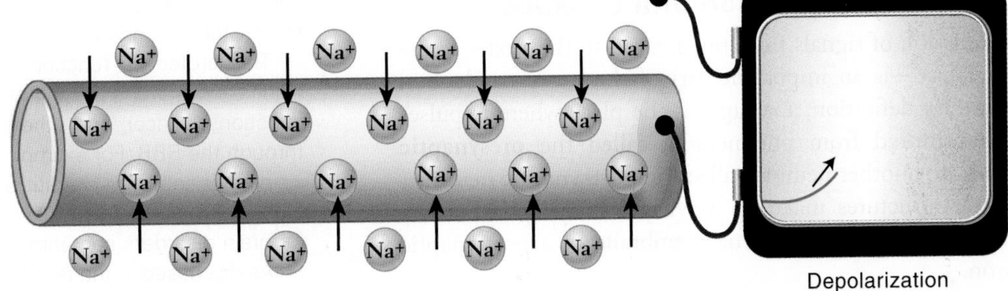

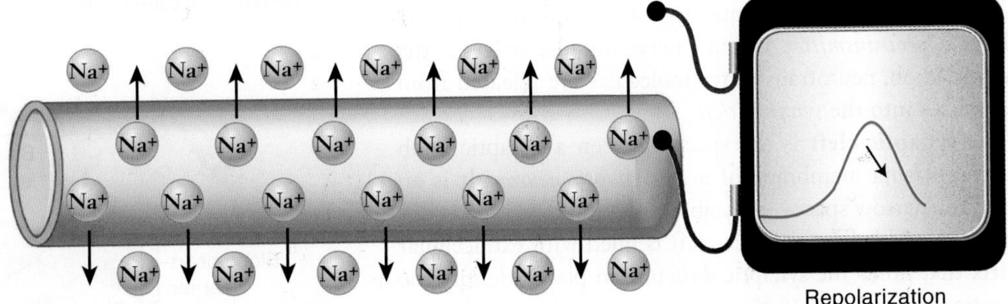

FIGURE 10-8 Mechanism of the nerve impulse. A voltmeter *(right)* shows how the charge difference across the membrane fluctuates as the balance of positive ions (Na^+) changes.

called *polarization*. This occurs because there is normally an excess of sodium ions (Na^+) on the outside of the membrane.

When a section of the membrane is stimulated, its Na^+ channels suddenly open, and Na^+ ions rush inward. The inside of the membrane temporarily becomes positive, and the outside becomes negative—a process called *depolarization*.

The depolarized section of the membrane then immediately recovers—a process called *repolarization*. However, the depolarization has already stimulated Na^+ channels in the next section of the membrane to open.

Conduction of Nerve Impulses

The impulse—or **action potential**—cannot go backward during the brief moment of repolarization and recovery of the previous section of membrane. Thus a self-propagating wave of electrical disturbance—a nerve impulse—travels continuously in one direction across the neuron's surface (**Figure 10-9**, *A*).

Nerve impulses are also called *action potentials* because each one is a difference in charge (called "electrical potential") that usually triggers an action by the cell—in this case, transmission of the impulse itself.

If the traveling impulse encounters a section of membrane covered with insulating myelin, it simply "jumps" around the myelin from one gap to the next. Called **saltatory conduction,** this type of impulse travel is much faster than is possible in nonmyelinated sections. Saltatory conduction is illustrated in **Figure 10-9**, *B*.

To learn more about nerve impulses, go to AnimationDirect online at *evolve.elsevier.com*.

Synapses

Structure and Function of a Synapse

Transmission of signals from one neuron to the next—across the synapse—is an important part of the nerve conduction process. By definition, a synapse is the place where impulses are transmitted from one neuron, called the **presynaptic neuron,** to another neuron, called the **postsynaptic neuron.**

Three structures make up a synapse: a synaptic knob, a synaptic cleft, and the plasma membrane of a postsynaptic neuron.

10

A **synaptic knob** is a tiny bulge at the end of a terminal branch of a presynaptic neuron's axon (**Figure 10-10**). Each synaptic knob contains many small sacs or vesicles. Each vesicle contains a very small quantity of a chemical compound called a *neurotransmitter*. When a nerve impulse arrives at the synaptic knob, neurotransmitter molecules are released from the vesicles into the *synaptic cleft*.

The **synaptic cleft** is the space between a synaptic knob and the plasma membrane of a *postsynaptic neuron*. It is an incredibly narrow space—only about two millionths of a centimeter in width. The synaptic cleft is filled with extracellular matrix that holds the synaptic structure in place. Identify the synaptic cleft in **Figure 10-10**.

The plasma membrane of a postsynaptic neuron has protein molecules embedded in it opposite each synaptic knob. These serve as receptors to which neurotransmitter molecules bind. This binding can initiate an impulse in the postsynaptic neuron by opening ion channels in the postsynaptic membrane.

After impulse conduction by postsynaptic neurons is initiated, neurotransmitter activity is rapidly terminated. Either one or both of two mechanisms cause this: reuptake or enzyme breakdown.

Some neurotransmitter molecules are transported out of the synaptic cleft and back into synaptic knobs in a process called *reuptake*. Like taking the key out of the ignition switch and returning it to your pocket, the receptor stops functioning when the neurotransmitters are taken back. There, they are repackaged into vesicles to be used again later—a mechanism called *recycling*.

The neurotransmitters may instead be broken apart by specific enzymes in the extracellular matrix of the synaptic cleft. The neurotransmitters may also be transported into a

CLINICAL APPLICATION

THE BLOOD-BRAIN BARRIER

Astrocytes have an important function other than supporting neurons and blood vessels. Notice in the figure that the "feet" of the astrocytes form a wall around the outside of blood vessels in the nervous system. This astrocyte wall, along with the vessel wall, forms a structure known as the **blood-brain barrier (BBB).**

The BBB allows water, oxygen, carbon dioxide, and a few other substances—such as alcohol—to move between the blood and the tissue of the brain. However, many toxins and pathogens that can enter other tissues through blood vessel walls cannot enter nervous tissue because of this barrier. This adaptation enhances survival because it protects vital brain and nerve tissues from damage.

This protective function of the BBB has great clinical significance. Drugs used in other parts of the body to treat infections, cancer, and other disorders often cannot pass through the BBB. For example, penicillin and other antibiotics cannot enter the interstitial fluid of brain tissue from the blood. Obviously, this makes development of treatments for brain disorders sometimes very difficult.

As discussed in the text (p. 259), parkinsonism resulting from a lack of dopamine in the brain cannot be treated with dopamine because it cannot cross the BBB. However, the dopamine precursor L-dopa *can* cross the BBB and be converted to dopamine in the brain in some patients.

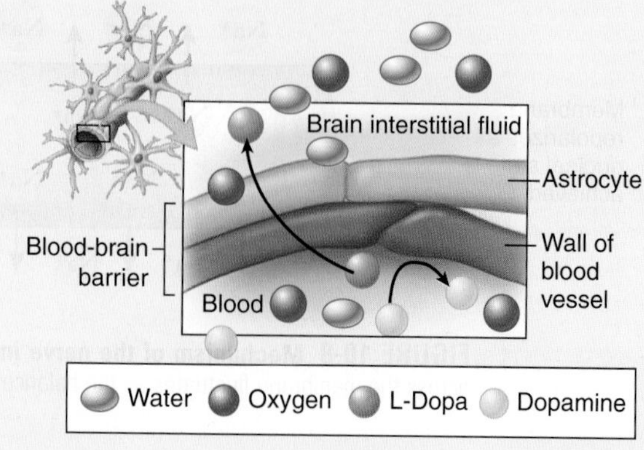

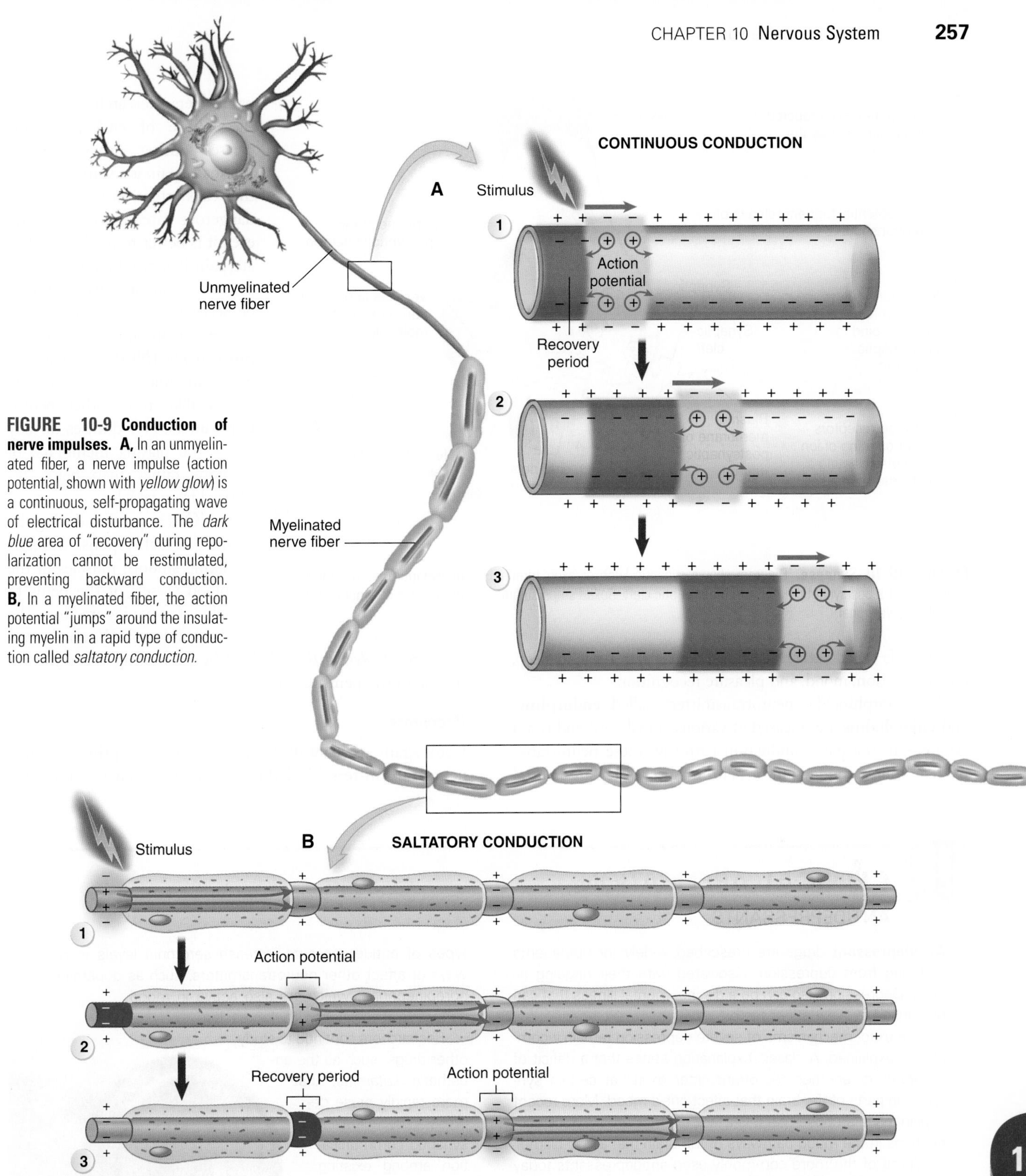

FIGURE 10-9 Conduction of nerve impulses. A, In an unmyelinated fiber, a nerve impulse (action potential, shown with *yellow glow*) is a continuous, self-propagating wave of electrical disturbance. The *dark blue* area of "recovery" during repolarization cannot be restimulated, preventing backward conduction. **B,** In a myelinated fiber, the action potential "jumps" around the insulating myelin in a rapid type of conduction called *saltatory conduction.*

nearby glial cell and broken apart by enzymes there. In either case, the "pieces" left after enzyme breakdown are returned to the presynaptic neuron to be recycled into new neurotransmitter molecules.

Neurotransmitters and Receptors

Neurotransmitters

Neurotransmitters are chemicals by which neurons communicate. As previously noted, at trillions of synapses in the CNS, presynaptic neurons release neurotransmitters that assist, stimulate, or inhibit postsynaptic neurons. At least 30 different compounds have been identified as neurotransmitters. They are not distributed randomly through the spinal cord and brain. Instead, specific neurotransmitters are localized in discrete groups of neurons and released in specific pathways.

For example, the substance named **acetylcholine (ACh)** is released at some of the synapses in the spinal cord and at neuromuscular (nerve-muscle) junctions. Other well-known neurotransmitters include **norepinephrine (NE), dopamine,** and **serotonin.** These three neurotransmitters belong to a group of

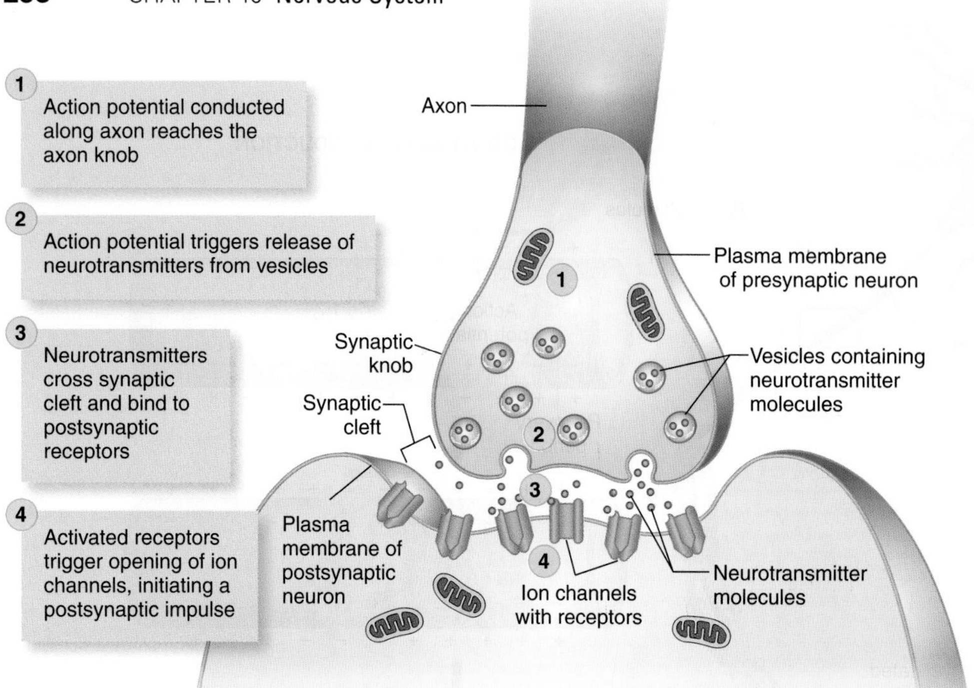

FIGURE 10-10 Synapse. Diagram shows synaptic knob or axon terminal of presynaptic neuron, the plasma membrane of a postsynaptic neuron, and a synaptic cleft. Boxed steps summarize the process of synaptic transmission of a nerve signal.

compounds called **catecholamines,** which play a role in sleep, motor function, mood, and pleasure recognition.

Two morphinelike neurotransmitters called **endorphins** and **enkephalins** are released at various spinal cord and brain synapses in the pain conduction pathway. These neurotransmitters inhibit conduction of pain impulses. They are natural pain killers. Research shows that the release of endorphins increases during heavy exercise. Normally, pain is a warning signal that calls attention to injuries or dangerous circumstances. However, it is better to inhibit moderate pain if it would stop us from continuing an activity that may be necessary for survival.

Very small molecules such as **nitric oxide (NO)** also have an important role as neurotransmitters. Unlike most other neurotransmitters, NO diffuses directly across the plasma membrane of neurons rather than being released from vesicles.

NO is important during the male sexual response in regulating smooth muscles in the blood vessels of the penis to allow for penile erection. The drug sildenafil (Viagra) treats *male erectile dysfunction (MED)* by promoting the same response in the penis as NO.

Receptors

Look again at **Figure 10-10** and locate the receptors to which neurotransmitters bind during synaptic transmission of a

CLINICAL APPLICATION

ANTIDEPRESSANTS

Antidepressant drugs are prescribed widely for outpatients suffering from depression associated with their ongoing illness. Depression also can be a distinct mental illness that can be caused by a variety of factors.

The exact mechanisms of severe depression remain incompletely explained. A classic explanation states that a deficit of *serotonin* or another neurotransmitter exists at certain synapses in parts of the brain that affect one's mood. More recent explanations propose that it is a lack of sufficient synaptic connections in the mood pathways of the brain that are to blame.

Some of the more commonly used antidepressants today are paroxetine (Paxil), fluoxetine (Prozac), and sertraline (Zoloft). These drugs produce their effects by blocking the uptake of serotonin back into presynaptic neurons. Drugs in this class of antidepressants are called *serotonin-specific reuptake inhibitors (SSRIs).*

Serotonin-uptake inhibition causes an increase in the amount of serotonin in the synapse, thereby reversing the serotonin deficit that may contribute to feelings of depression. However, it may be these drugs' ability to produce new neurons that produces most of the antidepressant effect. Other types of antidepressants increase serotonin levels in other ways or affect other neurotransmitters, such as dopamine or norepinephrine, which are also active in mood pathways associated with depression.

Researchers continue to study other drugs, such as the anesthetic *ketamine,* that more rapidly erase depression by increasing synaptic communication among existing neurons in the brain's mood pathways.

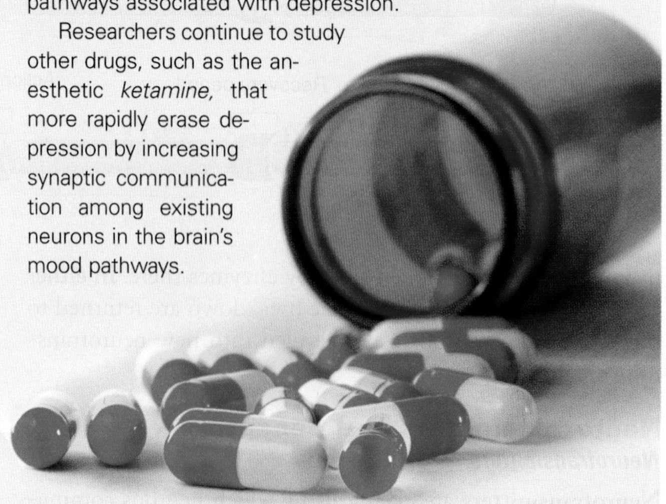

nerve signal. Notice that those shown are a component of the ion channels that produce a postsynaptic impulse when they open in response to activation of their receptors. Receptors can also be separate membrane structures that send a chemical signal to nearby ion channels to trigger their opening.

Each type of neurotransmitter can bind only to a receptor that chemically fits with it. Thus dopamine can only bind to a dopamine receptor and not to an acetylcholine receptor. However, there are slightly different kinds of dopamine receptors—each responding only to dopamine. The different kinds of dopamine receptors are found in different locations in the body.

The characteristics of neurotransmitter and receptors allow for very precise control of body functions. They also give an opportunity for developing effective drugs. For example, some asthma drugs target certain catecholamine receptors in the airways of lung and trigger them to dilate and allow a person to breathe more easily. However, such drugs can have unwanted side effects in other parts of the body that also have catecholamine receptors. So researchers attempt to find a drug that will trigger just the specific kind of catecholamine receptors found in the airways—and thus have fewer unwanted side effects.

To learn more about synapses, go to AnimationDirect online at *evolve.elsevier.com*.

Parkinson Disease

Parkinson disease (PD) is a chronic, progressive nervous disorder resulting from a deficiency of the neurotransmitter *dopamine* in certain parts of the brain.

The group of signs associated with this disorder, a syndrome called *parkinsonism,* includes rigidity and trembling of the head and extremities, a forward tilt of the trunk, and a shuffling manner of walking (**Figure 10-11**). You may have noticed these signs in former boxing champion Muhammad Ali, the actor Michael J. Fox, or in others you may know with PD. All of these characteristics result from lack of dopamine, leading to misinformation in the parts of the brain that normally prevents the skeletal muscles from being overstimulated.

Dopamine injections and dopamine pills are not effective treatments for PD because dopamine cannot cross the blood-brain barrier (see box on p. 256).

A breakthrough in the treatment of Parkinson disease came when the drug *levodopa* or L-dopa (Sinemet) was found to increase the dopamine levels in afflicted patients. Neurons use L-dopa, which can cross the blood-brain barrier, to make dopamine. For some reason, L-dopa does not always have the desired effects in individual patients or its effect may wear off over time, so a number of alternative treatments have been developed. For example, the drug apomorphine (Apokyn) has proved useful in treating individuals who no longer respond to L-dopa.

A treatment option that has shown some success is surgical grafting of normal dopamine-secreting neurons into the brains of individuals with PD. Another experimental option is an artificial implant that gives electrical stimulation to the brain, causing it to produce more dopamine.

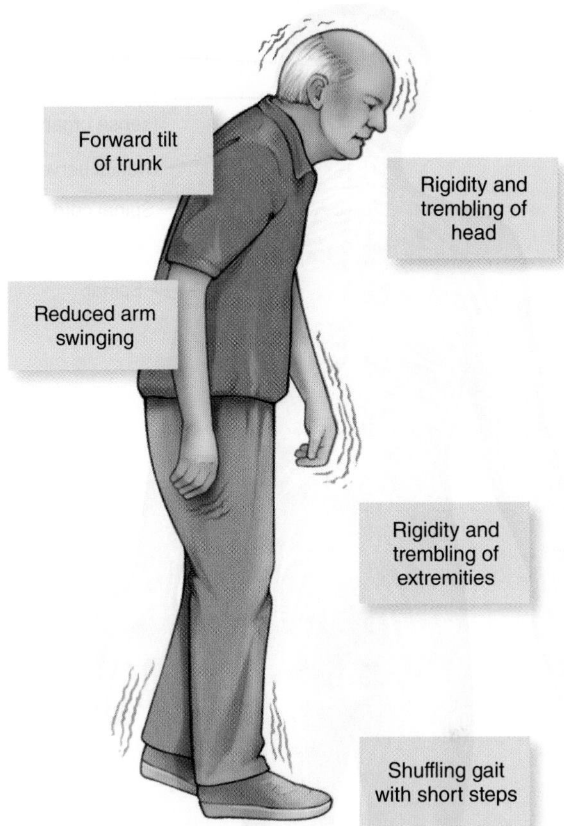

FIGURE 10-11 Parkinsonism. Parkinsonism is a syndrome typically found in individuals with Parkinson disease (PD). The signs include (but are not limited to) rigidity and trembling of the head and extremities, a forward tilt of the trunk, and a shuffling gait with short steps and reduced arm swinging.

QUICK CHECK

1. Why are nerve impulses often referred to as *action potentials*?
2. How does myelin increase the speed of nerve impulse conduction?
3. How do neurotransmitters transmit signals across the synapse?
4. What are the characteristics of *parkinsonism*?
5. How are the terms *reuptake* and *recycle* used when discussing the synapse?

10

Central Nervous System

The CNS, as its name implies, is centrally located. Its two major structures, the brain and spinal cord, are axial—found in the central axis of the body (**Figure 10-12**). The brain is protected in the cranial cavity of the skull, and the spinal cord is surrounded in the spinal cavity by the vertebral column. In addition, the brain and spinal cord are also covered by protective membranes called *meninges,* which are discussed in a later section of the chapter.

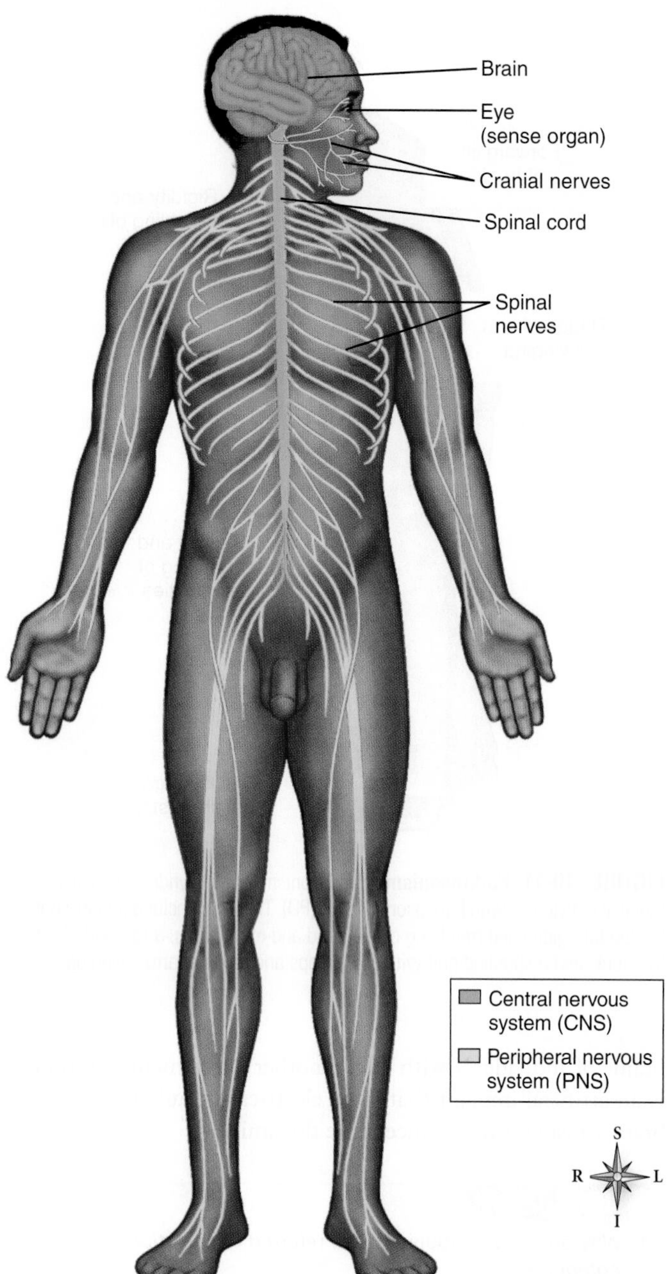

FIGURE 10-12 Nervous system. The brain and spinal cord *(highlighted green)* constitute the central nervous system (CNS), and the nerves *(yellow)* make up the peripheral nervous system (PNS).

10

Brain

Divisions of the Brain

The brain, one of our largest organs, consists of the following major divisions, named in ascending order beginning with the most inferior part:

I. Brainstem
   A. Medulla oblongata
   B. Pons
   C. Midbrain
II. Cerebellum
III. Diencephalon
   A. Hypothalamus
   B. Thalamus
   C. Pineal gland
IV. Cerebrum

Observe in Figure 10-13 the location and relative sizes of the different divisions of the brain.

Brainstem

The lowest part of the brainstem is the medulla oblongata. Immediately above the medulla lies the pons and above that the midbrain. Together these three structures are called the *brainstem* (see Figure 10-13).

The **medulla oblongata** is an enlarged, upward extension of the spinal cord. It lies just inside the cranial cavity above the large hole in the occipital bone called the *foramen magnum*. The **pons** bulges out a bit more than medulla, forming a bridge to the narrower midbrain.

In the brainstem, small bits of gray matter mix closely and intricately with white matter to form the **reticular formation** (*reticular* means "netlike"). In the spinal cord, gray and white matter do not intermingle; gray matter forms the interior core of the cord, and white matter surrounds it.

All three parts of the brainstem function as two-way conduction paths. Sensory fibers conduct impulses up from the spinal cord to other parts of the brain, and motor fibers conduct impulses down from the brain to the spinal cord.

In addition, many important reflex centers lie in the brainstem. The cardiac, respiratory, and vasomotor centers (collectively called the *vital centers*), for example, are located in the medulla. Impulses from these centers control heartbeat, respirations, and blood vessel diameter (which is important in regulating blood pressure).

Cerebellum

Structure

Look at Figure 10-13 to find the location, appearance, and size of the cerebellum.

The **cerebellum** is the second largest part of the human brain. It lies under the occipital lobe of the cerebrum. In the cerebellum, folded gray matter composes the thin outer layer, forming a large surface area of nervous connections that allow for a huge amount of information processing.

White matter tracts form most of the interior. Notice that these tracts branch in a treelike pattern called the *arbor vitae* (literally, "living tree").

Function

Most of our previous knowledge about cerebellar functions has come from observing patients who have some sort of disease of the cerebellum and from animals that have had the cerebellum removed. From such observations, we know that the cerebellum plays an essential part in the production of normal movements.

Perhaps a few examples will make this clear. A patient who has a tumor of the cerebellum frequently loses his balance and topples over; he may feel like a drunken man when he walks. He cannot coordinate his muscles normally. He may complain, for instance, that he is clumsy about everything he does—that he cannot even drive a nail or draw a straight line. With the loss of normal cerebellar functioning, he has lost the ability to make precise movements.

The most obvious functions of the cerebellum, then, are to produce smooth coordinated movements, maintain equilibrium, and sustain normal postures.

Recent studies using new brain imaging methods show that the cerebellum may have far more functions than earlier observed. The cerebellum may assist the cerebrum and other parts of the brain, perhaps having an overall coordinating function for the whole brain.

Diencephalon

The **diencephalon** is a small but important part of the brain located between the midbrain below and the cerebrum above. It consists of three major structures: the hypothalamus, thalamus, and pineal gland. Find these structures in **Figure 10-13** before reading further.

Hypothalamus

The **hypothalamus,** as its name suggests, is located below the thalamus. The posterior pituitary gland, the stalk that attaches it to the undersurface of the brain, and areas of gray matter located in the side walls of a fluid-filled space called the *third ventricle* are extensions of the hypothalamus. Identify the pituitary gland and the hypothalamus in **Figure 10-13**.

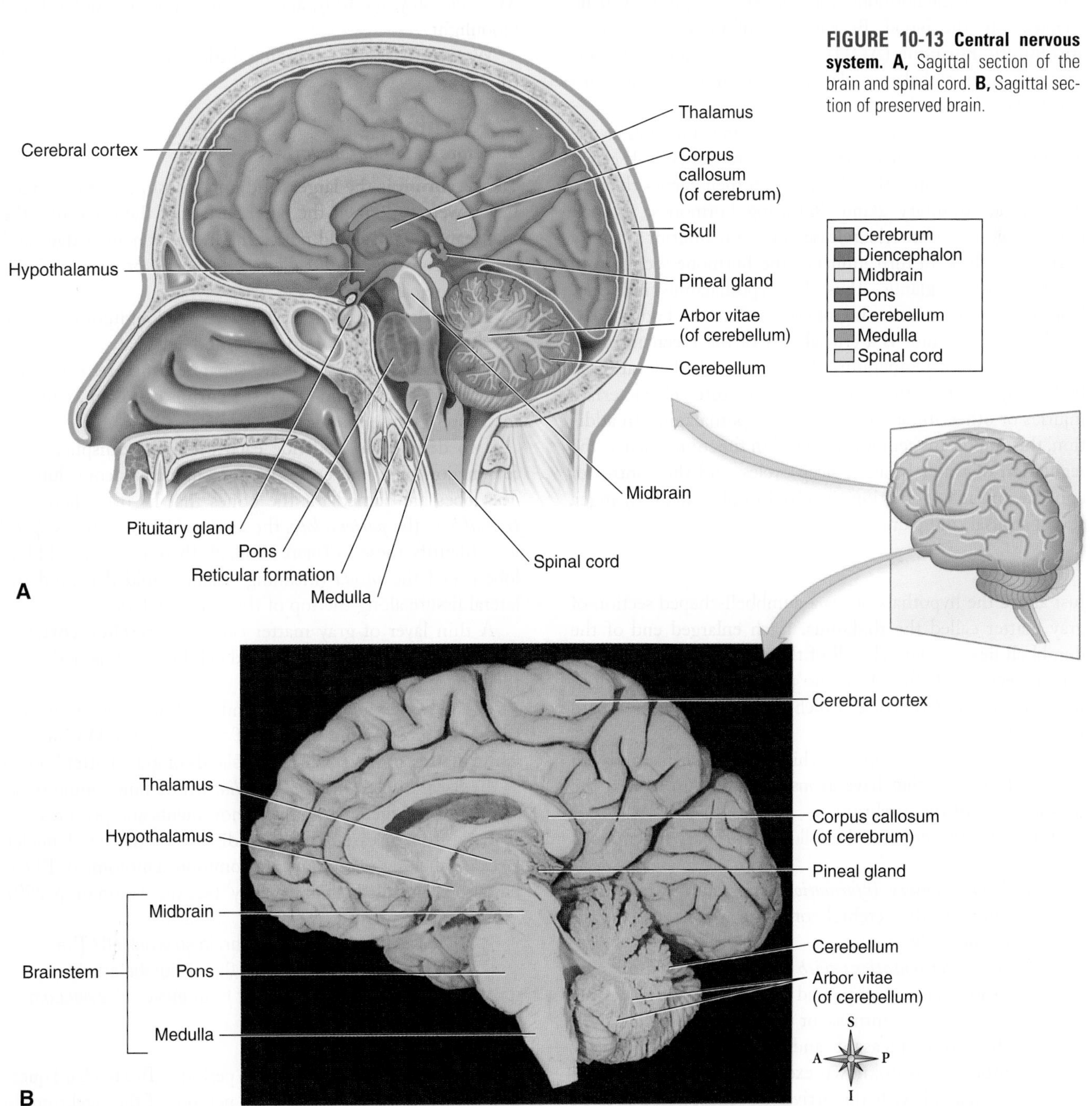

FIGURE 10-13 Central nervous system. A, Sagittal section of the brain and spinal cord. **B,** Sagittal section of preserved brain.

10

The old adage, "Don't judge by appearances," applies well to appraising the importance of the hypothalamus. Measured by size, it is one of the least significant parts of the brain, but measured by its contribution to healthy survival, it is one of the most important brain structures.

Impulses from neurons whose dendrites and cell bodies lie in the hypothalamus are conducted by their axons to neurons located in the spinal cord, and many of these impulses are then relayed to muscles and glands all over the body. Thus the hypothalamus exerts major control over virtually all internal organs. Among the vital functions that it helps control are the heartbeat, constriction and dilation of blood vessels, and contractions of the stomach and intestines.

Some neurons in the hypothalamus function in a surprising way; they make the hormones that the posterior pituitary gland secretes into the blood. Because one of these hormones—**antidiuretic hormone (ADH)**—affects the volume of urine excreted, the hypothalamus plays an essential role in maintaining the body's water balance.

Some of the neurons in the hypothalamus function as endocrine (ductless) glands. Their axons secrete chemicals called *releasing hormones* into the blood, which then carries them to the anterior pituitary gland. Releasing hormones, as their name suggests, control the release of certain anterior pituitary hormones. These in turn influence the hormone secretion of other endocrine glands. Thus the hypothalamus indirectly helps control the functioning of every cell in the body.

The hypothalamus is a crucial part of the mechanism for maintaining body temperature. Therefore marked elevation in body temperature in the absence of disease often characterizes injuries or other abnormalities of the hypothalamus. In addition, this important center is involved in functions such as the regulation of water balance, sleep cycles, and the control of appetite and many emotions involved in pleasure, fear, anger, sexual arousal, and pain.

Thalamus

Just above the hypothalamus is a dumbbell-shaped section of gray matter called the **thalamus.** Each enlarged end of the dumbbell lies in a lateral wall of the *third ventricle.* The thin center section of the thalamus passes from left to right through the third ventricle, which we discuss in more detail later in this chapter.

The thalamus is composed chiefly of dendrites and cell bodies of neurons that have axons extending up toward the sensory areas of the cerebrum.

The thalamus performs the following primary functions:

1. *Relays sensory information.* Its neurons relay impulses to the cerebral cortex from the sense organs of the body.
2. *Associates sensations with emotions.* Almost all sensations are accompanied by a feeling of some degree of pleasantness or unpleasantness. The way that these pleasant and unpleasant feelings are produced is unknown except that they seem to be associated with the arrival of sensory impulses in the thalamus.
3. *Regulates level of consciousness.* It plays a part in the so-called *arousal* or alerting mechanism that keeps us awake.
4. *Participates in motor reflexes.* It plays a role in mechanisms that produce complex reflex movements.

Pineal gland

Posterior to the thalamus is a tiny mass protruding from the back of the diencephalon called the **pineal gland** or *pineal body.* It resembles a small pine nut or kernel of corn.

The pineal gland receives sensory information about the strength of light seen by the eyes and adjusts its output of the hormone *melatonin.* Melatonin is known as "the timekeeping hormone" because it helps keep the body's clock "on time" with the daily, monthly, and seasonal cycles of sunlight and moonlight.

We return to this amazing little organ in Chapter 12 (p. 338).

Cerebrum

Structure of the Cerebrum

The **cerebrum** is the largest and uppermost part of the brain. If you were to look at the outer surface of the cerebrum, the first features you would notice might be its many ridges and grooves. The ridges are called *convolutions,* or **gyri,** and the grooves are called **sulci.**

The deepest sulci are called *fissures.* The longitudinal fissure divides the cerebrum into right and left halves or hemispheres. These halves are almost separate structures except for an inferior central band called the **corpus callosum,** which is made up of white matter tracts (see Figure 10-13).

Two deep sulci subdivide each cerebral hemisphere into four major lobes and each lobe into numerous convolutions. The lobes are named for the bones that lie over them: the *frontal lobe,* the *parietal lobe,* the *temporal lobe,* and the *occipital lobe.* Identify these in Figure 10-14, *A.* There is also a "hidden" lobe called the *insula* (meaning "island") folded behind the lateral fissure along the top of the temporal lobe.

A thin layer of gray matter called the **cerebral cortex** is made up of neuron dendrites and cell bodies; this makes up the surface of the cerebrum.

White matter, made up of bundles of nerve fibers (tracts), composes most of the interior of the cerebrum. Within this white matter, however, are a few islands of gray matter known as the **basal nuclei,** or *basal ganglia,* whose functioning is essential for producing automatic movements and postures.

Parkinson disease (PD) is a disease of the basal nuclei. Because shaking or tremors are common symptoms of PD, it has also been called "shaking palsy" (see discussion on p. 259).

Why is the surface of the brain so wrinkled? The answer may surprise you. Check out the article *Brain Wrinkles* at Connect It! at *evolve.elsevier.com.*

Function of the Cerebrum

What functions does the cerebrum perform? This is a hard question to answer briefly because the neurons of the cerebrum do not function alone. They function with many other neurons in

10

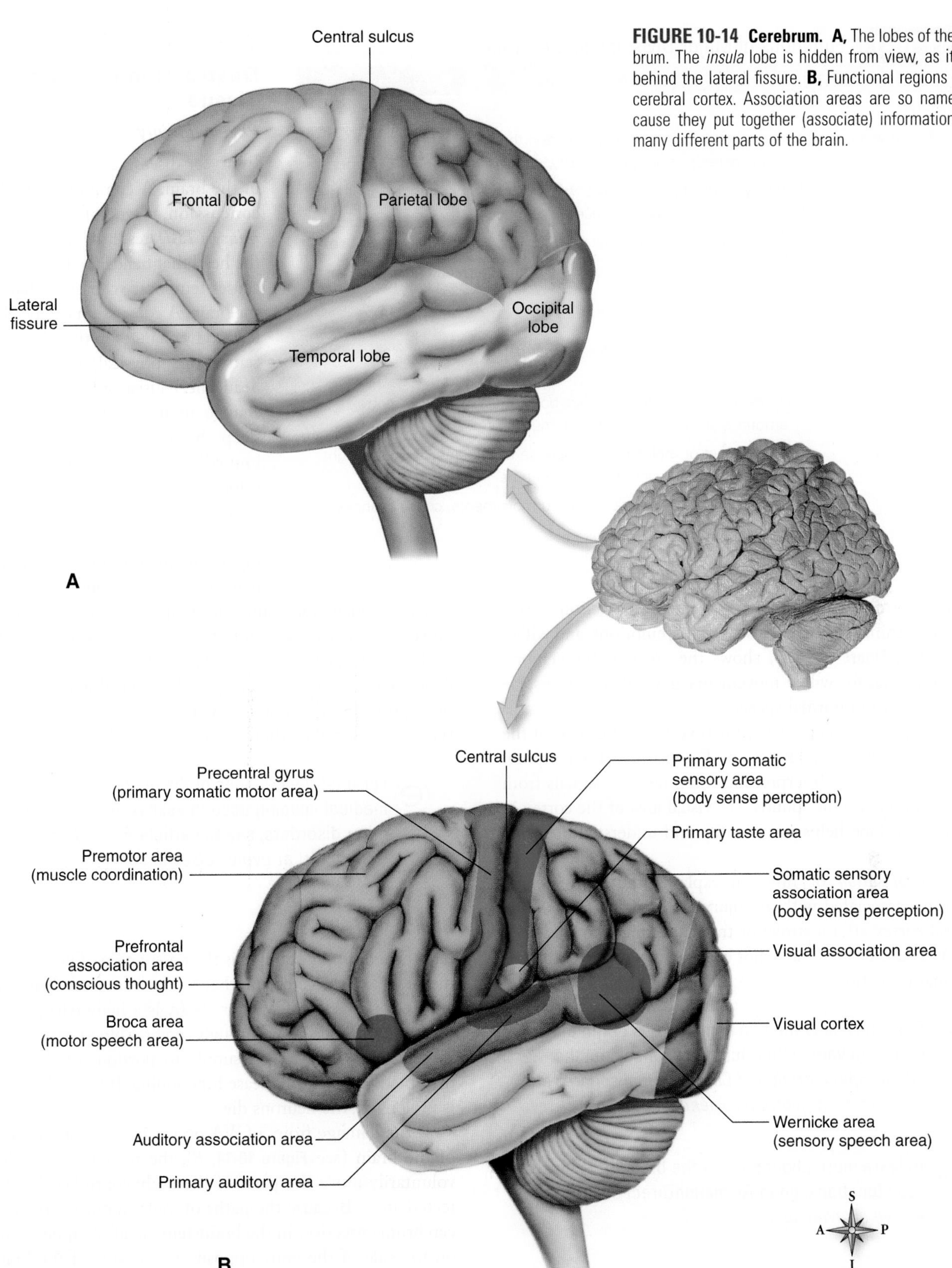

FIGURE 10-14 Cerebrum. A, The lobes of the cerebrum. The *insula* lobe is hidden from view, as it folds behind the lateral fissure. **B,** Functional regions of the cerebral cortex. Association areas are so named because they put together (associate) information from many different parts of the brain.

many other parts of the brain and in the spinal cord. Neurons of these various structures continually bring impulses to cerebral neurons and also continually transmit impulses away from them.

If all other neurons were functioning normally and only cerebral neurons were not functioning, here are some of the things that you could not do: You could not think or use your will. You could not remember anything that has ever happened to you. You could not decide to make the smallest movement, nor could you make it. You would not see or hear. You could not experience any of the sensations that make life so rich and varied. Nothing would anger or frighten you, and nothing would bring you joy or sorrow. You would, in short, be unconscious.

10

TABLE 10-1 Functions of Major Divisions of the Brain

BRAIN AREA	FUNCTION
Brainstem	
Medulla oblongata	Two-way conduction pathway between the spinal cord and higher brain centers; cardiac, respiratory, and vasomotor control center
Pons	Two-way conduction pathway between areas of the brain and other regions of the body; influences respiration
Midbrain	Two-way conduction pathway; relay for visual and auditory impulses
Cerebellum	Muscle coordination; maintenance of equilibrium and posture; assists cerebrum
Diencephalon	
Hypothalamus	Regulation of body temperature, water balance, sleep-cycle control, appetite, and sexual arousal
Thalamus	Sensory relay station from various body areas to cerebral cortex; emotions and alerting or arousal mechanisms
Pineal gland	Adjusts output of melatonin in response to changes in external light, to keep the body's internal clock on time
Cerebrum	Sensory perception, emotions, willed movements, consciousness, and memory

These terms sum up the major cerebral functions: consciousness, thinking, memory, sensations, emotions, and willed movements. **Figure 10-14**, *B*, shows the areas of the cerebral cortex essential for willed movements, general sensations, vision, hearing, and normal speech.

It is important to understand that very specific areas of the cortex have very specific functions. For example, the temporal lobe's auditory areas interpret incoming nervous signals from the ear as very specific sounds. The visual area of the cortex in the occipital lobe helps you identify and understand specific images.

The *localization* of function explains the very specific symptoms associated with an injury to localized areas of the cerebral cortex after a stroke or traumatic injury to the head. **Table 10-1** summarizes the major components of the brain and their main functions.

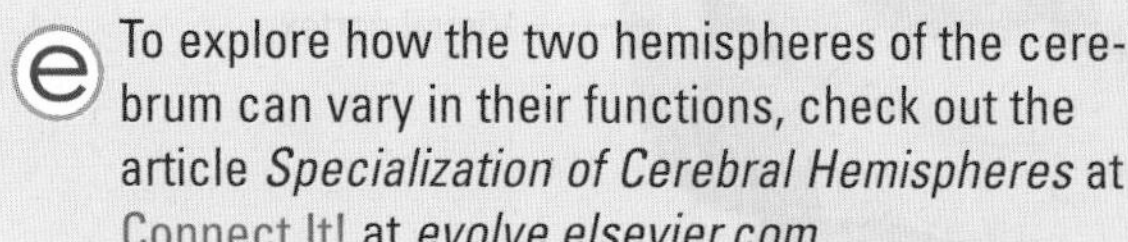
To explore how the two hemispheres of the cerebrum can vary in their functions, check out the article *Specialization of Cerebral Hemispheres* at Connect It! at *evolve.elsevier.com.*

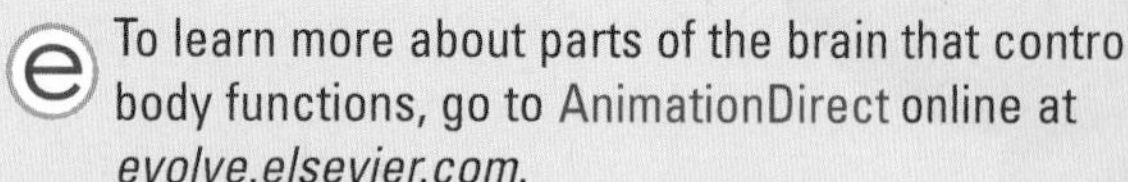
To learn more about parts of the brain that control body functions, go to AnimationDirect online at *evolve.elsevier.com.*

QUICK CHECK

1. What are the four main divisions of the brain?
2. What regions make up the *brainstem*?
3. What is the *corpus callosum*?
4. What structure in the diencephalon controls the hormone melatonin?
5. Why is the *hypothalamus* said to be a link between the nervous system and the endocrine system?

Brain Disorders

Destruction of Brain Tissue

Physical Injury

Injury or disease can destroy neurons in the brain. A common example is a **concussion,** a type of *traumatic brain injury (TBI)* resulting from a jolt to the head that bends the brainstem and causes temporary chemical changes in the brain. A concussion can be characterized by changes in thinking or concentration; physical symptoms like headache, nausea, or light sensitivity; a change in mood; and/or changes in sleep. Sometimes symptoms appear immediately but sometimes they develop over days or even months, as in *postconcussion syndrome.* If severe, or if a second jolt occurs, bleeding or swelling of brain also may occur, which may be life-threatening.

Most concussions (and the symptoms) are relatively mild, however, and with rest and other precautions may eventually heal without permanent effects. Studies show that the incidence of concussions can be greatly reduced with appropriate head protection, such as sports helmets, and avoiding certain types of movements that put the head at risk.

For brief descriptions of the major types of medical imaging used to assess brain damage and other disorders, see the article *Brain Studies* at Connect It! at *evolve.elsevier.com.*

Stroke

Another common example is the destruction of neurons of the motor area of the cerebrum that results from a **cerebrovascular accident (CVA).** A CVA, or *stroke,* is a hemorrhage from or cessation of blood flow through cerebral blood vessels. When this happens, the oxygen supply to portions of the brain is disrupted, and neurons cease functioning. If the lack of oxygen is prolonged, the neurons die.

If the damage from a CVA occurs in a motor control area of the brain (see **Figure 10-14**, *B*), the patient can no longer voluntarily move the parts of the body controlled by the affected areas. Because the paths of most motor neurons in the cerebrum cross over in the brainstem, paralysis appears mostly on the side of the body opposite to the side of the brain on which the CVA occurred.

Cerebral Palsy

One of the most common crippling diseases that can appear during childhood, **cerebral palsy (CP),** also results from damage to brain tissue. Cerebral palsy involves permanent, nonprogressive damage to motor control areas of the brain, which in turn causes abnormal muscle tension (spasticity) that

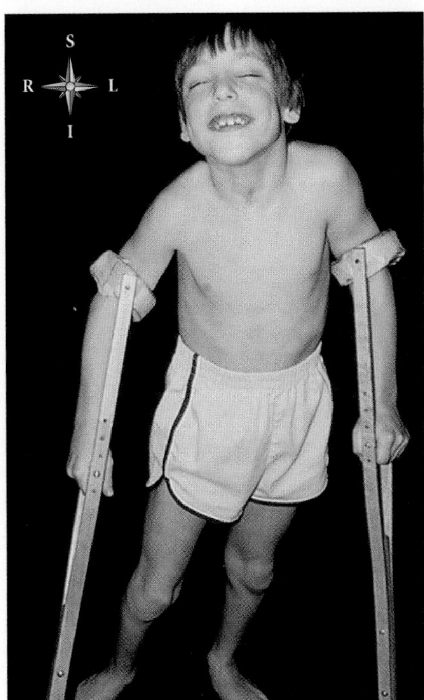

FIGURE 10-15 Cerebral palsy (CP). This patient requires crutches to walk because abnormal tension (spasticity) in muscles prevents normal walking movements.

hinders movement (**Figure 10-15**). Such damage is present at birth or occurs shortly after birth and remains throughout life.

Possible causes of brain damage include prenatal infections or diseases of the mother; mechanical trauma to the head before, during, or after birth; nerve-damaging poisons; and reduced oxygen supply to the brain. The resulting impairment to voluntary muscle control can manifest itself in a variety of ways. Many people with cerebral palsy exhibit **spastic paralysis,** a type of paralysis characterized by involuntary contractions of affected muscles. In cerebral palsy, spastic paralysis may affect most of one side of the body **(hemiplegia),** both legs **(paraplegia),** both legs and one arm **(triplegia),** or all four extremities **(quadriplegia).**

Degenerative Disease

A variety of degenerative diseases can result in destruction of neurons in the brain. This degeneration can progress to adversely affect memory, attention span, intellectual capacity, personality, and motor control. The general term for this syndrome is **dementia.**

Dementia is characteristic of **Alzheimer disease (AD).** Its characteristic lesions develop in the cortex during the middle to late adult years (**Figure 10-16**). Exactly what makes dementia-causing lesions develop in the brains of individuals with AD is not known. There is some evidence that this disease has a genetic basis—at least in some families. Other evidence indicates that environmental factors may play a role. These various triggers apparently cause accumulation of improperly folded proteins in brain cells, which in turn causes loss of function.

Once AD starts, the tangled proteins are able to move to nearby cells and trigger protein misfolding there. This is similar to the way *prions* can spread their damaging properties to other cells in the brain (see Chapter 6, p. 120). Because all the various mechanisms of AD are still not completely known, development of an effective treatment has proved difficult.

Currently, people diagnosed with AD are often treated with drugs such as donepezil (Aricept) for mild to moderate AD or with the drug memantine (Namenda) for moderate to advanced AD. These drugs cause therapeutic shifts in baseline neurotransmitter levels at synapses in the brain. In addition, treatment includes helping patients maintain their remaining mental abilities and looking after their hygiene, nutrition, and other aspects of personal health management.

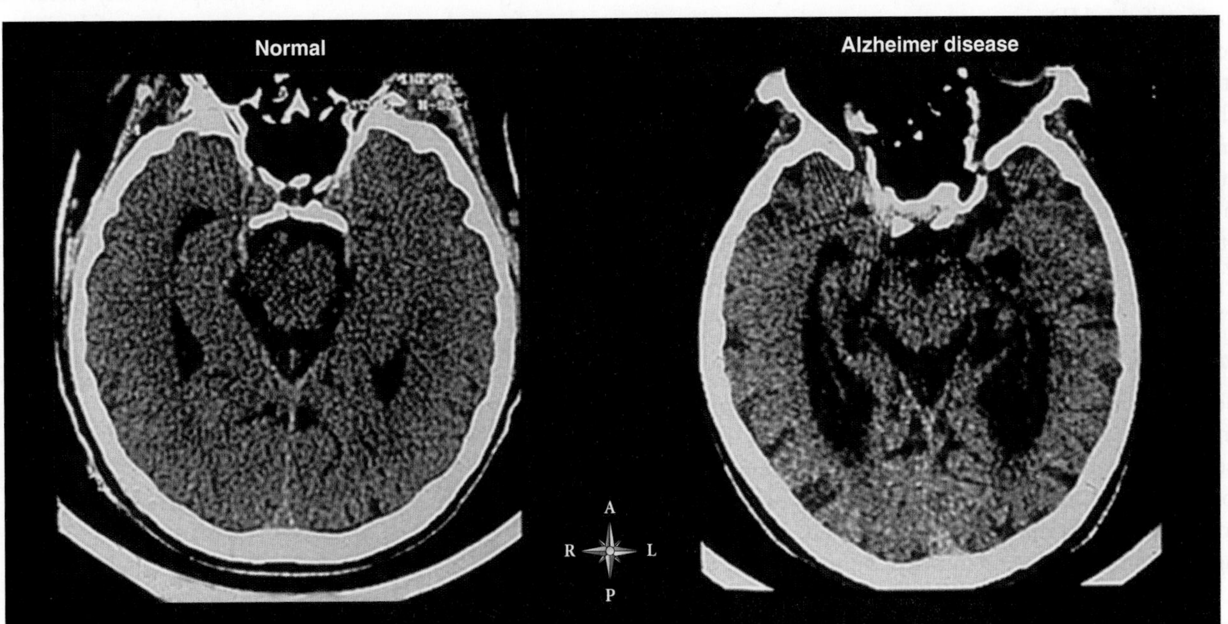

FIGURE 10-16 Alzheimer disease (AD). The CT scan on the left shows a horizontal section of a normal brain. In the CT scan on the right, however, you can see the dark patches in the cerebral cortex that show damage to brain tissue typical of AD.

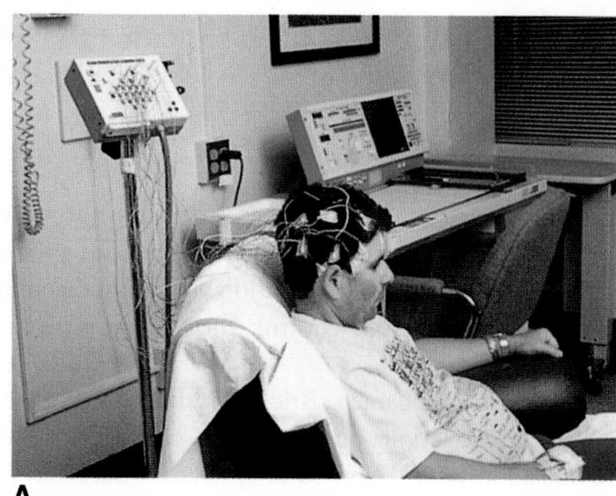

A

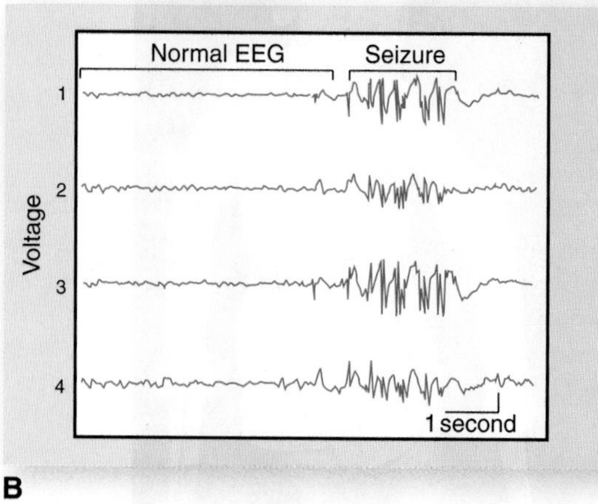

B

FIGURE 10-17 Electroencephalography. A, Photograph of a person with voltage-sensitive electrodes attached to his skull. Information from these electrodes is used to produce a graphic recording of brain activity—an electroencephalogram (EEG). **B,** An EEG tracing showing activity in four different places in the brain (obtained from four sets of electrodes). Compare the moderate chaotic activity identified as normal with the explosive activity that occurs during a seizure.

Chronic traumatic encephalopathy (CTE) similarly involves accumulation of abnormal proteins in the brain and memory loss. However, it is also characterized by parkinsonism (see **Figure 10-11** on p. 259), disordered thinking, and other neurological symptoms. As its name implies, CTE results from repeated trauma to the brain, including TBIs, as occurs in some sports, physical abuse, and those with seizure disorders or head-banging behavior.

Huntington disease (HD) is an inherited disease characterized by *chorea* (involuntary, purposeless movements) that progresses to severe dementia and death. The initial symptoms of this disease first appear between ages 30 and 40, with death occurring by age 55. Now that the gene responsible for Huntington disease has been located, researchers hope that an effective treatment will be found (see Chapter 25). The discovery of the HD gene poses an interesting question: if you could learn early in life that you will get HD, would you want to know?

The human immunodeficiency virus (HIV) that causes *acquired immunodeficiency syndrome (AIDS)* also can cause dementia. The immune deficiency characteristic of AIDS results from HIV infection of white blood cells critical to the proper function of the immune system (see Chapter 16). However, HIV also infects neurons and can cause progressive degeneration and shrinkage of the brain, resulting in dementia.

10

Seizure Disorders

Some of the most common nervous system abnormalities belong to the group of conditions called *seizure disorders.* These disorders are characterized by **seizures**—sudden bursts of abnormal neuron activity that result in temporary changes in brain function. Seizures may be very mild, causing subtle changes in the level of consciousness, motor control, or sensory perception. On the other hand, seizures may be quite severe, resulting in jerky, involuntary muscle contractions called *convulsions* or even unconsciousness.

Recurring or chronic seizure episodes constitute a condition called **epilepsy.** Although some cases of epilepsy can be traced to specific causes such as tumors, trauma, or chemical imbalances, most epilepsy is idiopathic (of unknown cause).

Those with epilepsy are often treated with well-known anticonvulsive drugs such as *phenobarbital, phenytoin (Dilantin),* or *valproic acid (Depakene)* that block neurotransmitters in affected areas of the brain. By thus blocking synaptic transmission, such drugs inhibit the explosive bursts of neuron activity associated with seizures.

Continuing research has resulted in the release of several new epilepsy drugs that have provided improved treatment options for many patients. Newer drugs include gabapentin (Neurontin) and lamotrigine (Lamictal). With proper medication, many people with epilepsy lead normal lives without the fear of experiencing uncontrollable seizures.

Diagnosis and evaluation of epilepsy or any seizure disorder often rely on a graphic representation of brain activity called an **electroencephalogram (EEG).** In **Figure 10-17**, *B,* a normal EEG shows the chaotic rise and fall of the electrical activity in different parts of the brain as a series of wavy lines (the so-called *brain waves*). A seizure manifests itself as an explosive increase in the size and frequency of waves—as seen on the right side of **Figure 10-17**, *B.* Classifications of epilepsy are based on the locations in the brain and the duration of these changes in brain activity.

QUICK CHECK

1. Identify the characteristics or symptoms of a *concussion.*
2. What is a *CVA*? How does it affect the brain?
3. What disorders are characterized by *dementia*?
4. What is *epilepsy*? How is epilepsy diagnosed and evaluated?

Spinal Cord

Structure

If you are of average height, your spinal cord is about 42 cm to 45 cm (17 or 18 inches) long (**Figure 10-18**). It lies inside the spinal column in the spinal cavity and extends from the occipital

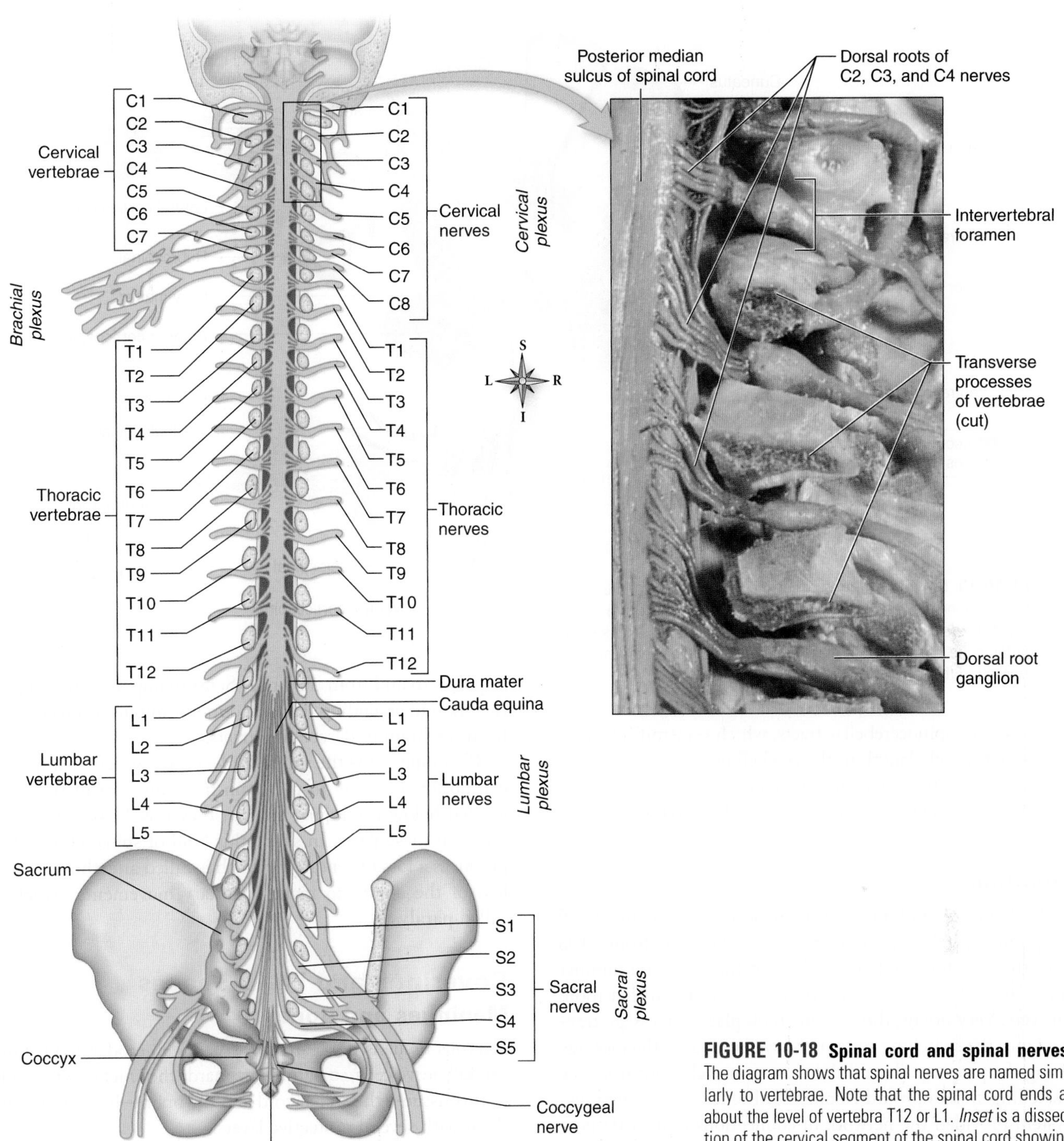

FIGURE 10-18 Spinal cord and spinal nerves. The diagram shows that spinal nerves are named similarly to vertebrae. Note that the spinal cord ends at about the level of vertebra T12 or L1. *Inset* is a dissection of the cervical segment of the spinal cord showing emerging cervical nerves. The spinal cord is viewed from behind (posterior aspect).

bone down to the bottom of the first lumbar vertebra. Place your hands on your hips, and they will line up with your fourth lumbar vertebra. Your spinal cord ends just above this level.

Look now at **Figure 10-19**. Notice the H-shaped core of the spinal cord. It consists of gray matter and thus is composed mainly of dendrites and cell bodies of neurons. This part of the spinal cord contains many synapses and interneurons, which enable it to be involved in many important reflex arcs.

Columns of white matter form the outer portion of the spinal cord, and bundles of myelinated nerve fibers—the **spinal tracts**—make up the white columns.

Spinal cord tracts provide two-way conduction paths to and from the brain. *Ascending tracts* conduct impulses up the spinal cord to the brain. *Descending tracts* conduct impulses down the spinal cord from the brain.

Tracts are functional organizations in that all axons composing one tract serve one general function. For instance, fibers of the spinothalamic tracts serve a sensory function. They carry impulses that produce sensations of crude touch, pain, and temperature. Other ascending tracts shown in **Figure 10-19** include the gracilis and cuneatus tracts, which transmit sensations of touch and pressure up to the brain, and the anterior

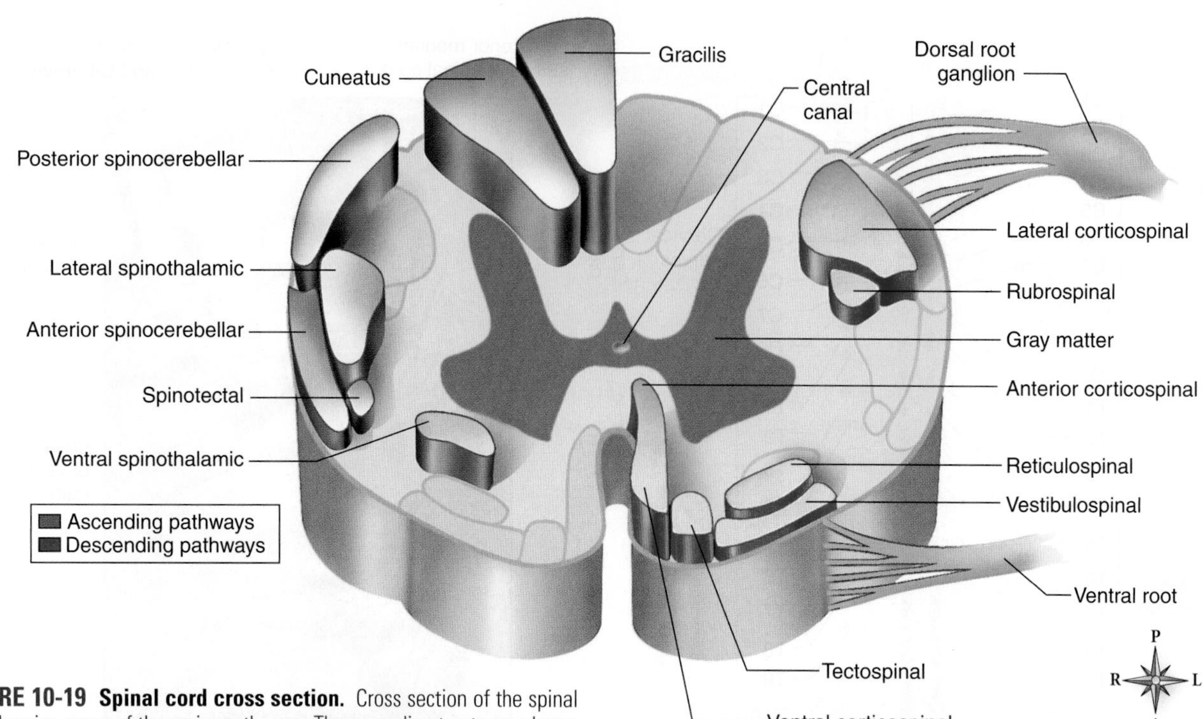

FIGURE 10-19 Spinal cord cross section. Cross section of the spinal cord showing some of the major pathways. The ascending tracts are shown in *blue* and the descending tracts are shown in *red*. You can also see the gray matter in the center of the spinal cord *(brown)* and the nerve roots *(yellow)* attached to the spinal cord.

and posterior spinocerebellar tracts, which transmit information about muscle length to the cerebellum.

Descending tracts include the lateral and ventral corticospinal tracts, which carry impulses controlling many voluntary movements.

Functions

To try to understand spinal cord functions, think about a hotel telephone switching system. Suppose a guest in Room 108 calls the switching system and keys in the extension number for Room 520, and in a second or so, someone in that room answers. Very briefly, three events took place: a message traveled into the switching system, the system routed the message along the proper path, and the message traveled out from the switching system toward Room 520. The telephone switching system provided the network of connections that made possible the completion of the call. We might say that the switching system transferred the incoming call to an outgoing line.

The spinal cord functions similarly. It contains the centers for thousands and thousands of reflex arcs. Look back at Figure 10-7. The interneuron shown there is an example of a spinal cord reflex center. It switches or transfers incoming sensory impulses to outgoing motor impulses, thereby making it possible for a reflex to occur.

Reflexes that result from conduction over arcs whose centers lie in the spinal cord are called *spinal cord reflexes*. Two common kinds of spinal cord reflexes are *withdrawal* and *jerk reflexes*. An example of a withdrawal reflex is pulling one's hand away from a hot surface. The familiar knee jerk is an example of a jerk reflex.

In addition to functioning as the primary reflex center of the body, the spinal cord tracts, as previously noted, carry impulses to and from the brain. Sensory impulses travel up to the brain in ascending tracts, and motor impulses travel down from the brain in descending tracts.

If an injury cuts the spinal cord all the way across, impulses can no longer travel to the brain from any part of the body located below the injury, nor can they travel from the brain down to these parts. In short, this kind of spinal cord injury produces a loss of sensation, which is called **anesthesia,** and a loss of the ability to make voluntary movements, which is called **paralysis.**

Coverings and Fluid Spaces

Meninges and Bone

Nervous tissue is not a sturdy tissue. Even moderate pressure can kill nerve cells, so nature safeguards the chief organs made of this tissue—the spinal cord and the brain—by surrounding them with several protective layers.

A tough, fluid-cushioned set of membranes called the **meninges** make up the inner layers of protection for the CNS.

The spinal meninges form a tubelike covering around the spinal cord and line the bony vertebral foramen of the vertebrae that surround the cord. Look at Figure 10-20, and you can identify the three layers of the spinal meninges. They are the **dura mater,** which is the tough outer layer that lines the vertebral canal, the **pia mater,** which is the innermost membrane covering the spinal cord itself, and the **arachnoid mater,** which is the membrane between the dura and the pia mater. The arachnoid mater resembles a cobweb with fluid in its spaces. The word *arachnoid* means "cobweblike." It comes from *arachne,* the Greek word for spider.

The meninges that form the protective covering around the spinal cord also extend up and around the brain to enclose it completely (peek ahead to Figure 10-22).

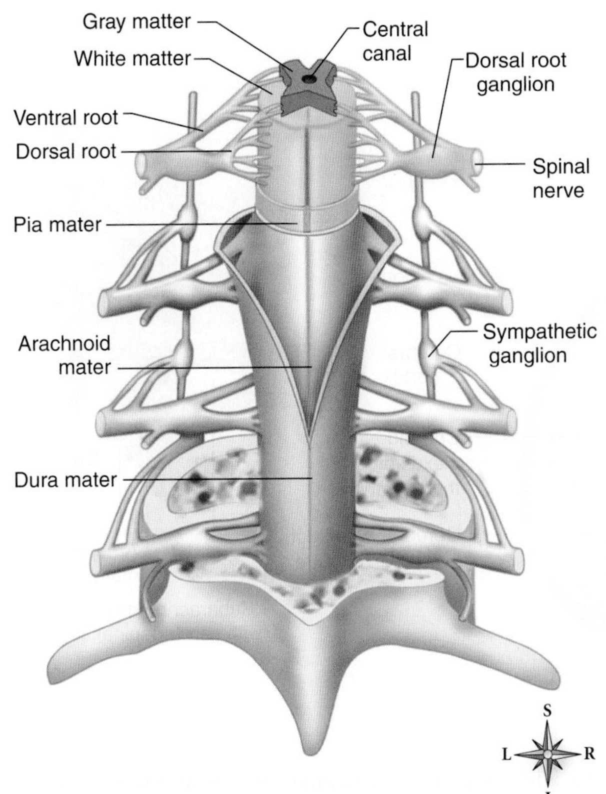

FIGURE 10-20 Spinal cord and its coverings. The meninges, spinal nerves, and sympathetic trunk are all depicted in a posterior view.

Infection or inflammation of the meninges is termed **meningitis.** This condition is most commonly caused by bacteria such as *Neisseria meningitidis* (meningococcus), *Streptococcus pneumoniae,* or *Haemophilus influenzae* (see Appendix A at *evolve.elsevier.com*). However, viral infections, mycoses (fungal infections), and tumors also may cause inflammation of the meninges.

Patients with meningitis usually complain of severe headaches and neck pain. Those experiencing symptoms should seek immediate attention to get the problem under control.

Depending on the primary cause, meningitis may be mild and self-limiting or may progress to a severe, perhaps fatal, condition. If only the spinal meninges are involved, the condition is called *spinal meningitis.*

The meninges are inner, soft coverings of the CNS. They are, in turn, surrounded by the hard bone of the skull and vertebrae—forming a highly protective shield from injury.

Cerebrospinal Fluid Spaces

Fluid fills the subarachnoid spaces between the pia mater and arachnoid in the brain and spinal cord. This fluid is called **cerebrospinal fluid (CSF).**

CSF also fills spaces in the brain called cerebral **ventricles.** In Figure 10-21, you can see the irregular shapes of the ventricles of the brain. These illustrations also can help you visualize the location of the ventricles if you remember that these large spaces lie deep inside the brain and that there are two lateral ventricles. One lies inside the right half of the cerebrum (the largest part of the human brain), and the other lies inside the left half.

CSF forms continually from fluid filtering out of the blood in a network of brain capillaries known as the **choroid plexus** and into the ventricles.

CSF is one of the body's circulating fluids. CSF seeps from the lateral ventricles into the third ventricle and flows down through the cerebral aqueduct (find this in Figures 10-21 and 10-22) into the fourth ventricle. Most of the CSF moves through tiny openings from the fourth ventricle into the subarachnoid space near the cerebellum. Some of it moves into the small, tubelike *central canal* of the spinal cord and then out into the subarachnoid spaces. Then CSF moves leisurely down and around the spinal cord and up and around the brain (in the subarachnoid spaces of their meninges) and returns to the blood (in the veins of the brain).

Remembering that this fluid forms continually from blood, circulates, and is resorbed into blood can be useful. It can help you understand certain abnormalities. Suppose a person has a brain tumor that presses on the cerebral aqueduct. This blocks the way for the return of CSF to the blood. Because the fluid continues to form but cannot drain away, it accumulates in the ventricles or in the meninges, creating enough pressure to damage or deform the nearby soft nervous tissue.

Other conditions can cause an accumulation of CSF in the ventricles. An example is **hydrocephalus,** or "water on the

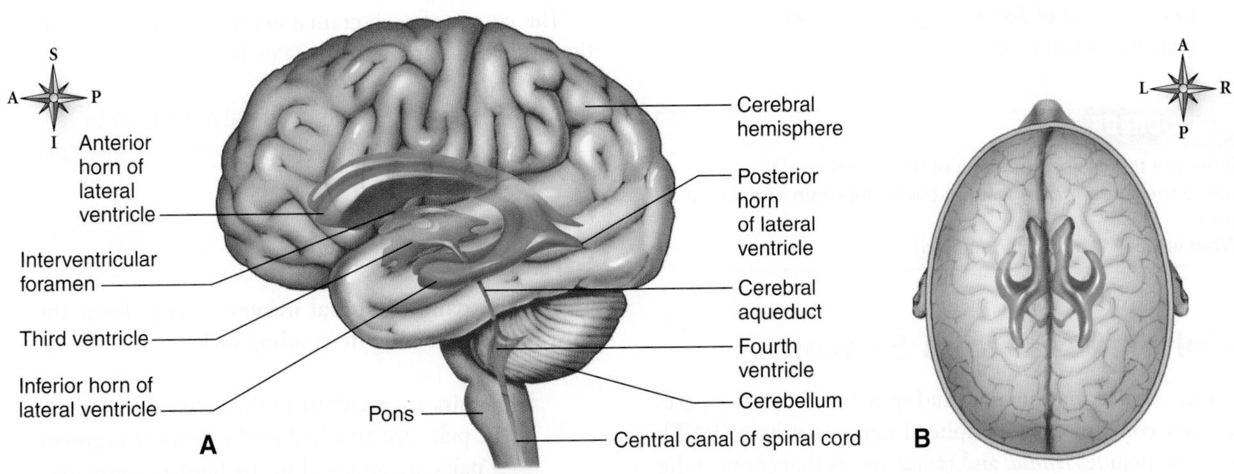

FIGURE 10-21 Fluid spaces of the brain. A, The ventricles are highlighted within the brain in a left lateral view. **B,** The ventricles shown from above.

10

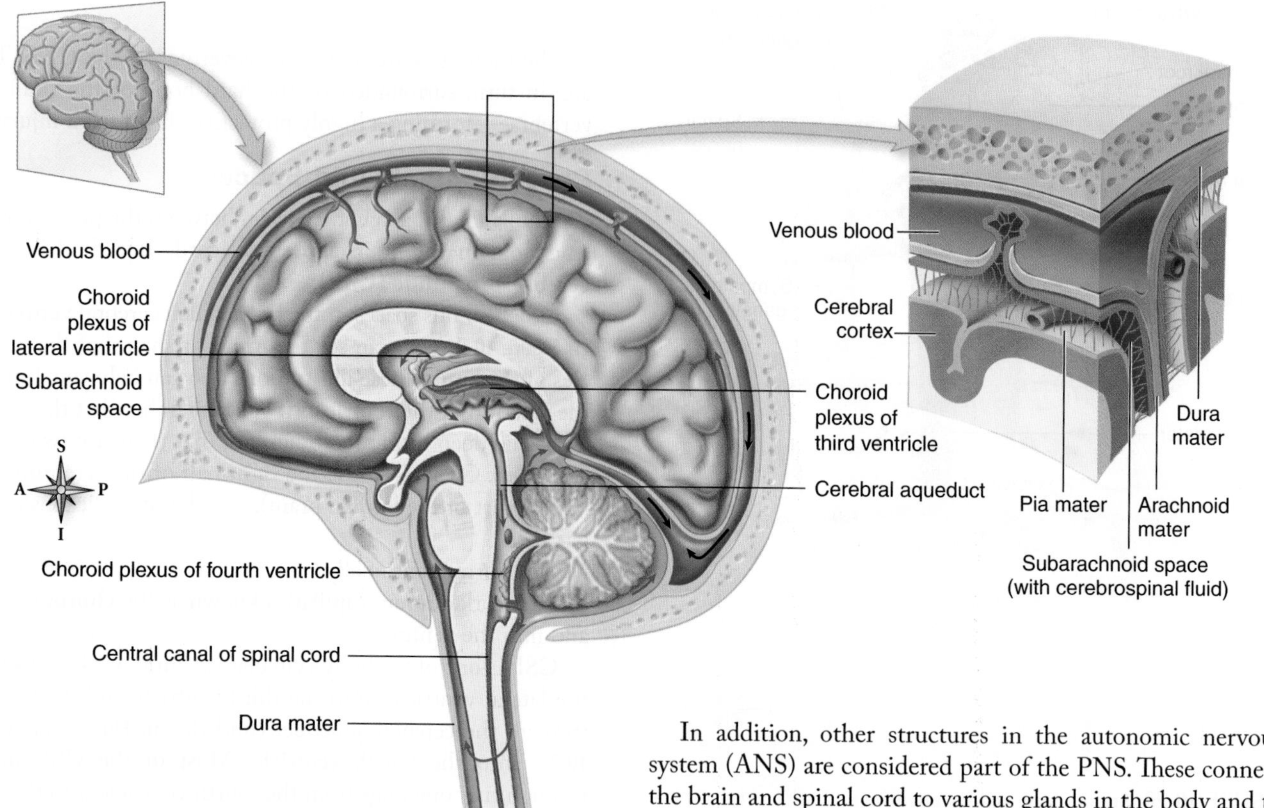

FIGURE 10-22 Flow of cerebrospinal fluid (CSF). The fluid produced by filtration of blood by the choroid plexus of each ventricle flows inferiorly through the lateral ventricles, interventricular foramen, third ventricle, cerebral aqueduct, fourth ventricle, and subarachnoid space and then to the blood.

brain." One form of treatment involves surgical placement of a hollow tube or catheter through the blocked channel so that CSF can drain into another location in the body (Figure 10-23).

The fluid spaces of the brain are better understood when one realizes that the CNS develops from a fluid-filled tube during embryonic development. To see a diagram and brief description of this process, check out the article *Embryonic Development of Tissues* at Connect It! at *evolve.elsevier.com*.

10

QUICK CHECK

1. What are the major functions of the spinal cord?
2. Name the three *meninges* that cover the brain and spinal cord.
3. What is *CSF*? What is its function?

Peripheral Nervous System

The nerves connecting the brain and spinal cord to other parts of the body constitute the peripheral nervous system (PNS). This system includes *cranial* and *spinal* nerves that connect the brain and spinal cord, respectively, to peripheral structures such as the skin surface and the skeletal muscles.

In addition, other structures in the autonomic nervous system (ANS) are considered part of the PNS. These connect the brain and spinal cord to various glands in the body and to the cardiac and smooth muscle in the thorax and abdomen.

Cranial Nerves

Twelve pairs of **cranial nerves (CN)** emerge from the undersurface of the brain. Figure 10-24 shows the attachments of these nerves to the brainstem and diencephalon. Their fibers conduct impulses between the brain and structures in the head and neck and in the thoracic and abdominal cavities.

For example, the second cranial nerve (CN II or *optic nerve*) conducts impulses from the eye to the brain, where these impulses produce vision. The third cranial nerve (CN III or *oculomotor nerve*) conducts impulses from the brain to muscles in the eye, where they cause contractions that move the eye. The tenth cranial nerve (CN X or *vagus nerve*) conducts impulses between the medulla oblongata and structures in the neck and thoracic and abdominal cavities.

The names of each cranial nerve and a brief description of their functions are listed in Table 10-2.

To learn more about cranial nerves, go to AnimationDirect online at *evolve.elsevier.com*.

Spinal Nerves

Structure

Thirty-one pairs of **spinal nerves** emerge from the spinal cord in the following descending order:

- 8 pairs are attached to the cervical segments
- 12 pairs are attached to the thoracic segments
- 5 pairs are attached to the lumbar segments
- 5 pairs are attached to the sacrospinal segments
- 1 pair is attached to the coccygeal segment

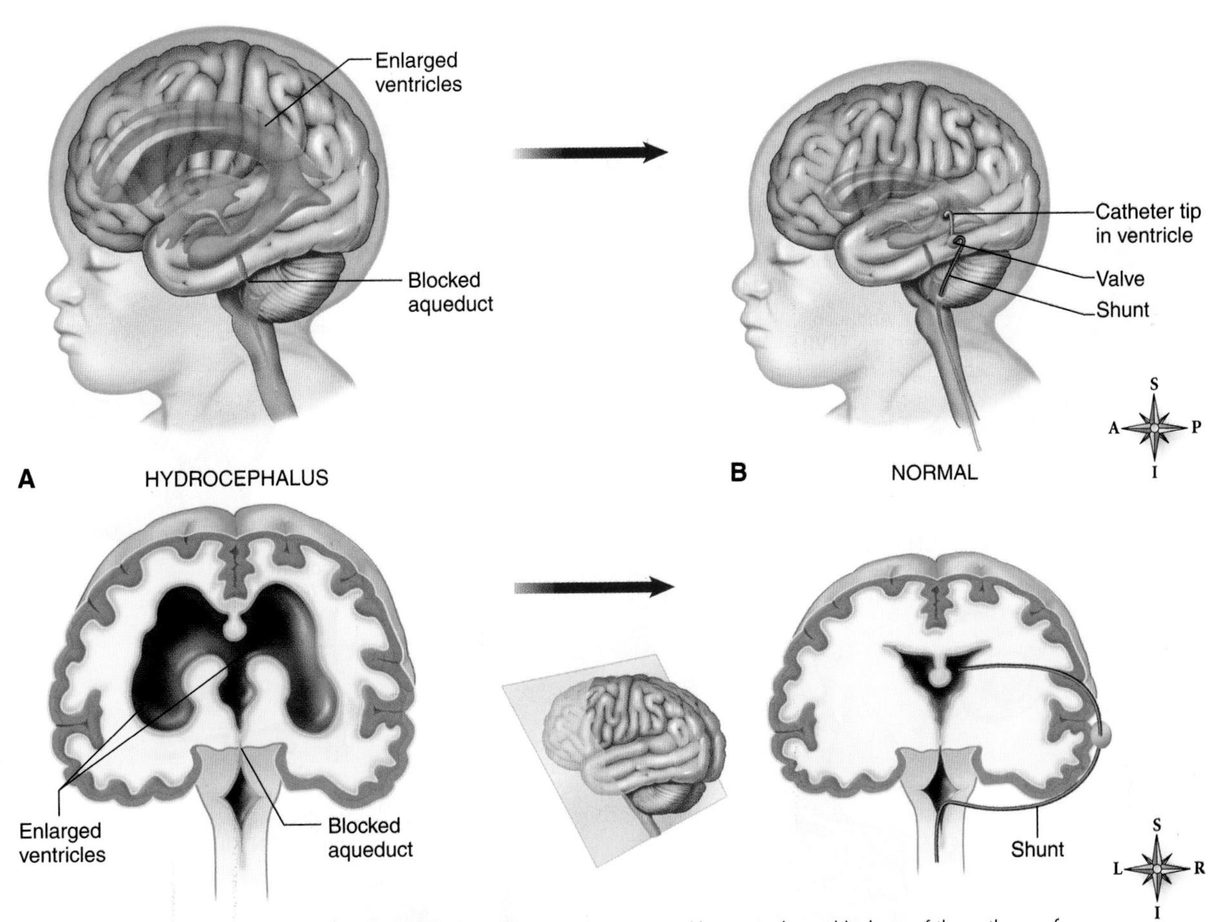

FIGURE 10-23 Hydrocephalus. A, Hydrocephalus is caused by narrowing or blockage of the pathways for CSF, causing the retention of CSF in the ventricles. **B,** This condition can be treated by surgical placement of a shunt or tube to drain the excess fluid. Notice in the cross sections of the brain how the ventricles and surrounding tissue return to their normal shapes and size after shunt placement.

TABLE 10-2 Cranial Nerves

NERVE*		IMPULSES	FUNCTIONS
I	Olfactory	From nose to brain	Sense of smell
II	Optic	From eye to brain	Vision
III	Oculomotor	From brain to eye muscles	Eye movements
IV	Trochlear	From brain to external eye muscles	Eye movements
V	Trigeminal	From skin and mucous membrane of head and from teeth to brain; also from brain to chewing muscles	Sensations of face, scalp, and teeth; chewing movements
VI	Abducens	From brain to external eye muscles	Eye movements
VII	Facial	From taste buds of tongue to brain; from brain to face muscles	Sense of taste; contraction of muscles of facial expression
VIII	Vestibulocochlear	From ear to brain	Hearing; sense of balance
IX	Glossopharyngeal	From throat and taste buds of tongue to brain; also from brain to throat muscles and salivary glands	Sensations of throat, taste, swallowing movements, secretion of saliva
X	Vagus	From throat, larynx, and organs in thoracic and abdominal cavities to brain; also from brain to muscles of throat and to organs in thoracic and abdominal cavities	Sensations of throat and larynx and of thoracic and abdominal organs; swallowing, voice production, slowing of heartbeat, acceleration of peristalsis (gut movements)
XI	Accessory	From brain to certain shoulder and neck muscles	Shoulder movements; turning movements of head
XII	Hypoglossal	From brain to muscles of tongue	Tongue movements

*The first letter of each word in the following sentence corresponds to the first letter of each of the cranial nerves, in ascending order from I to XII. Many anatomy students find that using this memory aid, or one like it, helps in memorizing the names and numbers of the cranial nerves. It is "On Old Olympus' Tiny Tops, A Friendly Viking Grew Vines And Hops."

10

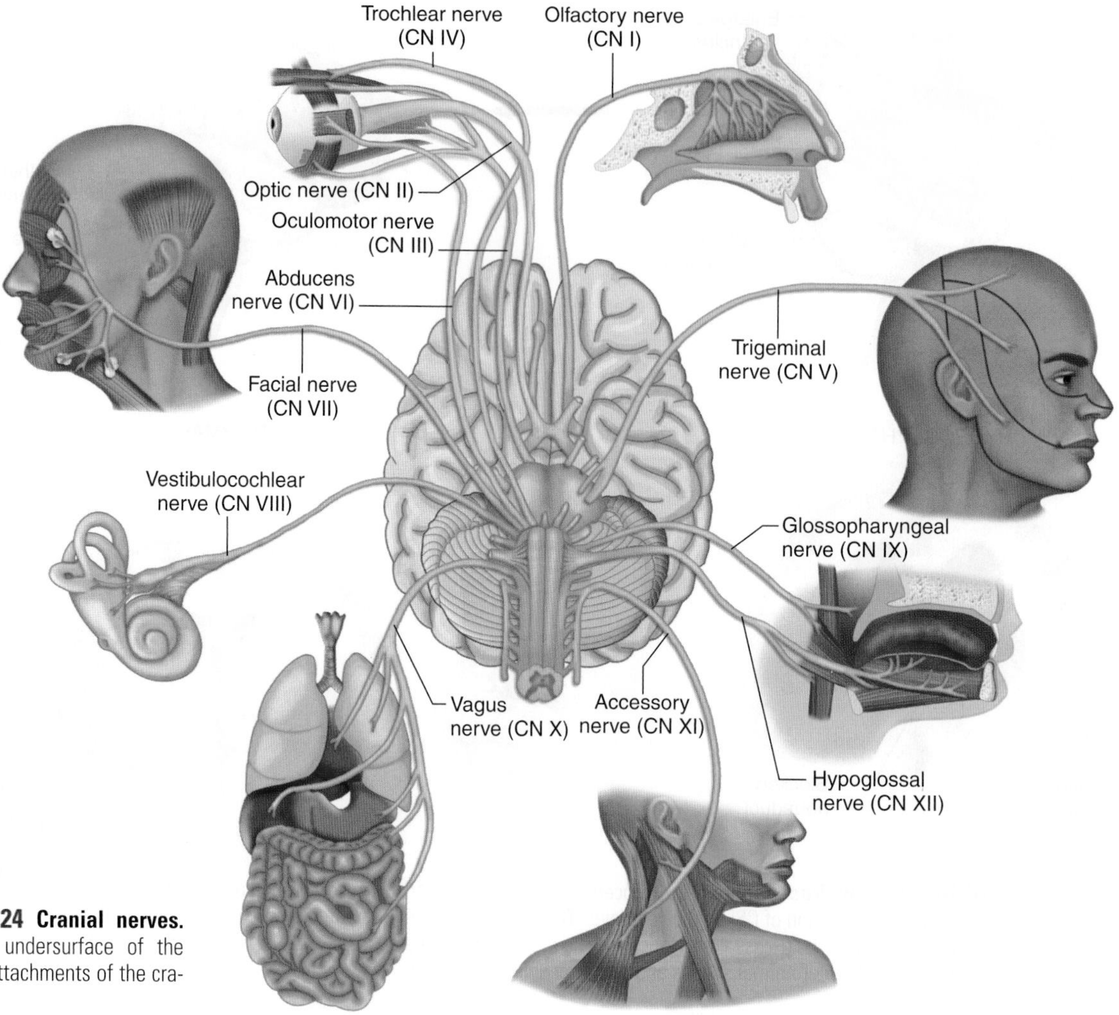

FIGURE 10-24 Cranial nerves. View of the undersurface of the brain shows attachments of the cranial nerves.

Unlike cranial nerves, spinal nerves have no special names. Instead, a letter and number identify each one (see **Figure 10-18**). *C1,* for example, indicates the pair of spinal nerves attached to the first segment of the cervical part of the cord, and *T8* indicates nerves attached to the eighth segment of the thoracic part of the spinal cord.

In **Figure 10-18** the cervical area of the spine has been dissected to show the emerging spinal nerves in that area. After spinal nerves exit from the spinal cord, they branch to form the many peripheral nerves of the trunk and limbs.

Sometimes, nerve fibers from several spinal nerves are reorganized to form a single peripheral nerve. This reorganization

CLINICAL APPLICATION

LUMBAR PUNCTURE

The meninges, the fluid-containing membranes surrounding the brain and spinal cord, extend beyond the spinal cord, an anatomical fact that is most convenient in regard to being able to perform lumbar punctures without putting the spinal cord at risk of injury. A **lumbar puncture,** or "spinal tap," is the withdrawal of some cerebrospinal fluid (CSF) from the subarachnoid space in the lumbar region of the spinal cord.

The physician inserts a needle just above or below the fourth lumbar vertebra, knowing that the spinal cord ends an inch or more above this level. The fourth lumbar vertebra can be easily located because it lies on a line with the iliac crest. Placing an adult patient on his side and having him arch his back by drawing the knees and chest together separates the vertebrae sufficiently to introduce the needle.

Lumbar punctures are often performed when CSF is needed for analysis or when it is necessary to reduce pressure caused by swelling of the brain or spinal cord after injury or disease. The normal sample of CSF from a lumbar puncture shown here is slightly yellowish and clear but the red color in the abnormal sample indicates bleeding (in this case, a hemorrhage in the subarachnoid space).

can be seen as a network of intersecting or "braided" branches called a **plexus.** Figure 10-18 shows several plexuses.

Functions

Spinal nerves conduct impulses between the spinal cord and the parts of the body not supplied by cranial nerves. The spinal nerves shown in Figure 10-18 contain, as do all spinal nerves, sensory and motor fibers. Spinal nerves therefore function to make possible a range of sensations and movements. A disease or injury that prevents conduction by a spinal nerve thus results in a loss of feeling and a loss of movement in the part supplied by that nerve.

Detailed mapping of the skin's surface reveals a close relationship between the source on the spinal cord of each spinal nerve and the part of the body that it innervates (Figure 10-25). Knowledge of the segmental arrangement of spinal nerves is useful to physicians. For instance, a neurologist can identify the site of a spinal cord or nerve abnormality by determining which area of the body is insensitive to a pinprick. Skin surface areas that are supplied by a single spinal nerve are called **dermatomes.** A dermatome "map" of the body is shown in Figure 10-25.

QUICK CHECK

1. What division of the nervous system includes the cranial nerves and spinal nerves?
2. Which cranial nerve conducts impulses from the brain to muscles in the eye?
3. What is a spinal nerve *plexus*?
4. What is a *dermatome*? What is the importance of a dermatome "map"?

Peripheral Nerve Disorders

Many afflictions of peripheral nerves, or their branches, involve inflammation—or **neuritis.** You may know someone who suffers from a form of neuritis called **sciatica.** This is a painful inflammation of the spinal nerve branch in the thigh called the *sciatic nerve*—the largest nerve in the body. This condition is characterized by nerve pain, or **neuralgia.** In some cases, this condition may lead to atrophy of the leg muscles.

Compression or degeneration of the fifth cranial nerve, the trigeminal nerve, may result in a condition called **trigeminal neuralgia,** or **tic douloureux.** This condition is characterized by recurring episodes of stabbing pain radiating from the angle of the jaw along a branch of the trigeminal nerve. Neuralgia of one branch occurs over the forehead and around the eyes. Pain along another branch is felt in the cheek, nose, and upper lip. Neuralgia of the third branch results in stabbing pains in the tongue and lower lip.

Compression, degeneration, or infection of CN VII, the *facial nerve,* may result in **Bell palsy.** Bell palsy is characterized by paralysis of some or all of the facial features innervated by the facial nerve, including the eyelids and mouth. This condition is often temporary but in some cases is irreversible. Plastic surgery is sometimes used to correct permanent disfigurement.

Herpes zoster, or **shingles,** is a unique viral infection that almost always affects the skin of a single dermatome (Figure 10-26). It is caused by a varicella zoster virus (VZV) of chickenpox. Nearly 15% of the population will suffer from shingles at least once by the time they reach the age of 80.

In most cases, shingles results from reactivation of the varicella virus. The virus probably travels through a cutaneous nerve and remains dormant in a dorsal root ganglion for years after an episode of the chickenpox. If the body's immunological protective mechanism becomes diminished in the elderly, after stress, or in individuals undergoing radiation therapy or taking immunosuppressive drugs, the virus may reactivate. If this occurs, the virus travels over the sensory nerve to the skin of a single dermatome. The result is a painful eruption of red, swollen plaques or vesicles that eventually rupture and crust before clearing in 2 to 3 weeks.

In severe cases of shingles, extensive inflammation, hemorrhagic blisters, and secondary bacterial infection may lead to permanent scarring. In most cases, the eruption of vesicles is preceded by 4 to 5 days of pre-eruptive pain, burning, and itching in the affected dermatome. Although an attack of herpes zoster does not confer lasting immunity, only 5% of cases are recurrences.

Some health officials are concerned about a possible shingles epidemic among adults caused by widespread use of chickenpox vaccines in children. Apparently, adults who have not had occasional immune-boosting exposures to children with chickenpox have an increased risk of developing shingles.

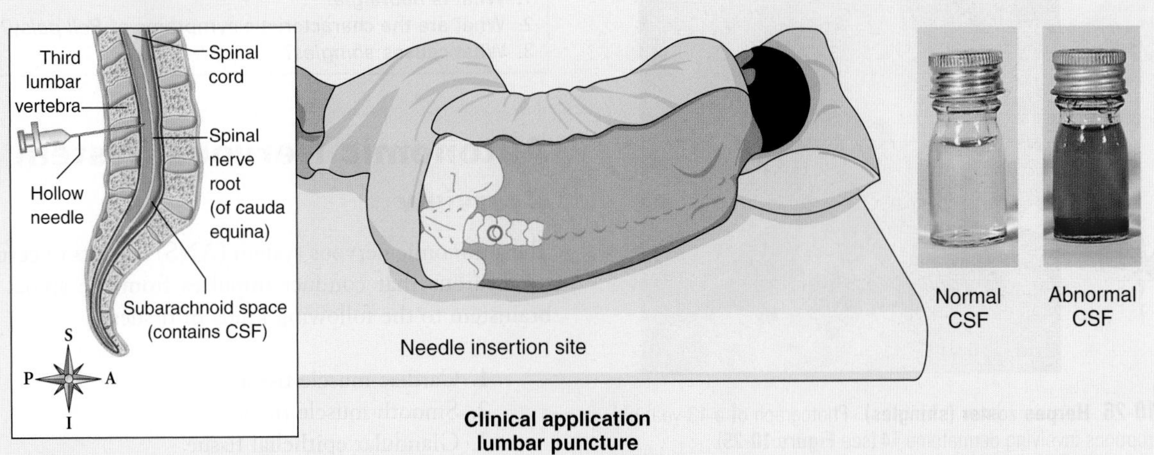

Clinical application lumbar puncture

10

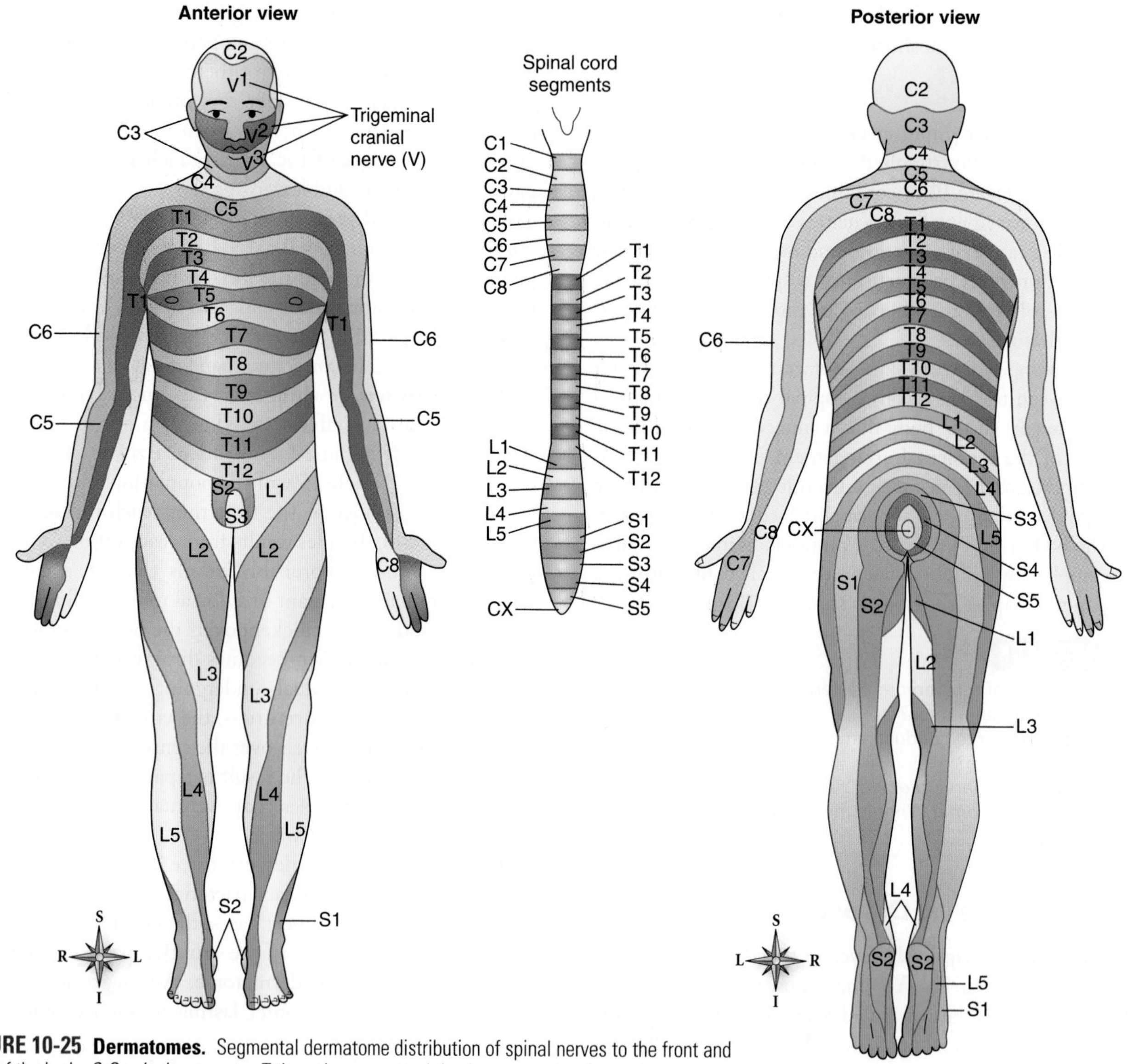

FIGURE 10-25 Dermatoms. Segmental dermatome distribution of spinal nerves to the front and back of the body. *C,* Cervical segments; *T,* thoracic segments; *L,* lumbar segments; *S,* sacral segments; *CX,* coccygeal segment.

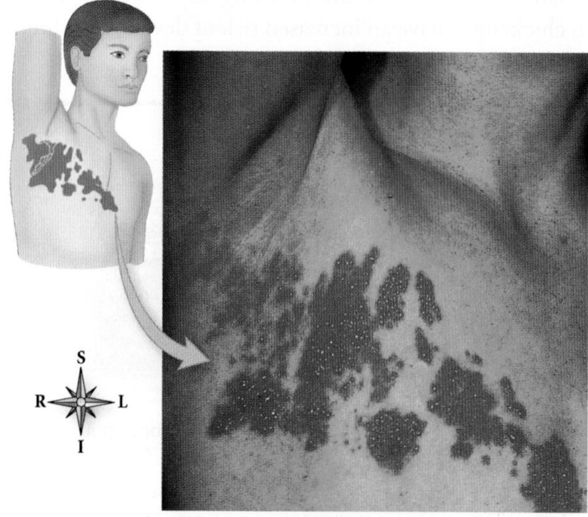

FIGURE 10-26 Herpes zoster (shingles). Photograph of a 13-year-old boy with eruptions involving dermatome T4 (see **Figure 10-25**).

A shingles vaccine is recommended for adults at risk, including people 60 and older who have had chickenpox.

QUICK CHECK

1. What is *neuralgia*?
2. What are the characteristic symptoms of *Bell palsy*?
3. What causes *shingles*?

Autonomic Nervous System

Overview

The autonomic nervous system (ANS) consists of certain motor neurons that conduct impulses from the spinal cord or brainstem to the following kinds of tissues:

1. Cardiac muscle tissue
2. Smooth muscle tissue
3. Glandular epithelial tissue

The ANS includes the parts of the nervous system that regulate involuntary functions (for example, the heartbeat, contractions of the stomach and intestines, and secretions by glands). On the other hand, motor nerves that control the voluntary actions of skeletal muscles are sometimes called the *somatic nervous system (SNS)*.

The autonomic nervous system consists of two main divisions: the **sympathetic division** and the **parasympathetic division** (Figure 10-27).

Functional Anatomy

Autonomic neurons are the motor neurons that make up the ANS. The dendrites and cell bodies of some autonomic neurons are located in the gray matter of the spinal cord or brainstem. Their axons extend from these structures and terminate in peripheral "junction boxes" called **ganglia.** These autonomic neurons are called **preganglionic neurons** because they conduct impulses between the spinal cord and a ganglion.

In the autonomic ganglia, the axon endings of preganglionic neurons synapse with the dendrites or cell bodies of postganglionic neurons. **Postganglionic neurons,** as their name suggests, conduct impulses from a ganglion to cardiac muscle, smooth muscle, or glandular epithelial tissue.

Autonomic effectors, or **visceral effectors,** are the tissues to which autonomic neurons conduct impulses. Specifically, a visceral effector is cardiac muscle that makes up the wall of the heart, smooth muscle that partially makes up the walls of blood vessels and other hollow internal organs, and glandular epithelial tissue that makes up the secreting part of glands.

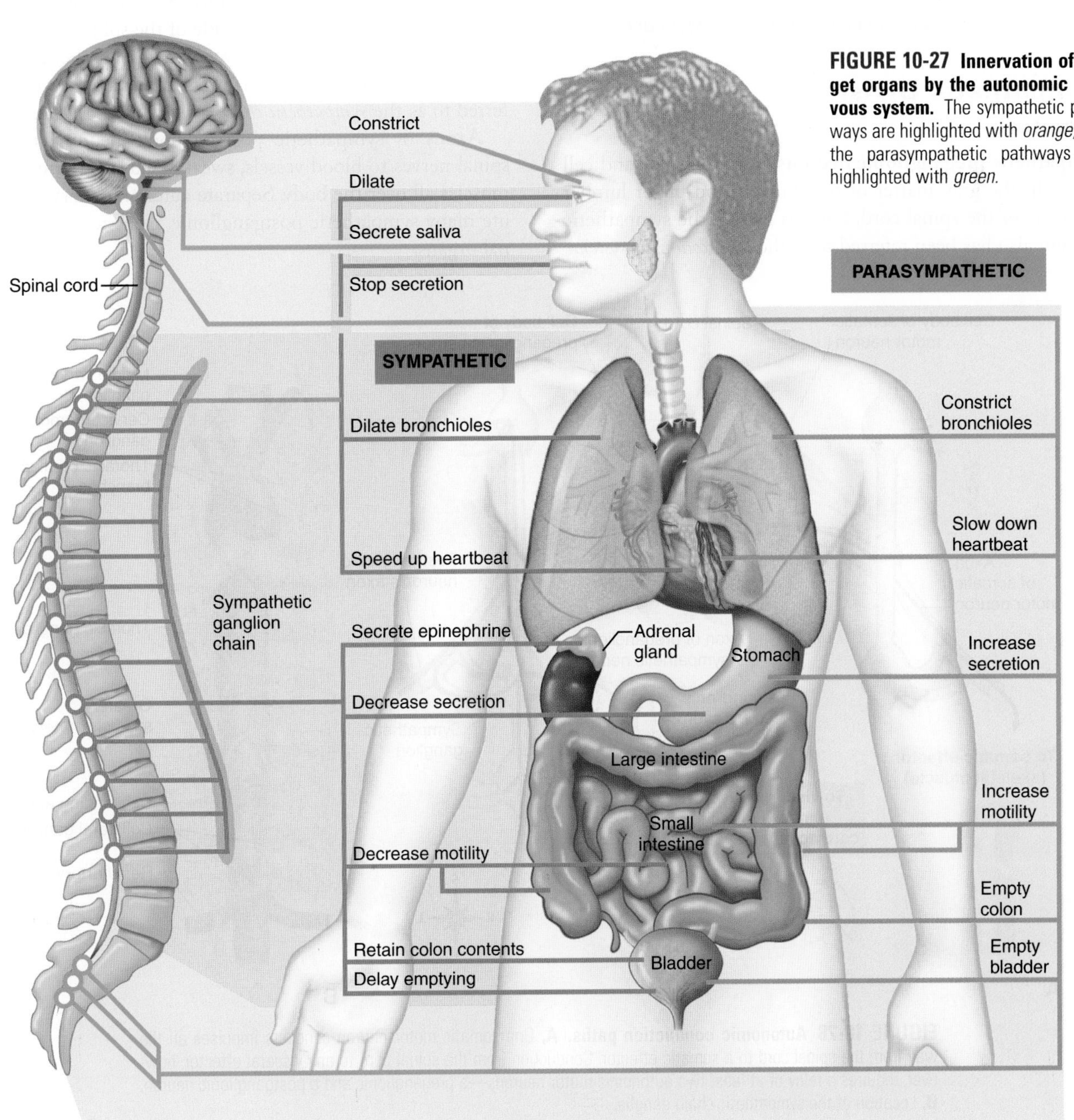

FIGURE 10-27 Innervation of target organs by the autonomic nervous system. The sympathetic pathways are highlighted with *orange,* and the parasympathetic pathways are highlighted with *green.*

10

Autonomic Conduction Paths

Conduction paths to visceral and somatic effectors from the CNS (spinal cord or brainstem) differ somewhat. Autonomic paths to visceral effectors, as the right side of **Figure 10-28**, *A* shows, consist of two-neuron relays. Impulses travel over preganglionic neurons from the spinal cord or brainstem to autonomic ganglia. There, they are relayed across synapses to postganglionic neurons, which then conduct the impulses from the ganglia to visceral effectors.

Compare the autonomic conduction path with the somatic conduction path illustrated on the left side of **Figure 10-28**, *A*. A single somatic motor neuron, like the one shown here, conducts impulses all the way from the spinal cord or brainstem to somatic effectors, with no intervening synapses.

To learn more about the difference between autonomic and somatic conduction paths, go to AnimationDirect online at *evolve.elsevier.com*.

Sympathetic Division

Structure

Sympathetic preganglionic neurons have dendrites and cell bodies in the gray matter of the thoracic and upper lumbar segments of the spinal cord. For this reason, the sympathetic division also has been referred to as the *thoracolumbar system*.

Look now at the right side of **Figure 10-28**, *A*. Follow the course of the axon of the sympathetic preganglionic neuron shown there. It leaves the spinal cord in the anterior (ventral) root of a spinal nerve. It next enters the spinal nerve but soon leaves it to extend to and through a sympathetic ganglion and terminate in a collateral ganglion. There, it synapses with several postganglionic neurons whose axons extend to terminate in visceral effectors.

Also shown in **Figure 10-28**, *A*, branches of the preganglionic axon may ascend or descend to terminate in ganglia above and below their point of origin. All sympathetic preganglionic axons therefore synapse with many postganglionic neurons, and these frequently terminate in widely separated organs. Hence sympathetic responses are usually widespread, involving many organs rather than just one.

Sympathetic postganglionic neurons have dendrites and cell bodies in sympathetic ganglia. Sympathetic ganglia are located anterior to and at each side of the spinal column. Because short fibers extend between the sympathetic ganglia, they look a little like two chains of beads and are often referred to as the *sympathetic chain ganglia* (**Figure 10-28**, *B*).

Axons of sympathetic postganglionic neurons travel in spinal nerves to blood vessels, sweat glands, and arrector hair muscles all over the body. Separate autonomic nerves distribute many sympathetic postganglionic axons to various internal organs.

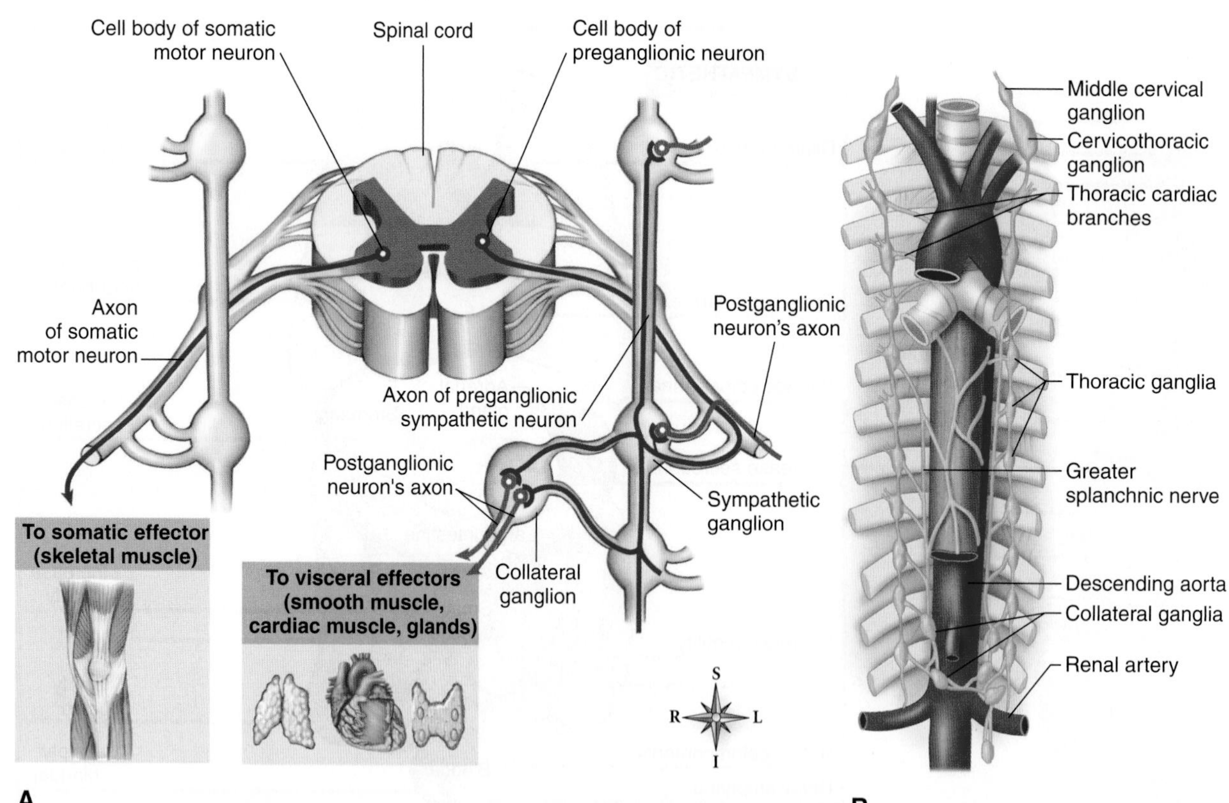

FIGURE 10-28 Autonomic conduction paths. A, One somatic motor neuron conducts impulses all the way from the spinal cord to a somatic effector. Conduction from the spinal cord to any visceral effector, however, requires a relay of at least two autonomic motor neurons—a preganglionic and a postganglionic neuron. **B,** Location of the sympathetic chain ganglia.

10

Function

The sympathetic division functions as an emergency system. Impulses over sympathetic fibers take control of many internal organs when we exercise strenuously and when strong emotions—anger, fear, hate, anxiety—are elicited. In short, when we must cope with stress of any kind, sympathetic impulses increase to many visceral effectors and rapidly produce widespread changes within our bodies.

The middle column of **Table 10-3** lists many of the possible sympathetic responses. The heart beats faster. Most blood vessels constrict, causing blood pressure to increase. Blood vessels in skeletal muscles dilate, supplying the muscles with more blood. Sweat glands and adrenal glands secrete more abundantly. Salivary and other digestive glands secrete more sparingly. Digestive tract contractions (peristalsis) become sluggish, hampering digestion.

Together, all these varied sympathetic responses make us ready for strenuous muscular work. We need such physiological preparation when facing a threat—we must be ready to either resist (fight) the threat or to avoid (fly from) the threat. Therefore, this group of changes induced by sympathetic control is known as the **fight-or-flight response.**

Parasympathetic Division

Structure

The dendrites and cell bodies of **parasympathetic preganglionic neurons** are located in the gray matter of the brainstem and the sacral segments of the spinal cord. For this reason, the parasympathetic division also has been referred to as the *craniosacral system.*

The parasympathetic preganglionic axons extend some distance before terminating in the parasympathetic ganglia located in the head and in the thoracic and abdominal cavities close to the visceral effectors that they control.

The dendrites and cell bodies of **parasympathetic postganglionic neurons** lie in these outlying parasympathetic ganglia, and their short axons extend into the nearby structures. Therefore, each parasympathetic preganglionic neuron synapses only with postganglionic neurons to a single effector.

For this reason, parasympathetic stimulation frequently involves response by only one organ. This is in stark contrast to sympathetic responses, which involve numerous organs.

Function

The parasympathetic system dominates control of many visceral effectors under normal, everyday conditions. Impulses carried by parasympathetic fibers, for example, tend to slow heartbeat, increase peristalsis, and increase secretion of digestive juices and insulin (see **Table 10-3**). Thus, we can think of parasympathetic function as counterbalancing sympathetic function.

e Recent evidence suggests that the sacral portion of the autonomic pathways may be sympathetic—not parasympathetic as first described over a century ago. Find out more at *New Model of ANS Pathways* at Connect It! at *evolve.elsevier.com.*

Autonomic Neurotransmitters

Turn your attention now to **Figure 10-29**. It illustrates information regarding autonomic neurotransmitters, the chemical compounds released from the axon terminals of autonomic neurons.

Observe that three of the axons shown in **Figure 10-29**—the sympathetic preganglionic axon, the parasympathetic

TABLE 10-3 Autonomic Functions

VISCERAL EFFECTORS	SYMPATHETIC CONTROL	PARASYMPATHETIC CONTROL
Heart muscle	Accelerates heartbeat	Slows heartbeat
Smooth muscle		
Of most blood vessels	Constricts blood vessels	None
Of blood vessels in skeletal muscles	Dilates blood vessels	None
Of the digestive tract	Decreases peristalsis; inhibits defecation	Increases peristalsis
Of the anal sphincter	Stimulates—closes sphincter	Inhibits—opens sphincter for defecation
Of the urinary bladder	Inhibits—relaxes bladder	Stimulates—contracts bladder
Of the urinary sphincters	Stimulates—closes sphincter	Inhibits—opens sphincter for urination
Of the eye		
Iris	Stimulates radial fibers—dilation of pupil	Stimulates circular fibers—constriction of pupil
Ciliary	Inhibits—accommodation for far vision (flattening of lens)	Stimulates—accommodation for near vision (bulging of lens)
Of hairs (pilomotor muscles)	Stimulates—"goose pimples"	No parasympathetic fibers
Glands		
Adrenal medulla	Increases epinephrine secretion	None
Sweat glands	Increases sweat secretion	None
Digestive glands	Decreases secretion of digestive juices	Increases secretion of digestive juices

preganglionic axon, and the parasympathetic postganglionic axon—release acetylcholine. These axons are therefore classified as **cholinergic fibers.**

Only one type of autonomic axon releases the neurotransmitter norepinephrine (noradrenaline). This is the axon of a sympathetic postganglionic neuron, and such neurons are classified as **adrenergic fibers.**

That each division of the ANS signals its effectors with a different neurotransmitter explains how an organ can tell which division is stimulating it. The heart, for example, responds to acetylcholine from the parasympathetic division by slowing down. The presence of norepinephrine in the heart, on the other hand, is a signal from the sympathetic division, and the response is an increase in heart activity.

Autonomic Nervous System as a Whole

The function of the autonomic nervous system is to regulate the body's automatic, involuntary functions in ways that maintain or quickly restore homeostasis.

Many internal organs are *dually innervated* by the ANS. In other words, they receive fibers from parasympathetic and sympathetic divisions. Parasympathetic and sympathetic impulses continually bombard them and, as **Table 10-3** indicates, influence their function in opposite or antagonistic ways.

For example, the heart continually receives sympathetic impulses that make it beat faster and parasympathetic impulses that slow it down. The ratio between these two antagonistic forces, determined by the ratio between the two different autonomic neurotransmitters, determines the actual heart rate.

The term *autonomic nervous system* is something of a misnomer. "Autonomy" seems to imply that this part of the nervous system is independent from other parts. But this is not true. Dendrites and cell bodies of preganglionic neurons are located, as observed in **Figure 10-29**, in the spinal cord and brainstem. They are continually influenced directly or indirectly by impulses from neurons located above them, notably by some in the hypothalamus and in the parts of the cerebral cortex called the **limbic system,** or *emotional brain.* Through conduction paths from these areas, emotions can produce widespread changes in the automatic functions of our bodies, in cardiac and smooth muscle contractions, and in secretion by glands. Anger and fear, for example, lead to increased sympathetic activity and the fight-or-flight response.

According to some physiologists, the slightly altered state of consciousness known as *meditation* leads to decreased sympathetic activity and a group of changes opposite those of the fight-or-flight response.

Disorders of the Autonomic Nervous System

Stress-Induced Disease

Considering the variety and number of effectors innervated by the autonomic nervous system, it is no wonder that

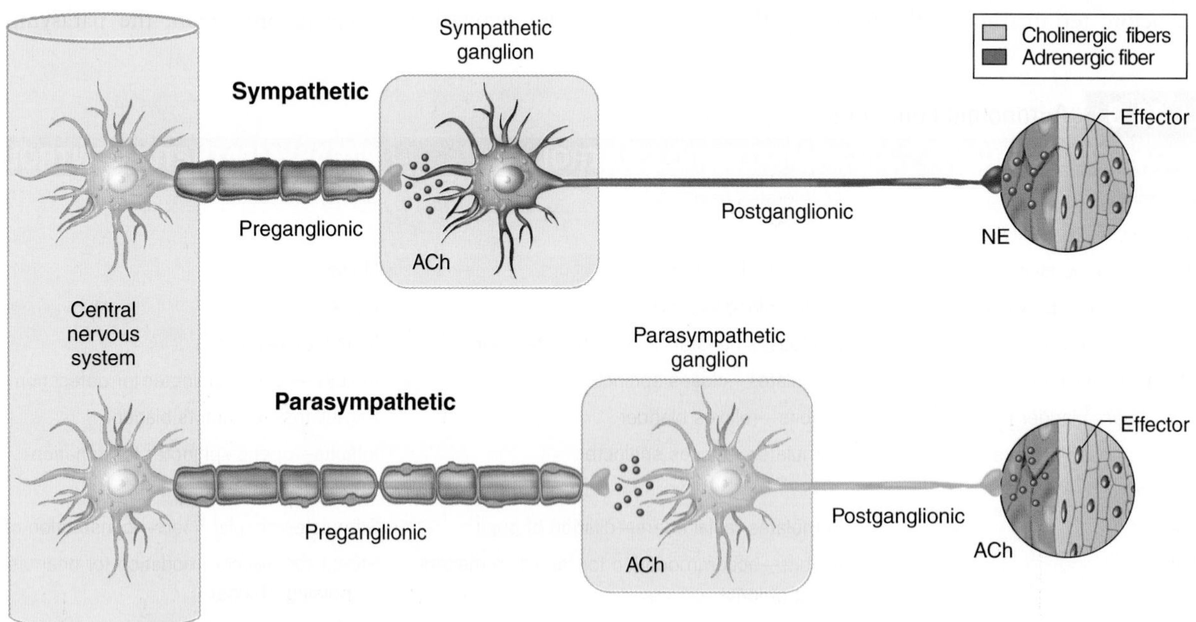

FIGURE 10-29 Autonomic neurotransmitters. Three of the four fiber types are cholinergic, secreting the neurotransmitter acetylcholine *(ACh)* into a synapse. Only the sympathetic postganglionic fiber is adrenergic, secreting norepinephrine *(NE)* into a synapse.

10

autonomic disorders have varied and far-reaching consequences. This is especially true of stress-induced diseases. Prolonged or excessive physiological response to stress, the fight-or-flight response, can disrupt normal functioning throughout the body.

Stress has been cited as an indirect cause or an important risk factor in a number of conditions. Only a few of those are listed here.

1. **Heart disease**—Although an extreme episode of stress can precipitate heart failure even in healthy people, chronic stress is known to increase the risk of certain heart disorders. One such condition is stress-induced high blood pressure, or *hypertension,* that can weaken the heart and blood vessels.
2. **Digestive problems**—*Colitis* (colon inflammation) and gastric ulcers, for example, may be precipitated by the changes in digestive secretion and movement, along with increased susceptibility to infection, that occur during prolonged or repeated stress responses.
3. **Reduced resistance to disease**—Hormones called *glucocorticoids* that are released by the adrenal glands during prolonged or repeated stress episodes depress the activity of the immune system. Depressed immune function leads to increased risk of infection and cancer.
4. **Spread of cancer**—The chronically elevated levels of norepinephrine caused by stress can speed up the flow of cancer cells out of tissues by way of the lymphatic system, possibly resulting in metastasis (see **Figure 6-11** on p. 130).

Because both the nervous system and endocrine system are involved in stress disorders, they are usually thought of as *neuroendocrine* disorders.

Neuroblastoma

Neuroblastoma is a malignant tumor of the sympathetic division. It most often occurs in the developing nervous systems of young children and metastasizes rapidly to other parts of the body. Symptoms often include exaggerated or inappropriate sympathetic effects, including increased heart rate, sweating, and high blood pressure. As with some other forms of cancer, spontaneous remissions may occur.

QUICK CHECK

1. What are the two main divisions of the ANS?
2. Describe the function of the parasympathetic nervous division.
3. Which two neurotransmitters are used by the autonomic nerve pathways?
4. What problems in the body arise from ANS malfunctions?

SCIENCE APPLICATIONS

NEUROSCIENCE

Otto Loewi (1873–1961)

The Austrian scientist Otto Loewi started his studies in the humanities, not science. Even after he did finally begin university studies in medicine, he often skipped his science classes to attend lectures in philosophy instead. But after Dr. Loewi turned his attention to human biology, his brilliance became evident. In 1921, while working to design an experiment that would unlock the mystery of how neurons communicate with other cells, he had a dream in which the answer was revealed to him. He rushed to his lab and performed a now famous experiment in which he discovered what we know as *acetylcholine.*

For his work that showed that it is neurotransmitters that carry signals from neurons, Loewi shared a Nobel Prize in 1936. Not surprisingly, Loewi later spent some of his time studying how dreams may help us understand subconscious thoughts.

Many professions depend on **neuroscientists** like Otto Loewi to provide information they need to help us improve our lives. For example, **neurologists, psychiatrists,** and other medical professionals use this information to treat disorders of the nervous system. **Pharmacologists** use these ideas to develop drug treatments that affect the nervous system—and **pharmacists** and **pharmacy technicians** supply these treatments.

Mental health professionals such as **psychologists** and *counselors* use concepts derived from neuroscience to better understand human emotions and behavior. Even people who specialize in *business* and *marketing* use some of the neuroscience discoveries—their focus is learning how to entice buyers to buy certain products or, perhaps, to predict the behavior of crowds.

LANGUAGE OF **SCIENCE** *(continued from p. 249)*

autonomic neuron
(aw-toh-NOM-ik NOO-ron)
[*auto-* **self,** *-nom-* **rule,** *-ic* **relating to,** *neuron* **nerve**]

axon
(AK-son)
[*axon* **axle**]

basal nuclei (basal ganglia)
(BAY-sal NOO-klee-aye [GANG-glee-ah])
sing., basal nucleus or ganglion
(BAY-sal NOO-klee-us [GANG-glee-un])
[*bas-* **foundation,** *-al* **relating to,** *nucle-* **nut or kernel** (*ganglion* **knot**)]

blood-brain barrier (BBB)
(blud brayn BAYR-ee-er [bee bee bee])

catecholamine
(kat-eh-KOHL-ah-meen)
[*catech-* **melt,** *-ol-* **alcohol,** *-amine* **ammonia compound**]

cell body
(sell BOD-ee)
[*cell* **storeroom**]

central nervous system (CNS)
(SEN-tral NER-vus SIS-tem [see en es])
[*centr-* **center,** *-al* **relating to,** *nerv-* **nerve,** *-ous* **relating to**]

cerebellum
(sayr-eh-BEL-um)
pl., cerebella or cerebellums
(sayr-eh-BEL-ah or sayr-eh-BEL-umz)
[*cereb-* **brain,** *-ell-* **small,** *-um* **thing**]

cerebral cortex
(seh-REE-bral KOR-teks)
[*cerebr-* **brain (cerebrum),** *-al* **relating to,** *cortex* **bark**]

cerebrospinal fluid (CSF)
(seh-ree-broh-SPY-nal FLOO-id [see es ef])
[*cerebr-* **brain,** *-spin-* **backbone,** *-al* **relating to**]

cerebrum
(SAYR-eh-brum)
[*cerebrum* **brain**]

cholinergic fiber
(koh-lin-ER-jik FYE-ber)
[*chole-* **bile,** *-erg-* **work,** *-ic* **relating to,** *fibr-* **thread**]

choroid plexus
(KOH-royd PLEK-sus)
pl., choroid plexuses
(KOH-royd PLEK-sus-ez)
[*chorio-* **skin,** *-oid* **like,** *plexus* **braid or network**]

corpus callosum
(KOHR-pus kah-LOH-sum)
pl., corpora callosa
(KOHR-por-ah kah-LOH-sah)
[*corpus* **body,** *callosum* **callous or tough**]

cranial nerve (CN)
(KRAY-nee-al nerv [see en])
[*crani-* **skull,** *-al* **relating to,** *nerv-* **string or nerve**]

dendrite
(DEN-dryte)
[*dendr-* **tree,** *-ite* **part (branch)**]

dermatome
(DER-mah-tohm)
[*derma-* **skin,** *-tome* **cut (segment)**]

diencephalon
(dye-en-SEF-ah-lon)
[*di-* **between,** *-en-* **within,** *-cephalon* **head**]

dopamine
(DOH-pah-meen)
[*dopa-* **amino acid,** *-amine* **ammonia**]

dura mater
(DOO-rah MAH-ter)
[*dura* **hard,** *mater* **mother**]

effector
(ef-FEK-tor)
[*effect-* **accomplish,** *-or* **agent**]

efferent neuron
(EF-fer-ent NOO-ron)
[*e-* **away,** *-fer-* **carry,** *-ent* **relating to,** *neur-* **string or nerve,** *-on* **unit**]

endoneurium
(en-doh-NOO-ree-um)
[*endo-* **inward,** *-neuri-* **nerve,** *-um* **thing**]

endorphin
(en-DOR-fin)
[*endo-* **within,** *-(m)orph-* **Morpheus (Roman god of dreams),** *-in* **substance**]

enkephalin
(en-KEF-ah-lin)
[*en-* **within,** *-kephalo-* **head,** *-in* **substance**]

epineurium
(ep-ih-NOO-ree-um)
[*epi-* **upon,** *-neuri-* **nerve,** *-um* **thing**]

fascicle
(FAS-ih-kul)
[*fasci-* **band or bundle,** *-cle* **small**]

fight-or-flight response
(fyte or flyte ree-SPAHNS)

ganglion
(GANG-lee-on)
pl., ganglia
(GANG-lee-ah)
[*gangli-* **knot,** *-on* **unit**]

glia
(GLEE-ah)
sing., glial cell
(GLEE-al sel)
[*glia* **glue**]

gray matter
(gray MAT-ter)

gyri
(JYE-rye)
sing., gyrus
(JYE-rus)
[*gyrus* **circle**]

hypothalamus
(hye-poh-THAL-ah-mus)
[*hypo-* **under or below,** *-thalamus* **inner chamber**]

interneuron
(in-ter-NOO-ron)
[*inter-* **between,** *-neuron* **nerve**]

limbic system
(LIM-bik SIS-tem)
[*limb-* **edge,** *-ic* **relating to**]

medulla oblongata
(meh-DUL-ah ob-long-GAH-tah)
[*medulla* **marrow or pith,** *oblongata* **oblong**]

meninges
(meh-NIN-jeez)
sing., meninx
(meh-NINKS)
[*meninx* **membrane**]

microglia
(my-KROG-lee-ah)
sing., microglial cell
(my-KROG-lee-al sel)
[*micro-* **small,** *-glia* **glue**]

midbrain
(MID-brayn)
[*mid-* **middle,** *-brain* **skull**]

motor neuron
(MOH-ter NOO-ron)
[*mot-* **movement,** *-or* **agent,** *neuron* **nerve**]

myelin
(MY-eh-lin)
[*myel-* **marrow,** *-in* **substance**]

myelinated fiber
(MY-eh-lih-nay-ted FYE-ber)
[*myel-* **marrow,** *-in-* **substance,** *-ate* **act of,** *fibr-* **thread**]

nerve
(nerv)
[*nerv-* **string or nerve**]

neurilemma
(noo-rih-LEM-mah)
[*neuri-* **neuron,** *-lemma* **sheath**]

neuroglia
(noo-ROH-glee-ah or noo-roh-GLEE-ah)
sing., neuroglial cell
(noo-ROH-glee-al sel)
[*neur-* **nerve,** *-glia* **glue**]

10

LANGUAGE OF **SCIENCE** *(continued from p. 280)*

neuron
(NOO-ron)
[*neur-* **string or nerve,** *-on* **unit**]

neurotransmitter
(noo-roh-trans-MIT-ter)
[*neuro-* **nerve,** *-trans-* **across,** *-mitt-* **send,** *-er* **agent**]

nitric oxide (NO)
(NYE-trik AWK-side [en oh])
[*nitr-* **nitrogen,** *-ic* **relating to,** *ox-* **oxygen,** *-ide* **chemical**]

node of Ranvier
(nohd ov rahn-vee-AY)
[*nod-* **knot,** *Louis A. Ranvier* **French pathologist**]

norepinephrine (NE)
(nor-ep-ih-NEF-rin [en ee])
[*nor-* **chemical prefix (unbranched C chain),** *-epi-* **upon,** *-nephr-* **kidney,** *-ine* **substance**]

oligodendrocyte
(ohl-ih-goh-DEN-droh-syte)
[*oligo-* **few,** *-dendr-* **part (branch) of,** *-cyte* **cell**]

parasympathetic division
(payr-ah-sim-pah-THET-ik dih-VIZH-un)
[*para-* **beside,** *-sym-* **together,** *-pathe-* **feel,** *-ic* **relating to**]

parasympathetic postganglionic neuron
(payr-ah-sim-pah-THET-ik post-gang-glee-ON-ik NOO-ron)
[*para-* **beside,** *-sym-* **together,** *-pathe-* **feel,** *-ic* **relating to,** *post-* **after,** *-ganglion-* **knot,** *-ic* **relating to,** *neur-* **string or nerve,** *-on* **unit**]

parasympathetic preganglionic neuron
(payr-ah-sim-pah-THET-ik pree-gang-glee-ON-ik NOO-ron)
[*para-* **beside,** *-sym-* **together,** *-pathe-* **feel,** *-ic* **relating to,** *pre-* **before,** *-ganglion-* **knot,** *-ic* **relating to,** *neur-* **string or nerve,** *-on* **unit**]

perineurium
(payr-ih-NOO-ree-um)
[*peri-* **around,** *-neur-* **nerve,** *-um* **thing**]

peripheral nervous system (PNS)
(peh-RIF-er-al NER-vus SIS-tem [pee en es])
[*peri-* **around,** *-phera-* **boundary,** *-al* **relating to,** *nerv-* **nerve,** *-ous* **relating to,** *system* **organized whole**]

pia mater
(PEE-ah MAH-ter)
[*pia* **tender,** *mater* **mother**]

pineal gland
(PIN-ee-al gland)
[*pine-* **pine,** *-al* **relating to,** *gland* **acorn**]

plexus
(PLEK-sus)
[*plexus* **braid or network**]
pl., plexuses
(PLEK-sus-ez)

pons
(ponz)
[*pons* **bridge**]

postganglionic neuron
(post-gang-glee-ON-ik NOO-ron)
[*post-* **after,** *-ganglion-* **knot,** *-ic* **relating to,** *neuron* **nerve**]

postsynaptic neuron
(post-sih-NAP-tik NOO-ron)
[*post-* **after,** *-syn-* **together,** *-apt-* **join,** *-ic* **relating to,** *neuron* **nerve**]

preganglionic neuron
(pree-gang-glee-ON-ik NOO-ron)
[*pre-* **before,** *-ganglion-* **knot,** *-ic* **relating to,** *neuron* **nerve**]

presynaptic neuron
(pree-sih-NAP-tik NOO-ron)
[*pre-* **before,** *-syn-* **together,** *-apt-* **join,** *-ic* **relating to,** *neuron* **nerve**]

receptor
(ree-SEP-tor)
[*recept-* **receive,** *-or* **agent**]

reflex
(REE-fleks)
[*re-* **again,** *-flex* **bend**]

reflex arc
(REE-fleks ark)
[*re-* **back or again,** *-flex* **bend,** *arc* **curve**]

reticular formation
(reh-TIK-yoo-lar for-MAY-shun)
[*ret-* **net,** *-ic-* **relating to,** *-ul-* **little,** *-ar* **characterized by,** *form-* **shape,** *-ation* **state**]

saltatory conduction
(SAL-tah-tor-ee kon-DUK-shun)
[*salta-* **leap,** *-ory* **relating to,** *con-* **with,** *-duct-* **lead,** *-ion* **process**]

Schwann cell
(shwon *or* shvon sell)
[*Theodor Schwann* **German anatomist,** *cell* **storeroom**]

sensory neuron
(SEN-sor-ee NOO-ron)
[*sens-* **feel,** *-ory* **relating to,** *neur-* **string or nerve,** *-on* **unit**]

serotonin
(sayr-oh-TOH-nin)
[*sero-* **watery body fluid,** *-ton-* **tension,** *-in* **substance**]

spinal nerve
(SPY-nal nerv)
[*spin-* **backbone,** *-al* **relating to,** *nerv-* **string or nerve**]

spinal tract
(SPY-nal trakt)
[*spin-* **backbone,** *-al* **relating to,** *trac-* **course or trail**]

sulci
(SUL-kye or SUL-kee)
sing., sulcus
[*sulcus* **furrow**]

sympathetic division
(sim-pah-THET-ik dih-VIZH-un)
[*sym-* **together,** *-pathe-* **feel,** *-ic* **relating to**]

sympathetic postganglionic neuron
(sim-pah-THET-ik post-gang-glee-ON-ik NOO-ron)
[*sym-* **together,** *-pathe-* **feel,** *-ic* **relating to,** *post-* **after,** *-ganglion-* **knot,** *-ic* **relating to,** *neur-* **string or nerve,** *-on* **unit**]

sympathetic preganglionic neuron
(sim-pah-THET-ik pree-gang-glee-ON-ik NOO-ron)
[*sym-* **together,** *-pathe-* **feel,** *-ic* **relating to,** *pre-* **before,** *-ganglion-* **knot,** *-ic* **relating to,** *neur-* **string or nerve,** *-on* **unit**]

synapse
(SIN-aps)
[*syn-* **together,** *-aps-* **join**]

synaptic cleft
(sin-AP-tik kleft)
[*syn-* **together,** *-apt-* **join,** *-ic* **relating to**]

synaptic knob
(sin-AP-tik nob)
[*syn-* **together,** *-apt-* **join,** *-ic* **relating to**]

thalamus
(THAL-ah-mus)
[*thalamus* **inner chamber**]

tract
(trakt)
[*trac-* **course or trail**]

ventricle
(VEN-trih-kul)
[*ventr-* **belly,** *-icle* **little**]

visceral effector
(VIS-er-al ee-FEK-tor)
[*viscer-* **internal organ,** *-al* **relating to,** *effect-* **accomplish,** *-or* **agent**]

white matter

withdrawal reflex
(with-DRAW-ul REE-fleks)
[*with-* **away,** *-draw-* **draw,** *-al* **relating to,** *re-* **again,** *-flex* **bend**]

LANGUAGE OF MEDICINE

Alzheimer disease (AD)
(AHLZ-hye-mer dih-ZEEZ [ay dee])
[*Alois Alzheimer* **German neurologist,** *dis-* **opposite of,** *-ease* **comfort**]

anesthesia
(an-es-THEE-zhah)
[*an-* **absence,** *-esthesia* **feeling**]

antidepressant
(an-tee-deh-PRESS-ant)
[*anti-* **against,** *-de-* **down,** *-press-* **press,** *-ant* **agent**]

Bell palsy
(bell PAWL-zee)
[*Charles Bell* **Scots anatomist,** *palsy* **paralysis**]

cerebral palsy (CP)
(seh-REE-bral PAWL-zee [see pee])
[*cerebr-* **brain,** *-al* **relating to,** *palsy* **paralysis**]

cerebrovascular accident (CVA)
(SAYR-eh-broh-VAS-kyoolar AK-sih-dent [see vee ay])
[*cerebr-* **brain,** *-vas-* **vessel,** *-cul-* **little,** *-ar* **relating to**]

chronic traumatic encephalopathy (CTE)
(KRON-ik traw-MAT-ik en-sef-al-OP-path-ee [see tee ee])
[*chron-* **time,** *-ic* **relating to,** *trauma-* **wound,** *-atic* **relating to,** *encephal-* **brain,** *-pathy* **disease**]

concussion
(kon-KUSH-in)
[*concuss-* **shake violently,** *-ion* **condition**]

dementia
(de-MEN-shah)
[*de-* **off,** *-ment-* **mind,** *-ia* **condition of**]

electroencephalogram (EEG)
(eh-lek-troh-en-SEF-uh-loh-gram [ee ee jee])
[*electro-* **electricity,** *-en-* **within,** *-cephal-* **head,** *-gram* **drawing**]

epilepsy
(EP-ih-lep-see)
[*epi-* **upon,** *-leps-* **seizure,** *-y* **state**]

glioma
(glee-OH-mah)
[*gli-* **neuroglia,** *-oma* **tumor**]

hemiplegia
(hem-ee-PLEE-jee-ah)
[*hemi-* **half,** *-pleg-* **stricken,** *-ia* **condition**]

herpes zoster
(HER-peez ZOS-ter)
[*herpe-* **creep,** *zoster* **belt or girdle**]

Huntington disease (HD)
(HUN-ting-ton dih-ZEEZ [aych dee])
[*George S. Huntington* **American physician,** *dis-* **opposite of,** *-ease* **comfort**]

hydrocephalus
(hye-droh-SEF-ah-lus)
[*hydro-* **water,** *-cephalus* **head**]

lumbar puncture
(LUM-bar PUNK-chur)
[*lumb-* **loin,** *-ar* **relating to**]

meningitis
(men-in-JYE-tis)
[*mening-* **membrane,** *-itis* **inflammation**]

multiple neurofibromatosis
(MUL-tih-pul noo-roh-fye-broh-mah-TOH-sis)
[*multi-* **many,** *-pl-* **fold,** *neuro-* **nerve,** *-fibr-* **fiber,** *-oma-* **tumor,** *-osis* **condition**]

multiple sclerosis (MS)
(MUL-tih-pul skleh-ROH-sis [em es])
[*multi-* **many,** *-pl-* **fold,** *scler-* **hard,** *-osis* **condition**]

neuralgia
(noo-RAL-jee-ah)
[*neur-* **nerve,** *-algia* **pain**]

neuritis
(noo-RYE-tis)
[*neur-* **nerve,** *-itis* **inflammation**]

neuroblastoma
(noo-roh-blas-TOH-mah)
[*neuro-* **nerve,** *-blast* **germ,** *-oma* **tumor**]

neurologist
(noo-ROL-uh-jist)
[*neuro-* **nerve,** *-log-* **words (study of),** *-ist* **agent**]

neuroma
(noo-ROH-mah)
[*neur-* **nerve,** *-oma* **tumor**]

neuroscientist
(noo-roh-SYE-en-tist)
[*neuro-* **nerve,** *-scien-* **knowledge,** *-ist* **agent**]

paralysis
(pah-RAL-ih-sis)
[*para-* **beyond,** *-lysis* **loosening**]

paraplegia
(payr-ah-PLEE-jee-ah)
[*para-* **beside,** *-pleg-* **stricken,** *-ia* **condition**]

Parkinson disease (PD)
(PARK-in-son dih-ZEEZ [pee dee])
[*James Parkinson* **English physician,** *dis-* **opposite of,** *-ease* **comfort**]

pharmacist
(FAR-mah-sist)
[*pharmac-* **drug,** *-ist* **agent**]

pharmacologist
(far-mah-KAHL-oh-jist)
[*pharmaco-* **drug,** *-log-* **words (study of),** *-ist* **agent**]

pharmacy technician
(FAR-mah-see tek-NISH-en)
[*pharmac-* **drug,** *-y* **location of activity,** *techn-* **art or skill,** *-ic* **relating to,** *-ian* **practitioner**]

psychiatrist
(sye-KYE-ah-trist)
[*psych-* **mind,** *-iatr-* **treatment,** *-ist* **agent**]

psychologist
(sye-KOL-uh-jist)
[*psych-* **mind,** *-log-* **words (study of),** *-ist* **agent**]

quadriplegia
(kwod-rih-PLEE-jee-ah)
[*quadri-* **fourfold,** *-pleg-* **stricken,** *-ia* **condition**]

sciatica
(sye-AT-ih-kah)
[*sciatica* **pain in the hip**]

seizure
(SEE-zhur)

shingles
(SHING-guls)
[from *cingulum* **belt or girdle**]

spastic paralysis
(SPAS-tik pah-RAL-ih-sis)
[*spast-* **pull,** *-ic* **relating to,** *para-* **beyond,** *-lysis* **loosening**]

tic douloureux
(tik doo-loo-ROO)
[*tic* **spasm,** *douloureux* **painful spasm**]

trigeminal neuralgia
(try-JEM-ih-nal noo-RAL-jee-ah)
[*tri-* **three,** *-gemina-* **twins or pair,** *-al* **relating to,** *neur-* **nerve,** *-algia* **pain**]

triplegia
(try-PLEE-jee-ah)
[*tri-* **three,** *-pleg-* **stricken,** *-ia* **condition**]

10

❑ OUTLINE SUMMARY

To download a digital version of the chapter summary for use with your device, access the ***Audio Chapter Summaries*** *online at evolve.elsevier.com.*

Scan this summary after reading the chapter to help you reinforce the key concepts. Later, use the summary as a quick review before your class or before a test.

Introduction

A. Both the nervous system and the endocrine system control various functions of the body by transmitting information

B. Homeostasis is possible only if control and integration systems function properly

Organization of the Nervous System
(**Figure 10-1**)

A. Central nervous system (CNS)—brain and spinal cord

B. Peripheral nervous system (PNS)—all nerves

C. Autonomic nervous system (ANS)

Cells of the Nervous System

A. Neurons
  1. Neuron structure
    a. Consist of three main parts—dendrites, cell body of neuron, and axon (**Figure 10-2**)
    b. Dendrites—branching projections that conduct impulses to cell body of neuron
    c. Axon—elongated projection that conducts impulses *away* from cell body of neuron
      (1) Myelin—white, fatty substance formed by glia, surrounding some axons as a sheath
      (2) Nodes of Ranvier—gaps in the myelin sheath
      (3) Neurilemma—outer layer of myelin sheath needed for repair of damaged axons
  2. Neuron types are classified according to function
    a. Sensory neurons: conduct impulses to the spinal cord and brain; also called *afferent neurons*
    b. Motor neurons: conduct impulses away from brain and spinal cord to muscles and glands; also called *efferent neurons*
    c. Interneurons: conduct impulses from sensory neurons to motor neurons or among a network of interneurons; also called *central* or *connecting neurons*

B. Glia (neuroglia)
  1. Function—support cells, bringing the cells of nervous tissue together structurally and functionally
  2. Central glia—three main types of glial cells of the CNS (**Figure 10-3**)
    a. Astrocytes—star-shaped cells that anchor small blood vessels to neurons
    b. Microglia—small cells that move in inflamed brain tissue carrying on phagocytosis
    c. Oligodendrocytes—form myelin sheaths on axons in the CNS (Schwann cells form myelin sheaths in PNS only)
  3. Peripheral glia—Schwann cells form myelin sheaths on axons of the PNS (**Figure 10-2**)

C. Disorders of nervous tissue
  1. Multiple sclerosis—characterized by myelin loss in central nerve fibers and resulting conduction impairments (**Figure 10-4**)
  2. Tumors
    a. General name for nervous system tumors is *neuroma*
    b. Most neuromas are gliomas (glial tumors)
    c. Multiple neurofibromatosis—characterized by numerous benign tumors (**Figure 10-5**)

Nerves and Tracts

A. Nerve—bundle of peripheral axons (**Figure 10-6**)
  1. Nerve coverings—fibrous connective tissue
    a. Endoneurium—surrounds individual fibers within a nerve
    b. Perineurium—surrounds a group (fascicle) of nerve fibers
    c. Epineurium—surrounds the entire nerve

B. Tract—bundle of central axons
  1. White matter—tissue composed primarily of myelinated axons (tracts)
  2. Gray matter—composed primarily of cell bodies and unmyelinated fibers

Nerve Signals

A. Reflex Arcs
  1. Nerve impulses are conducted from receptors to effectors over neuron pathways or reflex arcs; conduction by a reflex arc results in a reflex (that is, contraction by a muscle or secretion by a gland)
  2. The simplest reflex arcs are two-neuron arcs—consisting of sensory neurons synapsing in the spinal cord with motor neurons
  3. Three-neuron arcs consist of sensory neurons synapsing in the spinal cord with interneurons that synapse with motor neurons (**Figure 10-7**)

B. Nerve Impulses
1. Definition—self-propagating wave of electrical disturbance that travels along the surface of a neuron membrane (often called *action potentials*)
2. Mechanism (**Figure 10-8**)
   a. At rest, the neuron's membrane is slightly positive on the outside—polarized—from a slight excess of Na^+ on the outside
   b. A stimulus triggers the opening of Na^+ channels in the plasma membrane of the neuron
   c. Inward movement of Na^+ depolarizes the membrane by making the inside more positive than the outside at the stimulated point; this depolarization is a nerve impulse (action potential)
3. Conduction of nerve impulses (**Figure 10-9**)
   a. Continuous conduction—the stimulated section of membrane immediately repolarizes, but by that time the depolarization has already triggered the next section of membrane to depolarize, thus propagating a wave of electrical disturbances (depolarizations) all the way down the membrane
   b. Saltatory conduction—in myelinated fibers, conduction can "jump" from gap to gap and thus greatly speed up the rate of conduction

C. The Synapse
1. Definition—the place where impulses are transmitted from one neuron to another (the postsynaptic neuron) (**Figure 10-10**)
2. Synapse made of three structures—synaptic knob, synaptic cleft, and plasma membrane
3. Neurotransmitters bind to specific receptor molecules in the membrane of a postsynaptic neuron, opening ion channels and thereby stimulating impulse conduction by the membrane
4. Transmission stops when neurotransmitters are removed from receptors
   a. Reuptake of neurotransmitters into presynaptic neurons, with repackaging and recycling for later use
   b. Breakdown of neurotransmitters in extracellular matrix (ECM) of synaptic cleft
   c. Glia may break down neurotransmitters and return pieces to the presynaptic neuron for recycling
5. Examples of neurotransmitters—acetylcholine, catecholamines (norepinephrine, dopamine, and serotonin), endorphins, enkephalins, nitric oxide (NO), and other compounds
6. Receptors—specific to a particular neurotransmitter; possible to have several versions of receptors for a single neurotransmitter, each found in a different body location
7. Parkinson disease (PD)—characterized by abnormally low levels of dopamine in motor control areas of the brain; patients usually exhibit involuntary trembling and muscle rigidity (parkinsonism; **Figure 10-11**)

Central Nervous System

A. CNS is made up of the brain and spinal cord; central axis location; protected by hard bones and soft meninges (**Figure 10-12**)

B. Brain (**Figure 10-13** and **Table 10-1**)
1. Brainstem
   a. Consists of, named in ascending order, the medulla oblongata, pons, and midbrain
   b. Structure—white matter with bits of gray matter scattered through it
   c. Functions
      (1) All three parts of brainstem are two-way conduction paths
         (a) Sensory tracts in the brainstem conduct impulses to the higher parts of the brain
         (b) Motor tracts conduct from the higher parts of the brain to the spinal cord
      (2) Gray matter areas in the brainstem function as important reflex centers
2. Cerebellum
   a. Structure
      (1) Second largest part of the human brain
      (2) Gray matter outer layer is thin but highly folded, forming a large surface area for processing information
      (3) Arbor vitae—internal, treelike network of white matter tracts
   b. Function
      (1) Helps control muscle contractions to produce coordinated movements so that we can maintain balance, move smoothly, and sustain normal postures
      (2) Variety of additional coordinating effects, assisting the cerebrum and other regions of the brain
3. Diencephalon
   a. Hypothalamus
      (1) Consists mainly of the posterior pituitary gland, pituitary stalk, and gray matter
      (2) Acts as the major center for controlling the ANS; therefore helps control the functioning of most internal organs
      (3) Controls hormone secretion by anterior and posterior pituitary glands; therefore it indirectly helps control hormone secretion by most other endocrine glands
      (4) Contains centers for controlling appetite, wakefulness, pleasure, etc.
   b. Thalamus
      (1) Dumbbell-shaped mass of gray matter extending toward each cerebral hemisphere
      (2) Primary functions:
         (a) Relays sensory impulses to cerebral cortex sensory areas
         (b) In some way produces the emotions of pleasantness or unpleasantness associated with sensations

10

(c) Helps regulate level of consciousness by participating in arousal mechanism
(d) Plays a role in complex reflex movements

c. Pineal gland (pineal body)
(1) Small body resembling pine nut behind the thalamus
(2) Adjusts output of "timekeeping hormone" melatonin in response to changing levels of external light (sunlight and moonlight)

4. Cerebrum (**Figure 10-14**)
a. Largest part of the human brain
b. Outer layer of gray matter is the cerebral cortex; made up of lobes; composed mainly of dendrites and cell bodies of neurons
c. Interior of the cerebrum composed mainly of white matter
(1) Tracts—nerve fibers arranged in bundles
(2) Basal nuclei—islands of gray matter regulate automatic movements and postures
d. Functions of the cerebrum—mental processes of all types, including sensations, consciousness, memory, and voluntary control of movements; many functions are localized to specific areas of cortex

C. Brain disorders
1. Damage to brain tissue
a. Concussion—type of traumatic brain injury (TBI) caused by a jolt to the head and resulting changes in brain chemicals (or even bleeding or swelling) and characterized by changes in thinking, physical symptoms such as nausea, and mood or sleep changes; symptoms may develop months after an injury (postconcussion syndrome)
b. Cerebrovascular accident (CVA)—hemorrhage from or cessation of blood flow through cerebral blood vessels; a "stroke"
c. Cerebral palsy (CP)—condition in which damage to motor control areas of the brain before, during, or shortly after birth causes paralysis (usually spastic) of one or more limbs (**Figure 10-15**)
2. Dementia—syndrome that includes progressive loss of memory, shortened attention span, personality changes, reduced intellectual capacity, and motor control deficit
a. Alzheimer disease (AD)—brain disorder of the middle and late adult years characterized by dementia (**Figure 10-16**)
b. Chronic traumatic encephalopathy (CTE)—results from repeated head trauma and characterized by memory loss, parkinsonism, disordered thinking
c. Huntington disease (HD)—inherited disorder characterized by chorea (purposeless movement) progressing to severe dementia
d. HIV (also causes AIDS) can infect neurons and thus cause dementia
3. Seizure disorders
a. Seizure—sudden burst of abnormal neuron activity that results in temporary changes in brain function
b. Epilepsy—many forms, all characterized by recurring seizures
c. Electroencephalogram—graphic representation of voltage changes in the brain used to evaluate brain activity (**Figure 10-17**)

D. Spinal cord (**Figure 10-18**)
1. Columns of white matter, composed of bundles of myelinated nerve fibers, form the outer portion of the H-shaped core of the spinal cord; bundles of axons called tracts
2. Interior composed of gray matter made up mainly of neuron dendrites and cell bodies (**Figure 10-19**)
3. Spinal cord tracts provide two-way conduction paths—ascending and descending
4. Spinal cord functions as the primary center for all spinal cord reflexes; sensory tracts conduct impulses to the brain, and motor tracts conduct impulses from the brain

E. Coverings and fluid spaces of the brain and spinal cord
1. Meninges and bone (**Figure 10-20**)
a. Cerebral and spinal meninges
(1) Dura mater—tough outer membrane
(2) Arachnoid mater—cobweblike middle layer
(3) Pia mater—delicate inner layer; adheres to CNS tissue
b. Cranial bones and vertebrae form hard outer covering
2. Cerebrospinal fluid (CSF) spaces (**Figures 10-21** and **10-22**)
a. Subarachnoid spaces of meninges
b. Central canal inside cord
c. Ventricles in brain
d. Hydrocephalus can result from blocked CSF circulation (**Figure 10-23**)

Peripheral Nervous System

A. Cranial nerves (**Figure 10-24** and **Table 10-2**)
1. Twelve pairs—attached to undersurface of the brain
2. Connect brain with the neck and structures in the thorax and abdomen

B. Spinal nerves
1. Thirty-one pairs—contain dendrites of sensory neurons and axons of motor neurons
2. Conduct impulses necessary for sensations and voluntary movements
3. Dermatome—skin surface area supplied by a single cranial or spinal nerve (**Figure 10-25**)

C. Peripheral nerve disorders
1. Neuritis—general term referring to nerve inflammation
a. Sciatica—inflammation of the sciatic nerve that innervates the legs
b. Neuralgia, or muscle pain, often accompanies neuritis
2. Trigeminal neuralgia—recurring episodes of stabbing pain along one or more branches of the trigeminal (fifth cranial) nerve in the head
3. Bell palsy—paralysis of facial features resulting from damage to the facial (seventh cranial) nerve
4. Herpes zoster, or shingles (**Figure 10-26**)
a. Viral infection caused by chickenpox virus that has invaded the dorsal root ganglion and remained dormant until stress or reduced immunity precipitates an episode of shingles
b. Usually affects a single dermatome, producing characteristic painful plaques or vesicles

Autonomic Nervous System

A. Functional anatomy
1. Autonomic nervous system
a. Motor neurons that conduct impulses from the CNS to cardiac muscle, smooth muscle, and glandular epithelial tissue
b. Regulates the body's automatic, or involuntary, functions (**Figure 10-27**)
c. Distinct from the *somatic* nervous system, which instead regulates voluntary somatic effectors (skeletal muscles)
2. Autonomic neurons—preganglionic autonomic neurons conduct from spinal cord or brainstem to an autonomic ganglion; postganglionic neurons conduct from autonomic ganglia to cardiac muscle, smooth muscle, and glandular epithelial tissue
3. Autonomic or visceral effectors—tissues to which autonomic neurons conduct impulses (that is, cardiac and smooth muscle and glandular epithelial tissue)
4. Composed of two divisions—the sympathetic system and the parasympathetic system

B. Autonomic conduction paths (**Figure 10-28**)
1. Consist of two-neuron relays (that is, preganglionic neurons from the CNS to autonomic ganglia, synapses, postganglionic neurons from ganglia to visceral effectors)
2. In contrast, somatic motor neurons conduct all the way from the CNS to somatic effectors with no intervening synapses

C. Sympathetic division
1. Dendrites and cell bodies of sympathetic preganglionic neurons are located in the gray matter of the thoracic and upper lumbar segments of the spinal cord
2. Axons leave the spinal cord in the ventral roots of spinal nerves, extend to sympathetic, or collateral, ganglia and synapse with several postganglionic neurons whose axons extend to spinal or autonomic nerves to terminate in visceral effectors
3. A chain of sympathetic ganglia is in anterior and at each side of the spinal column
4. Functions of the sympathetic division
a. Serves as the emergency or stress system, controlling visceral effectors during strenuous exercise and when strong emotions (anger, fear, hate, or anxiety) are triggered
b. Group of changes induced by sympathetic control is called the *fight-or-flight response*

D. Parasympathetic division
1. Structure
a. Parasympathetic preganglionic neurons have dendrites and cell bodies in the gray matter of the brainstem and the sacral segments of the spinal cord
b. Parasympathetic preganglionic neurons terminate in parasympathetic ganglia located in the head and the thoracic and abdominal cavities close to visceral effectors
c. Each parasympathetic preganglionic neuron synapses with postganglionic neurons to only one effector
2. Function—dominates control of many visceral effectors under normal, everyday conditions; counterbalances sympathetic function

E. Autonomic neurotransmitters (**Figure 10-29**)
1. Cholinergic fibers—preganglionic axons of parasympathetic and sympathetic systems and parasympathetic postganglionic axons release acetylcholine
2. Adrenergic fibers—axons of sympathetic postganglionic neurons release norepinephrine (noradrenaline)

F. Autonomic nervous system as a whole
1. Regulates the body's automatic functions in ways that maintain or quickly restore homeostasis
2. Many visceral effectors are doubly innervated (that is, they receive fibers from parasympathetic and sympathetic divisions and are influenced in opposite ways by the two divisions)

G. Disorders of the autonomic nervous system
1. Stress-induced disease
a. Prolonged or excessive response to stress can disrupt normal functioning throughout the body
b. Examples of stress-induced conditions include heart disease, digestive problems, reduced resistance to disease, and spread of cancer
2. Neuroblastoma—highly malignant tumor of the sympathetic division, primarily affecting young children

❑ ACTIVE LEARNING

STUDY TIPS

Consider using these tips to achieve success in meeting your learning goals.

Review the synopsis of the nervous system in Chapter 5. The amount of material presented in Chapter 10 may be overwhelming at first, but it can be somewhat easier to learn if you divide the chapter into three parts: the microscopic structure and function of the nervous system, the central nervous system, and the peripheral nervous system.

1. Keep in mind that the nervous system functions as one organized system. The function of the nervous system is accomplished by two processes: conduction of nerve impulses and passing of the nerve impulse across a synapse. Nerve impulses are an exchange of ions between the interior and exterior of the neuron.
2. The synapse requires the production, release, and deactivation of neurotransmitters. Neurotransmitters function by stimulating receptors in the neuron on the other side of the synapse.
3. The material on the central nervous system can be learned best by using flash cards that match up the structure and function. Use online resources that provide tutorials and animations (for example, *getbodysmart.com*).
4. Make a chart showing the disorders of the nervous system. Organize them by type: myelin disorder, brain disorder, etc. Describe the damage done by the disorder and the effect it has on the body.
5. In your study group, you should go over the terms presented in the first part of the chapter. Review the Language of Science and Language of Medicine terms and their word origins to help you better understand the meaning of the nervous system terms. Discuss the processes of nerve impulse transmission and what occurs at the synapse. Review the flash cards with the names and functions of the parts of the central nervous system. Remember that most of the structures in the central nervous system have more than one function. Go over the disorder chart. If you remember the general functions of the sympathetic and parasympathetic divisions, the specific effects will be easier to remember. Also review the questions and outline summary at the end of the chapter.

Review Questions

Write out the answers to these questions after reading the chapter and reviewing the Chapter Summary. If you simply think through the answer without writing it down, you will not retain much of your new learning.

1. List the three types of neurons classified according to the direction in which the impulse is being transmitted. Define or explain each of them.
2. Define or explain the following terms: myelin, nodes of Ranvier, and neurilemma.
3. List and give the function of the three types of glia cells.
4. What occurs at the cellular level in multiple sclerosis? What effect does this have on the body?
5. Neuromas usually develop from what type of cells or tissues?
6. Define or explain the following terms: epineurium, perineurium, and endoneurium.
7. What causes gray matter to be gray and white matter to be white?
8. Explain how a reflex arc functions. What are two types of reflex arcs?
9. Explain what occurs during a nerve impulse. What is saltatory conduction?
10. Explain what occurs at a synapse. What are the two ways that neurotransmitter activity is terminated?
11. What is the cause of Parkinson disease? What are some treatment options?
12. Define dementia.
13. What is a seizure?
14. List two possible causes of Alzheimer disease.
15. List and describe the functions of the medulla oblongata.
16. List and describe the functions of the hypothalamus.
17. List and describe the functions of the thalamus.
18. Describe the function of the pineal gland (body).
19. List and describe the functions of the cerebellum.
20. List the general functions of the cerebrum. What are the specific functions of the occipital and temporal lobes?
21. What is a concussion? Describe its symptoms.
22. List and describe the functions of the spinal cord.
23. List and explain the three layers of the meninges.
24. What is the function of cerebrospinal fluid? Where and how is cerebrospinal fluid produced?
25. How many nerve pairs are generated from the spinal cord? How many nerve pairs are generated from each section of the spinal cord? How are these nerves named? What is a plexus?
26. Define neuritis and neuralgia.

27. What is the cause of tic douloureux? What is the cause of Bell palsy?
28. Explain the structure and function of the sympathetic nervous division.
29. Explain the structure and function of the parasympathetic nervous division.
30. Where is the "hidden lobe" located on the brain?

Critical Thinking

After finishing the Review Questions, write out the answers to these more in-depth questions to help you apply your new knowledge. Go back to sections of the chapter that relate to concepts that you find difficult.

31. Can you elaborate on why police officers use sobriety tests such as walking along a straight line, touching the tip of the nose with one finger, or maintaining balance with the eyes closed?
32. How would you explain why a person is more likely to survive damage to the cerebrum than damage to the brainstem?
33. There is a type of medication that inhibits the functioning of acetylcholinesterase (the enzyme that deactivates acetylcholine). Explain the effect this medication would have on the visceral effectors.
34. The body conserves everything it possibly can for later use. Each system attempts to be prudent with its resources. Can you give an example of how the nervous system demonstrates this concept with neurotransmitter molecules?

Chapter Test

Hint

After studying the chapter, test your mastery by responding to these items. Try to answer them without looking up the answers.

1. ________ is the name of the nervous system division that includes the nerves that extend to the outlying parts of the body.
2. ________ is the name of the nervous system division that includes the brain and spinal cord.
3. A group of peripheral axons bundled together in an epineurium is called a ________.
4. The two types of cells found in the nervous system are ________ and ________.
5. The knee-jerk reflex is a type of neural pathway called a ________.
6. A ________ is a self-propagating wave of electrical disturbance that travels along the surface of a neuron's plasma membrane.
7. ________ conduction is the term that describes the impulse as it "jumps" around the myelin.
8. The ________ is a place where impulses are passed from one neuron to another.
9. Acetylcholine and dopamine are examples of ________, which are chemicals used by neurons to communicate.
10. ________, ________, and ________ are the three membranes that make up the meninges.
11. When too much *CSF* accumulates in the ventricles, it may lead to a condition called ________ or "water on the brain."
12. The two hemispheres of the brain are almost separate structures except for their lower midportions, which are connected by a structure called the ________ ________.
13. There are ________ pairs of cranial nerves and ________ pairs of nerves that come from the spinal cord.
14. ________ are skin surface areas supplied by a single spinal nerve.
15. ________ is the part of the autonomic nervous system that regulates effectors during nonstress conditions.
16. ________ is the part of the autonomic nervous system that regulates effectors during the "fight-or-flight" response.
17. The preganglionic axons of the parasympathetic nervous system release the neurotransmitter ________. The postganglionic axons release ________.
18. The preganglionic axons of the sympathetic nervous system release the neurotransmitter ________. The postganglionic axons release ________.
19. After impulse conduction by postsynaptic neurons is initiated, neurotransmitter activity is rapidly terminated. Two mechanisms that cause this are ________ and ________.
20. The cerebrum has many ridges and grooves. The grooves are called ________.

10

Match each term in Column A with its corresponding function or description in Column B.

Column A	Column B
21. ________ dendrite	a. cells that make myelin for axons outside the CNS
22. ________ axon	b. glia cells that help form the blood-brain barrier
23. ________ myelin	c. a single projection that carries nerve impulses away from the cell body
24. ________ Schwann cells	d. cells that make myelin for axons inside the CNS
25. ________ astrocyte	e. a white fatty substance that surrounds and insulates the axon
26. ________ microglia	f. cells that act as microbe-eating scavengers in the CNS
27. ________ oligodendrocyte	g. a highly branched part of the neuron that carries impulses toward the cell body

Match each part of the central nervous system in Column A with its corresponding function in Column B.

Column A

28. ________ medulla oblongata
29. ________ pons
30. ________ midbrain
31. ________ hypothalamus
32. ________ thalamus
33. ________ cerebellum
34. ________ cerebrum
35. ________ spinal cord

Column B

a. part of the brainstem that is a conduction pathway between the brain and body; influences respiration
b. sensory relay station from various body areas to the cerebral cortex; also involved with emotion and alerting and arousal mechanisms
c. carries messages to and from the brain and the rest of the body; also mediates reflexes
d. part of the brainstem that contains cardiac, respiratory, and vasomotor centers
e. sensory perception, willed movements, consciousness, and memory are mediated here
f. regulates body temperature, water balance, sleep-wake cycle, and sexual arousal
g. regulates muscle coordination, maintenance of equilibrium and posture
h. part of the brainstem that contains relays for visual and auditory impulses

Match each disorder or disease in Column A with its description or cause in Column B.

Column A

36. ________ multiple sclerosis
37. ________ neuroma
38. ________ multiple neurofibromatosis
39. ________ Parkinson disease
40. ________ CVA
41. ________ dementia
42. ________ epilepsy
43. ________ meningitis
44. ________ tic douloureux
45. ________ Bell palsy
46. ________ neuroblastoma

Column B

a. inherited condition causing multiple benign tumors
b. cessation of blood flow to the brain; a stroke
c. syndrome that includes memory loss, short attention span, and reduced intellectual capacity
d. recurring or chronic seizure disorder
e. compression or degeneration of the seventh cranial nerve
f. disorder caused by the loss of myelin
g. a malignant tumor of the sympathetic nervous division
h. compression or degeneration of the fifth cranial nerve
i. infection or inflammation of the meninges
j. general term for a tumor in the nervous system
k. disease characterized by an abnormally low level of dopamine

Case Studies

To solve a case study, you may have to refer to the glossary or index, other chapters in this textbook, and other resources.

1. Tony's teachers describe him as a daydreamer. The teachers often find him staring off into space when they are trying to get his attention. When Tony's parents mentioned this to their family physician, the physician brings up the possibility of epilepsy. Could Tony's daydreaming be a sign of epilepsy? What test could help confirm such a diagnosis? What signs would one look for in such a test if epilepsy is present?
2. Over the last few years, your friend Angela has developed fibrous nodules in many areas of her skin. She recently confided that she has an inherited disorder of the nervous system that causes these bumps. What disease might Angela have? How can a nervous disorder cause skin lesions?
3. Baraka loves to dance. However, he and his friends notice that he misses easy steps more and more often. In fact, Baraka almost seems intoxicated because his coordination is so badly affected. Baraka's physicians tell him that he has a myelin disorder. How would such a disorder cause Baraka's symptoms? Name a specific myelin disorder and explain how this disease causes similar problems.
4. Matt has just returned home from his doctor's office. He has a kidney infection and his doctor has prescribed medication for it. Matt was concerned that he might not be able to drive while taking the medication, but his doctor reassured him that it would not "affect his brain." What did the doctor mean by that remark? Can you offer a more complete explanation?

Answers to Active Learning Questions can be found online at evolve.elsevier.com.

Senses

OUTLINE

Scan this outline before you begin to read the chapter, as a preview of how the concepts are organized.

OBJECTIVES

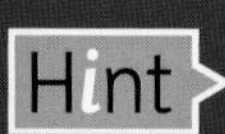

Before reading the chapter, review these goals for your learning.

After you have completed this chapter, you should be able to:

1. Classify sense organs as general or special and explain the basic differences between the two groups.
2. Discuss how a stimulus is converted into a sensation, as well as disorders involving general senses.
3. Discuss the general sense organs and their functions.
4. Describe the structure of the eye and the functions of its components.
5. Name and describe the major visual disorders.
6. Discuss the anatomy of the ear and its sensory function in hearing and equilibrium.
7. Name and describe the major forms of hearing impairment.
8. Describe the anatomy of the tongue and its sensory function in taste.
9. Describe the anatomy of the nasal cavity and its sensory function in smell.
10. Discuss how senses are integrated in the brain.

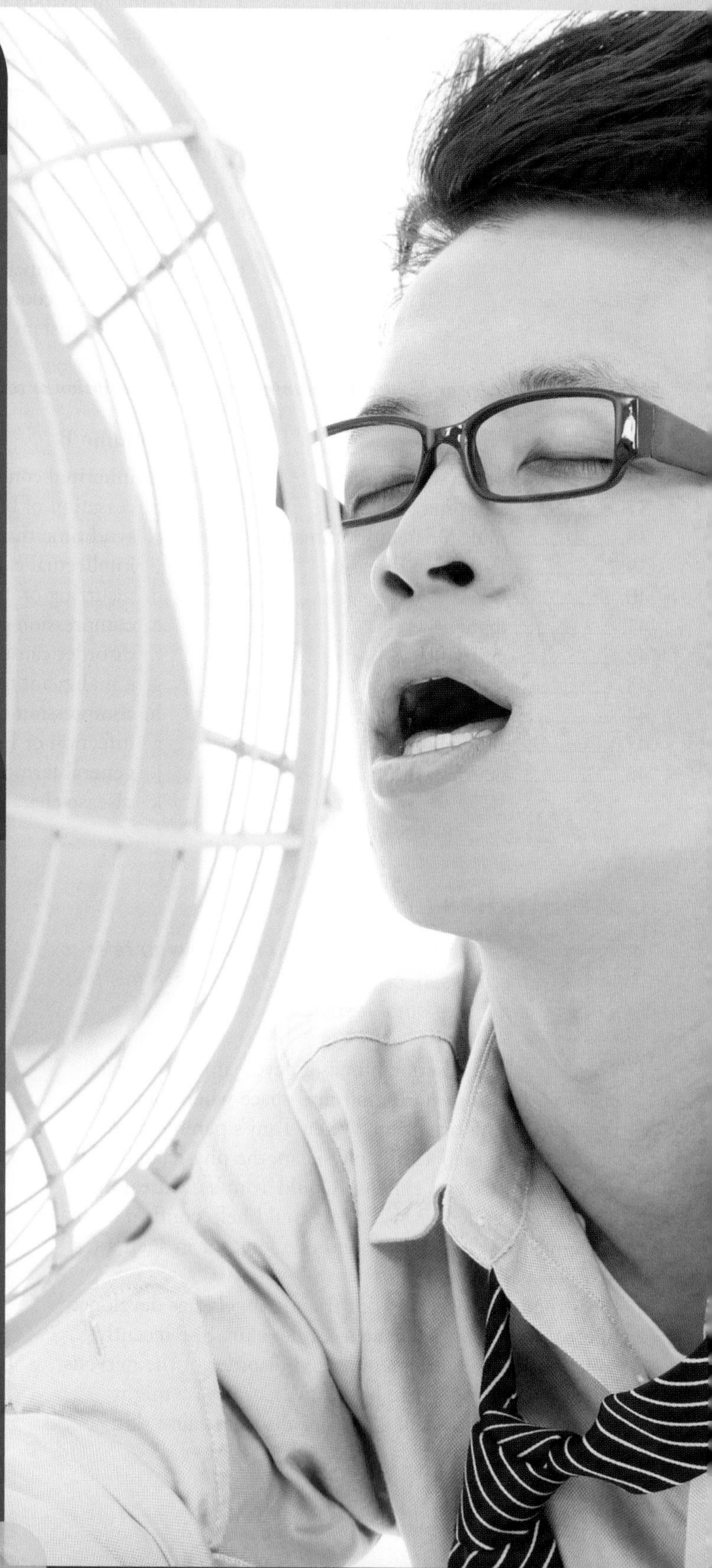

CHAPTER 11

If you were asked to name the sense organs, what organs would you name? Can you think of any besides the eyes, ears, nose, and taste buds? Actually there are millions of other sense organs throughout the body in our skin, internal organs, and muscles. They constitute the many **sensory receptors** that allow us to respond to stimuli such as touch, pressure, temperature, and pain. These microscopic receptors are located at the tips of dendrites of sensory neurons.

Our ability to detect changes in our external and internal environments is a requirement for maintaining homeostasis and for survival itself. We can initiate protective reflexes important to homeostasis only if we can *sense* a change or danger.

External dangers may be detected by sight or hearing. If the danger is internal, such as overstretching a muscle, detecting an increase in body temperature (fever), or sensing the pain caused by an ulcer, we have other receptors that make us aware of the problem, which makes it possible for us to then take appropriate action to maintain homeostasis.

Classification of Senses

The senses are often classified as either **general senses** or **special senses.**

General Senses

The *general senses* are those detected by rather simple, microscopic receptors that are widely distributed throughout the body in the skin, muscles, tendons, joints, and other internal organs of the body. They are responsible for such sensations as pain, temperature, touch, pressure, and body position.

LANGUAGE OF
SCIENCE

Hint Before reading the chapter, say each of these terms out loud. This will help you to avoid stumbling over them as you read.

adaptation
(ad-ap-TAY-shun)
[*adapt-* **fit to,** *-tion* **process**]

aqueous humor
(AY-kwee-us HYOO-mor)
[*aqu-* **water,** *-ous* **relating to,** *humor* **body fluid**]

auditory tube
(AW-dih-toh-ree toob)
[*audit-* **hear,** *-ory* **relating to**]

auricle
(AW-rih-kul)
[*auri-* **ear,** *-icle* **little**]

bony labyrinth
(BOHN-ee LAB-eh-rinth)
[labyrinth **maze**]

cerumen
(seh-ROO-men)
[*cer(a)-* **wax,** *-men* **formed of**]

ceruminous gland
(seh-ROO-mih-nus gland)
[*cer(a)-* **wax,** *-min-* **formed of,** *-ous* **relating to,** *gland* **acorn**]

chemoreceptor
(kee-moh-ree-SEP-tor)
[*chemo-* **chemical,** *-recept-* **receive,** *-or* **agent**]

choroid
(KOH-royd)
[*chor-* **skin,** *-oid* **like**]

ciliary muscle
(SIL-ee-ayr-ee)
[*ciliary* **eyelids or eyelashes,** *mus-* **mouse,** *-cle* **little**]

Continued on p. 310

Special Senses

The special senses are those detected by receptors that are grouped in specific areas and associated with complex structures that facilitate these senses. The senses of smell, taste, vision, hearing, and equilibrium are considered special senses because their receptors are grouped within distinct structures that enhance their function.

Sensory Receptor Types

Individual receptor cells are often identified structurally according to whether they are *encapsulated* or *unencapsulated*, that is, whether they are covered by some sort of capsule or are "free" or "naked" of any such covering.

Sensory receptor cells are also classified functionally by the types, or *modes*, of stimuli that activate them:

1. **Photoreceptors**—sensitive to change in intensity or color of light, as in vision
2. **Chemoreceptors**—sensitive to presence of certain chemicals, as in taste or smell
3. **Pain receptors**—sensitive to physical injury
4. **Thermoreceptors**—sensitive to changes in temperature
5. **Mechanoreceptors**—sensitive to mechanical stimuli that change their position or shape

Table 11-1 identifies the general sense organs as either free nerve endings or one of the six types of encapsulated nerve

TABLE 11-1 General Sense Organs

TYPE	MAIN LOCATIONS	GENERAL SENSES
Free Nerve Endings		
Naked nerve ending (several types exist)	Skin and mucosa (epithelial layers)	Pain, discriminative touch, tickle, and temperature
Encapsulated Nerve Endings		
Bulboid corpuscle (Krause corpuscle)	Skin (dermal layer), subcutaneous tissue, mucosa of lips and eyelids, and external genitals	Touch and possibly cold
Lamellar corpuscle (Pacini corpuscle)	Subcutaneous, submucous, and subserous tissues; around joints; in mammary glands and external genitals of both sexes	Pressure and high-frequency vibration
Tactile corpuscle (Meissner corpuscle)	Skin (in papillae of dermis) and fingertips and lips (numerous)	Fine touch and low-frequency vibration
Bulbous corpuscles (Ruffini corpuscle)	Skin (dermal layer) and subcutaneous tissue of fingers	Touch and pressure
Golgi tendon organ	Near junction of tendons and muscles	Proprioception (sense of muscle tension)
Muscle spindle (Intrafusal fibers)	Skeletal muscles	Proprioception (sense of muscle length)

11

TABLE 11-2 Special Sense Organs

SENSE ORGAN	SPECIFIC RECEPTOR	TYPE OF RECEPTOR	SENSE
Eye	Rods and cones	Photoreceptor	Vision
Ear	Organ of Corti (spiral organ)	Mechanoreceptor	Hearing
	Cristae ampullares	Mechanoreceptor	Dynamic equilibrium
	Maculae	Mechanoreceptor	Static equilibrium
Nose	Olfactory cells	Chemoreceptor	Smell
Taste buds	Gustatory cells	Chemoreceptor	Taste

endings, whereas Table 11-2 identifies the type of receptor cells in the special sense organs that are stimulated by specific types of stimuli.

Sensory Pathways

All sense organs, regardless of size, type, or location, have in common some important functional characteristics. First, they must be able to sense or detect a stimulus or a change in the quality or intensity of a particular stimulus in their environment. Next, detection of a stimulus must be converted into a nerve impulse. This signal is then conducted over a nervous system "pathway" to the brain, where the incoming information is filtered and sorted—often comparing it to information coming in along other sensory pathways. Only after all of this processing of information is the sensation actually perceived in the brain.

The sensory pathway for the general senses typically involves conduction of action potentials generated in the receptors through the spinal cord to the thalamus (cutaneous or skin receptors) or cerebellum (proprioceptors) where they synapse, and impulses are then relayed to specific areas of the cerebral cortex for conscious sensory interpretation.

The sensory pathways for the special senses are varied, but also ultimately end in specific sensory areas of the cerebral cortex.

General Senses

Distribution of General Sense Receptors

Microscopic general sense organ receptors are found in almost every part of the body, but they are most concentrated in the skin (Figure 11-1). However, these receptors are not evenly distributed over the body surface or in the internal organs. Also, they do not all respond to the same type of stimulus. To demonstrate this, try touching any point of your skin on a fingertip with the tip of a toothpick. You can hardly miss stimulating at least one receptor and almost instantaneously experiencing a sensation of touch.

The ability to distinguish one touch stimulus from two is called *two-point discrimination.* A neurological test that measures this function involves simultaneously touching two points on the skin over one area of the body to determine whether the ability to feel the two separate stimuli is present. The skin over different parts of the body will respond differently because of the differing numbers of touch receptors that are present.

Touch receptors are distributed closely together over the fingertips (2 to 8 mm apart), relatively close together over the palms (8 to 12 mm), and quite far apart over the back of the torso (40 to 60 mm). Lesions to the parietal lobe of the brain will impair two-point discrimination.

Modes of Sensation

Stimulation of general sensory receptors can lead to a variety of sensations. The difference in what kind of stimuli is detected is called the **mode** of the sensation. Different general sensory receptors can detect vibration, deep pressure, light pressure, pain, stretch, or temperature.

Examples of general sensory receptors of various modes are listed in Table 11-1 and illustrated in Figure 11-1.

Some general sensory receptors found near the point of junction between tendons and muscles and others found deep within skeletal muscle tissue are called **proprioceptors.** When stimulated by stretch, these mechanoreceptors provide us with information concerning the position or movement of the different parts of the body as well as the length and the extent of contraction of our muscles.

The Golgi tendon receptors and muscle spindles identified in Table 11-1 are important proprioceptors.

Many general sensory receptors are found in the skin, but some are present deep in the body. For example, there are stretch receptors in your stomach that signal you when it is

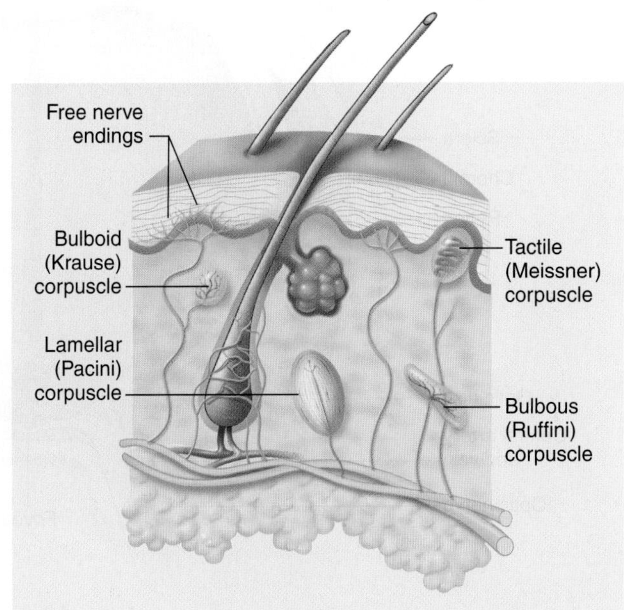

FIGURE 11-1 General sense receptors. This section of skin shows the placement of some of the receptors described in Table 11-1.

11

full. There are also stretch (pressure) receptors in most other hollow organs such as the stomach and intestines, arteries, vagina (birth canal), and urinary bladder that enable the normal functioning of those organs.

There are also important chemoreceptors in the aorta and other arteries that detect changes in pH and carbon dioxide levels in the blood—important information for regulating breathing and heart rate.

Disorders Involving General Senses

Disruption of general sense organs can occur by means of a variety of mechanisms. For example, third-degree burns can completely destroy general sense receptors throughout the affected area—resulting in loss of pain and touch sensations.

Temporary impairment of general sense receptors occurs when the blood flow to them is slowed. This commonly occurs when you put your legs in a position (such as crossing them above the knee or folding a leg under yourself as you sit) that causes pressure to be applied to your legs in a way that reduces blood flow. When you try to stand up, you cannot feel your legs because the general sense organs are temporarily impaired. You may not even be able to walk at first because you cannot tell where your legs are without looking at them. As blood flow returns, reactivation of the sense organs may produce a tingling sensation.

Disruption in the functioning of the general sense organs also can occur as a result of diabetes, cardiovascular disease, stroke, and spinal cord or brain injury or disease.

QUICK CHECK

1. What are two categories that the senses are often classified into when discussing the structure and function of the system?
2. Describe the sensory pathway of the general sense organs.
3. What is the function of a *proprioceptor*?
4. What is two-point *discrimination*?

Special Senses

Vision

Vision detects the color and intensity of light in our external environment. But when focused by the eyes and processed by the brain, it can do much more. For example, we can recognize the outlines and depth of objects, analyze movement, and determine distances. In this section, we discuss that complex and amazing tool of vision—the eye.

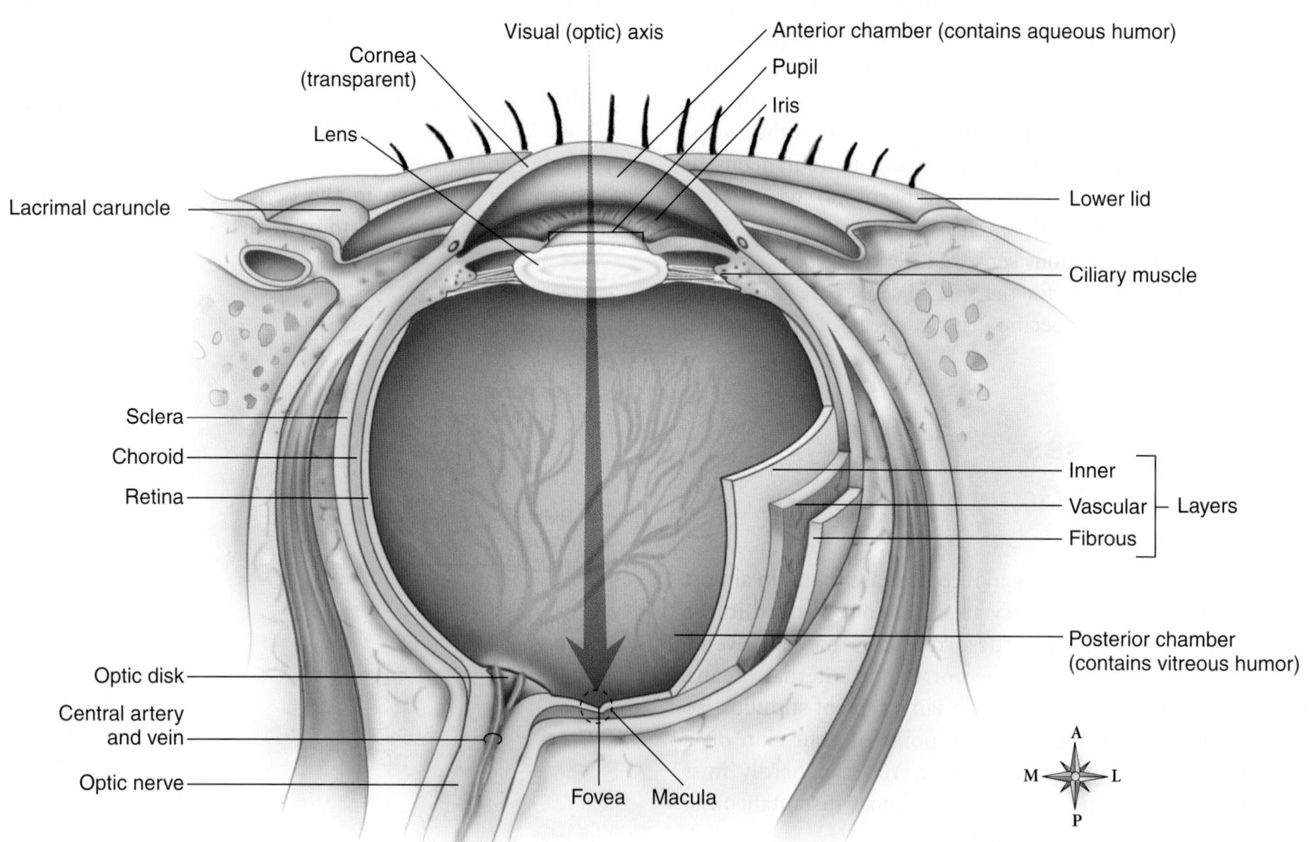

FIGURE 11-2 Eye. This transverse (horizontal) section through the left eyeball is shown as if viewed from above.

Structure and Function of the Eye

When you look at a person's eye, you see only a small part of the whole eye. As you can see in Figure 11-2, the eyeball is a fluid-filled sphere having a wall of three layers:

1. Fibrous layer
   - Sclera
   - Cornea
2. Vascular layer
   - Choroid
   - Ciliary muscle
   - Iris
   - Lens
3. Inner layer
   - Retina
   - Optic nerve
   - Retinal blood vessels

Fibrous Layer

The *fibrous layer* of the eyeball consists of tough fibrous tissue.

The "white" of the eye is part of the fibrous layer called the **sclera.** The sclera, made white by its dense bundles of collagen fibers, forms most of the fibrous layer.

The transparent circle on the anterior of the fibrous layer is called the **cornea.** The cornea is sometimes spoken of as the window of the eye because of its transparency.

Inflammation of the cornea is called *keratitis.* In addition to possible loss of transparency that may result from inflammation, any change in the shape of the cornea can dramatically change the ability of the eye to focus an image on the retina.

The fact that the shape of the cornea affects the eye's focus explains the popularity of surgical procedures that use lasers or other specialty instruments to "sculpt" and change the shape of the cornea. The result is improvement of many visual problems without the use of eyeglasses or contact lenses.

A mucous membrane known as the **conjunctiva** lines the eyelids and covers the fibrous layer in front. The blood vessels you see on the surface of the sclera actually belong to the conjunctiva. The conjunctiva is kept moist by tears secreted by the **lacrimal gland.**

Several strategies are available for replacing damaged corneas, as described in the article *Corneal Transplants* at Connect It! at *evolve.elsevier.com.*

Vascular Layer

The middle layer of the eyeball is called the *vascular layer* because it has a dense network of tiny blood vessels.

Most of the vascular layer is made up of the **choroid,** which contains a large amount of the dark pigment *melanin.* This almost-black layer absorbs light and thus helps prevent the scattering of incoming light rays, which could make it hard for the eye to focus an image.

Several involuntary muscles make up the anterior part of the choroid. Some are in the **iris,** the colored structure seen through the cornea. The iris may appear blue, green, brown, gray, or some combination of these colors when seen through the transparent cornea because of the pigments in this layer of the eyeball.

The black center of the iris is really a hole in this doughnut-shaped muscle—it is the **pupil** of the eye. Some of the fibers of the iris are arranged like spokes in a wheel. When they contract, the pupils dilate, letting in more light rays. Other fibers are circular. When they contract, the pupils constrict, letting in fewer light rays. Normally, the pupils constrict in bright light and dilate in dim light. Figure 11-3 shows how these muscles work under the control of autonomic nerves.

The **lens** of the eye lies directly behind the pupil. It is held in place by a ligament attached to an involuntary muscle called the **ciliary muscle** (see Figure 11-2) When we look at distant objects, the ciliary muscle is relaxed, and the lens has

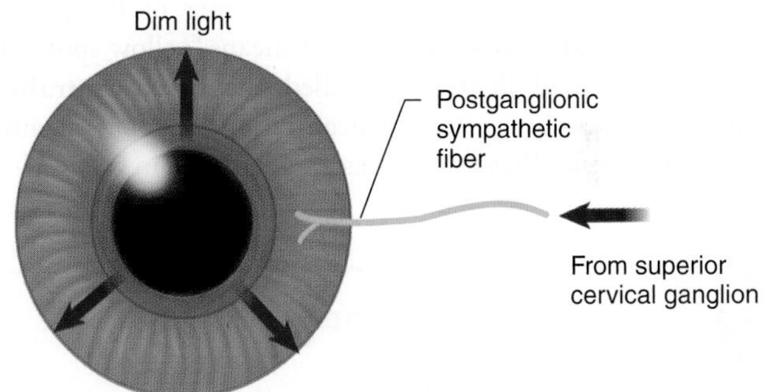

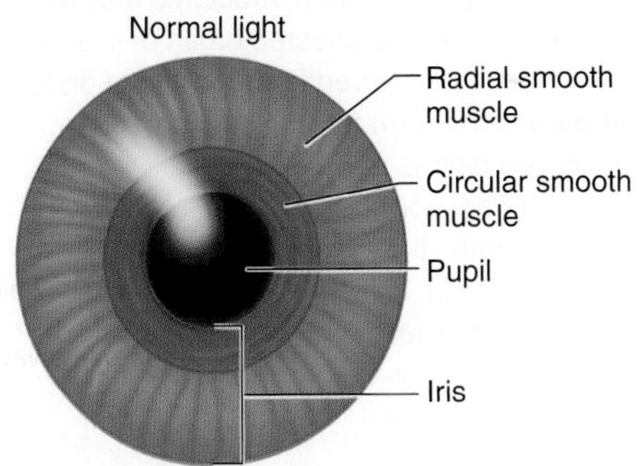

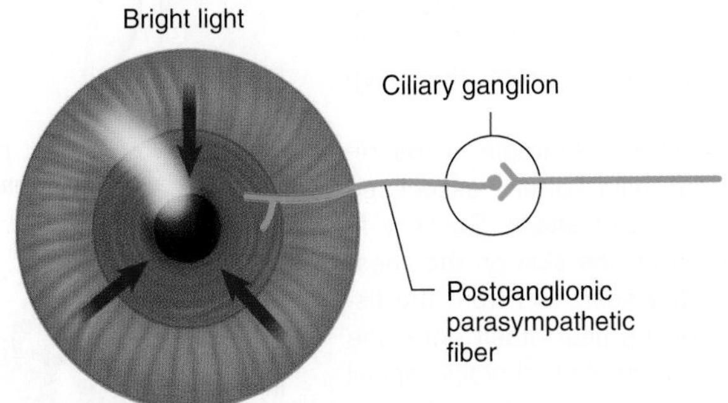

FIGURE 11-3 Control of pupil. This diagram of the muscular parts of the iris shows autonomic nerves stimulating radial muscles to dilate the pupil *(top)* and stimulating circular muscle to constrict the pupil *(bottom).*

only a slightly curved shape. To focus on near objects, the ciliary muscle must contract. As it contracts, it pulls the choroid coat forward toward the lens, thus causing the lens to bulge and curve even more.

Inner Layer

The **retina** makes up most of the *inner layer* of the eyeball. It contains microscopic photoreceptor cells to detect light (Figure 11-4). Most of these receptor cells are called *rods* and *cones* because of their shapes. Dim light of various wavelengths—or colors—can stimulate the rods, giving us monochrome (colorless) vision when lighting is low. However, fairly bright light is necessary to stimulate the cones. In other words, **rods** are the receptors for night vision and **cones** are the receptors for daytime vision.

There are three kinds of cones; each is sensitive to a different color: red, green, or blue. Scattered throughout the central portion of the retina, these three types of cones allow us to distinguish between different colors—but only in bright light.

There is a yellowish area near the center of the retina called the **macula lutea**—a term that means "yellow spot." It surrounds a small depression, called the **fovea centralis,** which contains the greatest concentration of cones of any area of the retina. These structures are identified in Figure 11-2 but can also be seen using a common medical device called an **ophthalmoscope,** shown in Figure 11-5.

In good light, greater visual acuity, or sharpness of visual perception, can be obtained if we look directly at an object and focus the image on the fovea. But in dim light or darkness, we see an object better if we look slightly to the side of it, thereby focusing the image nearer the periphery of the retina, where the rods are more plentiful.

Figure 11-4 also shows **ganglion cells,** which are also sensitive to light. Ganglion cells, like rods, are sensitive to various wavelengths (colors) of light, but they are not used to form visual images. Instead, information from ganglion cells helps the body determine whether it is day or night, as well as the level of moonlight (monthly phases). This helps our body's *internal clock* mechanisms synchronize themselves to the daily, monthly, and seasonal rhythms of our external environment.

Fluids of the Eyeball

Fluids fill the hollow spaces inside the eyeball. They maintain the normal shape of the eyeball and help refract light rays; that is, the fluids bend light rays to bring them to focus on the retina.

Aqueous humor is the name of the watery fluid in front of the lens (in the anterior chamber of the eye), and **vitreous**

CLINICAL APPLICATION

REFERRED PAIN

The stimulation of pain receptors in deep structures may be felt as pain in the skin that lies over the affected organ or in an area of skin on the body surface far removed from the site of disease or injury. **Referred pain** is the term for this phenomenon.

The cause of referred pain is related to a convergence of sensory nerve impulses from both the diseased organ and the skin in the area of referred pain. For example, pain originating in an organ deep in the abdominal cavity is often interpreted as coming from an area of skin whose sensory fibers enter the same segment of the spinal cord as the sensory fibers from the deep structure.

A classic example is the referred pain often associated with a heart attack. Sensory fibers from the skin on the chest over the heart and from the tissue of the heart itself enter the first to the fifth thoracic spinal cord segments and so do sensory fibers from the skin areas over the left shoulder and inner surface of the left arm. Part A of the figure shows the primary sensory fibers from both the skin and heart converging in the spinal cord. Sensory impulses from both these areas travel to the brain over a common pathway—the secondary sensory fiber. Thus the brain may locate the pain of a heart attack in the shoulder or arm (part B of the figure).

Misinterpretation in the brain in regard to the true location of sensory neurons being stimulated causes referred pain. In clinical medicine, an understanding of referred pain can be an im-

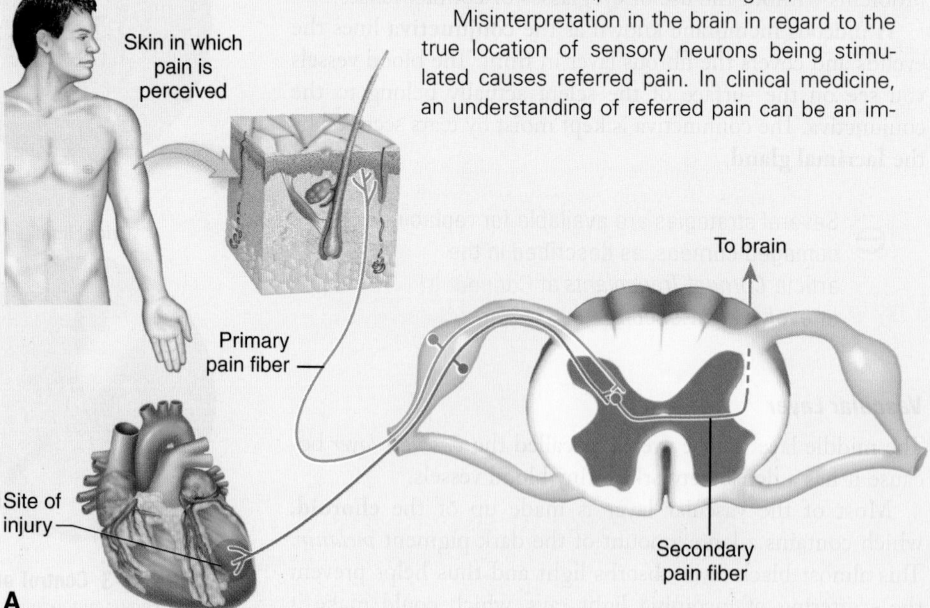

humor is the name of the jellylike fluid behind the lens (in the posterior chamber). Aqueous humor is constantly being formed, drained, and replaced in the anterior chamber. If drainage is blocked for any reason, the internal pressure within the eye will increase, and damage that could lead to blindness will occur. This condition is called *glaucoma,* which we discuss later in this chapter.

QUICK CHECK

1. Identify the three layers of tissues that form the eyeball.
2. How does the iris regulate the size of the pupil?
3. What is the "window of the eye"?
4. What is the function of *melanin* in the eye? Where is it located?
5. What are the *humors* of the eye? Describe the function of the *humors.*
6. How are rods and cones used in vision?

To learn more about the structures of the eye, go to AnimationDirect online at *evolve.elsevier.com.*

Visual Pathway

Light is the stimulus that results in vision (that is, our ability to see objects as they exist in our environment). Besides detecting intensity (brightness) and wavelength (color) of light, we can also perceive images and their movements.

Light enters the eye through the pupil and is *refracted,* or bent, so that it is focused on the retina. **Refraction** occurs as light passes through the cornea, the aqueous humor, the lens, and the vitreous humor on its way to the retina.

The innermost layer of the retina contains the rods and cones, which are the **photoreceptor cells** of the eye (see Figure 11-4). They respond to a light stimulus by producing a nervous impulse. The rod and cone photoreceptor cells synapse with neurons in the bipolar and ganglionic layers of the retina.

Nervous signals eventually leave the retina and exit the eye through the optic nerve on the posterior surface of the eyeball. No rods or cones are present in the area of the retina where the optic nerve fibers exit. The result is a "blind spot" known as the **optic disk** (see Figure 11-2).

After leaving the eye, the optic nerves enter the brain and travel to the visual cortex of the occipital lobe (Figure 11-6). Eventually, *visual interpretation* of the nervous impulses generated by light striking the retina results in "seeing."

portant determinant in whether the correct diagnosis of disease is made (see figure).

These concepts also relate to those explored in the article *Pain Control Areas* at Connect It! at *evolve.elsevier.com.*

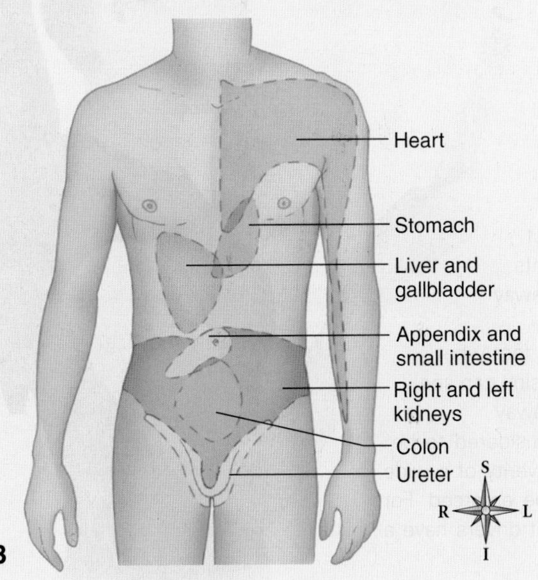

CLINICAL APPLICATION

FINDING YOUR BLIND SPOT

You can demonstrate the location of the blind spot in your visual field by covering your left eye and looking at the objects below. Begin by positioning your face about 35 cm (12 in) away from this page. Cover your left eye and stare continuously at the square with your right eye while slowly bringing your face closer and closer to the image. At one point, the circle will seem to disappear because its image has fallen on the blind spot.

Disorders of Vision

Healthy vision requires three basic processes: formation of an image on the retina (refraction), stimulation of rods and cones, and conduction of nerve impulses to the brain. Malfunction of any of these processes can disrupt this chain of processes, producing a visual disorder.

Refraction Disorders

Common Focusing Problems

Focusing a clear image on the retina is essential for good vision. In the normal eye, light rays enter the eye and are focused into a clear, upside-down image on the retina (Figure 11-7, *A*). The brain can easily right the upside-down image in our

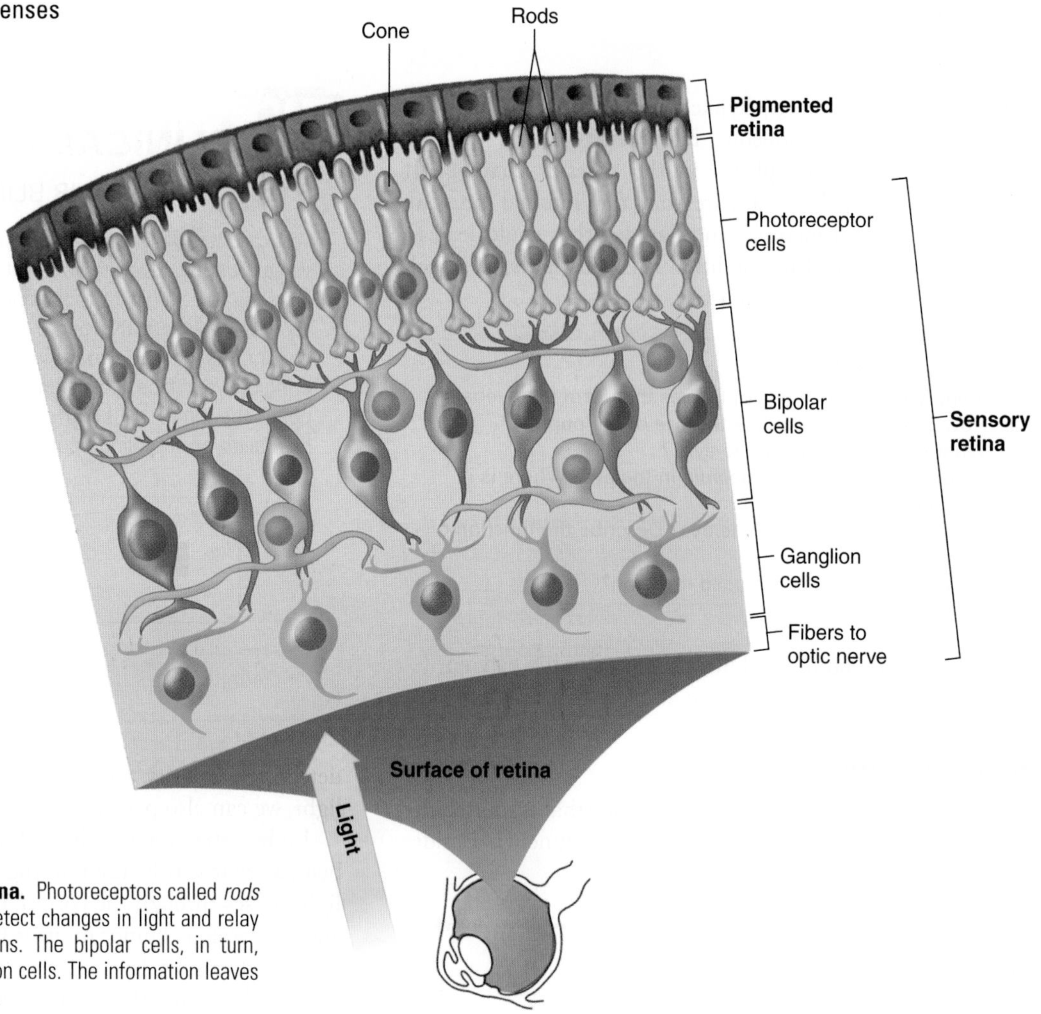

Cone
Rods
Pigmented retina
Photoreceptor cells
Bipolar cells
Ganglion cells
Fibers to optic nerve
Sensory retina
Surface of retina
Light

FIGURE 11-4 Cells of the retina. Photoreceptors called *rods* and *cones* (notice their shapes) detect changes in light and relay the information to bipolar neurons. The bipolar cells, in turn, conduct the information to ganglion cells. The information leaves the eye by way of the optic nerve.

CLINICAL APPLICATION

VISUAL ACUITY

Visual acuity is the clearness or sharpness of visual perception. Acuity is affected by our focusing ability, the efficiency of the retina, and the proper function of the visual pathway and processing centers in the brain.

One common way to measure visual acuity is to use the familiar test chart on which letters or other objects of various sizes and shapes are printed. The subject is asked to identify the smallest object that he or she can see from a distance of 20 feet (6.1 m). The resulting determination of visual acuity is expressed as a double number such as "20-20." The first number represents the distance (in feet) between the subject and the test chart—the standard being 20. The second number represents the number of feet a person with normal acuity would have to stand away from the chart to see the same objects clearly.

Thus a finding of 20-20 is normal because the subject can see at 20 feet what a person with normal acuity can see at 20 feet. A person with 20-100 vision can see objects at 20 feet that a person with normal vision can see at 100 feet away.

People whose acuity is worse than 20-200 after correction are considered to be *legally blind.* Legal blindness is the designation used to identify the severity of a wide variety of visual disorders so that laws that involve visual acuity can be enforced. For example, laws that govern the awarding of driving licenses require that drivers have a minimum level of visual acuity.

Smaller charts, such as the one shown in the figure, can be used to test near vision acuity.

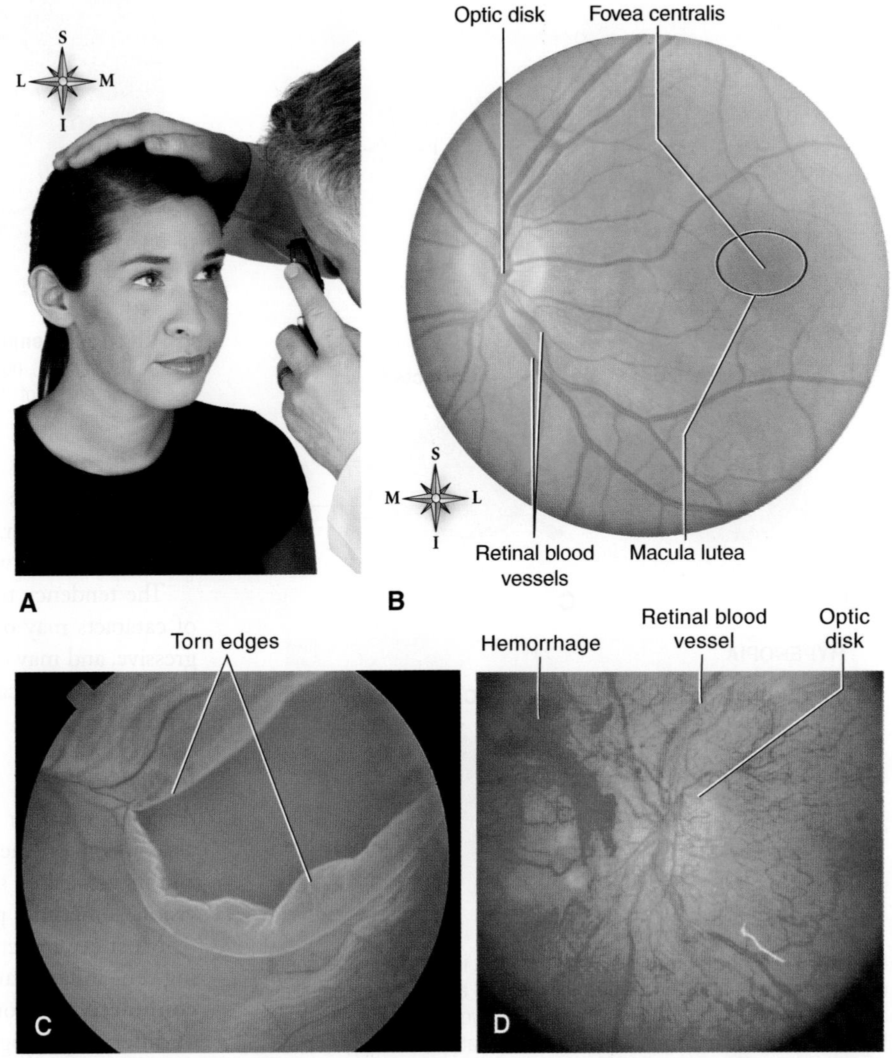

FIGURE 11-5 Examining the eye. A, Using an ophthalmoscope to view the retina. **B,** Ophthalmoscopic view of the retina, as seen through the pupil. **C,** A case of retinal tear and detachment. **D,** Diabetes can produce abnormal blood vessels and bleeding of the retina.

conscious perception but cannot correct an image that is not sharply focused.

If our eyeballs are elongated, the image focuses somewhere in front of the retina rather than directly on it. The retina receives only a fuzzy image. This condition, called **myopia** or *nearsightedness,* can be corrected by refractive eye surgery or by using contact lenses or glasses (**Figure 11-7**, *B* and *C*).

If our eyeballs are shorter (as measured from anterior to posterior) than normal, the image focuses behind the retina, also producing a fuzzy image. This condition, called **hyperopia** or *farsightedness,* also can be corrected by eye surgery or lenses (**Figure 11-7**, *D* and *E*).

An irregularity (unequal curvature) in the cornea or lens, a condition called **astigmatism** that distorts vision, also can be corrected with glasses or contact lenses.

For brief descriptions of some surgical options to treat refraction issues, such as **radial keratotomy (RK)** and **laser-assisted in situ keratomileusis (LASIK),** see the article *Refractive Eye Surgery* at Connect It! at *evolve.elsevier.com.*

In most young people, the lens is both transparent and somewhat elastic, making it capable of easily changing shape. As we grow older, however, most of us become more farsighted as we lose the ability to focus on close objects because our lenses lose their elasticity and can no longer bulge enough to bring near objects into focus. **Presbyopia,** literally "oldsightedness," is the name for this condition. Older individuals

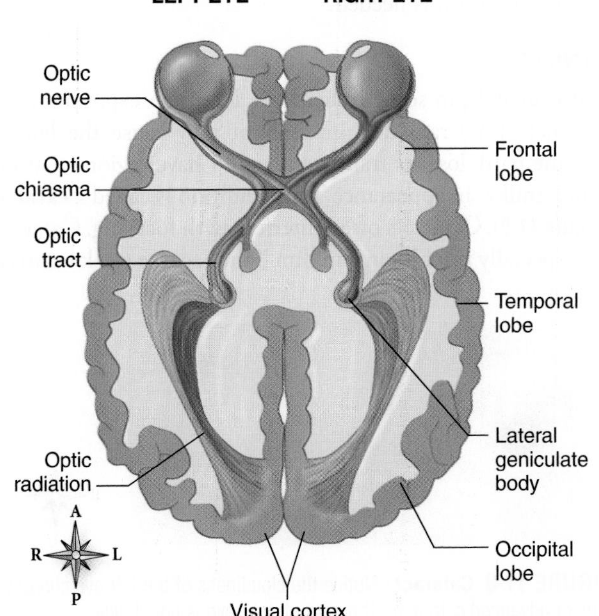

FIGURE 11-6 The visual pathway. Transverse section of the brain and eyes (inferior view). Note that the pathway from the retina of the left eye is color-coded *blue* and the pathway from the right eye is color-coded *green.* Identify the point at which half the information from each eye crosses over to the other side of the brain.

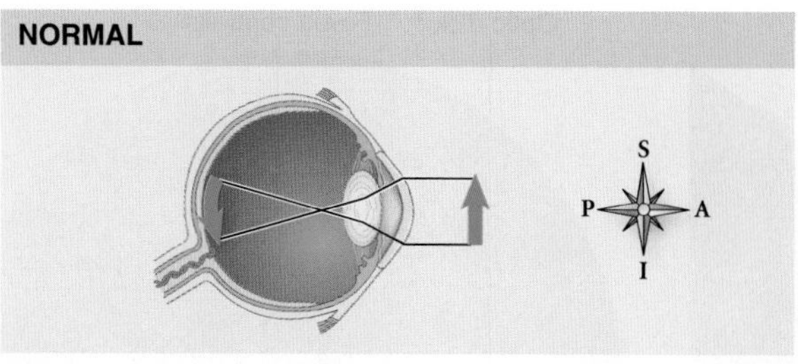

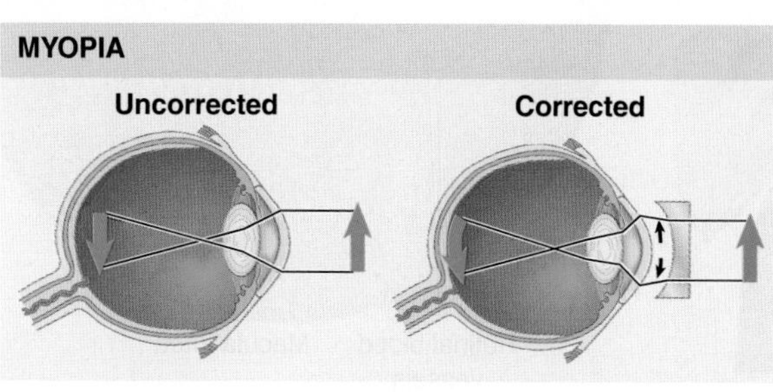

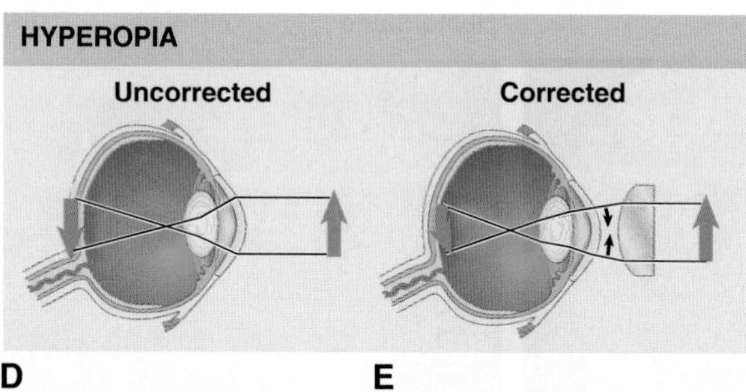

FIGURE 11-7 Refraction. A, Shows how light is refracted in the normal eye to form a well-focused image on the retina. **B** and **C,** The abnormal and corrected refraction observed in patients with myopia, or nearsightedness. **D** and **E,** Abnormal and corrected refraction in patients with hyperopia, or farsightedness.

can compensate for presbyopia by using "reading glasses" when near vision is needed.

Cataracts

Unfortunately, in some individuals, longtime exposure to ultraviolet (UV) radiation in sunlight may cause the lens to become hard, lose its transparency, and have regions that become "milky" in appearance. This condition is called a **cataract** (Figure 11-8). Cataracts often interfere with focusing. Cataracts are especially troublesome in dim light because weak beams of

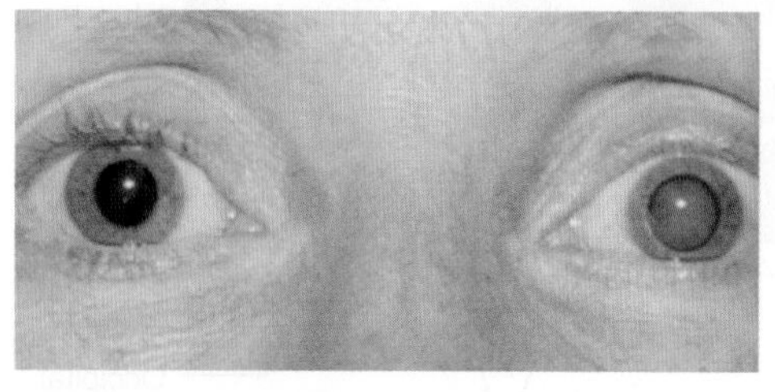

FIGURE 11-8 Cataract. Notice the cloudiness of the left eye characteristic of advanced cataracts. The normal right eye is not cloudy.

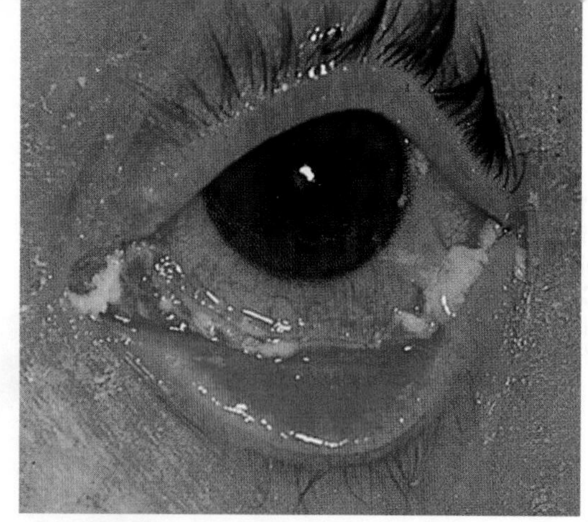

FIGURE 11-9 Conjunctivitis. In this case of acute bacterial infection, notice the discharge of mucous pus characteristic of this highly contagious infection of the conjunctiva.

light cannot pass through the cloudy spots the way some brighter light can. This fact accounts for the trouble many older adults have with their "night vision."

The tendency to develop cataracts is inherited. Formation of cataracts may occur in one or both eyes, tends to be progressive, and may eventually result in blindness. Cataracts can be removed surgically and the defective lens replaced with an artificial implant.

Conjunctivitis

A variety of other conditions can prevent the formation of a clear image on the retina. For example, infections of the eye and its associated structures also have the potential to impair vision, sometimes permanently.

Most eye infections start out in the conjunctiva, producing an inflammation response known as "pink eye" or **conjunctivitis.** You may recall from Chapter 6 that a variety of different pathogens can cause conjunctivitis. For example, the bacterium *Chlamydia trachomatis* that commonly infects the reproductive tract can cause a chronic infection called *chlamydial conjunctivitis* or **trachoma.** Because chlamydia and other pathogens often inhabit the birth canal, antibiotics are routinely applied to the eyes of newborns to prevent conjunctivitis.

Highly contagious *acute bacterial conjunctivitis,* characterized by drainage of a mucous pus (Figure 11-9), is most commonly caused by bacteria such as *Staphylococcus* and *Haemophilus.*

Conjunctivitis may produce lesions on the inside of the eyelid that can damage the cornea and thus impair vision. Occasionally infections of the conjunctiva spread to the tissues of the eye proper and cause permanent injury—even total blindness.

In addition to infection, conjunctivitis also may be caused by allergies. The red, itchy, watery eyes commonly associated with allergic reactions to pollen and other substances result from an allergic inflammation response of the conjunctiva. Allergy and hypersensitivity reactions are discussed further in Chapter 16. Trauma to the eye may cause bleeding below the conjunctiva resulting in a *subconjunctival hemorrhage.*

Strabismus

Yet another manner in which refraction can be disrupted occurs when one eye does not focus on the same object as the

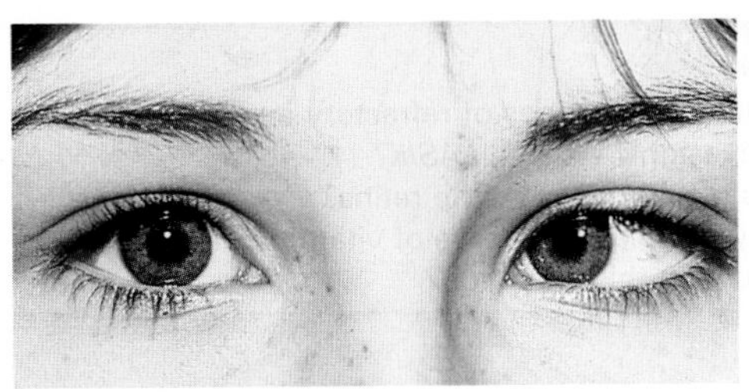

FIGURE 11-10 Strabismus. This child exhibits convergent left eye strabismus.

other eye. Normally, we use *binocular* ("two-eyed") vision in which both eyes aim toward the same object at the same time. Because the eyes are separated by a short distance, the images formed in each eye do not match exactly—a fact that allows *depth perception.*

If the positioning of the eyes cannot be coordinated, a condition called **strabismus** or "squint" results. In some cases, the eyes may diverge outward to the side, a condition called *divergent squint.* If one or both eyes converge toward the nose, the condition is called *convergent squint* or "cross-eye." The photo in Figure 11-10 shows a child with convergent left eye strabismus.

Usually the brain compensates for missing or unusual elements in the visual field. However, in severe cases of strabismus, the eyes may be so far off in their center of focus that the brain cannot mesh the two resulting images into a single picture.

Strabismus is usually caused by paralysis, weakness, or another abnormality affecting the external muscles of the eye. Whatever the cause, strabismus often can be corrected by treatment in early childhood that forces the eyes to focus together, by means of therapeutic training, corrective lenses, or by corrective surgery. If left untreated, by about 6 years of age the visual centers in the brain will learn to ignore information from one eye—causing a decrease in visual acuity and permanent blindness in the affected eye.

QUICK CHECK

1. How do *myopia* and *hyperopia* differ in terms of focusing an image on the retina?
2. What is *strabismus*? How is it treated?
3. What is *pink eye*?
4. What part of the eye is affected when an individual has *cataracts*?

Disorders of the Retina

Damage to the retina impairs vision because even a well-focused image cannot be perceived if some or all of the light receptors do not function properly.

Retinal Detachment

For example, in a condition called **retinal detachment,** part of the retina falls away from the tissue supporting it (see Figure 11-5, *C*). This condition often results from normal aging, eye tumors, or from sudden blows to the head—as in a sporting injury. Common warning signs include the sudden appearance of floating spots that may decrease over a period of weeks and odd "flashes of light" that appear when the eye moves. If left untreated, the retina may detach completely and cause total blindness in the affected eye.

A number of treatments are available for correcting retinal detachment. A traditional approach is the use of laser therapy. Another approach involves placing a tight collar around the eyeball to increase pressure within the eye. The high pressure of the jellylike vitreous humor holds the retina in place against the rear of the eyeball.

Diabetic Retinopathy

Diabetes mellitus, a disorder involving the hormone insulin, may cause a condition known as **diabetic retinopathy.** In this disorder, diabetes causes small hemorrhages in retinal blood vessels that disrupt the oxygen supply to the photoreceptors (see Figure 11-5, *D*). The eye responds by building new but abnormal vessels that block vision and may cause detachment of the retina.

Diabetic retinopathy is considered one of the leading causes of blindness in the United States. Fortunately, treatments developed over the last two decades have improved the outlook in this regard. For example, a type of laser therapy in which laser beams are used to seal off hemorrhaging retinal vessels has been used successfully in many cases.

Glaucoma

Another common condition that can damage the retina is **glaucoma.** Recall that glaucoma is excessive **intraocular pressure** caused by abnormal accumulation of aqueous humor. As fluid pressure against the retina increases above normal, blood flow through the retina slows. Reduced blood flow causes degeneration of the retina and thus a loss of vision.

Although acute forms of glaucoma can occur, most cases develop slowly over a period of years. This chronic form may not produce symptoms, especially in its early stages. For this reason, routine eye examinations typically include a screening test for glaucoma.

As chronic glaucoma progresses, damage first appears at the edges of the retina, causing a gradual loss of peripheral vision resulting in a condition known commonly as "tunnel vision." Blurred vision and headaches also may occur. As the damage becomes more extensive, "halos" are seen around bright lights. If untreated, glaucoma eventually produces total, permanent blindness.

Retinal Degeneration

Degeneration of the retina can cause difficulty seeing at night or in dim light. This condition is called **nyctalopia** or "night blindness."

Nyctalopia also can be caused by a deficiency of vitamin A. Vitamin A is needed to make *photopigment* in rod cells. **Photopigment** is a light-sensitive chemical that triggers stimulation of the visual nerve pathway. A lack of vitamin A may result in a lack of photopigment in rods, a condition that impairs dim-light vision.

11

The leading cause of permanent blindness in the elderly is progressive degeneration of the central part of the retina. Called **age-related macular degeneration (AMD),** this condition affects the part of the retina that is most essential to good vision—the central region called the *macula.* The exact cause of the degeneration is unknown, but the risk for developing this condition increases with age after reaching 50. Other known risk factors include cigarette smoking and a family history of the disorder.

Learn more about "wet" and "dry" forms of AMD and how lasers or injections may be used to treat them in the article *Therapy for Macular Degeneration* at Connect It! at *evolve.elsevier.com.*

Color Blindness

Retinal disorders are sometimes inherited. Most forms of **color blindness** are caused by genes on the X chromosome that produce abnormal photopigments in the cones. Each of three photopigments in cones is sensitive to one of the primary colors of light: green, blue, and red. In many color-blind individuals, the green-sensitive photopigment is missing or deficient; at other times, the red-sensitive photopigment is abnormal. Deficiency of the blue-sensitive photopigment is very rare.

Color-blind individuals see colors, but they cannot distinguish among all of them normally. Because color blindness is an X-linked genetic trait, more men than women have this condition (see Chapter 25). Although color blindness is an abnormality, it is not usually considered a clinical disease.

Colored figures are often used to screen individuals for color blindness (**Figure 11-11**). A person with red-green color blindness cannot see the number *74* in **Figure 11-11,** *A,* whereas a person with normal color vision can.

To determine which photopigment is deficient, red or green, a color-blind person may be evaluated using a figure similar to **Figure 11-11**, *B.* Persons with a deficiency of red-sensitive photopigment can distinguish only the number *2;* those deficient in green-sensitive photopigment can see only the number *4.*

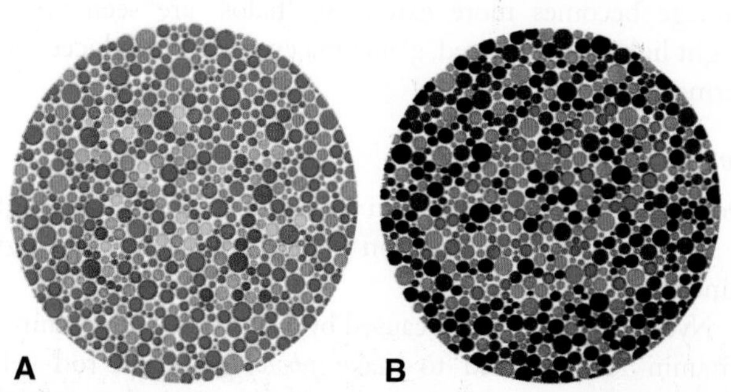

FIGURE 11-11 Color vision screening. A, People with normal color vision can see *74* in this mosaic; people with red-green color blindness cannot. **B,** This mosaic is used to classify the type of red-green color blindness a patient has. If the patient sees only the *2,* the red-sensitive cones are abnormal. If only the *4* is seen, the green-sensitive cones are abnormal.

QUICK CHECK

1. How do the various types of refractory eye surgery improve eyesight? What is *LASIK*?
2. How does glaucoma harm the retina?
3. What is *nyctalopia*? What type of vitamin deficiency is associated with this condition?

Disorders of the Visual Pathway

Damage or degeneration in the optic nerve, the brain, or any part of the visual pathway between them can impair vision (see **Figure 11-6**). For example, the pressure associated with glaucoma also can damage the optic nerve. Diabetes, already cited as a cause of retina damage, also can cause degeneration of the optic nerve.

Damage to the visual pathway does not always result in total loss of sight. Depending on where the damage occurs, only a part of the visual field may be affected. For example, a certain form of neuritis often associated with multiple sclerosis can cause loss of only the center of the visual field—a condition called **scotoma.**

A cerebrovascular accident (CVA), or stroke, can cause vision impairment when the resulting tissue damage occurs in one of the regions of the brain that process visual information (see **Figure 10-14**). For example, damage to an area that processes information about colors may result in a rare condition called *acquired cortical color blindness.* This condition is characterized by difficulty in distinguishing any color—not just one or two colors as in the more common inherited forms of retinal color blindness.

Hearing and Equilibrium

In addition to its role in hearing, the ear also functions as the sense organ of balance, or *equilibrium.* The stimulation, or "trigger," that activates receptors involved with hearing and equilibrium is mechanical, and the receptors themselves are called **mechanoreceptors.** In hearing, sound vibrations trigger nervous impulses that are eventually perceived in the brain as sound. In equilibrium, changes in position or movement of the body trigger impulses that lead to sensations of balance.

Ear

The ear is more than an appendage on the side of the head. A large part of the ear and its most important part lie hidden from view deep inside the temporal bone. The ear is divided into the following anatomical areas (**Figure 11-12**):

1. External ear
2. Middle ear
3. Inner (internal) ear

External Ear

The external ear has two parts: the **auricle,** or **pinna,** and the **external acoustic canal.** The auricle is the appendage on the side of the head surrounding the opening of the external acoustic canal. A number of the anatomical landmarks of the external ear are identified in **Figure 11-12.**

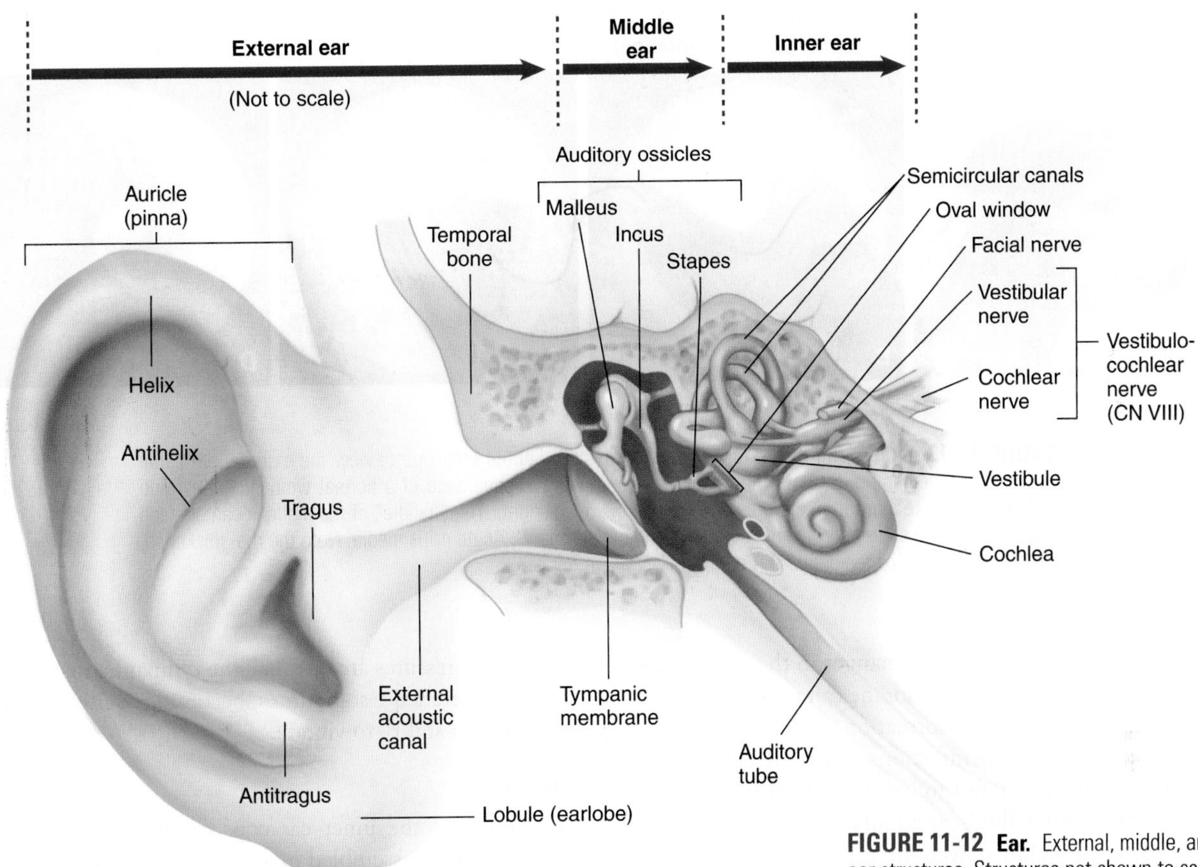

FIGURE 11-12 Ear. External, middle, and inner ear structures. Structures not shown to scale.

Because it lies exposed against the bony surface of the skull, the auricle is frequently injured by blunt trauma. Bruising may then cause an accumulation of blood and tissue fluid between the skin and underlying cartilage. If left untreated, the classic swelling of "cauliflower ear" may develop and become permanent. The skin behind the ear is rich in sebaceous glands. And, if they become infected, painful cysts develop that must be drained.

HEALTH AND WELL-BEING

SWIMMER'S EAR

External otitis, or *swimmer's ear,* is a common infection of the external ear in athletes. It can be bacterial or fungal in origin and is usually associated with prolonged exposure to water. The infection generally involves, at least to some extent, the acoustic canal and auricle. The ear as a whole is tender, red, and swollen. Treatment of swimmer's ear usually involves antibiotic therapy and prescription analgesics.

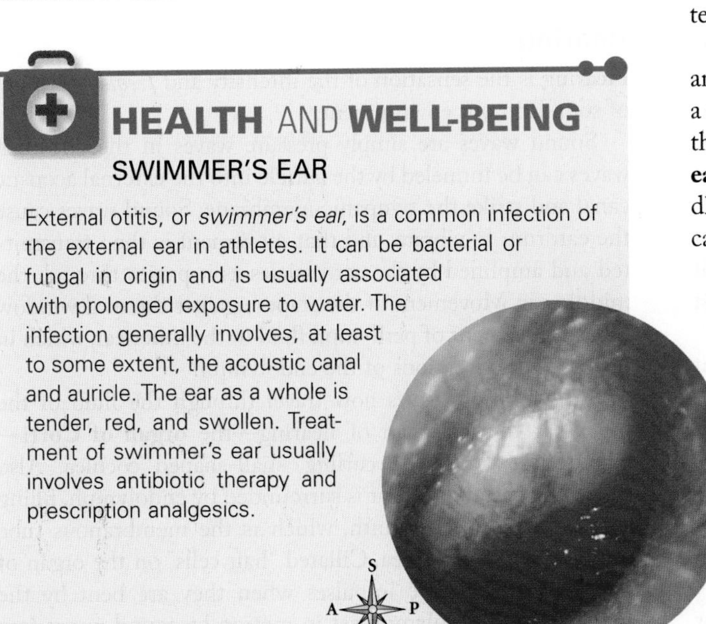

In people who suffer from gout, small nodules made up of urate crystals and called **tophi,** often appear on the upper edge of the **helix**. Another type of nodule that may appear on the helix is called a *Darwin tubercle.* It is considered a variation of normal and requires no treatment.

Although controversial, there is some evidence to suggest that the presence of oblique earlobe creases *(Frank's sign)* in individuals over the age of 50 may be related to coronary artery disease.

Note in **Figure 11-12** that the **tragus** of the auricle lies just anterior to the opening to the acoustic canal. The canal itself is a curving tube about 2.5 cm (1 inch) in length. It extends into the temporal bone and ends at the **tympanic membrane,** or **eardrum,** which is a partition between the external and middle ear. Sound waves traveling through the external acoustic canal strike the tympanic membrane and cause it to vibrate.

The skin of the acoustic canal, especially in its outer one third, contains many short hairs and **ceruminous glands** that produce a waxy substance called **cerumen.** Cerumen, or "earwax," helps keep the canal's skin from drying and flaking. It also traps dust that enters the canal and is then carried toward the exterior as the epithelium grows outward from the inner canal. However, cerumen may collect in excess and impair hearing by absorbing or blocking the passage of sound waves.

Just as an ophthalmoscope is used to view the interior of the eyeball and examine the retina, a lighted instrument called an **otoscope** is used to examine the external ear canal and outer surface of the tympanic

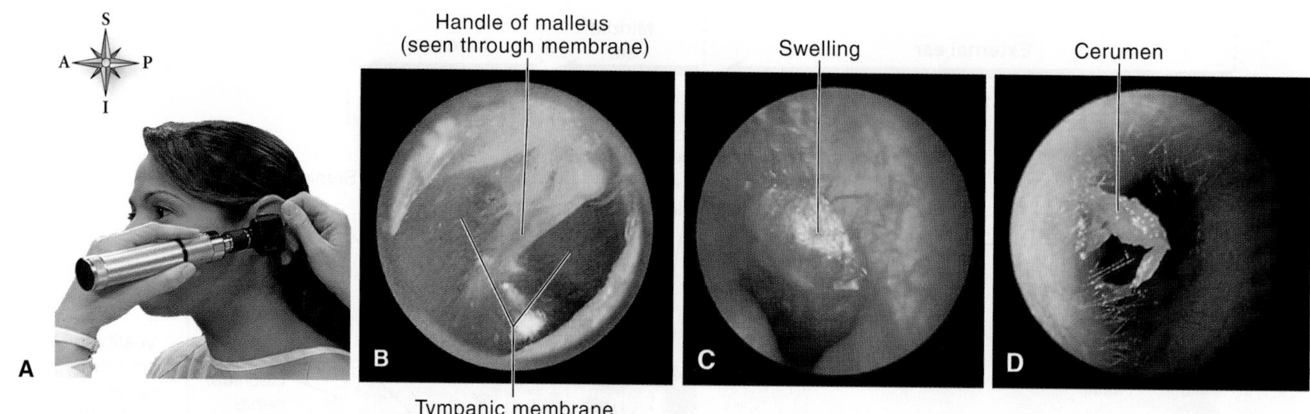

FIGURE 11-13 Examining the external ear. A, Using a lighted otoscope to view the external ear canal and tympanic membrane. **B,** Note the translucent, pearly-gray appearance of a normal tympanic membrane (with a bit of white glare from the otoscope light in the lower right). The "handle" of the malleus can be seen attaching near the center of the inner surface of the membrane. **C,** Acute otitis media. Note the red, thickened, and bulging tympanic membrane. **D,** Cerumen (earwax) in ear canal.

membrane (**Figure 11-13,** *A* and *B*). Changes in the appearance of the ear canal and tympanic membrane can provide a skilled observer with a great deal of information.

For example, middle ear infections cause the eardrum to become red and inflamed and to bulge outward into the ear canal as pus and other fluids accumulate in the middle ear (**Figure 11-13,** *C*). Foreign objects or excess cerumen in the ear canal, inflammation of the lining of the canal caused by prolonged exposure to moisture or bacterial infection (swimmer's ear), and perforation of the eardrum itself are also easily observed using an otoscope (**Figure 11-13,** *D*).

Middle Ear

The middle ear is a tiny and very thin epithelium-lined cavity hollowed out of the temporal bone (**Figure 11-12**). It is an air-filled space housing three very small bones. The names of these ear bones, called **ossicles,** describe their shapes—**malleus** (hammer), **incus** (anvil), and **stapes** (stirrup).

The "handle" of the malleus attaches to the inside of the tympanic membrane, and the "head" attaches to the incus (see **Figure 11-13,** *B*). The incus attaches to the stapes, and the stapes presses against a membrane that covers a small opening, the *oval window.* The oval window separates the middle ear from the inner ear.

When sound waves cause the eardrum to vibrate, that movement is transmitted and amplified by the ear ossicles as it passes through the middle ear. Movement of the stapes against the oval window causes movement of fluid in the inner ear.

A point worth mentioning, because it explains the frequent spread of infection from the throat to the ear, is that a tube—the **auditory tube,** or **eustachian tube**—connects the throat with the middle ear. The epithelial lining of the middle ears, auditory tubes, and throat are extensions of one continuous membrane. Consequently, infection causing a sore throat may spread to produce a middle ear infection called **otitis media** (see **Figure 11-13,** *C*).

The function of a healthy auditory tube is to equalize air pressure between the middle ear and the outside environment. When air pressures are unequal, the tympanic membrane may remain stretched—sometimes becoming quite painful and reducing its ability to vibrate.

Inner Ear

Anatomically, the inner ear consists of three spaces in the temporal bone, assembled in a complex maze called the **bony labyrinth.** This odd-shaped bony space is filled with a watery fluid called **perilymph** and is divided into the following parts: **vestibule, semicircular canals,** and **cochlea.** The vestibule is adjacent to the oval window between the semicircular canals and the cochlea (**Figure 11-14**).

Note in **Figure 11-14** that a balloonlike membranous sac is suspended in the perilymph and follows the shape of the bony labyrinth much like a "tube within a tube." This is the **membranous labyrinth,** and it is filled with a thicker fluid called **endolymph.**

Hearing

Hearing is the sensation of the intensity and *frequency* (tone) of sounds in our environment.

Sound waves are simply pressure waves in the air. Such waves can be funneled by the auricle into the external acoustic canal and strike the tympanic membrane. Sound waves cause the eardrum to vibrate, and that movement is then transmitted and amplified by the ear ossicles as it passes through the middle ear. Movement of the stapes against the oval window causes movement of perilymph fluid in the inner ear, which in turn triggers vibrations of the endolymph.

The vibration waves now travel through the fluid of the inner ear to the organ of hearing—the **organ of Corti**—which lies within the curling, snail-shaped cochlea. Also called the **spiral organ,** it is surrounded by endolymph, filling the membranous labyrinth, which is the membranous tube within the bony cochlea. Ciliated "hair cells" on the organ of Corti generate nerve impulses when they are bent by the movement of endolymph set in motion by sound waves (see **Figure 11-14** and **Figure 11-15**).

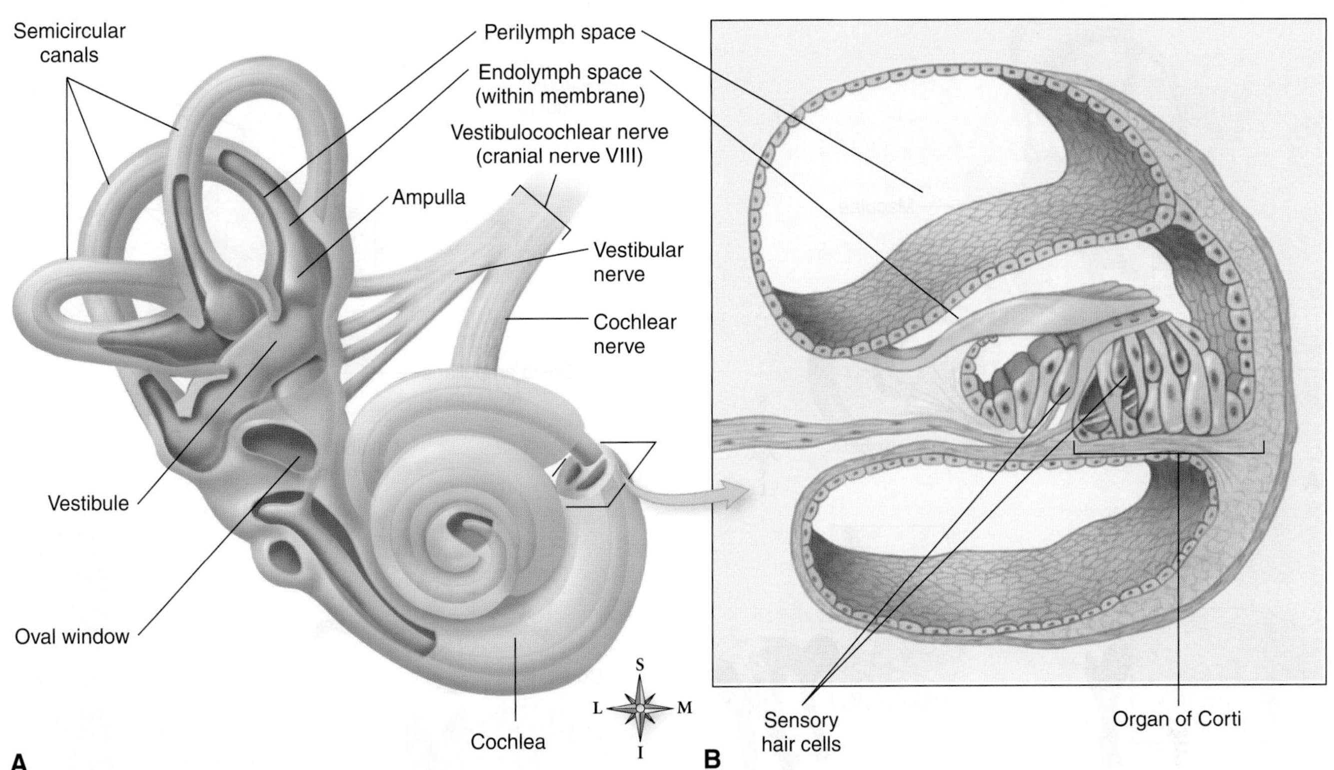

FIGURE 11-14 Inner ear. A, The bony labyrinth is the hard outer wall of the entire inner ear and includes semicircular canals, vestibule, and cochlea. Within the bony labyrinth is the membranous labyrinth *(purple),* which is surrounded by perilymph and filled with endolymph. Each ampulla in the vestibule contains a crista ampullaris that detects changes in head position and sends sensory impulses through the vestibular nerve to the brain. **B,** The *inset* shows a section of the membranous cochlea. Hair cells in the organ of Corti detect sound and send the information through the cochlear nerve. The vestibular and cochlear nerves join to form the vestibulocochlear nerve, or cranial nerve VIII.

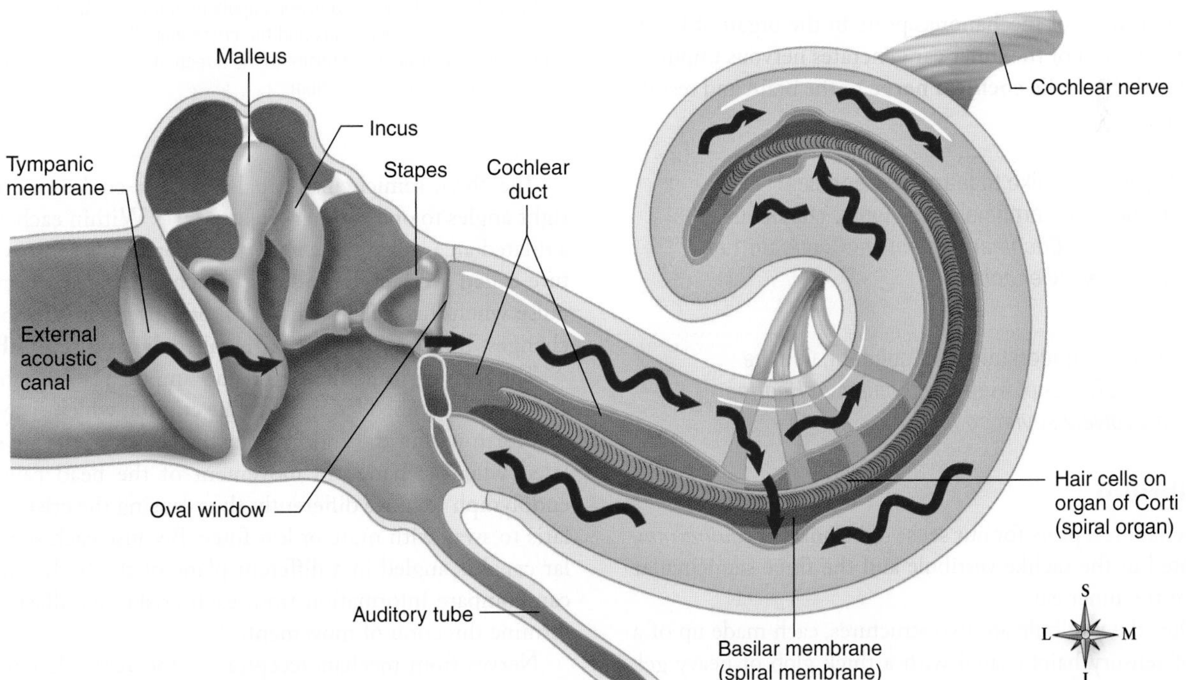

FIGURE 11-15 Effect of sound waves in the ear. Sound waves strike the tympanic membrane and cause it to vibrate. This vibration causes the membrane of the oval window to vibrate. Vibration of the oval window causes the perilymph in the bony labyrinth of the cochlea to move, which causes the endolymph in the membranous labyrinth of the cochlea or **cochlear duct** to move. This movement of endolymph stimulates hair cells on the organ of Corti (spiral organ) to generate a nerve impulse. The nerve impulse travels over the cochlear nerve, which becomes a part of cranial nerve VIII. Eventually, nerve impulses reach the auditory cortex and are interpreted as sound.

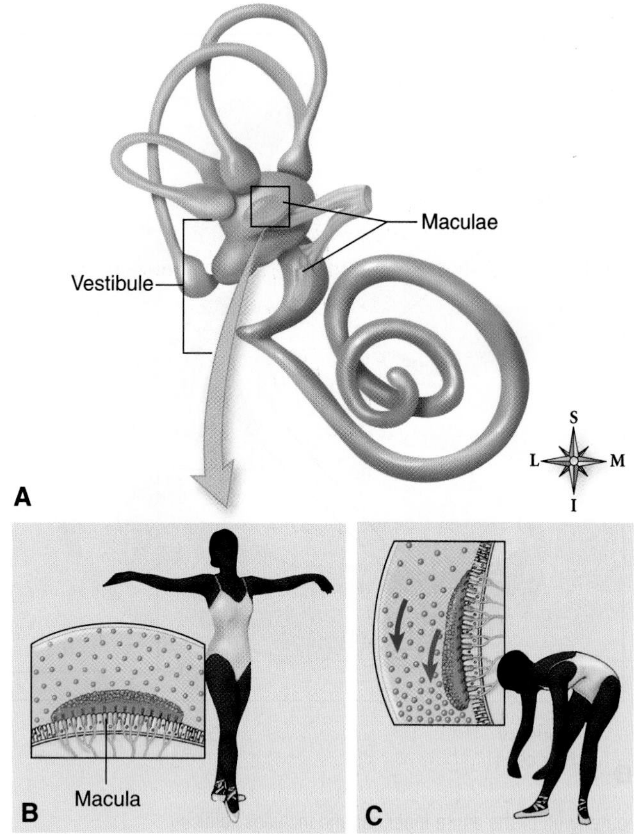

FIGURE 11-16 Static equilibrium. A, Structure of vestibule showing placement of the maculae, which have mechanoreceptors that detect our "sense of gravity" or static equilibrium. **B,** Macula stationary in upright position. **C,** Macula displaced by gravity as person bends over.

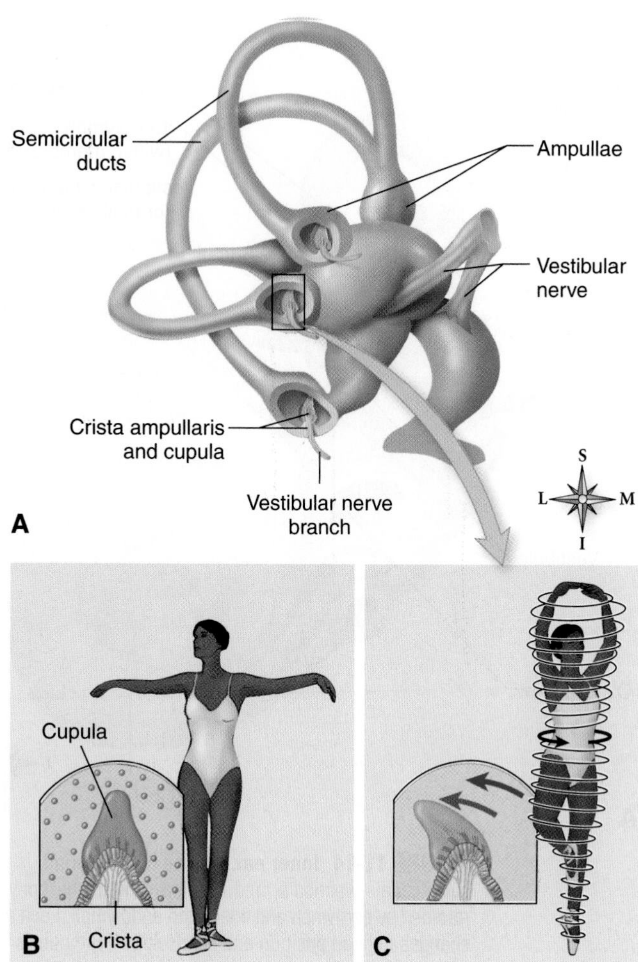

FIGURE 11-17 Dynamic equilibrium. A, Semicircular ducts showing location of the crista ampullaris and cupula in ampullae. **B,** When a person is at rest, the crista ampullaris and cupula do not move. **C,** As a person begins to spin, the cupula bends and the crista ampullaris is displaced by the endolymph in a direction opposite to the direction of spin. This produces the sensation of dynamic equilibrium.

This activation of mechanoreceptors in the organ of Corti inside the cochlea of the inner ear generates nervous impulses that travel through the **cochlear nerve** to the brain and results in hearing.

If you would like to learn how cochlear implants or "artificial ears" work, check out the article *Cochlear Implants* at Connect It! at *evolve.elsevier.com.*

To better understand this concept, use the Active Concept Map *Mechanism of Hearing* at *evolve.elsevier.com.*

Equilibrium

The mechanoreceptors for our sense of balance, or *equilibrium,* are located in the saclike vestibule and the three semicircular canals of the inner ear.

Within the vestibule are two structures, each made up of a patch of sensory hairs coated with a thick glob of heavy gel. Each of these structures is called a **macula.** When you bend your head, gravity acts on the heavy gel to pull it one way or the other (Figure 11-16). This, in turn, bends the cilia of the hair cells and thus produces a nerve signal. This signal is interpreted by the brain as our "sense of gravity" or **static equilibrium.**

The three semicircular canals are half-circles oriented at right angles to one another (Figure 11-17). Within each canal is a dilated area called the *ampulla* that contains a sensory structure called a **crista ampullaris,** which generates a nerve impulse when the speed or direction of movement of your head changes. This "sense of motion" is called **dynamic equilibrium.**

The sensory cells in the cristae ampullares have hairlike cilia that are embedded in a flaplike **cupula,** which sways back and forth within the endolymph. The sensory cells are stimulated when a change in movement of the head causes the endolymph to move differently, thus causing the crista ampullaris to sway with more or less force. Because each semicircular canal is angled in a different plane of the body, the brain can compare information from each crista ampullaris to determine direction of movement.

Nerves from mechanoreceptors in the vestibule join those from the semicircular canals to form the **vestibular nerve.** The vestibular nerve then joins with the cochlear nerve to form the *vestibulocochlear nerve (CN VIII)* (see Figure 11-12). Eventually, nervous impulses passing through this nerve reach the cerebellum and medulla—ultimately reaching the cerebral cortex.

QUICK CHECK

1. What structure in the ear vibrates when sound waves travel through the acoustic canal?
2. Why does an infection of the throat often spread and cause a middle ear infection called *otitis media*?
3. What fluid fills the membranous labyrinth? the organ of Corti?
4. Where are the mechanoreceptors for balance and equilibrium located?

To learn more about the process of hearing, go to AnimationDirect online at *evolve.elsevier.com*.

Hearing and Equilibrium Disorders

Hearing Disorders

Hearing problems can be divided into two basic categories: *conduction impairment* and *nerve impairment.* Conduction impairment refers to the blocking of sound waves as they travel through the external and middle ear to the sensory receptors of the inner ear (the conduction pathway). Nerve impairment results in insensitivity to sound because of inherited or acquired nerve damage.

The most obvious cause of conduction impairment is blockage of the external auditory canal. Waxy buildup of cerumen commonly blocks conduction of sound toward the tympanic membrane (see **Figure 11-13,** *D*). Foreign objects, tumors, and other matter can block conduction in the external or middle ear.

An inherited bone disorder called **otosclerosis** impairs conduction by causing structural irregularities in the stapes. Otosclerosis usually first appears during childhood or early adulthood as **tinnitus,** or "ringing in the ear."

Temporary conduction impairment often results from ear infection, or **otitis.** As stated earlier, the structure of the auditory tube makes the middle ear prone to bacterial or viral *otitis media.* Otitis media often produces swelling and pus formation that blocks the conduction of sound through the middle ear. Permanent damage to structures of the middle ear occasionally occurs in severe cases. Infectious organisms from a middle ear infection that invade the mastoid bone are often difficult to treat (see discussion in Chapter 8 on p. 185). They cause redness, inflammation, and swelling of the mastoid process that may push the auricle away from the skull. Hearing loss may be a complication in severe cases.

Hearing loss caused by nerve impairment is common in the elderly. Called **presbycusis,** this progressive hearing loss associated with aging results from degeneration of nerve tissue in the ear and the vestibulocochlear nerve.

A similar type of hearing loss occurs after chronic exposure to loud noises that damages receptors in the organ of Corti. Because different sound *frequencies* (tones) stimulate different regions of the organ of Corti, hearing impairment is limited to only frequencies associated with the damaged portion of the organ of Corti. For example, the portion of the organ of Corti that degenerates first in presbycusis is normally stimulated by high-frequency sounds. Thus the inability to hear high-pitched sounds is common among the elderly.

Whether a person is young or old, research shows that protecting oneself from loud noises and constant noises can reduce hearing loss over time.

Equilibrium Disorders

Equilibrium disorders are often characterized by **vertigo** (sensation of spinning), disorientation, falling (or feeling of falling), dizziness, and/or lightheadedness. Some anxiety disorders can also cause these symptoms, but are not true equilibrium disorders.

Some equilibrium disorders are caused by infection or inflammation of the inner ear, others by head injuries, nerve damage, or unknown causes. Temporary equilibrium impairment sometimes occurs when the brain receives conflicting sensory information about body movement from multiple senses (vision, balance, proprioception, etc.)—as in *motion sickness.*

Ménière disease is a chronic inner ear disease of unknown cause. Ménière disease is characterized by tinnitus, progressive nerve deafness, and vertigo.

QUICK CHECK

1. What is *tinnitus*?
2. What is the progressive hearing loss caused by nerve impairment that is common in the elderly?
3. What are the two basic categories of hearing problems?

Taste

Our sense of taste—or **gustation**—allows us to chemically analyze food before we bite or swallow it.

The **taste buds** are the sense organs of taste. They contain both supporting cells and chemoreceptors called **gustatory cells.** These cells generate the nervous impulses ultimately interpreted by the brain as taste (**Figure 11-18**).

Nervous impulses that are generated by stimulation of taste buds travel primarily through two cranial nerves (CN VII and CN IX) to end in the taste area of the cerebral cortex.

Although a few taste buds are located in the lining of the mouth and on the soft palate, most are located on the sides of much larger and differing shaped bumps scattered across the tongue called **papillae.** About 10 to 15 large **circumvallate papillae** form an inverted "V" pattern at the back of the tongue and contain the most taste buds.

Each taste bud, as you can see in **Figure 11-18,** *C,* opens through an opening into a trenchlike moat that surrounds the papilla and is filled with saliva. Chemicals dissolved in the saliva stimulate the chemoreceptor gustatory cells. All tastes can be detected in all areas of the tongue that contain taste buds.

Physiologists originally counted only four "primary" taste sensations—*sweet, sour, bitter,* and *salty*—that permit us to detect sugars, acids, alkalines, and sodium ions dissolved in our saliva. However, the list of "primary" taste sensations has expanded to include several others present in most individuals.

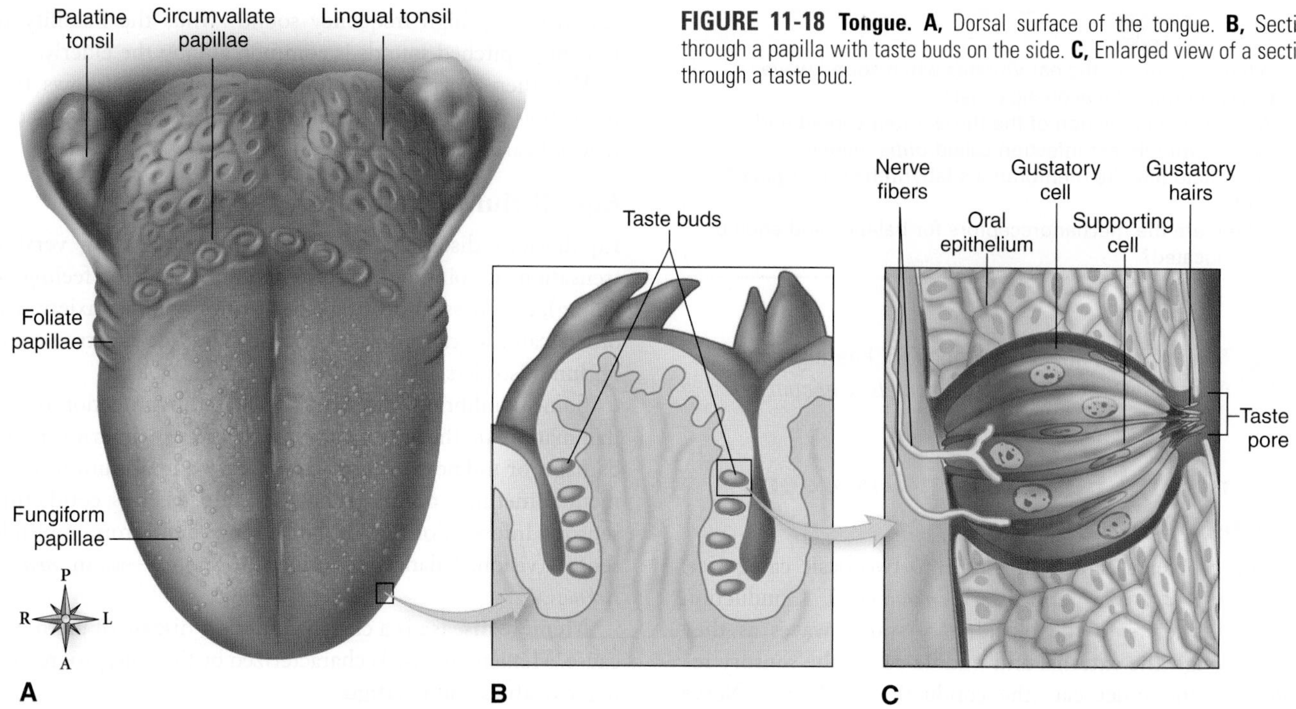

FIGURE 11-18 Tongue. A, Dorsal surface of the tongue. **B,** Section through a papilla with taste buds on the side. **C,** Enlarged view of a section through a taste bud.

Currently, **metallic** taste (to detect metal ions) and a savory, meaty taste called **umami** (to detect the amino acid glutamate) have been added to the list of primary tastes. The list continues to grow. Of course, some individuals are able to sense a larger number of tastes than others. Notable examples include "experts" and "supertasters" who, it is said, can detect literally dozens of discrete and different tastes in wine, coffee, tea, and other foods and beverages.

Smell

The chemoreceptors responsible for smell or **olfaction** are located in a small area of epithelial tissue in the upper part of the nasal cavity (Figure 11-19). The location of the **olfactory receptors** is somewhat hidden, and we often have to forcefully sniff air to smell delicate odors.

Each olfactory cell has a number of sensory cilia that detect different chemicals and cause the cell to respond by generating

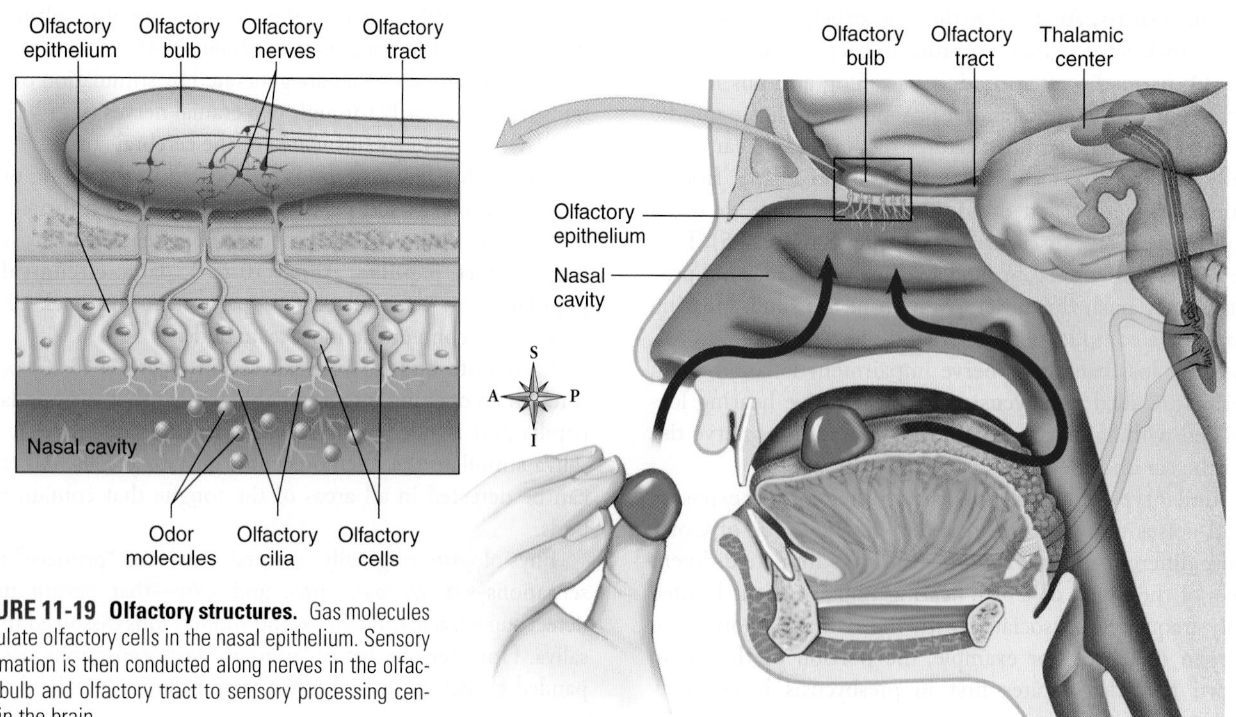

FIGURE 11-19 Olfactory structures. Gas molecules stimulate olfactory cells in the nasal epithelium. Sensory information is then conducted along nerves in the olfactory bulb and olfactory tract to sensory processing centers in the brain.

a nervous impulse. To be detected by olfactory receptors, chemicals must be dissolved in the watery mucus that lines the nasal cavity.

The olfactory receptors are extremely sensitive and respond quickly to even very slight odors. However, they are also easily fatigued—a fact that explains why odors that are at first very noticeable are not sensed at all after a short time. This decrease in receptor sensitivity is called **adaptation.**

After the olfactory cells are stimulated by odor-causing chemicals, the resulting nerve impulse travels through the olfactory nerves in the olfactory bulb and tract and then enters the thalamic and olfactory centers of the brain, where the nervous impulses are interpreted as specific odors. The pathways taken by olfactory nerve impulses and the areas where these impulses are interpreted are closely associated with areas of the brain important in memory and emotion. For this reason, we may retain vivid and long-lasting memories of particular smells and odors.

QUICK CHECK

1. Identify the primary tastes that humans can perceive.
2. Why are odors that are very noticeable at first, not sensed at all after a short time?
3. Trace the pathway of smell from the olfactory cells to the brain.

To learn more about how the brain interprets odors, go to AnimationDirect online at *evolve.elsevier.com*.

Integration of Senses

Looking at the "big picture" of sensation, we should remind ourselves that sensations are all perceived in the brain—not at the individual receptors scattered throughout the body. Some sensory signals never get to the brain, others are amplified or muffled in the brain. All incoming signals are integrated with other sensory signals and even memories to produce our perceptions, which are really a combined sensation of our world at that moment.

For example, most of what we think of as flavor sensations result from a combination of sensory stimuli detected by gustatory cells, olfactory receptors, and even touch and pain receptors. In other words, the myriad unique flavors we recognize are not just tastes alone but are a combined sensation based on tastes, odors, touch, temperature, and pain.

For this reason, severe nasal congestion can interfere with the stimulation of the olfactory receptors by odors from foods in the mouth, which can markedly dull flavor sensations (see **Figure 11-19**). Some foods seem to have a different flavor if they are crispy or warm or cold. And some spicy foods stimulate pain or temperature receptors to produce a "hot" flavor. Some mints can produce a sensation of coolness that adds to our experience of flavor.

Smell sensations, even more than other modes of sensation, are often powerful triggers of memory. Yet all sensations are compared to our learned memories, which help us accurately interpret what we are sensing at any one moment.

We often combine the senses of equilibrium with vision and proprioception to maintain our posture and balance—thus maintaining a safe body position under changing circumstances.

We should also remind ourselves that some sensory information is processed and perceived subconsciously. You cannot "feel" your blood pH go up or down, but your brain is constantly aware of such changes. Likewise, you cannot state your current blood oxygen saturation—but your subconscious mind is aware of the precise level.

With all the senses, advancing age often brings a structural degeneration that results in reduced function. Mechanoreceptors in our ears become less sensitive, our lenses become less able to adjust our visual focus, and we slowly lose taste and smell function. This may explain why some foods "just don't taste the same" as they did when we were younger. It is no wonder that some older adults become isolated and depressed when their contact with the outside world, through the senses, is gradually lost. Caring health professionals recognize these signs of aging and provide assistance needed by their aged patients to once again enjoy life.

SCIENCE APPLICATIONS

SENSES

Santiago Ramón y Cajal (1852-1934)

Santiago Ramón y Cajal is considered by many to be the originator of the modern view of the nervous system's organization. He not only uncovered much about sensory centers of the cortex and the structure of the retina but also made important discoveries about nearly every part of the nervous system. Most of this Spanish researcher's ideas about the nervous system are intact today. Although Santiago wanted to be an artist, his father convinced him to follow in his footsteps and become an anatomist—a choice that led to his receiving a Nobel Prize in 1906.

The study of the sensory aspect of the nervous system and its relationships with the rest of the body is useful in many different fields. For example, the ideas used by **optometrists** and **ophthalmologists, otologists** and **audiologists,** and other professionals who assess and treat sensory disorders are based on neuroscience.

Many other fields make indirect use of neuroscience as well. For example, *artists* use what we know of visual perception in creating their works, *musicians* and *architects* make use of our knowledge of sound perception when performing in or designing concert halls, and *aerospace professionals* can use what we know of equilibrium and how it is perceived in the brain to understand spatial orientation and motion sickness.

11

LANGUAGE OF SCIENCE (continued from p. 291)

circumvallate papilla
(ser-kum-VAL-ayt pah-PIL-ah)
pl., papillae
(pah-PIL-ee)
[*circum-* **around,** *-vall-* **post or stake,** *-ate* **relating to,** *papilla* **nipple**]

cochlea
(KOHK-lee-ah)
[*cochlea* **snail shell**]

cochlear duct
(KOHK-lee-ar dukt)
[*cochlea-* **snail shell,** *-ar* **relating to**]

cochlear nerve
(KOHK-lee-ar nerv)
[*cochlea-* **snail shell,** *-ar* **relating to,** *nerv-* **string or nerve**]

cone
(cohn)

conjunctiva
(kon-junk-TIH-vah)
[*con-* **together,** *-junct* **join**]

cornea
(KOR-nee-ah)
[*corn-* **horn,** *-a* **thing**]

crista ampullaris
(KRIS-tah am-pyoo-LAYR-is)
[*crista* **ridge,** *ampu-* **flask,** *-ulla-* **little,** *-aris* **relating to**]

cupula
(KYOO-pyoo-lah)
pl., cupulae
(KYOO-pyoo-lee)
[*cup-* **tub,** *-ula* **little**]

dynamic equilibrium
(dye-NAM-ik ee-kwi-LIB-ree-um)
[*dynam-* **moving force,** *-ic* **relating to,** *equi-* **equal,** *-libr-* **balance**]

eardrum
(EAR-drum)

endolymph
(EN-doh-limf)
[*endo-* **within,** *-lymph* **water**]

eustachian tube
(yoo-STAY-shun)
[*Bartolomeo Eustachio* **Italian anatomist,** *-an* **relating to**]

external acoustic canal
(eks-TER-nal ah-KOO-stik kah-NAL)
[*extern-* **outside,** *-al* **relating to,** *acoust-* **hearing,** *-ic* **relating to**]

fovea centralis
(FOH-vee-ah sen-TRAL-is)
[*fovea* **pit,** *centralis* **in the center**]

ganglion cell
(GANG-lee-on sell)
[*gangli-* **knot,** *-on* **unit,** *cell* **storeroom**]

general senses
(JEN-er-al SEN-sez)

gustation
(GUS-tay-shun)
[*gusta-* **taste,** *-tion* **process**]

gustatory cell
(GUS-tah-tor-ee sell)
[*gusta-* **taste,** *-ory* **relating to,** *cell* **storeroom**]

helix
(HEE-liks)
pl., helices
(HEE-lis-eez)
[*helix* **spiral**]

incus
(IN-kus)
[*incus* **anvil**]

intraocular pressure
(in-trah-OK-yoo-lar PRESH-ur)
[*intra-* **within,** *-ocul-* **eye,** *-ar* **relating to**]

iris
(AYE-ris)
[*iris* **rainbow**]

lacrimal gland
(LAK-rih-mal gland)
[*lacrima-* **tear,** *-al* **relating to,** *gland-* **acorn**]

lens
(lenz)
[*lens* **lentil**]

macula
(MAK-yoo-lah)
pl., maculae or maculas
(MAK-yoo-lee or MAK-yoo-lahz)
[*macula* **spot**]

macula lutea
(MAK-yoo-lah LOO-tee-ah)
[*macula* **spot,** *lutea* **yellow**]

malleus
(MAL-ee-us)
[*malleus* **hammer**]

mechanoreceptor
(mek-an-oh-ree-SEP-tor)
[*mechano-* **machine (mechanical),** *-recept-* **receive,** *-or* **agent**]

membranous labyrinth
(MEM-brah-nus LAB-eh-rinth)
[*membran-* **thin skin,** *-ous* **characterized by,** *labyrinth* **maze**]

metallic
(meh-TAL-ik)
[*metal-* **metal,** *-ic* **relating to**]

mode
(mohd)

olfaction
(ohl-FAK-shun)
[*olfact-* **smell,** *-ion* **process**]

olfactory receptor
(ohl-FAK-tor-ee ree-SEP-tor)
[*olfact-* **smell,** *-ory* **relating to,** *recept-* **receive,** *-or* **agent**]

optic disk
(OP-tic disk)
[*opti-* **vision,** *-ic* **relating to**]

organ of Corti
(OR-gan ov KOR-tee)
[*organ* **tool or instrument,** *Alfonso Corti* **Italian anatomist**]

ossicle
(OS-sih-kul)
[*os-* **bone,** *-icle* **little**]

pain receptor
(payn ree-SEP-tor)
[*recept-* **receive,** *-or* **agent**]

papilla
(pah-PIL-ah)
pl., papillae
(pah-PIL-ee)
[*papilla* **nipple**]

perilymph
(PAYR-ih-limf)
[*peri-* **around,** *-lymph* **water**]

photopigment
(foh-toh-PIG-ment)
[*photo-* **light,** *-pigment* **paint**]

photoreceptor
(FOH-toh-ree-sep-tor)
[*photo-* **light,** *-recept-* **receive,** *-or* **agent**]

pinna
(PIN-nah)
[*pinna* **wing or fin**]

proprioceptor
(proh-pree-oh-SEP-tor)
[*propri-* **one's own,** *-(re)cept-* **receive,** *-or* **agent**]

pupil
(PYOO-pill)
[*pup-* **doll,** *-il* **little**]

referred pain
(re-FERD payn)

refraction
(ree-FRAK-shun)
[*re-* **back or again,** *-fract-* **break,** *-tion* **process**]

retina
(RET-ih-nah)
[*ret-* **net,** *-ina* **relating to**]

LANGUAGE OF SCIENCE (continued from p. 310)

rod
(rod)

sclera
(SKLEH-rah)
[*sclera* **hard**]

semicircular canal
(sem-ih-SIR-kyoo-lar kah-nal)
[*semi-* **half,** *-circul-***circle,** *-ar* **relating to**]

sensory receptor
(SEN-soh-ree ree-sep-tohr)
[*sens-* **feel,** *-ory* **relating to,** *recept-* **receive,** *-or* **agent**]

special senses
(SPESH-ul SEN-sez)

spiral organ
(SPY-rel OR-gun)
[*spir-* **coiled,** *-al* **relating to,** *organ* **tool or instrument**]

stapes
(STAY-peez)
[*stapes* **stirrup**]

static equilibrium
(ee-kwih-LIB-ree-um)
[*stat-* **stand,** *-ic* **relating to,** *equi-* **equal,** *-libr-* **balance**]

taste bud
(tayst bud)

thermoreceptor
(ther-moh-ree-SEP-tor)
[*thermo-* **heat,** *-cept-* **receive,** *-or* **agent**]

tragus
(TRAY-gus)
[*tragus* **male goat**]

tympanic membrane
(tim-PAN-ik MEM-brayn)
[*tympan-* **drum,** *-ic* **relating to,** *membran-* **thin skin**]

umami
(oo-MAH-mee)
[*umami* **savory**]

vestibular nerve
(ves-TIB-yoo-lar nerv)
[*vestibul-* **entrance hall,** *-ar* **relating to,** *nerv-* **string or nerve**]

vestibule
(VES-tih-byool)
[*vestibul-* **entrance hall**]

vitreous humor
(VIT-ree-us HYOO-mer)
[*vitre-* **glassy,** *-ous* **of or like,** *humor* **fluid**]

LANGUAGE OF MEDICINE

age-related macular degeneration (AMD)
(ayj ree-LAYT-ed MAK-yoo-lar dih-jen-uh-RAY-shun)
[*macula-* **spot,** *-ar* **relating to,** *de-* **down,** *-generat-* **produce,** *-tion* **condition**]

astigmatism
(ah-STIG-mah-tiz-em)
[*a-* **not,** *-stigma-* **point,** *-ism* **condition**]

audiologist
(aw-dee-OL-oh-jist)
[*audio-* **hear,** *-log-* **words (study of),** *-ist* **agent**]

cataract
(KAT-ah-rakt)
[*cataract* **broken water**]

color blindness
(KUL-or BLIND-nes)

conjunctivitis
(kon-junk-tih-VYE-tis)
[*con-* **together,** *-junct-* **join,** *-iv-* **relating to,** *-itis* **inflammation**]

diabetic retinopathy
(dye-ah-BET-ik ret-in-AH-path-ee)
[*diabet-* **siphon (diabetes),** *-ic* **relating to,** *ret-* **net,** *-in-* **relating to,** *-path-* **disease,** *-y* **state**]

glaucoma
(glaw-KOH-mah)
[*glauco-* **gray or silver,** *-oma* **tumor (growth)**]

hyperopia
(hye-per-OH-pee-ah)
[*hyper-* **excessive or above,** *-op-* **vision,** *-ia* **condition**]

laser-assisted in situ keratomileusis (LASIK)
(LAY-zer ah-SIS-ted in- SYE-too kayr-at-oh-mil-YOO-sis)
[*laser-* **shortened "light amplification by stimulated emission of radiation",** *in* **in,** *situ* **place,** *kera-* **horn,** *-mileusis* **carving**]

Ménière disease
(men-ee-AYR dih-ZEEZ)
[*Prosper Ménière* **French physician,** *dis-* **opposite of,** *-ease* **comfort**]

myopia
(my-OH-pee-ah)
[*my-* **shut,** *-op-* **vision,** *-ia* **condition**]

nyctalopia
(nik-tah-LOH-pee-ah)
[*nyct-* **night,** *-op-* **vision,** *-ia* **condition**]

ophthalmologist
(off-thal-MOL-eh-jist)
[*oph-* **eye or vision,** *-thalm-* **inner chamber,** *-o-* **combining vowel,** *-log-* **words (study of),** *-ist* **agent**]

ophthalmoscope
(of-THAL-mah-skohp)
[*oph-* **eye or vision,** *-thalmo-* **inner chamber,** *-scop-* **see**]

optometrist
(op-TOM-eh-trist)
[*opti-* **vision,** *-metr-* **measure,** *-ist* **agent**]

otitis
(oh-TYE-tis)
[*ot-* **ear,** *-itis* **inflammation**]

otitis media
(oh-TYE-tis MEE-dee-ah)
[*ot-* **ear,** *-itis* **inflammation,** *media* **middle**]

otosclerosis
(oh-toh-skleh-ROH-sis)
[*oto-* **ear,** *-sclero-* **hard,** *-sis* **condition**]

otologist
(oh-TOL-o-jist)
[*oto-* **ear,** *-log-* **words (study of),** *-ist* **agent**]

otoscope
(OH-toh-skohp)
[*oto-* **ear,** *-scop-* **see**]

presbycusis
(pres-bih-KYOO-sis)
[*presby-* **elderly,** *-cusis* **hearing**]

presbyopia
(pres-bee-OH-pee-ah)
[*presby-* **aging,** *-op-* **vision,** *-ia* **condition**]

radial keratotomy (RK)
(RAY-dee-al KAR-ah-tah-toh-mee)
[*radi-* **ray,** *-al* **relating to,** *kera-* **horn,** *-tom-* **cut,** *-y* **action**]

11

LANGUAGE OF MEDICINE (continued from p. 311)

retinal detachment
(RET-in-al deh-TACH-ment)
[*ret-* **net,** *-in-* **relating to,** *-al* **relating to,** *de-* **remove,** *-tach-* **attach,** *-ment* **condition]**

scotoma
(skoh-TOH-mah)
[*scoto-* **darkness,** *-oma* **tumor]**

strabismus
(strah-BIS-mus)
[*strab-* **squinting,** *-ismus* **condition]**

tinnitus
(tih-NYE-tus or TIN-nit-us)
[*tinnitis* **ringing or tinkling]**

tophus
(TOH-fus)
pl., tophi
(TOH-fye)
[*tophus* **porous rock]**

trachoma
(trah-KOH-mah)
[*trach-* **rough,** *-oma* **tumor]**

vertigo
(VER-tih-go)
[*vertigo* **turning]**

❑ OUTLINE SUMMARY

To download a digital version of the chapter summary for use with your device, access the ***Audio Chapter Summaries*** *online at evolve.elsevier.com.*

Scan this summary after reading the chapter to help you reinforce the key concepts. Later, use the summary as a quick review before your class or before a test.

Classification of Senses

A. General senses
   1. Detected by sensory organs that exist as individual cells or receptor units (**Table 11-1**)
   2. Widely distributed throughout the body
B. Special senses (**Table 11-2**)
   1. Detected by large and complex organs, or localized grouping of sensory receptors
C. Sensory receptor types
   1. Classified by presence or absence of covering capsule
      a. Encapsulated
      b. Unencapsulated ("free" or "naked")
   2. Classified by type of stimuli (mode) required to activate receptors
      a. Photoreceptors (light)
      b. Chemoreceptors (chemicals)
      c. Pain receptors (injury)
      d. Thermoreceptors (temperature change)
      e. Mechanoreceptors (movement or shape change)

Sensory Pathways

A. All sense organs have common functional characteristics
   1. All are able to detect a particular stimulus
   2. A stimulus results in generation of a nerve impulse
   3. A nerve impulse is processed and perceived as a sensation in the central nervous system

General Senses

A. Distribution is widespread; single-cell receptors are common
B. Mode—the kind of stimulus or change a receptor or sense is able to detect
C. Examples of general sensory receptors and their modes (**Figure 11-1, Table 11-1**)
   1. Skin receptors
      a. Free nerve endings (several types)—pain, discriminative touch, tickle, and temperature
      b. Tactile (Meissner) corpuscle—fine touch and vibration
      c. Bulbous (Ruffini) corpuscle—touch and pressure
      d. Lamellar (Pacini) corpuscle—pressure and vibration
      e. Bulboid (Krause) corpuscle—touch
   2. Muscle receptors
      a. Golgi tendon receptor—proprioception
      b. Muscle spindle—proprioception
   3. Deep receptors
      a. Stretch (pressure) receptors in hollow organs
      b. Chemical receptors—detect pH, carbon dioxide, other chemicals

Special Senses

A. Vision
 1. Eye (Figure 11-2)
  a. Layers of eyeball
   (1) Fibrous layer—tough outer coat
    (a) Sclera—"white" of eye
    (b) Cornea—transparent part over iris
    (c) Conjunctiva—mucous membrane that covers front of fibrous layer and extends to inside of eyelids
    (d) Lacrimal gland—secretes tears that moisten conjunctiva
   (2) Vascular layer—has dense network of blood vessels
    (a) Choroid—pigmented, melanin-rich layer prevents scattering of light
    (b) Iris—the colored part of the eye; the pupil is the hole in the center of the iris; contraction of smooth muscle dilates or constricts pupil (Figure 11-3)
    (c) Lens—transparent body behind the pupil; focuses or refracts light rays on the retina
    (d) Ciliary muscle—near front of vascular layer, just outside the edge of the iris; contraction affects shape of lens just behind the iris, thus altering focus for near objects
   (3) Inner layer—innermost sensory layer
    (a) Retina—contains various kinds of photoreceptors (Figure 11-4 and Figure 11-5)
     i. Rods—receptors for night vision and peripheral vision
     ii. Cones—receptors for day vision and color vision
     iii. Ganglion cells—receptors for changing light patterns of days, months, seasons
  b. Eye fluids
   (1) Aqueous humor—in the anterior chamber in front of the lens
   (2) Vitreous humor—in the posterior chamber behind the lens
 2. Visual pathway
  a. Vision detects intensity (brightness) and wavelength (color) of light, as well as images and motion
  b. Light must be refracted (focused) by the eye to form a detectable image
  c. Innermost layer of retina contains rods and cones
  d. Impulse travels from the rods and cones through the bipolar and ganglionic layers of retina (Figure 11-4)
  e. Nerve impulse leaves the eye at the optic disk and continues through the optic nerve; the point of exit (optic disk) is free of receptors and is therefore also called a *blind spot* (Figure 11-5)
  f. Visual interpretation occurs in the visual cortex of the cerebrum (Figure 11-6)

B. Visual disorders
 1. Refraction disorders
  a. Common focusing problems (Figure 11-7)
   (1) Myopia (nearsightedness) is often caused by elongation of the eyeball
   (2) Hyperopia (farsightedness) is often caused by a shortened eyeball
   (3) Astigmatism is distortion caused by an irregularity of the cornea or lens
   (4) Presbyopia is farsightedness that occurs in late adulthood due to reduced elasticity
  b. Cataracts—lifetime exposure to UV radiation can trigger formation of hard, milky spots in the lens (Figure 11-8)
  c. Conjunctivitis (inflammation of the conjunctiva) can interfere with refraction
   (1) Trachoma—chronic chlamydial infection
   (2) Acute bacterial conjunctivitis—highly contagious infection that produces a discharge of mucous pus (Figure 11-9)
   (3) Conjunctivitis can be caused by allergies
  d. Strabismus—improper alignment of eyes (Figure 11-10)
   (1) Eyes can converge (cross) or diverge
   (2) If not corrected, can cause blindness
 2. Disorders of the retina
  a. Retinal detachment can be a complication of aging, eye tumors, or head trauma
  b. Diabetic retinopathy—damage to retina from hemorrhages and growth of abnormal vessels associated with diabetes mellitus
  c. Glaucoma—increased intraocular pressure decreases blood flow in retina and thus causes retinal degeneration
  d. Nyctalopia (night blindness) or the inability to see in dim light is caused by retinal degeneration or lack of vitamin A
  e. Macular degeneration—progressive degeneration of central part of retina; leading cause of permanent blindness in elderly
  f. Red-green color blindness is an X-linked genetic condition involving the inability to perceive certain colors; it is caused by an abnormality in the cones' photopigments (Figure 11-11)
 3. Disorders of the visual pathway
  a. Degeneration of the optic nerve resulting from diabetes, glaucoma, and other causes can impair vision
  b. Scotoma is the loss of only the central visual field when only certain nerve pathways are damaged
  c. Cerebrovascular accidents (CVAs) can damage visual processing centers; example is acquired cortical color blindness

C. Hearing and Equilibrium
 1. The ear functions in hearing and in equilibrium using receptors called *mechanoreceptors*

2. Ear (**Figure 11-12**)
   a. External ear
      (1) Auricle (pinna)
      (2) External acoustic canal
         (a) Curving canal 2.5 cm (1 inch) in length
         (b) Contains ceruminous glands
         (c) Ends at the tympanic membrane (**Figure 11-13**)
   b. Middle ear
      (1) Houses ear ossicles—malleus, incus, and stapes
      (2) Ends in the oval window
      (3) The auditory (eustachian) tube connects the middle ear to the throat
      (4) Inflammation called *otitis media*
   c. Inner ear (**Figure 11-14**)
      (1) Bony labyrinth filled with perilymph
      (2) Subdivided into the vestibule, semicircular canals, and cochlea
      (3) Membranous labyrinth filled with endolymph
3. Hearing (**Figure 11-15**)
   a. Hearing detects changes in intensity and frequency (tone) of sound waves, which are pressure waves
   b. Sound waves funneled by auricle into external acoustic canal and vibrate the tympanic membrane
   c. Vibrations of tympanic membrane are amplified by auditory ossicles and transmitted to the oval window
   d. Vibrations of the oval window trigger vibrations of perilymph, which in turn vibrates the endolymph
   e. Sensory hair cells on the organ of Corti (spiral organ) respond when bent by the movement of surrounding endolymph set in motion by sound waves; can become damaged by chronic exposure to loud noise
4. Equilibrium—two types of balance: static and dynamic
   a. Static equilibrium—sense of gravity (**Figure 11-16**)
      (1) Detected by ciliated hair cells (mechanoreceptors) of the two maculae in the vestibule
      (2) When the head tilts, gravity pulls the heavy gel of each macula, bending the sensory cilia and producing a nerve signal
   b. Dynamic equilibrium—sense of speed and direction of movement (**Figure 11-17**)
      (1) Detected by ciliated hair cells (mechanoreceptors) of the crista ampullaris (with flaplike cupula) in the ampulla of each semicircular canal
      (2) When speed or direction of movement of head changes, the flow of endolymph in semicircular canals is altered, which causes change in bending of sensory cilia (producing a nerve signal)
   c. Vestibular nerve carries nerve impulses from the equilibrium receptors of the vestibule; joins with cochlear nerve to form vestibulocochlear nerve (CN VIII)

D. Hearing and equilibrium disorders
   1. Hearing Disorders
      a. Conduction impairment
         (1) Can be caused by blockage of the external or middle ear (for example, by cerumen and tumors)
         (2) Otosclerosis—inherited bone disorder involving irregularity of the stapes; it first appears as tinnitus (ringing), then progresses to hearing loss
         (3) Otitis—ear inflammation caused by infection; can produce swelling and fluids that block sound conduction
      b. Nerve impairment
         (1) Presbycusis—progressive nerve deafness associated with aging
         (2) Progressive nerve deafness also can result from chronic exposure to loud noise
   2. Equilibrium disorders
      a. Characterized by vertigo, disorientation, falling (or falling feeling), dizziness, and/or lightheadedness
      b. Motion sickness—temporary equilibrium problem caused by mismatch of senses in the brain regarding body movement
      c. Ménière disease—chronic inner ear disorder of unknown cause; characterized by tinnitus, nerve deafness, and vertigo

E. Taste
   1. Sense of taste is also called gustation
   2. Receptors are chemoreceptors called gustatory cells, located in taste buds (**Figure 11-18**)
   3. Cranial nerves VII and IX carry gustatory impulses
   4. Primary taste modes
      a. Sweet—detects sugars
      b. Sour—detects acids
      c. Bitter—detects alkaline solutions
      d. Salty—detects sodium ions
      e. Metallic—detects metal ions
      f. Umami (savory)—detects glutamate (an amino acid)

F. Smell (olfaction)
   1. Olfactory receptors—sensory fibers of olfactory or cranial nerve I lie in olfactory mucosa of nasal cavity (**Figure 11-19**)
   2. Olfactory receptors are extremely sensitive but easily adapt (become fatigued)
   3. Odor-causing chemicals initiate a nervous signal that is interpreted as a specific odor by the brain
   4. Olfaction has a strong relationship with emotions and memory

11

Integration of Senses

1. All senses are processed and finally perceived in the brain (not receptors)
2. Sensory information is combined to form an overall sensory perception of our world
   a. Flavor
      (1) Combination of gustatory and olfactory senses; can be affected by other senses, such as touch, pain, or temperature
      (2) Nasal congestion interferes with stimulation of olfactory receptors and thereby dulls flavor sensations
   b. Posture and balance—both senses of equilibrium with vision and proprioception—combine to help us maintain a safe body position
3. Some sensory information is processed subconsciously
4. Our senses may decline as we age

❑ ACTIVE LEARNING

STUDY TIPS

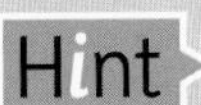

Consider using these tips to achieve success in meeting your learning goals.

Each of the body's senses must go through the following processes to perform its function: (1) detect the physical stimulus to which it responds and (2) convert that stimulus into a nerve impulse. For example, the eye must let light in and focus it on a specific point; the receptors convert that stimulus into a nerve impulse and send it to the brain.

1. When you study structures and their specific function in a sensory system, focus on how they contribute to one of these two processes. Use flash cards and other online resources to learn the specific structures and their functions in each sensory system. Link the disorders in the sensory system to their cause or mechanism. In the eye, these are refraction, retinal damage, or pathway damage. In the ear, conduction, infection, or nerve damage may cause disorders. A chart may be the best way to organize this information.
2. In your study group, discuss how each of the sensory systems detect and respond to a stimulus. Photocopy the figures of the sense organs, blacken out the labels, and quiz each other on the name, location, and function of each structure. Use online labeling exercises (www.getbodysmart.com) as a resource.
3. Review the Language of Science and Language of Medicine terms. Review the questions and outline summary at the end of the chapter and discuss possible test questions in your study group.

Review Questions

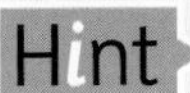

Write out the answers to these questions after reading the chapter and reviewing the Chapter Summary. If you simply think through the answer without writing it down, you will not retain much of your new learning.

1. Name the general senses found in the skin or subcutaneous tissue and list the type of stimuli to which each of them responds.
2. List the two general senses of proprioception and give the location of each.
3. With what type of information do proprioceptors provide us?
4. Describe how the iris changes the size of the pupil.
5. Explain how the ciliary muscles allow the eye to focus on near and far objects.
6. What is *presbyopia* and what is its cause?
7. List the three types of receptor cells in the retina. Explain the differences between the three receptors.
8. Define glaucoma and explain the cause of this condition.
9. What are cataracts? What causes cataracts and how can they be prevented?
10. What is meant by the *visual pathway?* Where is the *blind spot* and what causes it?
11. Explain how the disorder *strabismus* affects vision.
12. What causes diabetic retinopathy?
13. Briefly explain the structure of the external ear.
14. Explain how sound waves are transmitted through the middle ear.
15. Explain how sound waves are converted to an auditory impulse.
16. Explain how the structures in the inner ear help maintain balance or equilibrium.
17. What is Ménière disease?
18. Where are the gustatory cells located, and to what "primary" tastes do they respond?
19. Explain how the sense of smell is stimulated.
20. Describe how ganglion cells differ from rods and cones

11

Critical Thinking

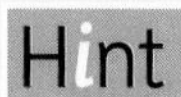

After finishing the Review Questions, write out the answers to these more in-depth questions to help you apply your new knowledge. Go back to sections of the chapter that relate to concepts that you find difficult.

21. Explain why food loses some of its taste when you have a bad cold or stuffy nose.
22. Explain why the longer you are in a newly painted room the less able you are to smell the paint.
23. Explain why the smell of a "doctor's office" or the smell of turkey cooking on Thanksgiving can easily generate an emotional response?
24. How is sensory response related to age?
25. Rock musicians sometimes lose their ability to hear high-frequency tones; explain how this can happen.

Chapter Test

After studying the chapter, test your mastery by responding to these items. Try to answer them without looking up the answers.

1. The eye can be classified as a photoreceptor. Taste and smell can be classified as ________, and Golgi tendon receptors and muscle spindles can be classified as ________.
2. The specific mechanoreceptor for balance is the ________.
3. The gustatory cells are involved with the sense of ________.
4. The four "primary" taste sensations that result from the stimulation of the taste buds are ________, ________, ________, and ________.
5. Taste buds can be found on much larger structures on the tongue called ________.
6. The chemoreceptors responsible for the sense of smell are the ________.
7. ________ is the gradual reduction in sensitivity to an odor after initial contact.
8. The ________ is also known as the *blind spot.*
9. ________ keeps the acoustic canal from drying and flaking.
10. ________ is used to examine the external ear canal and tympanic membrane

Match each structure of the eye in Column A with its corresponding function or description in Column B.

Column A	Column B
11. ________ sclera	a. gland in which tears are formed
12. ________ cornea	b. hole in the eye that lets light in
13. ________ iris	c. receptors for night vision or dim light
14. ________ pupil	d. thick jellylike fluid of the eye
15. ________ lacrimal	e. tough, white outer layer of the eye
16. ________ lens	f. receptors for red, blue, and green color vision
17. ________ rods	g. structure on which ciliary muscles pull to help the eye focus
18. ________ cones	h. dark pigmented middle layer of the eye that prevents the scattering of incoming light
19. ________ choroid coat	i. transparent part of the sclera, the window of the eye
20. ________ vitreous body	j. colored part of the anterior of the eye
21. ________ aqueous humor	k. thin, watery fluid of the eye

Match each structure of the ear in Column A with its corresponding function or description in Column B.

Column A	Column B
22. ________ tympanic membrane	a. tube connecting the middle ear and the throat
23. ________ ossicles	b. watery fluid that fills the bony labyrinth
24. ________ auditory tube	c. snail-shaped structure in the inner ear
25. ________ perilymph	d. organ of hearing
26. ________ endolymph	e. thick fluid in the membranous labyrinth
27. ________ cochlea	f. another term for eardrum
28. ________ organ of Corti	g. collective name for the incus, malleus, and stapes

Match each disorder in Column A with its description in Column B.

Column A	Column B
29. ________ myopia	a. caused by increased fluid pressure in the eye
30. ________ astigmatism	b. an inflammation of the conjunctiva, "pink eye"
31. ________ conjunctivitis	c. progressive degeneration of the central part of the retina
32. ________ strabismus	d. nearsightedness caused by the elongation of the eyeball
33. ________ diabetic retinopathy	e. an infection of the middle ear
34. ________ glaucoma	f. an improper alignment of the eyes; can cause them to converge (cross)
35. ________ age-related macular degeneration	g. chronic inner ear disorder of unknown cause; characterized by tinnitus, deafness, and vertigo
36. ________ color blindness	h. an X-linked genetic condition involving inability to perceive some colors
37. ________ otitis media	i. damage to the retina caused by hemorrhage and abnormal vessel growth
38. ________ Ménière disease	j. distortion of the image in the eye caused by irregularities of the cornea or lens

Case Studies

To solve a case study, you may have to refer to the glossary or index, other chapters in this textbook, and other resources.

1. Roger is legally blind. His vision impairment is a complication of diabetes mellitus. Can you describe what structural changes in Roger's eyes have caused his blindness? As you were helping him cross the street, a fellow pedestrian suddenly stumbled into your path. Without any signal from you, Roger jumped back to avoid hitting the other person. If Roger is blind, how could he have reacted this way?
2. As a child, Mrs. Stark was tested for color blindness and told that she had normal color vision. She never had any problems distinguishing colors until shortly after her retirement, when suddenly she lost her sense of color. She describes her perception of the world as being "like a black and white movie." She cannot distinguish yellows, oranges, blues, greens, reds, or any other colors. What might have caused Mrs. Stark's problems?
3. You have just been diagnosed with otitis media. Describe what has happened to your body to produce this condition. If left untreated, what are the possible outcomes of this condition?
4. Sharon went to the doctor complaining of abdominal pain in the right iliac region. After examination, the doctor advised Sharon that he suspected a spinal cord problem. He referred her to a neurologist and ordered preliminary tests of the spine to be sent to the doctor's office in preparation for her visit. Sharon is confused. Her pain is in her abdomen. Why is he sending her to a neurologist?

Answers to Active Learning Questions can be found online at evolve.elsevier.com.

Endocrine System

OUTLINE

Scan this outline before you begin to read the chapter, as a preview of how the concepts are organized.

OBJECTIVES

Before reading the chapter, review these goals for your learning.

After you have completed this chapter, you should be able to:

1. Do the following related to endocrine glands:
   - Distinguish between endocrine and exocrine glands.
   - Define the term *hormone.*
   - Identify the principal functions of each major endocrine hormone and describe the conditions that may result from hyposecretion or hypersecretion.
2. Describe the mechanisms of steroid and nonsteroid hormone action.
3. Explain how negative and positive feedback mechanisms regulate the secretion of endocrine hormones.
4. Explain the primary mechanisms of endocrine disorders.
5. Define and explain the importance of prostaglandins (PGs).
6. Identify and locate the primary endocrine glands, list the major hormones produced by each gland, and identify disorders of the endocrine system.

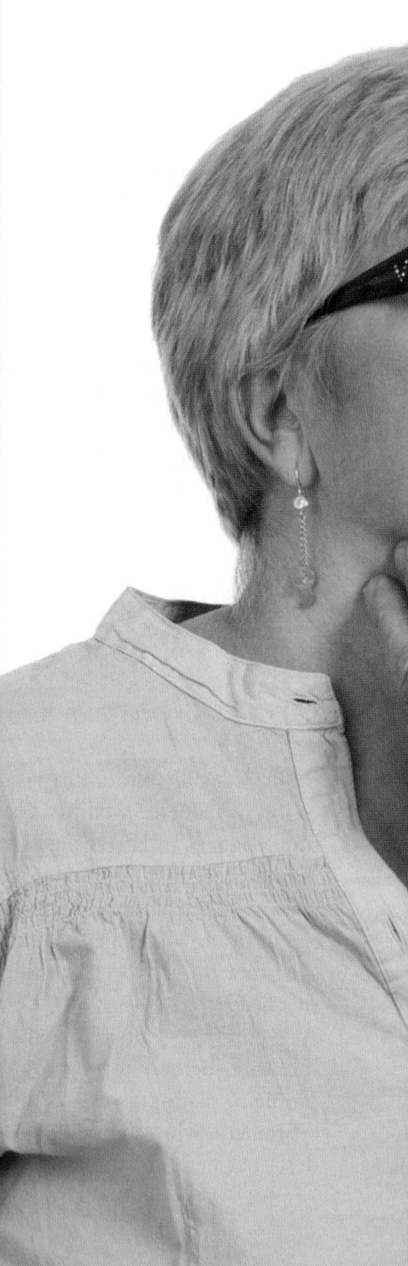

CHAPTER 12

Have you ever known anyone with thyroid problems or diabetes? Surely you have seen the dramatic changes that happen to a person's body as he or she goes through puberty. These are all proof of the importance of the endocrine system in normal development and health.

The **endocrine system** performs the same general functions as the nervous system: communication and control. The nervous system provides rapid, brief control by way of fast-traveling nerve impulses. The endocrine system provides slower but longer-lasting control by way of hormones (chemicals) secreted into and circulated by the blood.

The main organs of the endocrine system are located in widely separated parts of the body, as you can see in **Figure 12-1**.

In this chapter you will read about the functions of the main endocrine glands and discover why their importance is almost impossible to exaggerate. Hormones are the main regulators of metabolism, growth and development, reproduction, and many other body activities. They play important roles in maintaining homeostasis—fluid and electrolyte, acid-base, and energy balances, for example. Hormones make the difference between normalcy and

LANGUAGE OF SCIENCE

Hint Before reading the chapter, say each of these terms out loud. This will help you to avoid stumbling over them as you read.

adenohypophysis
(ad-eh-noh-hye-POF-ih-sis)
[*adeno-* **gland**, *-hypo-* **under or below**, *-physis* **growth**]

adrenal cortex
(ah-DREE-nal KOR-teks)
[*ad-* **toward**, *-ren-* **kidney**, *-al* **relating to**, *cortex* **bark**]

adrenal medulla
(ah-DREE-nal meh-DUL-ah)
[*ad-* **toward**, *-ren-* **kidney**, *-al* **relating to**, *medulla* **marrow or pith**]

adrenocorticotropic hormone (ACTH)
(ah-dree-noh-kor-teh-koh-TROH-pic HOR-mohn [ay see tee aych])
[*adreno-* **gland**, *-cortic-* **bark**, *-trop-* **nourish**, *-ic* **relating to**, *hormon-* **excite**]

aldosterone
(AL-doh-steh-rohn *or* al-DAH-stayr-ohn)
[*aldo-* **aldehyde**, *-stero-* **solid or steroid derivative**, *-one* **chemical**]

anabolic steroid
(an-ah-BOL-ik STAYR-oyd)
[*anabol-* **build up**, *-ic* **relating to**, *stero-* **solid**, *-oid* **like**]

anabolism
(ah-NAB-oh-liz-em)
[*anabol-* **build up**, *-ism* **action**]

androgen
(AN-droh-jen)
[*andro-* **male**, *-gen* **produce**]

Continued on p. 339

12

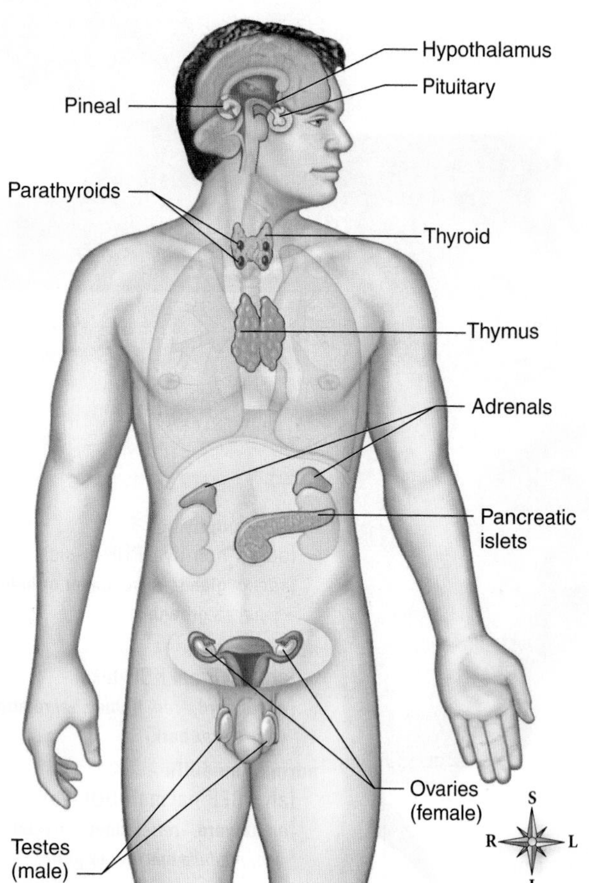

FIGURE 12-1 Location of endocrine glands. Some of the major endocrine structures are shown. Thymus gland is shown at maximum size at puberty.

many kinds of abnormalities such as dwarfism, gigantism, and sterility. They are important not only for the healthy survival of each one of us but also for the survival of the human species.

Endocrine Glands

All organs of the endocrine system are glands, but not all glands are organs of the endocrine system. Of the two types of glands in the body—**exocrine glands** and **endocrine glands**—only endocrine glands belong to this system.

You may recall from Chapters 4 and 5 that exocrine glands secrete their products into ducts that empty onto a surface or into a cavity. For example, sweat glands produce a watery secretion that empties onto the surface of the skin. Salivary glands are also exocrine glands, secreting saliva that flows into the mouth.

Endocrine glands, on the other hand, are ductless glands. They secrete chemicals known as **hormones** into intercellular spaces. From there, the hormones diffuse directly into the blood and are carried throughout the body. Each hormone molecule may then bind to a cell that has specific receptors for that hormone, triggering a reaction in the cell. Such a cell is called a **target cell.**

The list of endocrine glands and the organs in which their target cells are found *(target organs)* continues to grow. The names, locations, and functions of the well-known endocrine glands are given in **Figure 12-1** and **Table 12-1**.

To get an overview of the endocrine system, go to AnimationDirect online at *evolve.elsevier.com.*

Mechanisms of Hormone Action

A hormone causes its target cells to respond in particular ways; this has been the subject of intense interest and research. The two major classes of hormones—**nonsteroid hormones** and **steroid hormones**—differ in the mechanisms by which they influence target cells.

To better understand these concepts, use the Active Concept Map *Hormones* at *evolve.elsevier.com.*

Nonsteroid Hormones

Nonsteroid hormones are whole proteins, shorter chains of amino acids, or simply versions of single amino acids. Nonsteroid hormones typically work according to the **second-messenger mechanism.** According to this concept, a nonsteroid protein hormone, such as thyroid-stimulating hormone, acts as a "first messenger" (that is, it delivers its chemical message from the cells of an endocrine gland to highly specific

RESEARCH, ISSUES, AND TRENDS

SECOND-MESSENGER SYSTEMS

Rapid and revolutionary discoveries about how nonsteroid hormones act on their target cells began with the pioneering work of Earl Sutherland, who received the 1971 Nobel Prize for formulating the second-messenger hypothesis, and new discoveries continue to be made even today.

Later, the important role of the so-called **G protein** in getting the signal from the receptor to the enzyme that forms cyclic AMP (cAMP) was discovered. Look for the G protein in **Figure 12-2**. More recently, the role of nitric oxide (NO) in second-messenger systems has been worked out.

All of these discoveries resulted in Nobel Prizes, which shows the importance the scientific community has placed on them. Why? By working out the details of how hormones work, we can more clearly understand how and why things can go wrong that affect endocrine disorders. Perhaps we may even gain new knowledge about disorders that we previously did not even know involved hormone mechanisms.

Once the processes of disease mechanisms are figured out, we hope scientists will be able to design tests that can screen for such problems. Or perhaps they can develop drugs that will "fix" the broken mechanisms and cure the disease. After reading this chapter, you will discover that understanding how hormones act on target cells—the concept of **signal transduction**—will prepare you for the revolution in medicine that is now upon us.

membrane receptor sites on the cells of a target organ). This interaction between a hormone and its specific receptor site on the cell membrane of a target cell is often compared to the fitting of a unique key into a lock. This idea is the *lock-and-key model* of chemical activity.

After the hormone attaches to its specific receptor site, a number of chemical reactions occur. These reactions activate molecules within the cell called *second messengers.* One example of this mechanism occurs when the hormone-receptor interaction changes energy-rich ATP molecules inside the cell into **cyclic AMP** (adenosine monophosphate). Cyclic AMP serves as the second messenger, delivering information inside the cell that regulates the cell's activity. For example, cyclic AMP causes thyroid cells to respond to thyroid-stimulating hormone by secreting a thyroid hormone such as *thyroxine.* Cyclic AMP is only one of several second messengers that have been discovered.

In summary, nonsteroid hormones serve as first messengers, providing communication between endocrine glands and target organs. Another molecule, such as cyclic AMP, then acts as the second messenger, providing communication within a hormone's target cells. Figure 12-2 summarizes the mechanism of nonsteroid hormone action.

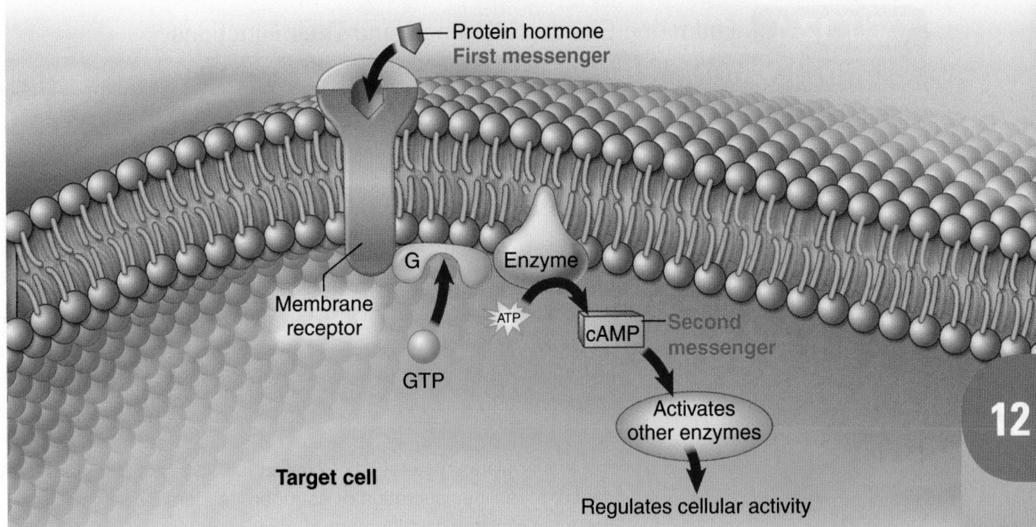

FIGURE 12-2 Mechanism of protein hormone action. The hormone acts as "first messenger," delivering its message via the bloodstream to a membrane receptor in the target cell much like a key fits into a lock. The "second messenger" is cyclic AMP (cAMP), which forms in response to the first messenger's signaling actions. cAMP causes the cell to respond and perform the function triggered by the hormone. Variations of this mechanism also exist.

Steroid Hormones

The primary actions of small, lipid-soluble steroid hormones such as estrogen do not occur by the second-messenger mechanism. Because they are lipid soluble, steroid hormones can pass intact directly through the cell membrane of the target cell.

Once inside the cell, steroid hormones pass through the cytoplasm and enter the nucleus, where they bind with a receptor (according to the lock-and-key model) to form a hormone-receptor complex. This complex acts on DNA, which ultimately causes the formation of a new protein in the cytoplasm that then produces specific effects in the target cell. In the case of estrogen, for example, that effect might be breast development in the female adolescent.

Figure 12-3 summarizes this mechanism of steroid hormone action. Because it takes some time to accomplish all of the steps illustrated in the diagram, steroid hormone responses typically are slow compared with responses triggered by nonsteroid hormones.

Besides the primary effects of steroids produced by the DNA-triggering mechanism just described, steroid hormones also may trigger membrane receptors to produce a variety of secondary effects. These secondary effects usually appear much more rapidly than do the primary steroid effects.

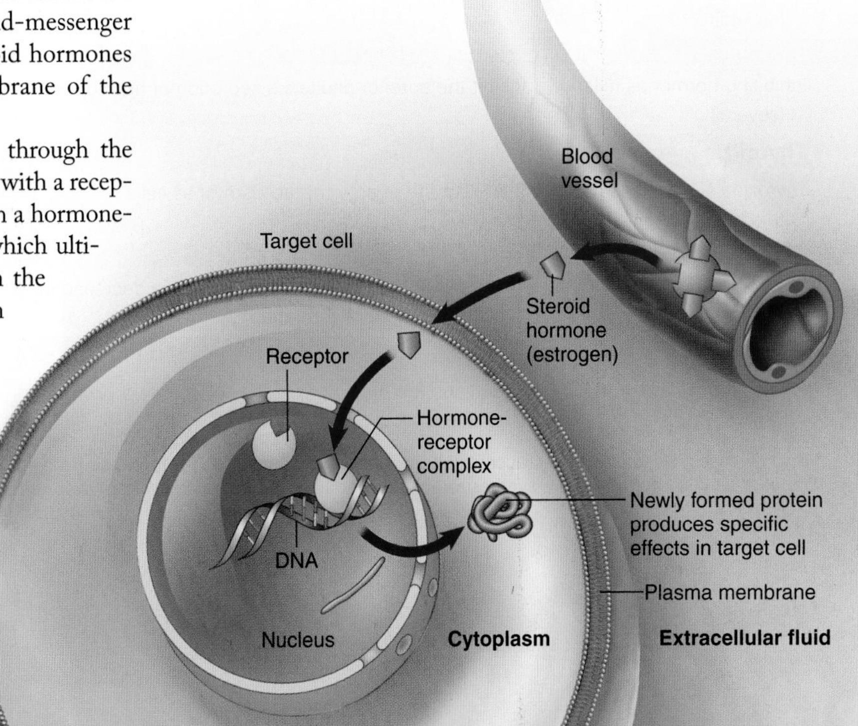

FIGURE 12-3 Mechanism of steroid hormone action. Steroid hormones pass through the plasma membrane and enter the nucleus to form a hormone receptor complex that acts on DNA. As a result, a new protein is formed in the cytoplasm that produces specific effects in the target cell.

TABLE 12-1 Endocrine Glands, Hormones, and Their Functions

GLAND/HORMONE	FUNCTION	DYSFUNCTION*
Anterior Pituitary		
Thyroid-stimulating hormone (TSH)	Tropic hormone Stimulates secretion of thyroid hormones	*Hypersecretion:* overstimulation of thyroid *Hyposecretion:* understimulation of thyroid
Adrenocorticotropic hormone (ACTH)	Tropic hormone Stimulates secretion of adrenal cortex	*Hypersecretion:* overstimulation of adrenal cortex hormones
Follicle-stimulating hormone (FSH)	Tropic hormone *Female:* stimulates development of ovarian follicles and secretion of estrogens *Male:* stimulates seminiferous tubules of testes to grow and produce sperm	*Hyposecretion:* understimulation of adrenal cortex hormones *Hyposecretion:* lack of sexual development and sterility
Luteinizing hormone (LH)	Tropic hormone *Female:* stimulates maturation of ovarian follicle and ovum; stimulates secretion of estrogen; triggers ovulation; stimulates development of corpus luteum (luteinization) *Male:* stimulates interstitial cells of the testes to secrete testosterone	*Hyposecretion:* lack of sexual development and sterility
Growth hormone (GH)	Stimulates growth in all organs; mobilizes nutrient molecules, causing an increase in blood glucose concentration	*Hypersecretion:* gigantism (pre-adult); acromegaly (mature adult) *Hyposecretion:* dwarfism (pre-adult)
Prolactin (PRL or lactogenic hormone)	Stimulates breast development during pregnancy and milk secretion (milk letdown) after pregnancy	*Hypersecretion:* inappropriate lactation in men or non-nursing women *Hyposecretion:* insufficient lactation in nursing women
Posterior Pituitary†		
Antidiuretic hormone (ADH)	Stimulates retention of water by the kidneys	*Hypersecretion:* abnormal water retention *Hyposecretion:* diabetes insipidus
Oxytocin (OT)	Stimulates uterine contractions at the end of pregnancy; stimulates the release of milk into the breast ducts	*Hypersecretion:* inappropriate ejection of milk in lactating women *Hyposecretion:* prolonged or difficult labor and delivery (uncertain)
Hypothalamus		
Releasing hormones (RHs) (several)	Stimulate the anterior pituitary to release hormones	*Hypersecretion:* hypersecretion by anterior pituitary *Hyposecretion:* hyposecretion by anterior pituitary
Inhibiting hormones (IHs) (several)	Inhibit the anterior pituitary's secretion of hormones	*Hypersecretion:* hyposecretion by anterior pituitary *Hyposecretion:* hypersecretion by anterior pituitary
Thyroid		
Thyroxine (T_4) and triiodothyronine (T_3)	Stimulate the energy metabolism of all cells	*Hypersecretion:* hyperthyroidism, Graves disease *Hyposecretion:* hypothyroidism, cretinism (pre-adult); myxedema (adult); goiter
Calcitonin (CT)	Inhibits the breakdown of bone; causes a decrease in blood calcium concentration	*Hypersecretion:* possible hypocalcemia *Hyposecretion:* possible hypercalcemia
Parathyroid		
Parathyroid hormone (PTH)	Stimulates the breakdown of bone; causes an increase in blood calcium concentration	*Hypersecretion:* possible hypercalcemia *Hyposecretion:* possible hypocalcemia

*In some cases, signs of hyposecretion result from target cell abnormality rather than from actual hyposecretion of a hormone.
†Posterior pituitary hormones are synthesized in the hypothalamus but released from axon terminals in the posterior pituitary.

TABLE 12-1 Endocrine Glands, Hormones, and Their Functions—cont'd

GLAND/HORMONE	FUNCTION	DYSFUNCTION*
Adrenal Cortex		
Mineralocorticoids: aldosterone	Regulate electrolyte and fluid homeostasis	*Hypersecretion:* increased water retention *Hyposecretion:* abnormal water loss (dehydration)
Glucocorticoids: cortisol (hydrocortisone)	Stimulate gluconeogenesis, causing an increase in blood glucose concentration; also have anti-inflammatory, anti-immunity, and antiallergy effects	*Hypersecretion:* Cushing syndrome *Hyposecretion:* Addison disease
Androgens	Stimulate sexual drive in the female but have negligible effects in the male	*Hypersecretion:* premature sexual (androgens) development; masculinization of female *Hyposecretion:* no significant effect
Adrenal/Medulla		
Epinephrine (Epi) and norepinephrine (NE)	Prolong and intensify the sympathetic nervous response during stress	*Hypersecretion:* stress effects (adrenaline) *Hyposecretion:* no significant effect
Pancreatic Islets		
Glucagon	Stimulates glycogenolysis in liver, causing an increase in blood glucose concentration	(Uncertain)
Insulin	Promotes glucose entry into all cells, causing a decrease in blood glucose concentration	*Hypersecretion:* severe hypoglycemia (insulin shock) *Hyposecretion:* diabetes mellitus
Ovary		
Estrogens	Promote development and maintenance of female sexual characteristics (see Chapter 23)	*Hypersecretion:* premature sexual development (female) and infertility *Hyposecretion:* lack of sexual development (female), infertility, and osteoporosis
Progesterone	Promotes conditions required for pregnancy (see Chapter 23)	*Hyposecretion:* sterility
Testis		
Testosterone	Promotes development and maintenance of male sexual characteristics	*Hypersecretion:* premature sexual development (male); muscle hypertrophy *Hyposecretion:* lack of sexual development (male)
Thymus		
Thymosin	Promotes development of immune-system cells	*Hyposecretion:* depression of immune system functions
Placenta		
Chorionic gonadotropin, estrogens, progesterone	Promote conditions required during early pregnancy	*Hyposecretion:* spontaneous abortion (miscarriage)
Pineal		
Melatonin	Inhibits tropic hormones that affect the ovaries; helps regulate the body's internal clock and sleep cycles	*Hypersecretion:* winter depression, sleep disorders, and other possible effects
Heart (Atria)		
Atrial natriuretic hormone (ANH)	Regulates fluid and electrolyte homeostasis	*Hyposecretion:* fluid or electrolyte imbalances; possible blood pressure problems
Gastrointestinal (GI) Tract		
Ghrelin	Affects energy balance (metabolism)	*Hypersecretion:* increase in hunger and suppression of fat utilization; possible obesity
Fat-Storing Cells		
Leptin	Controls how hungry or full we feel	*Hyposecretion:* possible obesity, other metabolic disorders

12

QUICK CHECK

1. What is the chemical messenger used by the endocrine system?
2. How do *nonsteroid* hormones and *steroid* hormones differ?
3. What is a *second-messenger* system?

Regulation of Hormone Secretion

Negative Feedback

Regulation of hormone levels in the blood depends on a homeostatic mechanism called *negative feedback*—a concept first introduced in Chapter 1.

The principle of **negative feedback** in the endocrine system can be illustrated by using the hormone insulin as an example. When released from endocrine cells in the pancreas, insulin lowers "blood sugar levels" or glucose concentration in the blood. Normally, elevated blood glucose occurs after a meal, after the absorption of glucose from the digestive tract takes place. The elevated blood glucose stimulates the release of insulin from the pancreas. Insulin then assists in the transfer of glucose from the blood into cells, causing blood glucose to drop back toward the normal set point.

As blood glucose concentration drops, the endocrine cells in the pancreas slow their production and release of insulin. These responses are *negative because they reverse the direction of a disturbance to the stability of the internal environment of the body*. Therefore, this homeostatic mechanism is called a *negative feedback control mechanism* because it reverses the change in blood glucose (**Figure 12-4**).

Positive Feedback

Positive feedback mechanisms, which are uncommon, amplify changes rather than reverse them. Usually, such amplification threatens homeostasis, but in some situations it can help the body maintain its stability.

For example, during labor, the muscle contractions that push the baby through the birth canal become stronger and stronger by means of a positive feedback mechanism that regulates secretion of the hormone oxytocin (see **Figure 1-12** on p. 16).

Levels of Regulation

The endocrine system provides a good example of the concept of different levels of homeostatic regulation. Regulating the secretion of a particular hormone is one level of control, but that in turn regulates specific functions in the target cells, which in turn changes some particular function of the body.

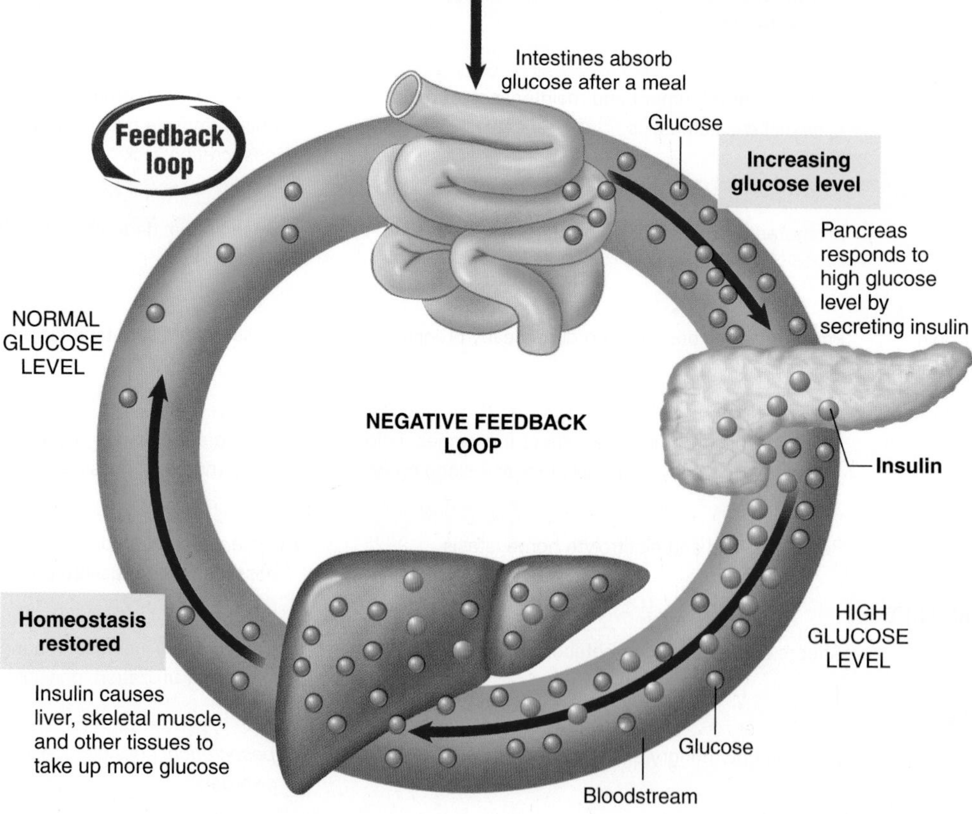

FIGURE 12-4 Negative feedback. The secretion of most hormones is regulated by negative feedback mechanisms that tend to reverse any deviations from normal. In this example, an increase in blood glucose triggers secretion of insulin. Because insulin promotes glucose uptake by cells, the blood glucose level is restored to its lower, normal level.

HEALTH AND WELL-BEING

STEROID ABUSE

Some steroid hormones are called **anabolic steroids** because they stimulate the building of large molecules (anabolism). Specifically, they stimulate the building of proteins in muscle and bone. Steroids such as testosterone and its synthetic derivatives are often abused by athletes and others who want to increase their athletic performance. The anabolic effects of the hormones increase the mass and strength of skeletal muscles.

Unfortunately, steroid abuse has other consequences that are not desirable. It disrupts the normal negative feedback control of hormones throughout the body and may result in tissue damage, sterility, mental imbalance, and many life-threatening metabolic problems. Abuse of steroids, other performance-enhancing drugs, or blood products in athletics is called *doping*. Such practices are outlawed in sports worldwide (see also the box "Blood Doping" on p. 354).

Typically, additional levels of control are involved in maintaining homeostasis. For example, feedback may trigger the secretion of a "releasing" hormone that targets another gland and triggers the secretion of that second gland's hormone.

Feedback may instead trigger autonomic nervous stimulation of a gland, which then secretes a releasing hormone. In turn, the releasing hormone triggers the release of another hormone that regulates its target cells, which change their functions to produce an effect that changes a variable to move back toward its set point.

Often, *all* the levels of control are receiving and reacting to feedback—thus providing extra efficiency and precision to the homeostatic control of body function.

Mechanisms of Endocrine Disease

Diseases of the endocrine system are numerous, varied, and sometimes catastrophic. Tumors or other abnormalities frequently cause the glands to secrete too much or too little of their hormones. Production of too much hormone by a diseased gland is called **hypersecretion.** If too little hormone is produced, the condition is called **hyposecretion.**

A variety of endocrine disorders that appear to result from hyposecretion are actually caused by a problem in the target cells. If the usual target cells of a particular hormone have damaged receptors, too few receptors, or some other abnormality, they will not respond to that hormone properly. In other words, lack of target cell response could be a sign of hyposecretion or a sign of target cell insensitivity. *Diabetes mellitus (DM),* for example, can result from insulin hyposecretion or from the target cells' insensitivity to insulin—or both.

Imbalances of one type of hormone often affect other hormones as well. Disorders that result from the hypersecretion or hyposecretion of several hormones are often called **polyendocrine disorders.**

Endocrinologists, scientists who specialize in endocrine function, or **endocrinology,** have developed a variety of strategies for treating endocrine disorders. Surgical or chemical treatment of tumors or damaged tissue is useful in some cases. Another common strategy is the use of pharmacological preparations of hormones. For example, insulin injections are used in treating some forms of diabetes mellitus.

The availability of synthetic hormones produced with genetic engineering technology has revolutionized the treatment of many endocrine disorders. Synthetic hormones are cheaper and more widely available than natural human hormones, and they do not carry the same risk of possible contamination with viruses or other dangerous substances (see box on p. 354).

Table 12-1 summarizes some of the major disorders of the endocrine system. Refer to this table often as you study the individual glands and hormones of the endocrine system. You may also want to refer to Appendix A at *evolve.elsevier.com*, which also contains a table listing major endocrine disorders.

Prostaglandins

Prostaglandins (PGs), or tissue hormones, are important and extremely powerful substances found in a wide variety of tissues. PGs are modified versions of fatty acids. PGs play an important role in communication and in the control of many body functions but do not meet the definition of a typical hormone.

The term *tissue hormone* is appropriate because in many instances a prostaglandin is produced in a tissue and then diffuses only a short distance to act on cells within that tissue. Typical hormones influence and control activities of widely separated organs; typical PGs influence activities of neighboring cells.

PGs, along with several other tissue hormones such as **leukotrienes** and **thromboxane,** are sometimes called **paracrine** agents. The term *paracrine* literally means "secrete beside"—an apt description for a regulatory agent released right next to its target cell.

The prostaglandins in the body can be divided into several groups. Three classes of prostaglandins—*prostaglandin A (PGA), prostaglandin E (PGE),* and *prostaglandin F (PGF)*—are among the best known.

PGs have profound effects on many body functions. They influence respiration, blood pressure, gastrointestinal secretions, inflammation, and the reproductive system. Researchers

RESEARCH, ISSUES, AND TRENDS

PROSTAGLANDIN THERAPY

Although much research is yet to be done, prostaglandins are already playing an important role in the treatment of diverse conditions such as glaucoma, high blood pressure, asthma, and ulcers. Because some prostaglandins have local muscle-relaxing effects, they can relax muscles in the walls of blood vessels to reduce blood pressure. In asthma, prostaglandins administered in a nebulizer (mist applicator) relax the muscles that constrict airflow during an asthma attack. Some gastric ulcers can be treated with prostaglandins that decrease stomach acid secretion.

Pharmacologists, scientists who study drug actions, or **pharmacology,** have discovered that prostaglandins are involved in some traditional therapies. For example, aspirin and its derivatives *(salicylates)* produce some of their effects by blocking prostaglandins involved in the inflammation response.

12

believe that most PGs regulate cells by influencing the production of cyclic AMP.

PGs are already playing an important role in the treatment of conditions such as glaucoma, high blood pressure, asthma, and ulcers—as described in the box above. In fact, many common treatments such as *aspirin* create their effects by altering the functions of PGs in the body.

QUICK CHECK

1. How does negative feedback affect hormone levels in the blood?
2. What may occur if the usual target cells of a particular hormone have damaged receptors, too few receptors, or some other abnormality?
3. Identify the strategies in treating endocrine disorders.
4. Why are prostaglandins called tissue *hormones*?

Pituitary Gland

Structure of the Pituitary Gland

The **pituitary gland** is a small but mighty structure. Although no larger than a pea, it is really two endocrine glands. One is called the anterior pituitary gland or **adenohypophysis,** and the other is called the posterior pituitary gland or **neurohypophysis.**

Differences between the two glands are indicated by their names—*adeno* means "gland," and *neuro* means "nervous." The adenohypophysis has the structure of an endocrine gland, whereas the neurohypophysis has the structure of nervous tissue. Hormones secreted by the adenohypophysis serve very different functions from those released from the neurohypophysis.

The protected location of this dual gland suggests its importance. The pituitary gland lies buried deep in the cranial cavity, in a well-protected location. It sits securely within a "seat" called the **sella turcica** formed by two bony projections at the top of the sphenoid bone (see **Figure 8-10,** *C,* on p. 184).

A stem-like structure, the *pituitary stalk,* attaches the gland to the undersurface of the brain. More specifically, the stalk attaches the pituitary body to the hypothalamus.

Anterior Pituitary Gland Hormones

The anterior pituitary gland secretes several major hormones. Each of the four hormones listed as a **tropic hormone** in **Table 12-1** stimulates another endocrine gland to grow and secrete its hormones.

Because the anterior pituitary gland exerts this tropic control over the structure and function of the thyroid gland, the adrenal cortex, the ovarian follicles, and the corpus luteum, in the past it was sometimes called the *master gland.* However, because its secretions are in turn controlled by the hypothalamus and other mechanisms, the anterior pituitary is hardly the master of body function it was once thought to be.

Thyroid-stimulating Hormone

Thyroid-stimulating hormone (TSH) acts on the thyroid gland. As its name suggests, it stimulates the thyroid gland to increase secretion of thyroid hormone.

Adrenocorticotropic Hormone

The **adrenocorticotropic hormone (ACTH)** acts on the adrenal cortex. It stimulates the adrenal cortex to increase in size and to secrete larger amounts of its hormones, especially larger amounts of cortisol (hydrocortisone).

Follicle-stimulating Hormone

Follicle-stimulating hormone (FSH) stimulates the primary ovarian follicles in an ovary to start growing and to continue developing to maturity (that is, to the point of ovulation). FSH also stimulates follicle cells to secrete estrogens. In the male, FSH stimulates the seminiferous tubules to grow and form sperm.

Luteinizing Hormone

Luteinizing hormone (LH) acts with FSH to perform several functions. It stimulates a follicle and ovum to complete their growth to maturity, it stimulates follicle cells to secrete estrogens, and it causes ovulation (rupturing of the mature follicle with expulsion of its ripe ovum). Because of this function, LH is sometimes called the *ovulating hormone.*

LH also stimulates the formation of a golden body, the corpus luteum, from the ruptured follicle. This process—called **luteinization**—is the one that earned LH its title of *luteinizing hormone.* As it promotes luteinization, LH stimulates the corpus luteum to produce the hormone progesterone.

The male pituitary gland also secretes LH. In males, LH stimulates interstitial cells in the testes to develop and secrete testosterone, the male sex hormone.

Growth Hormone

Another important hormone secreted by the anterior pituitary gland is **growth hormone (GH).** GH speeds up the movement of digested proteins (amino acids) out of the blood and into the cells, and this accelerates the cells' **anabolism** (building up) of amino acids to form tissue proteins (see Chapter 19). This anabolic action promotes normal growth.

Growth hormone also affects fat and carbohydrate metabolism. It accelerates fat catabolism (breakdown) but slows glucose catabolism. This means that less glucose leaves the blood to enter cells, and therefore the amount of glucose in the blood increases.

Thus growth hormone and insulin have opposite effects on blood glucose. Insulin decreases blood glucose, and growth hormone increases it. Too much insulin in the blood produces **hypoglycemia** (lower than normal blood glucose concentration). Too much growth hormone produces **hyperglycemia** (higher than normal blood glucose concentration).

Also called *human growth hormone (hGH),* this hormone is used by some people to keep themselves youthful or to boost athletic performance. These unapproved uses can have dangerous side effects by disrupting normal hormone balances in the body.

Hypersecretion of growth hormone during the early years of life produces a condition called **gigantism.** The name suggests the obvious characteristics of this condition. The child grows to a giant size.

If the anterior pituitary gland secretes too much growth hormone after the normal growth years, the disease called **acromegaly** develops. Characteristics of this disease are enlargement of the bones of the hands, feet, jaws, and cheeks. The facial appearance that is typical of acromegaly results from the combination of bone and soft tissue overgrowth. A prominent forehead and large nose are characteristic. In addition, patients with acromegaly may have enlarged skin pores and an overgrown mandible. **Figure 12-5** illustrates the major characteristics of gigantism and acromegaly.

Hyposecretion of growth hormone during the growth years often produces pituitary **dwarfism.** People with this condition usually have a body frame of normal proportions but are much smaller in overall size. Dwarfism caused by other conditions may produce an oddly proportioned body frame. Pituitary dwarfism can be treated with injections of synthetic growth hormone as the skeleton develops.

Prolactin

The anterior pituitary gland also secretes **prolactin (PRL),** or *lactogenic hormone.* During pregnancy, PRL stimulates the breast development necessary for eventual lactation (milk secretion). Also, soon after delivery of a baby, PRL stimulates the breasts to start secreting milk, a function suggested by prolactin's other name, *lactogenic hormone.*

One of the most common types of pituitary tumor is **prolactinoma,** a noncancerous adenoma that produces hypersecretion of PRL. Most prolactinomas are small and occur mostly in women. A patient may have symptoms typical of cranial tumors, such as headache, vision and other sensory changes, lethargy, and nausea. The excess PRL can cause changes in reproductive function including breast tenderness or enlargement, abnormal milk production, infertility, and loss of sexual interest or function.

For those who need treatment for prolactinoma, medications are usually effective. However, sometimes radiation therapy or surgery is required.

For a brief summary of anterior pituitary hormone target organs and functions, see **Figure 12-6**.

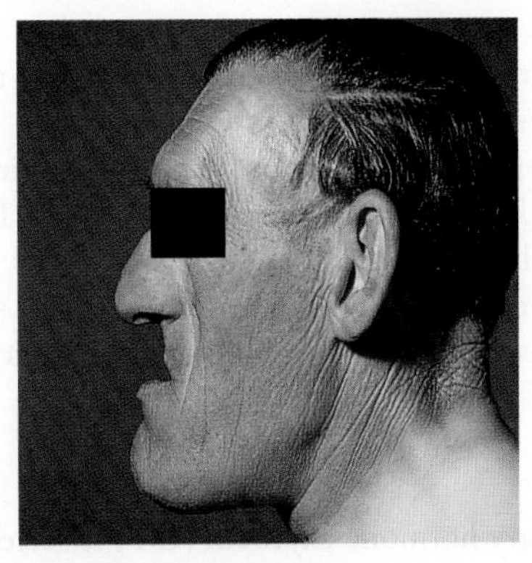

FIGURE 12-5 Growth hormone abnormalities. A, The 22-year-old man on the left with gigantism is much taller than his identical twin on the right. **B,** Acromegaly. Notice the large head, exaggerated projection of the lower jaw, and protrusion of the ridge above the eye orbits.

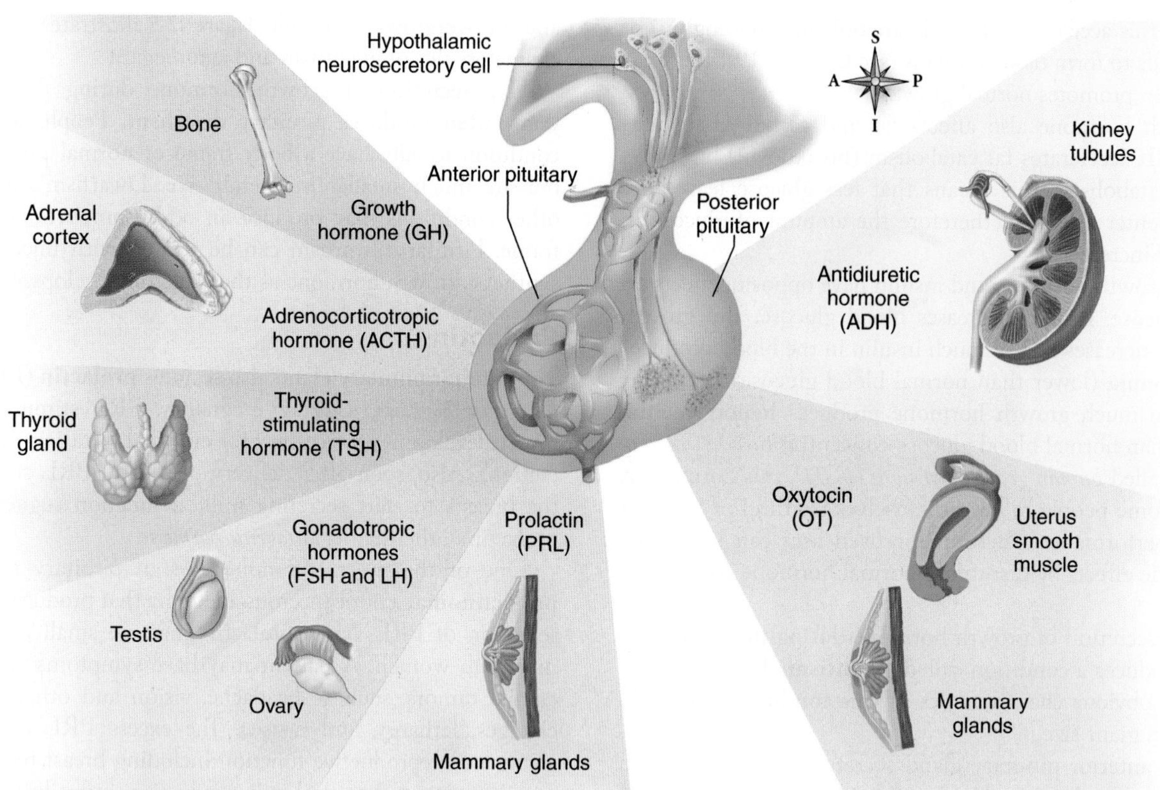

FIGURE 12-6 Pituitary hormones. Principal anterior and posterior pituitary hormones and their target organs.

Posterior Pituitary Gland Hormones

The posterior pituitary gland stores and releases two hormones—*antidiuretic hormone* and *oxytocin*. Both hormones are produced in cell bodies that are located in the hypothalamus but are released from the ends of axons that are located in the posterior pituitary.

Antidiuretic Hormone

Antidiuretic hormone (ADH) is a major regulator of fluid balance in the human body. ADH accelerates the reabsorption of water from urine in kidney tubules back into the blood when the body needs to conserve water. With more water moving out of the tubules into the blood, less water remains in the tubules, and therefore less urine leaves the body.

The term *antidiuretic* is appropriate because *anti-* means "against" and **diuretic** means "increasing the volume of urine excreted." Therefore, antidiuretic means "acting against an increase in urine volume"—in other words, ADH acts to decrease urine volume—and thus prevent dehydration.

Hyposecretion of ADH results in **diabetes insipidus,** a condition in which large volumes of urine are formed. Dehydration and electrolyte imbalances may cause serious problems. Although increased water intake can relieve mild symptoms, many cases also require administering a synthetic form of ADH.

Oxytocin

The posterior pituitary hormone **oxytocin (OT)** is secreted at high levels by a woman's body before and after she has a baby. Oxytocin stimulates contraction of the smooth muscle of the pregnant uterus and is believed to initiate and maintain labor. This is why physicians sometimes prescribe oxytocin injections to induce or increase labor.

Oxytocin also performs a function important to a newborn baby. It causes the glandular cells of the breast to release milk into ducts from which a baby can obtain it by sucking. In short, oxytocin stimulates "milk letdown."

Oxytocin is also thought to enhance social bonding—a function helpful in supporting the mother-infant bond.

The right side of **Figure 12-6** summarizes posterior pituitary functions. Disorders of the anterior and posterior pituitary are summarized in **Table 12-1**.

Hypothalamus

In discussing ADH and oxytocin, we noted that these hormones were *released* from the posterior lobe of the pituitary. As we also stated, actual production of these two hormones occurs in the hypothalamus. Two groups of secretory neurons in the hypothalamus synthesize the posterior pituitary hormones, which then pass down along axons into the pituitary gland. Release of ADH and oxytocin into the blood is controlled by nervous stimulation.

In addition to oxytocin and ADH, the hypothalamus also produces substances called **releasing hormones (RHs)** and **inhibiting hormones (IHs).** These substances are produced in the hypothalamus and then released directly into a connected blood capillary system. This system carries the hormones to the anterior pituitary gland, where they stimulate or inhibit the release of anterior pituitary hormones into the general blood circulation.

The combined nervous and endocrine functions of the hypothalamus allow the nervous system to influence many endocrine functions. Therefore, the hypothalamus plays a dominant role in the regulation of many body functions related to homeostasis. Examples include the regulation of body temperature, appetite, and thirst.

QUICK CHECK

1. How are the anterior pituitary and posterior pituitary different? How are they alike?
2. What makes a hormone a *tropic* hormone?
3. What are the effects of a prolactinoma?
4. How does the hypothalamus control the pituitary gland?

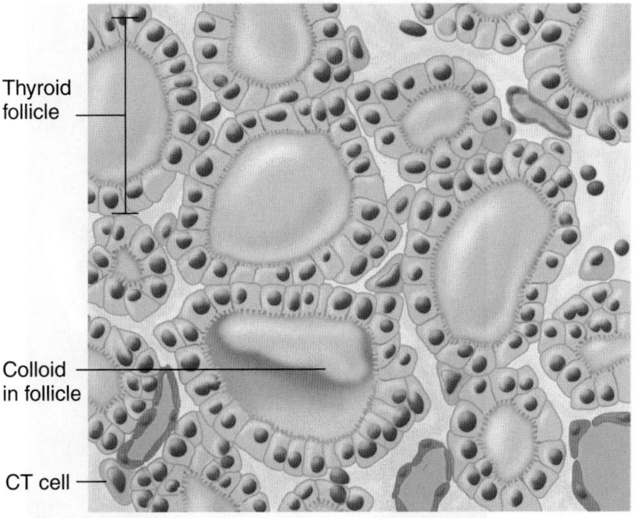

FIGURE 12-8 Thyroid gland tissue. Thyroid hormone is produced by follicular cells in the walls of the thyroid follicles. Note that each of the follicles is filled with colloid—a fluid with fine, suspended particles. The colloid serves as a storage medium for the thyroid hormones. Another type of thyroid gland cells, called *CT cells,* are outside the follicles and secrete calcitonin (CT).

12

Thyroid Gland

The thyroid gland lies in the neck just below the larynx (Figure 12-7). The thyroid gland secretes two thyroid hormones and the hormone **calcitonin (CT).**

As Figure 12-8 shows, thyroid tissue is organized into many chambers called **thyroid follicles.** Each thyroid follicle is filled with a thick fluid having many fine, suspended particles called *colloid.*

Thyroid Hormone

What was once thought of as a single thyroid hormone is actually two similar hormones: **triiodothyronine (T_3)** and **thyroxine (T_4).**

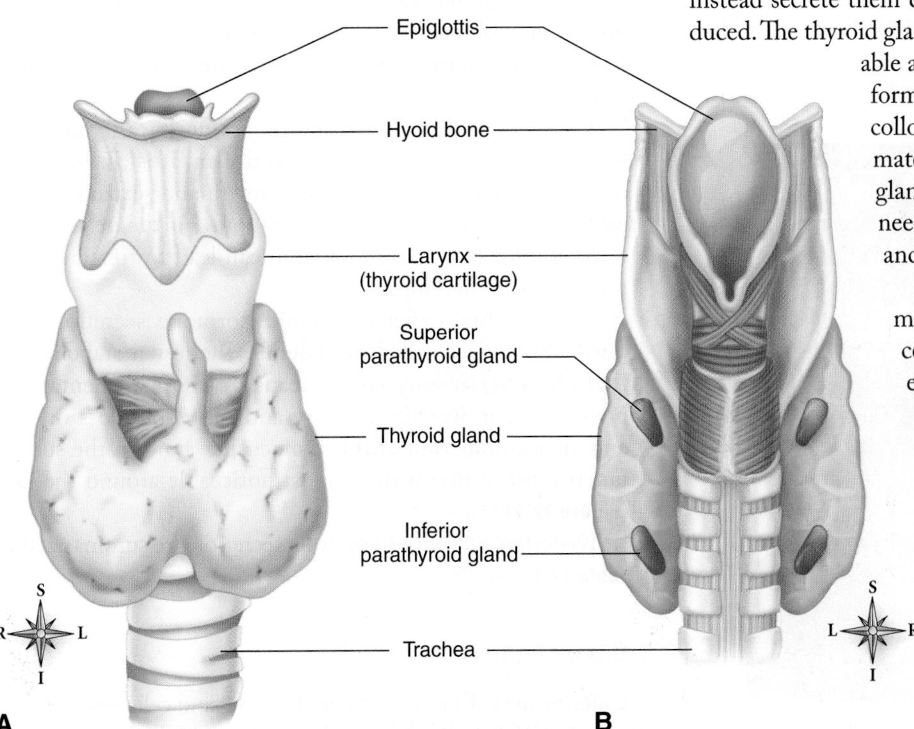

FIGURE 12-7 Thyroid and parathyroid glands. Note their relationship to each other and to the larynx (voice box) and trachea. **A,** Anterior view. **B,** Posterior view.

Of the two thyroid hormones, T_4 is the more abundant. However, T_3 is the more potent and is considered by physiologists to be the principal thyroid hormone. One molecule of T_4 contains four atoms of iodine, and one molecule of T_3, as its name suggests, contains three iodine atoms. For thyroid hormones to be produced in adequate amounts, the diet must contain sufficient iodine.

Most endocrine glands do not store their hormones but instead secrete them directly into the blood as they are produced. The thyroid gland is different in that it stores considerable amounts of the thyroid hormones in the form of molecules suspended in a fluid as a colloid, as seen in Figure 12-8. The colloid material is stored in the follicles of the gland, and when the thyroid hormones are needed, they are released from the colloid and secreted into the blood.

T_4 and T_3 are small, nonsteroid hormones that are able to enter their target cell to find their receptors. This is an exception to the general model of nonsteroid action requiring an internal second messenger.

T_4 and T_3 influence every one of the trillions of cells in our bodies. They make them speed up their release of energy from nutrients. In other words, these thyroid hormones stimulate cellular metabolism. This has far-reaching effects. Because all body functions depend on a normal supply of energy, they all depend on normal thyroid secretion. Even normal

12

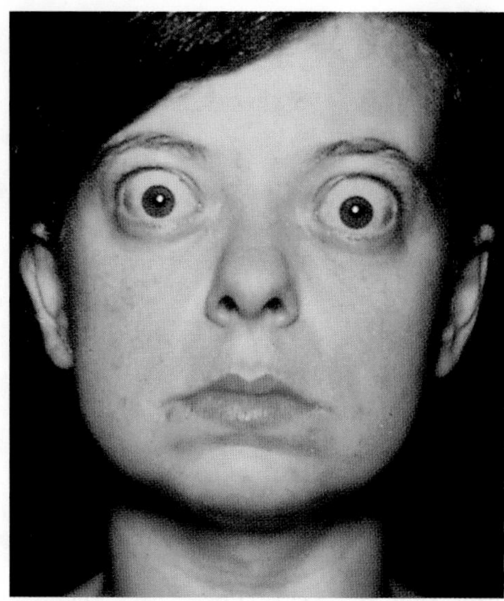

FIGURE 12-9 Hyperthyroidism. Note the prominent, protruding eyes (exophthalmos) of this woman with Graves disease.

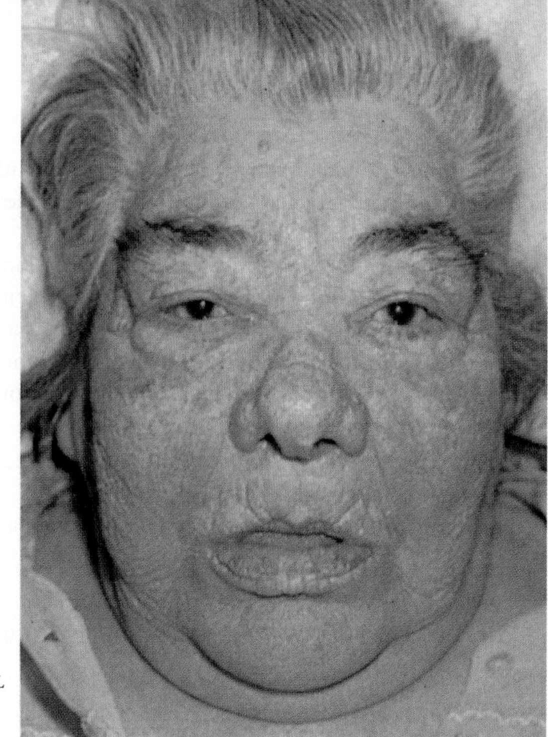

FIGURE 12-11 Myxedema. This condition results from hyposecretion of the thyroid gland during the adult years. Note the edema around the eyes, facial puffiness, prominent tongue, coarse hair, and dull yellowish skin.

mental and physical growth and development depend on normal thyroid functioning.

Hyperthyroidism, or oversecretion of the thyroid hormones, dramatically increases the metabolic rate. Nutrients are consumed by the cells at an excessive rate, and individuals who suffer from this condition lose weight, have an increased appetite, and show signs of nervous irritability. They appear restless, jumpy, and excessively active.

Many patients with hyperthyroidism also have very prominent, almost protruding eyes—a condition called **exophthalmos** (**Figure 12-9**). Hyperthyroidism with exophthalmos is characteristic of **Graves disease,** an inherited autoimmune condition that occurs five times more frequently in women than in men.

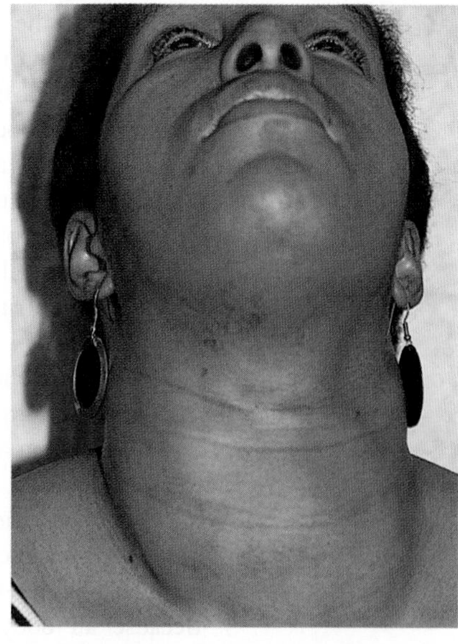

FIGURE 12-10 Goiter. The enlarged thyroid gland appears as a swelling of the neck. This condition—a *simple goiter*—results from a low dietary intake of iodine.

Hypothyroidism, or undersecretion of thyroid hormones, can be caused by and result in a number of different conditions. Low dietary intake of iodine causes a painless enlargement of the thyroid gland called a **simple goiter,** shown in **Figure 12-10**.

This condition was once common in areas of the United States where the iodine content of the soil and water was inadequate. The use of iodized salt has dramatically reduced the incidence of goiters caused by low iodine intake. To produce a **goiter,** the gland enlarges in an attempt to compensate for the lack of iodine in the diet necessary for the synthesis of thyroid hormones.

Hyposecretion of thyroid hormones during the formative years leads to a condition called **cretinism.** It is characterized by a low metabolic rate, retarded growth and sexual development, and often, mental retardation. Fortunately, health screening for low thyroid function can lead to treatment before cretinism develops.

Later in life, deficient thyroid hormone secretion produces the disease called **myxedema.** The low metabolic rate that characterizes myxedema leads to lessened mental and physical vigor, weight gain, dry and scaly skin, loss of hair, and an accumulation of thick, mucuslike fluid in the subcutaneous tissue that is often most noticeable around the face (**Figure 12-11**).

Disorders of thyroid hormone secretion are summarized in **Table 12-1**.

Calcitonin

Calcitonin (CT) is secreted by thyroid gland cells—sometimes called *CT cells*—that lie outside the thyroid follicles.

Calcitonin decreases the concentration of calcium in the blood by first acting on bone to inhibit its breakdown. With less bone being resorbed, less calcium moves out of bone into blood, and, as a result, the concentration of calcium in blood decreases.

An increase in calcitonin secretion quickly follows any increase in blood calcium concentration, even if it is a slight one. This causes blood calcium concentration to decrease to its normal level. Calcitonin thus helps maintain homeostasis of blood calcium. It prevents a harmful excess of calcium in the blood, a condition called **hypercalcemia,** from developing.

Parathyroid Glands

The **parathyroid glands** are small lumps of glandular epithelium. There are usually four of them, and they are found on the posterior surface of the thyroid gland (see **Figure 12-7**).

The parathyroid glands secrete **parathyroid hormone (PTH).**

PTH increases the concentration of calcium in the blood—the opposite effect of the thyroid gland's calcitonin. Whereas calcitonin acts to decrease the amount of calcium being dissolved and reabsorped from bone, PTH acts to increase it.

PTH stimulates mineral-dissolving osteoclast cells in bone tissue to increase their breakdown of bone's hard matrix, a process that frees the calcium stored in the matrix. The released calcium then moves out of bone into blood, and this in turn increases the blood's calcium concentration. PTH also promotes absorption of calcium from food and reduces loss of calcium in the urine.

Figure 12-12 provides a summary of the antagonistic effects of calcitonin and parathyroid hormone. This calcium-control mechanism is a matter of life-and-death importance because our cells are extremely sensitive to changing amounts of blood calcium. They cannot function normally with either too much or too little calcium.

For example, with too much blood calcium, brain cells and heart cells soon do not function normally; a person becomes mentally disturbed, and the heart may stop altogether. However, with too little blood calcium, nerve cells become overactive, sometimes to such a degree that they bombard muscles with so many impulses that the muscles go into spasms.

Disorders of parathyroid secretion are summarized in **Table 12-1**.

To learn more about the thyroid and parathyroid glands, go to AnimationDirect online at *evolve.elsevier.com*.

QUICK CHECK

1. Where are the thyroid and parathyroid glands located?
2. What is the danger of a *hyposecretion* of thyroid hormones during the formative years?
3. What is a *goiter* and how does it develop?
4. Calcitonin and parathyroid hormone both regulate the blood concentration of what important ion?

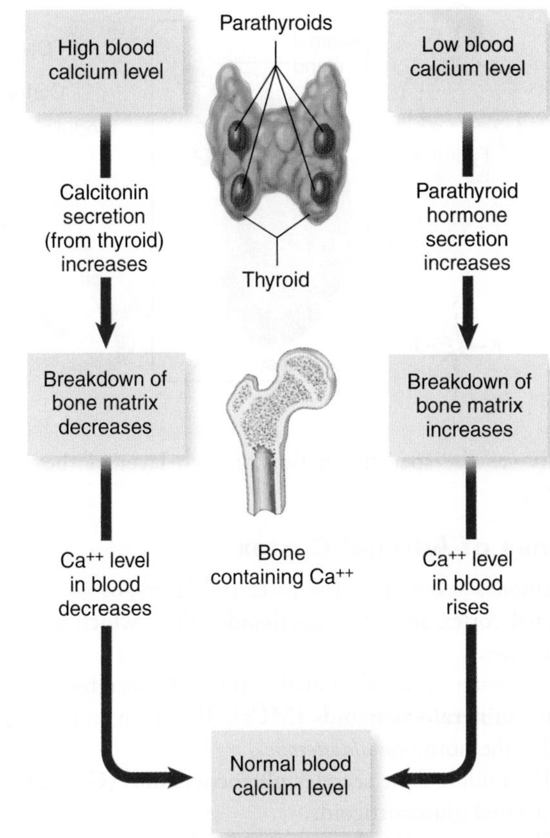

FIGURE 12-12 Regulation of blood calcium levels. Calcitonin and parathyroid hormones have antagonistic (opposite) effects on calcium concentration in the blood. Both are negative feedback effects because they reverse a trend away from normal blood calcium levels.

Adrenal Glands

Location of Adrenal Glands

As you can see in **Figure 12-1** and **Figure 12-13**, an adrenal gland curves over the superior surface of each kidney.

From the surface an adrenal gland appears to be only one organ, but it is actually two separate endocrine glands: the *adrenal cortex* and the *adrenal medulla.* Does this two-glands-in-one structure remind you of another endocrine organ? (See p. 326.)

The adrenal cortex is the outer part of an adrenal gland and is made up of glandular epithelium. The adrenal medulla is the inner part and it is made up of secretory nervous tissue—much like the secretory nervous tissue of the posterior pituitary. Each part releases a different set of hormones, as you might expect.

Adrenal Cortex

Three different zones or layers of cells make up the **adrenal cortex** as you can see in **Figure 12-13**. Follow this diagram carefully as you read the following paragraph, and you will

12

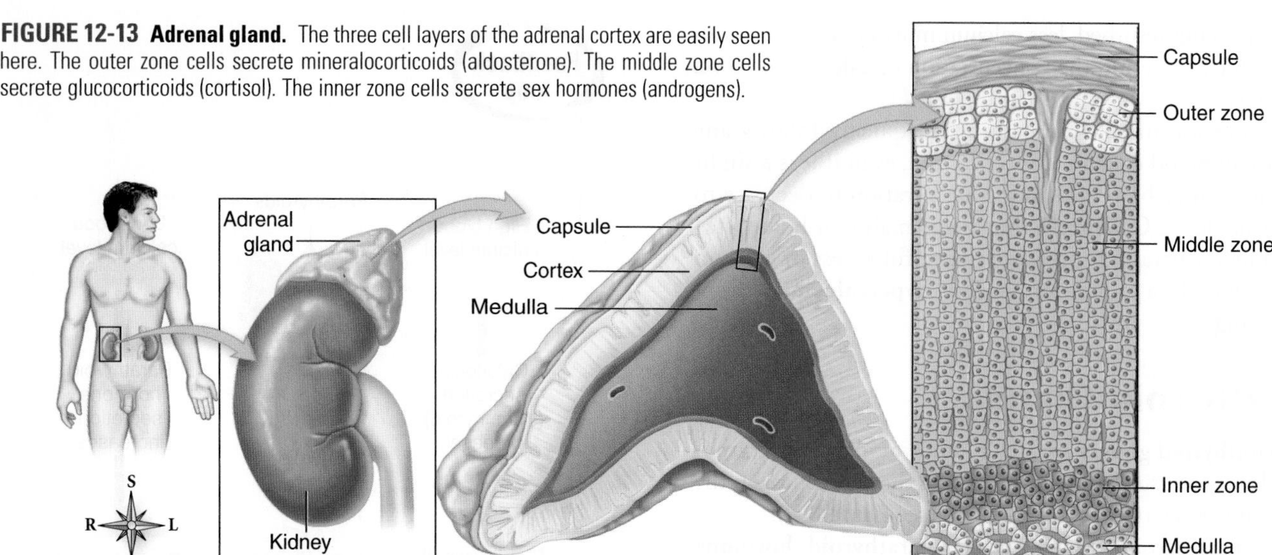

FIGURE 12-13 Adrenal gland. The three cell layers of the adrenal cortex are easily seen here. The outer zone cells secrete mineralocorticoids (aldosterone). The middle zone cells secrete glucocorticoids (cortisol). The inner zone cells secrete sex hormones (androgens).

easily see the special function of each layer of the adrenal cortex.

Zones of Adrenal Cortex

Hormones secreted by the three cell layers, or zones, of the adrenal cortex are called **corticoids,** all of which are steroid hormones.

The outer zone of adrenal cortex cells secretes hormones called **mineralocorticoids (MCs).** The main mineralocorticoid is the hormone *aldosterone.*

The middle zone secretes **glucocorticoids (GCs).** *Cortisol* is the chief glucocorticoid.

The innermost or deepest zone of the cortex secretes small amounts of **sex hormones.** Sex hormones secreted by the adrenal cortex resemble testosterone and are classified as *androgens* (male sex hormones).

We now discuss briefly the functions of the main cortical hormones.

Aldosterone

As their name suggests, *mineralocorticoids* help control the amount of certain mineral salts (mainly sodium chloride) in the blood.

Aldosterone is the chief mineralocorticoid. Remember its main functions—to increase the amount of sodium and decrease the amount of potassium in the blood—because these changes lead to other profound changes.

Aldosterone increases blood sodium and decreases blood potassium by influencing the kidney tubules. It causes the kidney tubules to speed up their reabsorption of sodium back into the blood so that less of it will be lost in the urine. At the same time, aldosterone causes the tubules to increase their secretion of potassium so that more of this mineral will be lost in the urine. The effects of aldosterone speed up kidney reabsorption of water.

Cortisol

An important function of glucocorticoids is to help maintain normal blood glucose concentration. **Cortisol**—called **hydrocortisone** when used as medical therapy—is the chief glucocorticoid produced by the adrenal cortex.

Cortisol and other glucocorticoids increase **gluconeogenesis,** a process in liver cells that converts amino acids or glycerol to glucose. Glucocorticoids act in several ways to increase gluconeogenesis. They promote the breakdown of tissue proteins to amino acids, especially in muscle cells. Amino acids thus formed move out of the tissue cells into blood and circulate to the liver. Liver cells then change them to glucose by the process of gluconeogenesis. The newly formed glucose leaves the liver cells and enters the blood. This action increases blood glucose concentration.

In addition to performing functions—which are necessary for maintaining normal blood glucose concentration—glucocorticoids such as cortisol also play an essential part in maintaining normal blood pressure. They act in a complicated way to make it possible for two other hormones secreted by the adrenal medulla to partially constrict blood vessels, a condition necessary for maintaining normal blood pressure.

Also, glucocorticoids act with these hormones from the adrenal medulla to produce an anti-inflammatory effect. They bring about a normal recovery from inflammations produced by many kinds of agents. The use of hydrocortisone to relieve skin rashes, for example, is based on the anti-inflammatory effect of glucocorticoids.

Review the basic concepts of the inflammatory response in the article *Inflammation* at Connect It! at *evolve.elsevier.com.*

Another effect produced by glucocorticoids is called their *anti-immunity, antiallergy effect.* Glucocorticoids bring about a decrease in the number of certain cells that produce antibodies, substances that make us immune to some factors and allergic to others.

When extreme stimuli act on the body, they produce an internal state or condition known as **stress.** Surgery, hemorrhage, infections, severe burns, and intense emotions are examples of extreme stimuli that bring on stress. The normal adrenal cortex responds to the condition of stress by quickly increasing its secretion of glucocorticoids.

Increased glucocorticoid secretion is only one of many ways in which the body responds to stress. However, it is one of the first stress responses and it brings about many of the other stress responses. Examine Figure 12-14 to discover some of the effects of glucocorticoids in the blood.

When resisting (or avoiding) a threat, the increased blood glucose can help improve our skeletal muscle function. Reduced inflammation may help keep us less swollen—and thus more mobile—while we deal with the threat. Decreased immunity may help us focus all our resources on the more immediate threat. Immunity resumes after a threatening encounter to deal with any damage.

Frequent or prolonged stress responses could cause metabolic problems by disturbing normal mechanisms keeping blood glucose and stored fats in balance. Chronic stress may also increase our susceptibility to cancer and infections by reducing our immunity. Prolonging the anti-inflammatory effects may cause constriction of blood vessels—possibly raising our blood pressure.

Adrenal Sex Hormones

The sex hormones that are secreted by the inner zone of the adrenal cortex are male hormones—**androgens**—similar to testosterone. These hormones are secreted in very small amounts in both adult males and adult females. However, they play an early role in the development of reproductive organs.

In women, these androgens may stimulate the female sexual drive. In men, so much testosterone is secreted by the testes that adrenal androgens are usually not very important in adults.

Adrenal Medulla

The **adrenal medulla,** or inner portion of the adrenal gland shown in Figure 12-13, secretes the hormones **epinephrine (Epi)** and **norepinephrine (NE).** Epinephrine is also known as *adrenaline.*

Our bodies have many ways to defend themselves against enemies that threaten their well-being. A physiologist might say that the body resists stress by producing a coordinated set of stress responses. We have just discussed increased glucocorticoid secretion. An even faster-acting stress response is increased hormone secretion by the adrenal medulla.

The adrenal medulla responds very rapidly to stress because nerve impulses conducted by sympathetic nerve fibers stimulate the adrenal medulla. When stimulated, it literally squirts epinephrine and norepinephrine into the blood. As with glucocorticoids, these hormones may help the body resist or avoid stress. In other words, these hormones produce the body's "fight-or-flight" response to danger (stress).

Suppose you suddenly face some threatening situation. Imagine encountering a large animal that is threatening you with bared teeth. Almost instantly, the medulla of each adrenal gland would be thrust into feverish activity. They would quickly secrete large amounts of epinephrine (adrenaline) into your blood. Many of your body functions would seem to be supercharged. Your heart would beat faster, your blood pressure would rise, more blood would be pumped to your skeletal muscles, your blood would contain increased glucose for more energy, and so on. In short, you would be geared for strenuous activity to either *resist* or *avoid* the animal attack—thus the phrase, "fight or flight."

Epinephrine prolongs and intensifies changes in body function brought about by the stimulation of the sympathetic

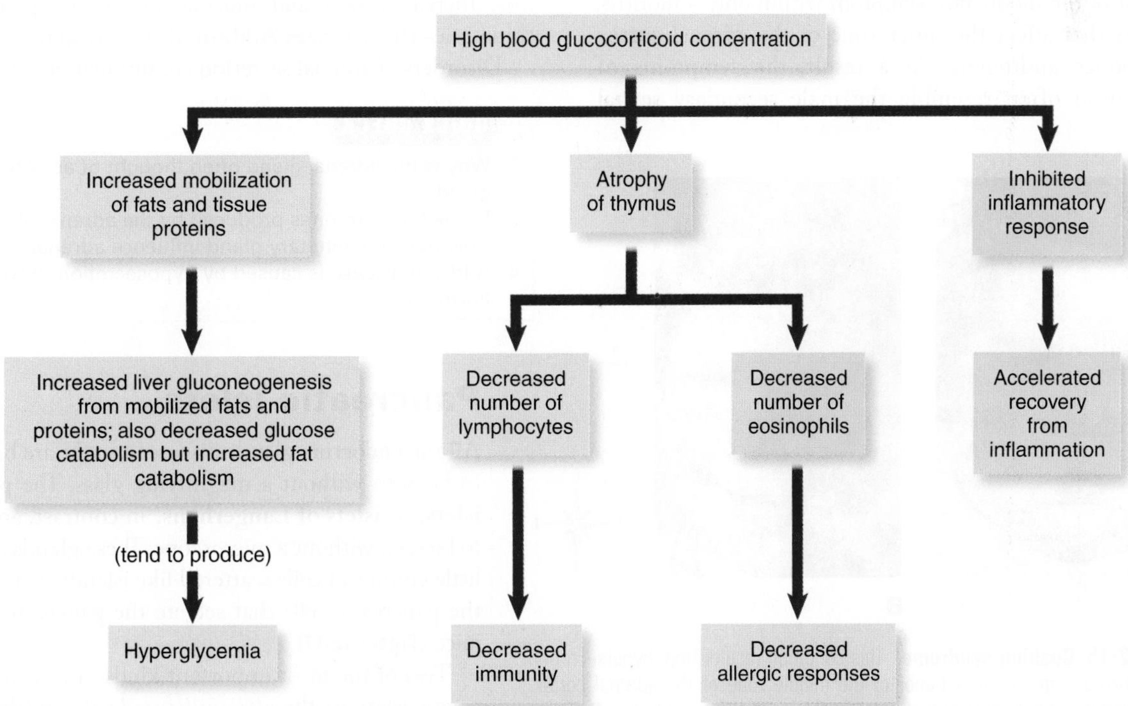

FIGURE 12-14 Stress responses. Stress may trigger elevated secretion of glucocorticoids (GCs) into the blood. This flow chart shows the possible effects induced by high blood GC concentration.

subdivision of the autonomic nervous system. Recall from Chapter 10 that sympathetic, or adrenergic, fibers release epinephrine and norepinephrine as neurotransmitters.

The close functional relationship between the nervous and the endocrine systems is perhaps most noticeable in the body's response to stress. In stress conditions, the hypothalamus acts on the anterior pituitary gland to cause the release of ACTH, which stimulates the adrenal cortex to secrete glucocorticoids. At the same time, the sympathetic subdivision of the autonomic nervous system is stimulated with the adrenal medulla, so the release of epinephrine and norepinephrine occurs to assist the body in responding to the stressful stimulus.

To learn more about adrenal function, go to AnimationDirect online at *evolve.elsevier.com.*

Adrenal Abnormalities

Injury, disease states, or malfunction of the adrenal glands can result in hypersecretion or hyposecretion of several different hormones.

Tumors of the adrenal cortex located in the middle zone often result in the production of abnormally large amounts of glucocorticoids. The medical name for the collection of symptoms that characterize hypersecretion of glucocorticoids is **Cushing syndrome.**

For some reason more women than men develop Cushing syndrome. Its most noticeable features are the so-called *moon face* (Figure 12-15) and the *buffalo hump* on the upper back that develop because of the redistribution of body fat. Individuals with Cushing syndrome also have elevated blood glucose levels and suffer frequent infections. Surgical removal of a glucocorticoid-producing tumor may result in dramatic improvement of the moon-face symptom within only 4 months.

Tumors that affect the inner zone of the adrenal cortex often produce androgens. As a result, the symptoms of hypersecretion often resemble the male secondary sexual characteristics such as beard growth, development of body hair, and increased muscle mass. If these masculinizing symptoms appear in a woman, the cause is frequently a **virilizing tumor** of the adrenal cortex. The term *virile* is from the Latin word *virilis* meaning "male" or "masculine."

Deficiency or hyposecretion of adrenal cortex hormones results in a condition called **Addison disease.** President John F. Kennedy suffered from Addison disease, which causes reduced cortical hormone levels that result in muscle weakness, reduced blood glucose, nausea, loss of appetite, and weight loss. Increased skin and mucous membrane pigmentation sometimes characterizes Addison disease (Figure 12-16).

Disorders of adrenal secretion are summarized in Table 12-1.

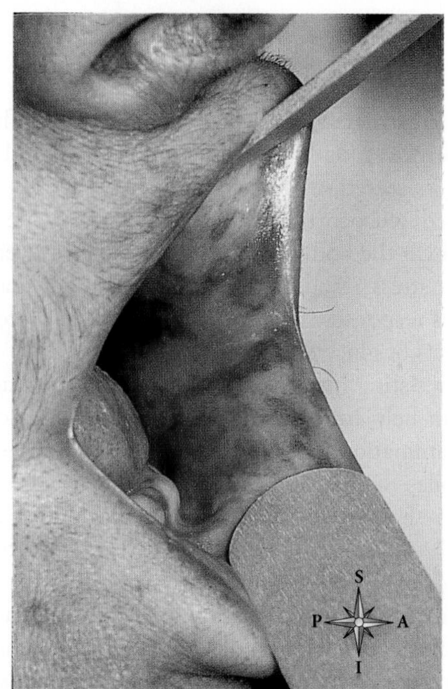

FIGURE 12-16 Addison disease. Addison disease may produce excess pigmentation of the skin and mucous membranes, as in the buccal (cheek) surface shown here.

QUICK CHECK

1. Why is the adrenal gland often thought of as two separate glands?
2. Name the hormones produced by the adrenal gland.
3. How does the pituitary gland influence adrenal function?
4. Addison disease is caused by hyposecretion of which hormone?

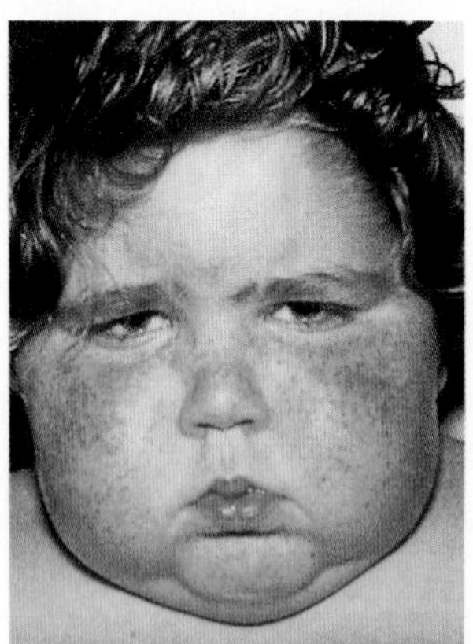

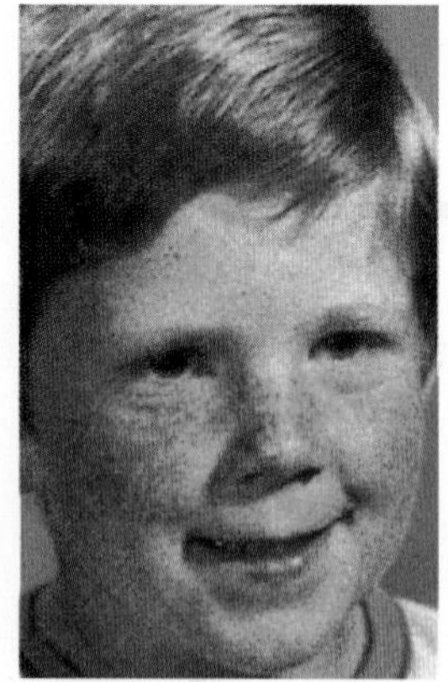

A B

FIGURE 12-15 Cushing syndrome. This condition results from hypersecretion of glucocorticoid hormone by a tumor of the middle zone of the adrenal cortex. **A,** Photo taken when patient was first diagnosed with Cushing syndrome. **B,** Taken 4 months after treatment.

Pancreatic Islets

All the endocrine glands discussed so far are big enough to be seen without a magnifying glass. The **pancreatic islets,** or **islets of Langerhans,** in contrast, are too tiny to be seen without a microscope. These glands are merely little clumps of cells scattered like islands in a sea among the pancreatic cells that secrete the pancreatic digestive juice (Figure 12-17).

Two of the most important kinds of cells in the pancreatic islets are the *alpha cells* (or *A cells*) and *beta cells* (or *B cells*). Alpha cells secrete a hormone called **glucagon,**

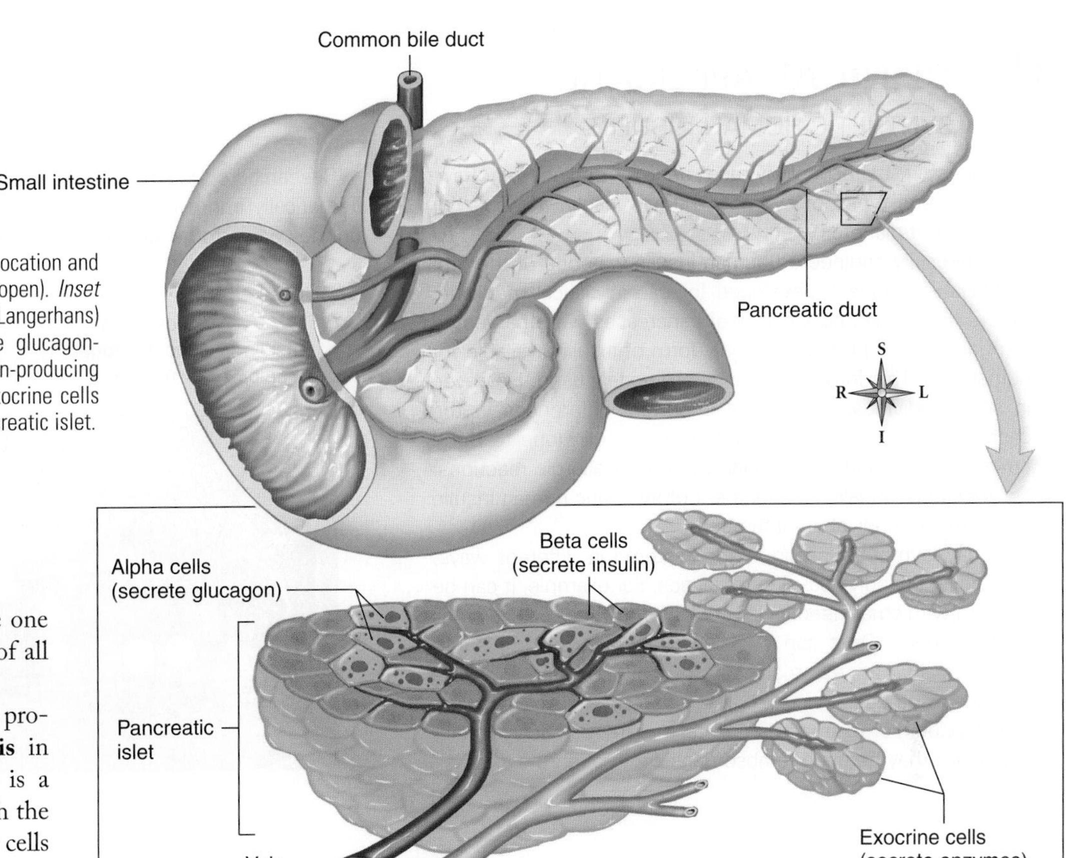

FIGURE 12-17 Pancreas. Location and structure of the pancreas (cut open). *Inset* shows a pancreatic islet (of Langerhans) in cross section, showing the glucagon-producing alpha cells and insulin-producing beta cells. Notice the many exocrine cells surrounding the endocrine pancreatic islet.

whereas beta cells secrete one of the most well-known of all hormones, **insulin.**

Glucagon accelerates a process called **glycogenolysis** in the liver. Glycogenolysis is a chemical process by which the glucose stored in the liver cells in the form of glycogen is converted to glucose. This glucose then leaves the liver cells and enters the blood. Glucagon therefore increases blood glucose concentration.

Insulin and glucagon are antagonists. In other words, insulin decreases blood glucose concentration; glucagon increases it. Insulin is the only hormone that can decrease blood glucose concentration. Several hormones, however, increase glucose concentration, including glucocorticoids, growth hormone, and glucagon. Insulin decreases blood glucose by accelerating its movement out of the blood, through cell membranes, and into cells. As glucose enters the cells at a faster rate, the cells increase their metabolism of glucose.

Briefly then, insulin decreases blood glucose and increases glucose metabolism.

If the pancreatic islets secrete a normal amount of insulin, glucose enters the cells easily, and a normal amount of glucose stays behind in the blood. "Normal" blood glucose is about 70-100 mg of glucose in every 100 mL (one deciliter or dL) of blood during fasting.

If the pancreatic islets secrete too much insulin, as they rarely do when a person has a tumor of the pancreas, more glucose than usual leaves the blood to enter the cells, and blood glucose decreases.

If the pancreatic islets secrete too little insulin, as they do in **type 1 diabetes mellitus,** less glucose leaves the blood to enter the cells, so the blood glucose increases, sometimes to even three or more times the normal amount.

Most cases of **type 2 diabetes mellitus** result from an abnormality of the insulin receptors or their signaling mechanism, preventing the normal effects of insulin on its target cells and thus also raising blood glucose levels.

Screening tests for all types of **diabetes mellitus (DM)** are based on the fact that the blood glucose level is elevated in this condition. Today, most screening is done with a simple test that requires only a drop of blood. Subjects with a high blood glucose level are suspected of having diabetes mellitus.

HEALTH AND WELL-BEING

EXERCISE AND DIABETES MELLITUS

Type 1 diabetes mellitus is characterized by high blood glucose concentration because the lack of sufficient insulin prevents glucose from entering cells. However, exercise physiologists have found that aerobic training increases the number of insulin receptors in target cells and the insulin affinity (attraction) of the receptors. Such training allows a small amount of insulin to have a greater effect than it would have otherwise had. Thus exercise reduces the severity of the diabetic condition.

All forms of diabetes benefit from properly planned exercise therapy. Not only is this form of treatment natural and cost-effective, but it also helps reduce or prevent other problems such as obesity and heart disease.

12

CLINICAL APPLICATION

SYNTHETIC HUMAN INSULIN

Advances in biotechnology and genetic engineering have made several types of synthetic human insulin available for treatment of diabetes.

Genetically engineered human insulin was one of the first "artificial hormones" developed for use in clinical medicine. Before its release, pork and beef pancreases were harvested from animals to produce the only form of insulin available for human use. This form of insulin is still widely used. However, in some individuals, small but important differences in the chemical makeup of animal and human insulin (one to three amino acids) results in immune system or allergic reactions. For these individuals, development of synthetic human insulin was a major medical breakthrough.

Today, synthetic insulin is administered in different ways, depending on individual circumstances. For example, it can be injected with small needles in various forms, such as in a small syringe "pen." Or it can be injected through an implanted needle by a timed, external insulin pump (pictured).

Other methods of insulin delivery also are available or being investigated. For example, so-called "smart" patches containing insulin may be used to absorb insulin through the skin.

Researchers also have developed insulin pills that could dramatically improve diabetes care for millions if in the long run they prove to be safe and effective.

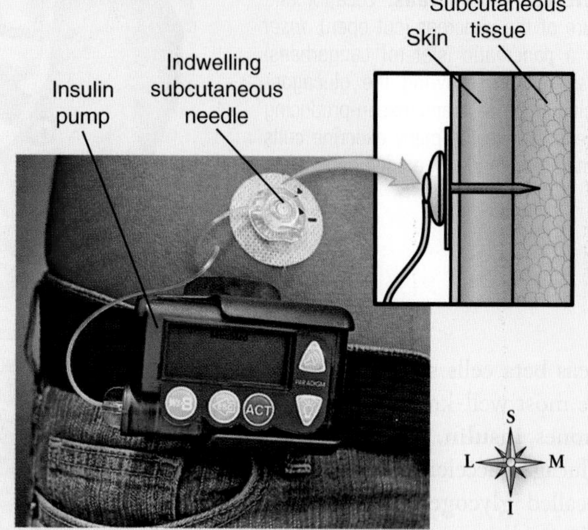

Testing for glucose in the urine is another common screening procedure. In diabetes mellitus, excess glucose is filtered out of the blood by the kidneys and lost in the urine, producing the condition **glycosuria.**

Figure 12-18 summarizes some of the many problems that can be caused by diabetes mellitus. A quick look at these problems underscores the importance of insulin and insulin receptors in healthy bodies.

Disorders of pancreatic islet secretion are summarized in **Table 12-1**.

Important concepts regarding DM are summarized in the article *Diabetes Mellitus* at Connect It! at *evolve.elsevier.com*.

To learn more about insulin function, go to AnimationDirect online at *evolve.elsevier.com*.

QUICK CHECK

1. Name the two primary hormones of the pancreatic islets.
2. What effect does insulin have on the blood's glucose concentration?
3. How does *type 1 diabetes* differ from *type 2 diabetes*?

Sex Glands

Female Sex Glands

A woman's primary sex glands are her two ovaries. Each ovary contains two different kinds of glandular structures: the *ovarian follicles* and the *corpus luteum.*

Ovarian follicles are little pockets in which egg cells, or **ova,** develop. Ovarian follicles also secrete **estrogen,** the "feminizing hormone." Estrogen is involved in the development and maturation of the breasts and external genitals. This hormone is also responsible for development of adult female body contours and initiation of the menstrual cycle.

The **corpus luteum** chiefly secretes **progesterone** but also some estrogen.

We shall save our discussion of the structure of these endocrine glands and the functions of their hormones for Chapter 23.

Male Sex Glands

Some of the cells of the testes produce the male sex cells called **sperm.** Other cells in the testes, male reproductive ducts, and glands produce the liquid portion of the male reproductive fluid called **semen.** The interstitial cells in the testes secrete the male sex hormone called **testosterone** directly into the blood. These cells of the testes are therefore the male endocrine glands.

Testosterone is the "masculinizing hormone." It is responsible for the maturation of the external genitals, beard growth, changes in voice at puberty, and for the muscular development and body contours typical of the male.

Chapter 23 contains more information about the structure of the testes and the functions of testosterone.

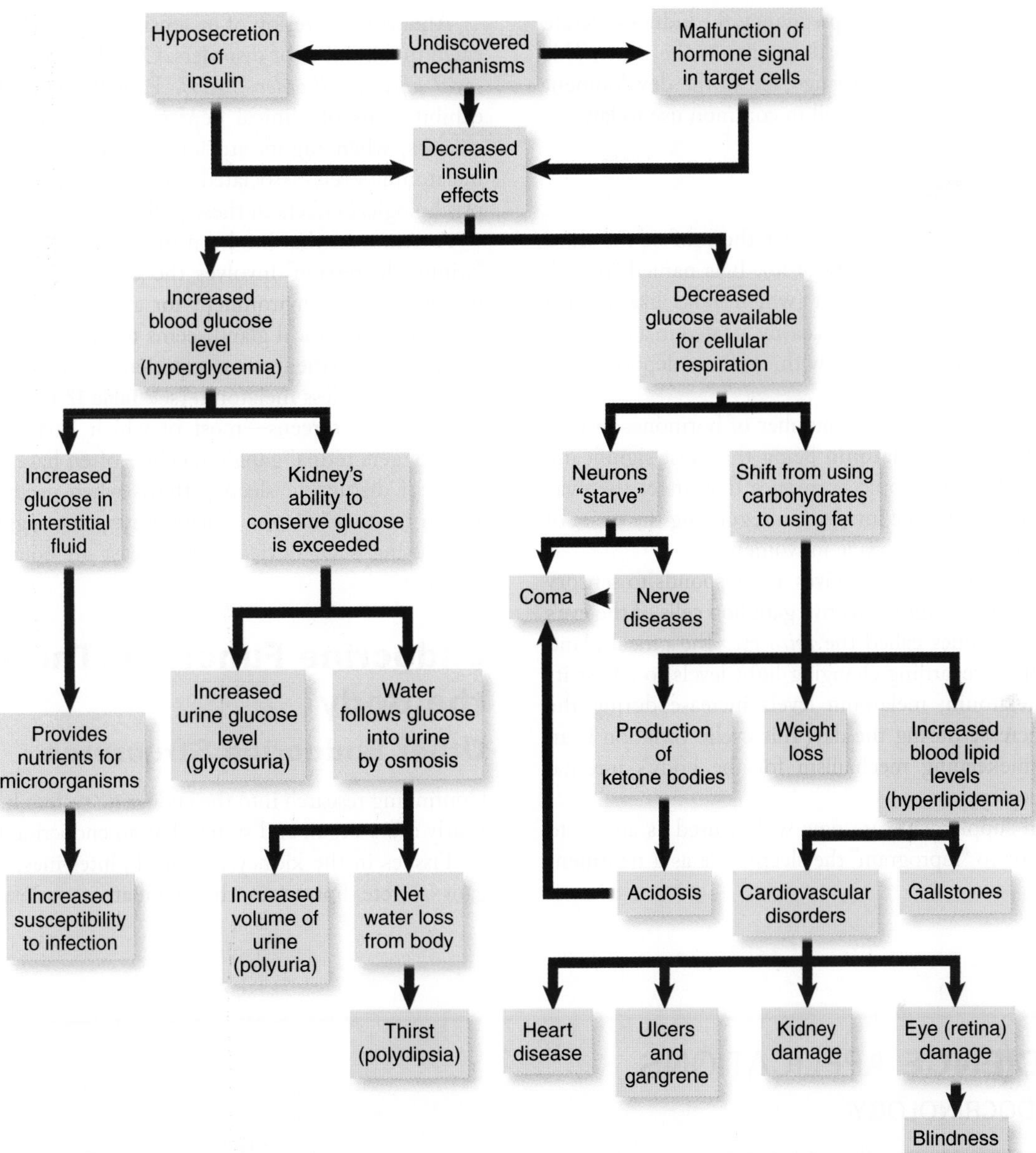

FIGURE 12-18 Diabetes mellitus. The signs and symptoms of this disorder *(highlighted in yellow)* all result from decreased insulin effects. Although this diagram may seem overwhelming at first glance, it is easy to follow if you trace each of the pathways step-by-step through to the end. By doing so, you will begin to appreciate how one event often triggers another in human physiology.

Thymus

The thymus is located in the mediastinum (see **Figure 12-1**), and in infants it may extend up into the neck as far as the lower edge of the thyroid gland. Like the adrenal gland, the thymus has a cortex and medulla. Both portions are composed largely of lymphocytes (white blood cells).

The thymus is the location where many of the body's cells of immunity develop. The hormone **thymosin** is actually a group of several hormones that together play an important role in regulating the development and function of T cells—an important category of immunity agents in the body. The function of T cells is discussed in detail in Chapter 16.

Although this structure is small—it weighs about 20 grams—it plays a critical part in the body's defenses against infections and cancer.

Placenta

The placenta functions as a temporary endocrine gland.

During pregnancy, it produces **chorionic gonadotropins,** so called because they are tropic hormones secreted by cells of the **chorion,** the outermost membrane that surrounds the embryo and fetus during development in the uterus. In addition to chorionic gonadotropins, the placenta also produces estrogen and progesterone.

During the earliest weeks of pregnancy, the kidneys excrete large amounts of chorionic gonadotropins in the urine. This fact, discovered nearly a century ago, led to the development of *early pregnancy tests* that are still in common use today.

Pineal Gland

The pineal gland is a small gland near the roof of the third ventricle of the brain (see **Figure 10-13**). It is named "pineal" because it resembles the pine nut (which looks like a small kernel of corn). The pineal gland is easily located in a child but becomes fibrous and encrusted with calcium deposits as a person ages.

The pineal gland produces a number of hormones in very small quantities, with **melatonin** being the most significant. Melatonin inhibits the tropic hormones that affect the ovaries, and it is thought to be involved in regulating the onset of puberty and the menstrual cycle in women.

Because the pineal gland receives and responds to sensory information from the light-sensitive ganglion cells of the eye's retina, it is sometimes called the *third eye.* The pineal gland uses information regarding changing light levels to adjust its output of melatonin; melatonin levels increase during the night and decrease during the day. This cyclic variation is an important timekeeping mechanism for the body's internal clock and sleep cycle.

Melatonin supplements are now widely used as an aid to induce sleep or to "reprogram" the sleep cycle as a treatment for jet lag.

Abnormal secretion of or sensitivity to melatonin is implicated in a number of disorders. One dramatic example is *seasonal affective disorder (SAD).* Patients with this condition exhibit signs of clinical depression only during the winter months, when nights are long. Apparently, unusually high melatonin levels associated with long winter nights cause psychological effects in these patients.

A treatment that has been successful in some cases of this "winter depression" involves the use of bright lights in the person's indoor environment for a few hours each day after sundown. The pineal gland seems to be tricked into responding as if the patient were experiencing a long summer day, thus secreting less melatonin (see **Table 12-1**).

Electronic screens—most of which have the bluish cast that triggers retinal ganglion cells—often have the unwanted effect of disrupting sleep patterns when used late at night. When treating insomnia, many physicians suggest refraining from using such devices for at least an hour or so before bedtime.

Endocrine Functions Throughout the Body

Other Endocrine Structures

Continuing research into the endocrine system has shown that nearly every organ and system has an endocrine function.

Tissues in the kidneys, stomach, intestines, and other organs secrete hormones that regulate a variety of essential

SCIENCE APPLICATIONS

ENDOCRINOLOGY

Frederick Banting (1891–1941)

The undisputed heroes of *endocrinology* are Canadian surgeon Frederick Banting and his assistant Charles Best. Until the early twentieth century, children with type 1 diabetes mellitus died a slow, horrible death as a result of their cells literally starving to death from lack of glucose. Acting on Banting's idea for removing insulin from the pancreatic islets of dogs, the two were the first to successfully isolate this important hormone. Chemist James Collip was able to purify the insulin sufficiently so that in 1921 their colleague, Scots physiologist John Macleod, could administer the insulin to a 14-year-old boy with diabetes. It worked! The treatment not only relieved the boy's suffering, but it also gave him a healthy, long life.

Their breakthrough, for which Banting and Macleod received the 1923 Nobel Prize, was the start of a century of rapid progress in understanding and treating endocrine disorders.

Charles Best (1899–1978)

Because hormones affect so many different body functions, nearly every kind of health professional, from *medical doctors* to *nurses* to *dietitians,* needs to be aware of their functions. Of course, hormones and chemicals that influence hormone actions are often used in treatments, so *pharmacologists* and *pharmacists* also must have an excellent knowledge of endocrinology.

Some scientists have applied principles of endocrinology in a variety of unexpected ways, including the development of early pregnancy test kits and ovulation test kits, to the use of synthetic hormones in healthy people to help them control their fertility.

human functions. For example, **ghrelin** is secreted by epithelial cells lining the stomach and boosts appetite, slows metabolism, and reduces fat burning. Ghrelin may, therefore, be involved in the development of obesity.

Another example is **atrial natriuretic hormone (ANH),** which is secreted by cells in the wall of the heart's atria (upper chambers). ANH is an important regulator of fluid and electrolyte homeostasis. ANH is an antagonist to aldosterone. Aldosterone stimulates the kidney to retain sodium ions and water, whereas ANH stimulates loss of sodium ions and water.

A more recently discovered hormone is **leptin,** which is secreted by fat-storing cells throughout the body. Leptin seems to regulate how hungry or full we feel and how fat is metabolized by the body. Researchers are now looking at how leptin works with other hormones in the hopes of finding ways to treat patients with obesity, diabetes mellitus, and other disorders involving fat storage.

In the *Clear View of the Human Body* (follows p. 8), try to find as many endocrine organs as you can and note their positions relative to other structures of the body.

Hormone Actions in Every Organ

This chapter has included a list of endocrine glands and hormones that may have seemed endless. Yet it is only a small fraction of the known hormones and hormone-producing cells.

We have mentioned the actions of hormones in previous chapters, and we will continue to discuss nearly all the hormones identified in this chapter as we proceed through the rest of this book. Why? Hormone actions are important regulators of homeostasis throughout the body. They play critical roles in the function of *every* organ of the body.

As you move forward in your course, always be on the lookout for the regulatory and coordinating roles of hormones. By doing so, you will have a more complete picture of whole-body function—a view that will serve you well in the future.

12

QUICK CHECK

1. Which hormones are produced by the male and female sex glands?
2. Why is the placenta considered to be a gland?
3. Why is the pineal gland sometimes called the "timekeeper" of the body?

LANGUAGE OF SCIENCE *(continued from p. 319)*

antidiuretic hormone (ADH)
(an-tee-dye-yoo-RET-ik HOR-mohn [ay dee aych])
[*anti-* **against,** *-dia-* **through,** *-uret-* **urination,** *-ic* **relating to,** *hormon-* **excite**]

atrial natriuretic hormone (ANH)
(AY-tree-al nay-tree-yoo-RET-ik HOR-mohn [ay en aych])
[*atria-* **entrance courtyard (atrium of heart),** *-al* **relating to,** *natri-* **natrium (sodium),** *-uret-* **urination,** *-ic* **relating to,** *hormon-* **excite**]

calcitonin (CT)
(kal-sih-TOH-nin [see tee])
[*calci-* **lime (calcium),** *-ton-* **tone,** *-in* **substance**]

chorion
(KOH-ree-on)
[*chorion* **skin**]

chorionic gonadotropin (hCG)
(koh-ree-ON-ik goh-nah-doh-TROH-pin [aych see jee])
[*chorion-* **skin,** *-ic* **relating to,** *gon-* **offspring,** *-ad-* **relating to,** *-trop-* **nourish,** *-in* **substance**]

corpus luteum
(KOHR-pus LOO-tee-um)
pl., corpora lutea
(KOHR-por-ah LOO-tee-ah)
[*corpus* **body,** *lute-* **yellow,** *-um* **thing**]

corticoid
(KOHR-tih-koyd)
[*cortic-* **cortex (bark),** *-oid* **like**]

cortisol
(KOHR-tih-sol)
[*cortis-* **cortex (bark),** *-ol* **alcohol**]

cyclic AMP (cAMP)
(SIK-lik ay em pee [see ay em pee])
[*cycl-* **circle,** *-ic* **relating to**]

endocrine gland
(EN-doh-krin gland)
[*endo-* **inward or within,** *-crin-* **secrete,** *gland* **acorn**]

endocrine system
(EN-doh-krin SIS-tem)
[*endo-* **inward or within,** *-crin-* **secrete,** *system* **organized whole**]

endocrinology
(en-doh-krin-OL-oh-jee)
[*endo-* **within,** *-crin-* **secrete,** *-o-* **combining vowel,** *-log-* **words (study of),** *-y* **activity**]

epinephrine (Epi)
(ep-ih-NEF-rin [EP-ee])
[*epi-* **upon,** *-nephr-* **kidney,** *-ine* **substance**]

estrogen
(ES-troh-jen)
[*estro-* **frenzy,** *-gen* **produce**]

exocrine gland
(EK-soh-krin gland)
[*exo-* **outside or outward,** *-crin-* **secrete,** *gland* **acorn**]

follicle-stimulating hormone (FSH)
(FOL-lih-kul-STIM-yoo-lay-ting HOR-mohn [ef es aych])
[*foll-* **bag,** *-icle* **little,** *stimulate-* **urge,** *-ing* **action,** *hormon-* **excite**]

G protein
(jee PROH-teen)
[*G* **for guanine-nucleotide binding,** *prote-* **first rank,** *-in* **substance**]

ghrelin
(GRAY-lin)
[*ghrel-* **grow (also acronym for "growth hormone releasing"),** *-in* **substance**]

glucagon
(GLOO-kah-gon)
[*gluca-* **sweet (glucose),** *-agon* **lead or bring**]

glucocorticoid (GC)
(gloo-koh-KOR-tih-koyd [jee see])
[*gluco-* **sweet (glucose),** *-cortic-* **cortex (bark),** *-oid* **like**]

gluconeogenesis
(gloo-koh-nee-oh-JEN-eh-sis)
[*gluco-* **sweet (glucose),** *-neo-* **new,** *-gen-* **produce,** *-esis* **process**]

glycogenolysis
(glye-koh-jeh-NOL-ih-sis)
[*glyco-* **sweet (glucose),** *-gen-* **produce,** *-o-* **combining vowel,** *-lysis* **loosening**]

growth hormone (GH or hGH)
(groth HOR-mohn [jee aych])
[*hormon-* **excite**]

Continued on p. 340

LANGUAGE OF SCIENCE (continued from p. 339)

hormone
(HOR-mohn)
[*hormon-* **excite**]

inhibiting hormone (IH)
(in-HIB-ih-ting HOR-mohn [aye aych])
[*inhib-* **restrain,** *-ing* **action,** *hormon-* **excite**]

insulin
(IN-suh-lin)
[*insul-* **island,** *-in* **substance**]

12

leptin
(LEP-tin)
[*lept-* **thin,** *-in* **substance**]

leukotriene
(loo-koh-TRY-een)
[*leuko-* **white,** *-tri-* **three,** *-ene* **chemical**]

luteinization
(loo-tee-in-ih-ZAY-shun)
[*lute-* **yellow,** *-ization* **process**]

luteinizing hormone (LH)
(loo-tee-in-AYE-zing HOR-mohn [el aych])
[*lute-* **yellow,** *-izing* **process,** *hormon-* **excite**]

melatonin
(mel-ah-TOH-nin)
[*mela-* **black,** *-ton-* **tone,** *-in* **substance**]

mineralocorticoid (MC)
(MIN-er-al-oh-KOR-tih-koyd [em see])
[*mineral-* **mine,** *-cortic-* **cortex (bark),** *-oid* **like**]

negative feedback
(NEG-ah-tiv FEED-bak)
[*negat-* **deny,** *-ive* **relating to,** *feedback* **information about the results of a process**]

neurohypophysis
(noo-roh-hye-POF-ih-sis)
[*neuro-* **nerve,** *-hypo-* **under or below,** *-physis* **growth**]

nonsteroid hormone
(non-STAYR-oyd HOR-mohn)
[*non-* **not,** *-stero-* **solid,** *-oid* **like,** *hormon-* **excite**]

norepinephrine (NE)
(nor-ep-ih-NEF-rin [en ar])
[*nor-* **chemical prefix (unbranched C chain),** *-epi-* **upon,** *-nephr-* **kidney,** *-ine* **substance**]

ova
(OH-vah)
sing., ovum
(OH-vum)
[*ovum* **egg**]

ovarian follicle
(oh-VAYR-ee-an FOL-ih-kul)
[*ov-* **egg,** *-arian* **relating to,** *foll-* **bag,** *-icle* **little**]

oxytocin (OT)
(ahk-see-TOH-sin [oh tee])
[*oxy-* **sharp or quick,** *-toc-* **birth,** *-in* **substance**]

pancreatic islet (islet of Langerhans)
(pan-kree-AT-ik aye-let) (AYE-let of lahn-GER-hans)
[*pan-* **all,** *-creat-* **flesh,** *-ic* **relating to,** *isl-* **island,** *-et* **little**] [*Paul Langerhans* **German pathologist**]

paracrine
(PAIR-ah-krin)
[*para-* **beside,** *-crin-* **secrete**]

parathyroid gland
(payr-ah-THYE-royd gland)
[*para-* **beside,** *-thyr-* **shield,** *-oid* **like,** *gland* **acorn**]

parathyroid hormone (PTH)
(payr-ah-THYE-royd HOR-mohn [pee tee aych])
[*para-* **besides,** *-thyr-* **shield,** *-oid* **like,** *hormon-* **excite**]

pituitary gland
(pih-TOO-ih-tayr-ee gland)
[*pituit-* **phlegm,** *-ary* **relating to,** *gland* **acorn**]

positive feedback
(POZ-it-iv FEED-bak)
[*posit-* **put or place,** *-ive* **relating to,** *feedback* **information about the results of a process**]

progesterone
(proh-JES-ter-ohn)
[*pro-* **before,** *-gester-* **bearing (pregnancy),** *-stero-* **solid or steroid derivative,** *-one* **chemical**]

prolactin (PRL)
(proh-LAK-tin [pee ar el])
[*pro-* **before,** *-lact-* **milk,** *-in* **substance**]

prostaglandin (PG)
(pros-tah-GLAN-din [pee jee])
[*pro-* **before,** *-stat-* **set or place (prostate),** *-gland-* **acorn (gland),** *-in* **substance**]

releasing hormone (RH)
(ree-LEE-sing HOR-mohn [ar aych])
[*hormon-* **excite**]

second-messenger mechanism
(SEK-und MES-en-jer MEK-an-is-em)

sella turcica
(SEL-lah TER-sih-kah)
[*sella* **saddle or seat,** *turcica* **Turkish**]

semen
(SEE-men)
[*semen* **seed**]

sex hormone
(seks HOR-mohn)
[*hormon-* **excite**]

signal transduction
(SIG-nul tranz-DUK-shen)
[*trans-* **across,** *-duc-* **transfer,** *-tion* **process**]

sperm
(sperm)
pl., sperms or sperm
[*sperm* **seed**]

steroid hormone
(STAYR-oyd HOR-mohn)
[*ster-* **sterol,** *-oid* **like,** *hormon-* **excite**]

stress
(stres)
[*stress* **tighten**]

target cell
(TAR-get sel)
[*cell* **storeroom**]

testosterone
(tes-TOS-teh-rohn)
[*testo-* **witness (testis),** *-stero-* **solid or steroid derivative,** *-one* **chemical**]

thromboxane
(throm-BOKS-ayne)
[*thrombo-* **clot,** *-oxa-* **oxygen,** *-ane* **chemical**]

thymosin
(THY-moh-sin)
[*thymos-* **thyme flower (thymus gland),** *-in* **substance**]

thyroid follicle
(THY-royd FOL-lih-kul)
[*thyro-* **shield (thyroid gland),** *-oid* **like,** *foll-* **bag,** *-icle* **little**]

thyroid-stimulating hormone (TSH)
(THY-royd STIM-yoo-lay-ting HOR-mohn [tee es aych])
[*thyro-* **shield,** *-oid* **like,** *stimulate-* **urge,** *-ing* **action,** *hormon-* **excite**]

thyroxine (T_4)
(thy-ROK-sin [tee for])
[*thyro-* **shield (thyroid gland),** *-ox-* **oxygen,** *-ine* **chemical**]

triiodothyronine (T_3)
(try-aye-oh-doh-THY-roh-neen [tee three])
[*tri-* **three,** *-iodo-* **violet (iodine),** *-thyro-* **shield (thyroid gland),** *-nine* **chemical**]

tropic hormone
(TROH-pik HOR-mohn)
[*trop-* **turn or change,** *-ic* **relating to,** *hormon-* **excite**]

LANGUAGE OF **MEDICINE**

acromegaly
(ak-roh-MEG-ah-lee)
[*acro-* **extremities,** *-mega-* **great,** *-aly* **state**]

Addison disease
(AD-ih-son dih-ZEEZ)
[*Thomas Addison* **English physician,** *dis-* **opposite of,** *-ease* **comfort**]

cretinism
(KREE-tin-iz-em)
[*cretin-* **idiot,** *-ism* **condition**]

Cushing syndrome
(KOOSH-ing SIN-drohm)
[*Harvey W. Cushing* **American neurosurgeon,** *syn-* **together,** *-drome* **running or (race) course**]

diabetes insipidus
(dye-ah-BEE-teez in-SIP-ih-dus)
[*diabetes* **siphon,** *insipidus* **without zest**]

diabetes mellitus (DM)
(dye-ah-BEE-teez MELL-ih-tus)
[*diabetes* **pass-through or siphon,** *mellitus* **honey-sweet**]

diuretic
(dye-yoo-RET-ik)
[*dia-* **through,** *-ure-* **urine,** *-ic* **relating to**]

dwarfism
(dwor-FIZ-em)
[*dwarf-* **something tiny,** *-ism* **condition**]

endocrinologist
(en-doh-krin-OL-oh-jist)
[*endo-* **within,** *-crin-* **secrete,** *-o-* **combining vowel,** *-log-* **words (study of),** *-ist* **agent**]

exophthalmos
(ek-sof-THAL-mus)
[*ex-* **outward,** *-oph-* **eye,** *-thalm-* **inner chamber,** *-os* **state**]

gigantism
(jye-GAN-tiz-em)
[*gigant-* **great,** *-ism* **condition**]

glycosuria
(glye-koh-SOO-ree-ah)
[*glyco-* **sweet (glucose),** *-ur-* **urine,** *-ia* **condition**]

goiter
(GOY-ter)
[*goiter* **throat**]

Graves disease
(gravz dih-ZEEZ)
[*Robert J. Graves* **Irish physician,** *dis-* **opposite of,** *-ease* **comfort**]

hydrocortisone
(hye-droh-KOHR-tih-zohn)
[*hydro-* **water,** *-cortisone* **cortex of adrenal gland**]

hypercalcemia
(hye-per-kal-SEE-mee-ah)
[*hyper-* **excessive,** *calc-* **lime (calcium),** *-emia* **blood condition**]

hyperglycemia
(hye-per-glye-SEE-mee-ah)
[*hyper-* **excessive,** *-glyc-* **sweet (glucose),** *-emia* **blood condition**]

hypersecretion
(hye-per-seh-KREE-shun)
[*hyper-* **excessive,** *-secret-* **separate,** *-tion* **process**]

hyperthyroidism
(hye-per-THYE-royd-iz-em)
[*hyper-* **excessive,** *-thyr-* **shield (thyroid gland),** *-oid-* **like,** *-ism* **condition**]

hypoglycemia
(hye-poh-glye-SEE-mee-ah)
[*hypo-* **under or below,** *-glyc-* **sweet (glucose),** *-emia* **blood condition**]

hyposecretion
(hye-poh-seh-KREE-shun)
[*hypo-* **under or below,** *-secret-* **separate,** *-tion* **process**]

hypothyroidism
(hye-poh-THYE-royd-iz-em)
[*hypo-* **under or below,** *-thyr-* **shield (thyroid gland),** *-oid-* **like,** *-ism* **condition**]

myxedema
(mik-seh-DEE-mah)
[*myx-* **mucus,** *-edema* **swelling**]

pharmacology
(farm-ah-KAHL-ah-jee)
[*pharmaco-* **medicine or poison,** *-log-* **words (study of),** *-y* **activity**]

polyendocrine disorder
(PAHL-ee-EN-doh-krin dis-OR-der)
[*poly-* **many,** *-endo-* **inward or within,** *-crin-* **secrete,** *dis-* **lack of,** *-order* **arrangement**]

prolactinoma
(proh-LAK-tih-noh-mah)
[*pro-* **before,** *-lact-* **milk,** *-in-* **substance,** *-oma* **tumor**]

simple goiter
(SIM-pel GOY-ter)
[*goiter* **throat**]

type 1 diabetes mellitus
(type won dye-ah-BEE-teez mel-AYE-tus)
[*diabetes* **siphon,** *mellitus* **honey sweet**]

type 2 diabetes mellitus
(type too dye-ah-BEE-teez mel-AYE-tus)
[*diabetes* **siphon,** *mellitus* **honey sweet**]

virilizing tumor
(VEER-il-aye-zing)
[*viril-* **male or masculine,** *-izing* **making,** *tumor* **swelling**]

12

❑ OUTLINE SUMMARY

To download a digital version of the chapter summary for use with your device, access the ***Audio Chapter Summaries*** *online at evolve.elsevier.com.*

Scan this summary after reading the chapter to help you reinforce the key concepts. Later, use the summary as a quick review before your class or before a test.

Endocrine Glands

A. Exocrine glands are ducted glands and are not included in the endocrine system
B. Endocrine glands are ductless glands that secrete chemicals (hormones) into the blood (see **Table 12-1**)
  1. Target cell—cell that has specific receptors for a particular hormone
  2. Endocrine glands are numerous and widespread in the body (**Figure 12-1**)

Mechanisms of Hormone Action

A. Endocrine glands secrete chemicals (hormones) into the blood (**Figure 12-1**)
B. Hormones perform general functions of communication and control but a slower, longer-lasting type of control than that provided by nerve impulses
C. Cells that respond to hormones are called *target cells* found within *target organs*
D. Nonsteroid hormones (first messengers) bind to receptors on the target cell membrane, triggering intracellular second messengers such as cyclic AMP to affect the cell's activities (**Figure 12-2**)
E. Steroid hormones
  1. Primary effects produced by binding to receptors within the target cell nucleus and influence cell activity by acting on DNA—a slower process than nonsteroid action (**Figure 12-3**)
  2. Secondary effects may occur when steroid hormones bind to membrane receptors to rapidly trigger functional changes in the target cell

Regulation of Hormone Secretion

A. Hormone secretion is controlled by homeostatic feedback
B. Negative feedback—mechanisms that reverse the direction of a change in a physiological system (**Figure 12-4**)
C. Positive feedback—(uncommon) mechanisms that amplify physiological changes
D. Levels of regulation—endocrine regulation of body function usually operates at multiple levels of control at the same time for better efficiency and precision

Mechanisms of Endocrine Disease (Table 12-1)

A. Hypersecretion—excess hormone secretion
B. Hyposecretion—insufficient hormone secretion
C. Polyendocrine disorders—hyper- or hyposecretion of more than one hormone
D. Target cell insensitivity produces results similar to hyposecretion
E. Endocrinologists have developed many different strategies for treatment (for example, surgery and hormone therapy)

Prostaglandins

A. Prostaglandins (PGs) are powerful lipid substances found in a wide variety of body tissues; PGs are modified fatty acids
B. PGs are typically produced in a tissue and diffuse only a short distance to act on cells in that tissue; often called *tissue hormones* or *paracrine* agents
C. Several classes of PGs include prostaglandin A (PGA), prostaglandin E (PGE), and prostaglandin F (PGF)
D. PGs influence many body functions, including respiration, blood pressure, gastrointestinal secretions, and reproduction

Pituitary Gland

A. Structure of the pituitary gland (**Figure 12-6**)
  1. Anterior pituitary—also called *adenohypophysis;* made up of glandular epithelium
  2. Posterior pituitary—also called *neurohypophysis;* made up of nervous tissue
  3. Location—in bony depression (sella turcica) of sphenoid bone in skull; connected to the hypothalamus by a pituitary stalk
B. Anterior pituitary gland (adenohypophysis)
  1. Names of major hormones
    a. Thyroid-stimulating hormone (TSH)
    b. Adrenocorticotropic hormone (ACTH)
    c. Follicle-stimulating hormone (FSH)
    d. Luteinizing hormone (LH)
    e. Growth hormone (GH)
    f. Prolactin (lactogenic hormone)
  2. Functions of major hormones
    a. TSH—stimulates growth of the thyroid gland; also stimulates it to secrete thyroid hormone
    b. ACTH—stimulates growth of the adrenal cortex and stimulates it to secrete glucocorticoids (mainly cortisol)
    c. FSH—initiates growth of ovarian follicles each month in the ovary and stimulates one or more follicles to develop to the stage of maturity and ovulation; FSH also stimulates estrogen secretion by developing follicles; stimulates sperm production in the male

d. LH—acts with FSH to stimulate estrogen secretion and follicle growth to maturity; causes ovulation; causes luteinization of the ruptured follicle and stimulates progesterone secretion by corpus luteum; causes interstitial cells in the testes to secrete testosterone in the male
e. GH—stimulates growth by accelerating protein anabolism; also accelerates fat catabolism and slows glucose catabolism; by slowing glucose catabolism, tends to increase blood glucose to higher than normal level (hyperglycemia)
   (1) Hypersecretion during childhood results in gigantism and during adulthood results in acromegaly (**Figure 12-5**)
   (2) Hyposecretion during childhood results in pituitary dwarfism
f. Prolactin (PRL) or lactogenic hormone—stimulates breast development during pregnancy and secretion of milk after the delivery of the baby
   (1) Prolactinoma—benign adenoma causing hypersecretion of PRL; occurs most frequently in females

C. Posterior pituitary gland (neurohypophysis) (**Figure 12-6**)
   1. Names of hormones
      a. Antidiuretic hormone (ADH)
         (1) Hyposecretion causes diabetes insipidus, characterized by excessive volume of urine
      b. Oxytocin (OT)
   2. Functions of hormones
      a. ADH—accelerates water reabsorption from urine in the kidney tubules into the blood, thereby decreasing urine secretion
      b. Oxytocin—stimulates the pregnant uterus to contract; may initiate labor; causes glandular cells of the breast to release milk into ducts

Hypothalamus

A. Produces posterior pituitary hormones
   1. Actual production of ADH and oxytocin occurs in the hypothalamus
   2. After production in the hypothalamus, hormones pass along axons into the pituitary gland
   3. The secretion and release of posterior pituitary hormones are controlled by nervous stimulation
B. Regulates anterior pituitary secretion
   1. Releasing hormones (RHs) and inhibiting hormones (IHs) control secretion by anterior pituitary
   2. RHs and IHs reach anterior pituitary through a direct capillary connection
C. The hypothalamus controls many body functions related to homeostasis (temperature, appetite, and thirst)

Thyroid Gland

A. Located in the neck, just inferior to the larynx (**Figure 12-7**)
B. Tissue made up of thyroid follicles filled with colloid (**Figure 12-8**)
C. Names of hormones
   1. Thyroid hormones—thyroxine (T_4) and triiodothyronine (T_3); produced by follicle cells and stored in colloid of follicles
   2. Calcitonin (CT)—made by CT cells outside the follicle walls
D. Functions of hormones
   1. Thyroid hormones—accelerate catabolism and energy production (increasing the body's metabolic rate)
   2. CT—decreases the blood calcium concentration by inhibiting breakdown of bone, which would release calcium into the blood
E. Hyperthyroidism (hypersecretion of thyroid hormones) increases metabolic rate
   1. Characterized by restlessness and exophthalmos (protruding eyes) (**Figure 12-9**)
   2. Graves disease is an inherited form of hyperthyroidism
F. Hypothyroidism (hyposecretion of thyroid hormones)
   1. May result from different conditions
   2. Simple goiter—painless enlargement of thyroid caused by dietary deficiency of iodine (**Figure 12-10**)
   3. Hyposecretion during early development may result in cretinism (retardation) and during adulthood in myxedema (characterized by edema, dry skin, and sluggishness; **Figure 12-11**)

Parathyroid Glands

A. Small lumps of glandular tissue located on the posterior surface of the thyroid gland (**Figure 12-7**)
B. Name of hormone—parathyroid hormone (PTH)
   1. Increases blood calcium concentration by increasing the breakdown of bone with the release of calcium into the blood; also promotes calcium absorption from food and reduces calcium loss in the urine
   2. PTH and CT have antagonistic effects that help maintain stable blood calcium concentrations needed for good health (**Figure 12-12**)

Adrenal Glands

A. Located on the superior end of each kidney; outer region is glandular and inner region is secretory nervous tissue (**Figure 12-13**)
B. Adrenal cortex
   1. Names of hormones (corticoids)
      a. Glucocorticoids (GCs)—chiefly cortisol (hydrocortisone)
      b. Mineralocorticoids (MCs)—chiefly aldosterone
      c. Sex hormones—small amounts of male hormones (androgens) secreted by adrenal cortex of both sexes
   2. Three cell layers (zones)
      a. Outer layer—secretes mineralocorticoids
      b. Middle layer—secretes glucocorticoids
      c. Inner layer—secretes sex hormones

3. Mineralocorticoids—increase blood sodium and decrease body potassium concentrations by accelerating kidney tubule reabsorption of sodium and excretion of potassium
4. Functions of glucocorticoids
   a. Help maintain normal blood glucose concentration by increasing gluconeogenesis—the formation of "new" glucose from amino acids produced by the breakdown of proteins, mainly those in muscle tissue cells; also the conversion to glucose of fatty acids produced by the breakdown of fats stored in adipose tissue cells
   b. Play an essential part in maintaining normal blood pressure—make it possible for epinephrine and norepinephrine to maintain a normal degree of vasoconstriction, a condition necessary for maintaining normal blood pressure
   c. Act with epinephrine and norepinephrine to produce an anti-inflammatory effect, to bring about normal recovery from inflammations of various kinds
   d. Produce anti-immunity, antiallergy effect; bring about a decrease in the number of lymphocytes and plasma cells and therefore a decrease in the amount of antibodies formed
   e. Secretion of glucocorticoid quickly increases when the body is thrown into a condition of stress; high blood concentration of glucocorticoids, in turn, brings about many other stress responses (**Figure 12-14**)
   f. Chronic stress can disturb the body's balance of metabolic and immune functions
5. Sex hormones—male androgens similar to testosterone are produced in both sexes; have a role in reproductive development

C. Adrenal medulla
1. Names of hormones—epinephrine (Epi), or adrenaline, and norepinephrine (NE)
2. Functions of hormones—help the body resist stress by intensifying and prolonging the effects of sympathetic stimulation; increased epinephrine secretion is the first endocrine response to stress

D. Adrenal abnormalities
1. Hypersecretion of glucocorticoids causes Cushing syndrome: moon face, hump on back, elevated blood glucose levels, frequent infections (**Figure 12-15**)
2. Hypersecretion of adrenal androgens may result from a virilizing tumor and cause masculinization of affected women
3. Hyposecretion of cortical hormones may result in Addison disease: increased pigmentation in skin and mucous membranes, muscle weakness, reduced blood glucose, nausea, loss of appetite, and weight loss (**Figure 12-16**)

Pancreatic Islets

A. Islands of endocrine tissue scattered within the exocrine tissue of the pancreas, a digestive gland near the junction of the stomach and small intestine (**Figure 12-17**)

B. Names of hormones
1. Glucagon—secreted by alpha cells
2. Insulin—secreted by beta cells

C. Functions of hormones
1. Glucagon increases the blood glucose level by accelerating glycogenolysis in liver (conversion of glycogen to glucose)
2. Insulin decreases the blood glucose by accelerating the movement of glucose out of the blood into cells, which increases glucose metabolism by cells

D. Diabetes mellitus (**Figure 12-18**)
1. Type 1 results from hyposecretion of insulin
2. Type 2 results from target cell insensitivity to insulin
3. Glucose cannot enter cells and thus blood glucose levels rise, producing glycosuria (glucose in the urine)

Sex Glands

A. Female sex glands
1. The ovaries contain two structures that secrete hormones—the ovarian follicles and the corpus luteum; see Chapter 23
2. Effects of estrogen (feminizing hormone)
   a. Development and maturation of breasts and external genitals
   b. Development of adult female body contours
   c. Initiation of menstrual cycle

B. Male sex glands
1. The interstitial cells of testes secrete the male hormone testosterone; see Chapter 23
2. Effects of testosterone (masculinizing hormone)
   a. Maturation of external genitals
   b. Beard growth
   c. Voice changes at puberty
   d. Development of musculature and body contours typical of the male

Thymus

A. Name of hormone—thymosin

B. Function of hormone—plays an important role in the development and function of T cells (agents of the body's immune system)

Placenta

A. Name of hormones—chorionic gonadotropins, estrogens, and progesterone

B. Functions of hormones—maintain the corpus luteum during pregnancy

Pineal Gland

A. A small gland near the roof of the third ventricle of the brain
1. Glandular tissue predominates in children and young adults
2. Becomes fibrous and calcified with age

B. Called *third eye* because its influence on secretory activity is related to the amount of light entering the eyes
C. Secretes melatonin, which
   1. Inhibits ovarian activity
   2. Regulates the body's internal clock
D. Abnormal secretion of (or sensitivity to) melatonin may produce seasonal affective disorder (SAD) or winter depression, a form of depression that occurs when exposure to sunlight is low and melatonin levels are high

Endocrine Functions Throughout the Body

A. Many organs (for example, the stomach, intestines, and kidneys) produce endocrine hormones
   1. Stomach lining produces ghrelin, which affects appetite and metabolism
   2. The atrial wall of the heart secretes atrial natriuretic hormone (ANH), which stimulates sodium loss from the kidneys
   3. Fat-storing cells secrete leptin, which controls how full or hungry we feel
B. Hormone actions occur in every organ of the body and are addressed throughout the rest of this book

12

❑ ACTIVE LEARNING

STUDY TIPS

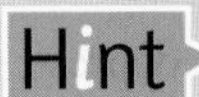

Consider using these tips to achieve success in meeting your learning goals.

Review the synopsis of the endocrine system in Chapter 5. The function of the endocrine system is the same as that of the nervous system. The differences are in the methods used and the extent of their effects. The endocrine system uses chemicals in the blood (hormones) rather than nerve impulses. Hormones can have a direct effect on almost every cell in the body, an impossible task for the nervous system. Steroid hormones can act directly because they can enter the cell; protein hormones cannot, so they need a second-messenger system.

1. Reviewing material from earlier chapters such as receptor proteins in the cell membrane, ATP, homeostasis, and negative feedback loops will help you understand the material in this chapter.
2. Use flash cards and online resources to learn the names of the hormones, what they do, and the names and locations of the glands that produce them. Develop a concept map that includes the gland, hormones, and their functions.
3. Remember that hormones released by the posterior pituitary gland are made in the hypothalamus.
4. When studying the disorders of the endocrine system, make a chart that identifies the disorders as a hyposecretion or hypersecretion of a specific gland. This may be more difficult than you think because many of the disorders are named after people, so the names themselves are not helpful in explaining the disorders. Usually, if you know the normal function of the hormone, you should be able to figure out what effect on the body hyposecretion or hypersecretion would have.
5. In your study group, discuss the hormone mechanisms and negative feedback loops involved in hormone regulation. Review the hormone flash cards. Use your electronic device to photocopy **Table 12-1**. The photocopy and the chapter outline summary at the end of the chapter would be a good way to organize almost all the information in the chapter. Go over the chart of endocrine disorders and the questions at the end of the chapter, and discuss possible test questions.

Review Questions

Write out the answers to these questions after reading the chapter and reviewing the Chapter Summary. If you simply think through the answer without writing it down, you will not retain much of your new learning.

1. Differentiate between endocrine and exocrine glands.
2. Define or explain the following terms: hormone, target organ, hypersecretion, and hyposecretion.
3. Explain the mechanism of action of nonsteroid hormones.
4. Explain the mechanism of action of steroid hormones.
5. Explain and give an example of a negative feedback loop for the regulation of hormone secretion.
6. Explain and give an example of a positive feedback loop for the regulation of hormone secretion.
7. Explain the difference between prostaglandins and hormones. List some of the body functions that can be influenced by prostaglandins.
8. Describe the structure of the pituitary gland and where it is located.
9. Name the four tropic hormones released by the anterior pituitary gland and briefly explain their functions.
10. Explain the function of growth hormone.

11. Gigantism and acromegaly have the same cause. What is the cause of these two endocrine disorders and how do they differ?
12. Explain the function of ADH.
13. What is the cause of diabetes insipidus? What are the signs and symptoms of the condition?
14. Explain the function of prolactin and oxytocin.
15. Explain the function of the hypothalamus in the endocrine system.
16. Explain the difference between T_3 and T_4. What is unique about the thyroid gland?
17. Describe the antagonistic effects of calcitonin and parathyroid hormone in the regulation of calcium.
18. Distinguish between cretinism and myxedema.
19. Name the hormones produced by the zones or areas of the adrenal cortex.
20. What are the signs and symptoms of Cushing syndrome? Of Addison disease?
21. Explain the function of aldosterone.
22. Explain the function of glucocorticoids.

Critical Thinking

After finishing the Review Questions, write out the answers to these more in-depth questions to help you apply your new knowledge. Go back to sections of the chapter that relate to concepts that you find difficult.

23. Explain why a secondary messenger system is needed for nonsteroid hormones but not for steroid hormones.
24. Why is the balance of blood calcium levels significant to homeostasis?
25. Why is a goiter usually more of a dietary problem rather than an endocrine problem?
26. A doctor discovered a patient had very low levels of thyroxine by noting high levels of TSH. Is the patient's problem in the thyroid gland or the pituitary gland? Explain your answer.
27. If a person diagnosed with diabetes mellitus were found to be producing a normal amount of insulin, what other cause could explain the diabetes?
28. Why is a program of regular exercise so important to a diabetic patient; especially someone with a diagnosis of type 1 diabetes?

Chapter Test

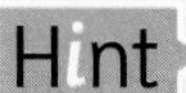

After studying the chapter, test your mastery by responding to these items. Try to answer them without looking up the answers.

1. ________ glands secrete their products into ducts that empty onto a surface or into a cavity.
2. ________ glands are ductless and secrete their products, called ________, into intercellular space, where they diffuse into the blood.
3. The two major classes of hormones are ________ and ________.
4. A cell or body organ that has receptors for a hormone that triggers a reaction is called a ________.
5. One example of a second-messenger system involves the conversion of ATP into ________.
6. The hormone receptors for a nonsteroid hormone are located in the ________, whereas the receptors for a steroid hormone are located in the ________.
7. "Tissue hormones" is another name for ________.
8. What part of the pituitary gland is made of nervous tissue? ________
9. What part of the pituitary gland is made of glandular tissue? ________
10. The hormone oxytocin is released by the ________ but is made in the ________.
11. The hormone hGH is a synonym for the________.
12. The ________ enters target cells and does *not* require a second-messenger action. It is an exception to the general model of nonsteroid action requiring an internal second messenger.
13. Patients with ________ exhibit signs of clinical depression only during the winter months, when nights are long.
14. A tropic hormone secreted by the anterior pituitary gland is:
    a. thyroid-stimulating hormone
    b. adrenocorticotropic hormone
    c. luteinizing hormone
    d. all of the above
15. Antidiuretic hormone (ADH):
    a. is made in the posterior pituitary gland
    b. accelerates water reabsorption in the kidney
    c. in high concentration causes diabetes insipidus
    d. all of the above
16. Which of the following hormones is released by the anterior pituitary gland and stimulates breast development during pregnancy necessary for eventual milk production?
    a. Estrogen
    b. Oxytocin
    c. Prolactin
    d. Progesterone
17. Which hormone released by the posterior pituitary gland stimulates the contraction of the pregnant uterus?
    a. Estrogen
    b. Oxytocin
    c. Prolactin
    d. Progesterone
18. A benign adenoma that causes a hypersecretion of prolactin (PRL) is called:
    a. virilizing tumor
    b. exophthalmos
    c. prolactinoma
    d. myxedema

19. What is the chemical process by which the glucose stored in the liver cells in the form of glycogen is converted to glucose?
    a. Glycolysis
    b. Gluconeogenesis
    c. Second-messenger mechanism
    d. Glycogenolysis

20. Which of the following is *not* a paracrine?
    a. Prostaglandin
    b. Leukotriene
    c. Thromboxane
    d. Ghrelin

Match each hormone in Column A with its function or source in Column B.

Column A	Column B
21. ______ parathyroid hormone	a. released by the adrenal medulla; prolongs the effect of the sympathetic nervous system
22. ______ mineralocorticoids	b. made in the heart; helps regulate blood sodium
23. ______ glucocorticoids	c. made in the pancreatic islets; decreases blood glucose levels
24. ______ epinephrine	d. has the opposite effect of calcitonin in the blood
25. ______ glucagon	e. made by the alpha cells in the pancreatic islets
26. ______ insulin	f. made in the outermost layer of the adrenal cortex
27. ______ chorionic gonadotropins	g. the most significant hormone released by the pineal gland
28. ______ melatonin	h. the hormone made in the placenta and detected by home pregnancy tests
29. ______ atrial natriuretic hormone	i. made by the middle layer of the adrenal cortex

Match each description or signs and symptoms in Column B with its corresponding endocrine disorder in Column A.

Column A	Column B
30. ______ gigantism	a. an inherited hyperthyroidism with exophthalmos
31. ______ acromegaly	b. hyposecretion of thyroid hormone in later life leading to lessened physical and mental vigor
32. ______ diabetes insipidus	c. hyposecretion of insulin causing an increased blood glucose level
33. ______ Graves disease	d. hypersecretion of growth hormone after the normal growth years
34. ______ myxedema	e. an enlarged thyroid gland as a result of dietary deficiency of iodine
35. ______ goiter	f. a condition caused by a hypersecretion of glucocorticoids
36. ______ cretinism	g. hypersecretion of growth hormone in the early years of life
37. ______ Cushing syndrome	h. hyposecretion of thyroid hormone in the formative years resulting in physical, mental, and sexual retardation
38. ______ diabetes mellitus	i. a condition caused by high secretions of melatonin, causing depression in winter
39. ______ seasonal affective disorder	j. hyposecretion of ADH, causing the production of a large volume of urine

Case Studies

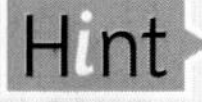

To solve a case study, you may have to refer to the glossary or index, other chapters in this textbook, and other resources.

1. George, the chief executive officer of a major institution, was jogging around his summer home when he became distressed at what seemed to be an irregularity of his heart rhythm. His assistants immediately rushed George to a hospital, where he was diagnosed as having atrial fibrillation (uncoordinated contractions of the upper heart chambers). George was even more distressed to hear that he had a specific heart condition, fearing it might disrupt his very active lifestyle. His physicians informed him that the overactivity of his heart—and perhaps other organs—was caused by hyperthyroidism. Explain how hyperthyroidism could cause George's problems. What strategies might his physicians have available for treating him?
2. In George's case (see preceding case study), the attending physicians chose to surgically remove part of the thyroid in an attempt to control George's hyperthyroidism. What precautions ought George's surgeons take in removing this tissue? (HINT: What anatomical structures in the thyroid area should they avoid cutting or removing?)
3. Your friend Lynn has type 1 diabetes mellitus. What therapy is likely to help her regain control of her metabolism and thus avoid possible tissue or organ damage? Lynn has told you that her condition, if untreated, results in "starvation" of cells in her body. This condition is characterized by hyperglycemia (elevated blood glucose), so you might wonder how the cells could starve if they have an excess of nutrients available. What is the explanation for this seemingly contradictory fact?

Answers to Active Learning Questions can be found online at evolve.elsevier.com.

Blood

OUTLINE

Scan this outline before you begin to read the chapter, as a preview of how the concepts are organized.

OBJECTIVES

Before reading the chapter, review these goals for your learning.

After you have completed this chapter, you should be able to:

1. Describe the primary functions and composition of blood, including the characteristics of blood tissue and plasma, and identify the most important function of each of the formed elements of blood.
2. Explain the mechanisms of blood disease.
3. Explain the structure and function of red blood cells, the purpose of an RBC count, the function of hemoglobin, and list RBC abnormalities.
4. Describe ABO and Rh systems for blood typing.
5. Identify and discuss common red blood cell disorders.
6. Describe the structure and function of white blood cells.
7. State the purpose of performing a WBC count, and identify WBC types.
8. Identify and discuss common white blood cell disorders.
9. Explain the steps involved in blood clotting and describe clotting disorders.

CHAPTER 13

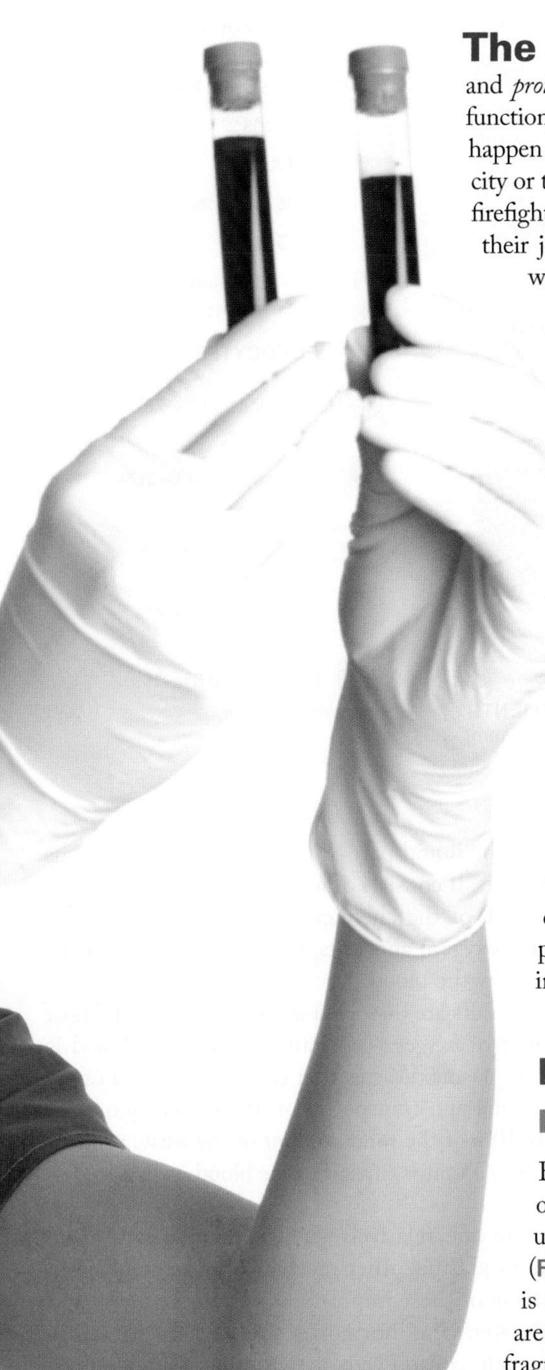

The next few chapters deal with *transportation* and *protection,* two of the body's most important functions. Have you ever thought of what would happen if the transportation ceased in your city or town? Or what would happen if the police, firefighters, and armed services stopped doing their jobs? Food would become scarce, garbage would pile up, and no one would protect you or your property. Stretch your imagination just a little, and you can imagine many disastrous results. Similarly, lack of transportation and protection for the cells—the "individuals" of the body—threatens the homeostasis of the body. The systems that provide these vital services for the body are the **cardiovascular system** *(circulatory system),* **lymphatic system,** and **immune system.**

In this chapter, we discuss the primary transportation fluid—blood. Blood not only performs vital pickup and delivery services but also provides much of the protection necessary to withstand foreign "invaders." The heart and blood vessels are discussed in Chapters 14 and 15. The lymphatic system and immunity are discussed in Chapter 16.

Blood Composition

Blood Tissue

Blood is a fluid tissue that has many kinds of chemicals dissolved in it and millions upon millions of cells floating in it (**Figure 13-1**). The liquid (extracellular) part is called **plasma.** Suspended in the plasma are many different types of cells and cell fragments that make up the **formed elements** of blood.

Many people are curious about just how much blood they have. The amount varies with

LANGUAGE OF **SCIENCE**

Hint Before reading the chapter, say each of these terms out loud. This will help you to avoid stumbling over them as you read.

ABO system
(ay bee oh SIS-tem)

agglutinate
(ah-GLOO-tin-ayt)
[*agglutin-* **glue,** *-ate* **process**]

agranular leukocyte
(ah-GRAN-yoo-lar LOO-koh-syte)
[*a-* **without,** *-gran-* **grain,** *-ul-* **little,** *-ar* **relating to,** *leuko-* **white,** *-cyte* **cell**]

albumin
(al-BYOO-min)
[*alb-* **white,** *-in* **substance**]

antibody
(AN-tih-bod-ee)
[*anti-* **against,** *-body* **main part**]

antigen
(AN-tih-jen)
[*anti-* **against,** *-gen* **produce**]

basophil
(BAY-soh-fil)
[*bas-* **base (high pH),** *-phil* **love**]

buffy coat
(BUF-ee koht)
[*buff-* **leather,** *-y* **characterized by**]

carbaminohemoglobin ($HbCO_2$)
(kar-bah-MEE-noh-hee-moh-GLOH-bin [aych bee see oh too])
[*carb-* **coal (carbon),** *-amino-* **ammonia compound (amino acid),** *-hemo-* **blood,** *-glob-* **ball,** *-in* **substance**]

cardiovascular system
(kar-dee-oh-VAS-kyoo-lar SIS-tem)
[*cardi-* **heart,** *-vas-* **vessel,** *-ular* **relating to**]

Continued on p. 368

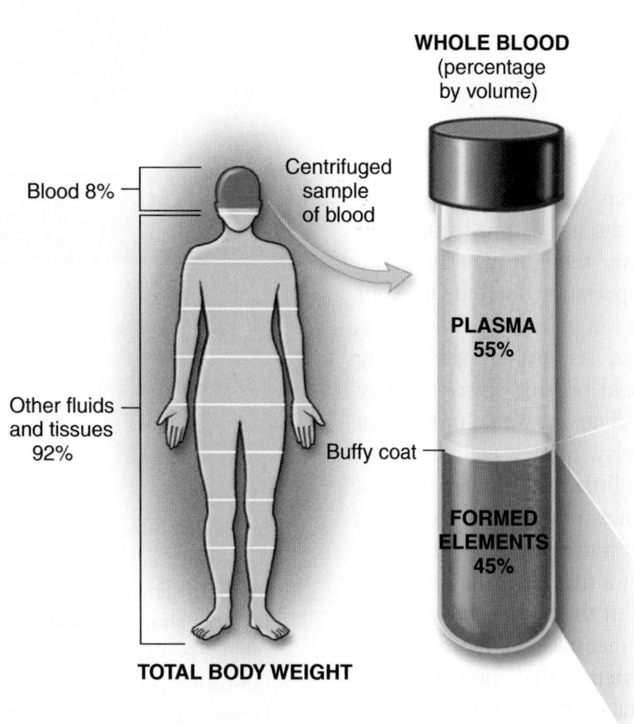

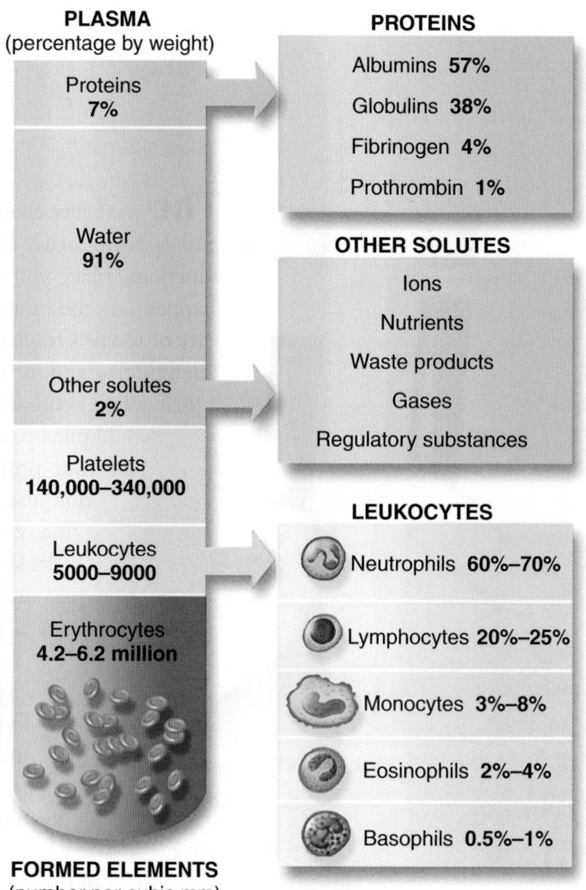

FIGURE 13-1 Blood components. Approximate values for the components of blood in a normal adult.

13

how big they are and whether they are male or female. A big person has more blood than a small person, and a man has more blood than a woman. But as a general rule, most adults probably have between 4 and 6 L of blood. It normally accounts for about 7% to 9% of the total body weight.

The volume of the plasma part of blood is usually a little more than half the entire volume of whole blood. An example of normal blood volumes for a person follows:

Plasma	2.6 L
+ Formed elements	2.4 L
Whole blood	5.0 L

Blood is slightly alkaline, with a pH between 7.35 and 7.45—always staying just above the chemically neutral point of 7.00 (see Chapter 2). If the alkalinity of your blood decreases toward neutral, you are a very sick person; in fact, you have **acidosis.** But even in this condition, blood almost never becomes the least bit acid; it just becomes less alkaline than normal.

QUICK CHECK

1. Name the liquid (extracellular) fraction of whole blood.
2. Blood accounts for what percent (%) of total body weight?
3. What is the normal blood pH?

Blood Plasma

Blood plasma is the liquid part of the blood, or blood minus its formed elements. It consists of water with many substances dissolved in it. All of the chemicals needed by cells to stay alive—nutrients, oxygen, and salts, for example—have to be brought to them by the blood.

Nutrients and salts are dissolved in plasma. About 1.5% of the total amount of *oxygen* (O_2) transported in the blood is also dissolved in plasma. Wastes that cells must get rid of are dissolved in plasma and transported to the excretory organs. Approximately 10% of the total amount of the waste product *carbon dioxide* (CO_2) that is carried in the blood is dissolved in the plasma.

In addition to the relatively small amounts of O_2 and CO_2 dissolved in plasma, other mechanisms are involved in the transportation of these important gases in the blood and are described later. The hormones and other regulatory chemicals that help control cells' activities are also dissolved in plasma.

As Figure 13-1 shows, the most abundant type of solute in the plasma is a group of **plasma proteins** that together make up about 7% of the plasma by weight. These proteins include **albumins,** which help retain water in the blood by osmosis. **Globulins,** which include the antibodies that help protect us from infections, circulate in the plasma. The plasma also carries *fibrinogen* and *prothrombin,* which are necessary for blood clotting.

Intravenous administration of albumin is sometimes used as a plasma volume expander in people with abnormally low blood volume. The injected albumin will draw about three to four times its volume of fluid into the blood through the process of osmosis. The result is an expansion of blood volume that can be lifesaving in cases of hemorrhage, severe burns, or kidney disease.

Blood **serum** is plasma minus its clotting factors, such as fibrinogen and prothrombin. Serum is obtained from whole blood by allowing the blood to first clot in the bottom of a tube; then the liquid serum that remains at the top is poured off. Serum still contains antibodies, so it can be used to treat patients who have a need for specific antibodies.

CLINICAL APPLICATION

CARDIAC BLOOD TESTS

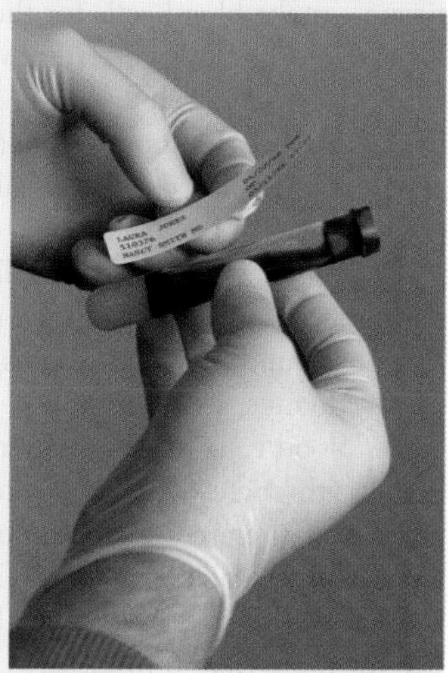

Sometimes blood plasma contains excesses of normal substances or low amounts of abnormal substances that may indicate disease. For example, when the heart muscle is damaged, enzymes contained within the muscle cells are released into the bloodstream, causing increased plasma levels of these enzymes.

Blood tests for a number of cardiac enzymes, including creatine kinase (CK), lactic dehydrogenase (LD), and serum glutamic-oxaloacetic transaminase (SGOT), are useful in confirming a *myocardial infarction (MI)*, or "heart attack."

The *troponins test* is another very valuable diagnostic aid. Rather than making an enzymatic determination, it identifies a specific biochemical marker present in cardiac disease. The presence of cardiac troponins is particularly useful in differentiating cardiac from noncardiac chest pain.

Formed Elements

There are three main types and several subtypes of formed elements:

1. Red blood cells (RBCs), or erythrocytes
2. White blood cells (WBCs), or leukocytes
   a. Granular leukocytes (have stained granules in their cytoplasm)
      (1) Neutrophils
      (2) Eosinophils
      (3) Basophils
   b. Agranular leukocytes (do not have stained granules in their cytoplasm)
      (1) Lymphocytes
      (2) Monocytes
3. Platelets or thrombocytes

Figure 13-1 shows the breakdown of numbers and percentages of the formed elements. **Table 13-1** lists the functions of these formed elements and shows what each looks like under the microscope.

It is difficult to believe just how many blood cells and cell fragments are in the human body. For instance, 5,000,000 RBCs, 7500 WBCs, and 300,000 platelets in 1 cubic millimeter (mm^3) of blood (a tiny fraction of a drop) would be considered normal counts. Because RBCs, WBCs, and platelets are continually being destroyed, the body also must continually make new ones to take their place at a really staggering rate; a few million RBCs are manufactured *each second!*

To learn more about the types of blood cells, go to AnimationDirect online at *evolve.elsevier.com*.

TABLE 13-1 Classes of Blood Cells

BODY CELL	FUNCTION
Erythrocyte	Oxygen and carbon dioxide transport
Neutrophil	Immune defense (phagocytosis)
Eosinophil	Defense against parasites
Basophil	Inflammatory response and heparin secretion
B lymphocyte	Antibody production (precursor of plasma cells)
T lymphocyte	Cellular immune response
Monocyte	Immune defenses (phagocytosis)
Thrombocyte	Blood clotting

Hematopoiesis

Recall from our previous discussion of bones in Chapter 8 that formation of new blood cells is called **hematopoiesis.** Two kinds of connective tissue—**myeloid tissue** and **lymphoid tissue**—make blood cells for the body.

Myeloid tissue is better known as *red bone marrow.* In the adult, it is found chiefly in the sternum, ribs, and coxal (hip) bones. A few other bones such as the vertebrae, clavicles, and cranial bones also contain small amounts of this important tissue.

Red bone marrow forms all types of blood cells except lymphocytes, which are formed in lymphoid tissue. Lymphoid tissue is found as white masses located chiefly in the lymph nodes, thymus, and spleen.

As blood cells mature, they move into the circulatory vessels. Erythrocytes circulate up to 4 months before they break apart and their components are removed from the bloodstream by the spleen and liver. Granular leukocytes often have a life span of only a few days, but agranular leukocytes may live for more than 6 months.

To see exactly where in the body hematopoiesis takes place, review the images in *Sites of Hematopoiesis* at Connect It! at *evolve.elsevier.com.*

13

QUICK CHECK

1. What is the most abundant type of solute in plasma? Name some examples.
2. Identify the formed elements of blood.
3. Name the two types of connective tissue that make blood cells for the body.

Mechanisms of Blood Disease

Most blood diseases are disorders of the formed elements. Thus it is not surprising that the basic mechanism of many blood diseases is the failure of the blood-producing myeloid and lymphoid tissues to form blood cells properly. In many cases, this failure is the result of damage by toxic chemicals or radiation. In other cases, it results from an inherited defect, viral infection, deficiency of nutrients, or even cancer.

If bone marrow failure is the suspected cause of a particular blood disorder, a sample of myeloid tissue may be drawn into a syringe from inside the pelvic bone (iliac crest) or the sternum. This procedure, called **aspiration biopsy cytology (ABC),** allows examination of the tissue that may help confirm or reject a tentative diagnosis.

If the bone marrow is severely damaged, the choice of a **bone marrow transplant** may be offered to the patient. In this procedure, myeloid tissue from a compatible donor is introduced into the recipient intravenously.

Transplantation also may involve infusion of *hematopoietic stem cells.* These blood-forming cells are harvested from the individual being treated, from a compatible donor, or from umbilical cord blood (see box on p. 659).

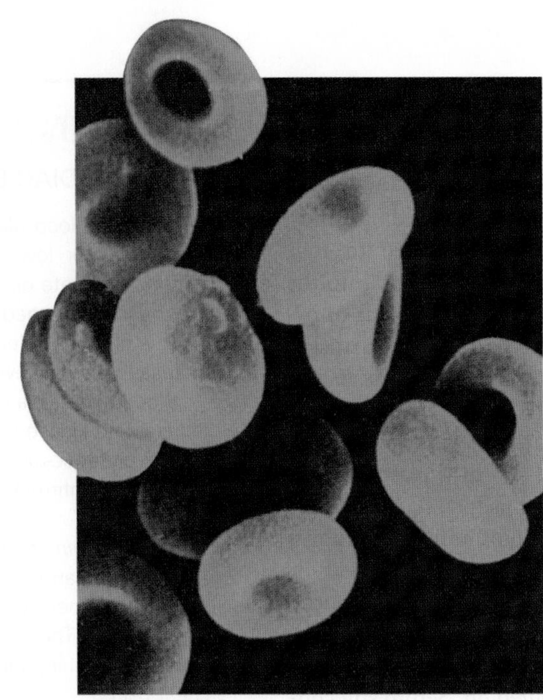

FIGURE 13-2 Red blood cells (RBCs). Color-enhanced scanning electron micrograph shows the detailed structure of normal RBCs. Note the biconcave shape of each RBC.

If the recipient's immune system does not reject the new tissue or stem cells, always a danger in transplant procedures, a new colony of healthy tissue may become established in the bone marrow. As a result, myeloid tissue destroyed by disease, high-dose irradiation, or chemotherapy will be replaced and begin again to produce normal, functioning blood cells.

QUICK CHECK

1. How does the procedure called *aspiration biopsy cytology (ABC)* differ from a bone marrow transplant?
2. What type of cell is involved in a bone marrow transplant?

Red Blood Cells

RBC Structure and Function

The **red blood cell (RBC)** is an elegant example of how structural adaptation can impact biological function. Note in **Figure 13-2** that the RBC, which is surrounded by a tough and flexible plasma membrane, is "caved in" on both sides so that each one has a thin center and thicker edges. This *biconcave* disk shape provides a large surface area for moving dissolved blood gases (O_2 and CO_2) and other solutes quickly in or out of the blood cell. It also helps keep the RBCs from spinning wildly as they flow through the bloodstream.

Mature RBCs have no nucleus or cytoplasmic organelles. Because of this they are unable to reproduce themselves or replace lost or damaged cellular components. The result is a relatively short life span of about 80 to 120 days.

However, the additional intracellular space that becomes available in each cell when the nucleus and cytoplasmic organelles are lost is filled to capacity with an important red pigment called **hemoglobin (Hb).** The unique chemical properties of hemoglobin permit the RBC to perform several critically important functions required for maintenance of homeostasis, such as carrying oxygen and buffering blood. Because RBCs are almost completely filled with hemoglobin, they are often called **erythrocytes** (literally, "red cells").

During hematopoiesis, RBCs lose their mitochondria. Thus, any oxygen carried in RBCs by Hb is not used up by mitochondria to generate adenosine triphosphate (ATP).

During its short life span, each RBC travels around the entire cardiovascular system more than 100,000 times! It is the flexible plasma membrane that permits each cell to "deform" and undergo drastic changes in shape as it repeatedly passes through capillaries whose lumen is smaller than the RBC cell's diameter. Because of the large numbers of RBCs and their unique biconcave shape, the total surface area available for them to perform their biological functions is enormous.

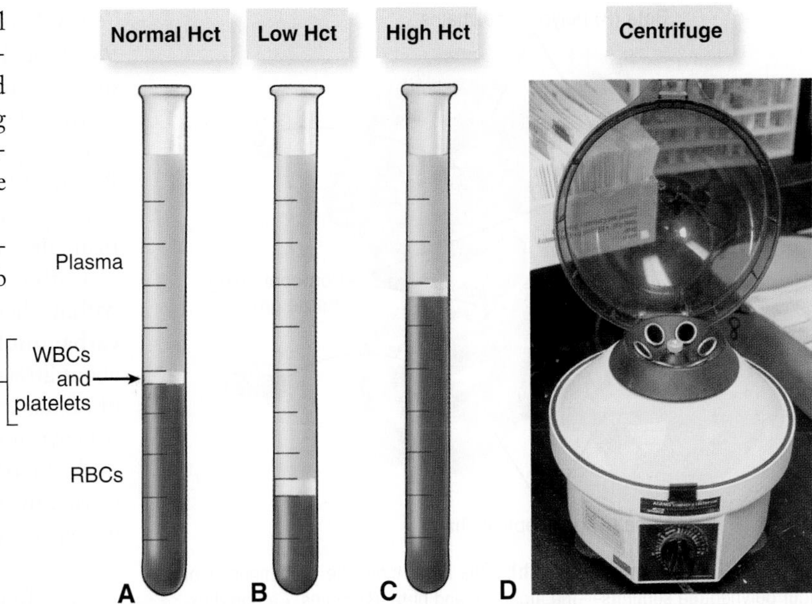

FIGURE 13-3 Hematocrit (Hct). Note the buffy coat located between the packed RBCs and the plasma. **A,** Normal blood with the typical percent of RBCs. **B,** Anemia (a low percent of RBCs). **C,** Polycythemia (a high percent of RBCs). **D,** Centrifuge that spins tubes of blood, causing RBCs to become densely packed into the bottom of the tubes.

13

RBC Count

The **CBC,** or **complete blood cell count,** is a battery of tests used to measure the amounts or levels of many blood constituents and is often ordered as a routine part of the physical examination (see box on this page).

Measuring the *numbers* of circulating RBCs per unit of blood volume is a valuable part of the CBC. Values listed in CBC results as "normal" will vary slightly between different laboratories and reference texts. For RBCs, a range of 4.2 to 6.2 million per cubic millimeter of blood (mm^3), with males generally having a higher number than females, is common. Normal deviations from average ranges often occur with age differences, level of hydration, altitude of residence, and other variables.

Originally RBC counts were done with a *hemocytometer,* a microscope slide with a counting grid etched on it. The current practice is to use a faster, more accurate automated blood cell counter.

The **hematocrit (Hct)** component of the CBC provides information about the *volume* of RBCs in a blood sample. If whole blood is placed in a special centrifuge tube and then "spun down," the heavier formed elements will quickly settle to the bottom of the tube.

During the procedure, RBCs are forced to the bottom of the tube first. The WBCs and platelets then settle out in a light-colored layer called the **buffy coat.** In Figure 13-3 the buffy coat can be seen between the packed RBCs on the bottom of the hematocrit tube and the liquid layer of plasma above.

The hematocrit test—also called *packed-cell volume (PCV)* test—gives an estimate of the proportion of RBCs to whole blood. Such information could help screen for dehydration, hemorrhaging, or other circumstances that affect the RBC ratio. Normally about 45% of the blood volume consists of RBCs (see Figure 13-1).

Another commonly used blood test focusing on RBCs is summarized in the article *Erythrocyte Sedimentation Rate* at Connect It! at *evolve.elsevier.com.*

QUICK CHECK

1. How does the *biconcave disk* shape of the red blood cell allow the red blood cell to carry out its functions?
2. What is the CBC, or *complete blood cell count*?
3. What is the importance of a *hematocrit* test?

CLINICAL APPLICATION

COMPLETE BLOOD CELL COUNT

One of the most useful and frequently performed clinical blood tests is called the *complete blood cell count,* or simply the CBC. The CBC is a collection of tests whose results, when interpreted as a whole, can yield an enormous amount of information regarding a person's health. Standard RBC, WBC, and thrombocyte counts, the differential WBC count, hematocrit, hemoglobin content, and other characteristics of the formed elements are usually included in this battery of tests. Normal ranges for blood values included in most CBC tests are found in Appendix C at *evolve.elsevier.com.*

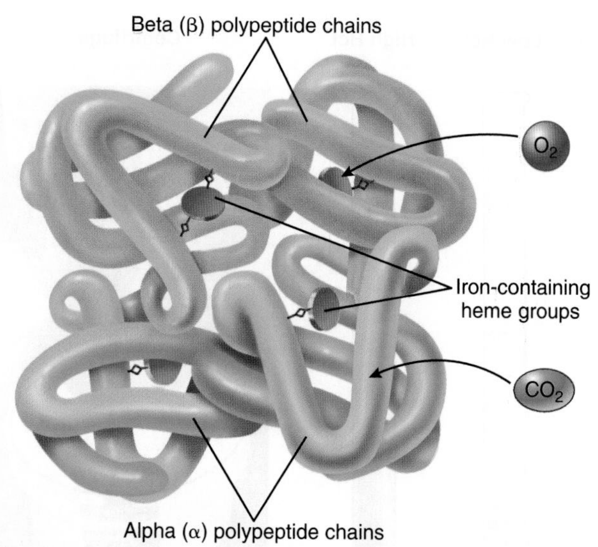

FIGURE 13-4 Hemoglobin (Hb). This large molecule is composed of four polypeptide subunits—the alpha (α) and beta (β) chains. Carbon dioxide may be carried on the amino acids of these chains. Each folded chain holds an iron-containing chemical group *(red)* at its core. The iron (Fe) gives hemoglobin its oxygen-carrying capacity.

13

Hemoglobin

The hemoglobin molecules that fill the millions of RBCs are critical in the transport and exchange of oxygen and carbon dioxide between the blood and the body's cells. They also play a key role in maintenance of acid-base balance in the body.

Hemoglobin is a quaternary protein made up of four folded polypeptide chains, two *alpha (α) chains* and two *beta (β) chains.* As you can see in **Figure 13-4**, there is a chemical structure called a *heme* group embedded within each folded chain. An iron (Fe) atom within each heme group attracts oxygen molecules to unite with hemoglobin and thus form an oxygen-hemoglobin complex called **oxyhemoglobin (HbO_2).**

Oxyhemoglobin makes possible the efficient transport of 98.5% of all of the oxygen required for the body cells (1.5% is dissolved in plasma).

Iron (Fe) is an essential nutrient needed to give hemoglobin its oxygen-carrying ability. Vitamin B_{12} and folate (also a B vitamin) are also among the critical nutrients needed by the red bone marrow to manufacture enough hemoglobin to maintain survival.

Carbon dioxide (CO_2) may attach to the amino acids within hemoglobin's alpha and beta chains to form **carbaminohemoglobin ($HbCO_2$).** This molecule transports about 20% of the carbon dioxide produced as a waste product of cellular metabolism to the lungs for disposal into the external environment. Recall that about 10% of CO_2 is transported in the blood dissolved in plasma. The majority of CO_2 (70%) carried in the blood is converted in RBCs to bicarbonate for its journey to the lungs for excretion (see Chapter 17).

To better understand these concepts, use the Active Concept Map *Transport of Oxygen and Carbon Dioxide in the Blood* at *evolve.elsevier.com.*

RBC Abnormalities

In peripheral blood smears, an RBC of normal size is about 7 to 9 μm in diameter and is called a *normocytic RBC* (*normo-* normal, *-cyte* cell). A normocytic RBC is approximately the same size as the nucleus of a small lymphocyte (**Figure 13-5**). Abnormally small RBCs are called *microcytes* (*micro-* small, *-cyte* cell) and larger RBCs are called *macrocytes* (*macro-* large, *-cyte* cell).

Figure 13-5 also compares the appearance of RBCs with normal amounts of the red pigment hemoglobin, called *normochromic RBCs* (*normo-* normal, *-chromic* color) with those that are deficient in hemoglobin, called *hypochromic RBCs* (*hypo-* low, *-chromic* color), and those that have an excess of hemoglobin, called *hyperchromic RBCs* (*hyper-* high, *-chromic* color).

HEALTH AND WELL-BEING

BLOOD DOPING

A number of athletes have reportedly improved their performance by a practice called *blood boosting* or **blood doping.**

A few weeks before an important event, an athlete has some blood drawn. The RBCs in this sample are separated and frozen. Just before competition, the RBCs are thawed and injected back into the athlete. The increased hematocrit (Hct) that results slightly improves the oxygen-carrying capacity of the blood, which theoretically improves performance. However, in practice the effects are slight.

In addition to blood transfusions, injection of substances such as hormones that increase RBC levels in an attempt to improve athletic performance also has been condemned by leading authorities in the area of sports medicine and by athletic organizations around the world. "Doping" with either the naturally occurring hormone **erythropoietin (EPO)** or with synthetic drugs that have similar biological effects—such as Epogen and Procrit—can result in devastating medical outcomes.

Blood doping is judged to be an unfair and unwise practice in athletics. Besides the significant health risks compared with only slight improvement in performance, there are the personal and professional risks of cheating. Widely reported blood doping scandals in professional cycling competitions and Olympic sports attest to the devastating effects of such scandal.

An alternative to blood doping is *high-altitude training.* Athletes who train at moderately high altitudes—usually at a reduced intensity—will naturally build up a slightly higher Hct in a negative-feedback response to the lower O_2 available in the atmosphere at those altitudes. This type of training must be managed carefully to avoid the risks associated with high Hct.

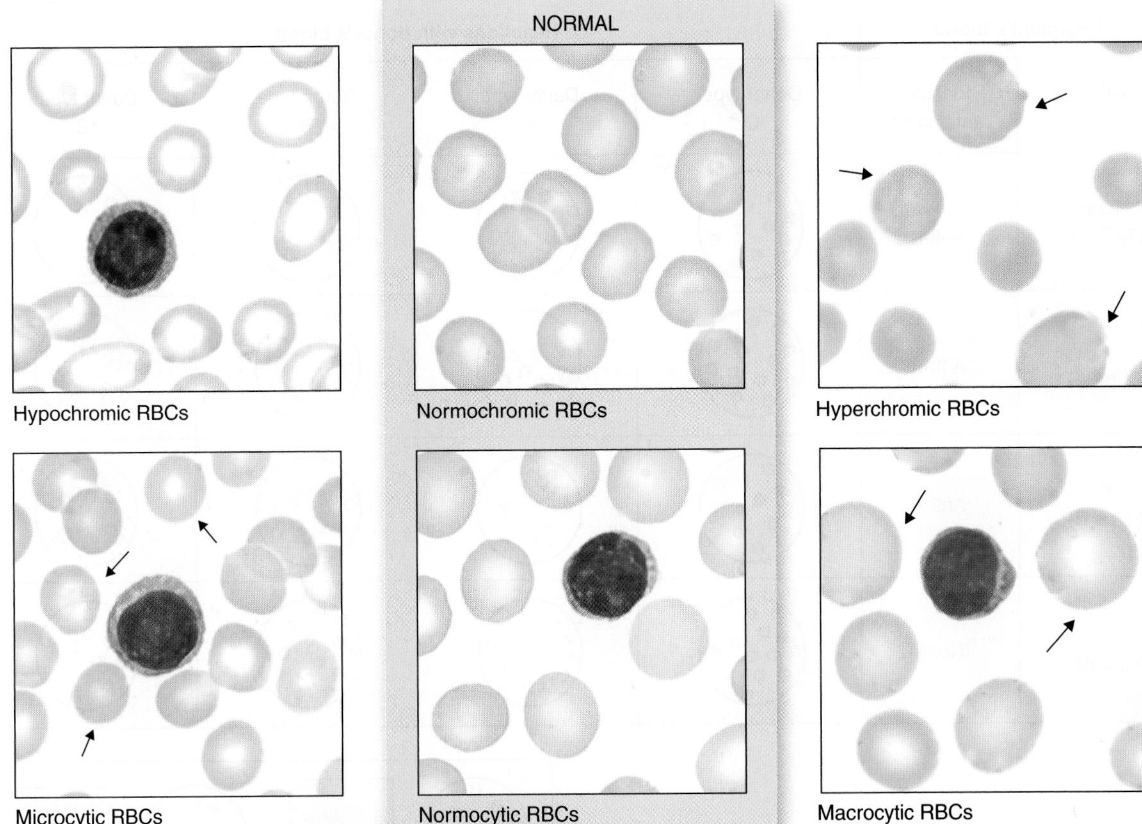

FIGURE 13-5 RBC abnormalities. Micrographs showing normal red blood cells (RBCs) in a smear compared to abnormal RBCs. The cells with a large, dark nucleus shown in some of the images are lymphocytes—a type of white blood cell similar in size to an RBC.

Production of *macrocytic hyperchromic RBCs* during periods of chronic blood loss is a good example of a *negative feedback* response that helps maintain homeostasis. Because the body is unable to produce adequate numbers of RBCs with normal levels of hemoglobin in each cell to replace those lost by hemorrhage, the body increases the size and amount of hemoglobin in those cells it can produce to help restore and maintain the oxygen-carrying capacity of the blood.

Blood Types

Systems of Blood Typing

Blood is often identified as a specific "type" by using the **ABO system** and **Rh system** of classification. Other blood types also occur, but usually do not have the significant clinical applications of the ABO and Rh systems.

Blood types are identified by certain antigens on the surfaces of RBCs (**Figure 13-6**). An **antigen** is a substance that can stimulate the body to make antibodies. Almost all substances that act as antigens are foreign proteins. That is, they are not the body's own natural proteins but instead are proteins that have entered the body from the outside by means of infection, transfusion, or some other method.

The word *antibody* can be defined in terms of what causes its formation or in terms of how it functions. Defined the first way, an **antibody** is a substance made by the body in response to stimulation by an antigen. Defined according to its functions, an antibody is a substance that reacts with the antigen that stimulated its formation.

Many antibodies react with their antigens to cause clumping—that is, they **agglutinate** the antigens. In other words, the antibodies cause their targeted antigens to stick together in little clusters, which disrupts the functions of agglutinated cells and makes them easy targets for the body's immune responses.

ABO System

Every person's blood is one of the following blood types in the ABO system of typing:

1. Type A
2. Type B
3. Type AB
4. Type O

Suppose that you have type A blood (as do about 41% of Americans). The letter *A* stands for a certain type of antigen in the plasma membrane of your RBCs that has been present since birth. Because you were born with type A antigen, your body does not form antibodies to react with it. In other words, your blood plasma contains no anti-A antibodies. It does, however, contain anti-B antibodies. For some unknown reason, these antibodies are present naturally in type A blood

Recipient's blood		Reactions with donor's blood			
RBC antigens	Plasma antibodies	Donor type O	Donor type A	Donor type B	Donor type AB
None (Type O)	Anti-A Anti-B				
A (Type A)	Anti-B				
B (Type B)	Anti-A				
AB (Type AB)	(None)				

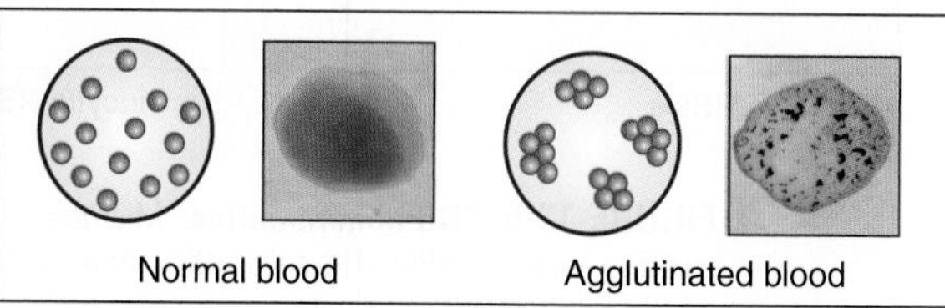

FIGURE 13-6 ABO blood typing. The left columns show the recipient's blood characteristics and the top row shows the donor's blood type. *Inset,* Photo showing samples of agglutinated and nonagglutinated blood.

plasma. The body did not form them in response to the presence of the B antigen—they are simply part of the body's genetic makeup.

In summary, in type A blood the RBCs contain type A antigen and the plasma contains anti-B antibodies.

Similarly, in type B blood, the RBCs contain type B antigen, and the plasma contains anti-A antibodies. In type AB blood, as its name indicates, the RBCs contain both type A and type B antigens, and the plasma contains neither anti-A nor anti-B antibodies.

The opposite is true of type O blood—its RBCs contain neither type A nor type B antigens, and its plasma contains both anti-A and anti-B antibodies.

Figure 13-6 shows the results of different combinations of donor and recipient blood.

Rh System

You may be familiar with the term **Rh-positive** blood. It means that the RBCs of this blood type contain an antigen called the *Rh factor*. If, for example, a person has type AB, Rh-positive blood, his red blood cells contain type A antigen, type B antigen, and the Rh factor antigen. The term *Rh* is used because this important blood cell antigen was first discovered in the blood of Rhesus monkeys.

In **Rh-negative** blood, the RBCs do not have the Rh antigens on their surfaces. Plasma never naturally contains anti-Rh antibodies. But if Rh-positive blood cells are introduced into an Rh-negative person's body, anti-Rh antibodies soon appear in the recipient's blood plasma.

Without appropriate precautions, there could be some danger for offspring born to an Rh-negative mother and an Rh-positive father. If the offspring inherits the Rh-positive trait from his father, the Rh factor on his RBCs may stimulate the mother's body to form anti-Rh antibodies. Then, if she later carries another Rh-positive fetus, it may develop a type of **hemolytic anemia** called *erythroblastosis fetalis,* caused by the mother's Rh antibodies reacting with the baby's Rh-positive cells (**Figure 13-7**).

All Rh-negative mothers who carry an Rh-positive offspring should be treated with an immunoglobulin (antibody) serum, widely marketed under the brand name **RhoGAM.** RhoGAM stops the mother's body from forming anti-Rh antibodies and thus prevents the possibility of harm to the next Rh-positive offspring.

Likewise, a person with Rh-negative blood who receives a transfusion of Rh-positive blood will also develop anti-Rh antibodies and be at risk of an immune reaction if exposed to Rh-positive blood again later.

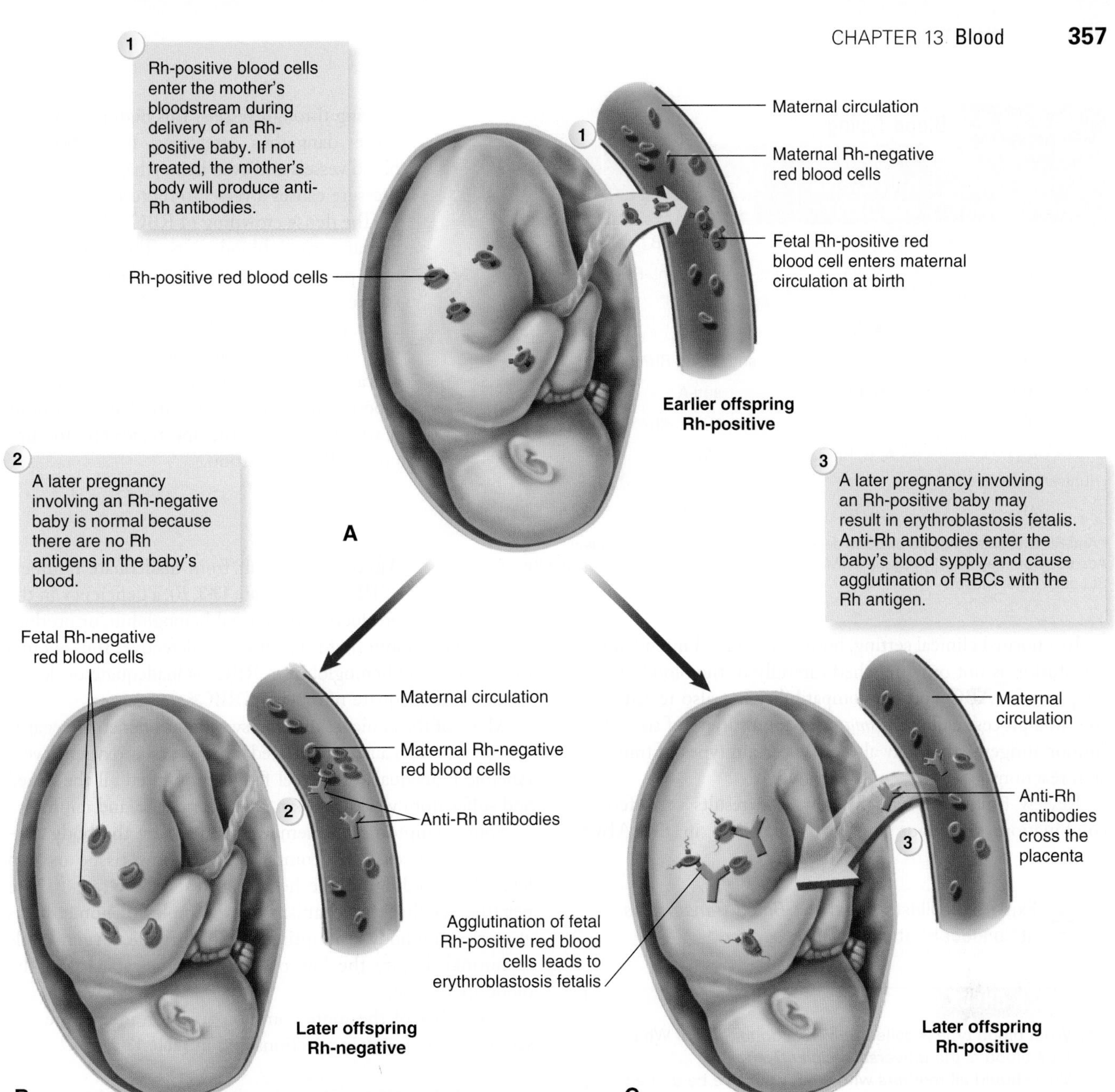

FIGURE 13-7 Erythroblastosis fetalis. Under certain conditions, anti-Rh antibodies may enter the offspring's blood supply and cause agglutination of RBCs with the Rh antigen.

Combined ABO-Rh System

Both the ABO and Rh systems are often used in combination to identify a person's blood type, as you can see in **Table 13-2**. For example, the blood type AB+ refers to the ABO type "AB" and the Rh-positive type. Likewise, O− identifies the blood type of a person with the "O" version of ABO type and the Rh-negative version of Rh type.

Knowing one's blood type can be lifesaving in a medical emergency or during surgery, when a blood transfusion may be needed to maintain the total blood volume. Harmful effects or even death can result from a blood transfusion reaction if the donor's RBCs become agglutinated by antibodies in the recipient's plasma.

If a donor's RBCs do not contain any A, B, or Rh antigen, they cannot be clumped by anti-A, anti-B, or anti-Rh antibodies. For this reason, the type of blood that contains no A, B, or Rh antigens, which is type O−, can be used in an emergency as donor blood. With type O−, there is no danger of anti-A, anti-B, or anti-Rh antibodies clumping its RBCs. Type O− blood has therefore been called **universal donor blood.**

Similarly, blood type AB+ has been called **universal recipient blood** because it contains no anti-A, anti-B, or anti-Rh antibodies in its plasma. Therefore, type AB+ blood does not clump any donor's RBCs containing A, B, or Rh antigens.

TABLE 13-2 Blood Typing

BLOOD TYPE (ABO and Rh)	ANTIGENS PRESENT*	ANTIBODIES PRESENT*	PERCENT OF GENERAL POPULATION
O+	Rh	anti-A, anti-B	35%
O− (universal donor)	None	anti-A, anti-B, anti-Rh?	7%
A+	A, Rh	anti-B	35%
A−	A	anti-B, anti-Rh?	7%
B+	B, Rh	anti-A	8%
B−	B	anti-A, anti-Rh?	2%
AB+ (universal recipient)	A, B, Rh	None	4%
AB−	A, B	anti-Rh?	2%

*Anti-Rh antibodies may be present, depending on exposure to Rh antigens.
Adapted from Pagana KD, Pagana TJ: *Mosby's manual of diagnostic and laboratory tests,* ed 5, St Louis, 2014, Mosby.

In a normal clinical setting, however, all blood intended for transfusion is not only matched carefully to the blood of the recipient for ABO and Rh compatibility but also tested further in a process called *crossmatching* for a variety of so-called "minor antigens" that may also cause certain types of transfusion reactions.

Review **Figure 13-6**, which shows the results of different combinations of donor and recipient blood in the ABO system.

Explore the illustrated article *Blood Transfusions* at Connect It! at *evolve.elsevier.com.*

QUICK CHECK

1. What blood type is called *universal donor blood*? What type blood is the *universal recipient*?
2. Why should all mothers who are Rh positive be treated with RhoGam?

Red Blood Cell Disorders

RBC disorders most often involve either overproduction of RBCs *(polycythemia)* or a condition that results in low oxygen-carrying capacity of the blood *(anemia).*

Polycythemia

Polycythemia is a serious blood disorder characterized by dramatic increases in RBC numbers. One cause of this disorder is cancer of cells in the red bone marrow. Blood doping also can cause temporary polycythemia (see box on p. 354).

In polycythemia, RBC counts may reach 10 million/mm^3 or higher. Such high numbers of RBCs thicken the blood. Thicker than normal blood resists flow—just as thicker than normal ketchup resists flow out of the bottle. This increased flow resistance *(viscosity)* often causes *hypertension* (high blood pressure) because more pressure is needed to keep thick blood moving through vessels. This high pressure can cause dangerous stretching (distention) of blood vessels and even hemorrhaging when the pressure causes a vessel to rupture.

The dense crowding of RBCs also can trigger abnormal blood clots. These conditions may put a person at risk for heart attack or stroke.

In polycythemia, the hematocrit (RBC %) may reach 60%—way above the normal 45% average (see **Figure 13-3**, *C*). Treatment involves blood removal (bleeding), irradiation of bone marrow, or chemotherapy treatment to suppress RBC production.

Anemia

Anemia can result from inadequate numbers of RBCs (see **Figure 13-3**, *B*), a deficiency in the production of normal hemoglobin, or production of hemoglobin that is in some way defective. Thus anemia can occur if the hemoglobin in RBCs is inadequate or defective, even if adequate numbers of RBCs are present.

Many of the clinical signs and symptoms of anemia, regardless of type or cause, are related to low tissue oxygen levels. Anemic individuals often feel fatigued or "tired all the time," and suffer from weakness, skin pallor, headache, and faintness.

Some symptoms of anemia are caused by the body's attempts to increase or "compensate" for low tissue oxygen levels by speeding up the heart and respiratory rates. These negative feedback mechanisms are examples of homeostasis at work—the body attempting to return tissue oxygen levels to normal despite the low oxygen-carrying capacity of the blood in anemia.

A number of the more common types of anemia are described in the following sections and in **Table 13-3**.

To learn more about the clinical manifestations of anemias, go to AnimationDirect online at *evolve.elsevier.com.*

Hemorrhagic Anemia

Hemorrhagic anemia is caused by an actual decrease in the number of circulating RBCs because of hemorrhage or bleeding. It is referred to as either *acute blood-loss anemia* resulting, for example, from extensive surgery or sudden trauma, or *chronic blood-loss anemia* caused by slow but continuous loss of blood over time from diseases such as cancer or ulcers.

As noted earlier, although fewer RBCs are present in circulating blood because of the hemorrhage, those RBCs that are produced during this time are both macrocytic and hyperchromic—an attempt to restore homeostasis by compensating for the low oxygen-carrying capacity caused by the hemorrhagic anemia. Once the actual bleeding is stopped, transfusion of whole blood or red cells and successful treatment of the underlying reason for chronic blood loss are curative.

Aplastic Anemia

Aplastic anemia is characterized by abnormally low RBC counts and destruction of bone marrow. The cause is often related to high-dose exposure to certain toxic chemicals such as benzene or mercury; irradiation; and in susceptible individuals, certain drugs including chloramphenicol.

Acute cases of aplastic anemia are very serious, with death rates reaching 70% at 3 or 4 months after diagnosis. Bone marrow or stem cell transplants provide the most effective treatment.

Deficiency Anemias

Reduction of Normal Hemoglobin

Deficiency anemias are caused by an inadequate supply of some substance, such as vitamin B_{12} or iron, required for red blood cell or hemoglobin production. In addition to adequate numbers of normally functioning RBCs, the amount and quality of hemoglobin are critical factors in maintaining the oxygen-carrying capacity of the blood.

Normal hemoglobin ranges from 12 to 14 grams per 100 milliliters (g/100 mL) of whole blood for adult females and from 14 to 17 g/100 mL for adult males. A hemoglobin value less than 9 g/100 mL indicates anemia.

Figure 13-4 shows the relationship of the four iron-containing groups and four protein subunits or chains in the hemoglobin molecule. Chemical changes in the hemoglobin molecule can result in development of defective hemoglobin and red cells and thus impair the ability of the blood to deliver adequate oxygen to the body tissues.

Pernicious Anemia

Pernicious anemia results from a dietary deficiency of vitamin B_{12} or from the failure of the stomach lining to produce **intrinsic factor**—the substance that allows vitamin B_{12} to be absorbed.

Genetics plays a role in development of pernicious anemia, and research evidence suggests that it is an autoimmune disease.

Vitamin B_{12} deficiency impairs the bone marrow and results in decreased RBC production, as well as a reduction in WBC and platelet numbers. The reduced number of red cells that do enter the circulation are macrocytes and are much larger than normal RBCs.

In addition to the classic symptoms of anemia caused by low oxygen delivery to tissues, patients with pernicious anemia develop numerous nervous system problems such as numbness, tingling, and burning in the feet and hands. Mental impairment, delusions, irritability, and depression are also common.

Pernicious anemia is successfully treated by repeated injections of vitamin B_{12}.

Folate Deficiency Anemia

Folate deficiency anemia is similar to pernicious anemia because it also causes a decrease in the RBC count resulting from a vitamin deficiency. In this condition, it is *folic acid* (vitamin B_9) that is deficient.

Folic acid deficiencies are common among individuals with alcoholism and other malnourished individuals. Treatment for acute folate deficiency anemia involves taking vitamin supplements until a balanced diet can be restored.

Iron Deficiency Anemia

Iron deficiency anemia, as the name suggests, is caused by a deficiency of iron, which is required for hemoglobin synthesis. Although the body carefully protects its iron reserves, they may be depleted through hemorrhage, increased requirements such as wound healing or pregnancy, or by low intake.

Unfortunately, iron deficiency is the most common nutritional deficiency in the world. The tragic result is that an estimated 10% of the population in some developed countries and up to 50% in developing countries suffer from iron deficiency anemia.

In most cases of iron deficiency anemia, the RBC numbers are only slightly below normal. However, the cells are small (microcytic) and appear pale due to the reduction in hemoglobin content (**Figure 13-8**).

TABLE 13-3 Laboratory Results for Types of Anemia

ANEMIA	FOLATE CONTENT	HEMOGLOBIN	HEMATOCRIT	IRON CONTENT	RBC SIZE (VOLUME)	VITAMIN B_{12} CONTENT
Aplastic anemia	Normal	Low to normal	Low to normal	High	Normal to slightly high	Normal
Pernicious anemia	Normal	Low	Low	High	High	Low
Hemorrhagic anemia						
Acute blood-loss anemia	Normal	Low to normal	Low to normal	Normal	Slightly low	Normal
Chronic blood-loss anemia	Normal	Low	Low	Low	Low to normal	Normal
Folate deficiency anemia	Low	Low	Low	High	High	Normal
Iron deficiency anemia	Normal	Low	Low	Low	Low	Normal
Hemolytic anemia (sickle cell anemia and thalassemia)	Normal	Low	Low	Normal to high	Normal to high	Normal

13

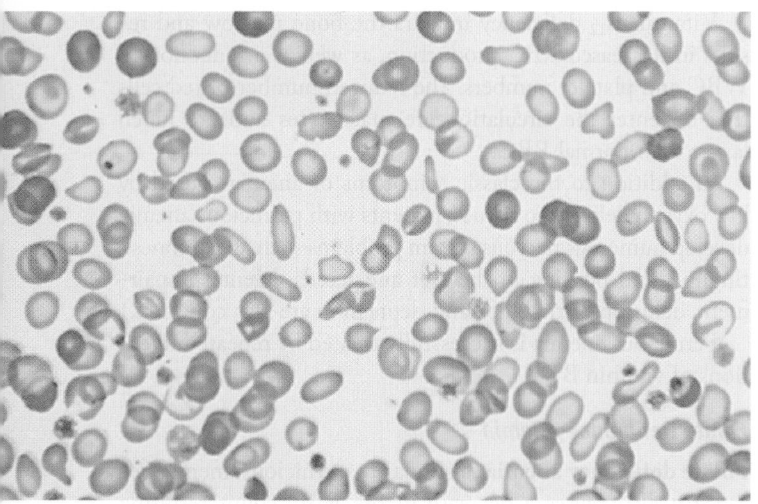

FIGURE 13-8 Iron deficiency anemia. Note the small (microcytic), pale (hypochromic) red blood cells (RBCs). Lack of adequate color in the RBCs is due to reduced hemoglobin content. Compare to RBCs in **Figure 13-5** on p. 355.

A low hematocrit value is common in iron deficiency anemia. Can you explain how this can be if the RBC numbers are near normal? The reason is that the *size* of the RBCs is small (microcytic) so the red cell *volume* and therefore the hematocrit value are both decreased.

Oral administration of iron-containing compounds, such as ferrous sulfate or ferrous gluconate, is very effective in treating the basic iron deficiency seen in the disease. The probability of a complete cure is excellent if, in addition to administration of iron, any underlying causes such as chronic bleeding or iron malabsorption problems are corrected.

Excess availability of iron in the blood can also cause health problems, as in cases of **hemochromatosis** or *iron storage disease.* Explore this condition in the article *Hemochromatosis* at Connect It! at *evolve.elsevier.com.*

Hemolytic Anemias

RBC Destruction

Hemolytic anemias as a group are all associated with a decreased RBC life span caused by an increased rate of destruction. Frequently, an abnormal hemoglobin will cause red blood cells to become distorted and easily broken.

The hemolytic anemias have some distinguishing symptoms in addition to those expected because of low oxygen delivery to tissues. Many are related to the fact that the body retains many of the breakdown products of the excess numbers of RBCs that are destroyed, including iron and pigments. The result may be *jaundice,* a yellowing of skin and other tissues caused by the breakdown of red hemoglobin into yellowish *bilirubin* in the liver.

Swelling of the spleen, problems associated with excess iron storage, and gallstone formation are also common. Some symptoms are unique to a particular type of hemolytic anemia, as discussed later.

Sickle Cell Anemia

Sickle cell anemia is a genetic disease that results in the formation of limited amounts of an abnormal type of hemoglobin called *sickle hemoglobin,* or *hemoglobin S (HbS).* The genetic defect produces an amino acid substitution in one of the beta (β) polypeptide chains (see **Figure 13-4**), causing the resulting HbS to be less stable and less soluble than normal hemoglobin. The defective hemoglobin forms crystals and causes the red cell to become fragile and assume a sickle (crescent) shape when the blood oxygen level is low (**Figure 13-9**).

A person who inherits only one defective gene develops only a small amount of HbS and has a form of the disease called **sickle cell trait.** Those with sickle cell trait most often have no symptoms at all. However, in some stressful or high-exertion situations, a person with sickle cell trait could become ill.

If two defective genes are inherited (one from each parent), then more HbS is produced and a much more severe condition called *sickle cell disease* develops. In addition to RBC sickling and rupture, high levels of HbS may cause reduction in blood flow; blood clotting; and in episodes of "crisis," severe pooling of red cells, particularly in the spleen, causing sudden death.

Treatment is primarily supportive because no effective anti-sickling drugs are currently available. However, patient education, early diagnosis, preventive measures to reduce dehydration and infection, and limited use of blood transfusions to treat episodes of crisis are improving survival rates.

Sickle cell anemia is found almost entirely in those of black African descent, and in the United States nearly 1 in every 600 African-American newborns is affected with sickle cell trait or disease.

Check out the brief, illustrated article *Sickle Cell Anemia* at Connect It! at *evolve.elsevier.com.*

Thalassemia

Thalassemia refers to a group of inherited hemolytic anemias. The most common type, which occurs most often in individuals of Mediterranean descent, is characterized by production of abnormal hemoglobin and inadequate numbers of small (microcytic) and often oddly shaped RBCs that are short lived.

Thalassemia, like sickle cell anemia, occurs in two forms, mild and severe. In both forms, flawed protein synthesis in the RBCs results in defective hemoglobin production and early hemolysis, or death, of defective red cells.

Thalassemia minor, or *thalassemia trait,* occurs when only one defective gene is

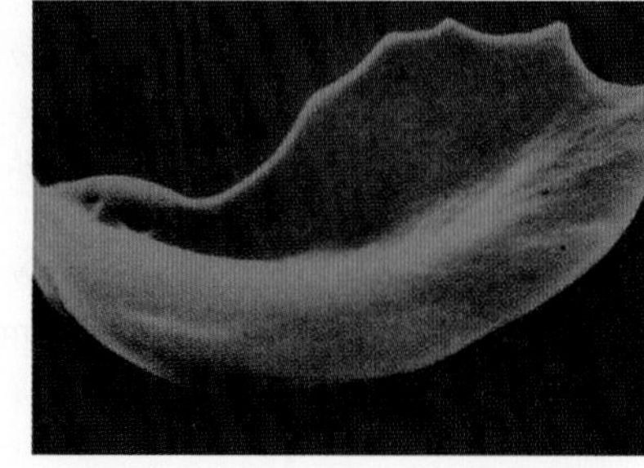

FIGURE 13-9 Sickle cell. A sickle-shaped red blood cell typical of sickle cell anemia.

inherited and is characterized by mild anemia, minimal RBC changes, and few symptoms.

Thalassemia major, which occurs when two defective genes are inherited, is a very serious and life-threatening hemolytic anemia. Red cells are quickly destroyed, hemoglobin levels often fall below 7 g/100 mL of blood, low blood and tissue oxygen levels cause multiple problems, bone marrow mass expands causing crippling and skeletal deformities, and swelling of the spleen and liver occurs.

If adequate and ongoing treatment is not initiated, iron released as a result of RBC hemolysis accumulates in pathological tissue deposits throughout the body.

Bone marrow and stem cell transplantation and experimental gene manipulation initiatives hold the most promise for long-term treatment success. Because thalassemia is a genetically transmitted disease, genetic counseling is appropriate.

Hemolytic Disease of the Newborn

Hemolytic disease of the newborn (HDN) begins during pregnancy if fetal RBCs of a different blood type (see p. 356) than the mother cross the placenta and enter the mother's circulation.

The most common HDN is called *fetal-maternal ABO incompatibility*. This can start during delivery as blood cells leak from the placenta as it pulls away from the lining of the uterus (womb). If this should occur, antibodies against them will be formed because antigens on the fetal RBCs are "foreign" to the mother. Problems for the fetus begin if the maternal antibodies against the "foreign" fetal RBCs cross the placental barrier and enter the fetal circulation. If this occurs, the maternal antibodies will attack and destroy the fetus's red cells, causing a hemolytic anemia to develop.

Rh incompatibility between an Rh-positive fetus and its Rh-negative mother results in hemolytic anemia called **erythroblastosis fetalis.** Review Figure 13-7 on p. 357.

Rh factor incompatibility is clinically more important than ABO incompatibility because the hemolytic response, although it occurs less frequently, is generally more severe. Fortunately, infant mortality caused by Rh incompatibility has been drastically reduced following introduction and widespread use of a product called RhoGAM in Rh-negative mothers (see p. 356).

HDN caused by either ABO or Rh incompatibility may occur early in pregnancy or become apparent only at birth. Red cell numbers and hemoglobin levels decline. Jaundice, intravascular coagulation, heart and lung damage, and swelling of the liver and spleen are common. If problems are detected by laboratory tests of amniotic fluid or from fetal or maternal blood sampling before birth, in utero exchange transfusions and early delivery may be needed to save the life of the infant.

QUICK CHECK

1. How does polycythemia differ from anemia?
2. Describe the clinical signs and symptoms of anemia.
3. Give examples of *hemorrhagic, deficiency,* and *hemolytic* types of anemias.
4. What is *hemolytic disease of the newborn*? Why does it develop? How can it be prevented?

To learn more about hemolytic disease of the newborn, go to AnimationDirect online at *evolve.elsevier.com*.

White Blood Cells

Introduction to WBCs

Recall from the listing of formed elements found in the blood (see p. 351) that the **white blood cells (WBCs)** are also called **leukocytes.**

The WBC, when stained on a microscope slide, shows a prominent and sometimes oddly shaped nucleus—far different in appearance than the RBC, which has no nucleus. WBCs have no hemoglobin and thus are translucent when unstained. A mass of WBCs looks whitish in appearance because of the diffusion of light, much as clear snowflakes appear white when found in a mass.

Different types of WBCs are categorized by the presence or absence of stained granules in their cytoplasm. **Granular leukocytes** *(granulocytes)* have stained granules and **agranular leukocytes** *(agranulocytes)* do not. The granulocytes include the *neutrophils, eosinophils,* and *basophils* (Figure 13-10 *A, B,* and *C*). The *lymphocytes* and *monocytes* (Figure 13-10, *D* and *E*) are agranulocytes.

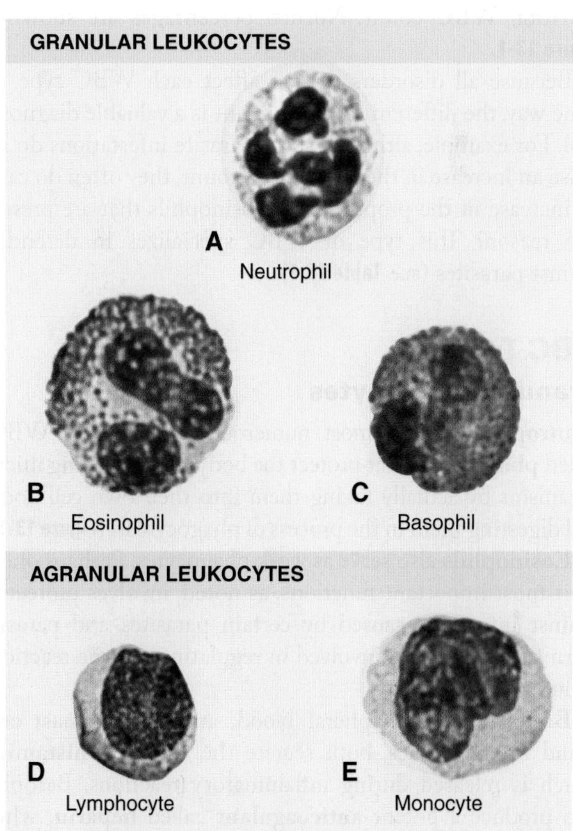

FIGURE 13-10 Leukocytes. A-E, Each light micrograph shows a different type of stained WBC.

13

All of the WBCs are involved in immunity. However, each type and subtype of WBC has its own unique roles to play in immunity, such as phagocytosis of foreign particles or virus-infected cells. Details of the roles of some WBCs are found in Chapter 16.

WBC Count

Normally, the total number of WBCs per cubic millimeter of whole blood (mm^3) ranges between 5000 and 10,000.

The term **leukopenia** is used to describe an abnormally *low* WBC count (less than 5000 WBCs/mm^3 of blood). Leukopenia does not occur often. However, malfunction of blood-forming tissues and cells and some diseases affecting the immune system, such as AIDS (discussed in Chapter 16), may lower WBC numbers.

Leukocytosis refers to an abnormally *high* WBC count (that is, more than 10,000 WBCs/mm^3 of blood). It is a much more common problem than leukopenia and almost always accompanies bacterial infections. In addition, leukocytosis is also seen in many forms of blood cancer (described later), which are often diagnosed when tremendous increases in WBC numbers are detected in blood tests.

13

A special type of white blood cell count called a **differential WBC count** reveals more information than simply counting the total number of all of the different types of WBCs in a blood sample. In a differential WBC count, a component test in the CBC (see box, p. 353), the *proportions* of each type of white blood cell are reported as percentages of the total WBC count. Normal percentages are shown in **Figure 13-1**.

Because all disorders do not affect each WBC type the same way, the differential WBC count is a valuable diagnostic tool. For example, although some parasite infestations do not cause an increase in the total WBC count, they often do cause an increase in the proportion of eosinophils that are present. The reason? This type of WBC specializes in defending against parasites (see **Table 13-1**).

WBC Types

Granular Leukocytes

Neutrophils are the most numerous of the active WBCs, called **phagocytes,** that protect the body from invading microorganisms by actually taking them into their own cell bodies and digesting them in the process of phagocytosis (**Figure 13-11**).

Eosinophils also serve as weak phagocytes. Perhaps one of their most important functions, as noted, involves protection against infections caused by certain parasites and parasitic worms. They also are involved in regulating allergic reactions, including asthma.

Basophils, in peripheral blood, and related **mast cells** found in the tissues, both secrete the chemical **histamine,** which is released during inflammatory reactions. Basophils also produce a potent **anticoagulant** called **heparin,** which helps prevent blood from clotting as it flows through the blood vessels of the body.

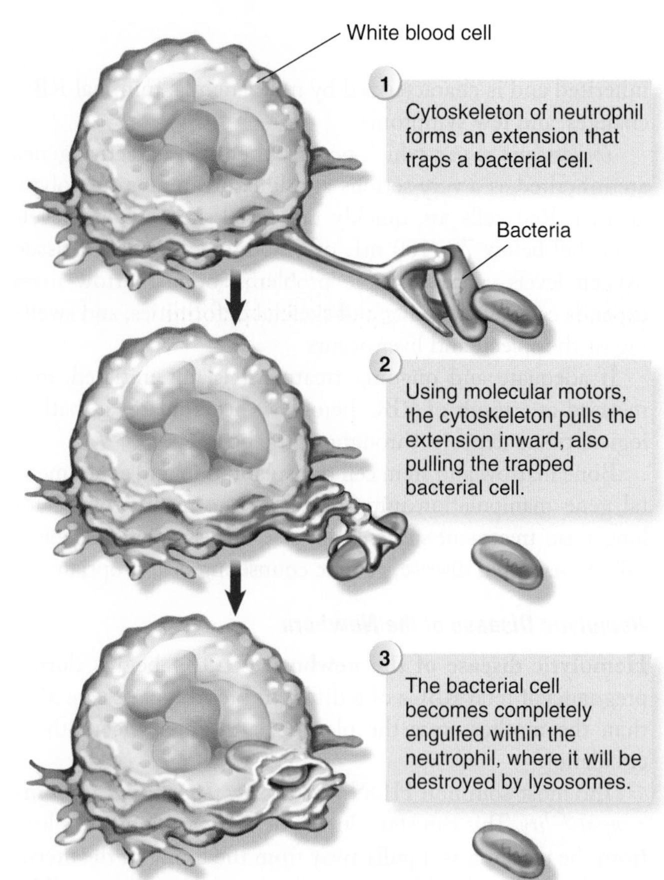

FIGURE 13-11 Phagocytosis. Diagrammatic representation of phagocytosis by a neutrophil (note the multilobed nucleus). Extension of cytoplasm envelopes the bacteria, which are drawn through the cell membrane and into the cytoplasm. (also see **Figure 16-13**).

Agranular leukocytes

Monocytes are the largest leukocytes. Like neutrophils, they are aggressive phagocytes. Because of their size, they are capable of engulfing larger bacterial organisms and cancerous cells. **Macrophages** (meaning "large eater") are monocytes that have grown to several times their original size after migrating out of the bloodstream. They are discussed further in Chapter 16.

Lymphocytes help protect us against infections, but they do it by a process different from phagocytosis. Lymphocytes function in the immune mechanism, the complex process that makes us immune to infectious diseases.

Lymphocytes called *B lymphocytes* develop within several lymphoid organs of the body. B lymphocytes (B cells) secrete plasma proteins called *antibodies* that attach to specific molecules related to bacteria, viruses, chemical toxins, or other foreign substances. Active B lymphocytes, called **plasma cells,** are formed in unusually large numbers in a type of bone marrow cancer called *multiple myeloma,* which is described later.

Other lymphocytes, called *T lymphocytes,* develop in the thymus. They do not secrete antibodies but instead protect us by directly attacking bacteria, virus-infected cells, or cancer cells.

Details of the role of lymphocytes in the immune system are discussed in Chapter 16.

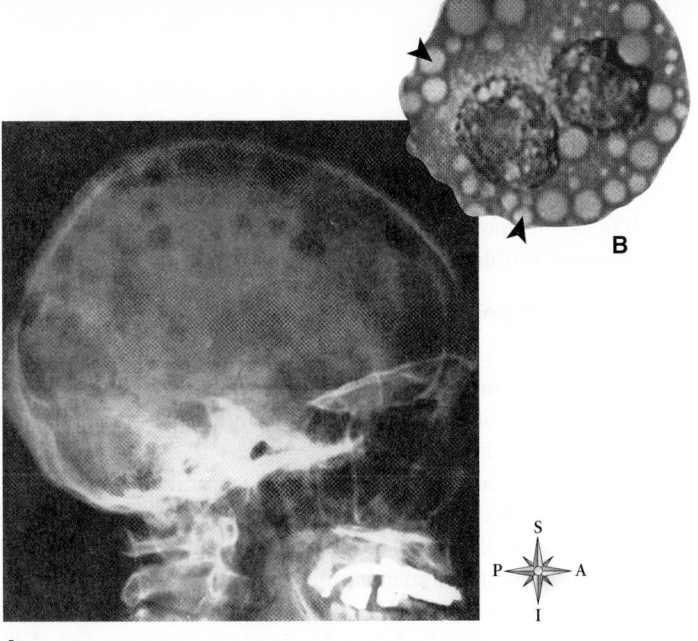

FIGURE 13-12 Multiple myeloma. A, X-ray film of skull showing "honeycomb" or "punched-out" bone defects caused by diseased antibody from plasma cells. **B,** Malignant plasma cell. Vacuoles *(arrowheads)* contain defective antibodies.

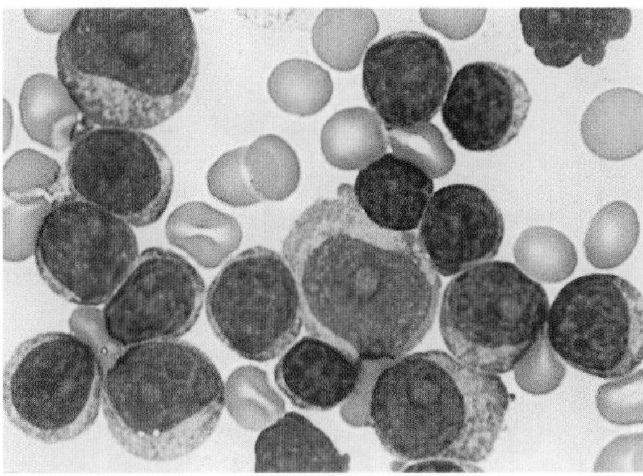

FIGURE 13-13 Chronic lymphocytic leukemia (CLL). Peripheral blood smear showing large numbers of diseased B lymphocytes.

White Blood Cell Disorders

Two major groups of disease conditions constitute a majority of WBC and blood-related cancers, or neoplasms. **Lymphoid neoplasms** arise from lymphoid precursor cells that normally produce B lymphocytes, T lymphocytes, or their descendant cell types. **Myeloid neoplasms** appear as a result of malignant transformation of myeloid precursor cells that normally produce granulocytic WBCs, monocytes, RBCs, and platelets (see **Table 13-1**).

Multiple Myeloma

Multiple myeloma is cancer of mature, antibody-secreting B lymphocytes called *plasma cells* (**Figure 13-12**). It is the most common and one of the most deadly forms of blood-related cancers in people older than 65 years of age.

The cancerous transformation of plasma cells results in impairment of bone marrow function, production of defective antibodies, recurrent infections, anemia, and the painful destruction and fracture of bones in the skull, vertebrae, and throughout the skeletal system. The x-ray photo in **Figure 13-12** shows typical "honeycomb-" or "punched-out"–appearing defects in skull bones caused by the defective myeloma antibody.

Treatment may lengthen life and help relieve symptoms but does not cure the disease. Chemotherapy, marrow and stem cell transplantation, and certain drug and antibody treatments are being used with varying degrees of success.

Leukemia

Leukemia is the term used to describe a number of blood cancers affecting the WBCs. In almost every form of leukemia, marked leukocytosis, or elevated WBC levels, occur. Leukocyte counts in excess of 100,000/mm^3 in circulating blood are common.

The different types of leukemia are identified as either *acute* or *chronic,* based on how quickly symptoms appear after the disease begins. Leukemias also can be identified as *lymphocytic* or *myeloid* depending on the cell type involved.

Four of the most common leukemias are briefly described subsequently.

13

Chronic Lymphocytic Leukemia

Chronic lymphocytic leukemia (CLL) most often affects older adults and is rare in individuals younger than 30 years of age. Average age of onset is about 65 years, appearing more often in men than in women. In those with CLL, malignant precursor B lymphocytes are produced in great numbers (**Figure 13-13**).

They are long lived but do not produce normal antibodies and, as a result, some increase in infections may occur. However, early in the disease few symptoms are apparent, and many patients are diagnosed inadvertently as part of routine physical examinations when results of blood tests become available. When symptoms do appear, they are often quite mild. Anemia, fatigue, and development of enlarged but generally painless lymph nodes are common.

Many patients with CLL live many years after diagnosis with little or no treatment. More severe cases often benefit from chemotherapy and irradiation.

Acute Lymphocytic Leukemia

Acute lymphocytic leukemia (ALL) is primarily a disease of children and constitutes the most common form of "blood cancer" in children between 3 and 7 years of age (**Figure 13-14**). Fully 80% of children who develop leukemia have this form of the disease. Although always a serious condition, it is highly curable in children but less so when it occurs in adults.

Onset of ALL is sudden and often marked by fever, bone pain, and increased rates of infection. Cancerous cells crowd out other bone marrow cells and decrease the production of RBCs, platelets, and other nonmalignant lymphocyte precursor cells, thereby producing anemia. As cancerous transformation of

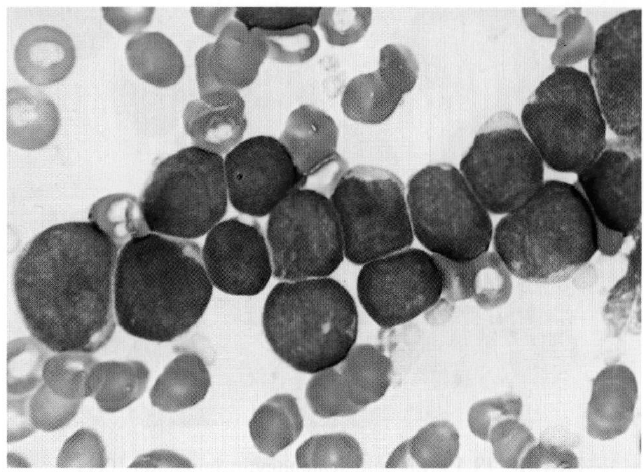

FIGURE 13-14 Acute lymphocytic leukemia (ALL). Appearance of B lymphocytes in acute lymphocytic leukemia.

B lymphocytes continues, their numbers increase in circulating blood and swelling often occurs in lymph nodes, spleen, and liver.

Treatment may involve chemotherapy, irradiation, and bone marrow or stem cell transplants.

Chronic Myeloid Leukemia

Chronic myeloid leukemia (CML) accounts for about 20% of all cases of leukemia and occurs most often in adults between 25 and 60 years of age. CML results from cancerous transformation of granulocytic (neutrophil, eosinophil, and basophil) precursor cells in the bone marrow.

Onset of CML is slow and early symptoms, such as fatigue, weakness, and weight loss, tend to be nonspecific. Once established, the disease progresses slowly. Diagnosis is often made by discovery of marked elevations of granulocytic WBCs in peripheral blood (**Figure 13-15**) and by extreme spleen enlargement.

Bone marrow transplants produce a cure in up to 70% of cases. At the beginning of this century, introduction of a new type of "rational" drug called *Gleevec* was a major advance in treatment of CML. It specifically seeks out and blocks the flawed signals in CML cancer cells that cause runaway proliferation.

Acute Myeloid Leukemia

The pathological transformation of myeloid stem cells resulting in **acute myeloid leukemia (AML)** accounts for 80% of all cases of acute leukemia in adults and 20% of acute leukemia in children.

As the name suggests, onset is sudden, and once symptoms appear, the disease progresses rapidly. Patients most often seek help because of fatigue, bone and joint pain, spongy bleeding gums, symptoms of anemia, and recurrent infections.

The prognosis in AML is poor, with only about 50% of children and 30% of adults achieving long-term survival. Advances in bone marrow and stem cell transplantation have increased cure rates in selected patients.

Infectious Mononucleosis

Infectious mononucleosis is a common noncancerous WBC disorder appearing most often in adolescents and young adults between 15 and 25 years of age.

The disease is usually caused by the *Epstein-Barr virus (EBV),* found in the saliva of infected individuals. "Mono" can be spread by kissing or any other direct contact with an infected person's saliva, such as sharing a straw, toothbrush, or eating utensil.

Leukocytosis is common early in the disease, with total WBC counts averaging between 12,000 to 18,000/mm^3. More than 60% of the leukocytes can be identified in a differential WBC count as large, *atypical* (abnormal) *lymphocytes* that have abundant cytoplasm and a large nucleus (**Figure 13-16**).

Symptoms of "mono" vary greatly but, in addition to the leukocytosis and atypical lymphocytes seen in peripheral blood, fever, sore throat, rash, severe fatigue, and enlargement of lymph nodes and the spleen are common findings.

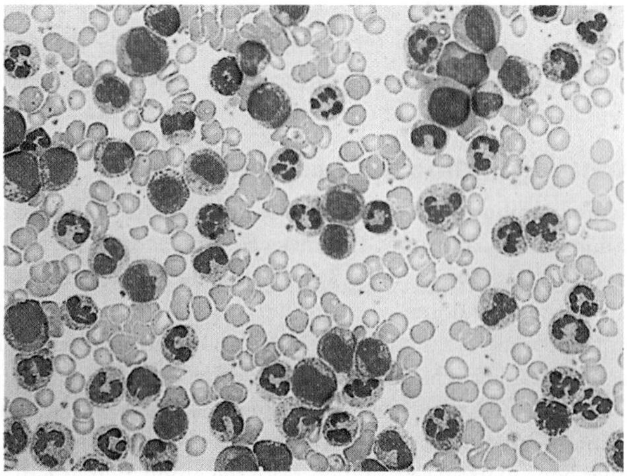

FIGURE 13-15 Chronic myeloid leukemia (CML). Severe granulocytic leukocytosis typical of CML.

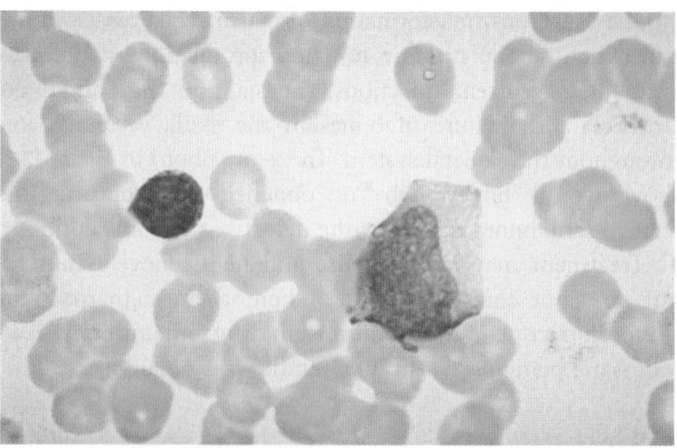

FIGURE 13-16 Infectious mononucleosis. The cell on the left is a typical small lymphocyte with its nucleus almost filling the cell. Often seen in infectious mononucleosis is the larger atypical lymphocyte on the right, which has much more cytoplasm and a larger nucleus.

Infectious mononucleosis is generally self-limited and resolves without complications in about 4 to 6 weeks, although fatigue may last for longer periods. Occasionally, severe complications affecting almost any body organ system may occur in individuals with weakened immune systems.

QUICK CHECK

1. What are the two categories of white blood cells? What determines how the white blood cells are categorized?
2. Define the terms *leukopenia* and *leukocytosis*.
3. Name one function for each of the leukocyte cell types.
4. How do *B lymphocytes* differ from *T lymphocytes*?
5. What type of leukemia is primarily a disease of children? Of older adults?

Platelets and Blood Clotting

Platelets

The **platelet,** the third main type of formed element, plays an essential part in blood clotting or *coagulation*. Your life might someday be saved just because your blood can clot. A clot plugs up torn or cut vessels and stops bleeding that otherwise might prove fatal. Platelets also are called **thrombocytes** from *thrombus* meaning "clot."

Much smaller than RBCs, platelets are tiny cell fragments that have broken away from a much larger precursor cell. Each tiny platelet is filled with chemicals necessary for triggering the formation of a blood clot.

Blood Clotting

The story of how we stop bleeding when an injury occurs—a process called **hemostasis**—is the story of a chain of rapid-fire reactions. All these reactions culminate in the formation of a blood clot.

When an injury occurs, smooth muscles around the wall of the vessel may reflexively contract and thereby constrict the diameter of the vessel—a process called **vasoconstriction.** The resulting pressure helps temporarily close any gaps in the vessel wall and reduces local blood flow until other measures come into play. Pressure applied from outside the wound by a first responder often enhances this effect.

As vessels constrict, damaged tissue cells release various clotting factors into the plasma. These factors rapidly react with other factors already present in the plasma to form **prothrombin activator.**

Normally the lining of blood vessels is extremely smooth, but an injury makes a rough spot with exposed collagen fibers. This attracts platelets to the site, which become "sticky" at the point of injury and rapidly accumulate near the break in the blood vessel, forming a soft, temporary **platelet plug.** As the platelets accumulate, they release additional clotting factors, forming even more prothrombin activator—a kind of self-amplifying, *positive-feedback* response.

If the normal amount of blood calcium is present, prothrombin activator triggers the next step of clotting by converting **prothrombin** (a protein in normal blood) to **thrombin.** In the last step, thrombin reacts with **fibrinogen** (a normal plasma protein) to change it to a fibrous gel called **fibrin.** Under the microscope, fibrin looks like a tangle of fine threads with RBCs caught in the tangle. Figure 13-17 illustrates the steps in the blood-clotting mechanism.

The clotting mechanism contains clues for ways to stop bleeding by speeding up blood clotting. For example, you might simply apply gauze to a bleeding surface. Its slight roughness would cause more platelets to stick together and release more clotting factors. These additional factors would then make the blood clot more quickly.

Physicians sometimes prescribe vitamin K before surgery to make sure that the patient's blood will clot fast enough to prevent hemorrhage. Vitamin K stimulates liver cells to increase the synthesis of prothrombin. More prothrombin in blood allows faster production of thrombin during clotting and thus faster clot formation.

To learn more about hemostasis, go to AnimationDirect online at *evolve.elsevier.com*.

Clotting Disorders

Abnormal Blood Clots

Types of Abnormal Clots

Unfortunately, clots sometimes form in unbroken blood vessels of the heart, brain, lungs, or some other organ—a dreaded thing because clots may produce sudden death by shutting off the blood supply to a vital organ. When a clot stays in the place where it formed, it is called a **thrombus** and the condition is spoken of as **thrombosis.**

If part of the clot dislodges and circulates through the bloodstream, the dislodged part is then called an **embolus,** and the condition is called an **embolism.** For example, a clot fragment that lodges in the lung is called a *pulmonary embolism*—a situation that may prove fatal (Figure 13-18).

Suppose that your doctor told you that you had a clot in one of your coronary arteries. Which diagnosis would she make—coronary thrombosis or coronary embolism—if she thought that the clot had formed originally in the coronary artery as a result of the accumulation of fatty material in the vessel wall?

Crushing injuries of skeletal muscle can cause widespread abnormal clotting. Review the article *Rhabdomyolysis* at Connect It! at *evolve.elsevier.com*.

Anticoagulant Therapy

A number of drugs are now available to help dissolve abnormal clots. *Streptokinase* and *recombinant* **tissue plasminogen activator (TPA or tPA)** are drugs frequently used in a variety of conditions, including treatment of clot-induced strokes, heart attacks, and other thrombus-induced and embolus-induced medical emergencies.

13

FIGURE 13-17 **Blood clotting.** **A,** The extremely complex clotting mechanism can be distilled into three basic steps outlined in the boxes. **B,** Color-enhanced scanning electron micrograph shows RBCs and WBCs entrapped in a fibrin *(yellow)* mesh during clot formation (platelets are *blue*).

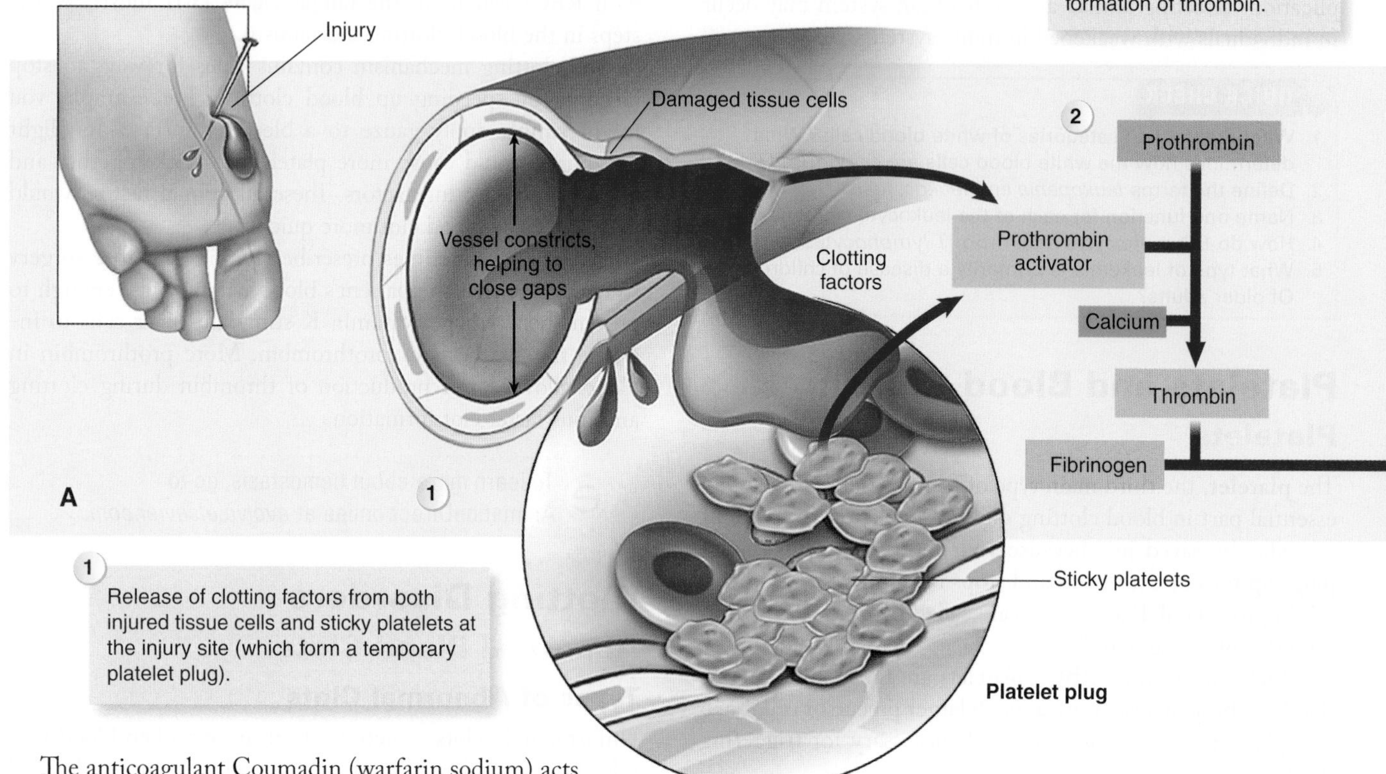

The anticoagulant Coumadin (warfarin sodium) acts by inhibiting the synthesis of prothrombin and other vitamin K–dependent clotting factors. By doing so, Coumadin decreases the ability of blood to clot and is effective in preventing repeat thromboses after a heart attack or the formation of clots after surgical replacement of heart valves.

Heparin can also be used to prevent excessive blood clotting. Heparin inhibits the conversion of prothrombin to thrombin, thus preventing formation of a thrombus.

The most widely used anticoagulant is low-dose (81-mg) aspirin. This readily available drug inhibits the formation of tiny platelet plugs and the subsequent formation of emboli, which may cause blockage of small blood vessels in the brain and lead to a stroke.

A laboratory test called the **prothrombin time (PT)** is often used to regulate dosage of anticoagulant drugs. In this test, thromboplastin (a blood clotting factor) and calcium are added simultaneously to a tube of the patient's plasma and a tube containing a normal control solution, and the time required for clot formation in both tubes is determined. A patient prothrombin time in excess of the standard control value (11 to 12.5 seconds) indicates the level of anticoagulant effect caused by the administered drug.

Unfortunately, PT test results may vary among different clinical laboratories. Variability is often caused by differing techniques or differences in the sensitivity of reagents used.

To minimize the effects of these and other variables and standardize the results of anticoagulation testing, a system called the **INR** (abbreviation for **I**nternational **N**ormalized **R**atio) has been developed. PT is reported in seconds. The INR is a mathematical calculation and is reported as a number. An INR of 0.8 to 1.2 is considered normal. In regulating anticoagulant therapy, keeping the INR between 1.5 and 3 will help ensure the prevention of unwanted blood coagulation in "at risk" individuals. By monitoring changes in the INR, a physician can adjust the dose of anticoagulant drug needed to maintain an appropriate level of anticoagulant effect.

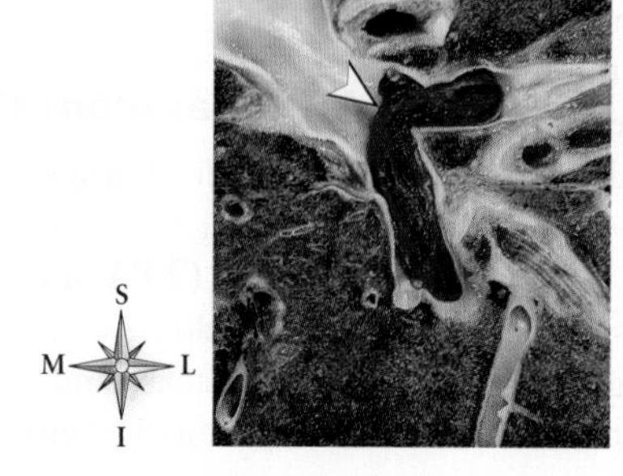

FIGURE 13-18 **Pulmonary embolism.** An embolus (clot fragment) that formed in the leg but broke away and now lodged in a branch of the pulmonary artery within the lung. *Arrowhead* shows the embolus blocking the artery, thus drastically reducing gas exchange between air and blood in the affected lung.

Hemophilia

Hemophilia is an X-linked inherited disorder (see Chapter 25, pp. 683-684) that affects more than 300,000 people around the world. Typically, it is transmitted from a symptom-free "carrier mother" to an affected son. Hemophilia is a

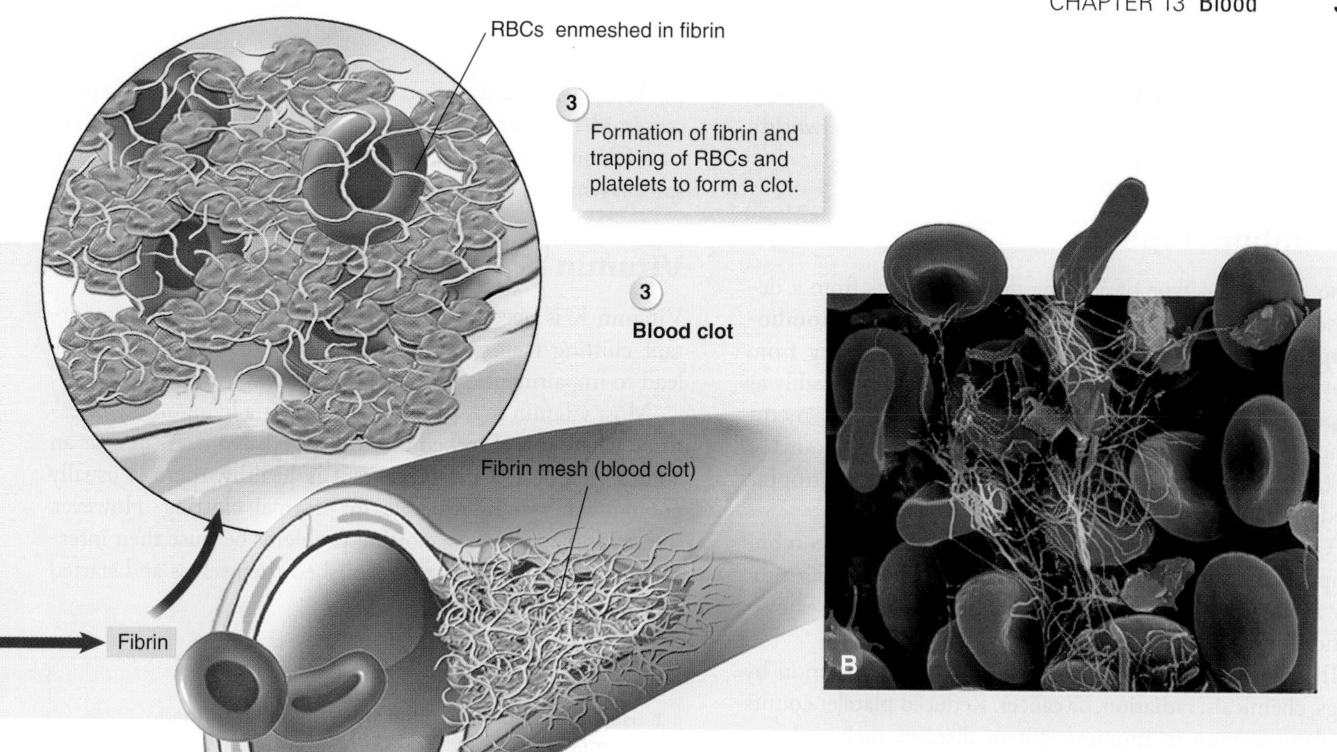

"bleeding disorder" that results from a failure to produce one or more plasma proteins responsible for blood clotting—a process illustrated in **Figure 13-17**.

The most common form of the disease, called *hemophilia A,* is caused by absence of normal **factor VIII**—one of the many clotting factors needed to form prothrombin activator. This serious coagulation disorder has plagued the royal families of Europe for hundreds of years and, as a result, its signs and symptoms and the genetics of its transmission are well known. Simply stated, people with hemophilia are relatively unable to form blood clots.

Because minor blood vessel injuries are common in ordinary life, hemophilia can be a life-threatening condition. Mild forms may not be apparent until the individual is subjected to surgery or trauma, whereas more severe cases may result in frequent and even spontaneous episodes of extensive bleeding.

The most common signs of the disease include easy bruising, deep muscle hemorrhage, nosebleeds, blood in the urine, and in severe cases, even bleeding into the brain. Perhaps the most characteristic sign is repeated episodes of bleeding into the joints—especially the elbows, knees, and ankles. The result is chronic pain and progressive joint deformity.

Treatment of hemophilia involves initiation of lifestyle changes that help prevent injury, prompt response to bleeding episodes, avoiding drugs such as aspirin that alter the clotting mechanism, and administration of factor VIII.

Historically, only small amounts of factor VIII could be obtained from large quantities of plasma obtained from many donors. Given the shortage of donated blood, this method could not meet demand as physicians prescribed more factor VIII to prevent as well as treat bleeding episodes.

Further, even with new and more effective blood banking safety precautions, purification methods and diagnostic tests, pooling and fractionating donated blood and plasma still involve some risk of disease transmission—especially HIV and viral hepatitis. Currently, recombinant methods eliminate

SCIENCE APPLICATIONS
HEMATOLOGY

Charles Richard Drew (1904–1950)

African-American physician Charles Richard Drew was a pioneer in **hematology,** the study of blood. During World War II, he developed the idea of blood banks and researched the best way to store blood for transfusions to wounded soldiers. In New York, he set up the first blood bank ever, in 1941. This blood bank served as the model for a network of blood banks opened by the American Red Cross.

Many *hematologists* continue in Drew's footsteps, refining and perfecting the practice of blood science. Many professions benefit from this research. **Phlebotomists** collect blood for testing or storage, *clinical laboratory technicians* analyze blood samples, and many different health professionals use blood analysis and blood transfusions to help their patients. Of course, military **medics** still rely on blood banking technology to provide immediate aid to wounded combat and terrorism victims.

these risks and are used to produce enough *recombinant antihemophilic factor VIII (rAHF)* to meet the needs of the world's hemophiliac population.

Thrombocytopenia

A more common type of clotting disorder results from a decrease in the platelet count—a condition called **thrombocytopenia.** This condition is characterized by bleeding from many small blood vessels throughout the body, most visibly as **purpura**—purple splotches in the skin and mucous membranes. If the number of thrombocytes falls to 20,000/mm^3 or less (normal range is 150,000 to 400,000/mm^3), catastrophic bleeding may occur.

A number of different mechanisms can result in thrombocytopenia. For example, platelet numbers below 50,000/mm^3 may result from mechanical destruction as blood passes over artificial heart valves.

The usual cause, however, is bone marrow destruction by drugs, chemicals, radiation, or cancer. Reduced platelet counts are also common in immune system diseases such as lupus and HIV/AIDS, in which a reduction in platelets occurs early in the course of infection.

13

Some drugs, such as aspirin, may cause thrombocytopenia as a side effect. In such cases, stopping the use of the drug usually solves the problem.

Active treatment options include administration of corticosteroid-type drugs, which increase platelet production, transfusion of platelets, and in severe cases, removal of the spleen, which is a major site of platelet destruction.

Vitamin K Deficiency

Vitamin K is needed by the body to produce several important clotting factors. Thus, a deficiency of this vitamin may lead to impairment of the clotting mechanism.

Most vitamin K is produced by bacteria living in the intestines, where it is absorbed into the bloodstream. As long as an adult's gastrointestinal (GI) tract is healthy, there is usually enough vitamin K available for normal clotting. However, infants sometimes have clotting problems because their intestinal bacteria have not yet established themselves and started producing vitamin K.

QUICK CHECK

1. How does a platelet differ from the other formed elements?
2. What is the role of *prothrombin, thrombin, fibrinogen,* and *fibrin* in the blood clotting mechanism?
3. What is the difference between a thrombus and an embolus?
4. Identify two types of clotting disorders.

LANGUAGE OF **SCIENCE** *(continued from p. 349)*

eosinophil
(ee-oh-SIN-oh-fil)
[*eos-* **dawn (red),** *-in-* **substance,** *-phil* **love**]

erythrocyte
(eh-RITH-roh-syte)
[*erythro-* **red,** *-cyte* **cell**]

erythropoietin (EPO)
(eh-RITH-roh-POY-eh-tin [ee pee oh])
[*erythro-* **red,** *-poiet-* **make,** *-in* **substance**]

factor VIII
(FAK-ter ayt)
[*VIII* **Roman numeral eight**]

fibrin
(FYE-brin)
[*fibr-* **fiber,** *-in* **substance**]

fibrinogen
(fye-BRIN-oh-jen)
[*fibr-* **fiber,** *-in-* **substance,** *-gen* **produce**]

formed element
(formd EL-eh-ment)

globulin
(GLOB-yoo-lin)
[*glob-* **ball,** *-ul-* **little,** *-in* **substance**]

granular leukocyte
(GRAN-yoo-lar LOO-koh-syte)
[*gran-* **grain,** *-ul-* **little,** *-ar* **relating to,** *leuko-* **white,** *-cyte* **cell**]

hematology
(hee-mah-TOL-oh-jee)
[*hema-* **blood,** *-to-* **combining syllable,** *-log-* **words (study of),** *-y* **activity**]

hematopoiesis
(hee-mat-oh-poy-EE-sis)
[*hemo-* **blood,** *-poiesis* **making**]

hemoglobin (Hb)
(hee-moh-GLOH-bin [aych bee])
[*hemo-* **blood,** *-glob-* **ball,** *-in* **substance**]

hemostasis
(hee-moh-STAY-sis)
[*hemo-* **blood,** *-stasis* **standing**]

heparin
(HEP-ah-rin)
[*hepar-* **liver,** *-in* **substance**]

histamine
(HIS-tah-meen)
[*hist-* **tissue,** *-amine* **ammonia compound**]

immune system
(ih-MYOON SIS-tem)
[*immun-* **free**]

intrinsic factor
(in-TRIN-sik)
[*intr-* **inside or within,** *-insic* **beside**]

leukocyte
(LOO-koh-syte)
[*leuko-* **white,** *-cyte* **cell**]

lymphatic system
(lim-FAT-ik SIS-tem)
[*lymph-* **water,** *-atic* **relating to**]

lymphocyte
(LIM-foh-syte)
[*lymph-* **water (lymphatic system),** *-cyte* **cell**]

lymphoid tissue
(LIM-foyd TISH-yoo)
[*lymph-* **water (lymphatic system),** *-oid* **like,** *tissu* **fabric**]

macrophage
(MAK-roh-fayj)
[*macro-* **large,** *-phag-* **eat**]

mast cell
(mast sel)
[*mast* **fattening,** *cell* **storeroom**]

monocyte
(MON-oh-syte)
[*mono-* **single,** *-cyte* **cell**]

LANGUAGE OF SCIENCE (continued from p. 368)

myeloid tissue
(MY-eh-loyd TISH-yoo)
[*myel-* **marrow**, *-oid* **of or like**, *tissu* **fabric**]

neutrophil
(NOO-troh-fil)
[*neutr-* **neither**, *-phil* **love**]

oxyhemoglobin (HbO_2)
(ahk-see-hee-moh-GLOH-bin [aych bee oh too])
[*oxy-* **sharp (oxygen)**, *-hemo-* **blood**, *-glob-* **ball**, *-in* **substance**]

phagocyte
(FAG-oh-syte)
[*phago-* **eat**, *-cyte* **cell**]

plasma
(PLAZ-mah)
[*plasma* **something molded or created**]

plasma cell
(PLAZ-mah sel)
[*plasma* **something molded or created (blood plasma)**, *cell* **storeroom**]

plasma protein
(PLAZ-mah PROH-teen)
[*plasma* **something molded or created (blood plasma)**, *prote-* **primary**, *-in* **substance**]

platelet
(PLAYT-let)
[*plate-* **flat**, *-let* **small**]

platelet plug
(PLAYT-let plug)
[*plate-* **flat**, *-let* **small**]

prothrombin
(proh-THROM-bin)
[*pro-* **before**, *-thromb-* **clot**, *-in* substance]

prothrombin activator
(proh-THROM-bin AK-tih-vay-tor)
[*pro-* **before**, *-thromb-* **clot**, *-in* **substance**, *act-* **perform**, *-iv-* **relating to**, *-at-* **process**, *-or* **condition**]

red blood cell (RBC)
(red blud sel [ar bee see])
[*cell* **storeroom**]

Rh system
(R H SIS-tem)
[*Rh* **short for "rhesus monkey"**]

serum
(SEER-um)
[*serum* **watery body fluid**]

thrombin
(THROM-bin)
[*thromb-* **clot**, *-in* **substance**]

thrombocyte
(THROM-boh-syte)
[*thromb-* **clot**, *-cyte* **cell**]

vasoconstriction
(vay-soh-kon-STRIK-shun)
[*vas-* **vessel**, *-constrict-* **draw tight**, *-tion* **state**]

white blood cell (WBC)
(whyte blud sel [DUB-el-yoo bee see])
[*cell* **storeroom**]

LANGUAGE OF MEDICINE

acidosis
(as-ih-DOH-sis)
[*acid-* **sour**, *-osis* **condition**]

acute lymphocytic leukemia (ALL)
(ah-KYOOT LIM-foh-sit-ik loo-KEE-mee-ah [ay el el])
[*acu-* **sharp**, *lymph-* **water (lymphatic system)**, *-cyt-* **cell**, *-ic* **relating to**, *leuk-* **white**, *-emia* **blood condition**]

acute myeloid leukemia (AML)
(ah-KYOOT MY-eh-loyd loo-KEE-mee-ah [ay em el])
[*acu-* **sharp**, *myel-* **marrow**, *-oid* **of or like**, *leuk-* **white**, *-emia* **blood condition**]

anemia
(ah-NEE-mee-ah)
[*an-* **without**, *-emia* **blood condition**]

anticoagulant
(an-tee-koh-AG-yoo-lant)
[*anti-* **against**, *-coagul-* **curdle**, *-ant* **agent**]

aplastic anemia
(a-PLAS-tik ah-NEE-mee-ah)
[*a-* **without**, *-plast-* **form**, *-ic* **relating to**, *an-* **without**, *-emia* **blood condition**]

aspiration biopsy cytology (ABC)
(as-pih-RAY-shun BYE-op-see sye-TOL-oh-jee [ay bee see])
[*a-* **act of**, *-spir-* **breathe**, *-ation* **process**, *bio-* **life**, *-ops-* **view**, *-y* **act of**, *cyt-* **cell**, *-o-* **combining form**, *-log-* **words (study of)**, *-y* **activity**]

blood doping
(blud DOH-ping)
[*dop-* **thick liquid (opium)**]

bone marrow transplant
(bohn MAYR-oh TRANS-plant)
[*trans-* **across**, *-plant* **sprout**]

chronic lymphocytic leukemia (CLL)
(KRON-ik LIM-foh-sit-ik loo-KEE-mee-ah [see el el])
[*chron-* **time**, *-ic* **relating to**, *lymph-* **water (lymphatic system)**, *-cyte* **cell**, *-ic* **relating to**, *leuk-* **white**, *-emia* **blood condition**]

chronic myeloid leukemia (CML)
(KRON-ik MY-eh-loyd loo-KEE-mee-ah [see em el])
[*chron-* **time**, *-ic* **relating to**, *myel-* **marrow**, *-oid* **like**, *leuk-* **white**, *-emia* **blood condition**]

complete blood cell count (CBC)
(kom-PLEET blud sel kount [see bee see])
[*cell* **storeroom**]

differential WBC count
(dif-er-EN-shal DUB-el-yoo bee see kownt)
[*different-* **difference**, *-al* **relating to**, *WBC* **white blood cell**]

embolism
(EM-boh-liz-em)
[*embol-* **plug**, *-ism* **condition**]

embolus
(EM-boh-lus)
[*embolus* **plug**]

erythroblastosis fetalis
(eh-rith-roh-blas-TOH-sis feh-TAL-is)
[*erythro-* **red**, *-blast-* **bud or sprout**, *-osis* **condition**]

folate deficiency anemia
(FOH-layt deh-FISH-en-see ah-NEE-mee-ah)
[*fol-* **leaf**, *-ate* **relating to**, *de-* **down**, *-fic-* **perform**, *-ency* **state**, *an-* **without**, *-emia* **blood condition**]

hematocrit (Hct)
(hee-MAT-oh-krit [aych see tee])
[*hemato-* **blood**, *-crit* **separate**]

LANGUAGE OF MEDICINE (continued from p. 369)

hemochromatosis
(hee-moh-kroh-mah-TOH-sis)
[*hemo-* **blood,** *-chroma-* **color,** *-osis* **condition**]

hemolytic anemia
(hee-moh-LIT-ik ah-NEE-mee-ah)
[*hemo-* **blood,** *-lyt-* **loosen,** *-ic* **relating to,** *an-* **without,** *-emia* **blood condition**]

hemolytic disease of the newborn (HDN)
(hee-moh-LIT-ik dih-ZEEZ ov thuh NOO-born)
[*hemo-* **blood,** *-lyt-* **loosen,** *-ic* **relating to,** *dis-* **opposite of,** *-ease* **comfort**]

hemophilia
(hee-moh-FIL-ee-ah)
[*hemo-* **blood,** *-phil-* **love,** *-ia* **condition**]

hemorrhagic anemia
(HEM-oh-raj-ick ah-NEE-mee-ah)
[*hemo-* **blood,** *-rrh(e)a-* **flow,** *-ag(e)-* **process or state,** *-ic* **relating to,** *an-* **without,** *-emia* **blood condition**]

infectious mononucleosis
(in-FEK-shuss mah-noh-noo-klee-OH-sis)
[*infect-* **stain,** *-ous* **relating to,** *mono-* **single,** *-nucle-* **nut,** *-osis* **condition**]

INR (international normalized ratio)
(aye en ar [in-ter-NASH-en-ul NOR-mah-lyzed RAY-shee-oh])

iron deficiency anemia
(AYE-ern deh-FISH-en-see ah-NEE-mee-ah)
[*iron* **element 26,** *de-* **down,** *-fic-* **perform,** *-ency* **state,** *an-* **without,** *-emia* **blood condition**]

leukemia
(loo-KEE-mee-ah)
[*leuk-* **white,** *-emia* **blood condition**]

leukocytosis
(loo-koh-SYE-toh-sis)
[*leuko-* **white,** *-cyt-* **cell,** *-osis* **condition**]

leukopenia
(loo-koh-PEE-nee-ah)
[*leuko-* **white,** *-penia* **lack**]

lymphoid neoplasm
(LIM-foyd NEE-oh-plaz-em)
[*lymph-* **water (lymphatic system),** *-oid* **like,** *neo-* **new,** *-plasm* **something molded or created**]

medic
(MED-ik)
[*med-* **heal,** *-ic-* **relating to**]

multiple myeloma
(MUL-tih-pul my-LOH-mah)
[*multi-* **many,** *-pl-* **fold,** *myel-* **marrow,** *-oma* **tumor**]

myeloid neoplasm
(MY-eh-loyd NEE-oh-plaz-em)
[*myel-* **marrow,** *-oid* **like,** *neo-* **new,** *-plasm* **something molded or created**]

pernicious anemia
(per-NISH-us ah-NEE-mee-ah)
[*pernici-* **destruction,** *-ous* **relating to,** *an-* **without,** *-emia* **blood condition**]

phlebotomist
(fleh-BOT-uh-mist)
[*phleb-* **vein,** *-tom-* **cut,** *-ist* **agent**]

polycythemia
(pahl-ee-sye-THEE-mee-ah)
[*poly-* **many,** *-cyt-* **cell,** *-emia* **blood condition**]

prothrombin time (PT)
(proh-THROM-bin tyme)
[*pro-* **before,** *-thromb-* **clot,** *-in* **substance**]

purpura
(PUR-pah-rah)
[*purpura* **purple**]

Rh-negative
(R H NEG-ah-tiv)
[*Rh* **short for "rhesus monkey,"** *negat-* **deny,** *-ive* **relating to**]

RhoGAM
(ROH-gam)
[**brand name derived from** *Rho* **17th letter of Greek alphabet (refers to Rh antigen),** *GAM* **from gamma-globulin (Rh antibody)**]

Rh-positive
(R H POZ-ih-tiv)
[*Rh* **short for "rhesus monkey,"** *posit-* **put or place,** *-ive* **relating to**]

sickle cell anemia
(SIK-ul sel ah-NEE-mee-ah)
[*sickle* **crescent-shaped tool,** *cell* **storeroom,** *an-* **without,** *-emia* **blood condition**]

sickle cell trait
(SIK-ul sel trayt)
[*sickle* **crescent,** *cell* **storeroom**]

thalassemia
(thal-ah-SEE-mee-ah)
[*thalas-* **sea,** *-emia* **blood condition**]

thrombocytopenia
(throm-boh-sye-toh-PEE-nee-ah)
[*thrombo-* **clot,** *-cyto-* **cell,** *-penia* **lack**]

thrombosis
(throm-BOH-sis)
[*thrombo-* **clot,** *-osis* **condition**]

thrombus
(THROM-bus)
[thrombus **clot**]

tissue plasminogen activator (TPA or tPA)
(TISH-yoo plaz-MIN-oh-jen AK-tih-vay-tor)
[*tissu* **fabric,** *plasm-* **something molded or created (plasma),** *-in-* **substance,** *-gen* **produce,** *act-* **perform,** *-ive* **relating to,** *-or* **agent**]

universal donor blood
(yoo-neh-ver-sal DOH-nor blud)
[*uni-* **one,** *-vers-* **turn (into),** *-al* **relating to,** *don-* **give,** *-or* **agent**]

universal recipient blood
(yoo-neh-ver-sal REE-sip-ee-ahnt blud)
[*uni-* **one,** *-vers-* **turn (into),** *-al* **relating to,** *recip-* **take,** *-ent* **agent**]

❑ OUTLINE SUMMARY

To download a digital version of the chapter summary for use with your device, access the ***Audio Chapter Summaries*** *online at evolve.elsevier.com.*

Scan this summary after reading the chapter to help you reinforce the key concepts. Later, use the summary as a quick review before your class or before a test.

Blood Composition

A. Blood tissue
   1. Blood tissue components
      a. Liquid fraction of whole blood (extracellular part) called *plasma* (**Figure 13-1**)
      b. Cellular components suspended in the plasma make up the formed elements
   2. Normal volumes of blood
      a. Plasma—2.6 L
      b. Formed elements—2.4 L
      c. Whole blood—4 to 6 L average or 7% to 9% of total body weight
   3. Blood pH
      a. Blood is alkaline—pH 7.35 to pH 7.45
      b. Blood pH decreased toward neutral creates a condition called *acidosis*

B. Blood plasma
   1. Liquid fraction of whole blood minus formed elements (**Figure 13-1**)
   2. Composition—water containing many dissolved substances including:
      a. Nutrients, salts
      b. About 1.5% of total O_2 transported in blood
      c. About 10% of total CO_2
      d. Most abundant solutes dissolved in plasma are plasma proteins
         (1) Albumins
         (2) Globulins
         (3) Fibrinogen
         (4) Prothrombin
   3. Plasma minus clotting factors is called *serum*
      a. Serum is liquid remaining after whole blood clots
      b. Serum contains antibodies

C. Formed elements
   1. Types (**Figure 13-1** and **Table 13-1**)
      a. Red blood cell; also called RBC or erythrocyte
      b. White blood cell; also called WBC or leukocyte
         (1) Granular leukocytes—neutrophils, eosinophils, and basophils
         (2) Agranular leukocytes—lymphocytes and monocytes
      c. Platelet; also called thrombocyte
   2. Number
      a. RBCs—4.2 to 6.2 million/mm^3 of blood
      b. WBCs—5000 to 10,000/mm^3 of blood
      c. Platelets—150,000 to 400,000/mm^3 of blood

D. Hematopoiesis—formation of new blood cells
   1. Myeloid tissue (red bone marrow)—forms all blood cells except some lymphocytes; found within bones
   2. Lymphoid tissue—forms additional white blood cells in the lymph nodes, thymus, and spleen
   3. RBCs live about 4 months; WBCs live for a few days (granular) to over 6 months (agranular)

Mechanisms of Blood Disease

A. Most blood diseases result from failure of myeloid and lymphoid tissues

B. Causes include toxic chemicals, radiation, inherited defects, nutritional deficiencies and cancers, including leukemia

C. Aspiration biopsy cytology (ABC) permits examination of blood-forming tissues to assist in diagnosis of blood diseases

D. Bone marrow, cord blood, and hematopoietic stem cell transplants may be used to replace diseased or destroyed blood-forming tissues

Red Blood Cells

A. RBC structure and function
   1. RBC offers excellent example of how structural adaptation affects biological function
   2. Tough and flexible plasma membrane deforms easily allowing RBCs to pass through small-diameter capillaries
   3. Biconcave disk shape (thin center and thicker edges) results in large cellular surface area (**Figure 13-2**)
   4. Absence of nucleus and cytoplasmic organelles
      a. Limits RBC life span to about 120 days
      b. Provides more cellular space for hemoglobin (Hb)
   5. Transport of respiratory gases (O_2 and CO_2)

B. RBC count
   1. Complete blood cell count (CBC)—battery of laboratory tests used to measure the amounts or levels of many blood constituents
   2. Hematocrit (Hct)
      a. Also called packed cell volume (PCV)
      b. Hct expressed as the percentage of whole blood that is RBCs (**Figure 13-3**)

C. Hemoglobin (Hb)
   1. Quaternary protein made up of four polypeptide chains, each with an oxygen-attracting heme group at center (**Figure 13-4**)

13

2. Iron (Fe), folate (a B vitamin), and vitamin B_{12} are among the critical nutrients needed to manufacture Hb
3. Transport of respiratory gases (O_2 and CO_2)
   a. Combined with hemoglobin
      (1) Oxyhemoglobin (Hb + O_2)
      (2) Carbaminohemoglobin (Hb + CO_2)
   b. CO_2 converted to bicarbonate by the RBCs
4. Important role in homeostasis of acid-base balance

D. RBC abnormalities (**Figure 13-5**)
1. Named according to size
   a. Normocytes—normal size (about 8 to 9 μm in diameter)
   b. Microcytic—small size
   c. Macrocytic—large size
2. Named according to hemoglobin content of cell
   a. Normochromic—normal Hb content
   b. Hypochromic—low Hb content
   c. Hyperchromic—high Hb content

E. Blood Types (**Table 13-2**)
1. ABO system (**Figure 13-6**)
   a. Antigen—substance that can activate immune system
   b. Antibody—substance made by body in response to stimulation by an antigen
   c. ABO blood types
      (1) Type A blood—type A self-antigens in RBCs; anti–B type antibodies in plasma
      (2) Type B blood—type B self-antigens in RBCs; anti–A type antibodies in plasma
      (3) Type AB blood—type A and type B self-antigens in RBCs; no anti-A or anti-B antibodies in plasma
      (4) Type O blood—no type A or type B self-antigens in RBCs; both anti-A and anti-B antibodies in plasma
2. Rh system
   a. Rh-positive blood—Rh factor antigen present in RBCs
   b. Rh-negative blood—no Rh factor present in RBCs; no anti-Rh antibodies present naturally in plasma; anti-Rh antibodies, however, appear in the plasma of Rh-negative persons if Rh-positive RBCs have been introduced into their bodies; an RH-negative person can generate anti-Rh antibodies following exposure to the Rh antigen
   c. Erythroblastosis fetalis—may occur when Rh-negative mother carries a second Rh-positive fetus; caused by mother's Rh antibodies reacting with the fetus's Rh-positive cells (**Figure 13-7**)
3. Combined ABO-Rh system
   a. Both systems commonly used together
   b. Type O−: universal donor blood
   c. Type AB+: universal recipient blood

13

Red Blood Cell Disorders

A. Most often related to either overproduction of RBCs—called *polycythemia;* or to low oxygen-carrying capacity of blood—called *anemia*

B. Polycythemia
1. Cause is generally cancerous transformation of red bone marrow
2. Dramatic increase in RBC numbers—often in excess of 10 million/mm^3 of blood; hematocrit may reach 60%
3. Signs and symptoms include:
   a. Increased blood viscosity or thickness
   b. Slow blood flow and coagulation problems
   c. Frequent hemorrhages
   d. Distention of blood vessels and hypertension
4. Treatment may include:
   a. Blood removal
   b. Irradiation and chemotherapy to suppress RBC production

C. Anemia (**Table 13-3**)
1. Caused by low numbers or abnormal RBCs or by low levels or defective types of hemoglobin
   a. Normal Hb levels 12 to 14 g/100 mL of blood
   b. Low Hb level (below 9 g/100 mL of blood) classified as anemia
2. Majority of clinical signs of anemia related to low tissue oxygen levels
   a. Fatigue; skin pallor
   b. Weakness; faintness; headache
   c. Compensation results in increased heart and respiratory rates
3. Hemorrhagic anemia
   a. Acute—blood loss is immediate (for example, surgery or trauma)
   b. Chronic—blood loss occurs over time (for example, ulcers or cancer)
4. Aplastic anemia
   a. Characterized by low RBC numbers and destruction of bone marrow
   b. Often caused by toxic chemicals, irradiation, or certain drugs
5. Deficiency anemias—caused by inadequate supply of some substance needed for RBC or hemoglobin production
   a. Pernicious anemia
      (1) Caused by vitamin B_{12} deficiency
      (2) Genetic-related autoimmune disease
      (3) Decreased RBC, WBC, and platelet numbers
      (4) RBCs are macrocytic
      (5) Classic symptoms of anemia coupled with central nervous system (CNS) impairment
      (6) Treatment is repeated vitamin B_{12} injections
   b. Folate deficiency anemia
      (1) Caused by folate (vitamin B_9) deficiency
      (2) Decreased RBC count
      (3) Common in alcoholism and malnutrition

c. Iron deficiency anemia
(1) Caused by deficiency of or inability to absorb iron needed for Hb synthesis (dietary iron deficiency is common worldwide)
(2) RBCs are microcytic and hypochromic (**Figure 13-8**)
(3) Hematocrit is decreased
(4) Treatment is oral administration of iron compounds

6. Hemolytic anemias
a. Caused by decreased RBC life span or increased RBC rate of destruction
b. Symptoms—such as jaundice, swelling of spleen, gallstone formation, and tissue iron deposits—are related to retention of RBC breakdown products
c. Sickle cell anemia (**Figure 13-9**)
(1) Genetic disease resulting in formation of abnormal hemoglobin (HbS); primarily found in African-American people
(2) RBCs become fragile and assume sickled shape when blood oxygen levels decrease
(3) Sickle cell trait—result of one defective gene; usually mild or no symptoms at all
(4) Sickle cell disease—result of two defective genes; more severe; causes blood stasis, clotting, and "crises" that may be fatal
(5) Affects 1 in every 600 African-American newborns
d. Thalassemia
(1) Group of inherited hemolytic anemias occurring primarily in people of Mediterranean descent
(2) RBCs microcytic and short lived
(3) Present as mild thalassemia trait and severe thalassemia major
(4) Hb levels often fall below 7 g/100 mL of blood in thalassemia major
(5) Classic symptoms of anemia coupled with skeletal deformities and swelling of spleen and liver
e. Hemolytic disease of the newborn (HDN)
(1) Caused by blood ABO or Rh factor incompatibility during pregnancy between developing offspring and mother
(2) Maternal antibodies against fetal RBCs of different blood type can cross placenta, enter the fetal circulation, and destroy some of the fetus's red cells—as in erythroblastosis fetalis (**Figure 13-7**)
(3) Symptoms in developing fetus related to decline in RBC numbers and Hb levels; jaundice, intravascular coagulation, and heart and lung damage are common
(4) Treatment may include in utero blood transfusions and premature delivery of the offspring
(5) Prevention of Rh factor incompatibility now possible by administration of RhoGAM to Rh-negative mothers

White Blood Cells

A. Introduction to WBCs
1. Categorized by presence of stained nuclei and granules in translucent cytoplasm (**Figure 13-10**)
a. Granulocytes—possess granules that stain
b. Agranulocytes—absence of stained granules
2. WBCs are all involved in immunity

B. WBC count
1. Complete WBC count—normal range is 5000 to 10,000/mm^3 of blood
2. Leukopenia—abnormally low WBC count (below 5000/mm^3 of blood)
a. Occurs infrequently
b. May occur with malfunction of blood-forming tissues or diseases affecting immune system, such as AIDS
3. Leukocytosis—abnormally high WBC count (over 10,000/mm^3 of blood)
a. Frequent finding in bacterial infections
b. Classic sign in blood cancers (leukemia)
4. Differential WBC count—component test in CBC; measures proportions of each type of WBC in blood sample (**Figures 13-1** and **13-10**)

C. WBC types
1. Granular leukocytes (granulocytes)
a. Neutrophils
(1) Most numerous type of phagocyte
(2) Numbers increase in bacterial infections
b. Eosinophils
(1) Weak phagocyte
(2) Active against parasites and parasitic worms
(3) Involved in allergic reactions
c. Basophils
(1) Related to mast cells in tissue spaces
(2) Both mast cells and basophils secrete histamine (related to inflammation)
(3) Basophils also secrete heparin (an anticoagulant)
2. Agranular leukocytes (agranulocytes)
a. Monocytes
(1) Largest leukocyte
(2) Aggressive phagocyte—capable of engulfing larger bacteria and cancer cells (**Figure 13-11**)
(3) Develop into much larger cells called *macrophages* after leaving blood to enter tissue spaces
b. Lymphocytes
(1) B lymphocytes involved in immunity against disease by secretion of antibodies
(2) Mature B lymphocytes called *plasma cells*
(3) T lymphocytes involved in direct attack on bacteria or cancer cells (not antibody production)

White Blood Cell Disorders

A. Two major types of WBC cancers or neoplasms
   1. Lymphoid neoplasms—result from B and T lymphocyte precursor cells or their descendant cell types
   2. Myeloid neoplasms—result from malignant transformation of precursor cells of granulocytic WBCs, monocytes, RBCs, and platelets
B. Multiple myeloma (**Figure 13-12**)
   1. Cancer of B lymphocytes called *plasma cells*
   2. Most deadly blood cancer in people older than age 65
   3. Causes bone marrow dysfunction and production of defective antibodies
   4. Characterized by:
      a. Recurrent infections and anemia
      b. Destruction and fracture of bones
   5. Treatment includes chemotherapy, drug and antibody therapy, and marrow and stem cell transplantation
C. Leukemias—WBC-related blood cancers
   1. Characterized by marked leukocytosis
   2. Identified as:
      a. Acute or chronic—based on timing of pathogenesis
         (1) Acute—rapid development of symptoms
         (2) Chronic—slow development of symptoms
      b. Lymphocytic or myeloid—based on origin tissue of involved cells
         (1) Lymphocytic—affects lymphocytes
         (2) Myeloid—affects granular leukocytes as they develop in the red bone marrow
   3. Chronic lymphocytic leukemia (CLL) (**Figure 13-13**)
      a. Average age at onset is 65; rare before age 30
      b. More common in men than in women
      c. Often diagnosed unexpectedly in routine physical exams with discovery of marked B lymphocytic leukocytosis
      d. Generally mild symptoms include anemia, fatigue, and enlarged—often painless—lymph nodes
      e. Most patients live many years following diagnosis
      f. Treatment of severe cases involves chemotherapy and irradiation
   4. Acute lymphocytic leukemia (ALL) (**Figure 13-14**)
      a. Primarily a disease of children between 3 and 7 years of age; 80% of children who develop leukemia have this form of the disease
      b. Highly curable in children but less so in adults
      c. Onset is sudden—marked by fever, leukocytosis, bone pain, and increases in infections
      d. Lymph node, spleen, and liver enlargement is common
      e. Treatment includes chemotherapy, irradiation, and bone marrow or stem cell transplantation
   5. Chronic myeloid leukemia (CML) (**Figure 13-15**)
      a. Accounts for about 20% of all cases of leukemia
      b. Occurs most often in adults between 25 and 60 years of age
      c. Caused by cancerous transformation of granulocytic precursor cells in the bone marrow
      d. Onset and progression of disease is slow with symptoms of fatigue, weight loss, and weakness
      e. Diagnosis often made by discovery of marked granulocytic leukocytosis and extreme spleen enlargement
      f. Treatment by new "designer drug" Gleevec or bone marrow transplantation is curative in more than 70% of cases
   6. Acute myeloid leukemia (AML)
      a. Accounts for 80% of all cases of acute leukemia in adults and 20% of acute leukemia in children
      b. Characterized by sudden onset and rapid progression
      c. Symptoms include leukocytosis, fatigue, bone and joint pain, spongy bleeding gums, anemia, and recurrent infections
      d. Prognosis is poor with only about 50% of children and 30% of adults achieving long-term survival
      e. Bone marrow and stem cell transplantation has increased cure rates in selected patients
D. Infectious mononucleosis (**Figure 13-16**)
   1. Noncancerous WBC disorder
   2. Highest incidence between 15 and 25 years of age
   3. Caused by virus present in saliva of infected individuals
   4. Leukocytosis of atypical lymphocytes with abundant cytoplasm and large nuclei
   5. Symptoms include fever, severe fatigue, sore throat, rash, and enlargement of lymph nodes and spleen
   6. Generally self-limited and resolves without complications in about 4 to 6 weeks

Platelets and Blood Clotting

A. Platelets—also called thrombocytes
   1. Tiny cell fragments filled with clot-triggering chemicals
   2. Play essential role in blood clotting
B. Blood clotting mechanism (**Figure 13-17**)
   1. Vasoconstriction of blood vessels helps close gaps in blood vessel wall and reduces local blood flow
   2. Blood vessel damage releases clotting factors that react with plasma factors to form prothrombin activator
   3. At the same time, platelets adhere to the break and form a "platelet plug" and release additional clotting factors promoting formation of prothrombin activator
   4. Prothrombin activator and calcium convert prothrombin to thrombin
   5. Thrombin reacts with fibrinogen to form fibrin
   6. Fibrin threads form a tangle to trap RBCs (and other formed elements) to produce a blood clot
C. Altering the blood clotting mechanism
   1. Application of gauze (rough surface) to wound causes platelet aggregation and release of clotting factors
   2. Administration of vitamin K will increase synthesis of prothrombin

Clotting Disorders

A. Abnormal blood clots
   1. Types of abnormal clots
      a. Thrombus—stationary blood clot
      b. Embolus—circulating blood clot (**Figure 13-18**)
   2. Anticoagulant therapy
      a. Drug called *tissue plasminogen activator [TPA or tPA]* used to dissolve clots that have already formed
      b. Drug warfarin sodium will delay clotting by inhibiting prothrombin synthesis
      c. Heparin delays clotting by inhibiting conversion of prothrombin to thrombin
      d. Laboratory tests used to monitor effectiveness of anticoagulant therapy include:
         (1) Prothrombin time—reported in seconds (7 to 10 seconds is normal)
         (2) INR (International Normalized Ratio)—a calculated value reported as a number (0.8 to 1.2 is normal)

B. Hemophilia—X-linked inherited disorder that results from inability to produce factor VIII (a plasma protein) responsible for blood clotting
   1. Most serious "bleeding disease" worldwide; hemophilia A most common form
   2. Characterized by easy bruising, deep muscle hemorrhage, blood in urine, and repeated episodes of bleeding into joints causing pain and deformity
   3. Treatment includes administration of factor VIII, injury prevention, and avoiding drugs such as aspirin that alter the clotting mechanism

C. Thrombocytopenia—caused by reduced platelet counts
   1. Characterized by bleeding from small blood vessels, most visibly as purpura (purplish spots) in the skin and mucous membranes
   2. Platelet count below 20,000/mm^3 may cause catastrophic bleeding (normal platelet count 150,000 to 400,000/mm^3)
   3. Most common cause is bone marrow destruction by drugs; chemicals; radiation; and diseases such as cancer, lupus, and HIV/AIDS
   4. Treatment may involve transfusion of platelets, corticosteroid-type drugs, or removal of the spleen

D. Vitamin K deficiency—can result in abnormally reduced clotting

13

❑ ACTIVE LEARNING

STUDY TIPS

Hint *Consider using these tips to achieve success in meeting your learning goals.*

Blood consists of a liquid portion, the plasma, and formed elements: red blood cells, white blood cells, and platelets. The function of the blood is to carry substances from one part of the body to another. Many transported materials are dissolved in the plasma, so the composition of the plasma varies based on what is going on in the body. Because of its function, the blood plays an important role in a number of other systems such as the respiratory, digestive, urinary, and immune systems. The material in this chapter shows up again in later chapters.

1. Flash cards and online tutorials will help you learn the names and functions of the various blood cells. Check out *www.getbodysmart.com* for tutorials and animations that illustrate a white blood cell differential test and blood typing.
2. The process of blood clot formation is important, and it is necessary that you know and understand the correct sequence of events. Develop a concept map that illustrates the sequence of the events that lead to a blood clot.
3. In order to understand the terminology, review the Language of Science and Language of Medicine terms.
4. When studying the blood disorders, make a chart that identifies the type of disorder: red blood cell, white blood cell, or clotting disorder. List the name and the specific cause of each disorder.
5. As you study the ABO blood typing system, be sure you give extra attention to learning what antigens are on the red blood cells and what antibodies are in the plasma. The antigens give a blood type its name, that is, type A blood has A antigens. Antibodies are the opposite of the blood type. Type A blood has anti-B antibodies. Type O has no antigens and both antibodies; type AB has both antigens and no antibodies.
6. In your study group, review the flash cards you made and online resources for studying the function of the blood cells. Discuss the process of blood clot formation. Go over the blood disorder chart. Review the antigens and antibodies of the various blood types. Go over the questions and the outline summary at the end of the chapter and discuss possible test questions.

Review Questions

Hint

Write out the answers to these questions after reading the chapter and reviewing the Chapter Summary. If you simply think through the answer without writing it down, you will not retain much of your new learning.

1. Name several substances found in blood plasma.
2. Explain the function of albumins, globulins, and fibrinogen.
3. What is the difference between serum and plasma?
4. What two types of connective tissue form blood cells? Where are they found and what do each of them form?
5. Describe the structure of a red blood cell. What advantage does this unique shape give the red blood cell that helps it perform its function?
6. Explain how type A blood differs from type B blood.
7. Explain the cause of erythroblastosis fetalis.
8. Both aplastic anemia and pernicious anemia are characterized by a low red blood cell count. How does the cause of each disease condition differ?
9. What is the buffy coat?
10. Explain the function of neutrophils and monocytes.
11. Explain the function of lymphocytes.
12. Explain the function of eosinophils and basophils.
13. Distinguish between leukopenia and leukocytosis.
14. How is hemophilia transmitted? What blood clotting factors can be affected?
15. Explain the process of blood clot formation.
16. Differentiate between a thrombus and an embolus.

Critical Thinking

After finishing the Review Questions, write out the answers to these more in-depth questions to help you apply your new knowledge. Go back to sections of the chapter that relate to concepts that you find difficult.

17. When a patient is being prepped for surgery, a physician will often prescribe vitamin K. What is the significance of administering vitamin K before surgery?
18. Explain how heparin inhibits blood clot formation.
19. You are a medical examiner, and a body is brought to you to determine the cause of death. You find a very large agglutination (not a clot) in a major vein. What judgment would you make regarding the cause of death?
20. Why is the first Rh-positive baby born to an Rh-negative mother usually unaffected?
21. Some athletes, seeking a competitive edge, may resort to blood doping. How would you explain blood doping? What information would you use to discourage athletes from participating in the practice of blood doping?
22. How do red blood cells confirm that "structure fits function"?

Chapter Test

After studying the chapter, test your mastery by responding to these items. Try to answer them without looking up the answers.

1. The liquid part of the blood is called ________.
2. Three important plasma proteins are ________, ________, and ________.
3. Blood plasma without the clotting factors is called ________.
4. The three formed elements of the blood are ________, ________, and ________.
5. The two types of connective tissue that make blood cells are ________ and ________.
6. The red pigment in red blood cells that carries oxygen is called ________.
7. An ________ is a foreign substance that can cause the body to produce an antibody.
8. A person with type AB blood has ________ antigens on the blood cell and ________ antibodies in the plasma.
9. A person with type B blood has ________ antigens on the blood cell and ________ antibodies in the plasma.
10. Type ________ blood is considered the universal donor.
11. Type ________ blood is considered the universal recipient.
12. A condition called ________ can develop if an Rh-negative mother produces antibodies against the blood of an Rh-positive fetus.
13. The term ________ is used to describe a number of disease conditions caused by the inability of red blood cells to carry a sufficient amount of oxygen.
14. If the body produces an excess of red blood cells, the condition is called ________.
15. Which white blood cells are the most numerous of the phagocytes? ________
16. Which white blood cells produce antibodies to fight microbes? ________
17. Prothrombin activator and the mineral ________ in the blood convert prothrombin into thrombin in the formation of a blood clot.
18. Thrombin converts the inactive plasma protein ________ into a fibrous gel called ________.
19. Vitamin ________ stimulates the liver to increase synthesis of prothrombin.
20. A ________ is a pathological rather than curative blood clot that stays in the place where it was formed.
21. If a part of a blood clot is dislodged and circulates through the bloodstream, it is called a/an________.
22. The ________ provides information about the volume of red blood cells in a blood sample. It is an estimate of the proportion of red blood cells to whole blood.
23. The ________ are the largest of the leukocytes and are very aggressive phagocytes.
24. A ________ provides the proportions of each type of white blood cell as percentages of the total white blood cell count. It is a valuable diagnostic tool.

Match each blood disorder in Column A with its corresponding description or cause in Column B.

Column A	Column B
25. ________ pernicious anemia	a. a type of inherited anemia that produces abnormal hemoglobin and red blood cell deformities
26. ________ sickle cell anemia	b. an abnormally low white blood cell count
27. ________ thalassemia	c. an inherited disorder in which a small amount of hemoglobin is produced; can be major or minor
28. ________ hemophilia	d. an inherited inability to form some blood clotting factors
29. ________ thrombocytopenia	e. an abnormally low number of platelets
30. ________ leukocytosis	f. a low number of red blood cells because of a lack of vitamin B_{12}
31. ________ leukopenia	g. a low number of red blood cells related to destruction of bone marrow
32. ________ aplastic anemia	h. an abnormally high white blood cell count

Case Studies

To solve a case study, you may have to refer to the glossary or index, other chapters in this textbook, and other resources.

1. Angela's physician suspects that Angela has just suffered a *myocardial infarction,* or heart attack. She tells Angela that she is going to take a blood sample so that the hospital lab can perform a test to confirm her diagnosis. What information can Angela's blood yield to help the physician?
2. Yvonne has just been told that she has a condition called *pernicious anemia.* She looked up the definition of this disease in a dictionary and learned that it is caused by a decreased availability of vitamin B_{12} needed for manufacturing RBCs. Yvonne promptly went to the local pharmacy to buy some vitamin B_{12} tablets to help her overcome this condition. Is this a wise course of action?
3. Your brother has just been diagnosed as having anemia, but your mother cannot seem to remember the specific type. She shows you a copy of your young brother's CBC results, but no diagnosis is stated. Based on the results given below, what would you guess is your brother's condition? (HINT: see **Table 13-3**.) Are you likely to develop this condition?

   Folate content: normal
   Hematocrit: low
   Hemoglobin content: low
   Iron content: slightly high
   RBC size (mean corpuscular volume): high
   Vitamin B_{12} content: normal

Answers to Active Learning Questions can be found online at evolve.elsevier.com.

13

Heart

OUTLINE

Hint *Scan this outline before you begin to read the chapter, as a preview of how the concepts are organized.*

OBJECTIVES

Before reading the chapter, review these goals for your learning.

After you have completed this chapter, you should be able to:

1. Discuss the location, size, and position of the heart in the thoracic cavity.
2. Identify and discuss the heart chambers, pericardium, heart valves, and major valve disorders.
3. Discuss the normal heart sounds and identify common abnormal heart sounds.
4. Trace blood through the heart, compare the functions of the heart chambers on the right and left sides, and explain how a myocardial infarction might occur.
5. Explain the cardiac cycle.
6. List the anatomical components of the heart conduction system and discuss the features of the normal electrocardiogram.
7. Describe the major types of cardiac dysrhythmia.
8. Explain how heart rate and stroke volume affect cardiac output.
9. List and describe the possible causes of heart failure.

CHAPTER 14

The system that supplies our body's transportation needs is the **cardiovascular system.** We need such a system to make sure that each cell is surrounded by fluid that is constantly replenished with oxygen, water, and nutrients as they are used up by a cell. In addition, we need the waste products in extracellular fluid to be continually removed as they are released from cells.

A circulating stream of blood can pick up substances from various parts of the body and deliver them to others—thus allowing our body to move substances around in a way that helps us maintain a relatively constant internal environment. Clearly, circulation of the blood is critical to maintaining the homeostatic balance of your body.

We begin the study of the cardiovascular system in this chapter with the heart—the pump that keeps blood moving through a closed circuit of blood vessels. Details related to heart structure will be followed by a discussion of how the heart functions. Chapter 15 then continues our story with a study of the vessels through which blood flows as a result of the pumping action of the heart.

Location of the Heart

No one needs to be told where the heart is or what it does. Everyone knows that the heart is in the chest, that it beats night and day to keep the blood flowing, and that if it stops, life stops.

Most of us probably think of the heart as being located on the left side of the body. As you can see in **Figure 14-1**, the heart is located between the lungs in the lower portion of the mediastinum. Draw an imaginary line through the middle of the trachea in **Figure 14-1** and continue the line down through the thoracic cavity to divide it into right and left halves. Note that about two-thirds of the mass of the heart is to the left of this line and one-third is to the right.

The heart is often described as a triangular organ, shaped and sized roughly like a closed fist. In **Figure 14-1** you can see that

LANGUAGE OF SCIENCE

Hint Before reading the chapter, say each of these terms out loud. This will help you to avoid stumbling over them as you read.

aorta
(ay-OR-tah)
[*aort-* **lifted**, *-a* **thing**]

aortic semilunar valve
(ay-OR-tik sem-ih-LOO-nar valv)
[*aort-* **lifted**, *-ic* **relating to**, *semi-* **half**, *-luna* **moon**, *-ar* **relating to**]

apex
(AY-peks)
pl., apices
(AY-pih-seez)
[*apex* **tip**]

atrioventricular bundle (AV bundle)
(ay-tree-oh-ven-TRIK-yoo-lar BUN-del [ay vee BUN-del])
[*atrio-* **entrance courtyard**, *-ventr-* **belly**, *-icul-* **little**, *-ar* **relating to**]

atrioventricular node (AV node)
(ay-tree-oh-ven-TRIK-yoo-lar nohd [ay vee nohd])
[*atrio-* **entrance courtyard**, *-ventr-* **belly**, *-icul-* **little**, *-ar* **relating to**, *nod-* **knot**]

atrioventricular valve (AV valve)
(ay-tree-oh-ven-TRIK-yoo-lar valv [ay vee valv])
[*atrio-* **entrance courtyard**, *-ventr-* **belly**, *-icul-* **little**, *-ar* **relating to**]

atrium
(AY-tree-um)
pl., atria
(AY-tree-ah)
[*atrium* **entrance courtyard**]

auricle
(AW-rih-kul)
[*auri-* **ear**, *-icle* **little**]

Continued on p. 395

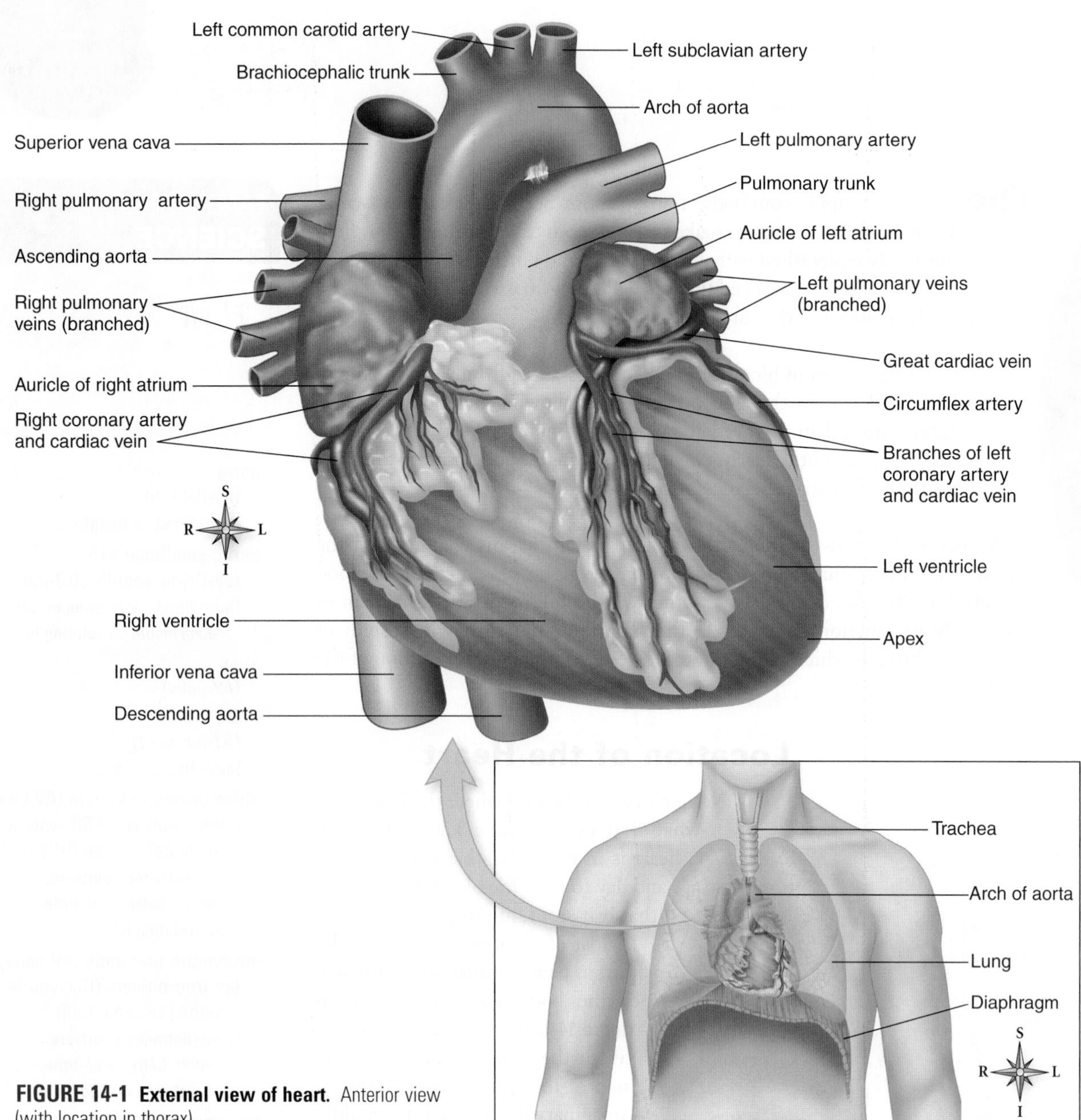

FIGURE 14-1 External view of heart. Anterior view (with location in thorax).

14

the **apex,** or blunt point, of the lower edge of the heart lies on the diaphragm, pointing toward the left.

Physicians and nurses often listen to the heart sounds by placing a stethoscope on the chest wall directly over the apex of the heart. Sounds of the so-called **apical heart beat** are easily heard in this area (that is, in the space between the fifth and sixth ribs on a line even with the midpoint of the left clavicle).

The heart is positioned in the thoracic cavity between the sternum in front and the bodies of the thoracic vertebrae behind. Because of this placement, it can be compressed or squeezed by application of pressure to the lower portion of the body of the sternum using the heel of the hand. Rhythmic compression of the heart in this way can maintain blood flow in cases of cardiac arrest and, if combined with effective artificial respiration, the resulting procedure, called **cardiopulmonary resuscitation (CPR),** can be lifesaving.

The exact procedures for CPR change frequently as new research data become available, so it is important for individuals certified in CPR to recertify on a regular basis.

Locate the heart and surrounding structures in the *Clear View of the Human Body* (follows p. 8).

Functional Anatomy of the Heart

Heart Chambers

If you cut open a heart, you can see many of its main structural features (**Figure 14-2**). This organ is hollow, not solid. A partition divides it into right and left sides. The heart contains four cavities, or hollow chambers. The two upper chambers are called

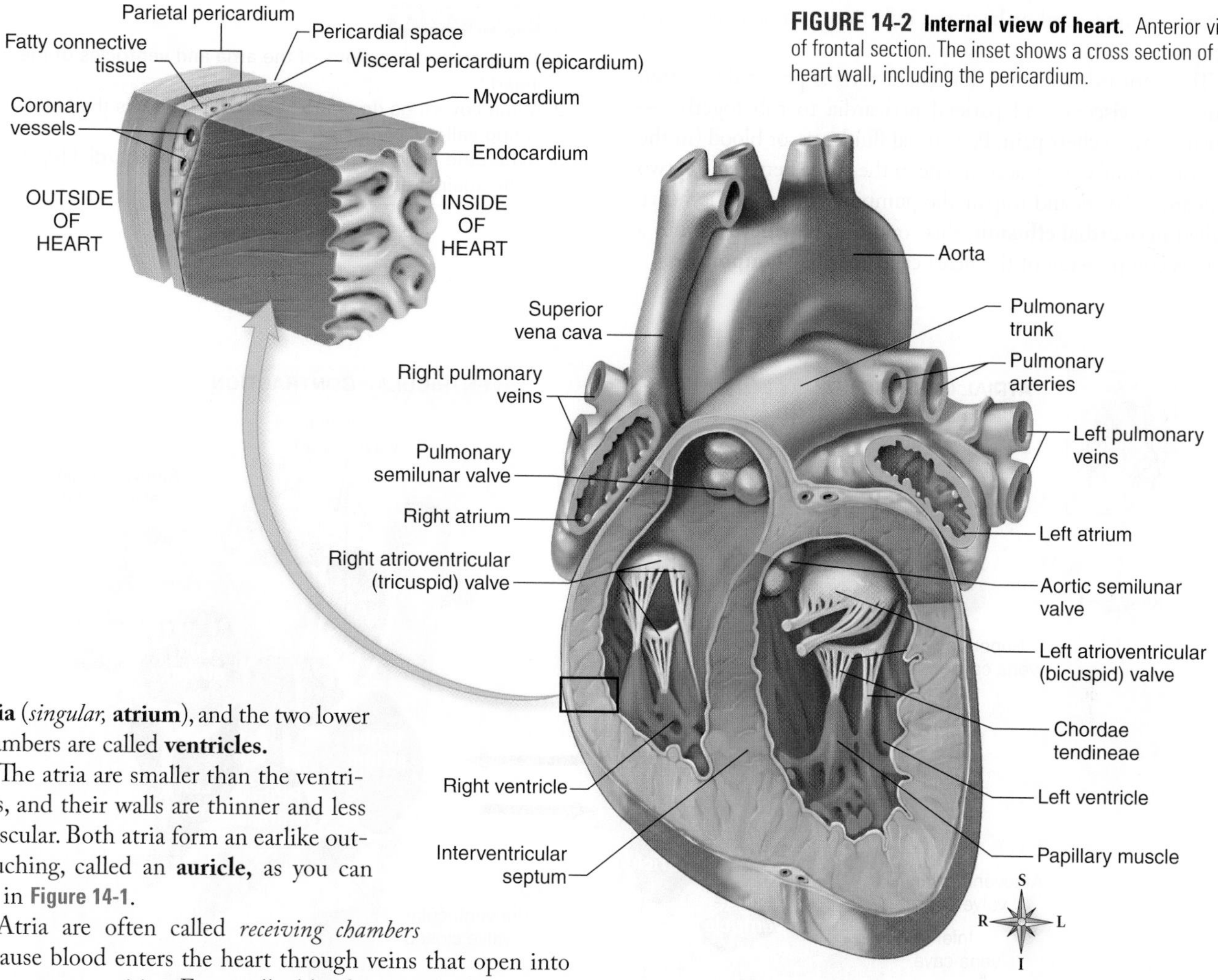

FIGURE 14-2 Internal view of heart. Anterior view of frontal section. The inset shows a cross section of the heart wall, including the pericardium.

atria (*singular,* **atrium**), and the two lower chambers are called **ventricles.**

The atria are smaller than the ventricles, and their walls are thinner and less muscular. Both atria form an earlike outpouching, called an **auricle,** as you can see in Figure 14-1.

Atria are often called *receiving chambers* because blood enters the heart through veins that open into these upper cavities. Eventually, blood is pumped from the heart into arteries that exit from the ventricles. The ventricles are therefore sometimes referred to as the *discharging chambers* of the heart.

Each heart chamber is named according to its location. Thus there are right and left atrial chambers above and right and left ventricular chambers below.

The wall of each heart chamber is composed of cardiac muscle tissue usually referred to as the **myocardium.** The septum between the atrial chambers is called the *interatrial septum.* The *interventricular septum* separates the ventricles.

Each chamber of the heart is lined by a thin layer of very smooth tissue called the **endocardium** (see Figure 14-2). Inflammation of this lining is referred to as **endocarditis.** If inflamed, the endocardial lining can become rough and abrasive to RBCs passing over its surface. Blood flowing over a rough surface is subject to clotting, and a **thrombus,** or clot, may form (see Chapter 13).

Unfortunately, rough spots caused by endocarditis or injuries to blood vessel walls often cause the release of platelet factors. The result is often the formation of a fatal blood clot.

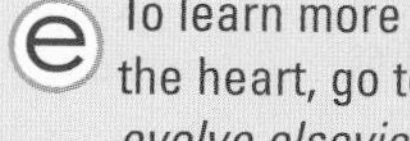
To learn more about the interior anatomy of the heart, go to AnimationDirect online at *evolve.elsevier.com.*

Pericardium

Coverings of the Heart

The heart's covering—the **pericardium**—consists of two layers of fibrous tissue with a small space in between them. The inner layer of the pericardium is called the **visceral pericardium,** or **epicardium.** It covers the heart the way an apple skin covers an apple. The outer layer of pericardium is called the **parietal pericardium.** It fits around the heart like a loose-fitting sack, allowing enough room for the heart to beat.

It is easy to remember the difference between the *endocardium,* which lines the heart chambers, and the *epicardium,* which covers the surface of the heart (see Figure 14-2), if you understand the meaning of the prefixes *endo-* and *epi-*. *Endo-* comes from the Greek word meaning "inside" or "within," and *epi-* comes from the Greek word meaning "upon" or "on."

The two pericardial layers slide over each other without friction when the heart beats because these are serous membranes with moist, slippery, surfaces. A thin film of pericardial fluid provides the lubricating moistness between the heart and its enveloping pericardial sac.

Pericarditis

If the pericardium becomes inflamed, a condition called **pericarditis** results. Pericarditis may be caused by a variety of factors: trauma, viral or bacterial infection, tumors, and other factors.

The pericardial edema that characterizes pericarditis often causes the visceral and parietal pericardia to rub together—causing severe chest pain. Pericardial fluid, pus, or blood (in the case of an injury) may accumulate in the space between the two pericardial layers and impair the pumping action of the heart. Called **pericardial effusion,** this condition may develop into a serious compression of the heart called **cardiac tamponade.**

QUICK CHECK

1. What are the functions of the atria and ventricles of the heart?
2. What coverings does the heart have? What is the heart's lining called?
3. When the heart beats, how do the two pericardial layers slide against each other without friction?

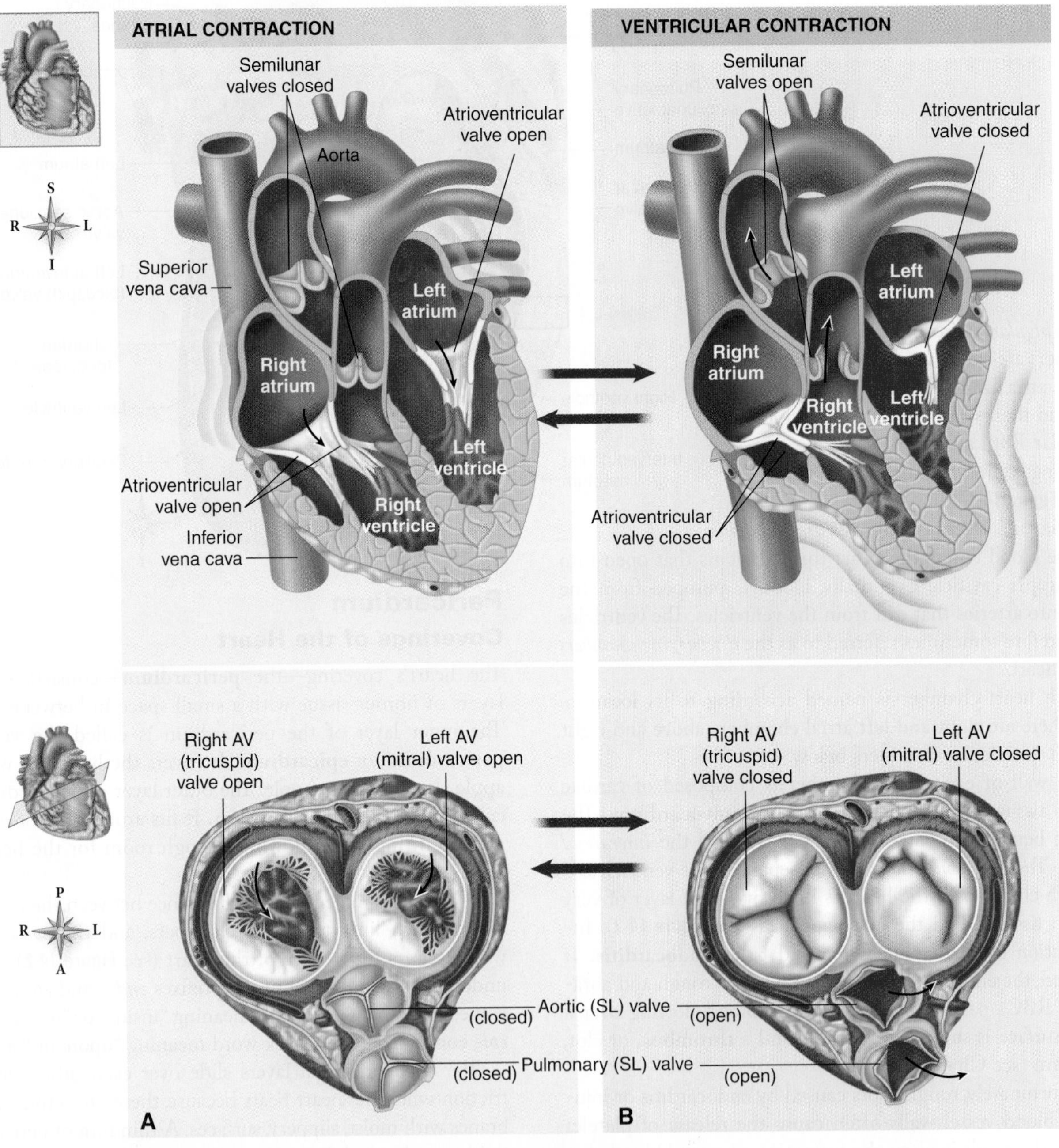

FIGURE 14-3 Heart action. A, During atrial systole (contraction), cardiac muscle in the atrial wall contracts, forcing blood through the atrioventricular (AV) valves and into the ventricles. Bottom illustration shows superior view of all four valves, with semilunar (SL) valves closed and AV valves open. **B,** During ventricular systole that follows, the AV valves close, and blood is forced out of the ventricles through the semilunar valves and into the arteries. Bottom illustration shows superior view of SL valves open and AV valves closed.

14

Heart Action

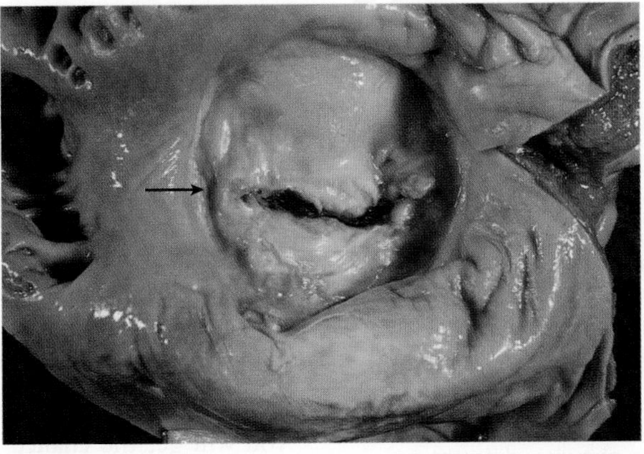

FIGURE 14-4 Mitral valve stenosis. Stenosed valves are valves that are narrower than normal when open, slowing blood flow from a heart chamber. Compare this valve *(arrow)* with the normal open valve shown in **Figure 14-3**, *A*.

The heart serves as a muscular pumping device for distributing blood to all parts of the body. Contraction of the heart is called **systole,** and relaxation is called **diastole.**

When the heart beats (that is, when it contracts), the atria contract first (atrial systole), forcing blood that has not yet leaked into the ventricles toward the ventricles. Once filled, the two ventricles contract (ventricular systole) and force blood out of the heart (**Figure 14-3**).

For the heart to be efficient in its pumping action, more than just the rhythmical contraction of its muscular fibers is required. The direction of blood flow must be directed and controlled. This is accomplished by four sets of valves located at the entrance and near the exit of the ventricles.

Heart Valves

Valve Structure and Function

The two valves that separate the atrial chambers above from the ventricles below are called **AV valves,** or **atrioventricular valves.** The left AV valve is also known as the **bicuspid valve,** or **mitral valve.** It is located between the left atrium and ventricle. The right AV valve is also known as the **tricuspid valve.** It is located between the right atrium and ventricle.

The AV valves prevent backflow of blood into the atria when the ventricles contract. Locate the AV valves in **Figures 14-2** and **14-3**. Note that a number of stringlike structures called **chordae tendineae** attach edges of the leaflets or "flaps" of the AV valves to fingerlike projections of cardiac muscle in the wall of the ventricles.

The **SL valves,** or **semilunar valves,** are located between each ventricular chamber and its large artery that carries blood away from the heart when contraction occurs (see **Figure 14-3**). The ventricles, like the atria, contract together. Therefore, the two semilunar valves open and close at the same time.

The **pulmonary semilunar valve** is located at the beginning of the pulmonary artery and allows blood going to the lungs to flow out of the right ventricle during systole but prevents it from flowing back into the ventricle during diastole. The **aortic semilunar valve** is located at the beginning of the aorta and allows blood to flow out of the left ventricle up into the aorta but prevents backflow into this ventricle.

Valve Disorders

Disorders of the cardiac valves can have several effects. For example, a congenital defect in valve structure can result in mild to severe pumping inefficiency. **Incompetent valves** leak, allowing some blood to flow back into the chamber from which it came. **Stenosed valves** are valves that are narrower than normal, slowing blood flow from a heart chamber (**Figure 14-4**).

Rheumatic heart disease is cardiac damage resulting from a delayed inflammatory response to streptococcal infection that occurs most often in children. A few weeks after an untreated or improperly treated streptococcal infection, the cardiac valves and other tissues in the body may become inflamed—a condition called *rheumatic fever.* If severe, the inflammation can result in stenosis or other deformities of the valves, chordae tendineae, or myocardium.

Mitral valve prolapse (MVP), a condition affecting the left AV valve (mitral valve) has a genetic basis in some cases but can result from rheumatic fever or other factors. A prolapsed mitral valve is one whose flaps extend back into the left atrium, causing incompetence (leaking) of the valve (**Figure 14-5**). This condition was once thought to be common, but recent studies show that many patients previously diagnosed with MVP have normal heart function.

Damaged or defective cardiac valves often can be replaced surgically. Animal valves and artificial valves made from synthetic materials are commonly used in valve replacement procedures.

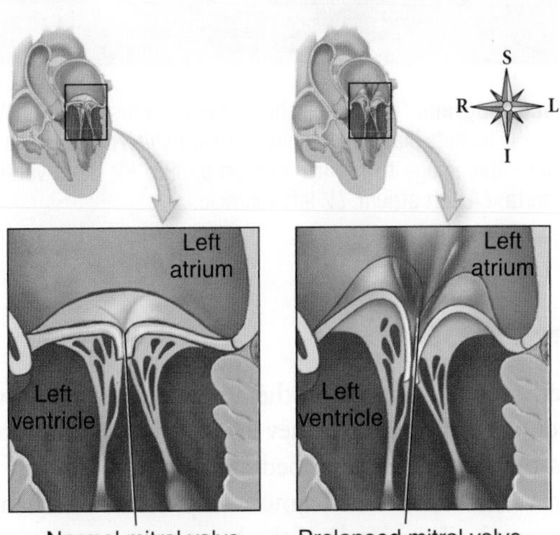

FIGURE 14-5 Mitral valve prolapse (MVP). The normal mitral valve *(upper left)* prevents backflow of blood from the left ventricle into the left atrium during ventricular systole (contraction). The prolapsed mitral valve *(right)* permits leakage because the valve flaps billow backward, parting slightly.

CLINICAL APPLICATION

ECHOCARDIOGRAPHY

Although the stethoscope is still the basic tool of the **cardiologist,** or heart specialist, more sophisticated methods for detecting abnormalities in heart valve function are available. One widely used technique is **echocardiography.** Ultrasound (extremely high-pitched sound) directed toward the heart is reflected back (echoed) by the tissues (see the box *Medical Imaging of the Body* on p. 132 and the *Cardiology* box on p. 392).

Like an airport's radar, a detector picks up the echoed ultrasound and produces an image showing different regions of blood and heart tissues. As the valves and other structures move during a series of heartbeats, the image changes. A cardiologist can examine a continuous video/audio recording called an **echocardiogram** (pictured) and determine the nature of a valve problem or other heart disorder.

To see a diagram that explains the concept of echocardiography, and additional echocardiogram examples, check out the article *Echocardiography* at Connect It! at *evolve.elsevier.com.*

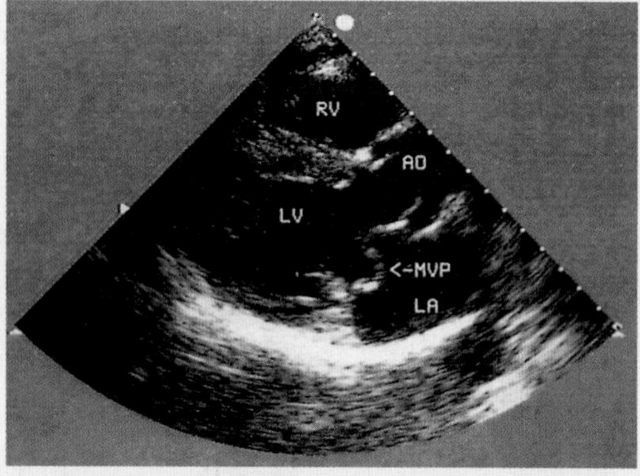

Echocardiogram. This image shows a mitral valve prolapse *(MVP)* where the leaflets (flaps) of the mitral valve *(white)* bulge noticeably. Compare this image to **Figure 14-5** on p. 383. *RV,* Right ventricle; *AO,* aorta; *LA,* left atrium; *LV,* left ventricle.

Heart Sounds

If a stethoscope is placed on the anterior chest wall, two distinct sounds can be heard. They are rhythmical and repetitive sounds that are often described as *lub dup.* Disorders of the cardiac valves are often diagnosed by detecting changes in these normal valve sounds of the heart.

The first, or *lub,* sound is caused by the vibration and abrupt closure of the AV valves as the ventricles contract. Closure of the AV valves prevents blood from rushing back up into the atria during contraction of the ventricles. This first sound is of longer duration and lower pitch than the second. The pause between this first sound and the *dup,* or second sound, is shorter than that after the second sound and the *lub dup* of the next systole.

The second heart sound is caused by the closing of both the semilunar valves when the ventricles undergo diastole (relax) (see **Figure 14-3**).

Abnormal heart sounds called **heart murmurs** are often caused by disorders of the valves. For example, incompetent valves may cause a swishing sound as a "lub" or "dup" ends. Stenosed valves, on the other hand, often cause swishing sounds just before a "lub" or "dup."

Later in the chapter, when you get to **Figure 14-10** (p. 388), you will get the chance to compare the timing of the heart sounds to other events of the cardiac pumping cycle. This will further clarify the clinical importance of heart sounds as indicators of heart function.

Blood Flow Through the Heart

When the heart "beats," first the atria contract simultaneously. This is *atrial systole.* After the ventricles fill with blood, they contract together during *ventricular systole.* Although the atria contract as a unit followed by the ventricles below, the right and left sides of the heart act as separate pumps. As we study the blood flow through the heart, the separate functions of the two pumps will become clearer.

Note in **Figure 14-6** that blood enters the right atrium through two large veins called the **superior vena cava** and **inferior vena cava.** The right heart pump receives oxygen-poor blood from the veins. After entering the right atrium, it flows through the right AV, or tricuspid, valve and enters the right ventricle. When the ventricles contract, blood in the right ventricle is pumped through the pulmonary semilunar valve into the **pulmonary artery** and eventually to the lungs, where oxygen is added and carbon dioxide is lost.

As you can see in **Figure 14-6,** oxygen-rich blood returns to the left atrium of the heart through four **pulmonary veins.** It then passes through the left AV, or bicuspid, valve into the left ventricle. When the left ventricle contracts, blood is forced through the aortic semilunar valve into the **aorta** and is distributed to the body as a whole.

As you can tell from **Figure 14-6,** the two sides of the heart actually pump blood through two separate "circulations" and function as two separate pumps:

- **Pulmonary circulation**—flow of blood from the right ventricle to the lungs and back to the left atrium (shaded with *blue* in **Figure 14-6**)
- **Systemic circulation**—flow of blood from the left ventricle throughout the body and back to the right atrium (shaded with *yellow* in **Figure 14-6**)

The vessels of the pulmonary and systemic circulations are discussed in Chapter 15.

One of the classic challenges often given to beginning students is to trace the path that a blood cell would take as it flows from the right atrium, all the way through both major circulatory routes, and back to the right atrium. Learning that

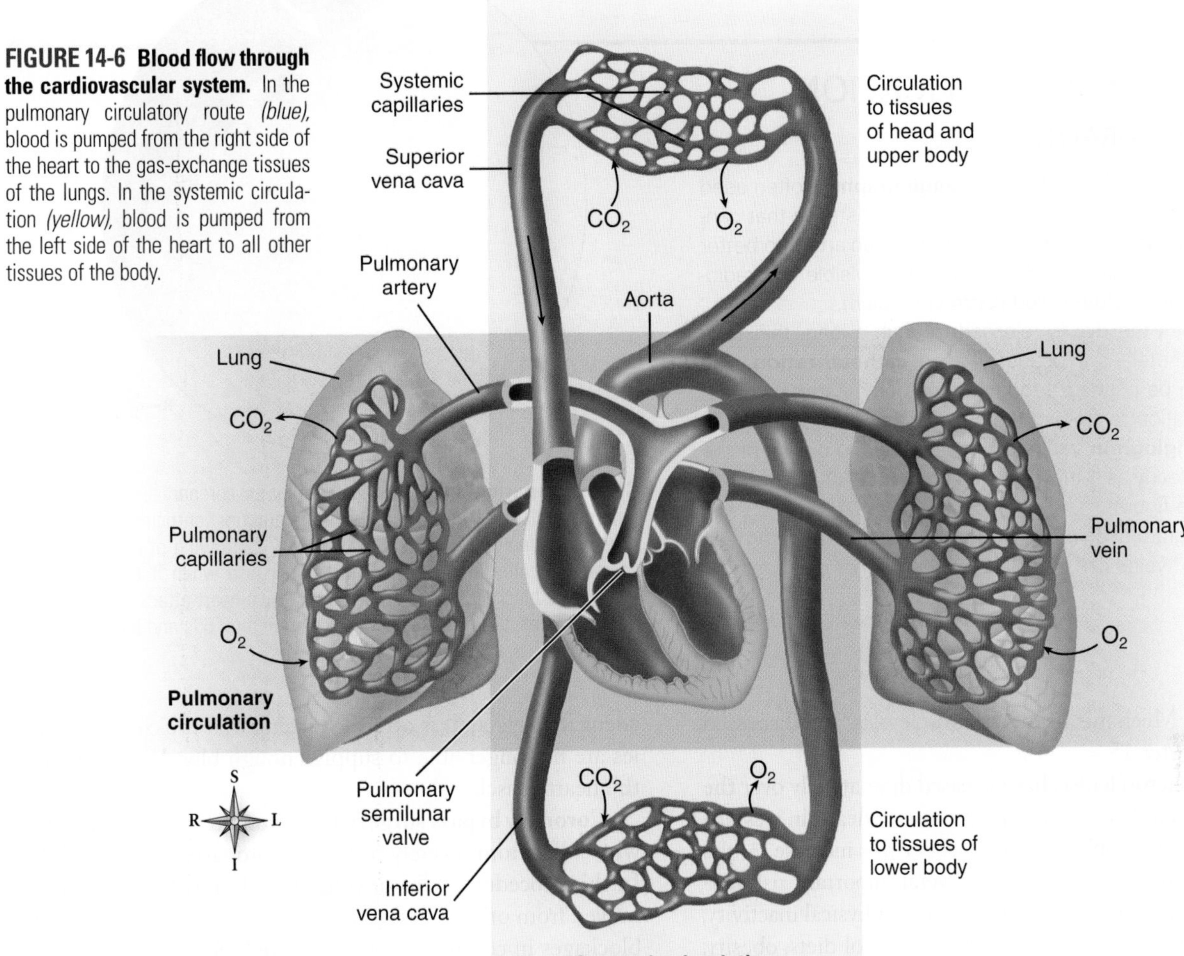

FIGURE 14-6 Blood flow through the cardiovascular system. In the pulmonary circulatory route *(blue)*, blood is pumped from the right side of the heart to the gas-exchange tissues of the lungs. In the systemic circulation *(yellow)*, blood is pumped from the left side of the heart to all other tissues of the body.

pathway will help in many of your future learning and clinical experiences.

To better understand this concept, use the Active Concept Map *Blood Flow Through the Heart* at *evolve.elsevier.com*.

Need help tracing the flow of blood through the major circulatory routes? Check out the article *How to Trace the Flow of Blood* at Connect It! at *evolve.elsevier.com*.

Blood Supply to Heart Muscle

To sustain life, the heart must pump blood throughout the body on a regular and ongoing basis. As a result, the heart muscle or myocardium requires a constant supply of blood containing nutrients and oxygen to function effectively. The delivery of oxygen and nutrient-rich arterial blood to cardiac muscle tissue and the return of oxygen-poor blood from this active tissue to the venous system is called the **coronary circulation** (**Figure 14-7**, *A*).

Blood flows into the heart muscle by way of two small vessels—the **right** and **left coronary arteries.** The coronary arteries are the aorta's first branches, as you can see in **Figure 14-7**, *A*.

Figure 14-7, *B*, shows that the openings into these small vessels lie behind the flaps of the aortic semilunar valves. During ventricular systole, the myocardium is contracting and putting pressure on the coronary arteries, so little blood can enter them. However, during ventricular diastole, blood that backs up behind the aortic SL valve can flow easily into the coronary arteries.

In both coronary thrombosis and coronary **embolism,** a blood clot occludes or plugs up some part of a coronary artery. Blood cannot pass through the occluded vessel and so cannot reach the heart muscle cells it normally supplies. Deprived of oxygen, these cells soon become damaged. In medical terms, **myocardial infarction (MI),** or tissue death, occurs.

An MI, also referred to as a "heart attack," is a common cause of death during middle and late adulthood. Recovery from a myocardial infarction is possible if the amount of heart tissue damaged was small enough so that the remaining undamaged heart muscle can still pump blood effectively enough to supply the needs of the rest of the heart and the body.

Coronary arteries also may become blocked as a result of **atherosclerosis,** a type of "hardening of the arteries" in which lipids and other substances build up on the inside wall of

CLINICAL APPLICATION

ANGIOGRAPHY

A special type of radiography called **angiography** is often used to visualize arteries. A *radiopaque dye*—a substance that cannot be penetrated by x-rays—is injected into an artery to better visualize vessels that would otherwise be invisible in a radiograph. This dye is often called *contrast medium*.

Sometimes the dye is released through a long, thin tube called a catheter—a procedure called **catheterization.** The catheter can be pushed through arteries until its tip is in just the right location to release the dye. As the dye begins to circulate, an **angiogram** (radiograph) will show the outline of the arteries as clearly as if they were made of bone or other dense material (see figure).

An angiogram of an artery is often called an *arteriogram*. An angiogram of veins can be called a *venogram* or *phlebogram*.

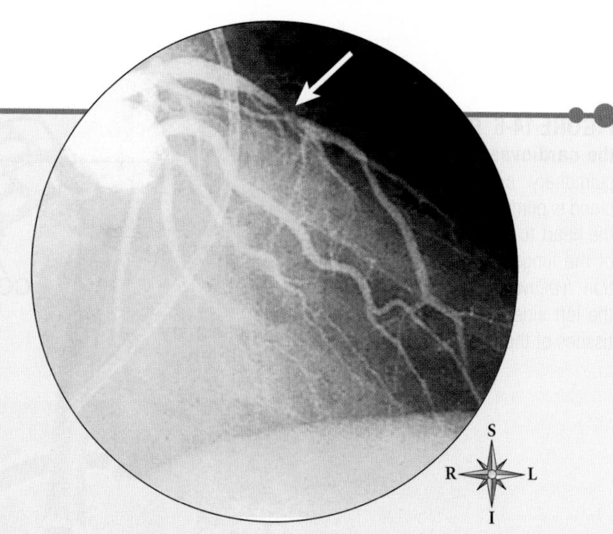

Coronary arteriogram. This angiogram of the coronary arteries shows a narrowing *(arrow)* of the channel in the anterior ventricular (left anterior descending or LAD) artery of the heart. Narrowing of this artery is sometimes called the "widow maker" because when it becomes inflamed and blocked by a blood clot, a massive heart attack and sudden death may result—often occurring in men in their 50s and 60s.

blood vessels. Mechanisms of atherosclerosis are discussed further in Chapter 15.

Coronary atherosclerosis has increased dramatically over the last few decades to become a leading cause of death in western countries. Many pathophysiologists believe this increase results from a change in lifestyle. They cite several important risk factors associated with coronary atherosclerosis: physical inactivity, cigarette smoking, high-fat and high-cholesterol diets, obesity, hypertension (high blood pressure), and diabetes.

The term **angina pectoris** is used to describe the severe chest pain that occurs when the myocardium is deprived of adequate oxygen. It is often a warning that the coronary arteries are no longer able to supply enough blood and oxygen to the heart muscle.

Coronary bypass surgery is a common treatment for those who suffer from severely restricted coronary artery blood flow. In this procedure, veins or other vessels are "harvested" or removed from other areas of the body and used to bypass partial blockages in coronary arteries (**Figure 14-8**).

A therapy called **coronary angioplasty,** a less invasive procedure, is often attempted first to treat blockages to coronary blood flow. Angioplasty, in which a device is inserted into a blocked artery to force open a channel for blood flow, is described in greater detail in Chapter 15, pp. 406-407.

14

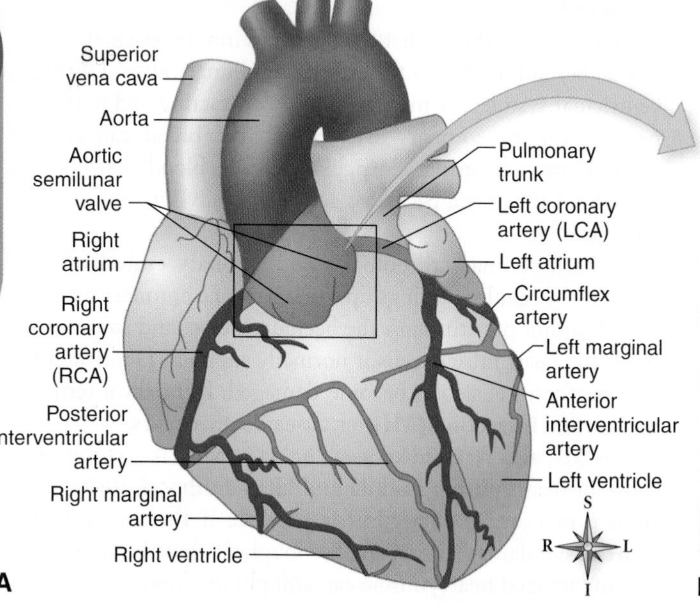

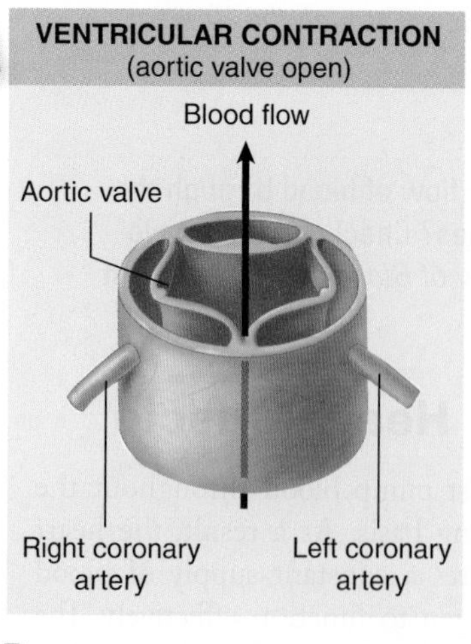

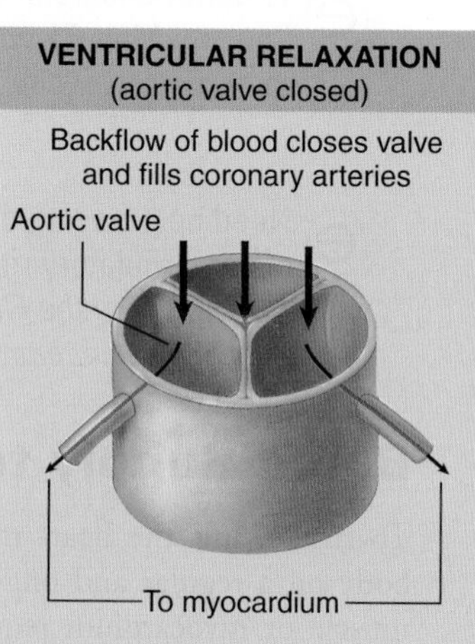

FIGURE 14-7 Coronary arteries. A, Diagram showing the major coronary arteries (anterior view). Clinicians often refer to the interventricular arteries as *descending* arteries. Thus a cardiologist would refer to the *left anterior descending (LAD) artery* and an anatomist would refer to the same vessel as the *anterior interventricular branch* or *artery.* **B,** The unusual placement of the coronary artery opening behind the leaflets of the aortic valve allows the coronary arteries to fill during ventricular relaxation.

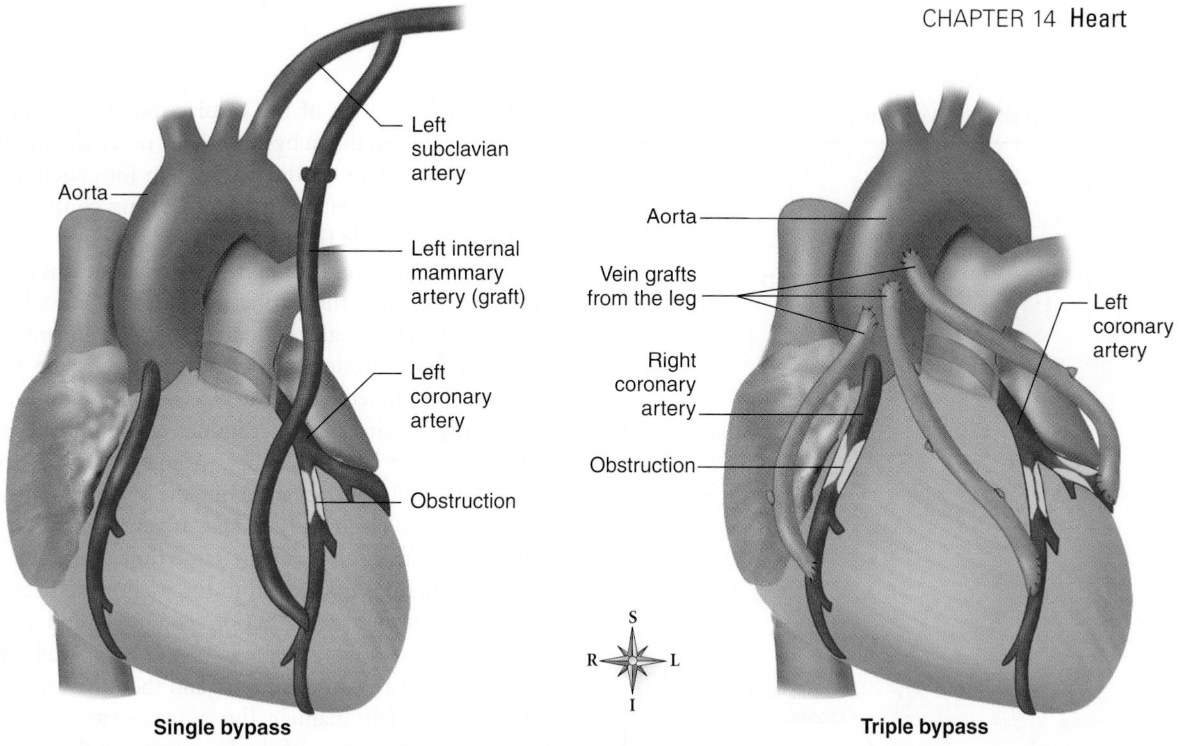

FIGURE 14-8 Coronary bypass. In coronary bypass surgery, blood vessels are "harvested" from other parts of the body and used to construct detours around blocked coronary arteries. Artificial vessels also can be used.

After blood has passed through the capillary beds in the myocardium, it flows into **cardiac veins,** which empty into the **coronary sinus** and finally into the right atrium. **Figure 14-9** shows how venous blood from the coronary circulation enters the right atrium through this "secret passage" rather than through the usual pathway through the superior or inferior vena cava.

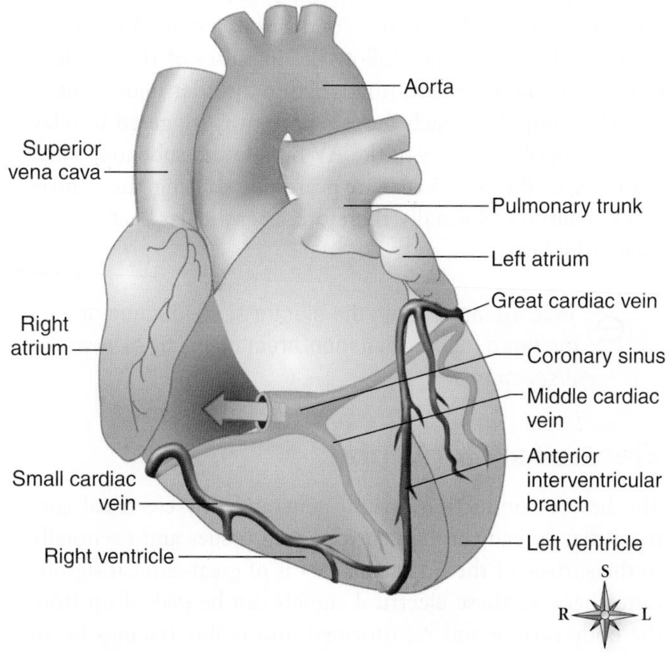

FIGURE 14-9 Coronary veins. Diagram showing the major veins of the coronary circulation (anterior view). Vessels near the anterior surface are more darkly colored than vessels of the posterior surface seen through the heart.

Cardiac Cycle

The beating of the heart is a regular and rhythmical process—what an engineer would call a *pumping cycle.* Each complete heartbeat is called a **cardiac cycle** and includes the contraction (systole) and relaxation (diastole) of atria and ventricles. Each cycle takes about 0.8 second to complete if the heart is beating at an average resting rate of about 72 beats per minute.

Figure 14-10 summarizes some of the important events of the cardiac pumping cycle. Although it looks overly complicated at first, a few minutes exploring this set of graphs can help make sense of all the processes you are learning about the rhythm of the heart's pumping cycle. For example, note that most of the atrial blood moves into the ventricles *before* the atria have a chance to contract.

Another oddity is that there is a brief period at the beginning of ventricular contraction where there is no change in volume. This occurs because it takes a moment for the ventricular pressure to overcome the force needed to open the semilunar valves. You also can see there is another period of constant volume as the ventricles begin to relax—before the mitral valves open and blood gushes rapidly in from the atria.

QUICK CHECK

1. What are *systole* and *diastole* of the heart?
2. What are the two major "circulations" of the body?
3. What causes each of the two major *heart sounds*?
4. Why is there a brief period at the beginning of ventricular contraction where there is no change in volume?
5. What is *angina pectoris*?

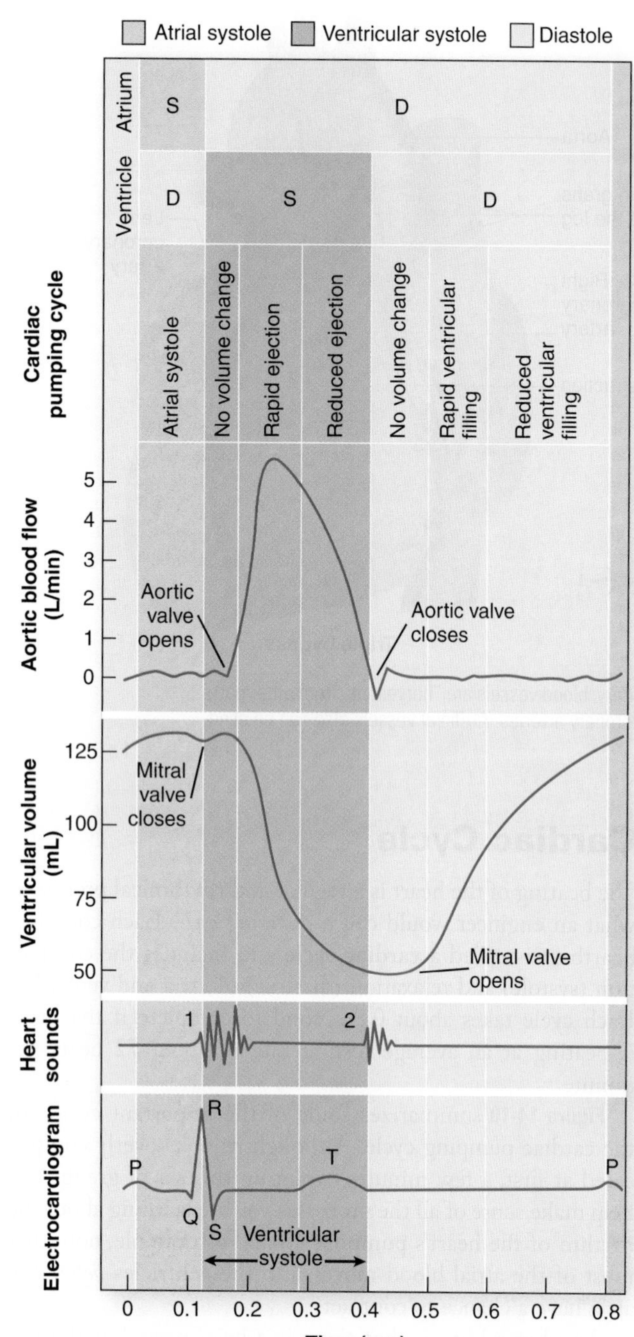

FIGURE 14-10 Composite chart of heart function. This chart is a composite of several diagrams of heart function during rest (72 beats/min). Along the top, *S* represents systole and *D* represents diastole of each heart chamber. Below that, details of the cardiac pumping cycle, aortic blood flow, ventricular volume, valve actions, heart sounds, and ECG are all adjusted to the same time scale. Although it appears daunting at first glance, this "stack of graphs" will be a valuable reference tool as you proceed through the rest of this chapter and try to "put it all together."

Electrical Activity of the Heart

Conduction System

Cardiac muscle fibers can contract rhythmically on their own. However, they must be coordinated by electrical signals (impulses) if the heart is to pump effectively.

Although the rate of the cardiac muscle's rhythm can be sped up or slowed down by autonomic nerve signals, the heart has its own built-in conduction system for generating action potentials spontaneously and coordinating contractions during the cardiac cycle (**Figure 14-11**).

The most important thing to realize about this conduction system is that all of the cardiac muscle fibers in each region of the heart are electrically linked together. The *intercalated disks* that were first introduced in Chapter 4 (see **Figure 4-17**, p. 83) are actually connections that electrically join muscle fibers into a single unit that can conduct an impulse through the entire wall of a heart chamber without stopping. Thus both atrial walls will contract at about the same time because all their fibers are electrically linked. Likewise, both ventricular walls will contract at about the same time.

Four structures embedded in the wall of the heart specialize in generating strong action potentials and conducting them rapidly to certain regions of the heart wall. Thus they make sure that the atria contract and then the ventricles contract in an efficient manner. The main structures that make up this conduction system of the heart are as follows:

1. **Sinoatrial node,** also called the SA node or the **pacemaker**
2. **Atrioventricular node,** or **AV node**
3. **AV bundle,** or **bundle of His**
4. **Subendocardial branches,** also called **Purkinje fibers**

Impulse conduction normally starts as a spontaneous action potential in the heart's pacemaker, namely, the SA node. From there, it spreads, as you can see in **Figure 14-11**, in all directions through the atria. Although each myocardial fiber can generate its own action potentials, because they are electrically linked together, they normally match the activity of the fibers that make up the conduction system. When the myocardial fibers do not follow the impulses of the conduction system, however, a cardiac rhythm disorder may result.

When impulses reach the AV node, it is triggered to relay its own impulses by way of the AV bundle and subendocardial branches to the ventricular myocardium, causing the ventricles to contract. Normally, therefore, a ventricular beat follows each atrial beat.

To learn more about the electrical conduction of the heart, go to AnimationDirect online at *evolve.elsevier.com*.

Electrocardiography

The heart's conduction system generates tiny electrical currents that spread through surrounding tissues and eventually to the surface of the body. This fact is of great clinical significance because these electrical signals can be picked up from the body surface and transformed into visible tracings by an instrument called an **electrocardiograph.**

The **electrocardiogram** is the graphic record of the heart's electrical activity obtained using an electrocardiograph apparatus. This graphic chart is also called an **ECG—**or **EKG**

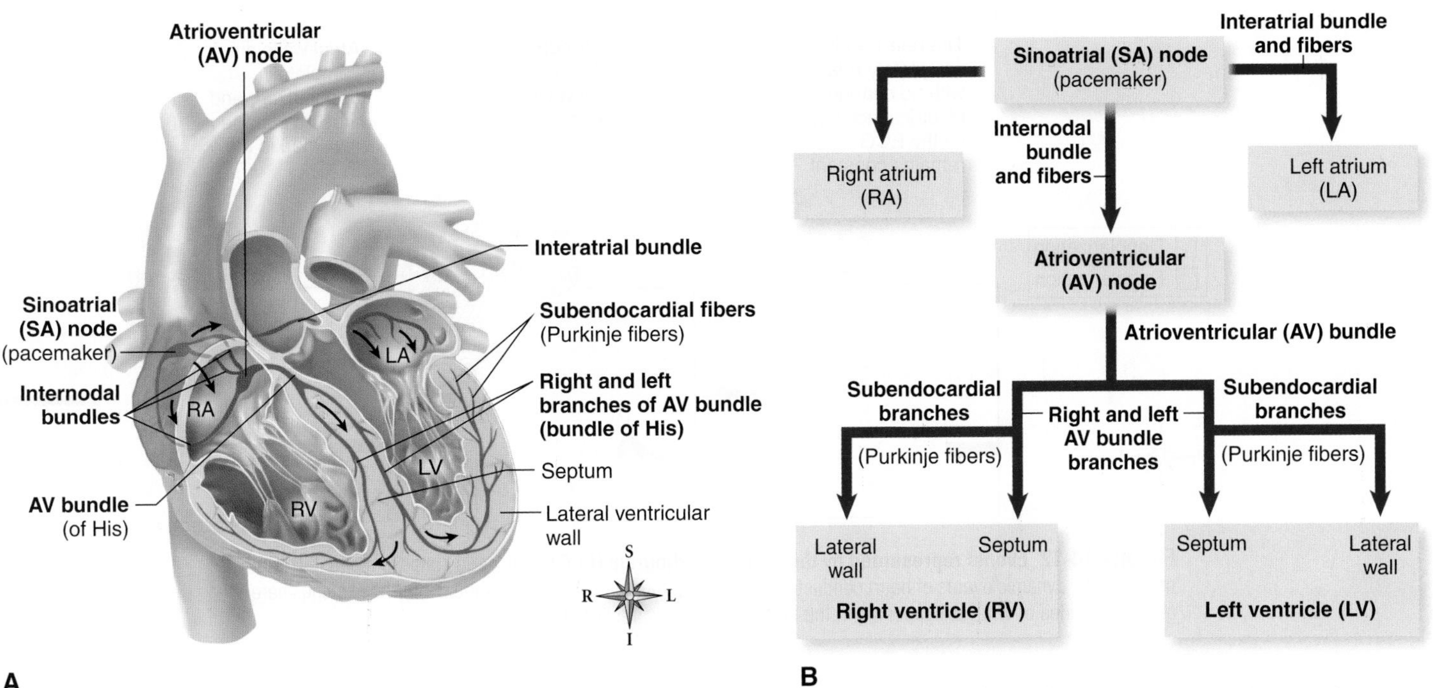

FIGURE 14-11 Conduction system of the heart. Specialized cardiac muscle cells *(red)* in the wall of the heart rapidly conduct an electrical impulse throughout the myocardium. The signal is initiated by the sinoatrial (SA) node (pacemaker) and spreads to the rest of the atrial myocardium and to the atrioventricular (AV) node. The AV node then initiates a signal that is conducted through the ventricular myocardium by way of the AV bundle (of His) and Purkinje fibers. Labels for parts of the heart's conduction system are highlighted in bold font.

when spoken aloud. Skilled interpretation of these ECG records may sometimes make the difference between life and death. A normal ECG tracing is shown in **Figure 14-12**.

A normal ECG tracing has three very characteristic deflections, or waves, called the **P wave,** the **QRS complex,** and the **T wave.** These deflections represent the electrical activity that regulates the contraction or relaxation of the atria or ventricles. The term **depolarization** describes the electrical activity that triggers contraction of the heart muscle. *Repolarization* begins just before the relaxation phase of cardiac muscle activity.

In the normal ECG shown in **Figure 14-12**, the small P wave occurs with depolarization of the atria. The QRS complex occurs as a result of depolarization of the ventricles, and the T wave results from electrical activity generated by repolarization of the ventricles. You may wonder why no visible record of atrial repolarization is noted in a normal ECG. The reason is simply that the deflection is very small and is hidden by the large QRS complex that occurs at the same time.

Now is a good time to go back and review **Figure 14-10** and explore the relationship between the ECG and other events of the cardiac cycle. By looking at changes in blood flow and volume, for example, you will discover that ECG deflections occur *before* myocardial contractions—not *during* these contractions. This occurs because depolarizations trigger contractions and a trigger always comes before the event that is triggered.

Damage to cardiac muscle tissue that is caused by a myocardial infarction or disease affecting the heart's conduction system results in distinctive changes in the ECG. Therefore ECG tracings are extremely valuable in the diagnosis and treatment of **heart disease.**

Check out the illustrated article *Electrocardiography* at Connect It! at *evolve.elsevier.com.*

To learn more about the electrocardiogram, go to AnimationDirect online at *evolve.elsevier.com.*

QUICK CHECK

1. List the structures that make up the conduction system of the heart.
2. Where does the impulse conduction in the heart begin?
3. What information is in an electrocardiogram?
4. Name the three characteristic deflections or waves of an ECG.

Cardiac Dysrhythmia

Various conditions such as endocarditis or myocardial infarction can damage the heart's conduction system and thereby disturb the rhythmic beating of the heart. The term **dysrhythmia** refers to an abnormality of heart rhythm. The term *arrhythmia* is still sometimes used to refer to rhythm abnormalities.

Heart Block

One kind of dysrhythmia is called a **heart block.** In *AV node block,* impulses are blocked from getting through to the ventricular myocardium, resulting in the ventricles contracting at a much slower rate than normal. On an ECG, there may be a large distance between the P wave and the R peak of the QRS complex. *Complete heart block* occurs when the P waves do not match up with the QRS complexes at all—as in an

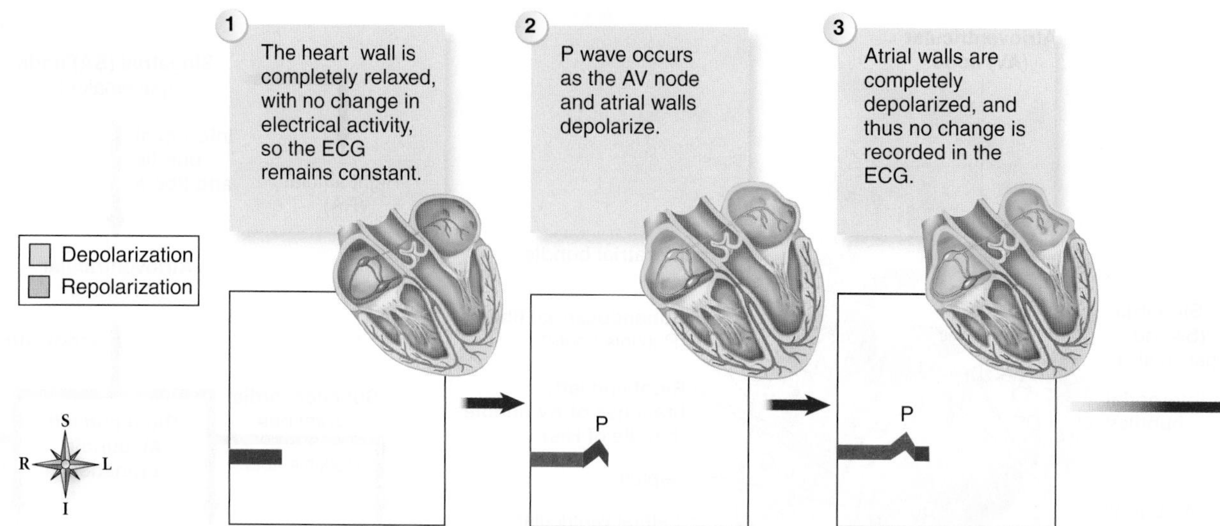

FIGURE 14-12 Events represented by the electrocardiogram (ECG). It is nearly impossible to illustrate the invisible, dynamic events of heart conduction in a few cartoon panels or "snapshots," but the sketches here give you an idea of what is happening in the heart as an ECG is recorded.

ECG that shows two or more P waves for every QRS complex (**Figure 14-13**, *A*).

A physician may treat heart block by implanting in the heart an **artificial pacemaker,** a battery-operated device implanted under the skin and connected by thin wires to the myocardium (**Figure 14-14**). This device stimulates the myocardium with timed electrical impulses that cause ventricular contractions at a rate fast enough to maintain an adequate circulation of blood.

To learn more about artificial pacemaker insertion, go to AnimationDirect online at *evolve.elsevier.com*.

14

Bradycardia

Bradycardia is a slow heart rhythm—less than 60 beats per minute (**Figure 14-13**, *B*). Slight bradycardia is normal during sleep and in conditioned athletes while they are awake (but at rest). Abnormal bradycardia can result from improper autonomic nervous control of the heart or from a damaged SA node.

If the problem is severe, artificial pacemakers can be used to increase the heart rate by taking the place of the SA node. For example, *demand pacemakers* take over SA node function only when the heart rate falls below a level programmed into the pacemaker by the physician.

Tachycardia

Tachycardia is a rapid heart rhythm—more than 100 beats per minute (**Figure 14-13**, *C*).

Tachycardia is normal during and after exercise and during the stress response. Abnormal tachycardia can result from improper autonomic control of the heart, blood loss or shock, the action of drugs and toxins, fever, and other factors.

Sinus Dysrhythmia

Sinus dysrhythmia is a variation in heart rate during the breathing cycle (**Figure 14-13**, *D*). Typically, the rate increases during inspiration and decreases during expiration. The causes of sinus dysrhythmia are not clear. This phenomenon is common in young people and does not require treatment.

Premature Contractions

Premature contractions, or *extrasystoles,* are contractions that occur before the next expected contraction in a series of cardiac cycles. For example, *premature atrial contractions (PACs)* may occur shortly after the ventricles contract—an early P wave on the ECG. *Premature ventricular contractions (PVCs)* occur when the electrical signal begins in the ventricle rather than in the SA node (**Figure 14-13**, *E*).

Premature contractions often occur with lack of sleep, anxiety, cold medications, too much caffeine or nicotine, alcoholism, or heart damage. PVCs can occur in otherwise healthy newborns, young children, and adult athletes following intense activity.

Fibrillation

Frequent premature contractions can lead to **fibrillation,** a condition in which cardiac muscle fibers contract out of step with each other. This event can be seen in an ECG as the absence of regular P waves or abnormal QRS and T waves. In fibrillation, the affected heart chambers do not effectively pump blood.

Atrial fibrillation (AF or A-fib) occurs commonly in mitral stenosis, rheumatic heart disease, and infarction of the atrial myocardium. **Figure 14-13**, *F*, shows an example of atrial fibrillation in an ECG strip.

Ventricular fibrillation (VF or V-fib) is an immediately life-threatening condition in which the lack of ventricular pumping suddenly stops the flow of blood to vital tissues. Unless ventricular fibrillation is corrected immediately by defibrillation or some other method, death may occur within minutes. **Figure 14-13**, *G*, shows an example of ventricular fibrillation in an ECG strip.

4
The QRS complex occurs as the atria repolarize and the ventricular walls depolarize.
P
Q
S
5
The atrial walls are now completely repolarized, the ventricular walls are now completely depolarized, and thus no change is seen in the ECG.
R
P
Q
S
6
The T wave appears on the ECG as the ventricular walls repolarize.
QRS complex
P wave
T wave
7
Once the ventricles are completely repolarized, we are back at the baseline of the ECG—essentially back where we began.
QRS complex
P wave
T wave

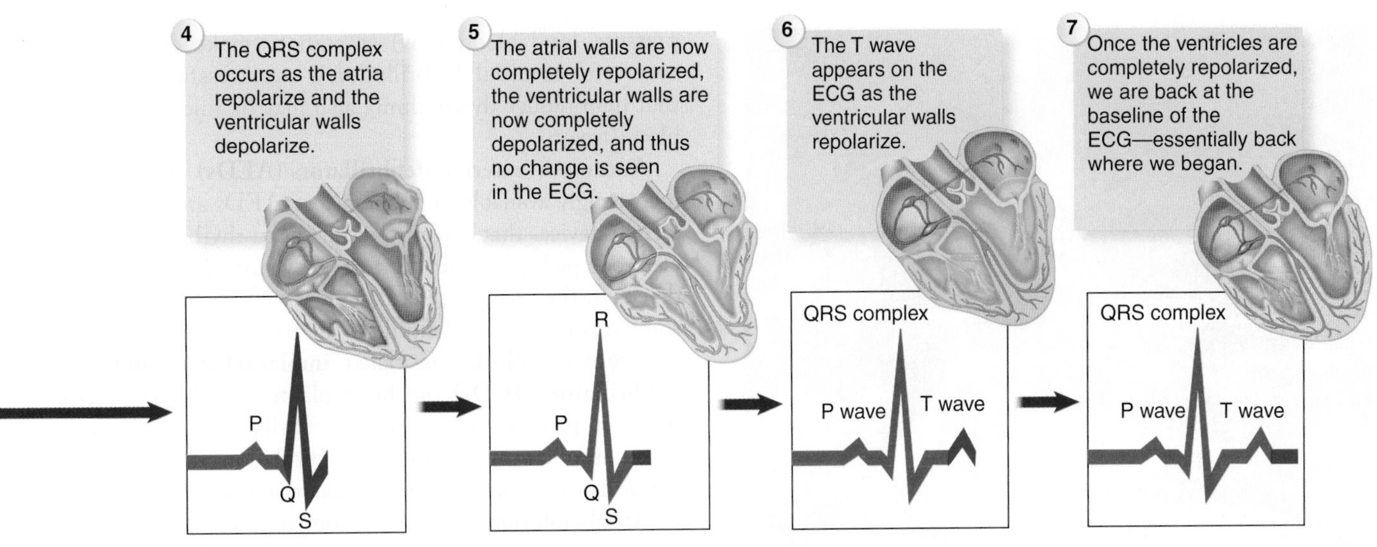

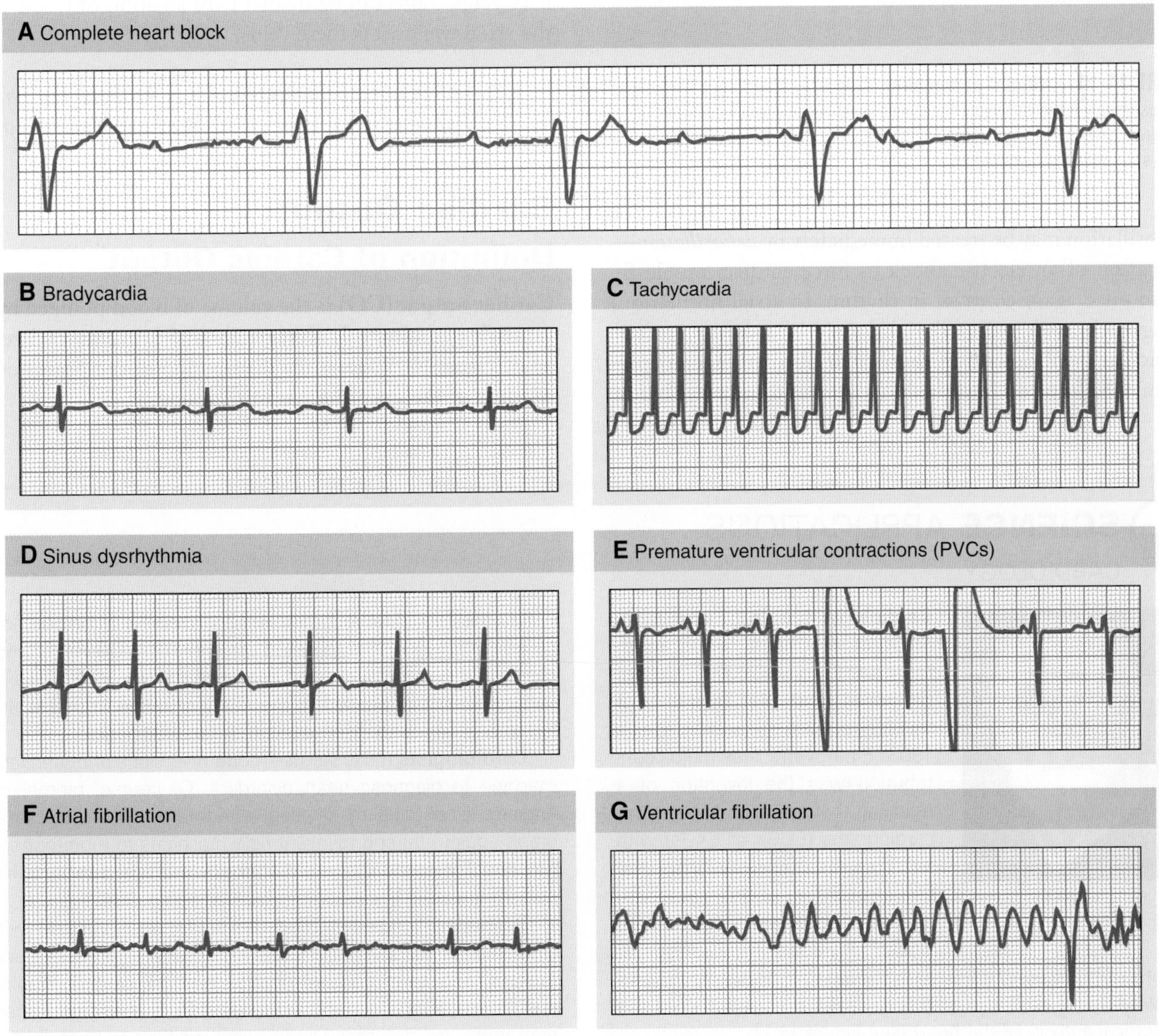

A Complete heart block
B Bradycardia
C Tachycardia
D Sinus dysrhythmia
E Premature ventricular contractions (PVCs)
F Atrial fibrillation
G Ventricular fibrillation

FIGURE 14-13 Dysrhythmia. Examples of different types of dysrhythmia are shown as they would appear in an electrocardiogram (ECG strip recording).

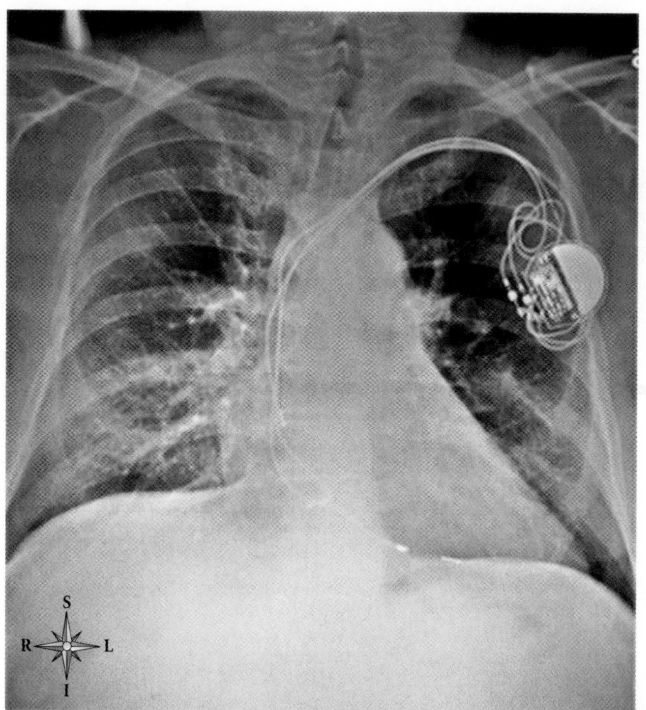

FIGURE 14-14 Artificial pacemaker. This x-ray photograph shows the stimulus generator in the subcutaneous tissue of the chest wall. Thin, flexible wires extend through veins to the heart, where timed electrical impulses stimulate the myocardium.

Fibrillation may be treated immediately by *defibrillation*—application of an electric shock to force cardiac muscle fibers to once again contract in rhythm. In atrial fibrillation, a drug such as digoxin (digitalis) may be used to prevent ventricular involvement. In ventricular fibrillation, epinephrine may be injected into the bloodstream to increase blood pressure (and blood flow) enough to make defibrillation successful. If initial defibrillation is unsuccessful, then drugs that help reduce dysrhythmia also may be injected into the bloodstream.

Automatic external defibrillators (AEDs) are becoming increasingly available in public places. AEDs are small, lightweight devices that detect a person's heart rhythm using small electrode pads placed on the torso. If ventricular fibrillation is detected, then a nonmedical rescuer will be talked through some simple steps to defibrillate the patient.

Internal defibrillators called **implantable cardioverter-defibrillators (ICDs)** can be implanted much like a pacemaker in patients prone to cardiac fibrillation or tachycardia. ICDs automatically monitor for fibrillation, then produce a defibrillating shock without any external intervention.

Atrial **ablation** is the intentional destruction of heart muscle in a specific location to treat atrial fibrillation by eliminating the pathway of abnormal electrical signals. **Atrial flutter (AFL)**—a rapid and irregular atrial rhythm often triggered by abnormal electrical signals from the nearby pulmonary veins—also can be treated with ablation of tissue where the atrial wall meets these veins.

Check out the illustrated article *Artificial Cardiac Pacemakers* at Connect It! at *evolve.elsevier.com*.

Cardiac Output

Definition of Cardiac Output

Cardiac output (CO) is the volume of blood pumped by one ventricle per minute. It averages about 5 L in a normal, resting adult. **Figure 14-15** shows the distribution of the heart's output to some of the major organs of the body.

SCIENCE APPLICATIONS

CARDIOLOGY

Willem Einthoven (1860–1927)

Cardiology, the study and treatment of the heart, owes much to Dutch physiologist Willem Einthoven and his invention of the modern electrocardiograph in 1903. Einthoven's first major contribution was the invention of a machine that could record electrocardiograms (ECGs or EKGs) with far greater sensitivity than the crude machines of the nineteenth century. Then, with the help of British physician Thomas Lewis, Einthoven demonstrated and named the P, Q, R, S, and T waves and proved that these waves precisely record the electrical activity of the heart (see **Figure 14-12**).

In 1905, Einthoven even invented a way that ECG data could be sent from a patient over the telephone line to his laboratory where they could be recorded and analyzed—a technique now called *telemetry*. His detailed studies of ECG recordings changed the practice of heart medicine forever. In fact, his invention was later applied to the study of nerve impulses and led to breakthrough discoveries in the neurosciences.

Cardiologists today still use modern versions of Einthoven's machine to diagnose heart disorders. Of course, biomedical engineers continue to develop refinements to electrocardiograph equipment and to invent new machines to monitor heart function. The photo shows a **diagnostic medical sonographer** performing an echocardiogram (see box on p. 384).

Biomedical engineers and designers have worked with cardiologists to develop artificial heart valves, artificial pacemakers, and even artificial hearts! With all of this medical equipment being used in cardiology, and in medicine in general, there are also many technicians working to keep it all in good repair.

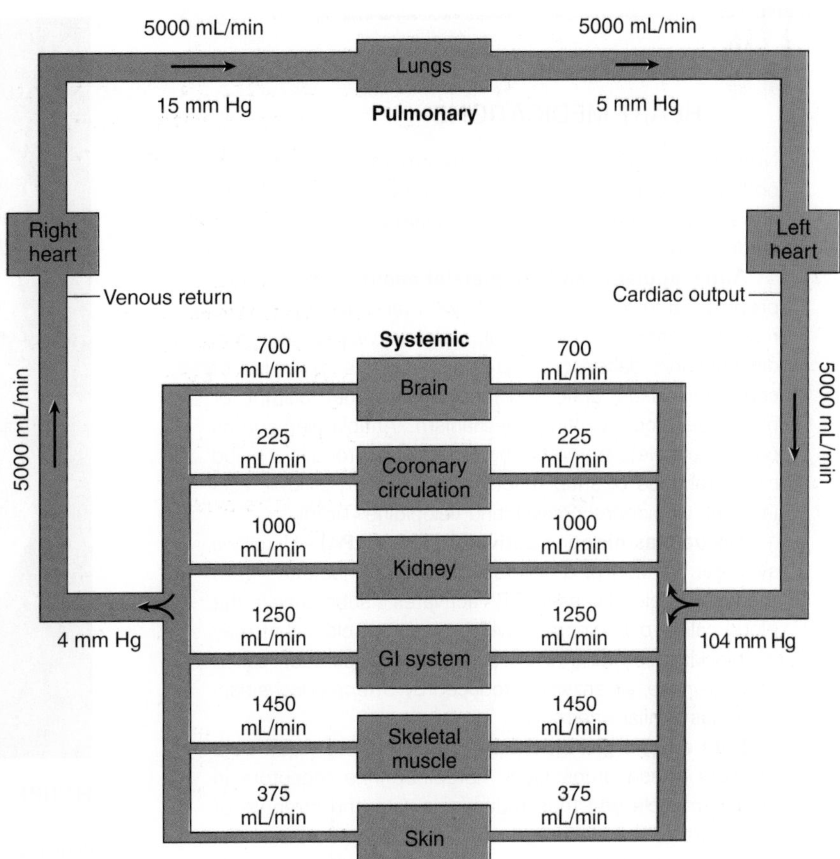

FIGURE 14-15 Cardiac output. This diagram shows that a typical resting cardiac output (CO) of 5000 mL/min (or 5 L/min) is distributed among the various systems and organs of the body. *GI*, Gastrointestinal.

The cardiac output is determined mainly by the **heart rate (HR)** and **stroke volume (SV).** *Heart rate* refers to the number of heart beats (cardiac cycles) per minute. The term *stroke volume* refers to the volume of blood ejected from the ventricle during each beat. The relationship is illustrated by this simple equation:

$$HR\left(\frac{\text{beat}}{\text{min}}\right) \times SV\left(\frac{\text{volume}}{\text{beat}}\right) = CO\left(\frac{\text{volume}}{\text{min}}\right)$$

The heart rate is determined mostly by the natural rhythm of the heart created by the heart's own conduction system (see **Figure 14-11** on p. 389). Abnormally decreased CO can result in fatigue or, with a significant drop in CO, even death.

Heart Rate

As you learned in Chapter 10, *the autonomic nervous system (ANS)* may alter the heart's rhythm to increase or decrease HR. **Figure 10-27** on p. 275 shows that the *sympathetic* division of the ANS increases HR. Neurons of the sympathetic *cardiac nerve* release the neurotransmitter *norepinephrine (NE),* which causes the SA node to increase its usual pace and thereby increase HR.

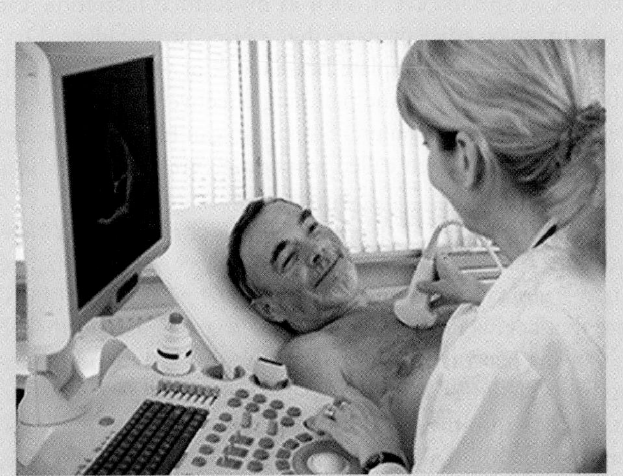

The same figure also shows that the parasympathetic division of the ANS slows down HR. This happens when neurons of the *vagus nerve (cranial nerve X)* release *acetylcholine (ACh)* to decrease the pace of the SA node.

The balance between the antagonistic influence of sympathetic and parasympathetic signals to the heart is tilted by a variety of factors. When blood CO_2 levels rise during exercise, for example, there is a reflexive rise in HR. This is an attempt by the body to restore homeostasis of blood gases.

A sudden drop in blood pressure can trigger a reflexive increase in HR as the body attempts to restore normal blood flow out of the heart. Stress—the recognition of a threat to homeostatic balance—also can cause a sudden increase in HR so that skeletal muscles will be ready to resist or avoid the stressor.

Various dysrhythmias may affect HR by disrupting the normal rhythm of the heart.

Stroke Volume

The volume of blood ejected by the ventricles is determined by the volume of blood returned to the heart by the veins, or **venous return** (see **Figure 14-15**). Generally, the higher the venous return, the higher the SV.

Venous return can change when the volume of the blood changes, as in dehydration or blood loss due to hemorrhage. Various hormones, many of which will be discussed in later chapters, can influence total blood volume and thus also affect

CLINICAL APPLICATION

HEART MEDICATIONS

Numerous drugs are used in the treatment of heart disease both in the critical care unit and in home health care. Here are a few examples of the basic pharmacological tools of cardiac care.

Anticoagulants and antiplatelet agents—These drugs prevent clot formation in patients with valve damage or who have experienced a myocardial infarction. Warfarin (Coumadin), heparin, dalteparin, and danaparoid are examples of commonly used anticoagulants—agents that disrupt or block the blood clotting mechanism. Antiplatelet agents prevent platelets from sticking together to produce a blood clot. Examples of antiplatelet drugs are acetylsalicylic acid (aspirin), clopidogrel (Plavix), and ticlopidine (Ticlid).

Tissue plasminogen activator (TPA or **tPA)**—Usually a synthetic version of a naturally occurring substance from the walls of blood vessels, TPA activates a substance in the blood called *plasminogen,* which dissolves clots that may be blocking coronary arteries. Another preparation called *streptokinase,* an enzyme produced by *Streptococcus* bacteria, has similar effects.

Beta-adrenergic blockers—Often referred to as *beta blockers,* these drugs block norepinephrine receptors in cardiac muscle and thus reduce the rate and strength of the heartbeat. Such drugs can help correct certain dysrhythmias, as well as reduce the amount of oxygen required by the myocardium. Propranolol (Inderal) and related drugs are beta-adrenergic blockers.

Calcium-channel blockers—These drugs block the flow of calcium ions (Ca^{++}) into cardiac muscle cells, which contribute to depolarization. Blocking Ca^{++} inflow thus reduces heart contractions. Calcium-channel blockers may be used in treating certain dysrhythmias and coronary heart disease. Some examples include diltiazem, verapamil, and nifedipine.

Digitalis—This drug slows and increases the strength of cardiac contractions. It plays an important part in the treatment of congestive heart failure and certain dysrhythmias. Digoxin is one of several commonly used digitalis preparations.

Nitroglycerin—This drug dilates (widens) coronary blood vessels, thus increasing the flow of oxygenated blood to the myocardium. Nitroglycerin is often used to prevent or relieve angina pectoris.

14

SV. Movement of skeletal muscles, including breathing, influences pressure on veins—which increases venous blood flow and thus increases the rate of venous return.

The strength of myocardial contraction also helps determine SV. Ion imbalances can affect muscle fiber function and thus impair contraction—thus also decreasing SV. Valve disorders, coronary artery blockage, or myocardial infarction can all decrease stroke volume and thus may decrease cardiac output.

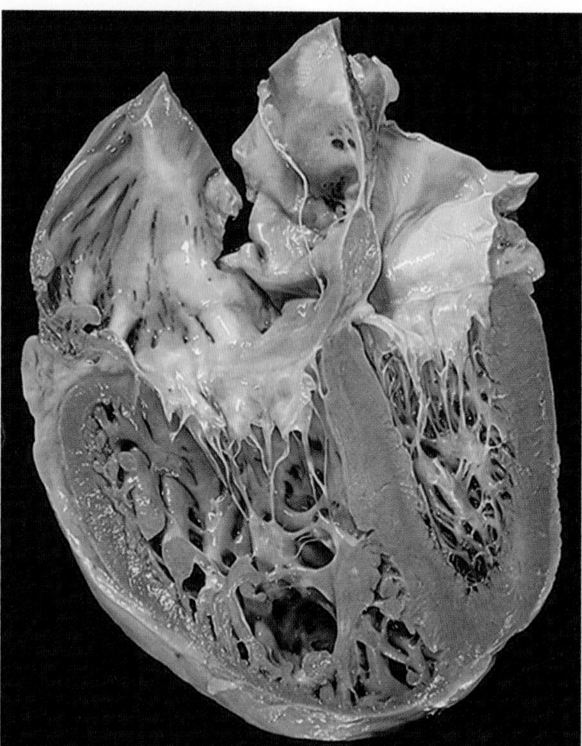

FIGURE 14-16 Cor pulmonale. When pulmonary blood backs up into the right side of the heart during right heart failure, it may stretch the right ventricle. Note in this photograph how large the right ventricle is in comparison with the left ventricle.

Heart Failure

Heart failure is the inability of the heart to pump enough blood to sustain life. Put another way, heart failure is a significant drop in cardiac output.

Heart failure can be the result of many different heart diseases. Valve disorders can reduce the pumping efficiency of the heart enough to cause heart failure. **Cardiomyopathy,** or disease of the myocardial tissue, may reduce pumping effectiveness. A specific event, such as myocardial infarction, can result in myocardial damage that causes heart failure. Dysrhythmias, such as complete heart block or ventricular fibrillation, also can impair the pumping effectiveness of the heart and thus cause heart failure. Stress also can trigger temporary cardiomyopathy, often called *stress cardiomyopathy* or "broken heart syndrome."

Failure of the right side of the heart, or *right heart failure,* accounts for about one-fourth of all cases of heart failure. Right heart failure often results from the progression of disease that begins in the left side of the heart. Failure of the left side of the heart results in reduced pumping of blood returning from the lungs. Blood backs up into the pulmonary circulation, then into the right heart—causing an increase in pressure that the right side of the heart simply cannot overcome.

Right heart failure also can be caused by lung disorders that obstruct normal pulmonary blood flow and thus overload the right side of the heart—a condition called **cor pulmonale** (**Figure 14-16**).

Congestive heart failure (CHF), or simply *left heart failure,* is the inability of the left ventricle to pump blood effectively. Most often, such failure results from myocardial infarction caused by coronary artery disease. It is called *congestive heart failure* because it decreases pumping pressure in the systemic circulation, which in turn causes the body to retain fluids. Portions of the systemic circulation thus become congested with extra fluid.

As stated above, left heart failure also causes congestion of blood in the pulmonary circulation, termed *pulmonary edema*—possibly leading to right heart failure.

Patients in danger of death because of heart failure may be candidates for heart *transplants* or heart *implants.* Heart transplants are surgical procedures in which healthy hearts from recently deceased donors replace the hearts of patients with heart disease (**Figure 14-17**). Unfortunately, a continuing problem with this procedure is the tendency of the body's immune system to reject the new heart as a foreign tissue. More details about the rejection of transplanted tissues are found in Chapter 16.

Heart implants are artificial hearts that are made of biologically inert synthetic materials. After decades of false starts and cumbersome implants with external pumps, the era of the "artificial heart" seems to have finally arrived. Although still not widely used, a number of small internal pumps have been successfully implanted in patients to take over the pumping duties of the heart.

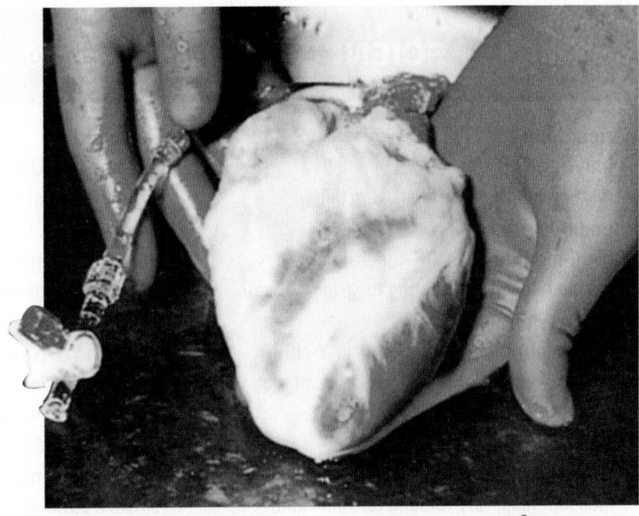

FIGURE 14-17 Heart transplant. Human heart prepared for transplantation into a patient.

QUICK CHECK

1. What does the term *dysrhythmia* mean?
2. What is the difference between *tachycardia* and *bradycardia*?
3. How is *fibrillation* corrected?
4. What is atrial ablation? In what instance is it used?
5. What is cardiac output and how is it determined?

LANGUAGE OF SCIENCE *(continued from p. 379)*

bicuspid valve
(bye-KUS-pid valv)
[*bi-* **double,** *-cusp-* **point,** *-id* **characterized by**]

bundle of His
(BUN-del of his)
[*Wilhelm His, Jr.* **Swiss cardiologist**]

cardiac cycle
(KAR-dee-ak SYE-kul)
[*cardi-* **heart,** *-ac* **relating to,** *cycl-* **circle**]

cardiac output (CO)
(KAR-dee-ak OUT-put [see oh])
[*cardi-* **heart,** *-ac* **relating to**]

cardiac vein
(KAR-dee-ak vayn)
[*cardi-* **heart,** *-ac* **relating to,** *cycle* **circle,** *vena* **blood vessel**]

cardiovascular system
(kar-dee-oh-VAS-kyoo-lar SIS-tem)
[*cardi-* **heart,** *-vas-* **vessel,** *-ular* **relating to**]

chordae tendineae
(KOR-dee ten-DIN-ee)
[*chorda* **string or cord,** *tendinea* **pulled tight**]

coronary artery
(KOHR-oh-nayr-ee AR-ter-ee)
[*corona-* **crown,** *-ary* **relating to,** *arteri-* **vessel**]

coronary circulation
(KOHR-oh-nayr-ee ser-kyoo-LAY-shun)
[*corona-* **crown,** *-ary* **relating to,** *circulat-* **go around,** *-tion* **process**]

coronary sinus
(KOR-oh-nayr-ee SYE-nus)
[*corona-* **crown,** *-ary* **relating to,** *sinus* **hollow**]

depolarization
(dee-poh-lar-ih-ZAY-shun)
[*de-* **opposite,** *-pol-* **pole,** *-ar-* **relating to,** *-ization* **process**]

diastole
(dye-AS-toh-lee)
[*dia-* **through,** *-stole* **contraction**]

endocardium
(en-doh-KAR-dee-um)
[*endo-* **within,** *-cardi-* **heart,** *-um* **thing**]

epicardium
(ep-ih-KAR-dee-um)
[*epi-* **on or upon,** *-cardi-* **heart,** *-um* **thing**]

heart rate (HR)
(hart rayt [aych ar])

inferior vena cava
(in-FEER-ee-or VEE-nah KAY-vah)
pl., venae cavae
(VEE-nee KAY-vee)
[*infer-* **lower,** *-or* **quality,** *vena* **vein,** *cava* **hollow**]

mitral valve
(MY-tral valv)
[*mitr-* **bishop's hat,** *-al* **relating to**]

myocardium
(my-oh-KAR-dee-um)
[*myo-* **muscle,** *-cardi-* **heart,** *-um* **thing**]

P wave
(pee wave)
[**named for letter of Roman alphabet**]

pacemaker
(PAYS-may-ker)

parietal pericardium
(pah-RYE-ih-tal payr-ih-KAR-dee-um)
[*pariet-* **wall,** *-al* **relating to,** *peri-* **around,** *-cardi-* **heart,** *-um* **thing**]

Continued on p. 396

14

LANGUAGE OF SCIENCE (continued from p. 395)

pericardium
(payr-ih-KAR-dee-um)
[*peri-* **around,** *-cardi-* **heart,** *-um* **thing**]

pulmonary artery
(PUL-moh-nayr-ee AR-ter-ee)
[*pulmon-* **lung,** *-ary* **relating to,** *arteri-* **vessel**]

pulmonary circulation
(PUL-moh-nayr-ee ser-kyoo-LAY-shun)
[*pulmon-* **lung,** *-ary* **relating to,** *circulat-* **go around,** *-tion* **process**]

pulmonary semilunar valve
(PUL-moh-nayr-ee sem-ee-LOO-nar valv)
[*pulmon-* **lung,** *-ary* **relating to,** *semi-* **half,** *-luna* **moon,** *-ar* **relating to**]

pulmonary vein
(PUL-moh-nayr-ee vayn)
[*pulmon-* **lung,** *-ary* **relating to,** *vena* **blood vessel**]

Purkinje fiber
(pur-KIN-jee FYE-ber)
[*Johannes E. Purkinje* **Czech physiologist,** *fiber* **thread**]

QRS complex
(kyoo ar es KOM-pleks)
[**named for letters of Roman alphabet**]

semilunar valve (SL valve)
(sem-ih-LOO-nar valv [es el valv])
[*semi-* **half,** *-luna* **moon,** *-ar* **relating to**]

sinoatrial node (SA node)
(sye-noh-AY-tree-al nohd [es ay nohd])
[*sin-* **hollow (sinus),** *-atri-* **entrance courtyard,** *-al* **relating to,** *nod-* **knot**]

stroke volume (SV)
(strohk VOL-yoom [es vee])
[*stroke* **a striking**]

subendocardial branch
(sub-en-doh-KAR-dee-al)
[*sub-* **under,** *-endo-* **within,** *-cardi-* **heart,** *-al* **relating to**]

superior vena cava
(soo-PEER-ee-or VEE-nah KAY-vah)
pl., venae cavae
(VEE-nee KAY-vee)
[*super-* **over or above,** *-or* **quality,** *vena* **vein,** *cava* **hollow**]

systemic circulation
(sis-TEM-ik ser-kyoo-LAY-shun)
[*system-* **body systems,** *-ic* **relating to,** *circulat-* **go around,** *-tion* **process**]

systole
(SIS-toh-lee)
[*systole* **contraction**]

T wave
(tee wave)
[**named for letter of Roman alphabet**]

tricuspid valve
(try-KUS-pid valv)
[*tri-* **three,** *-cusp-* **point,** *-id* **characterized by**]

venous return
(VEE-nus reh-TURN)
[*ven-* **vein,** *-ous* **relating to**]

ventricle
(VEN-trih-kul)
[*ventr-* **belly,** *-icle* **little**]

visceral pericardium
(VIS-er-al payr-ih-KAR-dee-um)
[*viscer-* **internal organ,** *-al* **relating to,** *peri-* **around,** *-cardi-* **heart,** *-um* **thing**]

LANGUAGE OF MEDICINE

14

ablation
(ab-LAY-shun)
[*ablat-* **take away,** *-tion* **process**]

angina pectoris
(an-JYE-nah PEK-tor-is)
[*angina* **strangling,** *pect-***breast,** *-oris* **relating to**]

angiogram
(AN-jee-oh-gram)
[*angi-* **vessel,** *-gram* **drawing**]

angiography
(an-jee-AH-graf-ee)
[*angi-* **vessel,** *-graph-* **draw,** *-y* **process**]

apical heart beat
(AY-pik-al hart beet)
[*apic-* **tip,** *-al* **relating to**]

artificial pacemaker
(ar-tih-FISH-al PAYS-may-ker)

atherosclerosis
(ath-er-oh-skleh-ROH-sis)
[*ather-* **porridge,** *-sclero-* **harden,** *-osis* **condition**]

atrial fibrillation (AF or A-fib)
(AY-tree-al fib-ril-LAY-shun [ay ef or AY-fib])
[*atri-* **entrance courtyard,** *-al* **relating to,** *fibr-* **thread or fiber,** *-illa-* **little,** *-ation* **process**]

atrial flutter (AFL)
(AY-tree-al FLUT-er [ay ef el])
[*atri-* **entrance courtyard,** *-al* **relating to**]

automatic external defibrillator (AED)
(aw-toh-MAT-ik eks-TERN-al dee-FIB-rih-lay-tor [ay ee dee])
[*auto-* **self,** *-mat-* **act,** *-ic* **relating to,** *extern-* **outside,** *-al* **relating to,** *de-* **off,** *-fibr-* **thread,** *-illa-* **little,** *-at-* **process,** *-or* **agent**]

bradycardia
(bray-dee-KAR-dee-ah)
[*brady-* **slow,** *-cardi-* **heart,** *-ia* **condition**]

cardiac tamponade
(KAR-dee-ak tam-poh-NOD)
[*cardi-* **heart,** *-ac* **relating to,** *tampon-* **plug,** *-ade* **condition**]

cardiologist
(kar-dee-AH-loh-jist)
[*cardi-* **heart,** *-o-* **combining vowel,** *-log-* **words (study of),** *-ist* **agent**]

cardiology
(kar-dee-AHL-uh-jee)
[*cardio-* **heart,** *-log-* **words (study of),** *-y* **activity**]

cardiomyopathy
(kar-dee-oh-my-OP-ah-thee)
[*cardi-* **heart,** *-myo-* **muscle,** *-path-* **disease,** *-y* **state**]

cardiopulmonary resuscitation (CPR)
(kar-dee-oh-PUL-moh-nayr-ree ree-sus-ih-TAY-shun [see pee ar])
[*cardio-* **heart,** *-pulmon-* **lung,** *-ary* **relating to,** *resuscitat-* **revive,** *-tion* **process**]

catheterization
(kath-eh-ter-ih-ZAY-shun)
[*cathe-* **send down,** *-er* **agent,** *-tion* **process**]

congestive heart failure (CHF)
(kon-JES-tiv hart FAYL-yoor [see aych ef])
[*congest-* **crowd,** *-ive* **relating to**]

cor pulmonale
(kohr pul-mah-NAL-ee)
[*cor-* **heart,** *pulmon-* **lung,** *-ale* **relating to**]

LANGUAGE OF **MEDICINE** *(continued from p. 396)*

coronary angioplasty
(KOHR-oh-nayr-ee AN-jee-oh-plas-tee)
[*corona-* **crown,** *-ary* **relating to,** *angio-* **vessel,** *-plasty* **surgical repair**]

coronary bypass surgery
(KOHR-oh-nayr-ee BYE-pass SER-jer-ee)
[*corona-* **crown,** *-ary* **relating to**]

diagnostic medical sonographer
(dye-ag-NOS-tik MED-ih-kul son-AH-graf-er)
[*dia-* **through,** *-gnos-* **knowledge,** *-ic* **relating to,** *medic-* **heal,** *-al* **relating to,** *sono-* **sound,** *-graph-* **draw,** *-er* **agent**]

dysrhythmia
(dis-RITH-mee-ah)
[*dys-* **disordered,** *-rhythm-* **rhythm,** *-ia* **condition**]

echocardiogram
(ek-oh-KAR-dee-oh-gram)
[*echo-* **reflect sound,** *-cardi-* **heart,** *-gram* **drawing**]

echocardiography
(ek-oh-kar-dee-OG-rah-fee)
[*echo-* **reflect sound,** *-cardi-* **heart,** *-graph-* **draw,** *-y* **activity**]

electrocardiogram (ECG or EKG)
(eh-lek-troh-KAR-dee-oh-gram [ee see jee or ee kay jee])
[*electro-* **electricity,** *-cardio-* **heart,** *-gram* **drawing**]

electrocardiograph
(eh-lek-troh-KAR-dee-oh-graf)
[*electro-* **electricity,** *-cardio-* **heart,** *-graph* **draw**]

embolism
(EM-boh-liz-em)
[*embol-* **plug,** *-ism* **condition**]

endocarditis
(en-doh-kar-DYE-tis)
[*endo-* **within,** *-cardi-* **heart,** *-itis* **inflammation**]

fibrillation
(fib-rih-LAY-shun)
[*fibr-* **fiber,** *-illa-* **little,** *-ation* **process**]

heart block
(hart blok)

heart disease
(hart dih-ZEEZ)
[*dis-* **opposite of,** *-ease* **comfort**]

heart failure
(hart FAYL-yoor)

heart murmur
(hart MUR-mur)
[*murmur* **mutter**]

incompetent valve
(in-KOM-peh-tent valv)
[*in-* **not,** *-compet-* **coincide or fit,** *-ent* **state**]

implantable cardioverter-defibrillator (ICD)
(im-PLAN-tah-bel KAR-dee-oh-vert-er dee-FIB-rih-lay-tor [aye see dee])
[*im-* **in,** *-planta-* **set or place,** *-able* **capable,** *cardio-* **heart,** *-vert-* **turn,** *-er* **agent,** *de-* **off,** *-fibr-* **thread,** *-illa-* **little,** *-at-* **process,** *-or* **agent**]

mitral valve prolapse (MVP)
(MY-tral valv PROH-laps [em vee pee])
[*mitr-* **bishop's hat,** *-al* **relating to,** *pro-* **forward,** *-laps-* **fall**]

myocardial infarction (MI)
(my-oh-KAR-dee-al in-FARK-shun [em aye])
[*myo-* **muscle,** *-cardi-* **heart,** *-al* **relating to,** *in-* **in,** *-farc-* **stuff,** *-tion* **process**]

pericardial effusion
(payr-ih-KAR-dee-al ef-FYOO-zhen)
[*peri-* **around,** *-cardi-* **heart,** *-al* **relating to,** *e(x)-* **outside,** *-fus-* **pour,** *-sion* **process**]

pericarditis
(payr-ih-kar-DYE-tis)
[*peri-* **around,** *-cardi-* **heart,** *-itis* **inflammation**]

premature contraction
(pree-mah-TUR kon-TRAK-shun)
[*pre-* **before,** *-mature* **ripen,** *con-* **together,** *-tract-* **draw,** *-tion* **process**]

rheumatic heart disease
(roo-MAT-ik hart dih-ZEEZ)
[*rheuma-* **flow,** *-ic* **relating to,** *dis-* **opposite of,** *-ease* **comfort**]

sinus dysrhythmia
(SYE-nus dis-RITH-mee-ah)
[*sinus* **hollow,** *dys-* **disordered,** *-rhythm-* **rhythm,** *-ia* **condition**]

stenosed valve
(steh-NOSD valv)
[*stenos-* **narrow**]

tachycardia
(tak-ih-KAR-dee-ah)
[*tachy-* **rapid,** *-cardi-* **heart,** *-ia* **condition**]

thrombus
(THROM-bus)
[*thrombus* **clot**]

ventricular fibrillation (VF or V-fib)
(ven-TRIK-yoo-lar fib-ril-LAY-shun [vee ef or VEE-fib])
[*ventr-* **belly,** *-icul-* **little,** *-ar* **relating to,** *fibr-* **thread or fiber,** *-illa-* **little,** *-ation* **process**]

14

❑ OUTLINE SUMMARY

To download a digital version of the chapter summary for use with your device, access the ***Audio Chapter Summaries*** *online at evolve.elsevier.com.*

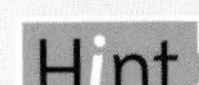

Scan this summary after reading the chapter to help you reinforce the key concepts. Later, use the summary as a quick review before your class or before a test.

Location of the Heart

A. Triangular organ located in mediastinum with two-thirds of the mass to the left of the body midline and one-third to the right; the apex is on the diaphragm; shape and size of a closed fist (**Figure 14-1**)

B. Cardiopulmonary resuscitation (CPR)—heart lies between the sternum in front and the bodies of the thoracic vertebrae behind; rhythmic compression of the heart between the sternum and vertebrae can maintain blood flow during cardiac arrest; when combined with an artificial respiration procedure, CPR can be lifesaving

Functional Anatomy of the Heart

A. Heart chambers (**Figure 14-2**)
   1. Two upper chambers are called *atria* (receiving chambers)—right and left atria
   2. Two lower chambers are called *ventricles* (discharging chambers)—right and left ventricles
   3. Wall of each heart chamber is composed of cardiac muscle tissue called *myocardium*
   4. Endocardium—smooth lining of heart chambers
      a. Inflammation of endocardium is called *endocarditis*
      b. Inflamed endocardium can become rough and abrasive and thereby cause a thrombus

B. The pericardium and pericarditis
   1. Pericardium—a two-layered fibrous sac with a lubricated space between the two layers
      a. Inner layer is called visceral pericardium, or epicardium
      b. Outer layer is called parietal pericardium
   2. Pericarditis—inflammation of the pericardium
   3. Cardiac tamponade—compression of the heart caused by fluid building up between the visceral pericardium and parietal pericardium

C. Heart action
   1. Contraction of the heart is called *systole*
   2. Relaxation of the heart is called *diastole*

D. Heart valves and valve disorders (**Figure 14-3**)
   1. Valves keep blood flowing through the heart; prevent backflow
   2. Atrioventricular (AV) valves
      a. Tricuspid valve—at the opening of the right atrium into the ventricle
      b. Bicuspid (mitral) valve—at the opening of the left atrium into the ventricle
      c. Chordae tendineae—stringlike structures that anchor the edges of AV valve leaflets to fingerlike projections of muscle in the heart wall
   3. Semilunar (SL) valves
      a. Pulmonary semilunar valve—at the beginning of the pulmonary artery
      b. Aortic semilunar valve—at the beginning of the aorta
   4. Valve disorders
      a. Incompetent valves "leak," allowing some blood to flow backward into the chamber from which it came
      b. Stenosed valves are narrower than normal, reducing blood flow (**Figure 14-4**)
      c. Rheumatic heart disease—cardiac damage resulting from a delayed inflammatory response to streptococcal infection
      d. Mitral valve prolapse (MVP)—incompetence of mitral valve caused by its edges extending back into the left atrium when the left ventricle contracts (**Figure 14-5**)

14

Heart Sounds

A. Two distinct heart sounds in every heartbeat, or cycle—"lub-dup" (**Figure 14-10**)
B. First sound (lub) is caused by the vibration and closure of AV valves during contraction of the ventricles
C. Second sound (dup) is caused by the closure of the semilunar valves during relaxation of the ventricles
D. Heart murmurs—abnormal heart sounds often caused by abnormal valves

Blood Flow Through the Heart

A. Heart acts as two separate pumps—the right atrium and ventricle performing different functions from the left atrium and ventricle (**Figure 14-6**)
B. Sequence of blood flow
   1. Systemic venous blood enters the right atrium through the superior and inferior venae cavae—passes from the right atrium through the tricuspid valve to the right ventricle
   2. From the right ventricle through the pulmonary semilunar valve to the pulmonary artery to the lungs
   3. Blood from the lungs through pulmonary veins to the left atrium, passing through the bicuspid (mitral) valve to left ventricle
   4. Blood in the left ventricle is pumped through the aortic semilunar valve into the aorta and is distributed to the body as a whole

Blood Supply to Heart Muscle

A. Blood, which supplies oxygen and nutrients to the myocardium of the heart, flows through the right and left coronary arteries (**Figure 14-7**)
   1. Blockage of blood flow through the coronary arteries can cause myocardial infarction (heart attack)
   2. Atherosclerosis
      a. Type of "hardening of arteries" in which lipids build up on the inside wall of blood vessels; can partially or totally block coronary blood flow
      b. Blocked areas may be opened with angioplasty or heart bypass surgery (**Figure 14-8**)
   3. Angina pectoris—chest pain caused by inadequate oxygen to the heart
B. Systemic venous blood from the myocardium is returned to the right atrium through a "back door" by way of the cardiac sinus (not by way of either vena cava) (**Figure 14-9**)

Cardiac Cycle

A. Heartbeat is regular and rhythmical—each complete beat called a *cardiac cycle*—average is about 72 beats per minute
B. Each cycle, about 0.8 second long, subdivided into systole (contraction phase) and diastole (relaxation phase)
C. Events of the cardiac cycle can be correlated with heart sounds, changes in blood flow and volume, and electrical activity of the heart (**Figure 14-10**)

Electrical Activity of the Heart

A. Conduction system of the heart (**Figure 14-11**)
   1. Intercalated disks electrically connect all the cardiac muscle fibers in a region together so that they receive impulses (action potentials), and thus contract, at about the same time
   2. Specialized conduction system structures generate and transmit the electrical impulses that result in contraction of the heart
      a. SA (sinoatrial) node, the pacemaker—located in the wall of the right atrium near the opening of the superior vena cava
      b. AV (atrioventricular) node—located in the right atrium along the lower part of the interatrial septum
      c. AV bundle (bundle of His)—located in the septum of the ventricle
      d. Subendocardial branches (Purkinje fibers)—located in the walls of the ventricles

B. Electrocardiography (**Figure 14-12**)
   1. The tiny electrical impulses traveling through the heart's conduction system can be picked up on the surface of the body and transformed into visible tracings by a machine called an *electrocardiograph*
   2. The visible tracing of these electrical signals is called an *electrocardiogram,* or ECG or EKG
   3. The normal ECG has three deflections, or waves
      a. P wave—associated with depolarization of the atria
      b. QRS complex—associated with depolarization of the ventricles
      c. T wave—associated with repolarization of the ventricles

C. Cardiac dysrhythmia—abnormality of heart rhythm (**Figure 14-13**)
   1. Heart block—conduction of impulses is blocked
      a. Complete heart block—impaired AV node conduction, producing complete dissociation of P waves from QRS complexes
      b. Can be treated by implanting an artificial pacemaker (**Figure 14-14**)
   2. Bradycardia—slow heart rate (less than 60 beats/min)
   3. Tachycardia—rapid heart rate (more than 100 beats/min)
   4. Sinus dysrhythmia—variation in heart rate during breathing cycle
   5. Premature contraction (extrasystole)—contraction that occurs sooner than expected in a normal rhythm
   6. Fibrillation—condition in which cardiac muscle fibers are "out of step," producing no effective pumping action
      a. Atrial fibrillation (AF or A-fib)—fibrillation in the atrial myocardium
      b. Ventricular fibrillation (VF or V-fib)—fibrillation in the ventricular myocardium
      c. Defibrillation (electrical shock) or ablation (destruction of myocardial tissue) are options for treating fibrillation

Cardiac Output

A. Cardiac output (CO)
   1. Amount of blood that one ventricle can pump each minute—average is about 5 L per minute at rest (**Figure 14-15**)
   2. Cardiac output is determined by heart rate (HR) and stroke volume (SV)—CO = HR × SV
      a. Heart rate is the number of beats (cardiac cycles) per minute
      b. Stroke volume is the volume of blood ejected from one ventricle with each beat (cycle)
      c. Any factor that affects HR or SV may thus also affect CO

B. Factors that affect heart rate
   1. HR determined mainly by heart's pacemaker
   2. Autonomic nervous system (ANS) can influence pacemaker
      a. Sympathetic cardiac nerve releases norepinephrine (NE) to increase HR
      b. Parasympathetic vagus nerve (cranial nerve X) releases acetylcholine (ACh) to decrease HR
      c. Exercise, change in blood pressure, stress, and dysrhythmias can cause changes in HR

C. Factors that affect stroke volume
   1. Venous return—volume of blood returned to the heart by veins (**Figure 14-15**)
      a. A high venous return results in a high SV
      b. Affected by total blood volume, which in turn can be affected by dehydration, hemorrhage, various hormones, activity of skeletal muscles
   2. Strength of myocardial contraction
      a. Impaired contractions reduce SV
      b. Can be influenced by ion imbalances, valve disorders, coronary artery blockage, or MI

Heart Failure

A. Heart failure—inability to pump enough returned blood to sustain life—can be caused by many different heart diseases
   1. Right heart failure—failure of the right side of the heart to pump blood, usually because the left side of the heart is not pumping effectively (**Figure 14-16**)
   2. Left heart failure (congestive heart failure, CHF)—inability of the left ventricle to pump effectively, resulting in congestion of the systemic and pulmonary circulations

B. Diseased hearts can be replaced by donated living hearts (transplants) or by artificial hearts (implants), although both procedures have yet to be perfected (**Figure 14-17**)

❑ ACTIVE LEARNING

STUDY TIPS

Consider using these tips to achieve success in meeting your learning goals.

Review the synopsis of the cardiovascular system in Chapter 5. Chapter 14 deals with the heart, the pump that moves the blood through the blood vessels.

1. Make flash cards and check out online resources to help you learn the structures of the heart. Review the Language of Science and Language of Medicine terms.
2. The location of the semilunar valves should be easy to remember because their names tell you where they are. It is harder to remember where the tricuspid and mitral valves are because their names do not give any clues to their locations. An easier way to remember them is to use their other names, the right and left atrioventricular valves. These names tell you exactly where they are, between the atria and ventricles on the right or left side.
3. Blood moves through the heart in one direction: from the right heart, to the lungs, to the left heart, and out the aorta. Electrical conduction through the heart wall may make more sense if you remember that atria contract from the top down, but ventricles must contract from the bottom up so the electrical impulse for contraction must be carried to the bottom of the ventricles before they start contracting. *my-ap.us/QmBof4* offers a tutorial on the electrical conduction pathways of the heart. The letters for the ECG waves do not stand for anything; they are arbitrary, but they do indicate a sequence of events (that is, the P tracing comes before the QRS complex). Check out *my-ap.us/L2JjzP* for a tutorial on ECG. Find additional online tips at *my-ap.us/LDwVq7*.
4. Make a chart of the disorders of the heart. Organizing them based on their source would be useful: the pericardium, the heart muscle, the heart valves, the conduction system, or general heart failure.
5. Bring photocopies of heart illustrations (for example, **Figures 14-1, 14-2, and 14-11**) to your study group. Blacken out the labels and use them as tools for learning the structures of the heart.
6. In your study group, discuss the flow of blood through the heart, the conduction system, the parts of the ECG, and the chart of the disorders of the heart. Go over the questions and the outline summary at the end of the chapter and discuss possible test questions.

Review Questions

Write out the answers to these questions after reading the chapter and reviewing the Chapter Summary. If you simply think through the answer without writing it down, you will not retain much of your new learning.

1. Describe the heart and its position in the body.
2. List the four chambers of the heart.
3. What is the *myocardium*? What is the *endocardium*?
4. Describe the two layers of the pericardium. What is the function of pericardial fluid?
5. Define or explain *pericarditis* and *pericardial effusion*.
6. What is *systole*? What is *diastole*?
7. List the four heart valves; identify where they are located.
8. Explain what is meant by a mitral valve prolapse.
9. Explain what occurs in a myocardial infarction.
10. Trace the flow of blood from the superior vena cava to the aorta.
11. Define *angina pectoris*.
12. Trace the path and name the structures involved in the conduction system of the heart.
13. What is *heart block*? What is *bradycardia*? What is *tachycardia*?
14. What is *fibrillation*? Which is more dangerous, atrial fibrillation or ventricular fibrillation?
15. Identify the factors that affect *heart rate* and *stroke volume*.
16. Differentiate between *stroke volume* and *cardiac output*.

Critical Thinking

After finishing the Review Questions, write out the answers to these more in-depth questions to help you apply your new knowledge. Go back to sections of the chapter that relate to concepts that you find difficult.

17. Explain how the tracings on an ECG relate to the electrical activity of the heart.
18. Explain how right heart failure is usually caused by left heart failure.
19. It has been determined that Danny has a heart rate of 72 beats per minute. His stroke volume is 70 mL. Calculate his cardiac output.
20. How does an angiography differ from a normal radiographic (x-ray) procedure?
21. Explain how blood that has provided oxygen and nutrients to the heart muscle returns to the right atrium.

Chapter Test

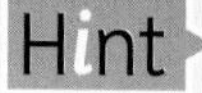

After studying the chapter, test your mastery by responding to these items. Try to answer them without looking up the answers.

1. ________ are the thicker chambers of the heart, sometimes called the discharging chambers.

2. ________ are the thinner chambers of the heart, sometimes called the receiving chambers.
3. Cardiac muscle tissue also may be called ________.
4. The ventricles of the heart are separated into right and left sides by a wall called the ________.
5. The thin layer of tissue lining the interior of each of the heart chambers is called the ________.
6. If the pericardium becomes inflamed, a condition called ________ results.
7. When the heart is contracting, it is said to be in ________.
8. A number of stringlike structures called ________ ________ attach the AV valves to the walls of the ventricles.
9. The heart valve located between the right atrium and right ventricle is called the ________ valve.
10. The ________ is the pacemaker of the heart and begins the contraction of the atria.
11. The ________ are extensions of the atrioventricular fibers and cause the contraction of the ventricles.
12. The ECG tracing that occurs when the ventricles are depolarizing is called the ________.
13. The term ________ refers to the volume of blood ejected from the ventricle during each beat.
14. The amount of blood that 1 ventricle can pump in 1 minute is called the ________ ________.
15. Blood returns from the lungs to the left ventricle through ________ pulmonary veins. HINT: How many?
16. The delivery of oxygen and nutrient-rich arterial blood to cardiac muscle tissue and the return of oxygen-poor blood from this active tissue to the venous system is called ________.
17. The first branches of the aorta are the ________.
18. The buildup of lipids and other substances on the inside wall of blood vessels is known as ________.
19. Contractions that occur before the next expected contraction in a series of cardiac cycles is known as ________.
20. An ________ detects a person's heart rhythm and enables nonmedical rescuers to defibrillate a patient.
21. An alternative to an AED, which is implanted in the patient and delivers a defibrillating shock without external intervention, is a(n) ________.
22. Place the following structures in their proper order in relation to blood flow through the heart. Put a 1 in front of the first structure the blood would pass through and a 10 in front of the last structure the blood would pass through.
    ________ a. left atrium
    ________ b. tricuspid valve (right atrioventricular valve)
    ________ c. right ventricle
    ________ d. pulmonary veins
    ________ e. aortic semilunar valve
    ________ f. mitral valve (left atrioventricular valve)
    ________ g. left ventricle
    ________ h. pulmonary artery
    ________ i. right atrium
    ________ j. pulmonary semilunar valve

Match each heart disorder in Column A with its corresponding description or cause in Column B.

Column A

23. ________ pericarditis
24. ________ mitral valve prolapse
25. ________ myocardial infarction
26. ________ angina pectoris
27. ________ heart block
28. ________ bradycardia
29. ________ tachycardia
30. ________ fibrillation
31. ________ congestive heart failure

Column B

a. damage to the heart cells due to a lack of blood flow
b. slow heart rhythm
c. a condition in which the cardiac muscles contract out of step with each other
d. rapid heart rhythm
e. also called left heart failure
f. inflammation of the pericardium
g. a condition in which contraction impulses are prevented from getting through to the ventricles
h. severe chest pain that occurs when the heart muscle is deprived of oxygen
i. a condition that allows blood to leak back into the left atrium when the left ventricle contracts

Case Studies

To solve a case study, you may have to refer to the glossary or index, other chapters in this textbook, and other resources.

1. You are visiting a friend in the hospital. Beside her bed is a video monitor that displays your friend's ECG. She asks you what the large spikes represent—can you tell her? Suddenly the ECG line becomes completely disorganized, with no discernible P, QRS, or T waves. What may have happened? What could be done for her? Or is treatment necessary?
2. Your classmate Vivian told you during lunchtime that she has been diagnosed with MVP. Describe this structural abnormality and its possible effects on heart function. Drawing on the information in your text, consider what might have caused her condition and some medical options for Vivian.
3. Uncle John is about to undergo coronary bypass surgery. His surgeon explained Uncle John's condition and the surgical procedure to correct it, but your uncle was too upset to pay close attention. Now that he is calmer, he realizes that he has very little idea of what "triple bypass surgery" is all about. Describe to your uncle the probable condition of his heart and explain the planned surgery.

Answers to Active Learning Questions can be found online at evolve.elsevier.com.

14

Circulation of Blood

OUTLINE

Scan this outline before you begin to read the chapter, as a preview of how the concepts are organized.

OBJECTIVES

Before reading the chapter, review these goals for your learning.

After you have completed this chapter, you should be able to:

1. Describe the structure and function of each major type of blood vessel: artery, vein, and capillary.
2. List the major disorders of blood vessels and explain how they develop.
3. Trace the path of blood through the systemic, pulmonary, hepatic portal, and fetal circulations.
4. Define hemodynamics, and identify and discuss the factors involved in the generation of blood pressure and how they relate to each other.
5. Define pulse and locate the major pulse points on the body.
6. Define hypertension and its associated risk factors and complications.
7. Explain what is meant by the term *circulatory shock* and describe the major types.

CHAPTER 15

In the previous chapter we discussed the basic structure and function of the circulatory system's pump: the heart. In this chapter, we continue our explanation of how blood circulates through the internal environment of the body.

First, the structure of blood vessels is discussed in some detail. Next, we explore the functions of the vessels in transporting vital substances and exchanging them with the body's tissues and the external environment. Then, we discuss how the vessels fit together into routes for the conduction of blood. The last part of this chapter deals with the driving force of blood circulation—blood pressure.

At every opportunity we also discuss major circulatory disorders to help clarify what happens when normal function fails.

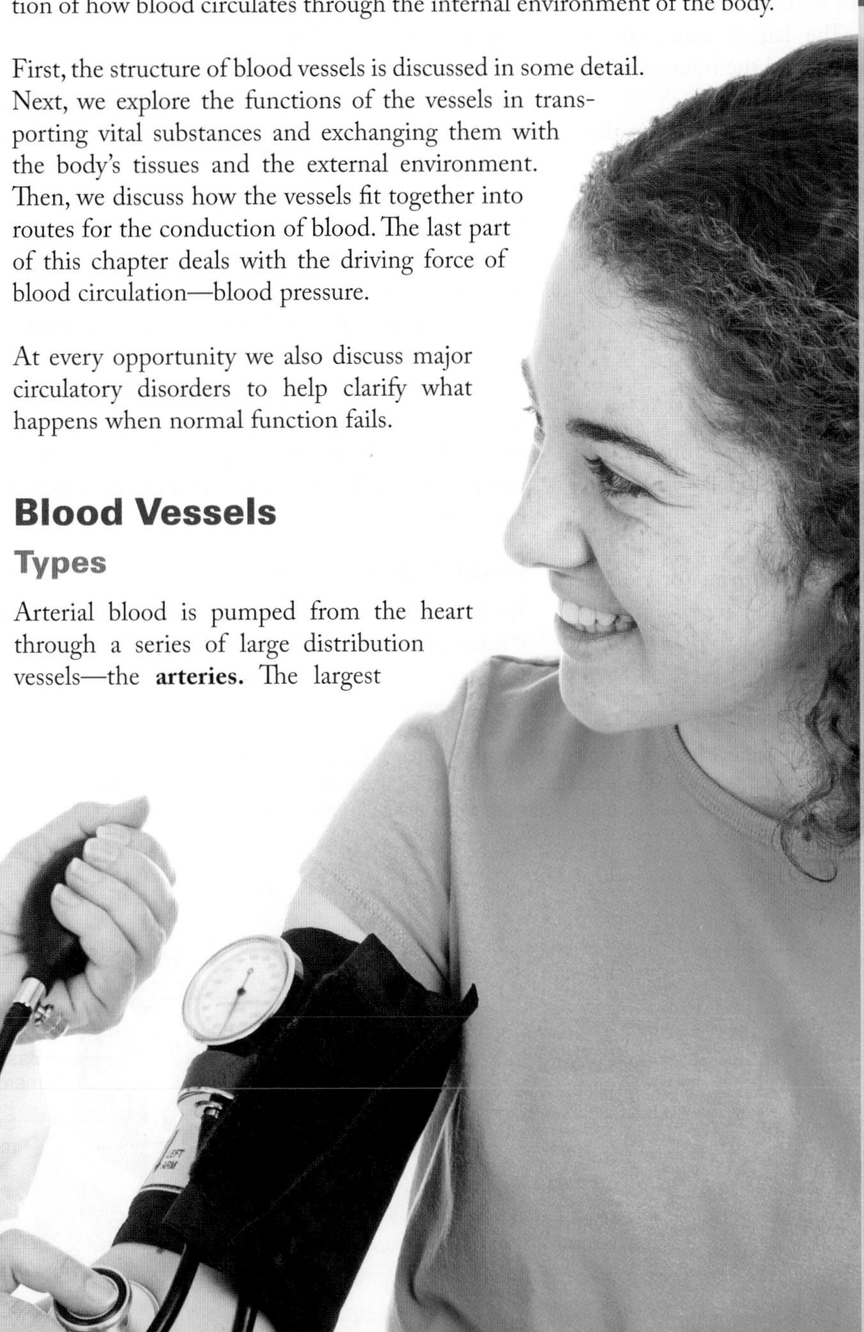

Blood Vessels

Types

Arterial blood is pumped from the heart through a series of large distribution vessels—the **arteries.** The largest

LANGUAGE OF **SCIENCE**

Hint Before reading the chapter, say each of these terms out loud. This will help you to avoid stumbling over them as you read.

arteriole
(ar-TEER-ee-ohl)
[*arteri-* **vessel,** *-ole* **little**]

artery
(AR-ter-ee)
[*arter-* **vessel,** *-y* **thing**]

blood pressure
(blud PRESH-ur)

blood pressure gradient
(blud PRESH-ur GRAY-dee-ent)
[*gradi-* **step,** *-ent* **state**]

capillary
(KAP-ih-layr-ee)
[*capill-* **hair of head,** *-ary* **relating to**]

cardiac output (CO)
(KAR-dee-ak OUT-put [see oh])
[*cardi-* **heart,** *-ac* **relating to**]

central venous pressure
(SEN-tral VEE-nus PRESH-ur)
[*centr-* **center,** *-al* **relating to,** *ven-* **vein,** *-ous* **relating to**]

diastolic blood pressure
(dye-ah-STOL-ik blud PRESH-ur)
[*dia-* **apart,** *-stol-* **position,** *-ic* **relating to**]

ductus arteriosus
(DUK-tus ar-teer-ee-OH-sus)
[*ductus* **duct,** *arteri-* **vessel,** *-osus* **relating to**]

ductus venosus
(DUK-tus veh-NOH-sus)
[*ductus* **duct,** *ven-* **vessel (vein),** *-osus* **relating to**]

endothelium
(en-doh-THEE-lee-um)
[*endo-* **within,** *-theli-* **nipple,** *-um* **thing**]

Continued on p. 421

artery in the body is the aorta. Arteries subdivide into vessels that become progressively smaller and finally become tiny **arterioles** that control the flow into microscopic exchange vessels called **capillaries.**

In the so-called *capillary beds,* the exchange of nutrients and respiratory gases occurs between the blood and tissue fluid around the cells.

Blood exits, or is drained, from the capillary beds and then enters the small **venules,** which join with other venules and increase in size, becoming **veins.** The largest veins, often called *sinuses,* are the superior vena cava and the inferior vena cava.

As noted in Chapter 14, arteries carry blood away from the heart and toward capillaries. Veins carry blood toward the heart and away from capillaries, and capillaries carry blood from the tiny arterioles into tiny venules. The aorta carries blood out of the left ventricle of the heart, and the venae cavae return blood to the right atrium after the blood has circulated through the body.

Structure

Arteries, veins, and capillaries differ in structure. Three coats or layers are found in both arteries and veins (Figure 15-1).

Outer Layer

The outermost layer is called the **tunica externa** (or *tunica adventitia*). The word *tunica* means "coat" and *externa* means "outside."

This outer layer is made of connective tissue fibers that reinforce the wall of the vessel so that it will not burst under pressure. The connective fibers also connect to the extracellular matrix of surrounding tissues to help hold the vessel in place.

Middle Layer

Figure 15-1 shows that smooth muscle tissue is found in the middle layer, or **tunica media,** of arteries and veins. The term *tunica media* means "middle coat."

This muscle layer is much thicker in arteries than it is in veins. Why is this important? Because the thicker muscle layer in the artery wall is able to resist great pressures generated by ventricular systole. In arteries, the tunica media plays a critical role in maintaining blood pressure and controlling blood distribution.

The tunica media—often simply called the *media*—is mostly smooth muscle, so it is controlled by the autonomic nervous system. The tunica media also sometimes includes a thin layer of elastic fibrous tissue.

Smooth muscle cells along the wall of arterioles are sometimes called **precapillary sphincters.** They encircle the arteriole walls and by contracting or relaxing, they regulate how much blood will flow into a capillary bed, as you can see in Figure 15-2.

Inner Layer

An inner layer of endothelial cells called the **tunica intima** ("innermost coat") lines arteries and veins.

FIGURE 15-1 Artery and vein. Schematic drawings of an artery and a vein show comparative thicknesses of the three layers: the outer layer or tunica externa, the muscle layer or tunica media, and the tunica intima made of endothelium. Note that the muscle and outer layers are much thinner in veins than in arteries and that veins have valves.

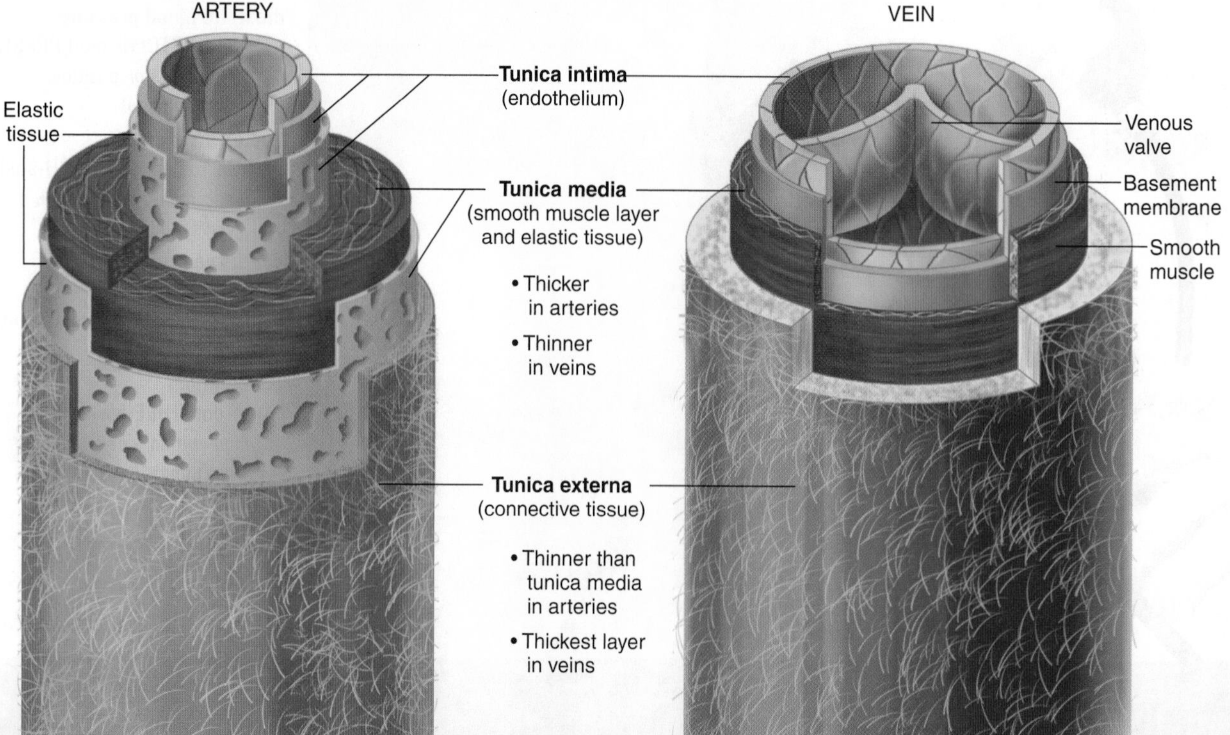

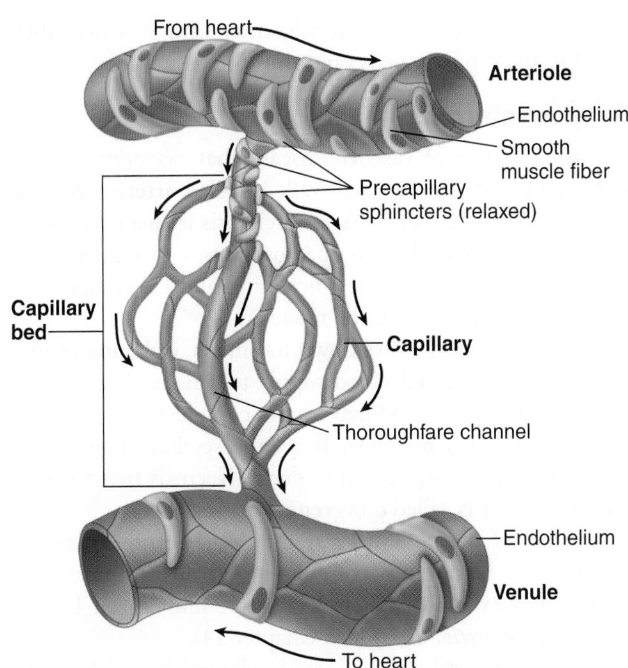

FIGURE 15-2 Capillaries. Capillaries are microscopic, thin-walled vessels that form networks joining arterioles to venules. Smooth muscle fibers (precapillary sphincters) around the arterioles can regulate how much blood flows into a capillary bed. Occasionally, these fibers wrap around the entrances to capillaries to more precisely control local blood flow.

The tunica intima is actually a single layer of squamous epithelial cells called **endothelium** that lines the inner surface of the entire cardiovascular system. This single layer of cells provides a very smooth lining that prevents the accidental formation of blood clots. The tunica intima also sometimes includes a thin layer of elastic fibrous tissue.

As you can see in Figure 15-1, many veins have a unique structural feature not present in arteries. A vein's tunica intima is equipped with pockets that act as one-way valves. These valves prevent the back-flow of blood—thus keeping blood flowing in one direction, back toward the heart.

These venous valves also allow veins to act as supplemental pumps that help maintain *venous return* of blood to the heart. Figure 15-3 shows how occasional activity of skeletal muscles surrounding the veins of the body creates pressure on blood that drives these venous pumps. This explains why stretching, walking, and other activities help improve blood circulation and prevent the formation of thrombi (abnormal clots) in the veins.

When a surgeon cuts into the body, only arteries, arterioles, veins, and venules can be seen. Capillaries cannot be seen because they are microscopic. The most important structural feature of capillaries is their extreme thinness—only one layer of flat, endothelial cells composes the capillary membrane. Instead of three layers or coats, the capillary wall is composed of only one—the tunica intima. Substances such as glucose, oxygen, carbon dioxide, hormones, and wastes can quickly pass through it on their way to or from cells.

Functions

Together, arteries, capillaries, and veins all conduct blood around the body's circulatory routes. However, each has its own unique roles to play.

Arteries and Arterioles

Arteries and arterioles distribute blood from the heart to capillaries in all parts of the body.

In addition, by constricting or dilating, arterioles help maintain arterial blood pressure at a normal level. As we discuss later in this chapter, arterial pressure is a major force in keeping blood flowing.

Capillary Exchange

Capillaries function as exchange vessels—thus carrying out a central function of the cardiovascular system. For example, glucose and oxygen move out of the blood in capillaries into interstitial fluid and then on into cells. Carbon dioxide and other substances move in the opposite direction (that is, into the capillary blood from the cells). Fluid is also exchanged between capillary blood and interstitial fluid (see Chapter 21).

Figure 15-4 illustrates the concept that two opposing forces influence capillary exchange. These forces include osmosis and filtration. Recall from Chapter 3 that *osmosis* is passive movement of water when some solutes cannot cross the membrane and *filtration* is passive movement of fluid resulting from a hydrostatic pressure gradient (see p. 51).

Figure 15-4 shows that the capillary exchange forces vary, depending on location. At the arterial end of a capillary, the outwardly directed forces are dominant and tend to move

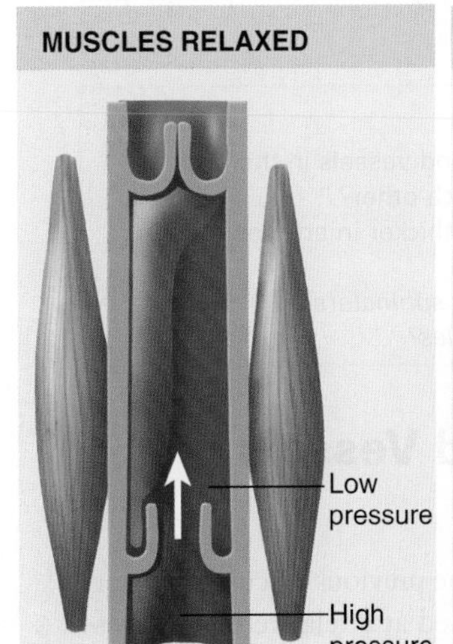

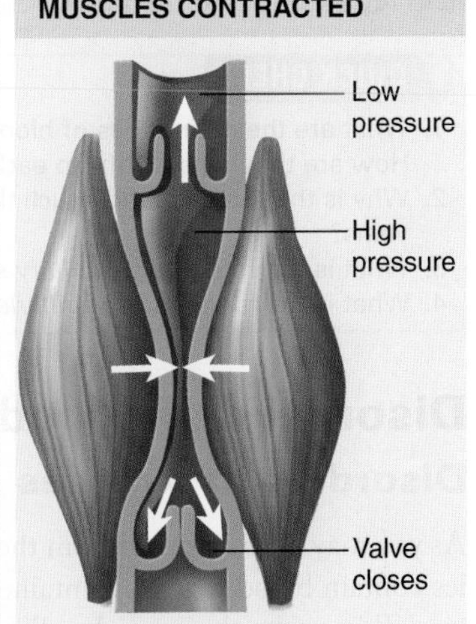

FIGURE 15-3 Venous valve function. Normal skeletal muscle contractions push on the walls of veins, which have one-way valves that allow the veins to act as pumps that push blood back toward the heart. This is similar to the action of the myocardium and heart valves acting together as a pump—except that this venous pump mechanism is not continuous and rhythmic.

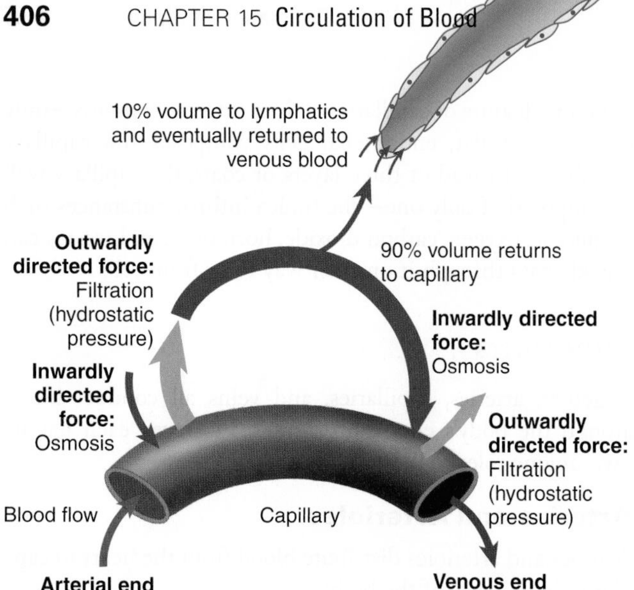

FIGURE 15-4 Capillary exchange. Osmosis (osmotic pressure) and filtration (hydrostatic pressure) are major forces that drive capillary exchange, tending to move fluids out of the capillary at the arterial end and into the capillary at the venous end. Excess tissue fluid can be collected by lymphatic vessels to be returned to the venous blood.

fluids from blood to tissue. At the venous end of a capillary, the inwardly directed forces are greater and thus tend to move fluids from tissue to blood. Excess tissue fluids not moved into the blood are collected by the lymphatic system to be eventually returned to venous blood (see Chapter 16).

Factors that affect osmotic pressure (such as plasma albumin levels) or the hydrostatic pressure (such as blood pressure) that drives filtration can disrupt capillary exchange—perhaps resulting in dehydration or overhydration of tissue (see Chapter 21).

Veins and Venules

Venules and veins collect blood from capillaries and return it to the heart.

The larger veins also serve as blood reservoirs because they carry blood under lower pressure (than arteries) and can expand to hold a larger volume of blood or constrict to hold a much smaller amount. As noted previously, external pressure can turn veins, which have one-way valves, into pumps that help return blood to the heart.

QUICK CHECK

1. What are the main types of blood vessels in the body? How are they different from each other?
2. Why is the *tunica media* much thicker in arteries than in veins?
3. What is the role of *precapillary sphincters*?
4. What is the function of *capillaries*?

15

Disorders of Blood Vessels

Disorders of Arteries

As you may have gathered from the previous discussion, arteries contain blood that is maintained at a relatively high pressure. This means the arterial walls must be able to withstand a great deal of force, or they will burst. The arteries must also stay free of obstruction; otherwise they cannot deliver their blood to the capillary beds (and thus the tissues they serve).

Arteriosclerosis

A common type of vascular disease that occludes (blocks) arteries and weakens arterial walls is called **arteriosclerosis,** or *hardening of the arteries.* Arteriosclerosis is characterized by thickening of arterial walls that progresses to hardening as calcium deposits form. The thickening and calcification reduce the flow of blood to the tissues.

If the blood flow slows down too much, **ischemia** results. Ischemia, or decreased blood supply to a tissue, involves the gradual death of cells and may lead to complete tissue death—a condition called **necrosis.** If a large section of tissue becomes necrotic, it may begin to decay. Necrosis that has progressed this far is called **gangrene.**

Because of the potential tissue damage involved, arteriosclerosis may be not only painful—it may be life-threatening as well. For example, ischemia of heart muscle can lead to *myocardial infarction (MI)* (see Chapter 14).

There are several types of arteriosclerosis, but perhaps the most well-known is *atherosclerosis,* described in Chapter 14 as the blockage of arteries by lipids and other matter (**Figure 15-5**).

Eventually, the fatty deposits in the arterial walls become fibrous and perhaps calcified—resulting in sclerosis (hardening). High blood levels of triglycerides and cholesterol, which may be caused by a high-fat, high-cholesterol diet, smoking, and a genetic predisposition, are associated with atherosclerosis. (See Chapter 2 for a discussion of triglycerides and cholesterol.)

In general, arteriosclerosis develops with advanced age, diabetes, high-fat and high-cholesterol diets, hypertension (high blood pressure), and smoking. Arteriosclerosis can be treated by drugs called **vasodilators** that trigger the smooth muscles of the arterial walls to relax, thus causing the arteries to dilate (widen).

Some cases of atherosclerosis are treated by mechanically opening the affected area of an artery, a type of procedure called **angioplasty.** In one such procedure, a deflated balloon attached to a long tube called a *catheter* is inserted into a partially blocked artery and then inflated (**Figure 15-6**). As the balloon inflates, the **plaque** (fatty deposits and tissue) is pushed outward, and the artery widens to allow near-normal blood flow.

In a similar procedure, metal springs or mesh tubes called **stents** are inserted in affected arteries to hold them open. Other types of angioplasty use lasers, drills, or spinning loops of wire to clear the way for normal blood flow. Severely affected arteries also can be surgically bypassed or replaced, as discussed in Chapter 14.

To learn more about altherosclerosis and angioplasty, go to AnimationDirect online at *evolve.elsevier.com.*

Aneurysm

Damage to arterial walls caused by arteriosclerosis or other factors may lead to the formation of an **aneurysm.** An aneurysm is a section of an artery that has become abnormally

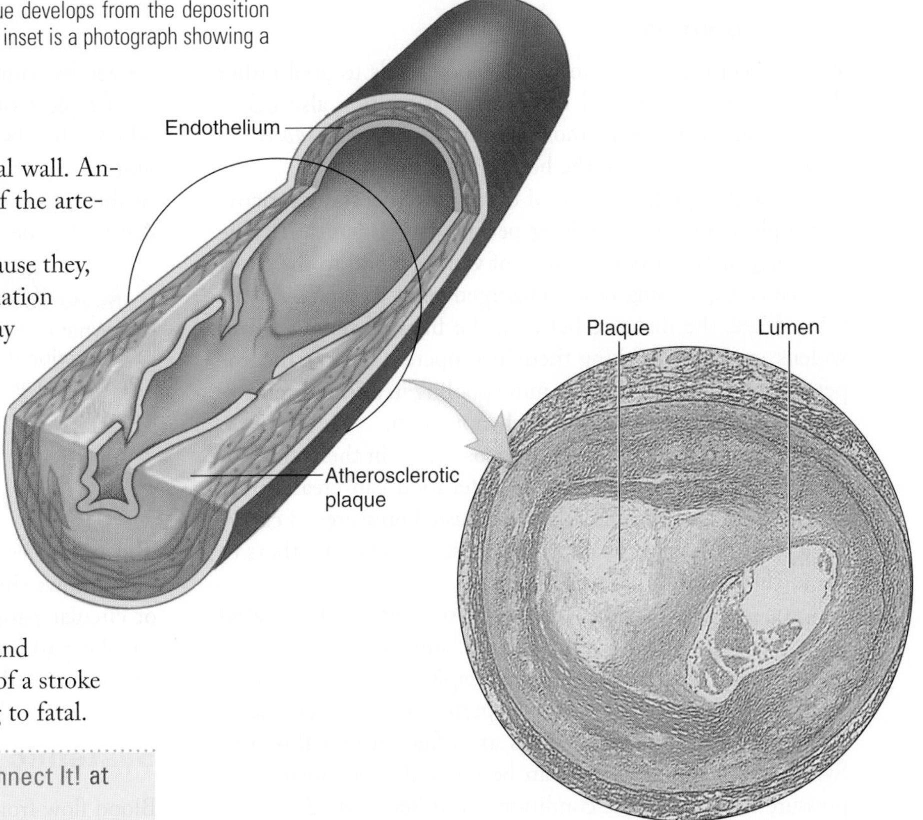

FIGURE 15-5 Atherosclerosis. Atherosclerotic plaque develops from the deposition of fats and other substances in the wall of the artery. The inset is a photograph showing a cross section of an artery partially blocked by plaque.

widened because of a weakening of the arterial wall. Aneurysms sometimes form a saclike extension of the arterial wall.

One reason aneurysms are dangerous is because they, like atherosclerotic plaques, promote the formation of thrombi (abnormal clots). A thrombus may cause an embolism (blockage) in the heart or some other vital tissue. Another reason aneurysms are dangerous is their tendency to burst, causing severe hemorrhaging that may result in death.

A brain aneurysm may lead to a *stroke,* or **cerebrovascular accident (CVA).** A stroke results from ischemia of brain tissue caused by an embolism or ruptured aneurysm. Depending on the amount of tissue affected and the place in the brain the CVA occurs, effects of a stroke may range from hardly noticeable to crippling to fatal.

Check out the article *Aneurysm* at Connect It! at *evolve.elsevier.com.*

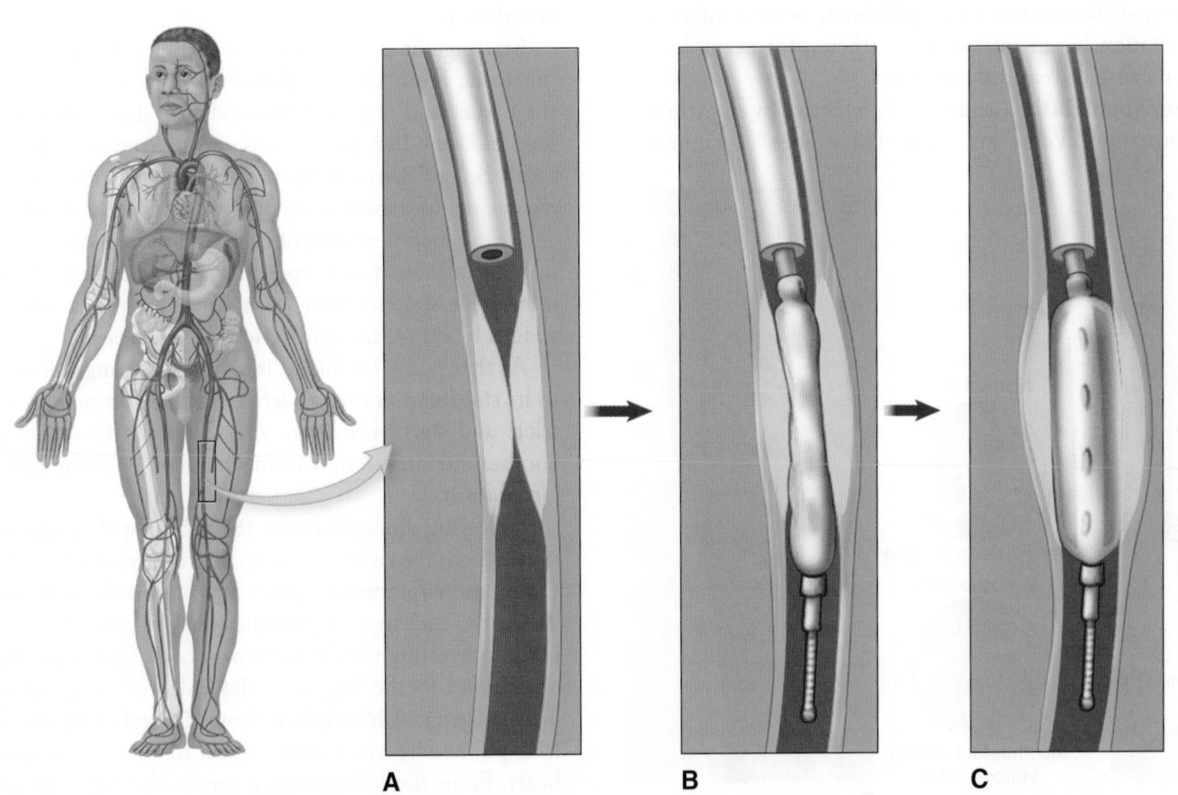

FIGURE 15-6 Balloon angioplasty. A, A catheter is inserted into the vessel until it reaches the affected region. **B,** A probe with a metal tip is pushed out the end of the catheter into the blocked region of the vessel. **C,** The balloon is inflated, pushing the walls of the vessel outward. Sometimes metal coils or tubes (stents) are inserted to keep the vessel open.

Disorders of Veins

Varicose Veins

Varicose veins are veins in which blood tends to pool rather than continue on toward the heart. Varicosities, also called **varices** (singular, *varix*), most commonly occur in *superficial veins* near the surface of the body (**Figure 15-7**).

The large superficial veins of the leg often become varicose in people who stand for long periods (see **Figure 15-7**). The force of gravity slows the return of venous blood to the heart in such cases, causing blood-engorged veins to dilate. As the veins dilate, the distance between the flaps of venous valves widens, eventually making them incompetent (leaky). Incompetence of valves causes even more pooling in affected veins—an abnormal positive-feedback phenomenon.

Hemorrhoids, or *piles,* are varicose veins in the rectum or anus. Excessive straining during defecation can create pressures that cause hemorrhoids. The unusual pressures of carrying a child during pregnancy predispose expectant mothers to hemorrhoids and other varicosities.

Varicose veins can be treated by supporting the dilated veins from the outside. For example, support stockings can reduce blood pooling in the great saphenous vein. Surgical removal of varicose veins can be performed in severe cases. Advanced cases of hemorrhoids are often treated this way. Symptoms of milder cases can be relieved by removing the pressure that caused the condition and other methods.

Phlebitis

A number of factors can cause **phlebitis,** or vein inflammation. Irritation by an intravenous catheter, for example, is a common cause of vein inflammation.

Thrombophlebitis is acute phlebitis caused by clot (thrombus) formation. Veins are more likely sites of thrombus formation than arteries because venous blood moves more slowly and is under less pressure. Thrombophlebitis is characterized by pain and discoloration of the surrounding tissue.

If a piece of a clot breaks free, it may cause an embolism when it blocks a blood vessel. **Pulmonary embolism,** for example, could result when an embolus lodges in the circulation of the lung (see **Figure 13-18** on p. 366). Pulmonary embolism can lead to death quickly if too much blood flow is blocked.

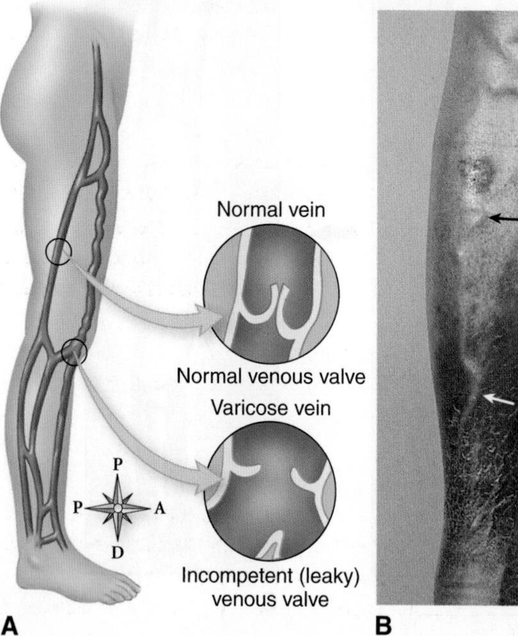

FIGURE 15-7 Varicose veins. A, Veins near the surface of the body—especially in the legs—may bulge and cause venous valves to leak. **B,** Photograph showing varicose veins on the surface of the leg.

QUICK CHECK

1. What is the medical term for *hardening of the arteries?* Describe this condition.
2. What is an *aneurysm?*
3. What causes *varicose veins?*

Routes of Circulation

The term *blood circulation* is self-explanatory, meaning that blood flows through vessels that comprise a complete circuit or circular pattern. A *route of circulation* is a particular set of circular pathways—such as from the heart to the lungs and back or from the heart to a particular organ and back.

Systemic and Pulmonary Circulation

Blood flow from the left ventricle of the heart through blood vessels to all parts of the body and back to the right atrium of the heart was described in Chapter 14 as the **systemic circulation.**

Starting our story at the left ventricle, blood is pushed into the aorta. From there, it flows into arteries that carry it into the tissues and organs of the body. As shown by the concept map in **Figure 15-8**, blood moves from arteries to arterioles to capillaries of systemic tissues. There, the vital two-way exchange of substances occurs between blood and cells.

Next, blood flows out of each organ's capillary beds by way of its venules and then its veins to drain eventually into the inferior or superior vena cava. These two great veins return venous blood to the right atrium of the heart.

At that point, the blood is short of coming full circle back to its starting point in the left ventricle. To reach the left ventricle and start on its way again, it must first flow through another circuit, referred to in Chapter 14 as the **pulmonary circulation.**

Following along in **Figure 15-8**, observe that venous blood moves from the right atrium to the right ventricle and then to the pulmonary artery to pulmonary arterioles and capillaries. There, the exchange of gases between the blood and air takes place, converting the deep crimson color typical of deoxygenated blood to the bright scarlet color of oxygenated blood. This oxygenated blood then flows through lung venules into four pulmonary veins and returns to the left atrium of the heart. From the left atrium, it enters the left ventricle, from which it will once again be pumped throughout the body in the systemic circulation.

Study **Figure 15-9** and **Table 15-1** to learn the names of the main systemic and pulmonary arteries of the body. Likewise,

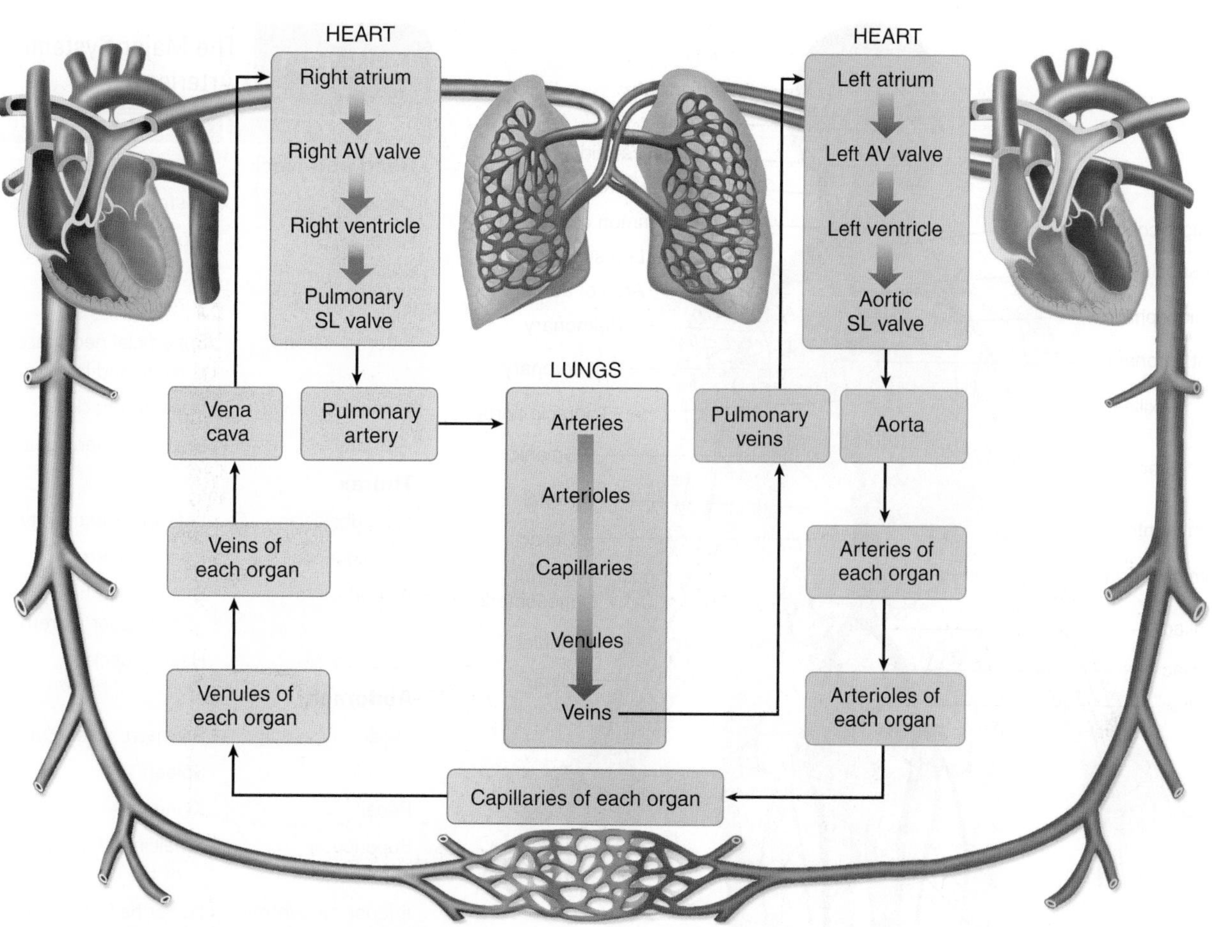

FIGURE 15-8 Diagram of blood flow in the cardiovascular system. Blood leaves the heart through arteries, then travels through arterioles, capillaries, venules, and veins before returning to the opposite side of the heart. *AV,* Atrioventricular; *SL,* semilunar.

study **Figure 15-10** and **Table 15-2** for the names of the main systemic and pulmonary veins. Explore the *Clear View of the Human Body* (follows p. 8) to see the locations of some of the major blood vessels relative to other organs.

To learn more about pulmonary circulation and systemic circulation, go to AnimationDirect online at *evolve.elsevier.com.*

To better understand these concepts, use the Active Concept Map *Blood Flow Through the Heart* at *evolve.elsevier.com.*

Hepatic Portal Circulation

The term **hepatic portal circulation** refers to the route of blood flow to and through the liver. The term *portal* means "doorway" and refers to a systemic circulatory route that is a doorway to a second set of systemic tissues.

Veins from the spleen, stomach, pancreas, gallbladder, and intestines do not pour their blood directly into the inferior vena cava as do the veins from other abdominal organs. Instead, blood flow from these organs is detoured to the liver by means of the hepatic portal vein (**Figure 15-11**). The blood then passes through the capillary beds and venous spaces of the liver before it reenters the more direct venous return pathway to the heart. Blood leaves the liver by way of the hepatic veins, which drain into the inferior vena cava.

As noted in **Figure 15-8**, most of the blood flows from arteries to arterioles to capillaries to venules to veins and back to the heart. Blood flow that is diverted to the hepatic portal circulation, however, does not follow this direct route. The diverted venous blood, instead of returning directly to the heart, is sent instead through a second capillary bed in the liver. The hepatic portal vein shown in **Figure 15-11** is located between two capillary beds—one located in the digestive organs and the other in the liver.

Once blood exits from the liver capillary beds, it returns to the systemic blood pathway, returning to the right atrium of the heart.

The detour of venous blood through a second capillary bed in the liver before its return to the heart serves some valuable purposes. For example, when nutrients from a meal are being absorbed, the blood in the portal vein contains a higher-than-normal concentration of glucose. Recall from Chapter 12 (see **Figure 12-4** on p. 324) that such high glucose levels trigger the secretion of insulin from pancreatic islets. Influenced by insulin, liver cells remove the excess glucose

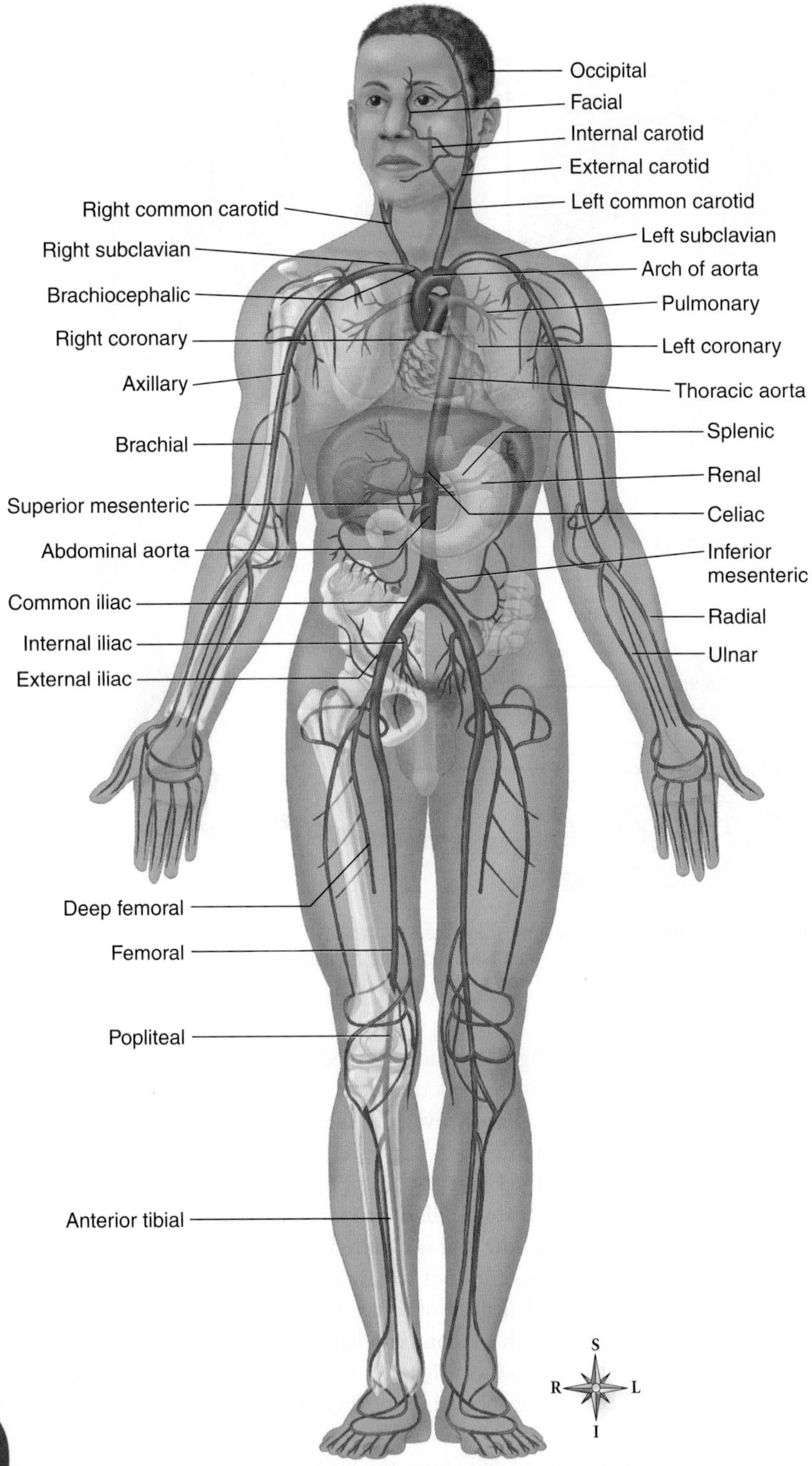

TABLE 15-1 The Major Systemic Arteries

ARTERY	TISSUES SUPPLIED
Head and Neck	
Occipital	Posterior head and neck
Facial	Mouth, pharynx, and face
Internal carotid	Anterior brain and meninges
External carotid	Superficial neck, face, eyes, and larynx
Common carotid	Head and neck
Vertebral	Brain and meninges
Thorax	
Left subclavian	Left upper extremity
Brachiocephalic	Head and arm
Arch of aorta	Branches to head, neck, and upper extremities
Coronary	Heart muscle
Abdomen	
Celiac	Stomach, spleen, and liver
Splenic	Spleen
Renal	Kidneys
Superior mesenteric	Small intestine; upper half of the large intestine
Inferior mesenteric	Lower half of the large intestine
Upper Extremity	
Axillary	Axilla (armpit)
Brachial	Arm
Radial	Lateral side of the hand
Ulnar	Medial side of the hand
Lower Extremity	
Internal iliac	Pelvic structures: viscera, genitalia, and rectum
External iliac	Lower trunk and lower extremities
Deep femoral	Deep thigh muscles
Femoral	Thigh
Popliteal	Knee and leg
Anterior tibial and posterior tibial	Leg

FIGURE 15-9 Principal arteries of the body.

15

and store it as glycogen. Therefore, blood leaving the liver usually has a lower blood glucose concentration than blood entering the liver.

Liver cells also remove and detoxify various poisonous substances that may be present in the blood. The hepatic portal circulation brings any new toxins absorbed from food directly to the liver where they can be detoxified.

The hepatic portal system is an excellent example of how "structure fits function" in helping the body maintain homeostasis.

To learn more about hepatic portal circulation, go to AnimationDirect online at *evolve.elsevier.com*.

TABLE 15-2 The Major Systemic Veins

VEIN	TISSUES DRAINED
Head and Neck	
Superior sagittal sinus	Brain
Facial and anterior facial	Anterior and superficial face
External jugular	Superficial tissues of the head and neck
Internal jugular	Sinuses of the brain
Thorax	
Brachiocephalic	Head, neck, and upper extremities
Subclavian	Upper extremities
Superior vena cava	Head, neck, and upper extremities
Right and left coronary	Heart
Inferior vena cava	Lower body
Abdomen	
Hepatic	Liver
Long thoracic	Abdominal and thoracic muscles
Hepatic portal	Intestines and nearby internal organs
Splenic	Spleen
Superior mesenteric	Small intestine and most of the colon
Inferior mesenteric	Descending colon and rectum
Upper Extremity	
Cephalic	Lateral arm
Axillary	Axilla and arm
Basilic	Medial arm
Median cubital	Cephalic vein (to basilic vein)
Radial	Lateral forearm
Ulnar	Medial forearm
Lower Extremity	
Internal iliac	Pelvic structures: viscera, genitals, rectum
External iliac	Lower limb
Femoral	Thigh
Great saphenous	Lower extremity
Small saphenous	Foot
Popliteal	Leg
Fibular (peroneal)	Foot
Anterior tibial	Deep anterior leg and dorsal foot
Posterior tibial	Deep posterior leg and plantar aspect

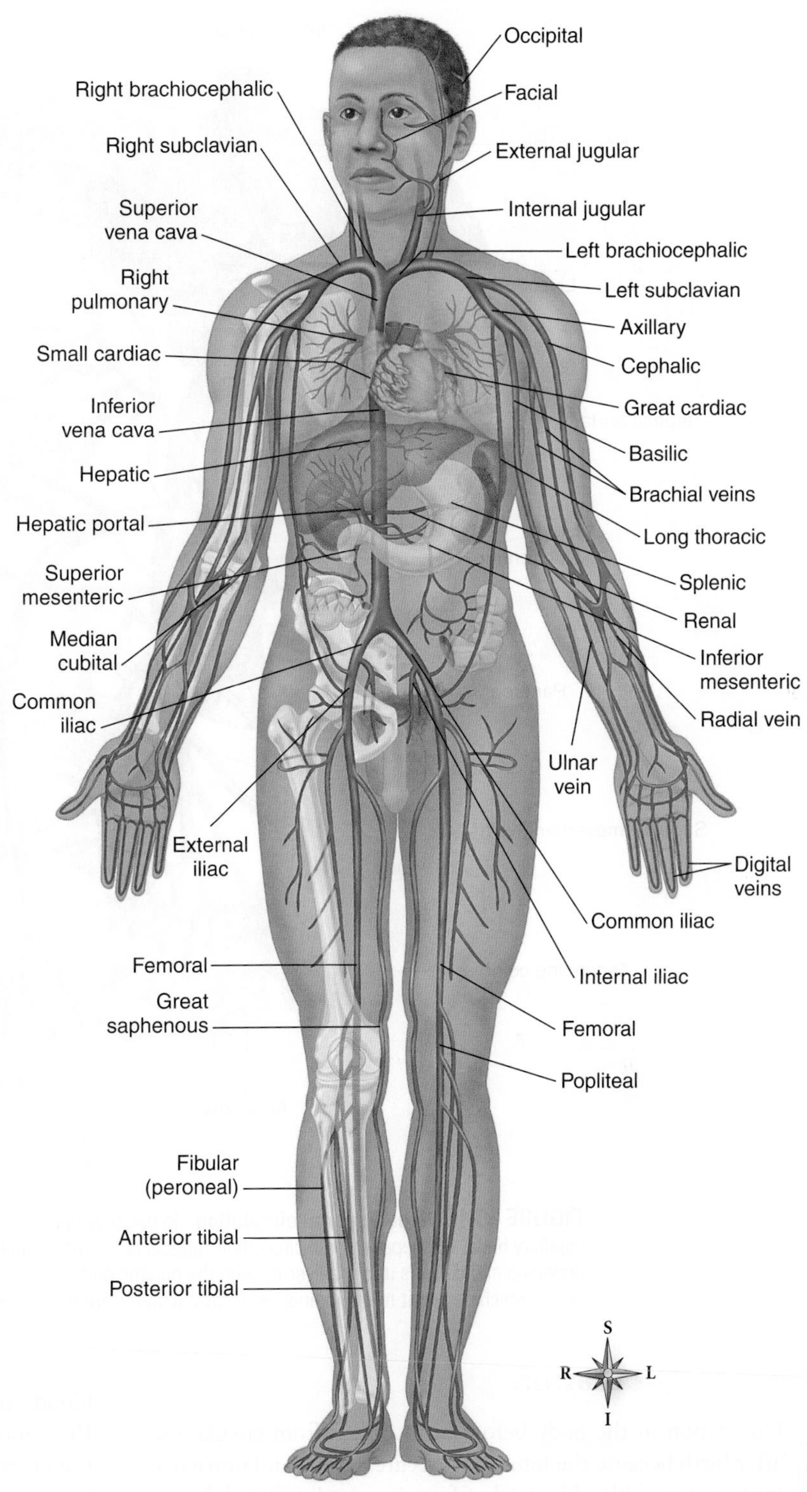

FIGURE 15-10 Principal veins of the body.

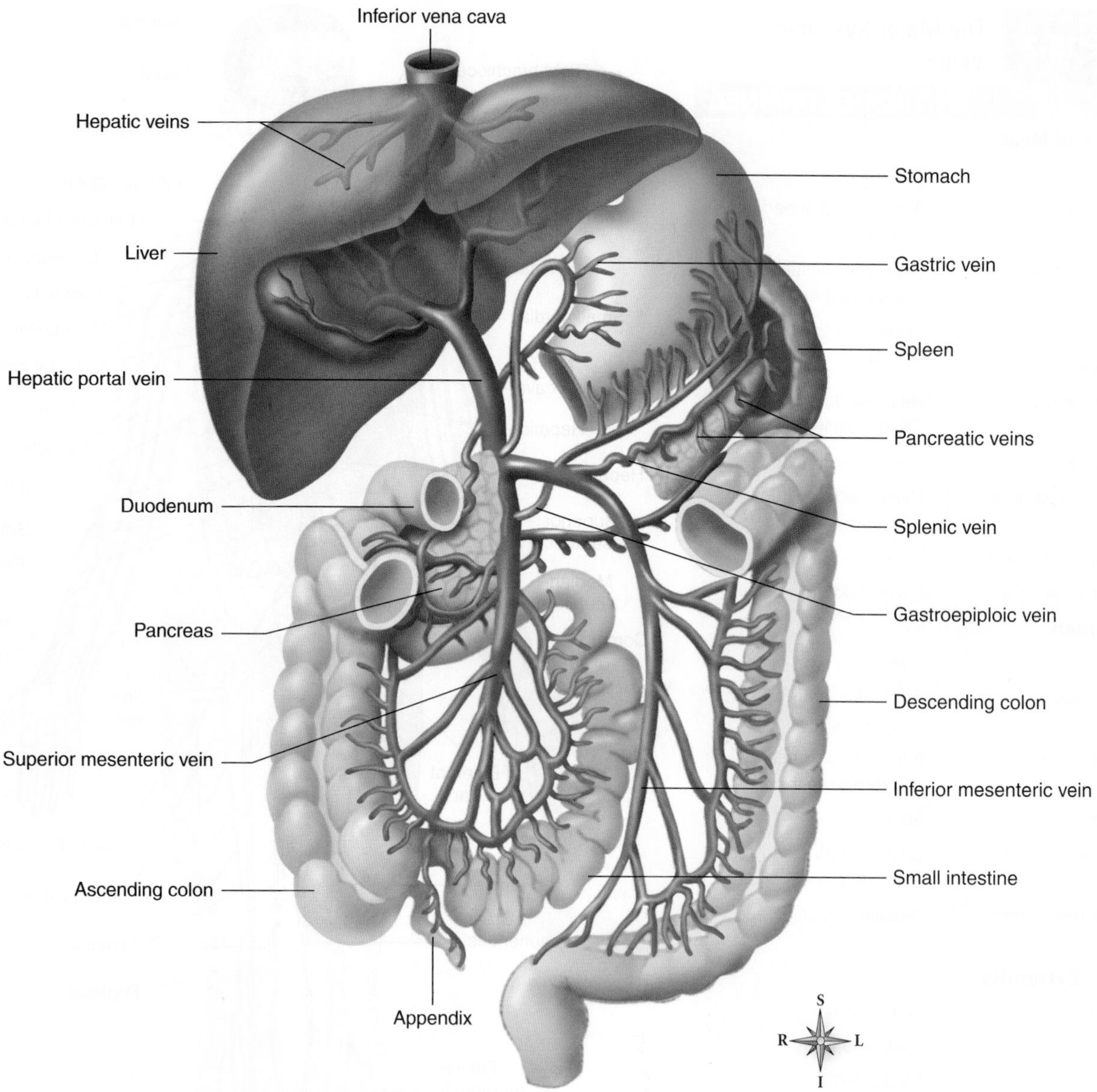

FIGURE 15-11 Hepatic portal circulation. In this very unusual circulation, a vein is located between two capillary beds. The hepatic portal circulation collects blood from capillaries in visceral structures located in the abdomen and delivers it to the liver through the hepatic portal vein. The blood leaves the liver through hepatic veins, which deliver it to the inferior vena cava. (Organs are not drawn to scale here.)

Fetal Circulation

Circulation in the body before birth differs from circulation after birth because the fetus must secure oxygen and nutrients from maternal blood instead of from its own lungs and digestive organs.

For the exchange of nutrients and oxygen to occur between fetal and maternal blood, blood vessels must carry the fetal blood to the **placenta,** where the exchange occurs, and then return it to the fetal body. Three vessels (shown in **Figure 15-12** as part of the **umbilical cord**) accomplish this purpose. They are the two small **umbilical arteries** and a single, much larger **umbilical vein.**

The movement of blood in the umbilical vessels may seem unusual at first in that the umbilical vein carries oxygenated blood, and the umbilical arteries carry oxygen-poor blood. Remember that arteries are vessels that carry blood away from the heart, whereas veins carry blood toward the heart, regardless of the oxygen content these vessels may have.

Umbilical hernia, which occurs when intestines protrude through the umbilical opening in the abdomen, are common in newborn infants. Review *Hernias* at Connect It! at *evolve.elsevier.com.*

Another structure unique to fetal circulation is called the **ductus venosus.** As you can see in **Figure 15-12**, it is actually a continuation of the umbilical vein. It serves as a shunt, allowing most of the blood returning from the placenta to bypass

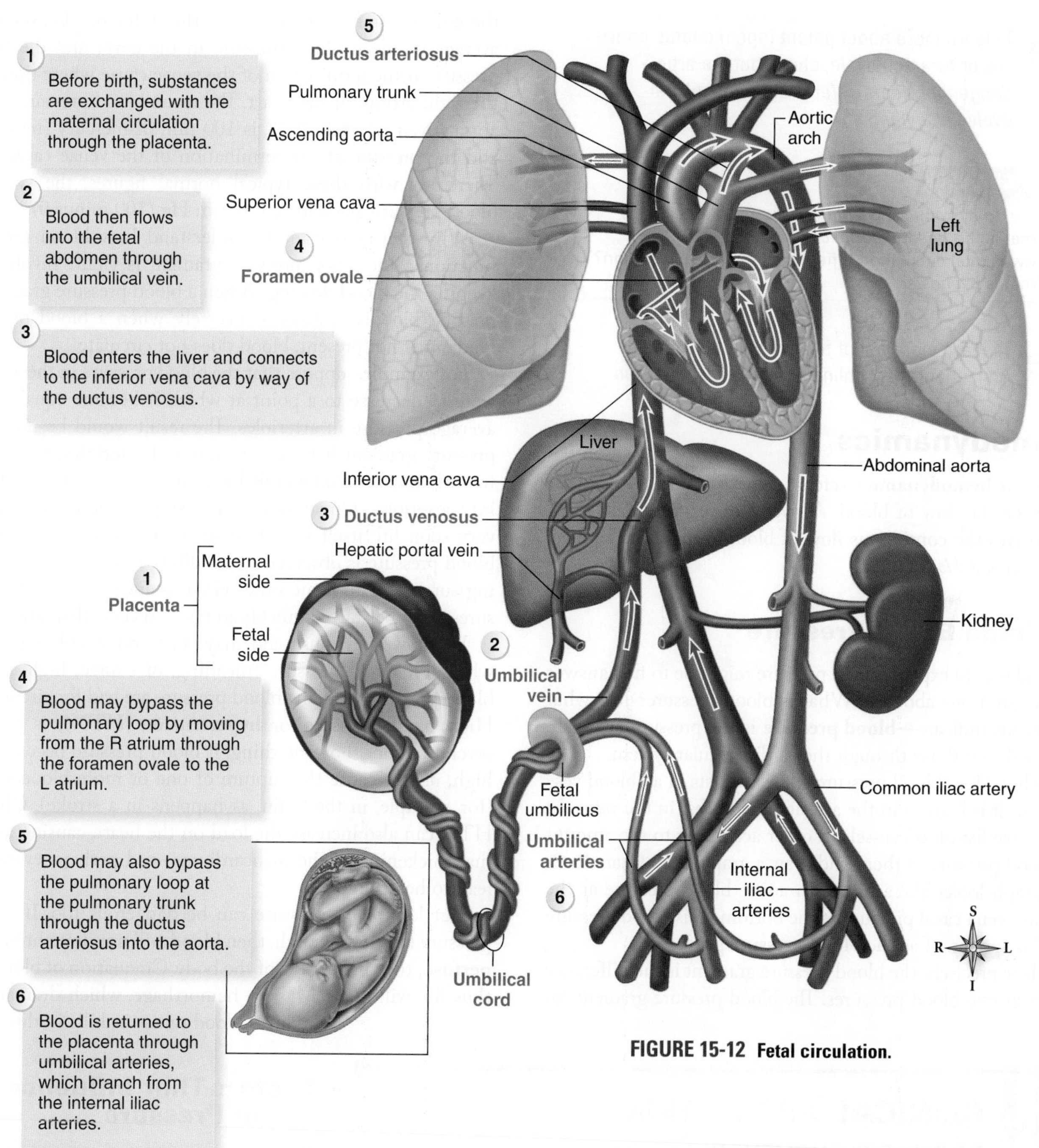

FIGURE 15-12 Fetal circulation.

the immature liver of the developing fetus and empty directly into the inferior vena cava.

Two other structures in the developing fetus allow most of the blood to bypass the developing lungs, which remain collapsed until birth. The **foramen ovale** shunts blood from the right atrium directly into the left atrium, and the **ductus arteriosus** connects the pulmonary artery to the aorta.

At birth, the infant's umbilical blood vessels and shunts must be rendered nonfunctional. When the newborn infant takes its first deep breaths, the cardiovascular system is subjected to increased pressure. The result is closure of the foramen ovale and rapid collapse of the umbilical blood vessels, the ductus venosus, and ductus arteriosus.

Several congenital disorders result from the failure of the cardiovascular system to shift from the fetal route of blood flow at the time of birth. The ductus arteriosus may fail to close, for example, and allow deoxygenated blood to bypass the lungs.

Similarly, the foramen ovale may fail to close and remain as a so-called *hole in the heart* that allows blood to bypass the pulmonary circulation. Defects in the septum between atria or between ventricles can also produce hole-in-the-heart conditions. Many people with small holes do not even know they have them, living relatively normal, healthy lives. In moderate to severe cases, however, a light-skinned baby may appear bluish because of the lack of oxygen in the systemic arterial blood. This condition of bluish tissue coloration is called **cyanosis.**

To learn more about patent (open) ductus arteriosus or foramen ovale, check out the article *Congenital Heart Defects* at Connect It! at *evolve.elsevier.com.*

QUICK CHECK

1. How do *systemic* and *pulmonary* circulations differ?
2. What is the *hepatic portal circulation*?
3. How is fetal circulation different from adult circulation?
4. What is *cyanosis*?

To learn more about fetal circulation, go to AnimationDirect online at *evolve.elsevier.com.*

Hemodynamics

The term **hemodynamics** refers to the set of processes that influence the flow of blood. As we shall see, the main force that drives the continuous flow of blood through its circulatory routes is *blood pressure.*

Defining Blood Pressure

A good way to explain blood pressure might be to first answer a few questions about it. What is blood pressure? Just what the words indicate—**blood pressure** is the pressure or "push" of blood as it flows through the cardiovascular system.

Where does blood pressure exist? It exists in all blood vessels, but it is highest in the arteries and lowest in the veins. In fact, if we list blood vessels in order according to the amount of blood pressure in them and draw a graph, as in Figure 15-13, the graph looks like a hill, with aortic blood pressure at the top and vena caval pressure at the bottom. This blood pressure "hill" is spoken of as the **blood pressure gradient.**

More precisely, the blood pressure gradient is the difference between two blood pressures. The blood pressure gradient for the entire systemic circulation is the difference between the average or mean blood pressure in the aorta and the blood pressure at the termination of the venae cavae where they join the right atrium of the heart. The mean blood pressure in the aorta, given in Figure 15-13, is 100 mm of mercury (mm Hg), and the pressure at the termination of the venae cavae is 0. Therefore, with these typical normal figures, the systemic blood pressure gradient is 100 mm Hg (100 minus 0).

Why is it important to understand how blood pressure functions? The blood pressure gradient is vitally involved in keeping the blood flowing. When a blood pressure gradient is present, blood circulates; conversely, when a blood pressure gradient is not present, blood does not circulate.

For example, suppose that the blood pressure in the arteries were to decrease to a point at which it became equal to the average pressure in arterioles. The result would be no blood pressure gradient between arteries and arterioles, and therefore no force would be available to move blood out of arteries into arterioles. Circulation would stop, in other words, and very soon life itself would cease. That is why when arterial blood pressure is observed to be falling rapidly, whether during surgery or in some other circumstance, emergency measures must be started quickly to try to reverse this fatal trend.

What we have just said may start you wondering about why high blood pressure (meaning, of course, high arterial blood pressure) and low blood pressure are bad for circulation. High blood pressure, or **hypertension (HTN),** is bad for several reasons. For one thing, if blood pressure becomes too high, it may cause the rupture of one or more blood vessels (for example, in the brain, as happens in a stroke). Chronic HTN can also increase the load on the heart, causing abnormal thickening of the myocardium—and perhaps eventually lead to heart failure.

But low blood pressure can be dangerous too. If arterial pressure falls low enough, then blood will not flow through, or **perfuse,** the vital organs of the body. Circulation of blood and thus life will cease. Massive hemorrhage, which dramatically reduces blood pressure, kills in this way.

CLINICAL APPLICATION

RAYNAUD PHENOMENON

A disorder characterized by sudden decreases in circulation in the digits (fingers or toes), often in response to stress or temperature change, is called **Raynaud phenomenon.** The decreased blood flow often causes pale discoloration of the affected digits (see figure), followed by numbness and cyanosis (blue discoloration) as oxygen levels drop. As blood flow returns to the digits, they may become dark and redder—often swelling and throbbing with pain. Symptoms range from mild to severe.

There is no known cause of Raynaud phenomenon, but some cases have been associated with inflammatory conditions such as scleroderma, rheumatoid arthritis, and lupus erythematosus.

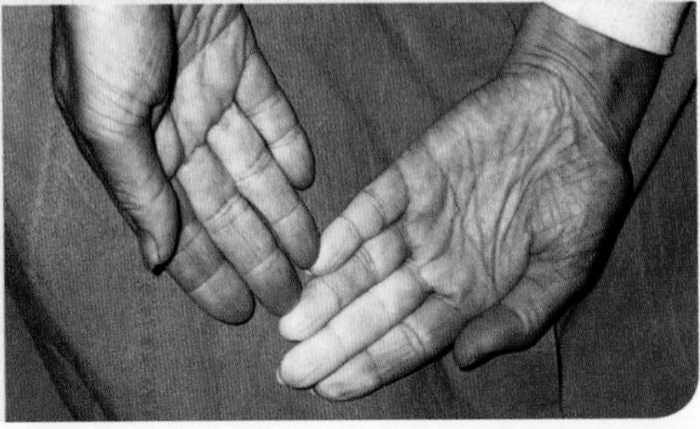

Factors That Influence Blood Pressure

What causes blood pressure, and what makes blood pressure change from time to time? Factors such as blood volume, the strength of each heart contraction, heart rate, the thickness of blood, and resistance to blood flow are all part of the answers to these questions. We explain further in the paragraphs that follow.

Blood Volume

The direct cause of blood pressure is the *volume* of blood in the vessels. The larger the volume of blood in the arteries, for example, the more pressure the blood exerts on the walls of the arteries, or the higher the arterial blood pressure.

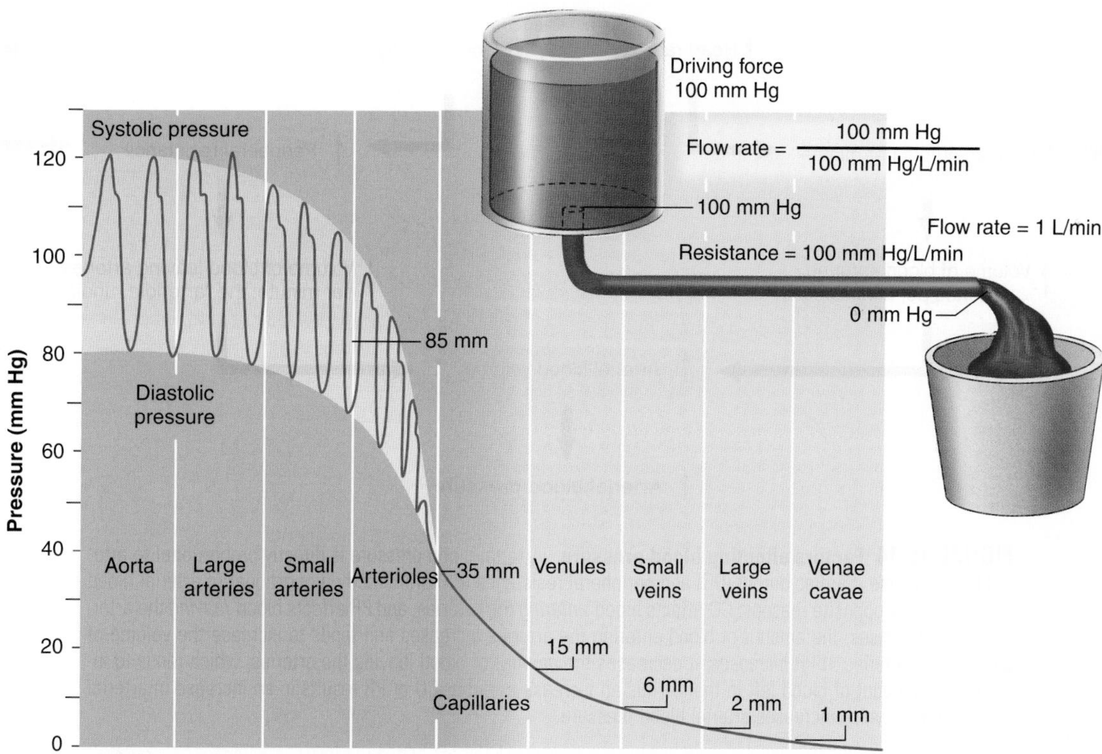

FIGURE 15-13 Pressure gradient that drives blood flow. Blood flows down a "blood pressure hill" from arteries, where blood pressure is highest, into arterioles, where it is somewhat lower, into capillaries, where it is lower still, and so on. All numbers on the graph indicate blood pressure measured in millimeters of mercury.

Conversely, the less blood in the arteries, the lower the blood pressure tends to be.

Hemorrhage demonstrates this relationship between blood volume and blood pressure. Hemorrhage is a pronounced loss of blood, and this decrease in the volume of blood causes blood pressure to drop. In fact, the major sign of hemorrhage is a rapidly falling blood pressure.

Another example is the fact that **diuretics**—drugs that promote water loss by increasing urine output—are often used to treat hypertension (high blood pressure). As water is lost from the body, blood volume decreases, and thus blood pressure decreases to a lower level.

The volume of blood in the arteries is determined by how much blood the heart pumps into the arteries and how much blood the arterioles drain out of them. The diameter of the arterioles plays an important role in determining how much blood drains out of arteries into arterioles.

Figure 15-14 summarizes some of the major factors that affect arterial blood volume, which influences arterial blood pressure, which is in turn the main factor driving continued blood flow in the body.

Strength of Heart Contractions

In the previous chapter, we discussed that the strength and the rate of the heartbeat affect **cardiac output (CO)** and therefore blood pressure.

Each time the left ventricle contracts, it squeezes a certain volume of blood (the **stroke volume [SV]**) into the aorta and on into other arteries. The stronger that each contraction is, the more blood it pumps into the aorta and arteries—that is, the SV is higher. Conversely, the weaker that each contraction is, the less blood it pumps—and the lower the stroke volume.

Suppose that one contraction of the left ventricle pumps 70 mL of blood into the aorta, and suppose that the heart beats 70 times a minute; 70 mL/beat × 70 beats/min equals 4900 mL/min. Almost 5 L of blood would enter the aorta and arteries every minute (the CO).

Now suppose that the heartbeat were to become weaker and that each contraction of the left ventricle pumps only 50 mL instead of 70 mL of blood into the aorta. If the heart still contracts just 70 times a minute, it will obviously pump much less blood into the aorta—only 3500 mL/min instead of the more normal 4900 mL/min. This decrease in the heart's CO decreases the volume of blood in the arteries, and the decreased arterial blood volume decreases arterial blood pressure.

In summary, the strength of the heartbeat affects blood pressure in this way: a stronger heartbeat increases blood pressure, and a weaker beat decreases it.

Heart Rate

The **heart rate (HR)** also may affect arterial blood pressure. You might reason that when the heart beats faster, more blood enters the aorta, and therefore the arterial blood volume and blood pressure would increase.

This is true only if the stroke volume does not decrease sharply when the heart rate increases. Often, however, when the heart beats faster, each contraction of the left ventricle

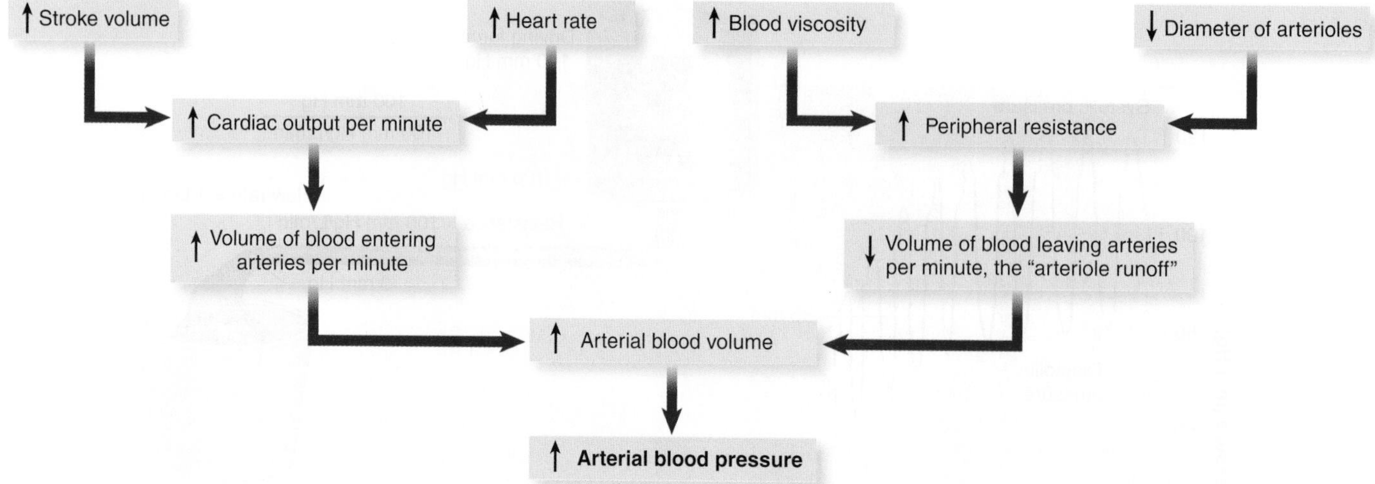

FIGURE 15-14 Factors affecting blood pressure. Arterial blood pressure is directly proportional to arterial blood volume. Cardiac output (CO) and peripheral resistance (PR) are directly proportional to arterial blood volume, but for opposite reasons: CO affects blood *entering* the arteries, and PR affects blood *leaving* the arteries. If CO increases, the amount of blood entering the arteries increases and tends to increase the volume of blood in the arteries. If PR increases, it decreases the amount of blood leaving the arteries, which tends to increase the amount of blood left in them. Thus an increase in either CO or PR results in an increase in arterial blood volume, which increases arterial blood pressure.

takes place so rapidly that it has little time to fill with blood and therefore squeezes out much less blood than usual into the aorta.

For example, suppose that the heart rate speeded up from 70 to 100 times per minute and that at the same time its stroke volume decreased from 70 mL to 40 mL. Instead of a CO of 70 × 70, or 4900 mL/min, the cardiac output would have changed to 100 × 40 or 4000 mL/min. Arterial blood volume decreases under these conditions, and therefore blood pressure also decreases, even though the heart rate has increased.

What generalization, then, can we make? We can say only that an increase in the rate of the heartbeat increases blood pressure, and a decrease in the rate decreases blood pressure. But whether a change in the heart rate actually produces a similar change in blood pressure depends on whether the stroke volume also changes and in which direction.

Blood Viscosity

Another factor that needs to be mentioned in connection with blood pressure is the *viscosity* of blood, or in plainer language, its thickness. The thicker the blood, the more resistance to flow there is—and the more blood pressure will build up. If blood becomes less viscous than normal, blood pressure decreases.

For example, if a person suffers a hemorrhage, fluid moves into the blood from the interstitial fluid. This dilutes the blood and decreases its viscosity, and blood pressure then falls because of the decreased viscosity. After hemorrhage, transfusion of whole blood or plasma is preferred to infusion of saline solution. The reason is that saline solution is not a viscous liquid and so cannot keep blood pressure at a normal level.

In a condition called **polycythemia,** discussed briefly in Chapter 13 (see **Figure 13-3** on p. 353), the number of red blood cells (RBCs) increases beyond normal and thus increases blood viscosity. This in turn increases blood pressure. An elevated RBC count also can occur when oxygen levels in the air decrease and the body attempts to increase its ability to attract oxygen to the blood—as happens when working at high altitude.

Resistance to Blood Flow

A factor that has a huge impact on local blood pressure gradients, and thus on blood flow, is any factor that changes the resistance to blood flow. The term **peripheral resistance (PR)** describes any force that acts against the flow of blood in a blood vessel. Viscosity of blood, for example, affects PR by influencing the ease with which blood flows through blood vessels.

Another factor that affects PR is the tension in smooth muscles of the blood vessel wall (**Figure 15-15**). When these muscles are relaxed, resistance is low and therefore blood pressure is low—thus blood may flow easily down its pressure gradient and into the vessel. When vessel wall muscles are contracted, however, resistance increases and therefore so does the blood pressure—thus the pressure gradient is reduced and blood will not flow so easily into the vessel.

Notice also in **Figure 15-15** that relatively minor changes in vessel diameter cause dramatic changes in blood flow. This fact means that with very slight adjustments of muscle tension in blood vessels, a wide range of different rates of blood flow can be achieved.

Such adjustment of muscle tension in vessel walls to control blood pressure, and therefore blood flow, is often called the **vasomotor mechanism.**

15

FIGURE 15-15 Vasomotor mechanism. Changes in smooth muscle tension in the wall of an arteriole influence the resistance of the vessel to blood flow. Relaxation of muscle results in decreased resistance; contraction of muscle results in increased resistance.

Decreased resistance

Diameter = 2

Flow = 256 mL/min

Smooth muscle relaxation

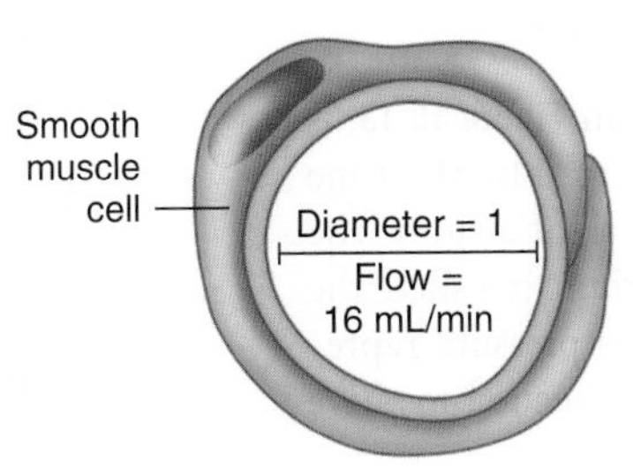

Normal resting tone

Increased resistance

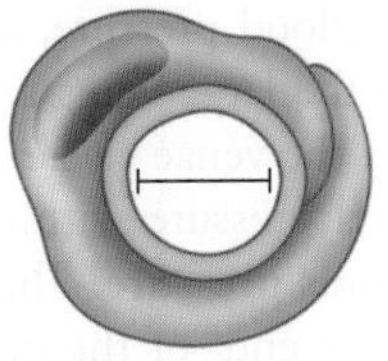

Diameter = 1/2

Flow = 1 mL/min

Smooth muscle contraction

Fluctuations in Arterial Blood Pressure

No one's blood pressure stays the same all the time. It fluctuates, even in a perfectly healthy individual. For example, it goes up when a person exercises strenuously. Not only is this normal, but the increased blood pressure serves a good purpose. It increases circulation to bring more blood to muscles each minute and thus supplies them with more oxygen and nutrients for more energy.

A normal, resting arterial blood pressure is below 120/80, or 120 mm Hg systolic pressure (maximum pressure) and

HEALTH AND WELL-BEING

CHANGES IN BLOOD FLOW DURING EXERCISE

Not only does the overall rate of blood flow increase during exercise, but also the relative blood flow through the different organs of the body changes. During exercise, blood is routed away from the kidneys and digestive organs and toward the skeletal muscles, cardiac muscle, and skin. Rerouting of blood is accomplished by contracting precapillary sphincters in some tissues (thus reducing blood flow) while relaxing precapillary sphincters in other tissues (thus increasing blood flow). How can homeostasis be better maintained as a result of these changes? One reason is that glucose and oxygen levels drop rapidly in muscles as they use up these substances to produce energy for exercising. Increased blood flow restores normal levels of glucose and oxygen more rapidly. Blood that has been warmed up in active muscles flows to the skin for cooling. This helps keep the body temperature from getting too high. Can you think of other ways this change in blood flow helps maintain homeostasis?

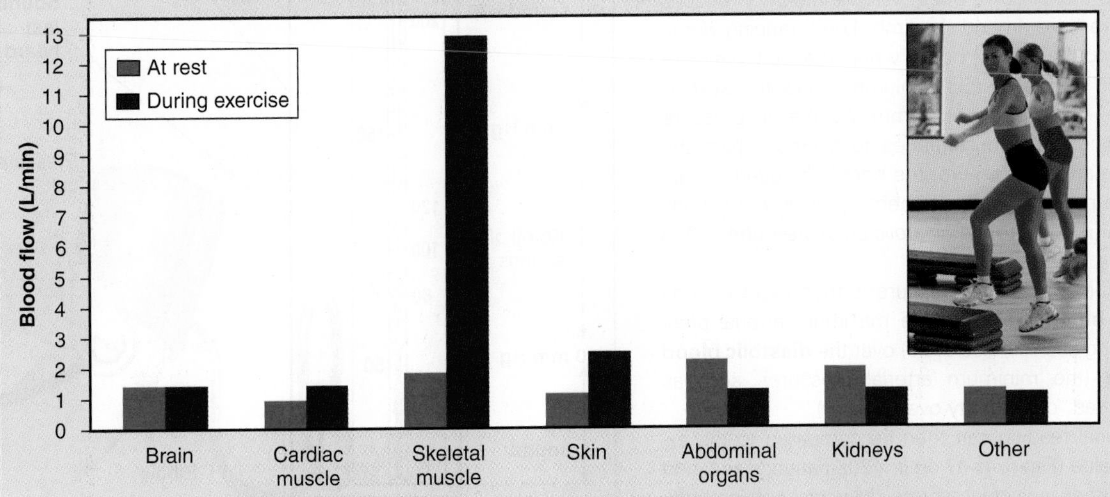

80 mm Hg diastolic pressure (minimum pressure). Remember, however, that what is "normal" varies somewhat among individuals and also varies with age.

Central Venous Blood Pressure

The venous blood pressure, as you can see in **Figure 15-13**, is very low in the large veins and falls almost to 0 by the time blood leaves the venae cavae and enters the right atrium. The venous blood pressure within the right atrium is called the **central venous pressure.** The central venous pressure represents the "low end" of the pressure gradient needed to drive blood flow all the way back to the heart.

The central venous pressure level is important because it influences the pressure that exists in the large peripheral veins. If the heart beats strongly, the central venous pressure is low as blood enters and leaves the heart chambers efficiently.

If the heart is weakened, however, central venous pressure increases, and the flow of blood into the right atrium is slowed. As a result, a person suffering heart failure, who is sitting at rest in a chair, often has distended external jugular veins as blood "backs up" in the venous network.

At least five mechanisms help to keep venous blood moving back through the cardiovascular system and back to the right atrium. They include the following:

1. Continued beating of the heart, which pumps blood through the entire cardiovascular system
2. Adequate blood pressure in the arteries, to push blood to and through the veins
3. Venous valves that ensure continued blood flow in one direction—toward the heart
4. Contraction of skeletal muscles, which squeeze veins, producing a kind of pumping action
5. Changing pressures in the chest cavity during breathing that produce a kind of pumping action in the veins in the thorax

QUICK CHECK

1. How does the blood pressure gradient explain what makes blood flow?
2. Name four factors that influence blood pressure.
3. Does a person's blood pressure stay the same all the time? Explain why this is so.

CLINICAL APPLICATION

BLOOD PRESSURE READINGS

A device called a **sphygmomanometer** is often used to measure blood pressures in both clinical and home health-care situations. The traditional sphygmomanometer is an inverted tube of mercury (Hg) with a balloonlike air cuff attached via an air hose.

The air cuff is placed around a limb, usually the subject's arm as shown in the figure. A stethoscope sensor is placed over a major artery (the brachial artery in the figure) to listen for the arterial pulse. A hand-operated pump fills the air cuff, increasing the air pressure and pushing the column of mercury higher.

While listening through the stethoscope, the operator opens the air cuff's outlet valve and slowly reduces the air pressure around the limb. Loud, tapping *Korotkoff sounds* suddenly begin when the cuff pressure measured by the mercury column equals the *systolic pressure*—often below 120 mm. As the air pressure surrounding the arm continues to decrease past the minimum arterial pressure, the Korotkoff sounds disappear. The pressure measurement at which the sounds disappear is equal to the *diastolic pressure*—often 70 to 80 mm.

The subject's blood pressure is then expressed as **systolic blood pressure** (the maximum arterial pressure during each cardiac cycle) over the **diastolic blood pressure** (the minimum arterial pressure), such as 120/80 (read "one-twenty over eighty").

The final reading can then be compared to the expected value (**Figure 15-17** on p. 420), patient's age, and various other individual factors. Mercury sphygmomanometers have been replaced in many clinical settings by nonmercury devices that similarly measure the maximum and minimum arterial blood pressures in mm Hg units. In home health-care settings, patients can often learn to monitor their own blood pressure.

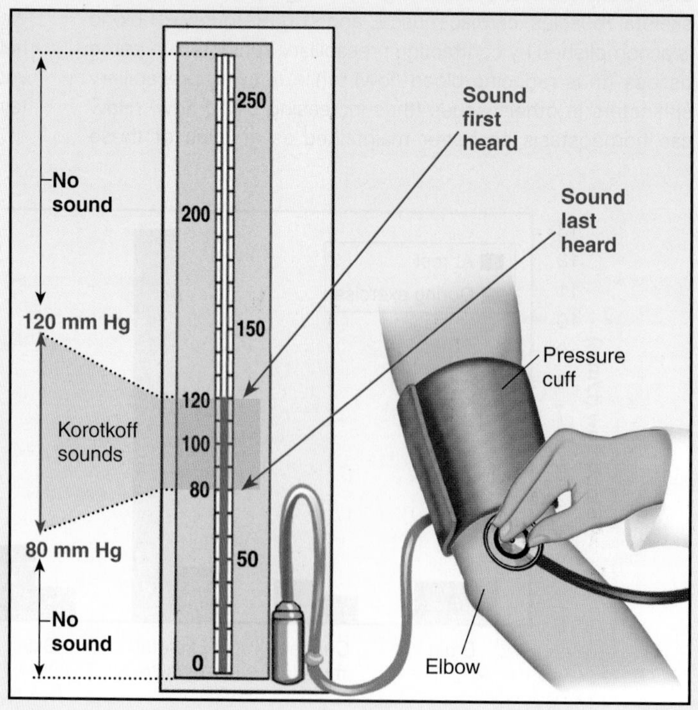

15

Pulse

What you feel when you take a **pulse** is an artery expanding and then recoiling alternately. The moving wave of expansion/recoil results from the changing arterial blood pressures that occur during the cardiac cycle. When the left ventricle ejects blood during contraction, expansion of the arterial wall results. When the aortic semilunar valve closes, and ejection ceases for a moment, the elastic arterial wall recoils.

To feel a pulse, you must place your fingertips over an artery that lies near the surface of the body and over a bone or other firm base. The pulse is a valuable clinical sign. It can provide information, for example, about the rate, strength, and rhythmicity of the heartbeat. It also can provide information about blood pressure.

Pulse is easily determined with little or no danger or discomfort. The major "pulse points" are named after the arteries over which they are felt. Locate each pulse point on **Figure 15-16** and on your own body.

Three pulse points are located on each side of the head and neck:

- Superficial temporal artery in front of the ear
- Common carotid artery in the neck along the front edge of the sternocleidomastoid muscle
- Facial artery at the lower margin of the mandible at a point below the corner of the mouth

A pulse is also detected at three points in the upper limb:

- Axillary artery in the armpit
- Brachial artery at the bend of the elbow along the inner or medial margin of the biceps brachii muscle
- Radial artery at the wrist

The so-called *radial pulse* is the most frequently monitored and easily accessible in the body.

The pulse also can be felt at four locations in the lower extremity:

- Femoral artery in the groin
- Popliteal artery behind and just proximal to the knee
- Posterior tibial artery just behind the medial malleolus (inner bump of the ankle)
- Dorsalis pedis artery on the front surface of the foot, just below the bend of the ankle joint

Hypertension

Definition

More visits to a physician's office are related to **hypertension (HTN),** or high blood pressure, than any other factor. More than 60 million cases of HTN have been diagnosed in the United States. This condition occurs when the force of blood exerted by the arterial blood vessel exceeds a blood pressure of 140/90 mm Hg.

Ninety percent of HTN cases are classified as *primary essential,* or idiopathic, with no single known causative etiology. Another classification, secondary HTN, is caused by kidney

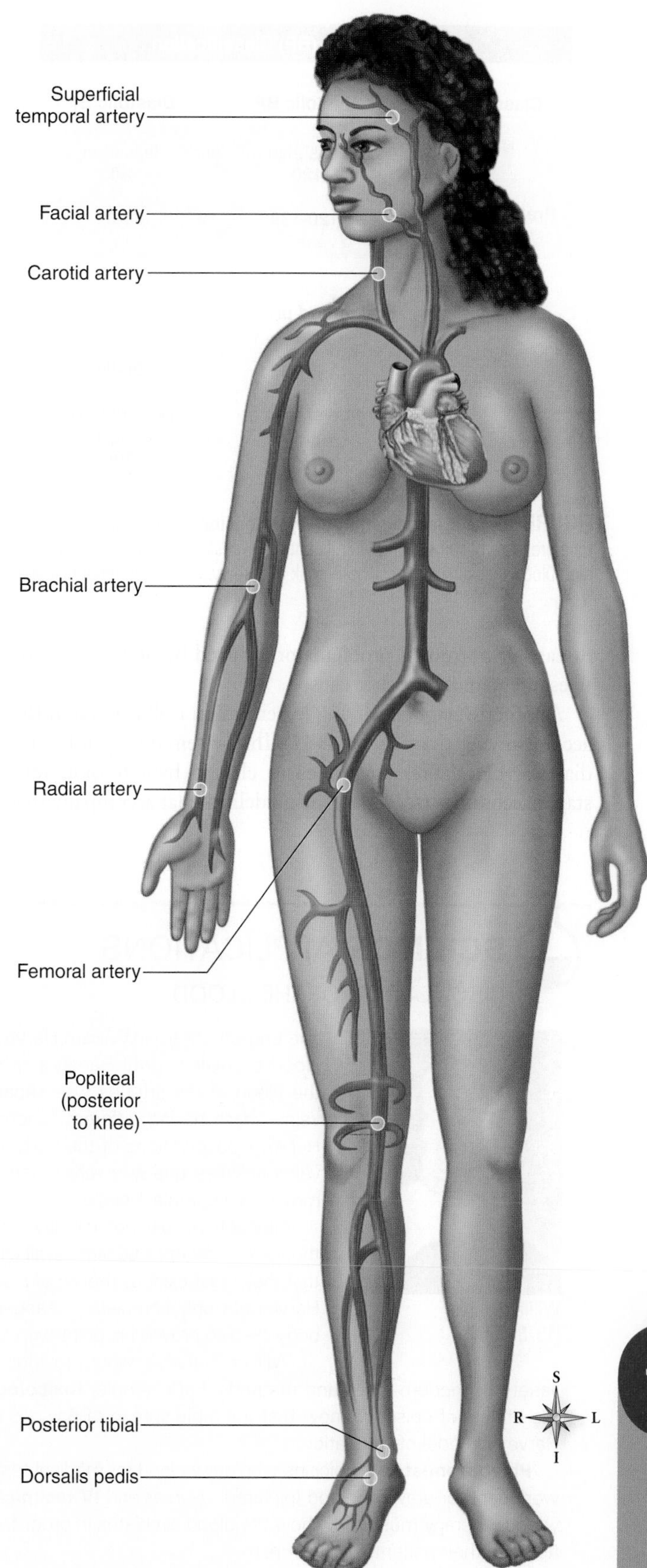

FIGURE 15-16 Pulse points. Each pulse point is named after the artery with which it is associated. External pressure applied to a pulse point can be used by first responders to slow bleeding from an injury distal to the pulse point or *pressure point.*

15

Blood Pressure (BP) Classification

Classification	Systolic BP		Diastolic BP
Normal	less than 120	and	less than 80
Prehypertension	120-139	or	80-89
High blood pressure			
Stage 1 hypertension	140-159	or	90-99
Stage 2 hypertension	greater than or equal to 160	or	greater than or equal to 100

FIGURE 15-17 Classification of hypertension. After age 50, the systolic pressure becomes more significant than diastolic pressure in assessing high blood pressure and associated risk of cardiovascular and renal disease.

disease or hormonal problems or induced by oral contraceptives, pregnancy, or other causes.

Another way of classifying hypertension is illustrated in the accompanying chart (**Figure 15-17**). This system uses systolic and diastolic blood pressure values to classify hypertension into stages according to severity. The guidelines that accompany this scheme emphasize the belief that there is no precise distinction between normal and abnormal values—thus even those in the high-normal range or the prehypertension range may be treated as having HTN.

Risk Factors

Many risk factors have been identified in the development of HTN. Genetic factors play a large role. There is an increased susceptibility or predisposition with a family history of HTN. Men experience higher rates of HTN at an earlier age than women, and HTN in African-Americans far exceeds that of Caucasians in the United States.

There is also a direct relationship between age and high blood pressure. This is because as age advances, the blood vessels become less compliant and there is a higher incidence of atherosclerotic plaque buildup. The hormone found in oral contraceptives can also cause HTN. Risk factors include high stress levels, obesity, calcium deficiencies, high levels of alcohol and caffeine intake, smoking, and lack of exercise.

Untreated HTN has many potential complications including ischemic heart disease and heart failure, kidney failure, and stroke. As many as 400,000 people per year experience a stroke. Because HTN manifests minimal or no overt signs, it is known as the "silent killer." Headaches, dizziness, and fainting have been reported but are not always symptomatic of HTN. Regular screenings at the worksite and screening

SCIENCE APPLICATIONS

CIRCULATION OF THE BLOOD

William Harvey (1578–1657)

The English physician William Harvey was the first to prove that blood circulates. Until Harvey's time, scientists believed that the blood of the arteries was separate from the blood of the veins—each having different functions in the body. However, Harvey's observations of the body, including his discovery that veins possess one-way valves, led him to discover that blood moves in a complete circle.

Although he did not directly observe the capillaries (even though microscopes became available in his day), Harvey proved that they must exist by means of a series of clever experiments. Harvey not only completely changed the way we think of the body, he also proved his point with logical experiments.

William Harvey's work provides the conceptual basis for a variety of modern ideas and methods. For example, **toxicologists** (scientists who study the effects of poisons) know that the rapid spread of poisons in the body is explained by Harvey's model of circulation.

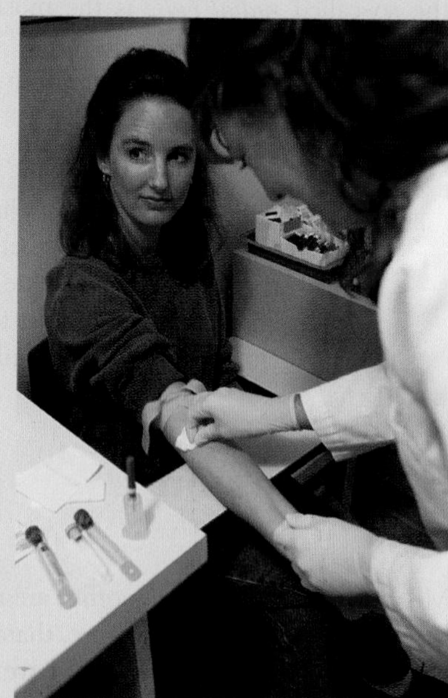

Phlebotomists, technicians who draw blood for medical tests, know which vessels will work best for drawing blood (pictured). *Nurses* and **IV technicians** that specialize in intravenous therapy must know how the blood circulates in order to effectively add therapeutic fluids to their patients' bloodstream.

Radiologists and *radiological technologists* must know which way blood flows in different vessels so that contrast dyes can be added to the bloodstream to help visualize structures of the body in an x-ray film.

Many different health professionals rely on their familiarity with the body's blood flow circuit when they measure blood pressure, inject intravenous drugs, perform surgeries, take pulses, attempt to stop bleeding after trauma to the body, and in many other medical procedures.

booths in malls and in hospitals often help identify asymptomatic HTN.

Circulatory Shock

The term **circulatory shock** refers to the failure of the circulatory system to adequately deliver oxygen to the tissues, resulting in the impairment of cell function throughout the body.

The body has a number of mechanisms that compensate for the changes that occur during shock. However, these mechanisms may fail to compensate for changes that occur in severe cases. If left untreated, circulatory shock may lead to death.

Circulatory failure has a variety of causes, all of which in some way reduce the flow of blood through the blood vessels of the body. Because of the variety of causes, circulatory shock is often classified as described in the following sections.

Cardiogenic Shock

Cardiogenic shock results from any type of heart failure, such as that after severe myocardial infarction (heart attack), heart infections, and other heart conditions. The term *cardiogenic* literally means "produced by the heart."

Because the heart can no longer pump blood effectively during heart failure, blood flow to the tissues of the body decreases or stops.

Hypovolemic Shock

Hypovolemic shock results from the loss of blood volume in the blood vessels. The term *hypovolemia* means "condition of low blood volume."

Reduced blood volume results in low blood pressure and reduced flow of blood to tissues. Hemorrhage is a common cause of blood volume loss leading to hypovolemic shock.

Hypovolemia also can be caused by loss of interstitial fluid, causing blood plasma to drain out of the vessels and into the tissue spaces. Loss of interstitial fluid is common in chronic diarrhea or vomiting, dehydration, intestinal blockage, severe or extensive burns, and some other conditions.

Neurogenic Shock

Neurogenic shock results from widespread dilation of blood vessels caused by an imbalance in autonomic stimulation of smooth muscles in vessel walls. The term *neurogenic* literally means "produced by nerves."

You may recall from Chapter 10 that autonomic effectors such as smooth muscle tissues are controlled by a balance of stimulation from the sympathetic and parasympathetic divisions of the autonomic nervous system. Normally, sympathetic stimulation maintains the muscle tone that keeps blood vessels at their usual diameter. If sympathetic stimulation is disrupted by an injury to the spinal cord or medulla, depressive drugs, emotional stress, or some other factor, blood vessels dilate significantly. Widespread vasodilation reduces blood pressure, thus reducing blood flow.

Anaphylactic Shock

Anaphylactic shock results from an acute allergic reaction called *anaphylaxis.* Anaphylaxis causes the same kind of blood vessel dilation characteristic of neurogenic shock.

Septic Shock

Septic shock results from complications of **septicemia,** a condition in which infectious agents release toxins into the blood. The toxins involved in septicemia often dilate blood vessels, thereby causing shock.

The situation is usually made worse by the damaging effects of the toxins on tissues combined with the increased cell activity caused by the accompanying fever.

One type of septic shock is *toxic shock syndrome (TSS),* which usually results from staphylococcal infections that begin in the vagina of menstruating women and spread to the blood (see Appendix A at *evolve.elsevier.com*).

QUICK CHECK

1. Where are the places on your body that you can likely feel your pulse?
2. As one gets older, why is the incidence of hypertension greater?
3. Describe four different causes of circulatory failure.

LANGUAGE OF SCIENCE *(continued from p. 403)*

foramen ovale
(foh-RAY-men oh-VAL-ee)
[*foramen* **opening,** *ovale* **egg shaped**]

heart rate (HR)
(hart rayt [aych ar])

hemodynamics
(hee-moh-dye-NAM-iks)
[*hemo-* **blood,** *-dynam-* **moving force,** *-ic* **relating to**]

hepatic portal circulation
(heh-PAT-ik POR-tall ser-kyoo-LAY-shun)
[*hepa-* **liver,** *-ic* **relating to,** *port-* **doorway,** *-al* **relating to,** *circulat-* **go around,** *-tion* **process**]

perfuse
(per-FYOOZ)
[*per-* **through,** *-fus-* **pour**]

peripheral resistance (PR)
(peh-RIF-er-al rih-ZIS-tens [pee ar])
[*peri-* **around,** *-pher-* **boundary,** *-al* **relating to,** *re-* **against,** *-sist-* **take a stand,** *-ance* **state**]

placenta
(plah-SEN-tah)
[*placenta* **flat cake**]

Continued on p. 422

LANGUAGE OF **SCIENCE** *(continued from p. 421)*

precapillary sphincter
(pree-CAP-pih-layr-ee SFINGK-ter)
[*pre-* **before,** *-capill-* **hair of head,** *-ary* **relating to,** *sphincter* **tight band**]

pulmonary circulation
(PUL-moh-nayr-ee ser-kyoo-LAY-shun)
[*pulmon-* **lung,** *-ary* **relating to,** *circulat-* **go around,** *-tion* **process**]

pulse
(puls)

stroke volume (SV)
(strowk VOL-yoom [es vee])
[*stroke* **a striking**]

systemic circulation
(sis-TEM-ik ser-kyoo-LAY-shun)
[*system-* **organized whole (body system),** *-ic* **relating to,** *circulat-* **go around,** *-tion* **process**]

systolic blood pressure
(sis-TOL-ik blud PRESH-ur)
[*sy(n)-* **together,** *-stol-* **position,** *-ic* **relating to**]

tunica externa
(TOO-nih-kah ex-TER-nah)
[*tunica* **tunic or coat,** *extern-* **outside**]

tunica intima
(TOO-nih-kah IN-tih-mah)
[*tunica* **tunic or coat,** *intima* **innermost**]

tunica media
(TOO-nih-kah MEE-dee-ah)
[*tunica* **tunic or coat,** *media* **middle**]

umbilical artery
(um-BIL-ih-kul AR-ter-ee)
[*umbilic-* **navel,** *-al* **relating to,** *arteri-* **vessel,** *-y* **thing**]

umbilical cord
(um-BIL-ih-kul)
[*umbilic-* **navel,** *-al* **relating to**]

umbilical vein
(um-BIL-ih-kul)
[*umbilic-* **navel,** *-al* **relating to,** *vein* **blood vessel**]

vasomotor mechanism
(vay-soh-MOH-tor MEK-ah-niz-em)
[*vas-* **vessel,** *-motor* **move**]

vein
(vayn)
[*vein* **blood vessel**]

venule
(VEN-yool)
[*ven-* **vessel (vein),** *-ule* **little**]

LANGUAGE OF **MEDICINE**

anaphylactic shock
(an-ah-fih-LAK-tik shok)
[*ana-* **without,** *-phylact-* **protection,** *-ic* **relating to,** *shock* **jolt**]

aneurysm
(AN-yoo-riz-em)
[*aneurysm* **widening**]

angioplasty
(AN-jee-oh-plas-tee)
[*angio-* **vessel,** *-plasty* **surgical repair**]

arteriosclerosis
(ar-tee-ree-oh-skleh-ROH-sis)
[*arteri-* **vessel (artery),** *-sclero-* **harden,** *-osis* **condition**]

cardiogenic shock
(kar-dee-oh-JEN-ik shok)
[*cardi-* **heart,** *-gen-* **produce,** *-ic* **relating to,** *shock* **jolt**]

cerebrovascular accident (CVA)
(SAYR-eh-broh-VAS-kyoo-lar)
[*cerebr-* **brain,** *-vas-* **vessel,** *-cul-* **little,** *-ar* **relating to**]

circulatory shock
(SER-kyoo-lah-tor-ee shok)
[*circulat-* **go around,** *-ory* **relating to,** *shock* **jolt**]

cyanosis
(sye-ah-NOH-sis)
[*cyan-* **blue,** *-osis* **condition**]

diuretic
(dye-yoo-RET-ik)
[*dia-* **through,** *-ure-* **urine,** *-ic* **relating to**]

gangrene
(GANG-green)
[*gangren-* **gnawing sore**]

hemorrhoid
(HEM-eh-royd)
[*hema-* **blood,** *-rrh-* **flow,** *-oid* **of or like**]

hypertension (HTN)
(hye-per-TEN-shun [aych tee en])
[*hyper-* **excessive,** *-tens-* **stretch or pull tight,** *-sion* **state**]

hypovolemic shock
(hye-poh-voh-LEE-mik shok)
[*hypo-* **under or below,** *-volem-* **volume,** *-ic* **relating to,** *shock* **jolt**]

IV (intravenous) technician
(aye-vee [in-trah-VEE-nus] tek-NISH-en)
[*intra-* **within,** *-ven-* **vein,** *-ous* **relating to,** *techn-* **art or skill,** *-ic* **relating to,** *-ian* **practitioner**]

ischemia
(is-KEE-mee-ah)
[*ische-* **hold back,** *-emia* **blood condition**]

necrosis
(neh-KROH-sis)
[*necr-* **death,** *-osis* **condition**]

neurogenic shock
(noo-roh-JEN-ik shok)
[*neuro-* **nerve,** *-gen-* **produce,** *-ic* **relating to,** *shock* **jolt**]

phlebitis
(fleh-BYE-tis)
[*phleb-* **vein,** *-itis* **inflammation**]

phlebotomist
(fleh-BOT-uh-mist)
[*phlebo-* **vein,** *-tom-* **cut,** *-ist* **agent**]

plaque
(plak)
[*plaque* **patch**]

polycythemia
(pahl-ee-sye-THEE-mee-ah)
[*poly-* **many,** *-cyt-* **cell,** *-emia* **blood condition**]

pulmonary embolism
(PUL-moh-nayr-ee EM-boh-liz-em)
[*pulmon-* **lung,** *-ary* **relating to,** *embol-* **plug,** *-ism* **condition**]

Raynaud phenomenon
(ray-NO fih-NOM-eh-nohn)
[*Maurice Raynaud* **French physician,** *phenomenon* **appearance**]

septic shock
(SEP-tik shok)
[*septi-* **putrid,** *-ic* **relating to,** *shock* **jolt**]

LANGUAGE OF **MEDICINE** *(continued from p. 422)*

septicemia
(sep-tih-SEE-mee-ah)
[*septic-* **putrid,** *-(h)em-* **blood,** *-ia* **condition**]

stent
(stent)
[*Charles Stent* **English dentist**]

sphygmomanometer
(sfig-moh-mah-NOM-eh-ter)
[*sphygmo-* **pulse,** *-mano-* **thin,** *-meter* **measure**]

thrombophlebitis
(throm-boh-fleh-BYE-tis)
[*thrombo-* **clot,** *-phleb-* **vein,** *-itis* **inflammation**]

toxicologist
(tok-sih-KOL-uh-jist)
[*toxic-* **poison,** *-log-* **words (study of),** *-ist* **agent**]

varice
(VAYR-ih-seez)
sing., varix
(VAYR-iks)
[*varic-* **swollen vein**]

varicose vein
(VAYR-ih-kohs vayn)
[*varic-* **swollen vein,** *-ose* **characterized by,** *vein* **blood vessel**]

vasodilator
(vay-so-DYE-lay-tor)
[*vaso-* **vessel or duct,** *-dilat-* **widen,** *-or* **agent**]

❑ OUTLINE SUMMARY

To download a digital version of the chapter summary for use with your device, access the ***Audio Chapter Summaries*** *online at evolve.elsevier.com.*

Scan this summary after reading the chapter to help you reinforce the key concepts. Later, use the summary as a quick review before your class or before a test.

Blood Vessels

A. Types
1. Arteries and arterioles—carry blood away from the heart and toward capillaries
2. Capillaries—carry blood from the arterioles to the venules
3. Veins and venules—carry blood toward the heart and away from capillaries

B. Structure (**Figure 15-1**)
1. Arteries
   a. Tunica intima—inner layer of endothelial cells
   b. Tunica media—smooth muscle, thick in arteries; some elastic tissue; important in blood pressure regulation
   c. Tunica externa—outer layer of fibrous connective elastic tissue, may have some elastic tissue
2. Capillaries—microscopic vessels (**Figure 15-2**)
   a. Only one layer thick—the tunica intima
   b. Precapillary sphincters in arterioles determine how much blood will flow into each bed of capillaries
3. Veins (**Figure 15-1**)
   a. Tunica intima—inner layer; venous valves prevent retrograde movement of blood (**Figure 15-3**)
   b. Tunica media—smooth muscle; thin in veins
   c. Tunica externa—heavy layer of fibrous connective tissue in many veins

C. Functions
1. Arteries and arterioles—distribute nutrients, gases, etc., carried in the blood by way of high pressure; assist in maintaining the arterial blood pressure and thus maintain blood flow
2. Capillaries—serve as exchange vessels for nutrients, wastes, gases, hormones, and fluids (a central cardiovascular function)
   a. Osmosis and filtration are major forces that drive capillary exchange (**Figure 15-4**)
   b. Outwardly directed forces are greater at arterial end of capillary, moving fluid from blood to tissue
   c. Inwardly directed forces are greater at venous end of capillary, moving fluid from tissue to blood
   d. Excess tissue fluid not returned to blood is collected by lymphatic system (see Chapter 16)
3. Veins and venules—collect blood for return to the heart; low-pressure flow of blood (compared to arteries); serve as blood reservoirs

Disorders of Blood Vessels

A. Disorders of arteries—arteries must withstand high pressure and remain free of blockage
1. Arteriosclerosis—hardening of arteries caused by calcification of fatty deposits on arterial walls (**Figure 15-5**)
   a. Thickening and calcification of arterial walls reduce flow of blood, possibly causing ischemia
   b. Ischemia may progress to necrosis (tissue death) and then gangrene

15

c. High blood levels of triglycerides and cholesterol, smoking, hypertension, advanced age, and genetic predisposition are associated factors
d. May be corrected by vasodilators (vessel-relaxing drugs) or angioplasty (mechanical widening of vessels, see **Figure 15-6**) or surgical replacement
2. Aneurysm—abnormal widening of arterial wall
a. Promotes formation of thrombi that may obstruct blood flow to vital tissues
b. Arterial walls may burst, resulting in life-threatening hemorrhaging
c. Cerebrovascular accident (CVA), or stroke—ischemia of brain tissue caused by embolism or hemorrhage

B. Disorders of veins—low-pressure vessels
1. Varicose veins (varices)—enlarged veins in which blood pools (**Figure 15-7**)
a. Hemorrhoids—varicose veins in the rectum
b. Treatments include supporting affected veins or surgical removal of veins
2. Phlebitis—vein inflammation; thrombophlebitis is accompanied by clot (thrombus) formation; may result in fatal pulmonary embolism

Routes of Circulation

A. Systemic and pulmonary circulation (**Figure 15-8**)
1. Blood circulation—refers to the flow of blood through all the vessels, which are arranged in a complete circuit or circular pattern; specific pathway to/from an organ called a *route of circulation*
2. Systemic circulation
a. Carries blood throughout the body
b. Path goes from left ventricle through aorta, smaller arteries, arterioles, capillaries, venules, venae cavae, to right atrium
3. Pulmonary circulation
a. Carries blood to and from the lungs
b. Arteries deliver deoxygenated blood to the lungs for gas exchange
c. Path goes from right ventricle through pulmonary arteries, lungs, pulmonary veins, to left atrium
4. Names of main arteries—see **Figure 15-9** and **Table 15-1**
5. Names of main veins—see **Figure 15-10** and **Table 15-2**

B. Special circulatory routes
1. Hepatic portal circulation (**Figure 15-11**)
a. Unique blood route through the liver
b. Vein (hepatic portal vein) exists between two capillary beds
c. Assists with homeostasis of blood glucose levels
2. Fetal circulation (**Figure 15-12**)
a. Refers to circulation before birth
b. Modifications required for fetus to efficiently secure oxygen and nutrients from the maternal blood
c. Unique structures include the placenta, umbilical arteries and vein, ductus venosus, ductus arteriosus, and foramen ovale
d. Failure of fetal circulation to shift to usual post-birth circulation may result in cyanosis caused by lack of oxygen

Hemodynamics

A. Defining blood pressure—push, or force, of blood in the blood vessels
1. Highest in arteries, lowest in veins (**Figure 15-13**)
2. Blood pressure gradient causes blood to circulate—liquids can flow only from areas of high pressure to areas of low pressure
3. Abnormally low blood pressure results in reduced blood flow to tissues
4. Hypertension (HTN)—high blood pressure
a. Can cause vessels to rupture
b. Can increase workload of heart, causing abnormally thickening of myocardium

B. Factors that influence blood pressure (**Figure 15-14**)
1. Blood volume
a. The larger the volume, the more pressure is exerted on vessel walls
b. Diuretics—drugs that promote water loss and thus loss of total blood volume
2. Strength of heart contractions—affects stroke volume (SV), which in turn affects cardiac output (CO); stronger heartbeat increases pressure; weaker beat decreases it
3. Heart rate (HR)—increased rate increases pressure; decreased rate decreases pressure
4. Blood viscosity (thickness)
a. Less-than-normal viscosity decreases pressure, more-than-normal viscosity increases pressure
b. Polycythemia—abnormally high hematocrit, which increases blood viscosity and thus increases blood pressure
5. Resistance to blood flow (peripheral resistance [PR])—affected by many factors, including the vasomotor mechanism (vessel muscle contraction/relaxation) (**Figure 15-15**)

C. Fluctuations in arterial blood pressure
1. Blood pressure varies within normal range
2. Normal systemic arterial blood pressure is below 120/80 at rest

D. Central venous pressure
1. Venous blood pressure within right atrium, the "low end" of the pressure gradient that drives blood flow
2. Venous return of blood to the heart depends on at least five mechanisms:
a. A strongly beating heart
b. An adequate arterial blood pressure
c. Valves in the veins
d. Pumping action of skeletal muscles as they contract
e. Changing pressures in the chest cavity caused by breathing

15

Pulse

A. Definition—alternate expansion and recoil of the blood vessel wall
B. Major pulse points named after arteries over which they are felt (**Figure 15-16**)

Hypertension (HTN)

A. Occurs when blood pressure exceeds 140/90 mm Hg (**Figure 15-17**)
B. Ninety percent of HTN cases are primary essential (idiopathic); secondary HTN can be caused by kidney disease or other causes
C. Many risk factors for HTN, including genetics, age, stress, obesity, and more
D. Untreated HTN may contribute to heart disease, kidney failure, and stroke

Circulatory Shock

A. Circulatory shock—failure of the circulatory system to deliver oxygen to the tissues adequately, resulting in cell impairment
B. When the cause is known, shock can be classified as follows:
   1. Cardiogenic shock—caused by heart failure
   2. Hypovolemic shock—caused by a drop in blood volume that causes blood pressure (and blood flow) to drop
   3. Neurogenic shock—caused by nerve condition that relaxes (dilates) blood vessels and thus reduces blood flow
   4. Anaphylactic shock—caused by a severe allergic reaction characterized by blood vessel dilation
   5. Septic shock—results from complications of septicemia (toxins in blood resulting from infection)

❑ ACTIVE LEARNING

STUDY TIPS

Consider using these tips to achieve success in meeting your learning goals.

The arteries and veins are composed of three different tissue layers. There is a difference in thickness in these layers because arteries carry blood under higher pressure. Arteries and veins carry blood in opposite directions: arteries away from the heart and veins toward the heart. Capillaries are the most important blood vessels in the system. The exchange of substances (e.g., O_2, CO_2, glucose) between the blood and the tissues, the function of the cardiovascular system, occurs in the capillaries. Because of this, the walls of the capillaries must be very thin.

1. If you are asked to learn the names and locations of specific blood vessels, make flash cards, use online resources, and use the figures in this chapter as learning tools.
2. The hepatic portal system makes more sense if you see it as a homeostatic mechanism. The liver helps keep the blood leaving the digestive system from having too high a concentration of various nutrients, such as glucose. It also has a protective function: detoxifying the blood.
3. Fetal circulation will make sense to you if you consider the environment in which the fetus is living. The blood sent to the fetus is already oxygenated and full of digested nutrients, so it does not have to go to the lungs or liver. **Figure 15-12** provides a visual that will help you remember the circulation route.
4. A liquid always moves from a higher pressure to a lower pressure, so pressure would be highest leaving the heart and lowest in the vena cava. Make a chart with the disorders of the vessels. It helps to organize them by whether they are arterial disorders or venous disorders.
5. Review the descriptions of circulatory shock found in the chapter.
6. In your study group, review the structure of the blood vessels and try to relate it to its function. Discuss hepatic portal circulation and fetal circulation in terms of their advantages or efficiencies. Go over the factors influencing blood pressure and the location of places where a pulse can be taken.
7. Refer to the Language of Science and Language of Medicine terms and review the questions and the outline summary at the end of the chapter and discuss possible test questions.

Review Questions

Write out the answers to these questions after reading the chapter and reviewing the Chapter Summary. If you simply think through the answer without writing it down, you will not retain much of your new learning.

1. List and describe the main types of blood vessels in the body.
2. Name the three tissue layers that make up arteries and veins.
3. What is arteriosclerosis?
4. Define ischemia and gangrene.
5. Describe the ways in which arteriosclerosis can be treated.
6. Define phlebitis.
7. Describe both systemic and pulmonary circulation.
8. Explain what would occur if the foramen ovale failed to close at the time of birth.
9. Name and briefly explain the four factors that influence blood pressure.
10. Identify where you could find a sinus in the blood vessels.
11. List five mechanisms that keep venous blood moving toward the right atrium.
12. What is circulatory shock? List the five types of circulatory shock.

Critical Thinking

After finishing the Review Questions, write out the answers to these more in-depth questions to help you apply your new knowledge. Go back to sections of the chapter that relate to concepts that you find difficult.

13. Explain how the formation of varicose veins is an example of a positive feedback mechanism.
14. Explain hepatic portal circulation. How is it different from normal circulation, and what advantages are gained from this type of circulation?
15. When nutrients from a meal are being absorbed, the blood in the portal vein contains a higher than normal concentration of glucose. Why does the blood, after it leaves the liver, usually have a somewhat normal blood glucose concentration?
16. Explain the differences between normal postnatal circulation and fetal circulation. Based on the environment of the fetus, explain how these differences make fetal circulation more efficient.
17. Explain the relationship between the central venous pressure and the pressure gradient.

Chapter Test

After studying the chapter, test your mastery by responding to these items. Try to answer them without looking up the answers.

1. The ________ are blood vessels that carry blood back to the heart.
2. The ________ are microscopic blood vessels where substances are exchanged between the blood and the tissues.
3. The innermost tissue layer in an artery is called the ________.
4. A (an) ________ is a section of an artery that has become abnormally widened because of a weakening of the arterial wall.
5. Systemic circulation involves moving blood throughout the body; ________ circulation involves moving blood to the lungs and back.
6. Medications that trigger smooth muscles of the arterial walls to relax and widen are called ________.
7. The two structures in the fetus that allow most of the blood to bypass the lungs are the ________ and the ________.
8. The strength of the heart contraction and blood volume are two factors that influence blood pressure. Two other factors are ________ and ________.
9. Two opposing forces influence capillary exchange. They are ________ and ________.
10. Varicose veins of the rectum are ________.
11. The term ________ refers to a systemic circulatory route that is a doorway to a second set of systemic tissues.
12. Other than the systemic and pulmonary circulation, name an example of a circulation that confirms the statement "structure fits function."

15

Match each disorder in Column A with its corresponding description or cause in Column B.

Column A

13. ________ arteriosclerosis
14. ________ necrosis
15. ________ aneurysm
16. ________ varicose veins
17. ________ hemodynamics
18. ________ cardiogenic shock
19. ________ hypovolemic shock
20. ________ septic shock
21. ________ anaphylactic shock
22. ________ neurogenic shock
23. ________ arterial pressure

Column B

a. cell death caused by ischemia
b. dilated, blood-engorged veins, usually found in the legs
c. circulatory shock caused by heart failure
d. circulatory shock that is a complication of septicemia
e. circulatory shock caused by an acute allergic reaction
f. also called "hardening of the arteries"
g. circulatory shock caused by autonomic stimulation of the smooth muscles in the blood vessels
h. circulatory shock due to loss of blood volume
i. set of processes that influence the flow of blood
j. a section of an artery that has widened due to a weakening of the arterial wall
k. major force in keeping blood flowing

Case Studies

To solve a case study, you may have to refer to the glossary or index, other chapters in this textbook, and other resources.

1. Kevin has just learned that he has hypercholesterolemia (high blood cholesterol). On the advice of his physician, he is starting a regular exercise program. How might this affect Kevin's cholesterol problem? (HINT: See Appendix A at *evolve.elsevier.com*.) What vascular disorder might Kevin develop if he is not able to correct his hypercholesterolemia?
2. Leo is a middle-aged man who has recently been experiencing pain in his legs, especially when he walks for even moderate distances. His physician tells him that he has atherosclerosis in a major artery in the affected leg. Why does this cause pain when Leo walks? What treatments might Leo's physician recommend to correct this problem? Explain how each will improve Leo's condition.
3. If balloon angioplasty is used to correct mitral valve stenosis (see Chapter 14), what route must the catheter travel if it enters at the femoral artery?
4. Jane is pregnant. Her last appointment with her obstetrician was stressful. The doctor noted that her baby's development was below average and it did not seem very active in the womb. He advised Jane that they were going to have to increase the nutrients and oxygen in the umbilical vein to ensure that her baby was receiving an adequate supply. Jane is confused. She thought that arteries not veins carried nutrients and oxygen. Could her doctor have been mistaken?

Answers to Active Learning Questions can be found online at evolve.elsevier.com.

15

Lymphatic System and Immunity

OUTLINE

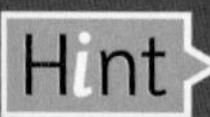

Scan this outline before you begin to read the chapter, as a preview of how the concepts are organized.

OBJECTIVES

Before reading the chapter, review these goals for your learning.

After you have completed this chapter, you should be able to:

1. Describe general functions of the lymphatic system and list the main lymphatic structures.
2. Compare innate and adaptive, inherited and acquired, and active and passive immunity.
3. Discuss the major types of immune system molecules and indicate how antibodies and complement proteins function.
4. Do the following related to immune system cells:
   - Name and describe the two primary cells of the immune systems, as well as types of each.
   - Discuss and contrast the development and functions of B and T cells.
5. Describe the mechanisms of allergy, autoimmunity, and alloimmunity.
6. List the major types of immune deficiencies and explain their causes.

CHAPTER 16

All of us live in a hostile and dangerous environment. Each day we are faced with potentially harmful toxins, disease-causing bacteria, viruses, and even cells from our own bodies that have been transformed into cancerous invaders. Fortunately, we are protected from this staggering variety of biological enemies by a remarkable set of defense mechanisms. We refer to this protective "safety net" as the **immune system.**

This system includes a set of fixed structural components, the lymphatic organs, along with a mobile group of defensive cells and molecules that protect us from infection and disease. This chapter begins with an overview of the lymphatic system, discussing a network of vessels that helps maintain fluid balance and lymphoid tissues that help defend the internal environment.

We then discuss the basic principles of immunity and the ways that highly specialized cells and molecules provide us with effective and very specific resistance to disease.

Lymphatic System

Organization of the Lymphatic System

Maintaining the constancy of the fluid around each body cell is possible only if numerous homeostatic mechanisms function together effectively in a controlled and integrated response to changing conditions. We know from Chapter 13 that the cardiovascular system plays a key role in bringing needed substances to cells and then removing the waste products that accumulate as a result of metabolism. This exchange of substances between blood and tissue fluid occurs in capillary beds. Many additional substances that cannot enter or return through the capillary walls, including excess fluid and protein molecules, are returned to the blood as **lymph.**

LANGUAGE OF **SCIENCE**

Hint Before reading the chapter, say each of these terms out loud. This will help you to avoid stumbling over them as you read.

adaptive immunity
(ah-DAP-tiv ih-MYOO-nih-tee)
[*adapt-* **fit to,** *-ive* **relating to,** *immun-* **free,** *-ity* **state**]

afferent lymphatic vessel
(AF-fer-ent lim-FAT-ik VES-el)
[*a[d]-* **toward,** *-fer-* **carry,** *-ent* **relating to,** *lymph-* **water,** *-atic* **relating to**]

agglutinate
(ah-GLOO-tin-ayt)
[*agglutin-* **glue,** *-ate* **process**]

antibody
(AN-tih-bod-ee)
[*anti-* **against**]

antibody-mediated immunity
(AN-tih-bod-ee-MEE-dee-ayt-ed ih-MYOO-nih-tee)
[*anti-* **against,** *-medi-* **middle,** *-ate* **process,** *immun-* **free,** *-ity* **state**]

antigen
(AN-tih-jen)
[*anti-* **against,** *-gen* **produce**]

antigen-presenting cell (APC)
(AN-tih-jen prih-ZENT-ing sel [ay pee see])
[*anti-* **against,** *-gen* **produce,** *present-* **place before,** *-ing* **action,** *cell* **storeroom**]

B cell
(bee sel)
[*B* **bursa-equivalent tissue,** *cell* **storeroom**]

B lymphocyte
(bee LIM-foh-syte)
[*B* **bursa-equivalent tissue,** *lympho-* **water (lymphatic system),** *-cyte* **cell**]

Continued on p. 449

16

Lymph is the excess fluid left behind by capillary exchange that drains from tissue spaces and is transported by way of **lymphatic vessels** to eventually reenter the bloodstream. Thus the lymphatic system is an important partner of the *cardiovascular system*—both vital components of the *circulatory system*.

In addition to lymph and the lymphatic vessels, the lymphatic system includes lymph nodes and lymphoid organs such as the thymus and spleen (**Figure 16-1**). Such lymphatic structures help to filter the body's fluids, removing harmful particles before they can cause significant damage to other parts of the body.

To learn more about the lymphatic system, go to AnimationDirect online at *evolve.elsevier.com*.

Lymph

Lymph forms in this way: blood plasma filters out of the capillaries into the microscopic spaces between tissue cells because of the hydrostatic pressure generated by the pumping

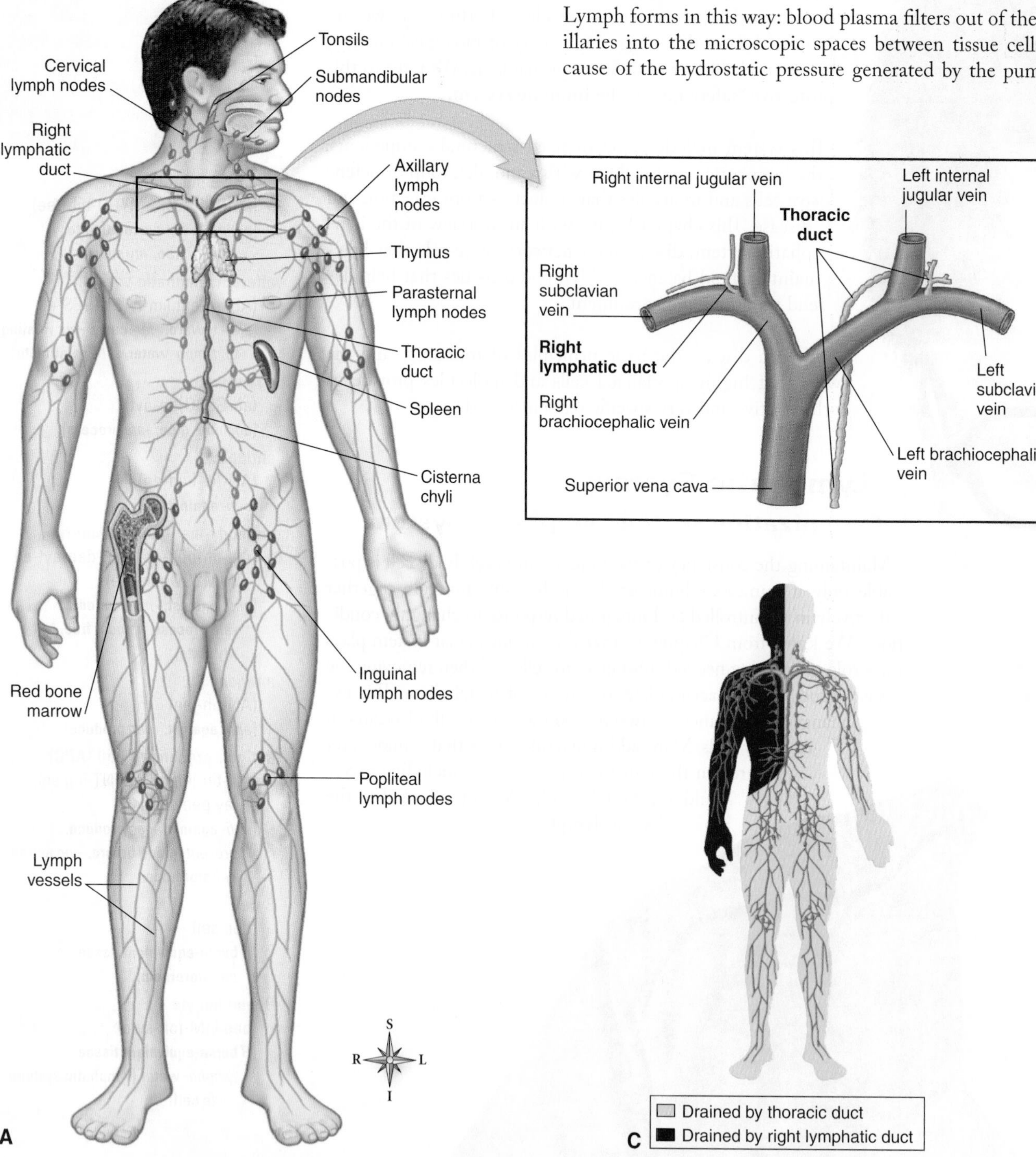

FIGURE 16-1 Lymphatic system. A, Principal organs of the lymphatic system. **B,** *Inset* showing the major lymphatic ducts draining lymphatic fluid into veins, just before systemic blood is returned to the heart. **C,** Lymph drainage. The right lymphatic duct drains lymph from the upper right quarter of the body into the right subclavian vein at its junction with the internal jugular vein. The thoracic duct drains lymph from the rest of the body into the left subclavian vein at its junction with the internal jugular vein.

action of the heart (see Figure 15-4 on p. 406). There, the liquid is called **interstitial fluid (IF),** or tissue fluid. Much of the interstitial fluid goes back into the blood by the same route it came out (that is, through the capillary membrane). The remainder of the interstitial fluid enters the lymphatic system before it returns to the blood.

The fluid, called *lymph* at this point, enters a network of tiny blind-ended tubes distributed in the tissue spaces. These tiny vessels, called **lymphatic capillaries,** permit excess tissue fluid along with some other substances such as dissolved protein molecules to leave the tissue spaces. Figure 16-2 shows how lymph forms as part of the process that maintains fluid homeostasis in the tissues of the body.

Lymphatic Vessels

Lymphatic and blood capillaries are similar in many ways. Both types of vessels are microscopic and both are formed from sheets consisting of a thin layer of simple squamous epithelium called *endothelium*. The flattened endothelial cells that form blood capillaries, however, fit tightly together so that large molecules cannot easily enter or exit from the vessel. The "fit" between endothelial cells forming the lymphatic capillaries is not as tight. As a result, they are more porous and allow larger molecules, including proteins and other substances, as well as the fluid itself, to enter the vessel and eventually return to the general circulation.

The movement of lymph in the lymphatic vessels is one way. Unlike blood, lymph does not flow over and over again through vessels that form a circular route. The lymphatic vessels often have a "beaded" appearance resulting from the presence of valves that assist in maintaining a one-way flow of lymph. These valves, similar to those in veins, sometimes cause lymph to back up behind them and cause swellings that look like beads.

Lymph flowing through the lymphatic capillaries next moves into successively larger and larger vessels sometimes called *lymphatic venules* and *lymphatic veins*. These lymphatic vessels eventually empty into one of two terminal vessels called the **right lymphatic duct** and the **thoracic duct,** which

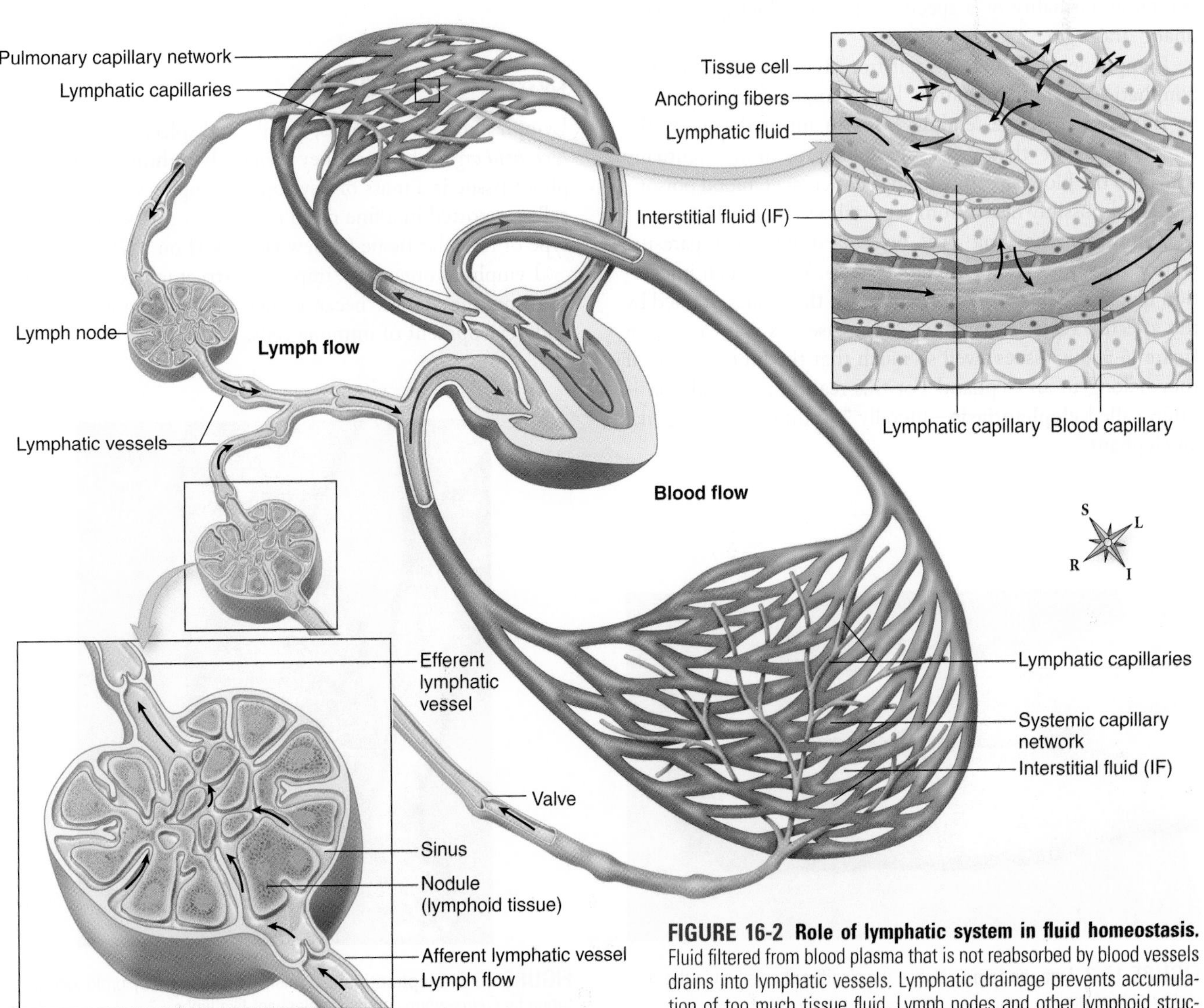

FIGURE 16-2 Role of lymphatic system in fluid homeostasis. Fluid filtered from blood plasma that is not reabsorbed by blood vessels drains into lymphatic vessels. Lymphatic drainage prevents accumulation of too much tissue fluid. Lymph nodes and other lymphoid structures filter the lymphatic fluid before it is returned to the bloodstream.

return their lymph into the blood in large veins of the neck region.

Lymph from about three-fourths of the body eventually drains into the thoracic duct, which is the largest lymphatic vessel in the body. Lymph from the right upper extremity and from the right side of the head, neck, and upper torso flows into the right lymphatic duct (see **Figure 16-1**).

Note in **Figure 16-1** that the thoracic duct in the abdomen has an enlarged pouchlike structure called the **cisterna chyli** that serves as a temporary holding area for lymph moving toward its point of entry into the veins.

Lymphatic capillaries in the wall of the small intestine are given the special name of **lacteals.** They transport fats obtained from digested food to the bloodstream and are discussed further in Chapter 18.

Lymphedema

Lymphedema is an abnormal condition in which tissues exhibit swelling (edema) because of the accumulation of lymph. Lymph may accumulate in tissue when the lymphatic vessels are partially blocked (**Figure 16-3**). This may result from a congenital abnormality or a specific injury or blockage of lymphatic drainage.

Lymphedema also may result from **lymphangitis,** that is, lymphatic vessel inflammation. Lymphangitis is characterized by thin, red streaks extending from an infected region. The infectious agent that causes lymphangitis may eventually spread to the bloodstream, causing septicemia ("blood poisoning") and possibly death from septic shock (**Figure 16-4**).

Rarely, lymphedema may be caused by small parasitic worms that infest the lymphatic vessels. When such infestation blocks the flow of lymph, edema of the tissues drained by the affected vessels occurs. In severe cases, as you can see in **Figure 16-5**, the tissues swell so much that the limbs look as if they belong to an elephant! For this reason, the condition is often called **elephantiasis**—literally "condition of being like an elephant."

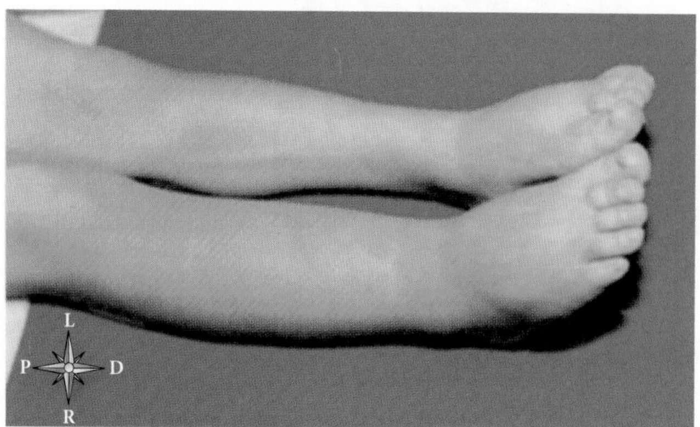

FIGURE 16-3 Lymphedema. Notice the significant swelling in the subject's right leg and foot.

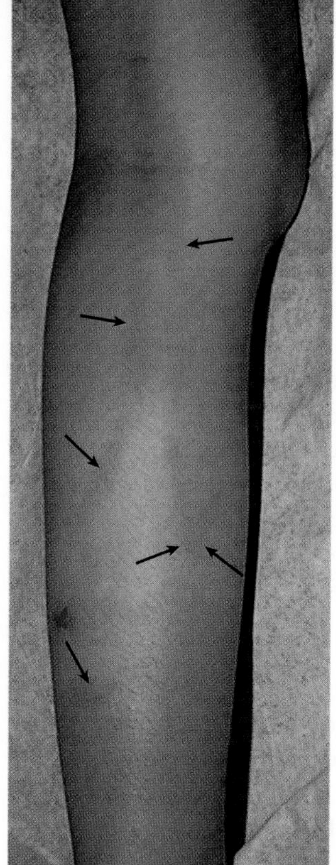

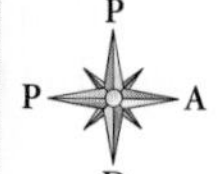

FIGURE 16-4 Lymphangitis. This condition is characterized by inflamed lymphatic vessels that appear as red streaks (highlighted by *arrows*) radiating from the source of infection.

Lymphoid Organs

Lymph nodes, the thymus, tonsils, and spleen are considered *lymphoid organs* because they contain **lymphoid tissue.** Lymphoid tissue is a mass of developing lymphocytes and related cells supported in a fine mesh of reticular fibers—making it a type of reticular tissue (review **Figure 4-11** on p. 79).

Lymphoid organs are important structural components of the immune system because they provide immune defense and development of immune cells.

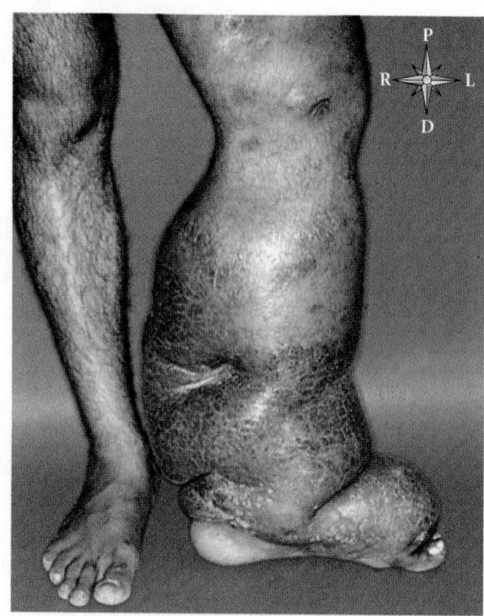

FIGURE 16-5 Elephantiasis. Lymphedema caused by prolonged infestation by *Filaria* worms produces elephant-like limbs.

To see an image of lymphoid tissue that includes the reticular fibers and developing lymphocytes in a lymph node, review *Sites of Hematopoiesis* at Connect It! at *evolve.elsevier.com*.

Lymph Nodes

Location

As lymph travels from its origin in the tissue spaces toward the thoracic or right lymphatic ducts and then into the venous blood, it is filtered by way of trickling through **lymph nodes,** which are located in clusters along the pathway of lymphatic vessels. Some of these nodes may be as small as a pinhead, and others may be as large as a lima bean.

With the exception of a relatively few single nodes, most of the larger lymph nodes occur in groups or clusters in certain areas. **Figure 16-1** shows the locations of the clusters of greatest clinical importance.

Figure 16-2 shows the structure of a typical lymph node. This structural pattern of a hollow capsule with nodules of lymphoid tissue suspended by reticular fibers is repeated in all the lymphoid organs.

To learn more about lymphatic vessels and lymph nodes, go to AnimationDirect online at *evolve.elsevier.com*.

Biological Filtration

In **Figure 16-6** a small node located next to an infected hair follicle is shown filtering bacteria from lymph. Lymph nodes perform biological filtration, a process in which cells (phagocytic cells in this case) alter the contents of the filtered fluid.

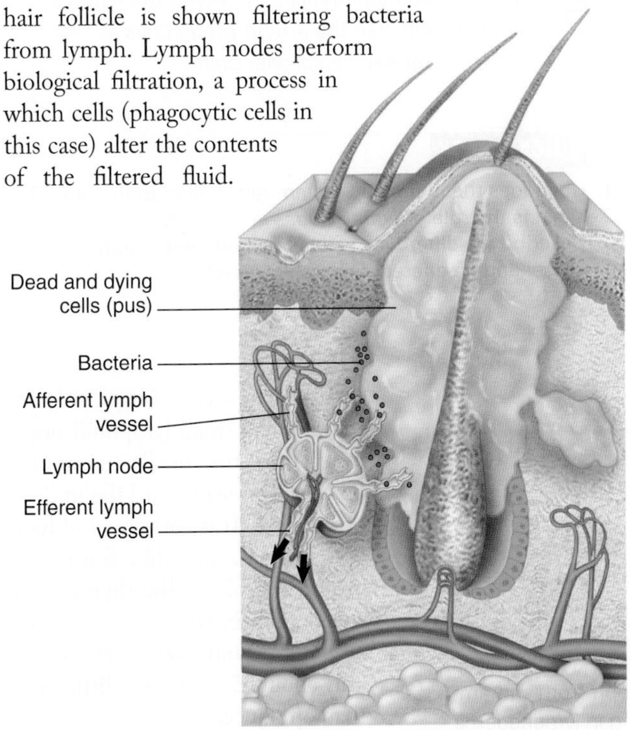

FIGURE 16-6 Lymph node function. Section of skin in which an infection surrounds a hair follicle. The *yellow areas* around the hair represent dead and dying cells (pus). The *black dots* around the yellow areas represent bacteria. Bacteria entering the node via the afferent lymphatics are filtered out.

Biological filtration of bacteria and other abnormal cells by phagocytosis helps prevent local infections from spreading.

Figure 16-2 shows that lymph enters the node through one or more **afferent lymphatic vessels.** The term *afferent* is from the Latin term for "carry toward." These vessels deliver lymph to the node.

Once lymph enters the node, it "percolates" slowly through spaces called *sinuses* that surround *nodules* found in the outer (cortex) and inner (medullary) areas of the node (see **Figure 16-2**). At the core of each nodule is a *germinal center* where new immune cells are produced.

Lymph exits from the node through one or more **efferent lymphatic vessels.** *Efferent* is from the Latin term for "carry away from."

In passing through the node, lymph is filtered so that bacteria, cancer cells, virus-infected cells, and damaged tissue cells are removed and prevented from entering the blood and circulating all over the body (see **Figure 16-6**). Lymph nodes accomplish this by a two-step process. First, debris is trapped by the web of reticular fibers that suspend the lymph nodules. Next, immune cells destroy and break apart the debris by phagocytosis and other biological processes.

Clusters of lymph nodes allow a very effective biological filtration of lymph flowing from specific body areas. **Figure 16-7** shows an x-ray image called a **lymphangiogram.** A special dye was injected into the soft tissues that drain the part of the lymphatic network that appears in the image. You can see that the dyed lymph appears in the vessels and nodes of the inguinal and pelvic regions.

Knowledge of lymph node location and function is important in clinical medicine. For example, a school nurse monitoring the progress of a child with an infected finger will watch the elbow and axillary regions for swelling and tenderness of the lymph nodes—a condition called **lymphadenitis.**

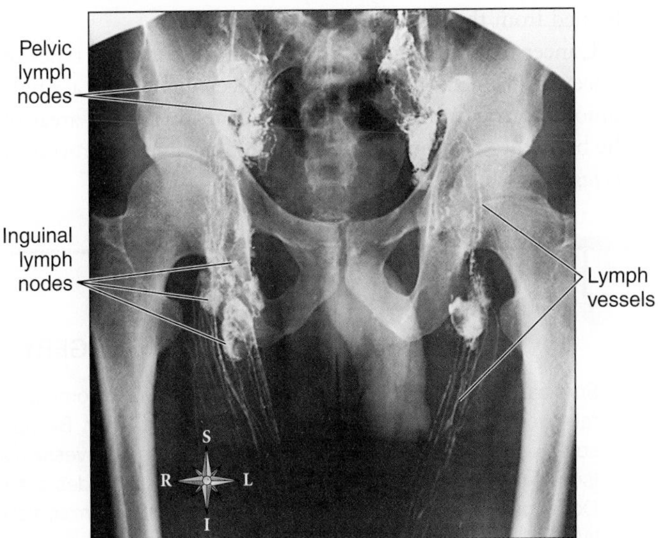

FIGURE 16-7 Lymphangiogram. A special dye that is opaque to x-rays is injected into the tissue fluids that drain into the inguinal and pelvic lymphatic pathways. Thus the outlines of the lymphatic vessels and lymph nodes can be visualized.

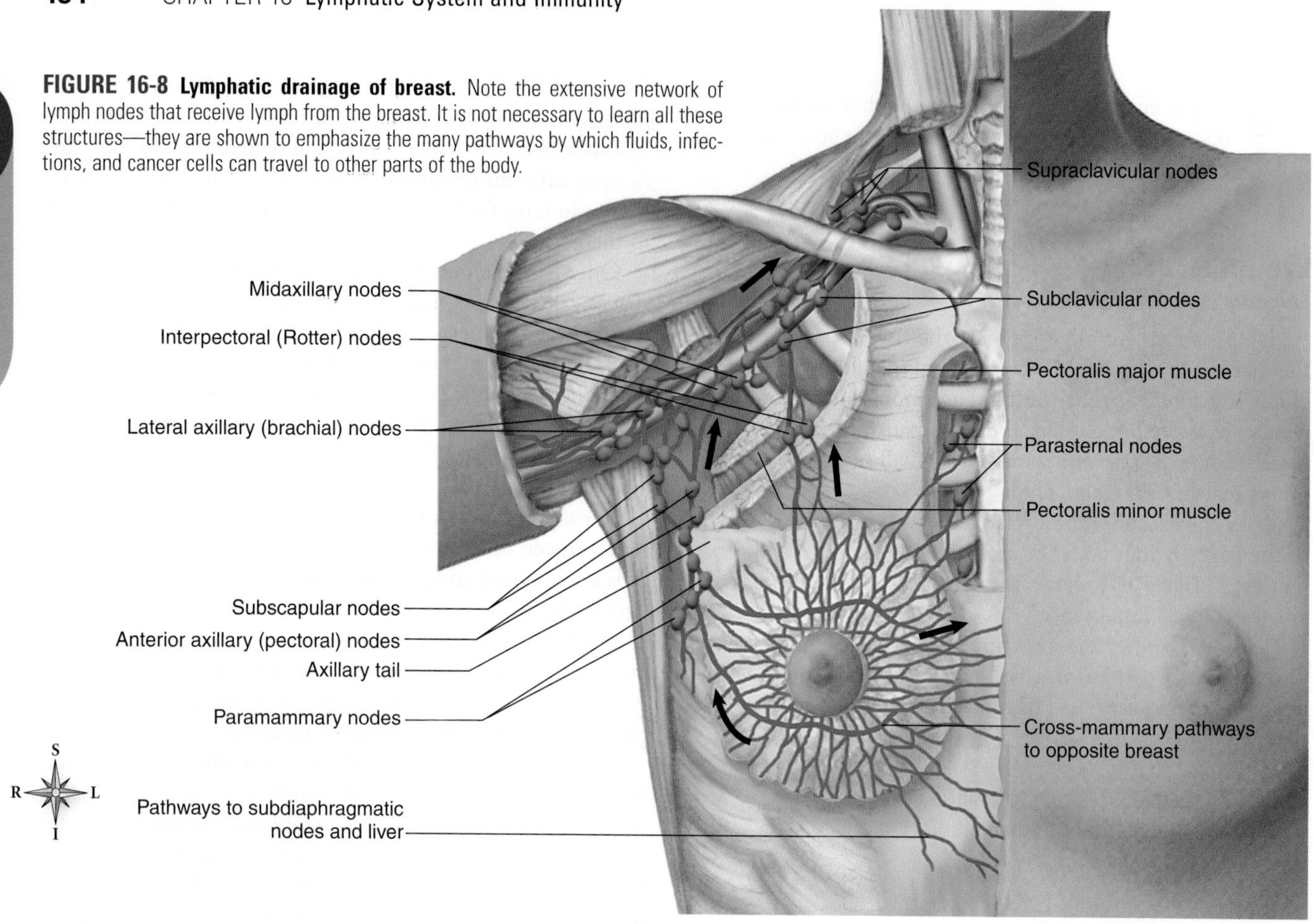

FIGURE 16-8 Lymphatic drainage of breast. Note the extensive network of lymph nodes that receive lymph from the breast. It is not necessary to learn all these structures—they are shown to emphasize the many pathways by which fluids, infections, and cancer cells can travel to other parts of the body.

These nodes filter lymph returning from the hand and may become infected by the bacteria they trap. As mentioned in the box *Lymphedema After Breast Surgery*, a surgeon uses knowledge of lymph node function when removing lymph nodes under the arms (axillary nodes) and in other nearby areas during an operation for breast cancer (**Figure 16-8**). These nodes may contain cancer cells filtered out of the lymph drained from the breast.

Cancer of the breast is one of the most common types of cancer in women. Unfortunately, cancer cells from a single tumorous growth in the breast often spread to other areas of the body through the lymphatic system during the process of *metastasis* (see **Figure 6-11** on p. 130).

For more about metastasis by way of the lymphatic system, see the article *Metastasis* at Connect It! at *evolve.elsevier.com*.

QUICK CHECK

1. How does the lymphatic system return fluid to the blood?
2. What is the role of *lymph nodes* in the body?
3. Describe the difference between an afferent lymphatic vessel and an efferent lymphatic vessel.

CLINICAL APPLICATION

LYMPHEDEMA AFTER BREAST SURGERY

Surgical procedures called **mastectomies,** in which some or all of the breast tissues are removed, are sometimes done to treat breast cancer. Because cancer cells can spread so easily through the extensive network of lymphatic vessels associated with the breast (see **Figure 16-8**), the lymphatic vessels and their nodes are sometimes also removed. Occasionally, such procedures interfere with the normal flow of lymph from the arm. When this happens, tissue fluid may accumulate in the arm—resulting in lymphedema.

In home health-care situations, the affected arm may be exercised and massaged to reduce swelling and encourage the growth of new lymphatic vessels. Some women wear an elastic sleeve that has a similar effect.

Thymus

As you can see in **Figure 16-1**, the **thymus** is a small lymphoid organ located in the mediastinum, extending upward toward the midline of the neck. It is composed of lymphocytes in a meshlike framework of reticular fibers. The thymus, also called the *thymus gland,* is largest at puberty and even then weighs only about 35 or 40 g—a little more than an ounce.

Although small in size, the thymus plays a central and critical role in the body's vital immunity mechanism. First, it is a source of lymphocytes before birth and is then especially important in the "maturation" or development of a type of lymphocyte that then leaves the thymus and circulates to the spleen, tonsils, lymph nodes, and other lymphoid tissues.

These **T lymphocytes,** or **T cells,** are critical to the functioning of the immune system and are discussed in more detail later. A group of hormones secreted by the thymus, called **thymosins,** influences the development of T cells.

The thymus appears to complete much of its work early in childhood, reaching its maximum size at puberty. The thymus tissue is then gradually replaced by fat and connective tissue, a process called *involution.* By age 60, the lymphoid tissue is about half its maximum size and is virtually gone by age 80 or so.

Tonsils

Masses of lymphoid tissue called *tonsils* are located in a protective ring under the mucous membranes in the mouth and throat (Figure 16-9). They help protect us against bacteria that may invade tissues in the area around the openings between the nasal and oral cavities.

The **palatine tonsils** are located on each side of the throat. The **pharyngeal tonsils,** known as **adenoids** when they become swollen, lie near the posterior opening of the nasal cavity. A third type of tonsil, the **lingual tonsils,** is found near the base of the tongue.

The tonsils serve as the first line of defense from the exterior and as such are subject to chronic infection, or **tonsillitis.** They may have to be removed surgically if antibiotic therapy is not successful at treating the chronic infection or if swelling impairs breathing or swallowing.

The protective lymphoid ring formed by the tonsils is one of many mechanisms that help protect the delicate tissues of the body's airways. To preview these strategies, check out the article *Protective Strategies of the Respiratory Tract* at Connect It! at *evolve.elsevier.com.*

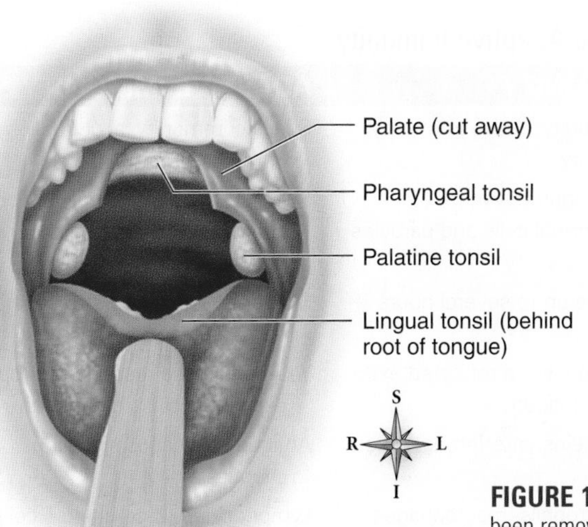

FIGURE 16-9 Location of tonsils. Small segments of the roof and floor of the mouth have been removed to show the protective ring of tonsils (lymphoid tissue) around the internal opening of the nose and throat.

Spleen

The spleen is the largest lymphoid organ in the body. As you can see in Figure 16-1 and Figure 1-6 (p. 10), it is located high in the upper left quadrant of the abdomen lateral to the stomach. Although the spleen is protected by the lower ribs, it can be injured by abdominal trauma.

The spleen has a very large network of reservoir veins and may contain more than 500 mL (about 1 pint) of blood. The spleen serves as a reservoir for blood that can be returned to the cardiovascular system when needed. If the spleen is damaged and bleeding, a surgical removal called a **splenectomy** may be required to stop the loss of blood and ensure survival.

After entering the spleen, blood flows through white, pulp-like accumulations of lymphoid tissue. As blood flows through this *white pulp,* the spleen removes bacteria and other debris by mechanical and biological filtration. The spleen also destroys worn out red blood cells (RBCs), which often fall apart when passing through the spleen's meshwork, and salvages the iron found in hemoglobin for future use.

The white pulp of the spleen also serves as a reservoir for monocytes, which can quickly leave the spleen to help repair damaged tissue anywhere in the body during an emergency.

Splenomegaly, or abnormal spleen enlargement, is observed in a variety of disorders. For example, infectious conditions such as scarlet fever, syphilis, and typhoid fever are characterized by splenomegaly. Spleen enlargement sometimes accompanies hypertension. Splenomegaly also accompanies some forms of hemolytic anemia in which red blood cells appear to be broken apart at an abnormally fast rate. Surgical removal of the spleen often cures such cases.

Although the spleen provides useful functions in maintaining the healthy stability of the body, we can survive without it if surgical removal is required to preserve our overall health.

Lymphoma

As you may recall from Chapter 6, *lymphoma* is a term that refers to lymphatic tumors. Lymphomas are most often malignant but in rare cases can be benign. The two principal categories of lymphoma are **Hodgkin disease** and **non-Hodgkin lymphoma.**

All types of lymphoma characteristically cause painless enlargements of the lymph nodes in the neck and other regions. This first sign is followed by anemia, weight loss, weakness, fever, and spread to other lymphoid tissues. In later stages, the lymphoma spreads to many other areas of the body.

When discovered early, lymphoma can be successfully treated with intensive radiation and chemotherapy. Lymphoma occurs more often in men than in women.

QUICK CHECK

1. Why is the *thymus* important for immunity?
2. What are *tonsils*? What is their function?
3. What is the role of the *spleen*?
4. Identify the two principal categories of lymphoma.

Immune System

Function of the Immune System

The body's defense mechanisms protect us from disease-causing microorganisms that invade our bodies, from foreign tissue cells that may have been transplanted into our bodies, and from our own cells when they have turned malignant or cancerous. The body's overall defense system is called the *immune system.* The immune system makes us immune—that is, able to either resist these threats to our health or free ourselves from them.

In the lymphatic system, we have seen many organs that help provide defense: lymph nodes, tonsils, thymus, and spleen. The immune system is not simply a small group of organs working together. Instead, it is an interactive network of many organs and billions of freely moving cells and trillions of free-floating molecules in many different areas of the body.

Before moving on, it is wise to remind ourselves that *leukocytes,* or *white blood cells (WBCs),* do much of the work of the immune system. Take a moment to go back and review WBC cell types in **Figure 13-1** on p. 350.

Innate Immunity

Overview

Innate immunity is maintained by mechanisms that attack any irritant or abnormal substance that threatens the internal environment. This type of immunity is called *innate* because we are born with these defenses, which do not require prior exposure to a harmful substance or threatening organism. Innate immunity is also sometimes called **nonspecific immunity** because it confers *general* protection rather than protection from specific kinds of threatening cells or chemicals.

As you can see in **Table 16-1**, the innate, nonspecific immune responses are more rapid than adaptive, specific immune responses—so they are often the "first responders" when threats occur in the body. Many of the innate immune mechanisms also trigger the specific immune mechanisms, which are slower to respond but have additional, complex strategies to help eliminate the threat.

There are many types of nonspecific immune defenses in the body, as you can see by scanning **Table 16-2**. The skin and mucous membranes, for example, are nonspecific mechanical barriers that prevent or slow entry into the body of bacteria and many other substances such as toxins and harmful chemicals. Tears and mucus also contribute to nonspecific immunity. Tears wash harmful substances from the eyes, and mucus traps foreign material that may enter through the various tracts of the body. Phagocytosis of bacteria by WBCs is also a nonspecific form of immunity.

To better understand the concept of innate immunity, use the Active Concept Map *Nonspecific Immunity* at *evolve.elsevier.com.*

Inflammatory Response

The **inflammatory response** is a set of innate responses that often occurs in the body. In the example shown in **Figure 16-10**, bacteria cause tissue damage that, in turn, triggers the release of chemical mediators from any of a variety of immune cells. Such signal molecules sent by cells are often called **cytokines.** Some of the cytokines attract WBCs to the area in the process of **chemotaxis.** Together, these factors produce the characteristic signs of inflammation: *heat, redness, pain,* and *swelling.*

These signs of inflammation are caused by increased blood flow (resulting in heat and redness) and vascular permeability (resulting in tissue swelling and the pain that it causes) in the affected region. Such changes help phagocytic WBCs and beneficial proteins reach the general area and enter the affected tissue.

Besides local inflammation, systemic inflammation may occur when the inflammation mediators—cytokines and other chemical signals—trigger responses that occur on a body-wide basis. A systemic (body-wide) inflammatory response may be manifested by a **fever**—a state of abnormally high body temperature. The elevated temperature of a low

TABLE 16-1 Innate and Adaptive Immunity

	INNATE IMMUNITY	ADAPTIVE IMMUNITY
Synonyms	Nonspecific immunity, native immunity, genetic immunity	Specific immunity, acquired immunity
Specificity	Not specific—recognizes variety of nonself or abnormal cells and particles	Specific—recognizes only specific antigens on certain cells or particles
Speed of reaction	Rapid—immediate up to several hours	Slower—several hours to several days
Memory	None—same response to repeated exposures to same antigen	Yes—enhanced response to repeated exposures to same antigen
Chemicals	Complement proteins, interferons, others	Antibodies, various signaling chemicals
Cells	Phagocytes (neutrophils, macrophages, dendritic cells)	Lymphocytes (B cells and T cells)

to moderate fever may facilitate some immune reactions and may also inhibit the reproduction of some bacteria. However, some immunologists still debate the role of fever in protecting the body.

A class of enzymes in blood plasma called *complement* can trigger a cascade of chemical reactions that literally punch holes in abnormal cells and regulate other immune mechanisms. Complement can be triggered in damaged or infected tissues by both innate, nonspecific mechanisms and by adaptive, specific immune mechanisms—as we discuss later in this chapter.

Adaptive Immunity

Overview

Adaptive immunity is able to adapt to newly encountered "enemies." It is also called **specific immunity** because it includes protective mechanisms that confer very specific protection against specific types of threatening microorganisms or other toxic materials. Adaptive immunity includes a long-term protective function called *immune memory,* which allows the immune system to effectively stop a second attack by the same specific pathogen.

In adaptive immunity, when the body is first attacked by particular bacteria or viruses, disease symptoms may occur as the body fights to destroy the threatening organism. However, if the body is exposed a second time to the same threatening organism, no serious symptoms occur because

TABLE 16-2 Mechanisms of Innate Defense

MECHANISM	DESCRIPTION
Mechanical and chemical barriers	Physical impediments to the entry of foreign cells or substances
Skin and mucous membranes	Forms a continuous wall that separates the internal environment from the external environment, preventing the entry of pathogens
Secretions	Secretions such as sebum, mucus, acids, and enzymes chemically inhibit the activity of pathogens
Inflammation	The inflammatory response isolates the pathogens and stimulates the speedy arrival of large numbers of immune cells
Fever	Fever may enhance immune reactions and inhibit pathogens
Phagocytosis	Ingestion and destruction of pathogens by phagocytic cells
Neutrophils	Granular leukocytes that are usually the first phagocytic cell to arrive at the scene of an inflammatory response
Macrophages	Monocytes that have enlarged to become giant phagocytic cells capable of consuming many pathogens; often called by more specific names when found in specific tissues of the body
Natural killer (NK) cells	Type of lymphocytes that kills any cell lacking a normal self-antigen
Complement	Group of plasma proteins (inactive enzymes) that produce a cascade of chemical reactions that ultimately causes lysis (rupture) of a foreign cell; the complement cascade can be triggered by adaptive or innate immune mechanisms
Interferon (IF)	Protein produced by cells after they become infected by a virus; inhibits the spread or further development of a viral infection (see box on p. 446)

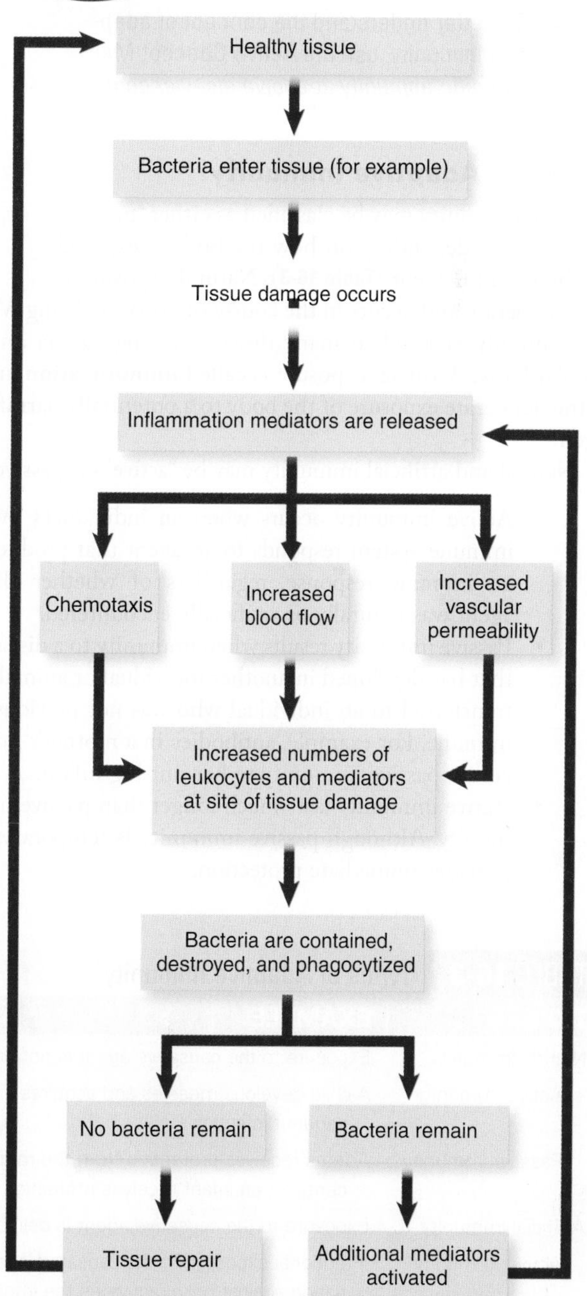

FIGURE 16-10 Inflammatory response. In this example, bacterial infection triggers a set of responses that tend to inhibit or destroy the bacteria.

16

the organism is destroyed quickly—the person is said to be *immune* to that particular organism. Such immunity is said to be specific because protection against one type of disease-causing bacteria or virus does not protect the body against others.

As **Table 16-1** shows, adaptive immune responses are slow compared to innate immune responses. However, adaptive immune responses have immune memory—the ability to produce a stronger, faster response to repeated exposure to the same antigen. **Table 16-1** summarizes other important features of both types of immunity, some of which are discussed later in this chapter.

To better understand the concept of adaptive immunity, use the Active Concept Map *Specific Immunity* at *evolve.elsevier.com.*

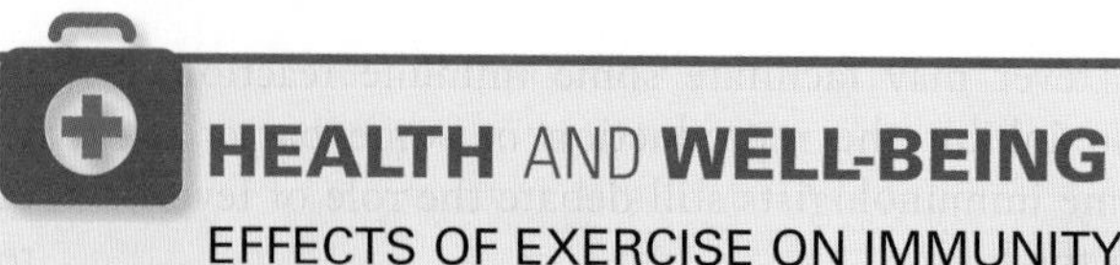

HEALTH AND WELL-BEING

EFFECTS OF EXERCISE ON IMMUNITY

Exercise physiologists have found that moderate exercise increases the number of white blood cells (WBCs), specifically granular leukocytes and lymphocytes. Not only is the number of circulating immune cells higher after exercise, but the activity of activated T cells is also increased. On the other hand, research also shows that strenuous exercise may actually inhibit immune function. Nevertheless, moderate exercise such as walking, when engaged in immediately after a trauma such as surgery, is often encouraged because of its immunity-strengthening effects.

Types of Adaptive Immunity

Specific immunity may be classified as either "natural" or "artificial" depending on how the body is exposed to the harmful agent (**Table 16-3**). Natural exposure is not deliberate and occurs in the course of everyday living. We are naturally exposed to many disease-causing agents on a regular basis. Artificial exposure is called **immunization** and is the deliberate exposure of the body to a potentially harmful agent.

Natural and artificial immunity may be "active" or "passive."

- Active immunity occurs when an individual's own immune system responds to an agent that produces an immune response, regardless of whether that agent was naturally or artificially encountered.
- Passive immunity results when immunity to a disease that has developed in another individual or animal is transferred to an individual who was not previously immune. For example, antibodies in a mother's milk confer passive immunity to her nursing infant.
- Active immunity lasts much longer than passive immunity. Although passive immunity is temporary, it provides immediate protection.

TABLE 16-3 Types of Adaptive Immunity

TYPE	EXAMPLE
Natural immunity	Exposure to the causative agent is not deliberate
Active immunity	A child develops measles and acquires an immunity to a subsequent infection
Passive immunity	A fetus receives protection from the mother through the placenta, or an infant receives protection via its mother's milk
Artificial immunity	Exposure to the causative agent is deliberate
Active immunity	Intentional exposure to the causative agent, such as a vaccination against polio, activates the immune system and thus confers immunity
Passive immunity	Injection of protective material (antibodies) that was developed by another individual's immune system

Table 16-3 lists the various forms of specific immunity and gives examples of each.

QUICK CHECK

1. What is the difference between *adaptive immunity* and *innate immunity*?
2. Outline the changes that occur in the body's *inflammatory response*.
3. Describe the difference between active immunity and passive immunity.

Immune System Molecules

The immune system functions because of adequate amounts of defensive protein molecules and protective cells. The protein molecules critical to immune system functioning include *cytokines, antibodies,* and *complement* proteins.

Cytokines

As mentioned earlier in this chapter, cytokines are chemicals released from cells to act as direct agents of innate, nonspecific immunity. They can also trigger or regulate many innate and adaptive immune responses. Often cytokines are critical to the cell-to-cell communication that is needed to coordinate the combined innate and adaptive actions that are unleashed during any immune response.

Many of the cytokines are proteins called **interleukins (ILs).** This name is apt for a substance used by WBCs to communicate between cells, because *inter-* means "between," *-leuk-* refers to leukocytes, and *-in* means "substance." ILs are often involved in signaling in both innate and adaptive immune mechanisms. For example, ILs are involved in producing a fever and in activating the cells of adaptive, specific immunity.

Antibodies

Definition

Antibodies are protein compounds that are normally present in the body. A defining characteristic of an antibody molecule is the uniquely shaped concave regions called **combining sites** on its surface. Another defining characteristic is the ability of an antibody molecule to combine with a specific molecule called an **antigen.**

All antigens are molecules that have small regions on their surfaces that are uniquely shaped to fit into the combining sites of a specific antibody molecule as precisely as a key fits into a specific lock. Antigens are often protein molecules imbedded in the surface membranes of threatening or diseased cells such as microorganisms or cancer cells.

Antibodies are also called **immunoglobulins (Igs)** and are grouped into classes named for letters of the alphabet, such as IgA, IgD, IgE, IgG, and IgM. For more discussion about immunoglobulin classes, see the article *Immunoglobulins* at Connect It! at *evolve.elsevier.com*.

Functions

In general, antibodies produce **humoral immunity,** or **antibody-mediated immunity,** by affecting the antigens in a way that prevents them from harming the body (**Figure 16-11**). To do this, an antibody must first bind to its specific antigen. This forms an *antigen-antibody complex.* The antigen-antibody complex then acts in one or more ways to make the antigen, or the cell on which it is present, harmless.

For example, if the antigen is a toxin, a substance poisonous to body cells, the toxin is neutralized or made nonpoisonous by becoming part of an antigen-antibody complex. Or if antigens are molecules in the surface membranes of threatening cells, when antibodies combine with them, the resulting antigen-antibody complexes may **agglutinate** the enemy cells (that is, make them stick together in clumps). Then macrophages or the other phagocytes can rapidly destroy them by ingesting and digesting large numbers of them at one time.

Another important function of antibodies is promotion and enhancement of phagocytosis. Certain antibody fractions help promote the attachment of phagocytic cells to the object they will engulf. As a result, the contact between the phagocytic cell and its target is enhanced, and the object is more easily ingested. This process contributes to the efficiency of immune system phagocytic cells, which is described on p. 440.

Probably the most important way in which antibodies act is a process called the **complement cascade.** Often, when antigens that are molecules on an antigenic or foreign cell's

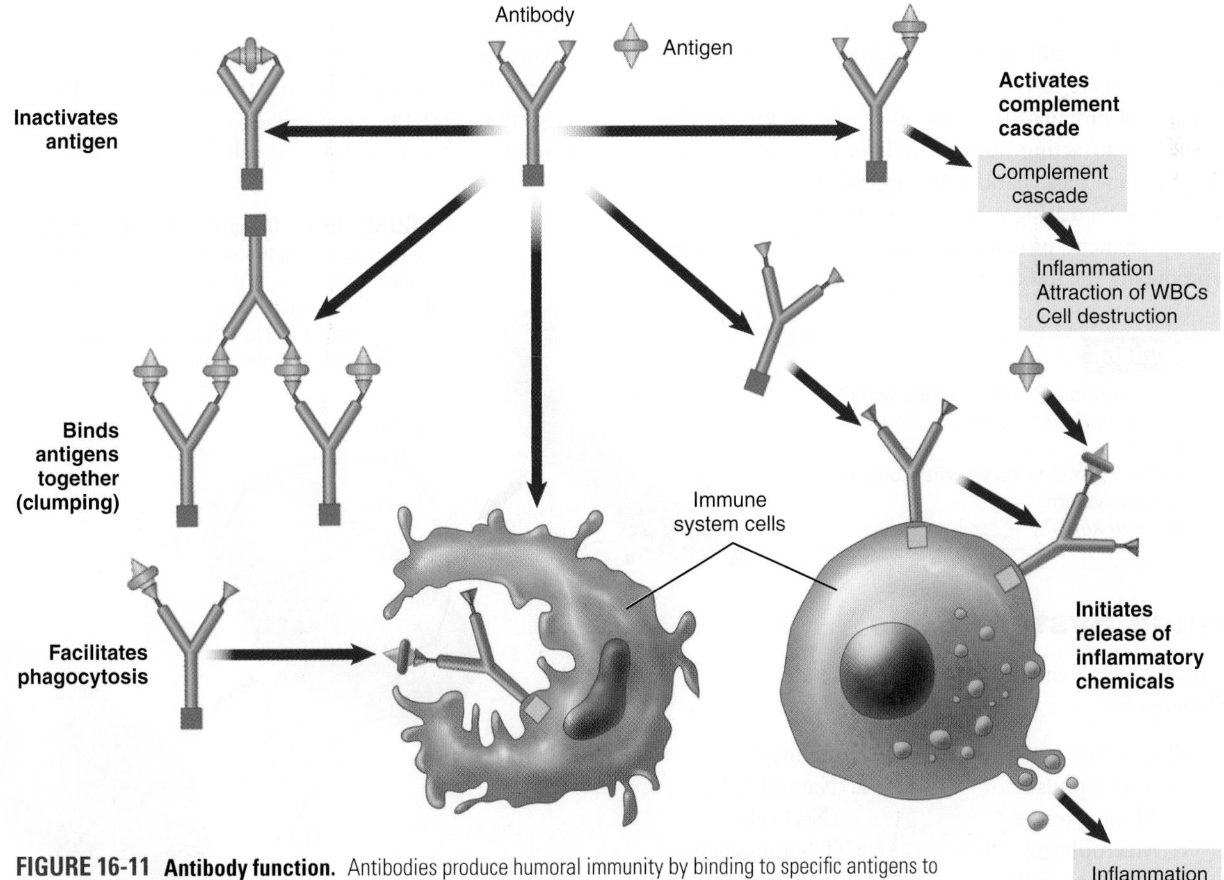

FIGURE 16-11 Antibody function. Antibodies produce humoral immunity by binding to specific antigens to form antigen-antibody complexes. These complexes produce a variety of changes that inactivate or kill threatening cells. *WBCs,* White blood cells.

surface combine with antibody molecules, they change the shape of the antibody molecule slightly, just enough to expose two previously hidden regions. These are called **complement-binding sites.** Their exposure initiates a series of events that eventually kill the cell. The next section describes these events.

Biologists can produce large quantities of pure and very specific antibodies called **monoclonal antibodies.** For details on how this medical advance works, see the article *Monoclonal Antibodies* at Connect It! at *evolve.elsevier.com.*

To learn more about antibodies and antigens, go to AnimationDirect online at *evolve.elsevier.com.*

Complement Proteins

Complement is the name used to describe a group of protein enzymes normally present in an inactive state in blood. These proteins may be activated by several triggers, including exposure of complement-binding sites on antibodies when they attach to antigens. The result is formation of highly specialized protein molecules that target foreign cells for destruction.

Recall that this process is a rapid-fire cascade or sequence of events collectively called the *complement cascade.* The end result of this process is that doughnut-shaped protein rings (complete with a hole in the middle) are formed and literally bore holes in the foreign cell!

At first, the tiny holes allow sodium to rapidly diffuse into the cell. Next, water follows sodium through the process of osmosis. The cell literally bursts as the internal osmotic pressure increases (**Figure 16-12**).

Complement proteins also serve other roles in the immune system, such as attracting immune cells to a site of infection, activating immune cells, marking foreign cells for destruction, and increasing permeability of blood vessels. Complement proteins also play a vital role in producing the inflammatory response.

QUICK CHECK

1. What are *antibodies*? How do they work?
2. What are *complement proteins*? How do they work?
3. Complement proteins serve what roles in the immune system.
4. What are *cytokines*?

Immune System Cells

The primary cells of the immune system include the following:

1. Phagocytes
   a. Neutrophils
   b. Monocytes
   c. Macrophages
   d. Dendritic cells (DCs)
2. Lymphocytes
   a. Natural killer (NK) cells
   b. T lymphocytes
   c. B lymphocytes

Phagocytes

Phagocytic WBCs are an important part of the immune system. In Chapter 13, phagocytes were described as bone marrow–derived cells that carry on phagocytosis, or ingestion and digestion, of foreign cells or particles (**Figure 16-13**).

Antibody molecules that bind to and coat certain foreign particles help macrophages function effectively. They serve as "flags" that alert the macrophage to the presence of foreign material, infectious bacteria, or cellular debris. They also help bind the phagocyte to the foreign material so that it can be engulfed more effectively.

Two important phagocytes are neutrophils and monocytes (see **Figure 13-10**, p. 361). These blood phagocytes migrate out of the blood and into the tissues in response to an infection.

Neutrophils are the most abundant immune cell in the body. Because neutrophils are needed only temporarily as phagocytes, they are short-lived in the tissues. The pus found at some infection sites is mostly dead neutrophils.

Once in the tissues, monocytes develop into phagocytic cells called **macrophages.** Most macrophages then "wander" throughout the tissues to engulf bacteria wherever they find them.

Another type of phagocytic cell is called the **dendritic cell (DC).** These highly branched (*dendrite* "branch") cells are produced in bone marrow and are released into the bloodstream (**Figure 16-14**). Some remain in the blood but many migrate to tissues in contact with the external environment—the skin, respiratory lining, digestive lining, and so on. Resident DCs in these barrier regions help protect us from threatening particles and cells.

Antibody molecules that bind to and coat certain foreign particles help macrophages function effectively. They serve as "flags" that alert the macrophage to the presence of foreign material, infectious bacteria, or cellular debris. They also help

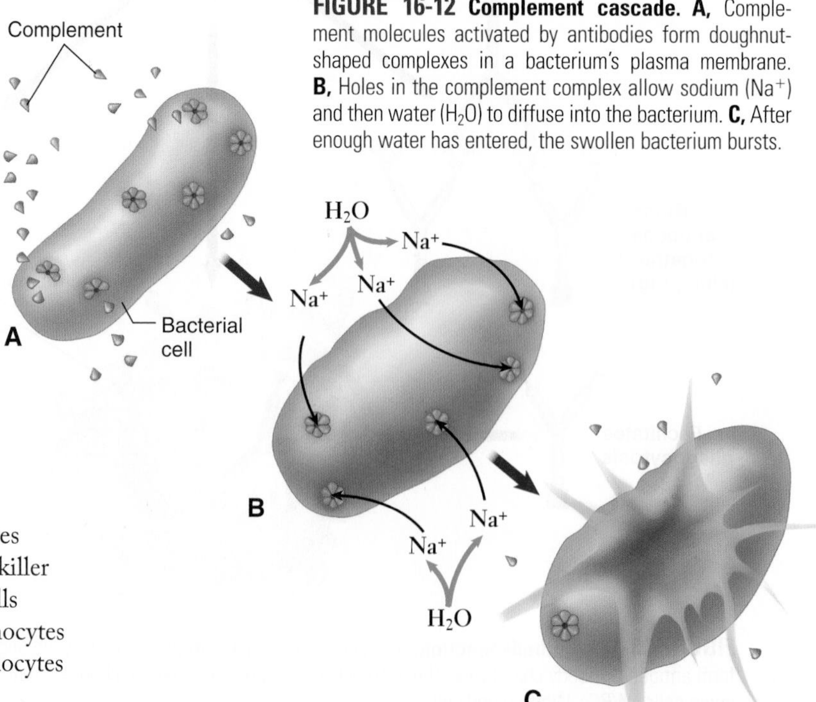

FIGURE 16-12 Complement cascade. A, Complement molecules activated by antibodies form doughnut-shaped complexes in a bacterium's plasma membrane. **B,** Holes in the complement complex allow sodium (Na^+) and then water (H_2O) to diffuse into the bacterium. **C,** After enough water has entered, the swollen bacterium bursts.

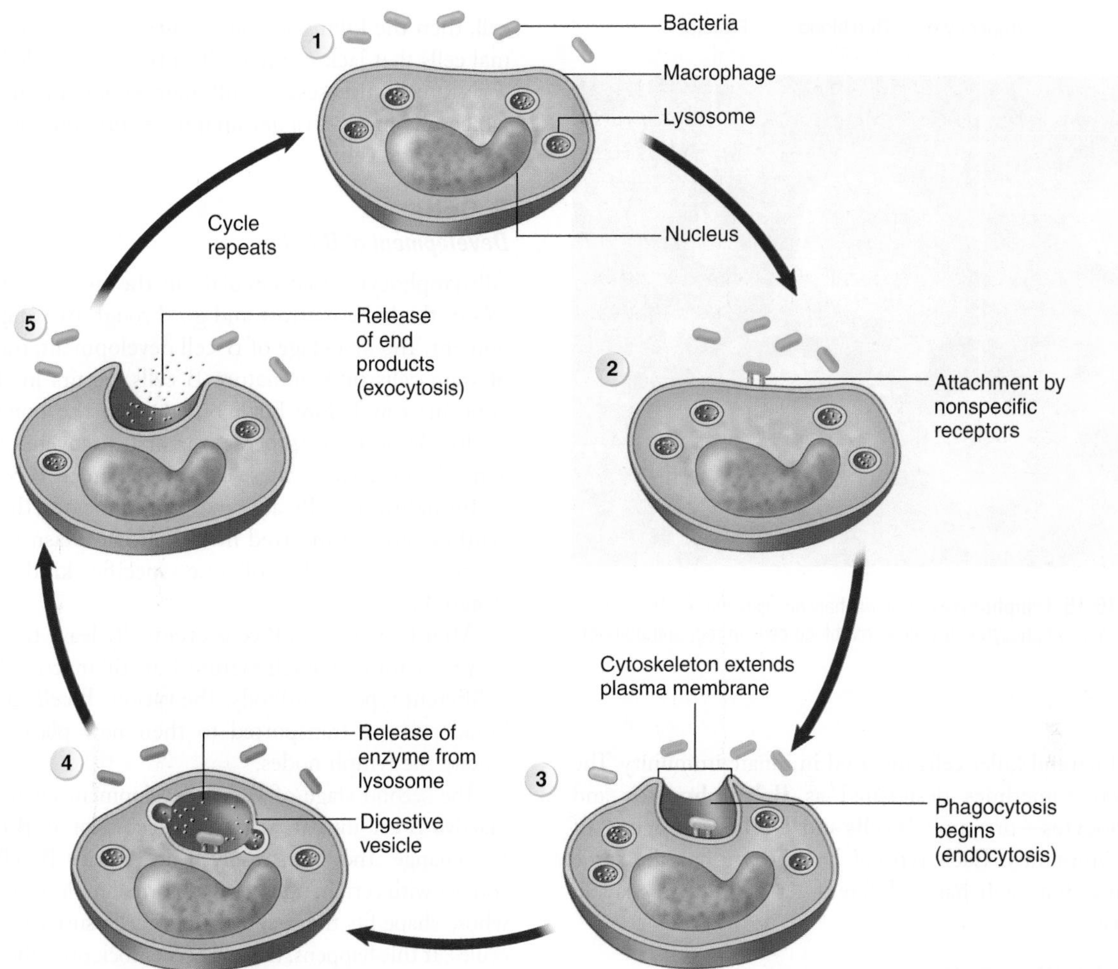

FIGURE 16-13 Phagocytosis. Drawing shows sequence of steps in phagocytosis of bacteria. The plasma membrane extends toward the bacterial cells, then envelops them. Once trapped, they are engulfed by the cell and destroyed by lysosomal enzymes.

bind the phagocyte to the foreign material so that it can be engulfed more effectively (see **Figure 16-13**).

Macrophages and DCs perform another important immune function besides destruction of threatening cells and particles. They also act as **antigen-presenting cells (APCs).** Macrophages and DCs ingest a cell or particle, remove its antigens, and display some of them on their cell surfaces. The displayed antigens can then be presented to other immune cells to trigger additional, adaptive immune responses.

FIGURE 16-14 Dendritic cell (DC). Scanning electron micrograph showing the detail of projections of the plasma membrane. DCs are phagocytic antigen-presenting cells (APCs) that are found in many areas of the body.

Lymphocytes

The most numerous cells of the immune system after neutrophils are the lymphocytes (**Figure 16-15**).

Lymphocytes are responsible for antibody production and other immune mechanisms. Lymphocytes circulate in the body's fluids. Huge numbers of them wander vigilantly throughout most of its tissues. Several million strong, lymphocytes continually patrol the body, searching out any enemy cells that may have entered or threatening virus-infected cells and cancer cells.

Developing and reserve lymphocytes densely populate the body's widely scattered lymph nodes and its other lymphoid tissues, especially the thymus gland in the chest and the spleen and liver in the abdomen.

We discuss three major types of lymphocytes here. One is the

16

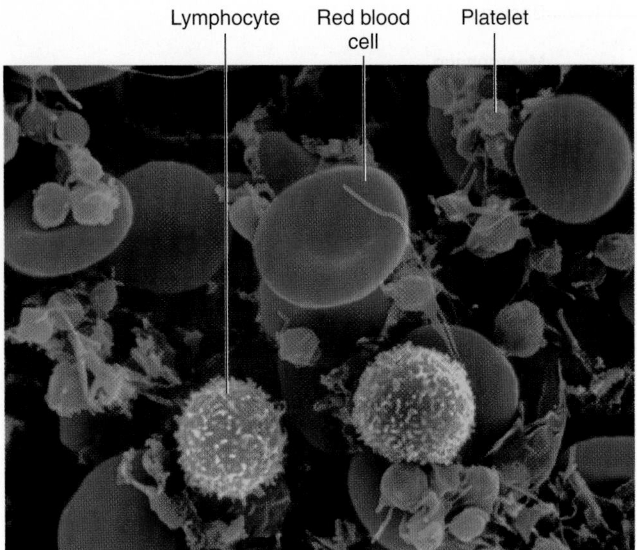

FIGURE 16-15 Lymphocytes. Color-enhanced scanning electron micrograph showing lymphocytes in yellow, red blood cells in red, and platelets in green.

so-called natural killer cells involved in innate immunity. The other two, sometimes designated as **B lymphocytes** and **T lymphocytes**—or simply **B cells** and **T cells**—are agents of adaptive immunity. Each type of lymphocyte has the same appearance, but each has a different set of roles to play in immunity.

Natural Killer Cells

The **natural killer (NK) cell** is a type of lymphocyte that kills many types of tumor cells and cells infected by different kinds of viruses. Because they have broad action, and do not require prior exposure to an antigen, NK cells qualify as agents of innate, nonspecific immunity.

NK cells recognize abnormal cells by using two different recognition receptors. One receptor acts as a "kill-activating" receptor and binds to any of several common surface molecules found on cells. Thus, it could potentially kill any cell in the body. However, the other receptor is the "kill-inhibiting" receptor. If it binds to a "self" antigen on a normal, healthy cell, then the killing action is "turned off." Thus, only abnormal cells that lack normal self-antigens are killed.

NK cells use several different killing methods, most of which chemically trigger apoptosis (programmed cell death) in the target cell.

B Cells

Development of B Cells

All lymphocytes that circulate in the tissues arise from *stem cells* in the bone marrow and go through two stages of development. The first stage of B-cell development, transformation of stem cells into immature B cells, occurs in the liver and bone marrow before birth but only in the bone marrow in adults. Because this process was first discovered in a bird organ called the *bursa,* these cells were named B cells.

Immature B cells are small lymphocytes that have each synthesized and inserted into their cytoplasmic membranes numerous molecules of one specific kind of antibody (**Figure 16-16**).

After they mature, B cells eventually leave the tissue where they were formed. Each mature, but still inactive, B cell carries a different type of antibody. The various B cells then enter the blood and are transported to their new place of residence, chiefly the lymph nodes.

The second stage of B-cell development changes a mature, inactive B cell into an activated B cell. Not all B cells undergo this change. They do so only if an inactive B cell comes into contact with certain nonself or abnormal molecules—antigens—whose shape fits the shape of the B cell's surface antibody molecules. If this happens, the antibodies lock onto the antigens and by so doing change the inactive B cell into an activated B cell. B-cell activation also requires a chemical signal (cytokine) from another immune cell—a type of T cell.

Then the activated B cell, by dividing rapidly and repeatedly, develops into groups or clones of many identical cells—all having the same type of antibody. A **clone** is a family of many identical cells, all descended from one cell.

Each clone of B cells is made up of two kinds of cells, **plasma cells** (also called **effector cells**) and **memory cells,** as you can see in **Figure 16-16**. Plasma cells secrete huge amounts of antibody into the blood—reportedly 2000 antibody molecules per second by each plasma cell for every second of the

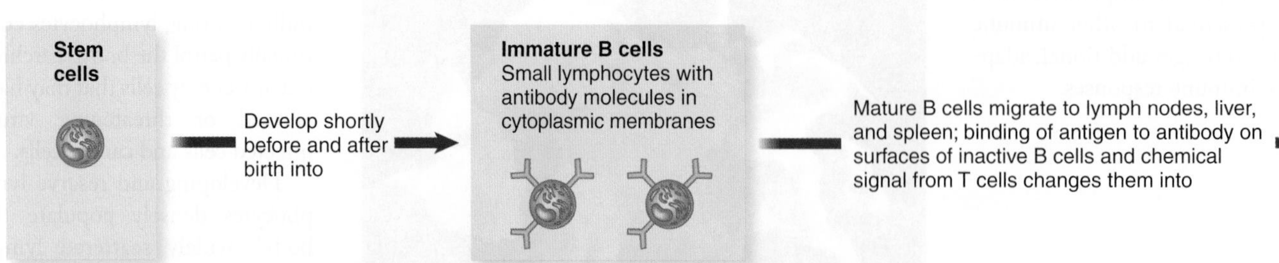

FIGURE 16-16 B-cell development. B-cell development takes place in two stages. First stage: Shortly before and after birth, stem cells develop into immature B cells, which then mature into inactive B cells that migrate to lymphoid organs. Second stage (occurs only when inactive B cell contacts its specific antigen): inactive B cell develops into activated B cell, which divides rapidly and repeatedly to form a clone of plasma cells and a clone of memory cells. Plasma cells secrete antibodies capable of combining with specific antigen that began the process. Stem cells maintain a constant population of newly differentiating inactive B cells.

few days that it lives. Antibodies circulating in the blood constitute an enormous, mobile, ever-on-duty army.

Memory cells can secrete antibodies but do not immediately do so. They remain in reserve in the lymph nodes until they are contacted by the same antigen that led to their formation. Then, the memory cells very quickly divide to produce clones of plasma cells—and more memory cells. The plasma cells secrete large amounts of antibody. Memory cells, in effect, seem to remember their ancestor activated-B cell's encounter with its appropriate antigen. They stand ready, at a moment's notice, to produce more plasma cells to release antibodies that will combine with this antigen.

Function of B Cells

B cells function indirectly to produce humoral immunity. Recall that *humoral immunity* is resistance to disease organisms produced by the actions of antibodies binding to specific antigens while circulating in body fluids. Activated B cells develop into plasma cells. Plasma cells secrete antibodies into the blood—thus serving as the "antibody factories" of the body. These antibodies, like other proteins manufactured for extracellular use, are formed in the endoplasmic reticulum of the cell.

T Cells

Development of T Cells

T cells are lymphocytes that have undergone their first stage of development in the thymus gland. Stem cells from the bone marrow seed the thymus, and shortly before and after birth, they develop into T cells. The newly formed T cells stream out of the thymus into the blood and migrate chiefly to the lymph nodes, where they take up residence.

Embedded in each T cell's cytoplasmic membrane are protein molecules shaped to fit only one specific kind of antigen molecule.

The second stage of T-cell development takes place when and if a T cell comes into contact with its specific antigen. If this happens, the antigen binds to the protein on the T cell's surface, thereby changing the T cell into an activated T cell (**Figure 16-17**).

As with B cells, T cells must also receive a chemical signal (cytokine) from another T cell to become activated. Likewise, activated T cells also produce a clone of identical cells, all able to react with the same antigen.

And as with B cells, T cells form a group of *effector cells* along with *memory cells*. The effector T cells immediately engage in immune responses, whereas the memory T cells do not. Later, if more effector T cells are needed, the memory T cells divide rapidly to produce additional clones that include more effector T cells.

Functions of T Cells

Activated T cells produce cell-mediated immunity. As the name suggests, **cell-mediated immunity** is resistance to disease organisms resulting from the actions of cells—chiefly activated T cells. One group of activated T cells kills infected cells and tumor cells directly. When bound to antigens on the abnormal cell's surface, these *cytotoxic T cells* release a substance that acts as a specific and lethal poison against the abnormal cell.

Activated T cells called *helper T cells* produce their deadly effects indirectly by means of chemical signals that they release into the area around enemy cells. Among these is a substance that attracts macrophages into the neighborhood of the enemy cells. The assembled macrophages then destroy the cells by phagocytosing (ingesting and digesting) them (**Figure 16-18**). Helper T cells also release the cytokines needed to help trigger the activation of B cells.

A third group of T cells called *regulatory T cells* helps shut down an immune reaction after the antigens have been destroyed and also helps prevent inappropriate immune reactions.

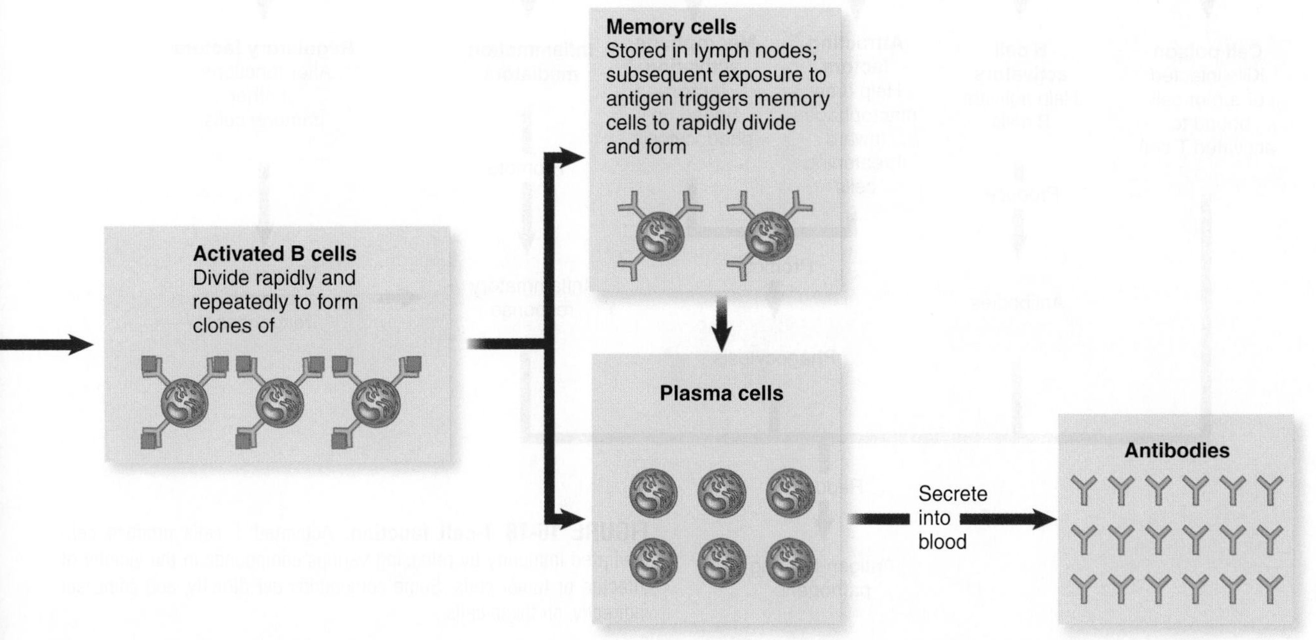

FIGURE 16-17 T-cell development. The first stage takes place in the thymus gland shortly before and after birth. Stem cells maintain a constant population of newly differentiating cells as they are needed. The second stage occurs only when a T cell contacts antigen, which combines with certain proteins on the T cell's surface.

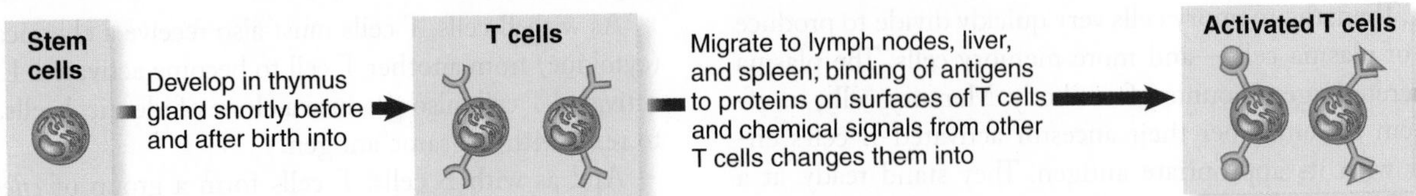

QUICK CHECK

1. What are *phagocytes*? How do they work?
2. What are *natural killer (NK) cells*?
3. What is the role of *B cells* in immunity?
4. What is the role of *T cells* in immunity?
5. What are *memory cells*? Describe their function.

Hypersensitivity of the Immune System

Hypersensitivity is an inappropriate or excessive response of the immune system. There are three types: allergy, autoimmunity, and alloimmunity.

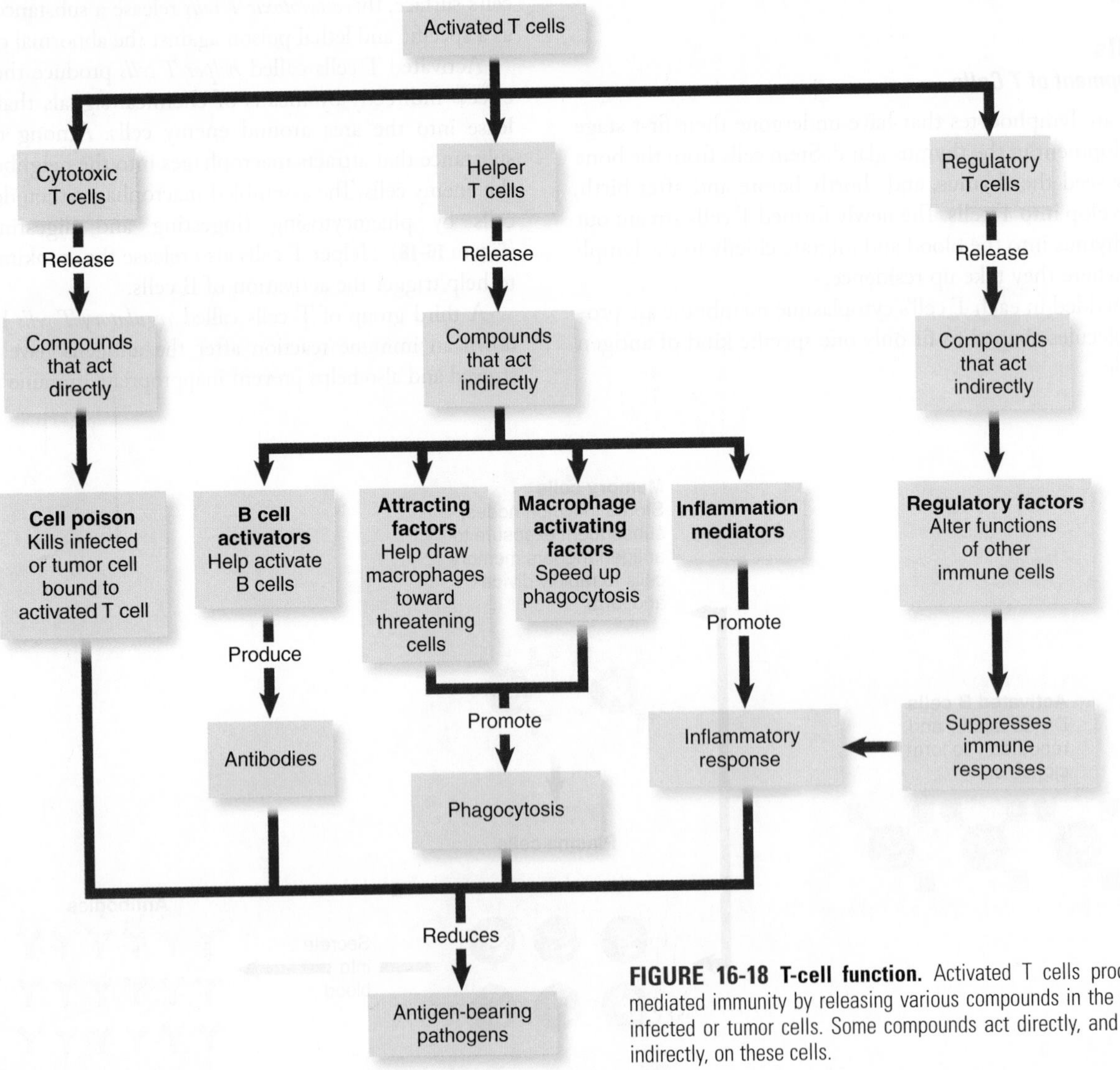

FIGURE 16-18 T-cell function. Activated T cells produce cell-mediated immunity by releasing various compounds in the vicinity of infected or tumor cells. Some compounds act directly, and some act indirectly, on these cells.

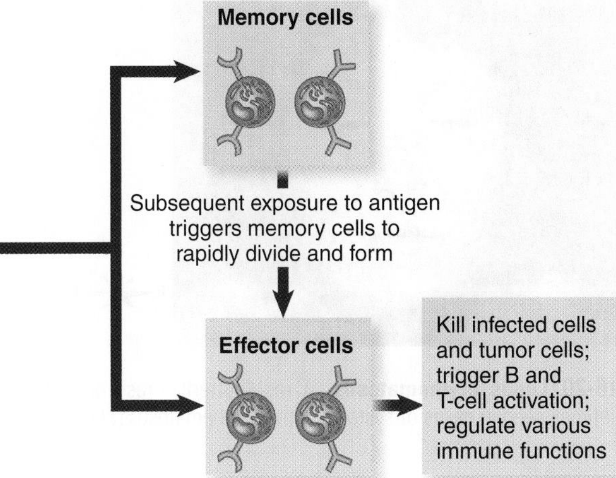

Allergy

The term **allergy** is used to describe hypersensitivity of the immune system to relatively harmless environmental antigens. Antigens that trigger an allergic response are often called **allergens.** One in six Americans has a genetic predisposition to exhibiting an allergy of some kind.

Immediate allergic responses involve antigen-antibody reactions. Before such a reaction occurs, a susceptible person must be exposed to an allergen repeatedly—triggering the production of antibodies. After a person is thus *sensitized,* exposure to an allergen causes antigen-antibody reactions that trigger the release of histamine, kinins, and other inflammatory substances. These responses usually cause typical allergy symptoms such as runny nose, conjunctivitis, and *urticaria* (hives).

In some cases of allergy, however, exposure to allergens may cause constriction of the airways, relaxation of blood vessels, and irregular heart rhythms that can progress to a life-threatening condition called **anaphylactic shock** (see Chapter 15). Drugs called *antihistamines* are sometimes used to relieve the symptoms of this type of allergy. *Epinephrine (Epi),* sometimes administered with a penlike injector carried by susceptible individuals, can also reduce severe allergic reactions.

Delayed allergic responses, on the other hand, involve cell-mediated immunity. In **contact dermatitis,** for example, T cells trigger events that lead to local skin inflammation a few hours or days after initial exposure to an antigen. Exposure to poison ivy, soaps, and certain cosmetics may cause contact dermatitis in this manner (**Figure 16-19**). Hypersensitive individuals may use *hypoallergenic* products (products without common allergens) to avoid such allergic reactions.

Autoimmunity

Autoimmunity is an inappropriate and excessive response to self-antigens. Disorders that result from autoimmune responses are called *autoimmune diseases.* Examples of autoimmune diseases are given in Table 9 of Appendix A at *evolve.elsevier.com.*

Self-antigens are molecules that are native to a person's body and that are used by the immune system to identify components of "self." Self-antigens can also be segments of a person's genetic material (DNA or RNA) or certain proteins or other chemicals made in the body. In autoimmunity, the immune system inappropriately attacks these antigens.

A common autoimmune disease is **systemic lupus erythematosus (SLE),** or simply *lupus.* Lupus is a chronic inflammatory disease that affects many tissues in the body: joints, blood vessels, kidneys, nervous system, and skin. The name *lupus erythematosus* refers to the red rash that sometimes develops on the face of those afflicted with SLE (**Figure 16-20**).

RESEARCH, ISSUES, AND TRENDS

MUCOSAL IMMUNITY

The *mucosal immune system* is a complex system of defense distinct from the systemic (internal) immune system that we have been discussing in most of this chapter. It is an innate (nonspecific) and adaptive (specific) system that is found within the mucous barriers of the body: digestive tract, urinary/reproductive tracts, respiratory tract, exocrine ducts, conjunctiva (eye covering), middle ear, and so on. The immune cells that make up the mucosal immune system are located mainly in or near *mucosa-associated lymphoid tissue (MALT).*

The main functions of the mucosal immune system involve preventing pathogens from colonizing the mucous surfaces of the body, preventing the accidental absorption of antigens from outside the body, and preventing inappropriate or intense responses of the systemic immune system to these external antigens.

Understanding the mucosal immune system and its cooperation with the systemic (internal) immune system promises to reveal new strategies of immunization. For example, researchers have found that immunizing through the bloodstream activates only the internal (systemic) B cells and T cells. Thus a pathogen would have to actually enter the internal environment before this type of adaptive immunity could protect us. Immunization of the mucosal lymphocytes, however, can activate both mucosal and systemic lymphocytes—providing a more thorough type of protection.

Another advantage of mucosal immunization is that it is easier to administer to patients than immunizations injected under the skin or into the bloodstream. For example, immunization can be delivered by nasal sprays or drops instead of "shots."

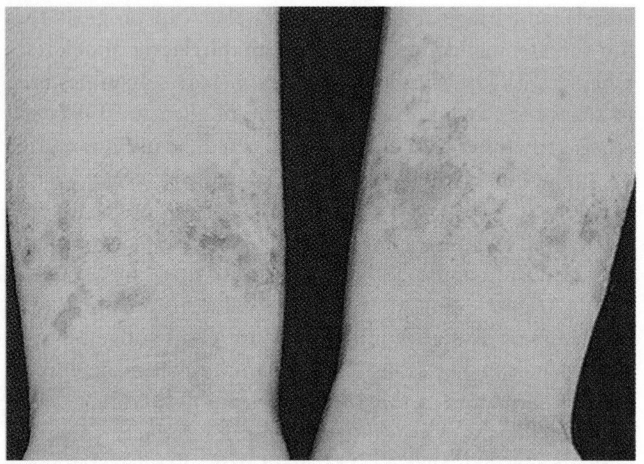

FIGURE 16-19 Contact dermatitis. Dermatitis, or skin inflammation, can result from contact with allergens—substances that trigger allergic responses in hypersensitive individuals.

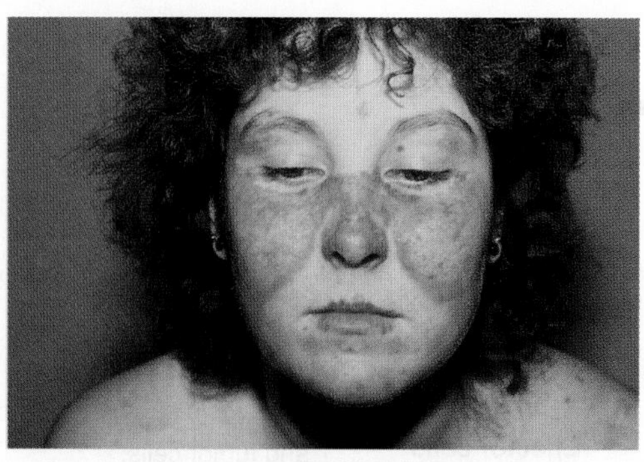

FIGURE 16-20 Lupus erythematosus. A red "butterfly" rash on the face is sometimes seen in cases of systemic lupus erythematosus (SLE).

The "systemic" part of the name comes from the fact that the disease affects many systems throughout the body. The systemic nature of SLE results from the production of antibodies against many different self-antigens.

Alloimmunity

Alloimmunity is excessive reaction of the immune system to antigens from a different individual of the same species. It is important in relation to pregnancy and tissue transplants. Alloimmunity is also sometimes called **isoimmunity.**

During pregnancy, antigens from the fetus may enter the mother's blood supply and sensitize her immune system. Antibodies that are formed as a result of this sensitization may enter the fetal circulation and cause an inappropriate immune reaction. One example, erythroblastosis fetalis, was discussed in Chapter 13.

Other pathological conditions also may be caused by damage to developing fetal tissues resulting from attack by the mother's immune system. Examples include congenital heart defects, Graves disease, and myasthenia gravis.

Tissue or organ **transplants** are medical procedures in which tissue from a donor is surgically *grafted* into the body. For example, skin grafts are often done to repair damage caused by burns. Donated whole blood tissue is often transfused into a recipient after massive hemorrhaging. A kidney is sometimes removed from a living donor and grafted into a person suffering

CLINICAL APPLICATION

INTERFERON

Interferons (IFs) are small proteins produced most often by body cells in response to viral infections. IFs play a significant role in producing innate (nonspecific) immunity to many viruses. Groups of IFs designated as alphas, betas, gammas, and omegas all have unique biological activity and are being used more and more often in clinical medicine.

One IF is produced by T cells within hours after they have been infected by a virus. This IF, when released from the T cells, protects other cells by interfering with the ability of the virus to reproduce as it moves from cell to cell. In the past, thousands of pints of blood had to be processed to harvest tiny quantities of leukocyte (T cell) IF for study.

Currently, different synthetic types of both human and non-naturally occurring IFs are being "manufactured" in bacteria as a result of gene-splicing techniques and are available in quantities sufficient for clinical use. Synthetic IF decreases the severity of many virus-related diseases including chickenpox, measles, and hepatitis. IF also shows promise as an anticancer agent. It has been shown to be effective in treating breast, skin, and other forms of cancer.

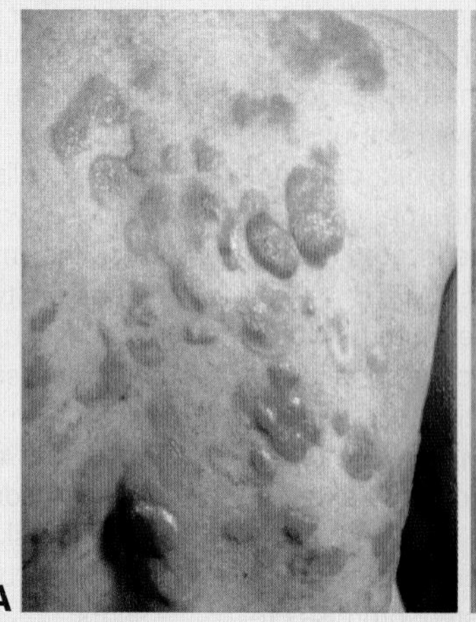

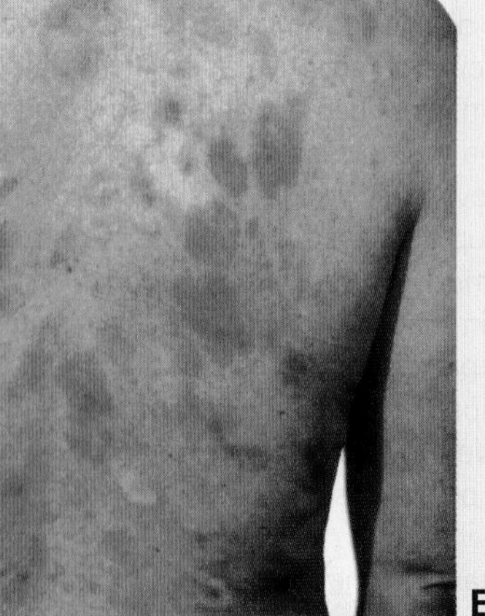

Interferon treatment. A, Cancerous skin tumors of Kaposi sarcoma (KS). **B,** After treatment with alpha interferon, the KS tumors have been reduced.

CLINICAL APPLICATION

IMMUNIZATION

Active immunity can be established artificially by using a technique called *vaccination*. The original vaccine was a live cowpox virus that was injected into healthy people to cause a mild cowpox infection. The term **vaccine** literally means "cow substance." Because the cowpox virus is similar to the deadly *smallpox* virus, vaccinated individuals developed antibodies that imparted immunity against both cowpox and smallpox viruses.

Modern vaccines work on a similar principle; substances that trigger the formation of antibodies against specific pathogens are introduced orally or by injection. Some of these vaccines are killed pathogens; some are live, attenuated (weakened) pathogens. Such pathogens still have their specific antigens intact, so they can trigger formation of the proper antibodies, but they are no longer *virulent* (able to cause disease). Although it is rare, these vaccines sometimes backfire and actually cause an infection. Many of the newer vaccines get around this potential problem by using only the part of the pathogen that contains antigens. Because the disease-causing portion is missing, such vaccines cannot cause infection. Rigorous studies have shown that vaccines are generally safe and effective.

The current recommended schedules for children and adults are available at *my-ap.us/VaccSched*

The amount of antibodies in a person's blood produced in response to vaccination or an actual infection is called the *antibody titer*. As you can see in the graph, the initial injection of vaccine triggers a rise in the antibody titer that gradually diminishes. Often, a *booster shot*, or second injection, is given to

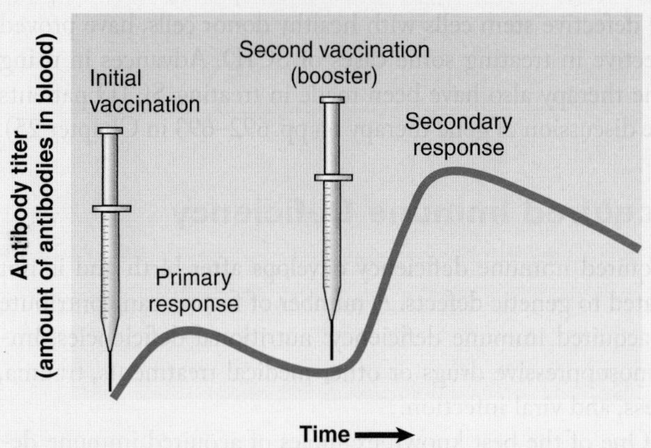

keep the antibody titer high or to raise it to a level that is more likely to prevent infection.

The secondary response is more intense than the primary response because memory B cells are standing ready to produce a large number of antibodies at a moment's notice. A later accidental exposure to the pathogen will trigger an even more intense response—thus preventing infection.

Toxoids are similar to vaccines but use an altered form of a bacterial toxin (poisonous chemical) to stimulate production of antibodies. Injection of toxoids imparts protection against toxins, whereas administration of vaccines imparts protection against pathogenic organisms and viruses.

To learn more about vaccination, go to AnimationDirect online at *evolve.elsevier.com*.

from kidney failure. Unfortunately, the immune system sometimes reacts against foreign antigens present in the grafted tissue, causing what is often called a *rejection syndrome*.

The antigens most commonly involved in transplant rejection are called **human lymphocyte antigens (HLAs)**.

Rejection of grafted tissues can occur in two ways. One is called *host-versus-graft rejection* because the recipient's immune system recognizes foreign HLAs and attacks them, destroying the donated tissue. The other is *graft-versus-host rejection* because the donated tissue (for example, bone marrow) attacks the recipient's HLAs, destroying tissue throughout the recipient's body. Graft-versus-host rejection may lead to death.

There are two ways to prevent rejection syndrome. One strategy is called *tissue typing*, in which HLAs and other antigens of a potential donor and recipient are identified. If they match, tissue rejection is unlikely to occur. Another strategy is the use of **immunosuppressive drugs** in the recipient. Immunosuppressive drugs such as *cyclosporine* and *prednisone* suppress the immune system's ability to attack the foreign antigens in the donated tissue.

QUICK CHECK

1. What is an *allergen*? How does it affect the immune system?
2. Why is *lupus* called an "autoimmune disorder"?
3. Why are immunosuppressive drugs given to organ transplant recipients?

Immune System Deficiency

Immune deficiency, or *immunodeficiency*, is the failure of immune system mechanisms in defending against pathogens. Immune system failure usually results from disruption of lymphocyte (B or T cell) function.

The chief characteristic of immune deficiency is the development of unusual or recurring severe infections or cancer. Although immune deficiency by itself does not cause death, the resulting infections or cancer can.

There are two broad categories of immune deficiencies, based on the mechanism of lymphocyte dysfunction: *congenital* and *acquired*.

Congenital Immune Deficiency

Congenital immune deficiency, which is rare, results from improper lymphocyte development before birth. Depending on the stage of development of stem cells (B or T cells) during which the defect occurs, different diseases can result.

For example, improper B-cell development can cause insufficiency or absence of antibodies in the blood. If stem cells are disrupted, a condition called **severe combined immune deficiency (SCID)** results. In most forms of SCID, humoral immunity and cell-mediated immunity are defective.

Temporary immunity can be imparted to children with SCID by injecting them with a preparation of antibodies (gamma globulin). Bone marrow transplants, which replace the defective stem cells with healthy donor cells, have proved effective in treating some cases of SCID. Advances in using gene therapy also have been made in treating SCID patients (see discussion of gene therapy on pp. 692–693 in Chapter 25).

Acquired Immune Deficiency

Acquired immune deficiency develops after birth and is not related to genetic defects. A number of factors can contribute to acquired immune deficiency: nutritional deficiencies, immunosuppressive drugs or other medical treatments, trauma, stress, and viral infection.

One of the best known examples of acquired immune deficiency is **acquired immunodeficiency syndrome (AIDS).** This syndrome (collection of symptoms) is caused by the **human immunodeficiency virus,** or **HIV.**

HIV, a retrovirus, contains RNA that undergoes reverse transcription inside affected cells to form its own DNA. The viral DNA often becomes part of the cell's DNA. When the viral DNA is activated, it directs the synthesis of its own RNA and protein coat, thus "stealing" raw materials from the cell. When this occurs in certain T cells, the cell is destroyed and immunity is impaired. As the T cell dies, it releases new retroviruses that can spread the HIV infection.

Although HIV can invade several types of cells, it has its most obvious effects in a certain type of T cell called a CD4+ T cell. When T-cell function is impaired, infectious organisms and cancer cells can grow and spread much more easily. Unusual conditions, such as *pneumocystosis* (a protozoan infection) and *Kaposi sarcoma,* or *KS* (a type of skin cancer caused by a herpes virus), also may appear. Because their immune systems are deficient, AIDS patients may eventually die from one of these infections or cancers.

After infection with HIV, an untreated person may not show signs of AIDS for months or years. This is because the immune system can hold the infection at bay for a long time before finally succumbing to it.

HIV infection has reached epidemic proportions in many countries, thus qualifying as a *pandemic.* There are several strategies for preventing the development of AIDS. Many agencies are trying to slow the spread of AIDS by educating people about how to avoid contact with the HIV retrovirus. HIV is spread by means of direct contact of body fluids, so preventing such contact reduces HIV transmission. Sexual relations, contaminated blood transfusions, and intravenous use of contaminated needles are common modes of HIV transmission. HIV can also be a **perinatal infection,** that is, an infection passing from mother to infant during birth.

Many researchers are working on HIV vaccines. Like many viruses, such as those that cause the common cold, HIV changes rapidly enough to make development of a vaccine difficult at best.

Another way to inhibit the progress of an HIV infection is by means of chemicals such as azidothymidine (AZT) and ritonavir (Norvir) that block HIV's ability to reproduce within infected cells. A "cocktail" of several antiviral drugs working together greatly reduces the number of virus particles in a patient's blood—thus reducing the effects of HIV infection. More than a hundred such compounds in various combinations are being evaluated for use in halting the progress of HIV infections.

QUICK CHECK

1. What is the difference between *congenital* and *acquired* immune deficiency?
2. What causes AIDS?
3. How can the progress of HIV infection be inhibited?

SCIENCE APPLICATIONS

VACCINES

Edward Jenner (1749–1823)

English surgeon Edward Jenner changed the world forever in 1789 when he inoculated his young son and two others against the terrible viral disease, smallpox. Using material from the blisters of a patient with the milder disease swinepox, he was able to trigger immunity to smallpox—the world's first vaccination. Later, in 1796, he found that vaccination with material from cowpox blisters worked even better in protecting people from smallpox.

A disease that had formerly killed millions upon millions of people worldwide eventually disappeared from the human population in the twentieth century because of Jenner's pioneering efforts.

In this century, interest in smallpox vaccinations has resurfaced because of the threat of smallpox as a weapon. Immunologists are at work improving on this important vaccine to protect people against such weapons. They also continue to work on vaccines for other infectious diseases such as HIV infection, new strains of influenza, Ebola, Zika, and even disorders such as heart disease and cancer.

Many health professionals use vaccines in their practice, of course, to boost the immune systems of their clients. Recently, misconceptions about the safety of vaccines threaten public health by reducing the number of children protected by vaccines that saved a whole generation from the devastating effects of measles, polio, smallpox, diphtheria, and more.

Many physicians also treat disorders of the immune system itself. For example, immune deficiencies such as AIDS, allergies such as "hay fever," and autoimmune disorders such as lupus and rheumatoid arthritis, are treated every day by physicians and other health professionals.

LANGUAGE OF **SCIENCE** *(continued from p. 429)*

cell-mediated immunity
(sel-MEE-dee-ayt-ed ih-MYOO-nih-tee)
[*cell* **storeroom,** *-medi-* **middle,** *-ate* **process,** *immun-* **free,** *-ity* **state**]

chemotaxis
(kee-moh-TAK-sis)
[*chemo-* **chemical,** *-taxis* **movement or reaction**]

cisterna chyli
(sis-TER-nah KYE-lee or KYE-lye)
[*cisterna* **vessel,** *chyli* **of juice**]

clone
(klohn)
[*clon-* **plant cutting**]

combining site
(kom-BYNE-ing syte)

complement
(KOM-pleh-ment)
[*comple-* **complete,** *-ment* **result of action**]

complement-binding site
(KOM-pleh-ment BIND-ing syte)
[*comple-* **complete,** *-ment* **result of action**]

complement cascade
(KOM-pleh-ment kas-KAYD)
[*comple-* **complete,** *-ment* **result of action,** *cascade* **waterfall**]

cytokine
(SYE-toh-kyne)
[*cyto-* **cell,** *-kine* **movement**]

dendritic cell (DC)
(DEN-drih-tik sel [dee see])
[*dendrit-* **tree branch,** *-ic* **relating to,** *cell* **storeroom**]

effector cell
(ef-FEK-tor sel)
[*effect-* **accomplish,** *-or* **agent,** *cell* **storeroom**]

efferent lymphatic vessel
(EF-fer-ent lim-FAT-ik VES-el)
[*e-* **away,** *-fer-* **carry,** *-ent* **relating to,** *lymph-* **water,** *-atic* **relating to**]

humoral immunity
(HYOO-mor-al ih-MYOO-nih-tee)
[*humor-* **liquid,** *-al* **relating to,** *immun-* **free,** *-ity* **state**]

immune system
(ih-MYOON)
[*immun-* **free (immunity)**]

immunoglobulin (Ig)
(ih-myoo-noh-GLOB-yoo-lin [aye jee])
[*immuno-* **free (immunity),** *-glob-* **ball,** *-ul-* **small,** *-in* **substance**]

inflammatory response
(in-FLAM-ah-toh-ree ree-SPONS)
[*inflam-* **set afire,** *-ory* **relating to**]

innate immunity
(IN-ayt ih-MYOO-nih-tee)
[*innat-* **inborn,** *immun-* **free,** *-ity* **state**]

interferon (IF)
(in-ter-FEER-on [aye ef])
[*inter-* **between,** *-fer-* **strike,** *-on* **substance**]

interleukin (IL)
(in-ter-LOO-kin [aye el])
[*inter-* **between,** *-leuk-* **white (blood cell),** *-in* **substance**]

interstitial fluid (IF)
(in-ter-STISH-al FLOO-id [aye ef])
[*inter-* **between,** *-stit-* **stand,** *-al* **relating to**]

lacteal
(LAK-tee-al)
[*lact-* **milk,** *-al* **relating to**]

lingual tonsil
(LING-gwal TAHN-sil)
[*lingua-* **tongue,** *-al* **relating to,** *tons-* **goiter,** *-il* **little**]

lymph
(limf)
[*lymph* **water**]

lymphatic capillary
(lim-FAT-ik KAP-ih-layr-ee)
[*lymph-* **water,** *-atic* **relating to,** *capill-* **hair,** *-ary* **relating to**]

lymphatic vessel
(lim-FAT-ik)
[*lymph-* **water,** *-atic* **relating to**]

lymph node
(limf nohd)
[*lymph* **water,** *nod-* **knot**]

lymphoid tissue
(LIM-foyd)
[*lymph-* **water (lymphatic system),** *-oid* **like,** *tissu* **fabric**]

macrophage
(MAK-roh-fayj)
[*macro-* **large,** *-phag-* **eat**]

memory cell
(MEM-oh-ree sel)
[*cell* **storeroom**]

natural killer cell (NK cell)
(NACH-er-ul KIL-er sel [en kay sel])
[*cell* **storeroom**]

nonspecific immunity
(non-speh-SIF-ik ih-MYOO-nih-tee)
[*non-* **not,** *-spec-* **form or kind,** *-ific* **relating to,** *immun-* **free,** *-ity* **state**]

palatine tonsil
(PAL-ah-tine TAHN-sil)
[*palat-* **palate,** *-ine* **relating to,** *tons-* **goiter,** *-il* **little**]

pharyngeal tonsil
(fah-RIN-jee-al TAHN-sil)
[*pharyng-* **throat,** *-al* **relating to,** *tons-* **goiter,** *-il* **little**]

plasma cell (effector B cell)
(PLAZ-mah sel [ef-FEK-tor bee sel])
[*plasma* **something molded (blood plasma),** *cell* **storeroom (***effect-* **accomplish,** *-or* **agent,** *B bursa-* **equivalent tissue)**]

right lymphatic duct
(ryte lim-FAT-ik dukt)
[*lymph-* **water,** *-atic* **relating to,** *duct* **lead**]

specific immunity
(speh-SIF-ik ih-MYOO-nih-tee)
[*spec-* **form or kind,** *-ific* **relating to,** *immun-* **free,** *-ity* **state**]

T cell
(T sel)
[*T* **thymus gland,** *cell*]

T lymphocyte
(tee LIM-foh-syte)
[*T* **thymus gland,** *lymph-* **water (lymphatic system),** *-cyte* **cell**]

thoracic duct
(thoh-RAS-ik)
[*thorac-* **chest (thorax),** *-ic* **relating to,** *duct* **lead**]

thymosin
(THY-moh-sin)
[*thymos-* **thyme flower (thymus gland),** *-in* **substance**]

thymus
(THY-mus)
pl., thymuses or thymi
(THY-mus-ez or THY-mye)
[*thymus* **thyme flower**]

LANGUAGE OF **MEDICINE**

acquired immunodeficiency syndrome (AIDS)
(ah-KWYERD IM-yoo-noh-deh-FISH-en-see SIN-drohm [ayds])
[*immuno-* **free**, *-defici-* **fail**, *-y* **state**, *syn-* **together**, *-drome* **running or (race) course**]

adenoid
(AD-eh-noyd)
[*adeno-* **gland**, *-oid* **like**]

allergen
(AL-er-jen)
[*all-* **other**, *-erg-* **work**, *-gen* **produce**]

allergy
(AL-er-jee)
[*all-* **other**, *-erg-* **work**, *-y* **state**]

alloimmunity
(al-oh-ih-MYOO-nih-tee)
[*allo-* **another or different**, *-immun-* **free**, *-ity* **state**]

anaphylactic shock
(an-ah-fih-LAK-tik)
[*ana-* **without**, *-phylact-* **protection**, *-ic* **relating to**]

autoimmunity
(aw-toh-ih-MYOO-nih-tee)
[*auto-* **self**, *-immun-* **free**, *-ity* **state**]

contact dermatitis
(KON-takt der-mah-TYE-tis)
[*derma-* **skin**, *-itis* **inflammation**]

elephantiasis
(el-eh-fan-TYE-ah-sis)
[*elephant-* **elephant**, *-iasis* **condition**]

fever
(FEE-ver)
[*fev-* **heat**]

Hodgkin disease
(HOJ-kin dih-ZEEZ)
[*Thomas Hodgkin* **English physician**, *dis-* **opposite of**, *-ease* **comfort**]

human immunodeficiency virus (HIV)
(ih-myoo-no-deh-FISH-en-see [aych aye vee])
[*immuno-* **free (immunity)**, *-de-* **down**, *-fic-* **perform**, *-ency* **state**, *virus* **poison**]

human lymphocyte antigen (HLA)
(HYOO-man LIM-foh-site AN-tih-jen [aych el ay])
[*lymph-* **water (lymphatic system)**, *-cyte* **cell**, *anti-* **against**, *-gen* **produce**]

hypersensitivity
(hye-per-SEN-sih-tiv-ih-tee)
[*hyper-* **excessive**, *sensitiv-* **able to feel**, *-ity* **state**]

immune deficiency
(ih-MYOON deh-FISH-en-see)
[*immun-* **free (immunity)**, *-defici-* **fail**, *-y* **state**]

immunization
(ih-myoo-nih-ZAY-shun)
[*immun-* **free (immunity)**, *-tion* **process**]

immunosuppressive drug
(ih-myoo-noh-soo-PRES-iv drug)
[*immuno-* **free (immunity)**, *-suppress-* **press down**, *-ive* **relating to**, *drug* **medicine**]

isoimmunity
(aye-soh-ih-MYOO-nih-tee)
[*iso-* **equal**, *-immun-* **free**, *-ity* **state**]

lymphadenitis
(lim-FAD-in-aye-tis)
[*lymph-* **water**, *-aden-* **gland**, *-itis* **inflammation**]

lymphangiogram
(lim-FAN-jee-oh-gram)
[*lymph-* **water**, *-angi-* **vessel**, *-gram* **drawing**]

lymphangitis
(lim-fan-JYE-tis)
[*lymph-* **water**, *-angi-* **vessel**, *-itis* **inflammation**]

lymphedema
(lim-fah-DEE-mah)
[*lymph-* **water**, *-edema* **swelling**]

mastectomy
(mas-TEK-toh-mee)
[*mast-* **breast**, *-ectomy* **surgical removal**]

monoclonal antibody
(mon-oh-KLONE-al AN-tih-bod-ee)
[*mono-* **single**, *-clon-* **plant cutting**, *-al* **relating to**, *anti-* **against**]

non-Hodgkin lymphoma
(non-HOJ-kin lim-FOH-mah)
[*non-* **not**, *Thomas Hodgkin* **English physician**, *lymph-* **water (lymphatic system)**, *-oma* **tumor**]

perinatal infection
(payr-ih-NAY-tal in-FEK-shun)
[*peri-* **around**, *-nat-* **birth or origin**, *-al* **relating to**, *infect-* **stain**, *-ion* **condition**]

severe combined immune deficiency (SCID)
(seh-VEER kom-BYNED ih-MYOON deh-FISH-en-see [skid])
[*immun-* **free (immunity)**, *-defici-* **fail**, *-y* **state**]

splenectomy
(spleh-NEK-toh-mee)
[*splen-* **spleen**, *-ec-* **out**, *-tom-* **cut**, *-y* **action**]

splenomegaly
(spleh-noh-MEG-ah-lee)
[*splen-* **spleen**, *-mega* **large**, *-ly* **relating to**]

systemic lupus erythematosus (SLE)
(sis-TEM-ik LOO-pus er-ih-them-ah-TOH-sus [es el ee])
[*system-* **organized whole**, *-ic* **relating to**, *lupus* **wolf**, *erythema-* **redness**, *-osus* **condition**]

tonsillitis
(tahn-sih-LYE-tis)
[*tonsil-* **tonsil**, *-itis* **inflammation**]

transplant
(trans-PLANT)
[*trans-* **across**, *-plant* **set or place**]

vaccine
(VAK-seen)
[*vaccin-* **cow (cowpox)**]

16

❑ OUTLINE SUMMARY

To download a digital version of the chapter summary for use with your device, access the ***Audio Chapter Summaries*** *online at evolve.elsevier.com.*

Hint *Scan this summary after reading the chapter to help you reinforce the key concepts. Later, use the summary as a quick review before your class or before a test.*

Lymphatic System

A. Organization of the lymphatic system—lymphatic fluid (lymph), lymphatic vessels, and many lymph organs make up this system (**Figure 16-1**)

B. Lymph—excess fluid left behind by capillary exchange that drains from tissue spaces and is transported by *lymphatic vessels* back to the bloodstream

C. Lymphatic vessels—permit only one-way movement of lymph
   1. Lymphatic capillaries—tiny blind-ended tubes distributed in tissue spaces (**Figure 16-2**)
      a. Microscopic in size
      b. Sheets consisting of one cell layer of simple squamous epithelium
      c. Poor "fit" between adjacent cells results in porous walls
      d. Called *lacteals* in the intestinal wall (fat transportation into bloodstream)
   2. Right lymphatic duct
      a. Drains lymph from the right upper extremity and right side of head, neck, and upper torso
   3. Thoracic duct
      a. Largest lymphatic vessel
      b. Has an enlarged pouch along its course, called *cisterna chyli*
      c. Drains lymph from about three-fourths of the body

D. Lymphedema—swelling (edema) of tissues caused by blockage of lymphatic vessels (**Figure 16-3**)
   1. Lymphangitis—inflammation of lymphatic vessels; may progress to septicemia (blood infection) (**Figure 16-4**)
   2. Elephantiasis—severe lymphedema of limbs resulting from parasite infestation of lymphatic vessels (**Figure 16-5**)

E. Lymphoid organs—have masses of developing WBCs (lymphoid tissue) and functions that include defense and WBC formation
   1. Lymph nodes
      a. Filter lymph (**Figure 16-6**)
      b. Located in clusters along the pathway of lymphatic vessels (**Figures 16-7** and **16-8**)
      c. Lymphoid tissue—mass of lymphocytes and related cells inside a lymphoid organ; provides immune function and development of immune cells
      d. Flow of lymph: to node via several afferent lymphatic vessels and drained from node by a single efferent lymphatic vessel
      e. Lymphadenitis—swelling and tenderness of lymph nodes
      f. Cancer cells can easily move through lymphatic vessels to other parts of the body in a process called metastasis
   2. Thymus
      a. Lymphoid tissue organ located in mediastinum
      b. Total weight of about 35 to 40 g—a little more than an ounce
      c. Plays a vital and central role in immunity
      d. Produces T lymphocytes, or T cells
      e. Secretes hormone called *thymosins,* which influence T cell development
      f. Lymphoid tissue is eventually replaced by fat (during childhood) in the process called *involution*
   3. Tonsils (**Figure 16-9**)
      a. Composed of three masses of lymphoid tissue around the openings of the mouth and throat
         (1) Palatine tonsils ("the tonsils")
         (2) Pharyngeal tonsils (also known as *adenoids*)
         (3) Lingual tonsils
      b. Subject to chronic infection
      c. Enlargement of pharyngeal tonsils may impair breathing
   4. Spleen
      a. Structure
         (1) Largest lymphoid organ in body
         (2) Located in upper left quadrant of abdomen
         (3) Often injured by trauma to abdomen
         (4) Surgical removal called *splenectomy*
      b. Functions
         (1) Phagocytosis of bacteria and old RBCs
         (2) Reservoir of monocytes, which are released for emergency tissue repairs elsewhere in the body
         (3) Acts as a blood reservoir
      c. Splenomegaly—enlargement of the spleen

5. Lymphoma—malignant tumor of lymph nodes
   a. Two principal types: Hodgkin disease and non-Hodgkin lymphoma
   b. All types cause painless enlargement of lymph nodes
   c. Can spread to many other areas of the body

Immune System

A. Protects body from pathological bacteria, foreign tissue cells, and cancerous cells
B. Made up of defensive cells and molecules
C. Two main strategies—innate (nonspecific) defenses and adaptive (specific) defenses (**Table 16-1**)
D. Innate immunity
   1. Called *innate* because we are born with it (no prior exposure needed)
   2. Also called *nonspecific immunity* because it includes mechanisms that act generally against any type of damage or threatening agent
   3. Many types of innate immunity occur in the body (**Table 16-2**)
      a. Nonspecific immunity is the rapid first response and often triggers slower specific responses
      b. Involves a variety of signaling chemicals called cytokines
   4. Skin—mechanical barrier to bacteria and other harmful agents
   5. Tears and mucus—wash eyes and trap and kill bacteria
   6. Inflammation
      a. Inflammatory response—attracts immune cells to site of injury, increases local blood flow, increases vascular permeability; promotes movement of WBCs to site of injury or infection (**Figure 16-10**)
      b. Fever—systemic effect of increased body temperature; may increase immune efficiency or inhibit infectious agents
   7. Complement—class of enzymes in blood plasma that can trigger a variety of immune responses; also involved in adaptive (specific) mechanisms
E. Adaptive immunity (**Table 16-1**)
   1. Adaptive because of its ability to recognize, respond to, and remember harmful substances or bacteria
   2. Also called *specific immunity* because it responds to particular antigens to which is has been exposed
   3. Types of adaptive immunity (**Table 16-3**)
      a. Natural immunity—exposure to causative agent is not deliberate
         (1) Active—active disease produces immunity
         (2) Passive—immunity passes from mother to fetus through placenta or from mother to child through mother's milk
      b. Artificial immunity—exposure to causative agent is deliberate
         (1) Active—vaccination results in activation of immune system and long-term protection
         (2) Passive—protective material developed in another individual's immune system and given to previously nonimmune individual, giving short-term protection

Immune System Molecules

A. Cytokines
   1. Cytokines are protein molecules that communicate among cells, coordinating immune responses
   2. Interleukins (ILs) are an example of cytokines
B. Antibodies (**Figure 16-11**)
   1. Protein molecules with specific combining sites
   2. Combining sites attach antibodies to specific antigens (foreign proteins), forming an antigen-antibody complex—this provides *humoral immunity (antibody-mediated immunity)*
   3. Antigen-antibody complexes may:
      a. Neutralize toxins
      b. Clump or agglutinate enemy cells
      c. Promote phagocytosis
C. Complement proteins
   1. Group of proteins normally present in blood in inactive state
   2. Complement cascade
      a. Important mechanism of action for antibodies
         (1) Complement-binding sites on antibody are exposed after attaching to antigen
         (2) Complement triggers a series (cascade) of reactions that produce tiny protein rings that create holes in the surface of a foreign cell
      b. Ultimately causes cell lysis by permitting entry of water through a defect created in the plasma membrane of the foreign cell (**Figure 16-12**)
   3. Also helps perform other functions—examples: attracting immune cells to a site of infection, activating immune cells, marking foreign cells for destruction, increasing permeability of blood vessels, the inflammatory response

Immune System Cells

A. Phagocytes
   1. Types
      a. Neutrophils—short-lived phagocytic cells; most abundant type of immune cell
      b. Monocytes—develop into phagocytic macrophages and migrate to tissues
      c. Dendritic cells (DCs)—often found at or near external surfaces (**Figure 16-14**)
   2. Ingest and destroy foreign cells or other harmful substances via phagocytosis (**Figure 16-13**)
   3. Macrophages and DCs act as antigen-presenting cells (APCs) by displaying ingested antigens on their outer surface to trigger specific immune cells
B. Lymphocytes
   1. Second most numerous of immune system cells, after neutrophils; include NK cells, B cells, and T cells (**Figure 16-15**)
   2. Natural killer (NK) cells—agents of innate immunity
      a. Can attach to any cell, but only kill cells lacking normal self-antigens
      b. Usually kill by triggering apoptosis (programmed cell death)
   3. B cells (B lymphocytes)—agents of adaptive immunity
      a. Development of B cells—primitive stem cells migrate from bone marrow and go through two stages of development (**Figure 16-16**)
         (1) First stage—stem cells develop into immature B cells
            (a) Takes place in the liver and bone marrow before birth and in the bone marrow only in adults
            (b) B cells are small lymphocytes with antibody molecules (which they have synthesized) in their plasma membranes
            (c) After they mature, inactive B cells migrate chiefly to lymph nodes
         (2) Second stage—inactive B cell develops into activated B cell
            (a) Initiated by inactive B cell's contact with antigens, which bind to its surface antibodies, plus cytokines (signal chemicals) from T cells
            (b) Activated B cell, by dividing repeatedly, forms two clones of cells—plasma (effector) cells and memory cells
            (c) Plasma cells secrete antibodies into blood; memory cells are stored in lymph nodes
            (d) If subsequent exposure to the specific antigen that activated the B cell occurs, memory cells become plasma cells and secrete antibodies in large quantity
      b. Function of B cells—indirectly, B cells produce humoral immunity
         (1) Activated B cells develop into plasma cells
         (2) Plasma cells secrete antibodies into the blood
         (3) Circulating antibodies produce humoral immunity (**Figure 16-16**)
   4. T cells (T lymphocytes)—agents of adaptive immunity
      a. Development of T cells—stem cells from bone marrow migrate to thymus gland (**Figure 16-17**)
         (1) First stage—stem cells develop into T cells
            (a) T cells mature in the thymus during few months before and after birth
            (b) Mature T cells migrate chiefly to lymph nodes
         (2) Second stage—T cells develop into activated T cells
            (a) Occurs when, and if, antigen binds to T cell's surface proteins and a cytokine (chemical signal) is received from another T cell
            (b) As with B cells, clones made up of effector cells and memory cells are formed
      b. Functions of T cells—produce cell-mediated immunity (**Figure 16-18**)
         (1) Cytotoxic T cells—kill infected or tumor cells by releasing a substance that poisons infected or tumor cells
         (2) Helper T cells—release cytokines that attract and activate macrophages to kill cells by phagocytosis; produce cytokines that help activate B cells
         (3) Regulatory T cells—release cytokines to suppress immune responses

Hypersensitivity of the Immune System

A. Hypersensitivity—inappropriate or excessive immune response (**Figure 16-19**)
B. Allergy—hypersensitivity to harmless environmental antigens (allergens)
   1. Immediate allergic responses usually involve humoral immunity
   2. Delayed allergic responses usually involve cell-mediated immunity
C. Autoimmunity—inappropriate, excessive response to self-antigens
   1. Causes autoimmune diseases
   2. Systemic lupus erythematosus (SLE)—chronic inflammatory disease caused by numerous antibodies attacking a variety of tissues (**Figure 16-20**)
D. Alloimmunity (isoimmunity)—excessive reaction to antigens from another human
   1. May occur between mother and fetus during pregnancy
   2. May occur in tissue transplants (causing rejection syndrome)

Immune System Deficiency

A. Congenital immune deficiency, or immunodeficiency (rare)
   1. Results from improper lymphocyte development before birth
   2. Severe combined immune deficiency (SCID)—caused by disruption of stem cell development
B. Acquired immune deficiency
   1. Develops after birth
   2. Acquired immunodeficiency syndrome (AIDS)—caused by HIV infection of T cells

❑ ACTIVE LEARNING

STUDY TIPS

Consider using these tips to achieve success in meeting your learning goals.

Review the synopsis of the lymphatic system in Chapter 5. The lymphatic system is partly a "sewer" system of the body. Plasma is pushed out of the capillaries and washes over the tissue cells. The interstitial fluid (IF) carries bacteria and other cellular debris, along with proteins and lipids, into blind-ended capillaries in the lymphatic system. The fluid is then called lymph. It is carried to the lymph node, where it is filtered, cleaned, and then carried by ducts back to the blood. Keep this process in mind when you study the structures of the lymphatic system.

1. There are several specific organs in the lymphatic system. Flash cards and online resources that include their names, locations, and functions will help you learn them.
2. A complex concept to understand is the inflammatory response. Use **Figure 16-10** or develop your own concept map to help you understand the steps in the inflammatory response.
3. Adaptive immunity can be classified as natural or artificial depending on how the body was exposed to the specific antigen, and active or passive depending on how involved the body's immune system was in developing the response.
4. The natural, active immune response is divided into humoral immunity and cell-mediated immunity. Humoral immunity is mediated by the B lymphocytes, or B cells. They stay in the lymph node and secrete antibodies into the blood (*humor* means "body fluid"). They also form memory cells, which give lifelong immunity. T lymphocytes, or T cells, provide cell-mediated immunity. They leave the lymph node and actively engage the antigen.
5. The best way to learn the disorders of the immune system is to make a chart organized by the mechanism or cause of the disorder: allergic reaction, autoimmunity, alloimmunity, and immune deficiencies.
6. In your study group, use flash cards or online resources to quiz each other on the terms and structures of the lymphatic and immune systems. Discuss the process of how lymph is formed, filtered, and returned to the blood. Discuss innate immunity, especially the inflammation response. Discuss the different types of adaptive immunity. Discuss the steps in humoral and cell-mediated immunity.
7. Refer to the Language of Science and Language of Medicine terms and review the disorder chart, chapter outline summary, and the questions at the end of the chapter and discuss possible test questions.

Review Questions

Write out the answers to these questions after reading the chapter and reviewing the Chapter Summary. If you simply think through the answer without writing it down, you will not retain much of your new learning.

1. Define *lymph* and explain its function.
2. Name the two major lymphatic ducts and the areas of the body each of them drains.
3. Describe the structure of a lymph node.
4. What is lymphedema? What is the cause of elephantiasis?
5. Explain the defense function of the lymph node.
6. Where is the thymus gland? What are the functions of the thymus gland?
7. Name the three pairs of tonsils and give the location of each.
8. Name the location and function of the spleen.
9. Explain the types of innate immunity.
10. Name and differentiate the four types of adaptive immunity.
11. What are antibodies? What are antigens?
12. Explain the role of complement in the immune system.
13. Explain the role of the macrophage in the immune system.
14. Explain the development and functioning of B cells.
15. Explain the development and functioning of T cells.
16. What is an allergy?
17. What is autoimmunity? Name an example of an autoimmune disease.
18. What is alloimmunity? Name an example of an alloimmunity disorder.
19. What are HLAs? How are they related to tissue typing?
20. What is SCID? What is its cause?
21. List three causes of acquired immunodeficiency syndrome.

Critical Thinking

After finishing the Review Questions, write out the answers to these more in-depth questions to help you apply your new knowledge. Go back to sections of the chapter that relate to concepts that you find difficult.

22. Differentiate between lymphatic capillaries and blood capillaries. Explain how the different structures relate to their function.
23. Explain the role of the lymph node in the spread of cancer.
24. Explain the difference in mechanisms in the development of the allergic reaction of runny nose and hives, and the allergic reaction to poison ivy.

16

Chapter Test

Hint *After studying the chapter, test your mastery by responding to these items. Try to answer them without looking up the answers.*

1. ________ is the fluid that leaves the blood capillaries and is not directly returned to the blood.
2. Lymph from about three-fourths of the body drains into the ________.
3. Lymph from the right upper extremity and the right side of the head drains into the ________.
4. An abnormal condition in which tissue swells because of accumulation of lymph is called ________.
5. The enlarged, pouchlike structure in the abdomen that serves as a storage area for lymph is called the ________.
6. The function of the ________ is to filter and clean the lymph.
7. The many lymphatic vessels that enter the lymph node are called ________ vessels; the single vessel leaving the lymph node is called the ________ vessel.
8. The ________ are white blood cells that mature in the thymus. The thymus also produces a hormone called ________.
9. The three pairs of tonsils are the ________, ________, and ________.
10. The largest lymphoid organ is the ________.
11. The signs of ________ are heat, redness, pain, and swelling.
12. ________ kills target cells by drilling holes in their plasma membrane, which disrupts the sodium and water balance.
13. Macrophages were ________ originally that migrated into the tissues.
14. A hypersensitivity of the immune system to a harmless environmental antigen is called a (an) ________.
15. An extreme allergic reaction causing life-threatening symptoms is called ________.
16. An inappropriate and excessive response to self-antigen is called ________.
17. Erythroblastosis is an example of what excessive immune reaction? ________
18. An attempt to identify and match HLAs between the organ donor and the recipient is called ________.
19. A congenital immune deficiency in which both humoral and cellular immunity are defective is called ________.
20. The cause of AIDS is ________.
21. Adaptive immunity is also known as ________.
22. The ________ of the spleen serves as a reservoir for monocytes that can quickly leave the spleen to help repair damaged tissue anywhere in the body during an emergency.
23. ________ are released from cells to act as direct agents of innate, nonspecific immunity.
24. ________ is sometimes administered with a penlike injector to reduce allergic reactions.
25. ________ cells recognize abnormal cells by using two different recognition receptors; a *kill activating* receptor and a *kill inhibiting* receptor.
26. The immunity that develops against polio after receiving a polio vaccination is an example of:
    a. active natural immunity
    b. passive natural immunity
    c. active artificial immunity
    d. passive artificial immunity
27. The immunity that is given to the fetus or newborn by the immune system of the mother is an example of:
    a. active natural immunity
    b. passive natural immunity
    c. active artificial immunity
    d. passive artificial immunity
28. The immunity that comes from the injection of antibodies made by another individual's immune system is an example of:
    a. active natural immunity
    b. passive natural immunity
    c. active artificial immunity
    d. passive artificial immunity
29. The immunity that develops after a person has had a disease is an example of:
    a. active natural immunity
    b. passive natural immunity
    c. active artificial immunity
    d. passive artificial immunity

For each of the following phrases, write a B in front of it if it describes the development or functioning of a B cell or write a T in front of it if it describes the development or functioning of a T cell.

30. ________ produces antibodies
31. ________ some develop into plasma cells
32. ________ the main cell involved in cell-mediated immunity
33. ________ the main cell involved in humoral immunity
34. ________ develops in the thymus gland
35. ________ moves to the site of the antigen and releases cell poison
36. ________ divides rapidly into clones once it is activated
37. ________ releases a substance that attracts macrophages
38. ________ some differentiate into memory cells
39. ________ mediates the contact dermatitis allergic response

Case Studies

To solve a case study, you may have to refer to the glossary or index, other chapters in this textbook, and other resources.

1. After having an infection in the groin, a young boy complains of painful swelling in his leg. On examination, you see that the entire leg is swollen, but the opposite leg is fine. The attending physician explains that this is a complication of the recent groin infection, which involved the lymph nodes in that area. Can you explain how this may have caused the boy's leg to swell? Why isn't the other leg affected?
2. Keith was sledding in the snow with his friends when he accidentally hit a tree. After examining Keith, the emergency room physician concluded that he had ruptured his spleen in the accident. How might a ruptured spleen be treated? What might happen if Keith's family or physician delays this treatment? Does Keith need his spleen to survive?
3. Many years ago there was a famous case of a boy born with severe combined immune deficiency (SCID). His physicians placed him in a pathogen-free chamber that resembled a giant glass bubble. What purpose was served by doing this? What treatments are available today that might have helped this boy?
4. Merrily recently had a mastectomy. When she was discharged, her doctor encouraged her to take a daily walk of 30 minutes and to swing her arms in an upward motion. Why did he add these orders to his discharge instructions?

Answers to Active Learning Questions can be found online at evolve.elsevier.com.

Respiratory System

OUTLINE

Scan this outline before you begin to read the chapter, as a preview of how the concepts are organized.

OBJECTIVES

Before reading the chapter, review these goals for your learning.

After you have completed this chapter, you should be able to:

1. List the major organs of the respiratory system and describe the function of each organ in relation to the major functions of the respiratory system.
2. Discuss the structures included in the upper respiratory tract, as well as identify and describe the major disorders of the upper respiratory tract.
3. Discuss the structures included in the lower respiratory tract, as well as identify and describe the major disorders of the lower respiratory tract.
4. Discuss respiration and pulmonary ventilation, including the mechanics of breathing and pulmonary volumes.
5. Describe the regulation of ventilation, and identify breathing patterns.
6. Compare, contrast, and explain the mechanism responsible for the exchange of gases that occurs during external and internal respiration.
7. Describe the transport of gases by blood.

CHAPTER 17

No one needs to be told how important the **respiratory system** is. The respiratory system serves the body much as a lifeline to an air pump serves a deep-sea diver. Think how panicked you would feel if suddenly your lifeline became blocked—if you could not breathe for a few seconds!

Of all the substances that cells, and therefore the body as a whole, must have to survive, oxygen is by far the most crucial. A person can live a few weeks without food, a few days without water, but only a few minutes without oxygen. Constant removal of carbon dioxide from the body is just as important for survival as a constant supply of oxygen.

The respiratory system ensures that oxygen is supplied to and carbon dioxide (a waste product) is removed from the body's cells. The process of respiration therefore is a vital **homeostatic mechanism.** By constantly supplying adequate oxygen and by removing carbon dioxide as it forms, the respiratory system helps maintain a constant internal environment that enables our body cells to function effectively.

To accomplish its functions, the respiratory system also effectively *filters, warms,* and *humidifies* the air we breathe. Respiratory organs such as vocal cords, sinuses, and specialized epithelium, also help produce speech and make possible the sense of smell, or **olfaction.** Even the primary function of gas exchange has a secondary effect—the removal of excess acid from the body. This pH-balancing function of the respiratory system is discussed further in Chapter 22.

Further discussion that relates to disorders of the upper and lower respiratory tracts emphasizes how disruptions in respiratory system homeostasis lead to both anatomical and physiological manifestations of disease.

LANGUAGE OF SCIENCE

Hint Before reading the chapter, say each of these terms out loud. This will help you to avoid stumbling over them as you read.

alveolar duct
(al-VEE-oh-lar dukt)
[*alve-* **hollow,** *-ol-* **little,** *-ar* **relating to**]

alveolar sac
(al-VEE-oh-lar sak)
[*alve-* **hollow,** *-ol-* **little,** *-ar* **relating to**]

alveoli
(al-VEE-oh-lye)
sing., alveolus
(al-VEE-oh-lus)
[*alve-* **hollow,** *-olus* **little**]

aortic body
(ay-OR-tik BOD-ee)
[*aort-* **lifted,** *-ic* **relating to**]

auditory tube
(AW-dih-toh-ree toob)
[*audit-* **listen,** *-or-* **agent,** *-y* **relating to**]

bicarbonate ion
(bye-KAR-boh-net)
[*bi-* **two,** *-carbon-* **coal (carbon),** *-ate* **oxygen compound**]

bronchi
(BRONG-kye)
sing., bronchus
(BRONG-kus)
[*bronchus* **windpipe**]

bronchiole
(BRONG-kee-ohl)
[*bronch-* **windpipe,** *-ol-* **little**]

carbaminohemoglobin ($HbCO_2$)
(kar-bah-MEE-noh-hee-moh-GLOH-bin [aych be see oh too])
[*carb-* **coal (carbon),** *-amino-* **ammonia compound (amino acid),** *-hemo-* **blood,** *-glob-* **ball,** *-in* **substance**]

Continued on p. 482

Structural Plan

Overview

Respiratory organs include the **nose, pharynx** *(throat),* **larynx** *(voice box),* **trachea** *(windpipe),* **bronchi** *(branches),* and **lungs.** The basic structural scheme of this organ system is that of a tube with many branches ending in millions of extremely tiny, very thin-walled sacs called **alveoli.** Figure 17-1 shows the extensive branching of the "respiratory tree" in both lungs. Think of this air distribution system as an "upside-down tree."

The trachea or windpipe then becomes the trunk and the bronchial tubes the branches. This idea is developed further when the types of bronchi and the alveoli are studied in more detail later in the chapter.

A network of capillaries fits like a hairnet around each microscopic alveolus. Incidentally, this is a good place for us to think again about a principle already mentioned several times, namely, that structure and function are intimately related. The function of alveoli—in fact, the function of the entire respiratory system—is to distribute air close enough to blood for a gas exchange to take place between air and blood. The passive transport process of **diffusion,** which was described in Chapter 3, is the mode for the exchange of gases that occurs in the respiratory system. You may want to review the discussion of diffusion on pp. 51–52 before you study the mechanism of gas exchange that occurs in the lungs and body tissues.

To learn more about respiratory system, go to AnimationDirect online at *evolve.elsevier.com.*

Respiratory Tract

The respiratory tract—the pathway of air flow—is often divided into upper and lower tracts, or divisions, to assist in the description of symptoms associated with common respiratory problems such as a cold. The organs of the upper respiratory tract are located outside of the thorax or chest cavity, whereas those in the lower tract, or division, are located almost entirely within it.

The **upper respiratory tract** is composed of the nose, pharynx, and larynx. The **lower respiratory tract,** or division, consists of the trachea, all segments of the bronchial tree, and the lungs.

The designation **upper respiratory infection (URI)** is often used by medical professionals to describe what many of us call a "head cold." Typically, the symptoms of an upper respiratory infection involve the sinuses,

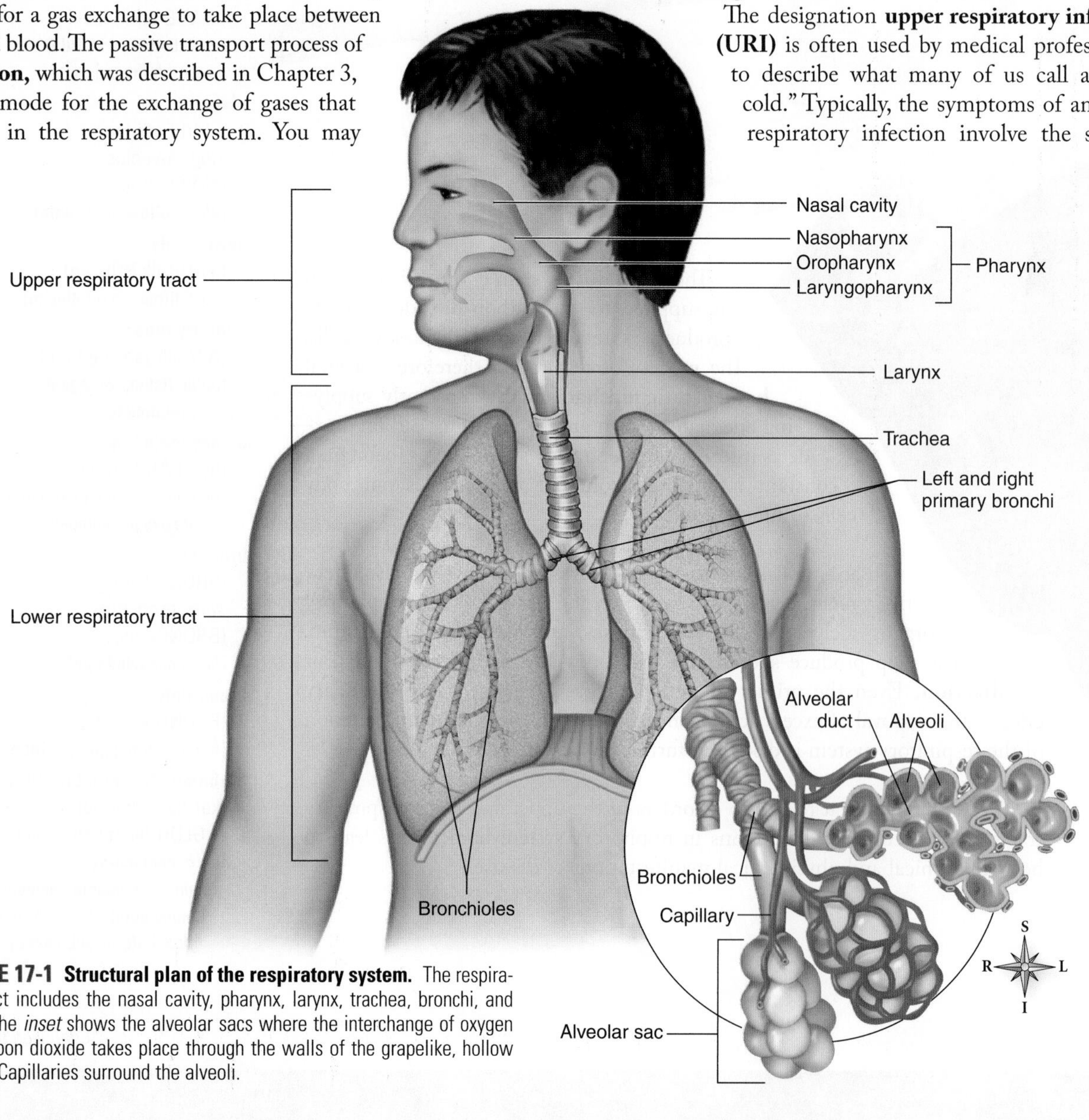

FIGURE 17-1 Structural plan of the respiratory system. The respiratory tract includes the nasal cavity, pharynx, larynx, trachea, bronchi, and lungs. The *inset* shows the alveolar sacs where the interchange of oxygen and carbon dioxide takes place through the walls of the grapelike, hollow alveoli. Capillaries surround the alveoli.

nasal cavity, pharynx, and/or larynx, whereas the symptoms of what is often referred to as a "chest cold" are similar to pneumonia and involve the organs of the lower respiratory tract.

Find the major structures of the respiratory tract in the *Clear View of the Human Body* (follows p. 8), noting their locations relative to other nearby body structures.

To learn more about respiratory mucosa, go to AnimationDirect online at *evolve.elsevier.com.*

Respiratory Mucosa

Structure

Before beginning the study of individual organs in the respiratory system, it is important to review the histology, or microscopic anatomy, of the **respiratory mucosa**—the membrane that lines most of the air distribution tubes in the system.

Respiratory mucosa is typically *ciliated pseudostratified epithelium,* as you can see in Figure 17-2. As its name implies, this type of tissue is covered with *cilia.* Look back at Figures 3-4 and 3-5 on p. 49 to see the structure of these tiny cell projections and how they can move fluids along the surfaces of a layer of cells. Figure 17-2 also shows the presence of *goblet cells,* which can produce and release huge amounts of **mucus.** Mucus varies in composition from very watery to very thick and sticky, depending on the specific location of the mucosa.

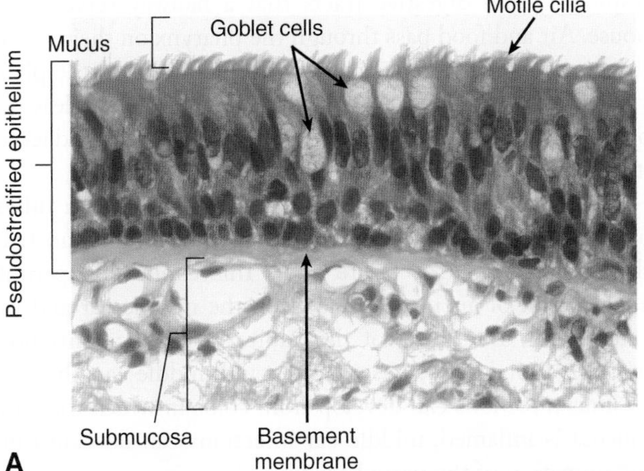

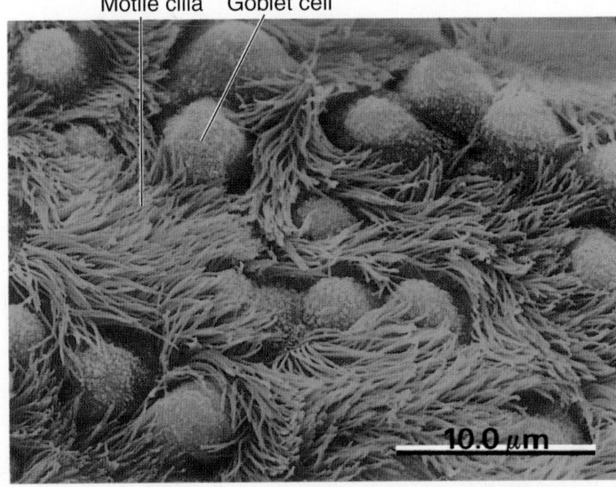

FIGURE 17-2 Respiratory mucosa. A, Light micrograph (×200) and **B,** scanning electron micrograph (×2000) of the ciliated pseudostratified epithelium typical of the respiratory lining. Note the numerous motile (moving) cilia and the mucus-producing goblet cells.

Although most of the respiratory passages are lined with ciliated pseudostratified epithelium, there are a few areas lined with other tissues. For example, protective *stratified squamous epithelium* is found just inside the nostrils, covering the vocal folds of the larynx, and lining the pharynx. Look back at Figure 4-4 on p. 74 to see the many layers of this thicker type of epithelium.

Simple squamous epithelium—an extremely thin tissue—lines the alveoli of the lungs where gas exchange occurs. Look back at Figure 4-3 on p. 74 to see the thinness of this type of epithelium.

Function

Recall that in addition to serving as air distribution passageways or gas exchange surfaces, the structures of the respiratory tract and lungs cleanse, warm, and humidify inspired air. Air entering the nose is generally contaminated with one or more common irritants; examples include some types of chemical air pollutants, dust, pollen, bacterial organisms, and even insects. A remarkably effective air purification mechanism removes almost every form of contaminant before inspired air reaches the alveoli, or terminal air sacs, in the lungs.

A layer of protective mucus, called a *mucous blanket,* covers nearly the entire ciliated pseudostratified epithelial lining of the air distribution tubes in the respiratory tree (see Figure 17-2). More than 125 mL of respiratory mucus is produced daily. It serves as the most important air purification mechanism. Air is purified when contaminants such as dust, pollen, and smoke particles stick to the mucus and become trapped.

Normally, the cleansing layer of mucus containing inhaled contaminants moves upward to the pharynx from the lower portions of the bronchial tree on the millions of hairlike cilia that beat or move only in one direction. This mechanism is often called the **ciliary escalator.**

Cigarette smoke and other irritants are detected by the cilia, which beat rapidly in response—an attempt to clear out the contaminants more efficiently. Prolonged exposure to cigarette smoke both increases production of mucus and eventually paralyzes cilia, thus causing accumulations of contaminated mucus to build up and remain in the respiratory passageways for longer periods of time. The result is a typical smoker's cough, which is the body's effort to clear these large quantities of contaminated mucus.

QUICK CHECK

1. What are the primary functions of the respiratory system?
2. Describe the characteristics of the *alveoli* that enable them to perform their function of gas exchange.
3. What distinguishes the *upper respiratory tract* from the *lower respiratory tract*?
4. What is the role of the *respiratory mucosa*?

17

17

Upper Respiratory Tract

Nose

Air enters the respiratory tract through the **external nares,** or nostrils. It then flows into the right and left **nasal cavities,** which are lined by respiratory mucosa. A partition called the **nasal septum** separates these two cavities.

The surface of the nasal cavities is moist from mucus and warm from blood flowing just under the epithelium. Nerve endings responsible for the sense of smell (olfactory receptors) are also located in the nasal mucosa.

Nasal polyps are painless, noncancerous tissue growths that may project from the nasal mucosa. They are frequently associated with chronic hay fever. Over time, nasal polyps may grow in size, partially obstruct the nasal passage, and impair breathing. In severe cases, surgical removal may be required.

Four **paranasal sinuses**—*frontal, maxillary, sphenoidal,* and *ethmoidal*—drain into the nasal cavities (Figure 17-3). The paranasal sinuses are lined with a mucous membrane that assists in the production of mucus for the respiratory tract. In addition, these hollow spaces help to lighten the skull bones and serve as resonant chambers that enhance the production of sound.

Because the mucosa that lines the sinuses is continuous with the mucosa that lines the nose, sinus infections, called **sinusitis,** often develop from colds in which the nasal mucosa is inflamed. Symptoms of sinusitis often include pressure; pain; headache; and external tenderness, swelling, and redness. In chronic cases, infection may spread to adjacent bone or into the cranial cavity inflaming meninges or brain tissue. Treatment includes decongestants, analgesics, antibiotics—and in rare cases, surgery to improve drainage.

Two ducts from the **lacrimal sacs** also drain into the nasal cavity, as Figure 17-3 shows. The lacrimal sacs collect tears from the corner of each eyelid and drain them into the nasal cavity. This arrangement explains why your nose may drip after producing excess tears, as in crying, allergies, or eye irritation.

Note in Figure 17-3 that three shelflike structures called **conchae** protrude into the nasal cavity on each side. The nasal conchae are sometimes called nasal **turbinates.** The mucosa-covered conchae greatly increase the surface over which air must flow as it passes through the nasal cavity. As air moves over the conchae and through the nasal cavities, it is warmed and humidified. This helps explain why breathing through the nose is more effective in humidifying inspired air than is breathing through the mouth.

If an individual who is ill requires supplemental oxygen, it is first bubbled through water to reduce the amount of moisture that would otherwise have to be removed from the lining of the respiratory tree to humidify it. Administration of "dry" oxygen pulls water from the mucosa and results in respiratory discomfort and irritation.

Pharynx

The **pharynx** is the structure that many of us call the throat. It is about 12.5 cm (5 inches) long and can be divided into three portions (Figure 17-4). The uppermost part of the tube just behind the nasal cavities is called the **nasopharynx.** The portion behind the mouth is called the **oropharynx.** The third or lowest segment is called the **laryngopharynx.**

The pharynx as a whole serves a similar purpose for the respiratory and digestive tracts that a hallway serves for a house. Air and food pass through the pharynx on their way to the lungs and the stomach, respectively. Air enters the pharynx from the two nasal cavities or the oral cavity and leaves it by way of the larynx. Food enters it from the mouth and leaves it by way of the esophagus.

The right and left **auditory tubes,** or **eustachian tubes,** open into the nasopharynx; they connect the middle ears with the nasopharynx (see Figure 17-4). This connection permits equalization of air pressure between the middle ear and the exterior ear. The lining of the auditory tubes is continuous with the lining of the nasopharynx and middle ear. Thus just as sinus infections can develop from colds in which the nasal mucosa is inflamed, middle ear infections can develop from inflammation of the nasopharynx.

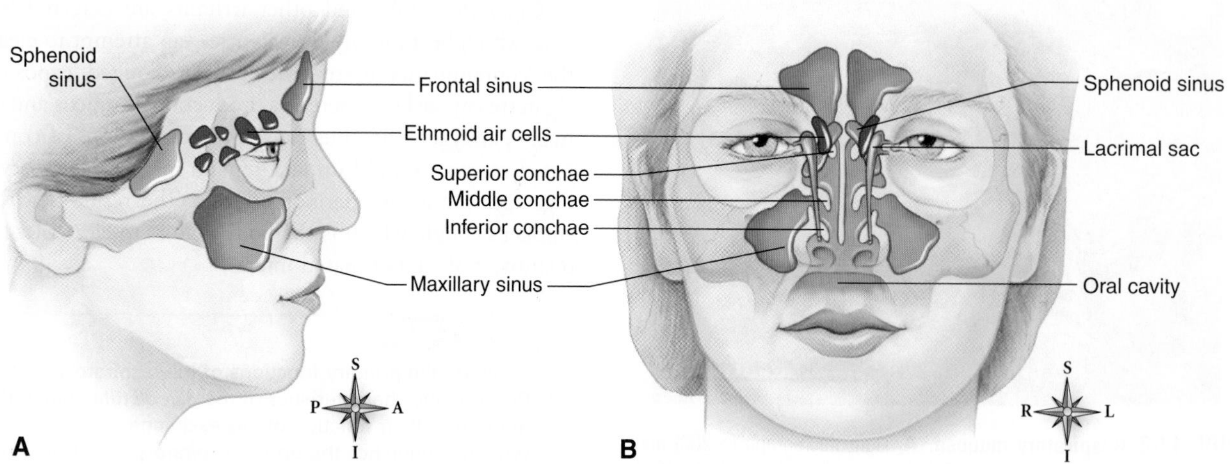

FIGURE 17-3 Paranasal sinuses. A, Lateral view of the position of the sinuses in adults. **B,** The anterior view shows the anatomical relationship of the paranasal sinuses to each other and to the nasal cavity.

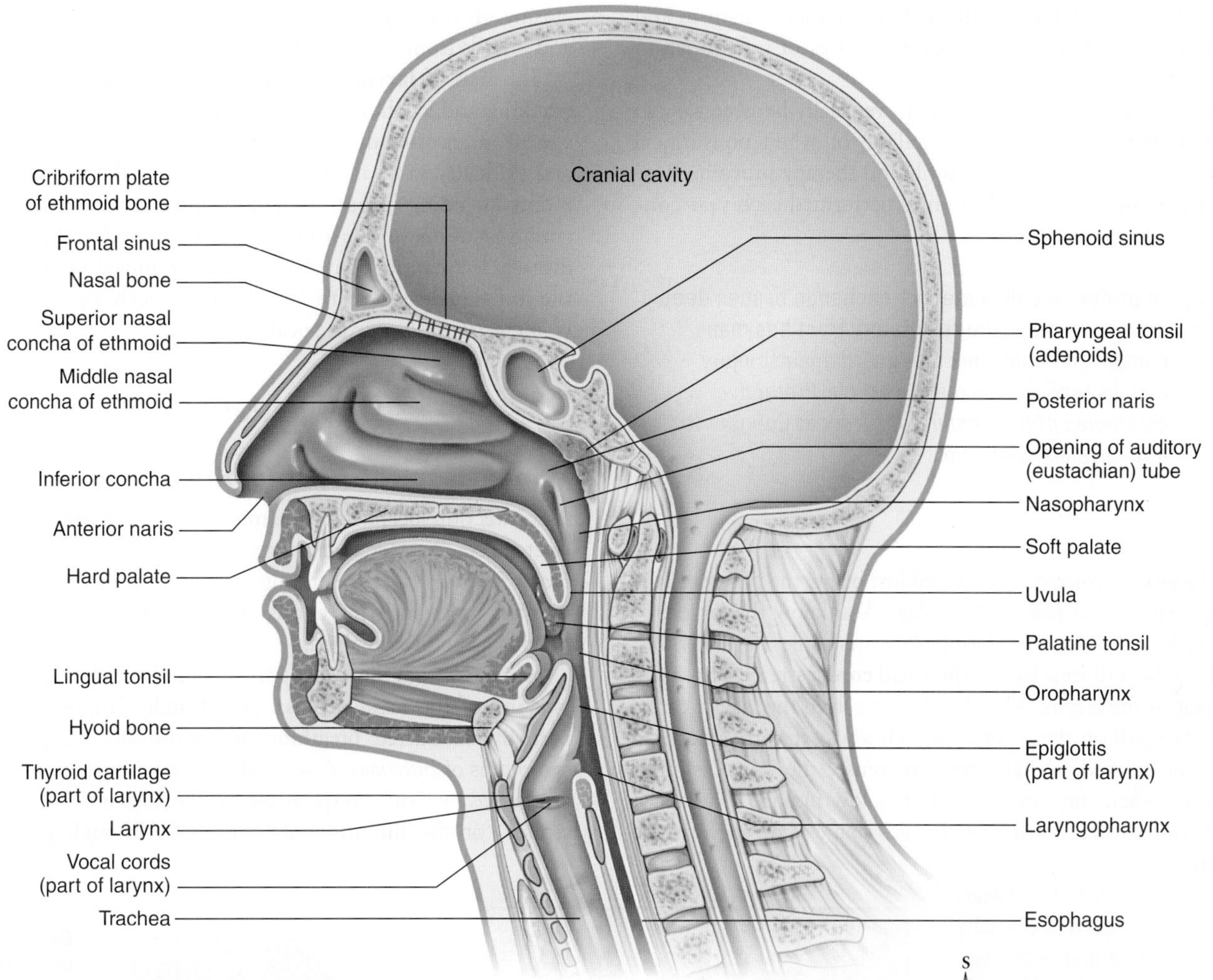

FIGURE 17-4 Head and neck. Sagittal section. The nasal septum has been removed, exposing the right lateral wall of the nasal cavity so that the nasal conchae can be seen. Note also the divisions of the pharynx and the position of the tonsils.

Masses of lymphoid tissue called **tonsils** are embedded in the mucous membrane of the pharynx. Recall the location of the tonsils from the previous chapter (see p. 435). The **lingual tonsils** and **palatine tonsils** are located in the oropharynx and the **pharyngeal tonsils,** also called the **adenoids** when swollen, are located in the nasopharynx.

As you read in Chapter 16, these tonsils form a ring of lymphoid tissue in the throat that provides immune protection at a critical boundary with the external environment. Although the tonsils usually protect us, they can also become infected and inflamed themselves—a condition called **tonsillitis** (Figure 17-5).

Swelling of the pharyngeal tonsils caused by infections may make it difficult or impossible for air to travel from the nose into the throat. In these cases the individual is forced to breathe through the mouth.

In a **tonsillectomy** both tonsils are generally removed by a surgeon. Once a very common surgical procedure, tonsillectomy, with its potentially serious complications—including severe hemorrhage—is now performed only after other options have been exhausted. Physicians now recognize the

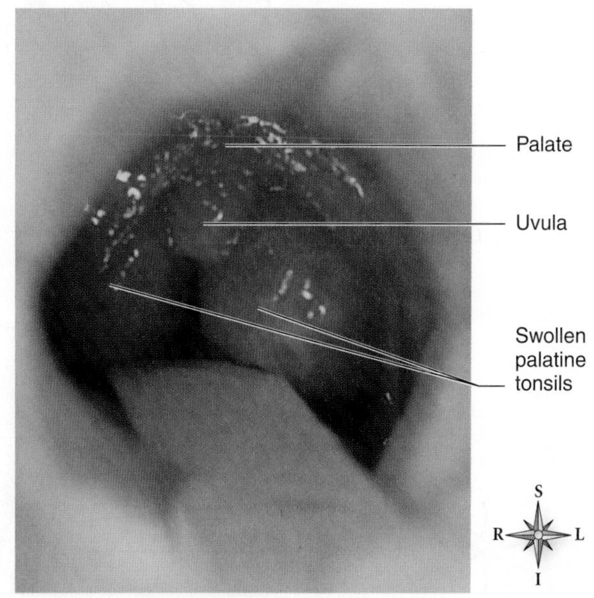

FIGURE 17-5 Tonsillitis. Enlarged palatine tonsils can be seen nearly meeting at the midline of the pharynx.

17

value of lymphoid tissue in the body's defense mechanism and delay removal of the tonsils—even in cases of inflammation (tonsillitis).

Although surgical removal may eventually be necessary in cases of repeated infections, swelling, or when nonsurgical treatments such as intensive antibiotic therapy prove ineffective, the number of tonsillectomies performed each year continues to decrease.

To protect the delicate gas-exchange tissues deep inside the lungs, the respiratory tract has many complex mechanisms that guard against injury and disease. Check out the article *Protective Strategies of the Respiratory Tract* at Connect It! at *evolve.elsevier.com.*

Larynx

The **larynx,** or voice box, is located just below the pharynx. It is composed of nine pieces of cartilage. You know the largest of these (the *thyroid cartilage*) as the "Adam's apple" (**Figure 17-6**).

Two short fibrous bands, the **vocal cords,** stretch across the interior of the larynx. Muscles that attach to the larynx cartilages can pull on these cords in such a way that they become tense or relaxed. When they are tense, the voice is high pitched; when they are relaxed, it is low pitched. The space between the vocal cords that changes shape as we speak is the **glottis.**

Another piece of cartilage, called the **epiglottis,** partially covers the opening of the larynx (see **Figure 17-6**). The epiglottis acts like a trapdoor, closing off the larynx during swallowing and preventing food and liquids from entering the trachea.

The risk of **laryngeal cancer** increases significantly with smoking and alcohol abuse. It occurs most often in men over age 50 and is often diagnosed because of persistent hoarseness and difficulty in swallowing. A number of therapeutic treatments including surgery, radiation, and chemotherapy can be curative but about one-third of those affected will die of the disease. If treatment involves surgical removal of the larynx, the individual must learn "esophageal speech" or use an electronic *artificial larynx* to speak.

Disorders of the Upper Respiratory Tract

Upper Respiratory Infection

Any infection localized in the mucosa of the upper respiratory tract (nose, pharynx, and larynx) can be called an *upper respiratory infection (URI).* Although the general designation URI is often used, such infections are sometimes named for the specific structure involved.

Rhinitis, from the Greek *rhinos,* "nose," is inflammation and swelling of the nasal mucosa. A red, itchy, "runny nose" and partially obstructed breathing are universally recognized as symptoms of *infectious rhinitis.* Most cases of infectious rhinitis are caused by viruses responsible for the common cold (rhinoviruses) or the flu (influenza viruses). Although potentially

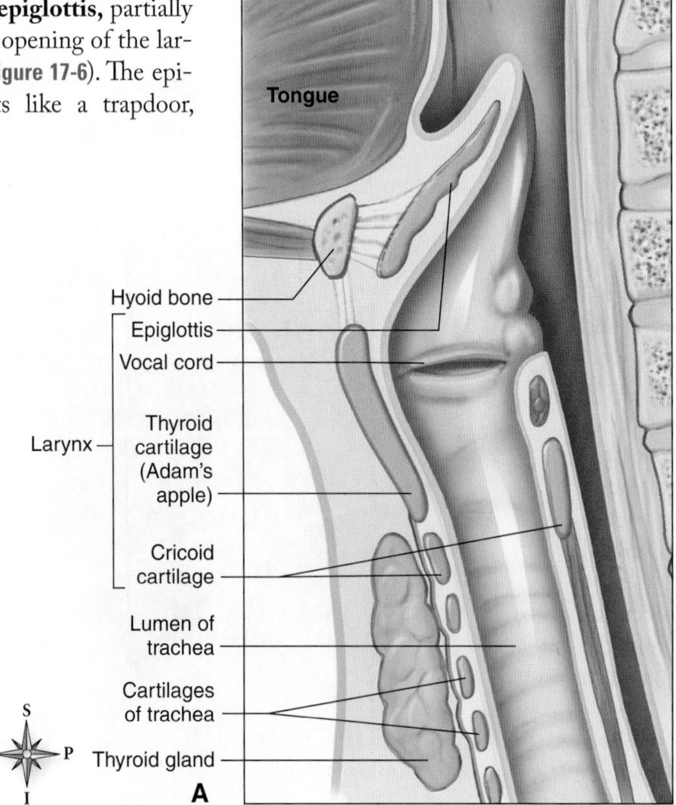

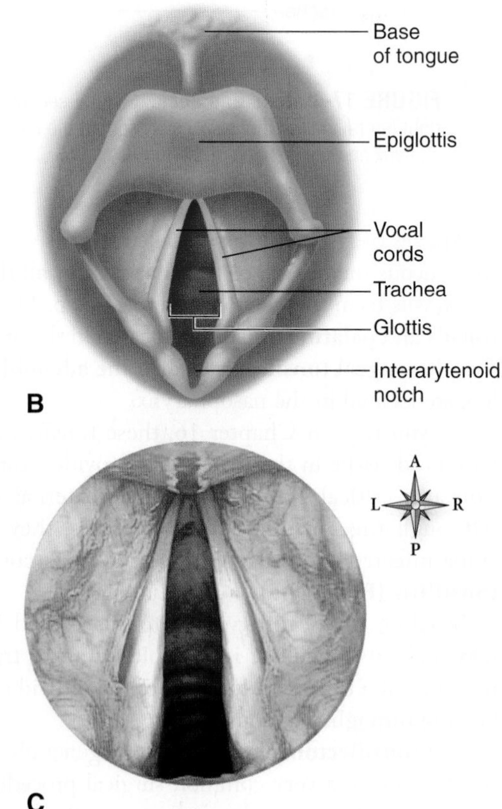

FIGURE 17-6 Larynx. A, Sagittal section of the larynx. **B,** Superior view of the larynx. **C,** Photograph of the larynx taken with an endoscope (optical device) inserted through the mouth and pharynx to the epiglottis.

CLINICAL APPLICATION

KEEPING THE TRACHEA OPEN

Often a tube is placed through the mouth, pharynx, and larynx into the trachea when patients need breathing support—especially when the airway may collapse or be obstructed. This procedure is called **endotracheal intubation.** The purpose of the tube is to ensure an open airway (see part *A* of the figure).

To ensure that the tube enters the trachea rather than the nearby esophagus (which leads to the stomach), anatomical landmarks such as the vocal folds are used. Likewise, the distinct feel of the V-shaped posterior groove called the **interarytenoid notch** (**Figure 17-6**, *B*) can help guide the proper insertion of the tube.

Another procedure done frequently in today's modern hospitals is a **tracheostomy.** This procedure involves the cutting of an opening into the trachea (part *B* of the figure). A surgeon may perform this procedure so that a suction device can be inserted to remove secretions from the bronchial tree or so that an *intermittent positive-pressure breathing (IPPB)* machine can be used to improve ventilation of the lungs.

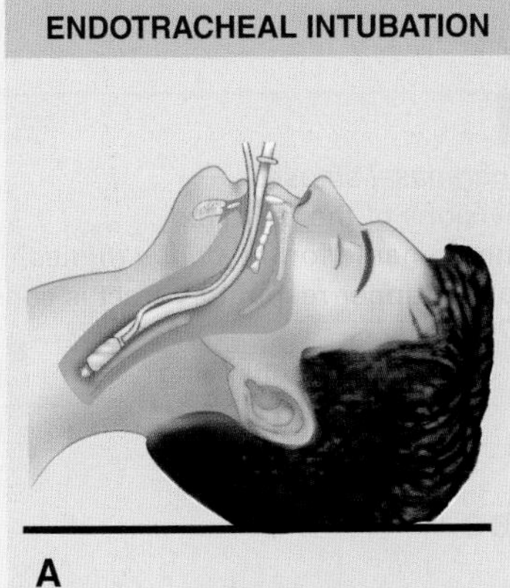

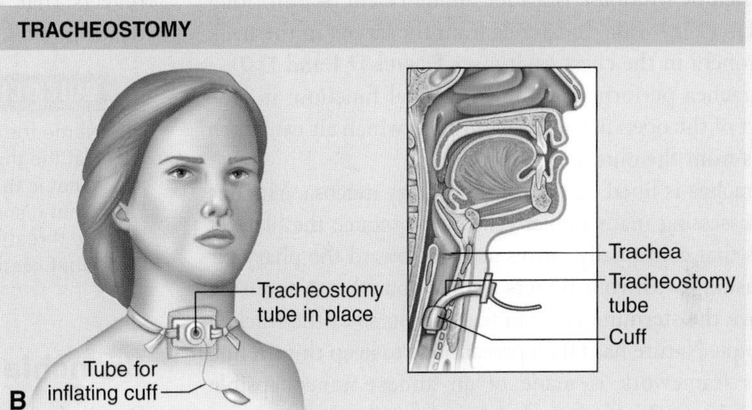

serious in susceptible individuals, most cases of infectious rhinitis resolve with (or without!) supportive treatment after about 7 to 10 days of misery.

The term *allergic rhinitis,* or "hay fever," is used to describe hypersensitivity-type reactions to many types of nasal irritants and airborne allergens including animal dander and plant pollens. Symptoms similar to infectious rhinitis may become chronic and result in formation of nasal polyps and secondary infections.

Pharyngitis, or *sore throat,* is inflammation or infection of the pharynx (throat). Pain, redness, and difficulty in swallowing are characteristic of pharyngitis. Pharyngitis may be caused by any of several pathogens, including the streptococcal bacteria that cause "strep throat" (see Appendix A at *evolve.elsevier.com*).

Laryngitis is inflammation of the mucous lining of the larynx. The inflammation is accompanied by edema of the laryngeal structures. If swelling of the vocal cords occurs, hoarseness or loss of voice results. The condition may be caused by bacteria, viruses, exposure to allergens, overuse of the voice, smoking, or other factors. Even a moderate amount of laryngeal swelling or edema, especially in a young child, can obstruct air flow and result in asphyxiation.

Epiglottitis is a life-threatening condition caused by *Haemophilus influenzae* type B *(Hib)* infection. Hib often struck children between 3 and 7 years of age a generation ago. However, introduction of Hib vaccines at the end of the twentieth century produced a 99% drop in the incidence of this infection, making this type of epiglottitis rare in our day.

The term **croup** is used to describe a non–life-threatening type of laryngitis generally seen in children younger than age 3. It is caused by the parainfluenza viruses. Symptoms include a harsh barklike cough and labored inspiration. Affected children often develop symptoms after going to sleep and awaken frightened and coughing but without a fever.

URIs are rather common, occurring several times a year in most individuals, because the upper respiratory tract is easily accessible to common airborne pathogens. Because the upper respiratory mucosa is continuous with the mucous lining of the sinuses, the eustachian tube and middle ear, and lower respiratory tract, URIs have an unfortunate tendency to spread. It is not unusual therefore to see a common cold progress to become sinusitis or *otitis media* (middle ear infection).

Anatomical Disorders

Deviated septum is a condition in which the nasal septum strays from the midline of the nasal cavity.

Nobody's nasal septum is *exactly* on the midsagittal plane, but most are fairly close. Some people, however, are born with a congenital defect of the septum that results in some degree of blockage to one or both sides of the nasal cavity. Others acquire a deviated septum after birth as a result of damage from an injury or infection. In either case, surgical correction of the anatomical abnormality often results in normal breathing through the nose.

Injury to the nose occurs relatively often because the nose projects some distance from the front of the head. Usually, common bumps and other injuries cause little if any serious damage. Occasionally, **epistaxis,** or nosebleed, occurs. The most common cause of nosebleed is a strong bump or blow, but it can result from severe inflammation or rubbing (as in rhinitis), hypertension, or even brain injury. Because of the rich blood supply close to the inside surface of the nasal cavity, even minor nosebleeds can produce a great deal of blood—causing them to appear to be a more serious injury than they are.

Lower Respiratory Tract

Trachea

17

The **trachea** or windpipe is a tube about 11 cm (4.5 in) long and 2.5 cm (1 in) wide. It extends from the larynx in the neck to the bronchi in the chest cavity (see **Figures 17-1** and **17-7**).

The trachea performs a simple but vital function: it provides part of the open passageway through which air can reach the lungs from the outside.

The trachea is lined by typical respiratory mucosa. Mucous glands possessing many goblet cells help produce the blanket of mucus that continually moves upward toward the pharynx.

By pushing with your fingers against your throat about an inch above the sternum, you can feel the shape of the trachea or windpipe. Nature has taken precautions to keep this lifeline open. Its framework is made of an almost noncollapsible material—15 to 20 C-shaped rings of cartilage placed one above the other with only a little soft tissue between them.

Figure 17-7, *B,* shows how the incomplete cartilage rings permit easy swallowing by allowing the esophagus (food tube) to stretch within the narrow space in the neck between the trachea and the vertebrae.

Despite the structural safeguard of cartilage rings, blockage of the trachea sometimes occurs. A tumor or an infection may enlarge the lymph nodes of the neck so much that they squeeze the trachea shut, or a person may aspirate (breathe in) a piece of food or something else that blocks the windpipe. Because air has no other way to get to the lungs, complete tracheal obstruction causes death in a matter of minutes.

Suffocation from all causes, including choking on food and other substances caught in the trachea, kills more than 4000 people each year—making it the fifth major cause of accidental deaths in the United States. Many experts recommend the *five-and-five* method (described in the box on p. 467 to free the trachea of ingested food or other foreign objects that would otherwise block the airway and cause death in choking cases.

QUICK CHECK

1. Name the four *paranasal sinuses.*
2. List the three divisions of the *pharynx.*
3. What is the common name for *infectious rhinitis?*
4. What disorder of the upper respiratory tract is considered life threatening?
5. What keeps the *trachea* from collapsing?

Bronchial Tree

Recall that one way to picture the thousands of air passages that make up the lungs is to think of an upside-down tree. The trachea is the main trunk of this tree; the right bronchus (the tube leading into the right lung) and the left bronchus (the tube leading into the left lung) are the trachea's first branches or **primary bronchi.**

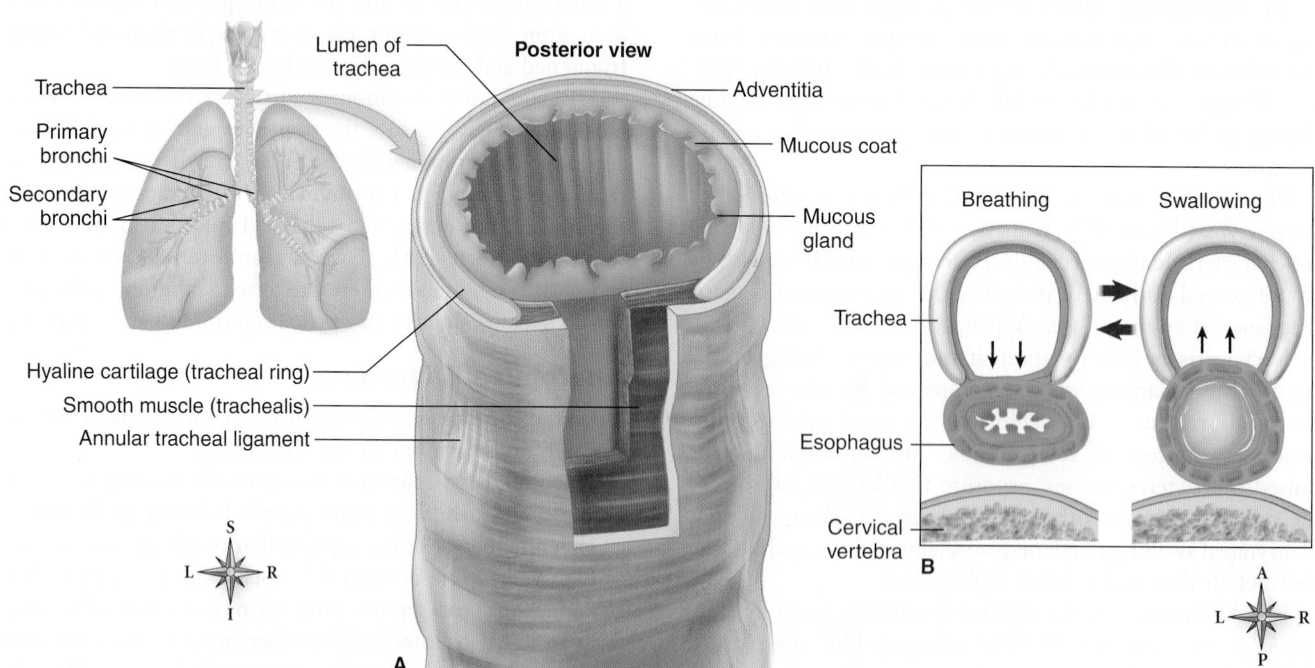

FIGURE 17-7 Trachea. A, Structure of trachea shown in a posterior view. *Inset* at top shows from where the transverse section was cut. **B,** Incomplete cartilage rings and elasticity of posterior tracheal wall allow the esophagus to expand during swallowing.

CLINICAL APPLICATION
FIVE-AND-FIVE RESCUE FOR CHOKING

Choking often occurs when something becomes lodged in the larynx and cannot be dislodged by normal coughing. Many experts recommend that a person who is choking should receive the *five-and-five* rescue method. Five blows to the back (between the scapulae) with the heel of the hand (Figure A) may be followed by five abdominal thrust maneuvers (Figure B). If needed, this sequence may be repeated until the object is dislodged.

Most accidental airway obstructions result from pieces of food aspirated during a meal—the condition is sometimes referred to as "café coronary." Other objects such as chewing gum or balloons are frequently the cause of obstructions in children. Individuals trained in emergency procedures must be able to tell the difference between airway obstruction and other conditions that produce similar symptoms, such as heart attacks. The key question they must ask the person who appears to be choking is, "Can you talk?" A person with an obstructed airway will not be able to speak, even while conscious.

In the five-and-five rescue, the five back blows help dislodge foreign material from the larynx. The abdominal thrust maneuver uses air already present in the lungs to expel the object obstructing the airway.

The abdominal thrust maneuver was formerly named after the physician who helped popularize it for general use in the 1970s, Henry Heimlich. More recently, the *eponym* (term that includes a person's name), that is, *Heimlich maneuver,* is dropping from common use and the technique is now often called simply **abdominal thrusts.**

The abdominal thrust maneuver, if the person is standing, consists of the rescuer grasping the person with both arms around the person's waist just below the rib cage and above the navel (Figure B). The rescuer makes a fist with one hand, grasps it with the other, and then delivers an upward thrust against the diaphragm just below the xiphoid process of the sternum. Air trapped in the lungs is pressurized, hopefully forcing the object that is choking the person out of the airway.

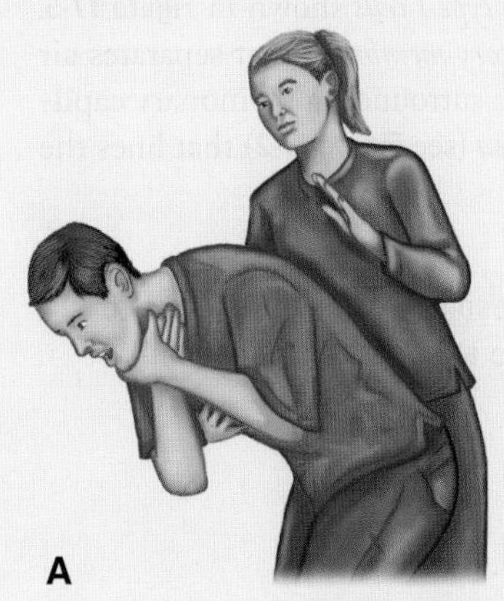
A

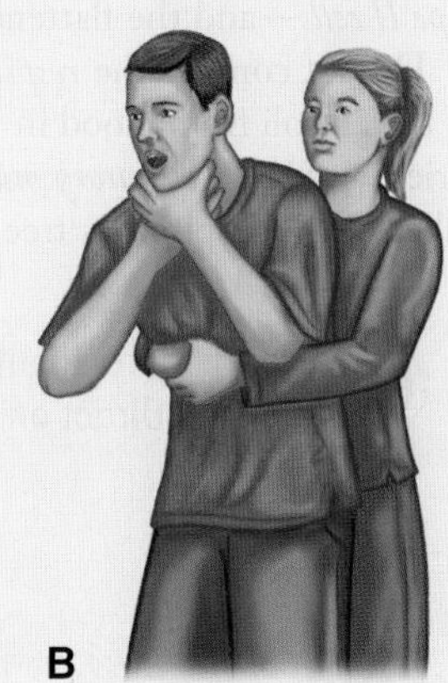
B

17

The right primary bronchus is more vertical and "in line" with the terminal trachea than is the left. As a result, aspirated objects that enter the trachea tend to enter and lodge in the right primary bronchus or lung more often than the left.

In each lung, the primary bronchi branch into smaller **secondary bronchi.** Cartilage rings in the wall of each secondary bronchus, like those of the trachea and primary bronchi, keep the air passages open. The secondary bronchi divide into smaller and smaller tubes, ultimately branching into tiny tubes whose walls contain only smooth muscle—no cartilage rings. These very small passageways are called **bronchioles.**

The bronchioles subdivide into microscopic tubes called **alveolar ducts,** which resemble the main stem of a bunch of grapes (**Figure 17-8**). Each alveolar duct ends in several **alveolar sacs,** each of which resembles a cluster of grapes, and the wall of each alveolar sac is made up of numerous **alveoli,** each of which resembles a single, hollow grape. Notice in the sectioned part of **Figure 17-8** that some of the alveoli are interconnected with each other. This anatomical feature allows more efficiency in ventilating all the alveoli equally—supporting the concept that *structure fits function*.

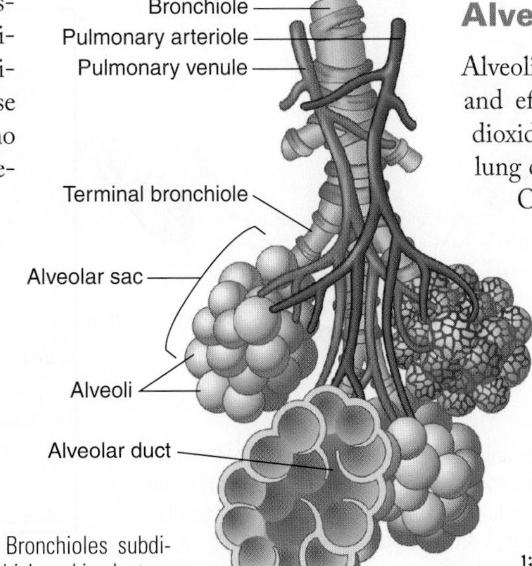

FIGURE 17-8 Bronchioles and alveoli. Bronchioles subdivide to form tiny tubes called *alveolar ducts,* which end in clusters of alveoli called *alveolar sacs.*

Alveoli

Alveoli are very effective in promoting the rapid and effective exchange of oxygen and carbon dioxide between blood circulating through the lung capillaries and alveolar air.

Once again, structure and function are closely related. Two aspects of the structure of alveoli assist in diffusion and make them able to perform this function admirably.

First, the wall of each alveolus is made up mostly of a single layer of simple squamous epithelial cells called *type I cells*. The walls of the capillaries that surround and lie in contact with them are also made of thin, flat endothelial cells. This means that, between the blood in the capillaries and the air in each alveolus, there is a barrier less than 1 micron thick. This extremely thin barrier is called the **respiratory membrane** (**Figure 17-9**).

Second, there are millions of alveoli. This means that together they create an enormous surface for gas exchange. The total surface area of all alveoli together is approximately 84 square meters (915 square feet)—about the size of a small home's floor plan. This huge surface area allows large amounts of oxygen and carbon dioxide to be rapidly exchanged.

The surface of the respiratory membrane inside each alveolus is covered by a substance called **surfactant.** Surfactant helps reduce *surface tension* or "stickiness" of the watery mucus lining the alveoli—keeping the alveoli from collapsing as air moves in and out during respiration. Note the difference in appearance between the surfactant-producing cells—called *type II cells*—and the flattened *type I cells* shown in **Figure 17-9**.

Do not confuse the *respiratory membrane* that separates air in the alveoli from blood in the surrounding pulmonary capillaries with the *respiratory mucosa* (see **Figure 17-2**) that lines the tubes of the respiratory tree.

To learn more about respiratory membrane, go to AnimationDirect online at *evolve.elsevier.com*.

Respiratory Distress

Respiratory distress results from the body's relative inability to inflate the alveoli of the lungs normally. **Respiratory distress syndrome (RDS)** is a condition most often caused by absence or impairment of the surfactant in the fluid that lines the alveoli.

Infant Respiratory Distress

Infant respiratory distress syndrome (IRDS) is a very serious, life-threatening condition that often affects premature infants of less than 37 weeks' gestation or those who weigh less than 2.2 kg (5 lb) at birth. IRDS is the leading cause of death among premature infants in the United States, claiming more than 5000 premature babies each year. The disease, characterized by a lack of surfactant in the alveolar air sacs, affects 50,000 babies annually.

Surfactant reduces the surface tension of the water fluid on the free surface of the alveolar walls and permits easy movement of air into and out of the lungs. The ability of the body to manufacture this important substance is not fully developed until shortly before birth—normally about 40 weeks after conception.

In newborn infants who are unable to manufacture surfactant, many air sacs collapse during expiration because of the

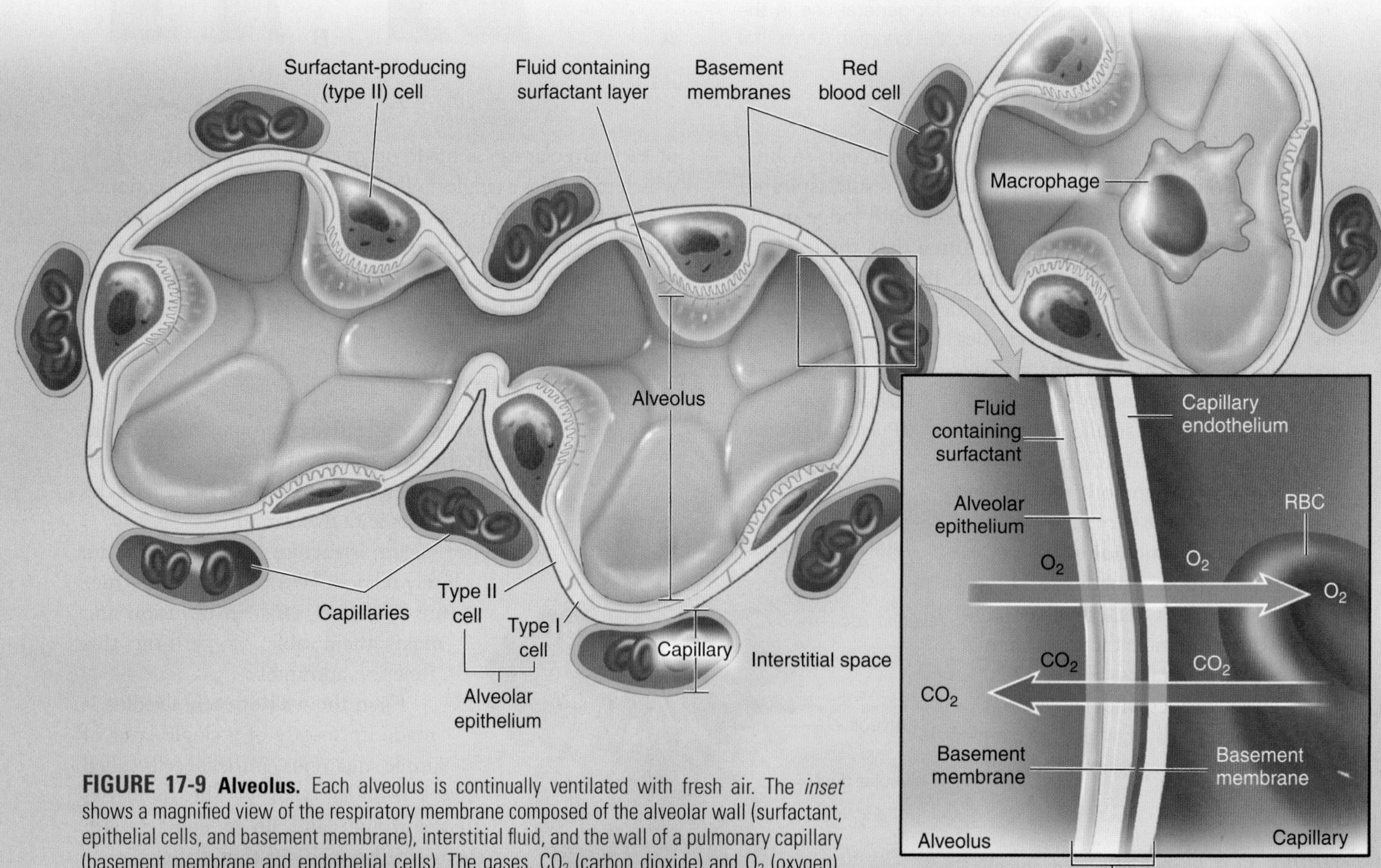

FIGURE 17-9 Alveolus. Each alveolus is continually ventilated with fresh air. The *inset* shows a magnified view of the respiratory membrane composed of the alveolar wall (surfactant, epithelial cells, and basement membrane), interstitial fluid, and the wall of a pulmonary capillary (basement membrane and endothelial cells). The gases, CO_2 (carbon dioxide) and O_2 (oxygen), diffuse across the respiratory membrane. *RBC*, Red blood cell.

increased surface tension. The effort required to reinflate these collapsed alveoli is much greater than that needed to reinflate normal alveoli with adequate surfactant. The baby soon develops labored breathing, and symptoms of respiratory distress appear shortly after birth.

Current treatment of IRDS usually involves delivering air under pressure and applying prepared surfactant directly into the baby's airways by means of a tube.

Adult Respiratory Distress

Adult respiratory distress syndrome (ARDS) is caused by impairment or removal of surfactant in the alveoli. For example, accidental inhalation of foreign substances such as water, vomit, smoke, or chemical fumes can cause ARDS. Edema of the alveolar tissue can impair surfactant and reduce the alveoli's ability to stretch, causing respiratory distress.

ARDS can be treated with supplemental oxygen supplied by nasal prongs or, in some cases, endotracheal intubation and a mechanical ventilator. The underlying cause of the condition is then assessed and treated.

Lungs

The **lungs** are fairly large organs. They develop to fill most of the thoracic cavity, leaving a small central space—the mediastinum—for the heart, large blood vessels, thymus, and esophagus. Note in **Figure 17-10** that deep grooves called *fissures* subdivide each lung into lobes. The right lung has three lobes and the left lung has two.

Figure 17-10 shows the relationship of the lungs to the rib cage at the end of a normal expiration. The narrow, superior portion of each lung, up under the collarbone, is the *apex*. The broad, inferior portion resting on the diaphragm is the *base*.

Each lung is made up of all the elements of the bronchial tree, alveoli, and pulmonary blood vessels—along with connective tissues, lymphatic vessels, and nerves. Each lung, therefore, is a combination of several kinds of structures that form a unit for respiration.

Pleurae

A **pleura** is a serous membrane that covers the outer surface of each lung and lines the inner surface of the rib cage. The pleura resembles other serous membranes in relation to its structure and function. Like the peritoneum or pericardium, the pleura is an extensive, thin, moist, slippery membrane. It lines a large, closed cavity of the body and covers the organs located within it.

The *parietal pleura* lines the walls of the thoracic cavity. The *visceral pleura* covers the lungs, and the intrapleural space lies between the two pleural membranes (**Figure 17-11**).

Normally the intrapleural space contains just enough pleural fluid to make both portions of the pleura moist and slippery and able to glide easily against each other as the lungs expand and deflate with each breath.

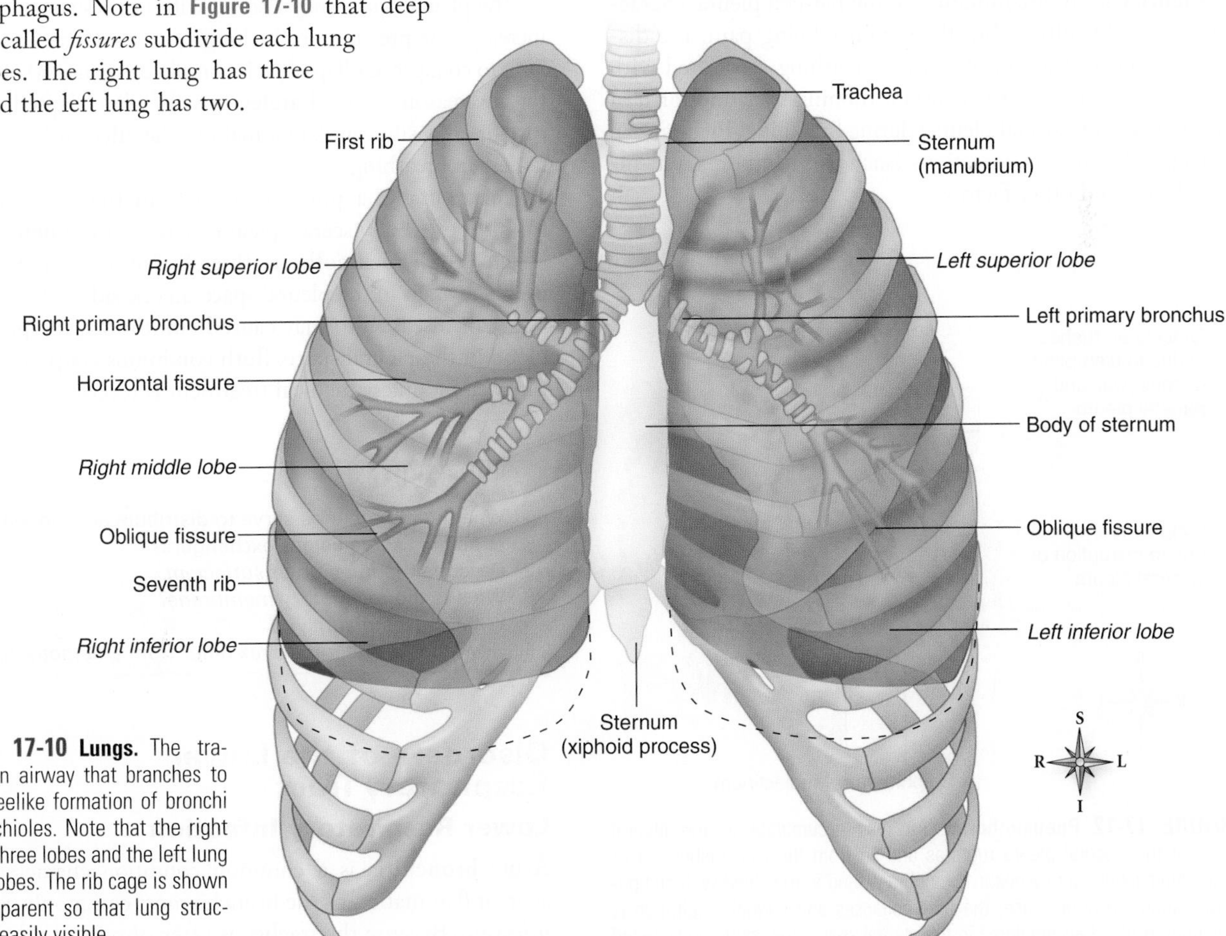

FIGURE 17-10 Lungs. The trachea is an airway that branches to form a treelike formation of bronchi and bronchioles. Note that the right lung has three lobes and the left lung has two lobes. The rib cage is shown semitransparent so that lung structures are easily visible.

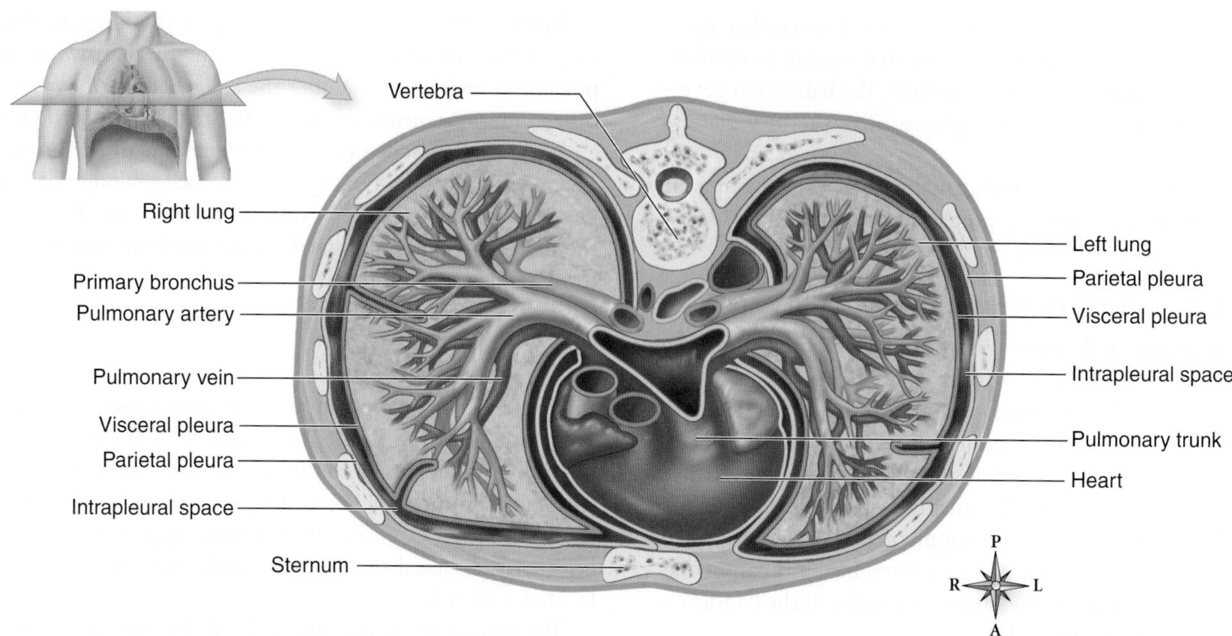

FIGURE 17-11 Lungs and pleura. The *inset* shows where the body was cut to show this transverse section of the thorax. A serous membrane lines the thoracic wall (parietal pleura) and then folds inward near the bronchi to cover the lung (visceral pleura). The intrapleural space contains a small amount of serous pleural fluid.

Pleurisy is an inflammation of the parietal pleura, characterized by difficulty in breathing and stabbing pain. The discomfort and restriction of normal breathing associated with pleurisy are caused by the constant rubbing back and forth of the visceral and parietal pleurae during breathing. Pleurisy can be caused by tumors, infections (such as pneumonia and tuberculosis), and other factors.

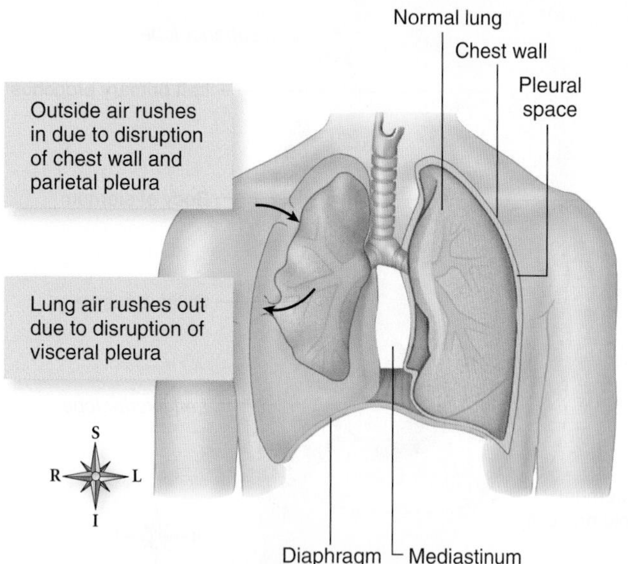

FIGURE 17-12 Pneumothorax. Air may accumulate in the pleural space if the visceral pleura ruptures and air from the lung rushes out or when atmospheric air rushes in through a wound in the chest wall and parietal pleura. In either case, the lung collapses and normal respiration is impaired. If blood accumulates in the pleural space, the condition is called *hemothorax*.

The pleural space may also fill with other substances, which increases the pressure on the lung's outer surface—causing the lung to collapse. Collapse (or incomplete expansion) of the lung for any reason is called **atelectasis.** While collapsed, the lung cannot be easily ventilated, making the affected lung virtually useless in breathing.

For example, a puncture wound to the chest wall or a rupture of the visceral pleura may cause **pneumothorax** (**Figure 17-12**). Pneumothorax (literally "air in the thorax") is the presence of air in the pleural space on one side of the chest. An injury or disease also may cause **hemothorax,** the presence of blood in the pleural space. Both conditions are potentially life threatening unless medical treatment is received.

QUICK CHECK

1. What lung structures serve to distribute air, and which structures serve as gas exchangers?
2. Describe the function of *surfactant.*
3. What is the *respiratory membrane*?
4. What causes *pleurisy*?
5. How does a *pneumothorax* differ from a *hemothorax*?

Disorders of the Lower Respiratory Tract

Lower Respiratory Infection

Acute **bronchitis** is a common condition characterized by acute inflammation of the bronchi, most commonly caused by infection. Because the trachea is often also involved, the condition may be called *tracheobronchitis.* This condition is often preceded by a URI that seems to move down into the trachea and bronchi after several days.

Acute bronchitis often starts with a nonproductive cough that progresses to a deep cough that produces **sputum** containing mucus and pus.

Pneumonia is an acute inflammation of the lungs in which the alveoli and bronchi become plugged with thick fluid (exudate).

The vast majority of pneumonia cases results from infection by *Streptococcus pneumoniae* bacteria, but it can be caused by several other bacteria, viruses, and fungi (see Appendix A at *evolve.elsevier.com*).

Pneumonia is characterized by a high fever, severe chills, headache, cough, and chest pain. The fact that each day more than 10,000 liters of potentially contaminated air enters the respiratory system helps explain why pneumonia is such a common illness—especially in individuals with lowered resistance or impaired immune systems.

Types of pneumonia include *lobar pneumonia,* which typically affects an entire lobe of the lung, and *bronchopneumonia* in which patches of infection are scattered along portions of the bronchial tree (**Figure 17-13**). The term *aspiration pneumonia* describes lung infections caused by inhalation of vomit or other infective material. It is common in acute alcohol intoxication and as a result of anesthesia.

Tuberculosis (TB) is a chronic bacillus infection caused by *Mycobacterium tuberculosis* (see Appendix A at *evolve.elsevier .com*). TB is a highly contagious disease transmitted through inhalation or swallowing of droplets contaminated with the TB bacillus. It usually affects the lungs and surrounding tissues but can invade any other tissue or organ as well.

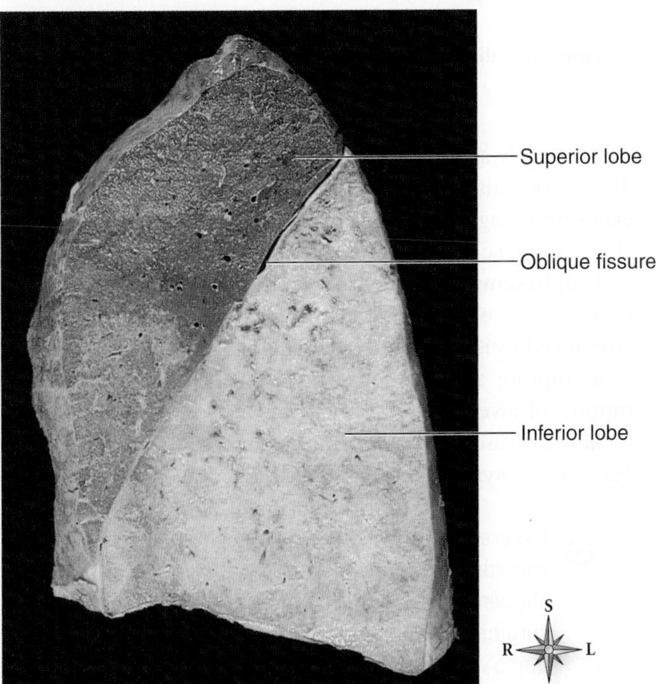

FIGURE 17-13 Lobar pneumonia. Exudate fills many of the alveoli and ducts of a single lobe of the left lung. Note the difference in texture and color of the affected inferior lobe. See **Figure 17-10**.

Early stages of TB are characterized by fatigue, chest pain, pleurisy, weight loss, and fever. As the disease progresses, lung hemorrhage and dyspnea may develop.

The name *tuberculosis* literally means "condition of having tubercles," which describes the protective capsules the body forms around colonies of TB bacilli. Successful treatment requires a combination of drugs and other therapies for an extended period—usually longer than a year. TB affects up to a third of the world's population and is a major cause of death in many poor, densely populated regions of the world. It has recently reemerged as a serious health problem in some major U.S. cities.

Any lung infection is usually treated primarily with antibiotic therapy directed at the specific type of pathogen suspected as the cause. Supportive therapy to maintain blood oxygen concentration and minimize patient discomfort is used alongside antibiotic therapy.

17

Restrictive Pulmonary Disorders

Restrictive pulmonary disorders involve restriction (reduced stretch) of the alveoli, as the name implies. Because inspiration requires that the lungs have the ability to stretch—a property called **compliance**—restrictive disorders reduce a person's ability to inhale normally. How such changes in pulmonary volumes are measured is described in a later section.

Some restrictive disorders arise in connective tissue of the lung itself. For example, inflammation or *fibrosis* (scarring) of lung tissue caused by exposure to asbestos, coal, or silicon dust can reduce compliance and thus restrict alveoli. Restriction of breathing also can be caused by the pain that accompanies pleurisy or mechanical injuries, such as rib fractures.

Another type of restrictive disorder is *cystic fibrosis (CF),* which was described in Chapter 3 (see p. 55). Recall that CF is characterized by thickened fluids in the lungs, which restricts compliance.

Obstructive Pulmonary Disorders

A number of different conditions may cause obstruction of the airways. For example, exposure to cigarette smoke and other common air pollutants can trigger a reflexive constriction of bronchial airways. Obstructive disorders may obstruct *inspiration* and *expiration,* whereas restrictive disorders mainly restrict *inspiration.* In obstructive disorders, the total lung capacity may be normal, or even high, but the time it takes to inhale or exhale maximally is significantly increased.

Some major obstructive disorders are summarized here and in **Figure 17-14**.

Acute obstruction of the airways, as when a piece of food blocks air flow, requires immediate action to avoid death from suffocation (see the box on p. 467).

Chronic obstructive pulmonary disease (COPD) is a broad term used to describe conditions of progressive irreversible obstruction of expiratory air flow. People with COPD have chronic difficulties with breathing, mainly emptying their lungs, and have visibly hyperinflated chests. Those with COPD often have a productive cough and intolerance of

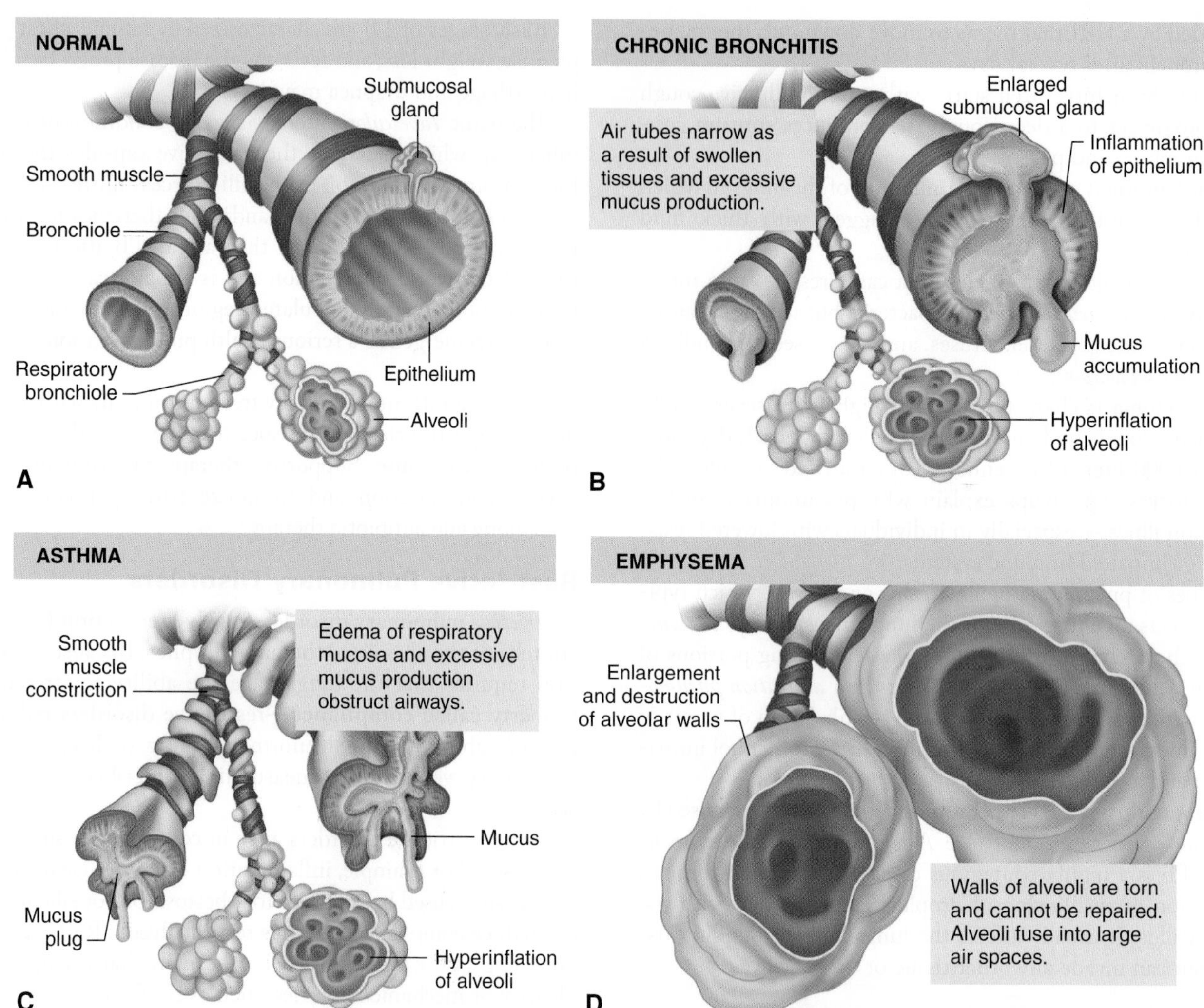

FIGURE 17-14 Major obstructive pulmonary disorders.

activity. The major disorders observed in people with COPD are chronic bronchitis and emphysema.

In North America, tobacco use is the primary cause of COPD, but air pollution, asthma, and respiratory infections also play a role. COPD is a leading cause of death, and the death rate from COPD is *increasing!* Until a few years ago, more men had COPD than women. However, the increase of smoking among women is thought to account for the fact that the occurrence rate for female COPD is growing rapidly.

Acute respiratory failure can occur when any of the disorders that produce COPD become intense. Heart failure resulting from the vascular resistance that develops with COPD is another possible outcome. Although there is no cure for chronic obstructive respiratory conditions, limiting symptoms can improve quality of life. Bronchodilators and corticosteroids have been used to relieve some of the airway obstruction involved in COPD.

Chronic bronchitis is a chronic inflammation of the bronchi and bronchioles. It is characterized by edema and excessive mucus production, which block air passages. People with chronic bronchitis have difficulty with exhaling and often cough deeply as they try to dislodge the accumulating mucus. The major cause of chronic bronchitis is cigarette smoking or exposure to cigarette smoke. Exposure to other air pollutants also may cause chronic bronchitis.

Emphysema may result from the progression of chronic bronchitis or other conditions as air becomes trapped within alveoli and causes them to enlarge. As the alveoli enlarge, their walls rupture and then fuse into large irregular spaces. The rupture of alveoli reduces the total surface area of the lung, making breathing difficult. Emphysema patients often develop *hypoxia,* or oxygen deficiency, in the internal environment.

Severe lung damage caused by emphysema is sometimes treated surgically. For micrographs showing emphysema damage and a description of surgical options, see the article *Lung Volume Reduction Surgery* at Connect It! at *evolve.elsevier.com.*

Asthma is an obstructive disorder characterized by recurring spasms of the smooth muscle in the walls of the bronchioles. The muscle contractions narrow the airways, making breathing difficult. Inflammation (edema and excessive mucus production) usually accompanies the spasms, further obstructing the airways. Asthma can be triggered by stress, heavy exercise, infection, or inhaling allergens or other irritants.

Lung Cancer

Lung cancer is a malignancy of pulmonary tissue that not only destroys the vital gas-exchange tissues of the lungs but, like other cancers, also may invade other parts of the body (metastasis).

Surgery is the most effective treatment for lung cancer known, but only half of the persons diagnosed as having lung cancer are good candidates for surgery because of extensive spread of the disease (metastasis) at the time of diagnosis. In a **lobectomy,** only the affected lobe of a lung is removed. **Pneumonectomy** is the surgical removal of an entire lung.

Review the article *Metastasis* at Connect It! at *evolve.elsevier.com.*

QUICK CHECK

1. What is the difference between *bronchopneumonia* and *lobar pneumonia*?
2. Do restrictive pulmonary disorders restrict mainly *inspiration* or *expiration*?
3. Give two examples of *chronic obstructive pulmonary disease (COPD).*
4. Why is surgery an effective treatment for only half of the persons diagnosed with lung cancer?

Respiration

Respiration means exchange of gases (oxygen and carbon dioxide) between a living organism and its environment. If the organism consists of only one cell, gases can move directly between it and the environment. If, however, the organism consists of billions of cells, as do our bodies, most of its cells are too far removed from the air source for a direct exchange of gases to occur. To overcome this obstacle, a pair of organs—the lungs—provides a place where air and a circulating fluid (blood) can come close enough to each other for oxygen to move out of the air into the blood while carbon dioxide moves out of the blood into the air.

Breathing, or **pulmonary ventilation,** is the process that moves air into and out of the lungs. It makes possible the exchange of gases between air in the lungs and in the blood. Together, these processes are often called **external respiration.**

In addition, exchange of gases occurs between the blood and the cells of the systemic tissues of the body, which then use the oxygen in the biochemical pathways that transfer energy from nutrient molecules to adenosine triphosphate (ATP). Together, these processes are called **internal respiration.** The term **cellular respiration** refers to the use of oxygen by cells in the process of metabolism, which is discussed further in Chapter 19.

All of these respiratory processes require transport of gases (oxygen and carbon dioxide) by the blood. And because a relatively constant setpoint concentration of these blood gases is required for survival, there are complex regulatory mechanisms that control them.

Figure 17-15 summarizes the major concepts of respiratory physiology, and we use it as the basis of discussion in the remaining sections of this chapter. Refer to this figure after reading each section to help you put your new learning into a useful "big picture" and deepen your understanding.

Pulmonary Ventilation

Mechanics of Breathing

Pulmonary ventilation, or breathing, has two phases. **Inspiration,** or *inhalation,* moves air into the lungs, and **expiration,** or *exhalation,* moves air out of the lungs.

The lungs are enclosed within the thoracic cavity. Thus changes in the shape and size of the thoracic cavity result in changes in the air pressure within that cavity and in the lungs. This difference in air pressure is the driving force of movement of air into and out of the lungs. Air moves from an area where pressure is high to an area where pressure is lower. Anything that flows—whether it is blood, lymph, or air—follows this primary principle: a fluid always flows down a pressure gradient.

For ventilation to occur, the tissues of the thorax and lungs must remain both *compliant* (able to stretch) and *elastic* (able to recoil after stretch). Respiratory muscles are responsible for the changes in the shape of the thoracic cavity that change the internal air pressures involved in breathing.

Inspiration

Inspiration occurs when the chest cavity enlarges. As the thorax enlarges, the lungs expand along with it, and air rushes into them and down into the alveoli. This happens because of a very important law of physics: the volume and pressure of a gas are inversely proportional. That means that when the volume of a gas goes up, as it does when we expand the thorax, then the pressure goes down. Thus, air pressure in the lungs decreases during inspiration. When air pressure in the lungs is less than atmospheric air pressure, air rushes down its pressure gradient into the lungs.

Muscles that increase the volume of the thorax are classified as **inspiratory muscles.** These include the *diaphragm* and the *external intercostal muscles.*

The **diaphragm** is the dome-shaped muscle separating the abdominal cavity from the thoracic cavity. The diaphragm flattens out when it contracts during inspiration. Instead of protruding up into the chest cavity, as it does at rest, it moves down toward the abdominal cavity as it contracts. Thus the contraction or flattening of the diaphragm makes the chest cavity longer from top to bottom. The diaphragm is the most important muscle of inspiration. Nerve impulses passing through both **phrenic nerves** stimulate the diaphragm to contract.

The external intercostal muscles are located between the ribs. When they contract, they pull the ribs upward and

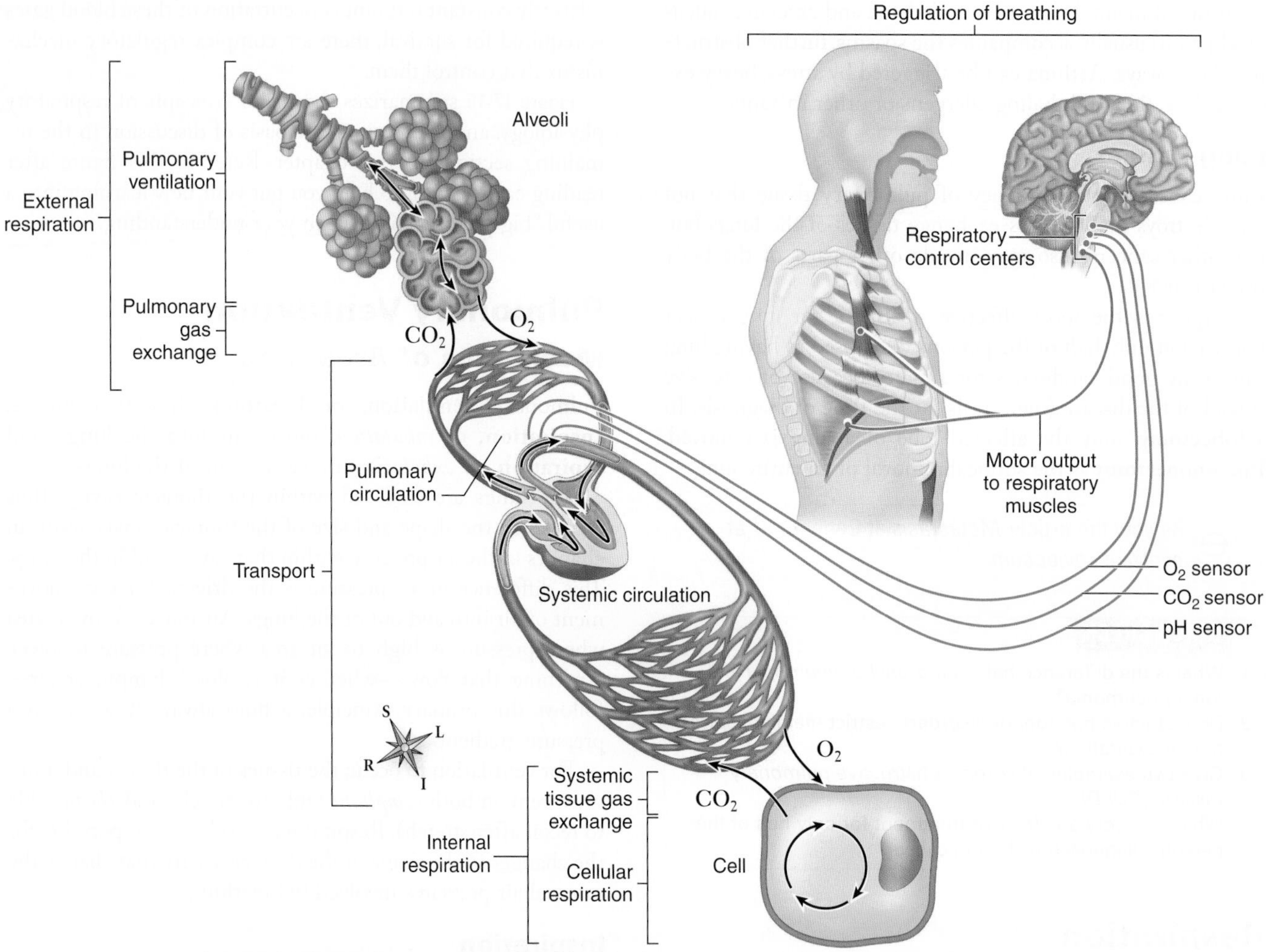

FIGURE 17-15 Overview of respiration. This chapter is organized around the principle that respiratory function includes external respiration (ventilation and pulmonary gas exchange), transport of gases by blood, and internal respiration (systemic tissue gas exchange and cellular respiration). Cellular respiration is discussed separately (see Chapter 19). Regulatory mechanisms centered in the brainstem use feedback from blood gas sensors to regulate ventilation.

outward. This enlarges the thorax by increasing the size of the cavity from anterior to posterior and from side to side.

Contraction of the inspiratory muscles increases the volume of the thoracic cavity and reduces lung air pressure below atmospheric air pressure, drawing air into the lungs (**Figure 17-16**).

Expiration

Quiet, resting expiration is ordinarily a passive process that begins when the inspiratory muscles relax and return to their resting length. The thoracic cavity then returns to its smaller size. The elastic nature of thoracic and lung tissue also causes these organs to "recoil" and decrease in size. Because volume and pressure are inversely proportional (one goes up as the other goes down), as lung volume decreases, the lung air pressure increases. As the lung air pressure rises above atmospheric air pressure, air flows down its pressure gradient and outward through the respiratory passageways.

When we speak, sing, or do heavy work, we may need more forceful expiration to increase the rate and depth of ventilation. During more forceful expiration, the **expiratory muscles** (internal intercostals and several abdominal muscles) contract.

When contracted, the internal intercostal muscles pull the rib cage inward and decrease the front-to-back size of the thorax. Contraction of the abdominal muscles pushes the abdominal organs against the underside of the diaphragm, pushing it farther upward into the thoracic cavity. As the thoracic cavity decreases in size, the air pressure within it increases above atmospheric air pressure and air flows out of the lungs (see **Figure 17-16**).

To learn more about pulmonary ventilation, go to AnimationDirect online at *evolve.elsevier.com*.

Pulmonary Volumes

A special device called a **spirometer** is used to measure the amount of air exchanged in breathing. **Figure 17-17** illustrates the various pulmonary volumes that can be measured as a subject breathes into a spirometer.

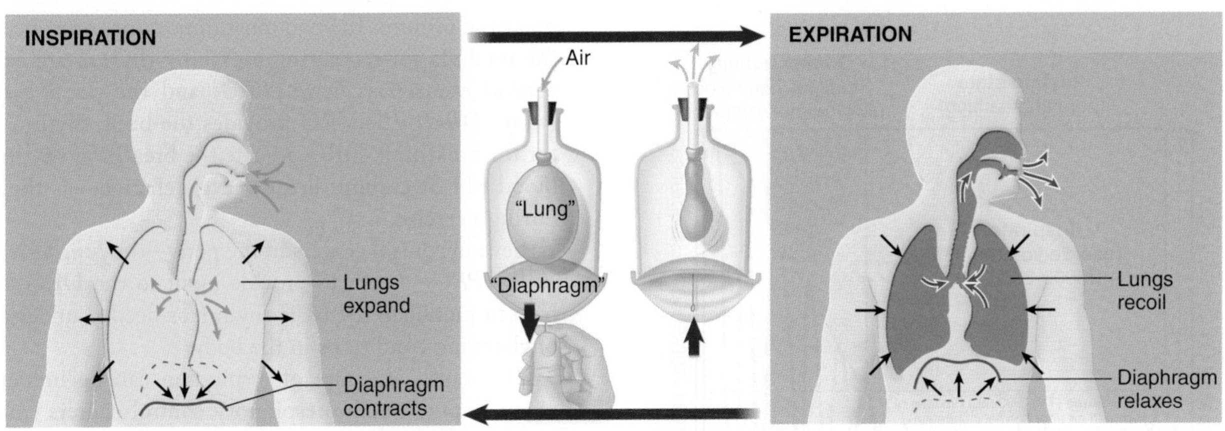

FIGURE 17-16 Mechanics of breathing. During *inspiration,* the diaphragm contracts, increasing the volume of the thoracic cavity. This increase in volume results in a decrease in pressure, which causes air to rush into the lungs. During *expiration,* the diaphragm returns to an upward position, reducing the volume in the thoracic cavity. Air pressure increases then, forcing air out of the lungs. The insets show the classic model in which a jar represents the rib cage, a rubber sheet represents the diaphragm, and a balloon represents the lungs.

We take approximately 500 mL (about a pint) of air into our lungs with each normal inspiration and expel it with each normal expiration. Because this amount comes and goes regularly like the tides of the sea, it is referred to as the **tidal volume (TV).**

The largest amount of air that we can breathe out in one expiration—by inhaling as deeply as possible, then exhaling fully—is known as the **vital capacity (VC).** In normal young men, this is about 4800 mL. Normal ranges for VC vary with age, gender, body size, fitness, and other variables.

Tidal volume and vital capacity are frequently measured in patients with lung or heart disease, conditions that often lead to abnormal volumes of air being moved in and out of the lungs.

Observe the area in **Figure 17-17** that represents the **expiratory reserve volume (ERV).** This is the amount of air that can be forcibly exhaled after expiring the tidal volume. Compare this with the area in **Figure 17-17** that represents the **inspiratory reserve volume (IRV).** The IRV is the amount of air that can be forcibly inspired over and above a normal inspiration. As the tidal volume increases, the ERV and IRV decrease.

Residual volume (RV) is simply the air that remains in the lungs after the most forceful expiration.

Note in **Figure 17-17** that vital capacity is the total of tidal volume, inspiratory reserve volume, and expiratory reserve volume—or expressed in another way: VC = TV + IRV + ERV. In pulmonary terminology, a *capacity* is a combination of two or more pulmonary volumes.

Clinically useful pulmonary volumes and are described in **Table 17-1**.

QUICK CHECK

1. What is *internal* respiration?
2. How does the diaphragm operate during inspiration? During expiration?
3. What nerves stimulate the diaphragm to contract?
4. What is the *vital capacity*? How is it measured?

Regulation of Ventilation

Homeostasis of Blood Gases

Although we may take only 12 to 18 breaths a minute at rest, we take considerably more than this when we are exercising. Not only do we take more breaths, but our tidal volume also increases with physical activity.

The reason our respiratory rate changes is because our body attempts to maintain a high setpoint level of oxygen and a low setpoint level of carbon dioxide in our blood. When our cells use up oxygen quickly during exercise, they draw more oxygen from the blood—reducing blood oxygen concentration below its set point. Likewise, cells release waste carbon dioxide into the blood more rapidly during exercise—thus raising the blood carbon dioxide concentration above its set point.

Various regulatory mechanisms respond to these changes through negative feedback loops that reverse blood gas concentrations back toward their setpoint values—by changing our respiratory rate and depth of breathing.

Brainstem Control of Respiration

Changes in respiration depend on proper functioning of the muscles of respiration. These muscles are stimulated by nervous impulses that originate in **respiratory control centers** located in the brainstem.

The brainstem centers are influenced by input from a number of sensory receptors located in different areas of the body. These receptors can sense the need for changing the rate or depth of respirations to maintain homeostasis. Certain receptors sense carbon dioxide or oxygen levels, whereas others sense blood acid levels or the amount of stretch in lung tissues.

A group of control centers in the medulla—the *medullary rhythmicity area*—seem to produce the basic rhythm of breathing. A normal resting breathing rate is about 12 to 18

17

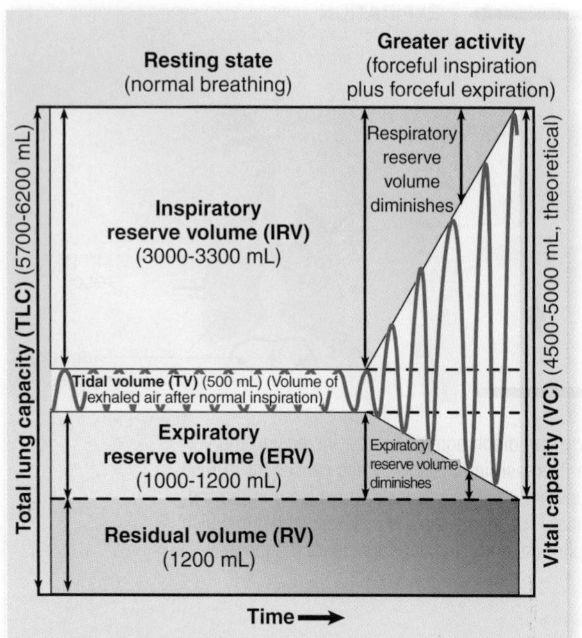

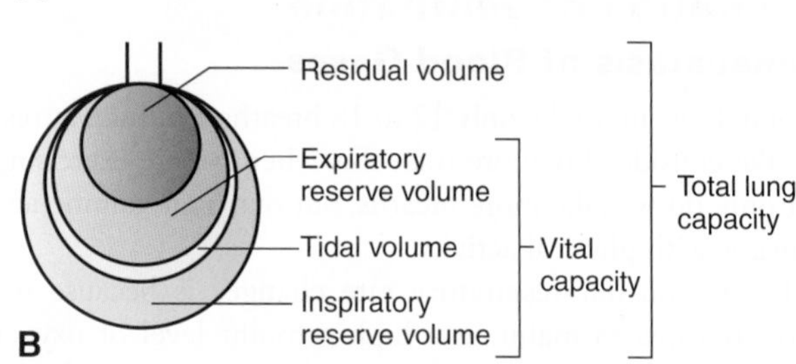

FIGURE 17-17 Pulmonary ventilation volumes. The chart in **A** shows a tracing like that produced with a spirometer. The diagram in **B** shows the pulmonary volumes as relative proportions of an inflated balloon (see **Figure 17-16**). During normal, quiet breathing, about 500 mL of air is moved into and out of the respiratory tract, an amount called the *tidal volume.* During forceful breathing (like that during and after heavy exercise), an extra 3300 mL can be inspired (the inspiratory reserve volume), and an extra 1000 mL or so can be expired (the expiratory reserve volume). The largest volume of air that can be moved in and out during ventilation is called the *vital capacity.* Air that remains in the respiratory tract after a forceful expiration is called the *residual volume.* See **Table 17-1**.

breaths a minute. The two most important control centers in the medulla for regulating breathing rhythm are called the *ventral respiratory group (VRG)* and the *dorsal respiratory group (DRG)*. The VRG provides the basic rhythm generator for breathing. DRG adjusts the breathing rhythm when blood pH or carbon dioxide levels change—as they would during exercise.

Several control centers in the pons—the *pontine respiratory group (PRG)*—seem to provide input to the DRG and thus help to modulate the basic rhythm as needed under a variety of changing conditions in the body.

The depth and rate of respiration can be influenced by many "inputs" to the respiratory control centers from other areas of the brain or from sensory receptors located outside of the central nervous system (**Figure 17-18**).

Cerebral Cortex Control of Respiration

The cerebral cortex can influence respiration by sending nerve signals that affect the function of the respiratory centers of the brainstem. In other words, an individual may voluntarily override the "automatic" brainstem rhythm of breathing and speed up or slow down the breathing rate—or greatly change the pattern of respiration during activities. This ability permits us to change respiratory patterns and even to hold our breath for short periods to accommodate activities such as speaking, eating, or swimming under water.

This voluntary control of respiration has limits. As indicated in a later section, other factors such as blood carbon dioxide levels are much more powerful in controlling respiration than conscious control. Regardless of cerebral intent to the contrary, we resume breathing when our bodies sense the need for more oxygen or if carbon dioxide levels increase to certain levels.

Respiratory Reflexes

Chemoreflexes

Chemoreceptors located in the **carotid** and **aortic bodies** are sensory receptors that are sensitive to increases in blood carbon dioxide level and decreases in blood oxygen level—both of which occur as cells do more work.

TABLE 17-1 Pulmonary Volumes and Capacities

VOLUME	DESCRIPTION	TYPICAL VALUE	CAPACITY	FORMULA	TYPICAL VALUE
Tidal volume (TV)	Volume moved into or out of the respiratory tract during a normal respiratory cycle	500 mL (0.5 L)	Vital capacity (VC)	TV + IRV +ERV	4500-5000 mL (4.5-5.0 L)
Inspiratory reserve volume (IRV)	Maximum volume that can be moved into the respiratory tract after a normal inspiration	3000-3300 mL (3.0-3.3 L)	Inspiratory capacity (IC)	TV + IRV	3500-3800 mL (3.5-3.8 L)
Expiratory reserve volume (ERV)	Maximum volume that can be moved out of the respiratory tract after a normal expiration	1000-1200 mL (1.0-1.2 L)	Functional residual capacity (FRC)	ERV + RV	2200-2400 mL (2.2-2.4 L)
Residual volume (RV)	Volume remaining in the respiratory tract after maximum expiration	1200 mL (1.2 L)	Total lung capacity (TLC)	TV + IRV + ERV + RV	5700-6200 mL (5.7-6.2 L)

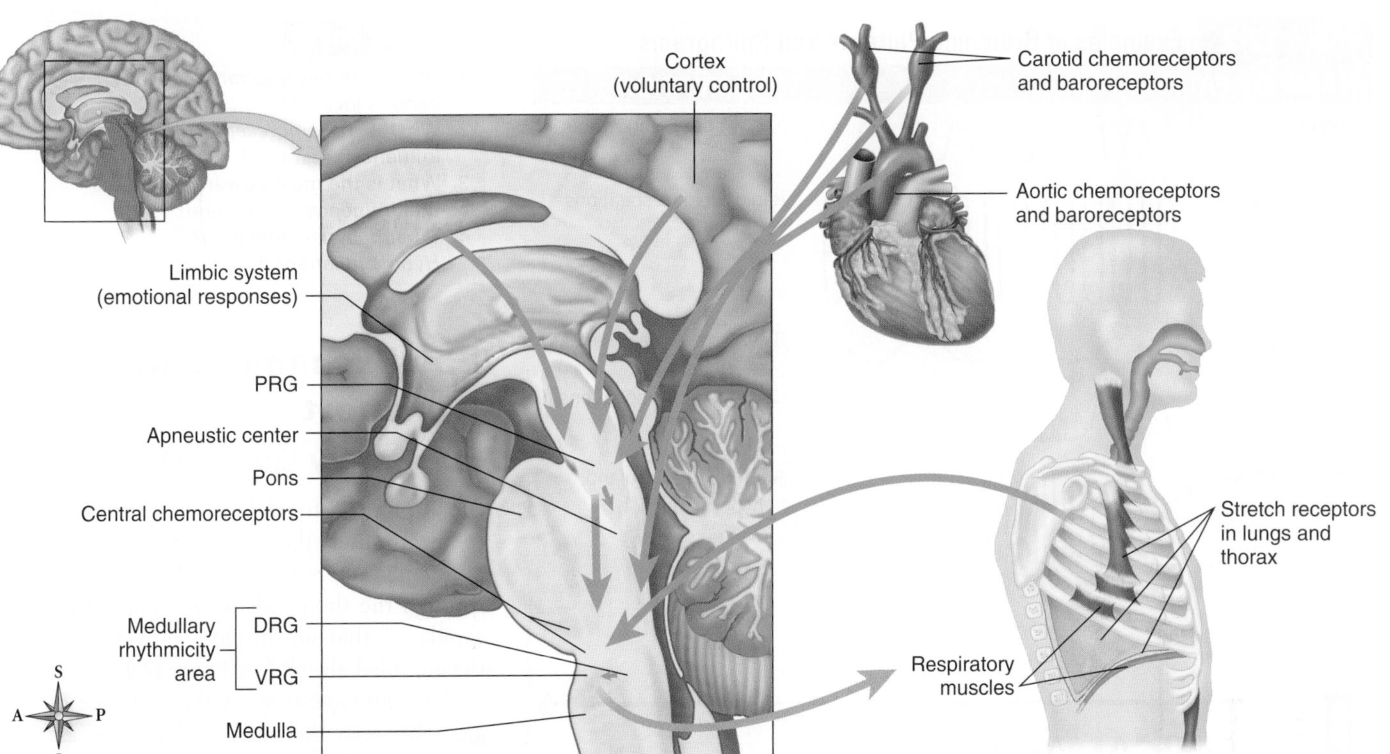

FIGURE 17-18 Regulation of respiration. Respiratory control centers in the brainstem control the basic rate and depth of breathing. The brainstem also receives input from other parts of the body; information from chemoreceptors and stretch receptors can alter the basic breathing pattern, as can emotional and sensory input. Despite these controls, the cerebral cortex can override the "automatic" control of breathing to some extent to accomplish activities such as singing or blowing up a balloon. *Green arrows* show the flow of regulatory information as it flows into the respiratory control centers. The *purple arrow* shows the flow of regulatory information from the control centers to the respiratory muscles that provide the power needed for breathing. *DRG,* Dorsal respiratory group; *PRG,* pontine respiratory group; *VRG,* ventral respiratory group.

17

Such chemoreceptors also sense and respond to increasing blood acid levels. Because carbon dioxide forms an acid in the blood plasma, blood pH goes down when work—and thus blood CO_2 levels—go up.

The carotid body receptors are found at the point where the common carotid arteries divide, and the aortic bodies are small clusters of chemosensitive cells that lie adjacent to the aortic arch near the heart (see **Figure 17-18**).

When stimulated by increasing levels of blood carbon dioxide, decreasing oxygen levels, or increasing blood acidity (lower plasma pH), these receptors send nerve impulses to the respiratory regulatory centers. This signal is interpreted as a change away from the set points for these physiological variables, so the control centers (integrators) modify respiratory rates to bring them back toward their set points.

The blood CO_2 level is the most powerful stimulus driving respiration. That is, the respiratory control centers seem to respond very quickly to even minor shifts in plasma CO_2.

Pulmonary Stretch Reflexes

Stretch receptors in the lungs are located throughout the pulmonary airways and in the alveoli (see **Figure 17-18**). Nerve impulses generated by these receptors influence the normal pattern of breathing and protect the respiratory system from excess stretching caused by harmful overinflation.

When the tidal volume of air has been inspired, the lungs are expanded enough to stimulate stretch receptors that then send inhibitory impulses to the inspiratory center. Relaxation of inspiratory muscles occurs, and expiration follows. After expiration, the lungs are sufficiently deflated to inhibit the stretch receptors, and inspiration is then allowed to start again.

Breathing Patterns

A number of clinical terms are used to describe breathing patterns. **Eupnea,** for example, refers to a normal respiratory rate. During eupnea, the need for oxygen and carbon dioxide exchange is being met, and the individual is usually not aware of their breathing pattern.

The terms **hyperventilation** and **hypoventilation** describe very rapid and deep or slow and shallow respirations, respectively. Hyperventilation sometimes results from a conscious voluntary effort preceding exertion. Or it can result from psychological factors—as in so-called "hysterical hyperventilation."

17

TABLE 17-2 Examples of Breathing Patterns and Spirograms

NAME OF PATTERN	DESCRIPTION
Eupnea	Normal breathing
Hyperventilation	Rapid, deep respirations
Hypoventilation	Slow, shallow respirations
Apnea	Cessation of respirations
Cheyne-Stokes respiration	Alternating apnea and hyperventilation

CLINICAL APPLICATION

SUDDEN INFANT DEATH SYNDROME

Sudden infant death syndrome (SIDS) is the third-ranking cause of infant death and accounts for about 1 in 9 of the nearly 30,000 infant deaths reported each year in the United States. Sometimes called "crib death," SIDS occurs most frequently in babies with no obvious medical problems who are younger than 3 months of age. The exact cause of death can seldom be determined even after extensive testing and autopsy.

SIDS occurs at a higher rate in African-American and Native-American babies than in white, Hispanic, or Asian infants, although the reasons remain a mystery. Regardless of infant ethnicity, recent data suggest that certain precautions, such as having babies sleep only on their backs and keeping cribs free of pillows or plush toys that might partially cover the nose or mouth, may reduce the incidence of SIDS. Also important is the elimination of smoking during pregnancy and protecting infants from exposure to "second-hand" cigarette smoke after birth.

Although the exact cause of SIDS remains unknown, abnormalities in the regulatory centers of the brainstem that control breathing may play a role in this tragic problem.

Dyspnea refers to labored or difficult breathing and is often associated with hypoventilation. Dyspnea that is relieved by moving into an upright or sitting position is called **orthopnea.**

If breathing stops completely for a brief period, regardless of cause, it is called **apnea.**

A series of cycles of alternating apnea and hyperventilation is called **Cheyne-Stokes respiration (CSR).** CSR occurs in critical diseases such as congestive heart failure, brain injuries, or brain tumors. CSR also may occur in the case of a drug overdose.

Failure to resume breathing after a period of apnea is called **respiratory arrest.**

Examples of breathing patterns are summarized in **Table 17-2**.

QUICK CHECK

1. Where are the *respiratory control centers* located?
2. What is a *chemoreceptor*? How does it influence breathing?
3. What is the most powerful stimulus that influences respiration?
4. What is *hyperventilation*? *Hypoventilation*?

Gas Exchange and Transport

Pulmonary Gas Exchange

Blood pumped from the right ventricle of the heart enters the pulmonary artery and eventually enters the lungs. It then flows through the thousands of tiny pulmonary capillaries that are in close proximity to the air-filled alveoli (see **Figure 17-9**).

External respiration, or the exchange of gases between the blood and alveolar air, occurs by diffusion. Diffusion is a passive process resulting in movement down a concentration gradient (see **Table 3-2** on p. 51). That is, substances move from an area of high concentration to an area of lower concentration of the diffusing substance.

The amounts or concentrations of some blood substances are measured in terms of weight. Reporting how many milligrams of a particular substance are present in 100 mL of blood (mg/dL) is one example. However, the concentration of a particular gas in air or within the blood is expressed as the *pressure* exerted by that gas and is reported in millimeters of mercury (mm Hg). Recall from Chapter 15 that blood pressure levels are also reported in mm Hg.

Several different gases are present in both air and blood. The total pressure of all gases present in an air or blood sample is, of course, the sum of the pressures exerted by each of the gases present. Because the pressure of the so-called *respiratory gases*—oxygen (O_2) and carbon dioxide (CO_2)—in air or blood constitutes only a part of the total pressure present, their concentration is reported as a **partial pressure (P).** The symbol used to designate partial pressure is the capital letter P preceding the chemical symbol for the gas. For respiratory gases the symbols PO_2 and PCO_2 are used.

Instead of referring directly to "concentration," respiratory physiologists state that blood gas particles diffuse from an area of high *partial pressure* to an area of lower *partial pressure.* Understanding the role of partial pressures of blood gases in normal gas exchange is important in the diagnosis and treatment of many disease conditions.

O_2 is continually removed from the blood and used by the body cells. By the time blood flows into the pulmonary capillaries, it has a Po_2 of about 40 mm Hg. Because alveolar air is rich in oxygen (Po_2 100 mm Hg), diffusion causes movement of oxygen from the area of high partial pressure (alveolar air) to the area of lower partial pressure (capillary blood). Put another way, oxygen diffuses "down" its partial pressure gradient.

Diffusion of carbon dioxide also occurs between blood in pulmonary capillaries and alveolar air. Blood flowing through the pulmonary capillaries is high in CO_2, having a Pco_2 of 46 mm Hg. The Pco_2 of alveolar air is about 40 mm Hg. Therefore, diffusion of carbon dioxide results in its movement from an area of high partial pressure in the pulmonary capillaries to an area of lower partial pressure in alveolar air. Then from the alveoli, carbon dioxide leaves the body in expired air (**Figure 17-19**).

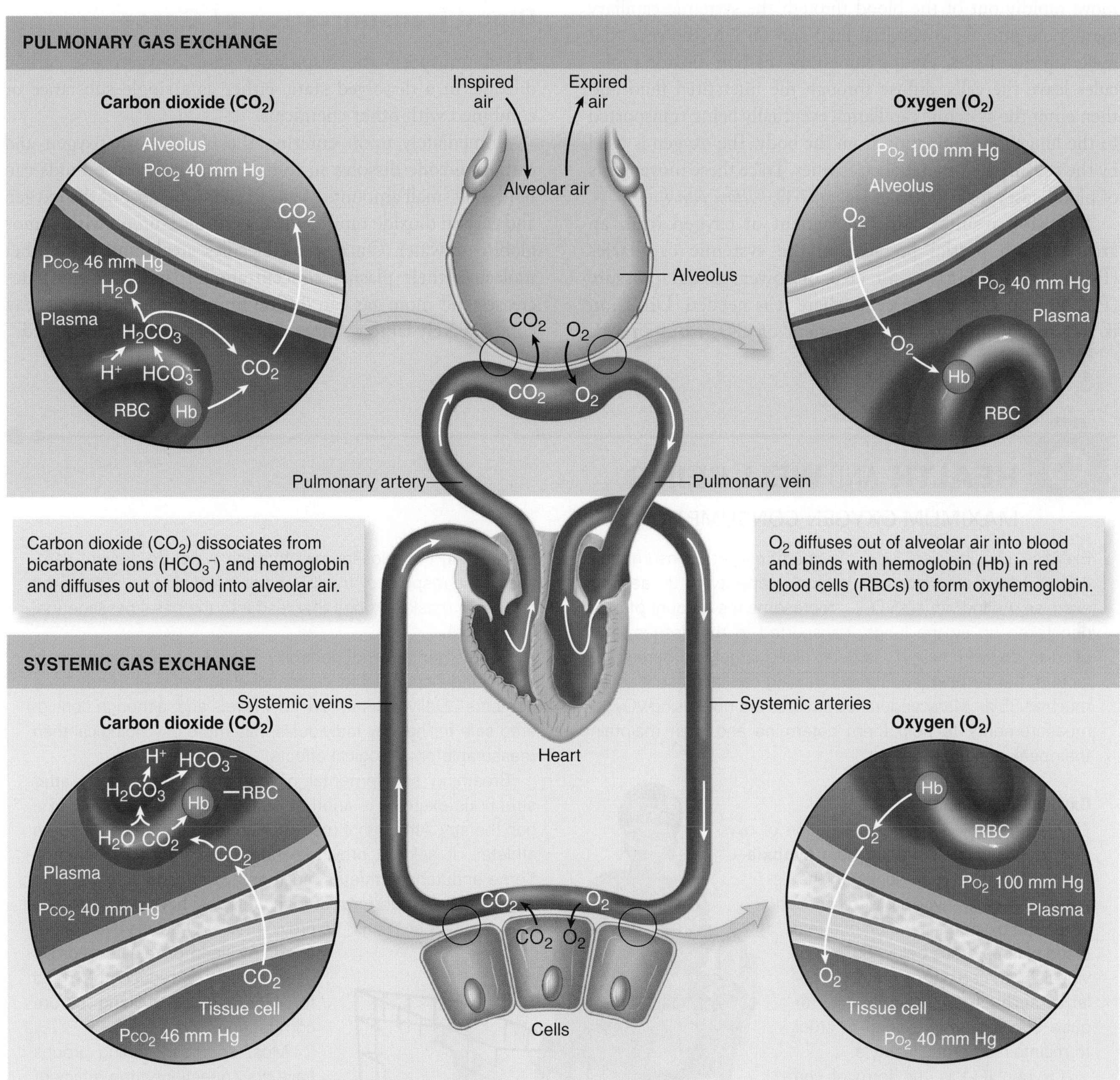

FIGURE 17-19 Exchange and transport of gases. The *top panel* of the diagram shows pulmonary gas exchange and the *bottom panel* shows systemic gas exchange. In each, the *left inset* shows the transport and movement of carbon dioxide (CO_2) and the *right inset* shows the transport and movement of oxygen (O_2).

Systemic Gas Exchange

The exchange of gases that occurs between blood in systemic capillaries and the body cells is called *internal respiration.*

As you would expect, the direction of movement of oxygen and carbon dioxide during internal respiration is just the opposite of that noted in the exchange that occurs during external respiration when gases are exchanged between the blood in the pulmonary capillaries and the air in alveoli.

During the process of internal respiration oxygen molecules move rapidly out of the blood through the systemic capillary membrane into the interstitial fluid and on into the cells that make up the tissues. At the same time, carbon dioxide molecules leave the cells, diffuse through the interstitial fluid and then enter the systemic capillaries, eventually being transported to the lungs for elimination from the body. The oxygen is used by the cells in their metabolic activities. Trace these movements of blood gases for yourself in **Figure 17-19**, *bottom panel.*

Diffusion results in the movement of oxygen from an area of high partial pressure in the systemic capillaries (Po_2 100 mm Hg) to an area of lower partial pressure (Po_2 40 mm Hg) in the cells where it is needed. Diffusion is also responsible for the movement of CO_2 from an area of high partial pressure in the cells (Pco_2 46 mm Hg) to an area of lower partial pressure (Pco_2 43 mm Hg) in the systemic capillaries.

Simply stated, during internal respiration oxygenated blood enters systemic capillaries and is changed into deoxygenated blood as it flows through them. In the process of losing oxygen, the waste product carbon dioxide is picked up and transported to the lungs for removal from the body.

Blood Transportation of Gases

Blood transports the respiratory gases, oxygen and carbon dioxide, in a dissolved state, either as a single substance or combined with other chemicals.

Immediately upon entering the blood, both oxygen and carbon dioxide dissolve in the plasma, but because fluids can hold only small amounts of gas in solution, most of the oxygen and carbon dioxide rapidly form a chemical union with hemoglobin or water. Once gas molecules are bound to another molecule, their plasma concentration (partial pressure) decreases and more gas can diffuse into the plasma. In this way, comparatively large volumes of the gases can be transported.

HEALTH AND WELL-BEING

MAXIMUM OXYGEN CONSUMPTION

Exercise physiologists use maximum oxygen consumption ($VO_{2\ max}$) as a predictor of a person's capacity to do aerobic exercise. An individual's $VO_{2\ max}$ represents the amount of oxygen taken up by the lungs, transported to the tissues, and used to do work. $VO_{2\ max}$ is determined largely by hereditary factors, but aerobic (endurance) training can increase it by as much as 35%. Many endurance athletes are now using $VO_{2\ max}$ measurements to help them determine and then maintain their peak condition.

Oxygen Supplements

Oxygen therapy is the administration of oxygen to individuals suffering from **hypoxia**—an insufficient oxygen supply to the tissues. Clinically, hypoxia is diagnosed when the oxygen saturation of arterial blood plasma drops below 80%. Individuals with certain respiratory problems, such as emphysema, may require supplemental oxygen in order to maintain a normal lifestyle.

Oxygen (O_2) in the form of compressed gas is commonly stored in and dispensed from small, green, metal cylinders or tanks (see the figure). Because the oxygen dispensed from such tanks is often cold and dry, it may need to be warmed and moistened, generally by bubbling the released gas through water, to prevent damage to the respiratory tract. Supplemental oxygen is delivered through a mask or tubes that lead into the nasal passage (nasal prongs).

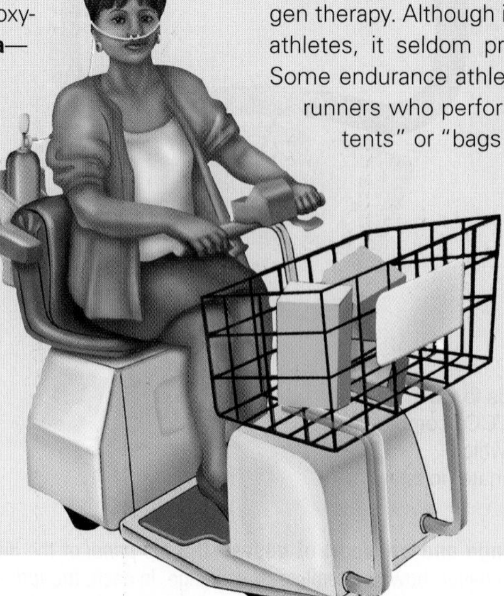

Supplemental (and generally very expensive) oxygen is now being dispensed for recreational purposes at trendy "oxygen bars." Delivery is at low flow levels and, although considered safe for healthy individuals, has more psychological than measurable physiological effects.

Breathing supplemental oxygen for short periods after strenuous exercise is another nonmedical application of oxygen therapy. Although it may shorten recovery times for some athletes, it seldom provides more than transitory benefits. Some endurance athletes, such as cyclists and long-distance runners who perform at high altitudes, have used "oxygen tents" or "bags" to provide lower O_2 levels over longer periods of sleep or rest to improve performance or reduce the need for high altitude training before competition.

Most sport-sanctioning groups have now questioned the ethics of this longer-term use of supplemental oxygen or have banned the practice outright as a form of "doping" (see box p. 354).

Transport of Oxygen

Only very limited amounts of oxygen can be dissolved in the blood. Of the total amount of oxygen that blood can transport, about 20.4 mL in 100 mL of blood, only about 1.5% or 0.3 mL is actually dissolved. Many times that amount, about 21.1 mL, combines with the hemoglobin (Hb) in 100 mL of blood to form **oxyhemoglobin (HbO_2)** so that it can be carried to the tissues and used by the body cells.

To combine with hemoglobin, oxygen must first diffuse into the red blood cells to form oxyhemoglobin. Hemoglobin molecules are large proteins that contain four iron-containing **heme** components, each of which is capable of combining with an oxygen molecule.

In many ways each hemoglobin molecule acts as the ultimate "oxygen sponge." Oxygen associates with hemoglobin rapidly—so rapidly, in fact, that about 97% of the blood's hemoglobin has united with oxygen, and become "oxygenated blood," by the time it leaves the pulmonary capillaries to return to the heart.

Oxygenated blood is found in the pulmonary veins and systemic arteries. Normally, oxygenated blood is 97% "saturated." So-called "deoxygenated blood," found in the systemic veins and pulmonary arteries, is about 75% saturated with oxygen. The difference in oxygen saturation results from the release of oxygen from oxyhemoglobin to supply the body cells. Therefore, the chemical combination of oxygen and hemoglobin is said to be "reversible" with oxyhemoglobin formation or oxygen release dependent on the partial pressure of oxygen driving the reaction.

Summing up, we can say that oxygen travels in two forms: (1) as simply dissolved O_2 in the plasma and (2) as a combination of O_2 and hemoglobin (oxyhemoglobin). Of these two forms of transport, oxyhemoglobin is the carrier of the vast majority of the total oxygen transported by the blood.

Unfortunately, other gases can also bind to hemoglobin, perhaps rendering the Hb incapable of transporting oxygen. For a common example of this, see the article *Carbon Monoxide Poisoning* at Connect It! at *evolve.elsevier.com*.

Transport of Carbon Dioxide

Carbon dioxide is a by-product of cellular metabolism and plays an important and necessary role in regulating the pH of body fluids. However, if it accumulates in the body beyond normal limits (40 to 50 mm Hg in venous blood), it can quickly become toxic. Elimination of excess CO_2 from the body occurs when it enters the alveoli and is expelled during expiration. For this to occur, CO_2 must be transported in the blood to the lungs in one of three forms, as described in the following sections.

Carbon Dioxide

About 10% of the total amount of carbon dioxide in blood is carried in the *dissolved form*. It is this dissolved CO_2 that produces the $P{CO_2}$ of blood plasma. However, all CO_2 in the blood must pass through the dissolved state before moving into or out of any of the states described in the following sections.

Carbaminohemoglobin

About 20% of the total CO_2 transported in the blood is in the form of **carbaminohemoglobin ($HbCO_2$).** $HbCO_2$ is formed by the binding of carbon dioxide to hemoglobin.

The formation of this compound is accelerated by an increase in $P{CO_2}$—as the extra dissolved CO_2 binds to hemoglobin. Likewise, formation of $HbCO_2$ is slowed—or even reversed—by a decrease in $P{CO_2}$.

Bicarbonate

About 70% of the total CO_2 transported in the blood is carried in the form of **bicarbonate ions (HCO_3^-).**

When CO_2 dissolves in water (as in blood plasma), some of the CO_2 molecules associate with water (H_2O) to form *carbonic acid (H_2CO_3)*. Once formed, some of the H_2CO_3 molecules dissociate to form hydrogen (H^+) and bicarbonate (HCO_3^-) ions. The speed of this process is quite slow when it occurs in the plasma, but the rate of reaction increases dramatically within RBCs because of the presence of an enzyme called **carbonic anhydrase (CA).** The reaction is summarized by the following chemical equation:

$$CO_2 + H_2O \rightleftharpoons H_2CO_3 \rightleftharpoons H^+ + HCO_3^-$$

Carbon dioxide, Water, Carbonic acid, Hydrogen ion, Bicarbonate ion

Note that the arrows go in both directions. This indicates that the reaction is *reversible*—it can go in either direction. If bicarbonate is being formed, CO_2 molecules entering into plasma are continually removed from the blood and travel to the lungs as bicarbonate. And, when the process is reversed in the lungs, CO_2 can be released from bicarbonate to enter the alveolar air and then be exhaled.

QUICK CHECK

1. When referring to respiratory gases, explain the use of *partial pressure (P)*.
2. Explain the process of *internal* respiration.
3. In what form does oxygen travel in the blood? What form does carbon dioxide travel in the blood?

To better understand these concepts, use the Active Concept Map *Transport of Oxygen and Carbon Dioxide in the Blood* at *evolve.elsevier.com*.

SCIENCE APPLICATIONS

RESPIRATORY MEDICINE

Christian Bohr (1855–1911)

The Danish physician Christian Bohr left a legacy of achievement in science in more ways than one. His son Niels Bohr (creator of the *Bohr model* of the atom seen in **Figure 2-2** on p. 26) and his grandson Aage Bohr both won Nobel Prizes in science, as did his student August Krogh. Christian Bohr's contributions to understanding respiration, however, have also left a lasting mark on respiratory physiology and medicine—resulting in three Nobel Prize nominations of his own.

Bohr's most famous discovery was the fact that a decrease in plasma pH or an increase in P_{CO_2} will decrease hemoglobin's binding affinity with oxygen. Called the *Bohr effect,* this phenomenon explains how hemoglobin so easily gives up its oxygen in very active tissues like muscles during exercise—where an increase in CO_2 and the accompanying acidity reflects the amount of cellular work and thus an increased use of oxygen.

The contributions of Bohr and many others to today's understanding of the relationship of respiration, blood gases, and pH continue to play a central role in health care. Today, countless *physicians, nurses,* **respiratory therapists, emergency medical technicians (EMTs),** and **paramedics,** continue to benefit from an understanding of these fundamental principles of physiology.

LANGUAGE OF SCIENCE (continued from p. 459)

carbonic anhydrase (CA)
(kar-BON-ik an-HYE-drayz [see ay])
[*carbo-* **coal,** *-ic* **relating to,** *a-* **without,** *-hydr-* **water,** *-ase* **enzyme**]

carotid body
(kah-ROT-id BOD-ee)
[*caro-* **heavy sleep,** *-id* **relating to**]

cellular respiration
(SEL-yoo-lar res-pih-RAY-shun)
[*cell* **storeroom,** *-ular* **relating to,** *re-* **again,** *-spir-* **breathe,** *-ation* **process**]

chemoreceptor
(kee-moh-ree-SEP-tor)
[*chemo-* **chemical,** *-recept-* **receive,** *-or* **agent**]

chemoreflex
(kee-moh-REE-fleks)
[*chemo-* **chemical,** *-re-* **back or again,** *-flex* **bend**]

ciliary escalator
(SIL-ee-ayr-ee ES-kuh-lay-ter)
[*cili-* **eyelash,** *-ary* **relating to,** *escalat-* **scale,** *-or* **agent**]

compliance
(kom-PLY-ans)
[*compli-* **fill up,** *-ance* **act of**]

conchae
(KONG-kee or KONG-kay)
sing., concha
(KONG-kah)
[*concha* **sea shell**]

diaphragm
(DYE-ah-fram)
[*dia-* **across,** *-phrag-* **enclose,** *-(u)m* **thing**]

diffusion
(dih-FYOO-zhen)
[*diffus-* **spread out,** *-sion* **process**]

epiglottis
(ep-ih-GLOT-is)
[*epi-* **upon,** *-glottis* **mouth of windpipe**]

eustachian tube
(yoo-STAY-shun toob)
[*Bartolomeo Eustachio,* **Italian anatomist,** *-an* **relating to**]

expiration (exhalation)
(eks-pih-RAY-shun [eks-huh-LAY-shun])
[*ex-* **out,** *-pir-* **breathe,** *-ation* **process** (*ex-* **out,** *-hal-* **breathe,** *-ation* **process**)]

expiratory muscle
(eks-PYE-rah-tor-ee MUS-el)
[*ex-* **out of,** *-[s]pir-* **breathe,** *-tory* **relating to,** *musc-* **mouse,** *-cle* **little**]

expiratory reserve volume (ERV)
(eks-PYE-rah-tor-ee ree-ZERV VOL-yoom [ee ar vee])
[*ex-* **out of,** *-[s]pir-* **breathe,** *-tory* **relating to**]

external nares
(eks-TER-nal NAY-reez)
sing., naris
(NAY-ris)
[*extern-* **outside,** *-al* **relating to,** *naris* **nostril**]

external respiration
(eks-TER-nal res-pih-RAY-shun)
[*extern-* **outside,** *-al* **relating to,** *re-* **again,** *-spir-* **breathe,** *-ation* **process of**]

glottis
(GLOT-is)
[*glottis* **mouth of windpipe**]

heme
(heem)
[*hem-* **blood**]

homeostatic mechanism
(hoh-mee-oh-STAT-ik MEK-ah-niz-em)
[*homeo-* **same or equal,** *-static* **relating to standing still**]

inspiration (inhalation)
(in-spih-RAY-shun [in-huh-LAY-shun])
[*in-* **in,** *-spir-* **breathe,** *-ation* **process**]

inspiratory muscle
(in-SPY-rah-tor-ee MUS-el)
[*in-* **in,** *-spir-* **breathe,** *-tory* **relating to,** *mus-* **mouse,** *-cle* **little**]

inspiratory reserve volume (IRV)
(in-SPY-rah-tor-ee ree-SERV VOL-yoom [aye ar vee])
[*in-* **in,** *-spir-* **breathe,** *-tory* **relating to**]

interarytenoid notch
(IN-ter-ar-ih-tee-noyd notch)
[*inter-* **among,** *-aryten-* **ladle,** *-oid* **like**]

internal respiration
(in-TER-nal res-pih-RAY-shun)
[*intern-* **inside,** *-al* **relating to,** *re-* **again,** *-spir-* **breathe,** *-ation* **process**]

lacrimal sac
(LAK-rih-mal sak)
[*lacrima-* **tear,** *-al* **relating to**]

laryngopharynx
(lah-ring-goh-FAYR-inks)
[*laryng-* **voice box (larynx),** *-pharynx* **throat**]

larynx
(LAYR-inks)
[*larynx* **voice box**]

LANGUAGE OF SCIENCE (continued from p. 482)

lingual tonsil
(LING-gwal TAHN-sil)
[*ling-* **tongue,** *-al* **relating to,** *tons-* **goiter,** *-il* **little**]

lower respiratory tract
(LOW-er RES-pih-rah-tor-ee trakt)
[*re-* **again,** *-spir-* **breathe,** *-tory* **relating to,** *tract* **trail**]

lung

mucus
(MYOO-kus)
[*mucus* **slime**]

nasal cavity
(NAY-zal KAV-ih-tee)
[*nas-* **nose,** *-al* **relating to,** *cav-* **hollow,** *-ity* **state**]

nasal septum
(NAY-zal SEP-tum)
[*nas-* **nose,** *-al* **relating to,** *septum* **wall**]

nasopharynx
(nay-zoh-FAYR-inks)
[*naso-* **nose,** *-pharynx* **throat**]

nose

olfaction
(ohl-FAK-shun)
[*olfact-* **smell,** *-tion* **condition**]

oropharynx
(or-oh-FAYR-inks)
[*oro-* **mouth,** *-pharynx* **throat**]

oxyhemoglobin (HbO_2)
(ahk-see-hee-moh-GLOH-bin [aych bee oh too])
[*oxy-* **sharp,** *-hemo-* **blood,** *-glob-* **ball,** *-in* **substance**]

palatine tonsil
(PAL-ah-tine TAHN-sil)
[*palat-* **palate,** *-ine* **relating to,** *tons-* **goiter,** *-il* **little**]

paranasal sinus
(payr-ah-NAY-zal SYE-nus)
[*para-* **beside,** *-nas-* **nose,** *-al* **relating to,** *sinus* **hollow**]

partial pressure (P)
(PAR-shal PRESH-ur [pee])

pharyngeal tonsil
(fah-RIN-jee-al TAHN-sil)
[*pharyng-* **throat,** *-al* **relating to,** *tons-* **goiter,** *-il* **little**]

pharynx
(FAYR-inks)
[*pharynx* **throat**]

phrenic nerve
(FREN-ik)
[*phren-* **mind,** *-ic* **relating to**]

pleura
(PLOO-rah)
pl., pleurae
(PLOO-ree)
[*pleura* **side of body (rib)**]

primary bronchi
(PRYE-mayr-ee BRONG-kye)
sing., bronchus
(BRONG-kus)
[*prim-* **first,** *-ary* **relating to,** *bronchus* **windpipe**]

pulmonary ventilation
(PUL-moh-nayr-ee ven-tih-LAY-shun)
[*pulmon-* **lung,** *-ary* **relating to,** *vent-* **fan or create wind,** *-tion* **process**]

residual volume (RV)
(reh-ZID-yoo-al VOL-yoom [ar vee])
[*residu-* **remainder,** *-al* **relating to**]

respiration
(res-pih-RAY-shun)
[*re-* **again,** *-spir-* **breathe,** *-ation* **process**]

respiratory control center
(RES-pih-rah-tor-ee kon-TROL SEN-ter)
[*re-* **again,** *-spir-* **breathe,** *-tory* **relating to**]

respiratory membrane
(RES-pih-rah-tor-ee MEM-brayn)
[*re-* **again,** *-spir-* **breathe,** *-tory* **relating to,** *membran-* **thin skin**]

respiratory mucosa
(RES-pih-rah-tor-ee myoo-KOH-sah)
[*re-* **again,** *-spir-* **breathe,** *-tory* **relating to,** *mucus* **slime**]

respiratory system
(RES-pih-rah-tor-ee SIS-tem)
[*re-* **again,** *-spir-* **breathe,** *-tory* **relating to**]

secondary bronchi
(SEK-on-dayr-ee BRONG-kye)
sing., bronchus
(BRONG-kus)
[*second-* **second,** *-ary* **relating to,** *bronchus* **windpipe**]

surfactant
(sur-FAK-tant)
[**combination of surf(ace) act(ive) a(ge)nt**]

tidal volume (TV)
(TYE-dal VOL-yoom [tee vee])
[*tid-* **time,** *-al* **relating to**]

tonsils
(TAHN-silz)
[*tons-* **goiter,** *-il* **little**]

trachea
(TRAY-kee-ah)
[*trachea* **rough duct**]

turbinate
(TUR-bih-nayt)
[*turbin-* **top (spinning toy),** *-ate* **of or like**]

upper respiratory tract
(UP-er RES-pih-rah-tor-ee trakt)
[*re-* **again,** *-spir-* **breathe,** *-tory* **relating to,** *tract* **trail**]

vital capacity (VC)
(VYE-tal kah-PAS-ih-tee [vee see])
[*vita-* **life,** *-al* **relating to,** *capac-* **hold,** *-ity* **state**]

vocal cords
(VOH-kull kordz)
[*voca-* **voice,** *-al* **relating to,** *cord-* **string**]

17

LANGUAGE OF MEDICINE

abdominal thrust
(ab-DOM-ih-nal thrust)
[*abdomin-* **belly,** *-al* **relating to**]

adenoid
(AD-eh-noyd)
[*adeno-* **gland,** *-oid* **like**]

adult respiratory distress syndrome (ARDS)
(ah-DULT RES-pih-rah-tor-ee dis-TRESS SIN-drohm [ardz or ay ar dee es])
[*re-* **again,** *-spir-* **breathe,** *-tory* **relating to,** *syn-* **together,** *-drome* **running or (race) course**]

apnea
(AP-nee-ah)
[*a-* **not,** *-pne-* **breathe,** *-a* **condition**]

asthma
(AZ-mah)
[*asthma* **panting**]

17

LANGUAGE OF MEDICINE (continued from p. 483)

atelectasis
(at-eh-LEK-tay-sis)
[*atele-* **incomplete,** *-ectasis* **extension**]

bronchitis
(brong-KYE-tis)
[*bronch-* **windpipe,** *-itis* **inflammation**]

chemoreceptor
(kee-moh-ree-SEP-tor)
[*chemo-* **chemical,** *-recept-* **receive,** *-or* **agent**]

Cheyne-Stokes respiration (CSR)
(chain stokes res-pih-RAY-shun [see es ar])
[*John Cheyne* **Scots physician,** *William Stokes* **Irish physician,** *re-* **again,** *-spir-* **breathe,** *-ation* **process of**]

chronic bronchitis
(KRON-ik brong-KYE-tis)
[*chron-* **time,** *-ic* **relating to,** *bronch-* **windpipe,** *-itis* **inflammation**]

chronic obstructive pulmonary disease (COPD)
(KRON-ik ob-STRUK-tiv PUL-moh-nayr-ee dih-ZEEZ [see oh pee dee])
[*chron-* **time,** *-ic* **relating to,** *obstruct-* **block,** *-ive* **relating to,** *pulmon-* **lung,** *-ary* **relating to,** *dis-* **opposite of,** *-ease* **comfort**]

croup
(kroop)
[*croup* **croak**]

deviated septum
(DEE-vee-ay-ted SEP-tum)
[*devia-* **turn aside,** *-ate* **process,** *septum* **partition**]

dyspnea
(DISP-nee-ah)
[*dys-* **painful,** *-pne-* **breathe,** *-a* **condition**]

emergency medical technician (EMT)
(eh-MER-jen-see MED-ih-kal tek-NISH-en [ee em tee])
[*medic-* **heal,** *-al* **relating to,** *techn-* **art or skill,** *-ic* **relating to,** *-ian* **practitioner**]

emphysema
(em-fih-SEE-mah)
[*em-* **in,** *-physema* **blowing or puffing up**]

endotracheal intubation
(en-doh-TRAY-kee-al in-too-BAY-shun)
[*endo-* **within,** *-trache-* **rough duct,** *-al* **relating to,** *in-* **within,** *-tub-* **tube,** *-ation* **process**]

epiglottitis
(ep-ih-glot-AYE-tis)
[*epi-* **upon,** *-glotti-* **mouth of windpipe,** *-itis* **inflammation**]

epistaxis
(ep-ih-STAK-sis)
[*epi-* **upon,** *-staxis* **drip**]

eupnea
(YOOP-nee-ah)
[*eu-* **easily,** *-pne-* **breathe,** *-a* **condition**]

hemothorax
(hee-moh-THOH-raks)
[*hemo-* **blood,** *-thorax* **chest**]

hyperventilation
(hye-per-ven-tih-LAY-shun)
[*hyper-* **excessive,** *-vent-* **fan or create wind,** *-tion* **process**]

hypoventilation
(hye-poh-ven-tih-LAY-shun)
[*hypo-* **under or below,** *-vent-* **fan or create wind,** *-tion* **process**]

hypoxia
(hye-POK-see-ah)
[*hypo-* **under or below,** *-ox-* **oxygen,** *-ia* **condition**]

infant respiratory distress syndrome (IRDS)
(IN-fant RES-pih-rah-toh-ree dih-STRESS SIN-drohm [irdz or aye ar dee es])
[*re-* **again,** *-spir-* **breathe,** *-tory* **relating to,** *syn-* **together,** *-drome* **course**]

laryngeal cancer
(lah-RIN-jee-al or lar-in-JEE-al KAN-ser)
[*laryng-* **voice box (larynx),** *-al* **relating to,** *cancer* **crab or malignant tumor**]

laryngitis
(lar-in-JYE-tis)
[*laryng-* **voice box (larynx),** *-itis* **inflammation**]

lobectomy
(loh-BEK-toh-mee)
[*lob-* **lobe,** *-ec-* **out,** *-tom-* **cut,** *-y* **action**]

nasal polyp
(NAY-zal PAH-lip)
[*nas-* **nose,** *-al* **relating to,** *polyp* **cuttlefish**]

orthopnea
(or-THOP-nee-ah)
[*ortho-* **straight or upright,** *-pne-* **breathe,** *-a* **condition**]

paramedic
(payr-ah-MED-ik)
[*para-* **beside,** *-med-* **heal,** *-ic-* **relating to**]

pharyngitis
(fayr-in-JYE-tis)
[*pharyng-* **throat (pharynx),** *-itis* **inflammation**]

pleurisy
(PLOOR-ih-see)
[*pleur-* **side of body (rib),** *-itis* **inflammation**]

pneumonectomy
(noo-moh-NEK-toh-mee)
[*pneumon-* **lung,** *-ec-* **out,** *-tom-* **cut,** *-y* **action**]

pneumonia
(noo-MOH-nee-ah)
[*pneumon-* **lung,** *-ia* **condition**]

pneumothorax
(noo-moh-THOH-raks)
[*pneumo-* **air or wind,** *-thorax* **chest**]

respiratory arrest
(RES-pih-rah-tor-ee ah-REST)
[*re-* **again,** *-spir-* **breathe,** *-tory* **relating to**]

respiratory distress syndrome (RDS)
(RES-pih-rah-tor-ee dih-STRESS SIN-drohm [ar dee es])
[*re-* **again,** *-spir-* **breathe,** *-tory* **relating to,** *syn-* **together,** *-drome* **running or (race) course**]

rhinitis
(rye-NYE-tis)
[*rhin-* **nose,** *-itis* **inflammation**]

respiratory therapist
(RES-pih-rah-tor-ee THAYR-ah-pist)
[*re-* **again,** *-spir-* **breathe,** *-tory* **relating to,** *therap-* **treatment,** *-ist* **agent**]

sinusitis
(sye-nyoo-SYE-tis)
[*sinus-* **hollow,** *-itis* **inflammation**]

spirometer
(spye-ROM-eh-ter)
[*spir-* **breathe,** *-meter* **measure**]

sputum
(SPYOO-tum)
[*sputum* **spit**]

sudden infant death syndrome (SIDS)
(SUD-den IN-fant deth SIN-drohm [sidz])
[*syn-* **together,** *-drome* **running or (race) course**]

tonsillectomy
(tahn-sih-LEK-toh-mee)
[*tonsil-* **tonsil,** *-ec-* **out,** *-tom-* **cut,** *-y* **action**]

tonsillitis
(tahn-sih-LYE-tis)
[*tonsil-* **tonsil,** *-itis* **inflammation**]

tracheostomy
(tray-kee-OS-toh-mee)
[*trache-* **rough duct (trachea),** *-os-* **mouth or opening,** *-tom-* **cut,** *-y* **action**]

tuberculosis (TB)
(too-ber-kyoo-LOH-sis [tee bee])
[*tuber-* **swelling,** *-cul-* **little,** *-osis* **condition**]

upper respiratory infection (URI)
(UP-er RES-pih-rah-tor-ee in-FEK-shun [yoo ar aye])
[*re-* **again,** *-spir-* **breathe,** *-tory* **relating to,** *infec-* **stain,** *-tion* **process**]

❑ OUTLINE SUMMARY

To download a digital version of the chapter summary for use with your device, access the ***Audio Chapter Summaries*** *online at evolve.elsevier.com.*

Scan this summary after reading the chapter to help you reinforce the key concepts. Later, use the summary as a quick review before your class or before a test.

Structural Plan

A. Overview
   1. Basic plan of respiratory system would be similar to an inverted tree if it were hollow; leaves of the tree would be comparable to alveoli, with the microscopic sacs enclosed by networks of capillaries (**Figure 17-1**)
   2. Diffusion is the mode for gas exchange that occurs in the respiratory mechanism

B. Divided into "upper" and "lower" to better describe locations in the air pathway of the respiratory system
   1. Upper respiratory tract—nose, pharynx, and larynx
   2. Lower respiratory tract—trachea, bronchial tree, and lungs

C. Respiratory mucosa
   1. Structure
      a. Mucous membrane that lines the air distribution tubes in the respiratory tree (**Figure 17-2**)
      b. Ciliated pseudostratified epithelium—lines most of tract; produces mucus
      c. Stratified squamous epithelium—lines nostrils, vocal folds, pharynx; protective function
      d. Simple squamous epithelium—lines alveoli; facilitates gas exchange
   2. Function
      a. More than 125 mL of mucus produced each day forms a "mucus blanket" over much of the respiratory mucosa
      b. Mucus serves as an air purification mechanism by trapping inspired irritants such as dust and pollen
      c. Ciliary escalator—cilia on mucosal cells beat in only one direction, moving mucus upward to pharynx for removal

Upper Respiratory Tract

A. Nose
   1. Structure
      a. Nasal septum separates interior of nose into two cavities
      b. Mucous membrane lines nose
      c. Nasal polyps—noncancerous growths that project from nasal mucosa (associated with chronic hay fever)
      d. Frontal, maxillary, sphenoid, and ethmoid sinuses drain into nose (**Figure 17-3**)
   2. Functions
      a. Warms and moistens inhaled air
      b. Contains sense organs of smell (olfactory receptors)

B. Pharynx
   1. Structure (**Figure 17-4**)
      a. Pharynx (throat) about 12.5 cm (5 inches) long
      b. Divided into nasopharynx, oropharynx, and laryngopharynx
      c. Two nasal cavities, mouth, esophagus, larynx, and auditory tubes all have openings into pharynx
      d. Tonsils—form ring of lymphoid tissue around throat
         (1) Pharyngeal tonsils and openings of auditory tubes open into nasopharynx
         (2) Lingual and palatine tonsils found in oropharynx
         (3) Tonsillitis—inflammation of tonsils; tonsillectomy is surgical removal of tonsils (**Figure 17-5**)
      e. Mucous membrane lines pharynx
   2. Functions
      a. Passageway for food and liquids
      b. Air distribution; passageway for air
      c. Tonsils—provide immune protection

C. Larynx
   1. Structure (**Figure 17-6**)
      a. Located just below pharynx; also referred to as the voice box
      b. Nine pieces of cartilage form framework
         (1) Thyroid cartilage (Adam's apple) is largest
         (2) Epiglottis partially covers opening into larynx
      c. Mucous lining
      d. Vocal cords stretch across interior of larynx; space between cords is the glottis
   2. Functions
      a. Air distribution; passageway for air to move to and from lungs
      b. Voice production
   3. Laryngeal cancer
      a. Incidence increases with age and alcohol abuse
      b. Occurs most often in men over age 50
      c. If larynx removed, "esophageal speech" or electric artificial larynx needed for speech

D. Disorders of the upper respiratory tract
   1. Upper respiratory infection (URI)
      a. Rhinitis—nasal inflammation, as in a cold, influenza, or allergy
         (1) Infectious rhinitis—common cold
         (2) Allergic rhinitis—hay fever
      b. Pharyngitis (sore throat)—inflammation or infection of the pharynx

17

c. Laryngitis—inflammation of the larynx resulting from infection or irritation
   (1) Epiglottitis—life-threatening
   (2) Croup—non–life-threatening
2. Anatomical disorders
   a. Deviated septum—septum that is abnormally far from the midsagittal plane (congenital or acquired)
   b. Epistaxis (bloody nose) can result from mechanical injuries to the nose, hypertension, or other factors

Lower Respiratory Tract

A. Trachea
1. Structure (**Figure 17-7**)
   a. Tube (windpipe) about 11 cm (4.5 inches) long that extends from larynx into the thoracic cavity
   b. Mucous lining
   c. C-shaped rings of cartilage hold trachea open, but allow for swallowing
2. Function—passageway for air to move to and from lungs
3. Obstruction
   a. Blockage of trachea occludes the airway, and if complete, causes death in minutes
   b. Tracheal obstruction causes more than 4000 deaths annually in the United States
   c. Five-and-five maneuver is a lifesaving technique used to free the trachea of obstructions; also see *abdominal thrusts* in box on p. 467
   d. Tracheostomy—surgical procedure in which a tube is inserted into an incision in the trachea so that a person with a blocked airway can breathe

B. Bronchial tree
1. Structure
   a. Trachea branches into right and left bronchi
      (1) Right primary bronchus more vertical than left
      (2) Aspirated objects most often lodge in right primary bronchus or right lung
   b. Each bronchus branches into smaller and smaller tubes (secondary bronchi), eventually leading to bronchioles
   c. Bronchioles end in clusters of microscopic alveolar sacs, the walls of which are made up of alveoli (**Figure 17-8**)
2. Function—air distribution; passageway for air to move to and from alveoli

C. Alveoli (**Figure 17-9**)
1. Respiratory membrane—thin wall that separates pulmonary blood from alveolar air, allowing diffusion of gases; flat type I cells form single, thin layer
2. Function—exchange of gases between air and blood
3. Surfactant—substance released by type II cells into alveoli to reduce surface tension and thus prevent collapse of alveoli

D. Respiratory distress—relative inability to inflate the alveoli
1. Infant respiratory distress syndrome (IRDS)—leading cause of death in premature infants, resulting from lack of surfactant production in alveoli
2. Adult respiratory distress syndrome (ARDS)—impairment of surfactant by inhalation of foreign substances or other conditions

E. Lungs
1. Structure (**Figure 17-10**)
   a. Size—large enough to fill the thoracic cavity, except for middle space (mediastinum) occupied by heart, large blood vessels, thymus, and esophagus
   b. Apex—narrow upper part of each lung, under collarbone
   c. Base—broad lower part of each lung; rests on diaphragm
2. Function—breathing (pulmonary ventilation)

F. Plurae
1. Thin membrane that lines thoracic cavity (parietal pleura) and covers outer surface of lungs (visceral pleura)
2. Moist, smooth, slippery serous membrane reduces friction between the lungs and chest wall during breathing (**Figure 17-11**)
3. Pleurisy—inflammation of the pleura
4. Atelectasis—collapse or incomplete expansion of the lung (alveoli) (**Figure 17-12**); can be caused by:
   a. Pneumothorax—presence of air in the pleural space
   b. Hemothorax—presence of blood in the pleural space

G. Disorders of the lower respiratory tract
1. Lower respiratory infection
   a. Acute bronchitis, or tracheobronchitis—inflammation of the bronchi or bronchi and trachea caused by infection (usually resulting from the spread of a URI)
   b. Pneumonia (**Figure 17-13**)—acute inflammation (infection) in which lung airways become blocked with thick exudates
      (1) Lobar pneumonia—affects entire lobe of lung
      (2) Bronchopneumonia—infection scattered along bronchial tree
   c. Tuberculosis (TB)—chronic, highly contagious lung infection characterized by tubercles in the lung; can progress to involve tissues outside the lungs and pleura
2. Restrictive pulmonary disorders reduce compliance (the ability of lung tissues to stretch), particularly during inspiration
   a. Factors inside the lungs, such as fibrosis (scarring) or inflammation, may restrict breathing
   b. Factors outside the lungs, such as pain from injury or pleurisy, may restrict breathing
   c. The thickened fluid in the lungs occurring in cystic fibrosis also restricts lung compliance

3. Obstructive pulmonary disorders
   a. Obstruct airways, thus obstructing inspiration and expiration
   b. Acute obstruction can be immediately life-threatening
   c. Chronic obstructive pulmonary disease (COPD) can develop from pre-existing obstructive conditions (**Figure 17-14**)
   d. Chronic bronchitis—chronic inflammation of the bronchial tree
   e. Emphysema—reduced surface area of lungs caused by rupture or other damage to alveoli
   f. Asthma—recurring spasms of the airways accompanied by edema and mucous production
4. Lung cancer—malignant tumor of the lungs, occasionally treatable with surgery, chemotherapy, and radiation

Respiration

A. Respiration involves several processes and mechanisms
   1. External respiration—pulmonary ventilation (breathing) and pulmonary gas exchange
   2. Transport of gases by blood and regulation of setpoint levels of blood gases
   3. Internal respiration—systemic gas exchange and cellular respiration
B. **Figure 17-15** summarizes all these processes and thus serves as a "big picture" view of respiration

Pulmonary Ventilation

A. Mechanics of breathing (**Figure 17-16**)
   1. Basic principles
      a. Pulmonary ventilation includes two phases called *inspiration* (movement of air into lungs) and *expiration* (movement of air out of lungs)
      b. Changes in size and shape of thorax cause changes in air pressure within that cavity and in the lungs because as volume changes, pressure changes in the opposite direction
      c. Pressure differences (gradients) cause air to move into or out of lungs; air moves from high air pressure toward low air pressure
      d. Thorax and lungs must remain:
         (1) Compliant—able to stretch
         (2) Elastic—able to recoil after stretch
   2. Inspiration (inhalation)
      a. Active process—air moves into lungs
      b. Inspiratory muscles include diaphragm and external intercostals
         (1) Diaphragm flattens when stimulated by phrenic nerves during inspiration—increases top-to-bottom length of thorax
         (2) External intercostal muscles contract and elevate the ribs—increases the size of the thorax from front to back and from side to side
      c. Increase in size of the thoracic cavity reduces pressure within it; air then enters the lungs by moving down its pressure gradient
   3. Expiration (exhalation)
      a. Quiet expiration is ordinarily a passive process
      b. During expiration, thorax returns to its resting size and shape
      c. Elastic recoil of lung tissues aids in expiration
      d. Expiratory muscles used in forceful expiration are internal intercostals and abdominal muscles
         (1) Internal intercostals—contraction depresses the rib cage and decreases the size of the thorax from front to back
         (2) Contraction of abdominal muscles elevates the diaphragm, thus decreasing size of the thoracic cavity from top to bottom
      e. Reduction in the size of the thoracic cavity decreases its volume and thus increases its pressure, so air moves down the pressure gradient and leaves the lungs
B. Pulmonary volumes (**Figure 17-17** and **Table 17-1**)
   1. Volumes of air exchanged (into and out of body) in breathing can be measured with a spirometer
   2. Tidal volume (TV)—amount normally breathed in or out with each breath
   3. Vital capacity (VC)—largest amount of air that one can breathe out in one expiration
   4. Expiratory reserve volume (ERV)—amount of air that can be forcibly exhaled after expiring the tidal volume
   5. Inspiratory reserve volume (IRV)—amount of air that can be forcibly inhaled after a normal inspiration
   6. Residual volume (RV)—air that remains in the lungs after the most forceful expiration
C. Regulation of ventilation
   1. Regulation of respiration permits the body to adjust to varying demands for oxygen supply and carbon dioxide removal by maintaining setpoint concentrations of blood gases
   2. Brainstem control of respiration (**Figure 17-18**)
      a. Most important central regulatory centers in brainstem are called *respiratory control centers*
      b. Medullary centers—under resting conditions, the medullary rhythmicity area produces a normal rate and depth of respirations (12 to 18 per minute)
      c. Pontine centers—as conditions in the body vary, these centers in the pons can alter the activity of the medullary rhythmicity area, thus adjusting breathing rhythm
      d. Brainstem centers are influenced by information from other parts of the brain and from sensory receptors located in other body regions
   3. Cerebral cortex—voluntary (but limited) control of respiratory activity
   4. Respiratory reflexes
      a. Chemoreflexes—chemoreceptors respond to changes in carbon dioxide, oxygen, and blood acid levels—receptors located in carotid and aortic bodies

b. Pulmonary stretch reflexes—respond to the stretch receptors in lungs, thus protecting respiratory organs from overinflation

D. Breathing patterns (**Table 17-2**)
1. Eupnea—normal breathing
2. Hyperventilation—rapid and deep respirations
3. Hypoventilation—slow and shallow respirations
4. Dyspnea—labored or difficult respirations
5. Orthopnea—dyspnea relieved by moving into an upright or sitting position
6. Apnea—stopped respiration
7. Cheyne-Stokes respiration (CSR)—cycles of alternating apnea and hyperventilation associated with critical conditions
8. Respiratory arrest—failure to resume breathing after a period of apnea

Gas Exchange and Transport

A. Pulmonary gas exchange—exchange of gases in lungs (**Figure 17-19**)
1. Carbaminohemoglobin breaks down into carbon dioxide and hemoglobin
2. Carbon dioxide moves out of lung capillary blood into alveolar air and out of body in expired air
3. Oxygen moves from alveoli into lung capillaries
4. Hemoglobin combines with oxygen, producing oxyhemoglobin

B. Systemic gas exchange—exchange of gases in tissues (**Figure 17-19**)
1. Oxyhemoglobin breaks down into oxygen and hemoglobin
2. Oxygen moves out of tissue capillary blood into tissue cells
3. Carbon dioxide moves from tissue cells into tissue capillary blood
4. Hemoglobin combines with carbon dioxide, forming carbaminohemoglobin

C. Blood transportation of gases
1. Transport of oxygen
   a. Only small amounts of oxygen (O_2) can be dissolved in blood
   b. Most oxygen combines with hemoglobin to form oxyhemoglobin (HbO_2) to be carried in blood
2. Transport of carbon dioxide
   a. Dissolved carbon dioxide (CO_2)—10%
   b. Carbaminohemoglobin ($HbCO_2$)—20%
   c. Bicarbonate ions (HCO_3^-)—70%

❑ ACTIVE LEARNING

STUDY TIPS

Consider using these tips to achieve success in meeting your learning goals.

Review the synopsis of the respiratory system in Chapter 5. The structures of the respiratory system can be described as tubes and bags. All the structures except the alveoli are tubes. Their job is to get air to and from the alveoli, where oxygen and carbon dioxide are exchanged in the blood.

1. Flash cards and online resources can be used to learn the names, locations, and functions of the structures of the respiratory system.
2. Lungs are passive organs. Remember that in order for air to be moved in and out of the lung, the pressure of the chest cavity must be raised or lowered. To lower the pressure, the volume must increase (Boyle's law); this is done by contracting the diaphragm, which causes air to enter the lung. When the diaphragm relaxes, the volume of the chest cavity decreases, the pressure goes up, and the air is pushed out of the lung.
3. When oxygen gets to the lung, it forms a weak bond with hemoglobin in the blood. When the blood gets to the tissue, it gives up the oxygen and takes on carbon dioxide. The blood carries carbon dioxide as bicarbonate ion or by combining it with hemoglobin. When the blood gets to the lung, the carbon dioxide dissociates and is exhaled. **Figure 17-19** shows CO_2 leaving the blood.
4. Review the Language of Science and Language of Medicine terms. Check out *my-ap.us/MOGBpB* for respiratory system tutorials.
5. The disorders of the respiratory system can be learned by making a chart of the various disorders. Organize the chart by mechanism or cause: upper respiratory infections, lower respiratory infections, restrictive disorders, and obstructive disorders.
6. In your study group, go over the flash cards of the structures of the respiratory system and pulmonary volumes. Discuss the processes of inspiration, expiration, and regulation of respiration. Discuss external and internal respiration. Go over the respiratory disorders chart, chapter outline summary, and the questions at the end of the chapter, and discuss possible test questions.

Review Questions

Write out the answers to these questions after reading the chapter and reviewing the Chapter Summary. If you simply think through the answer without writing it down, you will not retain much of your new learning.

1. Differentiate between the respiratory membrane and the respiratory mucosa.
2. Describe the ciliary escalator.
3. List the functions of the paranasal sinuses.
4. What is the function of the auditory tube?
5. What is the function of the epiglottis?
6. Describe, in decreasing order of size, the air tubes of the lung.
7. What constitutes an upper respiratory infection?
8. Describe rhinitis, pharyngitis, and laryngitis.
9. What is IRDS? What substance is missing from the lung that causes IRDS?
10. Describe the pleura. What is the function of pleural fluid?
11. What is atelectasis?
12. How is tuberculosis transmitted from person to person? What is the pathogen that causes tuberculosis?
13. What occurs to restrict breathing in asthma?
14. What is bronchitis?
15. What process in emphysema causes the reduction in lung surface area?
16. Distinguish between lobar pneumonia, bronchopneumonia, and aspiration pneumonia.
17. Define pulmonary compliance.
18. Name and explain the volumes that make up vital capacity.
19. Differentiate between external, internal, and cellular respiration.
20. Explain the mechanical process of inspiration.
21. Explain the mechanical process of expiration.
22. Explain the function of chemoreceptors in regulating breathing.
23. Explain the function of stretch receptors in regulating breathing.
24. Identify the two most important control centers in the medulla for regulating breathing rhythm.
25. Explain how the Po_2 and the Pco_2 impacts how gas is exchanged between the lung and the blood, and the blood and the tissues.
26. What is a spirometer?
27. Describe Cheyne-Stokes respiration.

Critical Thinking

After finishing the Review Questions, write out the answers to these more in-depth questions to help you apply your new knowledge. Go back to sections of the chapter that relate to concepts that you find difficult.

28. Explain the effect smoking has on the body's ability to move material trapped in the respiratory mucosa.
29. After strenuous exercise, inexperienced athletes will quite often attempt to recover and resume normal breathing by bending over or sitting down. Using the mechanics of ventilation, how would you modify the recovery practices of these athletes?
30. Calculate your vital capacity and total lung capacity if your pulmonary ventilation volumes were as follows: tidal volume—500 mL; inspiratory reserve volume—3200 mL; expiratory reserve volume—1100 mL; and residual volume—1150 mL.

17

Chapter Test

Hint *After studying the chapter, test your mastery by responding to these items. Try to answer them without looking up the answers.*

1. The organs of the respiratory system are designed to perform two basic functions: ______ and ______.
2. The upper respiratory tract consists of the ______, the ______, and the ______.
3. The membrane that separates the air in the alveoli from the blood in the surrounding capillaries is called the ______.
4. The membrane that lines most of the air distribution tubes of the respiratory system is called the ______.
5. Nerve endings responsible for the sense of ______ are located in the nasal mucosa.
6. The frontal, maxillary, sphenoidal, and ethmoidal cavities make up the ______.
7. The ______ sacs drain tears into the nasal cavity.
8. The ______ protrude into the nasal cavities and function to warm and humidify the air.
9. The ______ is the structure that can also be called the *throat.*
10. The ______ is also called the *voice box.*
11. The ______ is the large air tube in the neck.
12. The four progressively smaller air tubes that connect the trachea and the alveoli are the ______, ______, ______, and ______.
13. Exocrine mucous glands possessing many ______ cells help produce mucus.
14. ______ is a substance made by the lung to reduce the surface tension of water in the alveoli.
15. The right lung is made up of ______ lobes, and the left lung is made up of ______ lobes.
16. A collapse of the lung for any reason is called ______.
17. Air in the pleural space is called a ______.
18. A series of cycles of alternating apnea and hyperventilation is called ______ respiration.
19. The exchange of gases between the blood and the tissues is called ______.
20. The ______ is the most important muscle in respiration.
21. The exchange of gases by diffusion is accomplished by way of diffusion down their ______ ______ gradient.
22. Oxygen is carried in the blood as ______.
23. Carbon dioxide is carried in the blood as the ______ ion or combines with hemoglobin as ______.
24. The basic respiratory rhythm centers are located in the ______ of the brain.
25. ______ are the sensory receptors that help keep the lung from overexpanding.
26. ______ are the sensory receptors that help modify respiratory rates by detecting the amount of carbon dioxide, oxygen, or acid levels in the blood.
27. The amount of air that is moved in and out of the lung during normal, quiet breathing is called ______ volume.
28. The three volumes that make up vital capacity are ______, ______, and ______.
29. The volume included in total lung capacity, but not vital capacity, is ______ volume.

Match each disorder in Column A with its description or cause in Column B.

Column A

30. ______ rhinitis
31. ______ pharyngitis
32. ______ laryngitis
33. ______ deviated septum
34. ______ epistaxis
35. ______ IRDS
36. ______ pneumonia
37. ______ tuberculosis
38. ______ emphysema
39. ______ asthma

Column B

a. nosebleed
b. a condition in which the nasal septum strays from the midline of the nasal cavity
c. a chronic bacillus infection that usually affects the lung and is caused by a mycobacterium
d. an inflammation of the mucous lining of the larynx
e. a condition in which ruptured alveoli reduce the surface area of the lung, making breathing difficult
f. an inflammation of the nasal mucosa
g. an obstructive disorder characterized by recurring spasms of the smooth muscles of the bronchi
h. an acute inflammation of the lungs
i. an inflammation or infection of the pharynx
j. a disease characterized by a lack of surfactant in the alveoli; usually occurs in premature infants

Case Studies

To solve a case study, you may have to refer to the glossary or index, other chapters in this textbook, and other resources.

1. Curtis was having fun alongside a neighborhood swimming pool when he was accidentally pushed into the pool. Although he is a good swimmer, the suddenness of the fall caught him off guard and he inhaled some water before he was able to gain control of the situation. Luckily, a nearby swimmer assisted Curtis to the edge of the pool, but Curtis continued to have great difficulty in breathing. Can you name the syndrome that Curtis must be exhibiting? Explain what has happened to Curtis's lungs to cause his breathing difficulty.
2. Walter has aspergillosis in his lungs. This disease has caused a partial blockage of both of his bronchi. Does Walter have a restrictive or obstructive condition? What signs would you look for to confirm your diagnosis if Walter uses a spirometer to test his breathing? What type of pathogenic organism caused Walter's problem? (HINT: See Appendix A at *evolve.elsevier.com*.)
3. While you are chatting with your friend at an expensive restaurant, she suddenly stops in midswallow and looks panicked. When you ask what is wrong, she indicates that she can't speak and runs toward the restroom. What should you do? Assume that your first aid does not work—what procedure might emergency medical personnel use to help your friend?
4. Tom was an endurance runner. He was an average competitor but was hopeful that if he stuck with it, he might improve enough to earn a scholarship at a university. After a year on the team, he was discouraged because his performance did not improve enough to even be considered for any type of distance athletics. After talking to his coach, he was tempted to see an exercise physiologist to determine his maximum oxygen consumption ($VO_{2\ max}$). What type of information would this test provide to help Tom with future decisions regarding running for his high school team?

Answers to Active Learning Questions can be found online at evolve.elsevier.com.

17

Digestive System

OUTLINE

Scan this outline before you begin to read the chapter, as a preview of how the concepts are organized.

OBJECTIVES

Before reading the chapter, review these goals for your learning.

After you have completed this chapter, you should be able to:

1. List the main and accessory organs of the digestive system and discuss primary mechanisms of the digestive system.
2. List and describe the four layers of the digestive tract wall, and discuss the value of each layer to the digestive organs.
3. Discuss the structures of the mouth: the oral cavity, teeth, and salivary glands, as well as the disorders of the mouth.
4. Discuss the structure and function of the pharynx, esophagus, and stomach, as well as the disorders of the stomach.
5. Discuss the structure, function, and disorders of the small intestine.
6. Discuss the structure, function, and disorders of the liver, gallbladder, and pancreas.
7. Discuss the structure, function, and disorders of the large intestine, appendix, and peritoneum.
8. Define and contrast mechanical and chemical digestion.
9. Discuss the basics of carbohydrate, protein, and lipid digestion and give the end-products of each process.
10. Describe the process of absorption and how structural adaptations of the digestive tube affect the rate and efficiency of nutrient absorption.

CHAPTER 18

All of us enjoy a good meal! Food preferences differ widely among cultures and individuals, but there is no doubt that the sight, smell, taste, texture, and especially the nutrient content of the foods we eat contribute in many ways to our quality of life. Although we do not "live to eat," we certainly must "eat to live."

The ingestion of food is the first step in an important and complex biological process that begins when we consume a meal. Nutrients undergo three forms of processing in the body: **digestion, absorption,** and **metabolism.** Digestion and absorption are performed by the organs of the digestive system. Metabolism, on the other hand, is performed by all body cells. This chapter describes digestive organs, associated disease states, and the processes of digestion and absorption. The metabolism of nutrients after they have been absorbed is discussed in Chapter 19.

LANGUAGE OF SCIENCE

Hint Before reading the chapter, say each of these terms out loud. This will help you to avoid stumbling over them as you read.

absorption
(ab-SORP-shun)
[*absorp-* **swallow,** *-tion* **process**]

alimentary canal
(al-eh-MEN-tar-ee kah-NAL)
[*aliment-* **nourishment,** *-ary* **relating to**]

amylase
(AM-eh-lays)
[*amyl-* **starch,** *-ase* **enzyme**]

anal canal
(AY-nal kah-NAL)
[*an-* **ring (anus),** *-al* **relating to**]

anus
(AY-nus)
[*anus* **ring**]

ascending colon
(ah-SEND-ing KOH-lon)
[*a[d]-* **toward,** *-scend-* **climb,** *colon* **large intestine**]

bicuspid
(bye-KUS-pid)
[*bi-* **double,** *-cusp-* **point,** *-id* **characterized by**]

bile
(byle)
[*bil-* **liver secretion**]

body
(BOD-ee)
[*body* **main part**]

bolus
(BOW-lus)
[*bolus* **lump**]

canine
(KAY-nyne)
[*can-* **dog,** *-ine* **relating to**]

Continued on p. 520

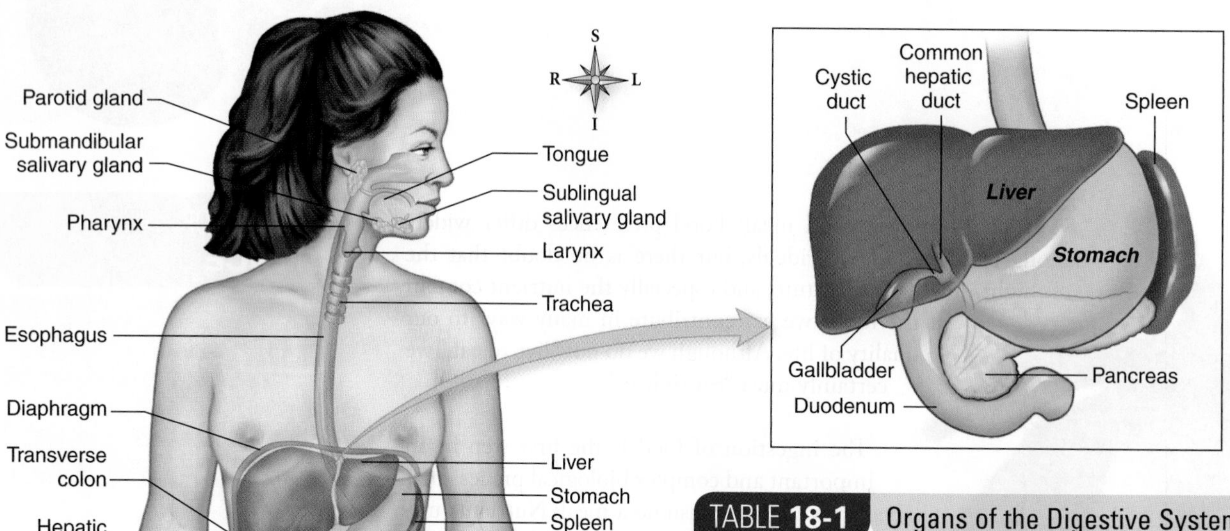

FIGURE 18-1 Digestive system.

TABLE 18-1 Organs of the Digestive System

MAIN ORGAN	ACCESSORY ORGAN
Mouth	Teeth and tongue Salivary glands Parotid Submandibular Sublingual
Pharynx (throat)	Tonsils
Esophagus	
Stomach	
Small intestine Duodenum Jejunum Ileum	Liver Gallbladder Pancreas
Large intestine Cecum Colon Ascending colon Transverse colon Descending colon Sigmoid colon Rectum Anal canal	Vermiform appendix

Many of the major structures of the digestive tract are shown in the layers of the *Clear View of the Human Body* (follows p. 8), which shows their relationships to other body structures.

Overview of Digestion

As you can see in **Figure 18-1**, the main organs of the **digestive system** form a continuous muscular tube of varying widths and open at both ends. This long, winding tube is called the **alimentary canal.** The term **gastrointestinal (GI) tract** technically refers only to the portion of the alimentary canal that includes the stomach and intestines, but it is often used to designate the entire *digestive tract.*

In the adult, this hollow digestive tube is about 9 meters (29 feet) long. Think of the tube as a passageway that extends through the body like a hallway through a building. Thus the food we eat and even the nutrient materials released by the digestive process are not truly "part of the body" until they have been absorbed through the wall of the GI tract and enter our body's internal environment.

Table 18-1 lists the organs of the digestive system. The *main organs* of digestion are those that make up the alimentary canal. The *accessory organs* of digestion are not segments of the tube, but are either within or surrounding the tube. For example, the teeth and tongue are inside the mouth and the various digestive glands surround the tract and pass their secretions through ducts into the lumen of the alimentary canal.

To accomplish the function of making nutrients available to each cell of the body, the digestive system uses various mechanisms (**Table 18-2**).

First, complex foods must be taken into the GI tract in a process called **ingestion.**

Then, the ingested food material must be broken down into simpler nutrients in the process that gives this system its name: **digestion.** The breakdown, or digestion, of food material is both *mechanical* and *chemical* in nature.

The teeth are used to physically break down large chunks of food before it is swallowed. The churning of material in the stomach then continues the mechanical digestive process. To physically break down large chunks of food matter into smaller bits and to move it along the tract, movement or **motility** of the GI wall is required.

In chemical digestion, large nutrient molecules are reduced to smaller molecules. This process requires **secretion** of

TABLE 18-2 Primary Mechanisms of the Digestive System

MECHANISM	DESCRIPTION
Ingestion	Process of taking food into the mouth, starting it on its journey through the digestive tract
Digestion	A group of processes that break complex nutrients into simpler ones, thus facilitating their absorption; *mechanical digestion* physically breaks large chunks into small bits; *chemical digestion* breaks molecules apart
Motility	Movement by the muscular components of the digestive tube, including processes of mechanical digestion; examples include *peristalsis* and *segmentation*
Secretion	Release of digestive juices (containing enzymes, acids, bases, mucus, bile, or other products that facilitate digestion); some digestive organs also secrete endocrine hormones that regulate digestion or metabolism of nutrients
Absorption	Movement of digested nutrients through the gastrointestinal (GI) mucosa and into the internal environment
Elimination	Excretion of the residues of the digestive process (feces) from the rectum, through the anus; defecation
Regulation	Coordination of digestive activity (motility, secretion, other digestive processes)

digestive enzymes and other products into the lumen of the GI tract. After the digestive processes have altered the physical and chemical composition of ingested food, the resulting nutrients are ready for the process of **absorption,** or movement through the GI mucosa into the internal environment.

Part of the digestive system, the large intestine, also serves as an organ of **elimination,** ridding the body of waste material, or **feces,** resulting from the digestive process.

To get an overview of the digestive system, go to AnimationDirect online at *evolve.elsevier.com.*

Wall of the Digestive Tract

The digestive tract has been described as a muscular tube that extends from the mouth to the anus. The wall of this digestive tube is fashioned of four layers of tissue (**Figure 18-2**). The inside, or hollow space within the tube, is called the **lumen.** The four layers, named from the inside coat to the outside of the tube, are as follows:

1. Mucosa or mucous membrane
2. Submucosa
3. Muscularis
4. Serosa

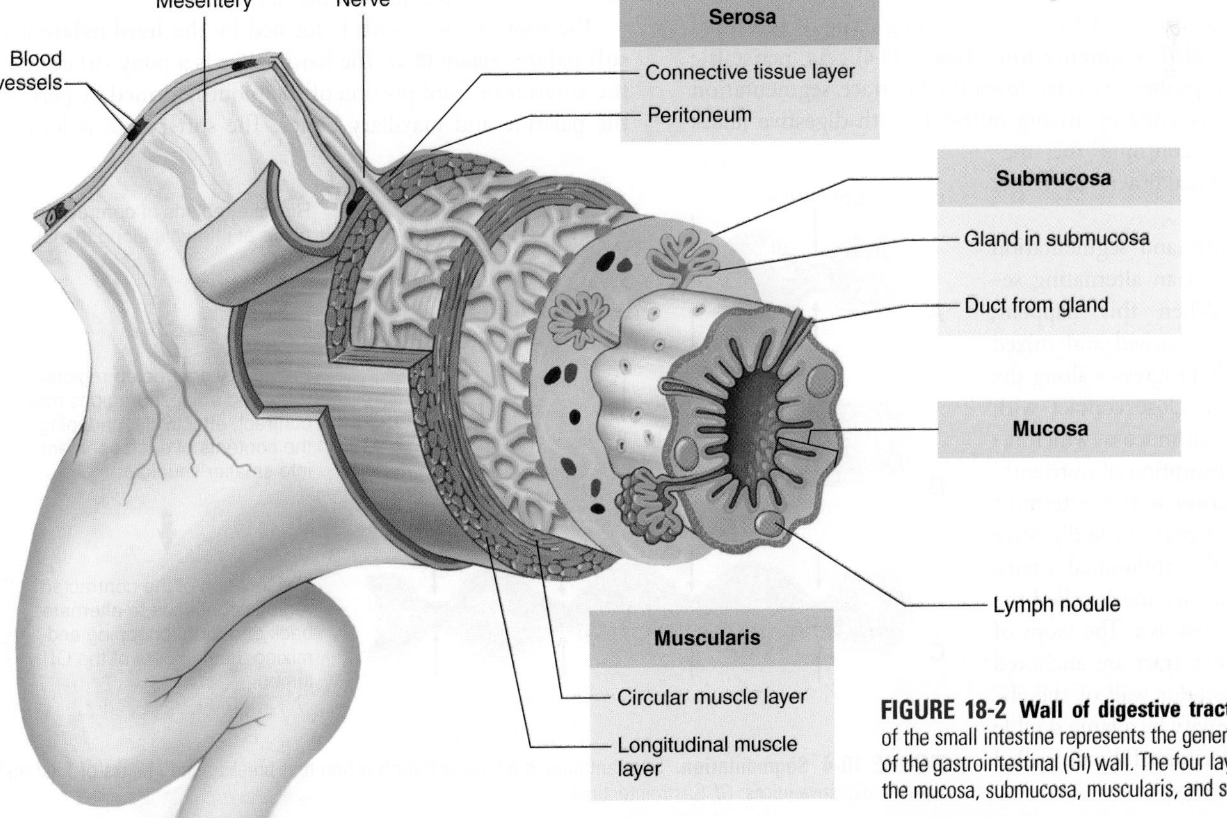

FIGURE 18-2 Wall of digestive tract. Segment of the small intestine represents the general structure of the gastrointestinal (GI) wall. The four layers include the mucosa, submucosa, muscularis, and serosa.

18

Although the same four tissue coats form every organ of the alimentary tract, their structure varies from organ to organ. The **mucosa** of the esophagus, for example, is composed of tough and abrasion-resistant stratified epithelium. The mucosa of the remainder of the tract is a delicate layer of simple columnar epithelium structured for absorption and secretion. The mucus produced by either type of epithelium coats the lining of the alimentary canal.

The **submucosa,** as the name implies, is a connective tissue layer that lies just below the mucosa. It contains many blood vessels, lymphatic vessels, and nerves.

Two or three layers of muscle tissue make up the **muscularis** (see **Figure 18-2**). These muscle layers have an important role to play in producing *motility* or movement of the GI tract during the digestive process.

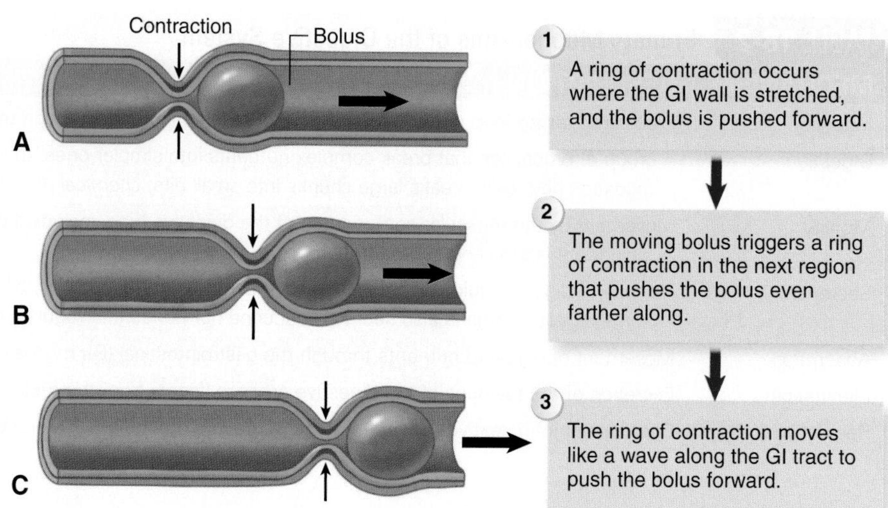

FIGURE 18-3 Peristalsis. Peristalsis is a progressive type of movement in which material is propelled from point to point along the gastrointestinal (GI) tract. Here, a ball or "bolus" of material is propelled by ringlike waves of smooth muscle contractions that produce peristalsis.

18

Peristalsis is a rhythmic, wavelike contraction of the gut wall caused by repeating waves of ringlike contraction along the circular muscle layer in the muscularis. This type of sequenced contraction squeezes and pushes ingested food material forward through the digestive tube's internal pathway—similar to how you might squeeze toothpaste out of its tube (**Figure 18-3**).

In addition to peristaltic contractions that cause material to move forward, alternating contraction of fibers of the muscularis within a single region, or segment, of the GI tract also produces a "back-and-forth" or "swishing" type of intestinal motility called **segmentation** (**Figure 18-4**). As peristaltic movement pushes material down the GI tract, segmentation contractions assist in mixing nutrients with digestive juices and helps continue the mechanical breakdown of larger particles.

Peristalsis and segmentation can occur in an alternating sequence. When this happens, material is churned and mixed as it slowly progresses along the GI tract in close contact with the intestinal mucosa, which facilitates absorption of nutrients.

The **serosa** is the outermost covering or coat of the digestive tube. In the abdominal cavity, this serosa covering is called the *visceral peritoneum.* The loops of the digestive tract are anchored to the posterior wall of the abdominal cavity by a large double fold of peritoneal tissue called the **mesentery.**

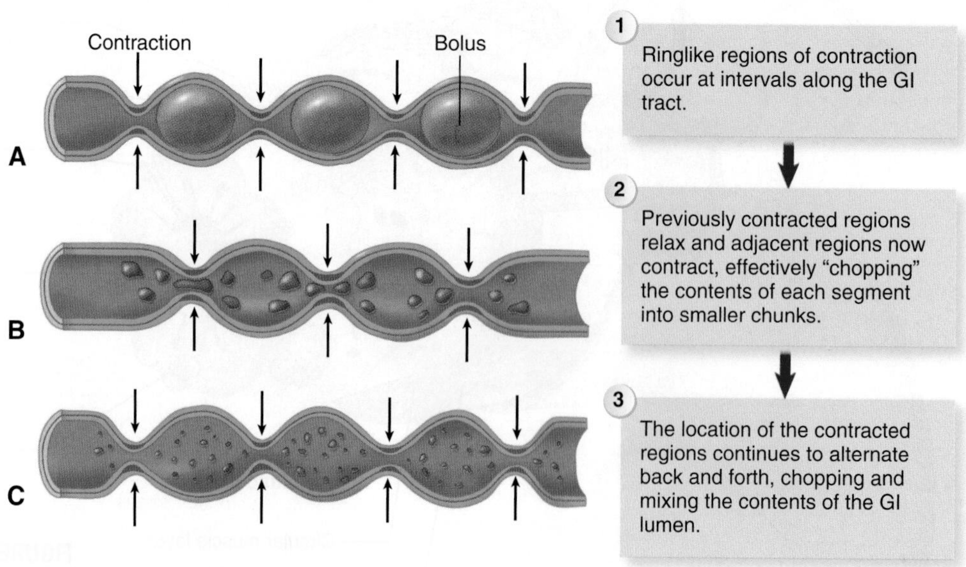

FIGURE 18-4 Segmentation. Segmentation is a back-and-forth action that breaks apart chunks of food and mixes in digestive juices. *GI,* Gastrointestinal.

To learn more about intestinal motility, go to AnimationDirect online at *evolve.elsevier.com.*

Mouth

Structure of the Oral Cavity

The **mouth,** or **oral cavity,** is a hollow chamber with a roof, a floor, and walls. Food enters, or is ingested, into the digestive tract through the mouth, and the process of digestion begins immediately. Like the remainder of the digestive tract, the mouth is lined with mucous membrane.

The roof of the mouth is formed by the **hard palate** and **soft palate** (**Figure 18-5**). The hard palate is a bony structure in the anterior or front portion of the mouth, formed by parts of the palatine and maxillary bones. The soft palate is located

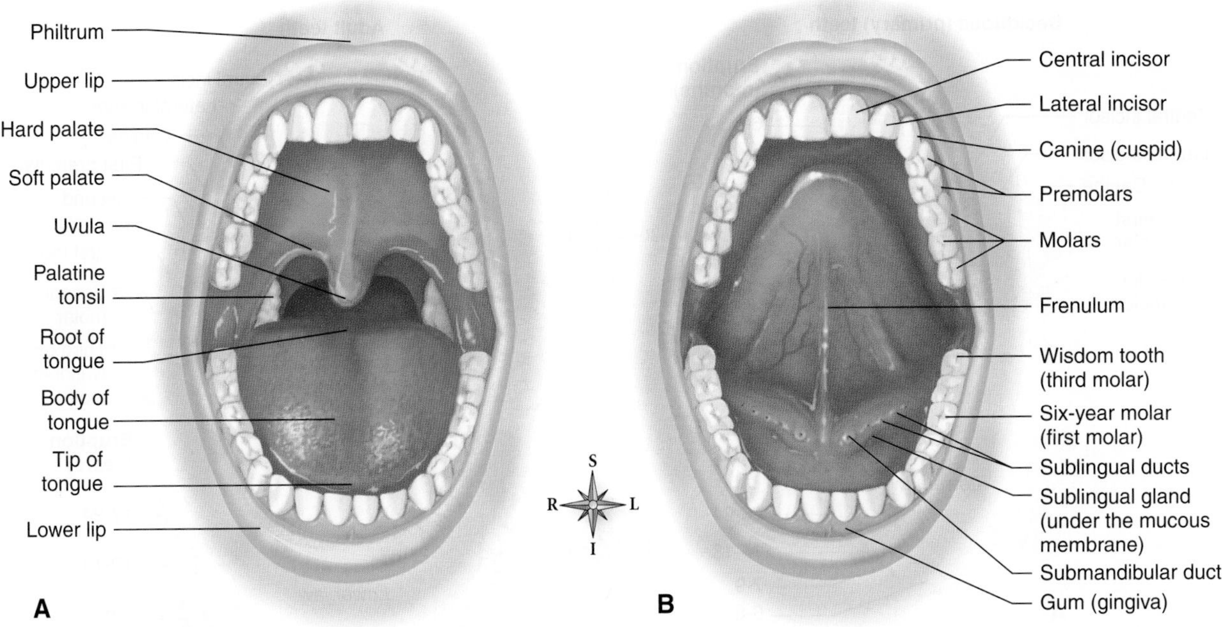

FIGURE 18-5 Mouth. A, Mouth cavity showing hard and soft palates, tongue surface, and uvula. **B,** Undersurface of tongue showing frenulum, sublingual gland, and opening of sublingual duct. *GI,* Gastrointestinal.

above the posterior or rear portion of the mouth. It is soft because it consists chiefly of muscle.

Hanging down from the center of the soft palate is an elongated process called the **uvula.** The uvula and the soft palate prevent any food and liquid from entering the nasal cavities above the mouth and also assist in speech and swallowing.

The floor of the mouth consists of the tongue and its muscles. The tongue is made of skeletal muscle covered with mucous membrane. It is anchored to bones in the skull and to the hyoid bone in the neck.

A thin membrane called the **frenulum** attaches the tongue to the floor of the mouth. Occasionally the frenulum is too short to allow free movements of the tongue and must be cut or surgically repaired to allow normal speech and swallowing.

Note in Figure 18-5, *A,* that the tongue can be divided into a blunt rear portion called the *root,* a pointed *tip*, and a central *body.*

The many small bumps on the surface of the tongue are called **papillae.** Recall from Chapter 11 that *taste buds*, which contain sensory receptors, are located on the sides of these papillae and chemically analyze food that may be swallowed. Figure 11-18 on p. 308 shows the detailed structure of these papillae.

Teeth

Types of Teeth

The shape and placement of the teeth assist them in their functions and are classified as one of four types:

1. Incisor
2. Canine
3. Premolar
4. Molar

The **incisors** are often called the "front teeth." Incisors have a sharp cutting edge (Figure 18-6) used to bite off or cut food into manageable portions to begin the process of **mastication,** or chewing of food.

The **canine teeth,** sometimes called **cuspids,** are usually more elongated and pointed in appearance and function to pierce or tear the food that is being eaten into smaller shreds. This tooth type is particularly large in meat-eating mammals such as dogs or cats.

The teeth referred to as **premolars** are also called **bicuspids** and are located just posterior to the canine teeth. They have two points called cusps that can saw through tough, fibrous food—especially when moved forward and backward against the premolars in the opposing jaw.

Behind the premolars are the **molars,** or **tricuspids.** The molars have comparatively larger surface areas with three grinding or crushing cusps on the surface.

The chewing made possible by the teeth begins the mechanical breakdown of food for digestion. After food has been chewed, it is formed into a small rounded mass called a **bolus** so that it can be swallowed.

By the time a baby is 2 years old, he or she probably has a full set of 20 *baby teeth*—the primary or **deciduous teeth.** By the time a young adult is somewhere between 17 and 24 years old, a full set of 32 **permanent teeth** is generally present.

The average age for "cutting" the first tooth (eruption of tooth through the gum) is about 6 months, and the average age for losing the first baby tooth and starting to cut the permanent teeth is about 6 years. Figure 18-6 gives the names of the teeth and shows which ones are lacking in the deciduous, or baby, set of teeth.

If an individual does not form wisdom teeth (third molars), then the number of adult teeth would be only 28. This is considered a normal variation and occurs most often in Asians

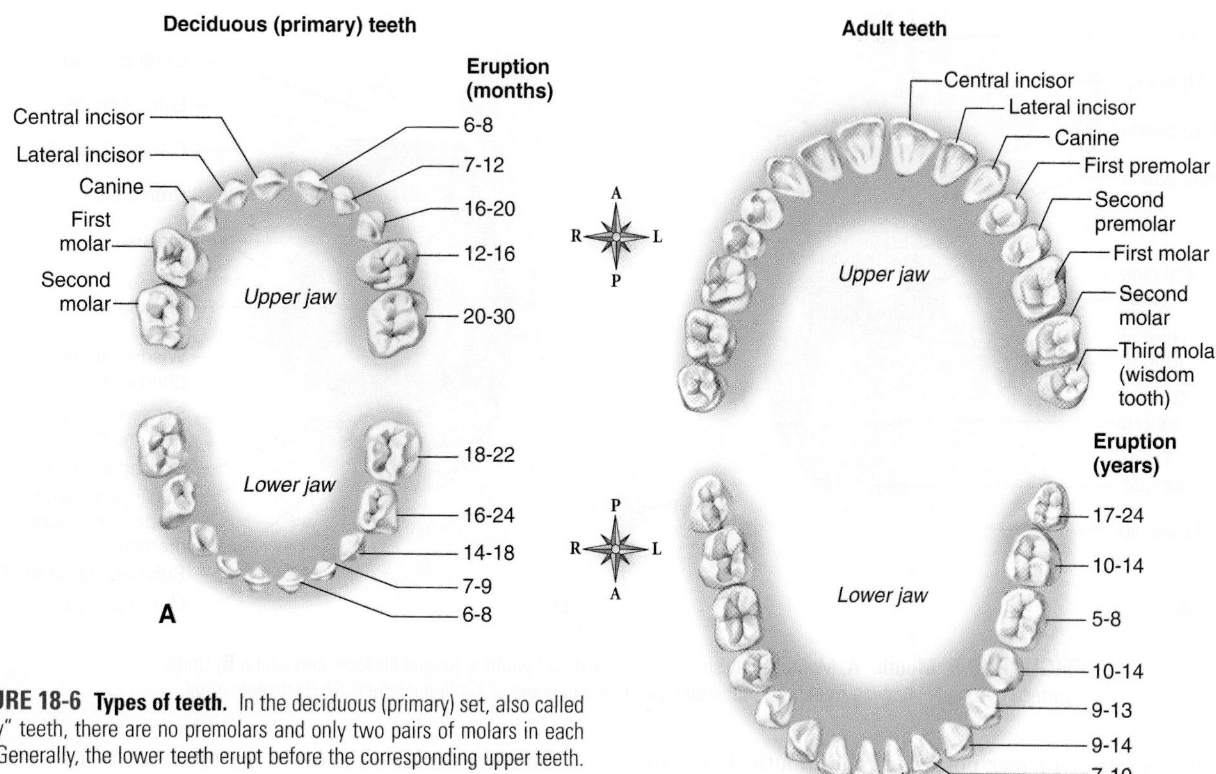

FIGURE 18-6 Types of teeth. In the deciduous (primary) set, also called "baby" teeth, there are no premolars and only two pairs of molars in each jaw. Generally, the lower teeth erupt before the corresponding upper teeth.

18

(30%), less frequently in Caucasians and Native Americans (about 12%), and only rarely in African-Americans (1% to 2%).

QUICK CHECK

1. Define the *alimentary canal* and name the four layers of the alimentary canal.
2. What three kinds of processing do nutrients undergo in the body?
3. Describe the interior of the mouth including the *hard* and *soft palate, uvula, frenulum, tongue* and *papillae.*
4. List the four major types of *teeth.*

Typical Tooth

A typical tooth can be divided into three main parts: crown, neck, and root (**Figure 18-7**).

The **crown** is the portion that is exposed and visible in the mouth. It is largely made of a bonelike material called **dentin** that is covered by **enamel.** Enamel is the hardest material made by the body and is ideally suited to withstand the grinding that occurs during the chewing of hard and brittle foods. The root and neck of each tooth are covered by **cementum.**

The center of the tooth contains a pulp cavity consisting of connective tissue, blood and lymphatic vessels, and sensory nerves.

The **neck** of a tooth is the narrow portion surrounded by the pinkish **gingiva** or *gum.* The neck joins the crown of the tooth to the root.

The **root** of the tooth fits into a socket of the upper or lower jaw bone—the maxilla or mandible (see **Figure 8-10** on p. 184). A fibrous **periodontal membrane** lines each tooth socket and anchors the tooth to the bone.

Salivary Glands

Three pairs of salivary glands—the parotids, submandibulars, and sublinguals—secrete most (about 1 L) of the watery saliva produced each day in the adult.

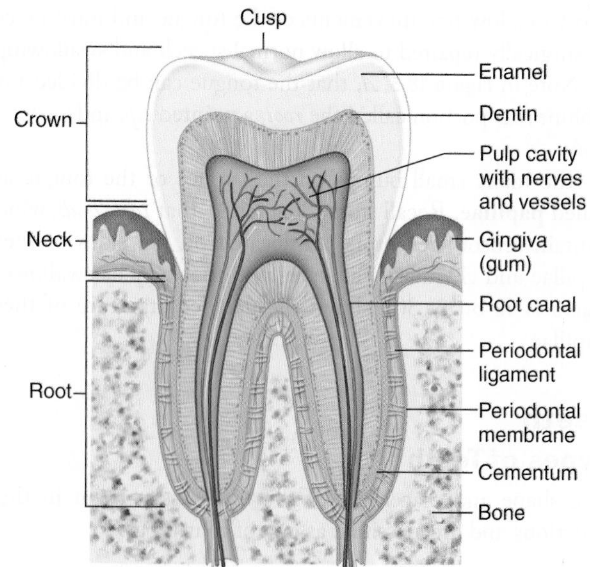

FIGURE 18-7 Typical tooth. A molar is sectioned here to show its bony socket and details of its three main parts: crown, neck, and root. Enamel (over the crown) and cementum (over the neck and root) surround the dentin layer. The pulp contains nerves and blood vessels.

CLINICAL APPLICATION
MALOCCLUSION

Malocclusion of the teeth occurs when missing teeth create wide spaces in the dentition, when teeth overlap, or when malposition of one or more teeth prevents correct alignment of the maxillary and mandibular dental arches (Figures *A* and *B*). Malocclusion that results in protrusion of the upper anterior teeth causing them to hang over the lower anterior teeth is called *overbite* (Figure *A*), whereas the positioning of the lower teeth outside the upper teeth is called *underbite* (Figure *B*).

Dental malocclusion may cause significant problems and chronic pain in the functioning of the temporomandibular joint, contribute to the generation of headaches, or complicate routine mastication of food. Fortunately, even severe malocclusion problems can be corrected by the use of braces and other dental appliances. **Orthodontics** is that branch of dentistry that deals with the prevention and correction of positioning irregularities of the teeth and malocclusion.

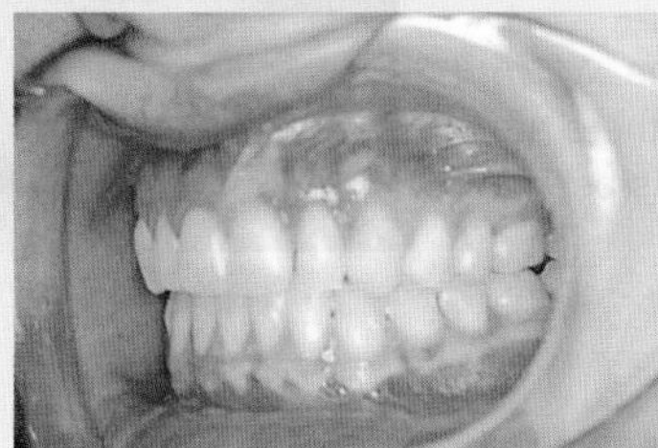
A

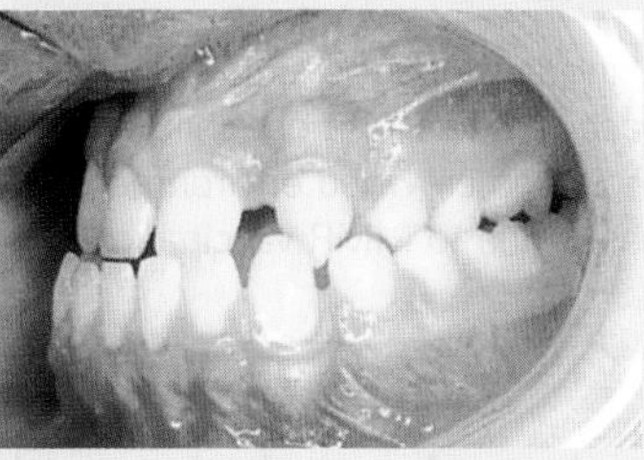
B

18

The salivary glands (**Figure 18-8**) are typical of the accessory glands associated with the digestive system because they are located outside the digestive tube itself and must convey their secretions by way of ducts into the tract. Because they have secretory ducts, they are considered *exocrine glands.*

The **parotid glands,** largest of the salivary glands, lie just inferior and anterior to each ear at the angle of the jaw. The parotid gland secretes a solution containing *sodium bicarbonate* ($NaHCO_3$), a base that helps neutralize bacterial acids. The openings of the parotid ducts can be found by looking in a mirror at the insides of the cheeks and next to the second molar tooth on either side of the upper jaw.

The ducts of the **submandibular glands** open into the mouth on either side of the lingual frenulum (**Figure 18-5**, *B*). The ducts of the **sublingual glands** open into the floor of the mouth.

Saliva also contains mucus and a digestive enzyme called salivary **amylase,** which begins the process of breaking down complex carbohydrates. Water and mucus moisten and lubricate the chewed food, allowing it to pass with less friction through the esophagus and on into the stomach.

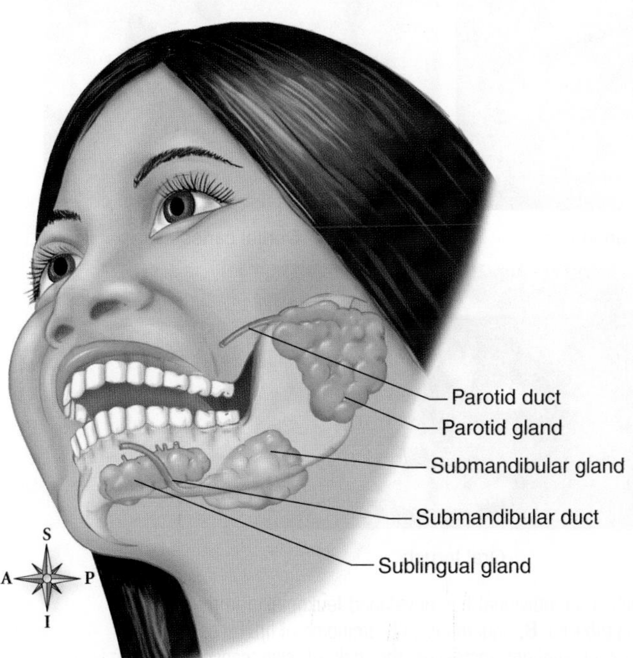

FIGURE 18-8 Salivary glands. The salivary glands and their associated ducts.

Disorders of the Mouth

Infections, cancer, congenital defects, and other disorders of the mouth and teeth can result in a variety of serious complications. Such conditions may cause pain or even damage to the mouth and teeth that makes chewing and swallowing difficult—perhaps causing a person to reduce the intake of food, thereby resulting in malnutrition. Mouth infections or cancer may spread to nearby tissues: the nasal cavity (then on to the sinuses, middle ear, and brain) or throat (and on to the esophagus, larynx, and thoracic organs).

Cancer

Cancer of the mouth may result from exposure to carcinogens found in tobacco smoke or in so-called smokeless tobacco (chewing tobacco), especially in combination with heavy alcohol consumption. Smokers may develop white patches, or **leukoplakia,** which may develop into malignant tumors.

Leukoplakia often develops in the fold between "cheek and gum" in users of smokeless tobacco (**Figure 18-9**, *A*). The

CLINICAL APPLICATION

MUMPS

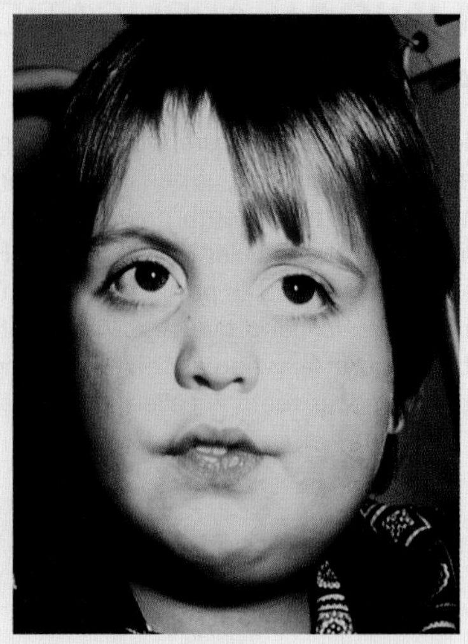

Mumps is an acute viral disease characterized by swelling of the parotid salivary glands, as you can see in the photo. Most of us think of mumps as a childhood disease because it most often affects children between the ages of 5 and 15 years of age. However, it can occur in adults—often producing a more severe infection.

The mumps infection can affect other tissues in addition to the parotid gland, including the joints, pancreas, myocardium, and kidneys. In about 25% of infected men, mumps causes inflammation of the testes, or *orchitis*. Orchitis resulting from mumps very rarely causes enough damage to render a man sterile.

Mumps is uncommon in developed countries because of the *MMR (mumps-measles-rubella) vaccine* given to most children. However, there are occasional outbreaks—especially in crowded environments. There has been a 99% decrease in mumps cases due to MMR vaccination.

condition, called **snuff dipper's pouch,** may lead to tooth and gum diseases and oral cancer.

Lip cancer may result from the carcinogenic effects of sunlight, which can be avoided by the use of lip balms containing sunscreen. The most common form of mouth cancer is squamous cell carcinoma (**Figure 18-9**, *B*).

18

Dental Conditions

Tooth decay, or **dental caries,** is one of the most common diseases in the developed world. It is a disease of the enamel, dentin, and cementum of teeth that results in the formation of a permanent defect called a *cavity* (**Figure 18-9**, *C*).

Many people living in the United States, Canada, and Europe are affected by the disease. Decay occurs on tooth surfaces where food debris, acid-secreting bacteria, and plaque accumulate. Sugar is the main ingredient in food that allows bacteria to produce the acid that damages the protective tooth enamel.

If the disease goes untreated, tooth decay results in infection, loss of teeth, and inflammation of the soft tissues in the

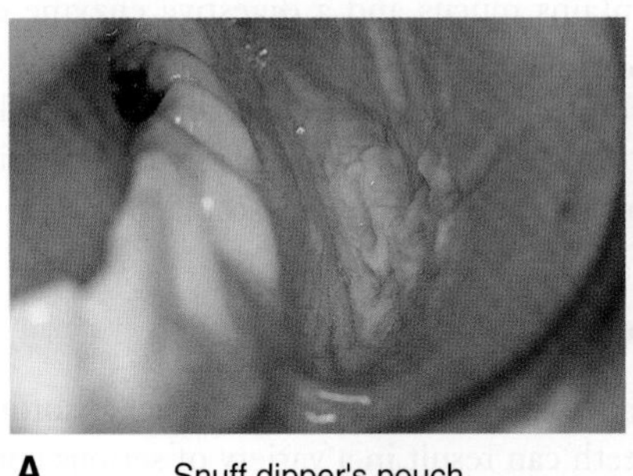

A Snuff dipper's pouch

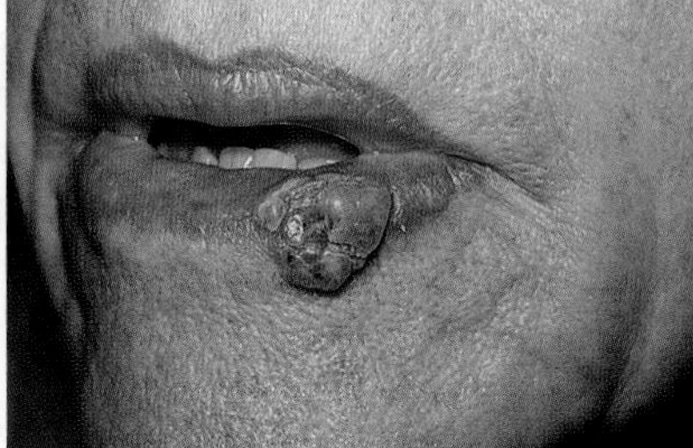

B Squamous cell carcinoma

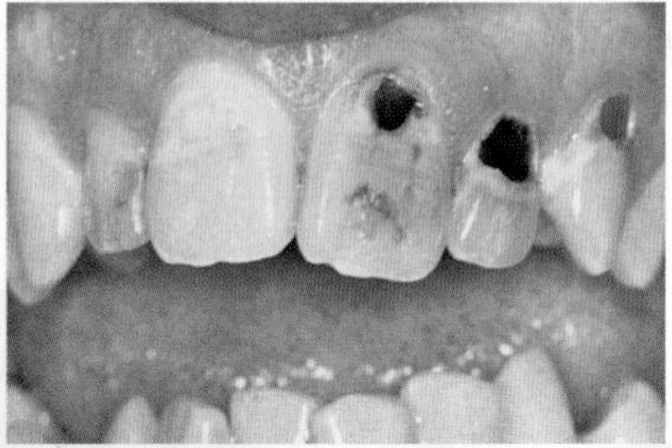

C Dental caries

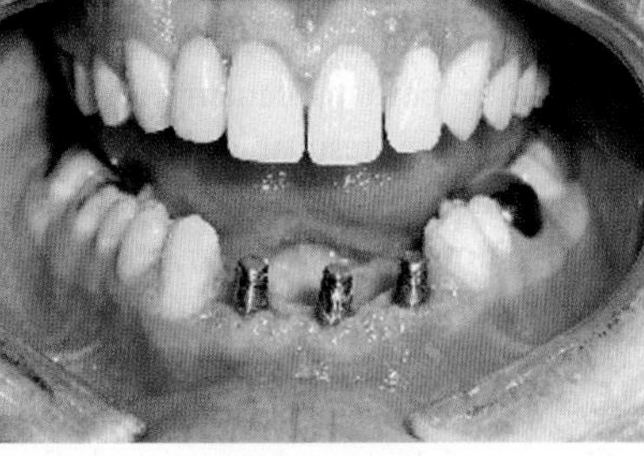

D Dental implants

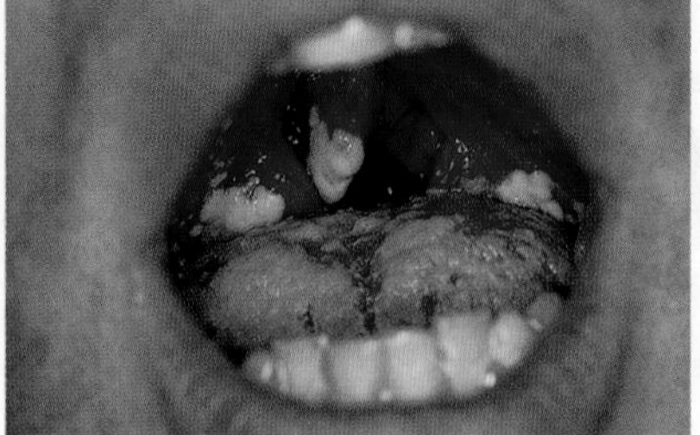

E Oral thrush

FIGURE 18-9 Mouth disorders. A, Snuff dipper's pouch. This individual has developed leukoplakia in the area between cheek and gum used for placement of chewing tobacco. **B,** Squamous cell carcinoma of lip. Excessive long-term exposure to ultraviolet light (UV) such as in sunlight increases the risk of skin cancer. **C,** Dental caries. These permanent defects, or cavities, are filled with decayed dental tissues. **D,** Dental implant. A permanent dental prosthesis will be affixed to the anchor after bone grows and healing has occurred. **E,** Oral thrush *(Candida albicans)*. Inflamed mucous membrane is covered with patches of creamy-white exudates.

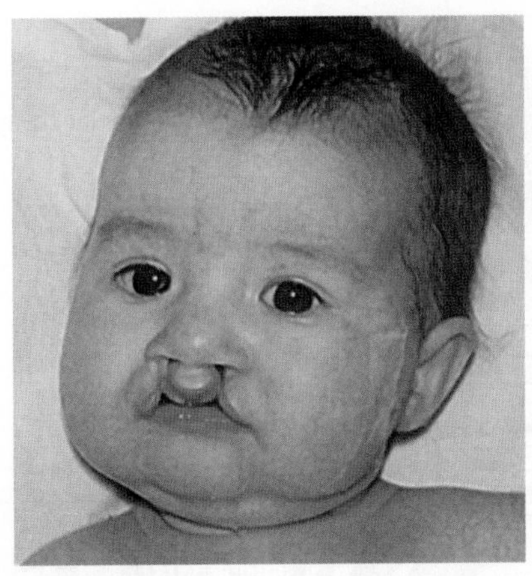

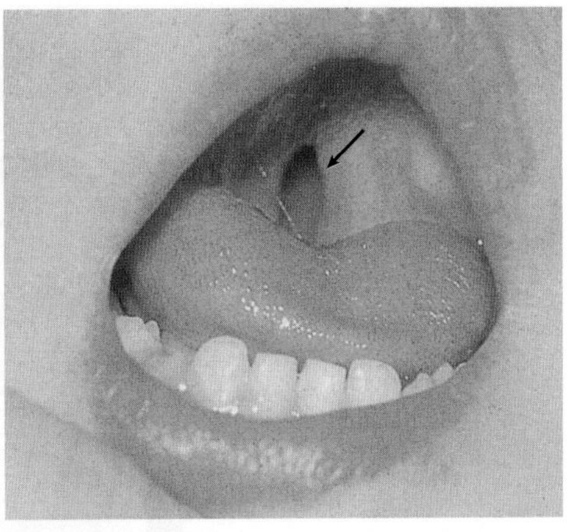

FIGURE 18-10 Congenital defects of the mouth. A, Bilateral cleft lip in an infant. **B,** Cleft palate.

mouth. Bacteria also may invade the paranasal sinuses or extend to the surface of the face and neck or enter the bloodstream, causing serious complications.

Dental caries are treated by removal of the decayed portion of the tooth followed by repair and filling of the defect using a variety of materials including porcelain and metal alloys.

If lost because of disease or trauma, teeth can be replaced with *dental appliances,* which include removable dentures and permanently fixed or implanted teeth. So called "dental implants" are anchors screwed into holes that have been drilled into a jawbone to support an artificial tooth or denture. About 6 months after insertion, new bone will have fused with and stabilized the anchor, permitting the attachment of the artificial tooth or dental appliance (Figure 18-9, *D*).

Gingivitis is the general term for inflammation or infection of the gums. Most cases of gingivitis result from poor dental hygiene—inadequate brushing and no flossing. Gingivitis also may be a complication of other conditions such as diabetes mellitus, vitamin deficiency, or pregnancy.

Periodontitis is the inflammation of the periodontal membrane, or *periodontal ligament,* which anchors the tooth to the bone of the jaw. Periodontitis is often a complication of advanced or untreated gingivitis and may spread to the surrounding bony tissue. Destruction of periodontal membrane and bone results in loosening and eventually complete loss of teeth. Periodontitis is the leading cause of tooth loss among adults.

Infection

Thrush, or **oral candidiasis,** is a mouth infection caused by a type of yeast known as *Candida* (see Figure 6-7 on p. 123). Candidiasis causes cream-colored "cheesy" patches of exudate to appear on an inflamed tongue and oral mucosa (Figure 18-9, *E*). The infection usually extends into the *oropharynx*—the region of the throat nearest the mouth.

Thrush is sometimes observed in otherwise healthy children but is most often seen in adults who are immunosuppressed, such as AIDS patients, or in individuals who have been on antibiotic therapy. The beneficial bacteria that are normal inhabitants of the microbiome of the mouth usually prevent the yeast population from expanding into an infection.

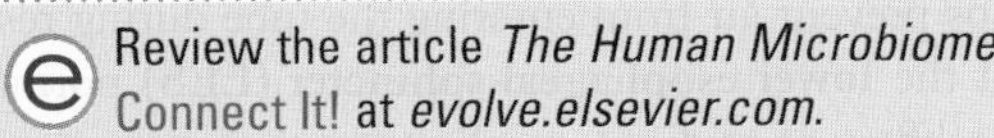
Review the article *The Human Microbiome* at Connect It! at *evolve.elsevier.com.*

Congenital Defects

Cleft lip and **cleft palate** (Figure 18-10) represent the most common forms of congenital defect affecting the mouth. They may occur alone or together and are caused by a failure of structures in the upper lip or palate to fuse or close properly during embryonic development. The condition may be inherited or be a spontaneous abnormality.

Cleft lip, which may occur on one or both sides, is generally repaired soon after birth. Surgical repair of cleft palate is usually done later in the first year of life. Modern reconstructive surgery techniques help minimize possible long-term complications that could include scarring, speech impairment, dental problems, and potential emotional maladjustment.

QUICK CHECK

1. What are the three main parts of a *typical tooth*?
2. What are the three pairs of *salivary glands*?
3. What is *snuff dipper's pouch*?
4. Distinguish between *dental caries, gingivitis,* and *periodontitis.*
5. Name the two most common forms of congenital defect affecting the mouth.

Pharynx

Structure

The **pharynx** is a tubelike structure made of muscle and lined with mucous membrane. Note its location in Figure 18-11. Because of its location behind the nasal cavities and mouth, it functions as part of both the respiratory and digestive systems. Air must pass through the pharynx on its way to the lungs, and a mass of chewed food must pass through it on its way to the stomach.

Recall that the pharynx as a whole is subdivided into three anatomical components: **nasopharynx, oropharynx,** and **laryngopharynx.** Also recall that the protective lymphoid ring formed by the three major pairs of *tonsils* in the pharynx guards against infections of the respiratory and digestive tracts.

18

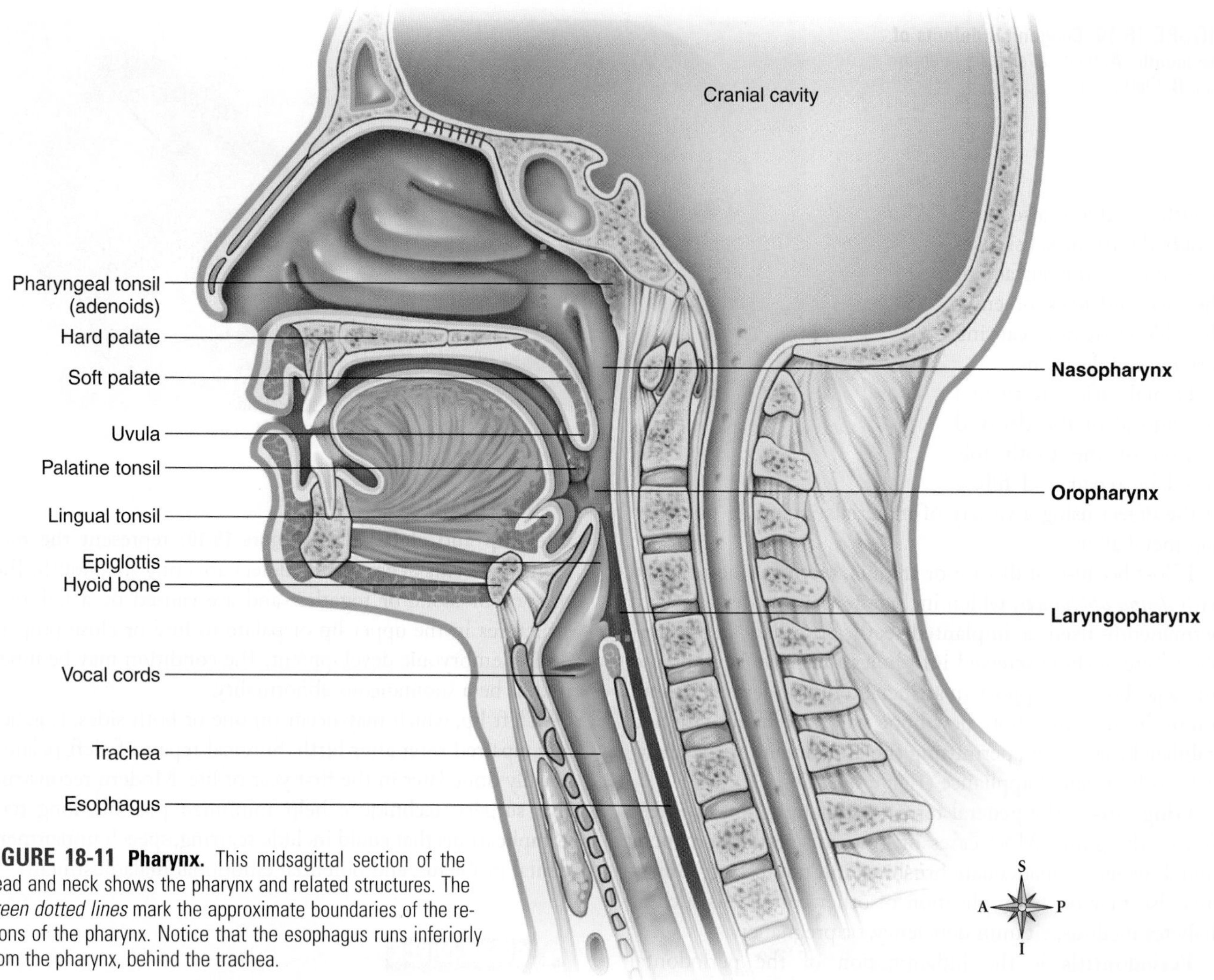

FIGURE 18-11 Pharynx. This midsagittal section of the head and neck shows the pharynx and related structures. The *green dotted lines* mark the approximate boundaries of the regions of the pharynx. Notice that the esophagus runs inferiorly from the pharynx, behind the trachea.

Review *Protective Strategies of the Respiratory Tract* at Connect It! at *evolve.elsevier.com*.

Function

Of the three anatomical divisions, the oropharynx is actively and most directly involved in the digestive process because of its important role in a specialized and coordinated type of GI tract motility involved in swallowing. The swallowing of a mass of chewed food is called **deglutition.**

First, mastication involves voluntary movements that result in formation of a ball or **bolus** of food in the mouth that is then moved involuntarily through the oropharynx and into the esophagus and, finally, into the stomach.

Swallowing is a complex process requiring the coordination of pharyngeal muscles and other muscles and structures in the head and neck. **Regulation** of voluntary swallowing movements is dependent on nervous impulses originating in the motor cortex of the cerebrum. Involuntary movements are regulated by impulses originating in the swallowing or "deglutition center" located in the medulla and pons of the brainstem (see **Figure 10-13** on p. 261).

To learn more about the pharynx and swallowing, go to AnimationDirect online at *evolve.elsevier.com*.

Esophagus

Structure and Function

The **esophagus** is a collapsible, muscular, mucus-lined tube about 25 cm (10 inches) long that extends from the pharynx to the stomach. It is the first segment of the digestive tube proper, and the four layers that form the wall of the GI tract can be easily identified when it is sectioned. Its muscular walls make it a dynamic passageway able to push food toward the stomach.

Each end of the esophagus is guarded by a muscular **sphincter.** Sphincters are valvelike rings of muscle tissue that often surround tubular structures or body openings. In the GI tract they normally act to keep ingested material moving in one direction down the tube. The **upper esophageal sphincter (UES)** helps prevent air from entering the tube during respiration, and the **lower esophageal sphincter (LES)** normally prevents backflow of acidic stomach contents.

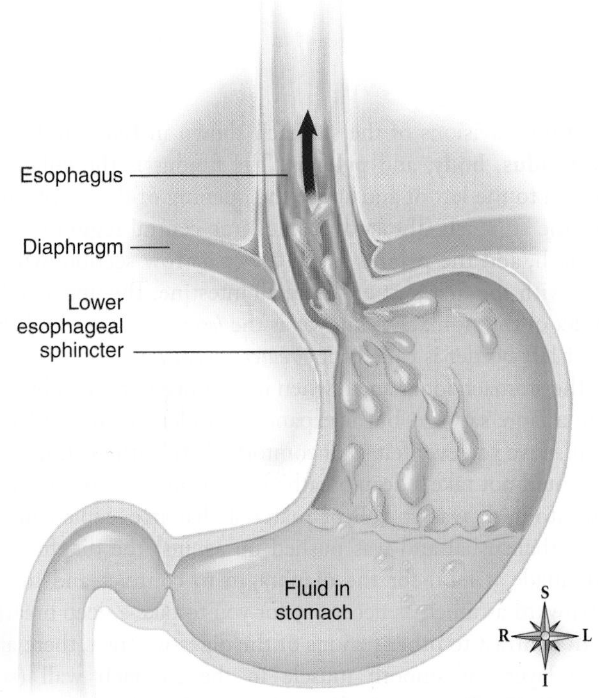

FIGURE 18-12 Reflux. In gastroesophageal reflux disease (GERD), reflux (backflow) of gastric acid up into the esophagus causes irritation of the lining of the esophagus.

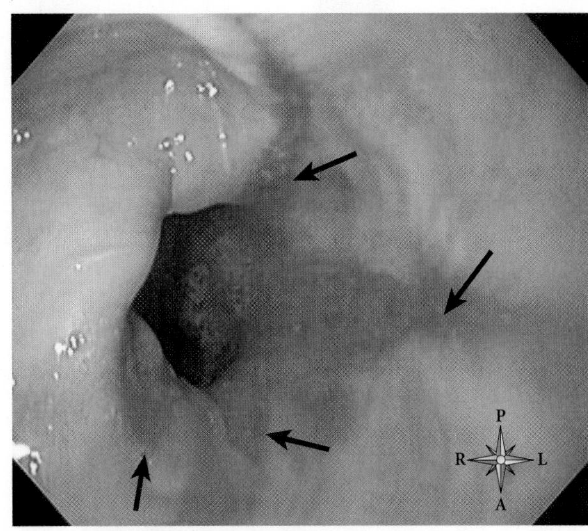

FIGURE 18-13 Esophageal inflammation. Chronic inflammation of the esophagus is characteristic of GERD (gastroesophageal reflux disease). *Arrows* show reddened, inflamed areas about midway along esophagus. This damage is caused by the frequent "splashing back" of acids from the stomach.

Reflux

The terms **heartburn** and *acid indigestion* are often used to describe a number of unpleasant symptoms experienced by more than 60 million Americans each month. Backward flow of stomach acid up into the esophagus causes these symptoms (**Figure 18-12**), which typically include burning and pressure behind the breastbone. The term **gastroesophageal reflux disease (GERD)** is used to better describe this very common and sometimes serious medical condition.

In its simplest form, GERD symptoms are mild and occur only infrequently (twice a week or less). In these cases, avoiding problem foods or beverages, stopping smoking, or losing weight if needed may solve the problem. Additional treatment with over-the-counter antacids or non-prescription-strength acid-blocking medications also may be helpful.

More severe and frequent episodes of GERD can trigger asthma attacks, cause severe chest pain, result in bleeding, or promote a narrowing (stricture) or chronic irritation of the esophagus (**Figure 18-13**). In these cases, more powerful inhibitors of stomach acid production may be prescribed. Other drugs that strengthen the LES and thus reduce backflow of stomach acid are also used in moderate to severe cases of GERD.

Several minimally invasive surgical procedures for treating serious cases of GERD are also available. In such procedures, which are done on an outpatient basis, a flexible tube called an **endoscope** is used to insert and then remove the necessary medical devices required for treatment. If GERD is left untreated, serious pathological (precancerous) changes in the esophageal lining may develop—a condition called **Barrett esophagus.**

A sign of Barrett esophagus is evident when the esophagus is viewed with an endoscope. The color of the esophageal mucosa changes from a pink to a reddish salmon color, indicating cellular changes that are caused by continual exposure of the esophagus to stomach acid. A sample of the lining of the esophagus is removed and viewed under the microscope to make the diagnosis of Barrett esophagus.

Other than heartburn, symptoms of Barrett's may include trouble swallowing, vomiting blood, and weight loss that results because eating is painful. Treatment may include medications that reduce stomach acid production and lifestyle changes aimed at preventing the reflux of acid into the esophagus. These changes include:

- Eating smaller and more frequent meals
- Avoid lying down until 3 hours after eating
- Elevating the head of the bed to prevent reflux during sleep
- Avoiding foods that can trigger heartburn, such as caffeine, nicotine, and alcohol
- Maintaining a healthy weight to decrease pressure on the stomach

Hiatal Hernia

GERD is a common symptom of **hiatal hernia.** A hernia results from an organ being pushed through a wall that normally acts as a barrier. In hiatal hernia, the stomach pushes through the gap, or hiatus, in the diaphragm that allows the esophagus to pass through it (**Figure 18-14**). Often the lower esophagus becomes enlarged, allowing acidic stomach contents to bypass the LES and flow upward into the esophagus.

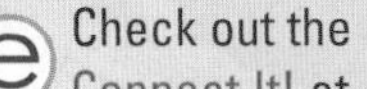

Check out the illustrated article *Hernias* at Connect It! at *evolve.elsevier.com*.

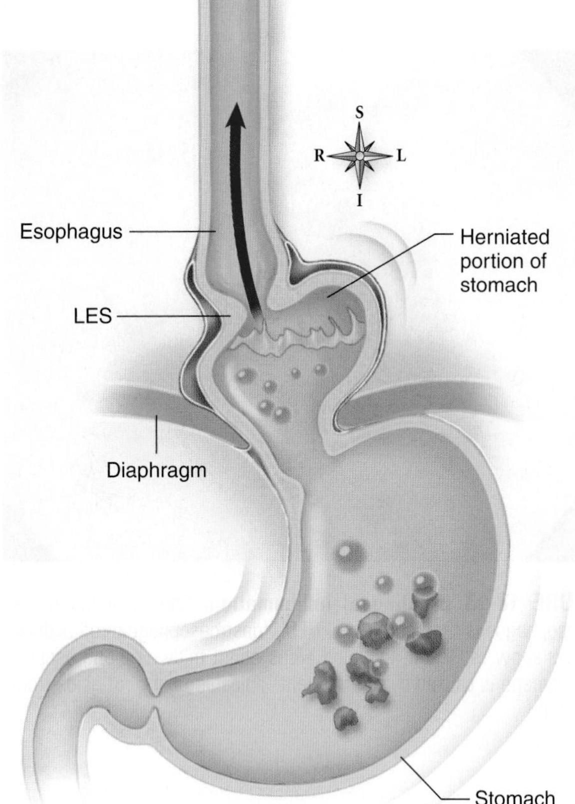

FIGURE 18-14 Hiatal hernia. Note herniated portion of stomach pushed through diaphragm. *LES,* Lower esophageal sphincter.

18

Stomach

The **stomach** (Figure 18-15) lies in the upper part of the abdominal cavity just under the diaphragm. It serves as a large pouch that ingested material enters after it has been chewed, swallowed, and passed through the esophagus.

Structure

The three divisions of the stomach shown in Figure 18-15 are the **fundus, body,** and **pylorus.** The fundus is the enlarged portion to the left of and above the opening of the esophagus into the stomach. The body is the large central region of the stomach, and the pylorus is the lower narrow section, which joins with the first part of the small intestine. The upper right border of the stomach is known as the *lesser curvature,* and the lower left border is called the *greater curvature.*

The stomach looks small when it is empty, not much bigger than a large sausage, but it expands considerably after a large meal. Have you ever felt so uncomfortably full after eating that you could not take a deep breath? If so, it probably meant that your stomach was so full of material that it occupied more space than usual and was pushed up against the diaphragm. This made it hard for the diaphragm to contract and move downward as much as necessary for you to take a deep breath.

In contrast to other regions of the digestive tract, there are three layers of smooth muscle in the stomach wall (see Figure 18-15). The muscle fibers that run lengthwise, around, and obliquely make the stomach one of the strongest internal organs—well able to break up chunks of ingested food into tiny particles and to mix them thoroughly with gastric juice to form a semisolid mixture called **chyme.** Stomach muscle contractions also result in peristalsis, which propels chyme down the digestive tract.

Mucous membrane lines the stomach, forming the *gastric mucosa.* It contains thousands of microscopic **gastric glands** that secrete *gastric juice* into the stomach. Cells in the stomach also secrete a chemical called *intrinsic factor* that protects vitamin B_{12} and saves it for its later absorption in the distal small intestine. Some individuals may require vitamin B_{12} injections after some stomach surgeries.

When the stomach is empty, its lining lies in folds called **rugae.**

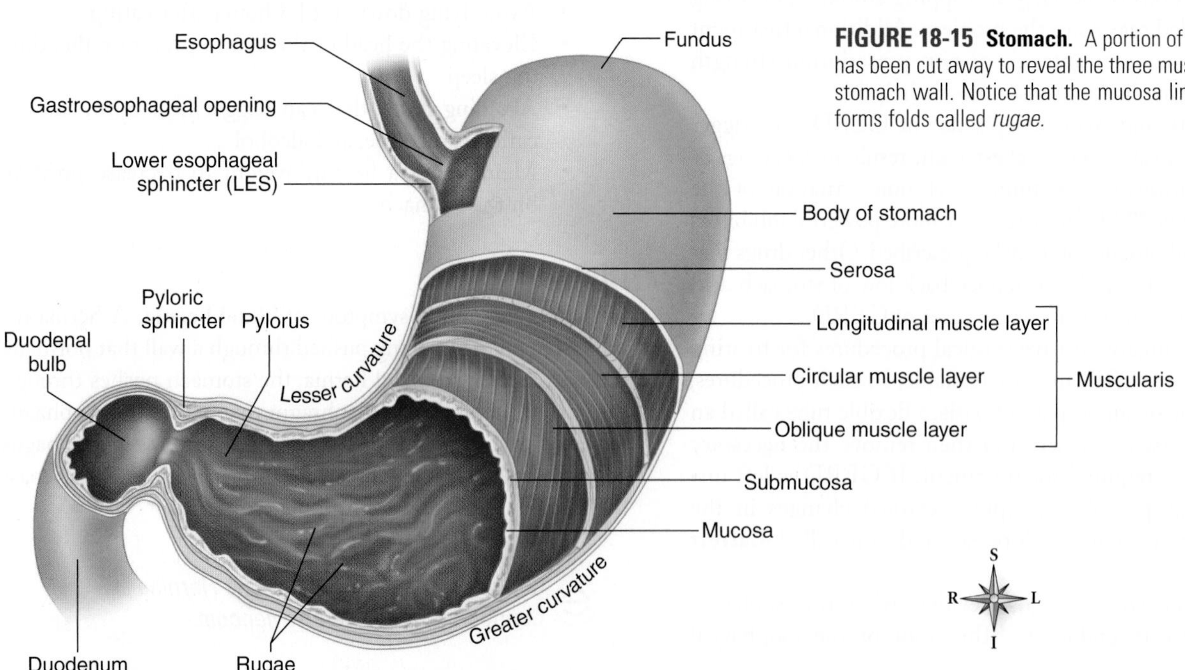

FIGURE 18-15 Stomach. A portion of the anterior wall has been cut away to reveal the three muscle layers of the stomach wall. Notice that the mucosa lining the stomach forms folds called *rugae.*

Function

After material has entered the stomach by passing through the muscular LES at the distal end of the esophagus, the digestive process continues.

Contraction of the stomach's muscular walls mixes the swallowed material thoroughly with the gastric juice and breaks it down into chyme, which eventually becomes more and more liquefied. This liquefaction process is a continuation of the mechanical digestion that begins in the mouth.

Mostly water, gastric juice also contains *hydrochloric acid (HCl)* that unfolds proteins by breaking hydrogen bonds. Then, enzymes also present in the gastric juice break apart some of the peptide bonds within protein molecules—all part of chemical digestion.

Partial digestion of proteins occurs after chyme is held in the stomach for some time by the **pyloric sphincter** muscle. The smooth muscle fibers of the sphincter stay contracted most of the time and thereby close off the opening of the pylorus into the small intestine. After chyme has been mixed in the stomach and protein digestion gets under way, it begins its passage through the pyloric sphincter into the first part of the small intestine.

QUICK CHECK

1. What are the three anatomical components of the *pharynx*?
2. What is a *hiatal hernia*?
3. What is *GERD*?
4. Name the three divisions of the *stomach*.

To learn more about the stomach, go to AnimationDirect online at *evolve.elsevier.com*.

Disorders of the Stomach

Signs and Symptoms

Gastroenterology is the study *(ology)* of the stomach *(gastro)* and intestines *(entero)* and their diseases. The stomach is the potential site of numerous diseases and conditions, some of which are briefly described in this section. Many of these disorders are characterized by one or more of these signs and symptoms:

1. **Gastritis**—stomach inflammation
2. **Anorexia**—chronic loss of appetite
3. **Nausea**—unpleasant feeling that often leads to vomiting
4. **Emesis**—vomiting

Pyloric Conditions

The pyloric sphincter is of clinical importance because **pylorospasm** is a fairly common condition in infants. The pyloric muscle fibers do not relax normally in infants with this condition. As a result, chyme is not able to leave the stomach, and the infant vomits nutrients instead of digesting and absorbing them. The condition is relieved by the administration of a drug that relaxes smooth muscles.

Another abnormality of the pyloric sphincter is called **pyloric stenosis**—an obstructive narrowing of its opening. This condition can be corrected surgically in infancy.

Gastric Ulcer

An **ulcer** is a craterlike wound or sore in a membrane caused by tissue destruction. Current statistics show that about 1 in 10 individuals in the United States will suffer from either a gastric (stomach) or duodenal ulcer in his or her lifetime (**Figure 18-16**, *A*).

Ulcers cause disintegration, loss, and death of tissue as they erode the layers of the wall of the stomach or duodenum. These craterlike lesions cause gnawing or burning pain and may ultimately result in hemorrhage, perforation, widespread inflammation, scarring, and other very serious medical complications. Usually perforation does not occur, but small, repeated hemorrhages over long periods can cause anemia.

Excessive acid secretion was thought for many years to be the primary cause of ulcers. It is now known that most gastric and duodenal ulcers result from infection with the ***Helicobacter pylori*** *(H. pylori)* bacterium (**Figure 18-16**, *B*). This is especially so if the infected individual has a genetic predisposition to ulcer development. The bacterium burrows through the protective mucus lining the GI tract and when it makes contact with the epithelium, it triggers immune reactions that include the inflammatory response. These responses impair the gastric lining's ability to produce acid-protective mucus. *H. pylori* infection is diagnosed by biopsy, breath, or blood antibody tests.

Long-term use of certain pain medications such as aspirin and ibuprofen, called *nonsteroidal antiinflammatory drugs (NSAIDs)* also can cause ulcers because they too decrease the secretion of mucus. NSAID-induced ulcers can be treated by stopping NSAID use and taking acid-reducing drugs until the ulcer heals.

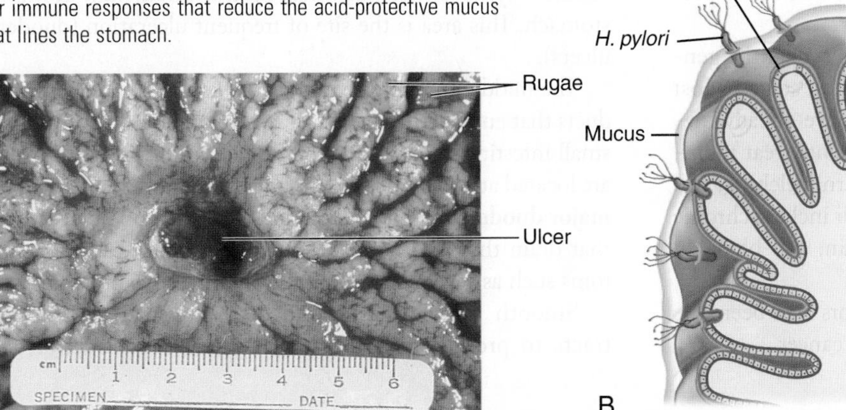

FIGURE 18-16 Disorders of the stomach. A, Gastric ulcer. **B,** *Helicobacter pylori* may infect the stomach mucosa and trigger immune responses that reduce the acid-protective mucus that lines the stomach.

CLINICAL APPLICATION

UPPER GASTROINTESTINAL X-RAY STUDY

The **upper GI (UGI)** study consists of a series of x-rays of the lower esophagus, stomach, and duodenum, produced with the aid of a contrast medium, usually barium sulfate. The test is used to detect ulcerations, tumors, inflammations, or anatomical malpositions such as *hiatal hernia* (protrusion of the stomach through the diaphragm). Obstruction of the upper GI tract is also easily detected with the UGI series.

First the patient drinks a flavored liquid containing barium sulfate. Then x-rays are taken periodically as the contrast medium travels through the system—the lower esophagus, gastric wall, pyloric channel, and duodenum; each structure is evaluated for defects.

Benign peptic ulcer is a common pathological condition that may affect these GI areas. Tumors, cysts, or enlarged organs near the stomach can also be identified when there is an anatomical distortion of the outline of the upper GI tract.

Compare the x-ray image shown here with the stomach structures depicted in **Figure 18-15**. Identify as many of these structures as you can in the x-ray. Can you locate the greater and lesser curvatures? Fundus, body, and pyloric areas? Location of the pyloric sphincter?

Note that although a majority of the barium contrast material has pooled in the stomach, some has passed through the pyloric sphincter, thereby outlining the duodenum.

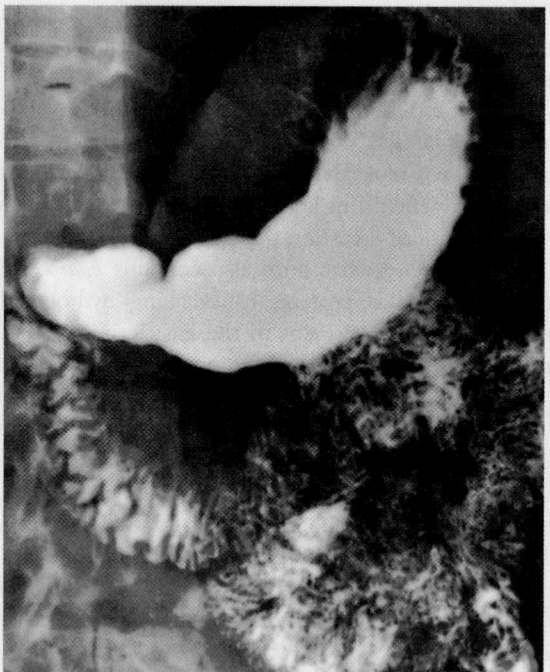

18

The discovery that most ulcers are caused by a bacterial organism led to development of a number of newer treatment programs. These treatments were designed to eradicate the bacteria by use of antibiotics while simultaneously blocking or reducing stomach acid secretion.

Currently, the standard antibiotic-based treatment used most often to both heal ulcers and prevent recurrences is called **triple therapy.** It requires that three medications be taken concurrently for about 2 weeks. More than one cycle may be required. Triple therapy combines a stomach-lining protector such as bismuth subsalicylate (Pepto-Bismol) and/or an acid reducer with two different antibiotics.

Stomach Cancer

Stomach cancer has been linked to excessive alcohol consumption, use of chewing tobacco, eating smoked or heavily preserved food, and to *H. pylori* infection.

Unfortunately, there is no practical way to screen the general population for stomach cancer in its earliest stages. Most stomach cancers, usually adenocarcinomas, have already metastasized before they are found because patients treat themselves for the early warning signs of heartburn, belching, and nausea. Later warning signs of stomach cancer include chronic indigestion, vomiting, anorexia, stomach pain, and blood in the feces.

Surgical removal of the malignant tumors has been the most successful method of treating stomach cancer.

Small Intestine

Structure

The **small intestine** is roughly 7 meters (20 feet) long. However, it is noticeably smaller in diameter than the large intestine, so in this respect its name is appropriate (**Figure 18-17**). Different names identify three different sections of the small intestine: **duodenum, jejunum,** and **ileum.**

To accommodate such a long tube within the relatively short abdominal cavity, it must be coiled into many loops. In this way, a small body cavity can contain a very long tube with a large surface area.

Most of the chemical digestion occurs in the first region of the small intestine, the duodenum. The duodenum is C-shaped and curves around the head of the pancreas (**Figure 18-18**). Acidic chyme enters the bulb of the duodenum from the stomach. This area is the site of frequent ulceration (duodenal ulcers).

The middle third of the duodenum contains the openings of ducts that empty pancreatic juice and bile from the liver into the small intestine. As you can see in **Figure 18-18**, the two openings are located at two bumps called the **minor duodenal papilla** and **major duodenal papilla.** Occasionally a gallstone blocks ducts that drain through the major duodenal papilla, causing symptoms such as severe pain, jaundice, and digestive problems.

Smooth muscle in the wall of the small intestine contracts to produce peristalsis, the wavelike contraction that

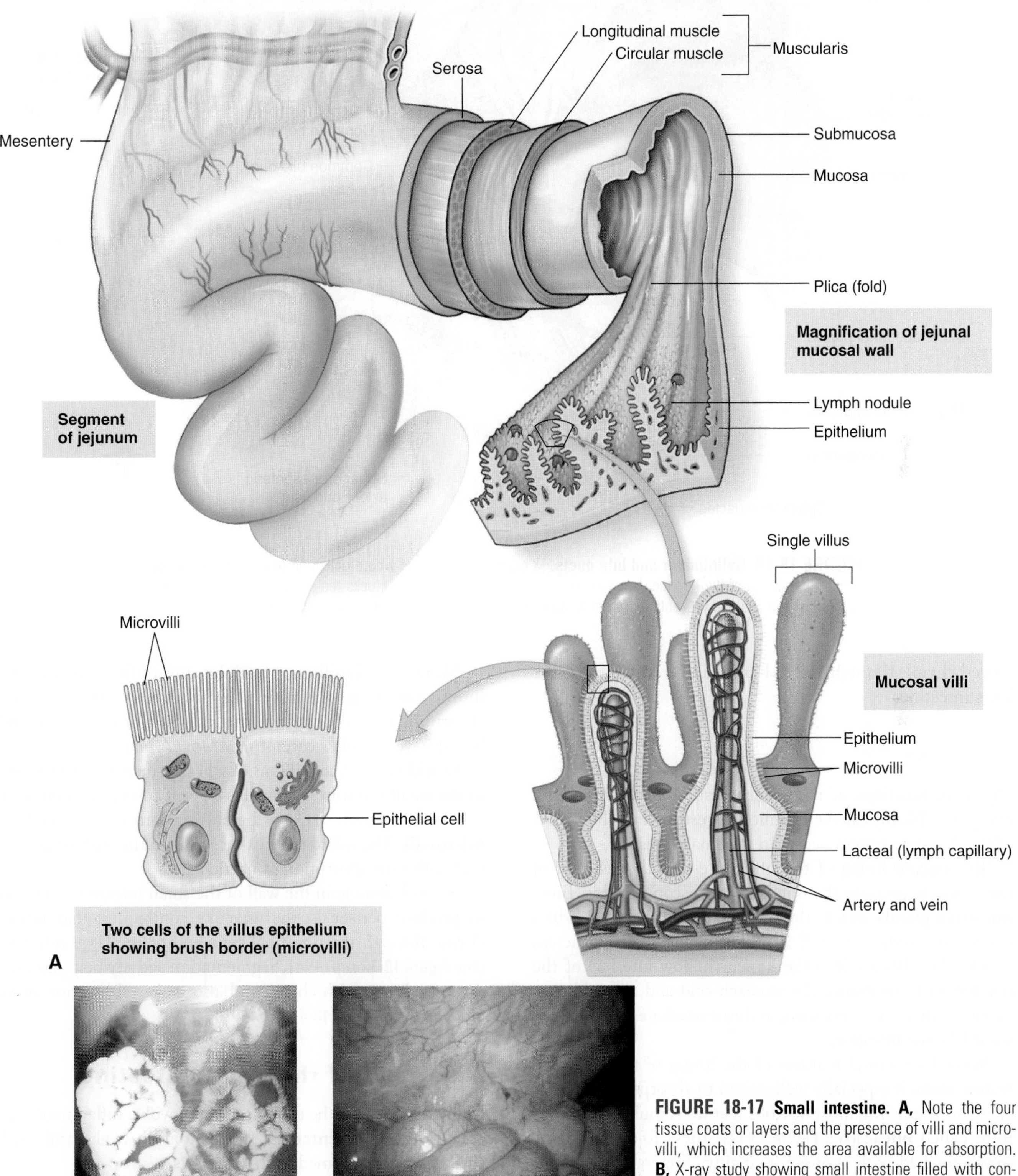

FIGURE 18-17 Small intestine. A, Note the four tissue coats or layers and the presence of villi and microvilli, which increases the area available for absorption. **B,** X-ray study showing small intestine filled with contrast medium to outline it clearly. **C,** Laparoscopic view of abdominopelvic cavity, with loops of small intestine visible.

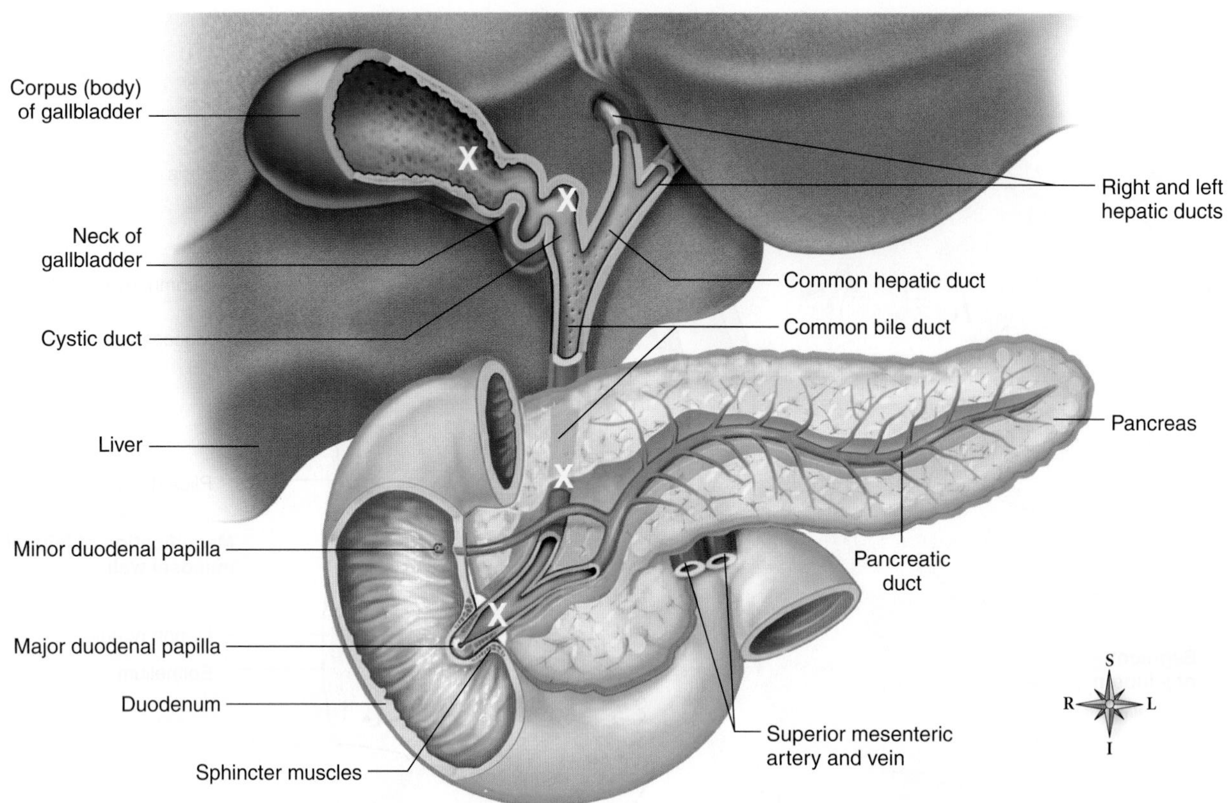

FIGURE 18-18 Gallbladder and bile ducts. *X* marks the locations where gallstone blockages commonly occur. Obstruction of the hepatic or common bile duct by stone or spasm blocks the exit of bile from the liver, where it is formed, and prevents bile from being ejected into the duodenum *(choledocholithiasis)*.

moves chyme through the GI tract and eventually to the large intestine.

Function

The main functions of the small intestine are *digestion* and *absorption.* Nearly all the chemical digestion and absorption of the digestive system occurs in the small intestine.

The mucous lining of the small intestine, as with that of the stomach, contains thousands of microscopic glands. These **intestinal glands** secrete the intestinal juice that is rich in a variety of enzymes as well as water and ions. The pancreas secretes bicarbonate into the lumen (hollow interior) of the duodenum to neutralize the stomach acid and also adds enzymes to digest fats, proteins, and carbohydrates that are absorbed by the intestine.

Several structural features of the lining of the small intestine make it especially well-suited to absorption of nutrients and water. The most obvious feature is multiple circular folds called **plicae** (see **Figure 18-17**). These folds are themselves covered with thousands of tiny "fingers" called **villi.** Under the microscope, the villi can be seen projecting into the lumen of the intestine. Inside each villus lies a rich network of blood capillaries that absorb the products of carbohydrate and protein digestion (sugars and amino acids). The villi also contain lymphatic capillaries called **lacteals** that absorb fats.

Millions and millions of villi jut inward from the mucous lining. This large absorptive surface area allows for faster absorption of nutrients from the intestine into the blood and lymph—yet another case of *structure fits function.*

In addition to the millions of villi that increase surface area in the small intestine, each villus is itself covered by epithelial cells, which each have a brushlike border composed of **microvilli.** The microvilli further increase the surface area of each villus for absorption of nutrients.

Smooth muscle in the wall of the small intestine contracts to produce peristalsis, the wavelike contraction that moves chyme through the intestinal tract and to the large intestine (see **Figure 18-3** on p. 496). Segmentation activity helps mix the digestive juices with chyme and also makes absorption more efficient (see **Figure 18-4** on p. 496).

Disorders of the Small Intestine

Many disorders of the small intestine involve inflammation, a condition termed **enteritis.** If the stomach is also inflamed, the condition is termed **gastroenteritis.**

Bacterial toxins or other irritants in the chyme, including stomach acid, can cause enteritis. Irritation or inflammation in the duodenum can produce a feeling of nausea that leads to emesis (vomiting). Because the duodenum may be emptied along with the stomach during vomiting, it is common to observe yellowish or brownish bile in the vomit.

X-ray studies of the small intestine, as well as direct viewing of either the inside lumen or exterior surface using an endoscope, are useful tools in both diagnosis and treatment of intestinal disease.

Malabsorption syndrome is a general term referring to a group of symptoms resulting from the failure of the small intestine to absorb nutrients properly. These symptoms include anorexia, weight loss, abdominal bloating, cramps, anemia, and fatigue. A number of underlying conditions can cause malabsorption syndrome, including mucosal changes due to surgery, blood flow changes, or disease.

Another disorder called **maldigestion** involves a deficit of digestive enzymes or bile salts. This reduces digestion and thereby reduces the amount of nutrients available for absorption.

Liver and Gallbladder

Structure

The **liver** is so large that it fills the entire upper right portion of the abdominal cavity and even extends partway into the left side (see Figure 18-1 and review Figure 1-6 on p. 10).

Because its cells secrete a substance called **bile** into ducts, the liver is classified as an exocrine gland. In fact, the liver is the largest gland in the body. Bile contains a mixture of substances, some of which have direct digestive functions described in the next section.

The liver also serves an excretory role, as it removes yellowish bile pigments formed by the breakdown of hemoglobin from old RBCs and puts them into the bile for elimination from the body. The liver has a wide variety of other metabolic functions that are discussed later in Chapter 19.

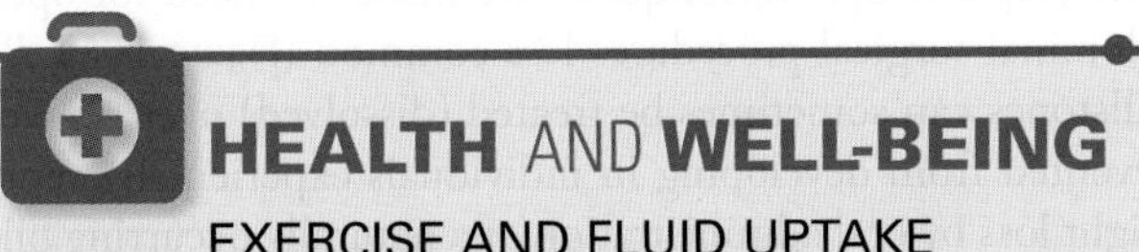

EXERCISE AND FLUID UPTAKE

Replacement of fluids lost during exercise, primarily through sweating, is essential for maintaining homeostasis. Nearly everyone increases his or her intake of fluids during and after exercise. The main limitation to efficient fluid replacement is how quickly fluid can be absorbed, rather than how much a person drinks. Very little water is absorbed until it reaches the intestines, where it is absorbed almost immediately. Thus the rate of *gastric emptying* into the intestine is critical.

Large volumes of fluid leave the stomach and enter the intestines more rapidly than small volumes. However, large volumes may create an uncomfortable feeling during exercise. Cool fluids (8° C to 13° C) empty more quickly than warm fluids. Fluids with a high solute concentration empty slowly and may cause nausea or stomach cramps. Thus large amounts of cool, dilute, or isotonic fluids are best for replacing fluids quickly during exercise.

The duration of exercise does not affect gastric emptying, but the intensity can. Strenuous exercise practically shuts down gastric emptying. Thus the harder you work, the harder it is to replace lost fluids.

Look again at Figure 18-18. First, identify the **hepatic ducts.** They drain bile out of the liver, a fact suggested by the name "hepatic," which comes from the Greek word for liver *(hepar)*. Next, notice the duct that drains bile into the small intestine (duodenum), the **common bile duct.**

The liver continuously secretes bile. If there is no chyme in the duodenum, then circular sphincter muscles within the duodenal papillae remain closed—and the bile backs up the common bile duct into the **cystic duct** that leads to the **gallbladder.** The folded lining of the gallbladder allows it to expand and thus act as an overflow reservoir for bile. The gallbladder also concentrates stored bile by reabsorbing water from bile back into the blood.

To learn more about bile ducts, go to AnimationDirect at *evolve.elsevier.com*.

Function

Chemically, bile contains significant quantities of cholesterol and substances *(bile salts)* that act as detergents to mechanically break up, or **emulsify,** fats. Because fats form large globules, they must be broken down, or emulsified, into smaller particles to increase the surface area to aid digestion.

In addition to emulsification of fats, bile that is eliminated from the body in the feces serves as a mechanism for excreting cholesterol from the body. Both emulsification of fats and elimination of cholesterol from the body are primary functions of bile.

When chyme containing lipid or fat enters the duodenum, it initiates a mechanism that contracts the gallbladder and forces bile into the small intestine. Fats in chyme trigger the secretion of the hormone **cholecystokinin (CCK)** from the intestinal mucosa of the duodenum. This hormone then travels through the bloodstream and promotes contraction of the gallbladder—and consequently bile flows into the duodenum. Secretion of CCK is a good example of a hormone acting to regulate GI motility.

QUICK CHECK

1. What bacterium is associated with ulcers?
2. Identify the different sections of the small intestine in the order in which chyme passes through them.
3. What is *gastroenteritis*?
4. What is the *gallbladder*? Where is *bile* formed, and what is its function?

Disorders of the Liver and Gallbladder

Gallstones and Jaundice

Gallstones are solid clumps of material (mostly cholesterol) that form in the gallbladder in 1 in 10 Americans. Some gallstones never cause problems and are called *silent gallstones,* whereas others produce painful symptoms or other medical complications and are called *symptomatic gallstones.*

Cholelithiasis literally means "condition of having bile (gall) stones" and often occurs in the presence of gallbladder inflammation, or **cholecystitis.**

18

Gallstones often form when the cholesterol concentration in bile becomes excessive, causing crystallization or precipitation to occur (**Figure 18-19**). Stone formation is much more likely to occur if the gallbladder does not empty regularly and chemically imbalanced or cholesterol-laden bile remains in the gallbladder for long periods of time.

When a gallstone blocks the common bile duct—a condition called **choledocholithiasis**—bile is not able to drain into the duodenum (see **Figure 18-18**). In such a case, feces then appear gray-white because the pigments from bile that normally give feces its characteristic color are absent. Often, pain accompanies this condition. The pain is called **biliary colic.**

Furthermore, because bile cannot be released into the digestive tract, excessive amounts of bile pigments are instead absorbed into the blood. A yellowish skin discoloration called **jaundice** results. Obstruction of the common hepatic duct also leads to jaundice because when bile cannot drain out of the liver, it is absorbed into the blood. Bile is not absorbed from the gallbladder so no jaundice occurs if only the cystic duct is blocked.

The relationship of dieting and weight loss to gallstone formation is under intense scrutiny. Physicians have known for years that in severely obese individuals (body mass index [BMI] over 40) the liver produces higher levels of cholesterol and the risk of developing gallstones is increased. However, only recently have scientists established with certainty that significant and rapid weight loss greatly increases the risk of symptomatic gallstone formation that may require surgery—a procedure called **cholecystectomy.**

Check out the article *Gallstones and Weight Loss* at Connect It! at *evolve.elsevier.com.*

Bariatrics (from Greek *baros,* weight) is a specialized field of medicine that deals with treatment of obesity. So-called *bariatric surgical procedures* used for producing weight loss, such as the *Lap Band Adjustable Gastric Banding System,* the more traditional restrictive gastric banding procedure (vertical-banded gastroplasty), or more extensive bypass operations (*RGB,* or *Roux-en-Y gastric bypass*), all reduce the size of the stomach and almost always result in rapid postsurgical weight loss, but more than one-third of these patients develop gallstones.

Unfortunately, individuals who choose nonsurgical approaches to achieve significant and rapid weight loss, such as very–low-calorie, ultra–low-fat or carbohydrate diets, also experience higher rates of gallstone formation. In these cases, stone formation is related to imbalances in bile chemistry and delayed emptying or incomplete gallbladder contractions.

If surgery is required for removal of symptomatic gallstones, laparoscopic techniques have made the need for open abdominal surgical procedures less common (**Figure 18-19,** *B*). Gallstones can sometimes be treated (dissolved) over time or prevented from developing in individuals experiencing rapid weight loss by oral administration of a naturally occurring bile constituent called *ursodeoxycholic acid* (Actigall).

A

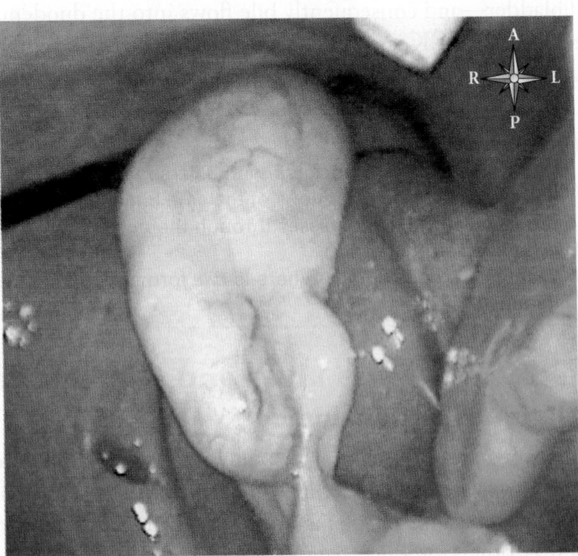

B

FIGURE 18-19 Gallstones. A, Inflamed gallbladder filled with yellow cholesterol gallstones. **B,** View of the gallbladder before removal using a laparoscope (viewing tube) inserted into the abdomen during surgery.

Hepatitis

Hepatitis is a general term referring to inflammation of the liver. Hepatitis is characterized by jaundice, liver enlargement, anorexia, abdominal discomfort, gray-white feces, and dark urine.

A number of different conditions can produce hepatitis. Alcohol, drugs, or other toxins may cause hepatitis. It may occur as a complication of bacterial or viral infection or parasite infestation. *Hepatitis A,* for example, results from infection by a virus that may be found in contaminated food.

Another viral hepatitis, *hepatitis B,* is more severe. It was historically called *serum hepatitis* because it is often transmitted by contaminated blood serum. Improperly sterilized tattooing needles contaminated with even trace amounts of hepatitis B–infected blood (0.004 mL), will cause disease. There are vaccines to prevent infection with both hepatitis A and hepatitis B viruses.

A

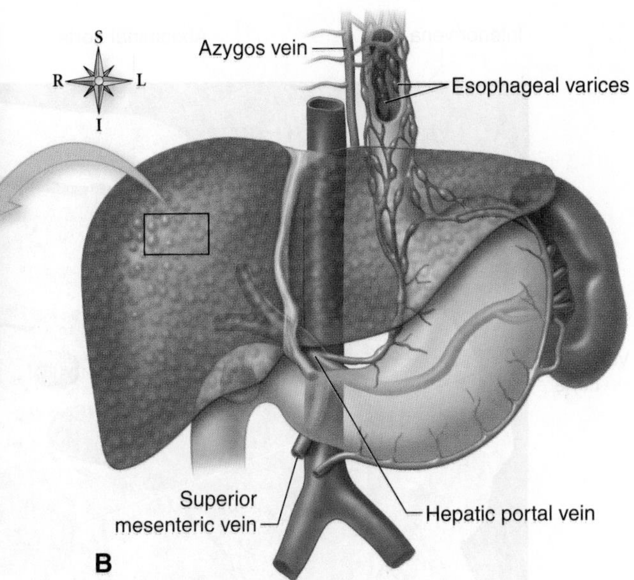

FIGURE 18-20 Liver damage. A, Alcoholic cirrhosis where liver surface is hard and covered with nodules that look like pebbles. **B,** Varicose veins (varices) of the esophagus caused by reduction of blood flow through liver with cirrhosis.

Hepatitis C is a form of liver inflammation caused by infection with the **hepatitis C virus (HCV).** It is most often associated with transfusion of contaminated blood or intravenous drug abuse. The disease may become chronic and result in cirrhosis (see later section) or liver cancer many months or even years after exposure. Oral drugs are now available that have a cure rate of nearly 100% when taken for 3 months. Vaccines for hepatitis C are also being developed.

Cirrhosis

Hepatitis, chronic alcohol abuse, malnutrition, or infection may lead to a degenerative liver condition known as **cirrhosis.**

The liver has a remarkable ability to regenerate its damaged tissue, but this process has its limits. For example, when the toxic effects of alcohol accumulate faster than the liver can regenerate itself, damaged tissue is replaced with a nodular, pebble-like fibrous or fatty tissue instead of normal tissue (Figure 18-20, *A*).

No matter what the cause of liver cirrhosis, the symptoms are the same: nausea, anorexia, gray-white stools, weakness, and pain. If the cause of cirrhosis is removed and high-protein foods are eaten, the liver may be able to repair itself—given enough time. If the damage is extensive, a liver transplant may be the only hope of saving someone with cirrhosis of the liver.

Acute or chronic liver disorders such as hepatitis or cirrhosis can block the flow of blood through the liver, thus causing it to back up into the *hepatic portal circulation* (Figure 15-11 on p. 412. As a result, the blood pressure in the hepatic circulation increases abnormally—a condition called **portal hypertension.**

To relieve the pressure of portal hypertension, the body forms additional new veins that connect to the exiting systemic veins (Figure 18-20, *B*). This often causes the veins lining the esophagus, stomach, and other organs to widen and become varicose. If these varicosities rupture after erosion by stomach acid, vomiting of blood occurs. This can lead to massive bleeding that may result in death.

Pancreas

Structure and Function

The pancreas lies behind the stomach in the concavity produced by the C shape of the duodenum (see Figure 18-18). It is an exocrine gland that secretes *pancreatic juice* into ducts and also an endocrine gland that secretes hormones into the blood. Locate the pancreas and nearby structures in Figure 18-21, which shows a transverse section of a human cadaver.

Pancreatic juice secreted into the duodenum contains enzymes that digest all three major kinds of energy-yielding nutrients—carbohydrates, proteins, and lipids. It also contains sodium bicarbonate, an alkaline substance that neutralizes the hydrochloric acid in the acidic chyme that enters the intestines from the stomach.

Pancreatic juice enters the duodenum of the small intestine at the same place that bile enters because both the common bile and pancreatic ducts open into the duodenum at the major duodenal papilla (see Figure 18-18).

Between the cells that secrete pancreatic juice into ducts lie clusters of cells that have no contact with any ducts. These are the **pancreatic islets (islets of Langerhans),** which secrete the hormones of the pancreas—mainly *insulin* and *glucagon*—described in Chapter 12.

To learn more about pancreatic ducts, go to AnimationDirect at *evolve.elsevier.com.*

Disorders of the Pancreas

Disorders of the pancreas include **diabetes mellitus (DM),** in those cases of DM involving the inability of the islet cells to make insulin.

Another pancreatic disorder is **pancreatitis,** which is inflammation of the pancreas. *Acute pancreatitis* may result from blockage of the common bile duct. The blockage causes

18

18

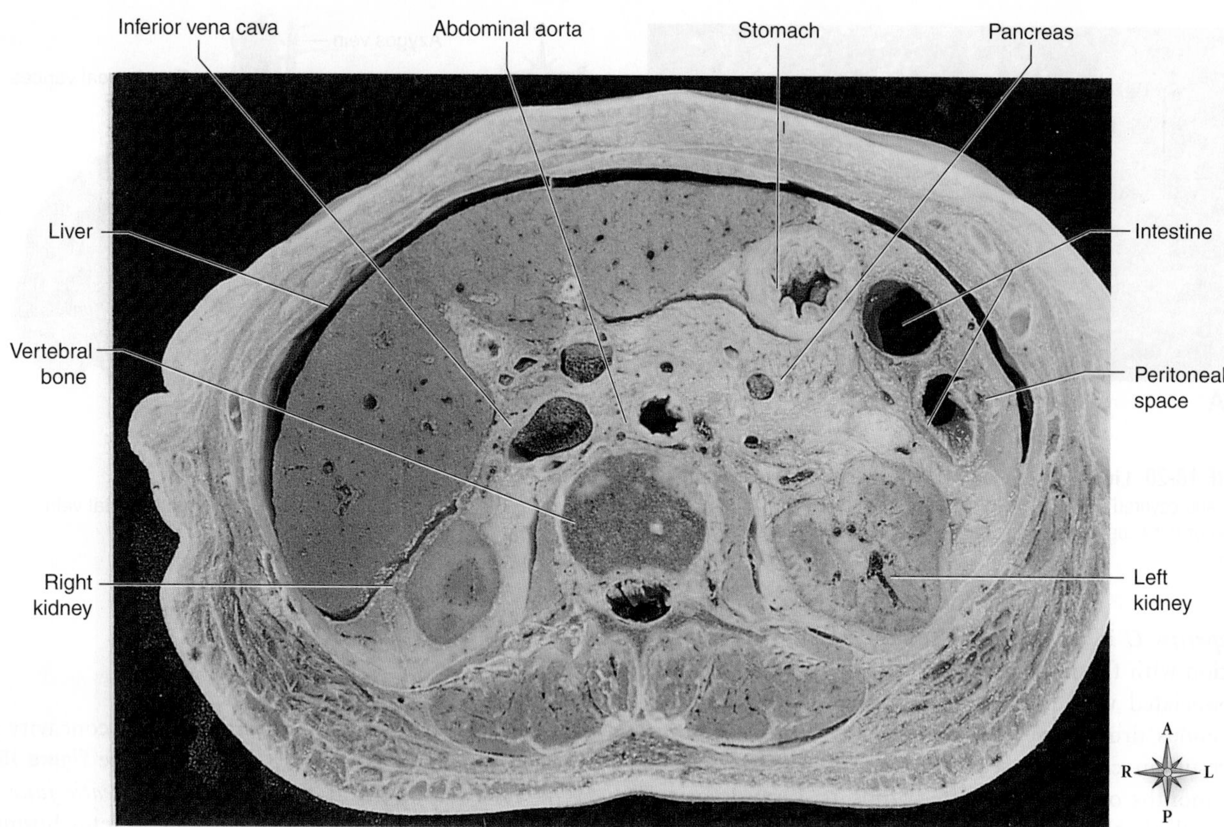

FIGURE 18-21 Abdominal organs. The photograph of a transverse section of a cadaver shows the relative position of some of the major digestive organs of the abdomen. Such a view is typical in imaging methods such as computed tomography (CT) scanning and magnetic resonance imaging (MRI).

pancreatic enzymes to "back up" into the pancreas and digest it. This is a very serious and potentially fatal condition.

Another condition that blocks the flow of pancreatic enzymes is **cystic fibrosis (CF),** which is an inherited disorder that disrupts cell membrane transport and causes exocrine glands to produce excessively thick secretions. Thick pancreatic secretions may build up and block pancreatic ducts, disrupting the flow of pancreatic enzymes and damaging the pancreas.

Another serious pancreatic disorder is *pancreatic cancer*. Usually a form of *adenocarcinoma,* advanced pancreatic cancer claims the lives of nearly all its patients within 5 years after diagnosis.

Large Intestine

Structure

The **large intestine** is only about 1.5 meters (5 feet) in length. It has a much larger diameter than the small intestine and forms the lower or terminal portion of the digestive tract.

Chyme containing undigested and unabsorbed material from ingested food enters the large intestine after passing through a sphincter called the **ileocecal valve** (**Figure 18-22**). Chyme, which has the consistency of soup, slowly changes to the more solid consistency of fecal matter as water and salts are reabsorbed during its passage through the small intestine. It is this pastelike material that passes through the valve into the large intestine.

The subdivisions of the large intestine are listed below in the order in which chyme or feces passes through them.

1. Cecum
2. Ascending colon
3. Transverse colon
4. Descending colon
5. Sigmoid colon
6. Rectum
7. Anal canal

Note in **Figure 18-22** that the ileocecal valve opens into a pouchlike area called the **cecum.** The opening itself is about 5 or 6 cm (2 inches) above the beginning of the large intestine. Material in the cecum flows upward to a region of the large intestine called the *colon.* Technically the colon does not include the entire large intestine, but often the terms *colon* and *large intestine* are used interchangeably. The colon is divided into three segments: *ascending, transverse,* and *descending* colon.

Material moves into the colon on the right side of the body—into the **ascending colon.** The **hepatic flexure** or **right colic flexure** is the bend between the ascending colon and the **transverse colon,** which extends across the front of the abdomen from right to left. The **splenic flexure** or **left colic flexure** marks the point where the **descending colon** turns downward on the left side of the abdomen. The **sigmoid colon** is the S-shaped segment that terminates in the **rectum.**

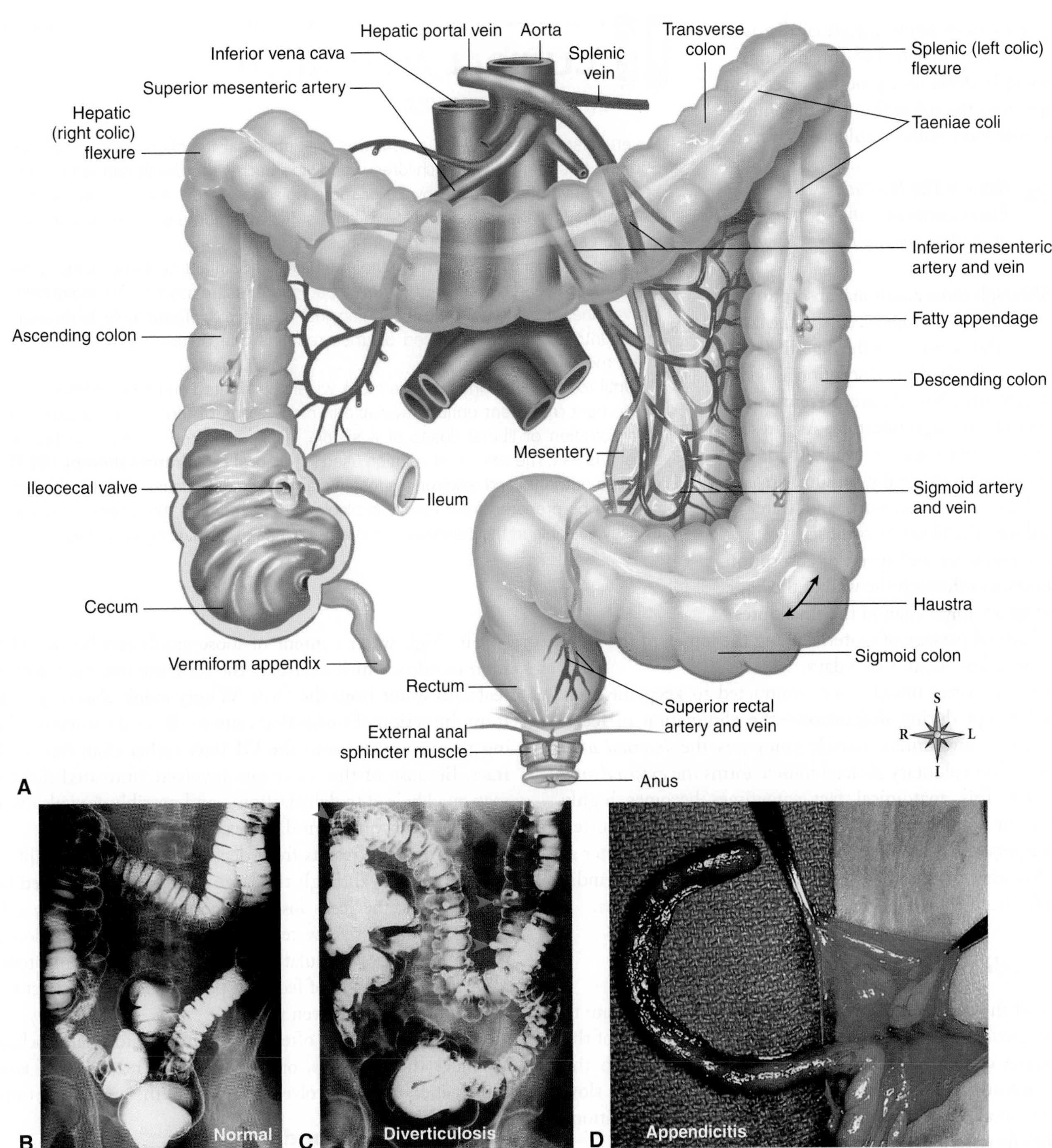

FIGURE 18-22 Large intestine. A, Artist's drawing of the large intestine. **B,** X-ray of large intestine and terminal ileum filled with barium contrast material (barium enema). **C,** X-ray of a barium enema showing diverticulosis *(arrowheads)*. **D,** Acute appendicitis. Appendix is gangrenous and showing signs of ischemia and putrefaction.

The terminal portion of the rectum is called the **anal canal,** which ends at the external opening, or **anus.**

Function

During its movement through the large intestine, material that remains after digestion in the small intestine is acted on by beneficial bacterial communities called the intestinal **microbiome** or *flora*. As a result of bacterial action, additional nutrients may be released from cellulose and other fibers and absorbed.

In addition to their digestive role, bacteria in the large intestine have other important functions. They are responsible for the synthesis of vitamin K needed for normal blood clotting and for the production of some of the B-complex vitamins. After they are formed, these vitamins are absorbed from the large intestine and enter the blood. The intestinal microbiome also plays a role in supporting immune functions that protect us from many intestinal diseases, some of which may be life-threatening.

Some bacteria also produce gases that escape from the colon through the anus—a phenomenon called flatulence or *flatus*.

Review *The Human Microbiome* at Connect It! at *evolve.elsevier.com.*

CLINICAL APPLICATION

INFANT DIARRHEA

Severe diarrhea caused by a *rotavirus,* an intestinal infection, kills more than 500,000 infants and young children worldwide each year. Death can result from severe dehydration caused by 20 or more episodes of diarrhea in a single day. Currently, more than 3 million U.S. children suffer symptoms of rotavirus intestinal infection annually and 55,000 require hospitalization.

Good medical care has limited the number of U.S. infant deaths caused by the disease each year to just a few. In developed countries, relatively new and promising vaccines provide good protection against the virus. Unfortunately, in developing countries, rotavirus-induced diarrhea remains one of the leading causes of infant mortality.

Until rotavirus vaccines can be more widely distributed and administered, one of the best treatment options available in many areas of the world involves oral administration of liberal doses of a simple, easily prepared solution containing sugar and salt. The salt-sugar solution used in this **oral rehydration therapy (ORT)** replaces nutrients and electrolytes lost in diarrheal fluid. Because the replacement fluid can be prepared from readily available and inexpensive ingredients, it is particularly valuable in the treatment of infant diarrhea in developing countries.

Although some additional absorption of water, salts, and vitamins occurs in the large intestine, this segment of the digestive tube is not as well suited for absorption as is the small intestine. No villi are present in the mucosa of the large intestine. As a result, much less surface area is available for absorption. Salts, especially sodium, are absorbed by active transport, and water is moved into the blood by osmosis.

The efficiency and speed of absorption of substances through the wall of the large intestine are lower than in the small intestine. Normal passage of material along the lumen of the large intestine takes about 3 to 5 days.

Two sphincter muscles stay contracted to keep the anus closed except during **defecation**—the elimination of feces. Smooth or involuntary muscle composes the *internal anal sphincter,* but voluntary skeletal muscle forms the *external anal sphincter.* This anatomical fact sometimes becomes highly important from a practical standpoint. For example, often after a person has had a stroke, the voluntary anal sphincter at first becomes paralyzed. This means, of course, that the individual has no control at that time over bowel movements.

Disorders of the Large Intestine

Most of the common disorders of the large intestine relate to *inflammation* or *abnormal motility,* or rate of movement of the intestinal contents. Abnormally rapid motility through the large intestine may result in *diarrhea,* and abnormally slow motility may result in *constipation.* These and other conditions are briefly described in this section.

Motility Disorders

Diarrhea usually occurs when the intestinal contents move so quickly that the resulting feces are more fluid than normal. Diarrhea is characterized by frequent passing of watery feces. In inflammatory conditions such as **dysentery,** the watery feces also may contain mucus, blood, or pus. Diarrhea is often accompanied by abdominal cramps—a symptom caused by excessive contractions of the intestinal muscles.

The increased intestinal motility that causes diarrhea often results from the presence of bacterial toxins, parasites, or other irritants. The intestines reflexively "speed up," a mechanism that quickly disposes of the irritant.

The high water content of loose stools can be caused by high motility, which decreases the time the intestines have to reabsorb water from the feces. Watery stools also may result from the action of toxins that causes cells in the intestinal lining to move water into the GI tract rather than out of the tract. Because of the water loss involved, untreated diarrhea may quickly lead to dehydration—and possibly convulsions or death (see box on infant diarrhea, above).

Constipation results from decreased intestinal motility. If passage of feces through the large intestine is prolonged beyond 5 days, the feces lose volume and become more solid because of excessive water reabsorption. This reduction in volume decreases stimulation of the bowel-emptying reflex, resulting in retention of feces—a positive-feedback affect that makes the condition even worse.

Acute constipation often results from intestinal blockage, low-fiber diets, tumors, or *diverticulitis.* Treatment of acute constipation usually involves treatment of the underlying cause.

Inflammatory Conditions

Diverticulitis is an inflammation of abnormal saclike pouches of the intestinal wall called *diverticula* (**Figure 18-22,** *C*). Diverticula often develop in adults older than 50 years of age who eat a low-fiber diet. As previously mentioned, diverticulitis can be a cause of constipation.

Colitis refers to any inflammatory condition of the large intestine. If present for prolonged periods of time, inflammatory bowel disease becomes a significant risk factor for colorectal cancer. Symptoms of colitis include diarrhea and abdominal cramps or constipation. Some forms of colitis also may produce bleeding and intestinal ulcers.

Colitis may be a result of emotional stress, as in *irritable bowel syndrome.* It also may result from an autoimmune disease, as in *ulcerative colitis.* Another type of autoimmune colitis is **Crohn disease,** which often also affects the small intestine.

If more conservative treatments fail, colitis may be corrected by surgical removal of the affected portions of the colon.

To learn more about a common medical imaging procedure used to assess the structure of the colon, check out the article *Barium Enema Study* at Connect It! at *evolve.elsevier.com.*

Colorectal Cancer

Colorectal cancer is a malignancy, usually *adenocarcinoma,* of the columnar epithelium that lines the lumen of the colon and/or rectum. Most colorectal cancers originate from nonmalignant *colonic polyps* that gradually undergo malignant transformation. Colorectal cancer occurs most often after age 50 and increases in incidence dramatically after age 75. The disease is slightly more common in men than in women and constitutes the second leading cause of death from cancer in the United States and the fourth most common type of cancer diagnosed after prostate, breast, and lung malignancy.

Diagnosis of colorectal cancer is made during a digital rectal examination or as a result of direct visualization of the rectum and lower colon during **sigmoidoscopy,** or of the entire colon during a **colonoscopy.** Other diagnostic tools include the use of barium enemas, ultrasound, various x-ray–based techniques, and magnetic resonance imaging.

Certain dietary habits, especially a high saturated fat intake, and genetic predisposition are known risk factors. The prognosis for recovery from colorectal cancer is based on a number of factors, including the degree of penetration (if any) of the tumor through the bowel wall, and the presence or absence of cancer cells in regional lymph nodes or distant body locations (metastases).

Early warning signs of this common type of cancer include changes in bowel habits (constipation or diarrhea), decreased stool diameter, rectal bleeding that may be obvious (gross or visible) or hidden (occult or microscopic), abdominal pain, unexplained anemia, weight loss, and fatigue. Large bowel obstruction is the most common complication of colon cancer.

Treatment requiring surgical removal of a tumor in the distal rectum also may require the creation of a **colostomy** (see box below). In addition to surgery, both rectal and colon cancer are often treated by use of chemotherapy. Radiation therapy is seldom used in treatment of colon cancer but has an important role in treatment of rectal malignancy.

Appendix

Structure and Function

The **vermiform appendix** (from *vermis* "worm" and *form* "shape") is, as the name implies, a wormlike, tubular structure. Note in **Figure 18-22** that the appendix is directly attached to the back of the cecum. The appendix contains a blind, tubelike interior lumen that communicates with the lumen of the large intestine 3 cm (1 inch) below the opening of the ileocecal valve into the cecum.

The appendix serves as a sort of incubator or "breeding ground" for the nonpathogenic intestinal bacteria normally residing in the colon. Maintaining a normal intestinal microbiome helps prevent pathogenic bacteria from becoming established. When the normal microbiome of the gut is disrupted, for example by infection or antibiotics, beneficial bacteria hidden away in the appendix can easily migrate into the colon to restore the normal ecological balance.

CLINICAL APPLICATION

COLOSTOMY

Colostomy is a surgical procedure in which an artificial anus is created on the abdominal wall by cutting the colon and bringing the cut end or ends out to the surface to form an opening called a **stoma** (see figure). This may be done during a surgery to remove a tumor or a section of the colon. After healing of the colon, the colostomy may be surgically reversed (removed).

Health-care workers help colostomy patients learn to accept the change in body image, which may cause emotional discomfort. The patient or caregiver is also trained in the regular changing of the disposable bag, including how to clean the stoma and how to prevent irritation, chapping, or infection. Irrigation of the colon with isotonic solutions is sometimes necessary. Deodorants may be added to the fresh bag to prevent unpleasant odors. Patients are also taught to manage their diet to include low-residue food and to avoid foods that produce gas or cause diarrhea. Fluid intake after colostomy is also carefully managed.

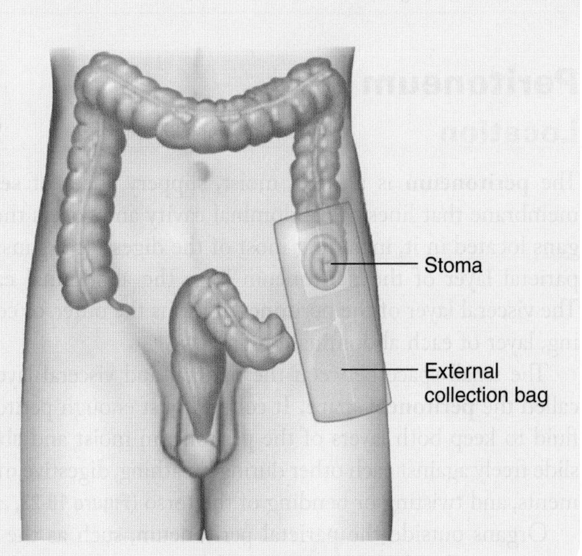

Review *The Human Microbiome* at Connect It! at *evolve.elsevier.com*.

Appendicitis

Inflammation of the appendix, or **appendicitis,** is a common and potentially very serious medical problem (**Figure 18-22**, *D*).

The opening between the lumen of the appendix and the cecum is quite large in children and young adults—a fact of great clinical significance because undigested chunks or fecal material trapped in the appendix may irritate and inflame its mucous lining, causing appendicitis. The opening between the appendix and the cecum is often completely obliterated in elderly persons, which explains the low incidence of appendicitis in this population.

A site on the surface of the anterior abdominal wall is often used to help in the diagnosis of appendicitis and to estimate the location of the appendix internally. It is called the *McBurney point* and is located in the right lower quadrant of the abdomen about a third of the way along a line from the right anterior superior iliac spine to the umbilicus. Extreme sensitivity and pain are common when the abdomen of persons with acute appendicitis is palpated over this point.

If infectious material becomes trapped in an inflamed appendix, the appendix may rupture and release the material into the abdominal cavity. Infection of the peritoneum and other abdominal organs can be life threatening. Appendicitis is the most common of the acute abdominal conditions that require surgery. It affects 7% to 12% of the population, generally before age 30.

QUICK CHECK

1. What is *cholelithiasis*?
2. What is *hepatitis C*? What are common causes of hepatitis C?
3. Name the seven subdivisions of the *large intestine*.
4. What is the most common acute abdominal condition requiring surgery?

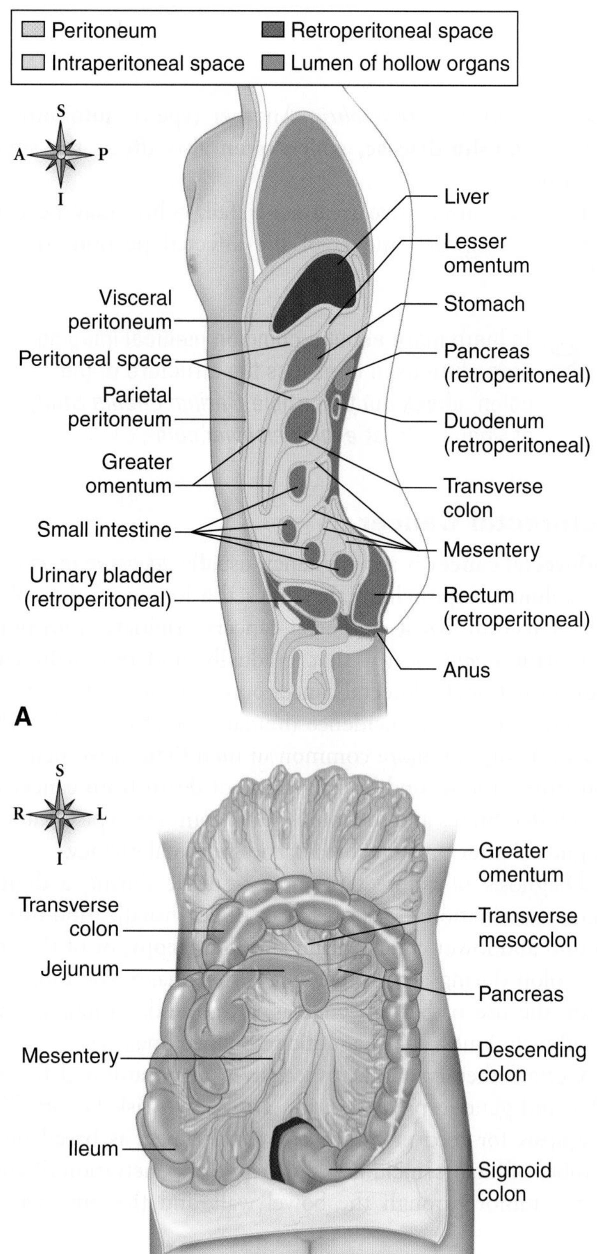

FIGURE 18-23 Peritoneum. A, The parietal layer of the peritoneum lines the abdominopelvic cavity and then extends as a series of mesenteries to form the visceral layer that covers abdominal organs. **B,** The transverse colon and greater omentum are raised and the small intestine is pulled to the side to show the mesentery.

Peritoneum

Location

The **peritoneum** is a large, moist, slippery sheet of **serous** membrane that lines the abdominal cavity and covers the organs located in it, including most of the digestive organs. The parietal layer of the peritoneum lines the abdominal cavity. The visceral layer of the peritoneum forms the outer, or covering, layer of each abdominal organ.

The small space between the parietal and visceral layers is called the **peritoneal space.** It contains just enough peritoneal fluid to keep both layers of the peritoneum moist and able to slide freely against each other during breathing, digestive movements, and twisting or bending of the torso (**Figure 18-23**, *A*).

Organs outside the parietal peritoneum, such as the kidneys, are said to be **retroperitoneal.**

Extensions

The two most prominent extensions of the peritoneum are the mesentery and the greater omentum.

The **mesentery** (**Figure 18-23**, *B*), an extension between the parietal and visceral layers of the peritoneum, is shaped like a giant, pleated fan. Its smaller edge attaches to the lumbar region of the posterior abdominal wall, and its long, loose outer edge encloses most of the small intestine, anchoring it to the posterior abdominal wall.

The **greater omentum** is a pouchlike extension of the visceral peritoneum from the lower edge of the stomach, part of the duodenum, and the transverse colon. Shaped like a large apron, it hangs down over the intestines, and because spotty deposits of fat give it a lacy appearance, it has been nicknamed the *lace apron.* It may envelop a badly inflamed appendix, walling the appendix off from the rest of the abdominal organs.

Peritonitis

Peritonitis is the inflammation of the peritoneum resulting from a bacterial infection or another irritating condition.

Peritonitis most commonly results from an infection that occurs after the rupture of the appendix or other abdominopelvic organ. It is characterized by abdominal distention, pain, nausea, vomiting, tachycardia (rapid heart rate), fever, dehydration, and other signs and symptoms. Circulatory shock progressing to heart failure may result.

Ascites

Ascites is the abnormal accumulation of fluid in the peritoneal space (**Figure 18-24**).

Fluid enters the peritoneal space from the blood because of local hypertension (high blood pressure) or an osmotic imbalance in the plasma (low plasma protein levels). This condition may be accompanied by abdominal swelling and decreased urinary output. It commonly occurs as a complication of cirrhosis, congestive heart failure, kidney disease, peritonitis, cancer, or malnutrition.

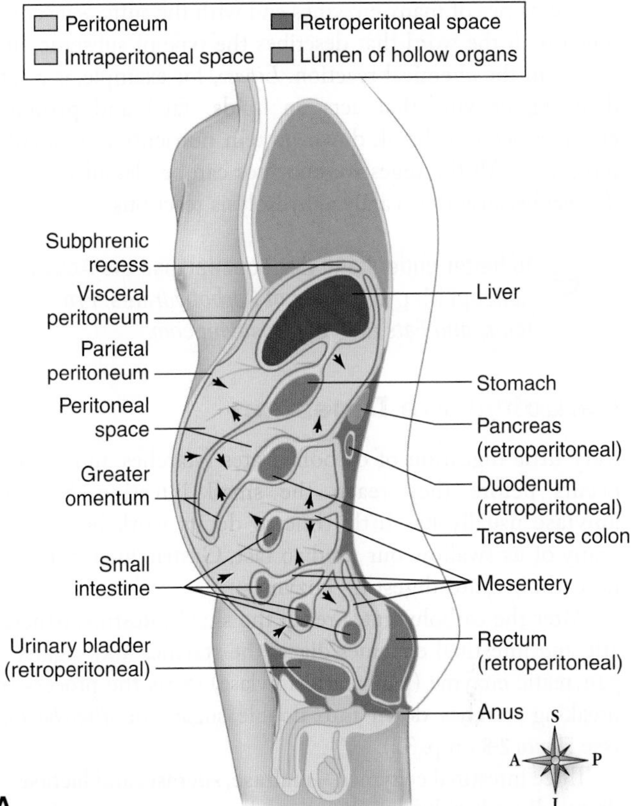

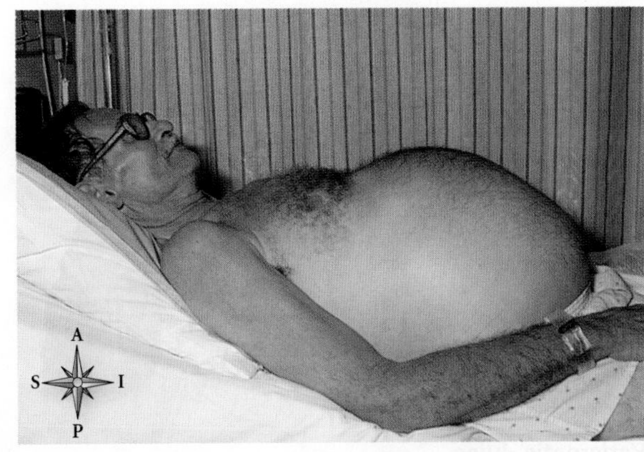

FIGURE 18-24 Ascites. Ascites results from an accumulation of fluid in the peritoneal space. **A,** The *arrows* indicate water filtering out of the peritoneal blood vessels, resulting from hypertension, or diffusing out of the vessels because of an osmotic imbalance in the blood. **B,** Abdominal bloating in ascites.

Digestion

Overview of Digestions

Digestion, a complex process that occurs in the alimentary canal, consists of physical and chemical changes that prepare nutrients for absorption.

Mechanical digestion breaks ingested food into tiny particles, mixes them with digestive juices, moves them along the alimentary canal, and finally eliminates the digestive wastes from the body. Chewing (mastication), swallowing (deglutition), peristalsis, and defecation are considered processes of mechanical digestion (see **Figure 18-3** and **Figure 18-4**).

Chemical digestion breaks down large, nonabsorbable nutrient molecules into smaller, absorbable molecules—molecules that are able to pass easily through the intestinal mucosa into blood and lymph. Chemical digestion consists of numerous chemical reactions catalyzed by enzymes in saliva, gastric juice, pancreatic juice, and intestinal juice.

Enzymes and Chemical Digestion

Enzymes are functional protein molecules that act as **catalysts.** That is, they speed up specific chemical reactions without themselves being changed or consumed during the reaction process.

During chemical digestion, certain enzymes very selectively speed up the breakdown of specific nutrient molecules and no others. Enzymes responsible for speeding up the breakdown of fats, for example, have no effect on carbohydrates or proteins.

The breakdown process facilitated by digestive enzymes is called **hydrolysis**—an important type of chemical reaction first discussed in Chapter 2. Recall that during hydrolysis, enzymes speed up reactions that add water *(hydro)* to chemically break up or split *(lysis)* larger molecules into smaller molecules (see **Figure 2-6** on p. 29).

18

The names of many enzymes end with the suffix *-ase* combined with the word that describes the type of substance involved in the chemical reaction. *Lipase,* for example, is a fat-digesting enzyme that acts on lipids (fats) and **protease** enzymes serve to break down protein nutrients into smaller molecules. All the digestive enzymes can be classified as *hydrolases* because they catalyze hydrolysis reactions.

To better understand this concept, use the Active Concept Map *Digestion of Carbohydrates, Proteins, and Fats* at *evolve.elsevier.com.*

Carbohydrate Digestion

Very little digestion of carbohydrates (starches and sugars) occurs before they reach the small intestine. Salivary **amylase** usually has little time to do its work because so many of us swallow our food so fast. Gastric juice contains no carbohydrate-digesting enzymes.

After the carbohydrates reach the small intestine, pancreatic and intestinal enzymes digest the starches and sugars. A pancreatic enzyme (pancreatic amylase) starts the process by breaking starches down into double sugars, or *disaccharides* (see **Figure 2-8** on p. 32).

18

Three intestinal enzymes—**maltase, sucrase,** and **lactase**—digest disaccharides by changing them into *monosaccharides* (simple sugars). *Maltase* digests **maltose** (malt sugar), *sucrase* digests **sucrose** (ordinary cane or table sugar), and *lactase* digests **lactose** (milk sugar).

Many adults and some children produce a low amount of lactase and therefore have difficulty digesting lactose—especially when consumed in large amounts, as in eating dairy products. This condition is called **lactose intolerance** and may produce digestive symptoms such as gas, bloating, cramps, or diarrhea. Lactose intolerance can be managed by avoiding foods high in lactose and/or taking a lactase supplement when eating dairy products.

The end products of carbohydrate digestion are monosaccharides, of which the most abundant is glucose.

Protein Digestion

Protein digestion starts in the stomach. **Hydrochloric acid (HCl)** in gastric juice helps unfold the large, complex protein shapes (see **Figure 2-12** on p. 34). This unfolding allows digestive enzymes to reach the *peptide bonds* that hold the *amino acids* together.

Pepsinogen, a protein in gastric juice, is converted into active **pepsin** enzyme by the HCl. Pepsin then begins breaking peptide bonds to form shorter and shorter chains of amino acids.

In the intestine, other enzymes—**trypsin** in pancreatic juice and **peptidases** in intestinal juice—finish the job of protein digestion.

TABLE 18-3 Chemical Digestion

DIGESTIVE JUICES AND ENZYMES	SUBSTANCE DIGESTED (OR HYDROLYZED)	RESULTING PRODUCT*
Saliva		
Salivary amylase	Starch (polysaccharide)	Maltose (a double sugar, or disaccharide)
Gastric Juice		
Protease (pepsin) plus hydrochloric acid	Proteins	Partially digested proteins
Pancreatic Juice		
Proteases (e.g., trypsin)†	Proteins (intact or partially digested)	Peptides and **amino acids**
Lipases	Fats emulsified by bile	**Fatty acids, monoglycerides,** and **glycerol**
Pancreatic amylase	Starch	Maltose
Intestinal Enzymes‡		
Peptidases	Peptides	**Amino acids**
Sucrase	Sucrose (cane sugar)	**Glucose** and **fructose**§ (simple sugars, or monosaccharides)
Lactase	Lactose (milk sugar)	**Glucose** and **galactose** (simple sugars)
Maltase	Maltose (malt sugar)	**Glucose**

*Substances in boldface type are end products of digestion (that is, completely digested nutrients ready for absorption).
†Secreted in inactive form (trypsinogen); activated by enterokinase, an enzyme in the intestinal brush border.
‡Brush-border enzymes.
§Glucose is also called *dextrose;* fructose is also called *levulose.*

When the protease enzymes have finally split up the large protein molecules into individual amino acids, protein digestion is completed. Hence the end product of protein digestion is amino acids.

Lipid Digestion

Just as with carbohydrates, very little fat and oil digestion occurs before they reach the small intestine. Most lipids are undigested until after being emulsified into tiny droplets by bile in the duodenum.

After the lipids are trapped inside tiny droplets, pancreatic **lipase** splits them into their components. Triglycerides and other large lipid molecules are thus broken down into *fatty acids* and *glycerol* (see **Figure 2-9** on p. 32), the end products of fat digestion.

End Products of Digestion

Table 18-3 and **Figure 18-25** summarize some key facts about chemical digestion.

When carbohydrate digestion has been completed, starches (polysaccharides) and double sugars (disaccharides) have been changed mainly to glucose, a simple sugar (monosaccharide). The end products of protein digestion, on the other hand, are amino acids. Fatty acid and glycerol are the end products of fat digestion.

Absorption

Mechanisms of Absorption

After food is digested, the resulting nutrients are absorbed and move through the mucous membrane lining of the small intestine into the blood and lymph (see **Figure 18-25**). In other words, nutrient **absorption** is the process by which molecules of amino acids, glucose, fatty acids, and glycerol go from the lumen of the intestines into the circulating fluids of the body.

Absorption of nutrients is just as essential as digestion of foods. The reason is fairly obvious. As long as food stays in the intestines, it cannot nourish the millions of cells that compose all other parts of the body. Their lives depend on the absorption

FIGURE 18-25 Digestion and absorption. Concept map summarizing how the digestive system breaks down major nutrients and moves them into the internal environment, where they can be used for metabolic function.

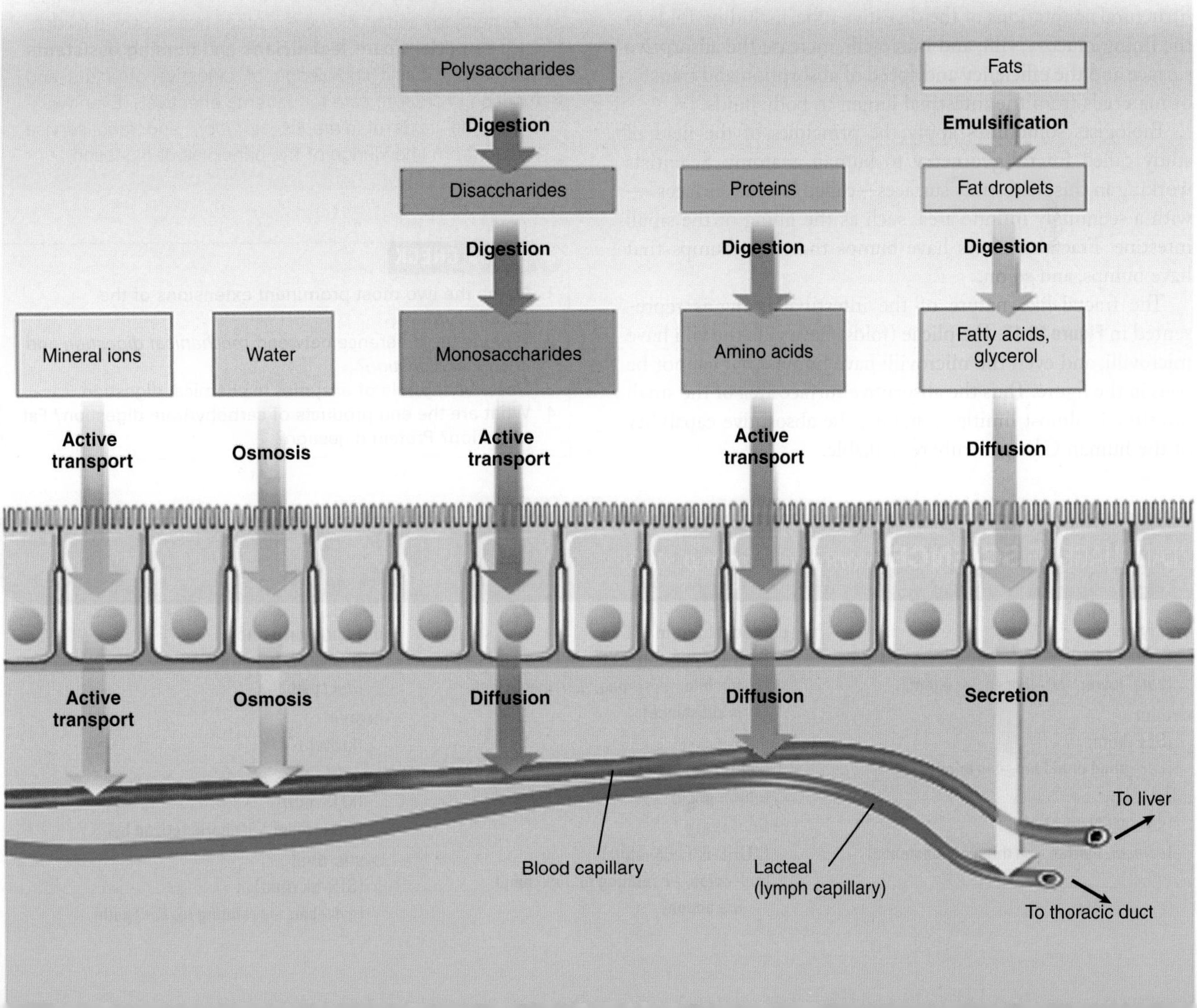

of digested food and its transportation to them by the circulating blood.

Many important minerals, such as sodium, are actively transported through the intestinal mucosa. Water follows by osmosis. Other nutrients, such as monosaccharides and amino acids, are also actively transported through the intestinal mucosa and then diffuse into the blood of capillaries in the intestinal villi.

Fatty acids and glycerol diffuse into the absorptive cells of the GI tract and then are "reassembled" into triglycerides. The triglycerides are then "repackaged" into phospholipid-coated spheres and secreted into the lacteals within intestinal villi. After eventually entering and traveling through the bloodstream, the triglycerides are broken down again in adipose and muscle tissue.

The "water-soluble vitamins" (vitamin C and the B vitamins) are dissolved in water and absorbed primarily from the small intestine. The "fat-soluble vitamins" (vitamins A, D, E, and K) are absorbed along with the end products of fat digestion in the small intestine and then pass into the lacteals. Bacterial action in the colon also produces some vitamin K that is absorbed through the lining of the large intestine.

18

Surface Area and Absorption

Structural adaptations of the digestive tube, including folds in the lining mucosa, villi, and microvilli, increase the absorptive surface and the efficiency and speed of absorption and transfer of materials from the intestinal lumen to body fluids.

Biologists sometimes apply the principles of the field of study called **fractal geometry** to human anatomy. Scientists working in this field study surfaces—called "fractal surfaces"—with a seemingly infinite area, such as the lining of the small intestine. Fractal surfaces have bumps that have bumps that have bumps, and so on.

The fractal-like nature of the intestinal lining is represented in **Figure 18-17**. The plicae (folds) have villi, the villi have microvilli, and even the microvilli have bumps that cannot be seen in the figure. Thus the absorptive surface area of the small intestine is almost limitless, making the absorptive capability of the human GI tract truly remarkable.

SCIENCE APPLICATIONS

GASTROENTEROLOGY

William Beaumont (1785–1853)

The word **gastroenterology** tells you by its parts that it is the study *(-ology)* of the stomach *(gastro-)* and the intestines *(-entero-)*.

One of the pioneering gastroenterologists was the American physician William Beaumont. In 1822, the young Québécois trapper Alexis St. Martin was shot with a musket near the Army hospital in Michigan where Beaumont was working. Beaumont treated his wound—although expecting St. Martin to die from the injury. However, St. Martin recovered and lived a long life even though the wound did not heal properly. For his entire life thereafter, an open hole remained in his abdomen that led directly into the stomach. Being grateful for his spared life and in need of income, St. Martin reluctantly allowed Beaumont to study gastric secretion through the opening.

Over many years, Beaumont made careful observations about how the stomach works. Many of his conclusions are still valid and serve as the foundation for modern gastroenterology.

Because of the complexity and importance of digestive processes, many *physicians* and *nurses* specialize in gastroenterology today. However, other health-care providers such as **patient care technicians** and **nursing assistants** still need a basic knowledge of digestive structure and function in order to care for patients effectively. Even workers in the fields of *dietetics, nutrition,* and *food service* benefit from knowledge of the principles of digestion.

QUICK CHECK

1. Name the two most prominent extensions of the *peritoneum*.
2. What is the difference between *mechanical digestion* and *chemical digestion*?
3. Describe the role of *enzymes* in chemical digestion.
4. What are the end products of carbohydrate digestion? Fat digestion? Protein digestion?

LANGUAGE OF SCIENCE *(continued from p. 493)*

catalyst
(KAT-ah-list)
[*cata-* **lower,** *-lys-* **loosen,** *-st* **agent**]

cecum
(SEE-kum)
[*cec-* **blind or hidden,** *-um* **thing**]

cementum
(see-MEN-tum)
[*cement-* **mortar,** *-um* **thing or substance**]

cholecystokinin (CCK)
(koh-lee-sis-toh-KYE-nin [see see kay])
[*chole-* **bile,** *-cyst-* **bag,** *-kin-* **movement,** *-in* **substance**]

chyme
(kyme)
[*chym-* **juice**]

colic flexure
(KOHL-ik FLEK-shur)
[*col-* **colon,** *-ic* **relating to,** *flex-* **bend,** *-ure* **action**]

common bile duct
(KOM-on byle dukt)
[*duct* **path**]

crown
(krown)

cuspid
(KUS-pid)
[*cusp-* **point,** *-id* **characterized by**]

cystic duct
(SIS-tik dukt)
[*cyst-* **sac,** *-ic* **relating to,** *duct* **path**]

LANGUAGE OF SCIENCE *(continued from p. 520)*

deciduous teeth
(deh-SID-yoo-us teeth)
[*decid-* **fall off,** *-ous* **relating to**]

defecation
(def-eh-KAY-shun)
[*de-* **remove,** *-feca-* **waste (feces),** *-tion* **process**]

deglutition
(deg-loo-TISH-un)
[*deglut-* **swallow,** *-tion* **process**]

dentin
(DEN-tin)
[*dent-* **tooth,** *-in* **substance**]

descending colon
(dih-SEND-ing KOH-lon)
[*de-* **down,** *-scend-* **climb,** *colon* **large intestine**]

digestion
(dye-JES-chun)
[*digest-* **break apart,** *-tion* **process**]

digestive system
(dye-JES-tiv SIS-tem)
[*digest-* **break apart,** *-tive* **relating to**]

duodenum
(doo-oh-DEE-num)
[*duodeni-* **12 fingers, shortened from** *intestinum duodenum digitorum* **intestine of 12 finger-widths**]

elimination
(eh-lim-uh-NAY-shun)
[*e-* **out,** *-limen-* **threshold,** *-ation* **process**]

emulsify
(eh-MUL-seh-fye)
[*e-* **out,** *-muls-* **milk,** *-i-* **combining vowel,** *-fy* **process**]

enamel
(ih-NA-mel)
[*en-* **in,** *-amel* **melt**]

enzyme
(EN-zyme)
[*en-* **in,** *-zyme* **ferment**]

esophagus
(ee-SOF-ah-gus)
[*eso-* **carry,** *-phagus* **food**]

feces
(FEE-seez)
[*feces* **waste**]

fractal geometry
(FRAK-tal jee-OM-eh-tree)
[*fract-* **break,** *-al* **relating to,** *geo-* **land,** *-metr-* **measure,** *-y* **activity**]

frenulum
(FREN-yoo-lum)
[*fren-* **bridle,** *-ul-* **little,** *-um* **thing**]

fundus
(FUN-dus)
[*fundus* **bottom**]

gallbladder
(GAWL-blad-er)
[*gall-* **bile**]

gastric gland
(GAS-trik gland)
[*gastr-* **stomach,** *-ic* **relating to,** *gland* **acorn**]

gastrointestinal tract (GI tract)
(gas-troh-in-TES-tih-nul trakt [jee aye trakt])
[*gastr-* **stomach,** *-intestin-* **intestine,** *-al* **relating to,** *tract* **trail**]

gingiva
(JIN-jih-vah)
[gingiva **gum**]

greater omentum
(GRAYT-er oh-MEN-tum)
[*omentum* **membrane covering intestines**]

hard palate
(hard PAL-et)

hepatic duct
(heh-PAT-ik dukt)
[*hepa-* **liver,** *-ic* **relating to,** *duct* **path**]

hepatic flexure
(heh-PAT-ik FLEK-sher)
[*hepa-* **liver,** *-ic* **relating to,** *flex-* **that may be bent,** *-ure* **action**]

hydrochloric acid (HCl)
(hye-droh-KLOR-ik AS-id [aych see el])
[*hydro-* **water,** *-chlor-* **green (chlorine),** *-ic* **relating to,** *acid* **sour**]

hydrolysis
(hye-DROHL-ih-sis)
[*hydro-* **water,** *-lysis* **loosening**]

ileocecal valve
(il-ee-oh-SEE-kal valv)
[*ileum* **groin or flank,** *cec-* **blind or hidden,** *-al* **relating to**]

ileum
(IL-ee-um)
[*ileum* **groin or flank**]

incisor
(in-SYE-zer)
[*in-* **into,** *-cis-* **cut,** *-or* **agent**]

ingestion
(in-JES-chun)
[*in-* **within,** *-gest-* **carry,** *-tion* **process**]

intestinal gland
(in-TES-tih-nal gland)
[*intestin-* **intestine,** *-al* **relating to,** *gland* **acorn**]

jejunum
(jeh-JOO-num)
[*jejun-* **empty,** *-um* **thing**]

lactase
(LAK-tayz)
[*lact-* **milk,** *-ase* **enzyme**]

lacteal
(LAK-tee-al)
[*lact-* **milk,** *-al* **relating to**]

lactose
(LAK-tohs)
[*lact-* **milk,** *-ose* **carbohydrate (sugar)**]

large intestine
(larj in-TES-tin)

laryngopharynx
(lah-ring-go-FAYR-inks)
[*laryng-* **voice box (larynx),** *-pharynx* **throat**]

lipase
(LYE-payz or LIP-ayz)
[*lip-* **fat,** *-ase* **enzyme**]

liver
(LIV-er)

lower esophageal sphincter (LES)
(LOH-er eh-sof-eh-JEE-al SFINGK-ter [el ee es])
[*eso-* **carry,** *-phag-* **food,** *-al* **relating to,** *sphinc-* **bind tight,** *-er* **agent**]

lumen
(LOO-men)
[*lumen* **light or window**]

major duodenal papilla
(MAY-jer doo-oh-DEE-nul [or doo-AH-de-nul] pah-PIL-ah)
[*major* **larger,** *duoden-* **12 fingers, shortened from** ***intestinum duodenum digitorum*** **intestine of 12 finger-widths,** *-al* **relating to,** *papilla* **nipple**]

maltase
(MAWL -tayz)
[*malt-* **grain,** *-ase* **enzyme**]

maltose
(MAWL-tohs)
[*malt-* **grain,** *-ose* **carbohydrate (sugar)**]

mastication
(mas-tih-KAY-shun)
[*mastica-* **chew,** *-ation* **process**]

mesentery
(MEZ-en-tayr-ee)
[*mes-* **middle,** *-enter-* **intestine,** *-y* **thing**]

metabolism
(meh-TAB-oh-liz-im)
[*meta-* **over,** *-bol-* **throw,** *-ism* **action**]

microbiome
(my-kroh-BYE-ohm)
[*micro-* **small,** *-bio-* **life,** *-ome* **entire collection**]

Continued on p. 522

18

LANGUAGE OF SCIENCE (continued from p. 521)

microvilli
(my-kroh-VIL-ee)
sing., microvillus
(my-kroh-VIL-us)
[*micro-* **small,** *-villi* **shaggy hairs**]

minor duodenal papilla
(MYE-ner doo-oh-DEE-nul [or doo-AH-de-nul] pah-PIL-ah)
[*minor* **smaller,** *duoden-* **12 fingers, shortened from** ***intestinum duodenum digitorum*** **intestine of 12 finger-widths,** *-al* **relating to,** *papilla* **nipple**]

molar
(MOHL-ar)
[*mol-* **millstone,** *-ar* **relating to**]

motility
(moh-TIL-ih-tee)
[*mot-* **move,** *-il-* **relating to,** *-ity* **state**]

mouth

mucosa
(myoo-KOH-sah)
pl., mucosae
(myoo-KOH-see)
[*muc-* **slime,** *-os-* **relating to,** *-a* **thing**]

muscularis
(mus-kyoo-LAYR-is)
[*mus-* **mouse,** *-cul-* **little,** *-ar-* **relating to,** *-is* **thing**]

nasopharynx
(nay-zoh-FAYR-inks)
[*naso-* **nose,** *-pharynx* **throat**]

neck
(nek)

oral cavity
(OR-al KAV-ih-tee)
[*or-* **mouth,** *-al* **relating to,** *cav-* **hollow,** *-ity* **state**]

oropharynx
(or-oh-FAYR-inks)
[*oro-* **mouth,** *-pharynx* **throat**]

pancreatic islet (islet of Langerhans)
(pan-kree-AT-ik aye-let) (AYE-let of LAHN-ger-hans)
[*pan-* **all,** *-creat-* **flesh,** *-ic* **relating to,** *isl-* **island,** *-et* **little**] [*Paul Langerhans* **German pathologist**]

papilla
(pah-PIL-ah)
pl., papillae
(pah-PIL-ee)
[*papilla* **nipple**]

parotid gland
(per-AH-tid gland)
[*par-* **beside,** *-ot-* **ear,** *-id* **relating to,** *gland* **acorn**]

periodontal membrane
(payr-ee-oh-DON-tull MEM-brayn)
[*peri-* **around,** *-dont-* **tooth,** *-al* **relating to,** *membrane* **thin skin**]

pepsin
(PEP-sin)
[*peps-* **digestion,** *-in* **substance**]

pepsinogen
(pep-SIN-oh-jen)
[*peps-* **digestion,** *-in* **substance,** *-o-* **combining form,** *-gen* **produce**]

peptidase
(PEP-tyd-ayz)
[*pept-* **digestion,** *-ide* **chemical,** *-ase* **enzyme**]

peristalsis
(payr-ih-STAL-sis)
[*peri-* **around,** *-stalsis* **contraction**]

peritoneal space
(payr-ih-toh-NEE-al)
[*peri-* **around,** *-tone-* **stretched,** *-al* **relating to**]

peritoneum
(payr-ih-toh-NEE-um)
[*peri-* **around,** *-tone-* **stretched,** *-um* **thing**]

permanent teeth
(PER-mah-nent teeth)

pharynx
(FAYR-inks)
[*pharynx* **throat**]

plica
(PLYE-kah)
pl., plicae
(PLYE-kee)
[*plica* **fold**]

premolar
(pree-MOHL-ar)
[*pre-* **before,** *-mola-* **millstone,** *-ar* **relating to**]

protease
(PROH-tee-ayz)
[*prote-* **protein,** *-ase* **enzyme**]

pyloric sphincter
(pye-LOR-ik SFINGK-ter)
[*pyl-* **gate,** *-or-* **to guard,** *-ic* **relating to,** *sphinc-* **bind tight,** *-er* **agent**]

pylorus
(pye-LOR-us)
[*pyl-* **gate,** *-orus* **guard**]

rectum
(REK-tum)
[*rect-* **straight,** *-um* **thing**]

regulation
(reg-yoo-LAY-shun)
[*regula-* **rule,** *-tion* **process**]

retroperitoneal
(reh-troh-pair-ih-toh-NEE-al)
[*retro-* **backward,** *peri-* **around,** *-tone-* **stretched,** *-al* **relating to**]

root

rugae
(ROO-gee)
sing., ruga
[*ruga* **wrinkle**]

secretion
(seh-KREE-shun)
[*secret-* **separate,** *-tion* **process**]

segmentation
(seg-men-TAY-shun)
[*segment-* **cut section,** *-ation* **process**]

serosa
(see-ROH-sah)
[*ser-* **watery fluid,** *-os-* **relating to,** *-a* **thing**]

serous
(SEE-rus)
[*ser-* **watery fluid,** *-ous* **relating to**]

sigmoid colon
(SIG-moyd KOH-lon)
[*sigm-* **sigma (Σ or σ) 18th letter of Greek alphabet (Roman S),** *-oid* **like,** *colon* **large intestine**]

small intestine
(smahl in-TEST-in)

soft palate
(PAL-et)
[*palat-* **roof of mouth**]

sphincter
(SFINGK-ter)
[*sphinc-* **bind tight,** *-er* **agent**]

splenic flexure
(SPLEN-ik FLEK-shur)
[*splen-* **spleen,** *-ic* **relating to,** *flex-* **bend,** *-ure* **action**]

stomach
(STUM-uk)
[*stomach* **mouth**]

sublingual gland
(sub-LING-gwall gland)
[*sub-* **under,** *-lingua-* **tongue,** *-al* **relating to,** *gland* **acorn**]

submandibular gland
(sub-man-DIB-yoo-lar gland)
[*sub-* **under,** *-mandibul-* **chew (mandible or jawbone),** *-ar* **relating to,** *gland* **acorn**]

submucosa
(sub-myoo-KOH-sah)
[*sub-* **under,** *-muc-* **slime,** *-os-* **relating to,** *-a* **thing**]

18

LANGUAGE OF **SCIENCE** *(continued from p. 522)*

sucrase
(SOO-krayz)
[*sucr-* **sugar,** *-ase* **enzyme**]

sucrose
(SOO-krohs)
[*sucr-* **sugar,** *-ose* **carbohydrate (sugar)**]

transverse colon
(tranz-VERS KOH-len)
[*trans-* **across,** *-vers-* **turn,** *colon* **large intestine**]

tricuspid
(try-KUS-pid)
[*tri-* **three,** *-cusp-* **point,** *-id* **characterized by**]

trypsin
(TRIP-sin)
[*tryps-* **pound,** *-in* **substance**]

upper esophageal sphincter (UES)
(UP-er eh-sof-eh-JEE-al SFINGK-ter [yoo ee es])
[*eso-* **carry,** *-phag-* **food (eat),** *-al* **relating to,** *sphinc-* **bind tight,** *-er* **agent**]

uvula
(YOO-vyoo-lah)
[*uva-* **grape (or bunch of grapes),** *-ula* **little**]

vermiform appendix
(VERM-ih-form ah-PEN-diks)
[*vermi-* **worm,** *-form* **shape,** *append-* **hang upon,** *-ix* **thing**]

villus
(VIL-us)
pl., villi
(VIL-aye)
[*villus* **shaggy hair**]

LANGUAGE OF **MEDICINE**

anorexia
(an-oh-REK-see-ah)
[*an-* **without,** *-orex-* **appetite,** *-ia* **condition**]

appendicitis
(ah-pen-dih-SYE-tis)
[*appendic-* **hang upon,** *-itis* **inflammation**]

ascites
(ah-SYE-tees)
[*ascit-* **bag,** *-es* **condition**]

bariatrics
(bayr-ee-AT-riks)
[*bari-* **weight,** *-iatr-* **physician,** *-ic* **relating to**]

Barrett esophagus
(BAHR-ett ee-SOF-ah-gus)
[*Norman R. Barrett* **English surgeon,** *eso-* **carry,** *-phag-* **food (eat)**]

biliary colic
(BIL-yah-ree KOL-ik)
[*bil-* **bile,** *-ary* **relating to,** *col-* **colon,** *-ic* **relating to**]

cholecystectomy
(kohl-eh-sis-TEK-toh-mee)
[*chole-* **bile,** *-cyst-* **bag,** *-ec-* **out,** *-tom-* **cut,** *-y* **action**]

cholecystitis
(koh-leh-sis-TYE-tis)
[*chole-* **bile,** *-cyst-* **bag,** *-itis* **inflammation**]

choledocholithiasis
(koh-LED-uh-koh-lih-THY-ah-sis)
[*chole-* **bile,** *-doch-* **containing,** *-lith-* **stone,** *-iasis* **condition**]

cholelithiasis
(koh-leh-lih-THY-ah-sis or koh-leh-lih-THEE-ah-sis)
[*chole-* **bile,** *-lith-* **stone,** *-iasis* **condition**]

cirrhosis
(sih-ROH-sis)
[*cirrhos-* **yellow-orange,** *-osis* **condition**]

cleft lip
(kleft lip)
[*cleft* **split**]

cleft palate
(kleft PAL-et)
[*cleft* **split,** *palate* **roof of mouth**]

colitis
(koh-LYE-tis)
[*col-* **colon,** *-itis* **inflammation**]

colonoscopy
(koh-lon-AH-skah-pee)
[*colon* **large intestine,** *-scop-* **see,** *-y* **activity**]

colorectal cancer
(koh-loh-REK-tal KAN-ser)
[*colo-* **colon,** *-rect-* **straight,** *-al* **relating to,** *cancer* **crab or malignant tumor**]

colostomy
(kah-LAH-stoh-mee)
[*colo-* **large intestine,** *-stom-* **mouth (opening),** *-y* **activity**]

constipation
(kon-stih-PAY-shun)
[*constipa-* **crowd together,** *-ation* **process**]

Crohn disease
(krohn dih-ZEEZ)
[*Burrill B. Crohn* **American physician,** *dis-* **opposite of,** *-ease* **comfort**]

cystic fibrosis (CF)
(SIS-tik fye-BROH-sis (see ef))
[*cyst-* **sac,** *-ic* **relating to,** *fibr-* **fiber,** *-osis* **condition**]

dental caries
(DENT-al KAYR-ees)
[*dent-* **tooth,** *-al* **relating to,** *caries* **decay**]

diabetes mellitus (DM)
(dye-ah-BEE-teez MELL-ih-tus) [dee em]
[*diabetes* **pass-through or siphon,** *mellitus* **honey-sweet**]

diarrhea
(dye-ah-REE-ah)
[*dia-* **through,** *-rrhea* **flow**]

diverticulitis
(dye-ver-tik-yoo-LYE-tis)
[*diverticul-* **bypath,** *-itis* **inflammation**]

dysentery
(DIS-en-tayr-ee)
[*dys-* **ill,** *-enter-* **intestines,** *-y* **activity**]

emesis
(EM-eh-sis)
[*emesis* **vomit**]

endoscope
(EN-doh-skohp)
[*endo-* **within,** *-scop-* **see**]

enteritis
(en-ter-AYE-tis)
[*enter-* **intestine,** *-itis* **inflammation**]

gallstone
(GAWL-stohn)
[*gall-* **bile**]

gastritis
(gas-TRY-tis)
[*gastr-* **stomach,** *-itis* **inflammation**]

gastroenteritis
(gas-troh-en-ter-AYE-tis)
[*gastr-* **stomach,** *-enter-* **intestine,** *-itis* **inflammation**]

Continued on p. 524

18

LANGUAGE OF MEDICINE (continued from p. 523)

gastroenterology
(gas-troh-en-ter-AHL-oh-jee)
[*gastr-* **stomach,** *-entero-* **intestine,** *-o-* **combining vowel,** *-log-* **words (study of),** *-y* **activity**]

gastroesophageal reflux disease (GERD)
(gas-troh-eh-sof-eh-JEE-al REE-fluks dih-ZEEZ [gerd])
[*gastro-* **stomach,** *-eso-* **carry,** *-phag-* **food (eat),** *-al* **relating to,** *re-* **again or back,** *-flux* **flow,** *dis-* **opposite of,** *-ease* **comfort**]

gingivitis
(jin-jih-VYE-tis)
[*gingiv-* **gum,** *-itis* **inflammation**]

heartburn
(hart-burn)

Helicobacter pylori
(HEEL-ih-koh-BAK-ter pye-LOH-ree)
[*Helic-* **helix,** *-bacter* **rod (bacterium),** *pyl-* **gate,** *-or-* **guard,** *-i* **of the**]

hepatitis
(hep-ah-TYE-tis)
[*hepat-* **liver,** *-itis* **inflammation**]

hepatitis C virus (HCV)
(hep-ah-TYE-tis see VYE-rus [aych see vee])
[*hepat-* **liver,** *-itis* **inflammation,** *C* **letter of Roman alphabet,** *virus* **poison**]

hiatal hernia
(hye-AY-tal HER-nee-ah)
[*hiat-* **gap,** *-al* **relating to,** *hernia* **rupture**]

jaundice
(JAWN-dis)
[*jaun-* **yellow,** *-ice* **state**]

lactose intolerance
(LAK-tohs in-TOL-er-ans)
[*lact-* **milk,** *-ose* **carbohydrate (sugar),** *in-* **not,** *-toler-* **bear,** *-ance* **state**]

leukoplakia
(loo-koh-PLAY-kee-ah)
[*leuko-* **white,** *-plak-* **flat area (tongue),** *-ia* **condition**]

malabsorption syndrome
(mal-ab-SORP-shun SIN-drohm)
[*mal-* **bad,** *-ab-* **from,** *-sorp-* **suck,** *-tion* **process,** *syn-* **together,** *-drome* **running or (race) course**]

maldigestion
(mal-dye-JES-chun)
[*mal-* **bad,** *-digest-* **break apart,** *-tion* **process**]

malocclusion
(mal-oh-CLEW-zhun)
[*mal-* **bad,** *-oc-* **against,** *-clu-* **shut or close,** *-sion* **state**]

nausea
(NAW-zee-ah)
[*naus-* **ship,** *-ea* **condition**]

nursing assistant
(NURS-ing ah-SIS-tent)
[*nurs-* **nourish or nurture,** *assist-* **help,** *-ant* **agent**]

oral candidiasis
(OR-al kan-dih-DYE-eh-sis)
[*or-* **mouth,** *-al* **relating to,** *candid-* **white,** *-iasis* **condition**]

oral rehydration therapy (ORT)
(OR-al ree-hye-DRAY-shun THAYR-ah-pee)
[*or-* **mouth,** *-al* **relating to,** *re-* **back again,** *-hydra-* **water,** *-ation* **process,** *therapy* **treatment**]

orthodontics
(or-thoh-DON-tiks)
[*ortho-* **straight or upright,** *-odont-* **tooth,** *-ic* **relating to**]

pancreatitis
(pan-kree-ah-TYE-tis)
[*pan-* **all,** *-creat-* **flesh,** *-itis* **inflammation**]

patient care technician
(PAY-shent kayr tek-NISH-en)
[*techn-* **art or skill,** *-ic* **relating to,** *-ian* **practitioner**]

periodontitis
(payr-ee-oh-don-TYE-tis)
[*peri-* **around,** *-odont-* **tooth,** *-itis* **inflammation**]

peritonitis
(payr-ih-toh-NYE-tis)
[*peri-* **around,** *-ton-* **stretch (peritoneum),** *-itis* **inflammation**]

portal hypertension
(PORT-al hye-per-TEN-shun)
[*port-* **doorway,** *-al* **relating to,** *hyper-* **excessive,** *-tens-* **stretch or pull tight,** *-sion* **state**]

pyloric stenosis
(pye-LOR-ik steh-NOH-sis)
[*pyl-* **gate,** *-or-* **guard,** *-ic* **relating to,** *stenos-* **narrow,** *-osis* **condition**]

pylorospasm
(pye-LOHR-oh-spaz-um)
[*pyl-* **gate,** *-or-* **guard,** *-spasm* **twitch or involuntary contraction**]

sigmoidoscopy
(SIG-moyd-os-koh-pee)
[*sigm-* **sigma (Σ or σ) 18th letter of Greek alphabet (Roman S),** *-oid-* **like,** *-scop-* **see,** *-y* **activity**]

snuff dipper's pouch
(snuf DIP-erz powch)
[*snuff* **powered tobacco**]

stoma
(STOH-mah)
pl., stomata
(STOH-mah-tuh or stoh-MAH-tuh)
[*stoma* **mouth**]

thrush
(thruhsh)
[*thrush* **throat**]

triple therapy
(TRIP-pul THAYR-ah-pee)
[*tri-* **three,** *-pl-* **plus or added,** *therapy* **treatment**]

ulcer
(UL-ser)
[*ulcer* **sore**]

upper GI (UGI)
(upper jee aye [yoo jee aye])
[*GI* **gastrointestinal**]

❑ OUTLINE SUMMARY

*To download a digital version of the chapter summary for use with your device, access the **Audio Chapter Summaries** online at evolve.elsevier.com.*

Scan this summary after reading the chapter to help you reinforce the key concepts. Later, use the summary as a quick review before your class or before a test.

Overview of Digestion

A. Alimentary canal, digestive tract, or gastrointestinal (GI) tract (**Figure 18-1**)
   1. Extends from mouth to anus—9 meters (29 feet) in length
   2. Involved in digestion and absorption of nutrients
   3. Main organs are part of the tract and accessory organs are inside or surrounding the tract (**Table 18-1**)

B. Primary mechanisms of the digestive system
   1. The digestive system uses many mechanisms (**Table 18-2**)
   2. Ingestion—complex foods taken into the GI tract
   3. Digestion—group of processes that break complex nutrients into simpler ones
      a. Mechanical digestion—breakup of large chunks of food into smaller bits
      b. Chemical digestion—breaks large molecules into smaller ones
   4. Motility—a number of GI movements resulting from muscular contraction
   5. Secretion—release of digestive juices and hormones that facilitate digestion
   6. Absorption—movement of digested nutrients into the internal environment of the body
   7. Elimination—movement of residues of digestion out of alimentary canal
   8. Regulation—neural, hormonal, and other mechanisms that regulate digestive activity

Wall of the Digestive Tract

A. Digestive tract described as muscular tube that extends from mouth to anus; the inner hollow space is called the *lumen*

B. Wall of the digestive tube is formed by four layers (**Figure 18-2**)
   1. Mucosa—type varies depending on GI location (tough and stratified or delicate and simple epithelium); mucous production
   2. Submucosa—connective tissue layer
   3. Muscularis—circular, longitudinal, and oblique (in stomach) layers of smooth muscle important in GI motility
      a. Peristalsis—"wavelike" movement pushes food down the tract (**Figure 18-3**)
      b. Segmentation—"back-and-forth" mixing movement (**Figure 18-4**)
   4. Serosa—serous membrane that covers the outside of abdominal organs; it attaches the digestive tract to the wall of the abdominopelvic cavity by forming folds called *mesenteries*

Mouth

A. Structure of oral cavity
   1. Roof—formed by hard palate (parts of maxillary and palatine bones) and soft palate, an arch-shaped muscle separating mouth from pharynx; uvula, a downward projection of soft palate helps in speech and swallowing (deglutition)
   2. Floor—formed by tongue and its muscles, lingual frenulum (fold of mucosa that helps anchor tongue) (**Figure 18-5**)

B. Teeth
   1. Types of teeth—incisors, canines (cuspids), premolars (bicuspids), and molars (tricuspids)
      a. Twenty teeth in deciduous or baby set; average age for cutting first tooth about 6 months; set complete at about 30 months of age
      b. Thirty-two teeth in permanent set; 6 years about average age for starting to cut first permanent tooth; set complete usually between ages of 17 and 24 years (**Figure 18-6**)
   2. Typical tooth (**Figure 18-7**)
      a. Three main parts—crown, neck, and root
      b. Enamel, which covers the crown, is hardest tissue in body

C. Salivary glands (**Figure 18-8**)
   1. Saliva
      a. Exocrine gland secretion flows through salivary ducts into the mouth
      b. Watery mixture contains enzymes (salivary amylase), sodium bicarbonate ($NaHCO_3$), and mucus
         (1) Begins digestion of carbohydrates
         (2) Lubricates food during mastication
         (3) Neutralizes bacterial acids in mouth
   2. Parotid glands
      a. Largest salivary glands
      b. Produces saliva containing $NaHCO_3$
   3. Submandibular glands—ducts open on either side of lingual frenulum
   4. Sublingual glands—multiple ducts open into floor of mouth

18

D. Disorders of the Mouth
1. Infections, cancer, congenital defects, and other disorders can cause serious complications including malnutrition; infections and cancer can spread to other structures
2. Cancer
a. Leukoplakia—precancerous condition of mouth tissue
(1) Snuff dipper's pouch—caused by use of chewing tobacco (**Figure 18-9**, *A*)
(2) Squamous cell carcinoma—most common form of mouth cancer (**Figure 18-9**, *B*)
3. Dental conditions
a. Dental caries (**Figure 18-9**, *C*)
(1) Tooth disease resulting in permanent defect called a "cavity"
(2) Infection may spread to other adjacent tissues or to blood
(3) Lost or diseased teeth may be replaced by dentures or implants (**Figure 18-9**, *D*)
b. Gingivitis—gum inflammation or infection
(1) Most cases result from poor oral hygiene
(2) Can be a complication of diabetes, vitamin deficiency, or pregnancy
c. Periodontitis—inflammation of periodontal membrane
(1) Often a complication of advanced or untreated gingivitis
(2) Leading cause of tooth loss among adults
4. Infection
a. Thrush, or oral candidiasis
(1) Caused by yeast organisms of the *Candida* group
(2) Patches of "cheesy"-looking exudate form over an inflamed tongue and oral mucosa, which itches and bleeds easily (**Figure 18-9**, *E*)
(3) Common in immunosuppressed individuals (with AIDS) or after antibiotic therapy
5. Congenital defects
a. Cleft lip and cleft palate—most common congenital defects of the mouth (**Figure 18-10**)
(1) May occur alone or together
(2) Caused by failure of mouth structures to fuse during embryonic development

Pharynx

A. Structure
1. Three divisions: nasopharynx, oropharynx, laryngopharynx (**Figure 18-11**)
2. Tonsils form lymphoid ring that prevents digestive tract infection
B. Function—deglutition (swallowing)
1. Oropharynx pharyngeal segment most involved in deglutition
2. Deglutition movements regulated by motor cortex of cerebrum (voluntary) and "deglutition center" of brainstem (involuntary)

Esophagus

A. Muscular, mucus-lined tube about 25 cm (10 inches) long
B. Connects pharynx to stomach
C. Dynamic passageway for food—muscular walls help push bolus toward stomach
D. Sphincters at each end of esophagus help keep ingested material moving in one direction down the tube
1. Upper esophageal sphincter (UES)
2. Lower esophageal sphincter (LES)
E. Reflux
1. Gastroesophageal reflux disease (GERD)
a. Backflow of acidic stomach contents into esophagus causes symptoms of heartburn and indigestion
b. Mild symptoms treated by nonsurgical measures include dietary changes, weight loss, acid blocking or buffering medications, and drugs that strengthen LES
c. Severe and frequent episodes of GERD can trigger asthma attacks, cause severe chest pain, bleeding, or narrowing and chronic irritation of esophagus (**Figure 18-12**)
d. Untreated GERD may result in a precancerous condition called Barrett esophagus (**Figure 18-13**)
2. Hiatal hernia
a. Stomach pushes through hiatus in the diaphragm (**Figure 18-14**)
b. Characterized by reflux (GERD); allows stomach acid to bypass LES

Stomach

A. Structure
1. Divisions—fundus (rounded base), body (main part), pylorus (apex)
2. Size—expands after large meal; about size of large sausage when empty (**Figure 18-15**)
3. Muscularis—many smooth muscle fibers in three layers; contractions produce churning movements (peristalsis)
4. Mucosa
a. Many microscopic gastric glands secrete gastric juice containing enzymes, hydrochloric acid, and intrinsic factor into stomach
b. Mucous membrane lies in folds (rugae) when stomach is empty
5. Pyloric sphincter muscle closes opening between pylorus (lower part of stomach) and duodenum
B. Function
1. Food enters stomach through LES and digestive process continues
2. Partial digestion of proteins occurs after chyme is held in the stomach for some time
C. Disorders of the stomach
1. Signs and symptoms
a. Gastroenterology—study of stomach and intestines and their diseases

b. Stomach is site of numerous possible diseases and conditions
c. Gastric diseases often exhibit these signs or symptoms: gastritis (inflammation), anorexia (appetite loss), nausea (upset stomach), and emesis (vomiting)
2. Pylorospasm—abnormal spasms of the pyloric sphincter
a. Common condition in infants
b. Pyloric stenosis is similar abnormality—obstructive narrowing of the pyloric opening
3. Ulcer—open wound caused by acid in gastric juice (**Figure 18-16**)
a. Often occurs in duodenum or stomach
b. Associated with infection by the bacterium *Helicobacter pylori* and use of NSAIDs
c. Current treatment involves triple therapy
4. Stomach cancer
a. Increased risk with consumption of alcohol, preserved food, use of chewing tobacco, and infection by *H. pylori*
b. No practical way to screen for early stages

Small Intestine

A. Structure
1. Size—about 7 meters (20 feet) long but only 2 cm or so in diameter (**Figure 18-17**)
2. Divisions
a. Duodenum
b. Jejunum
c. Ileum
3. Many coiled loops accommodate a long tube within the short abdominal cavity
4. Duodenum is site of much chemical digestion
a. Ducts from pancreas and liver enter tract here
b. Major and minor duodenal papillae are bumps where the secretions enter

B. Function
1. Main functions—digestion and absorption; small intestine does most of these functions for the digestive system
2. Intestinal secretions and digestions
a. Intestinal glands—many microscopic glands secrete intestinal juice (water, enzymes, ions)
b. Pancreatic and liver secretions
c. Most chemical digestion occurs in duodenum
3. Absorption
a. Huge absorptive surface area
(1) Circular folds (plicae)
(2) Intestinal villi and microvilli—microscopic finger-shaped projections
(3) Blood capillaries absorb carbohydrate and protein products (sugars; amino acids)
(4) Lacteals (lymph capillaries) absorb fats
4. Motility—smooth muscle fibers contract to produce movements
a. Peristalsis pushes chyme along, toward large intestine
b. Segmentation mixes digestive juices with chyme and helps with absorption

C. Disorders of the small intestine
1. Enteritis—intestinal inflammation
2. Gastroenteritis—inflammation of stomach and intestines
3. Malabsorption syndrome—group of symptoms resulting from failure to absorb nutrients properly (e.g., anorexia, weight loss, abdominal bloating, cramps, anemia, and fatigue)

Liver and Gallbladder

A. Structure (**Figure 18-18**)
1. Liver
a. Largest exocrine gland
b. Fills upper right section of abdominal cavity and extends over into left side (**Figure 18-1**)
c. Secretes bile, a mixture of chemicals in water
d. Excretory route for yellowish bile pigments from blood (from breakdown of old RBCs)
e. Many other metabolic functions (discussed in Chapter 19)
2. Gallbladder
a. Location—undersurface of the liver, sac with folded interior
b. Function—concentrates and stores bile produced in the liver
3. Ducts (**Figure 18-18**)
a. Hepatic—drains bile from liver
b. Cystic—duct by which bile enters and leaves gallbladder
c. Common bile—formed by union of hepatic and cystic ducts; drains bile from hepatic or cystic ducts into duodenum

B. Function
1. Bile contains bile salts that emulsify the fats in chyme
2. Bile contains cholesterol that can be eliminated from the body
3. CCK (cholecystokinin) is a hormone triggered by fat in chyme; CCK causes the gallbladder to contract and push stored bile into ducts leading to duodenum

C. Disorders of the liver and gallbladder
1. Gallstones—hard clumps made of cholesterol, crystallized bile pigments, and calcium salts
a. Cholelithiasis—condition of having gallstones (**Figure 18-19**)
b. Cholecystitis—inflammation of the gallbladder; may accompany cholelithiasis
c. Stones can obstruct bile canals—a condition called choledocholithiasis—causing jaundice
2. Hepatitis—liver inflammation
a. Characterized by liver enlargement, jaundice, anorexia, discomfort, gray-white feces, and dark urine
b. Caused by a variety of factors: toxins; bacteria; viruses; hepatitis A, B, and C; and parasites

18

3. Cirrhosis—degeneration of liver tissue involving replacement of normal (but damaged) tissue with fibrous and fatty tissue (**Figure 18-20,** *A*)
4. Portal hypertension—high blood pressure in the hepatic portal veins caused by obstruction of blood flow in a diseased liver; may cause varicosities of surrounding systemic veins (**Figure 18-20,** *B*)

Pancreas

A. Exocrine *and* endocrine gland that lies behind stomach (**Figure 18-18** and **Figure 18-21**)
B. Exocrine pancreatic cells secrete pancreatic juice
   1. Most important digestive juice, containing enzymes to digest carbohydrates, proteins, lipids; contains sodium bicarbonate that neutralizes stomach acid in chyme
   2. Secreted into pancreatic ducts; main duct empties into duodenum
C. Pancreatic islets (islets of Langerhans)—endocrine cells not connected with pancreatic ducts; secrete hormones glucagon and insulin into the blood
D. Pancreatic disorders
   1. Diabetes mellitus (DM)—some cases of DM result from failure of pancreatic islets to secrete sufficient insulin
   2. Pancreatitis—inflammation of pancreas
      a. Acute pancreatitis results from blocked ducts that force pancreatic juice to backflow
      b. Pancreatic enzymes digest the gland
   3. Cystic fibrosis—thick secretions block flow of pancreatic juice
   4. Pancreatic cancer—very serious; fatal in the majority of cases

18

Large Intestine

A. Structure (**Figure 18-22**)
   1. Cecum—blind-end pouch at beginning of large intestine; chyme enters cecum through ileocecal valve
   2. Colon—ascending, transverse, descending, and sigmoid segments
   3. Rectum—empties feces through anal canal and external opening called anus
B. Function
   1. Microbiome (flora)—microorganisms that help digest nutrients, produce vitamins, and support immune protection; produce gases (flatulence or flatus)
   2. Absorption of water, salts, vitamins
   3. Increased motility may produce diarrhea and decreased motility may result in constipation
   4. Defecation—elimination of feces; regulated by voluntary and involuntary anal sphincters
C. Disorders of the large intestine often relate to abnormal motility (rate of movement of contents)
   1. Motility disorders
      a. Diarrhea—results from abnormally increased intestinal motility; may result in dehydration or convulsions
      b. Constipation—results from decreased intestinal motility
   2. Inflammatory conditions
      a. Diverticulitis (inflammation of abnormal pouches called *diverticula*)—may cause constipation (**Figure 18-22,** *C*)
      b. Colitis—general name for any inflammatory condition of the large intestine
   3. Colorectal cancer—a common malignancy of the colon and rectum associated with colonic polyps; risk factors include advanced age, low-fiber and high-fat diets, and genetic predisposition

Appendix

A. Blind, worm-shaped tube off cecum (**Figure 18-22**)
B. Functions as an incubator for bacteria of the intestinal microbiome
C. Appendicitis—inflammation or infection of appendix
   1. If appendix ruptures, infectious material may spread to other organs (**Figure 18-22,** *D*)
   2. Most common acute abdominal condition requiring surgery
   3. Affects 7% to 12% of population younger than 30 years

Peritoneum

A. Location and description—large sheet of serous membrane (**Figure 18-23**)
   1. Parietal layer of peritoneum lines abdominal cavity
   2. Visceral layer of peritoneum covers abdominal organs
   3. Peritoneal space—lies between parietal and visceral layers; produces lubricating peritoneal (serous) fluid
   4. Retroperitoneal—describes structures outside the parietal peritoneum, such as kidneys
B. Extensions of peritoneum—largest are the mesentery and greater omentum (**Figure 18-23,** *B*)
   1. Mesentery—extension of parietal peritoneum, which attaches most of small intestine to posterior abdominal wall
   2. Greater omentum ("lace apron")—hangs down from lower edge of stomach and transverse colon in front of intestines
C. Peritonitis—inflammation of peritoneum resulting from infection or other irritant; often a complication of ruptured appendix
D. Ascites—abnormal accumulation of fluid in peritoneal space; often causes bloating of abdomen (**Figure 18-24**)

Digestion

A. Definition—process that transforms nutrients into a form that can be absorbed and used by cells (**Table 18-2**)
   1. Mechanical digestion—chewing, swallowing, and peristalsis break ingested material into tiny particles, mix them well with digestive juices, and move them along the digestive tract

2. Chemical digestion—breaks up large nutrient molecules into smaller molecules that can be easily absorbed; brought about by digestive enzymes

B. Enzymes and chemical digestion (**Table 18-3** and **Figure 18-25**)
1. Enzymes—protein molecules that act as catalysts, speeding up chemical reactions
2. Chemical digestion—specific enzymes speed up breakdown of specific molecules and no others
3. Hydrolysis—enzymatic reactions that add water to break large molecules into smaller molecules

C. Carbohydrate digestion—mainly in small intestine
1. Pancreatic amylase—breaks starches down into disaccharides
2. Intestinal juice enzymes
   a. Maltase—digests maltose
   b. Sucrase—digests sucrose
   c. Lactase—digests lactose; deficiency is called *lactose intolerance*

D. Protein digestion—starts in stomach; completed in small intestine
1. Hydrochloric acid in gastric juices unfolds large proteins and converts pepsinogen to active pepsin
2. Gastric juice enzyme pepsin partially digests proteins
3. Pancreatic enzyme trypsin digests proteins into smaller peptides
4. Intestinal peptidases break apart peptides into individual amino acids

E. Fat digestion
1. Bile contains no enzymes but emulsifies fats (breaks fat droplets into very small droplets)
2. Pancreatic lipase breaks down emulsified fats to fatty acids and glycerol in small intestine

Absorption

A. Mechanisms of absorption
1. Definition—process by which digested nutrients move from intestine into blood or lymph
2. Mechanisms include diffusion, osmosis, and active transport (**Figure 18-25**)
3. Nutrients and most water, minerals, and vitamins are absorbed from small intestine; some water and vitamin K also absorbed from large intestine

B. Surface area and absorption
1. Structural adaptations increase absorptive surface area
2. Fractal geometry—study of irregular "fragmented" geometric shapes such as those in lining of intestine that have almost unlimited surface area

❑ ACTIVE LEARNING

STUDY TIPS

Consider using these tips to achieve success in meeting your learning goals.

Review the synopsis of the digestive system in Chapter 5. The structures of the digestive system can be divided into two parts: the tube called the gastrointestinal tract and the accessory organs (organs that are not part of the tube). In most cases, the accessory organs produce substances that are released into the tube. The tube is composed of four layers of tissue. The actual process of digestion occurs inside this tube.

1. Make flash cards and check online resources to help you learn the name, location, and function of the organs of the gastrointestinal tract and the accessory organs.
2. Develop a concept map to outline the route nutrients take through the digestive system. Include the enzymes present in each organ where applicable. You also may wish to include what nutrients are digested and/or absorbed in each organ.
3. Make a chart of the disorders of the digestive system. Group them by the organ that is affected and the mechanism or cause of the disorder.
4. The process of digestion is explained in terms of what type of nutrient is being digested: carbohydrates, fats, or proteins. The chemical process of digestion uses some suffixes that can make the processes easier to learn. The suffix *-ose* indicates that the substance is a carbohydrate. The suffix *-ase* indicates the substance is an enzyme. In many cases, the first part of the enzyme's name tells you what substance is being digested. Malt*ose* is digested by the enzyme malt*ase*. If you know this general rule, remembering what digests what becomes a lot easier. The protein enzymes pepsin and trypsin are exceptions to this rule.
5. For a better understanding of the terms, refer to the Language of Science and the Language of Medicine sections.
6. In your study group, review your flash cards of the structures of the digestive system and your chart of the disorders of the digestive system. Use **Table 18-3** to quiz each other on the enzymes, the substances they digest, and the end products they produce. Review the chapter outline summary and the questions at the back of the chapter, and discuss possible test questions.

Review Questions

Write out the answers to these questions after reading the chapter and reviewing the Chapter Summary. If you simply think through the answer without writing it down, you will not retain much of your new learning.

1. Name and describe the four layers of the wall of the gastrointestinal tract.
2. Name the function of the uvula and soft palate.
3. Identify the function of the different types of teeth.
4. Describe the three main parts of the tooth.
5. What is leukoplakia? What could develop from it?
6. Distinguish between gingivitis and periodontitis.
7. Name the three pairs of salivary glands and describe where the duct from each enters the mouth.
8. Identify where carbohydrate digestion begins.
9. Define deglutition.
10. What is the function of the upper and lower esophageal sphincter muscles?
11. Define peristalsis.
12. Name the three parts of the triple therapy used to treat ulcers?
13. Explain how bile from the liver and gallbladder reaches the small intestine. What is the function of cholecystokinin?
14. What is the relationship between body weight and the formation of gallstones?
15. Define hepatitis. What are the signs and symptoms of hepatitis?
16. What is contained in pancreatic juice?
17. What do the bacteria in the large intestine contribute to the body?
18. List the seven subdivisions of the large intestine.
19. Describe the mesentery and the greater omentum.
20. Define the terms *peritonitis* and *ascites.*
21. Differentiate between mechanical digestion and chemical digestion.
22. Briefly describe the process of carbohydrate digestion.
23. Briefly describe the process of fat digestion.
24. Briefly describe the process of protein digestion.
25. Explain the process of absorption. What function do the lacteals have in absorption?

Critical Thinking

After finishing the Review Questions, write out the answers to these more in-depth questions to help you apply your new knowledge. Go back to sections of the chapter that relate to concepts that you find difficult.

26. What would happen if the small intestine had no plicae, villi, or microvilli?
27. Bile does not cause a chemical change; what is the effect of bile on fat, and why does this make fat digestion more efficient?
28. Some people are lactose intolerant. This means they are less able to fully digest lactose sugar. What enzyme is probably not functioning properly and what type of nutrients should these people try to avoid?

Chapter Test

After studying the chapter, test your mastery by responding to these items. Try to answer them without looking up the answers.

1. Nutrients undergo three kinds of processing in the body. All cells perform metabolism, but ________ and ________ are performed by the digestive system.
2. A thin membrane called the ________ attaches the tongue to the floor of the mouth.
3. The ________ and ________ prevent food and liquids from entering the nasal cavity above the mouth when food is swallowed.
4. The three main parts of the tooth are ________, ________, and ________.
5. The three salivary glands are the ________, ________, and ________.
6. The ________ layer of the wall of the gastrointestinal tract contains blood vessels and nerves.
7. The ________ is the innermost layer of the wall of the gastrointestinal tract.
8. The tube connecting the mouth and stomach is the ________.
9. The three divisions of the stomach are the ________, ________, and ________.
10. Gastric juice is composed of ________ and ________.
11. After the stomach mixes the swallowed bolus thoroughly with the gastric juice, the nutrients leave the stomach in a semisolid mixture called ________.
12. A condition during infancy when the pyloric fibers do not relax normally is called ________.
13. The three divisions of the small intestines are the ________, ________, and ________.
14. The tiny fingerlike projections covering the plicae of the small intestines are called ________.
15. The lymphatic vessels in the villi are called the ________.
16. The common bile duct is formed by the union of the ________ from the liver and the ________ from the gallbladder.
17. The pancreas secretes ________ which neutralizes the hydrochloric acid in the gastric juice that enters the small intestines.
18. The part of the large intestine between the ascending and descending colon is the ________.
19. The two most prominent extensions of the peritoneum are the ________ and the ________.
20. The process by which digested nutrients are moved from the digestive system to the circulating fluids is called ________.

Match each term in Column A with its corresponding statement in Column B.

Column A		Column B
21. ________	emulsification	a. enzyme that is made in the pancreas and digests fat
22. ________	amylase	b. enzyme that is made in the small intestine and digests protein
23. ________	pepsin	c. gland that produces bile
24. ________	cholecystokinin	d. the final end product of protein digestion
25. ________	peptidase	e. effect bile has on fat droplets
26. ________	cystic	f. the form that food takes in the mouth so that it can be swallowed easily
27. ________	bolus	g. the final end product of carbohydrate digestion
28. ________	simple sugars	h. one of the final end products of fat digestion
29. ________	amino acid	i. enzyme that is made in both the salivary glands and the pancreas and digests starch
30. ________	liver	j. enzyme that is made in the stomach and digests protein
31. ________	lipase	k. hormone that stimulates the contraction of the gallbladder
32. ________	glycerol	l. duct that connects the gallbladder to the common bile duct

Match each disorder in Column A with its cause or description in Column B.

Column A		Column B
33. ________	gingivitis	a. inflammation of the stomach and intestines
34. ________	periodontitis	b. a liver inflammation; B type is more serious than A type
35. ________	gastroenteritis	c. inflammation of abnormal outpouchings in the large intestine
36. ________	ulcer	d. inflammation of the periodontal membrane
37. ________	cholecystitis	e. abnormal accumulation of fluid in the peritoneal space
38. ________	hepatitis	f. a general term for the inflammation of the large intestine
39. ________	diverticulitis	g. a general term for gum inflammation or infection
40. ________	colitis	h. inflammation of the peritoneum
41. ________	peritonitis	i. inflammation of the gallbladder
42. ________	ascites	j. open wounds caused by gastric acid, often associated with *Helicobacter pylori*

Case Studies

To solve a case study, you may have to refer to the glossary or index, other chapters in this textbook, and other resources.

1. Because Alyse has failed to regularly brush or floss her teeth, she has been informed that she may eventually lose some of her teeth. Give the sequence of events in which a failure to brush and floss the teeth would lead to loss of teeth.
2. You have been diagnosed as having a gastric ulcer. Your physician prescribed the drug famotidine (Pepcid). This drug has an action similar to cimetidine, but it is much more potent. How will famotidine help your ulcer? Your brother also has an ulcer, but his physician has prescribed sucralfate (Carafate), a medication that has an antipepsin action and tends to adhere to membrane injuries. Explain how sucralfate might help your brother's ulcer.
3. Fred has been experiencing recurring episodes of heartburn, especially when he bends over to lace his shoes or when he lies down in bed. His physician believes Fred may have a hiatal hernia. To confirm this diagnosis, she has scheduled Fred for a "barium swallow." In this test, a barium-containing fluid that blocks x-rays is swallowed so that the stomach appears as a bright mass in a radiograph. If Fred has a hiatal hernia, what should the radiologist see on the radiograph? Can you explain what caused the symptoms that brought Fred in to see his physician?
4. Jim wants to begin an exercise program. He is 25 and has not been active in the past because he becomes dehydrated so rapidly and becomes ill. His doctor has offered some suggestions that might help. Circle what you would suggest and then offer a reason for his frequent dehydration.
   a. Drink (large or small) volumes of water
   b. Drink (cool or warm) liquids
   c. Drink fluids with (high or low) solute concentration
   d. (Mild or intense) exercise is best

Answers to Active Learning Questions can be found online at evolve.elsevier.com.

Nutrition and Metabolism

OUTLINE

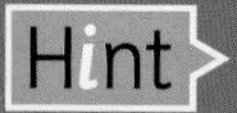

Hint *Scan this outline before you begin to read the chapter, as a preview of how the concepts are organized.*

OBJECTIVES

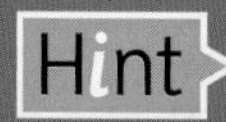

Hint *Before reading the chapter, review these goals for your learning.*

After you have completed this chapter, you should be able to:

1. Describe the role of the liver in metabolism.
2. Describe the metabolic roles of carbohydrates, fats, proteins, vitamins, and minerals.
3. Describe mechanisms that regulate food intake, and define basal metabolic rate, as well as list some factors that affect it.
4. List and describe three disorders associated with eating or metabolism.
5. Discuss the physiological mechanisms that regulate body temperature.

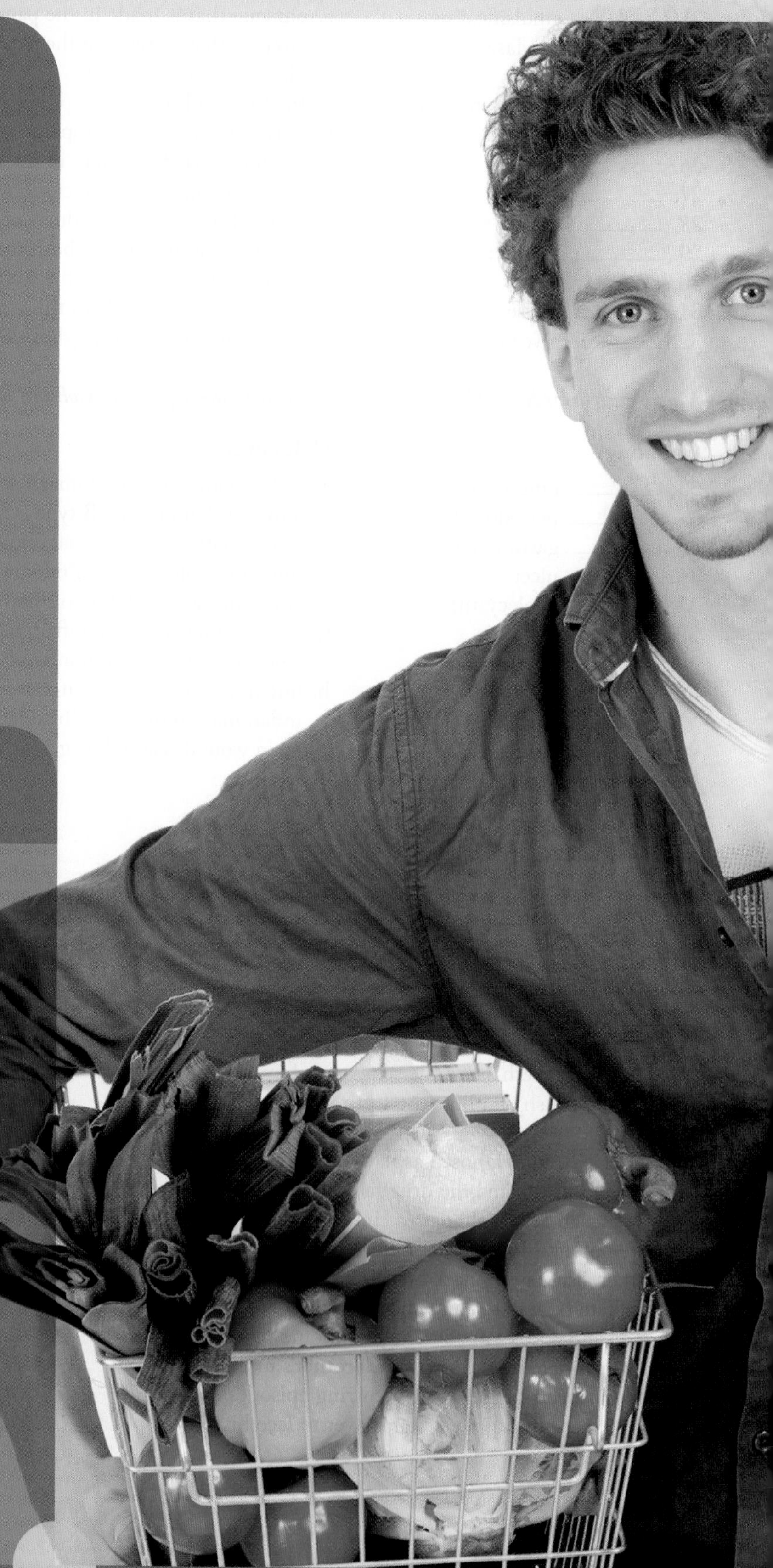

CHAPTER 19

Nutrition and metabolism are words that are often used together—but what do they mean? **Nutrition** is a term that refers to the foods that we eat and the nutrients they contain. Proper nutrition requires a balance of the three basic types of energy-yielding nutrients—*carbohydrates, fats,* and *proteins*—plus essential *vitamins* and *minerals.* Malnutrition is a deficiency or imbalance in the consumption of nutrients.

To promote good health, the United States, Canada, and other governments and organizations provide individualized online food guides that help a person determine the proper amounts and balance of nutrients (**Figure 19-1**).

A good phrase to remember in connection with the word **metabolism** is "use of nutrients" because basically this is what metabolism is—the use the body makes of nutrients after they have been digested, absorbed, and circulated to cells. It uses them in two ways: as an energy source and as building blocks for making complex chemical compounds.

Before they can be used in these two ways, nutrients have to be *assimilated.* **Assimilation** occurs when nutrient molecules enter cells and undergo many chemical changes there. All the chemical reactions that release energy from nutrient molecules make up the process of **catabolism,** a vital process because it is the only way that the body has of supplying itself with energy for doing any work. Catabolism breaks nutrient molecules down into smaller molecules and releases energy in the process. The many chemical reactions that build these smaller nutrient molecules into more complex chemical compounds constitute the process of **anabolism.**

Catabolism and anabolism together make up the process of metabolism. Metabolism is the ongoing remodeling and maintenance—breaking and building—that characterizes all life.

This chapter explores many of the basic ideas about why certain nutrients are necessary for survival, how they are used by the body, and what can go wrong in metabolic and eating disorders.

LANGUAGE OF SCIENCE

Hint Before reading the chapter, say each of these terms out loud. This will help you to avoid stumbling over them as you read.

acetyl CoA
(ah-SEET-il koh ay)
[*acetyl* **vinegar,** *CoA* **coenzyme A**]

aerobic
(ayr-OH-bik)
[*aer-* **air,** *-bi-* **life,** *-ic* **relating to**]

agricultural scientist
(ag-rih-KUL-cher-al SYE-en-tist)
[*agr-* **field,** *-cultur-* **tilling,** *-al* **relating to,** *scien-* **knowledge,** *-ist* **agent**]

amino acid
(ah-MEE-no AS-id)
[*amino* NH_2, *acid* **sour**]

anabolism
(ah-NAB-oh-liz-em)
[*ana-* **up,** *-bol-* **throw (build),** *-ism* **action**]

anaerobic
(an-ayr-OH-bik)
[*an-* **without,** *-aer-* **air,** *-bi-* **life,** *-ic* **relating to**]

antioxidant
(an-tee-OK-seh-dent)
[*anti-* **against,** *-oxi-* **sharp (oxygen),** *-ant* **agent**]

appetite center
(AP-ah-tyte SEN-ter)
[*a(d)-* **toward,** *-pet-* **seek out,** *-ite* **relating to**]

assimilation
(ah-sim-ih-LAY-shun)
[*assimila-* **make alike,** *-tion* **process**]

basal metabolic rate (BMR)
(BAY-sal met-ah-BAHL-ik rayt [bee em ar])
[*bas-* **basis,** *-al* **relating to,** *meta-* **over,** *-bol-* **throw,** *-ic* **relating to**]

Continued on p. 546

Metabolic Function of the Liver

As we discussed in Chapter 18, the liver plays an important role in mechanical digestion of lipids because it secretes *bile.* Recall that bile breaks large fat globules into smaller droplets of fat that are more easily digested by the enzyme lipase.

In addition, the liver performs many other functions necessary for healthy survival. The liver plays a major role in the metabolism of all three main types of nutrients.

For example, the liver helps maintain a normal blood glucose concentration for a few hours after a meal by storing glucose when it is overly abundant then releasing the glucose into the blood as needed. Many complex chemical reactions, regulated by many different hormones, assist in these storage and release processes.

Liver cells also carry out the first steps of protein and fat metabolism. Liver cells also synthesize several kinds of proteins. These proteins, when released into blood, are called the *blood proteins,* or **plasma proteins.** Prothrombin and fibrinogen, plasma proteins formed by liver cells, play essential parts in blood clotting (see p. 365). Another plasma protein made by liver cells, albumin, helps maintain normal blood volume.

The liver can also detoxify the body of toxic substances such as bacterial products and certain drugs. Once detoxified, these chemicals are often easier to excrete from the body. The liver can also store several substances, notably iron and vitamins A and D.

The liver is assisted by an interesting and unique structural pattern of the blood vessels that supply it. Recall from Chapter 15 that the hepatic portal vein delivers blood directly from the gastrointestinal tract to the liver (see **Figure 15-11**). This arrangement allows blood that has just absorbed nutrients and other substances to be processed by the liver before being distributed throughout the body. Thus excess nutrients and vitamins can be stored and toxins can be removed efficiently from blood arriving from the GI tract—before the blood reaches other areas of the body.

19

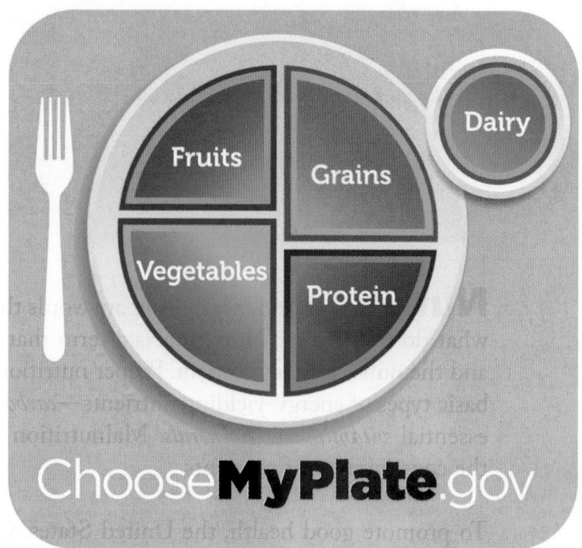

FIGURE 19-1 Food guide. Canada, the United States, and many other countries provide online, individualized food guides that help people determine proper amounts and a healthy balance of nutrients. The website *ChooseMyPlate.gov* is hosted by the U.S. Department of Agriculture (USDA).

QUICK CHECK

1. What are the three basic nutrient types?
2. What is *metabolism*?
3. Describe at least three functions of the *liver.*

Macronutrients

Dietary Sources of Nutrients

Put simply, the components of foods that are digested and absorbed by the body are called *nutrients.* The "big three" nutrients in our diets are carbohydrates ("carbs"), lipids (fats and oils), and proteins. Because they form the bulk of our diet, these three nutrients are sometimes called **macronutrients.** Vitamins and minerals, by contrast, are called **micronutrients** because they are needed in only very small quantities in our diet.

TABLE 19-1 Major Macronutrients

NUTRIENT	DIETARY SOURCES	FUNCTIONS
Carbohydrate		
Monosaccharide (glucose, galactose, fructose)	Fruit, honey, corn syrup	Used as a source of energy; used to build other carbohydrates
Disaccharide (sucrose, lactose, maltose)	Sugar, fruit, dairy, malted grain products	Source of monosaccharides (for energy)
Polysaccharide (starch, dietary fiber)	Grains, vegetables, nuts, fruits	Source of monosaccharides (for energy); dietary fiber promotes efficient digestive function
Fat (Lipid)		
Triglyceride (absorbed as fatty acids, glycerol)	Meat, vegetable oils	Provide energy Structural padding
Cholesterol	Meat, eggs, dairy	Transports lipids; stabilizes cell membranes; is basis of steroid hormones
Protein		
Many types (absorbed as individual amino acids)	Meat, eggs, dairy, vegetables, nuts	Form structures of the body (fibers such as keratin and collagen) Facilitate chemical reactions (enzymes) Send signals and regulate functions (neurotransmitters, nonsteroid hormones) Produce movement (myofilaments) May be used for energy

Table 19-1 summarizes the three macronutrients, their principal sources in our food, and their main functions in our body. The following sections explore some of these functions more deeply.

Carbohydrate Metabolism

Carbohydrates are the preferred energy nutrient of the body. The larger carbohydrate molecules are composed of smaller "building blocks," primarily *glucose* (see Chapter 2). Human cells catabolize (break down) glucose rather than other substances as long as enough glucose enters them to supply their energy needs.

Glucose Catabolism

Glucose catabolism involves three series of chemical reactions called "metabolic pathways," that occur in a precise sequence:

1. Glycolysis
2. Citric acid cycle (or Krebs cycle)
3. Electron transport system (ETS)

Glycolysis, the first step of glucose catabolism, occurs in the cytoplasm of each cell of the body. As **Figure 19-2** shows, glycolysis breaks down glucose (a six-carbon molecule) into two **pyruvic acids** (three carbon molecules). Glycolysis releases a small amount of energy—enough to generate two adenosine triphosphate (ATP) molecules—but requires no oxygen to do so. We thus say that it is an **anaerobic** process.

Each pyruvic acid molecule may then move into a *mitochondrion*—one of the cell's tiny "battery chargers" that transfers much more of the nutrient energy to ATP (see **Figure 3-2** on p. 45). After pyruvic acid is broken down into the 2-carbon acetyl molecule, a *coenzyme A (CoA)* then escorts it into the **citric acid cycle** or **Krebs cycle** as a molecule called **acetyl CoA.** The citric acid cycle releases high-energy electrons as it breaks down the acetyl CoA (two carbons) into two carbon dioxides (each having only one carbon) using enzymes located inside the mitochondrion.

The chemical reactions of glycolysis and the citric acid cycle release energy stored in the glucose molecule. More than half of the released energy is in the form of high-energy electrons. The **electron transport system,** embedded in the inner folds of the mitochondrion, transfers the energy from these electrons to molecules of *ATP.*

Up to 36 molecules of ATP can be generated in the mitochondrion for every original glucose molecule that enters this metabolic pathway. The rest of the energy originally stored in the glucose molecule is released as heat, which contributes to a person's body temperature.

The metabolic pathway inside the mitochondrion is, in contrast to glycolysis, an oxygen-using or **aerobic** process. A cell cannot operate the citric acid cycle or electron transport system—where most of the energy of glucose is released—without oxygen.

To better understand these concepts, use the Active Concept Map *Metabolism of Glucose to Generate ATP* at *evolve.elsevier.com.*

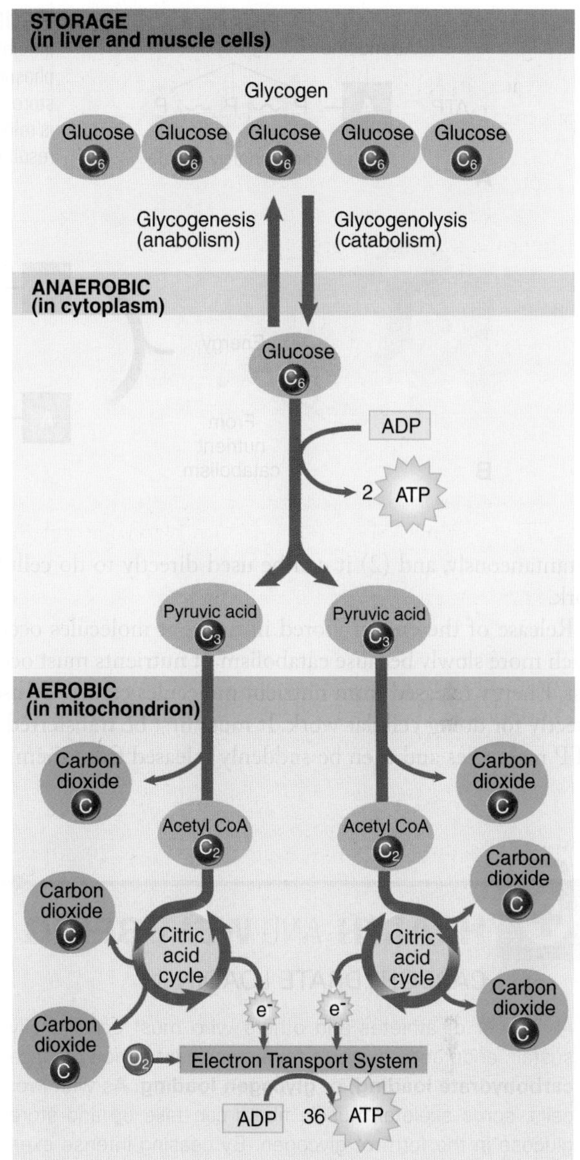

FIGURE 19-2 Metabolism of glucose. Glucose can be stored as subunits of glycogen in liver and muscle cells until needed to make adenosine triphosphate (ATP). After glycogen is split apart, each individual glucose molecule undergoes glycolysis in the cytoplasm. Glycolysis splits one molecule of glucose (six carbon atoms) into two molecules of pyruvic acid (three carbon atoms each) and produces enough energy to generate two ATPs. Each pyruvic acid is converted to the two-carbon acetyl molecule, which is escorted by *coenzyme A (CoA)* into the citric acid cycle in the mitochondrion as *acetyl CoA*. The citric acid cycle breaks apart each pyruvic acid molecule into three carbon dioxide molecules (one carbon atom each) and many high-energy electrons. The electron transport system (also in the mitochondrion) uses energy from these electrons to generate up to 36 ATPs in the presence of oxygen (O_2). *ADP,* Adenosine diphosphate.

ATP

ATP serves as the direct source of energy for doing cellular work in all kinds of living organisms from one-cell plants to trillion-cell animals, including humans. Among biological compounds, therefore, ATP ranks as one of the most important.

The energy transferred to ATP molecules differs in two ways from the energy stored in nutrient molecules: (1) the energy in ATP molecules is not stored but is released almost

19

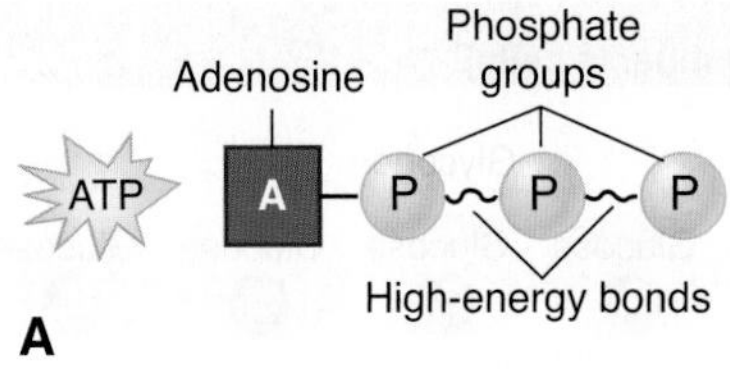

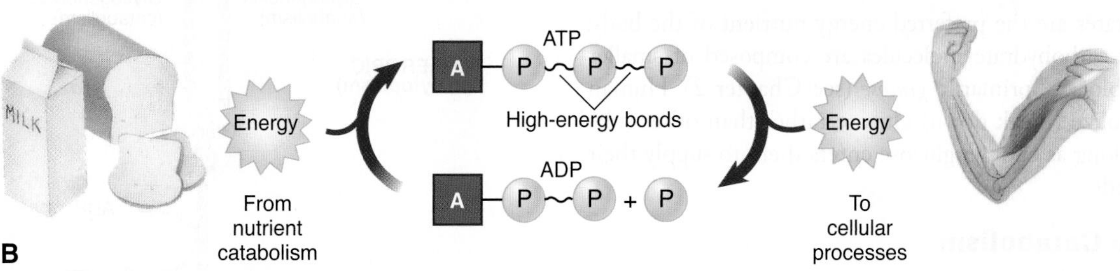

FIGURE 19-3 Adenosine triphosphate (ATP). A, The structure of ATP. A single adenosine group *(A)* has three attached phosphate groups *(P)*. The high-energy bonds between the phosphate groups can release chemical energy to do cellular work. **B,** ATP energy cycle. ATP stores energy in its last high-energy phosphate bond. When that bond is later broken, energy is released to do cellular work. The adenosine diphosphate (ADP) and phosphate groups that result can be resynthesized into ATP, capturing additional energy from nutrient catabolism.

instantaneously, and (2) it can be used directly to do cellular work.

Release of the energy stored in nutrient molecules occurs much more slowly because catabolism of nutrients must occur first. Energy released from nutrient molecules cannot be used directly for doing cellular work. It must first be transferred to ATP molecules and then be suddenly released from them.

HEALTH AND WELL-BEING

CARBOHYDRATE LOADING

A number of athletes and others who must occasionally sustain endurance exercise for a significant period practice **carbohydrate loading,** or **glycogen loading.** As with liver cells, some skeletal muscle fibers can take up and store glucose in the form of glycogen. By ceasing intense exercise and switching to a diet high in carbohydrates 2 or 3 days before an endurance event, an athlete can cause the skeletal muscles to store almost twice as much glycogen as usual. This allows the muscles to sustain aerobic exercise for up to 50% longer than usual. The concept of carbohydrate loading has been used to promote the use of "energy bar" sport snacks and some sports or "energy" drinks.

19

As **Figure 19-3** shows, ATP is made up of an adenosine group and three phosphate groups. The capacity of ATP to release large amounts of energy is found in the high-energy bonds that hold the phosphate groups (P) together, represented as curvy lines. When a phosphate group breaks off of the molecule, an *adenosine diphosphate (ADP)* molecule and free phosphate group result. Energy that had been holding the phosphate bond together is freed to do cellular work—muscle fiber contractions, for example.

As you can see in **Figure 19-3**, the ADP and phosphate are reunited by the energy produced by carbohydrate catabolism, making ATP a reusable energy-storage molecule. Only enough ATP for immediate cellular requirements is made at any one time. New ATP is constantly being made to meet changing cellular demands. Glucose that is not needed immediately for ATP production is built up (by anabolic processes) into larger molecules that are stored for later use.

Anabolism

Glucose anabolism is called **glycogenesis.** Carried on chiefly by liver and muscle cells, glycogenesis consists of a series of reactions that join glucose molecules together, like many beads in a necklace, to form **glycogen,** a compound sometimes called *animal starch.*

Later, when the glucose stored as glycogen is needed to make ATP, a process called **glycogenolysis** breaks down glycogen in the liver or muscle cells to release individual glucose molecules. Glycogenolysis is an example of catabolism.

Regulation of Carbohydrate Metabolism

Something worth noting is that the amount of glucose and other nutrients in the blood normally does not change very much, not even when we go without food for many hours, when we exercise and use a lot of nutrients for energy, or when we sleep and use few nutrients for energy. The amount of glucose in our blood, for example, usually stays at about 80 to 110 mg in 100 mL of blood when we are "fasting" between meals.

Several hormones help regulate carbohydrate metabolism to keep blood glucose at a normal level. **Insulin** is one of the most

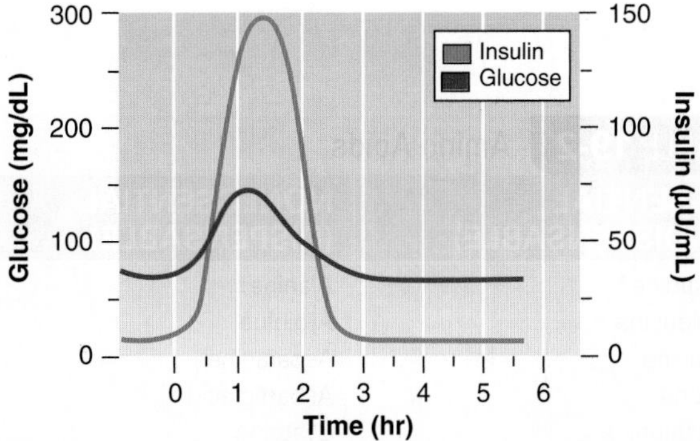

FIGURE 19-4 Role of insulin. Insulin operates in a negative feedback loop that prevents blood glucose concentration from increasing too far above the normal range. After a meal, intestinal absorption rises and hepatic portal blood glucose concentration increases—as shown by the *blue line* in the graph. Insulin secretion by the pancreatic islets increases in response *(orange line)*. Insulin promotes uptake of glucose (out of the blood) by liver cells. As blood glucose decreases to its setpoint level, feedback to the pancreatic islets reduces insulin secretion—thus maintaining normal blood glucose concentration. One expects to see a sharp rise in blood insulin levels shortly after a meal high in carbohydrates.

important of these. Although the exact details of its mechanism of action are still being worked out, insulin is known to accelerate glucose transport through cell membranes. As insulin secretion increases, more glucose leaves the blood and enters the cells—particularly the liver cells (see **Figure 12-4** on p. 324 and **Figure 19-4**).

Too little insulin secretion or resistance to insulin effects, such as occurs in people with various forms of *diabetes mellitus (DM),* produces the opposite effects. Less glucose leaves the blood and enters cells. More glucose therefore remains in the blood, and less glucose is metabolized by cells. In other words, high blood glucose (hyperglycemia) and a low rate of glucose metabolism characterize insulin deficiency or resistance.

Insulin is the only hormone that significantly lowers the blood glucose level. Several other hormones, on the other hand, can increase it. *Growth hormone* secreted by the anterior pituitary gland, *cortisone* secreted by the adrenal cortex, *epinephrine* secreted by the adrenal medulla, and *glucagon* secreted by the pancreatic islets are four of the most important hormones that increase blood glucose. More information about these hormones can be found in Chapter 12.

HEALTH AND WELL-BEING

LOW-CARB DIETS

Low-carbohydrate diets have become increasingly popular among those attempting weight loss. When carbohydrate catabolism equals energy needs, fats are not taken out of storage and catabolized. Low-carbohydrate diets are based on the rationale that when the body is not supplied with excess amounts of carbohydrates to meet its energy needs, it will not convert the surplus carbohydrates to fat and store it. Instead, the body relies on fat metabolism to supply energy needs between meals. This eventually reduces overall triglyceride stores in the body, and as a result, the person loses their excess weight. In addition, some research studies on these diets have demonstrated an improved plasma lipid profile.

However, there is still much controversy regarding the many types of low-carbohydrate diets and which are most safe and effective for those struggling with obesity, diabetes, and other disorders. Ultimately, the most successful weight-reducing diet may be the one that each person can stick to for the longest duration and produces the best long-term health effects.

To learn more about the citric acid cycle, go to AnimationDirect online at *evolve.elsevier.com.*

RESEARCH, ISSUES, AND TRENDS

MEASURING ENERGY

Physiologists studying metabolism must be able to express a quantity of energy in mathematical terms. The unit of energy measurement most often used is the calorie (cal). A **calorie** is the amount of energy needed to raise the temperature of 1 g of water 1° C. Because physiologists often deal with very large amounts of energy, the larger unit, **kilocalorie** (kcal) or **Calorie** (notice the uppercase C), is used. There are 1000 cal in 1 kcal or Calorie. Nutritionists in the United States prefer to use *Calorie* when they express the amount of energy stored in a nutrient.

Most physiologists in the United States—and most nutritionists outside the United States—prefer to use the metric unit **joule (J)** or *kilojoule (kJ)* instead of calorie-based units. A simple way to convert kilocalories to kilojoules is kcal × 4.2 kJ.

To learn more about measuring energy, including examples of the energy content of macronutrients and the energy cost of common activities, review the article *Measuring Energy* at Connect It! at *evolve.elsevier.com.*

Fat Metabolism

Lipids, like carbohydrates, are primarily energy nutrients. As cells begin to run low on adequate amounts of glucose to catabolize a few hours after a meal, they immediately shift to the catabolism of triglycerides—fat—for energy.

Fats are first broken down into fatty acids and glycerol in adipose tissue and released into the blood stream. In cells, fatty acids are broken down to form acetyl CoA, which then proceeds through the citric acid cycle (see **Figure 19-2**). Glycerol is converted to a compound that can enter the glycolysis pathway in a process is known as **gluconeogenesis.**

Gluconeogenesis, discussed in Chapter 12, is a process that is performed mainly by liver cells. By converting the components of fat into molecules that enter the pathway for glucose catabolism, this process allows energy to be transferred from a nutrient to ATP—which can be used directly for work in the cells of the body.

Fat catabolism happens normally when a person goes without carbohydrates for a few hours. It happens abnormally in individuals with untreated DM. Because of an insulin deficiency, too little glucose enters the cells of a diabetic person to supply all energy needs. The result is that the cells catabolize fats to make up the difference (**Figure 19-5**).

In all persons, fats not needed for catabolism are instead anabolized (built up) to form triglycerides and then are stored in adipose tissue, which contributes to weight gain.

Protein Metabolism

In a healthy person, proteins are catabolized to release energy only to a very small extent. When fat reserves are low, as they are in the starvation that accompanies certain eating disorders such as anorexia nervosa, the body can start to use more of its protein molecules as an energy source. Excess proteins in the diet can also be used for energy.

Specifically, the **amino acids** that make up proteins are each broken apart. The nitrogen group *amine* is removed by the liver and converted to the waste product **urea.** The rest of the molecule is converted by gluconeogenesis to a form that can enter the citric acid cycle—and thus release energy to "charge up" ATP (see **Figure 19-5**).

After a shift to reliance on protein catabolism as a major energy source occurs, death may quickly follow because vital proteins in the muscles and nerves are catabolized.

A more common situation in normal bodies is protein anabolism, the process by which the body builds amino acids into complex protein compounds (for example, enzymes and proteins that form the structure of the cell). Proteins are assembled from a pool of at least 20 different kinds of amino acids. If any one type of amino acid is deficient, vital proteins cannot be synthesized—a serious health threat.

One way your body maintains a constant supply of amino acids is by making them from other compounds already present in the body. Only about half of the required 20 types of amino acids can be made by the body, however. The remaining types of amino acids must be supplied in the diet. *Essential amino acids* are those that must be in the diet. *Nonessential amino acids* can be missing from the diet because they can be made by the body (**Table 19-2**).

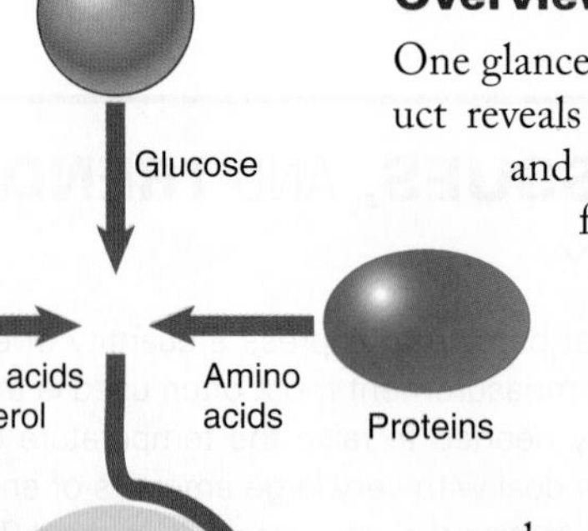

FIGURE 19-5 Catabolism of nutrients. Fats, carbohydrates, and proteins can be converted to products that enter the citric acid cycle to yield energy that is transferred to adenosine triphosphate (ATP).

TABLE **19-2** Amino Acids

ESSENTIAL (INDISPENSABLE)	NONESSENTIAL (DISPENSABLE)
Histidine*	Alanine
Isoleucine	Arginine
Leucine	Asparagine
Lysine	Aspartic acid
Methionine	Cysteine
Phenylalanine	Glutamic acid
Threonine	Glutamine
Tryptophan	Glycine
Valine	Proline
	Serine
	Tyrosine†

*Essential in infants and, perhaps, adult males.
†Can be synthesized from phenylalanine; therefore, is nonessential as long as phenylalanine is in the diet.

QUICK CHECK

1. Name three *macronutrients* and two *micronutrients.*
2. Identify the series of chemical reactions that make up the process of *glucose metabolism.*
3. How is energy transferred from glucose to ATP?
4. How are proteins used once they are absorbed into the body?
5. What are *essential amino acids*?

Micronutrients

Vitamins

Overview of Vitamins

One glance at the label of any packaged food product reveals the importance we place on vitamins and minerals. We know that carbohydrates, fats, and proteins are used by our bodies to build important molecules and to provide energy. So why do we need vitamins and minerals?

First, let's discuss the importance of vitamins. **Vitamins** are organic molecules needed in small quantities for normal metabolism throughout the body.

Most vitamin molecules attach to enzymes or *coenzymes* (molecules that assist enzymes) and help them work properly. Many enzymes are totally useless without the appropriate vitamins to activate them.

Some vitamins play other important roles in the body. For example, a form of vitamin A plays an important role in detecting light in the sensory cells of the retina. Vitamin D can be converted to a hormone that helps regulate calcium homeostasis in the body, and vitamin E acts as an **antioxidant** that prevents highly reactive oxygen

19

molecules called **free radicals** from damaging DNA and molecules in cell membranes.

Most vitamins cannot be made by the body, so we must eat them in our food. The body can store fat-soluble vitamins—A, D, E, and K—in the liver for later use. Because the body cannot store water-soluble vitamins such as B vitamins and vitamin C, they must be continually supplied in the diet. Vitamin deficiencies can lead to severe metabolic problems. **Table 19-3** lists some of the more well-known vitamins, their sources, functions, and symptoms of deficiency.

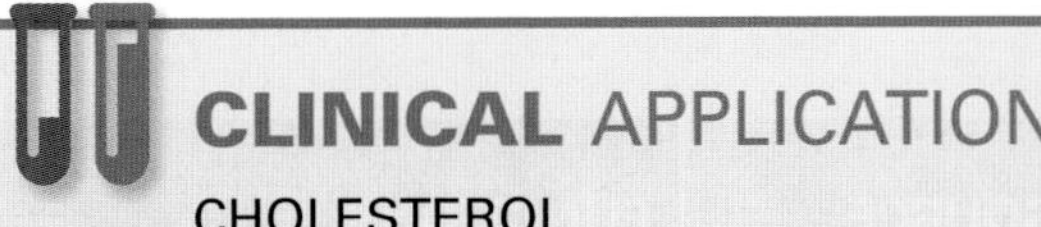

CLINICAL APPLICATION

CHOLESTEROL

Cholesterol is a type of lipid that has many uses in the body (see Chapter 2). The body derives steroid hormones from cholesterol (see Chapter 12) and uses cholesterol to stabilize the phospholipid bilayer that forms the plasma membrane and membranous organelles of all cells (see Chapter 3).

So why does such a useful substance have such a bad reputation? The reason lies in the fact that an *excess* of cholesterol in the blood, a condition called **hypercholesterolemia,** increases the risk of developing atherosclerosis (see *arrow* in figure). You may recall from Chapter 14 that *atherosclerosis* develops into a type of arteriosclerosis, or "hardening of the arteries," that can lead to heart disease, stroke, and other problems.

Hypercholesterolemia occurs most often in people with a genetic predisposition but is certainly also affected by other factors such as diet and exercise. People with hypercholesterolemia are encouraged to switch to diets low in cholesterol and saturated fats and to participate in aerobic exercise, both of which tend to lower blood cholesterol levels. Drugs such as **statins** are often used to control blood cholesterol when exercising and diet are not sufficient.

Chapter 2 discusses different types of cholesterol and their roles in health and disease.

Atherosclerotic plaque

Vitamin Imbalances

Vitamin deficiency, or **avitaminosis,** can lead to severe metabolic problems. For example, *avitaminosis C* (vitamin C deficiency) can lead to **scurvy** (**Figure 19-6**). Scurvy results from the inability of the body to manufacture and maintain collagen fibers. As you may have gathered from your studies thus far, collagen fibers are critical in many of the connective tissues that hold the body together. In scurvy, the body falls apart in the same way that a neglected house eventually falls apart.

More details about scurvy and other types of avitaminosis are given in Appendix A at *evolve.elsevier.com.*

Some forms of **hypervitaminosis**—or vitamin excess—can be just as serious as a deficiency of vitamins. For example, chronic *hypervitaminosis A* can occur if very large amounts of vitamin A are consumed daily over a period of 3 months or more. This condition first manifests itself with dry skin, hair loss, anorexia (appetite loss), and vomiting, but may progress to severe headaches and mental disturbances, liver enlargement, and occasionally cirrhosis. Acute hypervitaminosis A, characterized by vomiting, abdominal pain, and headache, can occur if a massive overdose is ingested.

Excesses of the fat-soluble vitamins (A, D, E, and K) are generally more serious than excesses of the water-soluble

19

SCIENCE APPLICATIONS

FOOD SCIENCE

George Washington Carver (1864–1943)

At the dawn of the twentieth century, one figure loomed large in the world of **food science**—George Washington Carver. Born a slave on a Missouri plantation during the Civil War, Carver overcame great obstacles to become one of the most admired American scientists of his era. Although talented in music and art, it was his knack for agriculture that led him to a long and successful career as a professor, researcher, and inventor in the agriculture department of Alabama's Tuskegee Institute.

At Tuskegee, his work resulted in the creation of 325 products made from peanuts, nearly 200 products from yams (sweet potatoes), and hundreds more from other plants native to the southern United States. Development of these new products helped poor farmers survive by allowing them to make money from a variety of crops that thrived on their land.

Today, breakthroughs continue to be made in the world of *agriculture* and food science. *Farmers* and *ranchers* work closely with **agricultural scientists** and *technicians* to improve food crops and to improve methods of raising livestock. As did Carver, they strive to work in ways that benefit the land and people. Of course, *nutritionists, dietitians, chefs,* and *food preparers* all play a role in getting these crops to our table in a healthy and appetizing way.

Food scientists and other *industrial scientists* work to develop technologies and methods for preparing, preserving, storing, and packaging foods.

TABLE 19-3 Major Vitamins

VITAMIN	DIETARY SOURCE	FUNCTIONS	CONSEQUENCES OF DEFICIENCY
Vitamin A	Green and yellow vegetables, dairy products, and liver	Maintains epithelial tissue and produces visual pigments	Night blindness and flaking skin
B-complex vitamins			
B_1 (thiamine)	Grains, meat, and legumes	Helps enzymes in the citric acid cycle	Nerve problems (beriberi), heart muscle weakness, and edema
B_2 (riboflavin)	Green vegetables, organ meat, eggs, and dairy products	Aids enzymes in the citric acid cycle	Inflammation of skin and eyes
B_3 (niacin)	Meat and grains	Helps enzymes in the citric acid cycle	Pellagra (scaly dermatitis and mental disturbances) and nervous disorders
B_5 (pantothenic acid)	Organ meat, eggs, and liver	Aids enzymes that connect fat and carbohydrate metabolism	Loss of coordination; decreased peristalsis (rare)
B_6 (pyridoxine)	Vegetables, meat, and grains	Helps enzymes that catabolize amino acids	Convulsions, irritability, and anemia
B_9 (folic acid)	Vegetables	Aids enzymes in amino acid catabolism and blood	Digestive disorders and anemia; neural defects in embryo or fetus
B_{12} (cyanocobalamin)	Meat and dairy products	Involved in blood production and other processes	Pernicious anemia
Biotin (vitamin H)	Vegetables, meat, and eggs	Helps enzymes in amino acid catabolism and fat and glycogen synthesis	Mental and muscle problems (rare)
Vitamin C (ascorbic acid)	Fruits and green vegetables	Helps in manufacture of collagen fibers	Scurvy and degeneration of skin, bone, and blood vessels
Vitamin D (calciferol)	Dairy products and fish liver oil	Aids in calcium absorption	Rickets and skeletal deformity
Vitamin E (tocopherol)	Green vegetables and seeds	Protects cell membranes from oxidation damage	Muscle and reproductive disorders (rare)
Vitamin K (group)	Mostly from intestinal bacteria Also spinach, other vegetables, and meat/dairy products	Needed to produce some clotting factors	Clotting disorders (most often in infants)

FIGURE 19-6 Scurvy. In scurvy, lack of vitamin C impairs the normal maintenance of collagen-containing connective tissues, causing bleeding and ulceration of the skin, gums, and other tissues, as these lesions on the skin show.

vitamins (B complex and C) because fat-soluble vitamins are stored, whereas excess water-soluble vitamins can be excreted.

Minerals

Minerals are just as essential for health as vitamins. Minerals are inorganic elements or salts found naturally in the earth and in many foods. As with vitamins, mineral ions can bind to enzymes and help them work effectively.

Minerals also function in a variety of other vital chemical reactions. For example, sodium, calcium, and other minerals are required for nerve conduction and for contraction in muscle fibers. Without these minerals, the brain, heart, and respiratory muscles would cease to function.

Information about some of the more important minerals is summarized in **Table 19-4**.

Like vitamins, minerals are beneficial only when taken in the proper amounts. Many of the minerals listed in **Table 19-4** are required only in trace amounts. Any intake of such minerals beyond the recommended trace amount may become toxic—perhaps even life threatening.

TABLE 19-4 Major Minerals

MINERAL	DIETARY SOURCE	FUNCTIONS	SYMPTOMS OF DEFICIENCY
Calcium (Ca)	Dairy products, legumes, and vegetables	Helps blood clotting, bone formation, and nerve and muscle function	Bone degeneration and nerve and muscle malfunction
Chlorine (Cl)	Salty foods	Aids in stomach acid production and acid-base balance	Acid-base imbalance
Cobalt (Co)	Meat	Helps vitamin B_{12} in blood cell production	Pernicious anemia
Copper (Cu)	Seafood, organ meats, and legumes	Involved in extracting energy from the citric acid cycle and in blood production	Fatigue and anemia
Iodine (I)	Seafood and iodized salt	Aids in thyroid hormone synthesis	Goiter (thyroid enlargement) and decrease of metabolic rate
Iron (Fe)	Meat, eggs, vegetables, and legumes	Involved in extracting energy from the citric acid cycle and in blood production	Fatigue and anemia
Magnesium (Mg)	Vegetables and grains	Helps many enzymes	Nerve disorders, blood vessel dilation, and heart rhythm problems
Manganese (Mn)	Vegetables, legumes, and grains	Helps many enzymes	Muscle and nerve disorders
Phosphorus (P)	Dairy products and meat	Aids in bone formation and is used to make ATP, DNA, RNA, and phospholipids	Bone degeneration and metabolic problems
Potassium (K)	Seafood, milk, fruit, and meat	Helps muscle and nerve function	Muscle weakness, heart problems, and nerve problems
Sodium (Na)	Salty foods	Aids in muscle and nerve function and fluid balance	Weakness and digestive upset
Zinc (Zn)	Many foods	Helps many enzymes	Metabolic problems

Foods that go beyond simply providing the nutrients needed for health and wellness because they have specific characteristics that prevent disease are often called functional foods. To learn more about this concept, review the article *Functional Foods* at Connect It! at *evolve.elsevier.com.*

Regulating Food Intake

Mechanisms for regulating food intake are still not clearly understood. That the *hypothalamus* in the diencephalon of the brain plays a part in these mechanisms, however, seems certain. There appears to be both an **appetite center** that promotes the feeling of hunger and a **satiety center** that promotes the feeling that we are satisfied or "full" in the hypothalamus. The balance of activity between these two centers appears to be the central mechanism that regulates food intake.

There are many factors that influence these hypothalamic centers and therefore influence the regulation of food intake. Among the many factors identified in affecting appetite are hormones, neurotransmitters, emotions, environmental cues, food sensations, habits, and more. Some examples of factors that affect the appetite-regulating centers of the hypothalamus are listed in **Table 19-5**. It is not important to memorize all these factors, but reading through them will help you understand the complexity of our body's regulation of food intake.

Metabolic Rates

The **basal metabolic rate (BMR)** is the rate at which nutrients are catabolized under basal conditions (that is, when the individual is resting but awake, is not digesting food, and is not adjusting to a cold external temperature). Or, stated differently, the BMR is the energy (measured in calories) that must be produced per hour by catabolic reactions just to keep the body alive, awake, and comfortably warm.

To provide energy for muscular work, as well as digestion and absorption of nutrients, an additional amount of energy must be produced by catabolism of nutrients. The amount of additional energy from nutrients needed depends mainly on how much work the individual does. The more active he or she is, the more nutrients the body must catabolize and the higher the total metabolic rate will be. The **total metabolic rate (TMR)** is the total amount of energy used by the body per day (**Figure 19-7**).

When the number of calories in your food intake equals your TMR, your weight remains constant (except for possible short-term variations resulting from water retention or water loss). When your food intake provides more calories than your TMR, you gain weight; when your food intake provides fewer calories than your TMR, you lose weight. Nature does not forget to count calories. A reducing diet must make use of this knowledge. The diet must include fewer calories than the TMR of the individual eating the diet, or no weight loss will be achieved.

19

TABLE 19-5 Factors That Influence Appetite*

FACTORS THAT STIMULATE APPETITE	FACTORS THAT INHIBIT APPETITE	SOURCE
Endogenous opioid peptides (EOP) Gamma-aminobutyric acid (GABA) Neuropeptide Y (NPY) Norepinephrine (NE) Orexins	Alpha-melanocyte–stimulating hormone (α-MSH) Cocaine- and amphetamine-regulated transcript (CART) Corticotropin-releasing hormone (CRH)	Hypothalamus
Emotions Environmental stimuli Food sensations (e.g., taste, smell, texture) Internal stimuli (e.g., blood temperature, glucose) Lifestyle choices and habits	Emotions Environmental stimuli Food sensations (e.g., taste, smell) Internal stimuli (e.g., blood temperature, glucose) Lifestyle choices and habits	Nervous system (outside hypothalamus)
Cortisol		Adrenal cortex
Ghrelin (GHRL)	Cholecystokinin (CCK) Glucagon-like peptide-1 (GLP-1)	Gastrointestinal (GI) tract
	Leptin Interleukin 18 (IL-18)	Adipose tissue
	Glucose	Liver
	Insulin Pancreatic polypeptide (PP)	Pancreas

*Factors that affect appetite-regulating centers in the hypothalamus.

Rapid, significant weight loss carries a risk of health risks. One such risk is outlined in the article *Gallstones and Weight Loss* at Connect It! at *evolve.elsevier.com.*

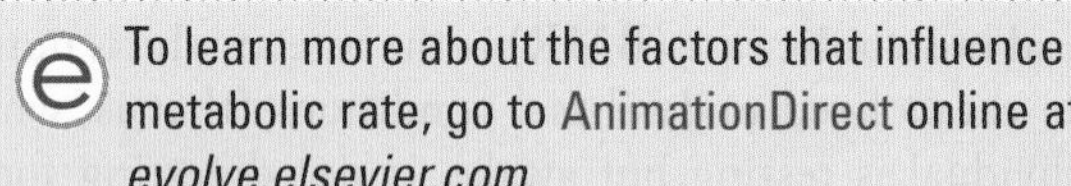

To learn more about the factors that influence metabolic rate, go to AnimationDirect online at *evolve.elsevier.com.*

QUICK CHECK

1. What is the overall job of *vitamins* in the body?
2. What is another name for the rate at which nutrients are catabolized under resting conditions?
3. How does the number of calories consumed relate to a person's body weight?
4. What is the difference between the *appetite center* and the *satiety center*? Where are these centers located?

Metabolic and Eating Disorders

Metabolic Imbalances

Disorders characterized by a disruption or imbalance of normal metabolism can be caused by several different factors.

For example, *inborn errors of metabolism* are a group of genetic conditions involving a deficiency or absence of a particular enzyme. Specific enzymes are required by cells to carry out each step of every metabolic reaction. Although an abnormal genetic code may affect the production of only a single enzyme, the resulting abnormal metabolism may have widespread effects. Specific diseases resulting from inborn errors of metabolism, such as **phenylketonuria (PKU),** are discussed in Chapter 25.

A number of metabolic disorders are complications of other conditions. For example, you may recall from Chapter 12 that both hyperthyroidism and hypothyroidism have profound effects on the BMR. Diabetes mellitus affects metabolism

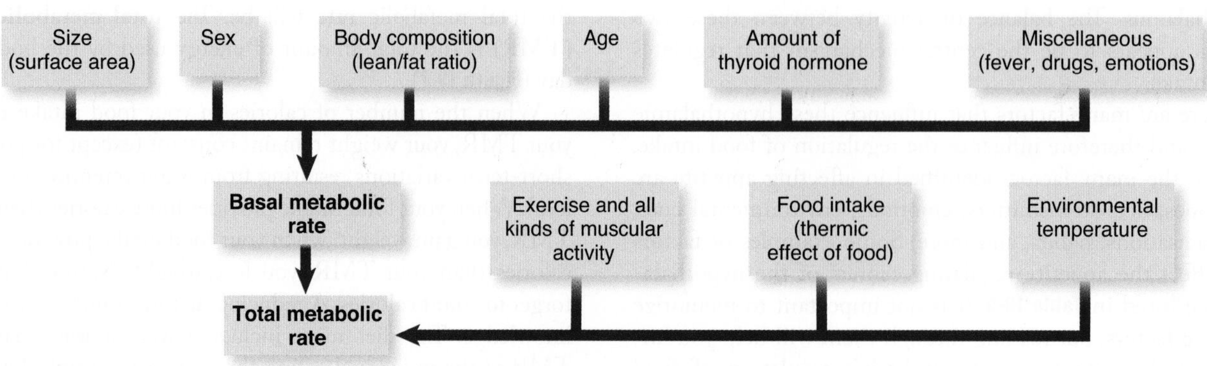

FIGURE 19-7 **Factors that determine the basal and total metabolic rates.**

throughout the body when an insulin deficiency limits the amount of glucose available for use by the cells.

Eating Disorders

Some metabolic disorders result from disruptions of normal mechanisms in the body that maintain homeostasis. For example, the body has several mechanisms that maintain a relatively constant level of glucose in the blood—glucose that is required by cells for life-sustaining catabolism. As mentioned earlier in this chapter, during *starvation* or in certain *eating disorders*, these mechanisms may become unbalanced as they attempt to maintain blood glucose homeostasis.

A few of the more well-known eating and nutrition disorders are briefly described in the following sections.

Anorexia Nervosa

A behavioral disorder characterized by chronic refusal to eat, **anorexia nervosa** often results from an abnormal fear of becoming obese. This condition is most commonly seen in teenage girls and young adult women and is often linked to emotional stress.

Treatment plans are usually directed at solving the resulting nutritional deficit, while at the same time dealing with the underlying behavioral problem.

Bulimia

Bulimia is a behavioral disorder characterized by insatiable craving for food alternating with periods of self-deprivation. The self-deprivation that follows a "food binge" is often accompanied by depression.

People with a form of this disorder called **bulimarexia** purposely induce the vomiting reflex to purge themselves of food they just ate. Excessive vomiting in this way can have a variety of consequences, including damage to the esophagus, pharynx, mouth, and teeth by stomach acid.

Obesity

Obesity is not an eating disorder itself but may be a result of chronic overeating behavior. Like anorexia nervosa and bulimia, eating disorders characterized by chronic overeating usually have an underlying emotional cause.

Obesity is defined as an abnormal increase in the proportion of fat in the body. Most of the excess fat is stored in the subcutaneous tissue and around the viscera. Obesity is a risk factor for a variety of chronic life-threatening diseases, including diabetes mellitus, many forms of cancer, and heart disease.

Protein-Calorie Malnutrition

Protein-calorie malnutrition (PCM) is an abnormal condition resulting from a deficiency of calories in general and protein in particular. PCM is likely to result from reduced intake of food but may also be caused by increased nutrient loss or increased use of nutrients by the body. **Table 19-6** summarizes a few of the wide variety of conditions that may lead to PCM.

Mild cases of PCM occur frequently during illness. As many as one in five patients admitted to the hospital is significantly malnourished. More severe cases of PCM are likely to occur in parts of the world where food, especially protein-rich food, is relatively unavailable.

There are two forms of advanced PCM: **marasmus** and **kwashiorkor** (**Figure 19-8**).

Marasmus results from an overall lack of calories and proteins, such as when sufficient quantities of food are not available. Marasmus is characterized by progressive wasting of muscle and subcutaneous tissue accompanied by fluid and electrolyte imbalances.

TABLE 19-6 Some Causes of Protein-Calorie Malnutrition

CONDITION	IMPACT ON NUTRIENTS
Conditions That Reduce Nutrient Intake	
Anorexia	Absence of appetite; reduced motivation to eat
Cachexia	Syndrome associated with cancer involves appetite loss, severe weight loss, and weakness
Dysphagia	Difficulty in swallowing; inhibition of normal eating
Gastrointestinal obstruction	Inability of food to be digested or absorbed
Nausea	Upset stomach; discomfort, which inhibits appetite
Pain	Discomfort, which discourages eating
Poverty	Inability to acquire proper nutrients
Social isolation	Absence of social cues or motivation for eating
Substance abuse	Reduction or replacement of the motivation to eat
Tooth problems	Difficulty in chewing, which discourages or prevents eating
Conditions That Increase Loss of Nutrients	
Diarrhea	Increased intestinal motility, which reduces absorption of nutrients
Glycosuria	Loss of glucose in the urine
Hemorrhage	Loss of blood and the nutrients it contains
Malabsorption	Failure to properly absorb nutrients, which causes nutrients to pass through the body unabsorbed
Conditions That Increase the Use of Nutrients by the Body	
Burns	Loss of nutrients from damaged tissues
Fever	Increased temperature and metabolic rate, which increase rate of nutrient catabolism
Infection	Increased immune activity and tissue repair, which increase the rate of nutrient use
Trauma and surgery	Increased immune activity, tissue repair, and homeostatic-compensating mechanisms, which increase the rate of nutrient use
Tumors	Increased tissue growth, which increases the rate of nutrient use

FIGURE 19-8 Protein-calorie malnutrition (PCM). A, Marasmus results from starvation. **B,** Kwashiorkor results from a diet sufficient in calories but deficient in protein. Note the abdominal bloating typical in kwashiorkor.

A

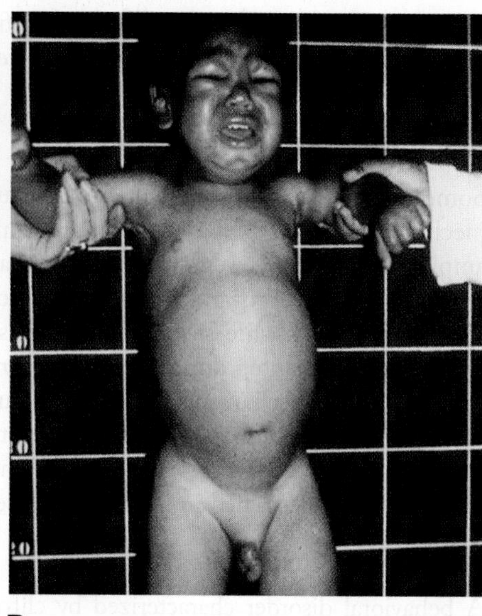
B

Kwashiorkor results from a diet that has sufficient calories, but is deficient in protein—as when a child is weaned from milk to low-protein foods. Kwashiorkor also causes wasting of tissues, but unlike marasmus, it is accompanied by pronounced ascites (abdominal bloating) and flaking dermatitis. The ascites results from a deficiency of plasma proteins, which changes the osmotic balance of the blood and thus promotes osmosis of water from the blood into the peritoneal space (see **Figure 18-24**).

Nutrition disorders, including many specific deficiency diseases, are summarized in Appendix A at *evolve.elsevier.com.*

QUICK CHECK

1. How do *anorexia nervosa* and *bulimia* differ? How are these conditions alike?
2. Obesity is a risk factor for which life-threatening diseases?
3. Name at least eight causes of *protein-calorie malnutrition.*

Body Temperature

Thermoregulation

Considering that more than 60% of the energy released from nutrient molecules during catabolism is converted to heat rather than being transferred to ATP, it is no wonder that maintaining a constant body temperature can be a challenge. Maintaining homeostasis of body temperature, or **thermoregulation,** is the function of the hypothalamus. The hypothalamus operates a variety of negative-feedback mechanisms that keep body temperature in its normal range (36.2° C to 37.6° C, or 97° F to 100° F).

The skin is often involved in negative-feedback loops that maintain body temperature. When the body is overheated, blood flow to the skin increases (see the box *Exercise and the Skin* on p. 156). Warm blood from the body's core can then be cooled by the skin, which acts as a radiator. At the skin, heat can be lost from blood by the following mechanisms, which are also illustrated in **Figure 19-9**:

1. **Radiation**—flow of heat waves from the blood and skin
2. **Conduction**—transfer of heat energy to the skin and then to cooler external environment
3. **Convection**—transfer of heat energy to cooler air that is continually flowing away from the skin
4. **Evaporation**—absorption of heat from blood and skin by water (sweat) vaporization

When necessary, heat can be conserved by reducing blood flow in the skin, as illustrated in Chapter 7 (see the box on p. 156). Heat can also be conserved by reducing any of the

RADIATION

A Flow of heat waves away from the blood and skin

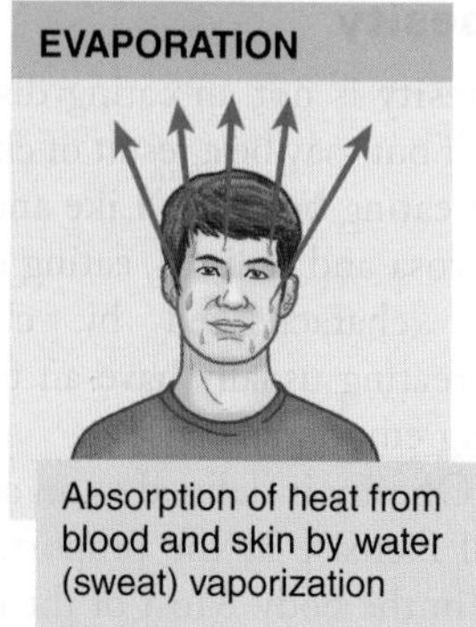

FIGURE 19-9 Mechanisms of heat loss. Heat can be lost from the blood and skin by means of radiation, conduction, convection, and evaporation. Heat can be conserved by altering blood flow in the skin or wearing warm clothing to block the mechanisms shown here.

four mechanisms described. For example, warm clothing can totally or partially block any of these mechanisms.

A number of other mechanisms can be called on to help maintain the homeostasis of body temperature. Heat-generating muscle activity such as shivering (short-term) and secretion of metabolism-regulating hormones (long-term) are two of the body's processes that can be altered to adjust the body's temperature (**Figure 19-10**, *A*). Conversely, an elevated body temperature can be reduced by increasing blood flow and sweating in the skin (**Figure 19-10**, *B*). The concept of using feedback control loops in homeostatic mechanisms was first introduced in Chapter 1.

Abnormal Body Temperature

Maintenance of body temperature within a narrow range is necessary for the normal functioning of the body. As **Figure 19-11** shows, straying too far out of the normal range of body temperatures can have very serious physiological consequences. The following sections identify a few of these important conditions related to body temperature.

Fever

Fever—A fever or febrile state is an unusually high body temperature associated with a systemic inflammation response. In the case of infections, chemicals called **pyrogens** (literally, "fire-makers") cause the thermostatic control centers of the hypothalamus to produce a fever. Because the body's "thermostat" is reset to a higher setting, a person feels a need to warm up to this new temperature and often experiences "chills" as the febrile state begins.

The high body temperature associated with infectious fever is thought to enhance the body's immune responses, helping to eliminate the pathogen. Strategies aimed at reducing the temperature of a febrile person are most often normally counteracted by the body's heat-generating mechanisms and thus have the effect of further weakening the infected person. Under ordinary circumstances, it is best to let the fever "break" on its own after the pathogen is destroyed.

Malignant Hyperthermia

Malignant hyperthermia (MH) is an inherited condition characterized by an abnormally increased body temperature

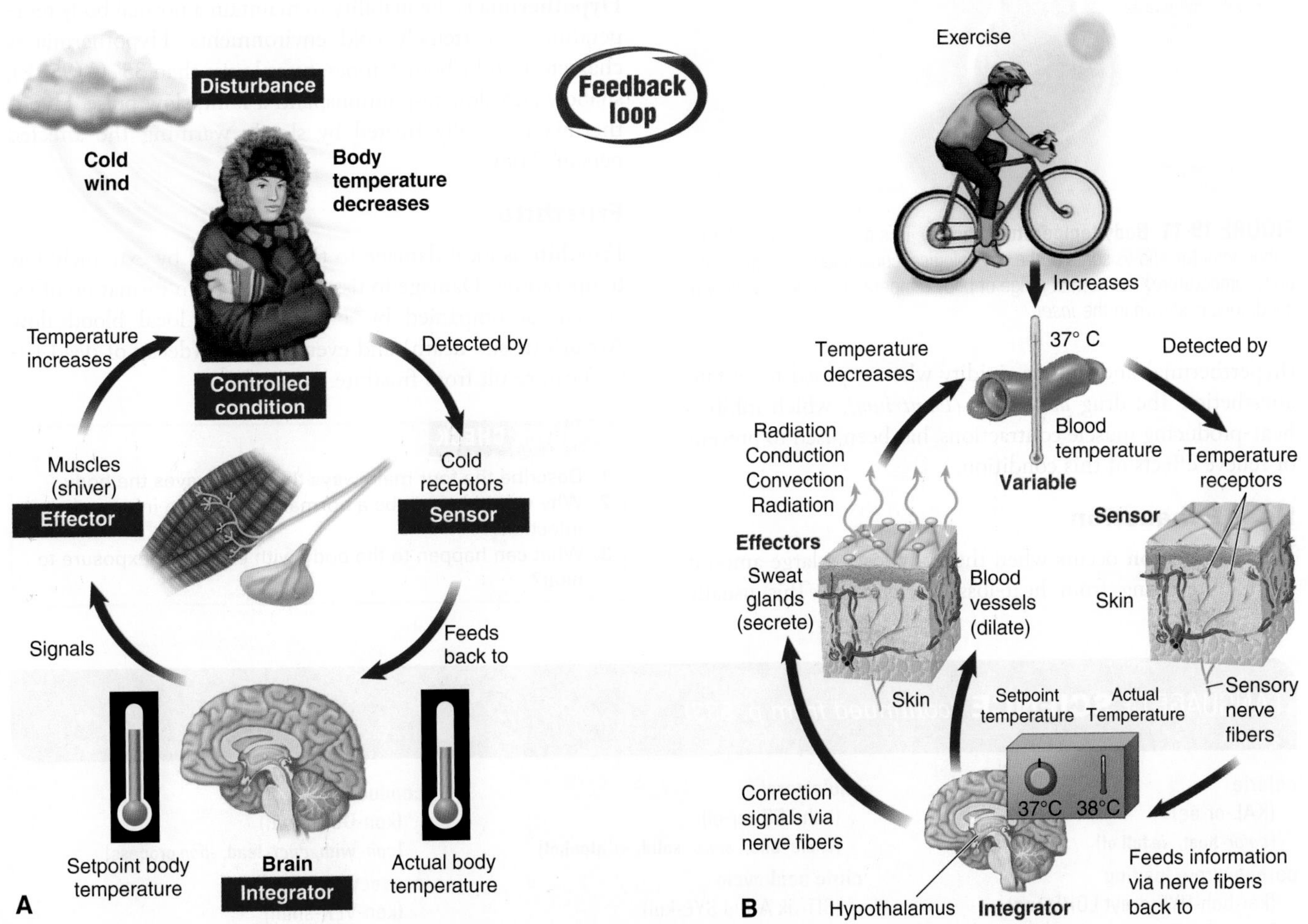

FIGURE 19-10 Feedback control of thermoregulation. A, When body temperature drops below its setpoint value, integrators in the hypothalamus of the brain trigger shivering of skeletal muscle effectors, which produces heat that raises body temperature back toward its set point. **B,** When body temperature rises above its setpoint value, the body's response is to increase sweating and blood flow to the skin, which facilitates loss of heat by several mechanisms—and brings the body temperature down toward its set point.

19

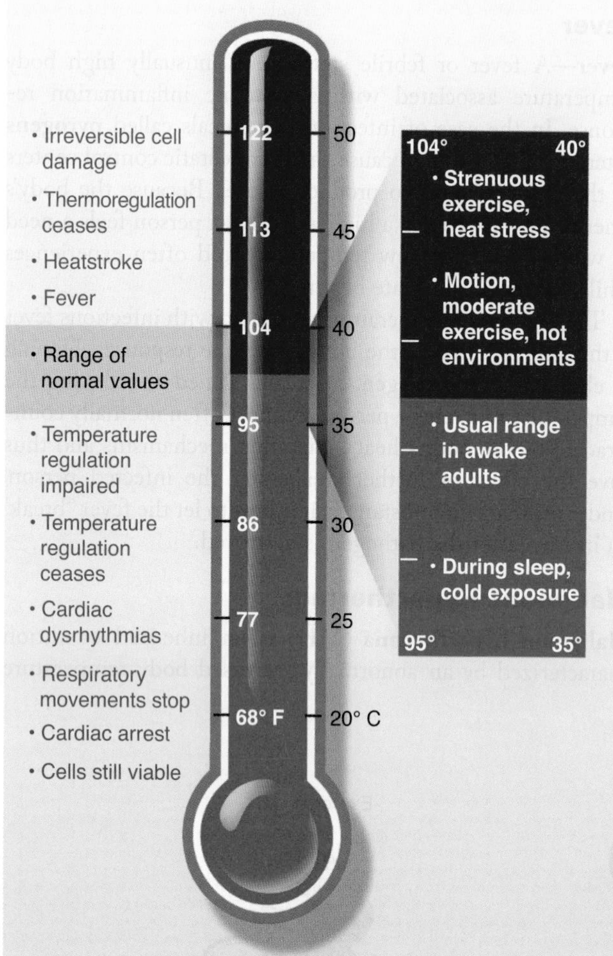

FIGURE 19-11 Body temperature range. This diagram, modeled after a thermometer, shows some of the physiological consequences of abnormal body temperature. The normal range of body temperature under a variety of conditions is shown in the *inset.*

19

(hyperthermia) and muscle rigidity when exposed to certain anesthetics. The drug *dantrolene (Dantrium),* which inhibits heat-producing muscle contractions, has been used to prevent or relieve effects of this condition.

Heat Exhaustion

Heat exhaustion occurs when the body loses a large amount of fluid resulting from heat-loss mechanisms. This usually happens when environmental temperatures are high. Although a normal body temperature is maintained, the loss of water and electrolytes can cause weakness, vertigo, nausea, and possible loss of consciousness. Heat exhaustion may also be accompanied by skeletal muscle cramps that are often called *heat cramps.* Heat exhaustion is treated with rest (in a cool environment) accompanied by fluid replacement.

Heatstroke

Also called *sunstroke,* **heatstroke** is a severe, sometimes fatal condition resulting from the inability to maintain a normal body temperature in an extremely warm environment. Such thermoregulatory failure may result from factors such as old age, disease, drugs that impair thermoregulation, or simply be caused by overwhelming elevated environmental temperatures.

Heatstroke is characterized by body temperatures of 41° C (105° F) or higher; tachycardia; headache; and hot, dry skin. Confusion, convulsions, or loss of consciousness may occur. Unless the body is cooled and body fluids are replaced immediately, death may result.

Hypothermia

Hypothermia is the inability to maintain a normal body temperature in extremely cold environments. Hypothermia is characterized by body temperatures lower than 35° C (95° F); shallow and slow respirations; and a faint, slow pulse. Hypothermia is usually treated by slowly warming the affected person's body.

Frostbite

Frostbite is local damage to tissues caused by extremely low temperatures. Damage to tissues results from formation of ice crystals accompanied by a reduction in local blood flow. *Necrosis* (tissue death) and even *gangrene* (decay of dead tissue) can result from frostbite.

QUICK CHECK

1. Describe the four main ways that heat leaves the body.
2. Why would a *fever* be a normal response to injury or infection?
3. What can happen to the body with excessive exposure to heat?

LANGUAGE OF **SCIENCE** *(continued from p. 533)*

calorie
(KAL-or-ee)
[*calor-* **heat,** *-ie* **full of**]

carbohydrate loading
(kar-boh-HYE-drayt LOHD-ing)
[*carbo-* **carbon,** *-hydr-* **hydrogen,** *-ate* **oxygen**]

catabolism
(kah-TAB-oh-liz-em)
[*cata-* **against,** *-bol-* **throw,** *-ism* **condition**]

cholesterol
(koh-LES-ter-ol)
[*chole-* **bile,** *-stero-* **solid,** *-ol* **alcohol**]

citric acid cycle
(SIT-rik AS-id SYE-kul)
[*citr-* **lemon,** *-ic* **relating to,** *acid* **sour,** *cycle* **circle**]

conduction
(kon-DUK-shun)
[*con-* **with,** *duct-* **lead,** *-tion* **process**]

convection
(kon-VEK-shun)
[*con-* **together,** *-vect-* **carry,** *-tion* **process**]

LANGUAGE OF SCIENCE *(continued from p. 546)*

electron transport system (ETS)
(eh-LEK-tron TRANZ-port SIS-tem [ee tee es])
[*electr-* **electricity,** *-on* **subatomic particle,** *trans-* **across,** *-port* **carry**]

evaporation
(ee-vap-oh-RAY-shun)
[*e-* **out from,** *-vapor* **steam,** *-ation* **process**]

food science
(food SYE-ens)
[*scienc-* **knowledge**]

free radical
(free RAD-ih-kal)
[*radic-* **root,** *-al* **relating to**]

gluconeogenesis
(gloo-koh-nee-oh-JEN-eh-sis)
[*gluco-* **sweet (glucose),** *-neo-* **new,** *-gen-* **produce,** *-esis* **process**]

glycogen
(GLYE-koh-jen)
[*glyco-* **sweet,** *-gen* **produce**]

glycogen loading
(GLYE-koh-jen LOHD-ing)
[*glyco-* **sweet (glucose),** *-gen* **produce**]

glycogenesis
(glye-koh-JEN-eh-sis)
[*glyco-* **sweet (glucose),** *-gen-* **produce,** *-esis* **process**]

glycogenolysis
(glye-koh-jeh-NOL-ih-sis)
[*glyco-* **sweet (glucose),** *-gen-* **produce,** *-o-* **combining vowel,** *-lysis* **loosening**]

glycolysis
(glye-KAHL-ih-sis)
[*glyco-* **sweet (glucose),** *-o-* **combining vowel,** *-lysis* **loosening**]

insulin
(IN-suh-lin)
[*insul-* **island,** *-in* **substance**]

joule (J or j)
(jool)
[*James Prescott Joule* **English physicist**]

kilocalorie (kcal; also calorie)
(KIL-oh-kal-oh-ree)
[*kilo-* **one thousand,** *-calor-* **heat,** *-ie* **full of**]

Krebs cycle
(krebz SYE-kul)
[*Sir Hans Adolf Krebs* **British biochemist,** *cycl-* **circle**]

macronutrient
(MAK-roh-NOO-tree-ent)
[*macro-* **large,** *-nutri-* **nourish,** *-ent* **agent**]

metabolism
(meh-TAB-oh-liz-em)
[*meta-* **over,** *-bol-* **throw,** *-ism* **action**]

micronutrient
(MY-kroh-NOO-tree-ent)
[*micro-* **small,** *-nutri-* **nourish,** *-ent* **agent**]

nutrition
(noo-TRIH-shun)
[*nutri-* **nourish,** *-tion* **process**]

plasma protein
(PLAZ-mah PROH-teen)
[*plasma* **substance,** *prote-* **primary,** *-in* **substance**]

pyrogen
(PYE-roh-jen)
[*pyro-* **heat,** *-gen* **produce**]

pyruvic acid
(pye-ROO-vik AS-id)
[*pyr-* **heat,** *-uv-* **grape,** *-ic* **relating to**]

radiation
(ray-dee-AY-shun)
[*radiat-* **send out rays,** *-ion* **process**]

satiety center
(sah-TYE-eh-tee SEN-ter)
[*sati-* **enough or full,** *-ety* **state**]

thermoregulation
(ther-moh-reg-yoo-LAY-shun)
[*therm-* **heat,** *-o-* **combining vowel,** *-regula-* **rule,** *-ation* **process**]

total metabolic rate (TMR)
(TOH-tal met-ah-BOL-ik rayt [tee em ar])
[*meta-* **over,** *-bol-* **throw,** *-ic* **relating to**]

urea
(yoo-REE-ah)
[*urea* **urine**]

vitamin
(VYE-tah-min)
[*vita-* **life,** *-amin* **ammonia compound**]

19

LANGUAGE OF MEDICINE

anorexia nervosa
(an-oh-REK-see-ah ner-VOH-sah)
[*an-* **without,** *-orex-* **appetite,** *-ia* **condition,** *nerv-* **nerve,** *-osa* **relating to**]

avitaminosis
(ay-vye-tah-mih-NOH-sis)
[*a-* **without,** *-vita-* **life,** *-amin-* **ammonia compound,** *-osis* **condition**]

bulimarexia
(boo-lee-mah-REK-see-ah)
[*bu-* **ox,** *-lim-* **hunger,** *-orex-* **appetite,** *-ia* **condition**]

bulimia
(boo-LEE-mee-ah)
[*bu-* **ox,** *-lim-* **hunger,** *-ia* **condition**]

fever
(FEE-ver)

frostbite
(FROST-byte)

heat exhaustion
(heet eg-ZAWS-chun)

heatstroke
(HEET-strohk)
[*stroke* **strike**]

hypercholesterolemia
(hye-per-koh-les-ter-ohl-EE-mee-ah)
[*hyper-* **excessive,** *-chole-* **bile,** *-stero-* **solid,** *-ol-* **alcohol,** *-emia* **blood condition**]

hypervitaminosis
(hye-per-vye-tah-mih-NOH-sis)
[*hyper-* **excessive,** *-vita-* **life,** *-amin-* **ammonia compound,** *-osis* **condition**]

hypothermia
(hye-poh-THER-mee-ah)
[*hypo-* **under or below,** *-therm-* **heat,** *-ia* **abnormal condition**]

kwashiorkor
(kwah-shee-OR-kor)
[*kwashiorkor* **one who is displaced (from the breast)**]

malignant hyperthermia (MH)
(mah-LIG-nant hye-per-THERM-ee-ah [em aych])
[*malign-* **bad,** *-ant* **state,** *hyper-* **excessive,** *-therm-* **heat,** *-ia* **abnormal condition**]

Continued on p. 548

LANGUAGE OF MEDICINE (continued from p. 547)

marasmus
(mah-RAZ-mus)
[*marasmus* **a wasting**]

obesity
(oh-BEES-ih-tee)
[*ob-* **over,** *-es-* **eat,** *-ity* **state**]

phenylketonuria (PKU)
(fen-il-kee-toh-NOO-ree-ah)
[*phen-* **shining (phenol),** *-yl-* **chemical,** *-keton-* **acetone,** *-ur-* **urine,** *-ia* **condition**]

protein-calorie malnutrition (PCM)
(PROH-teen-KAL-or-ee mal-noo-TRISH-un [pee see em])
[*prote-* **primary,** *-in* **substance,** *calor-* **heat,** *-ie* **full of,** *mal-* **bad,** *-nutri-* **nourish,** *-tion* **process**]

scurvy
(SKER-vee)
[*scur-* **sour milk,** *-vy* **a swelling**]

statin
(STAT-in)
[*stat-* **stand,** *-in* **substance**]

❑ OUTLINE SUMMARY

To download a digital version of the chapter summary for use with your device, access the ***Audio Chapter Summaries*** *online at evolve.elsevier.com.*

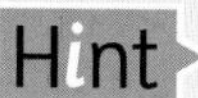

Scan this summary after reading the chapter to help you reinforce the key concepts. Later, use the summary as a quick review before your class or before a test.

Definitions

A. Nutrition—energy-yielding nutrients, vitamins, and minerals that are ingested and assimilated into the body (**Figure 19-1**)
B. Metabolism—process of using nutrient molecules as energy sources and as building blocks for our own molecules
C. Catabolism—process that breaks nutrient molecules down, releasing their stored energy; oxygen used in catabolism
D. Anabolism—process that builds nutrient molecules into complex chemical compounds

Metabolic Function of the Liver

A. Secretes bile to help mechanically digest lipids by emulsifying them
B. Processes blood immediately after it leaves the gastrointestinal tract
   1. Helps maintain normal blood glucose concentration
   2. Site of protein, carbohydrate, and fat metabolism
   3. Removes toxins from the blood
C. Synthesizes several kinds of plasma proteins, including albumins, fibrinogen, clotting factors, etc.
D. Stores useful substances, including glycogen, lipids, certain vitamins

Macronutrients

A. Dietary sources of nutrients
   1. Nutrients—food components digested and absorbed by the body
   2. Macronutrients—nutrients needed in large daily quantities (carbohydrates, fats, proteins) (**Table 19-1**)
   3. Micronutrients—nutrients needed in tiny daily quantities (vitamins and minerals) (**Table 19-3** and **Table 19-4**)
B. Carbohydrate metabolism
   1. Carbohydrates are the preferred energy nutrient of the body
   2. Three series of chemical reactions that occur in a precise sequence make up the process of glucose metabolism (**Figure 19-2**)
      a. Glycolysis—occurs in cytoplasm of the cell
         (1) Anaerobic process (uses no oxygen)
         (2) Changes glucose to pyruvic acid, which is then converted into acetyl CoA
         (3) Yields small amount of energy (transferred to ATP)
      b. Citric acid cycle (Krebs cycle)—occurs in the mitochondria
         (1) Aerobic process (requires oxygen)
         (2) Changes acetyl CoA to carbon dioxide
         (3) Most energy leaving the citric acid cycle is in the form of high-energy electrons
      c. Electron transport system (ETS)—occurs in the mitochondria
         (1) Transfers energy from high-energy electrons (from citric acid cycle) to ATP molecules
         (2) ATP serves as direct source of energy for cells (**Figure 19-3**)
   3. Adenosine triphosphate (ATP)—energy transferred to ATP differs from energy in nutrient molecules
      a. Not stored; released almost instantly
      b. Can be used directly to do cellular work

4. Anabolism and storage of glucose
   a. Glucose that is not needed immediately for making ATP is stored as glycogen (a long chain of glucose subunits) in liver and muscle cells
   b. Glycogenesis—anabolic process of joining glucose molecules together in a chain to form glycogen (storing glucose for later use)
   c. Glycogenolysis—catabolic process of breaking apart glycogen chains, releasing individual glucose molecules for use in making ATP
5. Blood glucose level—concentration of glucose in blood
   a. Normally maintained between about 80 and 110 mg per 100 mL of blood during fasting
   b. Insulin accelerates the movement of glucose out of the blood into cells, therefore decreasing blood glucose and increasing glucose catabolism (**Figure 19-4**)

C. Fat metabolism
1. Fats (triglycerides) are primarily an energy nutrient
2. Fatty acids and glycerol converted to forms of glucose by gluconeogenesis to be catabolized and energy transferred to ATP (**Figure 19-5**)
3. Excess fatty acids are anabolized to form triglycerides that are stored in adipose tissue

D. Protein metabolism
1. Proteins are catabolized for energy only after carbohydrate and fat stores are depleted; excess dietary proteins can also be catabolized for energy
2. Gluconeogenesis breaks apart amino acids to convert them to a form that enters the citric acid cycle to produce ATP; the nitrogenous waste product called *urea* is formed in this process (**Figure 19-5**)
3. *Essential amino acids* are those that must be in the diet because the body cannot make them (**Table 19-2**)

Micronutrients

A. Vitamins
1. Organic molecules that are needed in small amounts for normal metabolism (**Table 19-3**)
   a. May bind to enzymes and coenzymes to help them work effectively
   b. Vitamin A has role in vision
   c. Vitamin D converts to a hormone that regulates calcium homeostasis
   d. Vitamin E is an antioxidant that protects against free radicals
2. Vitamin imbalances
   a. Avitaminosis—deficiency of a vitamin
      (1) Can lead to severe metabolic problems
      (2) Avitaminosis C can lead to scurvy (**Figure 19-6**)
   b. Hypervitaminosis—excess of a vitamin
      (1) Can be just as serious as avitaminosis
      (2) May be chronic or acute

B. Minerals—inorganic molecules found naturally in the earth
1. Required by the body for normal function, including nerve conduction (**Table 19-4**)
2. May attach to enzymes to facilitate their work

Regulating Food Intake

A. Regulatory centers in the hypothalamus play a primary role in controlling food intake
1. Appetite center—produces feelings of hunger
2. Satiety center—produces feelings of satisfaction

B. Food intake regulation results from balance between hypothalamic control centers

C. Many diverse factors influence the hypothalamic control centers (**Table 19-5**)

Metabolic Rates

A. Basal metabolic rate (BMR)—rate of metabolism when a person is lying down but awake, not digesting food, and when the environment is comfortably warm

B. Total metabolic rate (TMR)—the total amounts of energy, expressed in calories, used by the body per day (**Figure 19-7**)

Metabolic and Eating Disorders

A. Disruption or imbalance of normal metabolism can be caused by several different factors
1. Inborn errors of metabolism—genetic conditions involving deficient or abnormal metabolic enzymes
2. Some metabolic disorders are complications of other conditions
   a. Hormonal imbalances and eating disorders

B. Eating disorders
1. Anorexia nervosa—characterized by chronic refusal to eat
2. Bulimia—an alternating pattern of craving of food followed by a period of self-denial; in bulimarexia, the self-denial triggers self-induced vomiting
3. Obesity—abnormally high proportion of body fat; may be a symptom of an eating disorder; risk factor for many chronic diseases
4. Protein-calorie malnutrition (PCM)—results from a deficiency of calories in general and proteins in particular; examples are marasmus and kwashiorkor (**Figure 19-8** and **Table 19-6**)

Body Temperature

A. Thermoregulation
1. Hypothalamus—regulates the homeostasis of body temperature (thermoregulation) through a variety of processes
2. Blood flow to the skin increases when body is overheated

3. Heat is lost through the skin by several mechanisms (**Figure 19-9**)
   a. Radiation—flow of heat waves from the blood and skin
   b. Conduction—transfer of heat energy to the skin and then to cooler external environment
   c. Convection—transfer of heat energy to cooler air that is continually flowing away from the skin
   d. Evaporation—absorption of heat from blood and skin by water (sweat) vaporization
4. The body can generate heat to maintain homeostasis over the short term (shivering) or the long term (changes in metabolic rates)
5. Heating and cooling of body is controlled by feedback loops that maintain a stable body temperature (**Figure 19-10**)

B. Abnormal body temperature can have serious physiological consequences (**Figure 19-11**)
   1. Fever (febrile state)—unusually high body temperature associated with systemic inflammation response
   2. Malignant hyperthermia (MH)—inherited condition that causes increased body temperature (hyperthermia) and muscle rigidity when exposed to certain anesthetics
   3. Heat exhaustion—results from loss of fluid as the body tries to cool itself; may be accompanied by heat cramps
   4. Heatstroke (sunstroke)—overheating of body resulting from failure of thermoregulatory mechanisms in a warm environment
   5. Hypothermia—reduced body temperature resulting from failure of thermoregulatory mechanisms in a cold environment
   6. Frostbite—local tissue damage caused by extreme cold; may result in necrosis or gangrene

❑ ACTIVE LEARNING

STUDY TIPS

Hint *Consider using these tips to achieve success in meeting your learning goals.*

This chapter begins by explaining the functions of the liver and the importance of the portal system, both of which were discussed in earlier chapters.

1. The process of metabolism refers to the body's use of nutrients. Fats and carbohydrates are used primarily for energy.
2. Glycolysis occurs in the cytoplasm of the cell; it requires no oxygen but produces very little energy. The end products of glycolysis enter the citric acid (or Krebs) cycle, which occurs in the mitochondria. This process requires oxygen and produces much more energy. Some of the end products of the citric acid cycle are high-energy molecules that are used to convert ADP into ATP; this is done in the electron transport system. Energy stored between the phosphates of the ATP molecules can be used for the needs of the cell. Fat and protein molecules can be modified so they can enter the citric acid cycle.
3. The term *nonessential amino acids* is somewhat misleading; it does not mean your body does not need them—it means that they can be made from other amino acids.
4. Vitamins and minerals assist in enzyme function. You can learn the names and functions of the vitamins and minerals from the tables in the text or by making flash cards and checking online resources. To better understand the terms in this chapter, refer to the Language of Science and the Language of Medicine sections.
5. Metabolic rates describe how quickly your body is using nutrients. The basal metabolic rate (BMR) is the amount of nutrients you burn just to stay alive and awake. Your total metabolic rate (TMR) depends on how active you are. Check out this website that allows you to calculate your BMR: *bmi-calculator.net/bmr-calculator/*.
6. Make a chart to help you learn the metabolic disorders. Organize the chart based on the mechanism or cause of each disorder: deficiency or excess of vitamins, nutrition disorders, and disorders of temperature regulation.
7. In your study group, review the flash cards for the vitamins and minerals or **Tables 19-2** and **19-3**. Discuss the processes of carbohydrate, protein, and fat metabolism. Discuss what constitutes basal and total metabolic rates and the ways heat can be lost from the body. Review the metabolic disorders chart, chapter outline summary and the questions at the end of the chapter, and discuss possible test questions.

Review Questions

Write out the answers to these questions after reading the chapter and reviewing the Chapter Summary. If you simply think through the answer without writing it down, you will not retain much of your new learning.

1. Define anabolism and catabolism.
2. Explain the function of the liver.
3. Compare macronutrients and micronutrients
4. Briefly explain the process of glycolysis.
5. Briefly explain the citric acid cycle.
6. What is the function of the electron transport system?
7. Explain the ways in which energy stored in ATP is different from energy stored in nutrient molecules.
8. List the primary hormones that tend to increase the amount of sugar in the blood.
9. Identify when fat catabolism usually occurs.
10. Identify when protein catabolism usually occurs.
11. Explain what is meant by a nonessential amino acid.
12. Explain the use of statin drugs.
13. Name three water-soluble and three fat-soluble vitamins.
14. Define avitaminosis. Name a disorder caused by avitaminosis. What vitamin deficiency causes this disorder?
15. Name the signs and symptoms of vitamin A hypervitaminosis.
16. Name three minerals needed by the body.
17. Briefly explain the function of vitamins and minerals in the body.
18. Locate the satiety center.
19. Differentiate between basal and total metabolic rate.
20. Distinguish between marasmus and kwashiorkor.
21. Name and explain four ways heat can be lost through the skin.
22. Explain the cause and symptoms of malignant hyperthermia.
23. Distinguish between heat exhaustion and heatstroke in terms of a person's body temperature.

Critical Thinking

After finishing the Review Questions, write out the answers to these more in-depth questions to help you apply your new knowledge. Go back to sections of the chapter that relate to concepts that you find difficult.

24. Differentiate between absorption and assimilation.
25. Explain the advantage the body gains by having the blood go through the hepatic portal system.
26. Diagram the ATP-ADP cycle. Include where the energy is added and where the energy is released.
27. A man went on a 10-day vacation. His total metabolic rate was 2600 calories a day. His total calorie intake was 3300 calories a day. He began the trip weighing 178 pounds. What did he weigh when he got back from vacation? (3500 excess calories = 1 pound)
28. Suppose your diet consisted of only protein. Is it possible for a person to keep from gaining weight by eating only protein? Explain.
29. Explain the process of gluconeogenesis and its function in fat metabolism.

Chapter Test

Hint *After studying the chapter, test your mastery by responding to these items. Try to answer them without looking up the answers.*

1. The process of ________ occurs when nutrient molecules enter the cell and undergo chemical change.
2. ________ is the term used to describe all the chemical processes that release energy from nutrients.
3. The plasma proteins ________ and ________ are made in the liver and are important in blood clot formation.
4. The vitamins ________ and ________ can be stored in the liver.
5. The hormone ________ is known to accelerate glucose transport through cell membranes.
6. An excess amount of cholesterol in the blood results in a condition called ________.
7. The B vitamins are ________ soluble, whereas vitamins K and E are ________ soluble.
8. ________ is the number of calories that must be used just to keep the body alive, awake, and comfortably warm.
9. In order to lose weight, your total caloric intake must be less than your ________.
10. One way heat can be lost by the skin is ________, which is the transfer of heat to the air that is continually flowing away from the skin.
11. One way heat can be lost by the skin is ________, which is the absorption of heat by water (sweat) vaporization.
12. ________ is the process used by the body as the second choice of energy metabolism.
13. In the healthy body, ________ is used almost exclusively for anabolism rather than catabolism.
14. ________ are amino acids needed by the body, but they can be made from other amino acids if they are not supplied directly by the diet.

Match each term in Column A with its corresponding description in Column B.

Column A	Column B
15. ________ glycolysis	a. part of the cell in which glycolysis occurs
16. ________ gluconeogenesis	b. the part of carbohydrate metabolism that does not require oxygen
17. ________ electron transport system	c. process that converts high-energy molecules from the citric acid cycle into ATP
18. ________ mitochondria	d. process that converts glycerol into a compound that can enter the glycolysis pathway
19. ________ cytoplasm	e. the body's direct source of energy
20. ________ ATP	f. molecule that results when adenosine triphosphate loses a phosphate group
21. ________ glycogenesis	g. the part of the cell in which the citric acid cycle takes place
22. ________ ADP	h. glucose anabolism

Match each disorder in Column A with its corresponding description or cause in Column B.

Column A	Column B
23. ________ avitaminosis	a. behavioral disorder characterized by a chronic refusal to eat
24. ________ hypervitaminosis	b. increased body temperature caused by exposure to anesthetics
25. ________ anorexia nervosa	c. type of malnutrition that results from an overall lack of calories
26. ________ bulimia	d. an overheating problem in which the body is dehydrated but the body temperature is normal
27. ________ marasmus	e. local tissue damage caused by ice crystals forming in the cells
28. ________ kwashiorkor	f. condition that results in the development of scurvy
29. ________ malignant hyperthermia	g. behavioral disorder characterized by insatiable craving for food alternating with self-deprivation; may include food binges
30. ________ heat exhaustion	h. a body temperature lower than 95° F
31. ________ heatstroke	i. overheating problem in which the body temperature can be as high as 105° F; potentially life-threatening
32. ________ hypothermia	j. type of malnutrition that results from a lack of protein with sufficient total calories
33. ________ frostbite	k. a vitamin excess, usually involving fat-soluble vitamins

Case Studies

To solve a case study, you may have to refer to the glossary or index, other chapters in this textbook, and other resources.

1. A friend of yours is helping you chop firewood on a hot day. She complains of muscle cramps and nausea but has a normal body temperature. What has happened to her? How would you help your friend?
2. While looking through an old family album, you can't help but notice that your great-great-great-grandfather's smile reveals that he has no teeth. When asked why this ancestor lost his teeth at an early age, your grandmother replies that he suffered from scurvy as a merchant marine and lost all his teeth as a result. Is this possible? Can you explain how scurvy can cause the loss of teeth?
3. Andrea is planning to adopt a totally vegetarian diet—a diet that includes no meats or animal products. Her friends have voiced some concern that her new diet may not contain certain essential amino acids. What is an essential amino acid? Why must her diet contain these nutrients?

Answers to Active Learning Questions can be found online at evolve.elsevier.com.

Urinary System

OUTLINE

Scan this outline before you begin to read the chapter, as a preview of how the concepts are organized.

OBJECTIVES

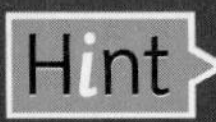

Before reading the chapter, review these goals for your learning.

After you have completed this chapter, you should be able to:

1. Do the following related to the kidneys:
   - Describe the location and structure of the kidneys and explain how they act as vital organs in maintaining homeostasis.
   - Identify the parts of a nephron and describe the role each component plays in the balancing of blood and formation of urine.
2. Explain the importance of filtration, tubular reabsorption, and tubular secretion in renal physiology.
3. Discuss the mechanisms that control urine volume.
4. Describe the structure and function of the ureters, urinary bladder, and urethra.
5. Describe the process of micturition and the control problems that frequently occur with this process.
6. Explain the purpose and importance of urinalysis.
7. List the major renal and urinary disorders and explain the mechanism of each disorder.

CHAPTER 20

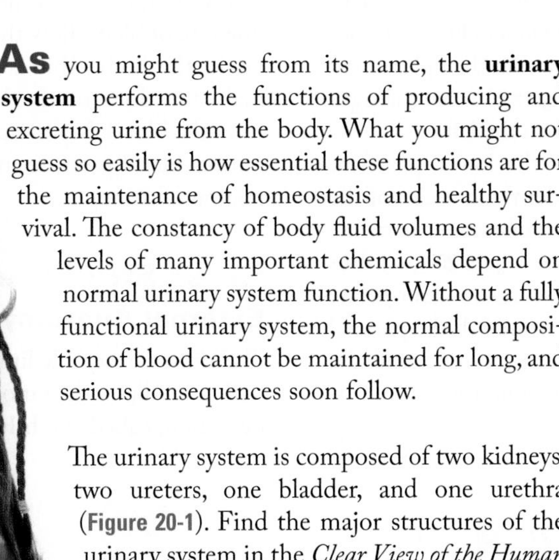

As you might guess from its name, the **urinary system** performs the functions of producing and excreting urine from the body. What you might not guess so easily is how essential these functions are for the maintenance of homeostasis and healthy survival. The constancy of body fluid volumes and the levels of many important chemicals depend on normal urinary system function. Without a fully functional urinary system, the normal composition of blood cannot be maintained for long, and serious consequences soon follow.

The urinary system is composed of two kidneys, two ureters, one bladder, and one urethra (**Figure 20-1**). Find the major structures of the urinary system in the *Clear View of the Human Body* (follows p. 8).

We begin our discussion with the kidneys. The kidneys "clear" or clean the blood of the many waste products continually produced as a result of metabolism of nutrients in body cells. As nutrients are burned for energy, the waste products produced must be removed from the blood, or they quickly accumulate to toxic levels—a condition called **uremia** or **uremic poisoning.**

The kidneys also play a vital role in maintaining electrolyte, water, and acid-base balances in the body. In this chapter, we discuss the structure and function of each organ of the urinary system. We also discuss disease conditions produced by abnormal functioning of the urinary system. In the following chapters, we continue the story by exploring fluid and electrolyte balance in Chapter 21 and acid-base balance in Chapter 22.

For an introduction to the urinary system, go to AnimationDirect online at *evolve.elsevier.com.*

LANGUAGE OF SCIENCE

Hint Before reading the chapter, say each of these terms out loud. This will help you to avoid stumbling over them as you read.

aldosterone
(AL-doh-steh-rohn *or* al-DAH-stayr-ohn)
[*aldo-* **aldehyde,** *-stero-* **solid or steroid derivative,** *-one* **chemical**]

antidiuretic hormone (ADH)
(an-tee-dye-yoo-RET-ik HOR-mohn [ay dee aych])
[*anti-* **against,** *-dia-* **through,** *-uret-* **urination,** *-ic* **relating to,** *hormon-* **excite**]

atrial natriuretic hormone (ANH)
(AY-tree-al nay-tree-yoo-RET-ik HOR-mohn [ay en aych])
[*atria-* **entrance courtyard (atrium of heart),** *-al* **relating to,** *natri-* **natrium (sodium),** *-uret-* **urination,** *-ic* **relating to,** *hormon-* **excite**]

Bowman capsule
(BOH-men KAP-sul)
[*William Bowman* **English anatomist,** *caps-* **box,** *-ule* **little**]

calyx
(KAY-liks)
pl., calyces
(KAY-lih-seez)
[*calyx* **seed pod or cup**]

collecting duct (CD)
(koh-LEK-ting dukt [see dee])
[*duct* **a leading**]

cortical nephron
(KOHR-tih-kal NEF-ron)
[*cortic-* **bark (cortex),** *-al* **relating to,** *nephro-* **kidney,** *-on* **unit**]

countercurrent mechanism
(KON-ter-ker-rent MEK-a-niz-em)
[*counter-* **against,** *-current* **flow**]

Continued on p. 574

Kidneys

Location of the Kidneys

To locate the kidneys on your own body, stand erect and put your hands on your hips with your thumbs meeting over your backbone. When you are in this position, your kidneys lie above your thumbs on either side of your spinal column, but their placement is higher than you might think. Note in **Figure 20-1** that the right kidney, which touches the liver, is lower than the left.

Both kidneys are protected a bit by the lower, posterior part of the rib cage. They are located under the muscles of the back and *behind* (outside) the parietal peritoneum—the membrane that lines the abdominal cavity (see **Figure 20-1**, *C*). Because of this **retroperitoneal** location, a surgeon can operate on a kidney from behind without cutting through the parietal peritoneum. Once the peritoneum has been cut or opened, the potential for spread of infection throughout the entire abdominal cavity increases.

A heavy cushion of fat—the *renal fat pad*—normally encases each kidney and helps hold it in place.

Note the relatively large diameter of the renal arteries in **Figure 20-1**, *A*. Normally, a little more than 20% of the total blood pumped by the heart each minute enters the kidneys. The rate of blood flow through this organ is among the highest in the body. This is understandable because one of the main functions of the kidney is to remove waste products from the blood. Maintenance of a high rate of blood flow and normal blood pressure in the kidneys is essential for the formation of urine.

Gross Structure of the Kidney

External Anatomy

The kidneys resemble lima beans in shape, that is, roughly oval with a medial indentation (see **Figure 20-1**, *A*). The medial indentation, called the **hilum,** is where vessels, nerves, and the

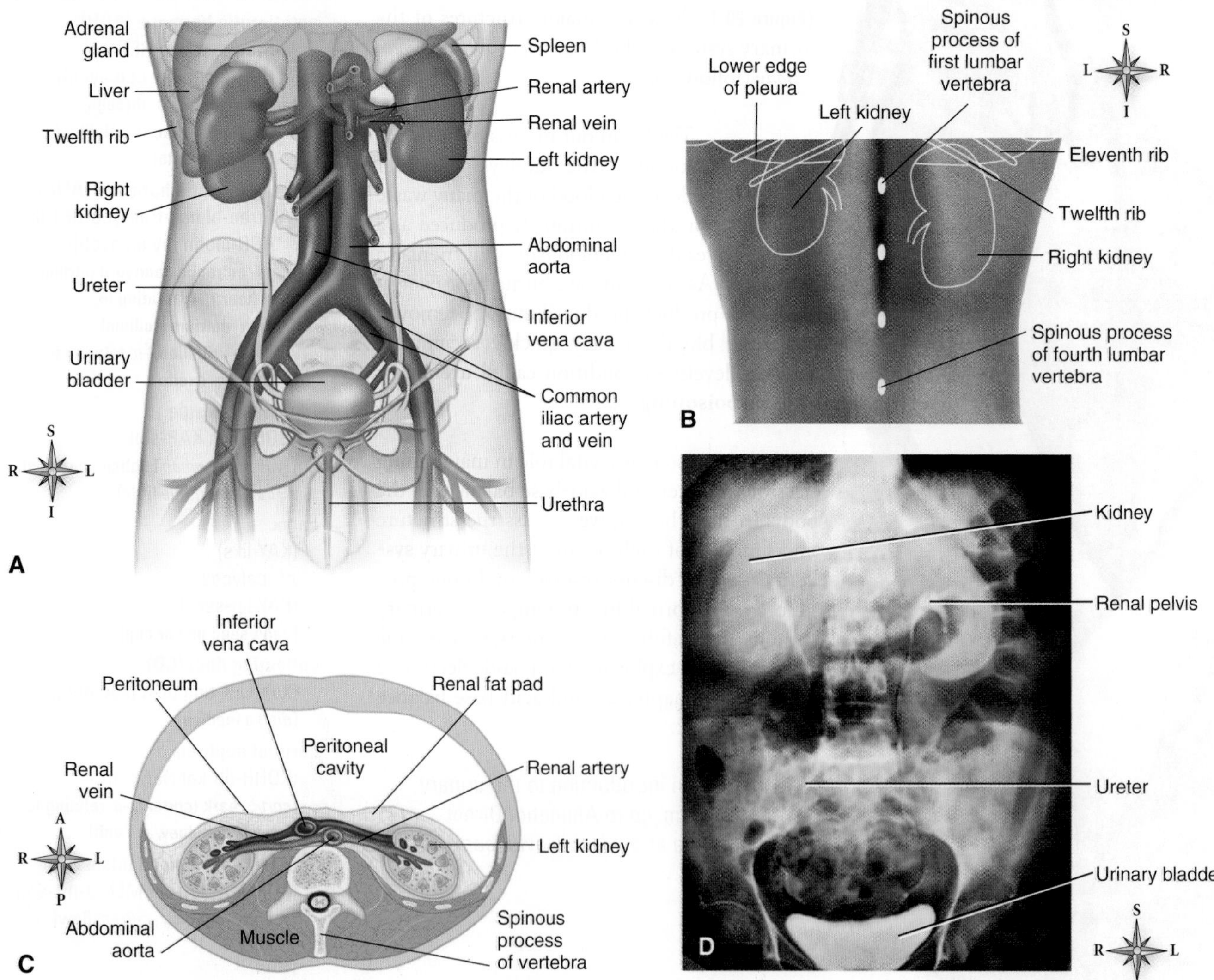

FIGURE 20-1 Urinary system. A, Anterior view of urinary organs. **B,** Surface markings of the kidneys, eleventh and twelfth ribs, spinous processes of L1 to L4, and lower edge of pleura viewed from behind. **C,** Horizontal (transverse) section of the abdomen showing the retroperitoneal position of the kidneys. **D,** X-ray film of the urinary organs.

20

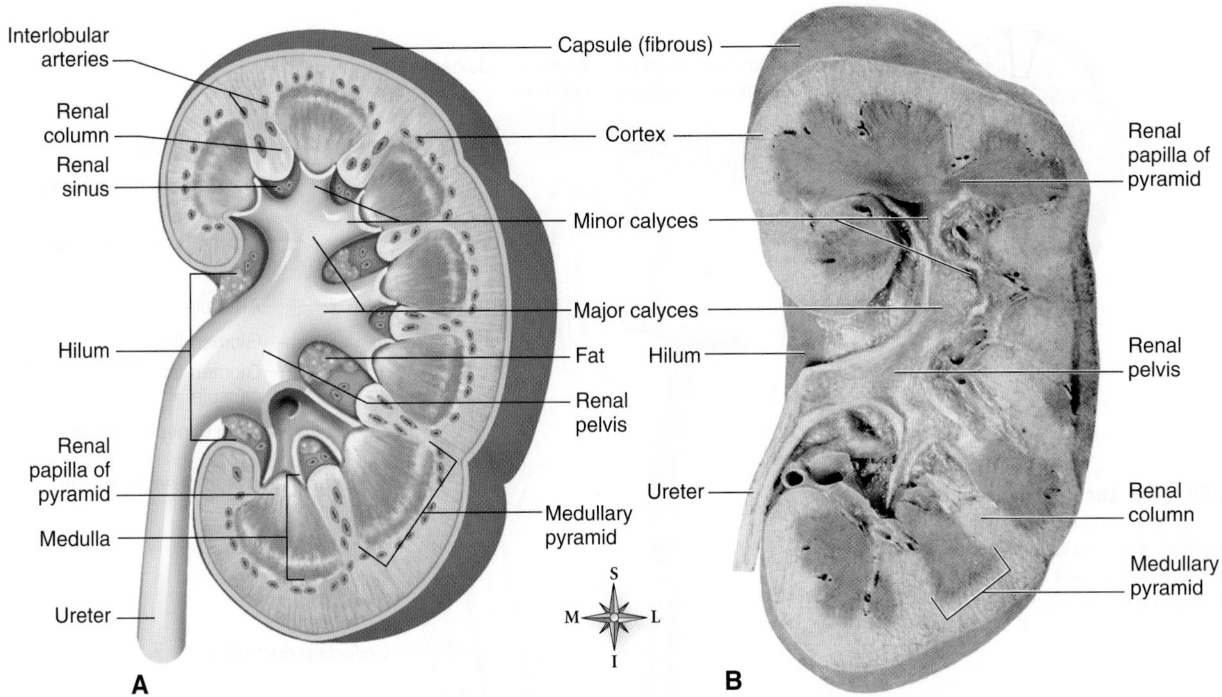

FIGURE 20-2 Kidney. Internal structure. **A,** Artist's rendering of a coronal section of the kidney. **B,** Photograph of coronal section of a preserved human kidney.

ureter connect with the kidney. An average-sized kidney measures approximately 11 × 7 × 3 cm (4.3 × 2.7 × 1.2 in).

There is a tough fibrous *capsule* that forms the exterior wall of the kidney.

Internal Anatomy

If you were to slice through a kidney from side to side and open it like the pages of a book (called a *coronal section*), you would see the structures shown in **Figure 20-2**. Identify each of the following parts:

1. **Renal cortex**—the outer part of the kidney (the word *cortex* comes from the Latin word for "bark," so the cortex of an organ is its outer layer).
2. **Renal medulla**—the inner portion of the kidney.
3. **Renal pyramids**—the triangular divisions of the medulla of the kidney. Extensions of cortical tissue that dip down into the medulla between the renal pyramids are called **renal columns.**
4. **Renal papilla** *(pl. papillae)*—narrow, innermost end of a pyramid.
5. **Renal pelvis**—(also called the *kidney pelvis*) an expansion of the upper end of a ureter (the tube that drains urine into the bladder).
6. **Calyx** *(pl. calyces)*—a division of the renal pelvis (the papilla of a pyramid opens into each calyx).

Microscopic Structure of the Kidney

More than a million microscopic units called **nephrons** make up each kidney's interior (**Figure 20-3**). The shape of a nephron is unique, unmistakable, and admirably suited to its function of producing urine. It looks a little like a tiny funnel with a very long stem, but it is an unusual stem in that it is highly convoluted—that is, it has many bends in it.

The nephron is composed of two principal components: the *renal corpuscle* and the *renal tubule.* The renal corpuscle can be subdivided still further into two parts and the renal tubule into four regions or segments. Identify each part of the renal corpuscle and renal tubule described in **Figure 20-4** and **Figure 20-5**.

A. **Renal corpuscle**
   1. **Glomerular capsule**—the cup-shaped top of a nephron. The hollow, saclike glomerular capsule surrounds the glomerulus. Also called **Bowman capsule.**
   2. **Glomerulus**—a network of blood capillaries tucked into the glomerular capsule. Note in **Figure 20-4,** *B* that the small artery *(afferent arteriole)* that delivers blood to the glomerulus is larger in diameter than the *efferent arteriole* that drains blood from the glomerulus and that it is relatively short. This partly explains the high blood pressure that exists in the glomerular capillaries. This high pressure is required to filter wastes from the blood.

B. **Renal tubule**
   1. **Proximal convoluted tubule (PCT)**—the first segment of a renal tubule. The PCT is called *proximal* because it lies nearest the tubule's origin from the glomerular capsule, and it is called *convoluted* because it has several bends.
   2. **Nephron loop (Henle loop)**—the extension of the proximal tubule into the renal medulla. Observe that the nephron loop consists of a straight

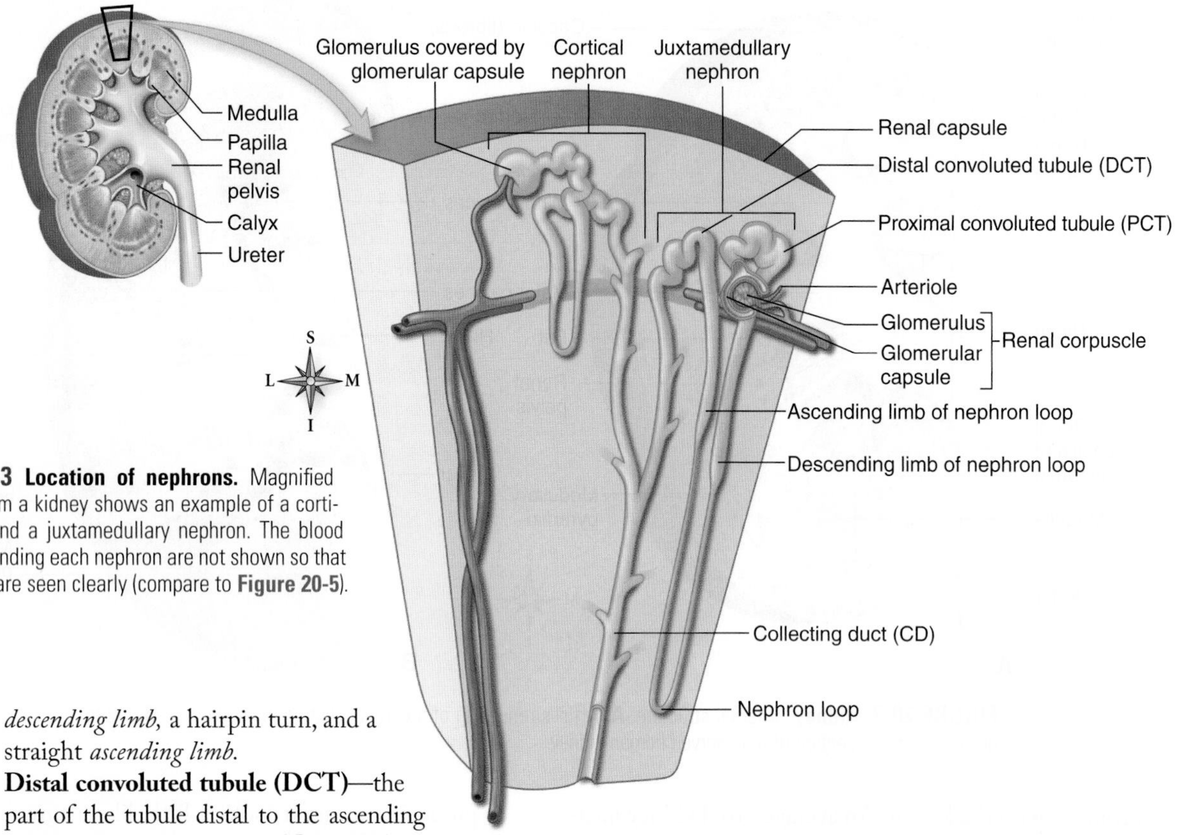

Medulla
Papilla
Renal pelvis
Calyx
Ureter
Glomerulus covered by glomerular capsule
Cortical nephron
Juxtamedullary nephron
Renal capsule
Distal convoluted tubule (DCT)
Proximal convoluted tubule (PCT)
Arteriole
Glomerulus
Glomerular capsule
Renal corpuscle
Ascending limb of nephron loop
Descending limb of nephron loop
Collecting duct (CD)
Nephron loop
S
L
M
I

FIGURE 20-3 Location of nephrons. Magnified wedge cut from a kidney shows an example of a cortical nephron and a juxtamedullary nephron. The blood vessels surrounding each nephron are not shown so that the nephrons are seen clearly (compare to **Figure 20-5**).

descending limb, a hairpin turn, and a straight *ascending limb.*

3. **Distal convoluted tubule (DCT)**—the part of the tubule distal to the ascending limb of the nephron loop. The DCT extends from the ascending limb to the collecting duct.
4. **Collecting duct (CD)**—a straight (that is, not convoluted) part of a renal tubule. Distal tubules of several nephrons join to form a single collecting duct.

Urine from the collecting ducts exits from the pyramid through the papilla and enters the calyx and renal pelvis before flowing into the ureter.

Look again at **Figure 20-3**. Notice the differing locations of the two nephrons in the illustration. One is located high in the cortex and is typical of about 85% of all nephrons. Nephrons in this group are located almost entirely in the renal

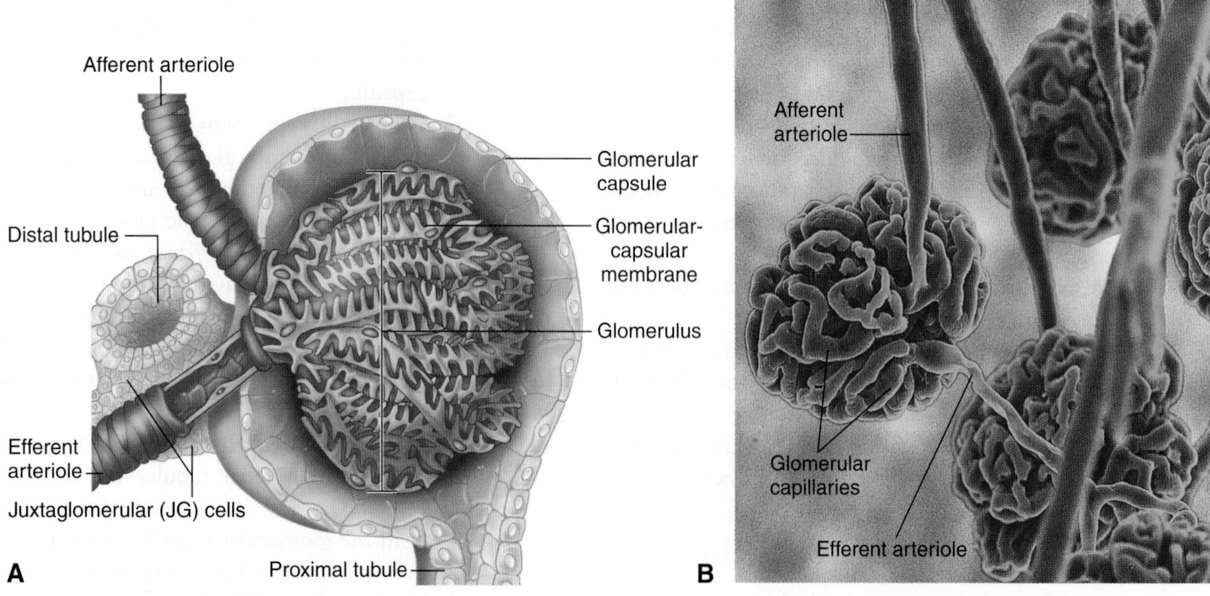

FIGURE 20-4 Renal corpuscle. A, Schematic showing relationship of glomerulus to glomerular capsule (Bowman capsule)—together called the *renal corpuscle*—and adjacent structures. **B,** Scanning electron micrograph showing several glomeruli and their associated blood vessels. The difference in diameters of afferent and efferent arterioles is clearly visible.

cortex and are called **cortical nephrons.** The remainder, called **juxtamedullary nephrons,** have their renal corpuscles near the junction (juxta) betweevn the cortex and medullary layers. These nephrons have nephron loops that dip far into the medulla. Juxtamedullary nephrons have an important role in concentrating urine.

Overview of Kidney Function

The kidneys are vital organs. The function they perform, that of forming urine, is essential for homeostasis and maintenance of life. Early in the process of urine formation, fluid, electrolytes, and wastes from metabolism are filtered from the blood and enter the nephron. Additional wastes may be secreted into the tubules of the nephron while substances useful to the body are reabsorbed into the blood.

Normally the kidneys balance the amount of many substances entering and leaving the blood over time so that normal concentrations can be maintained. In short, the kidneys adjust their output to equal the intake of the body. By eliminating wastes and adjusting fluid and electrolyte balance, the kidneys play an essential part in maintaining homeostasis of the whole body.

Homeostasis cannot be maintained—nor can life itself—if the kidneys fail and the condition is not soon corrected. Nitrogenous waste products accumulate as a result of protein breakdown and quickly reach toxic levels if not excreted. If kidney function is greatly reduced because of aging, injury, or disease, life can be maintained by using an artificial kidney to cleanse the blood of wastes.

Excretion of toxins and of nitrogen-containing waste products such as urea and ammonia represents only one of the important responsibilities of the kidney. The kidneys also play a key role in regulating the levels of many chemical substances in the blood such as chloride, sodium, potassium, and bicarbonate. The kidneys also regulate the proper balance between body water content and salt by selectively retaining or excreting both substances as requirements demand.

In addition, the cells of the **juxtaglomerular (JG) apparatus** (see **Figure 20-4**, *A*, and **Figure 20-5**) also function in blood volume

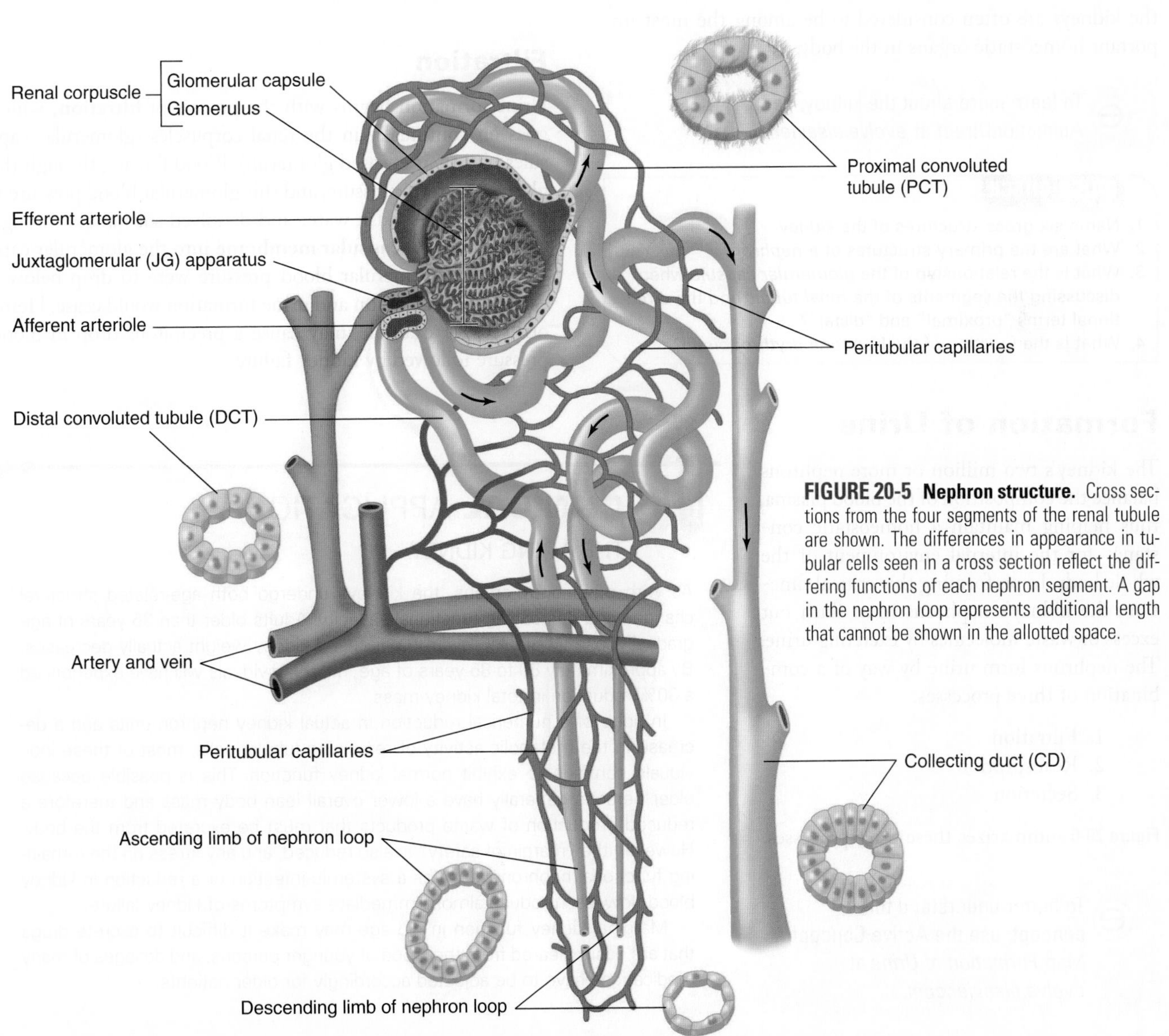

FIGURE 20-5 Nephron structure. Cross sections from the four segments of the renal tubule are shown. The differences in appearance in tubular cells seen in a cross section reflect the differing functions of each nephron segment. A gap in the nephron loop represents additional length that cannot be shown in the allotted space.

20

and blood pressure regulation. When blood pressure is low, which often occurs when blood plasma volume is low, these JG cells secrete an enzyme that triggers a system (discussed later in this chapter) to help restore normal blood volume and pressure.

Yet another important function of the kidney is secretion of the hormone **erythropoietin (EPO).** As a response to *hypoxia,* a deficiency of oxygen in the body, erythropoietin is released into the bloodstream. EPO travels in the bloodstream to the red bone marrow, where it stimulates the production of additional erythrocytes (red blood cells). The additional erythrocytes increase the ability of the blood to absorb and transport oxygen to oxygen-starved tissues.

EPO is sometimes used as a drug (one brand is Procrit) to treat anemia caused by critical illness such as cancer. EPO is sometimes abused by athletes attempting to improve their athletic performance by boosting hematocrit—thus increasing the oxygen-carrying capacity of their blood. (See *Blood Doping on p. 354.*)

As you probably guessed, kidney disease can cause anemia by reducing the body's ability to produce EPO when needed.

With all these vital functions, it is easy to understand why the kidneys are often considered to be among the most important homeostatic organs in the body.

To learn more about the kidney, go to AnimationDirect at *evolve.elsevier.com.*

QUICK CHECK

1. Name six gross structures of the *kidney.*
2. What are the primary structures of a *nephron*?
3. What is the relationship of the *glomerular capsule* when discussing the segments of the *renal tubule* and the directional terms "proximal" and "distal"?
4. What is the function of the hormone *erythropoietin*?

Glomerulus — Peritubular capillaries — Distal tubule (DCT) — Glomerular capsule — Proximal tubule (PCT) — Glucose — Nephron loop — Collecting duct (CD) — Na^+ — H_2O — K^+ — NH_3 — H^+

Filtration
Reabsorption
Secretion

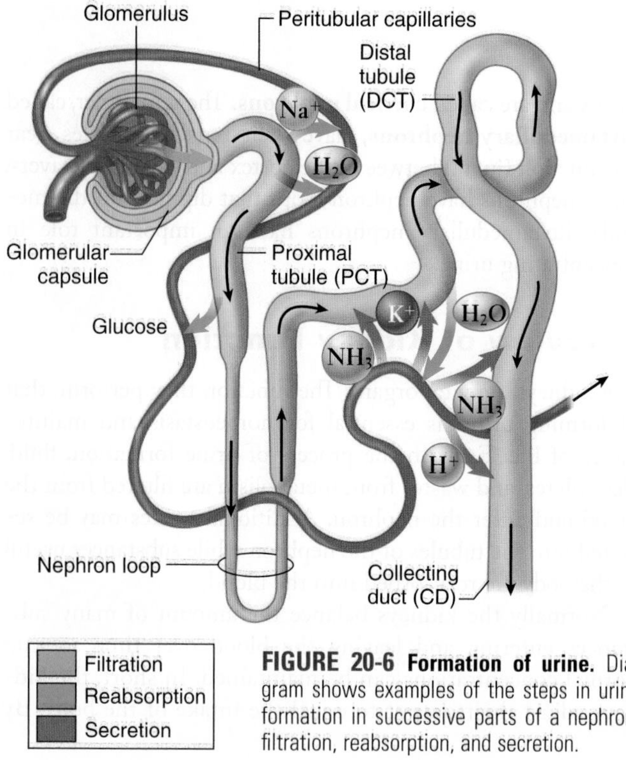

FIGURE 20-6 Formation of urine. Diagram shows examples of the steps in urine formation in successive parts of a nephron: filtration, reabsorption, and secretion.

Formation of Urine

The kidney's two million or more nephrons balance the composition of the blood plasma, thus helping maintain a homeostatic constancy for the internal environment of the whole body. In performing this critical function, the kidney's nephrons must flush out excess or waste molecules by excreting urine. The nephrons form urine by way of a combination of three processes:

1. Filtration
2. Reabsorption
3. Secretion

Figure 20-6 summarizes these three processes.

To better understand this concept, use the Active Concept Map *Formation of Urine* at *evolve.elsevier.com.*

Filtration

Urine formation begins with the process of **filtration,** which goes on continually in the renal corpuscles (glomerular capsules plus their encased glomeruli). Blood flowing through the glomeruli exerts pressure, and this glomerular blood pressure is high enough to push water and dissolved substances through the **glomerular-capsular membrane** into the glomerular capsule. If the glomerular blood pressure were to drop below a certain level, filtration and urine formation would cease. Hemorrhage, for example, may cause a precipitous drop in blood pressure followed by kidney failure.

CLINICAL APPLICATION

THE AGING KIDNEY

As with other body organs, the kidneys undergo both age-related structural changes and decreasing functional capacity. Adults older than 35 years of age gradually lose functional nephron units, and kidney weight actually decreases. By approximately 80 to 85 years of age, most individuals will have experienced a 30% reduction in total kidney mass.

In spite of a numerical reduction in actual kidney nephron units and a decrease in the metabolic activity of remaining tubular cells, most of these individuals continue to exhibit normal kidney function. This is possible because older persons generally have a lower overall lean body mass and therefore a reduced production of waste products that must be excreted from the body. However, the "margin of safety" is also reduced, and any stress on the remaining functional nephrons, such as a systemic infection or a reduction in kidney blood flow, can produce almost immediate symptoms of kidney failure.

Marginal kidney function in old age may make it difficult to excrete drugs that are easily cleared from the blood of younger persons, and dosages of many medications have to be adjusted accordingly for older patients.

Glomerular filtration normally occurs at the rate of 125 mL per minute. This is equivalent to about 180 L (nearly 50 gallons) of **glomerular filtrate** being produced by the kidneys every day.

Obviously no one ever excretes anywhere near 180 L of urine per day. Why? Because most of the fluid that leaves the blood by glomerular filtration, the first process in urine formation, returns to the blood by the second process—*reabsorption.*

Reabsorption

Reabsorption is the movement of substances out of the renal tubules into the blood capillaries located around the tubules (peritubular capillaries). Water, glucose, and other nutrients, as well as sodium and other ions, are substances that are reabsorbed. Reabsorption begins in the proximal convoluted tubules and continues in the nephron loop, distal convoluted tubules, and collecting ducts.

Large amounts of water—approximately 120 L per day—are reabsorbed by osmosis from the proximal tubules. In other words, roughly two-thirds of the 180 L of water that leaves the blood each day by glomerular filtration returns to the blood by proximal tubule reabsorption. The proximal tubules also reabsorb about two-thirds of most ions, as well as nearly all of the small organic molecules.

Smaller amounts of water and ions are later reabsorbed in the nephron loops, distal tubules, and collecting ducts.

Common table salt (NaCl) consumed in the diet or introduced by intravenous (IV) infusion of normal saline (0.9% NaCl) or other NaCl-containing fluids, provides the body with sodium ions (Na^+) and chloride ions (Cl^-). For the most part, sodium ions are actively transported back into blood from the tubular fluid in all segments of the kidney tubule except the collecting ducts.

Sodium reabsorption in the nephron loop is a special case. The nephron loop and its surrounding peritubular capillaries dip far into the medulla and then return back up in what is called *countercurrent* flow (see **Figure 20-5** on p. 559). This countercurrent flow—flow in opposite directions—of filtrate back up the nephron loop permits transport of large amounts of sodium and chloride into the interstitial fluid of the medulla. This makes the medulla very salty—or **hyperosmotic.** Hyperosmotic solutions are so named because they generally promote osmosis of water (into them), as do hypertonic solutions.

In addition, the countercurrent flow of blood in the peritubular capillaries surrounding the nephron loop fails to remove all of the excess sodium and chloride from the renal medulla.

Together, these **countercurrent mechanisms** maintain hyperosmotic conditions in the medulla. By maintaining a hyperosmotic medulla, the kidney is able to concentrate urine by reabsorbing more water by osmosis than otherwise possible. How the kidney thus regulates urine volume is covered subsequently.

The amount of sodium reabsorbed depends largely on the body's intake. In general the greater the amount of sodium intake, the less the amount reabsorbed and the greater the amount excreted in the urine. Also, the less sodium intake, the greater the reabsorption from kidney tubules and the less excreted in the urine.

Rather than being actively reabsorbed from renal tubules as are sodium ions (Na^+), chloride ions (Cl^-) passively move into blood because they carry a negative electrical charge. The positively charged sodium ions that have been reabsorbed and moved into the blood "attract" the negatively charged chloride ions from the tubule fluid into the peritubular capillaries.

Figure 20-7 explains the details of how sodium, chloride, and water are reabsorbed across the tubule wall and into the peritubular blood. Take a few minutes to review each step in the diagram.

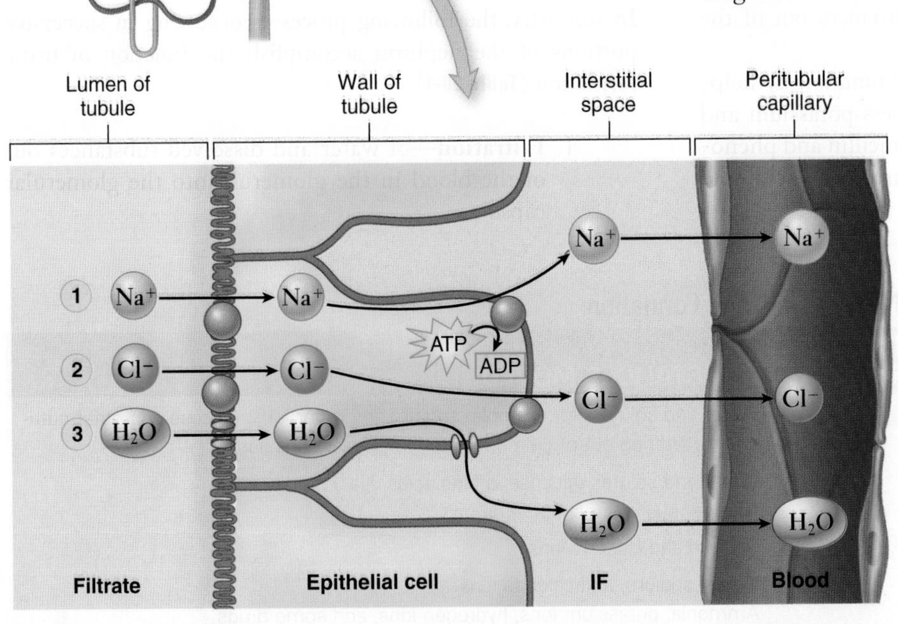

FIGURE 20-7 Reabsorption of ions and water. 1 Sodium ions (Na^+) are pumped from the tubule cell to interstitial fluid (IF), thereby increasing the interstitial Na^+ concentration to a level that drives diffusion of Na^+ into blood. As Na^+ is pumped out of the cell, more Na^+ passively diffuses in from the filtrate to maintain an equilibrium of concentration. Enough Na^+ moves out of the tubule and into blood that an electrical gradient is established (blood is positive relative to the filtrate). 2 Electrical attraction between oppositely charged particles drives diffusion of negative ions in the filtrate, such as chloride (Cl^-), into blood. 3 When the ion concentration in blood increases, osmosis of water from the tubule occurs. Thus active transport of sodium creates a situation that promotes passive transport of negative ions and water.

20

All of the filtered glucose is normally reabsorbed from the proximal tubules into peritubular capillary blood. None of this valuable nutrient is wasted by being lost in the urine. **Figure 20-8** shows how sodium-glucose carriers in the tubule wall allow glucose molecules to passively "tag along" as sodium is actively reabsorbed back into the blood.

The **transport maximum (T_{max})**—the largest amount of any substance that can be reabsorbed at one time—is determined mainly by the number of available transporters of that substance. The transport maximum of any substance helps determine the **renal threshold**—the amount of substance in the blood above which the kidney excretes the excess substance in the urine.

Unfortunately, sometimes not all the glucose in the tubule filtrate is recovered by the blood. For example, in *diabetes mellitus (DM),* if blood glucose concentration increases above the renal threshold, tubular filtrate then contains more glucose than kidney tubule cells can reabsorb. There are not enough sodium-glucose transporters to handle the excess glucose immediately. Some of the glucose therefore remains behind in the urine. Glucose in the urine—**glycosuria**—is a well-known sign of DM.

Retaining glucose in the urine also promotes other classic symptoms of DM. For example, high glucose concentration in the tubular filtrate means that less water will leave the tubules by osmosis. Thus urine is dilute and has a higher than normal volume—resulting in the excessive urination *(polyuria)* typical of untreated DM. This increased loss of water from the body, in turn, triggers thirst—called **polydipsia**—also a classic indicator of DM. Now is a good time to review the DM concept map shown in **Figure 12-18** on p. 337.

Secretion

Secretion is the process by which substances move into urine in the distal and collecting ducts from blood in the capillaries around these tubules. In this respect, secretion is reabsorption in reverse. Whereas reabsorption moves substances out of the urine into the blood, secretion moves substances out of the blood into the urine.

Tubular secretion serves an important function by helping to remove or "clear" the blood of excess potassium and hydrogen ions, certain drugs including penicillin and phenobarbital, and numerous wastes such as urea, uric acid, and creatinine.

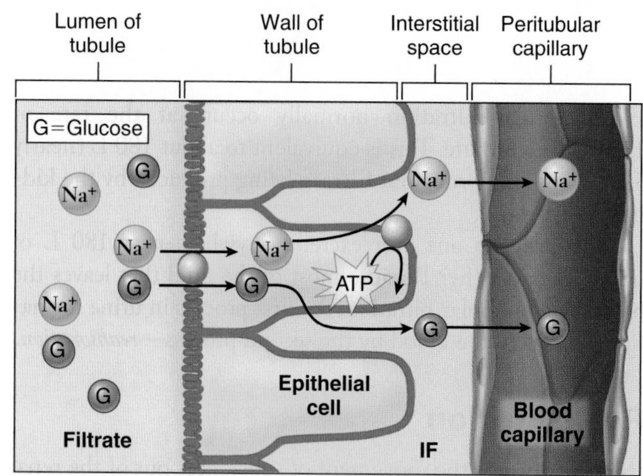

FIGURE 20-8 Reabsorption of glucose. The presence of sodium–glucose transporters provides a way for the active transport of sodium to also passively transport glucose across tubule cells and eventually back into the blood. The availability of these transporters may limit how much glucose can be reabsorbed at one time.

Many substances that are secreted from peritubular blood enter the filtrate primarily in the proximal tubule and, to a lesser extent, the distal convoluted tubule and collecting ducts.

The major exception to this "rule of thumb" is potassium ion, which is secreted primarily into the collecting ducts in an exchange with sodium. Urine volume of potassium ions (K^+) varies greatly with diet. Some **diuretic drugs,** which stimulate the production of urine (see the box on p. 591), are said to be "potassium wasting" because they increase *secretion* of potassium into tubular fluid and thus its excretion in the urine.

In the distal convoluted tubules and collecting ducts, sodium secretion is dependent on hormones that are also important in regulating urine volume, as discussed below. Ammonia is secreted passively by diffusion.

Kidney tubule secretion plays a crucial role in maintaining the body's fluid, electrolyte, and acid-base balance discussed in Chapters 21 and 22.

Summary of Urine Formation

In summary, the following processes occurring in successive portions of the nephron accomplish the function of urine formation (**Table 20-1**):

1. **Filtration**—of water and dissolved substances out of the blood in the glomeruli into the glomerular capsule.

20

TABLE 20-1 Functions of Parts of Nephron in Urine Formation

PART OF NEPHRON	PROCESS IN URINE FORMATION	SUBSTANCES MOVED
Glomerulus and glomerular capsule	Filtration	Water and solutes (for example, sodium and other ions, glucose and other nutrients filtering out of glomeruli into glomerular capsules)
Proximal tubule	Reabsorption Secretion	Water and solutes (glucose, amino acids, Na^+) Nitrogenous wastes, some drugs
Nephron loop	Reabsorption	Sodium and chloride ions
Distal and collecting tubules	Reabsorption Secretion	Water, sodium, and chloride ions Ammonia, potassium ions, hydrogen ions, and some drugs

2. **Reabsorption**—of water and dissolved substances out of kidney tubules back into blood. This prevents substances needed by the body from being lost in urine. Usually, up to 99% of water, sodium, and chloride filtered out of glomerular blood is retrieved from tubules—along with 100% of glucose and other small organic molecules.
3. **Secretion**—of hydrogen ions, potassium ions, and certain drugs from blood into kidney tubules.

To learn more about urine formation, go to AnimationDirect at *evolve.elsevier.com.*

QUICK CHECK

1. What are the three basic processes that occur in the nephron?
2. Where does *filtration* occur in the nephron?
3. Where does *reabsorption* occur in the nephron?

Control of Urine Volume

The body has ways to control the amount and composition of the urine that it excretes. It does this mainly by controlling the amount of water and dissolved substances that are reabsorbed by the kidney tubules.

Antidiuretic Hormone

An example of regulating water reabsorption in kidney tubules involves a hormone called **antidiuretic hormone (ADH)** secreted from the posterior pituitary gland. ADH decreases the amount of urine by making collecting ducts (CDs) permeable to water.

If no ADH is present, the CDs are practically impermeable to water, so little or no water is reabsorbed from them. When ADH appears in the blood, CDs become permeable to water and water is reabsorbed from them. As a result, less water is lost from the body as urine, and thus more water is retained by the body—think of it in whichever way you find it easier to remember. At any rate, for this reason ADH can accurately be described as the "water-retaining hormone" or the "urine-decreasing hormone."

Recall that the countercurrent mechanisms of the nephron loop and its capillaries maintain a hyperosmotic (salty) medulla. When filtrate moves down the collecting ducts, the action of ADH allows osmosis of water to equilibrate with the hyperosmotic interstitial fluid of the medulla—thus removing more water from the filtrate than would otherwise be possible. Maintaining a salty or hyperosmotic medulla allows ADH to have a pronounced effect in concentrating urine, thereby conserving the body's valuable water.

Aldosterone

The hormone **aldosterone,** secreted by the adrenal cortex, plays an important part in controlling the kidney tubules' reabsorption of sodium. Primarily, it stimulates the tubules to reabsorb sodium at a faster rate. Secondarily, aldosterone also increases tubular water reabsorption because "water always follows sodium" by osmosis whenever possible. The term *salt- and water-retaining hormone* therefore is an apt descriptive nickname for aldosterone. Like ADH, aldosterone reduces urine volume.

The kidney itself is responsible for triggering aldosterone secretion, a fact that illustrates the importance of the kidney in regulating overall fluid volume and blood pressure in the body. When blood volume and pressure drop below normal, this is sensed by cells in the JG apparatus. JG cells then release an enzyme called **renin** that initiates the **renin-angiotensin-aldosterone system (RAAS).** The RAAS eventually produces constriction of blood vessels and by doing so, raises blood pressure. The RAAS also triggers adrenal gland secretion of aldosterone, which promotes water retention and thus increases total blood volume—thus also contributing to a rise in blood pressure. Figure 20-9 illustrates the main events of the RAAS and how it acts to restore normal plasma volume and blood pressure. Aldosterone mechanisms are also discussed in the next chapter.

Atrial Natriuretic Hormone

Another hormone, **atrial natriuretic hormone (ANH)** secreted from the heart's atrial wall, has the opposite effect of aldosterone. ANH is the primary *atrial natriuretic peptide (ANP)* hormone in humans. ANH stimulates kidney tubules to secrete more sodium and thus lose more water. ANH is a *salt- and water-losing hormone.* Thus ANH increases urine volume while reducing blood volume.

The body secretes ADH, aldosterone, and ANH in different amounts, depending on the homeostatic balance of body fluids at any particular moment.

Abnormalities of Urine Volume

Sometimes the kidneys do not excrete normal amounts of urine as a result of kidney disease, endocrine imbalances, cardiovascular disease, stress, or a variety of other conditions. Here are some terms associated with abnormal amounts of urine:

1. **Anuria**—absence of urine
2. **Oliguria**—scanty amount of urine
3. **Polyuria**—unusually large amount of urine

Because a change in urine volume or output is a significant indicator of many types of fluid alterations and diseases, measurement of both fluid intake and fluid output (urine volume) over a specified period of time, often abbreviated as **I & O,** is a common practice in clinical medicine. The normal adult urine output is about 1500 to 1600 mL per day.

QUICK CHECK

1. What is the function of *ANH*?
2. How do *ADH* and *aldosterone* affect urine output?
3. How do *anuria* and *polyuria* differ?

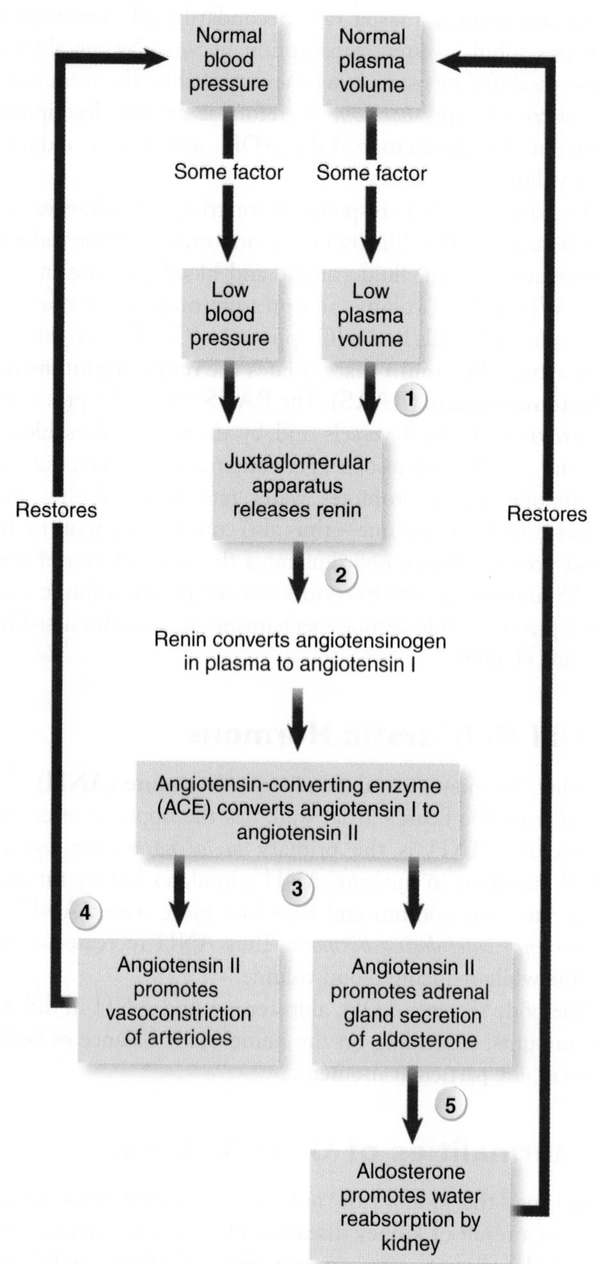

FIGURE 20-9 Renin-angiotensin-aldosterone system (RAAS).
1 Low plasma volume reduces blood pressure below normal, which is detected by juxtaglomerular (JG) cells in the juxtaglomerular apparatus of the kidney.
2 This triggers JG cells to release the enzyme renin, which converts angiotensinogen into angiotensin I.
3 The *angiotensin converting enzyme (ACE)* (found in various tissues) then converts angiotensin I to angiotensin II.
4 Angiotensin II stimulates constriction of arteriolar smooth muscles, increasing blood pressure back toward normal.
5 Angiotensin II also triggers adrenal gland secretion of aldosterone, which promotes water retention by the kidney and thus restoration of normal blood volume and pressure.

Elimination of Urine

Once urine is formed by the kidneys, it must be eliminated from the body. Our discussion now returns to a focus on anatomy as we discuss the "plumbing" needed to drain the urine away.

Ureters

Urine drains out of the collecting tubules of each kidney into the renal pelvis and down the ureter into the urinary bladder (see **Figure 20-1**). The **renal pelvis** is the basinlike upper end of the ureter located inside the kidney. **Ureters** are narrow tubes less than 6 millimeters (mm) (¼ inch) wide and 25 to 30 centimeters (cm) (10 to 12 inches) long.

Mucous membranes featuring easily stretchable *transitional epithelium* line both ureters and each renal pelvis. Note in **Figure 20-10** that the ureter has a thick, muscular wall. Contraction of the muscular coat produces peristaltic-type movements that assist in moving urine down the ureters into the bladder. The lining membrane of the ureters is richly supplied with sensory nerve endings.

Episodes of **renal colic**—pain caused by the passage of a kidney stone—have been described in medical writings since antiquity. Kidney stones cause intense pain if they have sharp edges or are large enough to distend the walls or cut the lining of the ureters or urethra as they pass from the kidneys to the exterior of the body. Some of the pain is caused by tearing or stretching of the urinary lining—along with the accompanying inflammation. However, much of the pain is associated with cramping of muscles that attempt to push a kidney stone forward. The term "colic" is used because of its similarity to

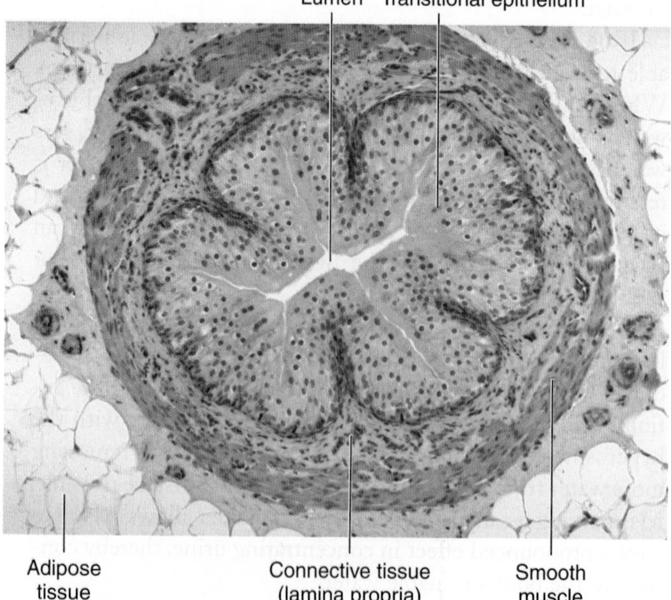

FIGURE 20-10 Ureter cross section. Note the many folds of the mucous lining (transitional epithelium) that permit stretching as urine passes through the tube. A thick muscular layer of smooth muscle helps "pump" urine toward the bladder. On its outer surface the ureter is covered by a tough fibrous connective tissue coat.

painful cramps sometimes experienced in the muscle layers of the colon.

Medical imaging techniques can clearly outline the segments of the urinary tract to show possible abnormalities. To see examples, check out the article *Visualizing the Urinary Tract* at Connect It! at *evolve.elsevier.com.*

Urinary Bladder

The empty urinary bladder lies in the pelvis just behind the pubic symphysis. When full of urine, it projects upward into the lower portion of the abdominal cavity. In women it sits in front of the uterus, whereas in men, it rests on the prostate.

Elastic fibers and involuntary muscle fibers in the wall of the urinary bladder make it well suited for expanding to hold variable amounts of urine and then contracting to empty itself. Mucous membrane containing transitional epithelium lines the urinary bladder (Figure 20-11). The lining is loosely attached to the deeper muscular layer so that the bladder is very wrinkled and lies in folds called **rugae** when it is empty. When the bladder is filled, its inner surface may stretch until it is smooth.

Note in Figure 20-11, *A,* that one triangular area on the back or posterior surface of the bladder is free of rugae. This area, called the **trigone,** is always smooth. There, the lining membrane is tightly fixed to the deeper muscle coat. The trigone extends between the openings of the two ureters above and the point of exit of the urethra below.

Urethra

To leave the body, urine passes from the bladder, down the **urethra,** and out of its external opening, the **urinary meatus.** In other words, the urethra is the lowest and last part of the urinary tract.

The same sheet of mucous membrane that lines each renal pelvis, the ureters, and the bladder extends down into the urethra, too. It is worth noting the continuity of the urinary mucous lining because it accounts for the fact that an infection of the urethra may spread upward through the urinary tract to cause **cystitis** (bladder infection).

The urethra is a narrow tube; it is only about 4 cm (1½ inches) long in a woman, but it is about 20 cm (8 inches) long in a man. In a man, the urethra has two functions: (1) it is the terminal portion of the urinary tract, and (2) it is the passageway for movement of the reproductive fluid (semen) from the body. In a woman, the urethra is a part of only the urinary tract.

Micturition

The terms **micturition, urination,** and **voiding** all refer to the passage of urine from the body or the emptying of the bladder.

Two **sphincters** (rings of muscle tissue) act as valves that guard the pathway leading from the bladder. The *internal urethral sphincter* is located at the bladder exit, and the *external urethral sphincter* circles the urethra just below the neck of the bladder (see Figure 20-11). When contracted, both sphincters seal off the bladder and allow urine to accumulate without leaking to the exterior. The internal urethral sphincter is involuntary, and the external urethral sphincter is composed of striated muscle and is under voluntary control.

The muscular wall of the bladder permits this organ to accommodate a considerable volume of urine with very little increase in pressure until a volume of 300 to 400 mL is reached. As the volume of urine increases, the need to void may be noticed at volumes of 150 mL, but micturition in adults does not normally occur much below volumes of 350 mL.

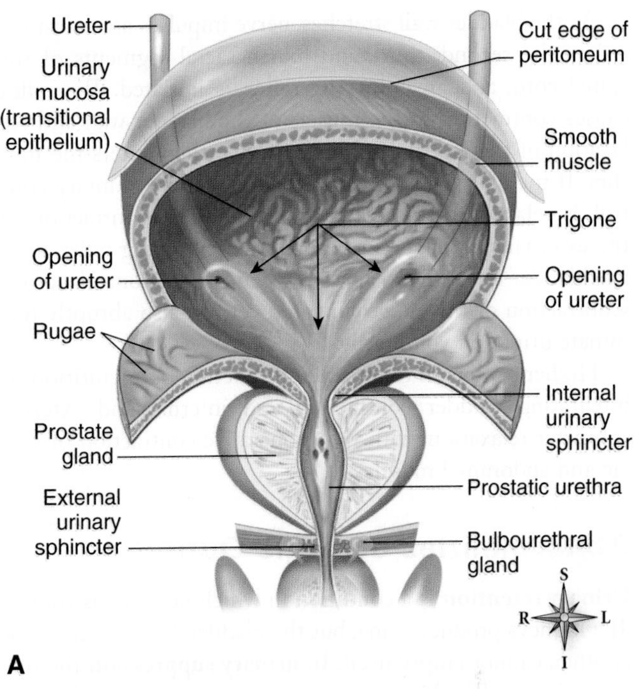

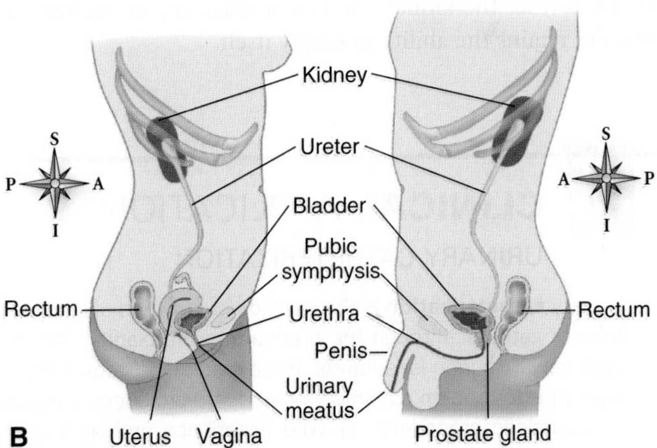

FIGURE 20-11 Urinary bladder. A, Frontal view of a fully distended male bladder dissected to show the interior. Note the relationship of the prostate gland, which surrounds the urethra as it exits the bladder. This relationship is discussed in Chapter 23. **B,** Sagittal section of the female urinary system *(left)* and the male urinary system *(right)* showing relationship of the bladder to other anatomical structures.

20

As the bladder wall stretches, nerve impulses are transmitted to the second, third, and fourth sacral segments of the spinal cord, and an **emptying reflex** is initiated. The reflex causes contraction of the muscle of the bladder wall and relaxation of the internal sphincter. Urine then enters the urethra. If the external sphincter, which is under voluntary control, is relaxed, micturition occurs. Voluntary contraction of the external sphincter can suppress the emptying reflex until the bladder is filled to capacity, when loss of control occurs. Contraction of this powerful sphincter can also abruptly terminate urination voluntarily.

Higher centers in the brain also function in micturition by integrating bladder contraction and internal and external sphincter relaxation, with the cooperative contraction of pelvic and abdominal muscles.

Abnormalities of Urine Output

Urinary retention is a condition in which no urine is voided. The kidneys produce urine, but the bladder, for one reason or another, cannot empty itself. In **urinary suppression** the opposite is true. The kidneys do not produce any urine, but the bladder retains the ability to empty itself.

Micturition is a complex body function. It requires control and integration of both voluntary and involuntary nervous system components acting on a variety of anatomic structures. Unfortunately, homeostatic control problems occur quite frequently in this complex system. In addition to the 15% to 20% of children who experience some degree of enuresis, voiding dysfunction affects nearly 15 million adult Americans. People over 60 are especially at risk, with elderly women affected almost twice as often as men.

Urinary incontinence or **enuresis** refers to involuntary voiding or loss of urine in an older child or adult. *Urge incontinence* is associated with smooth muscle overactivity in the bladder wall. The term *stress incontinence* is often used to describe the type of urine loss associated with laughing, coughing, or heavy lifting. It is a common problem in women with weakened pelvic floor muscles following pregnancy. So-called *overflow incontinence* is characterized by intermittent dribbling of urine. It results from urinary retention and an overdistended bladder—a common problem in men with an enlarged prostate gland (see Chapter 23).

Reflex incontinence occurs in the absence of any sensory warning or awareness. It is common in nervous system disorders such as stroke, parkinsonism, or spinal cord injury. If

CLINICAL APPLICATION

URINARY CATHETERIZATION

Urinary **catheterization** is the passage or insertion of a hollow tube or catheter through the urethra into the bladder for the withdrawal of urine (see figure). It is a medical procedure commonly performed on patients who undergo prolonged surgical or diagnostic procedures or who experience problems with urinary retention.

Correct catheterization procedures require aseptic techniques to prevent the introduction of infectious bacteria into the urinary system. Clinical studies have proved that improper catheterization techniques cause bladder infections—a condition called *cystitis*—and point out the need for extensive training of health professionals who perform catheterizations. To minimize the risk of infection, some facilities now use ultrasound imaging of the bladder to determine whether urine is being involuntarily retained in the bladder—replacing the former practice of catheterizing a patient at timed intervals.

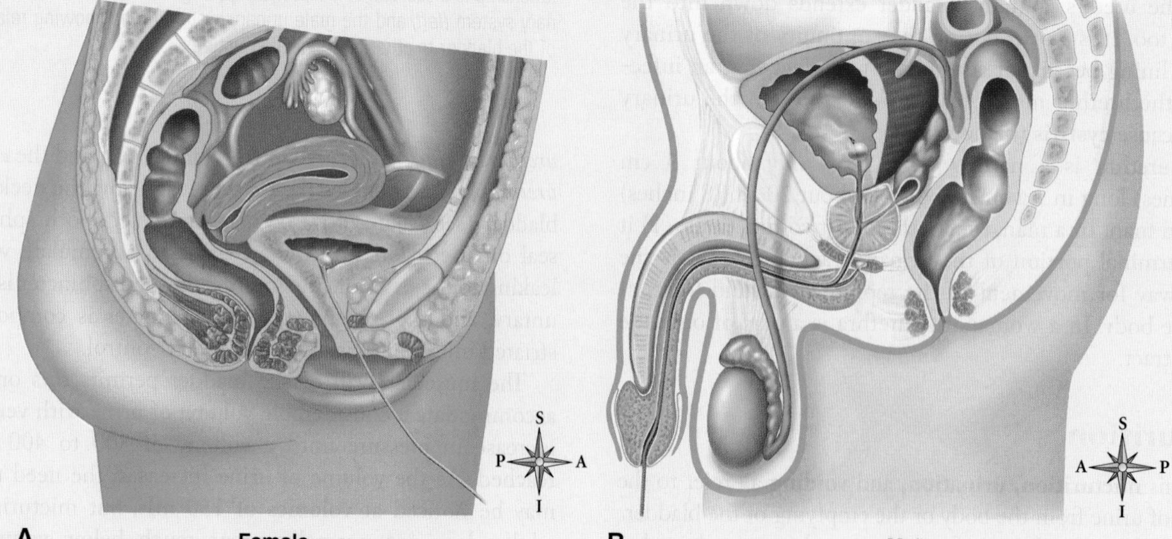

totally cut off from spinal innervation, the bladder musculature acquires some automatic action, and periodic but unpredictable voiding occurs—a condition called **neurogenic bladder.**

Bed wetting at night *(nocturnal enuresis)* often occurs in a child who is beyond the age when voluntary bladder control is expected. Incidence is higher in boys than in girls and is often due to maturational delay of the complex urinary reflexes needed for voluntary control of micturition.

QUICK CHECK

1. Through what tube does urine leave the kidney?
2. What structural characteristics of the bladder allow it to expand to hold urine?
3. Through what structure does urine pass from the bladder to the outside of the body?
4. What is *overflow incontinence*?

Urinalysis

The physical, chemical, and microscopic examination of urine is termed **urinalysis.** Like blood, urine is a fluid whose study can reveal much about the normal and abnormal functioning of the body. Changes in the normal characteristics of urine or the appearance of abnormal urine characteristics may be a sign of disease. Table 20-2 lists both the normal and abnormal characteristics of urine.

In clinical and laboratory situations, a standard urinalysis is often referred to as a "routine and microscopic" urinalysis, or simply an "R and M." The "routine" portion is a series of physical and chemical tests, whereas the "microscopic" part refers to the study of urine sediment with a microscope. This series of laboratory tests provides the variety of information often necessary for a physician to make a diagnosis.

Crushing injuries of skeletal muscle release intracellular contents into the bloodstream, which puts a heavy burden on the kidneys. Review the article *Rhabdomyolysis* and how it can be detected through urinalysis at Connect It! at *evolve.elsevier.com.*

HEALTH AND WELL-BEING

PROTEINURIA AFTER EXERCISE

Proteinuria is the presence of plasma proteins in the urine. Proteinuria is probably the most important indicator of renal disease (**nephropathy**) because only damaged nephrons consistently allow plasma protein molecules to leave the blood. However, intense exercise causes temporary proteinuria in many individuals. Some exercise physiologists believed that intense athletic activities cause kidney damage, but subsequent research has ruled out that explanation. One current hypothesis is that hormonal changes during strenuous exercise increase the permeability of the nephron's filtration membrane, allowing some plasma proteins to enter the filtrate. Some postexercise proteinuria is usually considered normal.

Renal and Urinary Disorders

You may have experienced the discomfort and painful, burning urination, called **dysuria,** associated with a bladder infection or know someone who has. Bladder infection is the most common urinary disorder, but it usually is not serious if promptly treated. A number of renal and urinary disorders are very serious, however. Any disorder that significantly reduces the effectiveness of the kidneys is immediately life threatening. In this section, we discuss some life-threatening kidney diseases, as well as a few of the less serious but more common disorders.

Obstructive Disorders

Obstructive urinary disorders are abnormalities that interfere with normal urine flow anywhere in the urinary tract. The severity of obstructive disorders depends on the location of the interference and the degree to which the flow of urine is impaired. Obstruction of urine flow usually results in "backing up" of the urine, perhaps all the way to the kidney itself.

The term **hydronephrosis** is used to describe pathological swelling or enlargement of the renal pelvis and calyces caused by blockage of urine outflow. The condition may be the result of congenital problems or be caused by blockage caused by stones, tumors, or inflammation. Regardless of cause, if left untreated, much of the internal structure of the kidney is lost as the cortex thins and medullary tissue is destroyed (Figure 20-12).

Some of the more important obstructive conditions are summarized in the following paragraphs.

Renal Calculi

Renal calculi, or *kidney stones,* are crystallized mineral chunks that develop in the renal pelvis or calyces (Figure 20-13). Calculi develop when calcium and other minerals, such as uric acid, crystallize on the renal papillae, then break off into the urine. *Staghorn calculi* are large, branched stones that form in, and take the shape of, the pelvis and branched calyces.

If the stones are small enough, they will simply flow through the ureters and eventually be voided with the urine. Larger stones may obstruct the ureters, causing an intense pain called *renal colic* as rhythmic muscle contractions of the ureter attempt to dislodge the stones. Hydronephrosis may occur if the stone does not move from its obstructing position.

Tumors

Tumors of the urinary system typically obstruct urine flow, possibly causing hydronephrosis in one or both kidneys. Most kidney tumors are malignant neoplasms called *renal cell carcinomas.* They usually occur only in one kidney. Renal cell carcinoma metastasizes most often to the lungs and bone tissue. *Bladder cancer* occurs about as frequently as renal cancer (each accounts for about two in every hundred cancer cases) and is often found in association with bladder stones.

TABLE 20-2 Characteristics of Urine

NORMAL CHARACTERISTICS	ABNORMAL CHARACTERISTICS
Color and Clarity	
Normal urine: Should be clear; color varies with specific gravity (Occasionally, normal urine may be cloudy because of high dietary levels of fat or phosphate.) Dilute urine: Light yellow to yellow Concentrated urine: Dark yellow to dark amber 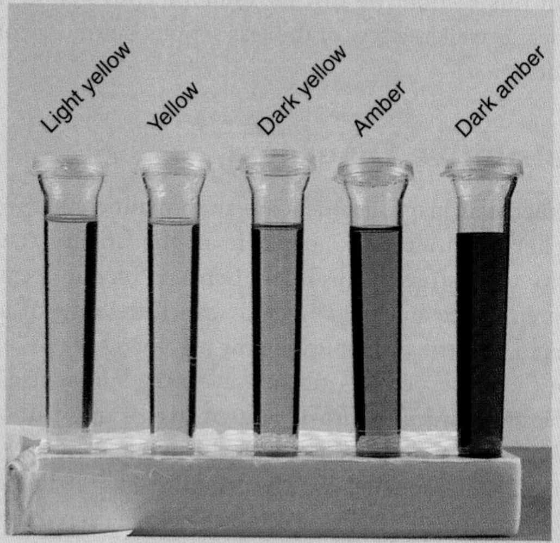 	Abnormally colored urine may result from: (1) pathological conditions, (2) certain foods, and (3) numerous drugs: 1. Pathological conditions (examples): Kidney cancer (hemorrhage)—red (RBCs) Bile duct obstruction (gallstones)—orange/yellow (bilirubin) *Pseudomonas* infection—green (bacterial toxins) Rhabdomyolysis—brown 2. Foods (examples): Beets—red Rhubarb—brown Carrots—dark yellow Vitamin supplements—bright yellow 3. Drugs (examples): Pyridium (urinary tract analgesic)—orange Dilantin (anticonvulsant)—pink/red brown Dyrenium (diuretic)—pale blue Cloudy urine may result from (examples): 1. Bacteria—active infection of urinary system organs 2. Blood cells RBCs—hemorrhage from kidney cancer WBCs—pus from urinary tract infection (UTI) 3. Casts—various types of tubelike clumps (blood cell, epithelial, hyaline, waxy, etc.) that form in diseased renal tubes 4. Proteinuria—(protein—usually albumin) in urine 5. Crystals—usually uric acid or phosphate/calcium oxalate in concentrated urine
Compounds	
Mineral ions (for example, Na^+, Cl^-, K^+) Nitrogenous wastes: Ammonia, creatinine, urea, uric acid Urine pigment: Urochrome (product of bilirubin metabolism)	Ketones—generally acetone Protein—generally albumin Glucose Crystals—generally uric acid and phosphate or calcium oxalate Pigments—abnormal levels of bilirubin metabolites
Odor	
Slight aromatic Some foods produce a characteristic odor (asparagus) Ammonialike odor on standing may result from decomposition in stored urine	Strong, sweet, fruity (acetone) odor—uncontrolled diabetes mellitus Foul odor—urinary tract infections (UTIs) Musty odor—phenylketonuria Maple syrup odor—congenital defect in protein metabolism
pH	
4.6-8.0 (average 6.0) Toward Low Normal: Some foods (meat and cranberries) and drugs (chlorothiazide diuretics) Toward High Normal: Some foods (citrus fruits, dairy products) and drugs (bicarbonate antacids)	High pH during alkalosis—kidneys compensate by excreting excess base Low pH during acidosis—kidneys compensate by excreting excess H^+
Specific Gravity	
Adult: 1.005-1.030 (usually, 1.010-1.025) Elderly: Values decrease with age Newborn: 1.001-1.020	Above normal limits—glycosuria, proteinuria, dehydration, high solute load (may result in precipitation of solutes and kidney stone formation) Below normal limits—chronic renal diseases (inability to concentrate urine), overhydration

20

Renal and bladder cancer have few symptoms early in their development, other than traces of blood in the urine, or **hematuria.** As the cancer develops, pelvic pain and symptoms of urinary obstruction may occur. Insertion of a **cystoscope** through the urethra and into the bladder permits direct inspection, biopsy, and surgical removal or treatment of bladder and other urinary tract lesions (**Figure 20-14,** *A*). The hollow tube allows passage of light, a viewing lens, and various catheters and

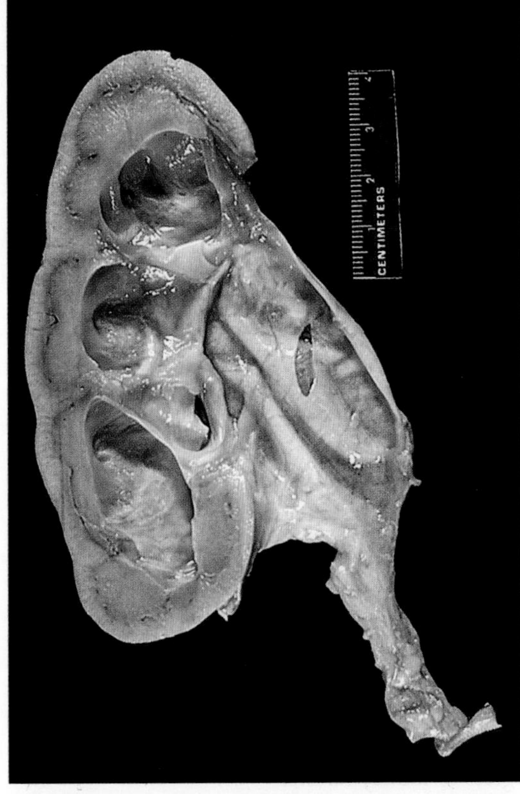

FIGURE 20-12 Hydronephrosis. Note the dramatic enlargement of the renal pelvis and calyces caused by blockage and "backing up" of urine.

operative devices. Figure 20-14, *B* shows the appearance of a malignant tumor on the bladder wall before its removal during a surgical cystoscopic procedure.

Renal tumors may require a biopsy to determine if they are cancerous. To learn more about how a **needle biopsy** can be used for this purpose, see the article *Kidney Biopsy* at Connect It! at *evolve.elsevier.com.*

Urinary Tract Infections

Most *urinary tract infections (UTIs)* are caused by bacteria, most often gram-negative types. UTIs can involve the urethra, bladder, ureter, and kidneys. Common types of urinary tract infections are summarized in the following paragraphs.

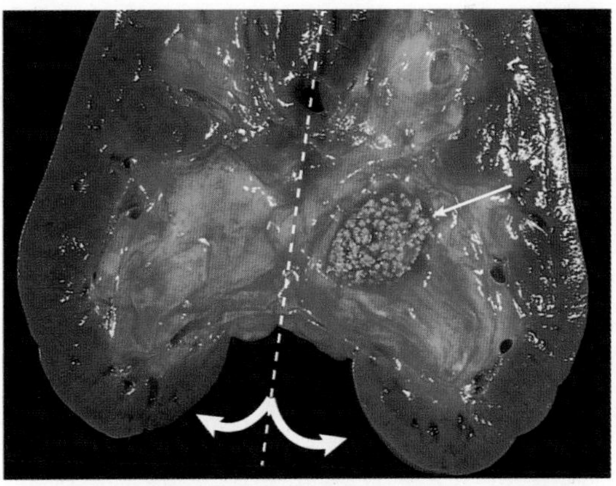

FIGURE 20-13 Renal calculi. Coronal (frontal) section of kidney, partially cut and opened like a book, showing a large stone *(arrow)* in the renal pelvis.

Urethritis

Urethritis is inflammation of the urethra that commonly results from bacterial infection, often *gonorrhea* (see Appendix A at *evolve.elsevier.com*). Nongonococcal urethritis is usually caused by *chlamydial* infection (see Appendix A). Males (particularly infants) suffer from urethritis more often than do females.

Cystitis

Cystitis is a term that refers to an inflammation of the bladder. Cystitis most commonly occurs as a result of infection but also can accompany calculi, tumors, or other conditions. Bacteria usually enter the bladder through the urethra. Cystitis occurs more often in women than in men because the female urethra is shorter and closer to the anus (a source of bacteria) than in the male. Bladder infections are characterized by pelvic pain, an urge to urinate frequently, painful urination (dysuria), and blood in the urine (hematuria).

If cystitis caused by bacterial infection becomes severe and chronic, the bladder epithelium may become ulcerated and covered with exudate. Extension of the infection may then inflame the ureters, renal pelvis, and kidney tissues. One common form of "nonbacterial" cystitis is *urethral syndrome.* Urethral

CLINICAL APPLICATION

REMOVAL OF KIDNEY STONES USING ULTRASOUND

Statistics suggest that approximately 1 in every 1000 adults in the United States suffers from kidney stones, or **renal calculi,** at some point in his or her life. Although symptoms of excruciating pain are common, many kidney stones are small enough to pass spontaneously out of the urinary system. If this is possible, no therapy is required other than treatment for pain and antibiotics if the calculi are associated with infection. Larger stones, however, may obstruct the flow of urine and therefore are much more serious and difficult to treat.

Formerly, only traditional surgical procedures were effective in removing relatively large stones that formed in the calyces and renal pelvis of the kidney. In addition to the risks that always accompany major medical procedures, surgical removal of stones from the kidneys frequently requires rather extensive hospital and home recovery periods, lasting 6 weeks or more.

A technique that uses ultrasound to pulverize the stones so that they can be flushed out of the urinary tract without surgery is now used in hospitals across the United States. The specially designed ultrasound generator required for the procedure is called a **lithotriptor.** Using a lithotriptor, physicians break up the stones with ultrasound waves—in a process called **lithotripsy**—without making an incision. Recovery time is minimal, and both patient risk and costs are reduced.

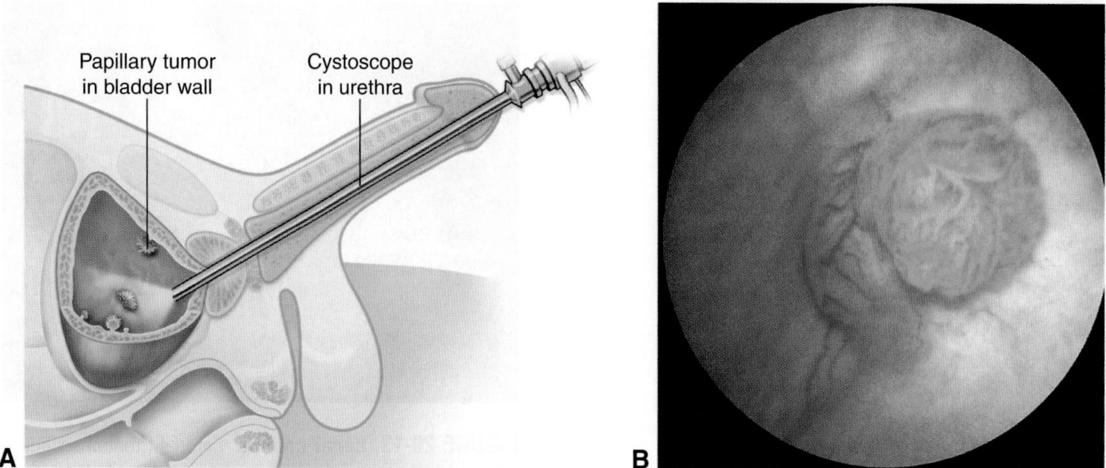

FIGURE 20-14 Imaging of bladder cancer. A, Cystoscope in male bladder. **B,** Cystoscopic view of a cancerous growth (a transitional cell carcinoma) on the bladder wall.

syndrome, which occurs most commonly in young women, has unknown causes but often develops into a bacterial infection.

The term **overactive bladder** refers to the need for frequent urination. The amounts voided are generally small, and feelings of extreme urgency and pain (dysuria) are common. The condition is called **interstitial cystitis** and is treated with drugs to decrease nervous stimulation and with physical distention of the bladder with fluid to increase capacity. The condition is another type of "nonbacterial" cystitis because symptoms occur without evidence of bacterial infection. Some clinicians believe it is autoimmune in origin because it is often associated with lupus (see Chapter 16).

Pyelonephritis

Nephritis is a general term referring to kidney disease, especially inflammatory conditions. **Pyelonephritis** is literally "pelvis nephritis" and refers to inflammation of the renal pelvis and connective tissues of the kidney. As with cystitis, pyelonephritis is usually caused by bacterial infection but also can result from viral infection, mycosis, calculi, tumors, pregnancy, and other conditions.

Acute pyelonephritis develops rapidly and is characterized by fever, chills, pain in the sides (flank), nausea, and an urge to urinate frequently. It often results from the spread of infection from the lower urinary tract or through the blood from other organs. *Chronic pyelonephritis* may be an autoimmune disease but is often preceded by a bacterial infection or urinary blockage.

Review the nature of the inflammatory response in the article *Inflammation* at Connect It! at *evolve.elsevier.com.*

20

Glomerular Disorders

Glomerulonephritis is a group of disorders that result from damage to the glomerular-capsular membrane. This damage can be caused by immune mechanisms, heredity, or bacterial infections. Without successful treatment, glomerular disorders can progress to kidney failure.

Nephrotic Syndrome

Nephrotic syndrome is a collection of signs and symptoms that accompany various glomerular disorders. This syndrome is characterized by the following:

1. **Proteinuria**—presence of proteins (especially *albumin*) in the urine. Protein, normally absent from urine, filters through damaged glomerular-capsular membranes and is not reabsorbed by the kidney tubules.
2. **Hypoalbuminemia**—low albumin concentration in the blood, resulting from the loss of albumin from the blood through holes in the damaged glomeruli. Albumin is the most abundant plasma protein. Because it normally cannot leave the blood vessels, it usually remains as a "permanent" solute in the plasma. This keeps plasma water concentration low and thus prevents osmosis of large amounts of water out of the blood and into tissue spaces. (Review the discussion of osmosis in Chapter 3 to help you understand this process.) In hypoalbuminemia, this osmotic function is lost, and fluid leaks out of blood vessels throughout the body and into tissue spaces, causing widespread edema.
3. **Edema**—general tissue swelling caused by accumulation of fluids in the tissue spaces. Edema associated with nephrotic syndrome is caused by loss of plasma protein (albumin) and the resulting osmosis of fluid from the blood.

Note that hematuria (blood in the urine) is not a feature of nephrotic syndrome.

Acute Glomerulonephritis

Acute glomerulonephritis is the most common form of kidney disease. It is caused by a delayed immune response to streptococcal infection—the same mechanism that causes

SCIENCE APPLICATIONS

FIGHTING INFECTION

Alexander Fleming (1881–1955)

Unfortunately, the structure of the urinary tract puts it at risk for infection by bacteria and other microorganisms. Because it is open to the external environment, bacteria can enter easily. In women, the short length of the urethra and its location close to the anus may further increase the risk of bacteria getting to the urinary bladder. Another risk factor is poor technique by healthcare workers when they insert catheters (tubes) into the urethras of patients who need help voiding their bladders of urine.

A breakthrough in the treatment of **urinary tract infections (UTIs)** came in 1928 in the laboratory of Scots researcher Alexander Fleming. Some mold spores accidentally contaminated one of the dishes in which Fleming was growing bacteria. He marveled at the fact that no bacteria could grow near the colony of mold. He isolated a substance from the mold that was responsible for this antibacterial effect and named it **penicillin.**

Earlier Fleming had discovered another natural antibiotic (lysozyme) that effectively attacked bacteria that did not often cause disease, so this concept was not totally new to him. However, through further experiments, Fleming showed that penicillin was effective against a variety of bacteria that do cause serious infections in humans, which made it an invaluable therapeutic tool.

Penicillin was touted as the first "miracle drug" and rapidly became the tool of choice in fighting bacteria. In 1943 another breakthrough came when laboratory worker Mary Hunt brought a moldy cantaloupe to work and researchers found that the new type of mold produced enough penicillin to make commercial production of the antibiotic possible.

Although forms of penicillin and other antibiotics derived from natural sources are still the weapon of choice in battling many infections, many bacteria are evolving into strains that resist common antibiotics. UTIs and other types of infections now require more powerful antibiotics and other special techniques to stop them. Unless current efforts to reverse this trend become effective, some scientists fear that the era of simple antibiotic therapy may be nearing an end.

Many professions are involved in the fight against infection. *Medical supply technicians* ensure that devices such as urethral catheters are sterile (free of microorganisms) before they are packaged and sent to hospitals and clinics (pictured).

Physicians, nurses, and others who deal directly with patients need to learn proper "sterile technique" to ensure that infections are not introduced by medical procedures. To help in this effort, most organizations designate an **infection control** manager—a health professional with the responsibility of preventing **nosocomial infections** (infections that begin in the hospital). Community health experts including **epidemiologists** and health service officers from the U.S. government's Centers for Disease Control and Prevention (CDC) also help prevent the spread of infection in local communities and worldwide. Of course, *pharmacology researchers,* **microbiologists,** and others continue in the quest to find newer and better treatments for UTIs and other infections that threaten human health.

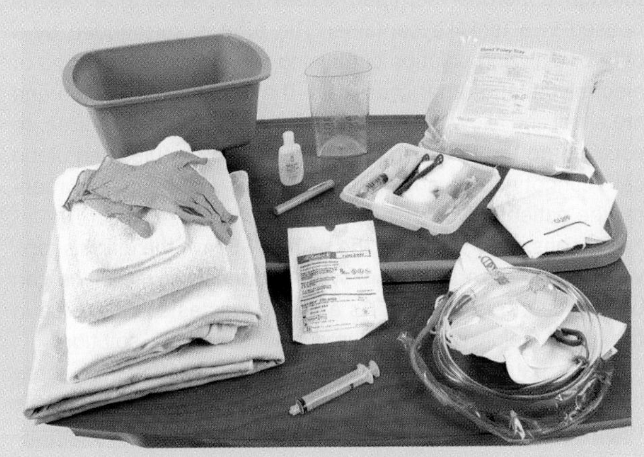

valve damage in rheumatic heart disease (see Chapter 14). For this reason, it is sometimes called *postinfectious glomerulonephritis.* Occurring 1 to 6 weeks after a streptococcal infection, this disorder is characterized by hematuria, oliguria, proteinuria, and edema.

Antibiotic therapy and bed rest are the usual treatments for acute glomerulonephritis. Recovery is often complete but may progress to a chronic form of glomerulonephritis.

Chronic Glomerulonephritis

Chronic glomerulonephritis is the general name for a variety of noninfectious glomerular disorders characterized by progressive kidney damage leading to renal failure. Early stages of chronic glomerulonephritis are asymptomatic. As this disorder progresses, hematuria, proteinuria, oliguria, and edema develop. Immune mechanisms are believed to be the major causes of chronic glomerulonephritis.

Kidney Failure

Kidney failure, or **renal failure,** is simply the failure of the kidney to properly process blood and form urine. Renal failure can be classified as either *acute* or *chronic.*

Acute Renal Failure

Acute renal failure is an abrupt reduction in kidney function characterized by oliguria and a sharp rise in nitrogenous compounds in the blood. The nitrogen-containing substances result from the breakdown of amino acids used for energy in cellular respiration.

The concentration of nitrogenous wastes in the blood is often assessed by the **blood urea nitrogen (BUN) test.** A high BUN result may indicate failure of the kidneys to remove urea from the blood.

Acute renal failure can be caused by a variety of factors that alter blood pressure or otherwise affect glomerular filtration. For example, hemorrhage, severe burns, acute glomerulonephritis or pyelonephritis, or obstruction of the lower urinary tract may progress to kidney failure. If the underlying cause of acute renal failure is successfully treated, recovery is usually rapid and complete.

Chronic Renal Failure

Chronic renal failure is a slow, progressive condition resulting from the gradual loss of nephrons. There are dozens of diseases that may result in the gradual loss of nephron function, including diabetes, infections, glomerulonephritis, tumors, systemic autoimmune disorders, and obstructive disorders.

Polycystic kidney disease (PKD) is one of the most common genetic disorders in humans. In PKD, large fluid-filled pockets (cysts) develop in the epithelium of the kidney tubules. In this condition, *primary cilia* (nonmoving cilia) on the epithelial cells that form kidney tubules fail to do their normal

CLINICAL APPLICATION

ARTIFICIAL KIDNEY

The artificial kidney is a mechanical device that uses the principle of dialysis to remove or separate waste products from the blood. In the event of kidney failure, the process, appropriately called **hemodialysis** (*hemo* "blood" and *lysis* "separate"), is implemented as a life-saving intervention for the patient.

During a hemodialysis treatment, a semipermeable membrane is used to separate large (nondiffusible) particles such as blood cells from small (diffusible) ones such as urea and other wastes. Figure **A** shows blood from the radial artery passing through a porous (semipermeable) cellophane tube that is housed in a tanklike container. The tube is surrounded by a bath or dialyzing solution containing varying concentrations of electrolytes and other chemicals. The pores in the membrane are small and allow only very small molecules, such as urea, to escape into the surrounding fluid. Larger molecules and blood cells cannot escape and are returned through the tube to reenter the patient via a wrist vein.

By constantly replacing the bath solution in the dialysis tank with freshly mixed solution, waste materials in the blood can be reduced to low levels. As a result, wastes such as urea in the blood rapidly pass into the surrounding wash solution.

For a patient with complete kidney failure, two or three hemodialysis treatments a week are required. These dialysis treatments are now being monitored and controlled by sophisticated computer components and software that have been integrated into the most current hemodialysis equipment. New and dramatic advances in both treatment techniques and equipment are expected in the future. Although most hemodialysis treatments are administered in hospital or clinical settings, equipment designed for use in the home is available and is appropriate for many individuals. Patients and their families using this equipment are initially instructed in its use and then monitored and supported on an ongoing basis by home health-care professionals.

Another technique used in the treatment of renal failure is called **continuous ambulatory peritoneal dialysis (CAPD).** In this procedure, 1 to 3 L of sterile dialysis fluid is introduced directly into the peritoneal cavity through an opening in the abdominal wall (Figure **B**). Peritoneal membranes in the abdominal cavity transfer waste products from the blood into the dialysis fluid, which is then drained back into a plastic container after about 2 hours. This technique is less expensive than hemodialysis and does not require the use of complex equipment. CAPD is the more frequently used home-based dialysis treatment for patients with chronic renal failure. CAPD is not practical for all patients, however, and successful long-term treatment is greatly enhanced by support from professionals trained in home health-care services.

To learn more about dialysis, go to AnimationDirect online at *evolve.elsevier.com.*

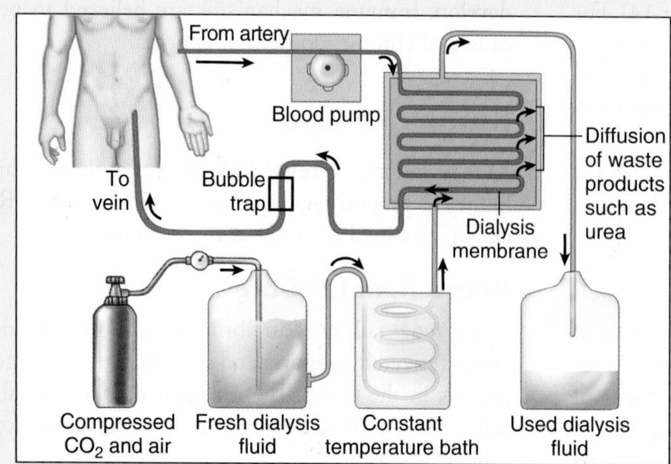

A

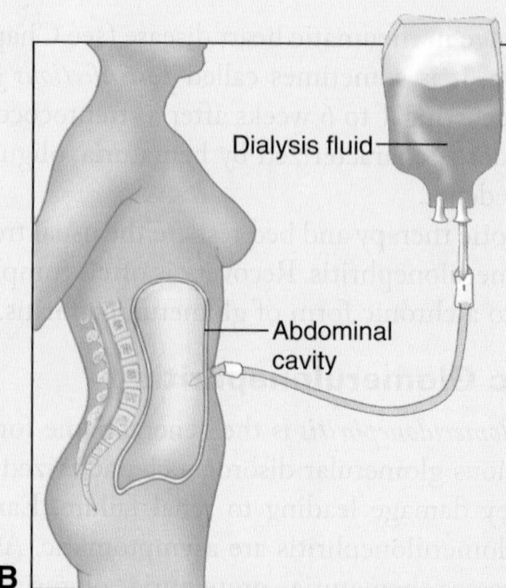

B

20

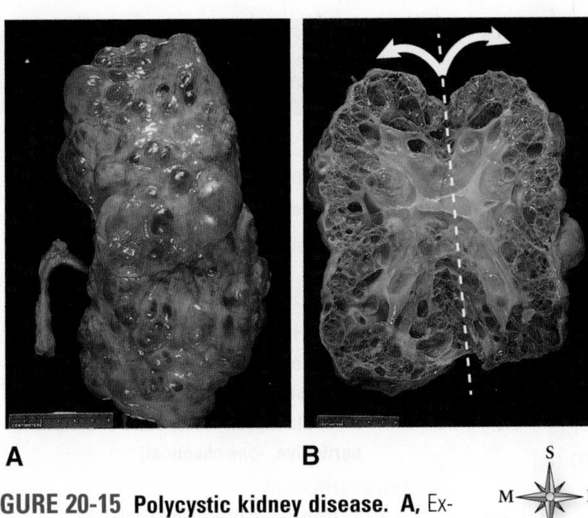

FIGURE 20-15 Polycystic kidney disease. A, External view shows characteristic cysts. **B,** Lateral view of a kidney partially cut along a frontal plane and then opened like a book to view the cysts inside the kidney.

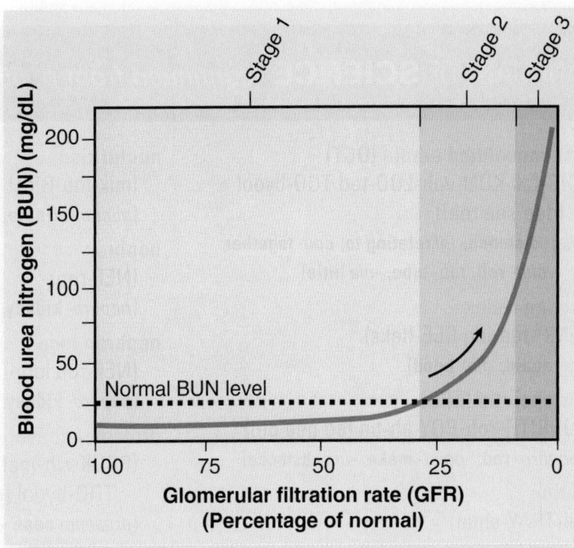

FIGURE 20-16 The three stages of chronic renal failure. *Stage 1:* As nephrons are lost (indicated by decreasing GFR), the remaining healthy nephrons compensate—keeping BUN values within the normal range. *Stage 2:* As more than 75% of kidney function is lost, BUN levels begin to climb. *Stage 3:* Uremia (elevated BUN) results from massive loss of kidney function.

job of regulating cell growth. The epithelial cells then overpopulate and obstruct the kidney tubules. The obstructions result in pockets of backed-up urine called *cysts*. Eventually, PKD leads to kidney failure.

One of the most common forms of PKD is called *adult polycystic kidney disease.* It is hereditary and appears in 1 in 500 to 1000 persons. Symptoms of the disease, which include flank pain and hematuria, generally appear after age 40 and progress slowly. Eventually, the kidneys achieve enormous size as they fill with growing numbers of cysts (**Figure 20-15**). Destruction of tissue results in progressive renal failure.

As kidney function is lost, the glomerular filtration rate (GFR) decreases, causing the blood urea nitrogen (BUN) and creatinine levels to climb. Chronic renal failure can be described as progressing through the three stages shown in **Figure 20-16** and described here:

1. **Stage 1**—During the first stage, some nephrons are lost but the remaining healthy nephrons compensate by enlarging and taking over the function of the lost nephrons. This stage is often asymptomatic and may last for years, depending on the underlying cause.
2. **Stage 2**—The second stage is often called *renal insufficiency.* During this stage, the kidney can no longer adapt to the loss of nephrons. The remaining healthy nephrons cannot handle the urea load, and BUN levels climb dramatically. Because the kidney's ability to concentrate urine is impaired, polyuria and dehydration may occur.
3. **Stage 3**—The final stage of chronic renal failure is called *uremia* or *uremic syndrome.* Uremia literally means "high blood urea" and is characterized by a very high BUN value caused by loss of kidney function. During this stage, low GFR causes low urine production and oliguria. Because fluids are retained by the body rather than being eliminated by the kidneys, edema and hypertension often occur. The uremic syndrome includes a long list of other symptoms caused directly or indirectly by the loss of kidney function. Unless an artificial kidney is used or a new kidney is transplanted, the progressive loss of kidney function eventually causes death.

QUICK CHECK

1. How do renal calculi develop?
2. Define nephrotic syndrome and list its characteristics.
3. What is *acute renal failure*? How is it assessed?

LANGUAGE OF **SCIENCE** *(continued from p. 555)*

distal convoluted tubule (DCT)
(DIS-tall KON-voh-LOO-ted TOO-byool [dee see tee])
[*dist-* **distance,** *-al* **relating to,** *con-* **together,** *-volut-* **roll,** *tub-* **tube,** *-ule* **little**]

emptying reflex
(EMP-tee-ing REE-fleks)
[*re-* **again,** *-flex* **bend**]

erythropoietin (EPO)
(eh-RITH-roh-POY-eh-tin [ee pee oh])
[*erythro-* **red,** *-poiet-* **make,** *-in* **substance**]

filtration
(fil-TRAY-shun)
[*filtr-* **strain,** *-ation* **process**]

glomerular-capsular membrane
(gloh-MER-yoo-lar KAP-sul-er MEM-brayne)
[*glomer-* **ball,** *-ul-* **little,** *-ar* **relating to,** *caps-* **box,** *-ul-* **little,** *-ar* **relating to,** *membrane* **thin skin**]

glomerular capsule
(gloh-MER-yoo-lar KAP-sul)
[*glomer-* **ball,** *-ul-* **little,** *-ar* **relating to,** *caps-* **box,** *-ule* **little**]

glomerular filtrate
(gloh-MER-yoo-lar fil-TRAYT)
[*glomer-* **ball,** *-ul-* **little,** *-ar* **relating to,** *filtr-* **strain,** *-ate* **result**]

glomerulus
(gloh-MAYR-yoo-lus)
pl., glomeruli
(gloh-MAYR-yoo-lye)
[*glomer-* **ball,** *-ul-* **little,** *-us* **thing**]

Henle loop
(HEN-lee loop)
[Friedrich Gustave Henle **German anatomist**]

hilum
(HYE-lum)
pl., hila
(HYE-lah)
[*hilum* **least bit**]

hyperosmotic
(hye-per-os-MOT-ik)
[*hyper-* **excessive or above,** *-osmo-* **push,** *-ic* **relating to**]

juxtaglomerular (JG) apparatus
(juks-tah-gloh-MER-yoo-lar [jay jee] ap-ah-RAT-us)
[*juxta-* **near or adjoining,** *-glomer-* **ball,** *-ul-* **little,** *-ar* **relating to,** *apparatus* **tool**]

juxtamedullary nephron
(jux-tah-MED-oo-lar-ee NEF-ron)
[*juxta-* **near or adjoining,** *-medulla-* **middle,** *-ary* **relating to,** *nephro-* **kidney,** *-on* **unit**]

micturition
(mik-too-RISH-un)
[*mictur-* **urinate,** *-tion* **process**]

nephron
(NEF-ron)
[*nephro-* **kidney,** *-on* **unit**]

nephron loop
(NEF-ron loop)
[*nephro-* **kidney,** *-on* **unit**]

proximal convoluted tubule (PCT)
(PROK-sih-mal KON-voh-LOO-ted TOO-byool [pee see tee])
[*proxima-* **near,** *-al* **relating to,** *con-* **together,** *-volut-* **roll,** *tub-* **tube,** *-ule* **little**]

reabsorption
(ree-ab-SORP-shun)
[*re-* **back again,** *-ab-* **from,** *-sorp-* **suck,** *-tion* **process**]

renal column
(REE-nall KOWL-um)
[*ren-* **kidney,** *-al* **relating to**]

renal corpuscle
(REE-nal KOR-pus-ul)
[*ren-* **kidney,** *-al* **relating to,** *corpus-* **body,** *-cle* **little**]

renal cortex
(REE-nal KOR-teks)
[*ren-* **kidney,** *-al* **relating to,** *cortex* **bark**]

renal medulla
(REE-nal meh-DUL-ah)
[*ren-* **kidney,** *-al* **relating to,** *medulla* **middle**]

renal papilla
(REE-nal pah-PIL-uh)
pl., papillae
(pah-PIL-ee)
[*ren-* **kidney,** *-al* **relating to,** *papilla* **nipple**]

renal pelvis
(REE-nal PEL-vis)
pl., pelves
(PEL-veez)
[*ren-* **kidney,** *-al* **relating to,** *pelvis* **basin**]

renal pyramid
(REE-nal PIR-ah-mid)
[*ren-* **kidney,** *-al* **relating to**]

renal threshold
(REE-nal THRESH-old)
[*ren-* **kidney,** *-al* **relating to**]

renal tubule
(REE-nal TOOB-yool)
[*ren-* **kidney,** *-al* **relating to,** *tub-* **pipe,** *-ule* **small**]

renin
(REE-nin)
[*ren-* **kidney,** *-in* **substance**]

renin-angiotensin-aldosterone system (RAAS)
(REE-nin–an-jee-oh-TEN-sin–al-DAH-stayr-ohn SYS-tem [ar ay ay es])
[*ren-* **kidney,** *-in* **substance,** *angio-* **vessel,** *-tens-* **pressure or stretch,** *-in* **substance,** *aldo-* **aldehyde,** *-stero-* **solid or steroid derivative,** *-one* **chemical**]

retroperitoneal
(reh-troh-payr-ih-toh-NEE-al)
[*retro-* **backward,** *-peri-* **around,** *-tone-* **stretched,** *-al* **relating to**]

rugae
(ROO-gee)
sing., ruga
[*ruga* **wrinkle**]

secretion
(seh-KREE-shun)
[*secret-* **separate,** *-tion* **process**]

sphincter
(SFINGK-ter)
[*sphinc-* **bind tight,** *-er* **agent**]

transport maximum (T_{max})
(TRANZ-port MAKS-im-um)
[*trans-* **across,** *-port* **carry,** *maximum* **greatest**]

trigone
(TRY-gohn)
[*tri-* **three,** *-gon-* **corner**]

ureter
(yoo-REE-ter or YOOR-eh-ter)
[*ure-* **urine,** *-ter* **agent or channel**]

urethra
(yoo-REE-thrah)
[*ure-* **urine,** *-thr-* **agent or channel**]

urinary meatus
(YOOR-ih-nayr-ee mee-AY-tus)
[*urin-* **urine,** *-ary* **relating to,** *meatus* **passage**]

urinary system
(YOOR-ih-nayr-ee SYS-tem)
[*urin-* **urine,** *-ary* **relating to**]

urination
(yoor-ih-NAY-shun)
[*urin-* **urine,** *-ation* **process**]

voiding
(VOYD-ing)
[*void-* **empty,** *-ing* **action**]

20

LANGUAGE OF MEDICINE

anuria
(ah-NOO-ree-ah)
[*a-* **not,** *-ur-* **urine,** *-ia* **condition**]

blood urea nitrogen (BUN) test
(blud yoo-REE-ah NYE-troh-jen [bun] test)
[*urea* **urine,** *nitro-* **soda,** *-gen* **produce**]

catheterization
(kath-eh-ter-ih-ZAY-shun)
[*cathe-* **send down,** *-er* **agent,** *-tion* **process of**]

continuous ambulatory peritoneal dialysis (CAPD)
(kon-TIN-yoo-us AM-byoo-lah-tor-ee payr-ih-toh-NEE-al dye-AL-ih-sis [see ay pee dee])
[*ambulat-* **walk,** *-ory* **relating to,** *peritone-* **peritoneum,** *-al* **relating to,** *dia-* **through,** *-lysis* **loosening**]

cystitis
(sis-TYE-tis)
[*cyst-* **bag,** *-itis* **inflammation**]

cystoscope
(SIS-toh-skohp)
[*cyst-* **bag,** *-scop-* **see**]

diuretic drug
(dye-yoo-RET-ik drug)
[*dia-* **through,** *-ure-* **urine,** *-ic* **relating to**]

dysuria
(dis-YOO-ree-ah)
[*dys-* **disordered,** *-ur-* **urine,** *-ia* **condition**]

edema
(eh-DEE-mah)
[*edema* **a swelling**]

enuresis
(en-yoo-REE-sis)
[*en-* **in,** *-uresis* **urinate**]

epidemiologist
(ep-ih-dee-mee-OL-uh-jist)
[*epi-* **upon,** *-dem-* **people,** *-log-* **words (study of),** *-ist* **agent**]

glomerulonephritis
(gloh-mer-yoo-loh-neh-FRY-tis)
[*glomer-* **ball,** *-ul-* **little,** *-nephr-* **kidney,** *-itis* **inflammation**]

glycosuria
(glye-koh-SOO-ree-ah)
[*glyco-* **sweet (glucose),** *-ur-* **urine,** *-ia* **condition**]

hematuria
(hem-ah-TOO-ree-ah)
[*hema-* **blood,** *-ur-* **urine,** *-ia* **condition**]

hemodialysis
(hee-moh-dye-AL-ih-sis)
[*hemo-* **blood,** *-dia-* **through or between,** *-lysis* **loosening**]

hydronephrosis
(hye-droh-neh-FROH-sis)
[*hydro-* **water,** *-nephr-* **kidney,** *-osis* **condition**]

hypoalbuminemia
(hye-poh-al-byoo-min-EE-mee-ah)
[*hypo-* **under or below,** *-alb-* **white,** *-emia* **blood condition**]

I & O
(aye and oh)
[*I* **input,** *&* **and,** *O* **output**]

infection control
(in-FEK-shun KON-trol)
[*infect-* **stain,** *-ion* **state**]

interstitial cystitis
(in-ter-STISH-al sis-TYE-tis)
[*inter-* **between,** *-stit-* **stand,** *-al* **relating to,** *cyst-* **bag,** *-itis* **inflammation**]

lithotripsy
(LITH-oh-trip-see)
[*litho-* **stone,** *-trips-* **pound,** *-y* **action**]

lithotriptor
(LITH-oh-trip-tor)
[*litho-* **stone,** *-trip-* **pound,** *-or* **agent**]

microbiologist
(my-kroh-bye-OL-uh-jist)
[*micro-* **small,** *-bio-* **life,** *-log-* **words (study of),** *-ist* **agent**]

needle biopsy
(NEE-dil BYE-op-see)
[*bio-* **life,** *-ops-* **view,** *-y* **action**]

nephritis
(neh-FRY-tis)
[*nephr-* **kidney,** *-itis* **inflammation**]

nephropathy
(neh-FROP-ah-thee)
[*nephro-* **kidney,** *-path* **disease,** *-y* **state**]

nephrotic syndrome
(neh-FROT-ik SIN-drohm)
[*nephr-* **kidney,** *-ic* **relating to,** *syn-* **together,** *-drome* **running or (race) course**]

neurogenic bladder
(noor-oh-JEN-ik BLAD-der)
[*neuro-* **nerves,** *-gen-* **produce,** *-ic* **relating to,** *bladder* **pimple**]

nosocomial infection
(no-zoh-KOHM-ee-al in-FEK-shun)
[*noso-* **disease,** *-com-* **care,** *-al* **relating to,** *in-* **into,** *-fec-* **put,** *-tion* **process**]

oliguria
(ohl-ih-GOO-ree-ah)
[*olig-* **few or little,** *-ur-* **urine,** *-ia* **condition**]

overactive bladder
(OH-ver-ak-tiv BLAD-der)

penicillin
(pen-ih-SIL-in)
[*penicill-* **brush (*Penicillum* mold),** *-in* **substance**]

polycystic kidney disease (PKD)
(pahl-ee-SIS-tik KID-nee dih-ZEEZ [pee kay dee])
[*poly-* **many,** *-cyst-* **bag,** *-ic* **relating to,** *dis-* **opposite of,** *-ease* **comfort**]

polydipsia
(pahl-ee-DIP-see-ah)
[*poly-* **many,** *-dips-* **thirst,** *-ia* **condition**]

polyuria
(pahl-ee-YOO-ree-ah)
[*poly-* **many,** *-ur-* **urine,** *-ia* **condition**]

proteinuria
(proh-teen-YOO-ree-ah)
[*prote-* **first rank,** *-in-* **substance,** *-ur-* **urine,** *-ia* **condition**]

pyelonephritis
(pye-eh-loh-neh-FRY-tis)
[*pyel-* **renal pelvis,** *-nephr-* **kidney,** *-itis* **inflammation**]

renal calculi
(REE-nal KAL-kyoo-lye)
sing., calculus
(KAL-kyoo-lus)
[*ren-* **kidney,** *-al* **relating to,** *calc-* **limestone,** *-ul-* **little,** *-i* **things**]

renal colic
(REE-nal KOL-ik)
[*ren-* **kidney,** *-al* **relating to,** *col-* **colon,** *-ic* **relating to**]

renal failure
(REE-nal FAIL-yoor)
[*ren-* **kidney,** *-al* **relating to**]

uremia
(yoo-REE-mee-ah)
[*ur-* **urine,** *-emia* **blood condition**]

uremic poisoning
(yoo-REE-mik POY-zun-ing)
[*ur-* **urine,** *-em-* **blood condition,** *-ic* **relating to**]

urethritis
(yoo-reh-THRY-tis)
[*ure-* **urine,** *-thr-* **agent or channel (urethra),** *-itis* **inflammation**]

urinalysis
(yoor-in-AL-is-is)
[*ur-* **urine,** *-in-* **chemical,** *-(an)a-* **apart,** *-lysis* **loosen or break**]

urinary incontinence
(YOOR-ih-nayr-ee in-KON-tih-nens)
[*urin-* **urine,** *-ary* **relating to,** *in-* **without,** *contin-* **contain,** *-ence* **ability**]

Continued on p. 576

LANGUAGE OF **MEDICINE** *(continued from p. 575)*

urinary retention
(YOOR-in-ayr-ee ree-TEN-shun)
[*urin-* **urine,** *-ary* **relating to,** *re-* **back,** *-ten-* **hold,** *-tion* **condition**]

urinary suppression
(YOOR-in-ayr-ee sup-PRESH-un)
[*urin-* **urine,** *-ary* **relating to,** *sup- (sub-)* **down,** *-press-* **press,** *-ion* **condition**]

urinary tract infection (UTI)
(YOOR-ih-nayr-ee trakt in-FEK-shun [yoo tee aye])
[*urin-* **urine,** *-ary* **relating to,** *tract* **trail,** *infect-* **stain,** *-tion* **condition**]

❑ OUTLINE SUMMARY

*To download a digital version of the chapter summary for use with your device, access the **Audio Chapter Summaries** online at evolve.elsevier.com.*

Scan this summary after reading the chapter to help you reinforce the key concepts. Later, use the summary as a quick review before your class or before a test.

Kidneys

A. Location—under back muscles, behind parietal peritoneum, just above waistline; right kidney usually a little lower than left (**Figure 20-1**)
B. Gross structure (**Figure 20-2**)
   1. External anatomy
      a. Kidney resembles a lima bean that is 11 cm × 7 cm × 3 cm
      b. Hilum—medial indentation where vessels, nerves, ureter connect
      c. Capsule—fibrous outer wall
   2. Internal anatomy
      a. Cortex—outer layer of kidney tissue
      b. Medulla—inner portion of kidney
      c. Pyramids—triangular divisions of tissue within the renal medulla
      d. Papilla—narrow, innermost end of a renal pyramid
      e. Pelvis—expansion of upper end of ureter; lies inside kidney
      f. Calyces—divisions of renal pelvis; each papilla faces a calyx
C. Microscopic structure of the kidney
   1. Interior of kidney composed of more than 1 million microscopic nephron units (**Figure 20-3**)
      a. Unique shape of nephron well suited to function
      b. Principal components are renal corpuscle and renal tubule
   2. Renal corpuscle (**Figure 20-4**)
      a. Glomerular capsule—cup-shaped top of nephron; also called *Bowman capsule*
      b. Glomerulus—network of blood capillaries surrounded by glomerular capsule
   3. Renal tubule (**Figure 20-5**)
      a. Proximal convoluted tubule (PCT)—first segment
      b. Nephron loop (Henle loop)—extension of proximal tubule; consists of descending limb, loop, and ascending limb
      c. Distal convoluted tubule (DCT)—extension of ascending limb of nephron loop
      d. Collecting duct (CD)—straight extension of distal tubule
   4. Location of nephrons
      a. Cortical nephrons—85% of total; most of nephron is located in renal cortex
      b. Juxtamedullary nephrons—have important role in concentrating urine; renal corpuscles are located near boundary between cortex and medulla
D. Kidney function
   1. Excrete toxins and nitrogenous wastes
   2. Regulate levels of many chemicals in blood
   3. Maintain water balance
   4. Help regulate blood pressure and volume
   5. Regulate red blood cell production by secreting erythropoietin (EPO)

Formation of Urine

A. Millions of nephrons balance blood and flush the excess/wastes as urine in a process that includes three functions: filtration, reabsorption, and secretion (**Figure 20-6** and **Table 20-1**)

20

B. Filtration
  1. Goes on continually in renal corpuscles
  2. Glomerular blood pressure causes water and dissolved substances to filter out of glomeruli into the glomerular capsule—across the glomerular-capsular membrane
  3. Normal glomerular filtration rate 125 mL per minute

C. Reabsorption
  1. Movement of substances out of renal tubules into blood in peritubular capillaries
  2. Water, nutrients, and ions are reabsorbed (**Figure 20-7**)
  3. Water is reabsorbed by osmosis from proximal tubules
  4. Countercurrent mechanisms in the nephron loop and surrounding peritubular capillaries concentrate sodium and chloride to make the renal medulla hyperosmotic, which helps concentrate urine (see *Control of Urine Volume* subsequently).
  5. Transport maximum (T_{max})—largest amount of substance that can be reabsorbed at one time
     a. Determined by the number of available transporters of the substance
     b. Determines the renal threshold—above this level, the kidney removes the substance from blood and excretes in urine
  6. All glucose is reabsorbed along with sodium, as long as glucose levels remain within a normal range and there are enough sodium-glucose transporters to accommodate all the glucose (**Figure 20-8**)

D. Secretion
  1. Movement of substances into urine in the distal tubule and collecting ducts from blood in peritubular capillaries
  2. Hydrogen ions, potassium ions, and certain drugs are secreted by active transport
  3. Ammonia is secreted by diffusion

Control of Urine Volume

A. Antidiuretic hormone (ADH)—secreted by posterior pituitary; promotes water reabsorption by collecting ducts; reduces urine volume

B. Hyperosmotic (salty) conditions in the renal medulla help ADH concentrate urine and thus conserve the body's water

C. Aldosterone—secreted by adrenal gland, triggered by the renin-angiotensin-aldosterone system (RAAS); promotes sodium and water reabsorption in nephron; reduces urine volume (**Figure 20-9**)

D. Atrial natriuretic hormone (ANH)—one of the peptide hormones (ANPs) secreted by atrial cells in heart; promotes loss of sodium and water into kidney tubules; increases urine volume

E. Abnormalities of urine volume
  1. Anuria—absence of urine
  2. Oliguria—scanty amount of urine
  3. Polyuria—unusually large amount of urine

Elimination of Urine

A. Ureters (**Figure 20-10**)
  1. Structure
     a. Long, narrow muscular tubes
     b. Lined with mucous membrane
     c. Expanded upper end forms renal pelvis located inside kidney
  2. Function—drain urine from renal pelvis to urinary bladder

B. Bladder
  1. Structure (**Figure 20-11**)
     a. Elastic muscular organ, capable of great expansion
     b. Lined with mucous membrane arranged in rugae, as is stomach mucosa
  2. Functions
     a. Storage of urine before voiding
     b. Voiding
  3. Cystitis—bladder infection

C. Urethra
  1. Structure (**Figure 20-11**)
     a. Narrow tube from urinary bladder to exterior
     b. Lined with mucous membrane
     c. Opening of urethra to the exterior called *urinary meatus*
  2. Functions
     a. Passage of urine from bladder to exterior of the body
     b. Passage of male reproductive fluid (semen) from the body

D. Micturition
  1. Passage of urine from body (also called *urination* or *voiding*)
  2. Regulatory sphincters
     a. Internal urethral sphincter (involuntary)
     b. External urethral sphincter (voluntary)
  3. Bladder wall expansion permits storage of urine with little increase in pressure
  4. Emptying reflex
     a. Initiated by stretch reflex in bladder wall
     b. Bladder wall contracts
     c. Internal sphincter (involuntary) relaxes
     d. External sphincter (voluntary) relaxes and urination occurs
     e. Enuresis—involuntary urination in young child
  5. Urinary retention—urine produced but not voided
  6. Urinary suppression—no urine produced but bladder is normal
  7. Urinary incontinence (enuresis)—urine is voided involuntarily
     a. Urge incontinence—associated with smooth muscle overactivity in the bladder wall
     b. Stress incontinence—associated with weakened pelvic floor muscles
     c. Overflow incontinence—associated with urinary retention and overdistended bladder

d. Reflex incontinence occurs in absence of any sensory warning or awareness—common following a stroke or spinal cord injury
e. Nocturnal enuresis—nighttime bed wetting
f. Neurogenic bladder—periodic but unpredictable voiding; related to paralysis or abnormal function of the bladder

Urinalysis

A. Examination of the physical, chemical, and microscopic characteristics of urine (**Table 20-2**)
B. May help determine the presence and nature of a pathological condition

Renal and Urinary Disorders

A. Obstructive disorders interfere with normal urine flow, possibly causing urine to back up and cause hydronephrosis or other kidney damage
   1. Hydronephrosis—enlargement of renal pelvis and calyces caused by blockage of urine flow (**Figure 20-12**)
   2. Renal calculi (kidney stones)—crystallized mineral chunks in renal pelvis or calyces; may block ureters, causing intense pain called *renal colic* (**Figure 20-13**)
   3. Tumors—renal cell carcinoma (kidney cancer) and bladder cancer (**Figure 20-14**); often characterized by hematuria (blood in the urine)
B. Urinary tract infections (UTIs) are often caused by gram-negative bacteria
   1. Urethritis—inflammation of the urethra
   2. Cystitis—inflammation or infection of the urinary bladder
   3. Pyelonephritis—inflammation of the renal pelvis and connective tissues of the kidney; may be acute (infectious) or chronic (autoimmune)
C. Glomerular disorders result from damage to the glomerular-capsular membrane of the renal corpuscles
   1. Nephrotic syndrome accompanies many glomerular disorders
      a. Proteinuria—protein in the urine
      b. Hypoalbuminemia—low plasma protein (albumin) level; caused by loss of proteins to urine
      c. Edema—tissue swelling caused by loss of water from plasma as a result of hypoalbuminemia
   2. Acute glomerulonephritis is caused by delayed immune response to a streptococcal infection
   3. Chronic glomerulonephritis is a slow inflammatory condition caused by immune mechanisms and often leads to renal failure
D. Kidney failure, or renal failure, occurs when the kidney fails to function
   1. Acute renal failure—abrupt reduction in kidney function that is usually reversible
   2. Chronic renal failure—slow, progressive loss of nephrons caused by a variety of underlying diseases
      a. Polycystic kidney disease (PKD)—numerous fluid-filled cysts destroy kidney tissue as they grow; hereditary (**Figure 20-15**)
      b. Progression of kidney failure (**Figure 20-16**)
         (1) Stage 1—early in this disorder, healthy nephrons often compensate for the loss of damaged nephrons
         (2) Stage 2—often called *renal insufficiency;* loss of kidney function ultimately results in uremia (high BUN levels) and its life-threatening consequences
         (3) Stage 3—called *uremia* or *uremic syndrome;* complete kidney failure results in death unless a new kidney is transplanted or an artificial kidney substitute is used

❑ ACTIVE LEARNING

STUDY TIPS

Hint *Consider using these tips to achieve success in meeting your learning goals.*

20

Review the synopsis of the urinary system in Chapter 5. The function of the urinary system is to maintain the homeostasis of the blood plasma.

1. The names, locations, and functions of the organs of the urinary system, the internal structure of the kidney, and the microscopic structures of the nephron all can be learned using flash cards and online resources. For a better understanding of the terms in this chapter, refer to the Language of Science and Language of Medicine sections.
2. The formation of urine involves three processes: filtration, reabsorption, and secretion. Filtration was discussed in Chapter 3. Reabsorption is the process of taking material out of the urine and returning it to the blood. Secretion is the process of taking material out of the blood and putting it into the urine.
3. Urine volume is controlled by three hormones, each produced by a different organ. Remember that the body cannot directly move water; it must first move solute by diffusion or active transport and then pull the water after it by osmosis. Make flash cards for each of the three hormones; include the name of the hormone, where it is made, its mechanism of action, and its overall effect on urine volume.

4. Make a chart of the disorders of the urinary system. Organize it based on the mechanism or cause of each disorder: obstructive disorders, urinary tract infections, glomerular disorders, and kidney failure.
5. The analogy of a combined waste-water treatment and garbage disposal facility linked to an incredibly efficient recycling center may help you to understand the big picture of urinary system function.
6. In your study group, review the material in this chapter using the flash cards and online resources. Use your cell phone and make photocopies of the figures of the organs of the urinary system, the internal structure of the kidney, and the microscopic structure of the nephron. Discuss how the kidney forms urine and the hormones involved in regulating urine volume. Make sure you know whether each of the hormones will increase or decrease urine volume. Go over the process of micturition, the chart of disorders of the urinary system, chapter outline summary and the questions at the end of the chapter; discuss possible test questions.

Review Questions

Write out the answers to these questions after reading the chapter and reviewing the Chapter Summary. If you simply think through the answer without writing it down, you will not retain much of your new learning.

1. Describe the location of the kidneys.
2. Name and describe the internal structures of the kidneys.
3. Name the principal components of the renal corpuscle and the renal tubule.
4. Define filtration, reabsorption, and secretion as they apply to kidney function.
5. Briefly explain the formation of urine.
6. Name several substances eliminated or regulated by the kidney.
7. Explain the function of the juxtaglomerular apparatus.
8. Identify the three hormones that regulate urine volume.
9. Describe the structure of the ureters.
10. Describe renal colic.
11. Describe the structure of the bladder. What is the trigone?
12. Describe the structure of the urethra.
13. Briefly explain the process of micturition.
14. Differentiate between retention and suppression of urine.
15. Define incontinence. What can cause incontinence?
16. Briefly explain what information a chemical urinalysis provides for medical caregivers.
17. Explain what casts are and why they are sometimes found in a urine sample.
18. Define hydronephrosis.
19. What is another term for renal calculi? What are they usually made of?
20. Name the most common urinary disorder.
21. Briefly explain the following disorders: urethritis, cystitis, and pyelonephritis.
22. Define proteinuria and hypoalbuminemia.
23. Briefly describe the three stages of chronic renal failure.

Critical Thinking

After finishing the Review Questions, write out the answers to these more in-depth questions to help you apply your new knowledge. Go back to sections of the chapter that relate to concepts that you find difficult.

24. Explain how salt and water balance is maintained by aldosterone and ADH.
25. Why is proper blood pressure necessary for proper kidney function?
26. If a person were doing strenuous work on a hot day and perspiring heavily, would there be a great deal of ADH in the blood or very little? Explain your answer.

20

Chapter Test

After studying the chapter, test your mastery by responding to these items. Try to answer them without looking up the answers.

1. The kidneys receive about ________% of the total amount of blood pumped by the heart each minute.
2. The renal corpuscle is made up of two structures, the ________ and the ________.
3. The two parts of the renal tubules that extend into the medulla of the kidney are the ________ and the ________.
4. The two parts of the renal tubules that are in the cortex of the kidney are the ________ and the ________.
5. A physical, chemical, and microscopic examination of the urine is called a ________.
6. ________ are crystallized mineral chunks that develop in the renal pelvis or calyces.
7. The involuntary muscle, ________, is at the exit of the bladder.
8. ________ is a condition in which the bladder is able to empty itself but no urine is being produced by the kidneys.
9. ________ is a condition in which a person voids urine involuntarily.
10. ________ is a condition in which the bladder is full and the kidney is producing urine but the bladder is unable to empty itself.
11. The nephrons form urine by way of a combination of three blood-balancing processes: ________, ________, and ________.
12. ________ mechanisms maintain hyperosmotic conditions in the medulla.
13. Some ________ drugs, which stimulate the production of urine, are said to be "potassium wasting" because they increase secretion of potassium.
14. The hormone ________ is released from the posterior pituitary gland and reduces the amount of water lost in the urine.
15. ________ is a hormone that is secreted by the adrenal cortex and plays a role in the reabsorption of sodium.
16. Juxtaglomerular cells release an enzyme that initiates the ________ system.
17. The cushion that normally encases each kidney is the ________.
18. The medial indentation on the kidney where vessels, nerves, and the ureter connect with the kidney is the ________.
19. Approximately 85% of all nephrons are located almost entirely in the renal cortex and are called cortical nephrons. The remainder of the nephrons are called ________.
20. ________ solutions promote osmosis of water into them, just as do hypertonic solutions. An example is the countercurrent mechanisms that maintain the ________ conditions in the medulla.

Match each term in Column A with its corresponding description in Column B.

Column A	Column B
21. ________ cortex	a. inner layer of the kidney
22. ________ medulla	b. expansion of the ureter in the kidney
23. ________ pyramid	c. cup-shaped part of the nephron that catches filtrate
24. ________ pelvis	d. tube leading from the bladder to outside of the body
25. ________ urethra	e. network of capillaries nestled within the glomerular capsule
26. ________ bladder	f. saclike structure used to hold urine until it is voided
27. ________ ureter	g. outer part of the kidney
28. ________ trigone	h. an area of the bladder that has openings for the two ureters and the urethra
29. ________ glomerular capsule	i. the part of the renal tubules that is located between the proximal and distal convoluted tubules
30. ________ glomerulus	j. tube connecting the kidney and bladder
31. ________ nephron loop	k. triangular division in the medulla of the kidney

Match each disorder in Column A with its corresponding description or cause in Column B.

Column A	Column B
32. ________ hydronephrosis	a. an inflammation of the urethra that commonly results from a bacterial infection
33. ________ renal calculi	b. another term for a kidney stone
34. ________ urethritis	c. protein, especially albumin, in the urine
35. ________ cystitis	d. condition caused by urine backing up into the kidney, causing swelling of the renal pelvis and calyces
36. ________ pyelonephritis	e. an inflammation of the bladder
37. ________ hypoalbuminemia	f. low albumin in the blood due to loss of albumin through damaged glomeruli
38. ________ proteinuria	g. inflammation of the renal pelvis and connective tissue of the kidney

Case Studies

To solve a case study, you may have to refer to the glossary or index, other chapters in this textbook, and other resources.

1. Sue has *anorexia nervosa* (see Chapter 19). Her body fat content has decreased to a level that is far below normal. How might the changing structure of the body affect the position of Sue's kidneys? How can a change in the position of one or both kidneys lead to kidney failure?
2. Drugs called *thiazide diuretics* are sometimes prescribed to control hypertension (high blood pressure). These drugs act on kidney tubules in a way that inhibits reabsorption of water. How does inhibition of water reabsorption by the kidney reduce high blood pressure? What effects would such drugs have on the volume of urine output?
3. Harriet is receiving *continuous ambulatory peritoneal dialysis (CAPD)*. As you may recall from the boxed essay earlier in this chapter, fluid is introduced into the peritoneal cavity and later withdrawn. Do you think that this dialysis fluid is hypertonic, isotonic, or hypotonic to normal blood plasma? Give reasons for your answer.
4. Jeff has learned from his urologist that his kidney disease has progressed to renal failure and that he must include dialysis in his treatment plan. Because you work on the dialysis unit of the hospital, Jeff has asked you to explain how the hemodialysis machine works. He also asks if the unit is portable so that he can use it at home, if necessary. What will you tell Jeff?

Answers to Active Learning Questions can be found online at evolve.elsevier.com.

Fluid and Electrolyte Balance

OUTLINE

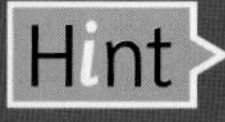

Hint *Scan this outline before you begin to read the chapter, as a preview of how the concepts are organized.*

OBJECTIVES

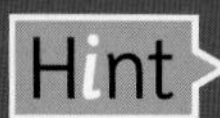

Hint *Before reading the chapter, review these goals for your learning.*

After you have completed this chapter, you should be able to:

1. Describe how body fluid volumes relate to age, gender, and body weight.
2. List, describe, and compare the body fluid compartments and their subdivisions.
3. Explain the mechanisms used by the body to maintain fluid balance, discussing the avenues by which water enters and leaves the body and the forces that move fluids into and out of the blood.
4. Describe examples of common fluid imbalances.
5. Discuss the nature and importance of electrolytes in body fluids.
6. Describe examples of common electrolyte imbalances.

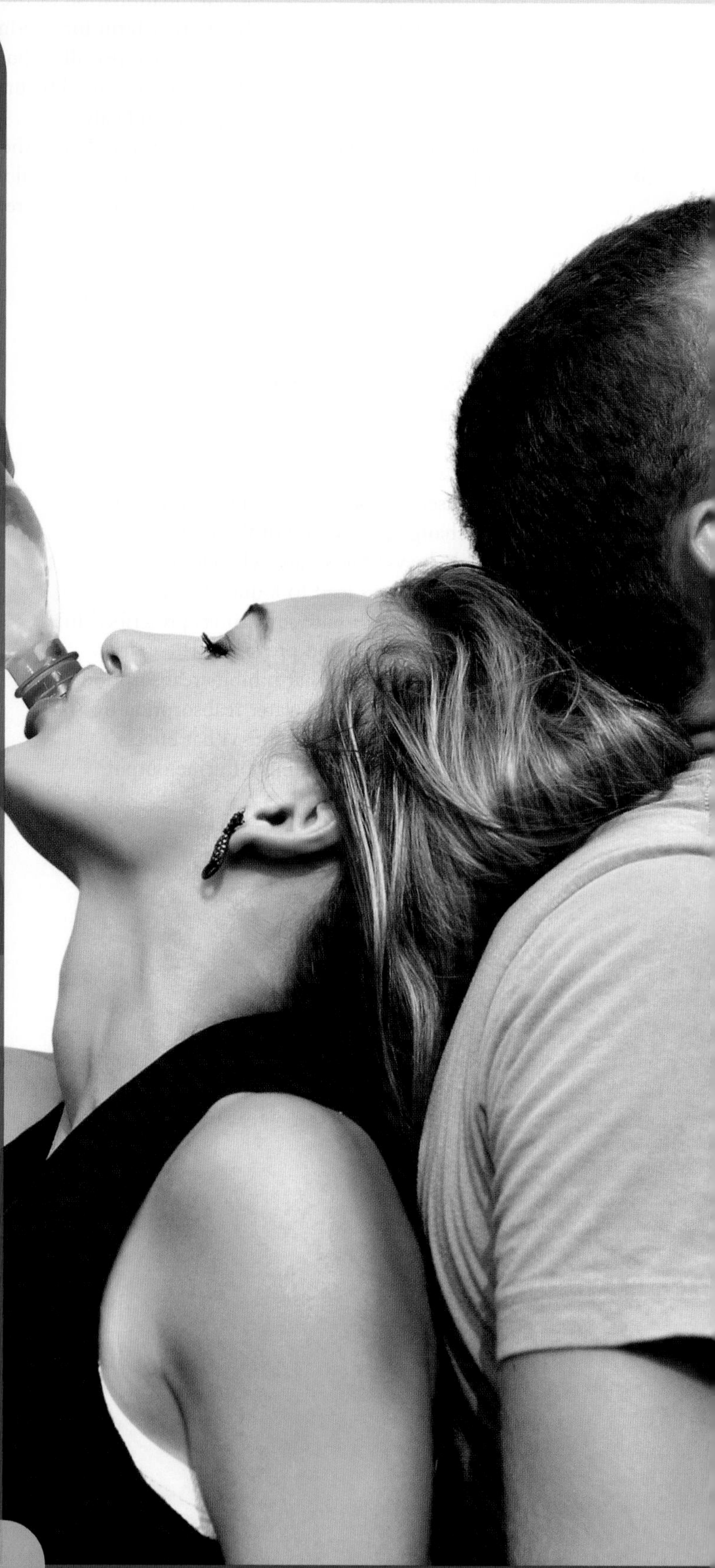

CHAPTER 21

Have you ever wondered why you sometimes excrete great volumes of urine and at other times excrete almost none at all? Why sometimes you feel so thirsty that you can hardly get enough to drink and other times you want no liquids at all? These conditions and many more relate to one of the body's most important functions—that of maintaining its *fluid and electrolyte balance.*

The phrase **fluid balance** implies homeostasis, or relative constancy of body fluid levels—a condition required for healthy survival. It means that both the total volume and distribution of water in the body remain normal and relatively constant. Body "input" or intake of water must be balanced by "output." If water in excess of requirements enters the body, it must be eliminated, and, if excess losses occur, prompt replacement is critical. Because fluid balance refers to normal homeostasis, fluid imbalance means that the total volume of water in the body or the amounts in one or more of its fluid compartments have increased or decreased beyond normal limits.

Electrolytes are substances such as salts that dissolve or break apart in water solution to form electrically charged atoms (or groups of atoms) called **ions. Electrolyte balance** refers to homeostasis or relative constancy of normal electrolyte levels in the body fluids.

The various types of body fluids serve differing functions in different areas of the body. To do so, each type of body fluid must maintain differing levels and types of electrolytes

LANGUAGE OF SCIENCE

Hint Before reading the chapter, say each of these terms out loud. This will help you to avoid stumbling over them as you read.

anion
(AN-aye-on)
[*ana-* **up,** *-ion* **to go (ion)**]

cation
(KAT-aye-on)
[*cat-* **down,** *-ion* **to go (ion)**]

dissociate
(dih-SOH-see-ayt)
[*dis-* **apart,** *-socia-* **unite,** *-ate* **action**]

electrolyte
(eh-LEK-troh-lyte)
[*electro-* **electricity,** *-lyt-* **loosening**]

electrolyte balance
(eh-LEK-troh-lyte BAL-ans)
[*electro-* **electricity,** *-lyt-* **loosening**]

extracellular fluid (ECF)
(eks-trah-SEL-yoo-lar FLOO-id [ee see ef])
[*extra-* **outside,** *-cell-* **storeroom,** *-ular* **relating to**]

fluid balance
(FLOO-id BAL-ans)

fluid compartment
(FLOO-id kom-PART-ment)

interstitial fluid (IF)
(in-ter-STISH-al FLOO-id [aye ef])
[*inter-* **between,** *-stit-* **stand,** *-al* **relating to**]

intracellular fluid (ICF)
(in-trah-SEL-yoo-lar FLOO-id [aye see ef])
[*intra-* **occurring within,** *-cell-* **storeroom,** *-ular* **relating to**]

ion
(AYE-on)
[*ion* **to go**]

Continued on p. 594

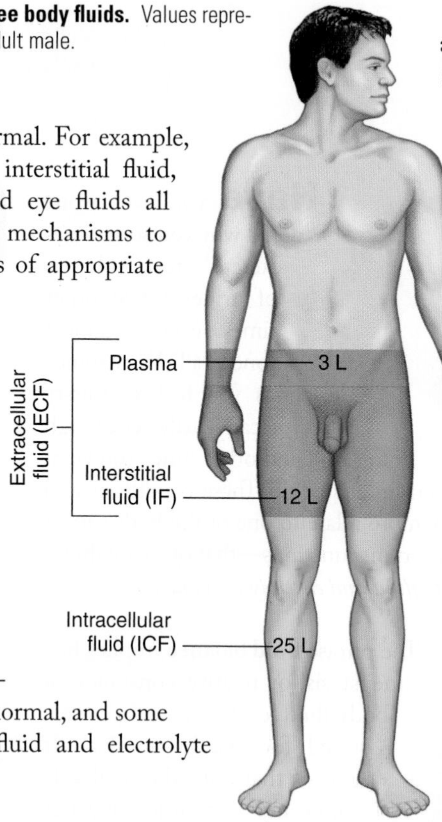

FIGURE 21-1 Relative volumes of three body fluids. Values represent typical fluid distribution in a young adult male.

within a very narrow range of normal. For example, blood, lymph, intracellular fluid, interstitial fluid, cerebrospinal fluid, and joint and eye fluids all depend on complex homeostatic mechanisms to adjust and maintain normal levels of appropriate electrolytes required for that particular type of body fluid to function as it should.

Health and sometimes even survival itself depend on maintaining the proper volume and distribution of body water and the appropriate levels and types of electrolytes within it. In this chapter you will find a discussion of body fluids and electrolytes, their normal values, the mechanisms that operate to keep them normal, and some of the more common types of fluid and electrolyte imbalances.

Body Fluid Volumes

Of the hundreds of compounds present in your body, the most abundant is water. Medical reference tables often refer to "average" fluid volumes based on healthy nonobese young adults. In such tables, males weighing 70 kg (154 pounds) will have on average about 60% of their body weight, nearly 40 L, as water (**Figure 21-1**). Young females average about 50% water.

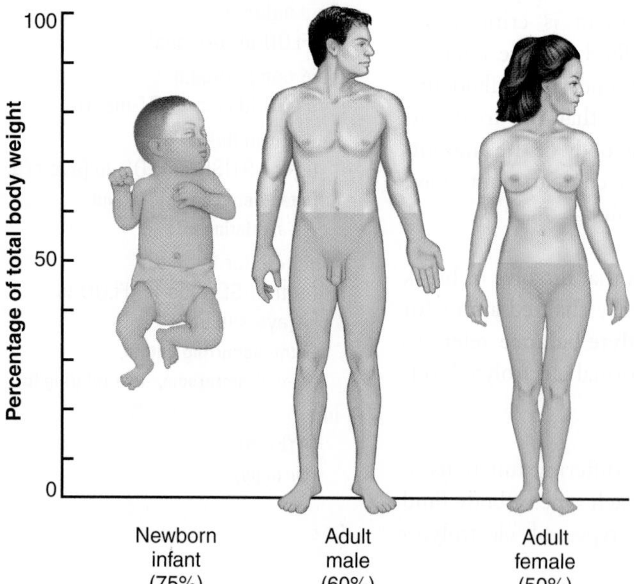

FIGURE 21-2 Water in the body. Proportion of body weight typically made up of water in infants, adult males, and adult females.

The reason fluid volume values in reference tables are based on nonobese individuals is that adipose, or fat tissue, contains the least amount of water of any body tissue. The more fat present in the body, the less the total water content per kilogram of body weight. Therefore, regardless of age, obese individuals, with their high body fat content, have less body water per kilogram of weight than slender people.

Although a nonobese male body typically consists of about 60% water, an obese male may consist of only 50% water or even less. The female body contains slightly less water per kilogram of weight because on average it contains slightly more fat than the male body.

Note in **Figure 21-2** that age, as well as gender, influences the amount of water in the body. Remember that body fluids are not all in a single, continuous space in the body—but often function as if they are.

Infants have more water as compared with body weight than adults of either sex. In a newborn, water may account for up to 80% of total body weight. The percentage of water is even higher in premature infants. The need for a high water content in the early stages of life is the reason fluid imbalances in infants caused by diarrhea, for example, can be so serious.

The percentage of body water decreases rapidly during the first 10 years of life and by adolescence, adult values are reached and gender differences, which account for about a 10% variation in body fluid volumes between the sexes, have developed.

In elderly individuals, the amount of water per kilogram of body weight decreases. One reason is that old age is often accompanied by a decrease in muscle mass (65% water) and an increase in fat (20% water). Certain drugs or toxins may have more potent effects in the elderly because they become more concentrated in the smaller volume of water present in the bodies of some elderly people. Of course, such drugs or toxins may have a reduced effect when diluted in the larger relative amount of water in a young person's body. In both cases, the key factor is the percentage of body weight represented by water.

Body Fluid Compartments

For the sake of discussion, the fluids of the body are thought of as being contained in theoretical "compartments." Each of these **fluid compartments** is actually a group of separated spaces in the body that in many ways function as if they are all in one compartment. Using this concept, total body fluid can be subdivided into two major fluid compartments called the *extracellular* and the *intracellular* fluid compartments. You can see the major fluid compartments illustrated in **Figure 21-3**.

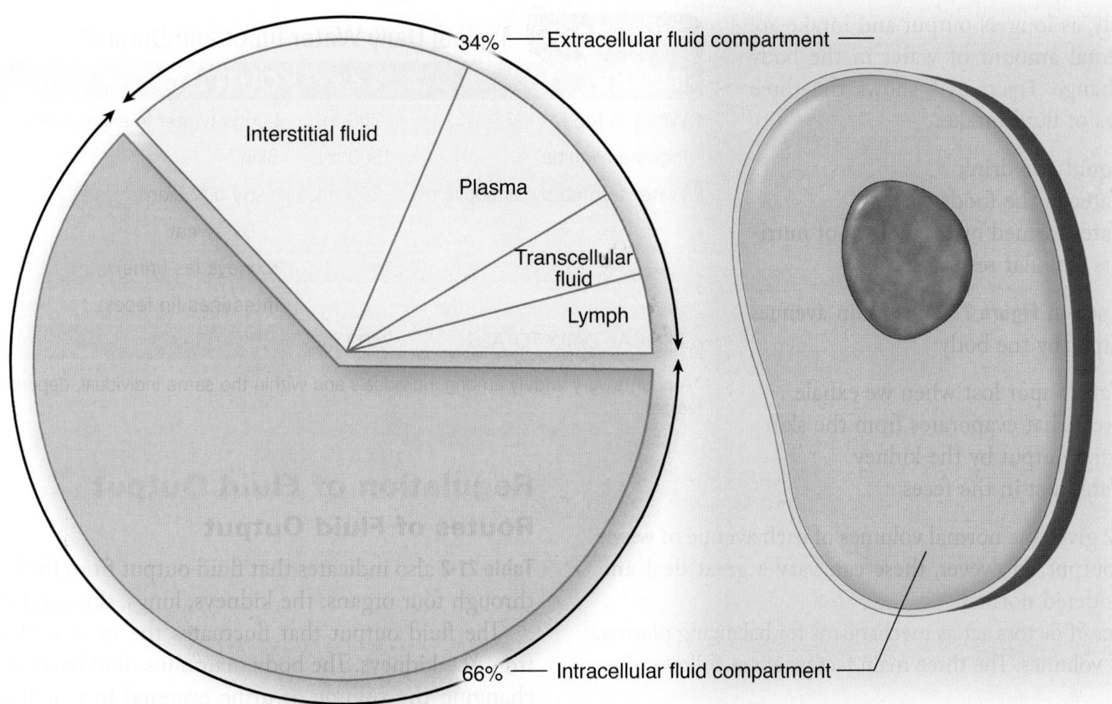

FIGURE 21-3 Distribution of total body water. The fluids of the body are separated by membranes into functional "compartments" of the body. The intracellular fluid (ICF) compartment includes all the fluids inside all the cells of the body. The extracellular fluid (ECF) compartment includes the interstitial fluid (IF) between cells of most tissues and the plasma of the blood tissue. ECF also includes lymph and transcellular fluids.

Extracellular Fluid

Extracellular fluid (**ECF**) consists mainly of the liquid part of whole blood called the **plasma,** found in the blood vessels, and the **interstitial fluid** (**IF**) that surrounds the cells.

In addition, a smaller volume of *lymph* and **transcellular fluids** are part of the extracellular fluid compartment. Transcellular fluids include cerebrospinal fluid (CSF), fluids of the eyeball, and the synovial joint fluids.

Table 21-1 lists typical percentage of body weight values for the extracellular fluid compartments. **Figure 21-3** shows the distribution of fluids in the extracellular fluid compartment as a percentage of total body water.

TABLE **21-1** Volumes of Body Fluid Compartments*

BODY FLUID	INFANT	ADULT MALE	ADULT FEMALE
Extracellular Fluid			
Plasma	4	4	4
Interstitial fluid, lymph, and transcellular fluids	26	16	11
Intracellular Fluid	45	40	35
TOTAL	75	60	50

*Percentage of body weight. Compare to volume in liters in **Figure 21-1** and percentage of total body water in **Figure 21-2**.

Intracellular Fluid

The term **intracellular fluid** (**ICF**) refers to the largest volume of body fluid by far. It is located inside all the cells of the body. Water has many functions inside the cell but mainly serves as a solvent in which important chemical reactions of the cell can occur.

QUICK CHECK

1. What are *electrolytes* and what is *electrolyte balance*?
2. What are the two main fluid compartments of the body?
3. What is meant by the term *fluid balance*?
4. What is the largest volume of body fluid?

Mechanisms That Maintain Fluid Balance

Overview of Fluid Balance

Under normal conditions, homeostasis of the total volume of water in the body is maintained or restored primarily by devices that adjust output (by adjusting urine volume) to intake and secondarily by mechanisms that adjust fluid intake. There is no question about which of the two mechanisms is more important; the body's chief mechanism, by far, for maintaining fluid balance is that of adjusting its fluid output so that it equals its fluid intake.

21

Obviously, as long as output and intake are equal, the total amount of water in the body does not change. **Figure 21-4** shows the three main sources of fluid intake:

1. Liquids we drink
2. Water in the foods we eat
3. Water formed by catabolism of nutrients (cellular respiration)

We also see in **Figure 21-4** the main avenues of water output by the body:

1. Water vapor lost when we exhale
2. Sweat that evaporates from the skin
3. Urine output by the kidney
4. Water lost in the feces

Table 21-2 gives the normal volumes of each avenue of water input and output. However, these can vary a great deal and still be considered normal.

A number of factors act as mechanisms for balancing plasma, IF, and ICF volumes. The three main factors are as follows:

1. Regulating fluid output
2. Regulating fluid input
3. Exchanging fluids between compartments and from place to place within the body

TABLE 21-2 Typical Daily Water Input and Output*

INTAKE		OUTPUT	
Water in foods	700 mL	Lungs (water in expired air)	350 mL
Ingested liquids	1500 mL	Skin	
Water formed by catabolism	200 mL	By diffusion	350 mL
		By sweat	100 mL
		Kidneys (as urine)	1400 mL
		Intestines (in feces)	200 mL
TYPICAL DAILY TOTALS	2400 mL		2400 mL

*Amounts vary widely among individuals and within the same individual, depending on many factors.

Regulation of Fluid Output

Routes of Fluid Output

Table 21-2 also indicates that fluid output from the body occurs through four organs: the kidneys, lungs, skin, and intestines.

The fluid output that fluctuates the most is that excreted from the kidneys. The body maintains fluid balance mainly by changing the volume of urine excreted to match changes in the volume of fluid intake. Everyone knows this from experience. The more liquid one drinks, the more urine one excretes. Conversely, the less the fluid intake, the less the urine volume. How changes in urine volume come about was discussed

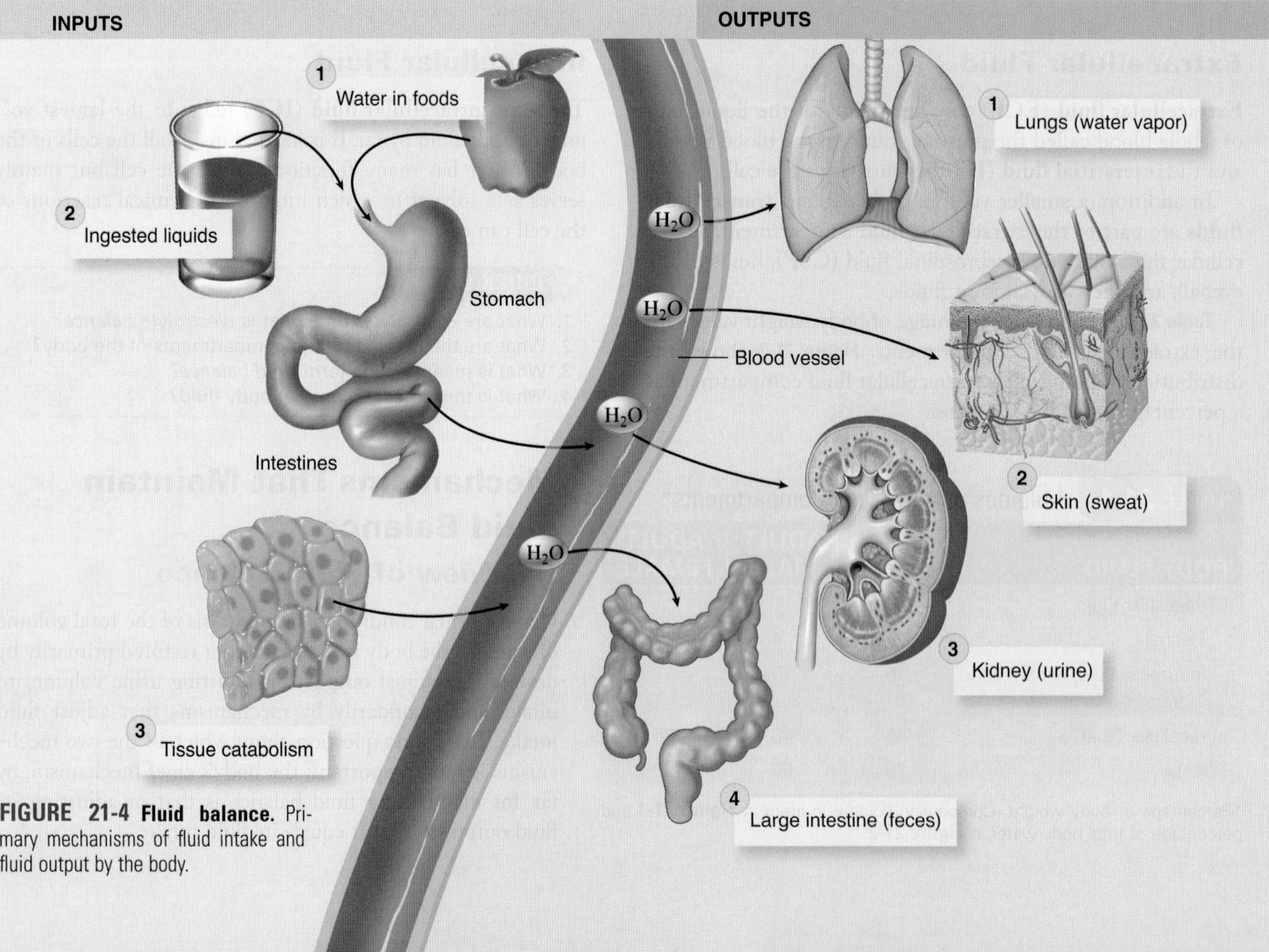

FIGURE 21-4 Fluid balance. Primary mechanisms of fluid intake and fluid output by the body.

21

on pp. 563-564. This would be a good time to review those paragraphs.

It is important to remember from your study of the urinary system that the rate of water and salt resorption by the renal tubules is the most important factor in determining urine volume. Urine volume is regulated chiefly by hormones that affect kidney tubule function.

ADH Mechanism

Antidiuretic hormone (ADH) release from the posterior pituitary increases as the ECF volume of the body decreases below normal. In Chapter 12, we learned that ADH promotes water reabsorption from the kidney tubule back into the blood. This reduces urine volume by retaining more water in the body. Thus, ADH reduces water output from the body.

Aldosterone Mechanism

Aldosterone from the adrenal cortex works with ADH to reduce water output even further. Aldosterone increases Na^+ reabsorption by the kidney tubules. Because water follows sodium, water reabsorption into the blood also increases. Thus, the body retains water that would otherwise be lost in the urine. Thus we see that ADH and aldosterone are water-conserving hormones.

Figure 21-5 traces the aldosterone mechanism in more detail. Begin in the upper right of the diagram and follow, in sequence, each step to see how the aldosterone mechanism helps maintain a constant volume of ECF in the body.

ANH Mechanism

Atrial natriuretic hormone (ANH) from the atrial wall of the heart, on the other hand, increases urine volume. ANH is released when blood volume is higher than normal, which stretches the atrium. ANH promotes sodium loss from the blood into kidney tubules. Because water follows sodium, water is also lost from the blood—thus increasing loss of water in the urine. Therefore, ANH is a water-loss hormone—or *diuretic* hormone.

Please review hormonal control of urine volume in Chapter 20 (pp. 563-564).

To learn more about the aldosterone regulation mechanism, go to AnimationDirect online at *evolve.elsevier.com*.

Regulation of Fluid Intake

Physiologists disagree about the details of the mechanism for controlling and regulating fluid intake to compensate for factors that would lead to dehydration.

In general the mechanism for regulating fluid intake appears to operate in the following ways. When dehydration starts to develop—that is, when fluid loss from the body exceeds fluid intake—changes occur in the ECF. The ECF volume decreases and the solute concentration (osmotic pressure) of the ECF increases.

Sensory receptors in the brain and elsewhere in the body detect the change in the volume and concentration of extracellular fluids caused by dehydration. They relay this information to the thirst centers of the hypothalamus. Signals from the hypothalamus cause water conservation throughout the body, including a decrease in salivary secretion. Decreased salivation produces a "dry-mouth feeling" that enhances a feeling of thirst. The dry mouth causes a person to "feel thirsty" and to drink water. Drinking water increases fluid intake and thereby compensates for previous fluid losses. This tends to restore fluid balance (**Figure 21-6**).

If an individual takes nothing by mouth for days, can his fluid output decrease to zero? The answer—no—becomes obvious after reviewing the information in **Table 21-2**. Despite

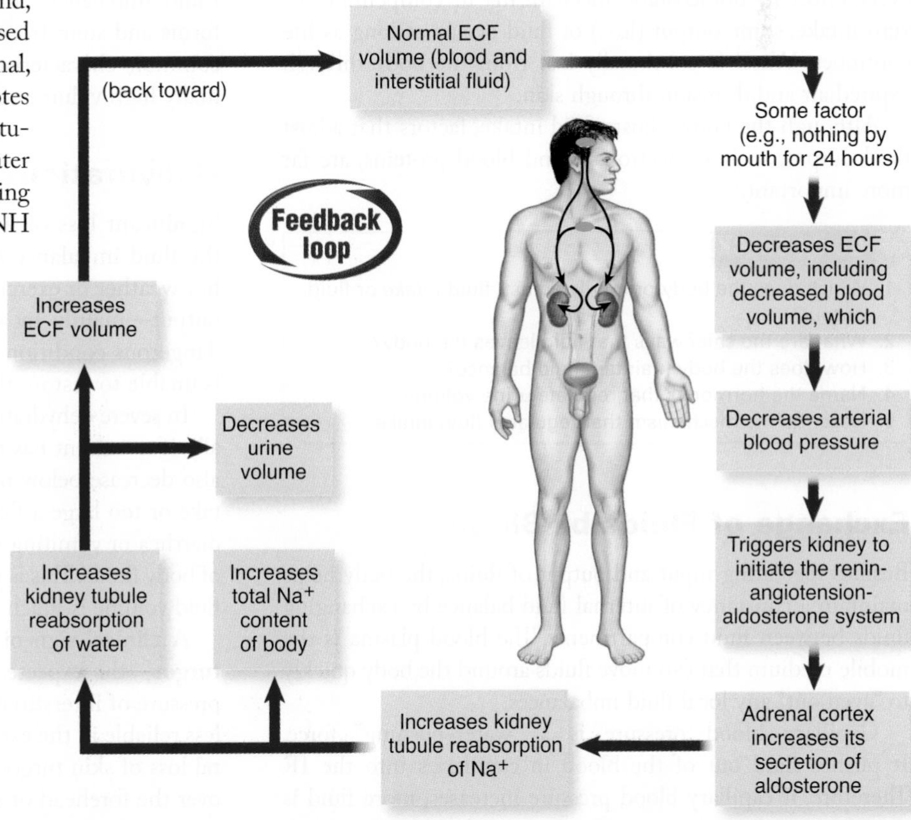

FIGURE 21-5 Aldosterone mechanism. Aldosterone restores normal extracellular fluid (ECF) volume when such levels decrease below normal. Excess aldosterone, however, leads to excess ECF volume—that is, excess blood volume (hypervolemia) and excess interstitial fluid volume (edema)—and also leads to an excess of the total Na^+ content of the body.

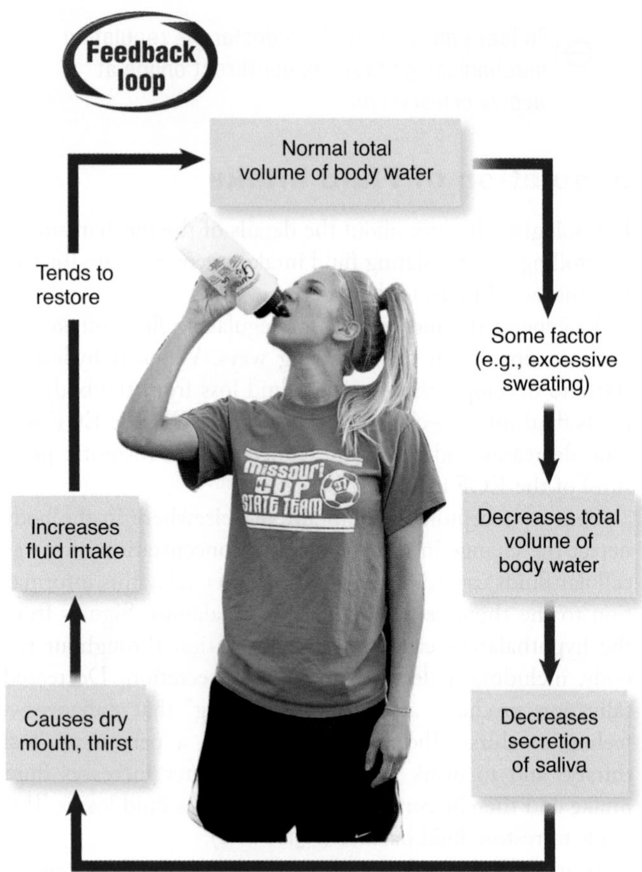

FIGURE 21-6 Thirst mechanism. A basic mechanism for adjusting intake to compensate for excess output of body fluid is diagrammed here.

every effort of homeostatic mechanisms to compensate for zero intake, some output (loss) of fluid occurs as long as life continues. Water is continually lost from the body through expired air and diffusion through skin.

Although the body adjusts fluid intake, factors that adjust fluid output, such as electrolytes and blood proteins, are far more important.

QUICK CHECK

1. Which does the body primarily adjust, fluid *intake* or fluid *output*?
2. What are the chief ways that fluid leaves the body?
3. How does the body maintain fluid balance?
4. Name the hormones that regulate urine volume.
5. Describe the mechanism that regulates fluid intake.

Exchange of Fluids by Blood

Besides regulating input and output of fluids, the body helps maintain a constancy of internal fluid balance by exchanging fluids between fluid compartments. The blood plasma is the mobile medium that can move fluids around the body quickly to "even out" any local fluid imbalances.

Capillary blood pressure is a "water-pushing" force. It pushes fluid out of the blood in capillaries into the IF. Therefore, if capillary blood pressure increases, more fluid is pushed—filtered—out of blood into the IF. The effect of an increase in capillary blood pressure, then, is to transfer fluid from blood to IF. In turn, this *fluid shift,* as it is called, changes blood and IF volumes. It decreases blood volume by increasing IF volume. If, on the other hand, capillary blood pressure decreases, less fluid filters out of blood into IF.

Water continually moves in both directions through the membranous walls of capillaries (see **Figure 21-4**). The amount that moves out of capillary blood into IF depends largely on capillary blood pressure, a water-pushing force. The amount that moves in the opposite direction (that is, into blood from IF) depends largely on the concentration of proteins in blood plasma. Review **Figure 15-4** on p. 406 to refresh your knowledge of these forces.

Plasma proteins contribute to osmotic pressure and thereby act as a water-pulling or water-holding force. They hold water in the blood and pull it into the blood from IF. If, for example, the concentration of proteins in blood decreases appreciably—as in protein deficiency—less water moves into blood from IF by osmosis (see **Figure 3-8** on p. 52). As a result, blood volume decreases and IF volume increases—causing edema.

Of the three main body fluids, IF volume varies the most. Plasma volume usually fluctuates only slightly and briefly. If a pronounced change in its volume occurs, adequate circulation cannot be maintained.

 To learn more about fluid shift, go to AnimationDirect online at *evolve.elsevier.com.*

Fluid Imbalances

Fluid imbalances are common ailments. They take several forms and stem from a variety of causes, but they all share a common characteristic—that of abnormally low or abnormally high volumes of one or more body fluids.

Dehydration

Significant loss of water from the body, or **dehydration,** is the fluid imbalance seen most often. **Figure 21-7** shows how hot weather or exercise can cause dramatic increases in water output—mainly by sweating. Dehydration is a potentially dangerous condition that can soon lead to death if a person is unable to restore the body's fluid volume.

In severe dehydration, IF volume decreases first, but eventually, if treatment has not been given, ICF and plasma volumes also decrease below normal levels. Either too small a fluid intake or too large a fluid output causes dehydration. Prolonged diarrhea or vomiting may result in dehydration due to the loss of body fluids. This is particularly true in infants where the total fluid volume is much smaller than it is in adults.

A clinical sign of dehydration is a decrease in the skin's **turgor**—the expected resiliency of skin due to the outward pressure of interstitial fluid (**Figure 21-8**). However, this sign is less reliable in the extremities of the elderly because of a natural loss of skin turgor due to aging. In aged patients, the skin over the forehead or sternum can be checked.

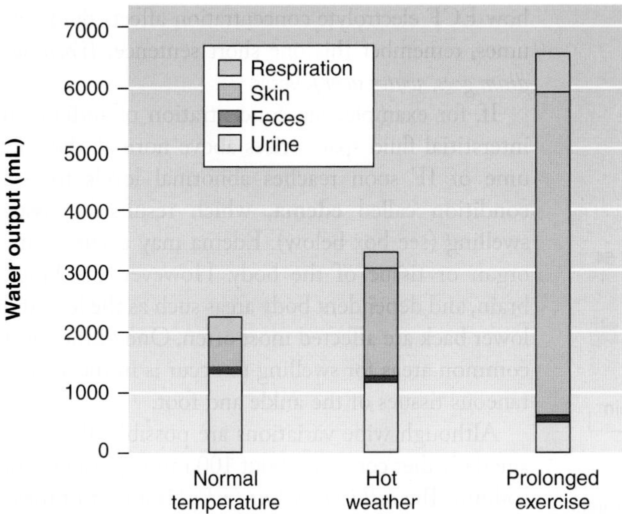

FIGURE 21-7 Water output by the body under varying conditions. Note that water loss from sweating (from the skin) increases total water loss by the body when the weather is hot and during prolonged exercise.

Overhydration

The condition of having more water in the body than needed for healthy survival is called **overhydration.** Although overhydration does occur, it is less common than dehydration—and is usually counteracted by a rapid loss of water in the urine.

One grave danger of giving intravenous fluids too rapidly or in too large of an amount is overhydration, which can put too heavy a burden on the heart by increasing the volume of blood to be pumped.

Water intoxication may result from rapidly drinking large volumes of water or giving hypotonic solutions to persons unable to dilute and excrete urine normally. This may occur in patients with kidney insufficiency or abnormal "thirst" mechanisms resulting from neurological disorders. Water content is elevated, and plasma sodium levels are diluted. Development of subtle mental changes such as confusion and lethargy occur. If intoxication is severe, stupor, seizures, and coma may result.

FIGURE 21-8 Testing for dehydration. A decrease of skin resiliency or turgor is a sign of dehydration. Skin that does not return quickly to its normal shape after being pinched (or "tented") indicates interstitial water loss. This sign is less reliable in the extremities of the elderly, who naturally experience loss of turgor due to aging.

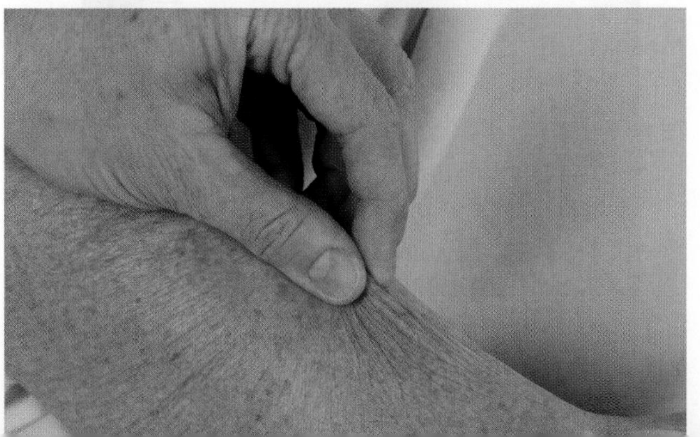

Correction of the neurological impairment along with water restriction can reverse the symptoms.

Water intoxication can happen in normal individuals if water intake is so rapid that the urinary mechanisms of water loss cannot keep up. Although this is unusual, it can happen—as witnessed by millions a few years ago when a radio station held a "water drinking race" on the air and a contestant died from the effects of severe water intoxication.

QUICK CHECK

1. How does an increase in capillary blood pressure cause fluid to move into the IF?
2. How do plasma proteins affect fluid balance?
3. What conditions might produce *dehydration*?
4. What is *water intoxication*?

Importance of Electrolytes in Body Fluids

Electrolytes and Nonelectrolytes

The bonds that hold together the molecules of certain organic substances such as glucose are such that they do not permit the compound to break up, or **dissociate,** in water solution. Such compounds are called **nonelectrolytes.** Crystals such as ordinary table salt, or sodium chloride (NaCl), that have *ionic bonds* that permit them to break up, or dissociate, in water solution into separate particles (Na^+ and Cl^-) are **electrolytes.**

Ions

The dissociated particles of an electrolyte are called **ions** and carry either a positive or negative electrical charge. As a group, all positively charged ions, such as Na^+, are called **cations.** All negatively charged ions, such as Cl^-, are called **anions.** Each of the body fluid compartments contains differing levels of many important ions—both positively charged cations and negatively charged anions. Sometimes the dissociated ions are themselves called *electrolytes.*

Important cations include sodium (Na^+), calcium (Ca^{++}), potassium (K^+), and magnesium (Mg^{++}). Important anions include chloride (Cl^-), bicarbonate (HCO_3^-), phosphates ($H_2PO_4^-$ and $HPO_4^=$), and many proteins. Proteins can be anionic when they contain negatively charged amino acids—amino acid side groups that have gained electrons to give them an electrical charge.

Figure 21-9 shows that although ECF contains a number of important ions, by far the most abundant are sodium (positive) and chloride (negative). However, in the ICF, we find mostly potassium (positive) and anionic proteins (negative).

Electrolyte Functions

A variety of electrolytes have important nutrient or regulatory roles in the body. Many ions are major or important "trace" elements in the body (see Appendix C at *evolve.elsevier.com*).

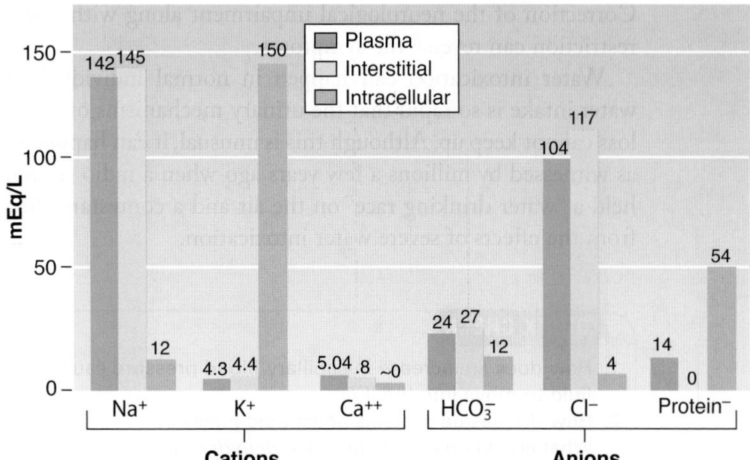

FIGURE 21-9 Electrolytes found in fluid compartments of the body. Note that sodium (Na^+) is the dominant positive ion and chloride (Cl^-) is the dominant negative ion in the extracellular fluid compartments (plasma and interstitial fluid). However, in the intracellular fluid compartment, potassium (K^+) and anionic proteins dominate. *mEq/L,* milliequivalent per liter.

Iron, for example, is required for hemoglobin production, and iodine must be available for synthesis of thyroid hormones. Electrolytes also are required for many cellular activities such as nerve conduction and muscle contraction.

In addition, electrolytes influence the movement of water among the three fluid compartments of the body. To remember how ECF electrolyte concentration affects fluid volumes, remember this one short sentence: *Where sodium goes, water soon follows.*

If, for example, the concentration of sodium in interstitial fluid spaces rises above normal, the volume of IF soon reaches abnormal levels too—a condition called **edema,** which results in tissue swelling (see box below). Edema may occur in any organ or tissue of the body. However, the lungs, brain, and dependent body areas such as the legs and lower back are affected most often. One of the most common areas for swelling to occur is in the subcutaneous tissues of the ankle and foot.

Although wide variations are possible, the average daily diet contains about 100 milliequivalents of sodium. The *milliequivalent (mEq)* is a unit of measurement related to ion reactivity.

In a healthy individual, sodium excretion from the body by the kidney is about the same as intake. The kidney acts as the chief regulator of sodium levels in body fluids. It is important to know that many electrolytes such as sodium not only pass into and out of the body but also move back and forth between a number of body fluids during each 24-hour period.

Figure 21-10 shows the large volumes of sodium-containing internal secretions produced each day. During a 24-hour period, more than 8 liters of fluid containing 1000 to 1300 mEq

CLINICAL APPLICATION

EDEMA

Edema may be defined as the presence of abnormally large amounts of fluid in the interstitial tissue spaces of the body. The term **pitting edema** is used to describe depressions in swollen subcutaneous tissue that do not rapidly refill after an examiner has exerted finger pressure (see photo). This type of edema is often a symptom in those with congestive heart failure.

The condition is a classic example of fluid imbalance and may be caused by disturbances in any factor that governs the interchange between blood plasma and IF compartments. Examples include the following:

1. **Retention of electrolytes (especially Na^+) in the extracellular fluid.** This can result from increased aldosterone secretion or can occur during serious kidney disease.
2. **An increase in capillary blood pressure.** Normally, fluid is drawn from the tissue spaces into the venous end of a tissue capillary because of the low venous pressure and the relatively high water-pulling force of the plasma proteins. This balance is upset by anything that increases the capillary hydrostatic pressure. The generalized venous congestion of heart failure is the most common cause of widespread edema. In patients with this condition, blood cannot flow freely through the capillary beds, and therefore the pressure will increase until venous return of blood improves.
3. **A decrease in the concentration of plasma proteins.** This decrease can be caused by "leakage" into the interstitial spaces of proteins normally retained in the blood. This may occur as a result of increased capillary permeability caused by infection, burns, or shock.

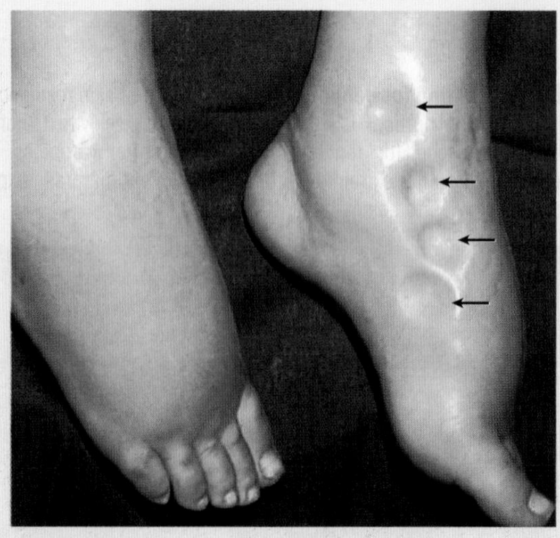

Pitting edema. Note the fingertip-shaped depressions *(arrows)* that do not rapidly refill after an examiner has exerted pressure.

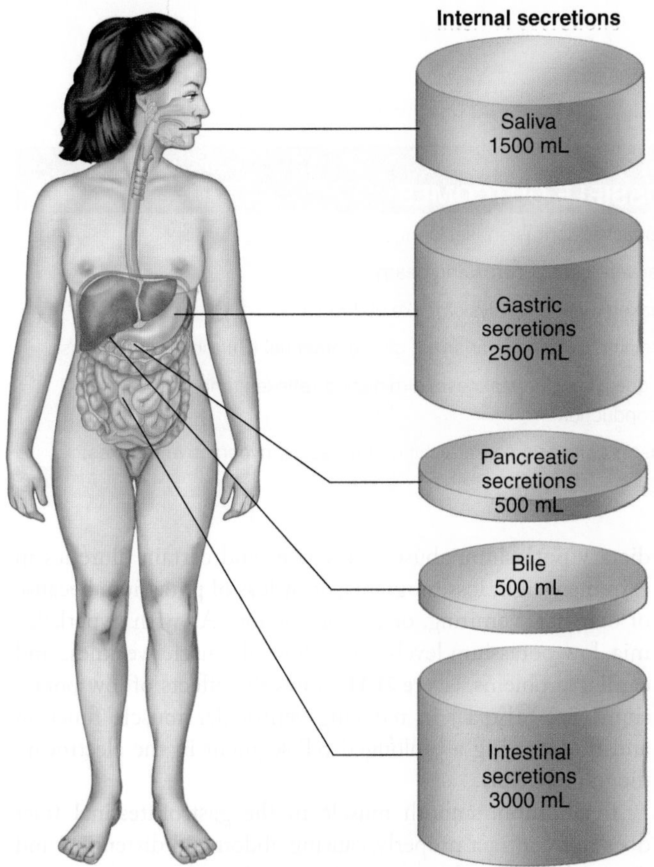

FIGURE 21-10 Sodium-containing internal secretions. The total volume of these secretions may reach 8000 mL or more in 24 hours.

of sodium are poured into the digestive system as part of saliva, gastric secretions, bile, pancreatic juice, and IF secretions. This sodium, along with most of that contained in the diet, is almost completely reabsorbed in the intestines. Very little sodium is lost in the feces. Precise regulation and control of sodium levels are required for survival.

QUICK CHECK

1. What is the difference between an *electrolyte* and a *nonelectrolyte*?
2. What are some of the major roles of ions in the body?
3. Identify the functions of electrolytes in the body.

Electrolyte Imbalances

Homeostasis of Electrolytes

Electrolyte balance, like fluid balance, is related to "intake" and "output" of specific electrolytes. Also important is the absorption of electrolytes that are ingested, their final distribution in the body fluids, and their "availability" for use by the body cells.

ECF normally contains differing levels of some electrolytes than does ICF. In order to maintain different concentrations of electrolytes in the different body fluids, differing homeostatic mechanisms that influence intake, absorption, distribution, and excretion of these electrolytes are needed.

Any disruption in a homeostatic mechanism that controls the level or normal chemical activity of a particular electrolyte in any of the different body fluids produces an *electrolyte imbalance.* Such imbalances are widespread and often very serious and sometimes fatal manifestations of disease.

CLINICAL APPLICATION

DIURETICS

The word **diuretic** is from the Greek word *diouretikos* meaning "causing urine." By definition a *diuretic drug* is a substance that promotes or stimulates the production of urine. Recall that an increase in urine volume represents a loss of water from the body.

As a group, diuretics are among the most commonly used drugs in medicine. They are used because of their role in influencing water and electrolyte balance, especially sodium, in the body. For example, diuretics can be used to remove fluid from the body to reduce blood pressure in patients with hypertension (HTN).

Diuretics have their effect on tubular function in the nephron, and the differing types of diuretics are often classified according to their major site of action. Examples would include (1) *proximal tubule diuretics* such as acetazolamide (Diamox), (2) *nephron loop diuretics* such as ethacrynic acid (Edecrin) or furosemide (Lasix), and (3) *distal tubule diuretics* such as chlorothiazide (Diuril).

Caffeine produces its mildly diuretic effects by inhibiting water reabsorption in the proximal tubules of the kidney.

Classification of diuretic drugs also can be made according to the effect the drug has on the level or concentration of sodium (Na^+), chloride (Cl^-), potassium (K^+), and bicarbonate (HCO_3^-) ions in the tubular fluid.

Alcohol is also a diuretic. It reduces secretion of ADH, which is a water-conserving hormone. Thus water that would have otherwise been conserved by the body is lost under the influence of alcohol.

Diuretics are sometimes abused by athletes to quickly reduce their weight just before an event or "weigh-in." Loss of water from the body does in fact reduce a person's weight, but it also reduces his or her athletic ability by creating the condition of dehydration. Diuretics (except for legitimate therapeutic use) are included on the Prohibited List by the World Anti-Doping Agency.

Nursing implications for caregivers monitoring patients receiving diuretics both in hospitals and in home health-care environments include (1) keeping a careful record of body weight and fluid intake and output and (2) assessing the patient for signs and symptoms of electrolyte and water imbalance. For example, diuretic-induced dehydration resulting in a loss of only 6% of initial body weight will cause tingling in the extremities, stumbling gait, headache, fever, and an increase in both pulse and respiratory rates.

Appendix C at *evolve.elsevier.com* lists the normal values of many important electrolytes and identifies disease states that may result in variations above or below normal levels. Electrolyte imbalances involving sodium, potassium, and calcium are common in clinical medicine and are described in the following sections.

Check out the article *Fluid and Electrolyte Therapy* at Connect It! at *evolve.elsevier.com.*

TABLE 21-3 Electrolyte Imbalances

ELECTROLYTE	IMBALANCE	BLOOD CONCENTRATION	POSSIBLE OUTCOMES
Sodium (Na^+)	Hypernatremia	>145 mEq/L	Headache; confusion; seizures
	Hyponatremia	<136 mEq/L	In severe cases: coma and death
Potassium (K^+)	Hyperkalemia	>5.1 mEq/L	Weakening and paralysis of skeletal muscle
	Hypokalemia	<3.5 mEq/L	Cardiac dysrhythmia or arrest gastrointestinal (GI) motility problems
Calcium (Ca^{++})	Hypercalcemia	>5.25 mEq/L	Fatigue, muscle weakness, diminished reflexes; impaired cardiac conduction
	Hypocalcemia	<4.2 mEq/L	Muscle cramping and twitching of muscles, hyperactive reflexes; cardiac dysrhythmia

Sodium Imbalance

The term *natrium* is the Latin word for sodium. The prefixes *hyper-* and *hypo-* refer to "above" and "below," respectively. Knowing this makes the terms *hypernatremia* and *hyponatremia* easier to understand and remember. **Hypernatremia** is used to describe a blood sodium level of more than 145 mEq/L. **Hyponatremia** occurs when blood sodium level is below 136 mEq/L (see Appendix C at *evolve.elsevier.com* and **Table 21-3**).

Hypernatremia may result from overuse of salt tablets, dehydration, or prolonged diarrhea. Regardless of cause, the condition is characterized by a relative deficit of water to salt in the ECF. Hyponatremia occurs when there is relatively too much water in the ECF compartment for the amount of sodium present. This can occur if excessive antidiuretic hormone is produced or after massive infusion of IV fluids, such as 5% dextrose in water, that do not contain sodium.

Hyponatremia also may be caused by excessive salt loss resulting from burns or certain diuretics.

Both of these conditions affect central nervous system (CNS) functioning and are characterized by headache; confusion; seizures; and, in the most severe cases, coma and death.

Potassium Imbalance

The normal range for potassium in blood is 3.5 to 5.1 mEq/L (see Appendix C at *evolve.elsevier.com*). Although most of the total body potassium is inside the cells, fluctuations or imbalance in the relatively small amounts present in the ECF will cause serious illness.

Hyperkalemia is the clinical term used to describe blood potassium levels of more than 5.1 mEq/L. (*Kalium* is the Latin word for potassium.) Elevation of potassium may be related to increased intake, a shift from the intracellular fluid into the blood—caused by tissue trauma or burns, or in cases of renal failure, by an inability of the kidneys to excrete excess potassium.

Many of the clinical manifestations of hyperkalemia are related to muscle malfunction (see **Table 21-3**). As potassium levels increase, skeletal muscles weaken and paralysis develops. Severe hyperkalemia results in cardiac arrest.

Hypokalemia refers to a low blood potassium level (below 3.5 mEq/L). It may be caused by fasting; fad diets low in dietary potassium; abuse of laxatives and certain diuretics in extreme weight loss programs; or by loss of potassium because of diarrhea, vomiting, or gastric suction. As with hyperkalemia, low potassium levels cause skeletal muscle weakness and cardiac problems. **Figure 21-11** shows the effects of low potassium (2.2 mEq/L) in reducing ventricular muscle function and thus causing a prolonged ST segment in the electrocardiogram (ECG).

In addition, smooth muscle in the gastrointestinal tract does not contract properly, causing abdominal distention and diminished rate of passage of intestinal contents.

Calcium Imbalance

Calcium is the most abundant mineral in the body. It serves as a basic structural building block in bone and teeth. In addition, it is essential for the maintenance of a normal heartbeat, for functioning of nerves and muscles, and for the role it plays in metabolism, blood coagulation, and in many other enzymatic reactions. Failure of homeostatic mechanisms that regulate levels of this important electrolyte can result in catastrophic illness.

The normal range for serum calcium is 8.4 to 10.5 mg/dL—or 4.2 to 5.25 mEq/L (see Appendix C at *evolve.elsevier.com* and **Table 21-3**).

Hypercalcemia occurs when blood calcium levels rise above normal limits. The condition may be caused by excessive input or by increased absorption that may occur following an overdose of vitamin D. Elevated levels also can result from

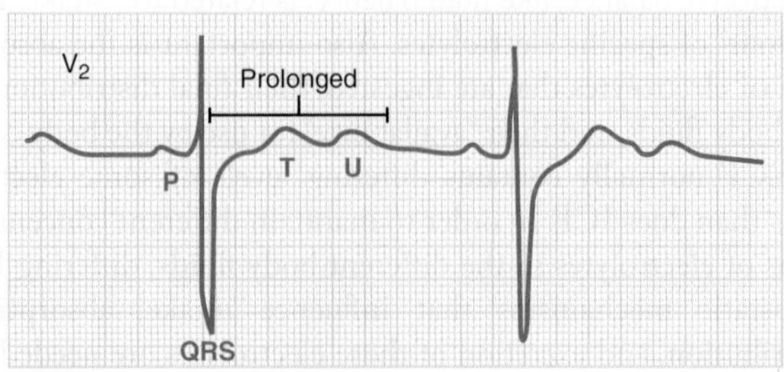

FIGURE 21-11 Hypokalemia effects on heart function. Low potassium levels (hypokalemia) can cause changes in heart function, including a prolonged ST segment caused by the presence of an extra wave called the *U wave*. Compare to the normal ECG (electrocardiogram) in **Figure 14-10** on p. 388.

shifts of calcium from bone into the ECF caused by Paget disease (see p. 202), other bone tumors, or hyperparathyroidism blood levels that will also increase if the kidney cannot normally excrete excess calcium in the urine—a side effect of certain diuretics.

Regardless of cause, hypercalcemia decreases neuromuscular ability to be stimulated—resulting in fatigue, muscle weakness, diminished reflexes, and delayed atrioventricular conduction in the heart.

Hypocalcemia may result from dietary calcium deficiency, decreased absorption or availability, and as a result of increased calcium excretion. Diseases such as pancreatitis, hypoparathyroidism, rickets, and osteomalacia and chronic renal insufficiency all lower blood calcium levels.

Clinical signs of hypocalcemia involve *increased* neuromuscular excitability, cramping and twitching of muscles, hyperactive reflexes, and abnormal cardiac rhythms characterized by impairment of myocardial contractility. For example, light taps on the cheek to stimulate the facial nerve (CN VII) may produce the **Chvostek sign**—an abnormal spasm of facial muscles—in hypocalcemic patients.

QUICK CHECK

1. What are the causes of hypernatremia? Hyponatremia?
2. Hypokalemia may cause what conditions?
3. Why is calcium a significant mineral in our body?

SCIENCE APPLICATIONS

THE CONSTANCY OF THE BODY

Claude Bernard (1813–1877)

In 1834, a young Claude Bernard left what he thought of at the time as his "boring job" as an apprentice apothecary (druggist) in Lyon, France, to make his fortune as a playwright in Paris. His plays were not appreciated in Paris, but he took a medical course while there and found that many of the doctors appreciated his research skills. Bernard went on to become one of the most important figures in the study of human physiology.

Bernard made groundbreaking discoveries in the functions of the pancreas and the liver, discovered the existence of muscles that control blood vessel dilation, and wrote a manual on experimental medicine that set the standard in research practice for a century. However, one of the most fundamental contributions he made to human physiology is the idea that the body is made up of cells living in an internal fluid environment.

Bernard stated that the internal fluid environment of the body is maintained in a relatively constant state—and that's what ensures the survival of the cells and therefore also ensures the survival of the whole body. Recall from Chapter 1 that we now call this concept *homeostasis* (see p. 14). It was Bernard who showed that the actions of hormones and other control mechanisms maintain constant conditions in the body's internal fluid environment. And it was Bernard who showed that nearly every function of the body somehow relates to the success of keeping body fluids constant.

Today, nearly every health-care professional uses concepts based on Bernard's original idea to help keep patients alive and healthy. Those who use these ideas most directly are the nurses, health technicians, **IV technicians** (pictured), and others who care for patients on an hour by hour basis. It is these professionals who must constantly assess the fluid balance of patients and possibly administer therapies to bring their fluids back into balance. Maintaining a healthy fluid and electrolyte balance is one of the key elements of successful patient care in the modern hospital and clinic.

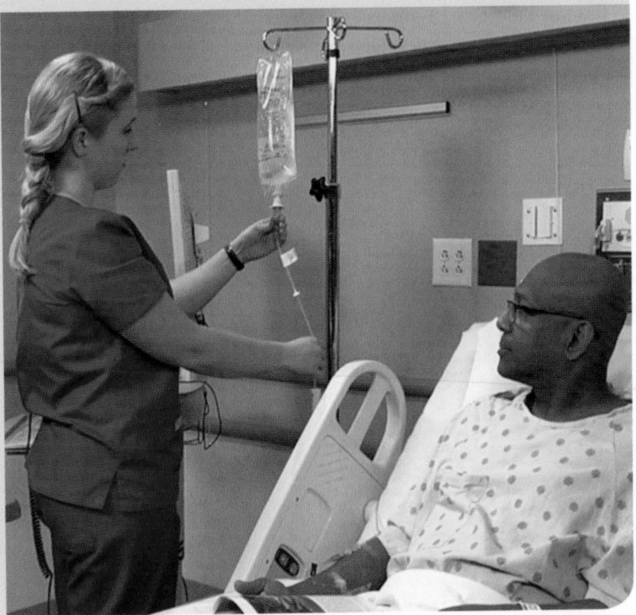

LANGUAGE OF **SCIENCE** *(continued from p. 583)*

nonelectrolyte
(non-ee-LEK-troh-lyte)
[*non-* **not,** *-electro-* **electricity,** *-lyt-* **loosening**]

plasma
(PLAZ-mah)
[*plasma* **substance**]

transcellular fluid
(tranz-SEL-yoo-lar)
[*trans-* **across,** *-cell-* **storeroom,** *-ular* **relating to**]

LANGUAGE OF **MEDICINE**

Chvostek sign
(ke-VOSH-tek syne)
[*Franz Chvostek* **Austrian surgeon**]

dehydration
(dee-hye-DRAY-shun)
[*de-* **remove,** *-hydro* **water,** *-ation* **process**]

diuretic
(dye-yoo-RET-ik)
[*dia-* **through,** *-ure-* **urine,** *-ic* **relating to**]

edema
(eh-DEE-mah)
[*edema* **swelling**]

hypercalcemia
(hye-per-kal-SEE-mee-ah)
[*hyper-* **excessive,** *-calc-* **lime (calcium),** *-emia* **blood condition**]

hyperkalemia
(hy-per-kal-EE-mee-ah)
[*hyper-* **excessive,** *-kal-* **kalium (potassium),** *-emia* **blood condition**]

hypernatremia
(hy-per-nah-TREE-mee-ah)
[*hyper-* **excessive,** *-natri-* **natrium (sodium),** *-emia* **blood condition**]

hypocalcemia
(hye-poh-kal-SEE-mee-ah)
[*hypo-* **under or below,** *-calc-* **lime (calcium),** *-emia* **blood condition**]

hypokalemia
(hye-poh-kal-EE-mee-ah)
[*hypo-* **under or below,** *-kal-* **kalium (potassium),** *-emia* **blood condition**]

hyponatremia
(hye-poh-nah-TREE-mee-ah)
[*hypo-* **under or below,** *-natri-* **natrium (sodium),** *-emia* **blood condition**]

IV (intravenous) technician
(aye-vee [in-trah-VEE-nus] tek-NISH-en)
[*intra-* **within,** *-ven-* **vein,** *-ous* **relating to,** *techn-* **art or skill,** *-ic* **relating to,** *-ian* **practitioner**]

overhydration
(oh-ver-hye-DRAY-shun)
[*over-* **above,** *-hydr-* **water,** *-ation* **process**]

pitting edema
(pit-ing eh-DEE-mah)
[*edema* **swelling**]

turgor
(TUR-ger)
[*turg-* **swell,** *-or* **condition**]

water intoxication
(WAH-ter in-TOK-sih-kay-shen)
[*in-* **in,** *-toxic-* **poison,** *-ation* **process**]

☐ OUTLINE SUMMARY

To download a digital version of the chapter summary for use with your device, access the ***Audio Chapter Summaries*** *online at evolve.elsevier.com.*

Hint *Scan this summary after reading the chapter to help you reinforce the key concepts. Later, use the summary as a quick review before your class or before a test.*

Body Fluid Volumes

A. Water is the most abundant body compound
   1. References to "average" body water volume in reference tables are based on a healthy, nonobese, 70-kg male
   2. Volume averages 40 L in a 70-kg male (**Figure 21-1**)
      a. Plasma (3 L)
      b. Interstitial fluid (12 L)
      c. Intracellular fluid (25 L)
   3. Water is 80% of body weight in newborn infants; 60% in adult males; 50% in adult females (**Figure 21-2**)
   4. Variation in total body water is related to:
      a. Total body weight of individual
      b. Fat content of body—the more fat in the body the less the total water content per kilogram of body weight (adipose tissue is low in water content)
      c. Gender—female body has about 10% less than male body (**Figure 21-2**)
      d. Age—in a newborn infant, water may account for 80% of total body weight. In the elderly, water per kilogram of weight decreases (muscle tissue—high in water—replaced by fat which is lower in water)

Body Fluid Compartments

A. The fluids of the body are contained within different "compartments" of the body (**Figure 21-3 and Table 21-1**)
B. Extracellular fluid (ECF)—called internal environment of body; surrounds cells and transports substances to and from them
   1. Plasma—liquid part of whole blood
   2. Interstitial fluid (IF)—surrounds the cells
   3. Transcellular—lymph; joint fluids; cerebrospinal fluid; eye humors
C. Intracellular fluid (ICF)—largest fluid compartment
   1. Located inside cells
   2. Serves as solvent to facilitate intracellular chemical reactions

Mechanisms That Maintain Fluid Balance

A. Sources of fluid intake (**Figure 21-4 and Table 21-2**)
   1. Liquids we drink
   2. Water in food we eat
   3. Metabolic water (from cellular respiration)
B. Sources of fluid output (**Figure 21-4 and Table 21-2**)
   1. Water vapor (during respiration)
   2. Sweating (from skin)
   3. Urine (from kidney)
   4. Water lost in the feces
C. Three main factors affect plasma, IF, and ICF volumes
   1. Regulating fluid output
   2. Regulating fluid input
   3. Exchanging fluid among compartments and around body
D. Regulation of fluid output
   1. Organs responsible for fluid output—lungs, skin, kidneys, and large intestine
   2. Fluid output, mainly urine volume, adjusts to fluid intake
   3. Antidiuretic hormone (ADH)
      a. ADH released from posterior pituitary gland when ECF volume is low
      b. ADH promotes water reabsorption from kidney tubules into blood
      c. Water is thus retained by body and less fluid is lost in urine
   4. Aldosterone (**Figure 21-5**)
      a. Aldosterone released from adrenal cortex.
      b. Aldosterone increases kidney tubule reabsorption of sodium in kidney tubules
      c. Water follows sodium from tubules into blood
      d. Water is retained by ECF (and total body fluid) by decreasing urine volume

5. Atrial natriuretic hormone (ANH)
   a. ANH is released from heart's atrial wall in response to high blood volume
   b. ANH promotes sodium loss from blood into kidney tubules
   c. Water follows sodium from blood, thus increasing loss of water in urine

E. Regulation of fluid intake (**Figure 21-6**)
   1. Sensory receptors detect change in volume and ECF concentration and send signals to the hypothalamus
   2. Signals from hypothalamus cause feeling of thirst, which triggers drinking of fluids to restore balance

F. Exchange of fluids by blood
   1. Constancy of internal fluid balance also maintained by exchanging fluids between fluid compartments
   2. Increased capillary blood pressure transfers fluid from blood to IF—a fluid shift
   3. Blood plasma protein concentration contributes to osmotic pressure, thus attracting water and holding it in the plasma

Fluid Imbalances

A. Dehydration—total volume of body fluids smaller than normal
   1. IF volume shrinks first, and then if treatment is not given, ICF volume and plasma volume decrease
   2. Dehydration occurs when fluid output exceeds intake for an extended period (**Figure 21-7** and **Figure 21-8**)

B. Overhydration—total volume of body fluids larger than normal
   1. Fluid intake exceeds output
   2. Excess volume burdens pumping action of heart

C. Water intoxication—possibly life-threatening neurological impairment caused by severe overhydration and accompanying electrolyte imbalance

Importance of Electrolytes in Body Fluids

A. Electrolytes and nonelectrolytes
   1. Nonelectrolytes—organic substances that do not break up or dissociate when placed in water solution (e.g., glucose)
   2. Electrolytes—compounds that break up or dissociate in water solution into separate particles called ions (e.g., ordinary table salt or sodium chloride)

B. Ions—the dissociated particles of an electrolyte that carry an electrical charge
   1. Cations are positively charged ions (e.g., potassium [K^+] and sodium [Na^+])
   2. Anions are negatively charged ions (e.g., chloride [Cl^-], bicarbonate [HCO_3^-], anionic proteins)

C. Electrolyte composition of body fluids (**Figure 21-9**)
   1. ECF dominated by sodium (positive) and chloride (negative)
   2. ICF dominated by potassium (positive) and anionic proteins (negative)

D. Edema—swelling caused by high IF volume

E. Sodium-containing internal secretions (**Figure 21-10**)

Electrolyte Imbalances (Table 21-3)

A. Related to "intake" and "output" of electrolytes and also absorption and distribution of electrolytes in body fluids and availability for use by body cells

B. Sodium imbalance
   1. Hypernatremia—blood sodium more than 145 mEq/L
   2. Characterized by relative deficit of water to salt in ECF
   3. Causes include overuse of salt tablets; dehydration; and prolonged diarrhea
   4. Hyponatremia—blood sodium less than 136 mEq/L
      a. Results when there is relatively too much water in the ECF for the amount of sodium present
      b. Causes include excessive secretion of antidiuretic hormone; massive infusion of sodium-free IV solution; burns; and prolonged use of certain diuretics
      c. Symptoms of both hyper- and hyponatremia are related to CNS malfunction and include headache, confusion, seizures, and coma

C. Potassium imbalance
   1. Hyperkalemia—blood potassium more than 5.1 mEq/L
      a. Causes include increased intake; shift of potassium from ICF to blood caused by tissue trauma and burns; renal failure

b. Clinical signs of hyperkalemia are related to muscle malfunction and include skeletal muscle weakness, paralysis, and cardiac arrest
2. Hypokalemia—blood potassium less than 3.5 mEq/L
a. Causes include fasting; diets low in potassium; abuse of laxatives and certain diuretics; diarrhea; vomiting; gastric suction
b. Clinical signs include skeletal muscle and cardiac problems; smooth muscle weakness causing abdominal distention; and slow rate of passage of GI contents (**Figure 21-11**)

D. Calcium imbalance
1. Hypercalcemia—blood calcium levels more than 10.5 mg/dL (5.25 mEq.L)
a. Caused by excessive input; increased absorption; shifts of calcium from bone to ECF; Paget disease and other bone tumors; hyperparathyroidism
b. Clinical signs related to decreased neuromuscular activity—fatigue; muscle weakness; diminished reflexes; cardiac problems
2. Hypocalcemia—blood calcium levels less than 8.4 mg/dL (4.2 mEq/L)
a. Caused by dietary deficiency, decreased absorption or availability, increased excretion, pancreatitis, hypoparathyroidism, rickets and osteomalacia, and renal insufficiency
b. Clinical signs related to increased neuromuscular irritability—cramping, muscle twitching; hyperactive reflexes; and abnormal cardiac rhythms

❑ ACTIVE LEARNING

STUDY TIPS

Consider using these tips to achieve success in meeting your learning goals.

Chapter 21 expands on some of the material from Chapter 20. A quick review of Chapter 20 will better prepare you for this chapter.

1. Make flash cards and check online resources to help you learn the terms in this chapter.
2. For a better understanding of the terms in this chapter, review the Language of Science and Language of Medicine sections.
3. Electrolytes are charged particles or ions. One of the functions of ions is to control water movement. The body cannot directly control water movement so it must move electrolytes and water will then follow.
4. The capillary pressure and blood protein mechanism regulates the movement of water between the blood and interstitial fluid. Blood pressure determines the amount of plasma that is pushed out into the interstitial fluid, and plasma proteins determine the amount of water that gets pulled back into the blood.
5. In your study group, review the flash cards with the terms. Discuss how electrolytes function in regulating water movement. Go over the aldosterone mechanism (see **Figure 21-5**). Discuss the plasma protein and capillary blood pressure mechanism for regulating the balance between blood plasma and interstitial fluid. Review the questions and chapter outline summary at the end of the chapter and discuss possible test questions.

Review Questions

Hint *Write out the answers to these questions after reading the chapter and reviewing the Chapter Summary. If you simply think through the answer without writing it down, you will not retain much of your new learning.*

1. Name and give the location of the three main fluid compartments of the body. Which of these make up ECF?
2. Explain *transcellular fluids* and list the *transcellular fluids* in the body.
3. List the factors that influence the percentage of water in the body. Explain the effect of each factor.
4. List the three sources of water for the body.
5. Identify the main factors that act as mechanisms for balancing plasma, IF, and ICF volumes.
6. List the four organs from which fluid output occurs.
7. Explain how aldosterone influences water movement between the kidney tubules and the blood.
8. Explain why the body is unable to reduce its fluid output to zero no matter how dehydrated it is.
9. Explain the role of capillary blood pressure in water movement between the plasma and interstitial fluid.
10. Explain the role of plasma proteins in water movement between the plasma and interstitial fluid.
11. Define dehydration and name a possible cause.
12. Define overhydration and name a possible cause.
13. Differentiate between an electrolyte and a nonelectrolyte.
14. Name three important anions.
15. Name three important cations.
16. What are the clinical manifestations of hyperkalemia?

Critical Thinking

After finishing the Review Questions, write out the answers to these more in-depth questions to help you apply your new knowledge. Go back to sections of the chapter that relate to concepts that you find difficult.

17. Regarding fluid and electrolyte balance, what would be the consequences of a large loss of skin (e.g., third-degree burns or scraping injuries)?
18. If a person rapidly drank a liter of distilled water, how would their ICF be affected?
19. Alcohol and caffeine are both diuretics. Explain how they affect diuresis in the body.

Chapter Test

After studying the chapter, test your mastery by responding to these items. Try to answer them without looking up the answers.

1. The extracellular fluid compartment is composed of ________ and ________.
2. The largest volume of water in the human body is contained in which fluid compartment? ________
3. The body's chief mechanism for maintaining fluid balance is to adjust its ________.
4. The body has three sources of fluid intake; the liquids we drink, the foods we eat, and ________.
5. The four organs from which fluid output occurs are the ________, ________, ________, and ________.
6. Urine volume is regulated by three hormones: ADH released from the pituitary gland, ________ released from the adrenal cortex, and ________ released from the heart.
7. When electrolytes dissociate in water, they form charged particles called ________.
8. The most abundant negatively charged particle in the blood is ________.
9. The most abundant positively charged particle in the blood is ________.
10. Depressions in swollen subcutaneous tissue that do not refill after an examiner has exerted finger pressure is referred to as ________.

***Indicate whether each of the next five questions is* true *or* false:**

11. ________ In general, an obese person has more water per pound of body weight than a slim person.
12. ________ In general, a man has less water per pound of body weight than a woman.
13. ________ In general, an infant has less water per pound of body weight than an adult.
14. ________ ANH is a diuretic hormone.
15. ________ The average diet contains 100 mEq of sodium.
16. When the blood level of aldosterone increases:
    a. sodium is moved from the blood to the kidney tubules
    b. sodium is moved from the kidney tubules to the blood
    c. more urine is formed
    d. ANH is released
17. Aldosterone causes:
    a. an increase in ICF
    b. a decrease in ICF
    c. an increase in ECF
    d. a decrease in ECF
18. Increased capillary pressure:
    a. moves fluid from the intracellular compartment to the extracellular compartment
    b. moves fluid from the plasma to the interstitial fluid
    c. moves fluid from the interstitial fluid to the plasma
    d. has no effect on fluid movement
19. Blood plasma proteins act to:
    a. move interstitial fluid into the plasma
    b. move plasma into the interstitial fluid
    c. move extracellular fluid into the intracellular space
    d. move interstitial fluid into the extracellular space
20. Signs of hypocalcemia may be manifested as a result of:
    a. pancreatitis
    b. hypoparathyroidism
    c. osteomalacia
    d. all of the above

Case Studies

To solve a case study, you may have to refer to the glossary or index, other chapters in this textbook, and other resources.

1. Like most people in the United States, Tom ingests 20 to 30 times more sodium each day than he needs for survival. How does Tom's body compensate for this imbalance to restore homeostasis?
2. Jo has not eaten anything all day but has consumed an excessive amount of distilled water. Will this affect her urine output? What unusual characteristics are likely to appear in a urinalysis of Jo's urine?
3. Jo is overhydrated (see Case Study number 2). This chapter states that *overhydration* can place a dangerously heavy burden on the heart. Explain how overhydration can tax the heart.
4. Liam and his friend love to sail in the gulf. They stay out for long periods at a time and are often on the boat for days. They try to conserve food and water when on trips but do not want to be foolish and put their health at risk. Can you offer a suggestion that might help Liam and his friend monitor their hydration while on extended trips?

Answers to Active Learning Questions can be found online at evolve.elsevier.com.

Acid-Base Balance

OUTLINE

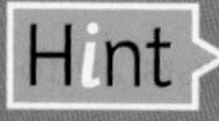

Hint *Scan this outline before you begin to read the chapter, as a preview of how the concepts are organized.*

OBJECTIVES

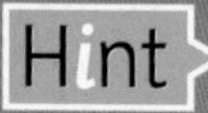

Hint *Before reading the chapter, review these goals for your learning.*

After you have completed this chapter, you should be able to:

1. Discuss the concept of pH, including:
   - Describe how to use the pH scale.
   - Discuss the pH unit.
   - Define the term *base* and explain the difference between strong and weak acids.
2. Do the following related to mechanisms that control pH of body fluids:
   - List the three mechanisms that control pH of body fluids.
   - Discuss buffers as a compensatory mechanism.
   - Compare and contrast the respiratory and urinary mechanisms of pH control.
3. Compare and contrast acidosis and alkalosis.
4. Discuss the two types of respiratory disturbances.

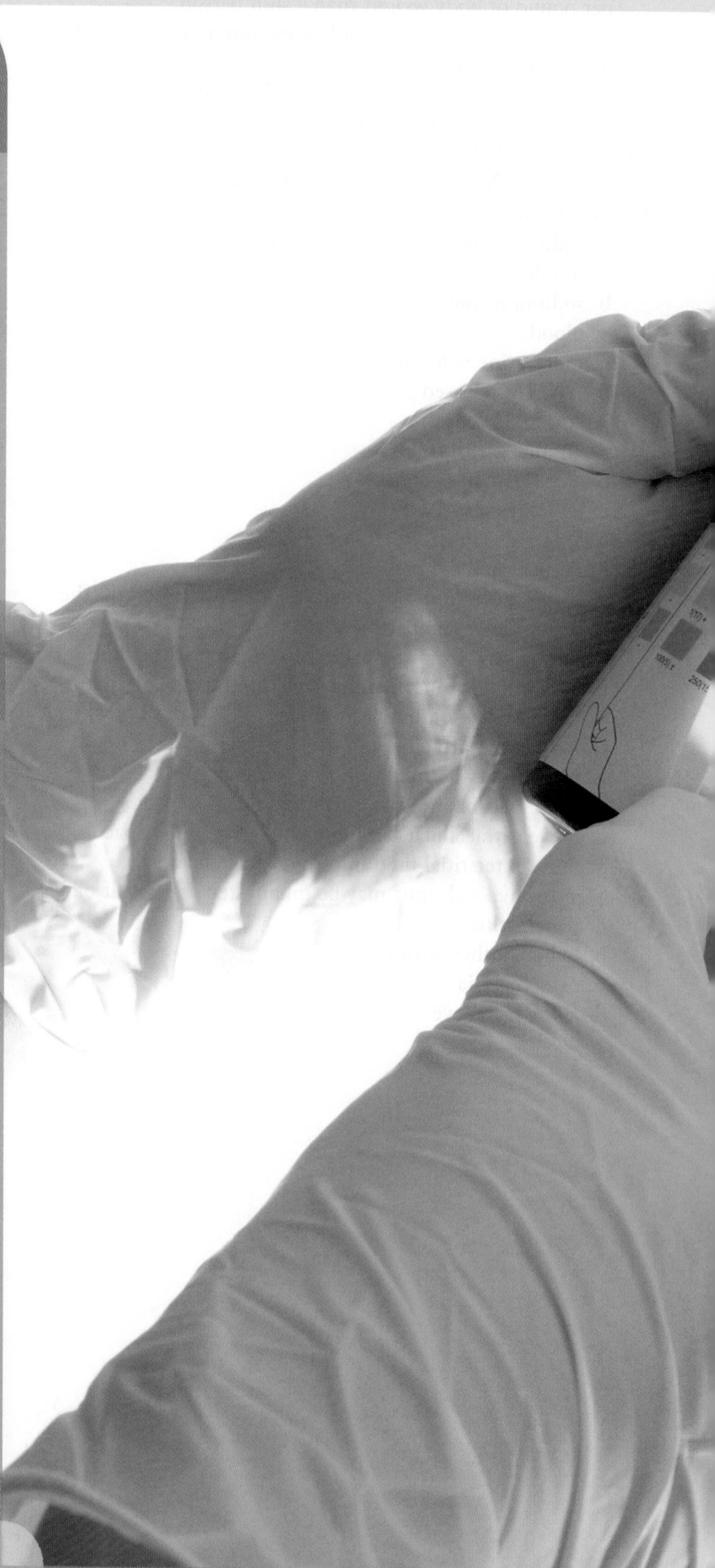

CHAPTER 22

Acid-base balance is one of the most important of the body's homeostatic mechanisms. Maintaining **acid-base balance** means keeping the concentration of hydrogen ions in body fluids relatively constant. Effective functioning of many important body proteins, such as cellular enzymes and hemoglobin, closely depends on maintaining precise regulation of hydrogen ion concentration. This is of vital importance. If the hydrogen ion concentration veers away from normal even slightly, serious illness or even death may occur. Healthy survival depends on the ability of the body to maintain, or quickly restore, the acid-base balance of its fluids if imbalances occur.

Acid-base regulation requires a series of coordinated homeostatic mechanisms that involve the blood and other body fluids, the lungs, and the kidneys. Ultimately all of these mechanisms are based on chemical processes. Recall that many important chemical principles related to the life process were covered in Chapter 2. You may wish to refer back to those principles of biochemistry as you study how the body so precisely regulates its acid-base balance.

pH of Body Fluids

As first outlined in Chapter 2, water and all water solutions contain **hydrogen ions (H^+)** and **hydroxide ions (OH^-).** pH is an acronym for "power of H^+." The term **pH** followed by a number indicates a solution's hydrogen ion concentration compared with hydroxide concentration.

Using the pH Scale

At pH 7.0 a solution contains an equal concentration of hydrogen and hydroxide ions. Therefore pH 7.0 also indicates that a fluid is **neutral** in reaction (that is, neither acidic nor alkaline) (**Figure 22-1**). The pH of pure water, for example, is 7.0.

LANGUAGE OF **SCIENCE**

Hint Before reading the chapter, say each of these terms out loud. This will help you to avoid stumbling over them as you read.

acidic
(ah-SID-ik)
[*acid-* **sour,** *-ic* **relating to**]

acid-base balance
(AS-id bays BAL-ans)
[*acid* **sour,** *bas-* **foundation,** *bal-* **twice,** *-lanc-* **dish (two scales)**]

alkaline
(AL-kah-lin)
[*alkal-* **ashes,** *-ine* **relating to**]

buffer
(BUF-er)
[*buffe-* **cushion,** *-er* **agent**]

buffer pair
(BUF-er payr)
[*buffe-* **cushion,** *-er* **agent**]

carbonic anhydrase (CA)
(kar-BON-ik an-HYE-drays [see ay])
[*carbo-* **coal,** *-ic* **relating to,** *an-* **without,** *-hydr-* **water,** *-ase* **enzyme**]

compensation
(kom-pen-SAY-shun)
[*compens-* **balance,** *-tion* **process**]

homeostasis
(hoh-mee-oh-STAY-sis)
[*homeo-* **same or equal,** *-stasis* **standing still**]

hydrogen ion (H^+)
(HYE-droh-jen AYE-on)
[*hydro-* **water,** *-gen* **produce,** *ion* **to go**]

hydroxide ion (OH^-)
(hye-DROK-side aye-on)
[*hydr-* **water (hydrogen),** *-ox-* **sharp (oxygen),** *-ide* **chemical,** *ion* **to go**]

Continued on p. 611

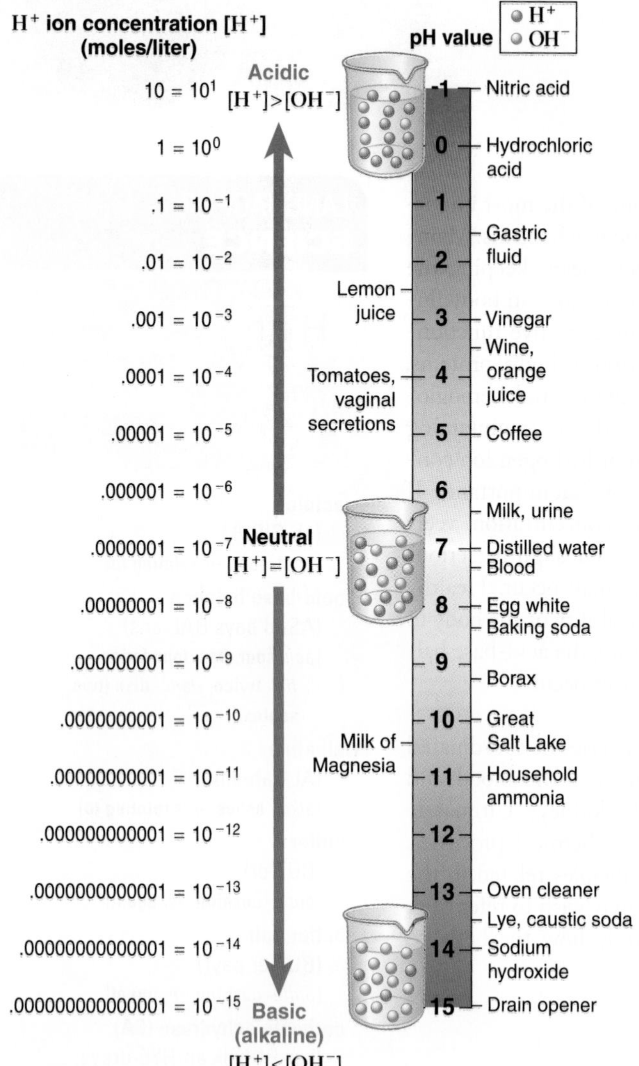

FIGURE 22-1 The pH range. The overall pH range is expressed numerically on what is called a *logarithmic scale* of 1 to 14. This means that a change of 1 pH unit represents a tenfold difference in actual concentration of hydrogen ions. Note that as the concentration of H^+ ions increases, the solution becomes increasingly acidic and the pH value decreases. As OH^- concentration increases, the pH value also increases, and the solution becomes more and more basic, or alkaline. A pH of 7 is neutral; a pH of 2 is very acidic, and a pH of 13 is very basic.

A pH higher than 7.0 indicates an **alkaline** solution (that is, one with a lower concentration of hydrogen than hydroxide ions). The more alkaline a solution, the higher is its pH value. Alkaline solutions are also called *basic* solutions.

A pH lower than 7.0 indicates an **acidic** solution (that is, one with a higher hydrogen ion concentration than hydroxide ion concentration). The higher the hydrogen ion concentration, the lower the pH and the more acidic a solution is.

With a pH as low as 1.6, gastric juice is the most acidic substance in the body. Saliva often has a pH of 7.7, on the alkaline side. Normally, the pH of arterial blood is about 7.4—with a normal upper limit of 7.45. The pH of venous blood, on the other hand, is a bit lower at 7.37—but normally not below 7.35. In clinical practice, the terms "arterial" and "venous" always refer to the systemic circulation—not the pulmonary circulation—unless stated otherwise.

By applying the information given in the previous paragraph, you can deduce the answers to the following questions. Is arterial blood slightly acidic or slightly alkaline? Is venous blood slightly acidic or slightly alkaline?

Arterial and venous blood are both slightly alkaline because both have a pH slightly higher than 7.0. Venous blood, however, is less alkaline than arterial blood because venous blood's pH of about 7.37 is slightly lower than arterial blood's pH of about 7.4.

To better understand this concept, use the Active Concept Map *Concept of pH* at *evolve.elsevier.com.*

The pH Unit

Looking at the left side of **Figure 22-1**, we see that the pH unit is based on exponents of 10 from one unit to the next. That means that on the pH scale moving from one unit to the next multiplies the relative H^+ concentration by 10 times. Thus the difference between pH 7 and pH 6 is a *tenfold* increase in H^+. Moving from pH 7 to pH 5 is a *hundredfold* increase in H^+ concentration.

This tenfold difference between pH units is important to remember when we look at the normal pH range of blood plasma—a key fluid compartment of the body. What may seem like a small change in acidity at first glance is really 10 times bigger than it looks!

A related mathematical quirk of the pH unit is that the difference between any two pH values below 7.4 is larger than it would be above 7.4.

Mechanisms That Control pH of Body Fluids

Overview of pH Control Mechanisms

The body has three mechanisms for regulating the pH of its fluids. They are

1. Buffer mechanism in blood
2. Respiratory mechanism
3. Urinary mechanism

Together, the listed processes constitute the complex pH homeostatic mechanism—the machinery that normally keeps blood slightly alkaline with a pH that stays remarkably constant. Its usual limits are very narrow, about 7.35 to 7.45.

The slightly lower pH of venous blood compared with arterial blood results primarily from carbon dioxide (CO_2) entering venous blood as a waste product of cellular metabolism. As carbon dioxide enters the blood, some of it combines with water (H_2O) and is converted into carbonic acid by **carbonic anhydrase (CA),** an enzyme found in red blood cells. The following chemical equation represents this reaction. If you need

SCIENCE APPLICATIONS

THE BODY IN BALANCE

Walter Bradford Cannon (1871–1945)

Keeping the pH of the body stable is but one aspect of maintaining health. The American physiologist Walter Cannon gave us a name for the principle of balance, or constancy, of the internal fluid environment of the body—**homeostasis.** In 1932, his popular book *The Wisdom of the Body* finally gave a name to the concept first explained by Claude Bernard seven decades earlier (see p. 593). However, Cannon did more than name the concept. In his book, Cannon explained the incredibly complex set of mechanisms that allows our bodies to adjust to tremendous internal and external fluctuations that would otherwise kill us.

Much of Cannon's thought came from his groundbreaking discoveries in how the body copes with stress. In examining the fight-or-flight response, the effects of emotional stimuli, the mechanisms of cardiovascular shock, and in developing the "case study" approach to learning about human health and disease, Walter Cannon developed a clear understanding of the interactive nature of the organs of the body. It was Cannon who led scientists to look at their work in this new framework that explains the "big picture" of human body function.

Cannon's explanation of homeostasis revolutionized the way we look at the body—and how we look at patient care. As with fluid and electrolyte balance, knowledge of the mechanisms of acid-base balance is critical in direct patient care. Therefore, many *physicians, nurses, respiratory therapists* (pictured), *IV technicians, first responders* (for example, *emergency medical technicians* and *paramedics*), and others need a basic knowledge of how the body maintains a constancy of pH in the blood.

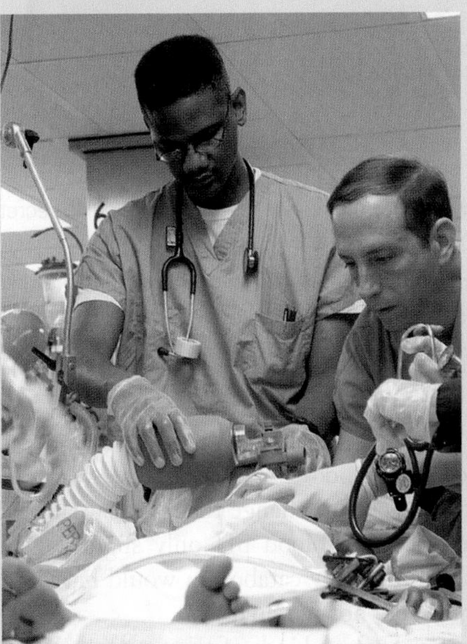

to review chemical formulas and equations, please refer to Chapter 2.

$$CO_2 + H_2O \xrightarrow{\text{carbonic anhydrase}} H_2CO_3$$

The lungs remove the equivalent of more than 30 L of carbonic acid each day from the venous blood by elimination of CO_2. This almost unbelievable quantity of acid is so well buffered that a liter of venous blood contains only about 1/100,000,000 grams more H^+ than does 1 liter of arterial blood. What incredible constancy! The pH homeostatic mechanism does indeed control effectively—astonishingly so.

To better understand this concept, use the Active Concept Map *Transport of Oxygen and Carbon Dioxide in the Blood* at *evolve.elsevier.com.*

Integration of pH Control

Integration of the three homeostatic mechanisms that act to maintain the pH of body fluids is illustrated in **Figure 22-2**.

Think of the circulating blood and RBCs as providing a *chemical pH control mechanism,* which is based on buffers (discussed later), and which acts immediately to help prevent harmful swings in pH when added acids or bases enter body fluids. If this immediate-acting chemical control mechanism is unable to stabilize the pH, the lungs and kidneys can both provide a *physiological pH control mechanism* to halt and reverse harmful pH shifts. The lungs respond in 1 to 2 minutes when the brainstem adjusts the respiratory rate (see **Figure 17-18**) and thus the adjustment of CO_2 is accomplished.

If the respiratory mechanism is unable to stop the pH shift, powerful but slower-acting renal mechanisms will be initiated within 24 hours. Details of each mechanism are discussed in the paragraphs that follow.

Buffers

Buffers are chemical substances that prevent a sharp change in the pH of a fluid when an acid or base is added to it. Strong acids and bases, if added to blood, would "dissociate" almost completely and release large quantities of H^+ or OH^- ions. The result would be drastic changes in blood pH. Survival itself depends on protecting the body from such drastic pH changes.

More acids than bases are usually added to body fluids. This is because catabolism, a process that goes on continually in every cell of the body, produces acids that enter blood as it flows through tissue capillaries. Almost immediately, one of

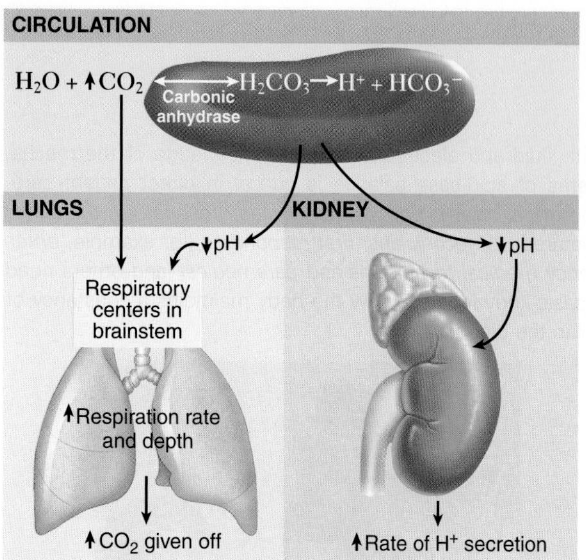

FIGURE 22-2 Integration of pH control mechanisms. Elevated CO_2 levels result in increased formation of carbonic acid in red blood cells. The resulting increase in hydrogen ions, coupled with elevated CO_2 levels, causes an increase in respiratory rate and secretion of hydrogen ions by the kidneys, thus helping to regulate the pH of body fluids.

the salts present in blood—a buffer, that is—reacts with these relatively strong acids to change them into weaker acids. The weaker acids decrease blood pH only slightly, whereas the stronger acids formed by catabolism would have decreased it greatly if they were not buffered.

Buffers consist of two kinds of substances and are therefore often called **buffer pairs.** One of the main blood buffer pairs is ordinary baking soda (sodium bicarbonate, or $NaHCO_3$) and carbonic acid (H_2CO_3).

Let us consider, as a specific example of buffer action, how the $NaHCO_3$–H_2CO_3 system works with a strong acid or base.

If a strong base, such as sodium hydroxide (NaOH), were added to this buffer system, the reaction shown in Figure 22-3 would take place. The H^+ of H_2CO_3 ($H \cdot HCO_3$), the weak acid of the buffer pair, combines with the OH^- of the strong base NaOH to form H_2O. Note what this accomplishes. It decreases the number of OH^- ions added to the solution, and this in turn prevents the drastic rise in pH that would occur without buffering.

Figure 22-4 shows how a buffer system works with a strong acid. Although useful in demonstrating the principles of buffer action, HCl or similar strong acids are never introduced directly into body fluids under normal circumstances. Instead, the $NaHCO_3$ buffer system is most often called on to buffer a number of weaker acids produced during catabolism. Lactic acid is a good example. As a weak acid, it does not "dissociate" as completely as HCl. Incomplete dissociation of lactic acid results in fewer hydrogen ions being added to the blood and a less drastic lowering of blood pH than would occur if HCl were added in an equal amount.

Without buffering, however, lactic acid buildup results in significant H^+ accumulation over time. The resulting decrease of pH can produce serious acidosis. Ordinary baking soda (sodium bicarbonate, or $NaHCO_3$) is one of the main buffers of the normally occurring "fixed" acids in blood. Lactic acid is one of the most abundant of the "fixed" acids (acids that do not break down to form a gas).

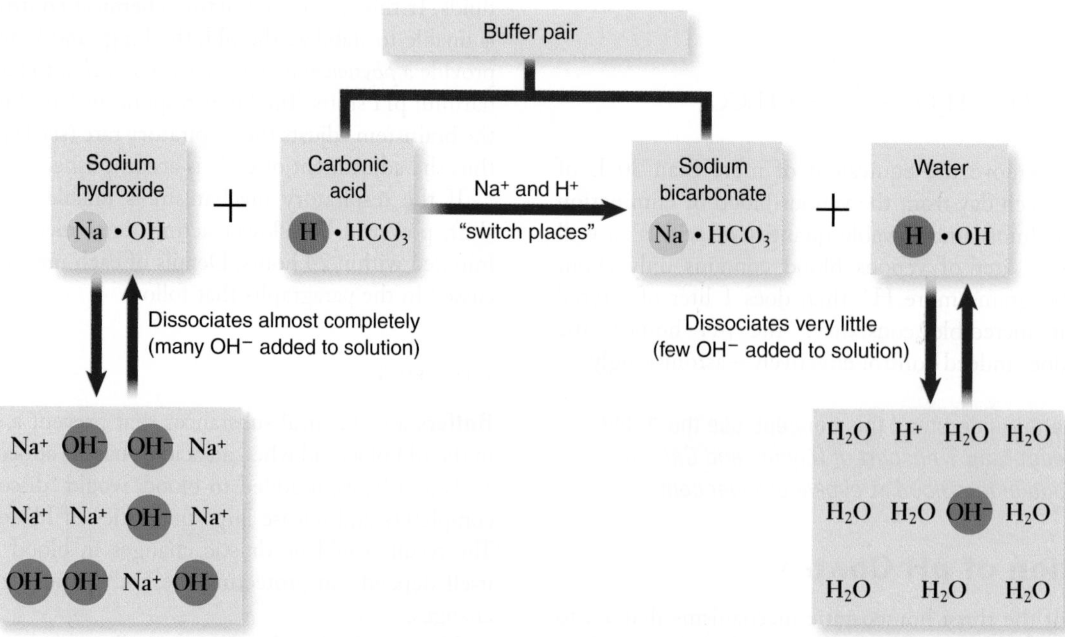

FIGURE 22-3 Buffering action of carbonic acid. Buffering of base NaOH by H_2CO_3. As a result of buffer action, the strong base (NaOH) is replaced by $NaHCO_3$ and H_2O. As a strong base, NaOH "dissociates" almost completely and releases large quantities of OH^-. Dissociation of H_2O is minimal. Buffering decreases the number of OH^- ions in the system.

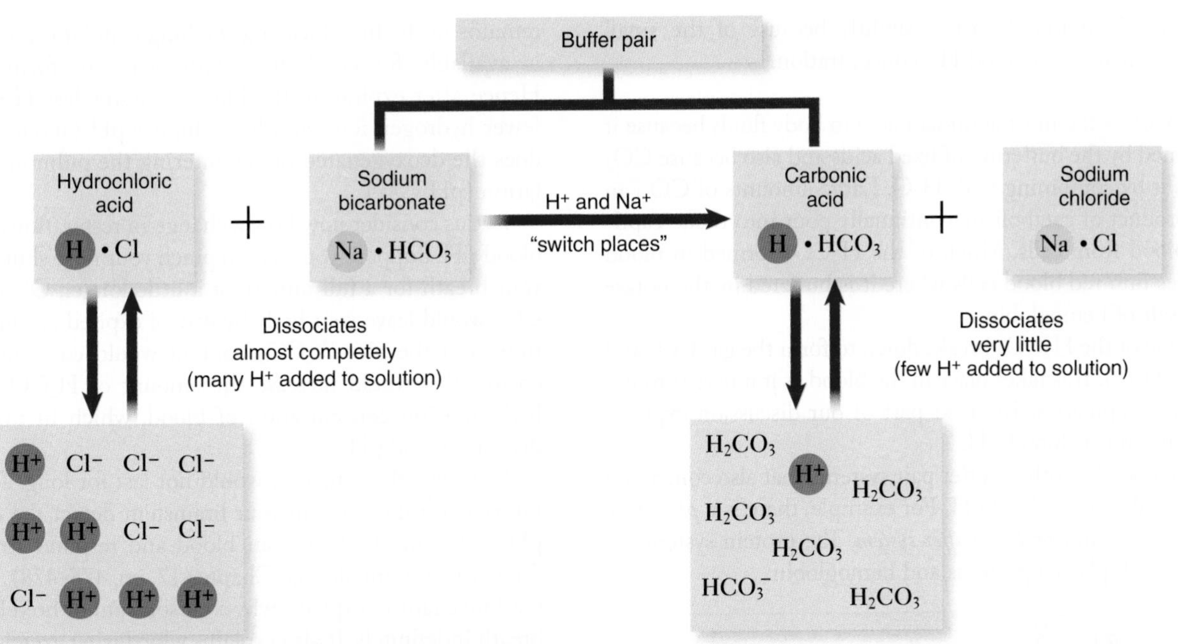

FIGURE 22-4 Buffering action of sodium bicarbonate. The acid HCl is buffered by $NaHCO_3$. As a result of the buffer action, the strong acid (HCl) is replaced by the weaker carbonic acid (H • HCO_3). Note that HCl, being a strong acid, "dissociates" almost completely and releases more H^+ than H_2CO_3. Buffering decreases the number of H^+ ions in the system.

Figure 22-5 shows the compounds formed by buffering of lactic acid (a "fixed" acid), produced by normal catabolism. The following changes in blood result from buffering of fixed acids in tissue capillaries:

1. The amount of H_2CO_3 in blood increases slightly because an acid (such as lactic acid) is converted to H_2CO_3.
2. The amount of bicarbonate in blood (mainly $NaHCO_3$) decreases because bicarbonate ions become part of the newly formed H_2CO_3. Normal arterial blood with a pH of 7.45 contains 20 times more $NaHCO_3$ than H_2CO_3. If this ratio decreases, blood pH decreases below 7.45.
3. The H^+ concentration of blood increases slightly. H_2CO_3 adds hydrogen ions to blood, but it adds fewer of them than lactic acid would have because it is a weaker acid than lactic acid. In other words, the buffering mechanisms do not totally prevent blood hydrogen ion concentration from increasing. It simply minimizes the increase.

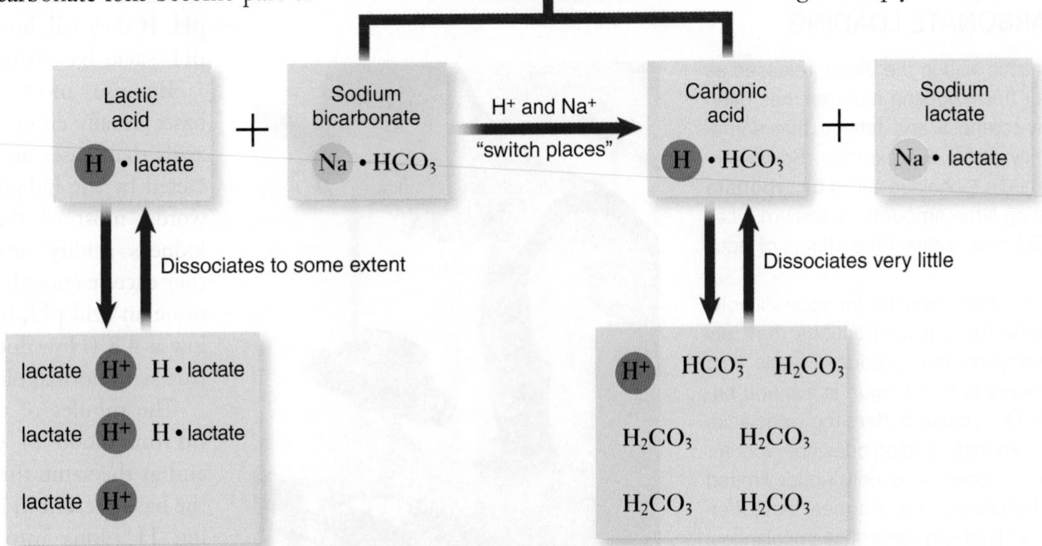

FIGURE 22-5 Lactic acid buffered by sodium bicarbonate. Lactic acid (H • lactate) and other "fixed" acids are buffered by $NaHCO_3$ in the blood. Carbonic acid (H • HCO_3, or H_2CO_3, a weaker acid than lactic acid) replaces lactic acid. As a result, fewer H^+ ions are added to blood than would be added if lactic acid were not buffered.

4. Blood pH decreases slightly because of the small increase in blood H^+ concentration.

H_2CO_3 is the most abundant acid in body fluids because it is formed by the buffering of fixed acids and also because CO_2 forms it by combining with H_2O. Large amounts of CO_2, an end product of catabolism, continually pour into tissue capillary blood from cells. Much of the H_2CO_3 formed in blood diffuses into red blood cells where it is buffered by the potassium salt of hemoglobin.

Some of the H_2CO_3 breaks down to form the gas CO_2 and water (H_2O). This takes place in the blood as it moves through the lung capillaries. The next part of our discussion explains how this affects blood pH.

The body has other buffer pair systems that also contribute to the stability of blood pH. For example, there is a *phosphate buffer system* and *protein buffer system*. The protein system includes both plasma proteins and hemoglobin.

QUICK CHECK

1. What three mechanisms does the body have for regulating pH of body fluids?
2. What is primarily responsible for the slightly lower pH in venous blood?
3. What are buffers?

Respiratory Mechanism of pH Control

Respirations play a vital part in controlling pH. With every expiration, CO_2 and H_2O leave the body in the expired air. The CO_2 has diffused out of the pulmonary blood as it moves through the lung capillaries. Less CO_2 therefore remains in the blood leaving the lung capillaries, so less of it is available for combining with water to form H_2CO_3. Hence after expiration the blood contains less H_2CO_3, has fewer hydrogen ions, and has a higher pH (about 7.4) than does the deoxygenated blood entering the pulmonary circulation (pH 7.37).

Let us consider now how a change in respirations can alter blood pH. Suppose you were to pinch your nose shut and hold your breath for a full minute or a little longer. Obviously, no CO_2 would leave your body by way of expired air during that time, and the blood's CO_2 content would consequently increase. This would increase the amount of H_2CO_3 and the hydrogen ion concentration of blood, which in turn would decrease blood pH.

However, this situation would not last for long. The respiratory control centers in your brainstem detect the dropping pH and rising CO_2 in your blood and respond strongly by forcing you to inhale (see Chapter 17, pp. 475-478). This survival mechanism explains why a person cannot hold his or her breath indefinitely. It also explains why during exercise, a drop in pH caused by increased muscle production of CO_2 triggers an increase in breathing rate. Of course, the opposite is true as well—when blood pH increases to or above normal, then the rate of breathing slows.

Urinary Mechanism of pH Control

Most people know that the kidneys are vital organs and that life soon ebbs away if they stop functioning. One reason is that the kidneys are the body's most effective regulators of blood pH. They can eliminate much larger amounts of acid than can the lungs and, if it becomes necessary, they also can excrete excess base. The lungs cannot. In short, the kidneys are the body's last and best defense against wide variations in blood pH. If they fail, homeostasis of pH—acid-base balance—fails.

Because more acids than bases usually enter blood, more acids than bases are usually excreted by the kidneys. In other words, most of the time the kidneys acidify urine; that is, they excrete enough acid to give urine an acid pH, frequently as low as 4.8. (How does this compare with normal blood pH?)

The tubules of the kidneys rid the blood of excess acid and at the same time conserve the base present in it by secreting H^+ ions into the urine while retaining HCO_3^- in the blood. Much of the excess H^+ is combined with the amine

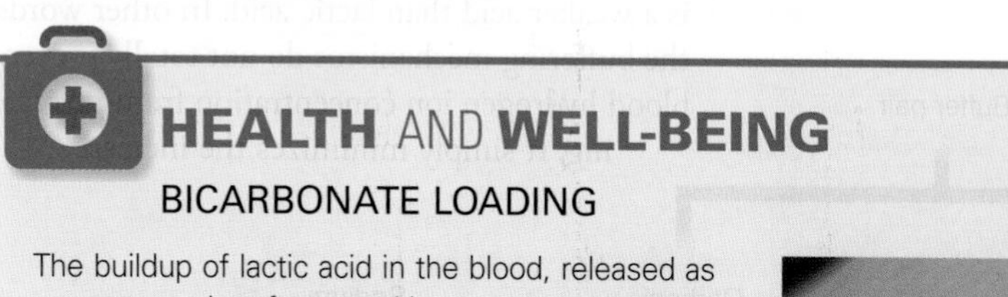

HEALTH AND WELL-BEING

BICARBONATE LOADING

The buildup of lactic acid in the blood, released as a waste product from working muscles, has been blamed for the soreness and fatigue that sometimes accompany strenuous exercise. Some athletes have adopted a technique called **bicarbonate loading,** ingesting large amounts of sodium bicarbonate ($NaHCO_3$) to counteract the effects of lactic acid buildup.

This practice is most popular for sports involving brief powerful muscle contractions that rely on aerobic respiration that produces lactic acid quickly. Their theory is that fatigue is avoided because the $NaHCO_3$, a base, buffers the lactic acid.

However, bicarbonate loading does not work for everyone. When it does, it is only under limited conditions. Unfortunately, the diarrhea that often results can trigger fluid and electrolyte imbalances. Long-term $NaHCO_3$ abuse can lead to disruption of acid-base balance and its disastrous effects.

CLINICAL APPLICATION

DIABETIC KETOACIDOSIS

An important part of home care for diabetics involves monitoring the level of glucose in the blood and, especially for patients taking insulin, carefully watching for the appearance of **ketone bodies** in the urine.

Accumulation of these acidic substances in the blood results from the excessive metabolism of fats, most often in those people with uncontrolled type 1 diabetes. Some type 2 diabetics may also develop ketoacidosis under severe conditions. These individuals have trouble metabolizing carbohydrates and instead burn fat as a primary energy source. Ketone bodies are one way that the body transports the fatty acids from stored fat to other cells of the body.

The accumulation of ketone bodies results in a condition called **diabetic ketoacidosis** that causes the blood to become dangerously acidic. As blood levels of ketones increase, they "spill over" into the urine and can be detected by use of appropriate reagent strips. Ketones also may give a "fruity" odor to the breath and urine. As the body compensates for the acidosis, rapid breathing may occur.

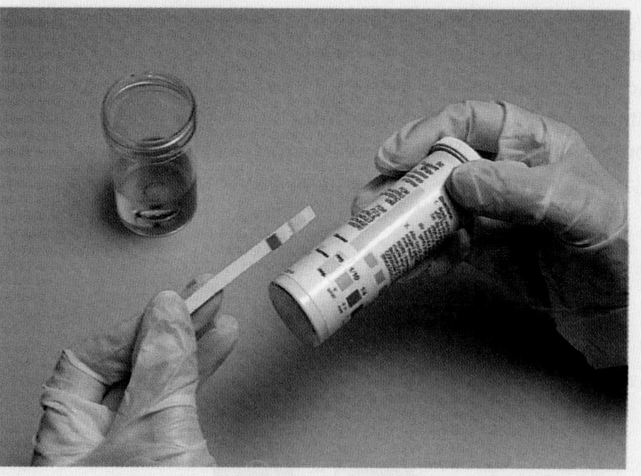

Ketonuria. Using a chemical test strip to check for the presence of ketone bodies in the urine of a diabetic patient.

group (NH_2) of an amino acid (glutamic acid) to form ammonia (NH_3) and ammonium ions (NH_4^+) before it is secreted into urine.

QUICK CHECK

1. How can breathing affect the pH of blood?
2. By what mechanism can the kidney change the pH of the blood?
3. What is the theory behind "bicarbonate loading," and what is the long term effect of this practice?

pH Imbalances

Acidosis and Alkalosis

The chemistry of life can operate only within the range of pH 6.8-8.0. The range of optimal human function is much narrower than that—pH 7.35 to 7.45. **Acidosis** and **alkalosis** are the two kinds of pH or acid-base imbalance that can threaten our health and survival.

Although any pH value above 7.0 is considered chemically basic, in clinical medicine the term *acidosis* is used to describe the condition that produces an arterial blood pH of less than 7.35 and *alkalosis* is used to describe the condition that produces an arterial blood pH greater than 7.45.

In acidosis the blood pH falls as H^+ ion concentration increases or bases are lost. Only rarely does it fall as low as 7.0 (neutrality), and almost never does it become even slightly acidic, because death usually occurs before the pH drops this much. In alkalosis, which develops less often than acidosis, the blood pH is higher than normal because of a loss of acids or an accumulation of bases.

From a clinical standpoint, disturbances in acid-base balance can be considered dependent on the relative quantities (ratio) of H_2CO_3 and $NaHCO_3$ in the blood. Components of this important buffer pair must be maintained at the proper ratio (20 times more $NaHCO_3$ than H_2CO_3) if acid-base balance is to remain normal. It is fortunate that the body can regulate both chemicals in the $NaHCO_3$–H_2CO_3 buffer system. Blood levels of $NaHCO_3$ can be regulated by the kidneys and H_2CO_3 levels by the respiratory system (lungs).

Metabolic and Respiratory Disturbances

Two types of disturbances, metabolic and respiratory, can alter the proper ratio of these components. Metabolic disturbances affect the bicarbonate ($NaHCO_3$) element of the buffer pair, and respiratory disturbances affect the H_2CO_3 element, as follows:

1. Metabolic disturbances
   a. **Metabolic acidosis** (bicarbonate deficit). Patients in metabolic acidosis with a bicarbonate deficit often suffer from renal disease, uncontrolled diabetes, prolonged diarrhea, or have ingested toxic chemicals such as antifreeze (ethylene glycol) or wood alcohol (methanol).
   b. **Metabolic alkalosis** (bicarbonate excess). The bicarbonate excess in metabolic alkalosis can result from diuretic therapy, loss of acid-containing gastric fluid caused by vomiting or suction, or from certain diseases such as Cushing syndrome.

CLINICAL APPLICATION

VOMITING

Vomiting, sometimes referred to as **emesis,** is the forcible emptying or expulsion of gastric and occasionally intestinal contents through the mouth. It can occur as a result of many stimuli, including foul odors or tastes, irritation of the stomach or intestinal mucosa caused by food poisoning, certain bacterial or viral infections, and alcohol intoxication.

A "vomiting center" in the brainstem regulates the many coordinated, but primarily involuntary, steps involved (see illustration). The pernicious vomiting of pregnancy and the severe and repetitive (cyclic) vomiting that sometimes occurs in childhood, especially with pyloric obstruction in infants, can be life threatening because of the fluid, electrolyte, and acid-base imbalances that may result.

One of the most frequent and serious complications of repetitive vomiting that continues over time is **metabolic alkalosis.** The bicarbonate excess of metabolic alkalosis results indirectly because of the massive loss of chloride. The lost chloride, which is a component of hydrochloric acid (HCl) in gastric secretions, is replaced by bicarbonate in the extracellular fluid. The result is metabolic alkalosis (see illustration).

The body "compensates" for the imbalance by suppressing respirations to increase blood CO_2 levels and, ultimately, levels of H_2CO_3 in the extracellular fluid. The kidneys also assist in the compensation process by conserving H^+ and eliminating additional HCO_3^- in an alkaline urine.

Therapy to actually *restore* the buffer pair ($NaHCO_3$ to H_2CO_3) ratio to normal includes intravenous administration of chloride-containing solutions such as **normal saline** (0.9% NaCl). The chloride ions of the solution replace the bicarbonate ions and thus help relieve the bicarbonate excess that is responsible for the imbalance.

Review the article *Fluid and Electrolyte Therapy* at Connect It! at *evolve.elsevier.com.*

1 Metabolic balance before onset of alkalosis
H_2CO_3 HCO_3^-
2 Metabolic alkalosis
H_2CO_3 HCO_3^-
HCO_3^- increases because of loss of chloride ions or excess ingestion of sodium bicarbonate
3 Body's compensation
Alkaline urine
$H^+ + HCO_3^-$
$CO_2 + H_2O$
CO_2 H_2CO_3 HCO_3^-
H^+ + HCO_3^-
CO_2
Breathing suppressed to hold CO_2
Kidneys conserve H^+ ions and eliminate HCO_3^- in alkaline urine
4 Therapy required to restore metabolic balance
H_2CO_3 HCO_3^- Cl^-
Chloride-containing solution
HCO_3^- ions replaced by Cl^- ions
A

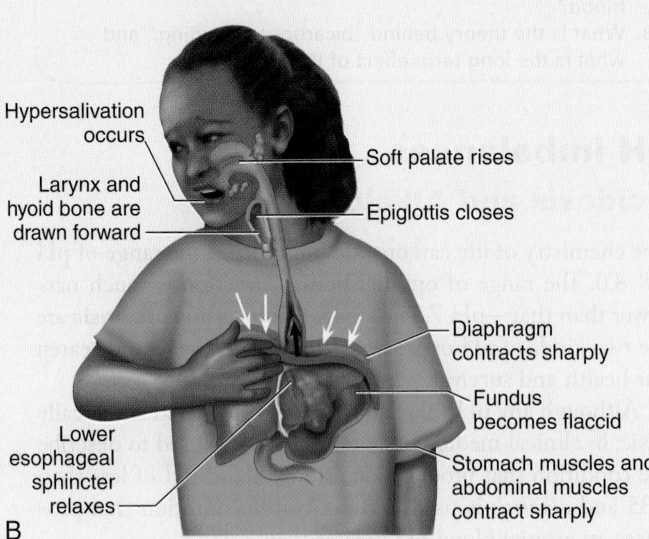

2. Respiratory disturbances
   a. **Respiratory acidosis** (H_2CO_3 excess). The increase in H_2CO_3 characteristic of respiratory acidosis is caused most frequently by slow breathing (hypoventilation), which results in excess CO_2 in the blood. Causes include depression of the respiratory center by drugs or anesthesia or by pulmonary diseases such as emphysema and pneumonia. Serious respiratory acidosis also follows recovery from cardiac arrest.

CLINICAL APPLICATION

CARDIAC ARREST AND RESPIRATORY ACIDOSIS

A cascade of rapid-fire catastrophic homeostatic failures follows *cardiac arrest*—the sudden cessation of blood pumping by the heart. One such failure involves almost immediate development of respiratory acidosis (carbonic acid excess) caused by retention of CO_2 in the body when respiration ceases and blood flow through the lung capillaries stops. Even if emergency CPR (cardiopulmonary resuscitation) measures can restore breathing and start the heart beating again, respiratory acidosis must be successfully treated and normal blood pH levels restored quickly in order to sustain life.

As in other types of pH imbalances, absolute changes in the amount or ratio of the bicarbonate-carbonic acid buffer pair components is the first line of defense to prevent massive changes in blood pH. Then, the body initiates both respiratory and renal compensatory mechanisms to help deal with the carbonic acid excess in severe respiratory acidosis.

The most important respiratory compensatory mechanism—increased breathing rate—does "blow off" some additional CO_2 but cannot significantly lower the very elevated carbonic acid buildup that follows cardiac arrest and remains after the blood buffers have been overwhelmed.

Finally, renal compensatory mechanisms that stabilize blood pH and help control many forms of respiratory acidosis are initiated after cardiac arrest. They include (1) decreasing the elimination of bicarbonate ions (HCO_3^-) and (2) increasing the elimination of hydrogen ions (H^+) in acidic urine. Although helpful in controlling chronic forms of respiratory acidosis that develop slowly over time, these slow-acting homeostatic compensatory mechanisms are unable to adequately address the serious, acute-onset acidosis that follows cardiac arrest. Medical intervention is required.

In the past, immediate intravenous (IV) infusion of bicarbonate- or lactate-containing solutions (lactate is converted to bicarbonate ions in the liver) was considered the emergency treatment of choice in treating respiratory acidosis after cardiac arrest—and these solutions are still used for that purpose. However, clinical studies have shown that aggressive treatment employing controlled ventilation to dramatically increase CO_2 elimination from the body may, in many cases, be more effective in restoring pH balance.

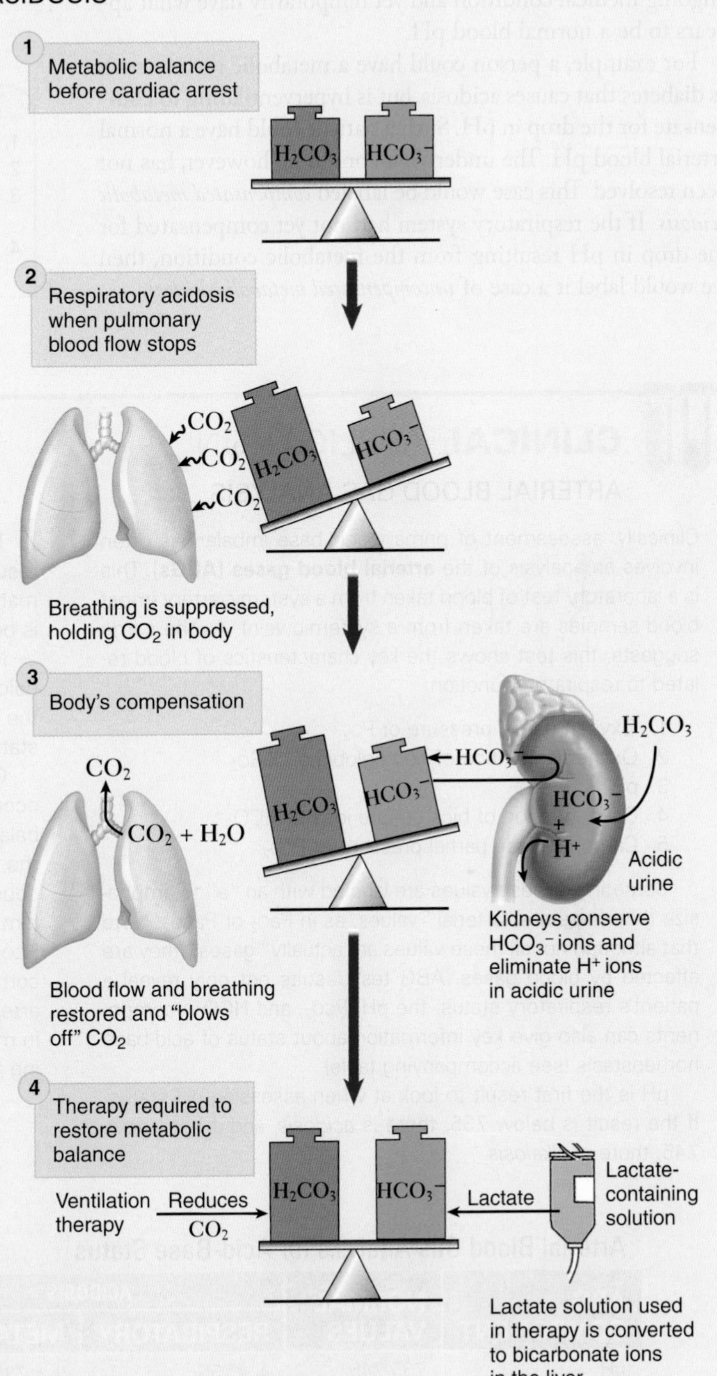

b. **Respiratory alkalosis** (H_2CO_3 deficit). Hyperventilation leads to a H_2CO_3 deficit caused by excessive loss of CO_2 in expired air. The result is respiratory alkalosis. Anxiety (hyperventilation syndrome), overventilation of patients on ventilators, or hepatic coma can all reduce H_2CO_3 and CO_2 to dangerously low levels.

Compensation for pH Imbalances

When acidosis or alkalosis occurs in the body, our various pH-balancing mechanisms—buffers and the respiratory and urinary mechanisms—try to restore balance as soon as possible. We often use the term **compensation** for this set of processes because the body is using means that "compensate" for the abnormal shift in pH.

Compensation is a clinically important concept. Because compensation mechanisms in the body can quickly counteract an abnormal shift in blood pH, a person may have a serious, ongoing medical condition and yet temporarily have what appears to be a normal blood pH.

For example, a person could have a metabolic disease such as diabetes that causes acidosis, but is hyperventilating to compensate for the drop in pH. Such a patient could have a normal arterial blood pH. The underlying condition, however, has not been resolved. This case would be labeled *compensated metabolic acidosis.* If the respiratory system had not yet compensated for the drop in pH resulting from the metabolic condition, then we would label it a case of *uncompensated metabolic acidosis.*

Crushing injuries of skeletal muscle can cause metabolic acidosis. Review the article *Rhabdomyolysis* at Connect It! at *evolve.elsevier.com.*

QUICK CHECK

1. What is *acidosis*? What is *alkalosis*?
2. What factors may cause a metabolic disturbance in pH?
3. What situations may cause a respiratory disturbance in pH?
4. How does vomiting sometimes create an acid-base imbalance?

22

CLINICAL APPLICATION

ARTERIAL BLOOD GAS ANALYSIS

Clinically, assessment of primary acid-base imbalances often involves an analysis of the **arterial blood gases (ABGs).** This is a laboratory test of blood taken from a systemic artery (most blood samples are taken from a systemic vein). As the name suggests, this test shows the key characteristics of blood related to respiratory function:

1. Oxygen partial pressure or Po_2
2. Oxygen saturation of hemoglobin or $\%So_2$
3. pH
4. Concentration of bicarbonate ions or HCO_3^-
5. Carbon dioxide partial pressure or Pco_2

Sometimes these values are labeled with an "a" to emphasize that these are "arterial" values, as in Pao_2 or $Paco_2$. Note that although not all these values are actually "gases," they are affected by blood gases. ABG test results not only reveal a patient's respiratory status, the pH, Pco_2, and HCO_3^- components can also give key information about status of acid-base homeostasis (see accompanying table).

pH is the first result to look at when assessing pH status. If the result is below 7.35, there is *acidosis*, and if it is above 7.45, there is *alkalosis*.

To determine the primary status, one next looks at the Pco_2 result. If the pH is low and Pco_2 is above 45 mm Hg, the primary status is *respiratory* acidosis, and if pH is high and Pco_2 is below 35 mm Hg, the status is *respiratory* alkalosis.

Next, look at the HCO_3^- result. If pH is low and HCO_3^- is below 22 mEq/L, the primary status is *metabolic* acidosis. If the pH is high and the HCO_3^- is above 26 mEq/L, the primary status is *metabolic* alkalosis.

One may then try to determine whether compensation is occurring in the body. This can be done by looking at the pH-balancing mechanism *not* directly involved in determining the primary status to see if it has changed in a way that counterbalances—or *compensates* for—the primary problem. For example, if the arterial pH is low (acidotic) and the Pco_2 (acid) is high, as in respiratory acidosis, the body may compensate by increasing the HCO_3^- (base). Likewise, if the arterial pH is high (alkalotic) and the HCO_3^- (base) is high, as in metabolic alkalosis, the body may compensate by increasing the Pco_2 (acid).

Arterial Blood Gas Analysis for Acid-Base Status

ABG COMPONENT	NORMAL VALUES	ACIDOSIS		ALKALOSIS	
		RESPIRATORY	METABOLIC	RESPIRATORY	METABOLIC
pH	7.35-7.45	↓ <7.35	↓ <7.35	↑ >7.45	↑ >7.45
Pco_2	35-45 mm Hg	↑ >45	Uncompensated: = 35-45 Compensated: ↓ <35	↓ <35	Uncompensated: = 35-45 Compensated: ↑ >45
HCO_3^-	22-26 mEq/L	Uncompensated: = 22-26 Compensated: ↑ >26	↓ <22	Uncompensated: = 22-26 Compensated: ↓ <22	↑ >26

↓, Decrease; ↑, increase; *ABG*, arterial blood gas; Pco_2, carbon dioxide pressure; $[HCO_3^-]$, bicarbonate concentration

LANGUAGE OF **SCIENCE** *(continued from p. 601)*

ketone body
(KEE-tohn BOD-ee)
[*keto-* **acetone,** *-one* **chemical**]

neutral
(NOO-trel)
[*neutr-* **neither,** *-al* **relating to**]

normal saline
(NOR-mall SAY-leen)
[*sal-* **salt,** *-ine* **relating to**]

pH
(pee aych)
[abbreviation for *potenz* **power,** *hydrogen* **hydrogen**]

22

LANGUAGE OF **MEDICINE**

acidosis
(as-ih-DOH-sis)
[*acid-* **sour,** *-osis* **condition**]

alkalosis
(al-kah-LOH-sis)
[*alkal-* **ashes,** *-osis* **condition**]

arterial blood gas (ABG)
(ar-TEER-ee-al blud gas)
[*arteri-* **airpipe (artery),** *-al* **relating to**]

bicarbonate loading
(bye-KAR-boh-net LOHD-ing)
[*bi-* **two,** *-carbon-* **coal (carbon),** *-ate* **oxygen**]

diabetic ketoacidosis
(dye-ah-BET-ik kee-toh-as-ih-DOH-sis)
[*diabet-* **siphon (diabetes mellitus),** *-ic* **relating to,** *keto-* **acetone,** *-acid-* **sour,** *-osis* **condition**]

emesis
(EM-eh-sis)
[*emesis* **vomiting**]

metabolic acidosis
(met-ah-BOL-ik as-ih-DOH-sis)
[*meta-* **over,** *-bol-* **throw,** *-ic* **relating to,** *acid-* **sour,** *-osis* **condition**]

metabolic alkalosis
(met-ah-BOL-ik al-kah-LOH-sis)
[*meta-* **over,** *-bol-* **throw,** *-ic* **relating to,** *alkal-* **ashes,** *-osis* **condition**]

respiratory acidosis
(RES-pih-rah-tor-ee as-ih-DOH-sis)
[*re-* **again,** *-spir-* **breathe,** *-tory* **relating to,** *acid-* **sour,** *-osis* **condition**]

respiratory alkalosis
(RES-pih-rah-tor-ee al-kah-LOH-sis)
[*re-* **again,** *-spir-* **breathe,** *-tory* **relating to,** *alkal-* **ashes,** *-osis* **condition**]

❑ OUTLINE SUMMARY

To download a digital version of the chapter summary for use with your device, access the ***Audio Chapter Summaries*** *online at evolve.elsevier.com.*

Scan this summary after reading the chapter to help you reinforce the key concepts. Later, use the summary as a quick review before your class or before a test.

pH of Body Fluids

A. pH—a number that indicates the relative hydrogen ion (H^+) concentration (compared with OH^-) of a fluid (**Figure 22-1**)
   1. pH 7.0 indicates neutrality (neutral solution)
   2. pH higher than 7.0 indicates alkalinity (alkaline or basic solution; base)
   3. pH less than 7.0 indicates acidity (acid solution)

B. Normal range of blood pH is approximately 7.35 to 7.45
   1. Arterial blood pH—about 7.45
   2. Venous blood pH—about 7.35

C. pH scale based on multiples of 10
   1. H^+ concentration changes by 10 times for each pH unit
   2. Large pH fluctuations may appear small

Mechanisms that Control pH of Body Fluids

A. pH homeostatic mechanism—three coordinated homeostatic mechanisms act to maintain the normal pH of body fluids and prevent pH swings when excess acids or bases are present (**Figure 22-2**)
   1. Chemical pH control mechanism—based on buffers in blood/RBCs/and body fluids—act immediately

2. Physiological pH control mechanisms
   a. Changes in pH regulated by changes in respiratory rate that result in changes in blood CO_2—act within minutes
   b. Changes in pH regulated by altered renal activity—act within hours

B. Buffers
   1. Definition—chemical substances that prevent a sharp change in the pH of a fluid when an acid or base is added to it (**Figure 22-3** and **Figure 22-4**)
   2. Buffers usually include two different chemicals—called a buffer pair
   3. "Fixed" acids are buffered mainly by sodium bicarbonate ($NaHCO_3$)
   4. Changes in blood produced by buffering of "fixed" acids in the tissue capillaries (**Figure 22-5**)
      a. Amount of carbonic acid (H_2CO_3) in blood increases slightly
      b. Amount of $NaHCO_3$ in blood decreases; ratio of amount of $NaHCO_3$ to the amount of H_2CO_3 does not normally change; normal ratio is 20:1
      c. H^+ concentration of blood increases slightly
      d. Blood pH decreases slightly below arterial level
   5. Body has other buffer pair systems
      a. Phosphates
      b. Proteins (plasma proteins and hemoglobin)

C. Respiratory mechanism of pH control
   1. Respirations remove some CO_2 from blood as blood flows through lung capillaries
   2. Amount of H_2CO_3 in blood is decreased and thereby its H^+ concentration is decreased; this in turn increases blood pH
   3. Respiratory control centers in brainstem react to dropping pH and promote increased respirations; when pH increases, then breathing slows

D. Urinary mechanism of pH control
   1. Kidneys are the body's most effective regulator of blood pH
   2. Usually urine is acidified by way of the distal tubules secreting hydrogen ions into the urine from blood in exchange for HCO_3^- being retained in the blood; much of the excess H^+ is secreted as ammonia (NH_3) and ammonium ions (NH_4^+)

pH Imbalances

A. Acidosis and alkalosis are the two kinds of pH, or acid-base, imbalances
   1. Disturbances in acid-base balance depend on relative quantities of $NaHCO_3$ and H_2CO_3 in the blood
   2. Body can regulate both of the components of the $NaHCO_3$–H_2CO_3 buffer system
      a. Blood levels of $NaHCO_3$ are regulated by kidneys
      b. H_2CO_3 levels are regulated by lungs

B. Metabolic and respiratory disturbances—both can alter the normal 20:1 ratio of $NaHCO_3$ to H_2CO_3 in blood
   1. Metabolic disturbances affect the $NaHCO_3$ levels in blood
      a. Metabolic acidosis—bicarbonate ($NaHCO_3$) deficit
      b. Metabolic alkalosis—bicarbonate ($NaHCO_3$) excess; complication of severe vomiting (see box, p. 608)
   2. Respiratory disturbances affect the H_2CO_3 levels in blood
      a. Respiratory acidosis (H_2CO_3 excess)
      b. Respiratory alkalosis (H_2CO_3 deficit)

C. Compensation for pH imbalances
   1. Compensated acidosis or alkalosis—occurs when the body's pH-balancing mechanisms temporarily counteract an abnormal shift in pH
   2. Uncompensated acidosis or alkalosis—occurs when the body's mechanisms have not yet normalized the pH

ACTIVE LEARNING

STUDY TIPS

Consider using these tips to achieve success in meeting your learning goals.

Learning the concepts presented in this chapter will be easier if you review a little basic chemistry.

1. The pH scale, acids, and bases are covered at the beginning of the chapter. If you need more of an explanation than that presented here, review Chapter 2 or check online resources.
2. Buffer systems can be thought of as hydrogen or hydroxide ion sponges. They remove those ions so they will have less of an effect on the pH of a solution, in this case, the blood. In the $NaHCO_3$–H_2CO_3 buffer system, the sodium bicarbonate can absorb hydrogen ions by having the hydrogen replace the sodium. The carbonic acid can give up one of its hydrogen atoms that then can react with a hydroxide ion to form water. In both cases the pH of the solution will change very little.
3. Blood carries carbon dioxide as carbonic acid. When the lungs exhale carbon dioxide, there is less carbonic acid in the blood and so the pH of the blood rises. The kidneys use a similar buffer system to secrete hydrogen ions.
4. The buffer system in the blood usually works well, but it can be overwhelmed. Acidosis is a condition in which the blood becomes too acidic, and alkalosis is a condition in which the blood becomes too basic. Develop a concept map of the respiratory and urinary mechanisms of control involved in maintaining normal pH of body fluids.
5. If you have difficulty with the chemistry in this chapter, discuss it in your study group. Someone in the group may have a stronger chemistry background. Review the Language of Science and Language of Medicine sections. Discuss the pH system. Carefully go over the diagrams of the blood and kidney buffer systems. Review the types of acidosis and alkalosis and what causes each of them. Go over the questions and the chapter outline summary at the end of the chapter and discuss possible test questions.

22

Review Questions

Write out the answers to these questions after reading the chapter and reviewing the Chapter Summary. If you simply think through the answer without writing it down, you will not retain much of your new learning.

1. Explain the relationship between pH and the relative concentration of hydrogen and hydroxide ions in a solution.
2. List the three primary mechanisms the body has for regulating the pH of its fluids.
3. Write out the chemical reaction formula that converts carbon dioxide and water to carbonic acid. What enzyme catalyzes this reaction?
4. Define buffer.
5. Explain *buffer pairs.*
6. Explain how a buffer pair would react if more hydrogen ions were added to the blood.
7. Explain how a buffer pair would react if more hydroxide ions were added to the blood.
8. List the four changes that occur in the blood as the result of buffering fixed acids.
9. Explain the respiratory mechanism of pH control.
10. Describe how changes in the respiration rate can affect blood pH.
11. Explain how the chemical reaction that occurs in the distal tubule of the kidney using NaH_2PO_4 removes hydrogen ions from the blood.
12. Define *acidosis* and *alkalosis.*
13. Explain metabolic disturbances of the buffer pair.
14. Explain respiratory disturbances of the buffer pair.

Critical Thinking

Hint *After finishing the Review Questions, write out the answers to these more in-depth questions to help you apply your new knowledge. Go back to sections of the chapter that relate to concepts that you find difficult.*

15. How would you explain pH in terms of the ions involved? What would be the hydrogen ion concentration of a solution with a pH of 4? With a pH of 6?
16. Explain how excessive vomiting causes metabolic alkalosis and explain why normal saline can be used to correct it.
17. What is the proper ratio of $NaHCO_3$ and H_2CO_3 in a buffer pair? Explain how the body can use this ratio to correct uncompensated metabolic acidosis.
18. The hormone aldosterone affects kidney tubule function. One of the effects is to increase H^+ secretion by the kidney tubules. What effect does this action have on the pH of the internal environment (blood plasma)? What might occur if there is hypersecretion of aldosterone? How would this change if there is hyposecretion of aldosterone?

Chapter Test

After studying the chapter, test your mastery by responding to these items. Try to answer them without looking up the answers.

1. The enzyme that converts carbon dioxide and water into carbonic acid is ______.
2. ______ are chemicals that prevent sharp changes in pH when an acid or base is added to a solution.
3. If a strong acid such as HCl were added to the buffer pair $NaHCO_3$ and H_2CO_3, the $NaHCO_3$ would become ______.
4. If a strong base such as NaOH were added to the buffer pair in question 3, the H_2CO_3 would become ______.
5. The part of the nephron that is important in regulation of blood pH is the ______.
6. When Na_2HPO_4 is used by the kidney to remove hydrogen ions from the blood, the end product that leaves the body in the urine is ______.
7. When ammonia is used by the kidney to remove hydrogen ions from the blood, the end product that leaves the body in the urine is ______.
8. The kidney is more effective in pH regulation than the lung because it can remove excess ______, which the lung cannot.
9. The condition in which the blood pH is higher than normal is called ______.
10. The condition in which the blood pH is lower than normal is called ______.
11. In order for the buffer pair to function correctly, the concentration of $NaHCO_3$ must be ______ times greater than the concentration of H_2CO_3.
12. Metabolic disturbances usually have an effect on the ______ part of the buffer pair.
13. Respiratory disturbances usually have an effect on the ______ part of the buffer pair.
14. Severe vomiting is a metabolic disturbance that can cause metabolic ______.

15. An acid solution has:
    a. a pH greater than 7.0
    b. a pH less than 7.0
    c. more hydroxide ions than hydrogen ions
    d. both b and c
16. An alkaline solution has:
    a. a pH greater than 7.0
    b. a pH less than 7.0
    c. more hydrogen ions than hydroxide ions
    d. both b and c
17. Which of the following statements is true?
    a. A solution with a pH of 5 has more hydrogen ions than a solution with a pH of 2.
    b. A solution with a pH of 9 is a base.
    c. The pH value increases as the number of hydrogen ions increases.
    d. Both a and c are true.
18. Arterial blood has a pH of 7.45, and venous blood has a pH of 7.35; therefore:
    a. arterial blood is slightly more acid than venous blood
    b. arterial blood is slightly more alkaline than venous blood
    c. venous blood is slightly more alkaline than arterial blood
    d. both a and c

For questions 19 through 24, fill in the blank with either *increases* or *decreases,* as appropriate.

19. When a fixed acid is buffered in the blood, the amount of $NaHCO_3$ in the blood ________.
20. When a fixed acid is buffered in the blood, the amount of hydrogen ions in the blood ________.
21. When a fixed acid is buffered in the blood, the amount of H_2CO_3 in the blood ________.
22. When a fixed acid is buffered in the blood, the pH of the blood ________.
23. Anything that causes an excessive increase in the respiration rate causes the pH of the blood to ________.
24. Anything that causes an appreciable decrease in the respiration rate causes the pH of the blood to ________.
25. IV normal saline is given to patients with severe emesis. This is done to replace the ________ ions and to restore homeostasis.

Case Studies

To solve a case study, you may have to refer to the glossary or index, other chapters in this textbook, and other resources.

1. Compensated respiratory acidosis is commonly found in persons with chronic bronchitis, an obstructive respiratory disorder discussed in Chapter 17. State what abnormal blood values a person should expect in such a case, and what factors produced them.
2. Larry is a diabetic who is suffering from metabolic acidosis. His breathing seems abnormally rapid. Is there a connection between Larry's acidosis and his rapid breathing? If so, explain the connection.
3. Elizabeth was advised by her doctor at her annual physical that her urine specimen indicated the presence of ketone bodies throughout the specimen. She asked what ketone bodies were and was told that they are acidic products of fat metabolism. He wanted to conduct some additional blood work to investigate the possibility of diabetes. Elizabeth is confused. Isn't diabetes a disorder involving carbohydrate metabolism? Why is the doctor concerned about fat metabolism and not concerned about "sugars"? Can you help Elizabeth understand ketone bodies, diabetic ketoacidosis, and why the doctor will be conducting additional blood work?

Answers to Active Learning Questions can be found online at evolve.elsevier.com.

22

Reproductive Systems

OUTLINE

Hint Scan this outline before you begin to read the chapter, as a preview of how the concepts are organized.

OBJECTIVES

Before reading the chapter, review these goals for your learning.

After you have completed this chapter, you should be able to:

1. Discuss the process of sexual reproduction, and describe the common structural and functional characteristics between the male and female systems.
2. Describe the structure and function of the male reproductive system, including the gross and microscopic structure of the gonads, the developmental steps in spermatogenesis, and the primary functions of the sex hormones.
3. List the major disorders of the male reproductive system and briefly describe each.
4. Describe the structure and function of the female reproductive system, including the gross and microscopic structure of the gonads and the functions of the sex hormones. Also identify and discuss the phases of the endometrial or menstrual cycle and correlate each phase with its occurrence in a typical 28-day cycle.
5. List the major disorders of the female reproductive system and briefly describe each.
6. Define the term *sexually transmitted disease* and describe the major types.

CHAPTER 23

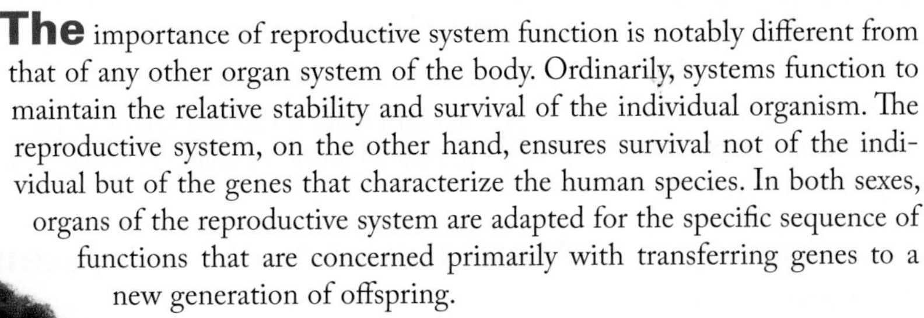

The importance of reproductive system function is notably different from that of any other organ system of the body. Ordinarily, systems function to maintain the relative stability and survival of the individual organism. The reproductive system, on the other hand, ensures survival not of the individual but of the genes that characterize the human species. In both sexes, organs of the reproductive system are adapted for the specific sequence of functions that are concerned primarily with transferring genes to a new generation of offspring.

This chapter explores the normal and pathological anatomy and physiology of the reproductive systems in both sexes and ends with a discussion of sexually transmitted diseases. After study of the reproductive system in both sexes, in Chapter 24 you will learn more about the topic of human development—a process extending from fertilization to death. Chapter 25 wraps up the story of the human body by examining the basic principles of genetics.

Sexual Reproduction

Producing Offspring

Sexual reproduction requires two parent organisms, a male and female, each of which contributes half of the nuclear chromosomes needed to form the first cell of an offspring organism. *Asexual reproduction,* on the other hand, requires only one parent who produces an offspring genetically identical to itself.

An advantage of sexual reproduction is that a new mixture of genes in each offspring increases the variety of genetic characteristics in the population. This variety of characteristics makes it more likely that in the case of environmental changes such as disease, natural disaster, or shifting climatic conditions, there will be at least some individuals likely to survive and carry on the reproductive line.

LANGUAGE OF **SCIENCE**

Hint Before reading the chapter, say each of these terms out loud. This will help you to avoid stumbling over them as you read.

accessory organ
(ak-SES-oh-ree OR-gun)
[*access-* **extra,** *-ory* **relating to,** *organ* **instrument**]

acrosome
(AK-roh-sohm)
[*acro-* **top or tip,** *-some* **body**]

andropause
(AN-droh-pawz)
[*andro-* **male,** *-paus-* **cease**]

antrum
(AN-trum)
[*antrum* **cave**]

areola
(ah-REE-oh-lah)
[*are-* **area or space,** *-ola* **little**]

Bartholin gland
(BAR-tul-in or bar-TOH-lin gland)
[*Caspar Bartholin* **Danish physician,** *gland* **acorn**]

body
(BOD-ee)
[*body* **main part**]

breast
(brest)

bulbourethral gland
(BUL-boh-yoo-REE-thral gland)
[*bulb-* **swollen root,** *-ure-* **urine,** *-thr-* **agent or channel (urethra),** *-al* **relating to,** *gland* **acorn**]

cervix
(SER-viks)
pl., cervices or cervixes
(SER-vis-eez or SER-viks-ez)
[*cervix* **neck**]

clitoris
(KLIT-oh-ris)
[*clitoris* **small key or latch**]

Continued on p. 641

The reproductive system of each parent produces the sex or reproductive cells called **gametes** needed to form the offspring. These gametes, called an **ovum** (from the female parent) and a **sperm** (from the male parent), fuse during the process of fertilization. The new offspring cell that results is called the **zygote.** After many complicated and amazing developmental stages, the zygote ultimately develops into the new individual.

Each reproductive system also produces hormones that regulate development of the secondary sex characteristics that promote successful reproduction. For example, hormones create structural and behavioral differences in the sexes that permit adults to recognize and form sexual attractions with the opposite sex. Reproductive hormones and other regulatory mechanisms give us the urge to have sex, which is often reinforced with the pleasant sensations that sexual activity can produce. This sex drive is essential to success in producing offspring.

Sexual maturity and the ability to reproduce are achieved by the end of **puberty.** The male reproductive system consists of organs whose functions are to produce, transfer, and ultimately introduce mature sperm into the female reproductive tract, where the nuclear chromosomes from each parent can unite to form a new offspring.

23

Male and Female Systems

Although the organs and specific functions of the male and female reproductive systems are discussed separately, it is important to understand that a common general structure and function can be identified between the systems in both sexes and that both sexes contribute in uniquely important ways to overall reproductive success.

In both men and women, the organs of the reproductive system are adapted for the specific sequence of functions that permit development of sperm or ova followed by successful fertilization and then the normal development and birth of a baby. In addition, production of hormones that permit development of secondary sex characteristics, such as breast development in women and beard growth in men, occurs as a result of normal reproductive system activity.

As you study the specifics of each system, keep in mind that the male organs function to produce, store, and ultimately introduce mature sperm into the female reproductive tract and that the female system is structured to produce ova, receive the sperm, and permit fertilization. In addition, the female reproductive system permits the fertilized ovum to develop and mature until birth.

The complex and cyclic control of reproductive functions in both men and women are particularly crucial to overall reproductive success in humans. The production of sex hormones—ultimately controlled by the brain—is required not only for development of the secondary sexual characteristics but also for normal reproductive functions in both sexes.

QUICK CHECK

1. What are *gametes*?
2. When do sexual maturity and the ability to reproduce occur?
3. What is the ultimate function of the reproductive systems?

To learn more about the reproductive systems, go to AnimationDirect online at *evolve.elsevier.com.*

Male Reproductive System

Structural Plan

Reproductive Tract

You may recall from Chapter 20 (see p. 565) that in males the urethra has a dual function. It serves as a passageway for *both* urine and semen from the body. The term *urogenital tract* is sometimes used in place of *reproductive tract* to describe this dual urinary and reproductive function.

So many organs make up the male reproductive system that we need to look first at the structural plan of the system as a whole. Reproductive organs can be classified as *essential* or *accessory.*

Essential Organs

The **essential organs** of reproduction in men and women are called the **gonads.** The gonads of men consist of a pair of main sex glands called the **testes.** We introduced the term *sex gland* in Chapter 12 (Endocrine System) to emphasize their role in producing *sex hormones.* The testes produce the male sex hormone *testosterone.* However, gonads have additional primary functions. The testes, for example, also produce the male sex cells—the *sperm* or **spermatozoa.**

Accessory Organs

The **accessory organs** of reproduction in men consist of the following structures:

1. A series of passageways or ducts that carry the sperm from the testes to the exterior
2. Additional sex glands that provide secretions that protect and nurture sperm
3. The external reproductive organs called the *external genitals*

TABLE 23-1 Male Reproductive Organs

ESSENTIAL ORGANS	ACCESSORY ORGANS
Gonads: testes (right testis and left testis)	Ducts: epididymis (two), vas deferens (two), ejaculatory duct (two), and urethra Supportive sex glands: seminal vesicles (two), bulbourethral (Cowper) glands (two), and prostate gland External genitals: scrotum and penis

Table 23-1 lists the names of the essential and accessory organs of reproduction in men, and **Figure 23-1** shows the location of most of them. The table and the illustration are included very early in the chapter to provide a preliminary but important overview. Refer back to this table and illustration frequently as you learn about each organ in the

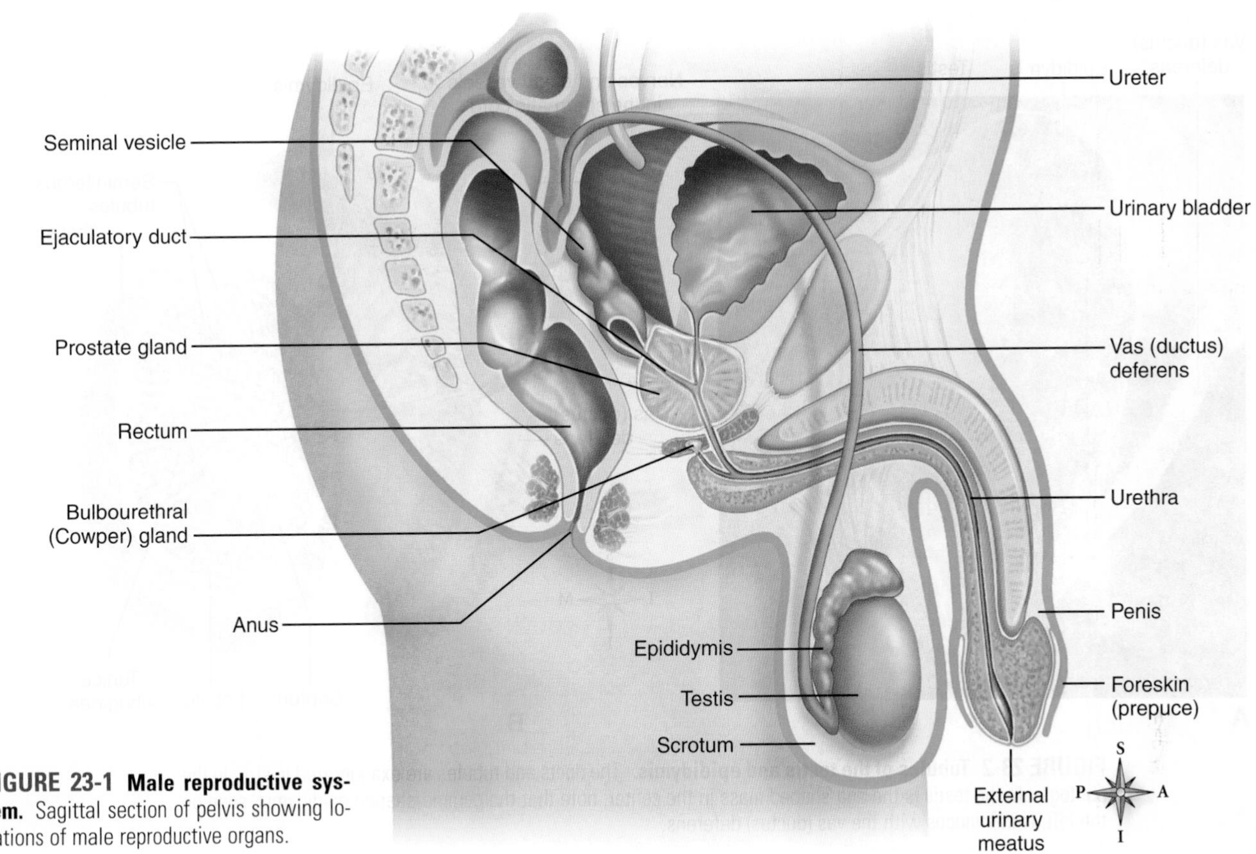

FIGURE 23-1 Male reproductive system. Sagittal section of pelvis showing locations of male reproductive organs.

pages that follow. Also find the major structures of the male reproductive system in the *Clear View of the Human Body* (follows p. 8).

Testes

Structure and Location

The paired **testes** are the gonads in males. They are located in the pouchlike **scrotum,** which is suspended outside of the body cavity behind the penis as you can see in **Figure 23-1**. This exposed location provides an environment about 1° C to 3° C cooler than normal core body temperature, an important requirement for the normal production and survival of sperm.

Each testis is a small, oval gland about 3.8 cm (1.5 inches) long and 2.5 cm (1 inch) wide. The testis is shaped like an egg that has been flattened slightly from side to side. Note in **Figure 23-2** that each testis is surrounded by a tough, whitish membrane called the **tunica albuginea.** This membrane covers the testis and then enters the gland to form the many septa that divide it into sections or lobules.

As you can see in **Figure 23-2**, each lobule consists of a narrow but long and coiled **seminiferous tubule.** These coiled structures form the bulk of the testicular tissue mass. Small endocrine cells lying near the septa that separate the lobules can be seen in **Figure 23-3**. These are the **interstitial cells** of the testes that secrete the male sex hormone **testosterone.**

Each seminiferous tubule is a long duct with a central lumen or passageway (see **Figure 23-3**). Sperm develop in the walls of the tubule and are then released into the lumen and begin their journey to the exterior of the body.

Testis Functions

Spermatogenesis

Sperm production is also called **spermatogenesis.** From puberty on, the seminiferous tubules continuously form spermatozoa, or sperm. Although the number of sperm produced each day diminish with increasing age, most men continue to produce significant numbers throughout life.

The testes prepare for sperm production before puberty by increasing the numbers of sperm precursor (stem) cells called **spermatogonia.** These cells are located near the outer edge of each seminiferous tubule (**Figure 23-4**, *A*). Before puberty, spermatogonia increase in number by the process of mitotic cell division, which was described in Chapter 3. Recall that mitosis results in the division of a "parent" cell into two "daughter" cells, each identical to the parent and each containing a complete copy of the genetic material represented in the normal number of 46 chromosomes.

The hypothalamus is a small but functionally important structure located near the base of the brain. One of its many functions, in both males and females, is to secrete **gonadotropin-releasing hormone (GnRH),** which then stimulates the anterior pituitary to secrete the gonadotropins **follicle-stimulating hormone (FSH)** and **luteinizing hormone (LH).** A *gonadotropin* is a hormone that has a stimulating effect on the *gonads*—the testes and ovaries.

You may want to review these roles of the hypothalamus and pituitary gland in Chapter 10 (p. 261) and Chapter 12 (p. 326). Also, peek ahead to **Figure 23-16**, where you will see the hypothalamus and pituitary depicted at the top of the diagram.

23

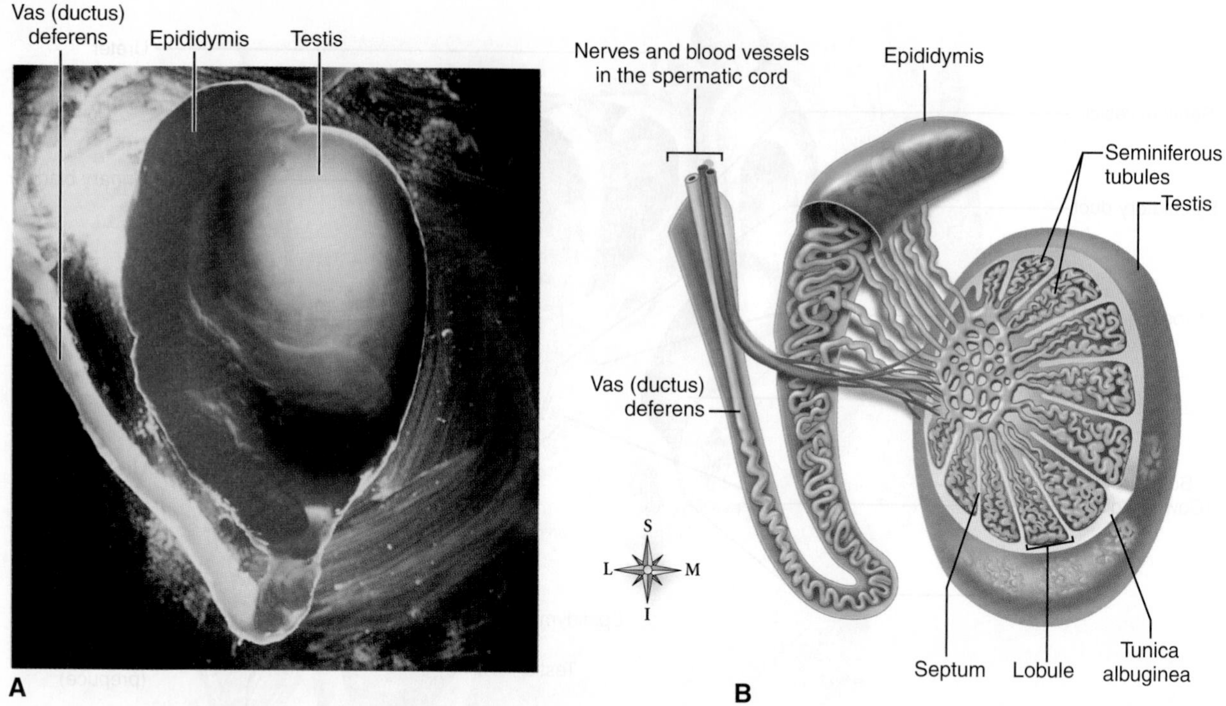

FIGURE 23-2 Tubules of the testis and epididymis. The ducts and tubules are exaggerated in size. In the photograph, the testis is the egg-shaped mass in the center; note that the comma-shaped epididymis, seen on the left, is continuous with the vas (ductus) deferens.

When a boy enters puberty, circulating levels of FSH cause a spermatogonium to undergo a unique series of cell divisions to produce sperm cells. When the spermatogonium undergoes cell division and mitosis under the influence of FSH, it produces two daughter cells. One of these cells remains as a spermatogonium and the other forms another type of cell called a **primary spermatocyte.** These primary spermatocytes then undergo another type of cell division characterized by **meiosis,** which ultimately results in sperm formation.

FIGURE 23-3 Testis tissue. Several seminiferous tubules surrounded by septa containing interstitial cells are shown.

Note in **Figure 23-4**, *B*, that in meiosis two cell divisions occur (not one as in mitosis) and that four daughter cells (not two as in mitosis) are formed. The daughter cells are called **spermatids.** Unlike the two daughter cells that result from mitosis, the four spermatids each have only half the genetic material in its nucleus and half of the nuclear chromosomes (23) of other body cells. These spermatids then develop into spermatozoa.

Look again at the diagram of meiosis in **Figure 23-4**, *B*. It shows that each primary spermatocyte ultimately produces four sperm cells. Note that, in the portion of a seminiferous tubule shown in **Figure 23-4**, spermatogonia are found at the outer surface of the tubule, primary and secondary spermatocytes lie deeper in the tubule wall, and mature but immotile sperm are seen about to enter the lumen of the tube and begin their journey through the reproductive ducts to the exterior of the body.

To learn more about spermatogenesis, go to AnimationDirect online at *evolve.elsevier.com.*

Sperm

Sperm are among the smallest and most unusual cells in the body (**Figure 23-5**, *A*). The term **sperm** comes from Latin *spermatozoan* meaning "seed animal." This is because, somewhat like a seed, each sperm cell is part of the reproductive process.

And each sperm cell has a tail and moves independently somewhat like a microscopic animal.

All of the characteristics that a baby will inherit from its father at fertilization are contained in the nuclear chromosomes found in each sperm head. However, this genetic information from the father will unite with chromosomes contained in the mother's ovum only if successful fertilization occurs.

The forceful ejection of fluid containing sperm, or **ejaculation,** into the female vagina during sexual intercourse is only one step in the long journey that these sex cells must make before they can meet and fertilize an ovum. To accomplish their task, these tiny packages of genetic information are equipped with tails for motility and enzymes to penetrate the outer membrane of the ovum when contact occurs with it.

The structure of a mature sperm is diagrammed in Figure 23-5, *B.* Note the sperm *head* containing the nucleus with its genetic material from the father. The sperm head is covered by the **acrosome**—a caplike structure containing enzymes that enable the sperm to break down the covering of the ovum and permit entry if contact occurs.

In addition to the head with its covering acrosome, each sperm has a *midpiece* and an elongated *tail.* Mitochondria in the midpiece release adenosine triphosphate (ATP) to provide energy for the tail movements required to propel the sperm and allow them to "swim" for relatively long distances through the female reproductive ducts. The tail is actually a *flagellum,* previously described in Chapter 3—see Figure 3-4 and Figure 3-5 (p. 49).

Production of Testosterone

In addition to spermatogenesis, the other function of the testes is to secrete the male hormone **testosterone.** This function is carried on by the **interstitial cells** of the testes, not by their seminiferous tubules. The gonadotropin LH stimulates interstitial cells to develop and produce testosterone.

Testosterone serves the following general functions:

1. Testosterone masculinizes. The various characteristics that we think of as "male" develop because of testosterone's influence. For instance, when a young boy's voice changes, it is testosterone that brings this about.

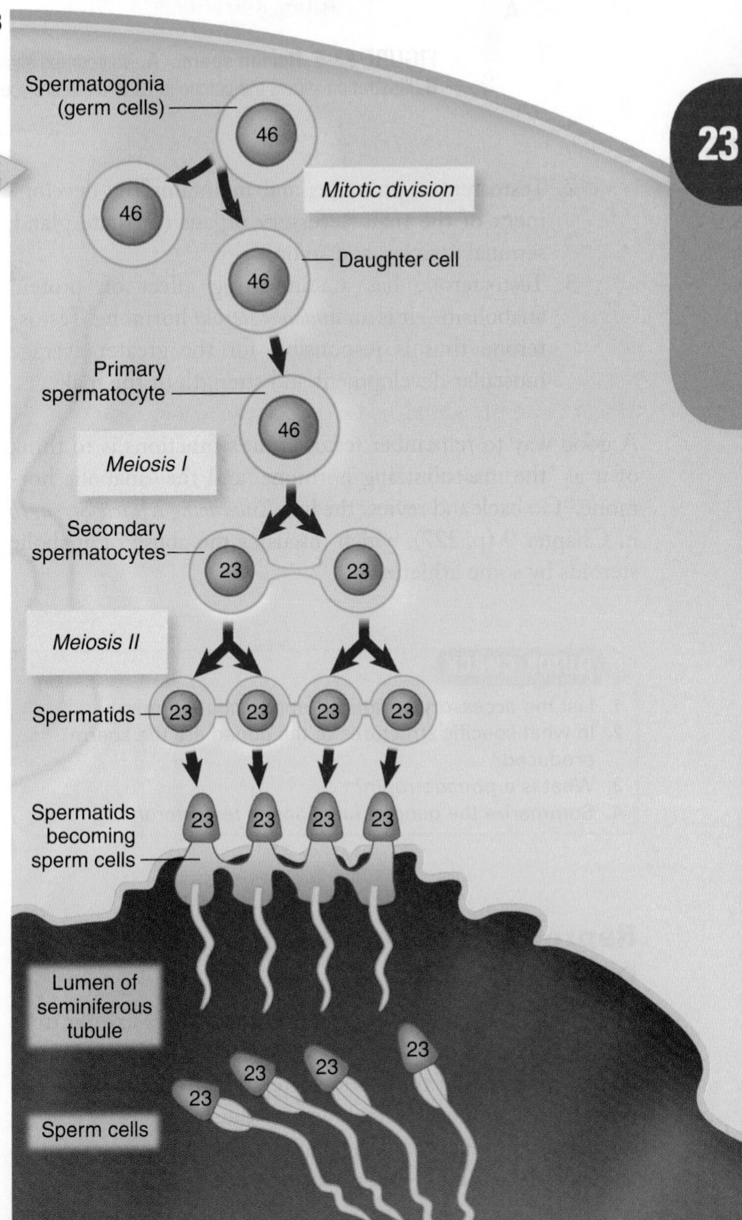

FIGURE 23-4 Spermatogenesis. A, Cross section of seminiferous tubule shows layers of cells undergoing the process of spermatogenesis. **B,** Diagram of spermatogenesis, including the role of meiosis in producing daughter sperm cells with half the number of nuclear chromosomes found in typical body cells.

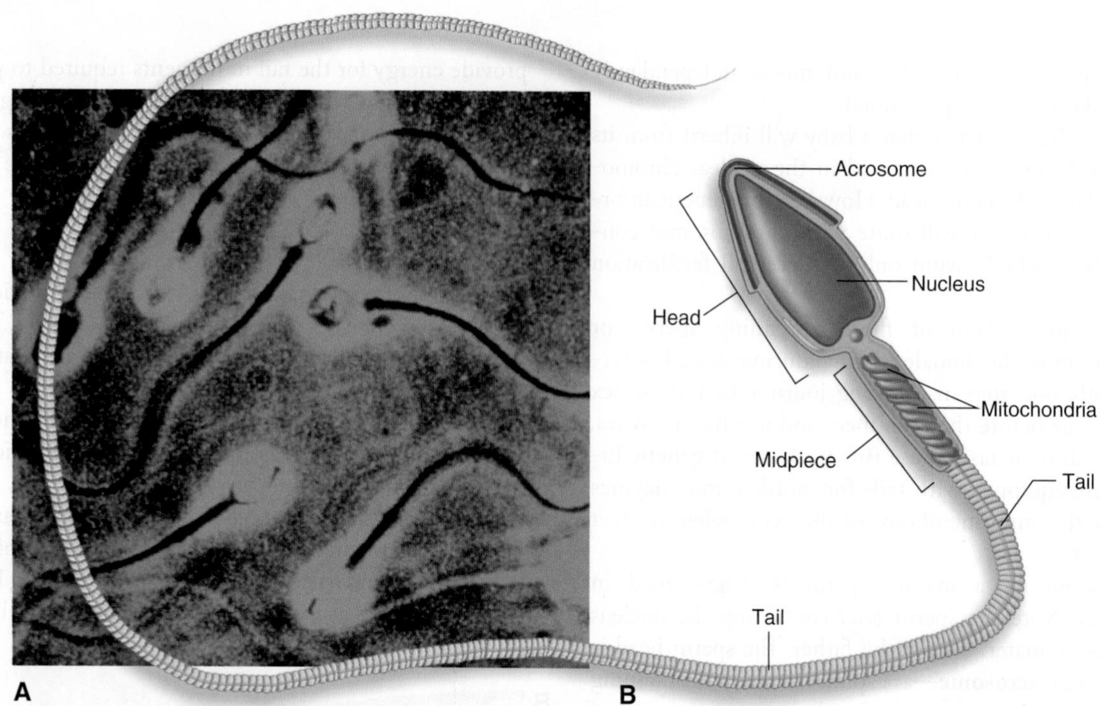

FIGURE 23-5 Human sperm. A, Micrograph shows the heads and long, slender tails of many spermatozoa. **B,** Illustration shows the components of a mature sperm cell and an enlargement of a sperm head and midpiece.

23

2. Testosterone promotes and maintains the development of the male accessory organs (prostate gland, seminal vesicles, and so on).
3. Testosterone has a stimulating effect on protein anabolism—it is an *anabolic steroid* hormone. Testosterone thus is responsible for the greater average muscular development and strength of the male.

A good way to remember testosterone's functions is to think of it as "the masculinizing hormone" and the "anabolic hormone." Go back and review the box *Enhancing Muscle Strength* in Chapter 9 (p. 227), which discusses the abuse of anabolic steroids by some athletes.

QUICK CHECK

1. List the *accessory organs* of reproduction in men.
2. In what specific structures of the gonad are the sperm produced?
3. What is a *gonadotropin*?
4. Summarize the general functions of *testosterone.*

Reproductive Ducts

Overview

The ducts through which sperm must pass after exiting from the testes until they reach the exterior of the body are important components of the accessory reproductive structures. The other two components included in the listing of accessory organs of reproduction in the male—the supportive sex glands and external genitals—are discussed separately here.

Sperm are formed within the walls of the seminiferous tubules of the testes. When they exit from these tubules within the testis, they enter and then pass, in sequence, through the epididymis, vas deferens (ductus deferens), ejaculatory duct, and the urethra on their journey out of the body.

Epididymis

Each **epididymis** consists of a single and very tightly coiled tube about 6 m (20 feet) in length. It is a comma-shaped structure (see **Figure 23-2**) that lies along the top and behind the testes inside the scrotum. Sperm mature and develop their ability to move, or swim, as they pass through the epididymis.

Cells lining the epididymis secrete nutrients for developing sperm and also remove substantial amounts of excess testicular fluid as the developing sex cells enter and eventually pass through the lumen of this highly coiled tube.

Epididymitis is a painful inflammation of the epididymis. (Recall that the suffix *-itis* signifies "inflammation of.") Epididymitis often occurs in association with sexually transmitted diseases, or STDs (see **Table 23-4**). The onset of pain is coupled with redness and swelling of the overlying scrotum, fever, and the appearance of white blood cells (WBCs) in the urine.

Vas Deferens

The **vas deferens,** or **ductus deferens,** is the tube that permits sperm to exit from the epididymis and pass from the scrotal sac upward into the pelvic cavity (see **Figure 23-1**). Each vas deferens is a thick, smooth, very muscular, and movable tube that can easily be felt or "palpated" through the thin skin of the scrotal wall. It passes through the inguinal canal into the

CLINICAL APPLICATION

VASECTOMY

Severing or clamping off the vas deferens—that is, a **vasectomy,** usually done through an incision in the scrotum—makes a man sterile. Why? Because it interrupts the route to the exterior from the epididymis. To leave the body, sperm must journey in succession through the epididymis, vas deferens, ejaculatory duct, and urethra.

Vasectomy is one of many types of reproductive planning called **contraception.** For more strategies, and how they help illustrate how reproduction functions, see the article *Contraception* at Connect It! at *evolve.elsevier.com.*

pelvic cavity as part of the *spermatic cord,* a connective tissue sheath that also encloses blood vessels and nerves.

Ejaculatory Duct and Urethra

Once in the pelvic cavity, the vas deferens extends over the top and down the posterior surface of the bladder, where it joins the duct from the seminal vesicle to form the **ejaculatory duct** (Figure 23-6).

Note in Figure 23-1 and Figure 23-6 that the ejaculatory duct passes through the substance of the prostate gland and permits sperm to empty into the **urethra,** which eventually passes through the penis and opens to the exterior at the **external urinary meatus.**

Accessory Glands

The term **semen,** or **seminal fluid,** is used to describe the mixture of sex cells or sperm produced by the testes and the secretions of the accessory or supportive sex glands. The accessory glands, which contribute more than 95% of the secretions to the gelatinous fluid part of the semen, include the two seminal vesicles, one prostate gland, and two bulbourethral (Cowper) glands. In addition to the production of sperm, the seminiferous tubules of the testes contribute somewhat less than 5% of the seminal fluid volume.

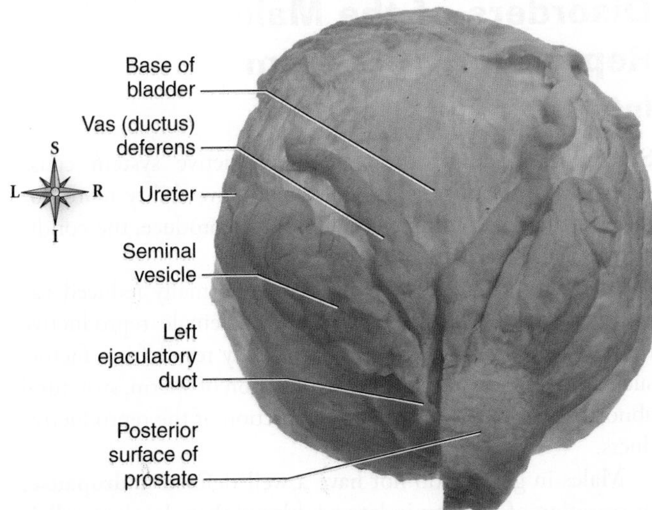

FIGURE 23-6 Male accessory glands. Dissection photo showing bladder, prostate, vas deferens, left ejaculatory duct, and seminal vesicles from behind.

Usually 3 to 5 mL (about 1 teaspoon) of semen is ejaculated at one time, and each milliliter normally contains about 20 million to 100 million sperm. These numbers vary considerably in healthy men, even from day to day. Semen is slightly alkaline and protects sperm from the acidic environment of the female reproductive tract.

Seminal Vesicles

The paired **seminal vesicles** (see Figure 23-1) are pouchlike glands that contribute about 60% of the seminal fluid volume. Their secretions are yellowish, thick, and rich in the sugar fructose. This fraction of the seminal fluid helps provide a source of energy for the highly motile sperm.

Prostate Gland

The **prostate gland** lies just below the bladder and is shaped like a doughnut. The urethra passes through the center of the prostate before traversing the penis to end at the external urinary orifice.

The prostate secretes a thin, milk-colored fluid that constitutes about 30% of the total seminal fluid volume. This portion of the ejaculate helps to activate the sperm and maintain their motility.

Bulbourethral Glands

Each of the two **bulbourethral glands** (also called **Cowper glands**) resembles a pea in size and shape. They are located just below the prostate gland and empty their secretions into the penile portion of the urethra. Because this fluid is often released just before the rest of the semen is ejaculated, it is called "pre-ejaculate."

The mucuslike secretions of the bulbourethral glands serve several functions. They neutralize any residue of sperm—damaging acidic urine in the urethra. They also lubricate the urethra to protect sperm from friction damage and add to the external lubrication of the penis needed for intercourse.

The bulbourethral glands contribute less than 5% of the seminal fluid volume ejaculated from the urethra.

External Genitals

The penis and scrotum constitute the external reproductive organs—sometimes called the **genitals** or **genitalia.**

The **penis** (Figure 23-7) is the organ that, when made stiff and erect by the filling of its spongy, or erectile, tissue components with blood during sexual arousal, can enter and deposit

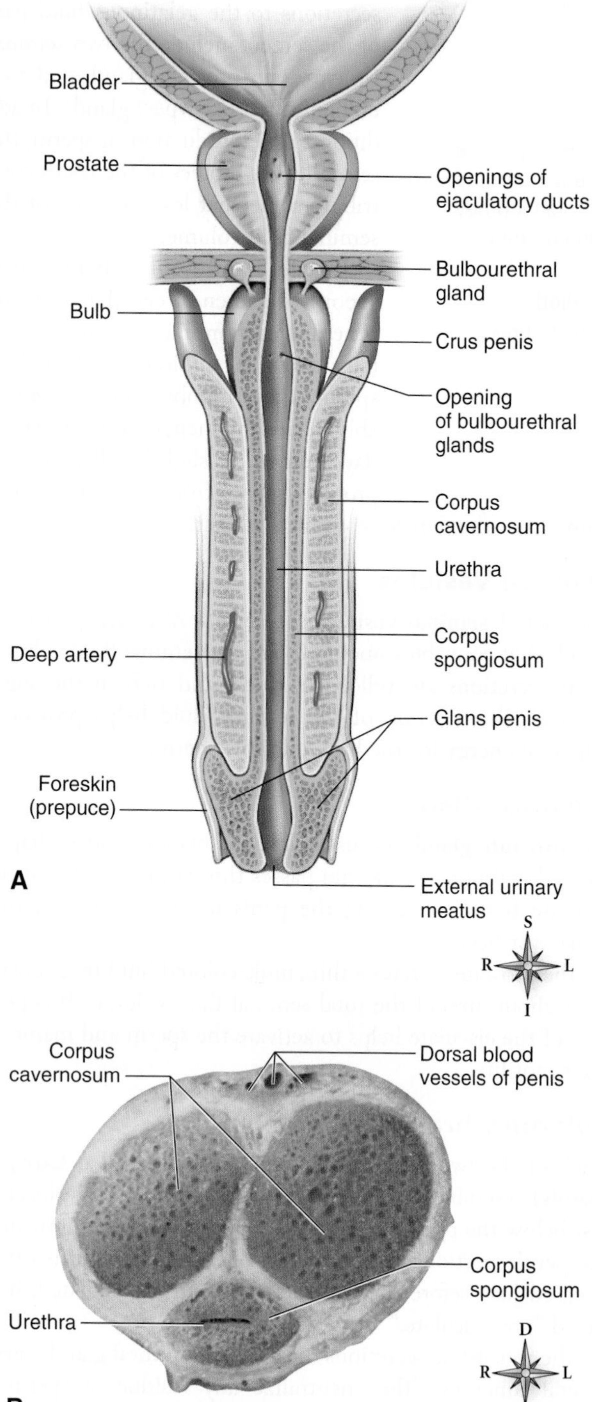

FIGURE 23-7 Penis. A, In this sagittal section of the penis viewed from above, the urethra is exposed throughout its length and can be seen exiting from the bladder and passing through the prostate gland before entering the penis to end at the external urinary meatus. **B,** Photograph of a cross section of the shaft of the penis showing the three columns of erectile or cavernous tissue. Note the urethra within the substance of the corpus spongiosum.

sperm in the vagina during intercourse. The penis has three separate columns of erectile tissue in its shaft: one **corpus spongiosum,** which surrounds the urethra, and two **corpora cavernosa,** which lie dorsally. The spongy nature of erectile tissue is apparent in **Figure 23-7**.

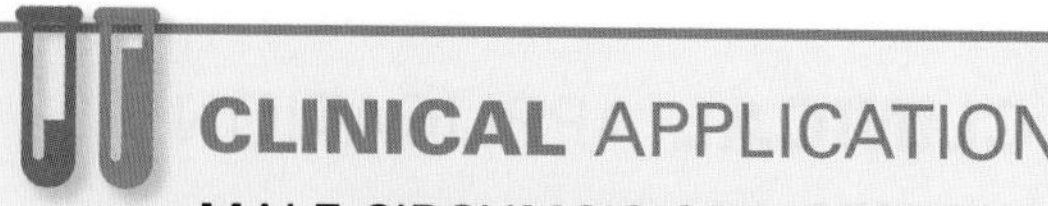

CLINICAL APPLICATION

MALE CIRCUMCISION

In recent decades, there has been disagreement among medical professionals regarding whether routine circumcision is justified in all male infants. Certainly, circumcision may be required if the foreskin fits so tightly over the glans that it cannot be retracted, a condition called **phimosis.**

Also, if the foreskin cannot be replaced to its usual position after it has been retracted behind the glans, a condition called **paraphimosis,** circumcision may be required.

After a few years of withholding the recommendation, most health experts are again recommending routine circumcision to reduce the spread of human immunodeficiency virus (HIV) and other *sexually transmitted infections (STIs)* based on extensive research findings. However, there are risks to circumcision, including loss of its normal protective and sensory role, and many question the ethics of a surgery in which the patient cannot give informed consent.

At the distal end of the shaft of the penis is the enlarged *glans penis,* or more simply **glans.** The glans is dense with sensory receptors that help stimulate the male sexual response. The *external urinary meatus* is the opening of the urethra at the tip of the glans.

The skin of the distal end of the penis is folded doubly to form a loose-fitting retractable, collar around the glans called the **foreskin,** or **prepuce.** Besides protecting the glans from abrasion, the foreskin is also dense with sensory receptors. Surgical removal of the foreskin is called **circumcision** (see Clinical Application box above).

The **scrotum** is a skin-covered pouch suspended from the groin. Internally, it is divided into two sacs by a septum; each sac contains a testis, epididymis, the lower part of the vas deferens, and the beginning of the spermatic cords.

Disorders of the Male Reproductive System

Infertility and Sterility

Several disorders of the male reproductive system cause **infertility.** Infertility is an abnormally low ability to reproduce. If there is a complete inability to reproduce, the condition is called **sterility.**

Infertility or sterility involves an abnormally reduced capacity to deliver healthy sperm to the female reproductive tract. Reduced reproductive capacity may result from factors such as a decrease in the testes' production of sperm, structural abnormalities in the sperm, or obstruction of the reproductive ducts.

Males in general do not have a well-defined **andropause,** or cessation of fertility, in late adulthood that closely parallels female menopause. However, sensitivity of the testis to LH may begin to decline after age 50, causing testosterone levels to drop. If it is a significant drop, it may be called "low T." Low testosterone production can cause sperm production to also decline somewhat. Even so, many men remain fertile throughout life.

Disorders of the Testes

Reduced Sperm Production

Disruption of the sperm-producing function of the seminiferous tubules can result in decreased sperm production, a condition called **oligospermia.** If the *sperm count* is too low, infertility may result. A large number of sperm is needed to ensure that many sperm will reach the ovum and dissolve its coating, allowing a single sperm to unite with the ovum.

Oligospermia can result from factors such as infection, fever, radiation, malnutrition, and high temperature in the testes. In some cases, oligospermia is temporary—as in some acute infections. Oligospermia is a leading cause of infertility. Of course, total absence of sperm production results in sterility.

Cryptorchidism

Early in fetal life the testes are located in the abdominopelvic cavity near the kidneys but normally descend into the scrotum about 2 months before birth. Occasionally a baby is born with undescended testes, a condition called **cryptorchidism,** which is readily detected by palpation of the scrotum at delivery. The word *cryptorchidism* is from the Greek words *kryptikos* (hidden) and *orchis* (testis).

Failure of the testes to descend may be caused by hormonal imbalances in the developing fetus or by a physical deficiency or obstruction. Regardless of cause, in the cryptorchid infant, the testes remain "hidden" in the abdominopelvic cavity.

Because the higher temperature inside the body cavity inhibits spermatogenesis, measures must be taken to bring the testes down into the scrotum to prevent permanent sterility. Early treatment of cryptorchidism by injection of testosterone, which stimulates the testes to descend, may result in normal testicular and sexual development. The condition may also be corrected surgically.

Testicular Cancer

Most *testicular tumors* are cancerous and arise from sperm-producing cells of the seminiferous tubules. Externally, they most often appear as a nontender mass fixed on the testis.

Malignancies of the testes are most common among men 15 to 30 years old. In addition to age group, this type of cancer is associated with genetic predisposition, trauma or infection of the testis, and cryptorchidism.

Treatment of testicular cancer is most effective when the diagnosis is made early in the development of the tumor. Many physicians encourage male patients to perform monthly self-examination of their testes, especially if they are in a high-risk group. The self-examination involves palpating each testis—preferably after a warm shower when the scrotum is relaxed and the testes are descended and accessible. Each testis should be palpated through the scrotal wall between the thumb and the index and middle fingers. They should feel firm, smooth, and rubbery but not hard. The examination may result in some tenderness but should not be painful. Any lump or change in texture should be reported to a physician for further assessment.

Disorders of the Prostate

Benign Prostatic Hypertrophy

A noncancerous condition called **benign prostatic hypertrophy (BPH)** is a common problem in older men. The condition is characterized by an enlargement or hypertrophy of the prostate gland. BPH is so common in late adulthood—more than 90% of men over 80—that it is considered a usual part of aging.

The fact that the urethra passes through the center of the prostate after exiting from the bladder is a matter of considerable clinical significance in this condition. As the prostate enlarges, it squeezes the urethra, possibly closing it so completely that urination becomes very difficult or even impossible.

In severe cases only, surgical removal of a part of the gland or the entire gland, a procedure called **prostatectomy,** may become necessary.

Prostate Cancer

Prostate cancer is the second leading cause of cancer deaths in men. Most are adenocarcinomas of the glandular tissue. Early diagnosis is critical for survival (see the Clinical Application box on page 626).

Once cancer is confirmed, treatment depends on the stage of cancer and the age and health of the patient. Watchful waiting is commonly recommended for some early stage prostate cancers, but more advanced cases may require prostatectomy—often in combination with other therapies. Options include systemic chemotherapy, cryotherapy (freezing) of prostatic tissue, microwave therapy, hormonal therapy, and various types of external beam x-ray radiation treatments.

One treatment protocol involves placing small radioactive "seeds" directly into the prostate tumor, where they give off very localized cancer-cell–destroying radiation for about a year. The treatment is called **brachytherapy**—from the Greek term *brachy*—meaning "short distance." The radiation is limited to a short distance because the radioactive seeds are placed in or near the tumor itself, thus reducing any radiation exposure to surrounding healthy tissue.

Disorders of the Penis and Scrotum

Penis Disorders

Infection, Cancer, and Structural Disorders

The penis is subject to *cancerous tumors* and is affected by numerous **sexually transmitted diseases,** or **STDs** (see Table 23-4). Development of herpes vesicles; genital warts; and various lesions of the foreskin, glans, and penile shaft are common.

Structural abnormalities such as *phimosis* and *paraphimosis,* discussed in the box on the opposite page, can obstruct the flow of urine or result in urinary tract infections.

The term **hypospadias** describes a congenital condition that is characterized by the opening of the urethral meatus on the underside of the glans or penile shaft. Surgical correction

CLINICAL APPLICATION

DETECTING PROSTATE CANCER

Many of the 32,000 men who die each year from *prostate cancer*—the most common nonskin form of cancer in American men and a leading cause of cancer deaths in men over 50—could be saved if the cancers were detected early enough to allow effective treatment. Several screening tests are now available for early detection of prostate cancer.

For example, physicians can sometimes detect prostate cancer early by palpating the prostate through the wall of the rectum using a gloved, lubricated finger. This is called a *digital rectal exam.*

The *prostate-specific antigen (PSA)* test is a blood test that assesses the level of a blood protein that may increase in prostate cancer. Although controversial when used alone for cancer screening because of potentially misleading results, it remains a valuable test for monitoring prostate health.

Because of the prevalence of prostate cancer, adult men are encouraged to have regular prostate examinations and to report any urinary or sexual difficulty to their physicians.

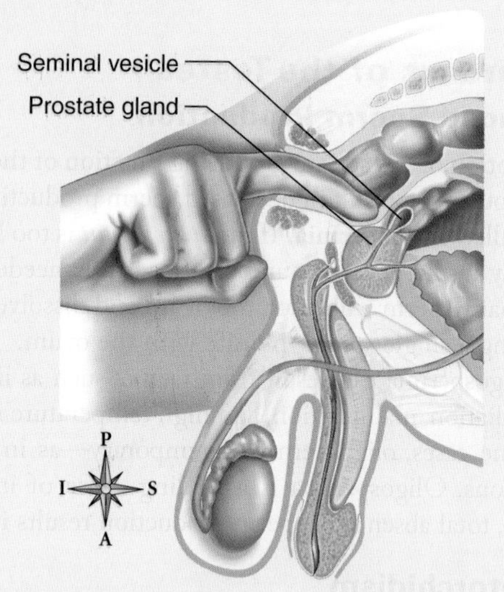

23

is performed if the defect is likely to cause urological or reproductive problems.

The term **epispadias** refers to a much less common congenital defect that involves the opening of the urethral meatus on the dorsal or top surface of the glans or penile shaft.

Erectile Dysfunction

Failure to achieve or maintain an erection of the penis adequate enough to permit sexual intercourse is called **erectile dysfunction (ED)** or *impotence.* ED affects men of all ages but is experienced most often after age 65. Impotence does not affect sperm production but infertility often results because normal intercourse may not be possible.

In the past, psychological problems such as anxiety, depression, and stress were often cited as the most important causes of impotence in sexually active men. There is no doubt that such conditions contribute to ED. However, current research suggests that purely psychological problems probably account for far fewer cases of impotence than previously thought.

We now know that ED is frequently caused by medical problems related to abnormal vascular or neural control of penile blood flow. Arteriosclerosis, diabetes, alcohol abuse, numerous medications, radiation therapy, tumors, spinal cord trauma, and surgery, especially if pelvic organs such as the prostate are involved, may all contribute to ED.

Treatment options for ED include use of drugs that increase blood flow to the spongy cavernous tissue of the penis causing it to stiffen and become erect. Oral medications such as Viagra (sildenafil), Levitra (vardenafil), Cialis (tadalafil), and Uprima (apomorphine) are generally preferred by men who do not have medical conditions that preclude their use.

A drug called Muse (alprostadil) is available as a tiny soft pellet that is inserted into the urethra using a small applicator. A similar drug available in solution, Caverject, is injected directly into the penis.

As a result of multiple options, even moderate to severe erectile dysfunction occurring in sexually active men can be treated with considerable success.

Scrotum Disorders

Swelling of the scrotum can be caused by a variety of conditions. One of the most common causes of scrotal swelling is an accumulation of fluid called a **hydrocele.** Hydroceles may be congenital, resulting from structural abnormalities present at birth.

In adults, hydrocele often occurs when fluid produced by the serous membrane lining the scrotum is not absorbed properly. The cause of adult hydrocele is not always known, but in some cases, it can be linked to trauma or infection.

Swelling of the scrotum may also occur when the intestines push through the weak area of the abdominal wall that separates the abdominopelvic cavity from the scrotum. This condition is a form of **inguinal hernia.** If the intestines protrude too far into the scrotum, the digestive tract may become obstructed, resulting in death.

In adults, inguinal hernia often occurs while lifting heavy objects, because of the high internal pressure generated by the contraction of abdominal muscles. Inguinal herniation also may be congenital (**Figure 23-8**).

Small inguinal hernias may be treated with external supports that prevent organs from protruding into the scrotum; more serious hernias must be repaired surgically.

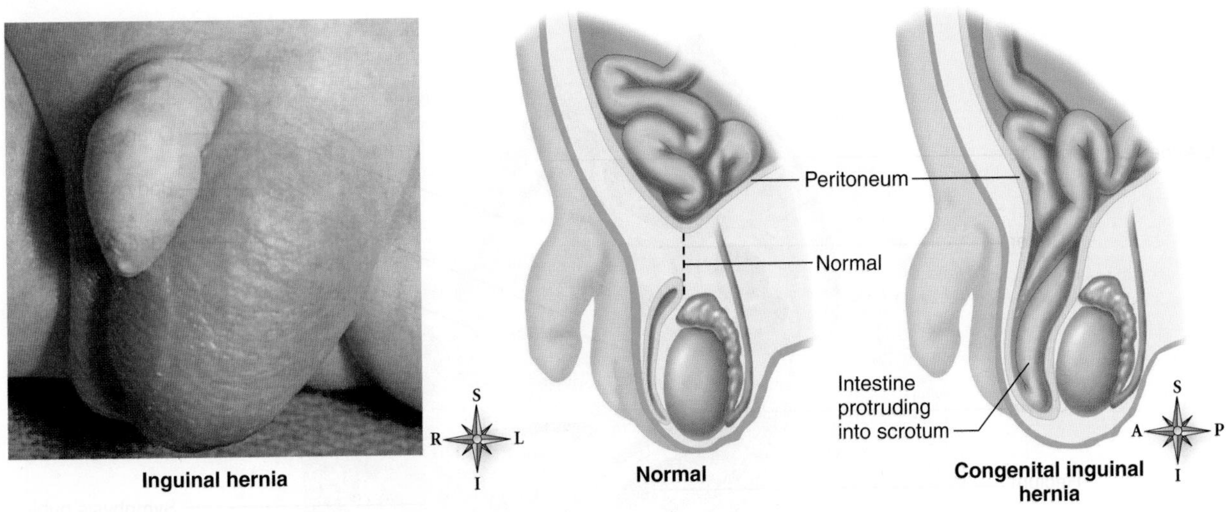

FIGURE 23-8 Inguinal hernia. Congenital inguinal hernia in infant male.

Although less common, inguinal hernias also occur in females. Review the article *Hernias* at Connect It! at *evolve.elsevier.com.*

QUICK CHECK

1. What duct leads from the epididymis?
2. Which organs produce the fluid in semen?
3. What is the function of the *erectile tissue*?
4. Identify the treatments for benign prostatic hypertrophy.

Female Reproductive System

Structural Plan

The structural plan of the reproductive system in both sexes is similar in that organs are characterized as *essential* or *accessory.*

Essential Organs

The essential organs of reproduction in women, the **gonads,** are the paired **ovaries.** The female sex cells, or **ova,** are produced in the ovaries. The ovaries also produce the hormones estrogen and progesterone.

Accessory Organs

The accessory organs of reproduction in women consist of the following structures:

1. A series of ducts or modified duct structures that extend from near the ovaries to the exterior
2. Additional sex glands, including the mammary glands, which have an important reproductive function only in women
3. The external reproductive organs or external genitals

Table 23-2 lists the names of the essential and accessory female organs of reproduction, and **Figure 23-9** shows the location of most of them. Refer back to this table and illustration as you read about each structure in the pages that follow. Also find the major structures of the female reproductive system in the *Clear View of the Human Body* (follows p. 8).

Ovaries

Structure and Location

The paired ovaries are the gonads of females. They have a puckered, uneven surface; each weighs about 3 g. The ovaries resemble large almonds in size and shape. They are attached to ligaments in the pelvic cavity on each side of the uterus.

Embedded in a connective tissue matrix just below the outer layer of each ovary in a newborn baby girl are about half a million **ovarian follicles.** Each follicle contains an **oocyte,** an immature stage of the female sex cell.

By the time a girl reaches puberty, however, further development has resulted in the formation of a reduced number (about 400,000) of what are then called **primary follicles.** Each primary follicle has a layer of **granulosa cells** around the oocyte.

The progression of development from primary follicle to ovulation is shown in **Figure 23-10**. As the thickness of the granulosa cell layer around the oocyte increases, a hollow chamber called an **antrum** appears, and a *secondary follicle* is formed.

During the reproductive lifetime of most women, only about 350 to 500 of these primary follicles fully develop into

TABLE 23-2 Female Reproductive Organs

ESSENTIAL ORGANS	ACCESSORY ORGANS
Gonads: ovaries (right ovary and left ovary)	Ducts: uterine tubes (two), uterus, vagina Accessory sex glands: vestibular glands (two pairs), breasts (two) External genitals: vulva

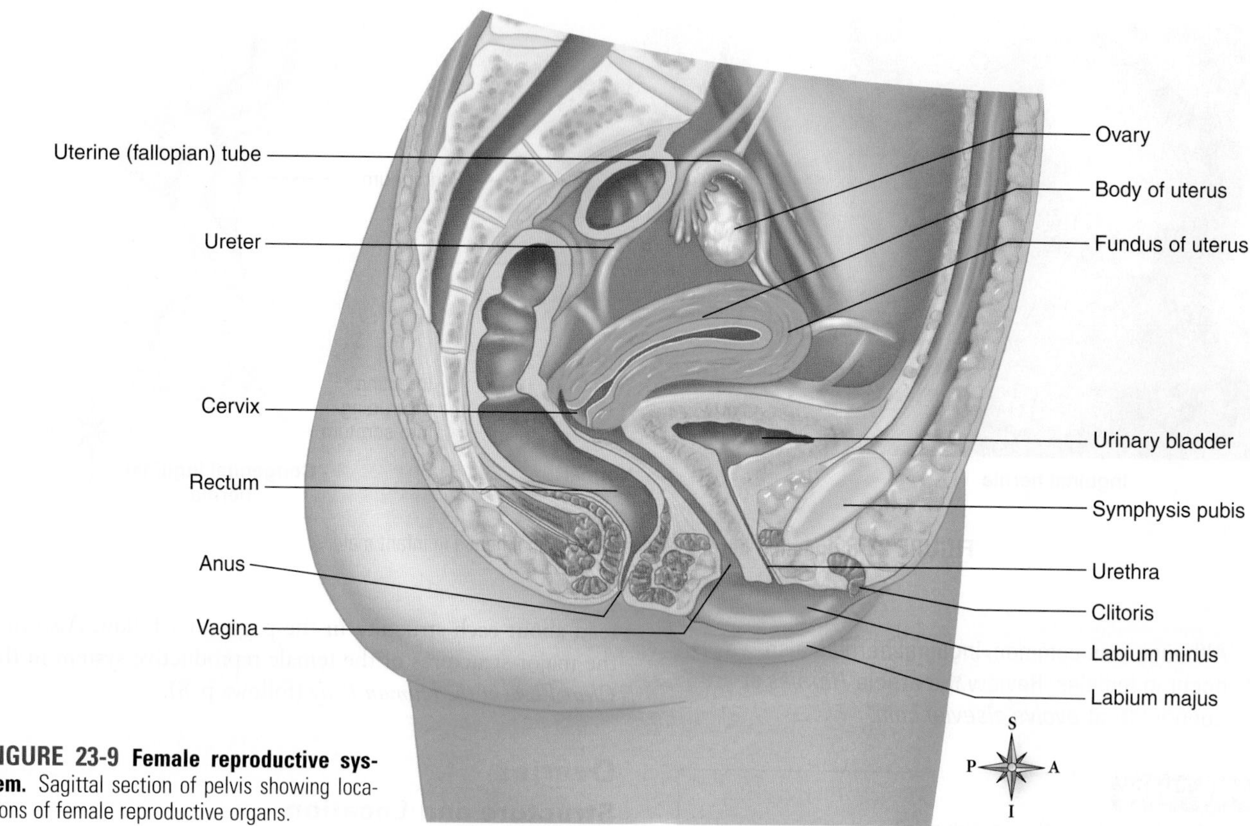

FIGURE 23-9 Female reproductive system. Sagittal section of pelvis showing locations of female reproductive organs.

mature follicles. It is the mature follicle that releases an ovum for potential fertilization—a process called **ovulation.** Follicles that do not mature degenerate and are reabsorbed into the ovarian tissue.

The sac containing a mature ovum is the *mature ovarian follicle*—often called a **graafian follicle,** in honor of the Dutch anatomist Regnier de Graaf who discovered it some 300 years ago.

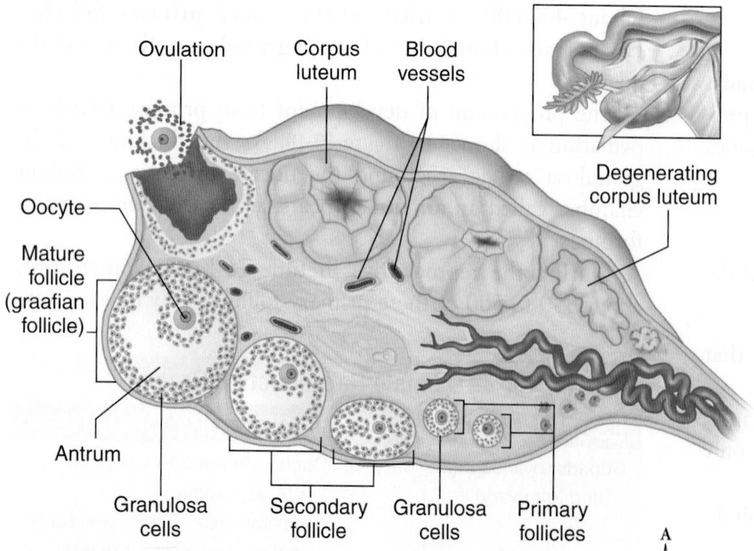

FIGURE 23-10 Ovary. Cross section of ovary shows successive stages of ovarian follicle and ovum development. Begin with the first stage (primary follicle) and follow around clockwise to the final state (degenerating corpus luteum).

After ovulation, the ruptured follicle is transformed into a hormone-secreting glandular structure called the **corpus luteum,** which is described later. *Corpus luteum* is a Latin phrase meaning "yellow body"—an appropriate name to describe the yellow appearance of this glandular structure.

Ovary Functions

Oogenesis

The production of female gametes, or sex cells, is called **oogenesis.**

The unusual form of cell division that results in sperm formation, meiosis, is also responsible for development of ova. During the developmental phases experienced by the female sex cell from its earliest stage to just after fertilization, two meiotic cell divisions occur.

As a result of meiosis in the female sex cell, the number of chromosomes is reduced equally in each daughter cell to half the number (23) found in other body cells (46).

However, the amount of cytoplasm is divided unequally among the daughter cells, as you can see in Figure 23-11. The result is formation of one large ovum and small daughter cells called **polar bodies** that degenerate. The ovum, with its large supply of cytoplasm, is one of the body's largest cells and is uniquely structured to provide nutrients for rapid development of the embryo until implantation in the uterus occurs.

At fertilization, the final phase of meiotic cell division in the ovum completes, and the last polar body is released.

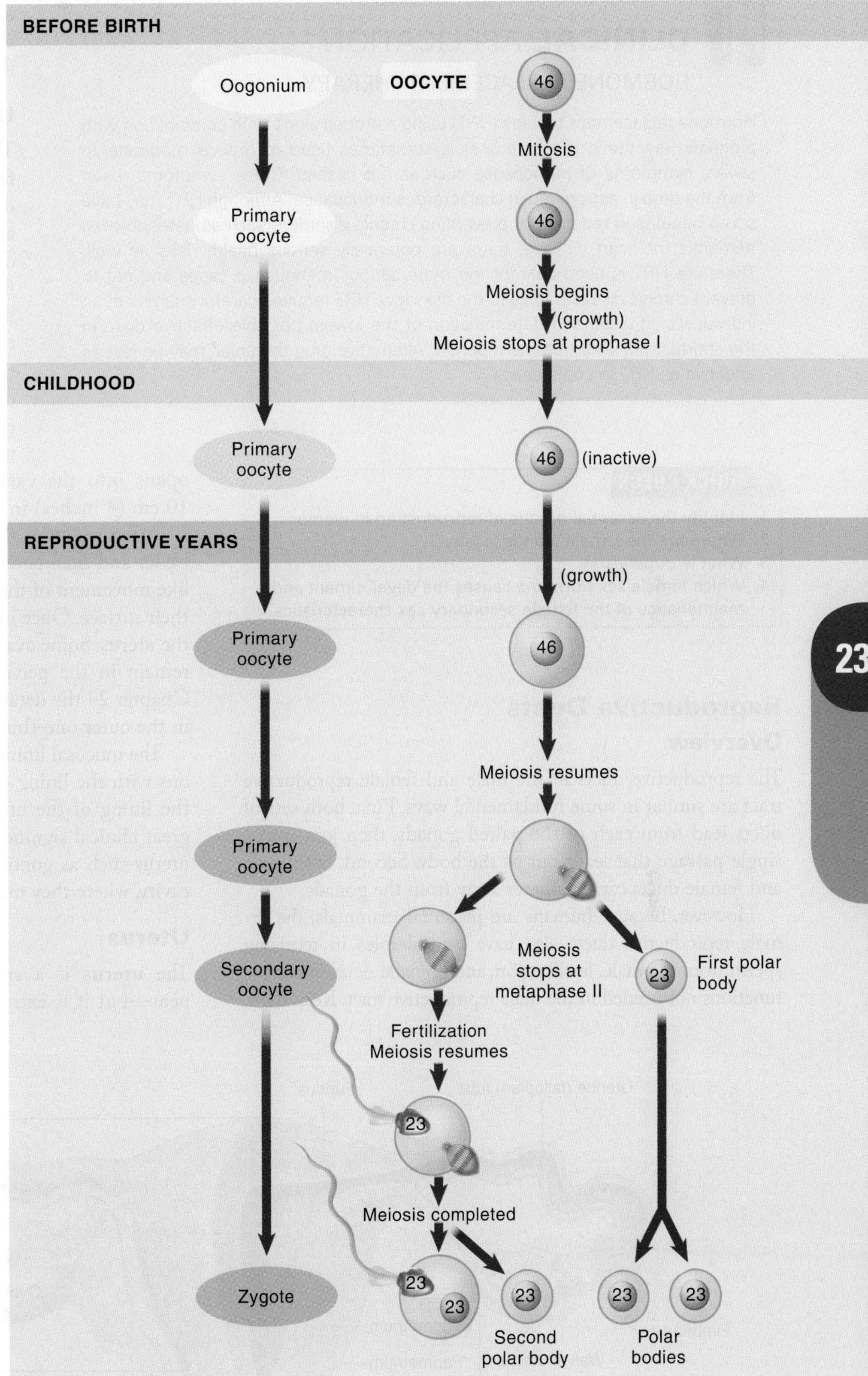

FIGURE 23-11 Oogenesis. Production of a mature ovum and subsequent fertilization are shown as a series of cell divisions. Notice that meiosis pauses in meiosis I before birth, then resumes in some primary oocytes beginning at puberty. Meiosis II does not complete until fertilization occurs.

The sex cells from both parents unite fully and the normal chromosome number (46) is achieved in the zygote that is formed.

Production of Estrogen and Progesterone

The second major function of the ovary, in addition to oogenesis, is secretion of the sex hormones, *estrogen* and *progesterone.* Hormone production in the ovary begins at puberty with the cyclic development and maturation of the ovum. The granulosa cells around the oocyte in the growing and mature follicle secrete estrogen. The corpus luteum, which develops after ovulation, chiefly secretes progesterone but also some estrogen.

Estrogen is the sex hormone that causes the development and maintenance of the female *secondary sex characteristics* and stimulates growth of the epithelial cells lining the uterus. Some of the actions of estrogen include the following:

1. Development and maturation of female reproductive organs, including the external genitals
2. Appearance of pubic hair and breast development
3. Development of female body contours by deposition of fat below the skin surface and in the breasts and hip region
4. Initiation of the first menstrual cycle

Progesterone is produced by the corpus luteum, which is a glandular structure that develops from a follicle that has just released an ovum. If stimulated by the appropriate anterior pituitary hormone, the corpus luteum produces progesterone for about 11 days after ovulation. Progesterone stimulates proliferation and vascularization of the epithelial lining of the uterus and acts with estrogen to initiate the menstrual cycle in girls entering puberty.

The surgical term **oophorectomy** is used to describe removal of the ovaries. If both ovaries are removed, sterility results and menopause follows.

CLINICAL APPLICATION

HORMONE REPLACEMENT THERAPY

Hormone replacement therapy (HRT) using estrogen alone or in combination with progestin (synthetic progesterone) is sometimes used to reduce moderate to severe symptoms of menopause such as hot flashes. These symptoms result from the drop in estrogen that characterizes menopause. Although HRT may have some benefits in reducing or preventing chronic disorders such as osteoporosis, dementia, or heart disease, there are potentially serious health risks as well. Therefore HRT is used only for the more serious menopause cases and not to prevent chronic disease. To keep the risks low, HRT requires careful analysis of an individual's situation and determination of the lowest possible effective dose in the shortest possible treatment period. Alternative drug therapies may be just as effective as HRT in some cases.

QUICK CHECK

1. Identify the essential organs of reproduction in women.
2. Where are the female glands located?
3. What is *oogenesis*?
4. Which female sex hormone causes the development and maintenance of the female secondary sex characteristics?

23

Reproductive Ducts

Overview

The reproductive ducts in the male and female reproductive tract are similar in some fundamental ways. First, both sets of ducts lead from each of the paired gonads, then join into a single passage that leads out of the body. Second, both male and female ducts carry gametes away from the gonads.

However, because humans are placental mammals, the female reproductive ducts also have central roles in receiving sperm from the male, fertilization, and prenatal development—functions not needed in the male reproductive tract. Keep these unique reproductive functions in mind as we now learn how *structure fits function* in the female reproductive tract.

Uterine Tubes

The two **uterine tubes,** also called **fallopian tubes** or **oviducts,** serve as ducts for the ovaries, even though they are not attached to them. The outer end of each tube terminates in an expanded, funnel-shaped structure that has fringe-like projections called **fimbriae** along its edge. This part of the tube curves over the top of each ovary (**Figure 23-12**) and opens into the pelvic cavity. The inner end of each uterine tube attaches to the uterus, and the cavity inside the tube opens into the cavity in the uterus. Each tube is about 10 cm (4 inches) in length.

After ovulation, the discharged ovum first enters the pelvic cavity and then enters the uterine tube assisted by the wave-like movement of the fimbriae and the beating of the cilia on their surface. Once in the tube, the ovum begins its journey to the uterus. Some ova never find their way into the oviduct and remain in the pelvic cavity where they are reabsorbed. In Chapter 24 the details of fertilization, which normally occurs in the outer one-third of the uterine tube, are discussed.

The mucosal lining of the uterine tubes is directly continuous with the lining of the pelvic cavity on one end and with the lining of the uterus and vagina on the other. This is of great clinical significance because infections of the vagina or uterus such as gonorrhea may pass into the abdominopelvic cavity, where they may become life threatening.

Uterus

The **uterus** is a small organ—only about the size of a pear—but it is extremely strong. It is almost all muscle, or

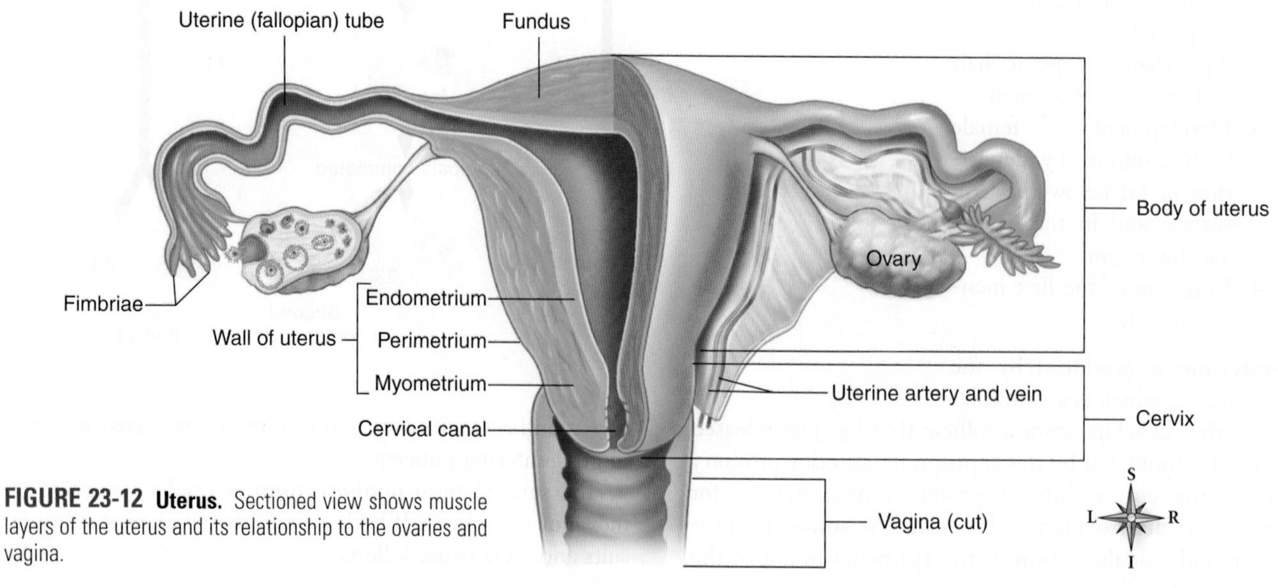

FIGURE 23-12 Uterus. Sectioned view shows muscle layers of the uterus and its relationship to the ovaries and vagina.

myometrium, with only a small cavity inside. During pregnancy the uterus grows many times larger so that it becomes big enough to hold a full-term fetus and a considerable amount of fluid.

The uterus is composed of several major regions. The upper portion of the uterus is the **body.** Just above the level where the uterine tubes attach to the body of the uterus, it rounds out to form a bulging prominence called the **fundus** (see Figure 23-12). The lower, narrow neck section is called the **cervix.**

Except during pregnancy, the uterus lies in the pelvic cavity just behind the urinary bladder. By the end of pregnancy, it has become large enough to extend up to the top of the abdominopelvic cavity. It then pushes the liver against the underside of the diaphragm—a fact that explains a comment such as "I can't seem to take a deep breath since I've gotten so big," made by many women late in their pregnancies.

Hysterectomy is surgical removal of the uterus. It may be excised and removed through a typical incision in the abdomen *(abdominal hysterectomy),* through the vagina *(vaginal hysterectomy),* or laparoscopically *(laparoscopic hysterectomy).* In *total hysterectomy* both the body and cervix are removed; in *subtotal hysterectomy* only the body of the uterus is removed, sparing the cervix.

The uterus functions in three processes—menstruation, pregnancy, and labor. The corpus luteum stops secreting progesterone and decreases its secretion of estrogens about 11 days after ovulation. About 3 days later, when the progesterone and estrogen concentrations in the blood are at their lowest, menstruation starts. Small pieces of the mucous membrane lining of the uterus, or the **endometrium** pull loose, leaving torn blood vessels underneath. Blood and bits of endometrium trickle out of the uterus into the vagina and out of the body.

Immediately after menstruation, the endometrium starts to repair itself. It again grows thick and becomes lavishly supplied with blood in preparation for pregnancy.

If fertilization does not take place, the uterus once more sheds the lining made ready for a pregnancy that did not occur. Because these changes in the uterine lining continue to repeat themselves, they are spoken of as the **menstrual cycle** (see pp. 633-635).

If fertilization occurs, pregnancy begins, and the endometrium remains intact. The events of pregnancy are discussed in Chapter 24.

Menstruation first occurs during puberty, often around the age of 12 to 13 years but sometimes even earlier. Normally it repeats itself about every 28 days or 13 times a year for some 30 to 40 years before it ceases at the time of **menopause,** when a woman is somewhere around the age of 50 years.

Vagina

The **vagina** is a distensible tube about 10 cm (4 inches) long made mainly of smooth muscle and lined with mucous membrane. It lies in the pelvic cavity between the urinary bladder and the rectum, as you can see in Figure 23-9. As the part of the female reproductive tract that opens to the exterior, the vagina is the organ that receives the penis during intercourse and through which sperm enter during their journey to meet an ovum.

The vagina is also the organ from which a baby emerges to meet its new world, and so it is also called the *birth canal.*

23

Accessory Glands

Vestibular Glands

Two pairs of exocrine glands lie imbedded in tissue to the left and right of the vaginal outlet and release mucous fluid into the vestibule of the *vulva* (described later in Figure 23-14).

One pair of these small glands are called the **greater vestibular glands,** and the other pair are called the **lesser vestibular glands.** The greater vestibular glands are also called **Bartholin glands,** and the lesser vestibular glands may be called **Skene glands** or *female prostate.*

Mucus from these glands may contribute to lubrication during sexual intercourse.

The vestibular glands have clinical importance because they may become infected. For example, *Neisseria gonorrhoeae*—the bacteria that cause gonorrhea—are often hard to eliminate once they infect a vestibular gland (see Table 23-4).

CLINICAL APPLICATION

ECTOPIC PREGNANCY

The term **ectopic pregnancy** is used to describe a pregnancy resulting from the implantation of a fertilized ovum in any location other than the uterus. Occasionally, because the outer ends of the uterine tubes open into the pelvic cavity and are not actually connected to the ovaries, an ovum does not enter an oviduct but becomes fertilized and remains in the pelvic cavity.

Although rare, if implantation occurs on the surface of an abdominal organ or on one of the mesenteries, development may continue to term. In such cases, delivery by cesarean section is required. Most ectopic pregnancies involve implantation in the uterine tube and are therefore called *tubal pregnancies.* If a tubal pregnancy is not terminated, catastrophic rupture of the uterine tube and death of both fetus and mother is likely to occur.

Breasts

The **breasts** lie over the pectoral muscles and are attached to them by *fibrous suspensory ligaments* (of Cooper). Breast size is determined more by the amount of fat around the glandular (milk-secreting) tissue than by the amount of glandular tissue itself. Hence the size of the breast has little to do with its ability to secrete adequate amounts of milk after the birth of a baby.

Each breast consists of 15 to 20 divisions or lobes that are arranged radially (Figure 23-13). Each lobe consists of several lobules, and each lobule consists of milk-secreting glandular cells. The milk-secreting cells are arranged in grapelike clusters

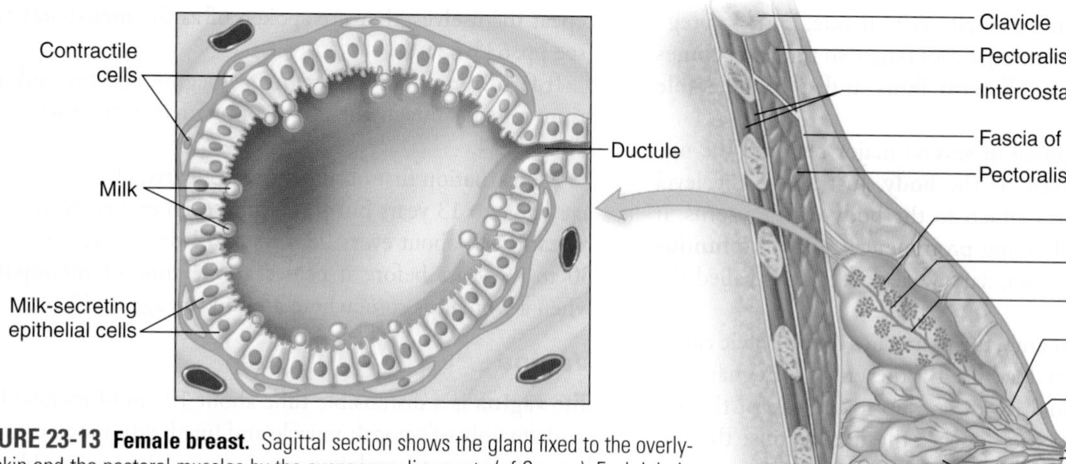

FIGURE 23-13 Female breast. Sagittal section shows the gland fixed to the overlying skin and the pectoral muscles by the suspensory ligaments (of Cooper). Each lobule of secretory tissue is drained by a lactiferous duct that opens through the nipple. The inset *(left)* shows one of the milk-producing alveoli of the mammary gland.

of small hollow chambers called *alveoli* (see **Figure 23-13**, *inset*). Small contractile cells surround the alveoli and push milk into ducts when stimulated by *oxytocin (OT)* released from the posterior pituitary gland—an event called "milk let-down."

Small **lactiferous ducts** drain the alveoli and converge toward the nipple like the spokes of a wheel. Only one lactiferous duct leads from each lobe to an opening in the *nipple.* Each lactiferous duct widens into a *lactiferous sinus* just before reaching the nipple. Each sinus acts like the bulb at the end of an eyedropper, pumping milk out of the nipple as an infant rhythmically squeezes its jaws as it nurses.

The colored surface area around the nipple is the **areola.** It contains many tiny bumps called *areolar glands.* Areolar glands are large sebaceous glands that secrete skin oils that condition the skin while nursing an infant. The areola also has a network of smooth muscles that contract to cause the nipple to become erect—which often helps an infant latch on to the breast at the most efficient location.

CLINICAL APPLICATION

FIBROCYSTIC DISEASE

The terms **fibrocystic disease** and **mammary dysplasia** are just two of the many names for a group of conditions characterized by benign lumps in one or both breasts. It is common in adult women before menopause, occurring in half of all women at some time, and is considered the most frequent breast lesion.

The lumps that characterize fibrocystic disease are often painful, especially during the secretory phase of the reproductive cycle. Treatment is usually aimed at relieving pain or tenderness that may occur. Although it is commonly called a *disease,* most experts agree that fibrocystic disease is simply a collection of normal variations in breast tissue. Even though the lumps associated with fibrocystic disease are benign, any suspicious lump or other change in breast tissue should be regarded as possibly cancerous until determined otherwise by a physician.

Cancerous cells from breast tumors often spread to other areas of the body through the lymphatic system. This lymphatic drainage is discussed in Chapter 16 (see also **Figure 16-8**).

Females and males both have breasts—and either can get breast cancer. For more information on male and female breast health, check out the articles *Male Breast Health* and *Breast Self-Examination* at Connect It! at *evolve.elsevier.com.*

To learn more about breast structure, go to AnimationDirect online at *evolve.elsevier.com.*

QUICK CHECK

1. What is another name for the *uterine tubes?*
2. What three major functions does the *uterus* perform?
3. What substance is conducted through *lactiferous ducts?*
4. Describe the function of the *areolar glands.*

External Genitals

The external **genitalia** of women consist of several structures collectively called the **vulva.** These include:

1. Mons pubis
2. Clitoris
3. External urinary meatus
4. Labia minora
5. Hymen
6. Openings of vestibular gland ducts
7. Orifice (opening) of vagina
8. Labia majora

The **mons pubis** is a skin-covered pad of fat over the symphysis pubis. Pubic hair appears on this mound of fat at puberty and persists throughout life.

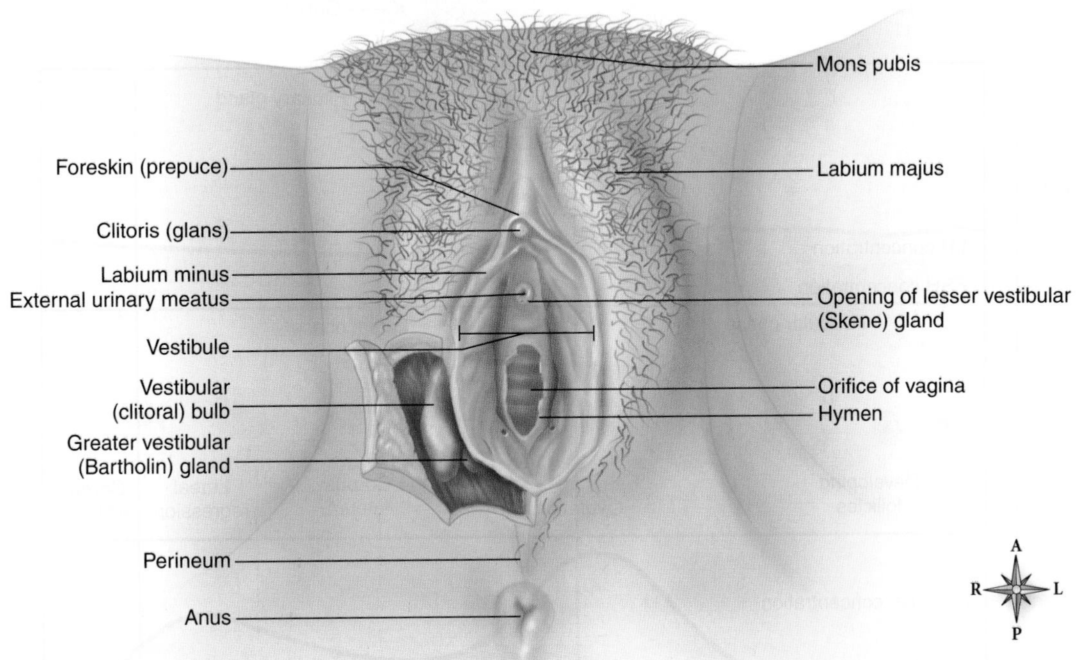

FIGURE 23-14 Vulva. External female genitals and related structures, shown from an inferior view.

Extending downward from the elevated mons pubis are the **labia majora,** literally "large lips." These elongated folds, which are composed mainly of fat and glands, are covered with pigmented skin and pubic hair on the outer surface and are smooth and free from hair on the inner surface. The **labia minora**—literally "small lips"—are nestled medially between the labia majora and are covered with thin skin. These two small lips join anteriorly at the midline.

The space between the labia minora is the **vestibule** (Figure 23-14). Several genital structures are located in the vestibule. The *glans* or head of the **clitoris,** which is composed of erectile tissue similar to that found in the penis, is located just behind the anterior junction of the labia minora. The deeper erectile tissue of the clitoris branches into two bulbs, one of which can be seen under a labium majus in the cut-away on right side of the specimen in Figure 23-14.

Situated between the glans clitoris and the vaginal opening is the *external urinary meatus.*

The vaginal orifice is bordered by a thin fold of mucous membrane called the **hymen.** Occasionally, the hymen partially blocks the vaginal opening. The ducts of the vestibular glands open on either side of the vaginal orifice, medial to the labia minora.

The term **perineum** is used to describe the area between the vaginal opening and anus. This area is sometimes cut in a surgical procedure called an **episiotomy** to prevent tearing of tissue during childbirth.

Menstrual Cycle

Overview

The menstrual cycle consists of many changes in the uterus, ovaries, and breasts and in the hypothalamus and anterior pituitary gland's secretion of hormones (Figure 23-15). In the majority of women, these changes occur with almost precise regularity throughout their reproductive years. The first indication of changes comes with the first menstrual period. The first *menses* or menstrual flow is referred to as the **menarche.**

A typical menstrual cycle covers a period of about 28 days. However, the length of the cycle varies among women. Some women, for example, may have a regular cycle that covers about 24 days. The length of the cycle also varies within one woman. Some women, for example, may have irregular cycles that range from 21 to 28 days, whereas others may be 2 to 3 months long.

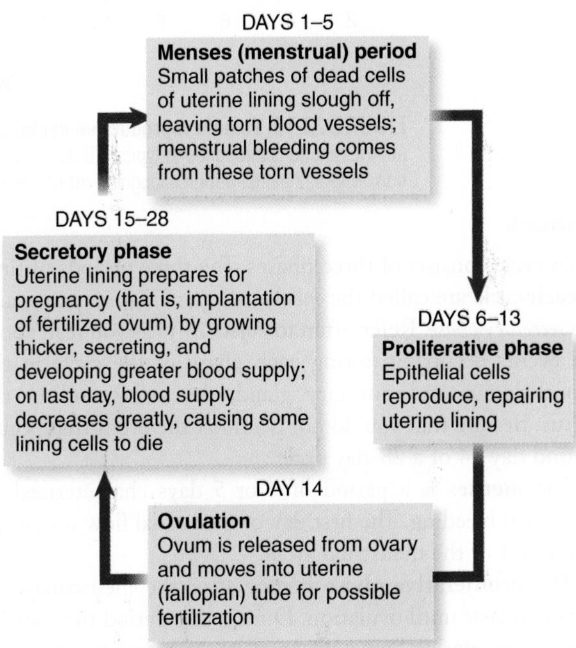

FIGURE 23-15 28-day menstrual cycle.

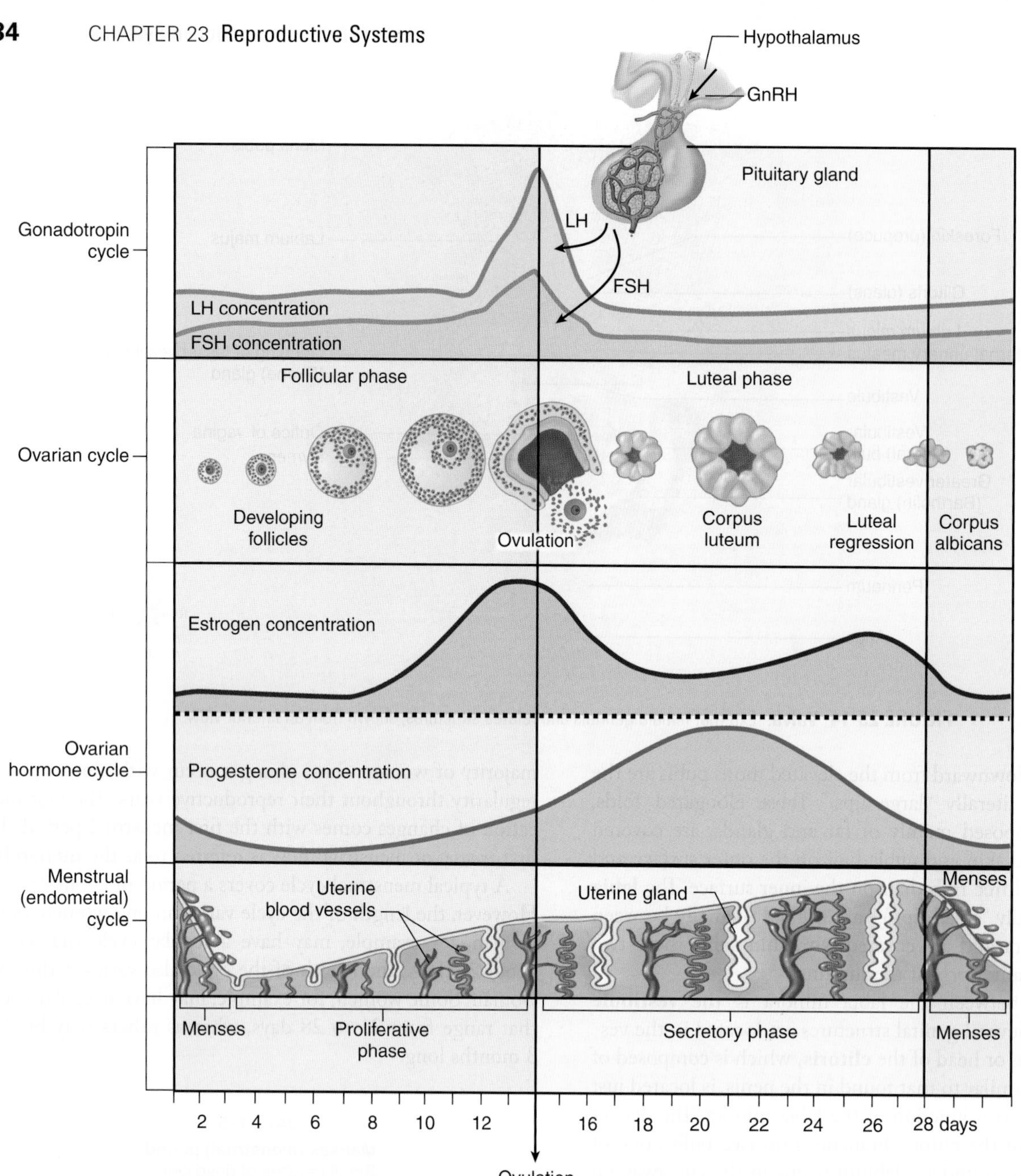

FIGURE 23-16 Female reproductive cycle. Diagram illustrates the interrelationship of pituitary, ovarian, and uterine functions throughout a typical 28-day cycle. A sharp increase in luteinizing hormone (LH) levels causes ovulation, whereas menstruation (sloughing off of the endometrial lining) is initiated by lower levels of progesterone.

Phases

Each cycle consists of three phases. The three periods of time in each cycle are called the *menses,* the *proliferative phase,* and the *secretory phase.* Refer often to **Figure 23-16** as you read about the events occurring during each phase of the cycle in the hypothalamus and pituitary gland, the ovary, and in the uterus. Be sure that you do not overlook the event that occurs around day 14 of a 28-day cycle.

The **menses** is a period of 4 or 5 days characterized by menstrual bleeding. The first day of menstrual flow is considered day 1 of the menstrual cycle.

The **proliferative phase** begins after the menstrual flow ends and lasts until ovulation. During this period the follicles mature, the uterine lining thickens (proliferates), and estrogen secretion increases to its highest level.

The **secretory phase** of the menstrual cycle begins at ovulation and lasts until the next menses begins. It is during this phase of the menstrual cycle that the uterine lining reaches its greatest thickness and the ovary secretes its highest levels of progesterone.

Ovulation

As a general rule, during the 30 or 40 years that a woman has periods, only one ovum matures each cycle. However, there are exceptions to this rule. Some cycles, more than one matures, and some cycles no ovum matures.

Ovulation occurs 14 days before the next menses begins. In a 28-day cycle, this means that ovulation occurs around day 14 of the cycle, as shown in **Figure 23-16.** (Recall that the first day of the menses is considered the first day of the cycle.) In a

30-day cycle, however, ovulation would not occur on the 14th cycle day, but instead on the 16th. And in a 25-day cycle, ovulation would occur on the 11th cycle day.

The time of ovulation has great practical importance because the possibility of fertilization—the fusion of a sperm and egg—can occur only during a short period of time during each menstrual cycle. Although a few "super" sperm may remain viable for up to 5 days, most sperm retain their fertilizing power for only 24 to 72 hours after being deposited in the female reproductive tract following ejaculation. And, the oocyte remains viable and capable of being fertilized for only about 12 to 24 hours after ovulation. A woman's fertile period therefore lasts only a few days each cycle—from between 3 to 5 days before, and no later than 24 hours after, ovulation.

To learn more about ovulation, go to AnimationDirect online at *evolve.elsevier.com*.

Control of the Menstrual Cycle

The anterior pituitary gland plays a critical role in regulating the cyclic changes that characterize the functions of the female reproductive system (see Chapter 12). As noted earlier, secretion of GnRH from the hypothalamus stimulates the anterior pituitary gland to secrete the gonadotropins FSH and LH. From day 1 to about day 7 of the menstrual cycle, GnRH selectively stimulates the anterior pituitary gland to secrete increasing amounts of FSH. A high blood concentration of FSH stimulates several immature ovarian follicles to start growing and secreting estrogen (see **Figure 23-16**).

Working together, increasing levels of estrogen and GnRH in blood stimulate the anterior pituitary gland to release increasing amounts of LH. LH causes maturing of a follicle and its ovum, ovulation (rupturing of mature follicle with ejection of ovum), and luteinization (formation of a yellow body, the corpus luteum, from the ruptured follicle).

Which hormone—FSH or LH—would you call the "ovulating hormone"? Do you think ovulation could occur if the blood concentration of FSH remained low throughout the menstrual cycle? If you answered LH to the first question and no to the second, you answered both questions correctly. Ovulation cannot occur if the blood level of FSH stays low because a high concentration of this hormone is essential to stimulation of ovarian follicle growth and maturation. With a low level of FSH, no follicles start to grow, and therefore none become ripe enough to ovulate. Ovulation is caused by the combined actions of FSH and LH. Birth control pills that contain estrogen substances suppress FSH secretion. This indirectly prevents ovulation.

Ovulation occurs, as we have said, because of the combined actions of the two anterior pituitary hormones, FSH and LH. The next question is: what causes menstruation? A brief answer is this: a sudden, sharp decrease in estrogen and progesterone secretion toward the end of the secretory phase causes the uterine lining to break down and another menstrual period to begin.

Disorders of the Female Reproductive System

Hormonal and Menstrual Disorders

Dysmenorrhea

Menstrual cramps, or **dysmenorrhea,** are terms used to describe the cramping, painful periods that affect 75% to 80% of women at some time during their reproductive years. For significant numbers of those affected, severe lower abdominal cramping and back pain accompanied by headache, nausea, and vomiting will disrupt their school, work, athletic, or other activities.

Primary dysmenorrhea is the most common type occurring in adolescents and young women. Symptoms, which can last from hours to days and vary in severity from cycle to cycle, are caused by overproduction of prostaglandins in the inner lining of the uterus. Prostaglandins cause spasms that decrease blood flow and oxygen delivery to uterine muscle resulting in pain. Fortunately, primary dysmenorrhea is not associated with pelvic disease, such as an infection or tumor, and generally can be treated effectively with over-the-counter anti-inflammatory drugs such as ibuprofen and naproxen, which decrease prostaglandin production. In severe cases a physician may prescribe more powerful anti-inflammatory drugs or certain hormones, including oral contraceptives, to alter menstrual cycle activity.

Secondary dysmenorrhea refers to menstrual-related pain caused by some type of pelvic pathology. The problem is generally a gynecological problem affecting one or more reproductive organ. Treatment of secondary dysmenorrhea involves treating the underlying disorder.

Amenorrhea

Amenorrhea is the absence of normal menstruation.

Primary amenorrhea is the failure of menstrual cycles to begin and may be caused by a variety of factors, such as hormone imbalances, genetic disorders, brain lesions, or structural deformities of the reproductive organs.

Secondary amenorrhea occurs when a woman who has previously menstruated slows to three or fewer cycles per year. Secondary amenorrhea may occur with weight loss, pregnancy, lactation, menopause, or disease of the reproductive organs.

Treatment of amenorrhea involves treating the underlying disorder or condition.

Dysfunctional Uterine Bleeding

Dysfunctional uterine bleeding (DUB) is irregular or excessive uterine bleeding that most often results from either a hormonal imbalance or some type of structural problem that causes a disruption of blood supply. DUB is a significant medical problem affecting nearly 2 million women in the United States each year. Excessive uterine bleeding over time can result in life-threatening anemia because of the chronic loss of blood.

CLINICAL APPLICATION

AMENORRHEA IN FEMALE ATHLETES

Failure to have a menstrual period is called **amenorrhea.** Amenorrhea occurs in some female athletes, probably resulting from a body fat composition that is too low to sustain normal reproductive function. Although it keeps the hematocrit (red blood cell level) higher than during menstruation, it is not considered a desirable condition. Besides infertility, amenorrhea may cause other problems. For example, the low blood levels of estrogen associated with long-term amenorrhea may cause osteoporosis (loss of bone mass).

To diagnose the cause of DUB, a physician may employ x-ray or ultrasound studies to look at the contours of the uterine cavity, look directly inside the uterus using a telescope-like instrument inserted through the vagina and cervix or examine tissue obtained by biopsy to exclude cancer.

If hormonal imbalance is the cause, it is the excessive growth and breakdown of delicate endometrial tissue that results in heavy bleeding. Structural problems, such as growth of a uterine malignancy, polyps, or fibroids (discussed on pp. 637–638), also may cause DUB by causing injury to the blood vessels of the uterine wall or lining.

Treatment of DUB generally begins with administration of nonsteroidal anti-inflammatory drugs and hormonal manipulation using low-dose birth control pills. If conservative treatment fails to stop the endometrial lining from hemorrhaging, hysterectomy remains one of the most effective curative options. Currently, about 20% of hysterectomies performed each year are for treatment of abnormal uterine bleeding.

Less invasive procedures, including **endometrial ablation** techniques, are now being used more frequently to destroy the endometrial lining and halt the bleeding. In *thermal ablation,* a balloon is inserted into the uterus and filled with fluid. A heat probe is then inserted into the balloon and the fluid is heated to a temperature that will destroy the endometrium. In *radiofrequency ablation,* a gold-plated mesh fabric is used to fill the uterine cavity and is then charged with radiofrequency energy that destroys the damaged and bleeding endometrial cells. Both procedures carry less risk and have shorter recovery periods than does a hysterectomy.

Premenstrual Syndrome

Premenstrual syndrome (PMS) is a condition that involves a collection of symptoms that regularly occur in some women during the secretory phase of their reproductive cycles. Symptoms include irritability, fatigue, nervousness, depression, and other problems that are often distressing enough to affect personal relationships. Because the cause of PMS is still unclear, current treatments focus on relieving the symptoms.

Infection and Inflammation

Infections

Infections of the female reproductive tract are often classified as *exogenous* or *endogenous.* Exogenous infections result from pathogenic organisms transmitted from another person, such as **sexually transmitted diseases (STDs).**

Endogenous infections result from pathogens that normally inhabit the intestines, vulva, or vagina. You may recall from Chapter 6 that many areas of the body are normally inhabited by pathogenic microbes but that they cause infection only when there is a change in conditions, or they are moved to a new area.

Pelvic Inflammatory Disease

Pelvic inflammatory disease (PID) occurs as either an acute or chronic inflammatory condition that can be caused by several different pathogens, which usually spread upward from the vagina. PID is a major cause of infertility and sterility and affects more than 800,000 women each year in the United States. It is a common complication following infection by gonococcal *(Neisseria gonorrhoeae)* and chlamydial microorganisms (see **Table 23-4**).

In PID any inflammation involving the uterus, uterine tubes, ovaries, and other pelvic organs often results in development of scar tissue and adhesions. As a result, serious complications, including infertility resulting from tubal obstruction or other damage to the reproductive tract, may occur. Uterine tube inflammation is termed **salpingitis** and inflammation of the ovaries is called **oophoritis.**

Laparoscopic examination is often used to make a definitive diagnosis or to determine the severity of the infection and the reproductive organs involved. Although some chlamydial infections may not cause symptoms, most cases of PID are accompanied by fever, pelvic tenderness, and pain. Unfortunately, because of scarring and adhesions, pain may continue even after antibiotic therapy has eliminated the active infection.

If left untreated, PID infections may spread to other tissues, including the blood, resulting in septic shock and death.

Vaginitis

Vaginitis is inflammation or infection of the vaginal lining.

Vaginitis most often results from STDs or from a yeast infection. Yeast infections are usually opportunistic infections of the fungus *Candida albicans,* producing vaginal **candidiasis** (see Appendix A at *evolve.elsevier.com*). Candidiasis infections are characterized by a whitish discharge—a symptom known as **leukorrhea.**

Tumors and Related Conditions

Benign Uterine Tumors

The terms **fibroid, myoma,** and **fibromyoma** are all words used to describe benign (noncancerous) tumors of uterine fibrous or smooth muscle tissue.

Individual fibroids may occur but multiple growths are not unusual. Fibroids are common in women during their reproductive years and develop most often in the myometrium of the uterine body and rarely in the cervix. The fact that they are seldom seen before puberty, increase in size during pregnancy, and tend to shrink in postmenopausal women suggests that age and estrogen levels may play a role in their development.

Fibroids range in size from small asymptomatic nodules to massive tumors that may be painful and exert pressure on other pelvic organs. Growth during pregnancy may result in placental hemorrhage or malpresentation of the fetus, complicating labor and delivery.

In addition to pain, symptoms of benign uterine tumors will vary depending on the size and location of the tumor. For example, if a large fibroid compresses the bladder and rectum, symptoms of urinary frequency and constipation may result. Even small tumors developing beneath the endometrium can cause severe hemorrhage (DUB).

Tumor size, location, and severity of symptoms determine treatment options. A relatively new technique, similar to a heart catheterization, called *uterine artery embolization* involves snaking a small catheter through an artery in the groin into the arterial vessel supplying blood to a fibroid. Tiny inert pellets are then injected into the artery, blocking the flow of blood. The procedure results in dramatic shrinkage of the treated fibroid and a reduction in symptoms, including hemorrhage.

Surgical removal of individual fibroids or, in more severe cases, hysterectomy may be indicated.

Ovarian Cysts

Ovarian cysts are very common fluid-filled cysts that develop either from follicles that fail to rupture completely *(follicular cysts)* or from corpora lutea that fail to degenerate *(luteal cysts).*

Most women develop a number of these cysts during their reproductive years and their presence does not represent a diagnosis of polycystic ovary syndrome. Although ovarian cysts are often multiple, they rarely become dangerous. However, on occasion they may become quite large and painful and be diagnosed by palpation or ultrasonography.

Luteal cysts are less common than follicular cysts but tend to cause more symptoms, such as pelvic pain and menstrual irregularities. Rarely, rupture of a large luteal cyst will result in internal bleeding that requires surgical intervention. The vast majority of all ovarian cysts will disappear within a few months of their appearance, most within 60 days.

Polycystic ovary syndrome (PCOS) is a condition that affects 10% of reproductive-age women but also can affect girls as young as 11 years old. It is characterized by enlarged ovaries that usually are studded with fluid-filled cysts about 0.5 to 1.5 cm in diameter (Figure 23-17). The cysts are found on both ovaries and develop from mature follicles that fail to rupture completely. Corpora lutea are generally absent.

Women with PCOS frequently have numerous endocrine abnormalities, including high levels of testosterone, infrequent menstrual cycles, and persistent anovulation. PCOS is the most common cause of female infertility.

Endometriosis

Endometriosis is the presence of functioning endometrial tissue outside the uterus. The displaced endometrial tissue can occur in many different places throughout the body but is most often found in or on pelvic and abdominal organs. The tissue reacts to ovarian hormones in the same way as the normal endometrium—exhibiting a cycle of growth and sloughing off.

Symptoms of endometriosis may include unusual bleeding, dysmenorrhea, and pain during intercourse. If symptoms are mild, pain medications are sometimes effective. Oral contraceptives, which alter the hormone levels that produce endometrial changes during the menstrual cycle, are effective in reducing the activity of endometriosis.

Cancer

Malignancies of reproductive and related organs, especially the breasts, account for the majority of cancer cases among women.

Breast Cancer

About 1 in 8 women eventually get breast cancer, often a form of adenocarcinoma. Treatment of breast cancer is often successful if the cancerous tumor is detected early. Because such tumors are often painless, most physicians recommend regular, frequent self-examination of breast tissue, as well as annual mammograms for women (see Chapter 6). Treatments often involve surgery, chemotherapy, and radiation therapy.

Breast surgeries can be very conservative, as in a simple lump removal or *lumpectomy.* If metastasis to surrounding tissue is suspected, a modified **radical mastectomy** may be performed. In this procedure the entire breast, with nearby lymph nodes, is removed.

FIGURE 23-17 Polycystic ovary syndrome (PCOS). The ovary is studded with fluid-filled cysts developed from follicles that have failed to rupture.

Just as lumpectomy results in less trauma than modified radical mastectomy, so called *limited-field radiation* can provide effective treatment for clearly defined early-stage cancers that have not spread. It does so with shorter treatment cycles and fewer side effects than whole-breast radiation.

In the past, after women had completed their initial treatment for breast cancer they had few options available to lessen the possibility of recurrence. For a number of years the drug *tamoxifen* has been used extensively to prevent the recurrence of breast cancer fueled by estrogen. It does so by blocking the estrogen receptor sites on the cancer cell membrane. Unfortunately, tamoxifen effectiveness is limited to about 5 years.

Newer drugs classified as *aromatase inhibitors* actually block estrogen production, instead of blocking estrogen receptors. This type of drug may replace tamoxifen or be prescribed for use after 5 years of tamoxifen therapy. Other "rational" drugs and other treatments are being developed to alter or block crucial metabolic pathways in treating breast and other forms of cancer.

Ovarian Cancer

Ovarian cancer is another malignancy that affects 1 in 70 women in America. Usually a type of adenocarcinoma, ovarian cancer is difficult to detect early and is often not easily apparent until it has grown into a large mass. Regular pelvic examinations that include palpation of the ovaries may result in earlier detection.

Risk factors for ovarian cancer include age (over 40), infertility, childlessness or few children, a history of miscarriages, and endometriosis.

Ovarian cancer is often treated by surgical removal of the ovaries combined with radiation therapy and chemotherapy.

Uterine Cancer

Cancer of the uterus can affect the body of the uterus or the cervix.

Cancers of the uterine body most often involve the endometrium *(endometrial cancer)* and mostly affect women beyond childbearing years; a common symptom is postmenopausal uterine bleeding. Risk factors for this type of cancer include obesity, prolonged estrogen therapy, and infertility.

Cervical cancer occurs most often in women between the ages of 30 and 50.

Cervical cancer is often diagnosed early, through screening tests such as the **Papanicolaou test,** or *Pap smear* (**Figure 23-18**). In this test, cells swabbed from the cervix are smeared on a glass slide, stained, and examined microscopically to determine whether any abnormalities exist. Current recommendations suggest two Pap smears 1 year apart beginning at age 21. If these two Pap smears are negative (that is, revealing no abnormalities), subsequent Pap smears should occur every 1 to 3 years thereafter.

Because early or frequent intercourse is a risk factor for cervical cancer, sexually active young women should have their first Pap smear much earlier—and have follow-ups done more often.

Because screening tests and other early detection methods have been so successful, the death rates for uterine cancers have dropped dramatically over the last few decades. Because *human papillomavirus (HPV)* infections dramatically increase the risk of developing cervical cancer, widespread use of HPV vaccines in both men and women have already begun to reduce the spread of HPV—thus reducing death rates from this type of cancer.

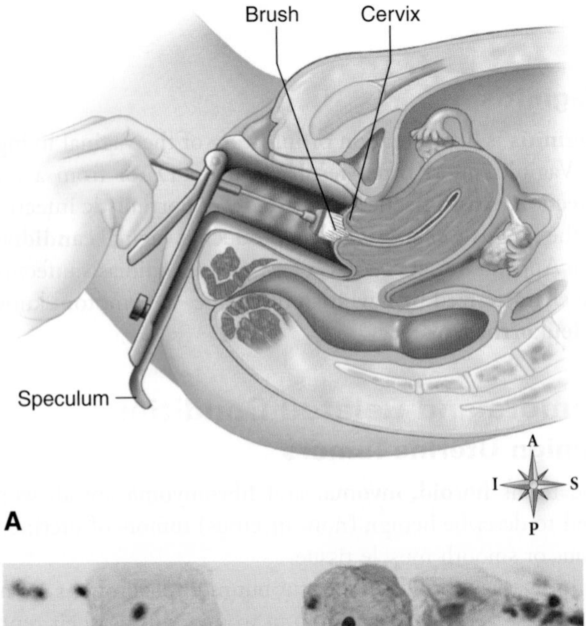

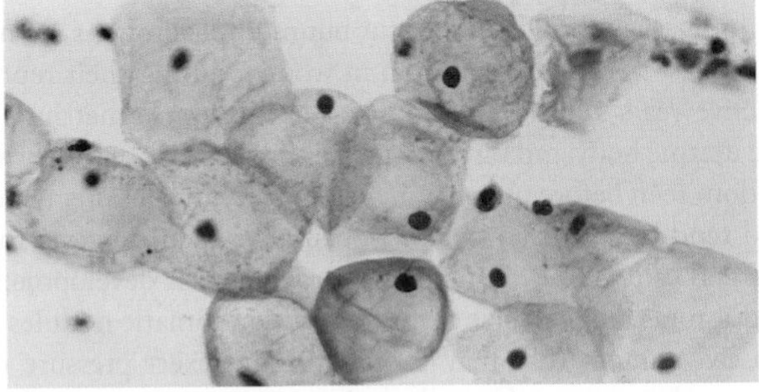

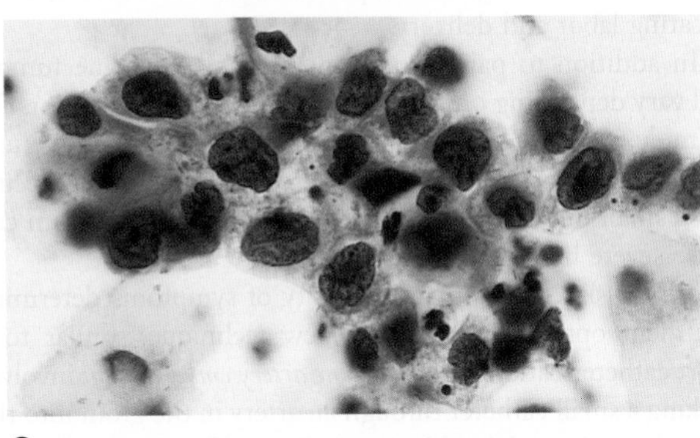

FIGURE 23-18 Papanicolaou (Pap) smear. A, Obtaining a Pap smear. **B,** Appearance of normal cervical epithelial cells in Pap smear. **C,** Appearance of cervical cancer cells in Pap smear. Note the reduction in cytoplasm and increased prominence of the nuclei compared with normal epithelial cells.

Infertility

Like in the male reproductive system, various disorders can disrupt normal function of the female reproductive tract so that successful reproduction is unlikely (infertility) or impossible (sterility). Infections, tumors, hormonal imbalances, and other factors can contribute to infertility or sterility in women.

SCIENCE APPLICATIONS
REPRODUCTIVE SCIENCES

William Masters (1915–2001)

The study of human reproduction, and especially sexual function, has many cultural implications. So it is no wonder that American researchers William Masters and Virginia Johnson encountered a great deal of controversy during their decades of pioneering work in the field of human sex and reproduction. They were the first to study human sexual physiology in the laboratory.

Dr. William Masters was a **gynecologist** (physician specializing in women's health) and Virginia Johnson was trained in psychology. In 1966, their book *Human Sexual Response* clearly explained the physiology of sex for the first time. Besides making discoveries in the physiology of human sex and reproduction, they also developed therapies for treating sex-related conditions, and they trained *therapists* from around the world.

In addition to the broad fields of *biology, medicine, psychology,* and the *behavioral sciences,* the pioneering work of Masters and Johnson paved the way for advances in such diverse and specialized areas of knowledge as *comparative neuroscience* and *social dynamics.*

Virginia Johnson (1925-2013)

Today, there are many opportunities to apply knowledge of reproductive science in a variety of professions. *Reproductive health nurses, gynecologists,* and **urologists** often provide primary reproductive care to adult men and women. Reproductive medicine clinical staff help couples improve fertility. *Psychologists* and *counselors* help patients struggling with various sexual concerns.

For example, inflammation or infection of the uterine tubes can result in scarring that blocks sperm from reaching the ovum or prevents the ovum from traveling to the uterus. Infections, cancer, or hormonal imbalances may inhibit the female reproductive cycle, preventing the production and release of a healthy ovum each cycle. Such conditions also may interfere with the development of the uterine lining that is essential for successful pregnancy.

Because sexual reproduction requires normal function of both male and female systems, infertility of a couple may result from the infertility of either partner. A couple is considered infertile if a pregnancy does not occur after a year of reasonably frequent sexual intercourse (without contraception). When couples seek help for infertility problems, one of the first steps in diagnosis is to determine whether there is a problem in the male partner or the female partner—or both.

For information on strategies available to couples to prevent pregnancy, see the article *Contraception* at Connect It! at *evolve.elsevier.com.*

Summary of Male and Female Reproductive Systems

The reproductive systems in both sexes revolve around the production of reproductive cells, or gametes (sperm and ova), as well as mechanisms to ensure union of these two cells; the fusion of these cells enables transfer of parental genetic information to the next generation.

Table 23-3 compares several analogous components of the reproductive systems in both sexes. You can see that men and women have similar structures to accomplish complementary functions. In addition, the female reproductive system permits development and birth of the offspring—the first subject of our next chapter.

Sexually Transmitted Diseases

Sexually transmitted diseases (STDs), formerly called *venereal diseases (VDs),* are infections caused by communicable pathogens such as viruses, bacteria, fungi, or protozoans (see Appendix A at *evolve.elsevier.com*).

TABLE 23-3 Analogous* Features of the Reproductive Systems

FEATURE	FEMALE	MALE
Essential organs	Ovaries	Testes
Sex cells	Ova (eggs)	Sperm
Hormones	Estrogen and progesterone	Testosterone
Hormone-producing cells	Granulosa cells and corpus luteum	Interstitial cells
Duct systems	Uterine (fallopian) tubes, uterus, and vagina	Epididymis, vas deferens, and urethra
External genitals	Clitoris and vulva	Penis and scrotum

*Resembling or similar in some respects.

The term **sexually transmitted infection (STI)** is sometimes used in place of the term STD but does not have exactly the same meaning. An STI is an infection that may or may not cause symptoms. An STD occurs when an STI progresses to actually produce symptoms that make a person feel sick—making STI a broader term than STD.

The factor that links all these infections or diseases and gives this category its name is the fact that they are transmitted by sexual contact. The term *sexual contact* refers to sexual intercourse in addition to any contact between the genitals of one person and the body of another person.

Diseases classified as STDs can be transmitted sexually, but that is not the only way to transmit them. For example, *human immunodeficiency virus (HIV)* infection is a viral condition that can be spread through sexual contact but is also spread by transfusion of infected blood and use of contaminated medical

TABLE 23-4 Examples of Sexually Transmitted Diseases (STDs)

DISEASE	PATHOGEN	DESCRIPTION
Acquired immunodeficiency syndrome (AIDS)	*Virus:* Human immunodeficiency virus (HIV)	HIV is transmitted by direct contact of body fluids, often during sexual contact. After a period that sometimes lasts many years, HIV infection produces the condition known as AIDS. AIDS is characterized by damage to lymphocytes (T cells), resulting in immune system impairment. Death results from secondary infections or tumors.
Candidiasis	*Fungus: Candida albicans*	This yeast infection is characterized by a white discharge (leukorrhea), peeling of skin, and bleeding. Although it can occur as an ordinary opportunistic infection, it may be transmitted sexually as well.
Chancroid	*Bacterium: Haemophilus ducreyi*	A highly contagious STI, chancroid is characterized by papules on the skin of the genitals that eventually ulcerate. About 90% of cases are reported by men.
Chlamydia	*Bacterium: Chlamydia trachomatis*	The most common STD, most infected people have no symptoms; symptoms include burning pain and discharge; easily treated with antibiotics; if not treated, may progress to PID.
Genital herpes	*Virus:* Herpes simplex virus (HSV)	HSV causes blisters on the skin of the genitals. The blisters may disappear temporarily but may reappear occasionally, especially as a result of stress.
Genital warts	*Virus:* Human papillomavirus (HPV-6, HPV-7)	Genital warts are nipplelike neoplasms of skin covering the genitals. An effective vaccination is available. HPV is known to cause cervical cancer.
Giardiasis	*Protozoan: Giardia lamblia*	This intestinal infection may be spread by sexual contact. Symptoms range from mild diarrhea to malabsorption syndrome, with about half of all cases being asymptomatic.
Gonorrhea	*Bacterium: Neisseria gonorrhoeae*	Gonorrhea primarily involves the genital and urinary tracts but can affect the throat, conjunctiva, or lower intestines. It may progress to PID.
Hepatitis	*Virus:* Hepatitis B virus (HBV)	This acute-onset liver inflammation may develop into a severe chronic disease, perhaps ending in death.
Lymphogranuloma venereum (LGV)	*Bacterium: Chlamydia trachomatis*	This chronic STD is characterized by genital ulcers, swollen lymph nodes, headache, fever, and muscle pain. *C. trachomatis* infection may cause a variety of other syndromes, including conjunctivitis, urogenital infections, and systemic infections. *C. trachomatis* infections constitute the most common STI in the United States (nongonorrheal urethritis) and often progress to PID.
Pubic (crab) lice	*Animal: Phthirius pubis*	Characterized by itching and the presence of visible nits (lice eggs on pubic hairs) or crawling lice, this infestation is treated by medication, manual removal of nits, and treatment of clothing, towels, bedding.
Scabies	*Animal: Sarcoptes scabiei*	Scabies is caused by infestation by the *itch mite,* which burrows into the skin to lay eggs. About 2 to 4 months after initial contact, a hypersensitivity reaction occurs, causing a rash along each burrow that itches intensely. Secondary bacterial infection is possible.
Syphilis	*Bacterium: Treponema pallidum*	Although transmitted sexually, syphilis can affect any system. *Primary syphilis* is characterized by chancre sores on exposed areas of the skin. If the primary infection goes untreated, *secondary syphilis* may appear 2 months after chancres disappear. The secondary stage occurs when the spirochete has spread throughout the body, presenting a variety of symptoms, and it is still highly contagious—even through kissing. *Tertiary syphilis* may appear years later, possibly resulting in death.
Trichomoniasis	*Protozoan: Trichomonas vaginalis*	This urological infection is asymptomatic in most women and nearly all men. Vaginitis may occur, characterized by itching or burning, and a foul-smelling discharge.

AIDS, Acquired immunodeficiency syndrome; *HIV,* human immunodeficiency virus; *PID,* pelvic inflammatory disease; *STI,* sexually transmitted infection.

instruments such as intravenous needles and syringes. Candidiasis, or yeast infection, is a common opportunistic infection, but it also can be transmitted through sexual contact. The *Zika virus* is widely transmitted by mosquito vectors, but can also be transmitted through sexual contact.

Sexually transmitted diseases are the most common of all communicable diseases. **Table 23-4** summarizes a few of the principal STDs.

For more information on STDs, including photographs of their typical signs, see the article *Sexually Transmitted Diseases* at Connect It! at *evolve.elsevier.com.*

QUICK CHECK

1. Which female structure is made of erectile tissue?
2. What is another term for *menses*?
3. Which hormone reaches a high peak just before ovulation?
4. Women with polycystic ovary syndrome frequently suffer from what conditions?

LANGUAGE OF SCIENCE *(continued from p. 617)*

corpora cavernosa
(KOHR-pohr-ah kav-er-NO-sah)
sing., corpus cavernosum
(KOHR-pus kav-er-NO-sum)
[*corpus* **body,** *cavern-* **large hollow,** *-os-* **relating to,** *-um* **thing**]

corpus luteum
(KOHR-pus LOO-tee-um)
pl., corpora lutea
(KOHR-pohr-ah LOO-tee-ah)
[*corpus* **body,** *lute-* **yellow,** *-um* **thing**]

corpus spongiosum
[*corpus* **body,** *spong-* **sponge,** *-os-* **relating to,** *-um* **thing**]
(KOHR-pus spun-jee-OH-sum)

Cowper gland
(KOW-per gland)
[*William Cowper* **English anatomist,** *gland* **acorn**]

ductus deferens
(DUK-tus DEF-er-enz)
[*ductus* **duct,** *de-* **away from,** *-fer-* **bear or carry**]

ejaculation
(ee-jak-yoo-LAY-shun)
[*e-* **out or away,** *-jacula-* **throw,** *-ation* **process**]

ejaculatory duct
(ee-JAK-yoo-lah-toh-ree dukt)
[*e-* **out or away,** *-jacula-* **throw,** *-ory* **relating to,** *duct* **a leading or path**]

endometrium
(en-doh-MEE-tree-um)
[*endo-* **within,** *-metr-* **womb,** *-um* **thing**]

epididymis
(ep-ih-DID-ih-mis)
[*epi-* **upon,** *-didymis* **pair**]

essential organ
(eh-SEN-shul OR-gun)
[*organ* **instrument**]

estrogen
(ES-troh-jen)
[*estro-* **frenzy,** *-gen* **produce**]

external urinary meatus
(eks-TER-nal YOOR-ih-nayr-ee mee-AY-tus)
pl., meatus or meatuses
(mee-AY-tus or mee-AY-tus-ez)
[*extern-* **outside,** *-al* **relating to,** *urin-* **urine,** *-ary* **relating to,** *meatus* **channel or passage**]

fallopian tube
(fal-LOH-pee-an toob)
[*Gabriele Fallopio* **Italian anatomist**]

fimbria
(FIM-bree-ah)
pl., fimbriae
(FIM-bree-yee)
[*fimbria* **fringe**]

follicle-stimulating hormone (FSH)
(FOL-lih-kul-STIM-yoo-lay-ting HOR-mohn [ef es aych])
[*foll-* **bag,** *-icle* **little,** *stimul-* **excite,** *-at-* **process,** *-ing* **action,** *hormon-* **excite**]

foreskin
(FORE-skin)
[*fore-* **front,** *-skin* **a hide**]

fundus
(FUN-dus)
[*fundus* **bottom**]

gamete
(GAM-eet)
[*gamet-* **sexual union or marriage partner**]

genital
(JEN-ih-tal)
pl., genitalia
(jen-ih-TAYL-yah)
[*gen-* **produce,** *-al* **relating to**]

glans
(glans)
[*glans* **acorn**]

gonad
(GOH-nad)
[*gon-* **offspring,** *-ad* **relating to**]

gonadotropin-releasing hormone (GnRH)
(goh-nah-doh-TROH-pin ree-LEES-ing HOR-mohn [jee en ar aych])
[*gon-* **offspring,** *-ad-* **relating to,** *-trop-* **nourish,** *-in* **substance,** *hormon-* **excite**]

graafian follicle
(GRAH-fee-en FOL-lih-kul)
[*Reijnier de Graaf* **Dutch physician,** *-an* **relating to,** *foll-* **bag,** *-icle* **little**]

granulosa cell
(gran-yoo-LOH-sah sel)
[*gran-* **grain,** *-ul-* **little,** *-osa* **relating to,** *cell* **storeroom**]

greater vestibular gland
(GRAYT-er ves-TIB-yoo-lar gland)
[*vestibul-* **entrance hall,** *-ar* **relating to,** *gland* **acorn**]

hymen
(HYE-men)
[*hymen* **membrane**]

interstitial cell
(in-ter-STISH-al sel)
[*inter-* **between,** *-stit-* **stand,** *-al* **relating to,** *cell* **storeroom**]

labia majora
(LAY-bee-ah mah-JOH-rah)
sing., labium majus
(LAY-bee-um MAY-jus)
[*labia* **lips,** *majora* **large**]

labia minora
(LAY-bee-ah mih-NO-rah)
sing., labium minus
(LAY-bee-um MYE-nus)
[*labia* **lips,** *minora* **small**]

lactiferous duct
(lak-TIF-er-us dukt)
[*lact-* **milk,** *-fer-* **bear or carry,** *-ous* **having to do with,** *duct* **a leading or path**]

23

Continued on p. 642

LANGUAGE OF **SCIENCE** *(continued from p. 641)*

lesser vestibular gland
(LES-er ves-TIB-yoo-lar gland)
[*vestibul-* **entrance hall,** *-ar* **relating to,** *gland* **acorn**]

luteinizing hormone (LH)
(loo-tee-in-AYE-zing HOR-mohn [el aych])
[*lute-* **yellow,** *-izing* **process,** *hormon-* **excite**]

meiosis
(my-OH-sis)
[*mei-* **smaller,** *-osis* **process**]

menarche
(meh-NAR-kee)
[*men-* **month,** *-arche* **beginning**]

menopause
(MEN-oh-pawz)
[*men-* **month,** *-paus-* **cease**]

menses
(MEN-seez)
[*menses* **months**]

menstrual cycle
(MEN-stroo-al SYE-kul)
[*mens-* **month,** *-al* **relating to,** *cycle* **circle**]

mons pubis
(monz PYOO-bis)
[*mons* **mountain,** *pubis* **groin**]

myometrium
(my-oh-MEE-tree-um)
[*myo-* **muscle,** *-metr-* **womb,** *-um* **thing**]

oocyte
(OH-oh-syte)
[*oo-* **egg,** *-cyte* **cell**]

oogenesis
(oh-oh-JEN-eh-sis)
[*oo-* **egg,** *-gen-* **produce,** *-esis* **process**]

ovarian follicle
(oh-VAYR-ee-an FOL-ih-kul)
[*ov-* **egg,** *-arian* **relating to,** *foll-* **bag,** *-icle* **little**]

ovary
(OH-var-ee)
[*ov-* **egg,** *-ar-* **relating to,** *-y* **location of process**]

oviduct
(OH-vih-dukt)
[*ovi-* **egg,** *- duct* **a leading or path**]

ovulation
(ov-yoo-LAY-shun)
[*ov-* **egg,** *-ation* **process**]

ovum
(OH-vum)
pl., ova
(OH-vah)
[*ovum* **egg**]

penis
(PEE-nis)
pl., penes or penises
[*penis* **male sex organ**]

perineum
(payr-ih-NEE-um)
[*peri-* **around or near,** *-ine-* **excrete or evacuate,** *-um* **thing**]

polar body
(POH-lar BOD-ee)
[*pol-* **pole,** *-ar* **relating to**]

prepuce
(PREE-pus)
[*pre-* **before,** *-puc-* **penis**]

primary follicle
(PRYE-mayr-ee FOL-ih-kul)
[*prim-* **first,** *-ary* **state,** *folli-* **bag,** *-cle* **small**]

primary spermatocyte
(PRYE-mayr-ee SPER-mah-toh-syte)
[*prim-* **first,** *-ary* **state,** *sperm-* **seed,** *-cyte* **cell**]

progesterone
(proh-JES-ter-ohn)
[*pro-* **provide for,** *-gester-* **bearing (pregnancy),** *-stero-* **solid or steroid derivative,** *-one* **chemical**]

proliferative phase
(PROH-lif-er-eh-tiv fayz)
[*proli-* **offspring,** *-fer-* **bear or carry,** *-at-* **process,** *-ive* **relating to**]

prostate gland
(PROS-tayt gland)
[*pro-* **before,** *-stat-* **set or place,** *gland* **acorn**]

puberty
(PYOO-ber-tee)
[*pubert-* **age of maturity,** *-y* **state**]

scrotum
(SKROH-tum)
[scrotum **bag**]

secretory phase
(SEEK-reh-toh-ree fayz)
[*secret-* **separate,** *-ory* **relating to**]

semen
(SEE-men)
[*semen* **seed**]

seminal fluid
(SEM-ih-nal FLOO-id)
[*semin-* **seed,** *-al* **relating to**]

seminal vesicle
(SEM-ih-nal VES-ih-kul)
[*semin-* **seed,** *-al* **relating to,** *vesic-* **blister,** *-cle* **little**]

seminiferous tubule
(seh-mih-NIF-er-us TOOB-yool)
[*semin-* **seed,** *-fer-* **bear or carry,** *-ous* **relating to,** *tub-* **tube,** *-ul-* **little**]

Skene gland
(skeen gland)
[*Alexander Johnston Chalmers Skene* **American gynecologist,** *gland* **acorn**]

sperm
(sperm)
pl., sperm
[*sperm* **seed**]

spermatid
(SPER-mah-tid)
[*sperm-* **seed,** *-id* **relating or belonging to**]

spermatogenesis
(sper-mah-toh-JEN-eh-sis)
[*sperm-* **seed,** *-gen-* **produce,** *-esis* **process**]

spermatogonia
(sper-mah-toh-GOH-nee-ah)
[*sperm-* **seed,** *-gonia* **offspring**]

spermatozoon
(sper-mah-tah-ZOH-on)
pl., spermatozoa
(sper-mah-tah-ZOH-ah)
[*sperm-* **seed,** *-zoon* **animal**]

testis
(TES-tis)
pl., testes
(TES-teez)
[*testis* **witness (male gonad)**]

testosterone
(tes-TOS-teh-rohn)
[*testo-* **witness (testis),** *-stero-* **solid or steroid derivative,** *-one* **chemical**]

tunica albuginea
(TOO-nih-kah al-byoo-JIN-ee-ah)
[*tunica* **tunic or coat,** *albuginea* **white**]

urethra
(yoo-REE-thrah)
[*ure-* **urine,** *-thr-* **agent or channel**]

uterine tube
(YOO-ter-in toob)
[*uter-* **womb,** *-ine* **relating to**]

uterus
(YOO-ter-us)
[*uterus* **womb**]

vagina
(vah-JYE-nah)
[*vagina* **sheath**]

vas deferens
(vas DEF-er-enz)
[*vas* **duct or vessel,** *de-* **away from,** *-fer-* **bear or carry**]

vestibule
(VES-tih-byool)
[*vestibul-* **entrance hall**]

vulva
(VUL-vah)
[*vulva* **wrapper**]

zygote
(ZYE-goht)
[*zygot-* **union or yoke**]

LANGUAGE OF **MEDICINE**

amenorrhea
(ah-men-oh-REE-ah)
[*a-* **without**, *-men-* **month**, *-rrhea* **flow**]

benign prostatic hypertrophy (BPH)
(be-NYNE proh-STAT-ik hye-PER-troh-fee [bee pee aych])
[*benign* **kind**, *pro-* **before**, *-stat-* **set or place**, *-ic* **relating to**, *hyper-* **excessive or above**, *-troph-* **nourishment**, *-y* **state**]

brachytherapy
(BRAKE-ee-THAYR-ah-pee)
[*brachy-* **short**, *-therapy* **curing**]

candidiasis
(kan-dih-DYE-eh-sis)
[*candid-* **white**, *-asis* **condition**]

circumcision
(ser-kum-SIH-zhun)
[*circum-* **around**, *-cision* **cutting**]

contraception
(kon-trah-SEP-shun)
[*contra-* **against**, *-cept-* **receive (conceive)**, *-tion* **process**]

cryptorchidism
(krip-TOR-kih-diz-em)
[*crypt-* **hidden**, *-orchid-* **testis**, *-ism* **condition**]

dysfunctional uterine bleeding (DUB)
(dis-FUNK-shun-al YOO-ter-in BLEED-ing [dee yoo bee])
[*dys-* **difficult**, *-function-* **performance**, *-al* **relating to**, *uter-* **womb**, *-ine* **relating to**]

dysmenorrhea
(dis-men-oh-REE-ah)
[*dys-* **painful**, *-men-* **month**, *-rhea* **flow**]

ectopic pregnancy
(ek-TOP-ik PREG-nan-see)
[*ec-* **out of**, *-top-* **place**, *-ic* **relating to**]

endogenous infection
(en-DOJ-en-us in-FEK-shun)
[*endo-* **within**, *-gen-* **produce**, *-ous* **relating to**, *in-* **into**, *-fec-* **put**, *-tion* **process**]

endometrial ablation
(en-doh-MEE-tree-al ab-LAY-shun)
[*endo-* **within**, *-metr-* **womb**, *-al* **relating to**, *ab-* **away from**, *-lat-* **carry**, *-tion* **process**]

endometriosis
(en-doh-mee-tree-OH-sis)
[*endo-* **within**, *-metr-* **womb**, *-osis* **condition**]

epididymitis
(ep-ih-did-ih-MY-tis)
[*epi-* **upon**, *-didymi-* **pair**, *-itis* **inflammation**]

episiotomy
(eh-piz-ee-OT-oh-mee)
[*episio-* **pubic region**, *-tom-* **cut**, *-y* **action**]

epispadias
(ep-is-PAY-dee-us)
[*epi-* **on or above**, *-spad-* **rip or split**, *-ias* **condition**]

erectile dysfunction (ED)
(eh-REK-tyle dis-FUNK-shun [ee dee])
[*erect-* **set up**, *-ile* **relating to**, *dys-* **bad or painful**, *-func-* **perform**, *-tion* **process**]

fallopian tube
(fal-LOH-pee-an toob)
[*Gabriele Fallopio* **Italian anatomist**]

fibroid
(FYE-broyd)
[*fibr-* **fiber**, *-oid* **of or like**]

fibrocystic disease
(fye-broh-SIS-tik dih-ZEEZ)
[*fibr-* **thread or fiber**, *cyst-* **sac**, *-ic* **relating to**, *dis-* **opposite of**, *-ease* **comfort**]

fibromyoma
(fye-broh-my-OH-mah)
[*fibro-* **fiber**, *-my-* **muscle**, *-oma* **tumor**]

gynecologist
(gye-neh-KOL-oh-jist)
[*gyneco-* **woman or female gender**, *-log-* **words (study of)**, *-ist* **agent**]

hydrocele
(HYE-droh-seel)
[*hydro-* **water**, *-cele* **tumor**]

hypospadias
(hye-poh-SPAY-dee-us)
[*hypo-* **under or below**, *-spad-* **rip or split**, *-ias* **condition**]

hysterectomy
(his-teh-REK-toh-mee)
[*hyster-* **uterus**, *-ec-* **out**, *-tom-* **cut**, *-y* **action**]

infertility
(in-fer-TIL-ih-tee)
[*in-* **not**, *-fertil-* **fruitful**, *-ity* **state**]

inguinal hernia
(ING-gwih-nal HER-nee-ah)
[*inguin-* **groin**, *-al* **relating to**, *hernia* **rupture**]

leukorrhea
(loo-koh-REE-ah)
[*leuko-* **white**, *-rrhea* **flow**]

mammary dysplasia
(MAM-mah-ree dis-PLAY-zhah)
[*mamma-* **breast**, *-ry* **relating to**, *dys-* **disordered**, *-plas(m)-* **substance or form**, *-ia* **condition**]

menstrual cramp
(MEN-stroo-al kramp)
[*mens-* **month**, *-al* **relating to**, *cramp* **bent**]

myoma
(my-OH-mah)
[*my-* **muscle**, *-oma* **tumor**]

oligospermia
(ol-ih-goh-SPER-mee-ah)
[*oligo-* **few or little**, *-sperm-* **seed**, *-ia* **condition**]

oophorectomy
(oh-of-eh-REK-toh-mee)
[*oophoro-* **ovary**, *-ec-* **out**, *-tom-* **cut**, *-y* **action**]

oophoritis
(oh-of-eh-RYE-tis)
[*oophoro-* **ovary**, *-itis* **inflammation**]

ovarian cyst
(oh-VAYR-ee-an SIST)
[*ov-* **egg**, *-arian* **relating to**, *cyst* **bag**]

oviduct
(OH-vih-dukt)
[*ovi-* **egg**, *-duct* **a leading or path**]

Papanicolaou (pap) test
(pah-peh-nik-oh-LAH-oo test)
[*George N. Papanicolaou* **Greek physician**]

paraphimosis
(para-fih-MOH-sis)
[*para-* **beside**, *-phim-* **muzzle**, *-osis* **condition**]

pelvic inflammatory disease (PID)
(PEL-vik in-FLAM-ah-tor-ee dih-ZEEZ [pee aye dee])
[*pelv-* **basin**, *-ic* **relating to**, *inflam-* **set afire**, *-ory* **relating to**, *dis-* **opposite of**, *-ease* **comfort**]

phimosis
(fih-MOH-sis)
[*phim-* **muzzle**, *-osis* **condition**]

polycystic ovary syndrome (PCOS)
(pahl-ee-SIS-tik OH-var-ee SIN-drohm [pee see oh es])
[*poly-* **many**, *-cyst-* **bag**, *-ic* **relating to**, *ov-* **egg**, *-ar-* **relating to**, *-y* **location of process**, *syn-* **together**, *-drome* **running or (race) course**]

premenstrual syndrome (PMS)
(pree-MEN-stroo-al SIN-drohm [pee em es])
[*pre-* **before**, *-mens-* **month**, *-al* **relating to**, *syn-* **together**, *-drome* **running or (race) course**]

prostate cancer
(PROS-tayt KAN-ser)
[*pro-* **before**, *-stat-* **set or place**, *cancer* **crab or malignant tumor**]

prostatectomy
(pros-tah-TEK-toh-mee)
[*pro-* **before**, *-stat-* **set or place (prostate gland)**, *-ec-* **out**, *-tom-* **cut**, *-y* **action**]

radical mastectomy
(RAD-ih-kal mas-TEK-toh-mee)
[*radic-* **root**, *-al* **relating to**, *mast-* **breast**, *-ec-* **out**, *-tom-* **cut**, *-y* **action**]

Continued on p. 644

LANGUAGE OF MEDICINE (continued from p. 643)

salpingitis
(sal-pin-JYE-tis)
[*salping-* **tube,** *-itis* **inflammation**]

sexually transmitted disease (STD)
(SEKS-yoo-al-ee trans-MIH-ted dih-ZEEZ [es tee dee])
[*dis-* **opposite of,** *-ease* **comfort**]

sexually transmitted infection (STI)
(SEKS-yoo-al-ee trans-MIH-ted in-FEK-shun [es tee aye])
[*in-* **into,** *-fec-* **put,** *-tion* **process**]

sterility
(steh-RIL-ih-tee)
[*steril-* **barren,** *-ity* **state**]

urologist
(yoo-ROL-uh-jist)
[*uro-* **urine,** *-log-* **words (study of),** *-ist* **agent**]

vaginitis
(vaj-ih-NYE-tis)
[*vagin-* **sheath (vagina),** *-itis* **inflammation**]

vasectomy
(va-SEK-toh-mee)
[*vas-* **vessel (vas deferens),** *-ec-* **out,** *-tom-* **cut,** *-y* **action**]

❑ OUTLINE SUMMARY

To download a digital version of the chapter summary for use with your device, access the ***Audio Chapter Summaries*** *online at evolve.elsevier.com.*

Scan this summary after reading the chapter to help you reinforce the key concepts. Later, use the summary as a quick review before your class or before a test.

Sexual Reproduction

A. Producing offspring
   1. Sexual reproduction involves two parents (unlike one-parent asexual reproduction); increases variation of genetic traits among offspring of same parents
   2. Gametes—sex cells that fuse at fertilization to form a one-celled zygote, the first cell of the offspring
      a. Sperm—gamete from the male parent
      b. Ovum—gamete from the female parent
   3. Reproductive hormones regulate sexual characteristics that promote successful reproduction
   4. Ability to reproduce begins at puberty

B. Male and female systems
   1. Common general structure and function can be identified between the systems in both sexes
   2. Systems adapted for development of sperm or ova followed by successful fertilization, development, and birth of offspring
   3. Sex hormones in both sexes are important in development of secondary sexual characteristics and normal reproductive system activity

Male Reproductive System

A. Structural plan of the reproductive tract (also called *urogenital tract*)
   1. Organs classified as *essential* or *accessory* (**Figure 23-1** and **Table 23-1**)
   2. Essential organs of reproduction are the gonads (testes in men), which produce sex cells (spermatozoa)
   3. Accessory organs of reproduction
      a. Ducts—passageways that carry sperm from testes to exterior
      b. Sex glands—produce protective and nutrient solution for sperm
      c. External genitals

B. Testes—the gonads of men
   1. Structure and location (**Figure 23-1** and **Figure 23-2**)
      a. Testes in scrotum—temperature is lower than inside body
      b. Covered by tunica albuginea, which divides testis into lobules containing seminiferous tubules
      c. Interstitial cells produce testosterone (**Figure 23-3**)
   2. Testis functions
      a. Spermatogenesis is process of sperm production (**Figure 23-4**)
         (1) FSH from pituitary stimulates spermatogenesis
         (2) Sperm precursor cells called *spermatogonia*
         (3) Meiosis produces primary spermatocyte, which forms four spermatids with 23 chromosomes
         (4) Spermatozoa—male reproductive cell (**Figure 23-5**)
            (a) Ejaculation—forceful ejection (from the penis) of fluid containing sperm
            (b) Head contains genetic material

(c) Acrosome contains enzymes to assist sperm in penetration of ovum
(d) Mitochondria provide energy for movement
b. Production of testosterone by interstitial cells
(1) LH from pituitary stimulates interstitial cells to develop and produce testosterone
(2) Testosterone function
(a) Testosterone "masculinizes" and promotes development of male accessory organs
(b) Stimulates protein anabolism (anabolic steroid) and development of muscle strength

C. Reproductive ducts—ducts through which sperm pass after exiting testes until they exit from the body
1. Epididymis—single, coiled tube about 6 m in length; lies along the top of and behind each testis in the scrotum
a. Sperm mature and develop the capacity for motility as they pass through the epididymis
b. Epididymitis—painful inflammation
2. Vas deferens—also called *ductus deferens*
a. Receives sperm from the epididymis and transports them from scrotal sac through the pelvic cavity
b. Passes through inguinal canal and then joins duct of seminal vesicle to form the ejaculatory duct (**Figure 23-6**)

D. Accessory glands—produce components of semen
1. Semen—also called *seminal fluid*
a. Mixture of sperm and secretions of accessory sex glands
b. Average of 3 to 5 mL per ejaculation, with each milliliter containing about 20 million to 100 million sperm (highly variable, even day to day)
2. Seminal vesicles
a. Pouchlike glands that produce about 60% of seminal fluid volume
b. Secretion is yellowish, thick, and rich in fructose to provide energy needed by sperm for motility
3. Prostate gland
a. Shaped like a doughnut and located below bladder
b. Urethra passes through the gland
c. Thin, milk-colored secretion that represents 30% of seminal fluid volume
d. Activates sperm and is needed for ongoing sperm motility
4. Bulbourethral glands—also called *Cowper glands*
a. Resemble peas in size and shape
b. Secrete mucuslike fluid (less than 5% of seminal fluid volume) that lubricates terminal portion of urethra

E. External genitals—also called *genitalia*
1. Penis and scrotum (**Figure 23-7**)
2. Penis has three columns of erectile tissue
a. Two dorsal columns called *corpora cavernosa*
b. One ventral column surrounding urethra called *corpus spongiosum*
3. Glans penis—distal end of penis
a. Covered by foreskin (prepuce)
b. Surgical removal of foreskin called *circumcision*
4. Scrotum—skin-covered pouch suspended from groin
a. Divided into two sacs by a septum
b. Each sac contains a testis, epididymis, part of vas deferens, and beginning of spermatic cords

Disorders of the Male Reproductive System

A. Infertility and sterility may result from reproductive disorders
1. Infertility—reduced reproductive ability
2. Sterility—total inability to reproduce
3. No distinct andropause (cessation of fertility) in late adulthood, as in female menopause
a. Testosterone production may decline (low T) in late adulthood
b. Low T may reduce fertility in some men; many men remain fertile for their lifetimes

B. Disorders of the testes
1. Oligospermia—low sperm production
2. Cryptorchidism—undescended testes
3. Testicular cancer—most common in males ages 15 to 30

C. Disorders of the prostate
1. Benign prostatic hypertrophy (BPH)—enlargement of prostate common in older men
2. Prostate cancer—malignancy of prostate tissue
a. Prostatectomy—surgical removal of part or all of the prostate
b. Brachytherapy—small radioactive "seeds" placed in prostate

D. Disorders of the penis and scrotum
1. Penis disorders
a. Phimosis—tight foreskin cannot be retracted over glans (Box, p. 624)
b. Paraphimosis—foreskin cannot be replaced to usual position after it has been retracted behind the glans (Box, p. 624)
c. Hypospadias—urethra opens on underside of glans or shaft
d. Epispadias—urethra opens on top of glans or shaft
e. Erectile dysfunction (ED)—failure to achieve or maintain erection of the penis

2. Scrotum disorders
   a. Hydrocele—accumulation of watery fluid in the scrotum
   b. Inguinal hernia—protrusion of abdominopelvic organs, possibly into the scrotum (**Figure 23-8**)

Female Reproductive System

A. Structural plan—organs classified as essential or accessory (**Figure 23-9** and **Table 23-2**)
   1. Essential organs are gonads (ovaries), which produce sex cells (ova)
   2. Accessory organs of reproduction
      a. Ducts or modified ducts—including oviducts, uterus, and vagina
      b. Sex glands—including those in the breasts
      c. External genitals

B. Ovaries
   1. Structure and location
      a. Paired glands weighing about 3 g each
      b. Resemble large almonds
      c. Attached to ligaments in pelvic cavity on each side of uterus
      d. Microscopic structure (**Figure 23-10**)
         (1) Ovarian follicles—contain oocyte, which is immature sex cell (about half a million at birth)
         (2) Primary follicles—about 400,000 at puberty—are covered with granulosa cells
         (3) About 350 to 500 mature follicles ovulate during the reproductive lifetime of most women—sometimes called *graafian follicles*
         (4) Secondary follicles have hollow chamber called *antrum*
         (5) Corpus luteum forms after ovulation
   2. Ovary functions
      a. Oogenesis (**Figure 23-11**)
         (1) Involves meiotic cell division (meiosis) that produces daughter cells with equal chromosome numbers (23) but unequal cytoplasm
         (2) Ovum is large; polar bodies are small and degenerate
      b. Production of estrogen and progesterone
         (1) Granulosa cells surrounding the oocyte in the mature and growing follicles produce estrogen
         (2) Corpus luteum produces progesterone
         (3) Estrogen causes development and maintenance of secondary sex characteristics
         (4) Progesterone stimulates secretory activity of uterine epithelium and assists estrogen in initiating menses
   3. Surgical removal called *oophorectomy*

C. Reproductive ducts
   1. Both male and female reproductive ducts carry gametes from each (of two) gonads, join into a single passage, and exit the body
   2. Only the female ducts also function in receiving sperm, fertilization, and prenatal development
   3. Uterine tubes—also called *fallopian tubes* or *oviducts*
      a. Extend about 10 cm from uterus into pelvic cavity
      b. Expanded distal end surrounded by fimbriae
      c. Mucosal lining of tube is directly continuous with lining of pelvic cavity
      d. Surgical removal called *hysterectomy*
   4. Uterus—composed of body, fundus, and cervix (**Figure 23-12**)
      a. Lies in pelvic cavity just behind urinary bladder
      b. Myometrium is muscle layer
      c. Endometrium lost in menstruation
      d. Menopause—end of repetitive menstrual cycles (about 45 to 50 years of age)
      e. Surgical removal called *hysterectomy*
         (1) Removal may be abdominal, vaginal, or laparoscopic
         (2) Total hysterectomy—removal of both body and cervix of uterus
         (3) Subtotal hysterectomy—removal of body of uterus only (cervix remains)
   5. Vagina
      a. Distensible tube about 10 cm long
      b. Located between urinary bladder and rectum in the pelvis
      c. Receives penis during sexual intercourse and is birth canal for normal delivery of baby at end of pregnancy

D. Accessory glands
   1. Greater and lesser vestibular glands
      a. Secrete mucous fluid that may lubricate during sexual intercourse
      b. Ducts open into vestibule (between labia minora)
      c. Clinically important when they become infected (e.g., *Neisseria gonorrhoeae* bacteria that cause gonorrhea)
   2. Breasts (**Figure 23-13**)
      a. Located over pectoral muscles of thorax
      b. Size determined by fat quantity more than amount of glandular (milk-secreting) tissue
      c. Lactiferous ducts drain at nipple, which is surrounded by pigmented areola
      d. Lymphatic drainage leads to spread of cancer cells to other body areas

E. External genitals (**Figure 23-14**)
   1. Vulva includes mons pubis, clitoris, external urinary meatus, openings of vestibular glands, orifice of vagina, labia minora and majora, and hymen

2. Perineum
   a. Area between vaginal opening and anus
   b. Surgical cut during birth called *episiotomy*

F. Menstrual cycle—involves many changes in the uterus, ovaries, vagina, and breasts (**Figure 23-15** and **Figure 23-16**)
   1. Length—about 28 days, varies from cycle to cycle among individuals and in the same individual
   2. Phases
      a. Menses—about the first 4 or 5 days of the cycle, varies somewhat
         (1) Characterized by sloughing of bits of endometrium (uterine lining) with bleeding
         (2) First day of flow is day 1 of menstrual cycle
      b. Proliferative phase—days between the end of menses and ovulation; varies in length
         (1) The shorter the cycle, the shorter the proliferative phase; the longer the cycle, the longer the proliferative phase
         (2) Characterized by proliferation (repair) of endometrium
      c. Secretory phase—days between ovulation and beginning of next menses
         (1) Characterized by further thickening of endometrium
         (2) Secretion by its glands in preparation for implantation of fertilized ovum
   3. Ovulation—typically one ovum released per cycle, 14 days before next menses; timing of ovulation is useful in timing sexual intercourse to maximize fertility
   4. Control—combined actions of the anterior pituitary hormones FSH and LH cause ovulation; sudden sharp decrease in estrogens and progesterone bring on menstruation if pregnancy does not occur

Disorders of the Female Reproductive System

A. Hormonal and menstrual disorders
   1. Dysmenorrhea (menstrual cramps)—painful menstruation
   2. Amenorrhea—absence of normal menstruation
   3. Dysfunctional uterine bleeding (DUB)—irregular or excessive bleeding resulting from a hormonal imbalance or pathology
   4. Premenstrual syndrome (PMS)—collection of symptoms that occur in some women before menstruation

B. Infection and inflammation
   1. Exogenous infections are often sexually transmitted; endogenous infections are caused by organisms already in or on the body
   2. Pelvic inflammatory disease (PID)—acute inflammatory condition of the uterus, uterine tubes, or ovaries caused by infection
   3. Vaginitis—infection of vaginal lining, it most often results from STDs or yeast infections

C. Tumors and related conditions
   1. Benign tumors of the uterus—myoma, fibroids, fibromyoma
   2. Ovarian cysts—fluid-filled enlargements; usually benign
      a. Follicular cysts—most common
      b. Luteal cysts—most symptomatic
      c. Most resolve in 60 days
   3. Polycystic ovary syndrome (PCOS)—enlarged ovaries with many fluid-filled cysts
      a. Affects 10% of reproductive-age women
      b. Most common cause of female infertility (**Figure 23-17**)
   4. Endometriosis—presence of functioning endometrial tissue outside the uterus
   5. Breast cancer is the most common type of cancer in women
   6. Cervical cancer is often detected by a Papanicolaou test (Pap smear) (**Figure 23-18**)

D. Infertility can result from factors such as infection and inflammation, tumors, and hormonal imbalances

23

Summary of Male and Female Reproductive Systems

A. In men and women the organs of the reproductive system are adapted for the specific sequence of functions that permit development of sperm or ova followed by the successful fertilization and then the normal development and birth of offspring

B. The male organs produce, store, and ultimately introduce mature sperm into the female reproductive tract

C. The female system produces ova, receives the sperm, and permits fertilization followed by fetal development and birth, with lactation afterward

D. Men and women have analogous reproductive structures (**Table 23-3**)

E. Production of sex hormones is required for development of secondary sex characteristics and for normal reproductive functions in both sexes

Sexually Transmitted Diseases

A. Sexually transmitted infections (STIs) can progress into STDs

B. STIs/STDs are transmitted sexually but can also be transmitted in other ways

C. STDs are the most common of all communicable diseases

D. STDs are caused by a variety of organisms (**Table 23-4**)

❑ ACTIVE LEARNING

STUDY TIPS

Hint *Consider using these tips to achieve success in meeting your learning goals.*

Review the synopsis of the male and female reproductive systems in Chapter 5.

1. Much of Chapter 23 deals with the names, locations, and functions of the structures of the male and female reproductive systems. Make flash cards and check online resources to help you learn this material. Review the Language of Science and Language of Medicine sections.
2. An effective way to understand the male reproductive ducts is to trace sperm in sequence from point of formation through the reproductive ducts to ejaculation.
3. The role of the male in reproduction is to produce as many sperm as possible, so four functional sperm are produced from meiosis. The female's role is the production of one egg containing 23 chromosomes; 69 of the original 92 chromosomes need to be "thrown away." This is the function of the polar bodies.
4. The responsibility of the female reproductive system is to produce an egg and prepare the body for a possible pregnancy. The reproductive cycle assists in doing this.
5. The reproductive cycle is regulated by four hormones: two from the pituitary gland and two from the ovary. Follicle-stimulating hormone does exactly what its name implies. *Luteinizing* hormone helps stimulate ovulation that causes the egg follicle to become the corpus *luteum*. Estrogen begins the initial preparation of the uterus. Progesterone prepares the uterus to receive a fertilized egg. Think of progesterone as "pro" (in favor of) "gesterone" (gestation); this may help you remember what it does. Construct a diagram that demonstrates hormonal integration in the menstrual cycle.
6. The disorders of the reproductive system can be put on a chart to help you learn them. Organize the disorders of the male reproductive system based on the specific structure of the male system that is affected. Disorders of the female system can be organized based on the specific structure of the female reproductive system or on menstruation disorders.
7. In your study group, go over the flash cards of the structures and photocopy figures to help you learn the locations. Discuss the process of meiosis and the differences between spermatogenesis and oogenesis. Discuss the reproductive cycle with emphasis on the function of the hormones. Go over the chart of the disorders, the chapter outline summary, and the questions at the end of the chapter, and discuss possible test questions.

Review Questions

Write out the answers to these questions after reading the chapter and reviewing the Chapter Summary. If you simply think through the answer without writing it down, you will not retain much of your new learning.

1. Describe the structure and location of the testes.
2. Describe the function of gonadotropin-releasing hormone.
3. Describe the structure of sperm.
4. List the functions of testosterone.
5. List and briefly describe the reproductive ducts of the male reproductive system.
6. List and briefly describe the glands of the male reproductive system. What does each gland contribute to the seminal fluid?
7. Distinguish between infertility, sterility, and impotence.
8. Describe circumcision and list the advantages and disadvantages of circumcision.
9. What is oligospermia? What is cryptorchidism?
10. Both a hydrocele and an inguinal hernia will produce swelling in the scrotum. Explain the difference between these two conditions.
11. Describe the structure and location of the ovaries.
12. Explain the development of an ovarian follicle from the primary follicle to the corpus luteum.
13. List the functions of estrogen.
14. List the functions of progesterone.
15. Describe the structure of the uterine tubes.
16. Describe the structure of the uterus.
17. Describe the structure of the vagina.
18. Describe the structure of the breast.
19. Explain what occurs during the proliferative phase of the reproductive cycle.

20. Explain what occurs during the secretory phase of the reproductive cycle.
21. Name the four hormones involved in the regulation of the reproductive cycle. Where is each made, and what is the function of each?
22. What is dysmenorrhea? What is amenorrhea?
23. Distinguish between salpingitis and oophoritis.
24. Describe endometriosis and identify some medical treatment strategies for this common disorder.
25. Distinguish between aromatase inhibitors and tamoxifen.

Critical Thinking

After finishing the Review Questions, write out the answers to these more in-depth questions to help you apply your new knowledge. Go back to sections of the chapter that relate to concepts that you find difficult.

26. Differentiate between spermatogenesis and oogenesis. How do these differences relate to the role of the male and female in reproduction?
27. How is the prostate gland related to the urethra? What problems can result from this relationship?
28. Why are the testes located outside the body cavity in the scrotum?
29. Female athletes may experience an undesirable condition called *amenorrhea*. What evidence can you find to support this statement?
30. Describe the characteristics of benign prostatic hypertrophy (BPH) and cancer of the prostate.

Chapter Test

After studying the chapter, test your mastery by responding to these items. Try to answer them without looking up the answers.

1. The essential organs of the male reproductive system are the ________.
2. The pouchlike sac where the male gonads are located is called the ________.
3. The membrane that covers the testis and also divides the interior into lobes is called the ________.
4. The ________ is a long duct in the testis where sperm develop.
5. The ________ are the cells in the testes that secrete testosterone.
6. The primary spermatocyte develops from a cell called the ________.
7. The primary spermatocyte forms sperm cells by undergoing a type of cell division called ________.
8. The sperm cell contains an ________, which contains an enzyme that can digest the covering of the ovum.
9. The ________ is a reproductive duct that consists of a tightly coiled tube that lies along the top of and behind the testis.
10. The ________ is a reproductive duct that permits the sperm to move out of the scrotum upward into the pelvic cavity.
11. The ________ is a gland that secretes a thin, milk-colored fluid that makes up about 30% of the seminal fluid.
12. The ________ are a pair of glands that produce a thick, yellowish, fructose-rich fluid that makes up about 60% of the seminal fluid.
13. The penis is composed of three columns of erectile tissue: one is called the corpus spongiosum, and the other two are called the ________.
14. ________ is the term used for a baby who is born with an undescended testicle.
15. The essential organs of the female reproductive system are the ________.
16. Another name for a mature ovarian follicle is a ________ follicle.
17. The process that produces the female gamete is called ________.
18. Meiosis in the female produces one large ovum and three small daughter cells called ________, which degenerate.
19. Meiosis is not fully complete in an ovum until ________ occurs.
20. The ________ are reproductive tubes connecting the ovary and the uterus.
21. The muscle layer of the uterus is called the ________.
22. The uterus is composed of several regions: the narrow lower part below the body is called the ________.
23. The innermost layer of the uterus, which is shed during menstruation, is called the ________.
24. The ________ is the part of the female reproductive system that opens to the exterior.
25. The ________ glands are glands that secrete a mucus-like fluid into the vestibule.
26. The milk-secreting glandular cells of the breast are arranged in grapelike structures called ________. These drain into ________ ducts that converge toward the nipple.
27. The ________ is composed of erectile tissue, similar to that found in the penis.
28. The first menses is known as the ________.
29. The term ________ is used to describe a pregnancy resulting from the implantation of a fertilized ovum in any location other than the uterus.

Match each disorder in Column A with its description or cause in Column B.

Column A	Column B
30. ________ cryptorchidism	a. a swelling of the scrotum caused by the overproduction of serous fluid
31. ________ benign prostatic hypertrophy	b. the lack of normal menstruation
32. ________ hydrocele	c. used to screen for cervical cancer
33. ________ inguinal hernia	d. the presence of functional endometrial tissue outside of the uterus
34. ________ dysmenorrhea	e. the inflammation of the uterine tubes
35. ________ amenorrhea	f. a sexually transmitted disease caused by a bacterium
36. ________ salpingitis	g. condition that occurs when a testis fails to descend into the scrotum
37. ________ Pap smear	h. painful menstruation syndrome
38. ________ myoma	i. swelling of the scrotum caused by intestine pushing through a weak part of the abdominal wall
39. ________ endometriosis	j. a sexually transmitted disease caused by a virus
40. ________ syphilis	k. a noncancerous condition characterized by enlargement of the prostate
41. ________ genital herpes	l. benign fibroid tumor of the uterus

Match each term in Column A with its corresponding description in Column B.

Column A	Column B
42. ________ FSH	a. term for the egg follicle after ovulation
43. ________ menstruation	b. ovarian hormone that reaches the highest concentration in the proliferative phase
44. ________ corpus luteum	c. caused by a rapid drop in the blood levels of estrogen and progesterone
45. ________ estrogen	d. phase of the reproductive cycle that begins after ovulation
46. ________ secretory phase	e. ovarian hormone that reaches its highest concentration during the secretory phase
47. ________ progesterone	f. term used to describe the egg being released from the ovary
48. ________ LH	g. phase of the reproductive cycle during which the uterine wall begins to thicken
49. ________ proliferative phase	h. pituitary hormone that stimulates the formation of an egg follicle
50. ________ ovulation	i. pituitary hormone that can be called the "ovulating hormone"

Case Studies

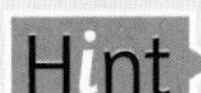

To solve a case study, you may have to refer to the glossary or index, other chapters in this textbook, and other resources.

1. As stated in the box on p. 626, one procedure that has been commonly used to screen for prostate cancer is palpation of the prostate through the wall of the rectum. Explain why digital (finger) palpation of the prostate is the only way to examine this gland from the outside without special equipment. What other prostate disorders might be detected this way?
2. Liz and Zeke have come to their physician with a problem: after 2 years of trying, they have not been able to conceive. According to the technical definition, is this couple infertile? On examination, Liz has been found to have pelvic inflammatory disease (PID). Could this condition have caused infertility?
3. Heather, age 22, briefly checks her breasts every few months and has never detected anything abnormal. However, a routine examination by her physician has revealed some small, tender lumps in both breasts. Does she have breast cancer? Can you think of any reason that Heather did not detect this condition herself?

Answers to Active Learning Questions can be found online at evolve.elsevier.com.

Growth, Development, and Aging

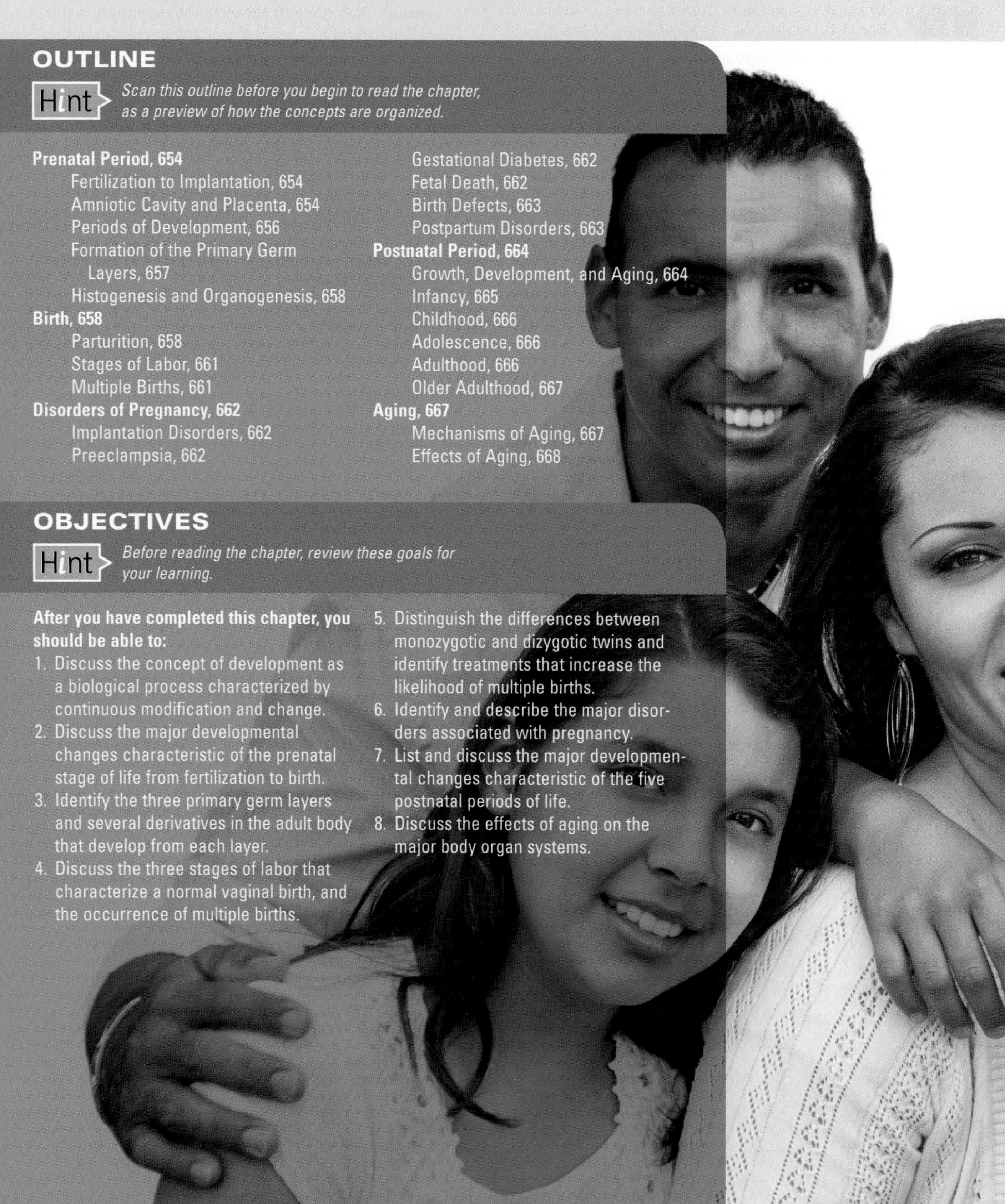

OUTLINE

Hint > *Scan this outline before you begin to read the chapter, as a preview of how the concepts are organized.*

OBJECTIVES

Hint > *Before reading the chapter, review these goals for your learning.*

After you have completed this chapter, you should be able to:

1. Discuss the concept of development as a biological process characterized by continuous modification and change.
2. Discuss the major developmental changes characteristic of the prenatal stage of life from fertilization to birth.
3. Identify the three primary germ layers and several derivatives in the adult body that develop from each layer.
4. Discuss the three stages of labor that characterize a normal vaginal birth, and the occurrence of multiple births.
5. Distinguish the differences between monozygotic and dizygotic twins and identify treatments that increase the likelihood of multiple births.
6. Identify and describe the major disorders associated with pregnancy.
7. List and discuss the major developmental changes characteristic of the five postnatal periods of life.
8. Discuss the effects of aging on the major body organ systems.

CHAPTER 24

Many of your fondest and most vivid memories are probably associated with your birthdays. The day of birth is an important milestone of life. Most people continue to remember their birthday in some special way each year; birthdays serve as pleasant and convenient reference points to mark periods of transition or change in our lives. The actual day of birth marks the end of one phase of life called the **prenatal period** and the beginning of a second called the **postnatal period.** The prenatal period begins at conception and ends at birth; the postnatal period begins at birth and continues until death.

Although important periods in our lives such as childhood and adolescence are often remembered as a series of individual and isolated events, they are in reality part of an ongoing and continuous process. In reviewing the many changes that occur during the cycle of life from conception to death, it is often convenient to isolate certain periods such as infancy or adulthood for study. It is important to remember, however, that life is not a series of stop-and-start events or individual and isolated periods of time. Instead,

LANGUAGE OF SCIENCE

Hint Before reading the chapter, say each of these terms out loud. This will help you to avoid stumbling over them as you read.

adolescence
(ad-oh-LES-ens)
[*adolesc-* **grow up,** *-ence* **state**]

adulthood
(ah-DULT-hood)

amniotic cavity
(am-nee-OT-ik KAV-ih-tee)
[*amnio-* **fetal membrane,** *-ic* **relating to,** *cav-* **hollow,** *-ity* **state**]

apoptosis
(ap-oh-TOH-sis *or* ap-op-TOH-sis)
[*apo-* **away,** *-ptosis* **falling**]

blastocyst
(BLAS-toh-sist)
[*blasto-* **bud,** *-cyst* **pouch**]

childhood
(CHILD-hood)

chorion
(KOH-ree-on)
[*chorion* **skin**]

chorionic villi
(koh-ree-ON-ik VIL-aye)
[*chorion-* **skin,** *-ic* **relating to**]

ectoderm
(EK-toh-derm)
[*ecto-* **outside,** *-derm* **skin**]

embryo
(EM-bree-oh)
[*em-* **in,** *-bryo* **fill to bursting**]

embryology
(em-bree-OL-oh-gee)
[*em-* **in,** *-bryo-* **fill to bursting,** *-log-* **words (study of),** *-y* **activity**]

embryonic phase
(em-bree-ON-ik fayz)
[*em-* **in,** *bryo* **fill to bursting,** *-ic* **relating to**]

Continued on p. 671

it is a biological process that is characterized by continuous modification and change.

This chapter discusses some of the events and changes that occur in the development of a human from conception to death. Study of development during the prenatal period is followed by a discussion of the birth process and a review of changes that occur during infancy and adulthood. Finally some important changes that occur in the individual organ systems of the body as a result of aging are discussed.

Prenatal Period

The *prenatal stage of development* begins at the time of conception or **fertilization** (that is, at the moment the female ovum and the male sperm cells unite) (**Figure 24-1**). The period of prenatal development continues until the birth of the child about 39 weeks later. The science of the development of the offspring before birth is called **embryology.** It is a story of biological marvels, describing the means by which a new human life is started and the steps by which a single microscopic cell is transformed into a complex human being.

FIGURE 24-1 Fertilization. Fertilization is a specific biological event. It occurs when the male and female sex cells fuse. After union between a sperm cell and the ovum has occurred, the cycle of life begins. The scanning electron micrograph shows spermatozoa attaching themselves to the surface of an ovum. Only one will penetrate and fertilize the ovum.

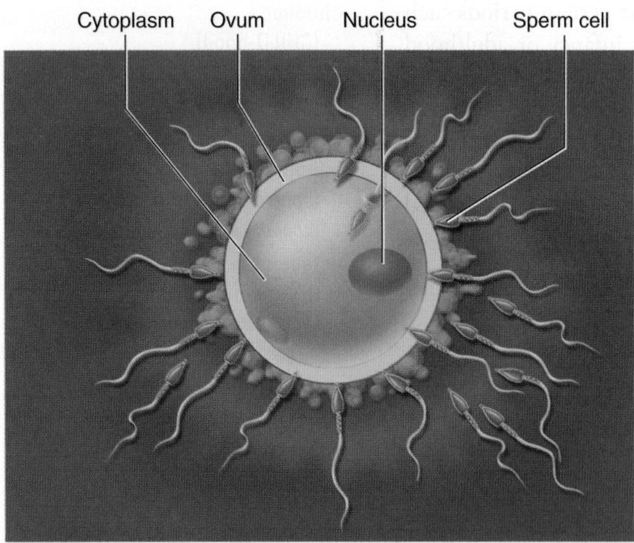

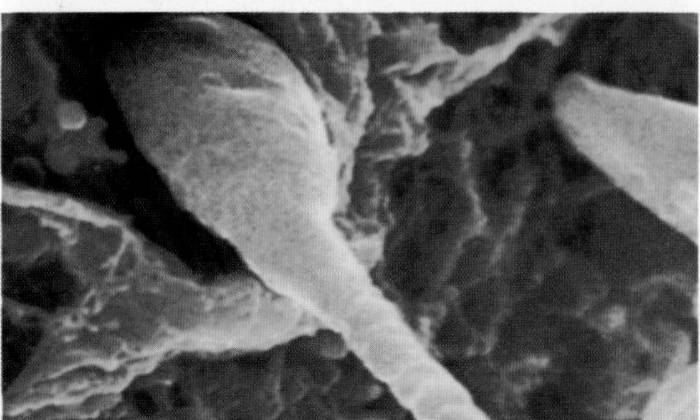

Fertilization to Implantation

After ovulation the discharged ovum first enters the pelvic cavity and then finds its way into the uterine (fallopian) tubes. Sperm cells swim up the uterine tube toward the ovum. Look at the relationship of the ovary, the two uterine tubes, and the uterus in **Figure 24-2**. Recall from Chapter 23 that each uterine tube extends outward from the uterus for about 10 cm. It then ends in the pelvic cavity near the ovary, as you can see in **Figure 24-2**, in an opening surrounded by fringelike processes, the *fimbriae.*

Sperm cells that are deposited in the vagina must enter and "swim" through the uterus and through the uterine tube to meet the ovum. Fertilization most often occurs in the outer one-third of the oviduct as shown in **Figure 24-2**.

The fertilized ovum, or **zygote,** is genetically complete—it is a new single-celled offspring. Time and nourishment are all that is needed for expression of characteristics such as sex, body build, and skin color that were determined at the time of fertilization. As you can see in the figure, the zygote immediately begins mitotic division, and in about 3 days a solid mass of cells called a **morula** is formed (see **Figure 24-2**). The cells of the morula continue to divide, and by the time the developing embryo reaches the uterus, it is a hollow ball of cells called a **blastocyst.**

During the 10 days from the time of fertilization to the time when the blastocyst completes **implantation** in the uterine lining, few nutrients from the mother are available. The rapid cell division taking place up to the blastocyst stage occurs with no significant increase in total mass compared with the zygote (**Figure 24-3**). One of the specializations of the ovum is its incredible store of nutrients that help support this embryonic development until implantation has occurred.

Amniotic Cavity and Placenta

Note in **Figure 24-4** that the blastocyst consists of an outer layer of cells and an inner cell mass. As the blastocyst develops, it forms a structure with two cavities, the **yolk sac** and **amniotic cavity.**

The yolk sac is most important in animals, such as birds, that depend heavily on yolk as the sole source of nutrients for the developing embryo. In these animals the yolk sac digests the yolk and provides the resulting nutrients to the embryo. Because uterine fluids provide nutrients to the developing human embryo until the placenta develops, the function of the yolk sac is not a nutritive one. Instead, it has other functions, including production of blood cells and stem cells that later form gametes in the gonads.

The yolk sac is an important site of hematopoiesis in early development. To see a diagram showing the shift in locations of blood-cell formation over the life span, review the article *Sites of Hematopoiesis* at Connect It! at *evolve.elsevier.com.*

The amniotic cavity becomes a fluid-filled, shock-absorbing sac, sometimes called the *bag of waters,* in which the embryo

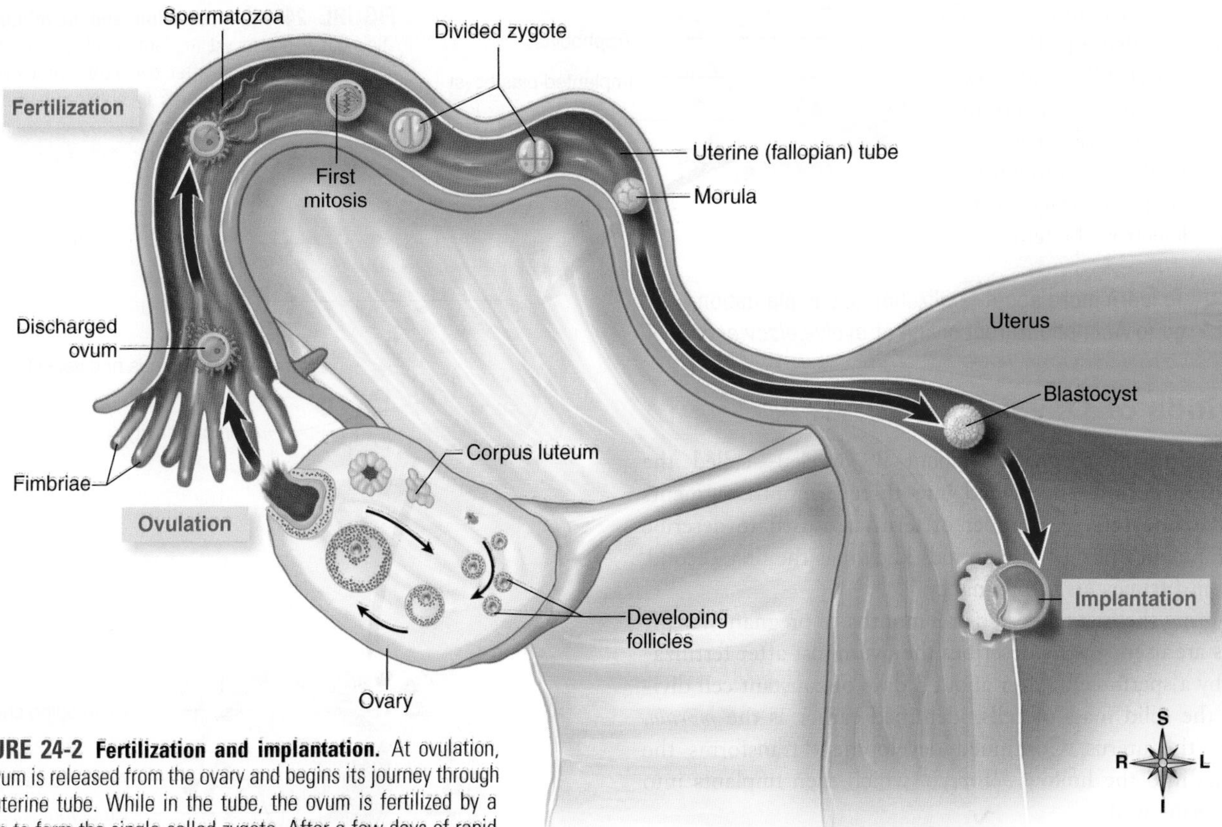

FIGURE 24-2 Fertilization and implantation. At ovulation, an ovum is released from the ovary and begins its journey through the uterine tube. While in the tube, the ovum is fertilized by a sperm to form the single-celled zygote. After a few days of rapid mitotic division, a ball of cells called a *morula* is formed. After the morula develops into a hollow ball called a *blastocyst,* implantation occurs.

floats during development. The **chorion,** shown in Figure 24-4 and Figure 24-5, develops into an important fetal membrane in the **placenta.** The **chorionic villi** shown in Figure 24-5 connect the blood vessels of the chorion to the rest of the placenta. The placenta (see Figure 24-5) anchors the developing fetus to the uterus and provides a "bridge" for the exchange of nutrients and waste products between mother and offspring.

The placenta is a unique structure that has a temporary but very important series of functions during pregnancy. It is composed of tissues from mother and child and functions not only as a structural "anchor" and nutritive bridge but also as an excretory, respiratory, and endocrine organ (see Figure 24-5).

Placental tissue normally separates the maternal blood, which fills the lacunae of the placenta, from the fetal blood so that no intermixing occurs. The very thin layer of placental tissue that separates maternal and fetal blood also serves as an effective "barrier" that can protect the developing offspring from many harmful substances that may enter the mother's bloodstream.

Unfortunately, toxic substances, such as alcohol and some infectious organisms, may nonetheless penetrate this protective

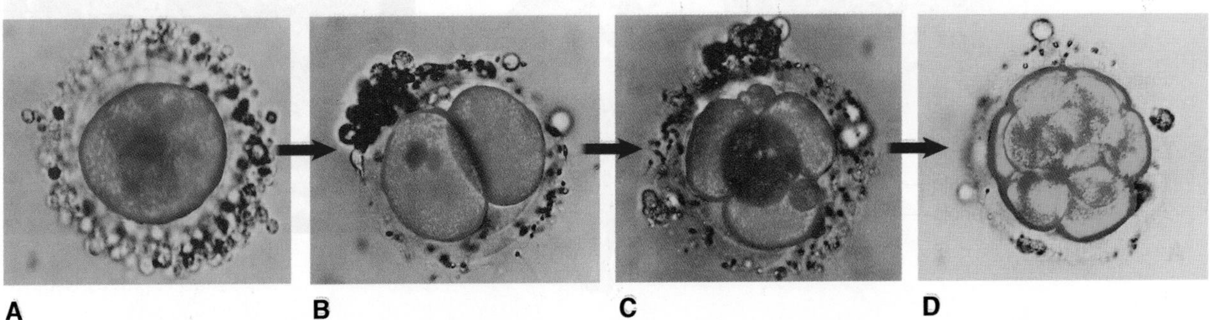

FIGURE 24-3 Early stages of development. A, Fertilized ovum or zygote. **B** to **D,** Early cell divisions produce more and more cells. The solid mass of cells shown in **D** forms the morula—an early stage in embryonic development.

placental barrier and injure the developing offspring. The *cytomegalovirus (CMV)*, the *Zika virus (ZV)*, or the bacterium that causes *syphilis*, for example, can easily pass through the placenta and cause tragic developmental defects in the fetus.

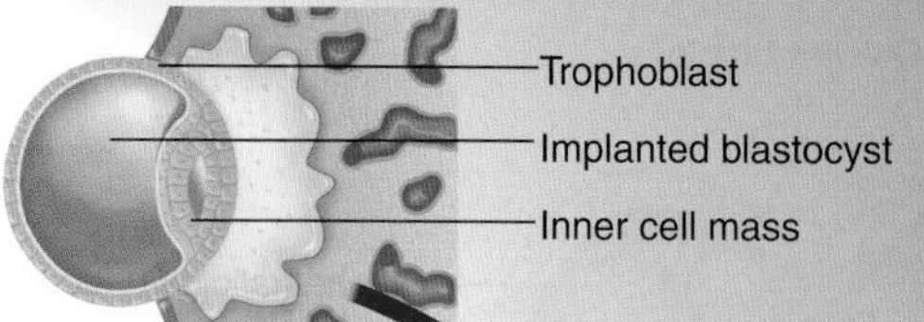

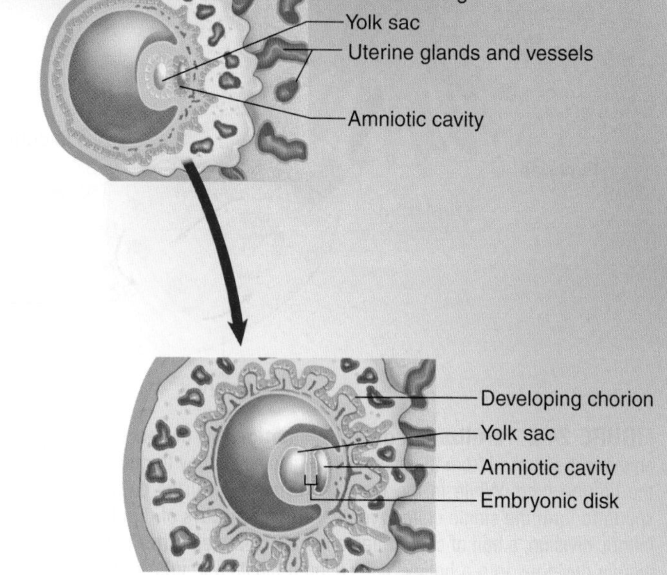

FIGURE 24-4 Implantation and development. The hollow blastocyst implants itself in the uterine lining about 10 days after ovulation. Until the placenta is functional, nutrients are obtained by diffusion from uterine fluids. Notice the developing chorion and how the blastocyst eventually forms a yolk sac and amniotic cavity.

To learn more about fertilization and implantation, go to AnimationDirect online at *evolve.elsevier.com.*

Periods of Development

The length of pregnancy (about 39 weeks)—called the **gestation period**—is divided into three 3-month segments called **trimesters.** A number of terms are used to describe development during these periods known as the first, second, and third trimesters of pregnancy.

During the first trimester, or 3 months, of pregnancy, many terms are used. *Zygote* describes the ovum just after fertilization by a sperm cell. After about 3 days of constant cell division, the solid mass of cells, identified earlier as the *morula,* enters the uterus. Continued development transforms the morula into the hollow *blastocyst,* which then implants into the uterine wall.

The **embryonic phase** of development extends from the third week after fertilization until the end of week 8 of **gestation.** During this period in the first trimester, the term **embryo** is used to describe the developing offspring. By day 35 of gestation (**Figure 24-6,** *A*), the heart is beating and, although the embryo is only 8 mm (about ⅜ inch) long, the eyes and so-called "limb buds," which ultimately form the arms and legs, are clearly visible.

The period of development extending from week 9 to week 39 is termed the **fetal phase.** During this period, the term *embryo* is replaced by **fetus.**

Figure 24-6, *C,* shows the stage of development of the fetus at the end of the first trimester of gestation. Body size is about 7 to 8 cm (3.2 inches) long. The facial features of the fetus are apparent, the limbs are complete, and gender can be identified. By month 4 (**Figure 24-6,** *D*) all organ systems are complete and in place.

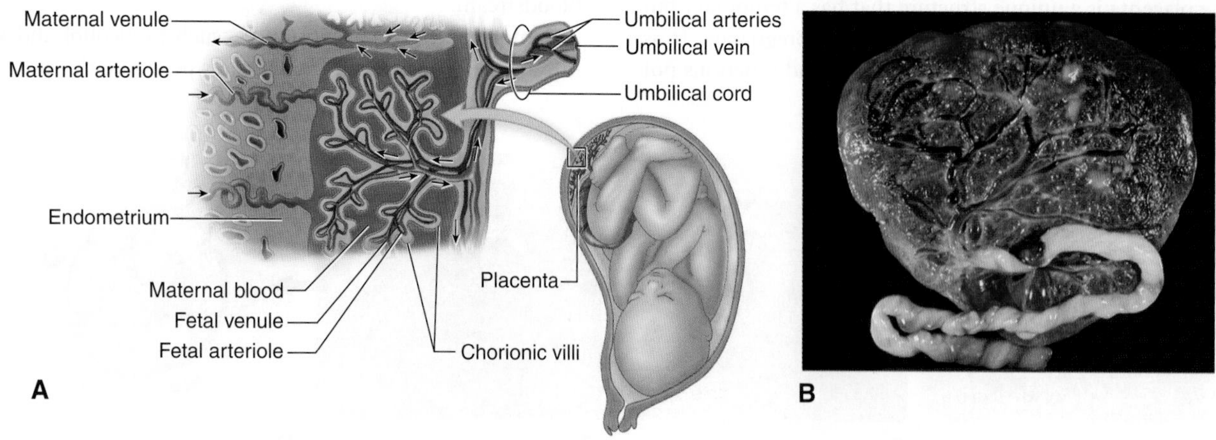

FIGURE 24-5 Placenta. The close placement of the fetal blood supply and the maternal blood in the placenta permits diffusion of nutrients and other substances. It also forms a thin barrier to prevent diffusion of most harmful substances. No mixing of fetal and maternal blood occurs. **A,** Diagram showing a cross section of the placental structure. **B,** Photograph of a normal, full-term placenta (fetal side) showing the branching of the placental blood vessels.

RESEARCH, ISSUES, AND TRENDS

IN VITRO FERTILIZATION

The Latin term *in vitro* means, literally, "within a glass." In the case of in vitro fertilization, it refers to the glass laboratory dish where an ovum and sperm are mixed and where fertilization occurs.

In the classic technique, the ovum is obtained from the mother by first inserting a fiber-optic viewing instrument called a **laparoscope** through a very small incision in the woman's abdomen. After it is in the pelvic cavity, the device allows the physician to view the ovary and then puncture and "suck up" an ovum from a mature follicle. Over the years, refinements to this technique have been made and less invasive procedures, such as the insertion of a needle through the vaginal wall, are currently being used. After the ovum is harvested, it is mixed with sperm so that fertilization occurs.

After about 2½ days' growth in a temperature-controlled environment, the developing zygote, which by then has reached the 8- or 16-cell stage, is returned by the physician to the mother's uterus. If implantation is successful, growth will continue and the subsequent pregnancy will progress. In the most successful fertility clinics in the United States, a normal term birth will occur in about 30% of in vitro fertilization attempts.

With the advent of genomics (see Chapter 25), we are now able to check embryos for genetic disorders before implanting them. However, the controversial practice of selecting embryos for gender has caused ethicists to debate whether the screening of embryos is appropriate medical practice.

To find out more about related **assisted reproductive technologies (ARTs)**, check out the article *In Vitro Fertilization* at Connect It! at *evolve.elsevier.com*.

Formation of the Primary Germ Layers

At the very beginning of the embryonic stage, all the cells are **stem cells.** Stem cells are unspecialized cells that reproduce to form specific lines of specialized cells. At this stage, they have their highest "stemness" or potency—that is, they are capable of producing many different kinds of cells in the body.

Adult stem cells remain after early development, but can only produce a few specialized kinds of cells in a particular tissue. We have already encountered these adult stem cells when we discussed **hematopoiesis**—formation of red blood cells (RBCs), white blood cells (WBCs), and platelets—in bone marrow. Other stem cells are found in the skin, many glands, muscles, nerve tissue, bone, and the gastrointestinal (GI) tract. Adult stem cells replace the specialized cells in a tissue and thus ensure stable, functional populations of the cell types needed for survival.

Early in the first trimester of pregnancy, three layers of stem cells develop that embryologists call the **primary germ layers** (**Table 24-1**). Each layer gives rise to definite structures such as the skin, nervous tissue, muscles, or digestive organs. **Table 24-1** lists a number of structures derived from each of the three primary germ layers:

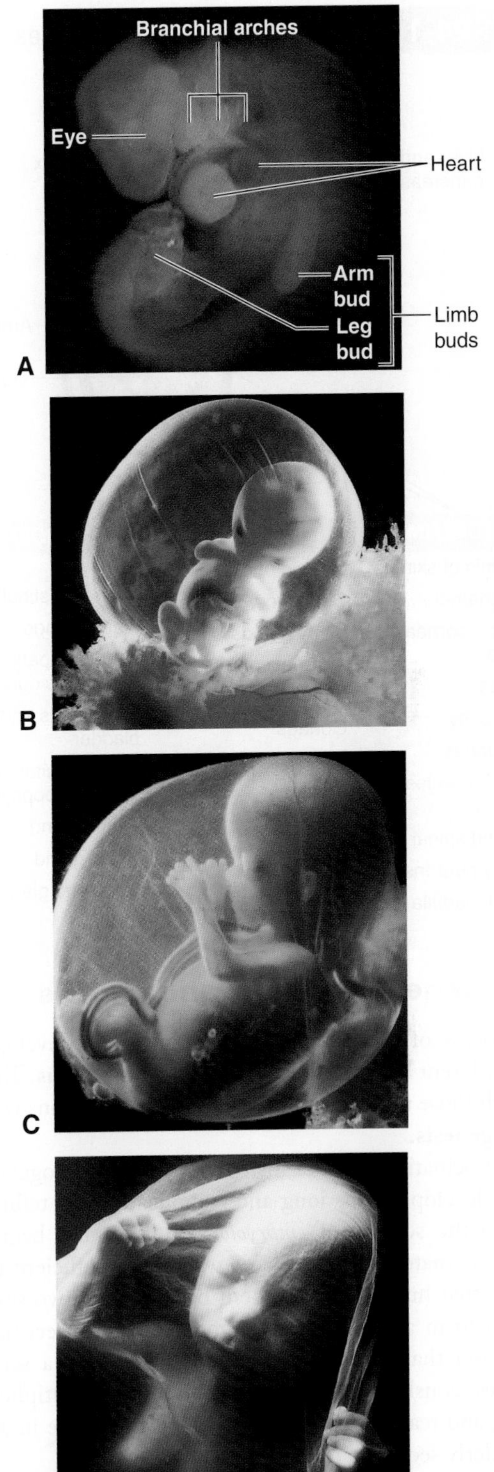

FIGURE 24-6 Embryos and fetuses. A, At 35 days. **B,** At 49 days. **C,** At the end of the first trimester. **D,** At 4 months.

1. **Endoderm**—inside layer
2. **Mesoderm**—middle layer
3. **Ectoderm**—outside layer

24

TABLE 24-1 Primary Germ Layer Derivatives

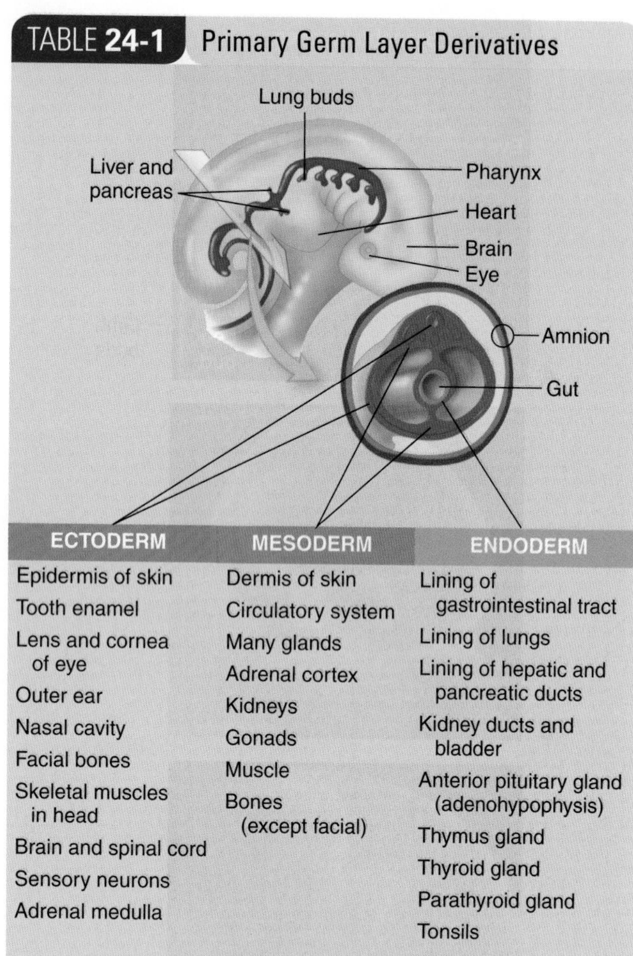

ECTODERM	MESODERM	ENDODERM
Epidermis of skin	Dermis of skin	Lining of gastrointestinal tract
Tooth enamel	Circulatory system	Lining of lungs
Lens and cornea of eye	Many glands	Lining of hepatic and pancreatic ducts
Outer ear	Adrenal cortex	Kidney ducts and bladder
Nasal cavity	Kidneys	Anterior pituitary gland (adenohypophysis)
Facial bones	Gonads	Thymus gland
Skeletal muscles in head	Muscle	Thyroid gland
Brain and spinal cord	Bones (except facial)	Parathyroid gland
Sensory neurons		Tonsils
Adrenal medulla		

Histogenesis and Organogenesis

The process of how the primary germ layers develop into many different kinds of tissues is called **histogenesis.** The way in which those tissues arrange themselves into organs is called **organogenesis.**

The fascinating story of histogenesis and organogenesis in human development is long and complicated—its telling belongs to the science of *embryology.* But for the beginning student of anatomy and physiology, it seems sufficient to appreciate that human development begins when two sex cells unite to form a single-celled zygote. It is also necessary to understand that the offspring's body evolves by a series of processes consisting of cell differentiation, multiplication, growth, and rearrangement, all of which take place in a definite, orderly sequence (**Figure 24-7**).

Development of structure and function go hand in hand, and from 4 months of gestation, when every organ system is complete and in place, until term (about 280 days), fetal development is mainly a matter of growth. **Figure 24-8**, *step 1,* shows the normal intrauterine placement of a fetus just before birth in a full-term pregnancy.

To learn more about how the various body tissues and organs develop, review the article *Embryonic Development of Tissues* at Connect It! at *evolve.elsevier.com.*

QUICK CHECK

1. What is the *postnatal period?* The *prenatal period?*
2. What is a *zygote?* How is it different from a morula or blastocyst?
3. Name and describe the three primary germ layers.
4. What are *stem cells?*
5. What is meant by the term *organogenesis?*

Birth

Parturition

The process of birth—sometimes called **parturition**—is the point of transition between the prenatal and postnatal periods of life.

As pregnancy draws to a close, the uterus becomes "irritable" and, ultimately, muscular contractions begin and cause the cervix to dilate or open, thus permitting the fetus to move from the uterus through the vagina, or "birth canal," to the exterior. When contractions occur, the amniotic sac or "bag of water" ruptures, and labor begins.

The process normally begins with the fetus taking a head-down position against the cervix (**Figure 24-8**, *1*). A *breech birth* is one in which the fetus fails to turn head downward and consequently the feet are born first. This condition usually requires the offspring to be born by **cesarean section.** Often called simply a *C-section,* it is a surgical procedure in which the newborn is delivered through an incision in the abdomen and uterine wall. The procedure may be done when abnormal conditions of the mother or fetus (or both) make normal vaginal delivery hazardous or impossible.

To learn more about cesarean section, go to AnimationDirect online at *evolve.elsevier.com.*

RESEARCH, ISSUES, AND TRENDS

HOW LONG DOES PREGNANCY LAST?

This seems like a silly question to most of us; the answer is 9 months, isn't it? Actually, the length of gestation (the amount of time one is pregnant) is defined in different ways in different situations and can vary from one pregnancy to another.

The average gestation in humans is 266 days, starting at the day of conception. But physicians instead usually count from the beginning of the woman's last menstrual period, for an average of 280 days. But these are only averages. What is normal in one case can be different from what is normal in another case. In practice, any pregnancy of less than 37 weeks (259 days) is said to be premature, and any lasting more than 42 weeks (294 days) is said to be postmature.

So as with many statistics regarding human function, what is "normal" can be spoken of only in generalities and averages.

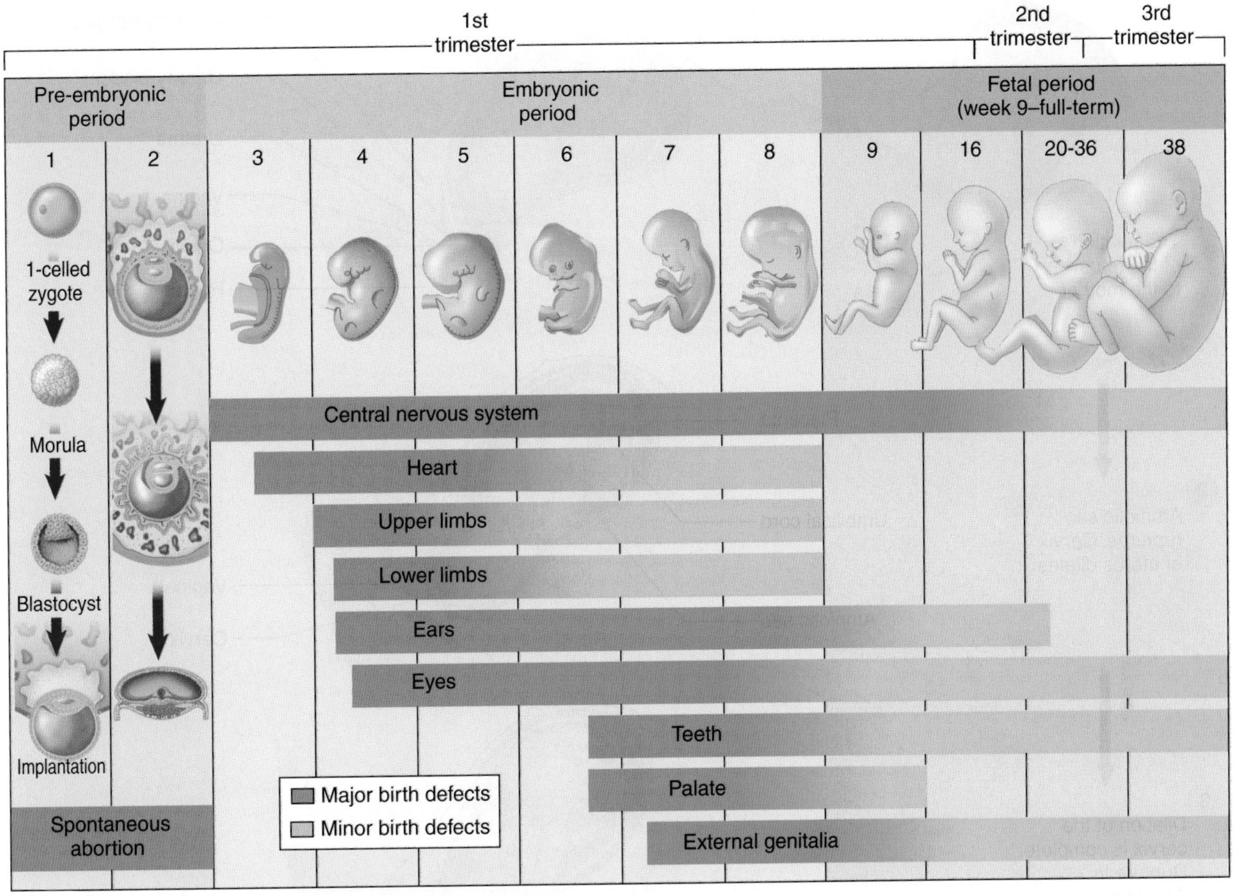

1st trimester
2nd trimester
3rd trimester

Pre-embryonic period
Embryonic period
Fetal period (week 9–full-term)

1 2 3 4 5 6 7 8 9 16 20-36 38

1-celled zygote
Morula
Blastocyst
Implantation

Central nervous system
Heart
Upper limbs
Lower limbs
Ears
Eyes
Teeth
Palate
External genitalia

Spontaneous abortion

Major birth defects
Minor birth defects

FIGURE 24-7 Critical periods of neonatal development. The *red areas* show when teratogens are most likely to cause major birth defects, and the *yellow areas* show when minor defects are more likely to arise. Numbers refer to weeks of gestation.

RESEARCH, ISSUES, AND TRENDS

FREEZING UMBILICAL CORD BLOOD

The concept of development of blood cells from red bone marrow, a process called **hematopoiesis,** was introduced in Chapter 13. Ultimately, the presence of **stem cells** is required for bone marrow to produce blood cells. The fact that umbilical cord blood is rich in these stem cells has great clinical significance.

In the past, if the stem cells in the bone marrow of a child were destroyed as a result of leukemia or by chemotherapy, death would result unless a bone marrow transplant was possible. Infusion of stored umbilical cord blood obtained from the child at the time of birth is an attractive alternative. The blood is rich in stem cells and can be obtained without risk. This procedure is much more cost-effective than a bone marrow transplant.

Removing and freezing umbilical cord blood at the time of birth may become a type of biological insurance against some types of leukemia that may affect a child later in life. Cord blood is readily available at birth and is a better source of stem cells than is bone marrow.

When the umbilical cord is cut after birth, the blood that remains in the cord is simply drained into a sterile bag (see photo), frozen, and then stored in liquid nitrogen in one of about a dozen cord-blood centers in the United States.

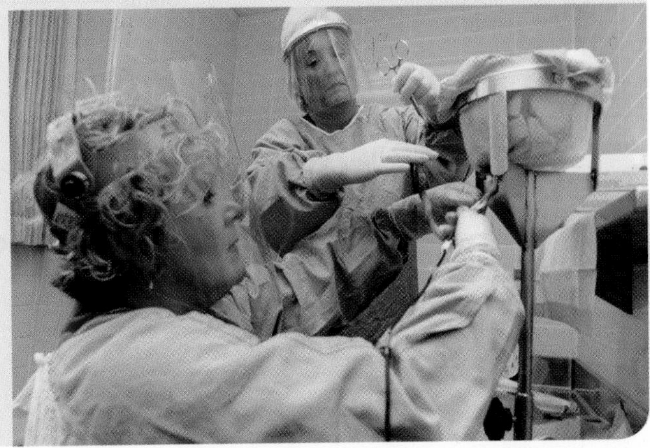

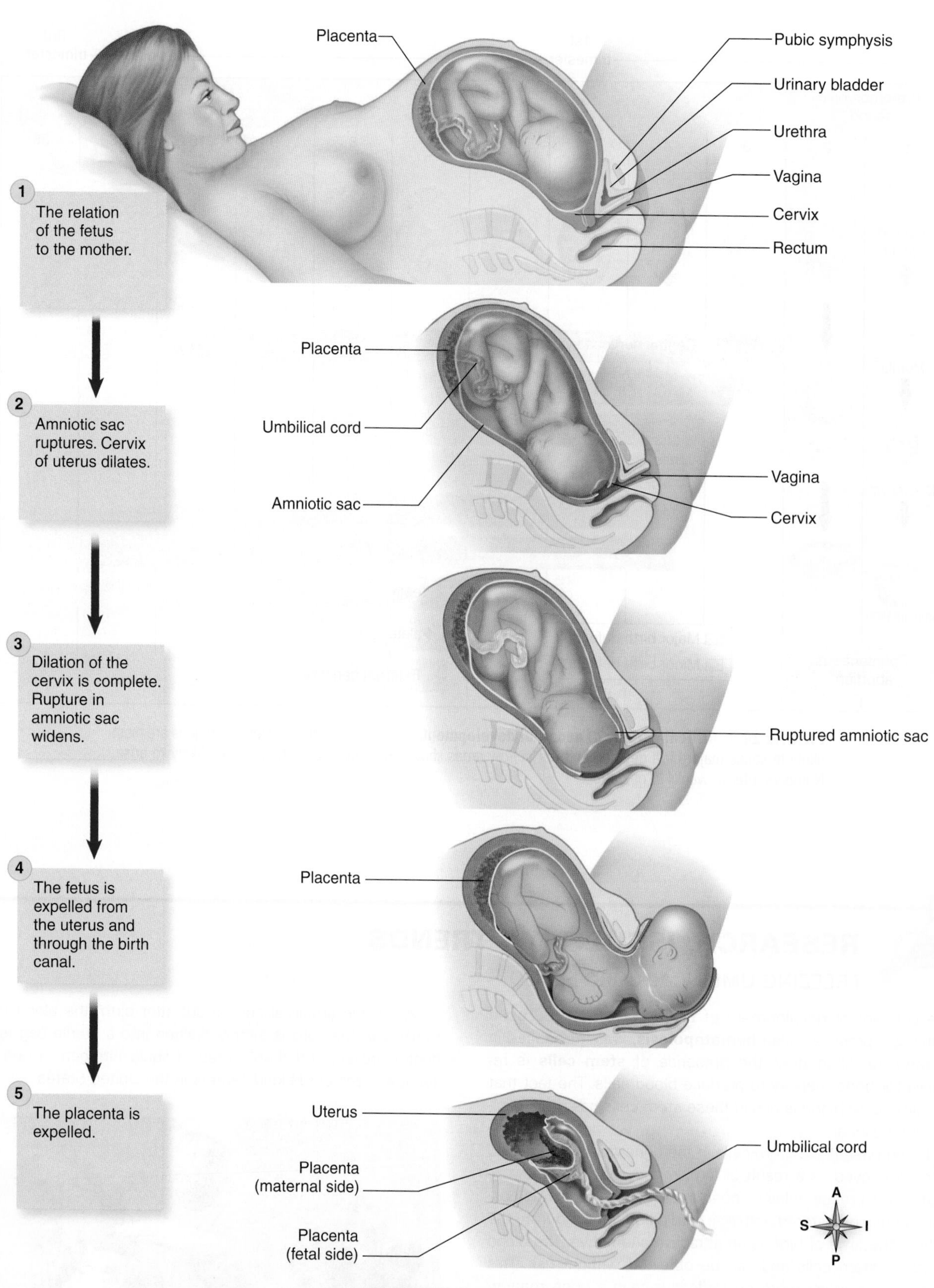

FIGURE 24-8 **Parturition.**

24

Stages of Labor

Labor is the process that results in the birth of an infant. It has three stages (**Figure 24-8**, *steps 2 to 5*):

- *Stage one*—period from onset of uterine contractions until dilation of the cervix is complete.
- *Stage two*—period from the time of maximal cervical dilation until the offspring exits through the vagina.
- *Stage three*—process of expulsion of the placenta through the vagina.

The time required for normal vaginal birth varies widely and may be influenced by many variables, including whether the woman has previously had a child. In most cases, stage one of labor lasts from 6 to 24 hours, and stage two lasts from a few minutes to an hour. Delivery of the placenta (stage three) normally occurs within 15 minutes after the birth of the infant.

Figure 1-12 on p. 16 illustrates the role of oxytocin (OT) in promoting a rapid delivery. A synthetic version of OT is sometimes given therapeutically if labor becomes dangerously slow.

To assess the general condition of a newborn, a system that evaluates five health criteria is often used. The criteria are heart rate (HR), respiration, muscle tone, skin color, and response to stimuli. Each aspect is scored as 0, 1, or 2—depending on the condition of the infant. The resulting total score is called the **Apgar score.** The Apgar score in a completely healthy newborn is 10.

To learn more about the three stages of birth, go to AnimationDirect online at *evolve.elsevier.com*.

Multiple Births

The term *multiple births* refers to the birth of two or more infants from the same pregnancy. The birth of twins is more common than the birth of triplets, quadruplets, or quintuplets.

Multiple-birth offspring are often born prematurely, so they are at a greater than normal risk of complications in infancy. However, premature infants that have modern medical care available have a much lower risk of complications than without such care.

Twinning, or double births, can result from at least two different natural processes, described in the following paragraphs.

Identical twins result from the splitting of embryonic tissue from the same zygote early in development. For this reason identical twins are also called **monozygotic twins.** As **Figure 24-9**, *A*, shows, identical twins usually share the same placenta but have separate umbilical cords.

Because they develop from the same fertilized egg, identical twins have the same genetic code. Despite this, identical twins are not absolutely identical in terms of structure and function. Different environmental factors and personal experiences lead to individuality even in genetically identical twins.

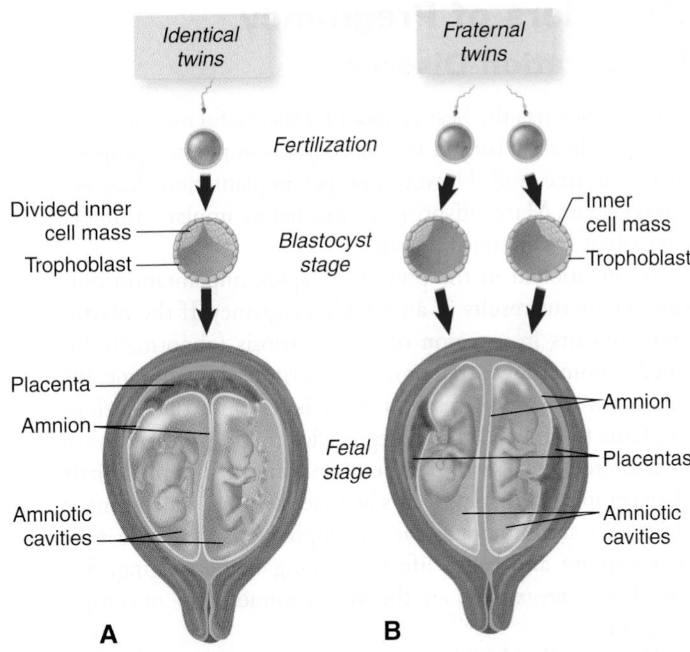

FIGURE 24-9 Multiple births. A, Identical (monozygotic) twins develop when embryonic tissue from a single zygote splits to form two individuals. Notice that the placenta and the part of the amnion separating the amniotic cavities are shared by the twins. **B,** Fraternal (dizygotic) twins develop when two ova are fertilized at about the same time, producing two separate zygotes. Notice that each fraternal twin has its own placenta and amnion.

Fraternal twins result from the fertilization of two different ova by two different spermatozoa—and are thus also called **dizygotic twins** (**Figure 24-9**, *B*). Dizygotic twinning requires the production of more than one mature ovum during a single menstrual cycle, a trait that is often inherited. Multiple ovulations also may occur in response to certain fertility drugs, especially the gonadotropin preparations.

Fraternal twins are no more closely related genetically than any other brother-sister relationship. Because two separate fertilizations must occur, it is even possible for fraternal twins to have different biological fathers. Triplets, quadruplets, and other multiple births may be identical, fraternal, or any combination.

More and more often, multiple births result from medical treatments. For example, fertility-enhancing drugs sometimes increase the number of eggs released at ovulation and thus increase the likelihood of multiple births. *In vitro* fertilization and other reproductive medical procedures often involve implantation of multiple embryos—increasing the odds of successful reproduction while also increasing the odds of multiple births (see the Research, Issues, and Trends box on p. 657).

QUICK CHECK

1. How does a *breech* birth differ from a normal birth?
2. Name and describe the three stages of *labor*?
3. How do *multiple births* occur?
4. Identify the primary difference between *identical* and *fraternal twins.*
5. What is an *Apgar score*?

24

Disorders of Pregnancy

Implantation Disorders

A pregnancy has the best chance of a successful outcome, the birth of a healthy baby, if the blastocyst is implanted properly in the uterine wall. However, proper implantation does not always occur. Many offspring are lost before implantation occurs, often for unknown reasons.

As mentioned in the previous chapter, implantation outside the uterus results in an ectopic pregnancy. If the blastocyst implants in a region of endometriosis (abnormally located endometrial tissue) or normal peritoneal membrane, the pregnancy may be successful if there is room for the developing fetus to grow. Ectopic pregnancies that do succeed must be delivered by C-section rather than by normal vaginal birth. If an ectopic pregnancy occurs in a uterine tube, which cannot stretch to accommodate the developing offspring, the tube may rupture and cause life-threatening hemorrhaging. So-called *tubal pregnancies* are the most common type of ectopic pregnancy.

Occasionally, the blastocyst implants in the uterine wall near the cervix. This in itself may present no problem, but if the placenta grows too closely to the cervical opening a condition called **placenta previa** results. The normal dilation and softening of the cervix that occurs in the third trimester often causes painless bleeding as the placenta near the cervix separates from the uterine wall. The massive blood loss that may result can be life threatening for both mother and offspring (**Figure 24-10**, *A*).

Separation of the placenta from the uterine wall can occur even when implantation occurs in the upper part of the uterus. When this occurs in a pregnancy of 20 weeks or more, the condition is called **abruptio placentae.** Complete separation of the placenta causes the immediate death of the fetus. The severe hemorrhaging that often results, sometimes hidden in the uterus, may cause circulatory shock and death of the mother within minutes. A cesarean section and perhaps also a hysterectomy must be performed immediately to prevent blood loss and death (**Figure 24-10**, *B*).

24

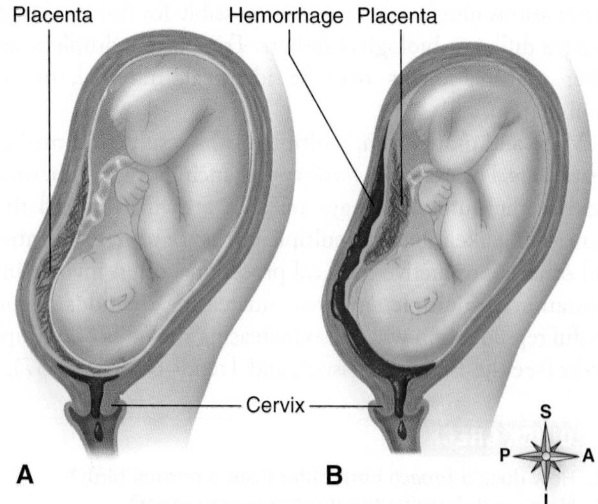

FIGURE 24-10 Implantation disorders. A, Placenta previa is the condition where the placenta grows too closely to the cervical opening. **B,** Abruptio placentae is the condition where there is complete separation of the placenta, causing the death of the fetus.

Preeclampsia

Preeclampsia, also called *toxemia of pregnancy,* is a serious disorder that occurs in about 1 in every 20 pregnancies. This disorder is characterized by the onset of acute hypertension after the twenty-fourth week, accompanied by proteinuria and edema.

The causes of preeclampsia are largely unknown, despite intense research efforts. Preeclampsia can result in complications such as abruptio placentae, stroke, hemorrhage, fetal malnutrition, and low birth weight.

Preeclampsia can progress to **eclampsia,** a life-threatening form of toxemia that causes severe convulsions, coma, kidney failure, and perhaps death of the fetus and mother.

Gestational Diabetes

The term **gestational diabetes mellitus (GDM)** is applied in cases of hyperglycemia (high blood glucose) that first occur during pregnancy. Some cases diagnosed as GDM actually may be previously undiagnosed cases of *type 1 DM* or *type 2 DM.* Now is a good time to review DM in Chapter 12 on pp. 335-337.

The conditions of pregnancy promote insulin resistance in cells. If insulin resistance is already an issue, then the insulin resistance is even more of a problem. GDM results from the failure of the pancreatic islets to produce additional insulin to compensate for this increased insulin resistance.

As **Figure 12-18** on p. 337 shows, the insulin resistance that occurs in DM produces a wide variety of changes in body function. Therefore, improperly managed GDM can have health effects in either or both the mother and fetus during pregnancy.

For example, high blood glucose in the fetus can lead to weight gain, which results in delivery of a large infant—a risk factor for complications during labor and delivery. GDM can also increase the risk of needing to deliver by C-section, rather than a vaginal delivery. GDM may increase the risk of hypertension that may progress to preeclampsia.

Fetal Death

A *miscarriage* is the loss of an embryo or fetus before the twentieth week (or a fetus weighing less than 500 g or 1.1 lb). Technically known as a **spontaneous abortion,** the most common cause of such a loss is a structural or functional defect in the developing offspring. Abnormalities of the mother, such as hypertension, uterine abnormalities, and hormonal imbalances, can also cause spontaneous abortions. After 20 weeks, delivery of a lifeless infant is termed a **stillbirth.**

Birth Defects

Developmental problems present at birth are often called **birth defects.** Such abnormalities may be structural or functional, perhaps even involving behavior and personality.

Birth defects may be caused by genetic factors such as abnormal genes or inheritance of an abnormal number of chromosomes. Birth defects also may be caused by exposure to environmental factors called **teratogens**—especially during organogenesis. Teratogens include:

- Radiation—for example, x-rays
- Toxic chemicals—for example, drugs, cigarette smoke, or alcohol
- Infections in the mother—for example, herpes virus, Zika virus, or cytomegalovirus

Some teratogens are also mutagens because they do their damage by changing the genetic code in cells of the developing embryo. Nutritional deficiencies during pregnancy also can lead to birth defects.

As Figure 24-7 shows, the period during the first trimester when the tissues are beginning to differentiate and the organs are just starting to develop is the time that teratogens are most likely to cause damage. In fact, teratogens can cause spontaneous abortion (miscarriage) if significant damage occurs during the pre-embryonic stage.

Umbilical hernia, which occurs when intestines protrude through the umbilical opening in the abdomen, is common in newborn infants. Review the article *Hernias* at Connect It! at *evolve.elsevier.com.*

Postpartum Disorders

Puerperal fever, or *childbed fever,* is a syndrome of postpartum mothers characterized by bacterial infection that progresses to septicemia (blood infection) and possibly death. Until the 1930s, puerperal fever was the leading cause of maternal death—claiming the lives of more than 20% of postpartum women. Modern antiseptic techniques prevent most postpartum infections now. Puerperal infections that do occur are usually treated successfully by an immediate and intensive program of antibiotic therapy.

After a child is born, it needs the nourishment of mother's milk to survive. However, a number of disorders of lactation (milk production) may occur to prevent a mother from nursing her infant. For example, anemia, malnutrition, emotional stress, and structural abnormalities of the breast can all interfere with normal lactation.

Mastitis, or breast inflammation, often caused by infection, can result in lactation problems or production of milk

RESEARCH, ISSUES, AND TRENDS

ANTENATAL DIAGNOSIS AND TREATMENT

Advances in **antenatal medicine** or "before-birth therapy" now permit extensive diagnosis and treatment of disease in the fetus much like any other patient. This new dimension in medicine began with techniques by which Rh fetuses could be given transfusions before birth.

Current procedures using images provided by ultrasound equipment (see figures) allow physicians to prepare for and perform, before birth, corrective surgical procedures such as bladder repair. These procedures also allow physicians to monitor the progress of other types of treatment on a developing fetus. Figure *A* shows placement of the ultrasound transducer on the abdominal wall. The resulting image is called an **ultrasonogram.** Figures *B* and *C* show two-dimensional and three-dimensional types of ultrasonogram of a 20-week and 21-week fetus (respectively).

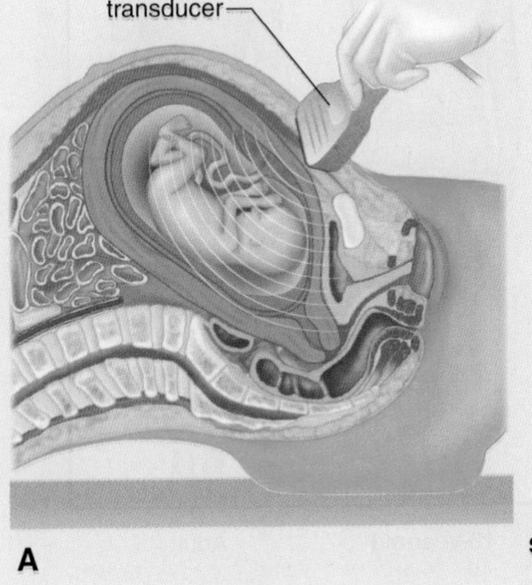

A

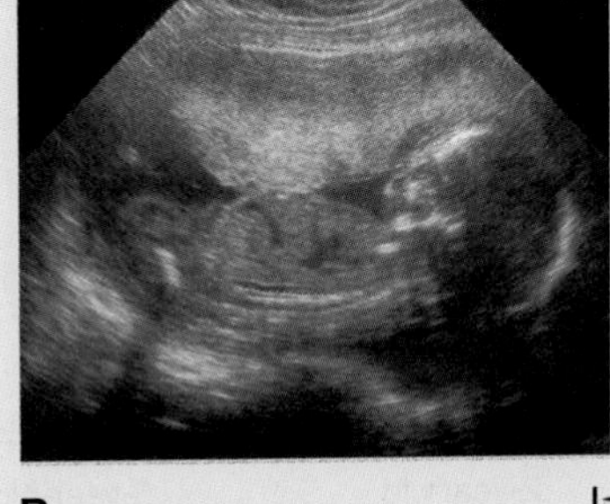

B

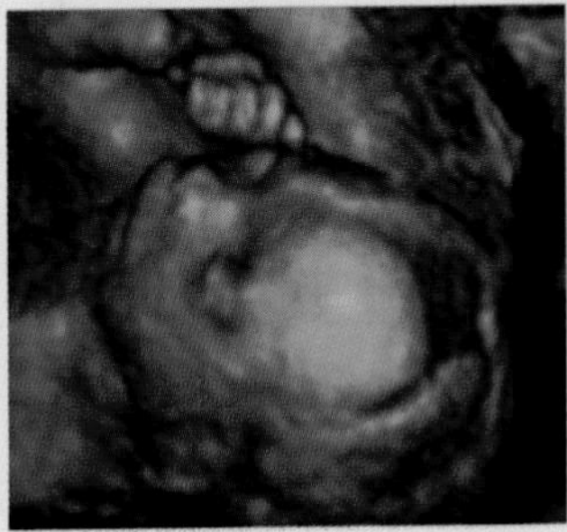

C

24

RESEARCH, ISSUES, AND TRENDS

FETAL ALCOHOL SYNDROME

Consumption of alcohol by the mother during pregnancy can have tragic effects on a developing fetus. Educational efforts to inform pregnant women about the dangers of alcohol are now receiving national attention. Even very limited consumption of alcohol during pregnancy poses significant hazards to the developing fetus because alcohol can easily cross the placental barrier and enter the fetal bloodstream.

When alcohol enters the fetal blood, the potential result, called **fetal alcohol syndrome (FAS),** can cause tragic **congenital** abnormalities such as "small head," or **microcephaly** (see figure), low birth weight, developmental disabilities such as mental retardation, and even fetal death.

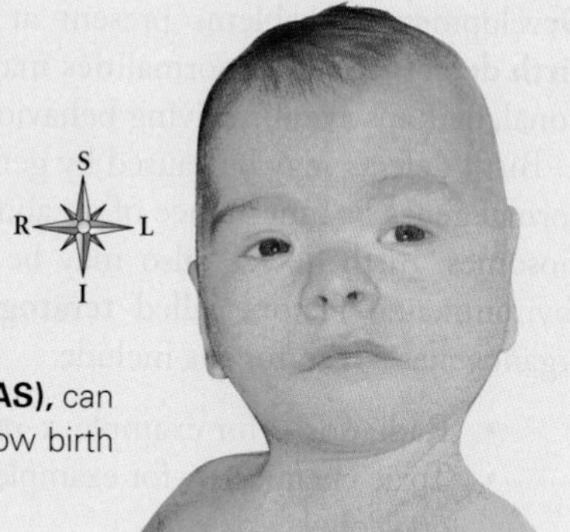

contaminated with pathogenic organisms. In many cultures, the availability of other nursing mothers or breast-milk substitutes allows proper nourishment of the infant when lactation problems develop. Most breast-milk substitutes are formulations of milk from another mammal, such as a cow.

Infants who lack the enzyme *lactase* may not be able to digest the lactose present in human or animal milk, resulting in a condition called **lactose intolerance.** Infants with lactose intolerance are sometimes given a lactose-free milk substitute made from soy beans or other plant products.

QUICK CHECK

1. What disorders can result from improper implantation in the uterine wall?
2. What complications may occur as a result of *preeclampsia*?
3. What is a *teratogen*? Give some examples of teratogens.
4. What is *gestational diabetes*?

Postnatal Period

Growth, Development, and Aging

The **postnatal period** begins at birth and lasts until death. Although it is often divided into major periods for study, we need to understand and appreciate that growth, development, and aging are continuous processes that occur throughout the life cycle.

Gradual changes in the physical appearance of the body as a whole and in the relative proportions of the head, trunk, and limbs are quite noticeable between birth and adolescence.

Note in **Figure 24-11** the obvious changes in the size of bones and in the proportionate sizes between different bones and body areas. The head, for example, becomes proportionately smaller. Whereas the infant head is approximately one-fourth of the total height of the body, the adult head is only about one-eighth of the total height. The facial bones also show several changes between infancy and adulthood. In an infant the

24

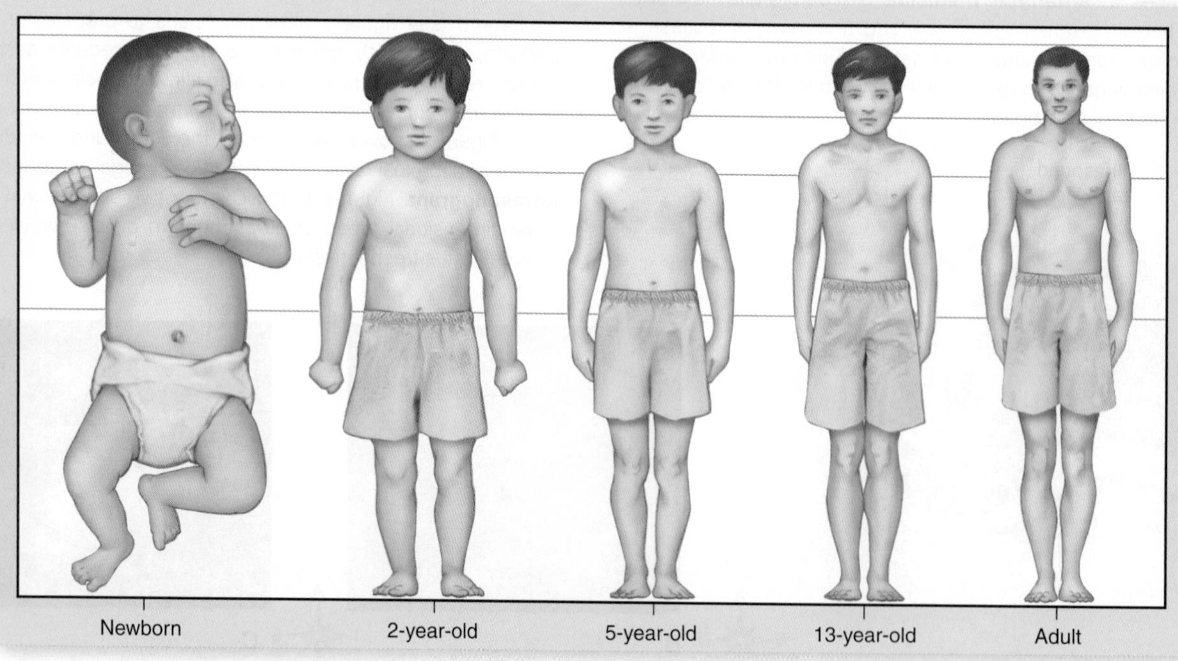

FIGURE 24-11 Developmental changes in body proportions. Note the dramatic differences in head size. As the individual grows, there is a gradual change in the relative proportions of the head, trunk, and limbs. The head becomes proportionately smaller, and the legs become proportionately longer and the trunk shorter.

TABLE 24-2 U.S. Population Projections by Age Group*

	2020	2030	2040	2050	2060	PERCENT CHANGE
Total population	334,503	359,402	380,219	398,328	416,795	25%
Infants and Toddlers						
0-4 years	20,568	21,178	21,471	22,147	22,778	11%
Children						
5-13 years	36,824	38,322	39,087	39,887	41,193	12%
Adolescents						
14-17 years	16,737	16,773	17,627	17,854	18,338	10%
Adults						
18-44 years	120,073	126,589	128,669	132,370	136,310	14%
45-64 years	83,861	82,434	91,021	98,074	100,013	19%
Older Adults						
65+ years	56,441	74,107	82,344	87,996	98,164	74%
85+ years	6,727	9,132	14,634	18,972	19,724	193%
100+ years	89	138	193	387	604	579%

*Numbers in thousands; life stages are "best fit" to available statistical age groups. (Source: U.S. Census Bureau.)

face is one-eighth of the skull surface, but in an adult the face is half of the skull surface.

Another change in proportion involves the trunk and lower extremities. The legs become proportionately longer and the trunk proportionately shorter. In addition, the thoracic and abdominal contours change, roughly speaking, from round to elliptical.

Such changes are good examples of the ever-changing and ongoing nature of growth, development, and aging. It is unfortunate that many of the changes that occur in the later years of life do not result in increased function. These degenerative changes are certainly important, however, and are discussed later in this chapter. The following are the most common postnatal periods: (1) *infancy,* (2) *childhood,* (3) *adolescence,* (4) *adulthood,* and (5) *older adulthood.*

Table 24-2 illustrates projected changes in U.S. population numbers in selected age-groups by decade through the year 2060. Notice that the older adulthood group is expected to have the largest percentage gain.

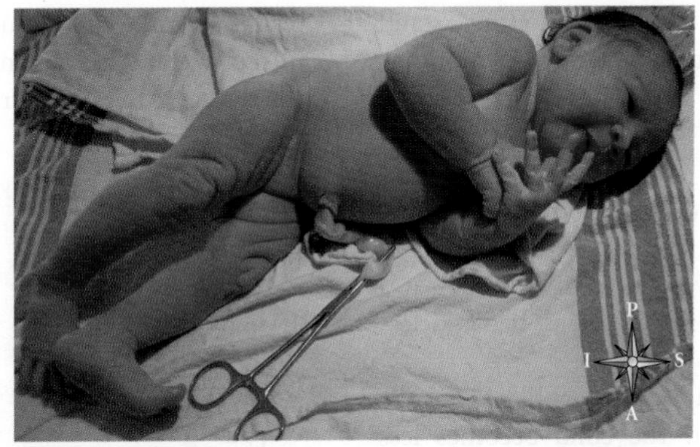

FIGURE 24-12 Neonate. The umbilical cord has been cut, separating the neonate from the placental blood supply, which requires the infant to breathe to get the oxygen once supplied by the mother.

Infancy

The period of **infancy** begins abruptly at birth and lasts about 18 months. The first 4 weeks of infancy are often referred to as the **neonatal period** (Figure 24-12). Dramatic changes occur at a rapid rate during this short but critical period.

Neonatology is the medical and nursing specialty concerned with the diagnosis and treatment of disorders of the newborn or **neonate.** Advances in this area have resulted in dramatically reduced infant mortality.

Many of the changes that occur in the cardiovascular and respiratory systems at birth are necessary for survival. Whereas the fetus totally depends on the mother for life support, the newborn infant must become totally self-supporting in terms of blood circulation and respiration immediately after birth. A baby's first breath is deep and forceful. The stimulus to breathe results primarily from the increasing amounts of carbon dioxide (CO_2) that accumulate in the blood after the umbilical cord is cut following delivery.

Many developmental changes occur between the end of the neonatal period and 18 months of age. Birth weight doubles during the first 4 months and then triples by 1 year. The baby also increases in length by 50% by the 12th month. The "baby fat" that accumulated under the skin during the first year begins to decrease, and the plump infant becomes leaner.

Early in infancy the baby has only one spinal curvature (Figure 24-13, *A*). The lumbar curvature appears between 12 and 18 months, and the once-helpless infant becomes a toddler who can stand (Figure 24-13, *B*).

One of the most striking changes to occur during infancy is the rapid development of the nervous and muscular systems. This permits the infant to follow a moving object with the eyes (2 months); lift the head and raise the chest (3 months); sit

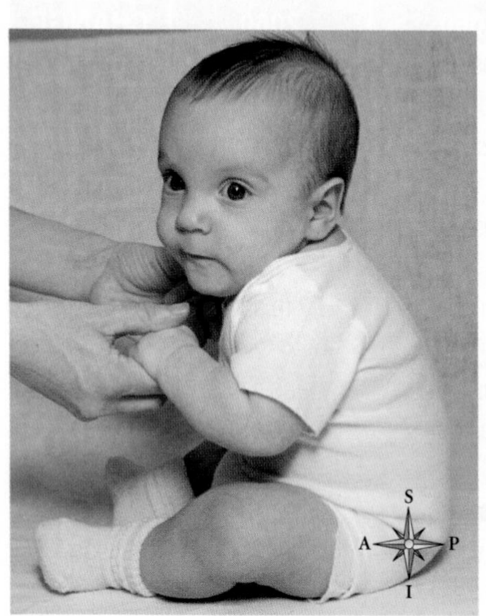

A

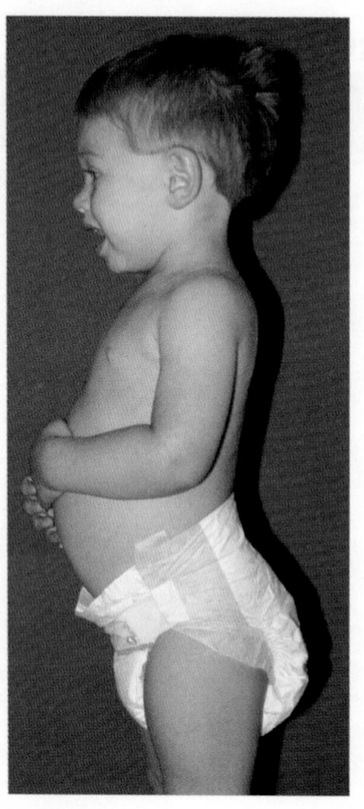

B

FIGURE 24-13 Spinal curvatures. A, Normal rounded curvature of the vertebral column in an infant. **B,** Normal vertebral curvature in a toddler. The *dark shadow* emphasizes the distinct lumbar curvature that develops with the ability to walk. Compare to adult curvatures in **Figure 8-14**, *A*, on p. 186.

when well supported (4 months); crawl (10 months); stand alone (12 months); and run, although a bit stiffly (18 months).

Childhood

Childhood extends from the end of infancy to sexual maturity or puberty—12 to 14 years in girls and 14 to 16 years in boys. Overall, growth during early childhood continues at a rather rapid pace, but month-to-month gains become less consistent.

By the age of 6 years, the child appears more like a preadolescent than an infant or toddler. The child becomes less chubby, the potbelly becomes flatter, and the face loses its babyish look. The nervous and muscular systems continue to develop rapidly during the middle years of childhood; by 10 years of age the child has developed numerous motor and coordination skills.

The *deciduous teeth,* which begin to appear at about 6 months of age, are lost during childhood, beginning at about 6 years of age. The *permanent teeth,* with the possible exception of the third molars, or wisdom teeth, all erupt by age 14.

Adolescence

The average age range of **adolescence** varies, but generally the teenage years (13 to 19) are called the adolescent years. This period is marked by rapid and intense physical growth, which ultimately results in sexual maturity.

Many of the developmental changes that occur during this period are controlled by the secretion of sex hormones and are classified as **secondary sex characteristics.** Breast development is often the first sign of approaching puberty in girls, beginning about age 10. Most girls begin to menstruate at 12 to 13 years of age, which is about 3 years earlier than a century ago. In boys the first sign of puberty is often enlargement of the testicles, which begins between 10 and 13 years of age.

Both sexes show a spurt in height during adolescence. In girls the spurt in height begins between the ages of 10 and 12 and is nearly complete by age 14 or 15. In boys the period of rapid growth begins between 12 and 13 and is generally complete by age 16. See **Figure 24-14.**

Adulthood

Many developmental changes that begin early in childhood are not completed until the early or middle years of **adulthood.** Examples include the maturation of bone, resulting in the full closure of the growth plates, and changes in the size and placement of other body components such as the sinuses. Many body traits do not become apparent for years

FIGURE 24-14 Growth in height. The figure shows typical patterns of gain in height to adulthood for girls and boys. Notice the rapid gain in height during the first few years, a period of slower growth, and then another burst of growth during adolescence—finally ending at the beginning of adulthood.

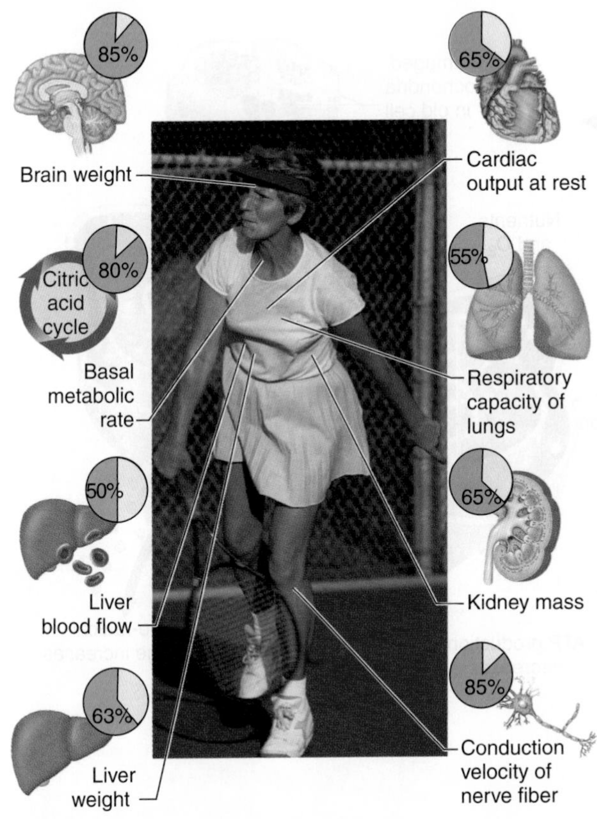

FIGURE 24-15 Changes in older adulthood. Insets show proportion of remaining function in the organs of a person in late adulthood compared with that of a 20-year-old. These are average numbers, so many individuals experience far different situations.

after birth. Normal balding patterns, for example, are determined at the time of fertilization by heredity but do not appear until maturity.

As a general rule, adulthood is characterized by maintenance of existing body tissues. With the passage of years, the ongoing effort of maintenance and repair of body tissues becomes more and more difficult. As a result, degeneration begins. It is the process of aging, and it culminates in death.

Older Adulthood

Most body systems are in peak condition and function at a high level of efficiency during the early years of adulthood. As a person grows older, a gradual but certain decline takes place in the functioning of every major organ system in the body. The study of aging is called **gerontology.**

Many of the biological changes associated with advancing age are shown in **Figure 24-15**. The illustration highlights the proportion of remaining function in a number of organs in **older adulthood** when compared with a 20-year-old person.

Aging

Mechanisms of Aging

Older adulthood is characterized by the processes of **senescence,** or degenerative aging. Unfortunately, the mechanisms and causes of aging are not well understood. A few of the more likely hypotheses are outlined here.

Some gerontologists believe that an important aging mechanism is the limit on cell reproduction. Laboratory experiments have shown that many types of human cells cannot reproduce more than 50 times, thus limiting the maximum life span. Cells die continually, in a process called **apoptosis,** no matter what a person's age, but in older adulthood many dead cells are not replaced, causing degeneration of tissues.

Perhaps the cells are not replaced because the surrounding cells have reached their limit of reproduction. Perhaps differences in each individual's aging process result from differences in the reproductive capacity of cells.

The cellular death mechanism seems to operate in individuals with **progeria,** a rare, genetic condition in which a person appears to age rapidly.

A variety of factors that affect the rates of cell death and cell reproduction have been cited as causes of aging. Some gerontologists believe that nutrition, injury, disease, and other environmental factors affect the aging process. A few have even proposed that aging results from cellular changes caused by slow-acting "aging" viruses found in all living cells. Other gerontologists have proposed that aging is caused by "aging" genes—genes in which aging is "preprogrammed."

Yet another proposed cause of aging is autoimmunity. You may recall from Chapter 16 that autoimmunity occurs when the immune system attacks a person's own tissues.

RESEARCH, ISSUES, AND TRENDS

PROGERIA

Progeria, also called *Hutchinson-Gilford progeria syndrome,* is a rare, fatal condition in which children appear to age rapidly.

In progeria, the reproductive capacity of cells seems to be diminished due to a toxic protein called *progerin,* which is also found in normal cells at much lower levels and increases as we age. Thus the tissues of children with progeria fail to maintain or repair themselves normally, and many of the degenerative conditions more commonly seen in elderly individuals appear.

Some of these conditions can be seen in this photograph of a boy with progeria: areas of tightened skin with stippled coloration; hair loss; loss of subcutaneous fat; and stiff, partially flexed joints. Children with progeria die of cardiovascular disease at an average age of 14 years.

24

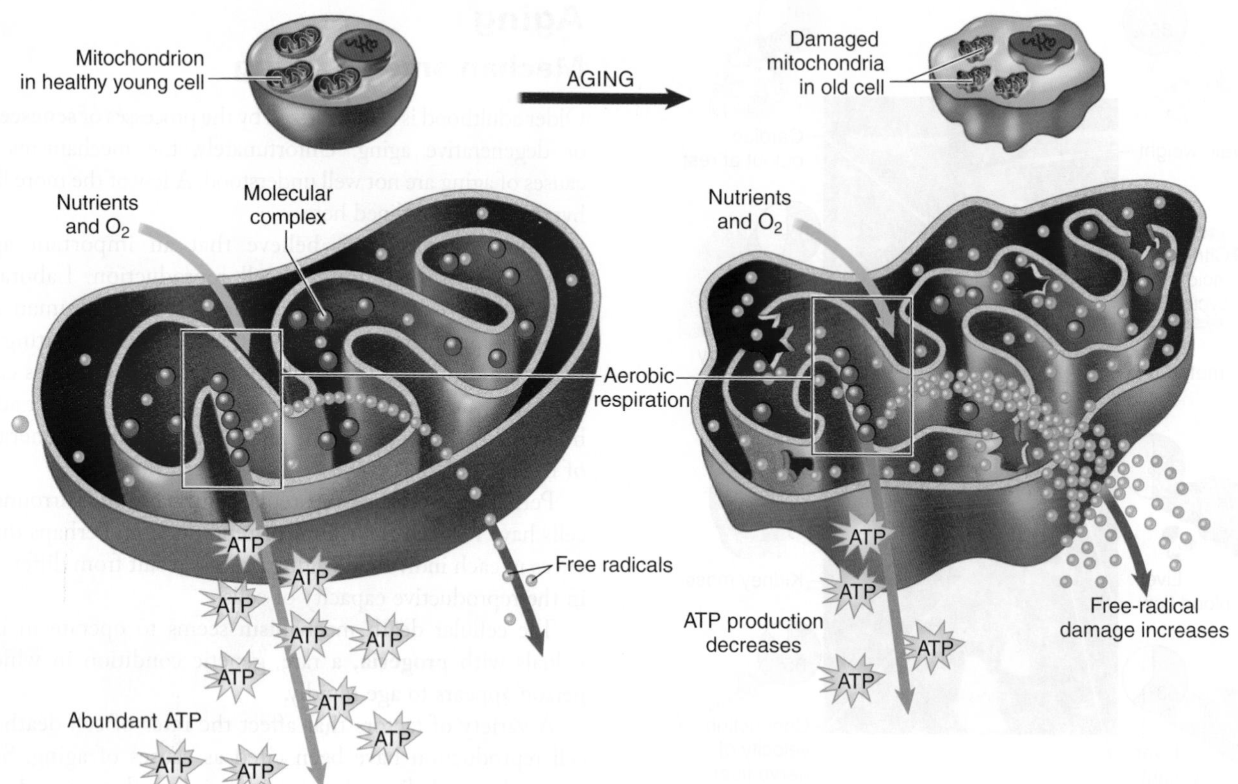

FIGURE 24-16 Free-radical theory of aging. Free-radical production by cells, one of many possible mechanisms of the aging processes, may increase as a person gets older, increasing the amount of cellular damage. Free radicals are highly reactive forms of oxygen that are normal by-products of cellular respiration in the mitochondria (shown) and other cell processes. As one ages, the number of free radicals increases as cellular efficiency decreases. Thus more cellular damage occurs, especially damage to cellular membranes, causing degeneration of the cell.

One popular theory of aging states that oxygen **free radicals** play a major role in cellular aging (**Figure 24-16**). Free radicals are highly reactive forms of oxygen that normally result from cellular activities, but may damage the cell. As a person's cells produce more and more free radicals during the later years, more damage occurs to cellular structures and functions.

Although the causes and basic mechanisms of aging are yet to be understood, at least many of the signs of aging are obvious. The remainder of this chapter deals with a number of the more common degenerative changes that frequently characterize processes of aging.

QUICK CHECK

1. Do the proportions of the human body change during postnatal development?
2. What is the *neonatal period* of development?
3. What biological changes happen during puberty?
4. What is *senescence*?
5. What are *free radicals,* and what role do they have in the aging process?

Effects of Aging

Although advanced age brings with it the higher risk of many disorders, it also brings some biological advantages. We explore just a few of the changes associated with aging here.

Skin

The skin (integumentary system) becomes dry, thin, and inelastic with advancing age. It "sags" on the body because of increased wrinkling and skin folds. Pigmentation changes and the thinning or loss of hair are also common conditions associated with the aging process.

Skeletal System

In older adulthood, bones undergo changes in texture, degree of calcification, and shape. Instead of clean-cut margins, older bones develop indistinct and shaggy-appearing margins with spurs—a process called *lipping*. This type of degenerative change restricts movement because of the piling up of bone tissue around the joints.

With advancing age, changes in calcification may result in reduction of bone size and in bones that are porous and subject to fracture. The lower cervical and thoracic vertebrae are the sites of frequent fractures. The result is curvature of the spine and the shortened stature so typical of late adulthood.

Degenerative joint diseases such as **osteoarthritis** are also common in elderly adults.

However, many of the aging effects seen in the skeleton can be lessened by physical activity—especially if exercise starts earlier in life. Loss of bone mass and reduced mobility can be avoided or reduced by an ongoing program of physical activity coupled with good nutrition.

RESEARCH, ISSUES, AND TRENDS

EXTENDING THE HUMAN LIFE SPAN

In the past few decades, the increased availability of better food, safer surroundings, and advanced medical care has extended quality living for many around the world. But even simple changes in lifestyle, regardless of modern medical wonders, can keep the effects of aging from creeping up too soon. Perhaps the three most important "low tech" methods for improving the quality of life as you age are healthful diet, exercise, and stress management.

A healthful diet is not available to some individuals, but it is available to most of us. We are learning more every day about what kind of diet is best, even to the point of being able to manage specific diseases through diet.

Exercise performed on a regular basis, even if light or moderate, cannot only keep our skeletal and muscular systems more fit but also can decrease aging's effects on the nervous system, endocrine system, digestive system, and immune system—the list seems endless.

And last, even ancient and simple techniques of stress management such as meditation and tai chi have been shown to help reduce the effects of aging and the diseases that often accompany aging such as heart disease and strokes.

In short, we can usually stay "young" longer if we *eat right, exercise,* and *relax.*

What roles do genes play in how long we live? Check out the article *Genes and Longevity* at Connect It! at *evolve.elsevier.com.*

Central Nervous System

Advancing age brings with it the risk of **dementia**—the loss of memory and other functions of conscious thinking—and other degenerative conditions that affect the central nervous system. For most of us, however, our memories remain mostly intact and have helped us develop a mature ability to reason and make decisions. Although some elderly individuals suffer from depression, especially when they become ill or separated from family, the average elderly person is happier than during early and middle adulthood.

Special Senses

The sense organs, as a group, all show a gradual decline in performance and capacity as a person ages.

Most people are farsighted by age 65 because eye lenses become hardened and lose elasticity; the lenses cannot be curved to accommodate for near vision. This hardening of the lens is called **presbyopia,** which means "old eye." Many individuals first notice the change at about 40 or 45 years of age, when it becomes difficult to do close-up work or read without holding printed material at arm's length. This explains the increased need, with advancing age, for bifocals, or glasses that incorporate two lenses, to assist the eye in accommodating for near and distant vision.

Loss of transparency of the lens or its covering capsule is another common age-related eye change. If the lens actually becomes cloudy and significantly impairs vision, it is called a **cataract** and must be removed surgically.

The incidence of **glaucoma,** the most serious age-related eye disorder, increases with age. Glaucoma causes an increase in the pressure within the eyeball and, unless treated, often results in blindness. The risk of retinal degeneration or detachment also increases with age.

In many elderly people a very significant loss of hair cells in the spiral organ of the inner ear causes a serious decline in the ability to hear certain frequencies. In addition, the eardrum and attached ossicles become more fixed and less able to transmit mechanical sound waves. Some degree of hearing impairment is universally present in the older adult.

The senses of smell and taste are also decreased. The resulting loss of appetite may be caused partly by the replacement of taste buds with connective tissue cells. Only about 40% of the taste buds present at age 30 remain in an individual at age 75.

Cardiovascular System

Degenerative heart and blood vessel disease are among the most common and serious effects of aging. Fatty deposits build up in blood vessel walls and narrow the passageway for the movement of blood, much as the build-up of scale in a water pipe decreases flow and pressure. The resulting condition, called **atherosclerosis,** often leads to eventual blockage of the coronary arteries and a "heart attack" (myocardial infarction [MI]).

If fatty accumulations or other substances in blood vessels calcify, actual hardening of the arteries, or **arteriosclerosis** occurs. Rupture of a hardened vessel in the brain (stroke or cerebrovascular accident [CVA]) is a frequent cause of serious disability or death in the older adult.

Hypertension (HTN), or high blood pressure, is also more common. Hardening of arteries reduces their elasticity, which can contribute to abnormally high arterial pressure.

Respiratory System

In older adulthood the costal cartilages that connect the ribs to the sternum become hardened or calcified. This makes it difficult for the rib cage to expand and contract as it normally does during inspiration and expiration. In time the ribs gradually become "fixed" to the sternum, and chest movements become difficult. When this occurs the rib cage remains in a more expanded position, respiratory efficiency decreases, and a condition called "barrel chest" results.

With advancing years, a generalized atrophy or wasting of muscle tissue takes place as the contractile muscle cells are

SCIENCE APPLICATIONS

EMBRYOLOGY

Rita Levi-Montalcini (1909–2012)

Rita Levi-Montalcini had just finished a medical degree in her native Italy when in 1938 the Fascist government under Mussolini barred all "non-Aryans" from working in academic and professional careers. Being Jewish, Levi-Montalcini was forced to move to Belgium to work. But when Belgium was about to be invaded by the Nazis, she decided to return home to Italy and work in secret.

Her home laboratory was very crude, but in it she made some important discoveries about how the nervous system develops during embryonic development. After World War II, she was invited to Washington University in St. Louis to work. There, she discovered the existence of *nerve growth factor (NGF)*, for which she later won the 1986 Nobel Prize. Her discovery of a chemical that regulates the growth of new nerves during early brain development has led to many different paths of investigation. For example, by learning more about growth regulators, we now know more about how the nervous system develops, as well as other tissues, organs, and systems of the body.

Today, many professions make use of the discoveries of **embryology**—the study of early development. Not only are these discoveries important for health professionals such as **obstetricians, obstetric nurses,** and others involved in *prenatal health care*, but they are also important in understanding adult medicine more fully. In fact, even **gerontology** (study of aging) and **geriatrics** (treatment of the aged) have benefited from embryological research. How? By providing insights on how tissue development is regulated in the embryo, scientists can better understand how to possibly stimulate damaged tissue in older adults to repair or regenerate itself.

replaced by connective tissue. This loss of muscle cells decreases the strength of the muscles associated with inspiration and expiration.

Urinary System

The number of nephron units in the kidney decreases by almost 50% between the ages of 30 and 75. Also, because less blood flows through the kidneys as an individual ages, there is a reduction in overall function and excretory capacity or the ability to produce urine. In the bladder, significant age-related problems often occur because of diminished muscle tone. Muscle atrophy (wasting) in the bladder wall results in decreased capacity and inability to empty or void completely.

Reproductive Systems

Although most men and women remain sexually active throughout their later years, mechanisms of the sexual response may change, and fertility declines. In men, erection may be more difficult to achieve and maintain. Urgency for sex may decline—perhaps from reduced testosterone or "low T." In women, lubrication of the vagina may decrease.

Men can continue to produce gametes as they age, but occasionally they exhibit a decline in testosterone severe enough to cause infertility—a process sometimes called *andropause.* Women experience a cessation of reproductive cycling between the ages of 45 and 60—**menopause.** Menopause results from a decrease in estrogen below that needed to sustain reproduction.

The process of menopause is not a disease condition and is considered a natural period of biological transition in a woman's life. However, in addition to the cessation of menstrual cycles, the decrease in blood estrogen levels during this period accounts for a number of common and often troubling symptoms which include hot flashes, sleep disturbances, and dryness and thinning of the vaginal wall in many women.

In the past, these menopause-related symptoms resulting from low estrogen levels were almost always treated with estrogen given as hormone replacement therapy (HRT) or more simply, hormone therapy (HT). In recent years this practice has been used more carefully because of the increased risk of some forms of cancer, stroke, blood clotting disorders, and other serious side effects in some older women who had been using HRT for long periods or had begun HRT well after the onset of menopause. HRT continues to be used in many younger women when they first enter menopause.

Fortunately, medications other than estrogen are also available to effectively treat or prevent most menopausal symptoms and other health problems, such as loss of bone mass, or osteoporosis (see Chapter 8), and heart disease, that increase in frequency in older women who have lower blood estrogen levels. As with any therapy, treatment for menopause-related symptoms requires careful, individualized risk-benefit analysis.

QUICK CHECK

1. What are some changes that occur in the skeleton as one ages?
2. How does one's eyesight change during late adulthood?
3. What changes in the cardiovascular system occur in older adults?
4. How is kidney function affected during old age?

24

LANGUAGE OF **SCIENCE** *(continued from p. 653)*

endoderm
(EN-doh-derm)
[*endo-* **within,** *-derm* **skin**]

fertilization
(FER-tih-lih-ZAY-shun)
[*fertil-* **fruitful,** *-ization* **process**]

fetal phase
(FEE-tal fayz)
[*fet-* **offspring,** *-al* **relating to**]

fetus
(FEE-tus)
[*fetus* **offspring**]

free radical
(free RAD-ih-kal)
[*radic-* **root,** *-al* **relating to**]

gestation
(jes-TAY-shun)
[*gesta-* **bear,** *-tion* **process**]

gestation period
(jes-TAY-shun PEER-ee-id)
[*gesta-* **bear,** *-tion* **process**]

hematopoiesis
(hee-mat-oh-poy-EE-sis)
[*hemo-* **blood,** *-poiesis* **making**]

histogenesis
(his-toh-JEN-eh-sis)
[*histo-* **tissue,** *-gen-* **produce,** *-esis* **process**]

implantation
(im-plan-TAY-shun)
[*im-* **in,** *-plant-* **set or place,** *-ation* **process**]

infancy
(IN-fan-see)
[*in-* **not,** *-fanc-* **speak,** *-y* **state**]

labor
(LAY-ber)

mesoderm
(MEZ-oh-derm)
[*meso-* **middle,** *-derm* **skin**]

morula
(MOR-yoo-lah)
[*mor-* **mulberry,** *-ula* **little**]

neonatal period
(nee-oh-NAY-tal PEER-ee-id)
[*neo-* **new,** *-nat-* **birth,** *-al* **relating to**]

neonate
(NEE-oh-nayt)
[*neo-* **new,** *-nat-* **born**]

older adulthood
(OLD-er ah-DULT-hood)

organogenesis
(or-gah-noh-JEN-eh-sis)
[*organ-* **instrument (organ),** *-gen-* **produce,** *-esis* **process**]

parturition
(pahr-too-RIH-shun)
[*parturi-* **give birth,** *-tion* **process**]

placenta
(plah-SEN-tah)
[*placenta* **flat cake**]

postnatal period
(POST-nay-tal PEER-ee-id)
[*post-* **after,** *-nat-* **birth,** *-al* **relating to**]

prenatal period
(PREE-nay-tal PEER-ee-id)
[*pre-* **before,** *-nat-* **birth,** *-al* **relating to**]

primary germ layer
(PRYE-mayr-ee jerm LAY-er)
[*prim-* **first,** *-ary* **state,** *germ* **sprout**]

secondary sex characteristic
(SEK-on-dayr-ee sex kayr-ak-ter-IS-tik)
[*second-* **second,** *-ary* **relating to**]

senescence
(seh-NES-enz)
[*senesc-* **grow old,** *-ence* **state**]

stem cell
[*stem* **stem of plant,** *cell* **storeroom**]

trimester
(TRY-mes-ter)
[*tri-* **three,** *-me(n)s-* **month,** *-ster* **thing**]

yolk sac
(yohk sak)
[*yolk* **yellow part**]

zygote
(ZYE-goht)
[*zygot-* **union or yoke**]

LANGUAGE OF **MEDICINE**

abruptio placentae
(ab-RUP-shee-oh plah-SEN-tay)
[*ab-* **away from,** *-ruptio* **rupture,** *placentae* **of flat cake (placenta)**]

antenatal medicine
(an-tee-NAY-tal MED-ih-sin)
[*ante-* **before,** *-nat-* **birth,** *-al* **relating to**]

Apgar score
(AP-gar skor)
[*Virginia Apgar* **American physician**]

arteriosclerosis
(ar-tee-ree-oh-skleh-ROH-sis)
[*arteri-* **vessel (artery),** *-sclero-* **harden,** *-osis* **condition**]

assisted reproductive technology (ART)
(ah-SIS-ted ree-proh-DUK-tiv tek-NOL-oh-jee [ay ar tee])
[*re-* **again,** *-pro-* **forward,** *-duct-* **bring or carry,** *-ive* **relating to,** *techn-* **art or skill,** *-log-* **words (study of),** *-y* **activity**]

atherosclerosis
(ath-er-oh-skleh-ROH-sis)
[*ather-* **porridge,** *-sclero-* **harden,** *-osis* **condition**]

birth defect
(berth DEE-fekt)

cataract
(KAT-ah-rakt)
[*cataract* **broken water**]

cesarean section
(seh-SAYR-ee-an SEK-shun)
[*Julius Caesar* **Roman emperor,** *-ean* **relating to,** *sect-* **cut,** *-ion* **condition**]

congenital
(kon-JEN-ih-tall)
[*con-* **with,** *-genit-* **born,** *-al* **relating to**]

dementia
(de-MEN-shah)
[*de-* **off,** *-mens-* **mind,** *-ia* **condition of**]

dizygotic twins
(dye-zye-GOT-ik twinz)
[*di-* **two,** *zygot-* **union or yoke,** *-ic* **relating to,** *twin* **twofold**]

eclampsia
(ee-KLAMP-see-ah)
[*ec-* **out,** *-lamp-* **shine,** *-sia* **condition**]

Continued on p. 672

24

LANGUAGE OF MEDICINE (continued from p. 671)

fetal alcohol syndrome (FAS)
(FEE-tal AL-koh-hol SIN-drohm [ef ay es])
[*fet-* **offspring,** *-al* **relating to,** *syn-* **together,** *-drome* **running or (race) course**]

fraternal twins
(frah-TERN-al twinz)
[*frater-* **brother,** *-al* **relating to,** *twin* **twofold**]

geriatrics
(jayr-ee-A-triks)
[*ger-* **old,** *-iatr-* **treatment,** *-ic* **relating to**]

gerontology
(jayr-on-TOL-uh-jee)
[*ger-* **old,** *-onto-* **age,** *-log-* **words (study of),** *-y* **activity**]

gestational diabetes mellitus (GDM)
(jes-TAY-shun-al dye-ah-BEE-teez MELL-ih-tus)
[*gesta-* **bear,** *-tion-* **process,** *-al* **relating to,** *diabetes* **pass-through or siphon,** *mellitus* **honey-sweet**]

glaucoma
(glaw-KOH-mah)
[*glauco-* **gray or silver,** *-oma* **tumor (growth)**]

hypertension (HTN)
(hye-per-TEN-shun [aych tee en])
[*hyper-* **excessive,** *-tens-* **stretch or pull tight,** *-sion* **state**]

identical twins
(aye-DEN-tih-kal twinz)
[*identic-* **sameness,** *-al* **relating to,** *twin* **twofold**]

lactose intolerance
(LAK-tohs in-TOL-er-ans)
[*lact-* **milk,** *-ose* **carbohydrate (sugar),** *in-* **not,** *-toler-* **bear,** *-ance* **state**]

laparoscope
(LAP-ah-roh-skope)
[*laparo-* **abdomen,** *-scop-* **see**]

mastitis
(mas-TYE-tis)
[*mast-* **breast,** *-itis* **inflammation**]

menopause
(MEN-oh-pawz)
[*men-* **month,** *-paus-* **cease**]

microcephaly
(my-kroh-SEF-ah-lee)
[*micro-* **small,** *-ceph-* **head,** *-al* **relating to,** *-y* **state**]

monozygotic twins
(mahn-oh-zye-GOT-ik twinz)
[*mono-* **one,** *zygot-* **union or yoke,** *-ic* **relating to,** *twin* **twofold**]

neonatology
(nee-oh-nay-TOL-oh-jee)
[*neo-* **new,** *-nat-* **born,** *-log-* **words (study of),** *-y* **activity**]

obstetric nurse
(ob-STET-rik nurs)
[*ob-* **in front,** *-stet-* **stand,** *-tric(s)* **female agent,** *nurs-* **nourish or nurture**]

obstetrician
(ob-steh-TRISH-an)
[*ob-* **in front,** *-stet-* **stand,** *-tric(s)* **female agent,** *-ian* **practitioner**]

osteoarthritis
(os-tee-oh-ar-THRY-tis)
[*osteo-* **bone,** *-arthr-* **joint,** *-itis* **inflammation**]

placenta previa
(plah-SEN-tah PREE-vee-ah)
[*placenta* **flat cake,** *previa* **gone before**]

preeclampsia
(pree-ee-KLAMP-see-ah)
[*pre-* **before,** *-ec-* **out,** *-lamp-* **shine,** *-sia* **condition**]

presbyopia
(pres-bee-OH-pee-ah)
[*presby-* **aging,** *-op-* **vision,** *-ia* **condition**]

progeria
(proh-JEER-ee-ah)
[*pro-* **early,** *-ger-* **old age,** *-ia* **condition**]

puerperal fever
(pyoo-ER-per-al FEE-ver)
[*puerp-* **childbirth,** *-al* **relating to**]

spontaneous abortion
(spon-TAY-nee-us ah-BOR-shun)
[*sponte-* **willingly,** *-ous* **relating to,** *ab-* **away from or amiss,** *-or-* **be born,** *-tion* **process**]

stillbirth
(STIL-berth)

teratogen
(TER-ah-toh-jen)
[*terato-* **monster,** *-gen* **produce**]

ultrasonogram
(ul-trah-SON-uh-gram)
[*ultra-* **beyond,** *-sono-* **sound,** *-gram* **drawing**]

24

❑ OUTLINE SUMMARY

To download a digital version of the chapter summary for use with your device, access the ***Audio Chapter Summaries*** *online at evolve.elsevier.com.*

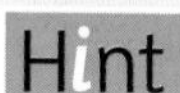

Scan this summary after reading the chapter to help you reinforce the key concepts. Later, use the summary as a quick review before your class or before a test.

Prenatal Period

A. Prenatal period begins at conception and continues until birth (about 39 weeks) (**Figure 24-1**)
B. Science of fetal growth and development is called *embryology*
C. Fertilization to implantation requires about 10 days
   1. Fertilization normally occurs in outer third of oviduct (uterine or fallopian tube) (**Figure 24-2**)
   2. Fertilized ovum called a *zygote;* zygote is genetically complete—all that is needed for expression of hereditary traits is time and nourishment

3. After 3 days of cell division, the zygote has developed into a solid cell mass called a *morula* (**Figure 24-3**)
4. Continued cell divisions of the morula produce a hollow ball of cells called a *blastocyst*
5. Blastocyst implants in the uterine wall about 10 days after fertilization
6. Blastocyst forms the amniotic cavity and chorion of the placenta (**Figure 24-4**)
7. Placenta provides for exchange of nutrients between the mother and fetus (**Figure 24-5**)

D. Periods of development
1. Length of pregnancy or gestation period is about 39 weeks
2. Embryonic phase extends from the third week after fertilization to the end of week 8 of gestation
3. Fetal phase extends from week 8 to week 39 of gestation
4. All organ systems are formed and functioning by month 4 of gestation (**Figure 24-6**)

E. Stem cells—unspecialized cells that reproduce to form specific lines of specialized cells

F. Three primary germ layers—appear in the developing embryo after implantation of the blastocyst (**Table 24-1**):
1. Endoderm—inside layer
2. Mesoderm—middle layer
3. Ectoderm—outside layer

G. Histogenesis and organogenesis (**Figure 24-7**)
1. Formation of new organs (organogenesis) and tissues (histogenesis) occurs from specific development of the primary germ layers
2. Each primary germ layer gives rise to definite structures such as the skin and muscles
3. Growth processes include cell differentiation, multiplication, growth, and rearrangement
4. From 4 months of gestation until delivery, the development of the offspring is mainly a matter of growth

Birth

A. Process of birth called *parturition* (**Figure 24-8**)
1. At the end of week 39 of gestation, the uterus becomes "irritable"
2. Fetus takes head-down position against the cervix
3. Muscular contractions begin, and labor is initiated
4. Amniotic sac ("bag of waters") ruptures
5. Cervix dilates
6. Fetus moves through vagina to exterior

B. Stages of labor
1. Stage one—period from onset of uterine contractions until dilation of the cervix is complete
2. Stage two—period from the time of maximal cervical dilation until the offspring exits through the vagina
3. Stage three—process of expulsion of the placenta through the vagina
4. Clinicians sometimes refer to the recovery period immediately following delivery of the placenta as the fourth stage of labor
5. Apgar score assesses general condition of a newborn infant
6. Cesarean section (C-section)—surgical delivery, usually through an incision in the abdomen and uterine wall

C. Multiple births—two or more infants from the same pregnancy (**Figure 24-9**)
1. Identical siblings result from the splitting of tissue from the same zygote, making them genetically identical
2. Fraternal siblings develop from different ova that are fertilized separately

Disorders of Pregnancy

A. Implantation disorders (**Figure 24-10**)
1. Ectopic pregnancy—implantation outside the uterus (e.g., tubal pregnancy)
2. Placenta previa—growth of the placenta at or near cervical opening, often resulting in separation of the placenta from the uterine wall
3. Abruptio placentae—separation of a normally placed placenta from the uterine wall

B. Preeclampsia (toxemia of pregnancy)
1. Syndrome of pregnancy that includes hypertension, proteinuria, and edema
2. May progress to eclampsia, a severe toxemia that may result in death

C. Gestational diabetes mellitus (GDM)
1. Insulin resistance increases during pregnancy, and if the pancreatic islets fail to compensate by increasing secretion of insulin, then GDM may result
2. May cause health issues in mother and/or fetus, including hypertension/preeclampsia, abnormal fetal weight gain (labor/delivery risk), and increased risk for C-section delivery

D. Fetal death
1. Spontaneous abortion (miscarriage)—loss before week 20 (or 500 g)
2. Stillbirth—loss after 20 weeks

E. Birth defects
1. May be inherited *(congenital abnormalities)* or acquired
2. Acquired defects are caused by teratogens (agents that disrupt normal development)

F. Postpartum disorders
1. Puerperal fever is caused by bacterial infection that may progress to septicemia and death; occurs in mothers after delivery (postpartum)
2. Lactation and thus infant nutrition can be disrupted by anemia, malnutrition, and other factors
   a. Mastitis—inflammation or infection of the breast
   b. Milk can be supplied by another nursing mother or by breast-milk substitutes
   c. Lactose intolerance results from an infant's inability to digest lactose present in human or animal milk

Postnatal Period

A. Postnatal period begins at birth and lasts until death
B. Divisions of postnatal period into isolated time frames can be misleading; life is a continuous process; growth and development are continuous (**Table 24-2**)
C. Obvious changes in the physical appearance of the body—in whole and in proportion—occur between birth and maturity (**Figure 24-11**)
D. Divisions of postnatal period
   1. Infancy
   2. Childhood
   3. Adolescence and adulthood
   4. Older adulthood
E. Infancy
   1. First 4 weeks called *neonatal period* (**Figure 24-12**)
   2. Neonatology—medical and nursing specialty concerned with the diagnosis and treatment of disorders of the newborn
   3. Many cardiovascular changes occur at the time of birth; fetus is totally dependent on mother, whereas the newborn must immediately become totally self-supporting (respiration and circulation)
   4. Respiratory changes at birth include a deep and forceful first breath
   5. Developmental changes between the neonatal period and 18 months include:
      a. Doubling of birth weight by 4 months and tripling by 1 year
      b. Fifty percent increase in body length by 12 months
      c. Development of normal spinal curvature by 15 months (**Figure 24-13**)
      d. Ability to raise head by 3 months
      e. Ability to crawl by 10 months
      f. Ability to stand alone by 12 months
      g. Ability to run by 18 months
F. Childhood
   1. Extends from end of infancy to puberty—13 years in girls and 15 in boys
   2. Overall rate of growth remains rapid but decelerates
   3. Continuing development of motor and coordination skills
   4. Loss of deciduous or baby teeth and eruption of permanent teeth
G. Adolescence
   1. Average age range of adolescence is from 13 to 19 years
   2. Period of rapid growth resulting in sexual maturity (adolescence)
   3. Appearance of secondary sex characteristics regulated by secretion of sex hormones
   4. Growth spurt typical of adolescence; begins in girls at about 10 and in boys at about 12 (**Figure 24-14**)
H. Adulthood
   1. Growth plates fully close in adult; other structures such as the sinuses acquire adult placement
   2. Adulthood characterized by maintenance of existing body tissues
   3. Degeneration of body tissue begins in adulthood
I. Older adulthood
   1. Degenerative changes characterize older adulthood, or aging (**Figure 24-15**)
   2. Gerontology—the study of aging
   3. Every organ system of the body undergoes degenerative changes, eventually culminating in death

Aging

A. Senescence—the process of degenerative aging
B. Mechanisms of aging—not well understood
   1. Cellular mechanism—limited cell reproduction and apoptosis (cell death) also limits the lifespan of tissues and the whole organism; progeria is a condition of rapid cell death that mimics senescence
   2. Free-radical theory—the number of oxygen free radicals increases as one ages, thus increasing the rate of cellular damage
   3. Other factors may play a role: nutrition, injury, disease, environmental and genetic factors, viruses, and autoimmunity
C. Effects of aging
   1. Integumentary system (skin)
      a. With age, skin "sags" and becomes thin, dry, and wrinkled (or inelastic)
      b. Pigmentation problems are common
      c. Frequent thinning or loss of hair occurs
   2. Skeletal system
      a. Aging causes changes in the texture, calcification, and shape of bones
      b. Bone spurs develop around joints
      c. Bones become porous and fracture easily
      d. Degenerative joint diseases such as osteoarthritis are common
      e. Physical activity can reduce loss of bone mass and mobility
   3. Central nervous system
      a. Increased risk of dementia
      b. Mature reasoning ability
   4. Special senses
      a. All sense organs show a gradual decline in performance with age
      b. Eye lenses become hard and cannot accommodate for near vision; result is farsightedness in many people by age 45 (presbyopia)
      c. Loss of transparency of lens or cornea is common (cataract)

d. Glaucoma (increase in pressure in eyeball) is often the cause of blindness in older adulthood
e. Loss of hair cells in inner ear produces frequency deafness in many older people
f. Decreased transmission of sound waves caused by loss of elasticity of eardrum and fixing of the bony ear ossicles is common in older adulthood
g. Some degree of hearing impairment is universally present in the aged
h. Ability to smell and taste may be reduced; only about 40% of the taste buds present at age 30 remain at age 75

5. Cardiovascular system
   a. Degenerative heart and blood vessel disease is among the most common and serious effects of aging
   b. Fat deposits in blood vessels (atherosclerosis) decrease blood flow to the heart and may cause complete blockage of the coronary arteries
   c. Hardening of arteries (arteriosclerosis) may result in rupture of blood vessels, especially in the brain (stroke)
   d. Hypertension or high blood pressure is common in older adulthood
6. Respiratory system
   a. Calcification of costal cartilages causes rib cage to remain in expanded position, resulting in barrel chest
   b. Wasting of respiratory muscles decreases respiratory efficiency
   c. Respiratory membrane thickens; movement of oxygen from alveoli to blood is slowed
7. Urinary system
   a. Nephron units decrease in number by 50% between ages 30 and 75
   b. Blood flow to kidney, and therefore ability to form urine, decreases
   c. Bladder problems such as inability to void completely are caused by muscle wasting in the bladder wall
8. Reproductive system
   a. Changes in the sexual response
      (1) Men—erection is more difficult to achieve and maintain; urgency for sex may decline due to declining testosterone
      (2) Women—lubrication during intercourse may decrease
   b. Changes in fertility
      (1) Men—may continue to be fertile throughout later adult years
      (2) Women—experience menopause (cessation of reproductive cycling) between the ages of 45 and 60

❑ ACTIVE LEARNING

STUDY TIPS

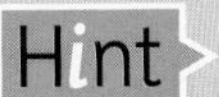

Consider using these tips to achieve success in meeting your learning goals.

Before studying this chapter, quickly review the reproductive systems in Chapter 23. Also review the synopsis of all the organ systems found in Chapter 5.

1. Make flash cards and check online resources to help you learn the early developmental stages. It would be helpful to include the order in which each developmental stage occurs. Remember to also include the functions of the amnion, chorion, and placenta.
2. The term *germ* in *primary germ layer* refers to "germinate." All of the structures of the body come from one of the primary germ layers. Each is named based on its location in the developing embryo. *Endoderm* means "inner skin," *mesoderm* means "middle skin," and *ectoderm* means "outer skin." Develop a chart or flash cards to help you learn to match up the primary germ layers and the structures that develop from each layer.
3. Review the Language of Science and Language of Medicine sections for a better understanding of the terminology.
4. The stages of labor, the important events in the postnatal periods, and the effects of old age on various organ systems also can be put on flash cards to facilitate learning.
5. Make a chart of the disorders of pregnancy; organize it by mechanism or cause: implantation disorders, preeclampsia, birth defects, and postpartum disorders.
6. In your study group, go over the flash cards of the stages of development, making sure you know the proper sequence. Go over your chart and flash cards for the primary germ layers and what organs come from each layer. Review the flash cards for the stages of labor, the postnatal periods, and the effects of aging, and review the chart of the disorders. Review the questions at the end of the chapter and the chapter outline summary and discuss possible test questions.

Review Questions

Write out the answers to these questions after reading the chapter and reviewing the Chapter Summary. If you simply think through the answer without writing it down, you will not retain much of your new learning.

1. Explain what occurs between the time of ovulation and the implantation of the fertilized egg into the uterus.
2. Explain the function of the chorion and placenta.
3. Explain the function of the yolk sac and amniotic cavity.
4. Name the three primary germ layers, and name three structures that develop from each layer.
5. Define histogenesis and organogenesis.
6. Describe and give the approximate length of each of the three stages of labor.
7. Describe a breech birth.
8. What is the difference between identical and fraternal twins?
9. What is an ectopic pregnancy? Where is it most likely to occur?
10. What is placenta previa? What is abruptio placentae?
11. What is preeclampsia?
12. What is a teratogen?
13. What is the stimulus for a baby's first breath?
14. Name three developmental changes that occur during infancy.
15. Briefly explain what biological developments occur during childhood.
16. Briefly explain what biological developments occur during adolescence.
17. Briefly explain what biological developments occur during adulthood.
18. What is progeria?
19. Explain the effects of aging on the skeletal system.
20. Explain the effects of aging on the respiratory system.
21. Explain the effects of aging on the cardiovascular system.
22. Explain the effects of aging on vision.

Critical Thinking

After finishing the Review Questions, write out the answers to these more in-depth questions to help you apply your new knowledge. Go back to sections of the chapter that relate to concepts that you find difficult.

23. Explain the procedure a physician might use if a normal vaginal delivery would be dangerous for the mother or offspring.
24. Why would your physician be reluctant to treat your menopause-related symptoms with hormone replacement therapy (HRT)?
25. Based upon what you know, explain how a sound exercise program can reduce some of the common effects of aging.

Chapter Test

After studying the chapter, test your mastery by responding to these items. Try to answer them without looking up the answers.

1. The science of the development of the offspring before birth is called ______.
2. ______ most often occurs in the outer one-third of the oviduct.
3. The fertilized ovum is called a ______.
4. After about 3 days of mitosis, the fertilized ovum forms a solid mass of cells called the ______.
5. Mitosis continues, and by the time the developing egg reaches the uterus, it has become a solid ball of cells called the ______.
6. At the very beginning of the embryonic stage, all the cells are ______.
7. The ______ anchors the developing fetus to the uterus and provides a bridge for the exchange of substances between the mother and offspring.
8. The ______ period lasts about 39 weeks and is divided into trimesters.
9. The three primary germ layers are the ______, the ______, and the ______.
10. ______ cells are unspecialized cells that are capable of producing many different kinds of cells.
11. The process by which the primary germ layers develop into tissues is called ______.
12. The process by which tissues develop into organs is called ______.
13. The process of birth is called ______.

14. Twins resulting from two different ova being fertilized by two different sperm are called ________ twins.
15. Twins resulting from the splitting of embryonic tissue from the same zygote are called ________ twins.
16. Environmental agents that can disrupt normal histogenesis and organogenesis are called ________.
17. After 20 weeks, delivery of a lifeless infant is termed a ________.
18. The first 4 weeks of infancy is referred to as the ________ period.
19. ________ is a syndrome of postpartum mothers characterized by bacterial infection that progresses to septicemia and possibly death.
20. ________ is a degenerative joint disease that is common to older adults.
21. ________ means "old eye" and causes older adults to be farsighted.
22. If the lens of the eye becomes cloudy and impairs vision, the condition is called a ________.
23. ________ is another name for "hardening of the arteries."
24. ________ are highly reactive forms of oxygen that normally result from cellular activities, but may damage the cell.

Match each term in Column A with its corresponding description in Column B.

Column A	Column B
25. ________ infancy	a. period during which the deciduous teeth are lost
26. ________ childhood	b. period during which closure of the bone growth plates occurs
27. ________ adolescence	c. period that begins at birth
28. ________ adulthood	d. senescence
29. ________ older adulthood	e. period during which the secondary sex characteristics usually begin to develop

Match each disorder in Column A with its corresponding description or cause in Column B.

Column A	Column B
30. ________ ectopic pregnancy	a. a condition in which the blastocyst implants too close to the cervical opening of the uterus
31. ________ abruptio placentae	b. inflammation of the breast
32. ________ placenta previa	c. a postpartum disorder characterized by a bacterial infection that progresses to septicemia
33. ________ preeclampsia	d. an inherited condition in which the person seems to age very rapidly
34. ________ puerperal fever	e. separation of the placenta from the uterine wall in a pregnancy of 20 weeks or longer
35. ________ mastitis	f. a disorder characterized by acute hypertension after 24 weeks of pregnancy
36. ________ progeria	g. the implantation of the blastocyst outside the uterus

Case Studies

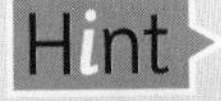

To solve a case study, you may have to refer to the glossary or index, other chapters in this textbook, and other resources.

1. Mary is pregnant with her first child and is touring the maternity floor of a local hospital in preparation for the upcoming birth. She keeps asking about their precautions regarding aseptic technique in the "home-style" birthing rooms. She is concerned because her own grandmother died of an infection after delivering Mary's mother. Why should Mary be concerned? What might happen if the hospital does not enforce aseptic technique in the birthing room?
2. Two-year-old Abe has always been small for his age and is now diagnosed as having developmental disabilities. Abe's grandmother believes that these disabilities are associated with the moderate drinking done by Abe's mother during her pregnancy. Is Abe's grandmother simply trying to justify her meddling—or could his mother's drinking have caused his problems?
3. Lactose intolerance is sometimes treated by removing lactose from the diet. For example, infants with lactose intolerance may be given a lactose-free milk substitute to avoid the indigestion, diarrhea, abdominal discomfort, and other symptoms of this condition. Older children and adults with lactose intolerance may avoid dairy products and other foods high in lactose. Your friend Aileen, who has lactose intolerance, takes a tablet with her favorite ice cream (chocolate fudge) that helps her avoid any problems. What might this tablet contain?

Answers to Active Learning Questions can be found online at evolve.elsevier.com.

24

Genetics and Genetic Diseases

OUTLINE

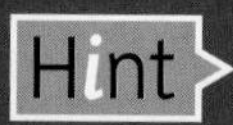

Hint *Scan this outline before you begin to read the chapter, as a preview of how the concepts are organized.*

OBJECTIVES

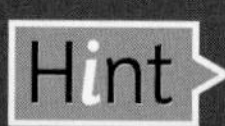

Hint *Before reading the chapter, review these goals for your learning.*

After you have completed this chapter, you should be able to:

1. Explain how genes can cause disease.
2. Distinguish between dominant and recessive genetic traits.
3. Describe sex-linked inheritance and explain how genetic mutations may occur.
4. Explain the mechanisms of genetic disease and list some important inherited diseases.
5. Describe how nondisjunction can result in trisomy or monosomy and list some disorders that result from it.
6. List some tools used in genetic counseling and explain how they are used to help clients.
7. Describe how genetic disorders can be treated.

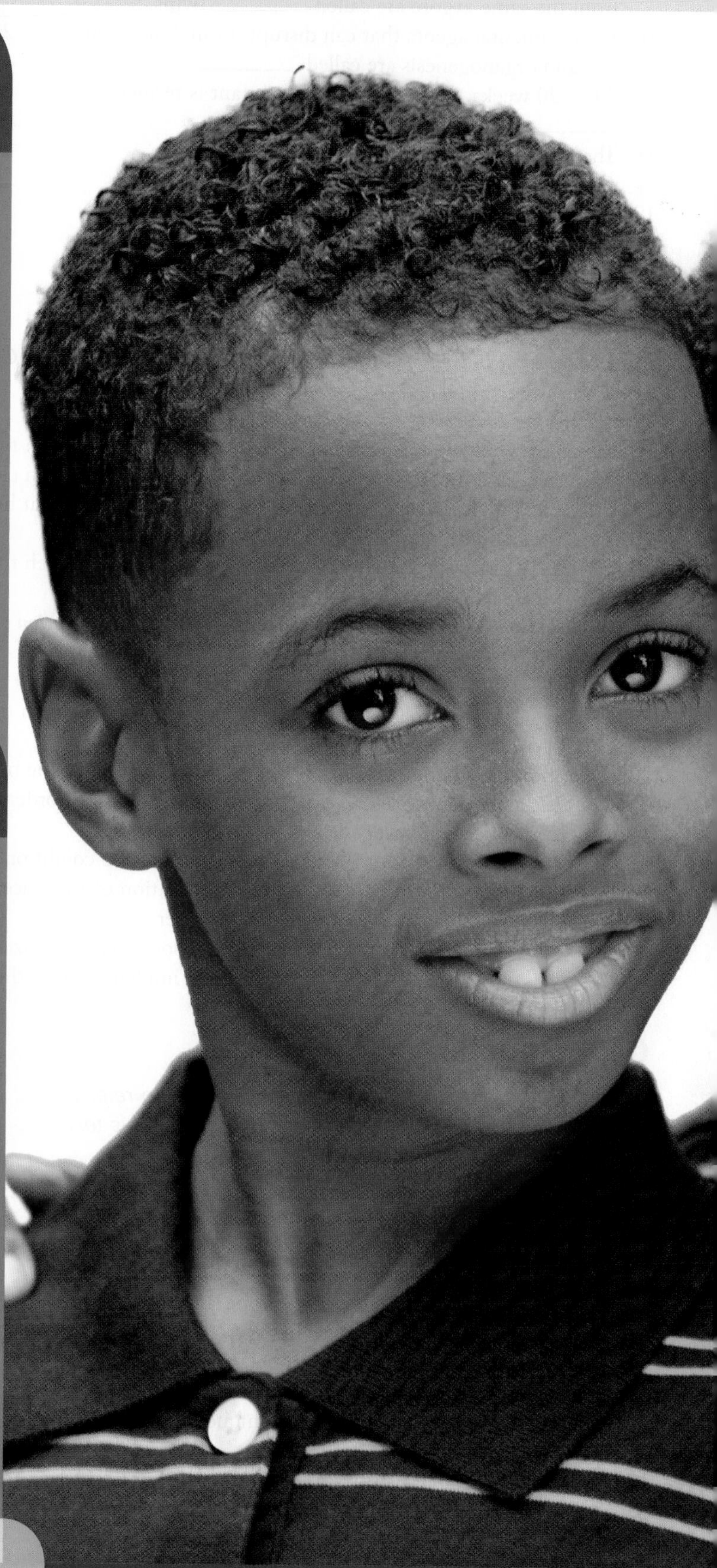

CHAPTER 25

It seems that today we are hearing more and more about the relationship of **genetics,** the scientific study of inheritance, and human disease. Popular news magazines are running story after story on the revolution in treating fatal inherited disorders by using something called **gene therapy.** Health and science blogs keep us informed of the latest discoveries of genes involved with disease, human behavior, and even longevity. Television programs outline the progress of the largest coordinated biological quest that anyone can remember: mapping the entire human genetic code, listing all the proteins encoded there, and working out how genes are regulated.

Clearly, a person cannot be informed about the mechanisms of human disease today without some knowledge of basic genetics. In this chapter, we briefly review the essential concepts of genetics and explain how misinformation in the genetic code can cause disease.

Genetics and Human Disease

History shows that humans have been aware of inheritance for thousands of years; however, it was not until the 1860s that the scientific

LANGUAGE OF **SCIENCE**

Hint Before reading the chapter, say each of these terms out loud. This will help you to avoid stumbling over them as you read.

autosome
(AW-toh-sohm)
[*auto-* **self,** *-som-* **body**]

carrier
(KARE-ee-er)

codominance
(koh-DOM-ih-nance)
[*co-* **together,** *-domina-* **rule,** *-ance* **state**]

crossing-over
(KROS-ing OH-ver)

dominant gene
(DOM-ih-nant jeen)
[*domina-* **rule,** *-ant* **state,** *gen-* **produce**]

epigenetics
(ep-ih-jeh-NET-iks)
[*epi-* **upon,** *gen-* **produce,** *-ic* **relating to**]

gamete
(GAM-eet)
[*gamet-* **sexual union or marriage partner**]

gene
(jeen)
[*gen-* **produce**]

genetic mutation
(jeh-NET-ik myoo-TAY-shun)
[*gene-* **produce,** *-ic* **relating to,** *muta-* **change,** *-ation* **process**]

genetics
(jeh-NET-iks)
[*gen-* **produce,** *-ic* **relating to**]

genome
(JEE-nohm)
[*gen-* **produce (gene),** *-ome* **entire collection**]

Continued on p. 694

study of inheritance—*genetics*—was born. At that time, a monk living in what is now part of Czechia discovered the basic mechanism by which traits are transmitted from parents to offspring. That man, Gregor Mendel, demonstrated that independent units (which we now call *genes*) are responsible for the inheritance of biological traits.

The science of genetics developed from Mendel's quest to explain how normal biological characteristics are inherited. As time went by, and more genetic studies were done, it became clear that certain diseases have a genetic basis. In Chapter 6, you learned that some diseases are inherited directly. For example, the blood-clotting disorder called *hemophilia* is inherited by children from a parent who has the genetic code for hemophilia.

Other diseases are only partly determined by genetics; that is, they involve genetic risk factors (see Chapter 6). For example, certain forms of skin cancer are thought to have a genetic basis. A person who inherits the genetic code associated with skin cancer will develop the disease only if the skin is also heavily exposed to the ultraviolet radiation in sunlight.

Chromosomes and Genes

Mechanisms of Gene Function

Mendel proposed that the genetic code is transmitted to offspring in discrete, independent units that we now call **genes.** In Chapter 3 we stated that each gene is a sequence of nucleotide bases in the deoxyribonucleic acid (DNA) molecule (also see Chapter 2). Each gene contains a genetic code that the cell transcribes to a messenger RNA (mRNA) molecule. Each mRNA molecule moves to a ribosome, where the code is translated to form a specific protein molecule.

Many of the protein molecules thus formed are *enzymes,* functional proteins that permit specific biochemical reactions to occur (see Chapter 2). Because enzymes regulate the biochemistry of the body, they regulate the entire structure and function of the body. Thus genes determine the structure and function of the human body by producing a set of specific regulatory enzymes.

As described in Chapter 3, genes are simply segments of a DNA molecule. While the genetic codes of its genes are being actively transcribed, the DNA is in a threadlike form called chromatin. During cell division, each replicated strand of chromatin coils to form a compact mass called a chromosome (**Figure 25-1**). Thus each DNA molecule can be called either a *chromatin strand* or a *chromosome.* Genes may be actively transcribed in the chromatin form of DNA but not in the chromosome form. To make things easy, however, we will simply use the term *chromosome* for DNA and the term *gene* for each distinct encoding segment within a DNA molecule.

25

Human Genome

The entire collection of genetic material in each typical cell of the human body is called the **genome.** The typical human genome includes 46 individual nuclear chromosomes and one mitochondrial chromosome.

In 2003, the **Human Genome Project (HGP)**—a publicly funded, worldwide collaboration to map all of the genes in the human genome—was completed. This landmark event coincided exactly with the fiftieth anniversary of the discovery of DNA. We now know that the human genome contains only about 19,000 or so genes. This is far less than originally estimated and among the smallest genomes of any animal!

We also know that less than 2% of the DNA carries genes that code for proteins. The rest has been called "junk DNA," or "noncoding DNA," because it is not used directly to make proteins.

A small portion of this noncoding DNA seems to be made up of broken bits of genes that are no longer functional—remnants of our evolutionary past called **pseudogenes.** According to the ongoing HGP offshoot *ENCODE (Encyclopedia of DNA Elements),* however, about 80% of the noncoding DNA is made up of regions that regulate the timely switching of genes on and off.

The current draft of the human genome shows us that most coding genes tend to lie in clusters, separated by long stretches of noncoding DNA. Hundreds of newly discovered genes in the human genome seem to be bacterial in origin, perhaps inserted there by bacteria in our distant ancestors.

Although we now seem to have the overall picture of the details of the human genome, much work still lies ahead in the field of **genomics,** the analysis of the genome's code. Besides filling in the remaining details of the rough draft, we have much work to do in discovering all the possible mutations that might exist (see the discussion later in this chapter) and all the proteins encoded by the genes that make up the human genome.

The quest to analyze the genome has generated another field called **proteomics,** the analysis of the proteins encoded by the genome. The entire group of proteins encoded by the human genome is called the human **proteome.** The ultimate goal of proteomics is to understand the role of each protein in the body. Understanding the roles of every single protein in the body will certainly go a long way toward improving our knowledge of the normal function of the body as well as the mechanisms for many diseases.

Information obtained about the human genome can be expressed in a variety of ways. As you can see in **Figure 25-1**, an **ideogram,** or simple cartoon of a chromosome, is often used in genomics to show the overall physical structure of a chromosome. The constriction in the ideogram shows the relative position of the chromosome's centromere. The shorter segment of the chromosome is called the **p-arm** and the longer segment is called the **q-arm.**

The bands in an ideogram of a chromosome show staining landmarks and help identify the regions of the chromosome. Sometimes physical maps of genes will show exact positions of individual genes on the p-arm and q-arm of a chromosome.

A more detailed representation of a gene would show the actual sequence of nucleotide bases, abbreviated here *a, c, g,* and *t* for *adenine, cytosine, guanine,* and *thymine,* as you see in **Figure 25-1**.

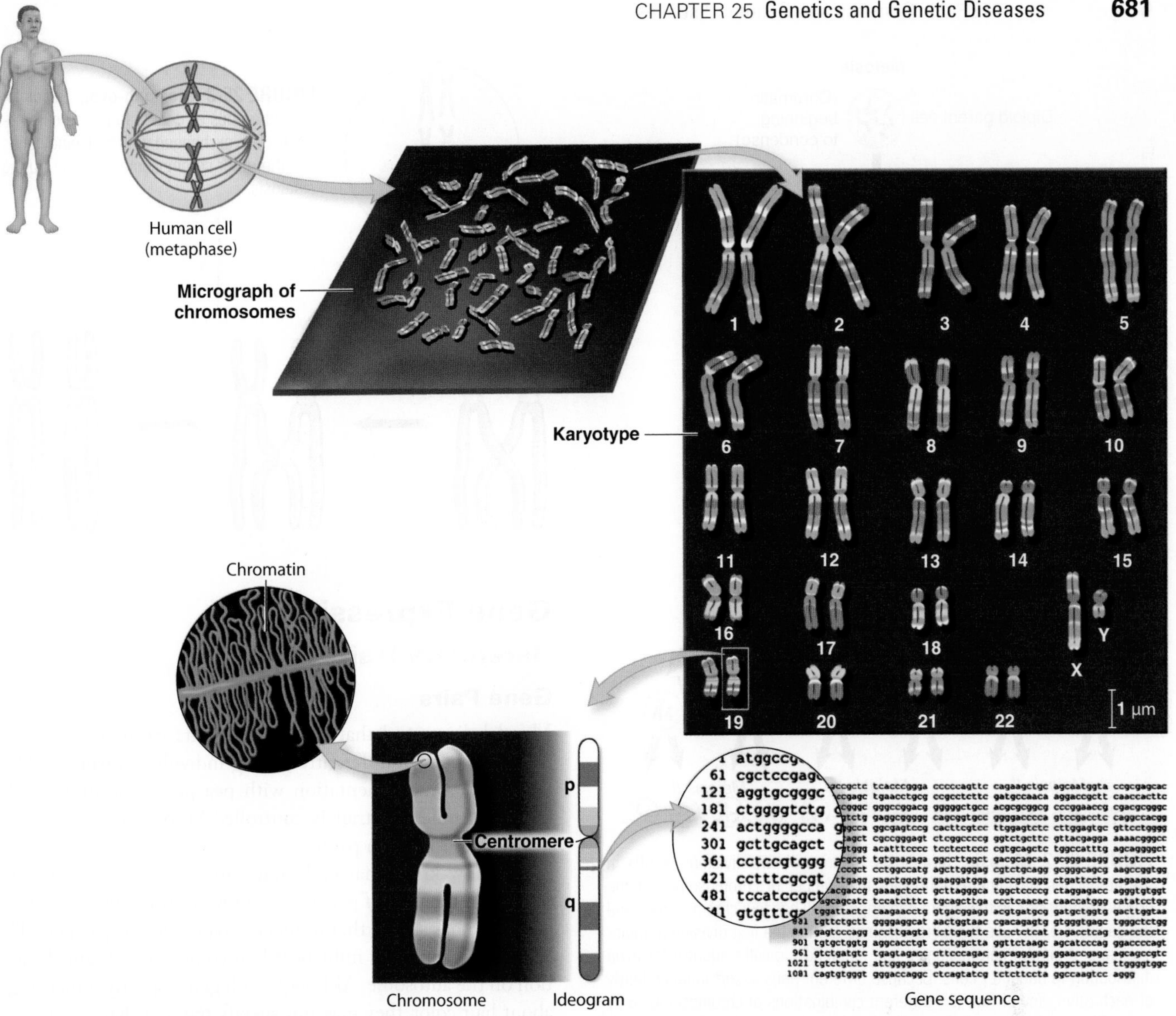

FIGURE 25-1 Human genome. A cell taken from the body is stained and photographed. A digital micrograph of nuclear chromosomes is then processed, sorting the 46 chromosomes into numbered pairs of decreasing size to form a chart called the *karyotype*. Each chromosome is a coiled mass of chromatin (DNA). In this figure, differentially stained bands in each chromosome appear as different bright colors. Such bands are useful as reference points when identifying the locations of specific genes within a chromosome. The staining bands are also represented on an ideogram, or simple graph, of the chromosome as reference points to locate specific genes. The genes themselves are usually represented as the actual sequence of nucleotide bases, abbreviated here as *a, c, g,* and *t.* In this figure, the sequence of one exon (segment) of a gene called *GPI* from chromosome 19 is shown. Each of these representations can be thought of as a type of "genetic map."

Distribution of Chromosomes to Offspring

Each cell of the human body contains 46 chromosomes. The only exceptions to this principle are the **gametes**—male *spermatozoa* and female *ova.* Chapter 23 introduced a special form of nuclear division called **meiosis** (Figure 25-2) that produces gametes with only 23 chromosomes—exactly half the usual number.

When a sperm (with its 23 chromosomes) unites with an ovum (with its 23 chromosomes) at conception, a *zygote* with 46 chromosomes is formed. Thus the zygote has the same number of chromosomes as each body cell in the parents.

As Figure 25-1 shows, the 46 human chromosomes can be arranged in 23 pairs according to size. One pair called the **sex chromosomes** may not match, but the remaining 22 pairs of **autosomes** always appear to be nearly identical to each other.

Because half of an offspring's chromosomes are from the mother and half are from the father, a unique blend of inherited traits is formed. According to principles first discovered by Mendel, each chromosome assorts itself independently during meiosis (see Figure 25-2). This means that as sperm are formed, chromosome pairs separate and the maternal and paternal chromosomes get mixed up and redistribute themselves independently of the other chromosome pairs. Thus each sperm is likely to have a *different* set of 23 chromosomes. Because ova are formed in the same manner, each ovum is likely to be genetically different from the ovum that preceded it.

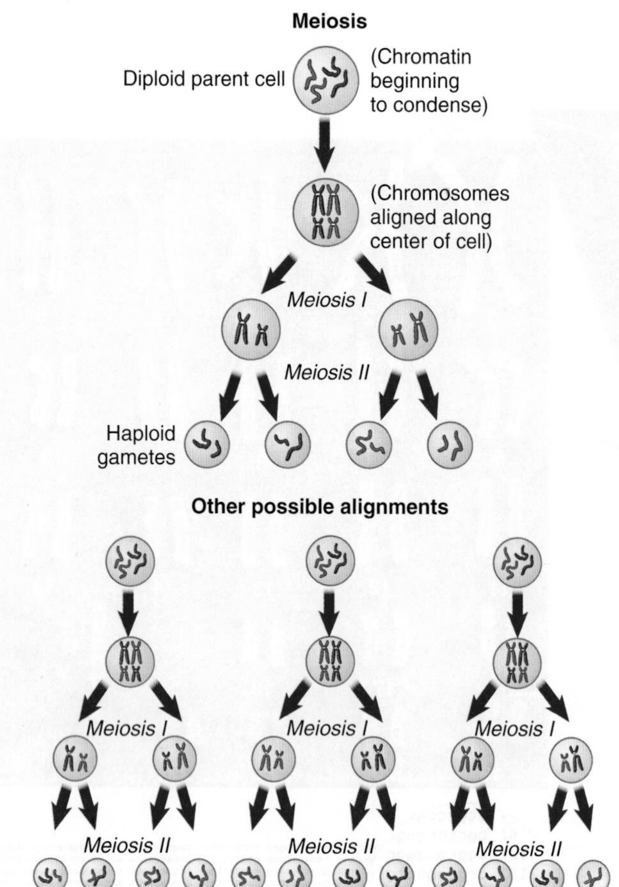

FIGURE 25-2 Meiosis. In meiosis, a series of two divisions results in the production of gametes with half the number of chromosomes of the original parent cell. In this figure, the original cell has four chromosomes and the gametes each have two chromosomes. During the first division of meiosis, pairs of similar chromosomes line up along the cell's equator for even distribution to daughter cells. Because different pairs assort independently of each other, any of four (2^2) different combinations of chromosomes may occur. Because human cells have 23 pairs of chromosomes, more than 8 million (2^{23}) different combinations are possible.

Independent assortment of chromosomes ensures that each offspring from a single set of parents is very likely to be genetically unique—a phenomenon known as *genetic variation.*

The **principle of independent assortment** also applies to individual genes or groups of genes. During one phase of meiosis, pairs of matching chromosomes line up along the equator of the cell and exchange genes. This process is called **crossing-over** because genes from a particular location cross over to the same location on the matching gene (**Figure 25-3**). Sometimes a whole group stays together and crosses over as a single unit—a phenomenon called **gene linkage.** Crossing-over introduces additional possibilities for genetic variation among offspring of a set of parents.

QUICK CHECK

1. How do genes produce biological traits?
2. Who might be considered the founder of the scientific study of genetics?
3. What is the difference between an *autosome* and a *sex chromosome*?
4. List some mechanisms that increase genetic variation among human offspring.

25

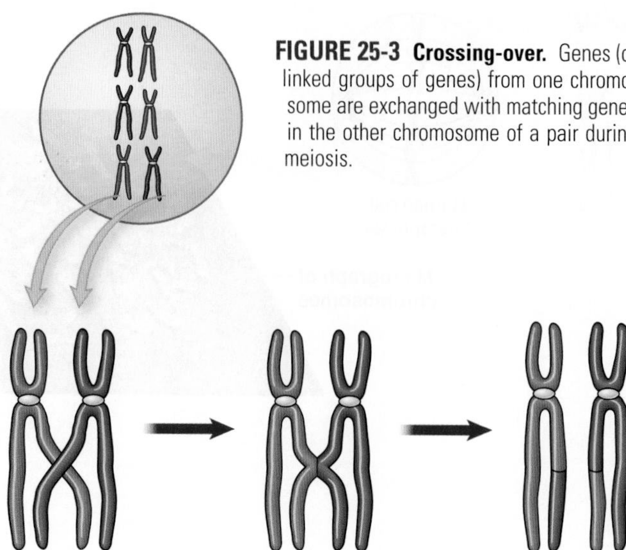

FIGURE 25-3 Crossing-over. Genes (or linked groups of genes) from one chromosome are exchanged with matching genes in the other chromosome of a pair during meiosis.

Gene Expression

Hereditary Traits

Gene Pairs

Mendel discovered that the genetic units we now call *genes* may be expressed differently among individual offspring. After rigorous experimentation with pea plants, he discovered that each inherited trait is controlled by two sets of similar genes, one from each parent.

We now know that each autosome in a pair matches its partner in the type of genes it contains. In other words, if one autosome has a gene that influences hair color, its partner will also have a gene that influences hair color—in the same location on the autosome. Although both genes specify something about hair color, they may not specify the *same* hair color.

Dominance and Recessiveness

Mendel also discovered that some genes and the traits they control are dominant and some are recessive. A **dominant gene** is one whose effects are seen and that is capable of masking the effects of a **recessive gene** for the same trait.

Consider the example of **albinism,** a lack of melanin pigment in the skin and eyes. Because they lack dark pigmentation, people with this condition have difficulty with seeing and protecting themselves from burns in direct sunlight. The genes that cause albinism are recessive; the genes that cause normal melanin production are dominant.

By convention, dominant genes are represented by uppercase letters and recessive genes by lowercase letters. One can represent the gene for albinism as *a* and the gene for normal skin pigmentation as *A.* A person with the gene combination *AA* has two dominant genes—and so will exhibit a normal skin color. Someone with the gene combination *Aa* will also have normal skin color because the normal gene *A* is dominant over the recessive albinism gene *a.* Only a person with the gene combination *aa* will have albinism because there is no dominant gene to mask the effects of the two recessive genes.

In the example of albinism, a person with the gene combination of *Aa* is said to be a genetic **carrier** of albinism. This

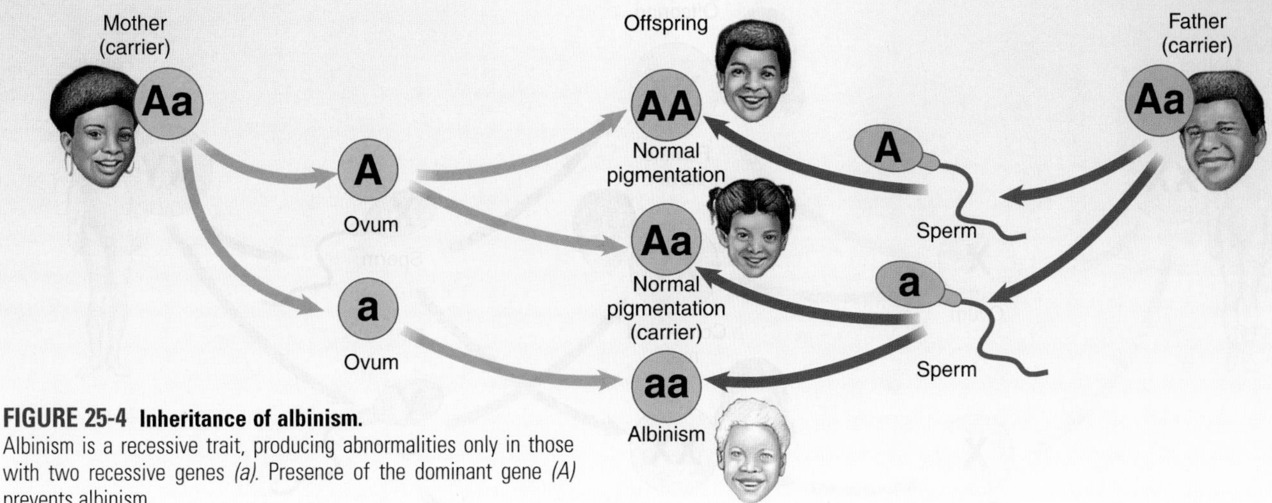

FIGURE 25-4 Inheritance of albinism. Albinism is a recessive trait, producing abnormalities only in those with two recessive genes *(a)*. Presence of the dominant gene *(A)* prevents albinism.

means that the person can transmit the albinism gene, *a*, to offspring. Thus two normal parents each having the gene combination *Aa* can produce both normal children and children that have albinism (**Figure 25-4**).

Codominance

What happens if two different dominant genes occur together? Suppose there is a gene A^1 for light skin and a gene A^2 for dark skin. In a form of dominance called **codominance,** they will simply have equal effects and a person with the gene combination A^1A^2 will exhibit a skin color that is something between light and dark.

We stated in Chapter 13 that the genes for sickle cell anemia behave this way. A person with two sickle cell genes will have *sickle cell anemia,* whereas a person with one normal gene and one sickle cell gene will have a much milder condition called *sickle cell trait.*

Sex-Linked Traits

We stated earlier that, in addition to the 22 pairs of autosomes, there is one pair of sex chromosomes. Notice in **Figure 25-1** that the chromosomes of this pair do not have matching structures. The larger sex chromosome is called the *X chromosome,* and the smaller one is called the *Y chromosome.* The *X* chromosome is sometimes called the *female chromosome* because it has genes that determine female sexual characteristics.

If a person has only *X* chromosomes, she is genetically a female. The *Y* chromosome is often called the *male chromosome* because anyone possessing a *Y* chromosome is genetically a male. Thus all normal females have the sex chromosome combination *XX* and all normal males have the combination *XY.* Because men produce both *X*-bearing and *Y*-bearing sperm, any two parents can produce male or female children (**Figure 25-5**).

The large *X* chromosomes contain many genes besides those needed for female sexual traits. Genes for producing certain clotting factors, the photopigments in the retina of the eye, and many other proteins are also found on the *X* chromosome. The tiny *Y* chromosome, on the other hand, contains few genes other than those that determine the male sexual characteristics.

Males and females need at least one normal *X* chromosome—otherwise, the genes for clotting factors and other essential proteins would be missing. Nonsexual traits carried on sex chromosomes are called sex-linked traits. Most **sex-linked**

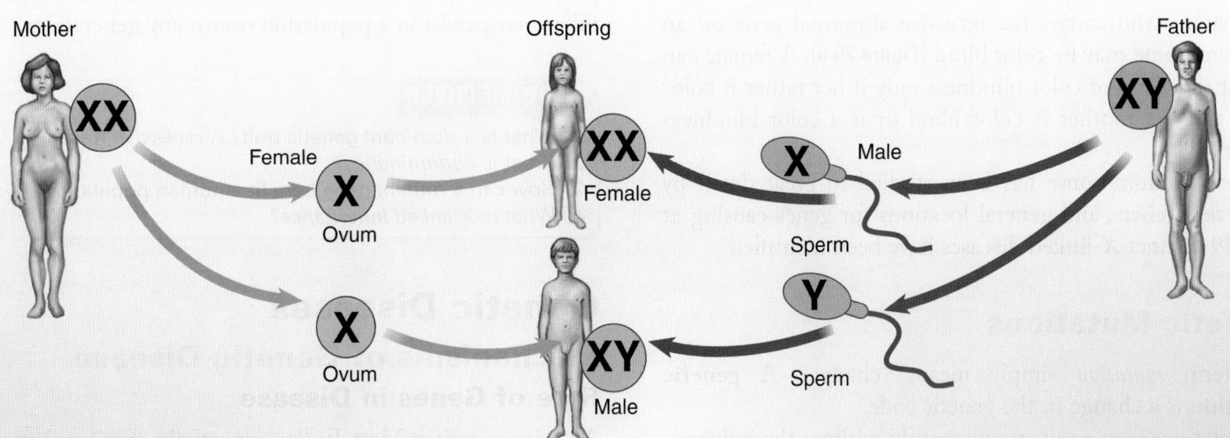

FIGURE 25-5 Sex determination. The presence of the *Y* chromosome specifies maleness. In the absence of a *Y* chromosome, an individual develops into a female.

25

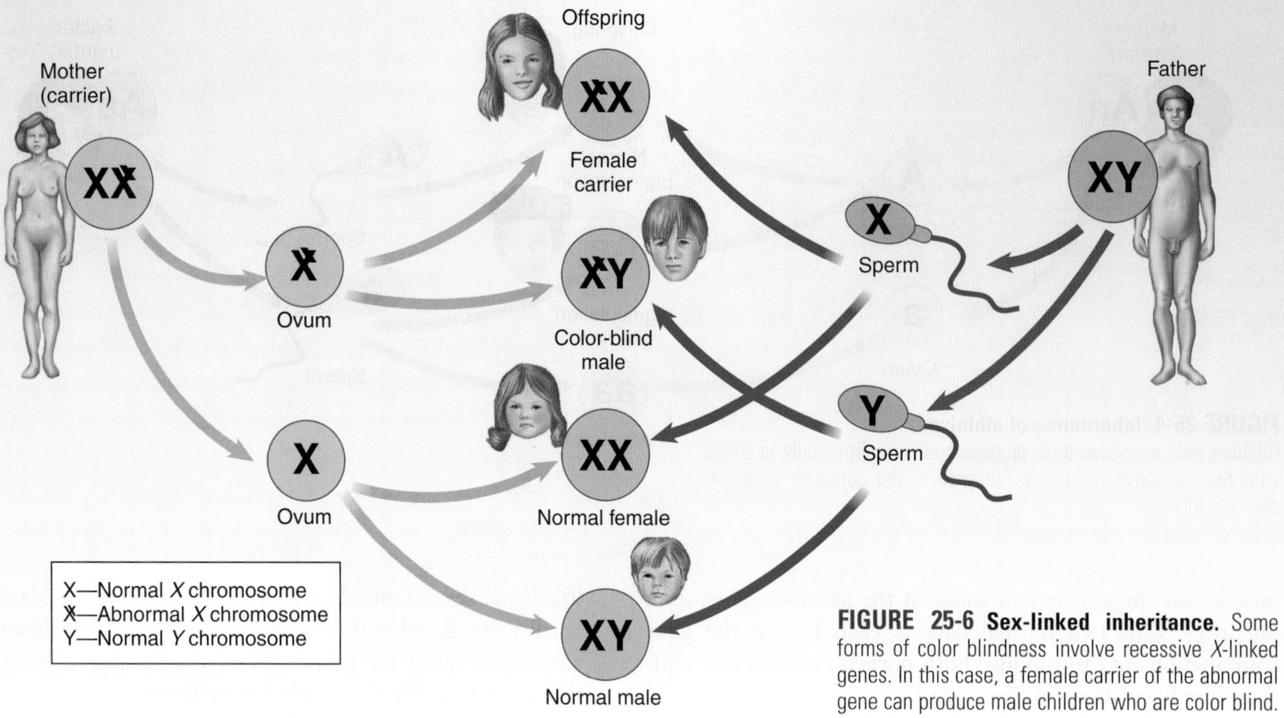

FIGURE 25-6 Sex-linked inheritance. Some forms of color blindness involve recessive *X*-linked genes. In this case, a female carrier of the abnormal gene can produce male children who are color blind.

traits are called *X-linked traits* because they are determined by the genes in the large *X* chromosome.

Dominant *X*-linked traits appear in each person as one would expect for any dominant trait. In females, recessive *X*-linked genes are masked by dominant genes in the other *X* chromosome. Only females with two recessive *X*-linked genes can exhibit the recessive trait.

Because males inherit only one *X* chromosome (from the mother), the presence of only one recessive *X*-linked gene is enough to produce the recessive trait. In short, in males, there are no matching genes in the *Y* chromosome to mask recessive genes in the *X* chromosome. For this reason, *X*-linked recessive traits appear much more commonly in males than in females.

An example of a recessive *X*-linked condition is *red-green color blindness,* which involves a deficiency of photopigments in the retina (see Chapter 11). In this condition, male children of a parent who carries the recessive abnormal gene on an *X* chromosome may be color blind (**Figure 25-6**). A female can inherit this form of color blindness only if her father is color blind and her mother is color blind or is a color blindness carrier.

The *X* chromosome has been studied in great detail by many researchers, and general locations for genes causing at least 59 distinct *X*-linked diseases have been identified.

Genetic Mutations

The term *mutation* simply means "change." A **genetic mutation** is a change in the genetic code.

Mutations may occur spontaneously, without the influence of factors outside the DNA itself. However, most genetic mutations are believed to be caused by **mutagens**—agents that cause mutations. Genetic mutagens include chemicals, some forms of radiation, and even viruses. Some mutations involve a change in the genetic code within a single gene, perhaps a slight rearrangement of the nucleotide sequence.

Other mutations involve damage to a portion of a chromosome or a whole chromosome. For example, a portion of a chromosome may completely break away.

Beneficial mutations allow organisms to adapt to their environments. Because such mutant genes benefit survival, they tend to spread throughout a population over the course of several generations. Harmful mutations inhibit survival, so they are not likely to spread through the population. Most harmful mutations kill the organism in which they occur—or at least prevent successful reproduction—and so are never passed to offspring. If a harmful mutation is only mildly harmful, it may persist in a population over many generations.

QUICK CHECK

1. What is a *dominant* genetic trait? A *recessive* trait?
2. What is *codominance?*
3. How can a *mutant* gene benefit a human population?
4. What is *X-linked inheritance?*

Genetic Diseases

Mechanisms of Genetic Disease

Role of Genes in Disease

As science writer Matt Ridley repeatedly and emphatically states in his best-selling book *Genome: The Autobiography of a*

Species in 23 Chapters, "GENES ARE NOT THERE TO CAUSE DISEASE."

Although we often hear of new "disease genes" being discovered and the pace is rapidly increasing—the function of these genes is not to cause disease any more than the function of an arm is to cause bone fractures. If you break an arm, a normal bone is broken and fails to serve its usual function. In genetic disorders, a normal gene or chromosome is broken (mutated) and fails to serve its usual function. Such a gene is sometimes called a "disease gene" because when it is broken, it is involved in the mechanism of a particular disease.

Keep this simple—but often overlooked—principle in mind as you read the following paragraphs.

Single-Gene Mechanisms

As we just stated, genetic diseases are diseases produced by an abnormality in the genetic code. Many genetic diseases are caused by individual mutant genes that are passed from one generation to the next, making them **single-gene diseases.**

The locations of some of the genes involved in selected single-gene diseases are shown in **Figure 25-7**. In single-gene diseases, the mutant gene may make an abnormal product that causes disease, or it may fail to make a product required for normal function. Some disease conditions result from the combined effects of inheritance and environmental factors.

We explore specific examples of single-gene diseases later.

Epigenetics

Epigenetics *(imprinting)* is the science that describes how environmental and behavioral factors may result in offspring with genetic traits that cannot be explained by genes alone.

For example, the diet of parents or grandparents may provide chemical groups to the chromosomes that may switch a certain gene on or off. Thus the genes may be imprinted or chemically marked, and certain diseases may or may not be seen in the offspring.

The effect of the environment on genes also explains why identical twins may not always share the same disease traits. Because they are not solely caused by genetic mechanisms, such conditions are not genetic diseases in the usual sense of the word—they are instead said to involve a *genetic predisposition.*

Learn more in the illustrated article *Epigenetics* at Connect It! at *evolve.elsevier.com.*

Chromosomal Mechanisms

Some genetic diseases do not result from an abnormality in a single gene. Instead, these diseases result from chromosome breakage or from the abnormal presence or absence of entire chromosomes.

For example, a condition called **trisomy** may occur in which there is a triplet of autosomes rather than a pair. Trisomy

RESEARCH, ISSUES, AND TRENDS

MITOCHONDRIAL INHERITANCE

Mitochondria are tiny, bacteria-like organelles present in every cell of the body (see **Figure 3-2**). Like a bacterium, each mitochondrion has its own circular DNA molecule, sometimes called **mitochondrial DNA (mDNA or mtDNA).** The figure shows an ideogram of the structure of a mitochondrial chromosome.

Inheritance of mDNA occurs only through one's mother because the few mitochondria that a sperm may contribute to the ovum during fertilization do not survive. Because mDNA contains the only genetic code for several important enzymes involved in the metabolic pathway that "recharges" adenosine triphosphate (ATP) in our cells, it has the potential for carrying mutations that produce disease.

Mitochondrial inheritance is now known to transmit genes for several degenerative nerve and muscle disorders. One such disease is **Leber hereditary optic neuropathy.** In this disease, young adults begin losing their eyesight as the optic nerve degenerates—resulting in total blindness by age 30.

Some medical researchers believe that at least some forms of several other diseases are associated with mDNA mutations. These diseases include *Parkinson disease, Alzheimer disease (AD), diabetes mellitus (DM)* with deafness, and maternally inherited forms of deafness, myopathy, and cardiomyopathy.

Researchers are developing experimental protocols for *in vitro fertilization (IVF)* in which mitochondrial inheritance of disease may be avoided by using an egg donor with normal mitochondria but using the egg nucleus from a mother with mutated mtDNA.

Review the article *In Vitro Fertilization* at Connect It! at *evolve.elsevier.com.*

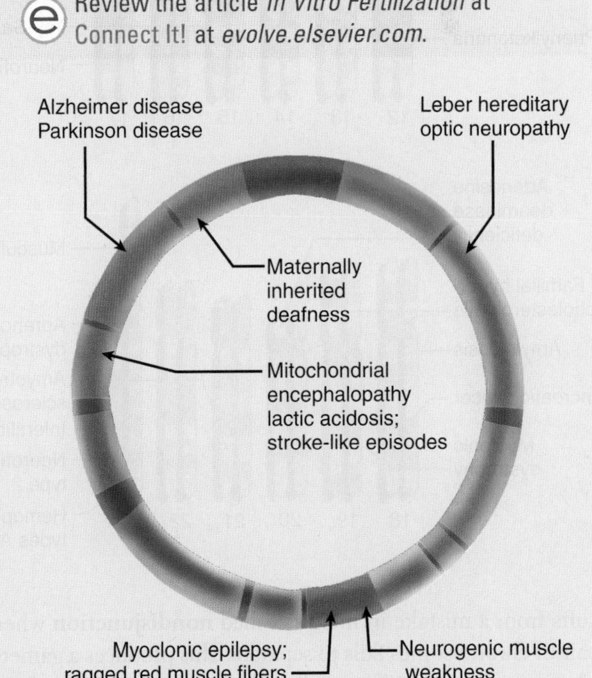

Map of mitochondrial DNA. Ideogram showing locations of some mtDNA genes associated with various diseases.

25

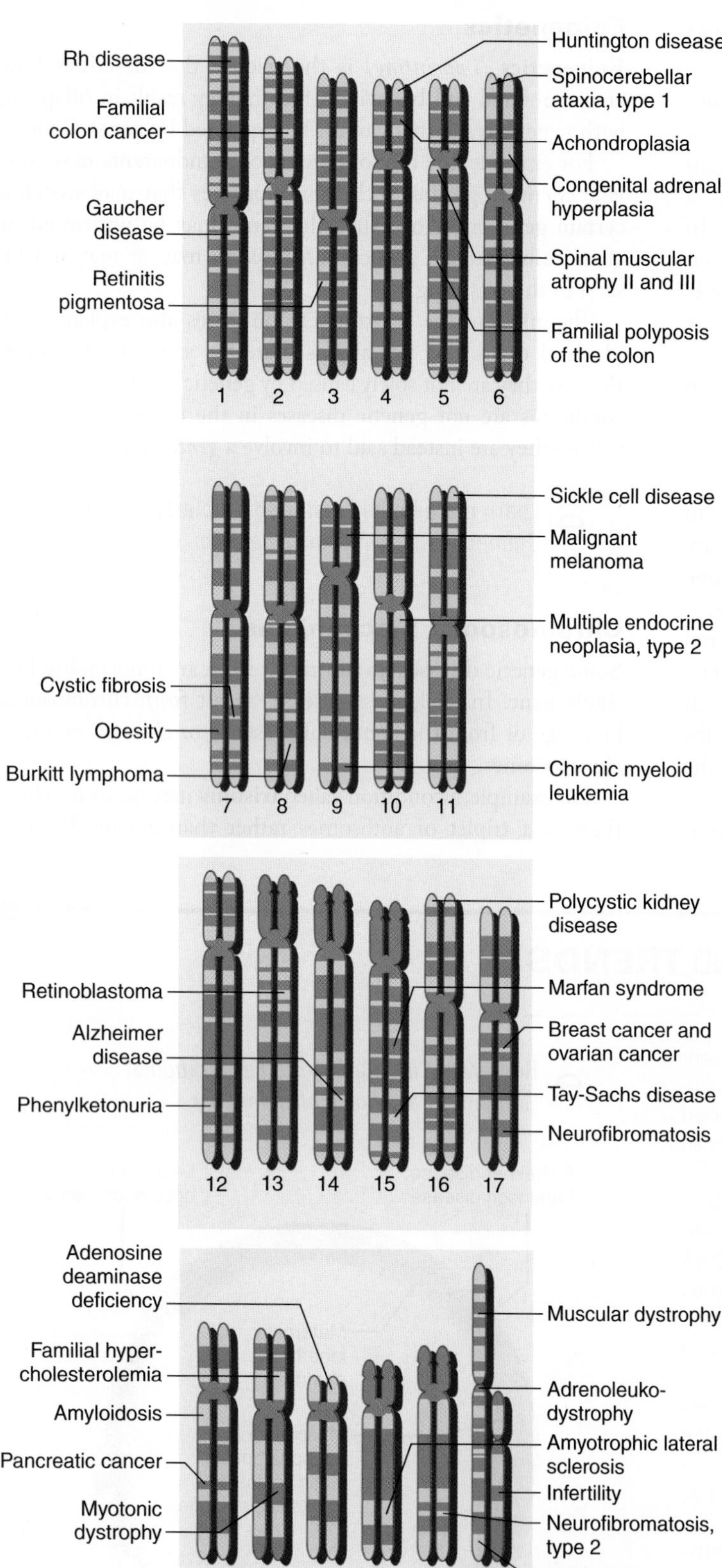

FIGURE 25-7 Location of genes involved in genetic diseases. The ideogram of each chromosome is labeled with the location of just one or two examples of the many genes associated with genetic disease.

results from a mistake in meiosis called **nondisjunction** when a pair of chromosomes fails to separate. This produces a gamete with two autosomes that are "stuck together" instead of the usual one. When this abnormal gamete joins with a normal gamete to form a zygote, the zygote has a triplet of autosomes (**Figure 25-8**).

Usually trisomy of any autosome pair is fatal. However, if trisomy occurs in autosome pair 13, 15, 18, 21, or 22, a person may survive for a time but not without profound developmental defects.

Monosomy, the presence of only one autosome instead of a pair, also may result from conception involving a gamete produced by nondisjunction (see **Figure 25-8**). Like trisomy, monosomy may produce life-threatening abnormalities.

Because most trisomic and monosomic individuals die before they can reproduce, these conditions are not usually passed from generation to generation.

Trisomy and monosomy are congenital conditions that are sometimes referred to as **chromosomal genetic diseases.** Specific examples are discussed later.

QUICK CHECK

1. How are *single-gene* diseases different from *chromosomal* conditions?
2. What is *nondisjunction*? How can it cause *trisomy*?

Single-Gene Diseases

There are many single-gene diseases. Only a few examples are discussed here and summarized in **Table 25-1**.

Cystic Fibrosis

Cystic fibrosis (CF) is caused by a recessive gene in chromosome 7 that codes for a protein called *CFTR (CF transmembrane conductance regulator). CFTR* normally regulates the transfer of sodium and chloride ions across cell membranes and serves as a chloride ion channel.

When this gene is missing a single codon, the abnormal version of *CFTR* causes impairment of ion transport across cell membranes. Such disruption causes exocrine cells to secrete thick mucus and sweat. The thickened mucus is especially troublesome in the gastrointestinal and respiratory tracts, where it can cause obstruction that leads to death.

A carrier of only one abnormal *CFTR* gene, however, may have improved resistance to diarrheal diseases such as cholera.

Advanced therapies for CF have greatly improved quality and length of life, but it is still a very serious condition.

Phenylketonuria

Phenylketonuria (PKU) is caused by a recessive gene that fails to produce the enzyme *phenylalanine hydroxylase.* This enzyme is needed to convert the amino acid phenylalanine into another amino acid, tyrosine. Thus phenylalanine absorbed from ingested

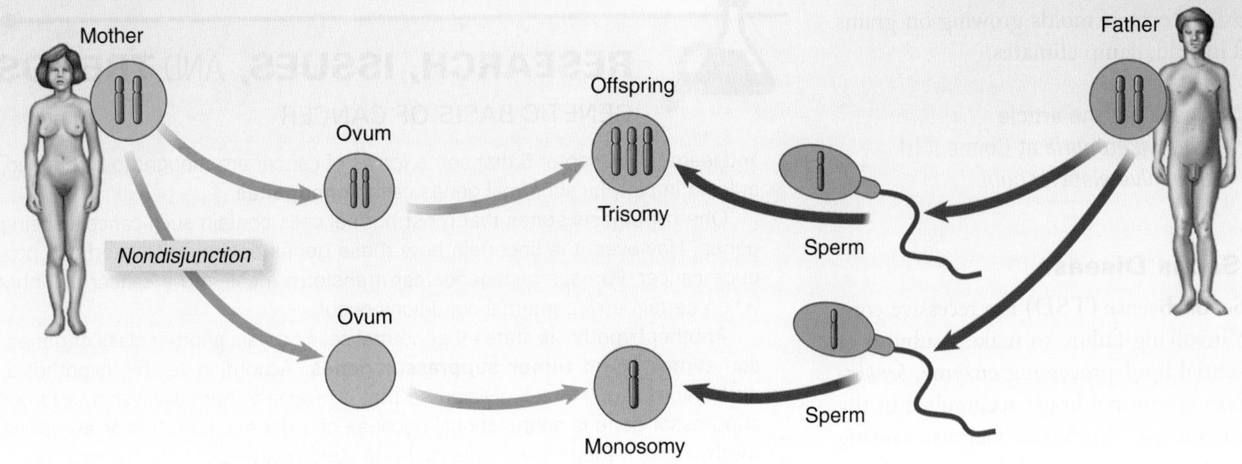

FIGURE 25-8 Effects of nondisjunction. Nondisjunction, failure of a chromosome pair to separate during gamete production, may result in trisomy or monosomy in the offspring.

food accumulates, resulting in the abnormal presence of phenylketone in the urine (hence the name *phenylketonuria*).

A high concentration of phenylalanine in the body destroys brain tissue, so babies born with this condition are at risk of progressive mental retardation and perhaps death.

Many PKU patients are identified at birth by state-mandated tests. Once identified, PKU victims are put on diets low in phenylalanine, thus avoiding a toxic accumulation of it.

You may be familiar with the printed warning for phenylketonurics commonly seen on products that contain aspartame (NutraSweet) or other substances made from phenylalanine.

The mutant PKU gene may have originated among the Celts in western Europe, where it offered protection against

TABLE 25-1 Examples of Single-Gene Disorders

DISORDER	DOMINANCE	DESCRIPTION
Hemophilia (some forms)	Recessive (*X*-linked)	Group of blood-clotting disorders caused by a failure to form clotting factors VIII, IX, or XI
Albinism	Recessive	Lack of the dark-brown pigment melanin in the skin and eyes, resulting in vision problems and susceptibility to sunburn and skin cancer
Sickle cell anemia and sickle cell trait	Codominant	Blood disorder in which abnormal hemoglobin is produced, causing red blood cells to deform into a sickle shape
Red-green color blindness (*X*-linked)	Recessive	Inability to distinguish red and green light, resulting from a deficiency of photopigments in the cone cells of the retina
Cystic fibrosis (CF)	Recessive	Condition characterized by excessive secretion of thick mucus and concentrated sweat, often causing obstruction of the gastrointestinal (GI) or respiratory tracts
Phenylketonuria (PKU)	Recessive	Excess of phenylketone in the urine, which is caused by accumulation of phenylalanine in the tissues and may cause brain injury and death
Tay-Sachs disease	Recessive	Fatal condition in which abnormal lipids accumulate in the brain and cause damage that leads to death by age 4
Osteogenesis imperfecta	Dominant	Group of connective tissue disorders characterized by imperfect skeletal development that produces brittle bones
Multiple neurofibromatosis	Dominant	Disorder characterized by multiple, sometimes disfiguring, benign tumors of the Schwann cells (neuroglia) that surround nerve fibers
Duchenne muscular dystrophy (DMD) (*X*-linked)	Recessive	Muscle disorder characterized by progressive atrophy of skeletal muscle without nerve involvement; caused by lack of normal dystrophin protein that supports muscle fiber structure
Hypercholesterolemia	Dominant	High blood cholesterol that may lead to atherosclerosis and other cardiovascular problems
Huntington disease (HD)	Dominant	Degenerative brain disorder characterized by chorea (purposeless movements), progressing to severe dementia and death by age 55
Severe combined immune deficiency (SCID)	Recessive	Failure of the lymphocytes to develop properly, in turn causing failure of the immune system's defense of the body; usually caused by adenosine deaminase (ADA) deficiency

the toxic effects of molds growing on grains stored in cold, damp climates.

Check out the article *Phenylketonuria* at Connect It! at *evolve.elsevier.com.*

Tay-Sachs Disease

Tay-Sachs disease (TSD) is a recessive condition involving failure to make a subunit of an essential lipid-processing enzyme, *hexosaminidase.* Abnormal lipids accumulate in the brain tissue of Tay-Sachs victims, causing severe retardation and death by 4 years of age. There is currently no specific therapy for this condition.

TSD is most prevalent among certain Jewish populations. Some epidemiologists believe that this ethnic distribution is related to the hypothesis that heterozygous carriers of the Tay-Sachs gene have a higher than normal resistance to tuberculosis (TB)—a potentially fatal disease that once killed millions in the crowded Jewish ghettos of many large cities. Residents of these TB-infested areas who carried the Tay-Sachs gene apparently survived longer—and reproduced more frequently—than noncarriers.

Tay-Sachs is also found in higher than average frequencies in French Canadians in southeastern Quebec and Cajun French families in southern Louisiana—probably due to the gene's presence in several founders of these family groups, rather than natural selection by the threat of TB.

RESEARCH, ISSUES, AND TRENDS

GENETIC BASIS OF CANCER

You learned in Chapter 6 that some forms of cancer are thought to be caused, at least in part, by abnormal genes called **oncogenes.**

One hypothesis states that most normal cells contain such cancer-causing genes. However, it is uncertain how these genes become activated and produce cancer. Perhaps oncogenes can transform a cell into a cancer cell only when certain environmental conditions occur.

Another hypothesis states that normal cells contain another class of genes, sometimes called **tumor suppressor genes.** According to this hypothesis, such genes regulate cell division so that it proceeds normally. When a tumor suppressor gene is nonfunctional because of a genetic mutation or excessive methylation, it then allows cells to divide abnormally—possibly producing cancer. If maternal and paternal tumor suppressor genes are both affected, the cancer risk would be more pronounced.

Another possible genetic basis for cancer relates to the genes that govern the cell's ability to repair damaged DNA. As mentioned in Chapter 7, a rare genetic disorder called **xeroderma pigmentosum** is characterized by the inability of skin cells to repair genetic damage caused by ultraviolet radiation in sunlight. Individuals with this condition almost always develop skin cancer when exposed to direct sunlight. The genetic abnormality does not cause skin cancer directly but inhibits the cell's cancer-preventing mechanisms.

Epigenetic Conditions

The study of how environment and behavior may influence the genes for both health and disease is rapidly growing. Some diseases are thought to be the result of epigenetic changes in DNA that alter gene activity that causes disorders. The number of epigenetic diseases may actually be huge.

Some cancers have been shown to be related to a reduction in chemical markers known as *methyl groups* ($—CH_3$). Other cancers are associated with too much methylation, which may inhibit genes that normally prevent cancer growth. Treatment of cancers with agents that either increase or decrease methylation of abnormal cells is being researched.

25

However, not all epigenetic conditions are associated with methylation. *Acetyl groups* ($—COCH_3$) or *ubiquitin* protein also may mark the DNA, while other mechanisms involve the RNA molecules that regulate the production of proteins in the cell.

Review the article *Epigenetics* at Connect It! at *evolve.elsevier.com.*

Fragile X syndrome (FXS) is a disease that is thought to be associated with overmethylation of a section of the DNA in the X chromosome. Methylation may occur when a string of repeating gene components, known as CGG in the normal protein gene, gets too long. The longer length of the repeating CGG nucleotides results in a greater severity of the syndrome.

Overmethylation of this section of DNA turns off the gene for a protein that normally prevents this genetic form of mental retardation.

FXS is more common in males because they only inherit one copy of the X chromosome. Females may have a milder form because they may inherit a normal X chromosome that is able to make some of the normal protein.

Some other epigenetic diseases may include type 2 diabetes mellitus (DM) and cardiovascular disease. Cancer risk may also have epigenetic components. The more we understand about the epigenetic basis of disease, the more likely it is that we will find effective treatments—or even cures.

Chromosomal Diseases

Some genetic disorders are not inherited but result instead from nondisjunction during gamete formation. As **Figure 25-8** shows, nondisjunction results in trisomy or monosomy.

Trisomy 21

The most well-known chromosomal disorder is *trisomy 21,* which produces a group of symptoms called **Down syndrome.** As shown in **Figure 25-9,** *A,* this condition is characterized by a triplet of chromosome 21 rather than the usual pair.

In the general population, trisomy 21 occurs in only 1 of every 600 or so live births. After age 35, however, a mother's chances of producing a trisomic child increase dramatically—to as high as 1 in 80 births by age 40. Down syndrome results

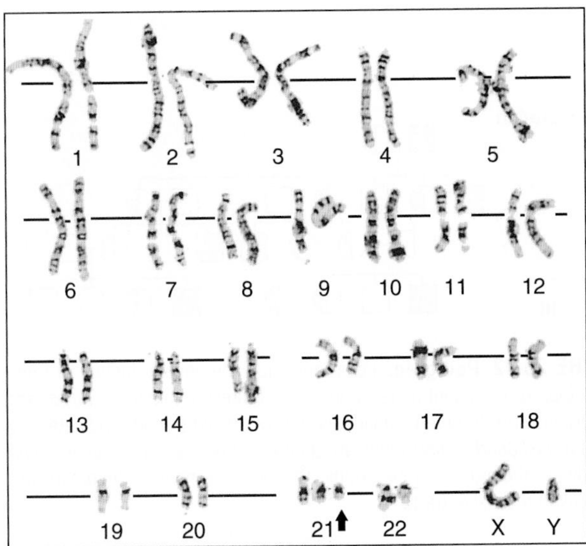

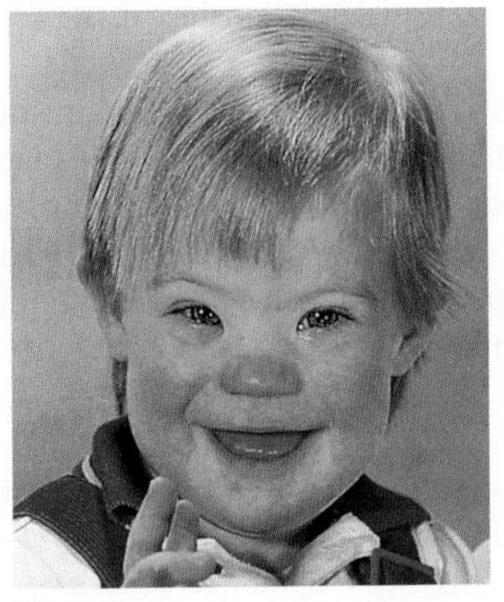

FIGURE 25-9 Down syndrome. A, Down syndrome is usually associated with trisomy of chromosome 21, as you can see in this karyotype. **B,** A child with Down syndrome. Notice the distinctive anatomical features: folds around the eyes, flattened nose, round face, and small hands with short fingers.

from trisomy 21 and rarely from other genetic abnormalities (which can be inherited).

Down syndrome is characterized by mental retardation (ranging from mild to severe) and multiple defects that include distinctive facial appearance (**Figure 25-9**, *B*), enlarged tongue, short hands and feet with stubby digits, congenital heart disease, and susceptibility to acute leukemia. People with Down syndrome have a shorter-than-average life expectancy but can survive to old age.

XXY

Klinefelter syndrome is another genetic disorder resulting from nondisjunction of chromosomes (**Figure 25-10**, *B*). This disorder occurs in males with a *Y* chromosome and at least two *X* chromosomes, typically the *XXY* pattern.

Characteristics of Klinefelter syndrome may include long legs, enlarged breasts, learning difficulties, small testes, sterility, and chronic pulmonary disease (**Figure 25-10**, *A*).

XO

Turner syndrome, sometimes called *XO syndrome,* occurs in females with a single sex chromosome, *X*. Like the conditions described earlier, it results from nondisjunction during gamete formation (**Figure 25-11**, *B*).

Turner syndrome is characterized by failure of the ovaries and other sex organs to mature (causing sterility), cardiovascular defects, dwarfism or short stature, a webbed neck, and possible learning disorders (see **Figure 25-11**, *A*). Symptoms of

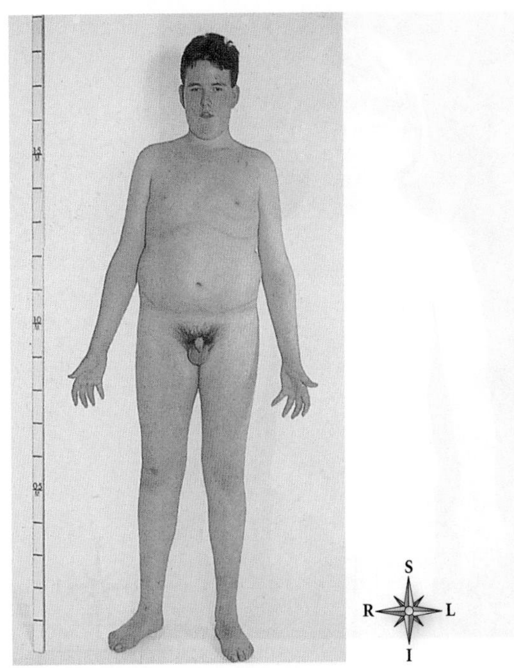

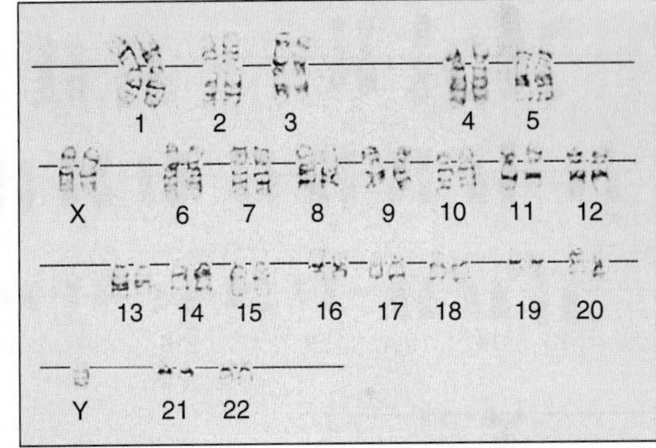

FIGURE 25-10 Klinefelter syndrome. A, This young man exhibits many of the characteristics of Klinefelter syndrome: small testes, some development of the breasts, sparse body hair, and long limbs. **B,** This syndrome results from the presence of two or more *X* chromosomes with a *Y* chromosome (genotypes XXY or XXXY, for example).

Turner syndrome can be reduced by hormone therapy using estrogens and growth hormone. Cardiovascular defects may be repaired surgically.

QUICK CHECK

1. How does avoidance of phenylalanine in the diet reduce the problems associated with *phenylketonuria (PKU)*?
2. What is *trisomy 21*?
3. What is an example of a female chromosomal disease that is caused by the monosomy condition XO?

Prevention and Treatment of Genetic Diseases

Genetic Counseling

The term **genetic counseling** refers to professional consultations with families regarding genetic diseases. Trained genetic counselors may help a family determine the risk of producing

25

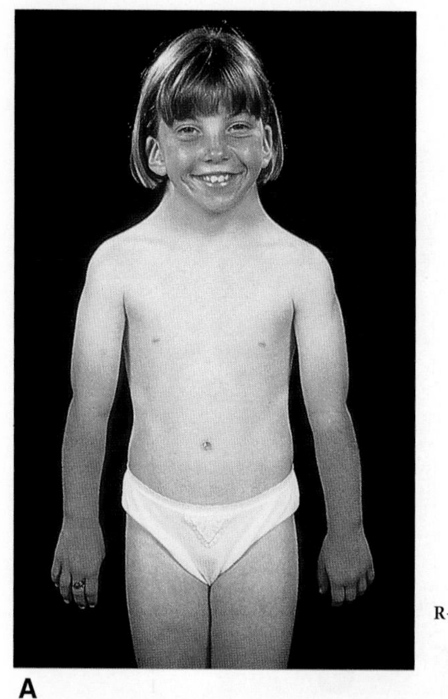

A

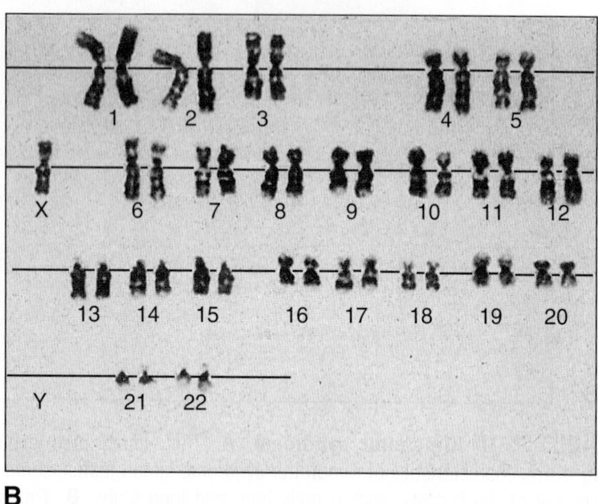

B

FIGURE 25-11 Turner syndrome. A, This woman exhibits many of the characteristics of Turner syndrome, including short stature, webbed neck, and sexual immaturity. **B,** As this karyotype shows, Turner syndrome results from monosomy of sex chromosomes (genotype XO).

children with genetic diseases. Parents with a high risk of producing children with genetic disorders may decide to avoid having children. Genetic counselors may also help evaluate whether any offspring already born have a genetic disorder and offer advice on treatment or care. A growing list of tools is available to genetic counselors, some of which are described below.

Pedigree

A **pedigree** is a chart that illustrates genetic relationships in a family over several generations (**Figure 25-12**). Using medical records and family histories, the genetic counselors assemble the chart, beginning with the client and moving backward through as many generations as are known.

Squares represent males; circles represent females. Fully shaded symbols represent affected individuals, whereas unshaded symbols represent normal individuals. Partially shaded symbols represent *carriers* of a recessive trait. A horizontal line between symbols designates a sexual relationship that produced offspring.

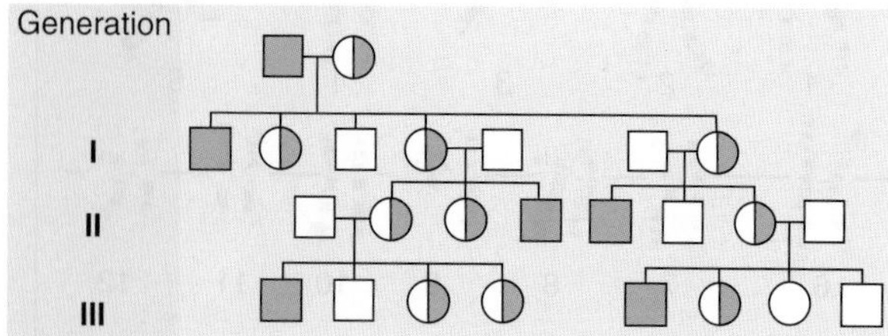

FIGURE 25-12 Pedigree. Pedigrees chart the genetic history of family lines. *Squares* represent males, and *circles* represent females. *Fully shaded symbols* indicate affected individuals; *partly shaded symbols* indicate carriers, and *unshaded symbols* indicate unaffected noncarriers. *Roman numerals* indicate the order of generations. This pedigree reveals the presence of an *X*-linked recessive trait.

The pedigree is useful in determining the possibility of producing offspring with certain genetic disorders. It also may tell a person whether he or she might have a genetic disorder that appears late in life, such as *Huntington disease (HD)*. In either case, a family can prepare emotionally, financially, and medically before a crisis occurs.

Punnett Square

The **Punnett square,** named after the English geneticist Reginald Punnett, is a grid used to help determine the *probability* of inheriting genetic traits.

As **Figure 25-13**, *A*, shows, genes in the mother's gametes are represented along one axis of the grid, and genes in the father's gametes are along the other axis. The ratio of different gene combinations in the offspring predicts their probability of occurrence in the next generation.

We can see in **Figure 25-13**, *A*, that offspring produced by two carriers of PKU (a recessive disorder) have a one in four (25%) chance of inheriting this recessive condition. There is a two in four (50%) chance that an individual child produced will be a PKU carrier.

Figure 25-13, *B*, shows that offspring between a carrier and a noncarrier cannot inherit PKU. What is the chance of an individual offspring being a PKU carrier in this case? **Figure 25-13**, *C*, shows the probability of producing an affected offspring when a PKU patient and a PKU carrier have children. **Figure 25-13**, *D*, shows the genetic probability when a PKU patient and a noncarrier produce children.

Karyotype

Disorders that involve trisomy (extra chromosomes), monosomy (missing chromosomes), and broken chromosomes can be detected after a **karyotype** is produced.

The first step in producing a karyotype is getting a sample of cells from the individual to be tested. This can be done by scraping cells from the lining of the cheek or from a blood sample containing white blood cells (WBCs).

Fetal tissue can be collected by **amniocentesis,** a procedure in which fetal cells floating in the amniotic fluid are collected with a syringe (**Figure 25-14**). **Chorionic villus sampling** is a

CLINICAL APPLICATION

DNA ANALYSIS

As a result of the intense efforts under way to map the entire human genome, new techniques have been developed to analyze the genetic makeup of individuals. Automated machines can now chemically analyze chromosomes and "read" their sequence of nucleotides—the genetic code.

One method by which this is done is called **electrophoresis,** which means "electric separation" *(A).* In electrophoresis, DNA fragments are chemically processed, then placed in a thick fluid or gel. An electric field in the gel causes the DNA fragments to separate into groups according to their relative sizes. The pattern *(B)* that results represents the sequence of codons in the DNA fragment.

This process is also the basis for so-called **DNA fingerprinting.** Like a fingerprint pattern, each person's DNA sequence is unique. After the exact sequences for specific diseases have been discovered, genetic counselors will be able to provide more details about the genetic makeup of their clients. This technique is also used in forensic science to show that an individual was at the scene of the crime and left their DNA.

The accuracy of DNA analysis using these techniques relies on several important factors. For example, contamination of samples must be avoided by using very strict laboratory protocols. The number of different DNA sequences analyzed, and the chromosomal locations of those sequences are also important factors.

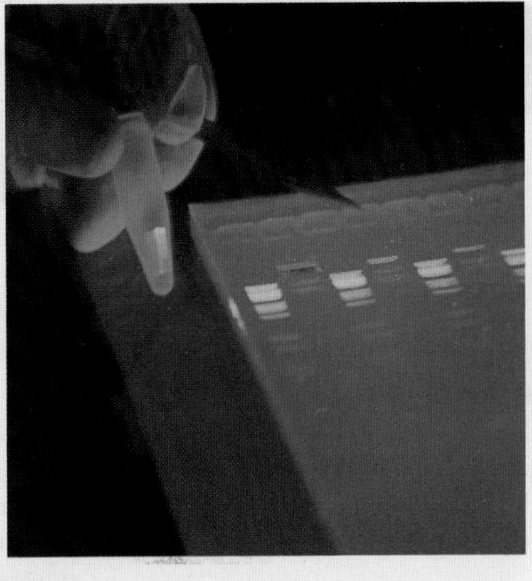

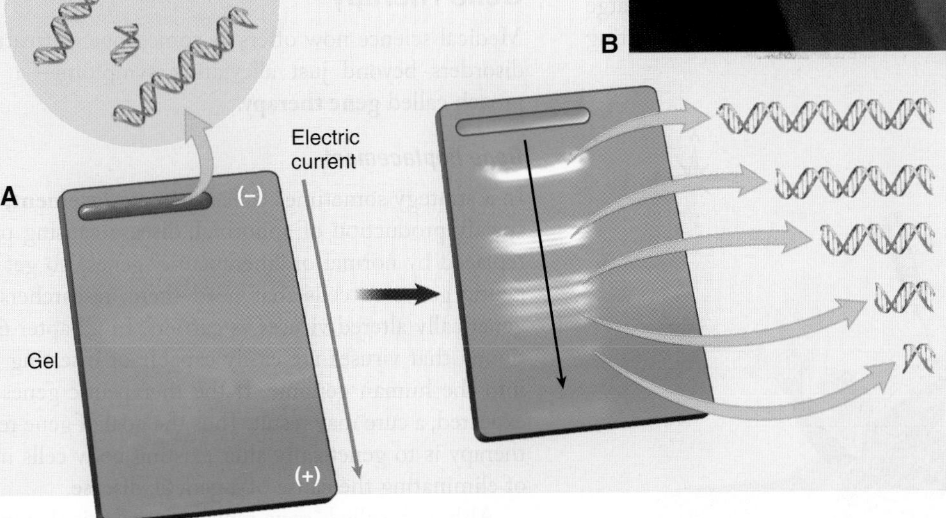

procedure in which cells from chorionic villi that surround a younger embryo (see Chapter 24) are collected through the opening of the cervix.

Newer, less invasive procedures that use fetal cells or fetal DNA recovered from maternal blood are in development and may soon be used widely. The newer tests have very low risks for complications compared with amniocentesis and chorionic villus sampling.

Collected fetal cells are grown in a special culture medium and allowed to reproduce. Cells in metaphase (when the chromosomes are most distinct) are stained and photographed using a microscope. The chromosomes are cut out of the photo and pasted on a chart in pairs according to size, as in **Figures 25-1** and **25-9**, *A*. Genetic counselors then examine the karyotype, looking for chromosome abnormalities. What chromosome abnormality is visible in **Figure 25-9**, *A*? Is this a male or female karyotype?

QUICK CHECK

1. What is *genetic counseling*?
2. How are *pedigrees* used by genetic counselors?
3. How is a *Punnett square* used to predict mathematical probabilities of inheriting specific genes?
4. How is a *karyotype* prepared? What is its purpose?

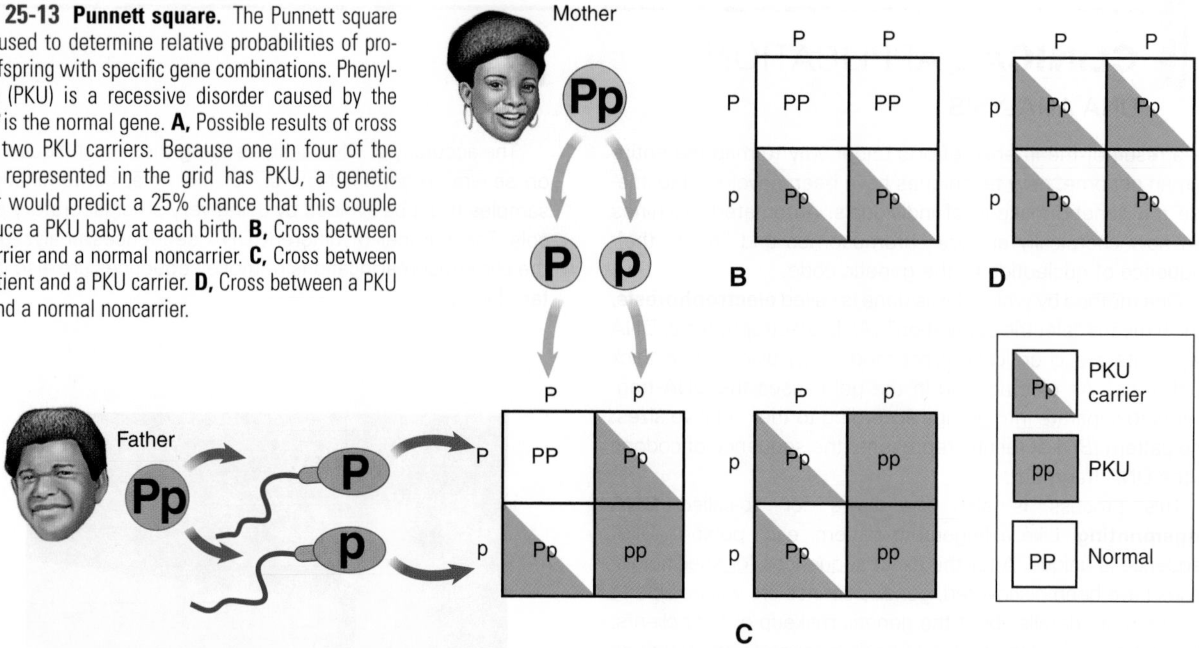

FIGURE 25-13 Punnett square. The Punnett square is a grid used to determine relative probabilities of producing offspring with specific gene combinations. Phenylketonuria (PKU) is a recessive disorder caused by the gene *p*. *P* is the normal gene. **A,** Possible results of cross between two PKU carriers. Because one in four of the offspring represented in the grid has PKU, a genetic counselor would predict a 25% chance that this couple will produce a PKU baby at each birth. **B,** Cross between a PKU carrier and a normal noncarrier. **C,** Cross between a PKU patient and a PKU carrier. **D,** Cross between a PKU patient and a normal noncarrier.

Treating Genetic Diseases

Treating Symptoms

Until this century, the only hope of treating any genetic disease was to treat the symptoms. In some diseases, such as PKU, this works well. If those affected by PKU simply avoid large amounts of phenylalanine in their diets, especially during critical stages of development, severe complications can be avoided. In Klinefelter and Turner syndromes, hormone therapy and surgery can alleviate some symptoms. However, there are no effective treatments for a vast majority of genetic disorders.

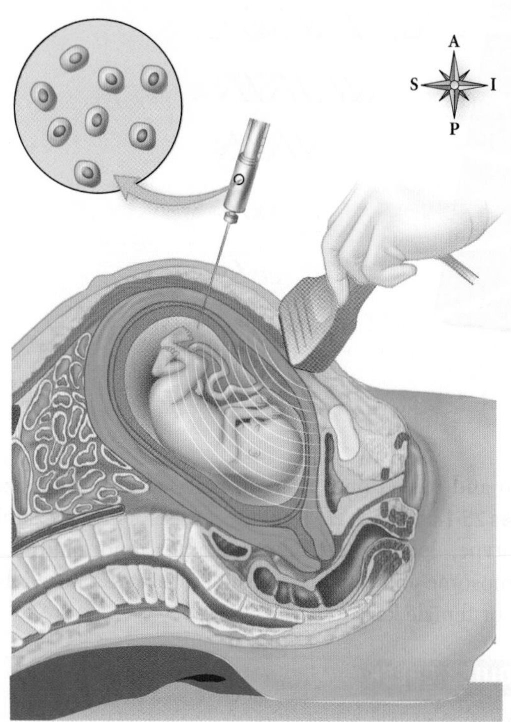

FIGURE 25-14 Amniocentesis. In amniocentesis, a syringe is used to collect amniotic fluid. Ultrasound imaging is used to guide the tip of the syringe needle to prevent damage to the placenta and fetus (see box on p. 663). Fetal cells in the collected amniotic fluid can then be chemically tested or used to produce a karyotype of the developing offspring.

Gene Therapy

Medical science now offers us some hope of treating genetic disorders beyond just alleviating symptoms—a newer approach called **gene therapy.**

Gene Replacement

In a strategy sometimes called **gene replacement,** genes that specify production of abnormal, disease-causing proteins are replaced by normal or "therapeutic" genes. To get the therapeutic genes to cells that need them, researchers are using genetically altered viruses as carriers. In Chapter 6, we mentioned that viruses are easily capable of inserting new genes into the human genome. If the therapeutic genes behave as expected, a cure may result. Thus the goal of gene replacement therapy is to genetically alter existing body cells in the hope of eliminating the cause of a genetic disease.

Although called "gene replacement," this therapy does not actually replace the defective genes—it instead inserts normal genes so that normal proteins can "replace" abnormal or missing proteins in the body's metabolic pathways.

Gene Augmentation

In a therapy called **gene augmentation,** normal genes are introduced with the hope that they will augment (add to) the production of the needed protein. In one form of gene augmentation, virus-altered cells are injected into the blood or implanted under the skin of a patient to produce adequate amounts of the missing protein.

Another approach is to use bacterial DNA rings called **plasmids** that have been altered by recombinant DNA techniques to carry the therapeutic gene(s). A more recent approach used the **human engineered chromosome (HEC).** In the HEC approach, a set of therapeutic genes is incorporated into a separate strand of DNA that is inserted into a cell's nucleus, thus acting like an extra, or forty-seventh, chromosome. Gene augmentation attempts to add genetically altered cells to the body, rather than to change existing body cells as in gene replacement therapy.

RNA Interference

RNA interference (RNAi) also may become a weapon against genetic disorders in an approach called **RNAi therapy.** RNAi is a method of *silencing* particular genes, thus rendering them unable to produce their encoded proteins. When harnessed in the laboratory, RNAi can turn off one gene at a time—greatly increasing the chances of figuring out which protein is encoded by that gene and what the function of that protein is. When used therapeutically, RNAi may be able to silence specific genes involved in disease mechanisms.

Potential of Gene Therapy

The use of genetic therapy began in 1990 with a group of young children having **adenosine deaminase (ADA) deficiency.** In this rare recessive disorder, the gene for producing the enzyme ADA is missing from both autosomes in pair 20. Deficiency of ADA results in **severe combined immune deficiency (SCID),** making its victims highly susceptible to infection (see Chapter 16).

As **Figure 25-15** shows, white blood cells from each patient were collected and infected with viruses carrying therapeutic genes. After reproducing 1000-fold, the genetically altered white blood cells were then injected into the patient. Because this treatment augments cells already present with genetically altered cells, it is a form of gene augmentation therapy.

Currently hundreds of gene therapy trials for diverse genetic disorders, cancer, and even aging are proposed or ongoing. Thousands of laboratory experiments in anticipation of human trials are also under way. Hurdles to overcome before we see widespread success of gene therapies include our lack of detailed knowledge of many of the "disease genes" and how multiple-gene diseases might be effectively treated—not to mention the high costs and risks involved. It is too early to say for sure, but there may soon come a time when many genetic diseases are treated—or even cured—with gene therapy.

To learn more about vector-mediated gene therapy, go to AnimationDirect online at *evolve.elsevier.com*.

QUICK CHECK

1. How are most genetic disorders treated today?
2. How does *gene replacement* therapy work?
3. What is the *human engineered chromosome (HEC)*?

FIGURE 25-15 Gene therapy. A, This method of gene augmentation therapy was used to treat children stricken with a form of severe combined immune deficiency syndrome (SCID). White blood cells taken from the patient were infected with viruses carrying the therapeutic gene. The altered cells were reproduced and injected into the bloodstream, thereby reducing the immunity-inhibiting effects of SCID. **B,** Gene augmentation therapy in this example uses plasmids containing the therapeutic gene for cystic fibrosis (CF) and delivers them to the lung tissues by means of a common inhaler.

SCIENCE APPLICATIONS

GENETICS AND GENOMICS

Gregor Mendel (1822–1884)

The Moravian-German Gregor Mendel was born to peasant farmers who taught him how plants and animals are bred for specific traits. Mendel's acceptance into a monastery allowed him to study the science that would later help him understand the mechanism of inheritance of biological traits.

Convinced that "particles" in the cells of the parents were responsible for inheritance of traits, Mendel carried out the now famous experiments with several generations of pea plants. In his report *Experiments with Plant Hybrids,* Mendel outlined what has become the foundation of the science of genetics. Not only did he reveal the presence of genetic particles (which are now called *genes*) and the basic patterns of how they are transmitted to offspring, he also set in motion two important movements in modern biology.

First, Mendel was among the first to use mathematical analysis to support his theory about inheritance. Mendel's work pioneered the systematic use of mathematics, quantified measurements, and applied statistics in biological research. Today, medical researchers often enlist the help of *statisticians, mathematicians, computer programmers,* and others in designing experiments, analyzing data, and interpreting results. In fact a whole field, sometimes called *biomathematics,* has now emerged to apply the principles of mathematics to biological study.

Second, Mendel was the first to discover how the biological mechanisms of inheritance worked in living organisms. This, of course, led to the science of *genetics.* Many disciplines have since grown from the study and application of genetics. For example, **genetic counselors** use principles of genetics to advise clients who wish to produce offspring but are worried about possible genetic disorders. *Agricultural scientists* use genetic principles in refining hybrid crop plants and livestock. **Genetic engineers** develop ways to manipulate the genetic code to produce a variety of therapies and enhanced biological characteristics of agricultural products. *Genomics scientists* analyze the genetic codes of organisms to help us better understand structure and function, which may lead to better treatments for genetic disorders.

LANGUAGE OF SCIENCE *(continued from p. 679)*

genomics
(jeh-NOH-miks)
[*gen-* **produce (gene),** *-om-* **entire collection,** *-ic* **relating to**]

meiosis
(my-OH-sis)
[*mei-* **smaller,** *-osis* **process**]

mitochondrial DNA (mDNA, mtDNA)
(my-toh-KON-dree-al dee en ay [em dee en ay, em tee dee en ay])
[*mito-* **thread,** *-chondrion-* **granule,** *-al* **relating to,** *DNA* **deoxyribonucleic acid**]

nondisjunction
(non-dis-JUNK-shun)
[*non-* **not,** *-dis-* **split in two,** *junc-* **join,** *-tion* **condition**]

p-arm
(pee arm)
[*p* **petite (small)**]

principle of independent assortment
(PRIN-suh-puhl ov in-dih-PEN-dent uh-SORT-ment)
[*princip-* **foundation,** *in-* **not,** *-de-* **upon,** *-pend-* **hang,** *-ent* **state,** *assort-* **match into groups,** *-ment* **process**]

proteome
(PROH-tee-ohm)
[*prote-* **first rank (protein),** *-ome* **entire collection**]

proteomics
(proh-tee-OH-miks)
[*prote-* **first rank (protein),** *-om-* **entire collection,** *-ic* **relating to**]

pseudogene
(SOOD-oh-jeen)
[*pseudo-* **false,** *gen-* **produce**]

q-arm
(kyoo arm)
[*q* **follows** *p* **in Roman alphabet**]

recessive gene
(ree-SES-iv jeen)
[*recess-* **retreat,** *-ive* **relating to,** *gen-* **produce**]

sex chromosome
(seks KROH-moh-sohm)
[*chrom-* **color,** *-som-* **body**]

sex-linked trait
(seks-linked trayt)

25

LANGUAGE OF **MEDICINE**

adenosine deaminase (ADA) deficiency
(ah-DEN-oh-seen dee-AM-ih-nayse [ay dee ay] dee-FISH-en-see)
[*adenosine* **blend of adenine and ribose,** *de-* **remove,** *-amin-* **ammonia,** *-ase* **enzyme,** *de-* **down,** *-fici-* **perform,** *-ency* **quality**]

albinism
(AL-bih-niz-em)
[*alb-* **white,** *-in-* **characterized by,** *-ism* **state**]

amniocentesis
(am-nee-oh-sen-TEE-sis)
[*amnio-* **birth membrane,** *-cent-* **prick,** *-esis* **process**]

chorionic villus sampling
(koh-ree-ON-ik VIL-lus SAM-pling)
[*chorion-* **skin,** *-ic* **relating to,** *villus* **shaggy hair**]

chromosomal genetic disease
(kroh-moh-SOH-mal jeh-NET-ik)
[*chrom-* **color,** *-soma-* **body,** *-al* **relating to,** *gen-* **produce,** *-ic* **relating to,** *dis-* **opposite of,** *-ease* **comfort**]

cystic fibrosis
(SIS-tik fye-BROH-sis)
[*cyst-* **sac,** *-ic* **relating to,** *fibr-* **fiber,** *-osis* **condition**]

DNA fingerprinting
(dee en ay FING-ger-print-ing)
[*DNA* **deoxyribonucleic acid**]

Down syndrome
(down SIN-drohm)
[*John L. Down* **English physician,** *syn-* **together,** *-drome* **running or (race) course**]

electrophoresis
(eh-lek-troh-foh-REE-sis)
[*electro-* **electricity,** *-phor-* **carry,** *-esis* **process**]

fragile X syndrome (FXS)
(FRAJ-il eks SIN-drohm [ef ex es])
[*fragil-* **frail,** *X* **sex chromosome X,** *syn-* **together,** *-drome* **running or (race) course**]

gene augmentation
(jeen awg-men-TAY-shun)
[*gen-* **produce or generate,** *aug-* **increase,** *-ment-* **state,** *-ation* **process**]

gene linkage
(jeen LINK-ej)
[*gen-* **produce or generate**]

gene replacement
(jeen ree-PLAYS-ment)
[*gen-* **produce or generate**]

gene therapy
(jeen THAYR-a-pee)
[*gen-* **produce or generate,** *therapy* **treatment**]

genetic counseling
(jeh-NET-ik KOWN-sel-ing)
[*gene-* **produce,** *-ic* **relating to,** *counsel-* **consult,** *-ing* **process**]

genetic counselor
(jeh-NET-ik KOWN-sel-er)
[*gene-* **produce,** *-ic* **relating to,** *counsel-* **consult,** *-or* **agent**]

genetic engineer
(jeh-NET-ik en-juh-NEER)
[*gene-* **produce,** *-ic* **relating to,** *engin-* **devise or design,** *-eer* **practitioner**]

human engineered chromosome (HEC)
(HYOO-man en-juh-NEERD KROH-meh-sohm [aych ee see])
[*engin-* **devise or design,** *-eer* **practitioner,** *-ed* **state,** *chrom-* **color,** *-som-* **body**]

Human Genome Project (HGP)
(HYOO-man JEE-nome PROJ-ekt [aych jee pee])
(*gen-* **to produce,** *-om(e)-* **whole collection**]

ideogram
(ID-ee-oh-gram)
[*ide-* **idea,** *-gram* **drawing**]

karyotype
(KAYR-ee-oh-type)
[*karyo-* **nucleus,** *-type* **kind**]

Klinefelter syndrome
(KLINE-fel-ter SIN-drohm)
[*Harry F. Klinefelter* **American physician,** *syn-* **together,** *-drome* **running or (race) course**]

Leber hereditary optic neuropathy
(LEE-ber heh-RED-ih-tayr-ee OP-tik noo-ROP-ah-thee)
[*Theodor von Leber* **German ophthalmologist,** *her-* **heir,** *-it(y)* **state,** *-ary* **relating to,** *opt-* **vision,** *-ic* **relating to,** *neuro-* **nerves,** *-path-* **disease,** *-y* **state**]

monosomy
(MON-oh-soh-mee)
[*mono-* **single,** *-som-* **body (chromosome),** *-y* **state**]

mutagen
(MYOO-tah-jen)
[*muta-* **change,** *-gen* **produce**]

oncogene
(ON-koh-jeen)
[*onco-* **swelling or mass (cancer),** *-gen-* **produce or generate**]

pedigree
(PED-ih-gree)
[**from** *pied de grue* **crane's foot pattern**]

phenylketonuria (PKU)
(fen-il-kee-toh-NOO-ree-ah [pee kay yoo])
[*phen-* **shining (phenol),** *-yl-* **chemical,** *-keton-* **acetone,** *-ur-* **urine,** *-ia* **condition**]

plasmid
(PLAS-mid)
[*plasm-* **formed substance,** *-id* **belonging to**]

Punnett square
(PUN-it skwayr)
[*Reginald C. Punnett* **English geneticist**]

RNA interference (RNAi)
(ar en ay in-ter-FEER-ens [ar en ay aye])
[*RNA* **ribonucleic acid,** *inter-* **between,** *-fere-* **strike,** *-ence* **process**]

RNAi therapy
(ar en ay aye THAYR-ah-pee)
[*RNA-* **ribonucleic acid,** *-i* **interference,** *therapy* **treatment**]

severe combined immune deficiency (SCID)
(seh-VEER kom-BYNED ih-MYOON deh-FISH-en-see [skid])
[*immun-* **free (immunity),** *-defici-* **fail,** *-y* **state**]

single-gene disease
(SING-jul jeen dih-ZEEZ)
[*gen-* **produce or generate,** *dis-* **opposite of,** *-ease* **comfort**]

Tay-Sachs disease (TSD)
(TAY-saks dih-ZEEZ [tee es dee])
[*Warren Tay* **English ophthalmologist,** *Bernard Sachs* **American neurologist,** *dis-* **opposite of,** *-ease* **comfort**]

trisomy
(TRY-soh-mee)
[*tri-* **three,** *-som-* **body (chromosome),** *-y* **state**]

tumor suppressor gene
(TOOM-er suh-PRES-er jeen)
[*tumor* **swelling,** *suppress-* **press down,** *-or* **agent,** *gen-* **produce or generate**]

Turner syndrome
(TUR-ner SIN-drohm)
[*Harry H. Turner* **American endocrinologist,** *syn-* **together,** *-drome* **running or (race) course**]

xeroderma pigmentosum
(zeer-oh-DER-mah pig-men-TOH-sum)
[*xero-* **dry,** *-derma* **skin,** *pigment-* **paint,** *-osum* **characterized by**]

❑ OUTLINE SUMMARY

*To download a digital version of the chapter summary for use with your device, access the **Audio Chapter Summaries** online at evolve.elsevier.com.*

Scan this summary after reading the chapter to help you reinforce the key concepts. Later, use the summary as a quick review before your class or before a test.

Genetics and Human Disease

A. Genetics—the scientific study of biological inheritance
B. Inherited traits can produce disease (see Chapter 6)

Chromosomes and Genes

A. Mechanisms of gene function
 1. Gene—independent genetic units (DNA segments) that carry the genetic code
 2. Genes dictate the production of enzymes and other molecules, which in turn dictate the structure and function of a cell
 3. Genes are active in the chromatin (strand) form and inactive when DNA is in the chromosome (compact) form (**Figure 25-1**)
B. The human genome (**Figure 25-1**)
 1. Genome—entire set of human chromosomes (46 in nucleus of each cell, 1 mitochondrial chromosome)
  a. Map of the entire human genome (nearly all nucleotides in sequence) was completed in 2003
  b. Contains about 19,000 or so genes and large amounts of noncoding DNA, including nonfunctioning pseudogenes
 2. Genomics—analysis of the sequence contained in the genome
 3. Proteomics—analysis of the entire group of proteins encoded by the genome, called the human *proteome*
 4. Genomic information can be expressed in various ways
  a. Ideogram—cartoon of a chromosome showing the centromere as a constriction and the short segment (p-arm) and long segment (q-arm)
  b. Genes are often represented as their actual sequence of nucleotide bases expressed by the letters *a, c, g,* and *t*
C. Distribution of chromosomes to offspring
 1. Meiotic cell division produces gametes with 23 chromosomes each (**Figure 25-2**)
 2. At conception, two gametes join and produce a zygote with 46 chromosomes—the complete human genome
 3. Twenty-two pairs of chromosomes are called *autosomes;* each member of a pair resembles its partner
 4. The chromosomes in the remaining pair (pair 23) are called *sex chromosomes*
 5. Genetic variation among offspring is increased by:
  a. Independent assortment of chromosomes during gamete formation (**Figure 25-2**)
  b. Crossing-over of genes or linked groups of genes between chromosome partners during meiosis (**Figure 25-3**)

Gene Expression

A. Hereditary traits
 1. Dominant genes have effects that appear in the offspring (dominant forms of a gene are often represented by uppercase letters)
  a. A genetic carrier is a person who carries a recessive gene but does not show its effects because of masking effect of a dominant gene
  b. Codominant genes are two or more genes that are all dominant and when they appear together produce a combined effect in offspring
 2. Recessive genes have effects that do not appear in the offspring when they are masked by a dominant gene (recessive forms of a gene are represented by lowercase letters)
B. Sex-linked traits (**Figures 24-5** and **24-6**)
 1. The large *X* chromosome ("female chromosome") contains genes for female sexual characteristics and many other traits
 2. The small *Y* chromosome ("male chromosome") contains only genes for male sexual characteristics
 3. Normal males have *XY* as pair 23; normal females have *XX* as pair 23
 4. Nonsexual traits carried on sex chromosomes are sex-linked traits; most are *X*-linked traits
C. Genetic mutations
 1. Can result in abnormalities in the genetic code that cause disease
 2. Most believed to be caused by mutagens

Genetic Diseases

A. Mechanisms of genetic disease
 1. Genes are not there to cause disease; malfunctions of genetic code may cause disease
 2. Single-gene diseases result from individual mutant genes (or groups of genes) that are passed from generation to generation (**Figure 25-7**)
 3. Epigenetics (imprinting) involves environmental factors that may result in offspring with genetic traits that cannot be explained by genes alone; may involve genetic predisposition for disease
 4. Chromosomal diseases result from chromosome breakage or from nondisjunction (failure of a

25

chromosome pair to separate during gamete formation) (**Figure 25-8**)
 a. Trisomy—a chromosome triplet (instead of the usual pair)
 b. Monosomy—a single chromosome (instead of a pair)

B. Examples of single-gene diseases (**Table 25-1**)
 1. Cystic fibrosis—recessive autosomal condition characterized by excessive secretion of mucus and sweat, often causing obstruction of the gastrointestinal or respiratory tracts
 2. Phenylketonuria (PKU)—recessive autosomal condition characterized by excess phenylketone in urine, caused by accumulation of phenylalanine in tissues; may cause brain injury and death
 3. Tay-Sachs disease (TSD) is a recessive condition involving failure to make a subunit of an essential lipid-processing enzyme

C. Examples of epigenetic conditions
 1. Result from abnormal addition of methyl groups, acetyl groups, or ubiquitin proteins that "mark" DNA and affect gene function
 2. Examples include fragile X syndrome (FXS), type 2 diabetes mellitus (DM), and cardiovascular disease

D. Examples of chromosomal diseases
 1. Down syndrome—usually caused by trisomy of chromosome 21; characterized by mental retardation and multiple structural defects (**Figure 25-9**)
 2. Klinefelter syndrome—caused by the presence of two or more *X* chromosomes in a male (usually trisomy *XXY*); characterized by long legs, enlarged breasts, low intelligence, small testes, sterility, chronic pulmonary disease (**Figure 25-10**)
 3. Turner syndrome—caused by monosomy of the *X* chromosome *(XO)*; characterized by immaturity of sex organs (resulting in sterility), short stature, webbed neck, cardiovascular defects, and learning disorders (**Figure 25-11**)

Prevention and Treatment of Genetic Diseases

A. Genetic counseling—professional consultations with families regarding genetic diseases
 1. Pedigree—chart illustrating genetic relationships over several generations (**Figure 25-12**)
 2. Punnett square—grid used to determine the probability of inheriting genetic traits (**Figure 25-13**)
 3. Karyotype—arrangement of chromosome photographs used to detect abnormalities
  a. Amniocentesis—involves collection of fetal cells floating in the amniotic fluid (**Figure 25-14**)
  b. Chorionic villus sampling—involves collection of embryonic cells from outside of chorionic tissue

B. Treating genetic diseases
 1. Most current treatments for genetic diseases are based on relieving or avoiding symptoms rather than attempting a cure
 2. Gene therapy—manipulates genes to cure genetic problems (**Figure 25-15**); most forms of gene therapy have just begun in humans
  a. Gene replacement therapy—abnormal genes in existing body cells are replaced by therapeutic genes
  b. Gene augmentation therapy—cells carrying normal genes are introduced into the body to augment production of a needed protein
  c. RNAi therapy—RNA interference—silences individual genes that cause disease

❑ ACTIVE LEARNING

STUDY TIPS

Hint *Consider using these tips to achieve success in meeting your learning goals.*

This chapter covers one of the most publicly discussed areas in biology. Stories on DNA fingerprinting, gene therapy, and genetically engineered medications and foods are often in the media. An understanding of the topics discussed in this chapter will allow you to better evaluate whether the stories you hear and read are based on good science.

1. The chapter requires an understanding of DNA; a review of the material in Chapter 3 may be helpful. It is important to understand how DNA controls the activity of the cell:

 DNA → mRNA → enzyme → biochemical reactions in the cell.

2. Make flash cards to review the Language of Science and the Language of Medicine sections for the genetic terms. Be sure to understand how the genetic makeup of the parents determines the probability for traits in the offspring in both autosomal and sex-linked traits.
3. The use of specific terms to designate the genetic constitution or gene combination on one hand, and the appearance or expression of that gene combination in the individual on the other, is often confusing. Repeated use of terms in the correct context and in conjunction with written expression of the appropriate genotype is helpful. Thus, individuals with genotypes for skin pigmentation *AA* and *Aa* have an identical phenotype (normal skin pigmentation), whereas a genotype of *aa* produces a different phenotype (albinism).
4. Make a chart of the genetic disorders and organize them based on the mechanism or cause, single-gene or chromosomal.

5. There are many online tutorials that cover pedigrees, Punnett squares, and karyotypes. You will also find several websites that provide genetic problems. Completing these problems will help you determine relative probabilities of producing offspring with specific gene combinations.
6. Be able to differentiate between gene replacement and gene augmentation therapy.
7. In your study group, review the genetic terms using flash cards. Discuss the relationship between the DNA sequence and the biochemical activity of the cell. Quiz each other on the probability of various traits in the offspring, based on the parental genes. Go over the disorders chart and the difference in the type of information gained by a pedigree, a Punnett square, and a karyotype. Go over the questions at the end of the chapter and the chapter outline summary and discuss possible test questions.

Review Questions

Write out the answers to these questions after reading the chapter and reviewing the Chapter Summary. If you simply think through the answer without writing it down, you will not retain much of your new learning.

1. Explain how the DNA code is able to regulate the biochemistry of the cell.
2. As they are used in this chapter, define *chromosome* and *gene.*
3. What is the human proteome?
4. What is meant by *independent assortment?*
5. Define or explain the terms *dominant, recessive,* and *codominant* in regard to genetics.
6. What is a sex-linked trait?
7. Define or explain the terms *nondisjunction, trisomy,* and *monosomy.*
8. What is a pedigree chart?
9. What is a Punnett square?
10. What is a karyotype? What are the two methods used to harvest cells for a karyotype?
11. Explain the difference between gene augmentation and gene replacement therapy.
12. Name and briefly describe the two single-gene diseases discussed in the chapter.
13. Name and briefly describe the three chromosomal diseases discussed in the chapter. Indicate whether the diseases are the result of trisomy or monosomy.

Critical Thinking

After finishing the Review Questions, write out the answers to these more in-depth questions to help you apply your new knowledge. Go back to sections of the chapter that relate to concepts that you find difficult.

14. How does crossing-over contribute to genetic variation?
15. Since all children inherit 50% of their genes from their mother and 50% of their genes from their father, why don't we all have characteristics halfway between those of our mother and those of our father?
16. Why must a boy always inherit an X-linked gene such as color blindness from his mother?
17. Which type of genetic mutation has the greatest long-term impact on the population, harmful or beneficial? Explain your answer.
18. If parents are concerned that their child might be born with Down syndrome, what would be the best way to determine this: a pedigree, a Punnett square, or a karyotype? Explain your answer.
19. Design a Punnett square for a mother and father who are both carriers for SCID. Use *S* for the normal gene and *s* for the recessive gene.
20. What is electrophoresis?

Chapter Test

Hint *After studying the chapter, test your mastery by responding to these items. Try to answer them without looking up the answers.*

1. ________ is the scientific study of inheritance.
2. The end product of protein synthesis is frequently an ________, which helps regulate the biochemistry of the body.
3. Most of the cells of the human body contain ________ chromosomes, but gametes contain ________ chromosomes.
4. A ________ gene is one whose effects are seen and is capable of masking a ________ gene for the same trait.

For questions 5, 6, and 7, let *A* stand for the dominant gene for normal skin pigment and let *a* stand for the recessive gene for albinism.

5. A father with *Aa* and a mother with *AA* have a ________% probability for having a child with albinism.
6. A father with *Aa* and a mother with *Aa* have a ________% probability for having a child with albinism.
7. A father with *Aa* and a mother with albinism have a ________% probability for having a child with albinism.
8. Nonsexual traits carried on the sex chromosome are called ________ traits.
9. Color blindness is carried on the *X* chromosome. If the father is color blind and the mother is not a carrier, the probability of having a color blind son would be ________%.
10. A change in the genetic code is called a ________.
11. A mistake in meiosis when a pair of chromosomes fails to separate is called ________.
12. ________ is a condition in which there is a triplet of autosomes rather than the normal pair.
13. ________ is a condition in which there is a single autosome rather than the normal pair.
14. A ________ is a chart that illustrates the genetic relationship in a family over several generations.
15. A ________ is a grid used to determine the probability of inheriting a genetic trait.
16. A ________ is a photograph of chromosomes arranged in pairs; amniocentesis can supply the cells for this.
17. Cancer is thought to be caused, at least in part, by abnormal genes called ________.
18. An ________, or simple cartoon of a chromosome, is often used in genomics to show the overall physical structure of a chromosome.
19. ________ is the science that describes how environmental and behavioral factors may result in offspring with genetic traits that can't be explained by genes alone.

Match each disorder in Column A with its corresponding cause or description in Column B.

Column A

20. ________ cystic fibrosis
21. ________ phenylketonuria
22. ________ Down syndrome
23. ________ Klinefelter syndrome
24. ________ Turner syndrome
25. ________ color blindness

Column B

a. a disorder caused by trisomy 21
b. a disorder caused by a recessive gene that fails to produce the enzyme phenylalanine hydroxylase
c. a disorder caused by the trisomy condition *XXY*
d. a condition caused by a recessive gene that causes an impairment in chloride ion transport across the cell membrane
e. a sex-linked trait that inhibits the production of certain photopigments
f. a disorder caused by the monosomy condition *XO*

Case Studies

To solve a case study, you may have to refer to the glossary or index, other chapters in this textbook, and other resources.

1. Quentin's family physician suspects that Quentin may have Klinefelter syndrome. Quentin's long limbs, small testes, and enlarged breasts seem to support the diagnosis. What test might the physician order to confirm the syndrome? What test results would be expected? What causes the genetic abnormality that produces Klinefelter syndrome?
2. A young infant just born at Memorial Hospital has parents that both report a history of Tay-Sachs disease in their families. Assuming both parents have the Tay-Sachs gene, what is the probability that this infant will develop this deadly disease?
3. Becky and Elliott have been told that Elliott is a carrier for PKU. Using *p* as the recessive gene and *P* as the normal gene, develop a Punnett square that will predict the probability of one of their children inheriting the genetic disease.

Answers to Active Learning Questions can be found online at evolve.elsevier.com.

GLOSSARY

A

abdomen (AB-doh-men) body area between the diaphragm and pelvis

abdominal (ab-DOM-ih-nal) relating to the abdomen

abdominal cavity (ab-DOM-ih-nal KAV-ih-tee) the cavity containing the abdominal organs

abdominal muscles (ab-DOM-ih-nal MUS-els) muscles supporting the anterior aspect of the abdomen

abdominal quadrants (ab-DOM-ih-nal KWOD-rants) four topographic subdivisions of the abdomen determined by two imaginary lines dividing the body through the navel—one vertical, one horizontal; health professionals use these designations to help locate specific internal organs

abdominal regions (ab-DOM-ih-nal REE-juns) nine topographic subdivisions of the abdomen determined by four imaginary lines configured in a tic-tac-toe pattern; anatomists use these named regions to identify the location of internal organs

abdominal thrust (ab-DOM-ih-nal thrust) emergency procedure in which sudden pressure on the abdomen of a person who is choking may dislodge material from the airway; formerly called *Heimlich maneuver*

abdominopelvic cavity (ab-DOM-ih-noh-PEL-vik KAV-ih-tee) the single cavity containing the abdominal and pelvic organs

abdominopelvic quadrant (ab-DOM-ih-noh-PEL-vik KWOD-rant) any of four regions formed by dividing the abdominopelvic cavity by an imaginary cross formed by a vertical and horizontal line

abdominopelvic region (ab-DOM-ih-noh-PEL-vik REE-jun) surface area of the body related to the abdominopelvic cavity

abduct (ab-DUKT) to move away from the midline of the body (or region)

abduction (ab-DUK-shun) moving away from the midline of the body (or region); opposite motion of **adduction**

ablation (ab-LAY-shun) destruction or cutting; for example, the intentional destruction of atrial muscle tissue to treat atrial fibrillation or atrial flutter

ABO system (ay bee oh SIS-tem) human blood classification system based on RBC antigens (A, B, AB, and O) and their corresponding antibodies

abruptio placentae (ab-RUP-shee-oh plah-SEN-tay) separation of normally positioned placenta from the uterine wall; may result in hemorrhage and death of the fetus and/or mother

absorption (ab-SORP-shun) passage of a substance through a membrane, such as skin or mucosa, into blood

accessory organ (ak-SES-oh-ree OR-gan) an organ that assists other organs in accomplishing their functions

acetabulum (as-eh-TAB-yoo-lum) socket in the hip bone (ox coxae or innominate bone) into which the head of the femur fits

acetylcholine (ACh) (as-ee-til-KOH-leen) chemical neurotransmitter

acid (AS-id) any substance that, when dissolved in water, contributes to an excess of H^+ ions (that is, a low pH)

acid-base balance (AS-id bays BAL-ans) maintaining the concentration of hydrogen ions in body fluids

acidic (ah-SID-ik) having the nature of an acid (a pH lower than 7.0)

acidosis (as-ih-DOH-sis) condition in which there is an excessive proportion of acid in the blood and thus an abnormally low blood pH; opposite of **alkalosis**

acne (AK-nee) a bacterial infection of the skin characterized by red pustules formed when hair follicles become infected

acne vulgaris (AK-nee vul-GAR-is) inflammatory skin condition affecting sebaceous gland ducts; *see* **comedones**

acquired immunity (ah-KWYERD ih-MYOO-nih-tee) immunity that is obtained after birth through the use of injections or exposure to a harmful agent

acquired immunodeficiency syndrome (AIDS) (ah-KWYERD IM-yoo-noh-deh-FISH-en-see SIN-drohm [aydz]) disease in which the human immunodeficiency virus attacks T cells, thereby compromising the body's immune system

acromegaly (ak-roh-MEG-ah-lee) condition caused by hypersecretion of growth hormone after puberty, resulting in enlargement of facial features (e.g., jaw, nose), fingers, and toes

acrosome (AK-roh-sohm) specialized structure covering the sperm head containing enzymes that break down the covering of the ovum to allow entry

actin (AK-tin) contractile protein found in the *thin* myofilaments of skeletal muscle

action potential (AK-shun poh-TEN-shal) nerve impulse

active transport (AK-tiv TRANS-port) movement of a substance into and out of a living cell requiring the use of cellular energy

acute (ah-KYOOT) intense; rapid onset, short in duration—as in acute disease

acute lymphocytic leukemia (ALL) (ah-KYOOT LIM-foh-sit-ik loo-KEE-mee-ah) type of acute (rapid onset and progression) blood cancer common in children 3 to 7 years of age; characterized by cancerous transformation and increased numbers of B lymphocytes

acute myeloid leukemia (AML) (ah-KYOOT MY-eh-loyd loo-KEE-me-ah) type of acute (rapid onset and progression) blood cancer most common in adults; characterized by cancerous transformation and increased numbers of myeloid precursor cells

adaptation (ad-ap-TAY-shun) condition of many sensory receptors in which the magnitude of a receptor potential decreases over a period of time in response to a continuous stimulus

adaptive immunity (ah-DAP-tiv ih-MYOO-nih-tee) a system of immunity with memory of a specific antigen and the ability to respond to that antigen, in contrast to the nonspecific nature of innate immunity; another name for **specific immunity**

Addison disease (AD) (AD-ih-son dih-ZEEZ) disease of the adrenal gland resulting in low blood glucose, weight loss, weakness, increase in blood sodium, and decrease in blood potassium

adduct (ad-DUKT) to move toward the midline of the body (or region)

adduction (ad-DUK-shun) moving toward the midline of the body (or region); opposite motion of **abduction**

adductor muscle (ad-DUK-tor MUS-el) any of several muscles that adduct a joint, moving a body part in from the side (lateral) and thus toward the midline (median or midsagittal plane) of the body or of the body region; for example, the pectoralis major and latissimus dorsi muscles contract together to pull the arm toward the trunk, thus *adducting* the shoulder joint

adenine (ADD-eh-een) one of several nitrogen-containing bases that make up nucleotides, which in turn make up nucleic acids such as DNA and RNA; in the cell, it can chemically bind to another nitrogenous base, thymine (T or t) or uracil (U or u), to form a more complex structure or in translating genetic codes; symbolized by the letter *A* or *a; see also* **guanine, cytosine, thymine, uracil**
adenocarcinoma (ad-eh-noh-kar-sih-NOH-mah) cancer of glandular epithelium
adenofibromas (ad-eh-noh-fye-BROH-mahs) benign neoplasms formed in epithelial and connective tissues
adenohypophysis (ad-eh-noh-hye-POF-ih-sis) anterior pituitary gland, which has the structure of an endocrine gland
adenoid (AD-eh-noyd) literally, glandlike; adenoids, or *pharyngeal tonsils,* are paired lymphoid structures in the nasopharynx; *see also* **tonsils**
adenoma (ad-eh-NOH-mah) benign tumor of glandular epithelium
adenosine deaminase (ADA) deficiency (ah-DEN-oh-seen dee-AM-ih-nayse dee-FISH-en-see) rare, inherited condition in which production of the enzyme adenosine deaminase is deficient, resulting in severe combined immune deficiency (SCID); first human disorder treated by gene therapy
adenosine diphosphate (ADP) (ah-DEN-oh-seen dye-FAHS-fayt) molecule similar to adenosine triphosphate but containing only two phosphate groups
adenosine triphosphate (ATP) (ah-DEN-oh-seen try-FAHS-fayt) chemical compound that provides energy for use by body cells
adipose (AD-ih-pohs) fat tissue; specialized tissue that stores lipids
adolescence (ad-oh-LES-ens) period of life between puberty and adulthood
adrenal cortex (ah-DREE-nal KOR-teks) outer portion of adrenal gland that secretes hormones called *corticoids*
adrenal gland (ah-DREE-nal) gland that rests on the top of the kidneys, made up of the cortex and medulla
adrenal medulla (ah-DREE-nal meh-DUL-ah) inner portion of adrenal gland that secretes epinephrine and norepinephrine
adrenergic fiber (ad-ren-ER-jik FYE-ber) any of the axons whose terminals release norepinephrine and epinephrine
adrenocorticotropic hormone (ACTH) (ah-dree-noh-kor-teh-koh-TROH-pic HOR-mohn) hormone that stimulates the adrenal cortex to secrete larger amounts of hormones
adult polycystic kidney disease (ah-DULT pah-lee-sis-tic KID-nee dih-ZEEZ) hereditary condition characterized by development of multiple cystic spaces in one or both kidneys that often fill with clear fluid or blood
adult respiratory distress syndrome (ARDS) (ah-DULT RES-pih-rah-tor-ee dis-TRES sin-drohm) relative inability to inflate alveoli normally; caused by impairment or removal of surfactant following accidental inhalation of destructive substances
adulthood (ah-DULT-hood) period of life after adolescence
aerobic (ayr-OH-bik) requiring oxygen
aerobic training (ayr-OH-bik TRAYN-ing) continuous vigorous exercise requiring the body to increase its consumption of oxygen and develop the muscles' ability to sustain activity over a long period
afferent (AF-fer-ent) carrying or conveying toward the center (e.g., an afferent neuron carries nerve impulses toward the central nervous system); opposite of **efferent**
afferent lymphatic vessel (AF-fer-ent lim-FAT-ik VES-el) any small lymphatic vessel that carries lymphatic fluid toward a lymph node; compare to **efferent lymphatic vessel**
afferent neuron (AF-fer-ent NOO-ron) neuron that conducts impulses toward the central nervous system; generally a sensory neuron
age (ayj) how old an organism is, usually measured from time of birth, hatching, or formation as an independent organism
age-related macular degeneration (AMD) (MAK-yoo-lar dih-jen-uh-RAY-shun) progressive deterioration of macula lutea of retina causing loss of central visual field
agglutinate (ah-GLOO-tin-ayt) antibodies causing antigens to clump or stick together
aging process (AYJ-ing PRAH-ses) the gradual degenerative changes that occur after young adulthood as a person ages
agranular leukocyte (ah-GRAN-yoo-lar LOO-koh-syte) class of white blood cell (leukocyte) that does not exhibit granules when stained; includes monocytes and lymphocytes; also called *non-granular leukocyte* or *agranulocyte*
agricultural scientist (ag-rih-KUL-cher-al SYE-en-tist) scientist who studies the growing of crops
AIDS-related complex (ARC) (AYDS ree-LAY-ted KOM-pleks) early manifestation of AIDS that produces fever, weight loss, and swollen lymph nodes in those whose immune systems are less deficient than those with full-blown AIDS
albinism (AL-bih-niz-em) recessive, inherited condition characterized by a lack of the dark brown pigment melanin in the skin and eyes, resulting in vision problems and susceptibility to sunburn and skin cancer; ocular albinism is a lack of pigment in the layers of the eyeball
albumin (al-BYOO-min) one of several types of proteins normally found in blood plasma; it helps thicken the blood
aldosterone (al-DOS-ter-own) hormone that stimulates the kidney to retain sodium ions and water
alimentary canal (al-eh-MEN-tar-ee kah-NAL) principal tubelike structure of the digestive system extending from mouth to anus—sometimes called the **gastrointestinal (GI) tract**
alkaline (AL-kah-lin) base; any substance that, when dissolved in water, contributes to an excess of OH^- ions (thus creating a high pH value)
alkaline phosphatase (AL-kah-lin FOS-fah-tays) enzyme present in blood plasma in high concentration during certain liver and malignant bone marrow disorders
alkalosis (al-kah-LOH-sis) condition in which there is an excessive proportion of alkali (base) in the blood; opposite of **acidosis**
allergen (AL-er-jen) harmless environmental antigen that stimulates an allergic reaction (hypersensitivity reaction) in a susceptible, sensitized person
allergy (AL-er-jee) hypersensitivity of the immune system to relatively harmless environmental antigens
all or none (all or nun) principle that a process will occur at its maximum or not at all, once it begins
allied health professions (AL-ayed helth proh-FESH-unz) fields of health-care work such as therapists, medical assistants, technicians, and others, who are not physicians or nurses
alloimmunity (al-oh-ih-MYOO-nih-tee) excessive reaction of the immune system to antigens from a different individual of the same species; sometimes called **isoimmunity**
alopecia (al-oh-PEE-sha) clinical term referring to hair loss
alpha cell (AL-fah sel) pancreatic cell that secretes glucagon
alveolar duct (al-VEE-oh-lar dukt) airway that branches from the smallest bronchioles; alveolar sacs arise from alveolar ducts
alveolar sac (al-VEE-oh-lar sak) sacs in the lungs that arise from the alveolar ducts and resemble a cluster of grapes

alveolus (al-VEE-oh-lus) (*pl.,* alveoli) literally, a small cavity; alveoli of lungs are microscopic saclike dilations of terminal bronchioles

Alzheimer disease (AD) (AHLZ-hye-mer dih-ZEEZ) brain disorder of the middle and late adult years characterized by loss of memory and dementia

amenorrhea (ah-men-oh-REE-ah) absence of normal menstruation

amino acid (ah-MEE-no AS-id) structural units from which proteins are built

amniocentesis (AM-nee-oh-sen-TEE-sis) procedure in which a sample of amniotic fluid is removed with a syringe for use in genetic testing, perhaps to produce a karyotype of the fetus; compare with **chorionic villus sampling**

amniotic cavity (am-nee-OT-ik KAV-ih-tee) cavity within the blastocyst that will become a fluid-filled sac in which the embryo will float during development

ameba (ah-MEE-bah) (*pl.,* amebas or amebae) protozoan of changing shape capable of causing infection

amphiarthrosis (am-fee-ar-THROH-sis) slightly movable joint such as the one joining the two pubic bones

amylase (AM-eh-lays) enzyme that digests carbohydrates; *see also* **salivary amylase**

anabolic steroid (an-ah-BOL-ik STAYR-oyd) a lipid molecule of the steroid variety that acts as a hormone to stimulate anabolism (specifically protein synthesis) in body tissues such as muscle (e.g., testosterone)

anabolism (ah-NAB-oh-liz-em) process in which cells make complex molecules (e.g., hormones) from simpler compounds (e.g., amino acids); opposite of **catabolism**

anaerobic (an-aXyr-OH-bik) requiring no oxygen

anal canal (AY-nal kah-NAL) terminal portion of the rectum

anaphase (AN-ah-fayz) stage of mitosis; duplicate chromosomes move to poles of dividing cell

anaphylactic shock (an-ah-fih-LAK-tik shok) circulatory failure (shock) caused by a type of severe allergic reaction characterized by blood vessel dilation; may be fatal

anaplasia (an-ah-PLAY-zhee-ah) growth of abnormal (undifferentiated) cells, as in a tumor or neoplasm

anatomical position (an-ah-TOM-ih-kal poh-ZISH-un) the standard neutral reference position for the body used to describe sites or motions of various body parts; gives meaning to directional terms

anatomist (ah-NAT-oh-mist) professional engaged in the study of the structure of an organism and the relationships of its parts

anatomy (ah-NAT-oh-mee) the study of the structure of an organism and the relationships of its parts

androgen (AN-droh-jen) male sex hormone

andropause (AN-droh-pawz) cessation of fertility in older adult males; not well-defined in humans

anemia (ah-NEE-mee-ah) deficient number of red blood cells or deficient hemoglobin

anesthesia (an-es-THEE-zhah) loss of sensation

aneurysm (AN-yoo-riz-em) abnormal widening of the arterial wall; aneurysms promote the formation of thrombi and also tend to burst

angina pectoris (an-JYE-nah PEK-tor-is) severe chest pain resulting when the myocardium is deprived of sufficient oxygen

angiogram (AN-jee-oh-gram) medical image of vessels produced by **angiography**

angiography (an-jee-AH-graf-ee) radiography in which radiopaque contrast medium is injected into a vessel to make it more visible in a medical image (angiogram); in arteries the image is called an *arteriogram;* in veins, a *venogram* or *phlebogram;* in lymphatic vessels, a *lymphangiogram*

angioplasty (AN-jee-oh-plas-tee) medical procedure in which vessels occluded by arteriosclerosis are opened (i.e., the channel for blood flow is widened)

Angstrom (ANG-strum) 0.1 mm (1/10,000,000,000 of a meter or about 1/250,000,000 of an inch); abbreviated Å

anion (AN-aye-on) negatively charged particle; a negative ion

anorexia (an-oh-REK-see-ah) loss of appetite (a symptom, rather than a distinct disorder)

anorexia nervosa (an-oh-REK-see-ah ner-VOH-sah) behavioral eating disorder characterized by chronic refusal to eat, often related to an abnormal fear of becoming obese

antagonist (an-TAG-oh-nist) any agent that has the opposite effect of the agent to which it is compared; for example, a hormone antagonist opposes the effect of the compared hormone

antagonist muscle (an-TAG-oh-nist MUS-el) a muscle having opposing actions to another muscle; for example, muscles that flex the arm are antagonists to muscles that extend it

antebrachial (an-tee-BRAY-kee-al) relating to the forearm

antenatal medicine (an-tee-NAY-tal MED-ih-sin) prenatal medicine

anterior (an-TEER-ee-or) front or ventral; opposite of **posterior** or **dorsal**

anthrax (AN-thraks) bacterial infection caused by *Bacillus anthracis,* ordinarily affecting herbivores (sheep, cattle, goats, antelope) and often killing them; rarely it occurs in humans through accidental or intentional exposure to bacterial spores through inhalation or skin contact; inhalation anthrax is life-threatening but can be treated successfully with medication; cutaneous anthrax is less serious, characterized by a reddish-brown patch on the skin that ulcerates and then forms a dark, nearly black scab, followed by muscle pain, internal hemorrhage (bleeding), headache, fever, nausea, and vomiting

anthropology (an-throh-POL-oh-jee) science of human origins, culture, characteristics, society, and beliefs

antibiotic (an-tih-by-OT-ik) compound usually produced by living organisms that destroys or inhibits microbes

antibody (AN-tih-bod-ee) substance produced by the body that destroys or inactivates a specific substance (antigen) that has entered the body

antibody-mediated immunity (AN-tih-bod-ee MEE-dee-ayt-ed ih-MYOO-nih-tee) immunity that is produced when antibodies make antigens unable to harm the body; also referred to as *humoral immunity*

anticoagulant (an-tee-koh-AG-yoo-lant) agent that opposes blood clotting

antidepressant (an-tee-deh-PRES-ant) drug that inhibits clinically significant feelings of depression or sadness

antidiuretic hormone (ADH) (an-tee-dye-yoo-RET-ik HOR-mohn) hormone produced in the posterior pituitary gland to regulate the balance of water in the body by accelerating the reabsorption of water

antigen (AN-tih-jen) substance that, when introduced into the body, causes formation of antibodies against it

antigen-presenting cell (APC) (AN-tih-jen prih-ZENT-ing sel) any of a variety of immune cells that present protein fragments (antigens) on their surface and thus allow recognition and reaction by other immune system cells; include macrophages, dendritic cells (DCs), and B cells

antihistamine (an-tih-HIS-tah-meen) agent that inhibits *histamine,* an inflammation agent

antioxidant (an-tee-OK-seh-dent) substance such as vitamin E that can inhibit free radicals (oxidants), which are highly reactive, electron-seeking molecules occurring normally in cells but which may damage electron-dense molecules such as DNA or molecules in cell membranes
antiplatelet agent (an-tee-PLAYT-let) drug therapy that inhibits platelets
antiviral drug (an-tee-VYE-ral [or an-tih-VYE-ral] drug) therapeutic agent that inhibits viral replication in body cells
antrum (AN-trum) cavity
anuria (ah-NOO-ree-ah) absence of urine
anus (AY-nus) distal end or outlet of the rectum
aorta (ay-OR-tah) main and largest artery in the body
aortic body (ay-OR-tik BOD-ee) small cluster of chemosensitive cells that respond to carbon dioxide and oxygen levels
aortic semilunar valve (ay-OR-tik sem-ih-LOO-nar valv) valve between the aorta and left ventricle that prevents blood from flowing back into the ventricle
apex (AY-peks) pointed end of a conical structure
Apgar score (AP-gar) system of assessing general health of newborn infant, in which heart rate, respiration, muscle tone, skin color, and response to stimuli are scored (a perfect total score is 10); named for the American physician Virginia Apgar
apical (AY-pik-al) relating to the apex (tip) of an organ, cell, or other structure; in a cell, often refers to the surface facing the lumen of the organ
apical heart beat (AY-pik-al hart beet) heart sound detected over the heart's apex in the space between the fifth and sixth ribs on a line even with the midpoint of the left clavicle
aplastic anemia (ay-PLAS-tik ah-NEE-mee-ah) blood disorder characterized by a low red blood cell count, caused by destruction of myeloid tissue in the bone marrow
apnea (AP-nee-ah) temporary cessation of breathing
apocrine (AP-oh-krin) relating to a category of exocrine gland that pinches off at its apical tip to release its secretion
apocrine sweat gland (AP-oh-krin swet gland) any of the sweat glands located in the axilla and genital regions; these glands enlarge and begin to function at puberty
apoptosis (ap-oh-TOH-si or ap-op-TOH-sis) programmed cell death by means of several biochemical processes built into each cell; apoptosis clears space for newer cells, as in early embryonic development or in tissue repair
appendage (ah-PEN-dij) something that is attached; for example, an attached body part such as an arm
appendicitis (ah-pen-dih-SYE-tis) inflammation of the vermiform appendix
appendicular (ah-pen-DIK-yoo-lar) relating to the upper and lower extremities of the body
appendicular skeleton (ah-pen-DIK-yoo-lar SKEL-eh-ton) the bones of the upper and lower extremities of the body
appendix (ah-PEN-diks) *see* **vermiform appendix**
appetite center (AP-ah-tyte SEN-ter) cluster of neurons in the hypothalamus whose impulses cause an increase in appetite
aqueous (AY-kwee-us) liquid mixture in which water is the solvent; for example, saltwater is an aqueous solution because water is the solvent
aqueous humor (AY-kwee-us HYOO-mor) watery fluid that fills the anterior chamber of the eye, in front of the lens
aqueous solution (AY-kwee-us suh-LOO-shun) a mixture made up of molecules dissolved in water
arachnoid mater (ah-RAK-noyd MAH-ter) delicate, weblike middle membrane covering the brain, the meninges
arch any structure resembling an arch or arc, as in the archlike arrangement of foot support structures; the process of forming an arch, as when flexing or extending the spine to form an arch
archaea (ark-EE-ah) type of microbe resembling bacteria but with different chemical makeup (especially in the cell wall) and different metabolic pathways; often capable of thriving in very harsh environments (very hot, very acid, very salty, etc.); not known to infect humans
areola (ah-REE-oh-lah) (*pl.,* areolae) small space; the pigmented ring around the nipple
areolar connective tissue (ah-REE-oh-lar koh-NEK-tiv TISH-yoo) a type of connective tissue consisting of fibers and a variety of cells embedded in a loose matrix of soft, sticky gel
arrector pili (ah-REK-tor PYE-lye) smooth muscles of the skin that are attached to hair follicles; when contraction occurs, the hair stands up, resulting in "goose flesh" or "goose bumps"
arrhythmia (ah-RITH-mee-ah) *see* **dysrhythmia**
arterial blood gas (ABG) (ar-TEER-ee-al blud gas) any of the blood characteristics related to respiratory gases normally measured in a lab analysis of arterial blood (Po_2, Pco_2, $\%SO_2$, pH, $[HCO_3^-]$)
arteriole (ar-TEER-ee-ohl) small branch of an artery
arteriosclerosis (ar-tee-ree-oh-skleh-ROH-sis) hardening of arteries; materials such as lipids (as in atherosclerosis) accumulate in arterial walls, often becoming hardened via calcification
artery (AR-ter-ee) vessel carrying blood away from the heart
arthritis (ar-THRY-tis) inflammatory joint disease, characterized by inflammation of the synovial membrane and a variety of systemic signs or symptoms
arthroplasty (AR-throh-plas-tee) the total or partial replacement of a diseased joint with an artificial device (prosthesis)
arthropod (AR-throh-pod) type of animal capable of infesting or parasitizing humans
arthroscopy (ar-THROS-skah-pee) process of viewing internal structures of a joint capsule using a lighted scope inserted through soft tissues
articular cartilage (ar-TIK-yoo-lar KAR-tih-lij) cartilage covering the joint ends of bones
articulation (ar-tik-yoo-LAY-shun) place of junction between two or more bones of the skeleton; also called a *joint*
artificial kidney (ar-tih-FISH-al KID-nee) mechanical device that removes wastes from the blood that would normally be removed by the kidney
artificial pacemaker (ar-tih-FISH-al PAYS-may-ker) an electrical device that is implanted into the heart to treat a heart block
ascending colon (ah-SEND-ing KOH-lon) portion of the colon extending from the cecum to the hepatic flexure
ascites (ah-SYE-tees) abnormal accumulation of fluid in intraperitoneal space
aseptic technique (ay-SEP-tik tek-NEEK) approach to limiting the spread of infection by preventing or reducing contacts with contaminated surfaces
asexual (ay-SEKS-yoo-al) one-celled plants and bacteria that do not produce specialized sex cells
aspiration biopsy cytology (as-pih-RAY-shun BYE-op-see sye-TOL-oh-jee) procedure that draws myeloid tissue into a syringe; allows for examination of tissue to confirm or reject diagnosis
assimilation (ah-sim-ih-LAY-shun) takes place when nutrient molecules enter the cell and undergo chemical changes
assisted reproductive technology (ART) (ah-SIS-ted ree-proh-DUK-tiv tek-NOL-oh-jee) any of several medical techniques used to enhance fertility

association area (ah-soh-see-AY-shun AYR-ee-ah) region of the cerebral cortex of the brain that functions to put together or "associate" information from many parts of the brain to help make sense of or analyze the information

asthma (AZ-mah) obstructive pulmonary disorder characterized by recurring spasms of muscles in bronchial walls accompanied by edema and mucous production, making breathing difficult

astigmatic keratotomy (AK) (AY-stig-mat-ic kayr-ah-TOT-ah-mee) type of refractory eye surgery for treatment of astigmatism that involves placement of transverse cuts across the corneal surface to alter its shape

astigmatism (ah-STIG-mah-tiz-em) irregular curvature of the cornea or lens that impairs refraction of a well-focused image in the eye

astrocyte (AS-troh-syte) a neuroglial cell

atelectasis (at-eh-LEK-tay-sis) total or partial collapse of the alveoli of the lung

atherosclerosis (ath-er-oh-skleh-ROH-sis) type of "hardening of the arteries" in which lipids and other substances build up on the inside wall of blood vessels

athletic trainer (ath-LET-ik TRAY-ner) health care professional who works with a physician and specializes in prevention, diagnosis, and therapy of sports-related injuries

atlas (AT-lis) another name for the first cervical vertebra (C1)

atom (AT-om) smallest particle of a pure substance (element) that still has the chemical properties of that substance; composed of protons, electrons, and neutrons (subatomic particles)

atomic mass (ah-TAH-mik MAS) combined total number of protons and neutrons in an atom

atomic number (ah-TAH-mik NUM-ber) total number of protons in an atom's nucleus; atoms of each element have a characteristic atomic number

atrial fibrillation (A-fib or AF) (AY-tree-al fib-ril-LAY-shun) frequent, chaotic premature contractions of the atrium

atrial flutter (AFL) (AY-tree-al FLUT-er) a rapid and irregular atrial rhythm often triggered by abnormal electrical signals from the nearby pulmonary veins

atrial natriuretic hormone (ANH) (AY-tree-al nay-tree-yoo-RET-ik HOR-mohn) hormone secreted by the heart cells that regulates fluid and electrolyte homeostasis

atrioventricular (AV) bundle (ay-tree-oh-ven-TRIK-yoo-lar BUN-del) bundle of rapidly conducting cardiac muscle fibers that extend from the AV node to the subendocardial branches (Purkinje fibers); involved in coordination of heart muscle contraction; also known as *bundle of His*

atrioventricular (AV) node (ay-tree-oh-ven-TRIK-yoo-lar nohd) a small mass of special impulse-generating cardiac muscle tissue near the junction of the left atrium and ventricle; part of the conduction system of the heart

atrioventricular (AV) valve (ay-tree-oh-ven-TRIK-yoo-lar valv) either of two valves that separate the atrial chambers from the ventricles

atrium (AY-tree-um) (*pl.*, atria) chamber or cavity; for example, atrium of each side of the heart

atrophy (AT-roh-fee) wasting away of tissue; decrease in size of a part; sometimes referred to as **disuse atrophy**

audiologist (aw-dee-OL-uh-jist) health-care professional who treats hearing disorders

auditory tube (AW-dih-toh-ree toob) tube extending from inside the middle ear to the throat to equalize air pressure; also called the **eustachian tube**

auricle (AW-rih-kul) part of the ear attached to the side of the head; earlike appendage of each atrium of the heart

autoimmunity (aw-toh-ih-MYOO-nih-tee) process in which a person's immune system attacks the person's own body tissues—the underlying cause of several diseases

automated lamellar keratoplasty (ALK) (AW-toh-may-ted lah-MEL-ahr kayr-AT-oh-plast-ee) type of refractory eye surgery that employs a microkeratome to cut off a cap of corneal tissue, which is replaced after the underlying tissue is reshaped

automatic external defibrillator (AED) (aw-toh-MAT-ik eks-TERN-al dee-FIB-rih-lay-tor) small, lightweight device that detects a person's heart rhythm using small electrode pads placed on the torso and, if ventricular fibrillation is detected, a nonmedical rescuer will be led through some simple steps to defibrillate the victim by applying brief electroshock to the heart

autonomic effector (aw-toh-NOM-ik ef-FEK-tor) tissues to which autonomic neurons conduct impulses

autonomic nervous system (ANS) (aw-toh-NOM-ik NER-vus SIS-tem) division of the human nervous system that regulates involuntary actions

autonomic neuron (aw-toh-NOM-ik NOO-ron) motor neurons that make up the autonomic nervous system

autopsy (AW-top-see) systematic dissection and analysis of a dead body, often for the purpose of discovering the cause of death and/or the presence of health conditions; also called *necropsy*

autosome (AW-toh-sohm) one of the 44 (22 pairs) chromosomes in the human genome other than the two sex chromosomes; means "same body," referring to the fact that members of a pair of autosomes match each other in size and other structural features

AV bundle (AV BUN-dul) fibers in the heart that relay a nerve impulse from the AV node to the ventricles; also known as the *bundle of His*

avitaminosis (ay-vye-tah-mih-NOH-sis) general name for any condition resulting from a vitamin deficiency

avulsion fracture (ah-VUL-shun FRAK-chur) fracture occurring when a powerful muscle contraction pulling on a ligamentous or tendinous attachment to a bone forcibly pulls a fragment of bone free from underlying osseous tissue

axial (AK-see-al) relating to the central axis of the body: head, neck, and torso or trunk

axial skeleton (AK-see-al SKEL-eh-ton) the bones of the head, neck, and torso

axilla (AK-sil-ah) relating to the armpit

axillary (AK-sih-layr-ee) relating to the area inside the shoulder joint or armpit

axon (AK-son) nerve cell process that transmits impulses away from the cell body

B

B cell (bee sel) a lymphocyte; activated B cells develop into plasma cells, which secrete antibodies into the blood

B lymphocyte (bee LIM-foh-syte) immune system cell that produces antibodies against specific antigens

bacillus (bah-SIL-us) (*pl.*, bacilli) rod-shaped bacterium

bacterium (bak-TEER-ee-um) microbe capable of causing disease; it is a primitive, single-celled organism without membranous organelles

Bard endoscopic suturing system (BARD en-doh-SKOP-ik SOO-chur-ing SIS-tem) use of an endoscope to place sutures in the lower esophageal sphincter to narrow the lumen

bariatrics (bayr-ee-AT-riks) field of medicine that deals with treatment of obesity

Barrett esophagus (BAHR-ett ee-SOF-ah-gus) precancerous condition of esophageal lining

Bartholin gland (BAR-toh-lin) exocrine mucous gland located on either side of the vaginal outlet; also known as **greater vestibular gland**

bartholinitis (bar-toh-lin-AYE-tis) inflammation of the Bartholin glands, accessory organs of the female reproductive tract

basal cell carcinoma (BAY-sal sel car-sih-NOH-mah) skin cancer, often occurring on upper face, with low potential for metastasizing

basal ganglia (BAY-sal GANG-glee-ah) *see* **basal nuclei** or **cerebral nuclei**

basal metabolic rate (BMR) (BAY-sal met-ah-BAHL-ik rayt) number of calories of heat that must be produced per hour by catabolism to keep the body alive, awake, and comfortably warm

basal nuclei (BAY-sal NOO-klee-aye) islands of gray matter located in the cerebral cortex that are responsible for automatic movements and postures; also called **basal ganglia** or **cerebral nuclei**

base 1. A chemical that, when dissolved in water, reduces the relative concentration of H^+ ions in the whole solution (sometimes by adding OH^- ions) 2. In the context of nucleic acids (DNA and RNA), *base* or *nitrogen base* refers to one part of a nucleotide (sugar, phosphate, and base) that is the basic building block of nucleic acid molecules; possible bases include adenine, thymine, guanine, cytosine, and uracil

basement membrane (BAYS-ment MEM-brayn) the connective tissue layer of the serous membrane that holds and supports the epithelial cells

basophil (BAY-soh-fil) white blood cell that stains readily with basic dyes

BBB *see* **blood-brain barrier**

Bell palsy (bell PAWL-zee) temporary or permanent paralysis of facial features caused by damage to cranial nerve VII (facial nerve)

benign (be-NYNE) refers to a tumor or neoplasm that does not metastasize or spread to different tissues

benign prostatic hypertrophy (BPH) (be-NYNE proh-STAT-ik hye-PER-troh-fee) benign enlargement of the prostate, a condition common in older males

benign tumor (be-NYNE TOO-mer) a noncancerous and generally harmless neoplasm

beta-adrenergic blocker (BAY-tah–ad-ren-ER-jik) drug that blocks beta-adrenergic receptors and therefore prevents dilation of blood vessels and increased contraction of heart muscle; also called *beta blocker*

beta cell (BAY-tah sel) pancreatic islet cell that secretes insulin

bicarbonate ion (bye-KAR-boh-nayt EYE-on) negative ion common in water solutions, including body fluids; HCO_3^-; often acts as a buffer to increase pH (reduce acidity) of a solution

bicarbonate loading (bye-KAR-boh-net LOHD-ing) ingesting large amounts of sodium bicarbonate to counteract the effects of lactic acid buildup, thereby reducing fatigue; however, there are potentially dangerous side effects

biceps brachii (BYE-seps BRAY-kee-aye) the primary flexor of the forearm

biceps femoris (BYE-seps FEM-oh-ris) powerful flexor of the leg

biconcave (bye-KON-kayv) depressed or "caved in" on two sides, as in the "pinched" disk shape of red blood cells

bicuspid (bye-KUS-pid) having two points; bicuspid tooth (also called **premolar**) has a large flat surface and two grinding cusps; *see also* **bicuspid valve**

bicuspid valve (bye-KUS-pid valv) one of the two AV valves, it is located between the left atrium and ventricle; also called the **mitral valve** or left **atrioventricular (AV) valve**

bilateral symmetry (bye-LAT-er-al SIM-eh-tree) concept of the right and left sides of the body being approximate mirror images of each other

bilayer (BYE-lay-er) double layer

bile (byle) substance that reduces large fat globules into smaller droplets of fat that are more easily broken down

bile duct (byle dukt) duct that drains bile into the small intestine and is formed by the union of the common hepatic and cystic ducts

biliary colic (BIL-yah-ree KOL-ik) pain that may occur when a gallstone blocks the common bile duct—a condition called *choledocholithiasis*

biochemist (bye-oh-KEM-ist) scientist who works primarily in the field of biochemistry; *see* **biochemistry**

biochemistry (bye-oh-KEM-is-tree) science of chemistry of living organisms

biological filtration (bye-EH-lah-jih-kal fil-TRAY-shun) process in which cells alter the contents of the filtered fluid

biomedical engineering (bye-oh-MED-ik-al en-juh-NEER-ing) field of machine design applied to therapeutic strategies; also called *bioengineering*

biopsy (BYE-op-see) procedure in which living tissue is removed from a patient for laboratory examination, as in determining the presence of cancer cells; see also **needle biopsy**

bioterrorism (bye-oh-TAYR-or-iz-em) unlawful release of biological agents (toxins or pathogens) for the purpose of intimidation

birth defect (DEE-fekt) any abnormality, whether caused by genetic or environmental factors, that exists at birth; *see* **teratogen**

blackhead description of sebum that accumulates, darkens, and enlarges some of the ducts of the sebaceous glands; also called a *comedo*

bladder (BLAD-der) a sac, usually referring to the urinary bladder

blastocyst (BLAS-toh-sist) postmorula stage of developing embryo; hollow ball of cells

blister (BLIS-ter) fluid-filled skin lesion; *see* **vesicle**

blood (blud) type of connective tissue characterized by a watery liquid matrix (blood plasma) and a variety of mobile cells that include red blood cells, white blood cells, and platelets

blood-brain barrier (BBB) (blud brayn BAYR-ee-er) structural and functional barrier formed by astrocytes and blood vessel walls in the brain; it prevents some substances from diffusing from the blood into brain tissue

blood doping (blud DOH-ping) a practice used to improve athletic performance by removing red blood cells weeks before an event and then reinfusing them just before competition to increase the oxygen-carrying capacity of the blood

blood pressure (blud PRESH-ur) pressure of blood in the blood vessels, expressed as systolic pressure over diastolic pressure (e.g., 120/80 mm Hg)

blood pressure gradient (blud PRESH-ur GRAY-dee-ent) the difference between any two blood pressures in the body; for example, the pressure difference between the blood in the left ventricle of the heart and the blood in the aorta is a pressure gradient

blood types (blud) the different types of blood that are identified by certain antigens in red blood cells (A, B, AB, O, and Rh-negative or Rh-positive)

blood urea nitrogen (BUN) test (yoo-REE-ah NYE-troh-jen) clinical laboratory measurement of the amount of nitrogen in urea present in the blood and used as a measure of the efficiency of the kidney's ability to clear urea from the body

body (BOD-ee) unified and complex assembly of structurally and functionally interactive components (as in *human body*); the main or central part of a structure (as in *cell body* or *body of an organ*)

body composition (BOD-ee com-poh-ZISH-un) assessment that identifies the percentage of the body that is lean tissue and the percentage that is fat

boil (BOY-el) *see* **furuncle**

bolus (BOW-lus) a small, rounded mass of masticated food ready to be swallowed

bond a chemical bond or union between two or more atoms to form a molecule; *see* **ionic bond** and **covalent bond**

bone (bohn) highly specialized connective tissue whose matrix is hard and calcified

bone marrow (bohn MAYR-oh) soft material that fills cavities of the bones; red bone marrow is vital to blood cell formation; yellow bone marrow is inactive fatty tissue

bone marrow transplant (bohn MAYR-oh TRANS-plant) treatment in which healthy blood-forming marrow tissue from a donor is intravenously introduced into a recipient

bony labyrinth (BOHN-ee LAB-eh-rinth) the fluid-filled complex maze of three spaces (the vestibule, semicircular canals, and cochlea) in the temporal bone

Bouchard node (boo-SHAR nohd) any of the abnormal enlargements seen at the proximal interphalangeal joints in people with osteoarthritis

bovine spongiform encephalopathy (BSE) (BOH-vyne SPUN-jeh-form en-sef-uh-LOP-uh-thee) also known as *mad cow disease;* a degenerative disease of the central nervous system caused by prions that convert normal proteins of the nervous system into abnormal proteins, causing loss of nervous system function; the abnormal form of the protein also may be inherited; *see also* **prion**

Bowman capsule (BOH-men KAP-sul) the cup-shaped beginning of a nephron that surrounds the glomerulus; also called *Bowman's capsule* or *glomerular capsule*

brachial (BRAY-kee-al) relating to the arm

brachialis (bray-kee-AL-is) skeletal muscle of the arm that flexes the forearm at the elbow

brachytherapy (brak-ih-THAYR-uh-pee) placement of radioactive "seeds" in close or direct contact with cancerous tissue

bradycardia (bray-dee-KAR-dee-ah) slow heart rhythm (below 60 beats/minute)

breast (brest) anterior aspect of the chest; in females, also an accessory sex organ

bronchiole (BRONG-kee-ohl) small branch of a bronchus

bronchitis (brong-KYE-tis) inflammation of the bronchi of the lungs, characterized by edema and excessive mucous production that causes coughing and difficulty in breathing (especially expiration); if the trachea is also inflamed, this condition may be referred to as *tracheobronchitis*

bronchus (BRONG-kus) (*pl.,* bronchi) the branches of the trachea

buccal (BUK-al) relating to the cheek

buffer (BUF-er) compound that combines with an acid or with a base to form a weaker acid or base, thereby lessening the change in hydrogen-ion concentration that would occur without the buffer

buffer pair (BUF-er payr) two kinds of chemical substances that together prevent a sharp change in the pH of a fluid; for example, sodium bicarbonate ($NaHCO_3$) and carbonic acid (H_2CO_3)

buffy coat thin layer of white blood cells (WBCs) and platelets located between red blood cells (RBCs) and plasma in a centrifuged sample of blood

bulboid corpuscle (BUL-boyd KOHR-pus-ul) mucous membrane receptor that detects sensations of touch and vibration; also known as **Krause end bulb**

bulbourethral gland (BUL-boh-yoo-REE-thral) small glands located just below the prostate gland whose mucuslike secretions lubricate the terminal portion of the urethra and contribute less than 5% of the seminal fluid volume; also known as *Cowper gland*

bulimarexia (boo-lee-mah-REK-see-ah) condition in which people purposely induce the vomiting reflex to purge themselves of food they just ate; an eating disorder

bulimia (boo-LEE-mee-ah) behavioral eating disorder characterized by an alternating pattern of overeating followed by self-denial (and perhaps purging of GI contents)

bundle of His (BUN-dul of his) *see* **AV bundle**

burn (bern) an injury to tissues resulting from contact with heat, chemicals, electricity, friction, or radiant and electromagnetic energy; classified into three categories, depending on the number of tissue layers involved

bursa (BER-sah) (*pl.,* bursae) small, cushionlike sacs found between moving body parts, which make movement easier

bursitis (ber-SYE-tis) inflammation of a bursa

C

cachexia (kah-KEES-ee-ah) syndrome associated with cancer and other chronic diseases that involves loss of appetite, weight loss, and general weakness

calcaneus (kal-KAY-nee-us) heel bone; largest tarsal in the foot

calcitonin (CT) (kal-sih-TOH-nin) a hormone secreted by the thyroid gland that decreases calcium in the blood

calcium-channel blocker (KAL-see-um CHAN-al) drug that inhibits the opening of calcium channels in cell membranes; for example, used to reduce heart muscle contractions

calculi (KAL-kyoo-lye) hard, crystalline stones that form in the lumen of hollow organs such as the gallbladder or liver (biliary calculi) or renal passages (renal calculi)

callus (KAL-us) bony tissue that forms a sort of collar around the broken ends of fractured bone during the healing process

calorie (c) (KAL-or-ee) heat unit; the amount of heat needed to raise the temperature of 1 g of water 1° C

Calorie (C) (KAL-or-ee) heat unit; kilocalorie; the amount of heat needed to raise the temperature of 1 kilogram of water 1° C

calyx (KAY-liks) cup-shaped division of the renal pelvis

canaliculi (kan-ah-LIK-yoo-lye) an extremely narrow tubular passage or channel in compact bone

cancellous bone (KAN-seh-lus) bone tissue containing tiny, branching trabeculae; also known as *spongy bone* or *trabecular bone*

cancer (KAN-ser) tumor (neoplasm) capable of metastasizing (spreading) to other parts of the body

candidiasis (kan-dih-DYE-eh-sis) infection caused by *Candida* yeast

canine (KAY-nyne) relating to a dog, as in the *canine tooth* with the longest crown and the longest root, which is located lateral to the second incisor that serves to pierce or tear food being eaten; the canine tooth is also called a **cuspid** tooth

capillary (KAP-ih-layr-ee or kap-IL-ah-ree) tiny vessels that connect arterioles and venules

capillary blood pressure (KAP-ih-layr-ee blud PRESH-ur) the blood pressure found in the capillary vessels

capsule (KAP-sul) hollowed out space found in diarthrotic joints, holds the bones of joints together while still allowing movement; made of fibrous connective tissue lined with a smooth, slippery synovial membrane

carbaminohemoglobin ($HbCO_2$) (kar-bah-MEE-noh-hee-moh-GLOH-bin) compound formed by the union of carbon dioxide with hemoglobin

carbohydrate (kar-boh-HYE-drayt) organic compounds containing carbon, hydrogen, and oxygen in certain specific proportions (C, H, O in a 1:2:1 ratio); for example, sugars, starches, and cellulose

carbohydrate loading (kar-boh-HYE-drayt LOHD-ing) a method used by athletes to increase the stores of muscle glycogen, allowing more sustained aerobic exercise; also called *glycogen loading*

carbon (KAR-bun) one of the chemical elements found in great quantity in the human body and always found in organic compounds; symbolized by C, as in CO_2 (carbon dioxide)

carbon dioxide (KAR-bun dye-AHK-syde) molecule made up of one carbon atom and two oxygen atoms; symbolized by the formula CO_2; produced by processes of cellular respiration as a waste product that must be excreted from the body through the respiratory system

carbonic anhydrase (CA) (kar-BON-ik an-HYE-drays) the enzyme that converts carbon dioxide into carbonic acid

carbuncle (KAR-bung-kul) a mass of connected boils, pus-filled lesions associated with hair follicle infections; *see* **furuncle**

carcinogen (kar-SIN-oh-jen) substance that promotes the development of cancer

carcinoma (kar-sih-NOH-mah) malignant tumor that arises from epithelial tissue

cardiac (KAR-dee-ak) relating to the heart

cardiac arrest (KAR-dee-ak ar-REST) abnormal condition in which the heart suddenly stops pumping blood, as after ventricular fibrillation

cardiac cycle (KAR-dee-ak SYE-kul) each complete heartbeat, including contraction and relaxation of the atria and ventricles

cardiac muscle (KAR-dee-ak MUS-el) the involuntary type of muscle tissue that makes up the heart wall

cardiac muscle tissue (KAR-dee-ak MUS-el TISH-yoo) *see* **cardiac muscle**

cardiac output (CO) (KAR-dee-ak OUT-put) volume of blood pumped by one ventricle per minute

cardiac sphincter (KAR-dee-ak SFINGK-ter) a ring of muscle between the stomach and esophagus that prevents food from reentering the esophagus when the stomach contracts

cardiac tamponade (KAR-dee-ak tam-poh-NOD) compression of the heart caused by fluid buildup in the pericardial space, as in pericarditis or mechanical damage to the pericardium

cardiac vein (KAR-dee-ak vayn) any vein that carries blood from the myocardial capillary beds to the coronary sinus and into the right ventricle

cardiogenic shock (kar-dee-oh-JEN-ik shok) circulatory failure (shock) caused by heart failure; literally "heart-caused" shock

cardiologist (kar-dee-AH-loh-jist) physician or researcher who specializes in the structure and function of the heart and associated structures

cardiology (kar-dee-OL-oh-jee) study and treatment of the heart and heart disease

cardiomyopathy (kar-dee-oh-my-OP-ah-thee) general term for disease of the myocardium (heart muscle)

cardiopulmonary resuscitation (CPR) (kar-dee-oh-PUL-moh-nayr-ree ree-sus-ih-TAY-shun) combined external cardiac (heart) massage and artificial respiration

cardiovascular (kar-dee-oh-VAS-kyoo-lar) relating to the heart and blood vessels

cardiovascular system (kar-dee-oh-VAS-kyoo-lar SIS-tem) the system that transports cells throughout the body by way of blood vessels; sometimes also called **circulatory system**

caries (KAYR-ees) decay of teeth or of bone; *see* **cavity**

carotid body (kah-ROT-id BOD-ee) chemoreceptor located in the carotid artery that detects changes in oxygen, carbon dioxide, and blood acid levels

carpal (KAR-pul) relating to the wrist

carpal tunnel syndrome (KAR-pul TUN-el SIN-drohm) muscle weakness, pain, and tingling in the radial side (thumb side) of the wrist, hand, and fingers—perhaps radiating to the forearm and shoulder; caused by compression of the median nerve within the carpal tunnel (a passage along the ventral concavity of the wrist)

carrier (KAYR-ee-er) in genetics, a person who possesses the gene for a recessive trait, but who does not actually exhibit the trait

cartilage (KAR-tih-lij) a specialized, fibrous connective tissue that has the consistency of a firm plastic or gristlelike gel

catabolism (kah-TAB-oh-liz-em) breakdown of nutrient compounds or cytoplasm into simpler compounds; opposite of **anabolism,** the other phase of metabolism

catalyst (KAT-ah-list) chemical that speeds up reactions without being changed itself

cataract (KAT-ah-rakt) opacity of the lens of the eye

catecholamine (kat-eh-KOHL-ah-meen) category of signaling molecule that includes norepinephrine and epinephrine

catheterization (kath-eh-ter-ih-ZAY-shun) passage of a flexible tube (catheter) into the bladder through the urethra for the withdrawal of urine (urinary catheterization)

cation (KAT-aye-on) positively charged particle; a positive ion

cavity (KAV-ih-tee) hollow place or space in a tooth resulting from decay; also referred to as *dental caries*

cecum (SEE-kum) blind pouch; the pouch at the proximal end of the large intestine

cell (sel) the basic biological and structural unit of the body consisting of a nucleus surrounded by cytoplasm and enclosed by a membrane

cell body (sel BOD-ee) the main part of a neuron from which the dendrites and axons extend

cell-mediated immunity (sel MEE-dee-ayt-ed ih-MYOO-nih-tee) resistance to disease organisms resulting from the actions of cells; chiefly T cells

cellular respiration (SEL-yoo-lar res-pih-RAY-shun) enzymes in the mitochondrial wall and matrix using oxygen to break down glucose and other nutrients to release energy needed for cellular work

cementum (see-MEN-tum) bonelike dental tissue covering the neck and root areas of teeth

centimeter (SEN-tih-mee-ter) $\frac{1}{100}$ of a meter; approximately 2.5 cm equal 1 inch

central canal (SEN-tral kah-NAL) longitudinal canal containing vascular elements and nervous tissue located in the center of an osteon, or Haversian system; central canal of any structure

central nervous system (CNS) (SEN-tral NER-vus SIS-tem) the brain and spinal cord

central venous pressure (SEN-tral VEE-nus PRESH-ur) venous blood pressure within the right atrium that influences the pressure in the large peripheral veins

centriole (SEN-tree-ohl) one of a pair of tiny cylinders in the centrosome of a cell; believed to be involved with the spindle fibers formed during mitosis

centromere (SEN-troh-meer) a beadlike structure that attaches one chromatid to another during the early stages of mitosis

centrosome (SEN-troh-sohm) area of the cytoplasm near the nucleus that coordinates the building and breaking up of microtubules in the cell

cephalic (seh-FAL-ik) relating to the head

cerebellum (sayr-eh-BEL-um) the second largest part of the human brain that plays an essential role in the production of normal movements

cerebral cortex (seh-REE-bral KOR-teks) a thin layer of gray matter made up of neuron dendrites and cell bodies that compose the surface of the cerebrum

cerebral nuclei (seh-REE-bral NOO-klee-aye) islands of gray matter located in the cerebral cortex that are responsible for autonomic movements and postures; also called **basal nuclei** or **basal ganglia**

cerebral palsy (CP) (seh-REE-bral PAWL-zee) abnormal condition characterized by permanent, nonprogressive paralysis (usually spastic paralysis) of one or more extremities caused by damage to motor control areas of the brain before, during, or shortly after birth

cerebrospinal fluid (CSF) (seh-ree-broh-SPY-nal FLOO-id) fluid that fills the subarachnoid space in the brain and spinal cord and in the cerebral ventricles

cerebrovascular accident (CVA) (seh-ree-broh-VAS-kyoo-lar accident) a hemorrhage or cessation of blood flow through cerebral blood vessels resulting in destruction of neurons; commonly called a *stroke*

cerebrum (SAYR-eh-brum) the largest and uppermost part of the human brain that controls consciousness, memory, sensations, emotions, and voluntary movements

cerumen (seh-ROO-men) ear wax

ceruminous gland (seh-ROO-mih-nus) gland that produces a waxy substance called **cerumen** (ear wax)

cervical (SER-vih-kal) relating to the neck

cervicitis (ser-vih-SYE-tis) inflammation of the cervix of the uterus

cervix (SER-viks) neck; any necklike structure

cesarean section (seh-SAYR-ee-an SEK-shun) surgical removal of a fetus, often through an incision of the skin and uterine wall; also called *C-section*

chemical level (KEM-ih-kal LEV-el) the level of the body's organization that includes atoms and molecules; the chemical substances that make up the body's structure

chemoreceptor (kee-moh-ree-SEP-tor) any receptor that responds to chemical changes; for example, receptors that detect the chemicals of taste and smell

chemoreflex (kee-moh-REE-fleks) any reaction triggered by a chemical change, as when the heart rate changes in response to shift in oxygen concentration in the blood

chemotaxis (kee-moh-TAK-sis) process in which white blood cells move toward the source of inflammation mediators

chemotherapy (kee-moh-THAYR-ah-pee) technique of using chemicals to treat disease (e.g., infections, cancer)

chest *see* **thorax**

Cheyne-Stokes respiration (CSR) (chain-stokes res-pih-RAY-shun) pattern of breathing associated with critical conditions such as brain injury or drug overdose and characterized by cycles of apnea and hyperventilation

childhood age period from infancy to puberty

chiropractic (kye-roh-PRAK-tik) system of therapy based on the principle that alignment of the skeleton promotes healing

chiropractor (KYE-roh-prak-ter) physician specializing in chiropractic therapy, which is based on the principle that alignment of the skeleton promotes healing

Chlamydia (klah-MID-ee-ah) small bacterium that infects human cells as an obligate parasite

cholangiography (kohl-an-jee-OG-rah-fee) specialized x-ray procedure used to visualize the gallbladder and the major bile and pancreatic ducts

cholecystectomy (kohl-eh-sis-TEK-toh-mee) surgical removal of the gallbladder

cholecystitis (koh-leh-sis-TYE-tis) inflammation of the gallbladder

cholecystokinin (CCK) (koh-lee-sis-toh-KYE-nin) hormone secreted from the intestinal mucosa of the duodenum that stimulates the contraction of the gallbladder, resulting in bile flowing into the duodenum

choledocholithiasis (koh-LED-uh-koh-lih-THY-ah-sis) condition of a gallstone blocking the common bile duct; a type of **cholelithiasis**

cholelithiasis (koh-leh-lih-THEE-ah-sis) condition of having gallstones (composed of cholesterol or bile salts), hard mineral deposits that may form and collect in the gallbladder

cholera (KAHL-er-ah) potentially fatal, infectious bacterial disease characterized by severe diarrhea, vomiting, cramps, dehydration; *see also* Appendix A, Table 3

cholesterol (koh-LES-ter-ol) steroid lipid found in many body tissues and in animal fat

cholinergic fiber (koh-lin-NER-jik FYE-ber) axon whose terminals release acetylcholine

chondrocyte (KON-droh-syte) cartilage cell

chondroma (kon-DROH-mah) benign tumor of cartilage

chondrosarcoma (kon-droh-sar-KOH-mah) cancer of cartilage tissue

chordae tendineae (KOR-dee ten-DIN-ee) stringlike structures that attach the AV valves to the wall of the heart

chorion (KOH-ree-on) structure that develops into an important fetal membrane in the placenta

chorionic gonadotropin (hCG) (koh-ree-ON-ik goh-nah-doh-TROH-pin) any of several hormones that are secreted as the uterus develops during pregnancy

chorionic villi (koh-ree-ON-ik VIL-aye) structures that connect the blood vessels of the chorion to the placenta

chorionic villus sampling (CVS) (koh-ree-ON-ik VIL-lus SAM-pling) procedure in which a tube is inserted through the (uterine) cervical opening and a sample of the chorionic tissue surrounding a developing embryo is removed for genetic testing; compare with **amniocentesis**

choroid (KOH-royd) middle layer of the eyeball that contains a dark pigment to prevent the scattering of incoming light rays

choroid plexus (KOH-royd PLEK-sus) a network of brain capillaries that are involved with the production of cerebrospinal fluid

chromatid (KROH-mah-tid) one of a pair of identical strands within a replicated chromosome

chromatin granule (KROH-mah-tin GRAN-yoo-ul) deep-staining, grainy-appearing substance in the nucleus of cells; condenses into distinct chromosomes during cell division

chromosomal genetic disease (kroh-moh-SOH-mal jeh-NET-ik) disease that results from chromosomal breakage or from abnormal presence or absence of entire chromosomes

chromosome (KROH-meh-sohm) DNA molecule that has coiled to form a compact mass during mitosis or meiosis; each chromosome is composed of regions called *genes,* each of which transmits hereditary information

chronic (KRON-ik) long-lasting, as in chronic disease

chronic bronchitis (KRON-ik brong-KYE-tis) chronic inflammation of the bronchi and bronchioles. It is characterized by edema and excessive mucous production, which often block air passages

chronic lymphocytic leukemia (CLL) (KRON-ik LIM-foh-sit-ik loo-KEE-mee-ah) type of chronic (slow onset and progression) blood cancer most common in older adults; characterized by cancerous transformation and increased numbers of B lymphocytes

chronic myeloid leukemia (CML) (KRON-ik MY-loyd loo-KEE-mee-ah) type of chronic (slow onset and progression) blood cancer characterized by cancerous transformation and increased numbers of granulocytic white blood cells (WBCs)

chronic obstructive pulmonary disease (COPD) (KRON-ik ob-STRUK-tiv PUL-moh-nayr-ee dih-ZEEZ) general term referring to a group of disorders characterized by progressive, irreversible obstruction of air flow in the lungs; *see* **bronchitis, emphysema**

chronic traumatic encephalopathy (CTE) (KRON-ik traw-MAT-ik en-sef-al-OP-path-ee) brain disorder resulting from repeated trauma to the brain that involves accumulation of abnormal proteins and is characterized by memory loss and parkinsonism

Chvostek sign (ke-VOSH-tek syne) abnormal spasms of facial muscles in hypocalcemic patients in response to light taps to stimulate the facial nerve (CN VII); named for Austrian surgeon Franz Chvostek

chyme (kyme) partially digested food mixture leaving the stomach

cilia (SIL-ee-ah) (*sing.*, cilium) tiny, hairlike projections of cells that detect changes outside the cell; some cilia can move, propelling mucus along a surface

ciliary escalator (SIL-ee-ayr-ee ES-kuh-lay-ter) process of cilia moving mucus and entrapped particles upward and out of the respiratory tract

ciliary muscle (SIL-ee-ayr-ee MUS-el) smooth muscle in the ciliary body of the eye that suspends the lens and functions in accommodation of focus for near vision

ciliate (SIL-ee-at) type of protozoan having cilia

cilium (SIL-ee-um) *see* **cilia**

circulatory shock (SER-kyoo-lah-tor-ee) failure of the circulatory (cardiovascular) system to deliver adequate oxygen to the tissues of the body

circulatory system (SER-kyoo-lah-tor-ee SIS-tem) *see* **cardiovascular system**

circumcision (ser-kum-SIH-zhun) surgical removal of the foreskin or prepuce on the penis or clitoris

circumduct (ser-kum-DUKT) to move a part so its distal end moves in a circle

circumduction (ser-kum-DUK-shun) moving of a part so its distal end moves in a circle

circumvallate (ser-kum-VAL-ayt) referring to anything encircled with a ridge or moat

circumvallate papilla (ser-kum-VAL-ayt pah-PIL-ah) any of the huge domelike bumps with central posts on the posterior surface of the tongue mucosa that form a transverse row; each one contains thousands of taste buds

cirrhosis (sih-ROH-sis) degeneration of liver tissue characterized by the replacement of damaged liver tissue with fibrous or fatty connective tissue

cisterna chyli (sis-TER-nah KYE-lee) an enlarged pouch on the thoracic duct that serves as a storage area for lymph moving toward its point of entry into the venous system

citric acid cycle (SIT-rik AS-id SYE-kul) the second series of chemical reactions in the process of glucose metabolism; it is an aerobic process; also referred to as the *Krebs cycle*

clavicle (KLAV-ih-kul) collar bone, connects the upper extremity to the axial skeleton

cleavage furrow (KLEE-vij FUR-oh) appears at the end of anaphase and begins to divide the cell into two daughter cells

cleft lip (kleft) congenital defect resulting in one or more clefts in the upper lip

cleft palate (kleft PAL-et) congenital defect resulting in a fissure of the palate in the roof of the mouth

clinical laboratory technician (KLIN-ih-kal LAB-rah-tor-ee tek-NISH-en) health-care worker who collects samples and scientifically analyzes tissues, body fluids, and other materials for medical purposes; also called medical laboratory technologist or technician

clitoris (KLIT-oh-ris) erectile tissue located within the vestibule of the vagina

clone (klohn) any of a family of many identical cells descended from a single "parent" cell

closed fracture (FRAK-chur) simple fracture; a bone fracture in which the skin is not pierced by bone fragments

coccus (KOK-us) (*pl.*, cocci) spherical bacterial cell

cochlea (KOHK-lee-ah) snail shell or structure of similar shape; relates to a structure within the inner ear

cochlear duct (KOHK-lee-ar dukt) membranous tube within the bony cochlea of the inner ear

cochlear implant (KOHK-lee-ar IM-plant) artificial hearing device that uses electronic circuits to perform the functions of the cochlea of the inner ear

cochlear nerve (KOHK-lee-ar nerv) part of vestibulocochlear nerve (cranial nerve VIII) attached to the cochlea; sensory nerve responsible for hearing

codominance (koh-DOM-ih-nance) in genetics, a form of dominance in which two dominant versions of a trait are both expressed in the same individual

codon (KOH-don) in RNA, a triplet of three base pairs that codes for a particular amino acid

coenzyme (koh-EN-zyme) molecule that assists an enzyme during metabolism, often by carrying a molecule (or molecule fragment) from one chemical pathway to another

colic flexure (left or right) (KOHL-ik FLEK-shur) bend of the colon; the *left colic flexure* is also called the splenic flexure and the *right colic flexure* is also called the **hepatic flexure**

colitis (koh-LYE-tis) any inflammatory condition of the colon and/or rectum

collagen (KAHL-ah-jen) principal organic constituent of connective tissue

collecting duct (CD) (koh-LEK-ting dukt) a straight part of a renal tubule formed by distal tubules of several nephrons joining together

colloid (KOL-oyd) dissolved particles with diameters of 1 to 100 millimicrons (1 millimicron equals about 1/25,000,000 inch)

colon (KOH-lon) *see* **intestine**

colonoscopy (koh-lon-AH-skah-pee) medical procedure in which the lining of the colon is checked for colorectal cancer or other abnormalities by inserting a flexible scope through the anus and into the colon

color blindness (KUL-or BLIND-nes) X-linked inherited condition in which one or more photopigments in the cones of the retina are abnormal or missing

colorectal cancer (kohl-oh-REK-tal KAN-ser) common form of cancer, usually adenocarcinoma, associated with advanced age, low-fiber/high-fat diet, and genetic predisposition

colostomy (kah-LAH-stoh-mee) surgical procedure in which an artificial anus is created on the abdominal wall by cutting the colon and bringing the cut ends out to the surface to form an opening called a **stoma**

columnar (koh-LUM-nar) cell shape in which cells are higher than they are wide

combining site (kom-BINE-ing syte) antigen-binding site; any of the antigen receptor regions on an antibody molecule; shape of each combining site is complementary to shape of a specific antigen

comedones (kom-eh-DOHNZ) (*sing.,* comedo) inflamed lesions associated with early stages of acne formed when sebaceous gland ducts become blocked

comminuted fracture (kom-ih-NOO-ted FRAK-chur) bone fracture characterized by many bone fragments

common bile duct (KOM-on byle dukt) duct from the liver that empties into the duodenum; made up of the merging of the hepatic duct with the cystic duct

communicable (koh-MYOO-nih-kah-bil) able to spread from one individual to another

compact bone (kom-PAKT) *see* **dense bone**

compensated metabolic acidosis (KOM-pen-say-ted met-ah-BOL-ik as-ih-DOH-sis) the body's successful adjustment of its body chemistry for the purpose of returning the blood pH value to near normal levels after metabolic acidosis has developed

compensation (kom-pen-SAY-shun) process by which the body attempts to counteract a shift away from homeostatic balance, thus compensating for the change

complement (KOM-pleh-ment) any of several inactive protein enzymes normally present in blood that when activated kill foreign cells by dissolving them

complement-binding sites (KOM-pleh-ment BIND-ing) locations on an antibody molecule that become available after exposure to an antigen and that bind to complement proteins in the blood plasma to trigger a complement cascade (immune system response) that harms the antigen-containing cell

complement cascade (KOM-pleh-ment kas-KAYD) rapid-fire series of chemical reactions involving proteins called *complements* (normally present in blood plasma) triggered by certain antibody-antigen reactions (and other stimuli) and resulting in the formation of tiny protein rings that create holes in a foreign cell and thus cause its destruction

complementary base pairing (kom-pleh-MEN-tah-ree bays PAYR-ing) bonding purines and pyrimidines in DNA; adenine always binds with thymine, and cytosine always binds with guanine

complete blood cell count (CBC) (kom-PLEET blud sel kownt) clinical blood test that usually includes standard red blood cell, white blood cell, thrombocyte counts, the differential white blood cell count, hematocrit, and hemoglobin content

complete fracture (kom-PLEET FRAK-chur) bone fracture characterized by complete separation of bone fragments

compliance (kom-PLY-ans) the ease of stretch of a material—as in lung compliance, the stretchability of the lung tissues

compound (KOM-pound) substance having more than one kind of element

computed tomography (CT) (kom-PYOO-ted toh-MOG-rah-fee) radiographic imaging technique in which a patient is scanned with x-rays and a computer constructs an image that appears to be a cut section of the person's body

concave (KON-kave) a rounded, somewhat depressed surface

concave curvature (kon-KAYV KUR-vah-chur) inward or *secondary* curvatures of the adult vertebral column in the cervical and lumbar regions

concentric contraction (kon-SENT-rik kon-TRAK-shen) type of isotonic muscle contraction in which a muscle's length decreases

concentric lamella (kon-SEN-trik lah-MEL-ah) ring of calcified matrix surrounding the central (Haversian) canal

concha (KONG-kah) (*pl.,* conchae) shell-shaped structure; for example, bony projections into the nasal cavity; also called *turbinate*

concussion (kon-KUSH-in) type of *traumatic brain injury (TBI)* resulting from a jolt to the head that bends the brainstem and causes temporary chemical changes in the brain, producing any of a variety of functional changes

conduction (kon-DUK-shun) in regard to body temperature regulation, transfer of heat energy to the skin and then the external environment

conductive keratoplasty (CK) (kon-DUK-tiv ker-ah-toh-PLAS-tee) therapy using radiofrequency (RF) energy to heat hair-thin probes that are then used to change the shape of the cornea to correct vision

condyloid joint (KON-dih-loyd joynt) ellipsoidal joint in which an oval process fits into an oval socket

cone receptor cell located in the retina that is stimulated by bright light; different types of cones are stimulated by different ranges of wavelengths (colors)

congenital (kon-JEN-ih-tall) term that refers to a condition present at birth; congenital conditions may be inherited or may be acquired in the womb or during delivery

congestive heart failure (CHF) (kon-JES-tiv hart FAYL-yoor) left heart failure; inability of the left ventricle to pump effectively, resulting in congestion in the systemic and pulmonary circulations

conjunctiva (kon-junk-TIH-vah) mucous membrane that lines the eyelids and covers the sclera (white portion)

conjunctivitis (kon-junk-tih-VYE-tis) inflammation of the conjunctiva, usually caused by irritation, infection, or allergy

connective tissue (koh-NEK-tiv TISH-yoo) most abundant and widely distributed tissue in the body and has numerous functions

connective tissue membrane (koh-NEK-tiv TISH-yoo MEM-brane) one of the two major types of body membranes; composed exclusively of various types of connective tissue

constipation (kon-stih-PAY-shun) condition caused by decreased motility of the large intestine, resulting in the formation of small, hard feces and difficulty in defecation

contact dermatitis (KON-takt der-mah-TYE-tis) a local skin inflammation that lasts a few hours or days and is initiated by the skin being exposed to an antigen

continuous ambulatory peritoneal dialysis (CAPD) (kon-TIN-yoo-us AM-byoo-lah-tor-ee payr-ih-toh-NEE-al dye-AL-ih-sis) an alternative form of treatment for renal failure that may be used instead of the more complex and expensive **hemodialysis**

contraception (kon-trah-SEP-shun) reproductive planning with a goal of avoiding pregnancy

contractile unit (kon-TRAK-til YOO-nit) the sarcomere, the basic functional unit of skeletal muscle

contractility (kon-trak-TIL-ih-tee) ability to contract a muscle

contraction (kon-TRAK-shun) ability of muscle cells to shorten or contract

control center (kon-TROHL SEN-ter) part of a homeostatic feedback loop that integrates (puts together) set point (preprogrammed) information with actual sensed information about a physiological variable and then possibly sends out a signal to an effector to change the variable

contusion (kon-TOO-zhun) local injury caused by mechanical trauma characterized by limited hemorrhaging under the skin, as in a muscle contusion or skin contusion caused by a blow to the body; a bruise

convection (kon-VEK-shun) transfer of heat energy to air that is flowing away from the skin

convex (KON-veks) a rounded, somewhat elevated surface

convex curvature (kon-VEKS KUR-vah-chur) the thoracic or sacral outward curving of the adult vertebral column; an infant has single *primary* outward curvature the length of its spine
cor pulmonale (kohr pul-mah-NAL-ee) failure of the right atrium and ventricle to pump blood effectively, resulting from obstruction of pulmonary blood flow
cornea (KOR-nee-ah) transparent, anterior portion of the sclera
corneal stem cell transplant (KOR-nee-al stem sel tranz-PLANT) procedure in which adult stem cells harvested from cadavers are transplanted into and around the edges of the corneas of a recipient to regrow a healthy cornea
coronal (koh-ROH-nal) literally "like a crown"; a coronal plane divides the body or an organ into anterior and posterior regions
coronary angioplasty (KOHR-oh-nayr-ee AN-jee-oh-plas-tee) medical procedure in which a device is inserted into a blocked coronary artery to force open a channel for blood flow through the myocardium of the heart
coronary artery (KOHR-oh-nayr-ee AR-ter-ee) the right and left coronary arteries are the first arteries to branch off the aorta; they supply blood to the myocardium (heart muscle)
coronary bypass surgery (KOHR-oh-nayr-ee BYE-pass SER-jer-ee) surgery to relieve severely restricted coronary blood flow; veins are taken from other parts of the body and then reattached where needed to bypass the partial blockage
coronary circulation (KOHR-oh-nayr-ee ser-kyoo-LAY-shun) delivery of oxygen and removal of waste product from the myocardium (heart muscle)
coronary embolism (KOHR-oh-nayr-ee EM-boh-liz-em) blocking of a coronary blood vessel by a clot
coronary heart disease (KOHR-oh-nayr-ee hart dih-ZEEZ) disease (blockage or other deformity) of the vessels that supply the myocardium (heart muscle); one of the leading causes of death among adults in the United States
coronary sinus (KOHR-oh-nayr-ee SYE-nus) area that receives deoxygenated blood from the coronary veins and empties it into the right atrium
coronary thrombosis (KOHR-oh-nayr-ee throm-BOH-sis) formation of a blood clot in a coronary blood vessel
coronary vein (KOHR-oh-nayr-ee vane) any vein that carries blood from the myocardial capillary beds to the coronary sinus
coronavirus (koh-ROHN-ah-vye-rus) category of RNA-containing viruses that infect humans and other vertebrate animals, sometimes causing severe respiratory infections (and sometimes intestinal infections and neurological syndromes); for example, SARS **(severe acute respiratory syndrome)** is caused by a type of coronavirus, **SARS-associated coronavirus** (SARS-CoV)
corpora cavernosa (KOHR-pohr-ah kav-er-NOH-sah) (*sing.*, corpus cavernosum) two columns of erectile tissue found in the shaft of the penis
corpus callosum (KOHR-pus kah-LOH-sum) brain structure at which the right and left cerebral hemispheres are joined
corpus luteum (KOHR-pus LOO-tee-um) a hormone-secreting glandular structure that is transformed after ovulation from a ruptured follicle; it secretes chiefly progesterone, with some estrogen secreted as well
corpus spongiosum (KOHR-pus spun-jee-OH-sum) a column of erectile tissue surrounding the urethra in the penis
cortex (KOHR-teks) outer part of an internal organ; for example, the outer part of the cerebrum and of the kidneys
cortical nephron (KOHR-tih-kahl NEF-ron) microscopic unit of the kidney that makes up 85% of all nephron units in the kidney; is located almost entirely in the renal cortex
corticoid (KOHR-tih-koyd) any of the hormones secreted by the three cell layers of the adrenal cortex
cortisol (KOHR-tih-sol) hormone secreted by the adrenal cortex to stimulate the availability of glucose in the blood; in large amounts, cortisol can depress immune functions, as when it is used as a drug treatment; *see* **hydrocortisone**
cosmetic surgery (koz-MET-ik SUR-jeh-ree) surgical medical specialty focused on improving one's appearance
cosmetician (koz-meh-TISH-un) worker who specializes in the manufacture, sale, or application of makeup or other products that affect one's appearance
cotransport (koh-TRANZ-port) active transport process in which two substances are moved together across a cell membrane; for example, sodium and glucose may be transported together across a membrane
countercurrent mechanism (KON-ter-ker-rent MEK-uh-niz-em) system in which renal tubule filtrate flows in opposite directions, maintaining a hyperosmotic medulla; facilitates urine concentration
covalent bond (koh-VAYL-ent) chemical bond formed by two atoms sharing one or more pairs of electrons
Cowper gland *see* **bulbourethral gland**
coxal bone (kok-SAL) the pelvic bone or hipbone (also known as the *os coxae* or the *innominate bone*); formed by fusion of three distinct bones (ilium, ischium, and pubis) during skeletal development
cramps (kramps) painful muscle spasms (involuntary twitches) that result from irritating stimuli, as in mild inflammation, or from ion imbalances
cranial (KRAY-nee-al) relating to (or toward) the head
cranial cavity (KRAY-nee-al KAV-ih-tee) space inside the skull that contains the brain
cranial nerve (CN) (KRAY-nee-al nerv) any of 12 pairs of nerves that attach to the undersurface of the brain and conduct impulses between the brain and structures in the head, neck, and thorax
craniosacral (kray-nee-oh-SAY-kral) relating to parasympathetic nerves
cranium (KRAY-nee-um) bony vault made up of eight bones that encases the brain
crenation (kreh-NAY-shun) abnormal notching in an erythrocyte caused by shrinkage after suspension in a hypertonic solution
cretinism (KREE-tin-iz-em) dwarfism caused by hyposecretion of the thyroid gland
crista ampullaris (KRIS-tah am-pyoo-LAYR-is) a specialized receptor located within the semicircular canals that detects head movements
Crohn disease (krohn dih-ZEEZ) chronic inflammatory bowel disease
crossing-over (KROS-ing OH-ver) phenomenon that occurs during meiosis when pairs of homologous chromosomes synapse and exchange genes
croup (kroop) non–life-threatening type of laryngitis generally seen in children less than 3 years of age; characterized by barklike cough and caused by parainfluenza viruses
crown (krown) topmost part of an organ or other structure, such as a tooth
cruciate ligament (KRU-shee-ayt LIG-uh-ment) either of two crossed ligaments inside the knee joint cavity that connect the tibia to the femur; the *anterior cruciate ligament (ACL)* and the *posterior cruciate ligament (PCL)*
crural (KROOR-al) relating to the leg (sometimes the entire lower extremity)
crust (krust) scab; area of the skin covered by dried blood or exudate

cryptorchidism (krip-TOR-kih-diz-em) undescended testicles
cubital (KYOO-bih-tall) relating to the elbow
cuboid (KYOO-boyd) resembling a cube
cuboidal (KYOO-boyd-al) cell shape resembling a cube
culture (KULT-chur) growth of microbes in a laboratory medium for the purpose of isolating and identifying pathogens from human body fluids
cupula (KYOO-pyoo-lah) the small cup-shaped, flaplike structure at the base of each semicircular canal of the ear that bends during movement of the head to facilitate the sense of dynamic equilibrium
Cushing syndrome (KOOSH-ing SIN-drohm) condition caused by the hypersecretion of glucocorticoids from the adrenal cortex
cuspid (KUS-pid) having cusps or points; for example, the **canine tooth** located lateral to the second incisor that serves to pierce or tear food being eaten is also called a *cuspid tooth*
cutaneous (kyoo-TAYN-ee-us) relating to the skin
cutaneous membrane (kyoo-TAYN-ee-us MEM-brane) primary organ of the integumentary system; the skin
cuticle (KYOO-tih-kul) skinfold covering the root of the nail
cyanosis (sye-ah-NOH-sis) condition in which light-skinned individuals exhibit a bluish coloration resulting from relatively low oxygen content in the arterial blood; literally "blue condition"
cyclic AMP (SIK-lik A M P) (adenosine monophosphate) one of several second messengers that delivers information inside the cell and thus regulates the cell's activity
cystic duct (SIS-tik dukt) joins with the common hepatic duct to form the common bile duct
cystic fibrosis (CF) (SIS-tik fye-BROH-sis) inherited disease involving abnormal chloride ion (Cl^-) transport; causes secretion of abnormally thick mucus and other problems
cystitis (sis-TYE-tis) inflammation or infection of the urinary bladder
cystoscope (SIS-toh-skohp) hollow instrument inserted through urethra into the bladder that permits passage of a light source and surgical instruments to be used for direct examination, biopsy, surgical removal, or treatment of bladder or other urinary tract lesions
cytokine (SYE-toh-kyne) chemical released from cells to trigger or regulate innate and adaptive immune responses
cytokinesis (sye-toh-kin-EE-sis) process by which a dividing cell splits its cytoplasm and plasma membrane into two distinct daughter cells; cytokinesis happens along with mitosis (or meiosis) during the cell division process
cytology (sye-TOL-oh-jee) study of cells
cytologist (SYE-TOL-uh-jist) scientist who studies cells
cytoplasm (SYE-toh-plaz-em) the gel-like substance of a cell exclusive of the nucleus and other organelles
cytosine (SYE-toh-seen) one of several nitrogen-containing bases that make up nucleotides, which in turn make up nucleic acids such as DNA and RNA; in the cell, it can chemically bind to another nitrogenous base, guanine (*G* or *g*), to form a more complex structure or in translating genetic codes; symbolized by the letter *C* or *c*; *see also* **guanine, adenine, thymine, uracil**
cytoskeleton (sye-toh-SKEL-eh-ton) cell's internal supporting, moving framework
cytotoxic T cell (sye-toh-TOK-sik) "cell killing" T cell

D

deciduous (deh-SID-yoo-us) temporary; shedding of structures at a certain stage of growth; for example, *deciduous teeth,* which are commonly referred to as *baby teeth,* are shed to make way for the permanent adult teeth
decubitus ulcer (deh-KYOO-bih-tus UL-ser) pressure sore that often develops over a bony prominence, such as the heel, when lying in one position for prolonged periods
deep in anatomy, a structure is deep to another structure when it is further inside the body; opposite of *superficial*
defecation (def-eh-KAY-shun) elimination of feces
defibrillation (deh-fib-rih-LAY-shun) electrical stimulation of the heart in order to restore normal heart rhythm (used when the heart fibrillates, or gets out of rhythm); *see* **ventricular fibrillation** and **automatic external defibrillator**
degeneration (dih-jen-uh-RAY-shun) a biological process, still somewhat puzzling to scientists, in which tissues break down as a normal consequence of aging; degeneration of one or more tissues resulting from disease, which can occur at any time
deglutition (deg-loo-TISH-un) swallowing
dehydration (dee-hye-DRAY-shun) clinical term that refers to an abnormal loss of fluid from the body's internal environment
dehydration synthesis (dee-hye-DRAY-shun SIN-the-sis) chemical reaction in which large molecules are formed by removing water from smaller molecules and joining them together
deltoid (DEL-toyd) having a triangular shape; for example, the deltoid muscle
dementia (deh-MEN-shah) syndrome of brain abnormalities that includes loss of memory, shortened attention span, personality changes, reduced intellectual capacity, and motor dysfunction
dendrite (DEN-dryte) branching or treelike; a nerve cell process that transmits impulses toward the body
dendritic cell (DC) (DEN-drih-tik) phagocytic cells of the immune system
dengue (DENG-gay or DENG-gee) serious viral infection caused by a type of flavivirus
dense bone bone that has a hard, dense outer layer; also called *compact bone*
dense fibrous connective tissue (dense FYE-brus koh-NEK-tiv TISH-yoo) connective tissue consisting of protein fibers packed densely in the matrix
dental caries (DENT-al KAYR-ees) *see* **caries**
dentin (DEN-tin) chief bonelike dental tissue covered by enamel in crown and by cementum in neck and root areas of tooth
deoxyribonucleic acid (DNA) (dee-ok-see-rye-boh-nook-LAY-ik AS-id) genetic material of the cell that carries the chemical "blueprint" of the body
depilatories (deh-PIL-ah-toh-rees) hair removers
depolarization (dee-poh-lar-ih-ZAY-shun) the electrical activity that triggers a contraction of the heart muscle
dermal-epidermal junction (DEJ) (DER-mal–EP-ih-der-mal JUNK-shun) junction between the thin epidermal layer of the skin and the dermal layer; provides support for the epidermis
dermal papilla (DER-mal pah-PIL-ah) (*pl.,* papillae) upper region of the dermis that forms part of the dermal-epidermal junction and forms the ridges and grooves of fingerprints
dermatitis (der-mah-TYE-tis) general term referring to any inflammation of the skin
dermatome (DER-mah-tohm) skin surface area supplied by a single spinal nerve
dermatosis (der-mah-TOH-sis) general term meaning "skin condition"
dermis (DER-mis) the deeper of the two major layers of the skin, composed of dense fibrous connective tissue interspersed with glands, nerve endings, and blood vessels; sometimes called the "true skin"

descending colon (dih-SEND-ing KOH-lon) portion of the colon that lies in the vertical position, on the left side of the abdomen; extends from below the stomach to the iliac crest

developmental process (dee-vel-op-MEN-tal PROS-es) changes and functions occurring during a human's early years as the body becomes more efficient and more effective

deviated septum (DEE-vee-ay-ted SEP-tum) abnormal condition in which the nasal septum (dividing wall between the two nasal air passages) is located far from its normal position, possibly obstructing normal nasal breathing

diabetes insipidus (dye-ah-BEE-teez in-SIP-ih-dus) condition resulting from hyposecretion of ADH in which large volumes of urine are formed and, if left untreated, may cause serious health problems

diabetes mellitus (dye-ah-BEE-teez mel-AYE-tus) a condition resulting when the pancreatic islets secrete too little insulin, resulting in increased levels of blood glucose

diabetic ketoacidosis (dye-ah-BET-ik kee-toh-as-ih-DOH-sis) low blood pH resulting from an accumulation of ketone bodies in the blood in diabetes mellitus

diabetic retinopathy (dye-ah-BET-ik ret-in-AH-path-ee) growth or hemorrhage of blood vessels caused by diabetes mellitus

diagnostic medical sonographer (dye-ag-NOS-tik MED-ih-kul son-AH-graf-er) health professional who uses sonography to examine internal body structures as a medical diagnostic strategy

dialysis (dye-AL-ih-sis) separation of smaller (diffusible) particles from larger (nondiffusible) particles through a semipermeable membrane

diaphragm (DYE-ah-fram) membrane or partition that separates one thing from another; the flat muscular sheet that separates the thorax and abdomen and is a major muscle of respiration

diaphysis (dye-AF-ih-sis) (*pl.*, diaphyses) shaft of a long bone

diarrhea (dye-ah-REE-ah) defecation of liquid feces

diarthrosis (dye-ar-THROH-sis) (*pl.*, diarthroses) freely movable joint

diastole (dye-AS-toh-lee) relaxation of the heart, interposed between its contractions; opposite of **systole**

diastolic blood pressure (dye-ah-STOL-ik blud PRESH-ur) blood pressure in arteries during diastole (relaxation) of the heart

diencephalon (dye-en-SEF-ah-lon) "between" brain; parts of the brain between the cerebral hemispheres and the mesencephalon (midbrain)

dietitian (dye-eh-TISH-en) person who works in nutrition science by developing healthful meals and dietary health strategies; also *dietician*

differential WBC count (dif-er-EN-shal WBC kownt) special type of white blood cell (WBC) count in which proportions of each type of WBC are reported as percentages of the total count

differentiate (dif-er-EN-shee-ayt) to become different in structure and function, as when some of the original cells of early developmental stages differentiate to become muscle cells and other cells become nerve cells, etc.

differentiation (dif-er-EN-shee-AY-shun) process by which daughter cells become different in structure and function (by using different genes from the genome, all cells of the body share), as when some of the original cells of early developmental stages differentiate to become muscle cells and other cells become nerve cells, and so on

diffusion (dih-FYOO-zhen) spreading; for example, scattering of dissolved particles

digestion (dye-JES-chun) the breakdown of food materials either mechanically (i.e., by chewing) or chemically (i.e., by action of digestive enzymes)

digestive system (dih-JES-tiv SIS-tem) organs that work together to ensure proper digestion and absorption of nutrients

digital (DIJ-ih-tal) in anatomy, relating to fingers and toes

digitalis (dij-ih-TAL-is) drug used to treat atrial fibrillation; digoxin is an example of a digitalis preparation

diploe (DIP-loh-EE) region of cancellous (spongy) bone within the wall of a flat bone of the cranium; also spelled *diploë*

directional term any term used to give direction in the body, such as left, right, anterior, posterior, superior, inferior, etc.

disaccharide (dye-SAK-ah-ryde) double sugar, such as sucrose, maltose, or lactose; type of carbohydrate made up of two saccharide groups (monosaccharides)

discharging chambers (dis-CHARJ-ing CHAYM-bers) the two lower chambers of the heart called *ventricles*

disease (dih-ZEEZ) any significant abnormality in the body's structure or function that disrupts a person's vital function or physical, mental, or social well-being

dislocation (dis-low-KAY-shun) abnormal movement of body parts, as in separation of bones of a joint; *see* **subluxation**

dissection (dye-SEK-shun) cutting technique used to separate body parts for study

dissociate (dih-SOH-see-ayt) to break apart a compound in solution

dissociation (dih-soh-see-AY-shun) separation of ions as they dissolve in water

distal (DIS-tal) toward the end of a structure; opposite of **proximal**

distal convoluted tubule (DCT) (DIS-tal KON-voh-loo-ted TOO-byool) the part of the tubule distal to the ascending limb of the nephron loop in the kidney

disuse atrophy (DIS-yoos AT-roh-fee) condition in which prolonged inactivity results in the muscles getting smaller in size; *see also* **atrophy**

diuretic (dye-yoo-RET-ik) referring to something that promotes the production of urine

diuretic drug (dye-yoo-RET-ik drug) therapeutic substance that promotes or stimulates the production of urine; diuretic drugs are among the most commonly used drugs in medicine

diverticulitis (dye-ver-tik-yoo-LYE-tis) inflammation of diverticula (abnormal outpouchings) of the large intestine, possibly causing constipation

dizygotic twins (dye-zye-GOT-ik twinz) *see* **fraternal twins**

DNA (dee en ay) *see* **deoxyribonucleic acid**

DNA fingerprinting (dee en ay FING-ger-print-ing) technique used to analyze the genetic makeup of individuals; compares nucleotide sequences using electrophoresis

DNA replication (dee en ay rep-lih-KAY-shun) the unique ability of DNA molecules to make copies of themselves

dominant gene (DOM-ih-nant jeen) in genetics, a *dominant gene* has effects that appear in the offspring (dominant forms of a gene are often represented by upper case letters); compare with **recessive** gene

dopamine (DOH-pah-meen) chemical neurotransmitter

doping (DOH-ping) the addition of blood (or blood products), steroids, or other performance-enhancing substances to the bloodstream, a practice performed by some athletes that can have serious (even fatal) side effects and is outlawed worldwide

dorsal (DOR-sal) referring to the back; opposite of **ventral;** in humans, the posterior is dorsal

dorsal body cavity (DOR-sal BOD-ee KAV-ih-tee) includes the cranial and spinal cavities

dorsal cavity *see* **dorsal body cavity**

dorsal root ganglion (DOR-sal root GANG-lee-on) ganglion of the dorsal root located near the spinal cord; where the neuron cell body of the dendrites of the sensory neuron is located

dorsiflex (dor-sih-FLEKS) to bend the foot with the toes pointing upward

dorsiflexion (dor-sih-FLEK-shun) movement in which the top of the foot is elevated (brought toward the front of the leg) with the toes pointing upward

double helix (HEE-lix) shape of DNA molecules; a double spiral

dowager's hump (DOW-ah-jerz) kyphosis (abnormal backward curvature of thoracic spine) caused by vertebral compression fractures in osteoporosis

Down syndrome (SIN-drohm) group of symptoms usually caused by trisomy of chromosome 21; characterized by mental retardation and multiple structural defects, including facial, skeletal, and cardiovascular abnormalities

Duchenne muscular dystrophy (DMD) (doo-SHEN MUS-kyoo-lar DIS-troh-fee) form of muscular dystrophy (abnormal muscle development in which normal muscle is replaced with fat and fibrous tissue) inherited on the X chromosome and characterized by mild leg muscle weakness that progresses rapidly to include the shoulder muscles and eventually death from cardiac or respiratory muscle weakness; also called **pseudohypertrophy** ("false muscle growth")

ductless gland (DUKT-les) specialized gland that secretes hormones directly into the blood; endocrine gland

ductus arteriosus (DUK-tus ar-teer-ee-OH-sus) connects the aorta and the pulmonary artery, allowing most blood to bypass the fetus's developing lungs

ductus deferens (DUK-tus DEF-er-enz) a thick, smooth, muscular tube that allows sperm to exit from the epididymis and pass from the scrotal sac into the abdominal cavity; also known as the *vas deferens*

ductus venosus (DUK-tus veh-NOH-sus) a continuation of the umbilical vein that shunts blood returning from the placenta past the fetus's developing liver directly into the inferior vena cava

duodenal papillae (doo-oh-DEE-nal pah-PIL-ee) ducts located in the middle third of the duodenum that empty pancreatic digestive juices and bile from the liver into the small intestine; there are two ducts, the major duodenal papilla and the minor papilla

duodenum (doo-oh-DEE-num or doo-AH-deh-num) the first subdivision of the small intestine where most chemical digestion occurs

dura mater (DOO-rah MAH-ter) literally "strong or hard mother"; outermost layer of the meninges

dust cells macrophages that ingest particulate matter in the small air sacs of the lungs

dwarfism (dwor-FIZ-em) condition of abnormally small stature, sometimes resulting from hyposecretion of growth hormone

dysentery (DIS-en-tayr-ee) inflammatory condition of colon characterized by frequent diarrhea that may contain blood or pus

dysfunctional uterine bleeding (DUB) (dis-FUNK-shun-al YOO-ter-in BLEED-ing) irregular or excessive bleeding from the uterus resulting from a hormonal imbalance

dysmenorrhea (dis-men-oh-REE-ah) painful menstruation

dyspnea (DISP-nee-ah) difficult or labored breathing

dysrhythmia (dis-RITH-mee-ah) any abnormality of cardiac rhythm

dysuria (dis-YOO-ree-ah) painful, burning urination

E

eardrum (EAR-drum) the tympanic membrane that separates the external ear and middle ear

eccentric contraction (ek-SENT-rik kon-TRAK-shun) type of isotonic muscle contraction in which a muscle's length increases under a load

eccrine (EK-rin) relating to an exocrine gland with secretory cells that release secretions by exocytosis, without losing part of the cell as in *apocrine* glands

eccrine sweat gland (EK-rin swet gland) small sweat glands distributed over the total body surface

echocardiogram (ek-oh-KAR-dee-oh-gram) medical image produced by a type of sonography called *echocardiography*

echocardiography (ek-oh-kar-dee-OG-rah-fee) heart imaging technique in which ultrasound waves echo back from heart tissues to form a continuous recording of heart structure movement during a series of cardiac cycles

eclampsia (eh-KLAMP-see-ah) potentially fatal condition associated with toxemia of pregnancy; characterized by convulsions and coma

ectoderm (EK-toh-derm) the innermost of the primary germ layers that develops early in the first trimester of pregnancy

ectopic pregnancy (ek-TOP-ik PREG-nan-see) a pregnancy in which the fertilized ovum implants someplace other than in the uterus

eczema (EK-zeh-mah) inflammatory skin condition associated with a variety of diseases and characterized by erythema, papules, vesicles, and crusts

edema (eh-DEE-mah) accumulation of fluid in a tissue, as in inflammation; swelling

effector (ef-FEK-tor) any organ that has an effect on the body's internal environment in response to feedback; for example, voluntary and involuntary muscle, the heart, and glands

effector B cell (ef-FEK-tor bee sel) cell that differentiates from a B cell; synthesizes and secretes huge amounts of antibodies

effector cell (ef-FEK-tor) any cell that has an effect in the body; a cell that acts as an effector; for example, the activated forms of B cells and T cells are called *effector cells*

effector T cell (ef-FEK-tor tee sel) cell that differentiates from a T cell; causes contact killing of a target cell

efferent (EF-fer-ent) carrying from, as in *efferent neurons* that transmit impulses from the central nervous system to the periphery; opposite of **afferent**

efferent lymphatic vessel (EE-fer-ent limf VES-el) any of the small lymphatic vessels that carry lymphatic fluid away from a lymph node; compare to *afferent lymphatic vessel*

efferent neuron (EF-fer-ent NOO-ron) neuron that transmits impulses away from the central nervous system and toward the periphery; generally a *motor neuron;* compare to **afferent neuron**

ehrlichiosis (ur-lik-ee-OH-sis) bacterial infection transmitted by ticks and similar to Lyme disease

ejaculation (ee-jak-yoo-LAY-shun) sudden discharging of semen from the body

ejaculatory duct (ee-JAK-yoo-lah-toh-ree dukt) duct formed by the joining of the vas deferens and the duct from the seminal vesicle that allows sperm to enter the urethra

elastin (e-LAS-tin) stretchy protein found in elastic fiber

electrocardiogram (ECG or EKG) (eh-lek-troh-KAR-dee-oh-gram) graphic record of the heart's action potentials

electrocardiograph (e-lek-troh-KAR-dee-oh-graf) machine that produces electrocardiograms, graphic records of the heart's electrical activity (voltage fluctuations)

electroencephalogram (EEG) (eh-lek-troh-en-SEF-uh-loh-gram) graphic representation of voltage changes in brain tissue used to evaluate nerve tissue function

electrolyte (eh-LEK-troh-lyte) substance that dissociates into ions in solution, rendering the solution capable of conducting an electric current

electrolyte balance (eh-LEK-troh-lyte BAL-ans) homeostasis of electrolytes

electron (eh-LEK-tron) small, negatively charged subatomic particle found in the outer regions of an atom

electron microscope (eh-LEK-tron MY-kroh-skope) a device that produces a greatly enlarged image of a tiny structure by using a beam of electrons focused by magnets (rather than a beam of light focused by glass lenses, as in a light microscope)

electron transport system (ETS) (eh-LEK-tron TRANZ-port SIS-tem) cellular process within mitochondria that transfers energy from high-energy electrons from glycolysis and the citric acid cycle to ATP molecules so that the energy is available to do work in the cell

electrophoresis (eh-lek-troh-foh-REE-sis) laboratory procedure in which different types of molecules are separated according to molecular weight by passing a weak electric current through their liquid medium

element (EL-eh-ment) pure substance, composed of only one type of atom

elephantiasis (el-eh-fan-TYE-ah-sis) extreme lymphedema (swelling due to lymphatic blockage) in the limbs caused by a parasitic worm infestation, so called because the limbs swell to "elephant proportions"

elimination (eh-lim-uh-NAY-shun) moving something out of the body, as in defecation

embolism (EM-boh-liz-em) obstruction of a blood vessel by foreign matter carried in the bloodstream

embolus (EM-boh-lus) a blood clot or other substance (such as a bubble of air) that is moving in the blood and may block a blood vessel

embryo (EM-bree-oh) animal in early stages of intrauterine development; in humans, the first 3 months after conception

embryology (em-bree-OL-oh-gee) study of the development of an individual from conception to birth

embryonic phase (em-bree-ON-ik fayz) the period extending from fertilization until the end of the eighth week of gestation; during this phase the term *embryo* is used

emesis (EM-eh-sis) vomiting

emphysema (em-fih-SEE-mah) abnormal condition characterized by trapping of air in alveoli of the lung that causes them to rupture and fuse to other alveoli

emptying reflex (EMP-tee-ing REE-fleks) the reflex that causes the contraction of the bladder wall and relaxation of the internal sphincter to allow urine to enter the urethra, which is followed by urination if the external sphincter is voluntarily relaxed

emulsify (eh-MUL-seh-fye) in digestion, when bile breaks up fats

enamel (ih-NA-mel) hard, mineralized connective tissue, harder than bone, forms hard covering of exposed tooth surfaces; hardest substance in body

endemic (en-DEM-ik) refers to a disease native to a local region of the world

endocarditis (en-doh-kar-DYE-tis) inflammation of the lining of the heart

endocardium (en-doh-KAR-dee-um) thin layer of very smooth tissue lining each chamber of the heart

endochondral ossification (en-doh-KON-dral os-ih-fih-KAY-shun) the process in which most bones are formed from cartilage models

endocrine (EN-doh-krin) secreting into the blood or tissue fluid rather than into a duct; opposite of exocrine

endocrine gland (EN-doh-krin gland) any ductless gland that is part of the endocrine system and secretes hormones into intercellular spaces and the blood

endocrine system (EN-doh-krin SIS-tem) the series of ductless glands that are found in the body

endocrinologist (en-doh-krin-OL-oh-jist) scientist or physician specializing in *endocrinology*

endocrinology (en-doh-krin-OL-oh-jee) study of endocrine glands and their functions

endoderm (EN-doh-derm) the outermost layer of the primary germ layers that develops early in the first trimester of pregnancy

endogenous infection (en-DOJ-en-us in-FEK-shun) infection caused by pathogens that normally inhabit the intestines, vulva, or vagina

endolymph (EN-doh-limf) clear potassium-rich fluid that fills the membranous labyrinth of the inner ear

endometrial ablation (en-doh-MEE-tree-al ab-LAY-shun) therapeutic destruction of endometrial tissue (the tissue that normally lines the uterus)

endometriosis (en-doh-mee-tree-OH-sis) presence of functioning endometrial tissue outside the uterus

endometrium (en-doh-MEE-tree-um) mucous membrane lining the uterus

endoneurium (en-doh-NOO-ree-um) the thin wrapping of fibrous connective tissue that surrounds each axon in a nerve

endoplasmic reticulum (ER) (en-doh-PLAZ-mik reh-TIK-yoo-lum) network of tubules and vesicles in cytoplasm; two types: *rough* and *smooth*

endorphin (en-DOR-fin) chemical in the central nervous system that influences pain perception; a natural painkiller

endoscope (EN-doh-skohp) flexible tube inserted through a small incision in order to view internal organs and sometimes to pass medical devices into the body or remove tissue from the body

endosteum (en-DOS-tee-um) a fibrous membrane that lines the medullary cavity

endothelium (en-doh-THEE-lee-um) squamous epithelial cells that line the inner surface of the entire cardiovascular system and the vessels of the lymphatic system

endotracheal intubation (en-doh-TRAY-kee-al in-too-BAY-shun) medical procedure in which a tube is placed through the mouth, pharynx, and larynx into the trachea to ensure an open airway

endurance training (en-DUR-ance TRAYN-ing) continuous vigorous exercise requiring the body to increase its consumption of oxygen and develop the muscles' ability to sustain activity over a prolonged period of time

energy level limited region surrounding the nucleus of an atom at a certain distance containing electrons; also called a *shell*

enkephalin (en-KEF-ah-lin) peptide chemical in the central nervous system that acts as a natural painkiller

enteritis (en-ter-AYE-tis) inflammation of the small intestine

enuresis (en-yoo-REE-sis) involuntary urination

environmental health (en-VYE-ron-ment-al helth) field of public health that focuses on the health effects of our surroundings (natural and artificial)

enzyme (EN-zyme) biochemical catalyst allowing chemical reactions to take place in a suitable time frame

eosinophil (ee-oh-SIN-oh-fil) white blood cell that is readily stained by eosin

epicardium (ep-ih-KAR-dee-um) the inner layer of the pericardium that covers the surface of the heart; it is also called the **visceral pericardium**

epidemic (ep-ih-DEM-ik) refers to a disease that occurs in many individuals at the same time

epidemiologist (ep-ih-dee-mee-OL-uh-jist) scientist engaged in the study, prevention, and treatment of the occurrence, distribution, and transmission of diseases in human populations

epidemiology (EP-ih-dee-mee-OL-oh-jee) study of the occurrence, distribution, and transmission of diseases in human populations

epidermis (ep-ih-DER-mis) "false" skin; outermost layer of the skin

epididymis (ep-ih-DID-ih-mis) one of two comma-shaped, long, tightly coiled tubes that carry sperm from testes to vas deferens

epididymitis (ep-ih-did-ih-MY-tis) inflammation of the epididymis

epigastric region (ep-ih-GAS-trik) the superior central region of the abdominopelvic cavity

epigenetics (ep-ih-jeh-NET-iks) any process of inheritance other than direct DNA inheritance, sometimes by adding a methyl group (or other chemical) to DNA, as in maternal/paternal imprinting of genes

epiglottis (ep-ih-GLOT-is) lidlike cartilage overhanging the entrance to the larynx

epiglottitis (EP-ih-glot-aye-tis) life-threatening type of laryngitis generally seen in children 3 to 7 years of age; characterized by laryngeal edema and high fever and caused by *Haemophilus influenzae* virus

epilepsy (EP-ih-lep-see) a seizure disorder characterized by recurring seizures

epinephrine (Epi) (ep-ih-NEF-rin) adrenaline; hormone secreted by the adrenal medulla

epineurium (ep-ih-NOO-ree-um) a tough fibrous sheath that covers the whole nerve

epiphyseal fracture (ep-ih-FEEZ-ee-al FRAK-chur) when the epiphyseal plate is separated from the epiphysis or diaphysis; this type of fracture can disrupt the normal growth of the bone

epiphyseal line (ep-ih-FEEZ-ee-al lyne) point of fusion seen in a mature bone that replaces the epiphyseal cartilage or growth plate that once separated the epiphysis and diaphysis of a growing bone

epiphyseal plate (ep-ih-FEEZ-ee-al) the cartilage plate that is between the epiphysis and the diaphysis and allows growth to occur; sometimes referred to as a *growth plate*

epiphysis (eh-PIF-ih-sis) (*pl.*, epiphyses) end of a long bone

episiotomy (eh-piz-ee-OT-oh-mee) a surgical procedure used during birth to prevent a laceration of the mother's perineum or the vagina

epispadias congenital defect in males characterized by opening of urethral meatus on dorsal (top) surface of glans or penile shaft

epistaxis (ep-ih-STAK-sis) clinical term referring to a bloody nose

epithelial membrane (ep-ih-THEE-lee-al MEM-brane) membrane composed of epithelial tissue with an underlying layer of specialized connective tissue

epithelial tissue (ep-ih-THEE-lee-al TISH-yoo) covers the body and its parts; lines various parts of the body; forms continuous sheets that contain no blood vessels; classified according to shape and arrangement

equilibration (ee-kwi-lib-RAY-shun) the state or act of coming into equilibrium or balanced state

equilibrium (ee-kwi-LIB-ree-um) a balanced state; a state in which two or more parts of a system remain in a relatively constant proportion to each other

erectile dysfunction (ED) (ee-REK-tyl dis-FUNK-shun) disorder in which the penis fails to become erect during the male sexual response, usually due to a lack of relaxation in smooth muscles in the wall of blood vessels in the penis; the drug Viagra (sildenafil) treats ED by promoting the same response in the penis as NO (nitric acid), which relaxes smooth muscles in the vessel walls

ergonomics (er-goh-NOM-iks) applied study of workers and their work environment

erythema (er-ih-THEE-mah) redness or inflammation of the skin or mucous membranes

erythroblastosis fetalis (eh-rith-roh-blas-TOH-sis fet-TAL-is) condition of a fetus or infant caused by the mother's Rh antibodies reacting with the baby's Rh-positive RBCs, characterized by massive agglutination of the blood and resulting in life-threatening circulatory problems

erythrocyte (eh-RITH-roh-syte) red blood cell; literally "red cell"

erythropoietin (EPO) (eh-RITH-roh-POY-eh-tin) glycoprotein secreted to increase red blood cell production in response to oxygen deficiency

esophagus (ee-SOF-ah-gus) the muscular, mucous-lined tube that connects the pharynx with the stomach; also known as the *foodpipe*

essential reproductive organ (ee-SEN-shal ree-proh-DUK-tiv OR-gan) reproductive organ that must be present for reproduction to occur; also known as **gonad**

estrogen (ES-troh-jen) sex hormone secreted by the ovary that causes the development and maintenance of the female secondary sex characteristics and stimulates growth of the epithelial cells lining the uterus

etiology (ee-tee-OHL-oh-jee) theory, or study, of the factors involved in causing a disease

eupnea (YOOP-nee-ah) normal respiration

eustachian tube (yoo-STAY-shun toob) *see* **auditory tube**

evaporation (ee-vap-oh-RAY-shun) heat being lost from the skin by sweat being vaporized

eversion (ee-VER-zhun) movement that turns a body part (such as the foot) outward

evert (ee-VERT) to turn outward

excimer laser surgery (EK-zim-er LAY-zer SIR-jer-ee) refractory eye surgery that uses an excimer or "cool" laser to vaporize corneal tissue in treating mild to moderate nearsightedness; also called **photorefractive keratectomy (PRK)**

excoriation (eks-koh-ree-AY-shun) skin lesion in which epidermis has been removed, as in a scratch wound

exercise physiologist (EK-ser-syze fiz-ee-OL-uh-jist) scientist who studies the process of muscular exercise and related phenomena

exhalation (eks-huh-LAY-shun) moving air out of the lungs; opposite of **inhalation,** or **inspiration;** also known as **expiration**

exocrine (EK-soh-krin) secreting into a duct; opposite of endocrine

exocrine gland (EK-soh-krin) glands that secrete their products into ducts that empty onto a surface or into a cavity; for example, sweat glands

exophthalmos (ek-sof-THAL-mus) condition of abnormally protruding eyeballs, occurring in a form of hyperthyroidism called **Graves disease;** also called *exophthalmia*

experimental control (eks-payr-ih-MEN-tel kon-TROL) any procedure within a scientific experiment that ensures that the test situation itself is not affecting the outcome of the experiment

experimentation (eks-payr-ih-men-TAY-shun) performing an experiment, which is usually a test of a tentative explanation of nature called a **hypothesis**

expiration (eks-pih-RAY-shun) moving air out of the lungs; opposite of **inhalation,** or **inspiration;** also known as **exhalation**

expiratory muscle (eks-PYE-rah-tor-ee MUS-el) any of the muscles that allow more forceful expiration to increase the rate and depth of ventilation; the internal intercostals and the abdominal muscles

expiratory reserve volume (ERV) (eks-PYE-rah-tor-ee ree-ZERV VOL-yoom) the amount of air that can be forcibly exhaled after expiring the tidal volume (TV)

extend (ek-STEND) to increase the angle between two bones at a joint; opposite of *flex*

extension (ek-STEN-shun) increasing the angle between two bones at a joint

external acoustic canal (eks-TER-nal ah-KOO-stik kah-NAL) a curved tube (approximately 2.5 cm long) extending from the auricle of the ear into the temporal bone, ending at the tympanic membrane; also *external auditory canal*

external ear (eks-TER-nal) the outer part of the ear that is made up of the auricle and the external auditory canal

external genitalia (eks-TER-nal jen-ih-TAYL-yah) external reproductive organs; also called *genitals* or simply *genitalia*

external intercostals (eks-TER-nal in-ter-KOS-talls) inspiratory muscles that enlarge the thorax, causing the lungs to expand and air to rush in

external nares (eks-TER-nal NAY-reez) (*sing.*, naris) nostrils

external oblique (eks-TER-nal oh-BLEEK) the outermost layer of the anterolateral abdominal wall

external otitis (eks-TER-nal oh-TYE-tis) a common infection of the external ear; also known as *swimmer's ear*

external respiration (eks-TER-nal res-pih-RAY-shun) the exchange of gases between air in the lungs and in the blood

external urinary meatus (eks-TER-nal YOOR-ih-nayr-ee mee-AY-tus) external opening of the urinary tract

extracellular fluid (ECF) (eks-trah-SEL-yoo-lar FLOO-id) the water found outside of cells located in two compartments between cells (interstitial fluid) and in the blood (plasma)

F

face (fays) anterior aspect of the head or skull; any flat surface of the external aspect of a structure

facial (FAY-shal) referring to the face

factor VIII (FAK-ter ayt) one of the blood clotting factors (coagulation factors)

fallen arch condition in which the tendons and ligaments of the foot weaken, allowing the normally curved arch to flatten out

fallopian tube (fal-LOH-pee-an toob) *see* **uterine tube**

false ribs the eighth, ninth, and tenth pairs of ribs, which are attached to the cartilage of the seventh ribs rather than the sternum

fascia (FASH-ee-ah) general name for the fibrous connective tissue masses located throughout the body

fascicle (FAS-ih-kul) small bundle of fibers, as in a small bundle of nerve fibers or muscle fibers

fasciculus (fah-SIC-yoo-lus) little bundle

fat one of the three basic nutrient types; primarily a source of energy

fatigue (fah-TEEG) loss of muscle power; weakness; state of exhaustion or tiredness

fat tissue (TISH-yoo) *see* **adipose**

fatty acid (FAT-tee AS-id) product of fat digestion; building block of fat molecules

febrile seizure (FEB-ril SEE-zhur) abnormal brain activity caused by a fever

feces (FEE-seez) waste material discharged from the intestines

feedback loop (FEED-bak loop) a highly complex and integrated communication control network, classified as negative or positive; negative feedback loops are the most important and numerous homeostatic control mechanisms

femoral (FEM-or-al) referring to the thigh

femur (FEE-mur) the thigh bone, which is the longest bone in the body

fertilization (FER-tih-lih-ZAY-shun) the action that takes place at the moment the female's ovum and the male's sperm cell unite

fetal alcohol syndrome (FAS) (FEE-tal AL-koh-hol SIN-drohm) a condition that may cause congenital abnormalities in a baby; it results from a woman consuming alcohol during pregnancy

fetal phase (FEE-tal fayz) period extending from the eighth to the thirty-ninth week of gestation; during this phase the term *fetus* is used

fetus (FEE-tus) unborn young, especially in the later stages; in human beings, from the third month of the intrauterine period until birth

fever (FEE-ver) elevated body temperature beyond the normal set point; usually triggered by the immune system in response to infection or injury

fiber (FYE-ber) threadlike structure; for example, nerve fiber or collagen fiber

fibrillation (fib-rih-LAY-shun) condition in which individual muscle fibers, or small groups of fibers, contract asynchronously (out of time) with other muscle fibers in an organ, producing no effective movement

fibrin (FYE-brin) insoluble protein in clotted blood

fibrinogen (fye-BRIN-oh-jen) soluble blood protein that is converted to insoluble fibrin during clotting

fibroid (FYE-broyd) *see* **fibromyoma**

fibromyoma (fye-broh-my-OH-mah) benign tumor of smooth muscle and fibrous connective tissue commonly occurring in the uterine wall, where it is often called a *fibroid; see also* **myoma**

fibromyositis (fye-broh-my-oh-SYE-tis) inflammation of muscle tissue accompanied by inflammation of nearby tendon tissue

fibrosarcoma (fye-broh-sar-KOH-mah) cancer of fibrous connective tissue

fibrosis (fye-BROH-sis) condition in which fibrous tissue replaces damaged tissues

fibrous connective tissue (FYE-brus koh-NEK-tiv TISH-yoo) strong, nonstretchable, white collagen fibers that compose tendons

fibula (FIB-yoo-lah) the slender non–weight-bearing bone located on the lateral aspect of the leg

fibularis (muscle) group (fib-YOO-lay-ris [MUS-el] groop) group of lateral muscles of the leg that act to pronate the foot, rotating it toward the midline, and plantar flex the foot, pulling it toes-downward; also called the **peroneus group**

fight-or-flight response (fyte or flyte) the changes produced by increased sympathetic impulses allowing the body to deal with any type of stress

filtration (fil-TRAY-shun) movement of water and solutes through a membrane by a higher hydrostatic pressure on one side

fimbria (FIM-bree-ah) (*sing.*, fimbriae) fringelike projection

first-degree burn minor burn with only minimal discomfort and no blistering; epidermis may peel but no dermal injury occurs; *see also* **partial-thickness burn**

fissure (FISH-ur) elongated break or groove

flagellate (FLAJ-eh-lat) protozoan possessing flagella

flagellum (flah-JEL-um) (*pl.*, flagella) single projection extending from the cell surface; only example in human is the "tail" of the male sperm

flat bone one of the four types of bone; the frontal bone is an example of a flat bone

flatfeet condition in which the tendons and ligaments of the foot are weak, allowing the normally curved arch to flatten out

flatulence (FLAT-yoo-lens) presence of air or other gases in the lumen of the gastrointestinal tract

flavivirus (FLAV-ih-vye-rus) category of RNA-containing viruses that typically require an insect vector to transmit them to humans; examples of flavivirus infections include yellow fever, dengue, St. Louis encephalitis, and West Nile virus (WNV)

flex (fleks) to bend; for example, to decrease the angle between two bones at the joint

flexion (FLEK-shun) act of bending; decreasing the angle between two bones at the joint

floating ribs (FLOW-ting) the eleventh and twelfth pairs of ribs, which are only attached to the thoracic vertebrae

fluid balance (FLOO-id BAL-ans) homeostasis of fluids; the volumes of interstitial fluid, intracellular fluid, and plasma and total volume of water remain relatively constant

fluid compartment (FLOO-id kom-PART-ment) any of the areas in the body where the fluid is located; for example, interstitial fluid

folate deficiency anemia (FOH-layt deh-FISH-en-see ah-NEE-mee-ah) blood disorder characterized by a decrease in the red blood cell count, caused by a deficiency of folic acid in the diet (as in malnourished individuals)

follicle (FOL-lih-kul) a pocket or bubble; for example, the pocket of skin from which a hair grows

follicle-stimulating hormone (FSH) (FOL-lih-kul STIM-yoo-lay-ting HOR-mohn) hormone present in males and females; in males, FSH stimulates the production of sperm; in females, FSH stimulates the ovarian follicles to mature and the follicle cells to secrete estrogen

fontanel (FON-tah-nel) any of the "soft spots" on an infant's head; incompletely ossified area in the infant skull

food science (food SYE-ens) study of the characteristics of food and effects of storing, handling, and preparing food

foramen (foh-RAY-men) small opening; for example, the vertebral foramen, which allows the spinal cord to pass through the vertebral canal

foramen ovale (foh-RAY-men oh-VAL-ee) shunts blood from the right atrium directly into the left atrium allowing most blood to bypass the baby's developing lungs; either of a pair of small oval openings for nerves in the sphenoid bone; literally *oval hole*

forensic science (foh-REN-zik SYE-ens) field of scientific investigation applied to legal questions, such as cause of death, crime scene investigation, and related matters

foreskin (FORE-skin) a loose-fitting, retractable casing located over the glans of the penis; also known as the **prepuce**

formed elements (formd EL-eh-mentz) cellular (RBC, WBC, and platelet) portion of blood of blood tissue, in contrast to the unformed (liquid) nature of blood plasma

formula (FOR-myoo-lah) shorthand notation for a chemical structure such as an atom or molecule, as in C for carbon and H_2O for water

fourth-degree burn complete destruction of epidermis, dermis, and subcutaneous tissue with additional damage below subcutaneous tissue to muscle and bone; *see* **full-thickness burn**

fovea centralis (FOH-vee-ah sen-TRAL-is) small depression in the macula lutea where cones are most densely packed; vision is sharpest where light rays focus on the fovea

fractal geometry (FRAK-tal jee-OM-eh-tree) the study of surfaces with a seemingly infinite area, such as the lining of the small intestine

fragile X syndrome (FXS) (FRAJ-il eks SIN-drohm) condition in which mental retardation results from breakage of X chromosome in males

fraternal twins (frah-TERN-al twinz) birth of two siblings at the same time that have developed from two separate zygotes (dizygotic); also called *dizygotic twins;* contrast with **identical (monozygotic) twins**

freckle (FREK-uhl) small brown or red macules that are a common genetic variant of normal skin pigmentation

free nerve ending (nerv END-ing) simple nerve receptor in the skin that responds to pain

free radical (RAD-ih-kal) highly reactive, electron-seeking molecules that occur normally in cells but may damage electron-dense molecules such as DNA or molecules in cell membranes; free radicals may be inhibited by antioxidants, such as vitamin E

frenulum (FREN-yoo-lum) the thin membrane that attaches the tongue to the floor of the mouth

frontal (FRON-tal) relating to the forehead or to the anterior aspect of a structure

frontal bone forehead bone

frontal muscle (FRUN-tall MUS-el) one of the muscles of facial expression; it moves the eyebrows and furrows the skin of the forehead

frontal plane (FRUN-tal playn) lengthwise plane running from side to side, dividing the body into anterior and posterior portions

frontal sinusitis (FRON-tall sye-nyoo-SYE-tis) inflammation in the frontal sinus

frostbite (FROST-byte) local tissue damage caused by extreme cold

full-thickness burn burn that (1) destroys epidermis, dermis, and subcutaneous tissue (*see* **third-degree burn**); and (2) extends below skin and subcutaneous tissue to reach muscle and bone (*see* **fourth-degree burn**)

functional protein (FUNK-shen-al PROH-teen) protein that has the role of regulating chemical reactions in the body, such as enzymes, some neurotransmitters, some hormones; compare to **structural protein**

fundus (of stomach) (FUN-dus) enlarged portion to the left of and above the opening of the esophagus into the stomach

fundus (of uterus) (FUN-duss of YOO-ter-us) bulging prominence above level where uterine tubes attach to the body of the uterus

fungus (FUNG-us) (*pl.,* fungi) organism similar to plants but lacking chlorophyll and capable of producing mycotic (fungal) infections

furuncle (FUR-un-kul) boil; a pus-filled cavity formed by some hair follicle infections

G

G protein a protein molecule usually embedded in a cell's plasma membrane that plays an important role in getting a signal from a receptor (also in the plasma membrane) to the inside of the cell

galactose (gah-LAK-tohs) simple sugar (monosaccharide) found in lactose (milk sugar)

gallbladder (GAWL-blad-er) hollow sac connected to the common bile duct and that stores and concentrates bile

gallstone (GAWL-stohn) solid concretions or stones, often composed of cholesterol or bile salts, found in the gallbladder; *see also* **cholelithiasis**

gamete (GAM-eet) either of the two sex cells, sperm (spermatozoa) and egg (ova), that have half the usual number of nuclear chromosomes (the haploid number)

ganglion (GANG-lee-on) (*pl.,* ganglia) a region of unmyelinated nerve tissue (usually this term is used only for regions in the peripheral nervous system [PNS])

ganglion cell (GANG-lee-on sel) photoreceptor cell of the eye's retina that does not help form an image but instead helps detect changes in environmental light to synchronize the body's internal clock to external daily, monthly, and seasonal cycles; compare to **rod** and **cone**

gangrene (GANG-green) tissue death (necrosis) that involves decay of tissue

gastric gland (GAS-trik) glands in stomach lining that secrete enzymes, mucus, or hydrochloric acid

gastritis (gas-TRY-tis) inflammation of the lining of the stomach

gastrocnemius (GAS-trok-NEE-mee-us) superficial muscle of the calf of the leg, connected (along with the soleus muscle) to the calcaneus bone of the foot by way of the Achilles (calcaneal) tendon; its action is to dorsiflex the foot, bending the toes upward

gastroenteritis (gas-troh-en-ter-EYE-tis) inflammation of the stomach and intestines

gastroenterology (gas-troh-en-ter-AHL-oh-jee) study and treatment of diseases of the gastrointestinal (GI) tract

gastroesophageal reflux disease (GERD) (gas-troh-eh-sof-eh-JEE-al REE-fluks dih-ZEEZ [gerd]) a set of symptoms resulting from a hiatal hernia that allows stomach (gastric) contents to flow back (reflux) into the esophagus; symptoms include heartburn or chest pain and coughing or choking during or just after a meal; also known as *GERD*

gastroesophageal sphincter (gas-troh-eh-sof-eh-JEE-al SFINK-ter) a ring of smooth muscle around the opening of the stomach at the lower end of the esophagus that acts as a valve to allow food to enter the stomach but prevents stomach contents from moving back into the esophagus

gastrointestinal (GI) tract (gas-troh-in-TES-tih-nul trakt) tubelike structure of the digestive system extending from stomach to anus—sometimes meant to include the entire **alimentary canal** (mouth to anus)

gene (jeen) one of many segments of a chromosome (DNA molecule); each gene contains the genetic code for synthesizing a protein molecule such as an enzyme or hormone

gene linkage (jeen LINK-ej) phenomenon that may occur during *crossing-over* of meiotic division in which a whole group of genes stays together and crosses over as a single unit

gene replacement therapeutic technique that replaces genes that specify production of abnormal proteins with normal genes

gene therapy (jeen THAYR-ah-pee) manipulation of genes to cure genetic problems; most forms of gene therapy have not yet been proven effective in humans

general senses (JEN-er-al SEN-sez) senses detected by simple, microscopic receptors widely distributed throughout the body (skin, muscles, tendons, joints, etc.) involving modes of pain, temperature, touch, pressure, or body position

genetic (jeh-NET-ik) relating to the genetic code and biological heredity; see also **genetics**

genetic counseling (jeh-NET-ik KOWN-se-ling) practice of consulting with families regarding genetic diseases

genetic counselor (jeh-NET-ik KOWN-se-lor) science professional who consults with families regarding genetic diseases

genetic engineer (jeh-NET-ik en-juh-NEER) someone who specializes in manipulating the genetic code

genetic mutation (jeh-NET-ik myoo-TAY-shun) change in the genetic material within a genome; may occur spontaneously or as a result of mutagens

genetics (jeh-NET-iks) scientific study of heredity and the genetic code

genital *see* **external genitalia**

genital ducts (jen-ih-tall dukts) tubelike structures in the embryo that develop into reproductive organs; also applies to adult reproductive ducts

genitalia (jen-ih-TAYL-yah) reproductive organs; *see* **external genitalia**

genome (JEE-nohm) entire set of chromosomes in a cell; the *human genome* refers to the entire set of human chromosomes

genomics (jeh-NOH-miks) field of endeavor involving the analysis of the genetic code contained in the human or other species' genome

geriatrics (jayr-ee-AT-riks) medical specialty that focuses on treatment of the elderly

gerontology (jayr-on-TOL-oh-jee) study of the aging process

gestation (jes-TAY-shun) pregnancy

gestation period (jes-TAY-shun PEER-ee-od) the time length or period of pregnancy, approximately 9 months in humans

gestational diabetes mellitus (GDM) (jes-TAY-shun-al dye-ah-BEE-teez MELL-ih-tus) condition characterized by a temporary decrease in blood levels of insulin during pregnancy

ghrelin (GRAY-lin) hormone secreted by epithelial cells lining the stomach; ghrelin boosts appetite, slows metabolism, and reduces fat burning; may be involved in the development of obesity

gigantism (jye-GAN-tiz-em) a condition produced by hypersecretion of growth hormone during the early years of life; results in a child who grows to gigantic size

gingiva (JIN-jih-vah) (*pl.,* gingivae) gums (of the mouth)

gingival (JIN-jih-val) relating to the gums (of the mouth)

gingivitis (jin-jih-VYE-tis) inflammation of the gum (gingiva), often as a result of poor oral hygiene

gland secreting structure

glandular epithelium (GLAN-dyoo-lar ep-ih-THEE-lee-um) cells that are specialized for secreting activity

glans (glanz) the sensitive distal end of the shaft of the penis or clitoris

glaucoma (glaw-KOH-mah) disorder characterized by elevated pressure in the eye

glia (GLEE-ah) supporting cells of nervous tissue; *see* **neuroglia**

gliding joint (GLY-ding joynt) type of diarthrotic joint formed by flat surfaces that glide past each other

glioma (glee-OH-mah) one of the most common types of brain tumors

globulin (GLOB-yoo-lin) a type of plasma protein that includes antibodies

glomerular capsule (gloh-MER-yoo-lar KAP-sul) *see* **Bowman capsule**

glomerular filtrate (gloh-MER-yoo-lar fil-TRAYT) watery fluid filtered from the plasma of renal blood capillaries into the glomerular capsule of the glomerulus

glomerular-capsular membrane (gloh-MER-yoo-lar KAP-soo-lahr) membrane made up of glomerular endothelium, basement membrane, and visceral layer of the glomerular (Bowman) capsule; function is filtration

glomerulonephritis (gloh-mer-yoo-loh-neh-FRY-tis) inflammatory disease of the glomerular-capsular membranes of the kidney

glomerulus (gloh-MAYR-yoo-lus) compact cluster; for example, capillaries in the kidneys

glottis (GLOT-is) the space between the vocal cords

glucagon (GLOO-kah-gon) hormone secreted by alpha cells of the pancreatic islets

glucocorticoid (GC) (gloo-koh-KOR-tih-koyd) hormone that influences nutrient metabolism; secreted by the adrenal cortex

gluconeogenesis (gloo-koh-nee-oh-JEN-eh-sis) formulation of glucose or glycogen from protein or fat compounds

glucose (GLOO-kohs) monosaccharide or simple sugar; the principal blood sugar

gluteal (GLOO-tee-al) of or near the buttocks

gluteus maximus (GLOO-tee-us MAX-ih-mus) major extensor of the thigh and also supports the torso in an erect position
glycerol (GLIS-er-ohl) product of fat digestion
glycogen (GLYE-koh-jen) polysaccharide made up of a chain of glucose (monosaccharide) molecules; animal starch
glycogen loading (GLYE-koh-jen LOHD-ing) *see* **carbohydrate loading**
glycogenesis (glye-koh-JEN-eh-sis) formation of glycogen from glucose or from other monosaccharides, fructose, or galactose
glycogenolysis (glye-koh-jeh-NOL-ih-sis) hydrolysis of glycogen to glucose-6-phosphate or to glucose
glycolysis (glye-KAHL-ih-sis) the first series of chemical reactions in glucose metabolism; changes glucose to pyruvic acid in a series of anaerobic reactions
glycosuria (glye-koh-SOO-ree-ah) glucose in the urine; a sign of diabetes mellitus
goblet cell (GOB-let sel) any of the specialized, goblet-shaped cells found in simple columnar epithelium that produce mucus
goiter (GOY-ter) enlargement of the thyroid gland
Golgi apparatus (GOL-jee ap-ah-RA-tus) small sacs stacked on one another near the nucleus that make carbohydrate compounds, combine them with protein molecules, and package the product in a globule
Golgi tendon receptors (GOL-jee TEN-don ree-SEP-tors) sensors that are responsible for proprioception
gonad (GOH-nad) essential organ of reproduction that produces gametes: testis in males; ovary in females
gonadotropin (goh-nah-doh-TROH-pin) any of the hormones (FSH and LSH) produced by the anterior pituitary or embryonic tissue (hCG) that stimulate growth and maintenance of the testes or ovaries
gonadotropin-releasing hormone (GnRH) (goh-nah-doh-TROH-pin ree-LEES-ing HOR-mohn) hormone released by the hypothalamus that stimulates the anterior pituitary gland to release its gonadotropin hormones
gout (gowt) abnormal condition in which excess uric acid is deposited in joints and other tissues as sodium urate crystals—the crystals produce inflammation or *gout arthritis*
gouty arthritis (gow-TEE ar-THRY-tis) metabolic disorder in which excess blood levels of uric acid are deposited within the synovial fluid of joints and other tissues
graafian follicle (GRAH-fee-en FOL-ih-kul) sac of the ovary that contains a mature ovum
gradient (GRAY-dee-ent) a slope or difference between two levels; for example, blood pressure gradient: a difference between the blood pressure in two different vessels
gram the unit of measure in the metric system on which mass is based (approximately 454 grams equals 1 pound)
Gram-staining technique (gram STAYN-ing tek-NEEK) process in microbiology in which a specific mixture of stains is used to distinguish between different categories of bacteria (gram-positive vs. gram-negative bacteria)
granular leukocyte (GRAN-yoo-lar LOO-koh-syte) white blood cell (leukocyte) with granules visible in cytoplasm when stained; also called *granulocyte*
granulosa cell (gran-yoo-LOH-sah sel) cell layer surrounding the oocyte
Graves disease (gravz dih-ZEEZ) inherited, possibly immune endocrine disorder characterized by hyperthyroidism accompanied by exophthalmos (protruding eyes)
gray matter (MAT-er) tissue in the central nervous system made up of cell bodies and unmyelinated axons and dendrites
greater omentum (GRAYT-er oh-MEN-tum) a pouchlike extension of the visceral peritoneum
greater vestibular gland (ves-TIB-yoo-lar gland) either of the exocrine mucous glands located on either side of the vaginal outlet; also known as **Bartholin gland**; *see also* **lesser vestibular gland**
growth hormone (GH or hGH) (HOR-mohn) hormone secreted by the anterior pituitary gland that controls the rate of skeletal and visceral growth
guanine (GWAH-neen) one of several nitrogen-containing bases that make up nucleotides, which in turn make up nucleic acids such as DNA and RNA; in the cell, it can chemically bind to another nitrogenous base, cytosine (*C* or *c*), to form a more complex structure or in translating genetic codes; symbolized by the letter *G* or *g*; *see also* **cytosine, adenine, thymine, uracil**
gustation (gus-TAY-shun) the process of tasting
gustatory cell (GUS-tah-tor-ee sel) cells of taste
gynecologist (gye-neh-KOL-uh-jist) physician specializing in medicine of the female reproductive system
gyrus (JYE-rus) (*pl.*, gyri) convoluted ridge found on the brain's surface

H

hair follicle (hayr FOL-ih-kul) a small tube where hair growth occurs
hair papilla (hayr pah-PIL-ah) a small, cap-shaped cluster of cells located at the base of the follicle where hair growth begins
hamstring muscle (HAM-string MUS-el) powerful flexor of the hip made up of the semimembranosus, semitendinosus, and biceps femoris muscles
Haversian canal (HAV-er-zhen or hah-VER-zhun kah-NAL) the central canal in the osteon (Haversian system) that contains a blood vessel; named for English physician Clopton Havers; also called **central canal** of the osteon
Haversian system (HAV-er-zhen or hah-VER-zhun SIS-tem) structural unit of compact bone tissue made up of concentric layers (lamellae) of hard bone matrix and bone cells (osteocytes); named for English physician Clopton Havers; also called **osteon**
health (helth) physical, mental, and social well-being; the absence of disease
heart block (hart blok) a blockage of impulse conduction from atria to ventricles so that the heart beats at a slower rate than normal
heart disease (hart dih-ZEEZ) any of a group of cardiac disorders that together constitute the leading cause of death in the United States
heart failure (hart FAYL-yoor) inability of the heart to pump returned blood sufficiently
heart murmur (hart MUR-mur) abnormal heart sound that may indicate valvular insufficiency (leaking) or stenosing (narrowing; blockage) of the valve
heart rate (HR) (hart rayt) heart beats (cardiac cycles) per unit of time; usually expressed as beats/min (beats per minute)
heartburn (HART-burn) esophageal pain caused by backflow of stomach acid into esophagus
heat exhaustion (heet eg-ZAWS-chun) condition caused by fluid loss resulting from activity of thermoregulatory mechanisms in a warm external environment
heatstroke (heet strohk) life-threatening condition characterized by high body temperature; failure of thermoregulatory mechanisms to maintain homeostasis in a very warm external environment
Heberden node (HEB-er-den nohd) any of the abnormal enlargements seen at the distal interphalangeal joints in osteoarthritis

Heimlich maneuver *see* **abdominal thrust**

Helicobacter pylori (HEEL-ih-koh-BAK-ter pye-LOH-ree) spiral-shaped bacterium known to be a major cause of gastric and duodenal ulcers

helix (HEE-lix) (*pl.,* helices) a spiral, as in the helix of the ear (a fold that spirals around the external ear)

helper T cell (TH cell) immune system cells that help B cells differentiate into antibody-secreting plasma cells; also help coordinate cellular immunity through direct contact with other immune cells

hematocrit (Hct) (hee-MAT-oh-krit) volume percent of blood cells in whole blood

hematology (hee-mah-TOL-oh-jee) study of the blood

hematopoiesis (hee-mah-toh-poy-EE-sis) blood cell formation

hematopoietic tissue (hee-mah-toh-poy-ET-ik TISH-yoo) specialized connective tissue that is responsible for the formation of blood cells and lymphatic system cells; found in red bone marrow, spleen, tonsils, and lymph nodes

hematuria (hem-ah-TOO-ree-ah) symptom of blood in the urine, often the result of a renal or urinary disorder

heme (heem) iron-containing chemical group found in hemoglobin

hemiplegia (hem-ee-PLEE-jee-ah) paralysis (lack of voluntary muscle control) of one entire side of the body (except the face)

hemochromatosis (hee-moh-kroh-mah-TOH-sis) condition characterized by excess availability of iron in the blood; also called *iron storage disease*

hemodialysis (hee-moh-dye-AL-ih-sis) use of dialysis to separate waste products from the blood

hemoglobin (Hb) (hee-moh-GLOH-bin) iron-containing protein in red blood cells

hemolytic anemia (hee-moh-LIT-ik ah-NEE-mee-ah) any of a group of blood disorders characterized by deficient or abnormal hemoglobin that causes deformation and fragility of red blood cells (e.g., sickle cell anemia, thalassemia)

hemolytic disease of the newborn (HDN) (hee-moh-LIT-ik dih-ZEEZ ov thuh NOO-born) condition caused by blood ABO or Rh factor incompatibility during pregnancy between developing offspring and mother

hemophilia (hee-moh-FIL-ee-ah) any of a group of X-linked inherited blood clotting disorders caused by a failure to form clotting factors VIII, IX, or XI

hemorrhagic anemia (HEM-oh-raj-ick ah-NEE-mee-ah) group of conditions characterized by low oxygen-carrying capacity of blood; caused by decreased red blood cell (RBC) life span and/or increased rate of RBC destruction

hemorrhoid (HEM-eh-royd) varicose vein in the rectum; hemorrhoids are also called *piles*

hemostasis (hee-moh-STAY-sis) stoppage of blood flow

hemothorax (hee-moh-THOH-raks) abnormal condition in which blood is present in the pleural space surrounding the lung, possibly causing collapse of the lung

Henle loop (HEN-lee loop) extension of the proximal tubule of the kidney; also called *nephron loop*

heparin (HEP-ah-rin) naturally occurring substance that inhibits formation of a blood clot; has been used as a drug to inhibit clotting

hepatic duct (heh-PAT-ik dukt) any of the liver ducts that drain bile out of the liver

hepatic flexure (heh-PAT-ik FLEK-sher) the bend between the ascending colon and the transverse colon; also called *hepatic colic flexure*

hepatic portal circulation (heh-PAT-ik POR-tall ser-kyoo-LAY-shun) the route of blood flow through the liver

hepatic portal vein (heh-PAT-ik POR-tall vane) delivers blood directly from the gastrointestinal tract to the liver

hepatitis (hep-ah-TYE-tis) inflammation of the liver due to viral or bacterial infection; injury; damage from alcohol, drugs, or other toxins; or other factors

hepatitis C virus (HCV) (hep-ah-TYE-tis see VYE-rus) one of several types of virus that causes liver inflammation and may eventually lead to cirrhosis or liver cancer if not treated

herniated ("slipped") disk (HER-nee-ayt-ed disk) rupture of a fibrocartilage intervertebral disk that permits the pulpy core of the disk to push against the spinal cord or spinal nerve roots, causing pain

herpes zoster (HER-peez ZOS-ter) viral infection that affects the skin of a single dermatome; commonly known as *shingles*

hiatal hernia (hye-AY-tal HER-nee-ah) a bulging out (hernia) of the stomach through the opening (hiatus) of the diaphragm through which the esophagus normally passes; this condition may prevent the valve between the esophagus and stomach from closing, thus allowing stomach contents to flow back into the esophagus; *see also* **gastroesophageal reflux disease**

hiccup (HIK-up) involuntary spasmodic contraction of the diaphragm

hilum (HYE-lum) (*pl.,* hila) small opening on the side of an organ (lung, kidney, lymph node) to allow vessels and nerves to enter/exit

hinge joint (hinj joynt) type of diarthrotic synovial joint that allows movement around a single axis in the manner of a hinge

hip the joint connecting the legs to the trunk

histamine chemical released by basophils and mast cells in allergic and inflammatory reactions; results in blood vessel vasodilation and bronchoconstriction

histogenesis (his-toh-JEN-eh-sis) formation of tissues from primary germ layers of embryo

histologist (hih-STOL-uh-jist) scientist that studies tissue structure and function

hives *see* **urticaria**

Hodgkin disease (HOJ-kin dih-ZEEZ) type of lymphoma (malignant lymph tumor) characterized by painless swelling of lymph nodes in the neck, progressing to other regions

homeostasis (hoh-mee-oh-STAY-sis) relative uniformity of the normal body's internal environment

homeostatic mechanism (hoh-mee-oh-STAT-ik MEK-ah-nih-zem) a system that maintains a constant environment enabling body cells to function effectively

hormone (HOR-mohn) substance secreted by an endocrine gland

human engineered chromosome (HEC) (HYOO-man en-juh-NEERD KROH-meh-sohm) gene augmentation procedure that inserts therapeutic genes into a separate strand of DNA that is inserted into nucleus of cell

Human Genome Project (HGP) (HYOO-man JEE-nome PROJ-ekt) a worldwide collaborative effort of scientists and others to map out the entire human genome and study the biological, medical, and ethical aspects of their discoveries; the HGP is largely funded by U.S. government sources such as the DOE (Department of Energy) and the NIH (National Institutes of Health); a currently active HGP offshoot is *ENCODE (The Encyclopedia of DNA Elements); see* **genome, genomics**

human immunodeficiency virus (HIV) (HYOO-man ih-myoo-noh-deh-FISH-en-see VYE-rus) the retrovirus that causes acquired immunodeficiency syndrome (AIDS)

human lymphocyte antigen (HLA) (HYOO-man LIM-foh-syte AN-tih-jen) type of self-antigen that the immune system uses to distinguish one's own tissue from that of a foreign entity

humerus (HYOO-mer-us) the second longest bone in the body; the long bone of the arm

humoral immunity (HYOO-mor-al ih-MYOO-nih-tee) *see* **antibody-mediated immunity**

Huntington disease (HD) (HUN-ting-ton dih-ZEEZ) degenerative, inherited brain disorder characterized by chorea (purposeless movements) progressing to severe dementia and death by age 55

hyaline cartilage (HYE-ah-lin KAR-tih-lij) most common type of cartilage; appears gelatinous and glossy

hybridoma (hye-brid-OH-mah) fused or hybrid cells that continue to produce the same antibody as the original lymphocyte

hydrocele (HYE-droh-seel) abnormal accumulation of watery fluid, as can occur in the scrotum

hydrocephalus (hye-droh-SEF-ah-lus) abnormal accumulation of cerebrospinal fluid; "water on the brain"

hydrochloric acid (HCl) (hye-droh-KLOR-ik AS-id) compound formed by the hydrogen ion and chloride ion, which releases the hydrogen ion in water to form an acid; produced in great quantity by gastric glands in the stomach

hydrocortisone (hye-droh-KOR-tih-zohn) a hormone ordinarily secreted by the adrenal cortex as cortisol, but as hydrocortisol it is used as a drug to reduce inflammation or other immune functions; *see* **cortisol**

hydrogen (HYE-droh-jen) one of the chemical elements found in great quantity in the human body; symbolized by H, as in H_2O (water); may form ions such as H^+ (hydrogen ion) or OH^- (hydroxide ion)

hydrogen bond (HYE-droh-jen bond) weak chemical bond that occurs between the partial positive charge on a hydrogen atom covalently bound to a nitrogen or oxygen atom and the partial negative charge of another polar molecule

hydrogen ion (HYE-droh-jen aye-on) found in water and water solutions; produces an acidic solution; symbol is H^+

hydrolysis (hye-DROHL-ih-sis) chemical reaction in which water is added to a large molecule, causing it to break apart into smaller molecules

hydronephrosis (hye-droh-neh-FROH-sis) pathological swelling or enlargement of renal pelvis or calyces caused by blockage of urine outflow

hydrostatic pressure (hye-droh-STAT-ik PRESH-ur) the force of a fluid pushing against some surface

hydroxide ion (hye-DROK-side aye-on) found in water and water solutions; produces an alkaline solution; symbol is OH^-

hydroxyurea (hye-DROK-see-yoo-REE-ah) an antineoplastic (antitumor) drug

hymen (HYE-men) Greek for "membrane"; mucous membrane that may partially or entirely occlude the vaginal outlet

hyoid bone (HYE-oyd bohn) U-shaped bone of the neck between the mandible and the larynx

hyperacidity (hye-per-ah-SID-ih-tee) excessive secretion of acid; an important factor in the formation of ulcers

hypercalcemia (hye-per-kal-SEE-mee-ah) a condition in which harmful excesses of calcium are present in the blood

hypercholesterolemia (hye-per-koh-les-ter-ohl-EE-mee-ah) condition of high blood cholesterol content

hyperglycemia (hye-per-glye-SEE-mee-ah) higher than normal blood glucose concentration

hyperkalemia (hy-per-kal-EE-mee-ah) abnormally high blood potassium level

hypernatremia (hy-per-nah-TREE-mee-ah) abnormally high blood sodium level

hyperopia (hye-per-OH-pee-ah) refractive disorder of the eye caused by a shorter than normal eyeball; farsightedness

hyperosmotic (hye-per-os-MOT-ik) relating to solutions that generally promote osmosis of water (into them)

hyperplasia (hye-per-PLAY-zee-ah) growth of an abnormally large number of cells at a local site, as in a neoplasm or tumor

hypersecretion (hye-per-seh-KREE-shun) too much of a substance is being secreted

hypersensitivity (hye-per-SEN-sih-tiv-ih-tee) inappropriate or excessive response of the immune system

hypertension (HTN) (hye-per-TEN-shun) abnormally high blood pressure

hyperthyroidism (hye-per-THYE-royd-iz-em) oversecretion of thyroid hormones, which increases metabolic rate resulting in loss of weight, increased appetite, and nervous irritability

hypertonic (hye-per-TON-ik) a solution containing a higher level of salt (NaCl) than is found in a living red blood cell (above 0.9% NaCl)

hypertrophy (hye-PER-troh-fee) increased size of a part caused by an increase in the size of its cells

hyperventilation (hye-per-ven-tih-LAY-shun) very rapid, deep respirations

hypervitaminosis (hye-per-vye-tah-mih-NOH-sis) general name for any condition resulting from an abnormally high intake of vitamins

hypoalbuminemia (hye-poh-al-byoo-min-EE-mee-ah) condition of low albumin (protein) in the blood plasma; it often results from renal disorders or malnutrition; loss of plasma protein usually causes edema of the tissue spaces

hypocalcemia (hye-poh-kal-SEE-mee-ah) abnormally low blood calcium level

hypochondriac region (hye-poh-KON-dree-ak REE-jun) the left and right upper regions of the abdominopelvic cavity, just under the lower part of the rib cartilage and on either side of the epigastric region; used when the abdominopelvic cavity is visualized as being subdivided into nine regions as in a tic-tac-toe grid

hypodermis (hye-poh-DER-mis) the loose ordinary (areolar) tissue just under the layers of skin and superficial to the muscles; made of loose connective tissue and fat; also called **subcutaneous tissue** or **superficial fascia**

hypogastric region (hye-poh-GAST-rik REE-jun) the central lower region of the abdominopelvic cavity, below the stomach and umbilicus (navel) and between the left and right iliac regions; used when the abdominopelvic cavity is visualized as being subdivided into nine regions as in a tic-tac-toe grid

hypoglycemia (hye-poh-glye-SEE-mee-ah) lower-than-normal blood glucose concentration

hypokalemia (hye-poh-kal-EE-mee-ah) abnormally low blood potassium level

hyponatremia (hye-poh-nah-TREE-mee-ah) abnormally low blood sodium level

hyposecretion (hye-poh-seh-KREE-shun) too little secretion of a substance

hypospadias (hye-poh-SPAY-dee-us) congenital defect in males characterized by opening of urethral meatus on underside of the glans or penile shaft

hypothalamus (hye-poh-THAL-ah-mus) portion of the floor and lateral wall of the third ventricle of the brain

hypothermia (hye-poh-THER-mee-ah) failure of thermoregulatory mechanisms to maintain homeostasis in a very cold external environment

hypothesis (hye-POTH-eh-sis) (*pl.,* hypotheses) a proposed explanation of an observed phenomenon
hypothyroidism (hye-poh-THYE-royd-iz-em) undersecretion of thyroid hormones; early in life results in cretinism; later in life results in myxedema
hypotonic (hye-poh-TON-ik) a solution containing a lower level of salt (NaCl) than is found in a living red blood cell (below 0.9% NaCl)
hypoventilation (hye-poh-ven-tih-LAY-shun) slow and shallow respirations
hypovitaminosis (hye-poh-VYTE-ah-min-oh-sis) condition of having too few vitamin molecules in the body for normal function
hypovolemic shock (hye-poh-voh-LEE-mik) circulatory failure (shock) caused by a drop in blood volume that causes blood pressure (and blood flow) to drop; literally "low volume" shock
hypoxia (hye-POCK-see-ah) abnormally low concentration of oxygen in the blood or tissue fluids
hysterectomy (his-teh-REK-toh-mee) surgical removal of the uterus

I

I & O measurement of both fluid intake and fluid output (urine volume) over a specified period of time; abbreviation of "input and output"
identical (monozygotic) twins birth of two siblings at the same time that have developed from a single zygote that splits into two offspring early during development; also called *monozygotic twins;* contrast with **fraternal (dizygotic) twins**
ideogram (ID-ee-oh-gram) a simple cartoon of a chromosome used in genomics to show the overall structure of the chromosome, including staining landmarks and the relative position of the centromere
idiopathic (id-ee-oh-PATH-ik) relating to a disease of undetermined cause
ileocecal valve (il-ee-oh-SEE-kal valv) the sphincterlike structure between the end of the small intestine and the beginning of the large intestine
ileum (IL-ee-um) the distal portion of the small intestine
iliac crest (IL-ee-ak krest) the superior edge of the ilium
iliac region (IL-ee-ak REE-jun) the left and right lower regions of the abdominopelvic cavity, near the iliac region of the pelvis and on either side of the hypogastric region; terminology used to describe the abdominopelvic cavity when it is visualized as being subdivided into nine regions as in a tic-tac-toe grid; also called *left and right inguinal regions*
iliopsoas (il-ee-oh-SOH-as) a flexor of the thigh and an important stabilizing muscle for posture
ilium (IL-ee-um) one of the three separate bones that fuse to form the os coxae or hip bone
immune deficiency (ih-MYOON deh-FISH-en-see) general term for complete or relative failure of the immune system to defend the internal environment of the body
immune system (ih-MYOON SIS-tem) the body's defense system against disease
immunization (ih-myoo-nih-ZAY-shun) deliberate artificial exposure to disease to produce acquired immunity in the body
immunoglobulin (Ig) (ih-myoo-noh-GLOB-yoo-lin) antibody
immunology (im-yoo-NOL-oh-jee) study of immune system functions and mechanisms
immunosuppressive drug (ih-myoo-noh-soo-PRES-iv drug) compound that suppresses, or reduces, the capacity of the immune system; such drugs are sometimes used to prevent rejection of transplanted tissues
immunotherapy (ih-myoo-noh-THAYR-ah-pee) therapeutic technique that bolsters a person's immune system in an attempt to control a disease
impacted fracture (im-PAK-ted FRAK-chur) fracture in which bone fragments are driven into each other
impetigo (im-peh-TYE-go) a highly contagious bacterial skin infection that occurs most often in children
implantable cardioverter-defibrillator (ICD) (im-PLAN-tah-bel KAR-dee-oh-vert-er dee-FIB-rih-lay-tor) surgically implanted medical device that automatically monitors for fibrillation, then produces a defibrillating shock without any external intervention.
implantation (im-plan-TAY-shun) occurs when a fertilized ovum implants in the uterus
impotence (IM-poh-tense) failure to achieve erection of the penis results in an inability to reproduce; also called *erectile dysfunction (ED)*
inborn immunity (IN-born ih-MYOO-nih-tee) immunity to disease that is inherited
incisor (in-SYE-zer) any of the front teeth, which are adapted for cutting
incompetent (cardiac) valve (in-KOM-peh-tent [KAR-dee-ak] valv) cardiac valve that "leaks," allowing some blood to flow back into the chamber from which it came
incomplete fracture (in-kom-PLEET FRAK-chur) bone fracture in which the bone fragments remain partially joined
incontinence (in-KON-tih-nens) condition in which an individual voids urine or feces involuntarily
incubation (in-kyoo-BAY-shun) early, latent stage of an infection, during which an infection has begun but signs or symptoms have not yet developed
incus (IN-kus) the anvil, the middle ear bone that is shaped like an anvil
induced abortion (in-DOOST ah-BOR-shun) purposeful termination of a pregnancy before the fetus is able to survive outside the womb
infancy (IN-fan-see) the age range from birth to about 18 months of age
infant respiratory distress syndrome (IRDS) (IN-fant RES-pih-rah-toh-ree dih-STRESS SIN-drohm) leading cause of death in premature babies, caused by a lack of surfactant in the alveolar air sacs
infection control (in-FEK-shun KON-trol) any practice intended to limit the spread of infection in a population
infectious (in-FEK-shus) describes any condition or substance that can induce an infection or is characterized by infection by a pathogen
infectious arthritis (in-FEK-shuss ar-THRY-tis) inflammation of joint tissues caused by a variety of pathogens (e.g., Lyme arthritis)
infectious mononucleosis (in-FEK-shuss mah-noh-noo-klee-OH-sis) a viral (noncancerous) white blood cell (WBC) disorder common in young adults; characterized by leukocytosis of atypical lymphocytes and severe fatigue
inferior (in-FEER-ee-or) lower; opposite of **superior**
inferior vena cava (in-FEER-ee-or VEE-nah KAY-vah) one of two large veins carrying blood into the right atrium

infertility (in-fer-TIL-ih-tee) lower-than-normal ability to reproduce
inflammation (in-flah-MAY-shun) group of responses to a tissue irritant marked by signs of redness, heat, swelling, and pain
inflammation mediator (in-flah-MAY-shun MEE-dee-ay-tor) chemical (e.g., prostaglandins, histamine, kinins) released by irritated tissues that promotes the events of the inflammation response
inflammatory (in-FLAM-ah-toh-ree) relating to inflammation, an immune response that often produces heat, swelling, redness, and pain
inflammatory exudate (in-FLAM-ah-toh-ree EK-soo-dayt) fluid that accumulates in inflamed tissues as a result of increased permeability of blood vessels
inflammatory response (in-FLAM-ah-toh-ree ree-SPONS) innate (nonspecific) immune process produced in response to injury and resulting in redness, pain, heat, and swelling and promoting movement of white blood cells to the affected area
ingestion (in-JES-chun) taking in of complex foods, usually by mouth
inguinal (ING-gwih-nal) of the groin
inguinal hernia (ING-gwih-nal HER-nee-ah) protrusion of abdominopelvic organs through the inguinal canal and into the scrotum
inhalation (in-hah-LAY-shun) breathing in; opposite of **exhalation,** or **expiration;** also called **inspiration**
inherited immunity (in-HAYR-ih-ted ih-MYOO-nih-tee) *see* **inborn immunity**
inhibiting hormone (IH) (in-HIB-ih-ting HOR-mohn) hormone produced by the hypothalamus that slows the release of anterior pituitary hormones
innate immunity (in-AYT ih-MYOON-ih-tee) *see* nonspecific immunity
inorganic compound (in-or-GAN-ik KOM-pownd) compound whose molecules do not contain carbon-carbon or carbon-hydrogen bonds
INR acronym for international normalized ratio
insertion (in-SER-shun) attachment of a muscle to the bone that it moves when contraction occurs (as distinguished from its origin)
inspiration (in-spih-RAY-shun) moving air into the lungs; opposite of **exhalation** or **expiration;** also referred to as **inhalation**
inspiratory muscle (in-SPY-rah-tor-ee MUS-el) the muscles that increase the size of the thorax, including the diaphragm and external intercostals, and allow air to rush into the lungs
inspiratory reserve volume (IRV) (in-SPY-rah-tor-ee ree-ZERV VOL-yoom) the amount of air that can be forcibly inspired over and above a normal respiration
insulin (IN-suh-lin) hormone secreted by the pancreatic islets
integument (in-TEG-yoo-ment) the skin
integumentary system (in-teg-yoo-MEN-tar-ee SIS-tem) the body system comprising only the skin; the skin is an organ and a system
interarytenoid notch (IN-ter-ar-ih-tee-noyd notch) V-shaped groove between the arytenoid cartilages of the larynx often used as a guide for inserting a tube safely into the airway
intercalated disk (in-TER-kah-lay-ted disk) any of the gap-junction connections that form between cardiac muscle fibers, visible as thin dark bands in stained microscopic specimens
intercostal muscle (in-ter-KOS-tal MUS-el) the respiratory muscles located between the ribs
interferon (IF) (in-ter-FEER-on) small proteins produced by the immune system that inhibit virus multiplication
interleukins (IL) (in-ter-LOO-kins) any of several intracellular signals (cytokines) released by white blood cells (leukocytes), usually involved in immune responses
internal oblique muscle (in-TER-nal oh-BLEEK MUS-el) muscle forming part of the middle layer of the anterolateral abdominal walls
internal respiration (in-TER-nal res-pih-RAY-shun) the exchange of gases that occurs between the blood and cells of the body
international normalized ratio (INR) (in-ter-NASH-en-ul NOR-mah-lyzed RAY-shee-oh) method of expressing the **prothrombin time** (time it takes for a blood sample to clot after tissue thromboplastin [prothrombin activator] is added) based on international standards
interneuron (in-ter-NOO-ron) nerve that conducts impulses from a sensory neuron to a motor neuron
interphalangeal joint (in-ter-fah-LAN-jee-al joynt) articulation that exists between the heads of the phalanges and the bases of the more distal phalanges
interphase (IN-ter-fayz) the phase immediately before the visible stages of cell division when the DNA of each chromosome replicates itself
interstitial (in-ter-STISH-al) in between; often used to describe the space or substance between cells
interstitial cell (in-ter-STISH-al sel) endocrine cells in the testes that secrete the male sex hormone, testosterone
interstitial cell-stimulating hormone (ICSH) (in-ter-STISH-al sel STIM-yoo-lay-ting HOR-mohn) the previous name for luteinizing hormone in males; causes testes to develop and secrete testosterone
interstitial cystitis (in-ter-STISH-al sis-TYE-tis) *see* **overactive bladder**
interstitial fluid (IF) (in-ter-STISH-al FLOO-id) fluid located in the microscopic spaces between the cells
intestinal gland (in-TES-tih-nal) thousands of glands found in the mucous membrane of the mucosa of the small intestines; secrete intestinal digestive juices
intestine (in-TES-tin) the part of the digestive tract through which food remains pass after leaving the stomach; separated into two segments, the small and the large
intracellular fluid (ICF) (in-trah-SEL-yoo-lar FLOO-id) fluid located within the cells; largest fluid compartment
intramembranous ossification (in-trah-MEM-brah-nus os-ih-fih-KAY-shun) process by which most flat bones are formed within connective tissue membranes
intramuscular injection (IM) (in-trah-MUS-kyoo-lar in-JEK-shun) administration of medication into the muscle
intraocular pressure (in-trah-OK-yoo-lar PRESH-ur) fluid pressure within the eyeball
intravenous (IV) (in-trah-VEE-nus) within, or into, a vein
intrinsic factor (in-TRIN-sik FAK-ter) substance that binds to molecules of vitamin B_{12}, protecting them from the acids and enzymes of the stomach; secreted by parietal cells of gastric glands
inversion (in-VER-zhun) movement that turns the sole of the foot inward, toward the median
invert (in-VERT) to move a part inward
in vitro (in VEE-troh) occurring in a test tube, dish, or other laboratory apparatus
involuntary muscle (in-VOL-un-tayr-ee MUS-el) smooth muscle that is not under conscious control and is found in organs such as

the stomach and small intestine; cardiac muscle is also an involuntary type of muscle; *see* **smooth muscle** and **cardiac muscle**

involution (in-voh-LOO-shun) return of an organ to its normal size after an enlargement; also after retrograde or degenerative change

ion (AYE-on) electrically charged atom or group of atoms

ionic bond (aye-ON-ik) chemical bond formed by the positive-negative attraction between two ions

iris (AYE-ris) circular, pigmented ring of muscle tissue behind the cornea; the center of the iris is perforated by the pupil

iron deficiency anemia (AYE-ern deh-FISH-en-see ah-NEE-mee-ah) condition in which there are inadequate levels of iron in the diet causing less hemoglobin to be produced; results in extreme fatigue

ischemia (is-KEE-mee-ah) reduced flow of blood to tissue resulting in impairment of cell function

ischium (IS-kee-um) one of three separate bones that forms the os coxae

islet of Langerhans *see* **pancreatic islet**

isoimmunity (aye-soh-ih-MYOO-nih-tee) immune response to antigens of another human, as in transplanted (grafted) tissues; also called **alloimmunity**; in some cases it is called *rejection syndrome*

isometric contraction (aye-soh-MET-rik) type of muscle contraction in which muscle does not shorten

isotonic (aye-soh-TON-ik) relating to equal or uniform pressures or tension

isotonic contraction (aye-soh-TON-ik) type of muscle contraction that maintains uniform tension or pressure

isotope (AYE-soh-tohp) atom with the same atomic number as another atom but with a different atomic weight (that is, with a different number of neutrons in the nucleus of the atom)

IV (intravenous) technician (aye-vee [in-trah-VEE-nus] tek-NISH-en) health-care professional specializing in preparation and administration of therapeutic fluids and medicines into veins

J

jaundice (JAWN-dis) abnormal yellowing of skin, mucous membranes, and white of eyes

jejunum (jeh-JOO-num) the middle third of the small intestine

joint (joynt) *see* **articulation**

joint capsule (joynt CAP-sool) fibrous connective tissue sleeve, lined with synovial membrane, that holds together opposing ends of articulating bones in a synovial joint

joule (J or j) (jool) unit of measuring energy; *see* **calorie**

juvenile rheumatoid arthritis (JRA) (JOO-veh-naye-il ROO-mah-toyd ar-THRY-tis) form of rheumatoid arthritis affecting people under 16 years of age; it may affect bone development

juxtaglomerular (JG) apparatus (jux-tah-gloh-MER-yoo-lar ap-ah-RAT-us) complex of cells in nephron near the glomerulus and adjacent to distal tubule and afferent arteriole; secretes enzyme (renin) important in regulation of blood pressure

juxtamedullary nephron (jux-tah-MED-oo-lar-ee NEF-ron) nephron units with renal corpuscles located near the junction between cortex and medullary layers of kidney; *see also* **nephron**

K

Kaposi sarcoma (KS) (KAH-poh-see sar-KOH-mah) a malignant neoplasm (cancer) of the skin caused by the Kaposi sarcoma–related herpes virus (KSHV), or human herpes virus 8 (HHV8), and characterized by purplish spots on the skin; is mainly found in certain ethnic groups and in those with immune deficiencies

karyotype (KER-ee-oh-type) ordered arrangement of photographs of chromosomes from a single cell used in genetic counseling to identify chromosomal disorders such as trisomy or monosomy

keloid (KEE-loyd) an unusually thick, irregularly shaped, progressively enlarging fibrous scar on the skin

keratin (KAYR-ah-tin) protein substance found in hair, nails, outer skin cells, and horny tissues

ketoacidosis (kee-toh-as-ih-DOH-sis) a condition of abnormally low blood pH (acidity) caused by the presence of an abnormally large number of ketone bodies or "keto acids" that are produced when fats are converted to forms of glucose to be used for cellular respiration; often occurs in those with diabetes mellitus, when it is more specifically called *diabetic ketoacidosis; see also* **acidosis**

ketone body (KEE-tohn BOD-ee) acidic product of lipid metabolism; they may accumulate abnormally in blood of individuals with uncontrolled type 1 diabetes

kidney (KID-nee) organ that cleanses the blood of waste products produced continually by metabolism

kidney dialysis (KID-nee dye-AL-ih-sis) therapy for kidney failure in which machines pump blood through permeable tubes in an external apparatus, allowing waste products to diffuse out of the blood and into a salt-water type of electrolyte fluid that surrounds the semipermeable dialysis tubes.

kilocalorie (Kcal) (KIL-oh-kal-oh-ree) 1000 calories

kinesthesia (kin-es-THEE-zee-ah) "muscle sense"; that is, sense of position and movement of body parts

Klinefelter syndrome (KLINE-fel-ter SIN-drohm) genetic disorder caused by the presence of two or more X chromosomes in a male (typically trisomy XXY); characterized by long legs, enlarged breasts, low intelligence, small testes, sterility, chronic pulmonary disease

Krause end bulb (krows) mucous membrane receptor that detects sensations of touch and vibration; also known as **bulboid corpuscle**

Krebs cycle *see* **citric acid cycle**

Kupffer cell (KOOP-fer sel) phagocytic cell found in spaces between liver cells

kwashiorkor (kwah-shee-OR-kor) nutritional disorder that results from a protein deficiency in the presence of sufficient calories

kyphosis (kye-FOH-sis) abnormally exaggerated thoracic curvature of the vertebral column

L

labia majora (LAY-bee-ah mah-JOH-rah) "large lips" of the vulva

labia minora (LAY-bee-ah mih-NOH-rah) "small lips" of the vulva

labor (LAY-ber) the process that results in the birth of the baby

laboratory technician (LAB-rah-tor-ee tek-NISH-en) a trained assistant in a medical or scientific laboratory

lacrimal gland (LAK-rih-mal) gland that produces tears; one gland located in the upper lateral portion of each eye orbit

lacrimal sac (LAK-rih-mal sak) widened upper part of nasolacrimal duct that conducts tears from the lacrimal glands

lactase (LAK-tayse) enzyme that breaks down lactose

lacteal (LAK-tee-al) a lymphatic vessel located in each villus of the intestine; serves to absorb fat materials from the chyme passing through the small intestine

lactiferous duct (lak-TIF-er-us dukt) the duct that drains the grape-like cluster of milk-secreting glands in the breast

lactogenic hormone (lak-toh-JEN-ik HOR-mohn) *see* **prolactin**
lactose (LAK-tohs) disaccharide sugar found in milk; also called *milk sugar*
lactose intolerance (LAK-tohs in-TOL-er-ans) lack of the enzyme lactase, resulting in an inability to digest lactose (a disaccharide present in milk and dairy products)
lacuna (lah-KOO-nah) (*pl.*, lacunae) space or cavity; for example, lacunae in bone contain bone cells
lambdoidal suture (LAM-doyd-al SOO-chur) the immovable joint formed by the parietal and occipital bones
lamella (lah-MEL-ah) (*pl.*, lamellae) thin layer, as of bone
lamellar corpuscle (lah-MEL-ar KOR-pus-ul) sensory receptor with a layered encapsulation found deep in the dermis that detects pressure on the skin surface; also known as *Pacini corpuscle*
lamina propria (LAM-in-ah PROH-pree-ah) fibrous connective tissue underlying the epithelium in mucous membranes
lanugo (lah-NOO-go) the extremely fine and soft hair found on a newborn infant
laparoscope (LAP-ah-roh-skope) specialized optical viewing tube
large intestine (in-TES-tin) part of GI tract that includes cecum; ascending, transverse, descending and sigmoid colons; rectum; and anal canal
laryngeal cancer (lah-RIN-jee-al or lar-in-JEE-al KAN-ser) malignancy of the voicebox (larynx)
laryngitis (lar-in-JYE-tis) inflammation of the mucous tissues of the larynx (voice box)
laryngopharynx (lah-ring-goh-FAYR-inks) the lowest part of the pharynx
larynx (LAYR-inks) the voice box located just below the pharynx; the largest piece of cartilage making up the larynx is the thyroid cartilage, commonly known as the *Adam's apple*
laser-assisted in situ keratomileusis (LASIK) (LAY-zer ah-SIS-ted in SYE-too kayr-at-oh-mil-YOO-sis) refractory eye surgery using a microkeratome to cut a corneal cap, which is replaced after an excimer laser is used to vaporize and reshape underlying corneal tissue
laser therapy (LAY-zer THAYR-ah-pee) use of laser (intense beams of light) to destroy a tumor, abnormal tissue, damaged tissue, or scars
laser thermal keratoplasty (LTK) (LAY-zer THER-mull kayr-AT-oh-plast-ee) refractory eye surgery employing ultrashort bursts (3 seconds) of laser energy to reshape the cornea
lateral (LAT-er-al) of or toward the side; opposite of **medial**
lateral longitudinal arch (LAT-er-al lawnj-ih-TOOD-in-al) outer lengthwise (anteroposterior) support structure of the foot
latissimus dorsi (lah-TIS-ih-mus DOR-sye) an extensor of the arm
law a scientific law is a theory, or explanation, of a scientific principle that is based on experimentation results and supported by scientists who have an extraordinarily high degree of confidence in its validity
Leber hereditary optic neuropathy (LEE-ber heh-RED-ih-tayr-ee OP-tik noo-ROP-ah-thee) inherited condition in which young adults begin losing their eyesight as the optic nerve degenerates—resulting in total blindness by age 30
lens (lenz) the refracting mechanism of the eye that is located directly behind the pupil
leptin (LEP-tin) hormone, secreted by fat-storing cells, that regulates how hungry or full we feel and how fat is metabolized by the body
lesion (LEE-zhun) any objective abnormality in a body structure
lesser vestibular gland (LES-er ves-TIB-yoo-lar gland) either of the exocrine mucous glands located on the sides of the urinary outlet in women; also known as *Skene gland; see also* **greater vestibular gland**
leukemia (loo-KEE-mee-ah) blood cancer characterized by an increase in white blood cells
leukocyte (LOO-koh-syte) white blood cells
leukocytosis (loo-koh-SYE-toh-sis) abnormally high white blood cell numbers in the blood
leukopenia (loo-koh-PEE-nee-ah) abnormally low white blood cell numbers in the blood
leukoplakia (loo-koh-PLAY-kee-ah) white patches in the mouth, commonly seen in chronic cigarette smokers; may lead to mouth cancer
leukorrhea (loo-koh-REE-ah) whitish discharge from the urogenital tract
leukotriene (loo-koh-TRY-een) cytokine compound that functions as an inflammation mediator
levels of organization (LEV-elz ov or-gan-ih-ZAY-shun) groupings of structural components from microscopic to gross, used as a manner of organizing concepts of biological scale
levodopa (LEV-oh-doh-pah) chemical manufactured by the brain cells and then converted into the neurotransmitter dopamine; has been used to treat disorders involving dopamine deficiencies such as Parkinson disease; also called *L-dopa*
lifestyle (LYFE-style) the mode of living of a person, including eating habits, activity, and choice of environment, which may not be completely voluntary
ligament (LIG-ah-ment) bond or band connecting two objects; in anatomy, a band of white fibrous tissue connecting bones
limbic system (LIM-bik) a collection of various small regions of the brain that act together to produce emotion and emotional response; sometimes called "the emotional brain"
linear fracture (LIN-ee-ar FRAK-chur) bone fracture characterized by a fracture line that is parallel to the bone's long axis
lingual tonsil (LING-gwal TAHN-sil) tonsil located at the base of the tongue
lipase (LYE-payse) fat-digesting enzymes
lipid (LIP-id) organic molecule usually composed of glycerol and fatty acid units; types include triglycerides, phospholipids, and cholesterol; a fat, wax, or oil
lipoma (lih-POH-mah) benign tumor of adipose (fat) tissue
lipoprotein (lip-oh-PROH-teen) substance that is part lipid and part protein; produced mainly in the liver
lithotripsy (lih-thoh-TRIP-see) use of ultrasound waves to break up kidney stones without making an incision
lithotriptor (LITH-oh-trip-tor) a specialized ultrasound generator that is used to pulverize kidney stones
liver (LIV-er) large, multilobed exocrine gland in the right upper abdominal quadrant, producing bile and having many metabolic functions
liver glycogenolysis (LIV-er glye-koh-jeh-NOL-ih-sis) chemical process by which liver glycogen is converted to glucose
lobectomy (loh-BEK-toh-mee) surgical removal of a single lobe of an organ, as in the removal of one lobe of a lung
lock-and-key model (lok and kee MAHD-el) concept that explains how molecules react when they fit together in a complementary way in the same manner that a key fits into a lock to cause the lock to open or close; the analogy is often used to explain the action of hormones, enzymes, and other biological molecules
longitudinal arch (lon-jih-TOO-dih-nal) two arches, the medial and lateral, that extend lengthwise in the foot
loop of Henle (loop of HEN-lee) *see* **Henle loop**
loose fibrous connective tissue *see* **areolar tissue**

lordosis (lor-DOH-sis) abnormally exaggerated lumbar curvature of the vertebral column; may also refer to normal concavity of the lumbar curvature

lower esophageal sphincter (LES) (LOH-er eh-sof-eh-JEE-al SFINGK-ter) ring of muscular tissue (sphincter) located between terminal esophagus and stomach

lumbar (LUM-bar) lower back, between the ribs and pelvis

lumbar puncture (LUM-bar PUNK-chur) when some cerebrospinal fluid is withdrawn from the subarachnoid space in the lumbar region of the spinal cord

lumbar region (LUM-bar REE-jun) the left and right middle regions of the abdominopelvic cavity, near the lumbar area of the vertebral column and on either side of the umbilical region; terminology used when the abdominopelvic cavity is visualized as being subdivided into nine regions as in a tic-tac-toe grid

lumen (LOO-men) (*pl.*, lumina or lumens) the hollow space within a tube

lung organ of respiration; the right lung has three lobes and the left lung has two lobes

lunula (LOO-nyoo-lah) crescent-shaped white area under the proximal nail bed

luteinization (loo-tee-in-ih-ZAY-shun) the process of development of a corpus luteum (golden body) in the ovary after an ovum is released from the follicle; stimulated by the action of luteinizing hormone (LH) from the anterior pituitary

luteinizing hormone (LH) (loo-tee-in-AYE-zing HOR-mohn) anterior pituitary hormone that stimulates the development of a corpus luteum (literally "yellow body") that secretes hormones at the surface of the ovary after a follicle has released its ovum; a tropic hormone also known as *LH*

Lyme arthritis (lyme ar-THRY-tis) infectious form of joint inflammation (arthritis) associated with Lyme disease; caused by a spirochete bacterium carried by deer ticks

lymph (limf) specialized fluid formed in the tissue spaces that returns excess fluid and protein molecules to the blood via lymphatic vessels

lymph node (limf) performs biological filtration of lymph on its way to the cardiovascular system

lymph nodule (limf NOD-yool) is a mass of lymphoid tissue (developing white blood cells) within a lymph node or making up a patch of lymph nodules (as in the tonsils)

lymphadenitis (lim-FAD-in-aye-tis) inflammation of a lymph node, usually caused by a bacterial infection or occasionally a neoplasm (benign or cancerous), and characterized by swelling and tenderness

lymphangiogram (lim-FAN-jee-oh-gram) radiograph (x-ray) of a part of the lymphatic network, which is produced by injecting a special dye that is opaque to x-rays into the soft tissues drained by the lymphatic network

lymphangitis (lim-fan-JYE-tis) inflammation of lymph vessels, usually caused by infection, characterized by fine red streaks extending from the site of infection; may progress to septicemia (blood infection)

lymphatic capillary (lim-FAT-ik CAP-ih-layr-ee) any of the tiny, blind-ended tubes for draining excess interstitial fluid distributed in the tissue spaces

lymphatic duct (lim-FAT-ik dukt) terminal vessel into which lymphatic vessels empty lymph; the duct then empties the lymph into the cardiovascular system

lymphatic system (lim-FAT-ik SIS-tem) a system that plays a critical role in the functioning of the immune system, moves fluids and large molecules from the tissue spaces and fat-related nutrients from the digestive system to the blood

lymphatic tissue (lim-FAT-ik TISH-yoo) *see* **lymphoid tissue**

lymphatic vessel (lim-FAT-ik VES-el) vessel that carries lymph to its eventual return to the cardiovascular system

lymphedema (lim-fah-DEE-mah) swelling (edema) of tissues caused by partial or total blockage of the lymph vessels that drain the affected tissue

lymphocyte (LIM-foh-syte) one type of white blood cell

lymphoid neoplasm (LIM-foyd NEE-oh-plaz-em) abnormal proliferation of lymphoid tissue or lymphoid precursor cells often associated with cancerous transformation

lymphoid tissue (LIM-foyd TISH-yoo) tissue that is responsible for manufacturing lymphocytes and monocytes; found mostly in the lymph nodes, thymus, and spleen

lymphoma (lim-FOH-mah) cancer of lymphoid tissue

lyse (lyze) disintegration of a cell

lysosome (LYE-soh-sohm) membranous organelles containing various enzymes that can dissolve most cellular compounds; thus called *digestive bags* or *suicide bags* of cells

M

macronutrient (MAK-roh-NOO-tree-ent) nutrient needed in large amounts; carbohydrates, fats, and proteins

macrophage (MAK-roh-fayj) phagocytic cells in the immune system

macula (MAK-yoo-lah) (*pl.*, maculae or maculas) strip of sensory epithelium in the utricle and saccule; provides information related to head position or acceleration [*macula* spot]

macula lutea (MAK-yoo-lah LOO-tee-ah) (*pl.*, maculae luteae) yellowish area near center of retina filled with cones permitting acute image formation and color vision

macular degeneration *see* **age-related macular degeneration (AMD)**

macule (MAK-yool) a flat skin lesion distinguished from the surrounding tissue by a difference in coloration

mad cow disease *see* **bovine spongiform encephalopathy**

magnetic resonance imaging (MRI) (mag-NET-ik REZ-ah-nans IM-ah-jing) scanning technique that uses a magnetic field to induce tissues to emit radio waves that can be used by computer to construct a sectional view of a patient's body

major duodenal papilla (MAY-jer doo-oh-DEE-nul [or doo-AH-de-nul] pah-PIL-ah) muscular bump in lining of duodenum where common bile duct enters; also called *greater duodenal papilla*

malabsorption syndrome (mal-ab-SORP-shun SIN-drohm) group of symptoms associated with the failure to absorb nutrients properly: anorexia, ascites, cramps, anemia, fatigue

maldigestion (mal-dye-JES-chun) failure to fully digest nutrients in the gut

malignant (mah-LIG-nant) referring to something harmful

malignant hyperthermia (MH) (mah-LIG-nant hye-per-THERM-ee-ah) inherited condition characterized by an abnormally increased body temperature (hyperthermia) and muscle rigidity when a person is exposed to certain anesthetics (e.g., succinylcholine)

malignant tumor (mah-LIG-nant TOO-mer) a tumor or neoplasm that is capable of metastasizing or spreading to new tissues (i.e., cancer)

malleus (MAL-ee-us) hammer; the tiny middle ear bone that is shaped like a hammer

malnutrition (mal-noo-TRISH-un) insufficient or imbalanced intake of nutrients, often causing any of a variety of diseases
malocclusion (mal-oh-CLEW-zhun) abnormal contact between the teeth of the upper and lower jaw
maltase (MAWL-tayz) enzyme that breaks apart maltose and thereby catalyzes the final steps of carbohydrate digestion
maltose (MAWL-tohs) disaccharide sugar formed by the breakdown of starch
mammary (MAM-mah-ree) relating to the breasts or milk-producing glands within the breasts
mammary dysplasia (MAM-mah-ree dis-PLAY-zhah) group of conditions characterized by benign lumps in one or both breasts; also called **fibrocystic disease**
mammary gland (MAM-mah-ree) milk-producing exocrine gland of the breasts; functionally classified as external accessory sex organ in females but is structurally part of the integumentary system
marasmus (mah-RAZ-mus) form of protein-calorie malnutrition; results from an overall lack of calories and protein
massage therapy (mah-SAHJ THAYR-ah-pee) pressing, rubbing, or other manipulation of muscle and other soft tissue to prevent or treat a variety of health conditions
masseter (mah-SEE-ter) large muscle of the cheek, used to lift the lower jaw (mandible) and thus provide chewing movement
mast cell immune system cell (related to the basophil) that secretes histamine and other inflammatory chemicals
mastectomy (mas-TEK-toh-mee) surgical removal of the breast
mastication (mas-tih-KAY-shun) chewing
mastitis (mas-TYE-tis) inflammation or infection of the breast
mastoiditis (mas-toyd-AYE-tis) inflammation of the air cells within the mastoid portion of the temporal bone; usually caused by infection
matrix (MAY-triks) the intracellular substance of a tissue; for example, the matrix of bone is calcified, whereas that of blood is liquid
matter any substance that occupies space and has mass
mature follicle (mah-CHUR FOL-lih-kul) *see* **graafian follicle**
maxilla (mak-SIH-lah) upper jaw bone
maximum oxygen consumption (VO_{2max}) (MAX-ih-mum OKS-ih-jen kon-SUMP-shun) the maximum amount of oxygen taken up by the lungs, transported to the tissues, and used to do work
mechanoreceptor (mek-an-oh-ree-SEP-tor) receptors that are mechanical in nature; for example, equilibrium and balance sensors in the ears
medial (MEE-dee-al) of or toward the middle; opposite of **lateral**
medial longitudinal arch (MEE-dee-al lon-jih-TOO-dih-nal) inner lengthwise (anteroposterior) support structure of the foot
mediastinum (MEE-dee-as-TYE-num) a subdivision in the midportion of the thoracic cavity
medic (MED-ik) member of a military medical corps
medicine (MED-ih-sin) practice of applying scientific principles to the prevention and treatment of health conditions
medulla (meh-DUL-ah) Latin for "marrow"; the inner portion of an organ in contrast to the outer portion or cortex
medulla oblongata (meh-DUL-ah ob-long-GAH-tah) the lowest part of the brainstem; an enlarged extension of the spinal cord; the vital centers are located within this area
medullary cavity (med-OO-layr-ee KAV-ih-tee) hollow area inside the diaphysis of the bone that contains yellow bone marrow
meibomian gland (my-BOH-mee-an gland) any of the small sebaceous glands along the edge (tarsus) of the eyelid; can become infected, resulting in a *sty*
meiosis (my-OH-sis) nuclear division in which the number of chromosomes are reduced to half their original number; produces gametes
Meissner corpuscle (MYZ-ner KOR-pus-ul) a sensory receptor located in the skin close to the surface that detects light touch; also known as **tactile corpuscle**
melanin (MEL-ah-nin) brown skin pigment
melanocyte (MEL-ah-noh-syte) specialized cells in the pigment layer that produce melanin
melanocyte-stimulating hormone (MSH) (MEL-ah-noh-syte STIM-yoo-lay-ting HOR-mohn) responsible for a rapid increase in the synthesis and dispersion of melanin granules in specialized skin cells
melanoma (mel-ah-NOH-mah) a malignant neoplasm (cancer) of the pigment-producing cells of the skin (melanocytes); also called *malignant melanoma*
melatonin (mel-ah-TOH-nin) important hormone produced by the pineal gland; believed to regulate the onset of puberty and the menstrual cycle; also referred to as the *third eye* because it responds to levels of light and is thought to be involved with the body's internal clock
membrane (MEM-brane) thin layer or sheet
membranous labyrinth (MEM-brah-nus LAB-eh-rinth) a membranous sac that follows the shape of the bony labyrinth and is filled with endolymph
memory cell (MEM-oh-ree sel) cell that remains in reserve in the lymph nodes until its ability to secrete antibodies is needed
menarche (meh-NAR-kee) beginnings of the menstrual function
Ménière disease (men-ee-AYR dih-ZEEZ) chronic inner ear disorder characterized by tinnitus, progressive nerve deafness, and vertigo
meninges (meh-NIN-jeez) (*sing.*, meninx) fluid-containing membranes surrounding the brain and spinal cord
meningitis (men-in-JYE-tis) inflammation of the meninges caused by a variety of factors including bacterial infection, mycosis, viral infection, and tumors
meniscus (meh-NIS-kus) (*pl.*, menisci) articular cartilage disk
menopause (MEN-oh-pawz) termination of menstrual cycles
menses (MEN-seez) menstrual flow
menstrual cramp (MEN-stroo-al kramp) painful contraction of the uterine muscle during menstruation
menstrual cycle (MEN-stroo-al SYE-kul) the cyclical changes in the uterine lining
mesentery (MEZ-en-tayr-ee) a large double fold of peritoneal tissue that anchors the loops of the digestive tract to the posterior wall of the abdominal cavity
mesoderm (MEZ-oh-derm) the middle layer of the primary germ layers
messenger RNA (mRNA) (MES-en-jer R N A) a duplicate copy of a gene sequence on the DNA that passes from the nucleus to the cytoplasm
metabolic (met-ah-BOL-ik) related to metabolism, the chemical reactions of the body
metabolic acidosis (met-ah-BOL-ik as-ih-DOH-sis) a disturbance affecting the bicarbonate element of the bicarbonate–carbonic acid buffer pair; bicarbonate deficit
metabolic alkalosis (met-ah-BOL-ik al-kah-LOH-sis) disturbance affecting the bicarbonate element of the bicarbonate–carbonic acid buffer pair; bicarbonate excess
metabolism (meh-TAB-oh-liz-em) complex process by which nutrients are used by a living organism

metacarpal (met-ah-KAR-pal) the part of the hand between the wrist and fingers
metallic (meh-TAL-ik) relating to metal, as in *metallic taste*
metaphase (MET-ah-fayz) second stage of mitosis, during which the nuclear membrane and nucleolus disappear
metastasis (meh-TAS-tah-sis) process by which malignant tumor cells fall off a primary tumor, then migrate to a new tissue to colonize a secondary tumor
metatarsal arch (met-ah-TAR-sal) the arch that extends across the ball of the foot; also called the *transverse arch*
metatarsals (met-ah-TAR-salz) any of the five bones that form the foot; articulate with tarsal bones proximally and the first row of toe phalanges distally
metazoan (met-ah-ZOH-an) (*pl.*, metazoa) animals (large multicellular organisms) that can sometimes cause or transmit disease
meter (MEE-ter) a measure of length in the metric system; equal to about 39.5 inches
methylation (meth-il-AY-shun) chemical process in which a methyl group (CH_3) is added to a molecule, as in adding methyl to DNA to regulate gene activity
microbe (MY-krobe) any microscopic organism
microbiologist (my-kroh-bye-OL-uh-jist) scientist specializing in the study of microorganisms such as bacteria
microbiome (my-kroh-BYE-ohm) all the interacting ecosystems of microbes (bacteria, fungi, etc.) that live on or in the human body; also called the *human microbiome* or *human microbial system*
microcephaly (my-kroh-SEF-ah-lee) a congenital abnormality in which an infant is born with a small head
microglia (my-KROG-lee-ah) one type of connective tissue found in the brain and spinal cord
micron (MY-kron) measurement that equals 1/1000 millimeter; 1/25,000 inch
micronutrient (MY-kroh-NOO-tree-ent) nutrient needed by the body in very small quantity, such as vitamins and minerals
microtubule (my-kroh-TOOB-yool) thick cell fiber (compared to microfilament); hollow tube responsible for movement of substances within the cell or movement of the cell itself
microvilli (my-kroh-VIL-ee) brushlike border made up of epithelial cells found on each villus in the small intestine and other areas of the body; increases the surface area (as for absorption of nutrients)
micturition (mik-too-RISH-un) urination, voiding (of bladder)
midbrain (MID-brayn) one of the three parts of the brainstem
middle ear (MID-ul eer) a tiny and very thin epithelium-lined cavity in the temporal bone that houses the ossicles; in the middle ear, sound waves are amplified
midsagittal plane (mid-SAJ-ih-tal) a cut or plane that divides the body or any of its parts into two equal halves
mineral (MIN-er-al) inorganic element or salt found naturally in the earth that may be vital to the proper functioning of the body
mineralocorticoid (MC) (min-er-al-oh-KOR-tih-koyd) hormone that influences mineral salt metabolism; secreted by adrenal cortex; aldosterone is the chief mineralocorticoid
minor duodenal papilla (MYE-ner doo-oh-DEE-nul [or doo-AH-de-nul] pah-PIL-ah) small muscular bump in lining of duodenum where the accessory pancreatic duct enters
mitochondria (my-toh-KON-dree-ah) plural form of **mitochondrion**
mitochondrial DNA (my-toh-KON-dree-al D N A) DNA located in the mitochondria of each cell, constituting a single chromosome; also called *mtDNA* or *mDNA*
mitochondrion (my-toh-KON-dree-on) organelle in which ATP generation occurs; often termed "powerhouse of cell"
mitosis (my-TOH-sis) indirect cell division involving complex changes in the nucleus
mitral valve (MY-tral valv) heart valve located between the left atrium and ventricle; also known as the **bicuspid valve**
mitral valve prolapse (MVP) (MY-tral valv PROH-laps) condition in which the bicuspid (mitral) valve extends into the left atrium, causing incompetence (leaking) of the valve
mode (mohd) category of sensation detected by a sensory receptor; also called *modality*
molar *see* **tricuspid**
mold large fungus (compared to a yeast, which is a small fungus)
molecule (MOL-eh-kyool) particle of matter composed of one or more smaller units called **atoms**
monoclonal antibody (mon-oh-KLONE-al AN-tih-bod-ee) specific antibody produced from a population of identical cells
monocyte (MON-oh-syte) largest type of white blood cell; a type of agranulocyte; often involved in phagocytosis of abnormal cells or particles
mononucleosis (MAHN-oh-NOO-klee-OH-sis) condition characterized by an increase in the number of mononuclear leukocytes; can be caused by the Epstein-Barr virus (EBV); also commonly called "*mono*"
monosaccharide (mon-oh-SAK-ah-ryde) a "simple sugar" composed of only a single saccharide group ($C_6H_{12}O_6$); examples include glucose, fructose, galactose
monosomy (MON-oh-soh-mee) abnormal genetic condition in which cells have only one chromosome where there should be a pair; usually caused by nondisjunction (failure of chromosome pairs to separate) during gamete production
monozygotic twins (mahn-oh-zye-GOT-ik twinz) twins that develop from a single zygote that has split during early development into two separate, but genetically identical, offspring; *see also* **identical twins**
mons pubis (monz PYOO-bis) skin-covered pad of fat over the symphysis pubis in the female
morbidity (mor-BID-ih-tee) illness or disease; the rate of incidence of a specific illness or disease in a specific population
mortality (mor-TAL-ih-tee) death; the rate of deaths caused by a specific disease within a specific population
morula (MOR-yoo-lah) a solid mass of cells formed by the divisions of a fertilized egg
motility (moh-TIL-ih-tee) ability to move
motor neuron (MOH-ter NOO-ron) neuron that transmits nerve impulses from the brain and spinal cord to muscles and glandular epithelial tissues
motor unit (MOH-ter YOO-nit) a single motor neuron along with the muscle cells it innervates
mouth (mowth) oral cavity
mucocutaneous junction (myoo-koh-kyoo-TAY-nee-us JUNK-shun) the transitional area where the skin and mucous membrane meet
mucosa (myoo-KOH-sah) mucous membrane
mucous membrane (MYOO-kus MEM-brane) epithelial membranes that line body surfaces opening directly to the exterior and secrete a thick, slippery material called **mucus**
mucus (MYOO-kus) thick, slippery material that is secreted by the mucous membrane and keeps the membrane moist
multiple myeloma (MUL-tih-pul my-LOH-mah) cancer of plasma cells
multiple neurofibromatosis (MUL-tih-pul noo-roh-fye-broh-mah-TOH-sis) disorder characterized by multiple, sometimes disfiguring, benign tumors of the Schwann cells (neuroglia) that surround nerve fibers

multiple sclerosis (MS) (MUL-tih-pul skleh-ROH-sis) the most common primary disease of the central nervous system; a myelin disorder
muscle fiber (MUS-el FYE-ber) the specialized contractile cells of muscle tissue that are grouped together and arranged in a highly organized way
muscle strain (MUS-el strayn) muscle injury resulting from overexertion or trauma and involving overstretching or tearing of muscle fibers
muscle tone (MUS-el tohn) tonic contraction; characteristic of muscle of a normal individual who is awake
muscular dystrophy (MUS-kyoo-lar DIS-troh-fee) a group of muscle disorders characterized by atrophy of skeletal muscle without nerve involvement; Duchenne muscular dystrophy (DMD) is the most common type
muscular system (MUS-kyoo-lar SIS-tem) the muscles of the body
muscularis (mus-kyoo-LAYR-is) two layers of muscle surrounding the digestive tube that produce wavelike, rhythmic contractions called **peristalsis**, which move food material
musculotendinous unit (mus-kyoo-loh-TEN-din-us YOO-nit) the functional unit formed by a skeletal muscle's muscle tissue, tendon, and the junction between the two tissues
mutagen (MYOO-tah-jen) agent capable of causing mutation (alteration) of DNA
myalgia (my-AL-jee-ah) general term referring to the symptom of pain in muscle tissue
myasthenia gravis (my-es-THEE-nee-ah GRAH-vis) autoimmune muscle disorder characterized by progressive weakness and chronic fatigue
mycotic infection (my-KOT-ik in-FEK-shun) fungal infection
myelin (MY-eh-lin) lipoid substance found in the myelin sheath around some nerve fibers
myelinated fiber (MY-eh-lih-nay-ted FYE-ber) axons outside the central nervous system that are surrounded by a segmented wrapping of myelin
myeloid (MY-eh-loyd) relating to bone marrow
myeloid neoplasm (MY-eh-loyd NEE-oh-plaz-em) abnormal proliferation of myeloid tissue or myeloid precursor cells often associated with cancerous transformation
myeloid tissue (MY-eh-loyd TISH-yoo) tissue that makes up bone marrow
myeloma (my-eh-LOH-mah) malignant tumor of bone marrow
myocardial infarction (MI) (my-oh-KAR-dee-al in-FARK-shun) death of cardiac muscle cells resulting from inadequate blood supply, as in coronary thrombosis
myocardium (my-oh-KAR-dee-um) muscle of the heart
myofilament (my-oh-FIL-ah-ment) any of the ultramicroscopic, threadlike structures found in myofibrils; two types: *thick* and *thin*
myoglobin (my-oh-GLOH-bin) a red, oxygen-storing protein pigment similar to hemoglobin found in muscle fibers
myoma (my-OH-mah) benign tumor of smooth muscle commonly occurring in the uterine wall; *see also* **fibromyoma**
myometrium (my-oh-MEE-tree-um) muscle layer in the uterus
myopathy (my-OP-ah-thee) general term referring to any muscle disease
myopia (my-OH-pee-ah) refractive disorder of the eye caused by an elongated eyeball; nearsightedness
myosin (MY-oh-sin) contractile protein found in the thick myofilaments of skeletal muscle
myositis (my-oh-SYE-tis) general term referring to muscle inflammation, as in infection or injury
myxedema (mik-seh-DEE-mah) condition caused by deficiency of thyroid hormone in adults

N

nail body (BOD-ee) the visible part of the nail
nail root the part of the nail hidden by the cuticle
nanometer (NAN-oh-mee-ter) a measure of length in the metric system; one billionth of a meter
nares (NAY-reez) (*sing.*, naris) nostrils
nasal (NAY-zal) relating to the nose
nasal cavity (NAY-zal KAV-ih-tee) the moist, warm cavities lined by mucosa located just beyond the nostrils; olfactory receptors are located in the mucosa
nasal polyp (NAY-zal PAH-lip) painless, noncancerous tissue growth that projects from nasal mucosa
nasal septum (NAY-zal SEP-tum) a partition that separates the right and left nasal cavities
nasopharynx (nay-zoh-FAYR-inks) the uppermost portion of the tube just behind the nasal cavities
natural killer cell (NK cell) (NACH-er-ul KIL-er sel) type of lymphocyte that kills many types of tumor cells
nausea (NAW-zee-ah) unpleasant sensation of the gastrointestinal tract that commonly precedes the urge to vomit; upset stomach
neck (nek) the body region connecting head to thorax; the narrow, connecting part of a structure, as in the region of the tooth that joins crown to root
necrosis (neh-KROH-sis) death of cells in a tissue, often resulting from ischemia (reduced blood flow)
needle biopsy (NEE-dil BYE-op-see) type of biopsy in which a specimen is withdrawn from the body through a hollow needle; *see* **biopsy**
negative feedback (NEG-ah-tiv FEED-bak) homeostatic control system in which information feeding back to the control center causes the level of a variable to be changed in the direction opposite to that of the initial stimulus
nematode (NEM-ah-tohd) roundworms—large parasites capable of infesting humans
neonatal period (nee-oh-NAY-tal PEER-ee-id) stage of early human development that corresponds to approximately the first 4 weeks after birth
neonate (NEE-oh-nayt) another name for an infant during the first 4 weeks after birth; *see* **neonatal period**
neonatology (nee-oh-nay-TOL-oh-jee) diagnosis and treatment of disorders of the newborn infant
neoplasm (NEE-oh-plaz-em) an abnormal mass of proliferating cells that may be either benign or malignant; a tumor
neoplastic (nee-oh-PLAS-tik) relating to tumors (neoplasms)
nephritis (neh-FRY-tis) general term referring to inflammatory or infectious conditions of renal (kidney) tissue
nephron (NEF-ron) anatomical and functional unit of the kidney, consisting of the renal corpuscle and the renal tubule
nephron loop (NEF-ron loop) extension of the proximal tubule of the kidney; also known as *loop of Henle* or *Henle loop*
nephropathy (neh-FROP-ah-thee) kidney disease
nephrotic syndrome (neh-FROT-ik SIN-drohm) group of symptoms and signs that often accompany glomerular disorders of the kidney: proteinuria, hypoalbuminemia, and edema
nerve (nerv) collection of nerve fibers
nerve impulse (nerv IM-puls) signals that carry information along the nerves

nervous system (NER-vus SIS-tem) body system made up of the brain, spinal cord, and nerves
nervous tissue (NER-vus TISH-yoo) consists of neurons and neuroglia and provides rapid communication and control of body function
neuralgia (noo-RAL-jee-ah) general term referring to nerve pain
neurilemma (noo-rih-LEM-mah) nerve sheath
neuritis (noo-RYE-tis) general term referring to nerve inflammation
neuroblastoma (noo-roh-blas-TOH-mah) malignant tumor of sympathetic nervous tissue, found mainly in young children
neurogenic bladder (noo-roh-JEN-ik BLAD-der) condition in which the nervous control of the urinary bladder is impaired, causing abnormal or obstructed flow of urine from the body
neurogenic shock (noo-roh-JEN-ik shok) circulatory failure (shock) caused by a nerve condition that relaxes (dilates) blood vessels and thus reduces blood flow; literally "nerve-caused" shock
neuroglia (noo-ROH-glee-ah) supporting cells of nervous tissue; also called simply *glia*
neurohypophysis (noo-roh-hye-POF-ih-sis) posterior pituitary gland
neurologist (noo-ROL-uh-jist) physician specializing in the treatment of nervous system disorders
neuroma (noo-ROH-mah) general term for nervous tissue tumors
neuromuscular junction (NMJ) (noo-roh-MUS-kyoo-lar JUNK-shun) the point of contact between the nerve endings and muscle fibers
neuron (NOO-ron) nerve cell, including its processes (axons and dendrites)
neuroscientist (noo-roh-SYE-en-tist) scientist specializing in research concerning the structure and function of the nervous system
neurotransmitter (noo-roh-trans-MIT-ter) chemicals by which neurons communicate
neutral (NOO-truhl) 1. relating to a solution having a pH of 7, being neither acid nor base; 2. having no electrical charge
neutron (NOO-tron) electrically neutral particle within the nucleus of an atom
neutrophil (NOO-troh-fil) white blood cell that stains readily with neutral dyes
nevus (NEE-vus) (*pl.,* nevi) small, pigmented benign tumor of the skin (e.g., a mole)
nitric oxide (NO) (NYE-trik AWK-side) compound composed of one nitrogen and one oxygen atom in each molecule, often acting as a small-molecule neurotransmitter
nitrogen (NYE-troh-jen) one of the chemical elements found in great quantity in the human body, especially in nucleic acids (DNA, RNA), proteins, and amino acids; symbolized by N, as in NH_3 (ammonia)
nitroglycerin (nye-troh-GLIS-eh-rin) heart medication that dilates coronary blood vessels thus improving supply of oxygen to myocardium
Nobel prize (noh-BEL) international award created by the late Alfred Nobel and awarded each year to up to three recipients in each of several categories such as chemistry, physics, and medicine or physiology (each Nobel laureate [prizewinner] receives a diploma, a medal, and a cash prize at a ceremony in Stockholm, Sweden)
nodes of Ranvier (rahn-vee-AY) indentations found between adjacent Schwann cells
nodule (NOD-yool) *see* **lymph nodule**
nondisjunction (non-dis-JUNK-shun) occurs during meiosis when a pair of chromosomes fails to separate
nonelectrolyte (non-ee-LEK-troh-lyte) compound that does not dissociate into ions in solution; for example, glucose
non-Hodgkin lymphoma (non-HOJ-kin lim-FOH-mah) type of lymphoma (malignant lymph tumor) characterized by swelling of lymph nodes and progressing to other areas
nonspecific immunity (non-spih-SIH-fik ih-MYOON-ih-tee) the protective mechanisms that provide immediate, generic protection against any bacteria, toxin, or other injurious particle; also called **innate immunity**
nonsteroid hormone (non-STAYR-oyd HOR-mohn) general type of hormone that does not have the lipid steroid structure (derived from cholesterol) but is instead a protein or protein derivative; also sometimes called **protein hormone**
norepinephrine (NE) (nor-ep-ih-NEF-rin) hormone secreted by adrenal medulla; released by sympathetic division; also known as *noradrenaline*
normal saline (SAY-leen) sodium chloride solution isotonic with body fluids
nose cartilaginous respiratory organ of the face
nosocomial infection (noh-zoh-KOHM-ee-al in-FEK-shun) infection that begins in the hospital or clinic
NSAID (EN-sayd) acronym for nonsteroidal anti-inflammatory drug, the term is applied to aspirin, ibuprofen, acetaminophen, and many other anti-inflammatory agents that do not contain steroid hormones or their derivatives
nuclear envelope (NOO-klee-ar AHN-vel-ohp) the boundary of a cell's nucleus, made up of a double layer of cellular membrane
nuclear medicine technologist (NOO-klee-ar MED-ih-sin tek-NOL-uh-jist) medical professional who prepares and administers radioactive drugs or other substances
nuclear membrane (NOO-klee-ar MEM-brane) membrane that surrounds the cell nucleus
nucleic acids (noo-KLAY-ik AS-ids) the two nucleic acids are ribonucleic acid (RNA), found in the cytoplasm, and deoxyribonucleic acid (DNA), found in the nucleus and mitochondrion; made up of units called **nucleotides** that each include a phosphate, a five-carbon sugar, and a nitrogen base
nucleolus (noo-KLEE-oh-lus) intercellular structure critical to protein formation because it "programs" the formation of ribosomes in the nucleus
nucleoplasm (NOO-klee-oh-plaz-im) a special type of cytoplasm found in the nucleus
nucleotide (NOO-klee-oh-tyde) molecule that connects to other nucleotides to form a nucleic acid such as DNA or RNA; each nucleotide has three parts: a phosphate group, a sugar (ribose or deoxyribose), and a nitrogenous base (adenine, thymine [or uracil], guanine, or cytosine)
nucleus (NOO-klee-us) spherical structure within a cell; a group of neuron cell bodies in the brain or spinal cord; central core of the atom, made up of protons and (sometimes) neutrons
nurse (nurs) health-care professional trained to care for the sick and injured
nursing assistant (NURS-ing ah-SIS-tent) health-care worker under the supervision of a nurse to care for patients **nutrition** (noo-TRIH-shun) food (nutrients), vitamins, and minerals that are ingested and assimilated into the body
nutritionist (noo-TRISH-en-ist) professional consultant specializing in diet and food
nyctalopia (nik-tah-LOH-pee-ah) condition caused by retinal degeneration or avitaminosis A and characterized by the relative inability to see in dim light; night blindness

O

obesity (oh-BEES-ih-tee) condition characterized by abnormally high proportion of body fat
oblique fracture (oh-BLEEK FRAK-chur) bone fracture characterized by a fracture line that is diagonal to the long axis of the broken bone
oblique plane (oh-BLEEK playn) imagined flat plane that runs diagonally to an axis of the body or one of its parts, producing a slanted, oblique section or cut
obstetric nurse (ob-STET-rik nurs) nurse specializing in pregnancy, labor, and delivery care
obstetrician (ob-steh-TRISH-an) physician specializing in pregnancy, labor, and delivery care
occipital (ok-SIP-it-al) relating to the area at the back of the lower skull
occipital bone (ok-SIP-it-al bohn) posterior and inferior bone of the skull
occupational therapist (ak-yoo-PAY-shun-al THAYR-ah-pist) health professional who treats injuries or disorders to develop or recover everyday living skills
old age life cycle phase after early and middle adulthood; *see* **senescence**
older adulthood *see* **old age** and **senescence**
olecranal (oh-LEK-rah-nal) relating to olecranon (back of elbow)
olecranon (oh-LEK-rah-non) the large bony process of the ulna; commonly referred to as the tip of the elbow; sometimes called *olecranon process*
olecranon fossa (oh-LEK-rah-non FOS-ah) a large depression on the posterior surface of the humerus
olfaction (ohl-FAK-shun) sense of smell
olfactory receptor (ohl-FAK-tor-ee ree-SEP-tor) chemical receptors responsible for the sense of smell; located in the epithelial tissue in the upper part of the nasal cavity
oligodendrocyte (ohl-ih-goh-DEN-droh-syte) a cell that holds nerve fibers together and produces the myelin sheath around axons in the CNS
oligospermia (ohl-ih-goh-SPER-mee-ah) low sperm production
oliguria (ohl-ih-GOO-ree-ah) scanty amounts of urine
oncogene (ON-koh-jeen) gene (DNA segment) thought to be responsible for the development of a cancer
oncology (ong-KOL-oh-jee) study of tumors and cancer; branch of medicine concerned with diagnosis and treatment of cancer
onycholysis (ahn-ik-oh-LYE-sis) separation of nail from the nail bed that begins at the distal or free edge of the affected nail
oocyte (OH-oh-syte) immature stage of the female sex cell
oogenesis (oh-oh-JEN-eh-sis) production of female gametes
oophorectomy (oh-of-eh-REK-toh-mee) surgical procedure to remove the ovaries
oophoritis (oh-of-eh-RYE-tis) inflammation of the ovaries
open fracture (OH-pen FRAK-chur) compound fracture; bone fracture in which bone fragments pierce the skin
ophthalmic (of-THAL-mik) relating to the eye
ophthalmologist (of-thal-MOL-eh-jist) physician specializing in treating disorders of the eye and vision
ophthalmoscope (of-THAL-mah-skohp) lighted instrument fitted with optical devices to permit examination of the retina and internal eye structures
opposition (op-oh-ZISH-un) moving the thumb to touch the tips of the fingers; the movement used to hold a pencil to write
optic disk (OP-tik disk) the area in the retina where the optic nerve fibers exit and there are no rods or cones; also known as a *blind spot*
oral (OR-al) relating to the mouth
oral candidiasis *see* **thrush**
oral cavity (OR-al KAV-ih-tee) mouth
oral rehydration therapy (ORT) treatment of infant diarrhea by the administration of a liberal dose of sugar and salt solution
orbicularis oculi (or-bik-yoo-LAYR-is OK-yoo-lye) facial muscle that causes a squint
orbicularis oris (or-bik-yoo-LAYR-is OR-is) facial muscle that puckers the lips
orbital (OR-bih-tal) relating to orbit of the eye (so-called "eye socket")
orchitis (or-KYE-tis) inflammation of the testes, often caused by infection
organ (OR-gan) group of several tissue types that performs a special function
organ of Corti (KOR-tee) *see* **spiral organ;** also *Corti organ*
organelle (or-gah-NELL) intercell organ; for example, the ribosome
organic compound (or-GAN-ik KOM-pownd) compound whose large molecules contain carbon and that include C—C bonds and/or C—H bonds
organism (OR-gah-niz-em) an individual living thing
organization (or-gan-ih-ZAY-shun) the characteristic of the body of being organized, that is, structured in different levels of complexity and coordinated in function; the human body is often said to be organized into different levels of organization: chemical, cell, tissue, organ, system, and body
organ of Corti (OR-gan ov KOR-tee) the organ of hearing located in the cochlea and filled with endolymph; also called *Corti organ* or *spiral organ*
organogenesis (or-gah-noh-JEN-eh-sis) formation of organs from the primary germ layers of the embryo
origin (OR-ih-jin) the attachment of a muscle to the bone, which does not move when contraction occurs, as distinguished from insertion
oropharynx (or-oh-FAYR-inks) the portion of the pharynx that is located behind the mouth
orthodontics (or-thoh-DON-tiks) dental specialty dealing with diagnosis and treatment of malocclusion of the teeth
orthopedic surgeon (or-thoh-PEE-dik SUR-jen) physician trained the medical specialty *orthopedics,* dealing with skeletal injury and disease
orthopnea (or-THOP-nee-ah) dyspnea (difficulty in breathing) that is relieved after moving into an upright or sitting position
os coxae (os KOK-see) hip bones; *see also* **coxal bone**
osmosis (os-MOH-sis) movement of a fluid through a semipermeable membrane
ossicle (OS-sih-kul) any of the little bones (malleus, incus, stapes) found in the ears
osteitis deformans (os-tee-AYE-tis deh-FOR-manz) *see* **Paget disease**
osteoarthritis (os-tee-oh-ar-THRY-tis) degenerative joint disease; a noninflammatory disorder of a joint characterized by degeneration of articular cartilage
osteoblast (OS-tee-oh-blast) bone-forming cell
osteoclast (OS-tee-oh-klast) bone-absorbing cell
osteocyte (OS-tee-oh-syte) bone cell
osteogenesis imperfecta (os-tee-oh-JEN-eh-sis im-per-FEK-tah) dominant, inherited disorder of connective tissue characterized by imperfect skeletal development, resulting in brittle bones
osteoma (os-tee-OH-mah) benign bone tumor

osteomalacia (os-tee-oh-mah-LAY-shah) bone disorder usually caused by vitamin D deficiency and characterized by loss of mineral in the bone matrix; the adult form of rickets

osteomyelitis (os-tee-oh-my-eh-LYE-tis) bacterial (usually staphylococcus) infection of bone tissue

osteon (AHS-tee-on) structural unit of compact bone tissue made up of concentric layers (lamellae) of hard bone matrix and bone cells (osteocytes); also called **Haversian system**

osteoporosis (os-tee-oh-poh-ROH-sis) bone disorder characterized by loss of minerals and collagen from bone matrix, reducing the volume and strength of skeletal bone

osteosarcoma (os-tee-oh-sar-KOH-mah) bone cancer

otitis (oh-TYE-tis) general term referring to inflammation or infection of the ear; otitis media involves the middle ear

otitis media (oh-TYE-tis MEE-dee-ah) a middle ear infection

otologist (oh-TOL-uh-jist) physician specializing in treating disorders of the ear

otosclerosis (oh-toh-skleh-ROH-sis) inherited bone disorder involving structural irregularities of the stapes in the middle ear and characterized by tinnitus progressing to deafness

otoscope (OH-toh-skohp) lighted device used to examine external ear canal and eardrum

ova (OH-vah) (*sing.*, ovum) female sex cells

oval window (OH-val WIN-doh) a small, membrane-covered opening that separates the middle and inner ear

ovarian cyst (oh-VAYR-ee-an SIST) smooth fluid-filled sac that forms in ovarian tissue

ovarian follicle (oh-VAYR-ee-an FOL-ih-kul) each contains an oocyte

ovary (OH-var-ee) female gonad that produces ova (female sex cells)

overactive bladder (OH-ver-ak-tiv BLAD-der) condition of frequent urination characterized by sensation of urgency and pain

overhydration (oh-ver-hye-DRAY-shun) too much fluid input in the body, which can put a burden on the heart

oviduct (OH-vih-dukt) also called *uterine tube* or *fallopian tube; see* **uterine tube** for definition

ovulation (ov-yoo-LAY-shun) release of an ovum from the ovary

ovum (OH-vum) (*pl.*, ova) egg; female sex cell (female gamete)

oxygen (O_2) (AHK-sih-jen) one of the chemical elements found in great quantity in the human body; symbolized by O, as in H_2O (water) or O_2 (oxygen gas)

oxygen concentrator (AHK-sih-jen KON-sen-tray-tor) a device used in health care that increases the proportion of oxygen gas in the air of the room in which it is placed; sometimes used in treatment of persons in respiratory distress and in other such conditions that produce hypoxia (low oxygen concentration in the blood)

oxygen debt (AHK-sih-jen det) continued increased metabolism that occurs in a cell to remove excess lactic acid that resulted from exercise

oxygen therapy (AHK-sih-jen) administration of oxygen gas to individuals suffering from low oxygen concentration in the blood (hypoxia)

oxyhemoglobin (HbO_2) (ahk-see-hee-moh-GLOH-bin) hemoglobin combined with oxygen

oxytocin (OT) (ahk-see-TOH-sin) hormone secreted by the posterior pituitary gland in a woman before and after she has delivered a baby; thought to initiate and maintain labor and also causes the release of breast milk into the mammary ducts to provide nourishment for the baby

P

P wave electrocardiogram deflection that represents depolarization of the atria

p-arm (PEE-arm) the short segment of a chromosome that is divided into two segments by a centromere

pacemaker (PAYS-may-ker) *see* **sinoatrial node**

Pacini corpuscle (pah-CHEE-nee KOHR-pus-ul) a receptor found deep in the dermis that detects pressure on the skin surface; also called *pacinian corpuscle* or *lamellar corpuscle*

Paget disease (PAJ-et dih-ZEEZ) osteitis deformans; a common, often mild bone disorder characterized by replacement of normal spongy bone with disorganized bone matrix

pain receptor (payn ree-SEP-tor) sensory neuron that detects painful stimuli; also called *nociceptor*

palate (PAL-let) the roof of the mouth; made up of the hard (anterior portion of the mouth) and soft (posterior portion of the mouth) palates

palatine tonsil *see* **tonsil**

paleontologist (pay-lee-un-TOL-uh-jist) scientist that studies organisms that lived in the ancient past

palmar (PAHL-mar) relating to the palm of the hand

palpable (PAL-pah-bul) can be felt or touched

palpebral fissure (PAL-peh-bral FISH-ur) opening between the two eyelids

pancreas (PAN-kree-as) endocrine gland located in the abdominal cavity; contains pancreatic islets that secrete glucagon and insulin

pancreatic islet (pan-kree-AT-ik eye-let) one of the microscopic, isolated endocrine portions of the pancreas; made up of alpha and beta cells, among others; also called *islet of Langerhans*

pancreatitis (pan-cree-ah-TYE-tis) inflammation of the pancreas

pandemic (pan-DEM-ik) refers to a disease that affects many people worldwide

Papanicolaou test (pah-peh-nik-oh-LAH-oo) cancer-screening test in which cells brushed off from the lining of the uterine cervix are smeared on a glass slide and examined for abnormalities; also called *Pap smear* or *Pap test*

papilla (pah-PIL-ah) (*pl.*, papillae) small, nipple-shaped elevation

papilloma (pap-ih-LOH-mah) benign skin tumor characterized by fingerlike projections (e.g., a wart)

papule (PAP-yool) raised, firm skin lesion less than 1 cm in diameter

paracrine [agent or hormone] (PAYR-ah-krin [AY-jent or HOHR-mohn]) hormone that regulates activity in nearby cells within the same tissue as their source

paralysis (pah-RAL-ih-sis) loss of the power of motion, especially voluntary motion

paramedic (payr-ah-MED-ik) health-care worker trained to assist a physician or to give care in the absence of a physician, often as part of a first-responder team

paranasal sinus (payr-ah-NAY-zal SYE-nus) four pairs of sinuses that have openings into the nose

paraphimosis (para-fih-MOH-sis) condition in which the male foreskin cannot be easily replaced after being retracted away from the glans penis

paraplegia (payr-ah-PLEE-jee-ah) paralysis (loss of voluntary muscle control) of both legs

parasite (PAYR-ah-syte) any organism that lives in or on another organism (a host) to obtain its nutrients; parasites may be harmless to the host, or they may disrupt normal body functions of the host and thus cause disease

parasympathetic division (payr-ah-sim-pah-THET-ik dih-VIZH-un) part of the autonomic nervous system that controls many visceral effectors under normal "maintenance" conditions; ganglia are connected to the brainstem and the sacral segments of the spinal cord (craniosacral segments)
parasympathetic postganglionic neuron (payr-ah-sim-pah-THET-ik post-gang-glee-ON-ik NOO-ron) ANS neuron in which dendrites and cell body are in a parasympathetic ganglion and axon travels to a variety of visceral effectors
parasympathetic preganglionic neuron (payr-ah-sim-pah-THET-ik pree-gang-glee-ON-ik NOO-ron) ANS neuron in which dendrites and cell body are located in the gray matter of the brainstem and sacral cord segments; axon terminates in a parasympathetic ganglion
parathyroid gland (payr-ah-THYE-royd) set of endocrine glands located in the neck on the posterior aspect of the thyroid gland; secrete parathyroid hormone (PTH)
parathyroid hormone (PTH) (payr-ah-THYE-royd HOR-mohn) hormone released by the parathyroid gland that increases the concentration of calcium in the blood
parenteral (pah-REN-ter-al) outside the intestinal tract; parenteral therapy is administration of nutrients, special fluids, and/or electrolytes by injection—thus bypassing intestinal absorption
parietal (pah-RYE-ih-tal) of the walls of an organ or cavity
parietal bone (pah-RYE-ih-tal) cranial bone of the top and side of the cranium
parietal pericardium (pah-RYE-ih-tal payr-ih-KAR-dee-um) pericardium surrounding the heart like a loose-fitting sack to allow the heart enough room to beat
parietal peritoneum (pah-RYE-ih-tal payr-ih-toh-NEE-um) serous membrane that lines and is adherent to the wall of the abdominal cavity
parietal pleura (pah-RYE-ih-tal PLOO-rah) serous membrane that lines and is adherent to the wall of the thoracic cavities
parietal portion (pah-RYE-ih-tal POR-shun) serous membrane that lines the walls of a body cavity
Parkinson disease (PD) (PARK-in-son dih-ZEEZ) a chronic disease of the nervous system characterized by a set of signs called *parkinsonism* that results from a deficiency of the neurotransmitter dopamine in certain regions of the brain that normally inhibit overstimulation of skeletal muscles; parkinsonism is characterized by muscle rigidity and trembling of the head and extremities, forward tilt of the body, and shuffling manner of walking
parotid duct (per-AH-tid dukt) either of the ducts of the parotid salivary glands; also known as **Stensen duct**
parotid gland (per-AH-tid) paired salivary gland that lies just below and in front of each ear at the angle of the jaw
partial pressure (P) (PAR-shal) pressure exerted by any one gas in a mixture of gases or in a liquid
partial-thickness burn term used to describe both minor burn injury and more severe burns that injure both epidermis and dermis (*see* **first-degree burn**) (*see* **second-degree burn**)
parturition (pahr-too-RIH-shun) act of giving birth
passive transport cellular process in which substances move through a cellular membrane with the energy supplied directly by the cell or its membrane
patella (pah-TEL-ah) small, shallow pan; the kneecap
pathogenesis (path-oh-JEN-eh-sis) pattern of a disease's development
pathologist (pah-THOL-uh-jist) scientist who studies disease processes
pathology (pah-THOL-oh-jee) the scientific study of disease
pathophysiology (path-oh-fiz-ee-OL-oh-jee) study of the underlying physiological aspects of disease
patient care technician (PAY-shent kayr tek-NISH-en) health-care worker who provides personal care to patients under the supervision of nurses, physicians, and other professionals
pectoral girdle (PEK-toh-ral GIRD-el) shoulder girdle; the scapula and clavicle, which connect the upper extremities to the axial skeleton
pectoralis major (pek-teh-RAH-liss MAY-jor) major flexor of the arm
pedal (PEED-al) relating to the foot
pedigree (PED-ih-gree) chart used in genetic counseling to illustrate genetic relationships over several generations
pelvic (PEL-vik) relating to the pelvis or hip bones, or to the nearby anatomical region
pelvic cavity (PEL-vik KAV-ih-tee) the inferior portion of the ventral cavity known as the abdominopelvic cavity
pelvic girdle (PEL-vik GIRD-el) ring of bone formed by the pelvic bones that connect the lower extremities to the axial skeleton
pelvic inflammatory disease (PID) (PEL-vik in-FLAM-ahtor-ee dih-ZEEZ) acute inflammatory condition of the uterus, fallopian tubes, and/or ovaries—usually the result of a sexually transmitted infection (STI)
pelvis (PEL-vis) basin- or funnel-shaped structure
penicillin (pen-ih-SIL-in) antibiotic derived from products of a specific type of mold; discovered in the lab of Alexander Fleming
penis (PEE-nis) (*pl.*, penes or penises) structure that forms part of the male genitalia; when sexually aroused, becomes stiff to enable it to enter and deposit sperm in the vagina
pepsin (PEP-sin) protein-digesting enzyme of the stomach
pepsinogen (pep-SIN-oh-jen) component of gastric juice that is converted into pepsin by hydrochloric acid
peptidase (PEP-tyd-ayz) intestinal enzyme that breaks apart peptide bonds in polypeptide strands that remain from protein digestion
peptide bond (PEP-tyde) covalent bond linking amino acids within a protein molecule
perfuse (per-FYOOZ) to flow through, as in flow of blood through a tissue
pericardial effusion (pair-ih-KAR-dee-all ef-FYOO-shen) accumulation of pericardial fluid, pus, or blood in the space between the two pericardial layers
pericarditis (payr-ih-kar-DYE-tis) condition in which the pericardium becomes inflamed
pericardium (payr-ih-KAR-dee-um) membrane that surrounds the heart
perilymph (PAYR-ih-limf) a watery fluid that fills the bony labyrinth of the ear
perinatal infection (payr-ih-NAY-tal in-FEK-shun) infection passed from a mother to an infant during the time of the birth process
perineal (payr-ih-NEE-al) relating to the area between the anus and genitals (the **perineum**)
perineum (payr-ih-NEE-um) the area between the anus and genitals
perineurium (payr-ih-NOO-ree-um) connective tissue that encircles a bundle (fascicle) of nerve fibers within a nerve
periodontal membrane (payr-ee-oh-DON-tull MEM-brayn) fibrous tissue that lines each tooth socket and serves to attach the tooth to underlying bone
periodontitis (payr-ee-oh-don-TYE-tis) inflammation of the periodontal membrane (periodontal ligament) that anchors teeth to jaw bone; common cause of tooth loss among adults

periosteum (payr-ee-OS-tee-um) tough, connective tissue covering the bone

peripheral (peh-RIF-er-al) relating to an outside surface

peripheral nervous system (PNS) (peh-RIF-er-al NER-vus SIS-tem) the nerves connecting the brain and spinal cord to other parts of the body

peripheral resistance (PR) (peh-RIF-er-al) resistance (blocked effort) to blood flow encountered in the peripheral arteries (arteries that branch off the aorta and pulmonary arteries)

peristalsis (payr-ih-STAL-sis) wavelike, rhythmic contractions of the stomach and intestines that move food material along the digestive tract

peritoneal space (payr-ih-toh-NEE-al) small, fluid-filled space between the visceral and parietal layers that allows the layers to slide over each other freely in the abdominopelvic cavity

peritoneum (payr-ih-toh-NEE-um) large, moist, slippery sheet of serous membrane that lines the abdominopelvic cavity (parietal layer) and its organs (visceral layer)

peritonitis (payr-ih-toh-NYE-tis) inflammation of the serous membranes in the abdominopelvic cavity; sometimes a serious complication of an infected appendix

permanent teeth (PER-mah-nent teeth) set of 32 teeth that replaces deciduous teeth; also called *adult teeth*

permeable membrane (PER-mee-ah-bul MEM-brayn) a membrane that allows passage of substances

permease system (PER-mee-ayz SIS-tem) a specialized cellular component that allows a number of active transport mechanisms to occur

pernicious anemia (per-NISH-us ah-NEE-mee-ah) deficiency of red blood cells caused by a lack of vitamin B_{12}

peroneal muscles (per-oh-NEE-al MUS-els) plantar flexors and evertors of the foot; the peroneus longus forms a support arch for the foot; *see* **peroneus (muscle) group**

peroneus (muscle) group (per-on-EE-uss groop) group of lateral muscles of the leg that act to pronate the foot, rotating it toward the midline, and plantar flex the foot, pulling it toes-downward; also called the **fibularis (muscle) group**

perspiration (per-spih-RAY-shun) transparent, watery liquid released by glands in the skin that eliminates ammonia and uric acid and helps maintain body temperature; also commonly known as **sweat**

pH (p H) mathematical expression of relative H^+ concentration (acidity); pH value higher than 7 is basic, pH value less than 7 is acidic, pH value equal to 7 is neutral

phagocyte (FAG-oh-syte) white blood cell that engulfs microbes and digests them

phagocytosis (fag-oh-sye-TOH-sis) ingestion and digestion of particles by a cell

phalanges (fah-LAN-jeez) the bones that make up the fingers and toes

pharmacist (FAR-mah-sist) health-care worker trained to dispense drugs and educate patients in their proper use

pharmacologist (far-mah-KAHL-uh-jist) scientist specializing in the study of drug actions

pharmacology (farm-ah-KAHL-ah-jee) study of drugs and their actions in the body

pharmacy technician (FAR-mah-see tek-NISH-en) health-care worker trained to dispense drugs under the supervision of a pharmacist

pharyngeal tonsil *see* **adenoid**

pharyngitis (fayr-in-JYE-tis) sore throat; inflammation or infection of the pharynx

pharynx (FAYR-inks) organ of the digestive and respiratory systems; commonly called the *throat*

phenylketonuria (PKU) (fen-il-kee-toh-NOO-ree-ah) recessive, inherited condition characterized by excess of phenylketone in the urine, caused by accumulation of phenylalanine (an amino acid) in the tissues; may cause brain injury and death if phenylalanine intake is not managed properly

phimosis (fih-MOH-sis) abnormal condition in which the prepuce (foreskin) fits tightly over the glans of the penis

phlebitis (fleh-BYE-tis) inflammation of a vein

phlebotomist (fleh-BOT-oh-mist) health-care worker specializing in drawing blood from veins for laboratory analysis or donation

phospholipid (fos-foh-LIP-id) phosphate-containing lipid (fat) molecule

photodynamic therapy (foh-toh-dye-NAM-ic) use of laser energy to trigger photosensitizing drugs in specialized treatment of superficial cancers and "wet" age-related macular degeneration

photopigments (foh-toh-PIG-ments) chemicals in retinal cells that are sensitive to light

photoreceptor (FOH-toh-ree-sep-tor) specialized nerve cell stimulated by light

photorefractive keratectomy (PRK) (FOH-toh-ree-frak-tiv kayr-ah-TEK-toh-mee) refractory eye surgery that uses an excimer or "cool" laser to vaporize corneal tissue in treating mild to moderate nearsightedness; also called **excimer laser surgery**

phrenic nerve (FREN-ik nerv) the nerve that stimulates the diaphragm to contract

physical education (FIS-ik-al ed-yoo-KAY-shun) teaching discipline that focuses on health, fitness, and sports

physical therapist (FIS-ik-al THAYR-ah-pist) health professional who helps patients improve body movements and manage pain

physician (fih-ZISH-en) health-care professional, usually holding a doctorate in medicine or related discipline, licensed to provide and supervise medical care

physiology (fiz-ee-OL-oh-jee) the study of body function

pia mater (PEE-ah MAH-ter) the vascular innermost covering (meninx) of the brain and spinal cord

pigment (PIG-ment) colored substance

pigment layer (PIG-ment LAY-er) the layer of the epidermis that contains the melanocytes that produce melanin to give skin its color; *stratum basale* or *basal layer*

pineal gland (PIN-ee-al gland) endocrine gland located in the third ventricle of the brain; produces melatonin; also known as *pineal body*

pinna (PIN-nah) flap of the external ear

pinocytosis (pin-oh-sye-TOH-sis) active transport mechanism used to transfer fluids or dissolved substances into cells

pitting edema (pit-ing eh-DEE-mah) depressions in swollen subcutaneous tissue that do not rapidly refill after exerted pressure is removed

pituitary gland (pih-TOO-ih-tayr-ee) endocrine gland located in the skull; made up of the adenohypophysis and the neurohypophysis

pivot joint (PIV-it joynt) type of diarthrotic synovial joint in which a projection from one bone articulates with a ring or notch in another bone, allowing rotational movement

placenta (plah-SEN-tah) anchors the developing fetus to the uterus and provides a "bridge" for the exchange of nutrients and waste products between the mother and developing baby

placenta previa (plah-SEN-tah PREE-vee-ah) abnormal condition in which a blastocyst implants in the lower uterus, developing a placenta that approaches or covers the cervical opening; placenta previa involves the risk of placental separation and hemorrhage

plane (of body) (playn) any completely flat cut through the body or any of its parts; a body plane can be oriented in any of several directions (e.g., sagittal, midsagittal, frontal [coronal], transverse [horizontal]) and is used to visualize the body from different perspectives; *see also* **section (of body)**

plantar (PLAN-tar) relating to the sole of the foot

plantar flex (PLAN-tar fleks) to move the ankle so that the bottom of the foot is directed inferiorly

plantar flexion (PLAN-tar FLEK-shun) movement in which the bottom of the foot is directed downward; this motion allows a person to stand on tiptoe

plaque (plak) raised skin lesion greater than 1 cm in diameter

plasma (PLAZ-mah) the liquid part of the blood

plasma cell (PLAZ-mah sel) cell that secretes copious amounts of antibody into the blood; also called **effector cell**

plasma membrane (PLAZ-mah MEM-brayn) membrane that separates the contents of a cell from the tissue fluid; encloses the cytoplasm and forms the outer boundary of the cell

plasma protein (PLAZ-mah PROH-teen) any of several proteins normally found in the plasma; includes albumins, globulins, and fibrinogen

plasmid (PLAS-mid) small circular ring of bacterial DNA

platelet (PLAYT-let) *see* **thrombocyte**

platelet plug (PLAYT-let) a temporary accumulation of platelets (thrombocytes) at the site of an injury; it precedes the formation of a blood clot

platyhelminth (plat-ih-HEL-minth) flatworm or fluke—animal parasite capable of infesting humans

pleura (PLOOR-ah) (*pl.*, pleurae) the serous membrane in the thoracic cavity

pleural (PLOOR-al) relating to the pleura or to the side of the thorax

pleural cavity (PLOOR-al KAV-ih-tee) a lateral subdivision of the thorax; cavity in which each lung is located

pleural space (PLOOR-al) the space between the visceral and parietal pleurae filled with just enough serous (pleural) fluid to allow them to glide effortlessly with each breath

pleurisy (PLOOR-ih-see) inflammation of the pleura

plexus (PLEK-sus) complex network formed by converging and diverging nerves, blood vessels, or lymphatic vessels

plica (PLYE-kah) (*pl.*, plicae) multiple circular folds

pneumocystosis (noo-moh-sis-TOH-sis) a protozoan infection, most likely to invade the body when the immune system has been compromised

pneumonectomy (noo-moh-NEK-toh-mee) surgical procedure in which an entire lung is removed

pneumonia (noo-MOH-nee-ah) abnormal condition characterized by acute inflammation of the lungs in which alveoli and bronchial passages become plugged with thick fluid (exudate)

pneumothorax (noo-moh-THOH-raks) abnormal condition in which air is present in the pleural space surrounding the lung, possibly causing collapse of the lung

podiatrist (poh-DYE-uh-trist) physician who specializes in health care of the foot, ankle, and leg

polar body (POH-lar BOD-ee) small, nonfunctional cell produced during meiotic cell divisions (meiosis) in the formation of female sex cells (gametes); incapable of being fertilized

poliomyelitis (pol-ee-oh-my-eh-LYE-tis) viral disease that damages motor nerves, often progressing to paralysis of skeletal muscles

polycystic kidney disease (PKD) (pahl-ee-SIS-tik KID-nee dih-ZEEZ) most common genetic disorder in humans, in which large fluid-filled pockets (cysts) develop in the epithelium of the kidney tubules

polycystic ovary syndrome (PCOS) (pahl-ee-SIS-tik OH-var-ee SIN-drohm) condition that is characterized by ovaries usually twice the normal size and that are studded with fluid-filled cysts

polycythemia (pahl-ee-sye-THEE-mee-ah) an excessive number of red blood cells

polydipsia (pahl-ee-DIP-see-ah) excessive and ongoing thirst

polyendocrine disorder (PAHL-ee-EN-doh-krin dis-OR-der) disorder caused by more than one endocrine malfunction or involving more than one hormone

polysaccharide (pahl-ee-SAK-ah-ryde) biomolecule made up of many saccharide sugars (monosaccharides)

polyuria (pol-ee-YOO-ree-ah) unusually large amounts of urine

pons (ponz) the part of the brainstem between the medulla oblongata and the midbrain

popliteal (pop-lih-TEE-al) relating to the area behind the knee

pore (por) pinpoint-size openings on the skin that are outlets of small ducts from the eccrine sweat glands

portal hypertension (PORT-al hye-per-TEN-shun) high blood pressure caused by blockage of blood flow in the liver (from cancer or cirrhosis)

port-wine stain pigmented, benign tumor of the skin present at birth and ranging in color from pale red to a deep reddish purple; also called *nevus flammeus* (*see* **nevus**)

positive feedback (POZ-it-iv FEED-bak) homeostatic control system in which information feeding back to the control center causes the level of a variable to be pushed farther in the direction of the original deviation, causing an amplification of the original stimulus; ordinarily this mechanism is used by the body to amplify a process and quickly finish it, as in labor contractions and blood clotting

posterior (pohs-TEER-ee-or) located behind; opposite of **anterior**

posterior pituitary gland (pohs-TEER-ee-or pih-TOO-ih-tayr-ee) neurohypophysis; produces hormones ADH and oxytocin

postganglionic neuron (post-gang-glee-ON-ik NOO-ron) autonomic neuron that conducts nerve impulses from a ganglion to cardiac or smooth muscle or glandular epithelial tissue

postherpetic neuralgia (post-her-PET-ik noo-RAL-jee-ah) pain (often severe) along nerve pathways previously affected by an outbreak of shingles (herpes zoster)

postnatal period (POST-nay-tal PEER-ee-id) the period beginning after birth and ending at death

postsynaptic neuron (post-sih-NAP-tik NOO-ron) a neuron situated distal to a synapse

posture (POS-chur) position of the body

precapillary sphincter (pree-CAP-pih-layr-ee SFINGK-ter) smooth muscle cells that guard the entrance to the capillary

preeclampsia (pree-ee-KLAMP-see-ah) syndrome of abnormal conditions in pregnancy of uncertain cause; syndrome includes hypertension, proteinuria, and edema; also called *toxemia of pregnancy*, it may progress to eclampsia—severe toxemia that may cause death

preexisting condition (pree-ig-ZIST-ing kon-DISH-un) disorder or health state that has become established before another condition occurs; also called a *primary* condition

preganglionic neurons (pree-gang-glee-ON-ik NOO-rons) autonomic neurons that conduct nerve impulses between the spinal cord and a ganglion

premature (cardiac) contraction (pree-mah-TUR [KAR-dee-ak] kon-TRAK-shun) contractions of the heart wall that occur before expected; extrasystoles

premenstrual syndrome (PMS) (pree-MEN-stroo-al SIN-drohm) syndrome of psychological changes (such as irritability) and physical changes (localized edema) that occur before menstruation in many women

premolar *see* **bicuspid**

prenatal period (PREE-nay-tal PEER-ee-id) the period after conception until birth

prepuce *see* **foreskin**

presbycusis (pres-bih-KYOO-sis) progressive hearing loss associated with advanced age

presbyopia (pres-bee-OH-pee-ah) farsightedness associated with advancing age

presynaptic neuron (pree-sih-NAP-tik NOO-ron) a neuron situated proximal to a synapse

primary bronchi (PRYE-mayr-ee BRAHN-kye) (*sing.,* bronchus) first branches of the trachea (right and left primary bronchi)

primary follicle (PRYE-mayr-ee FOL-ih-kul) ovarian follicle present at puberty; lined with granulosa cells

primary germ layer (PRYE-mayr-ee jerm LAY-er) any of the three layers of germ cells that give rise to definite structures as the embryo develops

primary protein (PRYE-mayr-ee PROH-teen) the preliminary structure of a protein: the sequence of amino acids held together with peptide bonds (this structure will then fold to become the secondary protein structure)

primary spermatocyte (SPER-mah-toh-syte) specialized cell that undergoes meiosis to ultimately form sperm

prime mover the muscle responsible for producing a particular movement

principle of independent assortment genetic principle that states as chromosome pairs separate, the maternal and paternal chromosomes redistribute themselves independently of the other chromosome pairs

prion (PREE-ahn) shortened form of the term "PROteinaceous INfectious particle"; pathogenic protein molecule that converts normal proteins of the body into abnormal proteins, causing abnormalities of function (the abnormal form of the protein also may be inherited by offspring of an affected person); *see also* **bovine spongiform encephalopathy**

product any substance formed as a result of a chemical reaction

progeria (proh-JEER-ee-ah) rare, genetic condition in which a person appears to age rapidly as a result of abnormal, widespread degeneration of tissues; adult and childhood forms exist, with the childhood form resulting in death by age 20 or so

progesterone (proh-JES-ter-ohn) hormone produced by the corpus luteum; stimulates secretion of the uterine lining; with estrogen, helps to initiate the menstrual cycle in girls entering puberty

prognosis (prog-NOH-sis) in medicine, the probable outcome of a disease

prolactin (PRL) (proh-LAK-tin) hormone secreted by the anterior pituitary gland during pregnancy to stimulate the breast development needed for lactation; also called *lactogenic hormone*

prolactinoma (proh-LAK-tih-noh-mah) benign adenoma (epithelial tumor) of the anterior pituitary, producing hypersecretion of prolactin (and exaggerated prolactin effects); is usually small and occurs most often in females

proliferative phase (PROH-lif-eh-rah-tiv faze) phase of menstrual cycle that begins after the menstrual flow ends and lasts until ovulation

pronate (PROH-nayt) to make a rotational movement of the forearm (turning the palm medially to face backward) or of the leg and ankle (turning the foot so toes point outward and the medial edge of the sole hits the ground); opposite of **supinate**

pronation (PROH-nay-shun) action in which the forearm or leg and ankle **pronates;** opposite of **supination**

prone used to describe the body lying in a horizontal position facing downward

prophase (PROH-fayz) first stage of mitosis during which chromosomes become visible

proprioceptor (proh-pree-oh-SEP-tor) stretch receptor located in the muscles, tendons, and joints; allows the body to recognize its position

prostaglandin (PG) (pros-tah-GLAN-din) any of a group of naturally occurring fatty acids that affect many body functions

prostatectomy (pros-tah-TEK-toh-mee) surgical removal of part or all of the prostate gland

prostate cancer (PROS-tayt KAN-ser) malignancy of the prostate gland

prostate gland (PROS-tayt) lies just below the bladder; secretes a fluid that constitutes about 30% of the seminal fluid volume; helps activate sperm and helps them maintain motility; also known simply as "the prostate"

prostate-specific antigen (PSA) (PROS-tayt–speh-SIF-ik AN-tih-jen) a protein (antigen) produced by prostate tissue that may be elevated in the blood of men with prostate cancer

prostatectomy (pros-tah-TEK-toh-mee) surgical removal of all or part of the prostate gland

prosthesis (pros-THEE-sis) an artificial body part or device that assists the functioning of a body part; used more narrowly, the term applies only to artificial limbs or limb extensions

protease (PROH-tee-ayz) protein-digesting enzyme

protein (PROH-teen) one of the basic nutrients needed by the body; a nitrogen-containing organic compound composed of a folded strand of amino acids

protein-calorie malnutrition (PCM) (PROH-teen-KAL-or-ee mal-noo-TRISH-un) abnormal condition resulting from a deficiency of calories in general and protein in particular; likely to result from reduced intake of food but may also be caused by increased nutrient loss or increased use of nutrients by the body

protein hormone (PROH-teen HOR-mohn) a nonsteroid; *see* **nonsteroid hormone**

proteinuria (proh-teen-YOO-ree-ah) presence of abnormally high amounts of plasma protein in the urine; usually an indicator of kidney disease

proteoglycan (PROH-tee-oh-GLYE-kan) large molecule made up of a protein strand that forms a backbone to which are attached many carbohydrate molecules

proteome (PROH-tee-ohm) the entire group of proteins encoded by the genome; *see* **genome**

proteomics (proh-tee-OH-miks) the endeavor that involves the analysis of the proteins encoded by the genome, with the ultimate goal of understanding the role of each protein in the body

prothrombin (proh-THROM-bin) a protein present in normal blood that is required for blood clotting

prothrombin activator (proh-THROM-bin AK-tih-vay-tor) combination of clotting factors and circulating plasma proteins that initiates conversion of prothrombin to thrombin in the clotting mechanism

prothrombin time (PT) (proh-THROM-bin tyme) time it takes for a blood sample to clot after tissue thromboplastin (prothrombin activator) is added—a way to assess efficiency of a person's extrinsic clotting mechanism; *see also* **international normalized ratio (INR)**

proton (PROH-ton) positively charged particle within the nucleus of an atom

protozoan (proh-toh-ZOH-an) (*pl.,* protozoa) single-celled organisms with nuclei and other membranous organelles that can infect humans

proximal (PROK-sih-mal) next or nearest; located nearest the center of the body or the point of attachment of a structure; opposite of **distal**

proximal convoluted tubule (PCT) (PROK-sih-mal kon-voh-LOO-ted TOOB-yool) the first segment of a renal tubule

pseudo (SOO-doh) false

pseudogene (SOOD-oh-jeen) possibly nonfunctional "broken" genetic code found in "junk DNA" located between the functioning, coding genes of a DNA molecule

pseudohypertrophy (soo-doh-hye-PER-troh-fee) literally, "false muscle growth"; another name for **Duchenne muscular dystrophy (DMD)**

pseudostratified epithelium (SOOD-oh-STRAT-ih-fyde ep-ih-THEE-lee-um) type of tissue similar to simple columnar epithelium that forms a membrane made up of a single layer of cells that are tall and narrow but that are squeezed together in a way that pushes the nuclei into two layers and thus gives an initial impression that it is stratified (having more than one layer of cells); compare to **simple columnar epithelium**

psoriasis (soh-RYE-ah-sis) chronic, inflammatory skin disorder characterized by cutaneous inflammation and scaly plaques

psychiatrist (sye-KYE-uh-trist) physician specializing in mental health

psychogenic (sye-koh-JEN-ik) relating to anything caused by psychological mechanisms; for example, psychogenic disorders are often caused by stress or other psychological trauma

psychologist (sye-KOL-uh-jist) someone who studies mental processes or treats mental conditions through counseling or related therapies

puberty (PYOO-ber-tee) stage of adolescence in which a person becomes sexually mature

pubis (PYOO-bis) joint in the midline between the two pubic bones

public health (PUB-lik helth) broad, interdisciplinary field aimed at promoting health and wellness of all people

puerperal fever (pyoo-ER-per-al FEE-ver) condition caused by bacterial infection in a woman after delivery of an infant, possibly progressing to septicemia and death; also called *childbed fever*

pulmonary artery (PUL-moh-nayr-ee AR-ter-ee) artery that carries deoxygenated blood from the right ventricle to the lungs

pulmonary circulation (PUL-moh-nayr-ee ser-kyoo-LAY-shun) venous blood flow from the right atrium to the lung and then to the left atrium

pulmonary embolism (PUL-moh-nayr-ee EM-boh-liz-em) blockage of the pulmonary circulation by a thrombus or other matter; may lead to death if blockage of pulmonary blood flow is significant

pulmonary semilunar valve (PUL-moh-nayr-ee sem-ih-LOO-nar valv) valve located at the beginning of the pulmonary artery

pulmonary vein (PUL-moh-nayr-ee vayn) any vein that carries oxygenated blood from the lungs to the left atrium

pulmonary ventilation (PUL-moh-nayr-ee ven-tih-LAY-shun) breathing; process that moves air in and out of the lungs

pulse (puls) alternating expansion and recoil of the arterial walls produced by the alternate contraction and relaxation of the ventricles; travels as a wave away from the heart

Punnett square (PUN-it skwayr) grid used in genetic counseling to determine the probability of inheriting genetic traits

pupil (PYOO-pill) the opening in the center of the iris that regulates the amount of light entering the eye

Purkinje fibers (pur-KIN-jee FYE-bers) specialized cells located in the walls of the ventricles; relay nerve impulses from the AV node to the ventricles, causing them to contract

purpura (PUR-pah-rah) condition in which small hemorrhages cause purplish discolorations in the skin, mucous membranes, and other body surfaces

pus accumulation of white blood cells, dead bacterial cells, and damaged tissue cells at the site of an infection

pustule (PUS-tyool) small, raised skin lesion filled with pus

P wave deflection on an ECG that occurs with depolarization of the atria

pyelonephritis (pye-eh-loh-neh-FRY-tis) infectious condition characterized by inflammation of the renal pelvis and connective tissues of the kidney

pyloric sphincter (pye-LOR-ik SFINGK-ter) sphincter that prevents food from leaving the stomach and entering the duodenum

pyloric stenosis (pye-LOR-ik steh-NOH-sis) anatomical abnormality in which the opening through the pylorus or pyloric sphincter is unusually narrow

pylorospasm (pye-LOHR-oh-spaz-um) spasm of pyloric sphincter of stomach

pylorus (pye-LOR-us) the small narrow section of the stomach that joins the first part of the small intestine

pyramids (PEER-ah-mids) triangular-shaped divisions of the medulla of the kidney; *see* **renal pyramid**

pyrogen (PYE-roh-jen) any systemic inflammatory chemical that causes the thermostatic control centers of the hypothalamus to produce a fever

pyruvic acid (pye-ROO-vik AS-id) product of the glycolysis of glucose, an energy-containing molecule that enters the mitochondrion for further catabolism and generation of ATP

Q

q-arm (KYU-arm) the long segment of a chromosome that is divided into two segments by a centromere

QRS complex (Q R S KOM-pleks) deflection on an ECG that occurs as a result of depolarization of the ventricles

quadrant (KWOD-runt) *see* **abdominal quadrants**

quadriceps femoris (KWOD-reh-seps feh-MOR-is) extensor muscle of the leg

quadriplegia (kwod-rih-PLEE-jee-ah) paralysis (loss of voluntary muscle control) in all four limbs

quaternary protein (KWAT-er-nayr-ee PROH-teen) the fourth level of structure in a protein formed when two or more tertiary (third-level) proteins unite to form a larger protein molecule

quickening (KWIK-en-ing) when a pregnant woman first feels recognizable movements of the fetus

R

radial keratotomy (RK) (RAY-dee-al KAR-ah-tah-toh-mee) surgical placement of six or more radial slits in a spokelike pattern around the cornea; flattens cornea and improves focus

radiation (ray-dee-AY-shun) flow of heat waves away from the blood

radiation sickness (ray-dee-AY-shun SIK-nes) illness caused by cell damage from high levels of radiation; symptoms may include diarrhea, headache, fever, dizziness, weakness, hair loss; also called *radiation poisoning* or *acute radiation syndrome*

radiation therapy (ray-dee-AY-shun THAYR-ah-pee) treatment often used for cancer in which high-intensity radiation is used to destroy cancer cells; also called *radiotherapy*

radical mastectomy (RAD-ih-kal mas-TEK-toh-mee) surgical procedure in which a cancerous breast is removed along with nearby muscle tissue and lymph nodes

radioactive isotope (ray-dee-oh-AK-tiv AYE-soh-tohp) form of an element in which atoms have a unique atomic number and also release particles or waves of radiation (*see also* **isotope**)

radiography (ray-dee-OG-rah-fee) imaging technique using x-rays that pass through some tissues more easily than others, allowing an image of tissues to form on a photographic plate or other sensitive surface; invented by Wilhelm Röntgen in 1895

radiological technologist (ray-dee-oh-LOJ-ih-kul tek-NOL-uh-jist) health-care worker who performs diagnostic imaging procedures, such as x-rays and CT or MRI scans

radiologist (ray-dee-AHL-uh-jist) physician who specializes in diagnosis using medical imaging such as x-rays and CT or MRI scans

radius (RAY-dee-us) one of the two bones in the forearm; located on the thumb side of the forearm

Raynaud phenomenon (ray-NO fih-NOM-eh-nohn) disorder characterized by sudden decreases in circulation in the digits (fingers or toes), often in response to stress or temperature change

reabsorption (ree-ab-SORP-shun) process of absorbing again that occurs in the kidneys

reactant (ree-AK-tant) any substance entering (and being changed by) a chemical reaction

receiving chambers (ree-SEE-ving CHAYM-bers) atria of the heart; receive blood from the superior and inferior venae cavae

receptor (ree-SEP-tor) peripheral beginning of a sensory neuron's dendrite

recessive (ree-SES-iv) in genetics, the term *recessive* refers to genes that have effects that do not appear in the offspring when they are masked by a dominant gene (recessive forms of a gene are represented by lowercase letters); compare with **dominant**

reconstructive surgery (ree-kon-STRUK-tiv SUR-jeh-ree) any surgical procedure in which anatomical structures are rebuilt to a different form

rectum (REK-tum) distal portion of the large intestine

rectus abdominis (REK-tus ab-DOM-ih-nus) muscle that runs down the middle of the abdomen; protects the abdominal viscera and flexes the spinal column

red blood cell (RBC) *see* **erythrocyte**

red bone marrow (red bohn MAR-oh) bone marrow (myeloid tissue) found in the ends of long bones and in flat bones; functions in the production of blood cells

referred pain (re-FERD payn) pain that originates in a different location in the body from where it is perceived by the brain

reflex (REE-fleks) involuntary action

reflex arc (REE-fleks ark) allows an impulse to travel in only one direction

reflux (REE-fluhks) backflow, as in flow of stomach contents back into esophagus

refraction (ree-FRAK-shun) bending of a ray of light as it passes from a medium of one density to one of a different density

regeneration (ree-jen-er-AY-shun) the process of replacing missing tissue with new tissue by means of cell division

regulation (reg-yoo-LAY-shun) process of control of body functions

regulator T cell (T-reg) T lymphocyte of the immune system that suppresses B-cell differentiation into plasma cells, allowing fine-tuning of antibody-mediated immune responses; also called *suppressor T cell*

rejection reaction (ree-JEK-shun re-AK-shun) immune responses to a donated or grafted tissue or organ; *see also* **alloimmunity**

releasing hormone (RH) (ree-LEE-sing HOR-mohn) hormone produced by the hypothalamus gland that causes the anterior pituitary gland to release its hormones

remission (ree-MISH-un) stage of a disease during which a temporary recovery from symptoms occurs

renal calculi (REE-nal KAL-kyoo-lye) kidney stones

renal colic (REE-nal KOL-ik) pain caused by the passage of a kidney stone

renal column (REE-nal KOL-um) extension of cortical tissue that dips down into the medulla of the kidney between the renal pyramids

renal corpuscle (REE-nal KOR-pus-ul) the part of the nephron located in the cortex of the kidney and made up of the glomerulus and glomerular (Bowman) capsule

renal cortex (REE-nal KOR-teks) (*pl.,* cortices) outer portion of the kidney

renal failure (REE-nal FAIL-yoor) acute or chronic loss of kidney function; acute kidney failure is often reversible, but chronic kidney failure slowly progresses to total loss of renal function (and death if kidney function is not restored through a kidney transplant or use of an artificial kidney)

renal medulla (REE-nal meh-DUL-ah) (*pl.,* medullae or medullas) inner portion of the kidney

renal papilla (REE-nal pah-PIL-ah) (*pl.,* papillae) nipplelike point of a renal pyramid, from which urine drips out of the kidney tubules

renal pelvis (REE-nal PEL-vis) basinlike upper end of the ureter that is located inside the kidney

renal ptosis (REE-nal TOH-sis) condition in which one or both kidneys descend, often because of loss of the fat pad that surrounds each kidney

renal pyramid (PIR-ah-mid) triangular-shaped division of the medulla of the kidney

renal threshold (REE-nal THRESH-hold) when the amount of a substance that is normally fully reabsorbed from tubular fluid (such as glucose) increases above this "threshold" level, the kidney tubules are unable to reabsorb all of it and the substances "spill over" into the urine

renal tubule (REE-nal TOOB-yool) one of the two principal parts of the nephron

renin (REE-nin) enzyme produced by the kidney that catalyzes the formation of angiotensin, a substance that increases blood pressure

renin-angiotensin-aldosterone system (RAAS) (REE-nin–an-jee-oh-TEN-sin–al-DAH-stayr-ohn SIS-tem) causes changes in blood plasma volume and blood pressure mainly by controlling aldosterone secretion

repolarization (ree-poh-lah-rih-ZAY-shun) begins just before the relaxation phase of cardiac muscle activity

reproductive system (ree-proh-DUK-tiv SIS-tem) produces hormones that permit the development of sexual characteristics and the propagation of the species

residual volume (RV) (reh-ZID-yoo-al VOL-yoom) the air that remains in the lungs after the most forceful expiration

respiration (res-pih-RAY-shun) breathing or pulmonary ventilation

respiratory acidosis (RES-pih-rah-tor-ee as-ih-DOH-sis) a respiratory disturbance that results in a carbonic acid excess

respiratory alkalosis (RES-pih-rah-tor-ee al-kah-LOH-sis) a respiratory disturbance that results in a carbonic acid deficit

respiratory arrest (RES-pih-rah-tor-ee ah-REST) cessation of breathing without resumption

respiratory control center (RES-pih-rah-tor-ee kon-TROL SEN-ter) nerve regulatory center located in the medulla and pons that stimulates the muscles of respiration

respiratory distress syndrome (RDS) (RES-pih-rah-tor-ee dih-STRESS SIN-drohm) difficulty in breathing caused by absence or failure of the surfactant in fluid lining the alveoli of the lung; IRDS is infant respiratory distress syndrome; ARDS is adult respiratory distress syndrome

respiratory membrane (RES-pih-rah-tor-ee MEM-brayn) the single layer of cells that makes up the wall of the alveoli

respiratory mucosa (RES-pih-rah-tor-ee myoo-KOH-sah) mucous-covered membrane that lines the tubes of the respiratory tree

respiratory muscle (RES-pih-rah-tor-ee MUS-el) any of the muscles responsible for the changing shape of the thoracic cavity that allows air to move in and out of the lungs

respiratory system (RES-pih-rah-tor-ee SIS-tem) the organs that allow the exchange of oxygen from the air with the carbon dioxide from the blood

respiratory therapist (RES-pih-rah-tor-ee THAYR-ah-pist) health professional who helps patients increase respiratory function and/or overcome or cope with the effects of respiratory conditions

respiratory tract (RES-pih-rah-tor-ee trakt) the two divisions of the respiratory system are the upper and lower respiratory tracts

reticular formation (reh-TIK-yoo-lar for-MAY-shun) located in the medulla where bits of gray and white matter mix intricately

reticular tissue (reh-TIK-yoo-lar) meshwork of netlike tissue that forms the framework of the spleen, lymph nodes, and bone marrow

retina (RET-ih-nah) innermost layer of the eyeball; contains rods and cones and continues posteriorly with the optic nerve

retinal detachment (RET-ih-nal) condition that occurs when part of the retina falls away from the tissue supporting it

retrograde endoscopic cholangiography (RET-roh-grayd en-doh-SKAHP-ik kohl-an-jee-OG-rah-fee) x-ray imaging of the bile duct system using an endoscope threaded through the major duodenal papilla to inject contrast material

retroperitoneal (reh-troh-payr-ih-toh-NEE-al) area outside of the peritoneum

retrovirus (ret-roh-VYE-rus) category of virus that uses its RNA to "transcribe backward" to produce the virus's primary genetic code and insert it into the host's DNA genome

Rh system classification of blood based on the presence (Rh^+) or absence (Rh^-) of a unique antigen on the surface of RBCs

rhabdomyolysis (RAB-doh-mye-OL-ih-sis) serious, acute condition resulting from damaged muscle fibers releasing their contents into the bloodstream

rheumatic heart disease (roo-MAT-ik hart dih-ZEEZ) cardiac damage (especially to the endocardium, including the valves) resulting from a delayed inflammatory response to streptococcal infection

rheumatoid arthritis (RA) (ROO-mah-toyd ar-THRY-tis) an autoimmune inflammatory joint disease characterized by synovial inflammation that spreads to other tissue

rhinitis (rye-NYE-tis) inflammation of the nasal mucosa often caused by nasal infections

Rh-negative (R H NEG-ah-tiv) red blood cells that do not contain the antigen called *Rh factor*

RhoGAM (ROH-gam) an injection of a special protein given to an Rh-negative woman who is pregnant to prevent her body from forming anti-Rh antibodies, which may harm an Rh-positive baby

Rh-positive (R H POZ-ih-tiv) red blood cells that contain an antigen called *Rh factor*

rib (rib) any of the 24 paired flat bones forming part of the framework of the thoracic wall

ribonucleic acid (RNA) (rye-boh-noo-KLAY-ik AS-id) a nucleic acid found in the cytoplasm that is crucial to protein synthesis

ribosomal RNA (rye-boh-SOHM-al R-N-A) also called *rRNA,* it is a form of RNA that makes up most of the structures (subunits) of the ribosome organelle of the cell

ribosome (RYE-boh-sohm) organelle in the cytoplasm of cells that synthesizes proteins; also known as a *protein factory*

rickets (RIK-ets) childhood form of osteomalacia, a bone-softening condition caused by vitamin D deficiency

Rickettsia (rih-KET-see-ah) small bacterium that infects human cells as an obligate parasite

right lymphatic duct (lim-FAT-ik) main lymphatic duct that drains lymph into the right subclavian vein

rigor mortis (RIG-or MOR-tis) literally "stiffness of death," the permanent contraction of muscle tissue after death caused by the depletion of ATP during the actin-myosin reaction—preventing myosin from releasing actin to allow relaxation of the muscle

risk factor (FAK-tor) predisposing condition; factor that puts one at a higher than usual risk for developing a particular disease

RNA (ar en ay) *see* **ribonucleic acid**

RNA interference (RNAi) (ar en ay in-ter-FEER-ens) a regulatory process of the cell in which a small molecule of dsRNA (double-stranded RNA) called *siRNA* (small interfering RNA) joins with a RISC (RNA-induced silencing complex) protein structure to break down a specific mRNA (messenger RNA) transcript and thus effectively silence the gene encoded by the mRNA; RNAi is a natural regulatory process thought to be involved with regulating gene expression, as in inhibiting viral infections, but is also used as a research technique to study the human genome (*RNA* ribonucleic acid)

RNAi therapy (ar en ay aye THAYR-ah-pee) any medical procedure in which RNAi techniques are used to silence (disable) the effects of a disease-causing gene; *see also* **RNA interference (RNAi)**

rod type of light receptor located in the retina responsible for monochrome, dim-light vision

root blunt tip of the tongue; portion of the tooth that fits into the socket of the alveolar process of either the upper or lower jaw

rotate (roh-TAYT) move in a circle around a central point

rotation (roh-TAY-shun) movement around a longitudinal axis; for example, shaking your head "no"

rugae (ROO-gee) (*sing.,* ruga) wrinkles or folds

rule of nines a frequently used method to determine the extent of a burn injury; the body is divided into 11 areas of 9% each to help estimate the amount of skin surface burned in an adult

S

sacrum (SAY-krum) bone of the lower vertebral column between the last lumbar vertebra and the coccyx, formed by the fusion of five sacral vertebrae

saddle joint (SAD-el joynt) type of diarthrotic joint formed by two saddle-shaped surfaces, allowing movement in two different axes

sagittal (SAJ-ih-tal) longitudinal; like an arrow
sagittal plane (SAJ-ih-tal playn) a longitudinal section or flat cut extending from front to back, dividing body or body part into right and left subdivisions
salivary amylase (SAL-ih-vayr-ee AM-ih-layz) digestive enzyme found in the saliva that begins the chemical digestion of carbohydrates (begins conversion of starch to smaller carbohydrate molecules)
salpingitis (sal-pin-JYE-tis) inflammation of the uterine (fallopian) tubes
salt compound formed when an acid and a base combine; sometimes specifically refers to the common salt, sodium chloride (NaCl)
saltatory conduction (SAL-tah-tor-ee kon-DUK-shun) when a nerve impulse encounters myelin and "jumps" from one node of Ranvier to the next
sarcoma (SAR-koh-mah) tumor of muscle tissue
sarcomere (SAR-koh-meer) contractile unit of muscle; length of a myofibril between two Z bands
SARS-associated coronavirus (SARS-CoV) (SARZ as-OH-see-ayt-ed koh-ROH-nah-vye-rus [SARZ koh-VEE]) a type of coronavirus shown to be the cause of severe acute respiratory syndrome (SARS); *see also* **severe acute respiratory syndrome (SARS)** and **coronavirus**
satiety center (sah-TYE-eh-tee SEN-ter) cluster of cells in the hypothalamus that send impulses to decrease appetite so that an individual feels satisfied
scabies (SKAY-bees) contagious skin condition caused by the itch mite *(Sarcoptes scabiei)*
scapula (SKAP-yoo-lah) shoulder blade
scar (skahr) thickened mass of tissue, usually fibrous connective tissue, that remains after a damaged tissue has been repaired
Schwann cell (shwon or shvon sel) large nucleated cell that forms myelin sheath around peripheral neurons
sciatica (sye-AT-ih-kah) neuralgia (pain) of the sciatic nerve
scientific method (sye-en-TIF-ik METH-odd) any logical and systematic approach to discovering principles of nature, often involving testing of tentative explanations called **hypotheses**
sclera (SKLEER-ah) white outer coat of the eyeball
scleroderma (skleer-oh-DER-mah) rare disorder affecting the vessels and connective tissue of skin and other tissues, characterized by tissue hardening
scoliosis (skoh-lee-OH-sis) abnormal lateral (side-to-side) curvature of the vertebral column
scotoma (skoh-TOH-mah) loss of the central visual field caused by nerve degeneration, it sometimes occurs with neuritis associated with multiple sclerosis
scrotum (SKROH-tum) pouchlike sac that contains the testes
scurvy (SKER-vee) condition caused by avitaminosis C (lack of vitamin C), which impairs the normal maintenance of collagen-containing connective tissues, causing bleeding and ulceration of the skin, gums, and other tissues
sebaceous gland (seh-BAY-shus) oil-producing glands found in the skin
sebum (SEE-bum) secretion of sebaceous glands
second-degree burn burn injury that is more severe than a first-degree burn and often involves damage to the dermis; *see also* **partial-thickness burn**
second messenger (SEK-und MES-en-jer) molecule that provides communication within the target cell of a chemical signal such as a hormone; for example, cyclic AMP
second-messenger mechanism (SEK-und MES-en-jer MEK-an-iz-em) a system of cellular communication (signal transduction) in which a molecule provides a communication link within the target cell of a chemical signal such as a hormone; for example, cyclic AMP links the external signal (arrival of the hormone or neurotransmitter) to the internal cellular processes that produce changes in the target cell
secondary bronchi (SEK-un-dayr-ee BRAHN-kye) (*sing.*, bronchus) smaller bronchial branches that result from division of the primary bronchi
secondary infection (SEK-on-dayr-ee in-FEK-shun) infection that occurs as a consequence of the weakened state of the body or damage caused by a previously existing disease
secondary protein (SEK-on-dayr-ee PROH-teen) second level of protein structure formed by the folding of the primary protein (string of amino acids) into helices (spirals) and pleated folds
secondary sex characteristic (SEK-on-dayr-ee seks kayr-ak-ter-IS-tik) any of the external physical characteristics of sexual maturity resulting from the action of the sex hormones; for example, growth of male and female patterns of body hair and fat distribution, external genital structures
secretion (seh-KREE-shun) in kidney function refers to active movement of substances such as electrolytes, waste products, or drugs through kidney tubule cells into the urine; in cells, secretion is the process of moving a substance out of the cell
secretory phase (SEEK-reh-toh-ree fayz) phase of menstrual cycle that begins at ovulation and lasts until the next menses begins
section (SEK-shun) a cut, ordinarily flat, through the body or any body part; *see also* **plane (of body)**
segmentation (seg-men-TAY-shun) occurs when digestive reflexes cause a forward-and-backward movement within a single region of the GI tract
seizure (SEE-zhur) sudden onset of abnormal body function, as in a brain seizure when a sudden disruption in the normal firing of neurons in the brain causes mild to severe neurological symptoms such as involuntary muscle spasms, changes in consciousness, or abnormal sensations
sella turcica (SEL-lah TER-sih-kah) small depression of the sphenoid bone that contains the pituitary gland
semen (SEE-men) male reproductive fluid or **seminal fluid**
semicircular canal (sem-ih-SIR-kyoo-lar kah-nal) any of three membranous, fluid-filled, curved tubes located in the inner ear; contains a sensory receptor called *crista ampullaris* that generates a nerve impulse on movement of the head
semilunar (SL) valve (sem-ih-LOO-nar valv) valve located between the two ventricular chambers and the large arteries that carries blood away from the heart; any of the valves found in the veins or lymphatic vessels
seminal fluid (SEM-ih-nal FLOO-id) *see* **semen**
seminal vesicle (SEM-ih-nal VES-ih-kul) paired, pouchlike glands that contribute about 60% of the seminal fluid volume; rich in fructose, which is a source of energy for sperm
seminiferous tubule (seh-mih-NIF-er-us TOOB-yool) long, coiled structure that forms the bulk of the testicular mass
senescence (seh-NES-enz) phase of human life cycle that constitutes older adulthood; process of aging
sense organ (sens OR-gan) any structure that detects changes in the internal or external environment of the body
sensor (SEN-sor) part of a homeostatic feedback loop that detects (senses) changes in the physiological variable that is regulated by the feedback loop
sensory neuron (SEN-sor-ee NOO-ron) neuron that transmits impulses to the spinal cord and brain from any part of the body

sensory receptor (SEN-soh-ree ree-sep-tohr) neuron in skin, internal organs, and muscles that allows body to detect various stimuli (changes)

septic shock (SEP-tik shok) circulatory failure (shock) resulting from complications of septicemia (toxins in blood resulting from infection)

serosa (see-ROH-sah) outermost covering of the digestive tract; composed of the parietal pleura in the abdominal cavity

serotonin (sayr-oh-TOH-nin) a neurotransmitter that belongs to a group of compounds called **catecholamines**

serous (SEE-rus) watery; refers to clear serous fluid or the type of membrane that produces it

serous membrane (SEE-rus MEM-brayn) a two-layer epithelial membrane that lines body cavities and covers the surfaces of organs

serum (SEER-um) blood plasma minus its clotting factors, still contains antibodies

severe acute respiratory syndrome (SARS) (seh-VEER ah-KYOOT res-pir-ah-TOR-ee SIN-drohm [sarz]) viral infection characterized by pneumonia and symptoms of fever, dry cough, dyspnea (shortness of breath), headache, hypoxia (low oxygen concentration in the blood), and sometimes progressing to death due to respiratory failure caused by damage to alveoli of the lungs

severe combined immune deficiency (SCID) (seh-VEER kom-BYNED ih-MYOON deh-FISH-en-see) nearly complete failure of the lymphocytes to develop properly, in turn causing failure of the immune system's defense of the body; very rare congenital immune disorder

sex chromosome (seks KROH-moh-sohm) either of a pair of chromosomes in the human genome that determine gender; normal males have one X chromosome and one Y chromosome (XY), whereas normal females have two X chromosomes (XX)

sex hormone (seks HOR-mohn) any hormone that has a direct effect on sexual structure or function, such as testosterone (male) and estrogens (female)

sex-linked trait (seks-linked trayt) nonsexual, inherited trait governed by genes located in a sex chromosome (X or Y); most known sex-linked traits are X-linked

sexually transmitted disease (STD) (SEKS-yoo-al-ee trans-MIH-ted dih-ZEEZ) any communicable disease that is commonly transmitted through sexual contact; compare to **sexually transmitted infection (STI)**

sexually transmitted infection (STI) (SEKS-yoo-al-ee trans-MIH-ted in-FEK-shun) any infection that is commonly transmitted through sexual contact and that may or may not produce symptoms; a sexually transmitted infection that produces symptoms (makes a person sick) may also be called a **sexually transmitted disease (STD)**

shaft *see* **diaphysis**

shingles (SHING-guls) *see* **herpes zoster**

sickle cell anemia (SIK-ul sel ah-NEE-mee-ah) severe, possibly fatal, hereditary disease caused by an abnormal type of hemoglobin

sickle cell trait (SIK-ul sel trayt) condition in which only one defective gene is inherited and only a small amount of hemoglobin that is less soluble than usual is produced

sigmoid colon (SIG-moyd KOH-lon) S-shaped segment of the large intestine that terminates in the rectum

sign (syne) objective deviation from normal (perceived by an examiner) that marks the presence of a disease

signal transduction (tranz-DUK-shen) term that refers to the whole process of getting a chemical signal (such as a hormone or neurotransmitter) to the inside of a cell; in a way, signal transduction is really "signal translation" by the cell

simple (SIM-pel) single, not complex

simple columnar epithelium (SIM-pel koh-LUM-nar ep-ih-THEE-lee-um) type of tissue that forms a membrane made up of a single layer of cells that are taller than they are wide

simple cuboidal epithelium (SIM-pel KYOO-boyd-al ep-ih-THEE-lee-um) type of tissue that forms a membrane made up of a single layer of cubelike cells

simple goiter (SIM-pel GOY-ter) condition in which the thyroid enlarges because iodine is lacking in the diet

simple squamous epithelium (SIM-pel SKWAY-mus ep-ih-THEE-lee-um) type of tissue that forms a membrane made up of a single layer of flattened cells

single-gene disease disease caused by individual mutant genes in nuclear DNA that pass from one generation to the next

sinoatrial (SA) node (sye-noh-AY-tree-al) the heart's pacemaker; where the impulse conduction of the heart normally starts; located in the wall of the right atrium near the opening of the superior vena cava

sinus (SYE-nus) a space or cavity inside some structures of the body, as inside the cranial bones (paranasal sinuses) and inside a lymph node; some large veins are also called *sinuses*

sinus dysrhythmia (SYE-nus dis-RITH-mee-ah) variation in the rhythm of heart rate during the breathing cycle (inspiration and expiration)

sinusitis (sye-nyoo-SYE-tis) sinus infection

skeletal muscle (SKEL-et-tal MUS-el) muscle tissue under willed or voluntary control; also known as *voluntary muscle*

skeletal muscle tissue *see* **skeletal muscle**

skeletal system (SKEL-et-tal SIS-tem) the bones, cartilage, and ligaments that provide the body with a rigid framework for support and protection

Skene gland (skeen gland) *see* **lesser vestibular gland**

skin *see* **cutaneous membrane**

skin graft surgical implantation of transplanted skin to replace burned or otherwise injured or removed skin

skull bony structure of the head

sliding filament model (SLY-ding FIL-ah-ment MAH-del) concept in muscle physiology describing the contraction of a muscle fiber in terms of the sliding of microscopic protein filaments past each other within the myofibrils in a manner that shortens the myofibrils and thus the entire muscle

small intestine (in-TEST-in) part of GI tract that includes duodenum, jejunum, and ileum

smooth muscle (MUS-el) muscle tissue that is not under conscious control; also known as *involuntary* or *visceral muscle;* forms the walls of blood vessels and hollow organs such as the stomach and small intestine

smooth muscle tissue *see* **smooth muscle**

snuff dipper's pouch precancerous leukoplakia (white patches) in fold between cheek and gum caused by use of smokeless tobacco

sodium-potassium pump (SOH-dee-um poh-TAS-ee-um) a system of coupled ion pumps that actively transports sodium ions out of a cell and potassium ions into the cell at the same time—found in all living cells

soft palate (soft PAL-let) soft, muscular posterior portion of the roof of the mouth

solute (SOL-oot) substance that dissolves into another substance; for example, in saltwater the salt is the solute dissolved in water

solution (suh-LOO-shun) liquid made up of a mixture of molecule types, usually made of solutes (solids) scattered in a solvent (liquid), such as salt in water

solvent (SOL-vent) substance in which other substances are dissolved; for example, in saltwater the water is the solvent for salt

somatic nervous system (soh-MAH-tik NER-vus SIS-tem) the motor neurons that control the voluntary actions of skeletal muscles

spastic paralysis (SPAS-tik pah-RAL-ih-sis) loss of voluntary muscle control characterized by involuntary contractions of affected muscles

special senses (SPESH-ul SEN-sez) senses detected by receptors in specific locations associated with complex structures and involve modes of smell, taste, vision, hearing, or equilibrium

specific immunity (spih-SIH-fik ih-MYOO-nih-tee) the protective mechanisms that provide specific protection against certain types of bacteria or toxins

sperm (*pl.,* sperms or sperm) the male spermatozoon; sex cell

spermatid (SPER-mah-tid) resulting daughter cell from the primary spermatocyte undergoing meiosis; a spermatid has only half the genetic material and half the chromosomes of regular body cells

spermatogenesis (sper-mah-toh-JEN-eh-sis) production of sperm cells

spermatogonia (sper-mah-toh-GOH-nee-ah) sperm precursor cells

spermatozoon (sper-mah-tah-ZOH-an) (*pl.,* spermatozoa) sperm cell or male sex cell (male gamete)

sphenoid bone (SFEE-noyd bohn) central keystone bone of the floor of the cranium; resembles a bat

sphincter (SFINGK-ter) ring-shaped muscle

sphygmomanometer (sfig-moh-mah-NOM-eh-ter) device for measuring blood pressure in the arteries of a limb

spinal cavity (SPY-nal KAV-ih-tee) the space inside the spinal column through which the spinal cord passes

spinal nerve (SPY-nal nerv) any of the nerves that connect the spinal cord to peripheral structures such as the skin and skeletal muscles

spinal tract (SPY-nal trakt) any of the white columns of the spinal cord that provide two-way conduction paths to and from the brain; *ascending tract* carries information to the brain, whereas *descending tracts* conduct impulses from the brain

spindle fiber (SPIN-dul FYE-ber) a network of microscopic tubules formed in the cytoplasm between the centrioles as they are moving away from each other during mitosis

spine (spyne) the vertebral column; a pointed ridge

spiral organ (SPY-rel OR-gun) the organ of hearing located in the cochlea with ciliated sensory receptor cells; also called *organ of Corti*

spirometer (spye-ROM-eh-ter) an instrument used to measure the amount of air exchanged in breathing

spleen largest lymphoid organ; filters blood, destroys worn-out red blood cells, salvages iron from hemoglobin, and serves as a blood reservoir

splenectomy (spleh-NEK-toh-mee) surgical removal of the spleen

splenic flexure (SPLEN-ik FLEK-shur) point at which the descending colon turns downward on the left side of the abdomen; also called *splenic colic flexure* or *left colic flexure*

splenomegaly (spleh-noh-MEG-ah-lee) condition of enlargement of the spleen

spongy bone (SPUN-jee) porous bone in the end of the long bone, which may be filled with marrow

spontaneous abortion (spon-TAY-nee-us ah-BOR-shun) miscarriage; loss of an embryo or fetus before the twentieth week of gestation (or fetus under a weight of 500 g)

spore (spor) nonreproducing form of a bacterium that resists adverse environmental conditions; spores revert to the active multiplying form when conditions improve

sporozoan (spor-oh-ZOH-an) (*pl.,* sporozoa) coccidia; parasitic protozoan that enters a host cell during one phase of a two-part life cycle

sprain (sprayn) an acute injury to soft tissues surrounding a joint, including muscle, tendon, and/or ligament

sputum (SPYOO-tum) material spit or coughed from the mouth

squamous (SKWAY-mus) scalelike

squamous cell carcinoma (SKWAY-mus sel kar-sih-NOH-mah) malignant tumor of the epidermis; slow-growing cancer that is capable of metastasizing; the most common type of skin cancer

squamous suture (SKWAY-mus SOO-chur) the immovable joint between the temporal bone and the sphenoid bone

stapes (STAY-peez) tiny, stirrup-shaped bone in the middle ear

staph (staf) a short word form for *Staphylococcus,* a category of bacteria that can infect the skin and other organs, sometimes seriously

static equilibrium (STAT-ik ee-kwih-LIB-ree-um) sense of the position of the body relative to gravity

statin (STAT-in) category of drugs that help control blood concentration of cholesterol

stem cell cell capable of dividing to produce new cell types

stenosed valve (steh-NOSD valv) cardiac valve that is narrower than normal, slowing blood flow from a heart chamber

Stensen duct (STEN-sen dukt) a duct of the parotid salivary glands; also known as **parotid duct**

stent (stent) metal spring or mesh tube inserted into an affected artery to keep it open

sterility (steh-RIL-ih-tee) as applied to humans, the inability to reproduce

sternoclavicular joint (ster-noh-klah-VIK-yoo-lar joynt) the direct point of attachment between the bones of the upper extremity and the axial skeleton

sternocleidomastoid (stern-oh-klye-doh-MAS-toyd) the diagonal "strap" muscle located on the anterior aspect of the neck

sternum (STER-num) breastbone

steroid (STAYR-oyd) any of a class of lipids related to sterols and forming numerous reproductive and adrenal hormones

steroid hormone (STAYR-oyd HOR-mohn) a type of lipid-soluble hormone that passes intact through the cell membrane of the target cell and influences cell activity by acting on specific genes

stillbirth (STIL-berth) delivery of a dead fetus (after twentieth week of gestation; before 20 weeks it is termed a **spontaneous abortion**)

stimulus (STIM-yoo-lus) (*pl.,* stimuli) agent that causes a change in the activity of a structure

stoma (STOH-mah) an opening, such as the opening created in a colostomy procedure

stomach (STUM-uk) an expansion of the digestive tract between the esophagus and small intestine

stork bite (stork byte) type of birthmark

strabismus (strah-BIS-mus) abnormal condition in which lack of coordination of, or weakness in, the muscles that control one or both eyes cause improper focusing of images on the retina, thus making depth perception difficult

strain (strayn) injury involving any component of the "musculotendinous unit"; although muscle is usually involved, the tendon, the junction between the two, as well as their attachments to bone, also may be involved

stratified (STRAT-ih-fyde) arranged in multiple layers

stratified squamous epithelium (STRAT-ih-fyde SKWAY-mus ep-ih-THEE-lee-um) type of tissue that forms a membrane made up of several layers of cells, with flattened cells in the surface layer(s)

stratified transitional epithelium (STRAT-ih-fyde tran-ZISH-en-al ep-ih-THEE-lee-um) type of tissue that forms a membrane made up of several layers of cells that can stretch out and flatten without breaking

stratum corneum (STRAH-tum KOR-nee-um) the tough outer layer of the epidermis; cells are filled with keratin

stratum germinativum (STRAH-tum jer-min-ah-TIV-um) the innermost layer of the tightly packed epithelial cells of the epidermis; cells in this layer are able to reproduce themselves

strawberry hemangioma (hem-an-jee-OH-mah) common pigmented and generally transient birthmark caused by a collection of dilated blood vessels

strength training (strengkth TRAYN-ing) contracting muscles against resistance to enhance muscle hypertrophy

stress (stres) actual or perceived threat to homeostatic balance or the physiological responses to such threat; pressure or load (mechanical stress)

Stretta procedure (STRET-ah proh-see-jur) procedure using an endoscope to deliver radiofrequency energy to burn, tighten, and reduce the size of the lumen of the lower esophageal sphincter in a person with gastroesophageal reflux disease (GERD)

striae (STRYE-ee) (*sing.,* stria) "stretch marks" caused by stretching of the skin beyond its ability to rebound

striated muscle (STRYE-ay-ted MUS-el) *see* **skeletal muscle**

stroke volume (SV) (strohk VOL-yoom) the amount of blood that is ejected from the ventricles of the heart with each beat; often expressed as mL/min (milliliters per minute)

structural protein (STRUK-shur-al PROH-teen) protein that has the role of building structures in the body, such as collagen fibers or keratin fibers; compare to **functional protein**

subcutaneous injection (sub-kyoo-TAY-nee-us in-JEK-shun) injection of liquid or pelleted material into the spongy and porous subcutaneous layer beneath the skin

subcutaneous tissue (sub-kyoo-TAY-nee-us TISH-yoo) *see* **hypodermis**

subendocardial branch (sub-en-doh-KAR-dee-al branch) *see* Purkinje fiber

sublingual gland (sub-LING-gwal gland) salivary gland that drains saliva into the floor of the mouth

subluxation (sub-luks-AY-shun) abnormal, partial separation of the bones in a joint; also called *incomplete dislocation*

submandibular gland (sub-man-DIB-yoo-lar gland) salivary glands that drain saliva into the mouth on either side of the lingual frenulum

submucosa (sub-myoo-KOH-sah) connective tissue layer containing blood vessels and nerves in the wall of the digestive tract

sucrase (soo-krays) enzyme that catalyzes the hydrolysis of sucrose into glucose and fructose; also called *invertase* or *saccharase*

sucrose (SOO-krohs) sugar made of a double saccharide molecule made up of a glucose unit and fructose unit

sudden infant death syndrome (SIDS) unexpected death of unknown origin in apparently normal infants; sometimes called "crib death"

sudoriferous gland (soo-doh-RIF-er-us) glands that secrete sweat; also referred to as *sweat glands*

sulcus (SUL-kus) (*pl.,* sulci) furrow or groove

superficial (soo-per-FISH-al) near the body surface

superficial fascia (soo-per-FISH-al FAH-shah) hypodermis; subcutaneous layer beneath the dermis

superior (soo-PEER-ee-or) higher; opposite of **inferior**

superior vena cava (soo-PEER-ee-or VEE-nah KAY-vah) one of two large veins returning deoxygenated blood to the right atrium

supinate (soo-pih-NAYT) to make a rotational movement of the forearm (turning the palm laterally to face forward) or of the leg and ankle (turning the foot so toes point inward and the lateral edge of the sole hits the ground); opposite of **pronate**

supination (soo-pih-NAY-shun) action in which the forearm or leg and ankle **supinates;** opposite of **pronation**

supine (SOO-pyne) description of the body lying in a horizontal position facing upward

supraclavicular (soo-prah-klah-VIK-yoo-lar) area above the clavicle

surfactant (sur-FAK-tant) a substance covering the surface of the respiratory membrane inside the alveolus; it reduces surface tension and prevents the alveoli from collapsing

suture (SOO-chur) immovable joint

sweat (swet) *see* **perspiration**

sweat gland (swet) *see* **sudoriferous gland**

sympathetic division (sim-pah-THET-ik dih-VIZH-un) part of the autonomic nervous system; ganglia are connected to the thoracic and lumbar regions of the spinal cord; functions as an emergency system

sympathetic postganglionic neurons (sim-pah-THET-ik post-gang-glee-ON-ik NOO-rons) dendrites and cell bodies are in sympathetic ganglia and axons travel to a variety of visceral effectors

sympathetic preganglionic neuron (sim-pah-THET-ik pree-gang-glee-ON-ik NOO-ron) nerve cell whose dendrites and cell bodies are located in the gray matter of the thoracic and lumbar segments of the spinal cord; leaves the cord through a ventral root of a spinal nerve and terminates in a collateral ganglion

symptom (SIMP-tum) subjective deviation from normal that marks the presence of a disease (perceived by a patient)

synapse (SIN-aps) junction between adjacent neurons

synaptic cleft (sin-AP-tik kleft) the space between a synaptic knob and the plasma membrane of a postsynaptic neuron

synaptic knob (sin-AP-tik nob) a tiny bulge at the end of a terminal branch of a presynaptic neuron's axon that contains vesicles with neurotransmitters

synarthrosis (sin-ar-THROH-sis) a joint in which fibrous connective tissue joins bones and holds them together tightly; commonly called **sutures**

syndrome (SIN-drohm) collection of signs or symptoms, usually with a common cause that defines or gives a clear picture of a pathological condition

synergist (SIN-er-jist) muscle that assists a prime mover

synovial fluid (sih-NOH-vee-al FLOO-id) the thick, colorless lubricating fluid secreted by the synovial membrane

synovial membrane (sih-NOH-vee-al MEM-brayn) connective tissue membrane lining the spaces between bones and joints that secretes synovial fluid

system (SIS-tem) group of organs arranged so that the group can perform a more complex function than any one organ can perform alone

systemic circulation (sis-TEM-ik ser-kyoo-LAY-shun) blood flow from the left ventricle to all parts of the body and back to the right atrium

systemic lupus erythematosus (SLE) (sis-TEM-ik LOO-pus er-ih-them-ah-TOH-sus) chronic inflammatory disease caused by widespread attack of self-antigens by the immune system (autoimmunity); characterized by a red rash on the face and other signs

systole (SIS-toh-lee) contraction of the heart muscle

systolic blood pressure (sis-TOL-ik blud PRESH-ur) force with which blood pushes against artery walls when ventricles contract

T

T cell (tee sel) *see* **T lymphocyte**

T lymphocyte (tee LIM-foh-syte) cells of the immune system that have undergone maturation in the thymus; produces cell-mediated immunity

T wave electrocardiogram deflection that reflects the repolarization of the ventricles

tachycardia (tak-ih-KAR-dee-ah) rapid heart rhythm (greater than 100 beats/minute)

tactile corpuscle (TAK-tyle KOR-pus-ul) a sensory receptor located in the skin close to the surface that detects light touch; also known as **Meissner corpuscle**

talus (TAY-lus) second largest tarsal (ankle) bone and articulates with the tibia at the ankle joint

target cell (TAR-get sel) cell that is acted on by a particular hormone (or other signaling molecule) and then responds to it

target organ (TAR-get OR-gan) organ containing *target cells* that are acted on by a particular hormone (or other signaling molecule) and then responds to it

tarsal (TAR-sal) relating to a flat plate or specifically to the heel; any of the seven bones of the heel and back part of the foot; the *calcaneus* is the largest tarsal bone

tarsal gland (TAR-sal gland) *see* **meibomian gland**

taste bud tongue structure housing chemical receptors that detect the sense of taste

Tay-Sachs disease (TAY-saks dih-ZEEZ) recessive, inherited condition in which abnormal lipids accumulate in the brain and cause tissue damage that leads to death by age 4

telemetry (tel-EM-eh-tree) technology by which data, such as heart activity monitored by an electrocardiograph, can be sent to a remote location through telephone wires, radio waves, or other communication pathway

telophase (TEL-oh-fayz or TEE-loh-fayz) last stage of mitosis in which the cell divides

temporal (TEM-poh-ral) relating to time or to the side of the head; muscle that assists the masseter in closing the jaw

temporal bone (TEM-poh-ral bohn) cranial bone located on lower side of cranium and part of its floor

tendon (TEN-don) band or cord of fibrous connective tissue that attaches a muscle to a bone or other structure

tendon sheath (TEN-don sheeth) tube-shaped structure lined with synovial membrane that encloses certain tendons

tenosynovitis (ten-oh-sin-oh-VYE-tis) inflammation of a tendon sheath

teratogen (TER-ah-toh-jen) any environmental factor that causes a birth defect (abnormality present at birth); common teratogens include radiation (e.g., x-rays), chemicals (e.g., drugs, cigarettes, or alcohol), and infections in the mother (e.g., herpes or rubella)

tertiary protein (TER-shee-ayr-ee PROH-teen) third level of protein structure formed by further folding of the secondary protein structure

testis (TES-tis) (*pl.,* testes) male gonad responsible for production of male sex cells or gametes (sperm) and testosterone

testosterone (tes-TOS-teh-rohn) male sex hormone produced by the interstitial cells in the testes; the "masculinizing hormone"

tetanic contraction (teh-TAN-ik kon-TRAK-shun) sustained contraction of muscle

tetanus (TET-ah-nus) state of sustained muscular contraction

thalamus (THAL-ah-mus) located just above the hypothalamus; its functions are to help produce sensations, associate sensations with emotions, and play a part in the arousal mechanism

thalassemia (thal-ah-SEE-mee-ah) any of a group of inherited hemoglobin disorders characterized by production of hypochromic, abnormal red blood cells

theory (THEE-ah-ree) an explanation of a scientific principle that has been tested experimentally and found to be true; compare to **hypothesis** and **law**

thermoreceptor (ther-moh-ree-SEP-tor) sensory receptor activated by heat or cold

thermoregulation (ther-moh-reg-yoo-LAY-shun) maintaining homeostasis of body temperature

third-degree burn involves complete destruction of both epidermis and dermis with injury extending into subcutaneous tissue; *see* **full-thickness burn**

thoracic (thoh-RAS-ik) relating to the chest area of the body (upper trunk)

thoracic cavity (thoh-RAS-ik KAV-ih-tee) organ-containing space inside the rib cage or chest of the body that includes the mediastinum and left and right pleural cavities

thoracic duct (thoh-RAS-ik dukt) largest lymphatic vessel in the body

thorax (THOR-aks) bony cage of the upper torso formed by 12 pairs of ribs, the sternum, and thoracic vertebrae; also called the *chest*

threshold stimulus (THRESH-hold STIM-yoo-lus) minimal level of stimulation required to cause a muscle fiber to contract

thrombin (THROM-bin) protein important in blood clotting

thrombocyte (THROM-boh-syte) type of blood cell that plays a role in blood clotting; also called *platelet*

thrombocytopenia (throm-boh-sye-toh-PEE-nee-ah) general term referring to an abnormally low blood platelet count

thrombophlebitis (throm-boh-fleh-BYE-tis) vein inflammation (phlebitis) accompanied by clot formation

thrombosis (throm-BOH-sis) formation of a clot in a blood vessel

thromboxane (throm-BOKS-ayne) prostaglandin-like substance in platelets that plays a role in hemostasis and blood clotting

thrombus (THROM-bus) stationary blood clot

thrush candidiasis of mouth (mouth infection) characterized by white, creamy patches of exudate on inflamed oral mucosa and tongue; caused by yeastlike fungal organism

thymine (THYE-meen) one of several nitrogen-containing bases that make up nucleotides, which in turn make up nucleic acids such as DNA (but not RNA); in the cell, it can bind to another nitrogenous base, adenine (*A* or *a*), to form a more complex structure or in translating genetic codes; symbolized by the letter *T* or *t*; *see also* **guanine, adenine, cytosine, uracil**

thymosin (THY-moh-sin) hormone produced by the thymus that is vital to the development and functioning of the body's immune system

thymus *see* **thymus gland**

thymus gland (THY-mus) endocrine gland located in the mediastinum; vital part of the body's immune system; often called simply "the thymus"

thyroid follicle (THY-royd FOL-lih-kul) pocket of thyroid colloid (suspended, stored form of thyroid hormone) in the thyroid gland

thyroid gland (THY-royd) endocrine gland located in the neck that stores its hormones until needed; thyroid hormones regulate cellular metabolism
thyroid-stimulating hormone (TSH) (THY-royd STIM-yoo-lay-ting HOR-mohn) a tropic hormone secreted by the anterior pituitary gland that stimulates the thyroid gland to increase its secretion of thyroid hormone
thyroxine (T_4) (thy-ROK-sin) thyroid hormone that stimulates cellular metabolism
tibia (TIB-ee-ah) shinbone
tibialis anterior (tib-ee-AL-is an-TEER-ee-or) dorsiflexor of the foot
tic douloureux (tik doo-loo-ROO) *see* **trigeminal neuralgia**
tidal volume (TV) (TYE-dal VOL-yoom) amount of air breathed in and out with each breath
tinea (TIN-ee-ah) fungal infection of the skin
tinea pedis (TIN-ee-ah PED-is) a fungal infection of the skin of the foot characterized by redness and itching; also called *athlete's foot*
tinnitus (tih-NYE-tus or TIN-it-us) abnormal sensation of ringing or buzzing in the ear
tissue (TISH-yoo) group of similar cells that perform a common function
tissue fluid (TISH-yoo FLOO-id) a dilute saltwater solution that bathes every cell in the body; also called *interstitial fluid*
tissue hormone (TISH-yoo HOR-mohn) prostaglandins; produced in a tissue and diffused only a short distance to act on cells within the tissue
tissue plasminogen activator (TPA or tPA) (TISH-yoo plaz-MIN-oh-jen AK-tih-vay-tor) naturally occurring substance that activates plasminogen and converts it to the active enzyme plasmin, which in turn dissolves fibrin blood clots
tissue typing (TISH-yoo TYE-ping) a procedure used to identify tissue compatibility before an organ transplant
T lymphocytes (T LIM-foh-sytes) cells that are critical to the function of the immune system; produce cell-mediated immunity
tone (tohn) tension; baseline contraction of any muscle
tonic contraction (TAHN-ik kon-TRAK-shun) sustained, baseline skeletal muscle contraction used to maintain posture
tonsil (TAHN-sil) mass of lymphoid tissue; protects against bacteria; three main sets: palatine tonsils, located on each side of the throat; pharyngeal tonsils (adenoids), near the posterior opening of the nasal cavity; and lingual tonsils, near the base of the tongue
tonsillectomy (tahn-sih-LEK-toh-mee) surgical procedure used to remove the tonsils
tonsillitis (tahn-sih-LYE-tis) inflammation of the tonsils, usually caused by infection
tophus (TOH-fus) (*pl.,* tophi) stonelike growths or deposits in tissues or around joints; may contain urate crystals in patients with gout
total metabolic rate (TMR) (TOH-tal met-ah-BOL-ik rayt) total amount of energy used by the body per day
toxicologist (tok-sih-KOL-uh-jist) scientist who studies the effects, treatments, and detection of poisons
trabecula (trah-BEK-yoo-lah) (*pl.,* trabeculae) branched, needlelike thread of tissue (for example, bone) that forms a network of spaces
trachea (TRAY-kee-ah) the windpipe; the tube extending from the larynx to the bronchi
tracheostomy (tray-kee-OS-toh-mee) medical procedure involving the cutting of an opening into the trachea
trachoma (trah-KOH-mah) chronic infection of the conjunctiva covering the eye caused by the bacterium *Chlamydia trachomatis;* also called *chlamydial conjunctivitis*
tract (trakt) a single nerve pathway made up of several bundles of axons and extending through the central nervous system; compare to **nerve**
tragus (TRAY-gus) prominent bump of skin-covered cartilage of the auricle (external ear) that lies just anterior the opening to the acoustic canal (ear canal)
transaminase (trans-AM-ih-nayz) enzyme released from damaged tissues; high blood concentration may indicate a heart attack or other pathological event
transcellular fluid (tranz-SEL-yoo-lar FLOO-id) part of the extracellular fluid that includes cerebrospinal fluid (CSF), fluids of the eyeball, and the synovial joint fluids (but not blood plasma or interstitial fluid)
transcription (trans-KRIP-shun) action that occurs when the double-stranded DNA molecule unwinds and becomes a template to form mRNA, thus making a copy of a gene
transfer RNA (TRANS-fer R N A) type of ribonucleic acid (RNA) that temporarily binds to specific amino acids and transfers them to specific sequences (codons) on a messenger RNA (mRNA) molecule; also known as *tRNA*
transitional (tranz-IH-shen-al) relating to a change or something capable of change, as in *transitional epithelium* (which can change cell shape as the tissue stretches)
transitional epithelium (tranz-IH-shen-al ep-ih-THEE-lee-um) type of epithelial tissue that forms membranes capable of stretching without breaking, as in the urinary bladder; cells in this type of tissue can stretch from a roughly columnar shape out to a flattened (squamous) shape and back again without sustaining damage; also called *stratified transitional epithelium*
translation (trans-LAY-shun) the synthesis of a protein by ribosomes (by translating genetic code)
transplant (trans-PLANT) tissue or organ graft; procedure in which a tissue (e.g., skin, bone marrow) or an organ (such as kidney, liver) from a donor is surgically implanted into a recipient
transport maximum (T_{max}) (TRANZ-port MAKS-im-um) the largest amount of any substance that can be moved by a cellular transporter (pump or carrier) at one time, determined mainly by the number of available transporters of that substance
transport process (TRANZ-port PROH-ses) process of carrying materials within the body, often across membranes and within fluids
transverse arch (TRANS-vers) *see* **metatarsal arch**
transverse canal (tranz-VERS kah-NAL) communicating canal between central (Haversian) canals that contains vessels to carry blood to the osteons; also carries nerves and lymphatic vessels; also called *Volkmann canal*
transverse colon (tranz-VERS KOH-len) division of the colon that passes horizontally across the abdomen
transverse fracture (TRANS-vers FRAK-chur) bone fracture characterized by a fracture line that is at a right angle to the long axis of the bone
transverse plane (tranz-VERS playn) a flat cut through the body (or a body part) that is horizontal or crosswise and thus divides the body (or body part) into upper and lower portions; *see also* **section (of body)**
transversus abdominis (trans-VER-sus ab-DAH-min-us) the innermost layer of the anterolateral abdominal wall
trapezium (trah-PEE-zee-um) the carpal bone of the wrist that forms the saddle joint that allows the opposition of the thumb
trapezius (trah-PEE-zee-us) triangular muscle in the back that elevates the shoulder and extends the head backward
traumatic (truh-MAT-ik) relating to injury (trauma)

triceps brachii (TRY-seps BRAY-kee-aye) extensor muscle of the elbow

tricuspid (try-KUS-pid) describes anything having three angles or points (cusps)

tricuspid tooth (try-KUS-pid) tooth with rather large flat surface with two or three grinding "cusps"; also called *molar*

tricuspid valve (try-KUS-pid valv) the valve located between the right atrium and ventricle

trigeminal neuralgia (try-JEM-ih-nal noo-RAL-jee-ah) pain in one or more (of three) branches of the fifth cranial nerve (trigeminal nerve) that runs along the face; also called *tic douloureux*

triglyceride (try-GLIH-ser-ayed) lipid that is synthesized from fatty acids and glycerol or from excess glucose or amino acids; stored mainly in adipose tissue cells

trigone (TRY-gon) triangular area on the wall of the urinary bladder

triiodothyronine (T_3) (try-aye-oh-doh-THY-roh-neen) thyroid hormone that stimulates cellular metabolism

trimester three-month segments of the gestation period (pregnancy)

triple therapy (TRIP-pul THAYR-ah-pee) treatment of ulcers using a combination of bismuth subsalicylate (Pepto-Bismol) and two antibiotics

triplegia (try-PLEE-jee-ah) paralysis (loss of voluntary muscle control) in three limbs, often two legs and one arm

trisomy (TRY-soh-mee) abnormal genetic condition in which cells have three chromosomes (a triplet) where there should be a pair; usually caused by nondisjunction (failure of chromosome pairs to separate) during gamete production

tropic hormone (TROH-pik HOR-mohn) hormone that stimulates another endocrine gland to grow and secrete its hormones

true ribs the first seven pairs of ribs, which are attached to the sternum

trypsin (TRIP-sin) protein-digesting enzyme (protease)

tubal pregnancy (TOO-bal PREG-nan-see) ectopic pregnancy that occurs in a uterine (fallopian) tube

tuberculosis (TB) (too-ber-kyoo-LOH-sis) chronic bacterial (bacillus) infection of the lungs or other tissues caused by *Mycobacterium tuberculosis* organisms

tumor (TOO-mer) growth of tissues in which cell proliferation is uncontrolled and progressive

tumor suppressor gene gene that works against the development of cancerous cells

tunica albuginea (TOO-nih-kah al-byoo-JIN-ee-ah) a tough, whitish membrane that surrounds each testis and enters the gland to divide it into lobules

tunica externa (TOO-nih-kah eks-TER-nah) the outermost layer found in blood vessels

tunica intima (TOO-nih-kah IN-tih-mah) endothelium that lines blood vessels; also called *tunica interna*

tunica media (TOO-nih-kah MEE-dee-ah) the muscular middle layer found in blood vessels; the tunica media of arteries is more muscular than that of veins

turbinate (TUR-bih-nayt) *see* **concha**

turgor (TUR-ger) resiliency or fluid pressure in the cells of the skin, often lost during dehydration

Turner syndrome (TUR-ner SIN-drohm) genetic disorder caused by monosomy of the X chromosome (XO) in females; characterized by immaturity of sex organs (causing sterility), webbed neck, cardiovascular defects, and learning disorders

T wave deflection on an electrocardiogram that occurs with repolarization of the ventricles

twitch a quick, jerky response to a single stimulus

tympanic (tim-PAN-ik) drumlike

tympanic membrane (tim-PAN-ik MEM-brayn) eardrum; membrane that separates external ear canal from middle ear

type 1 diabetes mellitus (dye-ah-BEE-teez mel-AYE-tus) a condition in which the pancreatic islets secrete too little insulin, resulting in increased levels of blood glucose; formerly known as *juvenile-onset diabetes* or *insulin-dependent diabetes mellitus*

type 2 diabetes mellitus (dye-ah-BEE-teez mel-AYE-tus) a condition in which cells of the body become less sensitive to the hormone insulin and perhaps the pancreatic islets secrete too little insulin, resulting in increased levels of blood glucose; formerly known as *maturity-onset diabetes* or *non–insulin-dependent diabetes mellitus*

U

ulcer (UL-ser) a necrotic open sore or lesion

ulna (UL-nah) one of the two forearm bones; located on the little finger side

ulnar deviation (UL-nur dee-vee-AY-shun) deformity of the hands as a result of rheumatoid arthritis

ultrasonogram (ul-trah-SOHN-oh-gram) a record obtained by using sound to produce images

ultrasonography (ul-trah-son-OG-rah-fee) an imaging technique in which high-frequency sound waves are reflected off tissue to form an image

umami (oo-MOM-ee) "meaty" or "savory" taste produced by glutamatic acid (an amino acid)

umbilical (um-BIL-ih-kul) relating to the navel or umbilicus, a structure made up of blood vessels connecting the developing fetus to the placenta

umbilical artery (um-BIL-ih-kul AR-ter-ee) two small arteries that carry oxygen-poor blood from the developing fetus to the placenta

umbilical cord (um-BIL-ih-kul) flexible structure connecting the fetus to the placenta, which allows the umbilical arteries and vein to pass

umbilical region (um-BIL-ih-kul REE-jun) the very center region of the abdominopelvic cavity, near the umbilicus (navel) and between the left and right lumbar regions; terminology used when the abdominopelvic cavity is visualized as being subdivided into nine regions as in a tic-tac-toe grid

umbilical vein (um-BIL-ih-kul vayn) a large vein carrying oxygen-rich blood from the placenta to the developing fetus

universal donor blood blood type O (or O−)

universal recipient blood blood type AB (or AB+)

upper esophageal sphincter (UES) (eh-sof-eh-JEE-al SFINGK-ter) ring of muscular tissue (sphincter) located between laryngopharynx and proximal end of esophagus

upper GI (UGI) (upper jee aye) the stomach and esophagus; an x-ray study of the lower esophagus, stomach, and duodenum, produced with the aid of a contrast medium and used to detect ulcerations, tumors, inflammations, or anatomical malpositions such as *hiatal hernia*

upper respiratory infection (URI) (RES-pih-rah-toree) infection localized in the mucosa of the upper respiratory tract (primarily the nose, nasal sinuses, and/or pharynx)

upper respiratory tract (UP-er res-PYE-rah-tor-ee trakt) division of respiratory tract outside the thorax that is composed of the nose, pharynx, and larynx

uracil (YOOR-ah-sil) one of several nitrogen-containing bases that make up nucleotides, which in turn make up nucleic acids such as

RNA (but not DNA); in the cell, it can chemically bind to another nitrogenous base, adenine (*A* or *a*), to form a more complex structure or in translating genetic codes; symbolized by the letter *U* or *u*; *see also* **guanine, adenine, thymine, cytosine**
urea (yoo-REE-ah) nitrogen-containing waste product
uremia (yoo-REE-mee-ah) condition in which blood urea concentration is abnormally elevated, expressed as a high BUN (blood urea nitrogen) value; uremia is often caused by renal failure; also called *uremic poisoning*
uremic poisoning (yoo-REE-mik POY-zon-ing) *see* **uremia**
ureter (YOO-reh-ter) muscular tube that conducts urine from the kidney to the urinary bladder
urethra (yoo-REE-thrah) passageway for elimination of urine; in males, also acts as a genital duct that carries sperm to the exterior
urethritis (yoo-reh-THRY-tis) inflammation or infection of the urethra
urinalysis (yoor-in-AL-is-is) clinical laboratory testing of urine samples
urinary bladder (YOOR-ih-nayr-ee BLAD-er) collapsible saclike organ that collects urine from the kidneys and stores it before elimination from the body
urinary incontinence (YOOR-ih-nayr-ee in-KON-tih-nens) condition in which an individual voids urine involuntarily; compare to **incontinence**
urinary meatus (YOOR-ih-nayr-ee mee-AY-tus) external opening of the urethra
urinary retention (YOOR-in-ayr-ee ree-TEN-shun) condition in which no urine is voided
urinary suppression (YOOR-in-ayr-ee sup-PRESH-un) condition in which kidneys do not produce urine
urinary system (YOOR-ih-nayr-ee SIS-tem) system responsible for excreting liquid waste from the body
urinary tract infection (UTI) (YOOR-ih-nayr-ee trakt in-FEK-shun) infection of the mucosa that lines the urinary tract
urination (yoor-ih-NAY-shun) passage of urine from the body; emptying of the bladder
urine (YOOR-in) fluid waste excreted by the kidneys
urologist (yoo-ROL-uh-jist) scientist or physician specializing in urine and the urogenital tract and its disorders
urticaria (er-tih-KAYR-ee-ah) an allergic or hypersensitivity response characterized by raised red lesions; also referred to as *hives*
uterine tube (YOO-ter-in toob) either of a pair of tubes that conduct the ovum from the ovary to the uterus; also called *fallopian tube* or *oviduct*
uterus (YOO-ter-us) hollow, muscular organ where a fertilized egg implants and grows
uvula (YOO-vyoo-lah) cone-shaped process hanging down from the soft palate that helps prevent food and liquid from entering the nasal cavities

V

vaccine (VAK-seen) application of killed or attenuated (weakened) pathogens (or portions of pathogens) to a patient to stimulate immunity against that pathogen
vagina (vah-JYE-nah) internal tube from uterus to vulva
vaginitis (vaj-ih-NYE-tis) inflammation of the vagina
variant Creutzfeldt-Jakob disease (vCJD) (VAYR-ee-ant KROYTS-felt YAH-kohb dih-ZEEZ) a degenerative disease of the central nervous system caused by prions (proteinaceous infectious particles) that convert normal proteins of the nervous system into abnormal proteins, causing loss of function; *see* **prion**
varicose vein (VAYR-ih-kohs vayn) enlarged vein in which blood pools; also called *varix*
varix (VAYR-iks) (*pl.*, varices) *see* **varicose vein**
vas deferens (vas DEF-er-enz) (*pl.*, vasa deferentia) *see* **ductus deferens**
vasectomy (va-SEK-toh-mee) surgical severing of the vas deferens to render a male sterile
vasoconstriction (vay-soh-kon-STRIK-shun) reduction in vessel diameter caused by increased contraction of the muscular coat (smooth muscle)
vasodilator (vay-so-DYE-lay-tor) class of drugs that trigger the smooth muscles of arterial walls to relax, causing the arteries to dilate
vasomotor mechanism (vay-soh-MOH-tor MEK-ah-niz-em) factors that control changes in the diameter of arterioles by changing the tension of smooth muscles in the vessel walls
vastus (VAS-tus) wide; of great size
vector (VEK-tor) arthropod that carries an infectious pathogen from one organism to another
vein (vayn) vessel carrying blood toward the heart
venous return (VEE-nus) amount of blood returned to the heart by the veins
ventral (VEN-tral) of or near the belly; in humans, front or **anterior;** opposite of **dorsal** or **posterior**
ventral body cavity (VEN-tral BOD-ee KAV-it-ee) organ-containing space in the anterior trunk of the body; for example, the thoracic and abdominopelvic cavities; compare with **dorsal body cavity**
ventricle (VEN-trih-kul) any small cavity or space
ventricular fibrillation (VF or V-fib) (ven-TRIK-yoo-lar fib-ril-LAY-shun) life-threatening condition in which the lack of ventricular pumping suddenly stops the flow of blood to vital tissues; unless ventricular fibrillation is corrected immediately by defibrillation or some other method, death may occur within minutes; *see also* **automatic external defibrillator**
venule (VEN-yool) small blood vessels that collect blood from the capillaries and join to form veins
vermiform appendix (VERM-ih-form ah-PEN-diks) a tubular structure attached to the cecum and composed of lymphoid tissue
vertebra (VER-teh-bra) (*pl.*, vertebrae) any of the bones that make up the spinal (vertebral) column
vertebral column (ver-TEE-bral KOL-um) the spinal column, made up of a series of separate vertebrae that form a flexible, curved rod
vertebroplasty (ver-tee-broh-PLAS-tee) orthopedic procedure used to treat the vertebral compression fractures that occur in osteoporosis; involves injecting bone cement, but without using a balloon
vertigo (VER-tih-go) abnormal sensation of spinning; dizziness
vesicle (VES-ih-kul) a clinical term referring to blisters, fluid-filled skin lesions
vestibular nerve (ves-TIB-yoo-lar nerv) a division of the vestibulocochlear nerve (the eighth cranial nerve)
vestibule (VES-tih-byool) located in the inner ear; the portion adjacent to the oval window between the semicircular canals and the cochlea
vestibule of the vulva (VES-tih-byool) the area between the labia minora; the clitoris and the orifice of the urethra are located in the vestibule
villus (VIL-us) (*pl.*, villi) any of the soft, fingerlike projections that cover the plicae (folds) of the small intestine

Vincent infection (VIN-sent in-FEK-shun) bacterial (spirochete) infection of the gum, producing gingivitis; also called *Vincent angina* and *trench mouth*

virilizing tumor (VEER-il-aye-zing TOOM-er) neoplasm of the adrenal cortex that stimulates overproduction of testosterone and therefore an increase in masculinization, even of women

virus (VYE-rus) microscopic, parasitic entity consisting of a nucleic acid bound by a protein coat and sometimes a lipoprotein envelope

visceral (VIS-er-al) relating to the viscera or internal organs

visceral effector (VIS-er-al ee-FEK-tor) any muscle or gland (effector) found within the cavities of the body and controlled by the autonomic nervous system; examples include cardiac muscle tissue, smooth muscle tissue, and internal glands

visceral muscle (VIS-er-al MUS-el) *see* **smooth muscle** and **involuntary muscle**

visceral pericardium (VIS-er-al payr-ih-KAR-dee-um) the pericardium that covers the heart; also called **epicardium**

visceral peritoneum (VIS-er-al payr-ih-TOHN-ee um) serous membrane that covers and is adherent to the abdominal viscera

visceral pleura (VIS-er-al PLOO-rah) serous membrane that covers and is adherent to the surface of the lungs

visceral portion (VIS-er-al POR-shun) serous membrane that covers the surface of organs found in the body cavity

vital capacity (VC) (VYE-tal kah-PAS-ih-tee) largest amount of air that can be moved in and out of the lungs in one inspiration and expiration

vitamin (VYE-tah-min) organic molecule needed in small quantities to help enzymes operate effectively or to otherwise regulate metabolism in the body

vitiligo (vit-ih-LYE-go) patchy areas of light skin caused by acquired loss of epidermal melanocytes

vitreous humor (VIT-ree-us HYOO-mor) the jellylike fluid found in the eye, posterior to the lens

vocal cords (VOH-kull kords) bands of tissue in larynx responsible for production of sound (speech)

voiding (VOYD-ing) emptying of the bladder

volar (VOH-lar) relating to the palm or sole

voluntary muscle (VOL-un-tayr-ee MUS-el) *see* **skeletal muscle**

vulva (VUL-vah) folds and other structures that together make up the external genitals of the female

vulvitis (vul-VYE-tis) inflammation of the vulva (the external female genitals)

W

wart raised bump that is a benign neoplasm (tumor) of the skin caused by viruses

water intoxication (WAH-ter in-TOK-sih-kay-shen) possibly life-threatening neurological impairment caused by severe overhydration and accompanying electrolyte imbalance

West Nile virus (WNV) (nyle VY-rus) sometimes fatal viral infection caused by a type of flavivirus transmitted to humans by an insect vector such as a mosquito, sand fly, or tick; characterized by sudden onset of fever and often accompanied by malaise, anorexia, nausea/vomiting, eye pain, headache, myalgia (muscle pain), rash, swollen lymph nodes, and sometimes progressing to severe neurological disease

wheal (weel) raised red skin lesion often associated with severe itching, as in hives

white blood cell (WBC) *see* **leukocyte**

white matter (MAT-ter) tissue made up of nerve tracts covered with white myelin

withdrawal reflex (with-DRAW-al REE-fleks) a reflex that moves a body part away from an irritating stimulus

X

xeroderma pigmentosum (zeer-oh-DER-mah pig-men-TOH-sum) rare genetic disorder characterized by the inability of skin cells to repair genetic damage caused by the ultraviolet (UV) radiation in sunlight

Y

yeast (yeest) single-celled fungus (compared to mold, which is a multicellular fungus)

yellow bone marrow (YEL-oh bohn MAYR-oh) fatty tissue found inside the medullary cavity of a long bone

yellow fever (YEL-oh FEE-ver) viral illness caused by a type of flavivirus (literally "yellow" virus) carried by mosquito vectors and characterized by fever

yolk sac (yohk sak) in humans, involved with the production of blood cells in the developing embryo

Z

Z disk disklike structure separating one structural unit (sarcomere) of the myofibril from the next unit of the myofibril, often seen as a dark band (Z line) in micrographs of the myofibrils of skeletal muscle fibers; also called Z *line*

Z line *see* **Z disk**

Zika virus (ZEE-kah VYE-rus) viral illness caused by a type of flavivirus carried by mosquito vectors or transmitted sexually and characterized by fever; known to cause birth defects such as microcephaly

zona fasciculata (ZOH-nah fas-sic-yoo-LAY-tah) middle zone of the adrenal cortex that secretes **glucocorticoids**

zona glomerulosa (ZOH-nah gloh-mayr-yoo-LOH-sah) outer zone of the adrenal cortex that secretes mineralocorticoids

zona reticularis (ZOH-nah reh-tik-yoo-LAYR-is) inner zone of the adrenal cortex that secretes small amounts of sex hormones

zygomatic (zye-goh-MAT-ik) cheek bone (malar bone) or related to the cheek bone and nearby area

zygomaticus (zye-goh-MAT-ik-us) muscle that elevates the corners of the mouth and lips; also known as the *smiling muscle*

zygote (ZYE-goht) a fertilized ovum

Illustration/Photo Credits

Chapter 1

1-2, 1-6, 1-7: Barbara Cousins; 1-4: Redrawn from Muscolino JE: *Know the body: muscle, bone, and palpation essentials,* St Louis, 2012, Mosby; Science Application box (Modern Anatomy, art): Joe Kulka; Science Application box (Modern Anatomy, photo): National Library of Medicine (National Institute of Health).

Chapter 2

2-1: From Sugimoto Y et al: Chemical identification of individual surface atoms by atomic force microscopy, *Nature* 446:64-67, 2007; Science Application box (Biochemistry, art): Joe Kulka.

Chapter 3

3-4A: Courtesy Charles Flickinger, University of Virginia; 3-4B: Lennart Nilsson, Albert Bonniers Forlag AB, Stockholm Sweden; 3-12: From Forbes CD, Jackson WF: *Color atlas and text of clinical medicine,* ed 3, London, 2003, Elsevier; 3-16: From Stevens A, Lowe J: *Pathology,* ed 2, St Louis, 2000, Mosby; Table 3-2: Network Graphics.

Chapter 4

4-1, Adapted from Fawcett DW: *Bloom and Fawcett's a textbook of histology,* ed 11, Philadelphia, 1986, Saunders, In Gartner LP, Hiatt JL: *Concise histology,* Philadelphia, 2011, Saunders; 4-2, 4-3B, 4-4B, 4-6B, 4-8B: Barbara Cousins; 4-3A, 4-4A, 4-6A,4-8A, 4-15A: From Gartner LP, Hiatt JL: *Color textbook of histology,* ed 3, Philadelphia, 2007, Saunders; 4-5: From Erlandsen SL, Magney J: *Color atlas of histology,* St Louis, 1992, Mosby; 4-7A, 4-10, 4-11, 4-12, 4-13, 4-14, 4-16A, 4-18A, 4-19: Dennis Strete; 4-17A: From Gartner LP: *Textbook of histology,* ed 4, Philadelphia, 2017, Elsevier; 4-20: From Callen JP, Paller AS, Greer KE, et al: *Color atlas of dermatology,* Philadelphia, 1993, Saunders; Science Application box (Microscopy, art): Joe Kulka; Science Application box (Microscopy, Modern Compound Light Microscope, art):Joe Kulka.

Chapter 5

5-2, 5-6: Barbara Cousins; 5-14: From Markstedt K, Athanasios M, Tournier I, et al: 3D Bioprinting human chondrocytes with nanocellulose-alginate bioink for cartilage tissue engineering applications, *Biomacromolecules* 16:1489-96, 2015; Science Application box (Radiography, drawing): Joe Kulka; Science Application box (Radiography, x-ray film): © Photo Researchers, Inc.

Chapter 6

6-1, 6-3: Centers for Disease Control and Prevention, Atlanta, GA; 6-4: Barbara Cousins; 6-5A: From Travis J: Drugs counter mad cow agent in cells, *Science News* 160(7):100, 2001, © Fred Cohen; 6-5B: Centers for Disease Control and Prevention/Teresa Hammett; 6-6, 6-7A-B: David M Phillips/Visual Unlimited; 6-8: From Schuster FL, Visvesvara GS: Free-living amoebae as opportunistic and non-opportunistic pathogens of humans and animals, *Int J Parasitol* 34(9):1001-1027, 2004; 6-9: Getty Images; 6-10, 6-11: Rolin Graphics; 6-12A: From LeTreut AL: *Mammography,* St Louis, 1991, Mosby; 6-12B: From Williams AL, Haughton VM: *Cranial computed tomography,* St Louis, 1985, Mosby; 6-12C: From Runge VM: *Enhanced magnetic resonance imaging,* St Louis, 1989, Mosby; 6-12D: From Hagen-Ansert SL: *Textbook of diagnostic ultrasonography,* ed 6, St Louis, 2006, Mosby; 6-13 inset: From Bolognia JL, Schaffer JV: *Dermatology essentials,* St Louis, 2014, Saunders; Table 6-7, Fig 1: Courtesy National Center for Veterinary Parasitology at Oklahoma State University; Table 6-7, Fig 2: Centers for Disease Control and Prevention/Dr. Amanda Loftis, Dr. William Nicholson, Dr. Will Reeves, Dr. Chris Paddock; Table 6-7, Figs 3, 4: Centers for Disease Control and Prevention/Anna Perez; Table 6-7, Fig 5: Centers for Disease Control and Prevention; Table 6-7, Fig 6: Centers for Disease Control and Prevention/Andrew J Brooks; Research, Issues, and Trends box (Centers for Disease Control and Prevention, photo): Centers for Disease control and Prevention: Public Health Image Library: Image ID: 19388 (http://phil.cdc.gov/phil/details.asp); Clinical Application box (Laboratory Identification of Pathogens, images): From Goering RV, Dockrell HM, Zuckerman M, et al: *Mim's Medical Microbiology,* ed 5, Elsevier, Ltd, 2012; Science Applications box (Public Health, art): Joe Kulka; Science Applications box (Public Health, photo): US Dept of Health & Human Services, http://www.usphs.gov/newsroom/gallery/adayinthelife.aspx; Research, Issues, and Trends box (Disease as a Weapon): From Beeching NJ, Nye FD: Diagnostic picture tests in clinical infectious diseases, St Louis, 1996, Mosby; Clinical Application box (Medical Imaging of the Body, A,B,C): © Photo Researchers, Inc.; Clinical Application box (Medical Imaging of the Body, D): From Ballinger PW et al: *Merrill's atlas of radiographic positions & radiologic procedures,* ed 10, St Louis, 2003, Mosby.

Chapter 7

7-4: Edward Reschke; 7-5: Courtesy James A Ischen, MD, Baylor College of Medicine; 7-6: Copyright Kevin Patton, Lion Den Inc, Weldon Spring, MO; 7-7, 7-21B: From Habif TP: *Clinical dermatology,* ed 6, St Louis, 2016, Mosby; 7-9: From Goldstein B, editor: *Practical dermatology,* ed 2, St Louis, 1997, Mosby; 7-11A, 7-12: From Habif T: *Clinical dermatology,* ed 3, St Louis, 1996, Mosby; 7-11B, 7-20B: From Habif TP: *Clinical dermatology,* ed 4, St Louis, 2004, Mosby; 7-13: From Callen JP, Paller AS, Greer KE, et al: *Color atlas of dermatology,* ed 2, Philadelphia, 2000, Saunders; 7-15: Hockenberry-Wilson: *Wong's nursing care of infants and children,* ed 9, St Louis, 2013, Mosby; 7-18: Courtesy Michael Peck, MD, University of North Carolina Burn Center, Chapel Hill, In Copstead-Kirkhorn L, Banasik J: *Pathophysiology,* ed 2, St Louis, 1999, Saunders; 7-20A: From Emond RTD, Welsby PD, Rowland HAK: *Colour atlas of infectious diseases,* ed 3, London, 1995, Mosby; 7-20C,D: Courtesy Jaime A Tschen, MD, Department of Dermatology, Baylor College of Medicine, Houston, TX; 7-21A: From Potter P, Perry A: *Fundamentals of nursing,* ed 7, St Louis, 2009, Mosby; 7-21C: From Habif TP: *Clinical dermatology,* ed 5, St Louis, 2010, Mosby; 7-20D: From Kumar V, Abbas A, Fausto N: *Robbins and Cotran pathologic basis of disease,* ed 8, Philadelphia, 2010, Saunders; 7-22A: From Goldman L, Schafer AI: *Goldman's Cecil medicine,* ed 24, Philadelphia, 2012, Saunders; 7-22B: From Noble J, Greene HL: *Textbook of primary care medicine,* St Louis, ed 3, Mosby,

2001; 7-22C: From Townsend C, Beauchamp RD, Evers BM, Mattox K: *Sabiston textbook of surgery,* ed 18, Philadelphia, 2008, Saunders; 7-22D: From Rakel R: *Textbook of family medicine,* ed 7, Philadelphia, 2007, Saunders; Science Application box (Secrets of the Skin, art): Joe Kulka; Science Application box (Secrets of the Skin, photo): Courtesy Angie's List, www.angieslist.com/articles/what-expect-laser-tattoo-removal.htm.

Chapter 8

8-4, 8-5: Dennis Strete; 8-6: Adapted from McCance K, Huether S: *Pathophysiology,* ed 7, St Louis, 2014, Mosby; 8-8: Network Graphics; 8-10B: Barbara Cousins; 8-12: Courtesy Dr. N. Blevins, New England Medical Center, Boston; 8-16: Adapted from Drake R, Vogl W, Mitchell AWM: *Gray's anatomy for students,* ed 3, Philadelphia, 2015, Churchill Livingstone; 8-17: From Hockenberry MJ: *Wong's essentials of pediatric nursing,* ed 9, St Louis, 2013, Mosby; 8-18D: From Barkauskas V, Baumann L, Stoltenberg-Allen K, Darling-Fisher C: *Health and physical assessment,* ed 2, St Louis, 1998, Mosby; 8-18E: Courtesy Nancy Lynch; 8-20C,F, 8-21B, 8-22B,D: From Vidic B, Suarez FR: *Photographic atlas of the human body,* St Louis, 1984, Mosby; 8-24: Yvonne Wylie Walston; 8-30A,B: Damjanov I, Linder J, Anderson WAD: *Anderson's pathology,* ed 10, St Louis, 1996, Mosby; 8-31: From Kumar V, Abbas A, Fausto N: *Robbins and Cotran pathologic basis of disease,* ed 7, Philadelphia, 2005, Saunders; 8-32, 8-35: From Kumar V: *Robbins basic pathology,* ed 9, Philadelphia, 2013, Saunders; 8-33: From Browner B, Jupiter J, Trafton P: *Skeletal trauma: basic science, management, and reconstruction,* ed 3, Philadelphia, 2003, Saunders; 8-34: From Canale ST: *Campbell's operative orthopaedics,* ed 9, St Louis, 1998, Mosby; 8-38: From Swartz MH: Textbook of physical diagnosis, ed 6, Philadelphia, 2010, Saunders; 8-39A: From Callen JP, Paller AS, Greer KE, et al: *Color atlas of dermatology,* ed 2, Philadelphia, 2000, WB Saunders; 8-39B: From Habif TP: *Clinical dermatology,* ed 5, St Louis, 2010, Mosby; Clinical Application box (Total Hip Replacement, fig): From Canale ST: *Campbell's operative orthopaedics,* ed 10, St Louis, 2003, Mosby; Health and Well-Being box (The Knee Joint, Fig A): Rolin Graphics; Science Application box (Bones and Joints, art): Joe Kulka; Science Application box (Bones and Joints, photo): From Cummings N, Stanley-Green S, Higgs P: *Perspectives in athletic training,* St Louis, 2009, Mosby; Clinical Application box (Arthroscopy, Fig A): From Miller MD, Howard RF, Plancher KD: *Surgical atlas of sports medicine,* Philadelphia, 2003, Saunders; Clinical Application box (Arthroscopy, Fig B): From Johnson LL: *Diagnostic and surgical arthroscopy,* ed 2, St Louis, 1981, Mosby.

Chapter 9

9-3C: Courtesy Dr. H.E. Huxley; 9-5B: Courtesy Dr. Paul C. Letourneau, Department of Anatomy, Medical School, University of Minnesota, MN; 9-10, 9-11, 9-12: John V. Hagen; 9-13 (Photo): Courtesy Kellie White; 9-14: CDC: Public Health Image Library: Image ID: 6373, http://phil.cdc.gov/phil/details.asp; Science Application box (Muscle Function, art): Joe Kulka; Clinical Application box (Intramuscular Injections, photo inset): From Perry AG, Potter PA, Ostendorf WR: *Clinical nursing skills and techniques,* ed 8, St Louis, 2014, Mosby.

Chapter 10

10-2C: Dennis Strete; 10-5: From Feldman M, Friedman L, Brandt L: *Sleisenger & Fordtran's gastrointestinal and liver disease,* ed 8, Philadelphia, 2006, Saunders; 10-13, 10-14 (photo): From Vidic B, Suarez FR: *Photographic atlas of the human body,* St Louis, 1984, Mosby; 10-15: From Zitelli BJ, Davis HW: *Atlas of pediatric physical diagnosis,* ed 5, Philadelphia, 2007, Mosby; 10-16: James King-Holmes and Science Photo Library; 10-17A: From Chipps EM, Clanin NJ, Campbell VG: *Neurologic disorders,* St Louis, 1992, Mosby; 10-26: From Habif TP: *Clinical dermatology,* ed 2, St Louis, 1990, Mosby; Clinical Application box (The Blood-Brain Barrier, fig): Courtesy John Nolte, PhD, University of Arizona College of Medicine; Clinical Application box (Antidepressants, photo): Copyright Adziohiciek (stock photo from Shutterstock.com); Clinical Application box (Lumbar Puncture, fig): From Forbes CD, Jackson WD: *Color atlas and text of clinical medicine,* ed 3, London, 2003, Mosby; Science Application box (Neuroscience, art): Joe Kulka.

Chapter 11

11-4: From Newell FW: *Ophthalmology: principles and concepts,* ed 7, St Louis, 1992, Mosby; 11-5C: From Swartz MH: *Textbook of physical diagnosis,* ed 7, Philadelphia, 2014, Saunders; 11-8: From Swartz MH: *Textbook of physical diagnosis,* ed 6, Philadelphia, 2010, Saunders; 11-9: From Newell FW: *Ophthalmology: principles and concepts,* ed 8, St Louis, 1996, Mosby; 11-10: Seidel H et al: *Mosby's guide to physical examination,* ed 3, St Louis, 2002, Mosby; 11-11: From *Ishihara's tests for colour deficiency,* Tokyo, Japan, 1973, Kanehara Trading Co, Copyright Isshinkai Foundation; 11-13A: From Ball JW, Dains JE, Flynn JA: *Seidel's guide to physical examination,* ed 8, St Louis, 2015, Mosby; 11-13C,D: From Bingham BJG, Hawke M, Kwok P: *Atlas of clinical otolaryngology,* St Louis, 1992, Mosby-Year Book; Health and Well-Being box (Swimmer's Ear, photo): From Zitelli, Davis: *Atlas of pediatric physical diagnosis,* ed 5, Philadelphia, 2007, Elsevier, Courtesy Michael Hawke, MD; Science Applications box (Senses, art): Joe Kulka.

Chapter 12

12-5A: Courtesy Robert F Gagel, MD, and Ian McCutcheon, MD, University of Texas M.D. Anderson Cancer Center, Houston, Texas, In Black JM, Hawks JH: *Medical-surgical nursing: clinical management for positive outcomes,* ed 8, St Louis, 2009, Mosby; 12-5B, 12-11, 12-16: From Forbes CD, Jackson WF: *Color atlas and text of clinical medicine,* ed 3, Edinburgh, 2003, Mosby; 12-9: From Seidel NM, Ball JW, Dains JE, Benedict GW: *Mosby's guide to physical examination,* ed 4, St Louis, 1999; 12-10: From Swartz MH: *Textbook of physical examination,* ed 6, Philadelphia, 2010, Saunders; 12-15: Courtesy Gower Medical Publishers; Clinical Application box (Synthetic Human Insulin, photo): Courtesy Medtronic, plc; Clinical Application box (Synthetic Human Insulin, art): From Cooper K, Gosnell K: Adult Health Nursing, ed 7, St Louis, 2015, Mosby; Science Applications box (Endocrinology, art): Joe Kulka.

Chapter 13

13-1: Barbara Cousins; 13-3A,B,C: From Bevelander G, Ramalay JA: *Essentials of histology,* ed 8, St Louis, 1979, Mosby; 13-3D: From Zakus SM: *Clinical procedures for medical assistants,* ed 3, St Louis, 1995, Mosby; 13-5: From Carr J, Rodak B: *Clinical hematology atlas,* ed 2, St Louis, 2004, Elsevier; 13-6 (photos): From Belcher AE: *Blood disorders,* St Louis, 1993, Mosby; 13-8: Courtesy of Dr. Robert W. McKenna, Department of Pathology, University of Texas Southwestern Medical School, Dallas, TX, In Kumar V, Abbas AK: *Robbins and Cotran pathologic basis of disease,* ed 8, Philadelphia, Saunders, 2010; 13-9: Dennis Strete; 13-12, 13-16: From Kumar V: *Robbins basic pathology,* ed 9, Philadelphia, 2013, Saunders; 13-13: Courtesy Dr. JV Melo; 13-14: Courtesy of Dr. Robert W McKenna, Department of Pathology, University of Texas Southwestern Medical School, Dallas, TX, In Kumar V et al: *Robbins basic pathology,* ed 9, Philadelphia, 2013, Saunders; 13-15: From Hoffbrand AV,

Pettit JE: *Clinical haematology illustrated,* Edinburgh/London, 1987, Churchill Livingstone/Gower Medical Publisher; 13-17B: Copyright Dennis Kunkel Microscopy, Inc; 13-18: From Kumar V, Abbas AK, Fausto N: *Robbins and Cotran pathologic basis of disease,* ed 8, Philadelphia, 2010, Elsevier; Clinical Application box (Cardiac Blood Tests, fig): From Warekois R, Robinson R: *Phlebotomy,* ed 2, St Louis, 2007, Saunders; Science Application box (Hematology, fig): Joe Kulka.

Chapter 14

14-4: From Kumar V, Abbas AK, Fausto N: *Robbins and Cotran pathologic basis of disease,* ed 8, Philadelphia, 2010, Elsevier; 14-12: Barbara Cousins; 14-14: From Libby P, Bonow RO, Mann DL, Zipes DP: *Braunwald's heart disease: a textbook of cardiovascular medicine,* ed 8, Philadelphia, 2008, Saunders; 14-17: Courtesy Patricia Kane, Indiana University Medical School; Clinical Application box (Echocardiography, photo): From Forbes, CD: *Color atlas and text of clinical medicine,* ed 3, Edinburgh, 2003, Mosby; Clinical Application box (Angiography, fig): From Goldman L, Ausiello D: *Cecil textbook of medicine,* ed 23, Philadelphia, 2008, Saunders; Science Application box (Cardiology, art): Joe Kulka; Unn 14-3B: From http://www.bls.gov/ooh/healthcare/diagnostic-medical-sonographers.htm#tab-3.

Chapter 15

Clinical Application box (Raynaud Phenomenon, photo): From Barkauskas VH, Baumann LC, Darling-Fisher CS: *Health and physical assessment,* ed 3, St Louis, 2002, Mosby; Science Applications box (Circulation of the Blood, art): Joe Kulka; Science Applications box (Circulation of the Blood, photo): From Warekois, R, Robinson, R: *Phlebotomy: worktext and procedures manual,* ed 3, St Louis, 2012, Saunders.

Chapter 16

16-1, inset: Drake RL et al: *Gray's anatomy for students,* ed 3, Philadelphia, 2015, Churchill-Livingstone; 16-3: Courtesy Walter Tunnesen, MD, The American Board of Pediatrics, Chapel Hill, NC; 16-4: From Zitelli B, McIntire SC, Nowalk AJ: *Atlas of pediatric physical diagnosis,* ed 6, Philadelphia, 2012, Mosby; 16-5: From Goldstein B, editor: *Practical dermatology,* ed 2, St Louis, 1997, Mosby; 16-7: From Ballinger P, Frank E: *Merrill's atlas of radiographic positions and radiologic procedures,* vol 1, ed 11, St Louis, 2007, Mosby; 16-8: From Ball JW, Dains JE, Flynn JA, et al: *Seidel's guide to physical examination,* ed 8, St Louis, 2015, Mosby; 16-14: Courtesy of Dr. Y-J Liu, MD, Anderson Cancer Center, Houston, TX, In Abbas A, Lichtman A: *Cellular and molecular immunology,* ed 8, Philadelphia, 2015, Saunders; 16-15: Copyright Dennis Kunkel Microscopy Inc.; 16-19: From Cerio R, Jackson WF: *Colour atlas of allergic skin disorders,* London, 1992, Mosby-Wolfe; 16-20: From Habif TP: *Clinical dermatology: a color guide to diagnosis and therapy,* ed 3, St Louis, 1996, Mosby; Clinical Application box (Interferon, photos): National Cancer Institute; Science Application box (Vaccines, art): Joe Kulka.

Chapter 17

17-1: Barbara Cousins; 17-6C: From Cox JD: *Radiation oncology,* ed 9, Philadelphia, 2010, Mosby; 17-8: Network Graphics; 17-13: From Kumar V, Abbas A, Fausto N: *Robbins and Cotran pathologic basis of disease,* ed 8, Philadelphia, 2010, Saunders; Clinical Application box (Keeping the Trachea Open, figs): Courtesy Andrew P. Evan, Indiana University School of Medicine; Science Application box (Respiratory Medicine, art): Joe Kulka.

Chapter 18

18-9A: From Regezi JA, Sciubba JJ, Pogrel MA: *Atlas of oral and maxillofacial pathology,* Philadelphia, 2000, Saunders; 18-9B: From Callen JP, Paller AS, Greer KE, et al: *Color atlas of dermatology,* ed 2, Philadelphia, 2000, WB Saunders; 18-9C: From Grundy JR, Jones JG: *A color atlas of clinical operative dentistry: crowns and bridges,* ed 2, London, 1993, Mosby-Wolfe; 18-9D: From Christensen GJ: *A consumer's guide to dentistry,* ed 2, St Louis, 2002, Mosby; 18-9E: From Emond RTD, Welsby PD, Rowland HAK: *Colour atlas of infectious diseases,* ed 4, Edinburgh, 2003, Mosby; 18-10A: From Wilson SF, Giddens JF: *Health assessment for nursing practice,* ed 2, St Louis, 2001, Mosby; 18-10B: From Greig JD, Garden OJ: *Color atlas of surgical diagnosis,* London, 1996, Times Mirror International Publishers; 18-13: Copyright Kevin Patton, Lion Den Inc, Weldon Spring, MO; 18-17B: From Weir J, Abrahams PH, Spratt JD, Salkowski LR: *Imaging atlas of human anatomy,* ed 4, Mosby Elsevier Ltd, 2011; 18-17C, 18-18 (x-ray only): From Abrahams PH, Boon JM, Spratt JD, et al: *McMinn's color atlas of human anatomy,* ed 6, Edinburgh, 2008, Mosby; 18-19: Courtesy Thompson JM, Wilson SF: *Health assessment for nursing practice,* St Louis 1996, Mosby; 18-20: From Kumar V, Abbas AK, Fausto N: *Robbins and Cotran pathologic basis of disease,* ed 6, Philadelphia, 1999, Elsevier; 18-21: From Vidic B, Suarez RF: *Photographic atlas of the human body,* St Louis, 1984, Mosby; 18-22B: From Abrahams PH, Boon JM, Spratt JD, et al: *McMinn's color atlas of human anatomy,* ed 6, Edinburgh, 2008, Mosby; 18-22C: From Heuman DM, Mills AS, McGuire HH: *Gastroenterology,* Philadelphia, 1997, Saunders; 18-22D: From Zitelli BJ, Davis HW: *Atlas of pediatric physical diagnosis,* ed 3, Philadelphia, 1997, Mosby; 18-24 (photo): From Swartz MH: *Textbook of physical diagnosis,* ed 6, Philadelphia, 2010, Saunders; 18-25: Barbara Cousins; Clinical Application box (Mumps, photo): Courtesy GDW McKendrick, In Zitelli BJ, Davis HW: *Atlas of pediatric physical diagnosis,* ed 5, Philadelphia, 2007, Mosby; Clinical Application box (Upper Gastrointestinal X-ray Study, photo): © Photo Researchers, Inc.; Science Application box (Gastroenterology, art): Joe Kulka.

Chapter 19

19-1: From the U.S. Department of Agriculture, Center for Nutrition Policy and Promotion, USDA's MyPlate (website): www.choosemyplate.gov, accessed August 29, 2016; 19-4: Adapted from Carroll R: *Elsevier's integrated physiology,* Philadelphia, 2007, Mosby; 19-6: From Morgan SL, Weinsier RL: *Fundamentals of clinical nutrition,* ed 2, St Louis, 1998, Mosby; 19-8A: Courtesy Dr. Jonathan Spector, Boston, MA, In Zitelli BJ, McIntire SC, Nowalk AJ: *Zitelli and Davis' atlas of pediatric physical diagnosis,* ed 6, Philadelphia, 2012, Mosby; 19-8B: From Zitelli BJ, McIntire SC, Nowalk AJ: *Zitelli and Davis' atlas of pediatric physical diagnosis,* ed 6, Philadelphia, 2012, Mosby; Health and Well-Being box (Carbohydrate Loading, photo): © i-stock; Science Application box (Food Science, art): Joe Kulka.

Chapter 20

20-1B: From Abrahams P, Hutchings RT, Marks SC: *McMinn's color atlas of human anatomy,* ed 5, St Louis, 2003, Mosby; 20-2B: From Abrahams P, Marks S, Hutchings R: *McMinn's color atlas of human anatomy,* ed 6, Philadelphia, 2008, Mosby; 20-4: Courtesy Andrew P. Evan, Indiana University School of Medicine; 20-10: From Telser A, Young J, Baldwin K: *Elsevier's integrated histology,* Philadelphia, 2008, Mosby; 20-12: From Kumar V, Abbas A, Fausto N: *Robbins and Cotran pathologic basis of disease,* ed 8, Philadelphia, 2010, Saunders; 20-13: Courtesy of Dr. E. Mosher, Brigham and Women's Hospital,

Boston, MA, In Kumar V, Abbas AK, Fausto N: *Robbins and Cotran pathologic basis of disease,* ed 8, Philadelphia, 2010, Elsevier; 20-14B: Courtesy of B. Bracken, MD, Cincinnati, Ohio, In Weiss MA, Mills SE: *Atlas of genitourinary tract disorders,* Philadelphia, 1988, J.B. Lippincott/Gower Medical Publishing; 20-15: From Kumar V, Abbas AK, Fausto N, Mitchell RN: *Robbin's basic pathology,* ed 7, Philadelphia, 2007, Saunders; Table 20-2 (photo): From Bonewit-West K: *Clinical procedures for medical assistants,* ed 8, St Louis, Saunders, 2011; Science Application box (Fighting Infection, art): Joe Kulka; Science Application box (Fighting Infection, photo): Copyright © Mosby's Clinical Skills: Essentials Collection.

Chapter 21

21-6: Copyright Kevin Patton, Lion Den Inc, Weldon Spring, MO; 21-8: From Fritz S: *Mosby's fundamentals of therapeutic massage,* ed 5, St Louis, 2013, Mosby; 21-11: Modified from Goldman L, Schafer AI: *Goldman's Cecil medicine,* ed 24, Philadelphia, 2012, Saunders; Clinical Application box (Edema, photo): From Bloom A, Ireland J: *Color atlas of diabetes,* ed 2, St Louis, 1992, Mosby; Science Applications box (The Constancy of the Body, art): Joe Kulka; Science Applications box (The Constancy of the Body, photo): Copyright © Mosby's Clinical Skills: Essentials Collection.

Chapter 22

Science Applications box (The Body in Balance, art): Joe Kulka; Science Applications box (The Body in Balance, photo): US Navy, www.navy.mil/view_image.asp?id=6952.

Chapter 23

23-2A: Lennart Nilsson, Albert Bonniers Forlag AB, Stockholm Sweden; 23-5A: Carolyn Coulam and John A McIntyre; 23-6: From Abraham PH, Boon JM, Spratt JD: *McMinn's clinical atlas of human anatomy,* ed 6, Edinburgh, 2008, Mosby; 23-7B: From Vidic B, Suarez FR: *Photographic atlas of the human body,* St Louis, 1984, Mosby; 23-8: From Zitelli BJ, McIntire SC, Nowalk AJ: *Atlas of pediatric diagnosis,* ed 6, Philadelphia, 2012, Mosby; 23-16: From Kumar V, Abbas AK, Fausto N: *Robbins and Cotran pathologic basis of disease,* ed 8, Philadelphia, 2010, Saunders; 23-17: Courtesy of Dr. Edmund S Cibas, Department of Pathology, Brigham & Women's Hospital, Boston, In Kumar V, Abbas AK, Fausto N: *Robbins and Cotran pathologic basis of disease,* ed 8, Philadelphia, 2010, Elsevier; Science Application box (Reproductive Sciences, art): Joe Kulka.

Chapter 24

24-1B, 24-6: Lennart Nilsson, Albert Bonniers Forlag AB, Stockholm Sweden; 24-3: Courtesy Lucinda L Veeck, Jones Institute for Reproductive Medicine, Norfolk, VA; 24-5B: From Kumar V, Abbas AK, Fausto N: *Robbins and Cotran pathologic basis of disease,* ed 7, Philadelphia, 2005, Elsevier; 24-12: Courtesy Karen Turner; 24-13A: From Hockenberry MJ, Wilson D: *Wong's essentials of pediatric nursing,* ed 8, St Louis, 2009, Mosby; 24-13B, Research, Issues, and Trends box (Antenatal Diagnosis and Treatment, figs A,B,C): Copyright Kevin Patton, Lion Den Inc, Weldon Spring, MO; 24-14: From Mahan LK, Escott-Stump S: *Krause's food & nutrition therapy,* ed 12, St Louis, 2008, Elsevier; Research, Issues, and Trends box (Freezing Umbilical Cord Blood, photo): Courtesy Craig Borck, St Paul Pioneer Press; Research, Issues, and Trends box (Fetal Alcohol, photo): Courtesy of Claus Simon/Michael Janner; Research, Issues, and Trends box (Progeria, photo): Courtesy of the Progeria Research Foundation; Peabody, Massachusetts http://www.progeriaresearch.org; Science Application box (Embryology, art): Joe Kulka.

Chapter 25

Quote (p. 685): Matt Ridley: *Genome: the autobiography of a species in 23 chapters;* 25-9A: Courtesy Lois McGavran, Denver Children's Hospital; 25-9B: From Zitelli BJ, Davis HW: *Atlas of pediatric physical diagnosis,* ed 5, Philadelphia, 2007, Mosby; 25-10, 25-11B: Courtesy Nancy S Wexler, PhD, Columbia University; 25-11A: From Llewellyn-Jones D, Oats J, Abraham S: *Fundamentals of obstetrics and gynaecology,* ed 9, Edinburgh, 2010, Mosby; Science Application box (Genetics and Genomics, art): Joe Kulka.

All chapter opening art used by permission of Shutterstock.com.

Index

A

Page numbers followed by *b*, *t*, and *f* indicate boxes, tables, and figures, respectively.

B

J

K

U

V